Achieve Your Potential

The authors J.S. Ratti and Marcus McWaters, have developed specific content in MyMathLab® to ensure you have many resources to help you achieve success in mathematics—and beyond!

The MyMathLab features described here will help you:

- **Brush up** on math skills and concepts you may have forgotten.
- **Retain** new concepts as you move through your math course.
- **Develop** skills that will help with your transition to college and to your profession.

Adaptive Study Plan

The Study Plan will help you study more efficiently and effectively.

Your performance and activity are ssessed continually in real time, providing a personalized experience based on your individual needs.

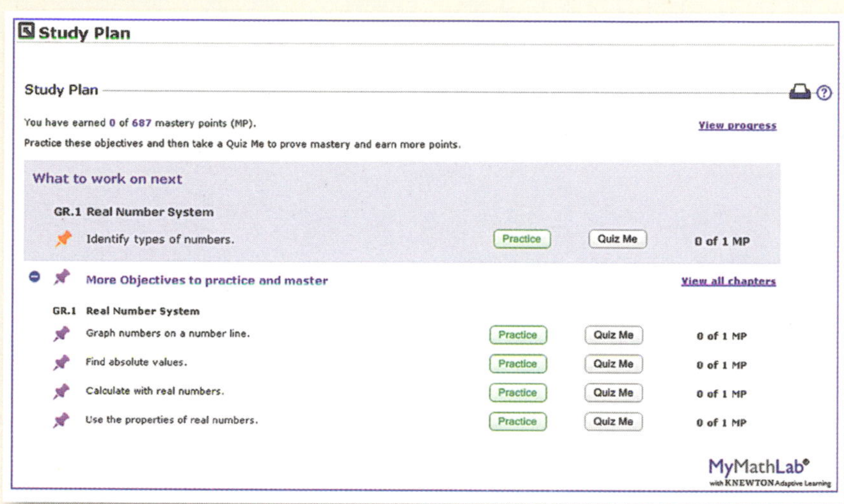

Skills for Success

The Skills for Success Module supports your continued success in college.

This module provides tutorials and guidance on a variety of topics, including the transition to college, online learning, time management, and more.

Additional content is provided to help with the development of professional skills such as resume writing and creation and interview preparation.

Getting Ready

Are you frustrated when you know you learned a math concept in the past but you can't quite remember the skill when it's time to use it? Don't worry!

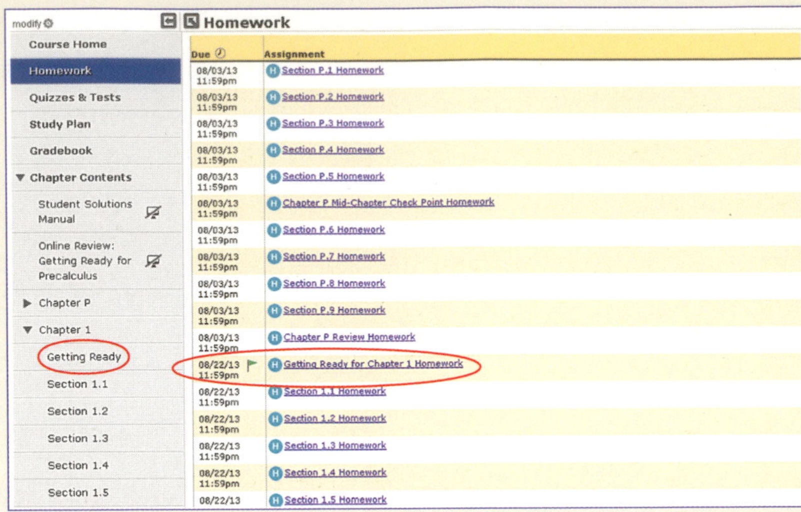

The authors have included Getting Ready material so that you can brush up on forgotten material efficiently by taking a quick skill review quiz to pinpoint the areas in which you need help.

Then a personalized homework assignment provides additional practice on those forgotten concepts, right when you need it.

Maintaining Skills

As you work through your math course, these MyMathLab® exercises support ongoing review to help you maintain essential skills.

The ability to recall important math concepts as you continually acquire new mathematical skills will help you be successful in this math course and in your future math courses.

Scan the code or visit **www.mymathlab.com/success-stories** to learn how students across the country have used MyMathLab to achieve their potential and be successful in their math courses.

College Algebra and Trigonometry

College Algebra and Trigonometry

Third Edition

J. S. Ratti
University of South Florida

Marcus McWaters
University of South Florida

PEARSON

Boston Columbus Indianapolis New York San Francisco Upper Saddle River
Amsterdam Cape Town Dubai London Madrid Milan Munich Paris Montréal Toronto
Delhi Mexico City São Paulo Sydney Hong Kong Seoul Singapore Taipei Tokyo

Editorial Director: Christine Hoag
Editor in Chief: Anne Kelly
Executive Content Editor: Christine O'Brien
Editorial Assistant: Judith Garber
Marketing Manager: Peggy Lucas
Marketing Assistant: Justine Goulart
Director of Production: Sarah McCracken
Senior Managing Editor: Karen Wernholm
Senior Production Project Manager: Beth Houston
Procurement Specialist: Debbie Rossi
Associate Design Director: Andrea Nix
Program Design Lead: Barbara Atkinson
Cover and Text Designer: Amanda/Kavanagh/Ark Design
Cover image: Waterscape Blue/White 1. Copyright Jim Nickelson Photography. All rights reserved.
Digital Assets Manager: Marianne Groth
Math XL Content Supervisor: Kristina Evans
Senior Content Developer: John Flanagan
Senior Author Support/Technology Specialist: Joe Vetere
Image Manager: Rachel Youdelman
Media Director: Ruth Berry
Media Producer: Tracy Menoza
Full-Service Project Management: PreMediaGlobal
Composition: PreMediaGlobal

Library of Congress Cataloging-in-Publication Data
Ratti, J. S., author.
College algebra and trigonometry / J.S. Ratti, University of South Florida, Marcus McWaters, University of South Florida. — 3 [edition].
pages cm
ISBN 0-321-86751-3 (student edition) —

1. Algebra—Textbooks. 2. Trigonometry—Textbooks. I. McWaters, Marcus, author. II. Title.
QA152.3.R38 2015
512′.13—dc23

2013028603

10 9 8 7 6 5 4 3 2 1-RRD-16 15 14 13

www.pearsonhighered.com

ISBN 13: 978-0-321-86751-3
ISBN 10: 0-321-86751-3

To Our Wives,

Lata and Debra

FOREWORD

We're pleased to present this new edition of our text for College Algebra and Trigonometry. Our experience teaching this material at the University of South Florida has been exceptionally rewarding. Because students are accustomed to information being delivered by electronic media, the introduction of MyMathLab® into our courses was seamless. We hope you will have a similar experience.

Many challenges face today's college algebra and trigonometry students and instructors. Students arrive with various levels of comprehension from their previous courses. Instead of really learning the concepts presented, students often resort to memorization to pass the course. As a result, a course needs to get students to a common starting point and engage them in becoming active learners, without sacrificing the solid mathematics essential for conceptual understanding. Instructors in this course are faced with the task of providing students with an understanding of college algebra and trigonometry, preparing them for the next step, and ensuring that they find mathematics useful and interesting. Our efforts, however, have been aided considerably by the many suggestions we have received from users of the first and second editions of this text.

In this text, there is a strong emphasis on both concept development and real-life applications. Topics such as functions, graphing, the difference quotient, and limiting processes provide thorough preparation for the study of calculus and will improve students' comprehension of algebra. Just-in-time review throughout the text ensures that all students are brought to the same level before being introduced to new concepts. Numerous applications are used to motivate students to apply the concepts and skills they learn in college algebra and trigonometry to other courses, including the physical and biological sciences, engineering, economics, and to on-the-job and everyday problem solving. Students are given ample opportunities throughout this course to think about important mathematical ideas and to practice and apply algebraic skills.

Throughout the text, we emphasize why the material being covered is important and how it can be applied. By thoroughly developing mathematical concepts with clearly defined terminology, students see the "why" behind those concepts, paving the way for a deeper understanding, better retention, less reliance on rote memorization, and ultimately more success. The level of exposition was selected so that the material is accessible to students and provides them with an opportunity to grow.

It is our hope that once you have read through our text, you will see that we were able to fulfill our initial goals of writing for today's students and for you, the instructor.

Marcus McWaters

(Marcus McWaters)

JS Ratti

(J. S. Ratti)

CONTENTS

Chapter P

Chapter 1

Chapter 2

Chapter 3

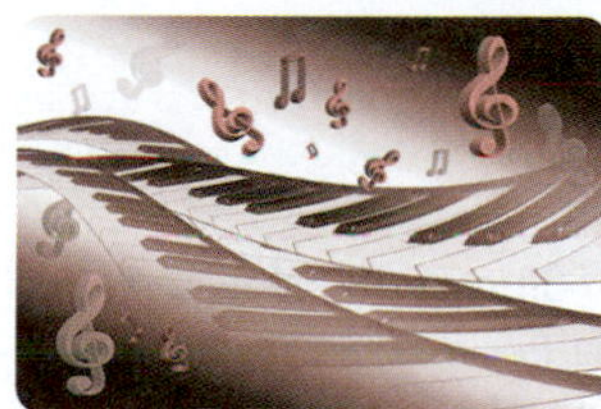

Chapter 4

Chapter 5

Chapter 6

Chapter 7

Chapter 8

Chapter 9

Chapter 10

Chapter 11

PREFACE

Students begin college algebra and trigonometry classes with widely varying backgrounds. Some haven't taken a math course in several years and may need to spend time reviewing prerequisite topics, while others are ready to jump right into new and challenging material. In Chapter P and in some of the early sections of other chapters, we have provided review material in such a way that it can be used or omitted as appropriate for your course. In addition, students may follow several paths after completing a college algebra and trigonometry course. Many will continue their study of mathematics in courses such as finite mathematics, statistics, and calculus. For others, college algebra and trigonometry may be their last mathematics course.

Responding to the current and future needs of all of these students was essential in creating this course. We introduce each exercise set with several concept exercises. These exercises consist of fill-in-the-blank and true–false exercises. They are not computation-reliant, but rather test whether students have absorbed the basic concepts and vocabulary of the section. Exercises asking students to extrapolate information from a given graph now appear in much greater number and depth throughout the course. We continue to present our content in a systematic way that illustrates how to study and what to review. We believe that if students use this material well, they will succeed in this course.

New To The Third Edition

EXERCISES

- We continue to improve the balance of exercises and have added more challenging exercises.
- Answers to the Practice Problems have been moved to the end of the section, just before the exercises.
- New **Maintaining Skills** exercises, placed at the end of each exercise set, help refresh important concepts learned in previous chapters and allow for practice of skills needed for more advanced concepts in upcoming chapters.
- Overall, approximately 20% of the exercises have been updated, and we've added over 1000 additional exercises.

CHAPTER 1

- Applications of linear equations have been relocated in Section 1.
- A separate section "Complex Numbers: Quadratic Equations with Complex Solutions" introduces complex numbers and discusses complex roots of quadratic equations.
- Polynomial and rational inequalities are now discussed in Section 5.

CHAPTER 2

- Discussion of the domain of composite functions has been expanded and includes more examples.

CHAPTER 3

- New examples of quadratic functions have been included.
- The procedure for graphing rational functions has been rewritten to identify the relation of the graph to horizontal and oblique asymptotes.

CHAPTER 4

- The number e is now introduced in Section 1.
- A new section on logarithmic scales has been added.

CHAPTER 7

- The ambiguous case is now included in Section 7.1.
- Section 7.3 has been expanded to include the areas of regular polygons.

CHAPTER 8

- Partial-Fraction Decomposition has been moved to immediately follow Section 2, "Systems of Linear Equations in Three Variables."

Features

CHAPTER OPENER Each chapter opener includes a description of applications relevant to the content of the chapter and the list of topics that will be covered in the chapter. In one page, students see what they are going to learn and why they are learning it.

APPLICATION The discussion in each section begins with a motivating anecdote or piece of information that is tied to an application problem. This problem is solved in an example later in the section, using the mathematics covered in the section.

SECTION OPENERS The application section openers lend continuity to the section and its content, utilizing material from a variety of fields: the physical and biological sciences (including health sciences), economics, art and architecture, the history of mathematics, and more. Of special interest are contemporary topics such as the greenhouse effect and global warming, CAT scans, and computer graphics.

REVIEW On the first page of each section is a list of topics that students should review prior to starting the chapter.

Section references with page numbers accompany the suggested review material so that students can readily find the material. The **Objectives** of the section are then clearly stated and numbered. Each numbered objective is paired with a similarly numbered subsection so that students can quickly find the section material for an objective.

DEFINITIONS AND THEOREMS All are boxed for emphasis and titled for ease of reference, as are lists of rules and properties.

FIGURES All figures are titled to make it easy to identify what is being illustrated.

EXAMPLES The examples include a wide range of computational, conceptual, and modern applied problems that are carefully selected to build confidence, competency, and understanding. Every example has a title that indicates its purpose. For every example, clarifying side comments are provided for each step in the detailed solution.

PRACTICE PROBLEM All examples are followed by a Practice Problem for students to try so that they can check their understanding of the concept covered. The answers to these problems are placed at the end of each section, just before the exercises.

PROCEDURE BOXES These boxes, interspersed throughout the text, present important procedures in numbered steps. Special Procedure in Action boxes present important multistep procedures, such as the steps for doing synthetic division, in a two-column format. The steps of the procedure are given in the left column, and an example is worked, following these steps, in the right column. This approach provides students with a clear model with which they can compare when encountering difficulty in their work. These boxes are a part of the numbered examples.

SUMMARY OF MAIN FACTS These boxes summarize information related to equations and their graphs, such as those of the conic sections.

HISTORICAL NOTES When appropriate, historical notes appear in the margin, giving students information on key people or ideas in the history and development of mathematics. This information is included to add flavor to the subject matter.

TECHNOLOGY CONNECTIONS Although the use of graphing calculators is optional in this course, Technology Connections give students tips on using calculators to solve problems, check answers, and reinforce concepts.

WARNINGS These boxes appear as appropriate throughout the text to let students know of common errors and pitfalls that can trip them up in their thinking or calculations.

RECALL Located in the margins, Recall Notes periodically remind students of a key idea they learned earlier in the text that will help them with the new problems at hand.

SIDE NOTES Students are given hints for handling newly introduced concepts.

DO YOU KNOW? These marginal notes provide students with additional interesting information on nonessential topics to keep them engaged in the mathematics presented.

EXERCISES The heart of any textbook is its exercises. Knowing this, we made sure that the quantity, quality, and variety of exercises meet the needs of all students. The problems in each exercise set are carefully graded to strengthen the skills developed in the associated section. Exercises are divided into three categories: **Basic Skills and Concepts** (developing fundamental skills), **Applying the Concepts** (using the section's material to solve real-world problems), and **Beyond the Basics** (providing more challenging exercises that give students an opportunity to reach beyond the material covered in the section). Exercises are paired so that the even-numbered Basic Skills and Concepts exercises closely follow the preceding odd-numbered exercises. All application exercises are titled and relevant to the topics of the section. Basic Skills and Concepts and Applying the Concepts exercises are intended for a typical student, whereas the Beyond the Basics exercises, generally more theoretical in nature, are suitable for honors students, special assignments, or extra credit. **Critical Thinking/Discussion/Writing** exercises appear in each exercise set and are designed to develop students' higher-level thinking skills and to reinforce and extend comprehension of the material. Calculator problems are included where appropriate.

END-OF-CHAPTER The chapter-ending material includes a **Summary of Definitions**, **Concepts**, **and Formulas**; **Review Exercises**; and **two Practice Tests**. The chapter summary, a brief description of key topics indicating where the material occurs in the text, encourages students to reread sections rather than memorize definitions out of context. The Review Exercises provide students with an opportunity to practice what they have learned in the chapter. (Students should then be prepared for Practice Test A in the usual open-ended format and Practice Test B, covering the same topics, in multiple-choice format.) All tests are designed to increase student comprehension and verify that students have mastered the skills and concepts in the chapter. Mastery of these materials should indicate a true comprehension of the chapter and the likelihood of success on the associated in-class examination.

THE NEW MAINTAINING SKILLS exercises, placed at the end of each exercise set, help refresh important concepts learned in previous chapters and allow for practice of skills needed for more advanced concepts in upcoming chapters.

CUMULATIVE REVIEW EXERCISES Starting with Chapter 2, these exercises appear at the end of every chapter to remind students that mathematics is not modular and that what is learned in the first part of the book will be useful in later parts of the book and on the final examination.

RESOURCES FOR SUCCESS

MyMathLab® Online Course (Access Code Required)

MyMathLab delivers **proven results** in helping individual students succeed. It provides **engaging experiences** that personalize, stimulate, and measure learning for each student. And, it comes from an **experienced partner** with educational expertise and an eye on the future. MyMathLab helps prepare students and gets them thinking more conceptually and visually through the following features:

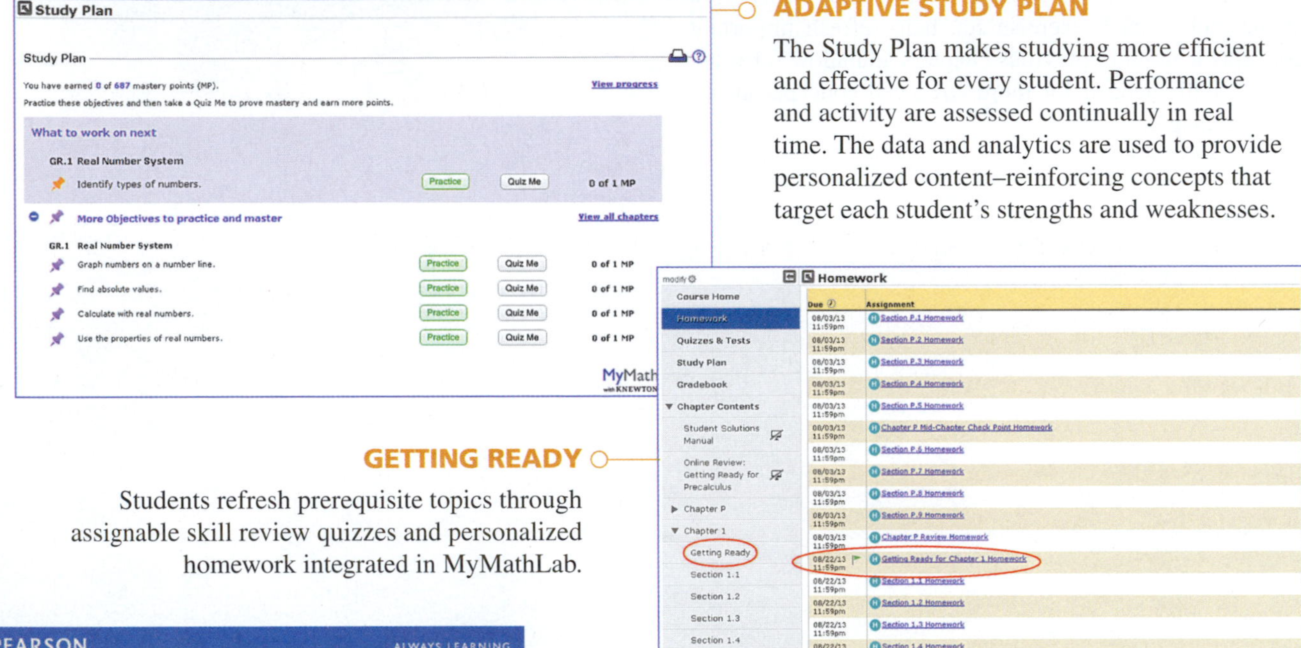

ADAPTIVE STUDY PLAN

The Study Plan makes studying more efficient and effective for every student. Performance and activity are assessed continually in real time. The data and analytics are used to provide personalized content–reinforcing concepts that target each student's strengths and weaknesses.

GETTING READY

Students refresh prerequisite topics through assignable skill review quizzes and personalized homework integrated in MyMathLab.

VIDEO ASSESSMENT

Video assessment is tied to key Example Solution videos to check student's conceptual understanding of important math concepts.

ENHANCED GRAPHING FUNCTIONALITY

Functionality allows graphing of 3-point quadratic functions, 4-point cubic graphs, and transformations in exercises.

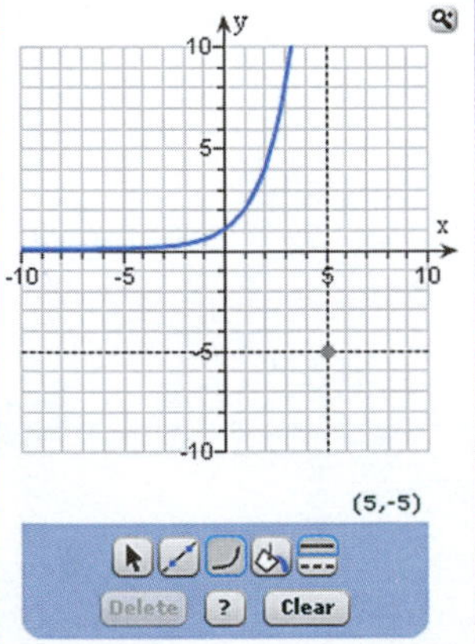

SKILLS FOR SUCCESS MODULE in MyMathLab helps students succeed in collegiate courses and prepare for future professions.

MAINTAINING SKILLS exercises help refresh important concepts and allows for practice of skills needed for more advanced topics. These exercises are also assignable in MyMathLab.

Instructor Resources

Additional resources can be downloaded from **www.pearsonhighered.com** or hardcopy resources can be ordered from your sales representative.

READY TO GO MYMATHLAB® COURSE

Now it is even easier to get started with MyMathLab. The Ready to Go MyMathLab course option includes author-chosen preassigned homework, integrated review, and more.

TESTGEN®

TestGen® (www.pearsoned.com/testgen) enables instructors to build, edit, print, and administer tests using a computerized bank of questions developed to cover all the objectives of the text.

POWERPOINT® LECTURE SLIDES

Fully editable slides that correlate to the textbook.

ANNOTATED INSTRUCTOR'S EDITION

This version of the text includes answers to all exercises presented in the book. Sample homework assignments, selected by the authors, are annotated with an underline.

INSTRUCTOR'S SOLUTIONS MANUAL

Includes fully worked solutions to all textbook exercises.

Student Resources

Additional resources to help student success.

VIDEO LECTURES

- Section Summary videos cover key definitions and procedures for most sections. Example Solution videos walk students through the detailed solution process for many examples in the textbook.
- There are over 20 hours of video instruction specifically filmed for this book, making it ideal for distance learning or supplemental instruction on a home computer or in a campus computer lab.
- Videos include optional subtitles in English and Spanish.

STUDENT'S SOLUTIONS MANUAL

Provides detailed worked-out solutions to odd-numbered exercises.

ACKNOWLEDGMENTS

We would like to express our gratitude to the reviewers of this and previous editions, who provided such invaluable insights and comments. Their contributions helped shape the development of the text and carry out the vision stated in the preface.

Reviewers

Alison Ahlgren, *University of Illinois at Urbana–Champaign*
Mohammad Aslam, *Georgia Perimeter College*
Mario Barrientos, *Angelo State University*
Ratan Barua, *Miami-Dade College*
Sam Bazzi, *Henry Ford Community College*
Diane Burleson, *Central Piedmont Community College*
Maureen T. Carroll, *University of Scranton*
Melissa Cass, *State University of New York–New Paltz*
Jason Cates, *Lone Star College–Montgomery*
Charles Conrad, *Volunteer State Community College*
Irene Coriette, *Cameron University*
Douglas Culler, *Midlands Technical College*
Baiqiao Deng, *Columbus State University*
Hussain Elaloui-Talibi, *Tuskegee University*
Kathleen M. Fick, *Methodist University-Fayetteville*
Junko Forbes, *El Camino College*
Gene Garza, *Samford University*
Steven Gonzales, *San Antonio College*
Bobbie Jo Hill, *Coastal Bend College*
Fran Hopf, *University of South Florida*
Natalie Hutchinson, *Old Dominion University*
Yvette Janecek, *Coastal Bend College*
Rose M. Jenkins, *Midlands Technical College*
Mohammad Kazemi, *University of North Carolina at Charlotte*
David Keller, *Kirkwood Community College*
Diana Klimek, *New Jersey Institute of Technology*
Mile Krajcevski, *University of South Florida*
Lance Lana, *University of Colorado–Denver*
Alexander Lavrentiev, *University of Wisconsin–Fox Valley*
Rebecca Leefers, *Michigan State University*
Paul Morgan, *College of Southern Idaho*
Kathy Nickell, *College of DuPage*
Catherine Pellish, *Front Range Community College*
Betty Peterson, *Mercer County Community College*
Edward K. Pigg, *Southern Union Community College*
Marshall Ransom, *Georgia Southern University*
Janice F. Rech, *University of Nebraska-Omaha*
Dr. Traci Reed, *St. Johns River Community College*
Linda Reist, *Macomb Community College*
Jeri Rogers, *Seminole Community College–Oviedo Campus*
Jason Rose, *College of Southern Idaho*
Mehdi Sadatmousavi, *Pima Community College–West*
Christy Schmidt, *Northwest Vista*
Nyeita Schult, *Coastal Carolina University*
Delphy Shaulis, *University of Colorado–Boulder*
Cindy Shaw, *Moraine Valley Community College*
Cynthia Sikes, *Georgia Southern University*
James Smith, *Columbia State Community College*
Linda Snellings-Neal, *Wright State University*
Jacqueline Stone, *University of Maryland–College Park*
Kay Stroope, *Phillips County Community College*
Robert J. Strozak, *Old Dominion University*
Helen Timberlake, *Rochester Institute of Technology*
Frances Tishkevich, *Massachusetts Maritime Academy*
Jo Tucker, *Tarrant County College–Southeast*
Roger Werbylo, *Pima Community College–West*
Tom Worthing, *Hutchinson Community College*
Vivian Zabrocki, *Montana State University–Billings*
Marti Zimmerman, *University of Louisville*

Our sincerest thanks go to the legion of dedicated individuals who worked tirelessly to make this book possible. We express special thanks to Elka Block and Frank Purcell of Twin Prime Editorial, Lenore Parens, Abby Tanenbaum, and Carrie Green for the excellent work they did as the development editors on the text and art development. We would also like to express our gratitude to our typist, Beverly DeVine-Hoffmeyer, for her amazing patience and skill. We must also thank Dr. Praveen Rohatgi, Dr. Nalini Rohatgi, and Dr. Bhupinder Bedi for the consulting they provided on all material relating to medicine. We particularly want to thank Professors Mile Krajcevski, Viktor Maymeskul, and Scott Rimbey for many helpful discussions and suggestions, particularly for improving the exercise sets. Further gratitude is due to Irena Andreevska, Gokarna Aryal, Ferene Tookos, and Christine Fitch for their assistance on the answers to the exercises in the text. In addition, we would like to thank Twin Prime Editorial, Viktor Maymeskul, Rhea Meyerholtz and Beverly Fusfield, for their meticulous accuracy in checking the text. Thanks are due as well to Erin Donahue and PrePress PMG for their excellent production work. Finally, our thanks are extended to the professional and remarkable staff at Pearson Education. In particular, we would like to thank Greg Tobin, President; Anne Kelly, Editor in Chief; Christine O'Brien, Project Editor; Judith Garber, Editorial Assistant; Beth Houston, Senior Production Project Manager; Peggy Sue Lucas, Senior Marketing Manager; Dona Kenly, Market Development Manager; Justine Goulart, Marketing Assistant; Barbara Atkinson, Cover Designer; Tracy Menoza, Media Producer; Karen Wernholm, Senior Managing Editor; and Joe Vetere, Senior Author Support/Technology Specialist.

We invite all who use this book to send suggestions for improvements to Marcus McWaters at mmm@cas.usf.edu.

ABOUT THE AUTHORS

J. S. Ratti

EDUCATION

B.Sc. Major: mathematics Minor: physics University of Bombay

M.Sc. Mathematics University of Bombay

PhD Mathematics Wayne State University

TEACHING

Wayne State University (teaching assistant and instructor)

University of Nevada at Las Vegas (instructor)

Oakland University (assistant professor)

University of South Florida (professor and past chair)

Undergraduate courses taught: college algebra and trigonometry, trigonometry, finite mathematics, calculus (all levels), set theory, differential equations, linear algebra

Graduate courses taught: number theory, abstract algebra, real analysis, complex analysis, graph theory

AWARDS

USF Research Council Grant

USF Teaching Incentive Program (TIP) Award

USF Outstanding Undergraduate Teaching Award

Academy of Applied Sciences grants

RESEARCH

Complex analysis, real analysis, graph theory, probability

PERSONAL INTERESTS

College and Professional Football, Traveling

Marcus McWaters

EDUCATION

BS Major: mathematics Minor: physics Louisiana State University New Orleans

PhD Major: mathematics Minor: philosophy University of Florida

TEACHING

Louisiana State University (teaching assistant)

University of Florida (teaching assistant and graduate fellow)

University of South Florida (associate professor and department chair)

Undergraduate courses taught: college algebra, trigonometry, finite mathematics, calculus, business calculus, life science calculus, advanced calculus, vector calculus, set theory, differential equations, linear algebra

Graduate courses taught: graph theory, number theory, discrete mathematics, geometry, advanced linear algebra, topology, mathematical logic, modern algebra, real analysis

AWARDS

Graduate Fellow, University of Florida (Center of Excellence Grant)

USF Research Council grant

USF Teaching Incentive Program (TIP) Award

Provost's Award

RESEARCH

Topology, algebraic topology, topological algebra

Founding member, USF Center for Digital and Computational Video

PERSONAL INTERESTS

Traveling with his wife and two daughters, theater, waterskiing, racquetball

CHAPTER P

Basic Concepts of Algebra

TOPICS

Many fascinating patterns in human and natural processes can be described in the language of algebra. We investigate events ranging from chirping crickets to the behavior of falling objects.

SECTION **P.1**

The Real Numbers and Their Properties

BEFORE STARTING THIS SECTION, REVIEW FROM YOUR PREVIOUS MATHEMATICS TEXTS

1 Arithmetic of signed numbers
2 Arithmetic of fractions
3 Long division involving integers
4 Decimals
5 Arithmetic of real numbers

OBJECTIVES

1 Classify sets of real numbers.
2 Use the ordering of the real numbers.
3 Specify sets of numbers in roster or set-builder notation.
4 Use interval notation.
5 Relate absolute value and distance on the real number line.
6 Identify the order of operations in arithmetic expressions.
7 Identify and use properties of real numbers.
8 Evaluate algebraic expressions.

◆ Cricket Chirps and Temperature

Crickets are sensitive to changes in air temperature; their chirps speed up as the temperature gets warmer and slow down as it gets cooler. It is possible to use the chirps of the male snowy tree cricket (*Oecanthus fultoni*), common throughout the United States, to gauge temperature. (The insect is found in every U.S. state except Hawaii, Alaska, Montana, and Florida.) By counting the chirps of this cricket, which lives in bushes a few feet from the ground, you can gauge temperature. Snowy tree crickets are more accurate than most cricket species; their chirps are slow enough to count, and they synchronize their singing. To convert cricket chirps to degrees Fahrenheit, count the number of chirps in 14 seconds and then add 40 to get the temperature. To convert cricket chirps to degrees Celsius, count the number of chirps in 25 seconds, divide by 3, and then add 4 to get the temperature. In Example 10, we evaluate algebraic expressions to learn the temperature from the number of cricket chirps.

1 Classify sets of real numbers.

Classifying Numbers

In algebra, we use letters such as a, b, x, y, and so on, to represent numbers. A letter that is used to represent one or more numbers is called a **variable**. A **constant** is a specific number such as 3 or $\frac{1}{2}$ or a letter that represents a fixed (but not necessarily specified) number. Physicists use the letter c as a constant to represent the speed of light ($c \approx 300{,}000{,}000$ meters per second).

We use two variables, a and b, to denote the results of the operations of addition $(a + b)$, subtraction $(a - b)$, multiplication $(a \times b$ or $a \cdot b)$, and division $\left(a \div b \text{ or } \frac{a}{b}\right)$. These operations are called **binary operations** because each is performed on two numbers.

SIDE **NOTE**

Here is one difficulty with attempting to divide by 0: If, for example, $\frac{5}{0} = a$, then $5 = a \cdot 0$. However, $a \cdot 0 = 0$ for all numbers a. So we would have $5 = 0$; this contradiction means that there is no appropriate choice for $\frac{5}{0}$.

We frequently omit the multiplication sign when writing a product involving two variables (or a constant and a variable) so that $a \cdot b$ and ab indicate the same product. Both a and b are called **factors** in the product $a \cdot b$. This is a good time to recall that we never divide by zero. For $\frac{a}{b}$ to represent a real number, b cannot be zero.

Equality of Numbers

The **equal sign**, $=$, is used much like we use the word *is* in English. The equal sign means that the number or expression on the left side is equal or equivalent to the number or expression on the right side. We write $a \neq b$ to indicate that a is not equal to b.

Classifying Sets of Numbers

The idea of a set is familiar to us. We regularly refer to "a set of baseball cards," a "set of CDs," or "a set of dishes." In mathematics, as in everyday life, a **set** is a collection of objects. The objects in the set are called the **elements**, or **members**, of the set. In the study of algebra, we are interested primarily in sets of numbers.

In listing the elements of a set, it is customary to enclose the listed elements in braces, $\{\ \}$, and separate them with commas.

We distinguish among various sets of numbers.

The numbers we use to count with constitute the set of **natural numbers**: $\{1, 2, 3, 4, \ldots\}$.

The three dots . . . may be read as "and so on" and indicate that the pattern continues indefinitely.

The **whole numbers** are formed by including the number 0 with the natural numbers to obtain the set $\{0, 1, 2, 3, 4, \ldots\}$.

The **integers** consist of the set of natural numbers together with their opposites and 0: $\{\ldots, -4, -3, -2, -1, 0, 1, 2, 3, 4, \ldots\}$.

Rational Numbers

The **rational numbers** consist of all numbers that *can* be expressed as the quotient or ratio, $\frac{a}{b}$ of two integers, where $b \neq 0$.

Examples of rational numbers are $\frac{1}{2}, \frac{5}{3}, \frac{-4}{17}$, and $0.07 = \frac{7}{100}$. Any integer a can be expressed as the quotient of two integers by writing $a = \frac{a}{1}$. Consequently, every integer is also a rational number. In particular, 0 is a rational number because $0 = \frac{0}{1}$.

The rational number $\frac{a}{b}$ can be written as a decimal by using long division. When any integer a is divided by an integer b, $b \neq 0$, the result is always a **terminating decimal** such as $\left(\frac{1}{2} = 0.5\right)$ or a **nonterminating repeating decimal** such as $\left(\frac{2}{3} = 0.666\ldots\right)$.

We sometimes place a bar over the repeating digits in a nonterminating repeating decimal. Thus, $\frac{2}{3} = 0.666\ldots = 0.\overline{6}$ and $\frac{141}{110} = 1.2818181\ldots = 1.2\overline{81}$.

EXAMPLE 1 **Converting Decimal Rationals to a Quotient**

Write the rational number $7.\overline{45}$ as the ratio of two integers in lowest terms.

Solution

Let $x = 7.454545\ldots$. Then

$$\begin{aligned} 100x &= 745.4545\ldots && \text{Multiply both sides by 100.} \\ \text{and} \quad x &= 7.4545\ldots \\ \hline 99x &= 738 && \text{Subtract } x \text{ from } 100x. \\ x &= \frac{738}{99} && \text{Divide both sides by 99.} \\ &= \frac{82 \times 9}{11 \times 9} && \text{Common factor} \\ x &= \frac{82}{11} && \text{Reduce to lowest terms.} \end{aligned}$$

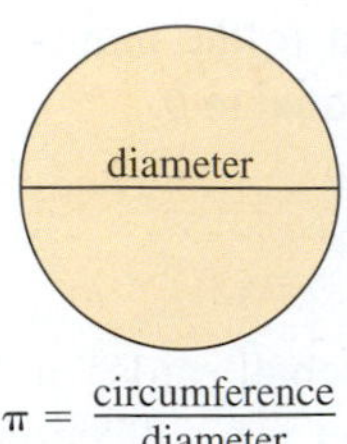

$\pi = \dfrac{\text{circumference}}{\text{diameter}}$

FIGURE P.1 Definition of π

Practice Problem 1 Repeat Example 1 for $2.132132132\ldots$.

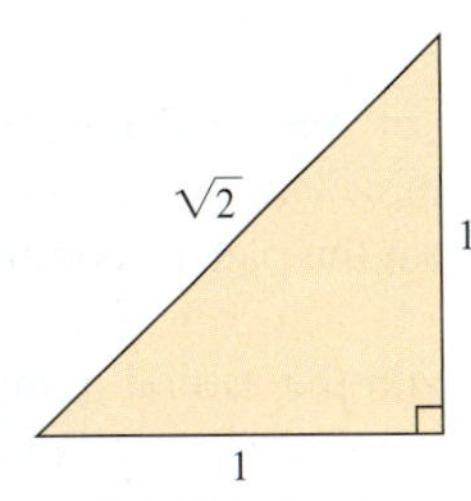

FIGURE P.2

Irrational Numbers

An **irrational number** is a number that cannot be written as a ratio of two integers. This means that its decimal representation must be nonrepeating and nonterminating. We can construct such a decimal using only the digits 0 and 1, such as 0.01001000100001. . . . Because each group of zeros contains one more zero than the previous group, no group of digits repeats. Other numbers such as π ("pi", see Figure P.1) and $\sqrt{2}$ (the square root of 2, see Figure P.2) can also be expressed as decimals that neither terminate nor repeat; so they are irrational numbers as well. We can obtain an approximation of an irrational number by using an initial portion of its decimal representation. For example, we can write $\pi \approx 3.14159$ or $\sqrt{2} \approx 1.41421$, where the symbol $\approx$ is read "is approximately equal to."

No familiar process, such as long division, is available for obtaining the decimal representation of an irrational number. However, your calculator can provide a useful approximation for irrational numbers such as $\sqrt{2}$. (Try it!) Because a calculator displays a fixed number of decimal places, it gives a **rational approximation** of an irrational number.

It is usually not easy to determine whether a specific number is irrational. One helpful fact in this regard is that *the square root of any natural number that is not a perfect square is irrational.* So $\sqrt{6}$ is irrational but $\sqrt{16} = \sqrt{4^2} = 4$ is rational.

RECALL

An integer is a *perfect square* if it is a product $a \cdot a$, where a is an integer. For example, $9 = 3 \cdot 3$ is a perfect square.

Because rational numbers have decimal representations that either terminate or repeat, whereas irrational numbers do not have such representations, *no number is both rational and irrational.*

The rational numbers together with the irrational numbers form the **real numbers**.

The diagram in Figure P.3 shows how various sets of numbers are related. For example, every natural number is also a whole number, an integer, a rational number, and a real number.

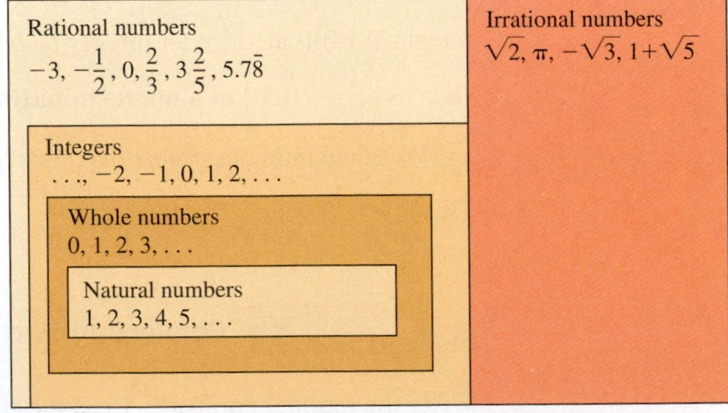

FIGURE P.3 Relationships among sets of real numbers

2 Use the ordering of the real numbers.

The Real Number Line

We associate the real numbers with points on a geometric line (imagined to be extended indefinitely in both directions) in such a way that each real number corresponds to exactly one point and each point corresponds to exactly one real number. The point is called the **graph** of the corresponding real number, and the real number is called the **coordinate** of the point. By agreement, ***positive numbers*** lie to the right of the point corresponding to 0 and ***negative numbers*** lie to the left of 0. See Figure P.4.

Notice that $\frac{1}{2}$ and $-\frac{1}{2}$, 2 and -2, and π and $-\pi$ correspond to pairs of points exactly the same distance from 0 but on opposite sides of 0.

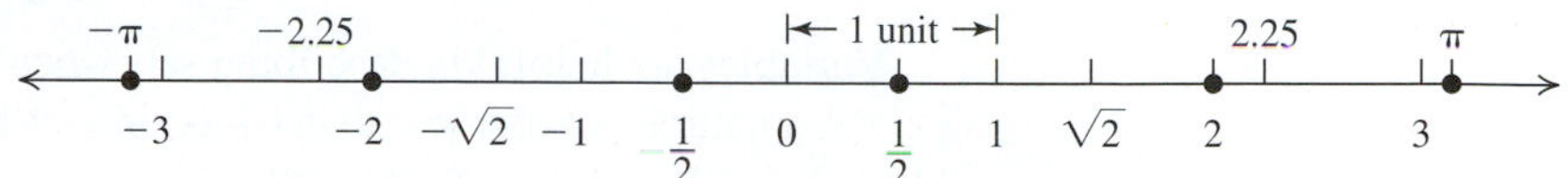

FIGURE P.4 The real number line

When coordinates have been assigned to points on a line in the manner just described, the line is called a **real number line**, a **coordinate line**, a **real line**, or simply a **number line**. The point corresponding to 0 is called the **origin**.

Inequalities

The real numbers are **ordered** by their size. We say that ***a* is less than *b*** and write $a < b$, provided that $b = a + c$ for some ***positive*** number c. We also write $b > a$, meaning the same thing as $a < b$, and say that ***b* is greater than *a***. On the real line, the numbers increase from left to right. Consequently, ***a is to the left of b on the number line when a < b***. Similarly, a is to the right of b on the number line when $a > b$. We sometimes want to indicate that at least one of two conditions is correct: Either $a < b$ or $a = b$. In this case, we write $a \le b$ or $b \ge a$. The four symbols $<$, $>$, $\le$, and $\ge$ are called **inequality symbols**.

EXAMPLE 2 **Identifying Inequalities**

Decide whether each of the following is true or false.

a. $5 > 0$ **b.** $-2 < -3$ **c.** $2 \le 3$ **d.** $4 \le 4$

SIDE NOTE

Notice that the inequality sign always points to the smaller number.

$2 < 7$, 2 is smaller.
$5 > 1$, 1 is smaller.

Solution

a. $5 > 0$ is true because 5 is to the right of 0 on the number line. See Figure P.5.

b. $-2 < -3$ is false because -2 is to the right of -3 on the number line.

c. $2 \le 3$ is true because 2 is to the left of 3 on the number line. (Recall that $2 \le 3$ is true if either $2 < 3$ or $2 = 3$.)

d. $4 \le 4$ is true because $4 \le 4$ is true if either $4 < 4$ or $4 = 4$.

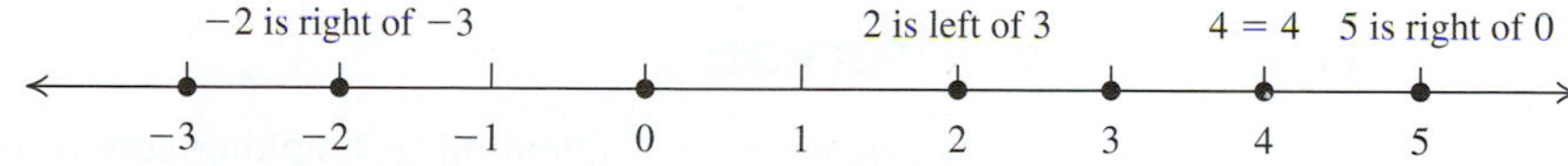

FIGURE P.5

Practice Problem 2 Decide whether each of the following is true or false.

a. $-2 < 0$ **b.** $5 \le 7$ **c.** $-4 > -1$

The following properties of inequalities for real numbers are used throughout this text.

Trichotomy Property: Exactly one of the following is true:

$$a < b, a = b, \text{ or } a > b$$

Transitive Property: If $a < b$ and $b < c$, then $a < c$.

The trichotomy property says that if two real numbers are not equal, then one is larger than the other. The transitive property says that "less than" works like "smaller than" or "lighter than." Frequently, we read $a > 0$ as "a is positive" instead of "a is greater than 0." We can also read $a < 0$ as "a is negative." If $a \geq 0$, then either $a > 0$ or $a = 0$, and we may say that "a is nonnegative."

3 Specify sets of numbers in roster or set-builder notation.

Sets

To specify a set, we do one of the following:

1. List the elements of the set (**roster method**)
2. Describe the elements of the set (often using **set-builder notation**)

Variables are helpful in describing sets when we use set-builder notation. The notation $\{x \mid x \text{ is a natural number less than } 6\}$ is in set-builder notation and describes the set $\{1, 2, 3, 4, 5\}$ using the roster method.

We read $\{x \mid x \text{ is a natural number less than } 6\}$ as "the set of all x such that x is a natural number less than six." Generally, $\{x \mid x \text{ has property } P\}$ designates the set of all x such that (the vertical bar is read "such that") x has property P.

It may happen that a description fails to describe any number. For example, consider $\{x \mid x < 2 \text{ and } x > 7\}$. Of course, no number can be simultaneously less than 2 and greater than 7; so this set has no members. We refer to a set with no elements as the **empty set**, or **null set**, and use the special symbol $\varnothing$ to denote it.

Definition of Union and Intersection

The **union** of two sets A and B, denoted $A \cup B$, is the set consisting of all elements that are in A *or* B (or both). See Figure P.6a. The **intersection** of A and B, denoted $A \cap B$, is the set consisting of all elements that are in both A *and* B. In other words, $A \cap B$ consists of the elements common to A and B. See Figure P.6b.

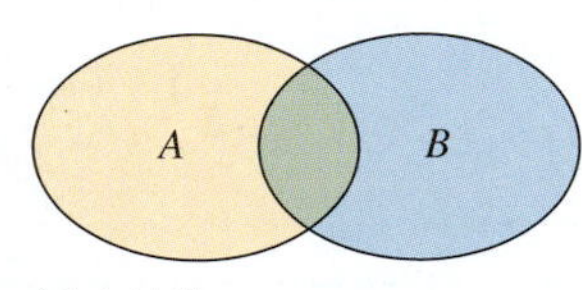

(a) $A \cup B$

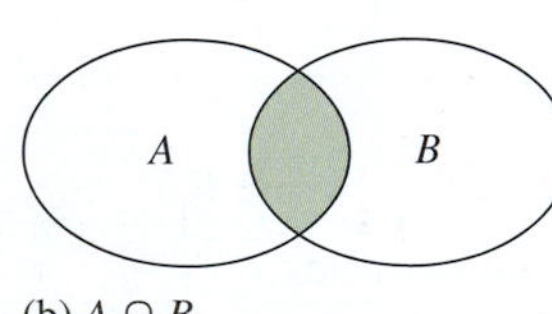

(b) $A \cap B$

FIGURE P.6 Picturing union and intersection

EXAMPLE 3 Forming Set Unions and Intersections

Find $A \cap B$, $A \cup B$, and $A \cap C$ if $A = \{-2, -1, 0, 1, 2\}$, $B = \{-4, -2, 0, 2, 4\}$, and $C = \{-3, 3\}$

Solution

$A \cap B = \{-2, 0, 2\}$, the set of elements common to both A and B.
$A \cup B = \{-4, -2, -1, 0, 1, 2, 4\}$, the set of elements that are in A or B (or both).
$A \cap C = \varnothing$.

Practice Problem 3 Find $A \cap B$ and $A \cup B$ if $A = \{-3, -1, 0, 1, 3\}$ and $B = \{-4, -2, 0, 2, 4\}$.

4 Use interval notation.

Intervals

We now turn our attention to graphing certain sets of numbers. That is, we graph each number in a given set. We are particularly interested in sets of real numbers, called **intervals**, whose graphs correspond to special sections of the number line.

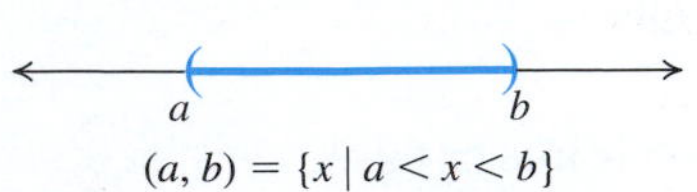

FIGURE P.7 An open interval

If $a < b$, then the set of real numbers between a and b, but not including either a or b, is called the **open interval** from a to b and is denoted by (a, b). See Figure P.7. Using set-builder notation, we can write

$$(a, b) = \{x \mid a < x < b\}.$$

We indicate graphically that the endpoints a and b are excluded from the open interval by drawing a left parenthesis at a and a right parenthesis at b. These parentheses enclose the numbers between a and b.

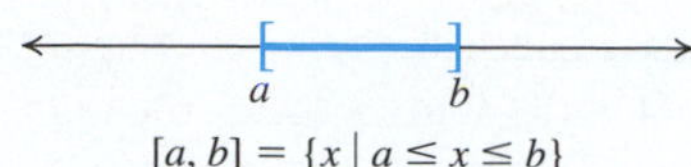

FIGURE P.8 A closed interval

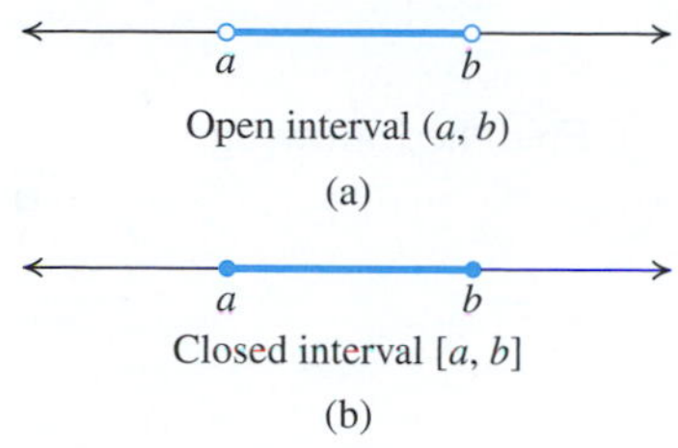

FIGURE P.9 Endpoint inclusion and exclusion

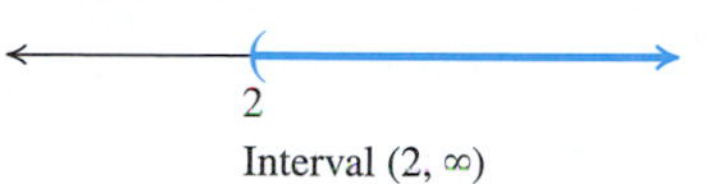

FIGURE P.10 An unbounded interval

The **closed interval** from a to b is the set

$$[a, b] = \{x | a \leq x \leq b\}.$$

The closed interval includes both endpoints a and b. We replace the parentheses with square brackets in the interval notation and on the graph. See Figure P.8. Sometimes we want to include only one endpoint of an interval and exclude the other. Table P.1 below shows how this is done.

An alternative notation for indicating whether endpoints are included uses open circles to show exclusion and closed circles to show inclusion. See Figure P.9.

If an interval extends indefinitely in one or both directions, it is called an **unbounded interval**. For example, the set of all numbers to the right of 2,

$$\{x | x > 2\},$$

is an unbounded interval denoted by $(2, \infty)$. See Figure P.10.

The symbol ∞ ("infinity") is not a number, but is used to indicate all numbers to the right of 2.

The symbol $-\infty$ is another symbol that does not represent a number. The notation $(-\infty, a)$ is used to indicate the set of all real numbers that are less than a. The notation $(-\infty, \infty)$ represents the set of all real numbers.

Table P.1 lists various types of intervals that we use in this book. In the table, when two points a and b are given, we assume that $a < b$. This is because if $a > b$, then (a, b) is the empty set.

SIDE NOTE

The symbols ∞ and $-\infty$ are always used with parentheses, not square brackets. Also note that $<$ and $>$ are used with parentheses and that $\leq$ and $\geq$ are used with square brackets.

TABLE P.1

Interval Notation	Set Notation	Graph
(a, b)	$\{x \| a < x < b\}$	
$[a, b]$	$\{x \| a \leq x \leq b\}$	
$(a, b]$	$\{x \| a < x \leq b\}$	
$[a, b)$	$\{x \| a \leq x < b\}$	
(a, ∞)	$\{x \| x > a\}$	
$[a, \infty)$	$\{x \| x \geq a\}$	
$(-\infty, b)$	$\{x \| x < b\}$	
$(-\infty, b]$	$\{x \| x \leq b\}$	
$(-\infty, \infty)$	$\{x \| x$ is a real number$\}$	

EXAMPLE 4 Union and Intersection of Intervals

Consider the two intervals $I_1 = (-3, 4)$ and $I_2 = [2, 6]$.

Find: **a.** $I_1 \cup I_2$ **b.** $I_1 \cap I_2$

Solution

a. From Figure P.11, we see that $I_1 \cup I_2 = (-3, 6]$. We note that every number in the interval $(-3, 6]$ is in either I_1 or I_2 or in both I_1 and I_2.

b. We see in Figure P.11 that $I_1 \cap I_2 = [2, 4)$. Every number in the interval $[2, 4)$ is in both I_1 and I_2. Notice that while 4 is in I_2, 4 is not in I_1.

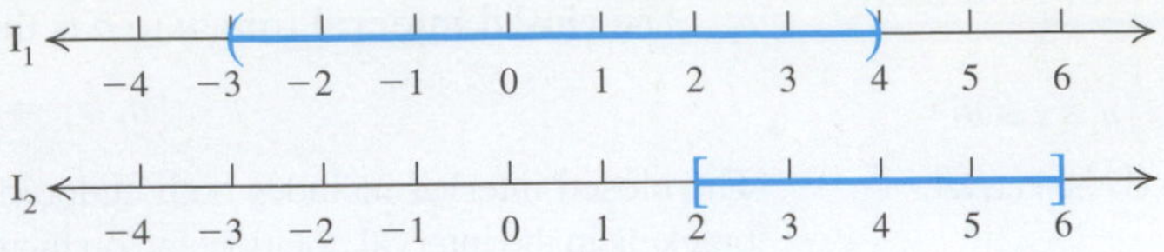

FIGURE P.11

Practice Problem 4 Let $I_1 = (-\infty, 5)$ and $I_2 = [-2, \infty)$. Find the following.

a. $I_1 \cup I_2$ **b.** $I_1 \cap I_2$

5 Relate absolute value and distance on the real number line.

Absolute Value

The ***absolute value*** of a number a, denoted by $|a|$, is the distance between the origin and the point on the number line with coordinate a. The point with coordinate -3 is 3 units from the origin, so we write $|-3| = 3$ and say that the absolute value of -3 is 3. See Figure P.12.

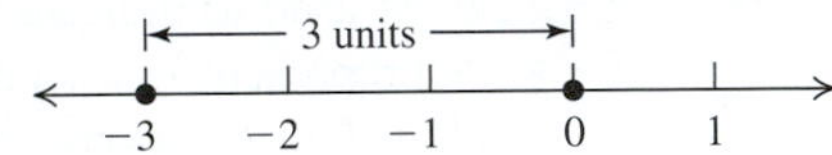

FIGURE P.12 Absolute value

Absolute Value

For any real number a, the **absolute value** of a, denoted $|a|$, is defined by

$$|a| = a \quad \text{if } a \geq 0 \qquad \text{and} \qquad |a| = -a \quad \text{if } a < 0.$$

SIDE NOTE

Finding the absolute value requires knowing whether the number or expression inside the absolute value bars is positive, zero, or negative. If it is positive or zero, you can simply remove the absolute value bars. If it is negative, you remove the bars and change the sign of the number or expression inside the absolute value bars.

EXAMPLE 5 Determining Absolute Value

Find the value of each of the following expressions.

a. $|4|$ **b.** $|-4|$ **c.** $|0|$ **d.** $|(-3) + 1|$

Solution

a. $|4| = 4$ Because the number inside the absolute value bars is 4 and $4 \geq 0$, just remove the absolute value bars.

b. $|-4| = -(-4) = 4$ Because the number inside the absolute value bars is -4 and $-4 < 0$, remove the bars and change the sign.

c. $|0| = 0$ Because the number inside the absolute value bars is 0 and $0 \geq 0$, just remove the absolute value bars.

d. $|(-3) + 1| = |-2| = -(-2) = 2$ Because the expression inside the absolute value bars is $(-3) + 1 = -2$ and $-2 < 0$, remove the bars and change the sign.

Practice Problem 5 Find the value of each of the following.

a. $|-10|$ **b.** $|3 - 4|$ **c.** $|2(-3) + 7|$

WARNING

The absolute value of a number represents a distance. Because distance can never be negative, the absolute value of a number is never negative; it is always positive or zero. However, if a is not 0, $-|a|$ is always negative. Thus, $-|5.3| = -5.3$, $-|-4| = -4$, and $-|1.\overline{18}| = -1.\overline{18}$.

Distance Between Two Points on a Real Number Line

The absolute value is used to define the distance between two points on a number line.

DISTANCE FORMULA ON A NUMBER LINE

If a and b are the coordinates of two points on a number line, then the distance between a and b, denoted by $d(a, b)$, is $|a - b|$. In symbols, $d(a, b) = |a - b|$.

TECHNOLOGY CONNECTION

The absolute value function on your graphing calculator will find the value of the expression entered and then compute its absolute value.

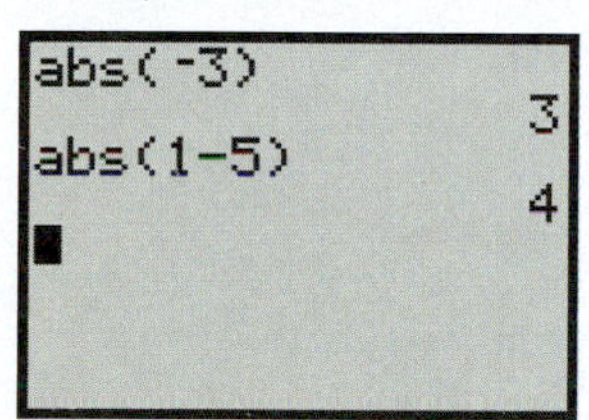
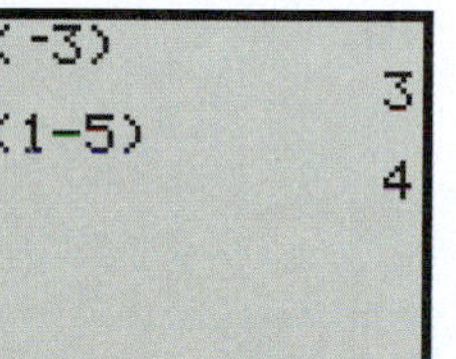

EXAMPLE 6 Finding the Distance Between Two Points

Find the distance between -3 and 4 on the number line.

Solution

Figure P.13 shows that the distance between -3 and 4 is 7 units. The distance formula gives the same answer.

$$d(-3, 4) = |-3 - 4| = |-7| = -(-7) = 7.$$

3 units | 4 units
−3 −2 −1 0 1 2 3 4
3 + 4 = 7 units

FIGURE P.13 Distance on the number line

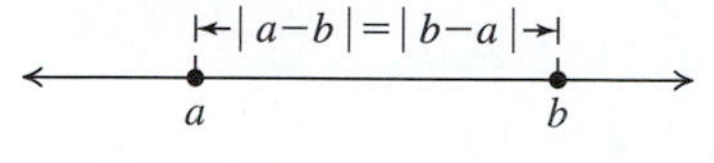

FIGURE P.14

Notice that reversing the order of -3 and 4 in this computation gives the same answer. That is, the distance between 4 and -3 is $|4 - (-3)| = |4 + 3| = |7| = 7$. It is always true that $|a - b| = |b - a|$. See Figure P.14.

Practice Problem 6 Find the distance between -7 and 2 on the number line.

We summarize the properties of absolute value next.

PROPERTIES OF ABSOLUTE VALUE

If a and b are any real numbers, the following properties apply.

	Property	Example
1.	$\|a\| \ge 0$	$\|-5\| = 5$ and $5 \ge 0$
2.	$\|a\| = \|-a\|$	$\|3\| = \|-3\|$
3.	$\|ab\| = \|a\|\|b\|$	$\|3(-5)\| = \|3\|\|-5\|$
4.	$\left\|\frac{a}{b}\right\| = \frac{\|a\|}{\|b\|}, b \neq 0$	$\left\|\frac{-7}{3}\right\| = \frac{\|-7\|}{\|3\|}$
5.	$\|a - b\| = \|b - a\|$	$\|2 - 7\| = \|7 - 2\|$
6.	$a \le \|a\|$	$-2 \le \|-2\|, 3 \le \|3\|$
7.	$\|a + b\| \le \|a\| + \|b\|$ (the triangle inequality)	$\|-2 + 5\| \le \|-2\| + \|5\|$

6 Identify the order of operations in arithmetic expressions.

Arithmetic Expressions

When we write numbers in a meaningful combination of the basic operations of arithmetic, the result is called an **arithmetic expression**. The real number that results from performing all operations in the expression is called the **value** of the expression. In arithmetic and algebra, parentheses ( ) are **grouping symbols** used to indicate which operations are to be performed first. Other common grouping symbols are square brackets [ ], braces { }, fraction bars — or /, and absolute value bars | |.

SIDE NOTE

You can remember the order of operations by the acronym PEMDAS (Please Excuse My Dear Aunt Sally):

Parentheses
Exponents
Multiplication
Division
Addition
Subtraction

Evaluating (finding the value of) arithmetic expressions requires carefully applying the following conventions for the order in which the operations are performed.

THE ORDER OF OPERATIONS

Whenever a fraction bar is encountered, work separately above and below it.

When parentheses or other grouping symbols are present, start inside the innermost pair of grouping symbols and work outward.

Step 1 Do operations involving exponents.

Step 2 Do multiplications and divisions in the order in which they occur, working from left to right.

Step 3 Do additions and subtractions in the order in which they occur, working from left to right.

TECHNOLOGY CONNECTION

The order of operations is built into your graphing calculator.

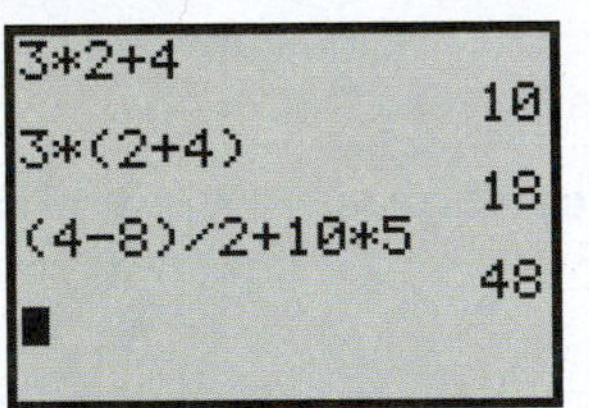

EXAMPLE 7 **Evaluating Arithmetic Expressions**

Use order of operations to evaluate each of the following.

a. $3 \cdot 2^5 + 4$ **b.** $5 - (3 - 1)^2$

Solution

a. $3 \cdot 2^5 + 4 = 3 \cdot 32 + 4$ Do operations involving exponents first, $2^5 = 32$.

$= 96 + 4$ Multiplication is done next, $3 \cdot 32 = 96$.

$= 100$ Addition is done next, $96 + 4 = 100$.

b. $5 - (3 - 1)^2 = 5 - (2)^2$ Work inside the parentheses first, $3 - 1 = 2$.

$= 5 - 4$ Next, do operations involving exponents, $2^2 = 4$.

$= 1$ Subtraction is done next, $5 - 4 = 1$.

Practice Problem 7 Evaluate.

a. $(-3) \cdot 5 + 20$ **b.** $5 - 12 \div 6 \cdot 2$ **c.** $\frac{9 - 1}{4} - 5 \cdot 7$ **d.** $-3 + (6 - 4)^2$

7 Identify and use properties of real numbers.

Properties of the Real Numbers

When doing arithmetic, we intuitively use important properties of the real numbers. We know, for example, that if we add or multiply two real numbers, the result is a real number. This fact is known as the **closure** property of real numbers. For each property listed below, *a, b, and c represent real numbers.*

PROPERTIES OF THE REAL NUMBERS

Name	Addition (A)	Multiplication (M)	Examples
Closure	$a + b$ is a real number	$a \cdot b$ is a real number	A: $1 + \sqrt{2}$ is a real number. M: $2 \cdot \pi$ is a real number.
Commutative	$a + b = b + a$	$a \cdot b = b \cdot a$	A: $4 + 7 = 7 + 4$ M: $3 \cdot 8 = 8 \cdot 3$
Associative	$(a + b) + c = a + (b + c)$	$(a \cdot b) \cdot c = a \cdot (b \cdot c)$	A: $(2 + 1) + 7 = 2 + (1 + 7)$ M: $(5 \cdot 9) \cdot 13 = 5 \cdot (9 \cdot 13)$
Identity	There is a unique real number 0, called the **additive identity**, such that $a + 0 = a$ and $0 + a = a$.	There is a unique real number 1, called the **multiplicative identity**, such that $a \cdot 1 = a$ and $1 \cdot a = a$.	A: $3 + 0 = 3$ and $0 + 3 = 3$ M: $7 \cdot 1 = 7$ and $1 \cdot 7 = 7$

Name	Addition (A)	Multiplication (M)	Examples
Inverse	There is a unique real number $-a$, called the **opposite of *a***, such that $a + (-a) = 0$ and $(-a) + a = 0$.	If $a \neq 0$, there is a unique real number $\frac{1}{a}$, called the **reciprocal of *a***, such that $a \cdot \frac{1}{a} = 1$ and $\frac{1}{a} \cdot a = 1$.	A: $5 + (-5) = 0$ and $(-5) + 5 = 0$ M: $4 \cdot \frac{1}{4} = 1$ and $\frac{1}{4} \cdot 4 = 1$
Distributive	Multiplication distributes over addition:	$a \cdot (b + c) = a \cdot b + a \cdot c$ $(a + b) \cdot c = a \cdot c + b \cdot c$	$3 \cdot (2 + 5) = 3 \cdot 2 + 3 \cdot 5$ $(2 + 5) \cdot 3 = 2 \cdot 3 + 5 \cdot 3$

TECHNOLOGY CONNECTION

On your graphing calculator, the ***opposite*** of a is denoted with a shorter horizontal hyphen than that used to denote subtraction. It is also placed somewhat higher than the subtraction symbol.

Two other useful properties of 0 involve multiplication rather than addition.

Zero-Product Properties

$$0 \cdot a = 0 \text{ and } a \cdot 0 = 0$$

$$\text{If } a \cdot b = 0, \text{ then } a = 0 \text{ or } b = 0.$$

Subtraction and Division of Real Numbers

You may have noticed that all of the properties discussed to this point apply to the operations of addition and multiplication. What about subtraction and division? We see next that subtraction and division can be defined in terms of addition and multiplication.

Subtraction of the number b from the number a is defined with the use of a and $-b$; to subtract b from a, add the opposite of b to a.

Subtraction

$$a - b = a + (-b)$$

If a and b are real numbers and $b \neq 0$, **division** of a by b is defined with the use of a and $\frac{1}{b}$; to divide a by b, multiply a by the reciprocal of b.

Division

$$\text{If } b \neq 0, \quad a \div b = \frac{a}{b} = a \cdot \frac{1}{b}$$

$$\frac{a}{0} \text{ is undefined.}$$

For $b \neq 0$, $\frac{a}{b}$ is called the **quotient**, "the **ratio** of a to b," or the **fraction** with **numerator** a and **denominator** b. Here are some useful properties involving opposites, subtraction, and division.

Throughout, we assume that the denominator of a fraction is nonzero.

PROPERTIES OF OPPOSITES

$(-1)a = -a$	$-(-a) = a$	$(-a)b = a(-b) = -(ab)$
$(-a)(-b) = ab$	$-(a + b) = -a - b$	$-(a - b) = b - a$
$\frac{-a}{b} = \frac{a}{-b} = -\frac{a}{b}$	$\frac{-a}{-b} = \frac{a}{b}$	$a(b - c) = ab - ac$

To combine real numbers by means of the division operation, we use the following properties. We write "$x \pm y$" as shorthand for "$x + y$ or $x - y$".

PROPERTIES OF FRACTIONS

All denominators are assumed to be nonzero.

$$\frac{a}{c} \pm \frac{b}{c} = \frac{a \pm b}{c} \qquad \frac{a}{b} \pm \frac{c}{d} = \frac{ad \pm bc}{bd}$$

$$\frac{a}{b} \cdot \frac{c}{d} = \frac{ac}{bd} \qquad \frac{a}{b} \div \frac{c}{d} = \frac{a}{b} \cdot \frac{d}{c}$$

$$\frac{ac}{bc} = \frac{a}{b} \qquad \frac{a}{b} = \frac{c}{d} \text{ means } a \cdot d = b \cdot c$$

The property $\frac{ac}{bc} = \frac{a}{b}$, equivalently $\frac{a}{b} = \frac{ac}{bc}$, can be used to produce a common denominator when adding or subtracting fractions.

EXAMPLE 8 **Adding, Subtracting, Multiplying, and Dividing Fractions**

Perform the indicated operations.

a. $\frac{5}{3} + \frac{3}{2}$ **b.** $\frac{5}{6} - \frac{2}{3}$ **c.** $\frac{4}{3} \cdot \frac{15}{8}$ **d.** $\dfrac{\frac{2}{5}}{\frac{9}{25}}$

SIDE NOTE

Part a of Example 8 can also be done as:

$$\frac{5}{3} + \frac{3}{2} = \frac{5 \cdot 2 + 3 \cdot 3}{3 \cdot 2}$$

using $\frac{a}{b} \pm \frac{c}{d} = \frac{ad \pm bc}{bd}$ directly. Part b can be done similarly.

Solution

a. $\frac{5}{3} + \frac{3}{2} = \frac{5 \cdot 2}{3 \cdot 2} + \frac{3 \cdot 3}{2 \cdot 3}$ $\qquad \frac{a}{b} = \frac{ac}{bc}$

$= \frac{5 \cdot 2 + 3 \cdot 3}{3 \cdot 2} = \frac{19}{6}$ $\qquad \frac{a}{c} + \frac{b}{c} = \frac{a + b}{c}$

b. $\frac{5}{6} - \frac{2}{3} = \frac{5}{6} - \frac{2 \cdot 2}{3 \cdot 2}$ $\qquad \frac{a}{b} = \frac{ac}{bc}$

$= \frac{5 - 2 \cdot 2}{6} = \frac{1}{6}$ $\qquad \frac{a}{c} - \frac{b}{c} = \frac{a - b}{c}$

c. $\frac{4}{3} \cdot \frac{15}{8} = \frac{4 \cdot 15}{3 \cdot 8}$ $\qquad \frac{a}{b} \cdot \frac{c}{d} = \frac{a \cdot c}{b \cdot d}$

$= \frac{\cancel{4} \cdot \cancel{3} \cdot 5}{\cancel{3} \cdot \cancel{4} \cdot 2} = \frac{5}{2}$ $\qquad \frac{a \cdot c}{b \cdot c} = \frac{a}{b}$

d. $\dfrac{\frac{2}{5}}{\frac{9}{25}} = \frac{2}{5} \cdot \frac{25}{9}$ $\qquad \frac{a}{b} \div \frac{c}{d} = \frac{a}{b} \cdot \frac{d}{c}$

$= \frac{2 \cdot 25}{5 \cdot 9}$ $\qquad \frac{a}{b} \cdot \frac{c}{d} = \frac{a \cdot c}{b \cdot d}$

$= \frac{2 \cdot \cancel{5} \cdot 5}{\cancel{5} \cdot 9} = \frac{10}{9}$ $\qquad \frac{a \cdot c}{b \cdot c} = \frac{a}{b}$

Practice Problem 8 Perform the indicated operations.

a. $\frac{7}{4} + \frac{3}{8}$ **b.** $\frac{8}{3} - \frac{2}{5}$ **c.** $\frac{9}{14} \cdot \frac{7}{3}$ **d.** $\dfrac{\frac{5}{8}}{\frac{15}{16}}$

8 Evaluate algebraic expressions.

Algebraic Expressions

In algebra, any number, constant, variable, or parenthetical group (or any product of them) is called a **term**. A combination of terms using the ordinary operations of addition, subtraction, multiplication, and division (as well as exponentiation and roots, which are discussed later in this chapter) is called an **algebraic expression**, or simply an **expression**. If we replace each variable in an expression with a specific number and get a real number, the resulting number is called the **value of the algebraic expression**. Of course, this value depends on the numbers we use to replace the variables in the expression. Here are some examples of algebraic expressions:

$$x - 2 \qquad \frac{1}{x} + \sqrt{7} \qquad \frac{10}{y+3} \qquad \sqrt{x} + |y| \div 5$$

WARNING

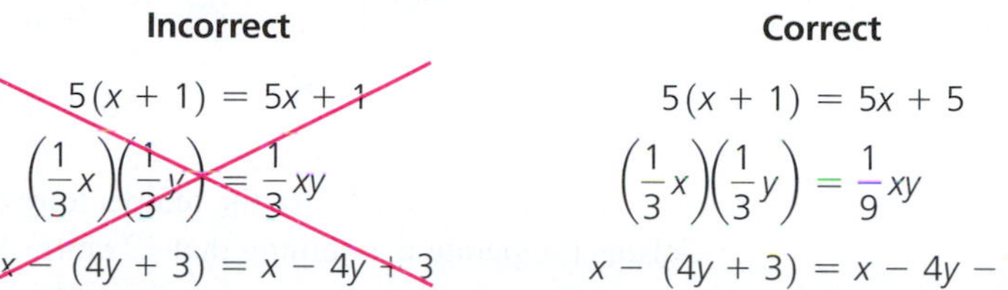

Incorrect	Correct
$5(x+1) = 5x + 1$	$5(x+1) = 5x + 5$
$\left(\frac{1}{3}x\right)\left(\frac{1}{3}y\right) = \frac{1}{3}xy$	$\left(\frac{1}{3}x\right)\left(\frac{1}{3}y\right) = \frac{1}{9}xy$
$x - (4y + 3) = x - 4y + 3$	$x - (4y + 3) = x - 4y - 3$

EXAMPLE 9 Evaluating an Algebraic Expression

Evaluate the following expressions for the given values of the variables.

a. $[(9 + x) \div 7] \cdot 3 - x$ for $x = 5$

b. $|x| - \frac{2}{y}$ for $x = 2$ and $y = -2$

Solution

a. $[(9 + x]) \div 7] \cdot 3 - x = [(9 + 5) \div 7] \cdot 3 - 5$ Replace x with 5.

$= [14 \div 7] \cdot 3 - 5$ Work inside the parentheses.

$= 2 \cdot 3 - 5$ Work inside the brackets.

$= 6 - 5$ Multiply.

$= 1$ The value of the expression

b. $|x| - \frac{2}{y} = |2| - \frac{2}{-2}$ Replace x with 2 and y with -2.

$= 2 - \frac{2}{-2}$ Eliminate the absolute value, $|2| = 2$.

$= 2 - (-1)$ $\frac{2}{-2} = -1$; division occurs before subtraction.

$= 3$ The value of the expression

Practice Problem 9 Evaluate.

a. $(x - 2) \div 3 + x$ for $x = 3$ **b.** $7 - \frac{x}{|y|}$ for $x = -1, y = 3$

EXAMPLE 10 Finding the Temperature from Cricket Chirps

Write an algebraic expression for converting cricket chirps to degrees Celsius. Assuming that you count 48 chirps in 25 seconds, what is the Celsius temperature?

Solution

Recall from the introduction to this section that to convert cricket chirps to degrees Celsius, you count the chirps in 25 seconds, divide by 3, and then add 4 to get the temperature.

$$\boxed{\text{Temperature in Celsius degrees}} = \left(\frac{\text{Number of chirps in 25 seconds}}{3}\right) + 4$$

If we let C = temperature in degrees Celsius and N = number of chirps in 25 seconds, we get the expression

$$C = \frac{N}{3} + 4.$$

Suppose we count 48 chirps in 25 seconds. The Celsius temperature is

$$C = \frac{48}{3} + 4$$
$$= 16 + 4 = 20 \text{ degrees.}$$

Practice Problem 10 Use the temperature expression in Example 10 to find the Celsius temperature assuming that 39 chirps are counted in 25 seconds.

Answers to Practice Problems

1. $\frac{710}{333}$ **2. a.** true **b.** true **c.** false

3. $A \cap B = \{0\}, A \cup B = \{-4, -3, -2, -1, 0, 1, 2, 3, 4\}$

4. a. $(-\infty, \infty)$ **b.** $[2, 5)$ **5. a.** 10 **b.** 1 **c.** 1

6. 9 **7. a.** 5 **b.** 1 **c.** −33 **d.** 1 **8. a.** $\frac{17}{8}$ **b.** $\frac{34}{15}$ **c.** $\frac{3}{2}$ **d.** $\frac{2}{3}$ **9. a.** $\frac{10}{3}$ **b.** $\frac{22}{3}$ **10.** 17°C

SECTION P.1 Exercises

Basic Concepts and Skills

1. Whole numbers are formed by adding the number ____________ to the set of natural numbers.
2. The number −3 is an integer, but it is also a(n) ____________ and a(n) ____________.
3. If $a < b$, then a is to the ____________ of b on the number line.
4. If a real number is not a rational number, it is a(n) ____________ number.
5. **True or False.** If $-x$ is positive, then $-x > 0$.
6. **True or False.** $\frac{-5}{2} < -2\frac{1}{2}$.

In Exercises 7–14, write each of the following rational numbers as a decimal and state whether the decimal is repeating or terminating.

7. $\frac{1}{3}$ **8.** $\frac{2}{3}$

9. $-\frac{4}{5}$ **10.** $-\frac{3}{12}$

11. $\frac{3}{11}$ **12.** $\frac{11}{33}$

13. $\frac{95}{30}$ **14.** $\frac{41}{15}$

In Exercises 15–22, convert each decimal to a quotient of two integers in lowest terms.

15. 3.75 **16.** −2.35 **17.** $-5.\overline{3}$

18. $9.\overline{6}$ **19.** $2.\overline{13}$ **20.** $3.\overline{23}$

21. $4.5\overline{23}$ **22.** $1.4\overline{235}$

In Exercises 23–30, classify each of the following numbers as rational or irrational.

23. −207 **24.** −114

25. $\sqrt{81}$ **26.** $-\sqrt{25}$

27. $\frac{7}{2}$ **28.** $-\frac{15}{12}$

29. $\sqrt{12}$ **30.** $\sqrt{3}$

In Exercises 31–40, use inequality symbols to write the given statements symbolically.

31. 3 is greater than −2.

32. −3 is less than −2.

33. $\frac{1}{2}$ is greater than or equal to $\frac{1}{2}$.

34. x is less than $x + 1$.

35. 5 is less than or equal to $2x$.

36. $x - 1$ is greater than 2.

37. $-x$ is positive.

38. x is negative.

39. $2x + 7$ is less than or equal to 14.

40. $2x + 3$ is not greater than 5.

In Exercises 41–44, fill in the blank with one of the symbols =, <, or > to produce a true statement.

41. $4 \_\_\_\_\_ \frac{24}{6}$

42. $-5 \_\_\_\_\_ -2$

43. $-4 \_\_\_\_\_ 0$

44. $-\frac{9}{2} \_\_\_\_\_ -4\frac{1}{2}$

In Exercises 45–52, find each set if $A = \{-4, -2, 0, 2, 4\}$, $B = \{-3, 0, 1, 2, 3, 4\}$, and $C = \{-4, -3, -2, -1, 0, 2\}$.

45. $A \cup B$

46. $A \cap B$

47. $A \cap C$

48. $B \cup C$

49. $(B \cap C) \cup A$

50. $(A \cup C) \cap B$

51. $(A \cup B) \cap C$

52. $(A \cup B) \cup C$

In Exercises 53–60, find the union and the intersection of the given sets.

53. $I_1 = (-2, 3]; I_2 = [1, 5)$

54. $I_1 = [1, 7]; I_2 = (3, 5)$

55. $I_1 = (-6, 2); I_2 = [2, 10)$

56. $I_1 = (-\infty, -3]; I_2 = (-3, \infty)$

57. $I_1 = (-\infty, 5); I_2 = [2, \infty)$

58. $I_1 = (-2, \infty); I_2 = (0, \infty)$

59. $I_1 = (-\infty, 3) \cup [5, \infty); I_2 = [-1, 7]$

60. $I_1 = (-\infty, 2] \cup (6, \infty); I_2 = [-3, 0]$

In Exercises 61–72, rewrite each expression without absolute value bars.

61. $|-4|$

62. $-|-17|$

63. $\left|\frac{5}{-7}\right|$

64. $\left|\frac{-3}{5}\right|$

65. $|5 - \sqrt{2}|$

66. $|\sqrt{2} - 5|$

67. $|\sqrt{3} - 2|$

68. $|3 - \pi|$

69. $\frac{8}{|-8|}$

70. $\frac{-8}{|8|}$

71. $|5 - |-7||$

72. $||4| - |7||$

In Exercises 73–80, use the absolute value to express the distance between the points with coordinates a and b on the number line. Then determine this distance by evaluating the absolute value expression.

73. $a = 3$ and $b = 8$

74. $a = 2$ and $b = 14$

75. $a = -6$ and $b = 9$

76. $a = -12$ and $b = 3$

77. $a = -20$ and $b = -6$

78. $a = -14$ and $b = -1$

79. $a = \frac{22}{7}$ and $b = -\frac{4}{7}$

80. $a = \frac{16}{5}$ and $b = -\frac{3}{5}$

In Exercises 81–88, graph each of the given intervals on a separate number line and write the inequality notation for each.

81. $(-3, 1]$

82. $[-6, -2)$

83. $[-3, \infty)$

84. $[0, \infty)$

85. $(-\infty, 5]$

86. $(-\infty, -1]$

87. $\left(-\frac{3}{4}, \frac{9}{4}\right)$

88. $\left(-3, -\frac{1}{2}\right)$

In Exercises 89–92, use the distributive property to write each expression without parentheses.

89. $4(x + 1)$

90. $(-3)(2 - x)$

91. $5(x - y + 1)$

92. $2(3x + 5 - y)$

In Exercises 93–96, find the additive inverse and reciprocal of each number.

Number	Additive Inverse	Reciprocal
93. 5		
94. $-\frac{2}{3}$		
95. 0		
96. 1.7		

In Exercises 97–108, name the property of real numbers that justifies the given equality. All variables represent real numbers.

97. $(-7) + 7 = 0$

98. $5 + (-5) = 0$

99. $(x + 2) = 1 \cdot (x + 2)$

100. $3a = 1 \cdot 3a$

101. $7(xy) = (7x)y$

102. $3 \cdot (6x) = (3 \cdot 6)x$

103. $\frac{3}{2}\left(\frac{2}{3}\right) = 1$

104. $2\left(\frac{1}{2}\right) = 1$

105. $(3 + x) + 0 = 3 + x$

106. $x(2 + y) + 0 = x(2 + y)$

107. $(x + 5) + 2y = x + (5 + 2y)$

108. $(3 + x) + 5 = x + (3 + 5)$

In Exercises 109–130, perform the indicated operations.

109. $\frac{3}{5} + \frac{4}{3}$

110. $\frac{7}{10} + \frac{3}{4}$

111. $\frac{6}{5} + \frac{5}{7}$

112. $\frac{9}{2} + \frac{5}{12}$

113. $\frac{5}{6} + \frac{3}{10}$

114. $\frac{8}{15} + \frac{2}{9}$

115. $\frac{5}{8} - \frac{9}{10}$

116. $\frac{7}{8} - \frac{1}{5}$

117. $\frac{5}{9} - \frac{7}{11}$

118. $\frac{5}{8} - \frac{7}{12}$

119. $\frac{2}{5} - \frac{1}{2}$

120. $\frac{1}{4} - \frac{1}{6}$

121. $\frac{3}{4} \cdot \frac{8}{27}$

122. $\frac{9}{7} \cdot \frac{14}{27}$

123. $\frac{\frac{8}{5}}{\frac{16}{15}}$

124. $\frac{\frac{5}{6}}{\frac{15}{6}}$

125. $\frac{\frac{7}{8}}{\frac{21}{16}}$

126. $\frac{\frac{3}{10}}{\frac{7}{15}}$

127. $5 \cdot \frac{3}{10} - \frac{1}{2}$

128. $2 \cdot \frac{7}{2} - \frac{3}{2}$

129. $3 \cdot \frac{2}{15} - \frac{1}{3}$

130. $2 \cdot \frac{5}{3} - \frac{3}{2}$

In Exercises 131–140, evaluate each expression for $x = 3$ and $y = -5$.

131. $2(x + y) - 3y$

132. $-2(x + y) + 5y$

133. $3|x| - 2|y|$

134. $7|x - y|$

135. $\frac{x - 3y}{2} + xy$

136. $\frac{y + 3}{x} - xy$

137. $\frac{2(1 - 2x)}{y} - (-x)y$

138. $\frac{3(2 - x)}{y} - (1 - xy)$

139. $\frac{\frac{14}{x} + \frac{1}{2}}{\frac{-y}{4}}$

140. $\frac{\frac{4}{-y} + \frac{8}{x}}{\frac{y}{2}}$

In Exercises 141–150, correct the error in each formula.

141. $\frac{x}{y} + \frac{x}{3} = \frac{x}{(y + 3)}$

142. $(x + 2)(x + 3) = x + 2x + 3$

143. $5(x + 3) = 5x + 3$

144. $(25x)(4x) = 100x$

145. $x - (3y + 2) = x - 3y + 2$

146. $2x - (4y - 5) = 2x - 4y - 5$

147. $\frac{x + y}{x} = 1 + y$

148. $\frac{x + y}{x + z} = 1 + \frac{y}{z}$

149. $(x + 1)(y + 1) = xy + 1$

150. $\frac{\frac{1}{x}}{\frac{1}{y}} = \frac{1}{xy}$

151. Let P and Q be two points on a number line with coordinates a and b, respectively. Show that the point M on the number line with coordinate $\frac{a + b}{2}$ is the midpoint of the line segment PQ. [*Hint:* Show that $d(P, M) = d(Q, M)$.]

152. Find 100 rational numbers between $-\frac{4}{13}$ and $\frac{7}{13}$.
(*Hint:* $-\frac{4}{13} = -\frac{40}{130}$ and $\frac{7}{13} = \frac{70}{130}$;
$-40 < -31 < -30 < \ldots < 0 < 1 < \ldots < 69$.)

Applying the Concepts

153. Media players. Let $A =$ the set of people who own MP3 players and $B =$ the set of people who own DVD players.
a. Describe the set $A \cup B$.
b. Describe the set $A \cap B$.

154. Standard car features. The table below indicates whether certain features are "standard" for each of three types of cars.

	Navigation System	Automatic Transmission	Leather Seats
2011 Lexus LS 460	yes	yes	yes
2011 Lincoln Town Car	no	yes	yes
2011 Audi A4	no	no	yes

Source: http://autos.yahoo.com

Use the roster method to describe each of the following sets.
a. $A =$ cars in which a navigation system is standard
b. $B =$ cars in which an automatic transmission is standard
c. $C =$ cars in which leather seats are standard
d. $A \cap B$
e. $B \cap C$ **f.** $A \cup B$ **g.** $A \cup C$

155. Blood pressure. A group of college students had systolic blood pressure readings that ranged from 119.5 to 134.5 inclusive. Let x represent the value of the systolic blood pressure readings. Use inequalities to describe this range of values and graph the corresponding interval on a number line.

156. Population projections. Population projections suggest that by 2050, the number of people 60 years and older in the United States will be about 107 million. In 1950, the number of people 60 years and older in the United States was 30 million. Let x represent the number (in millions) of people in the United States who are 60 years and older. Use inequalities to describe this population range from 1950 to 2050 and graph the corresponding interval on a number line.
Source: U.S. Census Bureau

157. Heart rate. For exercise to be most beneficial, the optimum heart rate for a 20-year-old person is 120 beats per minute. Use absolute value notation to write an expression that describes the difference between the heart rate achieved by each of the following 20-year-old people and the ideal exercise heart rate. Then evaluate that expression.
a. Latasha: 124 beats per minute
b. Frances: 137 beats per minute
c. Ignacio: 114 beats per minute

158. Downloading music. To download a 4 MB song with a 56 Kbs modem takes an average of 15 minutes. Use absolute value notation to write an expression that describes the difference between this average time and the actual time it took to download the following songs. Then evaluate that expression.
a. *Irreplaceable* (Beyoncé): 14 minutes
b. *Fallin'* (Alicia Keys): 17.5 minutes
c. *Your Song* (Elton John): 9 minutes

Negative calories. **In Exercises 159 and 160, use the fact that eating 100 grams of broccoli (a negative calorie food) actually results in a net *loss* of 55 calories.**

159. If a cheeseburger has 522.5 calories, how many grams of broccoli would a person have to consume to have a net gain of zero calories?

160. Carmen ate 600 grams of broccoli and now wants to eat just enough French fries so that she has a net calorie intake of zero. If a small order of fries has 165 calories, how many orders does she have to eat?

SECTION **P.2**

Integer Exponents and Scientific Notation

BEFORE STARTING THIS SECTION, REVIEW

1 Properties of opposites (Section P.1, page 11)
2 Variables (Section P.1, page 13)

OBJECTIVES

1 Use integer exponents.
2 Use the rules of exponents.
3 Simplify exponential expressions.
4 Use scientific notation.

Coffee and Candy Consumption in America

In 2012, Americans drank about 110 billion cups of coffee and spent more than \$29 billion on candy and other confectionery products. The American population at that time was about 314 million. Because our day-to-day activities bring us into contact with more manageable quantities, most people find large numbers like those just cited a bit difficult to understand. Using exponents, the method of scientific notation allows us to write large and small quantities in a manner that makes comparing such quantities fairly easy. In turn, these comparisons help us get a better perspective on these quantities. In Example 10 of this section, without relying on a calculator, we see that if we distribute the cost of the candy and other confectionery products used in 2012 evenly among all individuals in the United States, each person will spend over \$92. Distributing the coffee used in 2012 evenly among all individuals in the United States gives each person about 350 cups of coffee.

1 Use integer exponents.

Integer Exponents

The area of a square with side 5 feet is $5 \cdot 5 = 25$ square feet.
The volume of a cube that has sides of 5 feet each is $5 \cdot 5 \cdot 5 = 125$ cubic feet.

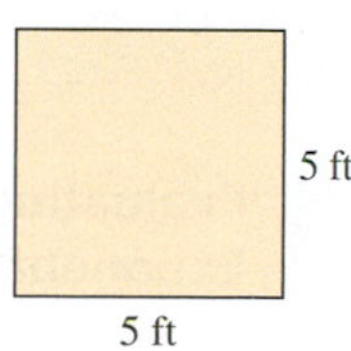

Area = 5·5 square feet

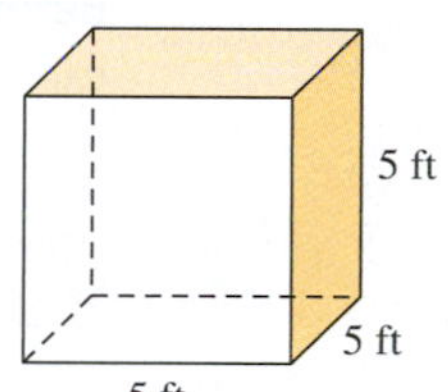

Volume = 5·5·5 cubic feet

A shorter notation for $5 \cdot 5$ is 5^2 and for $5 \cdot 5 \cdot 5$ is 5^3. The number 5 is called the *base* for both 5^2 and 5^3. The number 2 is called the *exponent* in the expression 5^2 and indicates that the base 5 appears as a factor twice.

Positive Integer Exponent

If a is a real number and n is a positive integer, then

$$a^n = \underbrace{a \cdot a \cdot \ldots \cdot a}_{n \text{ factors}}.$$

The number a^n is called the nth power of a and is read "a to the nth power", or "a to the n". The number a is called the **base**; n, the **exponent**. We adopt the convention that $a^1 = a$.

TECHNOLOGY CONNECTION

The notation for 5^3 on a graphing calculator is 5^3. Any expression on a calculator enclosed in parentheses and followed by ^n is raised to the nth power. A common error is to forget parentheses when computing an expression such as $(-3)^2$, with the result being -3^2.

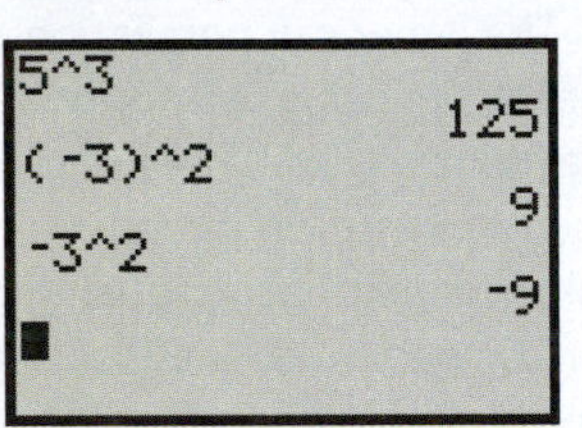

EXAMPLE 1 **Evaluating Expressions That Use Exponents**

Evaluate each expression.

a. 5^3 **b.** $(-3)^2$ **c.** -3^2 **d.** $(-2)^3$

Solution

a. $5^3 = 5 \cdot 5 \cdot 5 = 125$

b. $(-3)^2 = (-3)(-3) = 9$ $(-3)^2$ is the opposite of 3, squared.

c. $-3^2 = -3 \cdot 3 = -9$ Multiplication of $3 \cdot 3$ occurs first. (-3^2 is the opposite of 3^2.)

d. $(-2)^3 = (-2)(-2)(-2) = -8$

Practice Problem 1 Evaluate the following.

a. 2^3 **b.** $(3a)^2$ **c.** $\left(\frac{1}{2}\right)^4$

In Example 1, pay careful attention to the fact (from parts **b** and **c**) that $(-3)^2 \neq -3^2$. In $(-3)^2$, the parentheses indicate that the exponent 2 applies to the base -3, whereas in -3^2, the absence of parentheses indicates that the exponent applies only to the base 3. When n is even, $(-a)^n \neq -a^n$ for $a \neq 0$.

Zero and Negative Integer Exponents

For any nonzero number a and any positive integer n,

$$a^0 = 1 \text{ and } a^{-n} = \frac{1}{a^n}.$$

WARNING

Negative exponents indicate the ***reciprocal*** of a number. Zero cannot be used as a base with a negative exponent because zero does not have a reciprocal. Furthermore, **0^0 is not defined**. We assume throughout this book that the base is not equal to zero if any of the exponents are negative or zero.

EXAMPLE 2 **Evaluating Expressions That Use Zero or Negative Exponents**

Evaluate.

a. $(-5)^{-2}$ **b.** -5^{-2} **c.** 8^0 **d.** $\left(\frac{2}{3}\right)^{-3}$

Solution

a. $(-5)^{-2} = \frac{1}{(-5)^2} = \frac{1}{25}$ The exponent -2 applies to the base -5.

b. $-5^{-2} = -\frac{1}{5^2} = -\frac{1}{25}$ The exponent -2 applies to the base 5.

c. $8^0 = 1$ By definition

d. $\left(\frac{2}{3}\right)^{-3} = \frac{1}{\left(\frac{2}{3}\right)^3} = \frac{1}{\frac{8}{27}} = \frac{27}{8}$ $\left(\frac{2}{3}\right)^3 = \frac{2}{3}\cdot\frac{2}{3}\cdot\frac{2}{3} = \frac{8}{27}$

Practice Problem 2 Evaluate.

a. 2^{-1} **b.** $\left(\frac{4}{5}\right)^0$ **c.** $\left(\frac{3}{2}\right)^{-2}$

In Example 2, we see that parts **a** and **b** give different results. In part **a**, the base is -5 and the exponent is -2, whereas in part **b**, we evaluate the ***opposite*** of the expression with base 5 and exponent -2.

2 Use the rules of exponents.

Rules of Exponents

We now review the rules of exponents.

Let's see what happens when we multiply a^5 by a^3. We have

$$a^5 \cdot a^3 = \underbrace{(a \cdot a \cdot a \cdot a \cdot a)}_{5 \text{ factors}} \cdot \underbrace{(a \cdot a \cdot a)}_{3 \text{ factors}} = \underbrace{a \cdot a \cdot a \cdot a \cdot a \cdot a \cdot a \cdot a}_{5+3\,=\,8 \text{ factors}} = a^8.$$

Similarly, if you multiply m factors of a by n factors of a, you get $m + n$ factors of a.

RECALL

In equation (1), remember that a must be nonzero if the exponent is zero or negative.

PRODUCT RULE OF EXPONENTS

If a is a real number and m and n are integers, then

(1) $$a^m \cdot a^n = a^{m+n}.$$

To multiply exponential expressions with the same base, *keep the base and add exponents.*

EXAMPLE 3 **Using the Product Rule of Exponents**

Simplify. Use the product rule and (if necessary) the definition of a negative exponent or reciprocal to write each answer without negative exponents.

a. $2x^3 \cdot x^5$ **b.** $x^7 \cdot x^{-7}$ **c.** $(-4y^2)(3y^7)$

Solution

a. $2x^3 \cdot x^5 = 2x^{3+5} = 2x^8$ Add exponents: $3 + 5 = 8$.

b. $x^7 \cdot x^{-7} = x^{7+(-7)} = x^0 = 1$ Add exponents: $7+(-7) = 0$; simplify.

c. $(-4y^2)(3y^7) = (-4)3y^2y^7$ Group the factors with variable bases.

$= -12y^{2+7}$ Add exponents.

$= -12y^9$ $2 + 7 = 9$.

TECHNOLOGY CONNECTION

To evaluate an expression such as 5^{3+4} on a calculator, you must enclose $3 + 4$ in parentheses. Otherwise, 4 is added to the result of 5^3.

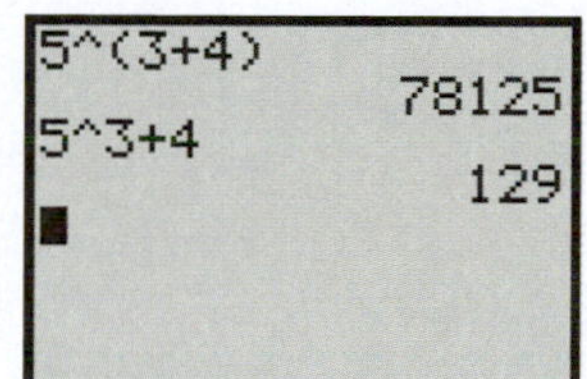

Practice Problem 3 Simplify. Write each answer without using negative exponents.

a. $x^2 \cdot 3x^7$ **b.** $(2^2x^3)(4x^{-3})$

Look what happens when we divide a^5 by a^3. We have

$$\frac{a^5}{a^3} = \frac{a \cdot a \cdot a \cdot a \cdot a}{a \cdot a \cdot a} = \frac{\cancel{a} \cdot \cancel{a} \cdot \cancel{a}}{\cancel{a} \cdot \cancel{a} \cdot \cancel{a}} \cdot \frac{a \cdot a}{1} = a \cdot a = a^2.$$

So $\frac{a^5}{a^3} = a^5 \cdot \frac{1}{a^3} = a^5 \cdot a^{-3} = a^{5-3} = a^2$. You subtract exponents because the three factors in the denominator cancel three of the factors in the numerator.

QUOTIENT RULE FOR EXPONENTS

If a is a nonzero real number and m and n are integers, then

$$\frac{a^m}{a^n} = a^{m-n}.$$

To divide two exponential expressions with the same base, *keep the base and subtract exponents*.

SIDE **NOTE**

Remember that if $a \neq 0$, then $a^0 = 1$; so it is okay for a denominator to contain a nonzero base with a zero exponent. For example, $\frac{3}{5^0} = \frac{3}{1} = 3$.

EXAMPLE 4 Using the Quotient Rule of Exponents

Simplify. Use the quotient rule to write each answer without negative exponents.

a. $\frac{5^{10}}{5^{10}}$ **b.** $\frac{2^{-1}}{2^3}$ **c.** $\frac{x^{-3}}{x^5}$

Solution

a. $\frac{5^{10}}{5^{10}} = 5^{10-10} = 5^0 = 1$

b. $\frac{2^{-1}}{2^3} = 2^{-1-3} = 2^{-4} = \frac{1}{2^4} = \frac{1}{16}$

c. $\frac{x^{-3}}{x^5} = x^{-3-5} = x^{-8} = \frac{1}{x^8}$

Practice Problem 4 Simplify. Write each answer without negative exponents.

a. $\frac{3^4}{3^0}$ **b.** $\frac{5}{5^{-2}}$ **c.** $\frac{2x^3}{3x^{-4}}$

To introduce the next rule of exponents, we consider $(2^3)^4$.

$$\begin{aligned}(2^3)^4 &= 2^3 \cdot 2^3 \cdot 2^3 \cdot 2^3 \qquad a^4 = a \cdot a \cdot a \cdot a;\text{ here } a = 2^3\\ &= (2 \cdot 2 \cdot 2)(2 \cdot 2 \cdot 2)(2 \cdot 2 \cdot 2)(2 \cdot 2 \cdot 2)\\ &= 2^{12}\end{aligned}$$

So $(2^3)^4 = 2^{3 \cdot 4} = 2^{12}$.
This suggests the following rule.

POWER-OF-A-POWER RULE FOR EXPONENTS

If a is a real number and m and n are integers, then

$$(a^m)^n = a^{mn}.$$

To find the power of a power, *keep the base and multiply exponents*.

EXAMPLE 5 Using the Power-of-a-Power Rule of Exponents

Simplify. Write each answer without negative exponents.

a. $(5^2)^0$ **b.** $[(-3)^2]^3$ **c.** $(x^3)^{-1}$ **d.** $(x^{-2})^{-3}$

Solution

a. $(5^2)^0 = 5^{2 \cdot 0} = 5^0 = 1$

b. $[(-3)^2]^3 = (-3)^{2 \cdot 3} = (-3)^6 = 729$ For $(-3)^6$, the base is -3.

c. $(x^3)^{-1} = x^{3(-1)} = x^{-3} = \dfrac{1}{x^3}$

d. $(x^{-2})^{-3} = x^{(-2)(-3)} = x^6$

Practice Problem 5 Simplify. Write each answer without negative exponents.

a. $(7^{-5})^0$ **b.** $(7^0)^{-5}$ **c.** $(x^{-1})^8$ **d.** $(x^{-2})^{-5}$

We now consider the power of a product.

$$(2 \cdot 3)^5 = (2 \cdot 3)(2 \cdot 3)(2 \cdot 3)(2 \cdot 3)(2 \cdot 3) \quad a^5 = a \cdot a \cdot a \cdot a \cdot a;\text{ here } a = 2 \cdot 3.$$
$$= (2 \cdot 2 \cdot 2 \cdot 2 \cdot 2)(3 \cdot 3 \cdot 3 \cdot 3 \cdot 3) \quad \text{Use associative and commutative properties.}$$
$$= 2^5 \cdot 3^5$$

So $(2 \cdot 3)^5 = 2^5 \cdot 3^5$.
This suggests the following rule.

POWER-OF-A-PRODUCT RULE

If a and b are real numbers and n is an integer, then

$$(a \cdot b)^n = a^n \cdot b^n.$$

EXAMPLE 6 Using the Power-of-a-Product Rule

Simplify. Use the power-of-a-product rule to write each expression without negative exponents.

a. $(3x)^2$ **b.** $(-3x)^{-2}$ **c.** $(-3^2)^3$ **d.** $(xy)^{-4}$ **e.** $(x^2y)^3$

Solution

a. $(3x)^2 = 3^2x^2 = 9x^2$

b. $(-3x)^{-2} = \dfrac{1}{(-3x)^2} = \dfrac{1}{(-3)^2x^2} = \dfrac{1}{9x^2}$ Negative exponents denote reciprocals.

c. $(-3^2)^3 = (-1 \cdot 3^2)^3 = (-1)^3(3^2)^3 = (-1)(3^6) = -729$

d. $(xy)^{-4} = \dfrac{1}{(xy)^4} = \dfrac{1}{x^4y^4}$ Negative exponents denote reciprocals.

e. $(x^2y)^3 = (x^2)^3y^3 = x^{2 \cdot 3}y^3 = x^6y^3$ Recall that $(x^2)^3 = x^{2 \cdot 3}$.

Practice Problem 6 Simplify. Write each answer without negative exponents.

a. $\left(\dfrac{1}{2}x\right)^{-1}$ **b.** $(5x^{-1})^2$ **c.** $(xy^2)^3$ **d.** $(x^{-2}y)^{-3}$

To see the last rule, we consider $\left(\frac{3}{2}\right)^5$. We have

$$\begin{aligned}\left(\frac{3}{2}\right)^5 &= \frac{3}{2}\cdot\frac{3}{2}\cdot\frac{3}{2}\cdot\frac{3}{2}\cdot\frac{3}{2}\\ &= \frac{3\cdot3\cdot3\cdot3\cdot3}{2\cdot2\cdot2\cdot2\cdot2}\\ &= \frac{3^5}{2^5}.\end{aligned}$$

More generally, we have the following rules.

POWER-OF-QUOTIENT RULES

If a and b are nonzero real numbers and n is an integer, then

(2) $\left(\frac{a}{b}\right)^n = \frac{a^n}{b^n}$ (3) $\left(\frac{a}{b}\right)^{-n} = \left(\frac{b}{a}\right)^n = \frac{b^n}{a^n}.$

EXAMPLE 7 Using the Power-of-Quotient Rules

Simplify. Use the power-of-quotient rules to write each answer without negative exponents.

a. $\left(\frac{3}{5}\right)^3$ **b.** $\left(\frac{2}{3}\right)^{-2}$

Solution

a. $\left(\frac{3}{5}\right)^3 = \frac{3^3}{5^3} = \frac{27}{125}$ Equation (2)

b. $\left(\frac{2}{3}\right)^{-2} = \left(\frac{3}{2}\right)^2 = \frac{3^2}{2^2} = \frac{9}{4}$ Equation (3) followed by equation (2)

Practice Problem 7 Simplify. Write each answer without negative exponents.

a. $\left(\frac{1}{3}\right)^2$ **b.** $\left(\frac{10}{7}\right)^{-2}$

WARNING

Incorrect	Correct
~~$(3x)^2 = 3x^2$~~	$(3x)^2 = 3^2x^2 = 9x^2$
~~$(x^2)^5 = x^7$~~	$(x^2)^5 = x^{2\cdot5} = x^{10}$
~~$x^3x^5 = x^{15}$~~	$x^3x^5 = x^{3+5} = x^8$

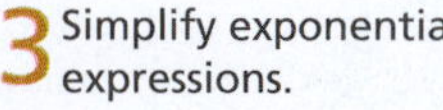

3 Simplify exponential expressions.

Simplifying Exponential Expressions

RULES FOR SIMPLIFYING EXPONENTIAL EXPRESSIONS

An exponential expression is considered **simplified** when

(i) each base appears only once,
(ii) each exponent is a positive number, and
(iii) no power is raised to a power.

EXAMPLE 8 **Simplifying Exponential Expressions**

Simplify the following.

a. $(-4x^2y^3)(7x^3y)$ **b.** $\left(\frac{x^5}{2y^{-3}}\right)^{-3}$

Solution

a. $(-4x^2y^3)(7x^3y) = (-4)(7)x^2x^3y^3y$ Group factors with the same base.

$= -28x^{2+3}y^{3+1}$ Apply the product rule to add exponents; remember that $y = y^1$.

$= -28x^5y^4$

b. $\left(\frac{x^5}{2y^{-3}}\right)^{-3} = \frac{(x^5)^{-3}}{(2y^{-3})^{-3}}$ The exponent -3 is applied to both the numerator and the denominator.

$= \frac{x^{5(-3)}}{2^{-3}(y^{-3})^{-3}}$ Multiply exponents in the numerator; apply the exponent -3 to each factor in the denominator.

$= \frac{x^{-15}}{2^{-3}y^{(-3)(-3)}}$ Power rule for exponents; $5(-3) = -15$.

$= \frac{x^{-15}}{2^{-3}y^9}$ $(-3)(-3) = 9$

$= \frac{x^{-15}x^{15}2^3}{x^{15}2^3 2^{-3}y^9}$ Multiply numerator and denominator by $x^{15}2^3$.

$= \frac{2^3}{x^{15}y^9}$ $x^{-15}x^{15} = x^0 = 1$; $2^3 2^{-3} = 2^0 = 1$

$= \frac{8}{x^{15}y^9}$ $2^3 = 8$

SIDE NOTE

There are many correct ways to simplify exponential expressions. The order in which you apply the rules for exponents is a matter of personal preference.

Practice Problem 8 Simplify each expression.

a. $(2x^4)^{-2}$ **b.** $\frac{x^2(-y)^3}{(xy^2)^3}$

4 Use scientific notation.

Scientific Notation

Scientific measurements and calculations often involve very large or very small positive numbers. For example, 1 gram of oxygen contains approximately

$$37{,}600{,}000{,}000{,}000{,}000{,}000{,}000 \text{ atoms}$$

and the mass of one oxygen atom is approximately

$$0.0000000000000000000000266 \text{ gram}.$$

Such numbers contain so many zeros that they are awkward to work with in calculations. Fortunately, scientific notation provides a better way to write and work with such large and small numbers.

Scientific notation consists of the product of a number less than 10, and greater than or equal to 1, and an integer power of 10. That is, scientific notation of a number has the form

$$c \times 10^n,$$

where c is a real number in decimal notation with $1 \le c < 10$ and n is an integer.

CONVERTING A DECIMAL NUMBER TO SCIENTIFIC NOTATION

1. Count the number, n, of places the decimal point in the given number must be moved to obtain a number c with $1 \le c < 10$.
2. If the decimal point is moved n places to the left, the scientific notation is $c \times 10^n$. If the decimal point is moved n places to the right, the scientific notation is $c \times 10^{-n}$.
3. If the decimal point does not need to be moved, the scientific notation is $c \times 10^0$.

EXAMPLE 9 Converting a Decimal Number to Scientific Notation

Write each decimal number in scientific notation.

a. 421,000 **b.** 10 **c.** 3.621 **d.** 0.000561

Solution

a. Because 421,000 = 421000.0, count five spaces to move the decimal point between the 4 and the 2 and produce the number 4.21.

$$421{,}000 = 4\,2\,1\,0\,0\,0.$$

5 places

Because the decimal point is moved five places to the *left*, the exponent is positive and we write

$$421{,}000 = 4.21 \times 10^5.$$

b. The decimal point for 10 is to the right of the units digit. Count one place to move the decimal point between 1 and 0 and produce the number 1.0.

$$10 = 1\,0\,.\,0$$

1 place

Because the decimal point is moved one place to the *left*, the exponent is positive and we write

$$10 = 1.0 \times 10^1.$$

c. The number 3.621 is already between 1 and 10, so the decimal does not need to be moved. We write

$$3.621 = 3.621 \times 10^0.$$

d. The decimal point in 0.000561 must be moved between the 5 and the 6 to produce the number 5.61. We count four places as follows:

$$0\,.\,0\,0\,0\,5\,6\,1$$

4 places

Because the decimal point is moved four places to the *right*, the exponent is negative and we write

$$0.000561 = 5.61 \times 10^{-4}.$$

Practice Problem 9 Write 732,000 in scientific notation.

TECHNOLOGY CONNECTION

To write numbers in scientific notation on a graphing calculator, you first must change the mode to *scientific*. Then on most graphing calculators, you will see the number displayed as a decimal number, followed by the letter *E*, followed by the exponent for 10. So 421000 is shown as 4.21E5 and 0.0018 is shown as 1.8E-3. Some calculators omit the *E*.

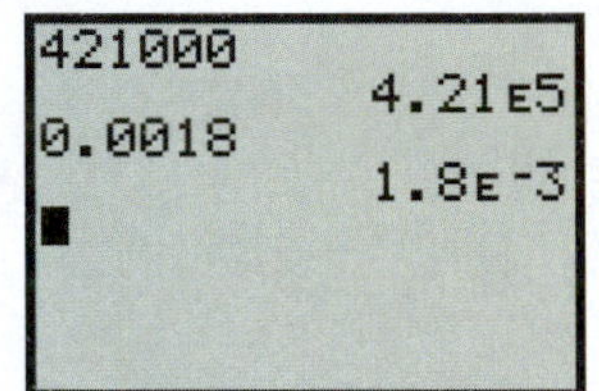

SIDE NOTE

Note that in scientific notation, "small" numbers (positive numbers less than 1) have negative exponents and "large" numbers (numbers greater than 10) have positive exponents.

EXAMPLE 10 Distributing Coffee and Candy in America

At the beginning of this section, we mentioned that in 2012, Americans drank about 110 billion cups of coffee and spent more than \$29 billion on candy and other confectionery products. To see how these products would be evenly distributed among the population, we first convert those numbers to scientific notation.

$$110 \text{ billion is } 110{,}000{,}000{,}000 = 1.1 \times 10^{11}$$
$$29 \text{ billion is } 29{,}000{,}000{,}000 = 2.9 \times 10^{10}.$$

The U.S. population in 2012 was about 314 million, and 314 million is $314{,}000{,}000 = 3.14 \times 10^8$.

To distribute the coffee evenly among the population, we divide:

$$\frac{1.1 \times 10^{11}}{3.14 \times 10^8} = \frac{1.10}{3.14} \times \frac{10^{11}}{10^8} = \frac{110}{314} \times 10^{11-8} \approx 0.35 \times 10^3 = 350 \text{ cups per person}$$

To distribute the cost of the candy evenly among the population, we divide:

$$\frac{2.9 \times 10^{10}}{3.14 \times 10^8} = \frac{2.90}{3.14} \times \frac{10^{10}}{10^8} = \frac{290}{314} \times 10^{10-8} \approx 0.92 \times 10^2 = 92,$$

or about \$92 per person

Practice Problem 10 If the amount spent on candy consumption remains unchanged when the U.S. population reaches 325 million, what is the cost per person when cost is distributed evenly throughout the population?

MAIN FACTS ABOUT EXPONENTS

$$a^n = \underbrace{a \cdot a \ldots a}_{n \text{ factors}} \text{ for } a \neq 0, \quad a^0 = 1 \quad \text{and} \quad a^{-n} = \frac{1}{a^n}$$

Product Rule: $a^m \cdot a^n = a^{m+n}$

Quotient Rule: $\dfrac{a^m}{a^n} = a^{m-n}$

Power-of-Power Rule: $(a^m)^n = a^{mn}$

Power-of-a-Product Rule: $(ab)^m = a^m b^m$

Power-of-Quotient Rules: $\left(\dfrac{a}{b}\right)^n = \dfrac{a^n}{b^n}$; $\left(\dfrac{a}{b}\right)^{-n} = \left(\dfrac{b}{a}\right)^n = \dfrac{b^n}{a^n}$, $b \neq 0$

Answers to Practice Problems

1. a. 8 **b.** $9a^2$ **c.** $\frac{1}{16}$ **2. a.** $\frac{1}{2}$ **b.** 1 **c.** $\frac{4}{9}$ **3. a.** $3x^9$ **b.** 16 **4. a.** 81 **b.** 125 **c.** $\frac{2x^7}{3}$ **5. a.** 1 **b.** 1 **c.** $\frac{1}{x^8}$ **d.** x^{10} **6. a.** $\frac{2}{x}$ **b.** $\frac{25}{x^2}$ **c.** x^3y^6 **d.** $\frac{x^6}{y^3}$ **7. a.** $\frac{1}{9}$ **b.** $\frac{49}{100}$ **8. a.** $\frac{1}{4x^8}$ **b.** $-\frac{1}{xy^3}$ **9.** 7.32×10^5 **10.** \$89 per person

SECTION P.2 Exercises

Basic Concepts and Skills

1. In the expression 7^2, the number 2 is called the ______________.
2. In the expression -3^7, the base is ______________.
3. The number $\frac{1}{4^{-2}}$ simplifies to the positive integer ______________.
4. The power-of-a-product rule allows us to rewrite $(5a)^3$ as ______________.
5. **True or False.** $(-11)^{10} = -11^{10}$.
6. **True or False.** The exponential expression $(x^2)^3$ is considered simplified.

In Exercises 7–16, name the exponent and the base.

7. 17^3
8. 10^2
9. 9^0
10. $(-2)^0$
11. $(-5)^5$
12. $(-99)^2$
13. -10^3
14. $-(-3)^7$
15. a^2
16. $(-b)^3$

In Exercises 17–46, evaluate each expression.

17. 6^1
18. 3^4
19. 7^0
20. $(-8)^0$
21. $(2^3)^2$
22. $(3^2)^3$
23. $(3^2)^{-2}$
24. $(7^2)^{-1}$
25. $(5^{-2})^3$
26. $(5^{-1})^3$
27. $(4^{-3}) \cdot (4^5)$
28. $(7^{-2}) \cdot (7^3)$
29. $3^0 + 10^0$
30. $5^0 - 9^0$
31. $3^{-2} + \left(\frac{1}{3}\right)^2$
32. $5^{-2} + \left(\frac{1}{5}\right)^2$
33. -2^{-3}
34. -3^{-2}
35. $(-3)^{-2}$
36. $(-2)^{-3}$
37. $\frac{2^{11}}{2^{10}}$
38. $\frac{3^6}{3^8}$
39. $\frac{(5^3)^4}{5^{12}}$
40. $\frac{(9^5)^2}{9^8}$
41. $\frac{2^5 \cdot 3^{-2}}{2^4 \cdot 3^{-3}}$
42. $\frac{4^{-2} \cdot 5^3}{4^{-3} \cdot 5}$
43. $\frac{-5^{-2}}{2^{-1}}$
44. $\frac{-7^{-2}}{3^{-1}}$
45. $\left(\frac{11}{7}\right)^{-2}$
46. $\left(\frac{13}{5}\right)^{-2}$

In Exercises 47–86, simplify each expression. Write your answers without negative exponents. Whenever an exponent is negative or zero, assume that the base is not zero.

47. x^4y^0
48. $x^{-1}y^0$
49. $x^{-1}y$
50. x^2y^{-2}
51. $x^{-1}y^{-2}$
52. $x^{-3}y^{-2}$
53. $(x^{-3})^4$
54. $(x^{-5})^2$
55. $(x^{-11})^{-3}$
56. $(x^{-4})^{-12}$
57. $-3(xy)^5$
58. $-8(xy)^6$
59. $4(xy^{-1})^2$
60. $6(x^{-1}y)^3$
61. $3(x^{-1}y)^{-5}$
62. $-5(xy^{-1})^{-6}$
63. $\frac{(x^3)^2}{(x^2)^5}$
64. $\frac{x^2}{(x^3)^4}$
65. $\left(\frac{2xy}{x^2}\right)^3$
66. $\left(\frac{5xy}{x^3}\right)^4$
67. $\left(\frac{-3x^2y}{x}\right)^5$
68. $\left(\frac{-2xy^2}{y}\right)^3$
69. $\left(\frac{-3x}{5}\right)^{-2}$
70. $\left(\frac{-5y}{3}\right)^{-4}$
71. $\left(\frac{4x^{-2}}{xy^5}\right)^3$
72. $\left(\frac{3x^2y}{y^3}\right)^5$
73. $\frac{x^3y^{-3}}{x^{-2}y}$
74. $\frac{x^2y^{-2}}{x^{-1}y^2}$
75. $\frac{27x^{-3}y^5}{9x^{-4}y^7}$
76. $\frac{15x^5y^{-2}}{3x^7y^{-3}}$
77. $\frac{1}{x^3}(x^2)^3x^{-4}$
78. $(8a^3b)^{-4}\left(\frac{2a}{b}\right)^{12}$
79. $(-xy^2)^3(-2x^2y^2)^{-4}$
80. $\left[\frac{(-x^2y)^3y^{-4}}{(xy)^5}\right]^{-2}$
81. $(4x^3y^2z)^2(x^3y^2z)^{-7}$
82. $\frac{(2xyz)^2}{(x^3y)^2(xz)^{-1}}$
83. $\frac{5a^{-2}bc^2}{a^4b^{-3}c^2}$
84. $\frac{(-3)^2a^5(bc)^2}{(-2)^3a^2b^3c^4}$
85. $\left(\frac{xy^{-3}z^{-2}}{x^2y^{-4}z^3}\right)^{-3}$
86. $\left(\frac{xy^{-2}z^{-1}}{x^{-5}yz^{-8}}\right)^{-1}$

In Exercises 87–94, write each number in scientific notation.

87. 125
88. 247
89. 850,000
90. 205,000
91. 0.007
92. 0.0019
93. 0.00000275
94. 0.0000038

Applying the Concepts

In Exercises 95 and 96, use the fact that when you uniformly stretch or shrink a three-dimensional object in every direction by a factor of *a*, the volume of the resulting figure is scaled by a factor of a^3. For example, when you uniformly scale (stretch or shrink) a three-dimensional object by a factor of 2, the volume of the resulting figure is scaled by a factor of 2^3, or 8.

95. A display in the shape of a baseball bat has a volume of 135 cubic feet. Find the volume of the display that results by uniformly scaling the figure by a factor of 2.

96. A display in the shape of a football has a volume of 675 cubic inches. Find the volume of the display that results by uniformly scaling the figure by a factor of $\frac{1}{3}$.

97. The area A of a square with side of length x is given by $A = x^2$. Use this relationship to

a. verify that doubling the length of the side of a square floor increases the area of the floor by a factor of 2^2.

b. verify that tripling the length of the side of a square floor increases the area of the floor by a factor of 3^2.

98. The area A of a circle with diameter d is given by $A = \pi\left(\frac{d}{2}\right)^2$. Use this relationship to

a. verify that doubling the length of the diameter of a circular skating rink increases the area of the rink by a factor of 2^2.

b. verify that tripling the length of the diameter of a circular skating rink increases the area of the rink by a factor of 3^2.

99. The cross section of one type of weight-bearing rod (the surface you would get if you sliced the rod perpendicular to its axis) used in the Olympics is a square, as shown in the accompanying figure. The rod can handle a stress of 25,000 pounds per square inch (psi). The relationship between the stress S the rod can handle, the load F the rod can carry, and the width w of the cross section is given by the equation $Sw^2 = F$. Assuming that $w = 0.25$ inch, find the load the rod can support.

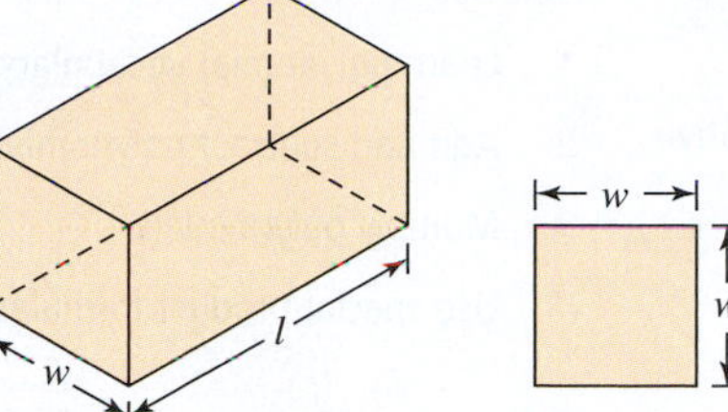

100. The cross section of one type of weight-bearing rod (the surface you would get if you sliced the rod perpendicular to its axis) used in the Olympics is a circle, as shown in the figure. The rod can handle a stress of 10,000 pounds per square inch (psi). The relationship between the stress S the rod can handle, the load F the rod can carry, and the diameter d of the cross section is given by the equation $S\pi(d/2)^2 = F$. Assuming that $d = 1.5$ inches, find the load the rod can support. Use $\pi \approx 3.14$ in your calculation.

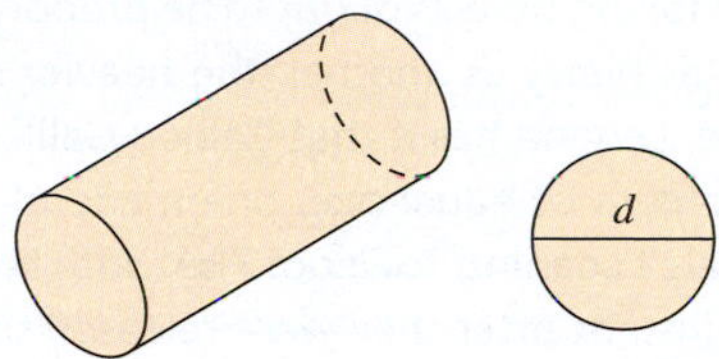

101. Complete the following table.

Celestial Body	Equatorial Diameter (km)	Scientific Notation
Earth	12,700	
Moon		3.48×10^3 (km)
Sun	1,390,000	
Jupiter		1.34×10^5 (km)
Mercury	4800	

In Exercises 102–107, express the number in each statement in scientific notation.

102. One gram of oxygen contains about 37,600,000,000,000,000,000,000 atoms.

103. One gram of hydrogen contains about 602,000,000,000,000,000,000 atoms.

104. One oxygen atom weighs about 0.0000000000000000000266 kilogram.

105. One hydrogen atom weighs about 0.00000000000000000000167 kilogram.

106. The distance from Earth to the moon is about 380,000,000 meters.

107. The mass of Earth is about 5,980,000,000,000,000,000,000,000 kilograms.

SECTION **P.3**

Polynomials

BEFORE STARTING THIS SECTION, REVIEW

1. Associative, commutative, and distributive properties (Section P.1, pages 10–11)
2. Properties of opposites (Section P.1, page 11)
3. Definition of subtraction (Section P.1, page 11)
4. Rules for exponents (Section P.2, page 25)

OBJECTIVES

1. Learn polynomial vocabulary.
2. Add and subtract polynomials.
3. Multiply polynomials.
4. Use special-product formulas.

Leaning Tower of Pisa

Galileo and Free-Falling Objects

At one time, it was widely believed that heavy objects should fall faster than lighter objects, or, more exactly, that the speed of falling objects ought to be proportional to their weights. So if one object is four times as heavy as another, the heavier object should fall four times as fast as the lighter one. Legend has it that Galileo Galilei (1564–1642) disproved this theory by dropping two balls of equal size, one made of lead and the other of balsa wood, from the top of Italy's Leaning Tower of Pisa. Both balls are said to have reached the ground at the same instant after they were released simultaneously from the top of the tower.

In fact, such experiments "work" properly only in a vacuum, where objects of different mass do fall at the same rate. In the absence of a vacuum, air resistance may greatly affect the rate at which different objects fall. Whether Galileo's experiment was actually performed, astronaut David R. Scott successfully performed a version of Galileo's experiment (using a feather and a hammer) on the surface of the moon on August 2, 1971. It turns out that on Earth (ignoring air resistance), regardless of the weight of the object we hold at rest and then drop, the expression $16t^2$ gives the distance in feet the object falls in t seconds. If we throw the object down with an initial velocity of v_0 feet per second, the distance the object travels in t seconds is given by the expression

$$16t^2 + v_0t.$$

In Example 1, we evaluate such expressions.

1 Learn polynomial vocabulary.

Polynomial Vocabulary

The height (in feet) of a golf ball above a driving range t seconds after being driven from the tee (at 70 feet per second) is given by the polynomial $70t - t^2$. After two seconds $(t = 2)$, the ball is $70(2) - (2)^2 = 136$ feet above the ground. *Polynomials* such as $70t - t^2$ appear frequently in applications.

We begin by reviewing the basic vocabulary of polynomials. A **monomial** is the simplest polynomial; it contains one term. In the variable x it has the form ax^k, where a is a constant and k is either a positive integer or zero. The constant a is called the **coefficient** of the monomial. For $a \neq 0$, the integer k is called the **degree** of the monomial. If $a = 0$, the monomial is 0 and has **no degree**. Any variable may be used in place of x to form a monomial in that variable.

The following are examples of monomials.

$2x^5$	The coefficient is 2, and the degree is 5.
$-3x^2$	The coefficient is -3, and the degree is 2.
-7	$-7 = -7(1) = -7x^0$. The coefficient is -7, and the degree is 0.
$8x$	The coefficient is 8, and the degree is 1 $(x = x^1)$.
x^{12}	The coefficient is 1 $(x^{12} = 1 \cdot x^{12})$, and the degree is 12.
$-x^4$	The coefficient is -1 $(-x^4 = (-1) \cdot x^4)$, and the degree is 4.

The expression $5x^{-3}$ is not a monomial because the exponent of x is negative.

Two monomials in the same variable with the same degree can be combined into one monomial using the distributive property. For example, $-3x^5$ and $9x^5$ can be added: $-3x^5 + 9x^5 = (-3 + 9)x^5 = 6x^5$. Or $-3x^5$ and $9x^5$ can be subtracted: $-3x^5 - 9x^5 = (-3 - 9)x^5 = -12x^5$. Two monomials in the same variable with the same degree are called **like terms**.

POLYNOMIALS IN ONE VARIABLE

A **polynomial** in x is any sum of monomials in x. By combining like terms, we can write any polynomial in the form

$$a_nx^n + a_{n-1}x^{n-1} + \cdots + a_2x^2 + a_1x + a_0,$$

where n is either a positive integer or zero and $a_n, a_{n-1}, \ldots a_1, a_0$ are constants, called the **coefficients** of the polynomial. If $a_n \neq 0$, then n, the largest exponent on x, is called the **degree** and a_n is called the **leading coefficient** of the polynomial. The monomials $a_nx^n, a_{n-1}x^{n-1}, \ldots, a_2x^2, a_1x$, and a_0 are the **terms** of the polynomial. The monomial a_nx^n is the **leading term** of the polynomial, and a_0 is the **constant term**.

The symbols $a_0, a_1, a_2, \ldots a_n$ in the general notation for a polynomial

$$a_nx^n + a_{n-1}x^{n-1} + \cdots + a_2x^2 + a_1x + a_0$$

are just constants; the numbers to the lower right of a are called subscripts. The notation a_2 is read "a sub 2", a_1 is read "a sub 1", and a_0 is read "a sub 0". This type of notation is used when a large or indefinite number of constants are required. The terms of the polynomial $5x^2 + 3x + 1$ are $5x^2$, $3x$, and 1; the coefficients are 5, 3, and 1; and the degree is 2.

Note that the terms in the general form for a polynomial are connected by the addition symbol +. How do we handle $5x^2 - 3x + 1$? Because subtraction is defined in terms of addition, we rewrite

$$5x^2 - 3x + 1 = 5x^2 + (-3)x + 1$$

and see that the terms of $5x^2 - 3x + 1$ are $5x^2$, $-3x$, and 1 and that the coefficients are 5, -3, and 1. (The degree is 2.) Once we recognize this fact, we no longer rewrite the polynomial with the + sign separating terms; we just remember that the "sign" is part of both the term and the coefficient. Polynomials and their properties will be discussed in more detail in Chapter 3.

Polynomials can be classified according to the number of terms they have. Polynomials with one term are called **monomials**, polynomials with two *unlike* terms are called **binomials**, and polynomials with three *unlike* terms are called **trinomials**. Polynomials with more than three unlike terms do not have special names.

By agreement, the only polynomial that has *no* degree is the **zero polynomial**, which results when all of the coefficients are 0. It is easiest to find the degree of a polynomial when it is written in **descending order**—that is, when the exponents decrease from left to right. In this case, the degree is the exponent of the leftmost term. A polynomial written in descending order is said to be in **standard form**.

EXAMPLE 1 Examining Free-Falling Objects

In our introductory discussion about Galileo and free-falling objects, we mentioned two well-known polynomials.

1. $16t^2$ gives the distance in feet a free-falling object falls in t seconds.
2. Letting $v_0 = 10$, $16t^2 + 10t$ gives the distance an object falls in t seconds when it is thrown down with an initial velocity of 10 feet per second. Use these polynomials to
   a. Find how far a wallet dropped from the 86th-floor observatory of the Empire State Building will fall in 5 seconds.
   b. Find how far a quarter thrown down with an initial velocity of 10 feet per second from a hot air balloon will travel after 5 seconds.

Solution

a. The value of $16t^2$ for $t = 5$ is $16(5)^2 = 400$. The wallet falls 400 feet.
b. The value of $16t^2 + 10t$ for $t = 5$ is $16(5)^2 + 10(5) = 450$. The quarter travels 450 feet on its downward path after five seconds.

Practice Problem 1 If $16t^2 + 15t$ gives the distance an object falls in t seconds when it is thrown down at an initial velocity of 15 feet per second, find how far a quarter thrown down with an initial velocity of 15 feet per second from a hot air balloon has traveled after 7 seconds.

2 Add and subtract polynomials.

Adding and Subtracting Polynomials

Like monomials, polynomials are added and subtracted by combining like terms. By convention, we write polynomials in standard form.

EXAMPLE 2 Adding Polynomials

Find the sum of the polynomials

$$-4x^3 + 5x^2 + 7x - 2 \quad \text{and} \quad 6x^3 - 2x^2 - 8x - 5.$$

Horizontal Method: Group like terms and then combine them.

$$\begin{aligned} &(-4x^3 + 5x^2 + 7x - 2) + (6x^3 - 2x^2 - 8x - 5) \\ &= (-4x^3 + 6x^3) + (5x^2 - 2x^2) + (7x - 8x) + (-2 - 5) && \text{Group like terms.} \\ &= 2x^3 + 3x^2 - x - 7 && \text{Combine like terms.} \end{aligned}$$

A second method for adding polynomials is to arrange them in columns so that like terms appear in the same column.

Column Method:

$$\begin{array}{rrrrr} & -4x^3 & + 5x^2 & + 7x & - 2 \\ (+) & 6x^3 & - 2x^2 & - 8x & - 5 \\ \hline & 2x^3 & + 3x^2 & - x & - 7 \end{array} \text{ (answer)}$$

Practice Problem 2 Find the sum of

$$7x^3 + 2x^2 - 5 \quad \text{and} \quad -2x^3 + 3x^2 + 2x + 1.$$

EXAMPLE 3 Subtracting Polynomials

Find the difference of the polynomials

$$9x^4 - 6x^3 + 4x^2 - 1 \quad \text{and} \quad 2x^4 - 11x^3 + 5x + 3.$$

Horizontal Method: First, using the definition of subtraction, change the sign of each term in the second polynomial. Then add the resulting polynomials by grouping like terms and combining them.

$$(9x^4 - 6x^3 + 4x^2 - 1) - (2x^4 - 11x^3 + 5x + 3)$$
$$= (9x^4 - 6x^3 + 4x^2 - 1) + (-2x^4 + 11x^3 - 5x - 3)$$
$$= (9x^4 - 2x^4) + (-6x^3 + 11x^3) + 4x^2 - 5x + (-1 - 3) \quad \text{Group like terms.}$$
$$= 7x^4 + 5x^3 + 4x^2 - 5x - 4 \quad \text{Combine like terms.}$$

The second method for finding the difference of polynomials requires arranging them in columns so that like terms appear in the same column.

Column Method: To find the difference, we again change the sign of each term in the second polynomial and then add.

$$\begin{array}{r} 9x^4 - 6x^3 + 4x^2 \qquad - 1 \\ (+) -2x^4 + 11x^3 \qquad - 5x - 3 \\ \hline 7x^4 + 5x^3 + 4x^2 - 5x - 4 \end{array} \quad \text{(answer)} \quad \text{Change signs and then add.}$$

Practice Problem 3 Find the difference of

$$3x^4 - 5x^3 + 2x^2 + 7 \quad \text{and} \quad -2x^4 + 3x^2 + x - 5.$$

3 Multiply polynomials.

Multiplying Polynomials

Previously, we multiplied two monomials, such as $-3x^2$ and $5x^3$, by using the product rule for exponents: $a^m \cdot a^n = a^{m+n}$. We multiply a monomial and a polynomial by combining the product rule and the distributive property.

EXAMPLE 4 Multiplying a Monomial and a Polynomial

Multiply $4x^3$ and $3x^2 - 5x + 7$.

$$4x^3(3x^2 - 5x + 7)$$
$$= (4x^3)(3x^2) + (4x^3)(-5x) + (4x^3)(7) \quad \text{Distributive property}$$
$$= (4 \cdot 3)x^{3+2} + 4(-5)x^{3+1} + (4 \cdot 7)x^3 \quad \text{Multiply monomials.}$$
$$= 12x^5 - 20x^4 + 28x^3 \quad \text{Simplify.}$$

Practice Problem 4 Multiply $-2x^3$ and $4x^2 + 2x - 5$.

We can use the distributive property repeatedly to multiply any two polynomials. As with addition, there are two methods.

EXAMPLE 5 Multiplying Polynomials

Multiply $4x^2 + 3x$ and $x^2 + 2x - 3$.

Horizontal Method: We treat the second polynomial as a single term and use the distributive property, $(a + b)c = ac + bc$, to multiply it by each of the two terms in the first polynomial.

$$(4x^2 + 3x)(x^2 + 2x - 3)$$
$$= 4x^2(x^2 + 2x - 3) + 3x(x^2 + 2x - 3) \quad \text{Distributive property}$$
$$= 4x^2(x^2) + 4x^2(2x) + 4x^2(-3) + 3x(x^2) + 3x(2x) + 3x(-3) \quad \text{Distributive property}$$
$$= 4x^4 + 8x^3 - 12x^2 + 3x^3 + 6x^2 - 9x \quad \text{Product rule}$$
$$= 4x^4 + (8x^3 + 3x^3) + (-12x^2 + 6x^2) - 9x \quad \text{Group like terms.}$$
$$= 4x^4 + 11x^3 - 6x^2 - 9x \quad \text{Combine like terms.}$$

The column method resembles the multiplication of two positive integers written in column form.

Column Method:

$$\begin{array}{rrrrl}
 & & x^2 & + 2x - 3 & \\
(*) & & 4x^2 & + 3x & \\
\hline
 & 3x^3 & + 6x^2 & - 9x & \text{Multiply by } 3x: (3x)(x^2 + 2x - 3). \\
4x^4 & + 8x^3 & - 12x^2 & & \text{Multiply by } 4x^2: (4x^2)(x^2 + 2x - 3). \\
\hline
4x^4 & + 11x^3 & - 6x^2 & - 9x \text{ (answer)} & \text{Combine like terms.}
\end{array}$$

Practice Problem 5 Multiply $5x^2 + 2x$ and $-2x^2 + x - 7$.

The basic multiplication rule for polynomials can be stated as follows.

MULTIPLYING POLYNOMIALS

To multiply two polynomials, multiply each term of one polynomial by every term of the other polynomial and combine like terms.

4 Use special-product formulas.

Special Products

Particular polynomial products called *special products* occur frequently enough to deserve special attention. We introduce a very useful method, called FOIL, for multiplying two binomials. Notice that

$$\begin{aligned}
(x + 2)(x - 7) &= x(x - 7) + 2(x - 7) && \text{Distributive property} \\
&= x^2 - 7x + 2x - 14 && \text{Distributive property} \\
&= x^2 - 5x - 14 && \text{Combine like terms.}
\end{aligned}$$

Now consider the relationship between the terms of the factors in $(x + 2)(x - 7)$ and the second line, $x^2 - 7x + 2x - 14$, in the computation of the product.

Product of First terms

$$(x + 2)(x - 7) = x^2 - 7x + 2x - 14 \qquad \text{First terms: } x \cdot x = x^2$$

Product of Outside terms

$$(x + 2)(x - 7) = x^2 - 7x + 2x - 14 \qquad \text{Outside terms: } x \cdot (-7) = -7x$$

Product of Inside terms

$$(x + 2)(x - 7) = x^2 - 7x + 2x - 14 \qquad \text{Inside terms: } 2 \cdot x = 2x$$

Product of Last terms

$$(x + 2)(x - 7) = x^2 - 7x + 2x - 14 \qquad \text{Last terms: } 2 \cdot (-7) = -14$$

FOIL METHOD FOR $(A + B)(C + D)$

$$(A + B)(C + D) = \overset{\text{F}}{A \cdot C} + \overset{\text{O}}{A \cdot D} + \overset{\text{I}}{B \cdot C} + \overset{\text{L}}{B \cdot D}$$

EXAMPLE 6 **Using the FOIL Method**

Use the FOIL method to find the following products.

a. $(2x + 3)(x - 1) = \overset{F}{(2x)(x)} + \overset{O}{(2x)(-1)} + \overset{I}{(3)(x)} + \overset{L}{(3)(-1)}$
$= 2x^2 - 2x + 3x - 3$
$= 2x^2 + x - 3$ Combine like terms.

b. $(3x - 5)(4x - 6) = \overset{F}{(3x)(4x)} + \overset{O}{(3x)(-6)} + \overset{I}{(-5)(4x)} + \overset{L}{(-5)(-6)}$
$= 12x^2 - 18x - 20x + 30$
$= 12x^2 - 38x + 30$ Combine like terms.

c. $(x + a)(x + b) = \overset{F}{x \cdot x} + \overset{O}{x \cdot b} + \overset{I}{a \cdot x} + \overset{L}{a \cdot b}$
$= x^2 + bx + ax + a \cdot b$ $xb = bx$ (commutative property)
$= x^2 + (b + a)x + ab$ Combine like terms.

Practice Problem 6 Use FOIL to find each product.

a. $(4x - 1)(x + 7)$ **b.** $(3x - 2)(2x - 5)$

Squaring a Binomial Sum or Difference

We can find a general formula for the square of any binomial sum $(A + B)^2$ by using FOIL.

$(A + B)^2 = (A + B)(A + B) = \overset{F}{A \cdot A} + \overset{O}{A \cdot B} + \overset{I}{B \cdot A} + \overset{L}{B \cdot B}$
$= A^2 + AB + AB + B^2$ $AB = BA$
$= A^2 + 2AB + B^2$ $AB + AB = 2AB$

SQUARING A BINOMIAL SUM

$$(A + B)^2 = A^2 + 2AB + B^2$$

A formula such as $(A + B)^2 = A^2 + 2AB + B^2$ can be used two ways.

EXAMPLE 7 **Finding the Square of a Binomial Sum**

Find $(2x + 3)^2$.

Solution

Pattern Method: In $(2x + 3)^2$, the first term is $2x$ and the second term is 3. Rewrite the formula as

Square of sum	=	Square of first term	+ 2	Product of first and second terms	+	Square of second term
$(2x + 3)^2$	=	$(2x)^2$	+	$2 \cdot (2x) \cdot 3$	+	3^2

$= 2^2x^2 + 2 \cdot 2 \cdot 3 \cdot x + 3^2$ Group numerical factors.
$= 4x^2 + 12x + 9$ Simplify.

Substitution Method: In the formula $(A + B)^2 = A^2 + 2AB + B^2$, substitute $2x = A$ and $3 = B$. Then

$(2x + 3)^2 = (2x)^2 + 2 \cdot (2x) \cdot 3 + 3^2$
$= 2^2x^2 + 2 \cdot 2 \cdot 3 \cdot x + 3^2$ Group numerical factors.
$= 4x^2 + 12x + 9$ Simplify.

Practice Problem 7 Find $(3x + 2)^2$.

The definition of subtraction allows us to write any difference $A - B$ as the sum $A + (-B)$. Because $A - B = A + (-B)$, we can use the formula for the square of a binomial sum to find a formula for the square of a binomial difference $(A - B)^2$.

$$\begin{aligned}(A - B)^2 &= [A + (-B)]^2 && \text{First term} = A\text{; second term} = -B\\ &= A^2 + 2A(-B) + (-B)^2 \\ &= A^2 - 2AB + B^2 && 2A(-B) = -2AB;\ (-B)^2 = B^2\end{aligned}$$

WARNING

Incorrect	Correct
~~$(x + y)^2 = x + y^2$~~	$(x + y)^2 = x^2 + 2xy + y^2$
~~$(x - y)^2 = x^2 + y^2$~~	$(x - y)^2 = x^2 - 2xy + y^2$
~~$(x + 4)(x + 5) = x + 4x + 5$~~	$(x + 4)(x + 5) = x^2 + 9x + 20$

The Product of the Sum and Difference of Terms

PRODUCTS GIVING A DIFFERENCE OF SQUARES

$$(A + B)(A - B) = A^2 - B^2$$

We leave it for you to use FOIL to verify the formula for finding the product of the sum and difference of terms.

EXAMPLE 8 Finding the Product of the Sum and Difference of Terms

Find the product $(3x + 4)(3x - 4)$.

Solution

We use the substitution method; substitute $3x = A$ and $4 = B$ in the formula $(A + B)(A - B) = A^2 - B^2$. Then

$$\begin{aligned}(3x + 4)(3x - 4) &= (3x)^2 - 4^2 \\ &= 9x^2 - 16 && (3x)^2 = 3^2x^2 = 9x^2\end{aligned}$$

Practice Problem 8 Find the product $(1 - 2x)(1 + 2x)$.

The special-products formulas, such as the formulas for squaring a binomial sum or difference, are used often and should be memorized. However, you should be able to derive them from FOIL, if you forget them. We list several of these formulas next.

SPECIAL-PRODUCT FORMULAS

A and B represent any algebraic expressions.

Formula	**Example**
Product giving a difference of squares	
$(A + B)(A - B) = A^2 - B^2$	$(5x + 2)(5x - 2) = (5x)^2 - 2^2 = 25x^2 - 4$
Squaring a binomial sum or difference	
$(A + B)^2 = A^2 + 2AB + B^2$	$(3x + 2)^2 = (3x)^2 + 2(3x)(2) + 2^2 = 9x^2 + 12x + 4$
$(A - B)^2 = A^2 - 2AB + B^2$	$(2x - 5)^2 = (2x)^2 - 2(2x)(5) + 5^2 = 4x^2 - 20x + 25$

Cubing a binomial sum or difference

$(A + B)^3 = A^3 + 3A^2B + 3AB^2 + B^3$ $\qquad$ $(x + 5)^3 = x^3 + 3x^2(5) + 3x(5)^2 + 5^3 = x^3 + 15x^2 + 75x + 125$

$(A - B)^3 = A^3 - 3A^2B + 3AB^2 - B^3$ $\qquad$ $(x - 4)^3 = x^3 - 3x^2(4) + 3x(4)^2 - 4^3 = x^3 - 12x^2 + 48x - 64$

Products giving a sum or difference of cubes

$(A + B)(A^2 - AB + B^2) = A^3 + B^3$ $\qquad$ $(x + 2)(x^2 - 2x + 4) = x^3 + 2^3 = x^3 + 8$

$(A - B)(A^2 + AB + B^2) = A^3 - B^3$ $\qquad$ $(x - 3)(x^2 + 3x + 9) = x^3 - 3^3 = x^3 - 27$

To this point, we have investigated polynomials only in a single variable. We next extend our previous vocabulary to discuss polynomials in more than one variable. Any product of a constant and two or more variables, each raised to whole-number powers, is a **monomial** in those variables. The constant is called the **coefficient** of the monomial, and the **degree** of the monomial is the sum of all the exponents appearing on its variables. For example, the monomial $5x^3y^4$ is of degree $3 + 4 = 7$ with coefficient 5.

EXAMPLE 9 Multiplying Polynomials in Two Variables

Multiply. **a.** $(7a + 5b)(7a - 5b)$ **b.** $(2x + 3y)^2$

Solution

a. $(7a + 5b)(7a - 5b) = (7a)^2 - (5b)^2$ $\qquad$ $(A + B)(A - B) = A^2 - B^2$

$= 49a^2 - 25b^2$ $\qquad$ Simplify.

b. $(2x + 3y)^2 = (2x)^2 + 2(2x)(3y) + (3y)^2$ $\qquad$ $(A + B)^2 = A^2 + 2AB + B^2$

$= 2^2x^2 + 2 \cdot 2 \cdot 3xy + 3^2y^2$ $\qquad$ Power-of-a-product rule

$= 4x^2 + 12xy + 9y^2$ $\qquad$ Simplify.

Practice Problem 9 Multiply $x^2(y + x)$.

Answers to Practice Problems

1. 889 ft **2.** $5x^3 + 5x^2 + 2x - 4$ **3.** $5x^4 - 5x^3 - x^2 - x + 12$ **4.** $-8x^5 - 4x^4 + 10x^3$ **5.** $-10x^4 + x^3 - 33x^2 - 14x$ **6. a.** $4x^2 + 27x - 7$ **b.** $6x^2 - 19x + 10$ **7.** $9x^2 + 12x + 4$ **8.** $1 - 4x^2$ **9.** $x^2y + x^3$

SECTION P.3 Exercises

Basic Concepts and Skills

1. The polynomial $-3x^7 + 2x^2 - 9x + 4$ has leading coefficient ______________ and degree ______________.
2. When a polynomial is written so that the exponents in each term decrease from left to right, it is said to be in ______________ form.
3. When a polynomial in x of degree 3 is added to a polynomial in x of degree 4, the resulting polynomial has degree ______________.
4. When a polynomial in x of degree 3 is multiplied by a polynomial in x of degree 4, the resulting polynomial has degree ______________.
5. **True or False.** When FOIL is used, the terms 7 and x are called the inside terms of the product $(4x + 7)(x - 2)$.
6. **True or False.** There are values of A and B for which $(A + B)^2 = A^2 + B^2$.

In Exercises 7–14, determine whether the given expression is a polynomial. If it is, write it in standard form.

7. $1 + x^2 + 2x$

8. $x - \frac{1}{x}$

9. $x^{-2} + 3x + 5$

10. $3x^4 + x^7 + 3x^5 - 2x + 1$

11. $x^2 - 2|x| + 4$

12. $x^3 + 5\sqrt{x} + 1$

13. $5x^3 - |\pi - 7|x + 11$

14. $2x^2 - \sqrt{3}x + 4$

In Exercises 15–18, find the degree and list the terms of the polynomial.

15. $7x + 3$

16. $-3x^2 + 7$

17. $x^2 - x^4 + 2x - 9$

18. $x + 2x^3 + 9x^7 - 21$

In Exercises 19–28, perform the indicated operations. Write the resulting polynomial in standard form.

19. $(x^3 + 2x^2 - 5x + 3) + (-x^3 + 2x - 4)$

20. $(x^3 - 3x + 1) + (x^3 - x^2 + x - 3)$

21. $(2x^3 - x^2 + x - 5) - (x^3 - 4x + 3)$

22. $(-x^3 + 2x - 4) - (x^3 + 3x^2 - 7x + 2)$

23. $(-2x^4 + 3x^2 - 7x) - (8x^4 + 6x^3 - 9x^2 - 17)$

24. $3(x^2 - 2x + 2) + 2(5x^2 - x + 4)$

25. $-2(3x^2 + x + 1) + 6(-3x^2 - 2x - 2)$

26. $2(5x^2 - x + 3) - 4(3x^2 + 7x + 1)$

27. $(3y^3 - 4y + 2) + (2y + 1) - (y^3 - y^2 + 4)$

28. $(5y^2 + 3y - 1) - (y^2 - 2y + 3) + (2y^2 + y + 5)$

In Exercises 29–62, perform the indicated operations.

29. $6x(2x + 3)$

30. $7x(3x - 4)$

31. $(x + 1)(x^2 + 2x + 2)$

32. $(x - 5)(2x^2 - 3x + 1)$

33. $(3x - 2)(x^2 - x + 1)$

34. $(2x + 1)(x^2 - 3x + 4)$

35. $(x + 1)(x + 2)$

36. $(x + 2)(x + 3)$

37. $(3x + 2)(3x + 1)$

38. $(x + 3)(2x + 5)$

39. $(-4x + 5)(x + 3)$

40. $(-2x + 1)(x - 5)$

41. $(3x - 2)(2x - 1)$

42. $(x - 1)(5x - 3)$

43. $(2x - 3a)(2x + 5a)$

44. $(5x - 2a)(x + 5a)$

45. $(x + 2)^2 - x^2$

46. $(x - 3)^2 - x^2$

47. $(x + 3)^3 - x^3$

48. $(x - 2)^3 - x^3$

49. $(4x + 1)^2$

50. $(3x + 2)^2$

51. $(3x + 1)^3$

52. $(2x + 3)^3$

53. $(5 - 2x)(5 + 2x)$

54. $(3 - 4x)(3 + 4x)$

55. $\left(x + \frac{3}{4}\right)^2$

56. $\left(x + \frac{2}{5}\right)^2$

57. $(2x - 3)(x^2 - 3x + 5)$

58. $(x - 2)(x^2 - 4x - 3)$

59. $(1 + y)(1 - y + y^2)$

60. $(y + 4)(y^2 - 4y + 16)$

61. $(x - 6)(x^2 + 6x + 36)$

62. $(x - 1)(x^2 + x + 1)$

In Exercises 63–72, perform the indicated operations.

63. $(x + 2y)(3x + 5y)$

64. $(2x + y)(7x + 2y)$

65. $(2x - y)(3x + 7y)$

66. $(x - 3y)(2x + 5y)$

67. $(x - y)^2(x + y)^2$

68. $(2x + y)^2(2x - y)^2$

69. $(x + y)(x - 2y)^2$

70. $(x - y)(x + 2y)^2$

71. $(x - 2y)^3(x + 2y)$

72. $(2x + y)^3(2x - y)$

In Exercises 73–78, use $a^2 + b^2 = (a + b)^2 - 2ab = (a - b)^2 + 2ab$.

73. If $x + y = 4$ and $xy = 3$, find the value of $x^2 + y^2$.

74. If $x - y = 3$ and $xy = 2$, find the value of $x^2 + y^2$.

75. If $x + \frac{1}{x} = 3$, find the value of

a. $x^2 + \frac{1}{x^2}$.

b. $x^4 + \frac{1}{x^4}$.

76. If $x - \frac{1}{x} = 2$, find the value of

a. $x^2 + \frac{1}{x^2}$.

b. $x^4 + \frac{1}{x^4}$.

77. If $3x + 2y = 12$ and $xy = 6$, find the value of $9x^2 + 4y^2$.

78. If $3x - 7y = 10$ and $xy = -1$, find the value of $9x^2 + 49y^2$.

Applying the Concepts

79. Ticket prices. Theater ticket prices (in dollars) x years after 2006 are described by the polynomial $-0.025x^2 + 0.44x + 4.28$. Find the value of this polynomial for $x = 6$ and state the result as the ticket price for a specific year.

80. Box-office grosses. Theater box-office grosses (in billions of dollars) x years after 2008 are described by the polynomial $0.035x^2 + 0.15x + 5.17$. Find the value of this polynomial for 2012 and state the result as box-office gross for a specific year.

81. Paper shredding. A business that shreds paper products finds that it costs $0.1x^2 + x + 50$ dollars to serve x customers. What does it cost to serve 40 customers?

82. Gift baskets. A company will produce x (for $x \geq 10$) gift baskets of wine, meat, cheese, and crackers at a cost of $(x - 10)^2 + 5x$ dollars per basket. What is the per basket cost for 15 baskets?

In Exercises 83 and 84, use the fact (from Example 1) that an object thrown down with an initial velocity of v_0 feet per second will travel $16t^2 + v_0t$ feet in t seconds.

83. Free fall. A sandwich is thrown from a helicopter with an initial downward velocity of 20 feet per second. How far has the sandwich fallen after five seconds?

84. Free fall. A ring is thrown off the Empire State Building with an initial downward velocity of 10 feet per second. How far has the ring fallen after two seconds?

85. Cruise ship revenue. A cruise ship decides to reduce ticket prices to its theater by x dollars from the current price of \$22.50.

a. Write a polynomial that gives the new price after the x-dollar deduction.

b. The revenue is the number of tickets sold times the price of each ticket. If 30 tickets were sold when the price was \$22.50 and 10 additional tickets are sold for each dollar the price is reduced (all tickets are sold at the same price), write a polynomial that gives the revenue in terms of x when the original price is reduced by x dollars.

86. Used-car rentals. A dealer who rents used cars wants to increase the monthly rent from the current \$250 in n increases of \$10.

a. Write a polynomial that gives the new price after n increases of \$10.

b. The revenue is the number of cars rented times the monthly rent for each car. If 50 cars can be rented at \$250 per month and 2 fewer cars can be rented for each \$10 rent increase, write a polynomial that gives the monthly revenue in terms of n.

SECTION P.4

Factoring Polynomials

BEFORE STARTING THIS SECTION, REVIEW

1. Exponents (Section P.2)
2. Polynomials (Section P.3)
3. Value of algebraic expressions (Section P.1, page 13)

OBJECTIVES

1. Identify and factor out the greatest common monomial factor.
2. Factor trinomials with a leading coefficient of 1.
3. Factor perfect-square trinomials.
4. Factor the difference of squares.
5. Factor the difference and sum of cubes.
6. Factor by grouping.
7. Factor trinomials.

◆ Profit From Research and Development

The finance department of a major drug company projects that an investment of x million dollars in research and development for a specific type of vaccine will return a profit (or loss) of $(0.012x^3 - 32.928)$ million dollars.

The company would have a "startup" cost of \$12 million and is prepared to invest up to \$24 million at a rate of \$1 million per year. You could probably stare at this polynomial for a long time without getting a clear picture of what kind of return to expect from different investments. In Example 7, we use the methods of this section to rewrite this profit–loss polynomial, making it easy to explain the result of investing various amounts.

1 Identify and factor out the greatest common monomial factor.

The Greatest Common Monomial Factor

The process of factoring polynomials that we study in this section "reverses" the process of multiplying polynomials that we studied in the previous section. To factor *any sum of terms* means to write it as a *product of factors*. Consider the product

$$(x + 3)(x - 3) = x^2 - 9.$$

The polynomials $(x + 3)$ and $(x - 3)$ are called *factors* of the polynomial $x^2 - 9$. When factoring, we start with one polynomial and write it as a product of other polynomials. The polynomials in the product are the **factors** of the given polynomial.

The first thing to look for when factoring a polynomial is a factor that is common to every term. This **common factor** can be "factored out" by the distributive property, $ab + ac = a(b + c)$.

Factoring Out a Monomial

In this section, we are concerned only with polynomials having integer coefficients. This restriction is called **factoring over the integers**.

Polynomial	Common Factor	Factored Form
$9 + 18y$	9	$9(1 + 2y)$
$2y^2 + 6y$	$2y$	$2y(y + 3)$
$7x^4 + 3x^3 + x^2$	x^2	$x^2(7x^2 + 3x + 1)$
$5x + 3$	No common factor	$5x + 3$

You can verify that any factorization is correct by multiplying the factors. We factored these polynomials by finding the *greatest common monomial factor* of their terms.

FINDING THE GREATEST COMMON MONOMIAL FACTOR

The term ax^n is the **greatest common monomial factor (GCF)** of a polynomial in x (with integer coefficients) if

1. a is the *greatest* integer that divides each of the polynomial coefficients.
2. n is the *least* exponent on x found in every term of the polynomial.

EXAMPLE 1 Factoring by Using the GCF

Factor. **a.** $16x^3 + 24x^2$ **b.** $5x^4 + 20x^2 + 25x$

Solution

a. The greatest integer that divides both 16 and 24 is 8. The exponents on x are 3 and 2, so 2 is the least exponent. So the **GCF** of $16x^3 + 24x^2$ is $8x^2$. We write

$$16x^3 + 24x^2 = 8x^2(2x) + 8x^2(3).$$ Write each term as the product of the GCF and another factor.

$$= 8x^2(2x + 3)$$ Use the distributive property to factor out the GCF.

We can check the results of factoring by multiplying.

Check: $8x^2(2x + 3) = 8x^2(2x) + 8x^2(3) = 16x^3 + 24x^2.$

b. The greatest integer that divides 5, 20, and 25 is 5. The exponents on x are 4, 2, and 1 $(x = x^1)$, so 1 is the least exponent. So the **GCF** of $5x^4 + 20x^2 + 25x$ is $5x$. We write

$$5x^4 + 20x^2 + 25x = 5x(x^3) + 5x(4x) + 5x(5) = 5x(x^3 + 4x + 5).$$

You should check this result by multiplying.

Practice Problem 1 Factor. **a.** $6x^5 + 14x^3$ **b.** $7x^5 + 21x^4 + 35x^2$

Polynomials that cannot be factored as a product of two polynomials (excluding the constant polynomials 1 and -1) are said to be **irreducible**. A polynomial is said to be **factored completely** when it is written as a product consisting of only irreducible factors. By agreement, the greatest integer common to all of the terms in the polynomial is not factored. We also allow repeated irreducible factors to be written as a power of a single factor.

2 Factor trinomials with a leading coefficent of 1.

Factoring Trinomials of the Form $x^2 + bx + c$

Recall that $(x + a)(x + b) = x^2 + (a + b)x + ab.$

We reverse the sides of this equation to get the form used for factoring

$$x^2 + (a + b)x + ab = (x + a)(x + b).$$

SIDE **NOTE**

To factor a trinomial in standard form with a leading coefficient of 1, we need two integers a and b whose product is the constant term and whose sum is the coefficient of the middle term.

EXAMPLE 2 Factoring $x^2 + bx + c$ by Using the Factors of c

Factor. **a.** $x^2 + 8x + 15$ **b.** $x^2 - 6x - 16$ **c.** $x^2 + x + 2$

Solution

a. We want integers a and b such that $ab = 15$ and $a + b = 8$.

Factors of 15	1, 15	−1, −15	3, 5	−3, −5
Sum of factors	16	−16	8	−8

Because the factors 3 and 5 of 15 in the third column have a sum of 8,

$$x^2 + 8x + 15 = (x + 3)(x + 5).$$

You should check this answer by multiplying.

b. We must find two integers a and b with $ab = -16$ and $a + b = -6$.

Factors of − 16	−1, 16	−2, 8	−4, 4	1, −16	2, −8	4, −4
Sum of factors	15	6	0	−15	−6	0

Because the factors 2 and −8 of −16 in the fifth column give a sum of −6,

$$x^2 - 6x - 16 = (x + 2)[x + (-8)] = (x + 2)(x - 8).$$

You should check this answer by multiplying.

c. We must find two integers a and b with $ab = 2$ and $a + b = 1$.

Factors of 2	1, 2	−1, −2
Sum of factors	3	−3

The coefficient of the middle term, 1, does not appear as the sum of any of the factors of the constant term, 2. Therefore, $x^2 + x + 2$ is irreducible.

Practice Problem 2 Factor. **a.** $x^2 + 6x + 8$ **b.** $x^2 - 3x - 10$ 

Factoring Formulas

We investigate how other types of polynomials can be factored by reversing the product formulas on pages 34 and 35.

FACTORING FORMULAS

A and B represent any algebraic expression.

$A^2 - B^2 = (A + B)(A - B)$	Difference of squares
$A^2 + 2AB + B^2 = (A + B)^2$	Perfect square, positive middle term
$A^2 - 2AB + B^2 = (A - B)^2$	Perfect square, negative middle term
$A^3 - B^3 = (A - B)(A^2 + AB + B^2)$	Difference of cubes
$A^3 + B^3 = (A + B)(A^2 - AB + B^2)$	Sum of cubes

3 Factor perfect-square trinomials.

Perfect-Square Trinomials

To factor a perfect square Trinomial, we use the special products:

$$A^2 + 2AB + B^2 = (A + B)^2 \quad \text{and} \quad A^2 - 2AB + B^2 = (A - B)^2$$

EXAMPLE 3 Factoring a Perfect-Square Trinomial

Factor. **a.** $x^2 + 10x + 25$ **b.** $16x^2 - 8x + 1$

Solution

a. The first term, x^2, and the third term, $25 = 5^2$, are perfect squares. Further, the middle term is twice the product of the terms being squared, that is, $10x = 2(5x)$.

$$x^2 + 10x + 25 = x^2 + 2 \cdot 5 \cdot x + 5^2 = (x + 5)^2$$

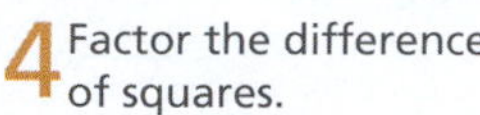

You should check your answer to any factoring problem by multiplying the factors to verify that the result is the original expression.

b. The first term, $16x^2 = (4x)^2$, and the third term, $1 = 1^2$, are perfect squares. Further, the middle term is the *negative* of twice the product of terms being squared, that is, $-8x = -2(4x)(1)$.

$$16x^2 - 8x + 1 = (4x)^2 - 2(4x)(1) + 1^2 = (4x - 1)^2$$

Practice Problem 3 Factor. **a.** $x^2 + 4x + 4$ **b.** $9x^2 - 6x + 1$

4 Factor the difference of squares.

Difference of Squares

To factor the difference of squares, we use the reverse of the special product $A^2 - B^2 = (A + B)(A - B)$.

EXAMPLE 4 Factoring the Difference of Squares

Factor. **a.** $x^2 - 4$ **b.** $25x^2 - 49$

Solution

a. We write $x^2 - 4$ as the difference of squares.

$$x^2 - 4 = x^2 - 2^2 = (x + 2)(x - 2)$$

b. We write $25x^2 - 49$ as the difference of squares.

$$25x^2 - 49 = (5x)^2 - 7^2 = (5x + 7)(5x - 7)$$

Practice Problem 4 Factor. **a.** $x^2 - 16$ **b.** $4x^2 - 25$

What about the *sum* of two squares? Let's try to factor $x^2 + 2^2 = x^2 + 4$. Notice that $x^2 + 4 = x^2 + 0 \cdot x + 4$ so that the middle term is $0 = 0 \cdot x$. We need two factors, a and b, of 4 whose sum is zero $(a + b = 0)$.

Factors of 4	1, 4	−1, −4	2, 2	−2, −2
Sum of factors	5	−5	4	−4

Because no sum of factors is 0, $x^2 + 4$ is irreducible. A similar result holds regardless of what integer replaces 2 in the process. Consequently, $x^2 + a^2$ is irreducible for any integer a.

SOME IRREDUCIBLE POLYNOMIALS

If a and c are integers *having no common factors,* then

$$ax + c \text{ is irreducible} \quad \text{and} \quad x^2 + a^2 \text{ is irreducible.}$$

EXAMPLE 5 Using the Difference of Squares Twice

Factor. $x^4 - 16$.

Solution

We write $x^4 - 16$ as the difference of squares.

$$x^4 - 16 = (x^2)^2 - 4^2 = (x^2 + 4)(x^2 - 4)$$

From Example 4, part **a**, we know that $x^2 - 4 = (x + 2)(x - 2)$. Consequently,

$$\begin{aligned} x^4 - 16 &= (x^2 + 4)(x^2 - 4) \\ &= (x^2 + 4)(x + 2)(x - 2) \quad \text{Replace } x^2 - 4 \text{ with } (x + 2)(x - 2). \end{aligned}$$

Because $x^2 + 4$, $x + 2$, and $x - 2$ are irreducible, $x^4 - 16$ is completely factored.

Practice Problem 5 Factor $x^4 - 81$.

5 Factor the difference and sum of cubes.

Difference and Sum of Cubes

We can also use the reverse of the special-product formulas to factor the difference and sum of cubes.

$$A^3 - B^3 = (A - B)(A^2 + AB + B^2) \quad \text{Difference of cubes}$$

(same sign: the sign between A^3 and B^3 and the sign in $(A - B)$; different sign: the sign before AB)

$$A^3 + B^3 = (A + B)(A^2 - AB + B^2) \quad \text{Sum of cubes}$$

(same sign: the sign between A^3 and B^3 and the sign in $(A + B)$; different sign: the sign before AB)

EXAMPLE 6 Factoring the Difference and Sum of Cubes

Factor. **a.** $x^3 - 64$ **b.** $8x^3 + 125$

Solution

a. We write $x^3 - 64$ as the difference of cubes.

$$x^3 - 64 = x^3 - 4^3 = (x - 4)(x^2 + 4x + 16)$$

b. We write $8x^3 + 125$ as the sum of cubes.

$$8x^3 + 125 = (2x)^3 + 5^3 = (2x + 5)(4x^2 - 10x + 25)$$

Practice Problem 6 Factor. **a.** $x^3 - 125$ **b.** $27x^3 + 8$

EXAMPLE 7 Determining Profit from Research and Development

In the beginning of this section, we introduced the polynomial $0.012x^3 - 32.928$ that gave the profit (or loss) a drug company would have after an investment of x million dollars. If the company had an initial investment of \$12 million and invests an additional \$1 million each year, determine when this investment returns a profit. What will the profit (or loss) be in six years?

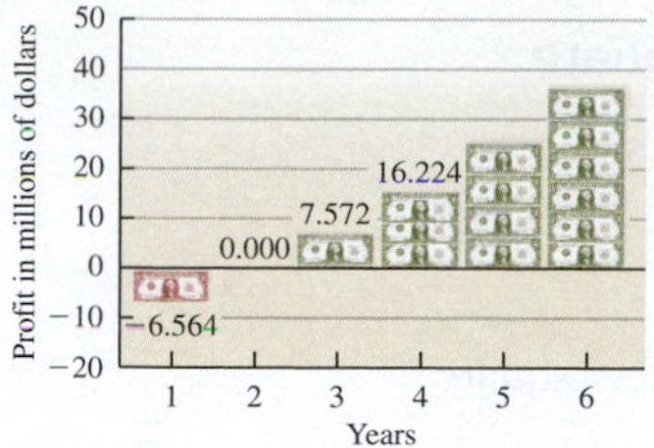

Solution

First, notice that 0.012 is a common factor of both terms of the polynomial because $32.928 = (0.012)(2744)$. We now factor $0.012x^3 - 32.928$ as follows:

$$\begin{aligned} 0.012x^3 - 32.928 &= 0.012x^3 - (0.012)(2744) \\ &= 0.012(x^3 - 2744) && \text{Factor out } 0.012. \\ &= 0.012(x^3 - 14^3) && 2744 = 14^3 \\ &= 0.012(x - 14)(x^2 + 14x + 196) && \text{Difference of cubes} \end{aligned}$$

This product has three factors, and the factor $x^2 + 14x + 196$ is positive for positive values of x because each term is positive. The sign of the factor $x - 14$ then determines the sign of the entire polynomial. The factor $x - 14$ is negative if x is less than 14, zero if x equals 14, and positive if x is greater than 14. Consequently, the company receives a profit when x is greater than 14. Because the initial investment is \$12 million, with an additional \$1 million invested each year, it will be two years before the company breaks even on this venture. Every year thereafter it will profit from the investment. In six years, the company will have invested the initial \$12 million plus an additional \$6 million (\$1 million per year). So the total investment is \$18 million. To find the profit (or loss) with \$18 million invested, we let $x = 18$ in the profit–loss polynomial: $0.012(18)^3 - 32.928 = \$37.056$ million.

The company will have made a profit of over \$37 million in six years.

Practice Problem 7 In Example 7, what will the profit be in four years?

6 Factor by grouping.

Factoring by Grouping

For polynomials having four terms and a GCF of 1, when none of the preceding factoring techniques apply, we can sometimes group the terms in such a way that each group has a common factor. This technique is called **factoring by grouping**.

EXAMPLE 8 **Factoring by Grouping**

Factor. **a.** $x^3 + 2x^2 + 3x + 6$ **b.** $6x^3 - 3x^2 - 4x + 2$ **c.** $x^2 + 4x + 4 - y^2$

Solution

a.
$$\begin{aligned} x^3 + 2x^2 + 3x + 6 &= (x^3 + 2x^2) + (3x + 6) && \text{Group terms so that each group can be factored.} \\ &= x^2(x + 2) + 3(x + 2) && \text{Factor the grouped binomials.} \\ &= (x + 2)(x^2 + 3) && \text{Factor out the common binomial factor } x + 2. \end{aligned}$$

b.
$$\begin{aligned} 6x^3 - 3x^2 - 4x + 2 &= (6x^3 - 3x^2) + (-4x + 2) && \text{Group terms so that each group can be factored.} \\ &= 3x^2(2x - 1) + (-2)(2x - 1) && \text{Factor the grouped binomials.} \\ &= (2x - 1)(3x^2 - 2) && \text{Factor out the common binomial factor } 2x - 1. \end{aligned}$$

c.
$$\begin{aligned} x^2 + 4x + 4 - y^2 &= (x^2 + 4x + 4) - y^2 && \text{Group terms so that each group can be factored.} \\ &= (x + 2)^2 - y^2 && \text{Factor the trinomial.} \\ &= (x + 2 - y)(x + 2 + y) && \text{Difference of squares.} \end{aligned}$$

Practice Problem 8 Factor.

a. $x^3 + 3x^2 + x + 3$ **b.** $28x^3 - 20x^2 - 7x + 5$ **c.** $x^2 - y^2 - 2y - 1$ 

EXAMPLE 9 **Factoring by Making a Perfect Square**

Factor: $x^4 + 3x^2 + 4$

Solution

Notice that a change in the middle term would give the perfect square $(x^2 + 2)^2 = x^4 + 4x^2 + 4$.

$$
\begin{aligned}
& x^4 + 3x^2 + 4 \\
&= x^4 + 4x^2 - x^2 + 4 && 3x^2 = 4x^2 - x^2 \\
&= x^4 + 4x^2 + 4 - x^2 && \text{Rewrite.} \\
&= (x^2 + 2)^2 - x^2 && x^4 + 4x^2 + 4 = (x^2 + 2)^2 \\
&= (x^2 + 2 + x)(x^2 + 2 - x) && \text{Difference of two squares} \\
&= (x^2 + x + 2)(x^2 - x + 2) && \text{Rewrite in standard form.}
\end{aligned}
$$

Practice Problem 9 Factor: $x^4 + 5x^2 + 9$

7 Factor trinomials.

Factoring Trinomials of the Form $Ax^2 + Bx + C$

To factor the trinomial $Ax^2 + Bx + C$ as $(ax + b)(cx + d)$, we use FOIL and combine like terms to get $Ax^2 + Bx + C = acx^2 + (ad + bc)x + bd$. Notice that $AC = acbd = (ad)(bc)$ and $B = ad + bc$. We can factor $Ax^2 + Bx + C$ if we can find integer factors of the product AC whose sum is B. We illustrate a procedure for doing this in the next example.

EXAMPLE 10 **Factoring Using FOIL and Grouping**

Factor. **a.** $6x^2 + 17x + 7$ **b.** $20x^2 - 7x - 6$

Solution

a. $Ax^2 + Bx + C = 6x^2 + 17x + 7$. We have $A \cdot C = 6 \cdot 7 = 42$. We are looking for two factors of 42 whose sum is 17. We have $42 = 3 \cdot 14$ and $3 + 14 = 17$. So

$$
\begin{aligned}
6x^2 + 17x + 7 &= 6x^2 + 3x + 14x + 7 && 17x = 3x + 14x \\
&= (6x^2 + 3x) + (14x + 7) && \text{Group terms.} \\
&= 3x(2x + 1) + 7(2x + 1) && \text{Factor out the common factor.} \\
&= (3x + 7)(2x + 1) && \text{Distributive property}
\end{aligned}
$$

b. $20x^2 - 7x - 6$. We have $(20)(-6) = -120$. The two factors of -120 whose sum is -7 are -15 and 8. So

$$
\begin{aligned}
20x^2 - 7x - 6 &= 20x^2 - 15x + 8x - 6 && -7x = -15x + 8x \\
&= (20x^2 - 15x) + (8x - 6) && \text{Group terms.} \\
&= 5x(4x - 3) + 2(4x - 3) && \text{Factor out the common factor.} \\
&= (5x + 2)(4x - 3) && \text{Distributive property}
\end{aligned}
$$

Practice Problem 10 Factor. **a.** $5x^2 + 11x + 2$ **b.** $4x^2 + x - 3$

Answers to Practice Problems

1. a. $2x^3(3x^2 + 7)$ **b.** $7x^2(x^3 + 3x^2 + 5)$
2. a. $(x + 4)(x + 2)$ **b.** $(x - 5)(x + 2)$
3. a. $(x + 2)^2$ **b.** $(3x - 1)^2$ **4. a.** $(x + 4)(x - 4)$
b. $(2x - 5)(2x + 5)$ **5.** $(x^2 + 9)(x - 3)(x + 3)$
6. a. $(x - 5)(x^2 + 5x + 25)$ **b.** $(3x + 2)(9x^2 - 6x + 4)$
7. \$16.224 million **8. a.** $(x + 3)(x^2 + 1)$
b. $(7x - 5)(2x + 1)(2x - 1)$ **c.** $(x + y + 1)(x - y - 1)$
9. $(x^2 + x + 3)(x^2 - x + 3)$ **10. a.** $(5x + 1)(x + 2)$
b. $(4x - 3)(x + 1)$

SECTION P.4 Exercises

Basic Concepts and Skills

1. The polynomials $x + 2$ and $x - 2$ are called ______________ of the polynomial $x^2 - 4$.
2. The polynomial $3y$ is the ______________ factor of the polynomial $3y^2 + 6y$.
3. The GCF of the polynomial $10x^3 + 30x^2$ is ______________.
4. A polynomial that cannot be factored as a product of two polynomials (excluding the constant polynomials ± 1) is said to be ______________.
5. **True or False.** When factoring a polynomial whose leading coefficient is positive, we need to consider only its positive factors.
6. **True or False.** The polynomial $x^2 + 7$ is irreducible.

In Exercises 7–18, factor each polynomial by removing any common monomial factor.

7. $8x - 24$
8. $5x + 25$
9. $-6x^2 + 12x$
10. $-3x^2 + 21$
11. $7x^2 + 14x^3$
12. $9x^3 - 18x^4$
13. $x^4 + 2x^3 + x^2$
14. $x^4 - 5x^3 + 7x^2$
15. $3x^3 - x^2$
16. $2x^3 + 2x^2$
17. $8ax^3 + 4ax^2$
18. $ax^4 - 2ax^2 + ax$

In Exercises 19–26, factor by grouping.

19. $x^3 + 3x^2 + x + 3$
20. $x^3 + 5x^2 + x + 5$
21. $x^3 - 5x^2 + x - 5$
22. $x^3 - 7x^2 + x - 7$
23. $6x^3 + 4x^2 + 3x + 2$
24. $3x^3 + 6x^2 + x + 2$
25. $12x^7 + 4x^5 + 3x^4 + x^2$
26. $3x^7 + 3x^5 + x^4 + x^2$

In Exercises 27–42, factor each trinomial or state that it is irreducible.

27. $x^2 + 7x + 12$
28. $x^2 + 8x + 15$
29. $x^2 - 6x + 8$
30. $x^2 - 9x + 14$
31. $x^2 - 3x - 4$
32. $x^2 - 5x - 6$
33. $x^2 - 4x + 13$
34. $x^2 - 2x + 5$
35. $2x^2 + x - 36$
36. $2x^2 + 3x - 27$
37. $6x^2 + 17x + 12$
38. $8x^2 - 10x - 3$
39. $3x^2 - 11x - 4$
40. $5x^2 + 7x + 2$
41. $6x^2 - 3x + 4$
42. $4x^2 - 4x + 3$

In Exercises 43–50, factor each polynomial as a perfect square.

43. $x^2 + 6x + 9$
44. $x^2 + 8x + 16$
45. $9x^2 + 6x + 1$
46. $36x^2 + 12x + 1$
47. $25x^2 - 20x + 4$
48. $64x^2 + 32x + 4$
49. $49x^2 + 42x + 9$
50. $9x^2 + 24x + 16$

In Exercises 51–60, factor each polynomial using the difference-of-squares pattern.

51. $x^2 - 64$
52. $x^2 - 121$
53. $4x^2 - 1$
54. $9x^2 - 1$
55. $16x^2 - 9$
56. $25x^2 - 49$
57. $x^4 - 1$
58. $x^4 - 81$
59. $20x^4 - 5$
60. $12x^4 - 75$

In Exercises 61–70, factor each polynomial using the sum-or-difference-of-cubes pattern.

61. $x^3 + 64$
62. $x^3 + 125$
63. $x^3 - 27$
64. $x^3 - 216$
65. $8 - x^3$
66. $27 - x^3$
67. $8x^3 - 27$
68. $8x^3 - 125$
69. $40x^3 + 5$
70. $7x^3 + 56$

In Exercises 71–78, factor by making a perfect square.

71. $x^4 + x^2 + 25$
72. $x^4 + x^2 + 1$
73. $x^4 + 15x^2 + 64$
74. $x^4 - 7x^2 + 9$
75. $x^4 - x^2 + 16$
76. $x^4 - 16x^2 + 36$
77. $x^4 + 4$
78. $x^4 + 64$

In Exercises 79–110, factor each polynomial completely. If a polynomial cannot be factored, state that it is irreducible.

79. $1 - 16x^2$
80. $4 - 25x^2$
81. $x^2 - 6x + 9$
82. $x^2 - 8x + 16$
83. $4x^2 + 4x + 1$
84. $16x^2 + 8x + 1$
85. $2x^2 - 8x - 10$
86. $5x^2 - 10x - 40$
87. $2x^2 + 3x - 20$
88. $2x^2 - 7x - 30$
89. $x^2 - 24x + 36$
90. $x^2 - 20x + 25$

91. $3x^5 + 12x^4 + 12x^3$

92. $2x^5 + 16x^4 + 32x^3$

93. $9x^2 - 1$

94. $16x^2 - 25$

95. $16x^2 + 24x + 9$

96. $4x^2 + 20x + 25$

97. $x^2 + 15$

98. $x^2 + 24$

99. $45x^3 + 8x^2 - 4x$

100. $25x^3 + 40x^2 - 9x$

101. $ax^2 - 7a^2x - 8a^3$

102. $ax^2 - 10a^2x - 24a^3$

103. $x^2 - 16a^2 + 8x + 16$

104. $x^2 - 36a^2 + 6x + 9$

105. $3x^5 + 12x^4y + 12x^3y^2$

106. $4x^5 + 24x^4y + 36x^3y^2$

107. $x^2 - 25a^2 + 4x + 4$

108. $x^2 - 16a^2 + 4x + 4$

109. $18x^6 + 12x^5y + 2x^4y^2$

110. $12x^6 + 12x^5y + 3x^4y^2$

Applying the Concepts

111. Garden area. A rectangular garden must have a perimeter of 16 feet. Write a completely factored polynomial whose values give the area of the garden, assuming that one side is x feet.

112. Serving tray dimensions. A rectangular serving tray must have a perimeter of 42 inches. Write a completely factored polynomial whose values give the area of the tray, assuming that one side is x inches.

113. Box construction. A box with an open top is to be constructed from a rectangular piece of cardboard with length 36 inches and width 16 inches by cutting out squares of equal side length x at each corner and then folding up the sides, as shown in the accompanying figure. Write a completely factored polynomial whose values give the volume of the box (volume = length × width × height).

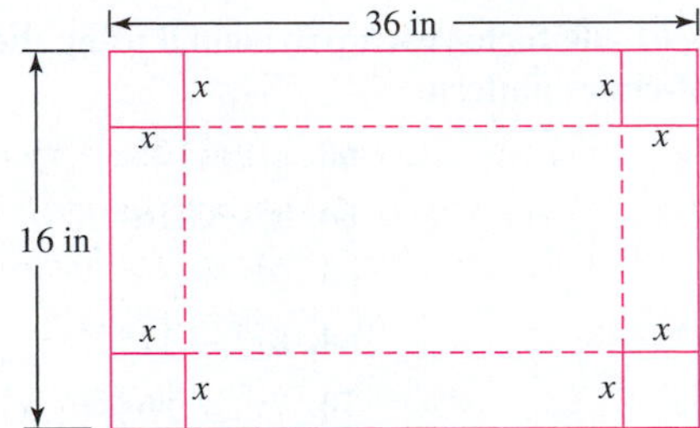

114. Surface area. For the box described in Exercise 113, write a completely factored polynomial whose value gives its surface area (surface area = sum of the areas of the five sides).

115. Area of a washer. As shown in the figure, a washer is made by removing a circular disk from the center of another circular disk of radius 2 cm. Assuming that the disk that is removed has radius x, write a completely factored polynomial whose values give the area of the washer.

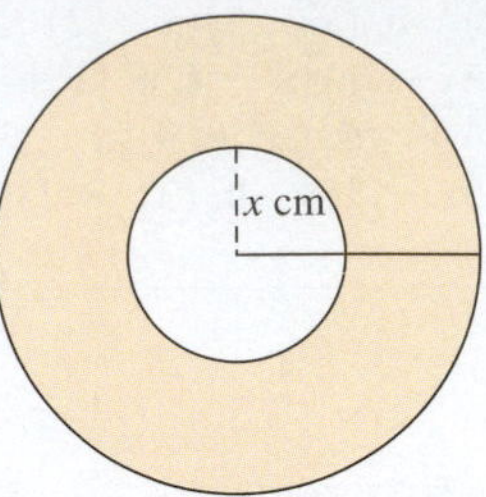

116. Insulation. Insulation must be put in the space between two concentric cylinders. Write a completely factored polynomial whose values give the volume between the inner and outer cylinders if the inner cylinder has radius 3 feet, the outer cylinder has radius x feet, and both cylinders are 8 feet high.

117. Grazing pens. A rancher has 2800 feet of fencing that will fence off a rectangular grazing pen for cattle. One edge of the pen borders a straight river and needs no fence. Assuming that the edges of the pen perpendicular to the river are x feet long, write a completely factored polynomial whose values give the area of the pen.

118. Centerpiece perimeter. A decorative glass centerpiece for a table has the shape of a rectangle with a semicircular section attached at each end, as shown in the figure. The perimeter of the figure is 48 inches. Write a completely factored polynomial whose values give the area of the centerpiece.

SECTION P.5

Rational Expressions

BEFORE STARTING THIS SECTION, REVIEW

1. Rational numbers (Section P.1, page 3)
2. Properties of fractions (Section P.1, page 12)
3. Factoring polynomials (Section P.4)
4. Irreducible polynomial (Section P.4, page 39)

OBJECTIVES

1. Reduce a rational expression to lowest terms.
2. Multiply and divide rational expressions.
3. Add and subtract rational expressions.
4. Identify and simplify complex fractions.

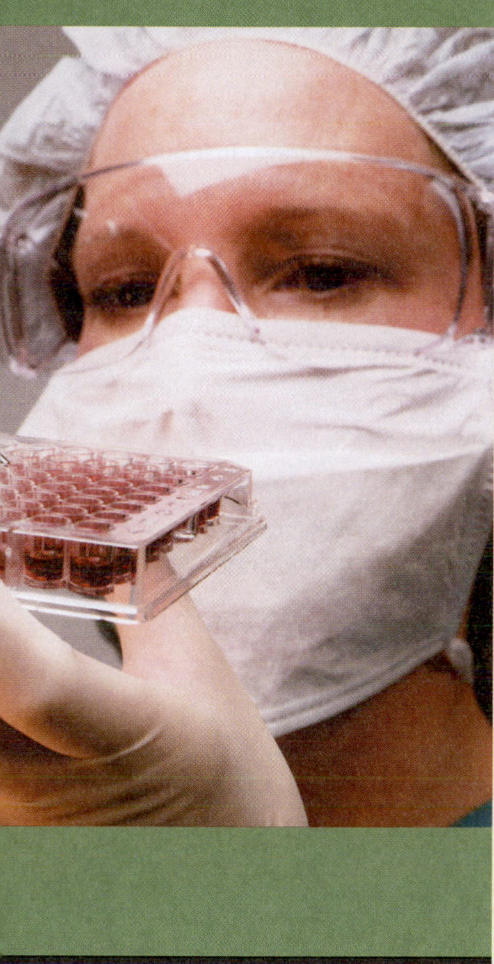

False Positive in Drug Testing

You are probably aware that the nonmedical use of steroids among adolescents and young adults is an ongoing national concern. In fact, many high schools have required students involved in extracurricular activities to submit to random drug testing. Many methods of drug testing are in use, and those who administer the tests are keenly aware of the possibility of what is termed a "false positive," which occurs when someone who is not a drug user tests positive for using drugs. How likely is it that a student who tests positive actually uses drugs? For a test that is 95% accurate, an expression for the likelihood that a student who tests positive but is not a drug user is

$$\frac{(0.05)(1 - x)}{0.95x + (0.05)(1 - x)},$$

where x is the percent of the student population that uses drugs. This is an example of a rational expression, which we study in this section. In Example 1, we see that in a student population among which only 5% of the students use drugs, the likelihood that a student who tests positive is not a drug user is 50% when the test used is 95% accurate.

Rational Expressions

Recall that the quotient of two integers, $\frac{a}{b}$ $(b \neq 0)$, is a rational number. The quotient of two polynomials is called a **rational expression**. Following are examples of rational expressions.

$$\frac{10}{17}, \quad \frac{x + 2}{3}, \quad \frac{x + 6}{(x - 3)(x + 4)}, \quad \frac{7}{x^2 + 1}, \quad \text{and} \quad \frac{x^2 + 2x - 3}{x^2 + 5x}$$

We use the same language to describe rational expressions that we use to describe rational numbers. For $\frac{x^2 + 2x - 3}{x^2 + 5x + 6}$, we call $x^2 + 2x - 3$ the **numerator** and $x^2 + 5x + 6$ the **denominator**. **The domain of a rational expression is the set of all real numbers except those that result in a zero denominator**. For example, the domain of $\frac{2}{x - 1}$ is all real

$x, x \neq 1$. We write $\frac{x + 6}{(x - 3)(x + 4)}, x \neq 3, x \neq -4$ to indicate that 3 and -4 are not in the domain of this rational expression.

EXAMPLE 1 **Calculating False Positives in Drug Testing**

With a test that is 95% accurate, an expression for the likelihood that a student who tests positive for drugs is a nonuser is

$$\frac{(0.05)(1 - x)}{0.95x + (0.05)(1 - x)},$$

where x is the percent (in decimal) of the student population that uses drugs. In a student population among which only 5% of the students use drugs, find the likelihood that a student who tests positive is a nonuser.

Solution

We convert 5% to a decimal to get $x = 0.05$, and we replace x with 0.05 in the given expression.

$$\frac{(0.05)(1 - x)}{0.95x + (0.05)(1 - x)} = \frac{(0.05)(1 - 0.05)}{0.95(0.05) + (0.05)(1 - 0.05)}$$

$$= 0.50 \qquad \text{Use a calculator.}$$

So the likelihood that a student who tests positive is a nonuser when the test used is 95% accurate is 50%. Because of the high rate of false positives, a more accurate test is needed.

Practice Problem 1 Rework Example 1 for a student population among which 10% of the students use drugs. 

1 Reduce a rational expression to lowest terms.

Lowest Terms for a Rational Expression

A rational expression is reduced to its lowest terms or **simplified** if its numerator and denominator have no common factor other than 1 or -1.

SIMPLIFYING RATIONAL EXPRESSION

To simplify a rational expression, we

1. Factor its numerator and denominator.
2. Divide the numerator and denominator by any common factors.

EXAMPLE 2 **Reducing a Rational Expression to Lowest Terms**

Simplify each expression.

a. $\frac{x^4 + 2x^3}{x + 2}, x \neq -2$ **b.** $\frac{x^2 - x - 6}{x^3 - 3x^2}, x \neq 0, x \neq 3$

Solution

Factor each numerator and denominator and divide out the common factors.

a. $\frac{x^4 + 2x^3}{x + 2} = \frac{x^3(x + 2)}{x + 2} = \frac{x^3\cancel{(x + 2)}}{\cancel{(x + 2)}} = x^3$

b. $\frac{x^2 - x - 6}{x^3 - 3x^2} = \frac{(x + 2)(x - 3)}{x^2(x - 3)} = \frac{(x + 2)\cancel{(x - 3)}}{x^2\cancel{(x - 3)}} = \frac{x + 2}{x^2}$

Practice Problem 2 Simplify each expression.

a. $\dfrac{2x^3 + 8x^2}{3x^2 + 12x}$ **b.** $\dfrac{x^2 - 4}{x^2 + 4x + 4}$

WARNING

Because x^2 is a *term*, not a *factor*, in both the numerator and denominator of $\dfrac{x^2 - 1}{x^2 + 2x + 1}$, you cannot remove x^2. You can remove only *factors* common to both the numerator and denominator of a rational expression.

2 Multiply and divide rational expressions.

Multiplication and Division of Rational Expressions

We use the same rules for multiplying and dividing rational expressions as we do for rational numbers.

SIDE NOTE

In working with the quotient of two rational expressions $\dfrac{\frac{A}{B}}{\frac{C}{D}} = \dfrac{A}{B} \cdot \dfrac{D}{C}$ three polynomials appear as denominators: B and D from $\dfrac{A}{B}$ and $\dfrac{C}{D}$ and C from $\dfrac{A}{B} \cdot \dfrac{D}{C}$

MULTIPLICATION AND DIVISION

For rational expressions $\dfrac{A}{B}$ and $\dfrac{C}{D}$,

$$\frac{A}{B} \cdot \frac{C}{D} = \frac{A \cdot C}{B \cdot D} \quad \text{and} \quad \frac{\frac{A}{B}}{\frac{C}{D}} = \frac{A}{B} \div \frac{C}{D} = \frac{A}{B} \cdot \frac{D}{C} = \frac{AD}{BC}$$

if $B \neq 0, D \neq 0$. if $B \neq 0, C \neq 0, D \neq 0$.

EXAMPLE 3 **Multiplying and Dividing Rational Expressions**

Multiply or divide as indicated. Simplify your answer and leave it in factored form.

a. $\dfrac{x^2 + 3x + 2}{x^3 + 3x} \cdot \dfrac{2x^3 + 6x}{x^2 + x - 2}, \quad x \neq 0, x \neq 1, x \neq -2$

b. $\dfrac{\dfrac{3x^2 + 11x - 4}{8x^3 - 40x^2}}{\dfrac{3x + 12}{4x^4 - 20x^3}}, \quad x \neq 0, x \neq 5, x \neq -4$

Solution

Factor each numerator and denominator and divide out the common factors.

a. $$\frac{x^2 + 3x + 2}{x^3 + 3x} \cdot \frac{2x^3 + 6x}{x^2 + x - 2} = \frac{(x + 2)(x + 1)}{x(x^2 + 3)} \cdot \frac{2x(x^2 + 3)}{(x - 1)(x + 2))}$$

$$= \frac{\cancel{(x + 2)}(x + 1)(2\cancel{x})\cancel{(x^2 + 3)}}{\cancel{x}\cancel{(x^2 + 3)}(x - 1)\cancel{(x + 2)}} = \frac{2(x + 1)}{x - 1}$$

b. $$\frac{\dfrac{3x^2 + 11x - 4}{8x^3 - 40x^2}}{\dfrac{3x + 12}{4x^4 - 20x^3}} = \frac{3x^2 + 11x - 4}{8x^3 - 40x^2} \cdot \frac{4x^4 - 20x^3}{3x + 12} = \frac{(x + 4)(3x - 1)}{8x^2(x - 5)} \cdot \frac{4x^3(x - 5)}{3(x + 4)}$$

$$= \frac{\cancel{(x + 4)}(3x - 1)\overset{x}{\cancel{(4x^3)}}\cancel{(x - 5)}}{\underset{2}{\cancel{8x^2}}\cancel{(x - 5)}(3)\cancel{(x + 4)}} = \frac{(3x - 1)x}{6} = \frac{x(3x - 1)}{6}$$

SIDE NOTE

The best strategy to use when multiplying and dividing rational expressions is to factor each numerator and denominator completely and then divide out the common factors.

Practice Problem 3 Multiply or divide as indicated. Simplify your answer.

$$\frac{\dfrac{x^2 - 2x - 3}{7x^3 + 28x^2}}{\dfrac{4x + 4}{2x^4 + 8x^3}} \qquad x \neq 0, x \neq -4, x \neq -1$$

3 Add and subtract rational expressions.

Addition and Subtraction of Rational Expressions

To add and subtract rational expressions, we use the same rules as for adding and subtracting rational numbers.

ADDITION AND SUBTRACTION OF RATIONAL EXPRESSIONS

For rational expressions $\frac{A}{D}$ and $\frac{C}{D}$ (same denominator $D \neq 0$),

$$\frac{A}{D} + \frac{C}{D} = \frac{A + C}{D} \quad \text{and} \quad \frac{A}{D} - \frac{C}{D} = \frac{A - C}{D}.$$

For rational expression $\frac{A}{B}$ and $\frac{C}{D}$ $(B \neq 0, D \neq 0)$,

$$\frac{A}{B} + \frac{C}{D} = \frac{AD + BC}{BD} \quad \text{and} \quad \frac{A}{B} - \frac{C}{D} = \frac{AD - BC}{BD}.$$

EXAMPLE 4 **Adding and Subtracting Rational Expressions with the Same Denominator**

Add or subtract as indicated. Simplify your answer and leave both the numerator and denominator in factored form.

a. $\dfrac{x - 6}{(x + 1)^2} + \dfrac{x + 8}{(x + 1)^2}, \qquad x \neq -1$

b. $\dfrac{3x - 2}{x^2 - 5x + 6} - \dfrac{2x + 1}{x^2 - 5x + 6}, \qquad x \neq 2, x \neq 3$

Solution

a. $$\frac{x - 6}{(x + 1)^2} + \frac{x + 8}{(x + 1)^2} = \frac{x - 6 + x + 8}{(x + 1)^2} = \frac{2x + 2}{(x + 1)^2}$$
$$= \frac{2\cancel{(x + 1)}}{(x + 1)\cancel{(x + 1)}} = \frac{2}{x + 1}$$

b. $$\frac{3x - 2}{x^2 - 5x + 6} - \frac{2x + 1}{x^2 - 5x + 6} = \frac{(3x - 2) - (2x + 1)}{x^2 - 5x + 6} = \frac{3x - 2 - 2x - 1}{x^2 - 5x + 6}$$
$$= \frac{x - 3}{(x - 2)(x - 3)} = \frac{\cancel{x - 3}}{(x - 2)\cancel{(x - 3)}} = \frac{1}{x - 2}$$

Practice Problem 4 Add or subtract as indicated. Simplify your answers.

a. $\dfrac{5x + 22}{x^2 - 36} + \dfrac{2(x + 10)}{x^2 - 36}, \qquad x \neq 6, x \neq -6$

b. $\dfrac{4x + 1}{x^2 + x - 12} - \dfrac{3x + 4}{x^2 + x - 12}, \qquad x \neq -4, x \neq 3$

When adding or subtracting rational expressions with different denominators, we proceed (as with fractions) by finding a common denominator. The one most convenient to use is called the **least common denominator (LCD)**, and it is the polynomial of least degree that contains each denominator as a factor. In the simplest case, the LCD is the product of the denominators.

EXAMPLE 5 Adding and Subtracting When Denominators Have No Common Factor

Add or subtract as indicated. Simplify your answer and leave it in factored form.

a. $\dfrac{x}{x+1} + \dfrac{2x-1}{x+3}, x \neq -1, x \neq -3$ **b.** $\dfrac{2x}{x+1} - \dfrac{x}{x+2}, x \neq -1, x \neq -2$

Solution

a. $$\frac{x}{x+1} + \frac{2x-1}{x+3} = \frac{x(x+3)}{(x+1)(x+3)} + \frac{(2x-1)(x+1)}{(x+1)(x+3)}$$ LCD = $(x+1)(x+3)$

$$= \frac{x(x+3)+(2x-1)(x+1)}{(x+1)(x+3)}$$

$$= \frac{x^2+3x+2x^2+2x-x-1}{(x+1)(x+3)}$$

$$= \frac{3x^2+4x-1}{(x+1)(x+3)}$$

b. $$\frac{2x}{x+1} - \frac{x}{x+2} = \frac{2x(x+2)}{(x+1)(x+2)} - \frac{x(x+1)}{(x+1)(x+2)}$$ LCD = $(x+1)(x+2)$

$$= \frac{2x(x+2)-x(x+1)}{(x+1)(x+2)} = \frac{2x^2+4x-x^2-x}{(x+1)(x+2)}$$

$$= \frac{x^2+3x}{(x+1)(x+2)} = \frac{x(x+3)}{(x+1)(x+2)}$$

Practice Problem 5 Add or subtract as indicated. Simplify your answers.

a. $\dfrac{2x}{x+2} + \dfrac{3x}{x-5}, \quad x \neq -2, x \neq 5$ **b.** $\dfrac{5x}{x-4} - \dfrac{2x}{x+3}, \quad x \neq 4, x \neq -3$

TO FIND THE LCD FOR RATIONAL EXPRESSIONS

1. Factor each denominator polynomial completely.
2. Form a product of the different irreducible factors of each polynomial. (Each distinct factor is used once.)
3. Attach to each factor in this product the greatest exponent that appears on this factor in *any* of the factored denominators.

EXAMPLE 6 Finding the LCD for Rational Expressions

Find the LCD for each pair of rational expressions.

a. $\dfrac{x+2}{x(x-1)^2(x+2)}, \dfrac{3x+7}{4x^2(x+2)^3}$ **b.** $\dfrac{x+1}{x^2-x-6}, \dfrac{2x-13}{x^2-9}$

Solution

a. **Step 1** The denominators are already completely factored.

Step 2 $4x(x-1)(x+2)$ Product of the different factors

Step 3 LCD $= 4x^2(x-1)^2(x+2)^3$ The greatest exponents are 2, 2, and 3.

b. **Step 1** $x^2 - x - 6 = (x + 2)(x - 3)$
$x^2 - 9 = (x + 3)(x - 3)$

Step 2 $(x + 2)(x - 3)(x + 3)$ Product of the different factors

Step 3 $\text{LCD} = (x + 2)(x - 3)(x + 3)$ The greatest exponent on each factor is 1.

Practice Problem 6 Find the LCD for each pair of rational expressions.

a. $\dfrac{x^2 + 3x}{x^2(x + 2)^2(x - 2)}, \dfrac{4x^2 + 1}{3x(x - 2)^2}$ **b.** $\dfrac{2x - 1}{x^2 - 25}, \dfrac{3 - 7x^2}{(x^2 + 4x - 5)}$

When adding or subtracting rational expressions with different denominators, the first step is to find the LCD. Here is the general procedure.

PROCEDURE FOR ADDING OR SUBTRACTING RATIONAL EXPRESSIONS WITH DIFFERENT DENOMINATORS

Step 1 Find the LCD.

Step 2 Using $\dfrac{A}{B} = \dfrac{AC}{BC}$, write each rational expression as a rational expression with the LCD as the denominator.

Step 3 Following the order of operations, add or subtract numerators, keeping the LCD as the denominator.

Step 4 Simplify.

EXAMPLE 7 Using the LCD to Add and Subtract Rational Expressions

Add or subtract as indicated. Simplify your answer and leave it in factored form.

a. $\dfrac{3}{x^2 - 1} + \dfrac{x}{x^2 + 2x + 1}, \quad x \neq 1, x \neq -1$

b. $\dfrac{x + 2}{x^2 - x} - \dfrac{3x}{4(x - 1)^2}, \quad x \neq 0, x \neq 1$

Solution

a. **Step 1** Note that $x^2 - 1 = (x - 1)(x + 1)$ and $x^2 + 2x + 1 = (x + 1)^2$; the LCD is $(x + 1)^2(x - 1)$.

Step 2 $\dfrac{3}{x^2 - 1} = \dfrac{3}{(x - 1)(x + 1)} = \dfrac{3(x + 1)}{(x - 1)(x + 1)^2}$ Multiply numerator and denominator by $x + 1$.

$\dfrac{x}{x^2 + 2x + 1} = \dfrac{x}{(x + 1)^2} = \dfrac{x(x - 1)}{(x + 1)^2(x - 1)}$ Multiply numerator and denominator by $x - 1$.

Steps 3–4 Add.

$$\frac{3}{x^2 - 1} + \frac{x}{x^2 + 2x + 1} = \frac{3(x + 1)}{(x - 1)(x + 1)^2} + \frac{x(x - 1)}{(x - 1)(x + 1)^2}$$

$$= \frac{3(x + 1) + x(x - 1)}{(x - 1)(x + 1)^2} = \frac{3x + 3 + x^2 - x}{(x - 1)(x + 1)^2}$$

$$= \frac{x^2 + 2x + 3}{(x - 1)(x + 1)^2}$$

b. Step 1 The LCD is $4x(x-1)^2$. $x^2 - x = x(x-1)$

Step 2

$$\frac{x+2}{x^2-x} = \frac{x+2}{x(x-1)} = \frac{(x+2)\cdot 4(x-1)}{x(x-1)\cdot 4(x-1)} = \frac{4(x+2)(x-1)}{4x(x-1)^2}$$ Multiply numerator and denominator by $4(x-1)$.

$$\frac{3x}{4(x-1)^2} = \frac{(3x)x}{4(x-1)^2 x} = \frac{3x^2}{4x(x-1)^2}$$ Multiply numerator and denominator by x.

Steps 3–4 Subtract.

$$\frac{x+2}{x^2-x} - \frac{3x}{4(x-1)^2} = \frac{4(x+2)(x-1)}{4x(x-1)^2} - \frac{3x^2}{4x(x-1)^2} = \frac{4(x+2)(x-1) - 3x^2}{4x(x-1)^2}$$
$$= \frac{4x^2 + 4x - 8 - 3x^2}{4x(x-1)^2} = \frac{x^2 + 4x - 8}{4x(x-1)^2}$$

Practice Problem 7 Add or subtract as indicated. Simplify your answers.

a. $\frac{4}{x^2 - 4x + 4} + \frac{x}{x^2 - 4}$, $x \neq 2, x = -2$

b. $\frac{2x}{3(x-5)^2} - \frac{6x}{2(x^2 - 5x)}$, $x \neq 0, x \neq 5$

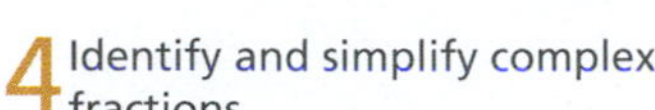

4 Identify and simplify complex fractions.

Complex Fractions

A rational expression that contains another rational expression in its numerator or denominator (or both) is called a **complex rational expression** or **complex fraction**. To **simplify** a complex fraction, we write it as a rational expression in lowest terms.

There are two effective methods of simplifying complex fractions.

PROCEDURES FOR SIMPLIFYING COMPLEX FRACTIONS

Method 1 Perform the operations indicated in both the numerator and denominator of the complex fraction. Then multiply the resulting numerator by the reciprocal of the denominator.

Method 2 Multiply the numerator and the denominator of the complex fraction by the LCD of all rational expressions that appear in either the numerator or the denominator. Simplify the result.

EXAMPLE 8 **Simplifying a Complex Fraction**

Simplify $\dfrac{\frac{1}{2} + \frac{1}{x}}{\frac{x^2 - 4}{2x}}$, $x \neq 0, x \neq 2, x \neq -2$, using each of the two methods.

Solution

Method 1

Reciprocal of the denominator

$$\frac{\frac{1}{2}+\frac{1}{x}}{\frac{x^2-4}{2x}} = \frac{\frac{x+2}{2x}}{\frac{x^2-4}{2x}} = \frac{x+2}{2x}\cdot\frac{2x}{x^2-4} = \frac{x+2}{2x}\cdot\frac{2x}{(x+2)(x-2)}$$

$$= \frac{(x+2)(2x)}{(2x)(x+2)(x-2)} = \frac{1}{x-2}$$

Method 2 The LCD of $\frac{1}{2}, \frac{1}{x}$, and $\frac{x^2-4}{2x}$ is $2x$.

$$\frac{\frac{1}{2}+\frac{1}{x}}{\frac{x^2-4}{2x}} = \frac{\left(\frac{1}{2}+\frac{1}{x}\right)(2x)}{\left(\frac{x^2-4}{2x}\right)(2x)}$$

Multiply numerator and denominator by the LCD.

$$= \frac{\frac{1}{2}(2x)+\frac{1}{x}(2x)}{\left[\frac{(x^2-4)}{2x}\right](2x)} = \frac{x+2}{x^2-4} = \frac{x+2}{(x-2)(x+2)} = \frac{1}{x-2}$$

Practice Problem 8 Simplify: $\dfrac{\frac{5}{3x}+\frac{1}{3}}{\frac{x^2-25}{3x}}, \quad x \neq 0, x \neq 5, x \neq -5$

EXAMPLE 9 Simplifying a Complex Fraction

Simplify: $\dfrac{x}{2x-\dfrac{7}{3-\frac{1}{2}}}, \quad x \neq \dfrac{7}{5}$

Solution

We use Method 1.

Reciprocal of the denominator

Begin with $\dfrac{7}{3-\frac{1}{2}} = \dfrac{7}{\frac{5}{2}} = 7\cdot\dfrac{2}{5} = \dfrac{14}{5}$. Replace $3-\frac{1}{2}$ with $\frac{5}{2}$.

Then $\dfrac{x}{2x-\dfrac{7}{3-\frac{1}{2}}} = \dfrac{x}{2x-\frac{14}{5}}$ Replace $\dfrac{7}{3-\frac{1}{2}}$ with $\frac{14}{5}$.

Reciprocal of the denominator

$$= \frac{x}{\frac{10x-14}{5}} = x\cdot\frac{5}{10x-14} = \frac{5x}{10x-14} = \frac{5x}{2(5x-7)}.$$

Practice Problem 9 Simplify: $\dfrac{5x}{3x-\dfrac{4}{2-\frac{1}{3}}}, \quad x \neq \dfrac{4}{5}$

Answers to Practice Problems

1. 0.32 **2. a.** $\frac{2x}{3}$ **b.** $\frac{x-2}{x+2}$ **3.** $\frac{x(x-3)}{14}$ **4. a.** $\frac{7}{x-6}$ **b.** $\frac{1}{x+4}$ **5. a.** $\frac{x(5x-4)}{(x+2)(x-5)}$ **b.** $\frac{x(3x+23)}{(x+3)(x-4)}$

6. a. $3x^2(x+2)^2(x-2)^2$ **b.** $(x-5)(x+5)(x-1)$ **7. a.** $\frac{x^2+2x+8}{(x+2)(x-2)^2}$ **b.** $\frac{45-7x}{3(x-5)^2}$ **8.** $\frac{1}{x-5}$ **9.** $\frac{25x}{15x-12}$

SECTION P.5 Exercises

Basic Concepts and Skills

1. The least common denominator for two rational expressions is the polynomial of least degree that contains ______________ as a factor.
2. The first step in finding the LCD for two rational expressions is to ______________ the denominators completely.
3. If the denominators for two rational expressions are $x^2 - 2x$ and $x^2 - x - 2$, the LCD is ______________.
4. A rational expression that contains another rational expression in its numerator or denominator is called a(n) ______________.
5. **True or False.** The expression $\frac{1}{2} + \frac{1}{x}$ is called a complex fraction.
6. **True or False.** A polynomial is also a rational expression.

In Exercises 7–22, reduce each rational expression to lowest terms. Specify the domain of the rational expression by identifying all real numbers that must be excluded from the domain.

7. $\frac{2x+2}{x^2+2x+1}$
8. $\frac{3x-6}{x^2-4x+4}$
9. $\frac{3x+3}{x^2-1}$
10. $\frac{10-5x}{4-x^2}$
11. $\frac{2x-6}{9-x^2}$
12. $\frac{15+3x}{x^2-25}$
13. $\frac{2x-1}{1-2x}$
14. $\frac{2-5x}{5x-2}$
15. $\frac{x^2-6x+9}{4x-12}$
16. $\frac{x^2-10x+25}{3x-15}$
17. $\frac{7x^2+7x}{x^2+2x+1}$
18. $\frac{4x^2+12x}{x^2+6x+9}$
19. $\frac{x^2-11x+10}{x^2+6x-7}$
20. $\frac{x^2+2x-15}{x^2-7x+12}$
21. $\frac{6x^4+14x^3+4x^2}{6x^4-10x^3-4x^2}$
22. $\frac{3x^3+x^2}{3x^4-11x^3-4x^2}$

In Exercises 23–40, multiply or divide as indicated. Simplify and leave the numerator and denominator in your answer in factored form. Assume all denominators are nonzero.

23. $\frac{x-3}{2x+4} \cdot \frac{10x+20}{5x-15}$
24. $\frac{6x+4}{2x-8} \cdot \frac{x-4}{9x+6}$
25. $\frac{2x+6}{4x-8} \cdot \frac{x^2+x-6}{x^2-9}$
26. $\frac{25x^2-9}{4-2x} \cdot \frac{4-x^2}{10x-6}$
27. $\frac{x^2-7x}{x^2-6x-7} \cdot \frac{x^2-1}{x^2}$
28. $\frac{x^2-9}{x^2-6x+9} \cdot \frac{5x-15}{x+3}$
29. $\frac{x^2-x-6}{x^2+3x+2} \cdot \frac{x^2-1}{x^2-9}$
30. $\frac{x^2+2x-8}{x^2+x-20} \cdot \frac{x^2-16}{x^2+5x+4}$
31. $\frac{2-x}{x+1} \cdot \frac{x^2+3x+2}{x^2-4}$
32. $\frac{3-x}{x+5} \cdot \frac{x^2+8x+15}{x^2-9}$
33. $\frac{x+2}{6} \div \frac{4x+8}{9}$
34. $\frac{x+3}{20} \div \frac{4x+12}{9}$
35. $\frac{x^2-9}{x} \div \frac{2x+6}{5x^2}$
36. $\frac{x^2-1}{3x} \div \frac{7x-7}{x^2+x}$
37. $\frac{x^2+2x-3}{x^2+8x+16} \div \frac{x-1}{3x+12}$
38. $\frac{x^2+5x+6}{x^2+6x+9} \div \frac{x^2+3x+2}{x^2+7x+12}$
39. $\left(\frac{x^2-9}{x^3+8} \div \frac{x+3}{x^3+2x^2-x-2}\right)\frac{1}{x^2-1}$
40. $\left(\frac{x^2-25}{x^2-3x-4} \div \frac{x^2+3x-10}{x^2-1}\right)\frac{x-2}{x-5}$

In Exercises 41–58, add and subtract as indicated. Simplify and leave the numerator and denominator in your answer in factored form.

41. $\frac{x}{5} + \frac{3}{5}$
42. $\frac{7}{4} - \frac{x}{4}$
43. $\frac{x}{2x+1} + \frac{4}{2x+1}$
44. $\frac{2x}{7x-3} + \frac{x}{7x-3}$

45. $\frac{x^2}{x+1} - \frac{x^2-1}{x+1}$

46. $\frac{2x+7}{3x+2} - \frac{x-2}{3x+2}$

47. $\frac{4}{3-x} + \frac{2x}{x-3}$

48. $\frac{-2}{1-x} + \frac{2-x}{x-1}$

49. $\frac{5x}{x^2+1} + \frac{2x}{x^2+1}$

50. $\frac{x}{2(x-1)^2} + \frac{3x}{2(x-1)^2}$

51. $\frac{7x}{2(x-3)} + \frac{x}{2(x-3)}$

52. $\frac{4x}{4(x+5)^2} + \frac{8x}{4(x+5)^2}$

53. $\frac{x}{x^2-4} - \frac{2}{x^2-4}$

54. $\frac{5x}{x^2-1} - \frac{5}{x^2-1}$

55. $\frac{x-2}{2x+1} - \frac{x}{2x-1}$

56. $\frac{2x-1}{4x+1} - \frac{2x}{4x-1}$

57. $\frac{-x}{x+2} + \frac{x-2}{x} - \frac{x}{x-2}$

58. $\frac{3x}{x-1} + \frac{x+1}{x} - \frac{2x}{x+1}$

In Exercises 59–66, find the LCD for each pair of rational fractions.

59. $\frac{5}{3x-6}, \frac{2x}{4x-8}$

60. $\frac{5x+1}{7+21x}, \frac{1-x}{3+9x}$

61. $\frac{3-x}{4x^2-1}, \frac{7x}{(2x+1)^2}$

62. $\frac{14x}{(3x-1)^2}, \frac{2x+7}{9x^2-1}$

63. $\frac{1-x}{x^2+3x+2}, \frac{3x+12}{x^2-1}$

64. $\frac{5x+9}{x^2-x-6}, \frac{x+5}{x^2-9}$

65. $\frac{7-4x}{x^2-5x+4}, \frac{x^2-x}{x^2+x-2}$

66. $\frac{13x}{x^2-2x-3}, \frac{2x^2-4}{x^2+3x+2}$

In Exercises 67–82, perform the indicated operations and simplify the result. Leave the numerator and denominator in your answer in factored form.

67. $\frac{5}{x-3} + \frac{2x}{x^2-9}$

68. $\frac{3x}{x-1} + \frac{x}{x^2-1}$

69. $\frac{2x}{x^2-4} - \frac{x}{x+2}$

70. $\frac{3x-1}{x^2-16} - \frac{2x+1}{x-4}$

71. $\frac{x-2}{x^2+3x-10} + \frac{x+3}{x^2+x-6}$

72. $\frac{x+3}{x^2-x-2} + \frac{x-1}{x^2+2x+1}$

73. $\frac{2x-3}{9x^2-1} + \frac{4x-1}{(3x-1)^2}$

74. $\frac{3x+1}{(2x+1)^2} + \frac{x+3}{4x^2-1}$

75. $\frac{x-3}{x^2-25} - \frac{x-3}{x^2+9x+20}$

76. $\frac{2x}{x^2-16} - \frac{2x-7}{x^2-7x+12}$

77. $\frac{3}{x^2-4} + \frac{1}{2-x} - \frac{1}{2+x}$

78. $\frac{2}{5+x} + \frac{5}{x^2-25} + \frac{7}{5-x}$

79. $\frac{x+3a}{x-5a} - \frac{x+5a}{x-3a}$

80. $\frac{3x-a}{2x-a} - \frac{2x+a}{3x+a}$

81. $\frac{1}{x+h} - \frac{1}{x}$

82. $\frac{1}{(x+h)^2} - \frac{1}{x^2}$

In Exercises 83–96, perform the indicated operations and simplify the result. Leave the numerator and denominator in your answer in factored form.

83. $\frac{\frac{2}{x}}{\frac{3}{x^2}}$

84. $\frac{\frac{-6}{x}}{\frac{2}{x^3}}$

85. $\frac{\frac{1}{x}}{1-\frac{1}{x}}$

86. $\frac{\frac{1}{x^2}}{\frac{1}{x^2}-1}$

87. $\frac{\frac{1}{x}-1}{\frac{1}{x}+1}$

88. $\frac{2-\frac{2}{x}}{1+\frac{2}{x}}$

89. $\frac{\frac{1}{x}-x}{1-\frac{1}{x^2}}$

90. $\frac{x-\frac{1}{x}}{\frac{1}{x^2}-1}$

91. $x - \frac{x}{x+\frac{1}{2}}$

92. $x + \frac{x}{x+\frac{1}{2}}$

93. $\frac{\frac{1}{x+h}-\frac{1}{x}}{h}$

94. $\frac{\frac{1}{(x+h)^2}-\frac{1}{x^2}}{h}$

95. $\frac{\frac{1}{x-a}+\frac{1}{x+a}}{\frac{1}{x-a}-\frac{1}{x+a}}$

96. $\frac{\frac{1}{x-a}+\frac{1}{x+a}}{\frac{x}{x-a}-\frac{a}{x+a}}$

Applying the Concepts

97. **Bearing length.** The length (in centimeters) of a bearing in the shape of a cylinder is given by $\frac{125.6}{\pi r^2}$, where r is the radius of the base of the cylinder. Find the length of a bearing with a diameter of 4 centimeters, using 3.14 as an approximation of π.

98. **Toy box height.** The height (in feet) of an open toy box that is twice as long as it is wide is given by $\frac{13.5 - 2x^2}{6x}$, where x is the width of the box. Find the height of a toy box that is 1.5 feet wide.

99. Diluting a mixture. A 100-gallon mixture of citrus extract and water is 3% citrus extract.

a. Write a rational expression in x whose values give the percentage (in decimal form) of the mixture that is citrus extract when x gallons of water are added to the mixture.

b. Find the percentage of citrus extract in the mixture assuming that 50 gallons of water are added to it.

100. Acidity in a reservoir. A half-full 400,000-gallon reservoir is found to be 0.75% acid.

a. Write a rational expression in x whose values give the percentage (in decimal form) of acid in the reservoir when x gallons of water are added to it.

b. Find the percentage of acid in the reservoir assuming that 100,000 gallons of water are added to it.

101. Diet lemonade. A company packages its powdered diet lemonade mix in containers in the shape of a cylinder. The top and bottom are made of a tin product that costs 5 cents per square inch. The side of the container is made of a cheaper material that costs 1 cent per square inch. The height of any cylinder can be found by dividing its volume by the area of its base.

a. Assuming that x is the radius of the base (in inches), write a rational expression in x whose values give the cost of each container, assuming that the capacity of the container is 120 cubic inches.

b. Find the cost of a container whose base is 4 inches in diameter, using 3.14 as an approximation of π.

102. Storage containers. A company makes storage containers in the shape of an open box with a square base. All five sides of the box are made of a plastic that costs 40 cents per square foot. The height of any box can be found by dividing its volume by the area of its base.

a. Assuming that x is the length of the base (in feet), write a rational expression in x whose values give the cost of each container, assuming that the capacity of the container is 2.25 cubic feet.

b. Find the cost of a container whose base is 1.5 feet long.

SECTION P.6

Rational Exponents and Radicals

BEFORE STARTING THIS SECTION, REVIEW

1. Absolute value (Section P.1, page 8)
2. Integer exponents (Section P.2, page 18)
3. Special-product formulas (Section P.3, pages 34 and 35)
4. Properties of fractions (Section P.1, page 12)

OBJECTIVES

1. Define and evaluate square roots.
2. Simplify square roots.
3. Define and evaluate *n*th roots.
4. Combine like radicals.
5. Rationalize denominators or numerators.
6. Define and evaluate rational exponents.

◆ Friction and Tap-Water Flow

Suppose your water company had a large reservoir or water tower 80 meters or more above the level of your cold-water faucet. If there was no friction in the pipes and you turned on your faucet, the water would come pouring out of the faucet at nearly 90 miles per hour! This fact follows from a result known as Bernoulli's equation, which predicts the velocity of the water (ignoring friction and related turbulence) in this situation to be equal to $\sqrt{2gh}$, where g is the acceleration due to gravity and h is the height of the water level above a hole from which the water escapes.

In Example 3, we use Bernoulli's equation to find the velocity of water.

1 Define and evaluate square roots.

Square Roots

When we raise the number 5 to the power 2, we write $5^2 = 25$. The reverse of this squaring process is called finding a **square root**. In the case of 25, we say that 5 is a square root of 25 and write $\sqrt{25} = 5$. In general, we say that b is a square root of a if $b^2 = a$. We can't call 5 *the* square root of 25 because it is also true that $(-5)^2 = 25$, so that -5 is *another* square root of 25.

The symbol $\sqrt{\ }$, called a **radical sign**, is used to distinguish between 5 and -5, the two square roots of 25. We write $5 = \sqrt{25}$ and $-5 = -\sqrt{25}$ and call $\sqrt{25}$ the *principal square root* of 25.

PRINCIPAL SQUARE ROOT

$\sqrt{a} = b$ means (1) $b^2 = a$ and (2) $b \geq 0$.

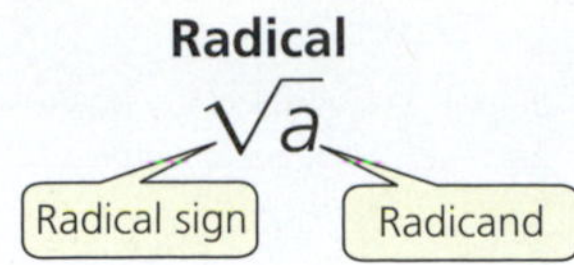

If $\sqrt{a} = b$, then $a = b^2 \geq 0$. Consequently, the symbol $\sqrt{a}$ denotes a real number only when $a \geq 0$. The number or expression under the radical sign is called the **radicand**, and the radicand together with the radical sign is called a **radical**.

According to the definition of the square root, two tests the number b must pass two tests for the statement $\sqrt{a} = b$ to be correct: (1) $b^2 = a$ and (2) $b \geq 0$. These tests are used to decide whether the statements below are true or false.

Statement $\sqrt{a} = b$	**Test (1)** $b^2 = a$?	**Test (2)** $b \geq 0$?	**Conclusion** True when b passes both tests
$\sqrt{4} = 2$	$2^2 = 4$ ✓	$2 \geq 0$ ✓	**True;** 2 passed both tests.
$\sqrt{4} = -2$	$(-2)^2 = 4$ ✓	$-2 \geq 0$ ✗	**False;** -2 failed Test (2).
$\sqrt{\frac{1}{9}} = \frac{1}{3}$	$(\frac{1}{3})^2 = \frac{1}{9}$ ✓	$\frac{1}{3} \geq 0$ ✓	**True;** $\frac{1}{3}$ passed both tests.
$\sqrt{7^2} = 7$	$7^2 = 7^2$ ✓	$7 \geq 0$ ✓	**True;** 7 passed both tests.
$\sqrt{(-7)^2} = -7$	$(-7)^2 = (-7)^2$ ✓	$-7 \geq 0$ ✗	**False;** -7 failed Test (2).

The last example (letting x represent -7) shows that $\sqrt{x^2}$ is not always equal to x. Now apply Test (1) and Test (2) to the statement $\sqrt{x^2} = |x|$, where x represents any real number.

Statement	**Test (1)**	**Test (2)**	**Conclusion**
$\sqrt{x^2} = |x|$	$|x|^2 = x^2$ ✓	$|x| \geq 0$ ✓	True; $|x|$ passed both tests.

For any real number x,

$$\sqrt{x^2} = |x|.$$

If $\sqrt{x} = b$, then by Test (1), $b^2 = x$ and by Test (2), b is nonnegative. Replacing b with $\sqrt{x}$ in the equation $b^2 = x$ yields the following result.

For any nonnegative real number x,

$$(\sqrt{x})^2 = x.$$

TECHNOLOGY CONNECTION

The square root symbol on a graphing calculator does not have a horizontal bar that goes completely over the number or expression whose square root is to be evaluated. You must enclose a rational number in parentheses before taking its square root and then use the "Frac" feature to have the answer appear as a fraction.

```
√(121)
                11
√(25/81)
       .5555555556
Ans▸Frac
               5/9
```

EXAMPLE 1 Finding the Square Roots of Perfect Squares and Their Ratios

Find. **a.** $\sqrt{121}$ **b.** $\sqrt{\frac{25}{81}}$ **c.** $\sqrt{\frac{16}{169}}$

Solution

Write each radicand as a perfect square.

a. $\sqrt{121} = \sqrt{11^2} = 11$

b. $\sqrt{\frac{25}{81}} = \sqrt{\frac{5^2}{9^2}} = \sqrt{\left(\frac{5}{9}\right)^2} = \frac{5}{9}$

c. $\sqrt{\frac{16}{169}} = \sqrt{\frac{4^2}{13^2}} = \sqrt{\left(\frac{4}{13}\right)^2} = \frac{4}{13}$

Practice Problem 1 Find.

a. $\sqrt{144}$ b. $\sqrt{\frac{1}{49}}$ c. $\sqrt{\frac{4}{64}}$

2 Simplify square roots.

Simplifying Square Roots

When working with square roots, we can simplify a radical expression by removing from under the radical sign perfect squares that are factors of the radicand. For example, to simplify $\sqrt{8}$, we write $\sqrt{8} = \sqrt{4 \cdot 2} = \sqrt{4}\sqrt{2} = 2\sqrt{2}$. We rely on the following rules.

PRODUCT AND QUOTIENT PROPERTIES OF SQUARE ROOTS

$$\sqrt{ab} = \sqrt{a}\sqrt{b} \qquad a \geq 0, b \geq 0$$

$$\sqrt{\frac{a}{b}} = \frac{\sqrt{a}}{\sqrt{b}} \qquad a \geq 0, b > 0$$

EXAMPLE 2 **Simplifying Square Roots**

Simplify: **a.** $\sqrt{117}$ **b.** $\sqrt{6}\sqrt{12}$ **c.** $\sqrt{48x^2}$

d. $\frac{\sqrt{36}}{\sqrt{3}}$ **e.** $\sqrt{\frac{25y^3}{9x^2}}, x \neq 0, y \geq 0$

Solution

a. $\sqrt{117} = \sqrt{9 \cdot 13} = \sqrt{9}\sqrt{13} = 3\sqrt{13}$

b. $\sqrt{6}\sqrt{12} = \sqrt{6 \cdot 12} = \sqrt{6 \cdot 6 \cdot 2} = \sqrt{6^2 \cdot 2} = \sqrt{6^2}\sqrt{2} = 6\sqrt{2}$

c. $\sqrt{48x^2} = \sqrt{48}\sqrt{x^2} = \sqrt{16 \cdot 3}\sqrt{x^2} = \sqrt{16}\sqrt{3}\sqrt{x^2} = 4\sqrt{3}|x|$

d. $\frac{\sqrt{36}}{\sqrt{3}} = \sqrt{\frac{36}{3}} = \sqrt{12} = \sqrt{4 \cdot 3} = \sqrt{4}\sqrt{3} = 2\sqrt{3}$

e. $\sqrt{\frac{25y^3}{9x^2}} = \frac{\sqrt{25y^3}}{\sqrt{9x^2}} = \frac{\sqrt{25 \cdot y^2 \cdot y}}{\sqrt{9 \cdot x^2}} = \frac{\sqrt{25}\sqrt{y^2}\sqrt{y}}{\sqrt{9}\sqrt{x^2}} = \frac{5|y|\sqrt{y}}{3|x|}$ $\quad x \neq 0, y \geq 0$

$= \frac{5y\sqrt{y}}{3|x|}$ $\quad |y| = y$

Practice Problem 2 Simplify.

a. $\sqrt{20}$ b. $\sqrt{6}\sqrt{8}$ c. $\sqrt{12x^2}$ d. $\sqrt{\frac{20y^3}{27x^2}}, x \neq 0, y \geq 0$

It is important to note that $\sqrt{a + b} \neq \sqrt{a} + \sqrt{b}$. For example, with $a = 1$ and $b = 4$, $\sqrt{1 + 4} \neq \sqrt{1} + \sqrt{4}$ because $\sqrt{1 + 4} = \sqrt{5}$, whereas $\sqrt{1} + \sqrt{4} = 1 + 2 = 3$ and $\sqrt{5} \neq 3$. Similarly, $\sqrt{a - b} \neq \sqrt{a} - \sqrt{b}$.

EXAMPLE 3 **Calculating Water Flow from a Punctured Water Tower**

From the introduction to this section, verify the claim that the water would come pouring out of the faucet at nearly 90 miles per hour.

Solution

We can use Bernoulli's equation to find the velocity of the water from the faucet. If we ignore friction, the velocity of water will be $\sqrt{2gh}$, where g is the acceleration due to gravity and h is the height of the water level above the faucet.

We replace h with 80 meters and roughly approximate g by 10 meters per second per second $(g \approx 10\,\text{m/s}^2)$ in the expression $\sqrt{2gh}$.

Thus, the velocity of the water is

$$
\begin{aligned}
&= \sqrt{2gh} \\
&\approx \sqrt{2(10\,\text{m/s}^2)(80\,\text{m})} = \sqrt{1600\,\text{m}^2/\text{s}^2} = 40 \text{ meters per second} \\
&= 40 \times 60 \times 60 \text{ meters per hour} && 60 \times 60 \text{ seconds each hour} \\
&= 144 \times 1000 \text{ meters per hour} \\
&\approx 144 \times 0.6214 \text{ mph} && 1000\,\text{m} \approx 0.6214 \text{ miles} \\
&= 89.4816 \text{ mph} \\
&\approx 90 \text{ miles per hour}
\end{aligned}
$$

Practice Problem 3 Repeat Example 3 if the water tower is 45 meters high.

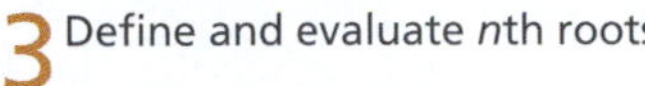
3 Define and evaluate nth roots.

Other Roots

If n is a positive integer, we say that b is an nth root of a if $b^n = a$. We next define the **principal nth root** of a real number a, denoted by the symbol $\sqrt[n]{a}$.

THE PRINCIPAL nth ROOT OF A REAL NUMBER

1. If a is positive $(a > 0)$, then $\sqrt[n]{a} = b$ provided that $b^n = a$ and $b > 0$.
2. If a is negative $(a < 0)$ and n is odd, then $\sqrt[n]{a} = b$ provided that $b^n = a$.
3. If a is negative $(a < 0)$ and n is even, then $\sqrt[n]{a}$ is not a real number.
4. If $a = 0$, then $\sqrt[n]{a} = 0$.

We use the same vocabulary for nth roots that we used for square roots. That is, $\sqrt[n]{a}$ is called a **radical expression**, $\sqrt{\ }$ is the **radical sign**, and a is the **radicand**. The positive integer n is called the **index**. The index 2 is not written; we write $\sqrt{a}$ rather than $\sqrt[2]{a}$ for the principal square root of a. It is also common practice to call $\sqrt[3]{a}$ the **cube root** of a.

EXAMPLE 4 **Finding Principal nth Roots**

Find each root. **a.** $\sqrt[3]{27}$ **b.** $\sqrt[3]{-64}$ **c.** $\sqrt[4]{16}$ **d.** $\sqrt[4]{(-3)^4}$ **e.** $\sqrt[8]{-46}$

Solution

a. The radicand, 27, is *positive*; so $\sqrt[3]{27} = 3$ because $3^3 = 27$ and $3 > 0$.
b. The radicand, -64, is *negative* and 3 is odd; so $\sqrt[3]{-64} = -4$ because $(-4)^3 = -64$.
c. The radicand, 16, is *positive*; so $\sqrt[4]{16} = 2$ because $2^4 = 16$ and $2 > 0$.
d. The radicand, $(-3)^4$, is *positive*; so $\sqrt[4]{(-3)^4} = 3$ because $3^4 = (-3)^4$ and $3 > 0$.
e. The radicand, -46, is *negative* and the index, 8, is even; so $\sqrt[8]{-46}$ is not a real number.

Practice Problem 4 Find each root.

a. $\sqrt[3]{-8}$ **b.** $\sqrt[5]{32}$ **c.** $\sqrt[4]{81}$ **d.** $\sqrt[6]{-4}$

The fact that the definition of the principal nth root depends on whether n is even or odd results in two rules to replace the square root rule $\sqrt{x^2} = |x|$.

SIMPLIFYING $\sqrt[n]{a^n}$

If n is *odd*, then $\sqrt[n]{a^n} = a$.

If n is *even*, then $\sqrt[n]{a^n} = |a|$.

The product and quotient rules for square roots can be extended to apply to nth roots as well.

PRODUCT AND QUOTIENT RULES FOR nth ROOTS

If a and b are real numbers and all indicated roots are defined, then

$$\sqrt[n]{ab} = \sqrt[n]{a} \cdot \sqrt[n]{b}$$

and

$$\sqrt[n]{\frac{a}{b}} = \frac{\sqrt[n]{a}}{\sqrt[n]{b}}.$$

EXAMPLE 5 Simplifying nth Roots

Simplify. **a.** $\sqrt[3]{135}$ **b.** $\sqrt[4]{162a^4}$

Solution

a. We find a factor of 135 that is a perfect cube. Here $135 = 27 \cdot 5$ and 27 is a perfect cube $(3^3 = 27)$. Then

$$\sqrt[3]{135} = \sqrt[3]{27 \cdot 5} = \sqrt[3]{27}\sqrt[3]{5} = 3\sqrt[3]{5}.$$

b. We find a factor of 162 that is a perfect fourth power. Here $162 = 81 \cdot 2$ and 81 is a perfect fourth power: $3^4 = 81$. Thus,

$$\sqrt[4]{162a^4} = \sqrt[4]{162}\sqrt[4]{a^4} = \sqrt[4]{81 \cdot 2}|a| = \sqrt[4]{81}\sqrt[4]{2}|a| = 3\sqrt[4]{2}|a|.$$

Practice Problem 5 Simplify.

a. $\sqrt[3]{72}$ **b.** $\sqrt[4]{48a^2}$

4 Combine like radicals.

Like Radicals

Radicals that have the same index and the same radicand are called **like radicals**. Like radicals can be combined with the use of the distributive property.

EXAMPLE 6 Adding and Subtracting Radical Expressions

Simplify: **a.** $\sqrt{45} + 7\sqrt{20}$ **b.** $5\sqrt[3]{80x} - 3\sqrt[3]{270x}$

Solution

a. We find perfect-square factors for both 45 and 20.

$$\begin{aligned}
\sqrt{45} + 7\sqrt{20} &= \sqrt{9 \cdot 5} + 7\sqrt{4 \cdot 5} && 45 = 9.5;\ 20 = 4.5 \\
&= \sqrt{9}\sqrt{5} + 7\sqrt{4}\sqrt{5} && \text{Product rule} \\
&= 3\sqrt{5} + 7 \cdot 2\sqrt{5} && \sqrt{9} = 3;\ \sqrt{4} = 2 \\
&= 3\sqrt{5} + 14\sqrt{5} && \text{Like radicals} \\
&= (3 + 14)\sqrt{5} && \text{Distributive property} \\
&= 17\sqrt{5}
\end{aligned}$$

b. This time we find perfect-cube factors for both 80 and 270.

$$\begin{aligned} 5\sqrt[3]{80x} - 3\sqrt[3]{270x} &= 5\sqrt[3]{8 \cdot 10x} - 3\sqrt[3]{27 \cdot 10x} && 80 = 8.10;\ 270 = 27.10 \\ &= 5\sqrt[3]{8} \cdot \sqrt[3]{10x} - 3\sqrt[3]{27} \cdot \sqrt[3]{10x} && \text{Product rule} \\ &= 5 \cdot 2 \cdot \sqrt[3]{10x} - 3 \cdot 3\sqrt[3]{10x} && \sqrt[3]{8} = 2;\ \sqrt[3]{27} = 3 \\ &= 10\sqrt[3]{10x} - 9\sqrt[3]{10x} && \text{Like radicals} \\ &= (10 - 9)\sqrt[3]{10x} && \text{Distributive property} \\ &= \sqrt[3]{10x} \end{aligned}$$

Practice Problem 6 Simplify.

a. $3\sqrt{12} + 7\sqrt{3}$ **b.** $2\sqrt[3]{135x} - 3\sqrt[3]{40x}$

WARNING

$\sqrt[n]{a} + \sqrt[n]{b}$ cannot be added if $a \neq b$.
$\sqrt[n]{a} + \sqrt[n]{a} = 2\sqrt[n]{a}$
$\sqrt[n]{a} + \sqrt[m]{a}$ cannot be added if $n \neq m$.

Radicals with Different Indexes

Suppose a is a positive real number and m, n, and r are positive integers. Then the property

$$\sqrt[n]{a^m} = \sqrt[nr]{a^{mr}}$$

is used to multiply radicals with different indexes.

EXAMPLE 7 **Multiplying Radicals with Different Indexes**

Write $\sqrt{2}\sqrt[3]{5}$ as a single radical.

Solution

Because $\sqrt{2}$ can also be written as $\sqrt[2]{2}$, we have

$$\sqrt{2} = \sqrt[2]{2} = \sqrt[2 \cdot 3]{2^3} = \sqrt[6]{2^3},$$

and similarly,

$$\sqrt[3]{5} = \sqrt[2 \cdot 3]{5^2} = \sqrt[6]{5^2}.$$

So

$$\sqrt{2}\sqrt[3]{5} = \sqrt[6]{2^3}\sqrt[6]{5^2} = \sqrt[6]{2^3 5^2} = \sqrt[6]{200}.$$

Practice Problem 7 Write $\sqrt{3}\sqrt[5]{2}$ as a single radical.

5 Rationalize denominators or numerators.

Rationalizing Radical Expressions

Removing radicals in the denominator or the numerator is called **rationalizing the denominator** or **rationalizing the numerator,** respectively. The procedure for rationalizing involves multiplying the fraction by 1, written in a special way, so as to obtain a perfect nth power.

EXAMPLE 8 **Rationalizing the Denominator**

Rationalize each denominator.

a. $\dfrac{\sqrt{5}}{\sqrt{2}}$ **b.** $\dfrac{\sqrt[3]{4}}{\sqrt[3]{9}}$

Solution

a. $\dfrac{\sqrt{5}}{\sqrt{2}} = \dfrac{\sqrt{5}}{\sqrt{2}} \cdot \dfrac{\sqrt{2}}{\sqrt{2}} = \dfrac{\sqrt{10}}{\sqrt{4}} = \dfrac{\sqrt{10}}{2}$

b. $\dfrac{\sqrt[3]{4}}{\sqrt[3]{9}} = \dfrac{\sqrt[3]{4}}{\sqrt[3]{9}} \cdot \dfrac{\sqrt[3]{3}}{\sqrt[3]{3}} = \dfrac{\sqrt[3]{12}}{\sqrt[3]{27}} = \dfrac{\sqrt[3]{12}}{3}$

Practice Problem 8 Rationalize each denominator.

a. $\dfrac{7}{\sqrt{8}}$ **b.** $\dfrac{\sqrt[3]{3}}{\sqrt[3]{4}}$

Conjugates

Pairs of expressions of the form $a\sqrt{x} + b\sqrt{y}$ and $a\sqrt{x} - b\sqrt{y}$ are called conjugates. We form the product of these two conjugates.

$$(a\sqrt{x} + b\sqrt{y})(a\sqrt{x} - b\sqrt{y}) = (a\sqrt{x})^2 - (b\sqrt{y})^2 \quad \text{Difference of two squares}$$
$$= a^2x - b^2y \quad (ab)^2 = a^2b^2; (\sqrt{r})^2 = r$$

We see that the product contains no radicals. So the product of conjugates can be used to rationalize the denominator or the numerator.

EXAMPLE 9 **Rationalizing the Numerator**

Rationalize each numerator.

a. $\dfrac{\sqrt{5} - \sqrt{2}}{4}$ **b.** $\dfrac{\sqrt{x} + \sqrt{3}}{x^2 - 9}$

Solution

a. $$\frac{\sqrt{5} - \sqrt{2}}{4} = \frac{\sqrt{5} - \sqrt{2}}{4} \cdot \frac{\sqrt{5} + \sqrt{2}}{\sqrt{5} + \sqrt{2}}$$
$$= \frac{(\sqrt{5})^2 - (\sqrt{2})^2}{4(\sqrt{5} + \sqrt{2})} \quad (a - b)(a + b) = a^2 - b^2$$
$$= \frac{5 - 2}{4(\sqrt{5} + \sqrt{2})} = \frac{3}{4(\sqrt{5} + \sqrt{2})}$$

b. $$\frac{\sqrt{x} + \sqrt{3}}{x^2 - 9} = \frac{\sqrt{x} + \sqrt{3}}{x^2 - 9} \cdot \frac{\sqrt{x} - \sqrt{3}}{\sqrt{x} - \sqrt{3}}$$
$$= \frac{x - 3}{(x^2 - 9)(\sqrt{x} - \sqrt{3})} \quad (\sqrt{x} + \sqrt{3})(\sqrt{x} - \sqrt{3}) = x - 3$$
$$= \frac{x - 3}{(x - 3)(x + 3)(\sqrt{x} - \sqrt{3})} \quad \text{Factor: } x^2 - 9 = (x - 3)(x + 3).$$
$$= \frac{1}{(x + 3)(\sqrt{x} - \sqrt{3})} \quad \text{Remove common factor.}$$

Practice Problem 9 Rationalize each denominator.

a. $\dfrac{2}{\sqrt{7} - \sqrt{5}}$ **b.** $\dfrac{x - 2}{\sqrt{x} + \sqrt{2}}$

6 Define and evaluate rational exponents.

Rational Exponents

Previously, we defined such exponential expressions as 2^3, but what do you think $2^{1/3}$ means? If we want to preserve the power rule of exponents, $(a^n)^m = a^{nm}$, for rational exponents such as $\frac{1}{3}$, we must have

$$(2^{1/3})^3 = 2^{1/3 \cdot 3} = 2^1 = 2.$$

Because $(2^{1/3})^3 = 2$, it follows that $2^{1/3}$ is the cube root of 2, that is, $2^{1/3} = \sqrt[3]{2}$.

Such considerations lead to the following definition.

THE EXPONENT $\frac{1}{n}$

For any real number a and any integer $n > 1$,

$$a^{1/n} = \sqrt[n]{a}.$$

When n is even and $a < 0$, $\sqrt[n]{a}$ and $a^{1/n}$ are not real numbers.
Notice that the denominator n of the rational exponent is the index of the radical.

EXAMPLE 10 Evaluating Expressions with Rational Exponents

Evaluate: **a.** $16^{1/2}$ **b.** $(-27)^{1/3}$ **c.** $\left(\frac{1}{16}\right)^{1/4}$ **d.** $32^{1/5}$ **e.** $(-5)^{1/4}$

Solution

a. $16^{1/2} = \sqrt{16} = 4$

b. $(-27)^{1/3} = \sqrt[3]{-27} = \sqrt[3]{(-3)^3} = -3$

c. $\left(\frac{1}{16}\right)^{1/4} = \sqrt[4]{\frac{1}{16}} = \sqrt[4]{\left(\frac{1}{2}\right)^4} = \frac{1}{2}$

d. $32^{1/5} = \sqrt[5]{32} = \sqrt[5]{2^5} = 2$

e. Because the base -5 is negative and $n = 4$ is even, $(-5)^{1/4}$ is not a real number.

Practice Problem 10 Evaluate.

a. $\left(\frac{1}{4}\right)^{1/2}$ **b.** $(125)^{1/3}$ **c.** $(-32)^{1/5}$

The box below shows a summary of our work.

THE EXPONENT $\frac{1}{n}$ AND RADICALS

If a is any real number and n is any integer greater than 1, then

	***n* even**	***n* odd**
If $a > 0$,	$\sqrt[n]{a} = a^{1/n}$ is positive.	$\sqrt[n]{a} = a^{1/n}$ is positive.
If $a < 0$,	$\sqrt[n]{a} = a^{1/n}$ is not a real number.	$\sqrt[n]{a} = a^{1/n}$ is negative.
If $a = 0$,	$\sqrt[n]{0} = 0^{1/n} = 0$.	$\sqrt[n]{0} = 0^{1/n} = 0$.

We have defined the expression $a^{1/n}$. How do we define $a^{m/n}$, where m and n are integers? If we insist that the power-of-a-product rule of exponents holds for rational numbers, then we must have

$$a^{m/n} = a^{1/n \cdot m} = (a^{1/n})^m = (\sqrt[n]{a})^m$$

and

$$a^{m/n} = a^{m \cdot 1/n} = (a^m)^{1/n} = \sqrt[n]{a^m}.$$

From this, we arrive at the following definition.

RATIONAL EXPONENTS

$$a^{m/n} = (\sqrt[n]{a})^m = \sqrt[n]{a^m},$$

provided that m and n are integers with no common factors, $n > 1$, and $\sqrt[n]{a}$ is a real number.

Note that for $a^{m/n}$ the numerator m of the exponent $\frac{m}{n}$ is the *exponent of the radical expression* and the denominator n is the *index of the radical* in $(\sqrt[n]{a})^m$. When n is even and $a < 0$, the symbol $a^{m/n}$ is not a real number.

When rational exponents with negative denominators such as $\frac{3}{-5}$ and $\frac{-2}{-7}$ are encountered, we use equivalent expressions such as $\frac{-3}{5}$ and $\frac{2}{7}$, respectively, before applying the definition. Similarly, we reduce the exponent $\frac{m}{n}$ to its lowest terms before applying the definition. Note that $(-8)^{2/6} \neq [(-8)^{1/6}]^2$ because $(-8)^{1/6} = \sqrt[6]{-8}$ is not a real number but $(-8)^{2/6}$ is a real number. Writing $\frac{2}{6} = \frac{1}{3}$, we have

$$(-8)^{2/6} = (-8)^{1/3} = \sqrt[3]{-8} = -2.$$

EXAMPLE 11 Evaluating Expressions Having Rational Exponents

Evaluate: **a.** $8^{2/3}$ **b.** $-16^{5/2}$ **c.** $100^{-3/2}$ **d.** $(-25)^{7/2}$

Solution

a. $8^{2/3} = (\sqrt[3]{8})^2 = 2^2 = 4$

b. $-16^{5/2} = -(\sqrt{16})^5 = -4^5 = -1024$

c. $100^{-3/2} = (100^{1/2})^{-3} = (\sqrt{100})^{-3} = \frac{1}{(\sqrt{100})^3} = \frac{1}{10^3} = \frac{1}{1000}$

d. $(-25)^{7/2} = [(-25)^{1/2}]^7$ Not a real number
$= (\sqrt{-25})^7$ because $\sqrt{-25}$ is not a real number

Practice Problem 11 Evaluate.

a. $(25)^{2/3}$ **b.** $-36^{3/2}$ **c.** $16^{-5/2}$ **d.** $(-36)^{-1/2}$

All of the properties of exponents we learned for integer exponents hold true for rational exponents. For convenience, we list these properties next.

PROPERTIES OF RATIONAL EXPONENTS

If r and s are rational numbers and a and b are real numbers, then

$$a^r \cdot a^s = a^{r+s} \qquad (a^r)^s = a^{rs} \qquad (ab)^r = a^r b^r$$

$$\frac{a^r}{a^s} = a^{r-s} \qquad \left(\frac{a}{b}\right)^r = \frac{a^r}{b^r} \qquad a^{-r} = \frac{1}{a^r} \qquad \left(\frac{a}{b}\right)^{-r} = \left(\frac{b}{a}\right)^r$$

provided that all of the expressions used are defined.

EXAMPLE 12 **Simplifying Expressions Having Rational Exponents**

Simplify. Express the answer with only positive exponents. Assume that x represents a positive real number.

a. $2x^{1/3} \cdot 5x^{1/4}$ **b.** $\dfrac{21x^{-2/3}}{7x^{1/5}}$ **c.** $(x^{3/5})^{-1/6}$

Solution

a. $2x^{1/3} \cdot 5x^{1/4} = 2 \cdot 5x^{1/3} \cdot x^{1/4}$ Group factors with the same base.

$= 10x^{1/3 + 1/4}$ Add exponents of factors with the same base.

$= 10x^{4/12 + 3/12} = 10x^{7/12}$

b. $\dfrac{21x^{-2/3}}{7x^{1/5}} = \left(\dfrac{21}{7}\right)\left(\dfrac{x^{-2/3}}{x^{1/5}}\right)$ Group factors with the same base.

$= 3x^{-2/3 - 1/5} = 3x^{-10/15 - 3/15} = 3x^{-13/15} = 3 \cdot \left(\dfrac{1}{x^{13/15}}\right) = \dfrac{3}{x^{13/15}}$

c. $(x^{3/5})^{-1/6} = x^{(3/5)(-1/6)} = x^{-1/10} = \dfrac{1}{x^{1/10}}$

Practice Problem 12 Simplify. Express your answer with only positive exponents.

a. $4x^{1/2} \cdot 3x^{1/5}$ **b.** $\dfrac{25x^{-1/4}}{5x^{1/3}}, x > 0$ **c.** $(x^{2/3})^{-1/5}, x \neq 0$

EXAMPLE 13 **Simplifying Radicals by Using Rational Exponents**

Assume that x represents a positive real number. Simplify.

a. $\sqrt[8]{x^2}$ **b.** $\sqrt[4]{9}\sqrt{3}$ **c.** $\sqrt{\sqrt[3]{x^8}}$

Solution

a. $\sqrt[8]{x^2} = (x^2)^{1/8} = x^{2/8} = x^{1/4} = \sqrt[4]{x}$

b. $\sqrt[4]{9}\sqrt{3} = 9^{1/4} \cdot 3^{1/2} = (3^2)^{1/4} \cdot 3^{1/2} = 3^{2/4} \cdot 3^{1/2} = 3^{1/2} \cdot 3^{1/2} = 3^{1/2+1/2} = 3^1 = 3$

c. $\sqrt{\sqrt[3]{x^8}} = \sqrt{(x^8)^{1/3}} = \sqrt{x^{8/3}} = (x^{8/3})^{1/2} = x^{8/3 \cdot 1/2} = x^{4/3} =$
$x^{1+1/3} = x^1 \cdot x^{1/3} = x\sqrt[3]{x}$

Practice Problem 13 Simplify.

a. $\sqrt[6]{x^4}$ **b.** $\sqrt[4]{25}\sqrt{5}$ **c.** $\sqrt{\sqrt[3]{x^{12}}}$

EXAMPLE 14 Simplifying an Expression Involving Negative Rational Exponents

Simplify $x(x + 1)^{-1/2} + 2(x + 1)^{1/2}$. Express the answer with only positive exponents and then rationalize the denominator. Assume that $(x + 1)^{-1/2}$ is defined.

Solution

$$x(x+1)^{-1/2} + 2(x+1)^{1/2} = \frac{x}{(x+1)^{1/2}} + \frac{2(x+1)^{1/2}}{1} \qquad (x+1)^{-1/2} = \frac{1}{(x+1)^{1/2}}$$

$$= \frac{x}{(x+1)^{1/2}} + \frac{2(x+1)^{1/2}(x+1)^{1/2}}{(x+1)^{1/2}}$$

$$= \frac{x + 2(x+1)}{(x+1)^{1/2}} = \frac{3x+2}{(x+1)^{1/2}} \qquad (x+1)^{1/2}(x+1)^{1/2} = x+1$$

To rationalize the denominator $(x + 1)^{1/2} = \sqrt{x + 1}$, multiply both numerator and denominator by $(x + 1)^{1/2}$.

$$\frac{3x+2}{(x+1)^{1/2}} = \frac{(3x+2)(x+1)^{1/2}}{(x+1)^{1/2}(x+1)^{1/2}}$$

$$= \frac{(3x+2)(x+1)^{1/2}}{x+1} \qquad (x+1)^{1/2}(x+1)^{1/2} = x+1$$

Practice Problem 14 Simplify $x(x + 3)^{-1/2} + (x + 3)^{1/2}$.

Answers to Practice Problems

1. a. 12 **b.** $\frac{1}{7}$ **c.** $\frac{1}{4}$ **2. a.** $2\sqrt{5}$ **b.** $4\sqrt{3}$ **c.** $2|x|\sqrt{3}$ **d.** $\frac{2y\sqrt{5y}}{3\sqrt{3}|x|}$ **3.** ≈67 mph **4. a.** −2 **b.** 2 **c.** 3 **d.** Not a real number **5. a.** $2\sqrt[3]{9}$ **b.** $2\sqrt[4]{3a^2}$ **c.** a^2 **6. a.** $13\sqrt{3}$ **b.** 0 **7.** $\sqrt[10]{972}$ **8. a.** $\frac{7\sqrt{2}}{4}$ **b.** $\frac{\sqrt[3]{6}}{2}$ **9. a.** $\sqrt{7} + \sqrt{5}$ **b.** $\sqrt{x} - \sqrt{2}$ **10. a.** $\frac{1}{2}$ **b.** 5 **c.** −2 **11. a.** $5\sqrt[3]{5}$ **b.** −216 **c.** $\frac{1}{1024}$ **d.** Not a real number **12. a.** $12x^{7/10}$ **b.** $\frac{5}{x^{7/12}}$ **c.** $\frac{1}{x^{2/15}}$ **13. a.** $\sqrt[3]{x^2}$ **b.** 5 **c.** x^2 **14.** $\frac{(2x+3)(x+3)^{1/2}}{x+3}$

SECTION P.6 Exercises

Basic Concepts and Skills

1. Any positive number has ______________ square roots.
2. To rationalize the denominator of an expression with denominator $2\sqrt{5}$, multiply the ______________ and ______________ by ______________.
3. Radicals that have the same index and the same radicand are called ______________.
4. The radical notation for $7^{1/3}$ is ______________.
5. **True or False.** For all real x, $\sqrt{x^2} = x$.
6. **True or False.** The radicals $-3\sqrt{5}$ and $12\sqrt{5}$ are "like radicals."

In Exercises 7–22, evaluate each root or state that the root is not a real number.

7. $\sqrt{64}$
8. $\sqrt{100}$
9. $\sqrt[3]{64}$
10. $\sqrt[3]{125}$
11. $\sqrt[3]{-27}$
12. $\sqrt[3]{-216}$
13. $\sqrt[3]{-\frac{1}{8}}$
14. $\sqrt[3]{\frac{27}{64}}$
15. $\sqrt{(-3)^2}$
16. $-\sqrt{4(-3)^2}$
17. $\sqrt[4]{-16}$
18. $\sqrt[6]{-64}$
19. $-\sqrt[5]{-1}$
20. $\sqrt[7]{-1}$
21. $\sqrt[5]{(-7)^5}$
22. $-\sqrt[5]{(-4)^5}$

In Exercises 23–46, simplify each expression using the product and quotient properties for square roots. Assume that x represents a positive real number.

23. $\sqrt{32}$
24. $\sqrt{125}$
25. $\sqrt{18x^2}$
26. $\sqrt{27x^2}$
27. $\sqrt{9x^3}$
28. $\sqrt{8x^3}$
29. $\sqrt{6x}\sqrt{3x}$
30. $\sqrt{5x}\sqrt{10x}$
31. $\sqrt{15x}\sqrt{3x^2}$
32. $\sqrt{2x^2}\sqrt{18x}$
33. $\sqrt[3]{-8x^3}$
34. $\sqrt[3]{64x^3}$
35. $\sqrt[3]{-x^6}$
36. $\sqrt[3]{-27x^6}$
37. $\sqrt{\dfrac{5}{32}}$
38. $\sqrt{\dfrac{3}{50}}$
39. $\sqrt[3]{\dfrac{8}{x^3}}$
40. $\sqrt[3]{\dfrac{7}{8x^3}}$
41. $\sqrt[4]{\dfrac{2}{x^5}}$
42. $-\sqrt[4]{\dfrac{5}{16x^5}}$
43. $\sqrt{\sqrt{x^8}}$
44. $\sqrt{\dfrac{4x^4}{25y^6}}$
45. $\sqrt{\dfrac{9x^6}{y^4}}$
46. $\sqrt[5]{x^{15}y^5}$

In Exercises 47–60, simplify each expression. Assume that all variables represent nonnegative real numbers.

47. $2\sqrt{3} + 5\sqrt{3}$
48. $7\sqrt{2} - \sqrt{2}$
49. $6\sqrt{5} - \sqrt{5} + 4\sqrt{5}$
50. $5\sqrt{7} + 3\sqrt{7} - 2\sqrt{7}$
51. $\sqrt{98x} - \sqrt{32x}$
52. $\sqrt{45x} + \sqrt{20x}$
53. $\sqrt[3]{24} - \sqrt[3]{81}$
54. $\sqrt[3]{54} + \sqrt[3]{16}$
55. $\sqrt[3]{3x} - 2\sqrt[3]{24x} + \sqrt[3]{375x}$
56. $\sqrt[3]{16x} + 3\sqrt[3]{54x} - \sqrt[3]{2x}$
57. $\sqrt{2x^5} - 5\sqrt{32x} + \sqrt{18x^3}$
58. $2\sqrt{50x^5} + 7\sqrt{2x^3} - 3\sqrt{72x}$
59. $\sqrt{48x^5y} - 4y\sqrt{3x^3y} + y\sqrt{3xy^3}$
60. $x\sqrt{8xy} + 4\sqrt{2xy^3} - \sqrt{18x^5y}$

In Exercises 61–74, rationalize the denominator of each expression.

61. $\dfrac{2}{\sqrt{3}}$
62. $\dfrac{10}{\sqrt{5}}$
63. $\dfrac{7}{\sqrt{15}}$
64. $\dfrac{-3}{\sqrt{8}}$
65. $\dfrac{1}{\sqrt{2} + x}$
66. $\dfrac{1}{\sqrt{5} + 2x}$
67. $\dfrac{3}{2 - \sqrt{3}}$
68. $\dfrac{-5}{1 - \sqrt{2}}$
69. $\dfrac{1}{\sqrt{3} + \sqrt{2}}$
70. $\dfrac{6}{\sqrt{5} - \sqrt{3}}$
71. $\dfrac{\sqrt{5} - \sqrt{2}}{\sqrt{5} + \sqrt{2}}$
72. $\dfrac{\sqrt{7} + \sqrt{3}}{\sqrt{7} - \sqrt{3}}$
73. $\dfrac{\sqrt{x + h} - \sqrt{x}}{\sqrt{x + h} + \sqrt{x}}$
74. $\dfrac{a\sqrt{x} + 3}{a\sqrt{x} - 3}$

In Exercises 75–82, rationalize the numerator of each expression.

75. $\dfrac{\sqrt{4 + h} - 2}{h}$
76. $\dfrac{\sqrt{y + 9} - 3}{y}$
77. $\dfrac{2 - \sqrt{4 - x}}{x}$
78. $\dfrac{\sqrt{x + 2} - \sqrt{2}}{x}$
79. $\dfrac{\sqrt{x} - 2}{x - 4}$
80. $\dfrac{\sqrt{x} - \sqrt{5}}{x^2 - 25}$
81. $\dfrac{\sqrt{x^2 + 4x} - x}{2}$
82. $\dfrac{\sqrt{x^2 + 2x + 3} - \sqrt{x^2 + 3}}{2}$

In Exercises 83–92, evaluate each expression without using a calculator.

83. $25^{1/2}$
84. $144^{1/2}$
85. $(-8)^{1/3}$
86. $(-27)^{1/3}$
87. $8^{2/3}$
88. $16^{3/2}$
89. $-25^{-3/2}$
90. $-9^{-3/2}$
91. $\left(\dfrac{9}{25}\right)^{-3/2}$
92. $\left(\dfrac{1}{27}\right)^{-2/3}$

In Exercises 93–104, simplify each expression, leaving your answer with only positive exponents. Assume that all variables represent positive numbers.

93. $x^{1/2} \cdot x^{2/5}$
94. $x^{3/5} \cdot 5x^{2/3}$
95. $x^{3/5} \cdot x^{-1/2}$
96. $x^{5/3} \cdot x^{-3/4}$
97. $(8x^6)^{2/3}$
98. $(16x^3)^{1/2}$
99. $(27x^6y^3)^{-2/3}$
100. $(16x^4y^6)^{-3/2}$
101. $\dfrac{15x^{3/2}}{3x^{1/4}}$
102. $\dfrac{20x^{5/2}}{4x^{2/3}}$
103. $\left(\dfrac{x^{-1/4}}{y^{-2/3}}\right)^{-12}$
104. $\left(\dfrac{27x^{-5/2}}{y^{-3}}\right)^{-1/3}$

In Exercises 105–114, convert each radical expression to its rational exponent form and then simplify. Assume that all variables represent positive numbers.

105. $\sqrt[4]{3^2}$
106. $\sqrt[4]{5^2}$
107. $\sqrt[3]{x^9}$
108. $\sqrt[3]{x^{12}}$
109. $\sqrt[3]{x^6y^9}$
110. $\sqrt[3]{8x^3y^{12}}$
111. $\sqrt[4]{9}\sqrt{3}$
112. $\sqrt[4]{49}\sqrt{7}$
113. $\sqrt{\sqrt[3]{x^{10}}}$
114. $\sqrt{\sqrt[3]{64x^6}}$

In Exercises 115–122, convert the given product to a single radical. Assume that all variables represent positive numbers.

115. $\sqrt[3]{2} \cdot \sqrt[5]{3}$

116. $\sqrt[3]{3} \cdot \sqrt[4]{2}$

117. $\sqrt[3]{x^2} \cdot \sqrt[4]{x^3}$

118. $\sqrt[5]{x^2} \cdot \sqrt[7]{x^5}$

119. $\sqrt[9]{2m^2n} \cdot \sqrt[3]{5m^5n^2}$

120. $\sqrt{3xy} \cdot \sqrt[3]{9x^2y^2}$

121. $\sqrt[5]{x^4y^3} \cdot \sqrt[6]{x^3y^5}$

122. $\sqrt[5]{3a^4b^2} \cdot \sqrt[6]{7a^2b^6}$

In Exercises 123–128, factor each expression. Use only positive exponents in your answer.

123. $\frac{4}{3}x^{1/3}(2x - 3) + 2x^{4/3}$

124. $\frac{4}{3}x^{1/3}(7x + 1) + 7x^{4/3}$

125. $4(3x + 1)^{1/3}(2x - 1) + 2(3x + 1)^{4/3}$

126. $4(3x - 1)^{1/3}(x + 2) + (3x - 1)^{4/3}$

127. $3x(x^2 + 1)^{1/2}(2x^2 - x) + 2(x^2 + 1)^{3/2}(4x - 1)$

128. $3x(x^2 + 2)^{1/2}(x^2 - 2x) + (x^2 + 2)^{3/2}(2x - 2)$

129. Show that $\frac{1}{3 - \sqrt{8}} - \frac{1}{\sqrt{8} - \sqrt{7}} + \frac{1}{\sqrt{7} - \sqrt{6}} - \frac{1}{\sqrt{6} - \sqrt{5}} + \frac{1}{\sqrt{5} - 2} = 5.$

130. Show that $\frac{1}{1 + \sqrt{2}} + \frac{1}{\sqrt{2} + \sqrt{3}} + \frac{1}{\sqrt{3} + \sqrt{4}} + \frac{1}{\sqrt{4} + \sqrt{5}} + \frac{1}{\sqrt{5} + \sqrt{6}} + \frac{1}{\sqrt{6} + \sqrt{7}} + \frac{1}{\sqrt{7} + \sqrt{8}} + \frac{1}{\sqrt{8} + \sqrt{9}} = 2.$

131. Let $x = 2 + \sqrt{3}$. Show that

a. $x + \frac{1}{x} = 4.$ **b.** $x^2 + \frac{1}{x^2} = 14.$

132. Let $x = 3 - 2\sqrt{2}$. Show that $x^2 + \frac{1}{x^2} = 34.$

Applying the Concepts

133. Sign dimensions. A sign in the shape of an equilateral triangle of area A has sides of length $\sqrt{\frac{4A}{\sqrt{3}}}$ centimeters. Find the length of a side assuming that the sign has an area of 692 square centimeters. (Use the approximation $\sqrt{3} \approx 1.73$.)

134. Return on investment. For an initial investment of P dollars to mature to S dollars after two years when the interest is compounded annually, an annual interest rate of $r = \sqrt{\frac{S}{P}} - 1$ is required. Find r assuming that $P = \$1{,}210{,}000$ and $S = \$1{,}411{,}344$.

135. Terminal speed. The terminal speed V of a steel ball that weighs w grams and is falling in a cylinder of oil is given by the equation $V = \sqrt{\frac{w}{1.5}}$ centimeters per second. Find the terminal speed of a steel ball weighing 6 grams.

136. Model airplane speed. A spring balance scale attached to a wire that holds a model airplane on a circular path indicates the force F on the wire. The speed of the plane V, in m/sec, is given by $V = \sqrt{(rF)/m}$, where m is the mass of the plane (in kilograms) and r is the length of the wire (in meters). What is the top speed of a 2-kilogram plane on an 18-meter wire if the force on the wire at top speed is 49 newtons?

137. Electronic game current. The current I (in amperes) in a circuit for an electronic game using W watts and having a resistance of R ohms is given by $I = \sqrt{\frac{W}{R}}$. Find the current in a game circuit using 1058 watts and having a resistance of 2 ohms.

138. Radius of the moon. The radius of a sphere having volume V is given by $r = \left(\frac{3V}{4\pi}\right)^{1/3}$. Assuming that we treat the moon as a sphere and are told that its volume is 2.19×10^{19} cubic meters, find its radius. Use 3.14 as an approximation for π.

139. Atmospheric pressure. The atmospheric pressure P, measured in pounds per square inch, at an altitude of h miles $(h < 50)$ is given by the equation $P = 14.7(0.5)^{h/3.25}$. Compute the atmospheric pressure at an altitude of 16.25 miles.

140. Force from constrained water. The force F that is exerted on the semicircular end of a trough full of water is approximately $F = 42r^{3/2}$ pounds, where r is the radius of the semicircular end. Find the force assuming that the radius is 3 feet.

SUMMARY Definitions, Concepts, and Formulas

P.1 The Real Numbers and Their Properties

i. Classifying numbers.

Natural numbers	$1, 2, 3, \ldots$
Whole numbers	$0, 1, 2, 3, \ldots$
Integers	$\ldots, -2, -1, 0, 1, 2, \ldots$
Rational numbers	Numbers that can be expressed as the quotient $\frac{a}{b}$ of two integers, where $b \neq 0$; a terminating or repeating decimal
Irrational numbers	Decimals that neither terminate nor repeat
Real numbers	All of the rational and irrational numbers

ii. Coordinate line. Real numbers can be associated with a geometric line called a number line or coordinate line. Numbers associated with points to the right (left) of zero are called positive (negative) numbers.

iii. Ordering numbers. $a < b$ means that $b = a + c$ for some positive number c.
Trichotomy property. For two numbers a and b, exactly one of the following is true: $a < b$, $a = b$, or $a > b$.
Transitive property. If $a < b$ and $b < c$, then $a < c$.

iv. Absolute value.

$$|a| = a \text{ if } a \geq 0, \text{ and } |a| = -a \text{ if } a < 0.$$

The **distance** between points a and b on a number line is $|a - b|$.

v. Algebraic properties.

Commutative	$a + b = b + a,$	$ab = ba$
Associative	$(a + b) + c = a + (b + c)$	
	$(ab)c = a(bc)$	
Distributive	$a \cdot (b + c) = a \cdot b + a \cdot c$	
	$(a + b) \cdot c = a \cdot c + b \cdot c$	
Identity	$a + 0 = 0 + a = a,$	$a \cdot 1 = 1 \cdot a = a$
Inverse properties	$a + (-a) = 0,$	$a \cdot \frac{1}{a} = 1,$ where $a \neq 0$
Zero-product properties	$0 \cdot a = 0 = a \cdot 0$	
	If $a \cdot b = 0$, then $a = 0$ or $b = 0$.	

P.2 Integer Exponents and Scientific Notation

i. If n is a positive integer, then

$$a^n = \underbrace{a \cdot a \cdots a}_{n \text{ factors}}$$

$$\left.\begin{aligned} a^0 &= 1 \\ a^{-n} &= \frac{1}{a^n} \end{aligned}\right\} \quad a \neq 0$$

ii. Rules of exponents. Denominators and exponents attached to a base of zero are assumed not to be zero.

Product Rule	$a^m \cdot a^n = a^{m+n}$
Quotient Rule	$\frac{a^m}{a^n} = a^{m-n}$
Power-of-a-power Rule	$(a^m)^n = a^{mn}$
Power-of-a-Product Rule	$(a \cdot b)^n = a^n \cdot b^n$
Power-of-a-Quotient Rules	$\left(\frac{a}{b}\right)^n = \frac{a^n}{b^n}$
	$\left(\frac{a}{b}\right)^{-n} = \left(\frac{b}{a}\right)^n = \frac{b^n}{a^n}$

P.3 Polynomials

An algebraic expression of the form

$$a_n x^n + a_{n-1}x^{n-1} + \cdots + a_2x^2 + a_1x + a_0,$$

where n is a nonnegative integer, is a polynomial in x. If $a_n \neq 0$, then n is called the degree of the polynomial. The term $a_n x^n$ is the **leading term**, and a_0 is the **constant term**.

P.4 Factoring Polynomials

Factoring formulas:

$$A^2 - B^2 = (A - B)(A + B)$$
$$A^2 + 2AB + B^2 = (A + B)^2$$
$$A^2 - 2AB + B^2 = (A - B)^2$$
$$A^3 - B^3 = (A - B)(A^2 + AB + B^2)$$
$$A^3 + B^3 = (A + B)(A^2 - AB + B^2)$$

P.5 Rational Expressions

The quotient of two polynomials is called a rational expression.

i. Multiplication and division.

$$\frac{A}{B} \cdot \frac{C}{D} = \frac{A \cdot C}{B \cdot D}$$

$$\frac{\frac{A}{B}}{\frac{C}{D}} = \frac{A}{B} \div \frac{C}{D} = \frac{A}{B} \cdot \frac{D}{C} = \frac{A \cdot D}{B \cdot C}$$

All of the denominators are assumed not to be zero.

ii. Addition and subtraction.

$$\frac{A}{B} \pm \frac{C}{D} = \frac{A \cdot D \pm B \cdot C}{B \cdot D}$$

The denominators are assumed not to be zero.

P.6 Rational Exponents and Radicals

i. If $a > 0$, then $\sqrt[n]{a} = b$ means that $b^n = a$ and $b > 0$.

ii. If $a < 0$ and n is odd, then $\sqrt[n]{a} = b$ means that $b^n = a$. If $a < 0$ and n is even, then $\sqrt[n]{a}$ is not defined.

iii. When all expressions are defined,

$$a^{m/n} = \sqrt[n]{a^m} = (\sqrt[n]{a})^m; \quad a^{-m/n} = \frac{1}{a^{m/n}}.$$

iv. $\sqrt{a^2} = |a|$

REVIEW EXERCISES

Basic Concepts and Skills

1. List all of the numbers in the set

$$\left\{4, \sqrt{7}, -5, \frac{1}{2}, 0.\overline{31}, 0, 0.2, \sqrt{12}\right\} \text{ that are}$$

a. natural numbers.
b. whole numbers
c. integers.
d. rational numbers.
e. irrational numbers.
f. real numbers.

In Exercises 2–5, state which property of the real numbers is being used.

2. $3x + 1 = 1 + 3x$
3. $\pi(x + y) = \pi x + \pi y$
4. $(x^2 + 1) \cdot 0 = 0 \cdot (x^2 + 1)$
5. $(x^2 + 1) \cdot 1 = x^2 + 1$

In Exercises 6 and 7, write each interval in inequality notation and then graph each interval.

6. $[-2, 3)$
7. $(-\infty, 1]$

In Exercises 8 and 9, write each inequality in interval notation and then graph the interval.

8. $1 \le x < 4$
9. $x > 0$

In Exercises 10–13, rewrite each expression without absolute value bars.

10. $\frac{-12}{|2|}$
11. $|2 - |-3||$
12. $|-5 - |-7||$
13. $|1 - \sqrt{15}|$

In Exercises 14–37, evaluate each expression.

14. -5^0
15. $(-4)^3$
16. $\frac{2^{18}}{2^{17}}$
17. $2^4 - 5 \cdot 3^2$
18. $(2^3)^2$
19. $-25^{1/2}$
20. $\left(\frac{1}{16}\right)^{1/4}$
21. $8^{4/3}$
22. $(-27)^{2/3}$
23. $\left(\frac{25}{36}\right)^{-3/2}$
24. $81^{-3/2}$
25. $2 \cdot 5^3$
26. $3^2 \cdot 2^5$
27. $\frac{21 \times 10^6}{3 \times 10^7}$
28. $\sqrt[4]{10{,}000}$
29. $\sqrt{5^2}$
30. $\sqrt{(-9)^2}$
31. $\sqrt[3]{-125}$
32. $3^{1/2} \cdot 27^{1/2}$
33. $64^{-1/3}$
34. $\sqrt{5}\sqrt{20}$
35. $(\sqrt{3} + 2)(\sqrt{3} - 2)$
36. $\left(\frac{1}{16}\right)^{-3/2}$
37. $\frac{3^{-2} \cdot 7^0}{18^{-1}}$

In Exercises 38–41, evaluate each expression for the given values of x and y.

38. $4(x + 7) + 2y;\ x = 3, y = 4$
39. $\frac{y - 6\sqrt{x}}{xy};\ x = 4, y = 2$
40. $\frac{|x|}{x} + \frac{|y|}{y};\ x = 6, y = -3$
41. $x^y;\ x = 16, y = -\frac{1}{2}$

In Exercises 42–49, simplify each expression. Write each answer with only positive exponents. Assume that all variables represent positive numbers.

42. $\frac{x^{-6}}{x^{-9}}$
43. $(x^{-2})^{-5}$
44. $\frac{x^{-3}}{y^{-2}}$
45. $\left(\frac{x^{-3}}{y^{-1}}\right)^{-2}$
46. $\left(\frac{xy^3}{x^5y}\right)^{-2}$
47. $(16x^{-2/3}y^{-4/3})^{3/2}$
48. $(49x^{-4/3}y^{2/3})^{-3/2}$
49. $\left(\frac{64y^{-9/2}}{x^{-3}}\right)^{-2/3}$

In Exercises 50–66, simplify each expression. Assume that all variables represent positive numbers.

50. $(2x^{2/3})(5x^{3/4})$
51. $(7x^{1/4})(3x^{3/2})$
52. $\frac{32x^{2/3}}{8x^{1/4}}$
53. $\frac{x^5(2x)^3}{4x^3}$
54. $\frac{(64x^4y^4)^{1/2}}{4y^2}$
55. $\left(\frac{x^2y^{4/3}}{x^{1/3}}\right)^6$
56. $7\sqrt{3} + 3\sqrt{75}$
57. $\frac{\sqrt{64}}{\sqrt{11}}$
58. $\sqrt{180x^2}$
59. $7\sqrt{6} - 3\sqrt{24}$
60. $7\sqrt[3]{54} + \sqrt[3]{128}$
61. $\sqrt{2x}\sqrt{6x}$
62. $\sqrt{75x^2}$
63. $\frac{\sqrt{100x^3}}{\sqrt{4x}}$
64. $5\sqrt{2x} - 2\sqrt{8x}$
65. $4\sqrt[3]{135} + \sqrt[3]{40}$
66. $y\sqrt[3]{56x} - \sqrt[3]{189xy^3}$

In Exercises 67–70, rationalize the denominator of each expression.

67. $\frac{7}{\sqrt{3}}$
68. $\frac{4}{9 - \sqrt{6}}$
69. $\frac{1 - \sqrt{3}}{1 + \sqrt{3}}$
70. $\frac{2\sqrt{5} + 3}{3 - 4\sqrt{5}}$
71. Write $3.7 \times (6.23 \times 10^{12})$ in scientific notation.
72. Write $\frac{3.19 \times 10^{-9}}{0.02 \times 10^{-3}}$ in decimal notation.

In Exercises 73–82, perform the indicated operations. Leave the resulting polynomial in standard form.

73. $(x^3 - 6x^2 + 4x - 2) + (3x^3 - 6x^2 + 5x - 4)$

74. $(10x^3 + 8x^2 - 7x - 3) - (5x^3 - x^2 + 4x - 9)$

75. $(4x^4 + 3x^3 - 5x^2 + 9) + (5x^4 + 8x^3 - 7x^2 + 5)$

76. $(8x^4 + 4x^3 + 3x^2 + 5) + (7x^4 + 3x^3 + 8x^2 - 3)$

77. $(x - 12)(x - 3)$

78. $(x - 7)^2$

79. $(x^5 - 2)(x^5 + 2)$

80. $(4x - 3)(4x + 3)$

81. $(2x + 5)(3x - 11)$

82. $(3x - 6)(x^2 + 2x + 4)$

In Exercises 83–104, factor each polynomial completely.

83. $x^2 - 3x - 10$

84. $x^2 + 10x + 9$

85. $24x^2 - 38x - 11$

86. $15x^2 + 33x + 18$

87. $x(x + 11) + 5(x + 11)$

88. $x^3 - x^2 + 2x - 2$

89. $x^4 - x^3 + 7x - 7$

90. $9x^2 + 24x + 16$

91. $10x^2 + 23x + 12$

92. $8x^2 + 18x + 9$

93. $12x^2 + 7x - 12$

94. $x^4 - 4x^2$

95. $4x^2 - 49$

96. $16x^2 - 81$

97. $x^2 + 12x + 36$

98. $x^2 - 10x + 100$

99. $64x^2 + 48x + 9$

100. $8x^3 - 1$

101. $8x^3 + 27$

102. $7x^3 - 7$

103. $x^3 + 5x^2 - 16x - 80$

104. $x^3 + 6x^2 - 9x - 54$

In Exercises 105–118, perform the indicated operation. Simplify your answer.

105. $\frac{4}{x - 9} - \frac{10}{9 - x}$

106. $\frac{2}{x^2 - 3x + 2} + \frac{6}{x^2 - 1}$

107. $\frac{3x + 5}{x^2 + 14x + 48} - \frac{3x - 2}{x^2 + 10x + 16}$

108. $\frac{x + 7}{x^2 - 7x + 6} - \frac{5 - 3x}{x^2 - 2x - 24}$

109. $\frac{x - 1}{x + 1} - \frac{x + 1}{x - 1}$

110. $\frac{x^3 - 1}{3x^2 - 3} \cdot \frac{6x + 6}{x^2 + x + 1}$

111. $\frac{x - 1}{2x - 3} \cdot \frac{4x^2 - 9}{2x^2 - x - 1}$

112. $\frac{x^2 + 2x - 8}{x^2 + 5x + 6} \cdot \frac{x + 2}{x + 4}$

113. $\frac{x^2 - 4}{4x^2 - 9} \cdot \frac{2x^2 - 3x}{2x + 4}$

114. $\frac{3x^2 - 17x + 10}{x^2 - 4x - 5} \cdot \frac{x^2 + 3x + 2}{x^2 + x - 2}$

115. $\frac{\frac{1}{x^2} - x}{\frac{1}{x^2} + x}$

116. $\frac{\frac{x}{x - 2} + x}{\frac{1}{x^2 - 4}}$

117. $\frac{\frac{x}{x - 3} + x}{\frac{x}{3 - x} - x}$

118. $\frac{x + 2 - \frac{18}{x - 5}}{x - 1 - \frac{12}{x - 5}}$

Applying the Concepts

119. Assuming that a right triangle has legs a and b of lengths 20 and 21, respectively, find the length of the hypotenuse.

120. One leg of a right triangle measures 5 inches. The hypotenuse is 1 inch longer than the other leg. Find the length of the hypotenuse.

121. Find the area A and the perimeter P of a right triangle with hypotenuse of length 20 and a leg of length 12.

122. Find the volume V of a box of length $\frac{1}{3}$, width 7, and height 60.

123. Find the area of a rectangular rug if its width is 6 feet and its diagonal measures 10 feet.

124. The amount of alcohol in the body of a person who drinks x grams of alcohol every hour over a relatively long period is $\frac{4.2x}{10 - x}$ grams. How many grams of alcohol will be in the body of a person who drinks 2 grams of alcohol every hour (over a relatively long period)?

125. Because of the earth's curvature, the maximum distance a person can see from a height h above the ground is $\sqrt{7920h + h^2}$, where h is given in miles. What is the maximum distance an airplane pilot can see when the plane is 0.7 mile above the ground?

126. A contaminated reservoir contains 2% arsenic. The percentage (in decimal) of arsenic in the reservoir can be reduced by adding water. Assuming that the reservoir contains a million gallons of water, write a rational expression whose values give the percentage of arsenic in the reservoir when x gallons of water are added to it.

127. The height (in feet) of an open fruit crate that is three times as long as it is wide is given by $\frac{36 - 3x^2}{8x}$, where x is the width of the box. Find the height of a fruit crate that is 2 feet wide.

128. An object thrown down with an initial velocity of v_0 feet per second will travel $16t^2 + v_0t$ feet in t seconds. How far has an orange that is dropped from a hot air balloon traveled after 9 seconds if it has an initial downward velocity of 25 feet per second?

PRACTICE TEST

Rewrite each expression without absolute value bars.

1. $|7 - |-3||$
2. $|\sqrt{2} - 100|$
3. Evaluate $\frac{x - 3y}{2} + xy$ for $x = 5$ and $y = -3$.
4. Determine the domain of the variable x in the expression $\frac{2x}{(x - 3)(x + 9)}$ and write it in interval notation.

Simplify each expression. Assume that all expressions containing variables are defined.

5. $\left(\frac{-3x^2y}{x}\right)^3$
6. $\sqrt[3]{-8x^6}$
7. $\sqrt{75x} - \sqrt{27x}$
8. $-16^{-3/2}$
9. $\left(\frac{x^{-2}y^4}{25x^3y^3}\right)^{-1/2}$
10. Rationalize the denominator of the expression $\frac{5}{1 - \sqrt{3}}$.

Perform the indicated operations. Write the resulting polynomial in standard form.

11. $4(x^2 - 3x + 2) + 3(5x^2 - 2x + 1)$
12. $(x - 2)(5x - 1)$
13. $(x^2 + 3y^2)^2$
14. A circular rug has a border of uniform width. The distance from the center of the rug to the inner and outer edges of the border are 5 feet and x feet respectively. Write a completely factored polynomial that gives the area of the border.

Factor each polynomial completely.

15. $x^2 - 5x + 6$
16. $9x^2 + 12x + 4$
17. $8x^3 - 27$

Perform the indicated operations and simplify the result. Leave your answer in factored form.

18. $\frac{6 - 3x}{x^2 - 4} - \frac{12}{2x + 4}$
19. $\frac{\frac{1}{x^2}}{9 - \frac{1}{x^2}}$
20. Find the length of the hypotenuse of a right triangle with legs a and b of lengths 35 and 12, respectively.

CHAPTER 1

Equations and Inequalities

TOPICS

Equations and inequalities are useful in analyzing designs found in the art and architecture of ancient civilizations, in understanding the dazzling geometric constructions that abound in nature, and in solving problems common to industry and daily life.

SECTION **1.1**

Linear Equations in One Variable

BEFORE STARTING THIS SECTION, REVIEW

1 Algebraic expressions (Section P.1, page 13)
2 Like terms (Section P.3, page 29)
3 Arithmetic of fractions, least common denominator (Section P.5, page 51)

OBJECTIVES

1 Learn the vocabulary and concepts used in studying equations.
2 Solve linear equations in one variable.
3 Solve formulas for a specific variable.
4 Use linear equations to solve applied problems.

◆ Bungee-Jumping TV Contestants

England's Oxford Dangerous Sports Club started the modern version of the sport of "bungee" (or bungy) jumping on April 1, 1978, from the Clifton Suspension Bridge in Bristol, England. Bungee cords consist of hundreds of continuous-length rubber strands encased in a nylon sheath. One end of the cord is tied to a solid structure such as a crane or bridge, and the other end is tied via padded ankle straps to the jumper's ankle. The jumper jumps, and if things go as planned, the cord extends like a rubber band and the jumper bounces up and down until coming to a halt. Suppose a television adventure program has its contestants bungee jump from a bridge 120 feet above the water. The bungee cord that is used has a 140%–150% elongation, meaning that its extended length will be the original length plus a maximum of an additional 150% of its original length. If the show's producer wants to make sure the jumper does not get closer than 10 feet to the water and if no contestant is more than 7 feet tall, how long can the bungee cord be? Using the information from this section, we learn in Example 11 that the cord should be about 41 feet long.

1 Learn the vocabulary and concepts used in studying equations.

Definitions

An **equation** is a statement that two mathematical expressions are equal. For example,

$$7 - 5 = 2$$

is an equation. In algebra, however, we are generally more interested in equations that contain variables.

An **equation in one variable** is a statement that two expressions, with at least one containing the variable, are equal. For example, $2x - 3 = 7$ is an equation in the variable x. The expressions $2x - 3$ and 7 are the *sides* of the equation. The **domain** of the variable in an equation is the set of all real numbers for which both sides of the equation are defined.

SIDE **NOTE**

When finding the domain of a variable, remember that

1. Division by 0 is undefined.
2. The square root (or any even root) of a negative number is not a real number.

EXAMPLE 1 **Finding the Domain of the Variable**

Find the domain of the variable x in each of the following equations.

a. $2 = \frac{5}{x - 1}$ **b.** $x - 2 = \sqrt{x}$ **c.** $2x - 3 = 7$

Solution

a. In the equation $2 = \frac{5}{x - 1}$, the left side 2 does not contain x; so it is defined for all values of x. Because division by 0 is undefined, $\frac{5}{x - 1}$ is not defined if $x = 1$. The domain of x is the set of all real numbers except the number 1. Frequently, when only a few numbers are excluded from the domain of the variable, we write the exceptions on the side of the equation. In this example, if we wanted to specify the domain of the variable, we would simply write $2 = \frac{5}{x - 1}, x \neq 1$. This domain can be written in interval notation as $(-\infty, 1) \cup (1, \infty)$.

b. The square root of a negative number is not a real number; so $\sqrt{x}$, the right side of the equation $x - 2 = \sqrt{x}$, is defined only when $x \geq 0$. The left side is defined for all real numbers. So the domain of x in this equation is $\{x | x \geq 0\}$, or in interval notation, $[0, \infty)$.

c. Because both sides of the equation $2x - 3 = 7$ are defined for all real numbers, the domain is $\{x | x \text{ is a real number}\}$, or in interval notation, $(-\infty, \infty)$.

Practice Problem 1 Find the domain of the variable x in each of the following equations.

a. $\frac{x}{3} - 7 = 5$ **b.** $\frac{3}{2 - x} = 4$ **c.** $\sqrt{x - 1} = 0$

When the variable in an equation is replaced with a specific value from its domain, the resulting statement may be true or false. For example, in the equation $2x - 3 = 7$, if we let $x = 1$, we obtain

$$2(1) - 3 = 7 \text{ or } -1 = 7,$$

which is obviously false. However, if we replace x with 5, we obtain $2(5) - 3 = 7$, a true statement. Those values (if any) of the variable that result in a true statement are called **solutions** or **roots** of the equation. Thus, 5 is a solution (or root) of the equation $2x - 3 = 7$. We also say that 5 **satisfies** the equation $2x - 3 = 7$. To **solve** an equation means to find all solutions of the equation; the set of all solutions of an equation is called its **solution set**. The process of finding the solution set of an equation is called **solving the equation**.

An equation that is satisfied by every real number in the domain of the variable is called an **identity**. The equations

$$2(x + 3) = 2x + 6 \quad \text{and} \quad x^2 - 9 = (x + 3)(x - 3)$$

are examples of identities. You should recognize that these equations are examples of the distributive property and the difference of squares, respectively.

An equation that is not an identity but is satisfied by at least one real number is called a **conditional equation**. To verify that an equation is a conditional equation, you must find at least one number that *is* a solution of the equation and at least one number that is *not* a solution of the equation. The equation $2x - 3 = 7$ is an example of a conditional equation: 5 satisfies the equation, but 1 does not.

An equation that *no* number satisfies is called an **inconsistent equation**. The solution set of an inconsistent equation is designated by the empty set symbol, $\varnothing$. The equation

$x = x + 5$ is an example of an inconsistent equation. It is inconsistent because no number is 5 more than itself.

In summary, there are three types of equations: *identities*, *conditional equations*, and *inconsistent equations*.

Equivalent Equations

Equations that have the same solution set are called **equivalent equations**. For example, equations $x = 4$ and $3x = 12$ are equivalent equations with solution set $\{4\}$.

In general, to solve an equation in one variable, we replace the given equation with a sequence of equivalent equations until we obtain an equation whose solution is obvious, such as $x = 4$. The following operations yield equivalent equations.

Generating Equivalent Equations

Operations	Given Equation	Equivalent Equation
Simplify expressions on either side by eliminating parentheses, combining like terms, etc.	$(2x - 1) - (x + 1) = 4$	$2x - 1 - x - 1 = 4$ or $x - 2 = 4$
Add (or subtract) the same expression (with the same domain) on *both* sides of the equation.	$x - 2 = 4$	$x - 2 + 2 = 4 + 2$ or $x = 6$
Multiply (or divide) *both* sides of the equation by the same *nonzero* number.	$2x = 8$	$\frac{1}{2} \cdot 2x = \frac{1}{2} \cdot 8$ or $x = 4$
Interchange the two sides of the equation.	$-3 = x$	$x = -3$

2 Solve linear equations in one variable.

Solving Linear Equations in One Variable

Linear Equations

A conditional linear equation in one variable, such as x, is an equation that can be written in the *standard form*

$$ax + b = 0,$$

where a and b are real numbers with $a \neq 0$.

Because the highest exponent on the variable in a linear equation is $1 (x = x^1)$, a linear equation is also called a *first-degree equation*. We can see that the linear equation $ax + b = 0$ with $a \neq 0$ has exactly one solution by using the following steps:

$$ax + b = 0 \quad \text{Original equation, } a \neq 0$$

$$ax = -b \quad \text{Subtract } b \text{ from both sides.}$$

$$x = -\frac{b}{a} \quad \text{Divide both sides by } a.$$

In the Procedure in Action box, we describe a step-by-step procedure for solving a linear equation in one variable.

PROCEDURE
IN ACTION

EXAMPLE 2 **Solving Conditional Linear Equations in One Variable**

OBJECTIVE

Solve a linear equation in one variable.

EXAMPLE

Solve $\frac{1}{3}x - \frac{2}{3} = \frac{1}{2}(1 - 3x)$.

Step 1 Eliminate fractions, if any. Multiply both sides of the equation by the least common denominator (LCD) of all of the fractions.

$\frac{1}{3}x - \frac{2}{3} = \frac{1}{2}(1 - 3x)$ Original equation

$6\left(\frac{1}{3}x - \frac{2}{3}\right) = 6 \cdot \frac{1}{2}(1 - 3x)$ Multiply both sides by 6 (LCD) to eliminate fractions.

Step 2 Simplify. Simplify both sides of the equation by removing parentheses and other grouping symbols (if any) and combining like terms.

$6 \cdot \frac{1}{3}x - 6 \cdot \frac{2}{3} = 3(1 - 3x)$ Distribute on the left side; simplify on the right.

$2x - 4 = 3 - 9x$ Simplify on the left side; distribute on the right.

Step 3 Isolate the variable term. Add appropriate expressions to both sides so that when both sides are simplified, the terms containing the variable are on one side and all constant terms are on the other side.

$2x - 4 + 9x + 4 = 3 - 9x + 9x + 4$ Add $9x + 4$ to both sides.

Step 4 Combine terms. Combine terms containing the variable to obtain one term that contains the variable as a factor.

$11x = 7$ Collect like terms and combine constants.

Step 5 Isolate the variable. Divide both sides by the coefficient of the variable to obtain the solution.

$x = \frac{7}{11}$ Divide both sides by 11.

Step 6 Check the solution. Substitute the solution into the original equation to check for computational errors.

Check:

Substitute $\frac{7}{11}$ for x in both sides of the original equation.

The result should be $-\frac{5}{11} = -\frac{5}{11}$.

Practice Problem 2 Solve $\frac{2}{3} - \frac{3}{2}x = \frac{1}{6} - \frac{7}{3}x$.

EXAMPLE 3 **Solving a Linear Equation That Is Inconsistent**

Solve $5x - [3x + 2(x - 1)] = 1$.

Solution

Step 2

$5x - [3x + 2(x - 1)] = 1$ Original equation

$5x - [3x + 2x - 2] = 1$ Remove innermost parentheses by distributing $2(x - 1) = 2x - 2$.

$5x - 3x - 2x + 2 = 1$ Remove square brackets and change the sign of each enclosed term.

$2 = 1$ Combine like terms.

Because $2 = 1$ is false, no number satisfies this equation. It is inconsistent.

Practice Problem 3 Solve $3x - [2x - 6(x + 1)] = 7x - 1$.

EXAMPLE 4 Solving an Equation That Is an Identity

Solve $1 - 5y + 2(y + 7) = 2y + 5(3 - y)$.

Solution

Step 2 $1 - 5y + 2(y + 7) = 2y + 5(3 - y)$ Original equation

$1 - 5y + 2y + 14 = 2y + 15 - 5y$ Distributive property

$15 - 3y = 15 - 3y$ Collect like terms and combine constants on each side.

Step 3 $15 - 3y + 3y = 15 - 3y + 3y$ Add $3y$ to both sides.

$15 = 15$ Simplify.

We have shown that $15 = 15$ is equivalent to the original equation. The equation $15 = 15$ is always true; its solution set is the set of all real numbers. Therefore, the solution set of the original equation is also the set of all real numbers. Thus, the original equation is an identity and the solution set is $(-\infty, \infty)$.

Practice Problem 4 Solve the equation $2(3x - 6) + 5 = 12 - (19 - 6x)$.

IDENTIFYING TYPES OF EQUATIONS

1. An equation that is satisfied by all values in the domain of the variable is an *identity*. For example, $2(x - 1) = 2x - 2$. When you try to solve an identity, you get a true statement such as $3 = 3$ or $0 = 0$.
2. An equation that is not an identity but is satisfied by at least one number in the domain of the variable is a *conditional* equation. For example, $2x = 6$ is a linear conditional equation; $x = 3$ is a solution, but $x = 0$ is not.
3. An equation that is not satisfied by any value of the variable is an *inconsistent* equation. For example, $x = x + 2$ is an inconsistent equation. Because no number is 2 more than itself, the solution set of the equation $x = x + 2$ is $\varnothing$. When you try to solve an inconsistent equation, you obtain a false statement such as $0 = 2$.

3 Solve formulas for a specific variable.

Formulas

Solving equations that arise from everyday experiences is one of the most important uses of algebra. An equation that expresses a relationship between two or more variables is called a **formula**.

EXAMPLE 5 Converting Temperatures

The formula for converting the temperature in degrees Celsius (C) to degrees Fahrenheit (F) is

$$F = \frac{9}{5}C + 32.$$

While planning a trip to the United States, a London resident notes that the local temperature will be 86° Fahrenheit. What is the temperature in degrees Celsius? (Temperatures are recorded in Celsius degrees in England.)

Solution

We first substitute 86 for F in the formula $F = \frac{9}{5}C + 32$

to obtain $86 = \frac{9}{5}C + 32.$

We now solve this equation for C.

$$5(86) = 5\left(\frac{9}{5}C + 32\right) \quad \text{Multiply both sides by 5.}$$
$$430 = 9C + 160 \quad \text{Simplify.}$$
$$270 = 9C \quad \text{Subtract 160 from both sides.}$$
$$\frac{270}{9} = C \quad \text{Divide both sides by 9.}$$
$$30 = C \quad \text{Simplify.}$$

Thus, 86° F converts to 30° C.

Practice Problem 5 Use the formula in Example 5 to find the temperature in degrees Celsius assuming that the temperature is 50° Fahrenheit.

In Example 5, we specified $F = 86$ and solved the equation for C. A more general result can be obtained by assuming that F is a fixed number, or constant, and solving for C. This process is called **solving for a specified variable**.

EXAMPLE 6 **Solving for a Specified Variable**

a. Solve $F = \frac{9}{5}C + 32$ for C. **b.** Solve $S = 2\ell w + 2\ell h + 2wh$ for h.

Solution

a.
$$F = \frac{9}{5}C + 32 \quad \text{Original equation}$$
$$F - 32 = \frac{9}{5}C \quad \text{Subtract 32 from both sides.}$$
$$\frac{5}{9}(F - 32) = C \quad \text{Multiply both sides by } \frac{5}{9};\ \frac{5}{9}\cdot\frac{9}{5} = 1$$
or $$C = \frac{5}{9}(F - 32)$$

b.
$$S = 2\ell w + 2\ell h + 2wh \text{ for } h \quad \text{Original equation}$$
$$S - 2\ell w = 2\ell h + 2wh \quad \text{Subtract } 2\ell w \text{ from both sides so that all terms with } h \text{ are on one side.}$$
$$S - 2\ell w = (2\ell + 2w)h \quad \text{Factor out } h.$$
$$\frac{S - 2\ell w}{2\ell + 2w} = h \quad \text{Divide both sides by } 2\ell + 2w.$$
or $$h = \frac{S - 2\ell w}{2\ell + 2w}$$

Practice Problem 6 Solve $P = 2\ell + 2w$ for w.

4 Use linear equations to solve applied problems.

Applications

Practical problems are often introduced through verbal descriptions. Frequently, they describe a situation, perhaps in a physical or financial setting, and ask for a value for some specific quantity that will favorably resolve the situation. The process of translating a practical problem into a mathematical one is called **mathematical modeling**. Generally, we use variables to represent the quantities identified in the verbal description and represent relationships between these quantities with algebraic expressions or equations. Ideally, we reduce the problem to an equation that, when solved, also solves the practical problem. Then we solve the equation and check our result against the practical situation presented in the problem.

Here is a general strategy for solving applied problems.

PROCEDURE FOR SOLVING APPLIED PROBLEMS

Step 1 Read the problem as many times as needed to understand it thoroughly. Pay close attention to the question asked to help you **identify the quantity the variable should represent.**

Step 2 **Assign a variable to represent the quantity you are seeking and when necessary, express all other unknown quantities in terms of this variable.** Frequently, it is helpful to draw a diagram to illustrate the problem or to set up a table to organize the information.

Step 3 Write an equation that describes the situation. **Look for a relationship between the unknown quantities in Step 2.**

Step 4 Solve the equation.

Step 5 Answer the question asked in the problem.

Step 6 Check the answer against the description of the original problem (not just the equation solved in Step 4).

Geometry

Common geometric figures and formulas are given on the final pages of the book.

EXAMPLE 7 Solving a Geometry Problem

The length of a rectangular soccer field is 30 yards less than twice its width. Find the dimensions of the field assuming that its perimeter is 360 yards.

Solution

Step 2 Let $x =$ the width of the field.
Then $2x - 30 =$ the length of the field. (See the following figure.)

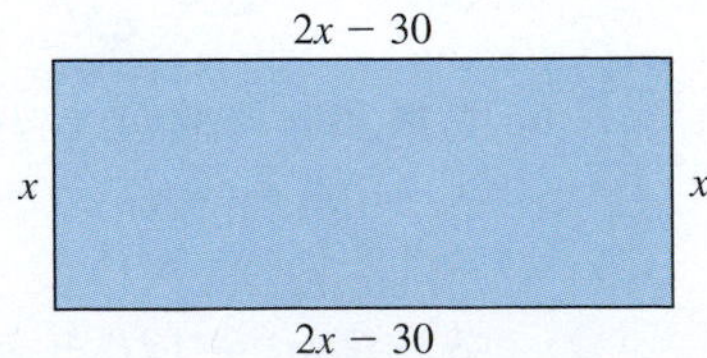

Step 3 Perimeter = sum of the lengths of the four sides of the rectangle

$$= (2x - 30) + x + (2x - 30) + x$$

$$\text{Perimeter} = 360 \text{ ft} \qquad \text{Given}$$

$$(2x - 30) + x + (2x - 30) + x = 360 \qquad \text{Replace the verbal description of perimeter with the algebraic expression.}$$

Step 4

$$6x - 60 = 360 \qquad \text{Combine terms.}$$

$$6x = 420 \qquad \text{Add 60 to both sides and simplify.}$$

$$x = \frac{420}{6} = 70 \qquad \text{Divide each side by 6.}$$

Step 5 The width of the field $= 70$ yards, and
the length of the field $= 2(70) - 30 = 140 - 30 = 110$ yards.
The dimensions of the soccer field are 110 by 70 yards.

Step 6 **Check:** The perimeter is $110 + 70 + 110 + 70 = 360$ yards.

Practice Problem 7 The length of a rectangle is 5 m more than twice its width. Find the dimensions of the rectangle assuming that its perimeter is 28 m.

Finance

EXAMPLE 8 **Analyzing Investments**

Tyrick invests \$15,000, some in stocks and the rest in bonds. If he invests twice as much in stocks as he does in bonds, how much does he invest in each?

Solution

Step 2 Let $x =$ the amount invested in stocks. The rest of the \$15,000 investment $(\$15{,}000 - x)$ is invested in bonds.

Step 3 We have one more important piece of information to use, as follows:

Amount invested in stocks, x	=	Twice the amount invested in bonds, $2(15{,}000 - x)$

$x = 2(15{,}000 - x)$ Replace the verbal descriptions with algebraic expressions.

Step 4

$x = 30{,}000 - 2x$ Distributive property

$3x = 30{,}000$ Add $2x$ to both sides.

$x = 10{,}000$ Divide both sides by 3.

Step 5 Tyrick invests \$10,000 in stocks and $\$15{,}000 - \$10{,}000 = \$5000$ in bonds.

Step 6 Check in the original problem. If Tyrick invests \$10,000 in stocks and \$5000 in bonds, then because $2(\$5000) = \$10{,}000$, he invested twice as much in stocks as he did in bonds.

Practice Problem 8 Repeat Example 8, but Tyrick now invests three times as much in stocks as he does in bonds.

Interest is money paid for the use of borrowed money. For example, while your money is on deposit at a bank, the bank pays you for the right to use the money in your account. The money you are paid by the bank is interest on your deposit. On any loan, the money borrowed for use (and on which interest is paid) is called the **principal**. The **interest rate** is the percent of the principal charged for its use for a specific time period. The most straightforward type of interest computation is described next.

Simple Interest

If a principal of P dollars is borrowed for a period of t years with interest rate r (expressed as a decimal) computed yearly, then the total interest paid at the end of t years is

$$I = Prt.$$

Interest computed with this formula is called **simple interest**.

When interest is computed yearly, the rate r is called an *annual* interest rate (or *per annum* interest rate).

EXAMPLE 9 **Solving Problems Involving Simple Interest**

Ms. Sharpy invests a total of $10,000 in blue-chip and technology stocks. At the end of a year, the blue chips returned 12% and the technology stocks returned 8% on the original investments. How much was invested in each type of stock if the total interest earned was $1060?

Step 1 We are asked to find two amounts: that invested in blue-chip stocks and that invested in technology stocks. If we know how much was invested in blue-chip stocks, then we know that the rest of the $10,000 was invested in technology stocks.

Step 2 Let x = amount invested in blue-chip stocks. Then

$10{,}000 - x$ = amount invested in technology stocks.

Step 3 We organize our information in the following table.

Investment	Principal P	Rate r	Time t	Interest $I = Prt$
Blue Chip	x	0.12	1	$0.12x$
Technology	$10{,}000 - x$	0.08	1	$0.08(10{,}000 - x)$

Interest from blue chip + Interest from technology = Total interest

Step 4

$$0.12x + 0.08(10{,}000 - x) = 1060$$ Replace descriptions with algebraic expressions.

$$12x + 8(10{,}000 - x) = 106{,}000$$ Multiply by 100 to eliminate decimals.

$$12x + 80{,}000 - 8x = 106{,}000$$ Distributive property

$$4x = 26{,}000$$ Combine like terms; subtract 80,000 from both sides.

$$x = 6500$$ Divide by 4 to get the amount in blue-chip stocks

$$10{,}000 - x = 10{,}000 - 6500$$ Replace x with 6500

$$= 3500$$ Substitute to get the amount in technology stocks

Step 5 Ms. Sharpy invests $3500 in technology stocks and $6500 in blue-chip stocks.

Step 6 Check in the original problem.

12% of $6500 is 0.12(6500) = $780

and 8% of $3500 is 0.08(3500) = $280.

Thus, the total interest earned is $780 + $280 = $1060.

Practice Problem 9 One-fifth of my capital is invested at 5%, one-sixth of my capital at 8%, and the rest at 10% per year. If the annual interest on my capital is $130, what is my capital?

Uniform Motion

If you drive 60 miles in two hours, your average speed is $\frac{60}{2} = 30$ miles per hour. If you multiply your average speed for the trip (30 miles per hour) by the time elapsed (two hours), you get the distance driven, 60 miles $(60 = 2 \cdot 30)$. When "rate" refers to the average speed of an object, use the following formula.

UNIFORM MOTION

If an object moves at a rate (average speed) r per unit time, then the distance d traveled in time t units is

$$d = rt.$$

WARNING

Care must be taken to make sure that units of measurement are consistent. If, for example, the rate is given in miles per hour, then the time should be given in hours. If the rate is given in miles per hour and the time is given as 15 minutes, change 15 minutes to $\frac{1}{4}$ hour before using the uniform-motion formula.

EXAMPLE 10 Solving a Uniform-Motion Problem

A motorcycle police officer is chasing a car that is speeding at 70 miles per hour. The police officer is 1 mile behind the car and is traveling 80 miles per hour. How long will it be before the officer overtakes the car?

Solution

Step 2 We are asked to find the amount of *time* before the officer overtakes the car. Draw a sketch to help visualize the problem. See Figure 1.1.

FIGURE 1.1

Let t = the time it takes the officer to overtake the car. Organize the information in a table.

Object	r (miles per hour)	t (hours)	$d = rt$ (miles)
Car	70	t	$70t$
Motorcycle	80	t	$80t$

Step 3 Because the distance the motorcycle travels is 1 mile more than the car travels, we have

$$80t = 70t + 1$$ The motorcycle's distance is 1 mile more than the car's distance.

Step 4

$$80t - 70t = 1$$ Subtract $70t$ from both sides.

$$10t = 1$$ Simplify.

$$t = \frac{1}{10}$$ Divide both sides by 10.

Step 5 The time required to overtake the car is

$$t = \frac{1}{10} \text{ hour} = 0.1 \text{ hour (or 6 minutes)}$$

Step 6 Check in the original problem. If the police officer travels for 0.1 of an hour at 80 miles per hour, he travels $d = (80)(0.1) = 8$ miles. The car traveling for this same 0.1 of an hour at 70 miles per hour travels $(70)(0.1) = 7$ miles. Because the officer travels 1 more mile during that time than the car does (the motorcycle started 1 mile behind the car), the officer does overtake the car.

Practice Problem 10 A train 130 meters long crosses a bridge in 21 seconds. The train is traveling at 25 meters per second. What is the length of the bridge?

EXAMPLE 11 Bungee-Jumping TV Contestants

The introduction to this section described a television show on which a bungee cord with 140%–150% elongation is used for contestants' bungee jumping from a bridge 120 feet above the water. The show's producer wants to make sure the jumper does not get closer than 10 feet to the water. How long can the bungee cord be given that no contestant is more than 7 feet tall?

Solution

Step 1 We want the extended cord plus the length of the jumper's body to be a minimum of 10 feet above the water. To be safe, we determine the maximum extended length of the cord, add the height of the tallest possible contestant, and ensure that this total length is 10 feet above the water.

Step 2 Let $x =$ length, in feet, of the cord to be used.

Step 3 Then $x + 1.5x = 2.5x$ is the maximum extended length (in feet) of the cord (with 150% elongation). We have

Extended cord length	+	Body length	+	10-foot buffer	=	Height of bridge above the water

$2.5x + 7 + 10 = 120$ Replace descriptions with arithmetic expressions.

Step 4

$2.5x + 17 = 120$ Combine constants.

$2.5x = 103$ Subtract 17 from both sides.

$x = 41.2$ feet Divide both sides by 2.5.

Step 5 If the length of the cord is less than 41.2 feet, all conditions are met.

Step 6 **Check:** $41.2 + 1.5(41.2) + 7 + 10 = 120$; so any length less than 41.2 ft will work.

Practice Problem 11 What length of cord should be used in Example 11 if the elongation is 200% instead of 140%–150%?

Answers to Practice Problems

1. a. $(-\infty, \infty)$ **b.** $(-\infty, 2) \cup (2, \infty)$ **c.** $[1, \infty)$ **2.** $\left\{-\frac{3}{5}\right\}$ **3.** $\varnothing$ **4.** $(-\infty, \infty)$ **5.** 10°C **6.** $w = \frac{P - 2\ell}{2}$ **7.** Width: 3 m; length: 11 m **8.** \$3750 in bonds and \$11,250 in stocks **9.** \$1500 **10.** 395 m **11.** 34.3 ft

SECTION 1.1 Exercises

Basic Concepts and Skills

1. The domain of the variable in an equation is the set of all real numbers for which both sides of the equation are ______.
2. Standard form for a linear equation in x is ______.
3. Two equations with the same solutions set are called ______.
4. A conditional equation is an equation that is not true for ______ values of the variables.
5. **True or False.** If \$100 is invested at 5% for three years, the interest $I = (100)(5)(3)$ dollars.
6. **True or False.** If an object is traveling at a uniform rate of 60 ft/sec, the distance covered in 15 minutes is $d = (60)(15)$ ft.

In Exercises 7–11, determine whether the given value of the variable is a solution of the equation.

7. $x - 2 = 5x + 6$
   a. $x = 0$ **b.** $x = -2$
8. $8x + 3 = 14x - 1$
   a. $x = -1$ **b.** $x = \frac{2}{3}$
9. $\frac{2}{x} = \frac{1}{3} + \frac{1}{x + 2}$
   a. $x = 4$ **b.** $x = 1$
10. $(x - 3)(2x + 1) = 0$
   a. $x = \frac{1}{2}$ **b.** $x = 3$
11. $2x + 3x = 5x$
   a. $x = 157$ **b.** $x = -2046$

In Exercises 12–16, find the domain of the variable in each equation. Write the answer in interval notation.

12. $(2 - x) - 4x = 7 - 3(x + 4)$
13. $\frac{y}{y - 1} = \frac{3}{y + 2}$
14. $\frac{1}{y} = 2 + \sqrt{y}$
15. $\frac{3x}{(x - 3)(x - 4)} = 2x + 9$
16. $\frac{1}{\sqrt{x}} = x^2 - 1$

In Exercises 17–20, determine whether the given equation is an identity. If the equation is not an identity, give a value that demonstrates that fact.

17. $2x + 3 = 5x + 1$
18. $3x + 4 = 6x + 2 - (3x - 2)$
19. $\frac{1}{x} + \frac{1}{2} = \frac{2 + x}{2x}$
20. $\frac{1}{x + 3} = \frac{1}{x} + \frac{1}{3}$

In Exercises 21–46, solve each equation.

21. $3x + 5 = 14$
22. $2x - 17 = 7$
23. $-10x + 12 = 32$
24. $-2x + 5 = 6$
25. $3 - y = -4$
26. $2 - 7y = 23$
27. $7x + 7 = 2(x + 1)$
28. $3(x + 2) = 4 - x$
29. $3(2 - y) + 5y = 3y$
30. $9y - 3(y - 1) = 6 + y$
31. $4y - 3y + 7 - y = 2 - (7 - y)$
32. $3(y - 1) = 6y - 4 + 2y - 4y$
33. $3(x - 2) + 2(3 - x) = 1$
34. $2x - 3 - (3x - 1) = 6$
35. $2x + 3(x - 4) = 7x + 10$
36. $3(2 - 3x) - 4x = 3x - 10$
37. $4[x + 2(3 - x)] = 2x + 1$
38. $3 - [x - 3(x + 2)] = 4$
39. $3(4y - 3) = 4[y - (4y - 3)]$
40. $5 - (6y + 9) + 2y = 2(y + 1)$
41. $2x - 3(2 - x) = (x - 3) + 2x + 1$
42. $5(x - 3) - 6(x - 4) = -5$
43. $\frac{2x + 1}{9} - \frac{x + 4}{6} = 1$
44. $\frac{2 - 3x}{7} + \frac{x - 1}{3} = \frac{3x}{7}$
45. $\frac{1 - x}{4} + \frac{5x + 1}{2} = 3 - \frac{2(x + 1)}{8}$
46. $\frac{x + 4}{3} + 2x - \frac{1}{2} = \frac{3x + 2}{6}$

In Exercises 47–58, solve each formula for the indicated variable.

47. $d = rt$ for r
48. $F = ma$ for a
49. $C = 2\pi r$ for r
50. $A = 2\pi rx + \pi r^2$ for x
51. $I = \frac{E}{R}$ for R
52. $A = P(1 + rt)$ for t
53. $A = \frac{(a + b)h}{2}$ for h
54. $T = a + (n - 1)d$ for d
55. $\frac{1}{f} = \frac{1}{u} + \frac{1}{v}$ for u
56. $\frac{1}{R} = \frac{1}{R_1} + \frac{1}{R_2}$ for R_2
57. $y = mx + b$ for m
58. $ax + by = c$ for y

In Exercises 59–62, write an algebraic expression for the specified quantity.

59. Leroy buys in-line skates for \$327.62, including sales tax of $6\frac{1}{2}$%. Let x = the price of in-line skates before tax. Write an algebraic expression in x for "the tax paid on the in-line skates."

60. A lamp manufacturing company produces 1600 reading lamps a day when it operates in two shifts. The first shift produces only $5/6$ as many lamps as the second shift. Let $x =$ the number of lamps produced per day by the second shift. Write an algebraic expression in x for the number of lamps produced per day by the first shift.

61. Kim invests \$22,000, some in stocks and the rest in bonds. Let $x =$ amount invested in stocks. Write an algebraic expression in x for "the amount invested in bonds."

62. An air-conditioning repair bill for \$229.50 showed a charge of \$72.00 dollars for parts, with the remainder of the charge for labor. Let $x =$ number of hours of labor it took to repair the air conditioner. Write an algebraic expression in x for the labor charge per hour.

Applying the Concepts

In Exercises 63–98, use mathematical modeling techniques to solve the problems.

63. **Swimming pool dimensions.** The volume of a swimming pool is 2808 cubic feet. Assuming that the pool is 18 feet long and 12 feet deep, find its width.

64. **Hole dimensions.** If a box-shaped hole 7 feet long and 3 feet wide holds 168 cubic feet of dirt, how deep must the hole be?

65. **Geometry.** The perimeter of a rectangle is 80 feet. Find its dimensions assuming that its length is 5 feet less than twice its width.

66. **Geometry.** The width of a rectangle is 3 feet more than one-half its length, and the perimeter of the rectangle is 36 feet. Find its dimensions.

67. **Geometry.** A circle has a circumference of 114π centimeters. Find the radius of the circle.

68. **Geometry.** A rectangle has a perimeter of 28 meters and a width of 5 meters. Find the length of the rectangle.

69. **Storage barrel dimensions.** A barrel has a surface area of 6π square meters and a radius of 1 meter. Find the height of the barrel.

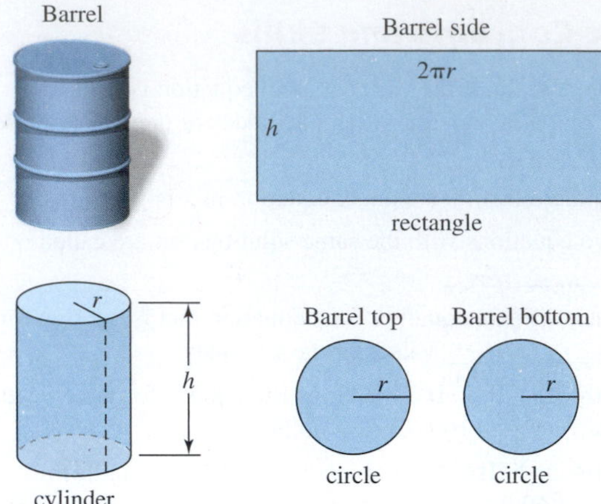

70. **Can dimensions.** A can has a volume of 148π cubic centimeters and a radius of 2 centimeters. Find the height of the can.

71. **Geometry.** A trapezoid has an area of 66 square feet and a height of 6 feet. If one base is 3 feet, what is the length of the other base?

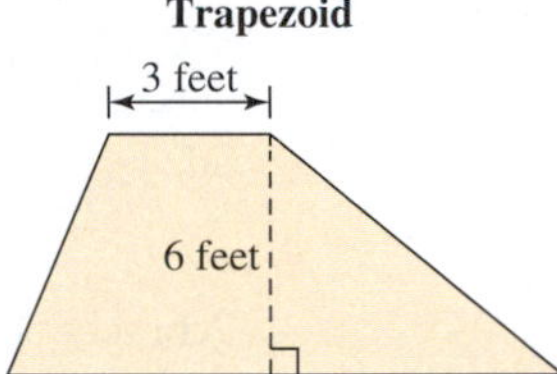

72. **Geometry.** A trapezoid has an area of 35 square centimeters. Assuming that one base is 9 centimeters and the other base is 11 centimeters, find the height of the trapezoid.

73. **Chain store properties.** A fast-food chain bought two pieces of land for new stores. The more expensive piece of land costs \$23,000 more than the less expensive piece of land, and the two pieces together cost \$147,000. How much did each piece cost?

74. **Administrative salaries.** The manager at a wholesale outlet earns \$450 more per month than the assistant manager. If their combined salary is \$3700 per month, what is the monthly salary of each?

75. **Lottery ticket sales.** The lottery ticket sales at Quick Mart for August were 10% above the sales for July. If Quick Mart sold a total of 1113 lottery tickets during July and August, how many tickets were sold each month?

76. **Sales commission.** Jan's commission for February was \$15 more than half her commission for March, and her total commission for the two months was \$633. Find her commission for each month.

77. **Inheritance.** An estate valued at \$225,000 is to be divided between two sons so that the older son receives four times as much as the younger son. Find each son's share of the estate.

78. **TV game show.** Kevin won \$735,000 on a TV game show. He decided to keep a certain amount for himself, give one-half the amount he kept for himself to his daughter, and give

one-fourth the amount he kept for himself to his dad. How much did each person get?

79. Grades. Your grades in the three tests in College Algebra are 87, 59, and 73.
  a. How many points do you need on the final to average 75?
  b. Find part (a) assuming that the final carries double weight.

80. Real estate investment. A real estate agent invested a total of \$4200. Part was invested in a high-risk real estate venture, and the rest was invested in a savings and loan. At the end of a year, the high-risk venture returned 15% on the investment and the savings and loan returned 8% on the investment. Find the amount invested in each assuming that the agent's total income from the investments was \$448.

81. Tax shelter. Mr. Mostafa received an inheritance of \$7000. He put part of it in a tax shelter paying 9% interest and part in a bank paying 6%. If his annual interest totals \$540, how much did he invest at each rate?

82. Interest. Ms. Jordan invested \$4900, part at 6% interest and the rest at 8%. If the yearly interest on each investment is the same, how much interest does she receive at the end of the year?

83. Interest. If \$5000 is invested in a bank that pays 5% interest, how much more must be invested in bonds at 8% to earn 6% on the total investment?

84. Retail profit. A retailer's cost for a microwave oven is \$480. If he wants to make a profit of 20% of the selling price, at what price should the oven be sold?

85. Profit from sales. The Beckly Company manufactures shaving sets for \$3 each and sells them for \$5 each. How many shaving sets must be sold for the company to recover an initial investment of \$40,000 and earn an additional \$30,000 as profit?

86. Jogging. Angelina jogs 100 meters in the same time that Harry bicycles 150 meters. If Harry bicycles 15 meters per minute faster than Angelina jogs, find the rate of each.

87. Overtaking a lead. A car leaves New Orleans traveling 50 kilometers per hour. An hour later a second car leaves New Orleans following the first car, traveling 70 kilometers per hour. How many hours after leaving New Orleans will it take the second car to overtake the first?

88. Separating planes. Two planes leave an airport traveling in opposite directions. One plane travels at 470 kilometers per hour and the other at 430 kilometers per hour. How long will it take them to be 2250 kilometers apart?

89. Overtaking a bike. Lucas is $\frac{1}{3}$ of a mile from home, bicycling at 20 miles per hour, when his brother leaves home on his bike to catch up. How fast must Lucas's brother go to catch Lucas a mile from home?

90. Overtaking a lead. Karen notices that her husband left for the airport without his briefcase. Her husband drives 40 miles per hour and has a 15-minute head start. If Karen (with the briefcase) drives 60 miles per hour and the airport is 45 miles away, will she catch her husband before he arrives at the airport?

91. Increasing distances. Two cars start out together from the same place. They travel in opposite directions, with one of them traveling 5 miles per hour faster than the other. After three hours, they are 405 miles apart. How fast is each car traveling?

92. Travel time. Aya rode her bicycle from her house to a friend's house, averaging 16 kilometers per hour. It turned dark before she was ready to return home, so her friend drove her home at a rate of 80 kilometers per hour. If Aya's total traveling time was three hours, how far away did her friend live?

93. Digital cameras. The price P (in dollars) for which a manufacturer will sell a digital camera is related to the number q of cameras ordered by the formula

$$P = 200 - 0.02q, \text{ for } 100 \le q \le 2000.$$

How many cameras must be ordered for a customer to pay \$170 per camera?

94. Quick ratio. The quick ratio Q of a business is related to its current assets A, its current inventory I, and its current liabilities L, by

$$Q = \frac{A - I}{L}.$$

Assuming that $Q = 37{,}000$, $L = \$1500$, and $I = \$3200$, find the current assets.

95. Numbers. The sum of two numbers is 28, and one number is three times the other. Find the numbers.

96. Numbers. The sum of three consecutive even integers is 42. Find the integers.

97. Numbers. The difference between two numbers is 14, and one number is twice the other. Find the numbers.

98. Numbers. One number is five more than a second number. The sum of the lesser number and twice the greater number is 49. Find the numbers.

Beyond the Basics

You can solve many problems by modeling procedures introduced in this section other than those presented in the examples in the section. Here are some types for you to try.

99. Explain whether the equations in each pair are equivalent.
  a. $x^2 = x, \quad x = 1$
  b. $x^2 = 9, \quad x = 3$
  c. $x^2 - 1 = x - 1, \quad x = 0$
  d. $\frac{x}{x - 2} = \frac{2}{x - 2}, \quad x = 2$

100. Find a value of k so that $3x - 1 = k$ and $7x + 2 = 16$ are equivalent.

101. Suppose you average 75 miles per hour over the first half of a drive from Denver to Las Vegas, but your average speed for the entire trip is 60 miles per hour. What was your average speed for the second half of the drive?

102. Davinder (D) and Mikhail (M) were 2 miles apart when they began walking toward each other. D walks at a constant rate of 3.7 miles per hour, and M walks at a constant rate of 4.3 miles per hour. When they started, D's dog, who runs at a constant rate of 6 miles per hour, ran to M and then turned back and ran to D. If the dog continued to run back and forth until D and M met and the dog lost no time turning around, how far did it run?

103. A mixture contains alcohol and water in the ratio 5:1. After the addition of 5 liters of water, the ratio of alcohol to water becomes 5:2. Find the quantity of alcohol in the original mixture.

104. An alloy contains zinc and copper in the ratio 5:8, and another alloy contains zinc and copper in the ratio 5:3. Equal amounts of both alloys are melted together. Find the ratio of zinc to copper in the new alloy.

105. Democritus has lived $\frac{1}{6}$ of his life as a boy, $\frac{1}{8}$ of his life as a youth, and $\frac{1}{2}$ of his life as a man and has spent 15 years as a mature adult. How old is Democritus?

106. A man and a woman have the same birthday. When he was as old as she is now, the man was twice as old as the woman. When she becomes as old as he is now, the sum of their ages will be 119. How old is the man now?

107. How many minutes is it before 6 P.M. if, 50 minutes ago, four times this number was the number of minutes past 3 P.M.?

108. Two pipes A and B can fill a tank in 24 minutes and 32 minutes, respectively. Initially, both pipes are opened simultaneously. After how many minutes should pipe B be turned off so that the tank is full in 18 minutes?

109. A small plane was scheduled to fly from Atlanta to Washington, D.C. The plane flies at a constant speed of 150 miles per hour. However, the flight was against a head wind of 10 miles per hour. The threat of mechanical failure forced the plane to turn back, and it returned to Atlanta with a tail wind of 10 miles per hour, landing 1.5 hours after it had taken off.

a. How far had the plane traveled before turning back?

b. What was the plane's average speed?

110. Three airports A, B, and C are located on an east-west line. B is 705 miles west of A, and C is 652.5 miles west of B. A pilot flew from A to B, had a stopover at B for three hours, and continued to C. The wind was blowing from the east at 15 miles per hour during the first part of the trip, but the wind changed to come from the west at 20 miles per hour during the stopover. The flight time between A and B and between B and C was the same. Find the airspeed of the plane.

Critical Thinking/Discussion/Writing

111. A pawnshop owner sells two watches for \$499 each. On one watch, he gained 10%, and on the other watch, he lost 10%. What is his percentage gain or loss?

a. No loss, no gain
b. 10% loss
c. 1% loss
d. 1% gain

112. The price of gasoline increased by 20% from July to August, while your consumption of gas decreased by 20% from July to August. What is the percentage change in your gas bill from July to August?

a. No change
b. 5% decrease
c. 4% increase
d. 4% decrease

Maintaining Skills

In Exercises 113–120, simplify each of the expressions.

113. $\sqrt{8}$

114. $\sqrt{27}$

115. $\sqrt{12}$

116. $\sqrt{45}$

117. $\frac{2 + 3\sqrt{8}}{2}$

118. $\frac{3 + \sqrt{18}}{3}$

119. $\frac{15 - \sqrt{75}}{5}$

120. $\frac{35 - 14\sqrt{12}}{7}$

In Exercises 121–132, factor each of the expressions.

121. $x^2 + x$

122. $2x^2 - 4x$

123. $x^2 - 4$

124. $x^2 - 25$

125. $x^2 + 4x + 4$

126. $x^2 - 6x + 9$

127. $x^2 - 8x + 7$

128. $x^2 + 2x - 15$

129. $6x^2 - x - 1$

130. $14x^2 + 17x - 6$

131. $-5x^2 + 3x + 2$

132. $-12x^2 + 9x + 3$

SECTION 1.2

Quadratic Equations

BEFORE STARTING THIS SECTION, REVIEW

1 Factoring (Section P.4)
2 Square roots (Section P.6, page 58)
3 Rationalizing (Section P.6, page 63)
4 Linear equations (Section 1.1, page 78)

OBJECTIVES

1 Solve a quadratic equation by factoring.
2 Solve a quadratic equation by the square root method.
3 Solve a quadratic equation by completing the square.
4 Solve a quadratic equation by using the quadratic formula.
5 Solve applied problems.

The Parthenon

The Pyramid of Giza

The Golden Rectangle

The ancient Greeks, in about the sixth century B.C., sought unifying principles of beauty and perfection, which they believed could be described mathematically. In their study of beauty, the Greeks used the term *golden ratio*. The golden ratio is frequently referred to as "phi," after the Greek sculptor Phidias, and is symbolized by the Greek letter Φ (phi). A geometric figure that is commonly associated with phi is the **golden rectangle**. (See Example 9.) This rectangle has sides of lengths a and b that are in proportion to the golden ratio. It has been said that the golden rectangle is the most pleasing rectangle to the eye. Numerous examples of art and architecture have employed the golden rectangle. Here are three of them:

The Mona Lisa

1. The exterior dimensions of the Parthenon in Athens, built about 440 B.C., form a perfect golden rectangle.
2. Leonardo Da Vinci called the golden ratio the "divine proportion" and featured it in many of his paintings, including the famous *Mona Lisa*. (Try drawing a rectangle around her face.)
3. The Great Pyramid of Giza is believed to be 4600 years old, a time long before the Greeks. Its dimensions are also based on the golden ratio.

In Example 9, we show that the numerical value of the golden ratio can be found by solving the quadratic equation $x^2 - x - 1 = 0$.

DO YOU **KNOW?**

The word *quadratic* comes from the Latin *quadratus*, meaning square.

QUADRATIC EQUATION

A quadratic equation in the variable x is an equation equivalent to the equation

$$ax^2 + bx + c = 0,$$

where a, b, and c are real numbers and $a \neq 0$.

A quadratic equation written in the form $ax^2 + bx + c = 0$ is said to be in **standard form**. Here are some examples of quadratic equations that are not in standard form.

$$2x^2 - 5x + 7 = x^2 + 3x - 1$$
$$3y^2 + 4y + 1 = 6y - 5$$
$$x^2 - 2x = 3$$

We discuss four methods for solving quadratic equations: (1) by factoring, (2) by taking square roots, (3) by completing the square, and (4) by using the quadratic formula.

We begin by studying real number solutions of quadratic equations.

1 Solve a quadratic equation by factoring.

Factoring Method

Some quadratic equations in the standard form $ax^2 + bx + c = 0$ can be solved by factoring and using the **zero-product property**.

ZERO-PRODUCT PROPERTY

Let A and B be two algebraic expressions. Then $AB = 0$ if and only if $A = 0$ or $B = 0$.

For a product of real numbers, a similar property was stated in Section P.1.

EXAMPLE 1 **Solving a Quadratic Equation by Factoring**

SIDE **NOTE**

To use the zero-product property to solve an equation, you must make sure that one side of the equation is 0.

Solve by factoring: $2x^2 + 5x = 3$

Solution

$2x^2 + 5x = 3$	Original equation
$2x^2 + 5x - 3 = 0$	Subtract 3 from both sides to get 0 on one side.
$(2x - 1)(x + 3) = 0$	Factor the left side.

We now set each factor equal to 0 and solve the resulting linear equations. The vertical line segment separates the computations leading to the two solutions.

$2x - 1 = 0$	$x + 3 = 0$	Zero-product property
$2x = 1$	$x = -3$	Isolate the x term on one side.
$x = \frac{1}{2}$	$x = -3$	Solve for x.

Check: You should check the solutions $x = \frac{1}{2}$ and $x = -3$ in the original equation.

The solution set is $\left\{\frac{1}{2}, -3\right\}$.

Practice Problem 1 Solve by factoring: $x^2 + 25x = -84$

EXAMPLE 2 Solving a Quadratic Equation by Factoring

Solve by factoring: $3t^2 = 2t$

Solution

$3t^2 = 2t$	Original equation
$3t^2 - 2t = 0$	Subtract $2t$ from both sides.
$t(3t - 2) = 0$	Factor.
$t = 0 \quad \vert \quad 3t - 2 = 0$	Zero-product property
$3t = 2$	Isolate the t terms on one side.
$t = \frac{2}{3}$	Solve for t.

Check: You should check the solutions $t = 0$ and $t = \frac{2}{3}$ in the original equation.

The solution set is $\left\{0, \frac{2}{3}\right\}$.

Practice Problem 2 Solve by factoring: $2m^2 = 5m$

SOLVING A QUADRATIC EQUATION BY FACTORING

Step 1 Write the given equation in standard form (so that one side is 0).

Step 2 Factor the nonzero side of the equation from Step 1.

Step 3 Set each factor obtained in Step 2 equal to 0.

Step 4 Solve the resulting equations in Step 3.

Step 5 Check the solutions obtained in Step 4 in the original equation.

The factoring method of solving an equation works *only when one of the sides of the equation is* 0. If neither side of an equation is 0, as in $(x + 1)(x - 3) = 10$, we cannot know the value of either factor. For example, we could have $x + 1 = 5$ and $x - 3 = 2$, or $x + 1 = 0.1$ and $x - 3 = 100$. There are countless possibilities for two factors whose product is 10.

EXAMPLE 3 Solving a Quadratic Equation with One Solution

Solve by factoring: $x^2 + 16 = 8x$

Solution

Step 1 Write the equation in standard form.

$x^2 + 16 = 8x$	Original equation
$x^2 + 16 - 8x = 0$	Subtract $8x$ from both sides.
$x^2 - 8x + 16 = 0$	Standard form

Step 2 Factor the left side of the equation.

$(x - 4)(x - 4) = 0$ Perfect-square trinomial

Steps 3 and 4 Set each factor equal to 0 and solve each resulting equation.

$x - 4 = 0$	$x - 4 = 0$	Zero-product property
$x = 4$	$x = 4$	

Step 5 Check the solution in the original equation.
Check: $x = 4$ because $x = 4$ is the only possible solution.

$$x^2 + 16 = 8x \qquad \text{Original equation}$$
$$(4)^2 + 16 \stackrel{?}{=} 8(4) \qquad \text{Replace } x \text{ with 4.}$$
$$32 = 32 \checkmark$$

Thus, 4 is the only solution of $x^2 + 16 = 8x$ and the solution set is $\{4\}$.

Practice Problem 3 Solve by factoring: $x^2 - 6x = -9$

In Example 3, because the factor $(x - 4)$ appears twice in the solution, the number 4 is called a **double root**, or a **root of multiplicity 2**, of the given equation.

2 Solve a quadratic equation by the square root method.

The Square Root Method

The solutions of an equation of the type $x^2 = d$ are obtained by the square root method. Suppose we are interested in solving the equation $x^2 = 3$, which is equivalent to the equation $x^2 - 3 = 0$. Applying the factoring method but *removing the restriction that coefficients and constants represent only integers*, we have

$$x^2 - 3 = 0$$
$$x^2 - (\sqrt{3})^2 = 0 \qquad \text{Because } 3 = (\sqrt{3})^2$$
$$(x + \sqrt{3})(x - \sqrt{3}) = 0 \qquad \text{Factor (using real numbers).}$$
$$x + \sqrt{3} = 0 \quad \text{or} \quad x - \sqrt{3} = 0 \qquad \text{Set each factor equal to zero.}$$
$$x = -\sqrt{3} \quad \text{or} \quad x = \sqrt{3} \qquad \text{Solve for } x.$$

Thus, the solutions of the equation $x^2 = 3$ are $-\sqrt{3}$ and $\sqrt{3}$. Because the solutions differ only in sign, we use the notation $\pm\sqrt{3}$ to write the two numbers $\sqrt{3}$ and $-\sqrt{3}$. In a similar way, we see that for any nonnegative real number d, the solutions of the quadratic equation $x^2 = d$ are $\pm\sqrt{d}$.

SIDE **NOTE**

If $d < 0$, then the equation $u^2 = d$ has no real solutions.

SQUARE ROOT PROPERTY

Suppose u is any algebraic expression and $d \geq 0$. If $u^2 = d$, then $u = \pm\sqrt{d}$.

EXAMPLE 4 **Solving an Equation by the Square Root Method**

Solve: $(x - 3)^2 = 5$

Solution

$$(x - 3)^2 = 5 \qquad \text{Original equation}$$
$$x - 3 = \pm\sqrt{5} \qquad \text{Square root property}$$
$$x = 3 \pm \sqrt{5} \qquad \text{Add 3 to both sides.}$$

The solution set is $\{3 - \sqrt{5}, 3 + \sqrt{5}\}$.

Practice Problem 4 Solve: $(x + 2)^2 = 5$

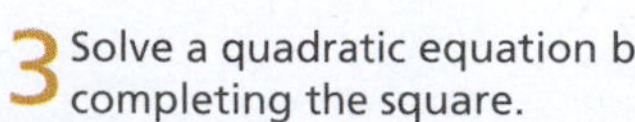

3 Solve a quadratic equation by completing the square.

Completing the Square

We use a method called **completing the square** to solve quadratic equations that cannot be solved by factoring and are not in the correct form to use the square root method. The method requires two steps. First, we write a given quadratic equation in the form $(x + k)^2 = d$. Then we solve the equation $(x + k)^2 = d$ by the square root method.

Recall that

$$(x + k)^2 = x^2 + 2kx + k^2.$$

Notice that the coefficient of x on the right side is $2k$, and half of this coefficient is k; that is, $\frac{1}{2}(2k) = k$. Notice that the constant term, k^2, in the trinomial $x^2 + 2kx + k^2$ is the *square of one-half the coefficient of x*.

PERFECT-SQUARE TRINOMIAL

A quadratic trinomial in x with coefficient of x^2 equal to 1 is a **perfect-square trinomial** if the constant term is the square of one-half the coefficient of x.

When the constant term is *not* the square of half the coefficient of x, we subtract the constant term from both sides and add a new constant term that will result in a perfect-square trinomial.

Here are some examples that show the number that should be added to $x^2 + bx$ to create a perfect-square trinomial.

$x^2 + bx$	b	$\frac{b}{2}$	Add $\left(\frac{b}{2}\right)^2$	Result is $\left(x + \frac{b}{2}\right)^2$
$x^2 + 6x$	6	3	$3^2 = 9$	$x^2 + 6x + 9 = (x + 3)^2$
$x^2 - 4x$	-4	-2	$(-2)^2 = 4$	$x^2 - 4x + 4 = (x - 2)^2$
$x^2 + 3x$	3	$\frac{3}{2}$	$\left(\frac{3}{2}\right)^2 = \frac{9}{4}$	$x^2 + 3x + \frac{9}{4} = \left(x + \frac{3}{2}\right)^2$
$x^2 - x$	-1	$-\frac{1}{2}$	$\left(-\frac{1}{2}\right)^2 = \frac{1}{4}$	$x^2 - x + \frac{1}{4} = \left(x - \frac{1}{2}\right)^2$

Thus, to make $x^2 + bx$ into a perfect square, we need to add

$$\left[\frac{1}{2}(\text{coefficient of } x)\right]^2 = \left[\frac{1}{2}(b)\right]^2 = \frac{b^2}{4}$$

so that

$$x^2 + bx + \frac{b^2}{4} = \left(x + \frac{b}{2}\right)^2.$$

We say that $\frac{b^2}{4}$ was added to $x^2 + bx$ to *complete the square*. See Figure 1.2.

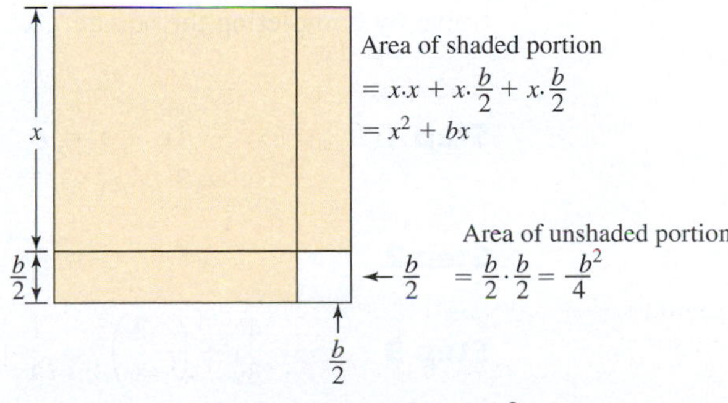

FIGURE 1.2 Area of a square $= \left(x + \frac{b}{2}\right)^2$

EXAMPLE 5 Solving a Quadratic Equation by Completing the Square

Solve by completing the square: $x^2 + 10x + 8 = 0$

Solution

First, isolate the constant on the right side.

$$x^2 + 10x + 8 = 0 \qquad \text{Original equation}$$
$$x^2 + 10x = -8 \qquad \text{Subtract 8 from both sides.}$$

Next, add $\left[\frac{1}{2}(10)\right]^2 = 5^2 = 25$ to both sides to complete the square on the left side.

$$x^2 + 10x + 25 = -8 + 25 \qquad \text{Add 25 to each side.}$$
$$(x + 5)^2 = 17 \qquad x^2 + 10x + 25 = (x + 5)^2$$
$$x + 5 = \pm\sqrt{17} \qquad \text{Square root property}$$
$$x = -5 \pm \sqrt{17} \qquad \text{Subtract 5 from each side.}$$

Thus, $x = -5 + \sqrt{17}$ or $x = -5 - \sqrt{17}$, and the solution set is $\{-5 + \sqrt{17}, -5 - \sqrt{17}\}$.

Practice Problem 5 Solve by completing the square: $x^2 - 6x + 7 = 0$

We summarize the procedure for solving a quadratic equation by completing the square.

METHOD OF COMPLETING THE SQUARE

Step 1 Rearrange the quadratic equation so that the terms in x^2 and x are on the left side of the equation and the constant term is on the right side.

Step 2 Make the coefficient of x^2 equal to 1 by dividing both sides of the equation by the original coefficient. (Steps 1 and 2 are interchangeable.)

Step 3 Add the square of one-half the coefficient of x to both sides of the equation.

Step 4 Write the equation in the form $(x + k)^2 = d$ using the fact that the left side is a perfect square.

Step 5 Take the square root of each side, prefixing $\pm$ to the right side.

Step 6 Solve the two equations resulting from Step 5.

EXAMPLE 6 Solving a Quadratic Equation by Completing the Square

Solve by completing the square: $3x^2 - 4x - 1 = 0$

Solution

Step 1 $3x^2 - 4x - 1 = 0$ Note that the coefficient of x^2 is 3.

$3x^2 - 4x = 1$ Add 1 to both sides.

Step 2 $x^2 - \frac{4}{3}x = \frac{1}{3}$ Divide both sides by 3.

Step 3 $x^2 - \frac{4}{3}x + \left(-\frac{2}{3}\right)^2 = \frac{1}{3} + \left(-\frac{2}{3}\right)^2$ Add $\left[\frac{1}{2}\left(-\frac{4}{3}\right)\right]^2 = \left(-\frac{2}{3}\right)^2$ to both sides.

Step 4 $\left(x - \frac{2}{3}\right)^2 = \frac{7}{9}$ $\quad x^2 - \frac{4}{3}x + \left(-\frac{2}{3}\right)^2 = \left(x - \frac{2}{3}\right)^2$

Step 5 $x - \frac{2}{3} = \pm\sqrt{\frac{7}{9}}$ Take the square root of both sides.

$x - \frac{2}{3} = \pm\frac{\sqrt{7}}{3}$ $\sqrt{\frac{7}{9}} = \frac{\sqrt{7}}{\sqrt{9}} = \frac{\sqrt{7}}{3}$

Step 6 $x = \frac{2}{3} \pm \frac{\sqrt{7}}{3}$ Add $\frac{2}{3}$ to both sides.

$x = \frac{2 \pm \sqrt{7}}{3}$ Add fractions.

The solution set is $\left\{\frac{2 - \sqrt{7}}{3}, \frac{2 + \sqrt{7}}{3}\right\}$.

Practice Problem 6 Solve by completing the square: $4x^2 - 24x + 25 = 0$

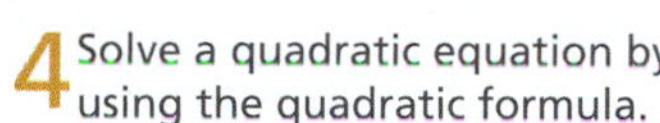

The Quadratic Formula

We can generalize the method of completing the square to derive a formula that gives a solution of *any* quadratic equation.

We solve a quadratic equation in standard form by completing the square.

$$ax^2 + bx + c = 0,\ a \neq 0$$

Step 1 $ax^2 + bx = -c$ Isolate the constant on the right side.

Step 2 $x^2 + \frac{b}{a}x = -\frac{c}{a}$ Divide both sides by a.

Step 3 $x^2 + \frac{b}{a}x + \left(\frac{b}{2a}\right)^2 = -\frac{c}{a} + \left(\frac{b}{2a}\right)^2$ Add the square of one-half the coefficient of x to both sides.

Step 4 $\left(x + \frac{b}{2a}\right)^2 = -\frac{c}{a} + \frac{b^2}{4a^2}$ The left side in Step 3 is a perfect square.

$\left(x + \frac{b}{2a}\right)^2 = \frac{b^2 - 4ac}{4a^2}$ Rearrange and combine fractions on the right side.

Step 5 $x + \frac{b}{2a} = \pm\sqrt{\frac{b^2 - 4ac}{4a^2}}$ Take the square roots of both sides.

Step 6 Solve the equations in Step 5.

$x + \frac{b}{2a} = \pm\sqrt{\frac{b^2 - 4ac}{4a^2}}$ Equations from Step 5

$x = -\frac{b}{2a} \pm \frac{\sqrt{b^2 - 4ac}}{2|a|}$ Add $-\frac{b}{2a}$ to both sides; $\sqrt{4a^2} = \sqrt{4}\sqrt{a^2} = 2|a|$.

$x = -\frac{b}{2a} \pm \frac{\sqrt{b^2 - 4ac}}{2a}$ $\pm 2|a| = \pm 2a$ because $|a| = a$ or $|a| = -a$.

$x = \frac{-b \pm \sqrt{b^2 - 4ac}}{2a}$ Combine fractions.

The last equation in Step 6 is called the **quadratic formula**.

SIDE NOTE

The quadratic formula shows that a quadratic equation can have, at most, two distinct solutions, and for the solutions to be real we must have $b^2 - 4ac \geq 0$.

QUADRATIC FORMULA

The solutions of the quadratic equation in the standard form $ax^2 + bx + c = 0$ with $a \neq 0$ are given by the formula

$$x = \frac{-b \pm \sqrt{b^2 - 4ac}}{2a}.$$

WARNING To use the quadratic formula, write the given quadratic equation in standard form. Then determine the values of a (coefficient of x^2), b (coefficient of x), and c (constant term).

EXAMPLE 7 Solving a Quadratic Equation by Using the Quadratic Formula

Solve by using the quadratic formula: $3x^2 = 5x + 2$

Solution

The equation $3x^2 = 5x + 2$ can be rewritten in the standard form by subtracting $5x + 2$ from both sides.

$$3x^2 - 5x - 2 = 0 \quad \text{Rewrite in standard form.}$$

$$\underset{\substack{\uparrow \\ a}}{3}x^2 + (\underset{\substack{\uparrow \\ b}}{-5})x + (\underset{\substack{\uparrow \\ c}}{-2}) = 0 \quad \text{Identify values of } a, b, \text{ and } c \text{ to be used in the quadratic formula.}$$

Then $a = 3$, $b = -5$, and $c = -2$, and we have

$$x = \frac{-b \pm \sqrt{b^2 - 4ac}}{2a} \quad \text{Quadratic formula}$$

$$x = \frac{-(-5) \pm \sqrt{(-5)^2 - 4(3)(-2)}}{2(3)} \quad \text{Substitute 3 for } a, -5 \text{ for } b, \text{ and } -2 \text{ for } c.$$

$$= \frac{5 \pm \sqrt{25 + 24}}{6} \quad \text{Simplify.}$$

$$= \frac{5 \pm \sqrt{49}}{6} = \frac{5 \pm 7}{6} \quad \text{Simplify.}$$

Then

$$x = \frac{5 + 7}{6} = \frac{12}{6} = 2 \text{ or } x = \frac{5 - 7}{6} = \frac{-2}{6} = -\frac{1}{3}, \text{ and the solution set is } \left\{-\frac{1}{3}, 2\right\}.$$

Now you should solve $3x^2 = 5x + 2$ by factoring to confirm that you get the same result.

Practice Problem 7 Solve $6x^2 - x - 2 = 0$ by using the quadratic formula.

5 Solve applied problems.

Applications

EXAMPLE 8 Partitioning a Building

A rectangular building whose depth (from the front of the building) is three times its frontage is divided into two parts by a partition that is 45 feet from and parallel to the front wall. Assuming that the rear portion of the building contains 2100 square feet, find the dimensions of the building.

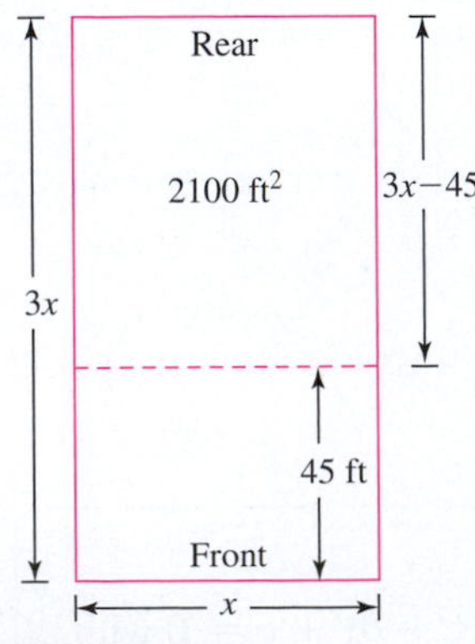

FIGURE 1.3

Solution

In solving problems of this type, it is advisable to draw a diagram such as Figure 1.3. We let

$x =$ the frontage of the building, in feet.

Then $3x =$ the depth of the building, in feet,

and $3x - 45 =$ the depth of the rear portion, in feet.

The area of the rear portion of the building is $x(3x - 45)$ square feet. This area is 2100 square feet, so

$$
\begin{aligned}
x(3x - 45) &= 2100 \\
3x^2 - 45x &= 2100 && \text{Distributive property} \\
3x^2 - 45x - 2100 &= 0 && \text{Subtract 2100 from both sides.} \\
x^2 - 15x - 700 &= 0 && \text{Divide both sides by 3.} \\
(x - 35)(x + 20) &= 0 && \text{Factor.} \\
x - 35 = 0 \quad \text{or} \quad x + 20 &= 0 && \text{Zero-product property} \\
x = 35 \quad \text{or} \quad x &= -20 && \text{Solve for } x.
\end{aligned}
$$

Because the dimensions of the building are positive numbers, we reject $x = -20$. Thus,

$$
\begin{aligned}
x &= 35 && \text{Frontage in feet} \\
3x &= 105 && \text{Depth in feet}
\end{aligned}
$$

Check: $3x - 45 = 3(35) - 45 = 105 - 45 = 60$ is the depth of the rear portion, and the area of the rear portion is $60 \cdot 35 = 2100$ square feet.

Practice Problem 8 In Example 8, find approximate dimensions of the building assuming that its depth is five times its frontage.

SIDE **NOTE**

Recall that two rectangles are similar if the lengths of corresponding sides are proportional.

Golden Rectangle

A rectangle of length p and width q, with $p > q$, is called a *golden rectangle* if you can divide the rectangle into a square with side of length q and a smaller rectangle that is similar to the original one. See Figure 1.4. The ratio of the longer side to the shorter side of a golden rectangle is called the *golden ratio*. In Figure 1.4, the golden ratio is $\frac{p}{q}$.

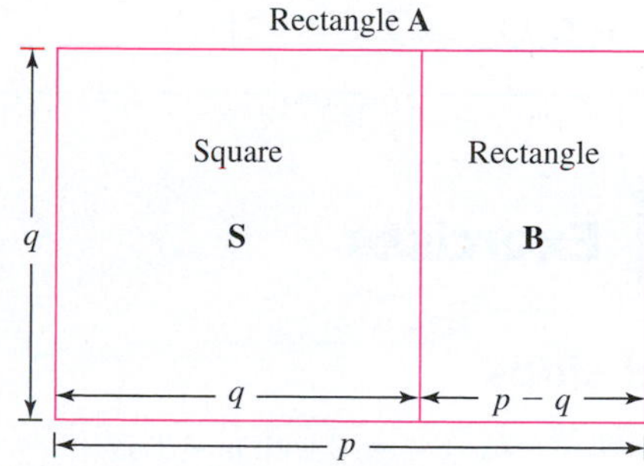

FIGURE 1.4 Golden Rectangle

EXAMPLE 9 Calculating the Golden Ratio

Use Figure 1.4 of a golden rectangle to calculate the golden ratio.

Solution

The rectangle **A** in Figure 1.4 is a golden rectangle. This means that we can divide rectangle **A** into a square **S** and a smaller rectangle **B** such that

$$\frac{\text{Longer side of } \mathbf{A}}{\text{Shorter side of } \mathbf{A}} = \frac{\text{Longer side of } \mathbf{B}}{\text{Shorter side of } \mathbf{B}}.$$

Thus, because rectangle **B** is the smaller rectangle, $p - q < q$, and we have

$$\frac{p}{q} = \frac{q}{p - q}$$

$$p(p - q) = q \cdot q \quad \text{Multiply both sides by } q(p - q)\text{; simplify.}$$

$$p^2 - pq = q^2 \quad \text{Distributive property}$$

$$\left(\frac{p}{q}\right)^2 - \frac{p}{q} = 1 \quad \text{Divide both sides by } q^2\text{; simplify.}$$

$$\Phi^2 - \Phi = 1 \quad \text{Replace } \frac{p}{q} \text{ with } \Phi\text{, the golden ratio.}$$

$$\Phi^2 - \Phi - 1 = 0 \quad \text{Write in standard form, treating } \Phi \text{ as the variable.}$$

$$\Phi = \frac{-(-1) \pm \sqrt{(-1)^2 - 4(1)(-1)}}{2(1)} \quad \text{Substitute 1 for } a \text{ and } -1 \text{ for both } b \text{ and } c.$$

$$= \frac{1 \pm \sqrt{5}}{2} \quad \text{Simplify.}$$

Because the ratio of two positive numbers is positive, we reject the negative root $\frac{1 - \sqrt{5}}{2}$.

Therefore, the golden ratio is

$$\Phi = \frac{1 + \sqrt{5}}{2} \approx 1.618.$$

Practice Problem 9 Find the length of a golden rectangle whose width is 36 feet.

Answers to Practice Problems

1. $\{-21, -4\}$ **2.** $\left\{0, \frac{5}{2}\right\}$ **3.** $\{3\}$ **4.** $\{-2 - \sqrt{5}, -2 + \sqrt{5}\}$ **5.** $\{3 - \sqrt{2}, 3 + \sqrt{2}\}$ **6.** $\left\{3 - \frac{\sqrt{11}}{2}, 3 + \frac{\sqrt{11}}{2}\right\}$ **7.** $\left\{-\frac{1}{2}, \frac{2}{3}\right\}$ **8.** 25.48 ft by 127.41 ft **9.** $18 + 18\sqrt{5} \approx 58.25$ ft

SECTION 1.2 Exercises

Basic Concepts and Skills

1. Any equation of the form $ax^2 + bx + c = 0$ with $a \neq 0$ is called a(n) ______________ equation.

2. If $P(x)$, $D(x)$, and $Q(x)$ are polynomials and $P(x) = D(x)Q(x)$, then the solutions of $P(x) = 0$ are the solutions of $Q(x) = 0$ together with the solutions of ______________ $= 0$.

3. From the square root property, we know that if $u^2 = 5$, then $u =$ ______________.

4. If you complete the square in the quadratic equation $ax^2 + bx + c = 0$, you get the quadratic formula for the solutions $x =$ ______________.

5. **True or False.** Every quadratic equation has, at most, two distinct solutions.

6. **True or False.** If we are given $x^2 + kx$, we form a perfect square by adding k^2.

In Exercises 7–14, show by substitution whether the number r is a solution of the corresponding quadratic equation.

7. $x^2 + 4x - 12 = 0;\quad r = -6$

8. $x^2 - 8x - 9 = 0;\quad r = 9$

9. $3x^2 + 7x - 6 = 0;\quad r = \frac{2}{3}$

10. $2x^2 - 5x - 3 = 0;\quad r = -\frac{1}{2}$

11. $x^2 - 4x + 1 = 0;\quad r = 2 - \sqrt{3}$

12. $x^2 - 6x + 1 = 0;\quad r = 3 + 2\sqrt{2}$

13. $4x^2 - 8x + 13 = 0;\quad r = 2 + \sqrt{3}$

14. $x^2 - 6x + 13 = 5;\quad r = 5 - \sqrt{2}$

15. Find k assuming that $x = 1$ is a solution of the equation $kx^2 + x - 3 = 0$.

16. Find k assuming that $x = \sqrt{7}$ is a solution of the equation $kx^2 + x - 3 = 0$.

In Exercises 17–22, solve each equation by factoring.

17. $x^2 - 5x = 0$

18. $x^2 - 5x + 4 = 0$

19. $x^2 + 5x = 14$

20. $x^2 - 11x = 12$

21. $x^2 = 5x + 6$

22. $x = x^2 - 12$

In Exercises 23–28, solve each equation by the square root method.

23. $3x^2 = 48$

24. $2x^2 = 50$

25. $x^2 + 1 = 5$

26. $2x^2 - 1 = 17$

27. $(x - 1)^2 = 16$

28. $(2x - 3)^2 = 25$

In Exercises 29–38, add a constant term to the expression to make it a perfect square.

29. $x^2 + 4x$

30. $y^2 + 10y$

31. $x^2 + 6x$

32. $y^2 - 8y$

33. $x^2 - 7x$

34. $x^2 - 3x$

35. $x^2 + \frac{1}{3}x$

36. $x^2 - \frac{3}{2}x$

37. $x^2 + ax$

38. $x^2 - \frac{2a}{3}x$

In Exercises 39–44, solve each equation by completing the square.

39. $x^2 + 2x - 5 = 0$

40. $x^2 + 6x = -7$

41. $x^2 - 3x - 1 = 0$

42. $x^2 - x - 3 = 0$

43. $2r^2 + 3r = 9$

44. $3k^2 - 5k + 1 = 0$

In Exercises 45–50, solve each equation by using the quadratic formula.

45. $x^2 + 2x - 4 = 0$

46. $m^2 + 3m + 2 = 0$

47. $6x^2 = 7x + 5$

48. $t^2 - 7 = 4t$

49. $3z^2 - 2z = 7$

50. $6y^2 + 11y = 10$

In Exercises 51–76, solve each equation by any method.

51. $2x^2 + 5x - 3 = 0$

52. $2x^2 - 9x + 10 = 0$

53. $(3x - 2)^2 - 16 = 0$

54. $(4x + 1)^2 - 25 = 0$

55. $5x^2 - 6x = 4x^2 + 6x - 3$

56. $x^2 + 7x - 5 = x - x^2$

57. $3p^2 + 8p + 4 = 0$

58. $x^2 = 5(x - 1)$

59. $3y^2 + 5y + 2 = 0$

60. $6x^2 + 11x + 4 = 0$

61. $5x^2 + 12x + 4 = 0$

62. $3x^2 - 2x - 5 = 0$

63. $5y^2 + 10y + 4 = 2y^2 + 3y + 1$

64. $3x^2 - 1 = 5x^2 - 3x - 5$

65. $2x^2 + x = 15$

66. $6x^2 = 1 - x$

67. $12x^2 - 10x = 12$

68. $-x^2 + 10x + 1200 = 0$

69. $(x + 13)(x + 5) = -2$

70. $3(x^2 + 1) = 2x^2 + 4x + 1$

71. $18x^2 - 45x = -7$

72. $18x^2 + 57x + 45 = 0$

73. $2t^2 - 5 = 0$

74. $3k^2 - 48 = 0$

75. $4x^2 - 10x - 750 = 0$

76. $12x^2 + 43x + 36 = 0$

In Exercises 77–80, write your answer rounded to two decimal places.

77. Find the length of the golden rectangle whose width is 14.72 in.

78. Find the length of the golden rectangle whose width is 18.63 ft.

79. Find the width of the golden rectangle whose length is 8.46 cm.

80. Find the width of the golden rectangle whose length is 4.68 m.

Applying the Concepts

81. Dimensions. The length of a rectangular plot is three times its width. Assuming that the area of the plot is 10,800 square feet, find the dimensions of the plot.

82. Dimensions. The diagonal of a square tile is 4 inches longer than its side. Find the length of the side.

83. Finding integers. Find two numbers whose sum is 28 and whose product is 147.

84. Finding integers. Find an integer such that the sum of twice the square of the integer and the integer itself is 55.

85. The sum of two numbers is 57, and their product is 782. Find the numbers.

86. The perimeter of a rectangle is 82 ft, and its area is 400 ft^2. Find dimensions of the rectangle.

87. Geometry. The length of a rectangle is 5 centimeters greater than its width. The area of the rectangle is 500 square centimeters. Find the dimensions of the rectangle.

88. Geometry. The sides of a rectangle are in the ratio 3:2. The area of the rectangle is 216 square centimeters. Find the dimensions of the rectangle.

89. Cutting a wire. A wire of length 16 inches is to be cut into two pieces; then each piece will be bent to form a square. Find the length of the two pieces assuming that the sum of the areas of the two squares is 10 square inches.

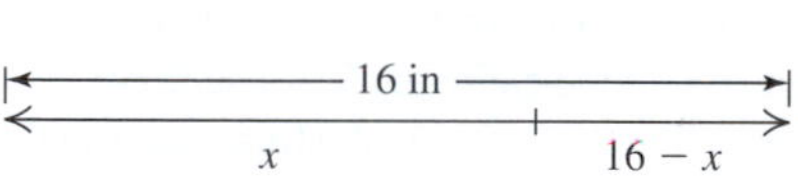

90. Cutting a wire. A piece of wire is 38 inches long. The wire is cut into two pieces; then each piece is bent to form a square. Find the length of each piece if three times the area of the larger square exceeds the area of the smaller square by 95.75 square inches.

91. Manufacturing. The surface area A of a cylinder with height h and radius r is given by the equation $A = 2\pi rh + 2\pi r^2$. A company makes soup cans by using 32π square inches of aluminum sheet for each can. Assuming that the height of the can is 6 inches, find the radius of the can.

92. Concrete patio. To make a rectangular concrete patio, a homeowner used 70 feet of perimeter, into which she poured 138 cubic feet of concrete to form a slab 6 inches thick. What were the dimensions of the patio?

93. Making a box. The volume of a box is given by the equation $V =$ length · width · height. A topless box is to be made from a square sheet of tin by cutting 5-inch squares from each of the four corners and then shaping the tin into an open box by turning up the sides. Assuming that the box is to hold 480 cubic inches, find the size of the piece of tin to be used.

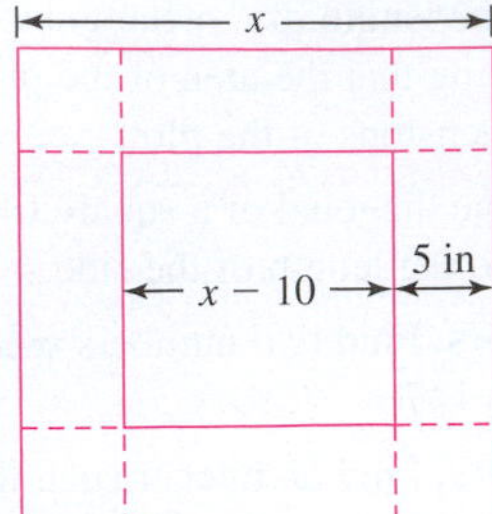

94. Making a box. A 2-inch square is cut from each corner of a rectangular piece of cardboard whose length exceeds the width by 4 inches. The sides are then turned up to form an open box. Assuming that the volume of the box is 64 cubic inches, find the dimensions of the box.

95. Bus travel. Two buses leave Atlanta at the same time, one traveling west at an average of 52 miles per hour and the other traveling north at an average of 39 miles per hour. When will they be 390 miles apart?

96. Plane travel. Two planes left San Francisco at 2 P.M. One flew east at a certain speed, and the other flew south at a speed that was 100 kilometers per hour more than the speed of the plane heading east. The planes were 2100 kilometers apart at 5 P.M. How fast was each plane flying?

97. Gardening. You want to expand your present 25-foot by 15-foot garden by planting a border of flowers. The border is to be of the same width around the entire garden. The flowers you bought will fill an area of 624 square feet. How wide should the border be?

98. Sky divers. Sky divers are in free fall from the time they jump out of a plane until they open their parachutes. A sky diver jumps from 5000 feet. The diver's height h above the ground t seconds after the jump is described by the equation $h = -16t^2 + 5000$. Find the time during which the diver is in free fall assuming that the parachute opens at 1000 feet.

99. Physics. Suppose you throw a ball straight up from the ground with a velocity of 112 feet per second. As the ball moves up, gravity slows it. Eventually, the ball begins to fall back to the ground. The height h of the ball after t seconds in the air is given by the equation $h = -16t^2 + 112t$.

a. Find the height of the ball after two seconds.

b. How long will it take the ball to reach a height of 96 feet?

c. How long after you throw the ball will it return to the ground?

100. Sound velocity. The speed of sound v (in ft/sec) depends on the temperature $T°$ Celsius and is given by
$$v = \frac{1087\sqrt{T + 273}}{16.25}.$$

a. Find the speed of sound if the air temperature is $T = 20°C$.

b. Find the air temperature if the speed of sound is 1150 ft/sec.

Height of a projectile. The height, h, of a projectile thrown straight up from the top of a 480-foot hill t seconds after being thrown upward with initial velocity v_0 ft/sec is given by $h = -16t^2 + v_0t + 480$. In Exercises 101–104, for a given value of v_0, find the time(s) when the projectile will (a) be at a height of 592 ft above the ground and (b) crash on the ground. Round your answers to two decimal places.

101. $v_0 = 96$ **102.** $v_0 = 112$

103. $v_0 = 256$ **104.** $v_0 = 64$

Beyond the Basics

In Exercises 105–110, find the values of k for which the given equation has equal roots.

105. $x^2 - kx + 3 = 0$ **106.** $x^2 + 3kx + 8 = 0$

107. $2x^2 + kx + k = 0$ **108.** $kx^2 + 2x + 6 = 0$

109. $x^2 + k^2 = 2(k + 1)x$ **110.** $kx^2 + (k + 3)x + 4 = 0$

111. Assuming that r and s are the roots of the quadratic equation $ax^2 + bx + c = 0$, show that
$$r + s = -\frac{b}{a} \quad \text{and} \quad r \cdot s = \frac{c}{a}.$$

112. Find the sum and the product of the roots of the following equations without solving the equation.

a. $3x^2 + 5x - 5 = 0$

b. $3x^2 - 7x = 1$

c. $\sqrt{3}x^2 = 3x + 4$

d. $(1 + \sqrt{2})x^2 - \sqrt{2}x - 5 = 0$

In Exercises 113 and 114, determine k so that the sum and the product of the roots are equal.

113. $2x^2 + (k - 3)x + 3k - 5 = 0$

114. $5x^2 + (2k - 3)x - 2k + 3 = 0$

115. Suppose r and s are the roots of the quadratic equation $ax^2 + bx + c = 0$. Show that you can write $ax^2 + bx + c = a(x - r)(x - s)$. Thus, the quadratic trinomial can be written in factored form. [*Hint:* From Exercise 111, $b = -a(r + s)$ and $c = ars$. Substitute in $ax^2 + bx + c$ and factor.]

116. Use Exercise 115 to factor each trinomial.

a. $4x^2 + 4x - 5$

b. $25x^2 + 40x + 11$

c. $25x^2 - 30x + 4$

d. $72x^2 + 95x - 1000$

117. If we know the roots r and s in advance, then by Exercise 115, a quadratic equation with roots r and s can be expressed in the form
$$a(x - r)(x - s) = 0.$$
Form a quadratic equation that has the listed numbers as roots.

a. $-3, 4$

b. $5, 5$

c. $3 + \sqrt{2}, 3 - \sqrt{2}$

118. Solve the equation $3x^2 - 4xy + 5 - y^2 = 0$
   a. for x in terms of y.
   b. for y in terms of x.

119. **How to construct a golden rectangle.** Draw a square $ABCD$ with side of length x, as in the accompanying diagram. Let M be the midpoint of DC. The segment MB sweeps out the arc BF so that F lies on the line through the segment DC. Show that the rectangle $AEFD$ is a golden rectangle.

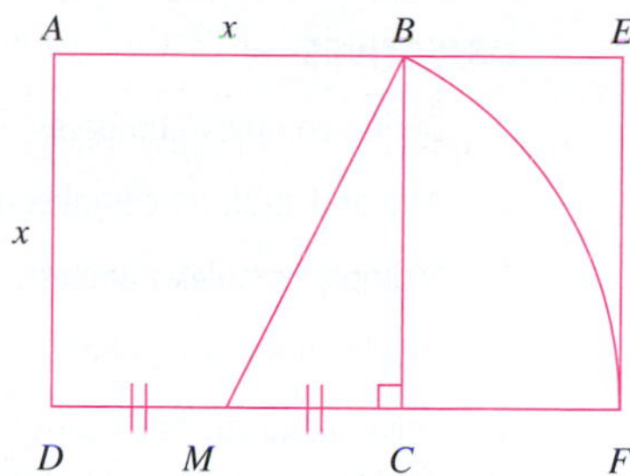

120. If the two roots of the quadratic equation

$$ax^2 + bx + c = 0$$

are reciprocals of each other, which of the following must be true?
   i. $b = c$ **ii.** $a = c$ **iii.** $a = 0$ **iv.** $b = 0$

121. Use Exercise 111 to show that a quadratic equation cannot have three distinct roots.

122. Suppose p is a positive integer greater than 1. Which of the following describes the roots of the equation $x^2 - (p + 1)x + p = 0$?
   i. real and equal **ii.** real and irrational
   iii. complex **iv.** integers and unequal

123. Amar, Akbar, and Anthony attempt to solve a barely legible quadratic equation written on an old paper. Amar misreads the constant term and finds the solutions 8 and 2. Akbar misreads the coefficient of x and finds the roots -9 and -1. Anthony solves the correct equation. What are Anthony's solutions of the equation?

In Exercises 124 and 125, choose the correct answer and explain why it is correct.

124. An equation equivalent to $x^2 + 4x - 5 = 0$ is
   a. $5x^2 - 4x + 1$ **b.** $x^2 - 4x + 5$
   c. $|x + 2| = 3$ **d.** $4x^2 + 5x - 1 = 0$

125. The equation $x^2 + 4x + k = 0$ has real roots. Then
   a. $k > 4$ **b.** $k \le 4$
   c. $k \le 0$ **d.** $k \ge 0$

Critical Thinking/Discussion/Writing

126. If a, b, and k are real numbers, show that the solutions of the equation $(x - a)(x - b) = k^2$ are real.

127. Show that for $0 < a < 4$, the equation $ax(1 - x) = 1$ has no real solutions.

Maintaining Skills

In Exercises 128–143, perform the indicated operations.

128. $(3 + \sqrt{2})(3 - \sqrt{2})$ **129.** $(1 - 2\sqrt{7})(1 + 2\sqrt{7})$

130. $(2 + \sqrt{5})^2$ **131.** $(\sqrt{2} - \sqrt{3})^2$

132. $\dfrac{5 + 5\sqrt{20}}{5}$ **133.** $\dfrac{6 + \sqrt{27}}{3}$

134. $\dfrac{16 - \sqrt{100}}{2}$ **135.** $\dfrac{18 - \sqrt{108}}{12}$

136. $(3x + 2) + (x - 7)$ **137.** $(5x - 9) + (6 - x)$

138. $(9x + 4) - (2x + 12)$ **139.** $(6x - 5) - (3x + 4)$

140. $(3x + 2)(x - 9)$ **141.** $(2x - 5)(3x + 4)$

142. $(x - 3)(x + 3)$ **143.** $(5x + 2)(5x - 2)$

In Exercises 144–147, solve each equation.

144. $x^2 + 4x + 1 = 0$ **145.** $x^2 + 5x + 5 = 0$

146. $5x^2 + 8x + 2 = 0$ **147.** $-2x^2 + 5x - 1 = 0$

SECTION 1.3

Complex Numbers: Quadratic Equations with Complex Solutions

BEFORE STARTING THIS SECTION, REVIEW

1 Properties of real numbers (Section P.1, pages 10 and 11)
2 Special products (Section P.3, pages 34 and 35)
3 Factoring (Section P.4)

OBJECTIVES

1 Define complex numbers.
2 Add and subtract complex numbers.
3 Multiply complex numbers.
4 Divide complex numbers.
5 Solve quadratic equations having complex solutions.

◆ Alternating Current Circuits

In the early days of the study of alternate current (AC) circuits, scientists concluded that AC circuits were somehow different from the battery-powered direct current (DC) circuits. However, both types of circuits obey the same physical and mathematical laws. In 1893, Charles Steinmetz provided the breakthrough in understanding AC circuits. He explained that in AC circuits, the voltage, current, and resistance (called *impedance* in AC circuits) were not scalars, but alternate in direction, and possessed frequency and phase shift that must be taken into account. He advocated the use of the polar form of complex numbers to provide a convenient method of symbolically denoting the magnitude, frequency, and phase shift simultaneously for the AC circuit quantities voltage, current, and impedance. Steinmetz eventually became known as "the wizard who generated electricity from the square root of minus one." In Example 8, we use complex numbers to compute the total impendance in an AC circuit.

1 Define complex numbers.

Complex Numbers

Because the square of a real number is nonnegative (that is, $x^2 \geq 0$ for any real number x), the equation $x^2 = -1$ has no solution in the set of real numbers. To solve such equations, we extend the real numbers to a larger set called the set of complex numbers. We first introduce a new number i whose square is -1.

Definition of *i*

The square root of -1 is called i.

$$i = \sqrt{-1} \text{ so that } i^2 = -1.$$

The number i is called the **imaginary unit**.

With the introduction of the number i, the equation $x^2 = -1$ has two solutions, $x = \pm\sqrt{-1} = \pm i$.

TECHNOLOGY CONNECTION

Most graphing calculators allow you to work with complex numbers by changing from *real* to $a + bi$ mode. Once the $a + bi$ mode is set, the $\sqrt{-1}$ is recognized as i. The i key is used to enter complex numbers such as $2 + 3i$.

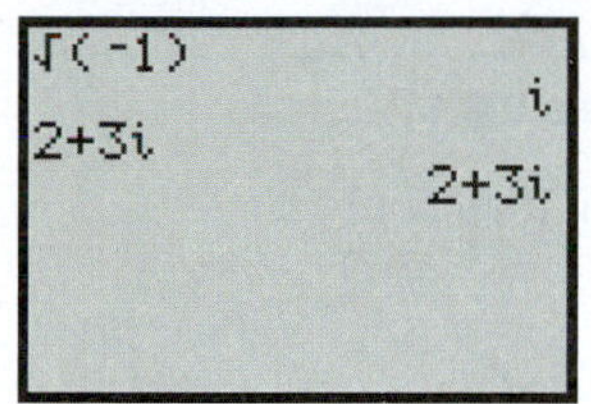

The real and imaginary parts of a complex number can then be found.

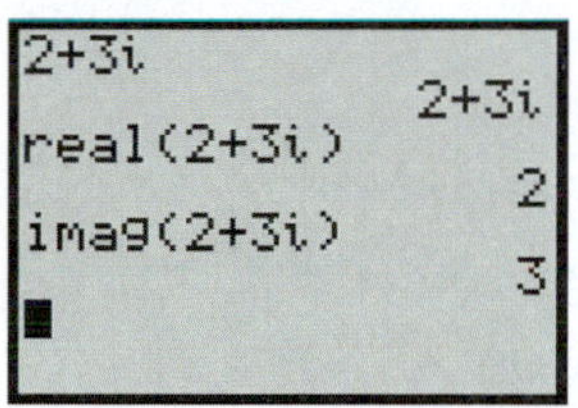

COMPLEX NUMBERS

A complex number is a number of the form

$$z = a + bi,$$

where a and b are **real numbers** and $i = \sqrt{-1}$, (so $i^2 = -1$).

The number a is called the **real part** of z, and we write $\text{Re}(z) = a$. The number b is called the **imaginary part** of z, and we write $\text{Im}(z) = b$.

A complex number z written in the form $a + bi$ is said to be in **standard form**. Where convenient, we use the form $a + ib$ as equivalent to $a + bi$. A complex number with $a = 0$ and $b \neq 0$, written as bi, is called a **pure imaginary number**. If $b = 0$, then the complex number $a + bi = a$ is a real number. So real numbers form a subset of the complex numbers (with imaginary part 0). Figure 1.5 shows how various sets of numbers are contained within larger sets.

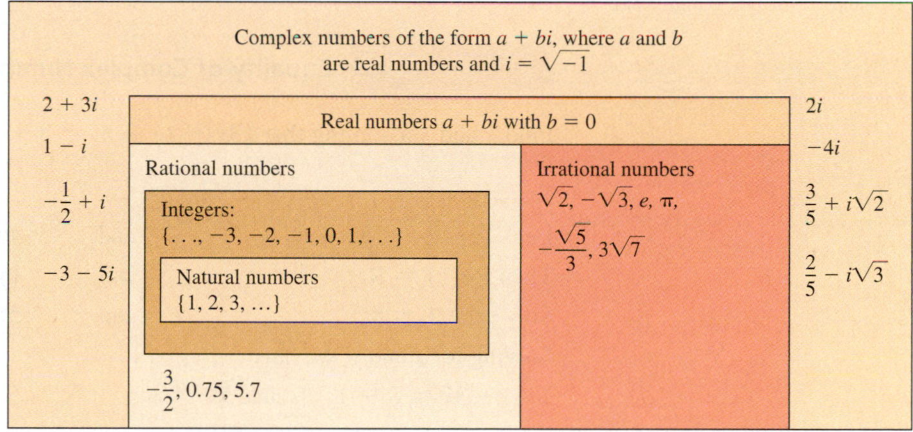

FIGURE 1.5 Complex numbers

DO YOU KNOW?

In 1777, the great Swiss mathematician Leonhard Euler introduced the i symbol for $\sqrt{-1}$. However, electrical engineers often use the letter j for $\sqrt{-1}$ because in electrical engineering problems, the symbol i is traditionally reserved for the electric current.

We can express the square root of any negative number as a product of a real number and i.

SQUARE ROOT OF A NEGATIVE NUMBER

For any positive number b,

$$\sqrt{-b} = (\sqrt{b})i = i\sqrt{b}.$$

EXAMPLE 1 Identifying the Real and Imaginary Parts of a Complex Number

Identify the real and imaginary parts of each complex number.

a. $2 + 5i$ **b.** $7 - \frac{1}{2}i$ **c.** $3i$ **d.** -9 **e.** 0 **f.** $3 + \sqrt{-25}$

Solution

To identify the real and imaginary parts, we express each number in the standard form $a + bi$.

a. $2 + 5i$ is already written as $a + bi$; real part 2, imaginary part 5

b. $7 - \frac{1}{2}i = 7 + \left(-\frac{1}{2}\right)i$; real part 7, imaginary part $-\frac{1}{2}$

c. $3i = 0 + 3i$; real part 0, imaginary part 3
d. $-9 = -9 + 0i$; real part -9, imaginary part 0
e. $0 = 0 + 0i$; real part 0, imaginary part 0
f. $3 + \sqrt{-25} = 3 + (\sqrt{25})\,i = 3 + 5i$; real part 3, imaginary part 5

Practice Problem 1 Identify the real and imaginary parts of each complex number.

a. $-1 + 2i$ **b.** $-\frac{1}{3} - 6i$ **c.** 8

EQUALITY OF COMPLEX NUMBERS

Two complex numbers $z = a + bi$ and $w = c + di$ are equal if and only if

$$a = c \quad \text{and} \quad b = d.$$

That is, $z = w$ if and only if $\text{Re}(z) = \text{Re}(w)$ and $\text{Im}(z) = \text{Im}(w)$.

EXAMPLE 2 Equality of Complex Numbers

Find x and y assuming that $(3x - 1) + 5i = 8 + (3 - 2y)i$.

Solution

Let $z = (3x - 1) + 5i$ and $w = 8 + (3 - 2y)i$.
Then $\text{Re}(z) = \text{Re}(w)$ and $\text{Im}(z) = \text{Im}(w)$.
So $3x - 1 = 8$ and $5 = 3 - 2y$.
Solving for x and y, we have

$3x = 8 + 1$ Isolate x. $\qquad 5 - 3 = -2y$ Isolate y.
$x = 3$ Solve for x. $\qquad -1 = y$ Solve for y.

So $x = 3$ and $y = -1$.

Practice Problem 2 Find a and b assuming that $(1 - 2a) + 3i = 5 - (2b - 5)i$.

2 Add and subtract complex numbers.

Addition and Subtraction

We find the sum or difference of two complex numbers by treating them as binomials in which i is the variable.

ADDITION AND SUBTRACTION OF COMPLEX NUMBERS

For real numbers a, b, c, and d, let $z = a + bi$ and $w = c + di$.

Sum: $z + w = (a + bi) + (c + di) = (a + c) + (b + d)i$

To add complex numbers, add their real parts, add their imaginary parts, and express the sum in standard form.

Difference: $z - w = (a + bi) - (c + di) = (a - c) + (b - d)i$

To subtract complex numbers, subtract their real parts, subtract their imaginary parts, and express the difference in standard form.

TECHNOLOGY CONNECTION

In $a + bi$ mode, most graphing calculators perform addition and subtraction of complex numbers using the ordinary addition and subtraction keys.

```
(3+7i)+(2-4i)
            5+3i
(5+9i)-(6-8i)
          -1+17i
```

EXAMPLE 3 Adding and Subtracting Complex Numbers

Write the sum or difference of two complex numbers in standard form.

a. $(3 + 7i) + (2 - 4i)$ **b.** $(5 + 9i) - (6 - 8i)$
c. $(2 + \sqrt{-9}) - (-2 + \sqrt{-4})$

Solution

a. $(3 + 7i) + (2 - 4i) = (3 + 2) + [7 + (-4)]i = 5 + 3i$

b. $(5 + 9i) - (6 - 8i) = (5 - 6) + [9 - (-8)]i = -1 + 17i$

c. $(2 + \sqrt{-9}) - (-2 + \sqrt{-4}) = (2 + 3i) - (-2 + 2i)$ $\quad \sqrt{-9} = 3i, \sqrt{-4} = 2i$

$$= (2 - (-2)) + (3 - 2)i$$
$$= 4 + i$$

Practice Problem 3 Write the following complex numbers in standard form.

a. $(1 - 4i) + (3 + 2i)$ **b.** $(4 + 3i) - (5 - i)$
c. $(3 - \sqrt{-9}) - (5 - \sqrt{-64})$

3 Multiply complex numbers.

Multiplying Complex Numbers

We multiply complex numbers by using FOIL (as we do with binomials; see page 32) and then replace i^2 with -1. For example,

$$\begin{aligned}(2 + 5i)(4 + 3i) &= \overset{F}{2 \cdot 4} + \overset{O}{2 \cdot 3i} + \overset{I}{5i \cdot 4} + \overset{L}{5i \cdot 3i} \\ &= 8 + 6i + 20i + 15i^2 \\ &= 8 + 26i + 15(-1) \qquad \text{Replace } i^2 \text{ with } -1. \\ &= (8 - 15) + 26i \\ &= -7 + 26i.\end{aligned}$$

You should always use the FOIL method to multiply complex numbers in standard form, but for completeness, we state the product rule formally.

MULTIPLYING COMPLEX NUMBERS

For all real numbers a, b, c, and d,

$$(a + bi)(c + di) = (ac - bd) + (ad + bc)i.$$

EXAMPLE 4 Multiplying Complex Numbers

Write the following products in standard form.

a. $(3 - 5i)(2 + 7i)$ **b.** $-2i(5 - 9i)$

Solution

a. $$\begin{aligned}(3 - 5i)(2 + 7i) &= \overset{F}{6} + \overset{O}{21i} - \overset{I}{10i} - \overset{L}{35i^2} \\ &= 6 + 11i + 35 \qquad \text{Because } i^2 = -1, -35i^2 = 35. \\ &= 41 + 11i \qquad \text{Combine terms.}\end{aligned}$$

b. $-2i(5 - 9i) = -10i + 18i^2$ — Distributive property

$= -10i - 18$ — Because $i^2 = -1$, $18i^2 = -18$.

$= -18 - 10i$

Practice Problem 4 Write the following products in standard form.

a. $(2 - 6i)(1 + 4i)$ **b.** $-3i(7 - 5i)$

WARNING

Recall from algebra that if a and b are positive real numbers, then

$$\sqrt{a}\sqrt{b} = \sqrt{ab}.$$

However, this property is not true for all complex numbers. For example,

$$\sqrt{-9}\sqrt{-9} = (3i)(3i) = 9i^2 = 9(-1) = -9,$$

but $\sqrt{(-9)(-9)} = \sqrt{81} = 9.$

Thus, $\sqrt{-9}\,\sqrt{-9} \neq \sqrt{(-9)(-9)}.$

When performing multiplication (or division) involving square roots of negative numbers (say, $\sqrt{-b}$ with $b > 0$), always write $\sqrt{-b} = i\sqrt{b}$ before performing the operation.

EXAMPLE 5 Multiplication Involving Square Roots of Negative Numbers

Perform the indicated operation and write the result in standard form.

a. $\sqrt{-2}\,\sqrt{-8}$ **b.** $\sqrt{-3}\,(2 + \sqrt{-3})$ **c.** $(-2 + \sqrt{-3})^2$

d. $(3 + \sqrt{-2})(1 + \sqrt{-32})$

Solution

a. $\sqrt{-2}\,\sqrt{-8} = i\sqrt{2}\cdot i\sqrt{8} = i^2\sqrt{2\cdot 8} = i^2\sqrt{16} = (-1)(4) = -4$

b. $\sqrt{-3}\,(2 + \sqrt{-3}) = i\sqrt{3}\,(2 + i\sqrt{3})$ — $\sqrt{-3} = i\sqrt{3}$

$= 2i\sqrt{3} + i^2\sqrt{9}$ — Distributive property

$= 2i\sqrt{3} + (-1)(3)$ — $i^2 = -1$, $\sqrt{9} = 3$

$= -3 + 2i\sqrt{3}$ — Standard form

c. $(-2 + \sqrt{-3})^2 = (-2 + i\sqrt{3})^2$ — $\sqrt{-3} = i\sqrt{3}$

$= (-2)^2 + 2(-2)(i\sqrt{3}) + (i\sqrt{3})^2$ — $(a + b)^2 = a^2 + 2ab + b^2$

$= 4 - 4i\sqrt{3} + 3i^2$ — $(i\sqrt{3})^2 = (i\sqrt{3})(i\sqrt{3}) = 3i^2$

$= 4 - 4i\sqrt{3} + 3(-1)$ — $i^2 = -1$

$= 1 - 4i\sqrt{3}$ — Simplify.

d. $(3 + \sqrt{-2})(1 + \sqrt{-32})$

$= (3 + i\sqrt{2})(1 + i\sqrt{32})$ — $\sqrt{-2} = i\sqrt{2}$, $\sqrt{-32} = i\sqrt{32}$

$= 3 + 3i\sqrt{32} + i\sqrt{2} + (i\sqrt{2})(i\sqrt{32})$ — FOIL

$= 3 + 3i(4\sqrt{2}) + i\sqrt{2} + i^2\sqrt{64}$ — $\sqrt{32} = \sqrt{16\cdot 2} = 4\sqrt{2}$

$= 3 + 12i\sqrt{2} + i\sqrt{2} + (-1)(8)$ — $i^2 = -1$

$= -5 + 13\sqrt{2}i$ — Add.

Practice Problem 5 Perform the indicated operation and write the result in standard form.

a. $(-3 + \sqrt{-4})^2$ **b.** $(5 + \sqrt{-2})(4 + \sqrt{-8})$

4 Divide complex numbers.

TECHNOLOGY CONNECTION

In $a + bi$ mode, most graphing calculators can compute the complex conjugate of a complex number.

```
conj(2+7i)
                2-7i
conj(5-3i)
                5+3i
```

Complex Conjugates and Division

To perform the division of complex numbers, it is helpful to learn about the *conjugate* of a complex number.

CONJUGATE OF A COMPLEX NUMBER

If $z = a + bi$, then the conjugate (or complex conjugate) of z is denoted by $\bar{z}$ and defined by $\bar{z} = \overline{a + bi} = a - bi$.

Note that the conjugate $\bar{z}$ of a complex number has the same real part as z does, but the sign of the imaginary part is changed. For example, $\overline{2 + 7i} = 2 - 7i$ and $\overline{5 - 3i} = \overline{5 + (-3)i} = 5 - (-3)i = 5 + 3i$.

EXAMPLE 6 Multiplying a Complex Number by Its Conjugate

Find the product $z\bar{z}$ for each complex number z.

a. $z = 2 + 5i$ **b.** $z = 1 - 3i$

Solution

a. If $z = 2 + 5i$, then $\bar{z} = 2 - 5i$.

$z\bar{z} = (2 + 5i)(2 - 5i) = 2^2 - (5i)^2$ Difference of squares

$= 4 - 25i^2$ $(5i)^2 = 5^2i^2 = 25i^2$

$= 4 - (-25)$ Because $i^2 = -1, 25i^2 = -25$.

$= 29$ Simplify.

b. If $z = 1 - 3i$, then $\bar{z} = 1 + 3i$.

$z\bar{z} = (1 - 3i)(1 + 3i) = 1^2 - (3i)^2$ Difference of squares

$= 1 - 9i^2$ $(3i)^2 = 3^2i^2 = 9i^2$

$= 1 - (-9)$ Because $i^2 = -1, 9i^2 = -9$.

$= 10$ Simplify.

Practice Problem 6 Find the product $z\bar{z}$ for each complex number z.

a. $1 + 6i$ **b.** $-2i$

The results in Example 6 correctly suggest the following theorem.

COMPLEX CONJUGATE PRODUCT THEOREM

If $z = a + bi$, then

$$z\bar{z} = a^2 + b^2.$$

To write the reciprocal of a nonzero complex number or the quotient of two complex numbers in the form $a + bi$, multiply the numerator and denominator by the conjugate of the denominator. By the Complex Conjugate Product Theorem, the resulting denominator is a real number.

DIVIDING COMPLEX NUMBERS

To write the quotient of two complex numbers w and $z\,(z \neq 0)$, write

$$\frac{w}{z} = \frac{w\bar{z}}{z\bar{z}}$$ Multiply numerator and denominator by $\bar{z}$.

and then write the right side in standard form.

TECHNOLOGY **CONNECTION**

In $a + bi$ mode, most graphing calculators perform division of complex numbers with the ordinary division key. The Frac option displays results using rational numbers.

```
1/(2+i)
              .4-.2i
Ans▸Frac
           2/5-1/5i
```

```
(4+√(-25))/(2-√(
-9))
-.5384615385+1....
Ans▸Frac
     -7/13+22/13i
```

The reciprocal of a complex number can also be found using the reciprocal key.

```
(2+i)⁻¹
              .4-.2i
Ans▸Frac
           2/5-1/5i
```

EXAMPLE 7 Dividing Complex Numbers

Write the following quotients in standard form.

a. $\frac{1}{2+i}$ **b.** $\frac{4+\sqrt{-25}}{2-\sqrt{-9}}$

Solution

a. The denominator is $2 + i$, so its conjugate is $2 - i$.

$$\frac{1}{2+i} = \frac{1(2-i)}{(2+i)(2-i)} \quad \text{Multiply numerator and denominator by } 2 - i.$$

$$= \frac{2-i}{2^2+1^2} \quad (2+i)(2-i) = 2^2 + 1^2$$

$$= \frac{2-i}{5} \quad \text{Simplify}$$

$$= \frac{2}{5} + \frac{-1}{5}i \quad \text{Standard form}$$

b. We write $\sqrt{-25} = 5i$ and $\sqrt{-9} = 3i$ so that $\frac{4+\sqrt{-25}}{2-\sqrt{-9}} = \frac{4+5i}{2-3i}$.

$$\frac{4+5i}{2-3i} = \frac{(4+5i)(2+3i)}{(2-3i)(2+3i)} \quad \text{Multiply numerator and denominator by } 2 + 3i.$$

$$= \frac{8 + 12i + 10i + 15i^2}{2^2 + 3^2} \quad \text{Use FOIL in the numerator; } (2-3i)(2+3i) = 2^2 + 3^2.$$

$$= \frac{-7+22i}{13} \quad 15i^2 = -15, 8 - 15 = -7$$

$$= -\frac{7}{13} + \frac{22}{13}i \quad \text{Standard form}$$

Practice Problem 7 Write the following quotients in standard form.

a. $\frac{2}{1-i}$ **b.** $\frac{-3i}{4+\sqrt{-25}}$

EXAMPLE 8 Using Complex Numbers in AC Circuits

In a parallel circuit, the total impendence Z_t is given by

$$Z_t = \frac{Z_1Z_2}{Z_1 + Z_2}.$$

Find Z_t assuming that $Z_1 = 2 + 3i$ ohms and $Z_2 = 3 - 4i$ ohms.

Solution

We first calculate $Z_1 + Z_2$ and Z_1Z_2.

$$Z_1 + Z_2 = (2+3i) + (3-4i) = 5 - i$$

$$Z_1Z_2 = (2+3i)(3-4i)$$

$$= 6 - 8i + 9i - 12i^2 \quad \text{FOIL}$$

$$= 18 + i \quad \text{Set } i^2 = -1 \text{ and simplify.}$$

$$Z_t = \frac{Z_1Z_2}{Z_1+Z_2} = \frac{18+i}{5-i}$$

$$= \frac{(18+i)(5+i)}{(5-i)(5+i)} \quad \text{Multiply numerator and denominator by conjugate of the denominator.}$$

$$= \frac{90 + 18i + 5i + i^2}{5^2 + 1^2} \qquad \text{Use FOIL in the numerator; } (5 - i)(5 + i) = 5^2 + 1^2.$$

$$= \frac{89 + 23i}{26} \qquad \text{Set } i^2 = -1 \text{ and simplify.}$$

$$= \frac{89}{26} + \frac{23}{26}i \qquad \text{Standard form}$$

Practice Problem 8 Repeat Example 8 assuming that $Z_1 = 1 + 2i$ and $Z_2 = 2 - 3i$.

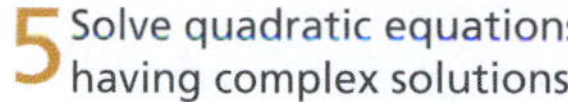

5 Solve quadratic equations having complex solutions.

Quadratic Equations with Complex Solutions

The methods of solving quadratic equations introduced in the previous section are also applicable to solving quadratic equations with complex solutions.

EXAMPLE 9 Solving Quadratic Equations with Complex Solutions

Solve each equation.

a. $x^2 + 4 = 0$ **b.** $x^2 + 2x + 2 = 0$

Solution

a. Use the square root method.

$$x^2 + 4 = 0 \qquad \text{Original equation}$$

$$x^2 = -4 \qquad \text{Add } -4 \text{ to both sides.}$$

$$x = \pm\sqrt{-4} = \pm(\sqrt{4})i = \pm 2i \qquad \text{Solve for } x.$$

The solution set is $\{-2i, 2i\}$. You should check these solutions.

b. $x^2 + 2x + 2 = 1 \cdot x^2 + 2x + 2 = 0$

$$x = \frac{-2 \pm \sqrt{2^2 - 4(1)(2)}}{2(1)} \qquad \text{Use the quadratic formula with } a = 1, b = 2, \text{ and } c = 2.$$

$$= \frac{-2 \pm \sqrt{-4}}{2} \qquad \text{Simplify.}$$

$$= \frac{-2 \pm 2i}{2} = \frac{\cancel{2}(-1 \pm i)}{\cancel{2}}$$

$$= -1 \pm i$$

The solution set is $\{-1 - i, -1 + i\}$. You should check these solutions.

Practice Problem 9 Solve.

a. $4x^2 + 9 = 0$ **b.** $x^2 = 4x - 13$

The Discriminant

If a, b, and c are real numbers and $a \neq 0$, we can predict the number and type of solutions that the equation $ax^2 + bx + c = 0$ will have without solving the equation. In the quadratic formula

$$x = \frac{-b \pm \sqrt{b^2 - 4ac}}{2a},$$

the quantity $b^2 - 4ac$ under the radical sign is called the **discriminant** of the equation. The discriminant reveals the type of solutions of the equation, as shown in the table on the next page.

Discriminant	Description of Solutions
$b^2 - 4ac > 0$	There are two unequal real solutions.
$b^2 - 4ac = 0$	There is one real solution.
$b^2 - 4ac < 0$	There are two nonreal complex solutions, and they are conjugates.

EXAMPLE 10 **Using the Discriminant**

Use the discriminant to determine the number and type of solutions of each quadratic equation.

a. $x^2 - 4x + 2 = 0$ **b.** $2t^2 + 2t + 19 = 0$ **c.** $4x^2 + 4x + 1 = 0$

Solution

Equation	$b^2 - 4ac$	Conclusion
a. $x^2 - 4x + 2 = 0$	$(-4)^2 - 4(1)(2) = 8 > 0$	Two unequal real solutions
b. $2t^2 + 2t + 19 = 0$	$(2)^2 - 4(2)(19) = -148 < 0$	Two nonreal complex solutions
c. $4x^2 + 4x + 1 = 0$	$(4)^2 - 4(4)(1) = 0$	Exactly one real solution

Practice Problem 10 Determine the number and type of solutions.

a. $9x^2 - 6x + 1 = 0$ **b.** $x^2 - 5x + 3 = 0$ **c.** $2x^2 - 3x + 4 = 0.$ 

Answers to Practice Problems

1. a. Real $= -1$, imaginary $= 2$ **b.** Real $= -\frac{1}{3}$, imaginary $= -6$ **c.** Real $= 8$, imaginary $= 0$
2. $a = -2; b = 1$ **3. a.** $4 - 2i$ **b.** $-1 + 4i$ **c.** $-2 + 5i$
4. a. $26 + 2i$ **b.** $-15 - 21i$ **5. a.** $5 - 12i$ **b.** $16 + 14\sqrt{2}i$
6. a. 37 **b.** 4
7. a. $1 + i$ **b.** $-\frac{15}{41} - \frac{12}{41}i$ **8.** $\frac{23}{10} + \frac{11}{10}i$
9. a. $\left\{-\frac{3}{2}i, \frac{3}{2}i\right\}$ **b.** $\{2 - 3i, 2 + 3i\}$
10. a. One real **b.** Two unequal real **c.** Two nonreal complex

SECTION 1.3 Exercises

Basic Concepts and Skills

1. We define $i =$ ____________ so that $i^2 =$ ____________.
2. A complex number in the form $a + bi$ is said to be in ____________.
3. For $b > 0$, $\sqrt{-b} =$ ____________.
4. The conjugate of $a + bi$ is ____________, and the conjugate of $a - bi$ is ____________.
5. **True or False.** The product of a complex number and its conjugate is a real number.
6. **True or False.** Division by a nonzero complex number z is accomplished by multiplying the numerator and denominator by $\bar{z}$.

In Exercises 7–10, use the definition of equality of complex numbers to find the real numbers x and y so that the equation is true.

7. $2 + xi = y + 3i$
8. $x - 2i = 7 + yi$
9. $x - \sqrt{-16} = 2 + yi$
10. $3 + yi = x - \sqrt{-25}$

In Exercises 11–34, perform each operation and write the result in the standard form $a + bi$.

11. $(5 + 2i) + (3 + i)$
12. $(6 + i) + (1 + 2i)$
13. $(4 - 3i) - (5 + 3i)$
14. $(3 - 5i) - (3 + 2i)$

15. $(-2 - 3i) + (-3 - 2i)$

16. $(-5 - 3i) + (2 - i)$

17. $3(5 + 2i)$

18. $4(3 + 5i)$

19. $-4(2 - 3i)$

20. $-7(3 - 4i)$

21. $3i(5 + i)$

22. $2i(4 + 3i)$

23. $4i(2 - 5i)$

24. $-3i(5 - 2i)$

25. $(3 + i)(2 + 3i)$

26. $(4 + 3i)(2 + 5i)$

27. $(2 - 3i)(2 + 3i)$

28. $(4 - 3i)(4 + 3i)$

29. $(3 + 4i)(4 - 3i)$

30. $(-2 + 3i)(-3 + 10i)$

31. $(\sqrt{3} - 12i)^2$

32. $(-\sqrt{5} - 13i)^2$

33. $(2 - \sqrt{-16})(3 + 5i)$

34. $(5 - 2i)(3 + \sqrt{-25})$

In Exercises 35–40, write the conjugate $\bar{z}$ of each complex number z. Then find $z\bar{z}$.

35. $z = 2 - 3i$

36. $z = 4 + 5i$

37. $z = \frac{1}{2} - 2i$

38. $z = \frac{2}{3} + \frac{1}{2}i$

39. $z = \sqrt{2} - 3i$

40. $z = \sqrt{5} + \sqrt{3}\,i$

In Exercises 41–54, write each quotient in the standard form $a + bi$.

41. $\frac{5}{-i}$

42. $\frac{2}{-3i}$

43. $\frac{-1}{1 + i}$

44. $\frac{1}{2 - i}$

45. $\frac{5i}{2 + i}$

46. $\frac{3i}{2 - i}$

47. $\frac{2 + 3i}{1 + i}$

48. $\frac{3 + 5i}{4 + i}$

49. $\frac{2 - 5i}{4 - 7i}$

50. $\frac{3 + 5i}{1 - 3i}$

51. $\frac{2 + \sqrt{-4}}{1 + i}$

52. $\frac{5 - \sqrt{-9}}{3 + 2i}$

53. $\frac{-2 + \sqrt{-25}}{2 - 3i}$

54. $\frac{-5 - \sqrt{-4}}{5 - \sqrt{-9}}$

In Exercises 55–64, solve each equation.

55. $x^2 + 5 = 1$

56. $4x^2 + 9 = 0$

57. $z^2 - 2z + 2 = 0$

58. $x^2 - 6x + 11 = 0$

59. $2x^2 - 20x + 49 = -7$

60. $4y^2 + 4y + 5 = 0$

61. $8(x^2 - x) = x^2 - 3$

62. $t(t + 1) = 3t^2 + 1$

63. $9k^2 + 25 = 0$

64. $3k^2 + 4 = 0$

Applying the Concepts

65. Series circuits. Assuming that the impedance of a resistor in a circuit is $Z_1 = 4 + 3i$ ohms and the impedance of a second resistor is $Z_2 = 5 - 2i$ ohms, find the total impedance of the two resistors when placed in series (sum of the two impedances).

Series circuit

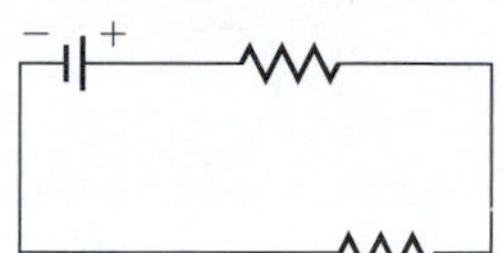

66. Parallel circuits. If the two resistors in Exercise 65 are connected in parallel, the total impedance is given by

$$\frac{Z_1Z_2}{Z_1 + Z_2}.$$

Find the total impedance assuming that the resistors in Exercise 65 are connected in parallel.

Parallel circuit

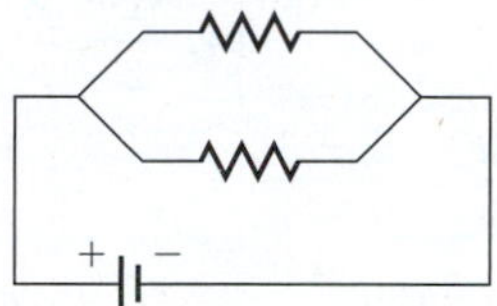

As with impedance, the current I and voltage V in a circuit can be represented by complex numbers. The three quantities (voltage, V; impedance, Z; and current, I) are related by the equation $Z = \frac{V}{I}$. Thus, if two of these values are given, the value of the third can be found from the equation.

In Exercises 67–72, use the equation $V = ZI$ to find the value that is not specified.

67. Finding impedance: $I = 7 + 5i \quad V = 35 + 70i$

68. Finding impedance: $I = 7 + 4i \quad V = 45 + 88i$

69. Finding voltage: $Z = 5 - 7i \quad I = 2 + 5i$

70. Finding voltage: $Z = 7 - 8i \quad I = \frac{1}{3} + \frac{1}{6}i$

71. Finding current: $V = 12 + 10i \quad Z = 12 + 6i$

72. Finding current: $V = 29 + 18i \quad Z = 25 + 6i$

Beyond the Basics

In Exercises 73–82, find each power of i and simplify the expression.

73. i^{17}

74. i^{125}

75. i^{-7}

76. i^{-24}

77. $i^{10} + 7$

78. $9 + i^3$

79. $3i^5 - 2i^3$

80. $5i^6 - 3i^4$

81. $2i^3(1 + i^4)$

82. $5i^5(i^3 - i)$

83. Prove that the reciprocal of $a + bi$, where both a and b are nonzero, is $\frac{a}{a^2 + b^2} - \frac{b}{a^2 + b^2}i$.

In Exercises 84–87, let $z = a + bi$ and $w = c + di$. Prove each statement.

84. $\text{Re}(z) = \frac{z + \bar{z}}{2}$

85. $\text{Im}(z) = \frac{z - \bar{z}}{2i}$

86. $\text{Re}\left(\frac{z}{z + w}\right) + \text{Re}\left(\frac{w}{z + w}\right) = 1$

87. The product $z\bar{z} = 0$ if and only if $z = 0$.

In Exercises 88–91, use the definition of equality of complex numbers to solve for x and y.

88. $\frac{x}{i} + y = 3 + i$

89. $x - \frac{y}{i} = 4i + 1$

90. $\frac{x + yi}{i} = 5 - 7i$

91. $\frac{5x + yi}{2 - i} = 2 + i$

92. Show that $\frac{1 - 2i}{5 - 5i} = \frac{2 - 3i}{9 - 7i}$.

93. Write in standard form:

$$\frac{1 + i}{1 - i} \div \frac{2 + i}{1 + 2i}$$

94. Solve for z:

$$(1 + 3i)z + (2 + 4i) = 7 - 3i$$

95. Let $z = 2 - 3i$ and $w = 1 + 2i$.

a. Show that $\overline{(zw)} = (\bar{z})(\bar{w})$.

b. Show that $\overline{\left(\frac{z}{w}\right)} = \frac{\bar{z}}{\bar{w}}$.

Critical Thinking/Writing/Discussion

96. State whether each of the following is true or false. Explain your reasoning.

a. Every real number is a complex number.

b. Every complex number is a real number.

c. Every complex number is an imaginary number.

d. A real number is a complex number whose imaginary part is 0.

e. The product of a complex number and its conjugate is a real number.

f. The equality $z = \bar{z}$ holds if and only if z is a real number.

97. Find the least positive n for which $\left(\frac{1 + i}{1 - i}\right)^n = 1$.

98. Show that the set of complex numbers does not have the ordering properties of the set of real numbers. [*Hint:* Assume that ordering properties hold. Then by the law of trichotomy, $i = 0$ or $i < 0$ or $i > 0$. Show that each leads to a contradiction.]

Maintaining Skills

In Exercises 99 and 100, factor each expression by grouping.

99. $x^3 - x^2 - 9x + 9$

100. $-x^3 - 2x^2 + 25x + 50$

In Exercises 101–106, find the LCD for each group of rational expressions.

101. $\frac{1}{x - 1}, \frac{1}{x}$

102. $\frac{3}{2 - x}, \frac{x}{x - 2}$

103. $\frac{15}{x + 3}, \frac{2x + 1}{x - 3}$

104. $\frac{3x}{x^2 - 25}, \frac{9}{5 - x}$

105. $\frac{x + 1}{(x - 3)(2 - x)}, \frac{12}{(x - 2)(x + 1)}$

106. $\frac{x}{4x^2 - 1}, \frac{7}{2x - 1}, \frac{1}{x}$

In Exercises 107–114, simplify each expression.

107. $27^{2/3}$

108. $8^{4/3}$

109. $4^{5/2}$

110. $25^{3/2}$

111. $(\sqrt{3 + x^2})^2$

112. $(1 + \sqrt{x + 1})^2$

113. $((3x^2 + 7)^{1/3})^3$

114. $((5x - 2)^{3/5})^{5/3}$

SECTION 1.4

Solving Other Types of Equations

Albert Einstein 1879–1955
Einstein began his schooling in Munich, Germany. Einstein renounced his German citizenship in 1896 and entered the Swiss Federal Polytechnic School in Zurich to be trained as a teacher in physics and mathematics. Unable to find a teaching job, he accepted a position as a technical assistant in the Swiss patent office. In 1905, Einstein earned his doctorate from the University of Zurich.

While at the patent office, he produced much of his remarkable work on the special theory of relativity. In 1909, he was appointed Professor Extraordinary at Zurich. In 1914, he was appointed professor at the University of Berlin. He immigrated to America to take the position of professor of physics at Princeton University. By 1949, Einstein was unwell. He drew up his will in 1950.

He left his scientific papers to the Hebrew University in Jerusalem. In 1952, the Israeli government offered the post of second president to Einstein. He politely declined the offer.

One week before his death, Einstein agreed to let Bertrand Russell place his name on a manifesto urging all nations to give up nuclear weapons.

BEFORE STARTING THIS SECTION, REVIEW

1 Factoring (Section P.4)
2 Radicals (Section P.6, page 59)
3 Quadratic equations (Sections 1.2 and 1.3)

OBJECTIVES

1 Solve equations by factoring.
2 Solve rational equations.
3 Solve equations involving radicals.
4 Solve equations with rational exponents.
5 Solve equations that are quadratic in form.

Time Dilation

Do you ever feel like time moves really quickly or really slowly? For example, don't the hours fly by when you are hanging out with your best friend and the seconds drag on endlessly when a dentist is working on your teeth? But you cannot really speed up or slow time down, right? It always "flows" at the same rate.

Einstein said "no." He asserted in his "Special Theory of Relativity" that time is relative. It speeds up or slows down depending on how fast one object is moving relative to another. He concluded that the closer we come to traveling at the speed of light, the more time will slow down for us relative to someone not moving. He called this slowing of time due to motion *time dilation*. Time dilation can be quantified by the equation

$$t_0 = t\sqrt{1 - \frac{v^2}{c^2}},$$

where $t_0 =$ the "proper time" measured in the moving frame of reference,

$t =$ the time observed in the *fixed* frame of reference,

$v =$ the speed of the moving frame of reference, and

$c =$ the speed of light.

We investigate this effect in Example 12.

1 Solve equations by factoring.

Solving Equations by Factoring

The method of factoring used in solving quadratic equations can also be used to solve certain nonquadratic equations. Specifically, we use the following strategy for solving those equations that can be expressed in factored form with 0 *on one side.*

PROCEDURE IN ACTION

EXAMPLE 1 **Solving Equations by Factoring**

OBJECTIVE

Use factoring techniques to solve an equation.

Step 1 Make one side zero. Move all nonzero terms in the equation to one side (say, the left side) so that the other side (the right side) is 0.

Step 2 Factor the left side.

Step 3 Use the zero-product property. Set each factor in Step 2 equal to 0 and then solve the resulting equations.

Step 4 Check your solutions.

EXAMPLE

Solve $x^3 + 3x^2 = 4x + 12$.

$x^3 + 3x^2 - 4x - 12 = 0$ Subtract $4x + 12$ from both sides.

$x^2(x + 3) - 4(x + 3) = 0$ Group for factoring.

$(x + 3)(x^2 - 4) = 0$ Factor out $x + 3$.

$(x + 3)(x + 2)(x - 2) = 0$ Factor $x^2 - 4$.

$x + 3 = 0$ or $x + 2 = 0$ or $x - 2 = 0$

$x = -3$ or $x = -2$ or $x = 2$

Check:

Check the solutions in the original equation.

For $x = -3$, $(-3)^3 + 3(-3)^2 \stackrel{?}{=} 4(-3) + 12$

$-27 + 27 = -12 + 12$ ✓

You should also verify that $x = -2$ *and* $x = 2$ *are solutions of* $x^3 + 3x^2 = 4x + 12$. *The solution set is* $\{-3, -2, 2\}$.

Practice Problem 1 Solve: $x^3 + 2x^2 - x - 2 = 0$

EXAMPLE 2 **Solving an Equation by Factoring**

Solve by factoring: $x^4 = 9x^2$

Solution

Step 1 **Move all terms to the left side.**

$x^4 = 9x^2$ Original equation

$x^4 - 9x^2 = 0$ Subtract $9x^2$ from both sides.

RECALL

$A^2 - B^2 = (A + B)(A - B)$

Difference of two squares, page 34

Step 2 **Factor.**

$x^2(x^2 - 9) = 0$ $x^4 - 9x^2 = x^2(x^2 - 9)$

$x^2(x + 3)(x - 3) = 0$ $x^2 - 9 = (x + 3)(x - 3)$

Step 3 **Set each factor equal to zero.**

$x^2 = 0$ or $x + 3 = 0$ or $x - 3 = 0$ Zero-product property

$x = 0$ or $x = -3$ or $x = 3$ Solve each equation for x.

Step 4 **Check:**

Let $x = 0$	Let $x = -3$	Let $x = 3$
$(0)^4 \stackrel{?}{=} 9(0)^2$	$(-3)^4 \stackrel{?}{=} 9(-3)^2$	$(3)^4 \stackrel{?}{=} 9(3)^2$
$0 = 0$ ✓	$81 = 81$ ✓	$81 = 81$ ✓

Thus, the solution set of the equation is $\{-3, 0, 3\}$. Note that 0 is a root of multiplicity 2.

Practice Problem 2 Solve: $x^4 = 4x^2$ 

WARNING

A common error when solving an equation such as $x^4 = 9x^2$ in Example 2 is to divide both sides by x^2, obtaining $x^2 = 9$ so that ± 3 are the two solutions. This leads to the loss of the solution 0. Remember, we can divide only by quantities that are not 0.

EXAMPLE 3 Solving an Equation by Factoring

Solve by factoring: $x^3 - 2x^2 = -x + 2$

Solution

Step 1 **Move all terms to the left side.**

$$x^3 - 2x^2 = -x + 2 \quad \text{Original equation}$$
$$x^3 - 2x^2 + x - 2 = -x + 2 + x - 2 \quad \text{Add } x - 2 \text{ to both sides.}$$
$$x^3 - 2x^2 + x - 2 = 0 \quad \text{Simplify.}$$

Step 2 **Factor.** Because there are four terms, we factor by grouping.

$$(x^3 - 2x^2) + (x - 2) = 0$$
$$x^2(x - 2) + 1 \cdot (x - 2) = 0 \quad \text{Factor } x^2 \text{ from the first two terms and rewrite } x - 2 \text{ as } 1 \cdot (x - 2).$$
$$(x - 2)(x^2 + 1) = 0 \quad \text{Factor out the common factor } (x - 2).$$

Step 3 **Set each factor equal to zero.**

$$x - 2 = 0 \qquad x^2 + 1 = 0$$
$$x = 2 \qquad x^2 = -1$$
$$x = \pm i$$

Step 4 **Check:**

Let $x = 2$.	Let $x = i$.	Let $x = -i$.
$(2)^3 - 2(2)^2 \stackrel{?}{=} -2 + 2$	$(i)^3 - 2(i)^2 \stackrel{?}{=} -(i) + 2$	$(-i)^3 - 2(-i)^2 \stackrel{?}{=} -(-i) + 2$
$8 - 8 \stackrel{?}{=} 0$	$-i + 2 = -i + 2$ ✓	$i + 2 = i + 2$ ✓
$0 = 0$ ✓		

Thus, the solution set of the equation is $\{2, i, -i\}$.

Practice Problem 3 Solve by factoring: $x^3 - 5x^2 = 4x - 20$

2 Solve rational equations.

Rational Equations

Recall that if at least one algebraic expression with the variable in the denominator appears in an equation, then the equation is a *rational equation*.

When we multiply a rational equation by an expression containing the variable, we may introduce a solution that satisfies the new equation but does not satisfy the original equation. Such a solution is called an **extraneous solution** or **extraneous root**. *So whenever we multiply an equation by an expression containing the variable, we must check all possible solutions obtained to reject extraneous solutions (if any).*

EXAMPLE 4 Solving a Rational Equation

Solve: $\frac{1}{6} + \frac{1}{x+1} = \frac{1}{x}$

Solution

Step 1 **Find the LCD.** The LCD from the denominators

$6, x + 1$, and x is $6x(x + 1)$.

Step 2 **Multiply both sides of the equation by the LCD and make the right side 0.**

$$6x(x+1)\left[\frac{1}{6} + \frac{1}{x+1}\right] = 6x(x+1)\left[\frac{1}{x}\right]$$ Multiply both sides by the LCD.

$$\frac{6x(x+1)}{6} + \frac{6x(x+1)}{x+1} = \frac{6x(x+1)}{x}$$ Distributive property

$$x(x+1) + 6x = 6(x+1)$$ Simplify.

$$x^2 + x + 6x = 6x + 6$$ Distributive property

$$x^2 + x - 6 = 0$$ Subtract $6x + 6$ from both sides.

Step 3 **Factor.**

$$x^2 + x - 6 = 0$$

$$(x+3)(x-2) = 0$$ Factor.

Step 4 **Set each factor equal to zero.**

$$x + 3 = 0 \quad \text{or} \quad x - 2 = 0$$

$$x = -3 \quad \text{or} \quad x = 2$$

Step 5 **Check:** You should check the solutions, -3 and 2, in the original equation. The solution set is $\{-3, 2\}$.

Practice Problem 4 Solve: $\frac{1}{x} - \frac{12}{5x+10} = \frac{1}{5}$

EXAMPLE 5 Solving a Rational Equation with an Extraneous Solution

Solve: $\frac{x}{x-1} - \frac{1}{x+1} = \frac{2x}{x^2-1}$

Solution

Step 1 The LCD of the denominators $(x-1), (x+1), (x^2-1) = (x-1)(x+1)$ is $(x-1)(x+1)$.

Step 2 $$(x-1)(x+1)\left[\frac{x}{x-1} - \frac{1}{x+1}\right]$$ Multiply both sides by the LCD.

$$= (x-1)(x+1)\left[\frac{2x}{x^2-1}\right]$$

$$(x-1)(x+1)\frac{x}{x-1} - (x-1)(x+1)\frac{1}{x+1}$$ Distributive property; $(x-1)(x+1) = x^2 - 1$

$$= (x^2-1)\cdot\frac{2x}{x^2-1}$$

$$(x+1)x - (x-1) = 2x$$ Simplify.

$$x^2 + x - x + 1 - 2x = 2x - 2x$$ Distributive property; add $-2x$ to both sides.

$$x^2 - 2x + 1 = 0$$ Simplify.

Step 3 $(x - 1)(x - 1) = 0$ — Factor.

Step 4 $x - 1 = 0$ or $x = 1$ — Solve for x.

Step 5 **Check:** Substitute $x = 1$ in the original equation.

$$\frac{x}{x-1} - \frac{1}{x+1} = \frac{2x}{x^2-1}$$

$$\frac{1}{1-1} - \frac{1}{1+1} \stackrel{?}{=} \frac{2(1)}{1^2-1}$$

The denominator of the first term is 0, so the equation is undefined for $x = 1$. Therefore, $x = 1$ is not a solution of the original equation; it is extraneous. The original equation has no solution.

Practice Problem 5 Solve: $\dfrac{x}{x-2} - \dfrac{2}{x+2} = \dfrac{4x}{x^2-4}$

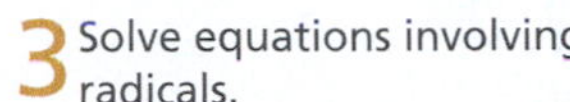

Equations Involving Radicals

If an equation involves radicals or rational exponents, the method of raising both sides to a positive integer power is often used to remove them. When this is done, the solution set of the new equation always contains the solutions of the original equation. However, in some cases, the new equation has *more* solutions than the original equation. For instance, consider the equation $x = 4$. If we square both sides, we get $x^2 = 16$. Notice that the given equation, $x = 4$, has only one solution (namely, 4), whereas the new equation, $x^2 = 16$, has two solutions: 4 and -4. Extraneous solutions may be introduced whenever both sides of an equation are raised to an *even* power, Thus, it is *essential that we check* all solutions obtained after raising both sides of an equation to any even power.

EXAMPLE 6 **Solving Equations Involving Radicals**

Solve: $x = \sqrt{6x - x^3}$

Solution

Because $(\sqrt{a})^2 = a$, we raise both sides of the equation to the second power to eliminate the radical sign. The solutions of the original equation are among the solutions of the equation

$$x^2 = (\sqrt{6x - x^3})^2 \qquad \text{Square both sides of the equation.}$$

$$x^2 = 6x - x^3 \qquad (\sqrt{6x - x^3})^2 = 6x - x^3$$

$$x^3 + x^2 - 6x = 0 \qquad \text{Add } x^3 - 6x \text{ to both sides.}$$

$$x(x^2 + x - 6) = 0 \qquad \text{Factor out } x \text{ from each term.}$$

$$x(x + 3)(x - 2) = 0 \qquad \text{Factor } x^2 + x - 6.$$

$$x = 0 \quad \text{or} \quad x + 3 = 0 \quad \text{or} \quad x - 2 = 0 \qquad \text{Zero-product property}$$

$$x = 0 \quad \text{or} \quad x = -3 \quad \text{or} \quad x = 2 \qquad \text{Solve each equation.}$$

Hence, the only possible solutions of $x = \sqrt{6x - x^3}$ are -3, 0, and 2.

Check:

Let $x = -3$.	Let $x = 0$.	Let $x = 2$.
$-3 \stackrel{?}{=} \sqrt{6(-3) - (-3)^3}$	$0 \stackrel{?}{=} \sqrt{6(0) - (0)^3}$	$2 \stackrel{?}{=} \sqrt{6(2) - (2)^3}$
$-3 \stackrel{?}{=} \sqrt{-18 + 27}$	$0 \stackrel{?}{=} \sqrt{0}$	$2 \stackrel{?}{=} \sqrt{12 - 8}$
$-3 \stackrel{?}{=} \sqrt{9}$	$0 \stackrel{?}{=} 0$ ✓	$2 \stackrel{?}{=} \sqrt{4}$
$-3 \neq 3$ ✗		$2 = 2$ ✓

Thus, -3 is an extraneous solution; 0 and 2 are the solutions of the equation $x = \sqrt{6x - x^3}$. The solution set is $\{0, 2\}$.

Practice Problem 6 Solve: $x = \sqrt{x^3 - 6x}$

EXAMPLE 7 **Solving an Equation Involving a Radical**

Solve: $\sqrt{2x + 1} + 1 = x$

Solution

Step 1 **We begin by isolating the radical to one side of the equation.**

$$\sqrt{2x + 1} + 1 = x \quad \text{Original equation}$$

$$\sqrt{2x + 1} = x - 1 \quad \text{Subtract 1 from both sides.}$$

Step 2 **Next, we square both sides and simplify.**

$$(\sqrt{2x + 1})^2 = (x - 1)^2 \quad \text{Square both sides.}$$

$$2x + 1 = x^2 - 2x + 1 \quad (\sqrt{a})^2 = a;\ (A - B)^2 = A^2 - 2AB + B^2$$

$$2x + 1 - (2x + 1) = x^2 - 2x + 1 - (2x + 1) \quad \text{Subtract } 2x + 1 \text{ from both sides.}$$

$$0 = x^2 - 4x \quad \text{Simplify.}$$

$$0 = x(x - 4) \quad \text{Factor.}$$

Step 3 **Set each factor equal to zero.**

$$x = 0 \quad \text{or} \quad x - 4 = 0$$

$$x = 0 \quad \text{or} \quad x = 4 \quad \text{Solve for } x.$$

Step 4 **Check:**

Let $x = 0$.	Let $x = 4$.
$\sqrt{2(0) + 1} + 1 \stackrel{?}{=} 0$	$\sqrt{2(4) + 1} + 1 \stackrel{?}{=} 4$
$\sqrt{1} + 1 \stackrel{?}{=} 0$	$\sqrt{9} + 1 \stackrel{?}{=} 4$
$1 + 1 \stackrel{?}{=} 0$	$3 + 1 \stackrel{?}{=} 4$
$2 \neq 0$	$4 = 4$ ✓

Thus, 0 is an extraneous root; the only solution of the given equation is 4, and the solution set is $\{4\}$.

Practice Problem 7 Solve: $\sqrt{6x + 4} + 2 = x$

Next, we illustrate a method for solving radical equations that contain two or more square root expressions.

EXAMPLE 8 **Solving an Equation Involving Two Radicals**

Solve: $\sqrt{2x - 1} - \sqrt{x - 1} = 1$

Solution

Step 1 **We first isolate one of the radicals.**

$$\sqrt{2x - 1} - \sqrt{x - 1} = 1 \quad \text{Original equation}$$

$$\sqrt{2x - 1} = 1 + \sqrt{x - 1} \quad \text{Add } \sqrt{x - 1} \text{ to both sides to isolate the radical } \sqrt{2x - 1}.$$

Step 2 **Square both sides of the last equation.**

$(\sqrt{2x - 1})^2 = (1 + \sqrt{x - 1})^2$ Square both sides.

$2x - 1 = 1^2 + 2 \cdot 1 \cdot \sqrt{x - 1} + (\sqrt{x - 1})^2$ $(A + B)^2 = A^2 + 2AB + B^2$

$2x - 1 = 1 + 2\sqrt{x - 1} + x - 1$ Simplify; use $(\sqrt{a})^2 = a$.

$2x - 1 = 2\sqrt{x - 1} + x$ Simplify.

Step 3 **The last equation still contains a radical. We repeat the process of isolating the radical expression.**

$x - 1 = 2\sqrt{x - 1}$ Subtract x from both sides and simplify to isolate $2\sqrt{x - 1}$.

$(x - 1)^2 = (2\sqrt{x - 1})^2$ Square both sides.

$x^2 - 2x + 1 = 4(x - 1)$ Use $(A - B)^2 = A^2 - 2AB + B^2$ and $(AB)^2 = A^2B^2$.

$x^2 - 2x + 1 = 4x - 4$ Distributive property

$x^2 - 6x + 5 = 0$ Add $-4x + 4$ to both sides; simplify.

$(x - 5)(x - 1) = 0$ Factor.

Step 4 **Set each factor equal to 0.**

$x - 5 = 0$ or $x - 1 = 0$

$x = 5$ or $x = 1$ Solve for x.

Step 5 **Check:** You should check that the solutions of the original equation are 1 and 5. The solution set is $\{1, 5\}$.

Practice Problem 8 Solve: $\sqrt{x - 5} + \sqrt{x} = 5$ 

SOLVING EQUATIONS CONTAINING SQUARE ROOTS

Step 1 Isolate one radical to one side of the equation.

Step 2 Square both sides of the equation in Step 1 and simplify.

Step 3 If the equation obtained in Step 2 contains a radical, repeat Steps 1 and 2 to get an equation that is free of radicals.

Step 4 Solve the equation obtained in Steps 1–3.

Step 5 Check the solutions in the original equation.

4 Solve equations with rational exponents.

Equations with Rational Exponents

If a given equation can be expressed in the form $u^{m/n} = k$, then we isolate u by raising both sides to the power $\frac{n}{m}\left(\text{the reciprocal of } \frac{m}{n}\right)$.

SOLVING EQUATIONS OF THE FORM $u^{m/n} = k$

Let m and n be positive integers, $k \geq 0$ a real number, and $\frac{m}{n}$ in lowest terms. Then if

***m* is odd**	***m* is even**
and $u^{m/n} = k$,	and $u^{m/n} = k$,
$u = k^{n/m}$	$u = \pm k^{n/m}$

EXAMPLE 9 **Solving Equations with Rational Exponents**

Solve.

a. $2(2x - 1)^{3/2} - 30 = 24$ **b.** $(3x - 1)^{2/3} + 5 = 9$

Solution

a.

$$2(2x - 1)^{3/2} - 30 = 24 \quad \text{Original equation}$$
$$2(2x - 1)^{3/2} = 24 + 30 \quad \text{Add 30 to both sides and simplify.}$$
$$= 54 \quad \text{Simplify.}$$
$$(2x - 1)^{3/2} = \frac{54}{2} = 27 \quad \text{Divide both sides by 2.}$$
$$((2x - 1)^{3/2})^{2/3} = (27)^{2/3} \quad \text{Raise both sides to the power } \frac{2}{3} \text{ with odd } m = 3.$$
$$2x - 1 = 9 \quad (27)^{2/3} = (\sqrt[3]{27})^2 = 3^2 = 9$$
$$2x = 10 \quad \text{Isolate } x.$$
$$x = 5 \quad \text{Solve for } x.$$

Check:

$$2(2x - 1)^{3/2} - 30 = 24$$
$$2(2 \cdot 5 - 1)^{3/2} - 30 \stackrel{?}{=} 24 \quad \text{Replace } x \text{ with 5.}$$
$$2(9)^{3/2} - 30 \stackrel{?}{=} 24 \quad 2 \cdot 5 - 1 = 9$$
$$2(27) - 30 \stackrel{?}{=} 24 \quad (9)^{3/2} = (\sqrt{9})^3 = (3)^3 = 27$$
$$54 - 30 \stackrel{?}{=} 24 \checkmark \quad \text{Yes}$$

b.

$$(3x - 1)^{2/3} + 5 = 9 \quad \text{Original equation}$$
$$(3x - 1)^{2/3} = 9 - 5 \quad \text{Subtract 5 from both sides and simplify.}$$
$$= 4 \quad \text{Simplify}$$
$$\left[(3x - 1)^{2/3}\right]^{3/2} = \pm(4)^{3/2} \quad \text{Because } m = 2 \text{ is even, we use the } \pm \text{ sign.}$$
$$3x - 1 = \pm 8 \quad (4)^{3/2} = (\sqrt{4})^3 = (2)^3 = 8$$
$$3x = 1 \pm 8 \quad \text{Add 1 to both sides.}$$

Then

$3x = 1 + 8$	or	$3x = 1 - 8$
$3x = 9$		$3x = -7$
$x = 3$		$x = -\frac{7}{3}$

Verify that both $x = 3$ and $x = -\frac{7}{3}$ satisfy the original equation.

The solution set is $\left\{-\frac{7}{3}, 3\right\}$.

Practice Problem 9

Solve: **a.** $3(x - 2)^{3/5} + 4 = 7$ **b.** $(2x + 1)^{4/3} - 7 = 9$

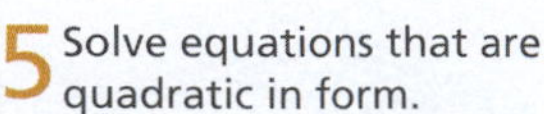
5 Solve equations that are quadratic in form.

Equations That Are Quadratic in Form

An equation in a variable x is **quadratic in form** if it can be written as

$$au^2 + bu + c = 0 \qquad (a \neq 0),$$

where u is an expression in the variable x. We solve the equation $au^2 + bu + c = 0$ for u. Then the solutions of the original equation can be obtained by replacing u with the expression in x that u represents.

EXAMPLE 10 Solving an Equation Quadratic in Form

Solve: $x^{2/3} - 5x^{1/3} + 6 = 0$

Solution

The given equation is not a quadratic equation. However, if we let $u = x^{1/3}$, then $u^2 = (x^{1/3})^2 = x^{2/3}$. The original equation becomes a quadratic equation if we replace $x^{1/3}$ with u.

$$x^{2/3} - 5x^{1/3} + 6 = 0 \quad \text{Original equation}$$
$$u^2 - 5u + 6 = 0 \quad \text{Replace } x^{1/3} \text{ with } u.$$
$$(u - 2)(u - 3) = 0 \quad \text{Factor.}$$
$$u - 2 = 0 \text{ or } u - 3 = 0 \quad \text{Zero-product property}$$
$$u = 2 \text{ or } u = 3 \quad \text{Solve for } u.$$

Because $u = x^{1/3}$, we find x from the equations $u = 2$ and $u = 3$.

$$x^{1/3} = 2 \text{ or } x^{1/3} = 3 \quad \text{Replace } u \text{ with } x^{1/3}.$$
$$x = 2^3 \text{ or } x = 3^3 \quad \text{Here } \frac{m}{n} = \frac{1}{3}, \text{ so } m = 1 \text{ is odd.}$$
$$x = 8 \text{ or } x = 27 \text{ are two apparent solutions.}$$

Check: You should check that both $x = 8$ and $x = 27$ satisfy the equation. The solution set is $\{8, 27\}$.

Practice Problem 10 Solve: $x^{2/3} - 7x^{1/3} + 6 = 0$

WARNING When u replaces an expression in the variable x in an equation that is quadratic in form, you are not finished once you have found values for u. You then must replace u with the expression in x and solve for x.

EXAMPLE 11 Solving an Equation That Is Quadratic in Form

Solve: $\left(x + \frac{1}{x}\right)^2 - 6\left(x + \frac{1}{x}\right) + 8 = 0$

Solution

We let $u = x + \frac{1}{x}$. Then $u^2 = \left(x + \frac{1}{x}\right)^2$. With this substitution, the original equation becomes a quadratic equation in u.

$$\left(x + \frac{1}{x}\right)^2 - 6\left(x + \frac{1}{x}\right) + 8 = 0 \quad \text{Original equation}$$
$$u^2 - 6u + 8 = 0 \quad \text{Replace } x + \frac{1}{x} \text{ with } u.$$
$$(u - 2)(u - 4) = 0 \quad \text{Factor.}$$
$$u - 2 = 0 \text{ or } u - 4 = 0 \quad \text{Zero-product property}$$
$$u = 2 \text{ or } u = 4 \quad \text{Solve for } u.$$

Replacing u with $x + \frac{1}{x}$ in these equations, we have

$$x + \frac{1}{x} = 2 \text{ or } x + \frac{1}{x} = 4.$$

Next, we solve each of these equations for x.

$$x + \frac{1}{x} = 2$$

$$x^2 + 1 = 2x \quad \text{Multiply both sides by } x \text{ (the LCD).}$$

$$x^2 - 2x + 1 = 0 \quad \text{Subtract } 2x \text{ from both sides; simplify.}$$

$$(x - 1)^2 = 0 \quad \text{Factor the perfect-square trinomial.}$$

$$x - 1 = 0 \quad \text{Square root property}$$

$$x = 1 \quad \text{Solve for } x.$$

Thus, $x = 1$ is a possible solution of the original equation.

$$x + \frac{1}{x} = 4$$

$$x^2 + 1 = 4x \quad \text{Multiply both sides by } x \text{ (the LCD).}$$

$$x^2 - 4x + 1 = 0 \quad \text{Subtract } 4x \text{ from both sides.}$$

$$x = \frac{-(-4) \pm \sqrt{(-4)^2 - 4(1)(1)}}{2(1)} \quad \text{Quadratic formula: } a = 1, b = -4, c = 1$$

$$x = \frac{4 \pm \sqrt{12}}{2} \quad \text{Simplify.}$$

$$x = \frac{4 \pm 2\sqrt{3}}{2} \quad \sqrt{12} = \sqrt{4 \cdot 3} = \sqrt{4}\sqrt{3} = 2\sqrt{3}$$

$$x = \frac{2(2 \pm \sqrt{3})}{2} \quad \text{Factor the numerator.}$$

$$x = 2 \pm \sqrt{3} \quad \text{Remove the common factor.}$$

Thus, $2 + \sqrt{3}$ and $2 - \sqrt{3}$ are also possible solutions of the original equation. Consequently, the possible solutions of the original equation are $1, 2 + \sqrt{3}$, and $2 - \sqrt{3}$.

You should verify that $1, 2 + \sqrt{3}$, and $2 - \sqrt{3}$ are solutions of the original equation. The solution set is $\{1, 2 - \sqrt{3}, 2 + \sqrt{3}\}$.

Practice Problem 11 Solve: $\left(1 + \frac{1}{x}\right)^2 - 6\left(1 + \frac{1}{x}\right) + 8 = 0$

The next example illustrates the paradoxical effect of time dilation (page 115) when one travels at speeds that are close to the speed of light.

EXAMPLE 12 Investigating Space Travel (Your Older Sister Is Younger Than You Are)

Your sister is five years older than you are. She decides that she has had enough of Earth and needs a vacation. She takes a trip to the Omega-One star system. Her trip to Omega-One and back in a spacecraft traveling at an average speed v took 15 years according to the clock and calendar on the spacecraft. But upon landing back on Earth, she discovers that her voyage took 25 years according to the time on Earth. This means that although you were five years younger than your sister before her vacation, you are five years older than her after her vacation! Use the time-dilation equation $t_0 = t\sqrt{1 - \frac{v^2}{c^2}}$ from the introduction to this section to calculate the speed of the spacecraft.

Solution

Substitute $t_0 = 15$ (moving-frame time) and $t = 25$ (fixed-frame time) to obtain

$$15 = 25\sqrt{1 - \frac{v^2}{c^2}}$$

$$\frac{3}{5} = \sqrt{1 - \frac{v^2}{c^2}} \qquad \text{Divide both sides by 25; } \frac{15}{25} = \frac{3}{5}.$$

$$\frac{9}{25} = 1 - \frac{v^2}{c^2} \qquad \text{Square both sides.}$$

$$\frac{v^2}{c^2} = 1 - \frac{9}{25} \qquad \text{Add } \frac{v^2}{c^2} - \frac{9}{25} \text{ to both sides.}$$

$$\left(\frac{v}{c}\right)^2 = \frac{16}{25} \qquad \text{Simplify.}$$

$$\frac{v}{c} = \pm\frac{4}{5} \qquad \text{Square root property}$$

$$\frac{v}{c} = \frac{4}{5} \qquad \text{Reject the negative value.}$$

$$v = \frac{4}{5}c = 0.8c \qquad \text{Multiply both sides by } c.$$

Thus, the spacecraft must have been traveling at 80% of the speed of light $(0.8c)$ when your sister found the "trip of youth."

Practice Problem 12 Suppose your sister's trip took 20 years according to the clock and calendar on the spacecraft, but 25 years judging by time on Earth. Find the speed of the spacecraft.

Answers to Practice Problems

1. $\{-2, -1, 1\}$ **2.** $\{-2, 0, 2\}$ **3.** $\{-2, 2, 5\}$ **4.** $\{-10, 1\}$ **5.** $\varnothing$ **6.** $\{0, 3\}$ **7.** $\{10\}$ **8.** $\{9\}$ **9. a.** $\{3\}$ **b.** $\left\{-\frac{9}{2}, \frac{7}{2}\right\}$ **10.** $\{1, 216\}$ **11.** $\left\{\frac{1}{3}, 1\right\}$ **12.** $0.6c$ or 60% of the speed of light

SECTION 1.4 Exercises

Basic Concepts and Skills

1. If an apparent solution does not satisfy the equation, it is called a(n) ____________ solution.
2. When solving a rational equation, we multiply both sides by the ____________.
3. If $x^{3/4} = 8$, then $x =$ ____________.
4. If $x^{4/3} = 16$, then $x =$ ____________ or $x =$ ____________.
5. **True or False.** Rational equations always have extraneous solutions.
6. **True or False.** If $x^{2/3} = k$, then $x = \pm\sqrt{k^3}$.

In Exercises 7–16, find the real solutions of each equation by factoring.

7. $x^3 = 2x^2$ **8.** $3x^4 - 27x^2 = 0$

9. $(\sqrt{x})^3 = \sqrt{x}$ **10.** $(\sqrt{x})^5 = 16\sqrt{x}$

11. $x^3 + x = 0$ **12.** $x^3 - 1 = 0$

13. $x^4 - x^3 = x^2 - x$ **14.** $x^3 - 36x = 16(x - 6)$

15. $x^4 = 27x$ **16.** $3x^4 = 24x$

In Exercises 17–26, solve each equation by multiplying both sides by the LCD.

17. $\frac{x+1}{3x-2} = \frac{5x-4}{3x+2}$

18. $\frac{x}{2x+1} = \frac{3x+2}{4x+3}$

19. $\frac{1}{x} + \frac{2}{x+1} = 1$

20. $\frac{x}{x+1} + \frac{x+1}{x+2} = 1$

21. $\frac{6x-7}{x} - \frac{1}{x^2} = 5$

22. $\frac{1}{x} + \frac{2}{x+1} + \frac{3}{x+2} = 0$

23. $\frac{1}{x} + \frac{1}{x-3} = \frac{7}{3x-5}$

24. $\frac{x+3}{x-1} + \frac{x+4}{x+1} = \frac{8x+5}{x^2-1}$

25. $\frac{5}{x+1} - \frac{4}{2x+2} + \frac{2}{2x-1} = \frac{13}{18}$

26. $\frac{6}{2x-2} - \frac{1}{x+1} = \frac{2}{2x+2} + 1$

In Exercises 27–34, solve each equation. Check your answers.

27. $\frac{x}{x-5} - \frac{5}{x+5} = \frac{10x}{x^2-25}$

28. $\frac{x}{x-4} - \frac{4}{x+4} = \frac{8x}{x^2-16}$

29. $\frac{x}{x-3} + \frac{3}{x+3} = \frac{6x}{x^2-9}$

30. $\frac{x}{x+7} + \frac{7}{x-7} = \frac{14x}{x^2-49}$

31. $\frac{1}{x-1} + \frac{x}{x+3} = \frac{4}{x^2+2x-3}$

32. $\frac{x}{x-2} + \frac{2}{x+3} = -\frac{10}{x^2+x-6}$

33. $\frac{2x}{x+3} - \frac{x}{x-1} = \frac{14}{x^2+2x-3}$

34. $\frac{2(x+1)}{x-2} - \frac{x}{x+1} = \frac{9}{x^2-x-2}$

In Exercises 35–56, solve each equation. Check your answers.

35. $\sqrt[3]{3x-1} = 2$

36. $\sqrt[3]{2x+3} = 3$

37. $\sqrt{x-1} = -2$

38. $\sqrt{3x+4} = -1$

39. $x + \sqrt{x+6} = 0$

40. $x - \sqrt{6x+7} = 0$

41. $\sqrt{y+6} = y$

42. $r + 11 = 6\sqrt{r+3}$

43. $\sqrt{6y-11} = 2y-7$

44. $\sqrt{3y+1} = y-1$

45. $t - \sqrt{3t+6} = -2$

46. $\sqrt{5x^2-10x+9} = 2x-1$

47. $\sqrt{x-3} = \sqrt{2x-5} - 1$

48. $x + \sqrt{x+1} = 5$

49. $\sqrt{2y+9} = 2 + \sqrt{y+1}$

50. $\sqrt{m} - 1 = \sqrt{m-5}$

51. $\sqrt{7z+1} - \sqrt{5z+4} = 1$

52. $\sqrt{3q+1} - \sqrt{q-1} = 2$

53. $\sqrt{2x+5} + \sqrt{x+6} = 3$

54. $\sqrt{5x-9} - \sqrt{x+4} = 1$

55. $\sqrt{2x-5} - \sqrt{x-3} = 1$

56. $\sqrt{3x+5} + \sqrt{6x+3} = 3$

In Exercises 57–60, solve each equation. Check your answers.

57. $(x-4)^{3/2} = 27$

58. $(x+7)^{3/2} = 64$

59. $(5x-3)^{2/3} - 5 = 4$

60. $(2x-6)^{2/3} + 9 = 13$

In Exercises 61–84, solve each equation by using an appropriate substitution. Check your answers.

61. $x - 5\sqrt{x} + 6 = 0$

62. $x - 3\sqrt{x} + 2 = 0$

63. $2y - 15\sqrt{y} = -7$

64. $y + 44 = 15\sqrt{y}$

65. $x^{-2} - x^{-1} - 42 = 0$

66. $x^{1/2} + 3 - 4x^{-1/2} = 0$

67. $x^{2/3} - 6x^{1/3} + 8 = 0$

68. $x^{2/5} + x^{1/5} - 2 = 0$

69. $2x^{1/2} + 3x^{1/4} - 2 = 0$

70. $2x^{1/2} - x^{1/4} - 1 = 0$

71. $x^4 - 13x^2 + 36 = 0$

72. $x^4 - 7x^2 + 12 = 0$

73. $2t^4 + t^2 - 1 = 0$

74. $81y^4 + 1 = 18y^2$

75. $p^2 - 3 + 4\sqrt{p^2-3} - 5 = 0$

76. $x^2 - 3 - 4\sqrt{x^2-3} - 12 = 0$

77. $(3t+1)^2 - 3(3t+1) + 2 = 0$

78. $(7z+5)^2 + 2(7z+5) - 15 = 0$

79. $(\sqrt{y}+5)^2 - 9(\sqrt{y}+5) + 20 = 0$

80. $(2\sqrt{t}+1)^2 - 2(2\sqrt{t}+1) - 3 = 0$

81. $(x^2-4)^2 - 3(x^2-4) - 4 = 0$

82. $(x^2+2)^2 - 5(x^2+2) - 6 = 0$

83. $(x^2-3x)^2 - 2(x^2-3x) - 8 = 0$

84. $(x^2-4x)^2 + 7(x^2-4x) + 12 = 0$

Applying the Concepts

85. **Fractions.** The numerator of a fraction is two less than the denominator. The sum of the fraction and its reciprocal is $\frac{25}{12}$. Find the numerator and denominator assuming that each is a positive integer.

86. **Fractions.** The numerator of a fraction is five less than the denominator. The sum of the fraction and six times its reciprocal is $\frac{25}{2}$. Find the numerator and denominator assuming that each is an integer.

87. **Investment.** Latasha bought some stock for \$1800. If the price of each share of the stock had been \$18 less, she could have bought five more shares for the same \$1800.
    a. How many shares of the stock did she buy?
    b. How much did she pay for each share?

88. **Bus charter.** A civic club charters a bus for one day at a cost of \$575. When two more people join the group, each person's cost decreases by \$2. How many people were in the original group?

89. **Depth of a well.** A stone is dropped into a well, and the time it takes to hear the splash is 4 seconds. Find the depth of the well (to the nearest foot). Assume that the speed of sound is 1100 feet per second and that $d = 16t^2$ is the distance d an object falls in t seconds.

90. **Estimating the horizon.** You are in a hot air balloon 2 miles above the ocean. (See the accompanying diagram.)

As far as you can see, there is only water. Given that the radius of the earth is 3960 miles, how far away is the horizon?

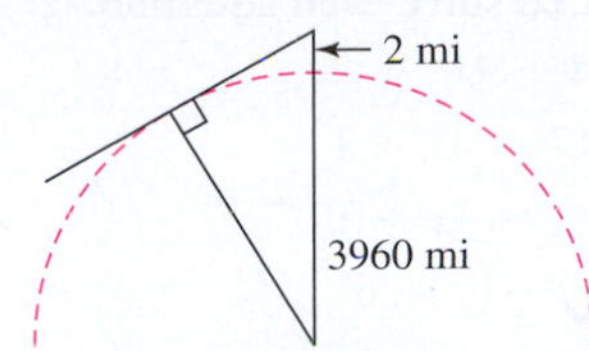

91. **Rate of current.** A motorboat is capable of traveling at a speed of 10 miles per hour in still water. On a particular day, the boat took 30 minutes longer to travel a distance of 12 miles upstream than it took to travel the same distance downstream. What was the rate of the current in the stream on that day?

92. **Train speed.** A freight train requires $2\frac{1}{2}$ hours longer to make a 300-mile journey than an express train does. If the express train averages 20 miles per hour faster than the freight train, how long does it take the express train to make the trip?

93. **Washing the family car.** A couple washed the family car in 24 minutes. Previously, when they each had washed the car alone, it took the husband 20 minutes longer to wash the car than it took the wife. How long did it take the wife to wash the car?

94. **Filling a swimming pool.** A small swimming pool can be filled by two hoses together in 1 hour and 12 minutes. The larger hose alone fills the pool in 1 hour less than the smaller hose. How long does it take the smaller hose to fill the pool?

95. **The area of a rectangular plot.** Together the diagonal and the longer side of a rectangular plot are three times the length of the shorter side. If the longer side exceeds the shorter side by 100 feet, what is the area of the plot?

96. **Emergency at sea.** You are at sea in your motorboat at a point A located 40 km from a point B that lies 130 kilometers from a hospital on a long, straight road. (See the accompanying figure.) While at point A, your companion feels sick and has to be taken to the hospital. You call the hospital on your cell phone to send an ambulance. Your motorboat can travel at 40 kilometers per hour, and the ambulance can travel at 80 kilometers per hour. How far from point B will you be when you and the ambulance arrive simultaneously at point C?

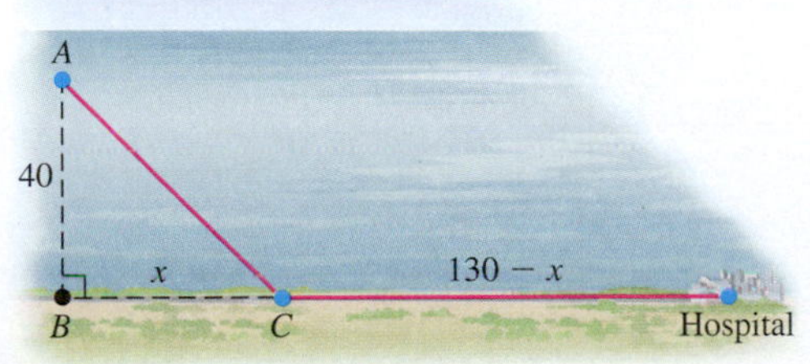

97. **Planning road construction.** Two towns A and B are located 8 and 5 miles, respectively, from a long, straight highway. The points C and D on the highway closest to the two towns are 18 miles apart. (Remember that the shortest distance from a point to a line is the length of the perpendicular segment from the point to the line.) Where should point E be located on the highway so that the sum of the lengths of the roads from A to E and E to B is 23 miles? (See the accompanying figure.)

[*Hint:* $205x^2 - 4392x + 18972 = (x - 6)(205x - 3162)$.]

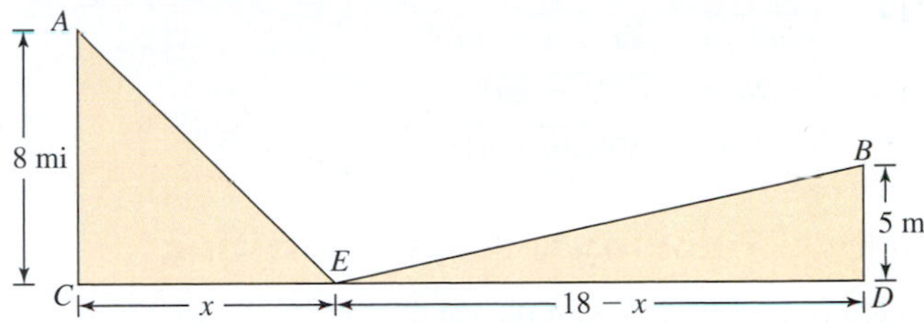

98. **Department store purchases.** A department store buyer purchased some shirts for \$540. Then she sold all but eight of them. On each shirt sold, a profit of \$6 (over the purchase price) was made. The shirts sold brought in revenue of \$780.

a. How many shirts were bought initially?
b. What was the purchase price of each shirt?
c. What was the selling price of each shirt?

Beyond the Basics

In Exercises 99–110, solve each equation by any method.

99. $\left(\dfrac{x^2 - 3}{x}\right)^2 - 6\left(\dfrac{x^2 - 3}{x}\right) + 8 = 0$

100. $\left(\dfrac{x + 2}{3x + 1}\right)^2 - 5\left(\dfrac{x + 2}{3x + 1}\right) + 6 = 0$

101. $\sqrt{\sqrt{3x - 2} + 2\sqrt{4x + 1}} = 2\sqrt{2}$

102. $\dfrac{5}{x - 1} - \dfrac{3x}{x + 1} = \dfrac{9}{x^2 - 1}$

103. $\dfrac{x + 2}{x} + \dfrac{2x + 6}{x^2 + 4x} = -\dfrac{1}{x + 4}$

104. $\left(x^2 + \dfrac{1}{x^2}\right) - 7\left(x - \dfrac{1}{x}\right) + 8 = 0$

$\left[\textit{Hint: } u = x - \dfrac{1}{x}, u^2 + 2 = x^2 + \dfrac{1}{x^2}.\right]$

105. $\left(\dfrac{t}{t + 2}\right)^2 + \dfrac{2t}{t + 2} - 15 = 0$

106. $6t^{-2/5} - 17t^{-1/5} + 5 = 0$

107. $2x^{1/3} + 2x^{-1/3} - 5 = 0$

108. $x^{4/3} - 4x^{2/3} + 3 = 0$

109. $x^{4/3} - 5x^{2/3} + 6 = 0$

110. $8\sqrt{\dfrac{x}{x + 3}} - \sqrt{\dfrac{x + 3}{x}} = 2$

111. Solve $x + \sqrt{x} - 42 = 0$ using these two methods.

a. Rationalize the equation and then solve it.
b. Use the substitution method.

112. Solve $\left(\frac{x}{x+1}\right)^2 - \frac{2x}{x+1} - 8 = 0$ using these two methods.

a. Use the substitution method.

b. Multiply by the LCD.

In Exercises 113–116, solve for x in terms of the other variables. (All letters denote positive real numbers.)

113. $\frac{x+b}{x-b} = \frac{x-5b}{2x-5b}$

114. $\frac{1}{x+a} + \frac{1}{x-2a} = \frac{1}{x-3a}$

115. $3x - 2a = \sqrt{a(3x-2a)}$

116. $3a - \sqrt{ax} = \sqrt{a(3x+a)}$

Critical Thinking/Discussion/Writing

In Exercises 117–119, find the value of x.

117. $x = \sqrt{1 + \sqrt{1 + \sqrt{1 + \cdots}}}$

[*Hint:* $x = \sqrt{1+x}$.]

118. $x = \sqrt{20 + \sqrt{20 + \sqrt{20 + \cdots}}}$

119. $x = \sqrt{n + \sqrt{n + \sqrt{n + \ldots}}}$, where n is a positive integer.

120. Assume that $x > 1$. Solve for x.

$\sqrt{x + 2\sqrt{x-1}} + \sqrt{x - 2\sqrt{x-1}} = 4$

[*Hint:* $x + 2\sqrt{x-1} = (x-1) + 1 + 2\sqrt{x-1} = (\sqrt{x-1}+1)^2$.]

Maintaining Skills

In Exercises 121–132, solve each equation.

121. $12x - 4 = 28 - 4x$

122. $2(x-3) + 17 = 4x + 1$

123. $2x + 3(x-4) = 7x + 10$

124. $2x - 3 - (3x-1) = 6$

125. $\frac{-2x}{3} = x + \frac{x}{2}$

126. $\frac{x-1}{3} + \frac{2x}{5} = x + 2$

127. $6x^2 - 102 = 0$

128. $(x+5)^2 = 27$

129. $25x^2 + 1 = 10x$

130. $x^2 - 10x + 25 = 0$

131. $3x^2 + 3x - 8 = 10$

132. $5x^2 - 6x = 4x^2 + 6x - 3$

In Exercises 133–136, perform the indicated operations, factoring all denominators.

133. $\frac{5}{2x} + 3$

134. $6x + \frac{5}{x-3}$

135. $\frac{3}{x+1} - \frac{x-2}{x+3}$

136. $\frac{x}{x^2+6x+9} - \frac{x-3}{x^2+5x+6}$

SECTION 1.5

Inequalities

BEFORE STARTING THIS SECTION, REVIEW

1. Inequalities (Section P.1, pages 6 and 7)
2. Intervals (Section P.1, page 7)
3. Linear equations (Section 1.1, page 78)

OBJECTIVES

1. Learn the vocabulary for discussing inequalities.
2. Solve and graph linear inequalities.
3. Solve and graph a compound inequality.
4. Solve polynomial and rational inequalities using test points.

The Bermuda Triangle

The Bermuda Triangle, or Devil's Triangle, is the expanse of the Atlantic Ocean between Florida, Bermuda, and Puerto Rico covering approximately 500,000 square miles of sea. This mysterious stretch of sea has an unusually high occurrence of disappearing ships and planes.

For example, on Halloween 1991, pilot John Verdi and his copilot were flying a Grumman Cougar jet over the triangle. They radioed the nearest tower to get permission to increase their altitude. The tower agreed and watched as the jet began the ascent and then disappeared off the radar. The jet did not fly out of range of the radar, did not descend, and did not radio a mayday (distress call). It just vanished, and the plane and crew were never recovered.

One explanation for such disappearances is based on the theory that strange compass readings occur in crossing the Atlantic due to confusion caused by the three north poles: magnetic (toward which compasses point), grid (the real North Pole, at 90 degrees latitude), and true or celestial north (determined by Polaris, the North Star). To get data to test this theory, a plane set out from Miami to Bermuda, a distance of 1035 miles, with the intention of setting the plane on automatic pilot once it reached a cruising speed of 300 miles per hour.

The plane was 150 miles along its path from Miami to Bermuda when it reached the 300-mile-per-hour cruising speed and was set on automatic pilot. After what length of time would you be confident in saying that the plane had encountered trouble? The answer to this question is found in Example 2, which uses a linear inequality, the topic discussed in this section.

1 Learn the vocabulary for discussing inequalities.

Inequalities

When we replace the equal sign $(=)$ in an equation with any of the four inequality symbols $<, \le, >,$ or $\ge$, the resulting expression is an *inequality*. Here are some examples.

Equation	Replace = with	Inequality
$x = 5$	$<$	$x < 5$
$3x + 2 = 14$	$\le$	$3x + 2 \le 14$
$5x + 7 = 3x + 23$	$>$	$5x + 7 > 3x + 23$
$x^2 = 0$	$\ge$	$x^2 \ge 0$

In general, an **inequality** is a statement that one algebraic expression is either less than or is less than or equal to another algebraic expression. The vocabulary for discussing inequalities is similar to that used for equations. The **domain** of a variable in an inequality is the set of all real numbers for which both sides of the inequality are defined. The real numbers that result in a true statement when those numbers are substituted for the variable in the inequality are called **solutions** of the inequality. Because replacing x with 1 in the inequality $3x + 2 \leq 14$ results in the true statement

$$3(1) + 2 \leq 14,$$

the number 1 is a solution of the inequality $3x + 2 \leq 14$. We also say that 1 *satisfies* the inequality $3x + 2 \leq 14$. To **solve** an inequality means to find all solutions of the inequality — that is, its solution set. The most elementary equations, such as $x = 5$, have only one solution. However, even the most elementary inequalities, such as $x < 5$, have infinitely many solutions. In fact, their solution sets are *intervals*, and we frequently graph the solution sets for inequalities in one variable on a number line. The graph of the inequality $x < 5$ is the interval $(-\infty, 5)$, shown in Figure 1.6.

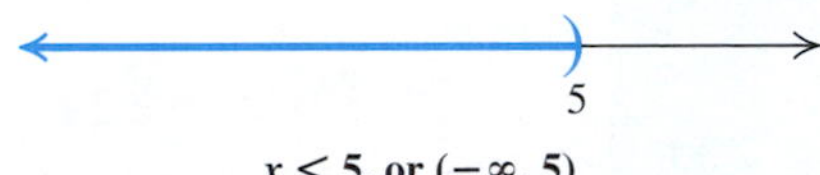

$x < 5$, or $(-\infty, 5)$

FIGURE 1.6 Graph of an inequality

Inequalities are classified the same way as equations: *conditional, inconsistent*, or *identities*. A **conditional inequality** such as $x < 5$ has in its domain at least one solution and at least one number that is not a solution. An inequality that *no* real number satisfies is called an **inconsistent inequality**, and an inequality that is satisfied by *every* real number in the domain of the variable is called an **identity**.

Because $x^2 = x \cdot x$ is the product of (1) two positive factors, (2) two negative factors, or (3) two zero factors, x^2 is either a positive number or zero. That is, x^2 is never negative, or is **nonnegative**. We call this fact the nonnegative identity.

NONNEGATIVE IDENTITY

$$x^2 \geq 0$$

for any real number x.

Two inequalities that have exactly the same solution set are called **equivalent inequalities**. The basic method of solving inequalities is similar to the method for solving equations: We replace a given inequality with a series of equivalent inequalities until we arrive at an equivalent inequality, such as $x < 5$, whose solution set we already know.

The following operations produce equivalent inequalities:

1. Simplifying one or both sides of an inequality by combining like terms and eliminating parentheses
2. Adding or subtracting the same expression on both sides of the inequality

Caution is required in multiplying or dividing both sides of an inequality by a real number or an expression representing a real number. Notice what happens when we multiply an inequality by -1.

We know that $2 < 3$, but how does $-2 = (-1)(2)$ compare with $-3 = (-1)(3)$? We have $-2 > -3$. See Figure 1.7.

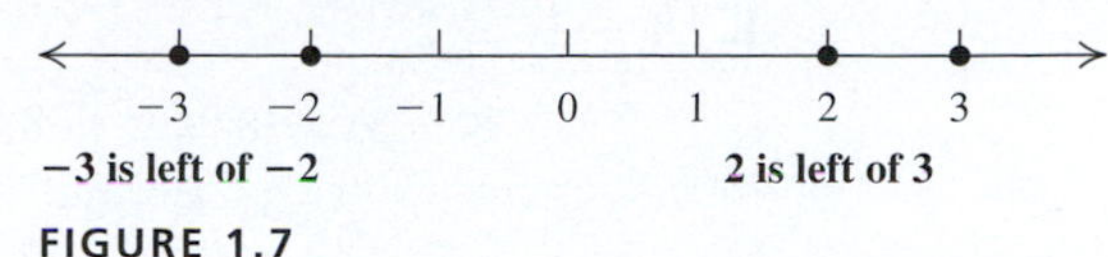

FIGURE 1.7

If we multiply (or divide) both sides of the inequality $2 < 3$ by -1, to get a correct result, we must exchange the $<$ symbol for the $>$ symbol. This exchange of symbols is called reversing the **sense** or the **direction** of the inequality.

The following chart describes the way multiplication and division affect inequalities.

If C represents a real number, then the following inequalities are all equivalent.

Sign of C	Inequality	Sense	Example
	$A < B$	$<$	$3x < 12$
C positive	$A \cdot C < B \cdot C$	Unchanged	$\frac{1}{3}(3x) < \frac{1}{3}(12)$
C positive	$\frac{A}{C} < \frac{B}{C}$	Unchanged	$\frac{3x}{3} < \frac{12}{3}$
C negative	$A \cdot C > B \cdot C$	Reversed	$-\frac{1}{3}(3x) > -\frac{1}{3}(12)$
C negative	$\frac{A}{C} > \frac{B}{C}$	Reversed	$\frac{3x}{-3} > \frac{12}{-3}$

Similar results apply when $<$ is replaced throughout with any of the symbols $\le$, $>$, or $\ge$.

2 Solve and graph linear inequalities.

Linear Inequalities

A **linear inequality in one variable** is an inequality that is equivalent to one of the forms

$$ax + b < 0 \quad \text{or} \quad ax + b \le 0,$$

where a and b represent real numbers and $a \ne 0$.

Inequalities such as $2x - 1 > 0$ and $2x - 1 \ge 0$ are linear inequalities because they are equivalent to $-2x + 1 < 0$ and $-2x + 1 \le 0$, respectively.

SIDE NOTE

A linear inequality becomes a linear equation when the inequality symbol is replaced with the equal sign (=).

EXAMPLE 1 Solving and Graphing Linear Inequalities

Solve each inequality and graph its solution set.

a. $7x - 11 < 2(x - 3)$ **b.** $8 - 3x \le 2$

Solution

a. $7x - 11 < 2(x - 3)$

$7x - 11 < 2x - 6$	Distributive property
$7x - 11 + 11 < 2x - 6 + 11$	Add 11 to both sides.
$7x < 2x + 5$	Simplify.
$7x - 2x < 2x + 5 - 2x$	Subtract $2x$ from both sides.
$5x < 5$	Simplify.
$\frac{5x}{5} < \frac{5}{5}$	Divide both sides by 5.
$x < 1$	Simplify.

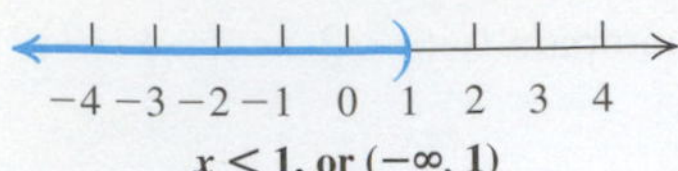

FIGURE 1.8

The solution set is $\{x | x < 1\}$, or in interval notation, $(-\infty, 1)$. The graph of the solution set is shown in Figure 1.8.

b. $8 - 3x \leq 2$

$$8 - 3x - 8 \leq 2 - 8 \quad \text{Subtract 8 from both sides.}$$
$$-3x \leq -6 \quad \text{Simplify.}$$
$$\frac{-3x}{-3} \geq \frac{-6}{-3} \quad \text{Divide both sides by } -3. \text{ (Reverse the direction of the inequality symbol.)}$$
$$x \geq 2 \quad \text{Simplify.}$$

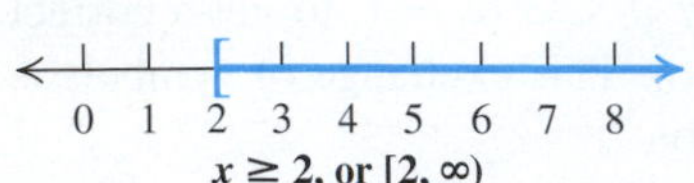

FIGURE 1.9

The solution set is $\{x | x \geq 2\}$, or in interval notation, $[2, \infty)$. The graph of the solution set is shown in Figure 1.9.

Practice Problem 1 Solve each inequality and graph the solution set.

a. $4x + 9 > 2(x + 6) + 1$ **b.** $7 - 2x \geq -3$

EXAMPLE 2 Calculating the Results of the Bermuda Triangle Experiment

In the introduction to this section, we discussed an experiment to test the reliability of compass settings and flight by automatic pilot along one edge of the Bermuda Triangle. The plane is 150 miles along its path from Miami to Bermuda, cruising at 300 miles per hour, when it notifies the tower that it is now set on automatic pilot.

The entire trip is 1035 miles, and we want to determine how much time we should let pass before we become concerned that the plane has encountered trouble.

Solution

Let t = time elapsed since the plane went on autopilot. Then

$$300t = \text{distance the plane has flown in } t \text{ hours}$$
$$150 + 300t = \text{plane's distance from Miami after } t \text{ hours.}$$

Concern is not warranted unless the Bermuda tower does not see the plane, which means that the following is true.

$$\boxed{\text{Plane's distance from Miami}} \geq \boxed{\text{Distance from Miami to Bermuda}}$$
$$150 + 300t \geq 1035 \quad \text{Replace the verbal description with an inequality.}$$
$$150 + 300t - 150 \geq 1035 - 150 \quad \text{Subtract 150 from both sides.}$$
$$300t \geq 885 \quad \text{Simplify.}$$
$$\frac{300t}{300} \geq \frac{885}{300} \quad \text{Divide both sides by 300.}$$
$$t \geq 2.95 \quad \text{Simplify.}$$

Because 2.95 is roughly three hours, the tower will suspect trouble if the plane has not arrived in Bermuda after three hours.

Practice Problem 2 How much time should pass in Example 2 if the plane was set on automatic pilot at 340 miles per hour when it was 185 miles from Miami?

If an inequality is true for all real numbers, its solution set is $(-\infty, \infty)$. If an inequality has no solution, its solution set is $\varnothing$.

EXAMPLE 3 **Solving an Inequality**

Write the solution set of each inequality.

a. $7(x + 2) - 20 - 4x < 3(x - 1)$ **b.** $2(x + 5) + 3x < 5(x - 1) + 3$

Solution

a.

$7(x + 2) - 20 - 4x < 3(x - 1)$	Original inequality
$7x + 14 - 20 - 4x < 3x - 3$	Distributive property
$3x - 6 < 3x - 3$	Combine like terms.
$-6 < -3$	Add $-3x$ to both sides and simplify.

The last inequality is equivalent to the original inequality and is always true. So the solution set of the original inequality is $(-\infty, \infty)$.

b.

$2(x + 5) + 3x < 5(x - 1) + 3$	Original inequality
$2x + 10 + 3x < 5x - 5 + 3$	Distributive property
$5x + 10 < 5x - 2$	Combine like terms.
$10 < -2$	Add $-5x$ and simplify.

The resulting inequality is equivalent to the original inequality and is always false. So the solution set of the original inequality is $\varnothing$.

Practice Problem 3 Write the solution set of each inequality.

a. $2(4 - x) + 6x < 4(x + 1) + 7$ **b.** $3(x - 2) + 5 \geq 7(x - 1) - 4(x - 2)$

3 Solve and graph a compound inequality.

Combining Two Inequalities

Sometimes we are interested in the solution set of two or more inequalities. The combination of two or more inequalities is called a **compound inequality**.

Suppose E_1 is an inequality with solution set (interval) I_1 and E_2 is another inequality with solution set I_2. Then the solution set of the compound inequality "E_1 and E_2" is $I_1 \cap I_2$ and the solution set of the compound inequality "E_1 or E_2" is $I_1 \cup I_2$.

EXAMPLE 4 **Solving and Graphing a Compound *OR* Inequality**

Graph and write the solution set of the compound inequality.

$$2x + 7 \leq 1 \quad \text{or} \quad 3x - 2 < 4(x - 1)$$

Solution

$2x + 7 \leq 1$	or	$3x - 2 < 4(x - 1)$	Original inequalities
$2x + 7 \leq 1$	or	$3x - 2 < 4x - 4$	Distributive property
$2x \leq -6$	or	$-x < -2$	Isolate variable term.
$x \leq -3$	or	$x > 2$	Solve for x; reverse second inequality symbol.

Graph each inequality and select the *union* of the two intervals.

$x \leq -3$

−5 −4 −3 −2 −1 0 1 2 3 4 5

$x > 2$

−5 −4 −3 −2 −1 0 1 2 3 4 5

$x \leq -3$ or $x > 2$

−5 −4 −3 −2 −1 0 1 2 3 4 5

The solution set of the compound inequality is $(-\infty, -3] \cup (2, \infty)$.

Practice Problem 4 Write the solution set of the compound inequality.

$$3x - 5 \geq 7 \quad \text{or} \quad 5 - 2x \geq 1$$

EXAMPLE 5 Solving and Graphing a Compound *AND* Inequality

Solve the compound inequality and graph the solution set.

$$2(x - 3) + 5 < 9 \quad \text{and} \quad 3(1 - x) - 2 \leq 7$$

Solution

$2(x - 3) + 5 < 9$	and	$3(1 - x) - 2 \leq 7$	Original inequalities
$2x - 6 + 5 < 9$	and	$3 - 3x - 2 \leq 7$	Distributive property
$2x < 10$	and	$-3x \leq 6$	Isolate variable term.
$x < 5$	and	$x \geq -2$	Solve for x; reverse second inequality symbol.

Graph each inequality and select the interval common to both inequalities.

$x < 5$

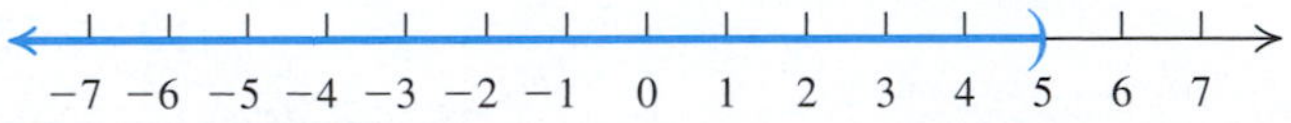

$x \geq -2$

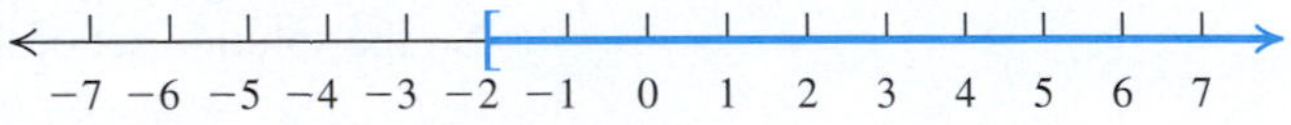

$x < 5$ and $x \geq -2$

The solution set of the compound inequality is the interval $[-2, 5)$.

Practice Problem 5 Solve and graph the compound inequality.

$$2(3 - x) - 3 < 5 \quad \text{and} \quad 2(x - 5) + 7 \leq 3$$

Sometimes we are interested in a joint inequality such as $-5 < 2x + 3 \leq 9$. This use of two inequality symbols in a single expression is shorthand for $-5 < 2x + 3$ *and* $2x + 3 \leq 9$. Fortunately, solving such inequalities requires no new principles, as we see in the next example.

EXAMPLE 6 Solving and Graphing a Compound Inequality

Solve the inequality $-5 < 2x + 3 \leq 9$ and graph its solution set.

Solution

We must find all real numbers that are solutions of *both* inequalities

$$-5 < 2x + 3 \quad \text{and} \quad 2x + 3 \leq 9.$$

We first solve these inequalities separately.

$-5 < 2x + 3$	$2x + 3 \leq 9$	Original inequalities
$-5 - 3 < 2x + 3 - 3$	$2x + 3 - 3 \leq 9 - 3$	Subtract 3 from both sides.
$-8 < 2x$	$2x \leq 6$	Simplify.
$\frac{-8}{2} < \frac{2x}{2}$	$\frac{2x}{2} \leq \frac{6}{2}$	Divide both sides by 2.
$-4 < x$	$x \leq 3$	Simplify.

The solution of the original pair of inequalities consists of all real numbers x such that $-4 < x$ *and* $x \leq 3$. This solution may be written more compactly as $\{x | -4 < x \leq 3\}$. In interval notation, we write $(-4, 3]$. The graph is shown in Figure 1.10.

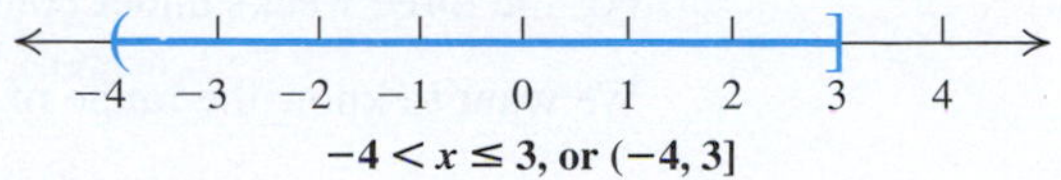

FIGURE 1.10

Notice that we did the same thing to both inequalities in each step of the solution process. We can accomplish this simultaneous solution more efficiently by working on both inequalities at the same time as follows.

$-5 < 2x + 3 \leq 9$	Original inequality
$-5 - 3 < 2x + 3 - 3 \leq 9 - 3$	Subtract 3 from each part.
$-8 < 2x \leq 6$	Simplify each part.
$\frac{-8}{2} < \frac{2x}{2} \leq \frac{6}{2}$	Divide each part by 2.
$-4 < x \leq 3$	Simplify each part.

The solution set is $(-4, 3]$, the same solution set we found previously.

Practice Problem 6 Solve and graph $-6 \leq 4x - 2 < 4$.

EXAMPLE 7 Finding the Interval of Values for a Linear Expression

If $-2 < x < 5$, find real numbers a and b so that $a < 3x - 1 < b$.

Solution

We start with the interval for x.

$-2 < x < 5$	
$3(-2) < 3x < 3(5)$	Multiply each part by 3 to get $3x$ in the middle.
$-6 < 3x < 15$	Simplify.
$-6 - 1 < 3x - 1 < 15 - 1$	Subtract 1 from each part to get $3x - 1$ in the middle.
$-7 < 3x - 1 < 14$	Simplify.

We have $a = -7$ and $b = 14$.

Practice Problem 7 Assuming that $-3 \leq x \leq 2$, find real numbers a and b so that $a \leq 3x + 5 \leq b$.

Our next example demonstrates a practical use of the method shown in Example 7.

EXAMPLE 8 Finding a Fahrenheit Temperature from a Celsius Range

The weather in London is predicted to range between 10° and 20° Celsius during the three-week period you will be working there. To decide what kind of clothes to take, you want to convert the temperature range to Fahrenheit temperatures. The formula for converting Celsius temperature C to Fahrenheit temperature F is $F = \frac{9}{5}C + 32$. What range of Fahrenheit temperatures might you find in London during your stay there?

Solution

First, we express the Celsius temperature range as an inequality. Let $C =$ temperature in Celsius degrees.

For the three weeks under consideration, $10 \le C \le 20$.

We want to know the range of $F = \frac{9}{5}C + 32$ when $10 \le C \le 20$.

$$10 \le C \le 20$$

$$\left(\frac{9}{5}\right)(10) \le \frac{9}{5}C \le \left(\frac{9}{5}\right)(20) \qquad \text{Multiply each part by } \frac{9}{5}.$$

$$18 \le \frac{9}{5}C \le 36 \qquad \text{Simplify.}$$

$$18 + 32 \le \frac{9}{5}C + 32 \le 36 + 32 \qquad \text{Add 32 to each part.}$$

$$50 \le \frac{9}{5}C + 32 \le 68 \qquad \text{Simplify.}$$

$$50 \le F \le 68 \qquad F = \frac{9}{5}C + 32$$

So the temperature range from 10° to 20° Celsius corresponds to a range from 50° to 68° Fahrenheit.

Practice Problem 8 What range in Fahrenheit degrees corresponds to the range of 15° to 25° Celsius?

4 Solve polynomial and rational inequalities using test points.

Using Test Points to Solve Inequalities

The **test-point** method, also known as the **sign-chart** method, involves writing an inequality (by rearranging if necessary) so that the expression on the left side of the inequality symbol is in factored form and the right side is 0. In Example 9, we illustrate this method to solve a **quadratic inequality**.

PROCEDURE **IN ACTION**

EXAMPLE 9 Using the Test-Point Method to Solve an Inequality

OBJECTIVE

Solve an inequality by using the test-point method.

Step 1 Rewrite (if necessary) the inequality so that **the right side of the inequality is 0.**

Simplify the left side.

Step 2 Factor the left side.

Step 3 Draw a number line. Find and plot the points where each factor is 0.

The n points so obtained divide the number line into $(n + 1)$ intervals.

EXAMPLE

Solve: $x^2 + 4x \ge 5x + 6$

1. $x^2 + 4x - 5x - 6 \ge 0$ Subtract $5x + 6$ from both sides.

$x^2 - x - 6 \ge 0$ Simplify.

2. $(x + 2)(x - 3) \ge 0$ Factor.

3. $x + 2 = 0$ for $x = -2$, and $x - 3 = 0$ for $x = 3$.

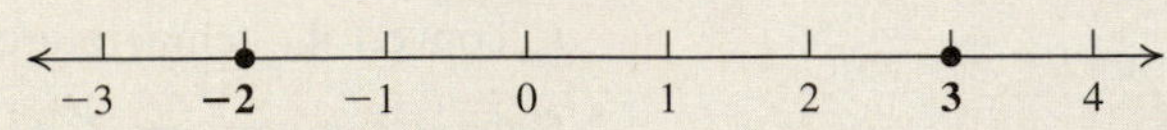

The three intervals determined by the two points -2 and 3 are $(-\infty, -2)$, $(-2, 3)$, and $(3, \infty)$.

(continued)

Step 4 The final expression on the left side of the inequality in Step 1 is always positive or always negative on each of the $(n + 1)$ intervals of Step 3. Select convenient "test points" in each interval to determine the sign of the inequality.

4.

Test Interval	Test Point	Sign of $(x + 2)(x - 3)$	Result
$(-\infty, -2)$	-3	$(-)(-)$	Positive
$(-2, 3)$	0	$(+)(-)$	Negative
$(3, \infty)$	4	$(+)(+)$	Positive

Step 5 **Draw a sign chart.** Show the information from Steps 3 and 4 on a number line.

5.

$x^2 - x - 6 = (x + 2)(x - 3)$

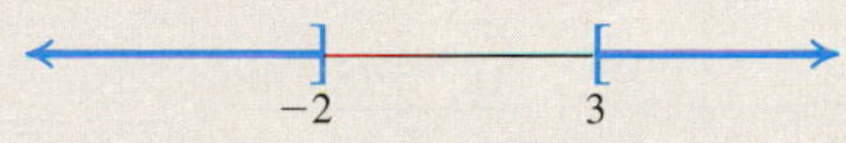

Step 6 From the sign chart in Step 5, write the solution set of the inequality. Graph the solution set.

6. To find where $x^2 - x - 6 \geq 0$ means to find where $x^2 - x - 6 > 0$ (indicated by + signs) or $x^2 - x - 6 = 0$ (indicated by 0) on the sign chart. The solution set of the inequality from the figure in step 5 is $(-\infty, -2] \cup [3, \infty)$. This set is shown below.

−2 3

Practice Problem 9 Solve: $x^2 + 2 < 3x + 6$

A **rational inequality** is an inequality that includes one or more rational expressions. We use the test-point method to solve a rational inequality.

EXAMPLE 10 Solving a Rational Inequality

Solve: $\dfrac{x^2 + 2x - 15}{x - 1} \geq 3$

Step 1

$$\frac{x^2 + 2x - 15}{x - 1} - 3 \geq 0 \qquad \text{Rearrange so that the right side is 0.}$$

$$\frac{x^2 + 2x - 15}{x - 1} - \frac{3(x - 1)}{x - 1} \geq 0 \qquad \text{Common denominator}$$

$$\frac{(x^2 + 2x - 15) - 3(x - 1)}{x - 1} \geq 0 \qquad \frac{a}{c} \pm \frac{b}{c} = \frac{a \pm b}{c}$$

$$\frac{x^2 + 2x - 15 - 3x + 3}{x - 1} \geq 0 \qquad \text{Distribute.}$$

$$\frac{x^2 - x - 12}{x - 1} \geq 0 \qquad \text{Simplify.}$$

Step 2 $\dfrac{(x + 3)(x - 4)}{x - 1} \geq 0$ Factor the numerator.

Step 3 $x + 3 = 0$ for $x = -3$

$x - 4 = 0$ for $x = 4$

$x - 1 = 0$ for $x = 1$ The rational expression is undefined at $x = 1$.

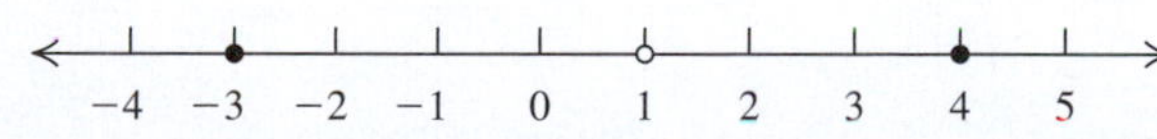

The four intervals determined by the three points -3, 1, and 4 are

$$(-\infty, -3),\ (-3, 1),\ (1, 4),\ \text{and}\ (4, \infty).$$

Step 4

Test Interval	Test Point	Sign of $\dfrac{(x+3)(x-4)}{(x-1)}$	Result
$(-\infty, -3)$	-4	$\dfrac{(-)(-)}{(-)}$	Negative
$(-3, 1)$	0	$\dfrac{(+)(-)}{(-)}$	Positive
$(1, 4)$	2	$\dfrac{(+)(-)}{(+)}$	Negative
$(4, \infty)$	5	$\dfrac{(+)(+)}{(+)}$	Positive

Step 5

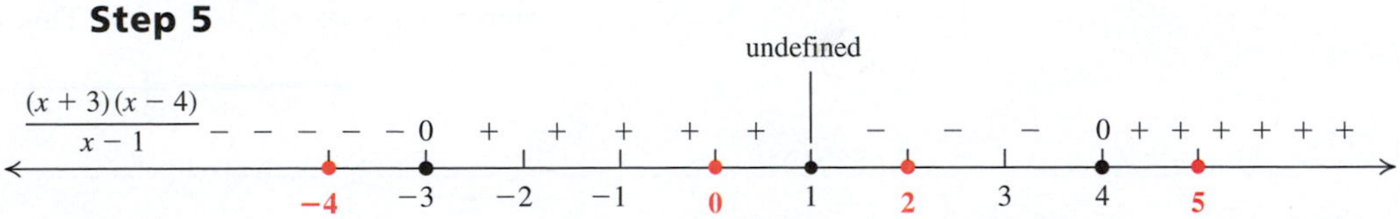

Step 6 $\dfrac{(x+3)(x-4)}{(x-1)} \geq 0$ on the set $[-3, 1) \cup [4, \infty)$, with the following graph:

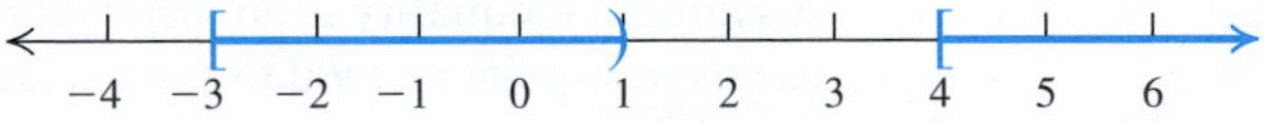

Practice Problem 10 Solve: $\dfrac{2x+5}{x-1} \leq 1$

WARNING You cannot solve a rational inequality by multiplying both sides by the LCD, as you would with a rational equation. Remember that you reverse the sense of an inequality when multiplying by a negative expression. However, when an expression contains a variable, you do not know whether the expression is positive or negative.

Answers to Practice Problems

1. a. $(2, \infty)$

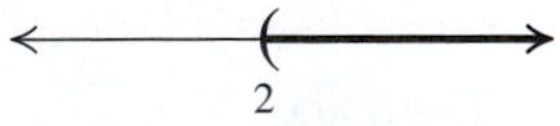

b. $(-\infty, 5]$

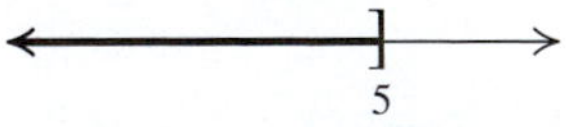

2. 2.5 hours **3. a.** $(-\infty, \infty)$ **b.** $\varnothing$ **4.** $(-\infty, 2] \cup [4, \infty)$

5. $(-1, 3]$

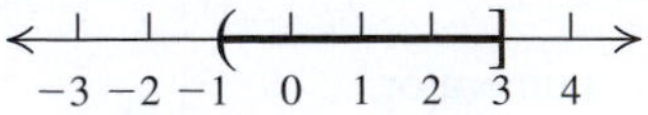

6. $\left[-1, \frac{3}{2}\right)$

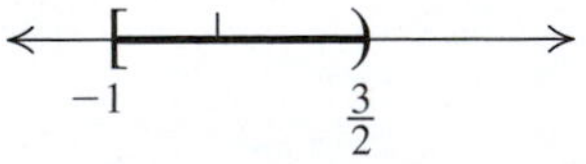

7. $a = -4, b = 11$ **8.** 59°F to 77°F

9. $(-1, 4)$ **10.** $[-6, 1)$

SECTION 1.5 Exercises

Basic Concepts and Skills

In Exercises 1–10, fill in the blank with the correct inequality symbol using the rules for producing equivalent inequalities.

1. If $x < 8$, then $x - 8$ ________ 0.

2. If $x \le -3$, then $x + 3$ ________ 0.

3. If $x \ge 5$, then $x - 5$ ________ 0.

4. If $x > 7$, then $x - 7$, ________ 0.

5. If $\frac{x}{2} < 6$, then x ________ 12.

6. If $\frac{x}{3} < 2$, then x ________ 6.

7. If $-2x \le 4$, then x ________ -2.

8. If $-3x > 12$, then x ________ -4.

9. If $x + 3 < 2$, then x ________ -1.

10. If $x - 5 \le 3$, then x ________ 8.

In Exercises 11–18, graph the solution set of each inequality and write it in interval notation.

11. $-2 < x < 5$ **12.** $-5 \le x \le 0$

13. $0 < x \le 4$ **14.** $1 \le x < 7$

15. $x \ge -1$ **16.** $x > 2$

17. $-5x \ge 10$ **18.** $-2x < 2$

In Exercises 19–40, solve each inequality. Write the solution in interval notation and graph the solution set.

19. $x + 3 < 6$ **20.** $x - 2 < 3$

21. $1 - x \le 4$ **22.** $7 - x > 3$

23. $2x + 5 < 9$ **24.** $3x + 2 \ge 7$

25. $3 - 3x > 15$ **26.** $8 - 4x \ge 12$

27. $3(x + 2) < 2x + 5$ **28.** $4(x - 1) \ge 3x - 1$

29. $3(x - 3) \le 3 - x$ **30.** $-x - 2 \ge x - 10$

31. $6x + 4 > 3x + 10$ **32.** $4(x - 4) > 3(x - 5)$

33. $8(x - 1) - x \le 7x - 12$

34. $3(x + 2) + 2x \ge 5x + 18$

35. $5(x + 2) \le 3(x + 1) + 10$

36. $x - 4 > 2(x + 8)$

37. $2(x + 1) + 3 \ge 2(x + 2) - 1$

38. $5(2 - x) + 4x \le 12 - x$

39. $2(x + 1) - 2 \le 3(2 - x) + 9$

40. $4(1 - x) + 2x > 5(2 - x) + 4x$

In Exercises 41–50, solve each rational inequality.

41. $9x - 6 \ge \frac{3}{2}x + 9$ **42.** $\frac{7x - 3}{2} < 3x - 4$

43. $\frac{x - 3}{3} \le 2 + \frac{x}{2}$ **44.** $\frac{2x - 3}{4} \ge 3 - \frac{x}{2}$

45. $\frac{3x + 1}{2} < x - 1 + \frac{x}{2}$ **46.** $\frac{2x - 1}{3} \ge \frac{x + 1}{4} + \frac{x}{12}$

47. $\frac{x - 3}{2} \ge \frac{x}{3} + 1$ **48.** $\frac{2x + 1}{3} < \frac{x - 1}{2} + \frac{1}{6}$

49. $\frac{3x + 1}{3} - \frac{x}{2} \le \frac{x + 2}{2}$ **50.** $\frac{x - 1}{3} + \frac{x + 1}{4} \le \frac{x}{2} + \frac{x}{12}$

In Exercises 51–58, solve each compound *or* inequality.

51. $2x + 5 < 1$ or $2 + x > 4$

52. $3x - 2 > 7$ or $2(1 - x) > 1$

53. $\frac{2x - 3}{4} \le 2$ or $\frac{4 - 3x}{2} \ge 2$

54. $\frac{5 - 3x}{3} \ge \frac{1}{6}$ or $\frac{x - 1}{3} \le 1$

55. $\frac{2x + 1}{3} \ge x + 1$ or $\frac{x}{2} - 1 > \frac{x}{3}$

56. $\frac{x + 2}{2} < \frac{x}{3} + 1$ or $\frac{x - 1}{3} > \frac{x + 1}{5}$

57. $\frac{x - 1}{2} > \frac{x}{3} - 1$ or $\frac{2x + 5}{3} \le \frac{x + 1}{6}$

58. $\frac{2x + 1}{3} \le \frac{x}{4} + 1$ or $\frac{3 - x}{2} > \frac{x}{3} - 1$

In Exercises 59–66, solve each compound *and* inequality.

59. $3 - 2x \le 7$ and $2x - 3 \le 7$

60. $6 - x \le 3x + 10$ and $7x - 14 \le 3x + 14$

61. $2(x + 1) + 3 \ge 1$ and $2(2 - x) > -6$

62. $3(x + 1) - 2 \ge 4$ and $3(1 - x) + 13 > 4$

63. $2(x + 1) - 3 > 7$ and $3(2x + 1) + 1 < 10$

64. $5(x + 2) + 7 < 2$ and $2(5 - 3x) + 1 < 17$

65. $5 + 3(x - 1) < 3 + 3(x + 1)$ and $3x - 7 \le 8$

66. $2x - 3 > 11$ and $5(2x + 1) < 3(4x + 1) - 2(x - 1)$

In Exercises 67–78, solve each compound inequality.

67. $3 < x + 5 < 4$ **68.** $9 \le x + 7 \le 12$

69. $-4 \le x - 2 < 2$ **70.** $-3 < x + 5 < 4$

71. $-9 \le 2x + 3 \le 5$ **72.** $-2 \le 3x + 1 \le 7$

73. $0 \le 1 - \frac{x}{3} < 2$ **74.** $0 < 5 - \frac{x}{2} \le 3$

75. $-1 < \frac{2x - 3}{5} \le 0$ **76.** $-4 \le \frac{5x - 2}{3} \le 0$

77. $5x \le 3x + 1 < 4x - 2$

78. $3x + 2 < 2x + 3 < 4x - 1$

In Exercises 79–84, find *a* and *b*.

79. If $-2 < x < 1$, then $a < x + 7 < b$.

80. If $1 < x < 5$, then $a < 2x + 3 < b$.

81. If $-1 < x < 1$, then $a < 2 - x < b$.

82. If $3 < x < 7$, then $a < 1 - 3x < b$.

83. If $0 < x < 4$, then $a < 5x - 1 < b$.

84. If $-4 < x < 0$, then $a < 3x + 4 < b$.

In Exercises 85–100, use the test-point method to solve each polynomial inequality.

85. $x^2 + 4x - 12 \le 0$

86. $x^2 - 8x + 7 > 0$

87. $6x^2 + 7x - 3 \ge 0$

88. $4x^2 - 2x - 2 < 0$

89. $(x + 3)(x + 1)(x - 1) \ge 0$

90. $(x + 4)(x - 1)(x + 2) \le 0$

91. $x^3 - 4x^2 - 12x > 0$

92. $x^3 + 8x^2 + 15x > 0$

93. $x^2 + 2x < -1$

94. $4x^2 + 12x < -9$

95. $x^3 - x^2 \ge 0$

96. $x^3 - 9x^2 \ge 0$

97. $x^2 \ge 1$

98. $x^4 \le 16$

99. $x^3 < -8$

100. $x^4 > 9$

In Exercises 101–120, solve each rational inequality.

101. $\frac{x + 2}{x - 5} < 0$

102. $\frac{x - 3}{x + 1} > 0$

103. $\frac{x + 4}{x} < 0$

104. $\frac{x}{x - 2} > 0$

105. $\frac{x + 1}{x + 2} \le 3$

106. $\frac{x - 1}{x - 2} \ge 3$

107. $\frac{(x - 2)(x + 2)}{x} > 0$

108. $\frac{(x - 1)(x + 3)}{x - 2} < 0$

109. $\frac{(x - 2)(x + 1)}{(x - 3)(x + 5)} \ge 0$

110. $\frac{(x - 1)(x - 3)}{(x + 2)(x + 4)} \ge 0$

111. $\frac{x^2 - 1}{x^2 - 4} \le 0$

112. $\frac{x^2 - 9}{x^2 - 64} \le 0$

113. $\frac{x + 4}{3x - 2} \ge 1$

114. $\frac{2x - 3}{x + 3} \le 1$

115. $3 \le \frac{2x + 6}{2x + 1}$

116. $\frac{x - 2}{2x + 1} < -1$

117. $\frac{x + 2}{x - 3} \ge \frac{x - 1}{x + 3}$

118. $\frac{x + 1}{x - 2} \ge \frac{x}{x - 1}$

119. $\frac{x - 1}{x + 1} \le \frac{x + 2}{x - 3}$

120. $\frac{x + 3}{x + 1} \le \frac{x - 1}{x - 2}$

Applying the Concepts

121. **Appliance markup.** The markup over the dealer's cost on a new refrigerator ranges from 15% to 20%. If the dealer's cost is \$1750, over what range will the selling price vary?

122. **Return on investment.** An investor has \$5000 to invest for a period of one year. Find the range of per annum simple interest rates required to generate interest between \$200 and \$275 inclusive.

123. **Hybrid car trip.** Sometime after passing a truck stop 300 miles from the start of her trip, Cora's hybrid car ran out of gas. Assuming that the tank could hold 12 gallons of gasoline and the hybrid car averaged 40 miles per gallon, find the range of gasoline (in gallons) that could have been in the tank at the start of the trip.

124. **Average grade.** Sean has taken three exams and earned scores of 85, 72, and 77 out of a possible 100 points. He needs an average of at least 80 to earn a B in the course. What range of scores on the fourth (and last) 100-point test will guarantee a B in the course?

125. **Butterfat content.** How much cream that is 30% butterfat must be added to milk that is 3% butterfat to have 270 quarts that are at least 4.5% butterfat?

126. **Pedometer cost.** A company produces a pedometer at a cost of \$3 each and sells the pedometer for \$5 each. If the company has to recover an initial expense of \$4000 before any profit is earned, how many pedometers must be sold to earn a profit in excess of \$3000?

127. **Car sales.** A car dealer has three times as many SUVs and twice as many convertibles as four-door sedans. How many four-door sedans does the dealer have if she has at least 48 cars of these three types?

128. **Temperature conversion.** The formula for converting Fahrenheit temperature F to Celsius temperature C is $C = \frac{5}{9}(F - 32)$. What range in Celsius degrees corresponds to a range of 68° to 86° Fahrenheit?

129. **Temperature.** The number N of water mites in a water sample depends on the temperature t in degrees Fahrenheit and is given by $N = 132t - t^2$. At what temperature will the number of mites exceed 3200?

130. Falling object. The height h of an object thrown from the top of a ski lift 1584 feet high after t seconds is $h = -16t^2 + 32t + 1584$. For what times is the height of the object at least 1200 feet?

131. Blackjack. A gambler playing blackjack in a casino using four decks noticed that 20% of the cards that had been dealt were jacks, queens, kings, or aces. After x cards had been dealt, he knew that the likelihood that the next card dealt would be a jack, a queen, a king, or an ace was $\frac{64 - 0.2x}{208 - x}$. For what values of x is this likelihood greater than 50%?

132. Area of a triangle. The base of a triangle is 3 centimeters greater than the height. Find the possible heights h so that the area of such a triangle will be at least 5 square centimeters.

Beyond the Basics

133. Find the numbers k for which the quadratic equation $2x^2 + kx + 2 = 0$ has two real solutions.

134. Find the numbers k for which the quadratic equation $2x^2 + kx + 2 = 0$ has no real solutions.

135. Find the numbers k for which the quadratic equation $x^2 + kx + k = 0$ has two real solutions.

136. Find the numbers k for which the quadratic equation $x^2 + kx + k = 0$ has no real solutions.

137. Solve: $\frac{x}{2x+1} \geq \frac{1}{4}$ *and* $\frac{6x}{4x-1} < \frac{1}{2}$

138. Solve: $\frac{2x-1}{x-7} > 1$ *and* $\frac{x-10}{x-8} > 2$

139. Find all values of c for which it is possible to find two numbers whose product is 36 and whose sum is c.

140. Find all values of c for which it is possible to find two numbers whose sum is 12 and the sum of whose squares is c.

141. An import firm pays a tax of \$10 on each radio it imports. In addition to this tax, a penalty tax must be paid if more than 1000 radios are imported. The penalty tax is computed by multiplying 5 cents by the number of radios imported in excess of 1000. This tax must be paid on each radio imported. If 1006 radios are imported, the penalty tax is $6 \cdot 5 = 30$ cents and is paid on each of the 1006 radios. If the firm wants to spend no more than a total of \$640,000 on import taxes, how many radios can it import?

142. A TV quiz program pays a contestant \$100 for each correct answer for ten questions. If all ten questions are answered correctly, bonus questions are asked. The reward for every correct answer is increased by \$50 for each bonus question that is correctly answered. Any incorrect answer ends the game. If two bonus questions are answered correctly, the contestant receives \$200 *for each of the 12 questions*. If a contestant won more than \$3500, how many questions must have been answered correctly?

Critical Thinking/Discussion/Writing

143. Give an example of a quadratic inequality with each of the following solution sets.

a. $(-4, 5)$
b. $[-2, 6]$
c. $(-\infty, \infty)$
d. $\varnothing$
e. $\{3\}$
f. $(-\infty, 2) \cup (2, \infty)$

144. Give an example of an inequality with each of the following solution sets.

a. $(-2, 4]$

b. $[3, 5)$

c. Is there a quadratic inequality whose solution set is $(2, 5]$?

Maintaining Skills

In Exercises 145–150, evaluate each expression.

145. $|-3|$ **146.** $|3 - 7|$
147. $|6 - 4|$ **148.** $|-\sqrt{2}|$
149. $|0|$ **150.** $|-15.8|$

In Exercises 151–154, find the distance between the given pair of points on a number line.

151. -2 and 5 **152.** -8 and -15
153. 2.3 and 5.7 **154.** -5 and 0

In Exercises 155–160, express each statement in algebraic notation using absolute values.

155. The distance from x to -2 is 5.

156. x is either 3 or -3.

157. x is at most 2 units from 4.

158. x is either to the left of -5 or to the right of 5 on a number line.

159. x is within 3 units of 5.

160. x is closer to 2 than to 6.

SECTION 1.6

Equations and Inequalities Involving Absolute Value

BEFORE STARTING THIS SECTION, REVIEW

1 Properties of absolute value (Section P.1, page 9)

2 Intervals (Section P.1, page 7)

3 Linear inequalities (Section 1.5, page 131)

OBJECTIVES

1 Solve equations involving absolute value.

2 Solve inequalities involving absolute value.

Rescuing a Downed Aircraft

Rescue crews arrive at an airport in Maine to conduct an air search for a missing Cessna with two people aboard. If the rescue crews locate the missing aircraft, they are to mark the global positioning system (GPS) coordinates and notify the mission base. The search planes normally average 110 miles per hour, but weather conditions can affect the average speed by as much as 15 miles per hour (either slower or faster). If a search plane has 30 gallons of fuel and uses 10 gallons of fuel per hour, what is its possible search range (in miles)? In Example 5, we use an inequality involving absolute value to find the plane's possible search range.

1 Solve equations involving absolute value.

Equations Involving Absolute Value

Recall that geometrically, the absolute value of a real number a is the distance from the origin on the number line to the point with coordinate a. The definition of absolute value is

$$|a| = a \text{ if } a \geq 0 \qquad \text{and} \qquad |a| = -a \text{ if } a < 0.$$

A summary of the most useful properties of absolute value can be found on page 9 in Section P.1.

Because the only two numbers on the number line that are exactly 2 units from the origin are 2 and -2, they are the only solutions of the equation $|x| = 2$. See Figure 1.11. This simple observation leads to the following rule for solving equations involving absolute value.

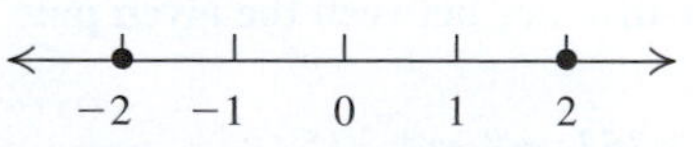

FIGURE 1.11 Distance from the origin

SOLUTIONS OF $|u| = a, a \geq 0$

If $a \geq 0$ and u is an algebraic expression, then

$$|u| = a \quad \text{is equivalent to} \quad u = a \text{ or } u = -a.$$
$$|u| = -a \quad \text{has no solution when } a > 0.$$

Note that if $a = 0$, then $u = 0$ is the only solution of $|u| = 0$.

EXAMPLE 1 **Solving an Equation Involving Absolute Value**

Solve each equation.

a. $|x + 3| = 0$ **b.** $|2x - 3| - 5 = 8$

Solution

a. Let $u = x + 3$. Then, $|u| = 0$ has only one solution: $u = 0$.

$$|x + 3| = 0 \quad |u| = 0$$
$$x + 3 = 0 \quad u = 0$$
$$x = -3 \quad \text{Solve for } x.$$

Now we check the solution in the original equation.

Check:

$$|x + 3| = 0 \quad \text{Original equation}$$
$$|-3 + 3| \stackrel{?}{=} 0 \quad \text{Replace } x \text{ with } -3.$$
$$|0| = 0 \checkmark \quad \text{Simplify.}$$

We have verified that -3 is a solution of the original equation.
The solution set is $\{-3\}$.

b. To use the rule for solving equations involving absolute value, we must first isolate the absolute value expression to one side of the equation.

$$|2x - 3| - 5 = 8$$
$$|2x - 3| = 13 \quad \text{Add 5 to both sides and simplify.}$$

$|u| = 13$ is equivalent to $u = 13$ or $u = -13$. Let $u = 2x - 3$; then $|2x - 3| = 13$ is equivalent to

$$2x - 3 = 13 \quad \text{or} \quad 2x - 3 = -13. \quad u = 13 \text{ or } u = -13$$
$$2x = 16 \quad | \quad 2x = -10 \quad \text{Add 3 to both sides of each equation.}$$
$$x = 8 \quad | \quad x = -5 \quad \text{Divide both sides by 2.}$$

We leave it to you to check these solutions.
The solution set is $\{-5, 8\}$.

Practice Problem 1 Solve each equation.

a. $|x - 2| = 0$ **b.** $|6x - 3| - 8 = 1$

With a little practice, you may find that you can solve these types of equations without actually writing u and the expression it represents. Then you can work with just the expression itself.

EXAMPLE 2 **Solving an Equation of the Form $|u| = |v|$**

Solve: $|x - 1| = |x + 5|$

Solution

If $|u| = |v|$, then u is equal to $|v|$ or u is equal to $-|v|$. Because $|v| = \pm v$ in every case, we have $u = v$ or $u = -v$. Thus,

$|u| = |v|$ is equivalent to $u = v$ or $u = -v$.

$|x - 1| = |x + 5|$ is equivalent to

$$x - 1 = x + 5 \quad \text{or} \quad x - 1 = -(x + 5) \quad u = x - 1, v = x + 5$$
$$-1 = 5 \text{ (False)} \quad | \quad x = -2 \quad \text{Solve for } x.$$

The only solution of $|x - 1| = |x + 5|$ is -2, so the solution set is $\{-2\}$. We leave it to you to check this solution.

Practice Problem 2 Solve: $|x + 2| = |x - 3|$

EXAMPLE 3 **Solving an Equation of the Form $|u| = v$**

Solve: $|2x - 1| = x + 5$

Solution

Letting $u = 2x - 1$ and $v = x + 5$, the equation $|u| = v$ is equivalent to

$u = v$	or $u = -v$	
$2x - 1 = x + 5$	$2x - 1 = -(x + 5)$	
$2x - 1 = x + 5$	$2x - 1 = -x - 5$	Distributive property
$x = 6$	$3x = -4$	Isolate x term.
$x = 6$	$x = -\frac{4}{3}$	Solve for x.

Check:

$x = 6$	or $x = -\frac{4}{3}$
$\lvert 2(6) - 1\rvert \stackrel{?}{=} 6 + 5$	$\left\lvert 2\left(-\frac{4}{3}\right) - 1\right\rvert \stackrel{?}{=} -\frac{4}{3} + 5$
$\lvert 11\rvert \stackrel{?}{=} 11$	$\left\lvert -\frac{8}{3} - 1\right\rvert \stackrel{?}{=} \frac{-4 + 15}{3}$
$\lvert 11\rvert \stackrel{?}{=} 11$ ✓ Yes	$\left\lvert -\frac{11}{3}\right\rvert \stackrel{?}{=} \frac{11}{3}$
	$\frac{11}{3} \stackrel{?}{=} \frac{11}{3}$ ✓ Yes

The solution set of the given equation is $\left\{-\frac{4}{3}, 6\right\}$.

Practice Problem 3 Solve $|3x - 4| = 2(x - 1)$.

2 Solve inequalities involving absolute value.

Inequalities Involving Absolute Value

The equation $|x| = 2$ has two solutions: 2 and -2. If x is any real number other than 2 or -2, then $|x| < 2$ or $|x| > 2$. Geometrically, this means that x is closer than 2 units to the origin if $|x| < 2$ or farther than 2 units from the origin if $|x| > 2$. Any number x that is closer than 2 units from the origin is in the interval $-2 < x < 2$. See Figure 1.12 (a). Any number x that is farther than 2 units from the origin is in the interval $(-\infty, -2)$ or the interval $(2, \infty)$, that is, either $x < -2$ or $x > 2$. See Figure 1.12 (b).

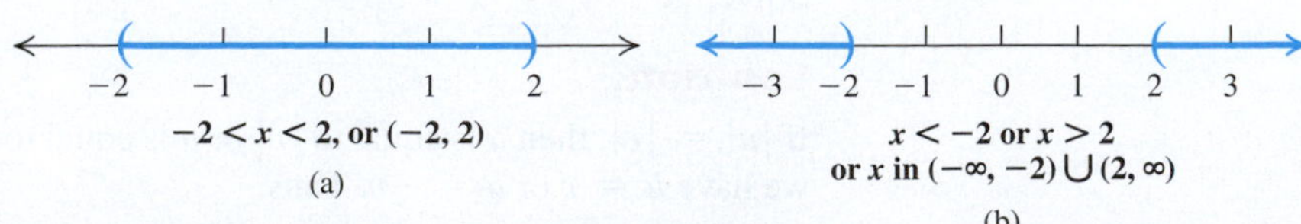

FIGURE 1.12 Graphs of inequalities involving absolute value

This discussion is equally valid when we use any positive real number a instead of the number 2, and it leads to the following rules for replacing inequalities involving absolute value with equivalent inequalities that do *not* involve absolute value.

RULES FOR SOLVING ABSOLUTE VALUE INEQUALITIES

If $a > 0$ and u is an algebraic expression, then

1. $|u| < a$ is equivalent to $-a < u < a$, or u in $(-a, a)$.
2. $|u| \le a$ is equivalent to $-a \le u \le a$, or u in $[-a, a]$.
3. $|u| > a$ is equivalent to $u < -a$ or $u > a$, or u in $(-\infty, -a) \cup (a, \infty)$.
4. $|u| \ge a$ is equivalent to $u \le -a$ or $u \ge a$, or u in $(-\infty, -a] \cup [a, \infty)$.

EXAMPLE 4 Solving an Inequality Involving Absolute Value

Solve the inequality $|4x - 1| \le 9$ and graph the solution set.

Solution

Rule 2 applies here with $u = 4x - 1$ and $a = 9$.

$|4x - 1| \le 9$ is equivalent to

$$-9 \le 4x - 1 \le 9 \quad \text{Rule 2, } -a \le u \le a$$

$$1 - 9 \le 4x - 1 + 1 \le 9 + 1 \quad \text{Add 1 to each part.}$$

$$-8 \le 4x \le 10 \quad \text{Simplify.}$$

$$-\frac{8}{4} \le \frac{4x}{4} \le \frac{10}{4} \quad \text{Divide by 4; the sense of the inequality is unchanged.}$$

$$-2 \le x \le \frac{5}{2} \quad \text{Simplify.}$$

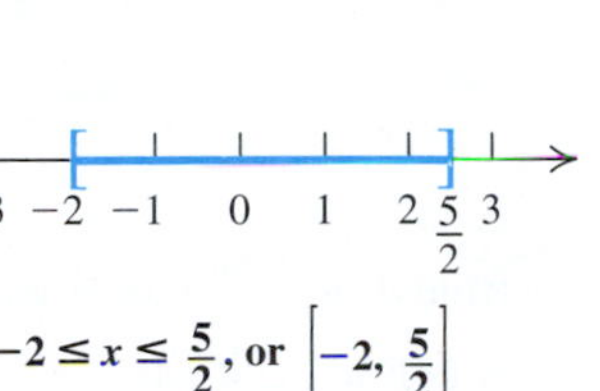

$-2 \le x \le \frac{5}{2}$, or $\left[-2, \frac{5}{2}\right]$

FIGURE 1.13

The solution set is $\left\{x \middle| -2 \le x \le \frac{5}{2}\right\}$; that is, the solution set is the closed interval $\left[-2, \frac{5}{2}\right]$. See Figure 1.13.

Practice Problem 4 Solve $|3x + 3| \le 6$ and graph the solution set.

Note the following techniques for solving inequalities similar to Example 4.

1. To solve $|4x - 1| < 9$, change $\le$ to $<$ throughout to obtain the solution $\left(-2, \frac{5}{2}\right)$.
2. To solve $|1 - 4x| \le 9$, note that $|1 - 4x| = |4x - 1|$; so the solution is the same as that for Example 4, $\left[-2, \frac{5}{2}\right]$.

EXAMPLE 5 Finding the Search Range of an Aircraft

In the introduction to this section, we wanted to find the possible search range (in miles) for a search plane that has 30 gallons of fuel and uses 10 gallons of fuel per hour. We were told that the search plane normally averages 110 miles per hour but that weather conditions can affect the average speed by as much as 15 miles per hour (either slower or faster). How do we find the possible search range?

Solution

To find the distance a plane flies (in miles), we need to know how long (time, in hours) it flies and how fast (speed, in miles per hour) it flies.

Let x = actual speed of the search plane, in miles per hour. We know that the actual speed is within 15 miles per hour of the average speed of 110 miles per hour. That is, |actual speed − average speed| ≤ 15 miles per hour.

$$|x - 110| \le 15 \quad \text{Replace the verbal description with an inequality.}$$

Solve this inequality for x.

$$-15 \le x - 110 \le 15 \qquad \text{Rewrite the inequality using Rule 2.}$$
$$110 - 15 \le x \le 110 + 15 \qquad \text{Add 110 to each part.}$$
$$95 \le x \le 125 \qquad \text{Simplify.}$$

The actual speed of the search plane is between 95 and 125 miles per hour.

Because the plane uses 10 gallons of fuel per hour and has 30 gallons of fuel, it can fly $\frac{30}{10}$, or 3, hours. Thus, the actual number of miles the search plane can fly is $3x$.

From $\qquad 95 \le x \le 125,$

we have $\qquad 3(95) \le 3x \le 3(125) \qquad$ Multiply each part by 3.

$$285 \le 3x \le 375 \qquad \text{Simplify.}$$

The search plane's range is between 285 and 375 miles.

Practice Problem 5 Repeat Example 5, but let the average speed of the plane be 115 miles per hour and suppose the wind speed can affect the average speed by 25 miles per hour.

EXAMPLE 6 Solving an Inequality Involving Absolute Value

Solve the inequality $|2x - 8| \ge 4$ and graph the solution set.

Solution

Rule 4 applies here with $u = 2x - 8$ and $a = 4$.

$|2x - 8| \ge 4$ is equivalent to

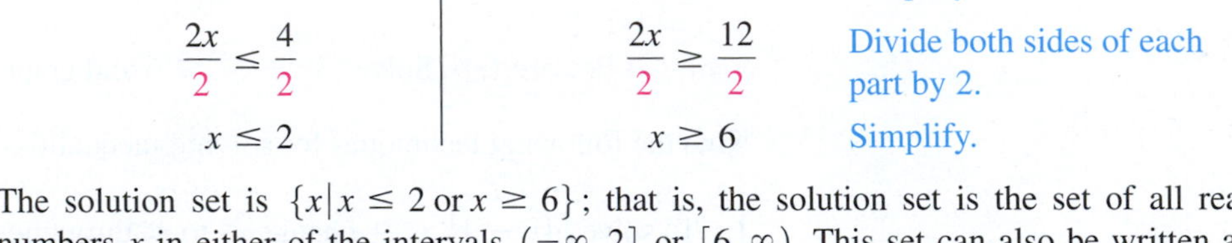

	or		
$2x - 8 \le -4$		$2x - 8 \ge 4$	Rule 4, $u \le -a$ or $u \ge a$
$2x - 8 + 8 \le -4 + 8$		$2x - 8 + 8 \ge 4 + 8$	Add 8 to each part.
$2x \le 4$		$2x \ge 12$	Simplify.
$\frac{2x}{2} \le \frac{4}{2}$		$\frac{2x}{2} \ge \frac{12}{2}$	Divide both sides of each part by 2.
$x \le 2$		$x \ge 6$	Simplify.

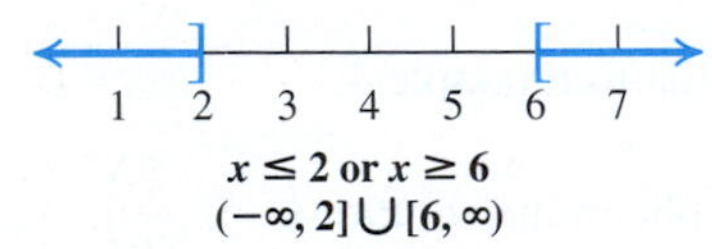

FIGURE 1.14

The solution set is $\{x | x \le 2 \text{ or } x \ge 6\}$; that is, the solution set is the set of all real numbers x in either of the intervals $(-\infty, 2]$ or $[6, \infty)$. This set can also be written as $(-\infty, 2] \cup [6, \infty)$. See Figure 1.14.

Practice Problem 6 Solve $|2x + 3| \ge 6$ and graph the solution set.

EXAMPLE 7 Solving Special Cases of Absolute Value Inequalities

Solve each inequality.

a. $|3x - 2| > -5$ $\qquad$ **b.** $|5x + 3| \le -2$

Solution

a. Because the absolute value is always nonnegative, $|3x - 2| > -5$ is true for all real numbers x. The solution set is the set of all real numbers; in interval notation, we write $(-\infty, \infty)$.

b. There is no real number with absolute value ≤ -2 because the absolute value of any number is nonnegative. The solution set for $|5x + 3| \le -2$ is the empty set, $\varnothing$.

Practice Problem 7 Solve.

a. $|5 - 9x| > -3$ $\qquad$ **b.** $|7x - 4| \le -1$

EXAMPLE 8 Solving an Inequality of the Form $|u| < |v|$

Solve: $|x + 1| < 3|x - 1|$

Solution

$$|x + 1| < 3|x - 1| \qquad \text{Original inequality}$$

$$\frac{|x + 1|}{|x - 1|} < 3 \qquad \text{For } x \neq 1\text{, divide both sides by } |x - 1|.$$

$$\left|\frac{x + 1}{x - 1}\right| < 3 \qquad \frac{|a|}{|b|} = \left|\frac{a}{b}\right|$$

$$-3 < \frac{x + 1}{x - 1} < 3 \qquad |u| < a \text{ implies } -a < u < a.$$

We solve the two inequalities:

$$-3 < \frac{x+1}{x-1} \quad \text{and} \quad \frac{x+1}{x-1} < 3$$

$$0 < \frac{x+1}{x-1} + 3 \qquad\qquad \frac{x+1}{x-1} - 3 < 0$$

$$0 < \frac{(x+1) + 3(x-1)}{x-1} \qquad\qquad \frac{(x+1) - 3(x-1)}{x-1} < 0 \qquad \text{Combine.}$$

$$0 < \frac{2(2x-1)}{x-1} \qquad\qquad \frac{-2(x-2)}{x-1} < 0 \qquad \text{Simplify.}$$

$$0 < \frac{2x-1}{x-1} \qquad\qquad \frac{-(x-2)}{x-1} < 0 \qquad \text{Divide by 2.}$$

$$\frac{2x-1}{x-1} > 0 \qquad\qquad \frac{x-2}{x-1} > 0 \qquad \text{Rewrite.}$$

The solution set of this inequality is S_1 (Figure 1.15). | The solution set of this inequality is S_2 (Figure 1.15).

SIDE NOTE

You can use the "test point" method to verify that the solution set of $\dfrac{2x-1}{x-1} > 0$ is S_1 and that of $\dfrac{x-2}{x-1} > 0$ is S_2 as shown in Figure 1.15.

$$S_1 = \left(-\infty, \frac{1}{2}\right) \cup (1, \infty)$$

$$S_2 = (-\infty, 1) \cup (2, \infty)$$

$$S_1 \cap S_2 = \left(-\infty, \frac{1}{2}\right) \cup (2, \infty)$$

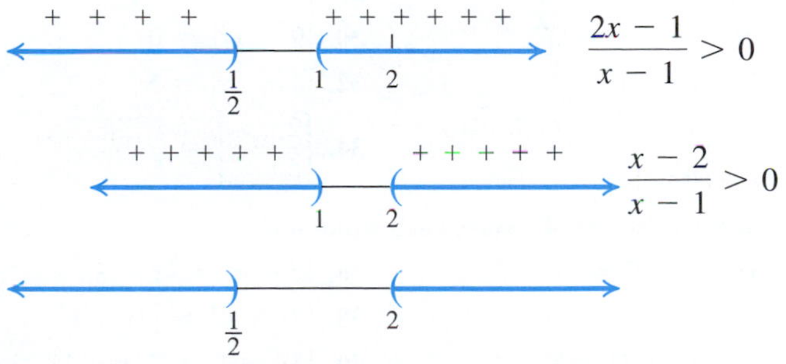

FIGURE 1.15

From Figure 1.15, we see that both inequalities are true on $S_1 \cap S_2$. So the solution set of the given inequality is $\left(-\infty, \frac{1}{2}\right) \cup (2, \infty)$.

Practice Problem 8 Solve: $|x - 2| < 4|x + 4|$

Answers to Practice Problems

1. a. $\{2\}$ **b.** $\{-1, 2\}$ **2.** $\left\{\frac{1}{2}\right\}$ **3.** $\left\{\frac{6}{5}, 2\right\}$

4. $[-3, 1]$

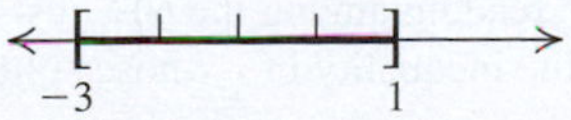

5. $[270, 420]$

6. $\left(-\infty, -\frac{9}{2}\right] \cup \left[\frac{3}{2}, \infty\right)$

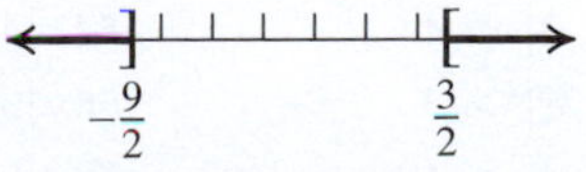

7. a. $(-\infty, \infty)$ **b.** $\varnothing$ **8.** $(-\infty, -6) \cup \left(-\frac{14}{5}, \infty\right)$

SECTION 1.6 Exercises

Basic Concepts and Skills

In Exercises 1–6, assume that $a > 0$.

1. The solution set of the equation $|x| = a$ is ____________.
2. The solution set of the inequality $|x| < a$ is ____________.
3. The solution set of the inequality $|x| \geq a$ is ____________.
4. The equation $|u| = |v|$ is equivalent to $u =$ ____________ or $u =$ ____________.
5. **True or False.** The solution set of $|3x - 2| < a$ is the same as the solution set of $|2 - 3x| < a$.
6. **True or False.** If $a > 0$, the solution set of $|3x - 2| \leq -a$ is $\varnothing$.

In Exercises 7–34, solve each equation.

7. $|3x| = 9$
8. $|4x| = 24$
9. $|-2x| = 6$
10. $|-x| = 3$
11. $|x + 3| = 2$
12. $|x - 4| = 1$
13. $|6 - 2x| = 8$
14. $|6 - 3x| = 9$
15. $|6x - 2| = 9$
16. $|6x - 3| = 9$
17. $|2x + 3| - 1 = 0$
18. $|2x - 3| - 1 = 0$
19. $\frac{1}{2}|x| = 3$
20. $\frac{3}{5}|x| = 6$
21. $\left|\frac{1}{4}x + 2\right| = 3$
22. $\left|\frac{3}{2}x - 1\right| = 3$
23. $6|1 - 2x| - 8 = 10$
24. $5|1 - 4x| + 10 = 15$
25. $2|3x - 4| + 9 = 7$
26. $9|2x - 3| + 2 = -7$
27. $|2x + 1| = -1$
28. $|3x + 7| = -2$
29. $|x^2 - 4| = 0$
30. $|9 - x^2| = 0$
31. $|1 - 2x| = 3$
32. $|4 - 3x| = 5$
33. $\left|\frac{1}{3} - x\right| = \frac{2}{3}$
34. $\left|\frac{2}{5} - x\right| = \frac{1}{5}$

In Exercises 35–44, solve each equation.

35. $|x + 3| = |x + 5|$
36. $|x + 4| = |x - 8|$
37. $|3x - 2| = |6x + 7|$
38. $|2x - 4| = |4x + 6|$
39. $|2x - 1| = x + 1$
40. $|3x - 4| = 2(x - 1)$
41. $|4 - 3x| = x - 1$
42. $|2 - 3x| = 2x - 1$
43. $|3x + 2| = 2(x - 1)$
44. $|4x + 7| = x + 1$

In Exercises 45–60, solve each inequality.

45. $|3x| < 12$
46. $|2x| \leq 6$
47. $|4x| > 16$
48. $|3x| > 15$
49. $|x + 1| < 3$
50. $|x - 4| < 1$
51. $|x| + 2 \geq -1$
52. $|x| + 2 > -7$
53. $|2x - 3| < 4$
54. $|4x - 6| \leq 6$
55. $|5 - 2x| > 3$
56. $|3x - 3| \geq 15$
57. $|3x + 4| \leq 19$
58. $|9 - 7x| < 23$
59. $|2x - 15| < 0$
60. $|x + 5| \leq -3$

In Exercises 61–70, solve each inequality.

61. $\left|\frac{x - 2}{x + 3}\right| < 1$
62. $\left|\frac{x + 3}{x - 1}\right| < 2$
63. $\left|\frac{2x - 3}{x + 1}\right| \leq 3$
64. $\left|\frac{2x - 1}{3x + 2}\right| \leq 1$
65. $\left|\frac{x - 1}{x + 2}\right| \geq 2$
66. $\left|\frac{x + 3}{x - 2}\right| \geq 3$
67. $\left|\frac{2x + 1}{x - 1}\right| > 4$
68. $\left|\frac{2x - 1}{3x + 2}\right| > 5$
69. $|x - 1| \leq 2|2x - 5|$
70. $2|x - 5| \leq |2x - 3|$

Applying the Concepts

71. **Varying temperatures.** The inequality $|T - 75| \leq 20$, where T is in degrees Fahrenheit, describes the daily temperature in Tampa during December. Give an interpretation for this inequality assuming that the high and low temperatures in Tampa during December satisfy the equation $|T - 75| = 20$.
72. **Scale error.** A butcher's scale is accurate to within ± 0.05 pound. A sirloin steak weighs 1.14 pounds on this scale. Let $x =$ actual weight of the steak. Write an absolute value inequality in x whose solution is the range of possible values for the actual weight of the steak.
73. **Company budget.** A company budgets \$700 for office supplies. The actual expense for budget supplies must be within $\pm$\$50 of this figure. Let $x =$ actual expense for the office supplies. Write an absolute value inequality in x whose solution is the range of possible amounts for the expense of the office supplies.
74. **National achievement scores.** Suppose 68% of the scores on a national achievement exam will be within ± 90 points of a score of 480. Let $x =$ a score among the 68% just described. Write an absolute value inequality in x whose solution is the range of possible scores within ± 90 points of 480.
75. **Blood pressure.** Suppose 60% of Americans have a systolic blood pressure reading of 120, plus or minus 6.75. Let $x =$ a blood pressure reading among the 60% just described. Write an absolute value inequality in x whose solution set is

the range of possible blood pressure readings within ± 6.75 points of 120.

76. Gas mileage. A motorcycle has approximately 4 gallons of gas, with an error margin of $\pm\frac{1}{4}$ of a gallon. If the motorcycle gets 37 miles per gallon of gas, how many miles can the motorcycle travel?

77. Event planning. An event planner expects about 120 people at a private party. She knows that this estimate could be off by 15 people (more or fewer). Food for the event costs \$48 per person. How much might the event planner's food expense be?

78. Company bonuses. Bonuses at a company are usually given to about 60 people each year. The bonuses are \$1200 each, and the estimate of 60 recipients may be off by 7 people (more or fewer). How much might the company spend on bonuses this year?

79. Fishing revenue. Sarah sells the fish she catches to a local restaurant for 60 cents a pound. She can usually estimate how much her catch weighs to within $\pm\frac{1}{2}$ pound. She estimates that she has 32 pounds of fish. How much might she get paid for her catch?

80. Ticket sales. Ticket sales at an amusement park average about 460 on a Sunday. If actual Sunday sales are never more than 25 above or below the average and if each ticket costs \$29.50, how much money might the park take in on a Sunday?

Beyond the Basics

In Exercises 81–86, solve each equation.

81. **a.** $|x^2 - 9| = x - 3$ **b.** $|x^2 - 8| = -2x$
c. $|x^2 - 5x| = 6$ **d.** $|x^2 + 3x - 2| = 2$

82. **a.** $|x^2 - 7| = |6x|$ **b.** $|x^2 - 2x| = |5x - 10|$
c. $|2x^2 - 3x + 5| = |x^2 - 4x + 7|$
d. $|x^2 + x + 3| = |x^2 + 5x + 1|$

83. $|2x - 3| + |x - 2| = 4$
[Consider three cases:
(i) $-\infty < x \le \frac{3}{2}$, (ii) $\frac{3}{2} < x \le 2$, (iii) $2 < x < \infty$.]

84. $2|2x - 3| - 3|x - 2| = 5$

85. $|x|^2 - 4|x| - 7 = 5$

86. $2|x|^2 - |x| + 8 = 11$

87. Show that if $0 < a < b$ and $0 < c < d$, then $ac < bd$.

88. Show that if $0 < a < b < c$ and $0 < e < f < g$, then $ae < bf < cg$.

89. Show that if $x^2 < a$ and $a > 0$, the solution set of the inequality $x^2 < a$ is $\{x | -\sqrt{a} < x < \sqrt{a}\}$, or in interval notation, $(-\sqrt{a}, \sqrt{a})$. [*Hint:* $a = (\sqrt{a})^2$; factor $x^2 - a$.]

90. Show that if $x^2 > a$ and $a > 0$, then the solution set of the inequality $x^2 > a$ is $(-\infty, -\sqrt{a}) \cup (\sqrt{a}, \infty)$.

In Exercises 91–102, write an absolute value inequality with variable x whose solution set is in the given interval(s).

91. $(1, 7)$ **92.** $[3, 8]$

93. $[-2, 10]$ **94.** $(-7, -1)$

95. $(-\infty, 3) \cup (11, \infty)$ **96.** $(-\infty, -1) \cup (5, \infty)$

97. $(-\infty, -5] \cup [10, \infty)$ **98.** $(-\infty, -3] \cup [-1, \infty)$

99. (a, b) **100.** $[c, d]$

101. $(-\infty, a) \cup (b, \infty)$ **102.** $(-\infty, c] \cup [d, \infty)$

103. Varying temperatures. The weather forecast predicts temperatures that will be between 8° of 70° Fahrenheit. Let $x =$ temperature in degrees Fahrenheit. Write an absolute value inequality in x whose solution is the predicted range of temperatures.

In Exercises 104–109, solve each inequality for x.

104. $|x - 1| \le 5$ and $|x| \ge 2$

105. $1 \le |x - 2| \le 3$

106. $|x - 1| + |x - 2| \le 4$

107. $|x - 2| + |x - 4| \ge 8$

108. $|x - 1| + |x - 2| + |x - 3| \le 6$

109. $\dfrac{|x - 1| - (x + 2)}{x + 2} \ge 0$

Critical Thinking/Discussion/Writing

110. For which values of x is $\sqrt{(x - 3)^2} = x - 3$?

111. For which values of x is
$\sqrt{(x^2 - 6x + 8)^2} = x^2 - 6x + 8$?

112. Solve $|x - 3|^2 - 7|x - 3| + 10 = 0$.
[*Hint:* Let $u = |x - 3|$.]

Maintaining Skills

In Exercises 113–120, simplify each expression.

113. $\dfrac{2 + 5}{2}$ **114.** $\dfrac{-3 + 7}{2}$

115. $\dfrac{-3-7}{2}$

116. $\dfrac{(a-b)-(a+b)}{2}$

117. $\sqrt{(5-2)^2+(3-7)^2}$

118. $\sqrt{(-8+3)^2+(-5-7)^2}$

119. $\sqrt{(2-5)^2+(8-6)^2}$

120. $\sqrt{(\sqrt{3}-\sqrt{12})^2+(\sqrt{2}+\sqrt{8})^2}$

In Exercises 121–126, add the appropriate term to each binomial so that it becomes a perfect square trinomial. Write and factor the trinomial.

121. $x^2 + 4x$

122. $x^2 - 6x$

123. $x^2 - 5x$

124. $x^2 + 7x$

125. $x^2 + \frac{3}{2}x$

126. $x^2 - \frac{4}{5}x$

SUMMARY Definitions, Concepts, and Formulas

1.1 Linear Equations in One Variable

i. An **equation** is a statement that two mathematical expressions are equal.

ii. The **solutions** or **roots** of an equation are those values (if any) of the variable that satisfy the equation.

iii. **Equivalent equations** are equations that have the same solution set.

iv. You can generate equivalent equations using the operations given on page 78.

v. The **standard form** of a conditional linear equation in x is $ax + b = 0, a \neq 0$.

vi. A **formula** is an equation that expresses a relationship between two or more variables.

1.2 Quadratic Equations

i. The equation $ax^2 + bx + c = 0, a \neq 0$, is the **standard form** of the quadratic equation in x.

ii. **Zero-product property:** Let A and B be two algebraic expressions. Then $AB = 0$ if and only if $A = 0$ or $B = 0$.

iii. **Square root method:** If $u^2 = d$, then $u = \pm\sqrt{d}$.

iv. **Method for completing the square:** See page 95.

v. **Quadratic formula:** $x = \dfrac{-b \pm \sqrt{b^2 - 4ac}}{2a}$

1.3 Complex Numbers: Quadratic Equations with Complex Solutions

i. Complex numbers are of the form $a + bi$, where a and b are real numbers and $i = \sqrt{-1}$; $i^2 = -1$.

ii. The number $a - bi$ is called the complex conjugate of $a + bi$.

iii. Operations with complex numbers can be performed as if they are binomials with the variable i. Set $i^2 = -1$ to simplify. Division is performed by first multiplying the numerator and the denominator by the complex conjugate of the denominator.

iv. The quantity $b^2 - 4ac$ in the quadratic formula is called the **discriminant** $(= D)$.

v. If $D > 0$, the quadratic equation has two distinct real roots.
If $D = 0$, there is one real solution.
If $D < 0$, there are two nonreal complex solutions, and they are conjugates.

1.4 Solving Other Types of Equations

Many types of equations can be solved by factoring. However, it is possible to introduce extraneous solutions when solving rational equations, equations involving radicals, or equations involving rational exponents. Remember to check the solutions. Extraneous solutions may be introduced when you

i. Multiply both sides by an LCD that contains a variable.

ii. Raise both sides to an even power.

1.5 Inequalities

i. An **inequality** is a statement that one algebraic expression is less than or is less than or equal to another algebraic expression.

ii. The real numbers that result in a true statement when those numbers are substituted for the variable in the inequality are called **solutions** of the inequality.

iii. A **linear inequality** is an inequality that can be written in the form $ax + b < 0$. The symbol $<$ can be replaced with $\leq, >$, or $\geq$.

iv. Test numbers can be used to solve polynomial or rational inequalities. See page 136.

1.6 Equations and Inequalities Involving Absolute Value

i. **Definition.** $|a| = a$ if $a \geq 0$ and $|a| = -a$ if $a < 0$.

ii. Let u be a variable or an algebraic expression and let $a > 0$. Then

a. $|u| < a$ if and only if $-a < u < a$.
b. $|u| > a$ if and only if $u < -a$ or $u > a$.

The properties in (a) and (b) are valid if $<$ and $>$ are replaced with $\leq$ and $\geq$, respectively.

REVIEW EXERCISES

Basic Concepts and Skills

In Exercises 1–18, solve each equation.

1. $5x - 4 = 11$

2. $12x + 7 = 31$

3. $3(2x - 4) = 9 - (x + 7)$

4. $4(x + 7) = 40 + 2x$

5. $3x + 8 = 3(x + 2) + 2$

6. $3x + 8 = 3(x + 1) + 4$

7. $x - (5x - 2) = 7(x - 1) - 2$

8. $7 + 4(3 + y) = 8(3y - 1) + 3$

9. $\dfrac{2}{x + 3} = \dfrac{5}{11x - 1}$

10. $\dfrac{7}{x + 2} = \dfrac{3}{x - 2}$

11. $\dfrac{y + 5}{2} + \dfrac{y - 1}{3} = \dfrac{7y + 3}{8} + \dfrac{4}{3}$

12. $\dfrac{y - 3}{6} - \dfrac{y - 4}{5} = -\dfrac{1}{6}$

13. $|2x - 3| = |4x + 5|$

14. $|5x - 3| = |x + 4|$

15. $|2x - 1| = |2x + 7|$

16. $|x - 1| = 2 - x$

17. $|3x - 2| = 2x + 1$

18. $|1 - 2x| = x + 5$

In Exercises 19–22, solve each equation for the indicated variable.

19. $p = k + gt$ for g

20. $RK = 4 + 3K$ for K

21. $T = \dfrac{2B}{B - 1}$ for B

22. $S = \dfrac{a}{1 - r}$ for r

In Exercises 23–46, solve each equation. (Include all complex solutions.)

23. $x^2 - 7x = 0$

24. $x^2 - 32x = 0$

25. $x^2 - 3x - 10 = 0$

26. $2x^2 - 9x - 18 = 0$

27. $(x - 1)^2 = 2x^2 + 3x - 5$

28. $(x + 2)^2 = x(3x + 2)$

29. $\dfrac{x^2}{4} + x = \dfrac{5}{4}$

30. $\dfrac{x^2}{4} + \dfrac{7}{16} = x$

31. $3x(x + 1) = 2x + 2$

32. $x^2 - x = 3(5 - x)$

33. $x^2 - 3x - 1 = 0$

34. $x^2 + 6x + 2 = 0$

35. $2x^2 + x - 1 = 0$

36. $x^2 + 4x + 1 = 0$

37. $3x^2 - 12x - 24 = 0$

38. $2x^2 + 4x - 3 = 0$

39. $2x^2 - x - 2 = 0$

40. $3x^2 - 5x + 1 = 0$

41. $x^2 - x + 1 = 0$

42. $3x^2 + 4x + 3 = 0$

43. $x^2 - 6x + 13 = 0$

44. $x^2 - 8x + 20 = 0$

45. $4x^2 - 8x + 13 = 0$

46. $3x^2 - 4x + 2 = 0$

In Exercises 47–50, find the discriminant and determine the number and type of roots.

47. $3x^2 - 11x + 6 = 0$

48. $x^2 - 14x + 49 = 0$

49. $5x^2 + 2x + 1 = 0$

50. $9x^2 = 25$

In Exercises 51–70, solve each equation.

51. $\sqrt{x^2 - 16} = 0$

52. $\sqrt{x + 6} = x$

53. $\sqrt{4 - 7x} = \sqrt{2x}$

54. $t - 2\sqrt{t} + 1 = 0$

55. $y - 2\sqrt{y} - 3 = 0$

56. $\sqrt{3x + 4} - \sqrt{x - 3} = 3$

57. $\sqrt{x} - 1 = \sqrt{5 + \sqrt{x}}$

58. $\dfrac{1}{(1 - x)^2} - \dfrac{7}{1 - x} = -10$

59. $(7x + 5)^2 + 2(7x + 5) - 15 = 0$

60. $(x^2 - 1)^2 - 11(x^2 - 1) + 24 = 0$

61. $x^{2/3} + 3x^{1/3} - 4 = 0$

62. $x^{-2/3} + x^{-1/3} - 6 = 0$

63. $(\sqrt{t} + 5)^2 - 9(\sqrt{t} + 5) + 20 = 0$

64. $3\left(\dfrac{y - 1}{6}\right)^2 - 7\left(\dfrac{y - 1}{6}\right) = 0$

65. $4x^4 - 37x^2 + 9 = 0$

66. $\dfrac{1}{x} + \dfrac{1}{x - 1} = \dfrac{5}{6}$

67. $\dfrac{2x + 1}{2x - 1} = \dfrac{x - 1}{x + 1}$

68. $6 - \dfrac{2}{x} = \dfrac{4}{x - 1}$

69. $\left(\dfrac{7x}{x + 1}\right)^2 - 3\left(\dfrac{7x}{x + 1}\right) = 18$

70. $\left(\dfrac{4x^2 - 3}{x}\right)^2 = 1$

In Exercises 71–74, solve each equation for x in terms of other variables.

71. $x^2 + 2yx - 3y^2 = 0$

72. $x^2 + (y - 2z)x - 2yz = 0$

73. $x^2 + (3 - 2y)x + y^2 - 3y + 2 = 0$

74. $x^2 + (1 - 2y)x + y^2 - y - 2 = 0$

In Exercises 75–90, solve each inequality. Write the solution in interval notation.

75. $x + 5 < 3$

76. $2x + 1 < 9$

77. $3(x - 3) \le 8$

78. $x + 5 \le 19 + 3x$

79. $x + 2 \ge \dfrac{2}{3}x - 2x$

80. $2x + 1 \ge \dfrac{5x - 6}{3}$

81. $\dfrac{1}{6} > \dfrac{4 - 3x}{3}$

82. $\dfrac{2}{5} > \dfrac{3 - 2x}{2}$

83. $\dfrac{x - 3}{3} - 2 \le \dfrac{x}{6} + \dfrac{1}{2}$

84. $\dfrac{x + 1}{2} - 3 \le \dfrac{x}{4} + \dfrac{1}{2}$

85. $\dfrac{3 - 2x}{4} + 1 > \dfrac{x - 5}{3}$

86. $\dfrac{5 - 3x}{5} - 2 > \dfrac{2 - x}{3}$

In Exercises 87–98, solve each compound inequality. Write the solution in interval notation.

87. $3x - 1 < 2$ or $11 - 2x < 5$

88. $3 - 2x > 5$ or $15 - 3x < 6$

89. $4x - 5 < 7$ and $7 - 3x < 1$

90. $2x - 1 < 3$ and $4 - 3x > 1$

91. $-3 \le 2x + 1 < 7$

92. $\dfrac{1}{6} \le \dfrac{3x - 4}{3} \le 4$

93. $-3 < 3 - 2x \le 97$

94. $-\dfrac{1}{2} < \dfrac{4 - 3x}{3} \le \dfrac{1}{2}$

95. $x^2 + x - 6 \ge 0$

96. $x^3 - 9x \le 0$

97. $\dfrac{(x - 1)(x + 3)}{(x + 2)(x + 5)} \ge 0$

98. $\dfrac{x^2 + 4x + 3}{x^2 + 6x + 8} \le 0$

In Exercises 99–108, solve each inequality. Write the solution set in interval notation.

99. $|3x + 2| \le 7$

100. $|x - 4| \ge 2$

101. $4|x - 2| + 8 > 12$

102. $3|x - 1| + 4 < 10$

103. $\left|\frac{4 - x}{5}\right| \ge 1$

104. $\left|\frac{1 - x}{6}\right| < 1$

105. $\left|\frac{x - 1}{x + 2}\right| \le 3$

106. $\left|\frac{x + 3}{x - 5}\right| < 4$

107. $\left|\frac{2x - 3}{x + 2}\right| \ge 2$

108. $\left|\frac{x + 1}{2x - 5}\right| > 3$

Applying the Concepts

109. A circular lens has a circumference of 22 centimeters. Find the radius of the lens.

110. A rectangle has a perimeter of 18 inches and a length of 5 inches. Find the width of the rectangle.

111. A trapezoid has an area of 32 square meters and a height of 8 meters. If one base is 5 meters, what is the length of the other base?

112. What principal must be deposited for four years at 7% annual simple interest to earn $354.20 in interest?

113. The volume of a box is 4212 cubic centimeters. If the box is 27 centimeters long and 12 centimeters wide, how high is it?

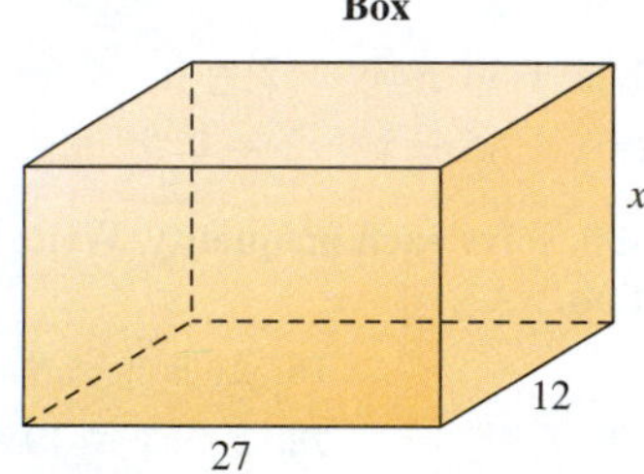

114. The ratio of the current assets of a business to its current liabilities is called the *current ratio* of the business. Assuming that the current ratio is 2.7 and the current assets total $256,500, find the current liabilities of the business.

115. A cylindrical shaft has a volume of 8750π cubic centimeters. If the radius is 5 centimeters, how long is the shaft?

116. A car dealer deducted 15% of the selling price of a car he had sold on consignment and gave the balance, $2210, to the owner. What was the selling price of the car?

117. The third angle in an isosceles triangle measures 40 degrees more than twice either base angle. Find the number of degrees in each angle of the triangle.

118. A 600-mile trip took 12 hours. Half of the time was spent traveling across flat terrain, and half was through hilly terrain. Assuming that the average rate over the hilly section was 20 miles per hour slower than the average rate over the flat section, find the two rates and the distance traveled at each rate.

119. A total of $30,000 is invested in two stocks. At the end of the year, one returned 6% and the other returned 8% on the original investment. How much was invested in each stock if the total profit was $2160?

120. The monthly note on a car that was leased for two years was $250 less than the monthly note on a car that was leased for a year and a half. The total income from the two leases was $21,300. Find the monthly note on each.

121. Two solutions, one containing $4\frac{1}{2}\%$ iodine and the other containing 12% iodine, are to be mixed to produce 10 liters of a 6% iodine solution. How many liters of each are required?

122. Two cars leave from the same place at the same time traveling in opposite directions. One travels 5 kilometers per hour faster than the other. After three hours, they are 495 kilometers apart. How fast is each car traveling?

123. A total of 28 handshakes were exchanged at the end of a party. Assuming that everyone shook hands with everyone else, how many people were at the party?

124. Find two positive numbers that differ by 7 and whose product is 408.

125. Lavina bought some shares of stock for $18,040. Later when the price had gone up $18 per share, she sold all but 20 of them for $20,000. How many shares had she bought?

126. Joann drives her car for 15 miles at a certain speed. She then increases her speed by 10 mph and drives for another 20 miles. If the total trip takes one hour, find her original speed.

127. One-fourth of a herd of wild horses is in a forest. Twice the square root of the number of horses in the herd has gone to the mountains, and the remaining 15 are on the bank of the river. What is the total number of horses in the herd?

128. A path of uniform width surrounds a rectangular swimming pool that is 30 ft long and 16 ft wide. Find the width of the path assuming that its area is 312 ft^2.

129. The Botany Club chartered a bus for $324 to go on a field trip, the cost to be divided equally among those attending. At the last minute, four more members decided to go, which reduced the cost by $0.90 per person. How many members went on the trip?

PRACTICE TEST A

1. Solve the equation $5x - 9 = 3x - 5$.
2. Solve the equation $\frac{7}{24} = \frac{x}{8} + \frac{1}{6}$.
3. Solve the equation $\frac{1}{x-2} - 5 = \frac{1}{x+2}$.
4. The width of a rectangle is 3 centimeters less than the length, and the area is 54 square centimeters. Find the dimensions of the rectangle.
5. Solve the equation $x^2 + 36 = -13x$.
6. If Fran invests \$8200 at 6% per year, how much additional money must she invest at 8% to ensure that the interest she receives each year is 7% of the total investment?
7. A frame of uniform width borders a painting that is 25 inches long and 11 inches high. If the area of the framed picture is 351 square inches, find the width of the border.
8. Solve $-6x - 15 = (2x + 5)^2$.
9. Determine the constant that should be added to the binomial

$$x^2 + \frac{2}{3}x$$

so that it becomes a perfect-square trinomial. Then write and factor the trinomial.

10. Solve $3x^2 - 5x - 1 = 0$.
11. A box with a square base and no top is made from a square piece of cardboard by cutting 2-inch squares from each corner and folding up the sides. What length of side must the original cardboard square have if the volume of the box is 50 cubic inches?
12. Solve $x^2 + 12x + 33 = 0$.
13. Solve the polynomial equation

$$3x^4 - 75x^2 = 0$$

by factoring and then using the zero-product property.

14. Solve the equation

$$3x - 2 - 5\sqrt{x} = 0$$

by making an appropriate substitution.

15. Solve the absolute value equation

$$\left|\frac{1}{3}x + 5\right| = \left|\frac{2}{3}x + 7\right|.$$

In Problems 16 and 17, solve the inequality. Express the solution set in interval notation.

16. $\frac{x}{2} - 5 \geq \frac{4x}{9}$
17. $-4 < 2x - 3 < 4$
18. Solve the inequality

$$\left|\frac{2}{3}x - 1\right| - 2 > \frac{1}{3}$$

by first rewriting it as an equivalent inequality without absolute value bars. Express the solution set in interval notation.

19. Solve the linear inequality $\frac{2}{5}y - 13 \leq -\left(7 + \frac{13}{5}y\right)$.
20. Solve the inequality $0 \leq 5x - 2 \leq 8$.

PRACTICE TEST B

1. Solve the equation $2x - 2 = 5x + 34$.
   a. $\left\{\frac{17}{4}\right\}$ **b.** $\{-18\}$ **c.** $\left\{\frac{15}{2}\right\}$ **d.** $\{-12\}$
2. Solve the equation $\frac{z}{2} = 2z + 35$.
   a. $\left\{\frac{47}{2}\right\}$ **b.** $\left\{-\frac{47}{2}\right\}$ **c.** $\{23\}$ **d.** $\left\{-\frac{70}{3}\right\}$
3. Solve the equation $\frac{1}{t-2} - \frac{1}{2} = \frac{-2t}{4t-1}$.
   a. $\left\{\frac{4}{9}\right\}$ **b.** $\left\{\frac{1}{2}\right\}$ **c.** $\left\{\frac{2}{3}\right\}$ **d.** $\{-2\}$
4. The length of a rectangle is 4 centimeters greater than the width, and the area is 77 square centimeters. Find the dimensions of the rectangle.
   a. 11 centimeters by 15 centimeters
   b. 5 centimeters by 9 centimeters
   c. 7 centimeters by 11 centimeters
   d. 9 centimeters by 13 centimeters
5. Solve the equation $x^2 + 12 = -7x$.
   a. $\{3, 4\}$ **b.** $\{-4, -3\}$
   c. $\{-3, 3, 4\}$ **d.** $\{4, 3\}$
6. If Rena invests \$7500 at 7% per year, how much additional money must she invest at 12% to ensure that the interest she receives each year is 10% of the total investment?
   a. \$11,250 **b.** \$12,375
   c. \$18,000 **d.** \$21,000
7. A frame of uniform width borders a painting that is 20 inches long and 13 inches high. Assuming that the area of the framed picture is 368 square inches, find the width of the border.
   a. 3.5 in. **b.** 3 in.
   c. 4 in. **d.** 1.5 in.
8. Solve $-6x - 2 = (3x + 1)^2$.
   a. $\varnothing$ **b.** $\left\{-\frac{1}{3}\right\}$
   c. $\left\{\frac{1}{3}, 1\right\}$ **d.** $\left\{-1, -\frac{1}{3}\right\}$

9. Determine the constant that should be added to the binomial

$$x^2 + \frac{1}{6}x$$

so that it becomes a perfect-square trinomial. Then write and factor the trinomial.

a. $\frac{1}{12}; x^2 + \frac{1}{6}x + \frac{1}{12} = \left(x + \frac{1}{6}\right)^2$

b. $\frac{1}{144}; x^2 + \frac{1}{6}x + \frac{1}{144} = \left(x + \frac{1}{12}\right)^2$

c. $\frac{1}{36}; x^2 + \frac{1}{6}x + \frac{1}{36} = \left(x + \frac{1}{6}\right)^2$

d. $144; x^2 + \frac{1}{6}x + 144 = (x + 12)^2$

10. Solve $7x^2 + 10x + 2 = 0$.

a. $\left\{\frac{-5 - \sqrt{11}}{7}, \frac{-5 + \sqrt{11}}{7}\right\}$

b. $\left\{\frac{-5 - \sqrt{39}}{7}, \frac{-5 + \sqrt{39}}{7}\right\}$

c. $\left\{\frac{-5 - \sqrt{11}}{14}, \frac{-5 + \sqrt{11}}{14}\right\}$

d. $\left\{\frac{-10 - \sqrt{11}}{7}, \frac{-10 + \sqrt{11}}{7}\right\}$

11. A box with a square base and no top is made from a square piece of cardboard by cutting 3-inch squares from each corner and folding up the sides. What length of side must the original cardboard square have if the volume of the box is 675 cubic inches?

a. 18 inches **b.** 21 inches
c. 15 inches **d.** 14 inches

12. Solve $x^2 + 14x + 38 = 0$.

a. $\{7 - \sqrt{38}, 7 + \sqrt{38}\}$
b. $\{14 + \sqrt{38}\}$
c. $\{7 + \sqrt{11}\}$
d. $\{-7 - \sqrt{11}, -7 + \sqrt{11}\}$

13. Solve the polynomial equation

$$5x^4 - 45x^2 = 0$$

by factoring and then using the zero-product property.

a. $\{-3, 0, 3\}$ **b.** $\{-3, 3\}$
c. $\{0\}$ **d.** $\{-3\sqrt{5}, 0, 3\sqrt{5}\}$

14. Solve the equation

$$x - 2048 - 32\sqrt{x} = 0$$

by making an appropriate substitution.

a. $\{4096\}$ **b.** $\{3072\}$
c. $\{8192\}$ **d.** $\{2048\}$

15. Solve the absolute value equation

$$\left|\frac{1}{2}x + 2\right| = \left|\frac{3}{4}x - 2\right|.$$

a. $\{12, 16\}$ **b.** $\varnothing$ **c.** $\{10\}$ **d.** $\{0, 16\}$

In Problems 16 and 17, solve the inequality. Express the solution set in interval notation.

16. $\frac{x}{6} - \frac{1}{3} \le \frac{x}{3} + 1$

a. $(-\infty, -8)$ **b.** $(-\infty, -8)$
c. $(-8, \infty)$ **d.** $[-8, \infty)$

17. $-13 \le -3x + 2 < -4$

a. $[-5, -2)$ **b.** $(2, 5]$
c. $(-5, -2]$ **d.** $[2, 5)$

18. Solve the inequality

$$8 + \left|1 - \frac{x}{2}\right| \ge 10$$

by first rewriting it as an equivalent inequality without absolute value bars. Express the solution set in interval notation.

a. $(-\infty, -6] \cup [2, \infty)$ **b.** $[-2, 6]$
c. $(-\infty, -2] \cup [6, \infty)$ **d.** $[-6, 2]$

19. Solve the linear inequality $\frac{2}{3}x - 2 < \frac{5}{3}x$.

a. $(-2, \infty)$ **b.** $(\infty, 2)$ **c.** $(-\infty, -2)$
d. $\{-2\}$ **e.** $(2, \infty)$

20. Solve the inequality $0 \le 7x - 1 \le 13$.

a. $\left[\frac{1}{7}, 2\right]$ **b.** $(-1, 2]$ **c.** $\frac{1}{7}$
d. $[-1, 2]$ **e.** $\left(-\infty, \frac{1}{7}\right]$

CHAPTER 2

Graphs and Functions

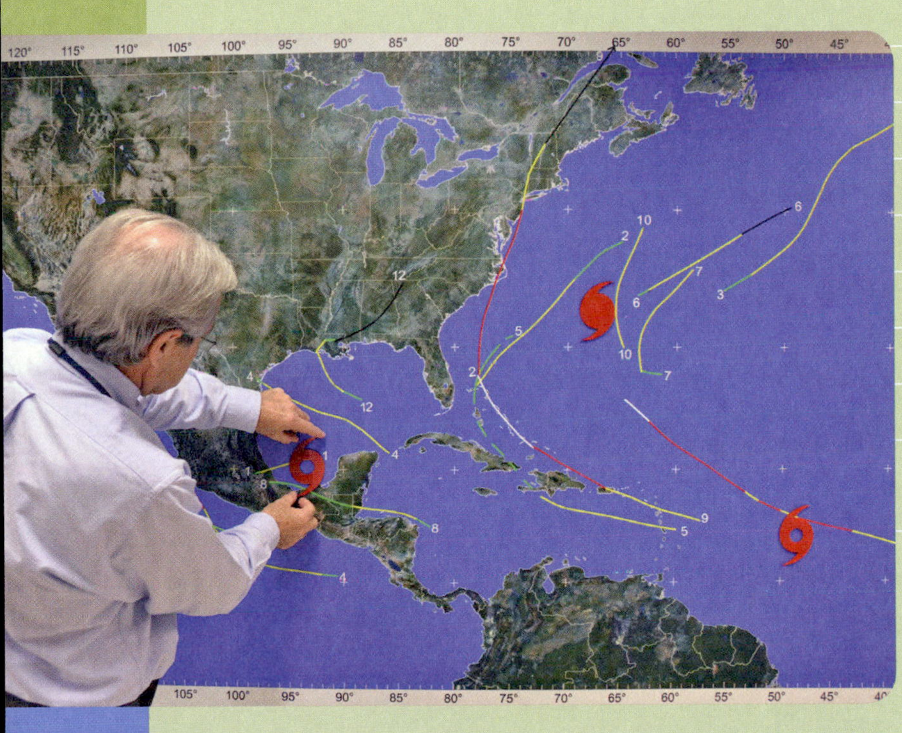

TOPICS

Hurricanes are tracked with the use of coordinate grids, and data from fields as diverse as medicine and sports are related and analyzed by means of functions. The material in this chapter introduces you to the versatile concepts that are the everyday tools of all dynamic industries.

SECTION 2.1

The Coordinate Plane

BEFORE STARTING THIS SECTION, REVIEW

1. The number line (Section P.1, page 5)
2. Equivalent equations (Section 1.1, page 78)
3. Completing squares (Section 1.2, page 94)
4. Interval notation (Section P.1, page 7)

OBJECTIVES

1. Plot points in the Cartesian coordinate plane.
2. Find the distance between two points.
3. Find the midpoint of a line segment.

◆ A Fly on the Ceiling

One day the French mathematician René Descartes noticed a fly buzzing around on a ceiling made of square tiles. He watched the fly and wondered how he could mathematically describe its location. Finally, he realized that he could describe the fly's position by its distance from the walls of the room. Descartes had just discovered the coordinate plane! In fact, the coordinate plane is sometimes called the Cartesian plane in his honor. The discovery led to the development of analytic geometry, the first blending of algebra and geometry.

Although the basic idea of graphing with coordinate axes dates all the way back to Apollonius in the second century B.C., Descartes, who lived in the 1600s, gets the credit for coming up with the two-axis system we use today. In Example 2, we will see how the Cartesian plane helps visualize data on smoking.

1 Plot points in the Cartesian coordinate plane.

The Coordinate Plane

A visually powerful device for exploring relationships between numbers is the Cartesian plane. A pair of real numbers in which the order is specified is called an **ordered pair** of real numbers. The ordered pair (a, b) has **first component** a and **second component** b. Two ordered pairs (x, y) and (a, b) are **equal**, and we write $(x, y) = (a, b)$ if and only if $x = a$ and $y = b$.

Just as the real numbers are identified with points on a line, called the ***number line*** or the ***coordinate line***, the sets of ordered pairs of real numbers are identified with points on a plane called the **coordinate plane** or the **Cartesian plane**.

We begin with two coordinate lines, one horizontal and one vertical, that intersect at their zero points. The horizontal line (with positive numbers to the right) is usually called the ***x*-axis**, and the vertical line (with positive numbers up) is usually called the ***y*-axis**. Their point of intersection is called the **origin**. The x-axis and y-axis are called **coordinate axes**, and the plane they form is sometimes called the ***xy*-plane**. The axes divide the plane into four regions called **quadrants**, which are numbered as shown in Figure 2.1. The points on the axes themselves do not belong to any of the quadrants.

SIDE NOTE

Although it is common to label the axes as x and y, other letters are also used, especially in applications. In a uv-plane or an st-plane, the first letter in the name refers to the horizontal axis; the second, to the vertical axis.

The notation $P(a, b)$, or $P = (a, b)$, designates the point P whose *first component* is a and whose *second component* is b. In an xy-plane, the first component, a, is called the ***x*-coordinate** of $P(a, b)$ and the second component, b, is called the ***y*-coordinate** of $P(a, b)$. The signs of the x- and y-coordinates for each quadrant are shown in Figure 2.1.

René Descartes (1596–1650)

Descartes was born at La Haye, near Tours in southern France. He is often called the father of modern science. Descartes established a new, clear way of thinking about philosophy and science by accepting only those ideas that could be proved by or deduced from first principles. He took as his philosophical starting point the statement **Cogito ergo sum**: "I think; therefore, I am." Descartes made major contributions to modern mathematics, including the Cartesian coordinate system and the theory of equations.

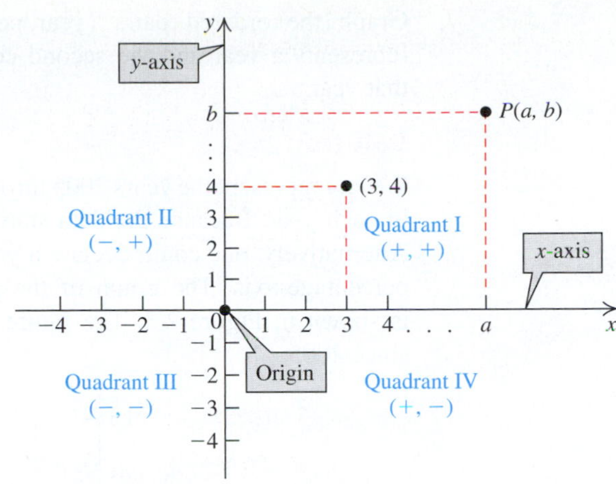

FIGURE 2.1 Quadrants in a plane

The point corresponding to the ordered pair (a, b) is called the **graph of the ordered pair** (a, b). However, we frequently ignore the distinction between an ordered pair and its graph.

EXAMPLE 1 Graphing Points

Graph the following points in the xy-plane:

$A(3, 1), B(-2, 4), C(-3, -4), D(2, -3)$, and $E(-3, 0)$.

Solution

Figure 2.2 shows a coordinate plane along with the graph of the given points. These points are located by moving left, right, up, or down starting from the origin $(0, 0)$.

$A(3, 1)$	3 units right, 1 unit up	$D(2, -3)$	2 units right, 3 units down
$B(-2, 4)$	2 units left, 4 units up	$E(-3, 0)$	3 units left
$C(-3, -4)$	3 units left, 4 units down		

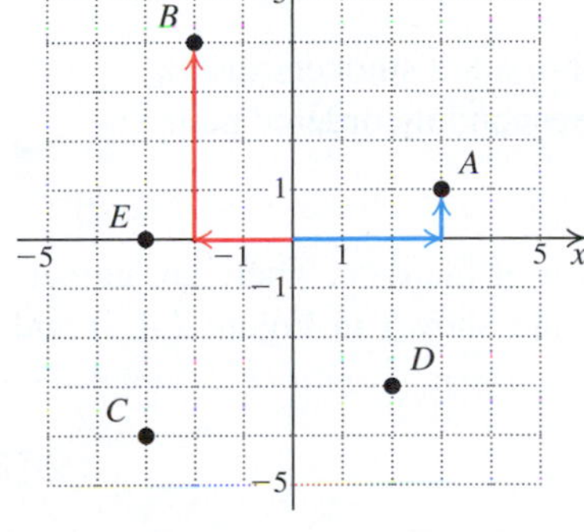

FIGURE 2.2 Graphing points

Practice Problem 1 Graph the following points in the xy-plane

$$P(-2, 2), Q(4, 0), R(5, -3), S(0, -3), \text{ and } T\left(-2, \frac{1}{2}\right).$$

EXAMPLE 2 Graphing Data on Adult Smokers in the United States

The data in Table 2.1 show the prevalence of smoking among adults 18 years and older in the United States over the years 2005–2012.

TABLE 2.1

Year	2005	2006	2007	2008	2009	2010	2011	2012
Percent of Adult Smokers	20.9	20.8	20.8	20.6	20.6	19.3	19.0	21.0

Source: Centers for Disease Control and Prevention, National Health Interview Survey.

Graph the ordered pairs (year, percent of adult smokers), where the first coordinate represents a year and the second coordinate represents the percent of adult smokers in that year.

Solution

We let t represent the years 2005 through 2012 and % represent the percent of adult smokers in each year. Because the data start from the year 2005, we show a break in the t-axis. Alternatively, one could declare a year—say, 2004—as 0. Similar comments apply to the percentage-axis. The graph of the points $(2005, 20.9), (2006, 20.8), \dots, (2012, 21.0)$ is shown in Figure 2.3. The figure depicts the percentage of adult smokers every year since 2005.

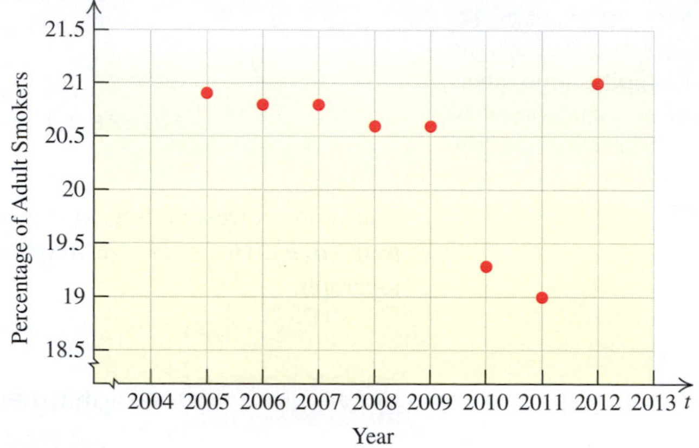

FIGURE 2.3 Percentage of adult smokers 2005–2012

Practice Problem 2 In Example 2, suppose the percent of adult smokers shown in Table 2.1 is decreased by 3 in each year. Write the corresponding ordered pairs (year, percent of adult smokers).

The display in Figure 2.3 is called a **scatter diagram** of the data. There are numerous other ways of displaying the data. Two such ways are shown in Figure 2.4(a) and Figure 2.4(b).

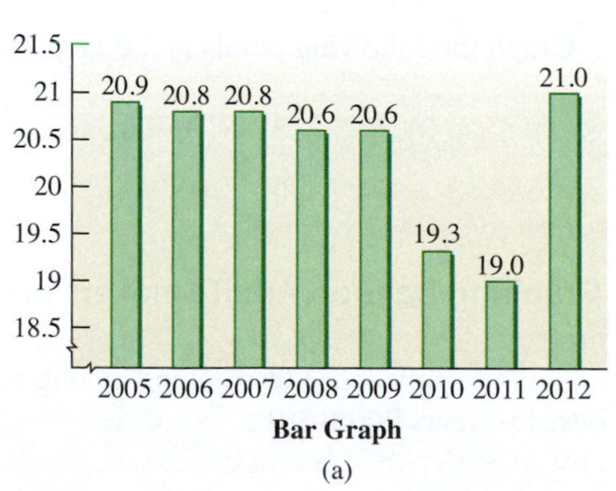

Bar Graph

(a)

y
21.5
21
20.5
20
19.5
19
18.5
2004 2005 2006 2007 2008 2009 2010 2011 2012 x

Line Graph

(b)

FIGURE 2.4 Two methods for displaying data

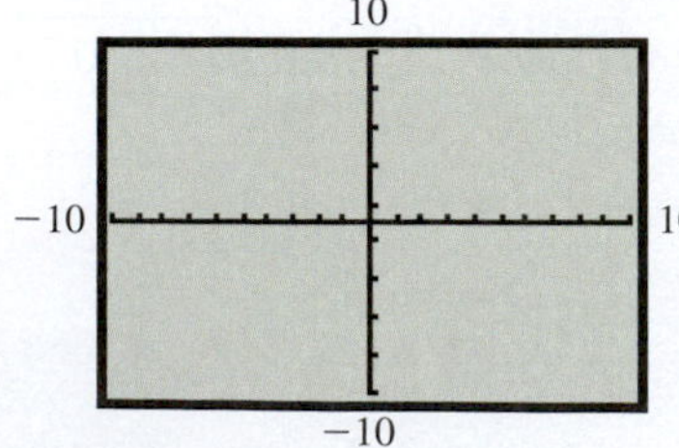

FIGURE 2.5 Viewing rectangle

Scales on a Graphing Utility

When drawing a graph, you can use different scales for the x- and y-axes. Similarly, the scale can be set separately for each coordinate axis on a graphing utility. Once scales are set, you get a **viewing rectangle**, where your graphs are displayed. For example, in Figure 2.5, the scale for the x-axis (the distance between each tick mark) is 1, whereas the scale for the y-axis is 2. Read more about viewing rectangles in your graphing calculator manual.

2 Find the distance between two points.

Distance Formula

If a Cartesian coordinate system has the same unit of measurement, such as inches or centimeters, on both axes, we can then calculate the distance between any two points in the coordinate plane in the given unit.

Recall that the Pythagorean theorem states that in a **right triangle** with hypotenuse of length c and the other two sides of lengths a and b,

$$a^2 + b^2 = c^2, \quad \text{Pythagorean theorem}$$

as shown in Figure 2.6.

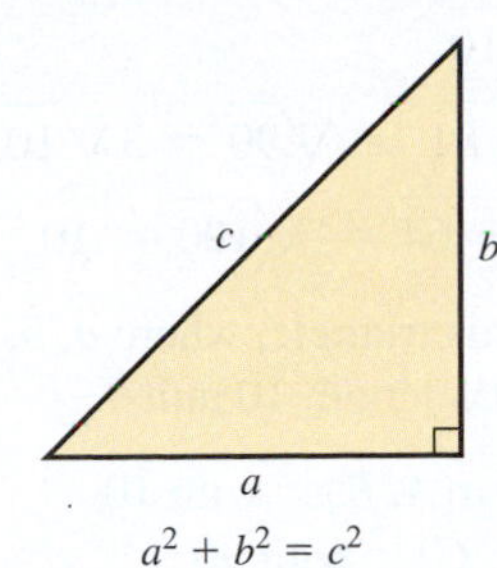

FIGURE 2.6 Pythagorean theorem

Suppose we want to compute the distance $d(P, Q)$ between the two points $P(x_1, y_1)$ and $Q(x_2, y_2)$. We draw a horizontal line through the point Q and a vertical line through the point P to form the right triangle PQS, as shown in Figure 2.7.

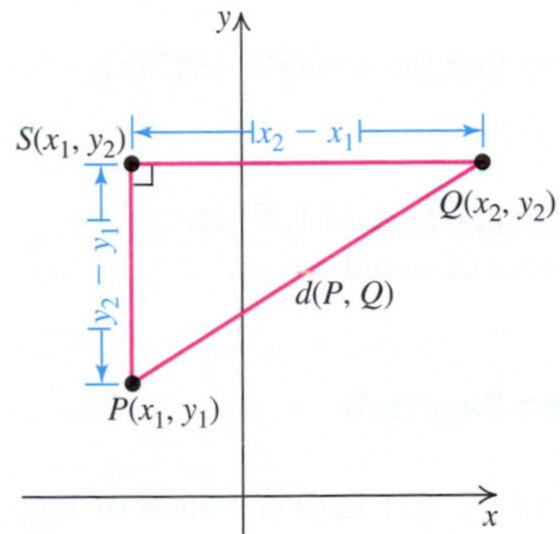

FIGURE 2.7 Visualizing the distance formula

The length of the horizontal side of the triangle is $|x_2 - x_1|$, and the length of the vertical side is $|y_2 - y_1|$. By the Pythagorean theorem, we have

$$[d(P, Q)]^2 = |x_2 - x_1|^2 + |y_2 - y_1|^2$$

$$d(P, Q) = \sqrt{|x_2 - x_1|^2 + |y_2 - y_1|^2} \quad \text{Take the square root of both sides.}$$

$$d(P, Q) = \sqrt{(x_2 - x_1)^2 + (y_2 - y_1)^2} \quad |a - b|^2 = (a - b)^2$$

DISTANCE FORMULA IN THE COORDINATE PLANE

Let $P = (x_1, y_1)$ and $Q = (x_2, y_2)$ be any two points in the coordinate plane. Then the distance between P and Q, denoted $d(P, Q)$, is given by the **distance formula**

$$d(P, Q) = \sqrt{(x_2 - x_1)^2 + (y_2 - y_1)^2}$$

EXAMPLE 3 Finding the Distance Between Two Points

Find the distance between the points $P(-2, 5)$ and $Q(3, -4)$.

Solution

Let $(x_1, y_1) = (-2, 5)$ and $(x_2, y_2) = (3, -4)$. Then

$$x_1 = -2, y_1 = 5, x_2 = 3, \text{ and } y_2 = -4.$$

$$d(P, Q) = \sqrt{(x_2 - x_1)^2 + (y_2 - y_1)^2} \quad \text{Distance formula}$$

$$= \sqrt{[3 - (-2)]^2 + (-4 - 5)^2} \quad \text{Substitute the values for } x_1, x_2, y_1, y_2.$$

$$= \sqrt{5^2 + (-9)^2} \quad \text{Simplify.}$$

$$= \sqrt{25 + 81} = \sqrt{106} \approx 10.3 \quad \text{Simplify; use a calculator.}$$

SIDE NOTE

Remember that in general $\sqrt{a^2 + b^2} \neq a + b$

Practice Problem 3 Find the distance between the points $(-5, 2)$ and $(-4, 1)$.

In the next example, we use the distance formula and the converse of the Pythagorean theorem to show that the given triangle is a right triangle.

EXAMPLE 4 Identifying a Right Triangle

RECALL

The converse of the Pythagorean Theorem states that: If in a triangle with side lengths a, b, and c, we have $a^2 + b^2 = c^2$, then the triangle is a right triangle.

Let $A(4, 3)$, $B(1, 4)$, and $C(-2, -5)$ be three points in the plane.

a. Sketch triangle ABC.

b. Find the length of each side of the triangle.

c. Show that ABC is a right triangle.

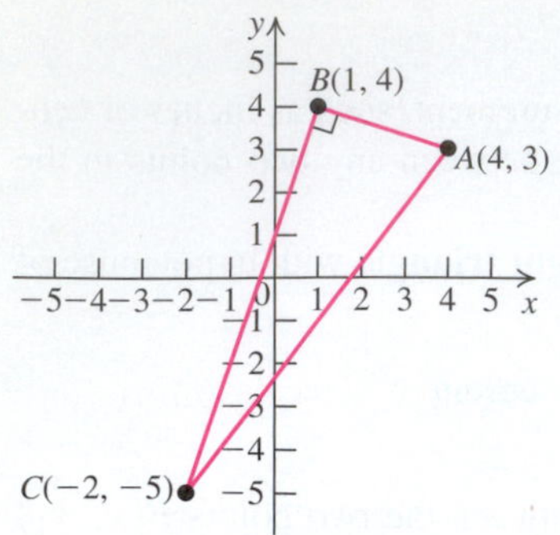

FIGURE 2.8

Solution

a. A sketch of the triangle formed by the three points A, B, and C is shown in Figure 2.8.

b. Using the distance formula, we have

$$d(A, B) = \sqrt{(4 - 1)^2 + (3 - 4)^2} = \sqrt{9 + 1} = \sqrt{10},$$
$$d(B, C) = \sqrt{[1 - (-2)]^2 + [4 - (-5)]^2} = \sqrt{9 + 81} = \sqrt{90} = 3\sqrt{10}$$
$$d(A, C) = \sqrt{[4 - (-2)]^2 + [3 - (-5)]^2} = \sqrt{36 + 64} = \sqrt{100} = 10.$$

c. We check whether the relationship $a^2 + b^2 = c^2$ holds in this triangle, where a, b, and c denote the lengths of its sides. The longest side, AC, has length 10 units.

$$[d(A, B)]^2 + [d(B, C)]^2 = 10 + 90 \quad \text{Replace } [d(A, B)]^2 \text{ with 10 and } [d(B, C)]^2 \text{ with 90.}$$
$$= (10)^2$$
$$= [d(A, C)]^2$$

It follows from the converse of the Pythagorean theorem that the triangle ABC is a right triangle.

Practice Problem 4 Is the triangle with vertices $(6, 2)$, $(-2, 0)$, and $(1, 5)$ an isosceles right triangle—that is, a right triangle with two sides of equal length?

EXAMPLE 5 Applying the Distance Formula to Baseball

The baseball diamond is in fact a square with a distance of 90 feet between each of the consecutive bases. Use an appropriate coordinate system to calculate the distance the ball travels when the third baseman throws it from third base to first base.

Solution

We can conveniently choose home plate as the origin and place the x-axis along the line from home plate to first base and the y-axis along the line from home plate to third base as shown in Figure 2.9. The coordinates of home plate (O), first base (A), second base (C), and third base (B) are shown in the figure.

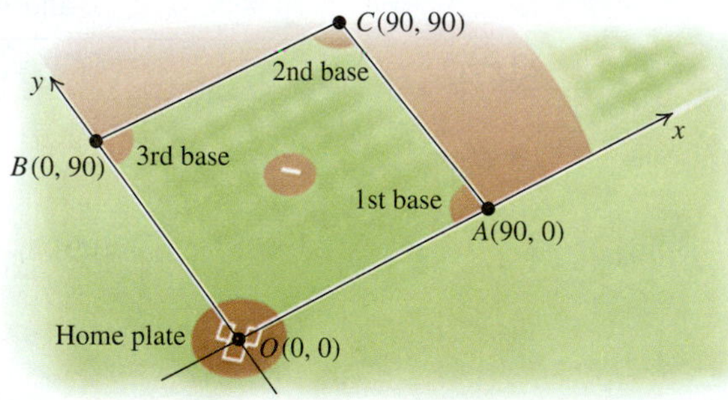

FIGURE 2.9

We are asked to find the distance between the points $A(90, 0)$ and $B(0, 90)$.

$$d(A, B) = \sqrt{(90 - 0)^2 + (0 - 90)^2} \quad \text{Distance formula}$$
$$= \sqrt{(90)^2 + (-90)^2}$$
$$= \sqrt{2(90)^2} \quad \text{Simplify.}$$
$$= 90\sqrt{2} \approx 127.28 \text{ feet} \quad \text{Simplify; use a calculator.}$$

Practice Problem 5 Young players might play baseball in a square "diamond" with a distance of 60 feet between consecutive bases. Repeat Example 5 for a diamond with these dimensions.

3 Find the midpoint of a line segment.

Midpoint Formula

Recall that a point M on the line segment PQ is its midpoint if $d(P, M) = d(M, Q)$. See Figure 2.10. We provide the midpoint formula in the xy-plane.

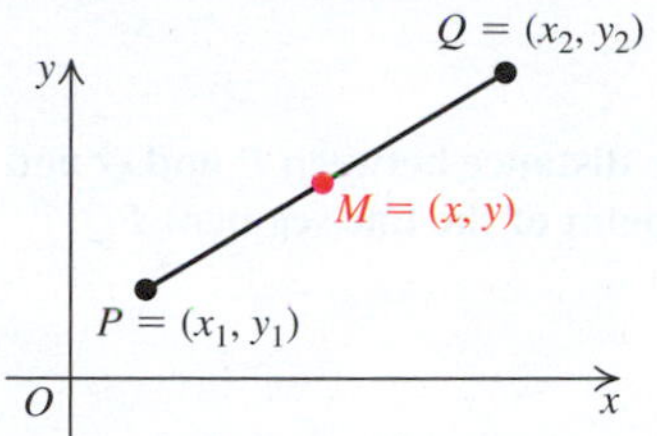

FIGURE 2.10 Midpoint of a segment

MIDPOINT FORMULA

The coordinates of the midpoint $M = (x, y)$ on the line segment joining $P = (x_1, y_1)$ and $Q = (x_2, y_2)$ are given by

$$M = (x, y) = \left(\frac{x_1 + x_2}{2}, \frac{y_1 + y_2}{2}\right).$$

We ask you to prove the midpoint formula in Exercise 68.

SIDE NOTE

The coordinates of the midpoint of a line segment are found by taking the average of the x-coordinates and the average of the y-coordinates of the endpoints.

EXAMPLE 6 Finding the Midpoint of a Line Segment

Find the midpoint of the line segment joining the points $P(-3, 6)$ and $Q(1, 4)$.

Solution

Let $(x_1, y_1) = (-3, 6)$ and $(x_2, y_2) = (1, 4)$. Then

$$x_1 = -3, y_1 = 6, x_2 = 1, \text{ and } y_2 = 4.$$

$$\text{Midpoint} = \left(\frac{x_1 + x_2}{2}, \frac{y_1 + y_2}{2}\right) \quad \text{Midpoint formula}$$

$$= \left(\frac{-3 + 1}{2}, \frac{6 + 4}{2}\right) \quad \text{Substitute values for } x_1, x_2, y_1, y_2.$$

$$= (-1, 5) \quad \text{Simplify.}$$

The midpoint of the line segment joining the points $P(-3, 6)$ and $Q(1, 4)$ is $M(-1, 5)$.

Practice Problem 6 Find the midpoint of the line segment whose endpoints are $(5, -2)$ and $(6, -1)$.

Answers to Practice Problems

1.

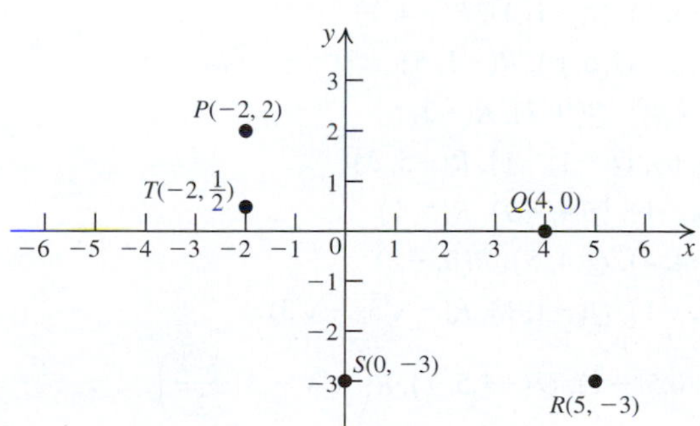

2. (2005, 17.9), (2006, 17.8), (2007, 17.8) (2008, 17.6), (2009, 17.6), (2010, 16.3), (2011, 16.0), and (2012, 18.0).

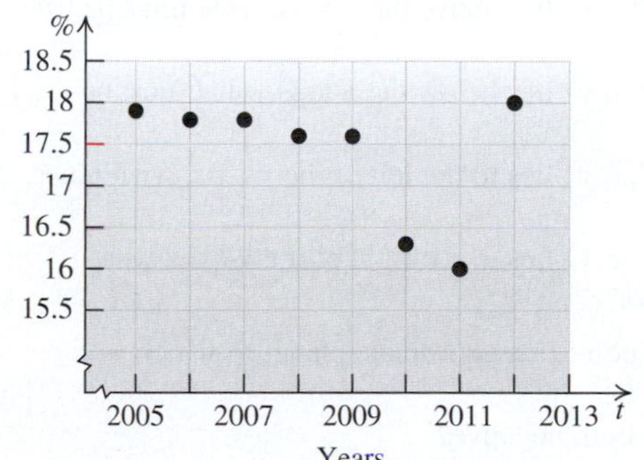

3. $\sqrt{2} \approx 1.4$ **4.** Yes **5.** $60\sqrt{2} \approx 84.85$ feet

6. $\left(\frac{11}{2}, -\frac{3}{2}\right)$

SECTION 2.1 Exercises

Basic Concepts and Skills

1. A point with a negative first coordinate and a positive second coordinate lies in the ______________ quadrant.

2. Any point on the x-axis has second coordinate ______________.

3. The distance between the points $P = (x_1, y_1)$ and $Q = (x_2, y_2)$ is given by the formula $d(P, Q) =$ ______________.

4. The coordinates of the midpoint $M = (x, y)$ of the line segment joining $P = (x_1, y_1)$ and $Q = (x_2, y_2)$ are given by $(x, y) =$ ______________.

5. True or False. For any points (x_1, y_1) and (x_2, y_2) $\sqrt{(x_1 - x_2)^2 + (y_1 - y_2)^2} = \sqrt{(x_2 - x_1)^2 + (y_2 - y_1)^2}$.

6. True or False. The point $(7, -4)$ is 4 units to the right and 2 units below the point $(3, 2)$.

7. Plot and label each of the given points in a Cartesian coordinate plane and state the quadrant, if any, in which each point is located. $(2, 2)$, $(3, -1)$, $(-1, 0)$, $(-2, -5)$, $(0, 0)$, $(-7, 4)$, $(0, 3)$, $(-4, 2)$

8. a. Write the coordinates of any five points on the x-axis. What do these points have in common?
b. Plot the points $(-2, 1)$, $(0, 1)$, $(0.5, 1)$, $(1, 1)$, and $(2, 1)$. Describe the set of all points of the form $(x, 1)$, where x is a real number.

9. a. If the x-coordinate of a point is 0, where does that point lie?
b. Plot the points $(-1, 1)$, $(-1, 1.5)$, $(-1, 2)$, $(-1, 3)$, and $(-1, 4)$. Describe the set of all points of the form $(-1, y)$, where y is a real number.

10. What figure is formed by the set of all points in a Cartesian coordinate plane that have
a. x-coordinate equal to -3?
b. y-coordinate equal to 4?

11. Let $P(x, y)$ be a point in a coordinate plane.
a. If the point $P(x, y)$ lies above the x-axis, what must be true of y?
b. If the point $P(x, y)$ lies below the x-axis, what must be true of y?
c. If the point $P(x, y)$ lies to the left of the y-axis, what must be true of x?
d. If the point $P(x, y)$ lies to the right of the y-axis, what must be true of x?

12. Let $P(x, y)$ be a point in a coordinate plane. In which quadrant does P lie
a. if x and y are both negative?
b. if x and y are both positive?
c. if x is positive and y is negative?
d. if x is negative and y is positive?

In Exercises 13–22, find (a) the distance between P and Q and (b) the coordinates of the midpoint of the line segment PQ.

13. $P(2, 1)$, $Q(2, 5)$
14. $P(3, 5)$, $Q(-2, 5)$
15. $P(-1, -5)$, $Q(2, -3)$
16. $P(-4, 1)$, $Q(-7, -9)$
17. $P(-1, 1.5)$, $Q(3, -6.5)$
18. $P(0.5, 0.5)$, $Q(1, -1)$
19. $P(\sqrt{2}, 4)$, $Q(\sqrt{2}, 5)$
20. $P(v - w, t)$, $Q(v + w, t)$
21. $P(t, k)$, $Q(k, t)$
22. $P(m, n)$, $Q(-n, -m)$

In Exercises 23–30, determine whether the given points are collinear. Points are *collinear* if they can be labeled P, Q, and R so that $d(P, Q) + d(Q, R) = d(P, R)$.

23. $(0, 0)$, $(1, 2)$, $(-1, -2)$
24. $(3, 4)$, $(0, 0)$, $(-3, -4)$
25. $(4, -2)$, $(-2, 8)$, $(1, 3)$
26. $(9, 6)$, $(0, -3)$, $(3, 1)$
27. $(-1, 4)$, $(3, 0)$, $(11, -8)$
28. $(-2, 3)$, $(3, 1)$, $(2, -1)$
29. $(4, -4)$, $(15, 1)$, $(1, 2)$
30. $(1, 7)$, $(-7, 8)$, $(-3, 7.5)$

31. Find the coordinates of the points that divide the line segment joining the points $P = (-4, 0)$ and $Q = (0, 8)$ into four equal parts.

32. Repeat Exercise 31 with $P = (-8, 4)$ and $Q = (16, -12)$.

In Exercises 33–40, identify the triangle PQR as *isosceles* (two sides of equal length), *equilateral* (three sides of equal length), or a *scalene* triangle (three sides of different lengths).

33. $P(-5, 5)$, $Q(-1, 4)$, $R(-4, 1)$
34. $P(3, 2)$, $Q(6, 6)$, $R(-1, 5)$
35. $P(-4, 8)$, $Q(0, 7)$, $R(-3, 5)$
36. $P(6, 6)$, $Q(-1, -1)$, $R(-5, 3)$
37. $P(0, -1)$, $Q(9, -9)$, $R(5, 1)$
38. $P(-4, 4)$, $Q(4, 5)$, $R(0, -2)$
39. $P(1, -1)$, $Q(-1, 1)$, $R(-\sqrt{3}, -\sqrt{3})$
40. $P(-0.5, -1)$, $Q(-1.5, 1)$, $R\left(\sqrt{3} - 1, \frac{\sqrt{3}}{2}\right)$

41. Show that the points $P(7, -12)$, $Q(-1, 3)$, $R(14, 11)$, and $S(22, -4)$ are the vertices of a square. Find the length of the diagonals.

42. Repeat Exercise 41 for the points $P(8, -10)$, $Q(9, -11)$, $R(8, -12)$, and $S(7, -11)$.

43. Find x such that the point $(x, 2)$ is 5 units from $(2, -1)$.

44. Find y such that the point $(2, y)$ is 13 units from $(-10, -3)$.

45. Find the point on the x-axis that is equidistant from the points $(-5, 2)$ and $(2, 3)$.

46. Find the point on the y-axis that is equidistant from the points $(7, -4)$ and $(8, 3)$.

Applying the Concepts

47. Population. The table shows the total population of the United States in millions. (205 represents 205,000,000.) The population is rounded to the nearest million. Plot the data in a Cartesian coordinate system.

Year	Total Population (millions)
1970	205
1975	216
1980	228
1985	238
1990	250
1995	267
2000	282
2005	296
2010	308
2012	314

Source: U.S. Census Bureau.

48. Use the midpoint formula for the years 2000 and 2012 to estimate the total population for 2006. What does this say about the line segment through the points (2000, 282) and (2012, 314)?

In Exercises 49–52, use the following vital statistics table.

	Rate per 1000 Population			
Year	**Births**	**Deaths**	**Marriages**	**Divorces**
1955	25.0	9.3	9.3	2.3
1960	23.7	9.5	8.5	2.2
1965	19.4	9.4	9.3	2.5
1970	18.4	9.5	10.6	3.5
1975	14.6	8.6	10.0	4.8
1980	15.9	8.8	10.6	5.2
1985	15.8	8.8	10.1	5.0
1990	16.7	8.6	9.4	4.7
1995	14.8	8.8	8.9	4.4
2000	14.4	8.7	8.5	4.2
2005	14.2	8.1	7.6	3.6
2008	14.0	8.1	7.1	3.5

Source: U.S. Census Bureau.

The table shows the rate per 1000 population of live births, deaths, marriages, and divorces from 1955 to 2008 at five-year intervals. Plot the data in a Cartesian coordinate system and connect the points with line segments.

49. Plot (year, births). **50.** Plot (year, deaths).

51. Plot (year, marriages). **52.** Plot (year, divorces).

In Exercises 53–55, assume that the graph through all of the data points is a line segment. Use the midpoint formula to find the estimate.

53. Murders in the United States. In the United States, 16,929 murders were committed in 2007 and 14,612 in 2011. Estimate the number of murders committed in 2009. (*Source:* www.disastercenter.com.)

54. Spending on prescription drugs. Americans spent \$228 billion on prescription drugs in 2007 and \$320 billion in 2011. Estimate the amount spent on prescription drugs during 2008, 2009, and 2010. (*Source:* www.cdc.gov)

55. Defense spending. The government of the United States spent \$548 billion on national defense in 2004 and \$925 billion in 2012. Estimate the amount spent on defense in each of the years 2005–2011. (*Source:* www.usgovernmentspending.com)

56. Length of a diagonal. The application of the Pythagorean theorem in three dimensions involves the relationship between the perpendicular edges of a rectangular block and the solid diagonal of the same block.

In the figure, show that $h^2 = a^2 + b^2 + c^2$.

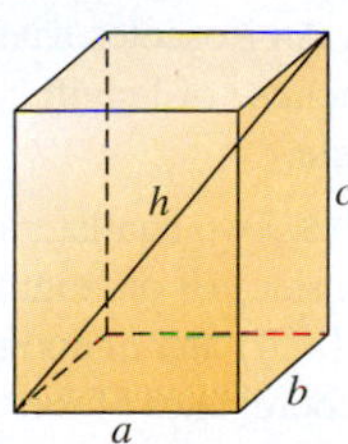

57. Distance. A pilot is flying from Dullsville to Middale to Pleasantville. With reference to an origin, Dullsville is located at $(2, 4)$, Middale at $(8, 12)$, and Pleasantville at $(20, 3)$, all numbers being in 100-mile units.

a. Locate the positions of the three cities on a Cartesian coordinate plane.

b. Compute the distance traveled by the pilot.

c. Compute the direct distance between Dullsville and Pleasantville.

58. Docking distance. A rope is attached to the bow of a sailboat that is 24 feet from the dock. The rope is drawn in over a pulley 10 feet higher than the bow at the rate of 3 feet per second. Find the distance from the boat to the dock after t seconds.

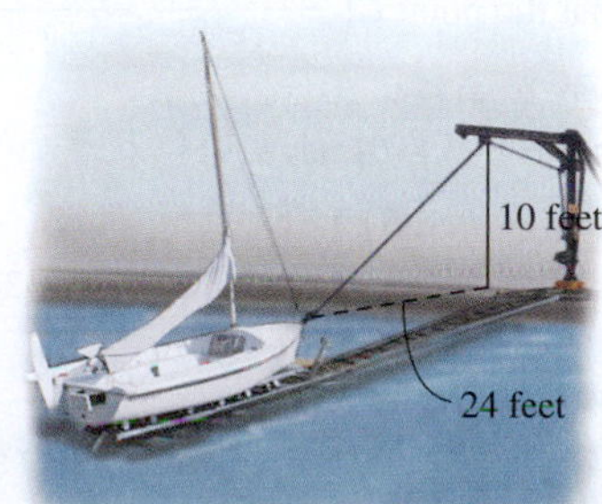

Beyond the Basics

59. Use coordinates to prove that the diagonals of a parallelogram bisect each other.
[*Hint:* Choose $(0, 0)$ $(a, 0)$, (b, c), and $(a + b, c)$ as the vertices of a parallelogram.]

60. Let $A(2, 3)$, $B(5, 4)$, and $C(3, 8)$ be three points in a coordinate plane. Find the coordinates of the point D such that the points A, B, C, and D form a parallelogram with
a. AB as one of the diagonals.
b. AC as one of the diagonals.
c. BC as one of the diagonals.
[*Hint:* Use Exercise 59.]

61. Prove that if the diagonals of a quadrilateral bisect each other, then the quadrilateral is a parallelogram. [*Hint:* Choose $(0, 0)$, $(a, 0)$, (b, c), and (x, y) as the vertices of a quadrilateral and show that $x = b - a$, $y = c$.]

62. a. Show that $(1, 2)$, $(-2, 6)$, $(5, 8)$, and $(8, 4)$ are the vertices of a parallelogram. [*Hint:* Use Exercise 61.]
b. Find (x, y) assuming that $(3, 2)$, $(6, 3)$, (x, y), and $(6, 5)$ are the vertices of a parallelogram.

63. Show that the sum of the squares of the lengths of the sides of a parallelogram is equal to the sum of the squares of the lengths of the diagonals. [*Hint:* Choose $(0, 0)$, $(a, 0)$, (b, c), and $(a + b, c)$ as the vertices of the parallelogram.]

64. Prove that in a right triangle, the midpoint of the hypotenuse is the same distance from each of the vertices.
[*Hint:* Let the vertices be $(0, 0)$, $(a, 0)$, and $(0, b)$.]

65. Isosceles triangle. An isosceles triangle ABC has right angle at C and hypotenuse of length c. Find the length of each leg of the triangle.

66. Equilateral triangles. Two equilateral triangles ABC and ABD have a common side AB of length $2a$. The side AB lies on the x-axis with $B = (a, 0)$ and midpoint at the origin O. (See the figure.) Find the coordinates of the vertices A, C, and D.

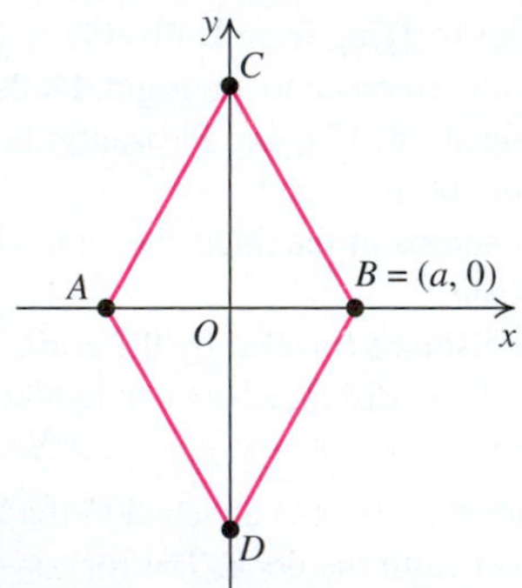

67. Trisecting a line segment. Let $A(x_1, y_1)$ and $B(x_2, y_2)$ be two points in a coordinate plane.
a. Show that the point $C\left(\frac{2x_1 + x_2}{3}, \frac{2y_1 + y_2}{3}\right)$ is one-third of the way from A to B.
[*Hint:* Show that $d(A, C) + d(C, B) = d(A, B)$ and that $d(A, C) = \frac{1}{3}d(A, B)$.]
b. Show that the point $D\left(\frac{x_1 + 2x_2}{3}, \frac{y_1 + 2y_2}{3}\right)$ is two-thirds of the way from A to B.
The points C and D in parts **a** *and* **b** *are called the points of trisection of segment AB.*
c. Find the points of trisection of the line segment joining $(-1, 2)$ and $(4, 1)$.

68. Midpoint formula. Use the notation in the accompanying figure to prove the midpoint formula

$$M(x, y) = \left(\frac{x_1 + x_2}{2}, \frac{y_1 + y_2}{2}\right).$$

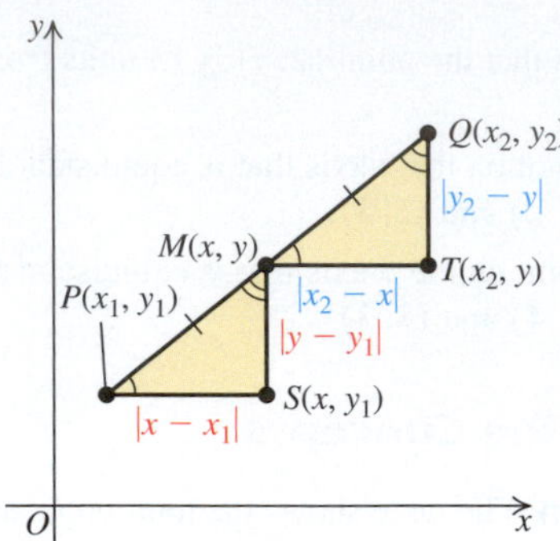

Critical Thinking/Discussion/Writing

In Exercises 69–73, describe the set of points $P(x, y)$ in the xy-plane that satisfies the given condition.

69. a. $x = 0$ **b.** $y = 0$
70. a. $xy = 0$ **b.** $xy \neq 0$
71. a. $xy > 0$ **b.** $xy < 0$
72. a. $x^2 + y^2 = 0$ **b.** $x^2 + y^2 \neq 0$
73. a. $x \geq 0$ **b.** $y \geq 0$

74. Describe how to determine the quadrant in which a point lies from the signs of its coordinates.

Maintaining Skills

In Exercises 75–78, evaluate each expression for the given values of x and y.

75. $x^2 + y^2$
a. $x = \frac{1}{2}, y = \frac{1}{2}$ **b.** $x = \frac{\sqrt{2}}{2}, y = \frac{\sqrt{2}}{2}$

76. $(x - 1)^2 + (y + 2)^2$
a. $x = -1, y = 1$ **b.** $x = 4, y = 2$

77. $\frac{x}{|x|} + \frac{|y|}{y}$
a. $x = 2, y = -3$ **b.** $x = -4, y = 3$

78. $\frac{|x|}{x} + \frac{|y|}{y}$
a. $x = -1, y = -2$ **b.** $x = 3, y = 2$

In Exercises 79–84, find the appropriate term that should be added so that the binomial becomes a perfect square trinomial.

79. $x^2 - 6x$ **80.** $x^2 - 8x$
81. $y^2 + 3y$ **82.** $y^2 + 5y$
83. $x^2 - ax$ **84.** $x^2 + xy$

SECTION **2.2**

Graphs of Equations

BEFORE STARTING THIS SECTION, REVIEW

1. How to plot points (Section 2.1, page 156)
2. Equivalent equations (Section 1.1, page 78)
3. Completing squares (Section 1.2, page 94)

OBJECTIVES

1. Sketch a graph by plotting points.
2. Find the intercepts of a graph.
3. Find the symmetries in a graph.
4. Find the equation of a circle.

◆ Doomsans Discover Deer

A herd of 400 deer was introduced onto a small island called Dooms. The natives both liked and admired these beautiful creatures. However, the Doomsans soon discovered that deer meat is excellent food. Also, a rumor spread throughout Dooms that eating deer meat prolonged one's life. The natives then began to hunt the deer. The number of deer, y, after t years from the initial introduction of deer into Dooms is described by the equation $y = -t^4 + 96t^2 + 400$. When does the population of deer become extinct in Dooms? (See Example 5.)

1 Sketch a graph by plotting points.

Graph of an Equation

We say that a *relation* exists between the two quantities when one quantity changes with respect to the other quantity. The two changing (or varying) quantities in a relation are often represented by *variables.* A relation may sometimes be described by an equation, which may be a formula. Following are examples of equations showing relationships between two variables.

$$y = 2x + 1, \quad x^2 + y^2 = 4, \quad y = x^2, \quad x = y^2, \quad F = \frac{9}{5}C + 32, \quad q = -3p^2 + 30$$

An ordered pair (a, b) is said to **satisfy** an equation with variables x and y if, when a is substituted for x and b is substituted for y in the equation, the resulting statement is true. For example, the ordered pair $(2, 5)$ satisfies the equation $y = 2x + 1$ because replacing x with 2 and y with 5 yields $5 = 2(2) + 1$, a true statement. The ordered pair $(5, -2)$ does not satisfy the equation $y = 2x + 1$ because replacing x with 5 and y with -2 yields $-2 = 2(5) + 1$, which is false. An ordered pair that satisfies an equation is called a **solution** of the equation.

In an equation involving x and y, if the value of y can be found given the value of x, then we say that y is the **dependent variable** and x is the **independent variable**. In the equation $y = 2x + 1$, there are infinitely many choices for x, each resulting in a corresponding value of y. Hence, we have infinitely many solutions of the equation $y = 2x + 1$. When these solutions are plotted as points in the coordinate plane, they constitute the *graph of the equation.* The graph of an equation is thus a geometric picture of its solution set.

GRAPH OF AN EQUATION

The **graph of an equation** in two variables, such as x and y, is the set of all ordered pairs (a, b) in the coordinate plane that satisfy the equation.

TECHNOLOGY CONNECTION

Calculator graph of $y = x^2 - 3$

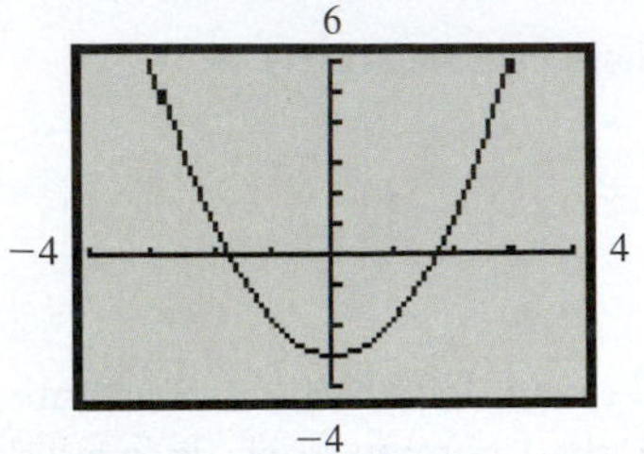

EXAMPLE 1 **Sketching a Graph by Plotting Points**

Sketch the graph of $y = x^2 - 3$.

Solution

There are infinitely many solutions of the equation $y = x^2 - 3$. To find a few, we choose integer values of x between -3 and 3. Then we find the corresponding values of y as shown in Table 2.2.

TABLE 2.2

x	$y = x^2 - 3$	(x, y)
-3	$y = (-3)^2 - 3 = 9 - 3 = 6$	$(-3, 6)$
-2	$y = (-2)^2 - 3 = 4 - 3 = 1$	$(-2, 1)$
-1	$y = (-1)^2 - 3 = 1 - 3 = -2$	$(-1, -2)$
0	$y = 0^2 - 3 = 0 - 3 = -3$	$(0, -3)$
1	$y = 1^2 - 3 = 1 - 3 = -2$	$(1, -2)$
2	$y = 2^2 - 3 = 4 - 3 = 1$	$(2, 1)$
3	$y = 3^2 - 3 = 9 - 3 = 6$	$(3, 6)$

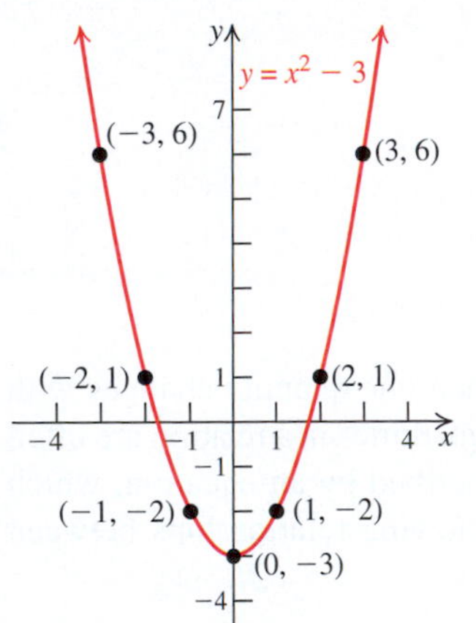

FIGURE 2.11 Graph of a parabola

We plot the seven solutions (x, y) and join them with a smooth curve as shown in Figure 2.11.

Practice Problem 1 Sketch the graph of $y = -x^2 + 1$.

The bowl-shaped curve sketched in Figure 2.11 is called a *parabola.* It is easy to find parabolas in everyday settings. For example, when you throw a ball, the path it travels is a parabola. Also, the reflector behind a car's headlight is parabolic in shape.

Example 1 suggests the following three steps for sketching the graph of an equation by plotting points.

SKETCHING A GRAPH BY PLOTTING POINTS

Step 1 Make a representative table of solutions of the equation.

Step 2 Plot the solutions as ordered pairs in the coordinate plane.

Step 3 Connect the solutions in Step 2 with a smooth curve.

Comment This point-plotting technique has obvious pitfalls. For instance, many different curves pass through the four points. See Figure 2.12. Assume that these points are solutions of a given equation. There is no way to guarantee that any curve we pass through the plotted points is the actual graph of the equation. However, in general, more plotted solutions, result in a more accurate graph of the equation.

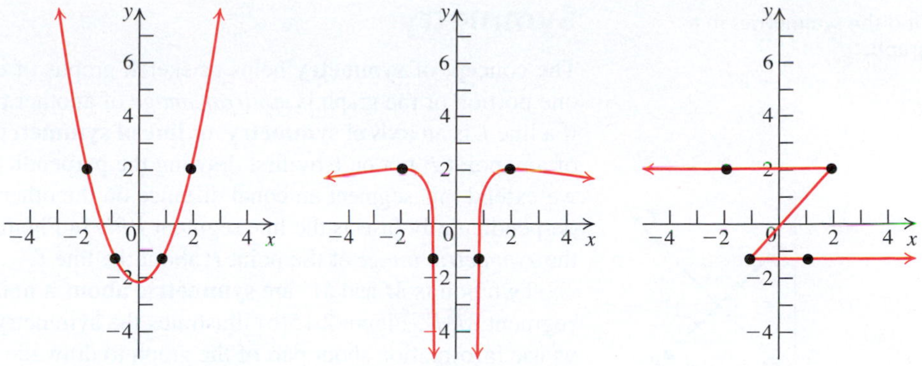

FIGURE 2.12 Several graphs through four points

We now discuss some special features of the graph of an equation.

2 Find the intercepts of a graph.

Intercepts

We examine the points where a graph intersects (crosses or touches) the coordinate axes. Because all points on the x-axis have a y-coordinate of 0, any point where a graph intersects the x-axis has the form $(a, 0)$. See Figure 2.13. The number a is called an ***x*-intercept** of the graph. Similarly, any point where a graph intersects the y-axis has the form $(0, b)$, and the number b is called a ***y*-intercept** of the graph.

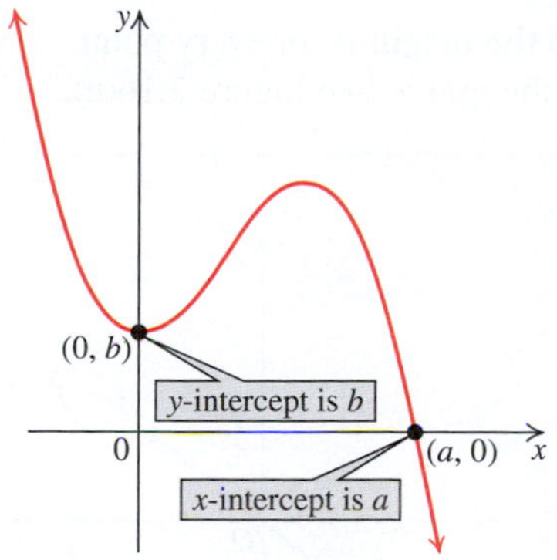

FIGURE 2.13 Intercepts of a graph

FINDING THE INTERCEPTS OF THE GRAPH OF AN EQUATION

Step 1 To find the x-intercepts of the graph of an equation, set $y = 0$ in the equation and solve for x.

Step 2 To find the y-intercepts of the graph of an equation, set $x = 0$ in the equation and solve for y.

Note that only real number solutions in Steps 1 and 2 correspond to the intercepts.

SIDE NOTE

Do not try to calculate the x-intercept by setting $x = 0$. The term *x-intercept* denotes the x-coordinate of the point where the graph touches or crosses the x-axis; so the y-coordinate must be 0.

EXAMPLE 2 Finding Intercepts

Find the x- and y-intercepts of the graph of the equation $y = x^2 - x - 2$.

Solution

Step 1 Set $y = 0$ in the equation and solve for x.

$$y = x^2 - x - 2 \quad \text{Original equation}$$
$$0 = x^2 - x - 2 \quad \text{Set } y = 0.$$
$$0 = (x + 1)(x - 2) \quad \text{Factor.}$$
$$x + 1 = 0 \quad \text{or} \quad x - 2 = 0 \quad \text{Zero-product property}$$
$$x = -1 \quad \text{or} \quad x = 2 \quad \text{Solve each equation for } x.$$

The x-intercepts are -1 and 2.

Step 2 Set $x = 0$ in the equation and solve for y.

$$y = x^2 - x - 2 \quad \text{Original equation}$$
$$y = 0^2 - 0 - 2 \quad \text{Set } x = 0.$$
$$y = -2 \quad \text{Solve for } y.$$

The y-intercept is -2.

The graph of the equation $y = x^2 - x - 2$ is shown in Figure 2.14.

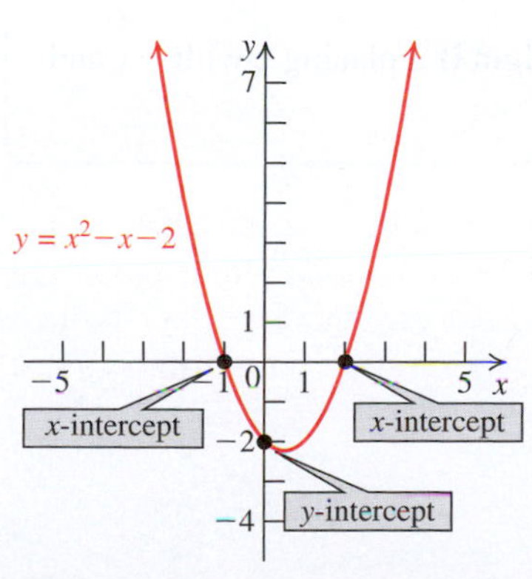

FIGURE 2.14 Intercepts of a graph

Practice Problem 2 Find the intercepts of the graph of $y = 2x^2 + 3x - 2$.

3 Find the symmetries in a graph.

Symmetry

The concept of **symmetry** helps us sketch graphs of equations. A graph has symmetry if one portion of the graph is a *mirror image* of another portion. As shown in Figure 2.15(a), if a line ℓ is an **axis of symmetry**, or **line of symmetry**, we can construct the mirror image of any point P not on ℓ by first drawing the perpendicular line segment from P to ℓ. Then we extend this segment an equal distance on the other side to a point P' so that the line ℓ perpendicularly bisects the line segment $\overline{PP'}$. In Figure 2.15(a), we say that the point P' is the *symmetric image* of the point P about the line ℓ.

Two points M and M' are **symmetric about a point** Q if Q is the midpoint of the line segment $\overline{MM'}$. Figure 2.15(b) illustrates the symmetry about the origin O. Symmetry lets us use information about part of the graph to draw the remainder of the graph.

The following three types of symmetries are frequently used.

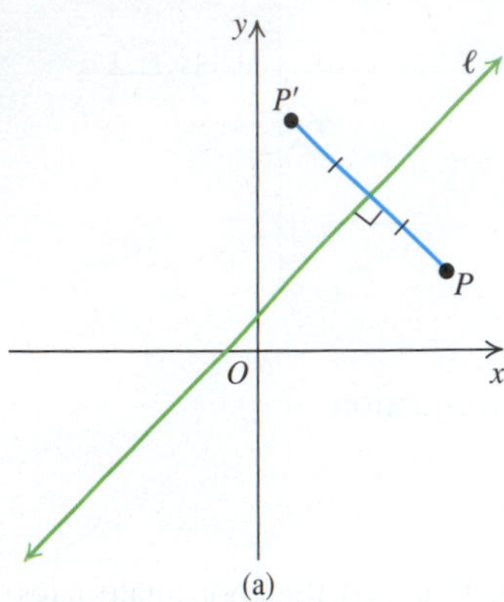

(a) Symmetric points about a line

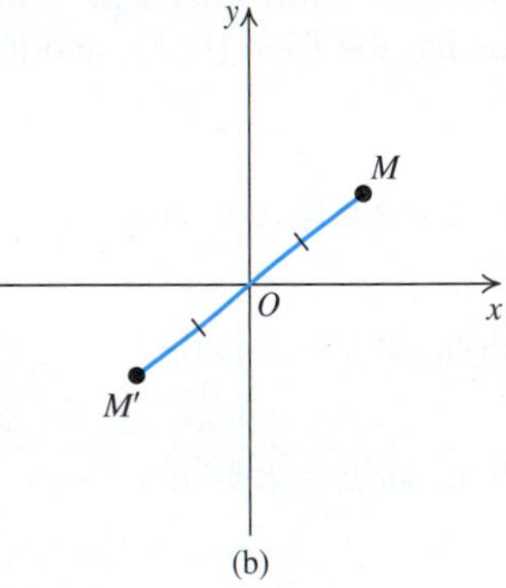

(b) Symmetric points about the origin

FIGURE 2.15

Symmetries

1. A graph is **symmetric with respect to (or about) the y-axis** if for every point (x, y) on the graph, the point $(-x, y)$ is also on the graph. See Figure 2.16(a).
2. A graph is **symmetric with respect to (or about) the x-axis** if for every point (x, y) on the graph, the point $(x, -y)$ is also on the graph. See Figure 2.16(b).
3. A graph is **symmetric with respect to (or about) the origin** if for every point (x, y) on the graph, the point $(-x, -y)$ is also on the graph. See Figure 2.16(c).

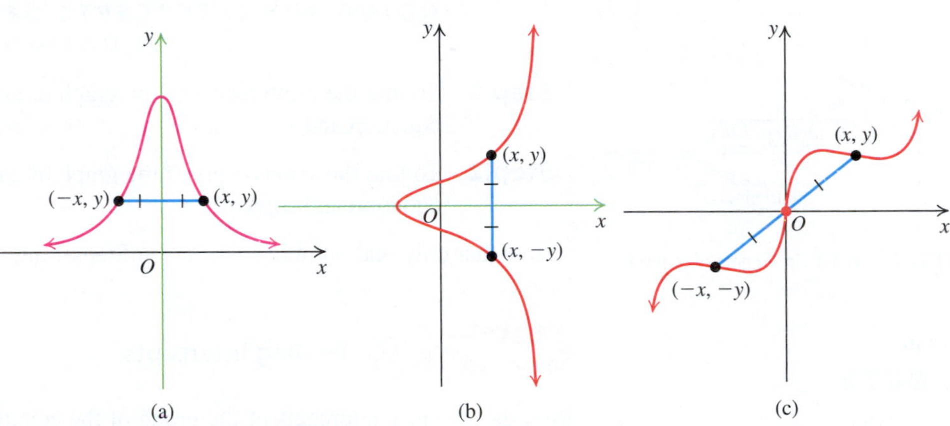

FIGURE 2.16 Three types of symmetry

TESTS FOR SYMMETRY

1. The graph of an equation is symmetric about the y-axis if replacing x with $-x$ results in an equivalent equation.
2. The graph of an equation is symmetric about the x-axis if replacing y with $-y$ results in an equivalent equation.
3. The graph of an equation is symmetric about the origin if replacing x with $-x$ and y with $-y$ results in an equivalent equation.

TECHNOLOGY CONNECTION

The calculator graph of $y = 1/(x^2 + 5)$ reinforces the result of Example 3.

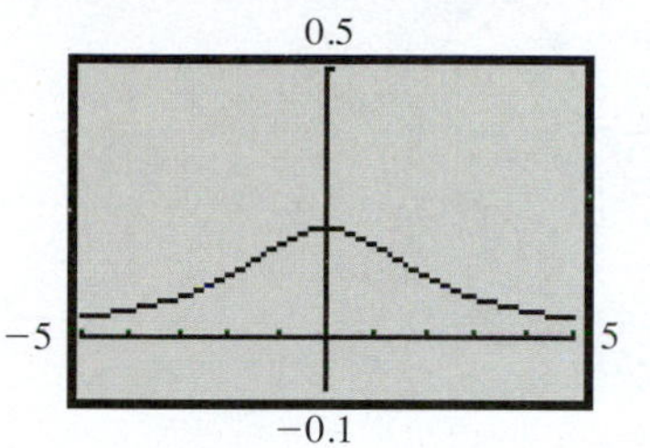

SIDE NOTE

Note that if *only* even powers of x appear in an equation, then the graph is symmetric with respect to the y-axis because for any integer n, $(-x)^{2n} = x^{2n}$.

EXAMPLE 3 Checking for Symmetry

Determine whether the graph of the equation $y = \dfrac{1}{x^2 + 5}$ is symmetric about the y-axis.

Solution

Replace x with $-x$ to see if $(-x, y)$ also satisfies the equation.

$$y = \frac{1}{x^2 + 5} \qquad \text{Original equation}$$

$$y = \frac{1}{(-x)^2 + 5} \qquad \text{Replace } x \text{ with } -x.$$

$$y = \frac{1}{x^2 + 5} \qquad \text{Simplify: } (-x)^2 = x^2.$$

Because replacing x with $-x$ gives us the original equation, the graph of $y = \dfrac{1}{x^2 + 5}$ is symmetric with respect to the y-axis.

Practice Problem 3 Check whether the graph of $x^2 - y^2 = 1$ is symmetric about the y-axis.

PROCEDURE IN ACTION

EXAMPLE 4 Sketching a Graph Using Symmetry

OBJECTIVE

Use symmetry to sketch the graph of an equation.

Step 1 Test for all three symmetries.
About the x-axis: Replace y with $-y$.
About the y-axis: Replace x with $-x$.
About the origin: Replace x with $-x$ and y with $-y$.

Step 2 Make a table of values using any symmetries found in Step 1.

Step 3 Plot the points from the table and draw a smooth curve through them.

EXAMPLE

Use symmetry to sketch the graph of $y = 4x - x^3$.

1.

x-axis	y-axis	origin
Replace y with $-y$	Replace x with $-x$	Replace x with $-x$ and y with $-y$
$-y = 4x - x^3$	$y = 4(-x) - (-x)^3$	$-y = 4(-x) - (-x)^3$
$y = -4x + x^3$	$y = -4x + x^3$	$-y = -4x + x^3$
		$y = 4x - x^3$
No	No	Yes

2. Origin symmetry: If (x, y) is on the graph, so is $(-x, -y)$. Use only positive x values in the table.

x	0	0.5	1	1.5	2	2.5
$y = 4x - x^3$	0	1.875	3	2.625	0	−5.625

3.

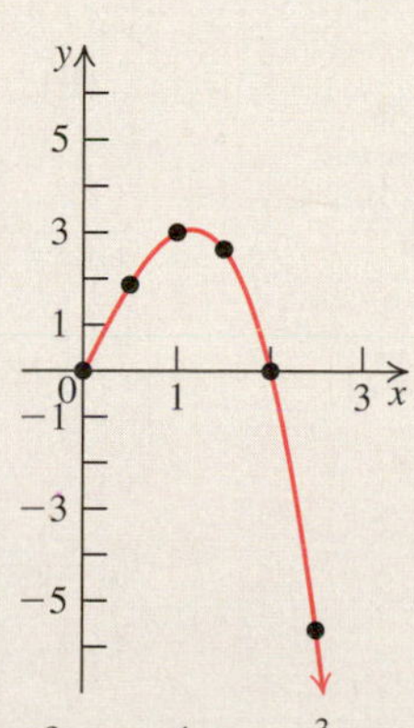

Graph of $y = 4x - x^3, x \geq 0$

(continued)

Step 4 Extend the portion of the graph found in Step 3, using symmetries.

4.

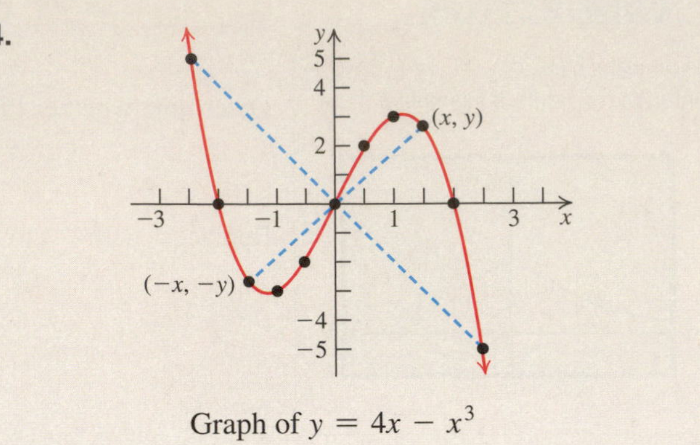

Graph of $y = 4x - x^3$

Practice Problem 4 Check whether the graph of $x^2 = y^3$ is symmetric about the x-axis, the y-axis, and the origin.

EXAMPLE 5 Sketching a Graph by Using Intercepts and Symmetry

Initially, 400 deer are on Dooms island. The number y of deer on the island after t years is described by the equation

$$y = -t^4 + 96t^2 + 400.$$

a. Sketch the graph of the equation $y = -t^4 + 96t^2 + 400$.

b. Adjust the graph in part **a** to account for only the physical aspects of the problem.

c. When do deer become extinct on Dooms?

Solution

a. (i) We find all intercepts. If we set $t = 0$ in the equation $y = -t^4 + 96t^2 + 400$, we obtain $y = 400$. Thus, the y-intercept is 400.

To find the t-intercepts, we set $y = 0$ in the given equation.

$$y = -t^4 + 96t^2 + 400 \quad \text{Original equation}$$
$$0 = -t^4 + 96t^2 + 400 \quad \text{Set } y = 0.$$
$$t^4 - 96t^2 - 400 = 0 \quad \text{Multiply both sides by } -1 \text{ and interchange sides.}$$
$$(t^2 + 4)(t^2 - 100) = 0 \quad \text{Factor.}$$
$$(t^2 + 4)(t + 10)(t - 10) = 0 \quad \text{Factor } t^2 - 100.$$
$$t^2 + 4 = 0 \quad \text{or} \quad t + 10 = 0 \quad \text{or} \quad t - 10 = 0 \quad \text{Zero-product property}$$
$$t = -10 \quad \text{or} \quad t = 10 \quad \text{Solve for } t\text{; there is no real solution of } t^2 + 4 = 0.$$

The t-intercepts are -10 and 10.

(ii) We check for symmetry. Note that t replaces x as the independent variable.

Symmetry about the t-axis:
Replacing y with $-y$ gives $-y = -t^4 + 96t^2 + 400$. The pair $(0, 400)$ is on the graph of $y = -t^4 + 96t^2 + 400$ but not of $-y = -t^4 + 96t^2 + 400$. Consequently, the graph is not symmetric about the t-axis.

Symmetry about the y-axis:
Replacing t with $-t$ in the equation $y = -t^4 + 96t^2 + 400$, we obtain $y = -(-t)^4 + 96(-t)^2 + 400 = -t^4 + 96t^2 + 400$, which is the original equation. Thus, $(-t, y)$ also satisfies the equation and the graph is symmetric about the y-axis.

Symmetry about the origin:
Replacing t with $-t$ and y with $-y$ in the equation $y = -t^4 + 96t^2 + 400$, we obtain $-y = -(-t)^4 + 96(-t)^2 + 400$, or $-y = -t^4 + 96t^2 + 400$. As we saw when we discussed symmetry in the t-axis, $(0, 400)$ is a solution but $(0, -400)$ is not a solution of this equation; so the graph is not symmetric with respect to the origin.

(iii) We sketch the graph by plotting points for $t \geq 0$ (see Table 2.3) and then using symmetry about the y-axis (see Figure 2.17).

TABLE 2.3

t	$y = -t^4 + 96t^2 + 400$	(t, y)
0	400	(0, 400)
1	495	(1, 495)
5	2175	(5, 2175)
7	2703	(7, 2703)
9	1615	(9, 1615)
10	0	(10, 0)
11	−2625	(11, −2625)

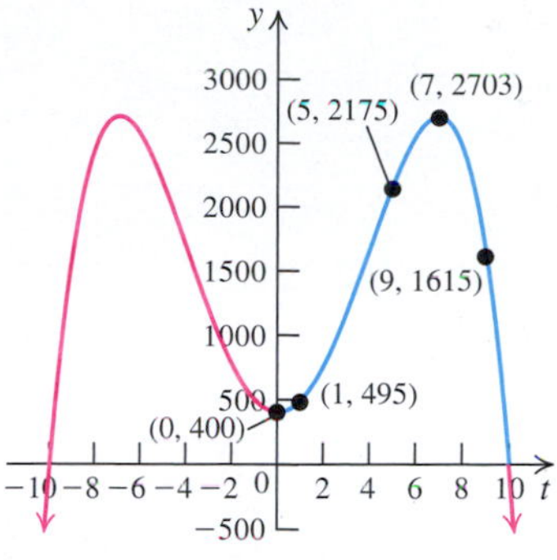

FIGURE 2.17

b. The graph pertaining to the physical aspects of the problem is the blue portion of the graph in Figure 2.17.

c. The positive t-intercept, 10, gives the time in years when the deer population of Dooms is 0; so deer are extinct after 10 years.

Practice Problem 5 Repeat Example 5 assuming that the initial deer population is 324 and the number of deer on the island after t years is given by the equation $y = -t^4 + 77t^2 + 324$.

4 Find the equation of a circle.

Circles

Sometimes a curve that is described geometrically can also be described by an algebraic equation. We illustrate this situation in the case of a circle.

Circle

A **circle** is a set of points in a Cartesian coordinate plane that are at a fixed distance r from a specified point (h, k). The fixed distance r is called the **radius** of the circle, and the specified point (h, k) is called the **center** of the circle.

Standard Form

A point $P(x, y)$ is on the circle if and only if its distance from the center $C(h, k)$ is r. Using the notation for the distance between the points P and C, we have

$$d(P, C) = r$$

$$\sqrt{(x - h)^2 + (y - k)^2} = r \quad \text{Distance formula}$$

$$(x - h)^2 + (y - k)^2 = r^2 \quad \text{Square both sides.}$$

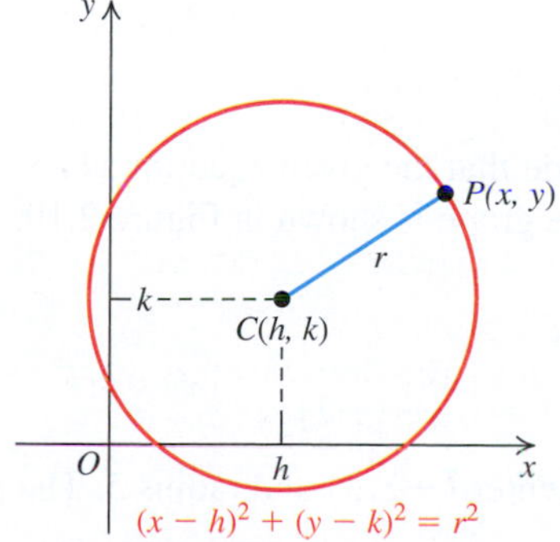

FIGURE 2.18

The equation $(x - h)^2 + (y - k)^2 = r^2$ is an equation of a circle with radius r and center (h, k). A point (x, y) is on the circle of radius r and center $C(h, k)$ if and only if it satisfies this equation. Figure 2.18 is the graph of a circle with center $C(h, k)$ and radius r.

The Standard Form for the Equation of a Circle

The equation of a circle with center (h, k) and radius r is

(1) $$(x - h)^2 + (y - k)^2 = r^2$$

and is called the **standard form** of an equation of a circle.

EXAMPLE 6 Finding the Equation of a Circle

Find the standard form of the equation of the circle with center $(7, -3)$ and that passes through the point $P = (5, -2)$.

Solution

$$(x - h)^2 + (y - k)^2 = r^2 \quad \text{Standard form}$$
$$(x - 7)^2 + (y - (-3))^2 = r^2 \quad \text{Replace } h \text{ with 7 and } k \text{ with } -3.$$
(2) $$(x - 7)^2 + (y + 3)^2 = r^2 \quad -(-3) = 3$$

Because the point $P = (5, -2)$ lies on the circle, its coordinates satisfy equation (2). So

$$(5 - 7)^2 + (-2 + 3)^2 = r^2 \quad \text{Replace } x \text{ with 5 and } y \text{ with } -2.$$
$$5 = r^2 \quad \text{Simplify.}$$

Replacing r^2 with 5 in equation (2) gives the required standard form

$$(x - 7)^2 + (y + 3)^2 = 5.$$

Practice Problem 6 Find the standard form of the equation of the circle with center $(3, -6)$ and radius 10.

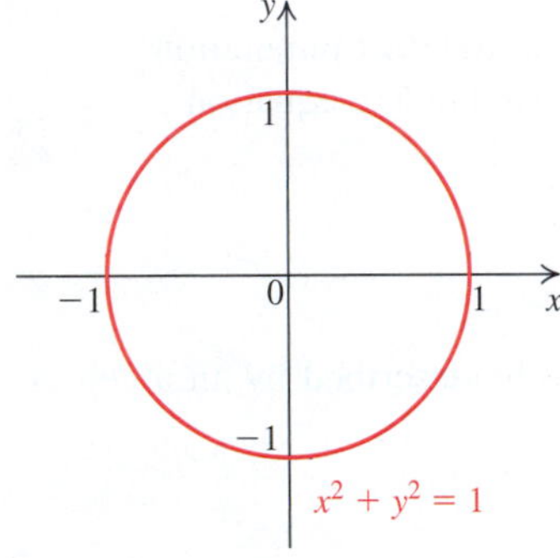

FIGURE 2.19 The unit circle

If an equation in two variables can be written in standard form (1), then its graph is a circle with center (h, k) and radius r.

EXAMPLE 7 Graphing a Circle

Specify the center and radius and graph each circle.

a. $x^2 + y^2 = 1$

b. $(x + 2)^2 + (y - 3)^2 = 25$

Solution

a. The equation $x^2 + y^2 = 1$ can be rewritten as

$$(x - 0)^2 + (y - 0)^2 = 1^2.$$

Comparing this equation with equation (1), we conclude that the given equation is an equation of a circle with center (0, 0) and radius 1. The graph is shown in Figure 2.19. This circle is called the **unit circle**.

b. Rewriting the equation $(x + 2)^2 + (y - 3)^2 = 25$ as

$$[x - (-2)]^2 + (y - 3)^2 = 5^2, \quad x + 2 = x - (-2)$$

we see that the graph of this equation is a circle with center $(-2, 3)$ and radius 5. The graph is shown in Figure 2.20.

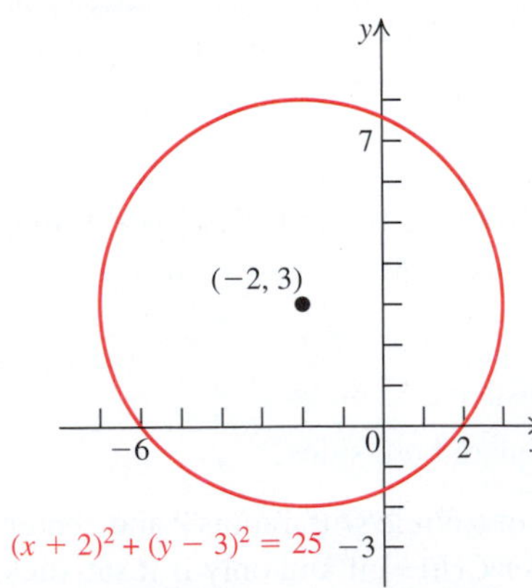

FIGURE 2.20 Circle with radius 5 and center at $(-2, 3)$

Practice Problem 7 Graph the equation $(x - 2)^2 + (y + 1)^2 = 36$.

Semicircles

Letting $h = 0$ and $k = 0$ in equation (1), we have

(3) $$x^2 + y^2 = r^2$$

Equation (3) is the standard form for a circle with center at the origin and radius r.

We solve equation (3) for y:

$$y^2 = r^2 - x^2 \quad \text{Subtract } x^2 \text{ from both sides.}$$
$$y = \pm\sqrt{r^2 - x^2} \quad \text{Square root property (page 58).}$$

or $$y = \sqrt{r^2 - x^2} \quad \text{and} \quad y = -\sqrt{r^2 - x^2}.$$

Similarly, solving equation (3) for x, we have two equations,

$$x = \sqrt{r^2 - y^2} \quad \text{and} \quad x = -\sqrt{r^2 - y^2}.$$

The graphs of these four equations are **semicircles** (half circles), shown in Figure 2.21.

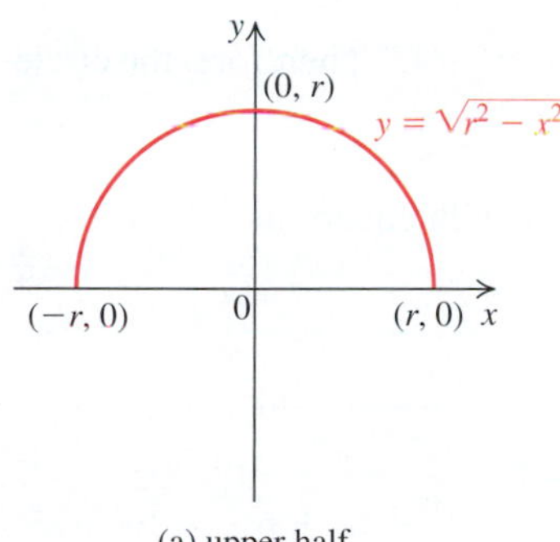

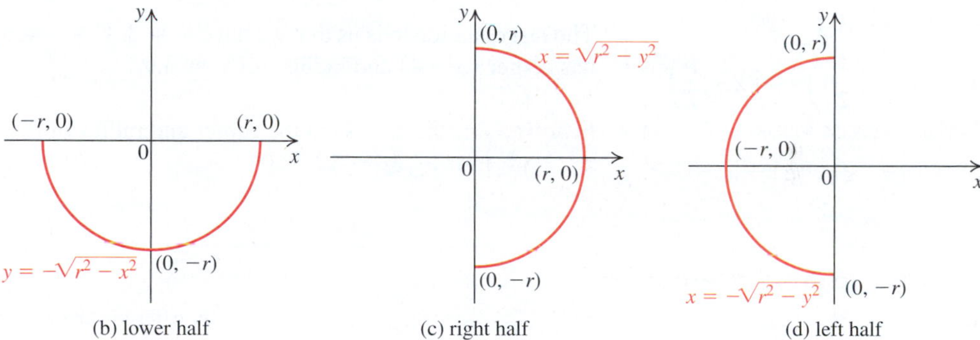

(a) upper half (b) lower half (c) right half (d) left half

FIGURE 2.21 Semicircles

TECHNOLOGY CONNECTION

To graph the equation $x^2 + y^2 = 1$ on a graphing calculator, we graph the two equations of the upper and lower semicircles:

$$Y_1 = \sqrt{1 - x^2} \quad \text{and}$$
$$Y_2 = -\sqrt{1 - x^2}$$

in the same window. The calculator graph does not look quite like a circle. We use the ZSquare option to make the display look like a circle.

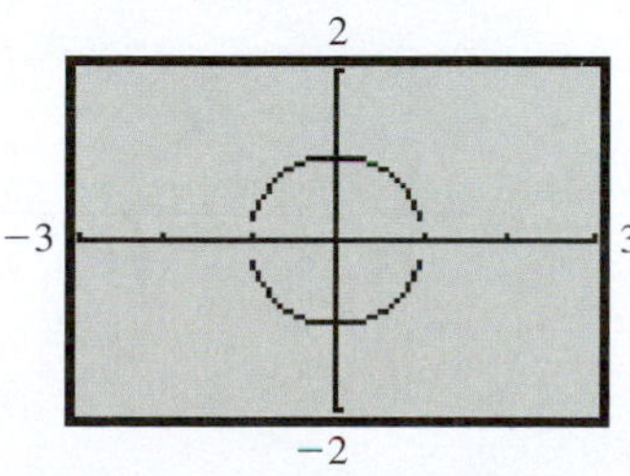

General Form

If we expand the squared expressions in the standard equation of a circle,

$$(x - h)^2 + (y - k)^2 = r^2,$$

and then simplify, we obtain an equation of the form

(4) $$x^2 + y^2 + ax + by + c = 0.$$

Equation (4) is called the *general form* of the equation of a circle. An equation $Ax^2 + By^2 + Cx + Dy + E = 0$ with $A \neq 0$ and $A = B$ can be converted to the general form of the equation of a circle by dividing both sides by A. See Exercise 83.

General Form of the Equation of a Circle

The **general form** of the equation of a circle is

$$x^2 + y^2 + ax + by + c = 0.$$

On the other hand, if we are given an equation in general form, we can convert it to standard form by completing the squares on the x- and y-terms. This gives

(5) $$(x - h)^2 + (y - k)^2 = d.$$

If $d > 0$, the graph of equation (5) is a circle with center (h, k) and radius $\sqrt{d}$. If $d = 0$, the graph of equation (5) is the point (h, k). If $d < 0$, there is no graph.

EXAMPLE 8 Converting the General Form to Standard Form

Find the center and radius of the circle with equation

$$x^2 + y^2 - 6x + 8y + 10 = 0.$$

Solution

Complete the squares on both the x-terms and y-terms to get standard form.

$$x^2 + y^2 - 6x + 8y + 10 = 0 \quad \text{Original equation}$$

$$(x^2 - 6x) + (y^2 + 8y) = -10 \quad \text{Group the } x\text{-terms and } y\text{-terms.}$$

$$(x^2 - 6x + 9) + (y^2 + 8y + 16) = -10 + 9 + 16 \quad \text{Complete the squares by adding 9 and 16 to both sides.}$$

$$(x - 3)^2 + (y + 4)^2 = 15 \quad \text{Factor and simplify.}$$

$$(x - 3)^2 + [y - (-4)]^2 = (\sqrt{15})^2$$

RECALL

To convert $x^2 + bx$ to a perfect square, add $\left(\frac{1}{2}b\right)^2$ to get

$$x^2 + bx + \left(\frac{1}{2}b\right)^2 = \left(x + \frac{b}{2}\right)^2.$$

This completes the square.

The last equation tells us that we have $h = 3$, $k = -4$, and $r = \sqrt{15}$. Therefore, the circle has center $(3, -4)$ and radius $\sqrt{15} \approx 3.9$.

Practice Problem 8 Find the center and radius of the circle with equation $x^2 + y^2 + 4x - 6y - 12 = 0$.

Answers to Practice Problems

1.

2. x-intercepts: $-2, \frac{1}{2}$; y-intercept: -2 **3.** Symmetric about the y-axis

4. Not symmetric about the x-axis, symmetric about the y-axis, not symmetric about the origin

5. a.

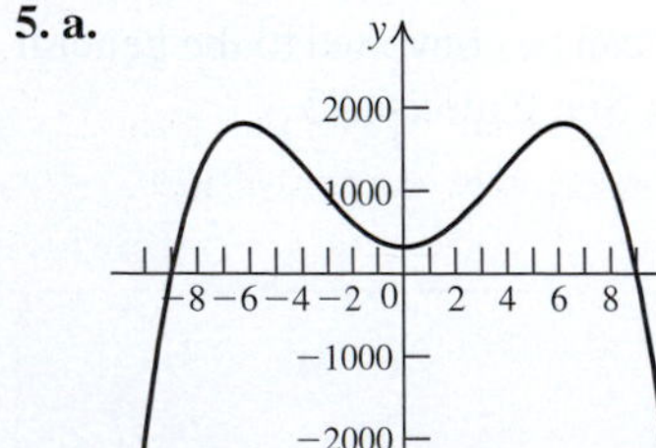

b.

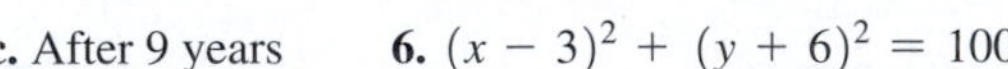

c. After 9 years **6.** $(x - 3)^2 + (y + 6)^2 = 100$

7.

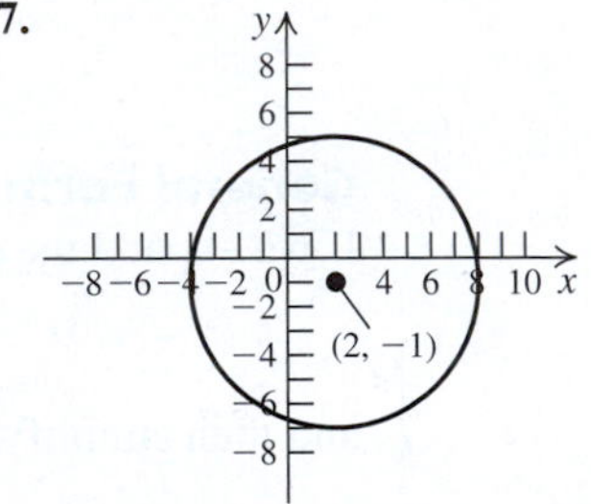

8. $(x + 2)^2 + (y - 3)^2 = 25$: $C = (-2, 3)$, $r = 5$

SECTION 2.2 Exercises

Basic Concepts and Skills

1. The graph of an equation in two variables such as x and y is the set of all ordered pairs (a, b) ______________.
2. If $(-2, 4)$ is a point on a graph that is symmetric with respect to the y-axis, then the point ______________ is also on the graph.
3. If $(0, -5)$ is a point of a graph, then -5 is a(n) ______________ intercept of the graph.
4. An equation in standard form of a circle with center $(1, 0)$ and radius 2 is ______________.
5. **True or False.** The graph of the equation $3x^2 - 2x + y + 3 = 0$ is a circle.
6. **True or False.** If a graph is symmetric about the x-axis, then it must have at least one x-intercept.

In Exercises 7–12, determine whether the given points are on the graph of the equation.

	Equation	Points
7.	$y = x - 1$	$(-3, -4), (1, 0), (4, 3), (2, 3)$
8.	$2y = 3x + 5$	$(-1, 1), (0, 2), \left(-\frac{5}{3}, 0\right), (1, 4)$
9.	$y = \sqrt{x + 1}$	$(3, 2), (0, 1), (8, -3), (8, 3)$
10.	$y = \frac{1}{x}$	$\left(-3, \frac{1}{3}\right), (1, 1), (0, 0), \left(2, \frac{1}{2}\right)$
11.	$x^2 - y^2 = 1$	$(1, 0), (0, -1), (2, \sqrt{3}), (2, -\sqrt{3})$
12.	$y^2 = x$	$(1, -1), (1, 1)\ (0, 0), (2, -\sqrt{2})$

In Exercises 13–24, find

a. the intercepts.

b. symmetries, about the x-axis, the y-axis, and the origin.

13.

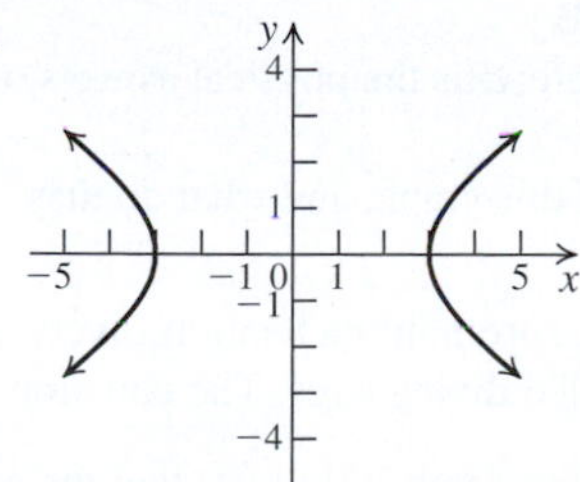

14.

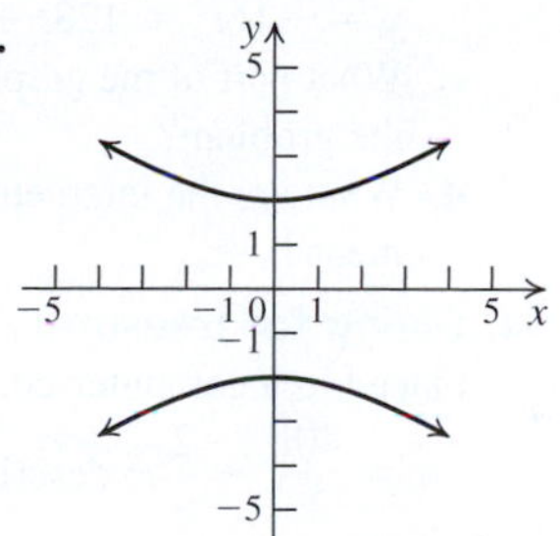

15.

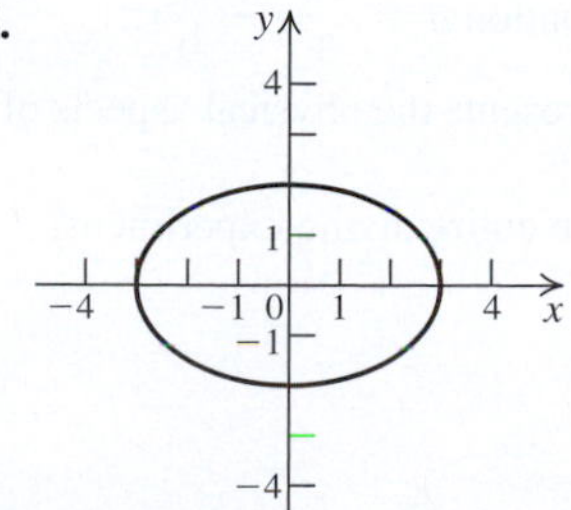

16.

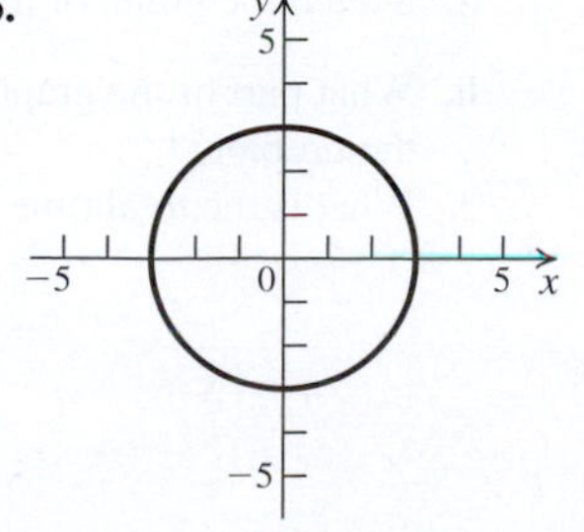

17.

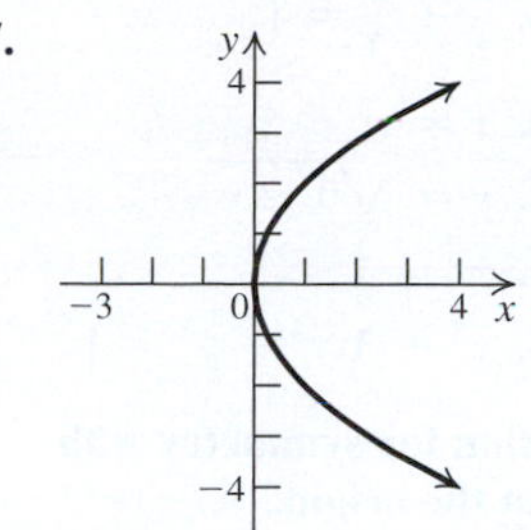

18.

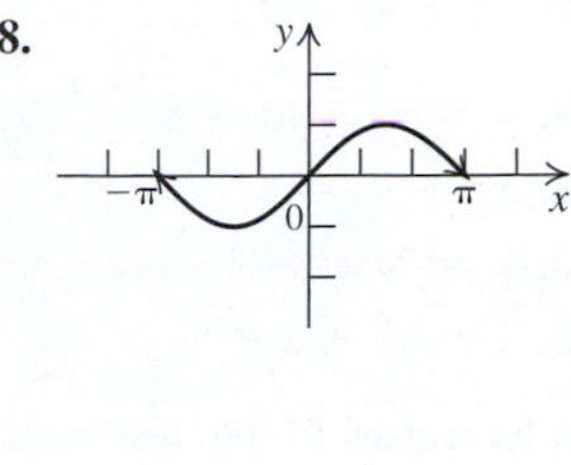

19.

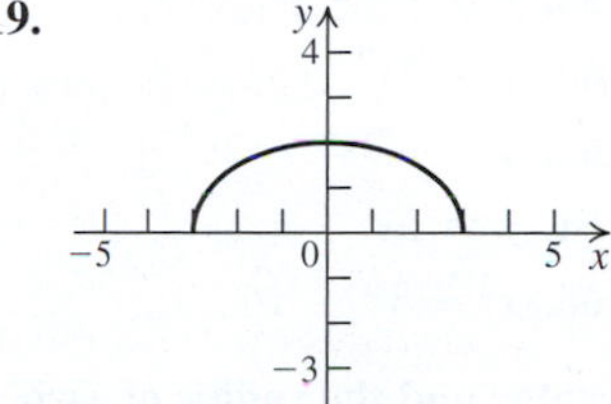

20.

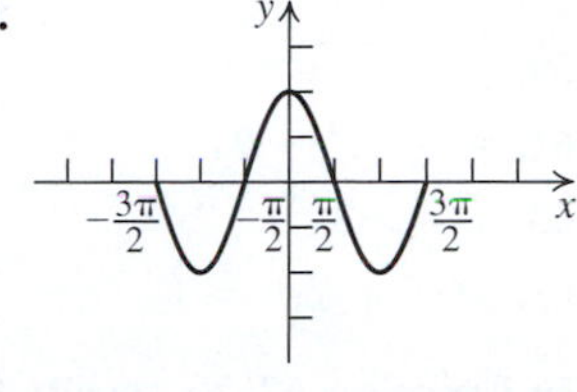

21.

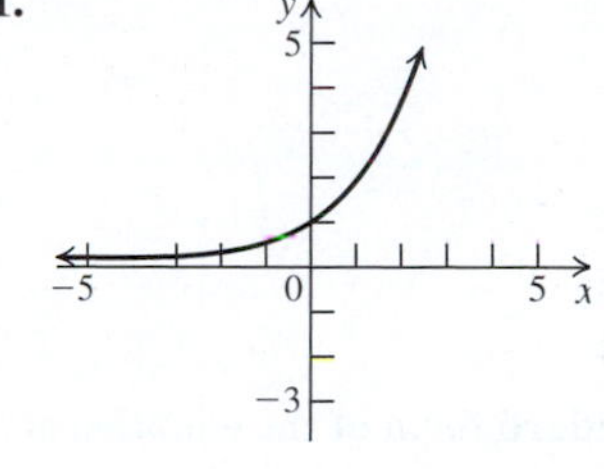

22.

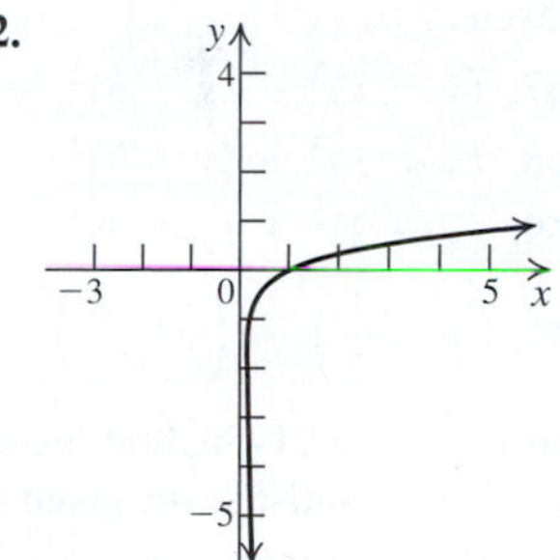

23.

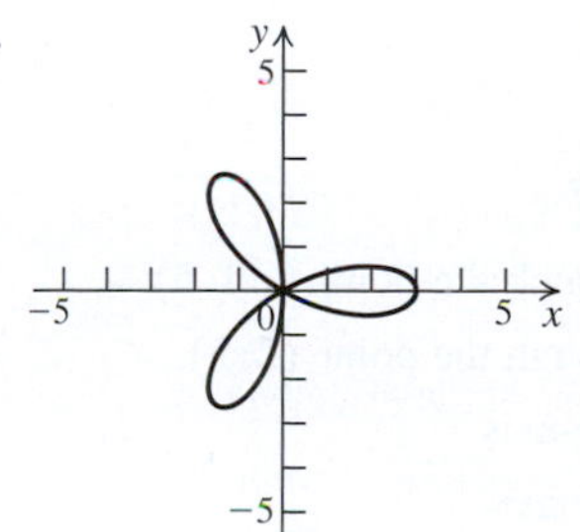

24.

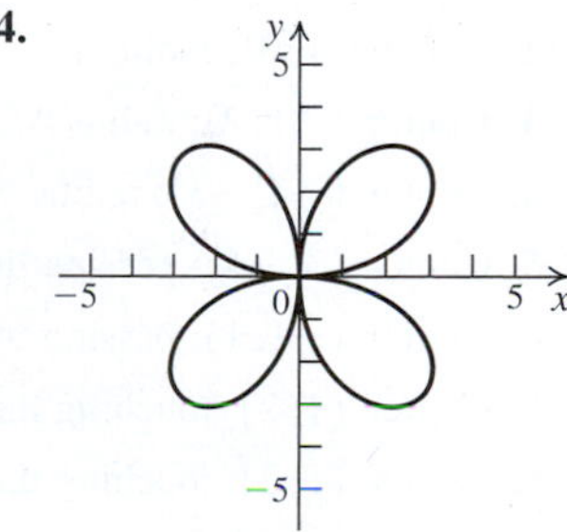

In Exercises 25–46, graph each equation by plotting points. Let $x = -3, -2, -1, 0, 1, 2$, and 3 where applicable.

25. $y = x + 1$ **26.** $y = x - 1$

27. $y = 2x$ **28.** $y = \frac{1}{2}x$

29. $y = x^2$ **30.** $y = -x^2$

31. $y = |x|$ **32.** $y = |x + 1|$

33. $y = |x| + 1$ **34.** $y = -|x| + 1$

35. $y = 4 - x^2$ **36.** $y = x^2 - 4$

37. $y = \sqrt{9 - x^2}$ **38.** $y = -\sqrt{9 - x^2}$

39. $y = x^3$ **40.** $y = -x^3$

41. $y^3 = x$ **42.** $y^3 = -x$

43. $x = |y|$ **44.** $|x| = |y|$

45. $y = |2 - x|$ **46.** $|x| + |y| = 1$

In Exercises 47–56, find the x- and y-intercepts of the graph of the equation.

47. $3x + 4y = 12$ **48.** $\frac{x}{5} + \frac{y}{3} = 1$

49. $y = x^2 - 6x + 8$ **50.** $x = y^2 - 5y + 6$

51. $x^2 + y^2 = 4$ **52.** $y = \sqrt{9 - x^2}$

53. $y = \sqrt{x^2 - 1}$ **54.** $xy = 1$

55. $y = x^2 + x + 1$ **56.** $x^3 + 3xy^2 + y^3 = 1$

In Exercises 57–66, test each equation for symmetry with respect to the x-axis, the y-axis, and the origin.

57. $y = x^2 + 1$ **58.** $x = y^2 + 1$

59. $y = x^3 + x$ **60.** $y = 2x^3 - x$

61. $y = 5x^4 + 2x^2$ **62.** $y = -3x^6 + 2x^4 + x^2$

63. $y = -3x^5 + 2x^3$ **64.** $y = 2x^2 - |x|$

65. $x^2y^2 + 2xy = 1$ **66.** $x^2 + y^2 = 16$

In Exercises 67–70, specify the center and the radius of each circle.

67. $(x - 2)^2 + (y - 3)^2 = 36$

68. $(x + 1)^2 + (y - 3)^2 = 16$

69. $(x + 2)^2 + (y + 3)^2 = 11$

70. $\left(x - \frac{1}{2}\right)^2 + \left(y + \frac{3}{2}\right)^2 = \frac{3}{4}$

In Exercises 71–80, find the standard form of the equation of a circle that satisfies the given conditions. Graph each equation.

71. Center $(0, 1)$; radius 2

72. Center $(1, 0)$; radius 1

73. Center $(-1, 2)$; radius $\sqrt{2}$

74. Center $(-2, -3)$; radius $\sqrt{7}$

75. Center $(3, -4)$; passing through the point $(-1, 5)$

76. Center $(-1, 1)$; passing through the point $(2, 5)$

77. Center $(1, 2)$; touching the x-axis

78. Center $(1, 2)$; touching the y-axis

79. Diameter with endpoints $(7, 4)$ and $(-3, 6)$

80. Diameter with endpoints $(2, -3)$ and $(8, 5)$

In Exercises 81–86, find

a. the center and radius of each circle.

b. the x- and y-intercepts of the graph of each circle.

81. $x^2 + y^2 - 2x - 2y - 4 = 0$

82. $x^2 + y^2 - 4x - 2y - 15 = 0$

83. $2x^2 + 2y^2 + 4y = 0$ **84.** $3x^2 + 3y^2 + 6x = 0$

85. $x^2 + y^2 - x = 0$ **86.** $x^2 + y^2 + 1 = 0$

Applying the Concepts

In Exercises 87–89, a graph is described *geometrically* as the path of a point $P(x, y)$ on the graph. Find an equation for the graph described.

87. Geometry. $P(x, y)$ is on the graph if and only if the distance from $P(x, y)$ to the x-axis is equal to its distance to the y-axis.

88. Geometry. $P(x, y)$ is the same distance from the two points $(1, 2)$ and $(3, -4)$.

89. Geometry. $P(x, y)$ is the same distance from the point $(2, 0)$ and the y-axis.

90. Spending on prescription drugs. Americans spent \$136 billion on prescription drugs in 2000 and \$234 billion in 2008. Assuming that this trend continued, estimate the amount spent on prescription drugs during 2002, 2004, and 2006. (*Source: U.S. Census Bureau.*)

91. Corporate profits. The equation $P = -0.5t^2 - 3t + 8$ describes the monthly profits (in millions of dollars) of ABCD Corp. for the year 2012, with $t = 0$ representing July 2012.

a. How much profit did the corporation make in March 2012?

b. How much profit did the corporation make in October 2012?

c. Sketch the graph of the equation.

d. Find the t-intercepts. What do they represent?

e. Find the P-intercept. What does it represent?

92. Female students in colleges. The equation

$$P = -0.002t^2 + 0.51t + 17.5$$

models the approximate number (in millions) of female college students in the United States for the academic years 2005–2009, with $t = 0$ representing 2005.

a. Sketch the graph of the equation.

b. Find the P-intercept. What does it represent? (*Source: Statistical Abstracts of the United States.*)

93. Motion. An object is thrown up from the top of a building that is 320 feet high. The equation $y = -16t^2 + 128t + 320$ gives the object's height (in feet) above the ground at any time t (in seconds) after the object is thrown.

a. What is the height of the object after 0, 1, 2, 3, 4, 5, and 6 seconds?

b. Sketch the graph of the equation $y = -16t^2 + 128t + 320$.

c. What part of the graph represents the physical aspects of the problem?

d. What are the intercepts of this graph, and what do they mean?

94. Diving for treasure. A treasure-hunting team of divers is placed in a computer-controlled diving cage. The equation $d = \frac{40}{3}t - \frac{2}{9}t^2$ describes the depth d (in feet) that the cage will descend in t minutes.

a. Sketch the graph of the equation $d = \frac{40}{3}t - \frac{2}{9}t^2$.

b. What part of the graph represents the physical aspects of the problem?

c. What is the total time of the entire diving experiment?

Beyond the Basics

In Exercises 95–98, sketch the complete graph by using the given symmetry.

95. **96.**

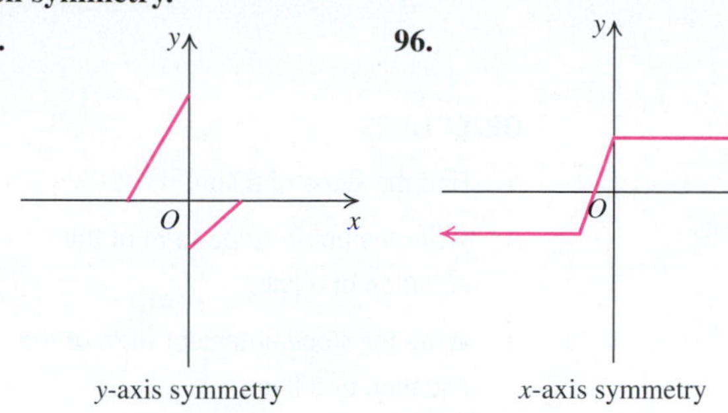

y-axis symmetry *x*-axis symmetry

97. **98.**

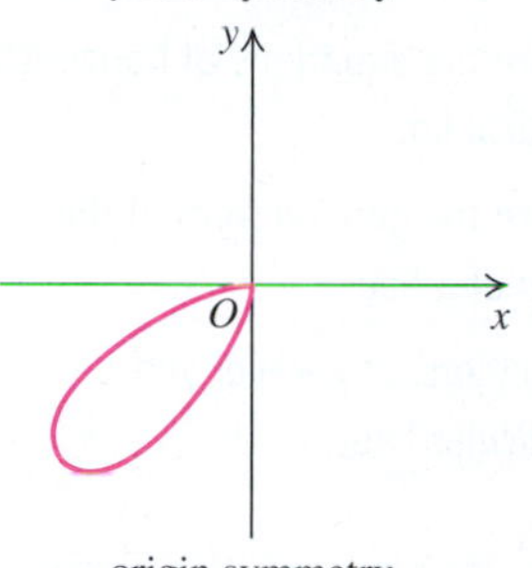

origin symmetry

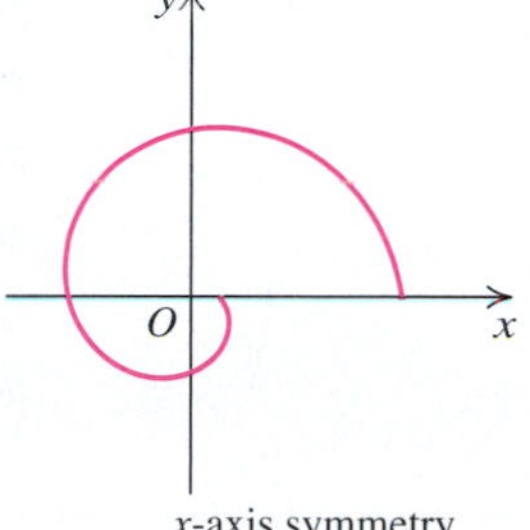

x-axis symmetry

99. In the same coordinate system, sketch the graphs of the two circles with equations $x^2 + y^2 - 4x + 2y - 20 = 0$ and $x^2 + y^2 - 4x + 2y - 31 = 0$ and find the area of the region bounded by the two circles.

100. Find the equation of a circle with radius 5 and x-intercepts -4 and 4.

(*Hint:* Center must be on the y-axis; there are two such circles.)

Critical Thinking/Discussion/Writing

101. Sketch the graph of $y^2 = 2x$ and explain how this graph is related to the graphs of $y = \sqrt{2x}$ and $y = -\sqrt{2x}$.

102. Show that a graph that is symmetric with respect to the x-axis and y-axis must be symmetric with respect to the origin. Give an example to show that the converse is not true.

103. **a.** Show that a circle with diameter having endpoints $A(0, 1)$ and $B(6, 8)$ intersects the x-axis at the roots of the equation $x^2 - 6x + 8 = 0$.

b. Show that a circle with diameter having endpoints $A(0, 1)$ and $B(a, b)$ intersects the x-axis at the roots of the equation $x^2 - ax + b = 0$.

c. Use graph paper, ruler, and compass to approximate the roots of the equation $x^2 - 3x + 1 = 0$.

104. The figure shows two circles each with radius r.

a. Write the coordinates of the center of each circle.

b. Find the area of the shaded region.

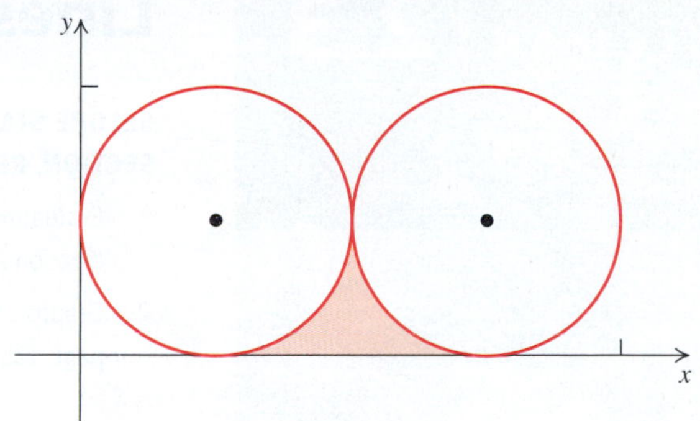

Maintaining Skills

In Exercises 105–110, perform the indicated operations.

105. $\dfrac{5 - 3}{6 - 2}$

106. $\dfrac{1 - 2}{-2 - 2}$

107. $\dfrac{2 - (-3)}{3 - 13}$

108. $\dfrac{3 - 1}{-2 - (-6)}$

109. $\dfrac{\frac{1}{2} - \frac{1}{4}}{\frac{3}{8} - \left(-\frac{1}{4}\right)}$

110. $\dfrac{\frac{3}{4} - 1}{\frac{1}{2} - \frac{1}{6}}$

In Exercises 111–116, solve each equation for the specified variable.

111. $A = P + Prt$ for P

112. $2x + 3y = 6$ for y

113. $\dfrac{x}{2} - \dfrac{y}{5} = 3$ for y

114. $y - 2 - \dfrac{2}{3}(x + 1) = 0$ for y

115. $ax + by + c = 0$ for y $(b \neq 0)$

116. $0.1x + 0.2y - 1 = 0$ for y

In Exercises 117–122, write the negative reciprocal of each number.

117. 2

118. -3

119. $-\dfrac{2}{3}$

120. $\dfrac{4}{3}$

121. $\dfrac{1 - \frac{1}{2}}{2 + \frac{3}{4}}$

122. $\dfrac{\frac{2}{3} - \frac{1}{4}}{\frac{5}{6} - \left(-\frac{3}{4}\right)}$

SECTION 2.3

Lines

BEFORE STARTING THIS SECTION, REVIEW

1. Evaluating algebraic expressions (Section P.1, page 13)
2. Graphs of equations (Section 2.2, page 165)

OBJECTIVES

1. Find the slope of a line.
2. Write the point–slope form of the equation of a line.
3. Write the slope–intercept form of the equation of a line.
4. Recognize the equations of horizontal and vertical lines.
5. Recognize the general form of the equation of a line.
6. Find equations of parallel and perpendicular lines.

A Texan's Tall Tale

Gunslinger Wild Bill Longley's *first* burial took place on October 11, 1878, after he was hanged before a crowd of thousands in Giddings, Texas.

Rumors persisted that Longley's hanging had been a hoax and that he had somehow faked his death and escaped execution. In 2001, Longley's descendants had his grave opened to determine whether the remains matched Wild Bill's description: a tall white male, age 27. Both the skeleton and some personal effects suggested that this was indeed Wild Bill. Modern science lent a hand too: The DNA of Wild Bill's sister's descendant Helen Chapman was a perfect match.

Now the notorious gunman could be buried back in the Giddings cemetery—for the *second* time. How did the scientists conclude from the skeletal remains that Wild Bill was approximately 6 feet tall? (See Example 8.)

1 Find the slope of a line.

Slope of a Line

In this section, we study various forms of first-degree equations in two variables. Because the graphs of these equations are straight lines, they are called **linear equations**. Just as we measure weight or temperature by a number, we measure the "steepness" of a line by a number called its **slope**.

Consider two points $P(x_1, y_1)$ and $Q(x_2, y_2)$ on a line, as shown in Figure 2.22. We say that the ***rise*** is the change in y-coordinates between the points (x_1, y_1) and (x_2, y_2) and that the ***run*** is the corresponding change in the x-coordinates. A positive run indicates change to the right, as shown in Figure 2.22; a negative run indicates change to the left. Similarly, a positive rise indicates upward change; a negative rise indicates downward change. A zero run or rise means no change in the corresponding coordinate.

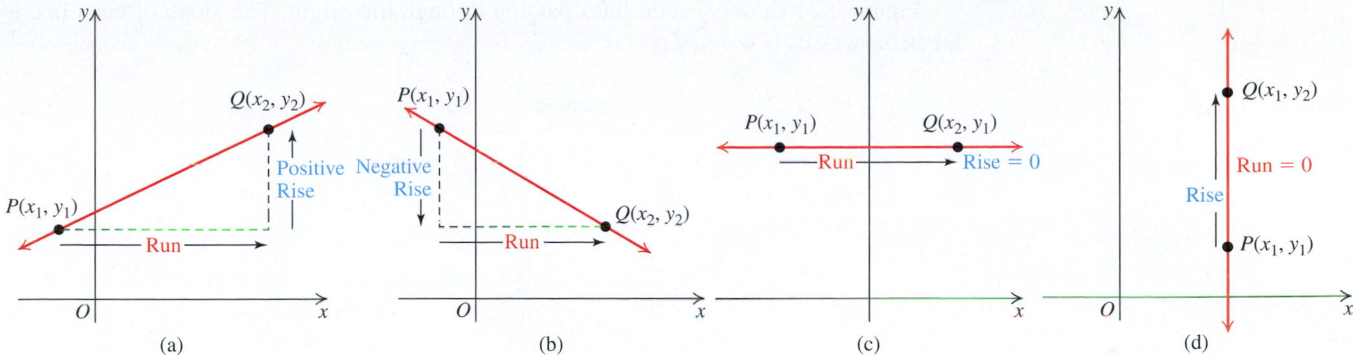

FIGURE 2.22 Slope of a line

Slope of a Line

The **slope** of a nonvertical line that passes through the points $P(x_1, y_1)$ and $Q(x_2, y_2)$ is denoted by m and is defined by

$$m = \frac{\text{rise}}{\text{run}} = \frac{y_2 - y_1}{x_2 - x_1}, \quad x_1 \neq x_2.$$

The slope of a vertical line is undefined.

Because $m = \frac{\text{rise}}{\text{run}} = \frac{\text{change in } y\text{-coordinates}}{\text{change in } x\text{-coordinates}}$, if the change in x is 1 unit to the right (x increases by 1 unit), then the change in y equals the slope of the line. The slope of a line is therefore the **change in *y* per unit change in *x***. In other words, the slope of a line measures the rate of change of y with respect to x. Roofs, staircases, graded landscapes, and mountainous roads all have slopes. For example, if the pitch (slope) of a section of the roof is 0.4, then for every horizontal distance of 10 feet in that section, the roof ascends 4 feet.

EXAMPLE 1 Finding and Interpreting the Slope of a Line

Sketch the graph of the line that passes through $P(1, -1)$ and $Q(3, 3)$. Find and interpret the slope of the line.

Solution

The graph of the line is sketched in Figure 2.23. The slope m of this line is given by

$$\begin{aligned} m &= \frac{\text{rise}}{\text{run}} = \frac{\text{change in } y\text{-coordinates}}{\text{change in } x\text{-coordinates}} \\ &= \frac{(y\text{-coordinate of } Q) - (y\text{-coordinate of } P)}{(x\text{-coordinate of } Q) - (x\text{-coordinate of } P)} \\ &= \frac{(3) - (-1)}{(3) - (1)} = \frac{3 + 1}{3 - 1} = \frac{4}{2} = 2. \end{aligned}$$

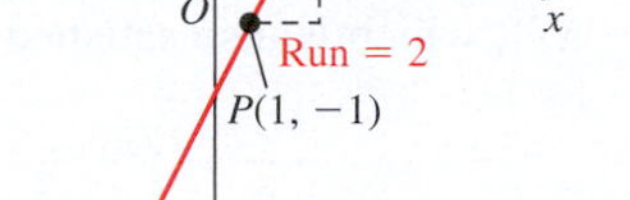

FIGURE 2.23 Interpreting slope

Interpretation

A slope of 2 means that the value of y increases 2 units for every 1 unit increase in the value of x.

Practice Problem 1 Find and interpret the slope of the line containing the points $(-7, 5)$ and $(6, -3)$.

Figure 2.24 shows several lines passing through the origin. The slope of each line is labeled.

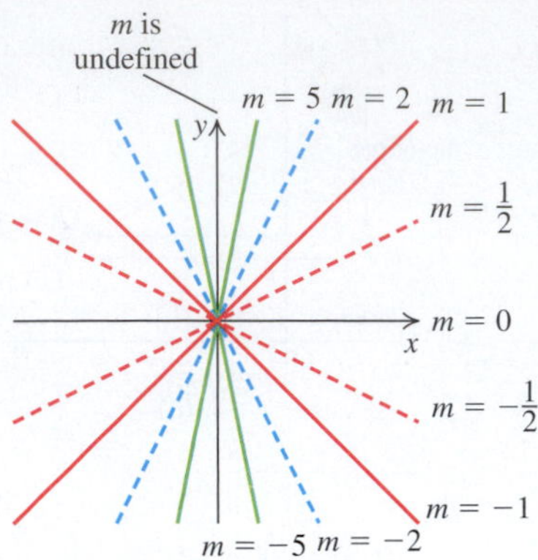

FIGURE 2.24 Slopes of lines

MAIN FACTS ABOUT SLOPES OF LINES

1. Scanning graphs from left to right, lines with positive slopes rise and lines with negative slopes fall.
2. The greater the absolute value of the slope, the steeper the line.
3. The slope of a vertical line is undefined.
4. The slope of a horizontal line is zero.

WARNING

Be careful when using the formula for finding the slope of a line joining the points (x_1, y_1). and (x_2, y_2). Be sure to subtract coordinates in the same order.

You can use either

$$m = \frac{y_2 - y_1}{x_2 - x_1} \quad \text{or} \quad m = \frac{y_1 - y_2}{x_1 - x_2},$$

but it is incorrect to use

$$m = \frac{y_1 - y_2}{x_2 - x_1} \quad \text{or} \quad m = \frac{y_2 - y_1}{x_1 - x_2}.$$

2 Write the point–slope form of the equation of a line.

Equation of a Line

We now find an equation of the line ℓ that passes through the point $A(x_1, y_1)$ and has slope m. Let $P(x, y)$, with $x \neq x_1$, be any point in the plane. Then $P(x, y)$ is on the line ℓ if and only if the slope of the line passing through $P(x, y)$ and $A(x_1, y_1)$ is m. This is true if and only if

$$\frac{y - y_1}{x - x_1} = m.$$

See Figure 2.25.

Multiplying both sides by $x - x_1$ gives $y - y_1 = m(x - x_1)$, which is also satisfied when $x = x_1$ and $y = y_1$.

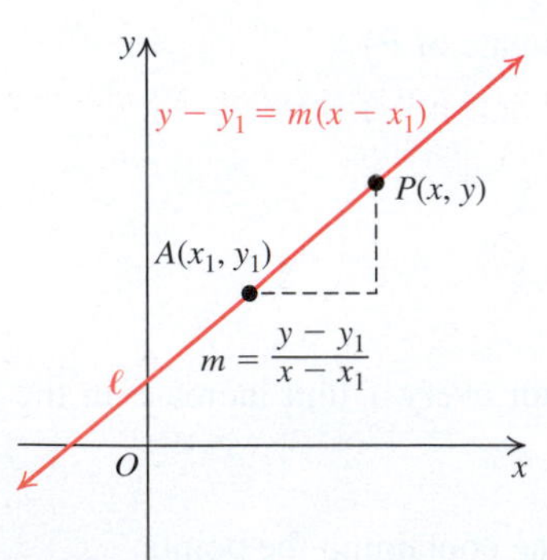

FIGURE 2.25 Point–slope form

The Point–Slope Form of the Equation of a Line

If a line has slope m and passes through the point (x_1, y_1), then the **point–slope form** of an equation of the line is

$$y - y_1 = m(x - x_1).$$

EXAMPLE 2 **Finding an Equation of a Line with Given Point and Slope**

Find the point–slope form of the equation of the line passing through the point $(1, -2)$ with slope $m = 3$. Then solve for y.

Solution

$$\begin{aligned} y - y_1 &= m(x - x_1) && \text{Point–slope form} \\ y - (-2) &= 3(x - 1) && \text{Substitute } x_1 = 1, y_1 = -2, \text{ and } m = 3. \\ y + 2 &= 3x - 3 && \text{Simplify.} \\ y &= 3x - 5 && \text{Solve for } y. \end{aligned}$$

Practice Problem 2 Find the point–slope form of the equation of the line passing through the point $(-2, -3)$ and with slope $-\frac{2}{3}$. Then solve for y.

EXAMPLE 3 **Finding an Equation of a Line Passing Through Two Given Points**

Find the point–slope form of an equation of the line ℓ passing through the points $(-2, 1)$ and $(3, 7)$. Then solve for y.

Solution

We first find the slope m of the line ℓ.

$$m = \frac{7 - 1}{3 - (-2)} = \frac{6}{3 + 2} = \frac{6}{5} \qquad m = \frac{y_2 - y_1}{x_2 - x_1}$$

We use $m = \frac{6}{5}$ and either of the two given points when substituting into the point–slope form:

$$y - y_1 = m(x - x_1)$$

With $(x_1, y_1) = (3, 7)$

$$y - 7 = \frac{6}{5}(x - 3)$$

$$y = \frac{6}{5}x + \frac{17}{5} \qquad \text{Simplify.}$$

With $(x_1, y_1) = (-2, 1)$

$$y - 1 = \frac{6}{5}(x - (-2))$$

$$y = \frac{6}{5}x + \frac{17}{5} \qquad \text{Simplify.}$$

Same result

Practice Problem 3 Find the point–slope form of an equation of the line passing through the points $(-3, -4)$ and $(-1, 6)$. Then solve for y.

3 Write the slope–intercept form of the equation of a line.

Slope–Intercept Form

If we know the slope m of a line and that a given point is the y-intercept, then we can use the point–slope form to create another form of an equation of this line.

EXAMPLE 4 Finding an Equation of a Line with a Given Slope and y-intercept

Find the point–slope form of the equation of the line with slope m and y-intercept b. Then solve for y.

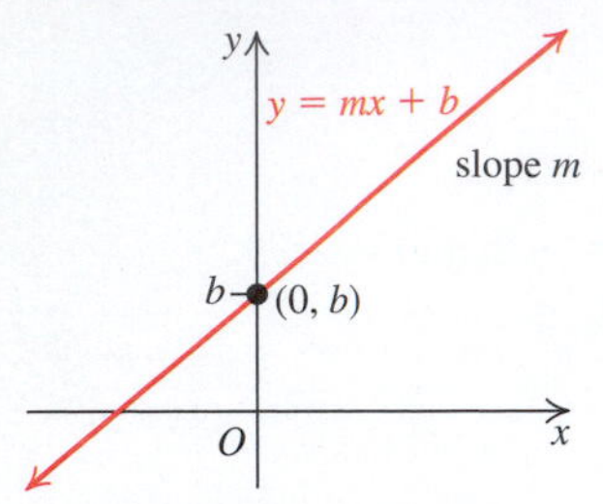

FIGURE 2.26 Slope–intercept form

Solution

Because the line has y-intercept b, the line passes through the point $(0, b)$. See Figure 2.26.

$$y - y_1 = m(x - x_1) \quad \text{Point–slope form}$$
$$y - b = m(x - 0) \quad \text{Substitute } x_1 = 0 \text{ and } y_1 = b.$$
$$y - b = mx \quad \text{Simplify.}$$
$$y = mx + b \quad \text{Solve for } y.$$

Practice Problem 4 Find the point–slope form of the equation of the line with slope 2 and y-intercept -3. Then solve for y.

Historical Note

No one is sure why the letter m is used for slope. Many people suggest that m comes from the French **monter** (to climb), but Descartes, who was French, did not use m. Professor John Conway of Princeton University has suggested that m could stand for "modulus of slope."

Slope–Intercept Form of the Equation of a Line

The **slope–intercept form** of an equation of the line with slope m and y-intercept b is

$$y = mx + b.$$

The linear equation in the form $y = mx + b$ displays the slope m (the coefficient of x) and the y-intercept b (the constant term). The number m tells which way and how much the line is tilted; the number b tells where the line intersects the y-axis.

EXAMPLE 5 Graph Using the Slope and y-intercept

Graph the line whose equation is $y = \frac{2}{3}x + 2$.

Solution

The equation

$$y = \frac{2}{3}x + 2$$

is in the slope–intercept form with slope $\frac{2}{3}$ and y-intercept 2. To sketch the graph, find two points on the line and draw a line through these two points. We use the y-intercept as one of the points and then use the slope to locate a second point. Because $m = \frac{2}{3}$, let 2 be the rise so that 3 is the run. From the point $(0, 2)$, move 3 units to the right (run) and 2 units up (rise). This gives $(0 + \text{run}, 2 + \text{rise}) = (0 + 3, 2 + 2) = (3, 4)$ as the second point.

The line we want joins the points $(0, 2)$ and $(3, 4)$. See Figure 2.27.

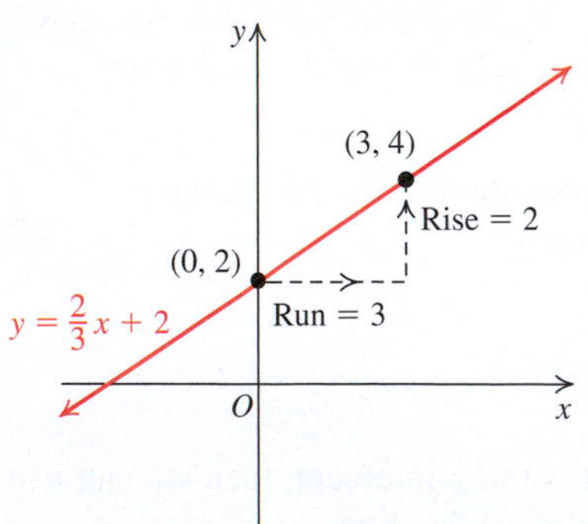

FIGURE 2.27 Locating a second point on a line

Practice Problem 5 Graph the line with slope $-\frac{2}{3}$ that contains the point $(0, 4)$.

4 Recognize the equations of horizontal and vertical lines.

Equations of Horizontal and Vertical Lines

Consider the horizontal line through the point (h, k). The y-coordinate of every point on this line is k. So we can write its equation as $y = k$. This line has slope $m = 0$ and y-intercept k. Similarly, an equation of the vertical line through the point (h, k) is $x = h$. This line has undefined slope and x-intercept h. See Figure 2.28.

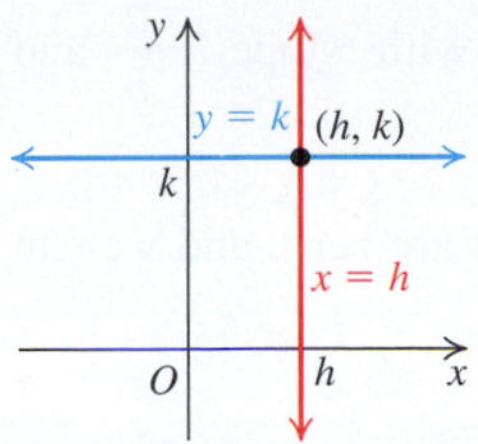

FIGURE 2.28 Horizontal and vertical lines

> **HORIZONTAL AND VERTICAL LINES**
>
> An equation of a horizontal line through (h, k) is $y = k$.
> An equation of a vertical line through (h, k) is $x = h$.

EXAMPLE 6 Recognizing Horizontal and Vertical Lines

Discuss the graph of each equation in the xy-plane.

a. $y = 2$ **b.** $x = 4$

Solution

a. The equation $y = 2$ may be considered as an equation in two variables x and y by writing

$$0 \cdot x + y = 2.$$

Any ordered pair of the form $(x, 2)$ is a solution of this equation. Some solutions are $(-1, 2)$, $(0, 2)$, $(2, 2)$, and $(7, 2)$. It follows that the graph is a line parallel to the x-axis and 2 units above it, as shown in Figure 2.29. Its slope is 0.

TECHNOLOGY CONNECTION

Calculator graph of $y = 2$

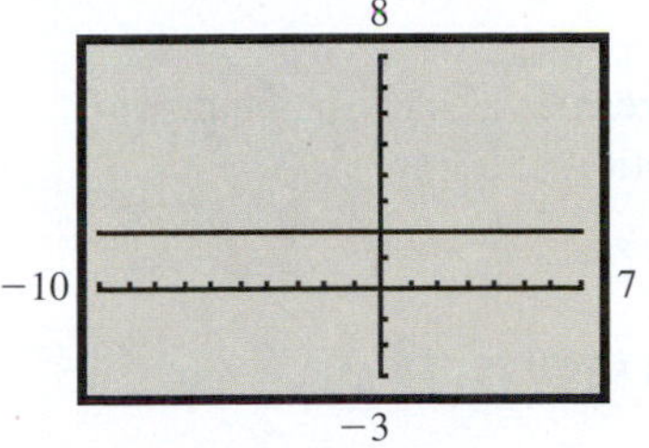

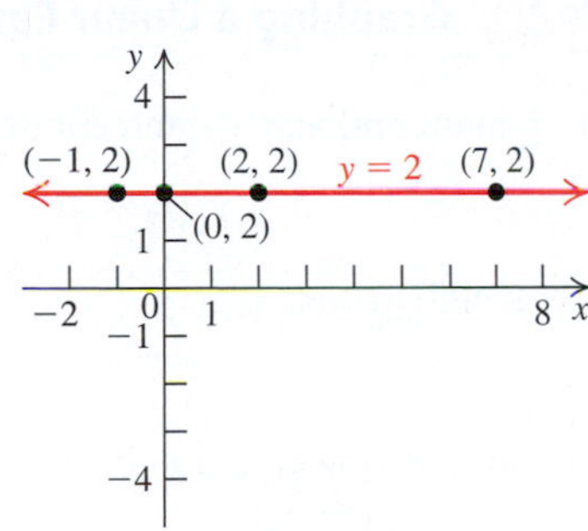

FIGURE 2.29 Horizontal line

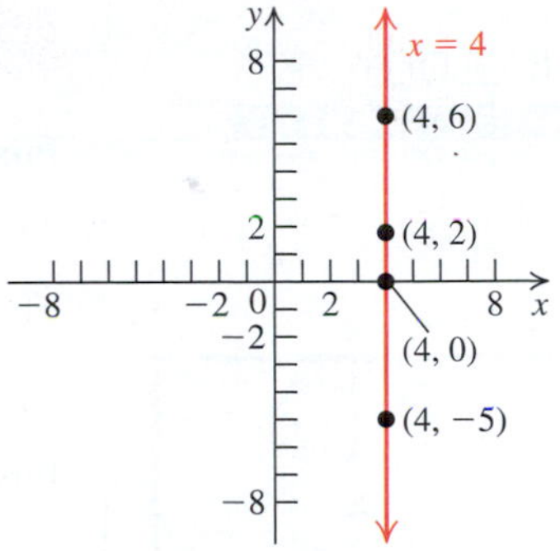

FIGURE 2.30 Vertical line

SIDE NOTE

Students sometimes find it easy to remember that if a linear equation
(1) does not have y in it, the graph is parallel to the y-axis or
(2) does not have x in it, the graph is parallel to the x-axis.

b. The equation $x = 4$ may be written as

$$x + 0 \cdot y = 4.$$

Any ordered pair of the form $(4, y)$ is a solution of this equation. Some solutions are $(4, -5)$, $(4, 0)$, $(4, 2)$, and $(4, 6)$. The graph is a line parallel to the y-axis and 4 units to the right of it, as shown in Figure 2.30. Its slope is undefined.

TECHNOLOGY CONNECTION

You cannot graph $x = 4$ using the $Y =$ key on your calculator.

Practice Problem 6 Sketch the graphs of the lines $x = -3$ and $y = 7$.

5 Recognize the general form of the equation of a line.

General Form of the Equation of a Line

An equation of the form

$$ax + by + c = 0,$$

where a, b, and c are constants and a and b are not both zero, is called the *general form* of a linear equation. Consider two possible cases: $b \neq 0$ and $b = 0$.

Suppose $b \neq 0$: We can isolate y on one side and rewrite the equation in slope–intercept form.

$ax + by + c = 0$	Original equation
$by = -ax - c$	Subtract $ax + c$ from both sides.
$y = -\frac{a}{b}x - \frac{c}{b}$	Divide by b.

The result is an equation of a line in slope–intercept form with slope $-\frac{a}{b}$ and y-intercept $-\frac{c}{b}$.

Suppose $b = 0$: We know that $a \neq 0$ because not both a and b are zero, and we can solve for x.

$$ax + 0 \cdot y + c = 0 \quad \text{Replace } b \text{ with } 0.$$
$$ax + c = 0 \quad \text{Simplify.}$$
$$x = -\frac{c}{a} \quad \text{Solve for } x.$$

The graph of the equation $x = -\frac{c}{a}$ is a vertical line.

General Form of the Equation of a Line

The graph of every linear equation

$$ax + by + c = 0,$$

where a, b, and c are constants and a and b are not both zero, is a line. The equation $ax + by + c = 0$ is called the **general form** of the equation of a line.

TECHNOLOGY CONNECTION

Calculator graph of $y = \frac{3}{4}x + 3$

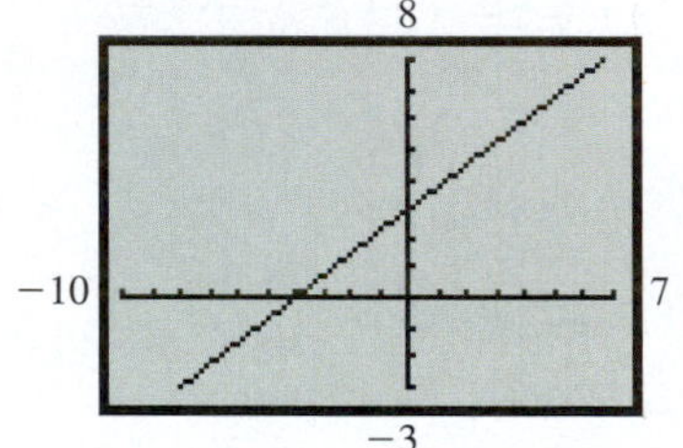

EXAMPLE 7 Graphing a Linear Equation Using Intercepts

Find the slope, y-intercept, and x-intercept of the line with equation

$$3x - 4y + 12 = 0.$$

Then sketch the graph.

Solution

First, solve for y to write the equation in slope–intercept form.

$$3x - 4y + 12 = 0 \quad \text{Original equation}$$
$$4y = 3x + 12 \quad \text{Add } 4y \text{ to both sides; interchange sides.}$$
$$y = \frac{3}{4}x + 3 \quad \text{Divide by 4.}$$

This equation tells us that the slope m is $\frac{3}{4}$ and the y-intercept is 3. To find the x-intercept, set $y = 0$ in the original equation and obtain $3x + 12 = 0$, or $x = -4$. So the x-intercept is -4.

We can sketch the graph of the equation if we can find two points on the graph. So we use the intercepts and sketch the line joining the points $(-4, 0)$ and $(0, 3)$. The graph is shown in Figure 2.31.

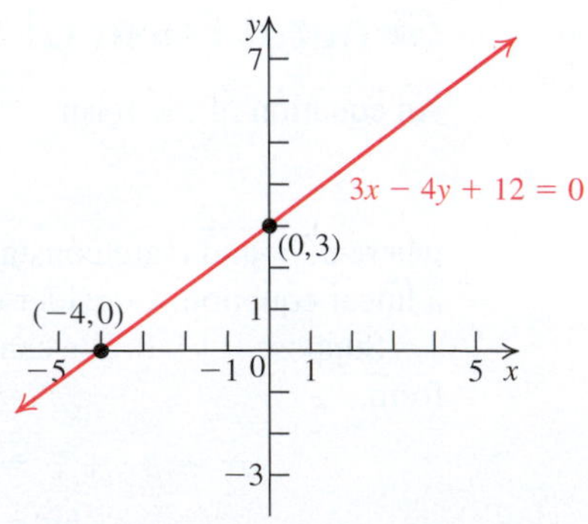

FIGURE 2.31 Graphing a line using intercepts

Practice Problem 7 Sketch the graph of the equation $3x + 4y = 24$.

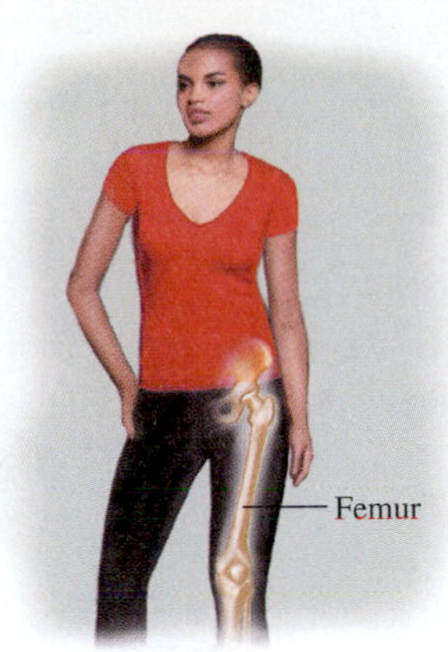

Forensic scientists use various clues to determine the identity of deceased individuals. Scientists know that certain bones serve as excellent sources of information about a person's height. The length of the *femur,* the long bone that stretches from the hip (pelvis) socket to the kneecap (patella), can be used to estimate a person's height. Knowing both the gender and race of the individual increases the accuracy of this relationship between the length of the bone and the height of the person.

EXAMPLE 8 Inferring Height from the Femur

The height H of a human male is related to the length x of his femur by the formula

$$H = 2.6x + 65,$$

where all measurements are in centimeters. The femur of Wild Bill Longley measured between 45 and 46 centimeters. Estimate the height of Wild Bill (in feet).

Solution

Substituting $x = 45$ and $x = 46$ into the preceding formula, we have possible heights

$$H_1 = (2.6)(45) + 65 = 182$$

$$\text{and} \quad H_2 = (2.6)(46) + 65 = 184.6.$$

Therefore, Wild Bill's height was between 182 and 184.6 centimeters.

1 foot $\approx$ 30.48 cm

Converting to feet, we conclude that his height was between $\dfrac{182}{30.48} \approx 5.97$ ft and $\dfrac{184.6}{30.48} \approx 6.06$ ft. So, he was approximately 6 feet tall.

Practice Problem 8 Suppose the femur of a person (male) measures between 43 and 44 centimeters. Estimate the height of the person in meters.

6 Find equations of parallel and perpendicular lines.

Parallel and Perpendicular Lines

We can use slope to decide whether two lines are parallel, perpendicular, or neither.

Parallel and Perpendicular Lines

Let ℓ_1 *and* ℓ_2 be two distinct lines with slopes m_1 *and* m_2, respectively. Then

ℓ_1 is parallel to ℓ_2 if and only if $m_1 = m_2$.

ℓ_1 is perpendicular to ℓ_2 if and only if $m_1 \cdot m_2 = -1$.

Any two vertical lines are parallel, as are any two horizontal lines. Any horizontal line is perpendicular to any vertical line.

In Exercises 113 and 114, you are asked to prove these assertions. Figure 2.32 shows examples of the slopes of a pair of parallel and a pair of perpendicular lines.

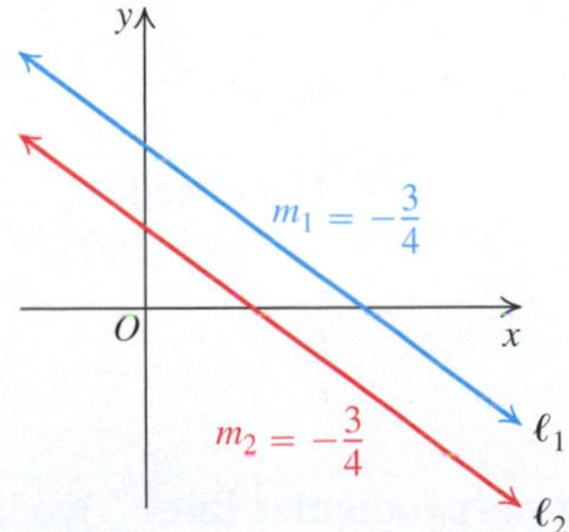

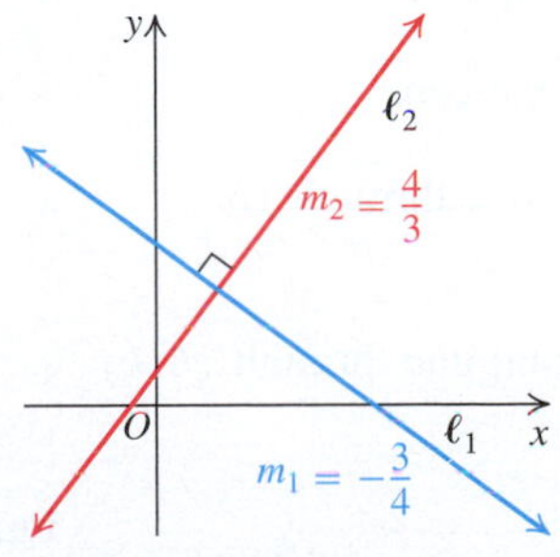

FIGURE 2.32 Parallel lines: $m_1 = m_2$ Perpendicular lines: $m_1 m_2 = -1$

PROCEDURE **IN ACTION**

EXAMPLE 9 Finding Equations of Parallel and Perpendicular Lines

OBJECTIVE

Let $\ell : ax + by + c = 0$. Find an equation of each line through the point (x_1, y_1):

a. ℓ_1 parallel to ℓ

b. ℓ_2 perpendicular to ℓ

Step 1 **Find the slope m of ℓ.**

Step 2 **Write slope of ℓ_1 and ℓ_2.**

The slope m_1 of ℓ_1 is m.

The slope m_2 of ℓ_2 is $-\frac{1}{m}$.

Step 3 **Write equations of ℓ_1 and ℓ_2.**

Use point–slope form to write equations of ℓ_1 and ℓ_2.

Simplify to write equations in the requested form.

EXAMPLE

Let $\ell : 2x - 3y + 6 = 0$. Find a general form of the equation of each line through the point $(2, 8)$:

(a) ℓ_1 parallel to ℓ

(b) ℓ_2 perpendicular to ℓ

1. $\ell : 2x - 3y + 6 = 0$ Original equation

$-3y = -2x - 6$ Subtract $2x + 6$.

$y = \frac{2}{3}x + 2$ Divide by -3.

The slope of ℓ is $m = \frac{2}{3}$.

2. $m_1 = \frac{2}{3}$ ℓ_1 parallel to ℓ

$m_2 = -\frac{3}{2}$ ℓ_2 perpendicular to ℓ

3. $\ell_1 : y - 8 = \frac{2}{3}(x - 2)$ Point–slope form

$2x - 3y + 20 = 0$ Simplify.

$\ell_2 : y - 8 = -\frac{3}{2}(x - 2)$ Point–slope form

$3x + 2y - 22 = 0$ Simplify.

Practice Problem 9 Find the general form of the equation of the line

a. Through $(-2, 5)$ and parallel to the line containing $(2, 3)$ and $(5, 7)$.

b. Through $(3, -4)$ and perpendicular to the line $4x + 5y + 1 = 0$.

SUMMARY OF **MAIN FACTS**

Form	Equation	Slope	Other Information
General	$ax + by + c = 0$ $a \neq 0$ or $b \neq 0$	$-\frac{a}{b}$ if $b \neq 0$ Undefined if $b = 0$	y-intercept is $-\frac{c}{b}$, $b \neq 0$. x-intercept is $-\frac{c}{a}$, $a \neq 0$.
Point–slope	$y - y_1 = m(x - x_1)$	m	Line passes through (x_1, y_1).
Slope–intercept	$y = mx + b$	m	x-intercept is $-\frac{b}{m}$ if $m \neq 0$; y-intercept is b.
Vertical line through (h, k)	$x = h$	Undefined	When $h \neq 0$, no y-intercept; x-intercept is h. When $h = 0$, the graph is the y-axis.
Horizontal line through (h, k)	$y = k$	0	When $k \neq 0$, no x-intercept; y-intercept is k. When $k = 0$, the graph is the x-axis.

Parallel and perpendicular lines: Two lines with slopes m_1 and m_2 are (a) parallel if $m_1 = m_2$ and (b) perpendicular if $m_1 \cdot m_2 = -1$.

7 Use linear regression.

Modeling Data Using Linear Regression

In many applications, the way in which one quantity depends on another can be modeled by a linear function. One technique for finding this linear function is called **linear regression**. To illustrate the method, we use Table 2.4, which gives the data for the number of registered motorcycles in the United States for the years 2006–2012.

TABLE 2.4

Year	2006	2007	2008	2009	2010	2011	2012
Registered motorcycles (millions)	6.7	7.1	7.8	7.9	8.4	8.8	9.5

The scatter diagram for the motorcycle data is shown in Figure 2.33(a), and the line used to model the data is added to the scatter diagram in Figure 2.33(b). The years 2006–2012 are represented by the numbers 0–6, respectively.

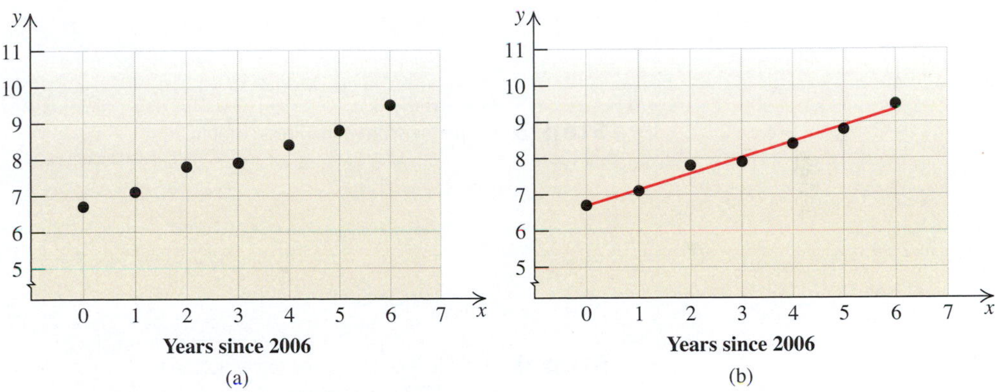

FIGURE 2.33

The line is called the **regression line** or the **least-squares line**. The slope–intercept form of the equation for the line in Figure 2.33(b) is $y = 0.44x + 6.70$, where x is the number of years after 2006, and the slope and intercept values are rounded to the nearest hundredth. The equation is found by using either a graphing calculator or statistical formulas.

EXAMPLE 10 Predictions Using Linear Regression

Use the equation $y = 0.44x + 6.70$ to predict the number of registered motorcycles in the United States for 2013.

Solution

Because 2013 is seven years after 2006, we set $x = 7$; then $y = 0.44(7) + 6.70 \approx 9.78$.

So according to the regression prediction, 9.78 million motorcycles will be registered in the United States for 2013. The quality of a prediction based on linear regression depends on how close the relationship between the data quantities is to being linear.

Practice Problem 10 Repeat Example 10 to predict the number of registered motorcycles in the United States for 2014.

The following Technology Connection shows how a graphing calculator displays the scatter diagram, regression line, and regression line equation.

TECHNOLOGY CONNECTION

You can use a graphing utility to investigate the linear regression model for data points that lie on or near a straight line. On the TI-84, the data for Example 10 are entered in five steps.

Step 1 Enter the data.
Press [STAT] 1 for STAT Edit. Enter the values of the independent variable *x* in L1 and the corresponding values of the dependent variable *y* in L2.

L1	L2	L3 2
0	6.7	------
1	7.1	
2	7.8	
3	7.9	
4	8.4	
5	8.8	
6	9.5	

L2(1)=6.7

Step 2 Turn on STAT Plot
Press [2nd] [Y =] 1 for Plot 1. Select *On*; then choose *Type*. [⋰]
Choose *X* List: L1; *Y* list: L2; and mark: [·].

Step 3 Graph the scatterplot. Press [Graph].

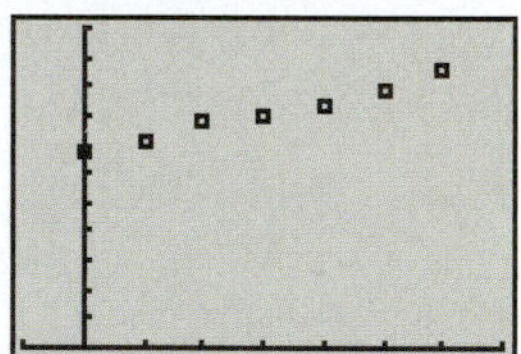

Step 4 Find the equation of the regression line.
Press [STAT] and highlight CALC. Press 4 for LinReg $(ax + b)$. Press [ENTER].

Step 5 Graph the regression line with the scatterplot.
Press [Y =].
Press [VARS] 5 for statistics. Highlight EQ and press [ENTER]. (You should see the regression equation in the $Y =$ window.)
Press [ZOOM] 9 for ZoomStat. (You should see the data graphed along with the regression line.)

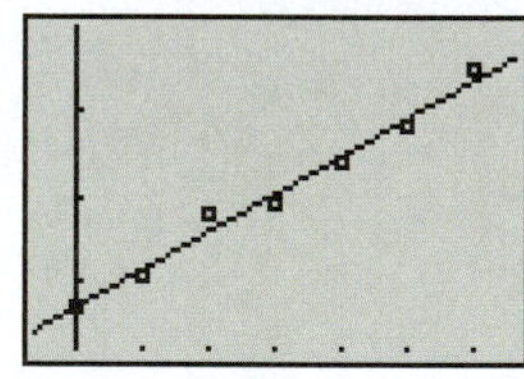

Answers to Practice Problems

1. $-\frac{8}{13}$; for every 13 units increase in x, the y-value decreases by 8 units.

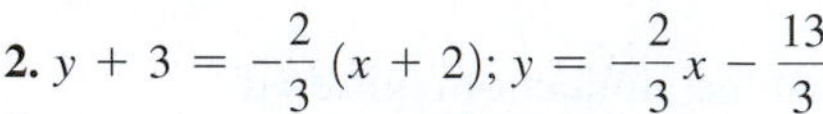

2. $y + 3 = -\frac{2}{3}(x + 2)$; $y = -\frac{2}{3}x - \frac{13}{3}$

3. $y + 4 = 5(x + 3)$; $y = 5x + 11$

4. $y + 3 = 2x$; $y = 2x - 3$

5.

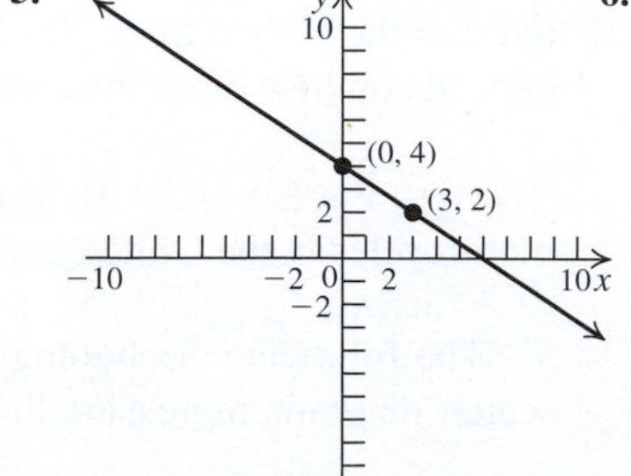

6.

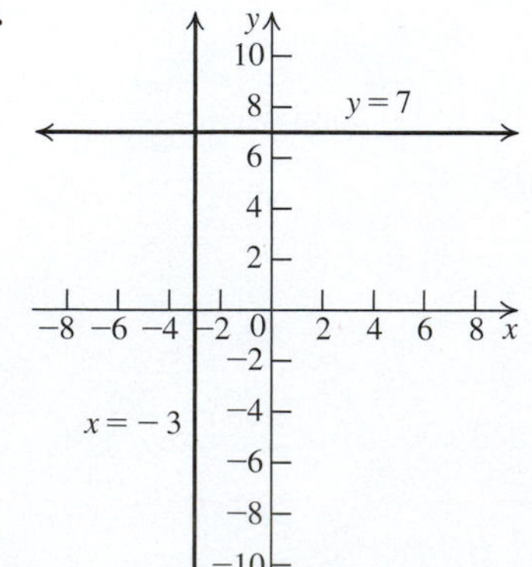

7.

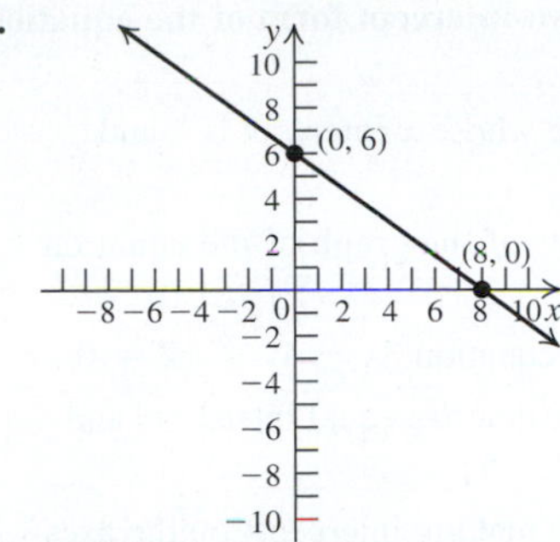

8. Between 1.77 and 1.79 meters

9. a. $4x - 3y + 23 = 0$

b. $5x - 4y - 31 = 0$

10. 10.22 million

SECTION 2.3 Exercises

Basic Concepts and Skills

1. The number that measures the "steepness" of a line is called the ______________.

2. In the slope formula $m = \frac{y_2 - y_1}{x_2 - x_1}$, the number $y_2 - y_1$ is called the ______________.

3. If two lines have the same slope, they are ______________.

4. The y-intercept of the line with equation $y = 3x - 5$ is ______________.

5. A line perpendicular to a line with slope $-\frac{1}{5}$ has slope ______________.

6. **True or False.** A line with undefined slope is perpendicular to every line with slope 0.

7. **True or False.** If $(1, 2)$ is a point on a line with slope 4, then so is the point $(2, 4)$.

8. **True or False.** If the points $(1, 3)$ and $(2, y)$ are on a line with negative slope, then $y > 3$.

In Exercises 9–16, find the slope of the line through the given pair of points. Without plotting any points, state whether the line is rising, falling, horizontal, or vertical.

9. $(1, 3), (4, 7)$

10. $(0, 4), (2, 0)$

11. $(3, -2), (6, -2)$

12. $(-3, 7), (-3, -4)$

13. $(0.5, 2), (3, -3.5)$

14. $(3, -2), (2, -3)$

15. $(\sqrt{2}, 1), (1 + \sqrt{2}, 5)$

16. $(1 - \sqrt{3}, 0), (1 + \sqrt{3}, 3\sqrt{3})$

17. In the figure, identify the line with the given slope m.

a. $m = 1$

b. $m = -1$

c. $m = 0$

d. m is undefined.

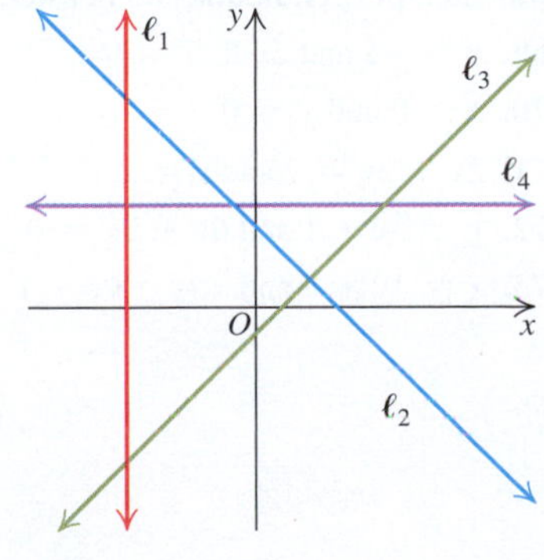

18. Find the slope of each line. (The scale is the same on both axes.)

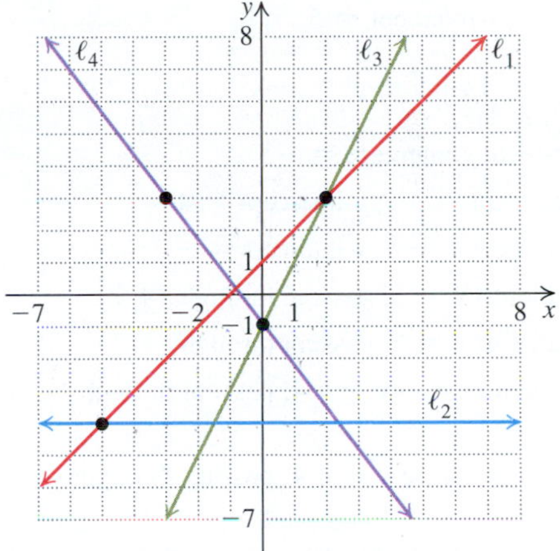

In Exercises 19 and 20, write an equation whose graph is described.

19. a. Graph is the x-axis.

b. Graph is the y-axis.

20. a. Graph consists of all points with a y-coordinate of -4.

b. Graph consists of all points with an x-coordinate of 5.

In Exercises 21–26, find an equation in slope–intercept form of the line that passes through the given point and has slope m. Also sketch the graph of the line by locating the second point with the rise-and-run method.

21. $(0, 4); m = \frac{1}{2}$

22. $(0, 4); m = -\frac{1}{2}$

23. $(2, 1); m = -\frac{3}{2}$

24. $(-1, 0); m = \frac{2}{5}$

25. $(5, -4); m = 0$

26. $(5, -4)$; m is undefined.

In Exercises 27–48, use the given conditions to find an equation in slope–intercept form of each nonvertical line. Write vertical lines in the form $x = h$.

27. Passing through $(0, 1)$ and $(1, 0)$

28. Passing through $(0, 1)$ and $(1, 3)$

29. Passing through $(-1, 3)$ and $(3, 3)$

30. Passing through $(-5, 1)$ and $(2, 7)$

31. Passing through $(-2, -1)$ and $(1, 1)$

32. Passing through $(-1, -3)$ and $(6, -9)$

33. Passing through $\left(\frac{1}{2}, \frac{1}{4}\right)$ and $(0, 2)$

34. Passing through $(4, -7)$ and $(4, 3)$

35. A vertical line through $(5, 1.7)$

36. A horizontal line through $(1.4, 1.5)$

37. A horizontal line through $(0, 0)$

38. A vertical line through $(0, 0)$

39. $m = 0$; y-intercept $= 14$

40. $m = 2$; y-intercept $= 5$

41. $m = -\frac{2}{3}$; y-intercept $= -4$

42. $m = -6$; y-intercept $= -3$

43. x-intercept $= -3$; y-intercept $= 4$

44. x-intercept $= -5$; y-intercept $= -2$

45. Parallel to $y = 5$; passing through $(4, 7)$

46. Parallel to $x = 5$; passing through $(4, 7)$

47. Perpendicular to $x = -4$; passing through $(-3, -5)$

48. Perpendicular to $y = -4$; passing through $(-3, -5)$

49. Let ℓ_1 be a line with slope $m = -2$. Determine whether the given line ℓ_2 is parallel to ℓ_1, perpendicular to ℓ_1, or neither.

a. ℓ_2 is the line through the points $(1, 5)$ and $(3, 1)$.

b. ℓ_2 is the line through the points $(7, 3)$ and $(5, 4)$.

c. ℓ_2 is the line through the points $(2, 3)$ and $(4, 4)$.

50. Find equations of the lines ℓ_1, ℓ_2, ℓ_3 and ℓ_4 of Exercise 18.

In Exercises 51–58, find the slope and intercepts from the equation of the line. Sketch the graph of each equation.

51. $x + 2y - 4 = 0$

52. $x = 3y - 9$

53. $3x - 2y + 6 = 0$

54. $2x = -4y + 15$

55. $x - 5 = 0$

56. $2y + 5 = 0$

57. $x = 0$

58. $y = 0$

59. Show that an equation of a line through the points $(a, 0)$ and $(0, b)$, with a and b nonzero, can be written in the form

$$\frac{x}{a} + \frac{y}{b} = 1.$$

This form is called the **two-intercept form** of the equation of a line.

In Exercises 60–64, use the two-intercept form of the equation of a line given in Exercise 59.

60. Find an equation of the line whose x-intercept is 4 and y-intercept is 3.

61. Find the x- and y-intercepts of the graph of the equation $2x + 3y = 6$.

62. Repeat Exercise 61 for the equation $3x - 4y + 12 = 0$.

63. Find an equation of the line that has equal intercepts and passes through the point $(3, -5)$.

64. Find an equation of the line making intercepts on the axes equal in magnitude but opposite in sign and passing through the point $(-5, -8)$.

65. Find an equation of the line passing through the points $(2, 4)$ and $(7, 9)$. Use the equation to show that the three points $(2, 4)$, $(7, 9)$, and $(-1, 1)$ are on the same line.

66. Use the technique of Exercise 65 to check whether the points $(7, 2)$, $(2, -3)$, and $(5, 1)$ are on the same line.

67. Write the slope–intercept form of the equation of the blue line shown in the figure that is perpendicular to the red line with x-intercept -4.

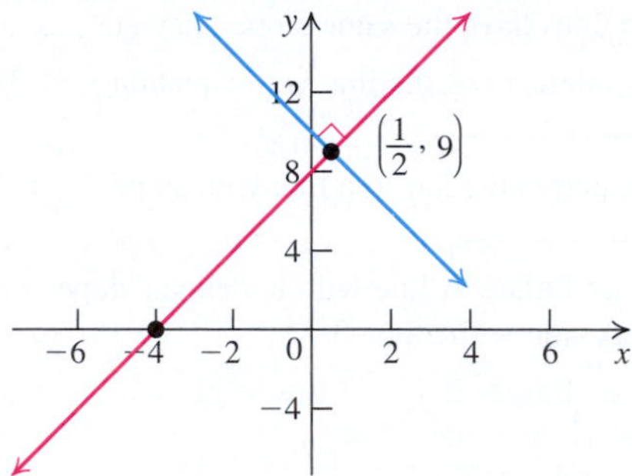

68. Write the point–slope form of the equation of the red line shown in the figure parallel to the blue line through the origin.

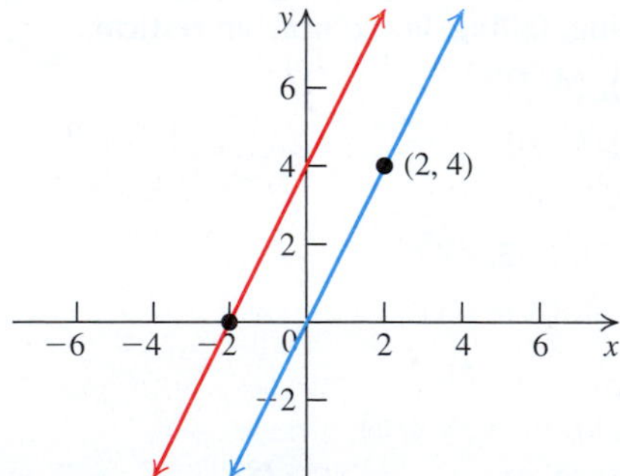

In Exercises 69–76, determine whether each pair of lines are parallel, perpendicular, or neither.

69. $x = -1$ and $2x + 7 = 9$

70. $x = 0$ and $y = 0$

71. $2x + 3y = 7$ and $y = 2$

72. $y = 3x + 1$ and $6y + 2x = 0$

73. $x = 4y + 8$ and $y = -4x + 1$

74. $4x + 3y = 1$ and $3 + y = 2x$

75. $3x + 8y = 7$ and $5x - 7y = 0$

76. $10x + 2y = 3$ and $y + 1 = -5x$

In Exercises 77–86, find the equation of the line in slope–intercept form satisfying the given conditions.

77. a. Passing through $(2, -3)$ and parallel to a line with slope 3

b. Passing through $(-1, 2)$ and perpendicular to a line with slope $-\frac{1}{2}$

78. a. Passing through $(-2, -5)$ and parallel to the line joining the points $(1, -2)$ and $(-3, 2)$

b. Passing through $(1, -2)$ and perpendicular to the line joining the points $(-3, 2)$ and $(-4, -1)$

79. Parallel to $x + y = 1$; passing through $(1, 1)$

80. Parallel to $y = 6x + 5$; y-intercept of -2

81. Perpendicular to $3x - 9y = 18$; passing through $(-2, 4)$

82. Perpendicular to $-2x + y = 14$; passing through $(0, 0)$

83. Perpendicular to $y = 6x + 5$; y-intercept of 4

84. Parallel to $-2x + 3y - 7 = 0$; passing through $(1, 0)$

85. Perpendicular bisector of the line segment with endpoints $A = (2, 3)$ and $B(-1, 5)$

86. Show that $y = x$ is the perpendicular bisector of the line segment with endpoints $A(a, b)$ and $B(b, a)$.

Applying the Concepts

87. Jet takeoff. Upon takeoff, a jet climbs to 4 miles as it passes over 40 miles of land below it. Find the slope of the jet's climb.

88. Road gradient. Driving down the Smoky Mountains in Tennessee, Samantha descends 2000 feet in elevation while moving 4 miles horizontally away from a high point on the straight road. Find the slope (gradient) of the road (1 mile = 5280 feet).

In Exercises 89–94, write an equation of a line in slope–intercept form. To describe this situation, interpret (a) the variables x and y and (b) the meaning of the slope and the y-intercept.

89. Christmas savings account. You currently have \$130 in a Christmas savings account. You deposit \$7 per week into this account. (Ignore interest.)

90. Golf club charges. Your golf club charges a yearly membership fee of \$1000 and \$35 per session of golf.

91. Wages. Judy's job at a warehouse pays \$11 per hour up to 40 hours per week. If she works more than 40 hours in a week, she is paid 1.5 times her usual wage for each hour after 40. [*Hint:* You will need two equations.]

92. Paying off a refrigerator. You bought a new refrigerator for \$700. You made a down payment of \$100 and promised to pay \$15 per month. (Ignore interest.)

93. Converting currency. In February 2013, according to the International Monetary Fund, 1 U.S. dollar equaled 53.87 Indian rupees. Convert currency from rupees to dollars.

94. Life expectancy. In 2010, the life expectancy of a female born in the United States was 80.8 years and was increasing at a rate of 0.17 years per year. (Assume that this rate of increase remains constant.)

95. Machine Value. The graph represents the value of a piece of factory machinery over a ten-year period. Each point (t, y) gives the machine's value y after t years.

a. What does the y-intercept represent?

b. What is the value of the machine after ten years?

c. How much does the machine depreciate in value each year?

d. Write a linear equation for the value of the machine after t years, for $0 \le t \le 10$.

e. What does the slope found in part d represent?

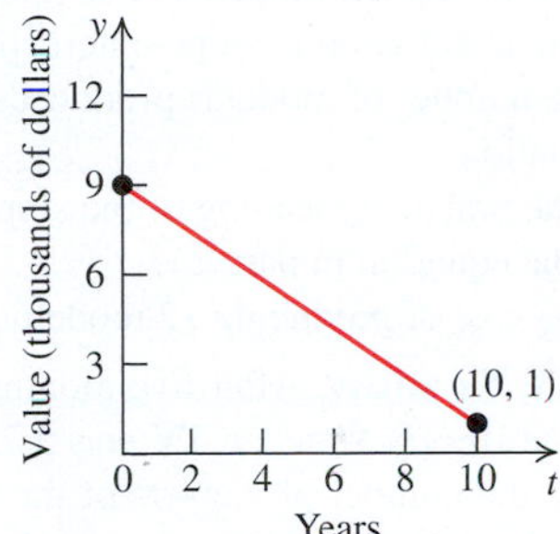

96. Depreciating a tractor. The value v of a tractor purchased for \$14,000 and depreciated linearly at the rate of \$1400 per year is given by $v = -1400t + 14{,}000$, where t represents the number of years since the purchase. Find the value of the tractor after (a) two years and (b) six years. When will the tractor have no value?

97. Playing for a concert. A famous band is considering playing a concert and charging \$40,000 plus \$5 per person attending the concert. Write an equation relating the income y of the band to the number x of people attending the concert.

98. Car rental. The U Drive car rental agency charges \$30.00 per day and \$0.25 per mile.

a. Find an equation relating the cost C of renting a car from U Drive for a one-day trip covering x miles.

b. Draw the graph of the equation in part (a).

c. How much does it cost to rent the car for one day covering 60 miles?

d. How many miles were driven if the one-day rental cost was \$47.75?

99. Female prisoners. The number of females in a state's prisons rose from 2425 in 2005 to 4026 in 2011.

a. Find a linear equation relating the number y of women prisoners to the year t. (Take $t = 0$ to represent 2005.)

b. Draw the graph of the equation from part (a).

c. How many women prisoners were there in 2008?

d. Predict the number of women prisoners in 2017.

100. **Fahrenheit and Celsius.** In the Fahrenheit temperature scale, water boils at 212°F and freezes at 32°F. In the Celsius scale, water boils at 100°C and freezes at 0°C. Assume that the Fahrenheit temperature F and Celsius temperature C are linearly related.
 a. Find the equation in the slope–intercept form relating F and C, with C as the independent variable.
 b. What is the meaning of slope in part (a)?
 c. Find the Fahrenheit temperatures, to the nearest degree, corresponding to 40°C, 25°C, -5°C, and -10°C.
 d. Find the Celsius temperatures, to the nearest degree, corresponding to 100°F, 90°F, 75°F, -10°F, and -20°F.
 e. The normal body temperature in humans ranges from 97.6°F to 99.6°F. Convert this temperature range to degrees Celsius.
 f. When is the Celsius temperature the same numerical value as the Fahrenheit temperature?

101. **Cost.** The cost of producing four modems is \$210.20, and the cost of producing ten modems is \$348.80.
 a. Write a linear equation in slope–intercept form relating cost to the number of modems produced. Sketch a graph of the equation.
 b. What is the practical meaning of the slope and the intercepts of the equation in part (a)?
 c. Predict the cost of producing 12 modems.

102. **Viewers of a TV show.** After five months, the number of viewers of *The Weekly Show* on TV was 5.73 million. After eight months, the number of viewers of the show rose to 6.27 million. Assume that the linear model applies.
 a. Write an equation expressing the number of viewers V after x months.
 b. Interpret the slope and the intercepts of the line in part (a).
 c. Predict the number of viewers of the show after 11 months.

103. **Health insurance coverage.** In Michigan, 12.70% of the population was not covered by health insurance in 2005. In 2008, the percentage of uninsured people dropped to 11.68%. Use the linear model to predict the percentage of people in Michigan that were not covered in 2013. (*Source:* Michigan Department of Community Health.)

104. **Oil consumption.** The total world consumption of oil increased from 82.7 million barrels per day in 2004 to 84.2 million barrels per day in 2009. Create a linear model involving a relation between the year and the daily oil consumption in that year. Let $t = 0$ represent the year 2004. (*Source:* www.cia.gov)

105. **Newspaper sales.** The table shows the number of newspapers (in thousands) that sell at a given price (in nickels) per copy for the college newspaper *Midstate Oracle* at Midstate College.

Price (nickels)	1	2	3	4	5
Numbers of newspapers sold (thousands)	11	8	6	4	3

 a. Use a graphing utility to find the equation of the line of regression relating the variables x (price in nickels) and y (number sold in thousands).
 b. Use a graphing utility to plot the points and graph the regression line.
 c. Approximately how many papers will be sold if the price per copy is 35¢?

106. **Sales and advertising.** Bildwell Computers keeps a careful record of its weekly advertising expenses (in thousands of dollars) and the number of computers (in thousands) it sells the following week. The table shows the data the company collected.

Weekly advertising expenses (in thousands of dollars)	100	200	300	400	500	600
Sales the following week (in thousands)	21	29	36	49	58	67

 a. Use a graphing utility to find the equation of the line of regression equation relating the variables x (weekly advertising expenses in thousands of dollars) and y (number of computers sold, in thousands).
 b. Use a graphing utility to plot the points and graph the regression line.
 c. Approximately how many thousands of computers will be sold if the company spends \$700,000 on advertising?

Beyond the Basics

In Exercises 107 and 108, show that the three given points are collinear by using (a) slopes and (b) the distance formula.

107. $(0, 1)$, $(1, 3)$, and $(-1, -1)$

108. $(1, 0.5)$, $(2, 0)$, and $(0.5, 0.75)$

In Exercises 109 and 110, find c so that the given three points are collinear.

109. $(-5, 1)$, $(1, c)$, and $(4, -2)$

110. $(2, 5), \left(c, \dfrac{11}{2}\right)$, and $(4, 6)$

111. Show that the triangle with vertices $A(1, 1)$, $B(-1, 4)$, and $C(5, 8)$ is a right triangle by using (a) slopes and (b) the converse of Pythagorean Theorem. See page 159.

112. Show that the four points, $A(-4, -1)$, $B(1, 2)$, $C(3, 1)$, and $D(-2, -2)$ are the vertices of a parallelogram.
[*Hint:* Opposite sides of a parallelogram are parallel.]

113. Show that if two nonvertical lines are parallel, their slopes are equal. Start by letting the two lines be show ℓ_1: $y = m_1x + b_1$ and ℓ_2: $y = m_2x + b_2$ $(b_1 > b_2)$. that if (x_1, y_1) and (x_2, y_2) are on ℓ_1, then $(x_1, y_1 - (b_1 - b_2))$ and $((x_2, y_2 - (b_1 - b_2))$ must be on ℓ_2. Then by computing m_1 and m_2, show that $m_1 = m_2$. (See the accompanying figure.)

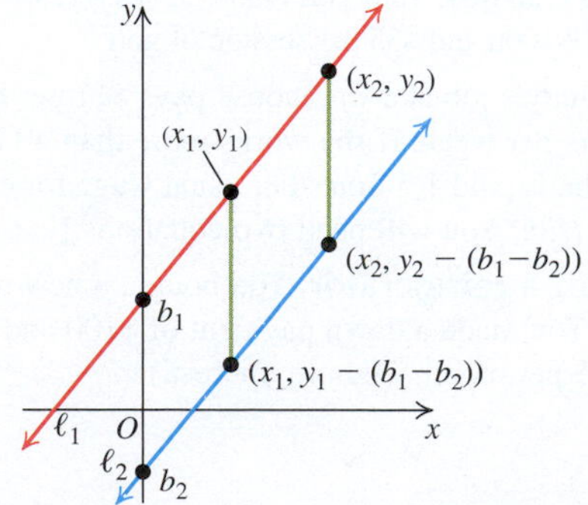

114. In the accompanying figure, note that lines ℓ_1 and ℓ_2 have slopes m_1 and m_2. Assume that ℓ_1 and ℓ_2 are perpendicular. Use the distance formula and the Pythagorean theorem to show that $m_1 \cdot m_2 = -1$.

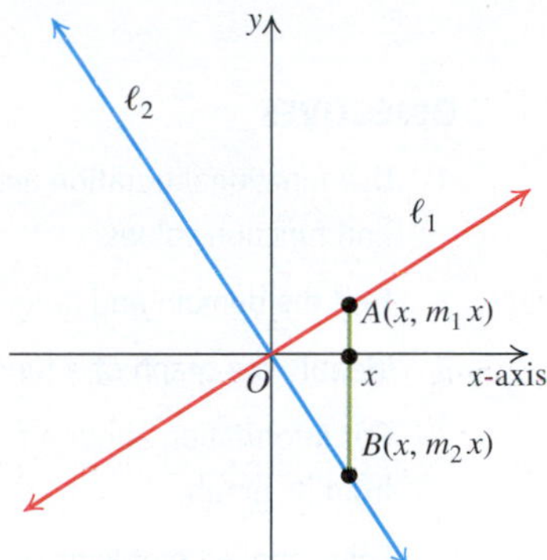

115. **Geometry.** Find the coordinates of the point B of a line segment AB, given that $A = (2, 2)$, $d(A, B) = 12.5$, and the slope of the line segment AB is $\frac{4}{3}$.

116. **Geometry.** Show that an equation of a circle with (x_1, y_1) and (x_2, y_2) as endpoints of a diameter can be written in the form

$$(x - x_1)(x - x_2) + (y - y_1)(y - y_2) = 0.$$

[*Hint:* Use the fact that when the two endpoints of a diameter and any other point on the circle are used as vertices for a triangle, the angle opposite the diameter is a right angle.]

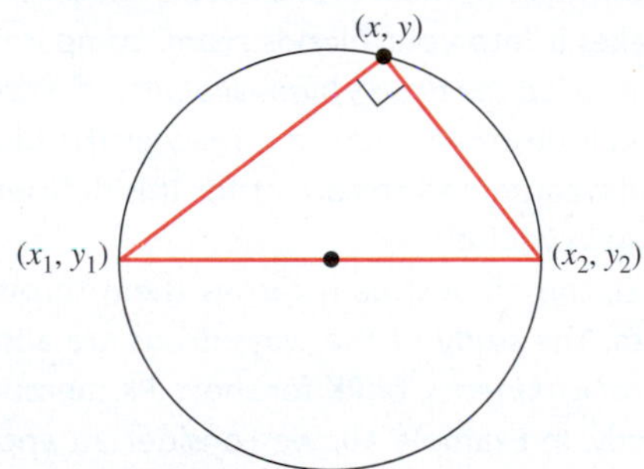

117. Find an equation of the circle the endpoints of whose diameter are the centers of the circles: $x^2 + y^2 + 6x - 14y - 1 = 0$ and $x^2 + y^2 - 4x + 10y - 2 = 0$. (*Hint:* Use Exercise 116.)

118. **Geometry.** Find an equation of the tangent line to the circle $x^2 + y^2 = 169$ at the point $(-5, 12)$. (See the accompanying figure.)

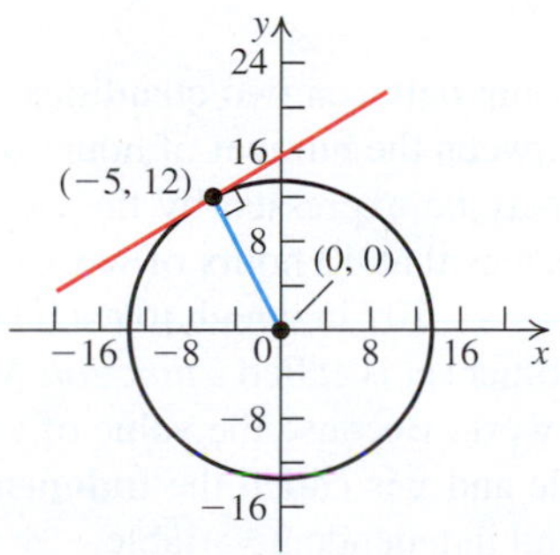

119. Find an equation of the tangent line to the circle $x^2 + y^2 = 25$ at the point $(4, -3)$.

120. Find an equation of the tangent line to the circle $x^2 + y^2 + 4x - 6y - 12 = 0$ at the point $(1, 7)$.

In Exercises 121–124, use a graphing device to graph the given lines on the same screen. Interpret your observations about the family of lines.

121. $y = 2x + b$ for $b = 0, \pm 2, \pm 4$

122. $y = mx + 2$ for $m = 0, \pm 2, \pm 4$

123. $y = m(x - 1)$ for $m = 0, \pm 3, \pm 5$

124. $y = m(x + 1) - 2$ for $m = 0, \pm 3, \pm 5$

Critical Thinking/Discussion/Writing

125. **a.** The equation $y + 2x + k = 0$ describes a family of lines for each value of k. Sketch the graphs of the members of this family for $k = -3, 0$, and 2. What is the common characteristic of the family?

b. Repeat part (a) for the family of lines $y + kx + 4 = 0$.

126. Explain how you can determine the sign of the x-coordinate of the point of intersection of the lines $y = m_1x + b_1$ and $y = m_2x + b_2$, if

a. $m_1 > m_2 > 0$ and $b_1 > b_2$.
b. $m_1 > m_2 > 0$ and $b_1 < b_2$.
c. $m_1 < m_2 < 0$ and $b_1 > b_2$.
d. $m_1 < m_2 < 0$ and $b_1 < b_2$.

Maintaining Skills

In Exercises 127–132, solve each equation or inequality for x.

127. $x^2 - x - 2 = 0$

128. $x^2 + 2x - 3 = 0$

129. $\frac{x^2 + 5x + 6}{x - 1} = 0$

130. $\frac{6x^2 - x - 1}{x^2 - x - 12} = 0$

131. $x^2 - 5x + 6 \geq 0$

132. $\frac{x - 1}{x + 2} + 1 \geq 0$

In Exercises 133 and 134, simplify each expression.

133. $\frac{1}{h}\left[\frac{1}{x + h} - \frac{1}{x}\right]$

134. $\frac{-(x + h)^2 + x^2}{h}$

In Exercises 135 and 136, rationalize the numerator.

135. $\frac{\sqrt{5} - \sqrt{2}}{3}$

136. $\frac{\sqrt{x + 2} - \sqrt{x}}{2}$

SECTION 2.4

Functions

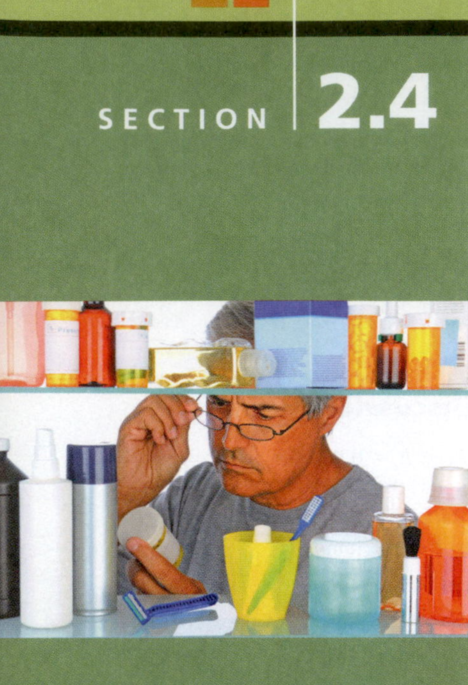

BEFORE STARTING THIS SECTION, REVIEW

1. Graph of an equation (Section 2.2, page 165)
2. Equation of a circle (Section 2.2, page 172)

OBJECTIVES

1. Use functional notation and find function values.
2. Find the domain and range of a function.
3. Identify the graph of a function.
4. Get information about a function from its graph.
5. Solve applied problems by using functions.

◆ The Ups and Downs of Drug Levels

Medicines in the form of a pill have a much harder time getting to work than you do. First, they are swallowed and dumped into a pool of acid in your stomach. Then they dissolve; leave the stomach; and begin to be absorbed, mostly through the lining of the small intestine. Next, the blood around the intestine carries the medicines to the liver, an organ designed to metabolize (break down) and remove foreign material from the blood. Because of this property of the liver, most drugs have a tough time getting past it. The amount of drug that makes it into your bloodstream, compared with the amount that you put in your mouth, is called the drug's *bioavailability*. If a drug has low bioavailability, then much of the drug is destroyed before it reaches the bloodstream. The amount of drug you take in a pill is calculated to correct for this deficiency so that the amount you need actually ends up in your blood.

Once the drugs are past the liver, the bloodstream carries them throughout the rest of the body in about one minute. The study of the ways drugs are absorbed and move through the body is called *pharmacokinetics,* or PK for short. PK measures the ups and downs of drug levels in your body. In Example 10, we consider an application of *functions* to the levels of drugs in the bloodstream.

(Adapted from *The Ups and Downs of Drug Levels,* by Bob Munk, www.thebody.com/content/treat/art1062.html)

1 Use functional notation and find function values.

Functions

There are many ways of expressing a relationship between two quantities. For example, if you are paid \$10 per hour, then the relation between the number of hours, x, that you work and the amount of money, y, that you earn may be expressed by the equation $y = 10x$. Replacing x with 40 yields $y = 400$ and indicates that 40 hours of work corresponds to a \$400 paycheck. A special relationship such as $y = 10x$ in which to each element x in one set there corresponds a unique element y in another set is called a *function*. We say that your pay is a function of the number of hours you work. Because the value of y depends on the value of x, y is called the **dependent variable** and x is called the **independent variable**. Any symbol may be used for the dependent and independent variable.

Function

A **function** from a set X to a set Y is a rule that assigns to each element of X exactly one element of Y. The set X is the **domain** of the function. The set of those elements of Y that correspond (are assigned) to the elements of X is the **range** of the function.

A rule of assignment may be described by a **correspondence diagram**, in which an arrow points from each domain element to the corresponding element or elements in the range.

To decide whether a correspondence diagram defines a function, we must check whether *any* domain element is paired with more than one range element. If this happens, that domain element does not have a *unique* corresponding element of Y and the correspondence is not a function. See the correspondence diagrams in Figure 2.34.

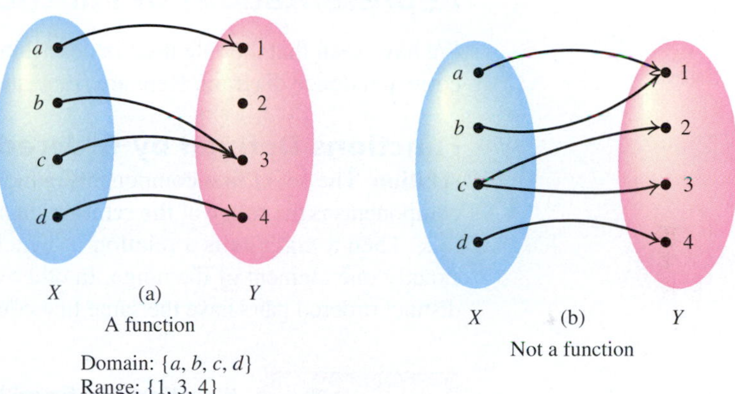

FIGURE 2.34 Correspondence diagrams

MAIN FACTS FROM THE DEFINITION OF A FUNCTION

Let f be a function from a set X to a set Y.

1. The domain of f is the entire set X. That is, *each* element of X must be assigned to some element of Y.
2. The range of f is the set of those elements in Y that are assigned to some element of X. So the range of f may not be the entire set Y.
3. It *is permissible* to assign the same element of Y to two or more elements of X. See Figure 2.34(a).
4. It *is not permissible* to assign an element of X to two or more elements of Y. See Figure 2.34(b).

Function Notation

DO YOU **KNOW?**

The functional notation $y = f(x)$ was first used by the great Swiss mathematician Leonhard Euler in the *Commentarii Academia Petropolitanae*, published in 1735.

We usually use single letters such as f, F, g, G, h, and H as the name of a function. Some special function names use more than one letter, such as *sin* (the sine function) and *log* (the logarithm function). If we choose f as the name of a function, then for each x in the domain of f, there corresponds a unique y in its range. The number y is denoted by $f(x)$, read as "f of x" or as "f at x." We call $f(x)$ the **value of f at the number x** and say that *f assigns* the value $f(x)$ to x. If the function shown in Figure 2.34(a) is named f, then the correspondence diagram may be described by:

$$f(a) = 1,\ f(b) = 3,\ f(c) = 3, \text{ and } f(d) = 4.$$

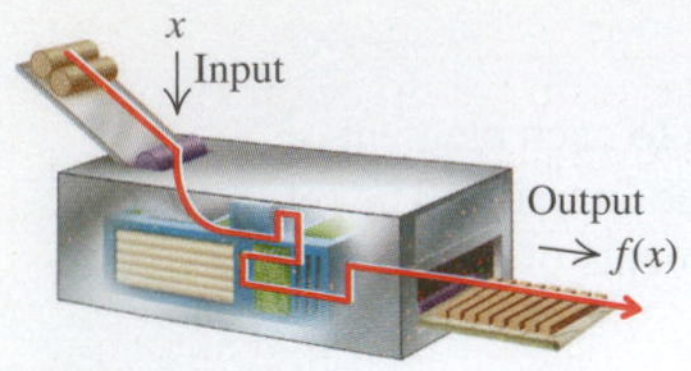

FIGURE 2.35 The function f, as a machine

Yet another way to illustrate the concept of a function is by a "machine," as shown in Figure 2.35. If x is in the domain of a function, then the machine accepts it as an "input" and produces the value $f(x)$ as the "output." In other words, **output is a function of the input**. A calculator is an example of a function machine. If an input number (say, 9) is entered and some key (such as $\sqrt{}$) is pressed, the display shows the function value (in this case, the output number 3).

WARNING

In writing $y = f(x)$, do not confuse f, which is the name of the function, with $f(x)$, which is a number in the range of f that the function assigns to x.
The symbol $f(x)$ does not mean "f times x."

Representations of Functions

We have seen that the rule used in describing a function may be expressed in the form of a correspondence diagram. Here are some additional forms.

Functions Defined by Ordered Pairs Any set of ordered pairs is called a **relation**. The set of first components is the **domain of the relation**, and the set of second components is the **range of the relation**. In the ordered pair (x, y), we say that y corresponds to x. Then a *function* is a relation in which each element of the domain corresponds to exactly one element of the range. In other words, a function is a relation in which no two distinct ordered pairs have the same first component.

EXAMPLE 1 Functions Defined by Ordered Pairs

Determine whether each relation defines a function.

a. $r = \{(-1, 2), (1, 3), (5, 2), (-1, -3)\}$
b. $s = \{(-2, 1), (0, 2), (2, 1), (-1, -3)\}$

Solution

a. The domain of the relation r is $\{-1, 1, 5\}$, and its range is $\{2, 3, -3\}$. It is not a function because the ordered pairs $(-1, 2)$ and $(-1, -3)$ have the same first component but the second components are different. See the correspondence diagram for the relation r in Figure 2.36.

b. The domain of the relation s is $\{-2, -1, 0, 2\}$, and its range is $\{-3, 1, 2\}$. The relation s is a function because no two ordered pairs of s have the same first component. Figure 2.37 shows the correspondence diagram for the function s.

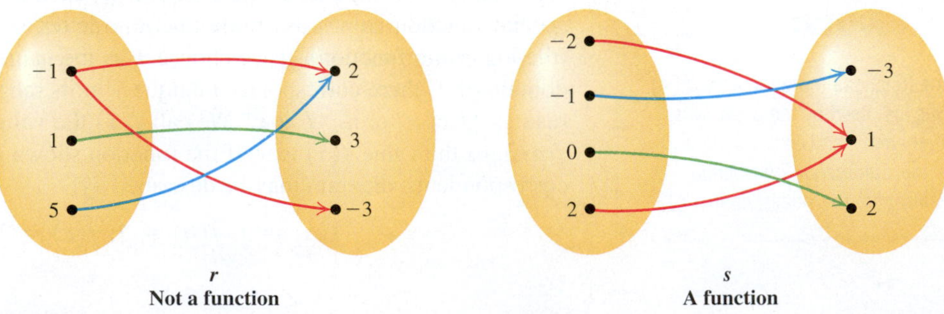

FIGURE 2.36 **FIGURE 2.37**

Practice Problem 1 Repeat Example 1 for each set of ordered pairs.

a. $R = \{(2, 1), (-2, 1), (3, 2)\}$
b. $S = \{(2, 5), (3, -2), (3, 5)\}$

TABLE 2.5

x	y
-1	2
1	3
5	2
-1	-3
Relation r	

TABLE 2.6

x	y
-2	1
-1	-3
0	2
2	1
Relation s	

Functions Defined by Tables and Graphs

Tables and graphs can be used to describe relations and functions. We can display in tabular form the list of ordered pairs (x, y) of a relation, where x is in the domain and the corresponding y is in the range. For example, the relations r and s of Example 1 described in tabular form are shown in Tables 2.5 and 2.6, respectively.

Alternatively, we can display geometrically the list of ordered pairs (x, y) in a relation by plotting each pair (x, y) as a point in the coordinate plane. The geometric display is the graph of the relation. Graphs of the relations r and s are given in Figure 2.38.

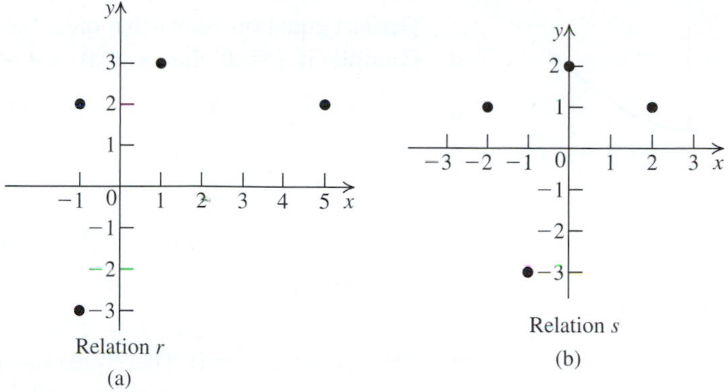

FIGURE 2.38 Graphs of relations

The graph of the discrete data points of a relation or function is called a **scatterplot**.

Functions Defined by Equations

When the relation defined by an equation in two variables is a function, we can often solve the equation for the dependent variable in terms of the independent variable. For example, the equation $y - x^2 = 0$ can be solved for y in terms of x:

$$y = x^2.$$

Now we can replace the dependent variable, in this case, y, with functional notation $f(x)$ and express the function as

$$f(x) = x^2 \quad \text{(read "}f\text{ of }x\text{ equals }x^2\text{")}.$$

Here $f(x)$ plays the role of y and is the value of the function f at x. For example, if $x = 3$ is an element (input value) in the domain of f, the corresponding element (output value) in the range is found by replacing x with 3 in the equation

$$f(x) = x^2$$

so that $\quad f(3) = 3^2 = 9. \quad$ Replace x with 3.

We say that the value of the function f at 3 is 9, or 9 is the evaluated function value $f(3)$. See Figure 2.39. In other words, the number 9 in the range corresponds to the number 3 in the domain and $(3, 9)$ is an ordered pair of the function f.

If a function g is defined by an equation such as $y = x^2 - 6x + 8$, the notations

$$y = x^2 - 6x + 8 \quad \text{and} \quad g(x) = x^2 - 6x + 8$$

define the same function.

SIDE NOTE

The definition of a function is independent of the letters used to denote the function and the variables. For example,

$$s(x) = x^2, g(y) = y^2, \text{ and } h(t) = t^2$$

represent the same function.

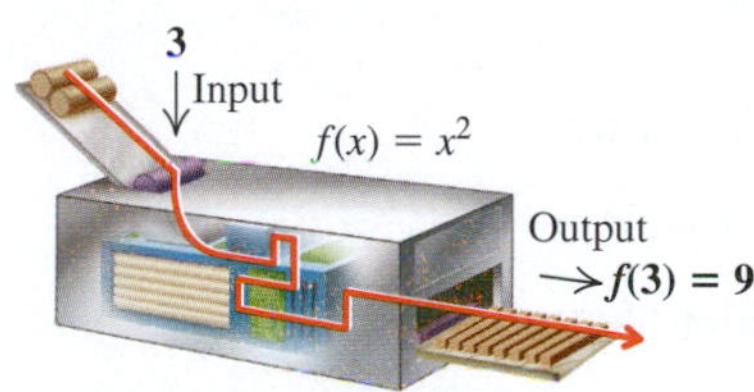

FIGURE 2.39 The function f, as a machine

EXAMPLE 2 Determining Whether an Equation Defines a Function

Determine whether y is a function of x for each equation.

a. $6x^2 - 3y = 12$ **b.** $y^2 - x^2 = 4$

Solution

Solve each equation for y in terms of x. If more than one value of y corresponds to the same value of x, then y is not a function of x.

a.

$6x^2 - 3y = 12$	Original equation
$6x^2 - 3y + 3y - 12 = 12 + 3y - 12$	Add $3y - 12$ to both sides.
$6x^2 - 12 = 3y$	Simplify.
$2x^2 - 4 = y$	Divide by 3.

The last equation shows that only one value of y corresponds to each value of x. For example, if $x = 0$, then $y = 0 - 4 = -4$. So y is a function of x.

b.

$y^2 - x^2 = 4$	Original equation
$y^2 - x^2 + x^2 = 4 + x^2$	Add x^2 to both sides.
$y^2 = x^2 + 4$	Simplify.
$y = \pm\sqrt{x^2 + 4}$	Square root property

The last equation shows that two values of y correspond to each value of x. For example, if $x = 0$, then $y = \pm\sqrt{0^2 + 4} = \pm\sqrt{4} = \pm 2$. Both $y = 2$ and $y = -2$ correspond to $x = 0$. Therefore, y is not a function of x. See Figure 2.40. However, the equation $y^2 - x^2 = 4$ defines two functions:

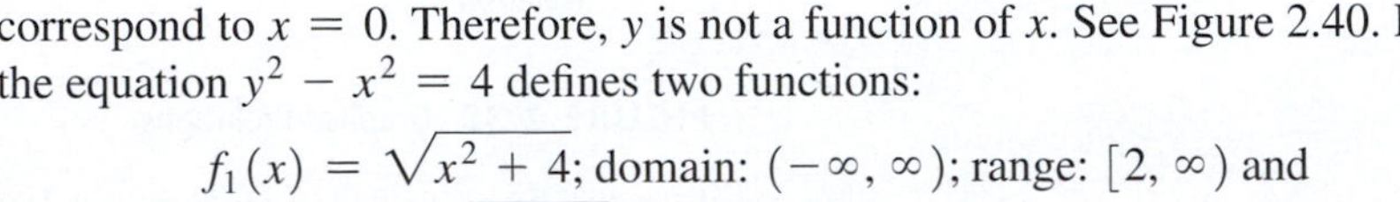

$$f_1(x) = \sqrt{x^2 + 4}; \text{ domain: } (-\infty, \infty); \text{ range: } [2, \infty) \text{ and}$$
$$f_2(x) = -\sqrt{x^2 + 4}; \text{ domain: } (-\infty, \infty); \text{ range } (-\infty, -2].$$

$y = f_1(x) = \sqrt{x^2 + 4}$

$y = f_2(x) = -\sqrt{x^2 + 4}$

FIGURE 2.40

Practice Problem 2 Determine whether y is a function of x for each equation.

a. $2x^2 - y^2 = 1$ **b.** $x - 2y = 5$

EXAMPLE 3 Evaluating a Function

Let g be the function defined by the equation

$$y = x^2 - 6x + 8.$$

Final each function value.

a. $g(3)$ **b.** $g(-2)$ **c.** $g\left(\frac{1}{2}\right)$ **d.** $g(a + 2)$ **e.** $g(x + h)$

Solution

Because the function is named g, we replace y with $g(x)$ and write

$$g(x) = x^2 - 6x + 8 \quad \text{Replace } y \text{ with } g(x).$$

In this notation, the independent variable x is a placeholder. We can write $g(x) = x^2 - 6x + 8$ as

$$g(\ ) = (\ )^2 - 6(\ ) + 8.$$

a.

$g(x) = x^2 - 6x + 8$	Given equation
$g(3) = 3^2 - 6(3) + 8$	Replace x with 3 at each occurrence of x.
$= 9 - 18 + 8 = -1$	Simplify.

The statement $g(3) = -1$ means that the value of the function g at 3 is -1.

Just as in part **a**, we can evaluate g at any value x in its domain.

$g(x) = x^2 - 6x + 8$	Original equation
b. $g(-2) = (-2)^2 - 6(-2) + 8 = 24$	Replace x with -2 and simplify.
c. $g\left(\frac{1}{2}\right) = \left(\frac{1}{2}\right)^2 - 6\left(\frac{1}{2}\right) + 8 = \frac{21}{4}$	Replace x with $\frac{1}{2}$ and simplify.

SIDE NOTE

Problems like Examples 3(d) and 3(e) will arise frequently. We emphasize that you must take whatever appears in the parentheses and substitute for the independent variable.

d. $g(a+2) = (a+2)^2 - 6(a+2) + 8$ Replace x with $(a+2)$ in $g(x)$.

$= a^2 + 4a + 4 - 6a - 12 + 8$ Recall: $(x+y)^2 = x^2 + 2xy + y^2$.

$= a^2 - 2a$ Simplify.

e. $g(x+h) = (x+h)^2 - 6(x+h) + 8$ Replace x with $(x+h)$ in $g(x)$.

$= x^2 + 2xh + h^2 - 6x - 6h + 8$ Simplify.

Practice Problem 3 Let g be the function defined by the equation $y = -2x^2 + 5x$. Find each function value.

a. $g(0)$ **b.** $g(-1)$ **c.** $g(x+h)$

EXAMPLE 4 Finding the Area of a Rectangle

In Figure 2.41, find the area of the rectangle *PLMN*.

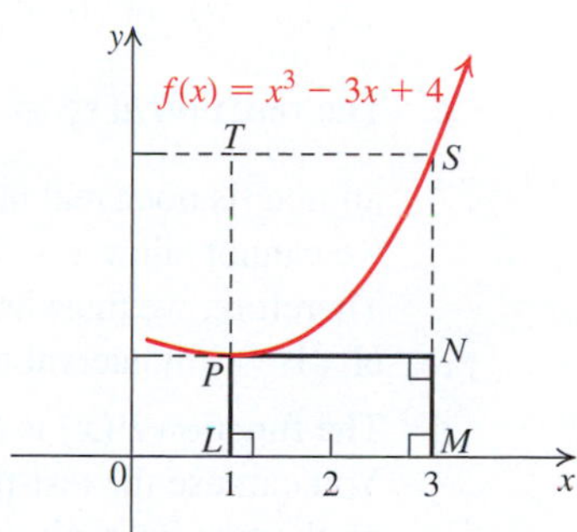

FIGURE 2.41

Solution

Area of the rectangle *PLMN* = (length)(height)

$= (3-1)(f(1))$ $P = (1, f(1))$

$= (2)(2)$ $f(1) = 1^3 - 3(1) + 4 = 2$

$= 4$ sq. units

Practice Problem 4 In Figure 2.41, find the area of the rectangle *TLMS*.

2 Find the domain and range of a function.

The Domain of a Function

Sometimes a function does not have a specified domain.

AGREEMENT ON DOMAIN

If the domain of a function that is defined by an equation is not specified, then we agree that the domain of the function is the largest set of real number inputs that result in real number outputs.

DO YOU KNOW?

The word *domain* comes from the Latin word *dominus*, or *master*, which in turn is related to *domus*, meaning "home." *Domus* is also the root for the words *domestic* and *domicile*.

When we use our agreement to find the domain of a function, we usually find the real values of the independent variable that do not result in real values of the dependent variable. Then we exclude those numbers from the domain. Remember that:

1. Division by zero is undefined.
2. The square root (or any even root) of a negative number is not a real number.

EXAMPLE 5 Finding the Domain of a Function

Find the domain of each function.

a. $f(x) = \dfrac{1}{1 - x^2}$ **b.** $g(x) = \sqrt{x}$

c. $h(x) = \dfrac{1}{\sqrt{x - 1}}$ **d.** $P(x) = \sqrt{x^2 - x - 6}$

Solution

a. The function f is not defined when the denominator $1 - x^2$ is 0. Because $1 - x^2 = 0$ if $x = 1$ or $x = -1$, the domain of f is the set $\{x | x \neq -1$ and $x \neq 1\}$, which in interval notation is written $(-\infty, -1) \cup (-1, 1) \cup (1, \infty)$.

b. Because the square root of a negative number is not a real number, negative numbers are excluded from the domain of g. The domain of $g(x) = \sqrt{x}$ is $\{x | x \geq 0\}$, or $[0, \infty)$ in interval notation.

c. The function $h(x) = \dfrac{1}{\sqrt{x - 1}}$ has *two* restrictions. The square root of a negative number is not a real number, so $\sqrt{x - 1}$ is a real number only if $x - 1 \geq 0$. However, we cannot allow $x - 1 = 0$ because $\sqrt{x - 1}$ is in the denominator and $\sqrt{0} = 0$. Therefore, we must have $x - 1 > 0$, or $x > 1$. The domain of h must be $\{x | x > 1\}$, or $(1, \infty)$ in interval notation.

d. The function $P(x)$ is defined when the expression under the radical sign is nonnegative. You can use the test-point method to see that $x^2 - x - 6 = (x + 2)(x - 3) \geq 0$ on the two intervals $(-\infty, -2]$ and $[3, \infty)$. So the domain of the function $P(x)$ in interval notation is $(-\infty, -2] \cup [3, \infty)$.

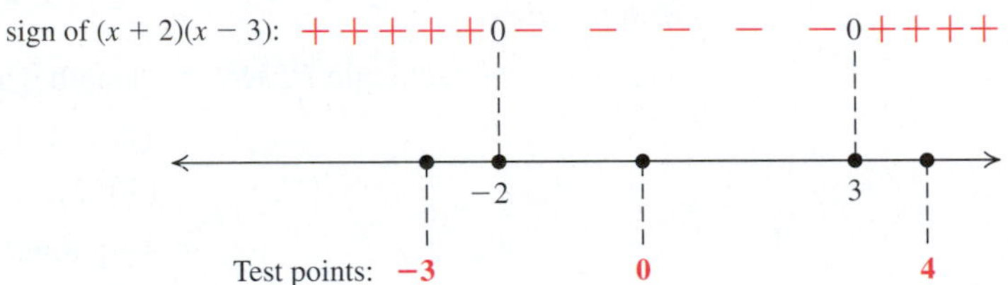

Practice Problem 5 Find the domain of each function.

a. $f(x) = \dfrac{1}{\sqrt{1 - x}}$. **b.** $g(x) = \sqrt{\dfrac{x + 2}{x - 3}}$

The Range of a Function

Suppose a function f is defined by an equation. A number y is in the range of f if the equation $f(x) = y$ has at least one solution.

EXAMPLE 6 Finding the Range of a Function

Let $f(x) = x^2$ with domain $X = [3, 5]$.

a. Is 10 in the range of f?

b. Is 4 in the range of f?

c. Find the range of f.

Solution

a. We find possible solutions of the equation

$$f(x) = 10.$$
$$x^2 = 10 \quad \text{Replace } f(x) \text{ with } x^2.$$
$$x = \pm\sqrt{10} \quad \text{Square root property}$$

Because $3 < \sqrt{10} < 5$, the number $x = \sqrt{10}$ is in the interval $X = [3, 5]$. So the equation $f(x) = 10$ has at least one solution in the domain of f. Therefore, 10 is in the range of f.

b. The solutions of the equation $x^2 = 4$ are $x = \pm 2$. Neither of these numbers is in the domain $X = [3, 5]$. So 4 is not in the range of f.

c. The range of f is the interval $[9, 25]$ because for each number y in the interval $[9, 25]$, the number $x = \sqrt{y}$ is in the interval $[3, 5]$ so that $f(x) = f(\sqrt{y}) = (\sqrt{y})^2 = y$.

Practice Problem 6 Repeat Example 6 for $f(x) = x^2$ with domain $X = [-3, 3]$.

Finding the range of a function such as $f(x) = \dfrac{x + 1}{x - 2}$ is a little more complicated. We discuss this problem in Section 2.9.

3 Identify the graph of a function.

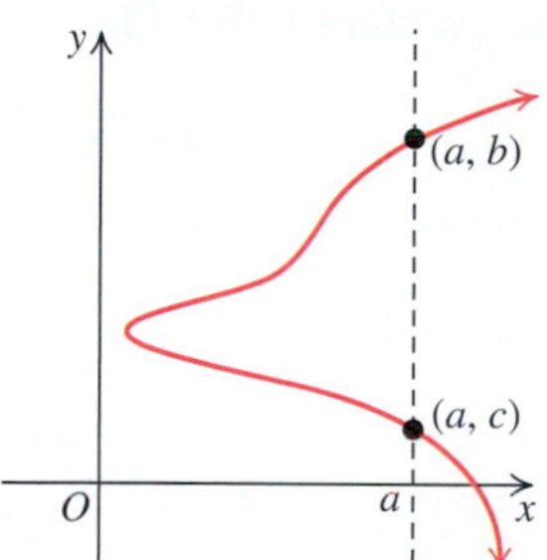

FIGURE 2.42 Graph does not represent a function

Graphs of Functions

The **graph of a function** f is the set of plots of ordered pairs $(x, f(x))$ such that x is in the domain of f. That is, the graph of f is the graph of the equation $y = f(x)$. We can sketch the graph of $y = f(x)$ by plotting a sufficiently large number of points and joining them with a smooth curve. The graph of a function provides valuable visual information about the function.

Figure 2.42 shows that not every curve in the plane is the graph of a function. In Figure 2.42, a vertical line $x = a$ intersects the curve at two distinct points (a, b) and (a, c). This curve cannot be the graph of $y = f(x)$ for any function f because having $f(a) = b$ and $f(a) = c$ means that f assigns two different range values to the same domain element a. We can express this statement, called the **vertical-line test**, as follows:

VERTICAL-LINE TEST

If no vertical line intersects the graph of a relation at more than one point, then the graph of the relation is the graph of a function.

EXAMPLE 7 **Identifying the Graph of a Function**

Determine which graphs in Figure 2.43 are graphs of functions.

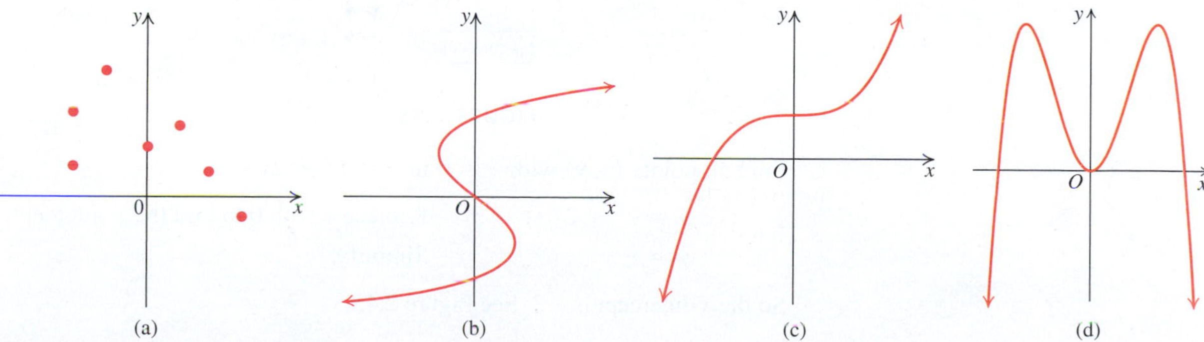

FIGURE 2.43 The vertical-line test

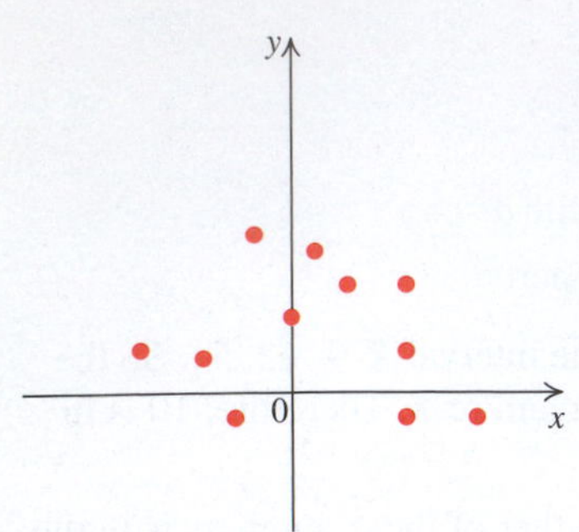

FIGURE 2.44

Solution

The graphs in Figures 2.43(a) and 2.43(b) are not graphs of functions because a vertical line can be drawn through the two points farthest to the left in Figure 2.43(a) and the y-axis is one of many vertical lines that contain more than one point on the graph in Figure 2.43(b). The graphs in Figures 2.43(c) and 2.43(d) are the graphs of functions because no vertical line intersects either graph at more than one point.

Practice Problem 7 Decide whether the graph in Figure 2.44 is the graph of a function.

EXAMPLE 8 Examining the Graph of a Function

Let $f(x) = x^2 - 2x - 3$.

a. Is the point $(1, -3)$ on the graph of f?

b. Find all values of x so that $(x, 5)$ is on the graph of f.

c. Find the y-intercept of the graph of f (if any).

d. Find all x-intercepts of the graph of f.

Solution

a. We check whether $(1, -3)$ satisfies the equation $y = x^2 - 2x - 3$.

$$-3 \stackrel{?}{=} (1)^2 - 2(1) - 3 \quad \text{Replace } x \text{ with 1 and } y \text{ with } -3.$$
$$-3 \stackrel{?}{=} -4 \quad \text{No!}$$

Note that $f(1) = 1^2 - 2(1) - 3 = -4$ so $(1, -4)$ is on the graph of f. But $(1, -3)$ is *not* on the graph of f. See Figure 2.45.

b. Substitute 5 for y in $y = x^2 - 2x - 3$ and solve for x.

$$5 = x^2 - 2x - 3$$
$$0 = x^2 - 2x - 8 \quad \text{Subtract 5 from both sides.}$$
$$0 = (x - 4)(x + 2) \quad \text{Factor.}$$
$$x - 4 = 0 \quad \text{or} \quad x + 2 = 0 \quad \text{Zero-product property}$$
$$x = 4 \quad \text{or} \quad x = -2 \quad \text{Solve for } x.$$

The point $(x, 5)$ is on the graph of f only when $x = -2$ or $x = 4$. Both points $(-2, 5)$ and $(4, 5)$ are on the graph of f. See Figure 2.45.

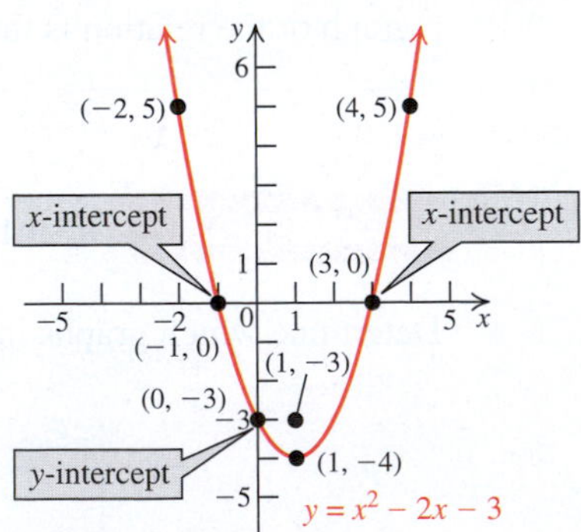

FIGURE 2.45

c. Find all points (x, y) with $x = 0$ in $y = x^2 - 2x - 3$.

$$y = 0^2 - 2(0) - 3 \quad \text{Replace } x \text{ with 0 to find the } y\text{-intercept.}$$
$$y = -3 \quad \text{Simplify.}$$

So the y-intercept is -3. See Figure 2.45.

d. Find all points (x, y) with $y = 0$ in $y = x^2 - 2x - 3$.

$$0 = x^2 - 2x - 3$$
$$0 = (x + 1)(x - 3) \quad \text{Factor.}$$
$$x + 1 = 0 \quad \text{or} \quad x - 3 = 0 \quad \text{Zero-product property}$$
$$x = -1 \quad \text{or} \quad x = 3 \quad \text{Solve for } x.$$

The x-intercepts of the graph of f are -1 and 3. See Figure 2.45.

Practice Problem 8 Let $f(x) = x^2 + 4x - 5$.

a. Is the point $(2, 7)$ on the graph of f?

b. Find all values of x so that $(x, -8)$ is on the graph of f.

c. Find the y-intercept of the graph of f.

d. Find any x-intercepts of the graph of f.

4 Get information about a function from its graph.

Function Information from Its Graph

Given a graph in the xy-plane, we use the vertical line test to determine whether the graph is that of a function. We can also obtain the following additional information from the graph of a function.

1. Point on a graph
A point (a, b) is on the graph of f means that a is in the domain of f and the value of f at a is b; that is, $f(a) = b$. We can visually determine whether a given point is on the graph of a function. In Figure 2.46, the point (a, b) is on the graph of f and the point (c, d) is not on the graph of f.

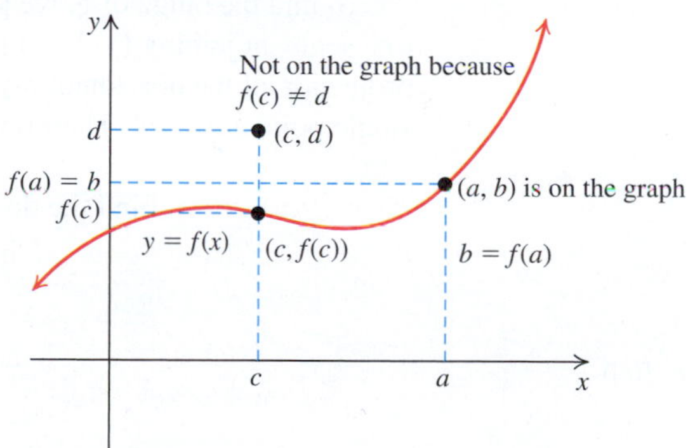

FIGURE 2.46

2. Domain and range from a graph
To determine the *domain* of a function, we look for the portion on the x-axis that is used in graphing f. We can find this portion by projecting (collapsing) the graph on the x-axis. This projection is the domain of f.

The range of f is the projection of its graph on the y-axis. See Figure 2.47.

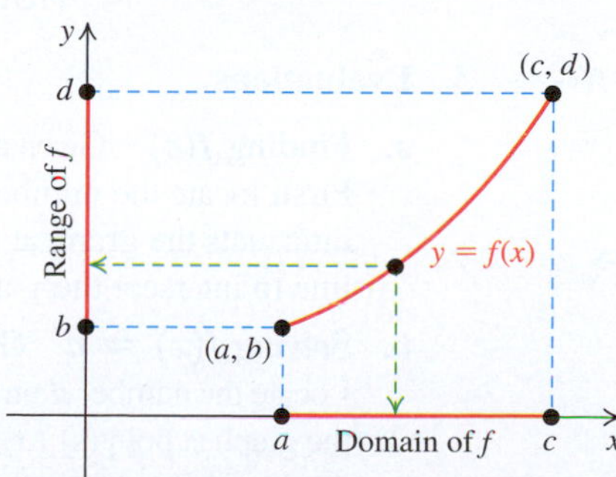

FIGURE 2.47 Domain and range of f

EXAMPLE 9 Finding the Domain and Range from a Graph

Use the graphs in Figure 2.48 to find the domain and the range of each function.

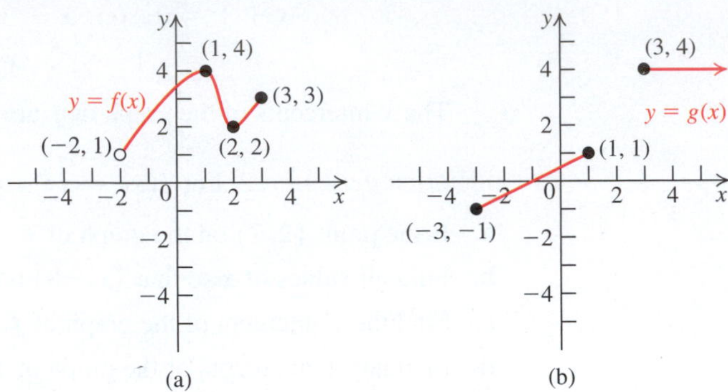

FIGURE 2.48 Domain and range from a graph

Solution

a. In Figure 2.48(a), the open circle at $(-2, 1)$ indicates that the point $(-2, 1)$ does not belong to the graph of f and the full circle at the point $(3, 3)$ indicates that the point $(3, 3)$ is part of the graph. When we project the graph of $y = f(x)$ onto the x-axis, we obtain the interval $(-2, 3]$. The domain of f in interval notation is $(-2, 3]$. Similarly, the projection of the graph of f onto the y-axis gives the interval $(1, 4]$, which is the range of f.

b. The projection of the graph of g in Figure 2.48(b) onto the x-axis is made up of two intervals $[-3, 1]$ and $[3, \infty)$. So the domain of g is $[-3, 1] \cup [3, \infty)$.

To find the range of g, we project its graph onto the y-axis. The projection of the line segment joining $(-3, -1)$ and $(1, 1)$ onto the y-axis is the interval $[-1, 1]$. The projection of the horizontal ray starting at the point $(3, 4)$ onto the y-axis is just a single point at $y = 4$. Therefore, the range of g is $[-1, 1] \cup \{4\}$.

Practice Problem 9 Find the domain and the range of $y = h(x)$ in Figure 2.49.

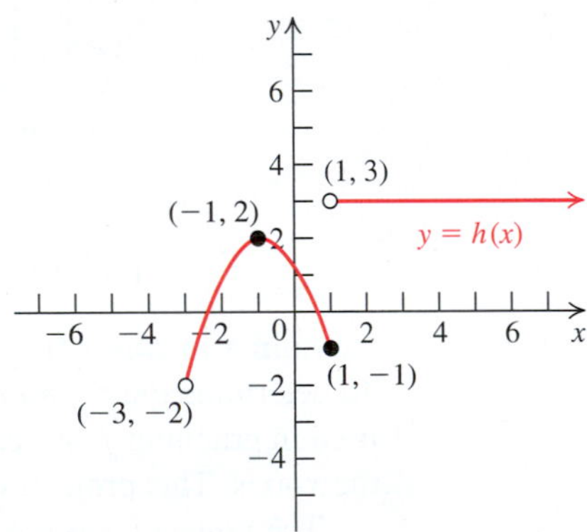

FIGURE 2.49

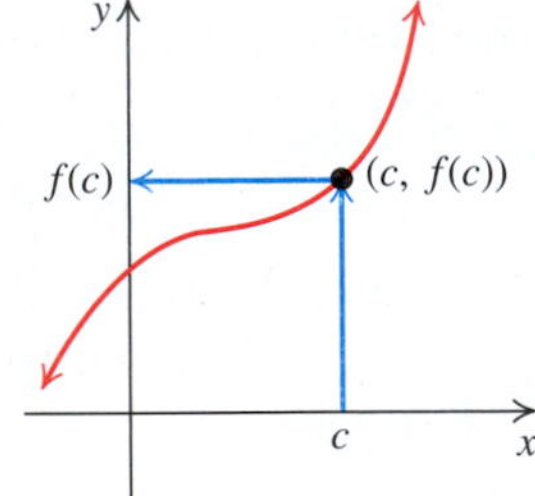

FIGURE 2.50 Locating $f(c)$ graphically

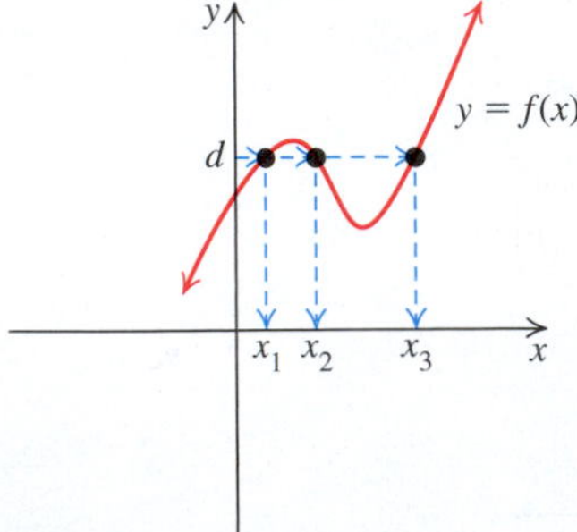

FIGURE 2.51 Graphically solving $f(x) = d$

3. Evaluations

a. Finding $f(c)$ Given a number c in the domain of f, we find $f(c)$ from the graph of f. First, locate the number c on the x-axis. Draw a vertical line through c. It intersects the graph at the point $(c, f(c))$. Through this point, draw a horizontal line to intersect the y-axis. Read the value $f(c)$ on the y-axis. See Figure 2.50.

b. Solving $f(x) = d$ Given d in the range of f, find values of x for which $f(x) = d$. Locate the number d on the y-axis. Draw a horizontal line through d. This line intersects the graph at point(s) (x, d). Through each of these points, draw vertical lines to intersect the x-axis. Read the values on the x-axis. These are the values of x for which $f(x) = d$. Figure 2.51 shows three solutions x_1, x_2, and x_3 of the equation $f(x) = d$.

SUMMARY OF **MAIN FACTS**

A function is usually described in one or more of the following six ways:

- A correspondence diagram
- A set of ordered pairs
- A table of values
- An equation or a formula
- A scatterplot or a graph
- In the words *y is a function of x*

Input	Output
x	y or $f(x)$
First coordinate	Second coordinate
Independent variable	Dependent variable
Domain is the set of all inputs.	Range is the set of all outputs.

5 Solve applied problems by using functions.

Building Functions

Level of Drugs in Bloodstream When you take a dose of medication, the drug level in the blood goes up quickly and soon reaches its *peak,* called C_{max} (the maximum concentration). As your liver or kidneys remove the drug, your blood's drug levels drop until the next dose enters your bloodstream. The lowest drug level is called the *trough,* or C_{min} (the minimum concentration). The ideal dose should be strong enough to be effective without causing too many side effects. We start by setting upper and lower limits on the drug's blood levels, shown by the two horizontal lines on a pharmacokinetics (PK) graph. See Figure 2.52. The upper line represents the drug level at which people start to develop serious side effects. The lower line represents the minimum drug level that provides the desired effect.

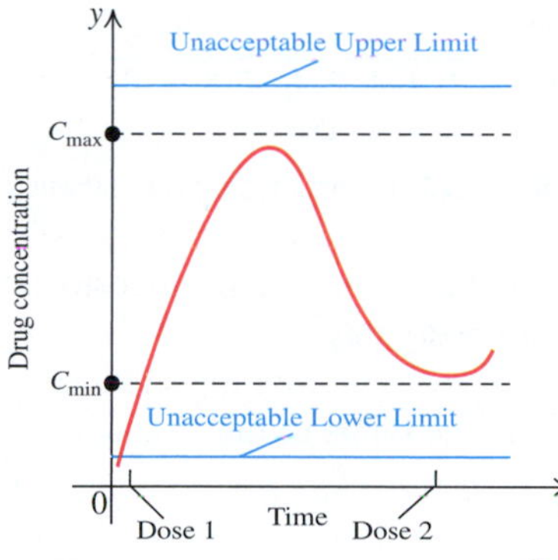

FIGURE 2.52 Drug levels

EXAMPLE 10 Cholesterol-Reducing Drugs

Many drugs used to lower high blood cholesterol levels are called *statins*. These drugs, along with proper diet and exercise, help prevent heart attacks and strokes. Recall from the introduction to this section that bioavailability is the amount of a drug you have ingested that makes it into your bloodstream. A statin with a bioavailability of 30% has been prescribed for Boris to treat his cholesterol levels. He is to take 20 milligrams daily. Every day his body filters out half the statin. Find the maximum concentration of the statin in the bloodstream on each of the first ten days of using the drug and graph the result.

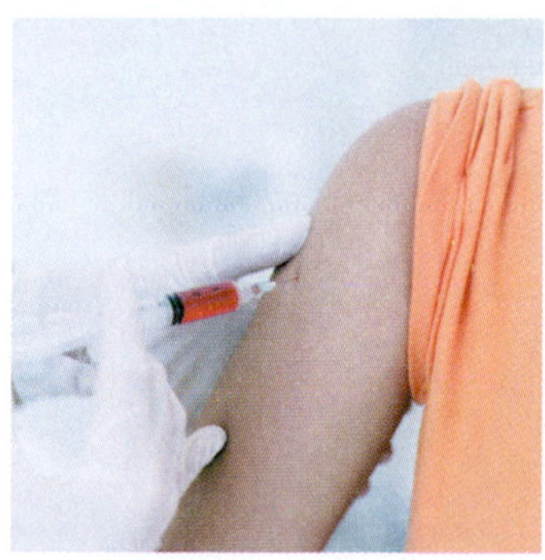

Solution

Because the statin has 30% bioavailability and Boris takes 20 milligrams per day, the maximum concentration in the bloodstream is 30% of 20 milligrams, or $20(0.3) = 6$ milligrams from each day's prescription. Because half the statin is filtered out of the body each day, the daily maximum concentration is

$$\frac{1}{2}(\text{previous days maximum concentration}) + 6.$$

So if $C(n)$ denotes the maximum concentration on the nth day, then

$$C(n) = \frac{1}{2}C(n-1) + 6, \quad C(0) = 0. \qquad (1)$$

Equation (1) is an example of a **recursive definition** of a function.

Table 2.7 shows the maximum concentration of the drug for each of the first ten days. After the first day, each number in the second column is computed (to three decimal places) by adding 6 to one-half the number above it. The graph of $C(t)$ is shown in Figure 2.53.

TABLE 2.7

Day	Maximum Concentration
1	6.000
2	9.000
3	10.500
4	11.250
5	11.625
6	11.813
7	11.907
8	11.954
9	11.977
10	11.989

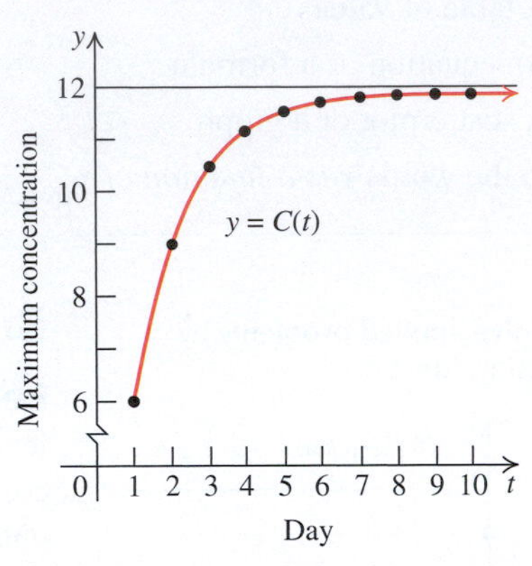

FIGURE 2.53 Maximum drug concentration

The graph shows that the maximum concentration of the statin in the bloodstream approaches 12 milligrams if Boris continues to take one pill every day.

Practice Problem 10 In Example 10, use Table 2.7 to find the range of the function $C(t)$. Also compute $C(11)$.

EXAMPLE 11 Cost of a Fiber-Optic Cable

Two points A and B are opposite each other on the banks of a straight river that is 500 feet wide. The point D is on the same side as B but is 1200 feet up the river from B. The local Internet cable company wants to lay a fiber-optic cable from A to D. The cost per foot of cable is \$5 per foot under water and \$3 per foot on land. To save money, the company lays the cable under water from A to P and then on land from P to D. In Figure 2.54, write the total cost C as a function of x.

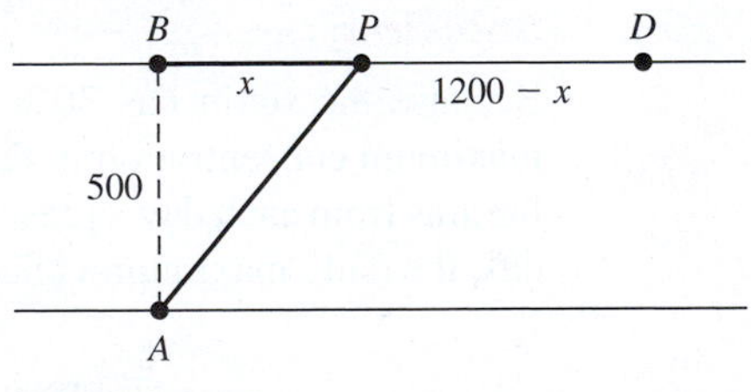

FIGURE 2.54

Solution

In Figure 2.54, we have

$$d(A, P) = AP = \sqrt{(500)^2 + x^2} \text{ feet} \qquad \text{Pythagorean theorem}$$
$$d(P, D) = PD = 1200 - x \text{ feet}$$

So the total cost C is given by

$$C = 5(AP) + 3(PD)$$
$$= 5\sqrt{(500)^2 + x^2} + 3(1200 - x).$$

Practice Problem 11 Repeat Example 11 assuming that the cost is 30% more under water than it is on land.

Functions in Economics

The important concepts of *breaking even, earning a profit,* or *taking a loss* in economics will be discussed throughout this book. Here we define some of the elementary functions in economics.

Let x represent the number of units of an item manufactured, and each item is sold at a price of p dollars. Then the cost $C(x)$ of producing x items includes the **fixed cost** (such as rent, insurance, product design, setup, promotion, and overhead) and the **variable cost** (which depends on the number of items produced at a certain cost per item).

Linear Cost Function A linear cost function $C(x)$ is given by

$$\begin{aligned} C(x) &= (\text{variable cost}) + (\text{fixed cost}) \\ &= ax + b, \end{aligned}$$

where the fixed cost is b and the **marginal cost** (cost of producing each item) is a dollars per item.

Average cost, denoted by $\overline{C}(x)$, is defined by $\overline{C}(x) = \dfrac{C(x)}{x}$.

Linear Price–Demand Function Suppose x items can be sold (demanded) at a price of p dollars per item. Then a linear demand function usually has the form

$$p(x) = -mx + d \qquad \text{Expressing } p \text{ as a function of } x$$

or

$$x(p) = -np + k \qquad \text{Expressing } x \text{ as a function of } p$$

Both expressions indicate that a higher price will result in fewer items sold. The constants $m, d, n,$ and k depend on the given situation, with $m > 0$ and $n > 0$. A **price–supply** function has the form $p = S(x)$ or $x = G(p)$, where a higher price results in a greater supply of items.

Revenue Function $R(x)$

Revenue = (Price per item) (Number of items sold)

$$R(x) = p \cdot x = (-mx + d)x \qquad p = p(x) = -mx + d$$

Profit Function $P(x)$

$$\begin{aligned} \text{Profit} &= \text{Revenue} - \text{Cost} \\ P(x) &= R(x) - C(x) \end{aligned}$$

A manufacturing company will

(i) Have a *profit* if $P(x) > 0$.

(ii) *Break even* if $P(x) = 0$.

(iii) Suffer a *loss* if $P(x) < 0$.

EXAMPLE 12 **Breaking Even**

Metro Entertainment Co. spent \$100,000 on production costs for its off-Broadway play *Bride and Prejudice*. Once the play runs, each performance costs \$1000 per show and the revenue from each show is \$2400. Using x to represent the number of shows,

a. Write the cost function $C(x)$.

b. Write the revenue function $R(x)$.

c. Write the profit function $P(x)$.

d. Determine how many showings of *Bride and Prejudice* must be held for Metro to break even.

Solution

a. $C(x) = (\text{Variable cost}) + (\text{Fixed cost})$
$= 1000x + 100{,}000$

b. $R(x) = (\text{Revenue per show})(\text{Number of shows})$
$= 2400x$

c. $P(x) = R(x) - C(x)$
$= 2400x - (1000x + 100{,}000)$
$= 2400x - 1000x - 100{,}000 = 1400x - 100{,}000$

d. Metro will break even when $P(x) = 0$.

$$1400x - 100{,}000 = 0$$
$$1400x = 100{,}000$$
$$x = \frac{100{,}000}{1400} \approx 71 \quad \text{Divide both sides by 1400.}$$

Only a whole number of shows is possible, so Metro needs 71 shows to break even.

Practice Problem 12 Suppose in Example 12 that once the play runs, each performance costs \$1200 and the revenue from each show is \$2500. How do the results in Example 12 change?

Answers to Practice Problems

1. a. Yes **b.** No **2. a.** No **b.** Yes **3. a.** 0 **b.** -7 **c.** $-2x^2 - 4hx + 5x - 2h^2 + 5h$ **4.** 44 sq. units **5. a.** $(-\infty, 1)$ **b.** $(-\infty, -2] \cup (3, \infty)$ **6. a.** No **b.** Yes **c.** $[0, 9]$ **7.** No **8. a.** Yes **b.** $-3, -1$ **c.** -5 **d.** $-5, 1$ **9.** Domain: $(-3, \infty)$; range: $(-2, 2] \cup \{3\}$ **10.** Range: $[6, 12)$; $C(11) = 11.994$ **11.** If c = cost per foot on land, then $C = c\left[1.3\sqrt{(500)^2 + x^2} + 1200 - x\right]$. **12. a.** $C = 1200x + 100{,}000$ **b.** $R = 2500x$ **c.** $P = 1300x - 100{,}000$ **d.** 77

SECTION 2.4 Exercises

Basic Concepts and Skills

1. In the functional notation $y = f(x)$, x is the ______________ variable.
2. If $f(-2) = 7$, then -2 is in the ______________ of the function f and 7 is in the ______________ of f.
3. If the point $(9, -14)$ is on the graph of a function f, then $f(9) =$ ______________.
4. If $(3, 7)$ and $(3, 0)$ are points on a graph, then the graph cannot be the graph of a(n) ______________.
5. To find the x-intercepts of the graph of an equation in x and y, we solve the equation ______________.
6. **True or False.** If a and b are in the domain of $f(x) = \frac{1}{x}$, then $a + b$ is also in the domain of f.
7. **True or False.** If $x = -7$, then $-x$ is in the domain of $f(x) = \sqrt{x}$.
8. **True or False.** The domain of $f(x) = \frac{1}{\sqrt{x + 2}}$ is all real x, $x \neq -2$.

In Exercises 9–14, determine the domain and range of each relation. Explain why each nonfunction is not a function.

9.
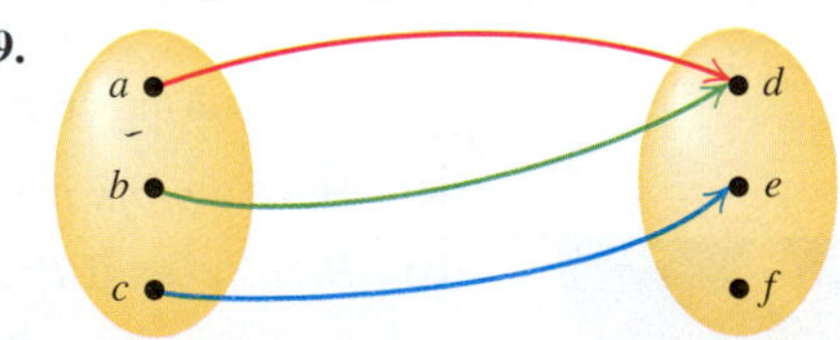

10.
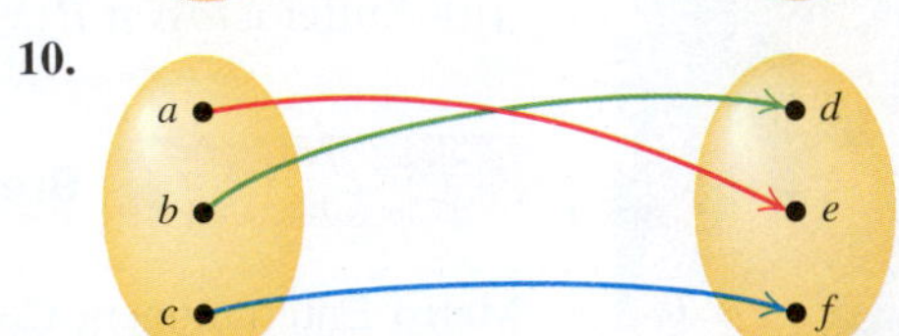

11.
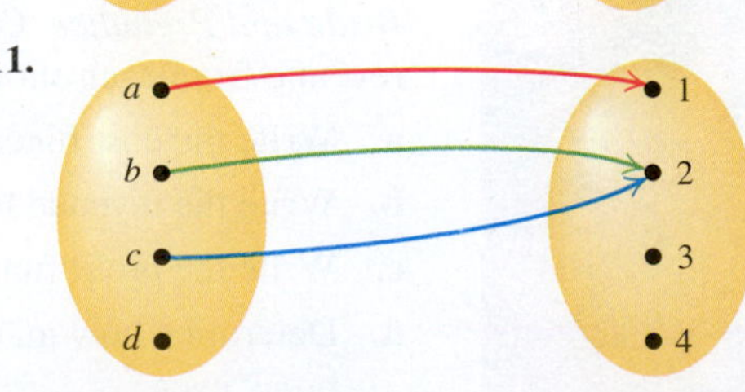

12.

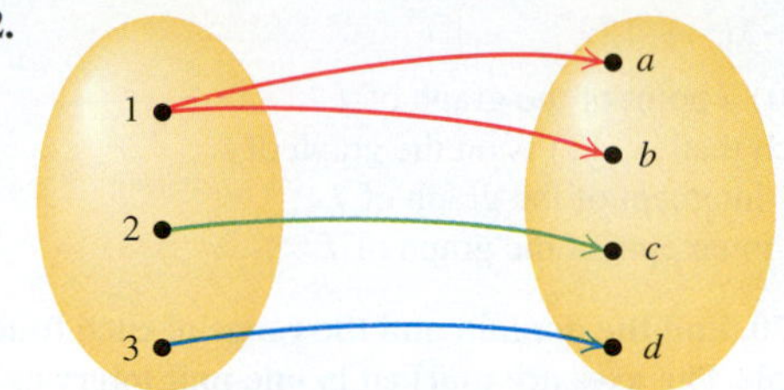

13.

x	0	3	8	0	3	8
y	-1	-2	-3	1	2	2

14.

x	-3	-1	0	1	2	3
y	-8	0	1	0	-3	-8

In Exercises 15–28, determine whether each equation defines y as a function of x.

15. $x + y = 2$
16. $x = y - 1$
17. $y = \frac{1}{x}$
18. $xy = -1$
19. $y^2 = x^2$
20. $x = |y|$
21. $y = \frac{1}{\sqrt{2x - 5}}$
22. $y = \frac{1}{\sqrt{x^2 - 1}}$
23. $2 - y = 3x$
24. $3x - 5y = 15$
25. $x + y^2 = 8$
26. $x = y^2$
27. $x^2 + y^3 = 5$
28. $x + y^3 = 8$

In Exercises 29–32, let $f(x) = x^2 - 3x + 1, g(x) = \frac{2}{\sqrt{x}}$, and $h(x) = \sqrt{2 - x}$.

29. Find $f(0), g(0), h(0), f(a)$, and $f(-x)$.
30. Find $f(1), g(1), h(1), g(a)$, and $g(x^2)$.
31. Find $f(-1), g(-1), h(-1), h(c)$, and $h(-x)$.
32. Find $f(4), g(4), h(4), g(2 + k)$, and $f(a + k)$.
33. Let $f(x) = \frac{2x}{\sqrt{4 - x^2}}$. Find each function value.
    a. $f(0)$
    b. $f(1)$
    c. $f(2)$
    d. $f(-2)$
    e. $f(-x)$
34. Let $g(x) = 2x + \sqrt{x^2 - 4}$. Find each function value.
    a. $g(0)$
    b. $g(1)$
    c. $g(2)$
    d. $g(-3)$
    e. $g(-x)$
35. In the figure for Exercise 35, find the sum of the areas of the shaded rectangles.

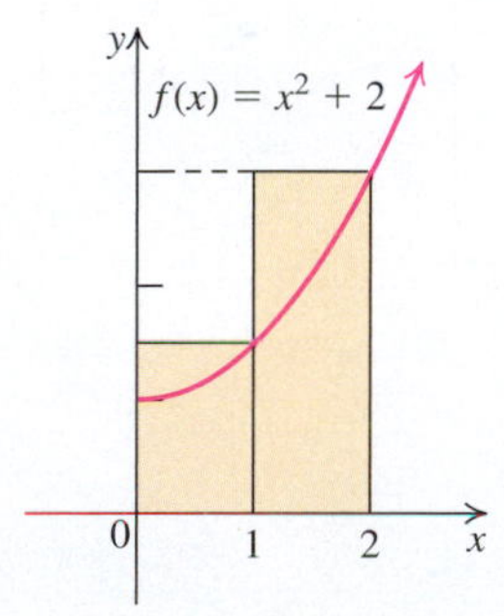

36. Repeat Exercise 35 in the figure for Exercise 36.

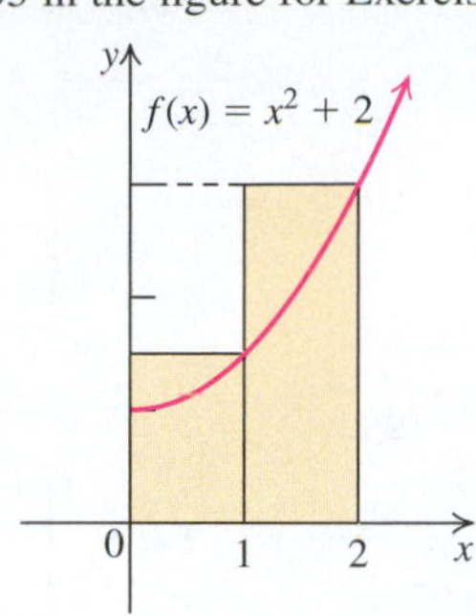

In Exercises 37–48, find the domain of each function.

37. $f(x) = -8x + 7$
38. $f(x) = 2x^2 - 11$
39. $f(x) = \frac{1}{x - 9}$
40. $f(x) = \frac{1}{x + 9}$
41. $h(x) = \frac{2x}{x^2 - 1}$
42. $h(x) = \frac{x - 3}{x^2 - 4}$
43. $G(x) = \frac{\sqrt{x - 3}}{x + 2}$
44. $f(x) = \frac{3}{\sqrt{4 - x}}$
45. $F(x) = \frac{x + 4}{x^2 + 3x + 2}$
46. $F(x) = \frac{1 - x}{x^2 + 5x + 6}$
47. $g(x) = \frac{\sqrt{x^2 + 1}}{x}$
48. $g(x) = \frac{1}{x^2 + 1}$

In Exercises 49–54, use the vertical-line test to determine whether the given graph represents a function.

49.

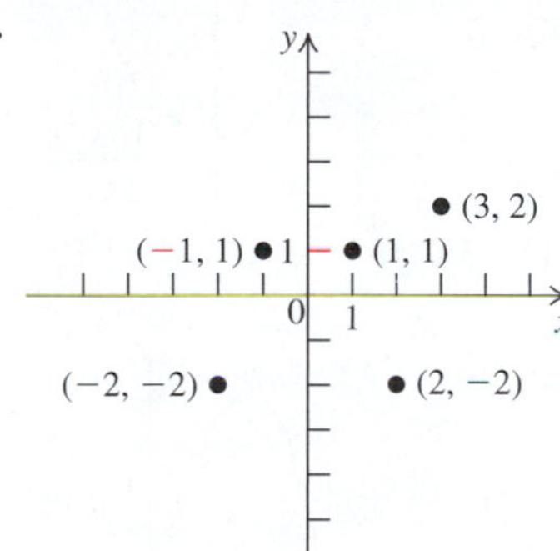

50.

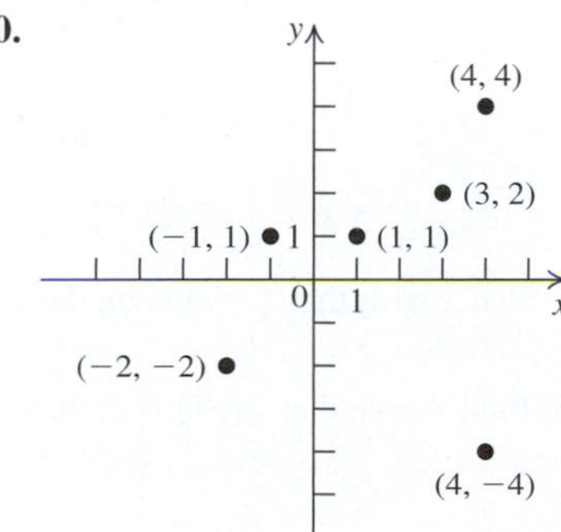

51.

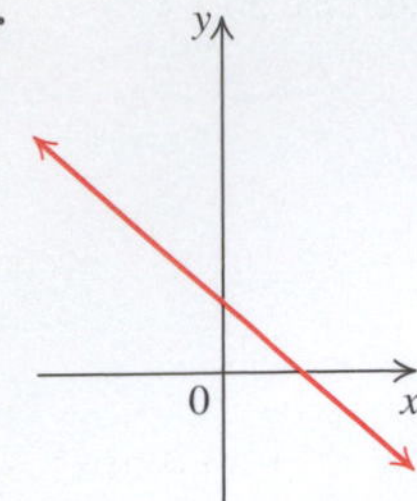

52.

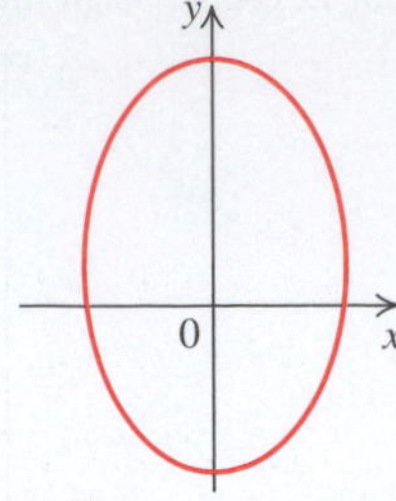

53.

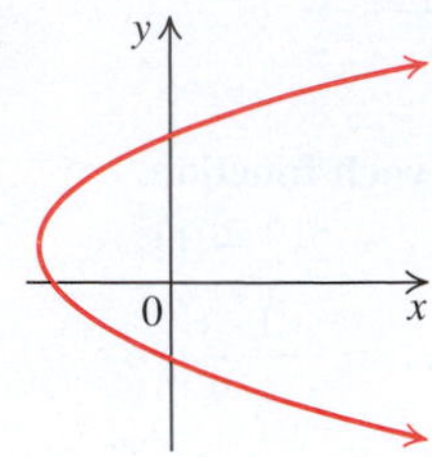

54.

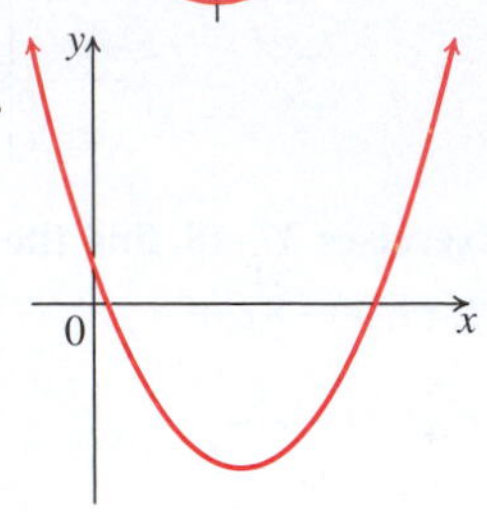

In Exercises 55–58, the graph of a function is given. Find the indicated function values.

55. $f(-4)$, $f(-1)$, $f(3)$, $f(5)$

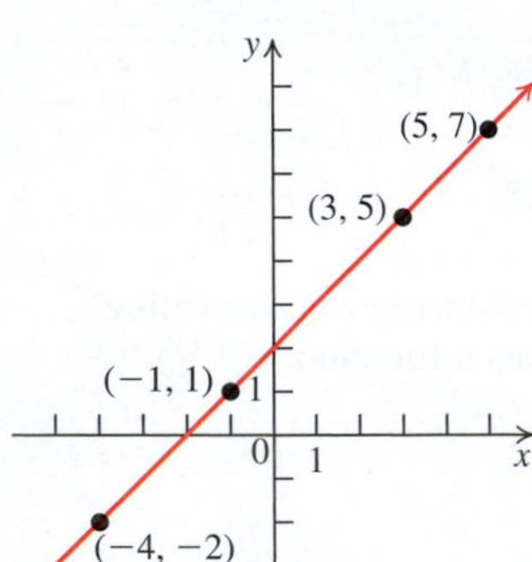

56. $g(-2)$, $g(1)$, $g(3)$, $g(4)$

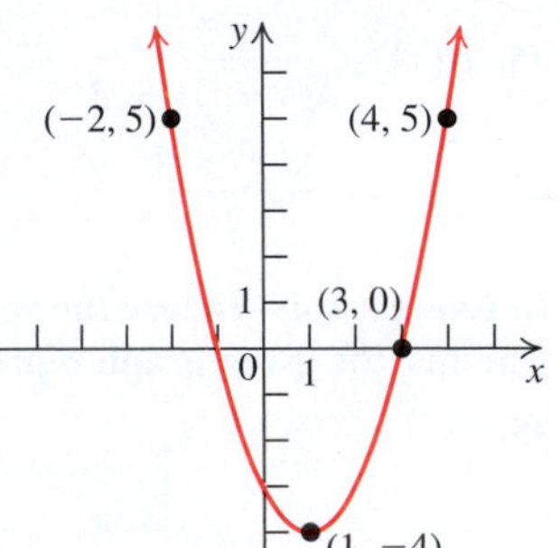

57. $h(-2)$, $h(-1)$, $h(0)$, $h(1)$

58. $f(-1)$, $f(0)$, $f(1)$

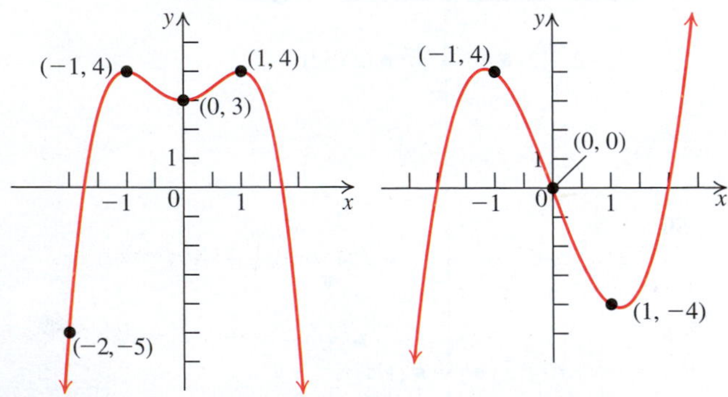

59. Let $h(x) = x^2 - x + 1$. Find x such that $(x, 7)$ is on the graph of h.

60. Let $H(x) = x^2 + x + 8$. Find x such that $(x, 7)$ is on the graph of H.

61. Let $f(x) = -2(x + 1)^2 + 7$.
   a. Is $(1, 1)$ a point of the graph of f?
   b. Find all x such that $(x, 1)$ is on the graph of f.
   c. Find all y-intercepts of the graph of f.
   d. Find all x-intercepts of the graph of f.

62. Let $f(x) = -3x^2 - 12x$.
   a. Is $(-2, 10)$ a point of the graph of f?
   b. Find x such that $(x, 12)$ is on the graph of g.
   c. Find the y-intercept of the graph of f.
   d. Find all x-intercepts of the graph of f.

In Exercises 63–70, find the domain and the range of each function from its graph. The axes are marked in one-unit intervals.

63.

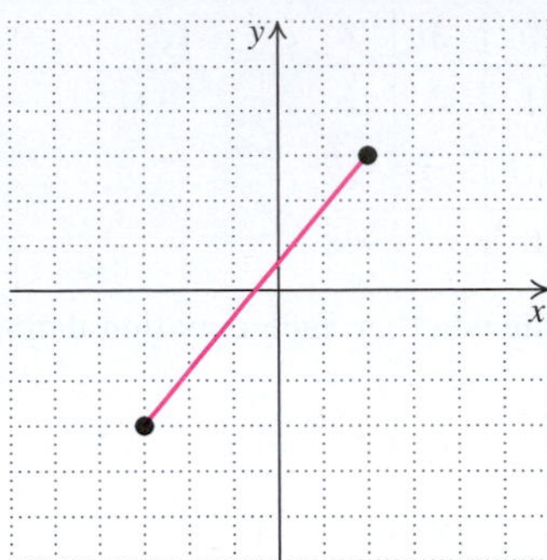

64.

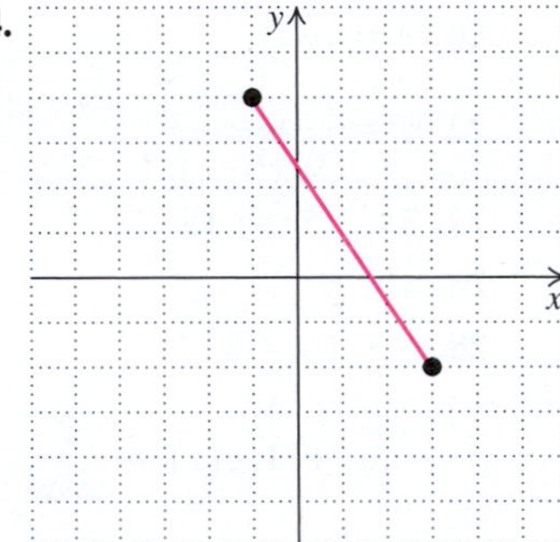

65.

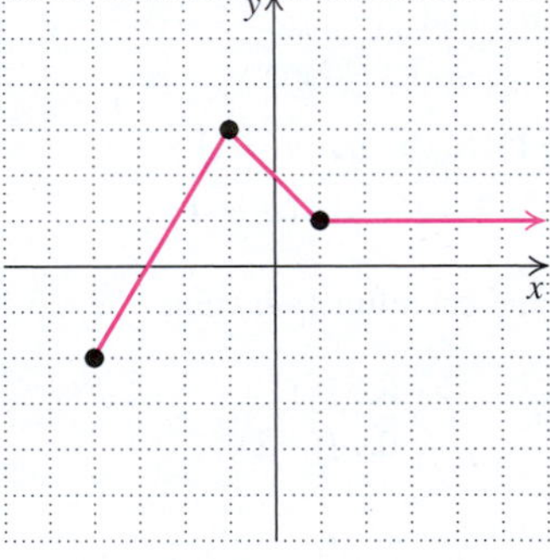

66.

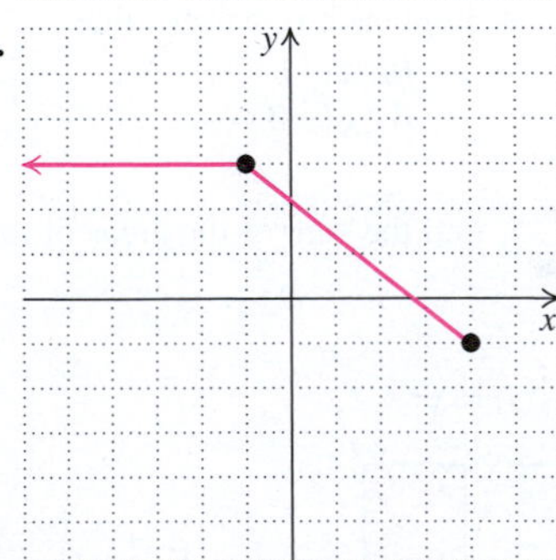

67.

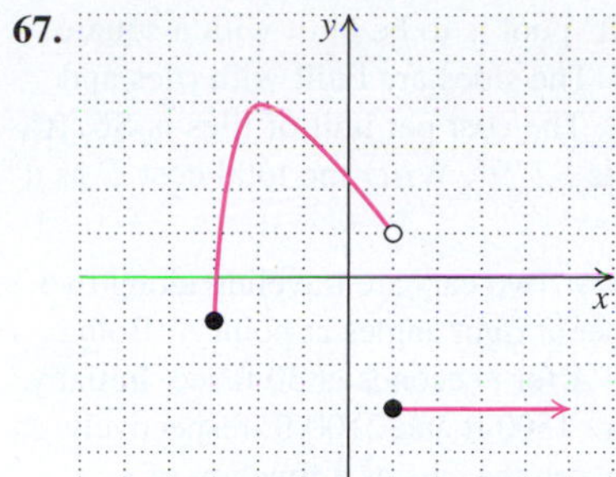

68.

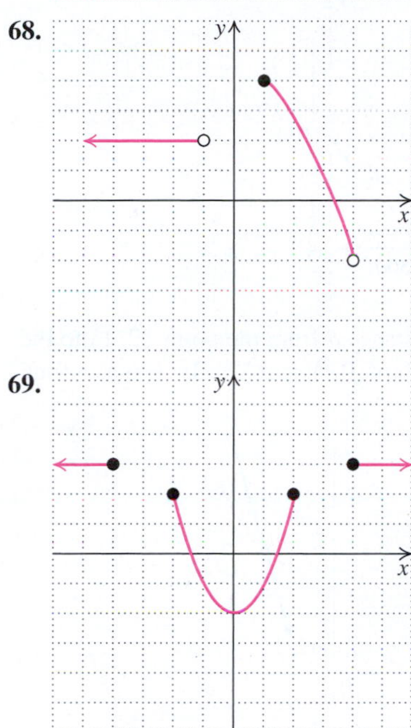

69.

70.

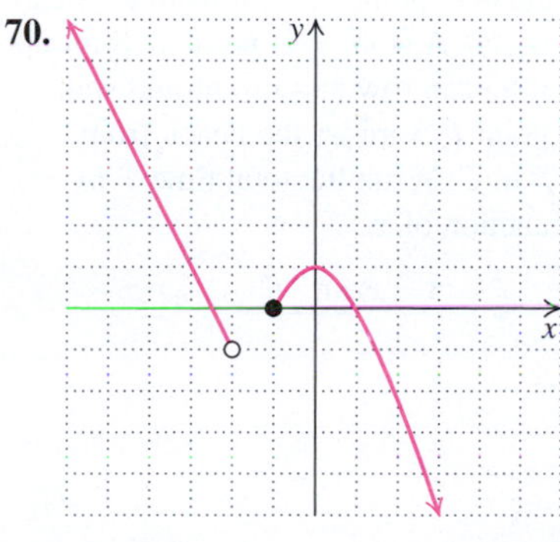

For Exercises 71–78, refer to the graph of $y = f(x)$ in the figure. The axes are marked off in one-unit intervals.

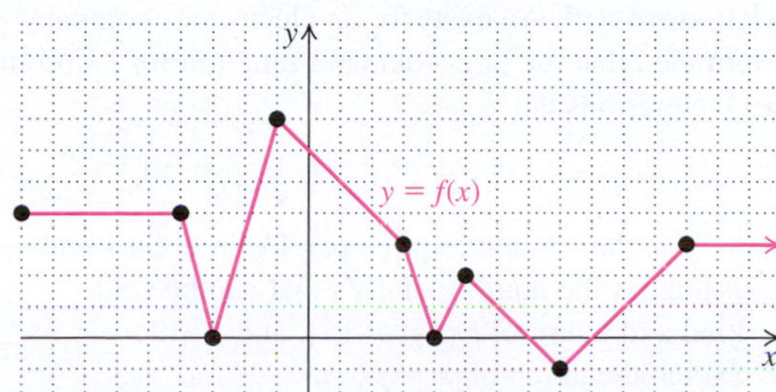

71. Find the domain of f.
72. Find the range of f.
73. Find the x-intercepts of f.
74. Find the y-intercepts of f.
75. Find $f(-7)$, $f(1)$, and $f(5)$.
76. Find $f(-4)$, $f(-1)$, and $f(3)$.
77. Solve the equation $f(x) = 3$.
78. Solve the equation $f(x) = -2$.

In Exercises 79–84, refer to the graph of $y = g(x)$ in the figure. The axes are marked off in one-unit intervals.

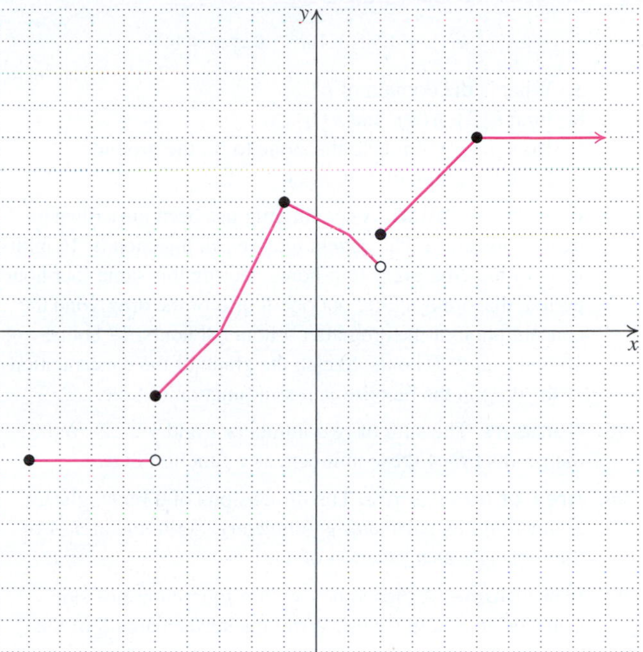

79. Find the domain of g.
80. Find the range of g.
81. Find $g(-4)$, $g(1)$, and $g(3)$.
82. Find $|g(-5)-g(5)|$.
83. Solve $g(x) = -4$.
84. Solve $g(x) = 6$.

Applying the Concepts

In Exercises 85–88, state whether the given relation is a function and explain why.

85. To each day of the year there corresponds the high temperature in your hometown on that day.
86. To each year since 1950 there corresponds the cost of a first-class postage stamp on January 1 of that year.
87. To each letter of the word *NUTS* there corresponds the states whose names start with that letter.
88. To each day of the week there corresponds the first names of people currently living in the United States born on that day of the week.
89. **Square tiles.** The area $A(x)$ of a square tile is a function of the length x of the square's side. Write a function rule for the area of a square tile. Find and interpret $A(4)$.
90. **Cube.** The volume $V(x)$ of a cube is a function of the length x of the cube's edge. Write a function rule for the volume of a cube. Evaluate the function for a cube with sides of length 3 inches.

91. Surface area. Is the total surface area S of a cube a function of the edge length x of the cube? If it is not a function, explain why not. If it is a function, write the function rule $S(x)$ and evaluate $S(3)$.

92. Measurement. One meter equals about 39.37 inches. Write a function rule for converting inches to meters. Evaluate the function for 59 inches.

93. Motion of a projectile. A stone thrown upward with an initial velocity of 128 feet per second will attain a height of h feet in t seconds, where

$$h(t) = 128t - 16t^2, \quad 0 \le t \le 8.$$

a. What is the domain of h?
b. Find $h(2)$, $h(4)$, and $h(6)$.
c. How long will it take the stone to hit the ground?
d. Sketch a graph of $y = h(t)$.

94. Drug prescription. A certain drug has been prescribed to treat an infection. The patient receives an injection of 16 milliliters of the drug every four hours. During this same four-hour period, the kidneys filter out one-fourth of the drug. Find the concentration of the drug after 4 hours, 8 hours, 12 hours, 16 hours, and 20 hours. Sketch the graph of the concentration of the drug in the bloodstream as a function of time.

95. Numbers. The sum of two numbers x and y is 28. Write the product P of these numbers as a function of x.

96. Area of a rectangle. The dimensions of a rectangle are x and y and its perimeter is 60 meters. Write the area A of the rectangle as a function of x.

97. Box volume. A closed box with a square base of side x inches is to hold 64 in^3. Write the surface area S of the box as a function of x.

98. Inscribed rectangle. In the figure, a rectangle is inscribed in a semicircle of diameter $2r$.
a. Write the perimeter P of the rectangle as a function of x.
b. Write the area A of the rectangle as a function of x.

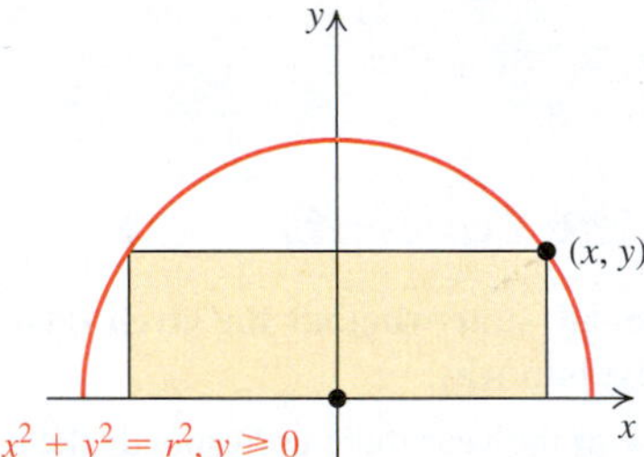

99. Piece of wire. A piece of wire 20 inches long is to be cut into two pieces. (See the figure.) The piece of length x is to form a circle, and the other is to form a square. Write A, the sum of the areas of the two figures, as a function of x.

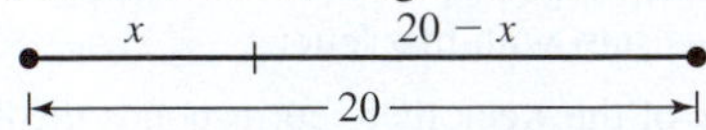

100. Area of metal. An open cylindrical tank with circular base of radius r is to be constructed of metal to contain a volume of 64 in^3. Write the surface area A of the metal as a function of r.

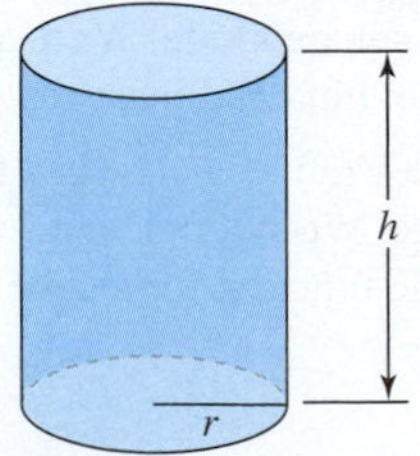

101. Cost of pool. A 288-ft^3 pool is to be built with a square bottom of side length x. The sides are built with tiles and the bottom with cement. The cost per unit of tiles is $\$6/\text{ft}^2$, and cost of the bottom is $\$2/\text{ft}^2$. Write the total cost C as a function of x.

102. Distance between cars. Two cars are traveling along two roads that cross each other at right angles at point A. Both cars are traveling toward A for t seconds at 30 ft/sec. Initially, their distances from A are 1500 ft and 2100 ft, respectively. Write the distance d between the cars as a function of t.

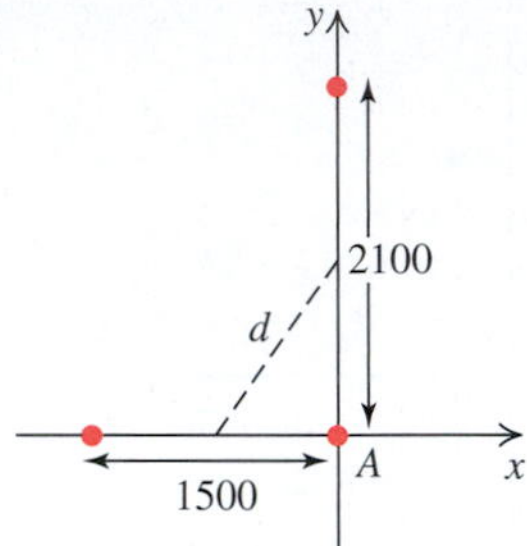

103. Distance. Write the distance d from the point $(2, 1)$ to the point (x, y) on the graph of $f(x) = x^3 - 3x + 6$ as a function of x.

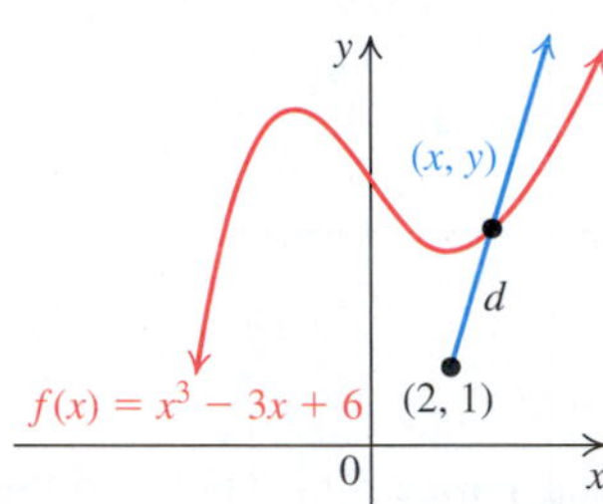

104. Time of travel. An island is at point A, 5 mi from the nearest point B on a straight beach. A store is at point C, 8 mi up the beach from point B. Julio can row at 4 mi/hr and walk 5 mi/hr. He rows to the point P, x mi up the beach from point B and walks from P to C. Write the total time T of travel from A to C as a function of x.

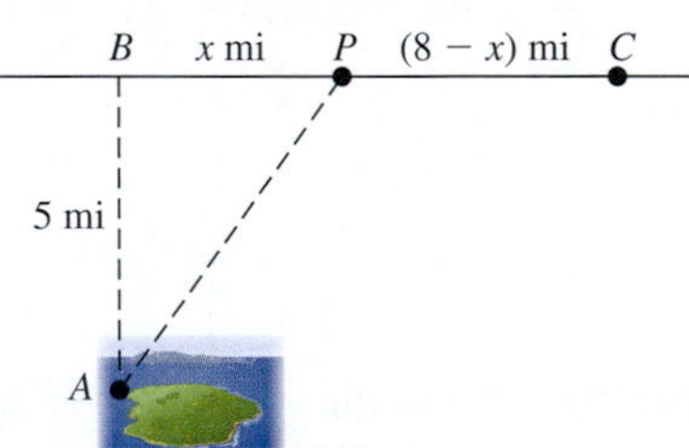

105. Price–demand. Analysts for an electronics company determined that the price–demand function for its 60-inch LED televisions is

$$p(x) = 1275 - 25x, \quad 1 \le x \le 30,$$

where p is the wholesale price per TV in dollars and x, in thousands, is the number of TVs that can be sold.
a. Find and interpret $p(5)$, $p(15)$, and $p(30)$.
b. Sketch the graph of $y = p(x)$.
c. Solve the equation $p(x) = 650$ and interpret your result.

106. Revenue.

a. Using the price–demand function $p(x) = 1275 - 25x$, $1 \le x \le 30$, of Exercise 105, write the company's revenue function $R(x)$ and state its domain.

b. Find and interpret $R(1)$, $R(5)$, $R(10)$, $R(15)$, $R(20)$, $R(25)$, and $R(30)$.

c. Using the data from part (b), sketch the graph of $y = R(x)$.

d. Solve the equation $R(x) = 4700$ and interpret your result.

107. Breaking even. Peerless Publishing Company intends to publish Harriet Henrita's next novel. The estimated cost is \$75,000 plus \$5.50 for each copy printed. The selling price per copy is \$15. The bookstores retain 40% of the selling price as commission. Let x represent the number of copies printed and sold.

a. Find the cost function $C(x)$.

b. Find the revenue function $R(x)$.

c. Find the profit function $P(x)$.

d. How many copies of the novel must be sold for Peerless to break even?

e. What is the company profit if 46,000 copies are sold?

108. Breaking even. Capital Records Company plans to produce a CD by the popular rapper Rapit. The fixed cost is \$500,000, and the variable cost is \$0.50 per CD. The company sells each CD to record stores for \$5. Let x represent the number of CDs produced and sold.

a. How many disks must be sold for the company to break even?

b. How many CDs must be sold for the company to make a profit of \$750,000?

Beyond the Basics

In Exercises 109–112, find an expression for $f(x)$ by solving for y and replacing y with $f(x)$. Then find the domain of f and compute $f(4)$.

109. $x = \dfrac{2}{y - 4}$

110. $xy - 3 = 2y$

111. $(x^2 + 1)y + x = 2$

112. $yx^2 - \sqrt{x} = -2y$

In Exercises 113–118, state whether f and g represent the same function. Explain your reasons.

113. $f(x) = 3x - 4, \quad 0 \le x \le 8, \quad g(x) = 3x - 4$

114. $f(x) = 3x^2, \quad g(x) = 3x^2, \quad -5 \le x \le 5$

115. $f(x) = x - 1, \quad g(x) = \dfrac{x^2 - 1}{x + 1}$

116. $f(x) = \dfrac{x + 2}{x^2 - x - 6}, \quad g(x) = \dfrac{1}{x - 3}$

117. $f(x) = x^2 + 1, \quad g(x) = 8x - 14$, both with domain $\{3, 5\}$

118. $f(x) = 2x^2 + 8, \quad g(x) = 3x + 7$, both with domain $\{1, 2\}$

119. Let $f(x) = ax^2 + ax - 3$. If $f(2) = 15$, find a.

120. Let $g(x) = x^2 + bx + b^2$. If $g(6) = 28$, find b.

121. Let $h(x) = \dfrac{3x + 2a}{2x - b}$. If $h(6) = 0$ and $h(3)$ is undefined, find a and b.

122. Let $f(x) = 2x - 3$. Find $f(x^2)$ and $[f(x)]^2$.

123. If $g(x) = x^2 - \dfrac{1}{x^2}$, show that $g(x) + g\left(\dfrac{1}{x}\right) = 0$.

124. If $f(x) = \dfrac{x - 1}{x + 1}$, show that $f\left(\dfrac{x - 1}{x + 1}\right) = -\dfrac{1}{x}$.

125. If $f(x) = \dfrac{x + 3}{4x - 5}$ and $t = \dfrac{3 + 5x}{4x - 1}$, show that $f(t) = x$.

Critical Thinking/Discussion/Writing

126. Write an equation of a function with each domain. Answers will vary.

a. $[2, \infty)$ **b.** $(2, \infty)$

c. $(-\infty, 2]$ **d.** $(-\infty, 2)$

127. Consider the graph of the function

$$f(x) = ax^2 + bx + c, a \ne 0.$$

a. Write an equation whose solution yields the x-intercepts.

b. Write an equation whose solution is the y-intercept.

c. Write (if possible) a condition under which the graph of $y = f(x)$ has no x-intercepts.

d. Write (if possible) a condition under which the graph of $y = f(x)$ has no y-intercepts.

128. Give an example (if possible) of a function matching each description. Some have many different correct answers.

a. Its graph is symmetric with respect to the y-axis.

b. Its graph is symmetric with respect to the x-axis.

c. Its graph is symmetric with respect to the origin.

d. Its graph consists of a single point.

e. Its graph is a horizontal line.

f. Its graph is a vertical line.

129. Let $X = \{a, b\}$ and $Y = \{1, 2, 3\}$

a. How many functions are there from X to Y?

b. How many functions are there from Y to X?

130. If a set X has m elements and a set Y has n elements, how many functions can be defined from X to Y? Justify.

Maintaining Skills

In Exercises 131–134, find the equation of the line passing through the given points in slope–intercept form.

131. $(0, 0)$ and $(2, -2)$

132. $(-1, 3)$ and $(4, -2)$

133. $(-3, 2)$ and $(1, 4)$

134. $(3, -5)$ and $(8, -3)$

In Exercises 135–138, find
a. $f(-x)$. **b.** $-f(x)$.

135. $f(x) = 2x^2 - 3x$

136. $f(x) = 3x^4 - 7x^2 + 5$

137. $f(x) = x^3 - 2x$

138. $f(x) = 2x^3 - 5x^2 + x$

SECTION 2.5

Properties of Functions

BEFORE STARTING THIS SECTION, REVIEW

1 Graph of a function (Section 2.4, page 201)

2 Evaluating a function (Section 2.4, page 198)

OBJECTIVES

1 Determine whether a function is increasing or decreasing on an interval.

2 Use a graph to locate relative maximum and minimum values.

3 Identify even and odd functions.

4 Find the average rate of change of a function.

Algebra in Coughing

Coughing is the body's way of removing foreign material or mucus from the lungs and upper airway passages or reacting to an irritated airway. When you cough, the windpipe contracts to increase the velocity of the outgoing air.

The average flow velocity, v, can be modeled by the equation

$$v = c(r_0 - r)r^2 \text{ mm/sec},$$

where r_0 is the rest radius of the windpipe in millimeters, r is its contracted radius, and c is a positive constant. In Example 3, we find the value of r that maximizes the velocity of the air going out. The diameter of the windpipe of a normal adult male varies between 25 mm and 27 mm, while that of a normal adult female varies between 21 mm and 23 mm.

1 Determine whether a function is increasing or decreasing on an interval.

Increasing and Decreasing Functions

Classifying functions according to how one variable changes with respect to the other variable can be very useful. Imagine a particle moving from left to right along the graph of a function. This means that the x-coordinate of the point on the graph is getting larger. If the corresponding y-coordinate of the point is getting larger, getting smaller, or staying the same, then the function is called *increasing, decreasing,* or *constant*, respectively.

Increasing, Decreasing, and Constant Functions

Let f be a function and let x_1 and x_2 be any two numbers in an open interval (a, b) contained in the domain of f. See Figure 2.55. The symbols a and b may represent real numbers, $-\infty$, or ∞. Then

(i) f is called an **increasing function** on (a, b) if $x_1 < x_2$ implies $f(x_1) < f(x_2)$.
(ii) f is called a **decreasing function** on (a, b) if $x_1 < x_2$ implies $f(x_1) > f(x_2)$.
(iii) f is **constant** on (a, b) if $x_1 < x_2$ implies $f(x_1) = f(x_2)$.

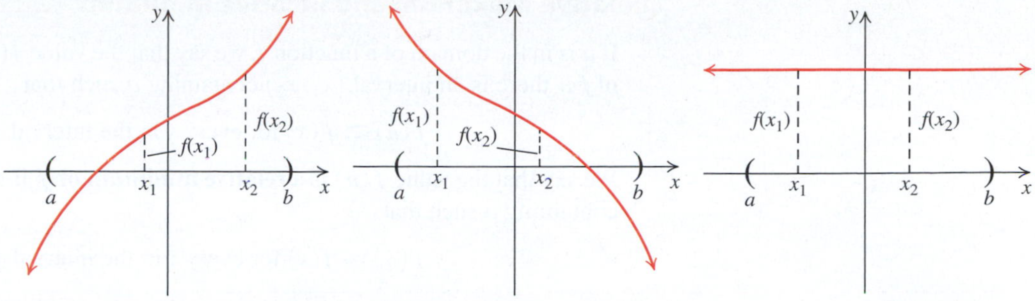

Increasing function on (a, b)
(a)

Decreasing function on (a, b)
(b)

Constant function on (a, b)
(c)

FIGURE 2.55 Increasing, decreasing, and constant functions

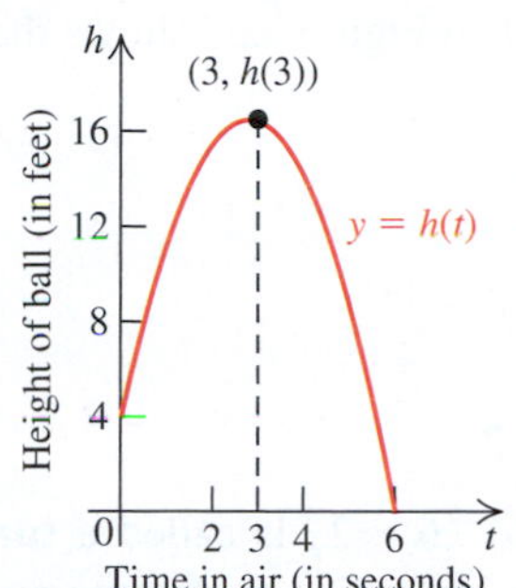

FIGURE 2.56 The function $f(t)$ increases on $(0, 3)$ and decreases on $(3, 6)$.

Geometrically, the graph of an increasing function on an open interval (a, b) rises as x-values increase on (a, b). This is because, by definition, as x-values increase from x_1 to x_2, the y-values also increase, from $f(x_1)$ to $f(x_2)$. Similarly, the graph of a decreasing function on (a, b) falls as x-values increase. The graph of a constant function is horizontal, or flat, over the interval (a, b).

Not every function always increases or always decreases. The function that assigns the height of a ball tossed in the air to the length of time it is in the air is an example of a function that increases, but also decreases, over different intervals of its domain. See Figure 2.56.

EXAMPLE 1 **Tracking the Behavior of a Function**

From the graph of the function g in Figure 2.57(a), find the intervals over which g is increasing, is decreasing, or is constant.

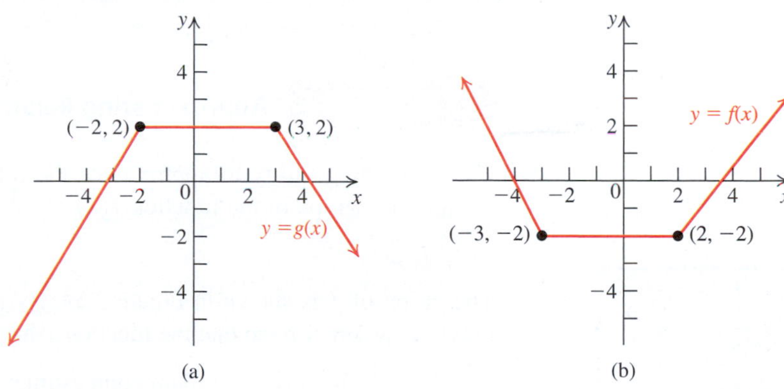

FIGURE 2.57

SIDE NOTE

When specifying the intervals over which a function f is increasing, decreasing, or constant, you need to use the intervals in the *domain* of f, *not* in the range of f.

Solution

The function g is increasing on the interval $(-\infty, -2)$. It is constant on the interval $(-2, 3)$. It is decreasing on the interval $(3, \infty)$.

Practice Problem 1 Using the graph in Figure 2.57(b), find the intervals over which f is increasing, is decreasing, or is constant.

2 Use a graph to locate relative maximum and minimum values.

Relative Maximum and Minimum Values

The y-coordinate of a point that is not lower than any nearby point on a graph is called a *relative maximum* value of the function. Such points are called the *relative maximum* points of the graph. Similarly, the y-coordinate of a point that is not higher than any nearby points on a graph is called a *relative minimum* value of the function. Such points are called *relative minimum* points of the graph. Relative maximum and minimum values are the y-coordinates of points corresponding to the peaks and troughs, respectively, of the graph.

Relative Maximum and Relative Minimum

If a is in the domain of a function f, we say that the value $f(a)$ is a **relative maximum of f** if there is an interval (x_1, x_2) containing a such that

$$f(a) \geq f(x) \text{ for every } x \text{ in the interval } (x_1, x_2).$$

We say that the value $f(a)$ is a **relative minimum of f** if there is an interval (x_1, x_2) containing a such that

$$f(a) \leq f(x) \text{ for every } x \text{ in the interval } (x_1, x_2).$$

The value $f(a)$ is called an **extreme value** of f if it is either a relative maximum value or a relative minimum value.

The graph of a function can help you find or approximate maximum or minimum values (if any) of the function. For example, the graph of a function f in Figure 2.58 shows that

- f has two relative maxima:
  (i) at $x = 1$ with relative maximum value $f(1) = 3$
  (ii) at $x = 4$ with relative maximum value $f(4) = 7$
- f has two relative minima:
  (i) at $x = 2$ with relative minimum value $f(2) = 1$
  (ii) at $x = 6$ with relative minimum value $f(6) = -2$

Points whose y-coordinates are relative maximum values

Points whose y-coordinates are relative minimum values

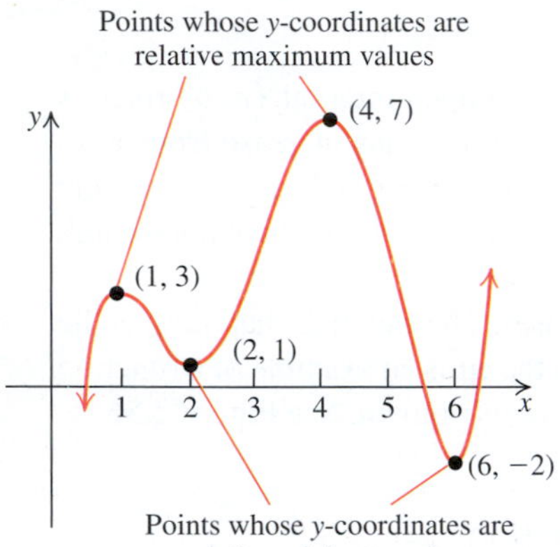

FIGURE 2.58 Relative maximum and minimum values

In Figure 2.58, each of the points $(1, 3)$, $(2, 1)$, $(4, 7)$, and $(6, -2)$ is called a **turning point**. At a turning point, a graph changes direction from increasing to decreasing or from decreasing to increasing. In calculus, you study techniques for finding the exact points at which a function has a relative maximum or a relative minimum value (if any). For the present, however, we settle for approximations of these points by using a graphing utility.

EXAMPLE 2 Approximating Relative Extrema

Use a graphing utility to approximate the relative maximum and the relative minimum point on the graph of the function $f(x) = x^3 - x^2$.

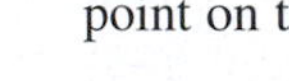

Solution

The graph of f is shown in Figure 2.59. By using the TRACE and ZOOM features of a graphing utility, we see that the function f has a

Relative minimum point estimated to be $(0.67, -0.15)$.

Relative maximum point estimated to be $(0, 0)$.

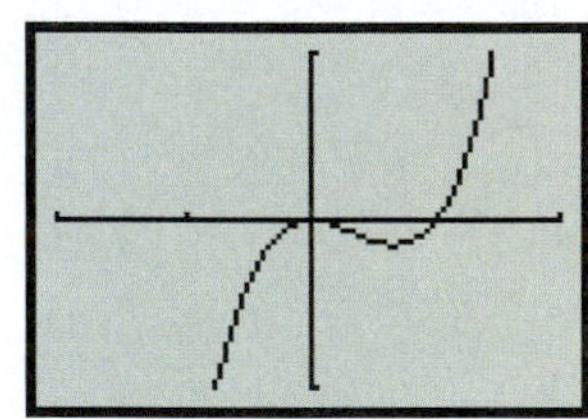

FIGURE 2.59

Practice Problem 2 Repeat Example 2 for $g(x) = -2x^3 + 3x^2$.

EXAMPLE 3 Algebra in Coughing

The average flow velocity, v, of outgoing air through the windpipe is modeled by (see introduction to this section)

$$v = c(r_0 - r)r^2 \text{ mm/sec} \qquad \frac{r_0}{2} \leq r \leq r_0,$$

where r_0 is the rest radius of the windpipe, r is its contracted radius, and c is a positive constant. For Mr. Osborn, assume that $c = 1$ and $r_0 = 13$ mm. Use a graphing utility to estimate the value of r that will maximize the airflow v when Mr. Osborn coughs.

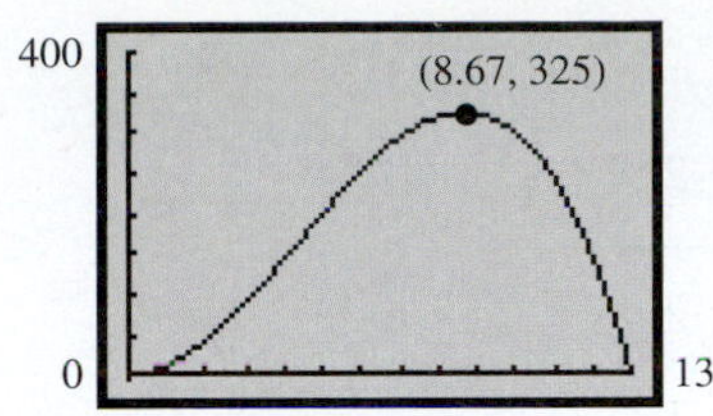

FIGURE 2.60 Flow velocity

Solution

The graph of v for $0 \leq r \leq 13$ is shown in Figure 2.60. By using the TRACE and ZOOM features, we see that the maximum point on the graph is estimated at $(8.67, 325)$. So Mr. Osborn's windpipe contracts to a radius of 8.67 mm to maximize the airflow velocity.

Practice Problem 3 Repeat Example 3 for Mrs. Osborn. Again, let $c = 1$ but $r_0 = 11$.

3 Identify even and odd functions.

Even–Odd Functions and Symmetry

This topic, *even–odd functions*, uses ideas of symmetry discussed in Section 2.2.

Even Function

A function f is called an **even function** if for each x in the domain of f, $-x$ is also in the domain of f and

$$f(-x) = f(x).$$

The graph of an even function is symmetric about the y-axis.

EXAMPLE 4 **Graphing the Squaring Function**

Show that the squaring function $f(x) = x^2$ is an even function and sketch its graph.

Solution

The function $f(x) = x^2$ is even because

$$\begin{aligned} f(-x) &= (-x)^2 && \text{Replace } x \text{ with } -x. \\ &= x^2 && (-x)^2 = x^2 \\ &= f(x) && \text{Replace } x^2 \text{ with } f(x). \end{aligned}$$

To sketch the graph of $f(x) = x^2$, we make a table of values. See Table 2.8(a). We then plot the points that are in the first quadrant and join them with a smooth curve. Next, we use y-axis symmetry to find additional points on the graph of f in the second quadrant, which are listed in Table 2.8(b). See Figure 2.61. Note that the scales on the x- and y-axes are different.

TABLE 2.8(a)

x	0	$\frac{1}{2}$	1	$\frac{3}{2}$	2	$\frac{5}{2}$	3
$f(x) = x^2$	0	$\frac{1}{4}$	1	$\frac{9}{4}$	4	$\frac{25}{4}$	9

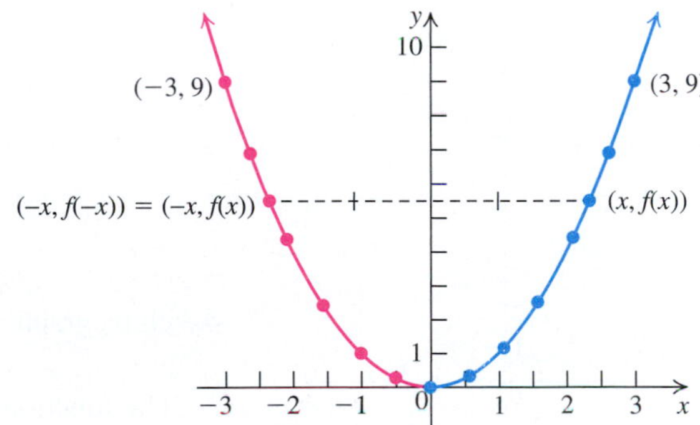

FIGURE 2.61 Graph of $f(x) = x^2$

SIDE NOTE

The graph of a nonzero function cannot by symmetric with respect to the x-axis because the two ordered pairs (x, y) and $(x, -y)$ would then be points of the graph on the same vertical line.

TABLE 2.8(b)

x	0	$-\frac{1}{2}$	-1	$-\frac{3}{2}$	-2	$-\frac{5}{2}$	-3
$f(x) = x^2$	0	$\frac{1}{4}$	1	$\frac{9}{4}$	4	$\frac{25}{4}$	9

Practice Problem 4 Show that the function $f(x) = -x^2$ is an even function and sketch its graph.

Odd Function

A function f is an **odd function** if for each x in the domain of f, $-x$ is also in the domain of f and

$$f(-x) = -f(x).$$

The graph of an odd function is symmetric about the origin.

EXAMPLE 5 Graphing the Cubing Function

Show that the cubing function defined by $g(x) = x^3$ is an odd function and sketch its graph.

Solution

The function $g(x) = x^3$ is an odd function because

$$\begin{aligned} g(-x) &= (-x)^3 && \text{Replace } x \text{ with } -x. \\ &= -x^3 && (-x)^3 = -x^3 \\ &= -g(x) && \text{Replace } x^3 \text{ with } g(x). \end{aligned}$$

To sketch the graph of $g(x) = x^3$, we plot points in the first quadrant and then use symmetry about the origin to extend the graph to the third quadrant. The graph is shown in Figure 2.62.

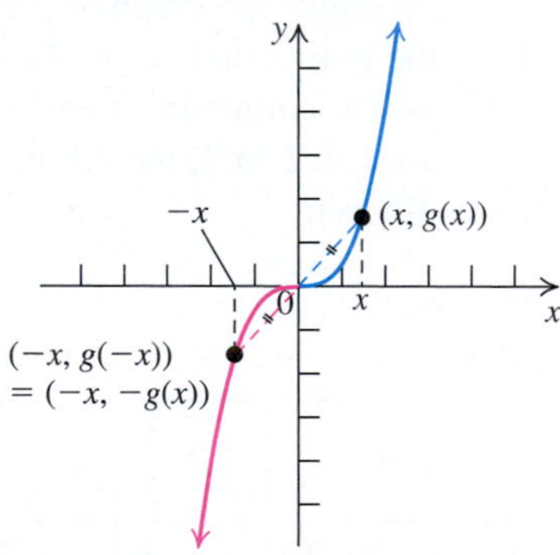

FIGURE 2.62 Graph of $g(x) = x^3$

Practice Problem 5 Show that the function $f(x) = -x^3$ is an odd function and sketch its graph.

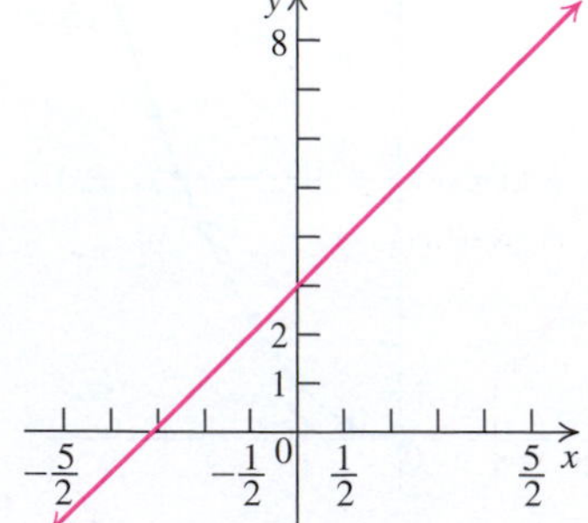

FIGURE 2.63 Graph of $h(x) = 2x + 3$

The function $h(x) = 2x + 3$, whose graph is sketched in Figure 2.63, is neither even nor odd, and its graph is neither symmetric with respect to the y-axis nor symmetric with respect to the origin. To see this algebraically, notice that $h(-x) = 2(-x) + 3 = -2x + 3$, which is not equivalent to $h(x)$ or $-h(x)$ (because $-h(x) = -2x - 3$).

EXAMPLE 6 Testing for Evenness and Oddness

Determine whether each function is even, odd, or neither.

a. $g(x) = x^3 - 4x$ **b.** $f(x) = x^2 + 5$ **c.** $h(x) = 2x^3 + x^2$

Solution

a. $g(x)$ is an odd function because

$$\begin{aligned} g(-x) &= (-x)^3 - 4(-x) && \text{Replace } x \text{ with } -x. \\ &= -x^3 + 4x && \text{Simplify.} \\ &= -(x^3 - 4x) && \text{Distributive property} \\ &= -g(x) \end{aligned}$$

b. $f(x)$ is an even function because

$$\begin{aligned} f(-x) &= (-x)^2 + 5 && \text{Replace } x \text{ with } -x. \\ &= x^2 + 5 && \text{Simplify.} \\ &= f(x) \end{aligned}$$

c. $$\begin{aligned} h(-x) &= 2(-x)^3 + (-x)^2 && \text{Replace } x \text{ with } -x. \\ &= -2x^3 + x^2 && \text{Simplify.} \end{aligned}$$

Comparing $h(x) = 2x^3 + x^2$, $h(-x) = -2x^3 + x^2$, and $-h(x) = -2x^3 - x^2$, we note that $h(-x) \neq h(x)$ and $h(-x) \neq -h(x)$. Hence, $h(x)$ is neither even nor odd.

Practice Problem 6 Repeat Example 6 for each function.

a. $g(x) = 3x^4 - 5x^2$ **b.** $f(x) = 4x^5 + 2x^3$ **c.** $h(x) = 2x + 1$

4 Find the average rate of change of a function.

Average Rate of Change

Suppose your salary increases from \$25,000 a year to \$40,000 a year over a five-year period. You then have the following:

Change in salary: \$40,000 − \$25,000 = \$15,000

Average rate of change in salary:

$$\frac{\$40{,}000 - \$25{,}000}{5} = \frac{\$15{,}000}{5} = \$3000 \text{ per year}$$

Regardless of when you actually received the raises during the five-year period, the final salary is the same as if you received your average annual increase of \$3000 *each* year.

The average rate of change can be defined in a more general setting.

Average Rate of Change of a Function

Let $(a, f(a))$ and $(b, f(b))$ be two points on the graph of a function f. Then the **average rate of change** of $f(x)$ as x changes from a to b is defined by

$$\frac{f(b) - f(a)}{b - a}, \quad a \neq b.$$

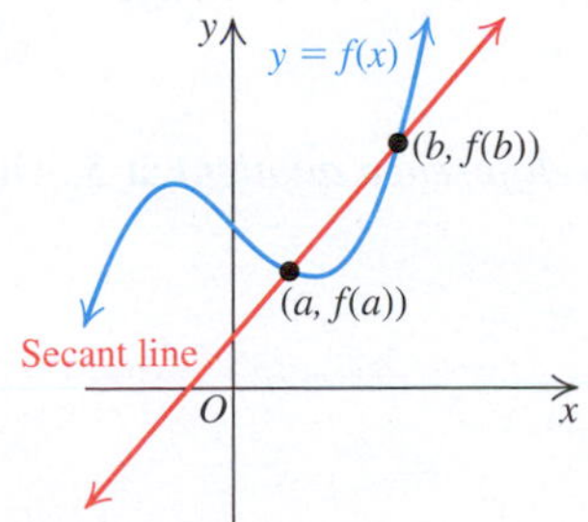

FIGURE 2.64

Note that the average rate of change of f as x changes from a to b is the slope of the **secant line** passing through the two points $(a, f(a))$ and $(b, f(b))$ on the graph of f. See Figure 2.64.

PROCEDURE IN ACTION

EXAMPLE 7 Finding the Average Rate of Change

OBJECTIVE	EXAMPLE
Find the average rate of change of a function f as x changes from a to b.	*Find the average rate of change of* $f(x) = 2 - 3x^2$ *as x changes from* $x = 1$ *to* $x = 3$.
Step 1 Find $f(a)$ and $f(b)$.	**1.** $f(1) = 2 - 3(1)^2 = -1 \quad a = 1$ $f(3) = 2 - 3(3)^2 = -25 \quad b = 3$
Step 2 Use the values from Step 1 in the definition of average rate of change.	**2.** $\frac{f(b) - f(a)}{b - a} = \frac{-25 - (-1)}{3 - 1} = -\frac{24}{2}$ $= -12$

Practice Problem 7 Find the average rate of change of $f(x) = 1 - x^2$ as x changes from 2 to 4.

EXAMPLE 8 Finding the Average Rate of Change

Find the average rate of change of $f(t) = 2t^2 - 3$ as t changes from $t = 5$ to $t = x$, $x \neq 5$.

Solution

$$\text{Average rate of change} = \frac{f(b) - f(a)}{b - a} \qquad \text{Definition}$$

$$= \frac{f(x) - f(5)}{x - 5} \qquad b = x, a = 5$$

$$\text{Average rate of change} = \frac{(2x^2 - 3) - (2 \cdot 5^2 - 3)}{x - 5} \qquad \text{Substitute for } f(x) \text{ and } f(5).$$

$$= \frac{2x^2 - 3 - 2 \cdot 5^2 + 3}{x - 5}$$

$$= \frac{2(x^2 - 5^2)}{x - 5} \qquad \text{Simplify.}$$

$$= \frac{2(x + 5)(x - 5)}{x - 5} \qquad \text{Difference of squares}$$

$$= 2x + 10 \qquad \text{Simplify: } x - 5 \neq 0.$$

Practice Problem 8 Find the average rate of change of $f(t) = 1 - t$ as t changes from $t = 2$ to $t = x$, $x \neq 2$.

The average rate of change calculated in Example 8 is a *difference quotient* at 5. The difference quotient is an important concept in calculus.

Difference Quotient

For a function f, the **difference quotient** is

$$\frac{f(x + h) - f(x)}{h}, \quad h \neq 0.$$

Equivalently, the difference quotient can be written in the form $\frac{f(x) - f(a)}{x - a}$, $x \neq a$.

EXAMPLE 9 **Evaluating and Simplifying a Difference Quotient**

Let $f(x) = 2x^2 - 3x + 5$. Find and simplify $\dfrac{f(x+h) - f(x)}{h}$, $h \neq 0$.

Solution

First, we find $f(x+h)$.

$$\begin{aligned} f(x) &= 2x^2 - 3x + 5 && \text{Original equation} \\ f(x+h) &= 2(x+h)^2 - 3(x+h) + 5 && \text{Replace } x \text{ with } (x+h). \\ &= 2(x^2 + 2xh + h^2) - 3(x+h) + 5 && (x+h)^2 = x^2 + 2xh + h^2 \\ &= 2x^2 + 4xh + 2h^2 - 3x - 3h + 5 && \text{Use the distributive property.} \end{aligned}$$

Then we substitute into the difference quotient.

$$\begin{aligned} \frac{f(x+h) - f(x)}{h} &= \frac{\overbrace{(2x^2 + 4xh + 2h^2 - 3x - 3h + 5)}^{f(x+h)} - \overbrace{(2x^2 - 3x + 5)}^{f(x)}}{h} \\ &= \frac{2x^2 + 4xh + 2h^2 - 3x - 3h + 5 - 2x^2 + 3x - 5}{h} && \text{Subtract.} \\ &= \frac{4xh - 3h + 2h^2}{h} && \text{Simplify.} \\ &= \frac{h(4x - 3 + 2h)}{h} && \text{Factor out } h. \\ &= 4x - 3 + 2h && \text{Divide out the common factor } h. \end{aligned}$$

Practice Problem 9 Repeat Example 9 for $f(x) = -x^2 + x - 3$.

Answers to Practice Problems

1. Increasing on $(2, \infty)$, decreasing on $(-\infty, -3)$, and constant on $(-3, 2)$ **2.** Relative minimum at $(0, 0)$; relative maximum at $(1, 1)$ **3.** Mrs. Osborn's windpipe should be contracted to a radius of 7.33 mm for maximizing the airflow velocity.

4. $f(-x) = -(-x)^2 = -x^2 = f(x)$; therefore, f is even;

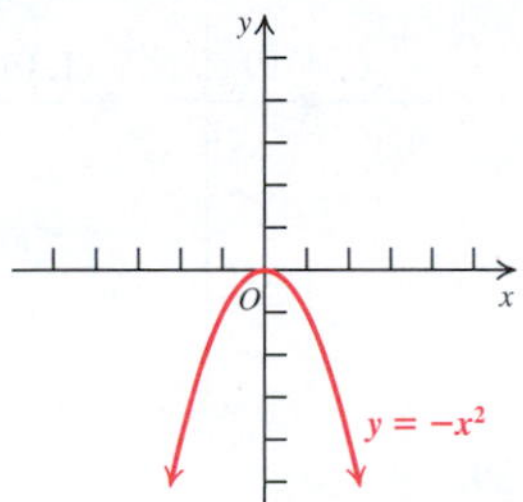

5. $f(-x) = -(-x^3) = x^3 = -f(x)$; therefore, f is odd;

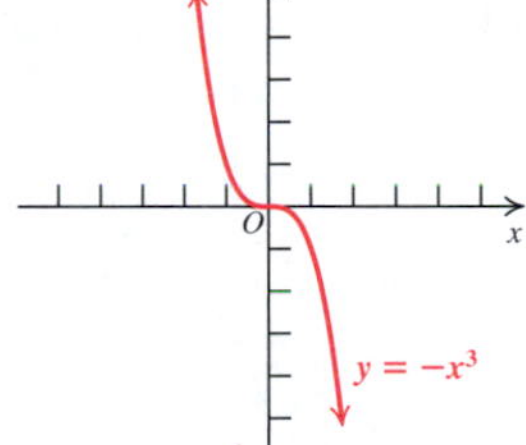

6. a. Even **b.** Odd **c.** Neither
7. -6 **8.** -1 **9.** $-2x + 1 - h$

SECTION 2.5 Exercises

Basic Concepts and Skills

1. A function f is decreasing if $x_1 < x_2$ implies that $f(x_1)$ __________ $f(x_2)$.
2. $f(a)$ is a relative maximum of f if there is an interval (x_1, x_2) containing a such that __________ for every x in the interval (x_1, x_2).
3. A function f is even if __________.
4. The average rate of change of f as x changes from $x = a$ to $x = b$ is __________, $a \neq b$.
5. **True or False.** The average rate of change of f as x changes from a to b is the slope of the line through the points $(a, f(a))$ and $(b, f(b))$.
6. **True or False.** A relative maximum or a relative minimum point on a graph is a turning point on the graph of a function.

In Exercises 7–14, the graph of a function is given. For each function, determine the intervals over which the function is increasing, decreasing, or constant.

7. **8.**

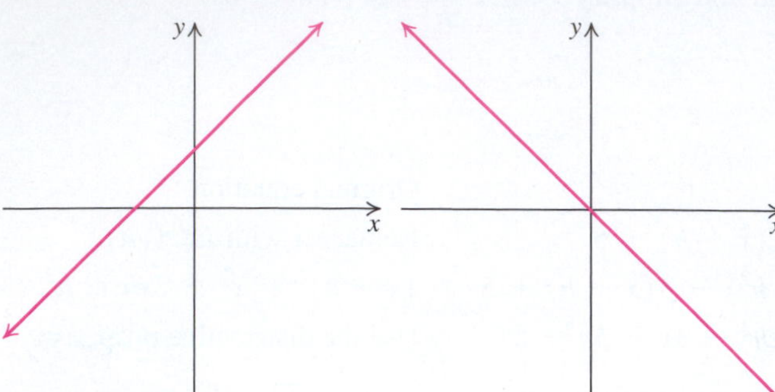

9.

10.

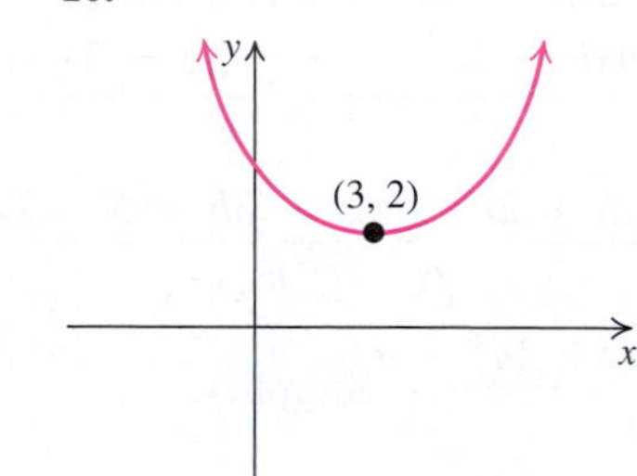

11. **12.**

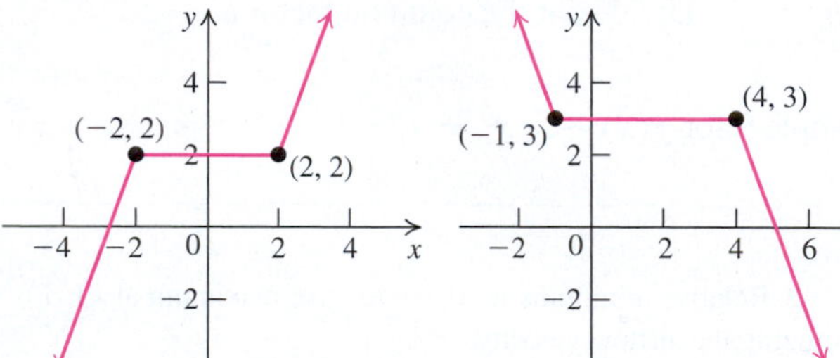

13. **14.**

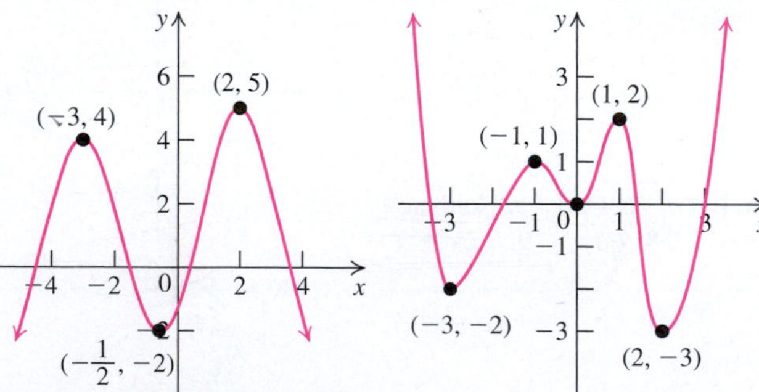

In Exercises 15–22, locate relative maximum and relative minimum points on the graph. State whether each relative extremum point is a turning point.

15. The graph of Exercise 7

16. The graph of Exercise 8

17. The graph of Exercise 9

18. The graph of Exercise 10

19. The graph of Exercise 11

20. The graph of Exercise 12

21. The graph of Exercise 13

22. The graph of Exercise 14

In Exercises 23–32, the graph of a function is given. State whether the function is odd, even, or neither.

23. **24.**

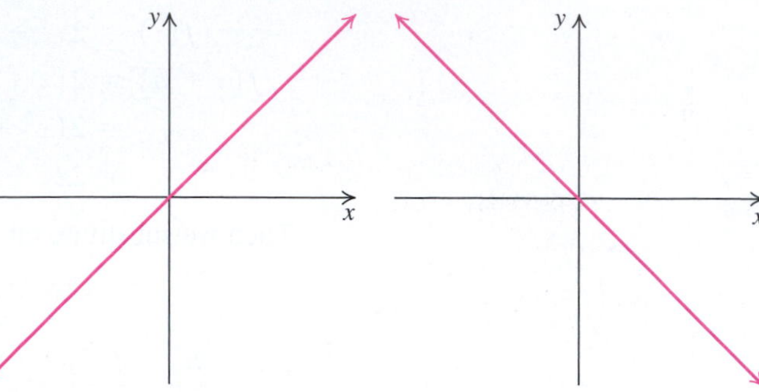

25. **26.**

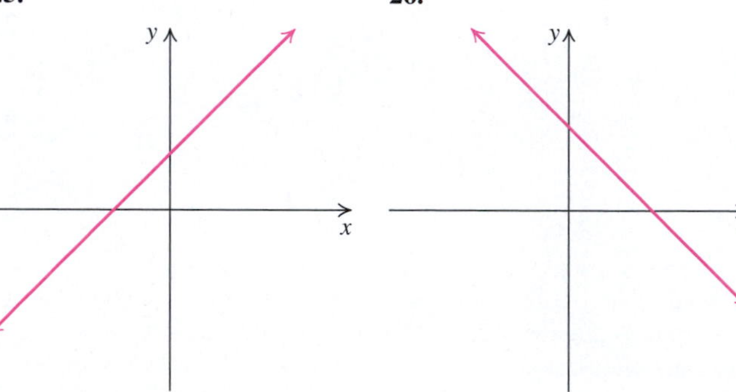

27. **28.**

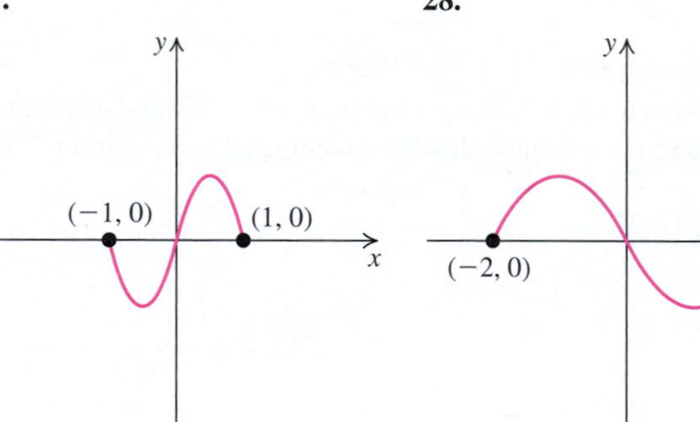

29. **30.**

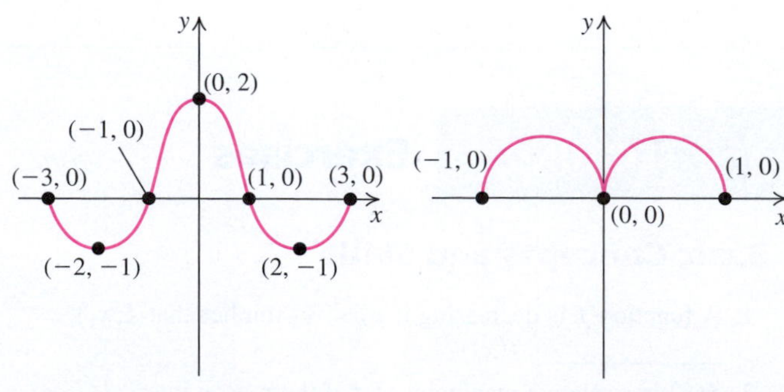

31. **32.**

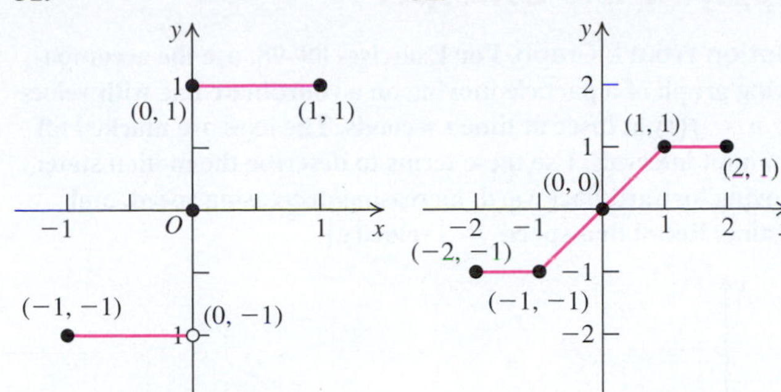

In Exercises 33–46, determine algebraically whether the given function is odd, even, or neither.

33. $f(x) = 2x^4 + 4$

34. $g(x) = 3x^4 - 5$

35. $f(x) = 5x^3 - 3x$

36. $g(x) = 2x^3 + 4x$

37. $f(x) = 2x + 4$

38. $g(x) = 3x + 7$

39. $f(x) = \dfrac{1}{x^2 + 4}$

40. $g(x) = \dfrac{x^2 + 2}{x^4 + 1}$

41. $f(x) = \dfrac{x^3}{x^2 + 1}$

42. $g(x) = \dfrac{x^4 + 3}{2x^3 - 3x}$

43. $f(x) = \dfrac{x}{x^5 - 3x^3}$

44. $g(x) = \dfrac{x^3 + 2x}{2x^5 - 3x}$

45. $f(x) = \dfrac{x^2 - 2x}{5x^4 + 7}$

46. $g(x) = \dfrac{3x^2 + 7}{x - 3}$

In Exercises 47–54, the graph of a function is given; each grid line represents 1 unit. Use the graph to find each of the following.

a. The domain and the range of the function
b. The intercepts, if any
c. The intervals on which the function is increasing, decreasing, or constant
d. The relative extrema points, if any
e. Whether the function is odd, even, or neither

47.

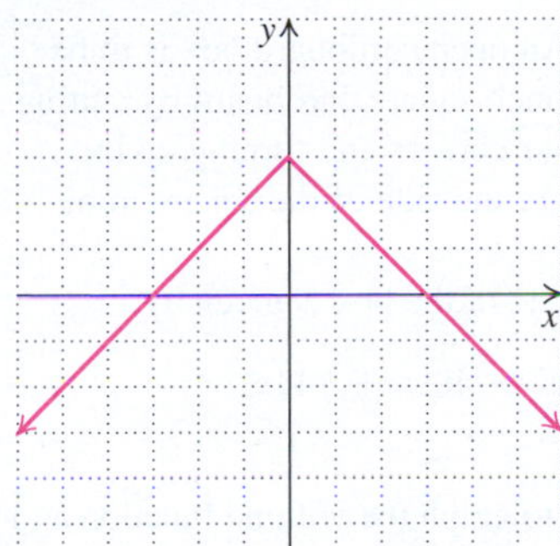

48.

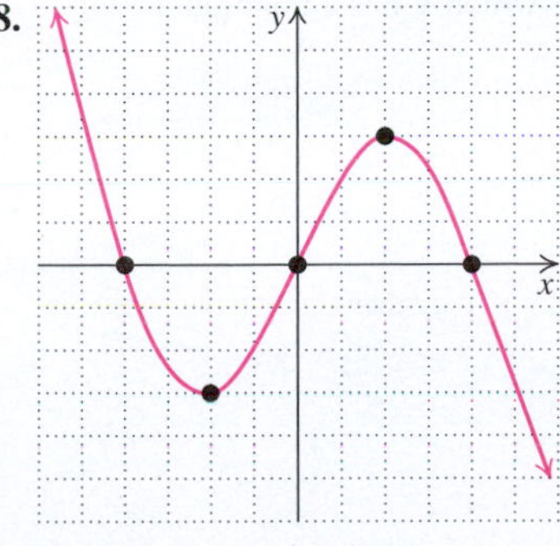

49.

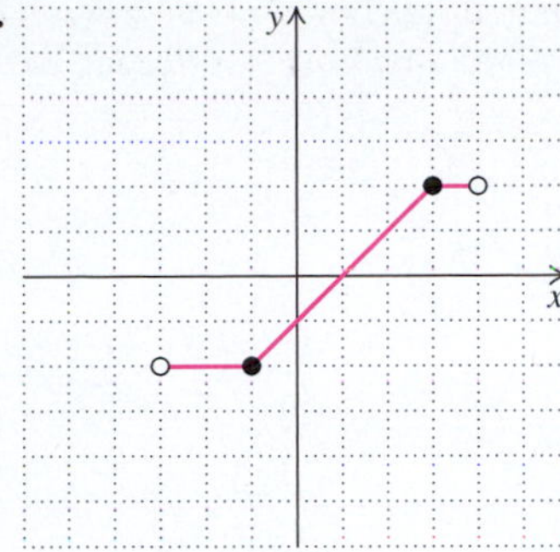

50.

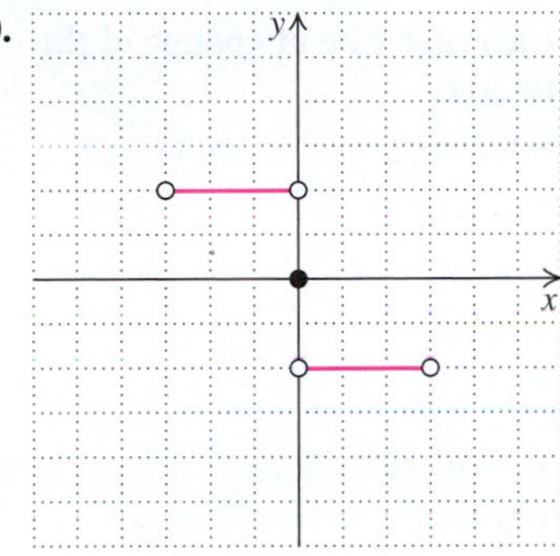

51.

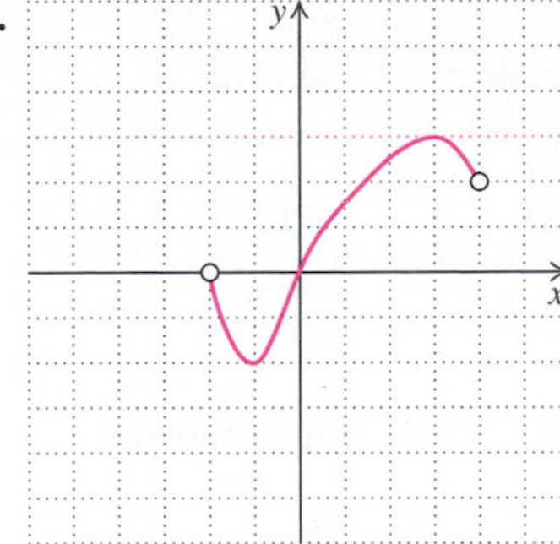

52.

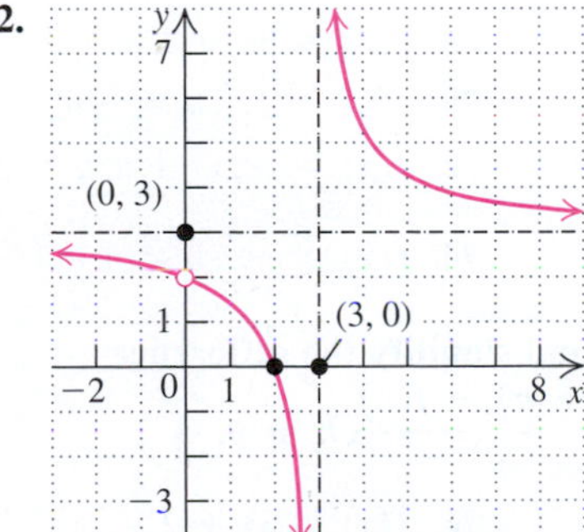

53.

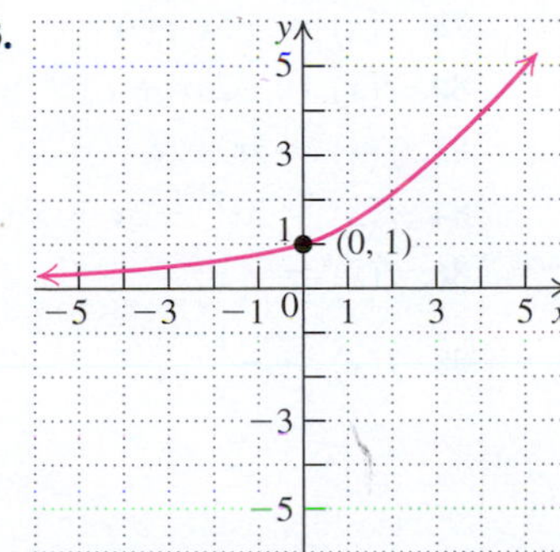

54. 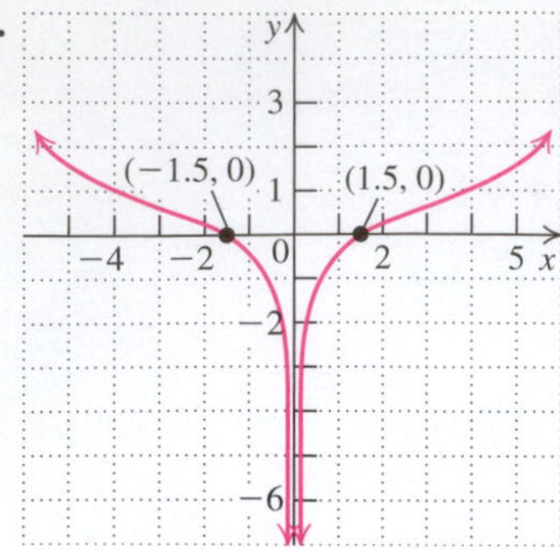

In Exercises 55–66, find the average rate of change of the function as x changes from a to b.

55. $f(x) = -2x + 7; a = -1, b = 3$
56. $f(x) = 4x - 9; a = -2, b = 2$
57. $f(x) = 3x + c; a = 1, b = 5$
58. $f(x) = mx + c; a = -1, b = 7$
59. $h(x) = x^2 - 1; a = -2, b = 0$
60. $h(x) = 2 - x^2; a = 3, b = 4$
61. $f(x) = (3 - x)^2; a = 1, b = 3$
62. $f(x) = (x - 2)^2; a = -1, b = 5$
63. $g(x) = x^3; a = -1, b = 3$
64. $g(x) = -x^3; a = -1, b = 3$
65. $h(x) = \frac{1}{x}; a = 2, b = 6$
66. $h(x) = \frac{4}{x + 3}; a = -2, b = 4$

In Exercises 67–74, find and simplify the difference quotient of the form $\frac{f(x) - f(a)}{x - a}, x \neq a$.

67. $f(x) = 2x, a = 3$
68. $f(x) = 3x + 2, a = 2$
69. $f(x) = -x^2, a = 1$
70. $f(x) = 2x^2, a = -1$
71. $f(x) = 3x^2 + x, a = 2$
72. $f(x) = -2x^2 + x, a = 3$
73. $f(x) = \frac{4}{x}, a = 1$
74. $f(x) = -\frac{4}{x}, a = 1$

In Exercises 75–88, find and simplify the difference quotient of the form $\frac{f(x + h) - f(x)}{h}, h \neq 0$.

75. $f(x) = x$
76. $f(x) = 3x + 2$
77. $f(x) = -2x + 3$
78. $f(x) = -5x - 6$
79. $f(x) = mx + b$
80. $f(x) = -2ax + c$
81. $f(x) = x^2$
82. $f(x) = x^2 - x$
83. $f(x) = 2x^2 + 3x$
84. $f(x) = 3x^2 - 2x + 5$
85. $f(x) = 4$
86. $f(x) = -3$
87. $f(x) = \frac{1}{x}$
88. $f(x) = -\frac{1}{x}$

Applying the Concepts

Motion From a Graph. **For Exercises 89–98, use the accompanying graph of a particle moving on a coordinate line with velocity $v = f(t)$ in ft/sec at time t seconds. The axes are marked off at 1-unit intervals. Use these terms to describe the motion state: moving forward/backward, increasing/decreasing speed, and resting. Recall that speed $= |\text{velocity}|$.**

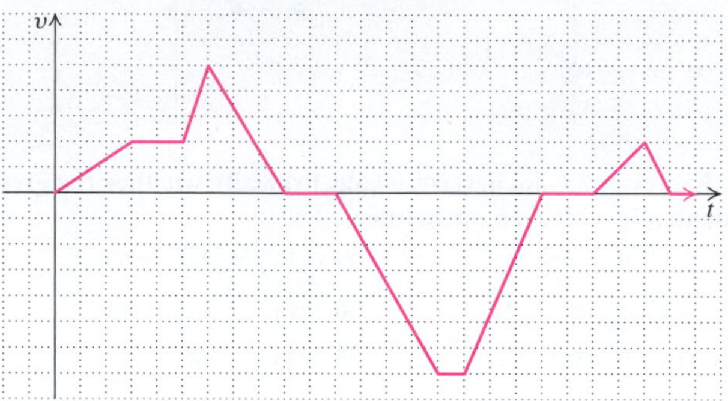

89. Find the domain of $v = f(t)$. Interpret.
90. Find the range of $v = f(t)$. Interpret.
91. Find the intervals over which the graph is above the t-axis. Interpret.
92. Find the intervals over which the graph is below the t-axis. Interpret.
93. Find the intervals over which $|v| = |f(t)|$ is increasing. Interpret for $v > 0$ and for $v < 0$.
94. Find the intervals over which $|v| = |f(t)|$ is decreasing. Interpret for $v > 0$ and for $v < 0$.
95. When does the particle have maximum speed?
96. When does the particle have minimum speed?
97. Describe the motion of the particle between $t = 5$ and $t = 6$.
98. Describe the motion of the particle between $t = 16$ and $t = 19$.
99. **Maximizing volume.** An open cardboard box is to be made from a piece of 12-inch square cardboard by cutting equal squares from the four corners and turning up the sides. Let x be the length of the side of the square to be cut out.
    a. Show that the volume, V, of the box is given by
    $$V = 4x^3 - 48x^2 + 144x.$$
    b. Find the domain of V.
     c. Use a graphing utility to graph the volume function and determine the range of V from its graph.
    d. Find the value of x for which V has a maximum value.

100. Minimizing the sum. Find two positive numbers whose product is 32 and whose sum is as small as possible.

a. Let one of the numbers be x, show that the sum $S = x + \frac{32}{x}$.

b. Use a graphing utility to graph the sum function and estimate the value of x for which S has a minimum value.

101. Cost. A computer notebook manufacturer has a daily fixed cost of \$10,500 and a marginal cost of \$210 per notebook.

a. Find the daily cost $C(x)$ of manufacturing x notebooks per day.

b. Find the cost of producing 50 notebooks per day.

c. Find the average cost of a computer notebook assuming that 50 notebooks are manufactured each day.

d. How many notebooks should be manufactured each day so that the average cost per notebook is \$315?

102. Geometry. Find an equation of the secant line to the graph of $f(x) = -2x^2 + 3x + 4$ passing through the points on the graph with x-coordinates 1 and 3.

103. Average velocity. A particle is moving in a straight line so that its position at time t in seconds is $f(t) = t^2 + 3t + 4$ feet to the right of a reference point for $t \geq 0$.

a. Find and interpret $f(0)$.

b. Find the distance traveled by the particle in four seconds.

c. Find the average velocity of the particle during the first three seconds.

d. Find its average velocity between the second and fifth seconds.

104. Population. The population of Sardonia (in millions) t years after 2000 is given by $P(t) = 0.01t^2 + 0.2t + 50$.

a. Find and interpret (*i*) $P(0)$, (*ii*) $P(4)$.

b. What was the average rate of growth of Sardonia's population between 2000 and 2010?

Beyond the Basics

105. Let $f(x) = \frac{x - 1}{x + 1}$. Show that $f(2x) = \frac{3f(x) + 1}{f(x) + 3}$.

106. What function is both odd and even?

107. Find algebraically the relative maximum point on the graph of $f(x) = -(x + 1)^2 + 5$. Does its graph have a relative minimum point?

108. Give an example to illustrate that a relative maximum point on a graph need not be a turning point on the graph.

In Exercises 109–112, find the difference quotient $\frac{f(x + h) - f(x)}{h}$, $h \neq 0$. Simplify with a rational numerator.

109. $f(x) = \sqrt{x}$

110. $f(x) = \sqrt{x - 1}$

111. $f(x) = -\frac{1}{\sqrt{x}}$

112. $f(x) = \frac{1}{x^2}$

Critical Thinking/Discussion/Writing

Let f be a function with domain $[a, b]$ so that a and b are endpoints of this domain.

113. Define relative maximum (minimum) at a.

114. Define relative maximum (minimum) at b.

115. Give an example (if possible) of a function f matching each description.

a. f has endpoint maximum and endpoint minimum.

b. f has endpoint minimum but no endpoint maximum.

c. f has endpoint maximum but no endpoint minimum.

d. f has neither endpoint maximum nor endpoint minimum

Maintaining Skills

In Exercises 116–118, find slope–intercept form of the equation of the line that passes through the given points P and Q.

116. $P = (-1, 3)$ and $Q = (4, -2)$

117. $P = (6, 2)$ and $Q = (7, -1)$

118. $P = (3, -5)$ and $Q = (6, -3)$

119. Let $f(x) = x^{3/2}$. Find (*i*) $f(2)$, (*ii*) $f(4)$, and (*iii*) $f(-4)$.

120. Let $f(x) = x^{2/3}$. Find (*i*) $f(2)$, (*ii*) $f(8)$, and (*iii*) $f(-8)$.

SECTION 2.6

A Library of Functions

BEFORE STARTING THIS SECTION, REVIEW

1. Equation of a line (Section 2.3)
2. Absolute value (Section P.1, page 8)
3. Graph of an equation (Section 2.2, page 165)

OBJECTIVES

1. Relate linear functions to linear equations.
2. Graph square root and cube root functions.
3. Evaluate and graph piecewise functions.
4. Graph additional basic functions.

◆ The Megatooth Shark

The giant "Megatooth" shark (*Carcharodon megalodon*), once estimated to be 100 to 120 feet in length, is the largest meat-eating fish that ever lived. Its actual length has been the subject of scientific controversy. Sharks first appeared in the oceans more than 400 million years ago, almost 200 million years before the dinosaurs. A shark's skeleton is made of cartilage, which decomposes rather quickly, making complete shark fossils rare. Consequently, scientists rely on calcified vertebrae, fossilized teeth, and small skin scales to reconstruct the evolutionary record of sharks.

The great white shark is the closest living relative to the giant "Megatooth" shark and has been used as a model to reconstruct the "Megatooth." John Maisey, curator at the American Museum of Natural History in New York City, used a partial set of "Megatooth" teeth to make a comparison with the jaws of the great white shark. In Example 2, we use a formula to calculate the length of the "Megatooth" on the basis of the height of the tooth of the largest known "Megatooth" specimen.

1 Relate linear functions to linear equations.

Linear Functions

We know from Section 2.3 that the graph of a linear equation $y = mx + b$ is a straight line with slope m and y-intercept b. A *linear function* has a similar definition.

Linear Functions

Let m and b be real numbers. The function $f(x) = mx + b$ is called a **linear function**. If $m = 0$, the function $f(x) = b$ is called a **constant function**. If $m = 1$ and $b = 0$, the resulting function $f(x) = x$ is called the **identity function**. See Figure 2.65.

The domain of a linear function is the interval $(-\infty, \infty)$ because the expression $mx + b$ is defined for any real number x. Figure 2.65 shows that the graph extends indefinitely to the left and right. The range of a nonconstant linear function also is the interval $(-\infty, \infty)$ because the graph extends indefinitely upward and downward. The range of a constant function $f(x) = b$ is the single real number b. See Figure 2.65(c).

GRAPHS OF $f(x) = mx + b$

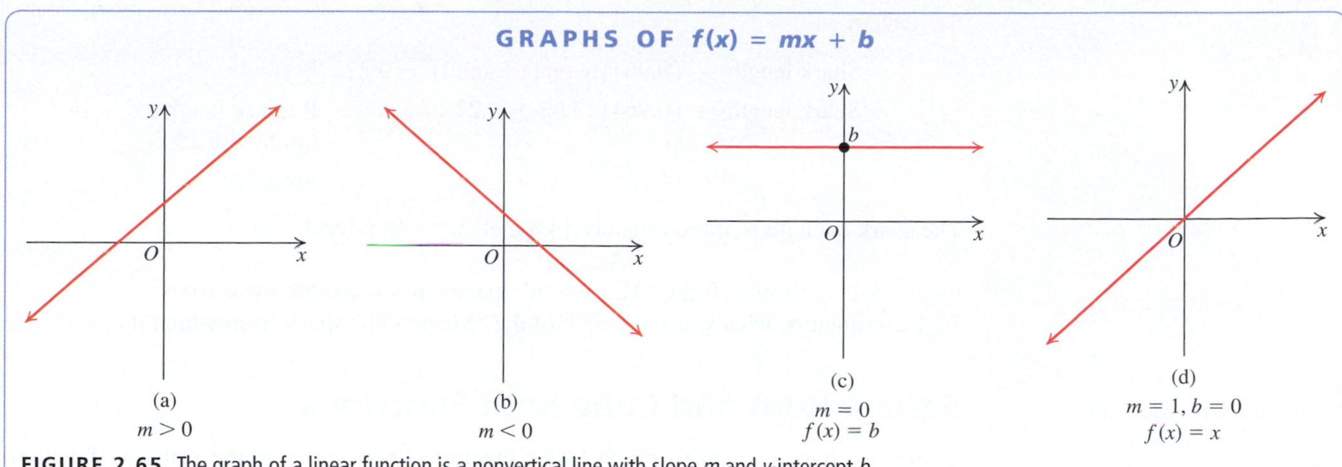

FIGURE 2.65 The graph of a linear function is a nonvertical line with slope m and y-intercept b.

EXAMPLE 1 Writing a Linear Function

Write a linear function g for which $g(1) = 4$ and $g(-3) = -2$.

Solution

We need to find the equation of the line passing through the two points

$$(x_1, y_1) = (1, 4) \text{ and } (x_2, y_2) = (-3, -2).$$

The slope of the line is given by

$$m = \frac{y_2 - y_1}{x_2 - x_1} = \frac{-2 - 4}{-3 - 1} = \frac{-6}{-4} = \frac{3}{2}.$$

Next, we use the point–slope form of a line.

$$y - y_1 = m(x - x_1) \quad \text{Point–slope form of a line}$$

$$y - 4 = \frac{3}{2}(x - 1) \quad \text{Substitute } x_1 = 1, y_1 = 4, \text{ and } m = \frac{3}{2}.$$

$$y - 4 = \frac{3}{2}x - \frac{3}{2} \quad \text{Distributive property}$$

$$y = \frac{3}{2}x + \frac{5}{2} \quad \text{Add } 4 = \frac{8}{2} \text{ to both sides and simplify.}$$

$$g(x) = \frac{3}{2}x + \frac{5}{2} \quad \text{Function notation}$$

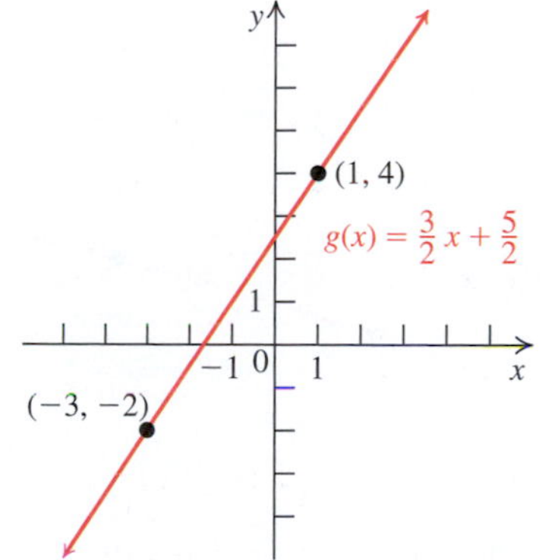

FIGURE 2.66 Point–slope form

The graph of the function $y = g(x)$ is shown in Figure 2.66.

Practice Problem 1 Write a linear function g for which $g(-2) = 2$ and $g(1) = 8$.

EXAMPLE 2 Determining the Length of the "Megatooth" Shark

The largest known "Megatooth" specimen is a tooth that has a total height of 15.6 centimeters. Calculate the length of the "Megatooth" shark from which it came by using the formula

$$\text{shark length} = (0.96)(\text{height of tooth}) - 0.22,$$

where shark length is measured in meters and tooth height is measured in centimeters.

Solution

Shark length $=$	$(0.96)(\text{height of tooth}) - 0.22$	Formula
Shark length $=$	$(0.96)(15.6) - 0.22$	Replace height of tooth with 15.6.
	$= 14.756$	Simplify.

The shark's length is approximately 14.8 meters ($\approx$48.6 feet).

Practice Problem 2 If the "Megatooth" specimen was a tooth measuring 16.4 centimeters, what was the length of the "Megatooth" shark from which it came? 

2 Graph square root and cube root functions.

Square Root and Cube Root Functions

So far, we have learned to graph linear functions: $y = mx + b$; the squaring function: $y = x^2$; and the cubing function: $y = x^3$. We now add some common functions to this list.

Square Root Function Recall (page 58) that for every nonnegative real number x, there is only one principal square root denoted by $\sqrt{x}$. Therefore, the equation $y = \sqrt{x}$ defines a function, called the **square root function**. The domain and range of the square root function is $[0, \infty)$.

EXAMPLE 3 Graphing the Square Root Function

Graph $f(x) = \sqrt{x}$.

Solution

To sketch the graph of $f(x) = \sqrt{x}$, we make a table of values. For convenience, we have selected the values of x that are perfect squares. See Table 2.9. We then plot the ordered pairs (x, y) and draw a smooth curve through the plotted points. See Figure 2.67.

TABLE 2.9

x	$y = f(x)$	(x, y)
0	$\sqrt{0} = 0$	$(0, 0)$
1	$\sqrt{1} = 1$	$(1, 1)$
4	$\sqrt{4} = 2$	$(4, 2)$
9	$\sqrt{9} = 3$	$(9, 3)$
16	$\sqrt{16} = 4$	$(16, 4)$

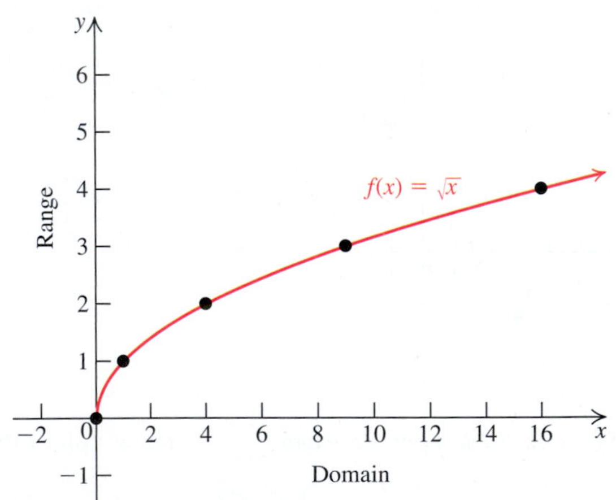

FIGURE 2.67 Graph of $y = \sqrt{x}$

The domain of $f(x) = \sqrt{x}$ is $[0, \infty)$, and its range also is $[0, \infty)$.

Practice Problem 3 Graph $g(x) = \sqrt{-x}$ and find its domain and range.

Cube Root Function There is only one cube root for each real number x. Therefore, the equation $y = \sqrt[3]{x}$ defines a function called the **cube root function**. The domain and range of the cube root function is $(-\infty, \infty)$.

EXAMPLE 4 **Graphing the Cube Root Function**

Graph $f(x) = \sqrt[3]{x}$.

Solution

We use the point plotting method to graph $f(x) = \sqrt[3]{x}$. For convenience, we select values of x that are perfect cubes. See Table 2.10. The graph of $f(x) = \sqrt[3]{x}$ is shown in Figure 2.68.

TABLE 2.10

x	$y = f(x)$	(x, y)
-8	$\sqrt[3]{-8} = -2$	$(-8, -2)$
-1	$\sqrt[3]{-1} = -1$	$(-1, -1)$
0	$\sqrt[3]{0} = 0$	$(0, 0)$
1	$\sqrt[3]{1} = 1$	$(1, 1)$
8	$\sqrt[3]{8} = 2$	$(8, 2)$

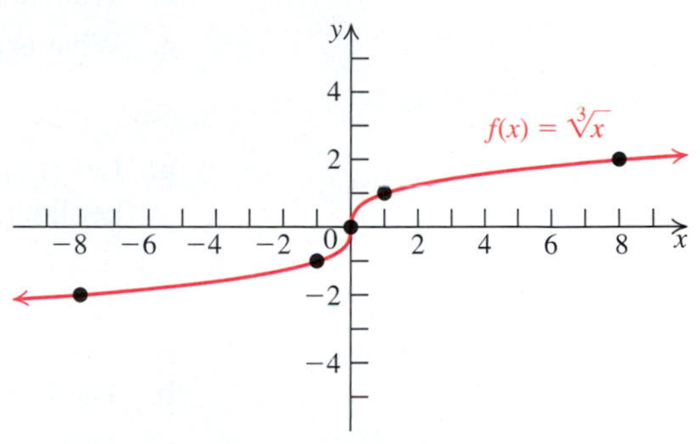

FIGURE 2.68 Graph of $y = \sqrt[3]{x}$

The domain of $f(x) = \sqrt[3]{x}$ is $(-\infty, \infty)$, and its range also is $(-\infty, \infty)$.

Practice Problem 4 Graph $g(x) = \sqrt[3]{-x}$ and find its domain and range.

3 Evaluate and graph piecewise functions.

Piecewise Functions

In the definition of some functions, different rules for assigning output values are used on different parts of the domain. Such functions are called **piecewise functions**; we illustrate the procedure for evaluating such functions next.

PROCEDURE IN ACTION

EXAMPLE 5 **Evaluating a Piecewise Function**

OBJECTIVE

Evaluate $F(a)$ for a piecewise function F.

Step 1 **Determine which line of the function applies to the number a.**

Step 2 **Evaluate $F(a)$ using the line chosen in Step 1.**

EXAMPLE

Let

$$F(x) = \begin{cases} x^2 & \text{if } x < 1 \\ 2x + 1 & \text{if } x \geq 2 \end{cases}$$

Find $F(0)$, $F(1)$ and $F(2)$.

1. Let $a = 0$. Because $0 < 1$, use the first line, $F(x) = x^2$.
Let $a = 1$, then since a is neither <1 nor ≥ 2 $a = 1$ is not in the domain of the F.
Let $a = 2$. Because $2 \geq 2$, use the second line, $F(x) = 2x + 1$.

2. $F(0) = (0)^2 = 0$, From $F(x) = x^2$
$F(1)$ is not defined.
$F(2) = 2(2) + 1 = 5$ From $F(x) = 2x + 1$

Practice Problem 5

Let $f(x) = \begin{cases} x^2 & \text{if } x \leq -1 \\ 2x & \text{if } x > -1 \end{cases}$. Find $f(-2)$ and $f(3)$.

EXAMPLE 6 **Finding and Evaluating a Piecewise Function**

In Peach County, Georgia, a section of the interstate highway has a speed limit of 55 miles per hour (mph). If you are caught speeding between 56 and 74 mph, your fine is \$50 plus \$3 for every mile per hour over 55 mph. For 75 mph and higher, your fine is \$150 plus \$5 for every mile per hour over 75 mph.

a. Find a piecewise function that gives your fine.

b. What is the fine for driving 60 mph?

c. What is the fine for driving 90 mph?

Solution

a. Let $f(x)$ be the piecewise function that represents your fine for speeding at x miles per hour. We express $f(x)$ as a piecewise function:

$$f(x) = \begin{cases} 50 + 3(x - 55), & 56 \le x < 75 \\ 150 + 5(x - 75), & x \ge 75 \end{cases}$$

b. The first line of the function means that if $56 \le x < 75$, your fine is

$$f(x) = 50 + 3(x - 55).$$

Suppose you are caught driving 60 mph. Then we have

$$\begin{aligned} f(x) &= 50 + 3(x - 55) && \text{Expression used for } 56 \le x < 75 \\ f(60) &= 50 + 3(60 - 55) && \text{Substitute 60 for } x. \\ &= 65 && \text{Simplify.} \end{aligned}$$

Your fine for speeding at 60 mph is \$65.

c. The second line of the function means that if $x \ge 75$, your fine is $f(x) = 150 + 5(x - 75)$. Suppose you are caught driving 90 mph. Then we have

$$\begin{aligned} f(x) &= 150 + 5(x - 75) && \text{Expression used for } x \ge 75 \\ f(90) &= 150 + 5(90 - 75) && \text{Substitute 90 for } x. \\ &= 225 && \text{Simplify.} \end{aligned}$$

Your fine for speeding at 90 mph is \$225.

Practice Problem 6 Repeat Example 6 if the speeding penalities are changed to \$50 plus \$4 for every mile per hour over 55 mph and \$200 plus \$5 for every mile per hour over 75 mph.

Graphing Piecewise Functions

Let's consider how to graph a piecewise function. The **absolute value function**, $f(x) = |x|$, can be expressed as a piecewise function by using the definition of absolute value:

$$f(x) = |x| = \begin{cases} -x \text{ if } x < 0 \\ x \text{ if } x \ge 0 \end{cases}$$

The first line in the function means that if $x < 0$, we use the equation $y = f(x) = -x$. So if $x = -3$, then

$$\begin{aligned} y = f(-3) &= -(-3) && \text{Substitute } -3 \text{ for } x \text{ in } f(x) = -x \\ &= 3 && \text{Simplify.} \end{aligned}$$

Thus, $(-3, 3)$ is a point on the graph of $y = |x|$. However, if $x \ge 0$, we use the second line in the function, which is $y = f(x) = x$. So if $x = 2$, then

$$y = f(2) = 2 \quad \text{Substitute 2 for } x \text{ in } f(x) = x$$

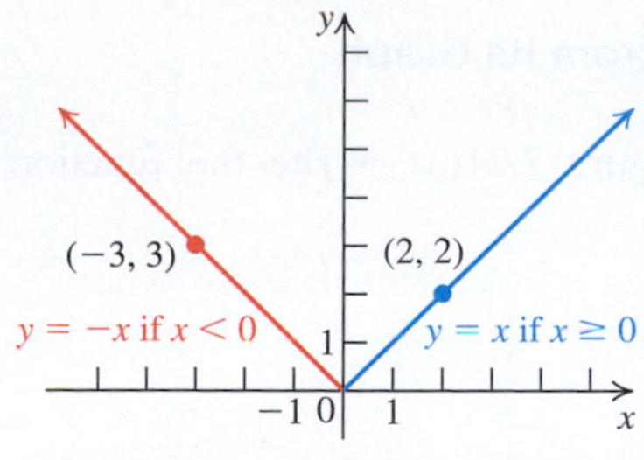

FIGURE 2.69 Graph of $y = |x|$

So $(2, 2)$ is a point on the graph of $y = |x|$. The two pieces $y = -x$ and $y = x$ are linear functions. We graph the appropriate parts of these lines ($y = -x$ for $x < 0$ and $y = x$ for $x \geq 0$) to form the graph of $y = |x|$. See Figure 2.69.

EXAMPLE 7 Graphing a Piecewise Function

Let

$$F(x) = \begin{cases} -2x + 1 & \text{if } x < 1 \\ 3x + 1 & \text{if } x \geq 1 \end{cases}.$$

Sketch the graph of $y = F(x)$.

TECHNOLOGY CONNECTION

To graph $F(x)$ on a TI-84 calculator:
$Y_1 = (x < 1)(-2x + 1) + (x \geq 1)(3x + 1)$

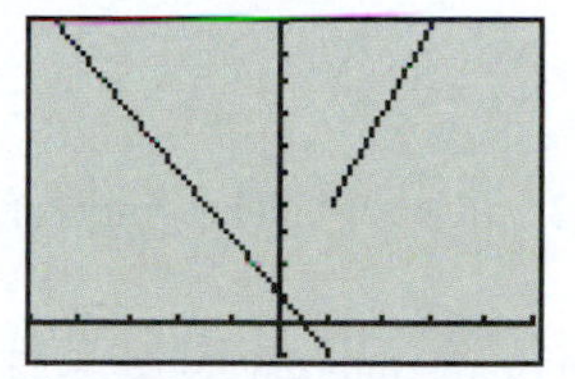

For the graph to appear correctly, change from *connected* to *dot* mode.

Solution

In the definition of F, the formula changes at $x = 1$. We call such numbers the **transition points** of the formula. For the function F, the only transition point is 1. Generally, to graph a piecewise function, we graph the function separately over the open intervals determined by the transition points and then graph the function at the transition points themselves. For the function $y = F(x)$, we graph the equation $y = -2x + 1$ on the interval $(-\infty, 1)$. See Figure 2.70(a). Next, we graph the equation $y = 3x + 1$ on the interval $(1, \infty)$ and at the transition point 1, where $y = F(1) = 3(1) + 1 = 4$. See Figure 2.70(b).

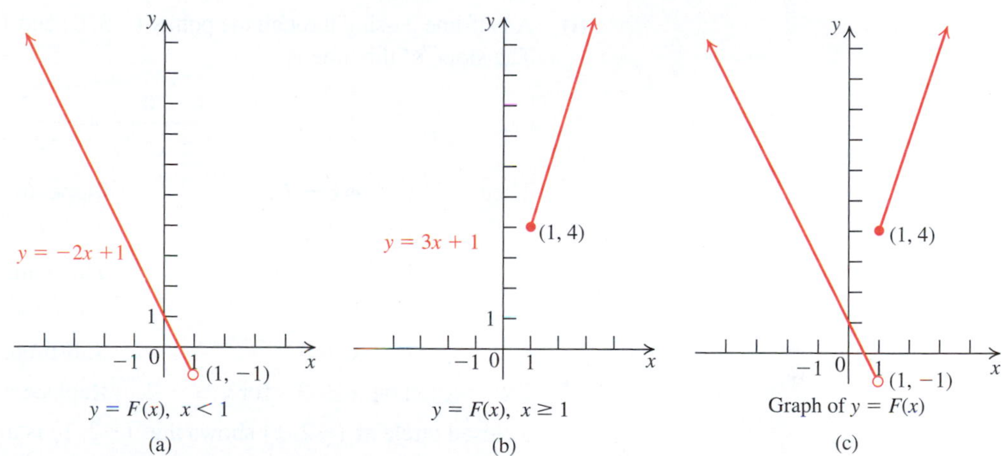

FIGURE 2.70 Steps for graphing a piecewise function

Combining these portions, we obtain the graph of $y = F(x)$, shown in Figure 2.70(c). When we come to the end of the part of the graph we are working with, we draw

(i) A closed circle if that point is included.

(ii) An open circle if the point is excluded.

You may find it helpful to think of this procedure as following the graph of $y = -2x + 1$ when x is less than 1 and then jumping to the graph of $y = 3x + 1$ when x is equal to or greater than 1.

Practice Problem 7 Let $F(x) = \begin{cases} -3x & \text{if } x \leq -1 \\ 2x & \text{if } x > -1 \end{cases}$. Sketch the graph of $y = F(x)$.

EXAMPLE 8 Writing a Piecewise Function from Its Graph

The graph of a piecewise function g is given in Figure 2.71(a). Write the function algebraically.

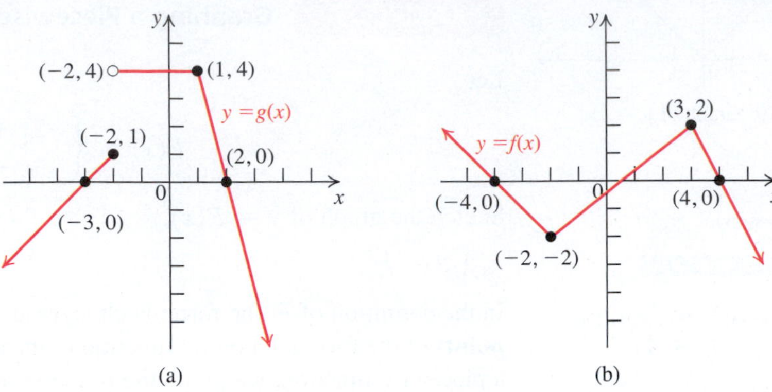

FIGURE 2.71

Solution

The domain of g is $(-\infty, \infty)$ with transition points at $x = -2$ and $x = 1$. The graph of g is made up of three parts.

(i) A half-line passing through the points $(-3, 0)$ and $(-2, 1)$ on the interval $(-\infty, -2]$. The slope of this line is:

$$m = \frac{1 - 0}{-2 - (-3)} = \frac{1}{-2 + 3} = 1.$$

Then	$y = mx + b$	Slope–intercept form
	$y = x + b$	$m = 1$
	$1 = -2 + b$	The point $(x, y) = (-2, 1)$ is on the line.
	$3 = b$	Solve for b.
	$y = x + 3$	Substitute $b = 3$ in $y = x + b$.
So	$g(x) = x + 3$ for $x \le -2$.	Replace y with $g(x)$.

a closed circle at $(-2, 1)$ shows that $(-2, 1)$ is a graph point.

(ii) A horizontal line segment passing through $(-2, 4)$ and $(1, 4)$; an open circle at $(-2, 4)$ shows that $(-2, 4)$ is not a graph point. We have

$$g(x) = 4 \quad \text{for } -2 < x \le 1.$$

A closed circle at $(1, 4)$ shows that $(1, 4)$ is a graph point.

(iii) A half-line passing through $(1, 4)$ and $(2, 0)$. As in **(i)**, we find $m = -4$ and $b = 8$. Because $(1, 4)$ is a graph point of the segment discussed in **(ii)** we need only define g for $x > 1$. We have

$$g(x) = -4x + 8 \quad \text{for } x > 1.$$

Combining **(i)**, **(ii)**, and **(iii)**, we have

$$g(x) = \begin{cases} x + 3 & \text{if } x \le -2 \\ 4 & \text{if } -2 < x \le 1 \\ -4x + 8 & \text{if } x > 1. \end{cases}$$

Practice Problem 8 Repeat Example 8 for the graph of $y = f(x)$ in Figure 2.71(b).

TECHNOLOGY CONNECTION

To graph the greatest integer function, enter $Y_1 = \text{int}(x)$. If you have your calculator set to graph in *connected* mode, you will get the following incorrect graph:

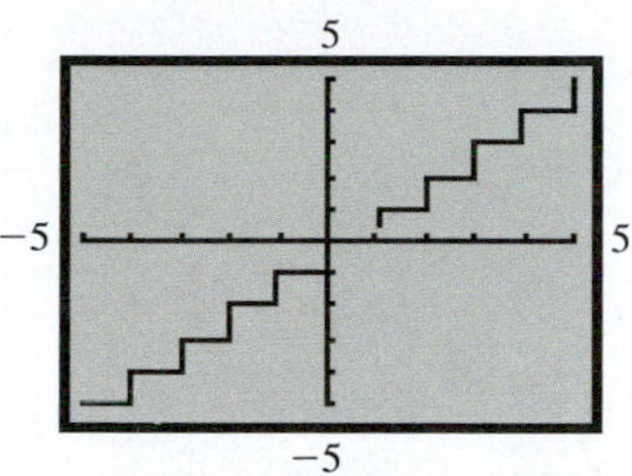

For the graph to appear correctly, change from *connected* to *dot* mode:

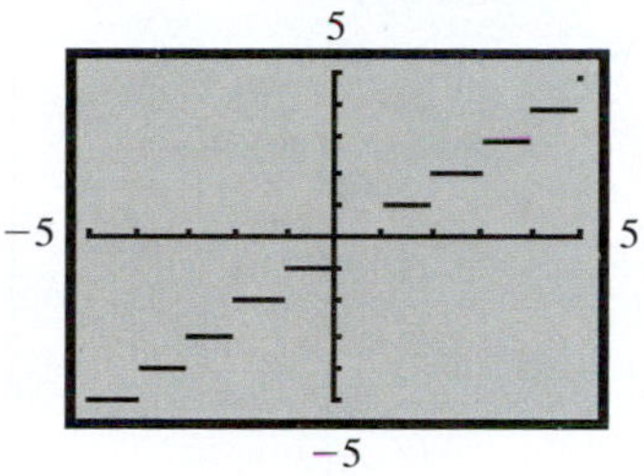

Note that the step from $0 \le x \le 1$ is obscured by the x-axis.

Some piecewise functions are called **step functions**. Their graphs look like the steps of a staircase.

The **greatest integer function** is denoted by $[\![x]\!]$, or $\text{int}(x)$, and is defined as the greatest integer less than or equal to x. For example,

$$[\![2]\!] = 2, \quad [\![2.3]\!] = 2, \quad [\![2.7]\!] = 2, \quad [\![2.99]\!] = 2$$

because 2 is the greatest integer less than or equal to 2, 2.3, 2.7, and 2.99. Similarly,

$$[\![-2]\!] = -2, \quad [\![-1.9]\!] = -2, \quad [\![-1.1]\!] = -2, \quad [\![-1.001]\!] = -2$$

because -2 is the greatest integer less than or equal to $-2, -1.9, -1.1$, and -1.001.

In general, if m is an integer such that $m \le x < m + 1$, then $[\![x]\!] = m$. In other words, if x is between two consecutive integers m and $m + 1$, then $[\![x]\!]$ is assigned the smaller integer m.

EXAMPLE 9 Graphing a Step Function

Graph the greatest integer function $f(x) = [\![x]\!]$.

Solution

Choose a typical closed interval between two consecutive integers—say, the interval $[2, 3]$. We know that between 2 and 3, the greatest integer function's value is 2. In symbols, if $2 \le x < 3$, then $[\![x]\!] = 2$. Similarly, if $1 \le x < 2$, then $[\![x]\!] = 1$. Therefore, the values of $[\![x]\!]$ are constant between each pair of consecutive integers and jump by 1 unit at each integer. The graph of $f(x) = [\![x]\!]$ is shown in Figure 2.72.

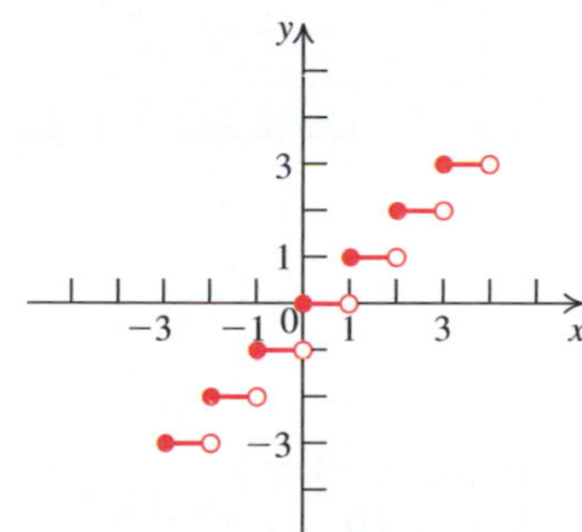

FIGURE 2.72 Graph of $y = [\![x]\!]$

The greatest integer function f can be interpreted as a piecewise function:

$$f(x) = [\![x]\!] = \begin{cases} \quad \vdots \\ -2 \text{ if } -2 \le x < -1 \\ -1 \text{ if } -1 \le x < 0 \\ \;\;0 \text{ if } \;\;0 \le x < 1 \\ \;\;1 \text{ if } \;\;1 \le x < 2 \\ \quad \vdots \end{cases}$$

Practice Problem 9 Find the values of $f(x) = [\![x]\!]$ for $x = -3.4$ and $x = 4.7$.

4 Graph additional basic functions.

Basic Functions

As you progress through this course and future mathematics courses, you will repeatedly come across a small list of basic functions. The following box lists some of these common algebraic functions, along with their properties. You should try to produce these graphs by plotting points and using symmetries. The unit length in all of the graphs shown is the same on both axes.

A LIBRARY OF BASIC FUNCTIONS

Constant Function

$f(x) = c$

Domain: $(-\infty, \infty)$
Range: $\{c\}$
Constant on $(-\infty, \infty)$
Even function (y-axis symmetry)

Identity Function

$f(x) = x$

Domain: $(-\infty, \infty)$
Range: $(-\infty, \infty)$
Increasing on $(-\infty, \infty)$
Odd function (origin symmetry)

Squaring Function

$f(x) = x^2$

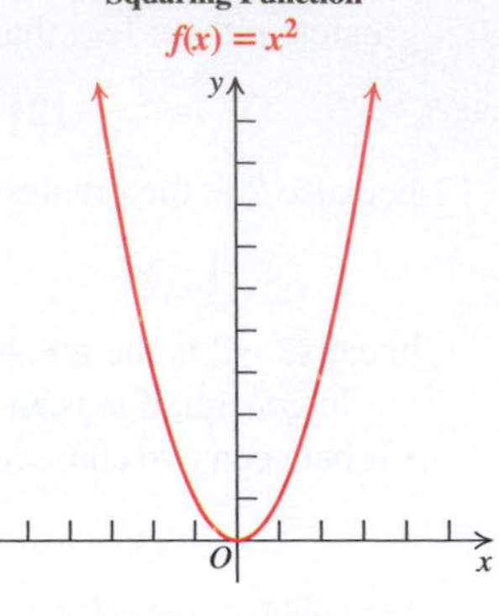

Domain: $(-\infty, \infty)$
Range: $[0, \infty)$
Decreasing on $(-\infty, 0)$
Increasing on $(0, \infty)$
Even function (y-axis symmetry)

Cubing Function

$f(x) = x^3$

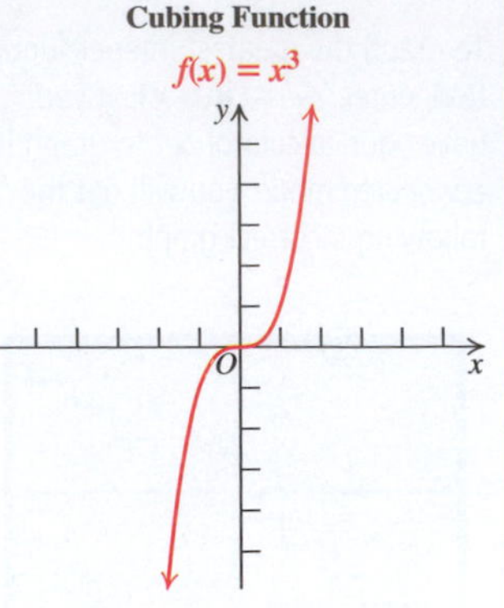

Domain: $(-\infty, \infty)$
Range: $(-\infty, \infty)$
Increasing on $(-\infty, \infty)$
Odd function (origin symmetry)

Absolute Value Function

$f(x) = |x|$

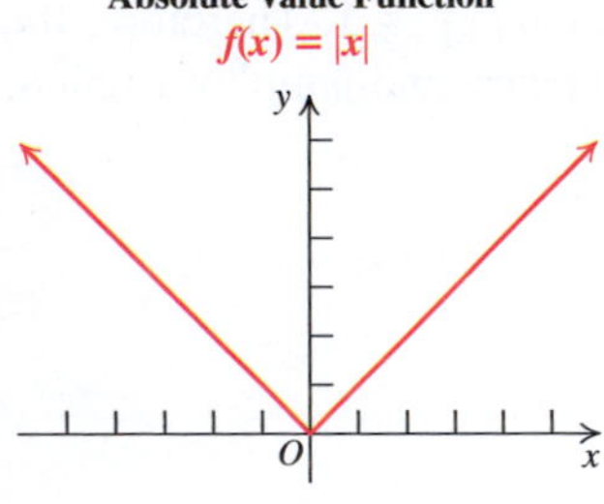

Domain: $(-\infty, \infty)$
Range: $[0, \infty)$
Decreasing on $(-\infty, 0)$
Increasing on $(0, \infty)$
Even function (y-axis symmetry)

Square Root Function

$f(x) = \sqrt{x} = x^{1/2}$

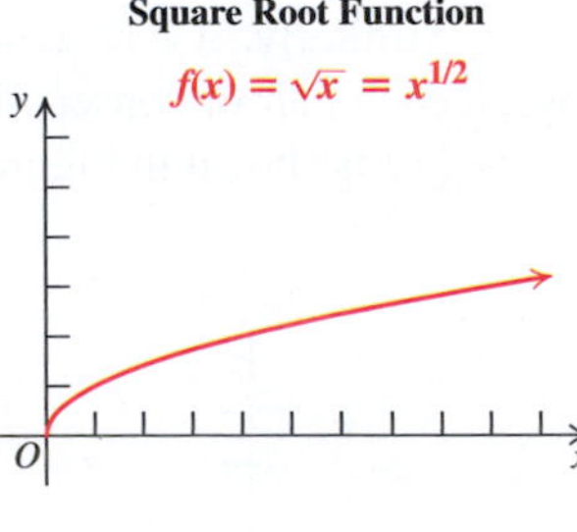

Domain: $[0, \infty)$
Range: $[0, \infty)$
Increasing on $(0, \infty)$
Neither even nor odd (no symmetry)

Cube Root Function

$f(x) = \sqrt[3]{x} = x^{1/3}$

Domain: $(-\infty, \infty)$
Range: $(-\infty, \infty)$
Increasing on $(-\infty, \infty)$
Odd function (origin symmetry)

Reciprocal Function

$f(x) = \frac{1}{x}$

Domain: $(-\infty, 0) \cup (0, \infty)$
Range: $(-\infty, 0) \cup (0, \infty)$
Decreasing on $(-\infty, 0) \cup (0, \infty)$
Odd function (origin symmetry)

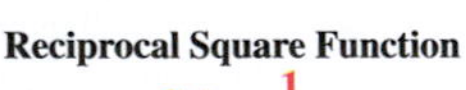

Reciprocal Square Function

$f(x) = \frac{1}{x^2}$

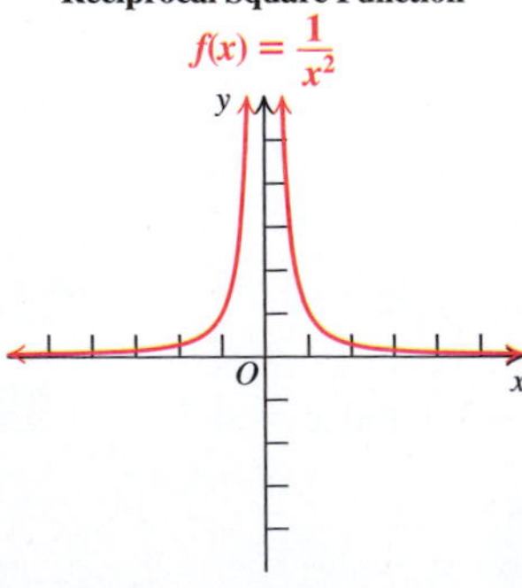

Domain: $(-\infty, 0) \cup (0, \infty)$
Range: $(0, \infty)$
Increasing on $(-\infty, 0)$
Decreasing on $(0, \infty)$
Even function (y-axis symmetry)

Rational Power Function

$f(x) = x^{\frac{3}{2}} = \left(x^{\frac{1}{2}}\right)^3$

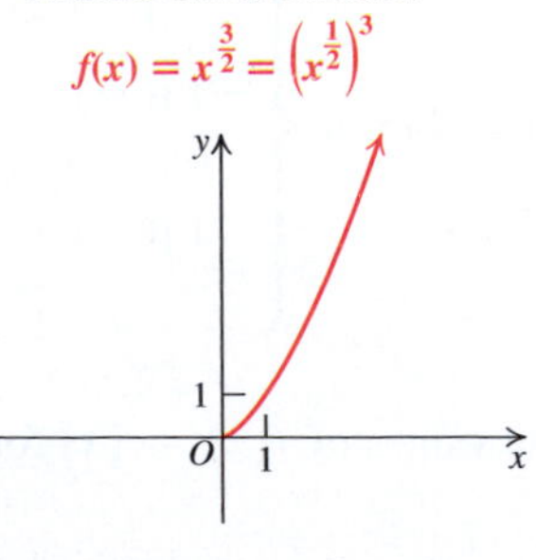

Domain: $[0, \infty)$
Range: $[0, \infty)$
Increasing on $(0, \infty)$
Neither even nor odd (no symmetry)

Rational Power Function

$f(x) = x^{\frac{2}{3}} = \left(x^{\frac{1}{3}}\right)^2$

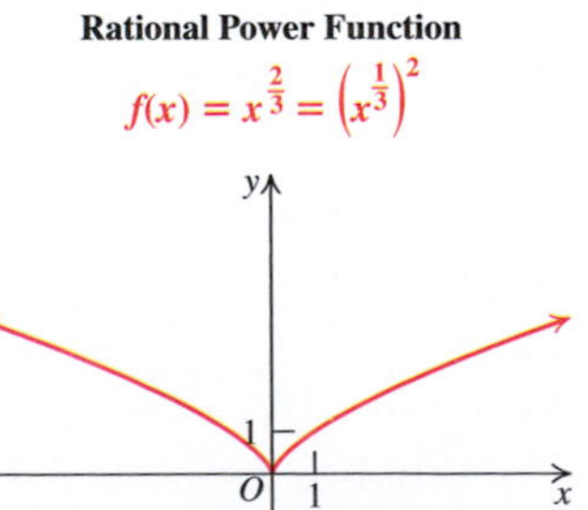

Domain: $(-\infty, \infty)$
Range: $[0, \infty)$
Decreasing on $(-\infty, 0)$
Increasing on $(0, \infty)$
Even function (y-axis symmetry)

Greatest–Integer Function

$f(x) = [\![x]\!]$

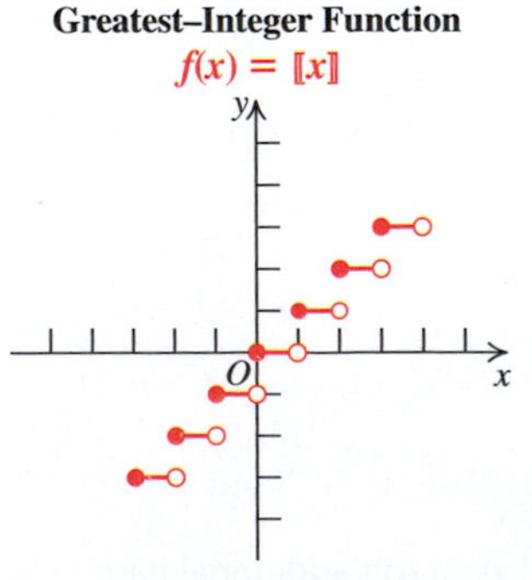

Domain: $(-\infty, \infty)$
Range: $\{\ldots, -3, -2, -1$ $0, 1, 2, 3, \ldots\}$
Neither even nor odd (no symmetry)

Answers to Practice Problems

1. $g(x) = 2x + 6$ **2.** 15.524 m

3. 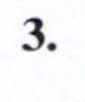

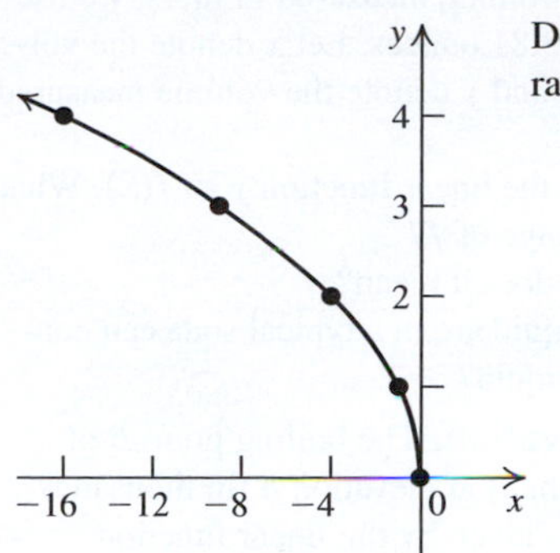

Domain: $(-\infty, 0]$; range: $[0, \infty)$

4. Domain: $(-\infty, \infty)$; range: $(-\infty, \infty)$

5. $f(-2) = 4, f(3) = 6$

6. a. $f(x) = \begin{cases} 50 + 4(x - 55), & 56 \le x < 75 \\ 200 + 5(x - 75), & x \ge 75 \end{cases}$ **b.** \$70

c. \$275 **7.** $f(x) = -3x, x \le -1 \quad f(x) = 2x, x > -1$

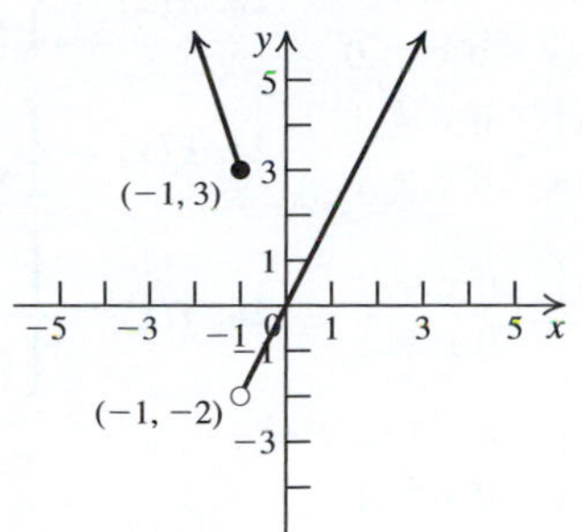

8. $f(x) = \begin{cases} -x - 4 & \text{if } x \le -2 \\ \frac{4}{5}x - \frac{2}{5} & \text{if } -2 < x < 3 \\ -2x + 8 & \text{if } x \ge 3 \end{cases}$

9. $[\![-3.4]\!] = -4; [\![4.7]\!] = 4$

SECTION 2.6 Exercises

Basic Concepts and Skills

1. The graph of the linear function $f(x) = b$ is a(n) ____________ line.
2. The absolute value function can be expressed as a piecewise function by writing $f(x) = |x| =$ ____________.
3. The graph of the function $f(x) = \begin{cases} x^2 + 2 \text{ if } x \le 1 \\ ax \quad\quad \text{if } x > 1 \end{cases}$ will have a break at $x = 1$ unless $a =$ ____________.
4. **True or False.** The function $f(x) = [\![x]\!]$ is increasing on the interval $(0, 3)$.

In Exercises 5–14, write a linear function f that has the indicated values. Sketch the graph of f.

5. $f(0) = 1, f(-1) = 0$
6. $f(1) = 0, f(2) = 1$
7. $f(-1) = 1, f(2) = 7$
8. $f(-1) = -5, f(2) = 4$
9. $f(1) = 1, f(2) = -2$
10. $f(1) = -1, f(3) = 5$
11. $f(-2) = 2, f(2) = 4$
12. $f(2) = 2, f(4) = 5$
13. $f(0) = -1, f(3) = -3$
14. $f(1) = \frac{1}{4}, f(4) = -2$

15. Let
$$f(x) = \begin{cases} x & \text{if } x \ge 2 \\ 2 & \text{if } x < 2 \end{cases}.$$
   a. Find $f(1), f(2)$, and $f(3)$.
   b. Sketch the graph of $y = f(x)$.

16. Let
$$g(x) = \begin{cases} 2x & \text{if } x < 0 \\ x & \text{if } x \ge 0 \end{cases}.$$
   a. Find $g(-1), g(0)$, and $g(1)$.
   b. Sketch the graph of $y = g(x)$.

17. Let
$$f(x) = \begin{cases} 1 & \text{if } x > 0 \\ -1 & \text{if } x < 0 \end{cases}.$$
   a. Find $f(-15)$ and $f(12)$.
   b. Sketch the graph of $y = f(x)$.
   c. Find the domain and the range of f.

18. Let
$$g(x) = \begin{cases} 2x + 4 & \text{if } x > 1 \\ x + 2 & \text{if } x \le 1 \end{cases}.$$
   a. Find $g(-3), g(1)$, and $g(3)$.
   b. Sketch the graph of $y = g(x)$.
   c. Find the domain and the range of g.

In Exercises 19–32, sketch the graph of each function and from the graph, find its domain and its range.

19. $f(x) = \begin{cases} -2x & \text{if } x < 0 \\ x^2 & \text{if } x \geq 0 \end{cases}$

20. $g(x) = \begin{cases} |x| & \text{if } x < 1 \\ x^2 & \text{if } x \geq 1 \end{cases}$

21. $g(x) = \begin{cases} \dfrac{1}{x} & \text{if } x < 0 \\ \sqrt{x} & \text{if } x \geq 0 \end{cases}$

22. $h(x) = \begin{cases} \sqrt[3]{x} & \text{if } x < 1 \\ \sqrt{x} & \text{if } x \geq 1 \end{cases}$

23. $f(x) = \begin{cases} [\![x]\!] & \text{if } x < 1 \\ \sqrt[3]{x} & \text{if } x \geq 1 \end{cases}$

24. $g(x) = \begin{cases} x^3 & \text{if } x < 0 \\ \sqrt{x} & \text{if } x \geq 0 \end{cases}$

25. $g(x) = \begin{cases} |x| & \text{if } x < 1 \\ x^3 & \text{if } x \geq 1 \end{cases}$

26. $f(x) = \begin{cases} \dfrac{1}{x} & \text{if } x < 0 \\ [\![x]\!] & \text{if } x \geq 0 \end{cases}$

27. $f(x) = \begin{cases} |x| & \text{if } -2 \leq x < 1 \\ \sqrt{x} & \text{if } x \geq 1 \end{cases}$

28. $g(x) = \begin{cases} \sqrt[3]{x} & \text{if } -8 \leq x < 1 \\ 2x & \text{if } x \geq 1 \end{cases}$

29. $g(x) = \begin{cases} [\![x]\!] & \text{if } x < 1 \\ -2x & \text{if } 1 \leq x \leq 4 \end{cases}$

30. $h(x) = \begin{cases} |x| & \text{if } x < 0 \\ \sqrt{x} & \text{if } 0 \leq x \leq 4 \end{cases}$

31. $f(x) = \begin{cases} 2x + 3 & \text{if } x < -2 \\ x + 1 & \text{if } -2 \leq x < 1 \\ -x + 3 & \text{if } x \geq 1 \end{cases}$

32. $g(x) = \begin{cases} -2x + 1 & \text{if } x \leq -1 \\ 2x + 1 & \text{if } -1 < x < 2 \\ x + 2 & \text{if } x \geq 2 \end{cases}$

In Exercises 33–36, write a piecewise function for the given graph.

33.

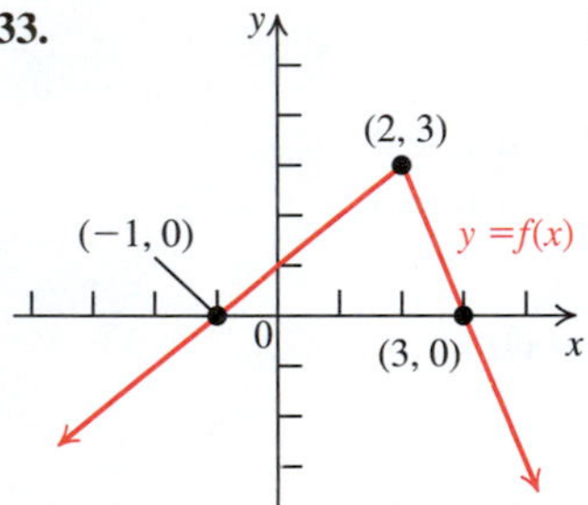

34.

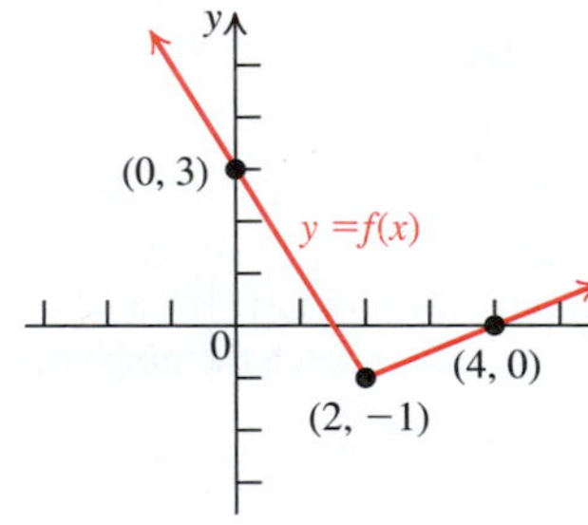

35.

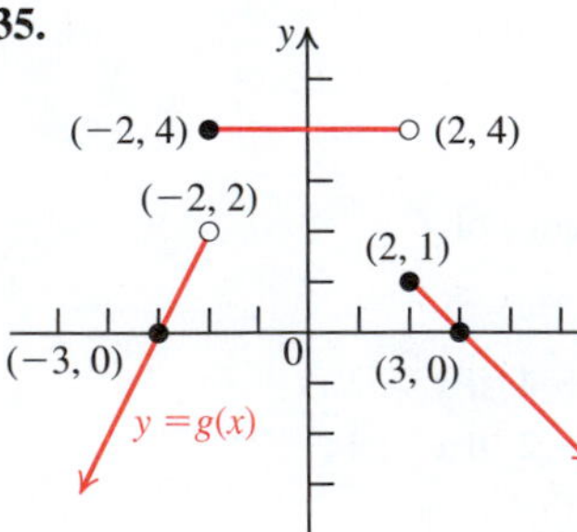

36.

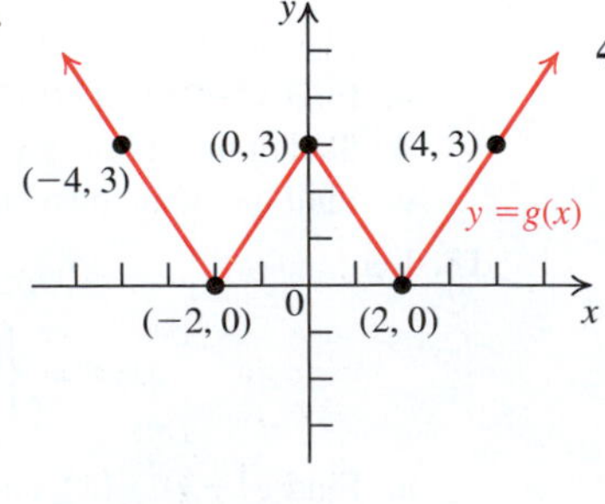

Applying the Concepts

37. Converting volume. To convert the volume of a liquid measured in ounces to a volume measured in liters, we use the fact that 1 liter $\approx$ 33.81 ounces. Let x denote the volume measured in ounces and y denote the volume measured in liters.

a. Write an equation for the linear function $y = f(x)$. What are the domain and range of f?

b. Compute $f(3)$. What does it mean?

c. How many liters of liquid are in a typical soda can containing 12 ounces of liquid?

38. Boiling point and elevation. The boiling point B of water (in degrees Fahrenheit) at elevation h (in thousands of feet) above sea level is given by the linear function $B(h) = -1.8h + 212$.

a. Find and interpret the intercepts of the graph of the function $y = B(h)$.

b. Find the domain of this function.

c. Find the elevation at which water boils at 98.6°F. Why is this height dangerous to humans?

39. Pressure at sea depth. The pressure P in atmospheres (atm) at a depth d feet is given by the linear function

$$P(d) = \frac{1}{33}d + 1.$$

a. Find and interpret the intercepts of $y = P(d)$.

b. Find $P(0)$, $P(10)$, $P(33)$, and $P(100)$.

c. Find the depth at which the pressure is 5 atmospheres.

40. Speed of sound. The speed V of sound (in feet per second) in air at temperature T (in degrees Fahrenheit) is given by the linear function $V(T) = 1055 + 1.1T$.

a. Find the speed of sound at 90°F.

b. Find the temperature at which the speed of sound is 1100 feet per second.

41. Manufacturer's cost. A manufacturer of printers has a total cost per day consisting of a fixed overhead of \$6000 plus product costs of \$50 per printer.

a. Express the total cost C as a function of the number x of printers produced.

b. Draw the graph of $y = C(x)$. Interpret the y-intercept.

c. How many printers were manufactured on a day when the total cost was \$11,500?

42. Supply function. When the price p of a commodity is \$10 per unit, 750 units of it are sold. For every \$1 increase in the unit price, the supply q increases by 100 units.

a. Write the linear function $q = f(p)$.

b. Find the supply when the price is \$15 per unit.

c. What is the price at which 1750 units can be supplied?

43. Apartment rental. Suppose a two-bedroom apartment in a neighborhood near your school rents for \$900 per month. If you move in after the first of the month, the rent is prorated; that is, it is reduced linearly.

a. Suppose you move into the apartment x days after the first of the month. (Assume that the month has 30 days.) Express the rent R as a function of x.

b. Compute the rent if you move in six days after the first of the month.

c. When did you move into the apartment if the landlord charged you rent of \$600?

44. College admissions. The average SAT scores (mathematics and critical reading) of college-bound seniors in the state of Florida was 995 in 2009; in 2011, it was 976. Assume that the scores have been changing linearly.

a. Express the average SAT score in Florida as a function of time since 2009.

b. If the trend continues, what will be the average SAT score of incoming students in Florida in 2013?

c. If the trend continues, when will the average SAT score be 900?

45. Breathing capacity. Suppose the average maximum breathing capacity for humans drops linearly from 100% at age 20 to 40% at age 80. What age corresponds to 50% capacity?

46. Drug dosage for children. If a is the adult dosage of a medicine and t is the child's age, then the *Friend's rule* for the child's dosage y is given by the formula $y = \frac{2}{25}ta$.

a. Suppose the adult dosage is 60 milligrams. Find the dosage for a 5-year-old child.

b. How old would a child have to be in order to be prescribed an adult dosage?

47. Air pollution. In Ballerenia, the average number y of deaths per month was observed to be linearly related to the concentration x of sulfur dioxide in the air. Suppose there are 30 deaths when $x = 150$ milligrams per cubic meter and 50 deaths when $x = 420$ milligrams per cubic meter.

a. Write y as a function of x.

b. Find the number of deaths when $x = 350$ milligrams per cubic meter.

c. If the number of deaths per month is 70, what is the concentration of sulfur dioxide in the air?

48. Child shoe sizes. Children's shoe sizes start with size 0, having an insole length L of $3\frac{11}{12}$ inches and each full size being $\frac{1}{3}$ inch longer.

a. Write the equation $y = L(S)$, where S is the shoe size.

b. Find the length of the insole of a child's shoe of size 4.

c. What size (to the nearest half size) shoe will fit a child whose insole length is 6.1 inches?

49. State income tax. Suppose a state's income tax code states that the tax liability T on x dollars of taxable income is as follows:

$$T(x) = \begin{cases} 0.04x & \text{if } 0 \le x < 20{,}000 \\ 800 + 0.06(x - 20{,}000) & \text{if } x \ge 20{,}000 \end{cases}$$

a. Graph the function $y = T(x)$.

b. Find the tax liability on each taxable income.

(i) \$12,000
(ii) \$20,000
(iii) \$50,000

c. Find your taxable income if you had each tax liability.

(i) \$600
(ii) \$1200
(iii) \$2300

50. Federal income tax. The tax table for the 2009 U.S. income tax for a single taxpayer is as follows.

If taxable income is over	But not over	The tax is
\$0	\$8350	10% of the amount over \$0.
\$8350	\$33,950	\$835 plus 15% of the amount over \$8350.
\$33,950	\$82,250	\$4675 plus 25% of the amount over \$33,950.
\$82,250	\$171,550	\$16,750 plus 28% of the amount over \$82,250.
\$171,550	\$372,950	\$41,754 plus 33% of the amount over \$171,550.
\$372,950	No limit	\$108,216 plus 35% of the amount over \$372,950.

Source: Internal Revenue Service.

a. Express the information given in the table as a six-part piecewise function f, using x as a variable representing taxable income.

b. Find the tax liability on each taxable income.

(i) \$35,000
(ii) \$100,000
(iii) \$500,000

c. Find your taxable income if you had each tax liability.

(i) \$3500
(ii) \$12,700
(iii) \$35,000

Beyond the Basics

51. Let $f(x) = \begin{cases} 3x + 5, & -3 \le x < -1 \\ 2x + 1, & -1 \le x < 2 \\ 2 - x, & 2 \le x \le 4 \end{cases}$.

a. Find the following.

(i) $f(-2)$
(ii) $f(-1)$
(iii) $f(3)$

b. Find x when $f(x) = 2$.

c. Sketch a graph of $y = f(x)$.

52. Let $g(x) = \begin{cases} 3, & -3 \le x < -1 \\ -6x - 3, & -1 \le x < 0 \\ 3x - 3, & 0 \le x \le 1 \end{cases}$.

a. Find the following.

(i) $g(-2)$
(ii) $g\left(\frac{1}{2}\right)$
(iii) $g(0)$
(iv) $g(-1)$

b. Find x when **(i)** $g(x) = 0$ and **(ii)** $2g(x) + 3 = 0$.

c. Sketch a graph of $y = g(x)$.

In Exercises 53 and 54, a function is given. For each function,

a. Find the domain and range.

b. Find the intervals over which the function is increasing, decreasing, or constant.

c. State whether the function is odd, even, or neither.

53. $f(x) = x - [\![x]\!]$

54. $f(x) = \dfrac{1}{[\![x]\!]}$

55. Let $f(x) = \dfrac{|x|}{x}, x \neq 0$. Find $|f(x) - f(-x)|$.

56. Windchill Index (WCI). Suppose the outside air temperature is T degrees Fahrenheit and the wind speed is v miles per hour. A formula for the WCI based on observations is as follows:

$$\text{WCI} = \begin{cases} T, & 0 \le v \le 4 \\ 91.4 + (91.4 - T)(0.0203v - 0.304\sqrt{v} - 0.474), & 4 < v < 45 \\ 1.6T - 55, & v \ge 45 \end{cases}$$

a. Find the WCI to the nearest degree if the outside air temperature is 40°F and

(i) $v = 2$ miles per hour.

(ii) $v = 16$ miles per hour.

(iii) $v = 50$ miles per hour.

b. Find the air temperature to the nearest degree if

(i) The WCI is -58°F and $v = 36$ miles per hour.

(ii) The WCI is -10°F and $v = 49$ miles per hour.

Critical Thinking/Discussion/Writing

57. Postage rate function. The U.S. Postal Service uses the function $f(x) = -[\![-x]\!]$, called the **ceiling integer function**, to determine postal charges. For instance, if you have an envelope that weighs 3.2 ounces, you will be charged the rate for $f(3.2) = -[\![-3.2]\!] = -(-4) = 4$ ounces. In 2009, it cost 44¢ for the first ounce (or a fraction of it) and 17¢ for each additional ounce (or a fraction of it) to mail a first-class letter in the United States.

a. Express the cost C (in cents) of mailing a letter weighing x ounces.

b. Graph the function $y = C(x)$.

c. What are the domain and range of $y = C(x)$?

58. The cost C of parking a car at the metropolitan airport is \$4 for the first hour and \$2 for each additional hour or fraction thereof. Write the cost function $C(x)$, where x is the number of parking hours, in terms of the greatest integer function.

59. Car rentals. The weekly cost of renting a compact car from U-Rent is \$150. There is no charge for driving the first 100 miles, but there is a 20¢ charge for each mile (or fraction thereof) driven over 100 miles.

a. Express the cost C of renting a car for a week and driving it for x miles.

b. Sketch the graph of $y = C(x)$.

c. How many miles were driven if the weekly rental cost was \$190?

60. Write as a piecewise function.

a. $f(x) = [\![x]\!] + [\![-x]\!]$

b. $g(x) = [\![x]\!] - [\![-x]\!]$

c. $h(x) = x - |x|$

d. $F(x) = x|x|$

e. $G(x) = |x - 1| + |x - 2|$

Maintaining Skills

For Exercises 61–64, recall that the graph of a function *f* is the graph of the set of ordered pairs $(x, f(x))$.

61. If we add 3 to each y-coordinate of the graph of f, we will obtain the graph of $y =$ ____________.

62. If we subtract 2 from each x-coordinate of the graph of f, we will obtain the graph of $y =$ ____________.

63. If we replace each x-coordinate with its opposite in the graph of f, we will obtain the graph of $y =$ ____________.

64. If we replace each y-coordinate with its opposite in the graph of f, we will obtain the graph of $y =$ ____________.

In Exercises 65–68, sketch the graphs of *f* and *g* on the same coordinate axes.

65. $f(x) = x^2$ and $g(x) = x^2 + 1$

66. $f(x) = |x|$ and $g(x) = |x + 2|$

67. $f(x) = \sqrt{x}$ and $g(x) = \sqrt{-x}$

68. $f(x) = x^2$ and $g(x) = -x^2$

SECTION 2.7

Transformations of Functions

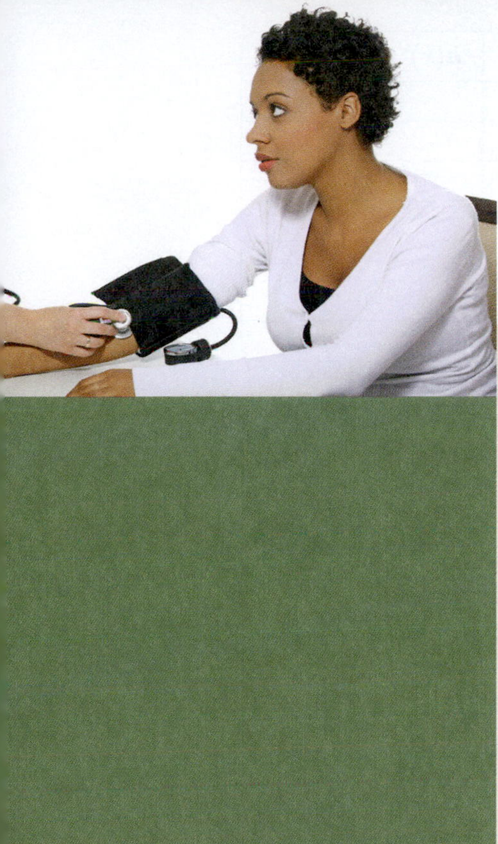

BEFORE STARTING THIS SECTION, REVIEW

1. Graphs of basic functions (Section 2.6, page 234)
2. Symmetry (Section 2.2, page 165)
3. Completing the square (Section 1.2, page 94)

OBJECTIVES

1. Define transformations of functions.
2. Use vertical or horizontal shifts to graph functions.
3. Use reflections to graph functions.
4. Use stretching or compressing to graph functions.
5. Use a sequence of transformations to graph functions.

Measuring Blood Pressure

Blood pressure is the force of blood per unit area against the walls of the arteries. Blood pressure is recorded as two numbers: the systolic pressure (as the heart beats) and the diastolic pressure (as the heart relaxes between beats). If your blood pressure is "120 over 80," it is rising to a maximum of 120 millimeters of mercury as the heart beats and is falling to a minimum of 80 millimeters of mercury as the heart relaxes.

The most precise way to measure blood pressure is to place a small glass tube in an artery and let the blood flow out and up as high as the heart can pump it. That was the way Stephen Hales (1677–1761), the first investigator of blood pressure, learned that a horse's heart could pump blood 8 feet 3 inches, up a tall tube. Such a test, however, consumed a great deal of blood. Jean Louis Marie Poiseuille (1797–1869) greatly improved the process by using a mercury-filled manometer, which allowed a smaller, narrower tube.

In Exercise 104, we investigate Poiseuille's Law for arterial blood flow.

1 Define transformations of functions.

Transformations

If a new function is formed by performing certain operations on a given function f, then the graph of the new function is called a **transformation** of the graph of f. For example, the graphs of $y = |x| + 2$ and $y = |x| - 3$ are transformations of the graph of $y = |x|$; that is, the graph of each is a special modification of the graph of $y = |x|$.

2 Use vertical or horizontal shifts to graph functions.

Vertical and Horizontal Shifts

A transformation that changes only the position of a graph but not its shape is called a **rigid transformation**. We first consider rigid transformations involving **vertical shifts** and **horizontal shifts** of a graph.

TECHNOLOGY CONNECTION

Many graphing calculators use abs (x) for $|x|$. The graphs of $y = |x|$, $y = |x| - 3$, and $y = |x| + 2$ illustrate that these graphs are identical in shape. They differ only in vertical placement.

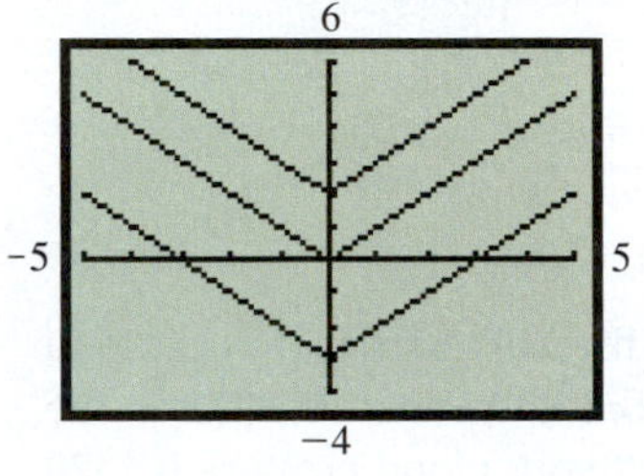

EXAMPLE 1 Graphing Vertical Shifts

Let $f(x) = |x|$, $g(x) = |x| + 2$, and $h(x) = |x| - 3$. Sketch the graphs of these functions on the same coordinate plane. Describe how the graphs of g and h relate to the graph of f.

Solution

Make a table of values and graph the equations $y = f(x)$, $y = g(x)$, and $y = h(x)$.

TABLE 2.11

x	$y = \|x\|$	$g(x) = \|x\| + 2$	$h(x) = \|x\| - 3$
−5	5	7	2
−3	3	5	0
−1	1	3	−2
0	0	2	−3
1	1	3	−2
3	3	5	0
5	5	7	2

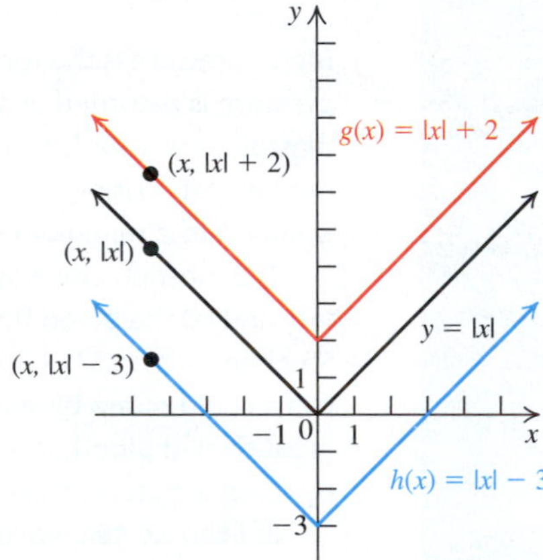

FIGURE 2.73 Vertical shifts of $y = |x|$

Notice that in Table 2.11 and Figure 2.73, each point $(x, |x|)$ on the graph of f has a corresponding point $(x, |x| + 2)$ on the graph of g and $(x, |x| - 3)$ on the graph of h. So the graph of $g(x) = |x| + 2$ is the graph of $y = |x|$ shifted 2 units up, and the graph of $y = |x| - 3$ is the graph of $y = |x|$ shifted 3 units down.

Practice Problem 1 Let

$$f(x) = x^3,\ g(x) = x^3 + 1, \text{ and } h(x) = x^3 - 2.$$

Sketch the graphs of all three functions on the same coordinate plane. Describe how the graphs of g and h relate to the graph of f.

Example 1 illustrates the concept of the vertical (up or down) shift of a graph.

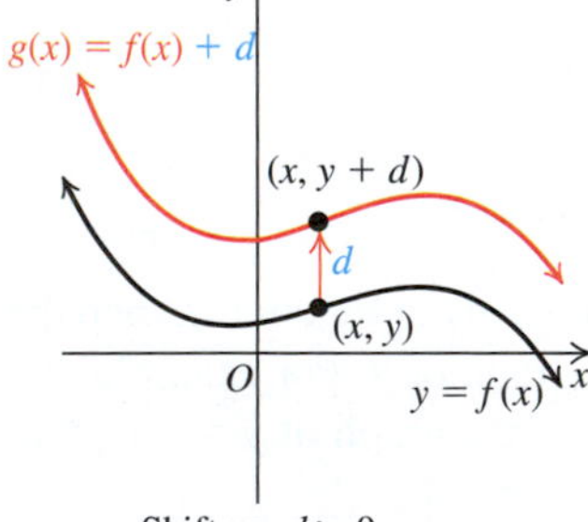

Shift up, $d > 0$

Shift down, $d > 0$

FIGURE 2.74 Vertical shifts

> **VERTICAL SHIFT**
>
> Let $d > 0$. The graph of $g(x) = f(x) + d$ is the graph of $y = f(x)$ shifted d units *up*, and the graph of $h(x) = f(x) - d$ is the graph of $y = f(x)$ shifted d units *down*. See Figure 2.74. If (x, y) is a point on the graph of $y = f(x)$, then the corresponding point $(x, y + d)$ is on the graph of $g(x) = f(x) + d$ and the point $(x, y - d)$ is on the graph of $h(x) = f(x) - d$.

Next, we consider the operation that shifts a graph horizontally.

EXAMPLE 2 **Writing Functions for Horizontal Shifts**

Let $f(x) = x^2$, $g(x) = (x - 2)^2$, and $h(x) = (x + 3)^2$. A table of values for *f*, *g*, and *h* is given in Table 2.12. The three functions *f*, *g*, and *h* are graphed on the same coordinate plane in Figure 2.75. Describe how the graphs of *g* and *h* relate to the graph of *f*.

TABLE 2.12

x	$y = x^2$	$y = (x - 2)^2$
−4	**16**	36
−3	**9**	25
−2	**4**	16
−1	**1**	9
0	**0**	4
1	**1**	1
2	4	0
3	9	1
4	16	4

(a)

x	$y = x^2$	$y = (x + 3)^2$
−4	16	1
−3	9	0
−2	4	1
−1	**1**	4
0	**0**	9
1	**1**	16
2	**4**	25
3	**9**	36
4	**16**	49

(b)

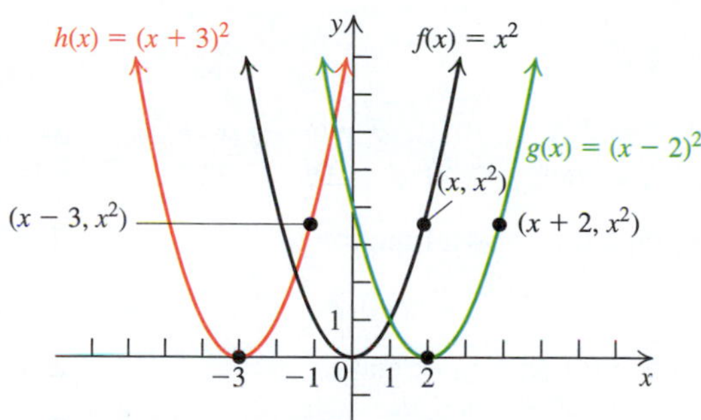

FIGURE 2.75 Horizontal shifts

Solution

First, notice that all three functions are squaring functions.

a. Each point (x, x^2) on the graph of f has a corresponding point $(x + 2, x^2)$ on the graph of g because $g(x + 2) = (x + 2 - 2)^2 = x^2$. So the graph of $g(x) = (x - 2)^2$ is just the graph of $f(x) = x^2$ shifted 2 units to the *right*. Noticing that $f(0) = 0$ and $g(2) = 0$ will help you remember which way to shift the graph. Table 2.12(a) and Figure 2.75 illustrate these ideas.

b. Each point (x, x^2) on the graph of f has a corresponding point $(x - 3, x^2)$ on the graph of h because $h(x - 3) = (x - 3 + 3)^2 = x^2$. So the graph of $h(x) = (x + 3)^2$ is just the graph of $f(x) = x^2$ shifted 3 units to the *left*. Noticing that $f(0) = 0$ and $h(-3) = 0$ will help you remember which way to shift the graph. Table 2.12(b) and Figure 2.75 confirm these considerations.

Practice Problem 2 Let

$$f(x) = x^3, g(x) = (x - 1)^3, \text{ and } h(x) = (x + 2)^3.$$

Sketch the graphs of all three functions on the same coordinate plane. Describe how the graphs of *g* and *h* relate to the graph of *f*.

HORIZONTAL SHIFTS

Let $c > 0$. The graph of $y = f(x - c)$ is the graph of $y = f(x)$ shifted c units to the right. The graph of $y = f(x + c)$ is the graph of $y = f(x)$ shifted c units to the left. See Figure 2.76. If (x, y) is a point on the graph of f, then the corresponding point $(x + c, y)$ is on the graph of $y = f(x - c)$ and the point $(x - c, y)$ is on the graph of $y = f(x + c)$.

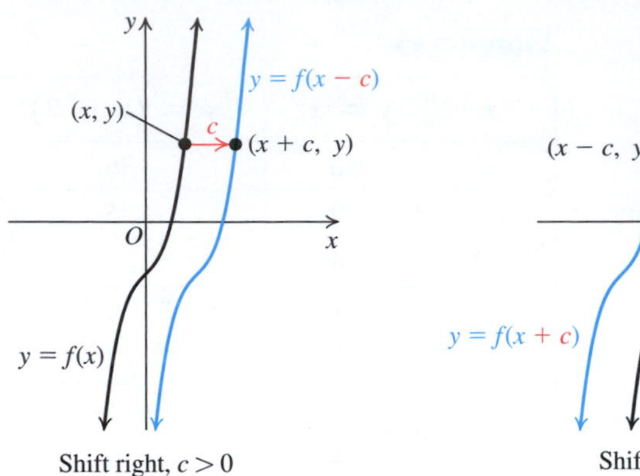

FIGURE 2.76 Horizontal shifts

PROCEDURE IN ACTION

EXAMPLE 3 **Graphing Combined Vertical and Horizontal Shifts**

OBJECTIVE

Sketch the graph of $g(x) = f(x \pm c) \pm d$, where f is a function whose graph is known, $c > 0$, $d > 0$.

Step 1 Identify and graph the known function f.

Step 2 Identify the *positive* constants c and d.

Step 3
(i) Graph $y = f(x - c)$ by shifting the graph of f horizontally c units to the right. Add c to the x-coordinate of each point.

(ii) Graph $y = f(x + c)$ by shifting the graph of f horizontally c units to the left. Subtract c from the x-coordinate of each point.

Step 4
(i) Graph $y = f(x \pm c) + d$ by shifting the graph of $y = f(x \pm c)$ vertically up d units. Add d to the y-coordinate of each point.

(ii) Graph $y = f(x \pm c) - d$ by shifting the graph of $y = f(x \pm c)$ vertically down d units. Subtract d from the y-coordinate of each point.

EXAMPLE

Sketch the graph of $g(x) = \sqrt{x + 2} - 3$.

1. Choose $f(x) = \sqrt{x}$. Domain: $[0, \infty)$; range: $[0, \infty)$

The graph of $y = \sqrt{x}$ is shown in Step 4.

2. $g(x) = \sqrt{x + 2} - 3$, so $c = 2$ and $d = 3$.

3. Because $c = 2 > 0$, the graph of $y = \sqrt{x + 2}$ is the graph of f shifted horizontally 2 units to the left. See the blue graph in Step 4.

4. Graph $y = \sqrt{x + 2} - 3$ by shifting the graph of $y = \sqrt{x + 2}$ down 3 units. Domain: $[-2, \infty)$; range: $[-3, \infty)$.

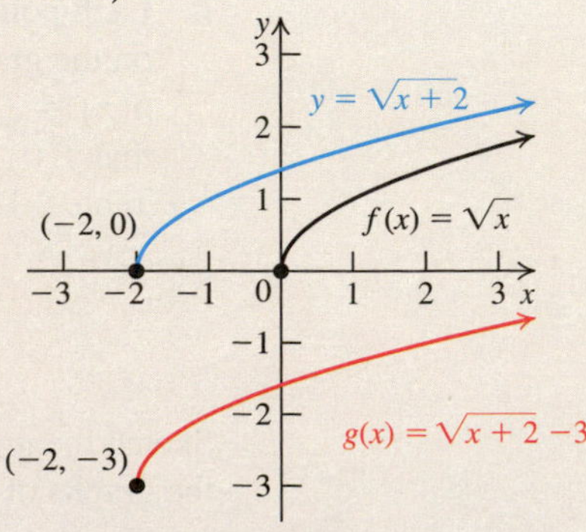

Horizontal and vertical shifts

Practice Problem 3 Sketch the graph of

$$f(x) = \sqrt{x - 2} + 3.$$

3 Use reflections to graph functions.

Reflections

We now consider rigid transformations that reflect a graph about a coordinate axis.

Comparing the Graphs of $y = f(x)$ and $y = -f(x)$

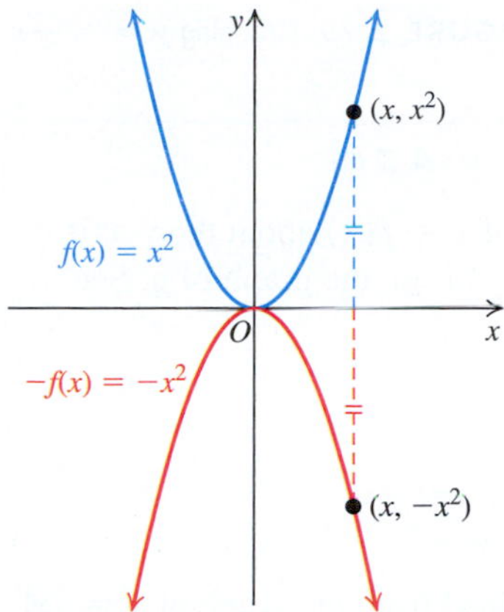

FIGURE 2.77 Reflection about the x-axis

Consider the graph of $f(x) = x^2$, shown in Figure 2.77. Table 2.13 gives some values for $f(x)$ and $g(x) = -f(x)$. Note that the y-coordinate of each point in the graph of $g(x) = -f(x)$ is the opposite of the y-coordinate of the corresponding point on the graph of $f(x)$. So the graph of $y = -x^2$ is the reflection of the graph of $y = x^2$ about the x-axis. This means that the points (x, x^2) and $(x, -x^2)$ are the same distance from, but on opposite sides of, the x-axis. See Figure 2.77.

TABLE 2.13

x	$f(x) = x^2$	$g(x) = -f(x) = -x^2$
−3	9	−9
−2	4	−4
−1	1	−1
0	0	0
1	1	−1
2	4	−4
3	9	−9

REFLECTION ABOUT THE x-AXIS

The graph of $g(x) = -f(x)$ is a reflection of the graph of $y = f(x)$ about the x-axis. If a point (x, y) is on the graph of f, then the point $(x, -y)$ is on the graph of g. See Figure 2.78.

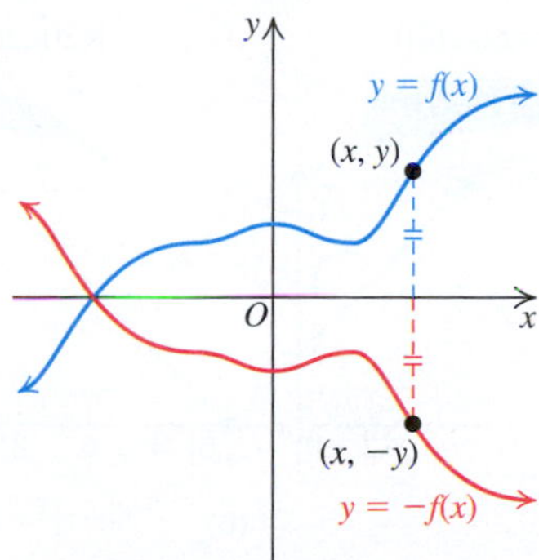

FIGURE 2.78 Reflection about the x-axis

Comparing the Graphs of $y = f(x)$ and $y = f(-x)$

To compare the graphs of $f(x)$ and $g(x) = f(-x)$, consider the graph of $f(x) = \sqrt{x}$. Then $f(-x) = \sqrt{-x}$. See Figure 2.79. Table 2.14 gives some values for $f(x)$ and $g(x) = f(-x)$. The domain of f is $[0, \infty)$, and the domain of g is $(-\infty, 0]$. Each point (x, y) on the graph of f has a corresponding point $(-x, y)$ on the graph of g. So the graph of $y = f(-x)$ is the reflection of the graph of $y = f(x)$ about the y-axis. This means that the points $(x, \sqrt{x})$ and $(-x, \sqrt{x})$ are the same distance from, but on opposite sides of, the y-axis. See Figure 2.79.

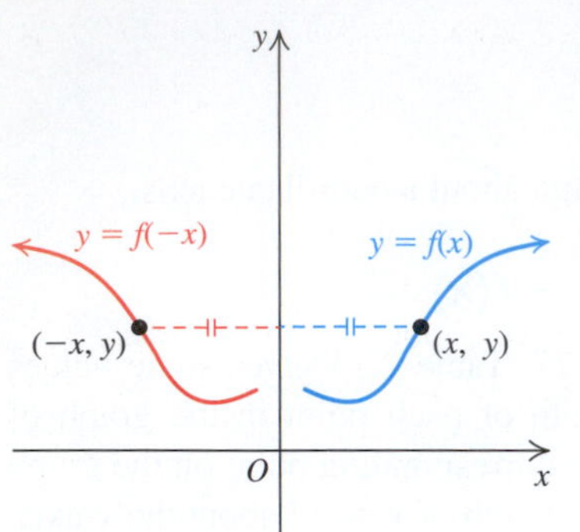

FIGURE 2.80 Reflection about the y-axis

TABLE 2.14

x	$f(x) = \sqrt{x}$	$-x$	$g(x) = \sqrt{-x}$
−4	Undefined	4	2
−1	Undefined	1	1
0	0	0	0
1	1	−1	Undefined
4	2	−4	Undefined

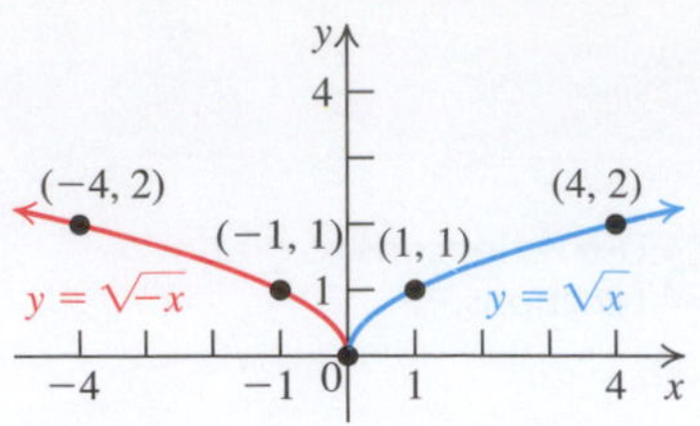

FIGURE 2.79 Graphing $y = \sqrt{-x}$

REFLECTION ABOUT THE y-AXIS

The graph of $g(x) = f(-x)$ is a reflection of the graph of $y = f(x)$ about the y-axis. If a point (x, y) is on the graph of f, then the point $(-x, y)$ is on the graph of g. See Figure 2.80.

EXAMPLE 4 Combining Transformations

Explain how the graph of $y = -|x - 2| + 3$ can be obtained from the graph of $y = |x|$.

Solution

Start with the graph of $y = |x|$. Follow the point $(0, 0)$ on $y = |x|$. See Figure 2.81(a).

Step 1 Shift the graph of $y = |x|$ to the right 2 units to obtain the graph of $y = |x - 2|$. Add 2 to the x-coordinate of each point. The point $(0, 0)$ moves to $(2, 0)$. See Figure 2.81(b).

Step 2 Reflect the graph of $y = |x - 2|$ about the x-axis to obtain the graph of $y = -|x - 2|$. Change the y-coordinate of each point to its opposite. The point $(2, 0)$ remains $(2, 0)$. See Figure 2.81(c).

Step 3 Finally, shift the graph of $y = -|x - 2|$ up 3 units to obtain the graph of $y = -|x - 2| + 3$. Add 3 to the y-coordinate of each point. The point $(2, 0)$ moves to $(2, 3)$. See Figure 2.81(d).

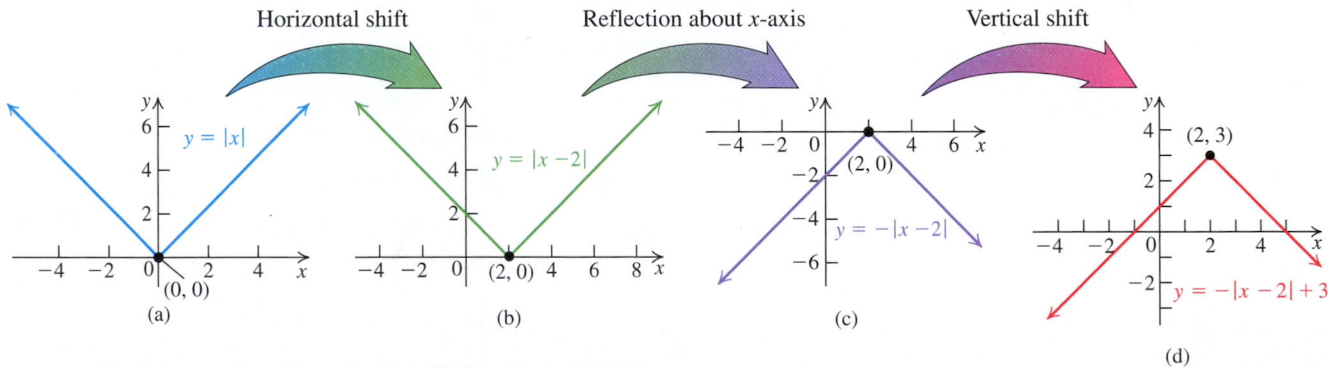

FIGURE 2.81 Transformations of $y = |x|$

Practice Problem 4 Explain how the graph of $y = -(x - 1)^2 + 2$ can be obtained from the graph of $y = x^2$.

EXAMPLE 5 Graphing $y = |f(x)|$

Use the graph of $f(x) = (x + 1)^2 - 4$ to sketch the graph of $y = |f(x)|$.

Solution

The graph of $f(x) = (x + 1)^2 - 4$ is obtained by shifting the graph of $y = x^2$ left by 1 unit and then shifting the resulting graph down by 4 units. See Figure 2.82(a). The function f has domain $(-\infty, \infty)$ and range $[-4, \infty)$.

We know that

$$|y| = \begin{cases} y & \text{if } y \geq 0 \\ -y & \text{if } y < 0. \end{cases}$$

This means that the portion of the graph on or above the x-axis $(y \geq 0)$ is unchanged and the portion of the graph below the x-axis $(y < 0)$ is reflected above the x-axis. See Figure 2.82(b). The function $|f|$ has domain $(-\infty, \infty)$ and range $[0, \infty)$.

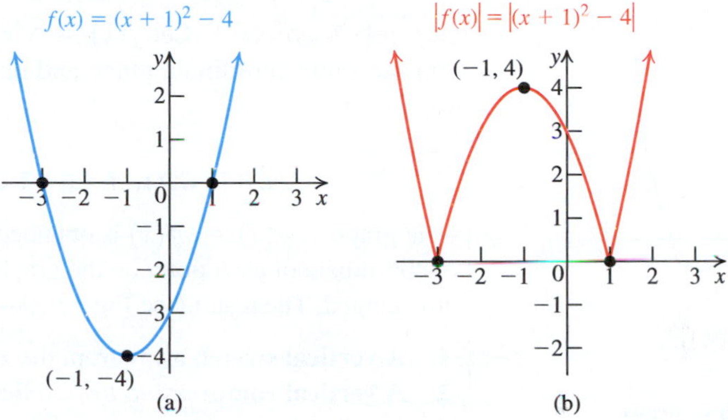

FIGURE 2.82 Graphing $y = |f(x)|$

Practice Problem 5 Use the graph of $f(x) = 2x - 4$ to sketch the graph of $y = |f(x)|$.

4 Use stretching or compressing to graph functions.

Stretching or Compressing

We now look at transformations that distort the shape of a graph, called **nonrigid transformations.** We consider the relationship of the graphs of $y = af(x)$ and $y = f(bx)$ to the graph of $y = f(x)$.

Comparing the Graphs of $y = f(x)$ and $y = af(x)$

EXAMPLE 6 **Stretching or Compressing a Graph Vertically**

Let $f(x) = |x|$, $g(x) = 2|x|$, and $h(x) = \frac{1}{2}|x|$. Sketch the graphs of f, g, and h on the same coordinate plane and describe how the graphs of g and h are related to the graph of f.

Solution

The graphs of $y = |x|$, $y = 2|x|$, and $y = \frac{1}{2}|x|$ are sketched in Figure 2.83. Table 2.15 gives some typical function values.

The graph of $y = 2|x|$ is the graph of $y = |x|$ vertically stretched (expanded) by multiplying each of its y-coordinates by 2. It is twice as high as the graph of $|x|$ at every real number x. The result is a taller V-shaped curve. See Figure 2.83.

The graph $y = \frac{1}{2}|x|$ is the graph of $y = |x|$ vertically compressed (shrunk) by multiplying each of its y-coordinates by $\frac{1}{2}$. It is half as high as the graph of $|x|$ at every real number x. The result is a flatter V-shaped curve. See Figure 2.83.

FIGURE 2.83 Vertical stretch and compression

TABLE 2.15

x	$f(x) = \lvert x\rvert$	$g(x) = 2\lvert x\rvert$	$h(x) = \frac{1}{2}\lvert x\rvert$
-2	2	4	1
-1	1	2	$\frac{1}{2}$
0	0	0	0
1	1	2	$\frac{1}{2}$
2	2	4	1

Practice Problem 6 Let $f(x) = \sqrt{x}$ and $g(x) = 2\sqrt{x}$. Sketch the graphs of f and g on the same coordinate plane and describe how the graph of g is related to the graph of f.

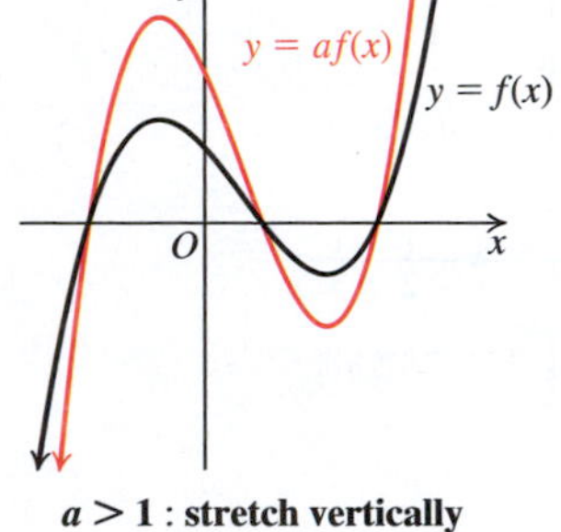

$a > 1$: stretch vertically

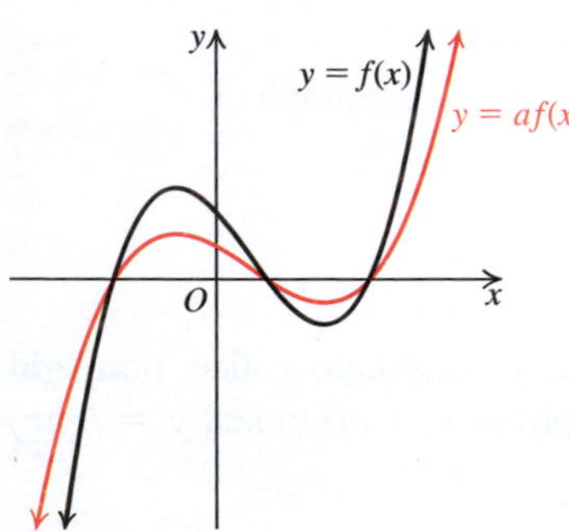

$0 < a < 1$: compress vertically

FIGURE 2.84 Vertical stretch or compression

VERTICAL STRETCHING OR COMPRESSING

The graph of $g(x) = af(x)$ is obtained from the graph of $y = f(x)$ by multiplying the y-coordinate of each point on the graph of $y = f(x)$ by a and leaving the x-coordinate unchanged. The result (see Figure 2.84) is as follows:

1. A **vertical stretch** away from the x-axis if $a > 1$
2. A **vertical compression** toward the x-axis if $0 < a < 1$

If $a < 0$, first graph $y = |a|f(x)$ by stretching or compressing the graph of $y = f(x)$ vertically. Then reflect the resulting graph about the x-axis.

So if (x, y) is a point on the graph of f, then the point (x, ay) is on the graph of g.

Comparing the Graphs of $y = f(x)$ and $y = f(bx)$

Given a function $y = f(x)$, we explore the effect of the constant b in graphing the function $y = f(bx)$. Consider the graphs of $y = f(x)$ and $y = f(2x)$ in Figure 2.85(a). Multiplying the *independent* variable x by 2 compresses the graph $y = f(x)$ horizontally toward the y-axis. Because the value $2x$ is twice the value of x, a point on the x-axis will be only half as far from the origin when $y = f(2x)$ has the same y value as $y = f(x)$. See Figure 2.85(a).

SIDE NOTE

Notice that replacing the variable x with $x \pm c$, or bx in $y = f(x)$ results in a horizontal change in the graphs. However, replacing the y value in $y = f(x)$ by $f(x) \pm d$, or $af(x)$, results in a vertical change in the graph of $y = f(x)$.

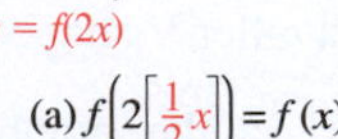

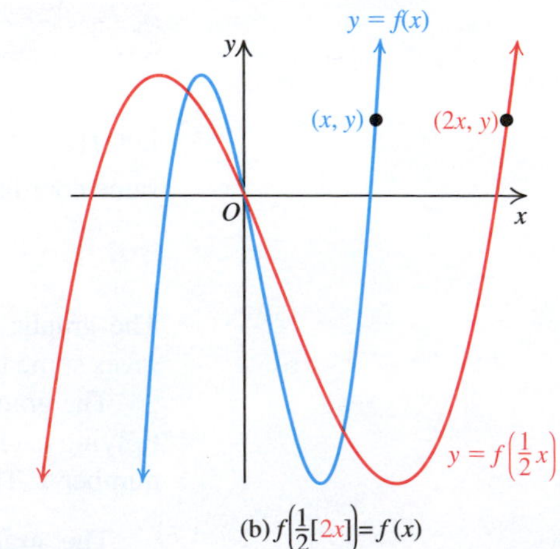

FIGURE 2.85 Horizontal stretch or compression

Now consider the graphs of $y = f(x)$ and $y = f\left(\frac{1}{2}x\right)$ in Figure 2.85(b). Multiplying the *independent* variable x by $\frac{1}{2}$ stretches the graph of $y = f(x)$ horizontally away from the

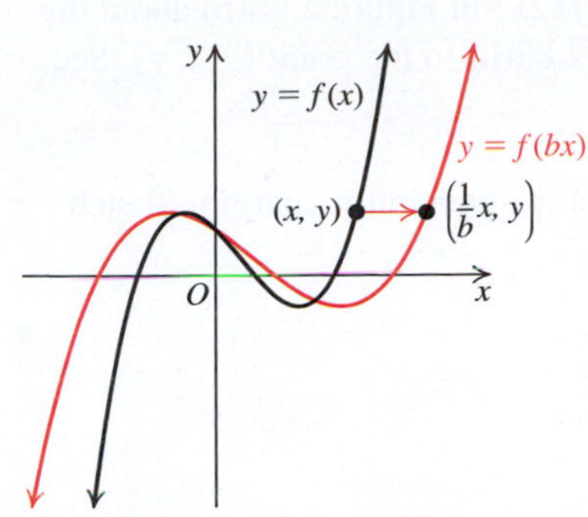

0 < b < 1 : stretch horizontally

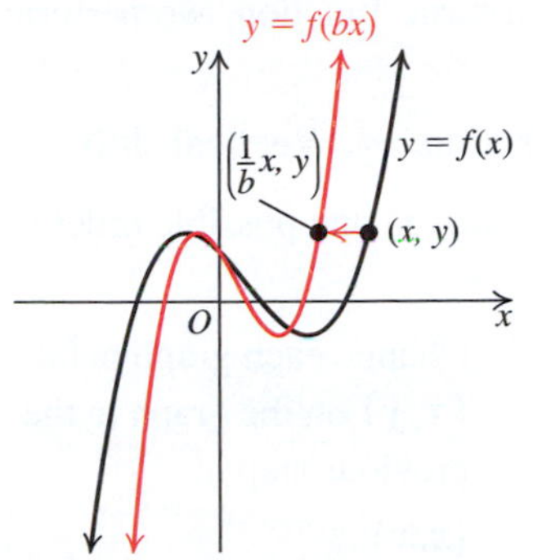

b > 1 : compress horizontally

FIGURE 2.86 Horizontal stretch and compression

y-axis. The value $\frac{1}{2}x$ is half the value of x so that a point on the x-axis will be twice as far from the origin for $y = f\left(\frac{1}{2}x\right)$ to have the same y value as in $y = f(x)$. See Figure 2.85(b).

HORIZONTAL STRETCHING OR COMPRESSING

Let $b > 0$. The graph of $g(x) = f(bx)$ is obtained from the graph of $y = f(x)$ by multiplying the x-coordinate of each point on the graph of $y = f(x)$ by $\frac{1}{b}$ and leaving the y-coordinate unchanged. The result (see Figure 2.86) is as follows:

1. A **horizontal stretch** away from the y-axis if $0 < b < 1$
2. A **horizontal compression** toward the y-axis if $b > 1$

If $b < 0$, first graph $f(|b|x)$ by stretching or compressing the graph of $y = f(x)$ horizontally. Then reflect the graph of $y = f(|b|x)$ about the y-axis.

If (x, y) is a point on the graph of $y = f(x)$, then the point $\left(\frac{1}{b}x, y\right)$ is on the graph of $g(x) = f(bx)$.

EXAMPLE 7 Stretching or Compressing a Function Horizontally

The graph of the function $y = f(x)$ whose equation is not given is shown in Figure 2.87. Sketch the following graphs.

a. $f\left(\frac{1}{2}x\right)$ **b.** $f(2x)$ **c.** $f(-2x)$

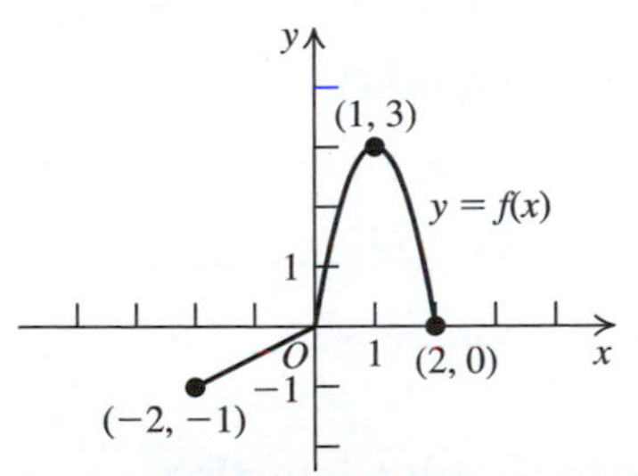

FIGURE 2.87

Solution

Note that the domain of f is $[-2, 2]$ and its range is $[-1, 3]$

a. To graph $y = f\left(\frac{1}{2}x\right)$, we stretch the graph of $y = f(x)$ horizontally by a factor of 2. In other words, we transform each point (x, y) in Figure 2.87 to the point $(2x, y)$. See Figure 2.88(a).

b. To graph $y = f(2x)$, we compress the graph of $y = f(x)$ horizontally by a factor of $\frac{1}{2}$. Therefore, we transform each point (x, y) in Figure 2.87 to $\left(\frac{1}{2}x, y\right)$. See Figure 2.88(b).

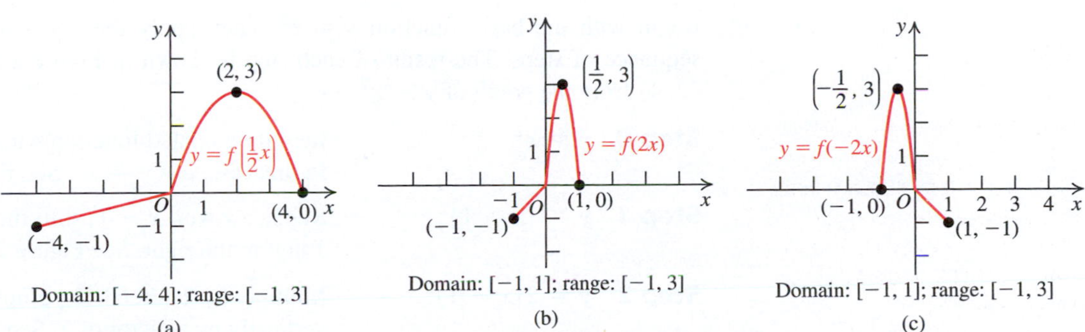

FIGURE 2.88

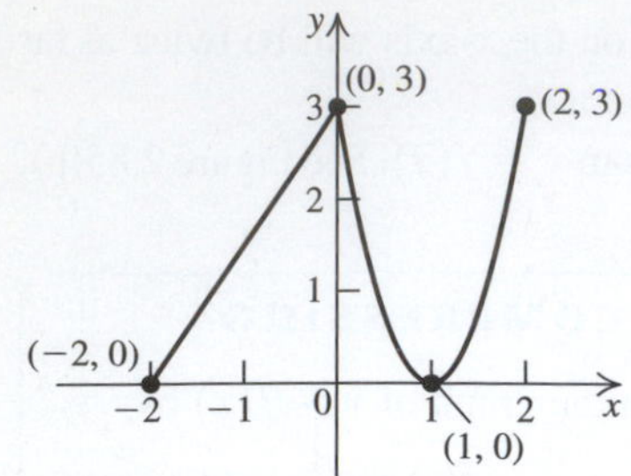

c. To graph $y = f(-2x)$, we reflect the graph of $y = f(2x)$ in Figure 2.88(b) about the y-axis. So we transform each point (x, y) in Figure 2.88(b) to the point $(-x, y)$. See Figure 2.88(c).

Practice Problem 7 The graph of a function $y = f(x)$ is given in the margin. Sketch the graphs of the following functions.

a. $f\left(\frac{1}{2}x\right)$ **b.** $f(2x)$

5 Use a sequence of transformations to graph functions.

Multiple Transformations in Sequence

When graphing requires more than one transformation of a basic function, we perform transformations in the following order:

1. Horizontal shift 2. Stretch or compression 3. Reflection 4. Vertical shift

To graph $y = af(bx - c) + d$ from the graph of $y = f(x)$, here is one possible order of transformations:

		Graph	Change each graph point (x, y) on the graph in the previous step to
Step 0	Given Graph	$y = f(x)$	(x, y)
Step 1	Horizontal Shift	$y = f(x - c)$	$(x + c, y)$
Step 2	Stretch/ Compression	$(i)\ y = f(\lvert b\rvert x - c)$	$\left(\frac{x}{\lvert b\rvert}, y\right)$
		$(ii)\ y = \lvert a\rvert f(\lvert b\rvert x - c)$	$(x, \lvert a\rvert y)$
Step 3	Reflections	$(i)\ y = \lvert a\rvert f(bx - c)$ for $b < 0$	$(-x, y)$
		$(ii)\ y = af(bx - c)$ for $a < 0$	$(x, -y)$
Step 4	Vertical Shift	$y = af(bx - c) + d$	$(x, y + d)$

EXAMPLE 8 **Using Multiple Transformations to Graph a Function**

Sketch the graph of the function $f(x) = -2(x - 1)^2 + 3$.

Solution

Begin with the basic function $y = x^2$. Then apply the necessary transformations in a sequence of steps. The result of each step is shown in Figure 2.89. We follow the point $(2, 4)$ from the graph of $y = x^2$.

Step 0	$y = x^2$	Identify a related function whose graph is familiar. In this case, use $y = x^2$. See Figure 2.89(a).
Step 1	$y = (x - 1)^2$	Replace x with $x - 1$; shift the graph of $y = x^2$, 1 unit to the right. See Figure 2.89(b).
Step 2	$y = 2(x - 1)^2$	Multiply by 2; stretch the graph of $y = (x - 1)^2$ vertically by a factor of 2. See Figure 2.89(c).
Step 3	$y = -2(x - 1)^2$	Multiply by -1. Reflect the graph of $y = 2(x - 1)^2$ about the x-axis. See Figure 2.89(d).
Step 4	$y = -2(x - 1)^2 + 3$	Add 3. Shift the graph of $y = -2(x - 1)^2$, up 3 units. See Figure 2.89(e).

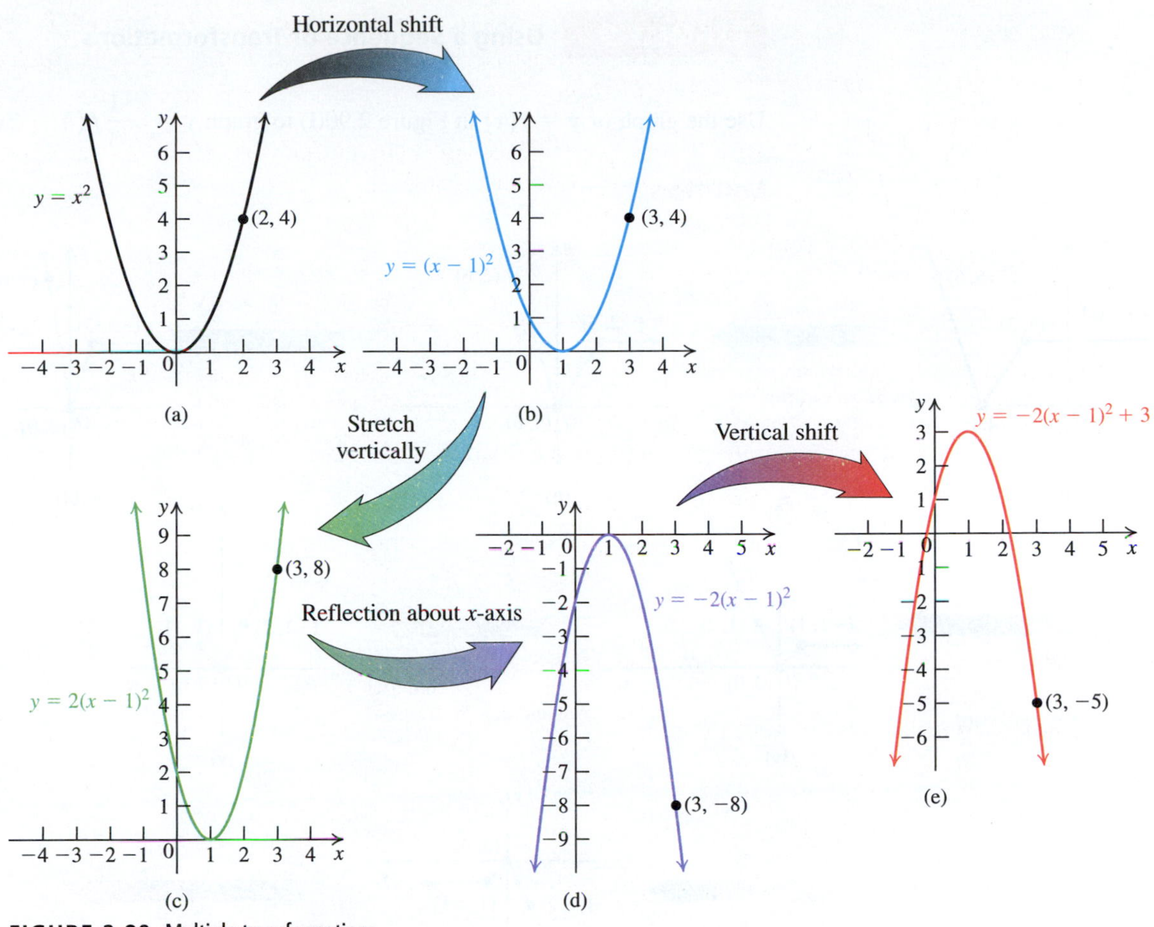

FIGURE 2.89 Multiple transformations

Practice Problem 8 Sketch the graph of the function

$$f(x) = 3\sqrt{x + 1} - 2.$$

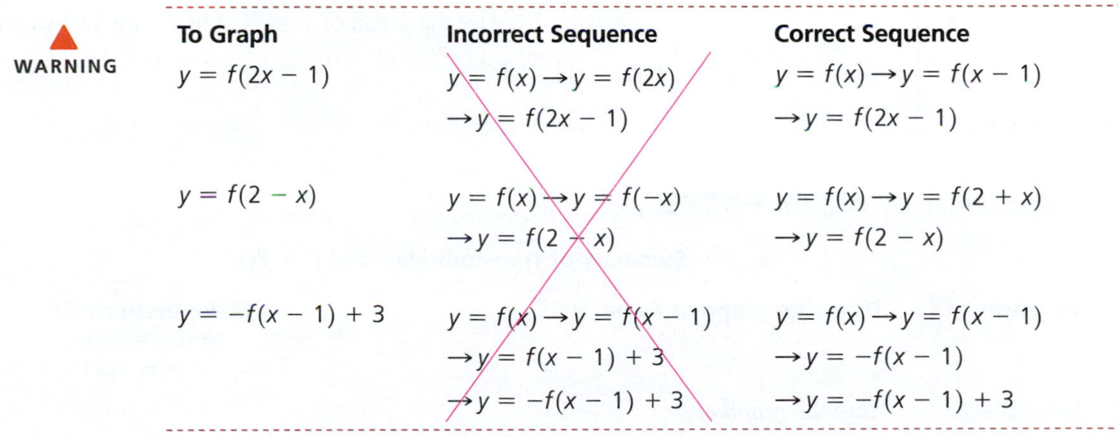

WARNING

To Graph	Incorrect Sequence	Correct Sequence
$y = f(2x - 1)$	$y = f(x) \rightarrow y = f(2x)$ $\rightarrow y = f(2x - 1)$	$y = f(x) \rightarrow y = f(x - 1)$ $\rightarrow y = f(2x - 1)$
$y = f(2 - x)$	$y = f(x) \rightarrow y = f(-x)$ $\rightarrow y = f(2 - x)$	$y = f(x) \rightarrow y = f(2 + x)$ $\rightarrow y = f(2 - x)$
$y = -f(x - 1) + 3$	$y = f(x) \rightarrow y = f(x - 1)$ $\rightarrow y = f(x - 1) + 3$ $\rightarrow y = -f(x - 1) + 3$	$y = f(x) \rightarrow y = f(x - 1)$ $\rightarrow y = -f(x - 1)$ $\rightarrow y = -f(x - 1) + 3$

EXAMPLE 9 Using a Sequence of Transformations

Use the graph of $y = f(x)$ in Figure 2.90(**i**) to graph $y = -\frac{1}{3}f(4 - 2x) + 5$.

Solution

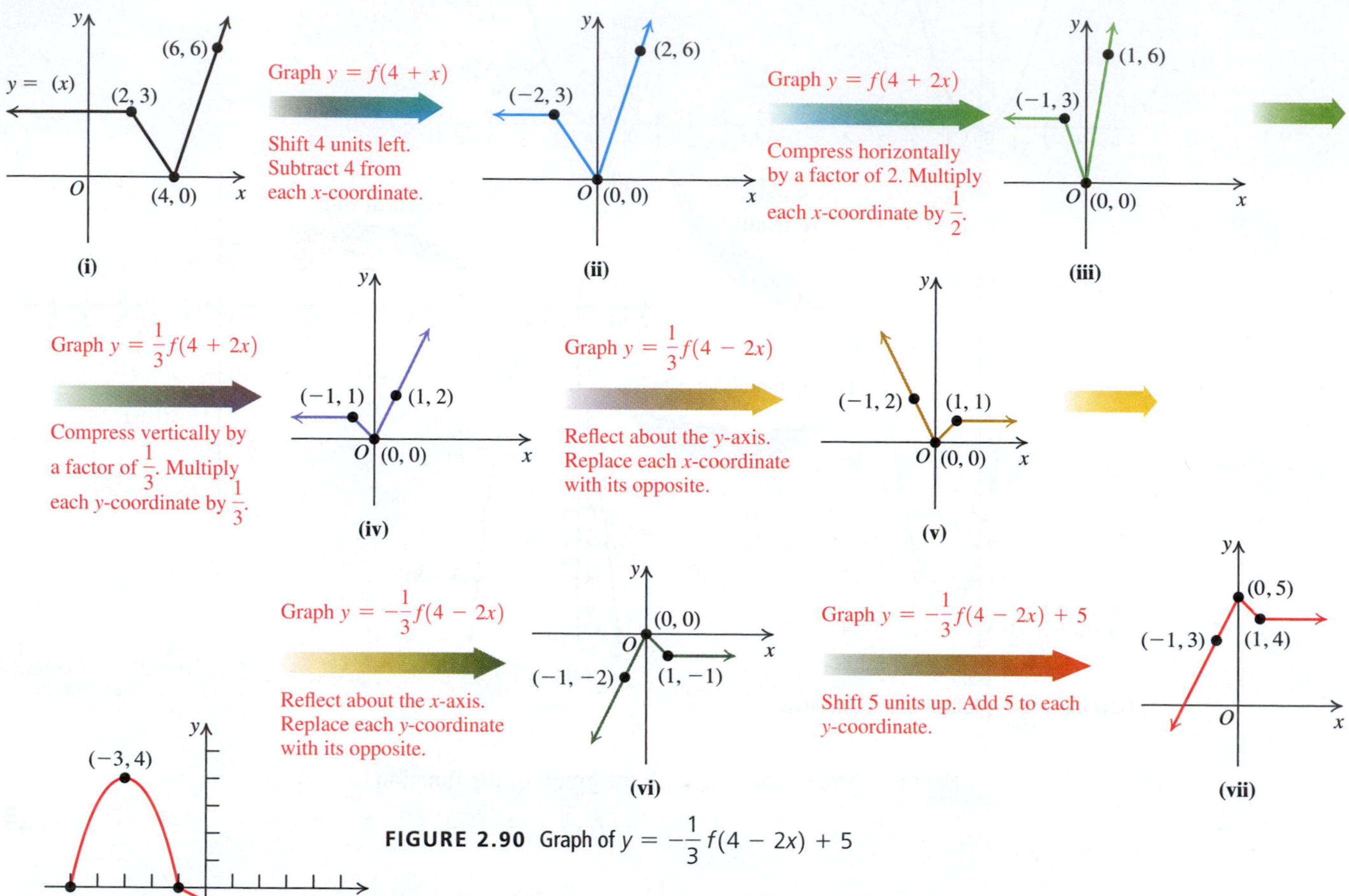

FIGURE 2.90 Graph of $y = -\frac{1}{3}f(4 - 2x) + 5$

FIGURE 2.91

Practice Problem 9 Use the graph of $y = f(x)$ in Figure 2.91 to graph $y = \frac{1}{2}f(2x - 1) + 3$.

SUMMARY OF MAIN FACTS

Summary of Transformations of $y = f(x)$

To graph	**Draw the graph of f and**	**Make these changes to the equation $y = f(x)$**	**Change the graph point (x, y) to**
Vertical shift (for $d > 0$)	*Shift the graph of f*		
$y = f(x) + d$	up d units.	Add d to $f(x)$.	$(x, y + d)$
$y = f(x) - d$	down d units.	Subtract d from $f(x)$.	$(x, y - d)$
Horizontal shift (for $c > 0$)	*Shift the graph of f*		
$y = f(x - c)$	to the right c units.	Replace x with $x - c$.	$(x + c, y)$
$y = f(x + c)$	to the left c units.	Replace x with $x + c$.	$(x - c, y)$

(continued)

To graph	Draw the graph of f and	Make these changes to the equation $y = f(x)$	Change the graph point (x, y) to
Reflection about the x-axis $y = -f(x)$	Reflect the graph of f about the x-axis.	Multiply $f(x)$ by -1.	$(x, -y)$
Reflection about the y-axis $y = f(-x)$	Reflect the graph of f about the y-axis.	Replace x with $-x$.	$(-x, y)$
Vertical stretching or compressing: $y = af(x)$	Multiply each y-coordinate of $y = f(x)$ by a. The graph of $y = f(x)$ is stretched vertically away from the x-axis if $a > 1$ and is compressed vertically toward the x-axis if $0 < a < 1$. If $a < 0$, sketch the graph of $y = \lvert a \rvert f(x)$, then reflect it about the x-axis.	Multiply $f(x)$ by a.	(x, ay)
Horizontal stretching or compressing: $y = f(bx)$	Multiply each x-coordinate of $y = f(x)$ by $\frac{1}{b}$. The graph of $y = f(x)$ is stretched away from the y-axis if $0 < b < 1$ and is compressed toward the y-axis if $b > 1$. If $b < 0$, sketch the graph of $y = f(\lvert b \rvert x)$, then reflect it about the y-axis.	Replace x with bx.	$\left(\frac{x}{b}, y\right)$

Summary for Combining Transformations in Sequence

To graph $y = af(bx + c) + d$, with $b > 0$ and $c > 0$

- Starting with the graph of $y = f(x)$, perform the indicated sequence of transformations. Here is one possible sequence:

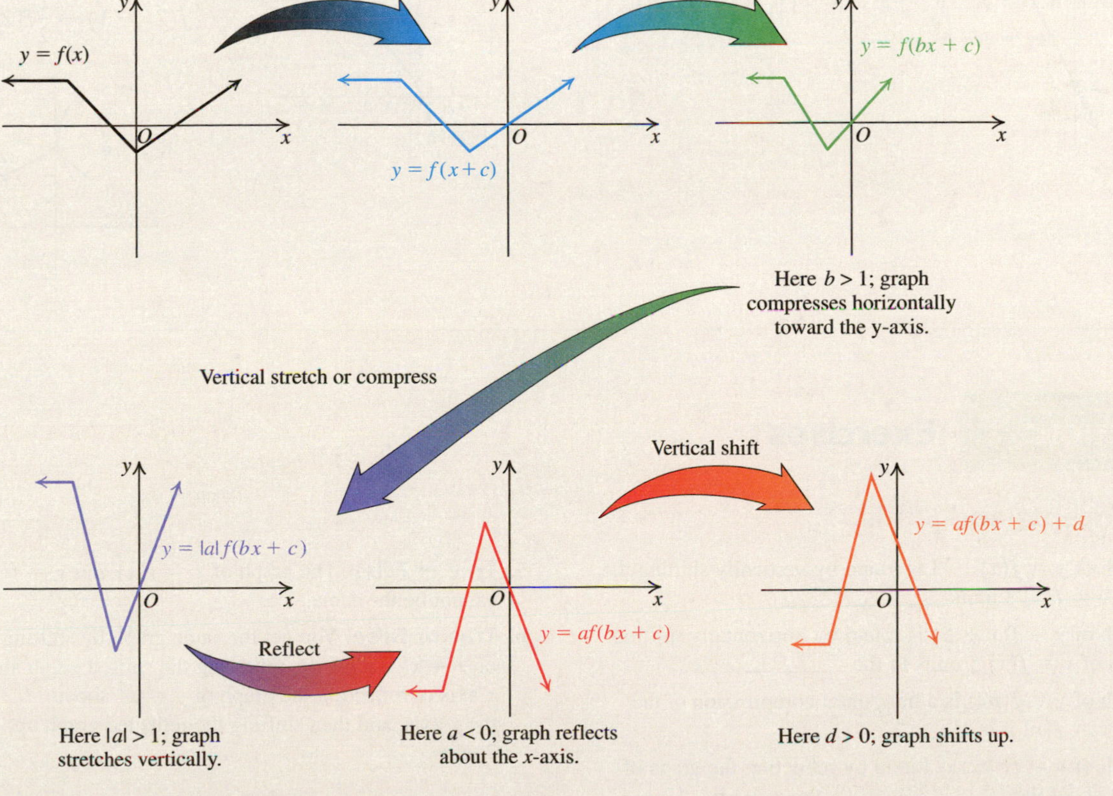

Note: If $c < 0$, then the horizontal shift should be $|c|$ units to the right, if $b < 0$, graph $y = af(|b|x + c) + d$, but reflect this graph about the y-axis before completing the last step, the vertical shift, d.

Answers to Practice Problems

1. The graph of g is the graph of f shifted 1 unit up; the graph of h is the graph of f shifted 2 units down.

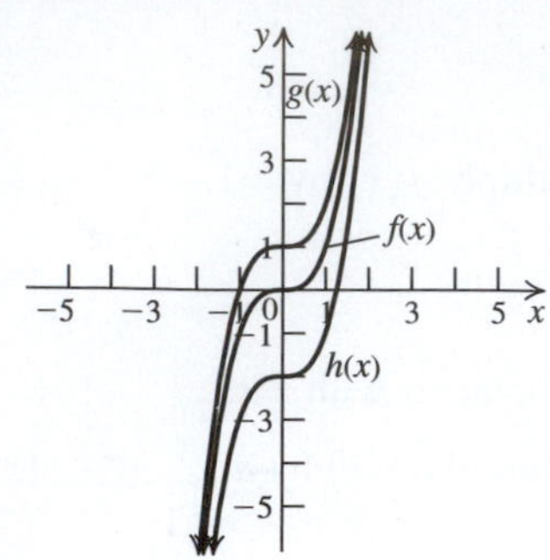

2. The graph of g is the graph of f shifted 1 unit to the right; the graph of h is the graph of f shifted 2 units to the left.

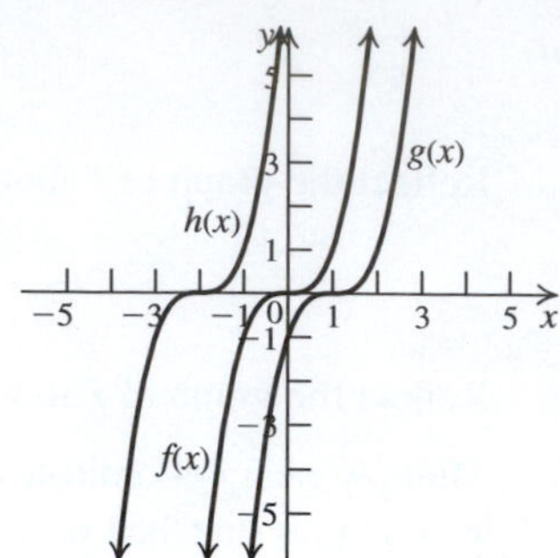

3.

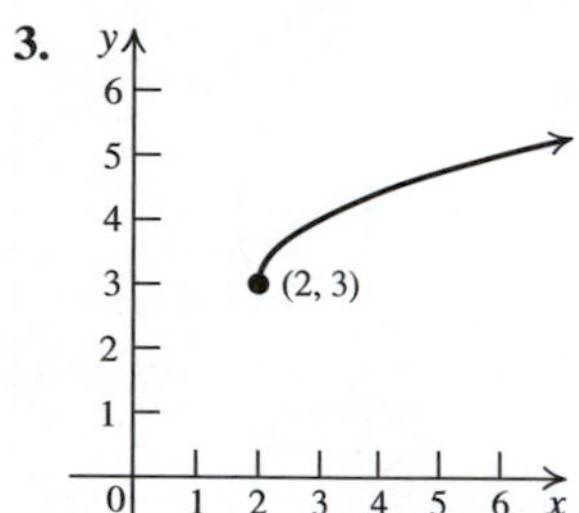

4. Shift the graph of $y = x^2$ to the right 1 unit, reflect the resulting graph about the x-axis, and shift it up 2 units.

5.

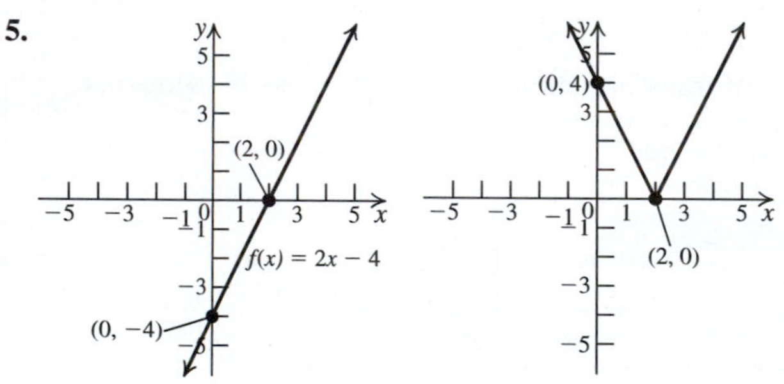

6. The graph of g is the graph of f vertically stretched by multiplying each of its y-coordinates by 2.

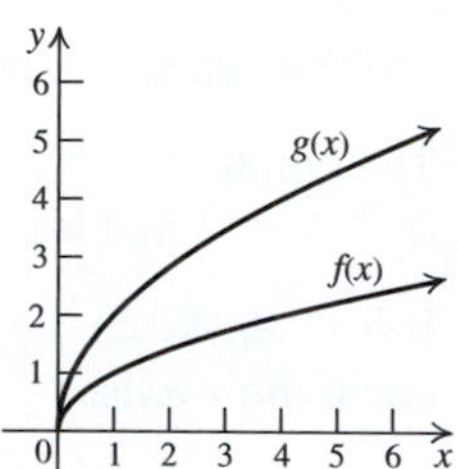

7. a. **b.**

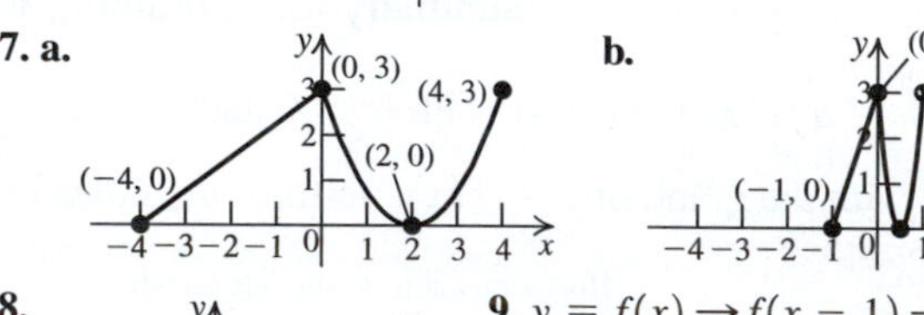

8.

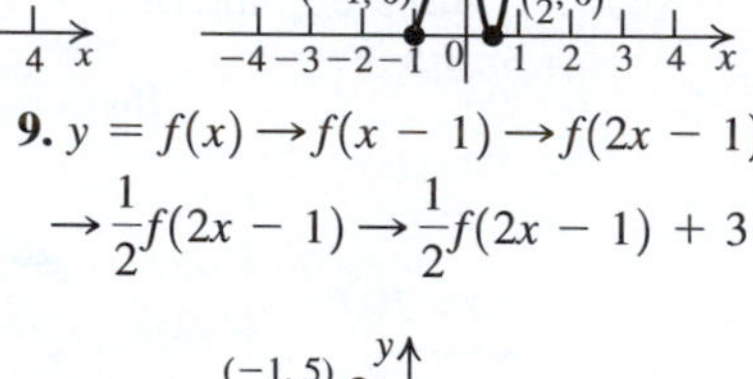

9. $y = f(x) \rightarrow f(x - 1) \rightarrow f(2x - 1)$
$\rightarrow \frac{1}{2}f(2x - 1) \rightarrow \frac{1}{2}f(2x - 1) + 3$

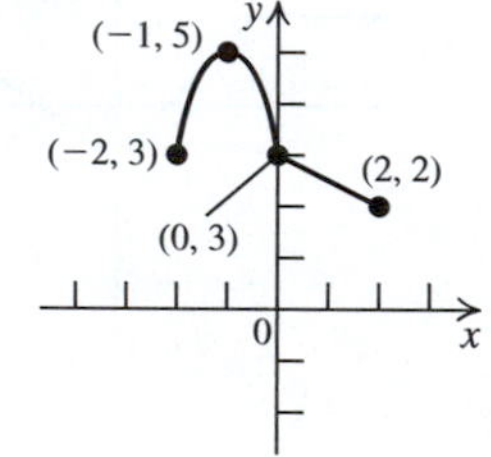

SECTION 2.7 Exercises

Basic Concepts and Skills

1. The graph of $y = f(x) - 3$ is found by vertically shifting the graph of $y = f(x)$ 3 units ______________.
2. The graph of $y = f(x + 5)$ is found by horizontally shifting the graph of $y = f(x)$ 5 units to the ______________.
3. The graph of $y = f(bx)$ is a horizontal compression of the graph of $y = f(x)$ if b ______________.
4. The graph of $y = f(-x)$ is found by reflecting the graph of $y = f(x)$ about the ______________.
5. **True or False.** The graph of $y = f(x)$ and $y = f(-x)$ cannot be the same.
6. **True or False.** You get the same graph by shifting the graph of $y = x^2$ 2 units up, reflecting the shifted graph about the x-axis or reflecting the graph of $y = x^2$ about the x-axis, and then shifting the reflected graph up 2 units.

In Exercises 7–20, describe the transformations that produce the graphs of *g* and *h* from the graph of *f*.

7. $f(x) = \sqrt{x}$
 a. $g(x) = \sqrt{x} + 2$
 b. $h(x) = \sqrt{x} - 1$

8. $f(x) = |x|$
 a. $g(x) = |x| + 1$
 b. $h(x) = |x| - 2$

9. $f(x) = x^2$
 a. $g(x) = (x + 1)^2$
 b. $h(x) = (x - 2)^2$

10. $f(x) = \dfrac{1}{x}$
 a. $g(x) = \dfrac{1}{x + 2}$
 b. $h(x) = \dfrac{1}{x - 3}$

11. $f(x) = \sqrt{x}$
 a. $g(x) = \sqrt{x + 1} - 2$
 b. $h(x) = \sqrt{x - 1} + 3$

12. $f(x) = x^2$
 a. $g(x) = -x^2$
 b. $h(x) = (-x)^2$

13. $f(x) = |x|$
 a. $g(x) = -|x|$
 b. $h(x) = |-x|$

14. $f(x) = \sqrt{x}$
 a. $g(x) = 2\sqrt{x}$
 b. $h(x) = \sqrt{2x}$

15. $f(x) = \dfrac{1}{x}$
 a. $g(x) = \dfrac{2}{x}$
 b. $h(x) = \dfrac{1}{2x}$

16. $f(x) = x^3$
 a. $g(x) = (x - 2)^3 + 1$
 b. $h(x) = -(x + 1)^3 + 2$

17. $f(x) = \sqrt{x}$
 a. $g(x) = -\sqrt{x} + 1$
 b. $h(x) = \sqrt{-x} + 1$

18. $f(x) = [\![x]\!]$
 a. $g(x) = [\![x - 1]\!] + 2$
 b. $h(x) = 3[\![x]\!] - 1$

19. $f(x) = \sqrt[3]{x}$
 a. $g(x) = \sqrt[3]{x} + 1$
 b. $h(x) = \sqrt[3]{x + 1}$

20. $f(x) = \sqrt[3]{x}$
 a. $g(x) = 2\sqrt[3]{1 - x} + 4$
 b. $h(x) = -\sqrt[3]{x - 1} + 3$

In Exercises 21–32, match each function with its graph (a)–(l).

21. $y = -|x| + 1$
22. $y = -\sqrt{-x}$
23. $y = \sqrt{x^2}$
24. $y = \dfrac{1}{2}|x|$
25. $y = \sqrt{x + 1}$
26. $y = 2|x| - 3$
27. $y = 1 - 2\sqrt{x}$
28. $y = -|x - 1| + 1$
29. $y = (x - 1)^2$
30. $y = -x^2 + 3$
31. $y = -2(x - 3)^2 - 1$
32. $y = 3 - \sqrt{1 - x}$

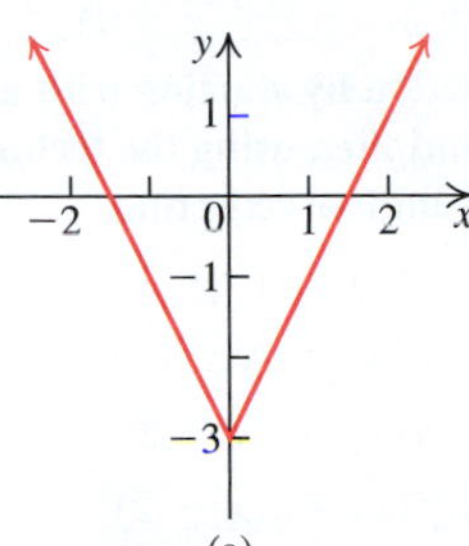

(a)

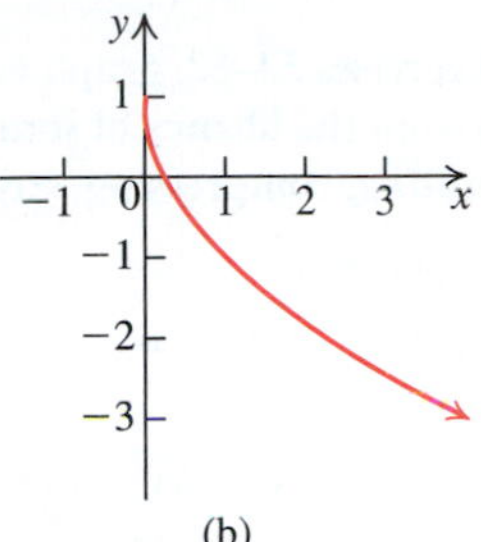

(b)

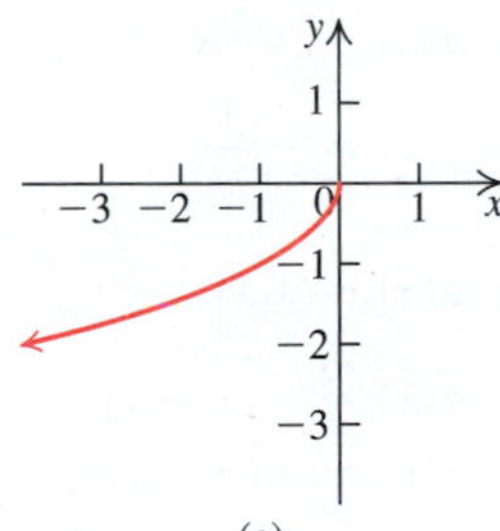

(c)

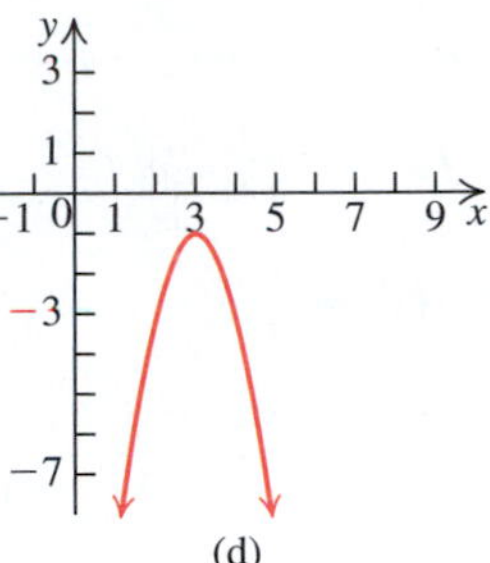

(d)

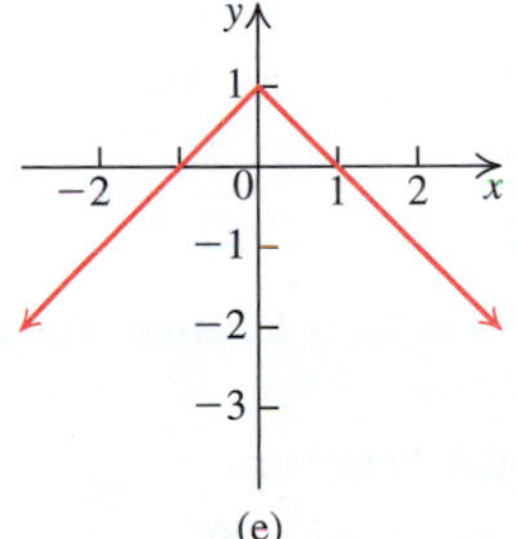

(e)

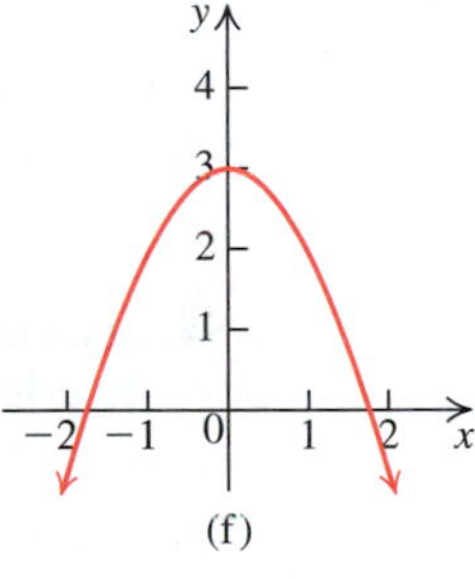

(f)

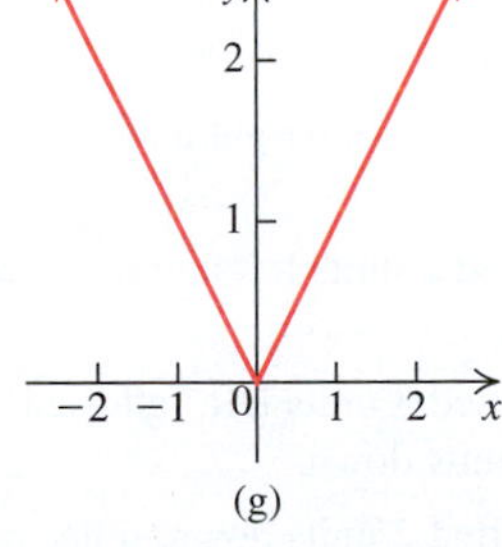

(g)

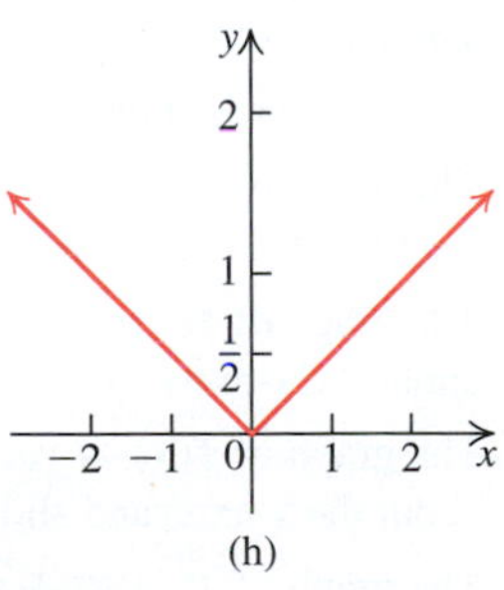

(h)

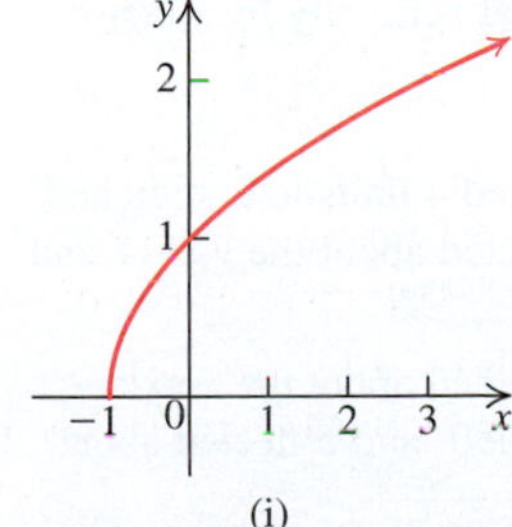

(i)

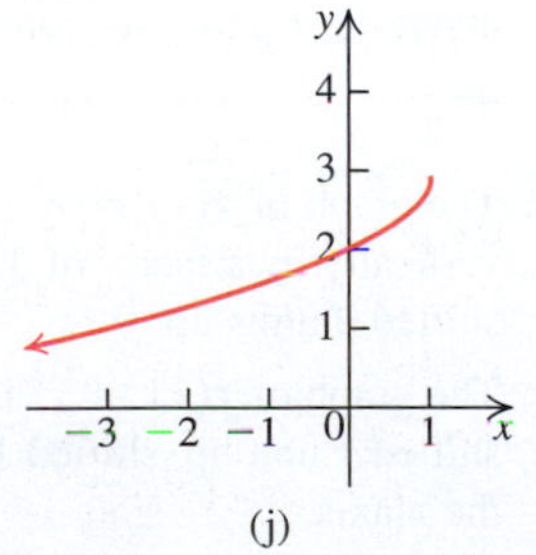

(j)

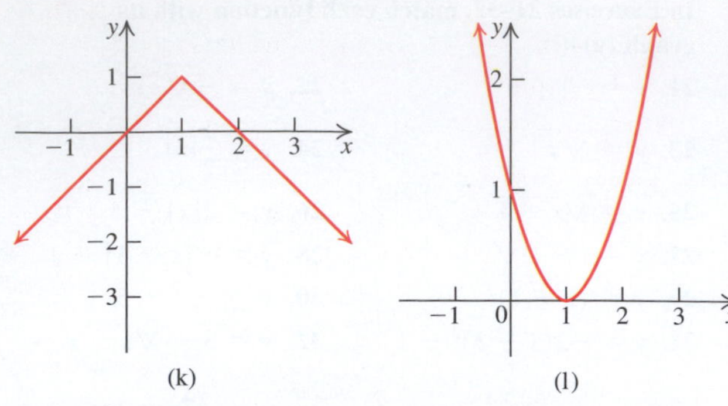

In Exercises 33–62, graph each function by starting with a function from the library of functions and then using the techniques of shifting, compressing, stretching, and/or reflecting.

33. $f(x) = x^2 - 2$

34. $f(x) = x^2 + 3$

35. $g(x) = \sqrt{x} + 1$

36. $g(x) = \sqrt{x} - 4$

37. $h(x) = |x + 1|$

38. $h(x) = |x - 2|$

39. $f(x) = (x - 3)^3$

40. $f(x) = (x + 2)^3$

41. $g(x) = (x - 2)^2 + 1$

42. $g(x) = (x + 3)^2 - 5$

43. $h(x) = -\sqrt{x}$

44. $h(x) = \sqrt{-x}$

45. $f(x) = -\frac{1}{x}$

46. $f(x) = -\frac{1}{2x}$

47. $g(x) = \frac{1}{2}|x|$

48. $g(x) = 4|x|$

49. $h(x) = -x^3 + 1$

50. $h(x) = -(x + 1)^3$

51. $f(x) = 2(x + 1)^2 - 1$

52. $f(x) = -(x - 1)^2$

53. $g(x) = 5 - x^2$

54. $g(x) = 2 - (x + 3)^2$

55. $h(x) = |1 - x|$

56. $h(x) = -2\sqrt{x - 1}$

57. $f(x) = -|x + 3| + 1$

58. $f(x) = 2 - \sqrt{x}$

59. $g(x) = -\sqrt{-x} + 2$

60. $g(x) = 3\sqrt{2 - x}$

61. $h(x) = 2[\![x + 1]\!]$

62. $h(x) = [\![-x]\!] + 1$

In Exercises 63–74, write an equation for a function whose graph fits the given description.

63. The graph of $f(x) = x^3$ is shifted 2 units up.

64. The graph of $f(x) = \sqrt{x}$ is shifted 3 units left.

65. The graph of $f(x) = |x|$ is reflected about the x-axis.

66. $f(x) = \sqrt{x}$ is reflected about the y-axis.

67. The graph of $f(x) = x^2$ is shifted 3 units right and 2 units up.

68. The graph of $f(x) = x^2$ is shifted 2 units left and reflected about the x-axis.

69. The graph of $f(x) = \sqrt{x}$ is shifted 3 units left, reflected about the x-axis, and shifted 2 units down.

70. The graph of $f(x) = \sqrt{x}$ is shifted 2 units down, reflected about the x-axis, and compressed vertically by a factor of $\frac{1}{2}$.

71. The graph of $f(x) = x^3$ is shifted 4 units left, stretched vertically by a factor of 3, reflected about the y-axis, and shifted 2 units up.

72. The graph of $f(x) = x^3$ is reflected about the x-axis, shifted 1 unit up, shifted 1 unit left, and reflected about the y-axis.

73. The graph of $f(x) = |x|$ is shifted 4 units right, compressed horizontally by a factor of 2, reflected about the x-axis, and shifted 3 units down.

74. The graph of $f(x) = |x|$ is shifted two units right, reflected about the y-axis, stretched horizontally by a factor of 2 and shifted three units down.

In Exercises 75–88, graph the function $y = g(x)$ given the following graph of $y = f(x)$.

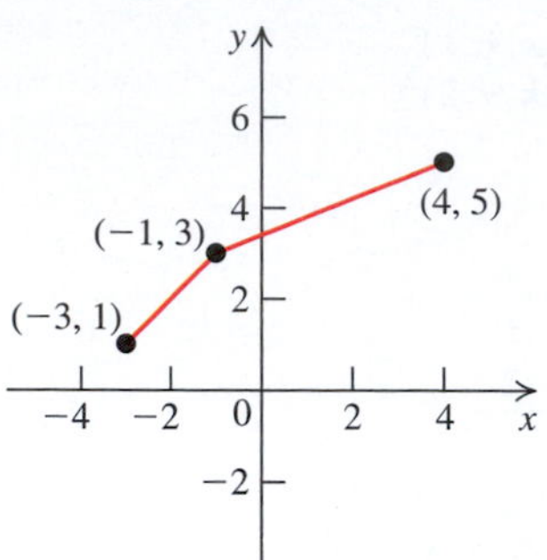

75. $g(x) = f(x) - 1$

76. $g(x) = f(x) + 3$

77. $g(x) = -f(x)$

78. $g(x) = f(-x)$

79. $g(x) = f(2x)$

80. $g(x) = f\left(\frac{1}{2}x\right)$

81. $g(x) = f(x + 1)$

82. $g(x) = f(x - 2)$

83. $g(x) = 2f(x)$

84. $g(x) = \frac{1}{2}f(x)$

85. $g(x) = f(x - 1) + 2$

86. $g(x) = -f(-2x) - 1$

87. $g(x) = f(1 - 2x)$

88. $g(x) = f(3 - 2x)$

In Exercises 89–96, graph (a) $y = g(x)$ and (b) $y = |g(x)|$, given the following graph of $y = f(x)$.

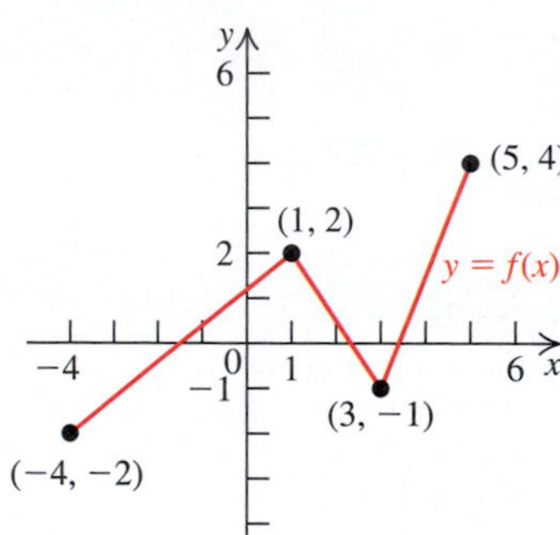

89. $g(x) = f(x) + 1$

90. $g(x) = -2f(x)$

91. $g(x) = f\left(\frac{1}{2}x\right)$

92. $g(x) = f(-2x)$

93. $g(x) = f(x - 1)$

94. $g(x) = f(2 - x)$

95. $g(x) = -2f(x + 1) + 3$

96. $g(x) = -f(-x + 1) - 2$

97. $g(x) = 2f(1 - 2x) - 3$

98. $g(x) = -f(-2 - 2x) + 1$

Applying the Concepts

In Exercises 99–102, let f be the function that associates the employee number x of each employee of the ABC Corporation with his or her annual salary $f(x)$ in dollars.

99. **Across-the-board raise.** Each employee was awarded an across-the-board raise of $800 per year. Write a function $g(x)$ to describe the new salary.

100. **Percentage raise.** Suppose each employee was awarded a 5% raise. Write a function $h(x)$ to describe the new salary.

101. **Across-the-board and percentage raises.** Suppose each employee was awarded a \$500 across-the-board raise and an additional 2% of his or her increased salary. Write a function $p(x)$ to describe the new salary.

102. **Across-the-board percentage raise.** Suppose the employees making \$30,000 or more received a 2% raise, while those making less than \$30,000 received a 10% raise. Write a piecewise function to describe these new salaries.

103. **Health plan.** The ABC Corporation pays for its employees' health insurance at an annual cost (in dollars) given by

$$C(x) = 5{,}000 + 10\sqrt{x - 1},$$

where x is the number of employees covered.

a. Use transformations of the graph of $y = \sqrt{x}$ to sketch the graph of $y = C(x)$.

b. Assuming that the company has 400 employees, find its annual outlay for the health coverage.

104. **Poiseuille's Law.** For an artery of radius R, the velocity v (in mm/min) of blood flow at a distance r from the center of the artery is given by

$$v = c(R^2 - r^2), 0 \le r \le R,$$

where c is a constant.

In an artery for Angie, $c = 1000$ and $R = 3$ mm. Find the velocity of blood flow in this artery at

a. The center of the artery.

b. The inner linings of the artery.

c. Midway between the center and the inner linings.

105. **Demand.** The weekly demand for paper hats produced by Mythical Manufacturers is given by

$$x(p) = 109{,}561 - (p + 1)^2,$$

where x represents the number of hats that can be sold at a price of p cents each.

a. Use transformations on the graph of $y = p^2$ to sketch the graph of $y = x(p)$.

b. Find the price at which 69,160 hats can be sold.

c. Find the price at which no hats can be sold.

106. **Revenue.** The weekly demand for cashmere sweaters produced by Wool Shop, Inc., is $x(p) = -3p + 600$. The revenue is given by $R(p) = -3p^2 + 600p$. Describe how to sketch the graph of $y = R(p)$ by applying transformations to the graph of $y = p^2$. [*Hint:* First, write $R(p)$ in the form $-3(p - h)^2 + k$.]

107. **Daylight.** At 60° north latitude, the graph of $y = f(t)$ gives the number of hours of daylight. (On the t-axis, note that 1 = January and 12 = December.)

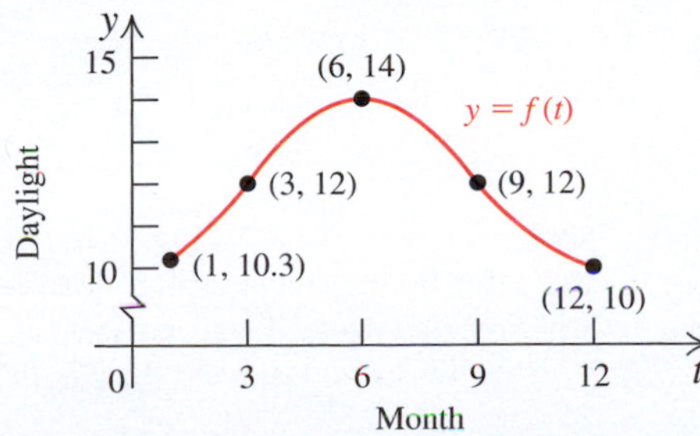

Sketch the graph of $y = f(t) - 12$.

108. Use the graph of $y = f(t)$ of Exercise 107 to sketch the graph of $y = 24 - f(t)$. Interpret the result.

Beyond the Basics

In Exercises 109 and 110, the graphs of $y = f(x)$ and $y = g(x)$ are given. Find an equation for $g(x)$ if the graph of g is obtained from the graph of f by a sequence of transformations.

109.

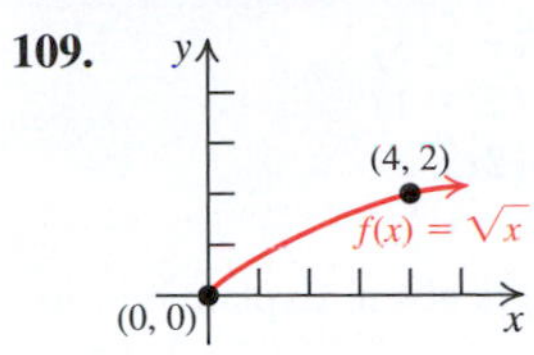

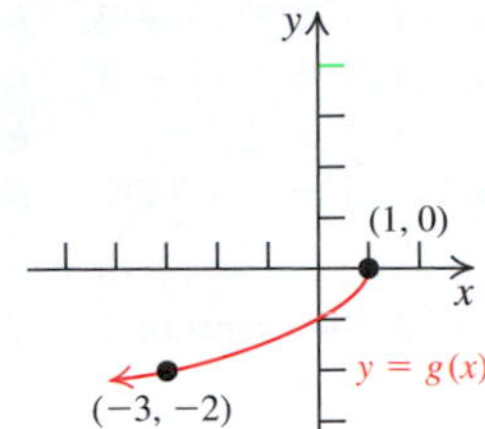

110.

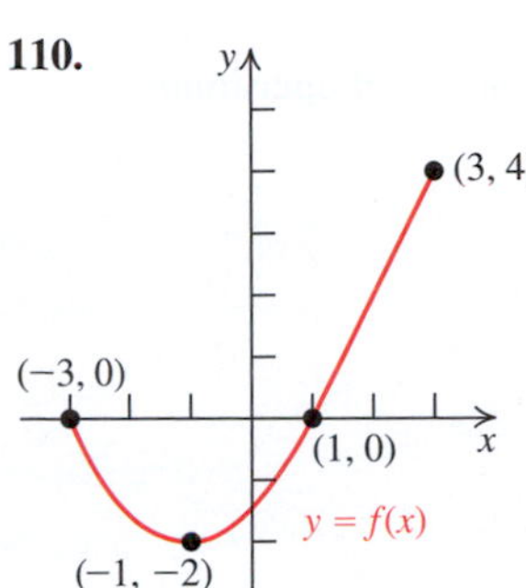

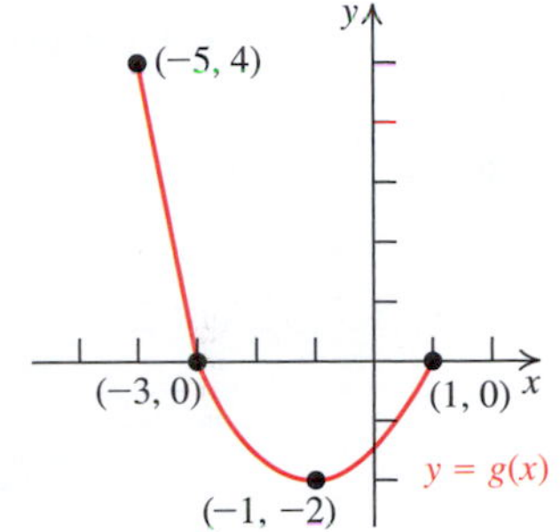

In Exercises 111–118, by completing the square on each quadratic expression, use transformations on $y = x^2$ to sketch the graph of $y = f(x)$.

111. $f(x) = x^2 + 4x$
[*Hint:* $x^2 + 4x = (x^2 + 4x + 4) - 4 = (x + 2)^2 - 4$.]

112. $f(x) = x^2 - 6x$

113. $f(x) = -x^2 + 2x$

114. $f(x) = -x^2 - 2x$

115. $f(x) = 2x^2 - 4x$

116. $f(x) = 2x^2 + 6x + 3.5$

117. $f(x) = -2x^2 - 8x + 3$

118. $f(x) = -2x^2 + 2x - 1$

In Exercises 119–124, sketch the graph of each function.

119. $y = |2x + 3|$

120. $y = |[\![x]\!]|$

121. $y = |4 - x^2|$

122. $y = \left|\frac{1}{x}\right|$

123. $y = \sqrt{|x|}$

124. $y = [\![|x|]\!]$

Critical Thinking/Discussion/Writing

125. If f is a function with x-intercept -2 and y-intercept 3, find the corresponding x-intercept and y-intercept, if possible, for

a. $y = f(x - 3)$.

b. $y = 5f(x)$.

c. $y = -f(x)$.

d. $y = f(-x)$.

126. If f is a function with x-intercept 4 and y-intercept -1, find the corresponding x-intercept and y-intercept, if possible, for

a. $y = f(x + 2)$.

b. $y = 2f(x)$.

c. $y = -f(x)$.

d. $y = f(-x)$.

127. Suppose the graph of $y = f(x)$ is given. Using transformations, explain how the graph of $y = g(x)$ differs from the graph of $y = h(x)$.

a. $g(x) = f(x) + 3, \quad h(x) = f(x + 3)$

b. $g(x) = f(x) - 1, \quad h(x) = f(x - 1)$

c. $g(x) = 2f(x), \quad h(x) = f(2x)$

d. $g(x) = -3f(x), \quad h(x) = f(-3x)$

128. Suppose the graph of $y = f(-4x)$ is given. Explain how to obtain the graph of $y = f(x)$.

Maintaining Skills

In Exercises 129–134, perform the indicated operations.

129. $(5x^2 + 5x + 7) + (x^2 + 9x - 4)$

130. $(x^2 + 2x) + (6x^3 - 2x + 5)$

131. $(5x^2 + 6x - 2) - (3x^2 - 9x + 1)$

132. $(x^3 + 2) - (2x^3 + 5x - 3)$

133. $(x - 2)(x^2 + 2x + 4)$

134. $(x^2 + x + 1)(x^2 - x + 1)$

In Exercises 135–138, find the domain of each function.

135. $f(x) = \dfrac{2x - 3}{x^2 - 5x + 6}$

136. $f(x) = \dfrac{x - 2}{x^2 - 4}$

137. $f(x) = \sqrt{2x - 3}$

138. $f(x) = \dfrac{1}{\sqrt{5 - 2x}}$

SECTION 2.8

Combining Functions; Composite Functions

BEFORE STARTING THIS SECTION, REVIEW

1. Domain of a function (Section 2.4, page 199)
2. Rational inequalities (Section 1.5, page 137)

OBJECTIVES

1. Define basic operations on functions.
2. Form composite functions.
3. Find the domain of a composite function.
4. Decompose a function.
5. Apply composition to practical problems.

BP Oil Spill

The Deepwater Horizon was a floating drilling rig that could operate in waters up to 10,000 feet deep. British Petroleum (BP) was the operator and principal developer of the exploratory well situated in the Gulf of Mexico about 41 miles off the Louisiana coast. On April 20, 2010 at approximately 9:45 PM, high-pressure methane gas from the well ignited and exploded, engulfing the platform of the rig. At the time, 126 crew members were on board. Of these, 11 workers were never found despite extensive search operations and are believed to have died in the explosion.

The total estimated volume of oil leaked was approximately 5 million barrels, making it the largest accidental oil spill in history. The spill directly impacted 68,000 square miles (about size of Oklahoma). By June 2010, oil had washed up on 1074 miles of the coasts of Louisiana, Mississippi, Alabama, and Florida.

On June 16, 2010, BP announced a $20 billion fund to settle claims arising from the disaster. The fund was to be used only for natural resource restoration and compensation for individual and business claimants. As of July, 2013, BP and its partners in the oil well, Transocean and Halliburton, are on trial to determine payouts and fines under the Clean Water Act and the Natural Resources Damages Assessment.

In Example 8, we calculate the area covered by an oil spill.

1 Define basic operations on functions.

Combining Functions

Just as numbers can be added, subtracted, multiplied, and divided to produce new numbers, so functions can be added, subtracted, multiplied, and divided to produce other functions. To add, subtract, multiply, or divide functions, we simply add, subtract, multiply, or divide their output values.

For example, if $f(x) = x^2 - 2x + 4$ and $g(x) = x + 2$, then

$$f(x) + g(x) = (x^2 - 2x + 4) + (x + 2) = x^2 - x + 6.$$

This new function $y = x^2 - x + 6$ is called the sum function $f + g$.

Sum, Difference, Product, and Quotient of Functions

Let f and g be two functions. The **sum** $f + g$, the **difference** $f - g$, the **product** fg, and the **quotient** $\frac{f}{g}$ are functions defined as follows:

- Sum $(f + g)(x) = f(x) + g(x)$
- Difference $(f - g)(x) = f(x) - g(x)$
- Product $fg(x) = f(x) \cdot g(x)$; also written as $(fg)(x)$
- Quotient $\frac{f}{g}(x) = \frac{f(x)}{g(x)}$, provided that $g(x) \neq 0$; also written as $\left(\frac{f}{g}\right)(x)$

The **domain** of each of the new functions consists of those values of x that are common to the domains of f and g, except that for $\frac{f}{g}$, all x for which $g(x) = 0$ must also be excluded.

EXAMPLE 1 **Combining Functions**

Let $f(x) = x^2 - 6x + 8$ and $g(x) = x - 2$. Find each of the following functions or function values.

a. $(f - g)(4)$ **b.** $\left(\frac{f}{g}\right)(3)$

c. $(f + g)(x)$ **d.** $(f - g)(x)$

e. $(fg)(x)$ **f.** $\left(\frac{f}{g}\right)(x)$

Solution

a.
$$f(4) = 4^2 - 6(4) + 8 = 0$$ Replace x with 4 in $f(x)$.
$$g(4) = 4 - 2 = 2$$ Replace x with 4 in $g(x)$.
$$(f - g)(4) = f(4) - g(4)$$ Definition of difference
$$= 0 - 2 = -2$$ Replace $f(4)$ with 0, $g(4)$ with 2.

b.
$$f(3) = 3^2 - 6(3) + 8 = -1$$ Replace x with 3 in $f(x)$.
$$g(3) = 3 - 2 = 1$$ Replace x with 3 in $g(x)$.
$$\left(\frac{f}{g}\right)(3) = \frac{f(3)}{g(3)} = \frac{-1}{1} = -1$$ Replace $f(3)$ with -1, $g(3)$ with 1.

c.
$$(f + g)(x) = f(x) + g(x)$$ Definition of sum
$$= (x^2 - 6x + 8) + (x - 2)$$ Add $f(x)$ and $g(x)$.
$$= x^2 - 5x + 6$$ Combine like terms.

d.
$$(f - g)(x) = f(x) - g(x)$$ Definition of difference
$$= (x^2 - 6x + 8) - (x - 2)$$ Subtract $g(x)$ from $f(x)$.
$$= x^2 - 7x + 10$$ Combine like terms.

e.
$$(fg)(x) = f(x) \cdot g(x)$$ Definition of product
$$= (x^2 - 6x + 8)(x - 2)$$ Multiply $f(x)$ and $g(x)$.
$$= x^2(x - 2) - 6x(x - 2) + 8(x - 2)$$ Distributive property
$$= x^3 - 2x^2 - 6x^2 + 12x + 8x - 16$$ Distributive property
$$= x^3 - 8x^2 + 20x - 16$$ Combine like terms.

f. $\left(\dfrac{f}{g}\right)(x) = \dfrac{f(x)}{g(x)}, \quad g(x) \neq 0$ Definition of quotient

$= \dfrac{x^2 - 6x + 8}{x - 2}, \quad x - 2 \neq 0$ Replace $f(x)$ with $x^2 - 6x + 8$, $g(x)$ with $x - 2$.

$= \dfrac{(x - 2)(x - 4)}{x - 2}, \quad x \neq 2$ Factor the numerator.

$= x - 4, \quad x \neq 2$

Because f and g are polynomials, the domain of both f and g is the set of all real numbers, or in interval notation, $(-\infty, \infty)$. So the domain for $f + g, f - g$, and fg is $(-\infty, \infty)$. However, for $\dfrac{f}{g}$, we must exclude 2 because $g(2) = 0$. The domain for $\dfrac{f}{g}$ is the set of all real numbers $x, x \neq 2$, or in interval notation, $(-\infty, 2) \cup (2, \infty)$.

Practice Problem 1 Let $f(x) = 3x - 1$ and $g(x) = x^2 + 2$. Find $(f + g)(x)$, $(f - g)(x)$, $(fg)(x)$, and $\left(\dfrac{f}{g}\right)(x)$.

WARNING

When rewriting a function such as

$$\left(\frac{f}{g}\right)(x) = \frac{(x - 2)(x - 4)}{x - 2}, \quad x \neq 2,$$

by dividing out the common factor, remember to specify "$x \neq 2$."

We identify the domain before we simplify and write

$$\left(\frac{f}{g}\right)(x) = x - 4, \quad x \neq 2.$$

EXAMPLE 2 Finding the Domain

Let $f(x) = \sqrt{x + 2}$ and $g(x) = \sqrt{5 - x}$. Find the domain of $f + g, f - g, fg$, and $\dfrac{f}{g}$.

Solution

Recall (see page 59) that $\sqrt{u}$ is defined only if $u \geq 0$.

So $f(x) = \sqrt{x + 2}$ is defined if $x + 2 \geq 0$ or $x \geq -2$. Sovle for x.

The domain of $f = D_1 = [-2, \infty)$ Interval notation

Similarly, $g(x) = \sqrt{5 - x}$ is defined only if $5 - x \geq 0$

or $5 \geq x$ Sovle for x.

or $x \leq 5$ Rewrite.

The domain of $g = D_2 = (-\infty, 5]$ Interval notation

To find the required domains, we first find $D_1 \cap D_2$ from the graphs of D_1 and D_2. See Figure 2.92.

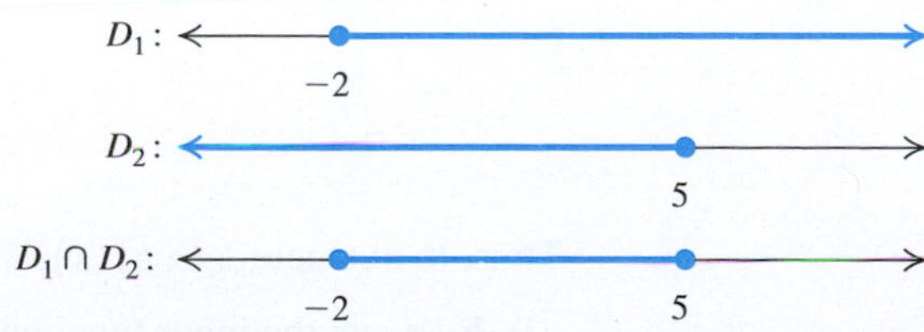

FIGURE 2.92

The domain of $f + g, f - g$, and fg is $D_1 \cap D_2 = [-2, 5]$. For the domain of $\frac{f}{g}$, we must exclude from $D_1 \cap D_2$ those numbers where $g(x) = \sqrt{5 - x} = 0$. Solving this equation for x, we have $x = 5$.

So the domain of $\frac{f}{g}$ is $[-2, 5)$.

Practice Problem 2 Let $f(x) = \sqrt{x - 1}$ and $g(x) = \sqrt{3 - x}$. Find the domain of $fg, \frac{f}{g}$, and $\frac{g}{f}$.

2 Form composite functions.

Composition of Functions

There is yet another way to construct a new function from two functions. Figure 2.93 shows what happens when you apply one function $g(x) = x^2 - 1$ to an input (domain) value, x, and then apply a second function $f(x) = \sqrt{x}$ to the output value, $g(x)$, from the first function. In this case, the output from the first function becomes the input for the second function.

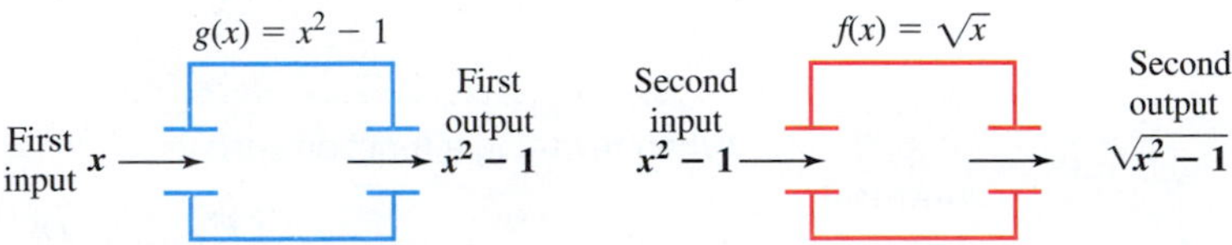

FIGURE 2.93 Composition of functions

Composition of Functions

If f and g are two functions, the composition of the function f with the function g is written as $f \circ g$ and is defined by the equation

$$(f \circ g)(x) = f(g(x))$$

We read $(f \circ g)(x)$ as "f composed with g of x" and $f(g(x))$ as "f of g of x." The domain of $f \circ g$ consists of those values x in the domain of g for which $g(x)$ is in the domain of f.

Figure 2.94 may help clarify the definition of $f \circ g$.

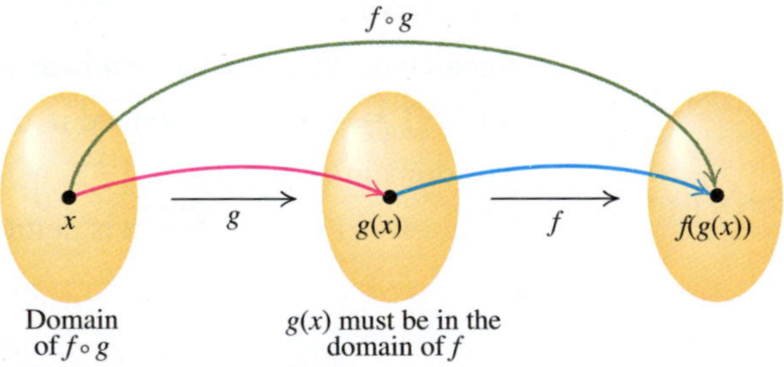

FIGURE 2.94 Composition of the function f with g

Evaluating $(f \circ g)(x)$ By definition,

inner function

$$(f \circ g)(x) = f(g(x))$$

outer function

Thus, to evaluate $(f \circ g)(x)$, we must

(i) Evaluate the inner function g at x.

(ii) Use the result $g(x)$ as the input for the outer function f.

EXAMPLE 3 **Evaluating a Composite Function**

Let $f(x) = x^3$ and $g(x) = x + 1$. Find each composite function value.

a. $(f \circ g)(1)$ **b.** $(g \circ f)(1)$ **c.** $(f \circ f)(-1)$

SIDE NOTE

When computing $(f \circ g)(x) = f(g(x))$, you must replace every x in $f(x)$ with $g(x)$.

Solution

a. $(f \circ g)(1) = f(g(1))$ Definition of $f \circ g$

$= [g(1)]^3$ Replace x with $g(1)$ in $f(x) = x^3$.

$= 2^3 = 8$ $g(1) = 1 + 1 = 2$; simplify.

b. $(g \circ f)(1) = g(f(1))$ Definition of $g \circ f$

$= f(1) + 1$ Replace x with $f(1)$ in $g(x) = x + 1$.

$= 1 + 1 = 2$ $f(1) = 1^3 = 1$; simplify.

c. $(f \circ f)(-1) = f(f(-1))$ Definition of $f \circ f$

$= [f(-1)]^3$ Replace x with $f(-1)$ in $f(x) = x^3$.

$= (-1)^3 = -1$ $f(-1) = (-1)^3 = -1$; simplify.

Practice Problem 3 Let $f(x) = -5x$ and $g(x) = x^2 + 1$. Find each composite function value.

a. $(f \circ g)(0)$ **b.** $(g \circ f)(0)$

EXAMPLE 4 **Finding Composite Functions**

Let $f(x) = 2x + 1$ and $g(x) = x^2 - 3$. Find each composite function.

a. $(f \circ g)(x)$ **b.** $(g \circ f)(x)$ **c.** $(f \circ f)(x)$

Solution

a. $(f \circ g)(x) = f(g(x))$ Definition of $f \circ g$

$= 2g(x) + 1$ Replace x with $g(x)$ in $f(x) = 2x + 1$.

$= 2(x^2 - 3) + 1$ Replace $g(x)$ with $x^2 - 3$.

$= 2x^2 - 5$ Simplify.

b. $(g \circ f)(x) = g(f(x))$ Definition of $g \circ f$

$= [f(x)]^2 - 3$ Replace x with $f(x)$ in $g(x) = x^2 - 3$.

$= (2x + 1)^2 - 3$ Replace $f(x)$ with $2x + 1$.

$= 4x^2 + 4x + 1 - 3$ $(A + B)^2 = A^2 + 2AB + B^2$

$= 4x^2 + 4x - 2$ Simplify.

c. $(f \circ f)(x) = f(f(x))$ Definition of $f \circ f$

$= 2f(x) + 1$ Replace x with $f(x)$ in $f(x) = 2x + 1$.

$= 2(2x + 1) + 1$ Replace $f(x)$ with $2x + 1$.

$= 4x + 3$ Simplify.

Practice Problem 4 Let $f(x) = 2 - x$ and $g(x) = 2x^2 + 1$. Find each composite function.

a. $(g \circ f)(x)$ **b.** $(f \circ g)(x)$ **c.** $(g \circ g)(x)$

Parts **a** and **b** of Example 3 illustrate that in general, $f \circ g \neq g \circ f$. In other words, the composition of functions is not commutative.

WARNING

Note that $f(g(x))$ is not $f(x) \cdot g(x)$. In $f(g(x))$, the output of g is *used as an input* for f; in contrast, in $f(x) \cdot g(x)$, the output $f(x)$ is *multiplied* by the output $g(x)$.

3 Find the domain of a composite function.

Domain of Composite Functions

Let f and g be two functions. The domain of the composite function $f \circ g$ consists of those values of x in the domain of g for which $g(x)$ is in the domain of f. So to find the domain of $f \circ g$, we first find the domain of g. Then from the domain of g, we exclude those values of x for which $g(x)$ is not in the domain of f.

PROCEDURE **IN ACTION**

EXAMPLE 5 Finding the Domain of a Composite Function

OBJECTIVE

Find the domain of $f \circ g$ from the defining equations for f and g.

EXAMPLE

Find the domain of $f \circ g$ if

$$f(x) = \frac{2}{x-3} \text{ and } g(x) = \frac{5}{x+1}.$$

Step 1 Let A be the domain of the inner function g. We may find A by first finding the values of x for which g is not defined.

1. $g(x) = \frac{5}{x+1}$ is not defined if the denominator $x + 1 = 0$, or $x = -1$. So $A = \{x | x \neq -1|\}$, or $A = (-\infty, -1) \cup (-1, \infty)$.

Step 2 Let B be the set of values of x for which $g(x)$ is in the domain of f. As before, we may do this by finding the values of x for which $g(x)$ is not in the domain of f.

2. $f(g(x)) = \frac{2}{g(x) - 3}$, so $f(g(x))$ is not defined if the denominator

$$g(x) - 3 = 0.$$

$$\frac{5}{x+1} - 3 = 0 \quad \text{Replace } g(x) \text{ with } \frac{5}{x+1}.$$

$$5 - 3(x+1) = 0 \quad \text{Multiply by } (x+1).$$

$$5 - 3x - 3 = 0 \quad \text{Simplify.}$$

$$x = \frac{2}{3} \quad \text{Solve for } x.$$

$$B = \left\{x | x \neq \frac{2}{3}\right\}, \text{ or } B = \left(-\infty, \frac{2}{3}\right) \cup \left(\frac{2}{3}, \infty\right)$$

Step 3 The domain of $f \circ g$ is $A \cap B$. The set $A \cap B$ is the values of x that are in the domain of g *and* for which $g(x)$ is in the domain of f.

3.

A: open point at -1

B: open point at $\frac{2}{3}$

$A \cap B$: open points at -1 and $\frac{2}{3}$

$A \cap B = (-\infty, -1) \cup \left(-1, \frac{2}{3}\right) \cup \left(\frac{2}{3}, \infty\right)$ is the domain of $f \circ g$.

Practice Problem 5 Let $f(x) = \sqrt{x+1}$ and $g(x) = \frac{2}{x-3}$. Find the domain of $f \circ g$.

EXAMPLE 6 **Finding the Domain of Composite Functions**

Let $f(x) = \sqrt{x - 2}$ and $g(x) = \sqrt{4 - x}$. Find the following functions and their domains using the procedure given in Example 5.

a. $f \circ g$ **b.** $g \circ f$

> SIDE **NOTE**
>
> If the function $f \circ g$ can be simplified, determine the domain *before* simplifying. For example, if $f(x) = x^2$ and $g(x) = \sqrt{x}$,
>
> $(f \circ g)(x) = f(g(x)) = f(\sqrt{x})$.
>
> Before simplifying:
>
> $$(f \circ g)(x) = (\sqrt{x})^2, x \geq 0$$
>
> The domain of f is $(-\infty, \infty)$, and $(f \circ g)(x)$ is defined on $[0, \infty)$.
> After simplifying:
>
> $$(f \circ g)(x) = x$$
>
> Here it appears that the domain of $f \circ g$ is $(-\infty, \infty)$. However this is not correct, it is $[0, \infty)$.

Solution

a. 1. $g(x) = \sqrt{4 - x}$ is defined if $4 - x \geq 0$ or $x \leq 4$. So $A = (-\infty, 4]$.

2. $f(g(x)) = \sqrt{g(x) - 2}$ Replace x with $g(x)$ in $f(x) = \sqrt{x - 2}$.

So, $g(x)$ is in the domine of f if:

$g(x) - 2 \geq 0$

$g(x) \geq 2$ Add 2 to both sides.

$\sqrt{4 - x} \geq 2$ Replace $g(x)$ with $\sqrt{4 - x}$.

$4 - x \geq 4$ Square both sides.

$-x \geq 0$ Subtract 4 from both sides.

$x \leq 0$ Solve for x.

So $B = (-\infty, 0]$.

3. $A \cap B = (-\infty, 4] \cap (-\infty, 0]$

$= (-\infty, 0]$ See Figure.

So, the domain of $f \circ g$ is $(-\infty, 0]$

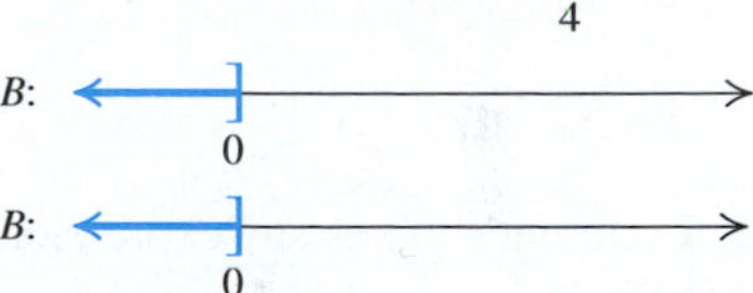

$A \cap B = (-\infty, 0]$

b. To find the domain of $g \circ f$, we interchange the role of g and f.

1. $f(x) = \sqrt{x - 2}$ is defined if $x - 2 \geq 0$ or $x \geq 2$. So $A = [2, \infty)$.

2. $g(f(x)) = \sqrt{4 - f(x)}$ Replace x with $f(x)$ in $g(x) = \sqrt{4 - x}$.

So, $f(x)$ is in the domine of g if:

$4 - f(x) \geq 0$

$4 \geq f(x)$ Add $f(x)$ to both sides.

$f(x) \leq 4$ Rewrite

$\sqrt{x - 2} \leq 4$ Replace $f(x)$ with $\sqrt{x - 2}$.

$x - 2 \leq 16$ Square both sides (if $0 < a \leq b$ then $a^2 \leq b^2$).

$x \leq 18$ Solve for x.

So $B = (-\infty, 18]$.

3. $A \cap B = [2, -\infty) \cap (-\infty, 18]$

$= [2, 18]$ See Figure.

So, the domain of $g \circ f = [2, 18]$.

$A \cap B = [2, 18]$

Practice Problem 6 Let $f(x) = \sqrt{x - 1}$ and $g(x) = \sqrt{4 - x^2}$. Find each of the following domains.

a. $f \circ g$ **b.** $g \circ f$

4 Decompose a function.

Decomposition of a Function

In the composition of two functions, we combine two functions and create a new function. However, sometimes it is more useful to use the concept of composition to *decompose* a function into simpler functions. For example, consider the function $H(x) = \sqrt{x^2 - 2}$. A way to write $H(x)$ as the composition of two functions is to let $f(x) = \sqrt{x}$ and $g(x) = x^2 - 2$ so that $f(g(x)) = f(x^2 - 2) = \sqrt{x^2 - 2} = H(x)$.

A function may be decomposed into simpler functions in several different ways by choosing various "inner" functions.

PROCEDURE **IN ACTION**

EXAMPLE 7 Decomposing a Function

OBJECTIVE

Write a function H as a composite of simpler functions f and g so that $H = f \circ g$.

EXAMPLE

Write

$$H(x) = \frac{1}{\sqrt{2x^2 + 1}}$$

as $f \circ g$ for some functions f and g.

Step 1 Define $g(x)$ as any expression in the defining equation for H.

1. Let $g(x) = 2x^2 + 1$.

Step 2 To get $f(x)$ from the defining equation for H, (a) replace the letter H with f and (b) replace the expression chosen for $g(x)$ with x.

2. $H(x) = \dfrac{1}{\sqrt{2x^2 + 1}}$

becomes

$$f(x) = \frac{1}{\sqrt{x}}$$

Replace the expression $2x^2 + 1$ from Step 1 with x and replace H with f.

Step 3 Now we have

$$H(x) = f(g(x)) = (f \circ g)(x).$$

3. $f(g(x)) = f(2x^2 + 1)$ $\qquad f(x) = \dfrac{1}{\sqrt{x}},\ g(x) = 2x^2 + 1$

$$= \frac{1}{\sqrt{2x^2 + 1}}$$

$$= H(x)$$

Note: Other choices include $g(x) = 2x^2$ and $f(x) = \dfrac{1}{\sqrt{x + 1}}$

Practice Problem 7 Repeat Example 7, letting $g(x) = \sqrt{2x^2 + 1}$. Find $f(x)$ and write $H(x) = (f \circ g)(x)$.

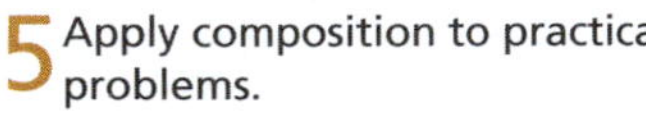

5 Apply composition to practical problems.

Applications of Composite Functions

EXAMPLE 8 Calculating the Area of an Oil Spill from a Tanker

Suppose oil spills from a tanker into the Pacific Ocean and the area of the spill is a perfect circle. The radius of this oil slick increases at the rate of 2 miles per hour.

a. Express the area of the oil slick as a function of time.

b. Calculate the area covered by the oil slick in six hours.

Solution

The area A of the oil slick is a function of its radius r.

$$A = f(r) = \pi r^2$$

The radius is a function of time. We know that r increases at the rate of 2 miles per hour. So in t hours, the radius r of the oil slick is given by the function

$$r = g(t) = 2t \qquad \text{Distance} = \text{Rate} \times \text{Time}$$

a. Therefore, the area A of the oil slick is a function of the radius, which is itself a function of time. It is a composite function given by

$$\begin{aligned} A &= f(g(t)) = f(2t) = \pi(2t)^2 \\ &= 4\pi t^2. \end{aligned}$$

b. Substitute $t = 6$ into A to find the area covered by the oil in six hours.

$$\begin{aligned} A &= 4\pi(6)^2 = 4\pi(36) \\ &= 144\pi \approx 452 \text{ square miles} \end{aligned}$$

Practice Problem 8 What would the result of Example 8 be if the radius of the oil slick increased at a rate of 3 miles per hour?

EXAMPLE 9 Applying Composition to Sales

A car dealer offers an 8% discount off the manufacturer's suggested retail price (MSRP) of x dollars for any new car on his lot. At the same time, the manufacturer offers a \$4000 rebate for each purchase of a new car.

a. Write a function $r(x)$ that represents the price after only the rebate.

b. Write a function $d(x)$ that represents the price after only the dealer's discount.

c. Write the functions $(r \circ d)(x)$ and $(d \circ r)(x)$. What do they represent?

d. Calculate $(d \circ r)(x) - (r \circ d)(x)$. Interpret this expression.

Solution

a. The MSRP is x dollars, and the manufacturer's rebate is \$4000 for each car; so

$$r(x) = x - 4000$$

represents the price of a car after the rebate.

b. The dealer's discount is 8% of x, or $0.08x$; therefore,

$$d(x) = x - 0.08x = 0.92x$$

represents the price of the car after the dealer's discount.

c. **(i)**
$$\begin{aligned} (r \circ d)(x) &= r(d(x)) && \text{Apply the dealer's discount first.} \\ &= r(0.92x) && d(x) = 0.92x \\ &= 0.92x - 4000 && \text{Evaluate } r(0.92x) \text{ using } r(x) = x - 4000. \end{aligned}$$

Then $(r \circ d)(x) = 0.92x - 4000$ represents the price of a car when the dealer's discount is applied first and then the manufacturer's rebate is applied.

(ii)
$$\begin{aligned} (d \circ r)(x) &= d(r(x)) && \text{Apply the manufacturer's rebate first.} \\ &= d(x - 4000) && r(x) = x - 4000 \\ &= 0.92(x - 4000) && \text{Evaluate } d(x - 4000) \text{ using } d(x) = 0.92x. \\ &= 0.92x - 3680 \end{aligned}$$

Thus, $(d \circ r)(x) = 0.92x - 3680$ represents the price of a car when the manufacturer's rebate is applied first and then the dealer's discount is applied.

d. $(d \circ r)(x) - (r \circ d)(x) = (0.92x - 3680) - (0.92x - 4000)$ Subtract function values.

$= \$320$

This equation shows that any car, regardless of its price, will cost \$320 more if you apply the rebate first and then the discount.

Practice Problem 9 Repeat Example 9 if the dealer offers a 6% discount and the manufacturer offers a \$4500 rebate.

Answers to Practice Problems

1. $(f+g)(x) = x^2 + 3x + 1$
$(f-g)(x) = -x^2 + 3x - 3$
$(fg)(x) = 3x^3 - x^2 + 6x - 2$
$\left(\frac{f}{g}\right)(x) = \frac{3x-1}{x^2+2}$

2. Domains: $fg = [1, 3]$; $\frac{f}{g} = [1, 3)$; $\frac{g}{f} = (1, 3]$

3. a. $(f \circ g)(0) = -5$ **b.** $(g \circ f)(0) = 1$

4. a. $(g \circ f)(x) = 2x^2 - 8x + 9$
b. $(f \circ g)(x) = 1 - 2x^2$
c. $(g \circ g)(x) = 8x^4 + 8x^2 + 3$

5. $(-\infty, 1] \cup (3, \infty)$ **6. a.** $[-\sqrt{3}, \sqrt{3}]$ **b.** $[1, 5]$

7. $f(x) = \frac{1}{x}$ **8. a.** $A = 9\pi t^2$ **b.** $A = 324\pi$ square miles ≈ 1018 square miles

9. a. $r(x) = x - 4500$ **b.** $d(x) = 0.94x$
c. (i) $(r \circ d)(x) = 0.94x - 4500$
(ii) $(d \circ r)(x) = 0.94x - 4230$
d. $(d \circ r)(x) - (r \circ d)(x) = 270$

SECTION 2.8 Exercises

Basic Concepts and Skills

1. If $f(2) = 12$ and $g(2) = 4$, then $(g - f)(2) =$ ________.

2. If $f(x) = 2x - 1$ and $f(x) + g(x) = 0$, then $g(x) =$ ________.

3. If $f(x) = x$ and $g(x) = 1$, then $(f \circ g)(x) =$ ________.

4. If $f(1) = 4$ and $g(4) = 7$, then $(g \circ f)(1) =$ ________.

5. **True or False.** If $f(x) = 2x$ and $g(x) = \frac{1}{2x}$, then $(f \circ g)(x) = x$.

6. **True or False.** It cannot happen that $f \circ g = g \circ f$.

In Exercises 7–10, functions f and g are given. Find each of the given values.

a. $(f+g)(-1)$ **b.** $(f-g)(0)$
c. $(f \cdot g)(2)$ **d.** $\left(\frac{f}{g}\right)(1)$

7. $f(x) = 2x$; $g(x) = -x$
8. $f(x) = 1 - x^2$; $g(x) = x + 1$
9. $f(x) = \frac{1}{\sqrt{x+2}}$; $g(x) = 2x + 1$
10. $f(x) = \frac{x}{x^2 - 6x + 8}$; $g(x) = 3 - x$

In Exercises 11–22, functions f and g are given. Find each of the following functions and state its domain.

a. $f + g$ **b.** $f - g$ **c.** $f \cdot g$
d. $\frac{f}{g}$ **e.** $\frac{g}{f}$

11. $f(x) = x - 3$; $g(x) = x^2$
12. $f(x) = 2x - 1$; $g(x) = x^2$
13. $f(x) = x^3 - 1$; $g(x) = 2x^2 + 5$
14. $f(x) = x^2 - 4$; $g(x) = x^2 - 6x + 8$
15. $f(x) = 2x - 1$; $g(x) = \sqrt{x}$
16. $f(x) = 1 - \frac{1}{x}$; $g(x) = \frac{1}{x}$
17. $f(x) = \frac{2}{x+1}$; $g(x) = \frac{x}{x+1}$
18. $f(x) = \frac{5x-1}{x+1}$; $g(x) = \frac{4x+10}{x+1}$
19. $f(x) = \frac{x^2}{x+1}$; $g(x) = \frac{2x}{x^2-1}$
20. $f(x) = \frac{x-3}{x^2-25}$; $g(x) = \frac{x-3}{x^2+9x+20}$
21. $f(x) = \frac{2x}{x^2-16}$; $g(x) = \frac{2x-7}{x^2-7x+12}$
22. $f(x) = \frac{x^2+3x+2}{x^3+4x}$; $g(x) = \frac{2x^3+8x}{x^2+x-2}$

In Exercises 23–26, find the domain of each function.

a. $f \cdot g$ **b.** $\dfrac{f}{g}$

23. $f(x) = \sqrt{x-1};\quad g(x) = \sqrt{5-x}$

24. $f(x) = \sqrt{x-2};\quad g(x) = \sqrt{x+2}$

25. $f(x) = \sqrt{x+2};\quad g(x) = \sqrt{9-x^2}$

26. $f(x) = \sqrt{x^2-4};\quad g(x) = \sqrt{25-x^2}$

In Exercises 27 and 28, use each diagram to evaluate $(g \circ f)(x)$. Then evaluate $(g \circ f)(2)$ and $(g \circ f)(-3)$.

27. $x \rightarrow$ $f(x) = x^2 - 1$ $\rightarrow$ $g(x) = 2x + 3$ $\rightarrow$

28. $x \rightarrow$ $f(x) = |x + 1|$ $\rightarrow$ $g(x) = 3x^2 - 1$ $\rightarrow$

In Exercises 29–36, let $f(x) = 2x + 1$ and $g(x) = 2x^2 - 3$. Evaluate each expression.

29. $(f \circ g)(2)$ **30.** $(g \circ f)(2)$

31. $(f \circ g)(-3)$ **32.** $(g \circ f)(-5)$

33. $(g \circ f)(a)$ **34.** $(g \circ f)(-a)$

35. $(f \circ f)(1)$ **36.** $(g \circ g)(-1)$

In Exercises 37–42, the functions f and g are given. Find $f \circ g$ and its domain.

37. $f(x) = \dfrac{2}{x+1};\ g(x) = \dfrac{1}{x}$

38. $f(x) = \dfrac{1}{x-1};\ g(x) = \dfrac{2}{x+3}$

39. $f(x) = \sqrt{x-3};\ g(x) = 2 - 3x$

40. $f(x) = \dfrac{x}{x-1};\ g(x) = 2 + 5x$

41. $f(x) = |x|;\ g(x) = x^2 - 1$

42. $f(x) = 3x - 2;\ g(x) = |x - 1|$

In Exercises 43–56, the functions f and g are given. Find each composite function and describe its domain.

a. $f \circ g$ **b.** $g \circ f$ **c.** $f \circ f$ **d.** $g \circ g$

43. $f(x) = 2x - 3;\ g(x) = x + 4$

44. $f(x) = x - 3;\ g(x) = 3x - 5$

45. $f(x) = 1 - 2x;\ g(x) = 1 + x^2$

46. $f(x) = 2x - 3;\ g(x) = 2x^2$

47. $f(x) = x^2;\ g(x) = \sqrt{x}$

48. $f(x) = x^2 + 2x;\ g(x) = \sqrt{x+2}$

49. $f(x) = \dfrac{1}{2x-1};\ g(x) = \dfrac{1}{x^2}$

50. $f(x) = x - 1;\ g(x) = \dfrac{x}{x+1}$

51. $f(x) = \sqrt{x-1};\ g(x) = \sqrt{4-x}$

52. $f(x) = x^2 - 4;\ g(x) = \sqrt{4-x^2}$

53. $f(x) = \sqrt{x^2-1};\ g(x) = \sqrt{4-x^2}$

54. $f(x) = \sqrt{x^2-9};\ g(x) = \sqrt{9-x^2}$

55. $f(x) = 1 + \dfrac{1}{x};\ g(x) = \dfrac{1+x}{1-x}$

56. $f(x) = \sqrt[3]{x+1};\ g(x) = x^3 + 1$

In Exercises 57–66, express the given function H as a composition of two functions f and g such that $H(x) = (f \circ g)(x)$.

57. $H(x) = \sqrt{x+2}$ **58.** $H(x) = |3x + 2|$

59. $H(x) = (x^2 - 3)^{10}$ **60.** $H(x) = \sqrt{3x^2 + 5}$

61. $H(x) = \dfrac{1}{3x-5}$ **62.** $H(x) = \dfrac{5}{2x+3}$

63. $H(x) = \sqrt[3]{x^2 - 7}$ **64.** $H(x) = \sqrt[4]{x^2 + x + 1}$

65. $H(x) = \dfrac{1}{|x^3 - 1|}$ **66.** $H(x) = \sqrt[3]{1 + \sqrt{x}}$

Applying the Concepts

67. Cost, revenue, and profit. A retailer purchases x shirts from a wholesaler at a price of \$12 per shirt. Her selling price for each shirt is \$22. The state has 7% sales tax. Interpret each of the following functions.

a. $f(x) = 12x$
b. $g(x) = 22x$
c. $h(x) = g(x) + 0.07g(x)$
d. $P(x) = g(x) - f(x)$

68. Cost, revenue, and profit. The demand equation for a product is given by $x = 5000 - 5p$, where x is the number of units produced and sold at price p (in dollars) per unit. The cost (in dollars) of producing x units is given by $C(x) = 4x + 12{,}000$. Express each of the following as a function of price.

a. Cost
b. Revenue
c. Profit

69. Cost and revenue. A manufacturer of radios estimates that his daily cost of producing x radios is given by the equation $C = 350 + 5x$. The equation $R = 25x$ represents the revenue in dollars from selling x radios.

a. Write and simplify the profit function
$$P(x) = R(x) - C(x).$$
b. Find $P(20)$. What does the number $P(20)$ represent?
c. How many radios should the manufacturer produce and sell to have a daily profit of \$500?
d. Find the composite function $(R \circ x)(C)$. What does this function represent?
[*Hint:* To find $x(C)$, solve $C = 350 + 5x$ for x.]

70. Mail order. You order merchandise worth x dollars from D-Bay Manufacturers. The company charges you sales tax of 4% of the purchase price plus shipping and handling fees of \$3 plus 2% of the after-tax purchase price.

a. Write a function $g(x)$ that represents the sales tax.
b. What does the function $h(x) = x + g(x)$ represent?
c. Write a function $f(x)$ that represents the shipping and handling fees.
d. What does the function $T(x) = h(x) + f(x)$ represent?

71. **Consumer issues.** You work in a department store in which employees are entitled to a 30% discount on their purchases. You also have a coupon worth \$5 off any item.
   a. Write a function $f(x)$ that models discounting an item by 30%.
   b. Write a function $g(x)$ that models applying the coupon.
   c. Use a composition of your two functions from (a) and (b) to model your cost for an item assuming that the clerk applies the discount first and then the coupon.
   d. Use a composition of your two functions from (a) and (b) to model your cost for an item assuming that the clerk applies the coupon first and then the discount.
   e. Use the composite functions from (c) and (d) to find how much more an item costs assuming that the clerk applies the coupon first.

72. **Consumer issues.** You work in a department store in which employees are entitled to a 20% discount on their purchases. In addition, the store is offering a 10% discount on all items.
   a. Write a function $f(x)$ that models discounting an item by 20%.
   b. Write a function $g(x)$ that models discounting an item by 10%.
   c. Use a composition of your two functions from (a) and (b) to model taking the 20% discount first.
   d. Use a composition of your two functions from (a) and (b) to model taking the 10% discount first.
   e. Use the composite functions from (c) and (d) to decide which discount you would prefer to take first.

73. **Test grades.** Professor Harsh gave a test to his college algebra class, and nobody got more than 80 points (out of 100) on the test. One problem worth 8 points had insufficient data, so nobody could solve that problem. The professor adjusted the grades for the class by (a) increasing everyone's score by 10% and (b) giving everyone 8 bonus points. Let x represent the original score of a student.
   a. Write statements (a) and (b) as functions $f(x)$ and $g(x)$, respectively.
   b. Find $(f \circ g)(x)$ and explain what it means.
   c. Find $(g \circ f)(x)$ and explain what it means.
   d. Evaluate $(f \circ g)(70)$ and $(g \circ f)(70)$.
   e. Does $(f \circ g)(x) = (g \circ f)(x)$?
   f. Suppose a score of 90 or better results in an A. What is the lowest original score that will result in an A if the professor uses
      (i) $(f \circ g)(x)$?
      (ii) $(g \circ f)(x)$?

74. **Sales commissions.** Henrita works as a salesperson in a department store. Her weekly salary is \$200 plus a 3% bonus on weekly sales over \$8000. Suppose her sales in a week are x dollars.
   a. Let $f(x) = 0.03x$. What does this mean?
   b. Let $g(x) = x - 8000$. What does this mean?
   c. Which composite function, $(f \circ g)(x)$ or $(g \circ f)(x)$, represents Henrita's bonus?
   d. What was Henrita's salary for the week in which her sales were \$17,500?
   e. What were Henrita's sales for the week in which her salary was \$521?

75. **Enhancing a fountain.** A circular fountain has a radius of x feet. A circular fence is installed around the fountain at a distance of 30 feet from its edge.
   a. Write the function $f(x)$ that represents the area of the fountain.
   b. Write the function $g(x)$ that represents the entire area enclosed by the fence.
   c. What area does the function $g(x) - f(x)$ represent?
   d. The cost of the fence was \$4200, installed at the rate of \$10.50 per running foot (per perimeter foot). You are to prepare an estimate for paving the area between the fence and the fountain at \$1.75 per square foot. To the nearest dollar, what should your estimate be?

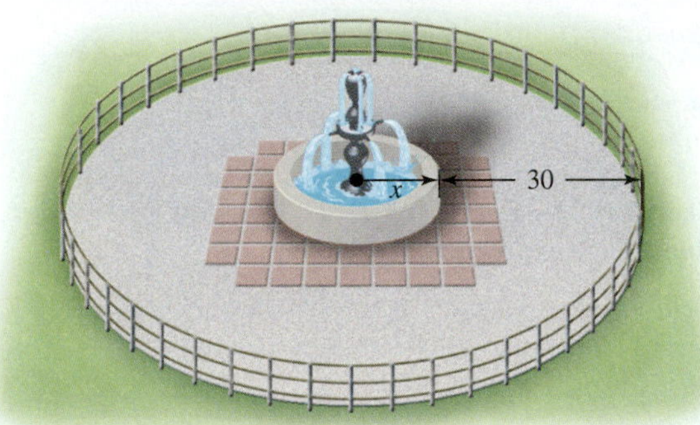

76. **Running track design.** An outdoor running track with semicircular ends and parallel sides is constructed. The length of each straight portion of the sides is 180 meters. The track has a uniform width of 4 meters throughout. The inner radius of each semicircular end is x meters.
   a. Write the function $f(x)$ that represents the area enclosed within the *outer* edge of the running track.
   b. Write the function $g(x)$ that represents the area of the field enclosed within the *inner* edge of the running track.
   c. What does the function $f(x) - g(x)$ represent?
   d. Suppose the inner perimeter of the track is 900 meters.
      (i) Find the area of the track.
      (ii) Find the outer perimeter of the track.

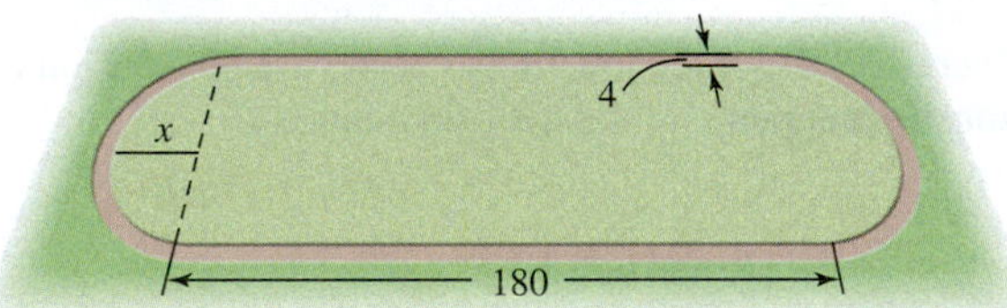

77. **Area of a disk.** The area A of a circular disk of radius r units is given by $A = f(r) = \pi r^2$. Suppose a metal disk is being heated and its radius r is increasing according to the equation $r = g(t) = 2t + 1$, where t is time in hours.
   a. Find $(f \circ g)(t)$.
   b. Determine A as a function of time.
   c. Compare parts (a) and (b).

78. **Volume of a balloon.** The volume V of a spherical balloon of radius r inches is given by the formula $V = f(r) = \frac{4}{3}\pi r^3$. Suppose the balloon is being inflated and its radius is increasing at the rate of 2 inches per second. That is, $r = g(t) = 2t$, where t is time in seconds.
   a. Find $(f \circ g)(t)$.
   b. Determine V as a function of time.
   c. Compare parts (a) and (b).

Beyond the Basics

79. Let $f = \{(-2, 3), (1, 2), (2, 1), (3, 0)\}$; $g = \{(-2, 0), (0, 2), (1, -2), (3, 2)\}$. Find each function.

a. $f + g$ **b.** fg **c.** $\frac{f}{g}$ **d.** $f \circ g$

80. Let

$$f(x) = \begin{cases} 2x & \text{if } -2 \le x \le 1 \\ x + 1 & \text{if } 1 < x \le 3 \end{cases};$$

$$g(x) = \begin{cases} x + 1 & \text{if } -3 \le x < 2. \\ 2x - 1 & \text{if } 2 \le x \le 3 \end{cases}$$

Find $f + g$ and its domain.

81. Let $h(x)$ be any function whose domain contains $-x$ whenever it contains x. Show that

a. $f(x) = h(x) + h(-x)$ is an even function.

b. $g(x) = h(x) - h(-x)$ is an odd function.

c. $h(x)$ can always be expressed as the sum of an even function and an odd function.

82. Use Exercise 81 to write each of the following functions as the sum of an even function and an odd function.

a. $h(x) = x^2 - 2x + 3$

b. $h(x) = [\![x]\!] + x$

83. Find the domain of $f(x) = \sqrt{\frac{1 - |x|}{2 - |x|}}$.

84. Let $f(x) = \begin{cases} -1 & \text{if } -2 \le x \le 0 \\ x - 1 & \text{if } 0 < x \le 2 \end{cases}$

Find $g(x) = f(|x|) + |f(x)|$

Critical Thinking/Discussion/Writing

85. Let $f(x) = \frac{1}{[\![x]\!]}$ and $g(x) = \sqrt{x(2 - x)}$. Find the domain of each of the following functions.

a. $f(x)$ **b.** $g(x)$

c. $f(x) + g(x)$ **d.** $\frac{f(x)}{g(x)}$

86. For each of the following functions, find the domain of $f \circ f$.

a. $f(x) = \frac{1}{\sqrt{-x}}$ **b.** $f(x) = \frac{1}{\sqrt{1 - x}}$

87. Even and odd functions. State whether each of the following functions is an odd function, an even function, or neither. Justify your statement.

a. The sum of two even functions

b. The sum of two odd functions

c. The sum of an even function and an odd function

d. The product of two even functions

e. The product of two odd functions

f. The product of an even function and an odd function

88. Even and odd functions. State whether the composite function $f \circ g$ is an odd function, an even function, or neither in the following situations.

a. f and g are odd functions.

b. f and g are even functions.

c. f is odd, and g is even.

d. f is even, and g is odd.

Maintaining Skills

89. a. Does the relation $R = \{(-3, 2), (-1, 1), (1, 3), (2, 1)\}$ define a function?

b. Construct a relation S by reversing the components of each ordered pair R. Does the relation S define a function?

90. Show that the line $y = x$ is perpendicular bisector of the line segment joining the points $P = (2, 5)$ and $P' = (5, 2)$. (This shows that the points P and P' are reflections of each other about the line $y = x$.)

91. Solve the equation $x = 2y + 3$ for y.

92. Solve the equation $x = y^2 + 1, y \ge 0$ for y.

93. Solve the equation $x^2 + y^2 = 4, x \le 0$ for x.

94. Solve the equation $2x - \frac{1}{y} = 3$ for y.

SECTION 2.9

Inverse Functions

BEFORE STARTING THIS SECTION, REVIEW

1. Vertical-line test (Section 2.4, page 201)
2. Domain and range of a function (Section 2.4, pages 199 and 200)
3. Composition of functions (Section 2.8, page 260)
4. Line of symmetry (Section 2.2, page 168)
5. Solving linear and quadratic equations (Sections 1.1–1.3)

OBJECTIVES

1. Define an inverse function.
2. Find the inverse function.
3. Use inverse functions to find the range of a function.
4. Apply inverse functions in the real world.

Water Pressure on Underwater Devices

The pressure at any depth in the ocean is the pressure at sea level (1 atm) plus the pressure due to the weight of the overlying water: The deeper you go, the greater the pressure. Pressure is a vital consideration in the design of all tools and vehicles that work beneath the water's surface. Every 11 feet of depth increases the pressure by approximately 5 pounds per square inch (1 atmosphere $\approx$ 15 psi). Computing the pressure precisely requires a somewhat complicated equation that takes into account the water's salinity, temperature, and slight compressibility. However, a basic formula is given by "pressure equals depth times 5, divided by 11 plus 15." Thus, we have pressure (psi) = depth (ft) $\times$ 5(psi per atm)/11(ft per atm) + 15, where *psi* means "pounds per square inch" and *atm* abbreviates "atmosphere."

For example, at 99 feet below the surface, the pressure is 60 psi. At 1 mile (5280 feet) below the surface, the pressure is $\frac{5280 \times 5}{11} + 15 = 2415$ psi. Suppose the pressure gauge on a diving bell breaks and shows a reading of 1800 psi. We can determine the bell's depth when the gauge failed by using an inverse function. We will return to this problem in Example 9.

1 Define an inverse function.

Inverses

Before we define inverse functions, we need to look at a special type of function called a one-to-one function.

One-to-One Function

A function is a **one-to-one function** if each y-value in its range corresponds to only one x-value in its domain.

Let f be a one-to-one function. Then the preceding definition says that for any two numbers x_1 and x_2 in the domain of f, if $x_1 \neq x_2$, then $f(x_1) \neq f(x_2)$, or equivalently, if $f(x_1) = f(x_2)$ then $x_1 = x_2$. That is, *f is a one-to-one function if different x-values correspond to different y-values.*

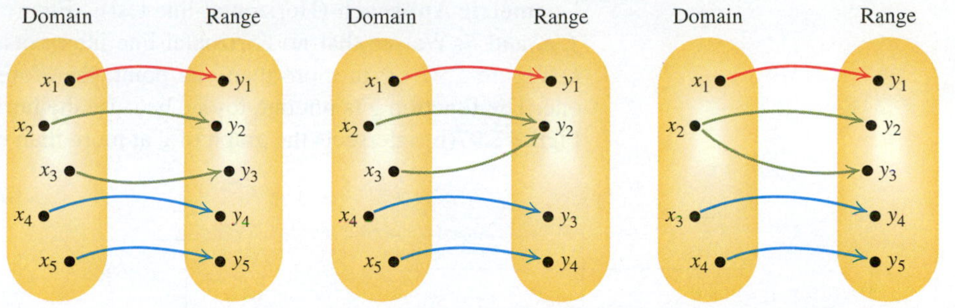

Domain Range Domain Range Domain Range

One-to-one function: Each y-value corresponds to only one x-value.
(a)

Not a one-to-one function: One y-value, y_2, corresponds to two different x-values.
(b)

Not a function: One x-value, x_2, corresponds to two different y-values.
(c)

FIGURE 2.95 Deciding whether a diagram respresents a function, a one-to-one function, or neither

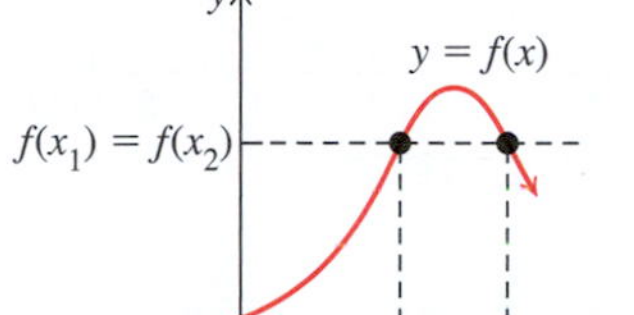

FIGURE 2.96 The function f is *not* one-to-one

Figure 2.95(a) represents a one-to-one function, and Figure 2.95(b) represents a function that is not one-to-one. The relation shown in Figure 2.95(c) is not a function.

Recall that with a one-to-one function, different x-values correspond to different y-values. Consequently, if a function f is *not* one-to-one, then there are at least two numbers x_1 and x_2 in the domain of f such that $x_1 \neq x_2$ and $f(x_1) = f(x_2)$. Geometrically, this means that the horizontal line passing through the two points $(x_1, f(x_1))$ and $(x_2, f(x_2))$ contains these *two* points on the graph of f. See Figure 2.96. The geometric interpretation of a one-to-one function is called the *horizontal-line test.*

Horizontal-Line Test

A function f is one-to-one if no horizontal line intersects the graph of f in more than one point.

EXAMPLE 1 Test for the One-to-One Property of a Function

Determine which of the following functions are one-to-one.

a. $f(x) = 2x + 5$ **b.** $g(x) = x^2 - 1$ **c.** $h(x) = 2\sqrt{x}$

Solution

Algebraic Approach Let x_1 and x_2 be two numbers in the domain of each function.

a. Suppose $f(x_1) = f(x_2)$. Then $2x_1 + 5 = 2x_2 + 5$

$$2x_1 = 2x_2 \quad \text{Subtract 5 from each side.}$$
$$x_1 = x_2 \quad \text{Divide both sides by 2.}$$

So $f(x_1) = f(x_2)$ implies $x_1 = x_2$; the function f is one-to-one.

b. Suppose $g(x_1) = g(x_2)$. Then $x_1^2 - 1 = x_2^2 - 1$

$$x_1^2 = x_2^2 \quad \text{Add 1 to both sides.}$$
$$x_1 = \pm x_2 \quad \text{Square-root property}$$

This means that for two distinct numbers, x_1 and $x_2 = -x_1$, $g(x_1) = g(x_2)$; so g is not a one-to-one function.

c. Suppose $h(x_1) = h(x_2)$. Then $2\sqrt{x_1} = 2\sqrt{x_2}$

$$\sqrt{x_1} = \sqrt{x_2} \quad \text{Divide both sides by 2.}$$
$$x_1 = x_2 \quad \text{Square both sides}$$

So $h(x_1) = h(x_2)$ implies $x_1 = x_2$; the function h is one-to-one.

Geometric Approach (Horizontal-line test) Figure 2.97 shows the graphs of functions f, g, and h. We see that no horizontal line intersects the graphs of f (Figure 2.97(a)) or h (Figure 2.97(c)) in more than one point; therefore, the functions f and h are one-to-one. The function g is not one-to-one because the horizontal line $y = 3$ (among others) in Figure 2.97(b) intersects the graph of g at more than one point.

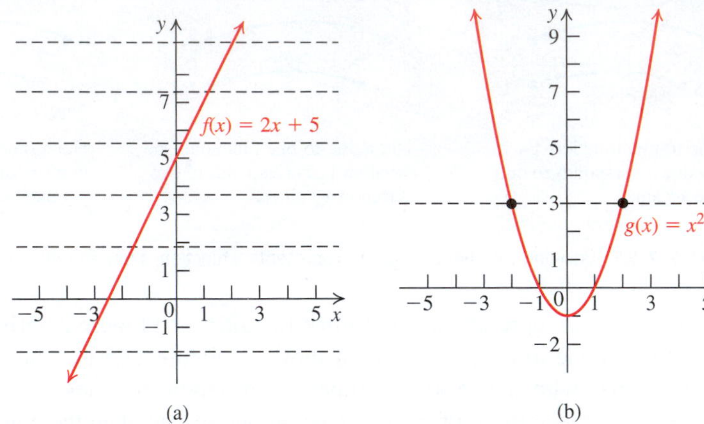

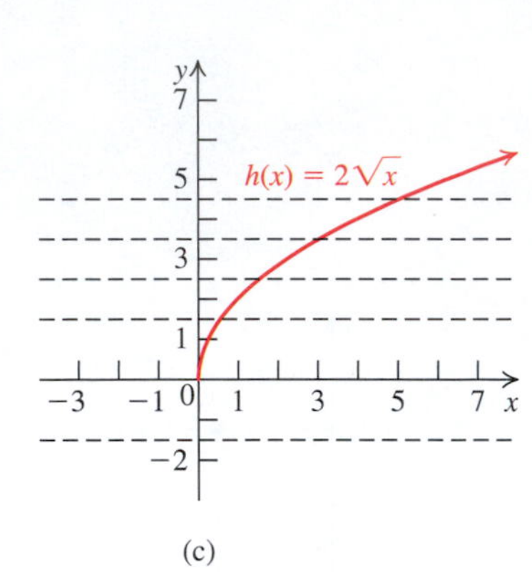

FIGURE 2.97 Horizontal-line test

Practice Problem 1 Determine wh1ether $f(x) = (x - 1)^2$ is a one-to-one function.

SIDE NOTE

The reason only a one-to-one function can have an inverse function is that if $x_1 \neq x_2$ but $f(x_1) = y$ and $f(x_2) = y$, then $f^{-1}(y)$ must be x_1 as well as x_2. This is not possible because $x_1 \neq x_2$.

Inverse Function

Let f represent a one-to-one function. Then if y is in the range of f, there is only one value of x in the domain of f such that $f(x) = y$. We define the inverse of f, called the **inverse function of f**, denoted f^{-1}, by $f^{-1}(y) = x$ if and only if $y = f(x)$.

From this definition we have the following:

Domain of f = Range of f^{-1} and Range of f = Domain of f^{-1}

WARNING

The notation $f^{-1}(x)$ does not mean $\frac{1}{f(x)}$. The expression $\frac{1}{f(x)}$ represents the reciprocal of $f(x)$ and is sometimes written as $[f(x)]^{-1}$.

EXAMPLE 2 Relating the Values of a Function and Its Inverse

Assume that f is a one-to-one function.

a. If $f(3) = 5$, find $f^{-1}(5)$. **b.** If $f^{-1}(-1) = 7$, find $f(7)$.

Solution

By definition, $f^{-1}(y) = x$ if and only if $y = f(x)$.

a. Let $x = 3$ and $y = 5$. Now reading the definition from right to left, $5 = f(3)$ if and only if $f^{-1}(5) = 3$. So $f^{-1}(5) = 3$.

b. Let $y = -1$ and $x = 7$. Now $f^{-1}(-1) = 7$ if and only if $f(7) = -1$. So $f(7) = -1$.

Practice Problem 2 Assume that f is a one-to-one function.

a. If $f(-3) = 12$, find $f^{-1}(12)$. **b.** If $f^{-1}(4) = 9$, find $f(9)$.

Consider the following input–output diagram and Figure 2.98 for $f^{-1} \circ f$.

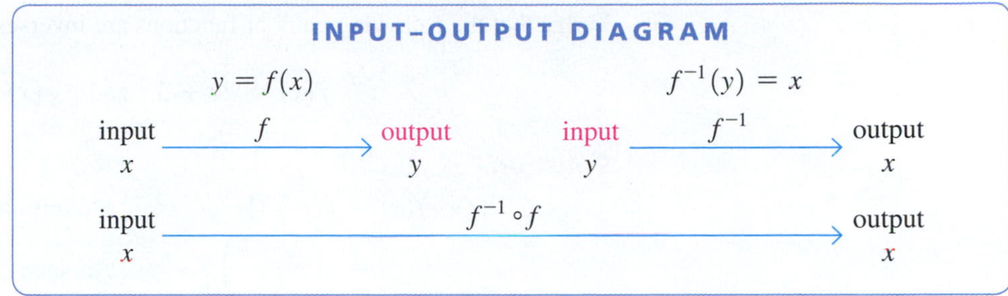

For the graphical interpretation of the input–output diagram, see Figure 2.98

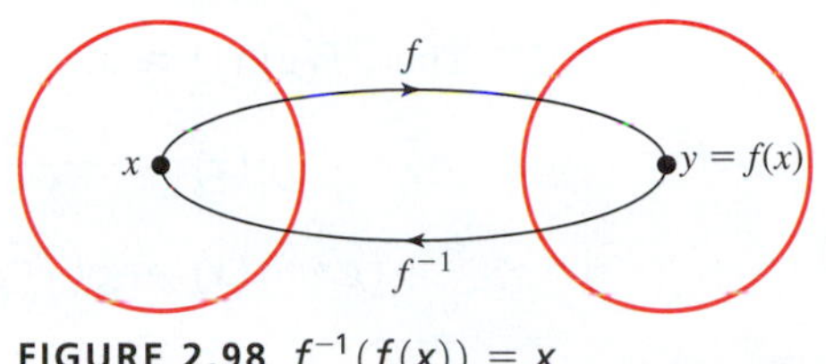

FIGURE 2.98 $f^{-1}(f(x)) = x$

Figure 2.98 suggests the following properties of inverse functions.

INVERSE FUNCTION PROPERTY

Let f denote a one-to-one function. Then

1. $f^{-1}(f(x)) = x$ for every x in the domain of f.
2. $f(f^{-1}(x)) = x$ for every x in the domain of f^{-1}.

Further, if g is any function such that (for all values of x in the domain of the inner function)

$$f(g(x)) = x \text{ and } g(f(x)) = x, \text{ then } g = f^{-1}.$$

One interpretation of the equation $(f^{-1} \circ f)(x) = x$ is that f^{-1} neutralizes the effect f on x. For example, let

$$f(x) = x + 2 \qquad f \text{ adds 2 to any input } x.$$

To undo what f does to x, we should subtract 2 from x. That is, the inverse of f should be

$$g(x) = x - 2 \qquad g \text{ subtracts 2 from } x.$$

Let's verify that $g(x) = x - 2$ is indeed the inverse function of x.

$$\begin{aligned}(f \circ g)(x) &= f(g(x)) && \text{Definition of } f \circ g\\ &= f(x - 2) && \text{Replace } g(x) \text{ with } x - 2.\\ &= (x - 2) + 2 && \text{Replace } x \text{ with } x - 2 \text{ in } f(x) = x + 2.\\ &= x && \text{Simplify.}\end{aligned}$$

We leave it for you to check that $(g \circ f)(x) = x$.

EXAMPLE 3 **Verifying Inverse Functions**

Verify that the following pairs of functions are inverses of each other:

$$f(x) = 2x + 3 \quad \text{and} \quad g(x) = \frac{x - 3}{2}$$

Solution

$$\begin{aligned}(f \circ g)(x) &= f(g(x)) && \text{Definition of } f \circ g\\ &= f\left(\frac{x-3}{2}\right) && \text{Replace } g(x) \text{ with } \frac{x-3}{2}.\\ &= 2\left(\frac{x-3}{2}\right) + 3 && \text{Replace } x \text{ with } \frac{x-3}{2} \text{ in } f(x) = 2x + 3.\\ &= x && \text{Simplify.}\end{aligned}$$

Thus, $f(g(x)) = x$.

$$\begin{aligned}g(x) &= \frac{x-3}{2} && \text{Given function } g\\ (g \circ f)(x) &= g(f(x)) && \text{Definition of } g \text{ of } f\\ &= g(2x + 3) && \text{Replace } f(x) \text{ with } 2x + 3.\\ &= \frac{(2x+3) - 3}{2} && \text{Replace } x \text{ with } 2x + 3 \text{ in } g(x) = \frac{x-3}{2}.\\ &= x && \text{Simplify.}\end{aligned}$$

Thus, $g(f(x)) = x$. Because $f(g(x)) = g(f(x)) = x$, f and g are inverses of each other.

Practice Problem 3 Verify that $f(x) = 3x - 1$ and $g(x) = \dfrac{x+1}{3}$ are inverses of 2each other.

In Example 3, notice how the functions f and g neutralize (undo) the effect of each other. The function f takes an input x, *multiplies* it by 2, and *adds* 3; g neutralizes (or undoes) this effect by *subtracting* 3 and *dividing* by 2. This process is illustrated in Figure 2.99. Notice that g reverses the operations performed by f *and* the order in which they are done.

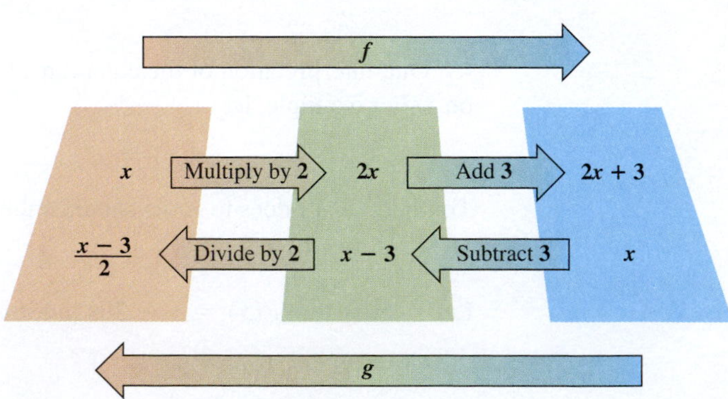

FIGURE 2.99 Function g undoes f

2 Find the inverse function.

Finding the Inverse Function

Let $y = f(x)$ be a one-to-one function; then f has an inverse function. Suppose (a, b) is a point on the graph of f. Then $b = f(a)$. This means that $a = f^{-1}(b)$; so (b, a) is a point on the graph of f^{-1}. The points (a, b) and (b, a) are symmetric about the line $y = x$, as shown in Figure 2.100. (See Exercises 85 and 86.) That is, if the graph paper is folded along the line $y = x$, the points (a, b) and (b, a) will coincide. Therefore, we have the following property.

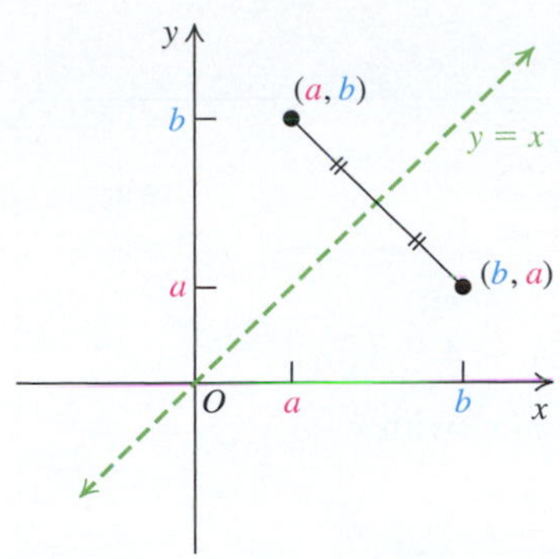

FIGURE 2.100 Relationship between points (a, b) and (b, a)

SYMMETRY PROPERTY OF THE GRAPHS OF *f* AND f^{-1}

The graph of a one-to-one function f and the graph of f^{-1} are symmetric about the line $y = x$.

EXAMPLE 4 Finding the Graph of f^{-1} from the Graph of f

The graph of a function f is shown in Figure 2.101. Sketch the graph of f^{-1}.

Solution

By the horizontal-line test, f is a one-to-one function; so it has an inverse function.

The graph of f consists of two line segments: one joining the points $(-3, -5)$ and $(-1, 2)$ and the other joining the points $(-1, 2)$ and $(4, 3)$.

The graph of f^{-1} is the reflection of the graph of f about the line $y = x$. The reflections of the points $(-3, -5)$, $(-1, 2)$, and $(4, 3)$ about the line $y = x$ are $(-5, -3)$, $(2, -1)$, and $(3, 4)$, respectively.

SIDE NOTE

It is helpful to notice that points on the x-axis, $(a, 0)$, reflect to points on the y-axis, $(0, a)$, and conversely. Also notice that points on the line $y = x$ are unaffected by reflection about the line $y = x$.

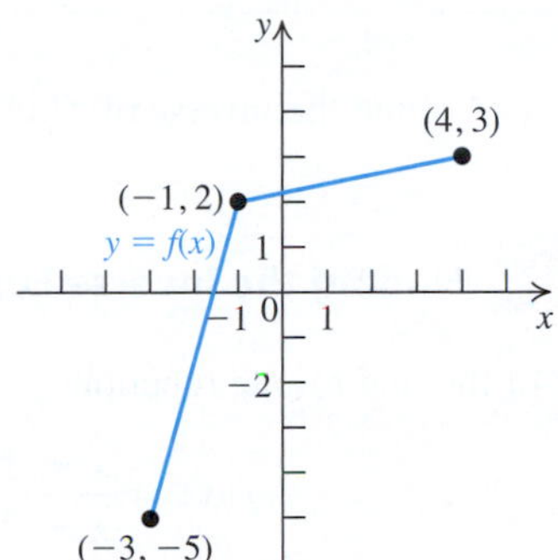

FIGURE 2.101 Graph of f

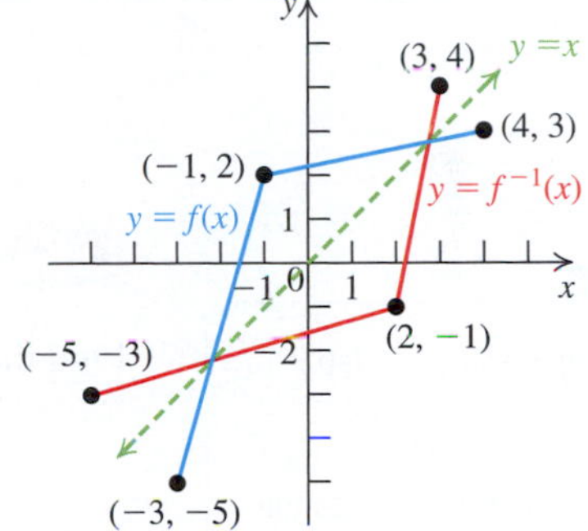

FIGURE 2.102 Graph of f^{-1}

The graph of f^{-1} consists of two line segments: one joining the points $(-5, -3)$ and $(2, -1)$ and the other joining the points $(2, -1)$ and $(3, 4)$. See Figure 2.102.

Practice Problem 4 Use the graph of a function f in Figure 2.103 to sketch the graph of f^{-1}.

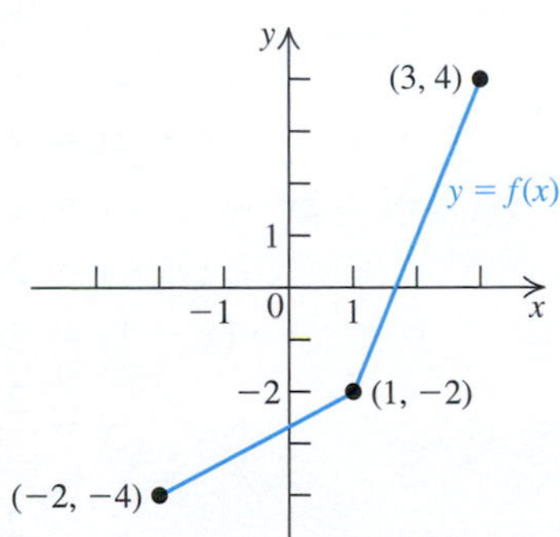

FIGURE 2.103 Graphing f^{-1} from the graph of f

The symmetry between the graphs of f and f^{-1} about the line $y = x$ tells us that we can find an equation for the inverse function $y = f^{-1}(x)$ from the equation of a one-to-one function $y = f(x)$ by interchanging the roles of x and y in the equation $y = f(x)$. This results in the equation $x = f(y)$. Then we solve the equation $x = f(y)$ for y in terms of x to get $y = f^{-1}(x)$.

PROCEDURE IN ACTION

EXAMPLE 5 Finding an Equation for f^{-1}

OBJECTIVE	EXAMPLE
Find the inverse of a one-to-one function f.	*Find the inverse of* $f(x) = 3x - 4$.
Step 1 Replace $f(x)$ with y in the equation defining $f(x)$.	**1.** $y = 3x - 4$ Replace $f(x)$ with y.
Step 2 Interchange x and y.	**2.** $x = 3y - 4$ Interchange x and y.
Step 3 Solve the equation in Step 2 for y.	**3.** $x + 4 = 3y$ Add 4 to both sides. $\frac{x+4}{3} = y$ Divide both sides by 3.
Step 4 Write $f^{-1}(x)$ for y.	**4.** $f^{-1}(x) = \frac{x+4}{3}$ We usually end with $f^{-1}(x)$ on the left.

Practice Problem 5 Find the inverse of $f(x) = -2x + 3$.

SIDE NOTE

Interchanging x and y in Step 2 is done so that we can write f^{-1} in the usual format with x as the independent variable and y as the dependent variable.

EXAMPLE 6 Finding the Inverse Function

Find the inverse of the one-to-one function

$$f(x) = \frac{x+1}{x-2}, \quad x \neq 2.$$

Solution

Step 1 $y = \frac{x+1}{x-2}$ Replace $f(x)$ with y.

Step 2 $x = \frac{y+1}{y-2}$ Interchange x and y.

Step 3 Solve $x = \frac{y+1}{y-2}$ for y. This is the most challenging step.

$x(y-2) = y + 1$ Multiply both sides by $y - 2$.

$xy - 2x = y + 1$ Distributive property

$xy - 2x + 2x - y = y + 1 + 2x - y$ Add $2x - y$ to both sides.

$xy - y = 2x + 1$ Simplify.

$y(x-1) = 2x + 1$ Factor out y.

$y = \frac{2x+1}{x-1}$ Divide both sides by $x - 1$ assuming that $x \neq 1$.

Step 4 $f^{-1}(x) = \frac{2x+1}{x-1}, x \neq 1$ Replace y with $f^{-1}(x)$.

To see if our calculations are accurate, we compute $f(f^{-1}(x))$ and $f^{-1}(f(x))$.

$$f^{-1}(f(x)) = f^{-1}\left(\frac{x+1}{x-2}\right) = \frac{2\left(\frac{x+1}{x-2}\right) + 1}{\frac{x+1}{x-2} - 1} = \frac{2x + 2 + x - 2}{x + 1 - x + 2} = \frac{3x}{3} = x$$

Multiply numerator and denominator by $x - 2$ and simplify.

You should also check that $f(f^{-1}(x)) = x$.

Practice Problem 6 Find the inverse of the one-to-one function $f(x) = \frac{x}{x + 3}, x \neq -3.$

3 Use inverse functions to find the range of a function.

Finding the Range of a One-to-One Function

It is not always easy to determine the range of a function that is defined by an equation. However, for a one-to-one function we can find the range of f by finding the domain of f^{-1}.

EXAMPLE 7 Finding the Domain and Range of a One-to-One Function

Find the domain and the range of the function $f(x) = \frac{x + 1}{x - 2}$ of Example 6.

RECALL

Remember that the domain of a function given by a formula is the largest set of real numbers for which the outputs are real numbers.

Solution

The domain of $f(x) = \frac{x + 1}{x - 2}$ is the set of all real numbers x such that $x \neq 2$. In interval notation, the domain of f is $(-\infty, 2) \cup (2, \infty)$.

From Example 6, $f^{-1}(x) = \frac{2x + 1}{x - 1}, x \neq 1$; therefore,

$$\text{Range of } f = \text{Domain of } f^{-1} = \{x | x \neq 1\}.$$

In interval notation, the range of f is $(-\infty, 1) \cup (1, \infty)$.

TECHNOLOGY CONNECTION 

With a graphing calculator, you can also see that the graphs of $g(x) = x^2 - 1, x \geq 0$ and $g^{-1}(x) = \sqrt{x + 1}$ are symmetric about the line $y = x$.

Enter Y_1, Y_2, and Y_3 as $g(x)$, $g^{-1}(x)$, and $f(x) = x$, respectively.

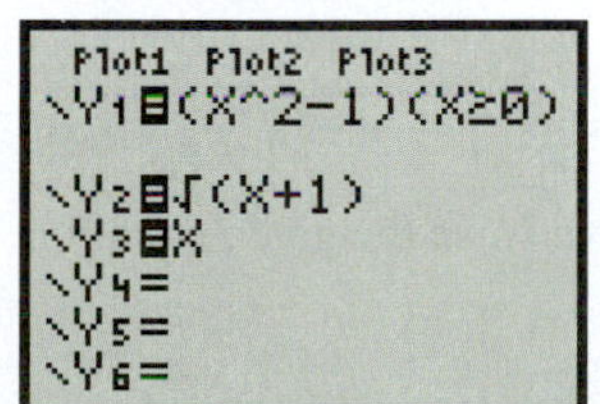

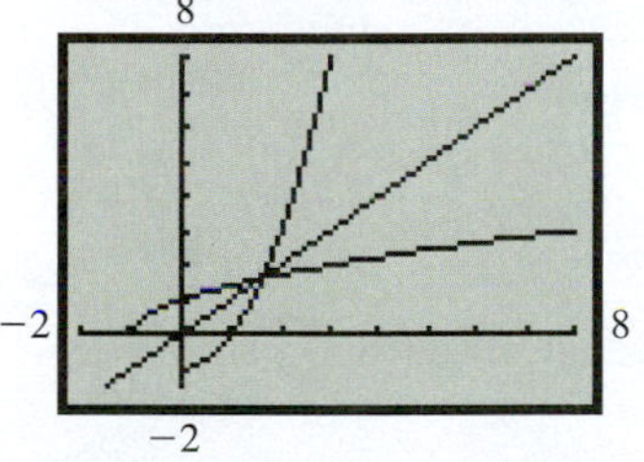

Practice Problem 7 Find the domain and the range of the function $f(x) = \frac{x}{x + 3}$.

If a function f is not one-to-one, then it does not have an inverse function. Sometimes by changing its domain, we can produce an interesting function that does have an inverse. (This technique is frequently used in trigonometry.) We saw in Example 1(b) that $g(x) = x^2 - 1$ is not a one-to-one function; so g does not have an inverse function. However, the horizontal-line test shows that the function

$$G(x) = x^2 - 1, \quad x \geq 0$$

with domain $[0, \infty)$ is one-to-one. See Figure 2.104. Therefore, G has an inverse function G^{-1}.

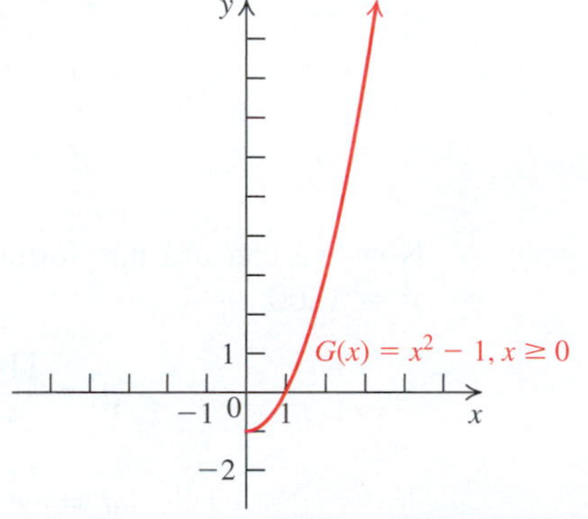

FIGURE 2.104 The function G has an inverse

EXAMPLE 8 Finding an Inverse Function

Find the inverse of $G(x) = x^2 - 1, x \geq 0$.

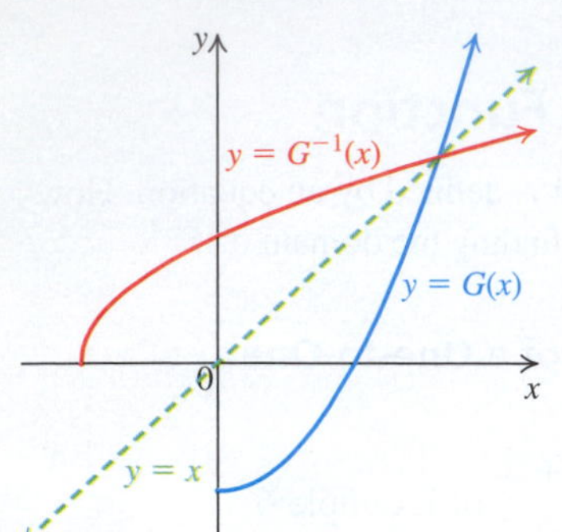

FIGURE 2.105 Graphs of G and G^{-1}

Solution

Step 1	$y = x^2 - 1, \quad x \geq 0$	Replace $G(x)$ with y.
Step 2	$x = y^2 - 1, \quad y \geq 0$	Interchange x and y.
Step 3	$x + 1 = y^2, \quad y \geq 0$	Add 1 to both sides.
	$y = \sqrt{x + 1}$, because $y \geq 0$	Solve for y.
Step 4	$G^{-1}(x) = \sqrt{x + 1}$	Replace y with $G^{-1}(x)$.

The graphs of G and G^{-1} are shown in Figure 2.105.

Practice Problem 8 Find the inverse of $G(x) = x^2 - 1, x \leq 0$.

4 Apply inverse functions in the real world.

Applications

Functions that model real-life situations are frequently expressed as formulas with letters that remind you of the variable they represent. In finding the inverse of a function expressed by a formula, interchanging the letters could be confusing. Accordingly, we omit Step 2, keep the same letters, and just solve the formula for the other variable.

EXAMPLE 9 Water Pressure on Underwater Devices

At the beginning of this section, we gave the formula for finding the pressure p (in pounds per square inch) at a depth d (in feet) below the surface of water. This formula can be written as $p = \frac{5d}{11} + 15$. Suppose the pressure gauge on a diving bell breaks and shows a reading of 1800 psi. How far below the surface was the bell when the gauge failed?

Solution

We want to find the unknown depth in terms of the known pressure. This depth is given by the inverse of the function $p = \frac{5d}{11} + 15$. To find the inverse, we solve the given equation for d.

$$p = \frac{5d}{11} + 15 \qquad \text{Original equation}$$

$$11p = 5d + 165 \qquad \text{Multiply both sides by 11.}$$

$$d = \frac{11p}{5} - 33 \qquad \text{Solve for } d.$$

Now we can use this formula to find the depth when the gauge read 1800 psi. We let $p = 1800$.

$$d = \frac{11p}{5} - 33 \qquad \text{Depth from pressure equation}$$

$$d = \frac{11(1800)}{5} - 33 \qquad \text{Replace } p \text{ with 1800.}$$

$$d = 3927$$

The device was 3927 feet below the surface when the gauge failed.

Practice Problem 9 In Example 9, suppose the pressure gauge showed a reading of 1650 psi. Determine the depth of the bell when the gauge failed.

Answers to Practice Problems

1. Not one-to-one; the horizontal line $y = 1$ intersects the graph at two different points.

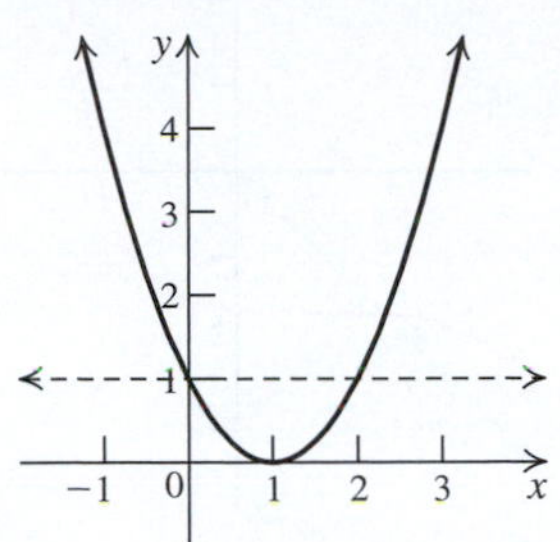

2. a. -3 **b.** 4

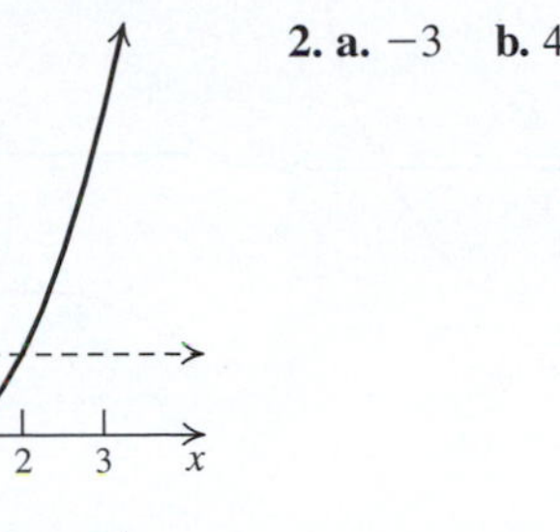

3. $(f \circ g)(x) = 3\left(\frac{x+1}{3}\right) - 1 = x + 1 - 1 = x$ and

$(g \circ f)(x) = \frac{3x - 1 + 1}{3} = \frac{3x}{3} = x$; therefore, f and g are inverses of each other.

4.

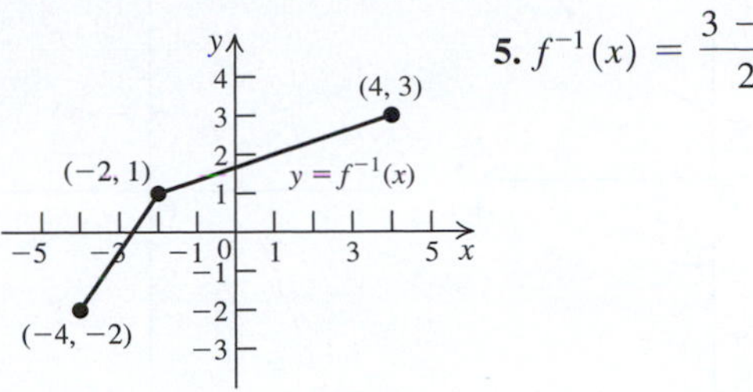

5. $f^{-1}(x) = \frac{3 - x}{2}$

6. $f^{-1}(x) = \frac{3x}{1 - x}, x \neq 1$ **7.** Domain: $(-\infty, -3) \cup (-3, \infty)$; range: $(-\infty, 1) \cup (1, \infty)$ **8.** $G^{-1}(x) = -\sqrt{x + 1}$ **9.** 3597 ft

SECTION 2.9 Exercises

Basic Concepts and Skills

1. If no horizontal line intersects the graph of a function f in more than one point, then f is a(n) ______________ function.
2. A function f is one-to-one when different x-values correspond to ______________.
3. If $f(x) = 3x$, then $f^{-1}(x) =$ ______________.
4. The graphs of a function f and its inverse f^{-1} are symmetric about the line ______________.
5. **True or False.** If a function f has an inverse, then the domain of the inverse function is the range of f.
6. **True or False.** It is possible for a function to be its own inverse, that is, for $f = f^{-1}$.

In Exercises 7–14, the graph of a function is given. Use the horizontal-line test to determine whether the function is one-to-one.

7. **8.**

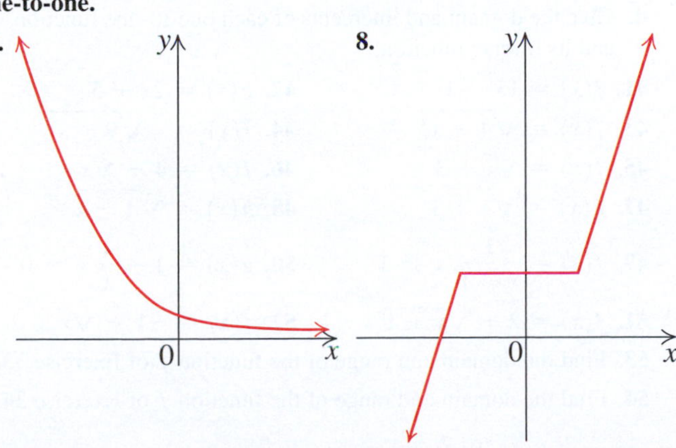

9.

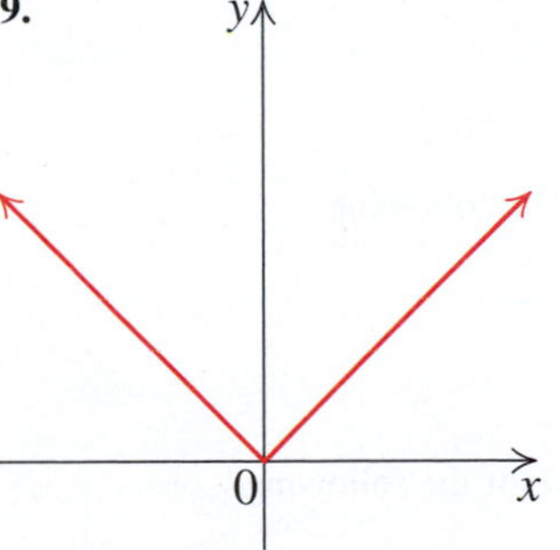

10.

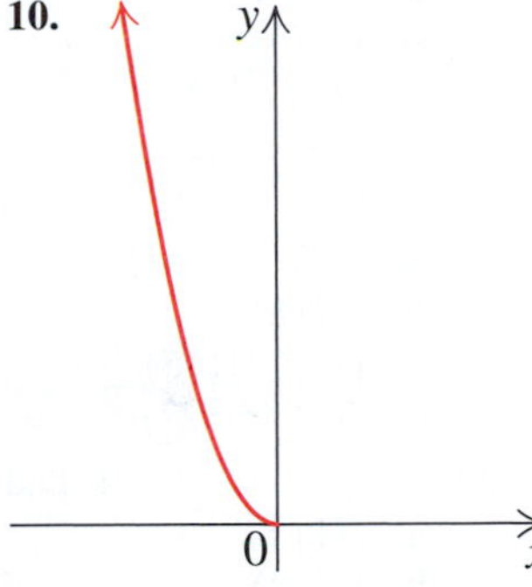

11.

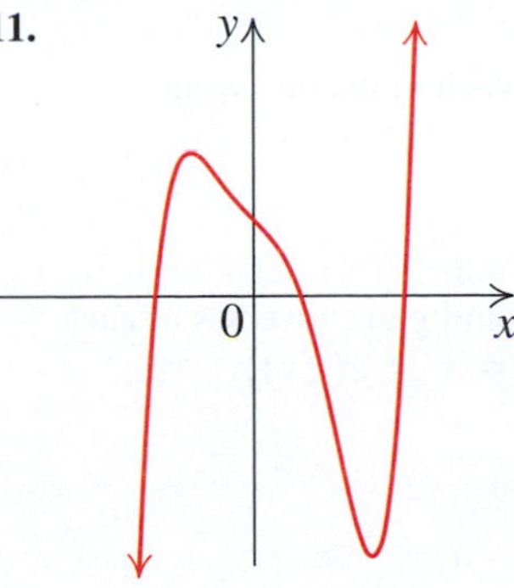

12.

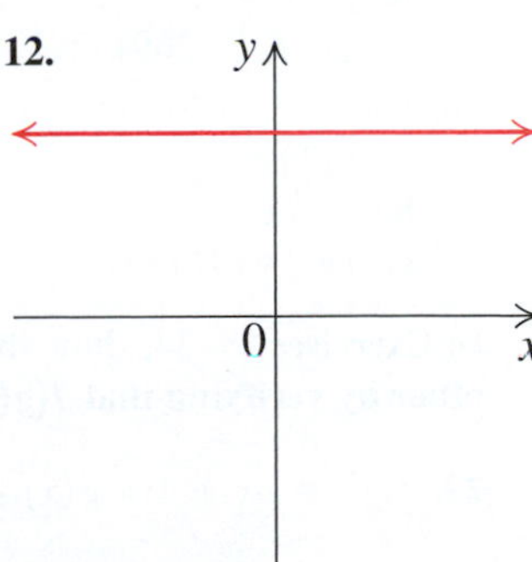

13.

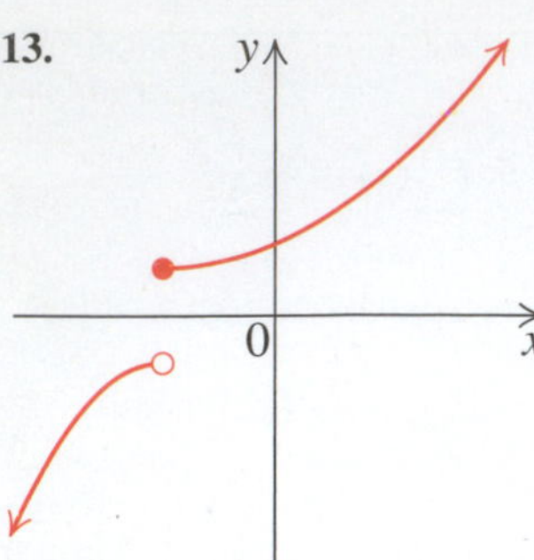

14.

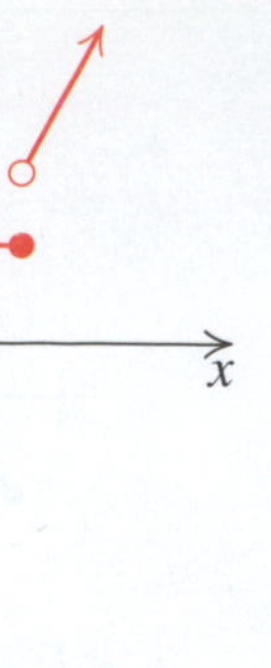

In Exercises 15–28, assume that the function f is one-to-one with domain: $(-\infty, \infty)$.

15. If $f(2) = 7$, find $f^{-1}(7)$.
16. If $f^{-1}(4) = -7$, find $f(-7)$.
17. If $f(-1) = 2$, find $f^{-1}(2)$.
18. If $f^{-1}(-3) = 5$, find $f(5)$.
19. If $f(a) = b$, find $f^{-1}(b)$.
20. If $f^{-1}(c) = d$, find $f(d)$.
21. Find $(f^{-1} \circ f)(337)$.
22. Find $(f \circ f^{-1})(25\pi)$.
23. Find $(f \circ f^{-1})(-1580)$.
24. Find $(f^{-1} \circ f)(9728)$.
25. For $f(x) = 2x - 3$, find each of the following.
    a. $f(3)$
    b. $f^{-1}(3)$
    c. $(f \circ f^{-1})(19)$
    d. $(f \circ f^{-1})(5)$
26. For $f(x) = x^3$, find each of the following.
    a. $f(2)$
    b. $f^{-1}(8)$
    c. $(f \circ f^{-1})(15)$
    d. $(f^{-1} \circ f)(27)$
27. For $f(x) = x^3 + 1$, find each of the following.
    a. $f(1)$
    b. $f^{-1}(2)$
    c. $(f \circ f^{-1})(269)$
28. For $g(x) = \sqrt[3]{2x^3 - 1}$, find each of the following.
    a. $g(1)$
    b. $g^{-1}(1)$
    c. $(g^{-1} \circ g)(135)$

In Exercises 29–34, show that f and g are inverses of each other by verifying that $f(g(x)) = x = g(f(x))$.

29. $f(x) = 3x + 1;\quad g(x) = \dfrac{x - 1}{3}$
30. $f(x) = 2 - 3x;\quad g(x) = \dfrac{2 - x}{3}$
31. $f(x) = x^3;\quad g(x) = \sqrt[3]{x}$
32. $f(x) = \dfrac{1}{x};\quad g(x) = \dfrac{1}{x}$
33. $f(x) = \dfrac{x - 1}{x + 2};\quad g(x) = \dfrac{1 + 2x}{1 - x}$
34. $f(x) = \dfrac{3x + 2}{x - 1};\quad g(x) = \dfrac{x + 2}{x - 3}$

In Exercises 35–40, the graph of a function f is given. Sketch the graph of f^{-1}.

35.

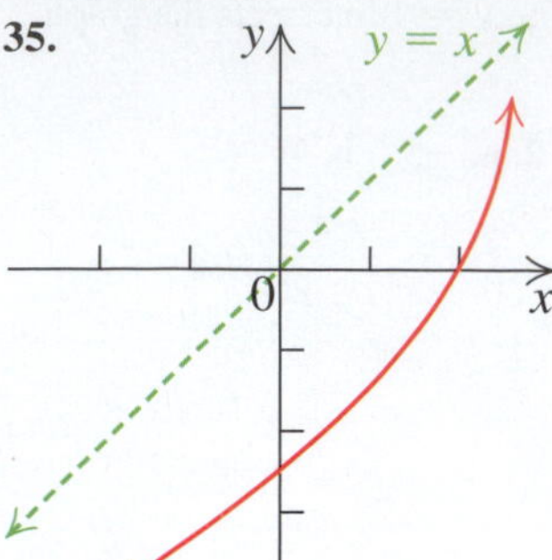

36.

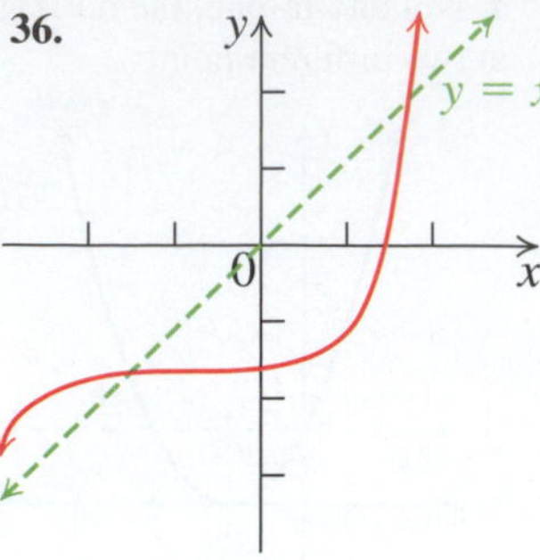

37.

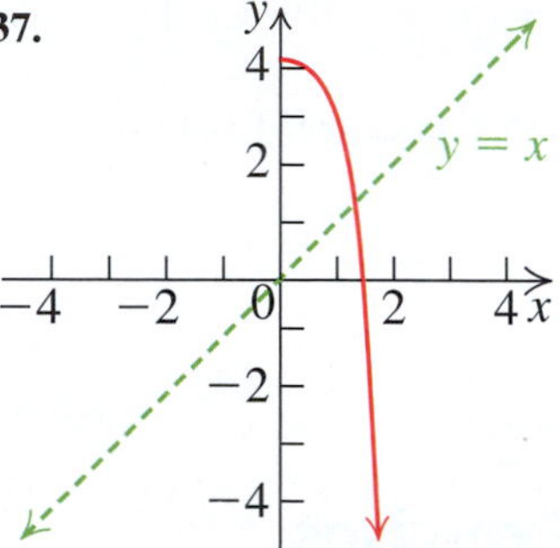

38.

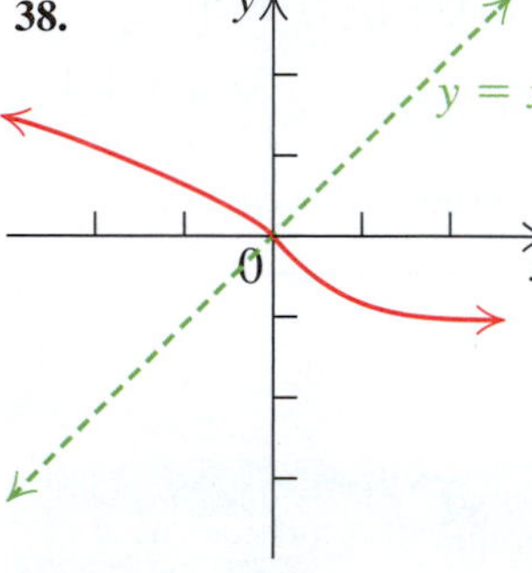

39.

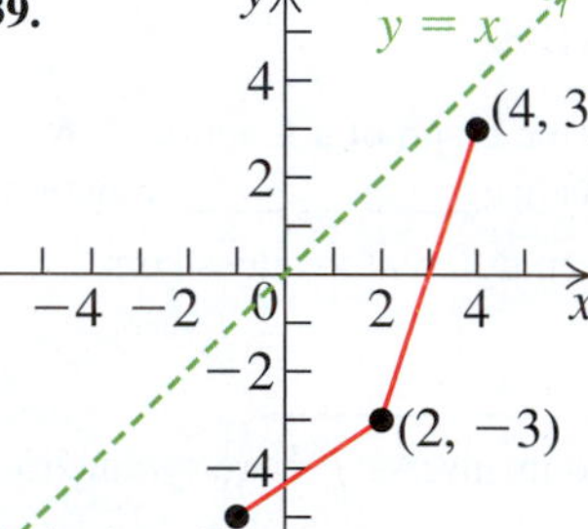

40.

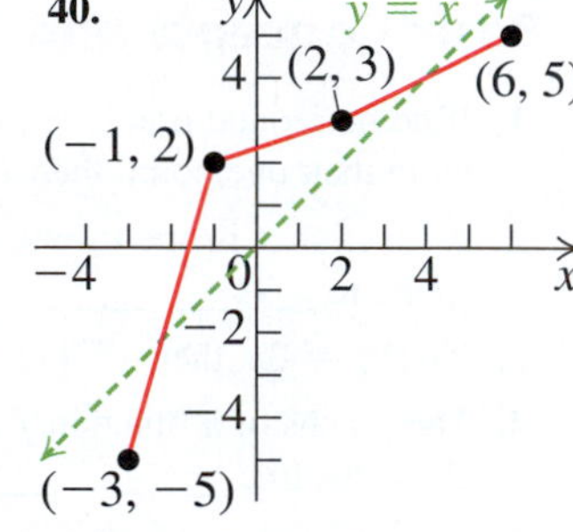

In Exercises 41–52,

a. Determine algebraically whether the given function is a one-to-one function.
b. If the function is one-to-one, find its inverse.
c. Sketch the graph of the function and its inverse on the same coordinate axes.
d. Give the domain and intercepts of each one-to-one function and its inverse function.

41. $f(x) = 15 - 3x$
42. $g(x) = 2x + 5$
43. $f(x) = \sqrt{4 - x^2}$
44. $f(x) = -\sqrt{9 - x^2}$
45. $f(x) = \sqrt{x} + 3$
46. $f(x) = 4 - \sqrt{x}$
47. $g(x) = \sqrt[3]{x + 1}$
48. $h(x) = \sqrt[3]{1 - x}$
49. $f(x) = \dfrac{1}{x - 1}, x \neq 1$
50. $g(x) = 1 - \dfrac{1}{x}, x \neq 0$
51. $f(x) = 2 + \sqrt{x + 1}$
52. $f(x) = -1 + \sqrt{x + 2}$
53. Find the domain and range of the function f of Exercise 33.
54. Find the domain and range of the function f of Exercise 34.

In Exercises 55–58, assume that the given function is one-to-one. Find the inverse of the function. Also find the domain and the range of the given function.

55. $f(x) = \frac{x+1}{x-2}, x \neq 2$ **56.** $g(x) = \frac{x+2}{x+1}, x \neq -1$

57. $f(x) = \frac{1-2x}{1+x}, x \neq -1$ **58.** $h(x) = \frac{x-1}{x-3}, x \neq 3$

In Exercises 59–66, find the inverse of each function and sketch the graph of the function and its inverse on the same coordinate axes.

59. $f(x) = -x^2, x \geq 0$ **60.** $g(x) = -x^2, x \leq 0$

61. $f(x) = |x|, x \geq 0$ **62.** $g(x) = |x|, x \leq 0$

63. $f(x) = x^2 + 1, x \leq 0$ **64.** $g(x) = x^2 + 5, x \geq 0$

65. $f(x) = -x^2 + 2, x \leq 0$ **66.** $g(x) = -x^2 - 1, x \geq 0$

Applying the Concepts

67. Temperature scales. Scientists use the **Kelvin** temperature scale, in which the lowest possible temperature (called absolute zero) is 0 K. (K denotes degrees Kelvin.)

The function $K(C) = C + 273$ gives the relationship between the Kelvin temperature (K) and Celsius temperature (C).

a. Find the inverse function of $K(C) = C + 273$. What does it represent?

b. Use the inverse function from (a) to find the Celsius temperature corresponding to 300 K.

c. A comfortable room temperature is 22°C. What is the corresponding Kelvin temperature?

68. Temperature scales. The boiling point of water is 373 K, or 212°F; the freezing point of water is 273 K, or 32°F. The relationship between Kelvin and Fahrenheit temperatures is linear.

a. Write a linear function expressing $K(F)$ in terms of F.

b. Find the inverse of the function in part (a). What does it mean?

c. A normal human body temperature is 98.6°F. What is the corresponding Kelvin temperature?

69. Composition of functions. Use Exercises 67 and 68 and the composition of functions to

a. Write a function that expresses F in terms of C.

b. Write a function that expresses C in terms of F.

70. Celsius and Fahrenheit temperatures. Show that the functions in (a) and (b) of Exercise 69 are inverse functions.

71. Currency exchange. Alisha went to Europe last summer. She discovered that when she exchanged her U.S. dollars for euros, she received 25% fewer euros than the number of dollars she exchanged. (She got 75 euros for every 100 U.S. dollars.) When she returned to the United States, she got 25% more dollars than the number of euros she exchanged.

a. Write each conversion function.

b. Show that in part (a), the two functions are not inverse functions.

c. Does Alisha gain or lose money after converting both ways?

72. Hourly wages. Anwar is a short-order cook in a diner. He is paid \$4 per hour plus 5% of all food sales per hour. His average hourly wage w in terms of the food sales of x dollars is $w = 4 + 0.05x$.

a. Write the inverse function. What does it mean?

b. Use the inverse function to estimate the hourly sales at the diner if Anwar averages \$12 per hour.

73. Hourly wages. In Exercise 72, suppose in addition that Anwar is guaranteed a minimum wage of \$7 per hour.

a. Write a function expressing his hourly wage w in terms of food sales per hour. [*Hint:* Use a piecewise function.]

b. Does the function in part (a) have an inverse? Explain.

c. If the answer in part (b) is yes, find the inverse function. If the answer is no, restrict the domain so that the new function has an inverse.

74. Simple pendulum. If a pendulum is released at a certain point, the **period** is the time the pendulum takes to swing along its path and return to the point from which it was released. The period T (in seconds) of a simple pendulum is a function of its length l (in feet) and is given by $T = 1.11\sqrt{l}$.

a. Find the inverse function. What does it mean?

b. Use the inverse function to calculate the length of the pendulum assuming that its period is two seconds.

c. The convention center in Portland, Oregon, has the longest pendulum in the United States. The pendulum's length is 70 feet. Find the period.

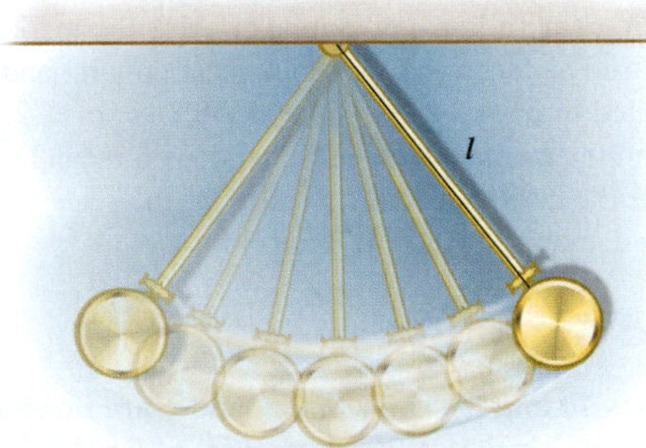

75. Water supply. Suppose x is the height of the water above the opening at the base of a water tank. The velocity V of water that flows from the opening at the base is a function of x and is given by $V(x) = 8\sqrt{x}$.

a. Find the inverse function. What does it mean?

b. Use the inverse function to calculate the height of the water in the tank when the flow is **(i)** 30 feet per second and **(ii)** 20 feet per second.

76. Physics. A projectile is fired from the origin over horizontal ground. Its altitude y (in feet) is a function of its horizontal distance x (in feet) and is given by

$$y = 64x - 2x^2.$$

a. Find the inverse function where the function is increasing.

b. Use the inverse function to compute the horizontal distance when the altitude of the projectile is **(i)** 32 feet, **(ii)** 256 feet, and **(iii)** 512 feet.

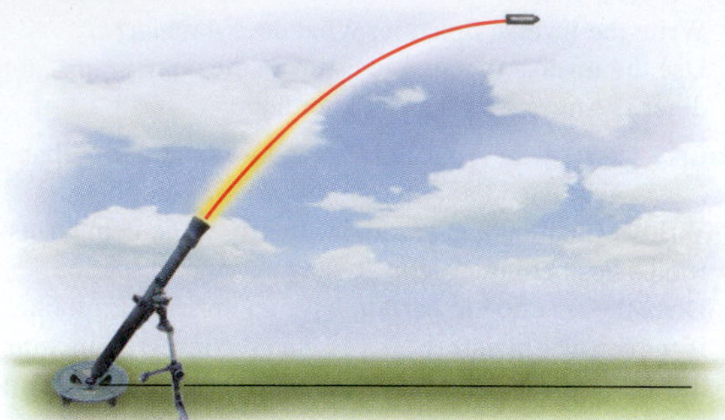

77. Loan repayment. Chris purchased a car at 0% interest for five years. After making a down payment, she agreed to pay the remaining $36,000 in monthly payments of $600 per month for 60 months.

a. What does the function $f(x) = 36{,}000 - 600x$ represent?

b. Find the inverse of the function in part (a). What does the inverse function represent?

c. Use the inverse function to find the number of months remaining to make payments if the balance due is $22,000.

78. Demand function. A marketing survey finds that the number x (in millions) of computer chips the market will purchase is a function of its price p (in dollars) and is estimated by $x = 8p^2 - 32p + 1200, 0 < p \le 2$.

a. Find the inverse function. What does it mean?

b. Use the inverse function to estimate the price of a chip if the demand is 1180.5 million chips.

Beyond the Basics

In Exercises 79 and 80, show that f and g are inverses of each other by verifying that $f(g(x)) = x = g(f(x))$.

79.

x	1	3	4
$f(x)$	3	5	2

x	3	5	2
$g(x)$	1	3	4

80.

x	1	2	3	4
$f(x)$	−2	0	−3	1

x	−2	0	−3	1
$g(x)$	1	2	3	4

81. Let $f(x) = \sqrt{4 - x^2}$.

a. Sketch the graph of $y = f(x)$.

b. Is f one-to-one?

c. Find the domain and the range of f.

82. Let $f(x) = \dfrac{[\![x]\!] + 2}{[\![x]\!] - 2}$, where $[\![x]\!]$ is the greatest integer function.

a. Find the domain of f.

b. Is f one-to-one?

83. Let $f(x) = \dfrac{x}{x + 1} = 1 - \dfrac{1}{x + 1}$.

a. Show that f is one-to-one.

b. Find the inverse of f.

c. Find the domain and the range of f.

84. Let $g(x) = \sqrt{1 - x^2}, 0 \le x \le 1$.

a. Show that g is one-to-one.

b. Find the inverse of g.

c. Find the domain and the range of g.

85. Let $P(3, 7)$ and $Q(7, 3)$ be two points in the plane.

a. Show that the midpoint M of the line segment PQ lies on the line $y = x$.

b. Show that the line $y = x$ is the perpendicular bisector of the line segment $\overline{PQ}$; that is, show that the line $y = x$ contains the midpoint found in part (a) and makes a right angle with $\overline{PQ}$. You will have shown that P and Q are symmetric about the line $y = x$.

86. Use the procedure outlined in Exercise 85 to show that the points (a, b) and (b, a) are symmetric about the line $y = x$. (See figure 2.100 on page 275.)

87. The graph of $f(x) = x^3$ is given in Section 2.5.

a. Sketch the graph of $g(x) = (x - 1)^3 + 2$ by using transformations of the graph of f.

b. Find $g^{-1}(x)$.

c. Sketch the graph of $g^{-1}(x)$ by reflecting the graph of g in part (a) about the line $y = x$.

88. Inverse of composition of functions. We show that if f and g have inverses, then

$$(f \circ g)^{-1} = g^{-1} \circ f^{-1}. \text{ (Note the order.)}$$

Let $f(x) = 2x - 1$ and $g(x) = 3x + 4$.

a. Find the following:

(i) $f^{-1}(x)$ **(ii)** $g^{-1}(x)$

(iii) $(f \circ g)(x)$ **(iv)** $(g \circ f)(x)$

(v) $(f \circ g)^{-1}(x)$ **(vi)** $(g \circ f)^{-1}(x)$

(vii) $(f^{-1} \circ g^{-1})(x)$ **(viii)** $(g^{-1} \circ f^{-1})(x)$

b. From part (a), conclude that

(i) $(f \circ g)^{-1} = g^{-1} \circ f^{-1}$. **(ii)** $(g \circ f)^{-1} = f^{-1} \circ g^{-1}$.

89. Repeat Exercise 88 for $f(x) = 2x + 3$ and $g(x) = x^3 - 1$.

90. Show that $f(x) = x^{2/3}, x \le 0$ is a one-to-one function. Find $f^{-1}(x)$ and verify that $f^{-1}(f(x)) = x$ for every x in the domain of f.

Critical Thinking/Discussion/Writing

91. Does every odd function have an inverse? Explain.

92. Is there an even function that has an inverse? Explain.

93. Does every increasing or decreasing function have an inverse? Explain.

94. A relation R is a set of ordered pairs (x, y). The inverse of R is the set of ordered pairs (y, x).

a. Give an example of a function whose inverse relation is not a function.

b. Give an example of a relation R whose inverse is a function.

Maintaining Skills

In Exercises 95–98, solve each quadratic equation.

95. $x^2 - 7x + 12 = 0$

96. $6x^2 + x - 2 = 0$

97. $12 - 3(x - 1)^2 = 0$

98. $x^2 - 4x + 1 = 0$

In Exercises 99 and 100, use transformations on the graph of $y = x^2$ to sketch graph.

99. $y = (x + 2)^2 - 3$

100. $y = -(x - 1)^2 + 2$

SUMMARY Definitions, Concepts, and Formulas

2.1 The Coordinate Plane

i. Ordered pair. A pair of numbers in which the order is specified is called an ordered pair of numbers.

ii. Distance formula. The distance between two points $P(x_1, y_1)$ and $Q(x_2, y_2)$, denoted by $d(P, Q)$, is given by

$$d(P, Q) = \sqrt{(x_2 - x_1)^2 + (y_2 - y_1)^2}.$$

iii. Midpoint formula. The coordinates of the midpoint $M(x, y)$ on the line segment joining $P(x_1, y_1)$ and $Q(x_2, y_2)$ are given by

$$(x, y) = \left(\frac{x_1 + x_2}{2}, \frac{y_1 + y_2}{2}\right).$$

2.2 Graphs of Equations

i. The graph of an equation in two variables, such as, x and y, is the set of all ordered pairs (a, b) in the coordinate plane that satisfies the equation. A graph of an equation, then, is a picture of its solution set.

ii. Sketching the graph of an equation by plotting points

Step 1 Make a representative table of solutions of the equation.

Step 2 Plot the solutions as ordered pairs in the coordinate plane.

Step 3 Connect the solutions in Step 2 with a smooth curve.

iii. Intercepts

1. The x-intercept is the x-coordinate of a point on the graph where the graph touches or crosses the x-axis. To find the x-intercepts, set $y = 0$ in the equation and solve for x.
2. The y-intercept is the y-coordinate of a point on the graph where the graph touches or crosses the y-axis. To find the y-intercepts, set $x = 0$ in the equation and solve for y.

iv. Symmetry A graph is symmetric with respect to

a. the y-axis if for every point (x, y) on the graph, the point $(-x, y)$ is also on the graph. That is, replacing x with $-x$ in the equation produces an equivalent equation.

b. the x-axis if for every point (x, y) on the graph, the point $(x, -y)$ is also on the graph. That is, replacing y with $-y$ in the equation produces an equivalent equation.

c. the *origin* if for every point (x, y) on the graph, the point $(-x, -y)$ is also on the graph. That is, replacing x with $-x$ and y with $-y$ in the equation produces an equivalent equation.

v. Circle A circle is the set of points in a Cartesian coordinate plane that are at a fixed distance r from a specified point (h, k). The fixed distance r is called the radius of the circle, and the fixed point (h, k) is called the center of the circle. The equation

$$(x - h)^2 + (y - k)^2 = r^2$$

is called the *standard form* of an equation of a circle.

2.3 Lines

i. The slope m of a nonvertical line through the points $P(x_1, y_1)$ and $Q(x_2, y_2)$ is

$$m = \frac{\text{rise}}{\text{run}} = \frac{(y\text{-coordinate of } Q) - (y\text{-coordinate of } P)}{(x\text{-coordinate of } Q) - (x\text{-coordinate of } P)}$$

$$= \frac{y_2 - y_1}{x_2 - x_1}.$$

The slope of a vertical line is undefined. The slope of a horizontal line is zero.

ii. Equation of a line

$y - y_1 = m(x - x_1)$	Point–slope form
$y = mx + b$	Slope–intercept form
$y = k$	Horizontal line
$x = h$	Vertical line
$ax + by + c = 0$	General form

iii. Parallel and perpendicular lines

Two distinct lines with respective slopes m_1 and m_2 are

1. Parallel if $m_1 = m_2$.
2. Perpendicular if $m_1 m_2 = -1$.

iv. Linear regression A method for modeling data using linear functions.

2.4 Functions

i. Any set of ordered pairs is a **relation**. The set of all first components is the **domain**, and the set of all second components is the **range**. The graph of all of the ordered pairs is the graph of the relation.

ii. A **function** is a relation in which each element of the domain corresponds to exactly one element in the range.

iii. If a function f is defined by an equation, then its domain is the largest set of real numbers for which $f(x)$ is a real number.

iv. **Vertical-line test.** If each vertical line intersects a graph at no more than one point, the graph is the graph of a function.

2.5 Properties of Functions

i. Let x_1 and x_2 be any two numbers in the interval (a, b).
- **a.** f is **increasing** on (a, b) if $x_1 < x_2$ implies $f(x_1) < f(x_2)$.
- **b.** f is **decreasing** on (a, b) if $x_1 < x_2$ implies $f(x_1) > f(x_2)$.
- **c.** f is **constant** on (a, b) if $x_1 < x_2$ implies $f(x_1) = f(x_2)$.

ii. Assume (x_1, x_2) is an interval in the domain of f containing a number a.
- **a.** $f(a)$ is **relative minimum of** f if $f(a) \leq f(x)$ for every x in (x_1, x_2).
- **b.** $f(a)$ is **relative maximum of** f if $f(a) \geq f(x)$ for every x in (x_1, x_2).
- **c.** A point (a, b) on the graph of f is a **turning point** if the graph of f changes direction at (a, b) from increasing to decreasing or from decreasing to increasing.

iii. Suppose that for each x in the domain of f, $-x$ is also in the domain of f.
- **a.** f is called an **even function** if $f(-x) = f(x)$. The graph of an even function is symmetric about the y-axis.
- **b.** f is called an **odd function** if $f(-x) = -f(x)$. The graph of an odd function is symmetric about the origin.

iv. The **average rate of change** of $f(x)$ as x changes from a to b is defined by $\dfrac{f(b) - f(a)}{b - a}$, $b \neq a$.

v. The expression $\dfrac{f(x) - f(a)}{x - a}$, $x \neq a$ is called a difference quotient at a.

The expression $\dfrac{f(x + h) - f(x)}{h}$, $h \neq 0$ is also called a difference quotient.

2.6 A Library of Functions

i. **Basic Functions**

Identity function	$f(x) = x$
Constant function	$f(x) = c$
Squaring function	$f(x) = x^2$
Cubing function	$f(x) = x^3$
Absolute value function	$f(x) = \lvert x \rvert$
Square root function	$f(x) = \sqrt{x}$
Cube root function	$f(x) = \sqrt[3]{x}$
Reciprocal function	$f(x) = \dfrac{1}{x}$
Reciprocal square function	$f(x) = \dfrac{1}{x^2}$
Greatest integer function	$f(x) = [\![x]\!]$

ii. For piecewise functions, different rules are used in different parts of the domain.

2.7 Transformations of Functions

i. **Vertical and horizontal shifts.** Let f be a function and c and d be positive numbers.

To Graph	Shift the graph of f
$f(x) + d$	up d units
$f(x) - d$	down d units
$f(x - c)$	right c units
$f(x + c)$	left c units

ii. **Reflections:**

To Graph	Reflect the graph of f in the
$-f(x)$	x-axis
$f(-x)$	y-axis

iii. **Stretching and compressing.** The graph of $g(x) = af(x)$ for $a > 0$ has the same shape as the graph of $f(x)$ and is stretched vertically away from the x-axis if $a > 1$ and compressed vertically towards the x-axis if $0 < a < 1$. If a is negative, the graph of $\lvert a \rvert f(x)$ is reflected about the x-axis.

The graph of $g(x) = f(bx)$ for $b > 0$ is obtained from the graph of f by stretching away from the y-axis if $0 < b < 1$ and compressing horizontally toward the y-axis if $b > 1$. If b is negative, the graph of $f(\lvert b \rvert x)$ is reflected about the y-axis.

2.8 Combining Functions; Composite Functions

i. Given two functions f and g, for all values of x for which both $f(x)$ and $g(x)$ are defined, the functions can be combined to form **sum**, **difference**, **product**, and **quotient** functions.

ii. The **composition** of f and g is defined by $(f \circ g)(x) = f(g(x))$.

The input of f is the output of g. The domain of $f \circ g$ is the set of all x in the domain of g such that $g(x)$ is in the domain of f.

In general $f \circ g \neq g \circ f$.

2.9 Inverse Functions

i. A function f is **one-to-one** if for any x_1 and x_2 in the domain of f, $f(x_1) = f(x_2)$ implies $x_1 = x_2$.

ii. **Horizontal-line test.** If each horizontal line intersects the graph of a function f in at most one point, then f is a one-to-one function.

iii. **Inverse function.** Let f be a one-to-one function. Then g is the inverse of f, and we write $g = f^{-1}$ if $(f \circ g)(x) = x$ for every x in the domain of g and $(g \circ f)(x) = x$ for every x in the domain of f. The graph of f^{-1} is the reflection of the graph of f in the line $y = x$.

REVIEW EXERCISES

Basic Concepts and Skills

In Exercises 1–8, state whether the given statement is true or false.

1. $(0, 5)$ is the midpoint of the line segment joining $(-3, 1)$ and $(3, 11)$.
2. The equation $(x + 2)^2 + (y + 3)^2 = 5$ is the equation of a circle with center $(2, 3)$ and radius 5.
3. If a graph is symmetric with respect to the x-axis and the y-axis, then it must be symmetric with respect to the origin.
4. If a graph is symmetric with respect to the origin, then it must be symmetric with respect to the x-axis and the y-axis.
5. In the graph of the equation of the line $3y = 4x + 9$, the slope is 4 and the y-intercept is 9.
6. If the slope of a line is 2, then the slope of any line perpendicular to it is $\frac{1}{2}$.
7. The slope of a vertical line is undefined.
8. The equation $(x - 2)^2 + (y + 3)^2 = -25$ is the equation of a circle with center $(2, -3)$ and radius 5.

In Exercises 9–14, find

a. The distance between the point P and Q.
b. The coordinates of the midpoint of the line segment PQ.
c. The slope of the line containing the points P and Q.

9. $P(3, 5), Q(-1, 3)$
10. $P(-3, 5), Q(3, -1)$
11. $P(4, -3), Q(9, -8)$
12. $P(2, 3), Q(-7, -8)$
13. $P(2, -7), Q(5, -2)$
14. $P(-5, 4), Q(10, -3)$
15. Show that the points $A(0, 5), B(-2, -3)$, and $C(3, 0)$ are the vertices of a right triangle.
16. Show that the points $A(1, 2), B(4, 8), C(7, -1)$, and $D(10, 5)$ are the vertices of a rhombus.
17. Which of the points $(-6, 3)$ and $(4, 5)$ is closer to the origin?
18. Which of the points $(-6, 4)$ and $(5, 10)$ is closer to the point $(2, 3)$?
19. Find a point on the x-axis that is equidistant from the points $(-5, 3)$ and $(4, 7)$.
20. Find a point on the y-axis that is equidistant from the points $(-3, -2)$ and $(2, -1)$.

In Exercises 21–24, specify whether the given graph has axis or origin symmetry (or neither).

21.

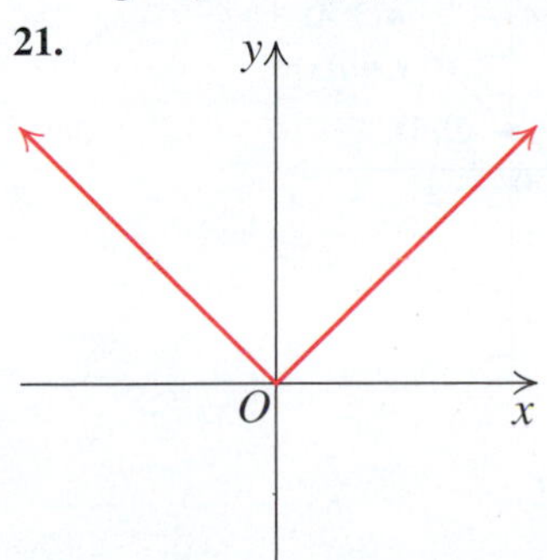

22.

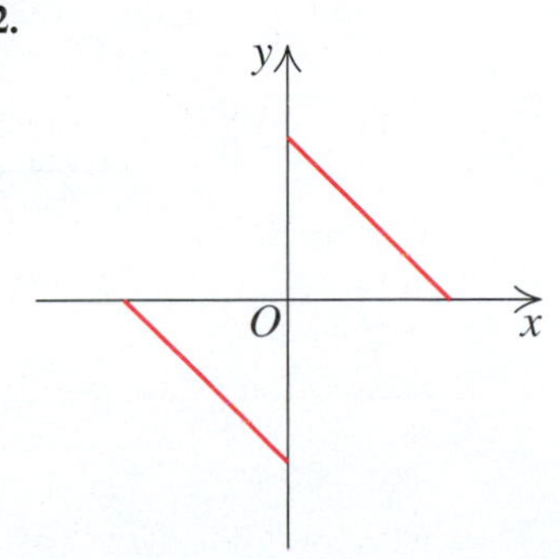

23. **24.**

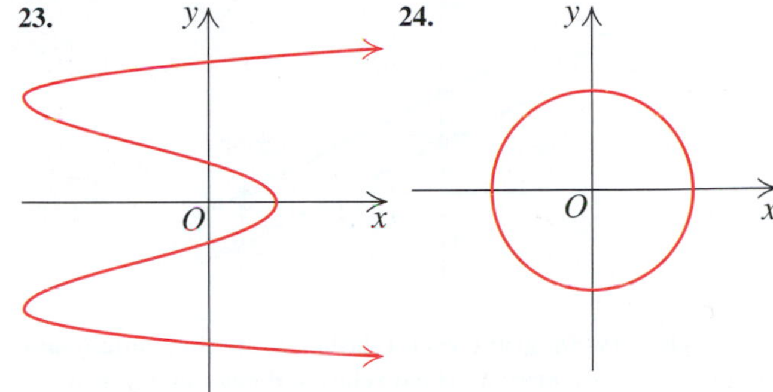

In Exercises 25–34, sketch the graph of the given equation. List all intercepts and describe any symmetry of the graph.

25. $x + 2y = 4$
26. $3x - 4y = 12$
27. $y = -2x^2$
28. $x = -y^2$
29. $y = x^3$
30. $x = -y^3$
31. $y = x^2 + 2$
32. $y = 1 - x^2$
33. $x^2 + y^2 = 16$
34. $y = x^4 - 4$

In Exercises 35–37, find the standard form of the equation of the circle that satisfies the given conditions.

35. Center $(2, -3)$, radius 5
36. Diameter with endpoints $(5, 2)$ and $(-5, 4)$
37. Center $(-2, -5)$, touching the y-axis

In Exercises 38–42, describe and sketch the graph of the given equation.

38. $2x - 5y = 10$
39. $\frac{x}{2} - \frac{y}{5} = 1$
40. $(x + 1)^2 + (y - 3)^2 = 16$
41. $x^2 + y^2 - 2x + 4y - 4 = 0$
42. $3x^2 + 3y^2 - 6x - 6 = 0$

In Exercises 43–47, find the slope–intercept form of the equation of the line that satisfies the given conditions.

43. Passing through $(1, 2)$ with slope -2
44. x-intercept 2, y-intercept 5
45. Passing through $(1, 3)$ and $(-1, 7)$
46. Passing through $(1, 3)$, perpendicular to the x-axis
47. Determine whether the lines in each pair are parallel, perpendicular, or neither.
    a. $y = 3x - 2$ and $y = 3x + 2$
    b. $3x - 5y + 7 = 0$ and $5x - 3y + 2 = 0$
    c. $ax + by + c = 0$ and $bx - ay + d = 0$
    d. $y + 2 = \frac{1}{3}(x - 3)$ and $y - 5 = 3(x - 3)$

48. Write the point–slope form of the equation of the line shown in the figure that is
   a. Parallel to the line in the figure with x-intercept 4.
   b. Perpendicular to the line in the figure with y-intercept 1.

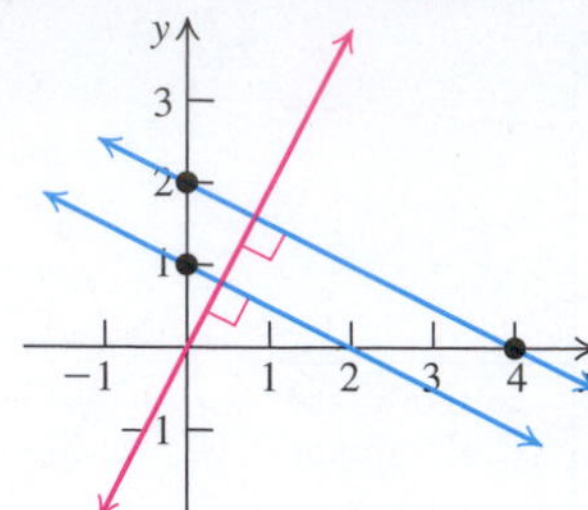

In Exercises 49–58, graph each relation. State the domain and range of each. Determine which relation determines y as a function of x.

49. $\{(0, 1), (-1, 2), (1, 0), (2, -1)\}$
50. $\{(0, 1), (0, -1), (3, 2), (3, -2)\}$
51. $x - y = 1$
52. $y = \sqrt{x - 2}$
53. $x^2 + y^2 = 0.04$
54. $x = -\sqrt{y}$
55. $x = 1$
56. $y = 2$
57. $y = |x + 1|$
58. $x = y^2 + 1$

In Exercises 59–76, let $f(x) = 3x + 1$ and $g(x) = x^2 - 2$. Find each of the following.

59. $f(-2)$
60. $g(-2)$
61. x if $f(x) = 4$
62. x if $g(x) = 2$
63. $(f + g)(1)$
64. $(f - g)(-1)$
65. $(f \cdot g)(-2)$
66. $(g \cdot f)(0)$
67. $(f \circ g)(3)$
68. $(g \circ f)(-2)$
69. $(f \circ g)(x)$
70. $(g \circ f)(x)$
71. $(f \circ f)(x)$
72. $(g \circ g)(x)$
73. $f(a + h)$
74. $g(a - h)$
75. $\dfrac{f(x + h) - f(x)}{h}$
76. $\dfrac{g(x + h) - g(x)}{h}$

In Exercises 77–82, graph each function and state its domain and range. Determine the intervals over which the function is increasing, decreasing, or constant.

77. $f(x) = -3$
78. $f(x) = x^2 - 2$
79. $g(x) = \sqrt{3x - 2}$
80. $h(x) = \sqrt{36 - x^2}$
81. $f(x) = \begin{cases} x + 1 & \text{if } x \geq 0 \\ -x + 1 & \text{if } x < 0 \end{cases}$
82. $g(x) = \begin{cases} x & \text{if } x \geq 0 \\ x^2 & \text{if } x < 0 \end{cases}$

In Exercises 83–86, use transformations to graph each pair of functions on the same coordinate axes.

83. $f(x) = \sqrt{x}, g(x) = \sqrt{x + 1}$
84. $f(x) = |x|, g(x) = 2|x - 1| + 3$
85. $f(x) = \dfrac{1}{x}, g(x) = -\dfrac{1}{x - 2}$
86. $f(x) = x^2, g(x) = (x + 1)^2 - 2$

In Exercises 87–92, state whether each function is odd, even, or neither. Discuss the symmetry of each graph.

87. $f(x) = x^2 - x^4$
88. $f(x) = x^3 + x$
89. $f(x) = |x| + 3$
90. $f(x) = 3x + 5$
91. $f(x) = \sqrt{x}$
92. $f(x) = \dfrac{2}{x}$

In Exercises 93–96, express each function as a composition of two functions.

93. $f(x) = \sqrt{x^2 - 4}$
94. $g(x) = (x^2 - x + 2)^{50}$
95. $h(x) = \sqrt{\dfrac{x - 3}{2x + 5}}$
96. $H(x) = (2x - 1)^3 + 5$

In Exercises 97–100, determine whether the given function is one-to-one. If the function is one-to-one, find its inverse and sketch the graph of the function and its inverse on the same coordinate axes.

97. $f(x) = x + 2$
98. $f(x) = -2x + 3$
99. $f(x) = \sqrt[3]{x - 2}$
100. $f(x) = 8x^3 - 1$

In Exercises 101 and 102, assume that f is a one-to-one function. Find f^{-1} and find the domain and range of f.

101. $f(x) = \dfrac{x - 1}{x + 2}, x \neq -2$
102. $f(x) = \dfrac{2x + 3}{x - 1}, x \neq 1$

103. For the graph of a function f shown in the figure,
   a. Write a formula for f as a piecewise function.
   b. Find the domain and range of f.
   c. Find the intercepts of the graph of f.
   d. Draw the graph of $y = f(-x)$.
   e. Draw the graph of $y = -f(x)$.
   f. Draw the graph of $y = f(x) + 1$.
   g. Draw the graph of $y = f(x + 1)$.
   h. Draw the graph of $y = 2f(x)$.
   i. Draw the graph of $y = f(2x)$.
   j. Draw the graph of $y = f\left(\dfrac{1}{2}x\right)$.
   k. Explain why f is one-to-one.
   l. Draw the graph of $y = f^{-1}(x)$.

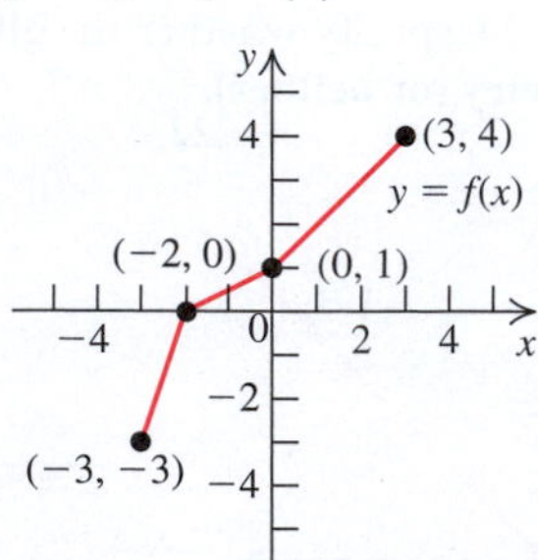

Applying the Concepts

104. Scuba diving. The pressure P on the body of a scuba diver increases linearly as she descends to greater depths d in seawater. The pressure at a depth of 10 feet is 19.2 pounds per square inch, and the pressure at a depth of 25 feet is 25.95 pounds per square inch.

a. Write the equation relating P and d (with d as independent variable) in slope–intercept form.

b. What is the meaning of the slope and the y-intercept in part (a)?

c. Determine the pressure on a scuba diver who is at a depth of 160 feet.

d. Find the depth at which the pressure on the body of the scuba diver is 104.7 pounds per square inch.

105. Waste disposal. The Sioux Falls, South Dakota, council considered the following data regarding the cost of disposing of waste material.

Year	1996	2000
Waste (in pounds)	87,000	223,000
Cost	\$54,000	\$173,000

Source: Sioux Falls Health Department.

Assume that the cost C (in dollars) is linearly related to the waste w (in pounds).

a. Write the linear equation relating C and w in slope–intercept form.

b. Explain the meaning of the slope and the intercepts of the equation in part (a).

c. Suppose the city has projected 609,000 pounds of waste for the year 2008. Calculate the projected cost of disposing of the waste for 2008.

d. Suppose the city's projected budget for waste collection in 2012 is \$1 million. How many pounds of waste can be handled?

106. Checking your speedometer. A common way to check the accuracy of your speedometer is to drive one or more measured (not odometer) miles, holding the speedometer at a constant 60 miles per hour and checking your watch at the start and end of the measured distance.

a. How does your watch show you whether the speedometer is accurate? What mathematics is involved?

b. Why shouldn't you use your odometer to measure the number of miles?

107. Playing blackjack. Chloe, a bright mathematician, read a book on counting cards in blackjack and went to Las Vegas to try her luck. She started with \$100 and discovered that the amount of money she had at time t hours after the start of the game could be expressed by the function $f(t) = 100 + 55t - 3t^2$.

a. What amount of money had Chloe won or lost in the first two hours?

b. What was the average rate at which Chloe was winning or losing money during the first two hours?

c. Did she lose all of her money? If so, when?

d. If she played until she lost all of her money, what was the average rate at which she was losing her money?

108. Volume discounting. Major cola distributors sell a case (containing 24 cans) of cola to a retailer at a price of \$4. They offer a discount of 20% for purchases over 100 cases and a discount of 25% for purchases over 500 cases. Write a piecewise function that describes this pricing scheme. Use x as the number of cases purchased and $f(x)$ as the price paid.

109. Air pollution. The daily level L of carbon monoxide in a city is a function of the number of automobiles in the city. Suppose $L(x) = 0.5\sqrt{x^2 + 4}$, where x is the number of automobiles (in hundred thousands). Suppose further that the number of automobiles in a given city is growing according to the formula $x(t) = 1 + 0.002t^2$, where t is time in years measured from now.

a. Form a composite function describing the daily pollution level as a function of time.

b. What pollution level is expected in five years?

110. Toy manufacturing. After being in business for t years, a toy manufacturer is making $x = 5000 + 50t + 10t^2$ GI Jimmy toys per year. The sale price p in dollars per toy has risen according to the formula $p = 10 + 0.5t$. Write a formula for the manufacturer's yearly revenue as

a. A function of time t.

b. A function of price p.

111. Height and weight of NBA players The table shows the roster for the National Basketball Association (NBA) team Boston Celtics for 2012.

Name	Height (in inches)	Weight (in pounds)
Leandro Barbosa	75	194
Brandon Bass	80	250
Avery Bradley	74	180
Jason Collins	84	255
Kevin Garnett	83	253
Jeff Green	81	235
Courtney Lee	77	200
Fab Melo	84	255
Paul Pierce	79	235
Rajon Rondo	73	186
Jared Sullinger	81	260
Jason Terry	74	180
Chris Wilcox	82	235

Source: http://www.nba.com/celtics/roster/2012–2013-roster.html

a. Use a graphing utility to find the line of regression relating the variables x (height in inches) and y (weight in pounds).

b. Use a graphing utility to plot the points and graph the regression line.

c. Use the line in part (a) to predict the weight of an NBA player whose height is 76 inches.

PRACTICE TEST A

1. Find the equation of a circle (in standard form) having a diameter with endpoints $(-2, 3)$ and $(-4, 5)$.
2. Determine which symmetries the graph of the equation $3x + 2xy^2 = 1$ has.
3. Find the x- and y-intercepts of the graph of $y = x^2(x - 3)(x + 1)$.
4. Graph the equation $x^2 + y^2 - 2x - 2y = 2$.
5. Write the slope–intercept form of the equation of the line with slope -1 passing through the point $(2, 7)$.
6. Write an equation of the line parallel to the line $8x - 2y = 7$ passing through $(2, -1)$.
7. Use $f(x) = -2x + 1$ and $g(x) = x^2 + 3x + 2$ to find $(fg)(2)$.
8. Use $f(x) = 2x - 3$ and $g(x) = 1 - 2x^2$ to evaluate $g(f(2))$.
9. Use $f(x) = x^2 - 2x$ to find $(f \circ f)(x)$.
10. If $f(x) = \begin{cases} x^3 - 2 & \text{if } x \le 0 \\ 1 - 2x^2 & \text{if } > 0 \end{cases}$, find (a) $f(-1)$, (b) $f(0)$, and (c) $f(1)$.
11. Find the domain of the function $f(x) = \dfrac{\sqrt{x}}{\sqrt{1 - x}}$.
12. Determine the average rate of change of the function $f(x) = 2x + 7$ between $x = 1$ and $x = 4$.
13. Determine whether the function $f(x) = 2x^4 - \dfrac{3}{x^2}$ is even, odd, or neither.
14. Find the intervals where the function whose graph is shown in the second column is increasing or decreasing.
15. Starting with the graph of $y = \sqrt{x}$, describe the sequence of transformations (in order) required to sketch the graph of $f(x) = 2\sqrt{x - 3} + 4$.
16. A ball is thrown upward from the ground. After t seconds, the height h (in feet) above the ground is given by $h = 25 - (2t - 5)^2$. How many seconds does the ball take to reach a height of 25 feet?

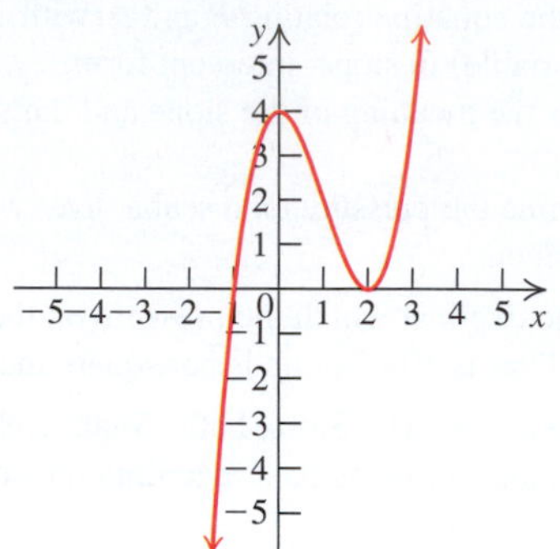

17. Assuming that f is a one-to-one function and $f(2) = 7$, find $f^{-1}(7)$.
18. Find the inverse function $f^{-1}(x)$ of the one-to-one function $f(x) = \dfrac{2x}{x - 1}$.
19. Now that Jo has saved \$1000 for a car, her dad takes over and deposits \$100 in her account each month. Write an equation that relates the total amount of money, A, in Jo's account to the number of months, x, that Jo's dad has been putting money in her account.
20. The cost C in dollars for renting a car for one day is a function of the number of miles traveled, m. For a car renting for \$30.00 per day and \$0.25 per mile, this function is given by

$$C(m) = 0.25m + 30.$$

   a. Find the cost of renting the car for one day and driving 230 miles.
   b. If the charge for renting the car for one day is \$57.50, how many miles were driven?

PRACTICE TEST B

1. The graph of the equation $|x| + 2|y| = 2$ is symmetric with respect to which of the following?
   I. the origin **II.** the x-axis **III.** the y-axis
   a. I only **b.** II only
   c. III only **d.** I, II, and III
2. Which are the x- and y-intercepts of the graph of $y = x^2 - 9$?
   a. x-intercept 3, y-intercept -9
   b. x-intercepts ± 3, y-intercept -9
   c. x-intercept 3, y-intercept 9
   d. x-intercepts ± 3, y-intercepts ± 9
3. The slope of a line is undefined if the line is parallel to
   a. The x-axis **b.** The line $x - y = 0$
   c. The line $x + y = 0$ **d.** The y-axis
4. Which is an equation of the circle with center $(0, -5)$ and radius 7?
   a. $x^2 + y^2 - 10y = 0$ **b.** $x^2 + (y - 5)^2 = 7$
   c. $x^2 + (y - 5)^2 = 49$ **d.** $x^2 + y^2 + 10y = 24$
5. Which is an equation of the line passing through $(2, -3)$ and having slope -1?
   a. $y + 3 = -(x + 2)$ **b.** $y - 3 = -(x - 2)$
   c. $y + 3 = -(x - 2)$ **d.** $y + 3 = -(x + 2)$

6. What is a second point on the line through $(3, 2)$ having slope $-\frac{1}{2}$?

a. $(2, 4)$ **b.** $(5, 1)$
c. $(7, 0)$ **d.** $(4, 4)$

7. Which is an equation of the line parallel to the line $6x - 3y = 5$ passing through $(-1, 2)$?

a. $y - 2 = -2(x + 1)$ **b.** $y - 2 = 6(x + 1)$
c. $y - 2 = -6(x + 1)$ **d.** $y - 2 = 2(x + 1)$

8. Use $f(x) = 3x - 5$ and $g(x) = 2 - x^2$ to find $(f \circ g)(x)$.

a. $-9x^2 + 30x - 23$ **b.** $-3x^2 + 1$
c. $3x^2 + 1$ **d.** $9x^2 - 30x + 23$

9. Use $f(x) = 2x^2 - x$ to find $(f \circ f)(x)$.

a. $8x^4 - 8x^3 + x$ **b.** $8x^4 + x$
c. $4x^4 - 4x^3 + x^2 - x$ **d.** $4x^4 - x^3 + x^2$
e. $-x^4 + 6x^2 - x$

10. Assuming that $g(t) = \frac{1 - t}{1 + t}$, find $g(a - 1)$.

a. $\frac{2 - a}{2 + a}$ **b.** $\frac{1 - a}{1 + a}$
c. $\frac{2 - a}{a}$ **d.** -1

11. The domain of the function $f(x) = \sqrt{x} + \sqrt{1 - x}$ is

a. $[0, 1]$ **b.** $(-\infty, -1] \cup [0, +\infty)$
c. $(-\infty, 1]$ **d.** $[0, +\infty)$

12. Let $f(x) = x^2 + 3x - 4$. Find all values of x for which $(x, 6)$ is on the graph of f.

a. $-5, 1$ **b.** $-2, 5$
c. $2, -4$ **d.** $2, -5$

13. Find the intervals where the function whose graph is shown here is increasing or decreasing.

a. Increasing on $(0, 3)$, decreasing on $(-3, 0) \cup (3, 4)$
b. Increasing on $(-1, 3)$, decreasing on $(2, -1) \cup (3, 1)$
c. Increasing on $(-3, 3)$, decreasing on $(2, -1) \cup (3, 1)$
d. Increasing on $(1.2, 4)$, decreasing on $(-3, -1.4)$

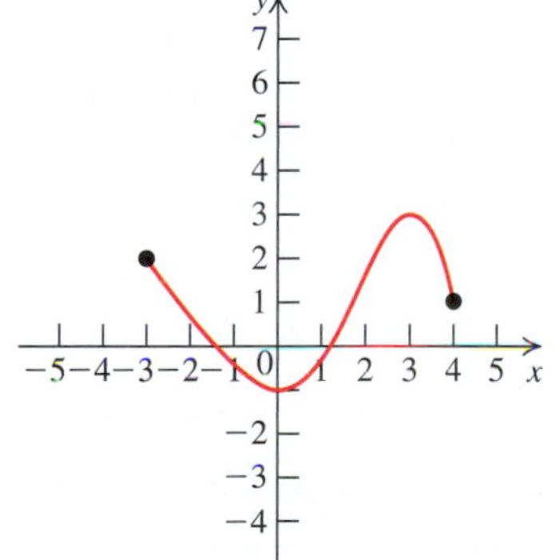

14. Suppose the graph of f is given. Describe how the graph of $y = f(x + 4)$ can be obtained from the graph of f.

a. By shifting the graph of f 4 units to the left
b. By shifting the graph of f 4 units down
c. By shifting the graph of f 4 units up
d. By shifting the graph of f 4 units to the right
e. By shifting the graph of f in the x-axis

15. Suppose the graph of f is given. Describe how the graph of $y = -2f(x - 4)$ can be obtained from the graph of f.

a. By shifting the graph of f 4 units to the left, shrinking vertically by a factor of $\frac{1}{2}$, and reflecting in the x-axis
b. By shifting the graph of f 4 units to the right, stretching vertically by a factor of 2, and reflecting in the x-axis
c. By shifting the graph of f 4 units to the left, shrinking vertically by a factor of $\frac{1}{2}$, and reflecting in the x-axis
d. By shifting the graph of f 4 units to the right, stretching vertically by a factor of 2, and reflecting in the y-axis

16. Which of the following is a one-to-one function?

a. $f(x) = |x + 3|$ **b.** $f(x) = x^2$
c. $f(x) = \sqrt{x^2 + 9}$ **d.** $f(x) = x^3 + 2$

17. If f is a one-to-one function and $f(3) = 5$, then $f^{-1}(5)$ equals

a. $\frac{1}{5}$ **b.** $\frac{1}{3}$
c. 3 **d.** -5

18. Find the inverse function $f^{-1}(x)$ of $f(x) = \frac{x}{3x + 2}$.

a. $\frac{3x}{x - 2}$ **b.** $\frac{2x}{3x - 1}$
c. $\frac{2x}{1 - 3x}$ **d.** $\frac{x}{x - 3}$

19. The ideal weight w (in pounds) for a man of height x inches is found by subtracting 190 from five times his height. Write a linear equation that gives the ideal weight of a man of height x inches. Then find the ideal weight of a man whose height is 70 inches.

a. $w = \frac{x + 190}{5}$, 160 lb **b.** $w = 5x - 190$, 160 lb
c. $w = 190 - \frac{x}{5}$, 176 lb **d.** $w = \frac{x + 190}{5}$, 52 lb

20. Cassie rented an intermediate sedan at \$25.00 per day plus \$0.20 per mile. How many miles can Cassie travel in one day for \$50.00?

a. 125 **b.** 1025
c. 250 **d.** 175

CUMULATIVE REVIEW EXERCISES CHAPTERS P–2

1. Simplify.

a. $\left(\frac{x^3}{y^2}\right)^2\left(\frac{y^2}{x^3}\right)^3$ **b.** $\frac{x^{-1}y^{-1}}{x^{-1} + y^{-1}}$

2. Factor.

a. $2x^2 + x - 15$ **b.** $x^3 - 2x^2 + 4x - 8$

3. Combine and simplify.

a. $\sqrt{75} + \sqrt{108} - \sqrt{192}$
b. $\frac{x - 1}{x + 1} - \frac{x - 2}{x + 2}$

4. Rationalize the denominator.

a. $\frac{1}{2 + \sqrt{3}}$ **b.** $\frac{1}{\sqrt{5} - 2}$

In Exercises 5–9, solve each equation for x.

5. **a.** $3x - 7 = 5$ **b.** $\frac{1}{x-1} = \frac{3}{x-1}$

6. **a.** $x^2 - 3x = 0$ **b.** $x^2 + 3x - 10 = 0$

7. **a.** $2x^2 - x + 3 = 0$ **b.** $4x^2 - 12x + 9 = 0$

8. **a.** $x - 6\sqrt{x} + 8 = 0$

b. $\left(x - \frac{1}{x}\right)^2 - 10\left(x - \frac{1}{x}\right) + 21 = 0$

9. **a.** $\sqrt{3x - 1} = 2x - 1$ **b.** $\sqrt{1 - x} = 2 - \sqrt{2x + 1}$

In Exercises 10–12, solve each inequality for x and write the solution in interval notation.

10. **a.** $2x - 5 < 11$ **b.** $-3x + 4 > -5$

11. **a.** $-3 < 2x - 3 < 5$ **b.** $5 \le 1 - 2x \le 7$

12. **a.** $|2x - 1| \le 7$ **b.** $|2x - 3| \ge 5$

13. Show that the triangle with vertices $A(5, -2)$, $B(6, 5)$, and $C(2, 2)$ is isosceles.

14. Sketch the graph of $4x^2 - 9y^2 = 0$. [*Hint:* $4x^2 - 9y^2 = (2x + 3y)(2x - 3y)$.]

15. Show that the points $(-3, -1)$, $(2, 4)$, $(5, 3)$, and $(6, 2)$ lie on a circle with center $(2, -1)$.

16. Find the center and radius of the circle with equation $x^2 + y^2 - 6x + 4y + 9 = 0$.

In Exercises 17–22, find the slope–intercept form of the equation of the line satisfying the given condition.

17. Slope $= -3$; y-intercept 5

18. Slope $= 2$; x-intercept 4

19. The line is perpendicular to $y = 2x + 3$ and passes through $(2, -1)$.

20. The line is parallel to $y = 2x + 3$ and passes through $(2, -1)$.

21. The line is the perpendicular bisector of the line segment joining $(3, -1)$ and $(5, 7)$.

22. The line is parallel to $x = -2$ and passes through $(5, 7)$.

23. Find x assuming that the line through $(x, 5)$ and $(5, 11)$ is parallel to a line with slope 2.

24. Find x assuming that the line through $(x, 3)$ and $(3, 7)$ is perpendicular to a line with slope $\frac{1}{2}$.

In Exercise 25–28, graph each equation.

25. $12x = 4y - 6$

26. $x^2 - 2x + y^2 - 4y - 4 = 0$

27. $f(x) = \sqrt{x + 2} + 3$

28. $f(x) = -(x - 1)^2 + 4$

29. Ms. Gutiérrez bought some used books for \$1650. She kept 16 of the books and sold the rest at a profit of \$10 each. If she recovered her original \$1650 from this sale, how many books did she purchase initially?

30. The monthly note on a car that was leased for two years was \$250 less than the monthly note on a car that was leased for a year and a half. The total expense for the two leases was \$21,000. Find the monthly note for each lease.

31. Let $f(x) = \sqrt{x + 1} - 3$.

a. Find the domain of f.

b. Find the intercepts.

c. Find $f(-1)$.

d. Solve $f(x) > 0$.

32. Let $f(x) = \begin{cases} -x & \text{if } x < 0 \\ x^2 & \text{if } x \ge 0 \end{cases}$.

a. Find $f(-2)$, $f(0)$, and $f(2)$.

b. Find the intervals on which f is increasing, decreasing, or constant.

33. Let $f(x) = \frac{1}{x - 2}$ and $g(x) = \frac{2}{x}$.

a. Find $(f \circ g)(x)$ and state its domain.

b. Find $(g \circ f)(x)$ and state its domain.

CHAPTER 3

Polynomial and Rational Functions

TOPICS

Polynomials and rational functions are used in navigation, in computer-aided geometric design, in the study of supply and demand in business, in the design of suspension bridges, in computations determining the force of gravity on planets, and in virtually every area of inquiry requiring numerical approximations. Their applicability is so widespread that it is hard to imagine an area in which they have not made an impact.

SECTION 3.1

Quadratic Functions

BEFORE STARTING THIS SECTION, REVIEW

1. Linear functions (Section 2.6, page 226)
2. Completing the square (Section 1.2, page 96)
3. Quadratic formula (Section 1.2, page 97)
4. Transformations (Section 2.7, page 250)

OBJECTIVES

1. Graph a quadratic function in standard form.
2. Graph any quadratic function.
3. Solve problems modeled by quadratic functions.

The Gateway Arch

The Gateway Arch, at 630-feet, is the world's tallest arch. Located on the west bank of the Mississippi River in St. Louis, Missouri, it stands as a monument to the westward expansion of the United States, begun 200 years ago by Lewis and Clark. Built in the form of a flattened catenary arch, it is the tallest man-made monument in the United States, and the hallmark of the Jefferson National Expansion Memorial. The arch was designed by Finnish-American architect Eero Saarinen and German-American structural engineer Hannskarl Bendel in 1947, and has become an internationally famous symbol of St. Louis. In Exercise 100 we investigate the dimensions of this famous structure.

1 Graph a quadratic function in standard form.

Quadratic Functions

Quadratic Function

A function of the form

$$f(x) = ax^2 + bx + c,$$

where a, b, and c, are real numbers with $a \neq 0$, is called a **quadratic function**.

Note that the squaring function $s(x) = x^2$ (whose graph is a parabola, see page 234) is the quadratic function $f(x) = ax^2 + bx + c$ with $a = 1$, $b = 0$, and $c = 0$. We will show that we can graph any quadratic function by transforming the graph of $s(x) = x^2$. Because the graph of $y = x^2$ is a parabola, the graph of any quadratic function is also a parabola.

First, we compare the graphs of $f(x) = ax^2$ for $a > 0$ and $g(x) = ax^2$ for $a < 0$. These are quadratic functions with $b = 0$ and $c = 0$.

$f(x) = ax^2, a > 0$	$g(x) = ax^2, a < 0$
The graph of $f(x) = ax^2$ is obtained from the graph of the squaring function $s(x) = x^2$ by multiplying the y-values by a, vertically stretching or compressing it. Here are three such graphs on the same axes for $a = 2, a = 1$, and $a = \frac{1}{2}$.	The graph of $g(x) = ax^2$ is the reflection of the graph of $f(x) = \lvert a \rvert x^2$ about the x-axis. Here are three graphs for $a = -2, a = -1$, and $a = -\frac{1}{2}$.

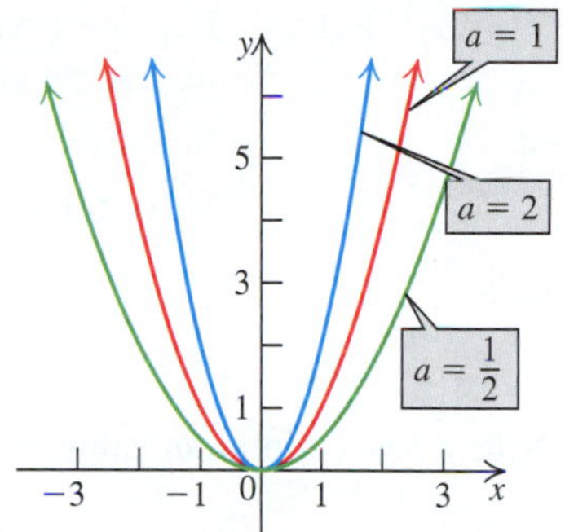

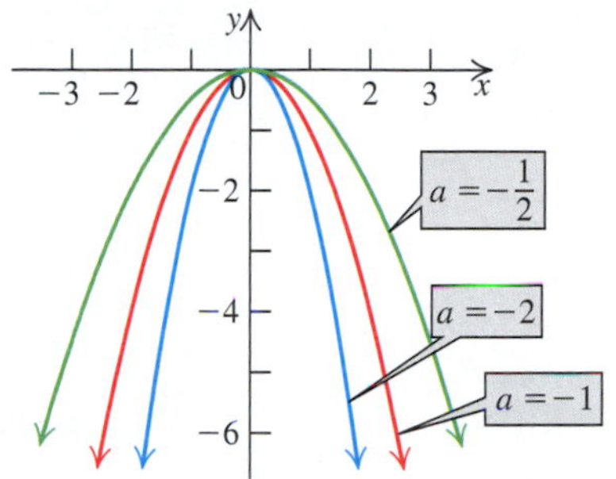

Standard Form of a Quadratic Function

Any quadratic function $f(x) = ax^2 + bx + c$ can be rewritten in the form $f(x) = a(x - h)^2 + k$, called the **standard form** of the quadratic function f. By using transformations (see Section 2.7), we can obtain the graph of $f(x) = a(x - h)^2 + k$ from the graph of the squaring function $s(x) = x^2$ as follows:

$s(x) = x^2$	The squaring function
$\downarrow$	
$g(x) = (x - h)^2$	Horizontal translation to the right if $h > 0$ and to the left if $h < 0$
$\downarrow$	
$G(x) = a(x - h)^2$	Vertical stretching or compressing. If $a < 0$, the graph is also reflected about the x-axis.
$\downarrow$	
$f(x) = a(x - h)^2 + k$	Vertical translation, up if $k > 0$ and down if $k < 0$

These transformations show that the graph of a quadratic function $f(x) = a(x - h)^2 + k$ is a parabola that opens up if $a > 0$ and down if $a < 0$. Its domain is $(-\infty, \infty)$. The parabola is symmetric about the vertical line $x = h$. The line of symmetry $x = h$ is called the ***axis*** (or ***axis of symmetry***) of the parabola. The point (h, k) where the axis meets the parabola is called the ***vertex*** of the parabola. The vertex of any parabola that opens up is the *lowest point* of the graph, and the vertex of any parabola that opens down is the *highest point* of the graph. Therefore, k is the ***minimum value*** for any parabola that opens up and k is the ***maximum value*** for any parabola that opens down. The standard form is particularly useful because it immediately identifies the vertex (h, k).

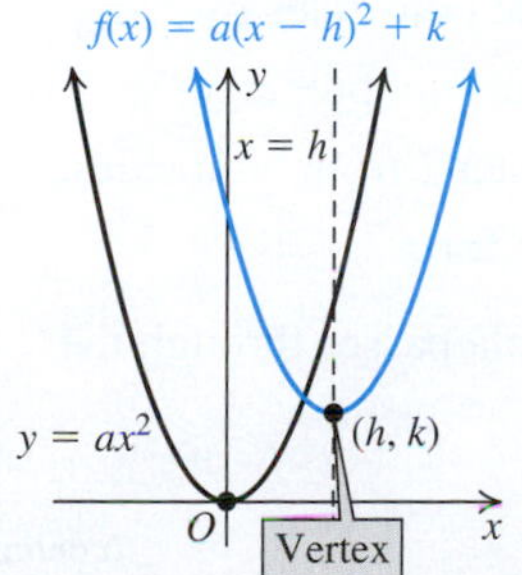

FIGURE 3.1 Quadratic functions

The Standard Form of a Quadratic Function

The quadratic function

$$f(x) = a(x - h)^2 + k, \ a \neq 0$$

is in **standard form**. The graph of f is a parabola with **vertex** (h, k). See Figure 3.1. The parabola is symmetric about the line $x = h$, called the **axis** of the parabola. If $a > 0$, the parabola opens up, and if $a < 0$, the parabola opens down. If $a > 0$, k is the **minimum value of** f, and if $a < 0$, k is the **maximum value of** f.

EXAMPLE 1 Writing the Equation of a Quadratic Function

Find the standard form of the quadratic function f whose graph has vertex $(-3, 4)$ and passes through the point $(-4, 7)$. Does f have a maximum or minimum value?

Solution

$f(x) = a(x - h)^2 + k$	Standard form of a quadratic function
$f(x) = a[x - (-3)]^2 + 4$	Replace h with -3 and k with 4.
$y = a(x + 3)^2 + 4$	Replace $f(x)$ with y, and simplify.
$7 = a(-4 + 3)^2 + 4$	Replace x with -4 and y with 7 because $(-4, 7)$ lies on the graph of $y = f(x)$.
$7 = a + 4$	Simplify.
$a = 3$	Solve for a.

The standard form is

$$f(x) = 3(x + 3)^2 + 4 \qquad a = 3, h = -3, k = 4$$

Because $a = 3 > 0$, f has a minimum value of 4 at $x = -3$.

Practice Problem 1 Find the standard form of the quadratic function f whose graph has vertex $(1, -5)$ and passes through the point $(3, 7)$. Does f have a maximum or minimum value?

PROCEDURE IN ACTION

EXAMPLE 2 Graphing a Quadratic Function in Standard Form

OBJECTIVE	EXAMPLE
Sketch the graph of $f(x) = a(x - h)^2 + k$.	*Sketch the graph of* $f(x) = -3(x + 2)^2 + 12$.
Step 1 The graph is a parabola because it has the form $f(x) = a(x - h)^2 + k$. Identify a, h, and k.	**1.** The graph of $f(x) = -3(x + 2)^2 + 12 = -3[x - (-2)]^2 + 12$ (with -3 labeled a, -2 labeled h, 12 labeled k) is a parabola; $a = -3$, $h = -2$, and $k = 12$.
Step 2 Determine how the parabola opens. If $a > 0$, the parabola opens *up*. If $a < 0$, it opens *down*.	**2.** Because $a = -3 < 0$, the parabola opens down.
Step 3 Find the vertex (h, k). If $a > 0$ (or $a < 0$), the function f has a minimum (or a maximum) value k at $x = h$.	**3.** The vertex $(h, k) = (-2, 12)$. Because the parabola opens down, the function f has a maximum value of 12 at $x = -2$.
Step 4 Find the x-intercepts (if any). Set $f(x) = 0$ and solve the equation $a(x - h)^2 + k = 0$ for x. If the solutions are real numbers, they are the x-intercepts. If not, the parabola lies above the x-axis (when $a > 0$) or below the x-axis (when $a < 0$).	**4.** $0 = -3(x + 2)^2 + 12$ — Set $f(x) = 0$. $3(x + 2)^2 = 12$ — Add $3(x + 2)^2$ to both sides. $(x + 2)^2 = 4$ — Divide both sides by 3. $x + 2 = \pm 2$ — Square root property $x = -2 \pm 2$ — Subtract 2 from both sides. $x = 0$ or $x = -4$ — Solve for x. The x-intercepts are 0 and -4. The parabola passes through the points $(0, 0)$ and $(-4, 0)$.

(continued)

Step 5 Find the y-intercept. Replace x with 0. Then $f(0) = ah^2 + k$ is the y-intercept.

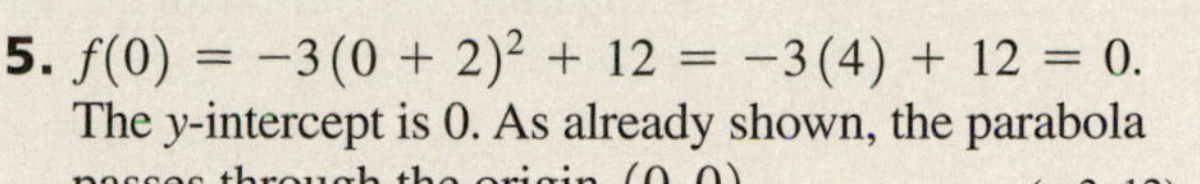

5. $f(0) = -3(0 + 2)^2 + 12 = -3(4) + 12 = 0.$
The y-intercept is 0. As already shown, the parabola passes through the origin $(0, 0)$.

Step 6 Sketch the graph. Plot the points found in Steps 3–5 and join them to form a parabola. Show the axis $x = h$ of the parabola by drawing a dashed vertical line.

If there are no x-intercepts, draw the half of the parabola that passes through the vertex and a second point, such as the y-intercept. Then use the axis of symmetry to draw the other half.

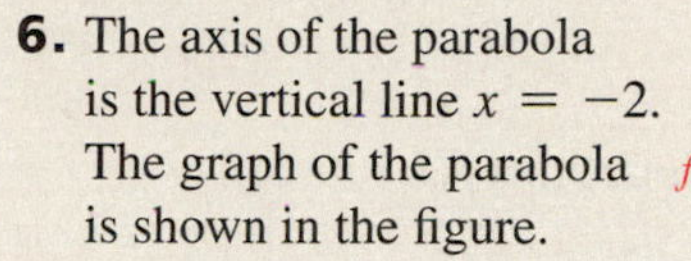

6. The axis of the parabola is the vertical line $x = -2$. The graph of the parabola is shown in the figure.

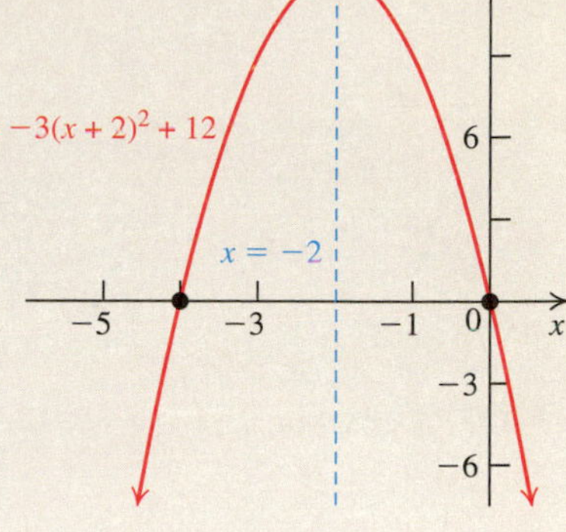

Practice Problem 2 Graph the quadratic function $f(x) = -2(x + 1)^2 + 3$. Find the vertex, maximum or minimum value, and x- and y-intercepts.

2 Graph any quadratic function.

Graphing a Quadratic Function $f(x) = ax^2 + bx + c$

A quadratic function $f(x) = ax^2 + bx + c$ can be changed to the standard form $f(x) = a(x - h)^2 + k$ by *completing the square.*

To write $x^2 - 4x + 3$ in standard form, we use $\left(\frac{-4}{2}\right)^2 = (-2)^2 = 4$, the square of one half the coefficient of the term $-4x$.

$$
\begin{aligned}
x^2 - 4x + 3 &= x^2 - 4x + 4 - 4 + 3 && \text{Add and subtract 4; } 4 = \left(\tfrac{-4}{2}\right)^2 \\
&= (x^2 - 4x + 4) - 4 + 3 && \text{Group the first three terms.} \\
&= (x - 2)^2 - 1 && x^2 - 4x + 4 = (x - 2)^2;\ -4 + 3 = -1
\end{aligned}
$$

To write $5x^2 - 20x + 8$ in standard form, we first group $5x^2 - 20x$.

$$
\begin{aligned}
5x^2 - 20x + 8 &= (5x^2 - 20x) + 8 && \text{Group } 5x^2 - 20x. \\
&= 5(x^2 - 4x) + 8 && \text{Factor out 5 from } 5x^2 - 20x. \\
&= 5(x^2 - 4x + 4 - 4) + 8 && \text{Add and subtract 4 within the parentheses; } 4 = \left(\tfrac{-4}{2}\right)^2 \\
&= 5(x^2 - 4x + 4) + 5(-4) + 8 && \text{Distributive property} \\
&= 5(x - 2)^2 - 12 && x^2 - 4x + 4 = (x - 2)^2\text{; simplify.}
\end{aligned}
$$

The general procedure for completing the square is given next.

Converting $f(x) = ax^2 + bx + c$ to Standard Form

$$
\begin{aligned}
f(x) &= ax^2 + bx + c && \text{Original function} \\
&= a\left(x^2 + \frac{b}{a}x\right) + c && \text{Factor out } a \text{ from } ax^2 + bx. \\
&= a\left(x^2 + \frac{b}{a}x + \frac{b^2}{4a^2} - \frac{b^2}{4a^2}\right) + c && \text{To complete the square, add and subtract } \left(\frac{1}{2}\cdot\frac{b}{a}\right)^2 = \frac{b^2}{4a^2} \text{ within the parentheses.} \\
&= a\left(x^2 + \frac{b}{a}x + \frac{b^2}{4a^2}\right) - a\cdot\frac{b^2}{4a^2} + c && \text{Distribute } a \text{ to } -\frac{b^2}{4a^2}\text{; regroup terms.} \\
&= a\left(x + \frac{b}{2a}\right)^2 + \left(c - \frac{b^2}{4a}\right) && \left(x + \frac{b}{2a}\right)^2 = x^2 + \frac{b}{a}x + \frac{b^2}{4a^2}
\end{aligned}
$$

Comparing this form with the standard form $f(x) = a(x - h)^2 + k$, we have

$$h = -\frac{b}{2a} \qquad \text{x-coordinate of the vertex}$$

$$\text{and} \quad k = c - \frac{b^2}{4a} \qquad \text{y-coordinate of the vertex}$$

Notice that $f(h) = a(h - h)^2 + k = k$ Replace x with h; $a(h - h)^2 = 0$.

Finding the Vertex of $f(x) = ax^2 + bx + c, a \neq 0$

To find the vertex (h, k) of $f(x) = ax^2 + bx + c$,

1. Find the x-coordinate $h = -\dfrac{b}{2a}$ of the vertex.

2. Calculate $k = f\left(-\dfrac{b}{2a}\right)$ to find its y-coordinate.

PROCEDURE **IN ACTION**

EXAMPLE 3 Graphing a Quadratic Function

OBJECTIVE

Graph $f(x) = ax^2 + bx + c, a \neq 0$.

Step 1 Identify a, b, and c.

Step 2 Determine how the parabola opens. If $a > 0$, the parabola opens *up*; if $a < 0$, the parabola opens *down*.

Step 3 Find the vertex (h, k). Use the following formulas:

$$h = -\frac{b}{2a}$$

$$k = f\left(-\frac{b}{2a}\right)$$

k is a minimum if $a > 0$.
k is a maximum if $a < 0$.

Step 4 Find the x-intercepts (if any). Let $f(x) = 0$ and solve $ax^2 + bx + c = 0$. If the solutions are real numbers, they are the x-intercepts. If not, the parabola lies entirely above the x-axis (when $a > 0$) or entirely below the x-axis (when $a < 0$).

EXAMPLE

Sketch the graph of $f(x) = 2x^2 + 8x - 10$.

1. In the equation $y = f(x) = 2x^2 + 8x - 10$, $a = 2$, $b = 8$, and $c = -10$.

2. Because $a = 2 > 0$, the parabola opens up.

3. $h = -\dfrac{b}{2a} = -\dfrac{8}{2(2)} = -2$ $a = 2, b = 8$

$$\begin{aligned} k &= f(h) = f(-2) \\ &= 2(-2)^2 + 8(-2) - 10 && \text{Replace } h \text{ with } -2. \\ &= -18 && \text{Simplify.} \end{aligned}$$

The vertex is $(-2, -18)$.

The function f has a minimum value of -18 at $x = -2$.

4.

$$\begin{aligned} 2x^2 + 8x - 10 &= 0 && \text{Set } f(x) = 0. \\ 2(x^2 + 4x - 5) &= 0 && \text{Factor out 2.} \\ 2(x + 5)(x - 1) &= 0 && \text{Factor.} \\ x + 5 = 0 \quad \text{or} \quad x - 1 &= 0 && \text{Zero-product property} \\ x = -5 \quad \text{or} \quad x &= 1 && \text{Solve for } x. \end{aligned}$$

The x-intercepts are at $x = -5$ and $x = 1$; the graph of f passes through the points $(-5, 0)$ and $(1, 0)$.

(continued)

Step 5 Find the y-intercept. Let $x = 0$. The result, $f(0) = c$, is the y-intercept.

5. Set $x = 0$ to obtain $f(0) = 2(0)^2 + 8(0) - 10 = -10$. The y-intercept is -10 and the graph passes through the point $(0, -10)$.

Step 6 The parabola is symmetric about its axis, $x = -\frac{b}{2a}$. Use this symmetry to find additional points.

6. The axis of symmetry is $x = -2$. The symmetric image of $(0, -10)$ about the axis $x = -2$ is $(-4, -10)$.

Step 7 Draw a parabola through the points found in Steps 3–6.

If there are no x-intercepts, draw the half of the parabola that passes through the vertex and a second point, such as the y-intercept. Then use the axis of symmetry to draw the other half.

7. The parabola passing through the points in Steps 3–6 is sketched in the figure.

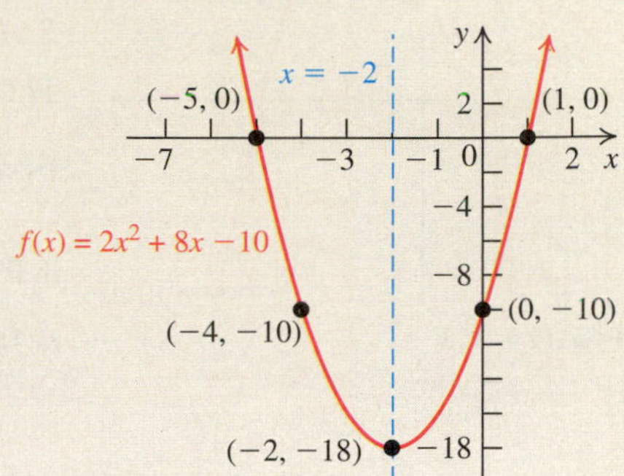

Practice Problem 3 Graph the quadratic function $f(x) = 3x^2 - 3x - 6$. Find the vertex, maximum or minimum value, and x- and y-intercepts.

As the graph suggests, when $a > 0$, every value of y, with $y \geq k$, is the second coordinate for some point on the graph of $f(x) = ax^2 + bx + c$. So when $a > 0$, the interval $[k, \infty)$ is the range of f. Similarly, when $a < 0$, the range of f is the interval $(-\infty, k]$.

Finding the Range of $f(x) = ax^2 + bx + c$ having Vertex (h, k)

1. If $a > 0$, the range is $[k, \infty)$.
2. If $a < 0$, the range is $(-\infty, k]$.

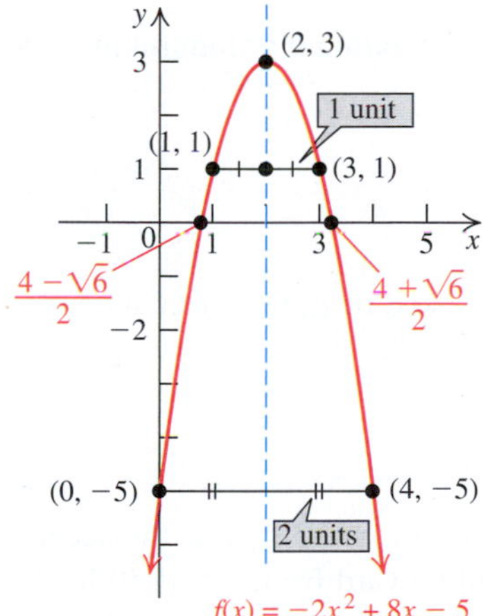

FIGURE 3.2

EXAMPLE 4 **Identifying the Characteristics of a Quadratic Function from Its Graph**

The graph of the function $f(x) = -2x^2 + 8x - 5$ is shown in Figure 3.2.

a. Find the domain and range of f.

b. Solve the inequality $-2x^2 + 8x - 5 > 0$.

Solution

The graph in Figure 3.2 is a parabola that opens down and whose vertex is $(2, 3)$.

a. The domain of f is $(-\infty, \infty)$. Because the parabola opens down, the maximum value of f is the y-coordinate, 3, of its vertex $(2, 3)$. So the range of f is $(-\infty, 3]$.

b. The x-intercepts of f are $\frac{4 - \sqrt{6}}{2} \approx 0.775$ and $\frac{4 + \sqrt{6}}{2} \approx 3.225$. The graph of f is above the x-axis between these intercepts. So $y = -2x^2 + 8x - 5 > 0$ for the x-values in the interval $\left(\frac{4 - \sqrt{6}}{2}, \frac{4 + \sqrt{6}}{2}\right)$. The solution set of the inequality $-2x^2 + 8x - 5 > 0$ is the interval $\left(\frac{4 - \sqrt{6}}{2}, \frac{4 + \sqrt{6}}{2}\right)$.

Practice Problem 4 Graph the function $f(x) = 3x^2 - 6x - 1$ and solve the inequality $3x^2 - 6x - 1 \leq 0$.

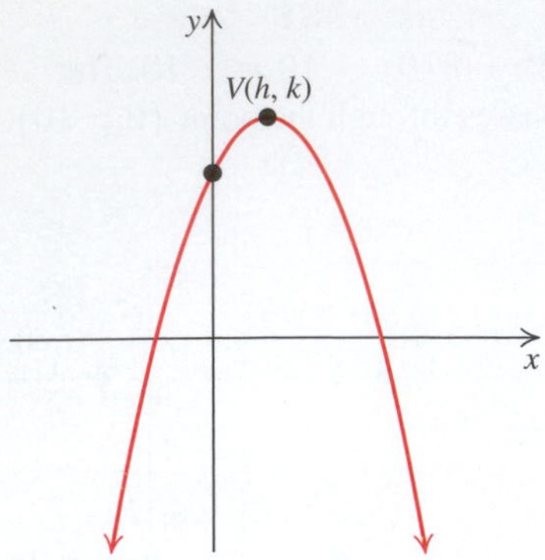

FIGURE 3.3

EXAMPLE 5 **Finding the Signs of a, b, and c from the Graph of $y = ax^2 + bx + c$**

The graph of $y = ax^2 + bx + c$ is given in Figure 3.3. Find the signs of a, b, and c and state which (if any) are zero.

Solution

The graph of the parabola opens down, so $a < 0$. The vertex $V(h, k)$ is in Q I, so $h > 0$. Because $h = -\frac{b}{2a}$ and $a < 0$, we must have $b > 0$. Otherwise, if b is negative, then $-b$ is positive and $\frac{-b}{2a} = \frac{+}{-}$ is negative. The y-intercept is positive, and the y-intercept is $a(0)^2 + b(0) + c = c$; so $c > 0$. So $a < 0$, $b > 0$, and $c > 0$.

Practice Problem 5 Repeat Example 5 for the graph in Figure 3.4.

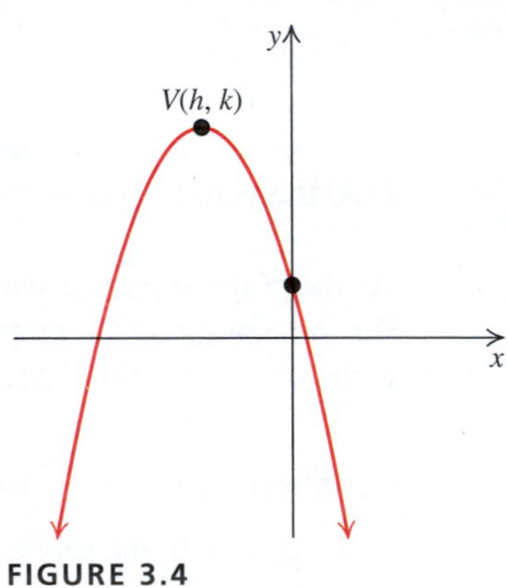

FIGURE 3.4

3 Solve problems modeled by quadratic functions.

Applications

Many applications of quadratic functions involve finding the maximum or minimum value of the function.

RECALL

For a quadratic function $f(x) = ax^2 + bx + c$, the vertex is $(h, k) = \left(-\frac{b}{2a}, f\left(-\frac{b}{2a}\right)\right)$. The function f has minimum value k if $a > 0$ and f has maximum value k if $a < 0$.

EXAMPLE 6 **Tracking a Moving Object**

The height h, in feet, of a ball thrown upward at an initial speed of v_0 ft/s from a point h_0 ft above the ground is given by the function

$$h(t) = -\frac{1}{2}g_E t^2 + v_0 t + h_0,$$

where t is the time in seconds after the ball was released and $g_E = 32 \text{ ft/s}^2$ is the acceleration due to gravity on Earth. If you have thrown the ball upward from a cliff 50 ft high and the maximum height that it reached was 99 ft,

a. Find the height of the ball as a function of t.

b. How long did it take the ball to reach its highest point?

Solution

a. With $g_E = 32$ and $h_0 = 50$, the height of the ball is given by the equation

(1) $$h(t) = -16t^2 + v_0 t + 50. \qquad -\frac{1}{2}g_E = -\frac{1}{2}(32) = -16$$

To find v_0, we observe that this function is a quadratic function with $a = -16 < 0$. It attains its maximum value at the point where

$$t = -\frac{v_0}{2(-16)} = \frac{v_0}{32}$$ Use the formula $t = -\frac{b}{2a}$ for the first coordinate of the vertex of a parabola.

Given that the maximum height of the ball was 99 ft and $h(t) = -16t^2 + v_0 t + 50$,

$$99 = h\left(\frac{v_0}{32}\right) = -16\left(\frac{v_0}{32}\right)^2 + v_0\left(\frac{v_0}{32}\right) + 50$$ Replace t with $\frac{v_0}{32}$ in $h(t)$.

Solving for v_0 yields

$$\frac{v_0^2}{64} + 50 = 99$$ $-16\left(\frac{v_0}{32}\right)^2 + v_0\left(\frac{v_0}{32}\right) = \frac{v_0^2}{64}$

$$\frac{v_0^2}{64} = 99 - 50 = 49$$ Subtract 50 from both sides.

$$v_0^2 = 64 \cdot 49$$ Multiply both sides by 64.

$$v_0 = \sqrt{64 \cdot 49} = 8 \cdot 7 = 56$$ v_0 is positive because the ball was thrown upward.

$$h(t) = -16t^2 + 56t + 50$$ Replace v_0 in equation (1) with 56.

b. We again use the formula for the first coordinate of the vertex of a parabola to conclude that the ball reached its highest point

$$t = -\frac{56}{2(-16)} = \frac{56}{32} = 1.75 \text{ seconds}$$

after it was released.

Practice Problem 6 Repeat Example 6 assuming that the height of the cliff was 100 ft and the ball went up to the maximum height of 244 ft above the ground.

EXAMPLE 7 Finding a Maximum Area

A farmer wants to build a rectangular grazing pen using a straight stone wall for one side. He has 100 feet of fence to use for the other three sides. What is the maximum area that can be enclosed? What are the dimensions of the pen?

Solution

Let $x =$ length of the side perpendicular to the wall (see Figure 3.5) and $y =$ length of the side parallel to the wall.

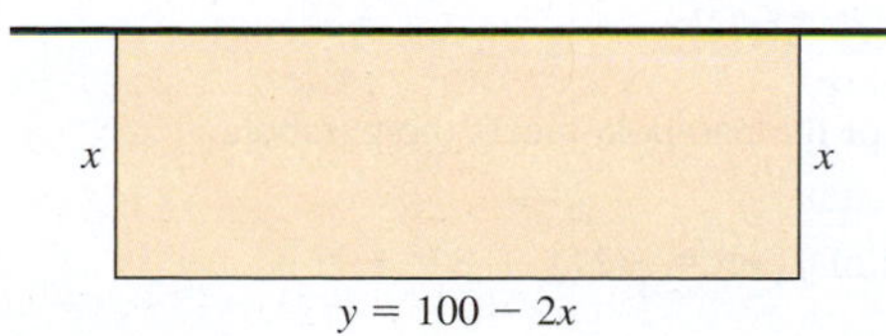

FIGURE 3.5

Then $2x + y = 100$ — The three sides use 100 feet of fence.

$y = 100 - 2x$ — Subtract $2x$ from both sides.

$A(x) =$ area of the pen $= xy$ — Area = (length)(width)

$A(x) = x(100 - 2x)$ — Replace y with $100 - 2x$.

The function $A(x) = x(100 - 2x) = -2x^2 + 100x$ gives the area of the pen, so we want to find the maximum value for $A(x)$. The vertex for the parabola is (h, k), where

$$h = -\frac{b}{2a} = \frac{-100}{2(-2)} = \frac{100}{4} = 25 \qquad A(x) = -2x^2 + 100x, \text{ so } a = -2, b = 100$$

$$k = A(25) = -2(25)^2 + 100(25) = 1250. \qquad k = \text{maximum value}$$

The maximum area is 1250 square feet, which occurs when $x = h = 25$.
The length of the side perpendicular to the wall is 25 feet.
The length of the side parallel to the wall is $100 - 2(25) = 50$ feet.

Practice Problem 7 A children's center receives a donation of 1000 ft of fence to enclose a rectangular playground. What is the maximum area that can be enclosed? What are the playground's dimensions?

Answers to Practice Problems

1. $y = 3(x - 1)^2 - 5$; f has minimum value -5.
2. The graph is a parabola. $a = -2, h = -1, k = 3$. It opens downward. Vertex $(-1, 3)$. It has maximum value 3; x-intercepts: $\pm\sqrt{\frac{3}{2}} - 1$; y-intercept: 1

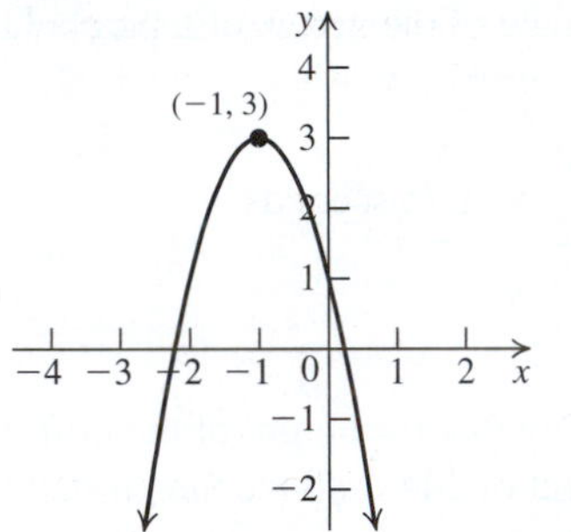

3. The graph is a parabola. It opens upward. Vertex $\left(\frac{1}{2}, \frac{-27}{4}\right)$; It has a minimum value $-\frac{27}{4}$; x-intercepts: $-1, 2$; y-intercept: -6

4.

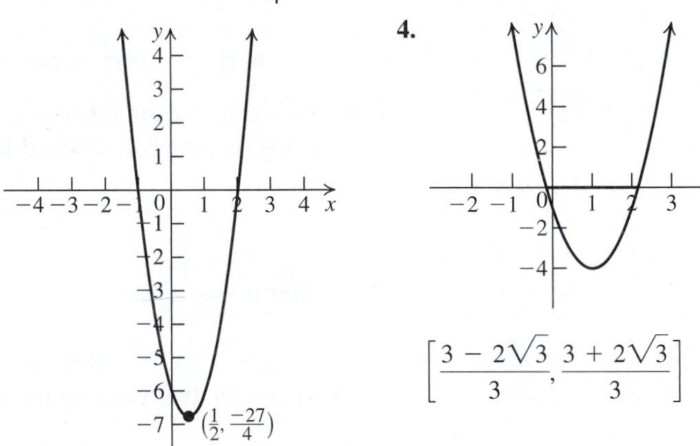

$$\left[\frac{3 - 2\sqrt{3}}{3}, \frac{3 + 2\sqrt{3}}{3}\right]$$

5. $a < 0, b < 0, c > 0$ **6. a.** $h(t) = -16t^2 + 96t + 100$ (feet) **b.** 3 seconds **7.** Maximum area is 62,500 sq ft. The playground is a square of side length 250 feet on each side.

SECTION 3.1 Exercises

Basic Concepts and Skills

1. A point where the axis of the parabola meets the parabola is called the ______________.
2. The vertex of the graph of $f(x) = -2(x + 3)^2 - 5$ is ______________.
3. **True or False.** The graph of the function in Exercise 2 opens down.
4. **True or False.** The graph of $f(x) = -2 - x + x^2$ opens down.
5. The x-coordinate of the vertex of the parabola of Exercise 4 is ______________.
6. **True or False.** The y-coordinate of the vertex of the parabola $f(x) = x^2 - 2x + 5$ is $f(1)$.
7. **True or False.** If $a > 0$, then the minimum value of $f(x) = ax^2 + bx + c$ is $f\left(-\frac{b}{2a}\right)$.
8. **True or False.** If $a < 0$, then $f(x) = ax^2 + bx + c$ has no minimum value.

In Exercises 9–16, match each quadratic function with its graph.

9. $y = -\frac{1}{3}x^2$ **10.** $y = -3x^2$

11. $y = -3(x + 1)^2$ **12.** $y = 2(x + 1)^2$

13. $y = (x - 1)^2 + 2$ **14.** $y = (x - 1)^2 - 3$

15. $y = 2(x + 1)^2 - 3$ **16.** $y = -3(x + 1)^2 + 2$

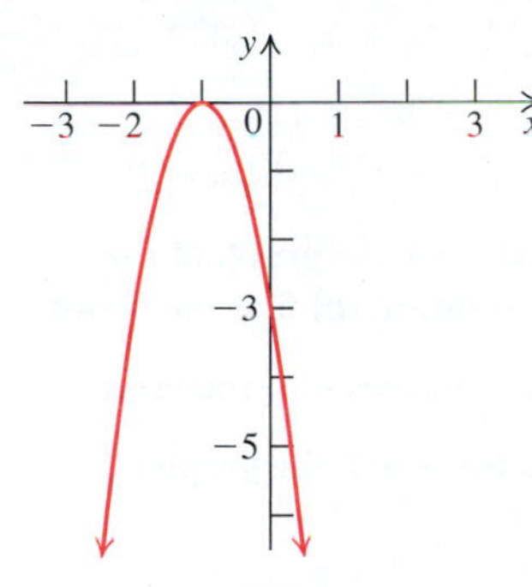
(a)

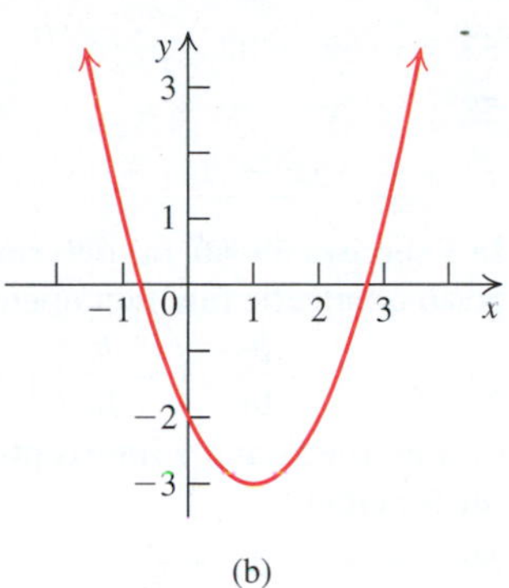
(b)

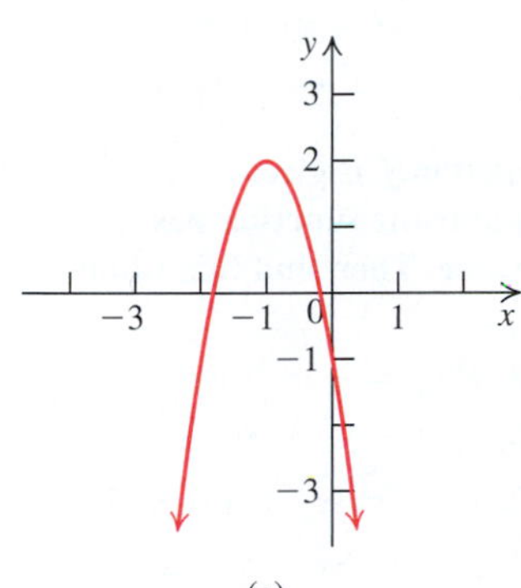
(c)

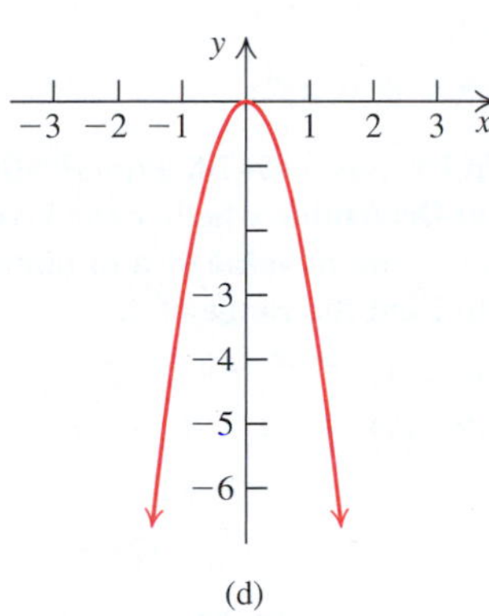
(d)

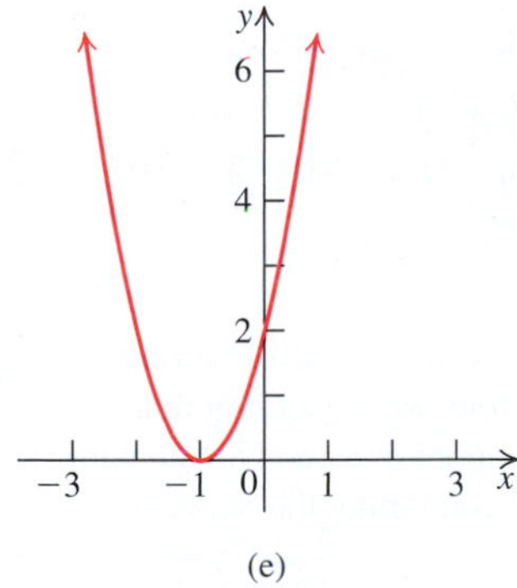
(e)

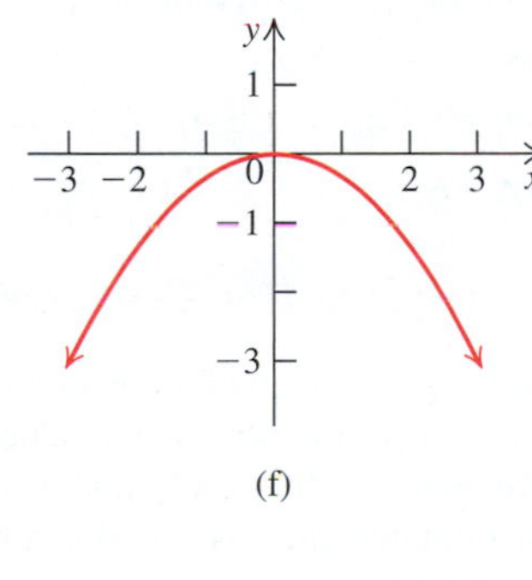
(f)

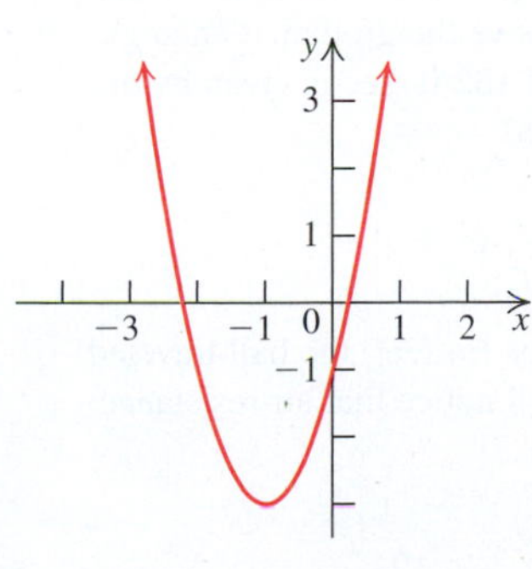
(g)

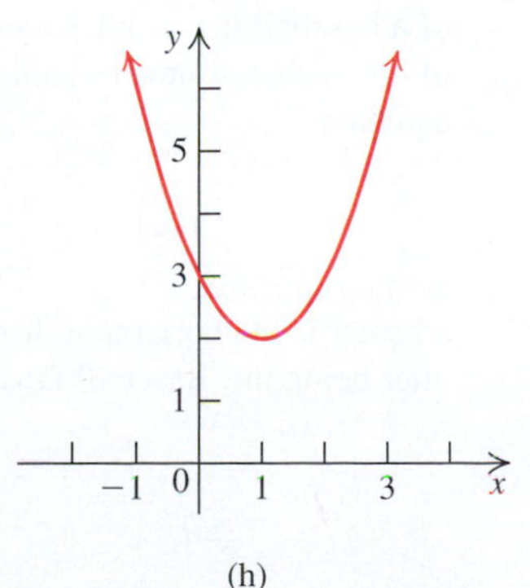
(h)

In Exercises 17–20, find a quadratic function of the form $y = ax^2$ that passes through the given point.

17. $(2, -8)$ **18.** $(-3, 3)$

19. $(2, 20)$ **20.** $(-3, -6)$

In Exercises 21–30, find the quadratic function $y = f(x)$ that has the given vertex and whose graph passes through the given point. Write the function in standard form.

21. Vertex $(0, 0)$; passing through $(-2, 8)$

22. Vertex $(2, 0)$; passing through $(1, 3)$

23. Vertex $(-3, 0)$; passing through $(-5, -4)$

24. Vertex $(0, 1)$; passing through $(-1, 0)$

25. Vertex $(2, 5)$; passing through $(3, 7)$

26. Vertex $(-3, 4)$; passing through $(0, 0)$

27. Vertex $(2, -3)$; passing through $(-5, 8)$

28. Vertex $(-3, -2)$; passing through $(0, -8)$

29. Vertex $\left(\frac{1}{2}, \frac{1}{2}\right)$; passing through $\left(\frac{3}{4}, -\frac{1}{4}\right)$

30. Vertex $\left(-\frac{3}{2}, -\frac{5}{2}\right)$; passing through $\left(1, \frac{55}{8}\right)$

In Exercises 31–34, the graph of a quadratic function $y = f(x)$ is given. Find the standard form of the function.

31.

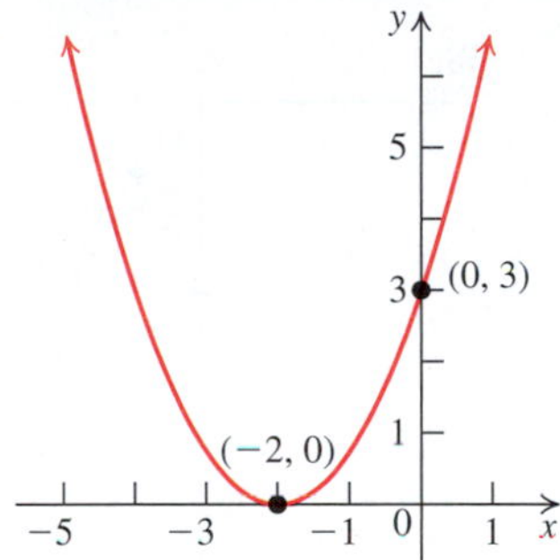

32.

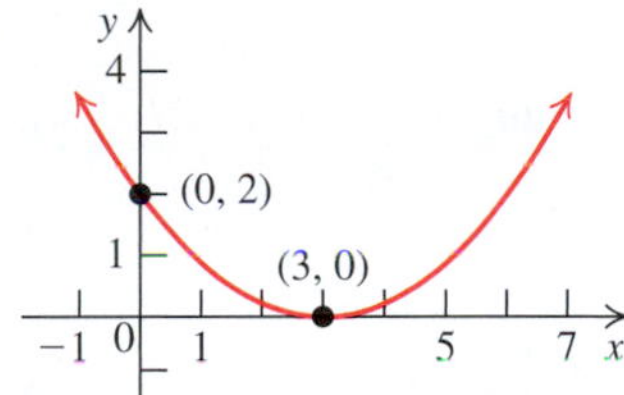

33.

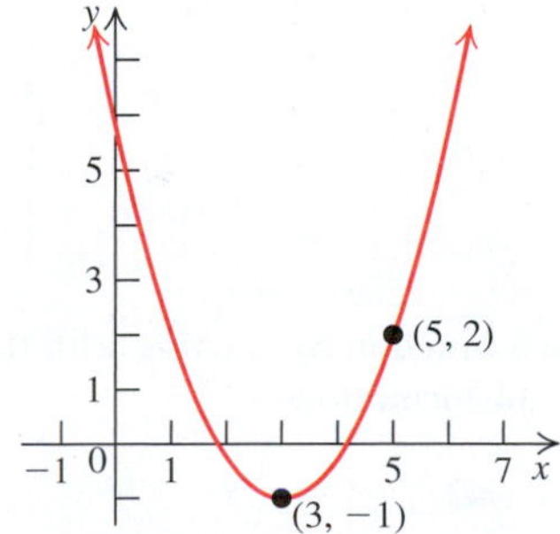

34.

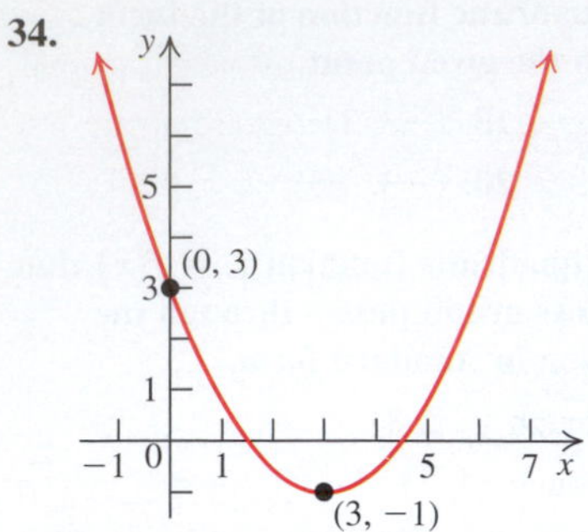

In Exercises 35–40, the graphs of $y = ax^2 + bx + c$ are given. Find the signs of a, b, and c and state which (if any) are zero.

35. 36.

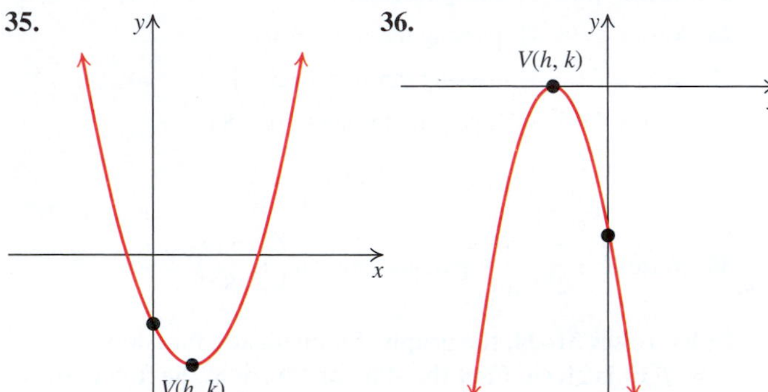

37. 38.

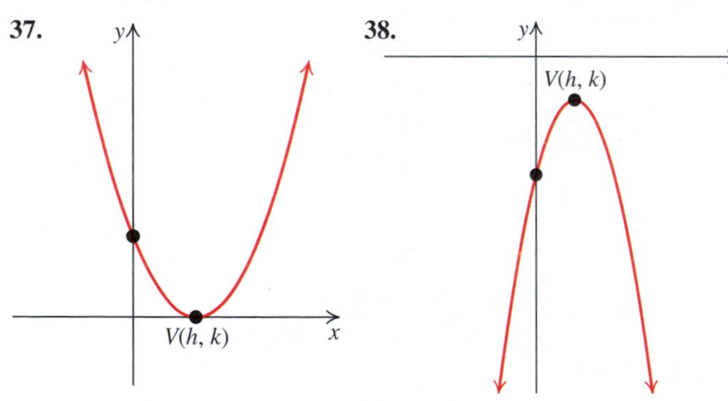

39. 40.

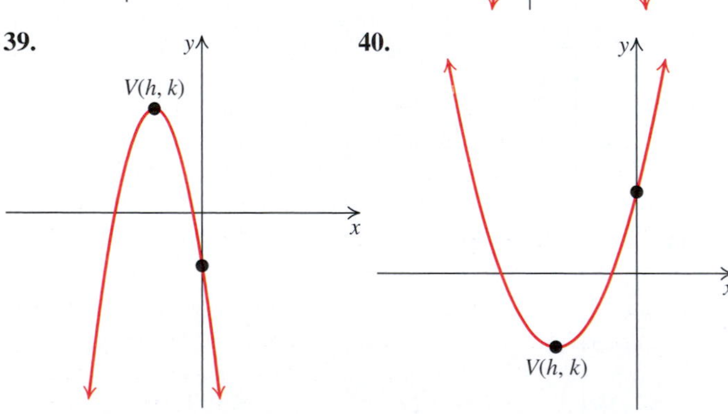

In Exercises 41–50, graph each function by starting with the graph of $y = x^2$ and using transformations.

41. $f(x) = 3x^2$

42. $f(x) = \frac{1}{3}x^2$

43. $g(x) = (x - 4)^2$

44. $g(x) = (x + 3)^2$

45. $f(x) = -2x^2 - 4$

46. $f(x) = -x^2 + 3$

47. $g(x) = (x - 3)^2 + 2$

48. $g(x) = (x + 1)^2 - 3$

49. $f(x) = -3(x - 2)^2 + 4$

50. $f(x) = -2(x + 1)^2 + 3$

In Exercises 51–58, graph the given function by writing it in the standard form $y = a(x - h)^2 + k$ and then using transformations on $y = x^2$. Find the vertex, the axis of symmetry, and the x- and y-intercepts.

51. $y = x^2 + 4x$

52. $y = x^2 - 2x + 2$

53. $y = 6x - 10 - x^2$

54. $y = 8 + 3x - x^2$

55. $y = 2x^2 - 8x + 9$

56. $y = 3x^2 + 12x - 7$

57. $y = -3x^2 + 18x - 11$

58. $y = -5x^2 - 20x + 13$

In Exercises 59–66, (a) determine whether the graph of the given quadratic function opens up or down, (b) find the vertex $(h, k) = \left(-\frac{b}{2a}, f\left(-\frac{b}{2a}\right)\right)$, (c) find the axis of symmetry, (d) find the x- and y-intercepts, and (e) sketch the graph of the function.

59. $y = x^2 - 8x + 15$

60. $y = x^2 + 8x + 13$

61. $y = x^2 - x - 6$

62. $y = x^2 + x - 2$

63. $y = x^2 - 2x + 4$

64. $y = x^2 - 4x + 5$

65. $y = 6 - 2x - x^2$

66. $y = 2 + 5x - 3x^2$

In Exercises 67–74, a quadratic function f is given. (a) Determine whether the given quadratic function has a maximum value or a minimum value. Then find this value. (b) Find the range of f.

67. $f(x) = x^2 - 4x + 3$

68. $f(x) = -x^2 + 6x - 8$

69. $f(x) = -4 + 4x - x^2$

70. $f(x) = x^2 - 6x + 9$

71. $f(x) = 2x^2 - 8x + 3$

72. $f(x) = 3x^2 + 12x - 5$

73. $f(x) = -4x^2 + 12x + 7$

74. $f(x) = 8x - 5 - 2x^2$

In Exercises 75–80, solve the given quadratic inequality by sketching the graph of the corresponding quadratic function.

75. $x^2 - 4 \le 0$

76. $x^2 + 4x + 5 < 0$

77. $x^2 - 4x + 3 > 0$

78. $x^2 + x - 2 > 0$

79. $-6x^2 + x + 7 > 0$

80. $-5x^2 + 9x - 4 \le 0$

Applying the Concepts

Note: The domain of a function that models a real-life problem may consist of only integer values. However, in solving these problems, we frequently find it convenient to extend the domain to a suitable interval of real numbers containing the relevant integers.

81. **Height of a baseball.** Ignoring air resistance, the height of a baseball that is hit 3 feet above the ground at an angle of 45° with an initial velocity of 132 ft/sec is given by the equation

$$h(x) = -\frac{32}{(132)^2}x^2 + x + 3,$$

where x is the horizontal distance (in feet) the ball traveled after being hit. Baseball fans will notice that air resistance severely distorts the actual distance a baseball travels. Use the graph to answer the following questions.

a. Approximately how many feet did the ball travel horizontally before hitting the ground?

b. Approximately how high did the ball go?

c. Use the function h to find answers to (a) and (b) rounded to the nearest integer.

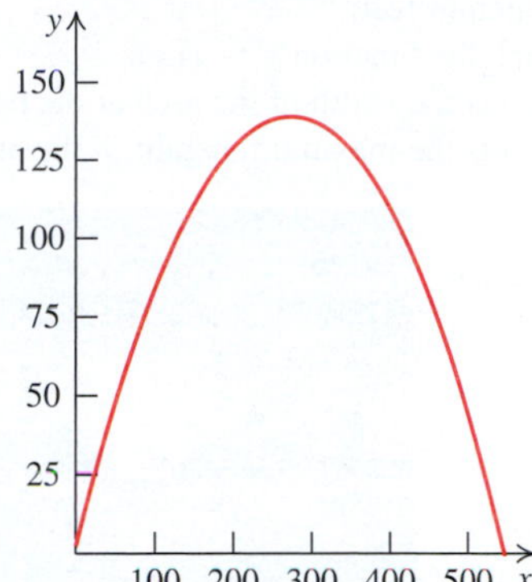

82. Apartment rentals. The number of apartments that can be rented in a 100-unit apartment complex decreases by one for each increase of \$25 in rent. The revenue generated from rent after x increases of \$25 is given by the equation $R(x) = -25x^2 + 1700x + 80{,}000$. Use the graph to answer the following questions.

a. Approximately what was the maximum revenue from the apartments?

b. Approximately how many \$25 increases generated the maximum revenue?

c. Use the function R to find answers to (a) and (b).

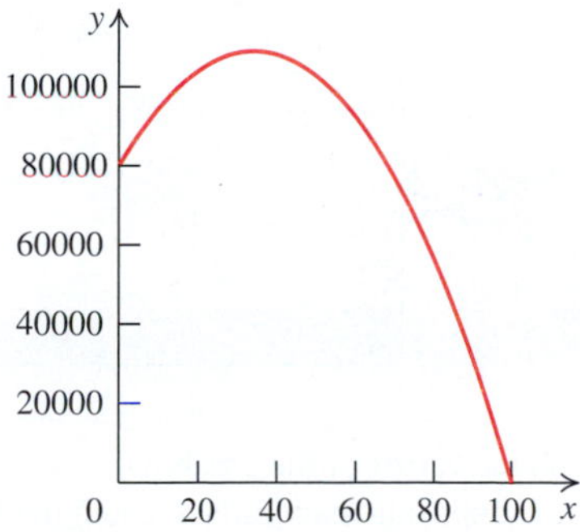

83. Maximizing revenue. A revenue function is given by $R(x) = 15 + 114x - 3x^2$, where x is the number of units produced and sold. Find the value of x for which the revenue is maximum.

84. Maximizing revenue. A demand function is given by $p = 200 - 4x$, where p is the price per unit and x is the number of units produced and sold (the demand). Find x for which the revenue is maximum. [Recall that $R(x) = x \cdot p$.]

85. Minimizing cost. The total cost of a product is $C = 200 - 50x + x^2$, where x is the number of units produced. Find the value of x for which the total cost is minimum.

86. Maximizing profit. A manufacturer produces x items of a product at a total cost of $(50 + 2x)$ dollars. The demand function for the product is $p = 100 - x$. Find the value of x for which the profit is maximum. How much is the maximum profit? (Recall that Profit = Revenue − Cost.)

87. Geometry. Find the dimensions of a rectangle of maximum area if the perimeter of the rectangle is 80 units. What is the maximum area?

88. Enclosing area. A rancher with 120 meters of fence intends to enclose a rectangular region along a river (which serves as a natural boundary requiring no fence). Find the maximum area that can be enclosed.

89. Enclosing playing fields. You have 600 meters of fencing, and you plan to lay out two identical playing fields. The fields can be side-by-side and separated by a fence, as shown in the figure. Find the dimensions and the maximum area of each field.

90. Enclosing area on a budget. You budget \$2400 for constructing a rectangular enclosure that consists of a high surrounding fence and a lower inside fence that divides the enclosure in half. The high fence costs \$8 per foot, and the low fence costs \$4 per foot. Find the dimensions and the maximum area of each half of the enclosure.

91. Maximizing apple yield. From past surveys and records, an orchard owner in northern Michigan has determined that if 26 apple trees per acre are planted, each tree yields 500 apples, on average. The yield decreases by ten apples per tree for each additional tree planted. How many trees should be planted per acre to achieve the maximum total yield?

92. Buying and selling beef. A steer weighing 300 pounds gains 8 pounds per day and costs \$1 a day to keep. You bought this steer today at a market price of \$1.50 per pound. However, the market price is falling 2¢ per pound per day. When should you sell the steer to maximize your profit?

93. Going on a tour. A group of students plans a tour. The charge per student is \$72 if 20 students go on the trip. If more than 20 students participate, the charge per student is reduced by \$2 times the number of students over 20. Find the number of students that will furnish the maximum revenue. What is the maximum revenue?

94. Computer disks. A 5-inch-radius computer disk contains information in units called *bytes* that are arranged in concentric tracks on the disk. Assume that m bytes per inch can be put on each track. The tracks are uniformly spread (that is, there is a fixed number, say, p, of tracks per inch, measured radially across the disk). If the number of bytes on each track must be the same, where should the innermost track be located to get the maximum number of bytes on the disk?

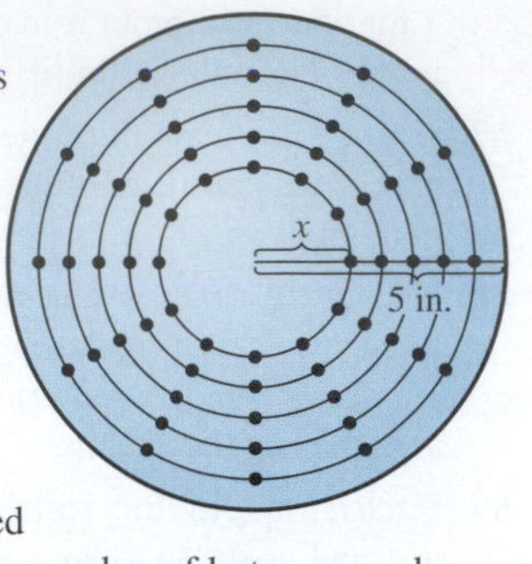

95. Motion on moon. On the Moon, the acceleration due to gravity is $g_M = 1.6 \text{ m/s}^2$. If an astronaut threw a stone from the 5 m high platform on his spacecraft directly upward and the stone went up to the maximum height of 25 m,

a. Find the height of the stone as a function of time t with $t = 0$ corresponding to the moment of releasing the stone.

b. How long did it take the stone to reach its highest point?

96. Repeat Exercise 95 assuming that the platform was 7 m high and the stone went up to the maximum height of 52 m.

97. Motion of a projectile. A projectile is fired straight up with a velocity of 64 ft/s. Its altitude (height) h after t seconds is given by

$$h(t) = -16t^2 + 64t.$$

a. What is the maximum height of the projectile?

b. When does the projectile hit the ground?

98. Motion of a projectile. A projectile is shot upward with a velocity of 64 feet per second. Its altitude (height) h after t seconds is given by

$$h(t) = -16t^2 + 64t + 336.$$

a. What is the maximum height of the projectile?

b. When does the projectile hit the ground?

99. Architecture. A window is to be constructed in the shape of a rectangle surmounted by a semicircle. If the perimeter of the window is 18 feet, find its dimensions for the maximum area.

100. Architecture. The shape of the Gateway Arch in St. Louis, Missouri, is a *catenary* curve, which closely resembles a parabola. The function

$$y = f(x) = -0.00635x^2 + 4x$$

models the shape of the arch, where y is the height in feet and x is the horizontal distance from the base of the left side of the arch in feet.

a. Graph the function $y = f(x)$.

b. What is the width of the arch at the base?

c. What is the maximum height of the arch?

101. Football. The altitude or height h (in feet) of a punted football can be modeled by the quadratic function

$$h = -0.01x^2 + 1.18x + 2,$$

where x (in feet) is the horizontal distance from the point of impact with the punter's foot. (See accompanying figure.)

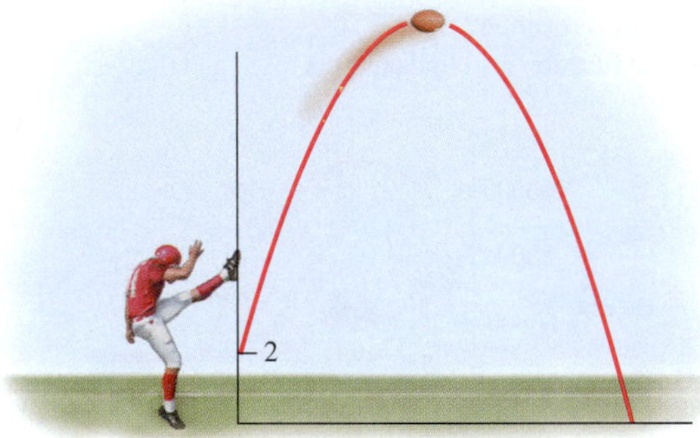

a. Find the vertex of the parabola.

b. If the opposing team fails to catch the ball, how far away from the punter does the ball hit the ground (to the nearest foot)?

c. What is the maximum height of the punted ball (to the nearest foot)?

d. The nearest defensive player is 6 feet from the point of impact. How high must the player reach to block the punt (to the nearest foot)?

e. How far downfield has the ball traveled when it reaches a height of 7 feet for the second time?

102. Field hockey. A *scoop* in field hockey is a pass that propels the ball from the ground into the air. Suppose a player makes a scoop with an upward velocity of 30 feet per second. The function

$$h = -16t^2 + 30t$$

models the altitude or height h in feet at time t in seconds. Will the ball ever reach a height of 16 feet?

Beyond the Basics

In Exercises 103–108, find a quadratic function of the form $f(x) = ax^2 + bx + c$ that satisfies the given conditions.

103. The graph of f is obtained by shifting the graph of $y = 3x^2$ two units horizontally to the left and three units vertically up.

104. The graph of f passes through the point $(1, 5)$ and has vertex $(-3, -2)$.

105. The graph of f has vertex $(1, -2)$ and y-intercept 4.

106. The graph of f has x-intercepts 2 and 6 and y-intercept 24.

107. The graph of f has x-intercept 7 and y-intercept 14, and $x = 3$ is the axis of symmetry.

108. The graph of f is obtained by shifting the graph of $y = x^2 + 2x + 2$ three units horizontally to the right and two units vertically down.

In Exercises 109–114, find two quadratic functions, one opening up and the other down, whose graphs have the given x-intercepts.

109. $-2, 6$

110. $-3, 5$

111. $-7, -1$

112. $2, 10$

113. Let $f(x) = 4x - x^2$. Solve $f(a + 1) - f(a - 1) = 0$ for a.

114. Let $f(x) = x^2 + 1$. Find $(f \circ f)(x)$.

Critical Thinking/Discussion/Writing

115. **Symmetry about the line $x = h$.** The graph of a polynomial function $y = f(x)$ is symmetric about the line $x = h$ if $f(h + p) = f(h - p)$ for every number p. For $f(x) = ax^2 + bx + c, a \neq 0$, set $f(h + p) = f(h - p)$ and show that $h = -\frac{b}{2a}$. This exercise proves that the graph of a quadratic function is symmetric about its axis: $x = -\frac{b}{2a}$.

116. In the graph of $y = 2x^2 - 8x + 9$, find the coordinates of the point symmetric to the point $(-1, 19)$ about the axis of symmetry.

117. Explain how to solve the inequality $ax^2 + bx + c > 0$ with $a \neq 0$ using only the sign of a and the x-intercepts of $f(x) = ax^2 + bx + c$.

118. Let $f(x) = a(x - h)^2 + k, a \neq 0$ and $g(x) = mx + b, m \neq 0$.

a. Show that the graph of $y = (f \circ g)(x)$ is a parabola and find its vertex.

b. Show that the graph of $y = (g \circ f)(x)$ is a parabola and find its vertex.

119. Let $f(x) = ax^2 + bx + c, a \neq 0$ with discriminant $b^2 - 4ac$ of the equation $f(x) = 0$. Discuss the relationship between the location of the vertex of $y = f(x)$ and the sign of the discriminant.

120. Show that a horizontal translation of the graph of a quadratic function is the composition of the quardratic function and a linear function.

Maintaining Skills

In Exercises 121–128, simplify each expression. Write your answers without negative exponents.

121. -5^0

122. $(-4)^{-2}$

123. -4^{-2}

124. $(2x^2)(-3x^4)$

125. $x^7 \cdot x^{-7}$

126. $(2x^2)(3x)(-x^3)$

127. $x^2\left(3 - \frac{3}{4}\right)$

128. $x^4\left(1 + \frac{3}{x^2} - \frac{5}{x^3}\right)$

In Exercises 129–136, factor each expression.

129. $4x^2 - 9$

130. $x^2 + 6x + 9$

131. $15x^2 + 11x - 12$

132. $14x^2 - 3x - 2$

133. $x^2(x - 1) - 4(x - 1)$

134. $9x^2(2x + 7) - 25(2x + 7)$

135. $x^3 + 4x^2 + 3x + 12$

136. $2x^3 - 3x^2 + 2x - 3$

SECTION 3.2 Polynomial Functions

BEFORE STARTING THIS SECTION, REVIEW

1. Graphs of equations (Section 2.2, page 66)
2. Solving linear equations (Section 1.1, page 78)
3. Solving quadratic equations (Section 1.2, page 52)
4. Transformations (Section 2.7, page 250)
5. Relative maximum and minimum values (Section 2.5, page 215)

OBJECTIVES

1. Learn properties of the graphs of polynomial functions.
2. Determine the end behavior of polynomial functions.
3. Find the zeros of a polynomial function by factoring.
4. Identify the relationship between degrees, real zeros, and turning points.
5. Graph polynomial functions.

Johannes Kepler (1571–1630)

Kepler was born in Weil in southern Germany and studied at the University of Tübingen. He studied with Michael Mastlin (professor of mathematics), who privately taught him the Copernican theory of the sun-centered universe while hardly daring to recognize it openly in his lectures. Kepler knew that to work out a detailed version of Copernicus's theory, he had to have access to the observations of the Danish astronomer Tycho Brahe (1546–1601). Kepler's correspondence with Brahe resulted in his appointment as Brahe's assistant. Brahe died about 18 months after Kepler's arrival. During those 18 months, Kepler learned enough about Brahe's work to create his own major project. He produced the famous three laws of planetary motion published in 1609 in his book *Astronomia Nova.*

Kepler's Wedding Reception

Kepler is best known as an astronomer who discovered the three laws of planetary motion. However, Kepler's primary field was not astronomy, but mathematics. Legend has it that at his own wedding reception, Kepler observed that the Austrian vintners could quickly and mysteriously compute the volumes of a variety of wine barrels. Each barrel had a hole, called a *bunghole*, in the middle of its side. The vintner would insert a rod, called the *bungrod*, into the hole until it hit the far corner. Then he would announce the volume. During the reception, Kepler occupied his mind with the problem of finding the volume of a wine barrel with a bungrod.

In Example 10, we show how he solved the problem for barrels that are perfect cylinders. But barrels are not perfect cylinders; they are wider in the middle and narrower at the ends and in between, the sides make a graceful curve that appears to be an arc of a circle. What could be the formula for the volume of such a shape? Kepler tackled this problem in his important book *Nova Stereometria Doliorum Vinariorum* (1615) and developed a related mathematical theory.

1 Learn properties of the graphs of polynomial functions.

Polynomial Functions

We review some basic terminology. A **polynomial function of degree n** is a function of the form

$$f(x) = a_n x^n + a_{n-1}x^{n-1} + \cdots + a_2x^2 + a_1x + a_0,$$

where n is a nonnegative integer and the *coefficients* $a_n, a_{n-1}, \ldots, a_2, a_1, a_0$ are real numbers with $a_n \neq 0$. The term a_nx^n is called the **leading term**, the number a_n (the coefficient of x^n) is called the **leading coefficient**, and a_0 is the **constant term**. A constant function $f(x) = a(a \neq 0)$, which may be written as $f(x) = ax^0$, for $x \neq 0$ is a polynomial of degree 0. Because the zero function $f(x) = 0$ has no nonzero coefficients, no degree is assigned to it.

In Section 2.6, we discussed linear functions, which are polynomial functions of degree 0 and 1. In Section 3.1, we studied quadratic functions, which are polynomial functions of degree 2. Polynomials of degree 3, 4, and 5 are also called **cubic, quartic**, and **quintic polynomials**, respectively. In this section, we concentrate on graphing polynomial functions of degree 3 or more.

Here are some common properties shared by all polynomial functions.

Common Properties of Polynomial Functions

1. The domain of a polynomial function is the set of all real numbers.
2. The graph of a polynomial function is a **continuous curve**. This means that the graph has no holes or gaps and can be drawn on a sheet of paper without lifting the pencil. See Figure 3.6.

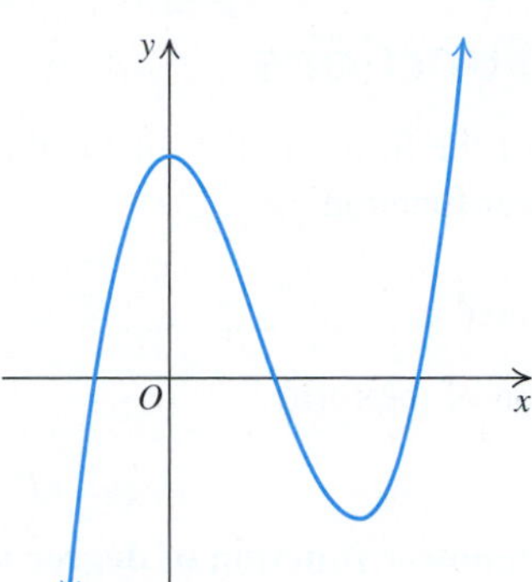

A continuous curve; can be the graph of a polynomial function

(a)

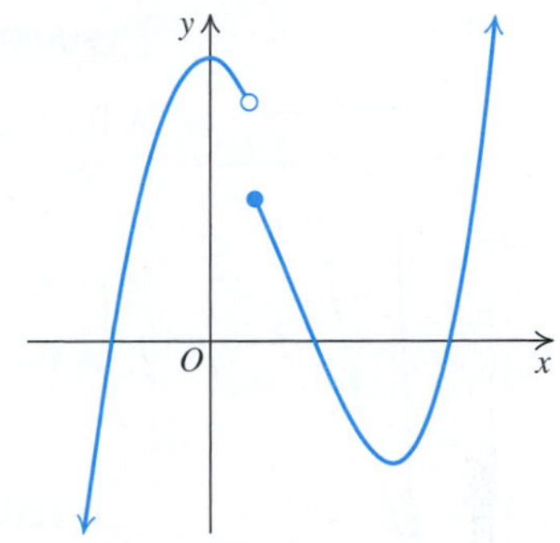

A discontinuous curve; cannot be the graph of a polynomial function

(b)

FIGURE 3.6

3. The graph of a polynomial function is a **smooth curve**. This means that the graph does not contain any sharp corners. See Figure 3.7.

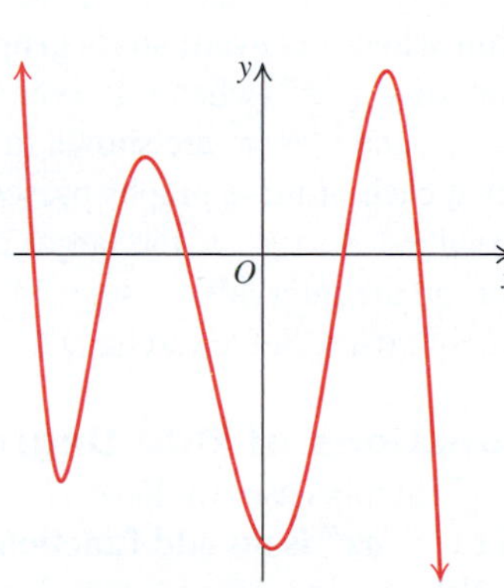

A smooth and continuous curve; can be the graph of a polynomial function

(a)

sharp corner

sharp corner

y

O

x

sharp corner

A continuous but not a smooth curve; cannot be the graph of a polynomial function

(b)

FIGURE 3.7

EXAMPLE 1 Polynomial Functions

State which functions are polynomial functions. For each polynomial function, find its degree, the leading term, and the leading coefficient.

a. $f(x) = 5x^4 - 2x + 7$ **b.** $g(x) = 7x^2 - x + 1,\ 1 \le x \le 5$

Solution

a. $f(x) = 5x^4 - 2x + 7$ is a polynomial function. Its degree is 4, the leading term is $5x^4$, and the leading coefficient is 5.

b. $g(x) = 7x^2 - x + 1, 1 \le x \le 5$ is not a polynomial function because its domain is not $(-\infty, \infty)$. However, $h(x) = 7x^2 - x + 1$ with no restrictions on the domain is a polynomial function of degree 2 with leading term $7x^2$ and leading coefficient 7.

Practice Problem 1 Repeat Example 1 for each function.

a. $f(x) = \dfrac{x^2 + 1}{x - 1}$ **b.** $g(x) = 2x^7 + 5x^2 - 17$

Power Functions

A function of the form $f(x) = ax^n$ is the simplest nth-degree polynomial function and is called a power function.

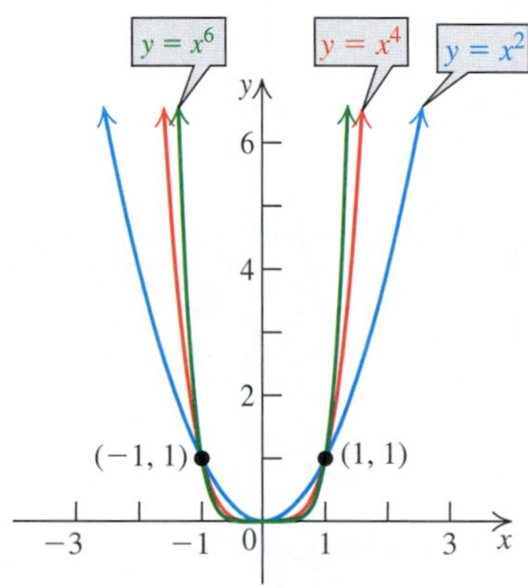

FIGURE 3.8 Power functions of even degree

Power Function

A function of the form

$$f(x) = ax^n$$

is called a **power function of degree n**, where a is a nonzero real number and n is a positive integer.

The graph of a power function can be obtained by stretching or shrinking the graph of $y = x^n$ (and reflecting it about the x-axis if $a < 0$). So the basic shape of the graph of a power function $f(x) = ax^n$ is similar to the shape of the graphs of the equations $y = \pm x^n$.

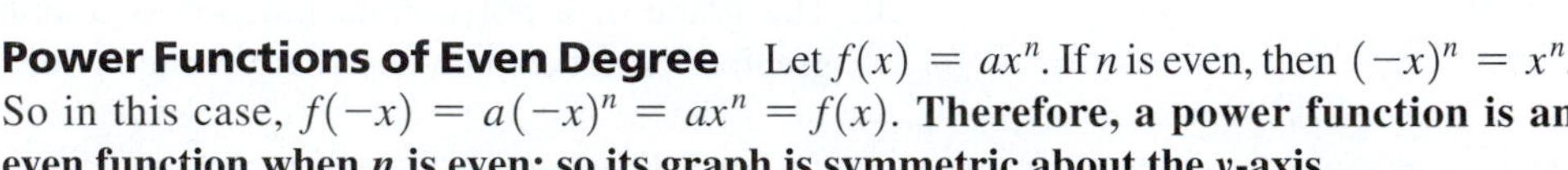

Power Functions of Even Degree Let $f(x) = ax^n$. If n is even, then $(-x)^n = x^n$. So in this case, $f(-x) = a(-x)^n = ax^n = f(x)$. **Therefore, a power function is an even function when n is even; so its graph is symmetric about the y-axis.**

The graph of $y = x^n$ (when n is even) is similar to the graph of $y = x^2$; the graphs of $y = x^2, y = x^4$, and $y = x^6$ are shown in Figure 3.8.

Notice that each of these graphs passes through the points $(-1, 1)$, $(0, 0)$, and $(1, 1)$. On the interval $-1 < x < 1$, the larger the exponent, the flatter the graph. See Exercise 91. However, on the intervals $(-\infty, -1)$ and $(1, \infty)$, the larger the exponent is, the more rapidly the graph rises. See Exercise 92.

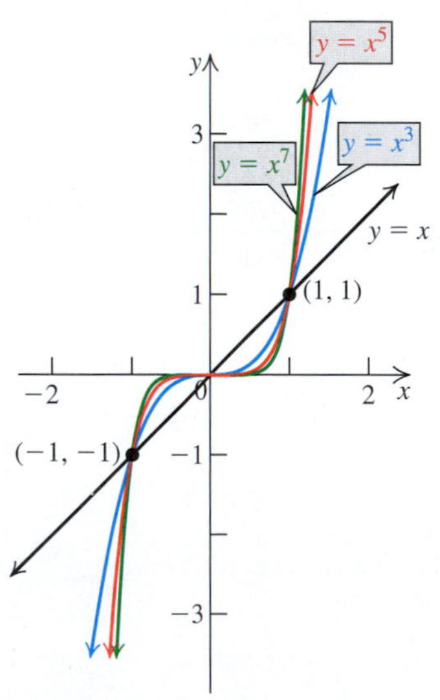

FIGURE 3.9 Power functions of odd degree

Power Functions of Odd Degree Again, let $f(x) = ax^n$. If n is odd, then $(-x)^n = -x^n$. In this case, we have $f(-x) = a(-x)^n = -ax^n = -f(x)$. **So the power function $f(x) = ax^n$ is an odd function when n is odd; its graph is symmetric about the origin.** When n is odd and $n \ge 3$, the graph of $f(x) = x^n$ is similar to the graph of $y = x^3$. The graphs of $y = x, y = x^3, y = x^5$, and $y = x^7$ are shown in Figure 3.9. If n is an odd positive integer, then the graph of $y = x^n$ passes through the points $(-1, -1)$, $(0, 0)$, and $(1, 1)$. The larger the value of n is, the flatter the graph is near the origin (in the interval $-1 < x < 1$) and the more rapidly it rises (or falls) on the intervals $(-\infty, -1)$ and $(1, \infty)$.

2 Determine the end behavior of polynomial functions.

End Behavior of Polynomial Functions

We investigate the behavior of the polynomial function $y = f(x)$ when the independent variable x is large in absolute value. The notation $x \to \infty$ (read "x approaches infinity") means that x gets larger and larger without bound: x can assume values greater than

SIDE NOTE

Here is a scheme for remembering the axis directions associated with the symbols $-\infty$ and ∞.

$x \to -\infty$ Left
$x \to \infty$ Right
$y \to -\infty$ Down
$y \to \infty$ Up

1, 10, 100, 1000, . . . , without end. Similarly, $x \to -\infty$ (read "x approaches negative infinity") means that x can assume values less than $-1, -10, -100, -1000, \ldots$, without end. Of course, x may be replaced by any other independent variable.

The behavior of a function $y = f(x)$ as $x \to \infty$ or $x \to -\infty$ is called the **end behavior** of the function. In the accompanying box, we describe the behavior of the power function $y = f(x) = ax^n$ as $x \to \infty$ or as $x \to -\infty$. **Every polynomial function exhibits end behavior similar to one of the four functions $y = x^2$, $y = -x^2$, $y = x^3$, or $y = -x^3$.**

END BEHAVIOR FOR THE GRAPH OF $y = f(x) = ax^n$

n Even

Case 1
$a > 0$

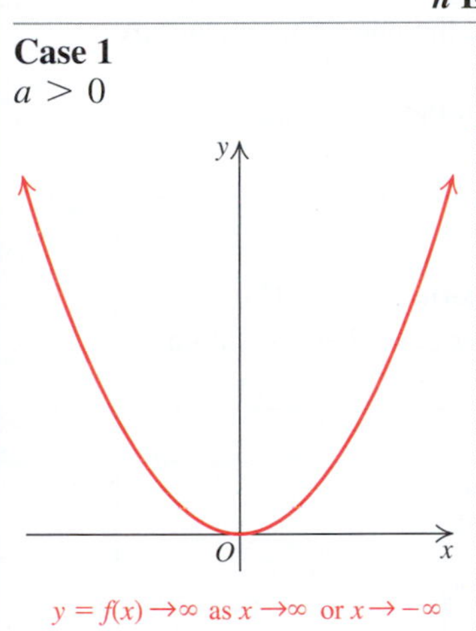

(a)

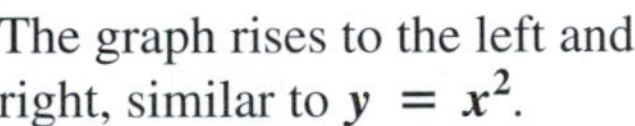

The graph rises to the left and right, similar to $y = x^2$.

Case 2
$a < 0$

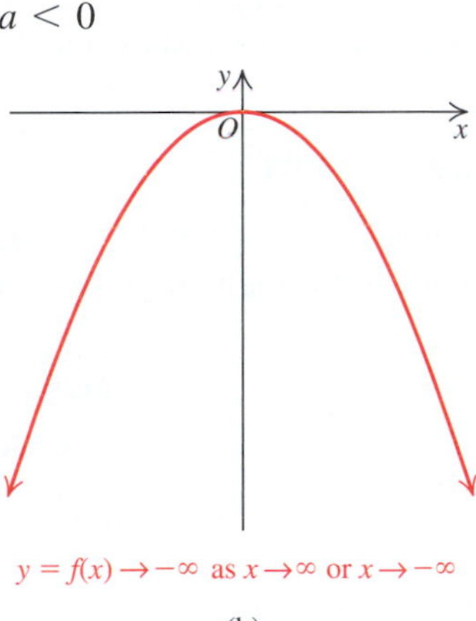

(b)

The graph falls to the left and right, similar to $y = -x^2$.

n Odd

Case 3
$a > 0$

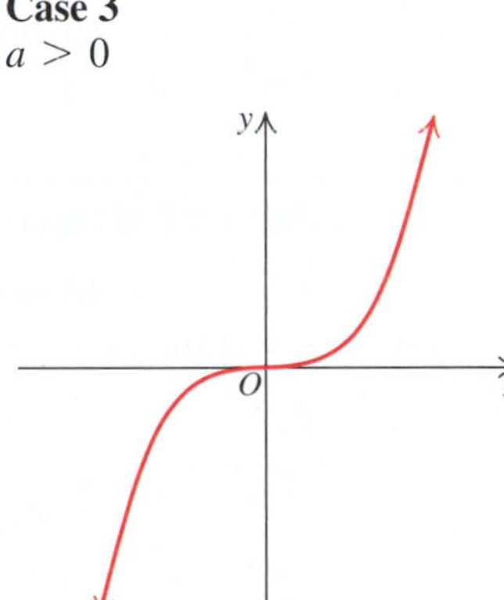

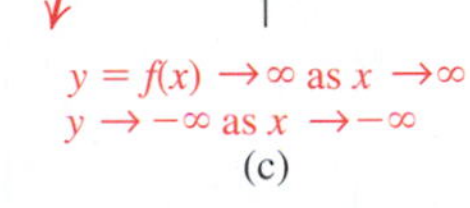

(c)

The graph rises to the right and falls to the left, similar to $y = x^3$.

Case 4
$a < 0$

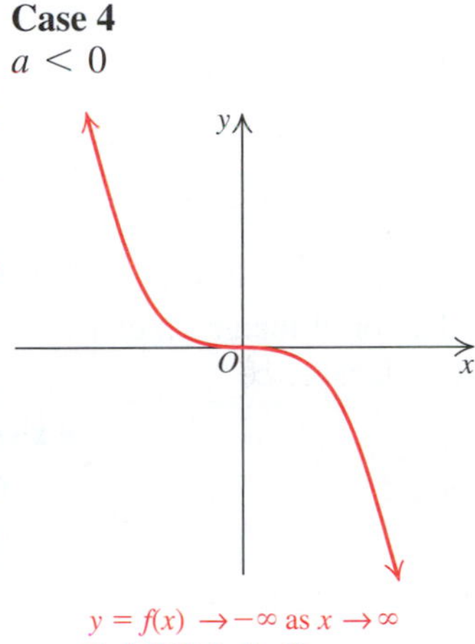

(d)

The graph rises to the left and falls to the right, similar to $y = -x^3$.

In the next example, we show that the end behavior of the given polynomial function is determined by its leading term. If a polynomial $P(x)$ has approximately the same values as $f(x) = ax^n$ when $|x|$ is very large, we write $\boldsymbol{P(x) \approx ax^n}$ when $|x|$ is very large.

EXAMPLE 2 Understanding the End Behavior of a Polynomial Function

Let $P(x) = 2x^3 + 5x^2 - 7x + 11$ be a polynomial function of degree 3. Show that $P(x) \approx 2x^3$ when $|x|$ is very large.

Solution

$$P(x) = 2x^3 + 5x^2 - 7x + 11 \qquad \text{Given polynomial}$$

$$= x^3\left(2 + \frac{5}{x} - \frac{7}{x^2} + \frac{11}{x^3}\right) \qquad \text{Distributive property}$$

TECHNOLOGY CONNECTION

Calculator graph of $P(x) = 2x^3 + 5x^2 - 7x + 11$

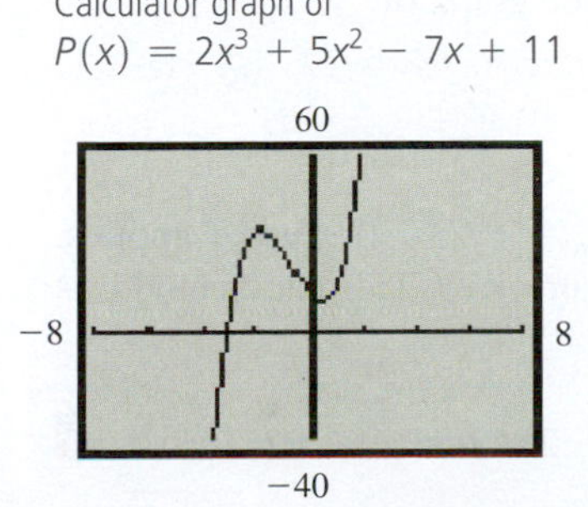

TABLE 3.1

x	$\frac{5}{x}$
10	0.5
10^2	0.05
10^3	0.005
10^4	0.0005

When $|x|$ is very large, the terms $\frac{5}{x}, \frac{7}{x^2}$, and $\frac{11}{x^3}$ are close to 0. Table 3.1 shows some values of $\frac{5}{x}$ as x increases. (You should compute the values of $\frac{7}{x^2}$ and $\frac{11}{x^3}$ for $x = 10, 10^2, 10^3$, and 10^4.) When $|x|$ is very large, we have

$$P(x) = x^3\left(2 + \frac{5}{x} - \frac{7}{x^2} + \frac{11}{x^3}\right) \approx x^3(2 + 0 - 0 + 0) = 2x^3$$

So the polynomial $P(x) \approx 2x^3$ when $|x|$ is very large.

Practice Problem 2 Let $P(x) = 4x^3 + 2x^2 + 5x - 17$. Show that $P(x) \approx 4x^3$ when $|x|$ is very large. 

It is possible to use the technique of Example 2 to show that the end behavior of any polynomial function is determined by its leading term.

The end behaviors of polynomial functions are shown in the following graphs.

THE LEADING-TERM TEST

Let $f(x) = a_nx^n + a_{n-1}x^{n-1} + \cdots + a_1x + a_0$ $(a_n \neq 0)$ be a polynomial function. Its leading term is a_nx^n. The behavior of the graph of f as $x \to \infty$ or $x \to -\infty$ is similar to one of the following four graphs and is described as shown in each case.

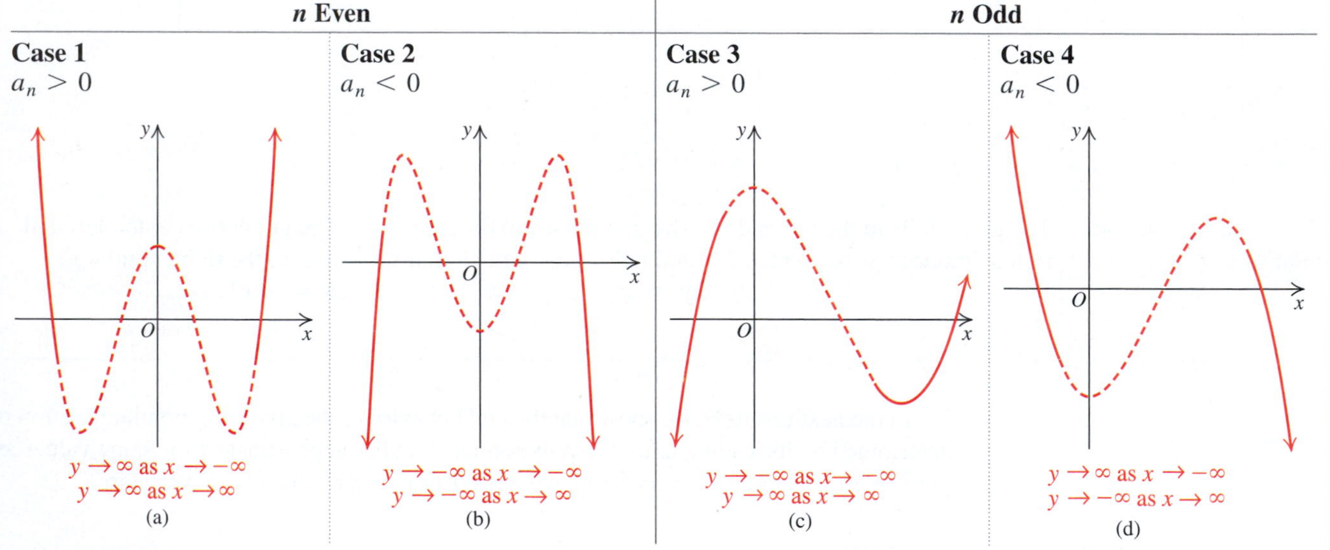

This test does not describe the middle portion of each graph, shown by the dashed lines.

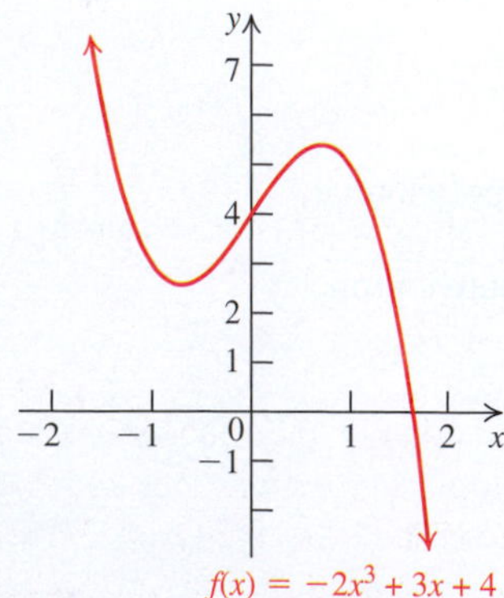

FIGURE 3.10 End behavior of $f(x) = -2x^3 + 3x + 4$

EXAMPLE 3 Using the Leading-Term Test

Use the leading-term test to determine the end behavior of the graph of

$$y = f(x) = -2x^3 + 3x + 4.$$

Solution

Here the degree $n = 3$ (odd) and the leading coefficient $a_n = -2 < 0$. Case 4 applies. The graph of f rises to the left and falls to the right. See Figure 3.10. This end behavior is described as $y \to \infty$ as $x \to -\infty$ and $y \to -\infty$ as $x \to \infty$.

Practice Problem 3 What is the end behavior of $f(x) = -2x^4 + 5x^2 + 3$? 

3 Find the zeros of a polynomial function by factoring.

Zeros of a Function

Let f be a function. An input c in the domain of f that produces output 0 is called a *zero* of the function f. For example, if $f(x) = (x - 3)^2$, then 3 is a zero of f because $f(3) = (3 - 3)^2 = 0$. In other words, a number c is a **zero** of f if $f(c) = 0$. The zeros of a function f can be obtained by solving the equation $f(x) = 0$. One way to solve a polynomial equation $f(x) = 0$ is to factor $f(x)$ and then use the zero-product property.

If c is a real number and $f(c) = 0$, then c is called a **real zero** of f. Geometrically, this means that the graph of f has an x-intercept at $x = c$.

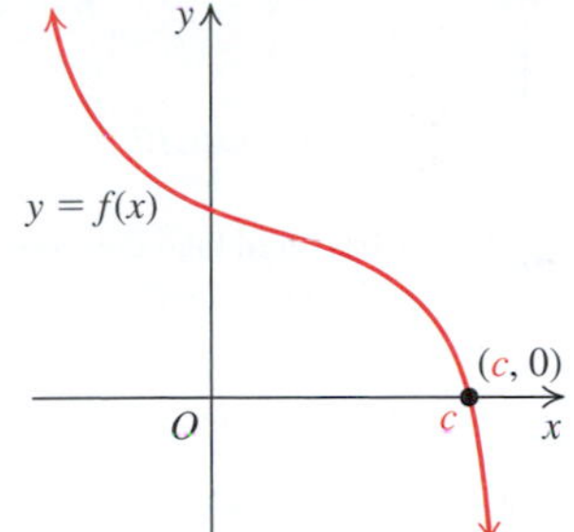

FIGURE 3.11 A zero at c of f

REAL ZEROS OF POLYNOMIAL FUNCTIONS

If f is a polynomial function and c is a real number, then the following statements are equivalent:

1. c is a **zero** of f.
2. c is a **solution** (or **root**) of the equation $f(x) = 0$.
3. c is an **x-intercept** of the graph of f. The point $(c, 0)$ is on the graph of f. See Figure 3.11.

$f(x) = x^4 - 2x^3 - 3x^2$

4
$Y1 = x^4 - 2x^3 - 3x^2$
−2 −15
$x = {}^-1$ $y = 0$
5

We can verify that the zeros of f are 0, −1, and 3 using the TRACE function.

$g(x) = x^3 - 2x^2 + x - 2$

has a real zero, 2.

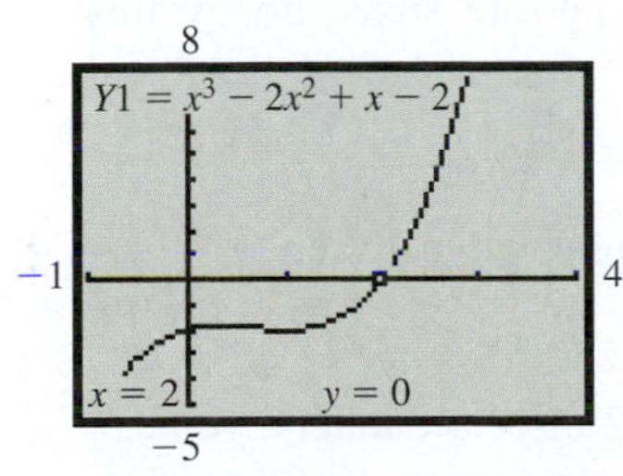

EXAMPLE 4 Finding the Zeros of a Polynomial Function

Find all real zeros of each polynomial function.

a. $f(x) = x^4 - 2x^3 - 3x^2$ **b.** $g(x) = x^3 - 2x^2 + x - 2$

Solution

a. We factor $f(x)$ and then solve the equation $f(x) = 0$.

$$
\begin{aligned}
f(x) &= x^4 - 2x^3 - 3x^2 && \text{Given function} \\
&= x^2(x^2 - 2x - 3) && \text{Factor out the common factor, } x^2. \\
&= x^2(x + 1)(x - 3) && \text{Factor } (x^2 - 2x - 3). \\
x^2(x + 1)(x - 3) &= 0 && \text{Set } f(x) = 0. \\
x^2 = 0, \quad x + 1 = 0, &\quad \text{or} \quad x - 3 = 0 && \text{Zero-product property} \\
x = 0, \quad x = -1, &\quad \text{or} \quad x = 3 && \text{Solve each equation.}
\end{aligned}
$$

The zeros of $f(x)$ are 0, −1, and 3. (See the calculator graph of f in the margin.)

b. By grouping terms, we factor $g(x)$ to obtain

$$
\begin{aligned}
g(x) &= x^3 - 2x^2 + x - 2 \\
&= x^2(x - 2) + 1 \cdot (x - 2) && \text{Group terms.} \\
&= (x - 2)(x^2 + 1) && \text{Factor out } x - 2. \\
0 &= (x - 2)(x^2 + 1) && \text{Set } g(x) = 0. \\
x - 2 = 0, &\quad \text{or} \quad x^2 + 1 = 0 && \text{Zero-product property}
\end{aligned}
$$

Solving for x, we obtain 2 as the only real zero of $g(x)$ because $x^2 + 1 > 0$ for all real numbers x. The calculator graph of g in the margin supports our conclusion.

Practice Problem 4 Find all real zeros of $f(x) = 2x^3 - 3x^2 + 4x - 6$.

Finding the zeros of a polynomial of degree 3 or more is one of the most important problems of mathematics. You will learn more about this topic in Sections 3.3 and 3.4. For now, we will approximate the zeros by using an *Intermediate Value Theorem*. Recall that the graph of a polynomial function $f(x)$ is a continuous curve. Suppose you locate two numbers a and b such that the function values $f(a)$ and $f(b)$ have opposite signs (one positive and the other negative). Then the continuous graph of f connecting the points

$(a, f(a))$ and $(b, f(b))$ must cross the x-axis (at least once) somewhere between a and b. In other words, the function f has a zero between a and b. See Figure 3.12.

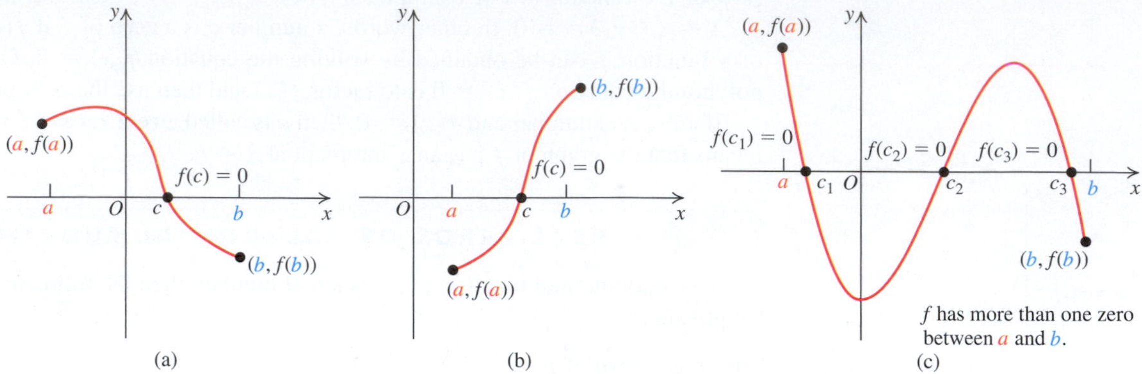

FIGURE 3.12 Each function has at least one zero between *a* and *b*

AN INTERMEDIATE VALUE THEOREM

Let a and b be two numbers with $a < b$. If f is a polynomial function such that $f(a)$ and $f(b)$ have opposite signs, then there is at least one number c, with $a < c < b$, for which $f(c) = 0$.

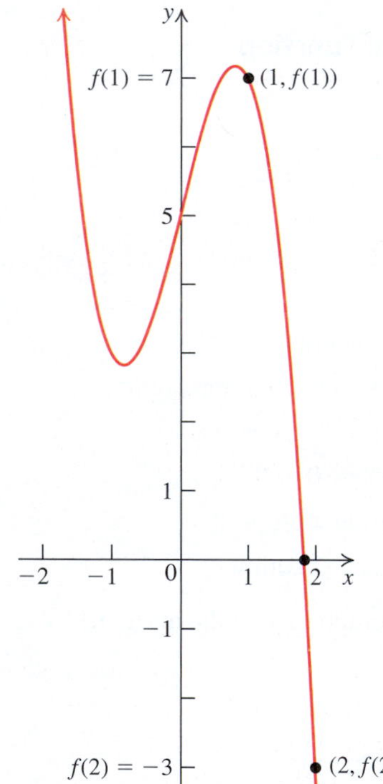

FIGURE 3.13 A zero lies between 1 and 2

Notice that Figure 3.12(c) shows more than one zero of f. That is why this Intermediate Value Theorem says that f has *at least one* zero between a and b.

EXAMPLE 5 Using an Intermediate Value Theorem

Show that the function $f(x) = -2x^3 + 4x + 5$ has a real zero between 1 and 2.

Solution

$$f(1) = -2(1)^3 + 4(1) + 5 \quad \text{Replace } x \text{ with1 in } f(x).$$
$$= -2 + 4 + 5 = 7 \quad \text{Simplify.}$$
$$f(2) = -2(2)^3 + 4(2) + 5 \quad \text{Replace } x \text{ with 2.}$$
$$= -2(8) + 8 + 5 = -3 \quad \text{Simplify.}$$

Because $f(1)$ is positive and $f(2)$ is negative, they have opposite signs. So by this Intermediate Value Theorem, the polynomial function $f(x) = -2x^3 + 4x + 5$ has a real zero between 1 and 2. See Figure 3.13.

Practice Problem 5 Show that $f(x) = 2x^3 - 3x - 6$ has a real zero between 1 and 2.

In Example 5, if you want to find the zero between 1 and 2 more accurately, you can divide the interval $[1, 2]$ into tenths and find the value of the function at each of the points 1, 1.1, 1.2, . . . , 1.9, and 2. You will find that

$$f(1.8) = 0.536 \quad \text{and} \quad f(1.9) = -1.118.$$

So by this Intermediate Value Theorem, f has a zero between 1.8 and 1.9. If you divide the interval $[1.8, 1.9]$ into tenths, you will find that $f(1.83) > 0$ and $f(1.84) < 0$; so f has a zero between 1.83 and 1.84. By continuing this process, you can find the zero of f between 1.8 and 1.9 to any desired degree of accuracy.

4 Identify the relationship between degrees, real zeros, and turning points.

Zeros and Turning Points

In Section 3.5, you will learn the Fundamental Theorem of Algebra. One of its consequences is the following fact:

REAL ZEROS OF A POLYNOMIAL FUNCTION

A polynomial function of degree n with real coefficients has, *at most*, n real zeros.

SIDE NOTE

A polynomial of degree n can have no more than n real zeros. In Example 6, parts **b** and **c**, we see two polynomials of degree 3 each having *fewer* than 3 zeros.

EXAMPLE 6 Finding the Number of Real Zeros

Find the number of distinct real zeros of the following polynomial functions of degree 3.

a. $f(x) = (x - 1)(x + 2)(x - 3)$ **b.** $g(x) = (x + 1)(x^2 + 1)$

c. $h(x) = (x - 3)^2(x + 1)$

Solution

a. $f(x) = (x - 1)(x + 2)(x - 3) = 0$ Set $f(x) = 0$.

$x - 1 = 0,$ or $x + 2 = 0,$ or $x - 3 = 0$ Zero-product property

$x = 1$ or $x = -2$ or $x = 3$ Solve for x.

So $f(x)$ has three real zeros: 1, -2, and 3.

b. $g(x) = (x + 1)(x^2 + 1) = 0$ Set $g(x) = 0$.

$x + 1 = 0$ or $x^2 + 1 = 0$ Zero-product property

$x = -1$ Solve for x.

Because $x^2 + 1 > 0$ for all real x, the equation $x^2 + 1 = 0$ has no real solution. So -1 is the only real zero of $g(x)$.

c. $h(x) = (x - 3)^2(x + 1) = 0$ Set $h(x) = 0$.

$(x - 3)^2 = 0,$ or $x + 1 = 0$ Zero-product property

$x = 3,$ or $x = -1$ Solve for x.

The real zeros of $h(x)$ are 3 and -1. Thus, $h(x)$ has two distinct real zeros.

FIGURE 3.14 Graph of $h(x) = (x - 3)^2(x + 1)$

Practice Problem 6 Find the distinct real zeros of

$$f(x) = (x + 1)^2(x - 3)(x + 5).$$

In Example 6c, the factor $(x - 3)$ appears twice in $h(x)$. We say that 3 is a zero repeated *twice*, or that 3 is a *zero of multiplicity* 2. Notice that the graph of $y = h(x)$ in Figure 3.14 *touches* the x-axis at $x = 3$, where h has a zero of multiplicity 2 (an even number). In contrast, the graph of h *crosses* the x-axis at $x = -1$, where it has a zero of multiplicity 1 (an odd number). We next consider the geometric significance of the multiplicity of a zero.

Multiplicity of a Zero

If c is a zero of a polynomial function $f(x)$ and the corresponding factor $(x - c)$ occurs exactly m times when $f(x)$ is factored as $f(x) = (x - c)^m q(x)$, with $q(c) \neq 0$, then c is called a **zero of multiplicity m**.

1. If m is odd, the graph of f crosses the x-axis at $x = c$.

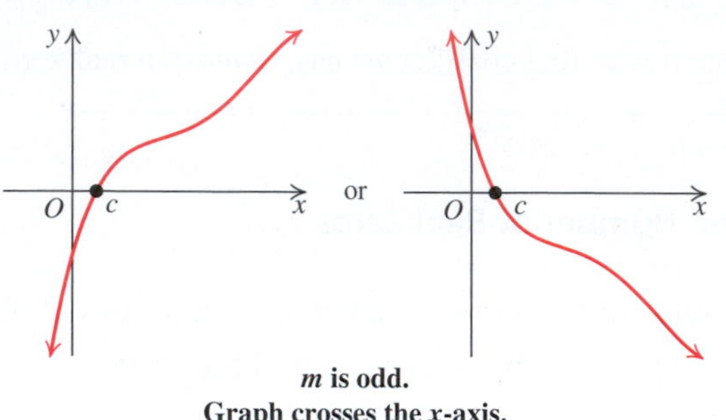

m is odd.
Graph crosses the x-axis.

2. If m is even, the graph of f touches but does not cross the x-axis at $x = c$.

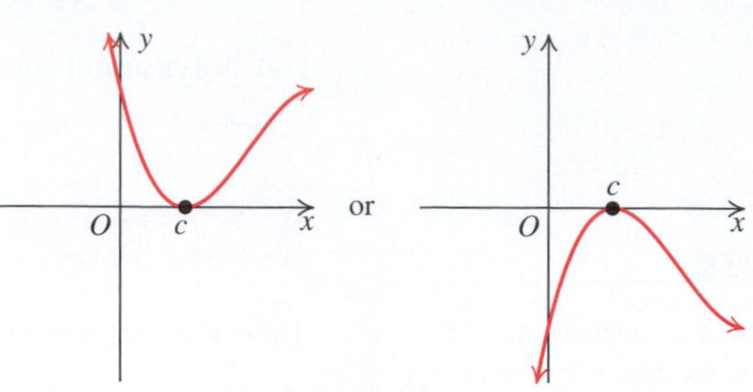

m is even.
Graph touches but does not cross the x-axis.

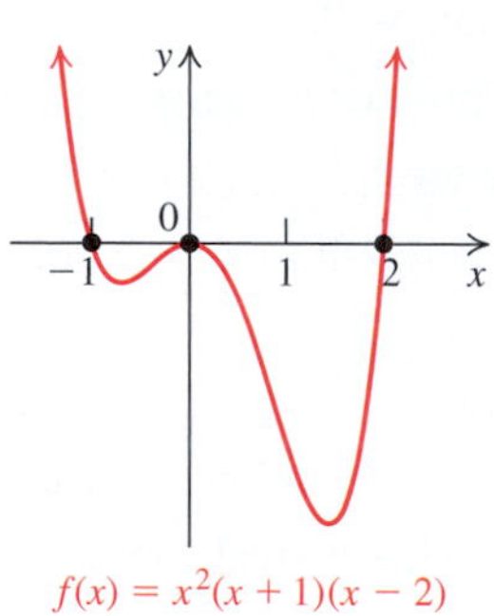

$f(x) = x^2(x + 1)(x - 2)$

FIGURE 3.15

EXAMPLE 7 Finding the Zeros and Their Multiplicities

Find the zeros of the polynomial function $f(x) = x^2(x + 1)(x - 2)$ and give the multiplicity of each zero.

Solution

The polynomial $f(x)$ is already in factored form.

$$f(x) = x^2(x + 1)(x - 2) = 0 \quad \text{Set } f(x) = 0.$$
$$x^2 = 0 \quad \text{or} \quad x + 1 = 0 \quad \text{or} \quad x - 2 = 0 \quad \text{Zero-product property}$$
$$x = 0 \quad \text{or} \quad x = -1 \quad \text{or} \quad x = 2 \quad \text{Solve for } x.$$

The function has three distinct zeros: 0, −1, and 2. The zero $x = 0$ has multiplicity 2, and each of the zeros −1 and 2 has multiplicity 1. The graph of f is given in Figure 3.15. Notice that the graph of f in Figure 3.15 crosses the x-axis at −1 and 2 (the zeros of odd multiplicity), while it touches but does not cross the x-axis at $x = 0$ (the zero of even multiplicity).

Practice Problem 7 Write the zeros of $f(x) = (x - 1)^2(x + 3)(x + 5)$ and give the multiplicity of each.

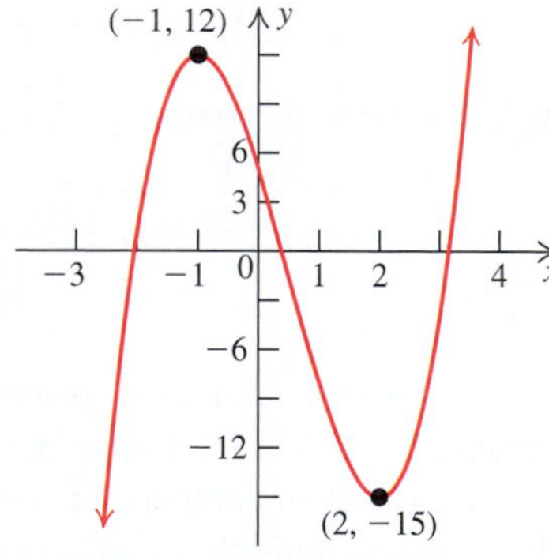

The graph of $f(x) = 2x^3 - 3x^2 - 12x + 5$ has turning points at (−1, 12) and (2, −15).

FIGURE 3.16 Turning points

Turning Points The graph of $f(x) = 2x^3 - 3x^2 - 12x + 5$ is shown in Figure 3.16. The value $f(-1) = 12$ is a **local** (or **relative**) **maximum** value of f. Similarly, the value $f(2) = -15$ is a **local** (or **relative**) **minimum** value of f. (See Section 2.5 for definitions.) The graph points corresponding to the local maximum and local minimum values of a polynomial function are the **turning points**. At each turning point, the graph changes direction from increasing to decreasing or from decreasing to increasing. One of the successes of calculus is finding the maximum and minimum values for many functions.

NUMBER OF TURNING POINTS

If $f(x)$ is a polynomial of degree n, then the graph of f has, *at most*, $(n - 1)$ turning points.

EXAMPLE 8 **Finding the Number of Turning Points**

SIDE NOTE

Notice that in Figure 3.17(c), the graph seems to "turn" but the graph is always increasing. There is no change from increasing to decreasing, so it has no turning points.

Use a graphing calculator and the window $-10 \le x \le 10$; $-30 \le y \le 30$ to find the number of turning points of the graph of each polynomial function.

a. $f(x) = x^4 - 7x^2 - 18$ **b.** $g(x) = x^3 + x^2 - 12x$

c. $h(x) = x^3 - 3x^2 + 3x - 1$

Solution

The graphs of f, g, and h are shown in Figure 3.17.

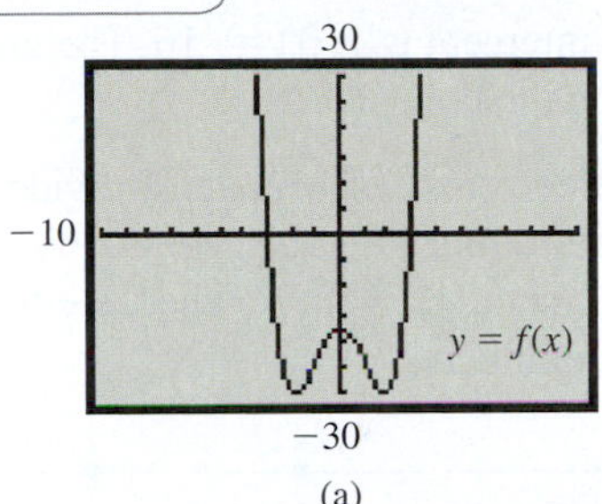

(a)

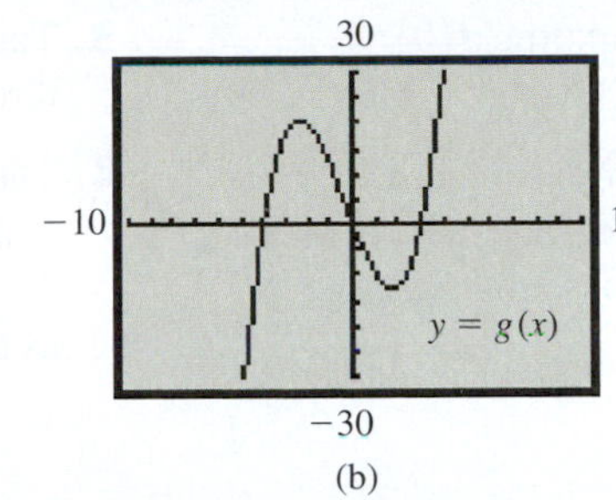

(b)

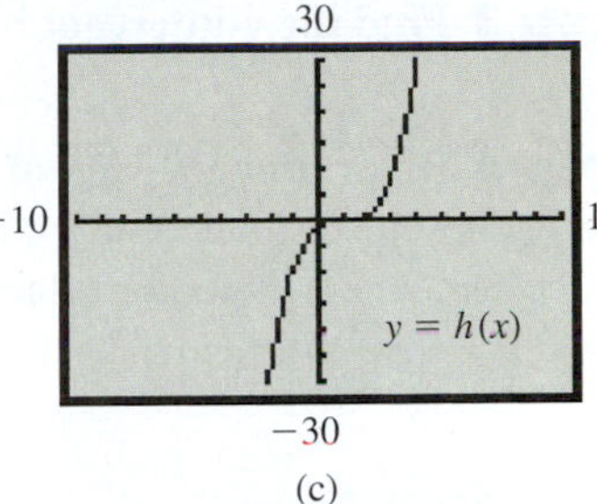

(c)

FIGURE 3.17 Turning points on a calculator graph

a. Function f has three turning points: two local minima and one local maximum.

b. Function g has two turning points: one local maximum and one local minimum.

c. Function h has no turning points: It is increasing on the interval $(-\infty, \infty)$.

Practice Problem 8 Find the number of turning points of the graph of $f(x) = -x^4 + 3x^2 - 2$.

5 Graph polynomial functions.

Graphing a Polynomial Function

You can graph a polynomial function by constructing a table of values for a large number of points, plotting the points, and connecting the points with a smooth curve. (In fact, a graphing calculator uses this method.) Generally, however, this process is a poor way to graph a function by hand because it is tedious, is error-prone, and may give misleading results. Similarly, a graphing calculator may give misleading results if the viewing window is not chosen appropriately. Next, we discuss a better strategy for graphing a polynomial function.

PROCEDURE IN ACTION

EXAMPLE 9 **Graphing a Polynomial Function**

OBJECTIVE

Sketch the graph of a polynomial function.

Step 1 Determine the end behavior. Apply the leading-term test.

EXAMPLE

Sketch the graph of $f(x) = -x^3 - 4x^2 + 4x + 16$.

1. Because the degree, 3, of f is odd and the leading coefficient, -1, is negative, the end behavior (Case 4, page 309) is similar to that of $y = -x^3$ as shown in this sketch.

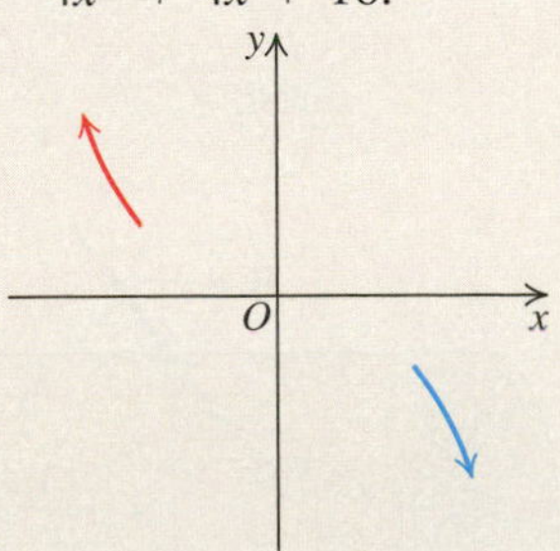

(continued)

Step 2 Find the real zeros of the polynomial function. Set $f(x) = 0$. Solve the equation $f(x) = 0$ by factoring. The real zeros with their multiplicities will give you the x-intercepts of the graph and determine whether the graph crosses the x-axis there.

2.
$$-x^3 - 4x^2 + 4x + 16 = 0$$
$$-(x + 4)(x + 2)(x - 2) = 0$$
$$x = -4, x = -2, \text{ or } x = 2$$

There are three zeros, each of multiplicity 1; so the graph crosses the x-axis at each zero.

Step 3 Find the y-intercept by computing $f(0)$.

3. The y-intercept is $f(0) = 16$. The graph passes through the point $(0, 16)$.

Step 4 Determine the sign of $f(x)$ by using arbitrarily chosen "test numbers" in the intervals defined by the x-intercepts. Find on which of these intervals the graph lies above or below the x-axis.

4. The three zeros -4, -2, and 2 divide the x-axis into four intervals: $(-\infty, -4)$, $(-4, -2)$, $(-2, 2)$, and $(2, \infty)$. We choose -5, -3, 0, and 3 as test numbers.

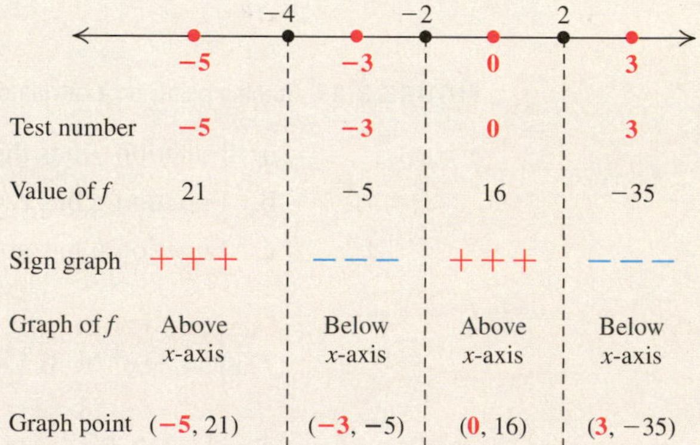

Test number	-5	-3	0	3
Value of f	21	-5	16	-35
Sign graph	$+++$	$---$	$+++$	$---$
Graph of f	Above x-axis	Below x-axis	Above x-axis	Below x-axis
Graph point	$(-5, 21)$	$(-3, -5)$	$(0, 16)$	$(3, -35)$

Step 5 Draw the graph using points from Steps 1–4. To check whether the graph is drawn correctly, use the fact that the number of turning points is less than the degree of the polynomial.

5. The graph connects the points we have found with a smooth curve.

The number of turning points is 2, which is less than 3, the degree of f.

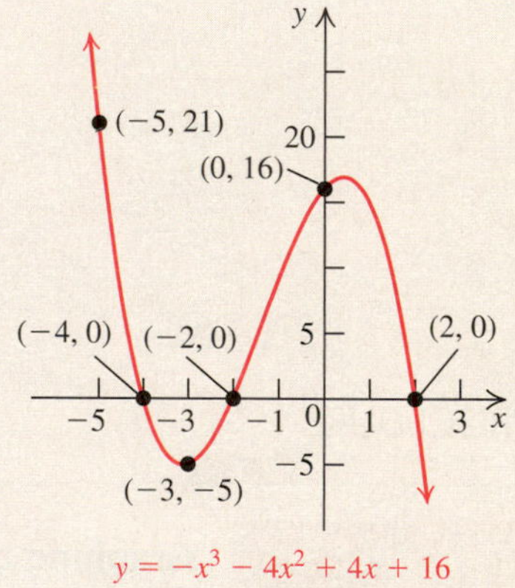

Practice Problem 9 Graph $f(x) = -x^4 + 5x^2 - 4$.

Volume of a Wine Barrel Kepler first analyzed the problem of finding the volume of a cylindrical wine barrel. (Review the introduction to this section and see Figure 3.18.) The volume V of a cylinder with radius r and height h is given by the formula $V = \pi r^2 h$. In Figure 3.18, $h = 2z$; so $V = 2\pi r^2 z$. Let x represent the value measured by the rod. Then by the Pythagorean theorem,

$$x^2 = 4r^2 + z^2. \qquad (2r)^2 = 4r^2$$

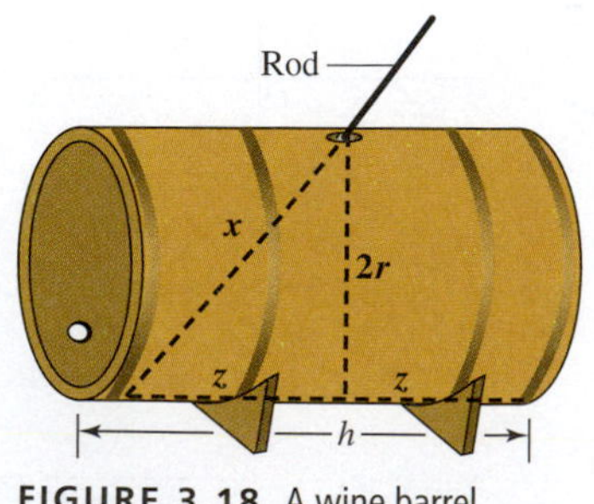

FIGURE 3.18 A wine barrel

Kepler's challenge was to calculate V, given only x—in other words, to express V as a function of x. To solve this problem, Kepler observed the key fact that Austrian wine barrels were all made with the same height-to-diameter ratio. Kepler assumed that the winemakers would have made a smart choice for this ratio, one that would maximize the volume for a fixed value of r. He calculated the height-to-diameter ratio to be $\sqrt{2}$.

EXAMPLE 10 Volume of a Wine Barrel

Express the volume V of the wine barrel in Figure 3.18 as a function of x. Assume that $\dfrac{\text{height}}{\text{diameter}} = \dfrac{2z}{2r} = \dfrac{z}{r} = \sqrt{2}$.

Solution

We have the following relationships:

$$V = 2\pi r^2 z \qquad \text{Volume with radius } r \text{ and height } 2z$$

$$\frac{z}{r} = \sqrt{2} \qquad \text{Given ratio}$$

$$x^2 = 4r^2 + z^2 \qquad \text{Pythagorean theorem}$$

$$\frac{x^2}{r^2} = 4 + \frac{z^2}{r^2} \qquad \text{Divide both sides by } r^2.$$

$$\frac{x^2}{r^2} = 4 + 2 = 6 \qquad \text{Because } \frac{z}{r} = \sqrt{2}, \frac{z^2}{r^2} = \left(\frac{z}{r}\right)^2 = (\sqrt{2})^2 = 2.$$

From $\dfrac{x^2}{r^2} = 6$, we have $r^2 = \dfrac{x^2}{6}$; so $r = \dfrac{x}{\sqrt{6}}$ $(r > 0)$. Using $\dfrac{z}{r} = \sqrt{2}$, we obtain

$$z = \sqrt{2}\,r = \sqrt{2}\,\frac{x}{\sqrt{6}} \qquad \text{Replace } r \text{ with } \frac{x}{\sqrt{6}}$$

$$= \frac{x}{\sqrt{3}} \qquad \frac{\sqrt{2}}{\sqrt{6}} = \frac{\sqrt{2}}{\sqrt{2}\cdot\sqrt{3}} = \frac{1}{\sqrt{3}}$$

Now

$$V = 2\pi r^2 z$$

$$= 2\pi\left(\frac{x^2}{6}\right)\left(\frac{x}{\sqrt{3}}\right) \qquad \text{Replace } z \text{ with } \frac{x}{\sqrt{3}} \text{ and } r^2 \text{ with } \frac{x^2}{6}.$$

$$= \frac{\pi}{3\sqrt{3}}x^3 \qquad \text{Simplify.}$$

$$\approx 0.6046x^3 \qquad \text{Use a calculator.}$$

Consequently, Austrian vintners could easily approximate the volume of a wine barrel by using the formula $V = 0.6046x^3$, where x was the length measured by the rod. For a quick estimate, use $V \approx 0.6x^3$.

Practice Problem 10 Use Kepler's formula to compute the volume of wine (in liters) in a barrel with $x = 70$ cm. Note: 1 decimeter (dm) = 10 cm and 1 liter = $(1 \text{ dm})^3$.

Answers to Practice Problems

1. a. Not a polynomial function **b.** A polynomial function; the degree is 7, the leading term is $2x^7$, and the leading coefficient is 2.

2. $P(x) = 4x^3 + 2x^2 + 5x - 17 = x^3\left(4 + \dfrac{2}{x} + \dfrac{5}{x^2} - \dfrac{17}{x^3}\right)$

When $|x|$ is large, the terms $\dfrac{2}{x}, \dfrac{5}{x^2}$, and $\dfrac{17}{x^3}$ are close to 0.

Therefore, $P(x) \approx x^3(4 + 0 + 0 - 0) = 4x^3$. **3.** $y \to -\infty$ as $x \to -\infty$ and $y \to -\infty$ as $x \to \infty$ **4.** $\dfrac{3}{2}$

5. $f(1) = -7, f(2) = 4$. Because $f(1)$ and $f(2)$ have opposite signs, by the Intermediate Value Theorem, f has a real zero between 1 and 2. **6.** $-5, -1$, and 3 **7.** -5 and -3 with multiplicity 1 and 1 with multiplicity 2 **8.** 3

9.

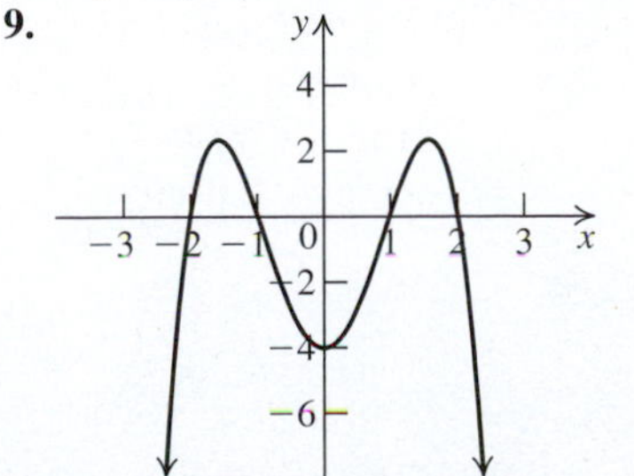

10. 207.378 liters

SECTION 3.2 Exercises

Basic Concepts and Skills

1. Consider the polynomial $2x^5 - 3x^4 + x - 6$. The degree of this polynomial is ____________, its leading term is ____________, its leading coefficient is ____________, and its constant term is ____________.
2. The domain of a polynomial function is the ____________.
3. The behavior of the function $y = f(x)$ as $x \to \infty$ or $x \to -\infty$ is called the ____________ of the function.
4. The end behavior of any polynomial function is determined by its ____________ term.
5. A number c for which $f(c) = 0$ is called a(n) ____________ of the function f.
6. If c is a zero of even multiplicity for a function f, then the graph of f ____________ the x-axis at c.
7. The graph of a polynomial function of degree n has, at most, ____________ turning points.
8. The zeros of a polynomial function $y = f(x)$ are determined by solving the equation ____________.

In Exercises 9–14, for each polynomial function, find the degree, the leading term, and the leading coefficient.

9. $f(x) = 2x^5 - 5x^2$
10. $f(x) = 3 - 5x - 7x^4$
11. $f(x) = \dfrac{2x^3 + 7x}{3}$
12. $f(x) = -7x + 11 + \sqrt{2}x^3$
13. $f(x) = \pi x^4 + 1 - x^2$
14. $f(x) = 5$

In Exercises 15–22, explain why the given function is not a polynomial function.

15. $f(x) = x^2 + 3|x| - 7$
16. $f(x) = 2x^3 - 5x, \quad 0 \le x \le 2$
17. $f(x) = \dfrac{x^2 - 1}{x + 5}$
18. $f(x) = \begin{cases} 2x + 3, & x \ne 0 \\ 1, & x = 0 \end{cases}$
19. $f(x) = 5x^2 - 7\sqrt{x}$
20. $f(x) = x^{4.5} - 3x^2 + 7$
21. $f(x) = \dfrac{x^2 - 1}{x - 1}, x \ne 1$
22. $f(x) = 3x^2 - 4x + 7x^{-2}$

In Exercises 23–28, explain why the given graph cannot be the graph of a polynomial function.

23.

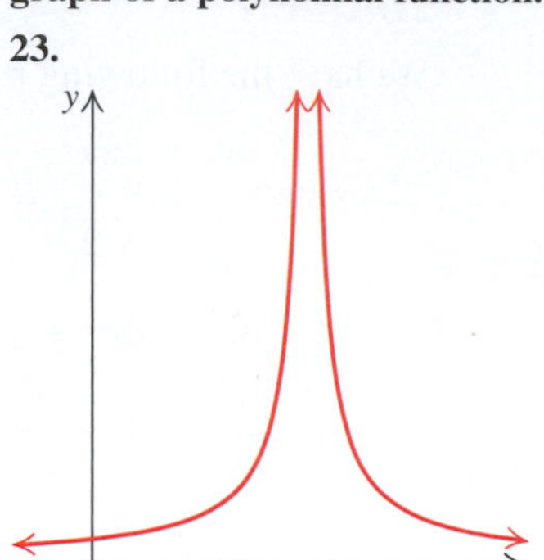

24.

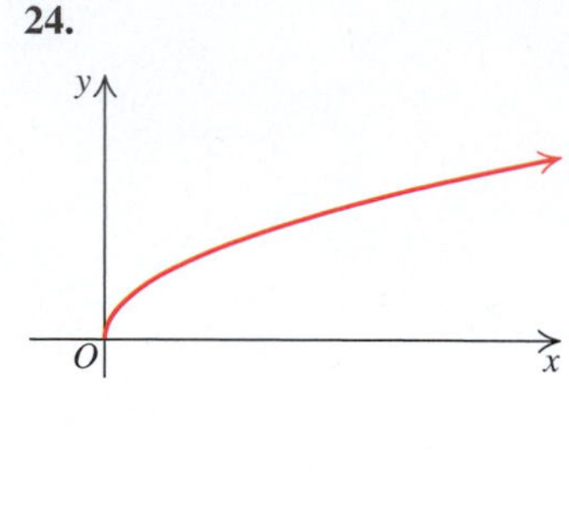

25.

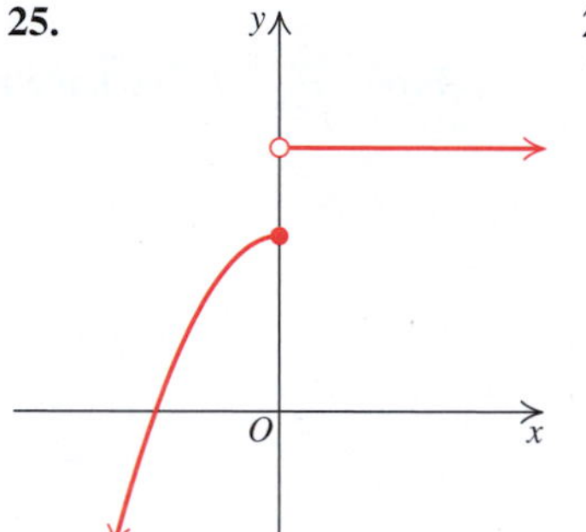

26.

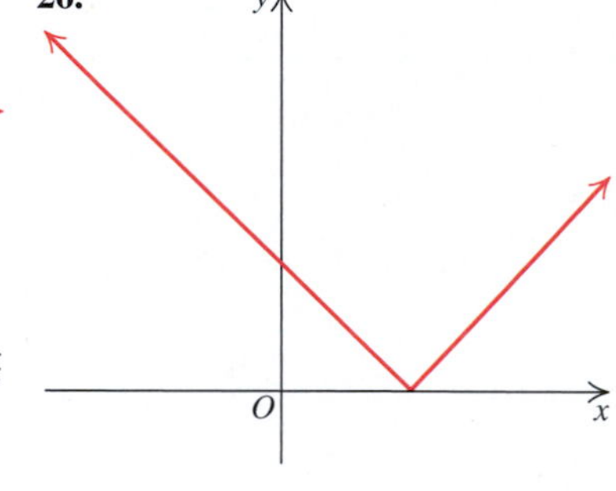

27.

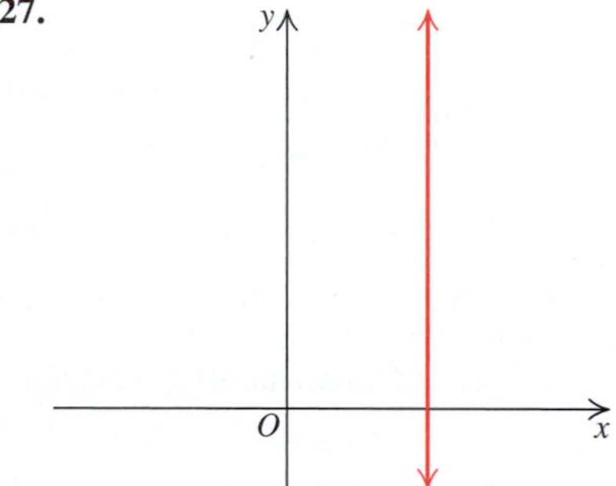

28.

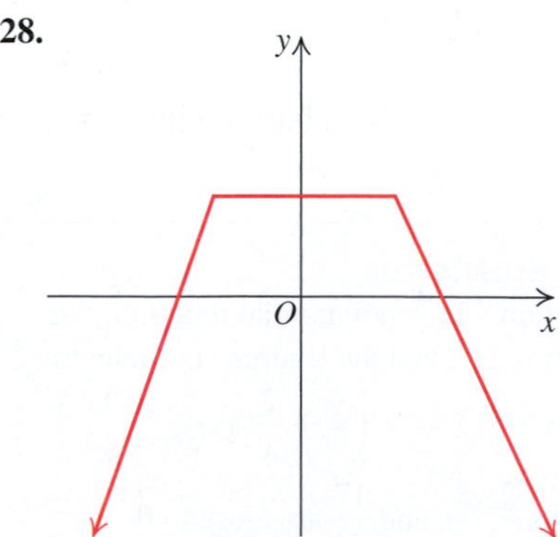

In Exercises 29–34, match the polynomial function with its graph. Use the leading-term test and the y-intercept.

29. $f(x) = -x^4 + 3x^2 + 4$

30. $f(x) = x^6 - 7x^4 + 7x^2 + 15$

31. $f(x) = x^4 - 10x^2 + 9$

32. $f(x) = x^3 + x^2 - 17x + 15$

33. $f(x) = x^3 + 6x^2 + 12x + 8$

34. $f(x) = -x^3 - 2x^2 + 11x + 12$

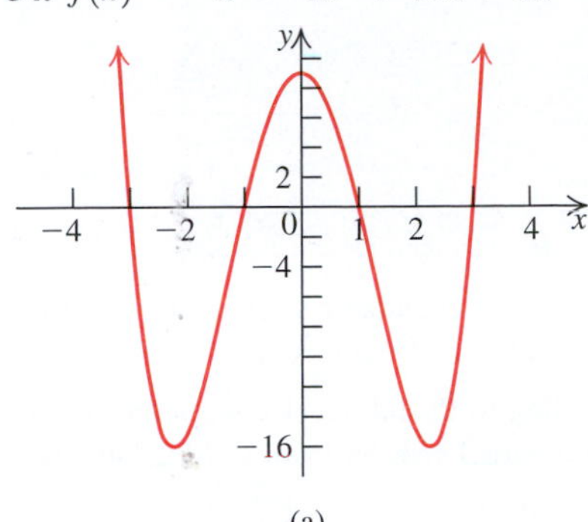

(a)

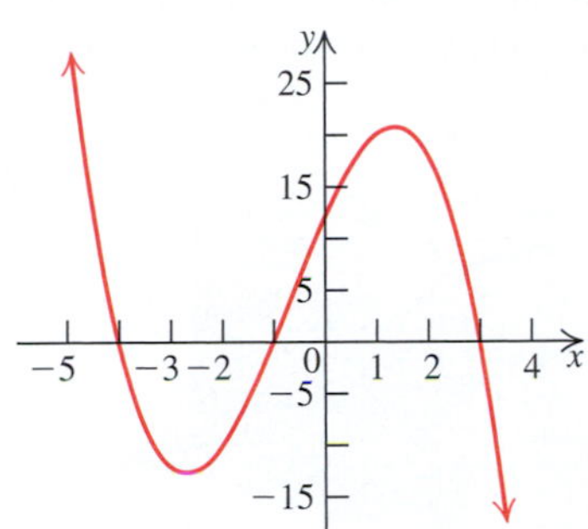

(b)

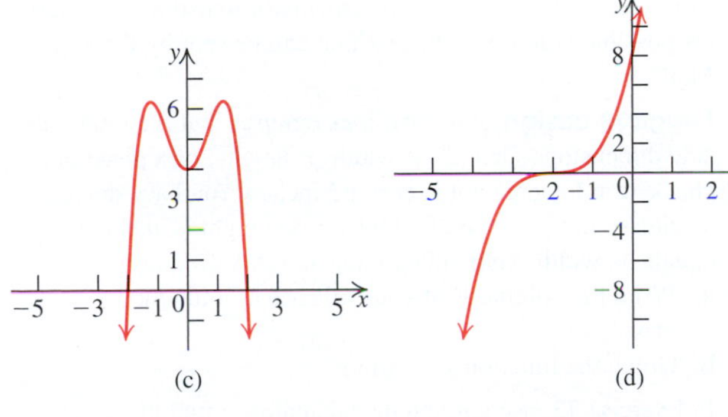

(c)

(d)

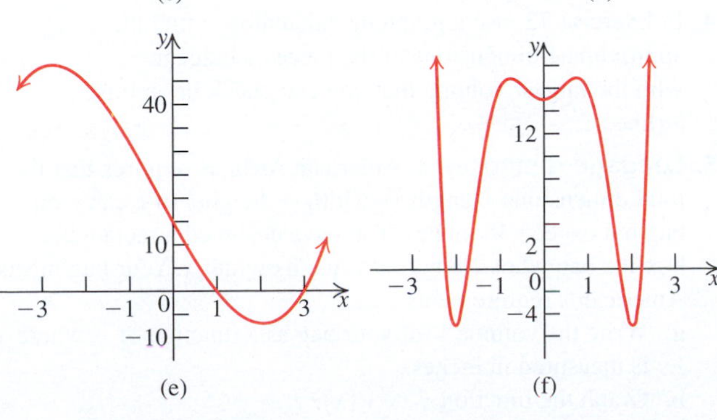

(e)

(f)

In Exercises 35–42, (a) find the real zeros of each polynomial function and state the multiplicity for each zero and (b) state whether the graph crosses or touches but does not cross the x-axis at each x-intercept. (c) What is the maximum possible number of turning points?

35. $f(x) = 5(x - 1)(x + 5)$

36. $f(x) = 3(x + 2)^2(x - 3)$

37. $f(x) = (x + 1)^2(x - 1)^3$

38. $f(x) = 2(x + 3)^2(x + 1)(x - 6)$

39. $f(x) = -5\left(x + \frac{2}{3}\right)\left(x - \frac{1}{2}\right)^2$

40. $f(x) = x^3 - 6x^2 + 8x$

41. $f(x) = -x^4 + 6x^3 - 9x^2$

42. $f(x) = x^4 - 5x^2 - 36$

In Exercises 43–46, use the Intermediate Value Theorem to show that each polynomial $P(x)$ has a real zero in the specified interval. Approximate this zero to two decimal places.

43. $P(x) = x^4 - x^3 - 10$; $[2, 3]$

44. $P(x) = x^4 - x^2 - 2x - 5$; $[1, 2]$

45. $P(x) = x^5 - 9x^2 - 15$; $[2, 3]$

46. $P(x) = x^5 + 5x^4 + 8x^3 + 4x^2 - x - 5$; $[0, 1]$

In Exercises 47–60, for each polynomial function f,

a. Describe the end behavior of f.

b. Find the real zeros of f. Determine whether the graph of f crosses or touches but does not cross the x-axis at each x-intercept.

c. Use the zeros of f and test numbers to find the intervals over which the graph of f is above or below the x-axis.

d. Determine the y-intercept.

e. Find any symmetries of the graph of the function.

f. Determine the maximum possible number of turning points.

g. Sketch the graph of f.

47. $f(x) = x^2 + 3$

48. $f(x) = x^2 - 4x + 4$

49. $f(x) = x^2 + 4x - 21$

50. $f(x) = -x^2 + 4x + 12$

51. $f(x) = -2x^2(x + 1)$

52. $f(x) = x - x^3$

53. $f(x) = x^2(x - 1)^2$

54. $f(x) = x^2(x + 1)(x - 2)$

55. $f(x) = (x - 1)^2(x + 3)(x - 4)$

56. $f(x) = (x + 1)^2(x^2 + 1)$

57. $f(x) = -x^2(x^2 - 1)(x + 1)$

58. $f(x) = -x^2(x^2 - 4)(x + 2)$

59. $f(x) = x(x + 1)(x - 1)(x + 2)$

60. $f(x) = x^2(x^2 + 1)(x - 2)$

Applying the Concepts

61. **Drug reaction.** Let $f(x) = 3x^2(4 - x)$.
  a. Find the zeros of f and their multiplicities.
  b. Sketch the graph of $y = f(x)$.
  c. How many turning points are in the graph of f?
  d. A patient's reaction $R(x)\,(R(x) \geq 0)$ to a certain drug, the size of whose dose is x, is observed to be $R(x) = 3x^2(4 - x)$. What is the domain of the function $R(x)$? What portion of the graph of $f(x)$ constitutes the graph of $R(x)$?

62. In Exercise 61, use a graphing calculator to estimate the value of x for which $R(x)$ is a maximum.

63. **Cough and air velocity.** The normal radius of the windpipe of a human adult at rest is approximately 1 cm. When you cough, your windpipe contracts to, say, a radius r. The velocity v at which the air is expelled depends on the radius r and is given by the formula $v = a(1 - r)r^2$, where a is a positive constant.
  a. What is the domain of v?
  b. Sketch the graph of $v(r)$.

64. In Exercise 63, use a graphing calculator to estimate the value of r for which v is a maximum.

65. **Maximizing revenue.** A manufacturer of slacks has discovered that the number x of khakis sold to retailers per week at a price p dollars per pair is given by the formula

$$p = 27 - \left(\frac{x}{300}\right)^2.$$

  a. Write the weekly revenue $R(x)$ as a function of x.
  b. What is the domain of $R(x)$?
  c. Graph $y = R(x)$.

66. a. In Exercise 65, use a graphing calculator to estimate the number of slacks sold to maximize the revenue.
  b. What is the maximum revenue from part (a)?

67. **Maximizing production.** An orange grower in California hires migrant workers to pick oranges during the season. He has 12 employees, and each can pick 400 oranges per hour. He has discovered that if he adds more workers, the production per worker decreases due to lack of supervision. When x new workers (above the 12) are hired, each worker picks $400 - 2x^2$ oranges per hour.
  a. Write the number $N(x)$ of oranges picked per hour as a function of x.
  b. Find the domain of $N(x)$.
  c. Graph the function $y = N(x)$.

68. a. In Exercise 67, use a graphing calculator to estimate the maximum number of oranges that can be picked per hour.
  b. Find the number of employees involved in part (a).

69. **Making a box.** From a rectangular 8×15 piece of cardboard, four congruent squares with sides of length x are cut out, one at each corner. (See the accompanying figure.) The resulting crosslike piece is then folded to form a box.
  a. Find the volume V of the box as a function of x.
  b. Sketch the graph of $y = V(x)$.

70. In Exercise 69, use a graphing calculator to estimate the largest possible value of the volume of the box.

71. **Mailing a box.** According to Postal Service regulations, the girth plus the length of a parcel sent by Priority Mail may not exceed 108 inches.
  a. Express, as a function of x, the volume V of a rectangular parcel with two square faces (with sides of length x inches) that can be sent by mail.
  b. Sketch a graph of $V(x)$.

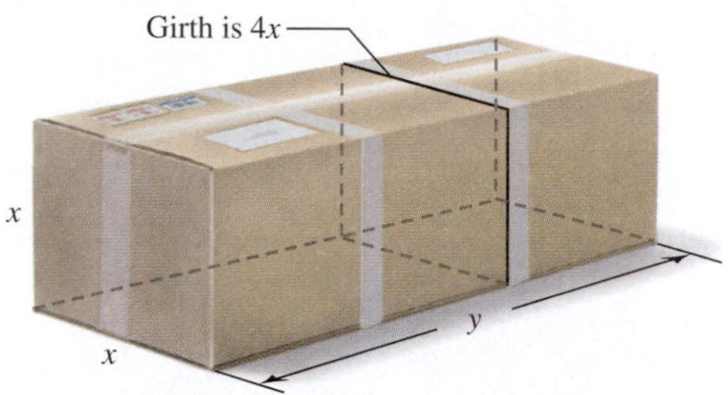

72. In Exercise 71, use a graphing calculator to estimate the largest possible volume of the box that can be sent by Priority Mail.

73. **Luggage design.** AAA Airlines requires that the total outside dimensions (length + width + height) of a piece of checked-in luggage not exceed 62 inches. You have designed a suitcase in the shape of a box whose height (x inches) equals its width. Your suitcase meets AAA's requirements.
  a. Write the volume V of your suitcase as a function of x.
  b. Graph the function $y = V(x)$.

74. In Exercise 73, use a graphing calculator to find the approximate dimensions of the piece of luggage with the largest volume that you can check in on this airline.

75. **Luggage dimensions.** American Airlines requires that the total dimensions (length + width + height) of a carry-on bag not exceed 45 inches. You have designed a rectangular boxlike bag whose length is twice its width x. Your bag meets American's requirements.
  a. Write the volume V of your bag as a function of x, where x is measured in inches.
  b. Graph the function $y = V(x)$.

76. In Exercise 75, use a graphing calculator to estimate the dimensions of the bag with the largest volume that you can carry on an American flight. (*Source:* www.aa.com)

77. Worker productivity. An efficiency study of the morning shift at an assembly plant indicates that the number y of units produced by an average worker t hours after 6 A.M. may be modeled by the formula $y = -t^3 + 6t^2 + 16t$. Sketch the graph of $y = f(t)$.

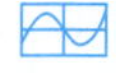
78. In Exercise 77, estimate the time in the morning when a worker is most efficient.

79. Deaths from FDA-approved drugs. The following table illustrates deaths (in thousands) that occurred in the United States during the 11 years from 2000 to 2010 as tabulated from the FDA's Adverse Event Reporting System for prescription drugs.

Year	Deaths
2000	19.445
2001	23.988
2002	28.181
2003	35.173
2004	34.928
2005	40.238
2006	37.465
2007	36.834
2008	49.958
2009	63.846
2010	82.724

Source: fda.gov/Drugs/GuidanceComplianceRegulatoryInformation/Surveillance/AdverseDrugEffects/ucm070461.htm

One of the simplest models that gives an approximation to these data is described by a cubic polynomial

$$D(t) = 0.205996t^3 - 2.526415t^2 + 11.156629t + 17.337104,$$

where $t = 0$ represents the year 2000.

Sketch the graph $y = D(t)$ and the actual data points on the same coordinate system and analyze the result.

80. Use your result in Exercise 79, to estimate the number of deaths in the United States from FDA-approved drugs in 2011. Explain your answer.

Beyond the Basics

In Exercises 81–90, use transformations of the graph of $y = x^4$ or $y = x^5$ to graph each function $f(x)$. Find the zeros of $f(x)$ and their multiplicity.

81. $f(x) = (x - 1)^4$

82. $f(x) = -(x + 1)^4$

83. $f(x) = x^4 + 2$

84. $f(x) = x^4 + 81$

85. $f(x) = -(x - 1)^4$

86. $f(x) = \frac{1}{2}(x - 1)^4 - 8$

87. $f(x) = x^5 + 1$

88. $f(x) = (x - 1)^5$

89. $f(x) = 8 - \frac{(x + 1)^5}{4}$

90. $f(x) = 81 + \frac{(x + 1)^5}{3}$

91. Prove that if $0 < x < 1$, then $0 < x^4 < x^2$. This shows that the graph of $y = x^4$ is flatter than the graph of $y = x^2$ on the interval $(0, 1)$.

92. Prove that if $x > 1$, then $x^4 > x^2$. This shows that the graph of $y = x^4$ is higher than the graph of $y = x^2$ on the interval $(1, \infty)$.

93. Sketch the graph of $y = x^2 - x^4$. Use a graphing calculator to estimate the maximum vertical distance between the graphs $y = x^2$ and $y = x^4$ on the interval $(0, 1)$. Also estimate the value of x at which this maximum distance occurs.

94. Repeat Exercise 93 by sketching the graph of $y = x^3 - x^5$.

In Exercises 95–98, write a polynomial with the given characteristics.

95. A polynomial of degree 3 that has exactly two x-intercepts

96. A polynomial of degree 3 that has three distinct x-intercepts and whose graph rises to the left and falls to the right

97. A polynomial of degree 4 that has exactly two distinct x-intercepts and whose graph falls to the left and right

98. A polynomial of degree 4 that has exactly three distinct x-intercepts and whose graph rises to the left and right

In Exercises 99–102, find the smallest possible degree of a polynomial with the given graph. Explain your reasoning.

99.

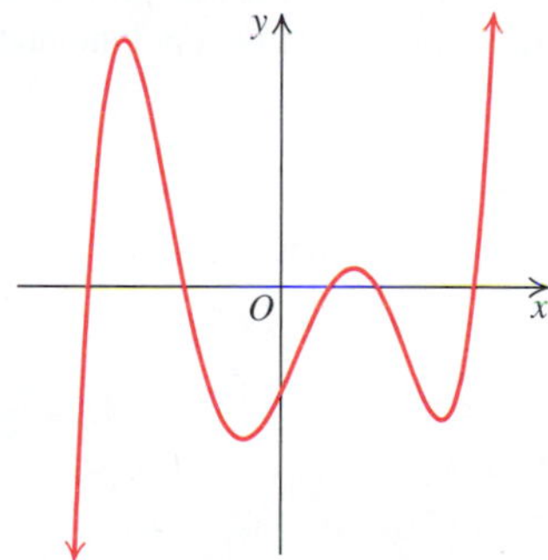

100.

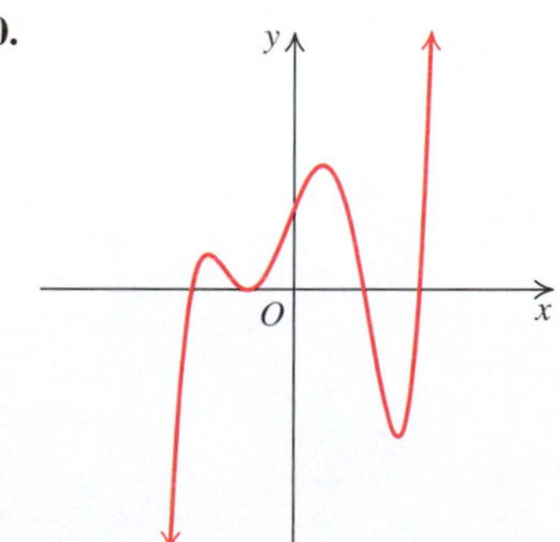

101.

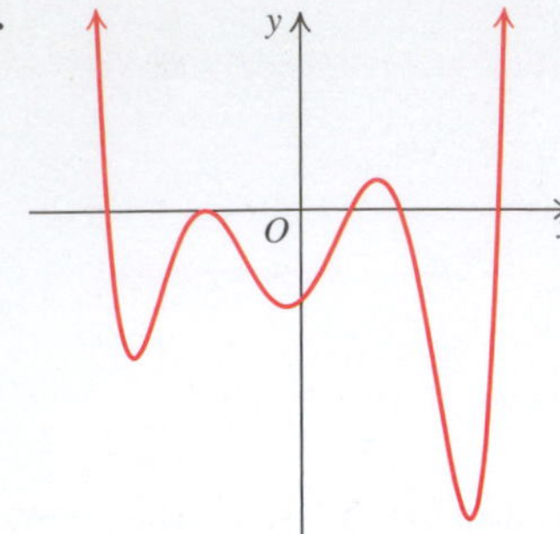

102.

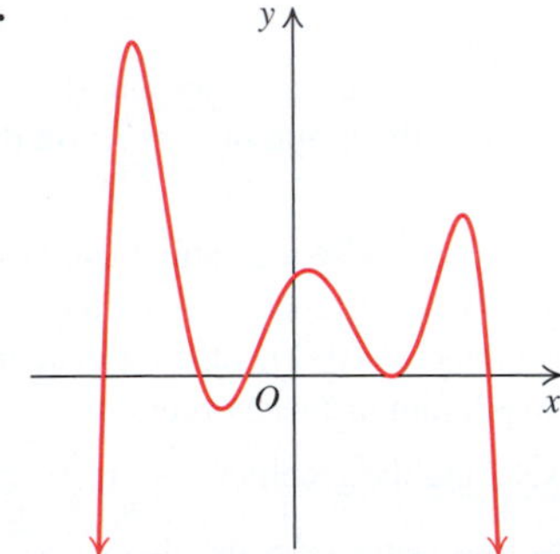

Critical Thinking/Discussion/Writing

103. Is it possible for the graph of a polynomial function to have no y-intercepts? Explain.

104. Is it possible for the graph of a polynomial function to have no x-intercepts? Explain.

105. Is it possible for the graph of a polynomial function of degree 3 to have exactly one local maximum and no local minimum? Explain.

106. Is it possible for the graph of a polynomial function of degree 4 to have exactly one local maximum and exactly one local minimum? Explain.

107. Let P and Q be polynomial functions of degree m and n, respectively. Is the composition $P \circ Q$ a polynomial function? If so, what is the degree of $P \circ Q$. If not, provide an example where the composition fails to be a polynomial.

Maintaining Skills

In Exercises 108–112, simplify each expression. (Do not use negative exponents.)

108. $\dfrac{12x^5}{-3x^3}$

109. $\dfrac{x^5}{x^7}$

110. $\dfrac{3x^4}{2x^2}$

111. $\dfrac{x^5 - 3x^2}{x^2}$

112. $\dfrac{5x^3 + 2x^2 - x}{x}$

In Exercises 113–115, recall that an integer a is a factor of an integer b is there is an integer c so that $b = ac$.

113. a. Show that 1 is a factor of 7.
b. Show that 7 is a factor of 7.
c. Show that -1 is a factor of 7.
d. Show that -7 is a factor of 7.

114. Find all integer factors of 7.

115. Find all integer factors of 120.

In Exercises 116–119, evaluate each function at the indicated values of x.

116. $f(x) = x^3 - 3x^2 + 2x - 9; x = -2$

117. $g(x) = x^4 + 2x^3 - 20x - 5; x = 3$

118. $g(x) = (-7x^5 + 2x)(x - 13); x = 13$

119. $f(x) = (x^8 - 15x^7 + 5x^3)(x + 12); x = -12$

In Exercises 120–123, factor each expression.

120. $x^2 - 3x - 10$

121. $-2x^2 + x + 1$

122. $(2x + 3)(x^2 + 6x - 7)$

123. $(x - 9)(6x^2 - 7x - 3)$

SECTION 3.3

Dividing Polynomials

BEFORE STARTING THIS SECTION, REVIEW

1 Multiplying polynomials (Section P.3, page 31)

2 Evaluating a function (Section 2.4, page 197)

3 Factoring trinomials (Section P.4, page 39)

OBJECTIVES

1 Learn the division algorithm.

2 Use synthetic division.

3 Use the Remainder and Factor Theorems.

Greenhouse Effect and Global Warming

The **greenhouse effect** refers to the fact that Earth is surrounded by an atmosphere that acts like the transparent cover of a greenhouse, allowing sunlight to filter through while trapping heat. It is becoming increasingly clear that we are experiencing *global warming*. This means that Earth's atmosphere is becoming warmer near its surface. Scientists who study climate agree that global warming is due largely to the emission of "greenhouse gases" such as carbon dioxide (CO_2), chlorofluorocarbons (CFCs), methane (CH_4), ozone (O_3), and nitrous oxide (N_2O). The temperature of Earth's surface increased by 1°F over the 20th century. The late 1990s and early 2000s were some of the hottest years ever recorded. Projections of future warming suggest a global temperature increase of between 2.5°F and 10.4°F by 2100.

Global warming will result in long-term changes in climate, including a rise in average temperatures, unusual rainfalls, and more storms and floods. The sea level may rise so much that people will have to move away from coastal areas. Some regions of the world may become too dry for farming.

The production and consumption of energy is one of the major causes of greenhouse gas emissions. In Example 8, we look at the pattern for the consumption of petroleum in the United States.

1 Learn the division algorithm.

The Division Algorithm

Dividing polynomials is similar to dividing integers. Both division processes are closely connected to multiplication. The fact that $10 = 5 \cdot 2$ is also expressed as $\frac{10}{5} = 2$. By comparison, in equation

(1) $$x^3 - 1 = (x - 1)(x^2 + x + 1),$$ Factor using difference of cubes.

dividing both sides by $x - 1$, we obtain

$$\frac{x^3 - 1}{x - 1} = x^2 + x + 1, \quad x \neq 1.$$

RECALL

A polynomial $P(x)$ is an expression of the form

$$a_n x^n + \cdots + a_1 x + a_0.$$

Polynomial Factor

A polynomial $D(x)$ is a **factor** of a polynomial $F(x)$ if there is a polynomial $Q(x)$ such that $F(x) = D(x) \cdot Q(x)$.

SIDE **NOTE**

It is helpful to compare the division process for integers with that of polynomials. Notice that

$$3\overline{)7}\quad 2,\; 6,\; 1$$

can also be written as

$$\frac{7}{3} = 2 + \frac{1}{3}$$

or

$$7 = 3 \cdot 2 + 1$$

Next, consider the equation

$$(2) \qquad x^3 - 2 = (x - 1)(x^2 + x + 1) - 1 \qquad \text{Subtract 1 from both sides of equation (1).}$$

Dividing both sides by $(x - 1)$, we can rewrite equation (2) in the form

$$\frac{x^3 - 2}{x - 1} = x^2 + x + 1 - \frac{1}{x - 1}, \quad x \neq 1.$$

Here we say that when the **dividend** $x^3 - 2$ is divided by the **divisor** $x - 1$, the **quotient** is $x^2 + x + 1$ and the **remainder** is -1. This result is an example of the *division algorithm.*

THE DIVISION ALGORITHM

If a polynomial $F(x)$ is divided by a polynomial $D(x)$ and $D(x)$ is not the zero polynomial, then there are unique polynomials $Q(x)$ and $R(x)$ such that

$$\frac{F(x)}{D(x)} = Q(x) + \frac{R(x)}{D(x)} \quad \text{or}$$

$$\underset{\text{Dividend}}{F(x)} = \underset{\text{Divisor}}{D(x)} \cdot \underset{\text{Quotient}}{Q(x)} + \underset{\text{Remainder}}{R(x)}$$

Either $R(x)$ is the *zero polynomial* or the degree of $R(x)$ is less than the degree of $D(x)$.

DO YOU **KNOW?**

An *algorithm* is a finite sequence of precise steps that leads to a solution of a problem.

The division algorithm says that the dividend is equal to the divisor times the quotient plus the remainder.

The expression $\frac{F(x)}{D(x)}$ is **improper** if degree of $F(x) \geq$ degree of $D(x)$. However, the expression $\frac{R(x)}{D(x)}$ is always **proper** because by the division algorithm, degree of $R(x) <$ degree of $D(x)$.

For an improper expression $\frac{F(x)}{D(x)}$, you may use the **long division** and the **synthetic division** processes reviewed in the next few examples.

PROCEDURE **IN ACTION**

EXAMPLE 1 **Long Division**

OBJECTIVE

Find the quotient and remainder when one polynomial is divided by another.

Step 1 Write the terms in the dividend and the divisor in descending powers of the variable.

EXAMPLE

Find the quotient and remainder when $2x^2 + x^5 + 7 + 4x^3$ *is divided by* $x^2 + 1 - x$.

1. *Dividend:* $x^5 + 4x^3 + 2x^2 + 7$ *Divisor:* $x^2 - x + 1$

(continued)

Step 2 Insert terms with zero coefficients in the dividend for any missing powers of the variable.

2. $x^5 + 0x^4 + 4x^3 + 2x^2 + 0x + 7$

Step 3 Divide the first term in the dividend by the first term in the divisor to obtain the first term in the quotient.

3. $\frac{x^5}{x^2} = x^3$ Divide first terms.

$$\begin{array}{r} x^3 \\ x^2 - x + 1 \overline{) x^5 + 0x^4 + 4x^3 + 2x^2 + 0x + 7} \end{array}$$

Step 4 Multiply the divisor by the first term in the quotient and subtract the product from the dividend.

4.

$$\begin{array}{r} x^3 \\ x^2 - x + 1 \overline{) x^5 + 0x^4 + 4x^3 + 2x^2 + 0x + 7} \\ (-)\underline{(x^5 - x^4 + x^3)} \\ x^4 + 3x^3 + 2x^2 + 0x + 7 \end{array}$$

Subtract. $(x^5 - x^4 + x^3) \leftarrow x^3(x^2 - x + 1)$; $x^4 + 3x^3 + 2x^2 + 0x + 7$ Remainder

Step 5 Treat the remainder obtained in Step 4 as a new dividend and repeat Steps 3 and 4. Continue this process until a remainder is obtained that is of lower degree than the divisor.

$\frac{x^4}{x^2} = x^2$ $\frac{4x^3}{x^2} = 4x$ $\frac{5x^2}{x^2} = 5$

5.

$$\begin{array}{r} x^3 + x^2 + 4x + 5 \\ x^2 - x + 1 \overline{) x^5 + 0x^4 + 4x^3 + 2x^2 + 0x + 7} \\ (-)\underline{(x^5 - x^4 + x^3)} \\ x^4 + 3x^3 + 2x^2 + 0x + 7 \\ (-)\underline{(x^4 - x^3 + x^2)} \\ 4x^3 + x^2 + 0x + 7 \\ (-)\underline{(4x^3 - 4x^2 + 4x)} \\ 5x^2 - 4x + 7 \\ (-)\underline{(5x^2 - 5x + 5)} \\ x + 2 \end{array}$$

$x^4 + 3x^3 + 2x^2 + 0x + 7 \leftarrow$ New dividend; $(x^4 - x^3 + x^2) \leftarrow x^2(x^2 - x + 1)$; $4x^3 + x^2 + 0x + 7$ Remainder; $(4x^3 - 4x^2 + 4x) \leftarrow 4x(x^2 - x + 1)$; $5x^2 - 4x + 7$ Remainder; $(5x^2 - 5x + 5) \leftarrow 5(x^2 - x + 1)$; $x + 2$ Remainder

Step 6 Write the quotient and the remainder.

6. Quotient $= x^3 + x^2 + 4x + 5$ Remainder $= x + 2$

Practice Problem 1 Find the quotient and remainder when $3x^3 + 4x^2 + x + 7$ is divided by $x^2 + 1$.

EXAMPLE 2 Using Long Division

Divide $x^4 - 13x^2 + x + 35$ by $x^2 - x - 6$.

Solution

Because the dividend does not contain an x^3 term, we use a zero coefficient for the missing term.

$$\begin{array}{r} x^2 + x - 6 \leftarrow \text{Quotient} \\ x^2 - x - 6 \overline{) x^4 + 0x^3 - 13x^2 + x + 35} \\ \underline{x^4 - x^3 - 6x^2} \\ x^3 - 7x^2 + x + 35 \\ \underline{x^3 - x^2 - 6x} \\ -6x^2 + 7x + 35 \\ \underline{-6x^2 + 6x + 36} \\ x - 1 \leftarrow \text{Remainder} \end{array}$$

The quotient is $x^2 + x - 6$, and the remainder is $x - 1$.

We can write this result in the form

$$\frac{x^4 - 13x^2 + x + 35}{x^2 - x - 6} = x^2 + x - 6 + \frac{x - 1}{x^2 - x - 6}.$$

Practice Problem 2 Divide $x^4 + 5x^2 + 2x + 6$ by $x^2 - x + 3$.

2 Use synthetic division.

Synthetic Division

You can decrease the work involved in finding the quotient and remainder when the divisor $D(x)$ is of the form $x - a$ by using the process known as *synthetic division*. The procedure is best explained by considering an example.

Long Division	Synthetic Division
$\begin{array}{r} 2x^2 + 1x + 6 \\ x - 3\overline{)2x^3 - 5x^2 + 3x - 14} \\ \underline{2x^3 - 6x^2} \qquad\qquad \\ 1x^2 + 3x - 14 \\ \underline{1x^2 - 3x} \qquad \\ 6x - 14 \\ \underline{6x - 18} \\ 4 \end{array}$	$\begin{array}{r\|rrrr} 3 & 2 & -5 & 3 & -14 \\ & & 6 & 3 & 18 \\ \hline & 2 & 1 & 6 & \| \; 4 \end{array}$ ← Remainder

Each circled term in the long division process is the same as the term above it. Furthermore, the boxed terms are terms in the dividend written in a new position. If these two sets of terms are eliminated in the synthetic division process, a more efficient format results.

The essential steps in the process can be carried out as follows:

(i) Bring down 2 from the first line to the third line.

$$\begin{array}{r|rrrr} 3 & 2 & -5 & 3 & -14 \\ & \downarrow & & & \\ \hline & 2 & & & \end{array}$$

(ii) Multiply 2 in the third line by the 3 in the divisor position and place the product, 6, under -5; then add vertically to obtain 1.

$$\begin{array}{r|rrrr} 3 & 2 & -5 & 3 & -14 \\ & & 6 & & \\ \hline & 2 & 1 & & \end{array}$$

(iii) Multiply 1 by 3 and place the product 3 under 3; then add vertically to obtain 6.

$$\begin{array}{r|rrrr} 3 & 2 & -5 & 3 & -14 \\ & & 6 & 3 & \\ \hline & 2 & 1 & 6 & \end{array}$$

(iv) Finally, multiply 6 by 3 and place the product, 18, under -14; then add to obtain 4.

$$\begin{array}{r|rrrr} 3 & 2 & -5 & 3 & -14 \\ & & 6 & 3 & 18 \\ \hline & 2 & 1 & 6 & 4 \end{array}$$

The first three terms, namely, 2, 1, and 6, are the coefficients of the quotient, $2x^2 + x + 6$, and the last number, 4, is the remainder. The procedure for synthetic division is given next.

TECHNOLOGY CONNECTION

Synthetic division can be done on a graphing calculator. In this problem, enter the first coefficient, 2. Then repeatedly multiply the answer by 3 and add the next coefficient.

```
                 2
Ans*3+-5
                 1
Ans*3+3
                 6
Ans*3+-14
                 4
```

PROCEDURE IN ACTION

EXAMPLE 3 Synthetic Division

OBJECTIVE

Divide a polynomial $F(x)$ by $x - a$.

EXAMPLE

Divide $2x^4 - 3x^2 + 5x - 63$ by $x + 3$.

Step 1 Arrange the coefficients of $F(x)$ in order of *descending powers of* x, supplying zero as the coefficient of each missing power.

1. $\underline{}|$ 2 0 −3 5 −63

Step 2 Replace the divisor $x - a$ with a. Place a in the position to the left of the coefficients.

2. Rewrite $x + 3 = x - (-3)$; $a = -3$

$\underline{-3}|$ 2 0 −3 5 −63

Step 3 Bring the first (leftmost) coefficient down below the line. Multiply it by a and write the resulting product one column to the right and above the line.

3. $\underline{-3}|$ 2 0 −3 5 −63
−6
2

⊗ means "multiply by −3."

Step 4 Add the product obtained in the previous step to the coefficient directly above it and write the resulting sum directly below it and below the line. This sum is the "newest" number below the line.

4. $\underline{-3}|$ 2 0 −3 5 −63
−6
2 −6

Step 5 Multiply the newest number below the line by a, write the resulting product one column to the right and above the line, and repeat Step 4.

5. $\underline{-3}|$ 2 0 −3 5 −63
−6 18
2 −6 15

Step 6 Repeat Step 5 until a product is added to the constant term. Separate the last number below the line with a short vertical line.

6. $\underline{-3}|$ 2 0 −3 5 −63
−6 18 −45 120
2 −6 15 −40 57

Step 7 The last number below the line is the remainder, and the other numbers, reading from left to right, are the coefficients of the quotient, which has degree one less than the dividend $F(x)$.

7. $\underline{-3}|$ 2 0 −3 5 −63 Dividend: degree 4
−6 18 −45 120
2 −6 15 −40 | 57 Quotient: degree 3

$\underbrace{2x^3 - 6x^2 + 15x - 40}_{\text{Quotient}}$ Remainder

Practice Problem 3 Divide $2x^3 - 7x^2 + 5$ by $x - 3$.

SIDE NOTE

A quick way to see if you remembered to put 0 coefficients in place of missing terms when dividing a polynomial of degree n by $x - a$ is to check that there are $n + 1$ numbers in the row next to $a\rfloor$.

EXAMPLE 4 Using Synthetic Division

Use synthetic division to divide $2x^4 + x^3 - 16x^2 + 18$ by $x + 2$.

Solution

Because $x - a = x + 2 = x - (-2)$, we have $a = -2$. Write the coefficients of the dividend in a line, supplying 0 as the coefficient of the missing x-term. Then carry out the steps of synthetic division.

$$\begin{array}{r|rrrrr} -2 & 2 & 1 & -16 & 0 & 18 \\ & & -4 & 6 & 20 & -40 \\ \hline & 2 & -3 & -10 & 20 & -22 \end{array}$$

The quotient is $2x^3 - 3x^2 - 10x + 20$, and the remainder is -22; so the result is

$$\frac{2x^4 + x^3 - 16x^2 + 18}{x + 2} = 2x^3 - 3x^2 - 10x + 20 + \frac{-22}{x + 2}$$

$$= 2x^3 - 3x^2 - 10x + 20 - \frac{22}{x + 2}.$$

Practice Problem 4 Use synthetic division to divide $2x^3 + x^2 - 18x - 7$ by $x + 3$.

3 Use the Remainder and Factor Theorems.

The Remainder and Factor Theorems

In Example 4, we discovered that when the polynomial

$$F(x) = 2x^4 + x^3 - 16x^2 + 18 \text{ is divided by } x + 2,$$

the remainder is -22, a constant.

Now evaluate F at -2. We have

$$\begin{aligned} F(-2) &= 2(-2)^4 + (-2)^3 - 16(-2)^2 + 18 && \text{Replace } x \text{ with } -2 \text{ in } F(x). \\ &= 2(16) - 8 - 16(4) + 18 = -22 && \text{Simplify.} \end{aligned}$$

We see that $F(-2) = -22$, the remainder we found when we divided $F(x)$ by $x + 2$. Is this result a coincidence? No, the following theorem asserts that such a result is always true.

REMAINDER THEOREM

If a polynomial $F(x)$ is divided by $x - a$, then the remainder R is given by

$$R = F(a).$$

Note that in the statement of the Remainder Theorem, $F(x)$ plays the dual role of representing a polynomial and a function defined by the same polynomial.

The Remainder Theorem has a very simple proof: The division algorithm says that if a polynomial $F(x)$ is divided by a polynomial $D(x) \neq 0$, then

$$F(x) = D(x) \cdot Q(x) + R(x),$$

where $Q(x)$ is the quotient and $R(x)$ is the remainder. Suppose $D(x)$ is a linear polynomial of the form $x - a$. Because the degree of $R(x)$ is less than the degree of $D(x)$, $R(x)$ is a constant. Call this constant R. In this case, by the division algorithm, we have

$$F(x) = (x - a)Q(x) + R.$$

Replacing x with a in $F(x)$, we obtain

$$F(a) = (a - a)Q(a) + R = 0 + R = R.$$

This proves the Remainder Theorem.

EXAMPLE 5 Using the Remainder Theorem

Find the remainder when the polynomial

$$F(x) = 2x^5 - 4x^3 + 5x^2 - 7x + 2$$

is divided by $x - 1$.

Solution

We could find the remainder by using long division or synthetic division, but a quicker way is to evaluate $F(x)$ when $x = 1$. By the Remainder Theorem, $F(1)$ is the remainder.

$$\begin{aligned} F(1) &= 2(1)^5 - 4(1)^3 + 5(1)^2 - 7(1) + 2 && \text{Replace } x \text{ with 1 in } F(x). \\ &= 2 - 4 + 5 - 7 + 2 = -2 \end{aligned}$$

The remainder is -2.

Practice Problem 5 Use the Remainder Theorem to find the remainder when

$$F(x) = x^{110} - 2x^{57} + 5 \text{ is divided by } x - 1.$$

Sometimes the Remainder Theorem is used in the other direction: to evaluate a polynomial at a specific value of x. This use is illustrated in the next example.

EXAMPLE 6 Using the Remainder Theorem

Let $f(x) = x^4 + 3x^3 - 5x^2 + 8x + 75$. Find $f(-3)$.

Solution

One way of solving this problem is to evaluate $f(x)$ when $x = -3$.

$$\begin{aligned} f(-3) &= (-3)^4 + 3(-3)^3 - 5(-3)^2 + 8(-3) + 75 && \text{Replace } x \text{ with } -3. \\ &= 6 && \text{Simplify.} \end{aligned}$$

Another way is to use synthetic division to find the remainder when the polynomial $f(x)$ is divided by $(x + 3)$.

$$\begin{array}{r|rrrrr} -3 & 1 & 3 & -5 & 8 & 75 \\ & & -3 & 0 & 15 & -69 \\ \hline & 1 & 0 & -5 & 23 & 6 \end{array}$$

The remainder is 6. By the Remainder Theorem, $f(-3) = 6$.

Practice Problem 6 Use synthetic division to find $f(-2)$, where

$$f(x) = x^4 + 10x^2 + 2x - 20.$$

Note that if we divide any polynomial $F(x)$ by $x - a$, we have

$$\begin{aligned} F(x) &= (x - a)Q(x) + R && \text{Division algorithm} \\ F(x) &= (x - a)Q(x) + F(a) && R = F(a), \text{ by the Remainder Theorem} \end{aligned}$$

The last equation says that if $F(a) = 0$, then

$$F(x) = (x - a)Q(x)$$

and $(x - a)$ is a factor of $F(x)$. Conversely, if $(x - a)$ is a factor of $F(x)$, then dividing $F(x)$ by $(x - a)$ gives a remainder of 0. Therefore, by the Remainder Theorem, $F(a) = 0$. This argument proves the Factor Theorem.

FACTOR THEOREM

A polynomial $F(x)$ has $(x - a)$ as a factor if and only if $F(a) = 0$.

The Factor Theorem shows that if $F(x)$ is a polynomial, then the following problems are equivalent:

1. Factoring $F(x)$.
2. Finding the zeros of the function $F(x)$ defined by the polynomial expression.
3. Solving (or finding the roots of) the polynomial equation $F(x) = 0$.

Note that if a is a zero of the polynomial function $F(x)$, then $F(x) = (x - a)Q(x)$. Any solution of the equation $Q(x) = 0$ is also a zero of $F(x)$. Because the degree of $Q(x)$ is less than the degree of $F(x)$, it is simpler to solve the equation $Q(x) = 0$ to find other possible zeros of $F(x)$. The equation $Q(x) = 0$ is called the **depressed equation** of $F(x)$.

The next example illustrates how to use the Factor Theorem and the depressed equation to solve a polynomial equation.

EXAMPLE 7 Using the Factor Theorem

Given that 2 is a zero of the function $f(x) = 3x^3 + 2x^2 - 19x + 6$, solve the polynomial equation $3x^3 + 2x^2 - 19x + 6 = 0$.

Solution

Because 2 is a zero of $f(x)$, we have $f(2) = 0$. The Factor Theorem tells us that $(x - 2)$ is a factor of $f(x)$. Next, we use synthetic division to divide $f(x)$ by $(x - 2)$.

$$\begin{array}{r|rrrr} 2 & 3 & 2 & -19 & 6 \\ & & 6 & 16 & -6 \\ \hline & 3 & 8 & -3 & 0 \end{array} \leftarrow \text{Remainder}$$

$\uparrow$ 3, 8, −3: Coefficients of the quotient

We now have the coefficients of the quotient $Q(x)$, with $f(x) = (x - 2)Q(x)$.

$$f(x) = 3x^3 + 2x^2 - 19x + 6 = (x - 2)\underbrace{(3x^2 + 8x - 3)}_{\text{Quotient}}$$

Any solution of the depressed equation $3x^2 + 8x - 3 = 0$ is a zero of f. Because this equation is of degree 2, any method of solving a quadratic equation may be used to solve it.

$$3x^2 + 8x - 3 = 0 \qquad \text{Depressed equation}$$

$$(3x - 1)(x + 3) = 0 \qquad \text{Factor}$$

$$3x - 1 = 0 \quad \text{or} \quad x + 3 = 0 \qquad \text{Zero-product property}$$

$$x = \frac{1}{3} \quad \text{or} \quad x = -3 \qquad \text{Solve each equation.}$$

The solution set is $\left\{-3, \frac{1}{3}, 2\right\}$.

Practice Problem 7 Solve the equation $3x^3 - x^2 - 20x - 12 = 0$ given that one solution is -2.

Application Energy is measured in British thermal units (BTU). One BTU is the amount of energy required to raise the temperature of 1 pound of water 1°F at or near 39.2°F; 1 million BTU is equivalent to the energy in 8 gallons of gasoline. Fuels such as coal, petroleum, and natural gas are measured in quadrillion BTU $(1 \text{ quadrillion} = 10^{15})$, abbreviated "quads."

EXAMPLE 8 Crude Oil and Petroleum Consumption

The total crude oil and petroleum products consumption C (in billion barrels) in the United States during 1993–2011 is given in the following table. (Data are rounded to the nearest tenth.)

Year	Consumption	Year	Consumption
1993	6.3	2003	7.3
1995	6.5	2005	7.6
1997	6.8	2007	7.6
1999	7.1	2009	6.9
2001	7.2	2011	6.9

Source: www.eia.gov

These data can be modeled by the function

$$C(t) = -0.0006t^3 + 0.00576t^2 + 0.10284t + 6.33488,$$

where $t = 0$ represents 1993.

The model indicates that $C(4) = 6.8$ billion barrels were consumed in 1997 $(t = 4)$. Find another year between 1997 and 2011 when the *model* estimates consumption of 6.8 billion barrels.

Solution

We are given that

$$C(4) = 6.8$$

$$C(t) = (t - 4)Q(t) + 6.8 \quad \text{Division Algorithm and Remainder Theorem}$$

$$C(t) - 6.8 = (t - 4)Q(t) \quad \text{Subtract 6.8 from both sides.}$$

Therefore, 4 is zero of

$$F(t) = C(t) - 6.8 = -0.0006t^3 + 0.00576t^2 + 0.10284t - 0.46512.$$

Finding another year between 1997 and 2011 when the crude oil and petroleum consumption was 6.8 billion barrels requires us to find another zero of $F(t)$ that is between 4 and 18. Because 4 is a zero of $F(t)$, we use synthetic division to find the quotient $Q(t)$. This yields

$$Q(t) = -0.0006t^2 + 0.00336t + 0.11628.$$

We now solve the depressed (quadratic) equation $Q(t) = 0$ using the quadratic formula.

$$Q(t) = -0.0006t^2 + 0.00336t + 0.11628 = 0$$

$$t = \frac{-0.00336 \pm \sqrt{(0.00336)^2 - 4(-0.0006)(0.11628)}}{2(-0.0006)} \quad \text{Quadratic formula}$$

$$t = -11.4 \quad \text{or} \quad t = 17 \quad \text{Use a calculator.}$$

Because t needs to be between $4(=1997 - 1993)$ and $18(=2011 - 1993)$, this model tells us that in 2010 $(t = 17)$, the crude oil and petroleum products consumption in the United States was also 6.8 billion barrels.

Practice Problem 8 The value of $C(x) = 0.23x^3 - 4.255x^2 + 0.345x + 41.05$ is 10 when $x = 3$. Find another positive number x for which $C(x) = 10$.

Answers to Practice Problems

1. Quotient: $3x + 4$; remainder $= -2x + 3$
2. $x^2 + x + 3 + \dfrac{2x - 3}{x^2 - x + 3}$
3. $2x^2 - x - 3 + \dfrac{-4}{x - 3}$
4. $2x^2 - 5x - 3 + \dfrac{2}{x + 3}$
5. 4
6. 32
7. $\left\{-2, -\dfrac{2}{3}, 3\right\}$
8. 18

SECTION 3.3 Exercises

Basic Concepts and Skills

1. In the division $\dfrac{x^4 - 2x^3 + 5x^2 - 2x + 1}{x^2 - 2x + 3} = x^2 + 2 + \dfrac{2x - 5}{x^2 - 2x + 3}$, the dividend is ____________, the divisor is ____________, the quotient is ____________, and the remainder is ____________.
2. If $P(x)$, $Q(x)$, and $F(x)$ are polynomials and $F(x) = P(x) \cdot Q(x)$, then the factors of $F(x)$ are ____________ and ____________.
3. The Remainder Theorem states that if a polynomial $F(x)$ is divided by $(x - a)$, then the remainder $R =$ ____________.
4. The Factor Theorem states that $(x - a)$ is a factor of a polynomial $F(x)$ if and only if ____________ $= 0$.
5. **True or False.** You cannot use synthetic division to divide a polynomial by $x^2 - 1$.
6. **True or False.** When a polynomial of degree $n + 1$ is divided by a polynomial of degree n, the remainder is a constant polynomial.

In Exercises 7–14, use long division to find the quotient and the remainder.

7. $\dfrac{6x^2 - x - 2}{2x + 1}$
8. $\dfrac{4x^3 - 2x^2 + x - 3}{2x - 3}$
9. $\dfrac{3x^4 - 6x^2 + 3x - 7}{x + 1}$
10. $\dfrac{x^6 + 5x^3 + 7x + 3}{x^2 + 2}$
11. $\dfrac{4x^3 - 4x^2 - 9x + 5}{2x^2 - x - 5}$
12. $\dfrac{y^5 + 3y^4 - 6y^2 + 2y - 7}{y^2 + 2y - 3}$
13. $\dfrac{z^4 - 2z^2 + 1}{z^2 - 2z + 1}$
14. $\dfrac{6x^4 + 13x - 11x^3 - 10 - x^2}{3x^2 - 5 - x}$

In Exercises 15–30, use synthetic division to find the quotient and the remainder when the first polynomial is divided by the second polynomial.

15. $x^3 - x^2 - 7x + 2; x - 1$
16. $2x^3 - 3x^2 - x + 2; x + 2$
17. $x^3 + 4x^2 - 7x - 10; x + 2$
18. $x^3 + x^2 - 13x + 2; x - 3$
19. $x^4 - 3x^3 + 2x^2 + 4x + 5; x - 2$
20. $x^4 - 5x^3 - 3x^2 + 10; x - 1$
21. $2x^3 + 4x^2 - 3x + 1; x - \dfrac{1}{2}$
22. $3x^3 + 8x^2 + x + 1; x - \dfrac{1}{3}$
23. $2x^3 - 5x^2 + 3x + 2; x + \dfrac{1}{2}$
24. $3x^3 - 2x^2 + 8x + 2; x + \dfrac{1}{3}$
25. $x^5 + x^4 - 7x^3 + 2x^2 + x - 1; x - 1$
26. $2x^5 + 4x^4 - 3x^3 - 7x^2 + 3x - 2; x + 2$
27. $x^5 + 1; x + 1$
28. $x^5 + 1; x - 1$
29. $x^6 + 2x^4 - x^3 + 5; x + 1$
30. $3x^5; x - 1$

In Exercises 31–34, use synthetic division and the Remainder Theorem to find each function value. Check your answer by evaluating the function at the given x-value.

31. $f(x) = x^3 + 3x^2 + 1$
    - **a.** $f(1)$
    - **b.** $f(-1)$
    - **c.** $f\left(\dfrac{1}{2}\right)$
    - **d.** $f(10)$
32. $g(x) = 2x^3 - 3x^2 + 1$
    - **a.** $g(-2)$
    - **b.** $g(-1)$
    - **c.** $g\left(-\dfrac{1}{2}\right)$
    - **d.** $g(7)$
33. $h(x) = x^4 + 5x^3 - 3x^2 - 20$
    - **a.** $h(1)$
    - **b.** $h(-1)$
    - **c.** $h(-2)$
    - **d.** $h(2)$
34. $f(x) = x^4 + 0.5x^3 - 0.3x^2 - 20$
    - **a.** $f(0.1)$
    - **b.** $f(0.5)$
    - **c.** $f(1.7)$
    - **d.** $f(-2.3)$

In Exercises 35–42, use the Factor Theorem to show that the linear polynomial is a factor of the second polynomial. Check your answer by synthetic division.

35. $x - 1; 2x^3 + 3x^2 - 6x + 1$

36. $x - 3; 3x^3 - 9x^2 - 4x + 12$

37. $x + 1; 5x^4 + 8x^3 + x^2 + 2x + 4$

38. $x + 3; 3x^4 + 9x^3 - 4x^2 - 9x + 9$

39. $x - 2; x^4 + x^3 - x^2 - x - 18$

40. $x + 3; x^5 + 3x^4 + x^2 + 8x + 15$

41. $x + 2; x^6 - x^5 - 7x^4 + x^3 + 8x^2 + 5x + 2$

42. $x - 2; 2x^6 - 5x^5 + 4x^4 + x^3 - 7x^2 - 7x + 2$

In Exercises 43–46, find the value of k for which the first polynomial is a factor of the second polynomial.

43. $x + 1; x^3 + 3x^2 + x + k$ [*Hint:* Let $f(x) = x^3 + 3x^2 + x + k$. Then $f(-1) = 0$.]

44. $x - 1; -x^3 + 4x^2 + kx - 2$

45. $x - 2; 2x^3 + kx^2 - kx - 2$

46. $x - 1; k^2 - 3kx^2 - 2kx + 6$

In Exercises 47–50, use the Factor Theorem to show that the first polynomial *is not* a factor of the second polynomial. (*Hint:* Show that the remainder is not zero.)

47. $x - 2; -2x^3 + 4x^2 - 4x + 9$

48. $x + 3; -3x^3 - 9x^2 + 5x + 12$

49. $x + 2; 4x^4 + 9x^3 + 3x^2 + x + 4$

50. $x - 3; 3x^4 - 8x^3 + 5x^2 + 7x - 3$

Applying the Concepts

51. Geometry. The area of a rectangle is $(2x^4 - 2x^3 + 5x^2 - x + 2)$ square centimeters. Its length is $(x^2 - x + 2)$ cm. Find its width.

52. Geometry. The volume of a rectangular solid is $(x^4 + 3x^3 + x + 3)$ cubic inches. Its length and width are $(x + 3)$ and $(x + 1)$ inches, respectively. Find its height.

53. Cell phone sales. The weekly revenue from one type of cell phone sold at a mall outlet is a quadratic function, R, of its price, x, in dollars. The weekly revenue is \$3000 if the phone is priced at \$40 or \$60 (fewer phones are sold at \$60) and is \$2400 if the phone is priced at \$30.

a. Find an equation for $R(x) - 3000$.

b. Find an equation for $R(x)$.

c. Find the maximum weekly revenue and the price that generates this revenue.

54. Ticket sales. The evening revenue from a theater is a quadratic function, R, of its ticket price, x, in dollars. The evening revenue is \$4725 if the tickets are priced at \$8 or \$12 (fewer tickets are sold at \$12) and is \$4125 if the tickets are priced at \$6.

a. Find an equation for $R(x) - 4725$.

b. Find an equation for $R(x)$.

c. Find the maximum weekly revenue and the price that generates this revenue.

55. Total energy consumption. The total energy consumption (in quads) in the United States from 1991–2011 can be approximated by the function

$$C(t) = -0.0006t^3 - 0.0613t^2 + 2.0829t + 82.904,$$

where $t = 0$ represents 1991. The model indicates that 97.6 quads of energy were consumed in 2002. Find another year after 2002 when the model estimates 97.6 quads of total energy consumption in the United States. Round your answer to the nearest year. (The model was modified from the data obtained from www.eia.gov.)

56. Lottery sales. In the lottery game called *Florida Lotto*, players should select 6 numbers out of 53. Various prizes are offered for matching three through six numbers drawn by the lottery. Florida Lotto sales (in millions of dollars) from 1998 through 2011 can be approximated by the function

$$s(t) = -0.1779t^3 - 2.8292t^2 + 42.0240t + 698.6831,$$

where $t = 0$ represents 1998. The model indicated that \$737.7 million in Florida Lotto sales occurred in 1999. Find another year after 1999 when the model indicates that \$737.7 million in Florida Lotto sales occurred. (The model was slightly modified from the data obtained from www.floridafiscalportal.state.fl.us.)

57. Marine Corps. The total number of U.S. Marine Corps, M (in thousands), during 1990–2010 can be modeled by the function

$$M(t) = -0.0027t^3 + 0.3681t^2 - 5.8645t + 195.2782,$$

where $t = 0$ represents 1990, $t = 1$ represents 1991, and so on. The model estimates the total number of U.S. Marine Corps in 1992 as 185,000. Find another year between 1992 and 2010 when the model indicates that U.S. Marine Corps were approximately 185,000. Round your answer to the nearest year. (The model was generated using slightly modified data from www.census.gov.)

58. Unemployment. U.S. unemployment, U (in percent), from 2002–2012 can be modeled by the function

$$U(t) = -0.0374t^3 + 0.5934t^2 - 2.0553t + 6.7478,$$

where $t = 0$ represents 2002, $t = 1$ represents 2003, and so on. The model indicates a 7.7% level of unemployment in 2008 ($t = 6$). Find another year after 2008 with about the same unemployment level. Round your answer to the nearest year. (The model was slightly modified from data obtained from www.multipl.com.)

Beyond the Basics

59. Show that $\frac{1}{2}$ is a root of multiplicity 2 of the equation

$$4x^3 + 8x^2 - 11x + 3 = 0.$$

60. Show that $-\frac{2}{3}$ is a root of multiplicity 2 of the equation

$$9x^3 + 3x^2 - 8x - 4 = 0.$$

61. Assuming that n is a positive integer, find the values of n for which each of the following is true.
a. $x + a$ is a factor of $x^n + a^n$.
b. $x + a$ is a factor of $x^n - a^n$.
c. $x - a$ is a factor of $x^n + a^n$.
d. $x - a$ is a factor of $x^n - a^n$.

62. Use Exercise 61 to prove the following:
a. $7^{11} - 2^{11}$ is divisible by 5.
b. $(19)^{20} - (10)^{20}$ is divisible by 261.

63. Synthetic division is a process of dividing a polynomial $F(x)$ by $x - a$. Modify the process to use synthetic division so that it finds the remainder when the first polynomial is divided by the second.
a. $2x^3 + 3x^2 + 6x - 2;\ \ 2x - 1$
$\left[\textit{Hint}: 2x - 1 = 2\left(x - \frac{1}{2}\right).\right]$
b. $2x^3 - x^2 - 4x + 1;\ 2x + 3$

64. Let $g(x) = 3x^3 - 5cx^2 - 3c^2x + 3c^3$. Use synthetic division to find each function value.
a. $g(-c)$
b. $g(2c)$

In Exercises 65–69, let $f(x) = ax^3 + bx^2 + cx + d, a \neq 0$ and $g(x) = kx^4 + lx^3 + mx^2 + nx + p, k \neq 0$. (*Hint:* Use the end behavior of polynomials and the Intermediate Value Theorem.)

65. Show that f has exactly one real zero or three real zeros counting multiplicities.

66. Show that if $kp < 0$, then g has exactly two real zeros or exactly four real zeros counting multiplicities.

67. Suppose $dp < 0$ and $kp < 0$. Show that the graphs of f and g must intersect.

68. Give an example to show that the results of Exercise 67 may not hold if $dp < 0$ and $kp > 0$.

69. Give an example of a fourth-degree polynomial that has no real zeros.

Critical Thinking/Discussion/Writing

70. a. Use the fact that $\frac{f(x)}{2x - 6} = \frac{1}{2}\left(\frac{f(x)}{x - 3}\right)$ to explain how to use synthetic division by $x - 3$ to find the quotient and remainder when $f(x)$ is divided by $2x - 6$.
b. Repeat part (a), replacing $2x - 6$ with $ax + b, a \neq 0$.

Maintaining Skills

In Exercises 71–74, use synthetic division to find each quotient and remainder.

71. $x^4 - x^3 + 2x^2 + x - 3$ is divided by $x - 1$.

72. $x^3 + x^2 - x + 2$ is divided by $x + 2$.

73. $-x^6 + 5x^5 + 2x^2 - 10x + 3$ is divided by $x - 5$.

74. $x^4 + 3x^3 + 3x^2 + 2x + 7$ is divided by $x + 1$.

In Exercises 75–78, find all of the factors of each integer.

75. 11

76. 18

77. 51

78. 72

In Exercises 79–82, identify the degree, leading coefficient, and constant term for each polynomial.

79. $x^2 + 2x - 1$

80. $-x^3 + x + 7$

81. $-7x^{10} + 3x^3 - 2x + 4$

82. $13x^9 + 2x^{15} + 3x - 11$

SECTION 3.4

The Real Zeros of a Polynomial Function

BEFORE STARTING THIS SECTION, REVIEW

1. Synthetic division (Section 3.3, page 326)
2. Zero-product property (Section 1.2, page 92)
3. Multiplicity of a zero (Section 3.2, page 314)

OBJECTIVES

1. Review real zeros of polynomials.
2. Use the Rational Zeros Theorem.
3. Find the possible number of positive and negative zeros of polynomials.
4. Find the bounds on the real zeros of polynomials.
5. Learn a procedure for finding the real zeros of a polynomial function.

◆ Dow Jones Industrial Average

The Dow Jones Industrial Average (DJIA) reflects the average cost per share of stock from 30 companies chosen by the board of the New York Stock Exchange as the most outstanding companies to provide a quantitative view of the stock market and, by extension, the U.S. economy. To calculate the DJIA on any day, we do not simply add the prices of the component stocks and divide by 30. The reason is that during the last 100-plus years, there have been many stock splits, spinoffs, and stock substitutions. Without making an adjustment to the divisor (30), the value of the DJIA would be distorted.

To understand this, consider a stock split. Suppose two stocks are trading at \$10 and \$40; the average of the two is \$25. If the company with \$40 stock has a two-for-one split, its shares are now priced at \$20 each. The average of the two stocks falls to \$15 $\left(\frac{10 + 20}{2} = 15\right)$. But due to the two-for-one split, the value of investment is unchanged. The \$40 stock simply sells for \$20 with twice as many shares available. So to keep the DJIA at \$25, we use the divisor x so that $\frac{10 + 20}{x} = 25$. Solving for x, we have $x = \frac{30}{25} = 1.2$.

Over time, the divisor 30 for the DJIA has been adjusted several times, mostly downward. It stood at 0.001960784 in September 2013. This explains why the DJIA can be reported as 15,300 even though the sum of the prices of all of its component stocks is a fraction of this number. In Exercise 83, we model an investment between 2004–2013 by a polynomial function.

1 Review real zeros of polynomials.

Real Zeros of a Polynomial Function

We first review what we have learned about the real zeros of a polynomial function.

1. **Definition** If c is a real number in the domain of a function f and $f(c) = 0$, then c is called a real zero of f.
   a. Geometrically, this means that the graph of f has an x-intercept at c.
   b. Algebraically, this means that $(x - c)$ is a factor of $f(x)$. In other words, we can write $f(x) = (x - c)g(x)$ for some polynomial function $g(x)$.

2. **Number of real zeros** A polynomial function of degree n has, at most, n real zeros.
3. **Polynomial in factored form** If a polynomial $F(x)$ can be written in factored form, we find its zeros by solving $F(x) = 0$ using the zero product property of real numbers.
4. **Intermediate Value Theorem** If a polynomial $F(x)$ cannot be factored, we find approximate values of the real zeros (if any) of $F(x)$ by using the Intermediate Value Theorem. See page 312.

In this section, we investigate other useful tools for finding the real zeros of a polynomial function.

2 Use the Rational Zeros Theorem.

Rational Zeros Theorem

We use the following test (see Exercise 101 for a proof) to find *possible* rational zeros of a polynomial function with integer coefficients. We then use the Factor Theorem to determine whether any of these numbers is, in fact, a zero of a polynomial function $F(x)$.

RECALL

The ratio $\frac{p}{q}$ of integers p and q is in lowest terms if p and q have no common factor other than ± 1.

RATIONAL ZEROS THEOREM

If $F(x) = a_n x^n + a_{n-1}x^{n-1} + \cdots + a_2 x^2 + a_1 x + a_0$ is a polynomial function with integer coefficients ($a_n \neq 0, a_0 \neq 0$) and $\dfrac{p}{q}$ is a rational number in lowest terms that is a zero of $F(x)$, then

1. p is a factor of the constant term a_0.
2. q is a factor of the leading coefficient a_n.

EXAMPLE 1 Using the Rational Zeros Theorem

Find all rational zeros of $F(x) = 2x^3 + 5x^2 - 4x - 3$.

Solution

First, we list all *possible* rational zeros of $F(x)$.

$$\text{Possible rational zeros} = \frac{p}{q} = \frac{\text{Factors of the constant term, } -3}{\text{Factors of the leading coefficient, } 2}$$

Factors of -3: $\pm 1, \pm 3$

Factors of 2: $\pm 1, \pm 2$

All combinations for the rational zeros can be found by using just the *positive* factors of the leading coefficient. The possible rational zeros are

$$\frac{\pm 1}{1}, \frac{\pm 1}{2}, \frac{\pm 3}{1}, \frac{\pm 3}{2}$$

or

$$\pm 1, \pm\frac{1}{2}, \pm\frac{3}{2}, \pm 3.$$

We use synthetic division to determine whether a rational zero exists among these eight candidates. We start by testing 1. If 1 is not a rational zero, we will test other possibilities.

$$\begin{array}{r|rrrr} 1 & 2 & 5 & -4 & -3 \\ & & 2 & 7 & 3 \\ \hline & 2 & 7 & 3 & 0 \end{array} \quad \checkmark \text{ Yes}$$

The zero remainder tells us that $(x - 1)$ is a factor of $F(x)$, with the other factor being $2x^2 + 7x + 3$. So

$$F(x) = (x - 1)(2x^2 + 7x + 3)$$
$$= (x - 1)(2x + 1)(x + 3) \quad \text{Factor } 2x^2 + 7x + 3.$$

We find the zeros of $F(x)$ by solving the equation $(x - 1)(2x + 1)(x + 3) = 0$.

$$x - 1 = 0 \quad \text{or} \quad 2x + 1 = 0 \quad \text{or} \quad x + 3 = 0 \qquad \text{Zero-product property}$$
$$x = 1 \quad \text{or} \quad x = -\frac{1}{2} \quad \text{or} \quad x = -3 \qquad \text{Solve each equation.}$$

The solution set is $\left\{1, -\frac{1}{2}, -3\right\}$. The rational zeros of F are -3, $-\frac{1}{2}$, and 1.

Practice Problem 1 Find all rational zeros of $F(x) = 2x^3 + 3x^2 - 6x - 8$.

Recall that the *zeros* of a polynomial function $y = F(x)$ are also the *roots* or the *solutions* of the polynomial equation $F(x) = 0$.

EXAMPLE 2 Solving a Polynomial Equation

Solve $3x^3 - 8x^2 - 8x + 8 = 0$.

Solution

$$\text{Possible rational roots} = \frac{\text{Factors of the constant term, 8}}{\text{Factors of the leading coefficient, 3}}$$
$$= \frac{\pm 1, \pm 2, \pm 4, \pm 8}{\pm 1, \pm 3}$$
$$= \pm 1, \pm 2, \pm 4, \pm 8, \pm\frac{1}{3}, \pm\frac{2}{3}, \pm\frac{4}{3}, \pm\frac{8}{3}$$

The graph of $y = 3x^3 - 8x^2 - 8x + 8$ is shown in Figure 3.19. The calculator graph is shown in the margin.

TECHNOLOGY CONNECTION

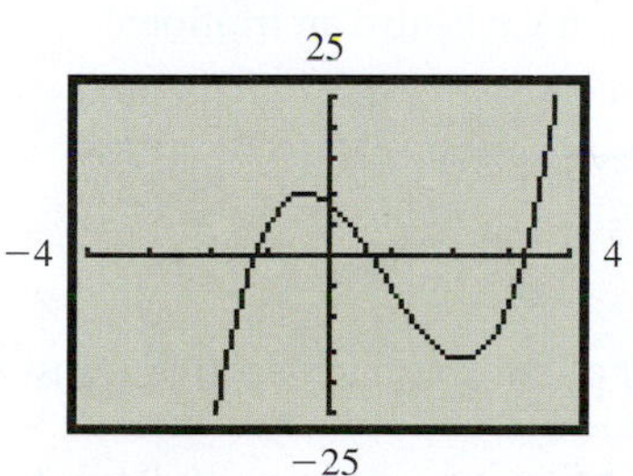

The calculator graph is helpful in anticipating where the rational zeros of $y = 3x^3 - 8x^2 - 8x + 8$ might be found.

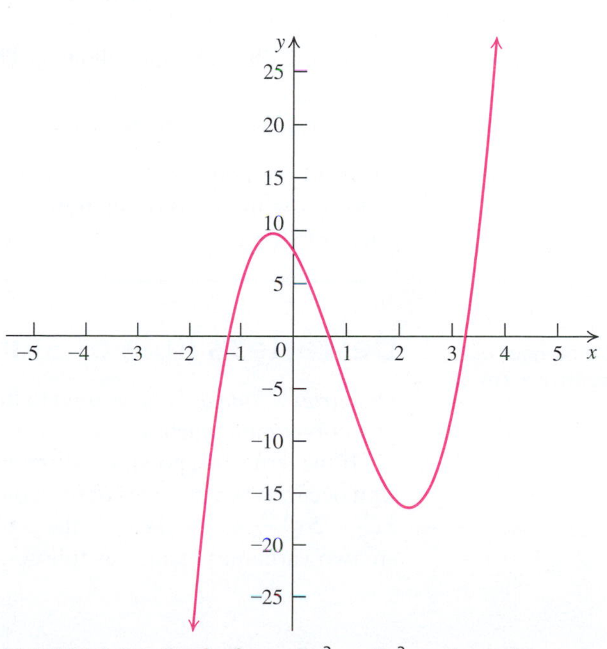

FIGURE 3.19 Graph of $y = 3x^3 - 8x^2 - 8x + 8$

From the graph in Figure 3.19, it appears that an x-intercept is between 0.5 and 1. We check whether $x = \frac{2}{3}$ is a root by synthetic division.

$$\begin{array}{r|rrrr} \frac{2}{3} & 3 & -8 & -8 & 8 \\ & & 2 & -4 & -8 \\ \hline & 3 & -6 & -12 & 0 \end{array} \quad \checkmark \text{ Yes}$$

Because the remainder in the synthetic division is 0, $\left(x - \frac{2}{3}\right)$ is a factor of the original equation with the depressed equation $3x^2 - 6x - 12 = 0$.

So

$$3x^3 - 8x^2 - 8x + 8 = 0 \qquad \text{Original equation}$$

$$\left(x - \frac{2}{3}\right)(3x^2 - 6x - 12) = 0 \qquad \text{Synthetic division}$$

$$\left(x - \frac{2}{3}\right) = 0 \quad \text{or} \quad 3x^2 - 6x - 12 = 0 \qquad \text{Zero-product property}$$

$$x = \frac{2}{3} \quad \Bigg| \quad x = \frac{-(-6) \pm \sqrt{(-6)^2 - 4(-12)(3)}}{2(3)} \qquad \text{Solve for } x \text{ using the quadratic formula}$$

$$= \frac{6 \pm \sqrt{36 + 144}}{6}$$

$$= \frac{6 \pm \sqrt{180}}{6} = \frac{6 \pm \sqrt{(36)(5)}}{6}$$

$$= \frac{6 \pm 6\sqrt{5}}{6} = 1 \pm \sqrt{5}$$

You can see from the graph in Figure 3.19 that $1 \pm \sqrt{5}$ are also the x-intercepts of the graph.

The three real roots of the given equation are

$$\frac{2}{3}, 1 + \sqrt{5}, \text{ and } 1 - \sqrt{5}.$$

Practice Problem 2 Solve $2x^3 - 9x^2 + 6x - 1 = 0$.

Example 2 illustrates the following Theorem.

CONJUGATE PAIRS THEOREM FOR IRRATIONAL ZEROS

If $P(x)$ is a polynomial with integer coefficients and $a + b\sqrt{c}$ (where a, b, and c are integers with $c > 0$) is an irrational zero of $P(x)$, then $a - b\sqrt{c}$ is also an irrational zero of $P(x)$.

3 Find the possible number of positive and negative zeros of polynomials.

Descartes's Rule of Signs

Descartes's Rule of Signs is used to find the possible number of positive and negative zeros of a polynomial function.

If the terms of a polynomial are written in descending order, we say that a **variation of sign** occurs when the signs of two consecutive terms differ. For example, in the polynomial $2x^5 - 5x^3 - 6x^2 + 7x + 3$, the signs of the terms are $+ \; - \; - \; + \; +$. Therefore, there are two variations of sign, as follows:

There are three variations of sign in

$$x^5 - 3x^3 + 2x^2 + 5x - 7.$$

DESCARTES'S RULE OF SIGNS

Let $F(x)$ be a polynomial function with real coefficients and with terms written in descending order.

1. The number of positive zeros of F is equal to the number of variations of sign of $F(x)$ or is less than that number by an even integer.
2. The number of negative zeros of F is equal to the number of variations of sign of $F(-x)$ or is less than that number by an even integer.

When using Descartes's Rule, a zero of multiplicity m should be counted as m zeros.

EXAMPLE 3 Using Descartes's Rule of Signs

Find the possible number of positive and negative zeros of

$$f(x) = x^5 - 3x^3 - 5x^2 + 9x - 7.$$

Solution

There are three variations of sign in $f(x)$.

$$f(x) = x^5 - 3x^3 - 5x^2 + 9x - 7$$

The number of positive zeros of $f(x)$ must be 3 or $3 - 2 = 1$. Furthermore,

$$\begin{aligned} f(-x) &= (-x)^5 - 3(-x)^3 - 5(-x)^2 + 9(-x) - 7 \\ &= -x^5 + 3x^3 - 5x^2 - 9x - 7. \end{aligned}$$

The polynomial $f(-x)$ has two variations of sign; therefore, the number of negative zeros of $f(x)$ is 2 or $2 - 2 = 0$.

Practice Problem 3 Find the possible number of positive and negative zeros of

$$f(x) = 2x^5 + 3x^2 + 5x - 1.$$

4 Find the bounds on the real zeros of polynomials.

Bounds on the Real Zeros

If a polynomial function $F(x)$ has no zeros greater than a number k, then k is called an **upper bound** on the zeros of $F(x)$. Similarly, if $F(x)$ has no zeros less than a number k, then k is called a **lower bound** on the zeros of $F(x)$. These bounds have the following rules.

RULES FOR BOUNDS

Let $F(x)$ be a polynomial function with real coefficients and a positive leading coefficient. Suppose $F(x)$ is synthetically divided by $x - k$.

1. If $k > 0$ and each number in the last row is zero or positive, then k is an upper bound on the zeros of $F(x)$.
2. If $k < 0$ and numbers in the last row alternate in sign, then k is a lower bound on the zeros of $F(x)$.

Zeros in the last row can be regarded as positive or negative.

EXAMPLE 4 **Finding the Bounds on the Zeros**

Find upper and lower bounds on the zeros of

$$F(x) = x^4 - x^3 - 5x^2 - x - 6.$$

Solution

By the Rational Zeros Theorem, the possible rational zeros of $F(x)$ are $\pm 1, \pm 2, \pm 3,$ and ± 6. There is only one variation of sign in $F(x) = x^4 - x^3 - 5x^2 - x - 6$, so $F(x)$ has only one positive zero. First, we try synthetic division by $x - k$ with $k = 1, 2, 3, \ldots$. The first integer that makes each number in the last row a 0 or a positive number is an upper bound on the zeros of $F(x)$.

$$\begin{array}{r|rrrrr} 1 & 1 & -1 & -5 & -1 & -6 \\ & & 1 & 0 & -5 & -6 \\ \hline & 1 & 0 & -5 & -6 & -12 \end{array}$$ Synthetic division with $k = 1$

$$\begin{array}{r|rrrrr} 2 & 1 & -1 & -5 & -1 & -6 \\ & & 2 & 2 & -6 & -14 \\ \hline & 1 & 1 & -3 & -7 & -20 \end{array}$$ Synthetic division with $k = 2$

$$\begin{array}{r|rrrrr} 3 & 1 & -1 & -5 & -1 & -6 \\ & & 3 & 6 & 3 & 6 \\ \hline & 1 & 2 & 1 & 2 & 0 \end{array}$$ Synthetic division with $k = 3$ ← All numbers are 0 or positive.

By the rule for bounds, 3 is an upper bound on the zeros of $F(x)$.

We now try synthetic division by $x - k$ with $k = -1, -2, -3, \ldots$. The first negative integer for which the numbers in the last row alternate in sign is a lower bound on the zeros of $F(x)$.

$$\begin{array}{r|rrrrr} -1 & 1 & -1 & -5 & -1 & -6 \\ & & -1 & 2 & 3 & -2 \\ \hline & 1 & -2 & -3 & 2 & -8 \end{array}$$ Synthetic division with $k = -1$

$$\begin{array}{r|rrrrr} -2 & 1 & -1 & -5 & -1 & -6 \\ & & -2 & 6 & -2 & 6 \\ \hline & 1 & -3 & 1 & -3 & 0 \end{array}$$ Synthetic division with $k = -2$ ← These numbers alternate in sign if we count 0 as positive.

By the rule of bounds, -2 is a lower bound on the zeros of $F(x)$.

SIDE NOTE

When checking the positive integers ($k = 1, 2, 3, \ldots$) for an upper bound on the zeros of a function, also check the last row for alternating signs indicating a lower bound. For example, for $f(x) = x^3 - 9x^2 + 26x - 24$, when checking $k = 1$, we have

$$\begin{array}{r|rrrr} 1 & 1 & -9 & 26 & -24 \\ & & 1 & -8 & 18 \\ \hline & 1 & -8 & 18 & -6 \end{array}$$

so 1 is a lower bound.

Practice Problem 4 Find upper and lower bounds on the zeros of

$$f(x) = 2x^3 + 5x^2 + x - 2.$$

An upper bound or a lower bound on the zeros of a polynomial function is not unique. For example, the zeros of $P(x) = x^3 + 2x^2 - x - 2$ are $-2, -1$, and 1. Any number greater than or equal to 1 is an upper bound on these zeros, and any number less than or equal to -2 is a lower bound on these zeros.

5 Learn a procedure for finding the real zeros of a polynomial function.

Find the Real Zeros of a Polynomial Function

We outline steps for gathering information and locating the real zeros of a polynomial function.

Step 1 Find the maximum number of real zeros by using the degree of the polynomial function.

Step 2 Find the possible number of positive and negative zeros by using Descartes's Rule of Signs.

Step 3 Write the set of possible rational zeros.

Step 4 Test the smallest positive integer in the set in Step 3, the next larger, and so on, until an integer zero or an upper bound of the zeros is found.

a. If a zero is found, use the function represented by the depressed equation in further calculations.

b. If an upper bound is found, discard all larger numbers in the set of possible rational zeros.

Step 5 Test the positive fractions that remain in the set in Step 3 after considering any bound that has been found.

Step 6 Use modified Steps 4 and 5 for negative numbers in the set in Step 3.

Step 7 If a depressed equation is quadratic, use any method (including the quadratic formula) to solve this equation.

SIDE **NOTE**

A graphing calculator may be used to obtain the approximate location of the real zeros. For a rational zero, verify by synthetic division.

EXAMPLE 5 Finding the Real Zeros of a Polynomial Function

Find the real zeros of $f(x) = 2x^5 - x^4 - 14x^3 + 7x^2 + 24x - 12$.

Solution

Step 1 The degree of f is 5; so f has, at most, 5 real zeros.

Step 2 $f(x) = 2x^5 - x^4 - 14x^3 + 7x^2 + 24x - 12$

There are three variations of sign in $f(x)$. So f has 3 or 1 positive zeros.

$$f(-x) = -2x^5 - x^4 + 14x^3 + 7x^2 - 24x - 12$$

There are two variations of signs in $f(-x)$. So f has 2 or 0 negative zeros.

Step 3 Possible rational zeros are

$$\left\{\pm 1, \pm 2, \pm 3, \pm 4, \pm 6, \pm 12, \pm\frac{1}{2}, \pm\frac{3}{2}\right\}.$$

Step 4 We test for zeros, 1, 2, 3, 4, 6, and 12 until a zero or an upper bound is found.

$$\begin{array}{r|rrrrrr} 1 & 2 & -1 & -14 & 7 & 24 & -12 \\ & & 2 & 1 & -13 & -6 & 18 \\ \hline & 2 & 1 & -13 & -6 & 18 & 6 \times \end{array} \qquad \begin{array}{r|rrrrrr} 2 & 2 & -1 & -14 & 7 & 24 & -12 \\ & & 4 & 6 & -16 & -18 & 12 \\ \hline & 2 & 3 & -8 & -9 & 6 & 0 \checkmark \end{array}$$

Because 2 is a zero, $(x - 2)$ is a factor of f. We have

$f(x) = (x - 2)(2x^4 + 3x^3 - 8x^2 - 9x + 6)$ Synthetic division

We consider the zeros of $Q_1(x) = 2x^4 + 3x^3 - 8x^2 - 9x + 6$.

The possible positive rational zeros of $Q_1(x)$ are $\left\{1, 2, 3, 6, \frac{1}{2}, \frac{3}{2}\right\}$. We discard 1 because 1 is not a zero of the original function. We test 2 because it may be a repeated zero of f.

$$\begin{array}{r|rrrrr} 2 & 2 & 3 & -8 & -9 & 6 \\ & & 4 & 14 & 12 & 6 \\ \hline & 2 & 7 & 6 & 3 & 12 \end{array} \leftarrow \text{Remainder} \neq 0$$

The last row shows that 2 is not a zero of $Q_1(x)$ but is an upper bound (all entries in the last row are positive) for the zeros of $Q_1(x)$, hence, also on the zeros of f.

Step 5 We test the remaining positive entries $\frac{1}{2}$ and $\frac{3}{2}$ in the list.

$$\begin{array}{r|rrrrr} \frac{1}{2} & 2 & 3 & -8 & -9 & 6 \\ & & 1 & 2 & -3 & -6 \\ \hline & 2 & 4 & -6 & -12 & 0 \checkmark \end{array} \qquad \begin{array}{r|rrrrr} \frac{3}{2} & 2 & 3 & -8 & -9 & 6 \\ & & 3 & 9 & \frac{3}{2} & -\frac{45}{4} \\ \hline & 2 & 6 & 1 & -\frac{15}{2} & -\frac{21}{4} \times \end{array}$$

Note that $\frac{1}{2}$ is a zero of $Q_1(x)$. Using the last row in synthetic division, we have

$$f(x) = (x-2)\left(x-\frac{1}{2}\right)(2x^3+4x^2-6x-12) = 0$$

or $(x-2)(2x-1)(x^3+2x^2-3x-6) = 0$ Simplify.

Step 6 We now find the negative zeros of $Q_2(x) = x^3 + 2x^2 - 3x - 6$. We test $-1, -2, -3,$ and -6.

$$\begin{array}{r|rrrr} -1 & 1 & 2 & -3 & -6 \\ & & -1 & -1 & 4 \\ \hline & 1 & 1 & -4 & -2 \times \end{array} \qquad \begin{array}{r|rrrr} -2 & 1 & 2 & -3 & -6 \\ & & -2 & 0 & 6 \\ \hline & 1 & 0 & -3 & 0 \checkmark \end{array}$$

Step 7 We can solve the depressed equation $x^2 - 3 = 0$ to obtain $x = \pm\sqrt{3}$. The real zeros of f are $-2, -\sqrt{3}, \frac{1}{2}, \sqrt{3},$ and 2. Because $f(x)$ has, at most, five real zeros, we have found all of them.

Practice Problem 5 Find the real zeros of $f(x) = 3x^4 - 11x^3 + 22x - 12$.

EXAMPLE 6 Graphing a Polynomial Function

Sketch the graph of $f(x) = 3x^3 - 2x^2 - 6x + 4$.

Solution

We use the procedure given in Section 3.2 on page 315.

Step 1 Because the degree, 3, is odd and the leading coefficient, 3, is positive, the end behavior is similar to that of $y = x^3$.

Step 2 Solve $3x^3 - 2x^2 - 6x + 4 = 0$ to find the real zeros. By the Rational Zeros Theorem, the possible rational roots are $\pm\frac{1}{3}, \pm\frac{2}{3}, \pm 1$. Trying each value, we find that $\frac{2}{3}$ is a real zero, as shown below. We solve the depressed equation $3x^2 - 6 = 0$ to find the remaining zeros, $x = \pm\sqrt{2}$. The y-intercept is $f(0) = 3(0)^3 - 2(0)^2 - 6(0) + 4 = 4$.

$$\begin{array}{r|rrrr} 2/3 & 3 & -2 & -6 & 4 \\ & & 2 & 0 & -4 \\ \hline & 3 & 0 & -6 & 0 \end{array}$$

Step 3 The three zeros, $-\sqrt{2}, \frac{2}{3}, \sqrt{2}$ divide the x-axis into four intervals.

Step 4 $(-\infty, -\sqrt{2}), \left(-\sqrt{2}, \frac{2}{3}\right), \left(\frac{2}{3}, \sqrt{2}\right)$, and, $(\sqrt{2}, \infty)$

Noting that $\sqrt{2} \approx 1.4$, we choose -2, 0, 1, and 2 as test numbers.

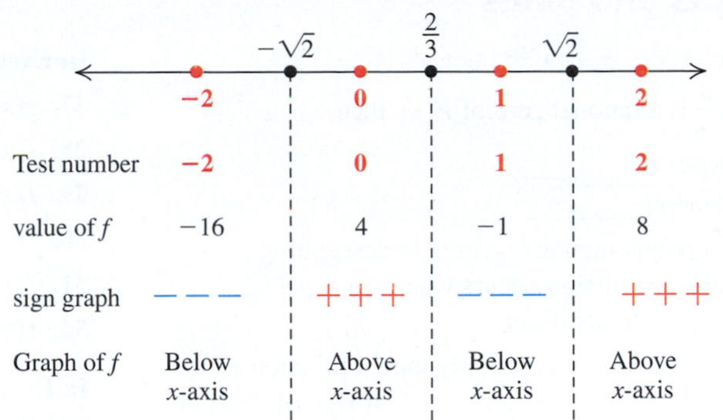

	$-\sqrt{2}$		$\frac{2}{3}$		$\sqrt{2}$	
Test number	-2		0		1		2
value of f	-16		4		-1		8
sign graph	− − −		+ + +		− − −		+ + +
Graph of f	Below x-axis		Above x-axis		Below x-axis		Above x-axis

Step 5 The graph is shown in Figure 3.20.

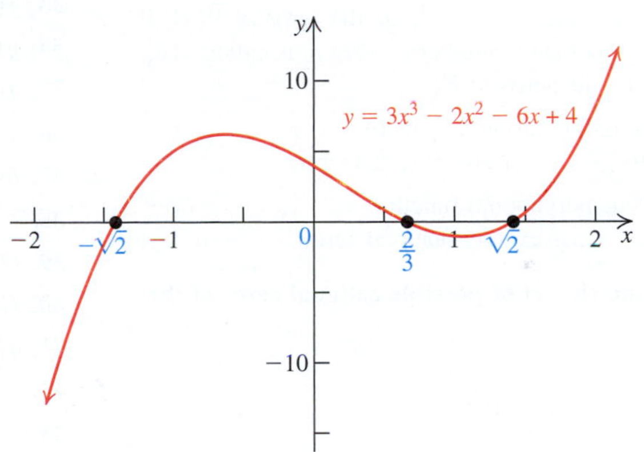

FIGURE 3.20

Practice Problem 6 Sketch the graph of $f(x) = 3x^3 - x^2 - 9x + 3$.

Answers to Practice Problems

1. $\{-2\}$ 2. $\frac{1}{2}, 2 + \sqrt{3}, 2 - \sqrt{3}$
3. Number of positive zeros: 1; number of negative zeros: 0 or 2
4. Upper bound: 1; lower bound: -3
5. $-\sqrt{2}, \frac{2}{3}, \sqrt{2}, 3$
6.

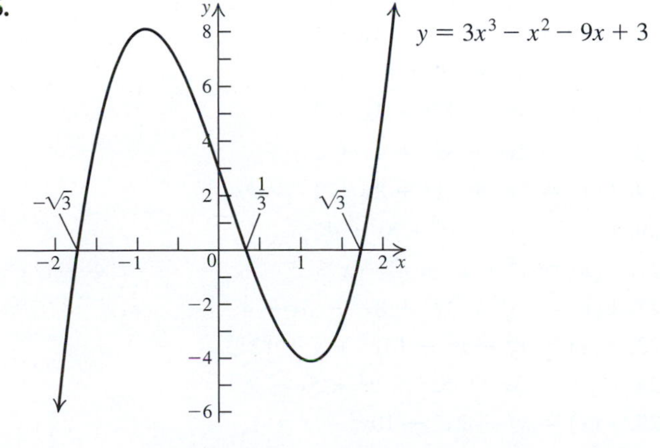

SECTION 3.4 Exercises

Basic Concepts and Skills

1. Let $P(x) = a_n x^n + \cdots + a_0$ with integer coefficients. If $\frac{p}{q}$ is a rational zero of $P(x)$, then $\frac{p}{q} = \frac{\text{possible factors of } \_\_\_\_\_}{\text{possible factors of } \_\_\_\_\_}$.
2. If the terms of a polynomial are written in descending order, then a variation of sign occurs when the signs of two __________ terms differ.
3. The number of positive zeros of a polynomial function $P(x)$ is equal to the number of __________ of sign of $P(x)$ or is less than that number by (a)n __________.
4. If a polynomial $P(x)$ has no zero greater than a number k, then k is called a(n) __________ on the zeros of $P(x)$. If $P(x)$ has no zero less than a number m, then m is called a(n) __________ on the zeros of $P(x)$.
5. **True or False.** Possible rational zeros of $P(x) = 9x^3 - 9x^2 - x + 1$ are $\pm 1, \pm 3, \pm 9$.
6. **True or False.** The polynomial function $P(x) = x^3 + x + 1$ has exactly one real zero.

In Exercises 7–10, find the set of possible rational zeros of the given function.

7. $f(x) = 3x^3 - 4x^2 + 5$
8. $g(x) = 2x^4 - 5x^2 - 2x + 1$
9. $h(x) = 4x^4 - 9x^2 + x + 6$
10. $F(x) = 6x^6 + 5x^5 + x - 35$

In Exercises 11–26, find all rational zeros of the given polynomial function.

11. $f(x) = x^3 - x^2 - 4x + 4$
12. $f(x) = x^3 - 4x^2 + x + 6$
13. $f(x) = x^3 + x^2 + 2x + 2$
14. $f(x) = x^3 + 3x^2 + 2x + 6$
15. $g(x) = 2x^3 + x^2 - 13x + 6$
16. $g(x) = 6x^3 + 13x^2 + x - 2$
17. $g(x) = 3x^3 - 2x^2 + 3x - 2$
18. $g(x) = 2x^3 + 3x^2 + 4x + 6$
19. $h(x) = 3x^3 + 7x^2 + 8x + 2$
20. $h(x) = 2x^3 + x^2 + 8x + 4$
21. $h(x) = x^4 - x^3 - x^2 - x - 2$
22. $h(x) = 2x^4 + 3x^3 + 8x^2 + 3x - 4$
23. $F(x) = x^4 - x^3 - 13x^2 + x + 12$
24. $G(x) = 3x^4 + 5x^3 + x^2 + 5x - 2$
25. $F(x) = x^4 - 2x^3 + 10x^2 - x + 1$
26. $G(x) = x^6 + 2x^4 + x^2 + 2$

In Exercises 27–32, graph each polynomial function.

27. $f(x) = 6x^3 + 13x^2 + x - 2$
28. $f(x) = 2x^3 - x^2 - 4x + 2$
29. $f(x) = 2x^3 - 3x^2 - 14x + 21$
30. $f(x) = 6x^3 + 17x^2 + x - 10$
31. $f(x) = 4x^4 + 4x^3 - 3x^2 - 2x + 1$
32. $f(x) = 9x^4 + 30x^3 + 13x^2 - 20x + 4$

In Exercises 33–46, determine the possible number of positive and negative zeros of the given function by using Descartes's Rule of Signs.

33. $f(x) = 5x^3 - 2x^2 - 3x + 4$
34. $g(x) = 3x^3 + x^2 - 9x - 3$
35. $f(x) = 2x^3 + 5x^2 - x + 2$
36. $g(x) = 3x^4 + 8x^3 - 5x^2 + 2x - 3$
37. $h(x) = 2x^5 - 5x^3 + 3x^2 + 2x - 1$
38. $F(x) = 5x^6 - 7x^4 + 2x^3 - 1$
39. $G(x) = -3x^4 - 4x^3 + 5x^2 - 3x + 7$
40. $H(x) = -5x^5 + 3x^3 - 2x^2 - 7x + 4$
41. $f(x) = x^4 + 2x^2 + 4$
42. $f(x) = 3x^4 + 5x^2 + 6$
43. $g(x) = 2x^5 + x^3 + 3x$
44. $g(x) = 2x^5 + 4x^3 + 5x$
45. $h(x) = -x^5 - 2x^3 + 4$
46. $h(x) = 2x^5 + 3x^3 + 1$

In Exercises 47–54, determine integer upper and lower bounds on the zeros of the given function.

47. $f(x) = 3x^3 - x^2 + 9x - 3$
48. $g(x) = 2x^3 - 3x^2 - 14x + 21$
49. $F(x) = 3x^3 + 2x^2 + 5x + 7$
50. $G(x) = x^3 + 3x^2 + x - 4$
51. $h(x) = x^4 + 3x^3 - 15x^2 - 9x + 31$
52. $H(x) = 3x^4 - 20x^3 + 28x^2 + 19x - 13$
53. $f(x) = 6x^4 + x^3 - 43x^2 - 7x + 7$
54. $g(x) = 6x^4 + 23x^3 + 25x^2 - 9x - 5$

In Exercises 55–74, find all rational zeros of f. Then (if necessary) use the depressed equation to find all roots of the equation $f(x) = 0$.

55. $f(x) = x^3 + 5x^2 - 8x + 2$
56. $f(x) = x^3 - 7x^2 - 5x + 3$
57. $f(x) = 2x^3 - x^2 - 6x + 3$
58. $f(x) = 3x^3 + 2x^2 - 15x - 10$
59. $f(x) = 2x^3 - 9x^2 + 6x - 1$
60. $f(x) = 2x^3 - 3x^2 - 4x - 1$
61. $f(x) = x^4 + x^3 - 5x^2 - 3x + 6$

62. $f(x) = x^4 - 2x^3 - 5x^2 + 4x + 6$

63. $f(x) = x^4 - 3x^3 + 3x - 1$

64. $f(x) = x^4 - 6x^3 - 7x^2 + 54x - 18$

65. $f(x) = 2x^4 - 5x^3 - 4x^2 + 15x - 6$

66. $f(x) = 3x^4 - 8x^3 - 18x^2 + 40x + 15$

67. $f(x) = 6x^4 - x^3 - 13x^2 + 2x + 2$

68. $f(x) = 6x^4 + x^3 - 19x^2 - 3x + 3$

69. $f(x) = x^5 - 2x^4 - 4x^3 + 8x^2 + 3x - 6$

70. $f(x) = x^5 + x^4 - 6x^3 - 6x^2 + 8x + 8$

71. $f(x) = 2x^5 + x^4 - 11x^3 - x^2 + 15x - 6$

72. $f(x) = 2x^5 - x^4 - 9x^3 + 10x + 4$

73. $f(x) = 2x^5 - 13x^4 + 27x^3 - 17x^2 - 5x + 6$

74. $f(x) = 3x^5 + x^4 - 9x^3 - 3x^2 + 6x + 2$

Applying the Concepts

75. Geometry. The length of a rectangle is $x^2 - 2x + 3$, and its width is $x - 2$. Find its dimensions assuming that its area is 306 square units.

76. Making a box. A square piece of tin 18 inches on each side is to be made into a box without a top by cutting a square from each corner and folding up the flaps to form the sides. What size corners should be cut so that the volume of the box is 432 cubic inches?

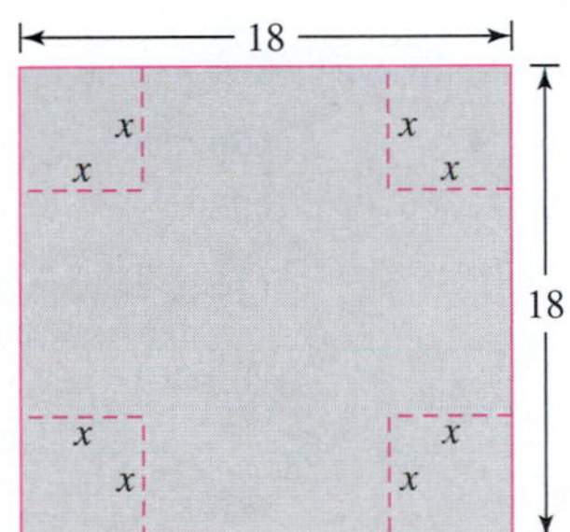

77. Cost function. The total cost of a product (in dollars) is given by $C(x) = 3x^3 - 6x^2 + 108x + 100$, where x is the demand for the product. Find the value of x that gives a total cost of \$628.

78. Profit. For the product in Exercise 77, the demand function is $p(x) = 330 + 10x - x^2$. Find the value of x that gives the profit of \$910.

79. Production cost. The total cost per month, in thousands of dollars, of producing x printers (with x measured in hundreds) is given by $C(x) = x^3 - 15x^2 + 5x + 50$. How many units must be produced so that the total monthly cost is \$125,000?

80. Break even. If the demand function for the printers in Exercise 79 is $p(x) = -3x + 3 + \frac{74}{x}$, find the number of printers that must be produced and sold to break even.

81. Oil import-export. A country maintains its normal oil needs by importing the deficit and exporting its surplus oil. The accompanying graph shows the country's balance of import and export (in million barrels) for 2004–2013, where $x = 0$ represents 2004.

a. How many years did the country export oil?

b. Construct a polynomial function of minimum degree that models this graph.

c. Use part (b) to estimate the oil import or export in 2009.

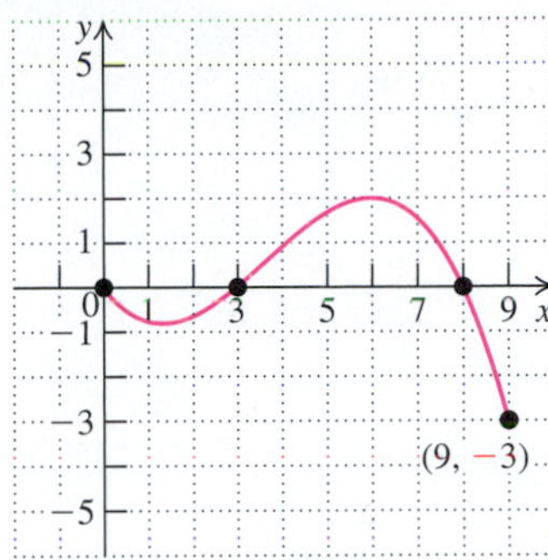

82. Number of tourists. The accompanying graph shows the number of tourists (in thousands) above or below normal during 2013, where $x = 1$ represents January.

a. Construct a polynomial function of minimum degree that models this graph.

b. Use part (a) to estimate the level of tourism in September 2013.

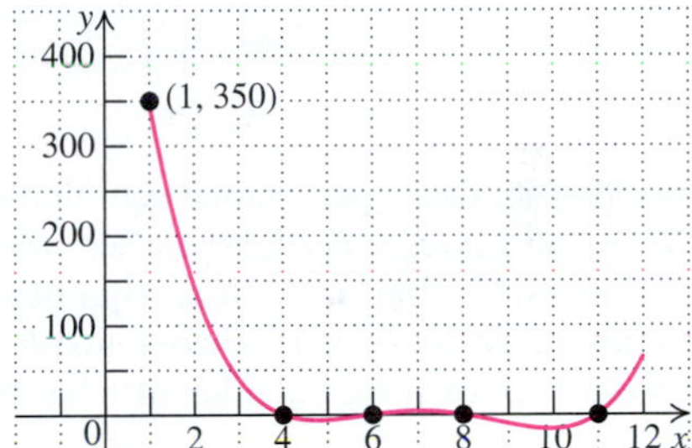

83. Investment. The accompanying graph shows the profit/loss (in thousand dollars) of Ms. Sharpy's investment in the stock market during 2004–2013, where $x = 0$ represents 2004.

a. Construct a polynomial of minimum degree that models the graph.

b. Use part (a) to estimate the profit/loss in 2008.

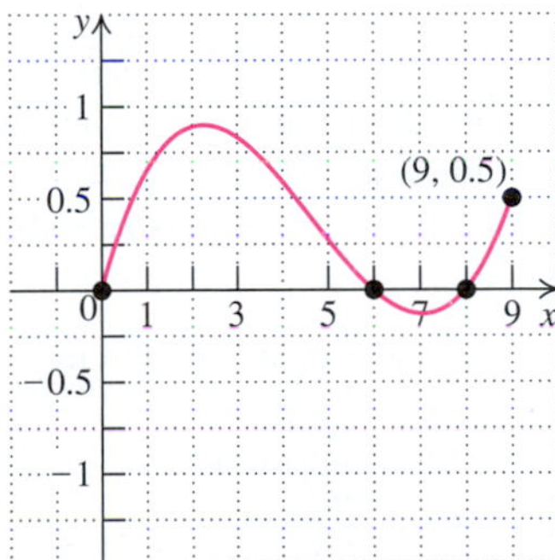

c. Use part (a) to determine the year(s) when Ms. Sharpy realized a profit of \$600.

d. Use part (a) to determine the maximum profit and the year it occurred.

84. County government. The accompanying graph shows the surplus/deficit (in million dollars) for a certain county government during 2004–2013, where $x = 0$ represents 2004.

a. Construct a polynomial function of minimum degree that models the graph.

b. Use part (a) to estimate the surplus/deficit in 2007 and 2011.

c. Use part (a) to determine the maximum amount of deficit.

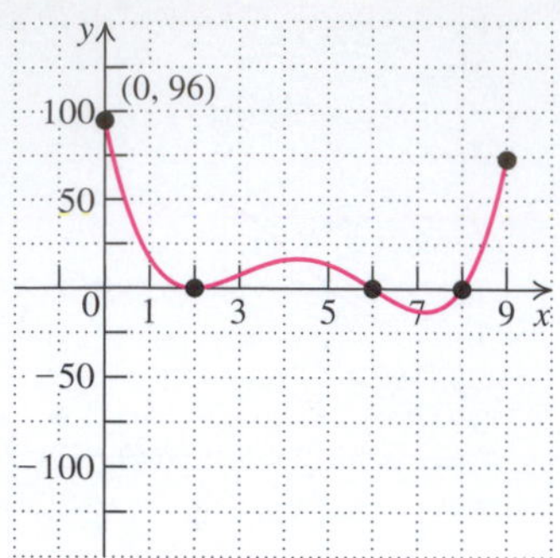

Beyond the Basics

In Exercises 85–88, show that the given number is irrational. [*Hint:* To show that $x = 1 + \sqrt{3}$ is irrational, form the equivalent equations $x - 1 = \sqrt{3}$, $(x - 1)^2 = 3$, and $x^2 - 2x - 2 = 0$ and show that the last equation has no rational roots.]

85. $\sqrt{3}$

86. $\sqrt[3]{4}$

87. $3 - \sqrt{2}$

88. $(9)^{\frac{2}{3}}$

In Exercises 89–100, find a polynomial equation of least possible degree with integer coefficient whose roots include the given numbers. Use the result: If an equation with integer coefficients has a root $(a + \sqrt{b})$ where a and b are rational but $\sqrt{b}$ is irrational, then the equation also has the root $(a - \sqrt{b})$.

89. $1, -1, 0$

90. $2, 1, -2$

91. $-\frac{1}{2}, 2, \frac{7}{3}$

92. $-\frac{1}{3}, -\frac{1}{2}, \frac{1}{6}$

93. $1 + \sqrt{2}, 3$

94. $3 - \sqrt{5}, 2$

95. $1 + \sqrt{3}, 3 - \sqrt{2}$

96. $2 - \sqrt{5}, 2 + \sqrt{3}$

97. $\sqrt{2} + \sqrt{3}$

98. $2 + \sqrt{3} + \sqrt{5}$

99. $3 - \sqrt{2} + \sqrt{5}$

100. $\sqrt{2} + \sqrt{3} + \sqrt{5}$

101. To prove the rational zeros test, we assume the **Fundamental Theorem of Arithmetic**, which states that every integer has a unique prime factorization. Supply reasons for each step in the following proof.

(i) $F\left(\frac{p}{q}\right) = 0$, and $\frac{p}{q}$ is in lowest terms.

(ii) $a_n\left(\frac{p}{q}\right)^n + a_{n-1}\left(\frac{p}{q}\right)^{n-1} + \cdots + a_1\left(\frac{p}{q}\right) + a_0 = 0$

(iii) $a_np^n + a_{n-1}p^{n-1}q + \cdots + a_1pq^{n-1} + a_0q^n = 0$

(iv) $a_np^n + a_{n-1}p^{n-1}q + \cdots + a_1pq^{n-1} = -a_0q^n$

(v) p is a factor of the left side of (iv).

(vi) p is a factor of $-a_0q^n$.

(vii) p is a factor of a_0.

(viii) $a_{n-1}p^{n-1}q + \cdots + a_1pq^{n-1} + a_0q^n = -a_np^n$

(ix) q is a factor of the left side of the equation in (viii).

(x) q is a factor of $-a_np^n$.

(xi) q is a factor of a_n.

102. Find all rational roots of the equation

$$2x^4 + 2x^3 + \frac{1}{2}x^2 + 2x - \frac{3}{2} = 0.$$

[*Hint:* Multiply both sides of the equation by the lowest common denominator of the coefficients.]

Critical Thinking/Discussion/Writing

103. True or False.

a. $\frac{1}{3}$ is a possible zero of $P(x) = x^5 - 3x^2 + x + 3$.

b. $\frac{2}{5}$ is a possible zero of $P(x) = 2x^7 - 5x^4 - 17x + 25$.

104. Find all the zeros of $f(x) = x^4 - 2x^3 - 4x^2 + 4x + 4$, if one of its zeros is $\sqrt{2}$.
(*Hint:* $f(x)$ has integer coefficients, so $-\sqrt{2}$ is also a zero.)

Maintaining Skills

In Exercises 105–114, simplify each expression.

105. $(-2i)^5$

106. $(x + i)(x - i)$

107. $(2x + i)(-2x + i)$

108. $(x + 3i)(x - 3i)$

109. $(5x + 2i)(5x - 2i)$

110. $(2x + 8i)(x - 4i)$

111. $(x - 1 + 2i)(x - 1 - 2i)$

112. $[x - (3 - i)][x - (3 + i)]$

113. $[x - (2 + 3i)][x - (2 - 3i)]$

114. $[x - (1 + \sqrt{3}i)][x - (1 - \sqrt{3}i)]$

SECTION 3.5

The Complex Zeros of a Polynomial Function

Gerolamo Cardano (1501–1576)
Cardano was trained as a physician, but his range of interests included philosophy, music, gambling, mathematics, astrology, and the occult sciences. Cardano was a colorful man, full of contradictions. His lifestyle was deplorable even by the standards of the times, and his personal life was filled with tragedies. His wife died in 1546. He saw one son executed for wife poisoning. He cropped the ears of a second son who attempted the same crime. In 1570, he was imprisoned for six months for heresy when he published the horoscope of Jesus. Cardano spent the last few years in Rome, where he wrote his autobiography, *De Vita Propria*, in which he did not spare the most shameful revelations.

BEFORE STARTING THIS SECTION, REVIEW

1. Long division (Section 3.3, page 324)
2. Synthetic division (Section 3.3, page 326)
3. Rational zeros theorem (Section 3.4, page 336)
4. Complex numbers (Section 1.3, page 104)
5. Conjugate of a complex number (Section 1.3, page 109)

OBJECTIVES

1. Learn basic facts about the complex zeros of polynomials.
2. Use the Conjugate Pairs Theorem to find zeros of polynomials.

◆ Cardano Accused of Stealing Formula

In 16th-century Italy, getting and keeping a teaching job at a university was difficult. University appointments were, for the most part, temporary. One way a professor could convince the administration that he was worthy of continuing in his position was by winning public challenges. Two contenders for a position would present each other with a list of problems, and later in a public forum, each person would present his solutions to the other's problems. As a result, scholars often kept secret their new methods of solving problems.

In 1535, in a public contest primarily about solving the cubic equation, Nicolo Tartaglia (1499–1557) won over his opponent Antonio Maria Fiore. Gerolamo Cardano (1501–1576), in the hope of learning the secret of solving a cubic equation, invited Tartaglia to visit him. After many promises and much flattery, Tartaglia revealed his method on the promise, probably given under oath, that Cardano would keep it confidential. When Cardano's book *Ars Magna* (The Great Art) appeared in 1545, the formula and the method of proof were fully disclosed. Angered at this apparent breach of a solemn oath and feeling cheated out of the rewards of his monumental work, Tartaglia accused Cardano of being a liar and a thief for stealing his formula. Thus began the bitterest feud in the history of mathematics, carried on with name-calling and mudslinging of the lowest order. As Tartaglia feared, the formula (see Group Project 1) has forever since been known as Cardano's formula for the solution of the cubic equation.

1 Learn basic facts about the complex zeros of polynomials.

RECALL

A number z of the form $z = a + bi$ is a **complex number**, where a and b are real numbers and $i = \sqrt{-1}$.

The Factor Theorem connects the concept of *factors* and *zeros* of a polynomial, and we continue to investigate this connection.

Because the equation $x^2 + 1 = 0$ has no real roots, the polynomial function $P(x) = x^2 + 1$ has no real zeros. However, if we replace x with the complex number i, we have

$$P(i) = i^2 + 1 = (-1) + 1 = 0.$$

Consequently, i is a complex number for which $P(i) = 0$; that is, i is a *complex zero* of $P(x)$. In Section 3.2, we stated that a polynomial of degree n can have, at most, n real zeros.

We now extend our number system to allow the variables and coefficients of polynomials to represent complex numbers. When we want to emphasize that complex numbers may be used in this way, we call the polynomial $P(x)$ a **complex polynomial**. If $P(z) = 0$ for a complex number z, we say that z is a zero or a **complex zero** of $P(x)$. In the complex number system, every nth-degree polynomial equation has exactly n roots and every nth-degree polynomial can be factored into *exactly* n linear factors. This fact follows from the Fundamental Theorem of Algebra.

FUNDAMENTAL THEOREM OF ALGEBRA

Every polynomial function

$$P(x) = a_n x^n + a_{n-1}x^{n-1} + \cdots + a_1 x + a_0 \quad (n \geq 1, a_n \neq 0)$$

with complex coefficients $a_n, a_{n-1}, \ldots, a_1, a_0$ has at least one complex zero.

RECALL

If $b = 0$, the complex number $a + bi$ is a real number. So the set of real numbers is a subset of the set of complex numbers.

This theorem was proved by the mathematician Karl Friedrich Gauss at age 20. The proof is beyond the scope of this book, but if we are allowed to use complex numbers, we can use the theorem to prove that every polynomial has a complete factorization.

FACTORIZATION THEOREM FOR POLYNOMIALS

If $P(x)$ is a complex polynomial of degree $n \geq 1$ with leading coefficient a, it can be factored into n (not necessarily distinct) linear factors of the form

$$P(x) = a(x - r_1)(x - r_2) \cdots (x - r_n),$$

where $a, r_1, r_2, \ldots, r_n$ are complex numbers.

SIDE **NOTE**

The Factor Theorem in the real number setting was proved in Section 3.3. It can be verified for the complex numbers by the same reasonings.

To prove the Factorization Theorem for Polynomials, we notice that if $P(x)$ is a polynomial of degree 1 or higher with leading coefficient a, then the Fundamental Theorem of Algebra states that there is a zero r_1 for which $P(r_1) = 0$. By the Factor Theorem, $(x - r_1)$ is a factor of $P(x)$ and

$$P(x) = (x - r_1)Q_1(x),$$

where the degree of $Q_1(x)$ is $n - 1$ and its leading coefficient is a. If $n - 1$ is positive, then we apply the Fundamental Theorem of Algebra to $Q_1(x)$; this gives us a zero, say, r_2, of $Q_1(x)$. By the Factor Theorem, $(x - r_2)$ is a factor of $Q_1(x)$ and

$$Q_1(x) = (x - r_2)Q_2(x),$$

where the degree of $Q_2(x)$ is $n - 2$ and its leading coefficient is a. Then

$$P(x) = (x - r_1)(x - r_2)Q_2(x).$$

This process can be continued until $P(x)$ is completely factored as

$$P(x) = a(x - r_1)(x - r_2) \cdots (x - r_n),$$

where a is the leading coefficient and $r_1, r_2, \ldots, r_n$ are zeros of $P(x)$. The proof is now complete.

In general, the numbers $r_1, r_2, \ldots, r_n$ may not be distinct. As with the polynomials we encountered previously, if the factor $(x - r)$ appears m times in the complete factorization of $P(x)$, then we say that r is a zero of *multiplicity* m. The polynomial $P(x) = (x - 2)^3(x - i)^2$ has zeros

$$2 \text{ (of multiplicity 3) and } i \text{ (of multiplicity 2)}.$$

Note that the polynomial $P(x) = (x - 2)^3(x - i)^2$ has only 2 and i as its distinct zeros, although it is of degree 5.

We have the following theorem.

NUMBER OF ZEROS THEOREM

Any polynomial of degree n has exactly n zeros, provided a zero of multiplicity m is counted m times.

EXAMPLE 1 Constructing a Polynomial Whose Zeros Are Given

Find a polynomial $P(x)$ of degree 4 with a leading coefficient of 2 and zeros $-1, 3, i$, and $-i$. Write $P(x)$

a. In completely factored form. **b.** By expanding the product found in part **a**.

Solution

a. Because $P(x)$ has degree 4, we write

$$P(x) = a(x - r_1)(x - r_2)(x - r_3)(x - r_4). \quad \text{Completely factored form}$$

Now replace the leading coefficient a with 2 and the zeros $r_1, r_2, r_3,$ and r_4 with $-1, 3, i,$ and $-i$ (in any order). Then

$$\begin{aligned} P(x) &= 2[x - (-1)](x - 3)(x - i)[x - (-i)] \\ &= 2(x + 1)(x - 3)(x - i)(x + i) \quad \text{Simplify.} \end{aligned}$$

b.
$$\begin{aligned} P(x) &= 2(x + 1)(x - 3)(x - i)(x + i) && \text{From part } \mathbf{a} \\ &= 2(x + 1)(x - 3)(x^2 + 1) && \text{Expand } (x - i)(x + i). \\ &= 2(x + 1)(x^3 - 3x^2 + x - 3) && \text{Expand } (x - 3)(x^2 + 1). \\ &= 2(x^4 - 2x^3 - 2x^2 - 2x - 3) && \text{Expand } (x + 1)(x^3 - 3x^2 + x - 3). \\ &= 2x^4 - 4x^3 - 4x^2 - 4x - 6 && \text{Distributive property} \end{aligned}$$

Practice Problem 1 Find a polynomial $P(x)$ of degree 4 with a leading coefficient of 3 and zeros $-2, 1, 1 + i, 1 - i$. Write $P(x)$

a. In completely factored form. **b.** By expanding the product found in part **a**.

2 Use the Conjugate Pairs Theorem to find zeros of polynomials.

Conjugate Pairs Theorem

In Example 1, we constructed a polynomial that had both i and its conjugate, $-i$, among its zeros. Our next result says that for all polynomials with real coefficients, nonreal zeros occur in conjugate pairs.

CONJUGATE PAIRS THEOREM

If $P(x)$ is a polynomial function whose coefficients are real numbers and if $z = a + bi$ is a zero of P, then its conjugate, $\bar{z} = a - bi$, is also a zero of P.

RECALL

The conjugate of a complex number $z = a + bi$ is $\bar{z} = a - bi$. For two complex numbers z_1 and z_2, we have (Section 1.3)

$$\overline{z_1 + z_2} = \overline{z_1} + \overline{z_2}$$
$$\overline{z_1 z_2} = \overline{z_1}\,\overline{z_2}.$$

To prove the Conjugate Pairs Theorem, let

$$P(x) = a_n x^n + a_{n-1}x^{n-1} + \cdots + a_1 x + a_0,$$

where $a_n, a_{n-1}, \ldots, a_1, a_0$ are real numbers. Now suppose that $P(z) = 0$. Then we must show that $P(\bar{z}) = 0$ also. We will use the fact that the conjugate of a real number a is identical to the real number a (because $\bar{a} = a - 0i = a$) and then use the conjugate sum and product properties $\overline{z_1 + z_2} = \overline{z_1} + \overline{z_2}$ and $\overline{z_1 z_2} = \overline{z_1}\,\overline{z_2}$.

$$\begin{aligned}
P(z) &= a_n z^n + a_{n-1} z^{n-1} + \cdots + a_1 z + a_0 && \text{Replace } x \text{ with } z.\\
P(\bar{z}) &= a_n(\bar{z})^n + a_{n-1}(\bar{z})^{n-1} + \cdots + a_1\bar{z} + a_0 && \text{Replace } z \text{ with } \bar{z}.\\
&= \bar{a}_n\overline{z^n} + \bar{a}_{n-1}\overline{z^{n-1}} + \cdots + \bar{a}_1\bar{z} + \bar{a}_0 && (\bar{z})^k = \overline{z^k}, \overline{a_k} = a_k\\
&= \overline{a_n z^n} + \overline{a_{n-1} z^{n-1}} + \cdots + \overline{a_1 z} + \bar{a}_0 && \bar{z}_1\,\bar{z}_2 = \overline{z_1 z_2}\\
&= \overline{a_n z^n + a_{n-1} z^{n-1} + \cdots + a_1 z + a_0} && \bar{z}_1 + \bar{z}_2 = \overline{z_1 + z_2}\\
&= \overline{P(z)} = \bar{0} = 0 && P(z) = 0
\end{aligned}$$

We see that if $P(z) = 0$, then $P(\bar{z}) = 0$, and the theorem is proved.

This theorem has several uses. If, for example, we know that $2 - 5i$ is a zero of a polynomial with real coefficients, then we know that $2 + 5i$ is also a zero. Because nonreal zeros occur in conjugate pairs, we know that there will always be an *even* number of nonreal zeros. Therefore, any polynomial of odd degree with real coefficients has at least one zero that is a real number.

ODD-DEGREE POLYNOMIALS HAVE A REAL ZERO

Any polynomial $P(x)$ of odd degree with real coefficients must have at least one real zero.

For example, the polynomial $P(x) = 5x^7 + 2x^4 + 9x - 12$ has at least one zero that is a real number because $P(x)$ has degree 7 (odd degree) and has real numbers as coefficients.

SIDE NOTE

In problems such as Example 2, once you determine all of the zeros, you can use the Factorization Theorem to write a polynomial with those zeros.

EXAMPLE 2 Using the Conjugate Pairs Theorem

A polynomial $P(x)$ of degree 9 with real coefficients has the following zeros: 2, of multiplicity 3; $4 + 5i$, of multiplicity 2; and $3 - 7i$. Write all nine zeros of $P(x)$.

Solution

Because complex zeros occur in conjugate pairs, the conjugate $4 - 5i$ of $4 + 5i$ is a zero of multiplicity 2 and the conjugate $3 + 7i$ of $3 - 7i$ is a zero of $P(x)$. The nine zeros of $P(x)$, including the repetitions, are

$$2, 2, 2, 4 + 5i, 4 + 5i, 4 - 5i, 4 - 5i, 3 + 7i, \text{ and } 3 - 7i.$$

Practice Problem 2 A polynomial $P(x)$ of degree 8 with real coefficients has the following zeros: -3 and $2 - 3i$, each of multiplicity 2, and i. Write all eight zeros of $P(x)$.

Every polynomial of degree n has exactly n zeros and can be factored into a product of n linear factors. If the polynomial has real coefficients, then, by the Conjugate Pairs Theorem, its nonreal zeros occur as conjugate pairs. Consequently, if $r = a + bi$ is a zero, then so is $\bar{r} = a - bi$, and both $x - r$ and $x - \bar{r}$ are linear factors of the polynomial. Multiplying these factors, we have

$$\begin{aligned}
(x - r)(x - \bar{r}) &= x^2 - (r + \bar{r})x + r\bar{r} && \text{Use FOIL.}\\
&= x^2 - 2ax + (a^2 + b^2) && r + \bar{r} = 2a;\ r\bar{r} = a^2 + b^2
\end{aligned}$$

The quadratic polynomial $x^2 - 2ax + (a^2 + b^2)$ has real coefficients and is *irreducible* (cannot be factored any further) *over the real numbers*. This result shows that each pair of nonreal conjugate zeros can be combined into one *irreducible* quadratic factor.

FACTORIZATION THEOREM FOR A POLYNOMIAL WITH REAL COEFFICIENTS

Every polynomial with real coefficients can be uniquely factored over the real numbers as a product of linear factors and/or irreducible quadratic factors.

EXAMPLE 3 Finding the Complex Zeros of a Polynomial Function

Given that $2 - i$ is a zero of $P(x) = x^4 - 6x^3 + 14x^2 - 14x + 5$, find the remaining zeros.

Solution

Because $P(x)$ has real coefficients, the conjugate $\overline{2 - i} = 2 + i$ is also a zero. By the Factorization Theorem, the linear factors $[x - (2 - i)]$ and $[x - (2 + i)]$ appear in the factorization of $P(x)$. Consequently, their product

$$\begin{aligned}
[x - (2 - i)][x - (2 + i)] &= (x - 2 + i)(x - 2 - i) \\
&= [(x - 2) + i][(x - 2) - i] && \text{Regroup.} \\
&= (x - 2)^2 - i^2 && (A + B)(A - B) = A^2 - B^2 \\
&= x^2 - 4x + 4 + 1 && \text{Expand } (x - 2)^2,\ i^2 = -1. \\
&= x^2 - 4x + 5 && \text{Simplify.}
\end{aligned}$$

is also a factor of $P(x)$. We divide $P(x)$ by $x^2 - 4x + 5$ to find the other factor.

$$\begin{array}{rl}
& x^2 - 2x + 1 \leftarrow \text{Quotient} \\
\text{Divisor} \rightarrow x^2 - 4x + 5 \overline{\smash{)}\,x^4 - 6x^3 + 14x^2 - 14x + 5} & \leftarrow \text{Dividend} \\
\underline{x^4 - 4x^3 + 5x^2} & \\
-2x^3 + 9x^2 - 14x & \\
\underline{-2x^3 + 8x^2 - 10x} & \\
x^2 - 4x + 5 & \\
\underline{x^2 - 4x + 5} & \\
0 & \leftarrow \text{Remainder}
\end{array}$$

RECALL

Dividend = Quotient · Divisor + Remainder

$$\begin{aligned}
P(x) &= (x^2 - 2x + 1)(x^2 - 4x + 5) && P(x) = \text{Quotient} \cdot \text{Divisor} \\
&= (x - 1)(x - 1)(x^2 - 4x + 5) && \text{Factor } x^2 - 2x + 1. \\
&= (x - 1)(x - 1)[x - (2 - i)][x - (2 + i)] && \text{Factor } x^2 - 4x + 5.
\end{aligned}$$

The zeros of $P(x)$ are 1 (of multiplicity 2), $2 - i$, and $2 + i$.

Practice Problem 3 Given that $2i$ is a zero of $P(x) = x^4 - 3x^3 + 6x^2 - 12x + 8$, find the remaining zeros.

In Example 3, because $x^2 - 4x + 5$ is irreducible over the real number system, the form $P(x) = (x - 1)(x - 1)(x^2 - 4x + 5)$ is a factorization of $P(x)$ into linear and irreducible quadratic factors. However, the form

$$P(x) = (x - 1)(x - 1)[x - (2 - i)][x - (2 + i)]$$

is a factorization of $P(x)$ into linear factors over the complex numbers.

EXAMPLE 4 Finding the Zeros of a Polynomial Function

Find all zeros of the polynomial function $P(x) = x^4 - x^3 + 7x^2 - 9x - 18$.

Solution

Because the degree of $P(x)$ is 4, $P(x)$ has four zeros. The Rational Zeros Theorem tells us that the possible rational zeros are

$$\pm 1, \pm 2, \pm 3, \pm 6, \pm 9, \pm 18.$$

Testing these possible zeros by synthetic division, we find that 2 is a zero.

$$\begin{array}{r|rrrrr} 2 & 1 & -1 & 7 & -9 & -18 \\ & & 2 & 2 & 18 & 18 \\ \hline & 1 & 1 & 9 & 9 & 0 \end{array}$$

Because $P(2) = 0$, $(x - 2)$ is a factor of $P(x)$ and the quotient is $x^3 + x^2 + 9x + 9$. We can solve the depressed equation $x^3 + x^2 + 9x + 9 = 0$ by factoring by grouping.

$x^2(x + 1) + 9(x + 1) = 0$	Group terms, distributive property
$(x^2 + 9)(x + 1) = 0$	Distributive property
$x^2 + 9 = 0$ or $x + 1 = 0$	Zero-product property
$x^2 = -9$ or $x = -1$	Solve each equation.
$x = \pm\sqrt{-9} = \pm 3i$	Solve $x^2 = -9$ for x.

The four zeros of $P(x)$ are -1, 2, $3i$, and $-3i$. The complete factorization of $P(x)$ is

$$P(x) = x^4 - x^3 + 7x^2 - 9x - 18 = (x + 1)(x - 2)(x - 3i)(x + 3i).$$

Practice Problem 4 Find all zeros of the polynomial

$$P(x) = x^4 - 8x^3 + 22x^2 - 28x + 16.$$

Answers to Practice Problems

1. a. $P(x) = 3(x + 2)(x - 1)(x - 1 - i)(x - 1 + i)$

b. $P(x) = 3x^4 - 3x^3 - 6x^2 + 18x - 12$

2. $-3, -3, 2 + 3i, 2 + 3i, 2 - 3i, 2 - 3i, i, -i$

3. The zeros are 1, 2, $2i$, and $-2i$.

4. The zeros are 2, 4, $1 + i$, and $1 - i$.

SECTION 3.5 Exercises

Basic Concepts and Skills

1. The Fundamental Theorem of Algebra states that a polynomial function of degree $n \geq 1$ has at least one ____________ zero.

2. The Number of Zeros Theorem states that a polynomial function of degree n has exactly ____________ zeros, provided a zero of multiplicity k is counted ____________ times.

3. If P is a polynomial function with real coefficients and if $z = a + bi$ is a zero of P, then ____________ is also a zero of P.

4. If $2 - 3i$ is a zero of a polynomial function with real coefficients, then so is ____________.

5. **True or False.** A polynomial function of degree $n \geq 1$ with real coefficients has at least one real zero and, at most, n real zeros.

6. **True or False.** A polynomial function $P(x)$ of degree $n \geq 1$ with real coefficients can be factored into n factors (not necessarily distinct) as

$$P(x) = a(x - r_1)(x - r_2)\ldots(x - r_n),$$

where $a, r_1, r_2, \ldots r_n$ are real numbers.

In Exercises 7–12, find all solutions of the equation in the complex number system.

7. $x^2 + 25 = 0$

8. $(x - 2)^2 + 9 = 0$

9. $x^2 + 4x + 4 = -9$

10. $x^3 - 8 = 0$

11. $(x - 2)(x - 3i)(x + 3i) = 0$

12. $(x - 1)(x - 2i)(2x - 6i)(2x + 6i) = 0$

In Exercises 13–18, find the remaining zeros of a polynomial $P(x)$ with real coefficients and with the specified degree and zeros.

13. Degree 3; zeros: 2, $3 + i$

14. Degree 3; zeros: 2, $2 - i$

15. Degree 4; zeros: 0, 1, $-5 + i$

16. Degree 4; zeros: $-i$, $1 - i$

17. Degree 6; zeros: 0, 5, i, $3i$

18. Degree 6; zeros: $2i$, $4 + i$, $i - 1$

In Exercises 19–22, find the polynomial $P(x)$ with real coefficients having the specified degree, leading coefficient, and zeros.

19. Degree 4; leading coefficient 2; zeros: $5 - i$, $3i$

20. Degree 4; leading coefficient -3; zeros: $2 + 3i$, $1 - 4i$

21. Degree 5; leading coefficient 7; zeros: 5 (multiplicity 2), 1, $3 - i$

22. Degree 6; leading coefficient 4; zeros: 3, 0 (multiplicity 3), $2 - 3i$

In Exercises 23–26, use the given zero to write $P(x)$ as a product of linear and irreducible quadratic factors.

23. $P(x) = x^4 + x^3 + 9x^2 + 9x$, zero: $3i$

24. $P(x) = x^4 - 2x^3 + x^2 + 2x - 2$, zero: $1 - i$

25. $P(x) = x^5 - 5x^4 + 2x^3 + 22x^2 - 20x$, zero: $3 - i$

26. $P(x) = 2x^5 - 11x^4 + 19x^3 - 17x^2 + 17x - 6$, zero: i

In Exercises 27–36, find all zeros of each polynomial function.

27. $P(x) = x^3 - 9x^2 + 25x - 17$

28. $P(x) = x^3 - 5x^2 + 7x + 13$

29. $P(x) = 3x^3 - 2x^2 + 22x + 40$

30. $P(x) = 3x^3 - x^2 + 12x - 4$

31. $P(x) = 2x^4 - 10x^3 + 23x^2 - 24x + 9$

32. $P(x) = 9x^4 + 30x^3 + 14x^2 - 16x + 8$

33. $P(x) = x^4 - 4x^3 - 5x^2 + 38x - 30$

34. $P(x) = x^4 + x^3 + 7x^2 + 9x - 18$

35. $P(x) = 2x^5 - 11x^4 + 19x^3 - 17x^2 + 17x - 6$

36. $P(x) = x^5 - 2x^4 - x^3 + 8x^2 - 10x + 4$

In Exercises 37–40, find an equation of a polynomial function of least degree having the given complex zeros, intercepts, and graph.

37. f has complex zero $3i$

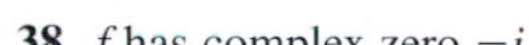

38. f has complex zero $-i$

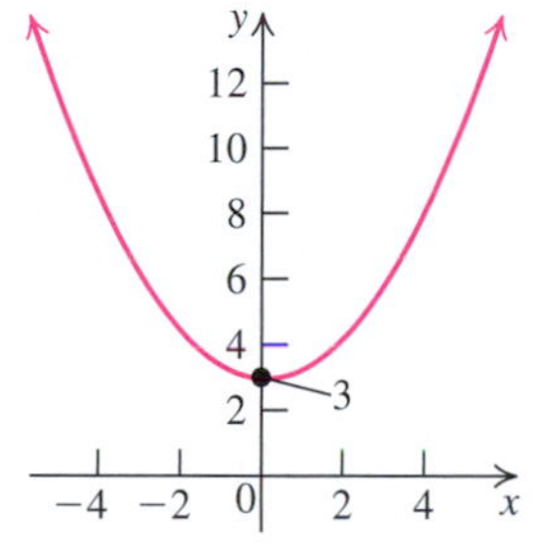

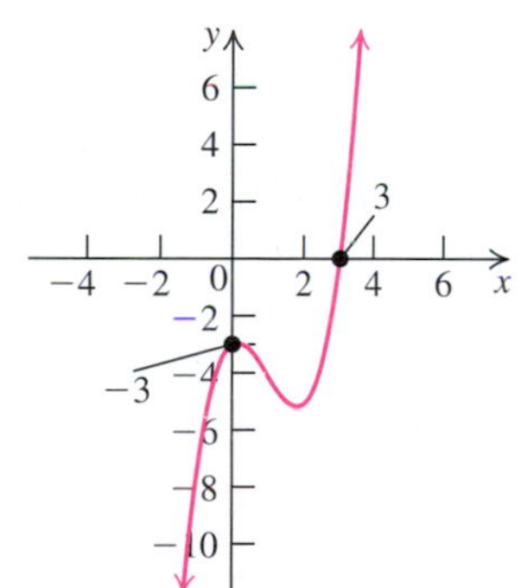

39. f has complex zeros i and $2i$

40. f has complex zero $-2i$

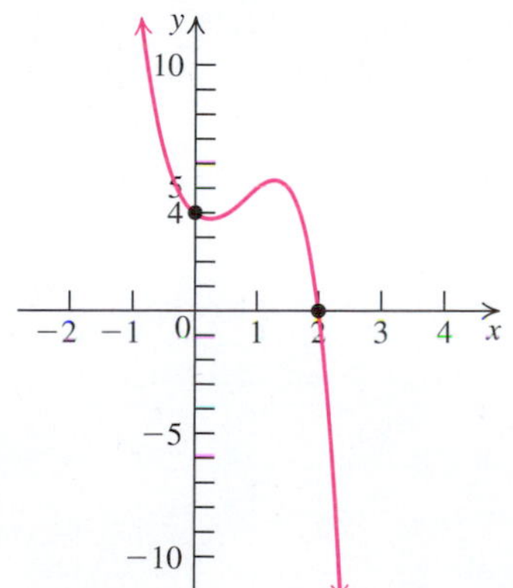

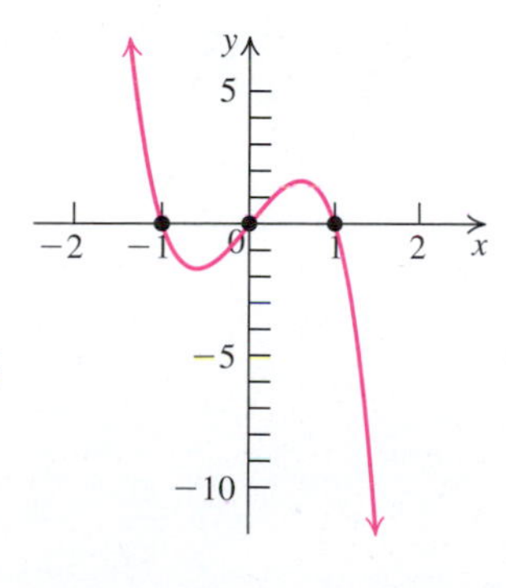

Beyond the Basics

41. The solutions -1 and 1 of the equation $x^2 = 1$ are called the square roots of 1. The solutions of the equation $x^3 = 1$ are called the cube roots of 1. Find the cube roots of 1. How many are there?

42. The solutions of the equation $x^n = 1$, where n is a positive integer, are called the nth roots of 1 or "nth roots of unity." Explain the relationship between the solutions of the equation $x^n = 1$ and the zeros of the complex polynomial $P(x) = x^n - 1$. How many nth roots of 1 are there?

In Exercises 43–46, use the given zero and synthetic division to determine all of the zeros of $P(x)$. Then factor the depressed equation to write $P(x)$ as a product of linear factors.

43. $P(x) = x^2 + (i - 2)x - 2i,\ x = -i$

44. $P(x) = x^2 + 3ix - 2,\ x = -2i$

45. $P(x) = x^3 - (3 + i)x^2 - (4 - 3i)x + 4i,\ x = i$

46. $P(x) = x^3 - (4 + 2i)x^2 + (7 + 8i)x - 14i,\ x = 2i$

In Exercises 47–50, find an equation with real coefficients of a polynomial function f that has the given characteristics. Then write the end behavior of the graph of $y = f(x)$.

47. Degree: 3; zeros: 2, $1 + 2i$; y-intercept: 40

48. Degree: 3; zeros: 1, $2 - 3i$; y-intercept: -26

49. Degree: 4; zeros: 1, -1, $3 + i$; y-intercept: 20

50. Degree: 4; zeros: $1 - 2i$, $3 - 2i$; y-intercept: 130

Critical Thinking/Discussion/Writing

51. Show that if $r_1, r_2, \ldots, r_n$ are the roots of the equation

$$a_n x^n + a_{n-1}x^{n-1} + \cdots + a_1 x + a_0 = 0 \ (a_n \neq 0),$$

then the sum of the roots satisfies

$$r_1 + r_2 + \cdots + r_n = -\frac{a_{n-1}}{a_n}$$

and the product of the roots satisfies

$$r_1 \cdot r_2 \cdot \cdots \cdot r_n = (-1)^n \frac{a_0}{a_n}.$$

(*Hint:* Factor the polynomial; then multiply it out using $r_1, r_2, \ldots, r_n$ and compare coefficients with those of the original polynomial.)

52. In his book *Ars Magna*, Cardano explained how to solve cubic equations. He considered the following example:

$$x^3 + 6x = 20$$

a. Explain why this equation has exactly one real solution.

b. Cardano explained the method as follows: "I take two cubes v^3 and u^3 whose difference is 20 and whose product is 2, that is, a third of the coefficient of x. Then, I say that $x = v - u$ is a solution of the equation. " Show that if $v^3 - u^3 = 20$ and $vu = 2$, then $x = v - u$ is indeed the solution of the equation $x^3 + 6x = 20$.

c. Solve the system

$$v^3 - u^3 = 20$$
$$uv = 2$$

to find u and v.

d. Consider the equation $x^3 + px = q$, where p is a positive number. Using your work in parts (a), (b), and (c) as a guide, show that the unique solution of this equation is

$$x = \sqrt[3]{\frac{q}{2} + \sqrt{\left(\frac{q}{2}\right)^2 + \left(\frac{p}{3}\right)^3}} - \sqrt[3]{-\frac{q}{2} + \sqrt{\left(\frac{q}{2}\right)^2 + \left(\frac{p}{3}\right)^3}}.$$

e. Consider an arbitrary cubic equation

$$x^3 + ax^2 + bx + c = 0.$$

Show that the substitution $x = y - \frac{a}{3}$ allows you to write the cubic equation as

$$y^3 + py = q.$$

53. Use Cardano's method to solve the equation $x^3 + 6x^2 + 10x + 8 = 0$.

Maintaining Skills

In Exercises 54–57, solve each equation.

54. $x^2 + 2x = 0$

55. $2x^2 + x - 3 = 0$

56. $x^3 + 3x^2 - 10x = 0$

57. $x^4 - x^3 - 12x^2 = 0$

In Exercises 58–61, find the quotient $Q(x)$ and remainder $R(x)$.

58. $\frac{x^3 - 3x + 1}{x}$

59. $\frac{x^2 - x - 4}{x - 2}$

60. $\frac{x^3 + 5}{x^2 + 3}$

61. $\frac{8x^4 + 6x^3 - 5}{2x^3 + x}$

In Exercises 62–65, evaluate each expression for the given value of the variable.

62. $\frac{-3}{2x + 1}; x = 2$

63. $\frac{7 - x}{2x^2 + 3x}; x = -1$

64. $\frac{2x + 3}{5 - 2x^2}; x = -3$

65. $\frac{x^2 + 4x - 1}{9 - x^3}; x = 2$

SECTION 3.6

Rational Functions

BEFORE STARTING THIS SECTION, REVIEW

1. Domain of a function (Section 2.4, 199)
2. Zeros of a function (Section 3.2, 311)
3. Quadratic formula (Section 1.2, 97)

OBJECTIVES

1. Define a rational function.
2. Define vertical and horizontal asymptotes.
3. Graph translations of $f(x) = \frac{1}{x}$.
4. Find vertical and horizontal asymptotes (if any).
5. Graph rational functions.
6. Graph rational functions with oblique asymptotes.
7. Graph a revenue curve.

Federal Taxes and Revenues

Federal governments levy income taxes on their citizens and corporations to pay for defense, infrastructure, and social services. The relationship between tax rates and tax revenues shown in Figure 3.21 is a "revenue curve."

All economists agree that a zero tax rate produces no tax revenues and that no one would bother to work with a 100% tax rate, so tax revenue would be zero. It then follows that in a given economy, there is some optimal tax rate T between zero and 100% at which people are willing to maximize their output and still pay their taxes. This optimal rate brings in the most revenue for the government. However, because no one knows the actual shape of the revenue curve, it is impossible to find the exact value of T.

Notice in Figure 3.21 that between the two extreme rates are two rates (such as a and b in the figure) that will collect the same amount of revenue. In a democracy, politicians can argue that taxes are currently too high (at some point b in Figure 3.21) and should therefore be reduced to encourage incentives and harder work (this is *supply-side economics*); at the same time, the government will generate more revenue. Others can argue that we are well to the left of T (at some point a in Figure 3.21), so the tax rate should be raised (for the rich) to generate more revenue. In Example 10, we sketch the graph of a revenue curve for a hypothetical economy.

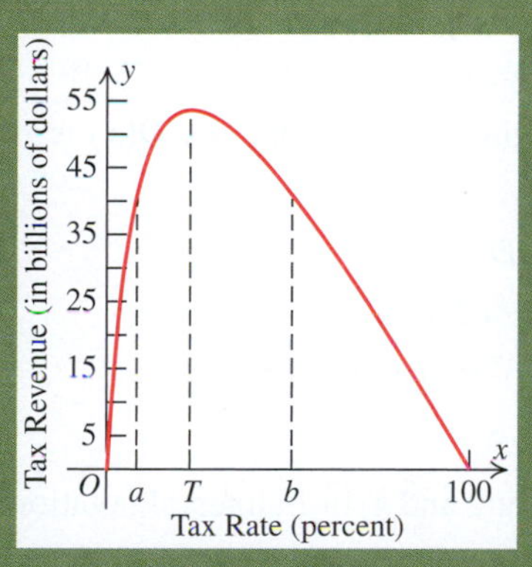

FIGURE 3.21 A revenue curve

1 Define a rational function.

Rational Functions

Recall that a sum, difference, or product of two polynomial functions is also a polynomial function. However, the quotient of two polynomial functions is generally *not* a polynomial function. In this case, we call the quotient of two polynomial functions a *rational function*.

Rational Function

A function f that can be expressed in the form

$$f(x) = \frac{N(x)}{D(x)}$$

where the numerator $N(x)$ and the denominator $D(x)$ are polynomials and $D(x)$ is not the zero polynomial, is called a **rational function**. The domain of f consists of all real numbers for which $D(x) \neq 0$.

Examples of rational functions are

$$f(x) = \frac{x-1}{2x+1}, \quad g(y) = \frac{5y}{y^2+1}, \quad h(t) = \frac{3t^2-4t+1}{t-2}, \quad \text{and} \quad F(z) = 3z^4 - 5z^2 + 1.$$

($F(z)$ has the constant polynomial function $D(z) = 1$ as its denominator.) By contrast, $S(x) = \dfrac{\sqrt{1-x^2}}{x+1}$ and $G(x) = \dfrac{|x|}{x}$ are not rational functions.

EXAMPLE 1 Finding the Domain of a Rational Function

Find the domain of each rational function.

a. $f(x) = \dfrac{3x^2-12}{x-1}$ **b.** $g(x) = \dfrac{x}{x^2-6x+8}$ **c.** $h(x) = \dfrac{x^2-4}{x-2}$

Solution

We eliminate all x for which $D(x) = 0$.

a. $D(x) = x - 1 = 0$, if $x = 1$. The domain of $f(x) = \dfrac{3x^2-12}{x-1}$ is the set of all real numbers x except $x = 1$, or in interval notation, $(-\infty, 1) \cup (1, \infty)$.

b. The domain of g is the set of all real numbers x for which the denominator $D(x)$ is nonzero.

$$\begin{aligned} D(x) = x^2 - 6x + 8 &= 0 && \text{Set } D(x) = 0. \\ (x-2)(x-4) &= 0 && \text{Factor } D(x). \\ x - 2 = 0 \quad \text{or} \quad x - 4 &= 0 && \text{Zero-product property} \\ x = 2 \quad \text{or} \quad x &= 4 && \text{Solve for } x. \end{aligned}$$

The domain of g is the set of all real numbers x except 2 and 4, or in interval notation, $(-\infty, 2) \cup (2, 4) \cup (4, \infty)$.

c. The domain of $h(x) = \dfrac{x^2-4}{x-2}$ is the set of all real numbers x except 2, or in interval notation, $(-\infty, 2) \cup (2, \infty)$.

Practice Problem 1 Find the domain of the rational function $f(x) = \dfrac{x-3}{x^2-4x-5}$.

RECALL

The graph of a polynomial function has no holes, gaps, or sharp corners.

In Example 1, part **c**, note that the functions $h(x) = \dfrac{x^2-4}{x-2} = \dfrac{(x-2)(x+2)}{x-2}$ and $H(x) = x + 2$ are not equal: The domain of $H(x)$ is the set of all real numbers, but the domain of $h(x)$ is the set of all real numbers x except 2. The graph of $y = H(x)$ is a line with slope 1 and y-intercept 2. The graph of $y = h(x)$ is the same line as $y = H(x)$, except that the point $(2, 4)$ is missing. See Figure 3.22.

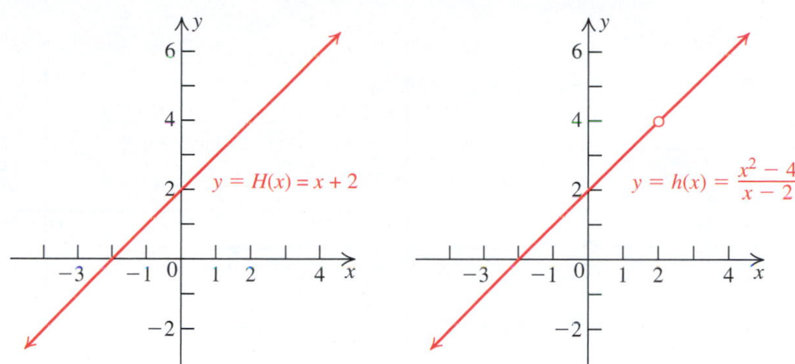

FIGURE 3.22 Functions $H(x)$ and $h(x)$ have different graphs

As with the quotient of integers, if the polynomials $N(x)$ and $D(x)$ have no common factors, then the rational function $f(x) = \frac{N(x)}{D(x)}$ is said to be in **lowest terms**. The zeros of $N(x)$ and $D(x)$ will play an important role in graphing the rational function f.

2 Define vertical and horizontal asymptotes.

Vertical and Horizontal Asymptotes

Consider the function $f(x) = \frac{1}{x}$, introduced in Section 2.6. We know that $f(0)$ is undefined, meaning that f does not assign any value to $x = 0$. We examine more closely the behavior of $f(x)$ "near" the excluded value $x = 0$. Consider the following table.

SIDE NOTE

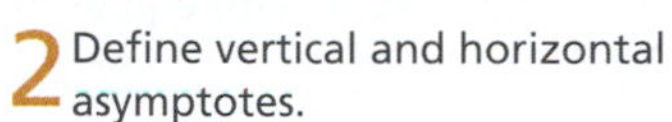

x	1	0.1	0.01	0.001	0.0001	10^{-9}	10^{-50}
$f(x) = \frac{1}{x}$	1	10	100	1000	10,000	10^{9}	10^{50}

We see that as x approaches 0 (with $x > 0$), $f(x)$ increases without bound. In symbols, we write "as $x \to 0^+, f(x) \to \infty$." This statement is read "As x approaches 0 *from the right,* $f(x)$ approaches infinity."

The symbol $x \to a^+$, read "x approaches a from the right," refers only to values of x near a that are greater than a. The symbol $x \to a^-$ is read "x approaches a from the left" and refers only to values of x near a that are less than a.

For the function $f(x) = \frac{1}{x}$, we can make another table of values that includes negative values of x near zero, such as $x = -1, -0.1, -0.001, \ldots, -10^{-9}$. We can conclude that as $x \to 0^-$, $f(x) \to -\infty$ and say that "as x approaches 0 *from the left,* $f(x)$ approaches negative infinity."

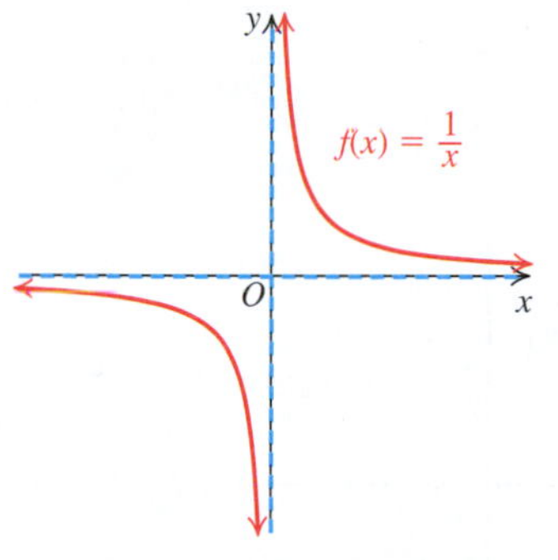

FIGURE 3.23 Reciprocal function, $f(x) = \frac{1}{x}$

The graph of $f(x) = \frac{1}{x}$ is the *reciprocal function* from Chapter 2. (See Figure 3.23). The vertical line $x = 0$ (the y-axis) is a *vertical asymptote*.

Vertical Asymptote

The line $x = a$ is called a **vertical asymptote** of the graph of a function f if $f(x) \to \infty$ as $x \to a^+$ or as $x \to a^-$ or if $f(x) \to -\infty$ as $x \to a^+$ or as $x \to a^-$.

This definition says that if the line $x = a$ is a vertical asymptote of the graph of a function f, then the graph of f *near* $x = a$ behaves like one of the graphs in Figure 3.24.

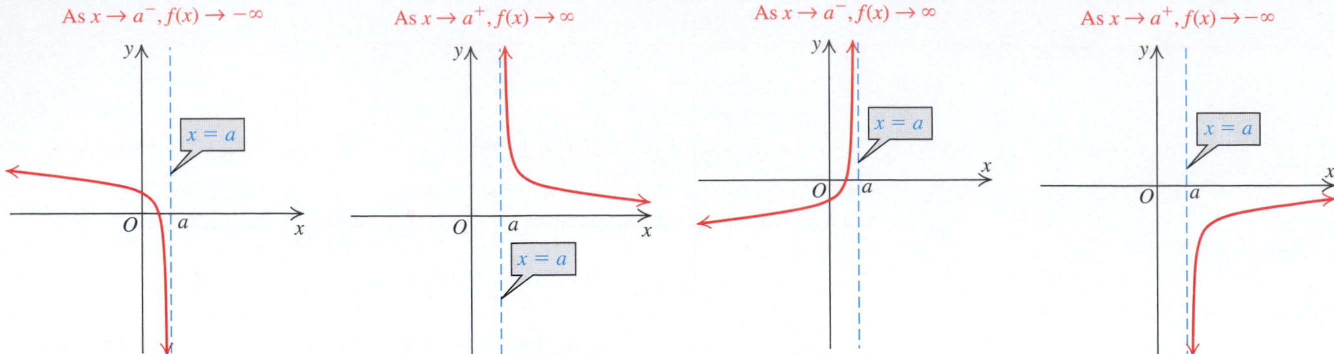

FIGURE 3.24 Behavior of a function near a vertical asymptote

If $x = a$ is a vertical asymptote for the graph of a rational function f, then a is not in its domain. So the graph of f *does not cross* the line $x = a$.

SIDE NOTE

Consider again the function $f(x) = \frac{1}{x}$. The larger we make x, the closer $\frac{1}{x}$ is to 0. So as $x \to \infty, f(x) = \frac{1}{x} \to 0$. Similarly, as $x \to -\infty, f(x) = \frac{1}{x} \to 0$. The x-axis, with equation $y = 0$, is a *horizontal asymptote* of the graph of $f(x) = \frac{1}{x}$. (See Figure 3.23).

Horizontal Asymptote

The line $y = k$ is a **horizontal asymptote** of the graph of a function f if $f(x) \to k$ as $x \to \infty$ or if $f(x) \to k$ as $x \to -\infty$.

This definition says that if the line $y = k$ is a horizontal asymptote of the graph of a function f, then the end behavior of the graph of f is similar to that of one of the graphs in Figure 3.25.

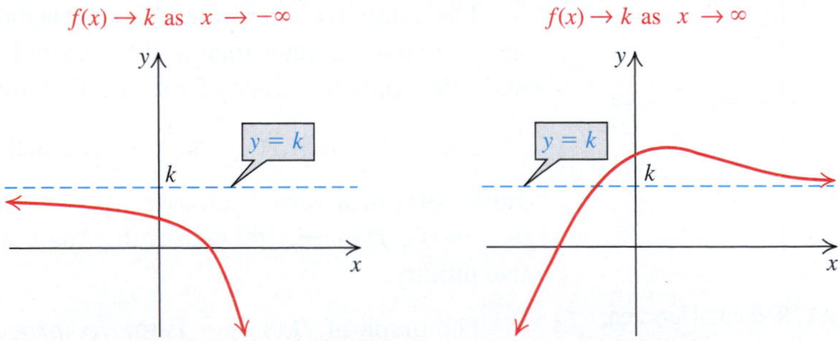

FIGURE 3.25 End behavior of a function near a horizontal asymptote

The graph of a function $y = f(x)$ can cross its horizontal asymptote (in contrast to a vertical asymptote). See Example 7, page 364.

3 Graph translations of $f(x) = \frac{1}{x}$.

Translations of $f(x) = \frac{1}{x}$

We can graph any function of the form

$$g(x) = \frac{ax + b}{cx + d}$$

starting with the graph of $f(x) = \frac{1}{x}$ and using the techniques of vertical stretching and compressing, shifting, and/or reflecting.

EXAMPLE 2 **Graphing Rational Functions Using Translations**

Graph each rational function, identify the vertical and horizontal asymptotes, and state the domain and range.

a. $g(x) = \frac{-2}{x + 1}$ **b.** $h(x) = \frac{3x - 2}{x - 1}$

Solution

a. Let $f(x) = \frac{1}{x}$. We can write $g(x)$ in terms of $f(x)$.

$$\begin{aligned} g(x) &= \frac{-2}{x + 1} \\ &= -2\left(\frac{1}{x + 1}\right) \\ &= -2f(x + 1) \quad \text{Replace } x \text{ with } x + 1 \text{ in } f(x) = \frac{1}{x}. \end{aligned}$$

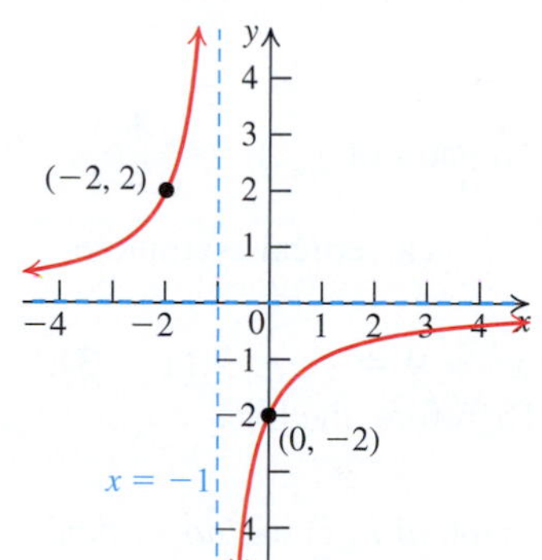

FIGURE 3.26 Graph of $g(x) = \frac{-2}{x + 1}$

The graph of $y = f(x + 1)$ is the graph of $y = f(x)$ shifted 1 unit to the left. This moves the vertical asymptote 1 unit to the left. The graph of $y = -2f(x + 1)$ is the graph of $y = f(x + 1)$ stretched vertically 2 units and then reflected about the x-axis. The graph is shown in Figure 3.26. The domain of g is $(-\infty, -1) \cup (-1, \infty)$, and the range is $(-\infty, 0) \cup (0, \infty)$. The graph has vertical asymptote $x = -1$ and horizontal asymptote $y = 0$ (the x-axis).

b. Using long division, we have

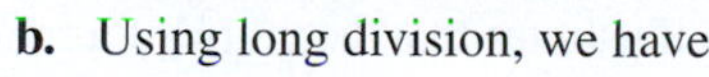

$$\begin{array}{r} 3 \\ x - 1\overline{)3x - 2} \\ \underline{3x - 3} \\ 1 \end{array} \qquad h(x) = 3 + \frac{1}{x - 1} = \frac{1}{x - 1} + 3$$

Then $h(x) = f(x - 1) + 3$ Replace x with $x - 1$ in $f(x) = \frac{1}{x}$.

We see that the graph of $y = h(x)$ is the graph of $f(x) = \frac{1}{x}$ shifted horizontally 1 unit to the right and then vertically up 3 units. The graph is shown in Figure 3.27. The domain of h is $(-\infty, 1) \cup (1, \infty)$, and the range is $(-\infty, 3) \cup (3, \infty)$. The graph of h has vertical asymptote $x = 1$ and horizontal asymptote $y = 3$.

FIGURE 3.27 Graph of $h(x) = \frac{3x - 2}{x - 1}$

Practice Problem 2 Repeat Example 2 with

a. $g(x) = \frac{3}{x - 2}$. **b.** $h(x) = \frac{2x + 5}{x + 1}$.

4 Find vertical and horizontal asymptotes (if any).

LOCATING VERTICAL ASYMPTOTES OF RATIONAL FUNCTIONS

If $f(x) = \dfrac{N(x)}{D(x)}$ is a rational function, where $N(x)$ and $D(x)$ do not have a common factor and a is a real zero of $D(x)$, then the line with equation $x = a$ is a vertical asymptote of the graph of f.

This means that the vertical asymptotes (if any) are found by locating the real zeros of the denominator.

TECHNOLOGY CONNECTION

Calculator graph for

$$h(x) = \frac{x^2 - 9}{x - 3}$$

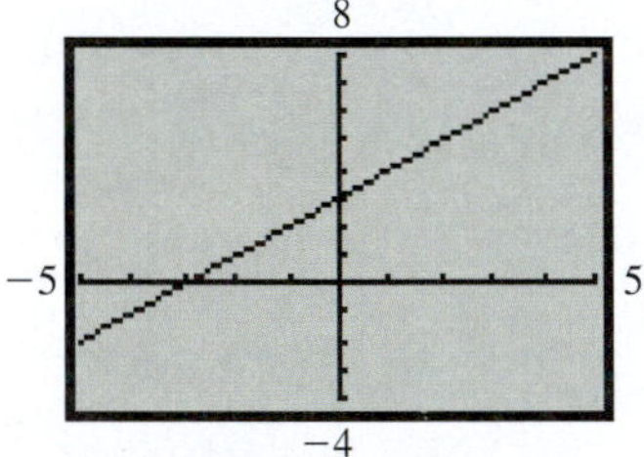

A calculator graph does not show the hole in the graph, but a calculator table explains the situation.

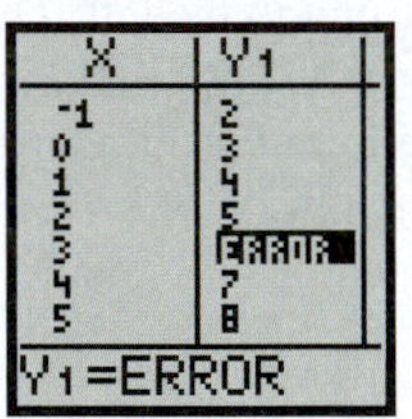

X	Y1
-1	2
0	3
1	4
2	5
3	ERROR
4	7
5	8

Y1=ERROR

EXAMPLE 3 Finding Vertical Asymptotes

Find all vertical asymptotes of the graph of each rational function.

a. $f(x) = \dfrac{1}{x - 1}$ **b.** $g(x) = \dfrac{1}{x^2 - 9}$ **c.** $h(x) = \dfrac{1}{x^2 + 1}$

Solution

a. There are no common factors in the numerator and denominator of $f(x) = \dfrac{1}{x - 1}$, and the only *zero* of the denominator is 1. Therefore, $x = 1$ is a vertical asymptote of $f(x)$.

b. The rational function $g(x)$ is in lowest terms. Factoring $x^2 - 9 = (x + 3)(x - 3)$, we see that the *zeros* of the denominator are -3 and 3. Therefore, the lines $x = -3$ and $x = 3$ are the two vertical asymptotes of $g(x)$.

c. Because the denominator $x^2 + 1$ has no real zeros, the graph of $h(x)$ has *no* vertical asymptotes.

Practice Problem 3 Find the vertical asymptotes of the graph of:

$$f(x) = \frac{x + 1}{x^2 + 3x - 10}.$$

The next example illustrates that the graph of a rational function may have gaps (missing points) with or without vertical asymptotes.

EXAMPLE 4 Rational Functions Whose Graphs Have a Hole

Find all vertical asymptotes of the graph of each rational function.

a. $h(x) = \dfrac{x^2 - 9}{x - 3}$ **b.** $g(x) = \dfrac{x + 2}{x^2 - 4}$

Solution

a.
$$\begin{aligned} h(x) &= \frac{x^2 - 9}{x - 3} && \text{Given function} \\ &= \frac{(x + 3)\cancel{(x - 3)}}{\cancel{x - 3}} && \text{Factor the numerator: } x^2 - 9 = (x + 3)(x - 3). \\ &= x + 3, \quad \text{if } x \neq 3 && \text{Simplify.} \end{aligned}$$

The graph of $h(x)$ is the line $y = x + 3$ with a gap (or hole) at $x = 3$. See Figure 3.28.

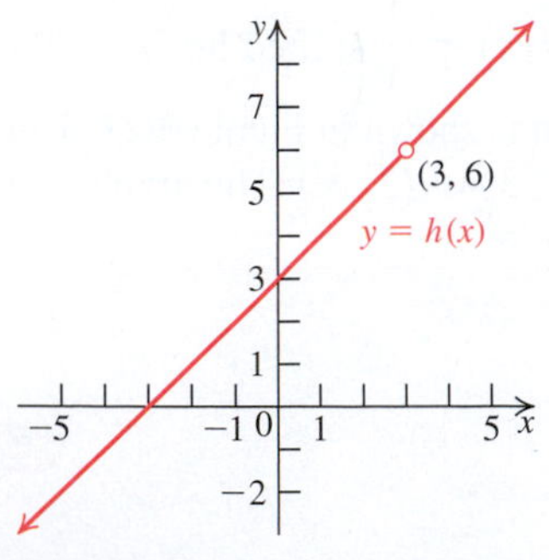

FIGURE 3.28 Graph with a hole

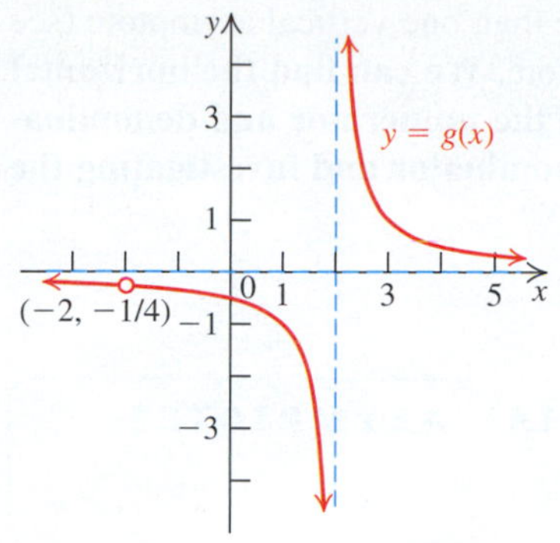

FIGURE 3.29 Graph with a hole and an asymptote

The graph of $h(x)$ has no vertical asymptote at $x = 3$, the zero of the denominator, because the numerator and the denominator have the common factor $(x - 3)$, both with multiplicity 1.

b. $g(x) = \dfrac{x+2}{x^2-4}$ Given function

$= \dfrac{\cancel{x+2}}{\cancel{(x+2)}(x-2)}$ Factor the denominator.

$= \dfrac{1}{x-2}, x \neq -2$ Simplify.

The graph of $g(x)$ has a hole at $x = -2$ and a vertical asymptote $x = 2$. See Figure 3.29.

Practice Problem 4 Find all vertical asymptotes of the graph of $f(x) = \dfrac{3-x}{x^2-9}$.

The graph of $g(x) = \dfrac{x+2}{x^2-4}$, shown in Figure 3.29, has a horizontal asymptote: $y = 0$. This can be verified algebraically. Divide the numerator and the denominator of g by x^2, the highest power of x in the denominator.

$$g(x) = \frac{x+2}{x^2-4} = \frac{\frac{x+2}{x^2}}{\frac{x^2-4}{x^2}}$$

$$= \frac{\frac{x}{x^2} + \frac{2}{x^2}}{\frac{x^2}{x^2} - \frac{4}{x^2}} \qquad \frac{a \pm b}{c} = \frac{a}{c} \pm \frac{b}{c}$$

$$= \frac{\frac{1}{x} + \frac{2}{x^2}}{1 - \frac{4}{x^2}} \qquad \text{Simplify each expression.}$$

As $x \to \pm\infty$, the expressions $\frac{1}{x}, \frac{2}{x^2}$, and $\frac{4}{x^2}$ all approach 0; so

$$g(x) \to \frac{0+0}{1-0} = \frac{0}{1} = 0.$$

Because $g(x) \to 0$ as $x \to \pm\infty$, the line $y = 0$ (the x-axis) is a horizontal asymptote.

TECHNOLOGY CONNECTION

Graphing calculators give very different graphs for

$$g(x) = \frac{1}{x^2-9}$$

depending on whether the mode is set to *connected* or *dot*.

In connected mode, the calculator connects the dots between plotted points, producing vertical lines at $x = -3$ and $x = 3$. Dot mode avoids graphing these vertical lines by plotting the same points but not connecting them. However, dot mode fails to display continuous sections of the graph.

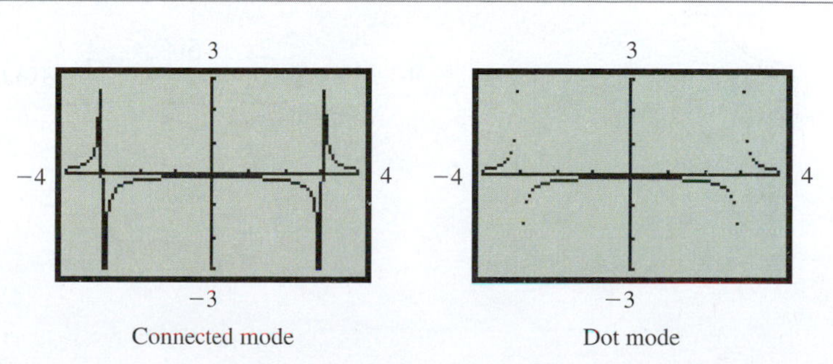

Note that although a rational function can have more than one vertical asymptote (see Example 6), it can have, at most, one horizontal asymptote. **We can find the horizontal asymptote (if any) of a rational function by dividing the numerator and denominator by the highest power of x that appears in the denominator and investigating the resulting expression as $x \to \pm\infty$.**

Alternatively, one may use the following rules.

RULES FOR LOCATING HORIZONTAL ASYMPTOTES

Let f be a rational function given by

$$f(x) = \frac{N(x)}{D(x)} = \frac{a_n x^n + a_{n-1}x^{n-1} + \cdots + a_2x^2 + a_1x + a_0}{b_m x^m + b_{m-1}x^{m-1} + \cdots + b_2x^2 + b_1x + b_0}, \; a_n \neq 0, b_m \neq 0.$$

To find whether the graph of f has one horizontal asymptote or no horizontal asymptote, we compare the degree of the numerator, n, with that of the denominator, m:

1. If $n < m$, then the x-axis $(y = 0)$ is the horizontal asymptote.
2. If $n = m$, then the line with equation $y = \dfrac{a_n}{b_m}$ is the horizontal asymptote.
3. If $n > m$, then the graph of f has *no* horizontal asymptote.

EXAMPLE 5 Finding the Horizontal Asymptote

Find the horizontal asymptote (if any) of the graph of each rational function.

a. $f(x) = \dfrac{5x + 2}{1 - 3x}$ **b.** $g(x) = \dfrac{2x}{x^2 + 1}$ **c.** $h(x) = \dfrac{3x^2 - 1}{x + 2}$

Solution

a. The numerator and denominator of $f(x) = \dfrac{5x + 2}{1 - 3x}$ are both of degree 1. The leading coefficient of the numerator is 5, and that of the denominator is -3. By Rule 2, the line $y = \dfrac{5}{-3} = -\dfrac{5}{3}$ is the horizontal asymptote of the graph of f.

b. For the function $g(x) = \dfrac{2x}{x^2 + 1}$, the degree of the numerator is 1 and that of the denominator is 2. By Rule 1, the line $y = 0$ (the x-axis) is the horizontal asymptote.

c. The degree of the numerator, 2, of $h(x)$ is greater than the degree of its denominator, 1. By Rule 3, the graph of h has no horizontal asymptote.

Practice Problem 5 Find the horizontal asymptote (if any) of the graph of each function.

a. $f(x) = \dfrac{2x - 5}{3x + 4}$ **b.** $g(x) = \dfrac{x^2 + 3}{x - 1}$ **c.** $h(x) = \dfrac{100x + 57}{0.01x^3 + 8x - 9}$

5 Graph rational functions.

Graphing Rational Functions

In the next procedure, we assume the rational function is in lowest terms. When this is not the case, reduce the function to the lowest terms, and plot the "hole" for the factor (or factors) that are divided out.

PROCEDURE IN ACTION

EXAMPLE 6 Graphing a Rational Function

OBJECTIVE

Graph $f(x) = \dfrac{N(x)}{D(x)}$.

EXAMPLE

Sketch the graph of $f(x) = \dfrac{2x^2 - 2}{x^2 - 9}$.

Step 0 **Plot a hole.** Remove any common factor(s) such as $x - a$ of $N(x)$ and $D(x)$.

0. There is no common factor of the form $x - a$ for $2x^2 - 2$ and $x^2 - 9$.

Step 1 **Find the intercepts.** Because f is in lowest terms, the x-intercepts are found by solving the equation $N(x) = 0$. The y-intercept, if there is one, is $f(0)$.

1.

$$2x^2 - 2 = 0 \qquad \text{Set } N(x) = 0.$$
$$2(x - 1)(x + 1) = 0 \qquad \text{Factor.}$$
$$x - 1 = 0 \quad \text{or} \quad x + 1 = 0 \qquad \text{Zero-product property}$$
$$x = 1 \quad \text{or} \quad x = -1 \qquad \text{Solve for } x.$$

The x-intercepts are -1 and 1, so the graph passes through the points $(-1, 0)$ and $(1, 0)$.

$$f(0) = \frac{2(0)^2 - 2}{(0)^2 - 9} \qquad \text{Replace } x \text{ with 0 in } f(x).$$
$$= \frac{2}{9} \qquad \text{Simplify.}$$

The y-intercept is $\frac{2}{9}$, so the graph of f passes through the point $\left(0, \frac{2}{9}\right)$.

Step 2 **Find the vertical asymptotes (if any).** Solve $D(x) = 0$ to find the vertical asymptotes of the graph. Sketch the vertical asymptotes.

2.

$$x^2 - 9 = 0 \qquad \text{Set } D(x) = 0.$$
$$(x - 3)(x + 3) = 0 \qquad \text{Factor.}$$
$$x - 3 = 0 \quad \text{or} \quad x + 3 = 0 \qquad \text{Zero-product property}$$
$$x = 3 \quad \text{or} \quad x = -3 \qquad \text{Solve for } x.$$

The vertical asymptotes of the graph of f are the lines $x = -3$ and $x = 3$.

Step 3 **Find the horizontal asymptote (if any).** Use the rules on page 362 for finding the horizontal asymptote.

3. Because $n = m = 2$, by Rule 2, the horizontal asymptote is

$$y = \frac{\text{Leading coefficient of } N(x)}{\text{Leading coefficient of } D(x)} = \frac{2}{1} = 2.$$

(continued)

Step 4 Locate the graph relative to the horizontal asymptote (if any). If $y = k$ is a horizontal asymptote, divide $N(x)$ by $D(x)$ and write $f(x) = \dfrac{N(x)}{D(x)} = k + \dfrac{R(x)}{D(x)}$. Use a sign graph for $f(x) - k = \dfrac{R(x)}{D(x)}$ and test numbers associated with the zeros of $R(x)$ and $D(x)$ to determine intervals where the graph of f is above the asymptote $y = k$, ($f(x) - k$ is *positive*) and where it is below $y = k$, ($f(x) - k$ is *negative*). The graph of f intersects the line $y = k$ if $f(x) - k = 0$.

4. So $f(x) = \dfrac{2x^2 - 2}{x^2 - 9}$ Using long division: $x^2 - 9\overline{)2x^2 - 2}$ gives quotient 2; $2x^2 - 18$; remainder 16

$= 2 + \dfrac{16}{x^2 - 9}$; $R(x) = 16$ has no zeros, and $D(x)$ has zeros -3 and 3. These zeros divide the x-axis into three intervals. See the figure. We choose test points -4, 0, and 4 to determine the sign of $\dfrac{16}{x^2 - 9}$.

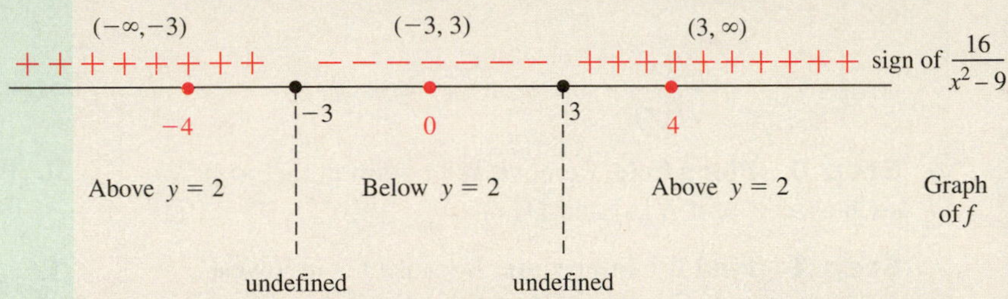

Step 5 Sketch the graph. Plot some points and graph the asymptotes found in Steps 1–4. Plot "holes" in the graph at $x = a$ from Step **0**.

5.

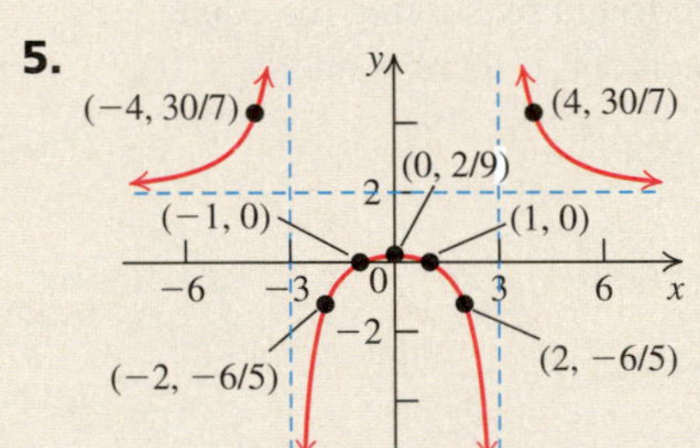

Practice Problem 6 Sketch the graph of $f(x) = \dfrac{2x}{x^2 - 1}$.

Notice that the rational function $g(x) = \dfrac{2x^3 - 2x}{x^3 - 9x}$ reduces to $f(x)$ of Example 6. The graph of g is identical to the graph of f with one exception: The graph of g has a "hole" at $(0, 2/9)$, because 0 is not in the domain of g.

EXAMPLE 7 Graphing a Rational Function

Sketch the graph of $f(x) = \dfrac{x^2 + 2}{(x + 2)(x - 1)}$.

Solution

Step 0 $f(x)$ is in lowest terms.

Step 1 Because $x^2 + 2 > 0$, the graph has no x-intercepts.

$$f(0) = \frac{0^2 + 2}{(0 + 2)(0 - 1)} \quad \text{Replace } x \text{ with 0 in } f(x).$$

$$= -1 \quad \text{Simplify.}$$

The y-intercept is -1.

Step 2 Set $(x + 2)(x - 1) = 0$. Solving for x, we have $x = -2$ or $x = 1$. The vertical asymptotes are the lines $x = -2$ and $x = 1$.

Step 3 We have $(x + 2)(x - 1) = x^2 + x - 2$, so $f(x) = \frac{x^2 + 2}{x^2 + x - 2}$. By Rule 2, page 362 , the horizontal asymptote is $y = \frac{1}{1} = 1$, because the leading coefficient for both the numerator and the denominator is 1.

$$\begin{array}{r} 1 \\ x^2 + x - 2\overline{)x^2 + 0x + 2} \\ \underline{x^2 + \ \ x - 2} \\ -x + 4 \end{array}$$

Step 4 By long division (see margin),

$f(x) = \frac{x^2 + 2}{(x + 2)(x - 1)} = 1 + \frac{4 - x}{(x + 2)(x - 1)}$; $R(x) = 4 - x$ has one zero, 4, and $D(x) = (x + 2)(x - 1)$ has two zeros, -2 and 1. These three zeros divide the x-axis into four intervals. See the figure. We choose test points -3, 0, 2, and 5 to determine the sign of $f(x) - 1 = \frac{4 - x}{(x + 2)(x - 1)}$.

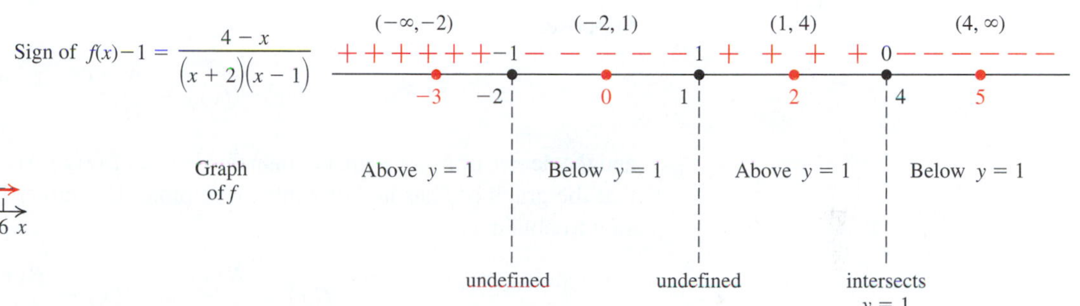

Step 5 The graph of f is shown in Figure 3.30.

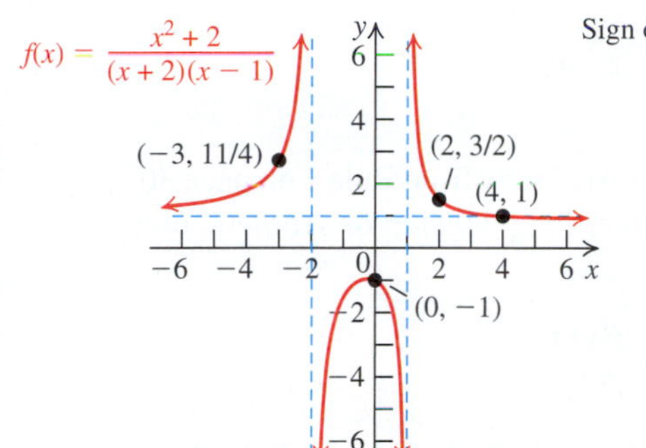

FIGURE 3.30 Graph crossing horizontal asymptote

Practice Problem 7 Sketch the graph of $f(x) = \frac{2x^2 - 1}{2x^2 + x - 3}$.

Graphing a rational function is more complicated than graphing a polynomial function. Plotting a larger number of points and checking to see if the function crosses its horizontal asymptote (if any) gives additional detail. Notice that in Figure 3.30 the graph of f crosses the horizontal asymptote $y = 1$. To find this point of intersection, set $f(x) = 1$ and solve for x.

$$\frac{x^2 + 2}{x^2 + x - 2} = 1 \qquad \text{Set } f(x) = 1;\ (x + 2)(x - 1) = x^2 + x - 2$$

$$x^2 + 2 = x^2 + x - 2 \qquad \text{Multiply both sides by } x^2 + x - 2.$$

$$x = 4 \qquad \text{Solve for } x.$$

The graph of f crosses the horizontal asymptote at the point $(4, 1)$.

We know that the graphs of polynomial functions are continuous (all in one piece). The next example shows that the graph of a rational function (other than a polynomial function) can also be continuous.

EXAMPLE 8 Graphing a Rational Function

Sketch a graph of $f(x) = \frac{x^2}{x^2 + 1}$.

Solution

Step 0 $f(x)$ is in lowest terms.

Step 1 Now $f(0) = 0$ and solving $f(x) = 0$ gives $x = 0$. Therefore, 0 is both the x-intercept and the y-intercept for the graph of f.

Step 2 Because $x^2 + 1 > 0$ for all x, there are no real zeros for the denominator, so there are no vertical asymptotes.

$$\begin{array}{r} 1 \\ x^2+1\overline{)x^2} \\ \underline{x^2+1} \\ -1 \end{array}$$

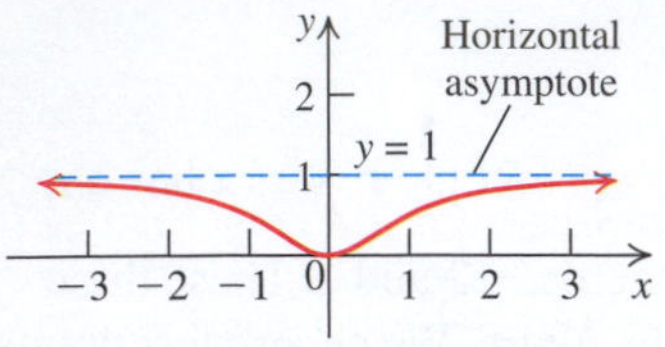

FIGURE 3.31 A continuous rational function

Step 3 By Rule 2 for locating horizontal asymptotes, the horizontal asymptote is $y = 1$.

Step 4 By long division (see margin), $f(x) = \dfrac{x^2}{x^2+1} = 1 + \dfrac{-1}{x^2+1}$. Neither $R(x) = -1$ nor $D(x) = x^2 + 1$ have real zeros. Because $\dfrac{-1}{x^2+1}$ is negative for all values of x, the graph of f is always below the line $y = 1$.

Step 5 The graph of $y = f(x)$ is shown in Figure 3.31.

Practice Problem 8 Sketch the graph of $f(x) = \dfrac{x^2+1}{x^2+2}$.

6 Graph rational functions with oblique asymptotes.

Oblique Asymptotes

Suppose

$$f(x) = \frac{N(x)}{D(x)}$$

and the degree of $N(x)$ is greater than the degree of $D(x)$. We know from Rule 3 on page 362 that the graph of f has no horizontal asymptote. Use either long division or synthetic division to obtain

$$f(x) = \frac{N(x)}{D(x)} = Q(x) + \frac{R(x)}{D(x)},$$

where the degree of $R(x)$ is less than the degree of $D(x)$. Rule 1 on page 362 tells us that as $x \to \infty$ or as $x \to -\infty$, the expression $\dfrac{R(x)}{D(x)} \to 0$.

Therefore, as $x \to \pm\infty$,

$$f(x) \to Q(x) + 0 = Q(x).$$

This means that as $|x|$ gets very large, the graph of f behaves like the graph of the polynomial $Q(x)$. If the degree of $N(x)$ is exactly one more than the degree of $D(x)$, then $Q(x)$ will have the linear form $mx + b$. In this case, the graph of f is said to have an **oblique** (or **slant**) asymptote. Consider, for example, the function

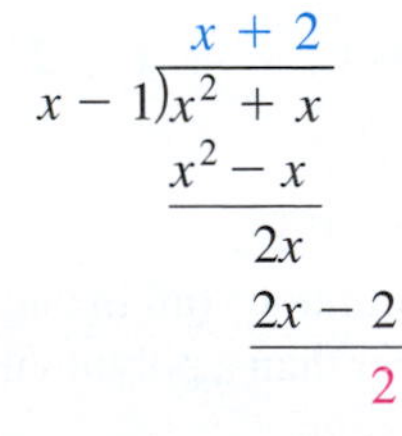

$$f(x) = \frac{x^2+x}{x-1} = x + 2 + \frac{2}{x-1}. \quad \text{Use long division; see margin.}$$

Now as $x \to \pm\infty$, the expression $\dfrac{2}{x-1} \to 0$ so that the graph of f approaches the graph of the oblique asymptote—the line $y = x + 2$, as shown in Figure 3.32. The graph of f is above the line $y = x + 2$ on $(1, \infty)$ and below it on $(-\infty, 1)$.

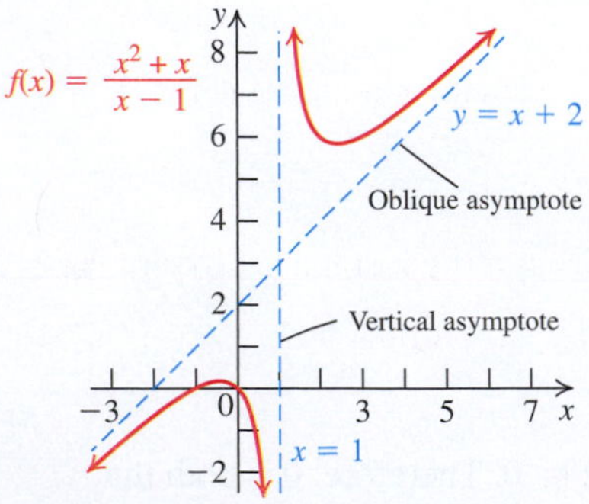

FIGURE 3.32 Graph with vertical and oblique asymptotes

EXAMPLE 9 Graphing a Rational Function with an Oblique Asymptote

Sketch the graph of $f(x) = \dfrac{x^2-4}{x+1}$.

Solution

Step 0 $f(x)$ is in lowest terms.

Step 1 **Intercepts:** $f(0) = \dfrac{0-4}{0+1} = -4$, so the y-intercept is -4. Set $x^2 - 4 = 0$. We have $x = -2$ and $x = 2$, so the x-intercepts are -2 and 2.

Step 2 **Vertical asymptotes:** Set $x + 1 = 0$. We get $x = -1$, so the line $x = -1$ is a vertical asymptote.

Step 3 **Asymptotes:** Since the degree of the numerator is greater than the degree of the denominator, $f(x)$ has no horizontal asymptote. However, by long division see margin,

$$\begin{array}{r} x - 1 \\ x + 1 \overline{) x^2 + 0x - 4} \\ \underline{x^2 + x } \\ -x - 4 \\ \underline{-x - 1} \\ -3 \end{array}$$

$$f(x) = \frac{x^2 - 4}{x + 1} = x - 1 + \frac{-3}{x + 1} = x - 1 - \frac{3}{x + 1}.$$

As $x \to \pm\infty$, the expression $\frac{3}{x + 1} \to 0$; so the line $y = x - 1$ is an oblique asymptote: The graph gets close to the line $y = x - 1$ as $x \to \infty$ and as $x \to -\infty$.

Step 4 **Location:** Use the intervals determined by the zeros of the numerator and of the denominator of $f(x) - (x - 1) = \frac{x^2 - 4}{x + 1} - (x - 1) = \frac{-3}{x + 1}$ to create a sign graph. The numerator, -3, has no zero, and the denominator, $x + 1$, has one zero, -1. The zero, -1, divides the x-axis into two intervals. See the figure. We choose test points -3 and 0 to determine the sign of $f(x) - (x - 1) = \frac{-3}{x + 1}$.

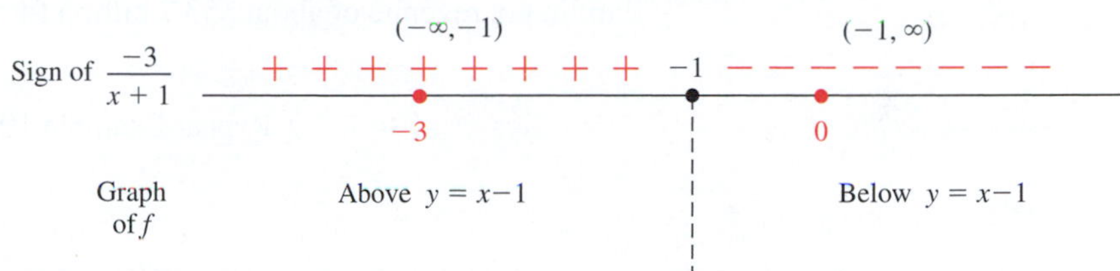

The graph is above the line $y = x - 1$ on $(-\infty, 1)$ and below it on $(-1, \infty)$.

Step 5 The graph of f is shown in Figure 3.33.

FIGURE 3.33

Practice Problem 9 Sketch the graph of $f(x) = \frac{x^2 + 2}{x - 1}$.

7 Graph a revenue curve.

Graph of a Revenue Curve

EXAMPLE 10 Graphing a Revenue Curve

The revenue curve for an economy of a country is given by

$$R(x) = \frac{x(100 - x)}{x + 10},$$

where x is the tax rate in percent and $R(x)$ is the tax revenue in billions of dollars.

a. Find and interpret $R(10)$, $R(20)$, $R(30)$, $R(40)$, $R(50)$, and $R(60)$.

b. Sketch the graph of $y = R(x)$ for $0 \le x \le 100$.

c. Use a graphing calculator to estimate the tax rate that yields the maximum revenue.

Solution

a. $R(10) = \frac{10(100 - 10)}{10 + 10} = \$45$ billion. This means that if the income is taxed at the rate of 10%, the total revenue for the government will be \$45 billion.

$$\begin{aligned} \text{Similarly, } R(20) &\approx \$53.3 \text{ billion,} \\ R(30) &= \$52.5 \text{ billion,} \\ R(40) &= \$48 \text{ billion,} \\ R(50) &\approx \$41.7 \text{ billion, and} \\ R(60) &\approx \$34.3 \text{ billion.} \end{aligned}$$

TECHNOLOGY CONNECTION

TRACE

$$y = \frac{100x - x^2}{x + 10}$$

to approximate the maximum revenue.

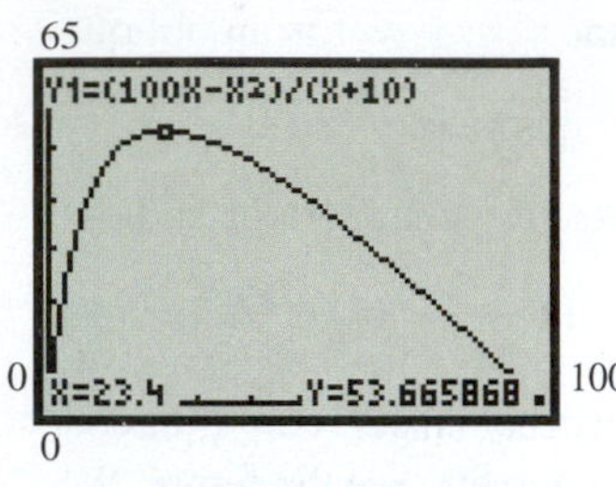

b. The graph of the function $y = R(x)$ for $0 \le x \le 100$ is shown in Figure 3.34.

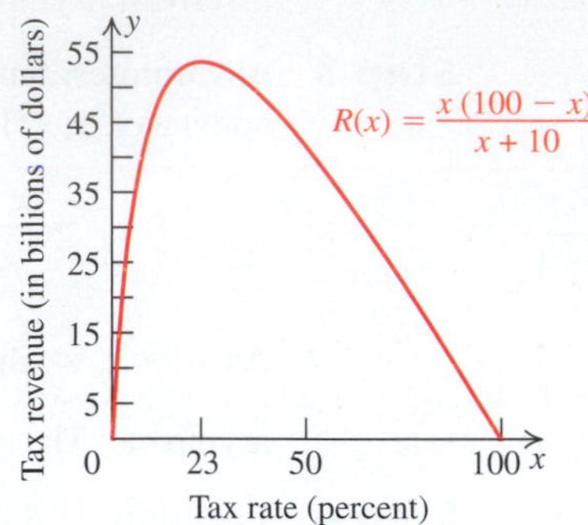

FIGURE 3.34

c. From the calculator graph of

$$Y = \frac{100x - x^2}{x + 10},$$

using the TRACE feature, you can see that the tax rate of about 23% produces the maximum tax revenue of about \$53.7 billion for the government.

Practice Problem 10 Repeat Example 10 for $R(x) = \dfrac{x(100 - x)}{x + 20}$.

Answers to Practice Problems

1. Domain: $(-\infty, -1) \cup (-1, 5) \cup (5, \infty)$

2. a. Domain: $(-\infty, 2) \cup (2, \infty)$; range: $(-\infty, 0) \cup (0, \infty)$; Vertical asymptote: $x = 2$; horizontal asymptote: $y = 0$

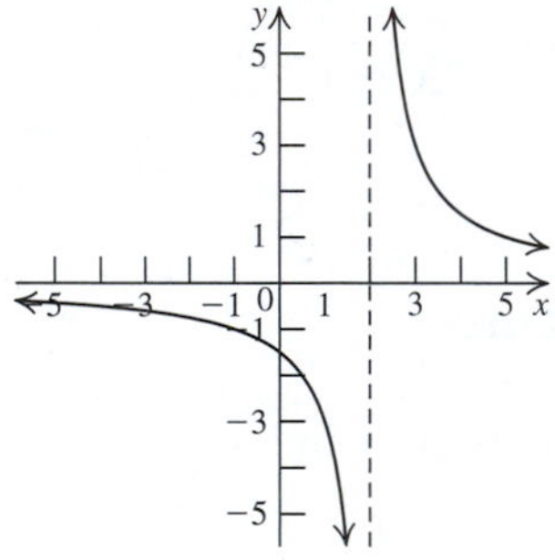

b. Domain: $(-\infty, -1) \cup (-1, \infty)$; range: $(-\infty, 2) \cup (2, \infty)$; vertical asymptote: $x = -1$; horizontal asymptote: $y = 2$

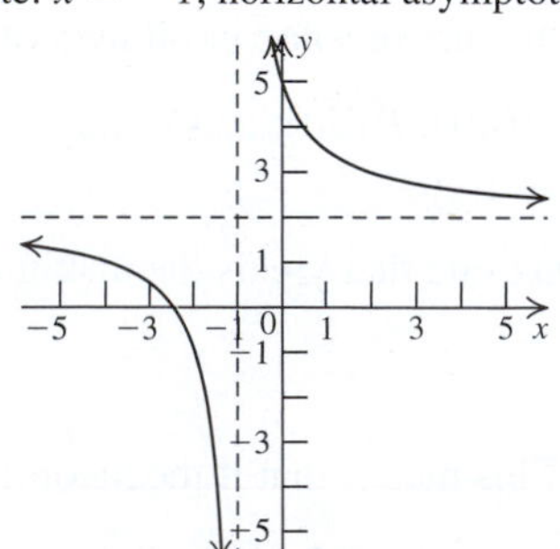

3. $x = -5$ and $x = 2$ **4.** $x = -3$ **5. a.** $y = \dfrac{2}{3}$

b. No horizontal asymptote **c.** $y = 0$

6. x-intercept: 0; y-intercept: 0; vertical asymptotes: $x = -1$ and $x = 1$; horizontal asymptote: x-axis. The graph is above the x-axis on $(-1, 0) \cup (1, \infty)$ and below the x-axis on $(-\infty, -1) \cup (0, 1)$.

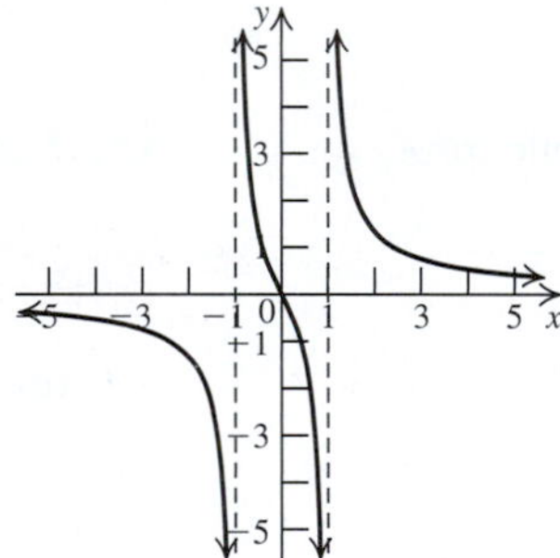

7. x-intercepts: $\pm\dfrac{\sqrt{2}}{2}$; y-intercept: $\dfrac{1}{3}$; vertical asymptotes: $x = -\dfrac{3}{2}$ and $x = 1$; horizontal asymptote: $y = 1$; The graph is above the line $y = 1$ on $\left(-\infty, -\dfrac{3}{2}\right) \cup (1, \infty)$ and below it on $\left(-\dfrac{3}{2}, 1\right)$. The graph crosses the horizontal $(2, 1)$.

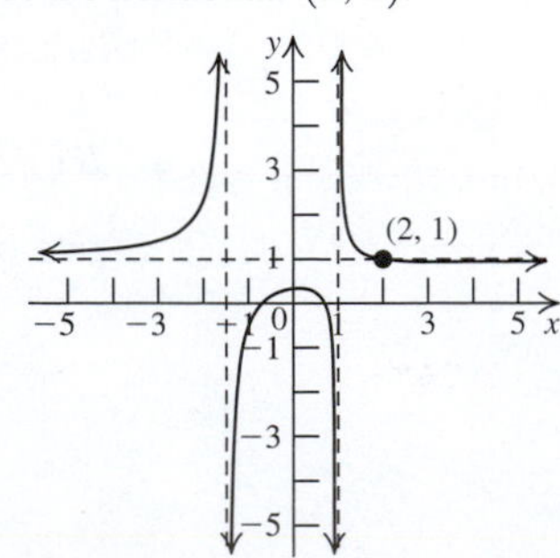

8. No x-intercept; y-intercept: $\frac{1}{2}$; no vertical asymptote; horizontal asymptote: $y = 1$. The graph is below the line $y = 1$ on $(-\infty, \infty)$.

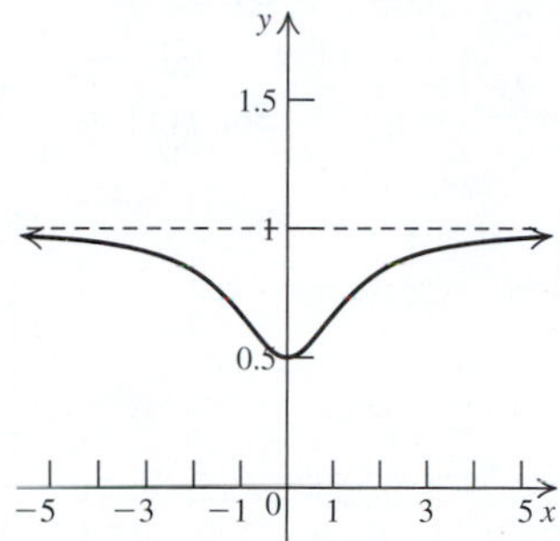

9. No x-intercept; y-intercept: -2; vertical asymptote: $x = 1$; no horizontal asymptote; $y = x + 1$ is an oblique asymptote. The graph is above the line $y = x + 1$ on $(1, \infty)$ and below it on $(-\infty, 1)$.

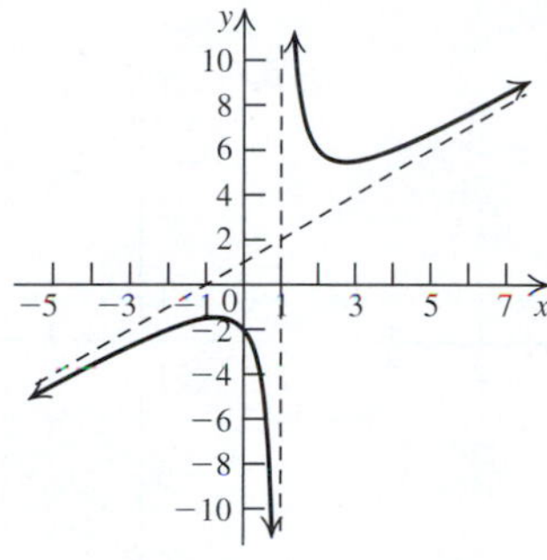

10. a. $R(10) = \$30$ billion
$R(20) = \$40$ billion
$R(30) = \$42$ billion
$R(40) = \$40$ billion
$R(50) = \$35.7$ billion
$R(60) = \$30$ billion

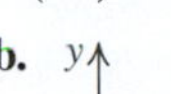

b.

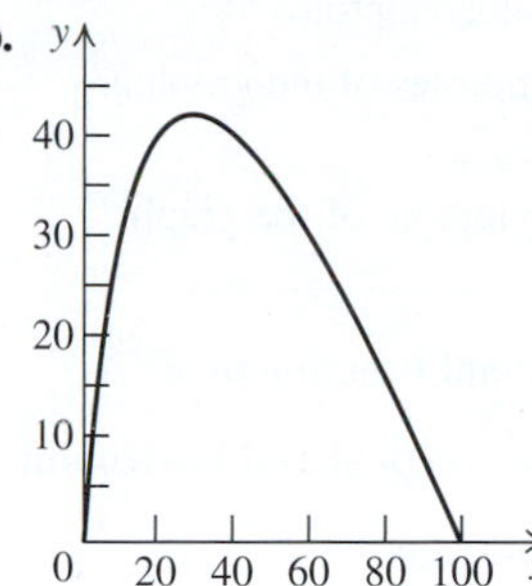

c. 29%

SECTION 3.6 Exercises

Basic Concepts and Skills

1. A rational function can be expressed in the form __________.
2. The line $x = a$ is a vertical asymptote of f if $|f(x)| \to \infty$ as $x \to$ __________ or as $x \to$ __________.
3. The line $y = k$ is a horizontal asymptote of f if $f(x) \to k$ as $x \to$ __________ or as $x \to$ __________.
4. If an asymptote is neither horizontal nor vertical, then it is called a(n) __________.
5. **True or False.** Every rational function has at least one vertical asymptote.
6. **True or False.** Every rational function has, at most, one horizontal asymptote.

In Exercises 7–14, find the domain of each rational function.

7. $f(x) = \dfrac{x - 3}{x + 4}$ **8.** $f(x) = \dfrac{x + 1}{x - 1}$

9. $g(x) = \dfrac{x - 1}{x^2 + 1}$ **10.** $g(x) = \dfrac{x + 2}{x^2 + 4}$

11. $h(x) = \dfrac{x - 3}{x^2 - x - 6}$ **12.** $h(x) = \dfrac{x - 7}{x^2 - 6x - 7}$

13. $F(x) = \dfrac{2x + 3}{x^2 - 6x + 8}$ **14.** $F(x) = \dfrac{3x - 2}{x^2 - 3x + 2}$

In Exercises 15–24, use the graph of the rational function $f(x)$ to complete each statement.

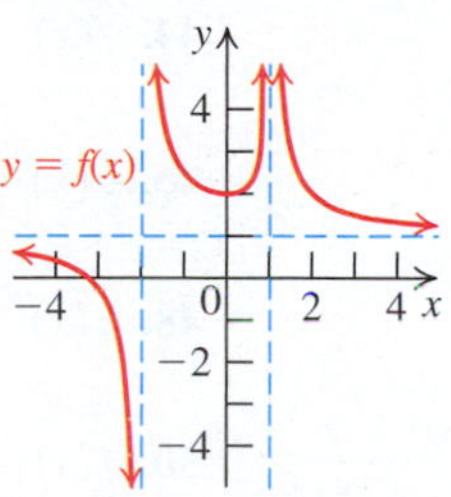

15. As $x \to 1^+$, $f(x) \to$ __________.

16. As $x \to 1^-$, $f(x) \to$ __________.

17. As $x \to -2^+, f(x) \to$ ____________.
18. As $x \to -2^-, f(x) \to$ ____________.
19. As $x \to \infty, f(x) \to$ ____________.
20. As $x \to -\infty, f(x) \to$ ____________.
21. The domain of $f(x)$ is ____________.
22. There are ____________ vertical asymptotes.
23. The equations of the vertical asymptotes of the graph are ____________ and ____________.
24. The equation of the horizontal asymptote of the graph is ____________.

In Exercises 25–32, graph each rational function as a translation of $f(x) = \frac{1}{x}$. Identify the vertical and horizontal asymptotes and state the domain and rage.

25. $f(x) = \dfrac{3}{x - 4}$
26. $g(x) = \dfrac{-4}{x + 3}$
27. $f(x) = \dfrac{-x}{3x + 1}$
28. $g(x) = \dfrac{2x}{4x - 2}$
29. $g(x) = \dfrac{-3x + 2}{x + 2}$
30. $f(x) = \dfrac{-x + 1}{x - 3}$
31. $g(x) = \dfrac{5x - 3}{x - 4}$
32. $f(x) = \dfrac{2x + 12}{x + 5}$

In Exercises 33–42, find the vertical asymptotes, if any, of the graph of each rational function.

33. $f(x) = \dfrac{x}{x - 1}$
34. $f(x) = \dfrac{x + 3}{x - 2}$
35. $g(x) = \dfrac{(x + 1)(2x - 2)}{(x - 3)(x + 4)}$
36. $g(x) = \dfrac{(2x - 1)(x + 2)}{(2x + 3)(3x - 4)}$
37. $h(x) = \dfrac{x^2 - 1}{x^2 + x - 6}$
38. $h(x) = \dfrac{x^2 - 4}{3x^2 + x - 4}$
39. $f(x) = \dfrac{x^2 - 6x + 8}{x^2 - x - 12}$
40. $f(x) = \dfrac{x^2 - 9}{x^3 - 4x}$
41. $g(x) = \dfrac{2x + 1}{x^2 + x + 1}$
42. $g(x) = \dfrac{x^2 - 36}{x^2 + 5x + 9}$

In Exercises 43–50, find the horizontal asymptote, if any, of the graph of each rational function.

43. $f(x) = \dfrac{x + 1}{x^2 + 5}$
44. $f(x) = \dfrac{2x - 1}{x^2 - 4}$
45. $g(x) = \dfrac{2x - 3}{3x + 5}$
46. $g(x) = \dfrac{3x + 4}{-4x + 5}$
47. $h(x) = \dfrac{x^2 - 49}{x + 7}$
48. $h(x) = \dfrac{x + 3}{x^2 - 9}$
49. $f(x) = \dfrac{2x^2 - 3x + 7}{3x^3 + 5x + 11}$
50. $f(x) = \dfrac{3x^3 + 2}{x^2 + 5x + 11}$

In Exercises 51–56, match the rational function with its graph.

51. $f(x) = \dfrac{2}{x - 3}$
52. $f(x) = \dfrac{x - 2}{x + 3}$
53. $f(x) = \dfrac{1}{x^2 - 2x}$
54. $f(x) = \dfrac{x}{x^2 + 1}$
55. $f(x) = \dfrac{x^2 + 2x}{x - 3}$
56. $f(x) = \dfrac{x^2}{x^2 - 4}$

a.
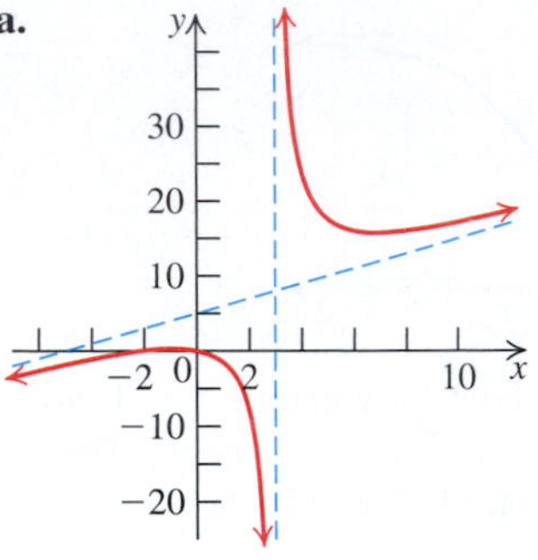

(a)

b.
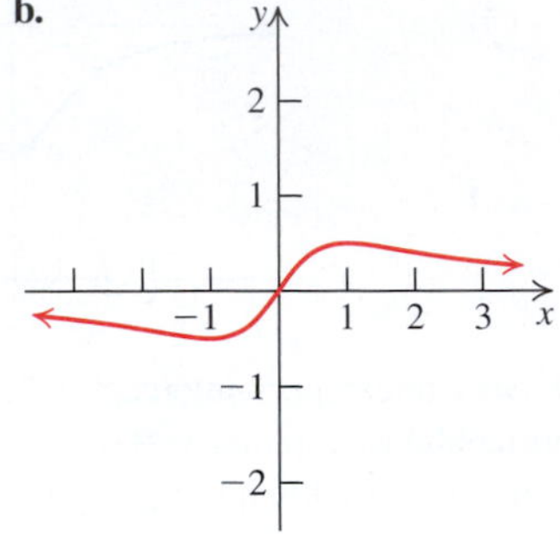

(b)

c. d.
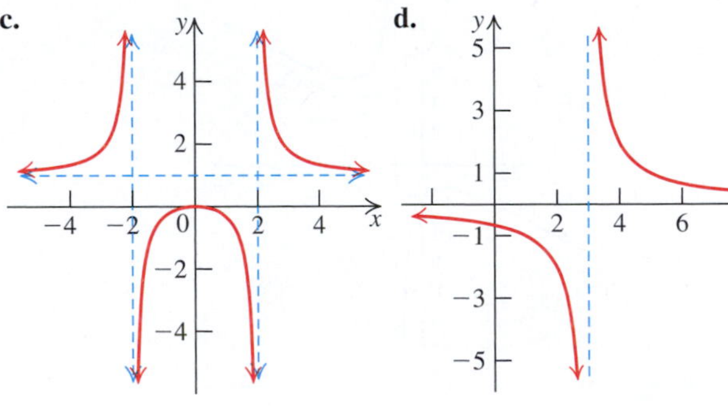

(c) (d)

e. f.
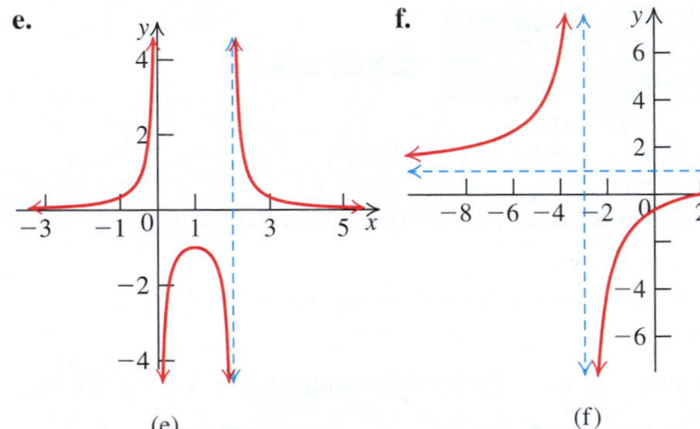

(e) (f)

In Exercises 57–72, use the six-step procedure on pages 363–364 to graph each rational function.

57. $f(x) = \dfrac{2x}{x - 3}$
58. $f(x) = \dfrac{-x}{x - 1}$
59. $f(x) = \dfrac{x}{x^2 - 4}$
60. $f(x) = \dfrac{x}{1 - x^2}$
61. $h(x) = \dfrac{-2x^2}{x^2 - 9}$
62. $h(x) = \dfrac{4 - x^2}{x^2}$

63. $f(x) = \dfrac{2}{x^2 - 2}$ 64. $f(x) = \dfrac{-2}{x^2 - 3}$

65. $g(x) = \dfrac{x + 1}{(x - 2)(x + 3)}$ 66. $g(x) = \dfrac{x - 1}{(x + 1)(x - 2)}$

67. $h(x) = \dfrac{x^2}{x^2 + 1}$ 68. $h(x) = \dfrac{2x^2}{x^2 + 4}$

69. $f(x) = \dfrac{x^3 - 4x}{x^3 - 9x}$ 70. $f(x) = \dfrac{x^3 + 32x}{x^3 + 8x}$

71. $g(x) = \dfrac{(x - 2)^2}{x - 2}$ 72. $g(x) = \dfrac{(x - 1)^2}{x - 1}$

In Exercises 73–76, find an equation of a rational function having the given asymptotes, intercepts, and graph.

73. **74.**

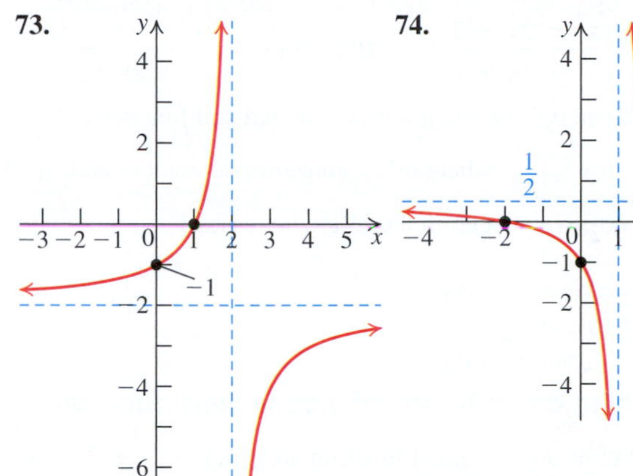

75. **76.**

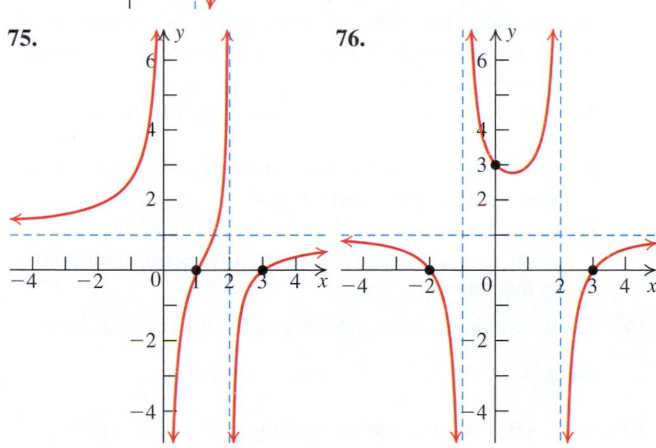

In Exercises 77–84, find the oblique asymptote and sketch the graph of each rational function.

77. $f(x) = \dfrac{2x^2 + 1}{x}$ 78. $f(x) = \dfrac{x^2 - 1}{x}$

79. $g(x) = \dfrac{x^3 - 1}{x^2}$ 80. $g(x) = \dfrac{2x^3 + x^2 + 1}{x^2}$

81. $h(x) = \dfrac{x^2 - x + 1}{x + 1}$ 82. $h(x) = \dfrac{2x^2 - 3x + 2}{x - 1}$

83. $f(x) = \dfrac{x^3 - 2x^2 + 1}{x^2 - 1}$ 84. $h(x) = \dfrac{x^3 - 1}{x^2 - 4}$

Applying the Concepts

For Exercises 85 and 86, use the following definition: Given a cost function C, the average cost of producing the first x items is found by the formula:

$$\overline{C}(x) = \frac{C(x)}{x}, x > 0.$$

85. Average cost. The Genuine Trinket Co. manufactures "authentic" trinkets for gullible tourists. Fixed daily costs are \$2000, and it costs \$0.50 to produce each trinket.

a. Write the cost function C for producing x trinkets.
b. Write the average cost $\overline{C}$ of producing x trinkets.
c. Find and interpret $\overline{C}(100)$, $\overline{C}(500)$, and $\overline{C}(1000)$.
d. Find and interpret the horizontal asymptote of the rational function $\overline{C}(x)$.

86. Average cost. The monthly cost C of producing x portable CD players is given by

$$C(x) = -0.002x^2 + 6x + 7000.$$

a. Write the average cost function $\overline{C}(x)$.
b. Find and interpret $\overline{C}(100)$, $\overline{C}(500)$, and $\overline{C}(1000)$.
c. Find and interpret the oblique asymptote of $\overline{C}(x)$.

87. Biology: birds collecting seeds. A bird is collecting seed from a field that contains 100 grams of seed. The bird collects x grams of seed in t minutes, where

$$t = f(x) = \frac{4x + 1}{100 - x}, 0 < x < 100.$$

a. Sketch the graph of $t = f(x)$.
b. How long does it take the bird to collect
 (i) 50 grams?
 (ii) 75 grams?
 (iii) 95 grams?
 (iv) 99 grams?
c. Complete the following statements if applicable:
 (i) As $x \to 100^-$, $f(x) \to$ ________.
 (ii) As $x \to 100^+$, $f(x) \to$ ________.
d. Does the bird ever collect all of the seed from the field?

88. Criminology. Suppose Las Vegas decides to be crime-free. The estimated cost of catching and convicting $x\%$ of the criminals is given by

$$C(x) = \frac{1000}{100 - x} \text{ million dollars.}$$

a. Find and interpret $C(50)$, $C(75)$, $C(90)$, and $C(99)$.
b. Sketch a graph of $C(x)$, $0 \le x < 100$.
c. What happens to $C(x)$ as $x \to 100^-$?
d. What percentage of the criminals can be caught and convicted for \$30 million?

89. Environment. Suppose an environmental agency decides to get rid of the impurities from the water of a polluted river. The estimated cost of removing $x\%$ of the impurities is given by

$$C(x) = \frac{3x^2 + 50}{x(100 - x)} \text{ billion dollars.}$$

a. How much will it cost to remove 50% of the impurities?
b. Sketch the graph of $y = C(x)$.
c. Estimate the percentage of the impurities that can be removed at a cost of \$30 billion.

90. Biology. The growth function $g(x)$ describes the growth rate of organisms as a function of some nutrient concentration x. Suppose

$$g(x) = a\frac{x}{k + x}, \quad x \geq 0,$$

where a and k are positive constants.

a. Find the horizontal asymptote of the graph of $g(x)$. Use the asymptote to explain why a is called the *saturation level* of the nutrient.

b. Show that k is the half-saturation constant; that is, show that if $x = k$, then $g(x) = \frac{a}{2}$.

91. Population of bacteria. The population P (in thousands) of a colony of bacteria at time t (in hours) is given by

$$P(t) = \frac{8t + 16}{2t + 1}, \quad t \geq 0.$$

a. Find the initial population of the colony. (Find the population at $t = 0$ hours.)

b. What is the long-term behavior of the population? (What happens when $t \to \infty$?)

92. Drug concentration. The concentration c (in milligrams per liter) of a drug in the bloodstream of a patient at time $t \geq 0$ (in hours since the drug was injected) is given by

$$c(t) = \frac{5t}{t^2 + 1}, \quad t \geq 0.$$

a. By plotting points or by using a graphing calculator, sketch the graph of $y = c(t)$.

b. Find and interpret the horizontal asymptote of the graph of $y = c(t)$.

 **c.** Find the approximate time when the concentration of drug in the bloodstream is maximal.

d. At what time is the concentration level equal to 2 milligrams per liter?

93. Book publishing. The printing and binding cost for a college algebra book is \$10. The editorial cost is \$200,000. The first 2500 books are samples and are given free to professors. Let x be the number of college algebra books produced.

a. Write a function f describing the average cost of saleable books.

b. Find the average cost of a saleable book if 10,000 books are produced.

c. How many books must be produced to bring the average cost of a saleable book under \$20?

d. Find the vertical and horizontal asymptotes of the graph of $y = f(x)$. How are they related to this situation?

94. A "phony" sale. A jewelry merchant at a local mall wants to have a "p percent off" sale. She marks up the stock by q percent to break even with respect to the prices before the sale.

a. Show that $q = \frac{p}{1 - p}, \quad 0 < p < 1$.

b. Sketch the graph of q in part (a).

c. How much should she mark up the stock to break even if she wants to have a "25% off" sale?

Beyond the Basics

In Exercises 95–104, use transformations of the graph of $y = \frac{1}{x}$ or $y = \frac{1}{x^2}$ to graph each rational function $f(x)$.

95. $f(x) = -\frac{2}{x}$

96. $f(x) = \frac{1}{2 - x}$

97. $f(x) = \frac{1}{(x - 2)^2}$

98. $f(x) = \frac{1}{x^2 + 2x + 1}$

99. $f(x) = \frac{1}{(x - 1)^2} - 2$

100. $f(x) = \frac{1}{(x + 2)^2} + 3$

101. $f(x) = \frac{1}{x^2 + 12x + 36}$

102. $f(x) = \frac{3x^2 + 18x + 28}{x^2 + 6x + 9}$

103. $f(x) = \frac{x^2 - 2x + 2}{x^2 - 2x + 1}$

104. $f(x) = \frac{2x^2 + 4x - 3}{x^2 + 2x + 1}$

105. Sketch and discuss the graphs of the rational functions of the form $y = \frac{a}{x^n}$, where a is a nonzero real number and n is a positive integer, in the following cases:

a. $a > 0$ and n is odd.

b. $a < 0$ and n is odd.

c. $a > 0$ and n is even.

d. $a < 0$ and n is even.

106. Graphing the reciprocal of a polynomial function. Let $f(x)$ be a polynomial function and $g(x) = \frac{1}{f(x)}$. Justify the following statements about the graphs of $f(x)$ and $g(x)$.

a. If c is a zero of $f(x)$, then $x = c$ is a vertical asymptote of the graph of $g(x)$.

b. If $f(x) > 0$, then $g(x) > 0$, and if $f(x) < 0$, then $g(x) < 0$.

c. The graphs of f and g intersect for those values (and only those values) of x for which $f(x) = g(x) = \pm 1$.

d. When f is increasing (decreasing, constant) on an interval of its domain, g is decreasing (increasing, constant) on that interval.

107. Use Exercise 106 to sketch the graphs of $f(x) = x^2 - 4$ and $g(x) = \frac{1}{x^2 - 4}$ on the same coordinate axes.

108. Use Exercise 106 to sketch the graphs of $f(x) = x^2 + 1$ and $g(x) = \frac{1}{x^2 + 1}$ on the same coordinate axes.

109. Recall that the inverse of a function $f(x)$ is written as $f^{-1}(x)$, while the reciprocal of $f(x)$ can be written as either $\frac{1}{f(x)}$ or $[f(x)]^{-1}$. The point of this exercise is to make clear that the reciprocal of a function has nothing to do with the inverse of a function. Let $f(x) = 2x + 3$. Find both $[f(x)]^{-1}$ and $f^{-1}(x)$. Compare the two functions. Graph all three functions on the same coordinate axes.

110. Let $f(x) = \frac{x - 1}{x + 2}$. Find $[f(x)]^{-1}$ and $f^{-1}(x)$. Graph all three functions on the same coordinate axes.

111. Sketch the graph of $g(x) = \dfrac{2x^3 + 3x^2 + 2x - 4}{x^2 - 1}$. Discuss the end behavior of $g(x)$.

112. Sketch the graph of $f(x) = \dfrac{x^3 - 2x^2 + 1}{x - 2}$. Discuss the end behavior of $f(x)$.

In Exercises 113 and 114, the graph of the rational function $f(x)$ crosses the horizontal asymptote. Graph $f(x)$ and find the point of intersection of the curve with its horizontal asymptote.

113. $f(x) = \dfrac{x^2 + x - 2}{x^2 - 2x - 3}$

114. $f(x) = \dfrac{4 - x^2}{2x^2 - 5x - 3}$

In Exercises 115–118, find an equation of a rational function f satisfying the given conditions.

115. Vertical asymptote: $x = 3$
Horizontal asymptote: $y = -1$
x-intercept: 2

116. Vertical asymptote: $x = -1, x = 1$
Horizontal asymptote: $y = 1$ x-intercept: 0

117. Vertical asymptote: $x = 0$
Oblique asymptote: $y = -x$
x-intercepts: $-1, 1$

118. Vertical asymptote: $x = 3$
Oblique asymptote: $y = x + 4$
y-intercept: 2; $f(4) = 14$

In Exercises 119–124, make up a rational function $f(x)$ that has all of the characteristics given in the exercise.

119. Has a vertical asymptote at $x = 2$, a horizontal asymptote at $y = 1$, and a y-intercept at $(0, -2)$

120. Has vertical asymptotes at $x = 2$ and $x = -1$, a horizontal asymptote at $y = 0$, a y-intercept at $(0, 2)$, and an x-intercept at 4

121. Has $f\left(\dfrac{1}{2}\right) = 0$; $f(x) \to 4$ as $x \to \pm\infty$, $f(x) \to \infty$ as $x \to 1^-$, and $f(x) \to \infty$ as $x \to 1^+$

122. Has $f(0) = 0$; $f(x) \to 2$ as $x \to \pm\infty$; has no vertical asymptotes and is symmetric about the y-axis

123. Has $y = 3x + 2$ as an oblique asymptote; has a vertical asymptote at $x = 1$

124. Is it possible for the graph of a rational function to have both a horizontal and oblique asymptote? Explain.

Critical Thinking/Discussion/Writing

125. Give an example of a rational function that intersects its horizontal asymptote at
a. No points.
b. One point.
c. Two points.

126. Give an example of a rational function that intersects its oblique asymptote at
a. No points.
b. One point.
c. Two points.

127. Construct a rational function $R(x)$ that has $y = x + 1$ as its oblique asymptote and the graph of $R(x)$ intersects the asymptote at the points $(2, R(2))$ and $(5, R(5))$.

128. Describe a rational function $R(x) = \dfrac{N(x)}{D(x)}$ that has asymptote $y = ax + b$ and $R(x)$ intersects the asymptote at n points whose x-coordinates are $c_1, c_2, \ldots, c_n$. Discuss the relationship between the degree of $D(x)$ and the number of points of intersection of $R(x)$ and $y = ax + b$.

Maintaining Skills

129. **a.** Find the equation of the line in slope–intercept form that passes through the point $(-5, 3)$ and has slope $-\dfrac{2}{3}$.
b. From the equation in part (a), find y when $x = 1$.

130. **a.** Find the equation of the line in slope–intercept form that passes through the point $(-2, -3)$ and is perpendicular to the line $4x + 5y = 6$.
b. From the equation in part (a), find y when $x = 10$.

131. Find c if the line $3x + cy = 11$ passes through the point $(1, 2)$.

132. Find k if the graph of the equation $y = kx^2 + 1$ passes through the point $(-2, 7)$. Then find y when $x = 2$.

133. Find k if the graph of the equation $y = \dfrac{k}{x^2}$ passes through the point $\left(\dfrac{1}{2}, 12\right)$. Then find y when $x = 2$.

SECTION 3.7

Variation

BEFORE STARTING THIS SECTION, REVIEW

1. Equation of a line (Section 2.3, page 186)
2. Solving linear equations (Section 1.1, page 78)

OBJECTIVES

1. Solve direct variation problems.
2. Solve inverse variation problems.
3. Solve joint and combined variation problems.

Sir Isaac Newton 1642–1727

Isaac Newton was a mathematician and physicist, one of the foremost scientific intellects of all time. He entered Cambridge University in 1661 and was elected a Fellow of Trinity College in 1667 and Lucasian Professor of Mathematics in 1669. He remained at the university, lecturing most years, until 1696.

As a firm opponent of the attempt by King James II to make the universities into Catholic institutions, Newton was elected Member of Parliament for the University of Cambridge to the Convention Parliament of 1689, and he sat again in 1701–1702.

Meanwhile, in 1696, he moved to London as warden of the Royal Mint. He became master of the Mint in 1699, an office he retained until his death. He was elected a Fellow of the Royal Society of London in 1671, and in 1703, he became president, being annually reelected for the rest of his life. One of his major works, *Opticks*, appeared the next year; he was knighted in Cambridge in 1705.

◆ Newton and the Apple

According to legend, Isaac Newton was sitting under an apple tree, and when an apple fell on his head, he suddenly thought of the Law of Universal Gravitation. As in many such legends, the story may not be true in detail, but it contains elements of the truth.

Before we tell you a more realistic version of this story, let's recall some terminology associated with the motion of objects.

$$\text{Recall that speed} = \frac{\text{distance}}{\text{time}}.$$

(i) The **velocity** of an object is the speed of an object in a specific direction.

(ii) **Acceleration**: If the velocity of an object is changing, we say that the object is accelerating. Acceleration is the rate of change of velocity.

(iii) **Force** = Mass × Acceleration

What really happened with the apple? Perhaps the correct version of the story is that upon observing an apple falling from a tree, Newton concluded that the apple accelerated because the velocity of the apple changed from when it began hanging on the tree (zero velocity) and then moved to the ground. He decided that a force must act on the apple to cause this acceleration. He called this force "gravity" and the associated acceleration (Force = Mass × Acceleration) the "acceleration due to gravity." He conjectured further that if the force of gravity reaches the top of the apple tree, it might reach even farther; might it reach all the way to the Moon? In that case, to keep the Moon moving in a circular orbit, rather than wandering into outer space, Earth must exert a gravitational force on the Moon. Newton realized that the force that brought the apple to the ground and the force that holds the Moon in orbit were the same. He concluded that any two objects in the universe exert gravitational attraction on each other, whereupon he gave us the *Law of Universal Gravitation*. See Example 6.

Some scientists believe that it was Kepler's Third Law of Planetary Motion (see Exercise 53), not an apple, that led Newton to his Law of Universal Gravitation.

1 Solve direct variation problems.

Direct Variation

Two types of relationships between quantities (variables) occur so frequently in mathematics that they are given special names: *direct variation* and *inverse variation*.

Direct variation describes any relationship between two quantities in which any increase (or decrease) in one causes a proportional increase (or decrease) in the other. For example, the sales tax in Hillsborough County, Florida, in 2012 was 7%, so that the sales tax on a \$1000 purchase was \$70. If we *double* the purchase to \$2000, the sales tax also *doubles*, to \$140. If we *reduce* the purchase by *half* to \$500, the sales tax is also *reduced* by half, to \$35. We say that the sales tax *varies directly* as the purchase price. The equation

$$\text{Sales tax} = (0.07)(\text{Purchase price})$$

expresses the fact that the sales tax was a constant (0.07) multiple of the purchase price.

SIDE NOTE

The statements

"y varies directly as x,"
"y varies as x,"
"y is directly proportional to x,"
"y is proportional to x,"
and "$y = kx, k \neq 0$"

have the same meaning.

Direct Variation

A quantity y is said to **vary directly** as the quantity x, or y is **directly proportional** to x, if there is a constant k such that $y = kx$. This constant k is called the **constant of variation** or the **constant of proportionality**.

PROCEDURE IN ACTION

EXAMPLE 1 Solving the Variation Problems

OBJECTIVE

Solve variation problems.

EXAMPLE

Suppose y varies as x and $y = 20$ when $x = \frac{4}{3}$. Find y when $x = 8$.

Step 1 Write the equation with the constant of variation, k.

1. $y = kx$ — y varies as x.

Step 2 Substitute the given values of the variables into the equation in Step 1 to find the value of the constant k.

2. $20 = k\left(\frac{4}{3}\right)$ — Replace y with 20 and x with $\frac{4}{3}$.

$20\left(\frac{3}{4}\right) = k\left(\frac{4}{3}\right)\left(\frac{3}{4}\right)$ — Multiply both sides by $\frac{3}{4}$.

$15 = k$ — Simplify to solve for k.

Step 3 Rewrite the equation in Step 1 with the value of k from Step 2.

3. $y = 15x$ — Replace k with 15.

Step 4 Use the equation from Step 3 to answer the question posed in the problem.

4. $y = 15(8)$ — Replace x with 8.

$y = 120$ — Simplify.

So $y = 120$ when $x = 8$.

Practice Problem 1 Suppose y varies directly as x. If y is 6 when x is 30, find y when $x = 120$.

EXAMPLE 2 Direct Variation in Electrical Circuits

The current in a circuit connected to a 220-volt battery is 50 amperes. If the current is directly proportional to the voltage of the attached battery, what voltage battery is needed to produce a current of 75 amperes?

Solution

Let I = current in amperes and V = voltage in volts of the battery.

Step 1 $I = kV$ — I varies directly as V.

Step 2 $50 = k(220)$ — Substitute $I = 50$, $V = 220$.

$\frac{50}{220} = k$ — Solve for k.

$\frac{5}{22} = k$ — Simplify.

Step 3 $I = \frac{5}{22}V$ — Replace k with $\frac{5}{22}$ in $I = kV$.

Step 4 $75 = \frac{5}{22}V$ — Substitute $I = 75$ in Step 3.

$\frac{22}{5} \cdot 75 = V$ — Solve for V.

$330 = V$ — Simplify.

A battery of 330 volts is needed to produce 75 amperes of current.

Practice Problem 2 In Example 2, if the current in the circuit is 60 amperes, what voltage battery is needed to produce a current of 75 amperes?

We can generalize the concept of direct variation to variation with powers.

Direct Variation with Powers

"A quantity y varies directly as the nth power of x" means

$$y = kx^n,$$

where k is a nonzero constant and $n > 0$.

Note that when $n = 1$, y varies directly as x. In the formula for the area of a circle, $A = \pi r^2$, A varies directly as the square (or second power) of the radius r, with π as the constant of variation.

EXAMPLE 3 Solving a Problem Involving Direct Variation with Powers

Suppose you had forgotten the formula for the volume of a sphere but were told that the volume V of a sphere varies directly as the cube of its radius r. In addition, you are given that $V = 972\pi$ when $r = 9$. Find V when $r = 6$.

Solution

Step 1 $V = kr^3$ — V varies directly as r^3.

Step 2 $972\pi = k(9)^3$ — Replace V with 972π and r with 9.

$972\pi = k(729)$ — $9^3 = 729$

$k = \frac{972\pi}{729}$ — Solve for k.

$= \frac{4}{3}\pi$ — Simplify.

Step 3 $V = \frac{4}{3}\pi r^3$ — Substitute $k = \frac{4}{3}\pi$ in Step 1.

Step 4 $V = \frac{4}{3}\pi(6)^3$ Replace r with 6.

$= 288\,\pi$ cubic units

Practice Problem 3 If y varies directly as the square of x and $y = 48$ when $x = 2$, find y when $x = 5$.

2 Solve inverse variation problems.

Inverse Variation

Inverse Variation

A quantity y **varies inversely** as the quantity x, or y is **inversely proportional** to x, if

$$y = \frac{k}{x}$$

where k is a nonzero constant.

Suppose a plane takes four hours to fly from Atlanta to Denver at an average speed of 300 miles per hour. If we *increase* the average speed to 600 miles per hour (twice the previous speed), the flight time *decreases* to two hours (half the previous time). For the Atlanta–Denver trip, the time of flight varies inversely as the speed of the plane.

$$\text{time} = \frac{\text{distance}}{\text{speed}}$$

$$= \frac{k}{\text{speed}}, \text{ where } k = \text{distance between Atlanta and Denver}$$

We see that the distance is the constant of variation when time and speed are the variables.

EXAMPLE 4 **Solving an Inverse Variation Problem**

Suppose y varies inversely as x and $y = 35$ when $x = 11$. Find y if $x = 55$.

Solution

Step 1 $y = \frac{k}{x}$ y varies inversely as x.

Step 2 $35 = \frac{k}{11}$ Substitute $y = 35, x = 11$.

$35 \cdot 11 = k$ Solve for k.

$385 = k$ Simplify.

Step 3 $y = \frac{385}{x}$ Replace k with 385 in $y = \frac{k}{x}$.

Step 4 $y = \frac{385}{55}$ Replace x with 55.

$y = 7$ Simplify.

Practice Problem 4 A varies inversely as B, and $A = 12$ when $B = 5$. Find A when $B = 3$.

Inverse Variation with Powers

A quantity y varies inversely as the nth power of x means that $y = \frac{k}{x^n}$, where k is a nonzero constant and $n > 0$.

DO YOU **KNOW?**

Light travels at a speed of 299,792,458 meters per second. The distance that light travels in a year is called a "light year" and is so large that it is a useful unit of distance in astronomy.

1. One light year is approximately 9.46×10^{15} m.
2. The nearest star (other than the Sun) is 4.3 light years away.
3. The Milky Way (our galaxy) is about 100,000 light years in diameter.
4. The most distant objects that astronomers can see are about 13 billion light years away. Thus, the light that we presently see from these objects began its journey to us about 13 billion years ago. Because this is close to the estimated age of the universe, that light is a kind of fossil record of the universe not long after its birth.

EXAMPLE 5 Solving a Problem Involving Inverse Variation with Powers

The intensity of light varies inversely as the square of the distance from the light source. If Rita doubles her distance from a lamp, what happens to the intensity of light at her new location?

Solution

Let I be the intensity of light at a distance d from the light source. Because I is inversely proportional to the square of d, we have

$$I = \frac{k}{d^2}.$$

If we replace d with $2d$, the intensity I_1 at the new location is given by

$$I_1 = \frac{k}{(2d)^2} \qquad \text{Replace } d \text{ with } 2d \text{ in } I = \frac{k}{d^2}.$$

$$I_1 = \frac{k}{4d^2} \qquad (2d)^2 = 2^2d^2 = 4d^2$$

Therefore,

$$I_1 = \frac{I}{4} \qquad \text{Replace } \frac{k}{d^2} \text{ with } I.$$

This equation tells us that if Rita doubles her distance from the light source, the intensity of light at the new location will be one-fourth the intensity at the original location.

Practice Problem 5 If y varies inversely as the square root of x and $y = \frac{3}{4}$ when $x = 16$, find x when $y = 2$.

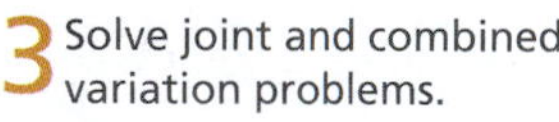

3 Solve joint and combined variation problems.

Joint and Combined Variation

Sometimes different types of variations occur in a problem that has more than one independent variable.

Joint and Combined Variation

The expression z *varies jointly as* x *and* y means that $z = kxy$ for some nonzero constant k. Similarly, if n and m are positive numbers, then the expression z *varies jointly as the nth power of* x *and mth power of* y means that $z = kx^ny^m$ for some nonzero constant k.

For example, the volume V of a right circular cylinder with radius r and height h is given by

$$V = \pi r^2 h.$$

Using the vocabulary of variation, we say, "The volume V of a right circular cylinder varies jointly as its height h and the square of its radius r." The constant of variation is π.

We can combine joint variation and inverse variation. For example, the relationship

$$w = \frac{kx^2y^3}{\sqrt{z}}$$

can be described as "w varies jointly as the square of x and cube of y and inversely as the square root of z." The same relationship also can be stated as "w varies directly as x^2y^3 and inversely as $\sqrt{z}$."

EXAMPLE 6 Newton's Law of Universal Gravitation

Newton's Law of Universal Gravitation says that every object in the universe attracts every other object with a force acting along the line of the centers of the two objects. Also, this attracting force is directly proportional to the product of the two masses and inversely proportional to the square of the distance between the two objects.

a. Write the law symbolically.

b. Estimate the value of g (the acceleration due to gravity) near the surface of Earth. Use the following estimates: radius of Earth $R_E = 6.38 \times 10^6$ meters, and mass of Earth $M_E = 5.98 \times 10^{24}$ kilograms.

Solution

a. Let m_1 and m_2 be the masses of the two objects and r be the distance between their centers. See Figure 3.35. Let F denote the gravitational force between the objects.

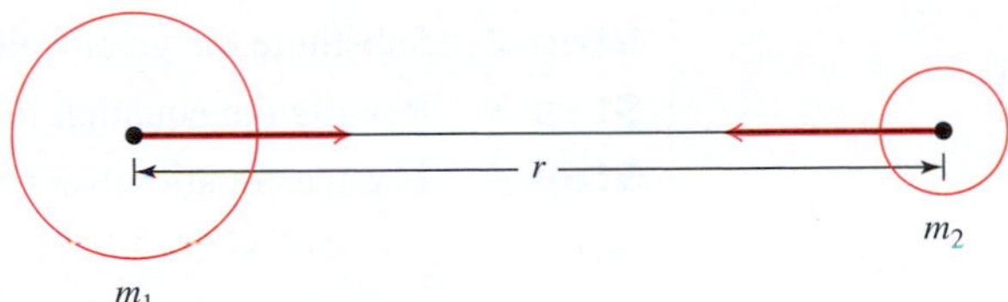

FIGURE 3.35 Gravitational attraction

Then according to Newton's Law of Universal Gravitation,

$$F = G \cdot \frac{m_1 m_2}{r^2}.$$

The constant of proportionality G is called the *universal gravitational constant*. It is "universal" because it is thought to be the same at all places and all times. If the masses m_1 and m_2 are measured in kilograms, r is measured in meters, and the force F is measured in newtons, then the value of G is approximately 6.67×10^{-11} meters cubed per kilogram per second squared. This value was experimentally verified by Henry Cavendish in 1795.

b. *Estimating the value of* g (acceleration due to gravity) *near the surface of Earth.* We use Newton's laws: Force = Mass × Acceleration = $m \cdot g$

$$\text{Force} = G \cdot \frac{m_1 m_2}{r^2}$$

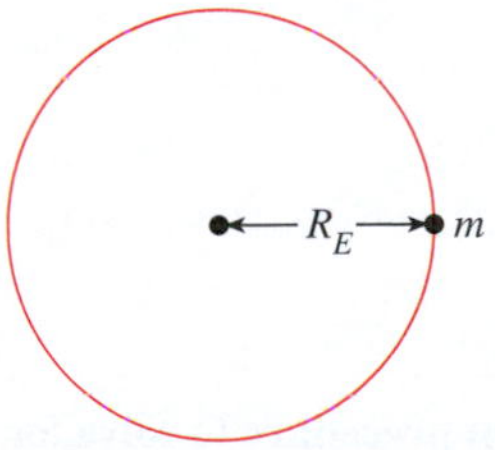

FIGURE 3.36 Finding g on Earth

For an object of mass m near the surface of Earth (see Figure 3.36):

$$\text{Force} = G \cdot \frac{m_1 m_2}{r^2}$$ Law of Universal Gravitation

$$\cancel{m} \cdot g = G \cdot \frac{\cancel{m} M_E}{R_E^2}$$ Force $= m \cdot g$; replace m_1 with m, m_2 with M_E, and r with R_E.

$$g = G \cdot \frac{M_E}{R_E^2}$$ Divide both sides by m.

$$= \frac{(6.67 \times 10^{-11}\ \text{m}^3/\text{kg}/\text{sec}^2) \cdot (5.98 \times 10^{24}\ \text{kg})}{(6.38 \times 10^6\ \text{m})^2}$$ Substitute appropriate values for G, M_E, and R_E.

$$\approx 9.8\ \text{m}/\text{sec}^2$$ Use a calculator.

Practice Problem 6 The mass of Mars is about 6.42×10^{23} kilograms, and its radius is about 3397 kilometers. What is the acceleration due to gravity near the surface of Mars?

We summarize the steps involved in solving most variation problems.

SOLVING VARIATION PROBLEMS

Step 1 Write the equation that models the problem.

a. y varies directly with x. $y = kx$

b. y varies with the nth power of x. $y = kx^n$

c. y varies inversely with x. $y = \frac{k}{x}$

d. y varies inversely with the nth power of x. $y = \frac{k}{x^n}$

e. z varies jointly with the nth power of x and the mth power of y. $z = kx^n y^m$

f. z varies directly with the nth power of x and inversely with the mth power of y. $z = \frac{kx^n}{y^m}$

Step 2 Substitute the given values in the equation in Step 1 and solve for k.

Step 3 Rewrite the equation in Step 1 with the value of k from Step 2.

Step 4 Use the equation from Step 3 to answer the question posed in the problem.

If a variation problem uses a more complicated relationship than those listed in Step 1, simply write the appropriate equation involving k and the variables and then proceed with Steps 2–4.

Answers to Practice Problems

1. $y = 24$ **2.** 275 **3.** $y = 300$ **4.** $A = 20$ **5.** $x = \frac{9}{4}$ **6.** $\approx 3.7 \text{ m/sec}^2$

SECTION 3.7 Exercises

Basic Concepts and Skills

In Exercises 1–8, write each statement as an equation.

1. P varies directly as T.
2. P varies inversely as V.
3. y varies as the square root of x.
4. F varies jointly as m_1 and m_2.
5. V varies jointly as the cube of x and fourth power of y.
6. w varies directly as the square root of x and inversely as y.
7. z varies jointly as x and u and inversely as square of v.
8. S varies jointly as the square root of x, the square of y, and the cube of z and inversely as the square of u.

In Exercises 9–24, use the four-step procedure to solve for the variable requested.

9. x varies directly as y, and $x = 15$ when $y = 30$. Find x if $y = 28$.
10. y varies directly as x, and $y = 3$ when $x = 2$. Find y when $x = 7$.
11. s varies directly as the square of t, and $s = 64$ when $t = 2$. Find s when $t = 5$.
12. y varies directly as the cube of x, and $y = 270$ when $x = 3$. Find x when $y = 80$.

13. r varies inversely as u, and $r = 3$ when $u = 11$. Find r if $u = \frac{1}{3}$.

14. y varies inversely as z, and $y = 24$ when $z = \frac{1}{6}$. Find y if $z = 1$.

15. B varies inversely with the cube of A. If $B = 1$ when $A = 2$, find B when $A = 4$.

16. y varies inversely with the cube root of x, and $y = 10$ when $x = 2$. Find x if $y = 40$.

17. z varies jointly as x and y, and $z = 42$ when $x = 2$ and $y = 3$. Find y if $z = 56$ and $x = 2$.

18. m varies directly as q and inversely as p, and $m = \frac{1}{2}$ when $p = 26$ and $q = 13$. Find m when $p = 14$ and $q = 7$.

19. z varies directly as the square of x, and $z = 32$ when $x = 4$. Find z if $x = 5$.

20. u varies inversely as the cube of t, and $u = 9$ when $t = 2$. Find u if $t = 6$.

21. P varies jointly as T and the square of Q, and $P = 36$ when $T = 17$ and $Q = 6$. Find P when $T = 4$ and $Q = 9$.

22. a varies jointly as b and the square root of c, and $a = 9$ when $b = 13$ and $c = 81$. Find a when $b = 5$ and $c = 9$.

23. z varies directly as the square root of x and inversely as the square of y, and $z = 24$ when $x = 16$ and $y = 3$. Find x when $z = 27$ and $y = 2$.

24. z varies jointly as u and the cube of v and inversely as the square of w; $z = 9$ when $u = 4$, $v = 3$, and $w = 2$. Find w when $u = 27$, $v = 2$, and $z = 8$.

In Exercises 25–28, solve for the variable requested without determining the constant of variation, k. Use the fact that if $x_1 = ky_1$ and $x_2 = ky_2$, then $\frac{x_1}{y_1} = k = \frac{x_2}{y_2}$ so that $\frac{x_1}{y_1} = \frac{x_2}{y_2}$.

25. If y is proportional to x and if $y = 12$ when $x = 16$, find y when $x = 8$.

26. If z varies directly as w and if $z = 17$ when $w = 22$, find z when $w = 110$.

27. If y is directly proportional to x and if $y = 100$ for the value x_0 of x, find y when x_0 is doubled—that is, when $x = 2x_0$.

28. If x varies directly as the square root of y and if $x = 2$ when $y = 9$, find y when $x = 3$.

Applying the Concepts

29. Hubble constant. The American astronomer Edwin Powell Hubble (1889–1953) is renowned for having determined that there are other galaxies in the universe beyond the Milky Way. In 1929, he stated that the galaxies observed at a particular time recede from each other at a speed that is directly proportional to the distance between them. The constant of proportionality is denoted by the letter H. Write Hubble's statement in the form of an equation.

30. Malthusian doctrine. The British scientist Thomas Robert Malthus (1766–1834) was a pioneer of population science and economics. In 1798, he stated that populations grow faster than the means that can sustain them. One of his arguments was that the rate of change, R, of a given population is directly proportional to the size P of the population. Write the equation that describes Malthus's argument.

31. Converting units of length. Use the fact that 1 foot $\approx$ 30.5 centimeters.

a. Write an equation that expresses the fact that a length measured in centimeters is directly proportional to the length measured in feet.

b. Convert the following measurements into centimeters.
  (i) 8 feet
  (ii) 5 feet 4 inches

c. Convert the following measurements into feet.
  (i) 57 centimeters
  (ii) 1 meter 24 centimeters

32. Converting weight. Use the fact that 1 kilogram $\approx$ 2.20 pounds.

a. Write an equation that expresses the fact that a weight measured in pounds is directly proportional to the weight measured in kilograms.

b. Convert the following measurements to pounds.
  (i) 125 grams
  (ii) 4 kilograms
  (iii) 2.4 kilograms

c. Convert the following measurements to kilograms.
  (i) 27 pounds
  (ii) 160 pounds

33. Protein from soybeans. The quantity P of protein obtained from dried soybeans is directly proportional to the quantity Q of dried soybeans used. If 20 grams of dried soybeans produces 7 grams of protein, how much protein can be produced from 100 grams of soybeans?

34. Wages. A person's weekly wages W are directly proportional to the number of hours h worked per week. If Samantha earned \$600 for a 40-hour week, how much would she earn if she worked only 25 hours in a particular week? What does the constant of proportionality mean here?

35. Physics. The distance d that an object falls varies directly as the square of the time t during which it is falling. If an object falls 64 feet in 2 seconds, how long will it take the object to fall 9 feet?

36. Physics. Hooke's Law states that the force F required to stretch a spring by x units is directly proportional to x. If a force of 10 pounds stretches a spring by 4 inches, find the force required to stretch a spring by 6 inches.

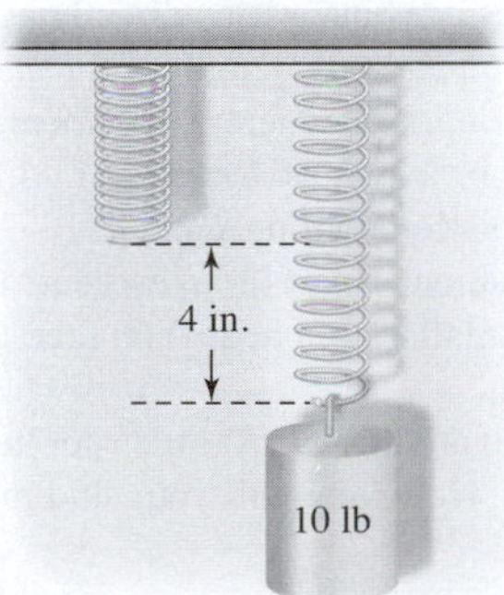

37. Chemistry. **Boyle's Law** states that at a constant temperature, the pressure P of a compressed gas is inversely proportional to its volume V. If the pressure is 20 pounds per square inch when the volume of the gas is 300 cubic inches, what is the pressure when the gas is compressed to 100 cubic inches?

38. Chemistry. In the Kelvin temperature scale, the lowest possible temperature (called absolute zero) is 0 K, where K denotes degrees Kelvin. The relationship between Kelvin temperature (T_K) and Celsius temperature (T_C) is given by $T_K = T_C + 273$.

The pressure P exerted by a gas varies directly as its temperature T_K and inversely as its volume V. Assume that at a temperature of 260 K, a gas occupies 13 cubic inches at a pressure of 36 pounds per square inch.

a. Find the volume of the gas when the temperature is 300 K and the pressure is 40 pounds per square inch.

b. Find the pressure when the temperature is 280 K and the volume is 39 cubic inches.

39. Weight. The weight of an object varies inversely as the square of the object's distance from the center of Earth. The radius of Earth is 3960 miles.

a. If an astronaut weighs 120 pounds on the surface of Earth, how much does she weigh 6000 miles above the surface of Earth?

b. If a miner weighs 200 pounds on the surface of Earth, how much does he weigh 10 miles below the surface of Earth?

40. Making a profit. Suppose you wanted to make a profit by buying gold by weight at one altitude and selling at another altitude for the same price per unit weight. Should you buy or sell at the higher altitude? (Use the information from Exercise 39.)

In Exercises 41 and 42, use Newton's Law of Universal Gravitation. (See Example 6.)

41. Gravity on the Moon. The mass of the Moon is about 7.4×10^{22} kilograms, and its radius is about 1740 kilometers. How much is the acceleration due to gravity on the surface of the Moon?

42. Gravity on the Sun. The mass of the Sun is about 2×10^{30} kilograms, and its radius is 696,000 kilometers. How much is the acceleration due to gravity on the surface of the Sun?

43. Illumination. The intensity I of illumination from a light source is inversely proportional to the square of the distance d from the source. Suppose the intensity is 320 candela at a distance of 10 feet from a light source.

a. What is the intensity at 5 feet from the source?

b. How far away from the source will the intensity be 400 candela?

44. Speed and skid marks. Police estimate that the speed s of a car in miles per hour varies directly as the square root of d, where d in feet is the length of the skid marks left by a car traveling on a dry concrete pavement. A car traveling 48 miles per hour leaves skid marks of 96 feet.

a. Write an equation relating s and d.

b. Use the equation in part (a) to estimate the speed of a car whose skid marks stretched (i) 60 feet, (ii) 150 feet, and (iii) 200 feet.

c. Suppose you are driving 70 miles per hour and slam on your brakes. How long will your skid marks be?

45. Simple pendulum. *Periodic motion* is motion that repeats itself over successive equal intervals of time. The time required for one complete repetition of the motion is called the *period*. The period of a simple pendulum varies directly as the square root of its length. What is the effect on the length if the period is doubled?

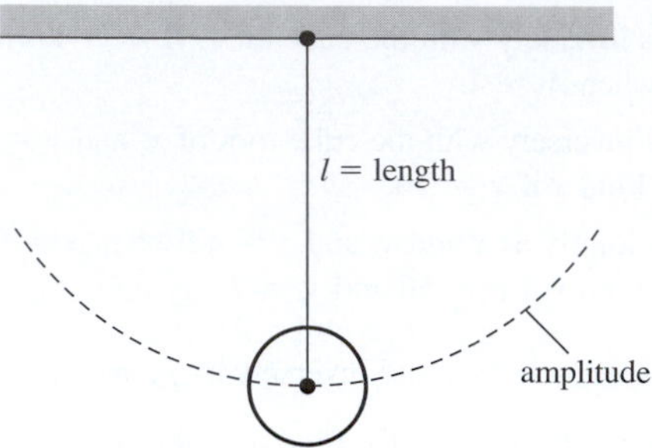

(*Note:* The amplitude of a pendulum has no effect on its period. This is what makes pendulums such good timekeepers. Because they invariably lose energy due to friction, their amplitude decreases but their period remains constant.)

46. Biology. The volume V of a lung is directly proportional to its internal surface area A. A lung with volume 400 cubic centimeters from a certain species has an average internal surface area of 100 square centimeters. Find the volume of a lung of a member of this species if the lung's internal surface area is 120 square centimeters.

47. Horsepower. The horsepower H of an automobile engine varies directly as the square of the piston radius R and the number N of pistons.

a. Write the given information in equation form.

b. What is the effect on the horsepower if the piston radius is doubled?

c. What is the effect on the horsepower if the number of pistons is doubled?

d. What is the effect on the horsepower if the radius of the pistons is cut in half and the number of pistons is doubled?

48. Safe load. The safe load that a rectangular beam can support varies jointly as the width and square of the depth of the beam and inversely as its length. A beam 4 inches wide, 6 inches deep, and 25 feet long can support a safe load of 576 pounds. Find the safe load for a beam that is of the same material but is 6 inches wide, 10 inches deep, and 20 feet long.

Beyond the Basics

49. Energy from a windmill. The energy E from a windmill varies jointly as the square of the length l of the blades and the cube of the wind velocity v.

a. Express the given information as an equation with k as the constant of variation.

b. If blades of length 10 feet and wind velocity 8 miles per hour generate 1920 watts of electric power, find k.

c. How much electric power would be generated if the blades were 8 feet long and the wind velocity was 25 miles per hour?

d. If the velocity of the wind doubles, what happens to E?

e. If the length of blades doubles, what happens to E?

f. What happens to E if both the length of the blades and the wind velocity are doubled?

50. Intensity of light. The intensity of light, I_d, at a distance d from the source of light varies directly as the intensity I of the source and inversely as d^2. It is known that at a distance of 2 meters, a 100-watt bulb produces an intensity of approximately 2 watts per square meter.

a. Find the constant of variation, k.

b. Suppose a 200-watt bulb is located on a wall 2 meters above the floor. What is the intensity of light at a point A that is 3 meters from the wall?

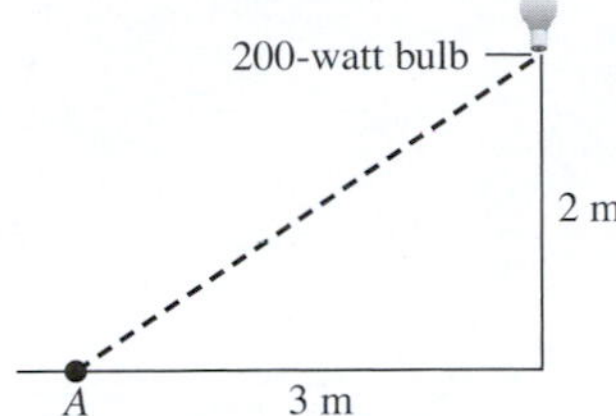

c. What is the intensity of illumination at the point A in the figure if the bulb is raised by 1 meter?

51. Metabolic rate. Metabolism is the sum total of all physical and chemical changes that take place within an organism. According to the laws of thermodynamics, all of these changes will ultimately release heat, so the metabolic rate is a measure of heat production by an animal. Biologists have found that the normal resting metabolic rate of a mammal is directly proportional to the $\frac{3}{4}$ power of its body weight.

The resting metabolic rate of a person weighing 75 kilograms is 75 watts.

a. Find the constant of proportionality, k.

b. Estimate the resting metabolic rate of a brown bear weighing 450 kilograms.

c. What is the effect on metabolic rate if the body weight is multiplied by 4?

d. What is the approximate weight of an animal with a metabolic rate of 250 watts?

52. Comparing gravitational forces. The masses of the Sun, Earth, and Moon are 2×10^{30} kilograms, 6×10^{24} kilograms, and 7.4×10^{22} kilograms, respectively. The Earth–Sun distance is about 400 times the Earth–Moon distance. Use Newton's Law of Universal Gravitation to compare the gravitational attraction between the Sun and Earth with that between Earth and the Moon.

53. Kepler's Third Law. Suppose an object of mass M_1 orbits around an object of mass M_2. Let r be the average distance in meters between the centers of the two objects and let T be the *orbital period* (the time in seconds the object completes one orbit). Kepler's Third Law states that T^2 is directly proportional to r^3 and inversely proportional to $M_1 + M_2$. The constant of proportionality is $\frac{4\pi^2}{G}$.

a. Write the equation that expresses Kepler's Third Law.

b. Earth orbits the Sun once each year at a distance of about 1.5×10^8 kilometers. Find the mass of the Sun. Use these estimates:
Mass of Earth + Mass of Sun $\approx$ Mass of Sun,
$G = 6.67 \times 10^{-11}$, 1 year $= 3.15 \times 10^7$ seconds

54. Use Kepler's Third Law. The Moon orbits Earth in about 27.3 days at an average distance of about 384,000 kilometers. Find the mass of Earth. [*Hint:* Convert the orbital period, defined in Exercise 53, to seconds; also use Mass of Earth + Mass of Moon $\approx$ Mass of Earth.]

55. Spreading a rumor. Let P be the population of a community. The rate R (per day) at which a rumor spreads in the community is jointly proportional to the number N of people who have already heard the rumor and the number $(P - N)$ of people who have not yet heard the rumor.

A rumor about the president of a college is spread in his college community of 10,000 people. Five days after the rumor started, 1000 people had heard it, and it was spreading at the rate of 45 additional people per day.

a. Write an equation relating R, P, and N.

b. Find the constant k of variation.

c. Find the rate at which the rumor was spreading when one-half the college community had heard it.

d. How many people had heard the rumor when it was spreading at the rate of 100 people per day?

Critical Thinking/Discussion/Writing

56. Electricity. The current I in an electric circuit varies directly as the voltage V and inversely as the resistance R. If the resistance is increased by 20%, what percent increase must occur in the voltage to increase the current by 30%?

57. Precious stones. The value of a precious stone is proportional to the square of its weight.

a. Calculate the loss incurred by cutting a diamond worth \$1000 into two pieces whose weights are in the ratio 2:3.

b. A precious stone worth \$25,000 is accidentally dropped and broken into three pieces, the weights of which are in the ratio 5:9:11. Calculate the loss incurred due to breakage.

c. A diamond breaks into five pieces, the weights of which are in the ratio 1:2:3:4:5. If the resulting loss is \$85,000, find the value of the original diamond. Also calculate the value of a diamond whose weight is twice that of the original diamond.

58. Bus service. The profit earned in running a bus service is jointly proportional to the distance and the number of passengers over a certain fixed number. The profit is \$80 when 30 passengers are carried a distance of 40 km and is \$180 when 35 passengers are carried 60 km. What is the minimum number of passengers that results in no loss?

59. Weight of a sphere. The weight of a sphere is directly proportional to the cube of its radius. A metal sphere has a hollow space about its center in the form of a concentric sphere, and its weight is $\frac{7}{8}$ times the weight of a solid sphere of the same radius and material. Find the ratio of the inner to the outer radius of the hollow sphere.

60. Train speed. A locomotive engine can go 24 miles per hour, and its speed is reduced by a quantity that varies directly as the square root of the number of cars it pulls. Pulling four cars, its speed is 20 miles per hour. Find the greatest number of cars the engine can pull.

Maintaining Skills

In Exercises 61–70, simplify each expression.

61. 5^0
62. 2^3
63. 3^{-2}
64. $\left(\frac{1}{2}\right)^3$
65. $\left(\frac{1}{2}\right)^{-4}$
66. $2^{x-1} \cdot 2^{3-x}$
67. $5^{2x-3} \cdot 5^{3-x}$
68. $\frac{2^{3x-2}}{2^{x-5}}$
69. $(2^{x-1})^x$
70. $\sqrt[3]{\sqrt[4]{2}}$

In Exercises 71–76, solve each equation for the requested variable.

71. $y = mx + b$ for m
72. $ax + by = 3$ for x
73. $A = B(1 + C)$ for C
74. $A = B(1 + C)^3$ for C
75. $A = B \cdot 10^{-n}$ for B
76. $A = B \cdot 10^m + C$ for B

SUMMARY Definitions, Concepts, and Formulas

3.1 Quadratic Functions

i. A **quadratic function** f is a function of the form

$$f(x) = ax^2 + bx + c, a \neq 0.$$

ii. The **standard form** of a quadratic function is

$$f(x) = a(x - h)^2 + k, a \neq 0.$$

iii. The graph of a quadratic function is a transformation of the graph of $y = x^2$.

iv. The graph of a quadratic function is a parabola with vertex

$$(h, k) = \left(-\frac{b}{2a}, f\left(-\frac{b}{2a}\right)\right).$$

v. The maximum (if $a < 0$) or minimum (if $a > 0$) value of a quadratic function $f(x) = ax^2 + bx + c$ occurs at the y-coordinate of the vertex of the parabola.

3.2 Polynomial Functions

i. A function f of the form

$$f(x) = a_n x^n + a_{n-1}x^{n-1} + \cdots a_2 x^2 + a_1 x + a_0, a_n \neq 0$$

is a polynomial function of degree n.

ii. The graph of a polynomial function is smooth and continuous.

iii. The end behavior of the graph of a polynomial function depends on the sign of the leading coefficient and the degree (even-odd) of the polynomial.

iv. A real number c is a **zero** of a function f if $f(c) = 0$. Geometrically, c is an x-intercept of the graph of $y = f(x)$.

v. If in the factorization of a polynomial function $f(x)$ the factor $(x - a)$ occurs exactly m times, then a is a zero of **multiplicity** m. If m is odd, the graph of $y = f(x)$ crosses the x-axis at a; if m is even, the graph touches but does not cross the x-axis at a.

vi. If the degree of a polynomial function $f(x)$ is n, then $f(x)$ has, at most, n real zeros and the graph of $f(x)$ has, at most, $(n - 1)$ turning points.

vii. **Intermediate Value Theorem:** Let $f(x)$ be a polynomial function and a and b be two numbers such that $a < b$. If $f(a)$ and $f(b)$ have opposite signs, then there is at least one number c, with $a < c < b$, for which $f(c) = 0$.

viii. See page 315 for graphing a polynomial function.

3.3 Dividing Polynomials

i. **Division algorithm:** If a polynomial $F(x)$ is divided by a polynomial $D(x) \neq 0$, there are unique polynomials $Q(x)$ and $R(x)$ such that $F(x) = D(x)Q(x) + R(x)$, where either $R(x) = 0$ or deg $R(x) <$ deg $D(x)$. In words, "The dividend equals the product of the divisor and the quotient plus the remainder."

ii. Synthetic division is a shortcut for dividing a polynomial $F(x)$ by $x - a$.

iii. **Remainder Theorem:** If a polynomial $F(x)$ is divided by $(x - a)$, the remainder is $F(a)$.

iv. **Factor Theorem:** A polynomial function $F(x)$ has $(x - a)$ as a factor if and only if $F(a) = 0$.

3.4 The Real Zeros of a Polynomial Function

i. **Rational Zeros Theorem:** If $\frac{p}{q}$ is a rational zero in lowest terms for a polynomial function with integer coefficients, then p is a factor of the constant term and q is a factor of the leading coefficient.

ii. **Descartes's Rule of Signs.** Let $F(x)$ be a polynomial function with real coefficients.

a. The number of positive zeros of F is equal to the number of variations of signs of $F(x)$ or is less than that number by an even integer.

b. The number of negative zeros of F is equal to the number of variations of sign of $F(-x)$ or is less than that number by an even integer.

iii. **Rules for Bounds on the Zeros.** Suppose a polynomial $F(x)$ is synthetically divided by $x - k$.

a. If $k > 0$ and each number in the last row is zero or positive, then k is an upper bound on the zeros of $F(x)$.

b. If $k < 0$ and the numbers in the last row alternate in sign, then k is a lower bound on the zeros of $F(x)$.

3.5 The Complex Zeros of a Polynomial Function

i. **Fundamental Theorem of Algebra.** An nth-degree polynomial equation has at least one complex zero.

ii. **Factorization Theorem for Polynomials.** If $P(x)$ is a polynomial of degree $n \geq 1$, it can be factored into

n (not necessarily distinct) linear factors of the form $P(x) = a(x - r_1)(x - r_2) \cdots (x - r_n)$, where a, $r_1, r_2, \ldots, r_n$ are complex numbers.

iii. **Number of Zeros Theorem.** A polynomial of degree n has exactly n complex zeros, provided a zero of multiplicity k is counted k times.

iv. **Conjugate Pairs Theorem.** If $a + bi$ is a zero of the polynomial function P (with real coefficients), then $a - bi$ is also a zero of P.

3.6 Rational Functions

i. A function $f(x) = \frac{N(x)}{D(x)}$, where $N(x)$ and $D(x)$ are polynomials and $D(x) \neq 0$, is called a rational function. The domain of f is the set of all real numbers except the real zeros of $D(x)$.

ii. The line $x = a$ is a vertical asymptote of the graph of f if $|f(x)| \to \infty$ as $x \to a^+$ or as $x \to a^-$.

iii. If $\frac{N(x)}{D(x)}$ is in lowest terms, then the graph of $F(x)$ has vertical asymptotes at the real zeros of $D(x)$.

iv. The line $y = k$ is a horizontal asymptote of the graph of f if $f(x) \to k$ as $x \to \infty$ or as $x \to -\infty$.

v. A procedure for graphing rational functions is given on page 363.

3.7 Variation

k is a nonzero constant called the constant of variation.

Variation	Equation
y varies directly with x.	$y = kx$
y varies with the nth power of x.	$y = kx^n$
y varies inversely with x.	$y = \frac{k}{x}$
y varies inversely with the nth power of x.	$y = \frac{k}{x^n}$
z varies jointly with the nth power of x and the mth power of y.	$z = kx^n y^m$
z varies directly with the nth power of x and inversely with the mth power of y.	$z = \frac{kx^n}{y^m}$

REVIEW EXERCISES

Basic Skills and Concepts

In Exercises 1–10, graph each quadratic function by finding (i) whether the parabola opens up or down and by finding (ii) its vertex, (iii) its axis, (iv) its x-intercepts, (v) its y-intercept, and (vi) the intervals over which the function is increasing and decreasing.

1. $y = (x - 1)^2 + 2$ **2.** $y = (x + 2)^2 - 3$

3. $y = -2(x - 3)^2 + 4$ **4.** $y = -\frac{1}{2}(x + 1)^2 + 2$

5. $y = -2x^2 + 3$ **6.** $y = 2x^2 + 4x - 1$

7. $y = 2x^2 - 4x + 3$ **8.** $y = -2x^2 - x + 3$

9. $y = 3x^2 - 2x + 1$ **10.** $y = 3x^2 - 5x + 4$

In Exercises 11–14, determine whether the given quadratic function has a maximum or a minimum value and then find that value.

11. $f(x) = 3 - 4x + x^2$ **12.** $f(x) = 8x - 4x^2 - 3$

13. $f(x) = -2x^2 - 3x + 2$ **14.** $f(x) = \frac{1}{2}x^2 - \frac{3}{4}x + 2$

In Exercises 15–18, graph each polynomial function by using transformations on the appropriate function $y = x^n$.

15. $f(x) = (x + 1)^3 - 2$ **16.** $f(x) = (x + 1)^4 + 2$

17. $f(x) = (1 - x)^3 + 1$ **18.** $f(x) = x^4 + 3$

In Exercises 19–24, for each polynomial function f,

(i) Determine the end behavior of f.
(ii) Determine the zeros of f. State the multiplicity of each zero. Determine whether the graph of f crosses or only touches the axis at each x-intercept.
(iii) Find the x- and y-intercepts of the graph of f.
(iv) Use test numbers to find the intervals over which the graph of f is above or below the x-axis.
(v) Sketch the graph of $y = f(x)$.

19. $f(x) = x(x - 1)(x + 2)$ **20.** $f(x) = x^3 - x$

21. $f(x) = -x^2(x - 1)^2$ **22.** $f(x) = -x^3(x - 2)^2$

23. $f(x) = -x^2(x^2 - 1)$

24. $f(x) = -(x - 1)^2(x^2 + 1)$

In Exercises 25–28, divide by using long division.

25. $\frac{6x^2 + 5x - 13}{3x - 2}$ **26.** $\frac{8x^2 - 14x + 15}{2x - 3}$

27. $\frac{8x^4 - 4x^3 + 2x^2 - 7x + 165}{x + 1}$

28. $\frac{x^3 - 3x^2 + 4x + 7}{x^2 - 2x + 6}$

In Exercises 29–32, divide by using synthetic division.

29. $\dfrac{x^3 - 12x + 3}{x - 3}$

30. $\dfrac{-4x^3 + 3x^2 - 5x}{x - 6}$

31. $\dfrac{2x^4 - 3x^3 + 5x^2 - 7x + 165}{x + 1}$

32. $\dfrac{3x^5 - 2x^4 + x^2 - 16x - 132}{x + 2}$

In Exercises 33–36, a polynomial function $f(x)$ and a constant c are given. Find $f(c)$ by (i) evaluating the function and (ii) using synthetic division and the Remainder Theorem.

33. $f(x) = x^3 - 3x^2 + 11x - 29; c = 2$

34. $f(x) = 2x^3 + x^2 - 15x - 2; c = -2$

35. $f(x) = x^4 - 2x^2 - 5x + 10; c = -3$

36. $f(x) = x^5 + 2; c = 1$

In Exercises 37–40, a polynomial function $f(x)$ and a constant c are given. Use synthetic division to show that c is a zero of $f(x)$. Use the result to final all zeros of $f(x)$.

37. $f(x) = x^3 - 7x^2 + 14x - 8; c = 2$

38. $f(x) = 2x^3 - 3x^2 - 12x + 4; c = -2$

39. $f(x) = 3x^3 + 14x^2 + 13x - 6; c = \dfrac{1}{3}$

40. $f(x) = 4x^3 + 19x^2 - 13x + 2; c = \dfrac{1}{4}$

In Exercises 41 and 42, use the Rational Zeros Theorem to list all possible rational zeros of $f(x)$.

41. $f(x) = x^4 + 3x^3 - x^2 - 9x - 6$

42. $f(x) = 9x^3 - 36x^2 - 4x + 16$

In Exercises 43–50, use Descartes's Rule of Signs and the Rational Zeros Theorem to find all real zeros of each polynomial function.

43. $f(x) = 5x^3 + 11x^2 + 2x$

44. $f(x) = x^3 + 2x^2 - 5x - 6$

45. $f(x) = x^3 + 3x^2 - 4x - 12$

46. $f(x) = 2x^3 - 9x^2 + 12x - 5$

47. $x^3 - 4x^2 - 5x + 14$

48. $x^3 - 5x^2 + 3x + 1$

49. $2x^3 - 5x^2 - 2x + 2$

50. $3x^3 - 29x^2 + 29x + 13$

In Exercises 51–56, find all of the zeros of $f(x)$, real and nonreal.

51. $f(x) = x^3 - 7x + 6$; one zero is 2.

52. $f(x) = x^4 + x^3 - 3x^2 - x + 2$; 1 is a zero of multiplicity 2.

53. $f(x) = x^4 - 2x^3 + 6x^2 - 18x - 27$; two zeros are -1 and 3.

54. $f(x) = 4x^3 - 19x^2 + 32x - 15$; one zero is $2 - i$.

55. $f(x) = x^4 + 2x^3 + 9x^2 + 8x + 20$; one zero is $-1 + 2i$.

56. $f(x) = x^5 - 7x^4 + 24x^3 - 32x^2 + 64$; $2 + 2i$ is a zero of multiplicity 2.

In Exercises 57–68, solve each equation in the complex number system.

57. $x^3 - x^2 - 4x + 4 = 0$

58. $2x^3 + x^2 - 12x - 6 = 0$

59. $4x^3 - 7x - 3 = 0$

60. $x^3 + x^2 - 8x - 6 = 0$

61. $x^3 - 8x^2 + 23x - 22 = 0$

62. $x^3 - 3x^2 + 8x + 12 = 0$

63. $3x^3 - 5x^2 + 16x + 6 = 0$

64. $2x^3 - 9x^2 + 18x - 7 = 0$

65. $x^4 - x^3 - x^2 - x - 2 = 0$

66. $x^4 - x^3 - 13x^2 + x + 12 = 0$

67. $2x^4 - x^3 - 2x^2 + 13x - 6 = 0$

68. $3x^4 - 14x^3 + 28x^2 - 10x - 7 = 0$

In Exercises 69 and 70, show that the given equation has no rational roots.

69. $x^3 + 13x^2 - 6x - 2 = 0$

70. $3x^4 - 9x^3 - 2x^2 - 15x - 5 = 0$

In Exercises 71 and 72, use the Intermediate Value Theorem to find the value of the real root between 1 and 2 of each equation to two decimal places.

71. $x^3 + 6x^2 - 28 = 0$

72. $x^3 + 3x^2 - 3x - 7 = 0$

In Exercises 73–80, graph each rational function by following the five-step procedure outlined in Section 3.6.

73. $f(x) = 1 + \dfrac{1}{x}$

74. $f(x) = \dfrac{2 - x}{x}$

75. $f(x) = \dfrac{x}{x^2 - 1}$

76. $f(x) = \dfrac{x^2 - 9}{x^2 - 4}$

77. $f(x) = \dfrac{x^3}{x^2 - 9}$

78. $f(x) = \dfrac{x + 1}{x^2 - 2x - 8}$

79. $f(x) = \dfrac{x^4}{x^2 - 4}$

80. $f(x) = \dfrac{x^2 + x - 6}{x^2 - x - 12}$

Applying the Concepts

81. **Variation.** Assuming that y varies directly as x and $y = 12$ when $x = 4$, find y when $x = 5$.

82. **Variation.** Assuming that p varies inversely as q and $p = 4$ when $q = 3$, find p when $q = 4$.

83. **Variation.** Assuming that s varies directly as the square of t and $s = 20$ when $t = 2$, find s when $t = 3$.

84. **Variation.** Assuming that y varies inversely as x^2 and $y = 3$ when $x = 8$, find x when $y = 12$.

85. Missile path. A missile fired from the origin of a coordinate system follows a path described by the equation $y = -\frac{1}{10}x^2 + 20x$, where x is in yards and the x-axis is on the ground. Sketch the missile's path and determine what its maximum altitude is and where it hits the ground.

86. Minimizing area. Suppose a wire 20 centimeters long is to be cut into two pieces, each of which will be formed into a square. Find the size of each piece that minimizes the total area.

87. A farmer wants to fence off three identical adjoining rectangular pens, each 400 square feet in area. See the figure. What should the width and length of each pen be so that the least amount of fence is used?

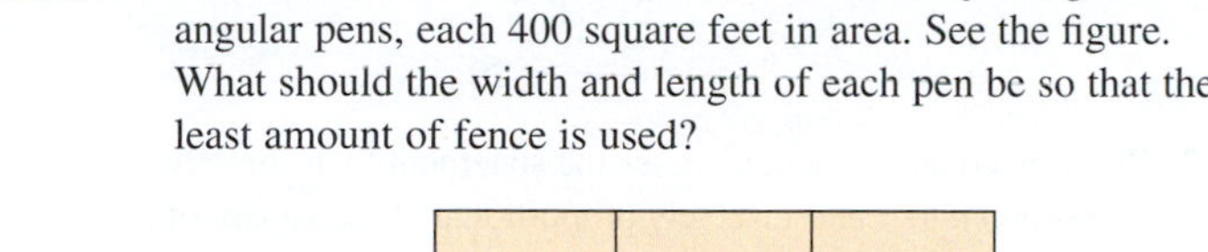

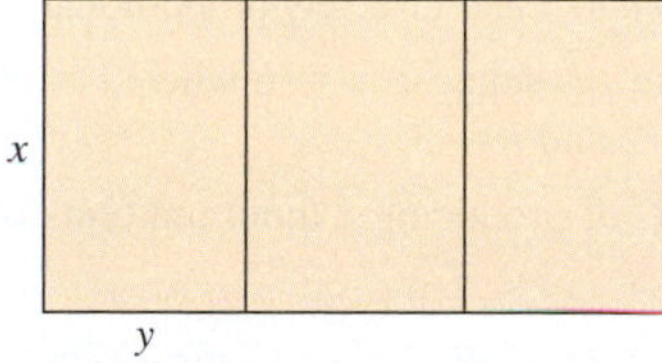

88. Suppose the outer boundary of the pens in Exercise 87 requires heavy fence that costs \$5 per foot and two internal partitions cost \$3 per foot. What dimensions x and y will minimize the cost?

89. Electric circuit. In the circuit shown in the figure, the voltage $V = 100$ volts and the resistance $R = 50$ ohms. We want to determine the size of the remaining resistor (x ohms). The power absorbed by the circuit is given by

$$p(x) = \frac{V^2 x}{(R + x)^2}.$$

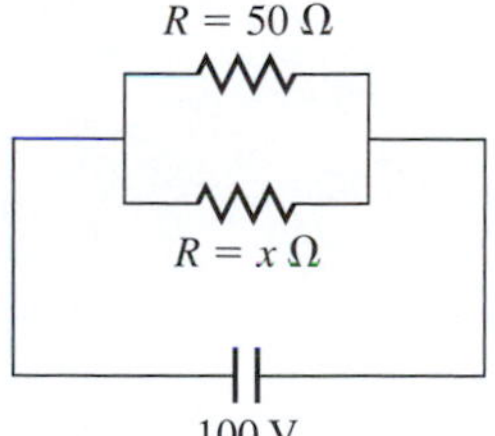

a. Graph the function $y = p(x)$.

b. Use a graphing calculator to find the value of x that maximizes the power absorbed.

90. Maximizing area. A sheet of paper for a poster is 18 square feet in area. The margins at the top and bottom are 9 inches each, and the margin on each side is 6 inches. What should the dimensions of the paper be if the printed area is to be a maximum?

91. Maximizing profit. A manufacturer makes and sells printers to retailers at \$24 per unit. The total daily cost C in dollars of producing x printers is given by

$$C(x) = 150 + 3.9x + \frac{3}{1000}x^2.$$

a. Write the profit P as a function of x.

b. Find the number of printers the manufacturer should produce and sell to achieve maximum profit.

c. Find the average cost $\overline{C}(x) = \frac{C(x)}{x}$. Graph $y = \overline{C}(x)$.

92. Meteorology. The function $p = \frac{69.1}{a + 2.3}$ relates the atmospheric pressure p in inches of mercury to the altitude a in miles from the surface of Earth.

a. Find the pressure on Mount Kilimanjaro at an altitude of 19,340 feet.

b. Is there an altitude at which the pressure is 0?

93. Wages. An employee's wages are directly proportional to the time he or she has worked. Sam earned \$280 for 40 hours. How much would Sam earn if he worked 35 hours?

94. Car's stopping distance. The distance required for a car to come to a stop after its brakes are applied is directly proportional to the square of its speed. If the stopping distance for a car traveling 30 miles per hour is 25 feet, what is the stopping distance for a car traveling 66 miles per hour?

95. Illumination. The amount of illumination from a source of light varies directly as the intensity of the source and inversely as the square of the distance from the source. At what distance from a light source of intensity 300 candlepower will the illumination be one-half the illumination 6 inches from the source?

96. Chemistry. *Charles's Law* states that at a constant pressure, the volume V of a gas is directly proportional to its temperature T (in Kelvin degrees). If a bicycle tube is filled with 1.2 cubic feet of air at a temperature of 295 K, what will the volume of the air in the tube be if the temperature rises to 310 K while the pressure stays the same?

97. Electric circuits. The current I (measured in amperes) in an electric circuit varies inversely as the resistance R (measured in ohms) when the voltage is held constant. The current in a certain circuit is 30 amperes when the resistance is 300 ohms.

a. Find the current in the circuit if the resistance is decreased to 250 ohms.

b. What resistance will yield a current of 60 amperes?

98. Safe load. The safe load that a circular column can support varies directly as the square root of its radius and inversely as the square of its length. A pillar with radius 4 inches and length 12 feet can safely support a 20-ton load. Find the load that a pillar of the same material with diameter 6 inches and length 10 feet can safely support.

99. Spread of disease. An infectious cold virus spreads in a community at a rate R (per day) that is jointly proportional to the number of people who are infected with the virus and the number of people in the community who are not infected yet. After the tenth day of the start of a certain infection, 15% of the total population of 20,000 people of Pollutville had been infected and the virus was spreading at the rate of 255 people per day.

a. Find the constant of proportionality, k.

b. At what rate is the disease spreading when one-half of the population is infected?

c. Find the number of people infected when the rate of infection reached 95 people per day.

100. Coulomb's Law. The electric charge is measured in coulombs. (The charges on an electron and a proton, which are equal and opposite, are approximately 1.602×10^{-19} coulomb.) Coulomb's Law states that the force F between two particles is jointly proportional to their charges q_1 and q_2 and inversely proportional to the square of the distance between the two particles. Two charges are acted upon by a repulsive force of 96 units. What is the force if the distance between the particles is quadrupled?

PRACTICE TEST A

1. Find the x-intercepts of the graph of $f(x) = x^2 - 6x + 2$.
2. Graph $f(x) = 3 - (x + 2)^2$.
3. Find the vertex of the parabola described by $y = -7x^2 + 14x + 3$.
4. Find the domain of the function
$$f(x) = \frac{x^2 - 1}{x^2 + 3x - 4}.$$
5. Find the quotient and remainder of
$$\frac{x^3 - 2x^2 - 5x + 6}{x + 2}.$$
6. Graph the polynomial function $P(x) = x^5 - 4x^3$.
7. Find all of the zeros of $f(x) = 2x^3 - 2x^2 - 8x + 8$, given that 2 is one of the zeros.
8. Find the quotient of $\dfrac{-6x^3 + x^2 + 17x + 3}{2x + 3}$.
9. Use the Remainder Theorem to find the value $P(-2)$ of the polynomial $P(x) = x^4 + 5x^3 - 7x^2 + 9x + 17$.
10. Find all rational roots of the equation $x^3 - 5x^2 - 4x + 20 = 0$ and then find the irrational roots if there are any.
11. Find the zeros of the polynomial function $f(x) = x^4 + x^3 - 15x^2$.
12. For $P(x) = 2x^{18} - 5x^{13} + 6x^3 - 5x + 9$, list all possible rational zeros found by the Rational Zeros Test, but do not check to see which values are actually zeros.
13. Describe the end behavior of $f(x) = (x + 3)^3(x - 5)^2$.
14. Find the zeros and the multiplicity of each zero for $f(x) = (x^2 - 4)(x + 2)^2$.
15. Determine how many positive and how many negative real zeros the polynomial function $P(x) = 3x^6 + 2x^3 - 7x^2 + 8x$ can have.
16. Find the horizontal and vertical asymptotes of the graph of
$$f(x) = \frac{2x^2 + 3}{x^2 - x - 20}.$$
17. Write an equation that expresses the statement "y is directly proportional to x and inversely proportional to the square of t."
18. In Problem 17, suppose $y = 6$ when $x = 8$ and $t = 2$. Find y if $x = 12$ and $t = 3$.
19. The cost C of producing x thousand units of a product is given by
$$C = x^2 - 30x + 335 \text{ (dollars)}.$$
Find the value of x for which the cost is minimum.
20. From a rectangular 8×17 piece of cardboard, four congruent squares with sides of length x are cut out, one at each corner. The sides can then be folded to form a box. Find the volume V of the box as a function of x.

PRACTICE TEST B

1. Find the x-intercepts of the graph of $f(x) = x^2 + 5x + 3$.

   a. $\dfrac{-5 \pm i\sqrt{13}}{2}$ **b.** $\dfrac{-5 \pm \sqrt{13}}{2}$

   c. $\dfrac{-5 \pm i\sqrt{37}}{2}$ **d.** $\dfrac{-5 \pm \sqrt{37}}{2}$

2. Which is the graph of $f(x) = 4 - (x - 2)^2$?

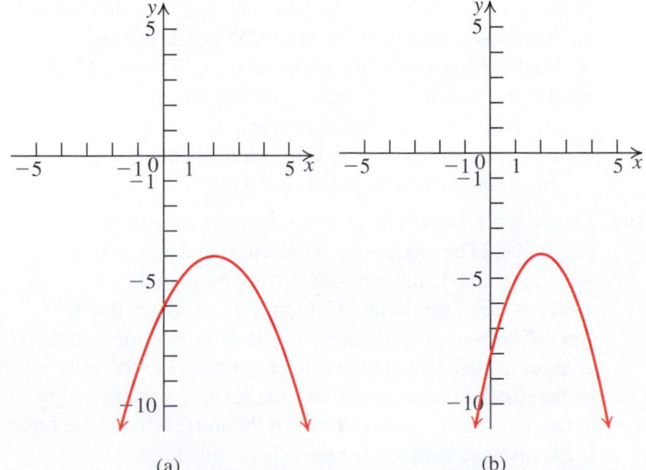

(a) (b)

(c) (d)

3. Find the vertex of the parabola described by $y = 6x^2 + 12x - 5$.

   a. $(-1, -11)$ **b.** $(1, -5)$

   c. $(-1, 13)$ **d.** $(1, 13)$

4. Which of the following is *not* in the domain of the function $f(x) = \dfrac{x^2}{x^2 + x - 6}$?

I. -3
II. 0
III. 2

a. I and II **b.** I and III
c. II and III **d.** I only

5. Find the quotient and remainder when $x^3 - 8x + 6$ is divided by $x + 3$.

a. $x^2 - 8; 2$ **b.** $x^2 - 8; 0$
c. $x^2 - 3x + 1; x + 3$ **d.** $x^2 - 3x + 1; 3$

6. Which is the graph of the polynomial $P(x) = x^4 + 2x^3$?

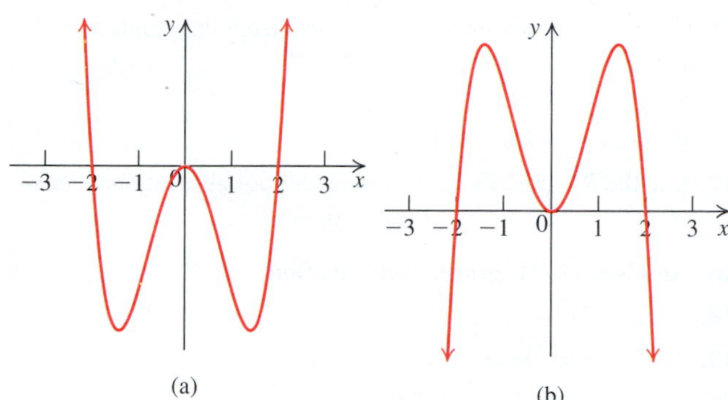

(a) (b)

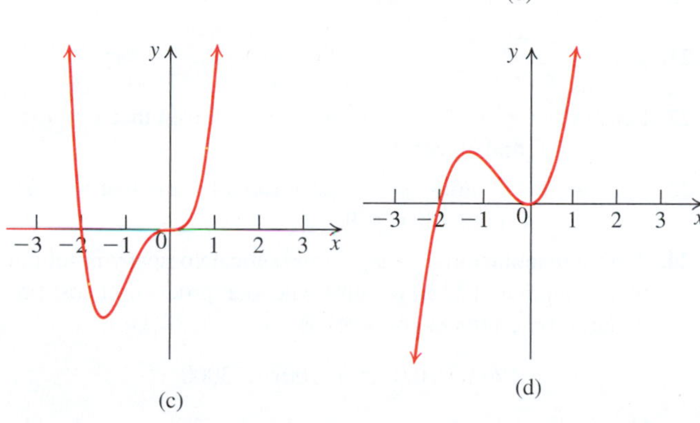

(c) (d)

7. Find all of the zeros of $f(x) = 3x^3 - 26x^2 + 61x - 30$ given that 3 is one of the zeros (that is, $f(3) = 0$).

a. $3, -5, -\frac{2}{3}$ **b.** $3, 2, \frac{5}{3}$
c. $3, 5, \frac{2}{3}$ **d.** $3, -2, \frac{5}{3}$

8. Find the quotient $\dfrac{-10x^3 + 21x^2 - 17x + 12}{2x - 3}$.

a. $x^2 - 3x + 4$ **b.** $-5x^2 + 3x - 4$
c. $x^2 + 3x - 4$ **d.** $-5x^2 - 4$

9. Use the Remainder Theorem to find the value $P(-3)$ of the polynomial $P(x) = x^4 + 4x^3 + 7x^2 + 10x + 15$.

a. 13 **b.** 15
c. 21 **d.** 6

10. Find all rational roots of the equation $-x^3 + x^2 + 8x - 12 = 0$ and then find the irrational roots if there are any.

a. -3 and 2 **b.** -3 and $\sqrt{2}$
c. $-1, -2,$ and 3 **d.** -3 and $\sqrt{3}$

11. Find the zeros of the polynomial function $f(x) = x^3 + x^2 - 30x$.

a. $x = -6, x = 5, x = 0$ **b.** $x = 0, x = -6$
c. $x = 4, x = 5$ **d.** $x = 0, x = 4, x = 5$

12. For $P(x) = x^{30} - 4x^{25} + 6x^2 + 60$, list all possible rational zeros found by the Rational Zeros Test, but do not check to see which values are actually zeros.

a. $\pm 2, \pm 3, \pm 4, \pm 5, \pm 6, \pm 8, \pm 10, \pm 12, \pm 15, \pm 20, \pm 30$, and ± 60
b. $\pm 1, \pm 2, \pm 3, \pm 4, \pm 5, \pm 6, \pm 10, \pm 12, \pm 15, \pm 18, \pm 20, \pm 24, \pm 30$, and ± 60
c. $\pm 1, \pm 3, \pm 4, \pm 5, \pm 6, \pm 12, \pm 15, \pm 20, \pm 30$, and ± 60
d. $\pm 1, \pm 2, \pm 3, \pm 4, \pm 5, \pm 6, \pm 10, \pm 12, \pm 15, \pm 20, \pm 30$, and ± 60

13. Which of the following correctly describes the end behavior of $f(x) = (x + 1)^2 (x - 2)^2$?

a. $\begin{cases} y \to \infty \text{ as } x \to -\infty \\ y \to -\infty \text{ as } x \to \infty \end{cases}$ **b.** $\begin{cases} y \to -\infty \text{ as } x \to -\infty \\ y \to -\infty \text{ as } x \to \infty \end{cases}$

c. $\begin{cases} y \to \infty \text{ as } x \to -\infty \\ y \to \infty \text{ as } x \to \infty \end{cases}$ **d.** $\begin{cases} y \to -\infty \text{ as } x \to -\infty \\ y \to \infty \text{ as } x \to \infty \end{cases}$

14. Find the zeros and the multiplicity of each zero for $f(x) = (x^2 - 1)(x + 1)^2$.

a. $\begin{cases} \text{zero } 1, & \text{multiplicity } 1 \\ \text{zero } -1, & \text{multiplicity } 2 \end{cases}$

b. $\begin{cases} \text{zero } 1, & \text{multiplicity } 1 \\ \text{zero } -1, & \text{multiplicity } 3 \end{cases}$

c. $\begin{cases} \text{zero } 1, & \text{multiplicity } 2 \\ \text{zero } -1, & \text{multiplicity } 2 \end{cases}$

d. $\begin{cases} \text{zero } i, & \text{multiplicity } 2 \\ \text{zero } -1, & \text{multiplicity } 2 \end{cases}$

15. Determine how many positive and how many negative real zeros the polynomial $P(x) = x^5 - 4x^3 - x^2 + 6x - 3$ can have.

a. positive 3, negative 2
b. positive 2 or 0, negative 2 or 0
c. positive 3 or 1, negative 2 or 0
d. positive 2 or 0, negative 3 or 1

16. The horizontal and the vertical asymptotes of the graph of $f(x) = \dfrac{x^2 + x - 2}{x^2 + x - 12}$ are

a. $x = -2, y = 3, y = -4.$
b. $x = 1, y = 3, y = -4.$
c. $y = 1, x = 3, x = -4.$
d. $y = -2, x = 3, x = -4.$

17. Write an equation that expresses the statement that S is directly proportional to the square of t and inversely proportional to the cube of x.

a. $S = \dfrac{kt^3}{\sqrt[3]{x}}$ **b.** $S = kt^2x^3$

c. $S = \dfrac{kt^3}{x^2}$ **d.** $S = \dfrac{kt^2}{x^3}$

18. In Problem 17, suppose $S = 27$ if $t = 3$ and $x = 1$. If $t = 6$ and $x = 3$, then what is S?

a. 54 **b.** 4
c. $\frac{4}{3}$ **d.** 9

19. The cost C of producing x thousand units of a product is given by

$$C = x^2 - 24x + 319 \text{ (dollars)}.$$

Find the value of x for which the cost is minimum.

a. 319 **b.** 12 **c.** 175 **d.** 24

20. From a rectangular 10×12 piece of cardboard, four congruent squares with sides of length x are cut out, one at each corner. The sides can then be folded to form a box. Find the volume V of the box as a function of x.

a. $V = 2x(10 - x)(12 - x)$
b. $V = 2x(10 - 2x)(12 - 2x)$
c. $V = x(10 - x)(12 - x)$
d. $V = x(10 - 2x)(12 - 2x)$

CUMULATIVE REVIEW EXERCISES CHAPTERS P–3

1. Find the distance between the points $P(-1, 3)$ and $Q(2, 5)$.

2. Find the midpoint of the line segment joining $P(2, -5)$ and $Q(-8, -3)$.

3. Find the x- and y-intercepts of the graph of the equation $y = x^2 - 2x - 8$ and sketch the graph.

In Exercises 4 and 5, find the slope and intercepts of the line and sketch the graph.

4. $x + 3y - 6 = 0$

5. $x = 2y - 6$

6. Find an equation in standard form of the circle with center $(2, -3)$ and radius 4.

7. Find the center and radius of the circle with equation

$$x^2 + y^2 + 2x - 4y - 4 = 0.$$

In Exercises 8 and 9, find the slope–intercept form of the line satisfying the given condition.

8. The line has slope 3 and passes through $(1, -2)$.

9. The line is parallel to $2x + 3y = 5$ and passes through $(1, 3)$.

In Exercises 10 and 11, find the domain of each function.

10. $f(x) = \dfrac{1}{2x + 3}$ **11.** $p(x) = \dfrac{1}{\sqrt{4 - 2x}}$

12. Let $f(x) = x^2 - 2x + 3$. Find $f(-2), f(3), f(x + h)$, and $\dfrac{f(x + h) - f(x)}{h}$.

13. Let $f(x) = \sqrt{x}$ and $g(x) = x^2 + 1$. Find each of the following.

a. $f(g(x))$ **b.** $g(f(x))$
c. $f(f(x))$ **d.** $g(g(x))$

14. Let $f(x) = \begin{cases} 3x + 2 & \text{if } x \le 2 \\ 4x - 1 & \text{if } 2 < x \le 3. \\ 6 & \text{if } x > 3 \end{cases}$

a. Find $f(1)$, $f(3)$, and $f(4)$.
b. Sketch the graph of $y = f(x)$.

15. Let $f(x) = 2x - 3$. Find $f^{-1}(x)$.

16. Use transformations on $y = \sqrt{x}$ to sketch the graph of each function.

a. $f(x) = \sqrt{x + 2}$
b. $g(x) = -2\sqrt{x + 1} + 3$

17. Use the Rational Zeros Test to list all possible rational zeros of $f(x) = 2x^4 - 3x^2 + 5x - 6$.

In Exercises 18–21, graph each function.

18. $f(x) = 2x^2 - 4x + 1$

19. $f(x) = -x^2 + 2x + 3$

20. $f(x) = (x - 1)^2(x + 2)$

21. $f(x) = \dfrac{x^2 - 1}{x^2 - 4}$

22. Let $f(x) = x^4 - 3x^3 + 2x^2 + 2x - 4$. Given that $1 + i$ is a zero of f, find all zeros of f.

23. Suppose that y varies as the square root of x and that $y = 6$ when $x = 4$. Find y if $x = 9$.

24. A drug manufactured by a pharmaceutical company is sold in bulk at a price of \$150 per unit. The total production cost (in dollars) for x units in one week is

$$C(x) = 0.02x^2 + 100x + 3000.$$

How many units of the drug must be manufactured and sold in a week to maximize the profit? What is the maximum profit?

25. The profit (in dollars) for a product is given by

$$P(x) = 0.02x^3 + 48.8x^2 - 2990x + 25{,}000,$$

where x is the number of units produced and sold. One break-even point occurs when $x = 10$. Use synthetic division to find another break-even point for the product.

CHAPTER 4

Exponential and Logarithmic Functions

TOPICS

The world of finance, the growth of many populations, musical pitch, the molecular activity that results in an atomic explosion, and countless everyday events are modeled with exponential and logarithmic functions.

SECTION 4.1

Exponential Functions

BEFORE STARTING THIS SECTION, REVIEW

1. Integer exponents (Section P.2, page 18)
2. Rational exponents (Section P.6, page 66)
3. Graphing and transformations (Section 2.7, page 250)
4. One-to-one functions (Section 2.9, page 270)
5. Increasing and decreasing functions (Section 2.5, page 214)

OBJECTIVES

1. Define an exponential function.
2. Graph exponential functions.
3. Develop formulas for simple and compound interest.
4. Understand the number *e*.
5. Graph the natural exponential function.
6. Model using exponential functions.

TABLE 4.1 Grains of Wheat on a Chessboard

Square Number	Grains Placed on this Square
1	$1\ (=2^0)$
2	$2\ (=2^1)$
3	$4\ (=2^2)$
4	$8\ (=2^3)$
5	$16\ (=2^4)$
6	$32\ (=2^5)$
.	.
.	.
.	.
63	2^{62}
64	2^{63}

◆ Fooling a King

One night in northwest India, a wise man named Shashi invented a new game called "Shatranj" (chess). The next morning he took it to King Rai Bhalit, who was so impressed that he said to Shashi, "Name your reward." Shashi merely requested that 1 grain of wheat be placed on the first square of the chessboard, 2 grains on the second, 4 on the third, 8 on the fourth, and so on, for all 64 squares. The king agreed to his request, thinking the man was an eccentric fool for asking for only a few grains of wheat when he could have had gold, jewels, or even his daughter's hand in marriage.

You may know the end of this story. Much more wheat was needed to satisfy Shashi's request. The king could never fulfill his promise to Shashi; instead, he had him beheaded.

The details of this rapidly growing wheat phenomenon are given in the margin. Table 4.1 shows the number of grains of wheat that need to be stacked on each square of the chessboard.

These data can be modeled by the function

$$g(n) = 2^{n-1},$$

where $g(n)$ is the number of grains of wheat and n is the number of the square (or the nth square) on the chessboard. Each square has twice the number of grains as the previous square. The function g is an example of an *exponential function* with base 2, with n restricted to $1, 2, 3, \ldots, 64$—the number of the chessboard squares. The name *exponential function* comes from the fact that the variable n occurs in the exponent. The base 2 is the *growth factor*.

When the growth rate of a quantity is directly proportional to the existing amount, the growth can be modeled by an exponential function. For example, exponential functions are often used to model the growth of investments and populations, the cell division of living organisms, and the decay of radioactive material. Example 12 discusses the growth of the world population.

1 Define an exponential function.

Exponential Functions

Exponential Function

A function f of the form

$$f(x) = a^x, a > 0, \text{ and } a \neq 1,$$

is called an **exponential function with base a and exponent x**. Its domain is $(-\infty, \infty)$.

In the definition of the exponential function, we rule out the base $a = 1$ because in this case, the function is simply the constant function $f(x) = 1^x$, or $f(x) = 1$. We exclude negative bases so that the domain includes all real numbers. For example, a cannot be -2 because $(-2)^{1/2} = \sqrt{-2}$ is not a real number. Other functions with constant base and variables in the exponent, such as $f(x) = 4 \cdot 3^x$, $g(x) = -5 \cdot 4^{3x-2}$, and $h(x) = c \cdot a^x$ ($a > 0$ and $a \neq 1$), are also called exponential functions.

Evaluate Exponential Functions

We can evaluate exponential functions by using the laws of exponents and/or calculators, as in the next example. (See Section P.2 for the definitions and the laws of exponents involving a^x when x is a rational number.)

TECHNOLOGY CONNECTION

You can use either the [^] or [x^y] key on your calculator to evaluate exponential functions. For example, to evaluate $4^{3.2}$, press

4 [^] 3.2 [ENTER]

or

4 [x^y] 3.2 [=]. You will see the following display:

Here the displayed number is an approximate value, so we should write $4^{3.2} \approx 84.44850629$.

EXAMPLE 1 Evaluating Exponential Functions

a. Let $f(x) = 3^{x-2}$. Find $f(4)$.

b. Let $g(x) = -2 \cdot 10^x$. Find $g(-2)$.

c. Let $h(x) = \left(\frac{1}{9}\right)^x$. Find $h\left(-\frac{3}{2}\right)$.

d. Let $F(x) = 4^x$. Find $F(3.2)$.

Solution

a. $f(4) = 3^{4-2} = 3^2 = 9$

b. $g(-2) = -2 \cdot 10^{-2} = -2 \cdot \frac{1}{10^2} = -2 \cdot \frac{1}{100} = -0.02$

c. $h\left(-\frac{3}{2}\right) = \left(\frac{1}{9}\right)^{-\frac{3}{2}} = (9^{-1})^{-\frac{3}{2}} = 9^{\frac{3}{2}} = (\sqrt{9})^3 = 27$

d. $F(3.2) = 4^{3.2} \approx 84.44850629$ Use a calculator.

Practice Problem 1 Let $f(x) = \left(\frac{1}{4}\right)^x$. Find $f(2), f(0), f(-1), f\left(\frac{5}{2}\right)$, and $f\left(-\frac{3}{2}\right)$.

The domain of an exponential function is $(-\infty, \infty)$. In Example 1, we evaluated exponential functions at some rational numbers. But what is the meaning of a^x when x is an irrational number? For example, what do expressions such as $3^{\sqrt{2}}$ and 2^π mean? It turns out that the definition of a^x (with $a > 0$ and x irrational) requires methods discussed in calculus. However, we can understand the basis for the definition from the following discussion. Suppose we want to define the number 2^π. We use several numbers that approximate π. A calculator shows that $\pi \approx 3.14159265\ldots$. We successively approximate 2^π by using the rational powers shown in Table 4.2.

TABLE 4.2 Values of $f(x) = 2^x$ for rational values of x that approach π

x	3	3.1	3.14	3.141	3.1415
2^x	$2^3 = 8$	$2^{3.1} = 8.5\ldots$	$2^{3.14} = 8.81\ldots$	$2^{3.141} = 8.821\ldots$	$2^{3.1415} = 8.8244\ldots$

It can be shown that the powers $2^3, 2^{3.1}, 2^{3.14}, 2^{3.141}, 2^{3.1415}, \ldots$ approach exactly one number. We define 2^π as that number. Table 4.2 shows that $2^\pi \approx 8.82$ (correct to two decimal places).

For our work with exponential functions, we need the following facts:

1. Exponential functions $f(x) = a^x$ are defined for all real numbers x.
2. The graph of an exponential function is a continuous (unbroken) curve.
3. The rules of exponents hold for all real number exponents and positive bases.

RULES OF EXPONENTS

Let a, b, x, and y be real numbers with $a > 0$ and $b > 0$.

$$a^x \cdot a^y = a^{x+y} \qquad (a^x)^y = a^{xy}$$

$$\frac{a^x}{a^y} = a^{x-y} \qquad a^0 = 1$$

$$(ab)^x = a^x b^x \qquad a^{-x} = \frac{1}{a^x} = \left(\frac{1}{a}\right)^x$$

EXAMPLE 2 Using Rules of Exponents

Use the rules of exponents to simplify each expression.

a. $6^{\sqrt{27}} \cdot 6^{\sqrt{12}}$ **b.** $(6^{\sqrt{3}})^{\sqrt{3}}$

Solution

a.
$$\begin{aligned} 6^{\sqrt{27}} \cdot 6^{\sqrt{12}} &= 6^{\sqrt{27}+\sqrt{12}} && a^x \cdot a^y = a^{x+y} \\ &= 6^{3\sqrt{3}+2\sqrt{3}} && \sqrt{27} = \sqrt{9.3} = 3\sqrt{3};\ \sqrt{12} = \sqrt{4 \cdot 3} = 2\sqrt{3} \\ &= 6^{5\sqrt{3}} && 3\sqrt{3} + 2\sqrt{3} = (3 + 2)\sqrt{3} = 5\sqrt{3} \\ &= (6^5)^{\sqrt{3}} \\ &= 7776^{\sqrt{3}} \end{aligned}$$

b.
$$\begin{aligned} (6^{\sqrt{3}})^{\sqrt{3}} &= 6^{\sqrt{3}\cdot\sqrt{3}} && (a^x)^y = a^{x \cdot y} \\ &= 6^3 && \sqrt{3} \cdot \sqrt{3} = \sqrt{3 \cdot 3} = \sqrt{9} = 3 \\ &= 216 \end{aligned}$$

Practice Problem 2 Simplify each expression.

a. $3^{\sqrt{8}} \cdot 3^{\sqrt{2}}$ **b.** $(a^{\sqrt{8}})^{\sqrt{2}}$

2 Graph exponential functions.

Graphing Exponential Functions

Let's see how to sketch the graph of an exponential function. Although the domain is the set of all real numbers, we usually evaluate the functions only for the integer values of x (for ease of computation). To evaluate a^x for noninteger values of x, use your calculator. Recall that the graph of a function f is the graph of the equation $y = f(x)$. We can use either $f(x) = a^x$ or $y = a^x$ to represent a given function.

EXAMPLE 3 Graphing an Exponential Function with Base $a > 1$

Graph the exponential function $f(x) = 3^x$.

Solution

First, make a table of a few values of x and the corresponding values of y.

x	-3	-2	-1	0	1	2	3
$y = 3^x$	$\frac{1}{27}$	$\frac{1}{9}$	$\frac{1}{3}$	1	3	9	27

FIGURE 4.1 The graph of $y = a^x$ with $a > 1$

Next, plot the points and draw a smooth curve through them. See Figure 4.1.

Note that the x-axis is the horizontal asymptote of the graph of $y = a^x$.

Practice Problem 3 Sketch the graph of

$$f(x) = 2^x.$$

We now sketch the graph of an exponential function $f(x) = a^x$ when $0 < a < 1$.

EXAMPLE 4 Graphing an Exponential Function $f(x) = a^x$, with $0 < a < 1$

Sketch the graph of $y = \left(\frac{1}{2}\right)^x$ by making a table of values.

Solution

Make a table similar to the one in Example 3.

x	-3	-2	-1	0	1	2	3
$y = \left(\frac{1}{2}\right)^x$	8	4	2	1	$\frac{1}{2}$	$\frac{1}{4}$	$\frac{1}{8}$

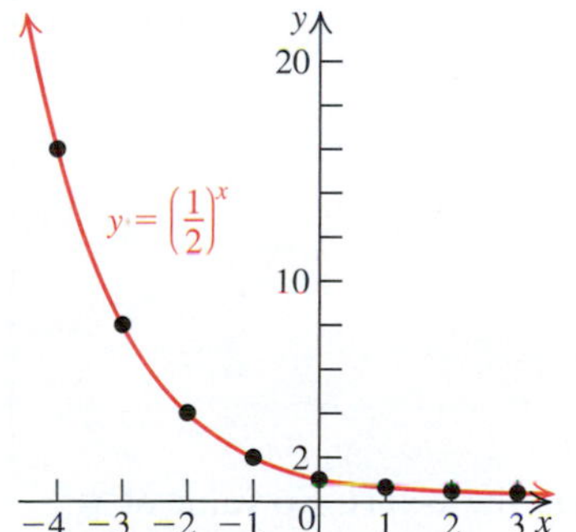

FIGURE 4.2 An exponential function $y = a^x$ with $0 < a < 1$

Plotting these points and drawing a smooth curve through them, we get the graph of $y = \left(\frac{1}{2}\right)^x$, shown in Figure 4.2. In this graph, as x increases in the positive direction, $y = \left(\frac{1}{2}\right)^x$ decreases toward 0. The x-axis is its horizontal asymptote.

Practice Problem 4 Sketch the graph of

$$f(x) = \left(\frac{1}{3}\right)^x.$$

In general, graphs of exponential functions have two basic shapes determined by the base a. If $a > 1$, the graph of $y = a^x$ is rising as in Figure 4.1. If $0 < a < 1$, the graph is falling as in Figure 4.2. To get the correct shape for the graph of $f(x) = a^x$, it is helpful to graph the three points:

$$\left(-1, \frac{1}{a}\right), (0, 1), \text{ and } (1, a).$$

RECALL

The graph of $y = f(-x)$ is the reflection of the graph of $y = f(x)$ about the y-axis.

Because $y = \left(\frac{1}{2}\right)^x = (2^{-1})^x = 2^{-x}$, the graph of $y = \left(\frac{1}{2}\right)^x$ can also be obtained by reflecting the graph of $y = 2^x$ about the y-axis. See Figure 4.3.

In Figure 4.3, we sketch the graphs of four exponential functions whose bases are pairwise reciprocals on the same set of axes. These graphs illustrate some general properties of exponential functions.

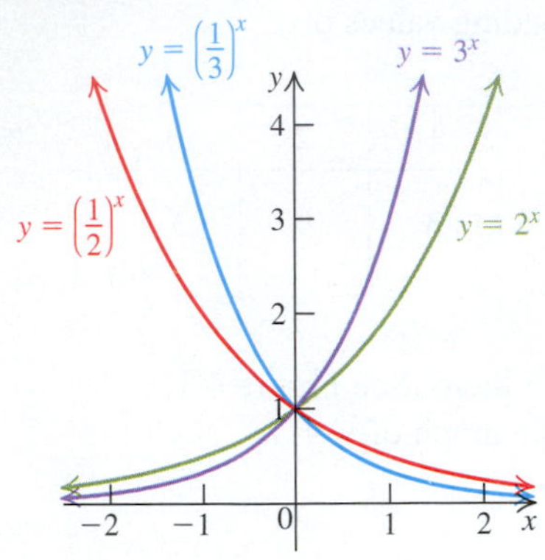

FIGURE 4.3 Graphs of exponential functions with reciprocal bases.

PROPERTIES OF EXPONENTIAL FUNCTIONS

Let $y = f(x) = a^x, a > 0, a \neq 1.$

1. The domain of f is $(-\infty, \infty)$.
2. The range of f is $(0, \infty)$: The entire graph lies above the x-axis.
3. Note that $f(x + 1) = a^{x+1} = a \cdot a^x = af(x)$. This means that the y-values change by a factor of a for each unit increase in x.
4. For $a > 1$, the **growth factor** is a. Because the y-values *increase* by a factor of a for each unit increase in x,
   (i) f is a increasing function. That is, if $x_1 < x_2$ then $a^{x_1} < a^{x_2}$; so the graph rises to the right.
   (ii) $y = a^x$ is $\begin{cases} <1 & \text{if} \quad x < 0 \\ =1 & \text{if} \quad x = 0 \\ >1 & \text{if} \quad x > 0 \end{cases}$
   (iii) As $x \to \infty, y \to \infty$.
   (iv) As $x \to -\infty, y \to 0$.
5. For $0 < a < 1$, the **decay factor** is a. Because the y-values *decrease* by a factor of a for each unit increase in x,
   (i) f is a decreasing function. That is, if $x_1 < x_2$ then $a^{x_1} > a^{x_2}$; so the graph falls to the right.
   (ii) $y = a^x$ is $\begin{cases} >1 & \text{if} \quad x < 0 \\ =1 & \text{if} \quad x = 0 \\ <1 & \text{if} \quad x > 0 \end{cases}$
   (iii) As $x \to -\infty, y \to \infty$.
   (iv) As $x \to \infty, y \to 0$.
6. Each exponential function f is one-to-one. So
   (i) If $a^m = a^n$, then $m = n$.
   (ii) f has an inverse.
7. The graph of f has no x-intercepts, so it never crosses the x-axis. No value of x will cause $f(x) = a^x$ to equal 0.
8. The graph of f is a smooth and continuous curve, and it passes through the points $\left(-1, \frac{1}{a}\right)$, $(0, 1)$, and $(1, a)$.
9. The x-axis is a horizontal asymptote for the graph of every exponential function of the form $f(x) = a^x$.
10. The graph of $y = a^{-x} = \left(\frac{1}{a}\right)^x$ is the reflection about the y-axis of the graph of $y = a^x$.
11. Let $1 < a < b$.
   (i) The graph of $f(x) = a^x$ lies below the graph of $g(x) = b^x$ on the interval $(0, \infty)$. This means that
   $$2^x < 3^x < 4^x < \ldots \text{ for all } x > 0.$$
   (ii) The graph of $f(x) = a^x$ lies above the graph of $g(x) = b^x$ on the interval $(-\infty, 0)$. This means that
   $$2^x > 3^x > 4^x > \ldots \text{ for all } x < 0.$$
   (iii) The graphs of f and g intersect at $(0, 1)$ because $f(0) = g(0) = 1$.

RECALL

A function f is one-to-one if $f(x_1) = f(x_2)$ implies $x_1 = x_2$.

RECALL

A line $y = k$ is a horizontal asymptote for the graph of f if $f(x) \to k$ as $x \to \infty$ or as $x \to -\infty$.

EXAMPLE 5 Finding Exponential Functions

Find the exponential function of the form $f(x) = ca^x$ that satisfies the given conditions.

a. $f(0) = 3$ and $f(2) = 75$.

b. The graph of f contains the points $(-1, 18)$ and $\left(4, \frac{2}{27}\right)$.

Solution

a. Because $f(0) = 3$, we have $3 = c \cdot a^0 = c \cdot 1 = c$. We now write

$$\begin{aligned} f(x) &= 3a^x && c = 3 \\ 75 &= 3a^2 && \text{Given } f(2) = 75 \\ 25 &= a^2 && \text{Divide both sides by 3.} \\ \pm 5 &= a && \text{Solve for } a. \\ a &= 5 && \text{Reject } a = -5 \text{, because the base } a \text{ is positive.} \\ \text{So } f(x) &= 3 \cdot 5^x && \text{Replace } c \text{ with 3 and } a \text{ with 5 in } f(x) = ca^x. \end{aligned}$$

b. The given condition means that $f(-1) = 18$ and $f(4) = 27$.

$$\begin{aligned} 18 &= ca^{-1} && (-1, 18) \text{ is on the graph of } y = ca^x. \\ 18a &= c && \text{Multiply both sides by } a\text{; simplify.} \\ \frac{2}{27} &= ca^4 && \left(4, \frac{2}{27}\right) \text{ is on the graph of } y = ca^x. \\ \frac{2}{27} &= 18aa^4 && \text{Replace } c \text{ with } 18a \text{ in the previous equation.} \\ \frac{2}{27} &= 18a^5 && a \cdot a^4 = a^5 \\ a^5 &= \frac{1}{243} && \text{Divide both sides by 18. Simplify.} \\ a &= \sqrt[5]{\frac{1}{243}} = \frac{1}{3} && \text{Solve for } a. \end{aligned}$$

To find c, we substitute $a = \frac{1}{3}$ in the equation $18a = c$. So

$$18\left(\frac{1}{3}\right) = c \quad \text{or} \quad c = 6.$$

Substituting $c = 6$ and $a = \frac{1}{3}$ in $f(x) = ca^x$, we have

$$f(x) = 6\left(\frac{1}{3}\right)^x.$$

Practice Problem 5 Repeat Example 5b with the graph points:

a. $(0, 1)$ and $(2, 49)$ **b.** $(-2, 16)$ and $\left(3, \frac{1}{2}\right)$

3 Develop formulas for simple and compound interest.

Simple Interest

We first review some terminology concerning interest.

Simple Interest

A fee charged for borrowing a lender's money is the **interest**, denoted by I.

The original, or initial, amount of money borrowed is the **principal**, denoted by P.

The period of time during which the borrower pays back the principal plus the interest is the **time**, denoted by t.

The **interest rate** is the percent charged for the use of the principal for the given period. The interest rate, denoted by r, is expressed as a decimal. Unless stated otherwise, the period is assumed to be one year; that is, r is an *annual* rate.

The amount of interest computed only on the principal is called **simple interest**.

RECALL

$$8\% = 8 \text{ percent} = 8 \text{ per hundred} = \frac{8}{100} = 0.08$$

When money is deposited with a bank, the bank becomes the borrower. For example, depositing \$1000 in an account at 8% interest means that the principal P is \$1000 and the interest rate r is 0.08 for the bank as the borrower.

SIMPLE INTEREST FORMULA

The simple interest I on a principal P at an annual rate of r (expressed as a decimal) per year for t years is

$$I = Prt. \quad (1)$$

EXAMPLE 6 Calculating Simple Interest

Juanita has deposited \$8000 in a bank for five years at a simple interest rate of 6%.

a. How much interest will she receive?

b. How much money will be in her account at the end of five years?

Solution

a. $P = \$8000$, $r = 0.06$, and $t = 5$.

$$\begin{aligned} I &= Prt && \text{Equation (1)} \\ &= \$8000\,(0.06)(5) && \text{Substitute values.} \\ &= \$2400 && \text{Simplify.} \end{aligned}$$

b. In five years, the amount A she will receive is the principal plus the interest earned:

$$\begin{aligned} A &= P + I \\ &= \$8000 + \$2400 \\ &= \$10{,}400 \end{aligned}$$

Practice Problem 6 Find the amount that will be in a bank account if \$10,000 is deposited at a simple interest rate of 7.5% for two years.

Given P and r, the amount $A(t)$ due in t years and calculated at simple interest is found by using the formula

$$A(t) = P + Prt. \quad (2)$$

Equation (2) is a linear function of t. Simple interest problems are examples of **linear growth**, which take place when the growth of a quantity occurs at a constant rate and so can be modeled by a linear function.

Compound Interest

In the real world, simple interest is rarely used for periods of more than one year. Instead, we use **compound interest**—the interest paid on both the principal and the accrued (previously earned) interest.

To illustrate compound interest, suppose \$1000 is deposited in a bank account paying 4% annual interest. At the end of one year, the account will contain the original \$1000 plus the 4% interest earned on the \$1000:

$$\$1000 + (0.04)(\$1000) = \$1040$$

Similarly, if P represents the initial amount deposited at an interest rate r (expressed as a decimal) per year, then the amount A_1 in the account after one year is

$$\begin{aligned} A_1 &= P + rP && \text{Principal } P \text{ plus interest earned on } P \\ &= P(1 + r) && \text{Factor out } P. \end{aligned}$$

During the second year, the account earns interest on the new principal A_1. The amount A_2 in the account after the second year will be equal to A_1 plus the interest on A_1.

$$\begin{aligned} A_2 &= A_1 + rA_1 && A_1 \text{ plus interest earned on } A_1 \\ &= A_1(1 + r) && \text{Factor out } A_1. \\ &= P(1 + r)(1 + r) && A_1 = P(1 + r) \\ &= P(1 + r)^2 \end{aligned}$$

The amount A_3 in the account after the third year is

$$\begin{aligned} A_3 &= A_2 + rA_2 && A_2 \text{ plus interest earned on } A_2 \\ &= A_2(1 + r) && \text{Factor out } A_2. \\ &= P(1 + r)^2(1 + r) && A_2 = P(1 + r)^2 \\ &= P(1 + r)^3 && (1 + r)^2(1 + r) = (1 + r)^3 \end{aligned}$$

In general, the amount A in the account after t years is given by

$$A = P(1 + r)^t. \quad (3)$$

We say that this type of interest is **compounded annually** because it is paid once a year.

EXAMPLE 7 **Calculating Compound Interest**

Juanita deposits \$8000 in a bank at the interest rate of 6% compounded annually for five years.

a. How much money will she have in her account after five years?

b. How much interest will she receive?

Solution

a. Here $P = \$8000$, $r = 0.06$, and $t = 5$; so

$$\begin{aligned} A &= P(1 + r)^t \\ &= \$8000(1 + 0.06)^5 && P = \$8000, r = 0.06, t = 5 \\ &= \$8000(1.06)^5 \\ &= \$10{,}705.80 && \text{Use a calculator.} \end{aligned}$$

b. Interest $= A - P = \$10{,}705.80 - \$8000 = \$2705.80$

Practice Problem 7 Repeat Example 7 assuming that the bank pays 7.5% interest compounded annually.

Comparing Examples 6 and 7, we see that compounding Juanita's interest made her money grow faster. We expect this because the function $A(t) = P(1 + r)^t$ is an exponential function with base $(1 + r)$.

Banks and other financial institutions usually pay savings account interest more than once a year. They pay a smaller amount of interest more frequently. Suppose the quoted annual interest rate r (also called the **nominal rate**) is compounded n times per year (at equal intervals) instead of annually. Then for each period, the interest rate is $\frac{r}{n}$, and there are $n \cdot t$ periods in t years. Accordingly, we can restate the formula $A(t) = P(1 + r)^t$ as follows:

COMPOUND INTEREST FORMULA

$$A = P\left(1 + \frac{r}{n}\right)^{nt} \qquad (4)$$

A = amount after t years
P = principal
r = annual interest rate (expressed as a decimal number)
n = number of times interest is compounded each year
t = number of years

The total amount accumulated after t years, denoted by A, is also called the **future value** of the investment.

EXAMPLE 8 Using Different Compounding Periods to Compare Future Values

If \$100 is deposited in a bank that pays 5% annual interest, find the future value A after one year if the interest is compounded

(i) Annually.
(ii) Semiannually.
(iii) Quarterly.
(iv) Monthly.
(v) Daily.

DO YOU KNOW?

Compounding that occurs 1, 2, 4, 12, and 365 times a year is known as compounding annually, semiannually, quarterly, monthly, and daily, respectively.

Solution

In the following computations, $P = \$100$, $r = 0.05$, and $t = 1$. Only n, the number of times interest is compounded each year, changes. Because $t = 1$, $nt = n(1) = n$.

(i) Annual Compounding: $A = P\left(1 + \frac{r}{n}\right)^{nt}$ $\quad n = 1; t = 1$

$$A = \$100(1 + 0.05) = \$105.00$$

(ii) Semiannual Compounding: $A = P\left(1 + \frac{r}{2}\right)^{2}$ $\quad n = 2; t = 1$

$$A = \$100\left(1 + \frac{0.05}{2}\right)^{2}$$

$$\approx \$105.06 \qquad \text{Use a calculator.}$$

(iii) Quarterly Compounding: $A = P\left(1 + \frac{r}{4}\right)^{4}$ $\quad n = 4; t = 1$

$$A = \$100\left(1 + \frac{0.05}{4}\right)^{4}$$

$$\approx \$105.09 \qquad \text{Use a calculator.}$$

(iv) Monthly Compounding: $A = P\left(1 + \frac{r}{12}\right)^{12}$ $\quad n = 12; t = 1$

$$A = \$100\left(1 + \frac{0.05}{12}\right)^{12}$$

$$\approx \$105.12 \quad \text{Use a calculator.}$$

(v) Daily Compounding: $A = P\left(1 + \frac{r}{365}\right)^{365}$ $\quad n = 365; t = 1$

$$A = \$100\left(1 + \frac{0.05}{365}\right)^{365}$$

$$\approx \$105.13 \quad \text{Use a calculator.}$$

SIDE NOTE

You may make errors such as forgetting parentheses when entering complicated expressions in your calculator. Look at the answer to see whether it is reasonable and makes sense.

Practice Problem 8 Repeat Example 8 assuming that \$5000 is deposited at a 6.5% annual rate.

The next example shows that we can solve equation (4) for the interest rate r.

EXAMPLE 9 Computing Interest Rate

Carmen has \$9000 to invest. She needs \$20,000 at the end of 8 years. If the interest is compounded quarterly, find the rate r needed.

Solution

Have $P = \$9000$, $A = \$20{,}000$, $t = 8$, and $n = 4$. Substituting these values in equation (4), we have

$$\$20{,}000 = \$9000\left(1 + \frac{r}{4}\right)^{4 \cdot 8}$$

or

$$\left(1 + \frac{r}{4}\right)^{32} = \frac{20{,}000}{9000} \quad \text{Rewrite, with } 4 \cdot 8 = 32 \text{ and divide both sides by 9000.}$$

$$= \frac{20}{9} \quad \text{Simplify the right side.}$$

Taking the 32nd root or $\frac{1}{32}$ power of both sides, we have

$$1 + \frac{r}{4} = \left(\frac{20}{9}\right)^{1/32}$$

$$\frac{r}{4} = \left(\frac{20}{9}\right)^{1/32} - 1 \quad \text{Subtract 1 from both sides.}$$

$$r = 4\left[\left(\frac{20}{9}\right)^{1/32} - 1\right] \quad \text{Multiply both sides by 4.}$$

$$\approx 0.1010692264 \quad \text{Use a calculator.}$$

$$\approx 10.107\%$$

Carmen needs an interest rate of about 10.107%.

Check: $9000\left(1 + \frac{0.10107}{4}\right)^{32} = 20{,}000.12073 \approx 20{,}000.$

Practice Problem 9 Repeat Example 9 if the interest is compounded monthly.

4 Understand the number e.

Leonhard Euler (1707–1783)

Leonhard Euler was the son of a Calvinist minister from Switzerland. At 13, Euler entered the University of Basel, pursuing a career in theology. At the university, Euler was tutored by Johann Bernoulli, of the famous Bernoulli family of mathematicians. Euler's interest and skills led him to abandon his theological studies and take up mathematics. In 1741, Euler moved to the Berlin Academy, where he stayed until 1766. He then returned to St. Petersburg, where he remained for the rest of his life.

Euler contributed to many areas of mathematics, including number theory, combinatorics, and analysis, as well as applications to areas such as music and naval architecture. He wrote over 1100 books and papers and left so much unpublished work that it took 47 years after he died for all of his work to be published. During Euler's life, his papers accumulated so quickly that he kept a large pile of articles awaiting publication. The Berlin Academy published the papers on top of this pile, so later results often were published before results they depended on or superseded. The project of publishing his collected works, undertaken by the Swiss Society of Natural Science, is still going on and will require more than 75 volumes.

TECHNOLOGY CONNECTION

Graphs of $y_1 = e$ and $y_2 = \left(1 + \frac{1}{x}\right)^x$

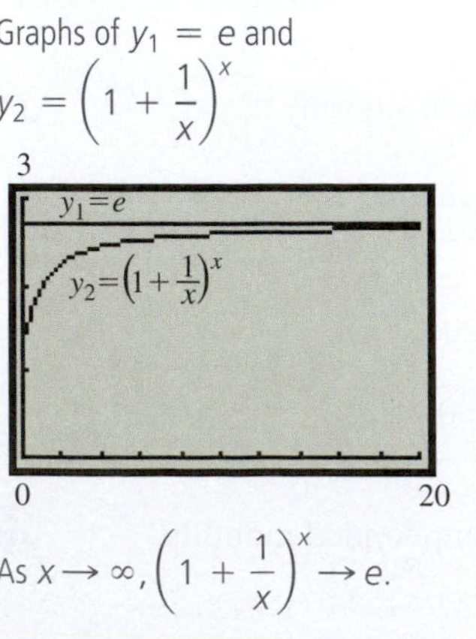

As $x \to \infty$, $\left(1 + \frac{1}{x}\right)^x \to e$.

Continuous Compound Interest Formula

Notice that in Example 8 the future value A increases with n, the number of compounding periods. (Of course P, r, and t are fixed.) The question is: If n increases indefinitely (100, 1000, 10,000 times, and so on), does the amount A also increase indefinitely? Let's see why the answer is "no". Let $h = \frac{n}{r}$. Then we have

$$\begin{aligned} A &= P\left(1 + \frac{r}{n}\right)^{nt} && \text{Equation (4)} \\ &= P\left[\left(1 + \frac{r}{n}\right)^{n/r}\right]^{rt} && nt = \frac{n}{r} \cdot rt \\ &= P\left[\left(1 + \frac{1}{h}\right)^{h}\right]^{rt} && h = \frac{n}{r}, \text{ so } \frac{1}{h} = \frac{r}{n} \end{aligned}$$

Table 4.3 shows the expression $\left(1 + \frac{1}{h}\right)^h$ as h takes on increasingly larger values.

TABLE 4.3

h	$\left(1 + \frac{1}{h}\right)^h$
1	2
2	2.25
10	2.59374
100	2.70481
1000	2.71692
10,000	2.71815
100,000	2.71827
1,000,000	2.71828

Table 4.3 suggests that as h gets larger and larger, $\left(1 + \frac{1}{h}\right)^h$ gets closer and closer to a fixed number. This observation can be proven, and the fixed number is denoted by e in honor of the famous mathematician Leonhard Euler (pronounced "oiler"). The number e, an irrational number, is sometimes called the **Euler number**.

> The value of e to 15 decimal places is
>
> $$e \approx 2.718281828459045.$$

We sometimes write

$$e = \lim_{h \to \infty}\left(1 + \frac{1}{h}\right)^h,$$

which means that when h is very large, $\left(1 + \frac{1}{h}\right)^h$ has a value very close to e. Note that as n gets very large, the quantity $h = \frac{n}{r}$ also gets very large because r is fixed. Therefore, the compounded amount $A = P\left(1 + \frac{r}{n}\right)^{nt} = P\left[\left(1 + \frac{r}{n}\right)^{n/r}\right]^{rt}$ approaches Pe^{rt}.

Continuous Compounding When interest is compounded continuously, the amount A after t years is given by the following formula:

CONTINUOUS COMPOUND INTEREST FORMULA

$$A = Pe^{rt} \quad (5)$$

A = amount after t years
P = principal
r = annual interest rate (expressed as a decimal number)
t = number of years

TECHNOLOGY CONNECTION

Most calculators have the $\boxed{e^x}$ or $\boxed{\text{EXP}}$ key for calculating e^x for a given value of x.

On a TI-83 calculator, to evaluate a number such as $\boxed{e^5}$, you press $\boxed{e^x}$ 5 $\boxed{\text{ENTER}}$ and read the display 148.4131591.

EXAMPLE 10 Calculating Continuous Compound Interest

Find the amount when a principal of \$8300 is invested at a 7.5% annual rate of interest compounded continuously for eight years and three months.

Solution

We use formula (5), with $P = \$8300$ and $r = 0.075$. We convert eight years and three months to 8.25 years.

$$A = \$8300e^{(0.075)(8.25)} \quad \text{Use the formula } A = Pe^{rt}.$$
$$\approx \$15{,}409.83 \quad \text{Use a calculator.}$$

Practice Problem 10 Repeat Example 10 assuming that \$9000 is invested at a 6% annual rate.

5 Graph the natural exponential functions.

The Natural Exponential Function

The exponential function

$$f(x) = e^x$$

with base e is so prevalent in the sciences that it is often referred to as *the* exponential function or the **natural exponential function**. We use a calculator to find e^x to two decimal places for $x = -2, -1, 0, 1,$ and 2 in Table 4.4.

TABLE 4.4

x	e^x
-2	0.14
-1	0.37
0	1
1	2.72
2	7.39

The graph of $f(x) = e^x$ is sketched in Figure 4.4 by using the ordered pairs in Table 4.4.

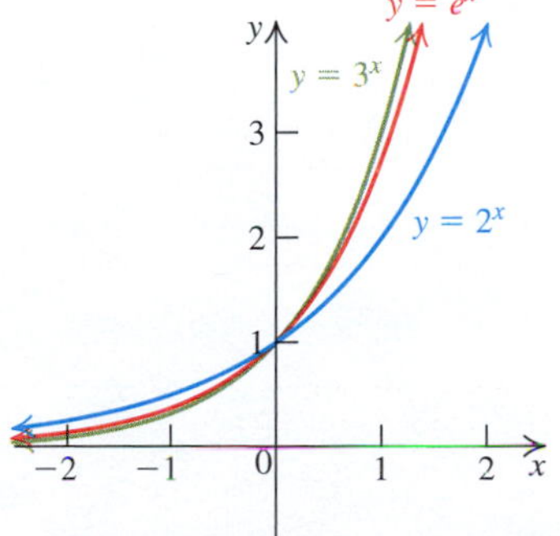

FIGURE 4.4 The graph of $y = e^x$ is between the graphs $y = 2^x$ and $y = 3^x$

DO YOU KNOW?

Transcendental numbers and transcendental functions.

Numbers that are solutions of polynomial equations with rational coefficients are called **algebraic:** -2 is algebraic because it satisfies the equation $x + 2 = 0$, and $\sqrt{3}$ is algebraic because it satisfies the equation $x^2 - 3 = 0$. Numbers that are not algebraic are called **transcendental**, a term coined by Euler to describe numbers such as e and π that appear to "transcend the power of algebraic methods."

There is a somewhat similar distinction between functions. The exponential and logarithmic functions, which are the main subject of this chapter, are transcendental functions.

Because $2 < e < 3$, the graph of $y = e^x$ lies between the graphs $y = 2^x$ and $y = 3^x$. The function $f(x) = e^x$ has all the properties of exponential functions with base $a > 1$ listed on page 396.

We can apply the transformations from Section 2.7 to the natural exponential function.

EXAMPLE 11 Using Transformations on $f(x) = e^x$

Use transformations to sketch the graph of

$$g(x) = e^{x-1} + 2.$$

Solution

We start with the natural exponential function $f(x) = e^x$ and shift its graph 1 unit right to obtain the graph of $y = e^{x-1}$.

We then shift the graph of $y = e^{x-1}$ up 2 units to obtain the graph of $g(x) = e^{x-1} + 2$. See Figure 4.5.

SIDE NOTE

Alternatively, we can rewrite $g(x)$ as:

$$g(x) = \frac{1}{e}e^x + 2$$

so that the horizontal shift can be replaced by shrinking vertically by a factor of $\frac{1}{e}$.

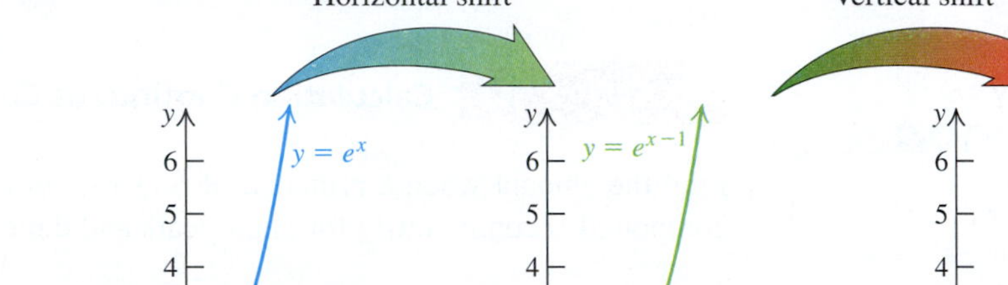

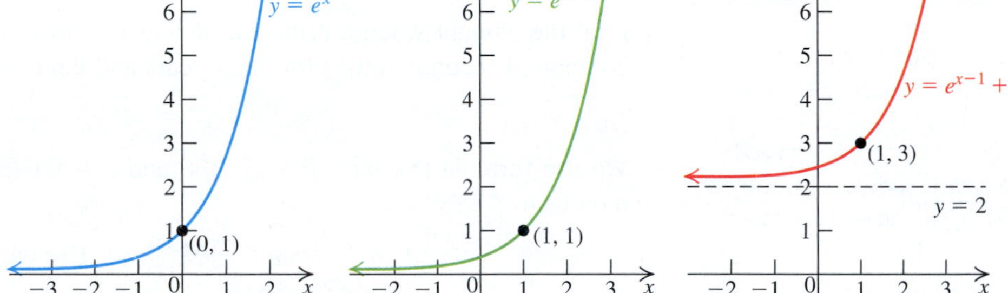

FIGURE 4.5 Graphing $g(x) = e^{x-1} + 2$

Practice Problem 11 Sketch the graph of $g(x) = -e^{x-1} - 2$.

6 Model using exponential functions.

Natural Exponential Growth and Decay

Numerous applications of the exponential function are based on the fact that many different quantities have a growth (or decay) rate proportional to their size. Using calculus, we can show that in such cases, we obtain the following mathematical models:

MODELS FOR NATURAL EXPONENTIAL GROWTH AND DECAY

For $k > 0$,

Exponential growth: $A(t) = A_0 e^{kt}$, and
Exponential decay: $A(t) = A_0 e^{-kt}$, where

$A(t)$ = the amount at time t
$A_0 = A(0)$, the initial amount (the amount at $t = 0$)
k = relative rate of growth or decay
t = time

EXAMPLE 12 Exponential Growth

In the year 2000, the human population of the world was approximately 6.08 billion. Assume the annual rate of growth from 1990 onwards at 1.5%. Using the exponential growth model, estimate the population of the world in the following years.

a. 2030 **b.** 1990

Solution

a. The year 2000 corresponds to $t = 0$. So $A_0 = 6.08$ billion, $k = 0.015,$ and 2030 corresponds to $t = 30$.

$$A(30) = 6.08\, e^{(0.015)(30)} \qquad A(t) = A_0 e^{kt}$$
$$A(30) \approx 9.54 \qquad \text{Use a calculator.}$$

Thus, the model predicts that if the rate of growth is 1.5% per year, over 9.54 billion people will be in the world in 2030.

b. The year 1990 corresponds to $t = -10$ (because 1990 is ten years prior to 2000). We have

$$A(-10) = 6.08\, e^{(0.015)(-10)} \qquad A(t) = A_0 e^{kt}$$
$$\approx 5.23 \qquad \text{Use a calculator.}$$

Thus, the model estimates that the world had over 5.23 billion people in 1990 assuming the growth rate had been 1.5% per year. (The actual population in 1990 was 5.28 billion.)

Practice Problem 12 Repeat Example 12 assuming that the annual rate of growth at 1.6%.

EXAMPLE 13 Exponential Decay

You buy a fishing boat for \$22,000. Your boat depreciates (exponentially) at the annual rate of 15% of its value. Find the depreciated value of your boat at the end of 5 years.

Solution

We use the exponential decay model: $A(t) = A_0 e^{-kt}$, where $A(t)$ represents the value of the boat after t years.

Here $A_0 = 22{,}000$, $k = 0.15$, and $t = 5$. So

$$A(t) = 22{,}000\, e^{-kt}$$
$$A(5) = 22{,}000\, e^{-(0.15)(5)} \qquad \text{Replace } k \text{ with 0.15 and } t \text{ with 5.}$$
$$\approx 10{,}392.06416 \qquad \text{Use a calculator.}$$

So to the nearest cent, the value of your boat at the end of 5 years will be \$10,392.06.

Practice Problem 13 Repeat Example 13 with $k = 18\%$ and $t = 6$ years.

Answers to Practice Problems

1. $f(2) = \frac{1}{16}$
$f(0) = 1$
$f(-1) = 4$
$f\left(\frac{5}{2}\right) = \frac{1}{32}$
$f\left(-\frac{3}{2}\right) = 8$

2. a. $3^{3\sqrt{2}} = 27^{\sqrt{2}}$ **b.** a^4

3.

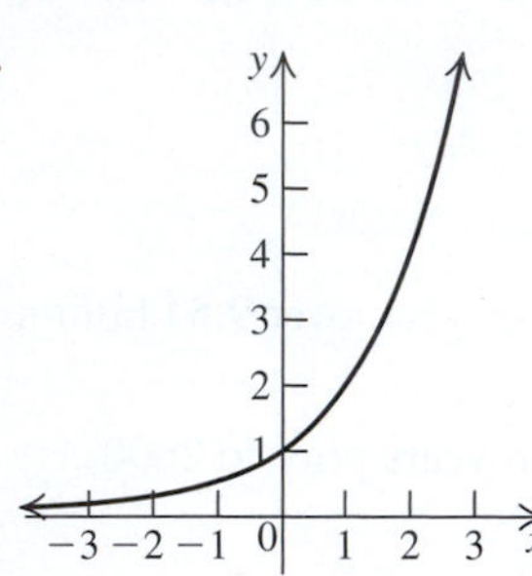

4.

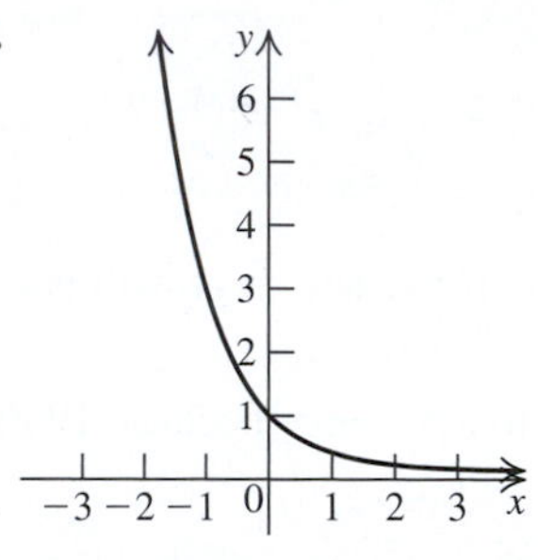

5. a. 7^x **b.** $4\left(\frac{1}{2}\right)^x$ **6.** \$11,500 **7. a.** $A =$ \$11,485.03 **b.** $I =$ \$3,485.03 **8. (i)** \$5325.00 **(ii)** \$5330.28 **(iii)** \$5333.01 **(iv)** \$5334.86 **(v)** \$5335.76 **9.** 10.023% **10.** \$14,764.48

11. $y = -e^{x-1} - 2$

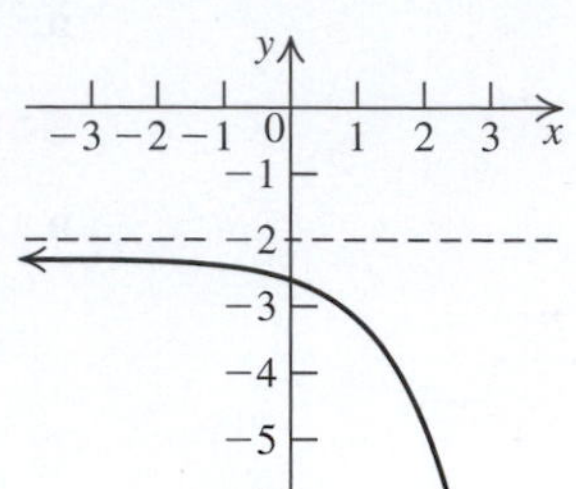

12. a. 9.82 **b.** 5.18 **13.** \$7471.10

SECTION 4.1 Exercises

Basic Concepts and Skills

1. For the exponential function $f(x) = ca^x$, $a > 0$, $a \neq 1$, the domain is ______________ and for $c > 0$, the range is ______________.
2. The graph of $f(x) = 3^x$ has y-intercept ______________ and has ______________ x-intercept.
3. The horizontal asymptote of the graph of $y = \left(\frac{1}{3}\right)^x$ is the ______________.
4. The exponential function $f(x) = a^x$ is increasing if ______________ and is decreasing if ______________.
5. The formula for compound interest at rate r compounded n times per year is $A =$ ______________.
6. The formula for continuous compound interest is $A =$ ______________.
7. **True or False.** The graphs of $y = 2^x$ and $y = \left(\frac{1}{2}\right)^x$ are symmetric with respect to the x-axis.
8. **True or False.** For the equation $y = a^x$ $(a > 0, a \neq 1)$, $y \to \infty$ as $x \to \infty$.
9. **True or False.** Let $y = e^{-x}$, then $y \to 0$ as $x \to \infty$.
10. **True or False.** The graph of $y = e^x + 1$ is obtained by shifting the graph of $y = e^x$ horizontally 1 unit to the left.

In Exercises 11–18, explain whether the given equation defines an exponential function. Give reasons for your answers. Write the base for each exponential function.

11. $y = x^3$
12. $y = 4^x$
13. $y = 2^{-x}$
14. $y = (-5)^x$
15. $y = x^x$
16. $y = 4^2$
17. $y = 0^x$
18. $y = (1.8)^x$

In Exercises 19–24, evaluate each exponential function for the given value(s). (Use a calculator if necessary.)

19. $f(x) = 5^{x-1}, f(0)$
20. $f(x) = -2^{x+1}, f(-2)$
21. $g(x) = 3^{1-x}$; $g(3.2), g(-1.2)$
22. $g(x) = \left(\frac{1}{2}\right)^{x+1}$; $g(2.8), g(-3.5)$
23. $h(x) = \left(\frac{2}{3}\right)^{2x-1}$; $h(1.5), h(-2.5)$
24. $f(x) = 3 - 5^x$; $f(2), f(-1)$

In Exercises 25–32, simplify each expression. Write the answer in the form a^x.

25. $3^{\sqrt{2}} \cdot 3^{\sqrt{2}}$
26. $7^{\sqrt{3}} \cdot 49^{\sqrt{12}}$
27. $8^{\pi} \div 4^{\pi}$
28. $9^{\sqrt{5}} \div 27^{\sqrt{5}}$
29. $(3^{\sqrt{2}})^{\sqrt{3}}$
30. $(2^{\sqrt{3}})^{\sqrt{5}}$
31. $(a^{\sqrt{3}})^{\sqrt{12}}$
32. $a^{\sqrt{2}} \cdot (a^2)^{\sqrt{8}}$

In Exercises 33–36, find the function of the form $f(x) = ca^x$ that contains the two given graph points.

33. a. $(0, 1)$ and $(2, 16)$ **b.** $(0, 1)$ and $\left(-2, \frac{1}{9}\right)$

34. a. $(0, 3)$ and $(2, 12)$ **b.** $(0, 5)$ and $(1, 15)$

35. a. $(1, 1)$ and $(2, 5)$ **b.** $(1, 1)$ and $\left(2, \frac{1}{5}\right)$

36. a. $(1, 5)$ and $(2, 125)$ **b.** $(-1, 4)$ and $(1, 16)$

In Exercises 37–44, sketch the graph of the given function by making a table of values. (Use a calculator if necessary.)

37. $f(x) = 4^x$ **38.** $g(x) = 10^x$

39. $g(x) = \left(\frac{3}{2}\right)^{-x}$ **40.** $h(x) = 7^{-x}$

41. $h(x) = \left(\frac{1}{4}\right)^x$ **42.** $f(x) = \left(\frac{1}{10}\right)^x$

43. $f(x) = (1.3)^{-x}$ **44.** $g(x) = (0.7)^{-x}$

Match each exponential function given in Exercises 45–48 with one of the graphs labeled (a), (b), (c), and (d).

45. $f(x) = 5^x$ **46.** $f(x) = -5^x$

47. $f(x) = 5^{-x}$ **48.** $f(x) = 5^{-x} + 1$

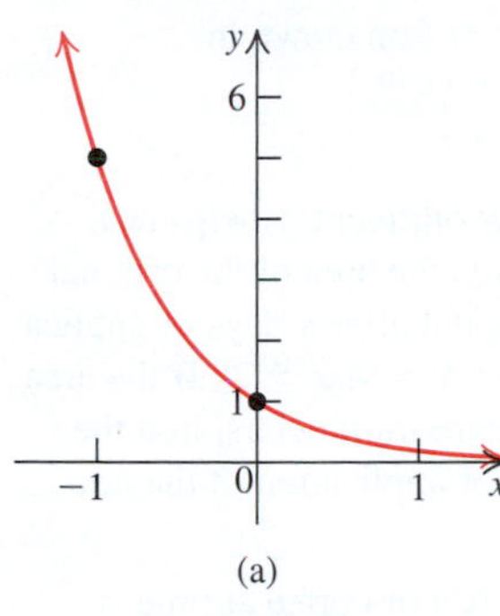
(a)

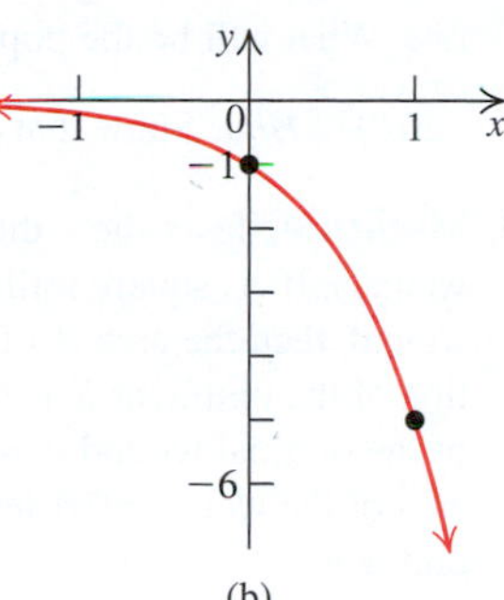
(b)

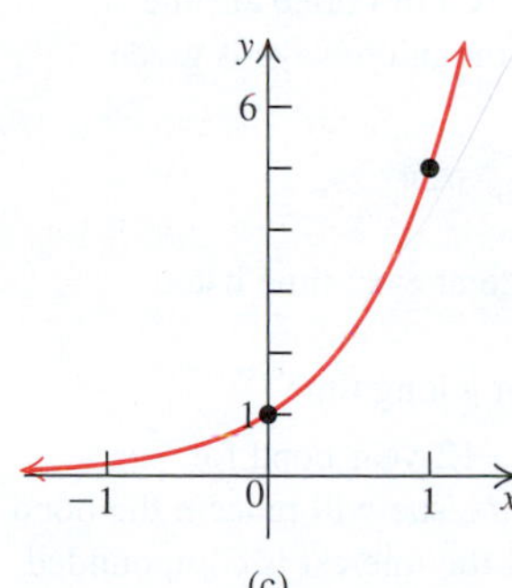
(c)

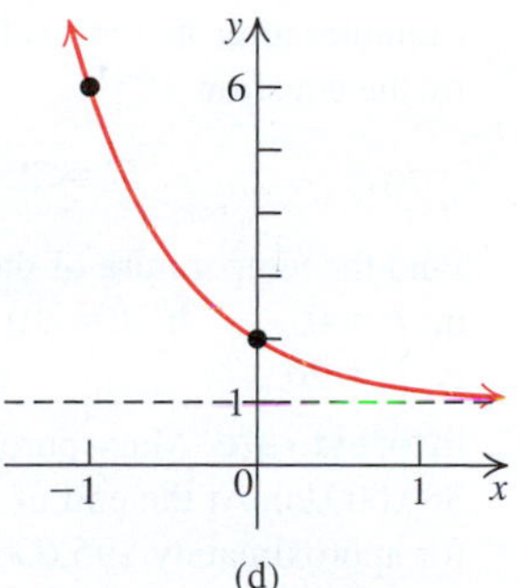
(d)

In Exercises 49–56, start with the graph of the appropriate basic exponential function f and use transformations to sketch the graph of the function g. State the domain and range of g and the horizontal asymptote of its graph.

49. $g(x) = 3^{x-1}$ **50.** $g(x) = 3^x - 1$

51. $g(x) = 4^{-x}$ **52.** $g(x) = -4^x$

53. $g(x) = -2 \cdot 5^{x-1} + 4$ **54.** $g(x) = \frac{1}{2} \cdot 5^{1-x} - 2$

55. $g(x) = -e^{x-2} + 3$ **56.** $g(x) = 3 + e^{2-x}$

In Exercises 57–60, write an equation of the form $f(x) = a^x + b$ from the given graph. Then compute $f(2)$.

57.

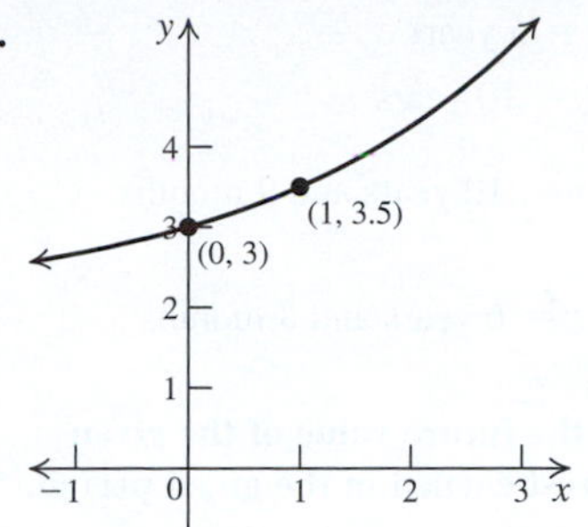

58.

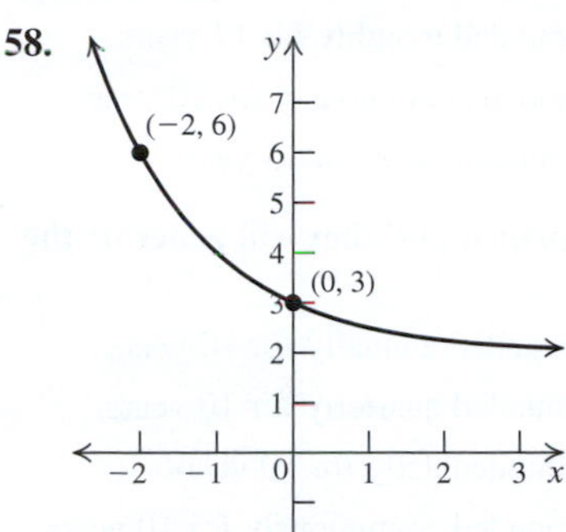

59.

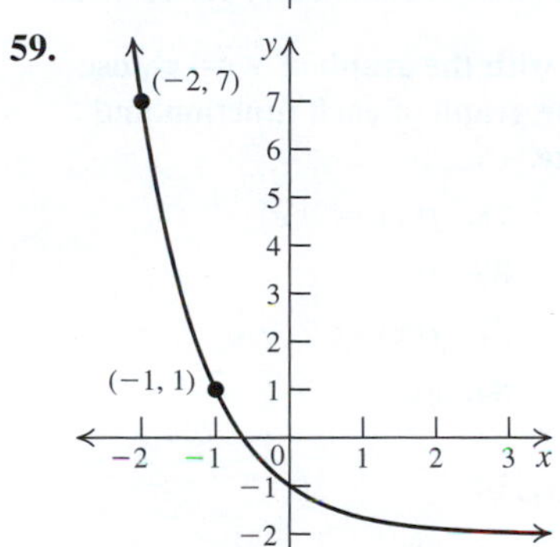

60.

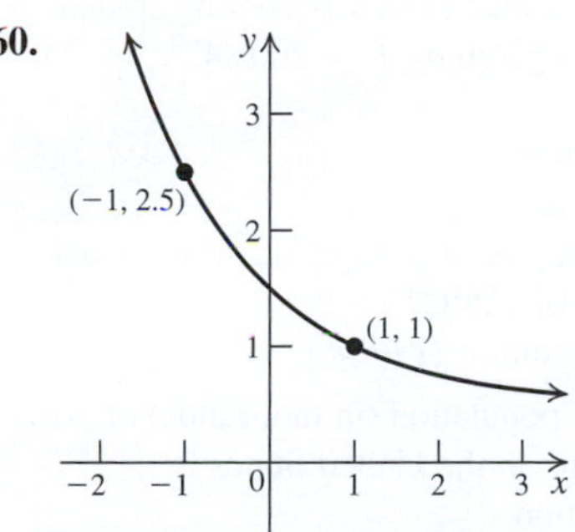

In Exercises 61–64, write an equation of each graph in the final position.

61. The graph of $y = 2^x$ is shifted 2 units left and then 5 units up.

62. The graph of $y = 3^x$ is shifted 3 units right and then is reflected about the y-axis.

63. The graph of $y = \left(\frac{1}{2}\right)^x$ is stretched vertically by a factor of 2 and then is shifted 5 units down.

64. The graph of $y = 2^{-x}$ is reflected about the x-axis and then is shifted 3 units up.

In Exercises 65–68, find the simple interest for each value of principal P, rate r per year, and time t.

65. $P = \$5000$, $r = 10\%$, $t = 5$ years

66. $P = \$10{,}000$, $r = 5\%$, $t = 10$ years

67. $P = \$7800$, $r = 6\frac{7}{8}\%$, $t = 10$ years and 9 months

68. $P = \$8670$, $r = 4\frac{1}{8}\%$, $t = 6$ years and 8 months

In Exercises 69–72, find (a) the future value of the given principal P and (b) the interest earned in the given period.

69. $P = \$3500$ at 6.5% compounded annually for 13 years

70. $P = \$6240$ at 7.5% compounded monthly for 12 years

71. $P = \$7500$ at 5% compounded continuously for 10 years

72. $P = \$8000$ at 6.5% compounded daily for 15 years

In Exercises 73–76, find the principal P that will generate the given future value A.

73. $A = \$10{,}000$ at 8% compounded annually for 10 years

74. $A = \$10{,}000$ at 8% compounded quarterly for 10 years

75. $A = \$10{,}000$ at 8% compounded daily for 10 years

76. $A = \$10{,}000$ at 8% compounded continuously for 10 years

In Exercises 77–84, starting with the graph of $y = e^x$, use transformations to sketch the graph of each function and state its horizontal asymptote.

77. $f(x) = e^{-x}$

78. $f(x) = -e^x$

79. $f(x) = e^{x-2}$

80. $f(x) = e^{2-x}$

81. $f(x) = 1 + e^x$

82. $f(x) = 2 - e^{-x}$

83. $f(x) = -e^{x-2} + 3$

84. $g(x) = 3 + e^{2-x}$

Applying the Concepts

85. Metal cooling. Suppose a metal block is cooling so that its temperature T (in °C) is given by $T = 200 \cdot 4^{-0.1t} + 25$, where t is in hours.

a. Find the temperature after
(i) 2 hours. **(ii)** 3.5 hours.

b. How long has the cooling been taking place if the block now has a temperature of 125°C?

c. Find the eventual temperature $(t \to \infty)$.

86. Ethnic population. The population (in thousands) of people of East Indian origin in the United States is approximated by the function

$$p(t) = 1600(2)^{0.1047t},$$

where t is the number of years since 2010.

a. Find the population of this group in
(i) 2010. **(ii)** 2018.

b. Predict the population in 2025.

87. Price appreciation. In 2010, the median price of a house in Miami was \$190,000. Assuming a rate of increase of 3% per year, what can we expect the price of such a house to be in 2015?

88. Investment. How much should a mother invest at the time her son is born to provide him with \$80,000 at age 21? Assume that the interest is 7% compounded quarterly.

89. Depreciation. Trans Trucking Co. purchased a truck for \$80,000. The company depreciated the truck at the end of each year at the rate of 15% of its current value. What is the value of the truck at the end of the fifth year?

90. Manhattan Island purchase. In 1626, Peter Du Minuit purchased Manhattan Island from the Native Americans for 60 Dutch guilders (about \$24). Suppose the \$24 was invested in 1626 at a 6% rate. How much money would that investment be worth in 2006 if the interest was

a. Simple interest.

b. Compounded annually.

c. Compounded monthly.

d. Compounded continuously.

91. Investment. Ms. Ann Scheiber retired from government service in 1941 with a monthly pension of \$83 and \$5000 in savings. At the time of her death in January 1995 at the age of 101, Ms. Scheiber had turned the \$5000 into \$22 million through shrewd investments in the stock market. She bequeathed all of it to Yeshiva University in New York City. What annual rate of return compounded annually would turn \$5000 to a whopping \$22 million in 54 years?

92. Population. The population of Sometown, USA, was 12,000 in 2000 and grew to 15,000 in 2010. Assume that the population will continue to grow exponentially at the same constant rate. What will be the population of Sometown in 2020? $\left[\textit{Hint: } \text{Show that } e^k = \left(\frac{5}{4}\right)^{1/10}.\right]$

93. Medicine. Tests show that a new ointment X helps heal wounds. If A_0 square millimeters is the area of the original wound, then the area A of the wound after n days of application of the ointment X is given by $A = A_0e^{-0.43n}$. If the area of the original wound was 10 square millimeters, find the area of the wound after ten days of application of the new ointment.

94. Cooling. The temperature T (in °C) of coffee at time t minutes after its removal from the microwave is given by the equation

$$T = 25 + 73e^{-0.28t}.$$

Find the temperature of the coffee at each time listed.

a. $t = 0$ **b.** $t = 10$

c. $t = 20$ **d.** after a long time

95. Interest rate. Mary purchases a 12-year bond for \$60,000.00. At the end of 12 years, she will redeem the bond for approximately \$95,600.00. If the interest is compounded quarterly, what was the interest rate on the bond?

96. Best deal. Fidelity Federal offers three types of investments: (i) 9.7% compounded annually, (ii) 9.6% compounded monthly, and (iii) 9.5% compounded continuously. Which investment is the best deal?

97. Paper stacking. Suppose we have a large sheet of paper 0.015 centimeter thick and we tear the paper in half and put the pieces on top of each other. We keep tearing and stacking in this manner, always tearing each piece in half. How high will the resulting pile of paper be if we continue the process of tearing and stacking

a. 30 times?

b. 40 times?

c. 50 times? [*Hint:* Use a calculator.]

98. Doubling. A jar with a volume of 1000 cubic centimeters contains bacteria that doubles in number every minute. If the jar is full in 60 minutes, how long did it take for the container to be half full?

Beyond the Basics

99. Let $f(x) = e^x$. Show that

a. $\dfrac{f(x+h) - f(x)}{h} = e^x\left(\dfrac{e^h - 1}{h}\right)$.

b. $f(x+y) = f(x)f(y)$.

c. $f(-x) = \dfrac{1}{f(x)}$.

100. Let $f(x) = 3^x + 3^{-x}$ and $g(x) = 3^x - 3^{-x}$. Find each of the following:

a. $f(x) + g(x)$

b. $f(x) - g(x)$

c. $[f(x)]^2 - [g(x)]^2$

d. $[f(x)]^2 + [g(x)]^2$

101. You need the following definition for this exercise: Let $0! = 1$; $1! = 1$; $2! = 2 \cdot 1$; and, in general, $n! = n(n-1)(n-2) \cdots 3 \cdot 2 \cdot 1$. The symbol $n!$ is read "n factorial." Now let

$$S_n = 2 + \frac{1}{2!} + \frac{1}{3!} + \frac{1}{4!} + \cdots + \frac{1}{n!}.$$

Compute S_n for $n = 5$, 10, and 15. Compare the resulting values with the value of e on page 402.

102. The **hyperbolic cosine function** (abbreviated as **cosh *x***) and hyperbolic sine function (abbreviated as **sinh *x***) are defined by $\cosh x = \dfrac{e^x + e^{-x}}{2}$ and $\sinh x = \dfrac{e^x - e^{-x}}{2}$.

a. Is $\cosh x$ odd, even, or neither?

b. Is $\sinh x$ odd, even, or neither?

c. Show that $\cosh x + \sinh x = e^x$.

d. Show that $(\cosh x)^2 - (\sinh x)^2 = 1$.

103. If an investment of P dollars returns A dollars after one year, the **effective annual interest rate** or **annual yield** y is defined by the equation

$$A = P(1 + y).$$

Show that if the sum of P dollars is invested at a nominal rate r per year, compounded m times per year, the effective annual yield y is given by the equation

$$y = \left(1 + \frac{r}{m}\right)^m - 1.$$

104. The relative rate of growth of $A(t) = Pe^{kt}$ in the interval $[t, t+h]$ is defined by $\dfrac{A(t+h) - A(t)}{A(t)}$.

a. Show that $\dfrac{A(t+1) - A(t)}{A(t)} = e^k - 1$.

b. Show that $\dfrac{A(t+h) - A(t)}{A(t)} = e^{kh} - 1$.

In Exercises 105–108, describe an order in which a sequence of transformations is applied to the graph of $y = e^x$ to obtain the graph of the given equation.

105. $y = 3e^{2x-1}$

106. $y = -2e^{3x+2}$

107. $y = 5e^{2-3x} + 4$

108. $y = -2e^{3-4x} - 1$

Critical Thinking/Discussion/Writing

109. Discuss why we do not allow a to be 1, 0, or a negative number in the definition of an exponential function of the form $f(x) = a^x$.

110. Consider the function $g(x) = (-3)^x$, which is not an exponential function. What are the possible rational numbers x for which $g(x)$ is defined.

111. Discuss the end behavior of the graph of $y = \dfrac{b}{1 + ca^x}$, where $a, b, c > 0$ and $a \neq 1$. (Consider the cases $a < 1$ and $a > 1$.)

112. Write a summary of the types of graphs (with sketches) for the exponential functions of the form $y = ca^x$, $c \neq 0$, $a > 0$, and $a \neq 1$.

113. Give a convincing argument to show that the equation $2^x = k$ has exactly one solution for every $k > 0$. Support your argument with graphs.

114. Find all solutions of the equation $2^x = 2x$. How do you know you found all solutions?

Maintaining Skills

In Exercises 115–122, simplify each expression.

115. 10^0

116. 10^{-1}

117. $(-8)^{1/3}$

118. $(25)^{1/2}$

119. $\left(\dfrac{1}{7}\right)^{-2}$

120. $\left(\dfrac{1}{9}\right)^{-1/2}$

121. $(18)^{1/2}$

122. $\left(\dfrac{1}{12}\right)^{-1/2}$

In Exercises 123–128, find the inverse of each function. Then find the domain and the range of f^{-1}.

123. $f(x) = 3x + 4$

124. $f(x) = \dfrac{1}{2}x - 5$

125. $f(x) = \sqrt{x}, x \geq 0$

126. $f(x) = x^2 + 1, x \geq 0$

127. $f(x) = \dfrac{1}{x - 1}$

128. $f(x) = \sqrt[3]{x}$

SECTION 4.2

Logarithmic Functions

BEFORE STARTING THIS SECTION, REVIEW

1. Finding the inverse of a one-to-one function (Section 2.9, page 276)
2. Polynomial and rational inequalities (Section 1.5)
3. Graphing and transformations (Section 2.7)

OBJECTIVES

1. Define logarithmic functions.
2. Evaluate logarithms.
3. Derive basic properties of logarithmic functions.
4. Find the domains of logarithmic functions.
5. Graph logarithmic functions.
6. Use natural logarithms in applications.

◆ Detecting Possible Fraud

Listed below is the size (in square kilometers) of the first 50 countries of the world in two columns. One of the columns contains the real data, and the other column contains fictitious data. Can you guess which column contains the real data? In Example 8, we show how to detect the real data.

Country	Size I	Size II	Country	Size I	Size II
Afghanistan	608,239	645,807	Bermuda	54	53
Albania	29,562	28,748	Bhutan	47,560	46,650
Algeria	2,481,743	2,381,741	Bolivia	998,582	1,098,581
American Samoa	201	197	Bosnia and Herzegovina	49,857	51,129
Andorra	465	464	Botswana	601,780	581,730
Angola	997,852	1,246,700	Bouvet Island	48	49
Anguilla	98	96	Brazil	8,645,521	8,544,418
Antarctica	15,207,012	13,209,000	British Virgin Islands	201	151
Antigua & Barbuda	445	442	Brunei Darussalam	6,123	5,765
Antilles, Netherlands	801	800	Bulgaria	99,949	110,994
Arabia, Saudi	2,250,790	2,149,690	Burkina Faso	269,013	267,950
Argentina	2,875,510	2,777,409	Burundi	27,735	27,834
Armenia	29,851	29,743	Cambodia	201,137	181,035
Aruba	201	193	Cameroon	485,447	475,442
Australia	7,801,663	7,682,557	Canada	9,976,231	9,976,137
Austria	83,760	83,858	Cape Verde	4,121	4,033
Azerbaijan	86,640	86,530	Cayman Islands	260	259
Bahamas	20,112	13,962	Central African Republic	623,541	622,436
Bahrain	695	694	Chad	999,494	1,284,000
Bangladesh	99,763	142,615	Chile	756,589	755,482
Barbados	423	431	China	9,806,441	9,806,391
Belarus	207,599	207,600	Christmas Island	241	135
Belgium	31,618	30,518	Cocos (Keeling) Islands	21	14
Belize	23,105	22,966	Colombia	986,779	1,141,748
Benin	99,879	112,622	Comoros	2,001	1,862

1 Define logarithmic functions.

Logarithmic Functions

Recall from Section 4.1 that every exponential function

$$f(x) = a^x, a > 0, a \neq 1,$$

is a *one-to-one* function and therefore has an inverse function. Let's find the inverse of the function $f(x) = 3^x$. If we think of an exponential function f as "putting an exponent on" a particular base, then the inverse function f^{-1} must "lift the exponent off" to undo the effect of f, as illustrated below:

$$x \xrightarrow{f} 3^x \xrightarrow{f^{-1}} x$$

Let's try to find the inverse of $f(x) = 3^x$ by the procedure outlined in Section 2.9.

Step 1 Replace $f(x)$ in the equation $f(x) = 3^x$ with y: $y = 3^x$.

Step 2 Interchange x and y in the equation in Step 1: $x = 3^y$.

Step 3 Solve the equation for y in Step 2.

In Step 3, we need to solve the equation $x = 3^y$ for y. Finding y is easy when $3 = 3^y$ or $27 = 3^y$, but what if $5 = 3^y$? There actually is a real number y that solves the equation, $5 = 3^y$, but we have not introduced the name for this number yet. (We were in a similar position concerning the solution of $y^3 = 7$ before we introduced the name for the solution, namely, $\sqrt[3]{7}$.) We can say that y is "the exponent on 3 that gives 5." Instead, we use the word *logarithm* (or *log* for short) to describe this exponent. We use the notation $\log_3 5$ (read as "log of 5 with base 3") to express the solution of the equation $5 = 3^y$. This log notation is a new way to write information about exponents: $\boldsymbol{y = \log_3 x}$ means $\boldsymbol{x = 3^y}$. In other words, $\log_3 x$ *is the exponent to which 3 must be raised to obtain x*. The inverse function of the exponential function $f(x) = 3^x$ is the *logarithmic function* with base 3. So

$$f^{-1}(x) = \log_3 x \quad \text{or} \quad y = \log_3 x, \text{ where } y = f^{-1}(x).$$

The previous discussion could be repeated for any base a instead of 3 provided that $a > 0$ and $a \neq 1$.

Logarithmic Function

For $x > 0$, $a > 0$, and $a \neq 1$,

$$y = \log_a x \quad \text{if and only if} \quad x = a^y.$$

The function $f(x) = \log_a x$ is called the **logarithmic function with base a.**

The definition of the logarithmic function says that the two equations

$$y = \log_a x \ (\textbf{logarithmic form}) \text{ and}$$
$$x = a^y \quad (\textbf{exponential form})$$

are equivalent. For instance, $\log_3 81 = 4$ because 3 must be raised to the fourth power to obtain 81.

EXAMPLE 1 Converting from Exponential to Logarithmic Form

Write each exponential equation in logarithmic form.

a. $4^3 = 64$ **b.** $\left(\frac{1}{2}\right)^4 = \frac{1}{16}$ **c.** $a^{-2} = 7$

Solution

a. $4^3 = 64$ is equivalent to $\log_4 64 = 3$.

b. $\left(\frac{1}{2}\right)^4 = \frac{1}{16}$ is equivalent to $\log_{1/2} \frac{1}{16} = 4$.

c. $a^{-2} = 7$ is equivalent to $\log_a 7 = -2$.

SIDE NOTE

To change from exponential form to logarithmic form, use the pattern

$a^u = v$ means $\log_a v = u$.

The exponent u is the logarithm.

Practice Problem 1 Write each exponential equation in logarithmic form.

a. $2^{10} = 1024$ **b.** $9^{-1/2} = \frac{1}{3}$ **c.** $p = a^q$

EXAMPLE 2 Converting from Logarithmic Form to Exponential Form

SIDE NOTE

To change from logarithmic form to exponential form, use

$\log_a v = u$ means $v = a^u$.

Write each logarithmic equation in exponential form.

a. $\log_3 243 = 5$ **b.** $\log_2 5 = x$ **c.** $\log_a N = x$

Solution

a. $\log_3 243 = 5$ is equivalent to $243 = 3^5$.

b. $\log_2 5 = x$ is equivalent to $5 = 2^x$.

c. $\log_a N = x$ is equivalent to $N = a^x$.

Practice Problem 2 Write each logarithmic equation in exponential form.

a. $\log_2 64 = 6$ **b.** $\log_v u = w$

2 Evaluate logarithms.

Evaluating Logarithms

The technique of converting from logarithmic form to exponential form can be used to evaluate some logarithms by inspection.

EXAMPLE 3 Evaluating Logarithms

Find the value of each of the following logarithms.

a. $\log_5 25$ **b.** $\log_2 16$ **c.** $\log_{1/3} 9$

d. $\log_7 7$ **e.** $\log_6 1$ **f.** $\log_4 \frac{1}{2}$

Solution

Logarithmic Form	Exponential Form	Value
a. $\log_5 25 = y$	$25 = 5^y$ or $5^2 = 5^y$	$y = 2$
b. $\log_2 16 = y$	$16 = 2^y$ or $2^4 = 2^y$	$y = 4$
c. $\log_{1/3} 9 = y$	$9 = \left(\frac{1}{3}\right)^y$ or $3^2 = 3^{-y}$	$y = -2$
d. $\log_7 7 = y$	$7 = 7^y$ or $7^1 = 7^y$	$y = 1$
e. $\log_6 1 = y$	$1 = 6^y$ or $6^0 = 6^y$	$y = 0$
f. $\log_4 \frac{1}{2} = y$	$\frac{1}{2} = 4^y$ or $2^{-1} = 2^{2y}$	$y = -\frac{1}{2}$

Practice Problem 3 Evaluate.

a. $\log_3 9$ **b.** $\log_9 \frac{1}{3}$ **c.** $\log_{1/2} 32$

3 Derive basic properties of logarithmic functions.

Basic Properties of Logarithms

We use the definition of the logarithm as an exponent to derive two basic properties for $a > 0$ and $a \neq 1$,

1. Because $a^1 = a$, we have $\log_a a = 1$.
2. Because $a^0 = 1$, we have $\log_a 1 = 0$.

The next two properties follow from the fact that a logarithmic function is the inverse of the appropriate exponential function.

RECALL

If f^{-1} is the inverse of f, then

$$f^{-1}(f(x)) = x$$

and

$$f(f^{-1}(x)) = x.$$

Let $f(x) = a^x$ and $f^{-1}(x) = \log_a x$; then

3. $x = f^{-1}(f(x)) = f^{-1}(a^x) = \log_a a^x$

and

4. $x = f(f^{-1}(x)) = f(\log_a x) = a^{\log_a x}$.

We summarize these basic properties of logarithms.

Basic Properties of Logarithms

For any base $a > 0$, with $a \neq 1$,

1. $\log_a a = 1$.
2. $\log_a 1 = 0$.
3. $\log_a a^x = x$ for any real number x.
4. $a^{\log_a x} = x$ for any $x > 0$.

EXAMPLE 4 **Using Basic Properties of Logarithms**

Evaluate.

a. $\log_3 3$ **b.** $5^{\log_5 7}$

Solution

a. Because $\log_a a = 1$ (property (1)), we have $\log_3 3 = 1$.

b. Because $a^{\log_a x} = x$ (property (4)), we have $5^{\log_5 7} = 7$.

Practice Problem 4 Evaluate. **a.** $\log_5 1$ **b.** $\log_3 3^5$ **c.** $7^{\log_7 5}$ 

4 Find the domains of logarithmic functions.

Domains of Logarithmic Functions

Because the exponential function $f(x) = a^x$ has domain $(-\infty, \infty)$ and range $(0, \infty)$, its inverse function **$y = \log_a x$ has domain $(0, \infty)$ and range $(-\infty, \infty)$**. Therefore, the logarithms of 0 and of negative numbers are not defined; so expressions such as $\log_a(-2)$ and $\log_a(0)$ are meaningless.

EXAMPLE 5 **Finding the Domain**

Find the domain.

a. $f(x) = \log_3(2 - x)$. **b.** $g(x) = \log_2\left(\dfrac{x+1}{x-3}\right)$

Solution

a. Because the domain of a logarithmic function is $(0, \infty)$, the expression $(2 - x)$ must be positive. The domain of f is the set of all real numbers x, where

$$2 - x > 0$$

$$2 > x \quad \text{Solve for } x.$$

So, the domain of f is $(-\infty, 2)$.

b. The domain of g consists of all real numbers x for which $\dfrac{x+1}{x-3} > 0$. We can solve this inequality using the test-point method (page 136).

Sign of $\dfrac{x+1}{x-3}$ ++ ++ | − − − − − − | + + + + +

-1, 3; test points: -2: $\dfrac{(-)}{(-)} = +$; 0: $\dfrac{(+)}{(-)} = -$; 4: $\dfrac{(+)}{(+)} = +$

FIGURE 4.6 Solving $\dfrac{x+1}{x-3} > 0$ by the test-point method.

From Figure 4.6, we see that $\dfrac{x+1}{x-3} > 0$ on $(-\infty, -1) \cup (3, \infty)$. So the domain of g is $(-\infty, -1) \cup (3, \infty)$.

Practice Problem 5 Find the domain of $f(x) = \log_{10}\sqrt{1-x}$.

5 Graph logarithmic functions.

Graphs of Logarithmic Functions

We now look at other properties of logarithmic functions and begin with an example showing how to graph a logarithmic function.

EXAMPLE 6 **Sketching a Graph**

Sketch the graph of $y = \log_3 x$.

RECALL

Remember that "$\log_3 x$" means the exponent on 3 that gives x.

Plotting points (Method 1) To find selected ordered pairs on the graph of $y = \log_3 x$, we choose the x-values to be powers of 3. We can easily compute the logarithms of these values by using Property 3, namely, $\log_3 3^x = x$. We compute the y values in Table 4.5.

TABLE 4.5

x	$y = \log_3 x$	(x, y)	
$\frac{1}{27} = 3^{-3}$	$y = \log_3 \frac{1}{27} = \log_3 3^{-3} = -3$	$\left(\frac{1}{27}, -3\right)$	
$\frac{1}{9} = 3^{-2}$	$y = \log_3 \frac{1}{9} = \log_3 3^{-2} = -2$	$\left(\frac{1}{9}, -2\right)$	
$\frac{1}{3} = 3^{-1}$	$y = \log_3 \frac{1}{3} = \log_3 3^{-1} = -1$	$\left(\frac{1}{3}, -1\right)$	
$1 = 3^0$	$y = \log_3 1 = 0$	$(1, 0)$	Property (2)
$3 = 3^1$	$y = \log_3 3 = 1$	$(3, 1)$	Property (3)
$9 = 3^2$	$y = \log_3 9 = \log_3 3^2 = 2$	$(9, 2)$	

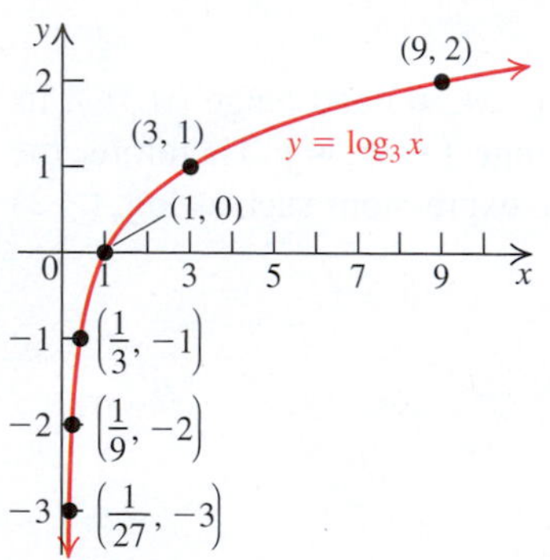

FIGURE 4.7 Graph $y = \log_3 x$ by plotting points

Plotting these ordered pairs and connecting them with a smooth curve gives us the graph of $y = \log_3 x$, shown in Figure 4.7.

Using the inverse function (Method 2) Because $y = \log_3 x$ is the inverse of the function $y = 3^x$, we first graph the exponential function $y = 3^x$, then reflect the graph of $y = 3^x$ about the line $y = x$. Both graphs are shown in Figure 4.8. Note that the horizontal asymptote $y = 0$ of the graph of $y = 3^x$ is reflected as the vertical asymptote $x = 0$ for the graph of $y = \log_3 x$.

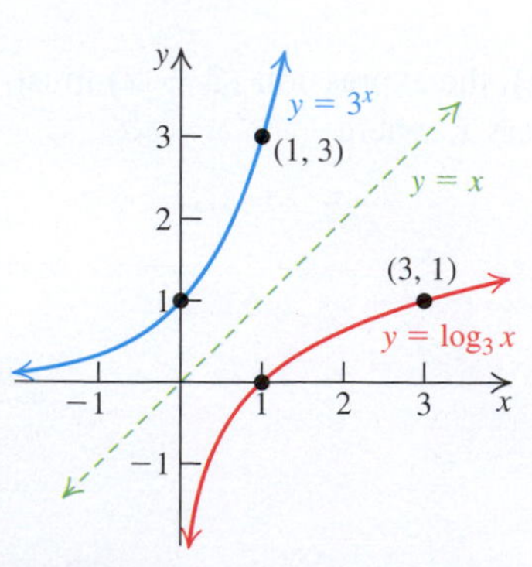

FIGURE 4.8 Graph $y = \log_3 x$ by using the inverse function $y = 3^x$

Practice Problem 6 Sketch the graph of

$$y = \log_2 x.$$

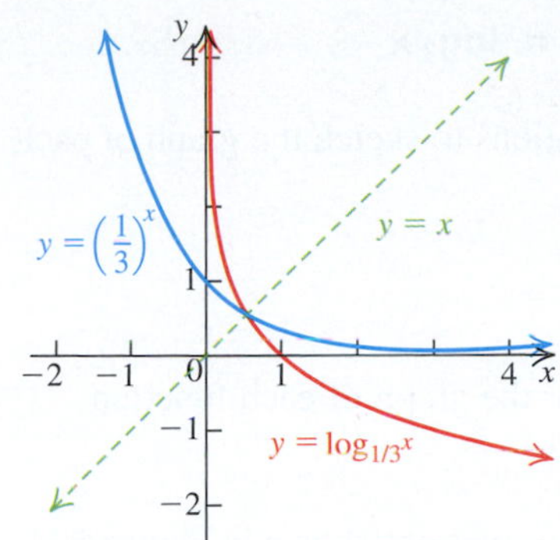

FIGURE 4.9 Graph of $y = \log_{1/3} x$ by using its inverse $y = \left(\frac{1}{3}\right)^x$

We can graph $y = \log_{1/3} x$ either by plotting points or by using the graph of its inverse. See Figure 4.9. The graph of $y = \log_3 x$ shown in Figure 4.8 is rising. The graph of $y = \log_{1/3} x$ shown in Figure 4.9, on the other hand, is falling.

In general, graphs of logarithmic functions have two basic shapes, determined by the base a. See Figure 4.10.

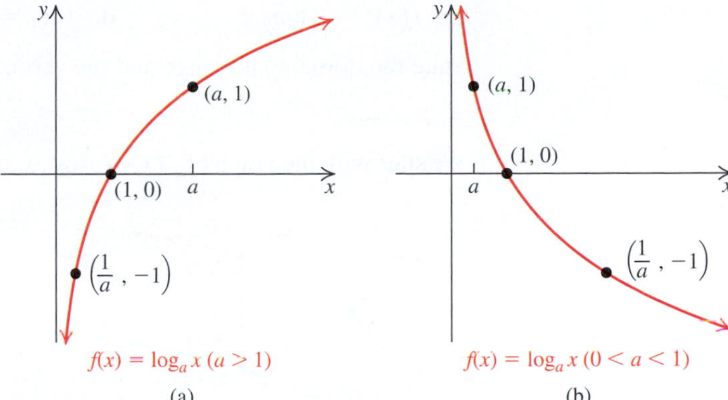

FIGURE 4.10 Basic shapes of $y = \log_a x$

SIDE **NOTE**

To get the correct shape for the graph of a logarithmic function $y = \log_a x$, it is helpful to graph the three points

$$(a, 1), (1, 0), \text{ and } \left(\frac{1}{a}, -1\right).$$

1. For $a > 1$:

(i) Figure 4.10(a) represents a typical shape for the graph of $y = \log_a x$.

(ii) $f(x) = \log_a x$ is an increasing function. That is, if $0 < x_1 < x_2$, then $\log_a x_1 < \log_a x_2$.

(iii) The values of $y = \log_a x$ are $\begin{cases} < 0 & \text{for} \quad 0 < x < 1 \\ = 0 & \text{for} \quad x = 1. \\ > 0 & \text{for} \quad x > 1 \end{cases}$

2. For $0 < a < 1$:

(i) Figure 4.10(b) represents a typical shape for the graph of $y = \log_a x$.

(ii) $f(x) = \log_a x$ is a decreasing function. That is, if $0 < x_1 < x_2$, then $\log_a x_1 > \log_a x_2$.

(iii) The values of $y = \log_a x$ are $\begin{cases} > 0 & \text{for} \quad 0 < x < 1 \\ = 0 & \text{for} \quad x = 1. \\ < 0 & \text{for} \quad x > 1 \end{cases}$

Next, we summarize important properties of the exponential and logarithmic functions.

Properties of Exponential and Logarithmic Functions

Exponential Function $y = a^x$	**Logarithmic Function $y = \log_a x$**
1. The domain is $(-\infty, \infty)$, and the range is $(0, \infty)$.	The domain is $(0, \infty)$, and the range is $(-\infty, \infty)$.
2. The y-intercept is 1, and there is no x-intercept.	The x-intercept is 1, and there is no y-intercept.
3. The x-axis $(y = 0)$ is the horizontal asymptote.	The y-axis $(x = 0)$ is the vertical asymptote.
4. The function is one-to-one; that is, $a^u = a^v$ if and only if $u = v$.	The function is one-to-one; that is, $\log_a u = \log_a v$ if and only if $u = v$.
5. The function is increasing if $a > 1$ and decreasing if $0 < a < 1$.	The function is increasing if $a > 1$ and decreasing if $0 < a < 1$.

Once we know the shape of a logarithmic graph, we can shift it vertically or horizontally, stretch it, compress it, check answers with it, and interpret the graph.

EXAMPLE 7 Using Transformations on $f(x) = \log_3 x$

Start with the graph of $f(x) = \log_3 x$ and use transformations to sketch the graph of each function.

a. $f(x) = \log_3 x + 2$ **b.** $f(x) = \log_3 (x - 1)$

c. $f(x) = -\log_3 x$ **d.** $f(x) = \log_3 (-x)$

State the domain and range and the vertical asymptote for the graph of each function.

Solution

We start with the graph of $f(x) = \log_3 x$ and use the transformations shown in Figure 4.11.

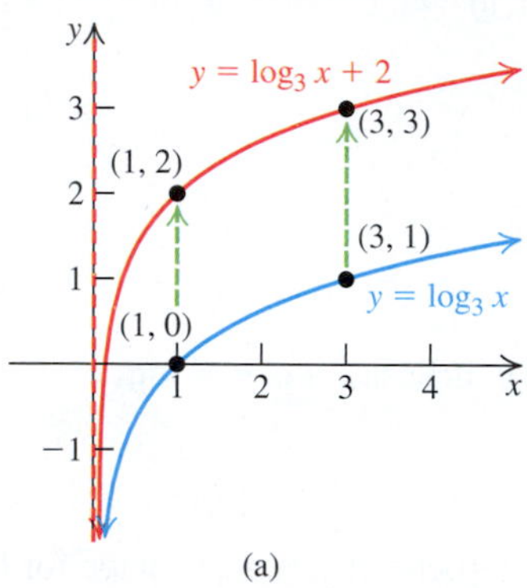

(a)

a. $f(x) = \log_3 x + 2$

Adding 2 to $\log_3 x$ shifts the graph 2 units up. Domain: $(0, \infty)$; range: $(-\infty, \infty)$; vertical asymptote: $x = 0$ (y-axis)

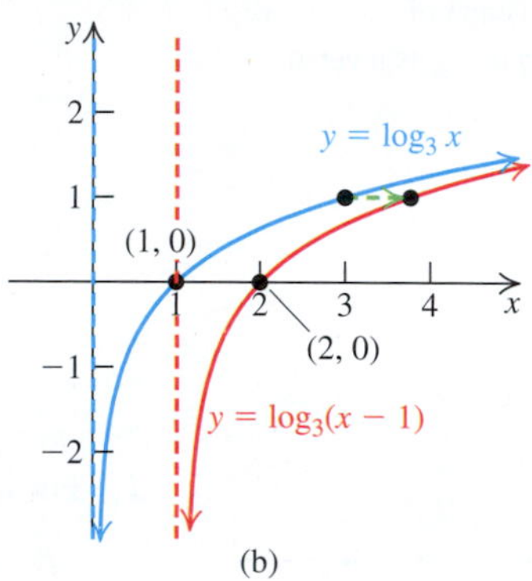

(b)

b. $f(x) = \log_3 (x - 1)$

Replacing x with $x - 1$ in $\log_3 x$ shifts the graph 1 unit right.
Domain: $(1, \infty)$; range $(-\infty, \infty)$; vertical asymptote: $x = 1$

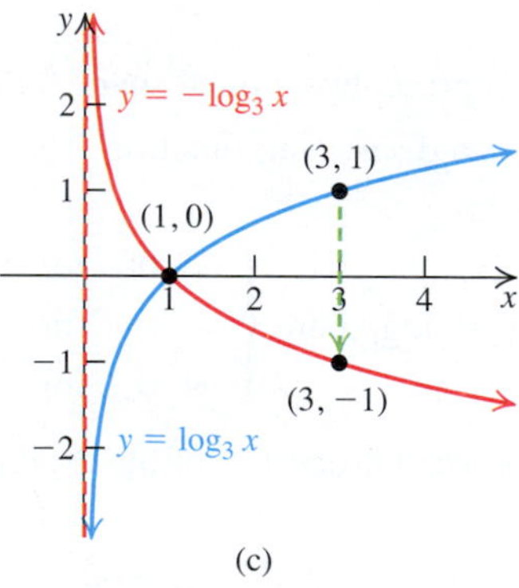

(c)

c. $f(x) = -\log_3 (x)$

Multiplying $\log_3 x$ by -1 reflects the graph about the x-axis.
Domain: $(0, \infty)$; range: $(-\infty, \infty)$; vertical asymptote: $x = 0$

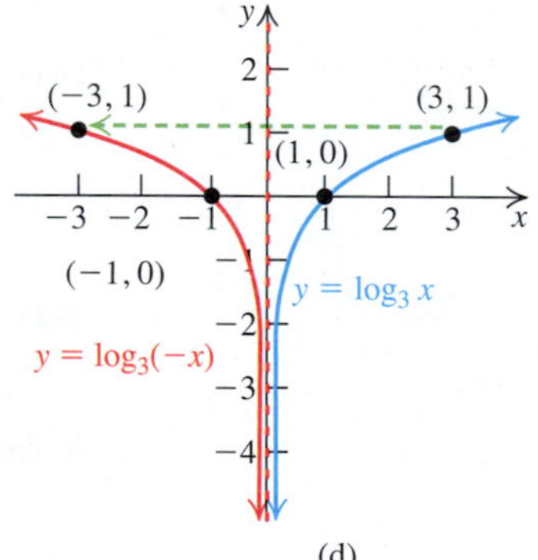

(d)

d. $f(x) = \log_3 (-x)$

Replacing x with $-x$ in $\log_3 x$ reflects the graph about the y-axis.
Domain: $(-\infty, 0)$; range: $(-\infty, \infty)$; vertical asymptote: $x = 0$

FIGURE 4.11 Transformations on $y = \log_3 x$

Practice Problem 7 Use transformations of the graph $f(x) = \log_2 x$ to sketch the graph of

$$y = -\log_2 (x - 3).$$

TECHNOLOGY CONNECTION

You can find the common logarithms of any positive number with the LOG key on your calculator. The exact key sequence will vary with different calculators. On most calculators, if you want to find log 3.7, press

3.7 LOG ENTER

or

LOG 3.7 ENTER

and see the display 0.5682017.

Although the display reads 0.5682017, this number really is an approximate value; so you should write $\log_{10} 3.7 \approx 0.5682017$

Common Logarithm

The logarithm with base 10 is called the **common logarithm** and is denoted by omitting the base, so

$$\log x = \log_{10} x.$$

$$y = \log x \ (x > 0) \quad \text{if and only if} \quad x = 10^y.$$

Applying the basic properties of logarithms (see page 413) to common logarithms, we have the following:

1. $\log 10 = 1$
2. $\log 1 = 0$
3. $\log 10^x = x$, x any real number
4. $10^{\log x} = x, x > 0$

We use property (3) to evaluate the common logarithms of numbers that are powers of 10. For example,

$$\log 1000 = \log 10^3 = 3$$
$$\text{and} \quad \log 0.01 = \log 10^{-2} = -2.$$

We use a calculator to find the common logarithms of numbers that are not powers of 10 by pressing the LOG key.

The graph of $y = \log x$ is similar to the graph in Figure 4.10(a) because the base for $\log x$ is $10 > 1$.

Frank Benford (1883–1948)

Frank Benford was an electrical engineer and a physicist who is best known for the probability distribution (Benford's Law) about the occurrence of digits in lists of real data sets. Auditors often use the law to check for fraudulent data. Benford graduated from the University of Michigan in 1910 and worked for General Electric until he retired in July 1948. He died suddenly at his home on December 4, 1948.

EXAMPLE 8 Benford's Law

Benford's Law states that the chances of the digit m occurring as the first decimal digit (from 1 to 9) from many real-life data sources is

$$P_m = \log(m + 1) - \log m.$$

This means that about $100P_m\%$ of the data can be expected to have m as the first digit.

a. Find P_1. Interpret your result.

b. Use part **a** to decide which column (size I or size II) represents the actual data for the size of countries listed on page 410.

Solution

a. $P_m = \log(m + 1) - \log m$ — Benford's law

$P_1 = \log 2 - \log 1$ — $m = 1$

$= \log 2$ — $\log 1 = \log_{10} 1 = 0$

≈ 0.301 — Use a calculator.

This means that about 30.1% of the data is expected to have 1 as the first digit.

b. We count the number of countries having 1 as the first digit in the columns with headings Size I and Size II. We find the following:

Number of countries with 1 as the first digit:

$$\text{Size I column} = 1$$
$$\text{Size II column} = 16$$

30% of 50 countries $= (0.30)(50) = 15$. So according to Benford's Law, size II column represents the actual data.

Practice Problem 8 Find P_2 and interpret your result.

TECHNOLOGY CONNECTION

To evaluate natural logarithms, we use the [LN] key on a calculator. To find ln (2.3) on most calculators, press

2.3 [LN] [ENTER]

or

[LN] 2.3 [ENTER]

and see the display 0.832909123.

Natural Logarithm

In most applications in calculus and the sciences, the convenient base for logarithms is the number e. The logarithm with base e is called the **natural logarithm** and is denoted by $\ln x$ (read "*ell en x*") so that

$$\ln x = \log_e x.$$

$$y = \ln x \ (x > 0) \quad \text{if and only if} \quad x = e^y.$$

Applying the basic properties of logarithms (see page 413) to natural logarithms, we have the following:

1. $\ln e = 1$
2. $\ln 1 = 0$
3. $\ln e^x = x$, x any real number
4. $e^{\ln x} = x, x > 0$

We can use property (3) to evaluate the natural logarithms of powers of e.

EXAMPLE 9 **Evaluating the Natural Logarithm Function**

Evaluate each expression.

a. $\ln e^4$ **b.** $\ln \frac{1}{e^{2.5}}$ **c.** $\ln 3$

Solution

a. $\ln e^4 = 4$ — Property (3)

b. $\ln \frac{1}{e^{2.5}} = \ln e^{-2.5} = -2.5$ — Property (3)

c. $\ln 3 \approx 1.0986123$ — Use a calculator.

Practice Problem 9 Evaluate each expression.

a. $\ln \frac{1}{e}$ **b.** $\ln 2$

Summary of Basic Properties of Logarithms

Base $a > 0, a \neq 1$	Base 10	Base e
1. $\log_a a = 1$	$\log 10 = 1$	$\ln e = 1$
2. $\log_a 1 = 0$	$\log 1 = 0$	$\ln 1 = 0$
3. $\log_a a^x = x$	$\log 10^x = x$	$\ln e^x = x$
4. $a^{\log_a x} = x$	$10^{\log x} = x$	$e^{\ln x} = x$

6 Use natural logarithms in applications.

Investments

We recall (page 403), the continuous compound interest formula:

$$A = Pe^{rt} \qquad (5)$$

We express equation (5) in logarithmic form:

$$\frac{A}{P} = e^{rt} \qquad \text{Divide both sides of (5) by } P.$$

$$\ln\frac{A}{P} = rt \qquad \text{Logarithmic form}$$

CONTINUOUS COMPOUND INTEREST

Exponential Form	Logarithmic Form
$A = Pe^{rt}$	$\ln\frac{A}{P} = rt$

The exponential form is used when we need to find A or P; logarithmic form is useful in calculating r or t.

EXAMPLE 10 Doubling Your Money

a. How long will it take to double your money if it earns 6.5% compounded continuously?

b. At what rate of return, compounded continuously, would your money double in 5 years?

Solution

If P dollars is invested and you want to double it, then the final amount $A = 2P$.

a.

$$\ln\left(\frac{A}{P}\right) = rt \qquad \text{Continuous compounding formula, logarithmic form}$$

$$\ln\left(\frac{2P}{P}\right) = 0.065t \qquad A = 2P,\ r = 0.065$$

$$\ln 2 = 0.065t \qquad \frac{2P}{P} = 2$$

$$\frac{\ln 2}{0.065} = t \qquad \text{Divide both sides by 0.065.}$$

$$t \approx 10.66 \qquad \text{Use a calculator.}$$

It will take approximately 11 years to double your money.

b.

$$\ln\left(\frac{A}{P}\right) = rt \qquad \text{Continuous compounding formula, logarithmic form}$$

$$\ln\left(\frac{2P}{P}\right) = r(5) \qquad A = 2P,\ t = 5$$

$$\ln 2 = 5r \qquad \frac{2P}{P} = 2$$

$$\frac{\ln 2}{5} = r \qquad \text{Divide both sides by 5.}$$

$$r \approx 0.1386 \qquad \text{Use a calculator.}$$

Your investment will double in 5 years at the approximate rate of 13.86%.

Practice Problem 10 Repeat Example 10 for tripling ($A = 3P$) your money.

Newton's Law of Cooling

When a cool drink is removed from a refrigerator, the drink warms to the temperature of the room. When removed from the oven, a pizza baked at a high temperature cools to the temperature of the room.

In situations such as these, the rate at which an object's temperature changes at any given time is proportional to the difference between its temperature and the temperature of the surrounding medium. This observation is called *Newton's Law of Cooling*, although, as in the case of a cool drink, it applies to warming as well.

Newton's Law of Cooling states that

$$T(t) = T_s + (T_0 - T_s)e^{-kt} \qquad (6)$$

where $T(t)$ is the temperature of the object at time t, T_s is the surrounding temperature, T_0 is the value of $T(t)$ at $t = 0$, and k is a positive constant that depends on the object.

We can express equation (6) in logarithmic form:

$$T(t) - T_s = (T_0 - T_s)e^{-kt} \qquad \text{Subtract } T_s \text{ from both sides of equation (6).}$$

$$\frac{T(t) - T_s}{T_0 - T_s} = e^{-kt} \qquad \text{Divide both sides by } T_0 - T_s.$$

$$\ln\left(\frac{T(t) - T_s}{T_0 - T_s}\right) = -kt \qquad \text{Logarithmic form}$$

NEWTON'S LAW OF COOLING

Exponential Form	Logarithmic Form
$T(t) = T_s + (T_0 - T_s)e^{-kt}$	$\ln\left(\frac{T(t) - T_s}{T_0 - T_s}\right) = -kt$

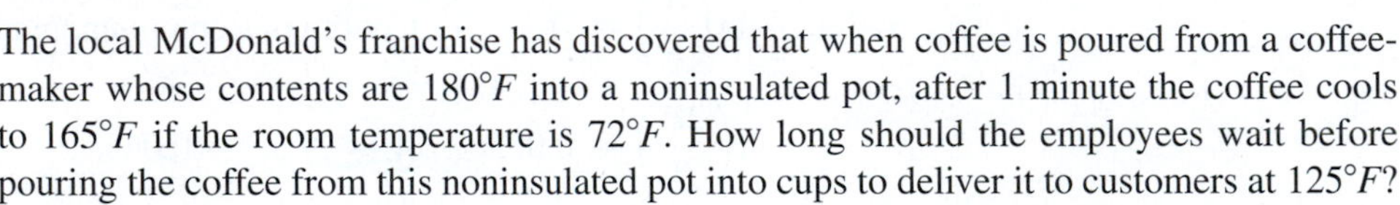

EXAMPLE 11 **McDonald's Hot Coffee**

The local McDonald's franchise has discovered that when coffee is poured from a coffee-maker whose contents are 180°F into a noninsulated pot, after 1 minute the coffee cools to 165°F if the room temperature is 72°F. How long should the employees wait before pouring the coffee from this noninsulated pot into cups to deliver it to customers at 125°F?

Solution

From the given data, we have $T = 165$ when $t = 1$, $T_0 = 180$, and $T_s = 72$.

$$\ln\left(\frac{165 - 72}{180 - 72}\right) = \ln\left(\frac{93}{108}\right) = -k \cdot 1 \qquad \text{Substitute given data in the logarithmic form.}$$

$$k = -\ln\left(\frac{93}{108}\right) \qquad \text{Solve for } k.$$

With this value of k, we find t when $T = 125$.

$$\ln\left(\frac{125 - 72}{180 - 72}\right) = \ln\left(\frac{53}{108}\right) = -\left(-\ln\left(\frac{93}{108}\right)\right) \cdot t \qquad \text{Substitute data in logarithmic form.}$$

$$\left[\frac{1}{\ln\left(\frac{93}{108}\right)}\right]\ln\left(\frac{53}{108}\right) = t \qquad \text{Solve for } t.$$

$$t \approx 4.76 \qquad \text{Use a calculator.}$$

The employees should wait approximately 4.76 minutes (realistically, about 5 minutes) to deliver the coffee at 125°F to the customers.

Practice Problem 11 Repeat Example 11 assuming that the coffee is to be delivered to the customers at 120°F.

Answers to Practice Problems

1. a. $\log_2 1024 = 10$ **b.** $\log_9\left(\frac{1}{3}\right) = -\frac{1}{2}$ **c.** $\log_a p = q$
2. a. $2^6 = 64$ **b.** $v^w = u$ **3. a.** 2 **b.** $-1/2$ **c.** -5
4. a. 0 **b.** 5 **c.** 5 **5.** $(-\infty, 1)$
6.

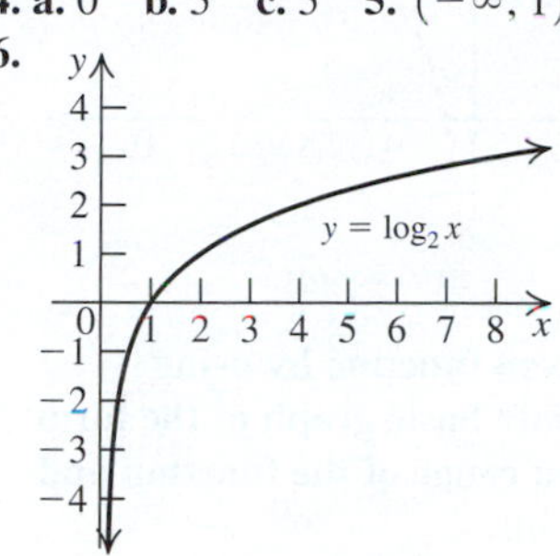

7. Shift the graph of $y = \log_2 x$ 3 units right and reflect the graph about the x-axis.

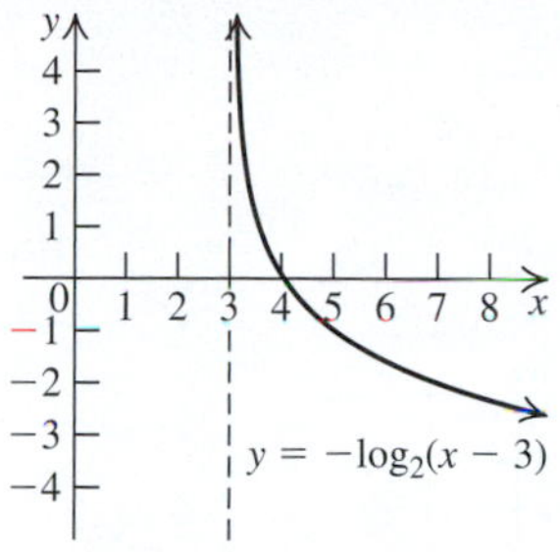

8. $P_2 = \log 3 - \log 2 \approx 0.176$. About 17.6% of the data is expected to have 2 as the first digit. **9. a.** -1 **b.** ≈ 0.693
10. a. ≈ 17 years **b.** 21.97%
11. ≈ 5.4 min

SECTION 4.2 Exercises

Basic Concepts and Skills

1. The domain of the function $y = \log_a x$ is ____________, and its range is ____________.
2. The logarithmic form $y = \log_a x$ is equivalent to the exponential form ____________.
3. The logarithm with base 10 is called the ____________ logarithm, and the logarithm with base e is called the ____________ logarithm.
4. $a^{\log_a x} =$ ____________, and $\log_a a^x =$ ____________.
5. **True or False.** The graph of $y = \log_a x$, $a > 0$, $a \neq 1$, is the graph of an increasing function.
6. **True or False.** The graph of $y = \log_a x$, $a > 0$, and $a \neq 1$, has no horizontal asymptote.

In Exercises 7–18, write each exponential equation in logarithmic form.

7. $5^2 = 25$
8. $(49)^{-1/2} = \frac{1}{7}$
9. $\left(\frac{1}{16}\right)^{-1/2} = 4$
10. $(a^2)^2 = a^4$
11. $10^0 = 1$
12. $10^4 = 10{,}000$
13. $(10)^{-1} = 0.1$
14. $3^x = 5$
15. $a^2 + 2 = 7$
16. $a^e = \pi$
17. $2a^3 - 3 = 10$
18. $5 \cdot 2^{ct} = 11$

In Exercises 19–30, write each logarithmic equation in exponential form.

19. $\log_2 32 = 5$
20. $\log_7 49 = 2$
21. $\log_{10} 100 = 2$
22. $\log_{10} 10 = 1$
23. $\log_{10} 1 = 0$
24. $\log_a 1 = 0$
25. $\log_{10} 0.01 = -2$
26. $\log_{1/5} 5 = -1$
27. $3 \log_8 2 = 1$
28. $1 + \log 1000 = 4$
29. $\ln 2 = x$
30. $\ln \pi = a$

In Exercises 31–40, evaluate each expression without using a calculator.

31. $\log_5 125$
32. $\log_9 81$
33. $\log 10{,}000$
34. $\log_3 \frac{1}{3}$
35. $\log_2 \frac{1}{8}$
36. $\log_4 \frac{1}{64}$
37. $\log_3 \sqrt{27}$
38. $\log_{27} 3$
39. $\log_{16} 2$
40. $\log_5 \sqrt{125}$

In Exercises 41–52, evaluate each expression.

41. $\log_3 1$
42. $\log_{1/2} 1$
43. $\log_7 7$
44. $\log_{1/9} \frac{1}{9}$
45. $\log_6 6^7$
46. $\log_{1/2}\left(\frac{1}{2}\right)^5$
47. $3^{\log_3 5}$
48. $7^{\log_7 \frac{1}{2}}$
49. $2^{\log_2 7} + \log_5 5^{-3}$
50. $3^{\log_3 5} - \log_2 2^{-3}$
51. $4^{\log_4 6} - \log_4 4^{-2}$
52. $10^{\log x} - e^{\ln y}$

In Exercises 53–60, find the domain of each function.

53. $f(x) = \log_2 (x + 1)$
54. $g(x) = \log_3 (x - 8)$
55. $f(x) = \log_3 \sqrt{x - 1}$
56. $g(x) = \log_4 \sqrt{3 - x}$
57. $f(x) = \log(x - 2) + \log(2x - 1)$
58. $g(x) = \ln\sqrt{x + 5} - \ln(x + 1)$

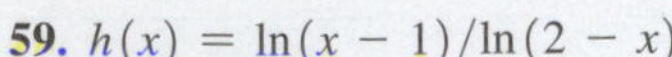

59. $h(x) = \ln(x - 1)/\ln(2 - x)$

60. $f(x) = \ln(x - 3) + \ln(2 - x)$

61. Match each logarithmic function with one of the graphs labeled **a–f**.

a. $f(x) = \log x$
b. $f(x) = -\log|x|$
c. $f(x) = -\log(-x)$
d. $f(x) = \log(x - 1)$
e. $y = (\log x) - 1$
f. $f(x) = \log(-1 - x)$

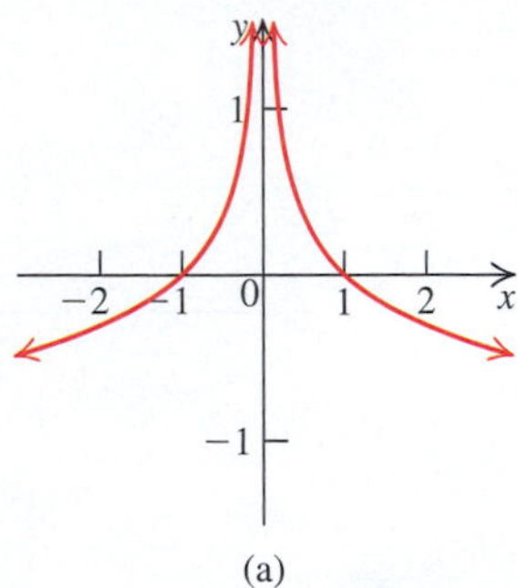
(a)

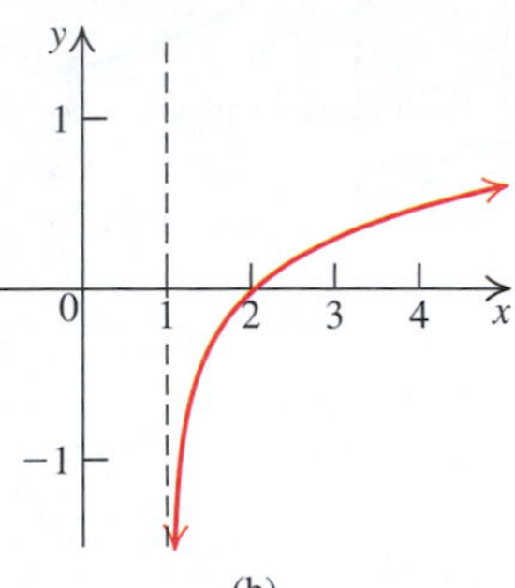
(b)

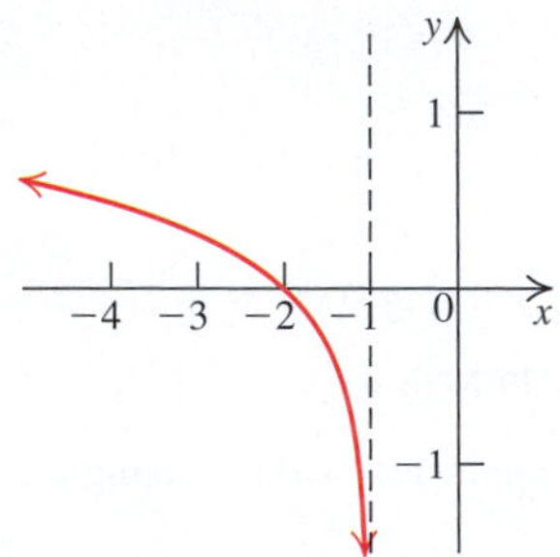
(c)

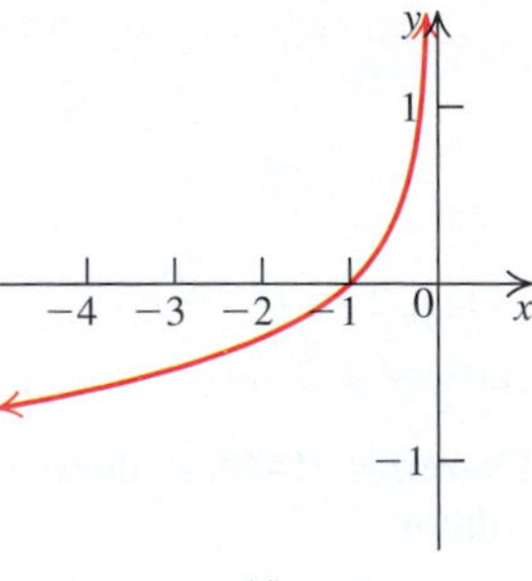
(d)

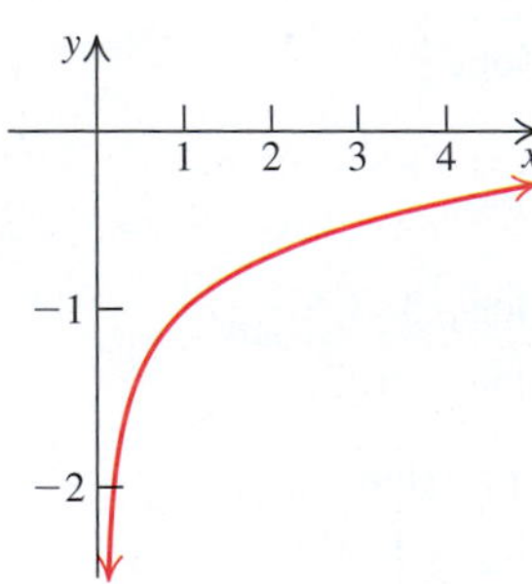
(e)

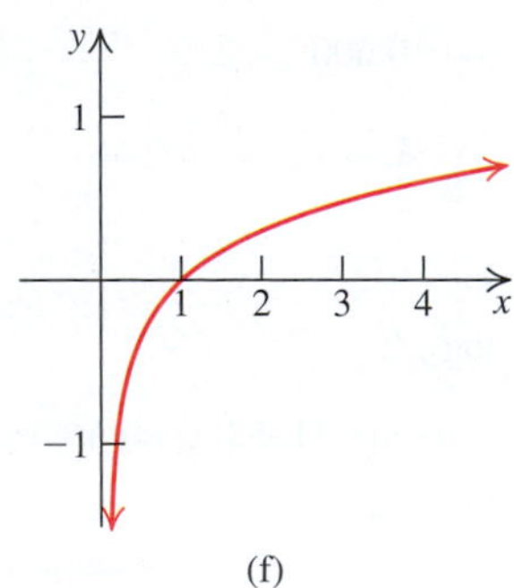
(f)

62. For each given graph, find a function of the form $f(x) = \log_a x$ that represents the graph.

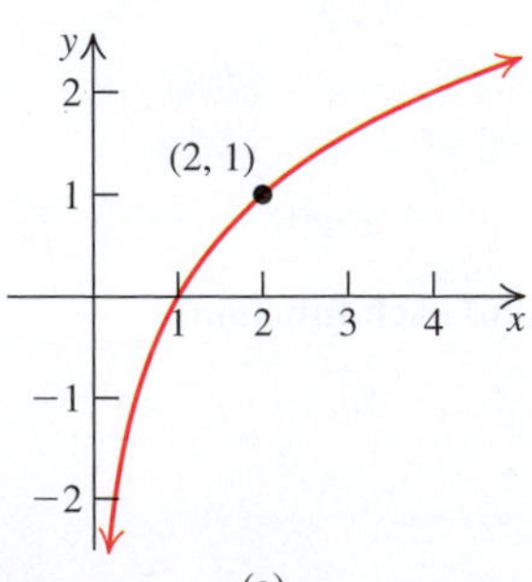

(a)

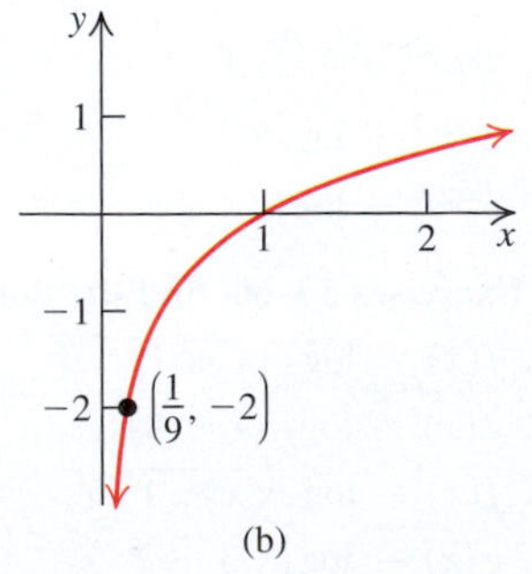

(b)

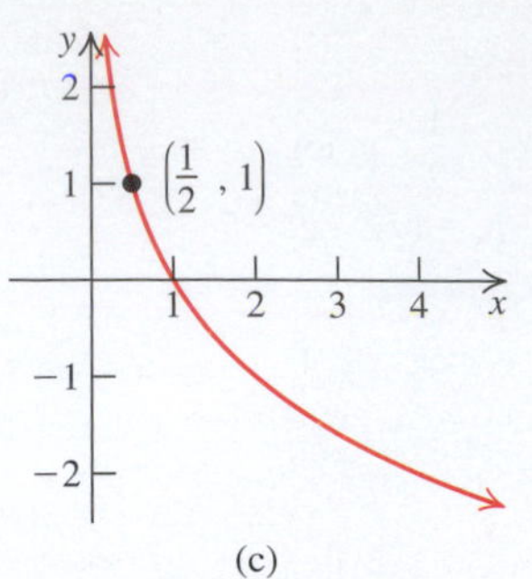

(c)

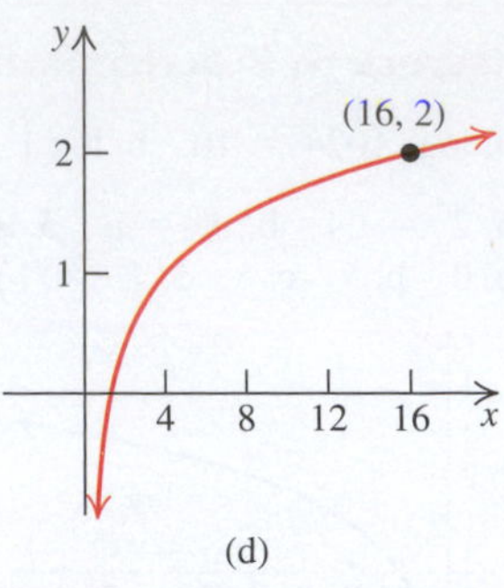

(d)

In Exercises 63–70, graph the given function by using transformations on the appropriate basic graph of the form $y = \log_a x$. State the domain and range of the function and the vertical asymptote of the graph.

63. $f(x) = \log_4 (x + 3)$
64. $g(x) = \log_{1/2} (x - 1)$
65. $y = -\log_5 x$
66. $y = 2 \log_7 x$
67. $y = \log_{1/5} (-x)$
68. $y = 1 + \log_{1/5} (-x)$
69. $y = |\log_3 x|$
70. $y = \log_3 |x|$

In Exercises 71–78, begin with the graph of $f(x) = \log_2 x$ and use transformations to sketch the graph of each function. Find the domain and range of the function and the vertical asymptote of the graph.

71. $y = \log_2(x - 1)$
72. $y = \log_2 (-x)$
73. $y = \log_2(3 - x)$
74. $y = \log_2 x$
75. $y = 2 + \log_2(3 - x)$
76. $y = 4 - \log_2 (3 - x)$
77. $y = \log_2 |x|$
78. $y = \log_2 x^2$

In Exercises 79–84, evaluate each expression.

79. $\log_4 (\log_3 81)$
80. $\log_4 [\log_3 (\log_2 8)]$
81. $\log_{\sqrt{2}} 2$
82. $\log_{2\sqrt{2}} 8$
83. $\log_{\sqrt{2}} 4$
84. $\log_{\sqrt{3}} 27$

In Exercises 85–88, find the value of x in each equation.

85. $\log x = 2$
86. $\log (x - 1) = 1$
87. $\ln x = 1$
88. $\ln x = 0$

In Exercises 89–94, begin with the graph of $y = \ln x$ and use transformations to sketch the graph of each of the given functions.

89. $y = \ln (x + 2)$
90. $y = \ln (2 - x)$
91. $y = -\ln (2 - x)$
92. $y = 2 \ln x$
93. $y = 3 - 2 \ln x$
94. $y = 1 - \ln (1 - x)$

Applying the Concepts

In Exercises 95–108, use the model $A = A_0 e^{kt}$ or $\ln \dfrac{A}{A_0} = kt$.

95. Doubling your money. How long would it take to double your money if you invested P dollars at the rate of 8% compounded continuously?

96. Investment goal. How long would it take to grow your investment from \$10,000 to \$120,000 at the rate of 10% compounded continuously?

97. Rate for doubling your money. At what annual rate of return, compounded continuously, would your investment double in six years?

98. Rate-of-investment goal. At what annual rate of return, compounded continuously, would your investment grow from \$8000 to \$50,000 in 25 years?

99. Population of the United States. The U.S. population in 2010 was 308 million. The U.S. Census Bureau reported that the population grew 9.7% since 2000 (the slowest growth rate since the depression).

a. What was the population of the United States in 2000?

b. What was the annual rate of growth between 2000 and 2010?

c. Assuming the same rate of growth, determine the year when the population of the United States will be 400 million.

100. Population of Canada. The population of Canada in 2010 was 35 million, and it grew 12.33% since 2000.

a. What was the population of Canada in 2000?

b. What was the annual rate of growth between 2000 and 2010?

c. Assuming the same rate of growth, determine the time from 2010 until Canada's population doubles.

d. Find the time from 2010 until Canada's population reaches 50 million.

101. Newton's Law of Cooling. A thermometer is taken from a room at 75°F to the outdoors, where the temperature is 20°F. The reading on the thermometer drops to 50°F after one minute.

a. Find the reading on the thermometer after

(i) Five minutes.

(ii) Ten minutes.

(iii) One hour.

b. How long will it take for the reading to drop to 22°F?

102. Newton's Law of Cooling. The last bit of ice in a picnic cooler has melted ($T_0 = 32$°F). The temperature in the park is 85°F. After 30 minutes, the temperature in the cooler is 40°F. How long will it take for the temperature inside the cooler to reach 50°F?

103. Cooking salmon. A salmon filet initially at 50°F is cooked in an oven at a constant temperature of 400°F. After ten minutes, the temperature of the filet rises to 160°F. How long does it take until the salmon is medium rare at 220°F?

104. The time of murder. A forensic specialist took the temperature of a victim's body lying in a street at 2:10 A.M. and found it to be 85.7°F. At 2:40 A.M., the temperature of the body was 84.8°F. When was the murder committed if the air temperature during the night was 55°F? [Remember, normal body temperature is 98.6°F.]

105. Water contamination. A chemical is spilled into a reservoir of pure water. The concentration of chemical in the contaminated water is 4%. In one month, 20% of the water in the reservoir is replaced with clean water.

a. What will be the concentration of the contaminant one year from now?

b. For water to be safe for drinking, the concentration of this contaminant cannot exceed 0.01%. How long will it be before the water is safe for drinking?

106. Toxic chemicals in a lake. In a lake, one-fourth of the water is replaced by clean water every year. Sixteen thousand cubic meters of soluble toxic chemical spill takes place in the lake. Let $T(n)$ represent the amount of toxin left after n years.

a. Find a formula for $T(n)$.

b. How much toxin will be left after 12 years?

c. When will 80% of the toxin be eliminated?

107. Sheep population. A herd of sheep doubles in size every three years. There are now 1500 sheep in the herd.

a. Find an exponential function of the form $P = P_0e^{rt}$ to model the herd's growth.

b. How many sheep will be in the herd seven years from now (round your answer)?

c. In how many years will the herd have 15,000 sheep?

108. Shark population. A school of sharks is losing one-ninth of its population every two years. The colony currently has 150 sharks.

a. Find an exponential function of the form $P = P_0e^{-rt}$ to model the school's decline.

b. How many sharks will be in the school 4 years from now?

c. In how many years (to the nearest tenth) will the colony have 35 sharks?

109. Benford's Law. The law states that the probability that the first decimal-digit of a raw data sample (from 1 to 9) is given by $P_m = \log(m + 1) - \log m$. That is, about $(100P_m)$% of the data can be expected to have m as the first digit.

a. What percent of the data can be expected to have 3 as the first digit?

b. Find $P_1 + P_2 + \cdots + P_9$. Interpret your result.

110. Prime Number Theorem. A natural number $p \ge 2$ is *prime* if the only positive divisors of p are 1 and p. An important topic in number theory involves counting prime numbers $\{2, 3, 5, 7, 11, 13, \ldots\}$. For any natural number $n \ge 2$, the quantity of prime numbers less than or equal to n is denoted by $\pi(n)$ (read "pie of n"). For example, $\pi(2) = 1$, $\pi(3) = 2$, $\pi(4) = 2$, $\pi(5) = 3$, and $\pi(6) = 3$. The *Prime Number Theorem* states that for large n, $\pi(n)$ can be approximated by $\dfrac{n}{\ln(n)}$.

a. Find $\pi(20)$ by listing the primes that are less than or equal to 20.

b. Find $\pi(50)$ by listing primes and compare your answer with $\dfrac{50}{\ln(50)}$.

c. Estimate $\pi(1{,}000{,}000)$ by using the Prime Number Theorem.

Beyond the Basics

111. Find the domain of $f(x) = \log_3\left(\dfrac{x - 2}{x + 1}\right)$.

112. Find the domain of $g(x) = \log_2\left(\dfrac{x + 3}{x - 2}\right)$.

113. Finding the domain. Find the domain of each function.

a. $f(x) = \log_2(\log_3 x)$

b. $f(x) = \log(\ln(x - 1))$

c. $f(x) = \ln(\log(x - 1))$

d. $f(x) = \log(\log(\log(x - 1)))$

114. Finding the inverse. Find the inverse of each function in Exercise 113.

In Exercises 115 and 116, describe an order for the sequence of transformations on the graph of $y = \log x$ to produce the graph of the given equation.

115. $y = -3\log\left(\frac{1}{2}x - 2\right) + 4$

116. $y = -4\log(2 - 3x) + 1$

117. Present value of an investment. Recall that if P dollars is invested in an account at an interest rate r compounded continuously, then the amount A (called the *future value of P*) in the account t years from now will be $A = Pe^{rt}$. Solving the equation for P, we get $P = Ae^{-rt}$. In this formulation, P is called the present value of the investment.

a. Find the present value of \$100,000 at 7% compounded continuously for 20 years.

b. Find the interest rate r compounded continuously that is needed to have \$50,000 be the present value of \$75,000 in ten years.

118. Your uncle is 40 years old, and he wants to have an annual pension of \$50,000 each year at age 65. What is the present value of his pension if the money can be invested at all times at a continuously compounded interest rate of

a. 5%?

b. 8%?

c. 10%?

Critical Thinking/Discussion/Writing

In Exercises 119 and 120 evaluate each expression without using a calculator.

119. $2^{\log_2 3} - 3^{\log_3 2}$

120. $(\log_3 4 + \log_2 9)^2 - (\log_3 4 - \log_2 9)^2$

121. Solve for x: $\log_3[\log_4(\log_2 x)] = 0$

122. Write as a piecewise function.

a. $f(x) = |\log x|$

b. $g(x) = |\ln(x - 1)| + |\ln(x - 2)|$

123. Inequalities involving logarithms.

a. If a is positive, is the statement "$a < b$ if and only if $\log a < \log b$" always true?

b. What property of logarithmic functions is used whenever the statement in part (a) is true?

124. Write a summary of the types of graphs (with sketches) for the logarithmic functions of the form $y = c\log_a x$, $c \neq 0$, $a > 0$, and $a \neq 1$.

Maintaining Skills

In Exercises 125–130 write each expression in the form a^n where n is a positive integer.

125. $a^2 \cdot a^7$

126. $(a^2)^3$

127. $\sqrt{a^8}$

128. $\sqrt[3]{a^6}$

129. $\left(\frac{243}{32}\right)^{4/5}$

130. $\sqrt[5]{(32)^{-3}}$

In Exercises 131 and 132, perform the indicated operations and write the answer in scientific notation.

131. $(4.7 \times 10^7)(8.1 \times 10^5)$

132. $\frac{7.2 \times 10^6}{2.4 \times 10^{-3}}$

In Exercises 133–136, verify that each equation is true by evaluating each side without using a calculator.

133. $\log_3 81 = \log_3 3 + \log_3 27$

134. $\log_2 8 = \log_2 128 - \log_2 16$

135. $\log_2 16 = 2\log_2 4$

136. $\log_4 64 = \frac{\log_2 64}{\log_2 4}$

SECTION 4.3

Rules of Logarithms

BEFORE STARTING THIS SECTION, REVIEW

1. Definition of logarithm (Section 4.2, page 411)
2. Basic properties of logarithms (Section 4.2, page 413)
3. Rules of exponents (Section 4.1, page 394)

OBJECTIVES

1. Learn the rules of logarithms.
2. Estimate a large number.
3. Change the base of a logarithm.
4. Apply logarithms to growth and decay.

King Tut's Golden Mask

The Boy King Tut

Today the most famous pharaoh of Ancient Egypt is King Nebkheperure Tutankhamun, popularly called "King Tut." He was only 9 years old when he became a pharaoh. In 2005, a team of Egyptian scientists headed by Dr. Zahi Hawass determined that the pharaoh was 19 years old when he died (around 1346 B.C.).

Tutankhamun was a short-lived boy king who, unlike the great Egyptian kings Khufu (builder of the Great Pyramid), Amenhotep III (builder of temples throughout Egypt), and Ramesses II (prolific builder and usurper), accomplished nothing significant during his reign. In fact, little was known about him prior to Howard Carter's discovery of his tomb (and the amazing treasures it held) in the Valley of the Kings on November 4, 1922. Carter, with his benefactor Lord Carnarvon at his side, entered the tomb's burial chamber on November 26, 1922. Lord Carnarvon died seven weeks after entering the burial chamber, giving rise to the theory of the "curse" of King Tut.

Work on the tomb continued until 1933. The tomb contained a pristine mummy of an Egyptian king, lying intact in his original burial furniture. He was accompanied by a small slice of the royal world of the pharaohs: golden chariots, statues of gold and ebony, a fleet of miniature ships to accommodate his trip to the hereafter, his throne of gold, bottles of perfume, precious jewelry, and more. The "Treasures of Tutankhamun" exhibition, first shown at the British Museum in London in 1972, traveled to many countries, including the United States.

In Example 8, we discuss the age of a work of art found in King Tut's tomb.

1 Learn the rules of logarithms.

Rules of Logarithms

Recall the basic properties of logarithms from Section 4.2 for $a > 0, a \neq 1$:

$$\log_a a = 1 \qquad (7)$$
$$\log_a 1 = 0 \qquad (8)$$
$$\log_a a^x = x, x \text{ real} \qquad (9)$$
$$a^{\log_a x} = x, x > 0 \qquad (10)$$

In this section, we will discuss some important rules of logarithms that are helpful in calculations, simplifications, and applications.

RULES OF LOGARITHMS

Let M, N, and a be positive real numbers with $a \neq 1$ and let r be any real number.

Rule	Description	Examples
1. *Product Rule:* $\log_a (MN) = \log_a M + \log_a N$	The logarithm of the product of two (or more) numbers is the sum of the logarithms of the numbers.	$\ln (5 \cdot 7) = \ln 5 + \ln 7$ $\log (3x) = \log 3 + \log x$ $\log_2 (5 \cdot 17) = \log_2 5 + \log_2 17$
2. *Quotient Rule:* $\log_a \left(\frac{M}{N}\right) = \log_a M - \log_a N$	The logarithm of the quotient of two numbers is the difference of the logarithms of the numbers.	$\ln \frac{5}{7} = \ln 5 - \ln 7$ $\log \frac{5}{x} = \log 5 - \log x$
3. *Power Rule:* $\log_a M^r = r \log_a M$	The logarithm of a number to the power r is r times the logarithm of the number.	$\ln 5^7 = 7 \ln 5$ $\log 5^{3/2} = \frac{3}{2} \log 5$ $\log_2 7^{-3} = -3 \log_2 7$

Rules 1, 2, and 3 follow from the corresponding rules of the exponents:

$$a^x \cdot a^y = a^{x+y} \qquad \text{Product rule}$$

$$\frac{a^x}{a^y} = a^{x-y} \qquad \text{Quotient rule}$$

$$(a^x)^r = a^{xr} \qquad \text{Power rule}$$

For example, to prove the product rule for logarithms, we let

$$\log_a M = x \quad \text{and} \quad \log_a N = y.$$

The corresponding exponential forms of these equations are

$$M = a^x \quad \text{and} \quad N = a^y.$$

Then

$$MN = a^x \cdot a^y$$

$$MN = a^{x+y} \qquad \text{Product rule of exponents}$$

$$\log_a MN = x + y \qquad \text{Logarithmic form}$$

$$\log_a MN = \log_a M + \log_a N \qquad \text{Replace } x \text{ with } \log_a M \text{ and } y \text{ with } \log_a N.$$

This proves the product rule of logarithms. In Exercise 114, you are asked to prove the quotient rule and the power rule similarly by using the corresponding rules for exponents. The following table compares the rules of exponents and logarithms.

COMPARE THE RULES OF EXPONENTS AND LOGARITHMS

	Exponents	Logarithms
1.	$a^x \cdot a^y = a^{x+y}$	$\log_a xy = \log_a x + \log_a y$
2.	$\frac{a^x}{a^y} = a^{x-y}$	$\log_a \frac{x}{y} = \log_a x - \log_a y$
3.	$(a^x)^y = a^{xy}$	$\log_a x^y = y \log_a x$

EXAMPLE 1 Using Rules of Logarithms to Evaluate Expressions

RECALL

Radicals can also be written as exponents. Recall that

$$\sqrt{x} = x^{1/2}$$

and $\sqrt[3]{x} = x^{1/3}$.

In general,

$$\sqrt[m]{x^n} = x^{n/m}.$$

Given that $\log_5 z = 3$ and $\log_5 y = 2$, evaluate each expression.

a. $\log_5 (yz)$ **b.** $\log_5 (125y^7)$ **c.** $\log_5 \sqrt{\frac{z}{y}}$ **d.** $\log_5 (z^{1/30}y^5)$

Solution

a. $\log_5 (yz) = \log_5 y + \log_5 z$ Product rule

$= 2 + 3 = 5$ Use the given values and simplify.

b. $\log_5 (125y^7) = \log_5 125 + \log_5 y^7$ Product rule

$= \log_5 5^3 + \log_5 y^7$ $125 = 5^3$

$= 3 + 7 \log_5 y$ Power rule and $\log_a a = 1$

$= 3 + 7(2) = 17$ Use the given values and simplify.

c. $\log_5 \sqrt{\frac{z}{y}} = \log_5 \left(\frac{z}{y}\right)^{1/2}$ Rewrite radical as exponent.

$= \frac{1}{2} \log_5 \frac{z}{y}$ Power rule

$= \frac{1}{2}(\log_5 z - \log_5 y)$ Quotient rule

$= \frac{1}{2}(3 - 2) = \frac{1}{2}$ Use the given values and simplify.

d. $\log_5 (z^{1/30}y^5) = \log_5 z^{1/30} + \log_5 y^5$ Product rule

$= \frac{1}{30} \log_5 z + 5 \log_5 y$ Power rule

$= \frac{1}{30}(3) + 5(2)$ Use the given values.

$= 0.1 + 10 = 10.1$ Simplify.

Practice Problem 1 Evaluate each expression for the given values of logarithms in Example 1.

a. $\log_5 (y/z)$ **b.** $\log_5 (y^2z^3)$

In many applications in more advanced mathematics courses, the rules of logarithms are used in both directions; that is, the rules are read from left to right and from right to left. For example, by the product rule of logarithms, we have

$$\log_2 3x = \log_2 3 + \log_2 x.$$

The expression $\log_2 3 + \log_2 x$ is the *expanded form* of $\log_2 3x$, while $\log_2 3x$ is the *condensed form,* or the *single logarithmic form* of $\log_2 3 + \log_2 x$.

EXAMPLE 2 Writing Expressions in Expanded Form

Write each expression in expanded form.

a. $\log_2 \frac{x^2(x - 1)^3}{(2x + 1)^4}$ **b.** $\log_c \sqrt{x^3y^2z^5}$

Solution

a. $\log_2 \dfrac{x^2(x-1)^3}{(2x+1)^4} = \log_2 x^2(x-1)^3 - \log_2 (2x+1)^4$ Quotient rule

$= \log_2 x^2 + \log_2 (x-1)^3 - \log_2 (2x+1)^4$ Product rule

$= 2\log_2 x + 3\log_2 (x-1) - 4\log_2 (2x+1)$ Power rule

b. $\log_c \sqrt{x^3y^2z^5} = \log_c (x^3y^2z^5)^{1/2}$ $\sqrt{a} = a^{1/2}$

$= \dfrac{1}{2}\log_c (x^3y^2z^5)$ Power rule

$= \dfrac{1}{2}[\log_c x^3 + \log_c y^2 + \log_c z^5]$ Product rule

$= \dfrac{1}{2}[3\log_c x + 2\log_c y + 5\log_c z]$ Power rule

$= \dfrac{3}{2}\log_c x + \log_c y + \dfrac{5}{2}\log_c z$ Distributive property

Practice Problem 2 Write each expression in expanded form.

a. $\ln \dfrac{2x-1}{x+4}$ **b.** $\log \sqrt{\dfrac{4xy}{z}}$

EXAMPLE 3 Writing Expressions in Condensed Form

Write each expression in condensed form.

a. $\log 3x - \log 4y$

b. $2\ln x + \dfrac{1}{2}\ln (x^2+1)$

c. $2\log_2 5 + \log_2 9 - \log_2 75$

d. $\dfrac{1}{3}[\ln x + \ln (x+1) - \ln (x^2+1)]$

Solution

a. $\log 3x - \log 4y = \log\left(\dfrac{3x}{4y}\right)$ Quotient rule

b. $2\ln x + \dfrac{1}{2}\ln (x^2+1) = \ln x^2 + \ln (x^2+1)^{1/2}$ Power rule

$= \ln (x^2\sqrt{x^2+1})$ Product rule; $\sqrt{a} = a^{1/2}$

c. $2\log_2 5 + \log_2 9 - \log_2 75 = \log_2 5^2 + \log_2 9 - \log_2 75$ Power rule

$= \log_2 (25 \cdot 9) - \log_2 75$ $5^2 = 25$; Product rule

$= \log_2 \dfrac{25 \cdot 9}{75}$ Quotient rule

$= \log_2 3$ $\dfrac{25 \cdot 9}{75} = \dfrac{9}{3} = 3$

d. $\dfrac{1}{3}[\ln x + \ln (x+1) - \ln (x^2+1)] = \dfrac{1}{3}[\ln x(x+1) - \ln (x^2+1)]$ Product rule

$= \dfrac{1}{3}\ln\left[\dfrac{x(x+1)}{x^2+1}\right]$ Quotient rule

$= \ln \sqrt[3]{\dfrac{x(x+1)}{x^2+1}}$ Power rule; $a^{1/3} = \sqrt[3]{a}$

Practice Problem 3 Write the expression in condensed form:
$\dfrac{1}{2}[\log (x+1) + \log (x-1)]$.

Be careful when using the rules of logarithms to simplify expressions. For example, there is no property that allows you to rewrite $\log_a (x + y)$.

In general,

$$\log_a (x + y) \neq \log_a x + \log_a y. \quad (11)$$

To illustrate this statement, let $x = 100$, $y = 10$, and $a = 10$. Then the value of the left side of (11) is

$$\log (100 + 10) = \log (110) \approx 2.0414. \quad \text{Use a calculator.}$$

The value of the right side of (11) is

$$\begin{aligned} \log 100 + \log 10 &= \log 10^2 + \log 10 \\ &= 2 \log 10 + \log 10 && \text{Power rule} \\ &= 2(1) + 1 = 3. \end{aligned}$$

Therefore, $\log (100 + 10) \neq \log 100 + \log 10$.

WARNING

To avoid common errors, be aware that in general,

$$\begin{aligned} &\log_a (M + N) \neq \log_a M + \log_a N. \\ &\log_a (M - N) \neq \log_a M - \log_a N. \\ &(\log_a M)(\log_a N) \neq \log_a MN. \\ &\log_a \left(\frac{M}{N}\right) \neq \frac{\log_a M}{\log_a N}. \\ &\frac{\log_a M}{\log_a N} \neq \log_a M - \log_a N. \\ &(\log_a M)^r \neq r \log_a M. \end{aligned}$$

2 Estimate a large number.

Number of Digits

In estimating the magnitude of a positive number K, we look for the number of digits needed to express K. If K has a fractional part, then "number of digits" means the number of digits to the left of the decimal point. For example, 357.29 and 231.4796 are viewed as 3-digit numbers in this context.

A number K written in the form $K = s \times 10^n$, where $1 \leq s < 10$ and n is an integer, is said to be in scientific notation. For example, $432.1 = 4.321 \times 10^2$ and $0.56 = 5.6 \times 10^{-1}$ are both in scientific notation.

The exponent n represents the magnitude of K and is closely related to the common logarithm. In fact, we have

$$\begin{aligned} \log K &= \log(s \times 10^n) && \text{Take log of both sides of } K = s \times 10^n. \\ &= \log s + \log 10^n && \text{Product rule} \\ &= \log s + n \log 10 && \text{Power rule} \\ &= \log s + n && \log 10 = \log_{10} 10 = 1 \end{aligned}$$

Because $1 \leq s < 10$, we have

$$\begin{aligned} \log 1 \leq \log s &< \log 10 && y = \log_{10} x \text{ is an increasing function.} \\ 0 \leq \log s &< 1 && \log 1 = 0 \text{ and } \log 10 = 1 \end{aligned}$$

This says that the exponent n in $K = s \times 10^n$ is the largest integer less than or equal to $\log K$. We note that a number m is a 3-digit number if

$$\begin{aligned} & 100 = 10^2 \leq m < 10^3 = 1000 \\ \text{or}\quad & \log 10^2 \leq \log m < \log 10^3 && y = \log_{10} x \text{ is an increasing function.} \\ & 2 \leq \log m < 3 && \log 10^x = \log_{10} 10^x = x \end{aligned}$$

Similarly, a number K has $(n + 1)$ digits if $n \leq \log K < n + 1$.

DIGITS AND COMMON LOGARITHMS

A positive number K has $(n + 1)$ digits if and only if $\log K$ is in the interval $[n, n + 1)$.

For example, the number 8765.43, which is a 4 digit number, has its common logarithm between 3 and 4. Conversely, if $\log N$ is approximately 57.3 then N is a 58-digit number.

EXAMPLE 4 **Estimating a Large Number**

Write an estimate of the number $K = e^{700}$ in scientific notation.

Solution

$$\begin{aligned} K &= e^{700} \\ \log K &= \log(e^{700}) \\ &= 700 \log e \\ &\approx 304.0061373 && \text{Use a calculator.} \end{aligned}$$

Because $\log K$ lies between the integers 304 and 305, the number K requires 305 digits to the left of the decimal point. Also, by definition of the common logarithm, we have

$$\begin{aligned} K &\approx 10^{304.0061373} && \text{Exponential form} \\ &= 10^{0.0061373} \times 10^{304} && a^{m+n} = a^m \times a^n \\ &\approx 1.014232 \times 10^{304} && \text{Use a calculator.} \end{aligned}$$

Practice Problem 4 Repeat Example 4 for $K = (234)^{567}$.

3 Change the base of a logarithm.

Change of Base

Calculators usually come with two types of log keys: a key for natural logarithms (base e, LN) and a key for common logarithms (base 10, LOG). These two types of logarithms are frequently used in applications. Sometimes, however, we need to evaluate logarithms for bases other than e and 10. The *change-of-base formula* helps us evaluate these logarithms.

Suppose we are given $\log_b x$ and we want to find an equivalent expression in terms of logarithms with base a. We let

$$u = \log_b x. \qquad (12)$$

In exponential form, we have

$$x = b^u. \qquad (13)$$

Then

$$\begin{aligned} \log_a x &= \log_a b^u && \text{Take } \log_a \text{ of each side of (13).} \\ \log_a x &= u \log_a b && \text{Power rule} \\ u &= \frac{\log_a x}{\log_a b} && \text{Solve for } u. \\ \log_b x &= \frac{\log_a x}{\log_a b} \quad (14) && \text{Substitute for } u \text{ from (12).} \end{aligned}$$

TECHNOLOGY CONNECTION

To graph a logarithmic function whose base is not e or 10, use the change-of-base formula. For example, graph $y = \log_5 x$ by entering $Y_1 = \text{LN}(X)/\text{LN}(5)$ or $Y_1 = \log(X)/\log(5)$.

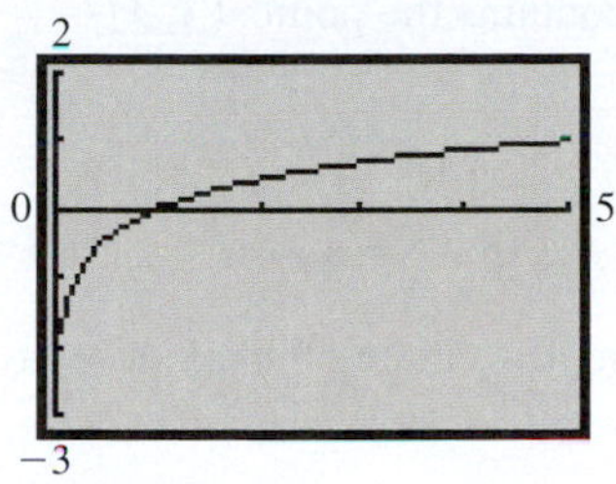

Equation (14) is the change-of-base formula. By choosing $a = 10$ and $a = e$, we can state the change-of-base formula as follows:

CHANGE-OF-BASE FORMULA

Let a, b, and x be positive real numbers with $a \neq 1$ and $b \neq 1$. Then $\log_b x$ can be converted to a different base as follows:

$$\log_b x = \underset{(\text{base } a)}{\frac{\log_a x}{\log_a b}} = \underset{(\text{base } 10)}{\frac{\log x}{\log b}} = \underset{(\text{base } e)}{\frac{\ln x}{\ln b}}$$

EXAMPLE 5 Using a Change of Base to Compute Logarithms

Compute $\log_5 13$ by changing to **a.** common logarithms and **b.** natural logarithms.

Solution

a. $\log_5 13 = \dfrac{\log 13}{\log 5}$ Change to base 10.

≈ 1.59369 Use a calculator.

b. $\log_5 13 = \dfrac{\ln 13}{\ln 5}$ Change to base e.

≈ 1.59369 Use a calculator.

Practice Problem 5 Find $\log_3 15$.

EXAMPLE 6 Modeling with Logarithmic Functions

Find the following.

a. An equation of the form $y = c + b \log x$ whose graph contains the points $(2, 3)$ and $(4, -5)$.

b. An equation of the form $y = c + b \log_a x$ whose graph contains the points $(2, 3)$ and $(4, -5)$ where a, b, and c are integers.

Solution

Substitute $(2, 3)$ and $(4, -5)$ in the equation $y = c + b \log x$ to obtain:

$$3 = c + b \log 2 \qquad (15)$$
$$-5 = c + b \log 4 \qquad (16)$$

a. Find b: Subtract equation (16) from equation (15) to obtain

$$8 = b \log 2 - b \log 4 = b(\log 2 - \log 4) = b\left(\log \frac{2}{4}\right) = b \log\left(\frac{1}{2}\right)$$
$$8 = b \log\left(\frac{1}{2}\right) = b \log 2^{-1} = -b \log 2$$

$b = -\dfrac{8}{\log 2}$ Solve for b.

Find c: $c = 3 - b \log 2$ From equation (15)

$= 3 - \left(-\dfrac{8}{\log 2}\right)\log 2$ Replace b with $-\dfrac{8}{\log 2}$

$= 3 + 8$ Simplify.

So, $y = 11 - \dfrac{8}{\log 2} \log x$ Substitute for b and c.

b. Use the change-of-base formula to find a, b, and c with integer values.

$$y = 11 - \frac{8}{\log 2}(\log x) = 11 - 8\left(\frac{\log x}{\log 2}\right) \quad \text{Answer to Part } \mathbf{a}$$

$$y = 11 - 8\log_2 x \quad \log_2 x = \frac{\log x}{\log 2}$$

Then $a = 2$, $b = -8$, and $c = 11$.

Practice Problem 6 Repeat Example 6 for the graph that contains the points $(3, 3)$ and $(9, 1)$.

4 Apply logarithms to growth and decay.

Growth and Decay

We express the exponential growth (or decay) formula (page 404) in an equivalent logarithmic form.

GROWTH AND DECAY FORMULA

Exponential Form	Logarithmic Form
$A(t) = A_0 e^{kt}$	$\ln\left(\frac{A(t)}{A_0}\right) = kt$

where $A(t)$ = the amount of substance (or population) at time t,
$A_0 = A(0)$ is the initial amount, and
t = time.

Growth occurs when $k > 0$ and decay occurs when $k < 0$.

Half-Life

The **half-life** of any quantity whose value decreases with time is the time required for the quantity to decay to half its initial value.

Radioactive substances undergo exponential decay. Krypton-85, for example, has a half-life of about ten years; this means that any initial quantity of krypton-85 will dissipate or decay to half that amount in about ten years.

If h is the half-life of a substance undergoing exponential decay at the rate k, then any mass A_0 decays to $\frac{1}{2}A_0$ in time h. See Figure 4.12. This fact leads to a half-life formula.

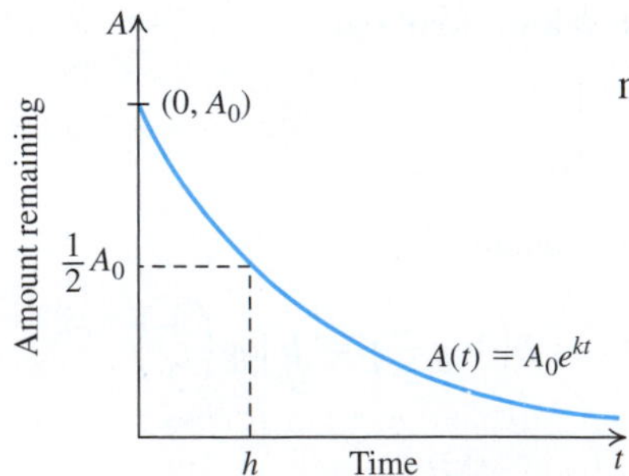

FIGURE 4.12 Time h is the half-life for the substance with this decay curve

$$\ln\left(\frac{\frac{1}{2}A_0}{A_0}\right) = kh \quad t = h \text{ and } A(h) = \frac{1}{2}A_0 \text{ in the logarithmic form}$$

$$\ln\left(\frac{1}{2}\right) = kh \quad \text{Simplify by dividing out } A_0.$$

$$-\ln 2 = kh \quad \ln\left(\frac{1}{2}\right) = \ln(2^{-1}) = -\ln 2$$

$$h = -\frac{\ln 2}{k} \quad \text{Solve for } h.$$

HALF-LIFE FORMULA

The half-life h of a substance undergoing exponential decay at a rate k $(k < 0)$ is given by the formula

$$h = -\frac{\ln 2}{k}.$$

EXAMPLE 7 Finding the Half-Life of a Substance

In an experiment, 18 grams of the radioactive element sodium-24 decayed to 6 grams in 24 hours. Find its half-life to the nearest hour.

Solution

$$\ln\left(\frac{A(t)}{A_0}\right) = kt \qquad \text{Logarithmic growth and decay model}$$

$$\ln\left(\frac{6}{18}\right) = k(24) \qquad A_0 = 18, A(24) = 6$$

$$\ln\left(\frac{1}{3}\right) = 24\,k \qquad \text{Simplify.}$$

$$-\ln 3 = 24k \qquad \ln\left(\frac{1}{3}\right) = \ln\left(3^{-1}\right) = -\ln 3$$

$$k = -\frac{\ln 3}{24} \qquad \text{Solve for } k.$$

To find the half-life of sodium-24, we use the formula

$$h = -\frac{\ln 2}{k} \qquad \text{Half-life formula}$$

$$h = -\frac{\ln 2}{\left(-\frac{\ln 3}{24}\right)} \qquad \text{Replace } k \text{ with } -\frac{\ln 3}{24}.$$

$$h = \frac{24 \ln 2}{\ln 3} \approx 15 \text{ hours} \qquad \text{Simplify; use a calculator.}$$

Practice Problem 7 In an experiment, 100 grams of the radioactive element strontium-90 decayed to 66 grams in 15 years. Find its half-life to the nearest year.

Willard Frank Libby (1908–1980)

Libby was born in Grand Valley, Colorado. In 1927, he entered the University of California at Berkeley, where he earned his BSc and PhD degrees in 1931 and 1933, respectively. He then became an instructor in the Department of Chemistry at the University of California (Berkeley) in 1933. He was awarded a Guggenheim Memorial Foundation Fellowship in 1941, but the fellowship was interrupted when Libby went to Columbia University on the Manhattan District Project upon America's entry into World War II.

Libby was a physical chemist and a specialist in radiochemistry, particularly in hot-atom chemistry, tracer techniques, and isotope tracer work. He became well known at the University of Chicago for his work on natural carbon-14 (a radiocarbon) and its use in dating archaeological artifacts and on natural tritium and its use in hydrology and geophysics. He was awarded the Nobel Prize in Chemistry in 1960.

Radiocarbon Dating

Ordinary carbon, called carbon-12 (^{12}C), is stable and does not decay. However, carbon-14 (^{14}C), is a form of carbon that decays radioactively with a half-life of 5700 years. Sunlight constantly produces carbon-14 in Earth's atmosphere. When a living organism breathes or eats, it absorbs carbon-12 and carbon-14. After the organism dies, no more carbon-14 is absorbed; so the age of its remains can be calculated by determining how much carbon-14 has decayed. The method of radiocarbon dating was developed by the American scientist W. F. Libby.

EXAMPLE 8 King Tut's Treasure

In 1960, a group of specialists from the British Museum in London investigated whether a piece of art containing organic material found in Tutankhamun's tomb had been made during his reign or (as some historians claimed) whether it belonged to an earlier period. We know that King Tut died in 1346 B.C. and ruled Egypt for ten years. What percent of the amount of carbon-14 originally contained in the object should be present in 1960 if the object was made during Tutankhamun's reign?

Solution

Because the half-life h of carbon-14 is approximately 5700 years, we have

$$5700 = -\frac{\ln 2}{k} \quad \text{Half-life formula with}$$

$$k = -\frac{\ln 2}{5700} \quad \text{Solve for } k.$$

$$\approx -0.0001216 \quad \text{Use a calculator.}$$

Substituting this value of k in the exponential form, we obtain

$$A(t) = A_0 e^{-0.0001216t} \qquad (17)$$

The time t that elapsed between King Tut's death and 1960 is

$$t = 1960 + 1346 \quad \text{1346 B.C. to A.D. 1960}$$

$$= 3306.$$

$$A(3306) = A_0 e^{-0.0001216(3306)} \quad \text{Replace } t \text{ with 3306 in (17).}$$

$$\approx 0.66897A_0.$$

Therefore, the percent of the original amount of carbon-14 remaining in the object (after 3306 years) is 66.897%.

Because King Tut began ruling Egypt ten years before he died, the time t_1 that elapsed from the beginning of his reign to 1960 is 3316 years.

$$A(3316) = A_0 e^{-0.0001216(3316)} \quad \text{Replace } t \text{ with 3316 in (17).}$$

$$\approx 0.66816A_0.$$

Therefore, a piece of art made during King Tut's reign would have between 66.816% and 66.897% of carbon-14 remaining in 1960.

Practice Problem 8 Repeat Example 8 assuming that the art object was made 194 years before King Tut's death.

SUMMARY OF PROPERTIES AND RULES OF LOGARITHMS

$\log_a 1 = 0$ $\quad \log_a a = 1$ $\quad \log_a a^x = x$ $\quad a^{\log_a x} = x$

$\log_a (MN) = \log_a M + \log_a N$ $\quad \log_a\left(\frac{M}{N}\right) = \log_a M - \log_a N$

$\log_a M^r = r \log_a M$ $\quad \log_b M = \frac{\log_a M}{\log_a b}$

Answers to Practice Problems

1. a. -1 **b.** 13

2. a. $\ln(2x - 1) - \ln(x + 4)$

b. $\log 2 + \frac{1}{2}\log x + \frac{1}{2}\log y - \frac{1}{2}\log z$

3. $\log\sqrt{x^2 - 1}$

4. $2.215088 \times 10^{1343}$

5. $\frac{\log 15}{\log 3} = \frac{\ln 15}{\ln 3} \approx 2.46497$

6. a. $y = 5 - \frac{2}{\log 3}\log x$ **b.** $y = 5 - 2\log_3 x$

7. ≈ 25 years **8.** $\approx 65.34\%$

SECTION 4.3 Exercises

Basic Concepts and Skills

1. $\log_a MN =$ ________ + ________
2. ________ $= \log_a M - \log_a N$
3. $\log_a M^r =$ ________
4. The change-of-base formula using base e is $\log_a M =$ ________.
5. **True or False.** $\log_a (u + v) = \log_a u + \log_a v$
6. **True or False.** $\log \dfrac{x}{10} = \log x - 1$

In Exercises 7–18, given that $\log x = 2$, $\log y = 3$, $\log 2 \approx 0.3$, and $\log 3 \approx 0.48$, evaluate each expression without using a calculator.

7. $\log 6$
8. $\log 4$
9. $\log 5 \left[\textit{Hint: } 5 = \dfrac{10}{2} \right]$
10. $\log (3x)$
11. $\log \left(\dfrac{2}{x} \right)$
12. $\log x^2$
13. $\log (2x^2 y)$
14. $\log xy^3$
15. $\log \sqrt[3]{x^2 y^4}$
16. $\log (\log x^2)$
17. $\log \sqrt[3]{48}$
18. $\log_2 3$

In Exercises 19–32, write each expression in expanded form. Assume that all expressions containing variables represent positive numbers.

19. $\ln [x(x - 1)]$
20. $\ln \dfrac{x(x + 1)}{(x - 1)^2}$
21. $\log_a \sqrt{xy^3}$
22. $\log_a \dfrac{3x^2}{\sqrt{y}}$
23. $\log_a \sqrt[3]{\dfrac{x}{y}}$
24. $\log_a \sqrt[3]{\dfrac{x^2}{y^5}}$
25. $\log_2 \sqrt[4]{\dfrac{xy^2}{8}}$
26. $\log \sqrt[3]{\dfrac{x^2 y}{100}}$
27. $\log \dfrac{\sqrt{x^2 + 1}}{x + 3}$
28. $\log_4 \left(\dfrac{x^2 - 9}{x^2 - 6x + 8} \right)^{2/3}$
29. $\log_b x^2 y^3 z$
30. $\log_b \sqrt{xyz}$
31. $\ln \left[\dfrac{x\sqrt{x - 1}}{x^2 + 2} \right]$
32. $\ln \left[\dfrac{\sqrt{x - 2}\,\sqrt[3]{x + 1}}{x^2 + 3} \right]$
33. $\ln \left[\dfrac{(x + 1)^2}{(x - 3)\sqrt{x + 4}} \right]$
34. $\ln \left[\dfrac{2x + 3}{(x + 4)^2 (x - 3)^4} \right]$
35. $\ln \left[(x + 1)\sqrt{\dfrac{x^2 + 2}{x^2 + 5}} \right]$
36. $\ln \left[\dfrac{\sqrt[3]{2x + 1}\,(x + 1)}{(x - 1)^2 (3x + 2)} \right]$
37. $\ln \left[\dfrac{x^3 (3x + 1)^4}{\sqrt{x^2 + 1}\,(x + 2)^{-5} (x - 3)^2} \right]$
38. $\ln \left[\dfrac{(x + 1)^{1/2} (x^2 - 2)^{2/5}}{(2x - 1)^{3/2} (x^2 + 2)^{-4/5}} \right]$

In Exercises 39–48, write each expression in condensed form.

39. $\log_2 x + \log_2 7$
40. $\log_2 x - \log_2 3$
41. $\dfrac{1}{2} \log x - \log y + \log z$
42. $\dfrac{1}{2} (\log x + \log y)$
43. $\dfrac{1}{5} (\log_2 z + 2 \log_2 y)$
44. $\dfrac{1}{3} (\log x - 2 \log y + 3 \log z)$
45. $\ln x + 2 \ln y + 3 \ln z$
46. $2 \ln x - 3 \ln y + 4 \ln z$
47. $2 \ln x - \dfrac{1}{2} \ln (x^2 + 1)$
48. $2 \ln x + \dfrac{1}{2} \ln (x^2 - 1) - \dfrac{1}{2} \ln (x^2 + 1)$

In Exercises 49–52, write an estimate of each number in scientific notation.

49. e^{500}
50. π^{650}
51. 324^{756}
52. 723^{416}
53. Which is larger, 234^{567} or 567^{234}?
54. Which is larger, 4321^{8765} or 8765^{4321}?
55. Find the number of digits in $17^{200} \cdot 53^{67}$.
56. Find the number of digits in $67^{200} \div 23^{150}$.

In Exercises 57–64, use the change-of-base formula and a calculator to evaluate each logarithm.

57. $\log_2 5$
58. $\log_4 11$
59. $\log_{1/2} 3$
60. $\log_{\sqrt{3}} 12.5$
61. $\log_{\sqrt{5}} \sqrt{17}$
62. $\log_{15} 123$
63. $\log_2 7 + \log_4 3$
64. $\log_2 9 - \log_{\sqrt{2}} 5$

In Exercises 65–72, find the value of each expression without using a calculator.

65. $\log_3 \sqrt{3}$
66. $\log_{1/4} 4$
67. $\log_3 (\log_2 8)$
68. $2^{\log_2 2}$
69. $5^{2 \log_5 3 + \log_5 2}$
70. $e^{3 \ln 2 - 2 \ln 3}$
71. $\log 4 + 2 \log 5$
72. $\log_2 160 - \log_2 5$

In Exercises 73–80, find an equation of the form $y = c + b\log_a x$ (a, b, and c integers) whose graph contains the two given points.

73. $(10, 1)$ and $(1, 2)$
74. $(4, 10)$ and $(2, 12)$
75. $(e, 1)$ and $(1, 2)$
76. $(3, 1)$ and $(9, 2)$
77. $(5, 4)$ and $(25, 7)$
78. $(4, 3)$ and $(8, 5)$
79. $(2, 4)$ and $(4, 9)$
80. $(1, 1)$ and $(5, 7)$

In Exercises 81–84, determine the half-life of each substance that decays from A_0 to A in time t.

81. $A_0 = 50$, $A = 23$, $t = 12$ years
82. $A_0 = 200$, $A = 65$, $t = 10$ years
83. $A_0 = 10.3$, $A = 3.8$, $t = 15$ hours
84. $A_0 = 20.8$, $A = 12.3$, $t = 40$ minutes

Applying the Concepts

In Exercises 85–88, assume the exponential growth model $A(t) = A_0e^{kt}$ or $\ln\left(\frac{A(t)}{A_0}\right) = kt$ and a world population of 7 billion in 2011.

85. **Rate of population growth.** If the world adds about 90 million people in 2012, what is the rate of growth?

86. **Population.** Use the rate of growth from Exercise 85 to estimate the year in which the world population will be
 a. 12 billion.
 b. 20 billion.

87. **Population growth.** If the population must stay below 20 billion during the next 100 years, what is the maximum acceptable annual rate of growth? (Round your answer to six decimal places.)

88. **Population decline.** If the world population must decline below 5 billion during the next 25 years, what must be the annual rate of decline? (Round your answer to six decimal places.)

89. **Continuous compounding.** If \$1000 is deposited in a bank at 10% interest compounded continuously, how many years will it take for the money to grow to \$3500?

90. **Continuous compounding.** At what rate of interest compounded continuously will an investment double in six years?

91. **Half-life.** Iodine-131, a radioactive substance that is effective in locating brain tumors, has a half-life of only eight days. A hospital purchased 20 grams of the substance but had to wait five days before it could be used. How much of the substance was left after five days?

92. **Half-life.** Plutonium-241 has a half-life of 13 years. A laboratory purchased 10 grams of the substance but did not use it for two years. How much of the substance was left after two years?

93. **Half-life.** Tritium is used in nuclear weapons to increase their power. It decays at the rate of 5.5% per year. Calculate the half-life of tritium.

94. **Half-life.** Sixty grams of magnesium-28 decayed into 7.5 grams after 63 hours. What is the half-life of magnesium-28?

Effective medicine dosage. Use the following model in Exercises 95–98: The concentration $C(t)$ of a drug administered to a patient intravenously jumps to its highest level almost immediately. The concentration subsequently decays exponentially according to the law

$$C(t) = C_0e^{-kt}, \text{ or equivalently, } \ln\left(\frac{C(t)}{C_0}\right) = -kt,$$

where $C(0) = C_0$. The physician administering the drug would like to have

$$m < C(t) < M,$$

where m is the concentration below which the drug is ineffective and M is the concentration above which the drug is dangerous.

95. **Drug concentration.** Suppose that immediately after a certain drug is administered, the concentration is 5 milligrams per milliliter. Ten hours later the concentration drops to 1.5 milligrams per milliliter. Determine the value of k for this drug.

96. **Drug concentration.** For the drug of Exercise 95, suppose the maximum safe level is $M = 12$ milligrams per milliliter and the minimum effective level is $m = 3$ milligrams per milliliter. If the initial concentration is M, find the maximum possible time between doses of this drug.

97. **Half-life.** The half-life of Valium is 36 hours. Suppose a patient receives 16 milligrams of the drug at 8 A.M. How much Valium is in the patient's blood at 4 P.M. the same day?

98. **Half-life.** The half-life of aspirin in the blood is 12 hours. Estimate the time required for the aspirin to decay to 90% of the original amount ingested.

In Exercises 99–102, use the following amortization formula:

$$P = \frac{r \cdot M}{1 - \left(1 + \frac{r}{n}\right)^{-nt}} \div n$$

where P = the payment, r = the annual interest rate, M = the mortgage amount, t = the number of years, and n = the number of payments per year.

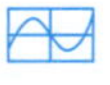
99. **Finding *P*.** What is the monthly payment on a mortgage of \$120,000 with a 6% interest rate for 20 years? How much interest will be paid over 20 years?

100. **Finding *t*.** The Garcia family wanted to take out a mortgage for \$150,000 at 8% interest with monthly payments. The family can afford monthly payments of \$1200. How long would they have to make payments to pay off their mortgage, and how much interest would they be paying?

101. **Finding *M*.** The First National Bank offers Andy an 8.5% interest rate on a 30-year mortgage to be paid back in monthly payments. The most Andy can afford to pay in monthly payments is \$850. What mortgage amount can Andy afford?

 **102. Finding r.** Lisa is 40 years old and needs to take out a mortgage of \$120,000. She can afford monthly payments up to \$850. She wants her mortgage paid off when she retires at age 65. What interest rate on the mortgage will satisfy all her requirements? [*Hint*: Start with $r = 7\%$ and compute the monthly payment, increasing r by 0.001 each time.]

Beyond the Basics

103. Show that
$\log_b(\sqrt{x^2+1} - x) + \log_b(\sqrt{x^2+1} + x) = 0.$

104. Show that
$\log_b(\sqrt{x+1} + \sqrt{x}) = -\log_b(\sqrt{x+1} - \sqrt{x}).$

105. Show that $(\log_b a)(\log_a b) = 1.$

106. Show that $\log \frac{a}{b} + \log \frac{b}{a} = 0.$

 **107.** Write $\log_2 x = \frac{\ln x}{\ln 2}$ by the change-of-base formula.

Then use a graphing calculator to sketch the graph of $y = \log_2 x$.

 **108.** Sketch the graph of $y = \log_5(x+3)$.

109. Simplify each expression.

a. $\log\left(\frac{a}{b}\right) + \log\left(\frac{b}{a}\right) + \log\left(\frac{c}{a}\right) + \log\left(\frac{a}{c}\right)$

b. $\log\left(\frac{a^2}{bc}\right) + \log\left(\frac{b^2}{ca}\right) + \log\left(\frac{c^2}{ab}\right)$

c. $\log_2 3 \cdot \log_3 4$

d. $\log_a b \cdot \log_b c \cdot \log_c a$

110. Find the number of digits in N if $\log_2(\log_2 N) = 4$.

111. Find the domain of
$f(x) = \log_4(\log_5(\log_3(18x - x^2 - 77)))$

112. Let $f(x) = \log_a x$. Show that

a. $f\left(\frac{1}{x}\right) = -f(x) = \log_{1/a} x.$

b. $\frac{f(x+h) - f(x)}{h} = \log_a\left(1 + \frac{h}{x}\right)^{1/h}, h \neq 0.$

113. State whether each of the following is true or false for all permissible values of the variable.

a. $\log(x+2) = \log x + \log 2$

b. $\log 2x = \log x + \log 2$

c. $\log_2 3x = 3\log_2 x$

d. $(\log_3 2x)^4 = 4\log_3 2x$

e. $\log_5 x^2 = 2\log_5 |x|$

f. $\log \frac{x}{3} = \log x - \log 3$

g. $\log \frac{x}{4} = \frac{\log x}{\log 4}$

h. $\ln\left(\frac{1}{x}\right) = -\ln x$

i. $\log_2 x^2 = (\log_2 x)(\log_2 x)$

j. $\log|10x| = 1 + \log|x|$

114. a. Prove the quotient rule for logarithms:
$\log_a \frac{M}{N} = \log_a M - \log_a N.$

b. Prove the power rule for logarithms: $\log_a M^r = r\log_a M.$

115. If $\log\left(\frac{a+b}{3}\right) = \frac{1}{2}(\log a + \log b)$, show that
$a^2 + b^2 - 7ab = 0.$

116. Evaluate $\frac{1}{1+\log_c ab} + \frac{1}{1+\log_a bc} + \frac{1}{1+\log_b ca}.$

[*Hint:* Convert to common logarithms and then simplify.]

Critical Thinking/Discussion/Writing

117. What went wrong? Find the error in the following argument:

$$3 < 4$$
$$3\log\frac{1}{2} < 4\log\frac{1}{2}$$
$$\log\left(\frac{1}{2}\right)^3 < \log\left(\frac{1}{2}\right)^4$$
$$\log\frac{1}{8} < \log\frac{1}{16}$$
$$\frac{1}{8} < \frac{1}{16}$$
$$2 < 1$$

118. By the power rule for logarithms, $\log x^2 = 2\log x$. However, the graphs of $f(x) = \log x^2$ and $g(x) = 2\log x$ are not identical. Explain why.

119. Prime Number. As of 2008, the largest known prime number was $2^{43112609} - 1$. How many digits does this prime number have?

[*Hint*: First, argue that if $p = 2^m - 1$ is a prime number, then p and 2^m have the same number of digits.]

120. Find the domain of $g(x) = \frac{1}{\log(1-x)} + \frac{1}{\ln(x+2)}.$

Maintaining Skills

In Exercises 121–124, simplify each expression and write your result without exponents.

121. $11 \cdot 3^0$

122. $-4 \cdot 5^x \cdot 5^{-x}$

123. $4^x \cdot 2^{-2x+1}$

124. $(7^x)^2 \cdot (7^2)^{-x}$

In Exercises 125–128, let $t = 5^x$ and write each expression in the form $at^2 + bt + c = 0$, where a, b, and c are real numbers.

125. $5^{2x} - 5^x = -1$

126. $3 \cdot 5^{2x} - 2 \cdot 5^x = 7$

127. $\frac{5^x + 3 \cdot 5^{-x}}{5^x} = \frac{1}{4}$

128. $\frac{5^{-x} + 5^x}{5^{-x} - 5^x} = \frac{1}{2}$

In Exercises 129–132, solve for x.

129. $2x - (11 + x) = 8x + (7 + 2x)$

130. $4x - 1 + (6x - 2) = 3 - (5x + 1)$

131. $x^2 + 3x - 1 = 3$

132. $2x^2 - 7x = 3x + 48$

In Exercises 133–136, solve for x.

133. $2x < 7 + x$

134. $5x - 2 \le 19 - (1 - x)$

135. $12x > 30 - 3x$

136. $4x - 17 \ge 6 - (5 + 2x)$

SECTION 4.4

Exponential and Logarithmic Equations and Inequalities

BEFORE STARTING THIS SECTION, REVIEW

1. One-to-one property of exponential functions (Section 4.1, page 396)
2. One-to-one property of logarithmic functions (Section 4.2, page 415)
3. Rules of logarithms (Section 4.3, page 426)

OBJECTIVES

1. Solve exponential equations.
2. Solve applied problems involving exponential equations.
3. Solve logarithmic equations.
4. Use the logistic growth model.
5. Use logarithmic and exponential inequalities.

◆ Logistic Growth Model

In Sections 4.1 and 4.3, we discussed the population growth formula

$$A(t) = A_0 e^{kt}. \qquad (18)$$

Equation (18) is a realistic model for a biological population that has plenty of food, enough space to grow, and no threat from predators.

However, most populations are constrained by limitations on resources such as the plants that populations eat. In the 1830s, the Belgian scientist P. F. Verhulst added another constraint to the population growth model: *There is a limited sustainable maximum population M of a species that the habitat can support*. In population biology, M is called the *carrying capacity*.

Verhulst replaced equation (18) with

$$P(t) = \frac{M}{1 + ae^{-kt}}, \qquad (19)$$

where k is the rate of growth and a is a positive constant. In 1840, using the data from the first five U.S. censuses, he made a prediction of the U.S. population for the year 1940. As things turned out, it was off by less than 1%.

The mathematical model represented by equation (19) is called the *logistic growth model* or the *Verhulst model*. Verhulst introduced the term *logistique* to refer to the "loglike" qualities of the S-shaped graph of equation (19). See Figure 4.13.

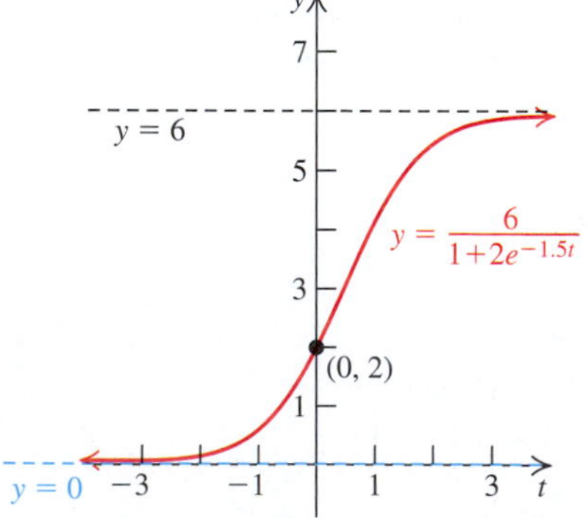

FIGURE 4.13 A logistic curve

To sketch the graph of equation (19), we notice that the y-intercept is

$$P(0) = \frac{M}{1 + ae^0} = \frac{M}{1 + a}.$$

As $t \to \infty$, the quantity $ae^{-kt} \to 0$ (because k is positive). Consequently, the denominator $1 + ae^{-kt} \to 1$; therefore, $P(t) \to M$ and the line $y = M$ is a horizontal asymptote.

Similarly, as $t \to -\infty$, $e^{-kt} \to \infty$. In this case, the denominator $1 + ae^{-kt} \to \infty$ and $P(t) \to 0$. The x-axis, or $y = 0$, is another horizontal asymptote. See the graph of $y = \frac{6}{1 + 2e^{-1.5t}}$ in Figure 4.13.

In Example 9 we use the logistic function defined by equation (19).

1 Solve exponential equations.

RECALL

The one-to-one property of exponential and logarithmic functions states that

- if $a^u = a^v$, then $u = v$.
- if $\log_a u = \log_a v$, then $u = v$.

Solving Exponential Equations

An equation containing terms of the form $a^x (a > 0, a \neq 1)$ is called an **exponential equation**. Here are some examples of simple exponential equations:

$$2^x = 15$$
$$9^x = 3^{x+1}$$
$$7^x = 13^{2x}$$
$$5 \cdot 2^{x-3} = 17$$

If both sides of an equation can be expressed as a power of the same base, then we can use the one-to-one property of exponential functions to solve the equation.

EXAMPLE 1 Solving an Exponential Equation

Solve each equation.

a. $25^x = 125$ **b.** $9^x = 3^{x+1}$

Solution

a.

$(5^2)^x = 5^3$	Rewrite 25 as 5^2 and 125 as 5^3.
$5^{2x} = 5^3$	Power rule for exponents
$2x = 3$	One-to-one property
$x = \dfrac{3}{2}$	Solve for x.

b.

$(3^2)^x = 3^{x+1}$	Rewrite 9 as 3^2.
$3^{2x} = 3^{x+1}$	Power rule of exponents
$2x = x + 1$	One-to-one property
$x = 1$	Solve for x.

Practice Problem 1 Solve each equation.

a. $3^x = 243$ **b.** $8^x = 4$

We can use logarithms to solve those equations for which both sides cannot be readily expressed with the same base. For example, the fact that $\log 2^x = x \log 2$ allows us to solve the equation $2^x = 9$ by first taking the log of both sides.

$2^x = 9$	
$\log 2^x = \log 9$	Take the log of both sides.
$x \log 2 = \log 9$	Power rule for logarithms.
$x = \dfrac{\log 9}{\log 2}$	Solve for x.

An approximation $x \approx 3.1699$ is found by using a calculator.

The general procedure for solving exponential equations using logarithms is given next.

PROCEDURE IN ACTION

EXAMPLE 2 Solving Exponential Equations Using Logarithms

OBJECTIVE	EXAMPLE	
Solve exponential equations when both sides are not expressed with the same base.	Solve for x: $5 \cdot 2^{x-3} = 17$	
Step 1 Isolate the exponential expression on one side of the equation.	**1.** $2^{x-3} = \frac{17}{5} = 3.4$	Solve for 2^{x-3}.
Step 2 Take the common or natural logarithm of both sides.	**2.** $\ln 2^{x-3} = \ln (3.4)$	Take ln of both sides.
Step 3 Use the power rule, $\log_a M^r = r \log_a M$.	**3.** $(x - 3) \ln 2 = \ln (3.4)$	Power Rule of logarithms
Step 4 Solve for the variable.	**4.** $x - 3 = \frac{\ln (3.4)}{\ln 2}$	Divide both sides by ln 2.
	$x = \frac{\ln (3.4)}{\ln 2} + 3$	Solve for x; Exact answer:
	≈ 4.766	Use a calculator.

Practice Problem 2 Solve for x: $7 \cdot 3^{x+1} = 11$.

EXAMPLE 3 Solving an Exponential Equation with Different Bases

Solve the equation $5^{2x-3} = 3^{x+1}$ and approximate the answer to three decimal places.

Solution

When different bases are involved, we begin with Step 2 of the procedure above.

$\ln 5^{2x-3} = \ln 3^{x+1}$	Take the natural log of both sides.
$(2x - 3)\ln 5 = (x + 1)\ln 3$	Power rule of logarithms
$2x \ln 5 - 3 \ln 5 = x \ln 3 + \ln 3$	Distributive property
$2x \ln 5 - x \ln 3 = \ln 3 + 3 \ln 5$	Collect terms containing x on one side.
$x(2 \ln 5 - \ln 3) = \ln 3 + 3 \ln 5$	Factor out x on the left side.
$x = \frac{\ln 3 + 3 \ln 5}{2 \ln 5 - \ln 3}$	Solve for x to obtain an *exact* solution.
≈ 2.795	Use a calculator.

SIDE NOTE

When using a calculator to evaluate such expressions, make sure you enclose the numerator and denominator in parentheses.

Practice Problem 3 Solve the equation $3^{x+1} = 2^{2x}$.

EXAMPLE 4 An Exponential Equation of Quadratic Form

Solve the equation $3^x - 8 \cdot 3^{-x} = 2$.

Solution

$3^x(3^x - 8 \cdot 3^{-x}) = 2(3^x)$	Multiply both sides by 3^x.
$3^{2x} - 8 \cdot 3^0 = 2 \cdot 3^x$	Distributive property; $3^x \cdot 3^{-x} = 3^0$
$3^{2x} - 8 = 2 \cdot 3^x$	$3^0 = 1$
$3^{2x} - 2 \cdot 3^x - 8 = 0$	Subtract $2 \cdot 3^x$ from both sides.

The last equation is quadratic in form. To see this, we let $y = 3^x$; then $y^2 = (3^x)^2 = 3^{2x}$. So

$$\begin{aligned}
3^{2x} - 2\cdot 3^x - 8 &= 0 \\
y^2 - 2y - 8 &= 0 && \text{Substitute: } y = 3^x, y^2 = 3^{2x}. \\
(y + 2)(y - 4) &= 0 && \text{Factor.} \\
y + 2 = 0 \quad \text{or} \quad y - 4 &= 0 && \text{Zero-product property} \\
y = -2 \quad \text{or} \quad y &= 4 && \text{Solve for } y. \\
3^x = -2 \quad \text{or} \quad 3^x &= 4 && \text{Replace } y \text{ with } 3^x.
\end{aligned}$$

But $3^x = -2$ is not possible because $3^x > 0$ for all real numbers x. Solve $3^x = 4$.

$$\begin{aligned}
\ln 3^x &= \ln 4 && \text{Take the natural log of both sides.} \\
x \ln 3 &= \ln 4 && \text{Power rule of logarithms} \\
x &= \frac{\ln 4}{\ln 3} && \text{An exact solution} \\
&\approx 1.262 && \text{Use a calculator.}
\end{aligned}$$

Practice Problem 4 Solve the equation $e^{2x} - 4e^x - 5 = 0$.

2 Solve applied problems involving exponential equations.

Applications of Exponential Equations

EXAMPLE 5 Solving a Population Growth Problem

The following table shows the approximate population and annual growth rate of the United States and Pakistan in 2010.

Country	Population	Annual Population Growth Rate
United States	308 million	0.9%
Pakistan	185 million	2.1%

Source: www.cia.gov

Use the alternate population model $P(t) = P_0(1 + r)^t$, where P_0 is the initial population and t is the time in years since 2010. Assume that the growth rate for each country stays the same.

a. Estimate the population of each country in 2020.

b. In what year will the population of the United States be 350 million?

c. In what year will the population of Pakistan be the same as the population of the United States?

Solution

The given model is

$$P(t) = P_0(1 + r)^t.$$

a. In 2010, the initial year, the U.S. population was $P_0 = 308$. In ten years, 2020, the population of the United States will be

$$308(1 + 0.009)^{10} \approx 336.87 \text{ million} \qquad P_0 = 308, r = 0.009, t = 10$$

In the same way, the population of Pakistan in 2020 will be

$$185(1 + 0.021)^{10} \approx 227.73 \text{ million} \qquad P_0 = 185, r = 0.021, t = 10$$

b. To find the year when the population of the United States will be 350 million, we solve for t in the equation

$$350 = 308(1 + 0.009)^t \qquad P(t) = 350,\ P_0 = 308,\ r = 0.009$$

$$\frac{350}{308} = (1.009)^t \qquad \text{Divide both sides by 308.}$$

$$\ln\left(\frac{350}{308}\right) = \ln(1.009)^t \qquad \text{Take natural logs of both sides.}$$

$$\ln\left(\frac{350}{308}\right) = t\ln(1.009) \qquad \text{Power rule for logarithms}$$

$$t = \frac{\ln\left(\frac{350}{308}\right)}{\ln(1.009)} \approx 14.27 \qquad \text{Solve for } t \text{ and use a calculator.}$$

The population of the United States will be 350 million approximately 14.27 years after 2010—that is, sometime in 2025.

c. To find when the population of the two countries will be the same, we solve for t in the equation

$$308(1.009)^t = 185(1.021)^t \qquad \text{Equate populations for both countries.}$$

$$\frac{308}{185} = \left(\frac{1.021}{1.009}\right)^t \qquad \text{Divide both sides by } 185(1.009)^t \text{ and rewrite.}$$

$$\ln\left(\frac{308}{185}\right) = \ln\left(\frac{1.021}{1.009}\right)^t \qquad \text{Take natural logs of both sides.}$$

$$\ln\left(\frac{308}{185}\right) = t\ln\left(\frac{1.021}{1.009}\right) \qquad \text{Power rule for logarithms}$$

$$t = \frac{\ln\left(\frac{308}{185}\right)}{\ln\left(\frac{1.021}{1.009}\right)} \qquad \text{Solve for } t \text{ to get } \textit{exact} \text{ solution.}$$

$$\approx 43.12 \text{ years} \qquad \text{Use a calculator.}$$

Therefore, the two populations will be equal in about 43.12 years, that is, during 2054.

Practice Problem 5 Repeat Example 5 assuming that the rates of growth for the United States and Pakistan are 1.1% and 3.3%, respectively.

3 Solve logarithmic equations.

Solving Logarithmic Equations

Equations that contain terms of the form $\log_a x$ are called **logarithmic equations**. Here are some examples of logarithmic equations.

$$\log_2 x = 4$$
$$\log_3(2x - 1) = \log_3(x + 2)$$
$$\log_2(x - 3) + \log_2(x - 4) = 1$$

If a logarithmic equation can be expressed in the form $\log_a N = c$, we can solve it by writing in the equivalent exponential form $N = a^c$.

For example, to solve an equation such as $\log_2 x = 4$, we rewrite it in the equivalent exponential form:

$$\log_2 x = 4 \text{ means } x = 2^4 = 16$$

So $x = 16$ is a solution of the equation $\log_2 x = 4$. Because the domain of logarithmic functions consists of positive numbers, we must check apparent solutions of logarithmic equations in the original equation.

EXAMPLE 6 Solving a Logarithmic Equation

Solve $4 + 3\log_2 x = 1$.

Solution

$$\begin{aligned} 4 + 3\log_2 x &= 1 && \text{Original equation} \\ 3\log_2 x &= 1 - 4 && \text{Subtract 4 from both sides.} \\ 3\log_2 x &= -3 && \text{Simplify.} \\ \log_2 x &= -1 && \text{Divide both sides by 3.} \\ x &= 2^{-1} && \text{Exponential form} \\ x &= \frac{1}{2} && a^{-n} = \frac{1}{a^n} \end{aligned}$$

RECALL

Remember that $\log_2 x$ means "the exponent on 2 that gives x."

Check Substitute $x = \frac{1}{2} = 2^{-1}$ into $4 + 3\log_2 x = 1$.

$$\begin{aligned} 4 + 3\log_2 2^{-1} &\stackrel{?}{=} 1 && \\ 4 + 3(-1)\log_2 2 &\stackrel{?}{=} 1 && \text{Power rule for logarithms} \\ 4 - 3 &\stackrel{?}{=} 1 && \log_2 2 = 1 \\ 1 &\stackrel{?}{=} 1 && \text{Yes} \end{aligned}$$

This check shows that the solution set is $\left\{\frac{1}{2}\right\}$.

Practice Problem 6 Solve $1 + 2\ln x = 4$.

EXAMPLE 7 Using the Product and Quotient Rules

Solve the following equations.

a. $\log_2(x - 3) + \log_2(x - 4) = 1$

b. $\log_2(x + 4) - \log_2(x + 3) = 1$

Solution

a.
$$\begin{aligned} \log_2(x - 3) + \log_2(x - 4) &= 1 && \text{Original equation} \\ \log_2[(x - 3)(x - 4)] &= 1 && \text{Product rule for logarithms} \\ (x - 3)(x - 4) &= 2^1 && \text{Exponential form} \\ x^2 - 7x + 10 &= 0 && \text{Write in standard form.} \\ (x - 2)(x - 5) &= 0 && \text{Factor.} \\ x - 2 = 0 \quad \text{or} \quad x - 5 &= 0 && \text{Zero-product property} \\ x = 2 \quad \text{or} \quad x &= 5 && \text{Solve for } x. \end{aligned}$$

We check the possible solutions in the original equation.

Check $x = 2$

$$\begin{aligned} \log_2(2 - 3) + \log_2(2 - 4) &\stackrel{?}{=} 1 \\ \log_2(-1) + \log_2(-2) &\stackrel{?}{=} 1 \quad \text{No} \end{aligned}$$

Because logarithms are not defined for negative numbers, $x = 2$ is an extraneous solution of the original equation.

Check $x = 5$

$$\begin{aligned} \log_2(5 - 3) + \log_2(5 - 4) &\stackrel{?}{=} 1 \\ \log_2 2 + \log_2 1 &\stackrel{?}{=} 1 \\ 1 + 0 &\stackrel{?}{=} 1 \\ 1 &\stackrel{?}{=} 1 \quad \text{Yes} \end{aligned}$$

The solution set is $\{5\}$.

b. $\log_2 (x + 4) - \log_2 (x + 3) = 1$ — Original equation

$\log_2 \frac{x + 4}{x + 3} = 1$ — Quotient rule

$\frac{x + 4}{x + 3} = 2^1$ — Exponential form

$x + 4 = 2(x + 3)$ — Multiply both sides by $x + 3$.

$(x + 2) = 0$ — Simplify.

$x = -2$ — Solve for x.

We check the possible solution in the original equation.

Check $x = -2$

$$\log_2 (-2 + 4) - \log_2 (-2 + 3) \stackrel{?}{=} 1$$
$$\log_2 2 - \log_2 1 \stackrel{?}{=} 1$$
$$1 - 0 \stackrel{?}{=} 1 \quad \text{Yes}$$

The solution set is $\{-2\}$.

Practice Problem 7 Solve $\log_3 (x - 8) + \log_3 x = 2$.

If each side of an equation can be expressed as a single logarithm with the same base, then we can use the one-to-one property of logarithms to solve the equation.

EXAMPLE 8 Using the One-to-One Property of Logarithms

Solve $\log_4 x + \log_4 (x + 1) = \log_4 (x - 1) + \log_4 6$.

Solution

$\log_4 x + \log_4 (x + 1) = \log_4 (x - 1) + \log_4 6$ — Original equation

$\log_4 [x(x + 1)] = \log_4 [6(x - 1)]$ — Product rule for logarithms

$x(x + 1) = 6(x - 1)$ — One-to-one property

$x^2 + x = 6x - 6$ — Distributive property

$x^2 - 5x + 6 = 0$ — Add $-6x + 6$ to both sides.

$(x - 2)(x - 3) = 0$ — Factor.

$x - 2 = 0$ or $x - 3 = 0$ — Zero-product property

$x = 2$ or $x = 3$ — Solve for x.

RECALL

The product rule says that $\log_a M + \log_a N = \log_a MN$.

Now we check these possible solutions in the original equation.

Check $x = 2$

$$\log_4 2 + \log_4 (2 + 1) \stackrel{?}{=} \log_4 (2 - 1) + \log_4 6$$
$$\log_4 2 + \log_4 3 \stackrel{?}{=} \log_4 1 + \log_4 6$$
$$\log_4 (2 \cdot 3) \stackrel{?}{=} \log_4 (1 \cdot 6) \quad \text{Yes}$$

Check $x = 3$

$$\log_4 3 + \log_4 (3 + 1) \stackrel{?}{=} \log_4 (3 - 1) + \log_4 6$$
$$\log_4 3 + \log_4 4 \stackrel{?}{=} \log_4 2 + \log_4 6$$
$$\log_4 (3 \cdot 4) \stackrel{?}{=} \log_4 (2 \cdot 6) \quad \text{Yes}$$

The solution set is $\{2, 3\}$.

Practice Problem 8 Solve $\ln (x + 5) + \ln (x + 1) = \ln (x - 1)$.

4 Use the logistic growth model.

Pierre François Verhulst (1804–1849)

Pierre Verhulst was born and educated in Brussels, Belgium. He received his PhD from the University of Ghent in 1825. Influenced by Lambert Quetelet, he became interested in social statistics. Verhulst's research on the law of population growth is important. Before Quetelet and Verhulst, scientists believed that an increasing population as a function of time was given by

$$P(t) = P_0(1 + k)^t \quad \text{or}$$
$$P(t) = P_0 e^{kt}.$$

Verhulst showed that forces tend to prevent the population growth according to these laws. He discovered the model given by equation (2).

EXAMPLE 9 **Using Logarithms in the Logistic Growth Model**

Suppose the carrying capacity M of the human population on Earth is 35 billion. In 1987, the world population was about 5 billion. Use the logistic growth model of P. F. Verhulst to calculate the average rate, k, of growth of the population given that the population was about 6 billion in 2003.

$$P(t) = \frac{M}{1 + ae^{-kt}} \quad (19) \qquad \text{From the section introduction}$$

Solution

If 1987 represents $t = 0$, then $P(0) = 5$ and $M = 35$; so equation (19) becomes

$$5 = \frac{35}{1 + ae^{-k(0)}}$$

$$5 = \frac{35}{1 + a}$$

$$5(1 + a) = 35 \qquad \text{Multiply both sides by } 1 + a.$$

$$a = 6 \qquad \text{Solve for } a.$$

Equation (19), with $M = 35$ and $a = 6$, becomes

$$P(t) = \frac{35}{1 + 6e^{-kt}} \quad (20)$$

We now solve equation (20) for k given that $t = 16$ (for 2003) and $P(16) = 6$.

$$6 = \frac{35}{1 + 6e^{-16k}}$$

$$6 + 36e^{-16k} = 35 \qquad \text{Multiply both sides by } 1 + 6e^{-16k} \text{ and distribute.}$$

$$e^{-16k} = \frac{29}{36} \qquad \text{Isolate } e^{-16k}.$$

$$-16k = \ln\left(\frac{29}{36}\right) \qquad \text{Logarithmic form}$$

$$k = -\frac{1}{16}\ln\left(\frac{29}{36}\right) \qquad \text{Solve for } k.$$

$$k \approx 0.0135 = 1.35\% \qquad \text{Use a calculator.}$$

The average growth rate of the world population was approximately 1.35%.

Practice Problem 9 Repeat Example 9 assuming that the world population in 2005 was about 6.5 billion.

5 Use logarithmic and exponential inequalities.

Logarithmic and Exponential Inequalities

The properties of logarithms and exponents, together with the same techniques we used to solve polynomial inequalities, can be used to solve inequalities involving logarithmic and exponential expressions.

The exponential and logarithmic functions with base $a > 1$ are increasing functions. That is, if $x_1 < x_2$, then $a^{x_1} < a^{x_2}$, and if $0 < x_1 < x_2$, then $\log_a x_1 < \log_a x_2$. As their graphs suggest, it is also true for base $a > 1$ that if $a^{x_1} < a^{x_2}$, then $x_1 < x_2$ and if $\log_a x_1 < \log_a x_2$, then $x_1 < x_2$. See Figures 4.14 and 4.15.

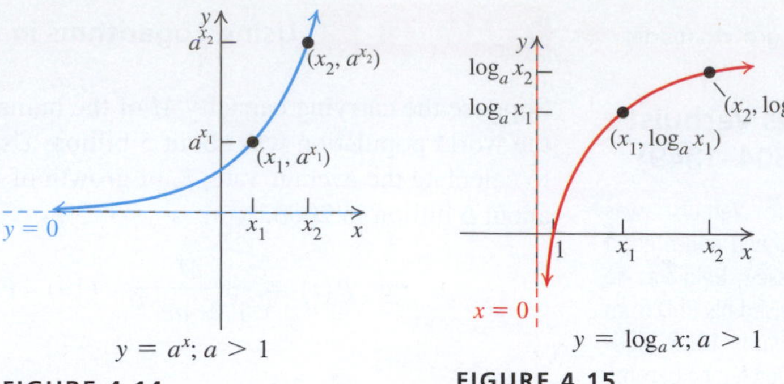

$y = a^x; a > 1$

FIGURE 4.14

$y = \log_a x; a > 1$

FIGURE 4.15

When working with inequalities, it is important to remember that $y = \log_a x, a > 1$, is negative for $0 < x < 1$.

EXAMPLE 10 Solving an Inequality Involving an Exponential Expression

Solve $5(0.7)^x + 3 < 18$.

Solution

$5(0.7)^x + 3 < 18$	Given inequality
$(0.7)^x < 3$	Isolate $(0.7)^x$ on one side.
$\ln(0.7)^x < \ln 3$	Take the natural log of both sides.
$x \ln(0.7) < \ln 3$	Power rule for logarithms
$x > \dfrac{\ln 3}{\ln(0.7)} \approx -3.080$	Divide both sides by $\ln(0.7)$; reverse the sense of the inequality because $\ln(0.7) < 0$.

RECALL

If $A < B$ and $C < 0$, then $AC > BC$.

Practice Problem 10 Solve $3(0.5)^x + 7 > 19$.

EXAMPLE 11 Solving an Inequality Involving a Logarithmic Expression

Solve $\log(2x - 5) \le 1$.

Solution

We first identify the domain for $\log(2x - 5)$. Because $2x - 5$ must be positive, we solve $2x - 5 > 0$; so $2x > 5$ or $x > \frac{5}{2}$.

Then

$\log(2x - 5) \le 1$	Given inequality
$10^{\log(2x-5)} \le 10^1$	If $u < v$, then $10^u < 10^v$.
$2x - 5 \le 10$	$a^{\log_a x} = x$
$x \le \dfrac{15}{2}$	Solve for x.

For $\log(2x - 5)$ to be a real number, we must have $x > \frac{5}{2}$. Combining this with the fact that $x \le \frac{15}{2}$, we find that the solution set for $\log(2x - 5) \le 1$ is the interval $\left(\frac{5}{2}, \frac{15}{2}\right]$.

Practice Problem 11 Solve $\ln(1 - 3x) > 2$.

Answers to Practice Problems

1. a. $x = 5$ **b.** $x = \dfrac{2}{3}$ **2.** $x = \dfrac{\ln\left(\frac{11}{7}\right)}{\ln 3} - 1 \approx -0.589$

3. $x = \dfrac{\ln 3}{2\ln 2 - \ln 3} \approx 3.819$ **4.** $x = \ln 5 \approx 1.609$

5. a. United States: 343.61 million; Pakistan: 255.96 million

b. Sometime in 2022 (after 11.69 yr) **c.** In 23.68 yrs (sometime in 2033) **6.** $x = e^{3/2}$ **7.** $x = 9$ **8.** $\varnothing$

9. The average growth rate was approximately 1.74%.

10. $x < -2$ **11.** $x < \dfrac{1 - e^2}{3}$

SECTION 4.4 Exercises

Basic Concepts and Skills

1. An equation that contains terms of the form a^x is called a(n) ______________ equation.

2. An equation that contains terms of the form $\log_a x$ is called a(n) ______________ equation.

3. The equation $y = \dfrac{M}{1 + ae^{-bx}}$ represents a(n) ______________ model.

4. True or False. A logistic curve always has two horizontal asymptotes.

5. True or False. Because the domain of $f(x) = \log_a x$ is the interval $(0, \infty)$, a logarithmic equation cannot have a negative solution.

6. True or False. The equation $8^{2x} = 4^{3x}$ is true for all real values of x.

In Exercises 7–22, solve each equation.

7. $2^x = 16$ **8.** $3^x = 243$

9. $8^x = 32$ **10.** $5^{x-1} = 1$

11. $4^{|x|} = 128$ **12.** $9^{|x|} = 243$

13. $5^{-|x|} = 625$ **14.** $3^{-|x|} = 81$

15. $\ln x = 0$ **16.** $\ln(x - 1) = 1$

17. $\log_2 x = -1$ **18.** $\log_2(x + 1) = 3$

19. $\log_3 |x| = 2$ **20.** $\log_2 |x + 1| = 3$

21. $\dfrac{1}{2}\log x - 2 = 0$ **22.** $\dfrac{1}{3}\log(x + 1) - 1 = 0$

In Exercises 23–52, solve each exponential equation. Write the exact answer with natural logarithms and then approximate the result correct to three decimal places.

23. $2^x = 3$ **24.** $3^x = 5$

25. $2^{2x+3} = 15$ **26.** $3^{2x+5} = 17$

27. $5 \cdot 2^x - 7 = 10$ **28.** $3 \cdot 5^x + 4 = 11$

29. $3 \cdot 4^{2x-1} + 4 = 14$ **30.** $2 \cdot 3^{4x-5} - 7 = 10$

31. $5^{1-x} = 2^x$ **32.** $3^{2x-1} = 2^{x+1}$

33. $2^{1-x} = 3^{4x+6}$ **34.** $5^{2x+1} = 3^{x-1}$

35. $2 \cdot 3^{x-1} = 5^{x+1}$ **36.** $5 \cdot 2^{2x+1} = 7 \cdot 3^{x-1}$

37. $(1.065)^t = 2$ **38.** $(1.0725)^t = 2$

39. $2^{2x} - 4 \cdot 2^x = 21$ **40.** $4^x - 4^{-x} = 2$

41. $9^x - 6 \cdot 3^x + 8 = 0$

42. $\dfrac{3^x + 5 \cdot 3^{-x}}{3} = 2$

43. $3^{3x} - 4 \cdot 3^{2x} + 2 \cdot 3^x = 8$

44. $2^{3x} + 3 \cdot 2^{2x} - 2^x = 3$

45. $\dfrac{3^x - 3^{-x}}{3^x + 3^{-x}} = \dfrac{1}{4}$ **46.** $\dfrac{e^x - e^{-x}}{e^x + e^{-x}} = \dfrac{1}{3}$

47. $\dfrac{4}{2 + 3^x} = 1$ **48.** $\dfrac{7}{2^x - 1} = 3$

49. $\dfrac{17}{5 - 3^x} = 7$ **50.** $\dfrac{15}{3 + 2 \cdot 5^x} = 4$

51. $\dfrac{5}{2 + 3^x} = 4$ **52.** $\dfrac{7}{3 + 5 \cdot 2^x} = 4$

In Exercises 53–70, solve each logarithmic equation.

53. $3 + \log(2x + 5) = 2$

54. $1 + \log(3x - 4) = 0$

55. $\log(x^2 - x - 5) = 0$

56. $\log(x^2 - 6x + 9) = 0$

57. $\log_4(x^2 - 7x + 14) = 1$

58. $\log_4(x^2 + 5x + 10) = 1$

59. $\ln(2x - 3) - \ln(x + 5) = 0$

60. $\log(x + 8) + \log(x - 1) = 1$

61. $\log x + \log(x + 9) = 1$

62. $\log_5(3x - 1) - \log_5(2x + 7) = 0$

63. $\log_a(5x - 2) - \log_a(3x + 4) = 0$

64. $\log(x - 1) + \log(x + 2) = 1$

65. $\log_6(x + 2) + \log_6(x - 3) = 1$

66. $\log_2(3x - 2) - \log_2(5x + 1) = 3$

67. $\log_3(2x - 7) - \log_3(4x - 1) = 2$

68. $\log_4 \sqrt{x + 3} - \log_4 \sqrt{2x - 1} = \dfrac{1}{4}$

69. $\log_7 3x + \log_7(2x - 1) = \log_7(16x - 10)$

70. $\log_3(x + 1) + \log_3 2x = \log_3(3x + 1)$

In Exercises 71–78, find a and k and then evaluate the function. Round your answer to three decimal places.

71. Let $f(x) = 20 + a(2^{kx})$ with $f(0) = 50$ and $f(1) = 140$. Find $f(2)$.

72. $f(x) = 40 + a(4^{kx})$ with $f(0) = -216$ and $f(2) = 39$. Find $f(1)$.

73. $f(x) = 16 + a(3^{kx})$ with $f(0) = 21$ and $f(4) = 61$. Find $f(2)$.

74. $f(x) = 50 + a(2^{kx})$ with $f(0) = 34$ and $f(4) = 46$. Find $f(2)$.

75. Let $f(x) = \dfrac{10}{3 + ae^{kx}}$ with $f(0) = 2$ and $f(1) = \dfrac{1}{2}$. Find $f(2)$.

76. Let $f(x) = \dfrac{6}{a + 2e^{kx}}$ with $f(0) = 1$ and $f(1) = 0.8$. Find $f(2)$.

77. Let $f(x) = \dfrac{4}{a + 4e^{kx}}$ with $f(0) = 2$ and $f(1) = 9$. Find $f(2)$.

78. Let $f(x) = \dfrac{a}{1 + 3e^{kx}}$ with $f(0) = -1$ and $f(2) = -2$. Find $f(1)$.

In Exercises 79–88, solve each inequality.

79. $5(0.3)^x + 1 \le 11$

80. $(0.1)^x - 4 > 15$

81. $-3(1.2)^x + 11 \ge 8$

82. $-7(0.4)^x + 19 < 5$

83. $\log(5x + 15) < 2$

84. $\log(2x + 0.9) > -1$

85. $\ln(x - 5) \ge 1$

86. $\ln(4x + 10) \le 2$

87. $\log_2(3x - 7) < 3$

88. $\log_2(5x - 4) > 4$

Applying the Concepts

89. Investment. Find the time required for an investment of \$10,000 to grow to \$18,000 at an annual interest rate of 6% if the interest is compounded
- **a.** Yearly.
- **b.** Quarterly.
- **c.** Monthly.
- **d.** Daily.
- **e.** Continuously.

90. Investment. How long will it take for an investment of \$100 to double in value if the annual rate of interest is 7.2% compounded
- **a.** Yearly.
- **b.** Quarterly.
- **c.** Monthly.
- **d.** Continuously.

91. Inflation. In the fictional land of Sardonia, the price of a car costing \$20,000 in U.S. currency will double in eight years. What will the car cost in five years? What is the annual rate of inflation?

92. Doubling time. An amount P dollars is deposited in a regular bank account. How long does it take to double the initial investment (called the **doubling time**) if
- **a.** The interest rate is r compounded annually.
- **b.** The interest rate is r (per year) compounded continuously.

93. Light absorption in physics. The light intensity I (in lumens) at a depth of x feet in Lake Elizabeth is given by

$$\log\left(\frac{I}{12}\right) = -0.025x.$$

- **a.** Find the light intensity at a depth of 30 feet.
- **b.** At what depth is the light intensity 4 lumens?

94. Rule of 70. Bankers use the **rule of 70** to estimate the doubling time for money invested at different rates. The rule of 70 states that

$$\text{Doubling time} \approx \frac{70}{100r} \text{ years},$$

where r is the annual interest rate (expressed as a decimal number). Explain why this formula works.

95. Epidemic outbreak. The number of people in a community who became infected during an epidemic t weeks after its outbreak is given by the function

$$f(t) = \frac{20{,}000}{1 + ae^{-kt}},$$

where 20,000 people of the community are susceptible to the disease. Assuming that 1000 people were infected initially and 8999 had been infected by the end of the fourth week,
- **a.** Find the number of people infected after 8 weeks.
- **b.** After how many weeks will 12,400 people be infected?

96. Spreading a rumor. A jealous student started a malicious rumor on an isolated college campus of 5000 students. The number of people who know of the rumor within t days after it was started is given by the function

$$R(t) = \frac{5000}{1 + ae^{-kt}}.$$

Half the students had known of the rumor within ten days.
- **a.** Find a and k and graph the function $y = R(t)$.
- **b.** How many students would have known of the rumor within 15 days after it started?
- **c.** How many students altogether will know the rumor?

97. Biological growth. In his laboratory experiment in 1934, G. F. Gause placed paramecia (unicellular microorganisms) in 5 cubic centimeters of a saline (salt) solution with a constant amount of food and measured their growth on a daily basis. He found that the population $P(t)$ after t days was approximated by

$$P(t) = \frac{4490}{1 + e^{5.4094 - 1.0255t}}, \quad t \ge 0.$$

- **a.** What was the initial population of the paramecia?
- **b.** What was the carrying capacity of the medium?
- **c.** Graph the equation $y = P(t)$.

98. Facebook users. In 2004, there were 1 million Facebook users, and in 2012, there were 845 million users. Use the model $f(t) = \frac{20{,}000}{20 + ae^{-kt}}$, where t is the number of years since 2004 and $f(t)$ is the number of users in millions, to:

a. Find a and k.

b. Estimate the number of users in 2015.

c. Estimate the year in which the number of users will be 2 billion.

(*Source*: www.benposter.com/facebook-user-growth-chart-2004-2010)

Beyond the Basics

99. Solve for t: $P = \frac{M}{1 + e^{-kt}}$.

100. Find k so that

a. $2^x = e^{kx}$.

b. $e^x = 2^{kx}$.

101. If $\frac{\log x}{2} = \frac{\log y}{3} = \frac{\log z}{5} (=k)$, show that

a. $xy = z$. **b.** $y^2z^2 = x^8$.

[*Hint:* First, express x, y, and z in terms of k.]

102. The present value P of an annuity with payment size R, periodic rate i, and term of n payments is given by the formula

$$P(1 + i)^n = R\frac{(1 + i)^n - 1}{i}.$$

Solve this equation for n.

In Exercises 103–110, solve each equation for x.

103. $(\log x)^2 = \log x$

104. $(\log_2 x)(\log_2 8x) = 10$

105. $(\log_3 x)(\log_3 3x) = 2$

106. $\log_2 x + \log_4 x = 6$

107. $\log_4 x^2(x - 1)^2 - \log_2(x - 1) = 1$

108. $\frac{\log(7x - 12)}{\log x} = 2$

109. $\frac{\log(3x - 5)}{2} = \log x$

110. $\log(x - 4) - \log x = \log\left(\frac{1}{10 - x}\right)$

In Exercises 111–118, use the four-step procedure (page 276) to find $f^{-1}(x)$ for the given function $f(x)$.

111. $f(x) = 3^x + 5$

112. $f(x) = 2^{-x} + 4$

113. $f(x) = 3 \cdot 4^x + 7$

114. $f(x) = 2 \cdot 3^{x-1} - 5$

115. $f(x) = 1 + \log_2(x - 1)$

116. $f(x) = 7 + 2\log_6(3x - 1)$

117. $f(x) = \frac{1}{2}\ln\left(\frac{x - 1}{x + 1}\right)$

118. $f(x) = \frac{1}{2}\log\left(\frac{1 + x}{1 - x}\right)$

In Exercises 119–122, find the smallest integer n for which the given property holds.

119. $7^n > 43^{67}$

120. $9^n > e^{321}$

121. $8^{1/n} < 1.01$

122. $12^{1/n} < 1.001$

123. Find an integer n so that 31^n has 567 digits. [*Hint*: $566 \le \log 31^n < 567$.]

124. Find an integer n so that n^{123} has 456 digits.

125. Evaluate $\frac{x^{\log y}}{x^{\log z}} \cdot \frac{y^{\log z}}{y^{\log x}} \cdot \frac{z^{\log x}}{z^{\log y}}$.

126. Evaluate $\frac{1}{\log_{xy} xyz} + \frac{1}{\log_{yz} xyz} + \frac{1}{\log_{zx} xyz}$.

Critical Thinking/Discussion/Writing

127. Logistic function. For $f(t) = \frac{P}{1 + ae^{-kt}}$, it can be shown that the maximum rate of growth occurs when $f(t) = \frac{P}{2}$.

Find the time when the rate of growth is maximum.

128. Solve each of the following equations for x.

a. $\log_4(x - 1)^2 = 3$

b. $2\log_4(x - 1) = 3$

c. $2\log_4|x - 1| = 3$

Explain why these three equations do not have identical solutions.

Maintaining Skills

In Exercises 129–132, convert each equation from exponential to logarithmic form.

129. $9^2 = 81$

130. $3^{-2} = \frac{1}{9}$

131. $2 \cdot 10^x + 1 = 7$

132. $3e^{2x} + 5 = 17$

In Exercises 133–136, convert each equation from logarithmic to exponential form.

133. $\log_2 64 = 6$

134. $\log_{\frac{1}{2}} 16 = -4$

135. $\log\left(\frac{A}{2}\right) = 3$

136. $\ln\left(\frac{A}{P}\right) = kt$

In Exercises 137–144, solve each equation for x.

137. $\log_5 x = -3$

138. $\log_3 \frac{1}{27} = x - 1$

139. $\log_x 1000 = 3$

140. $\log_2(x^2 - 6x + 10) = 1$

141. $2^{x+1} = 32$

142. $3^{2x-1} = 7$

143. $3^{x+1} = 5^{2x-3}$

144. $\log(x^2 + x) = \log(3x + 3)$

SECTION 4.5

Logarithmic Scales

BEFORE STARTING THIS SECTION, REVIEW

1. Exponential form for logarithms (Section 4.2, page 411)
2. Properties of logarithms (Section 4.2, page 413)
3. Rules for logarithms (Section 4.3, page 426)

OBJECTIVES

1. Define pH.
2. Define the Richter scale for measuring earthquake intensity.
3. Define scales for measuring sound.
4. Define magnitude of star brightness.

A **logarithmic scale** is a scale in which logarithms are used in the measurement of quantities. Suppose we have a quantity that has a small positive range of variation (for example, from $0.00001 = 10^{-5}$ to $0.0001 = 10^{-4}$); then the common logarithms of such numbers would be between -5 and -4. Similarly, if a quantity has a large positive range of variation from $10{,}000 = 10^4$ to $100{,}000 = 10^5$, the common logarithms of these numbers would be between 4 and 5. So a logarithmic scale serves to make data more manageable by expanding small variations and compressing large ones.

In Example 2 we will see how a small difference on a logarithmic scale results in an impressive increase in the acidity of acid rain compared to an ordinary rain.

1 Define pH.

pH Scale

In chemistry, acidity of a substance is related to the presence of positively charged hydrogen ions. pH represents the effective concentration (activity) of hydrogen ions H^+ in water. Because H^+ concentrations are much smaller than other dissolved species in water, the activity of hydrogen ions is expressed most conveniently in logarithmic units.

pH SCALE FORMULA

We define the pH scale of a solution by the formula

$$\text{pH} = -\log\,[H^+],$$

where $[H^+]$ is the concentration of H^+ ions in moles per liter. (A mole is a unit of measurement, equal to 6.023×10^{23} atoms.)

The pH value is a measure of the acidity or alkalinity of a solution. The pH value for pure water at $25°C\ (77°F)$ is 7.

Søren Sørensen (1868–1939)

The term *pH* was originally derived from the French term *pouvoir hydrogène.* In English, this means "hydrogen power." The term *pH* is always written with a lowercase p and an uppercase H. The pH scale was defined by the Danish biochemist Sören Sörensen in 1909.

Because pH is defined as $-\log[H^+]$, pH decreases as $[H^+]$ increases (which will happen if acid is added to water). *Acidic* solutions have pH values less than 7, and *alkaline* (also called *basic*) solutions have pH values greater than 7. The smaller the number on the pH scale, the more acidic the substance is.

EXAMPLE 1 Calculating pH and $[H^+]$

a. Calculate to the nearest tenth the pH value of grapefruit juice if $[H^+]$ of grapefruit juice is 6.32×10^{-4}.

b. Find the hydrogen concentration $[H^+]$ in beer if its pH value is 4.82.

Solution

a.

$$\begin{aligned} \text{pH} &= -\log[H^+] && \text{Definition of pH} \\ &= -\log(6.32 \times 10^{-4}) && \text{Given } [H^+] = 6.32 \times 10^{-4} \\ &= -(\log 6.32 + \log 10^{-4}) && \text{Product rule} \\ &= -\log 6.32 + 4 && -\log 10^{-4} = -(-4)\log 10 = 4 \\ &\approx -0.8007 + 4 && \text{Use a calculator.} \\ &= 3.1993 \approx 3.2 && \text{Simplify.} \end{aligned}$$

b.

$$\begin{aligned} 4.82 &= -\log[H^+] && \text{Replace pH with 4.82.} \\ \log[H^+] &= -4.82 && \text{Rewrite equation.} \\ [H^+] &= 10^{-4.82} && \text{Exponential form} \\ &= 10^{0.18} \times 10^{-5} && -4.82 = -5 + 0.18 \\ &\approx 1.51 \times 10^{-5} && \text{Use a calculator.} \end{aligned}$$

Practice Problem 1

a. The $[H^+]$ of a solution is 2.68×10^{-6}. Find its pH value.

b. Find $[H^+]$ of seawater if its pH is 8.47.

EXAMPLE 2 Acid Rain

How much more acidic is acid rain with a pH value of 3 than an ordinary rain with a pH value of 6?

Solution

We have

$$3 = \text{pH}_{\text{acid rain}} = -\log[H^+]_{\text{acid rain}} \qquad \text{Definition of pH}$$

so that

$$[H^+]_{\text{acid rain}} = 10^{-3} \qquad \text{Exponential form of } \log[H^+] = -3$$

Similarly,

$$[H^+]_{\text{ordinary rain}} = 10^{-6}.$$

Therefore,

$$\frac{[H^+]_{\text{acid rain}}}{[H^+]_{\text{ordinary rain}}} = \frac{10^{-3}}{10^{-6}} = 10^{-3+6} = 10^3 = 1000.$$

So the hydrogen ion concentration in this acid rain is 1000 times greater than that in ordinary rain. That is, this acid rain is 1000 times more acidic than the ordinary rain.

Practice Problem 2 How much more acidic is an acid rain with a pH value of 2.8 than an ordinary rain with a pH value of 6.2?

DO YOU KNOW?

Acid rain is rain, snow, or fog that is polluted by acid in the atmosphere and damages the environment. Two culprits that acidify rain are sulfur dioxide (SO_2) and nitrogen oxide N_2O. Rain measuring between 0 and 5 on the pH scale is acidic and therefore called "acid rain."

SIDE NOTE

Notice that each one point increase in pH value corresponds to a factor of 10 increase in H^+ concentration.

Table 4.6 lists a few typical pH values.

DO YOU **KNOW?**

An acid is a substance with sour taste, and it reacts with a base to form a salt. Acids turn blue litmus paper (also called pH paper) red. Strong acids can burn your skin. Common antacid tablets (for indigestion) are basic and work by neutralizing stomach acid.

TABLE 4.6

Solution	pH Value
Battery Acid	1
Lemon Juice	2
Stomach Acid	2–3
Vinegar	3
Milk	6–7
Baking Soda, Seawater	8–9
Milk of Magnesia	9–10
Ammonia	10–11
Drain opener	10–12
Lye	13

Maximum acidity — Neutral — Maximum alkalinity

0 — 7 — 14 — x

$0 =$ Maximum acidity

$7 =$ Neutral point

$14 =$ Maximum alkalinity (opposite of acidity)

2 Define the Richter scale for measuring earthquake intensity.

Earthquake Intensity

An earthquake is the vibration, sometimes violent, of Earth's surface that follows a release of energy in Earth's crust. Earth's crust is composed of about 20 huge plates that float on the molten material beneath the crust. These plates slowly move over, under, and past each other. Sometimes the movement is gradual. At other times, the plates are locked together, unable to release the accumulating energy. When this energy grows strong enough, the plates break free and vibrations called "seismic waves" or earthquakes are generated.

The Richter scale was invented in the 1930s by the American scientist Dr. Charles Richter to measure the magnitude of an earthquake. It is based on the idea of comparing the intensity of an earthquake with that of a zero-level earthquake. He defined the zero-level earthquake as an earthquake whose seismographic reading measures 1 micron (1000 microns $=$ 1 millimeter, or 1,000,000 microns $=$ 1 meter) at a distance of 100 kilometers (about 62 miles) from the epicenter of the earthquake.

Let us denote the intensity of a zero-level earthquake by I_0. A zero-level earthquake is just noticeable and is the *threshold level* below which we would not be aware of a quake.

Charles Francis Richter (1900–1985)

Charles Richter was born in Hamilton, Ohio, and received his PhD from California Institute of Technology in 1928. He developed his scale to measure the strength of earthquakes in 1935. Earlier scales had been developed by de Rossi in the 1880s and by Giuseppe Mercalli in 1902, but both used a descriptive scale defined in terms of damage to buildings and the behavior and response of the population. Richter's scale is an absolute one, based on the amplitude of the waves produced by the earthquake. He defined the magnitude of an earthquake as the logarithm to the base 10 of the maximum amplitude of the waves, measured in microns. This means that waves whose amplitudes differ by a factor of 100 will differ by 2 points on the Richter scale.

RICHTER SCALE

The magnitude M of an earthquake is a function of its intensity I and is defined by

$$M = \log\left(\frac{I}{I_0}\right),$$

where I_0 is the intensity of the zero-level earthquake.

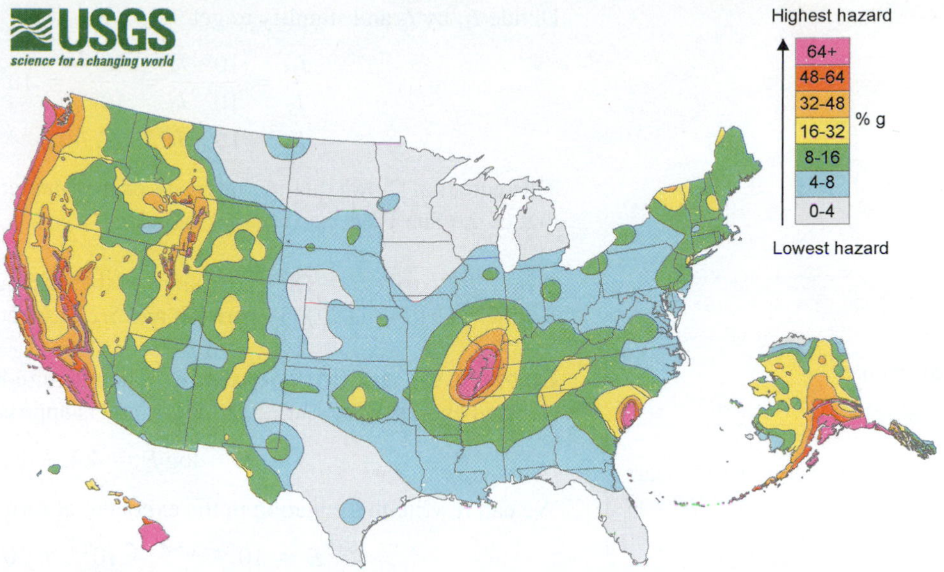

Source: US Geological Survey (USGS).

EXAMPLE 3 Intensity of an Earthquake

The magnitude of an earthquake is 4.0 on the Richter scale. What is the intensity of this earthquake?

Solution

$$M = \log\left(\frac{I}{I_0}\right) \qquad \text{Definition of magnitude}$$

$$4 = \log\left(\frac{I}{I_0}\right) \qquad \text{Replace } M \text{ with 4.}$$

$$\frac{I}{I_0} = 10^4 \qquad \text{Exponential form}$$

$$I = 10^4 I_0 = 10{,}000\, I_0 \qquad \text{Multiply both sides by } I_0.$$

The intensity of this earthquake is 10,000 times the intensity of I_0 (the zero-level earthquake). Notice that a 1 point increase on the Richter scale corresponds to a factor of 10 increase in intensity.

DO YOU KNOW?

To get some idea of the destructive power of an earthquake, the energy released at the epicenter of a zero-level earthquake is equivalent to more than 10,000 atomic bombs of the kind that leveled Hiroshima, Japan, in 1945 exploding simultaneously.

Practice Problem 3 The magnitude of an earthquake is 6.5 on the Richter scale. What is the intensity of this earthquake?

EXAMPLE 4 Comparing Two Earthquakes

Compare the intensity of the Mexico City earthquake of 1985, which registered 8.1 on the Richter scale, to that of the San Francisco area earthquake of 1989, which measured 6.9 on the Richter scale.

Solution

Let I_M and I_S denote the intensities of the Mexico City and the San Francisco earthquakes, respectively. We have

$$8.1 = \log\left(\frac{I_M}{I_0}\right) \quad \Big| \quad 6.9 = \log\left(\frac{I_S}{I_0}\right) \qquad \text{Definition of magnitude}$$

$$\frac{I_M}{I_0} = 10^{8.1} \quad \Big| \quad \frac{I_S}{I_0} = 10^{6.9} \qquad \text{Exponential form}$$

$$I_M = 10^{8.1} I_0 \quad \Big| \quad I_S = 10^{6.9} I_0. \qquad \text{Multiply both sides by } I_0.$$

Divide I_M by I_S and simplify to get

$$\frac{I_M}{I_S} = \frac{10^{8.1} I_0}{10^{6.9} I_0} = \frac{10^{8.1}}{10^{6.9}} = 10^{8.1-6.9} = 10^{1.2}$$
$$I_M = 10^{1.2} I_S \approx 16\, I_S.$$

This equation shows that the intensity of the Mexico City earthquake was about 16 times that of the San Francisco area earthquake.

Practice Problem 4 The magnitudes of the earthquakes of Mozambique (2006) and Southern California (2005) were 7.0 and 5.2, respectively. Compare their intensities. 

BY THE **WAY**

A joule is a unit of energy. 100 joules of energy could light up a 100-watt lightbulb for a second.

Energy of an Earthquake The magnitude M of an earthquake is related to its released energy E (measured in joules) and is approximated by the equation

$$\log E = 4.4 + 1.5\,M.$$

We can rewrite this equation in the exponential form as

$$E = 10^{4.4+1.5M} = 10^{0.4} \times 10^4 \times 10^{1.5M}$$

or

$$E \approx (2.5 \times 10^4) \times 10^{1.5M} \qquad 10^{0.4} \approx 2.5.$$

ENERGY OF AN EARTHQUAKE

The energy E (in joules) released by an earthquake of magnitude M (Richter scale) is given by

$$\log E = 4.4 + 1.5M$$

or

$$E \approx (2.5 \times 10^4) \times 10^{1.5M}.$$

The energy E_0 released by a zero-level earthquake can be calculated by substituting $M = 0$ in the formula. We have

$$E_0 = 2.5 \times 10^4 \text{ joules.}$$

TABLE 4.7 Major Earthquakes ($M \geq 7.5$) of the 20th & 21st Century

Date	Location	Magnitude
April 1906	San Francisco	7.8
December 1908	Messina, Italy	7.5
December 1920	Gansu, China	8.6
September 1923	Sagami Bay, Japan	8.3
February 1931	New Zealand	7.9
August 1950	Assam, India	8.7
July 1976	Tangshan, China	8.0
September 1985	Mexico City	8.1
June 1990	Iran	7.7
May 1997	Iran	7.5
December 2004	Sumatra, Indonesia	9.0
March 2005	Sumatra, Indonesia	8.6
February 2010	Maule, Chile	8.8
March 2011	Tohoku, Japan	9.0

Source: National Earthquake Information Center, U.S. Geological Survey.

If two earthquakes have magnitude M_1 and M_2, we can compute the ratio of their corresponding energies E_1 and E_2 as follows:

$$\frac{E_1}{E_2} = \frac{(2.5 \times 10^4) \times 10^{1.5M_1}}{(2.5 \times 10^4) \times 10^{1.5M_2}} = \frac{10^{1.5M_1}}{10^{1.5M_2}} = 10^{1.5M_1 - 1.5M_2} = 10^{1.5(M_1 - M_2)}.$$

EXAMPLE 5 **Comparing Two Earthquakes**

Compare the estimated energies released by the San Francisco earthquake of 1906 and the Northridge earthquake of 1994 with magnitude of 6.7.

Solution

Let M_{1906} and E_{1906} represent the magnitude and energy released by the 1906 quake of San Francisco. Attach similar meaning to M_{1994} and E_{1994}.

With $M_{1906} = 7.8$ (Table 4.7) and $M_{1994} = 6.7$ and the equation $\frac{E_1}{E_2} = 10^{1.5(M_1 - M_2)}$, we have

$$\frac{E_{1906}}{E_{1994}} = 10^{1.5(7.8-6.7)} = 10^{1.5(1.1)} \approx 45 \qquad \text{Use a calculator.}$$

The 1906 earthquake released more than 45 times as much energy as the 1994 earthquake.

Practice Problem 5 Compare the energies released by the Japan earthquake of 2011 and Iran earthquake of 1997 (see Table 4.7).

3 Define scales for measuring sound.

Loudness of Sound

The *intensity* of a sound wave is defined as the amount of power the wave transmits through a given area. The intensity of a sound is measured in watts per square meters (abbreviated W/m^2). The faintest sound that the human ear can detect has an intensity of 10^{-12} W/m^2. This faintest sound is known as the *threshold of hearing* (TOH). The most intense sound that the human ear can safely detect without suffering any physical damage is more than one billion times the TOH.

Because the range of intensities that the human ear can detect is so large, a logarithmic scale called *decibels* (abbreviated dB) is used to measure the loudness of a sound. A decibel (as the name suggests), is one-tenth of a *bel*, a unit named after the telephone inventor Alexander Graham Bell.

LOUDNESS OF SOUND AND DECIBELS

The *loudness* (or relative intensity) L of a sound measured in decibels is related to its intensity I by the formula

$$L = 10 \log\left(\frac{I}{I_0}\right) \qquad \text{(logarithmic form)}$$

or

$$I = I_0 \times 10^{L/10} \qquad \text{(exponential form)}$$

where $I_0 = 10^{-12}$ W/m^2 is the intensity of TOH.

If we substitute $I = I_0$ in the formula for loudness, we obtain

$$\text{Loudness of TOH} = 10 \log \frac{I_0}{I_0} = 10 \log 1 = 10(0) = 0.$$

So the threshold of hearing has 0 dB loudness.

Table 4.8 lists approximate loudness of some common sounds.

TABLE 4.8 Approximate loudness of common sounds

Source	Intensity in W/m^2	Loudness in dB
Threshold of hearing	10^{-12}	0
Rustling leaves	10^{-11}	10
Whisper	10^{-10}	20
Background noise in average home	10^{-8}	40
Normal conversation	10^{-6}	60
Busy street traffic	10^{-5}	70
Vacuum cleaner	10^{-4}	80
Portable audio player at maximum level	10^{-2}	100
Front rows of rock concert	10^{-1}	110
Threshold of pain	10	130
Jet aircraft 50 m away	10^{2}	140

EXAMPLE 6 Computing Loudness

Find the decibel level of a TV that has an intensity of 250×10^{-7} W/m^2.

Solution

We are given $I = 250 \times 10^{-7}$ W/m^2, and we know that $I_0 = 10^{-12}$ W/m^2. We have

$$\begin{aligned}
L &= 10 \log\left(\frac{I}{I_0}\right) && \text{Formula for loudness} \\
&= 10 \log\left(\frac{250 \times 10^{-7}}{10^{-12}}\right) && I = 250 \cdot 10^{-7} \text{ and } I_0 = 10^{-12} \\
&= 10 \log(250 \times 10^{5}) && \frac{10^{-7}}{10^{-12}} = 10^{-7+12} = 10^{5} \\
&= 10[\log 250 + \log 10^{5}] && \text{Product rule for logarithms} \\
&= 10[\log 250 + 5 \log 10] && \text{Power rule for logarithms} \\
&= 10[\log 250 + 5] && \log 10 = \log_{10} 10 = 1 \\
&\approx 74 && \text{Use a calculator, rounding.}
\end{aligned}$$

So the decibel level of this TV is approximately 74 dB.

Practice Problem 6 Find the decibel level of a TV that has intensity of 200×10^{-7} W/m^2.

EXAMPLE 7 **Comparing Intensities**

How much more intense is a 65 dB sound than a 42 dB sound?

Solution

$$
\begin{aligned}
I &= I_0 \times 10^{\frac{L}{10}} && \text{Formula for intensity, exponential form}\\
I_{65} &= I_0 \times 10^{\frac{65}{10}} && I_{65} \text{ is the intensity of a 65 dB sound.}\\
I_{42} &= I_0 \times 10^{\frac{42}{10}} && I_{42} \text{ is the intensity of a 42 dB sound.}\\
\frac{I_{65}}{I_{42}} &= \frac{I_0 \times 10^{\frac{65}{10}}}{I_0 \times 10^{\frac{42}{10}}} && \text{Divide } I_{65} \text{ by } I_{42}.\\
&= \frac{I_0 \times 10^{6.5}}{I_0 \times 10^{4.2}}\\
&= 10^{6.5-4.2} && \text{Remove } I_0 \text{ and use the quotient rule for exponents.}\\
&= 10^{2.3} && \text{Simplify.}\\
&\approx 199.53 && \text{Use a calculator.}\\
&\approx 200
\end{aligned}
$$

So a sound of 65 dB is about 200 times more intense than a sound of 42 dB.

Practice Problem 7 How much more intense is a 75 dB sound than a 55 dB sound?

EXAMPLE 8 **Computing Sound Intensity**

Calculate the intensity in watts per square meter of a sound of 73 dB.

Solution

$$
\begin{aligned}
I &= I_0 \times 10^{\frac{L}{10}} && \text{Formula for intensity, exponential form}\\
I &= 10^{-12} \times 10^{\frac{73}{10}} && \text{Replace } I_0 \text{ with } 10^{-12} \text{ and } L \text{ with 73.}\\
&= 10^{-12} \times 10^{7.3}\\
&= 10^{-4.7} && \text{Simplify.}\\
&= 10^{0.3} \times 10^{-5} && -4.7 = 0.3 - 5\\
&\approx 1.995 \times 10^{-5} && \text{Use a calculator.}\\
&\approx 2 \times 10^{-5}\ \text{W/m}^2.
\end{aligned}
$$

Practice Problem 8 What is the intensity of a 48 dB sound?

Musical Pitch

Pitch of a sound is determined by its frequency. For example, on a piano, the pitch of the note A above middle C may be denoted by A440. This means that A vibrates at 440 Hertz (cycles per second). In music, an interval whose higher note frequency is twice that of its lower note is an **octave**. Two notes that are an octave apart sound essentially "the same," but one has a higher pitch. For this reason, notes an octave apart are given the same name. An octave jump is usually divided into 12 approximately equal *semitones* on a log scale. In an equal tempered scale, a semitone is further divided into 100 *cents*. So there are 1200 cents in an octave.

CHANGE IN PITCH

If f is a frequency measured in Hertz (Hz) and f_0 is a reference frequency, then the change in pitch $P(f)$ in cents from f_0 to f is given by:

$$P(f) = 1200 \log_2\left(\frac{f}{f_0}\right).$$

We note that the pitch difference of 1 octave means $f = 2f_0$, so

$$P(2f_0) = 1200 \log_2\left(\frac{2f_0}{f_0}\right) = 1200 \log_2 2 = 1200 \text{ cents},$$

which agrees with our previous assertion.

Most people are not sensitive to pitch differences of less than about 2 cents.

EXAMPLE 9 Equal Tempered Scale

In an "equal tempered" scale for a given reference frequency f_0 of a note, find the frequency of the next higher-pitched semitone.

Solution

For the given reference frequency f_0, let f be the frequency of the next higher-pitched semitone. Because in an equal tempered scale every semitone is 100 cents, we have $P(f) = 100$.

$$P(f) = 1200 \log_2\left(\frac{f}{f_0}\right) \qquad \text{Change in pitch formula}$$

$$100 = 1200 \log_2\left(\frac{f}{f_0}\right) \qquad \text{Replace } P(f) \text{ with 100.}$$

$$\frac{1}{12} = \log_2\left(\frac{f}{f_0}\right) \qquad \text{Divide by 1200 and simplify.}$$

$$\frac{f}{f_0} = 2^{\frac{1}{12}} = \sqrt[12]{2} \qquad \text{Exponential form}$$

$$f = \sqrt[12]{2}\, f_0 \qquad \text{Multiply by } f_0.$$

Practice Problem 9 In an equal tempered scale, the notes are denoted as follows:

A A# B C C# D D# E F F# G G# A

If the frequency of A is 440 Hz, estimate the frequency of B.

The next example shows that an equal tempered scale does not always produce the best harmonics.

EXAMPLE 10 Harmony in Music

A jump of one *perfect fifth* gives a frequency increase of 50%. An equal tempered *fifth* is 700 cents. Find the difference in cents (rounded to the nearest cent). Is the difference noticeable?

Solution

If the reference frequency is f_0 and it increases by 50%, then $f = f_0 + 0.5f_0 = (1 + 0.5)f_0 = 1.5f_0$. So

$$P(f) = 1200 \log_2(1.5f_0/f_0) \qquad \text{Replace } f \text{ with } 1.5f_0$$

$$= 1200 \log_2(1.5) \qquad \text{Simplify.}$$

$$= 1200\frac{\log(1.5)}{\log 2} \qquad \text{Change to base 10.}$$

$$\approx 701.955 \qquad \text{Use a calculator.}$$

$$\approx 702$$

Because a fifth on an equal tempered scale is 700 cents, the difference is a little less than 2 cents. This difference is barely noticeable to most listeners.

Practice Problem 10 Repeat Example 10 if a perfect *major* third gives a frequency increase of 25% and an equal tempered *major* third is 400 cents.

4 Define magnitude of star brightness.

Star Brightness

Two thousand years ago Greek astronomers Hipparchus and Ptolemy created a system to quantify the *apparent brightness* (wattage received on the surface of the Earth) of stars. They classified star brightness on a scale from magnitude 1 for the brightest visible star to magnitude 6 for the dimmest visible star. Other stars were assigned magnitudes based on their brightness compared with those of magnitude 1 or 6. This system was refined considerably, leading to a much larger range of magnitude values and the following definition.

APPARENT MAGNITUDE

If two stars of magnitudes m_1 and m_2 have apparent brightness b_1 and b_2, respectively, then

$$m_2 - m_1 = 2.5 \log\left(\frac{b_1}{b_2}\right).$$

Although this definition of apparent magnitude was originally intended for comparing the brightness of stars, it can also be used for objects such as planets and their moons. The Hubble telescope can detect stars with apparent magnitude 30.

EXAMPLE 11 Comparing Star Brightness

a. Compare the brightness of a magnitude 1 star with that of a magnitude 6 star.

b. Find the magnitude m of a star that is 650 times as bright as one of magnitude 7.25.

Solution

a. $m_2 - m_1 = 2.5 \log\left(\frac{b_1}{b_2}\right)$ Apparent magnitude formula

$6 - 1 = \frac{5}{2} \log\left(\frac{b_1}{b_2}\right)$ $m_2 = 6, m_1 = 1$, and $2.5 = \frac{5}{2}$

$2 = \log\left(\frac{b_1}{b_2}\right)$ Multiply both sides by $\frac{2}{5}$ and simplify.

$\frac{b_1}{b_2} = 10^2$ Exponential form

$b_1 = 10^2 b_2 = 100b_2$ Multiply both sides by b_2.

So a star of magnitude 1 is 100 times brighter than a star of magnitude 6.

b. Letting $m_2 = m$, $m_1 = 7.25$ and $b_2 = 650b_1$ in the formula, we have

$$m - 7.25 = 2.5 \log\left(\frac{b_1}{650b_1}\right)$$

$$m - 7.25 = 2.5 \log\left(\frac{1}{650}\right) = 2.5 \log(650)^{-1}$$

$m - 7.25 = -2.5 \log(650)$ Power rule for logs

$m = 7.25 - 2.5 \log(650)$ Add 7.25 to both sides.

$m \approx 0.2177$ Use a calculator.

Practice Problem 11

a. Compare the brightness of a magnitude 0 star with that of a magnitude 2 star.

b. Find the magnitude m of a star that is 50% brighter than a star of magnitude 4.6. 

Answers to Practice Problems

1. a. 5.57 **b.** $3.39 \cdot 10^{-9}$ **2.** ≈ 2512
3. $3{,}162{,}278 I_0$ **4.** $I_{\text{Moz}} \approx 63 I_{\text{Cal}}$ **5.** 178 times
6. 73 dB **7.** 100 times more
8. $6.3 \cdot 10^{-8}$ W/m² **9.** ≈ 494 hertz
10. $\approx 400 - 386 = 14$ (cents)
11. a. $10^{0.8} \approx 6.3$ times brighter **b.** ≈ 4.1598

SECTION 4.5 Exercises

Basic Concepts and Skills

1. If the pH value of a solution is less than 7, the solution is __________.

2. As $[H^+]$ increases, the pH value of a substance __________.

3. On the Richter scale, the magnitude of an earthquake $M =$ __________.

4. The energy E released by an earthquake of magnitude M is given by $\log E =$ __________.

5. The loudness L of a sound of intensity I is given by $L =$ __________, and $I_0 = 10^{-12}$ W/m².

6. If f_0 is a reference frequency and f is the frequency of a note, then the change in pitch $P(f) =$ __________.

7. If two stars have magnitudes m_1 and m_2 with apparent brightness b_1 and b_2, respectively, then $m_2 - m_1 =$ __________.

8. True or False. The acidity of a solution increases as its pH value increases.

9. True or False. For a 1-unit increase in the magnitude of an earthquake, its intensity increases tenfold.

10. True or False. The apparent brightness of a star increases if its magnitude decreases.

In Exercises 11–14, the concentration $[H^+]$ of a substance is given. Calculate the pH value of each substance and classify each substance as an acid or a base.

11. $[H^+] = 10^{-8}$

12. $[H^+] = 10^{-4}$

13. $[H^+] = 2.3 \times 10^{-5}$

14. $[H^+] = 4.7 \times 10^{-9}$

In Exercises 15–18, use the following information. Negative ions, designated by the notation $[OH^-]$, are always present in any acid or base. The concentration, $[OH^-]$, of these ions is related to $[H^+]$ by the equation $[OH^-] \cdot [H^+] = 10^{-14}$ moles per liter. Find the concentrations of $[OH^-]$ and $[H^+]$ in moles per liter for the substances with the following pH values.

15. pH = 6

16. pH = 8

17. pH = 9.5

18. pH = 3.7

In Exercises 19–22, the magnitude M of an earthquake is given.
a. Find the earthquake intensity I in terms of the zero-level earthquake intensity I_0.
b. Find the energy released by the earthquake.

19. $M = 5$

20. $M = 2$

21. $M = 7.8$

22. $M = 3.7$

In Exercises 23–26, the energy E released by an earthquake is given.
a. Find the earthquake's magnitude.
b. Find the earthquake's intensity.

23. $E = 10^{13.4}$ joules

24. $E = 10^{10.4}$ joules

25. $E = 10^{12}$ joules

26. $E = 10^{9}$ joules

In Exercises 27–30, the intensity I of a sound is given. Find the loudness L of the sound ($I_0 = 10^{-12}$ W/m²).

27. $I = 10^{-8}$ W/m²

28. $I = 10^{-10}$ W/m²

29. $I = 3.5 \times 10^{-7}$ W/m²

30. $I = 2.37 \times 10^{-5}$ W/m²

In Exercises 31–34, a loudness L is given. Find the intensity of the sound.

31. $L = 80$ dB

32. $L = 90$ dB

33. $L = 64.7$ dB

34. $L = 37.4$ dB

In Exercises 35–38, use the equal tempered scale with the twelve semitones A, A#, B, C, C#, D, D#, E, F, F#, G, and G# with the frequency of A440 Hz. On a piano keyboard, the distance between two white keys that are side-by-side is a whole note if there is a black key between them and is a semitone if there is no black key between them. A black key (such as the one between C and D) is called a sharp. The black key between C and D is denoted by C#.

Remember that the change in frequency $P(f)$ between any two consecutive semitones is 100 cents.

35. Find the frequencies of A# and C.

36. Find the frequencies of D and E.

37. A whole tone is a frequency ratio of 10:9. That is, $\frac{f}{f_0} = \frac{10}{9}$. An equal tempered whole tone is 200 cents. Find the difference in cents between a whole tone and an equal tempered whole tone. Is the difference noticeable?

38. Repeat Exercise 37 if a whole tone is a frequency ratio of 9:8.

39. The magnitudes of two stars, A and B, are 4 and 20, respectively. Compare the brightness of these stars.

40. Repeat Exercise 39 for two stars with magnitudes -2 and 5.

41. The magnitude of star A is 2 more than that of star B. How is their corresponding brightness related?

42. The magnitude of star A is 1 more than that of star B. How is their corresponding brightness related?

Applying the Concepts

43. pH of human blood. The hydrogen ion concentration $[H^+]$ of blood is approximately 3.98×10^{-8} moles per liter. Find the pH value of human blood to the nearest tenth. Is human blood acidic or basic?

44. pH of milk. The hydrogen ion concentration $[H^+]$ of milk is approximately 3.16×10^{-7} moles per liter. Find the pH value of milk to the nearest tenth. Is milk acidic or basic?

45. pH of some common substances. The pH value of a sample of each substance is given. Find $[H^+]$ of the substance.

a. Wine grapes: pH = 3.15
b. Water: pH = 7.2
c. Eggs: pH = 7.78
d. Vinegar: pH = 3

46. pH of some common substances. The pH value of a sample of each substance is given. Find $[H^+]$ of the substance.

a. Battery acid: pH = 1.0
b. Baking soda: pH = 8.7
c. Ammonia: pH = 10.6
d. Stomach acid: pH = 2.3

47. Acid rain. In the Netherlands, the average pH value of the rainfall is 3.8. Find the average concentration of hydrogen ions $[H^+]$ in the rainfall. How much more acidic is this rain than ordinary rain with a pH value of 6?

48. pH of a solution. Suppose the pH value of a solution A is 1.0 more than the pH value of a solution B. How are the concentrations of hydrogen ions in the two solutions related?

49. Comparing pH of solutions. Suppose the hydrogen ion concentration in a solution A is 100 times that in solution B. How are the pH values of the two solutions related?

50. pH of a solution. Suppose the pH value of a solution is increased by 1.5. How much change does this represent in the hydrogen ion concentration of this solution? Does the increase in pH make the solution more acidic or more basic?

51. pH of a solution. Suppose the hydrogen ion concentration of a solution is increased 50 times. How much change does this represent in the pH value of this solution? Does this increase in $[H^+]$ make the solution more acidic or more basic?

52. China earthquake. The Great China Earthquake of 1920 registered 8.6 on the Richter scale.

a. What was the intensity of this earthquake?
b. How many joules of energy were released?

53. San Francisco earthquake. Repeat Exercise 52 for the Great San Francisco Earthquake of 1906, which registered 7.8 on the Richter scale.

54. Comparing earthquakes. Table 4.7 lists two earthquakes in Japan.

a. Compare the intensities of these earthquakes.
b. Compare the energies released by these quakes.

55. Comparing earthquakes. Suppose earthquake A registers 1 more point on the Richter scale than earthquake B.

a. How are their corresponding intensities related?
b. How are their released energies related?

56. Comparing earthquakes. Repeat Exercise 55, if earthquake A registers 1.5 more points on the Richter scale than earthquake B.

57. Comparing magnitudes. If one earthquake is 150 times as intense as another, what is the difference in the Richter scale readings of the two earthquakes?

58. Comparing magnitudes. If the energy released by one earthquake is 150 times that of another, what is the difference in the Richter scale readings of the two earthquakes?

In Exercises 59–66, use Table 4.8 as necessary.

59. dB and intensity. What is the loudness, in decibels, of a radio with intensity 5.2×10^{-5} W/m^2?

60. dB of a jet. What is the loudness of a jet aircraft with intensity 2.5×10^2 W/m^2?

61. Comparing two sounds. How much more intense is a sound of the threshold of pain (130 dB) than a conversation at 65 dB?

62. Comparing two sounds. How much more intense is a 75 dB sound than a 62 dB sound?

63. Comparing two sounds. Suppose a sound is 1000 times as intense as one at the threshold of pain for the human ear. Find the loudness of this sound in decibels.

64. Comparing intensities. Show that if two sounds of intensity I_1 and I_2 register decibel levels of L_1 and L_2, respectively, then $\frac{I_2}{I_1} = 10^{0.1(L_2 - L_1)}$.

65. Comparing intensities. Suppose one sound registers 1 decibel more than another. How are the intensities of the two sounds related?

66. Comparing sounds. The noise level of a street sound outside a new office building in downtown Chicago is measured to be approximately 70 dB. With special insulation material, the noise level inside the office building is reduced to 29 dB. Compare the intensities of sounds outside and inside the building.

67. Violin tuning. A violinist is tuning her A-string to match concert pitch. The concert's A is a frequency of 440 Hz. Her A-string is improperly tuned to 441 Hz. Find the difference in cents.

68. Violin tuning. Repeat Exercise 67 if her A-string is improperly tuned to 438 Hz.

In Exercises 69–74, use the following table of approximate apparent magnitudes of some celestial objects.

Object	The Sun	Full Moon	Venus	Sirius	Vega	Saturn	Neptune
Magnitude	−27	−13	−4	−1	0.03	1.47	7.8

69. How many times brighter is the Sun than the full Moon?

70. Moon-Vega. How many times brighter is the full Moon than Vega?

71. Sun-Venus. How many times brighter is the Sun than Venus?

72. Venus-Neptune. How many times brighter is Venus than Neptune?

73. Saturn-Star. What is the magnitude of a star that is 560 times brighter than Saturn?

74. Neptune-Star. What is the magnitude of a star that is 1 billion times brighter than Neptune?

Beyond the Basics

75. Decibel levels. The decibel level of Professor Stout's voice decreases with the distance from the professor according to the formula

$$L = 10 \log\left(\frac{4 \times 10^4}{r^2}\right),$$

where L is the decibel level and r is the distance in feet from the professor to the listener.

a. Find the decibel levels at distances of 10, 25, 50, and 100 feet.

b. How far must a listener be from the professor so that the decibel level drops to 0?

c. Express L in the form $L = a + b \log r$, for suitable constants a and b.

d. Express r as a function of L.

76. Sound intensity. The intensity of sound is inversely proportional to the square of the distance. That is, $I = \frac{K}{r^2}$, where I is the intensity at a distance r from the source and K is a constant.

a. Show that if L_1 and L_2 are decibel readings of a sound at distances r_1 and r_2 (in meters), respectively, then $L_1 - L_2 = 20 \log\left(\frac{r_2}{r_1}\right)$.

b. A military jet at 30 meters has 140 decibel level of loudness. How loud will you hear this jet sound at a distance of 100 meters? 300 meters?

c. Express $\frac{r_2}{r_1}$ in the exponential form.

77. Nuclear bomb. Compare the energy released by a 1 megaton nuclear bomb (about 5×10^{15} joules) to the energy released by a great earthquake of magnitude 8.

78. Adrian was calculating the pitch of a sound and mistakenly transposed the frequency f of the sound with the reference frequency f_0. The correct answer is 40 cents. What was her answer?

79. If the A-string of a violin is tuned to 880 Hz with an error of ± 10 cents, what is the frequency range of the A-string?

80. Show that the formula for star brightness can be written as $\frac{b_1}{b_2} = 10^{0.4(m_2 - m_1)}$.

81. A star of magnitude m is 176 times brighter than a star of magnitude 3.42. Find m.

82. Star luminosity. The luminosity L of a star is the power output at a star's surface (measured in watts). The star's apparent brightness b is the wattage received on Earth's surface. The apparent magnitude m of a star depends on its absolute magnitude M (related to L) and its distance D in parsecs (1 parsec $\approx$ 3.26 light years) by the formula

$$m - M = 2.5 \log \frac{L}{b} = 5 \log \frac{D}{10}.$$

a. Given m and M for a star, show that its distance D from Earth is $D = 10^{1+(m-M)/5}$ parsecs.

b. For the star Alpha Centauri, $m = 0$ and $M = 4.39$. How far is it from Earth?

c. Compare the luminosity and brightness of Alpha Centauri.

Maintaining Skills

In Exercises 83–86, find the slope and the y-intercept of the equation of each line.

83. $3x - 2y = 12$

84. $3x + 5y = 15$

85. $e^{1.3}x + e^{2.5}y = e^{4.7}$

86. $x \log 2 - y \log 4 = \log 8$

In Exercises 87–94, find the center and radius of each circle.

87. $x^2 + (y - 2)^2 = 9$

88. $(x + 1)^2 + y^2 = 5$

89. $(x - 3)^2 + (y + 1)^2 = 7$

90. $(x + 2)^2 + (y - 3)^2 = 16$

91. $x^2 + y^2 - 2x - 3 = 0$

92. $x^2 + y^2 + 4y = 0$

93. $x^2 + y^2 - 2x + 4y - 4 = 0$

94. $x^2 + y^2 + 4x - 6y - 3 = 0$

SUMMARY Definitions, Concepts, and Formulas

4.1 Exponential Functions

i. A function $f(x) = a^x$, with $a > 0$ and $a \neq 1$, is called an exponential function with base a and exponent x.

Rules of exponents: $a^x a^y = a^{x+y}$, $\dfrac{a^x}{a^y} = a^{x-y}$,

$$(a^x)^y = a^{xy},\ a^0 = 1,\ a^{-x} = \frac{1}{a^x}$$

ii. Exponential functions are one-to-one: If $a^u = a^v$, then $u = v$.

iii. If $a > 1$, then $f(x) = a^x$ is an increasing function; $f(x) \to \infty$ as $x \to \infty$ and $f(x) \to 0$ as $x \to -\infty$.

iv. If $0 < a < 1$, then $f(x) = a^x$ is a decreasing function; $f(x) \to 0$ as $x \to \infty$ and $f(x) \to \infty$ as $x \to -\infty$.

v. The graph of $f(x) = a^x$ has y-intercept 1, and the x-axis is a horizontal asymptote.

vi. **Simple interest formula.** If P dollars is invested at an interest rate r per year for t years, then the simple interest is given by the formula $I = Prt$. The future value $A(t) = P + Prt$.

vii. **Compound interest.** P dollars invested at an annual rate r compounded n times per year for t years amounts to

$$A(t) = P\left(1 + \frac{r}{n}\right)^{nt}.$$

viii. The Euler constant $e = \lim\limits_{h \to \infty}\left(1 + \dfrac{1}{h}\right)^h \approx 2.718.$

ix. **Continuous compounding.** P dollars invested at an annual rate r compounded continuously for t years amounts to $A = Pe^{rt}$.

x. The function $f(x) = e^x$ is the natural exponential function.

4.2 Logarithmic Functions

i. For $x > 0$, $a > 0$, and $a \neq 1$, $y = \log_a x$ if and only if $x = a^y$.

ii. Basic properties: $\log_a a = 1$, $\log_a 1 = 0$,

$$\log_a a^x = x, \quad a^{\log_a x} = x \qquad \text{Inverse properties}$$

iii. The domain of $\log_a x$ is $(0, \infty)$, the range is $(-\infty, \infty)$, and the y-axis is a vertical asymptote. The x-intercept is 1.

iv. Logarithmic functions are one-to-one: If $\log_a x = \log_a y$, then $x = y$.

v. If $a > 1$, then $f(x) = \log_a x$ is an increasing function; $f(x) \to \infty$ as $x \to \infty$ and $f(x) \to -\infty$ as $x \to 0^+$.

vi. If $0 < a < 1$, then $f(x) = \log_a x$ is a decreasing function; $f(x) \to -\infty$ as $x \to \infty$ and $f(x) \to \infty$ as $x \to 0^+$.

vii. The common logarithmic function is $y = \log x$ (base 10); the natural logarithmic function is $y = \ln x$ (base e).

4.3 Rules of Logarithms

i. **Rules of logarithms:**

$$\log_a (MN) = \log_a M + \log_a N \qquad \text{Product rule}$$

$$\log_a\left(\frac{M}{N}\right) = \log_a M - \log_a N \qquad \text{Quotient rule}$$

$$\log_a M^r = r \log_a M \qquad \text{Power rule}$$

ii. **Change-of-base formula:**

$$\log_b x = \frac{\log_a x}{\log_a b} = \frac{\log x}{\log b} = \frac{\ln x}{\ln b}$$

$$(\text{base } a) \quad (\text{base } 10) \quad (\text{base } e)$$

4.4 Exponential and Logarithmic Equations and Inequalities

An *exponential equation* is an equation in which a variable occurs in one or more exponents.

A *logarithmic equation* is an equation that involves the logarithm of a function of the variable.

Exponential and logarithmic equations are solved by using some or all of the following techniques:

i. Using the one-to-one property of exponential and logarithmic functions

ii. Converting from exponential to logarithmic form or vice versa

iii. Using the product, quotient, and power rules for exponents and logarithms

We use similar techniques to solve logarithmic and exponential inequalities

4.5 Logarithmic Scales

A *logarithmic scale* is a scale in which logarithms are used in the measurement of quantities.

i. $\text{pH} = -\log[\text{H}^+]$, where $[\text{H}^+]$ is the concentration of H^+ ions in moles per liter.

ii. The Richter scale is used to measure the *magnitude* M of an earthquake.

iii. $M = \log\left(\dfrac{I}{I_0}\right)$, where I is the intensity of the earthquake and I_0 is the zero-level earthquake.

iv. The *energy* E released by an earthquake of magnitude M is given by $\log E = 4.4 + 1.5M$.

v. The *intensity* of a sound wave is defined as the amount of power the wave transmits through a given area. The *loudness* L of a sound of intensity I is given by $L = 10 \log\left(\dfrac{I}{I_0}\right)$, where $I_0 = 10^{-12}\ \text{W/m}^2$ is the intensity of the threshold of hearing. Equivalently, $I = 10^{L/10} I_0$.

vi. If f is a frequency and f_0 is a reference frequency, then the change in pitch $P(f)$ in cents is given by $P(f) = 1200 \log_2\left(\frac{f}{f_0}\right)$.

vii. If two stars of magnitudes m_1 and m_2 have apparent brightness b_1 and b_2, respectively, then $m_2 - m_1 = 2.5 \log\left(\frac{b_1}{b_2}\right)$.

REVIEW EXERCISES

Basic Concepts and Skills

In Exercises 1–10, state whether the given statement is true or false.

1. The function $f(x) = a^x$ is exponential if $a > 0$.

2. The graph of $f(x) = 4^x$ approaches the x-axis as $x \to -\infty$.

3. The domain of $f(x) = \log(2 - x)$ is $(-\infty, 2]$.

4. The equation $u = 10^v$ means that $\log u = v$.

5. The inverse of $f(x) = \ln x$ is $g(x) = e^x$.

6. The graph of $y = a^x$ $(a > 0, a \neq 1)$ always contains the points $(0, 1)$ and $(1, a)$.

7. $\ln(M + N) = \ln M + \ln N$

8. $\log \sqrt{300} = 1 + \frac{1}{2} \log 3$

9. The functions $f(x) = 2^{-x}$ and $g(x) = \left(\frac{1}{2}\right)^x$ have the same graph.

10. $\ln u = \frac{\log u}{\log e}$

In Exercises 11–18, match the function with its graph in (a)–(h).

11. $f_1(x) = \log_2 x$

12. $f_2(x) = 2^x$

13. $f_3(x) = \log_2(3 - x)$

14. $f_4(x) = \log_{1/2}(x - 1)$

15. $f_5(x) = -\log_2 x$

16. $f_6(x) = \left(\frac{1}{2}\right)^x$

17. $f_7(x) = 3 - 2^{-x}$

18. $f_8(x) = \frac{6}{1 + 2e^{-x}}$

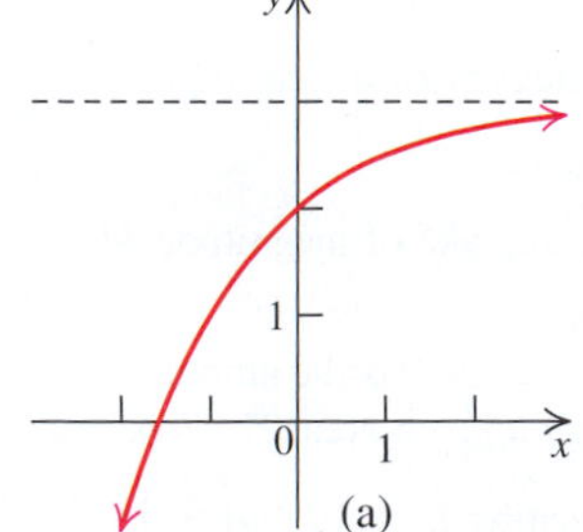

(a)

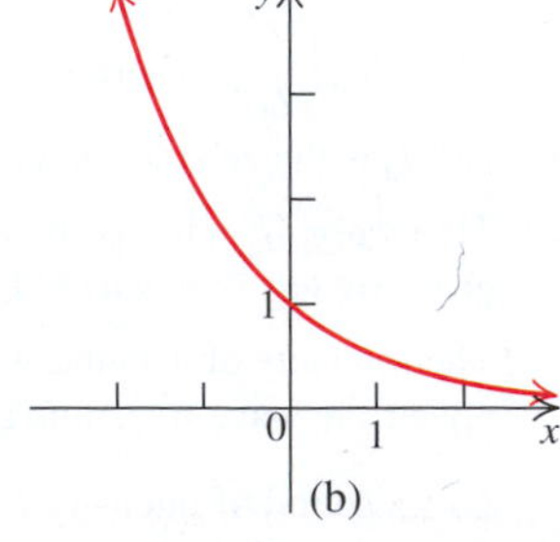

(b)

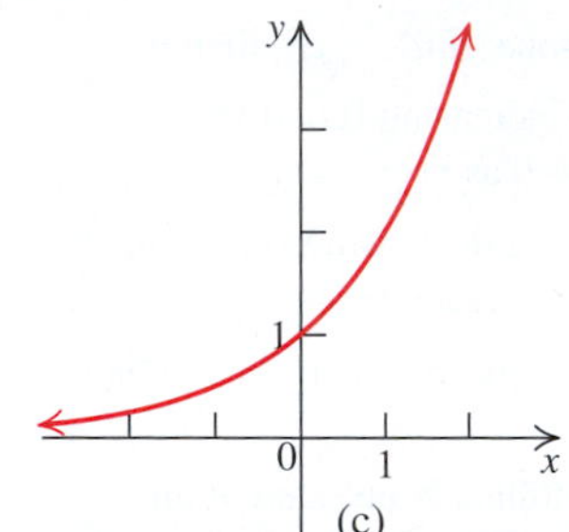

(c)

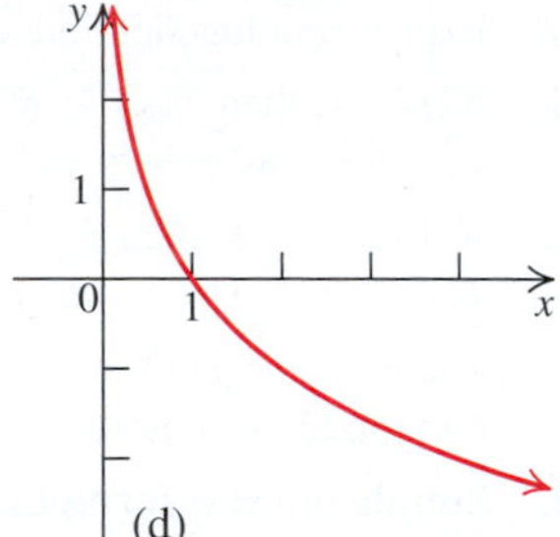

(d)

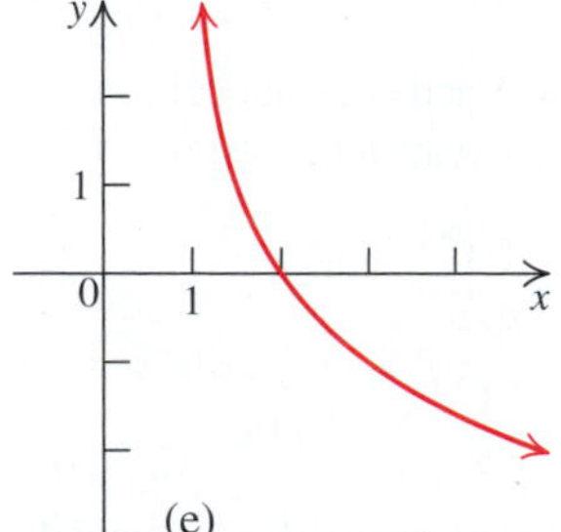

(e)

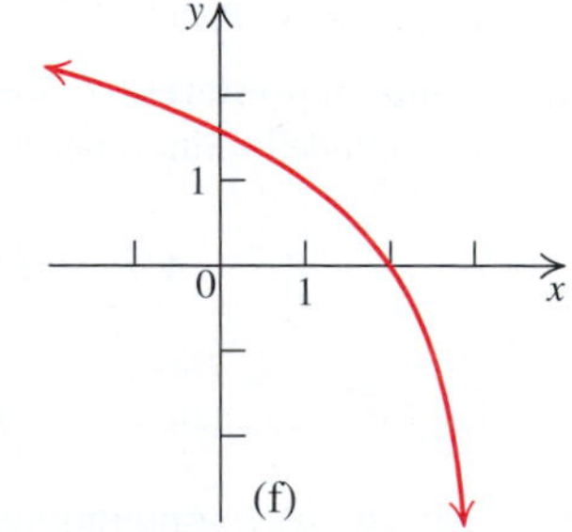

(f)

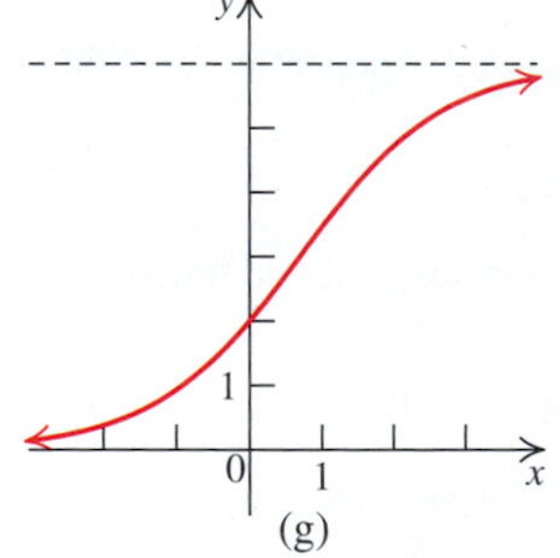

(g)

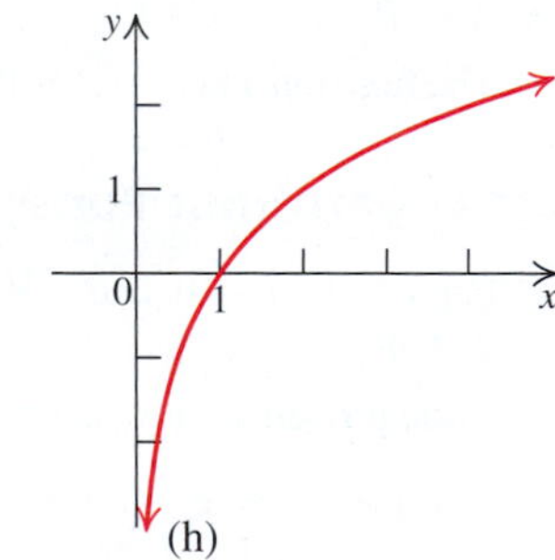

(h)

In Exercises 19–30, graph each function using transformations on an appropriate graph. Determine the domain, range, and asymptotes (if any).

19. $f(x) = 2^{-x}$

20. $g(x) = 2^{-0.5x}$

21. $h(x) = 3 + 2^{-x}$

22. $f(x) = 5^{|x|}$

23. $g(x) = 5^{-|x|}$

24. $h(x) = e^{-x+1}$

25. $f(x) = \ln(-x)$

26. $g(x) = 2 \ln |x|$

27. $h(x) = 2 \ln(x - 1)$

28. $f(x) = 2 - \left(\frac{1}{2}\right)^x$

29. $g(x) = 2 - \ln(-x)$

30. $h(x) = 3 + 2 \ln(5 + x)$

In Exercises 31–34, sketch the graph of the given function using these two steps:

a. Find the intercepts. **b.** Find the end behavior of f.

31. $f(x) = 3 - 2e^{-x}$

32. $f(x) = \dfrac{5}{2 + 3e^{-x}}$

33. $f(x) = e^{-x^2}$

34. $f(x) = 3 - \dfrac{6}{1 + 2e^{-x}}$

In Exercises 35–38, find *a* and *k* and then evaluate the function.

35. Let $f(x) = a(2^{kx})$ with $f(0) = 10$ and $f(3) = 640$. Find $f(2)$.

36. Let $f(x) = 50 - a(5^{kx})$ with $f(0) = 10$ and $f(2) = 0$. Find $f(1)$.

37. Let $f(x) = \dfrac{4}{1 + ae^{-kx}}$ with $f(0) = 1$ and $f(3) = \dfrac{1}{2}$. Find $f(4)$.

38. Let $f(x) = 6 - \dfrac{3}{1 + ae^{-kx}}$ with $f(0) = 5$ and $f(4) = 4$. Find $f(10)$.

In Exercises 39 and 40, find an exponential function of the form $f(x) = ca^x$ with the given graph.

39. **40.**

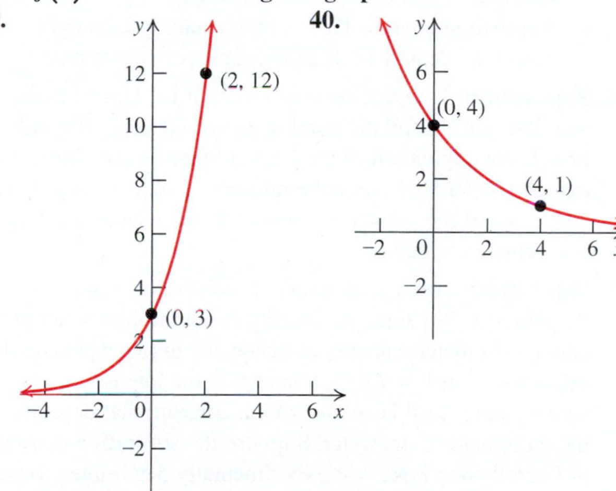

In Exercises 41 and 42, find a logarithmic function of the form $y = \log_a(x - c)$ with the given graph.

41.

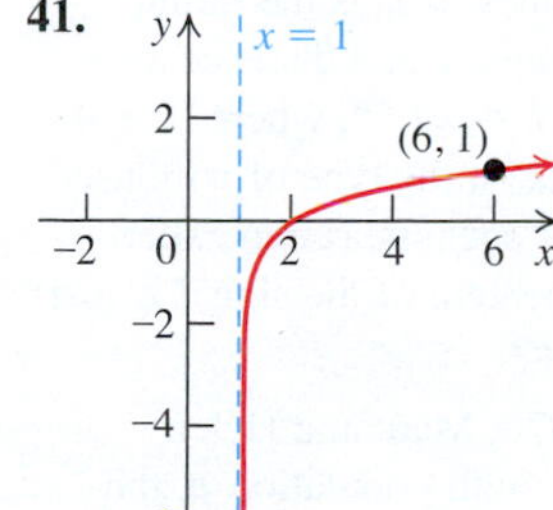

42.

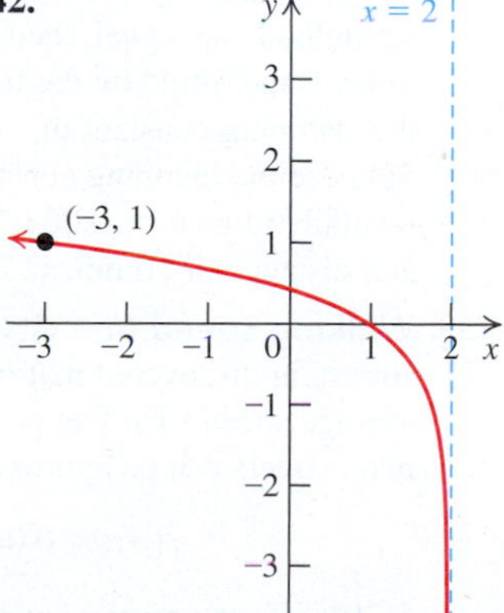

43. Start with the graph of $y = 2^x$. Find an equation of the graph that results when you

a. Shift the graph right 1 unit, shift up 3 units, and reflect about the x-axis.

b. Shift the graph right 1 unit, reflect about the x-axis, and shift up 3 units.

44. Start with the graph of $y = \ln x$. Find an equation of the graph that results when you

a. Shift the graph left 1 unit, stretch horizontally by a factor of 2, and compress vertically by a factor of 3.

b. Compress horizontally to $\frac{1}{3}$, shift left by 1 unit, and reflect about the y-axis.

In Exercises 45–48, write each logarithm in expanded form.

45. $\ln(xy^2z^3)$

46. $\log(x^3\sqrt{y - 1})$

47. $\ln\left[\dfrac{x\sqrt{x^2 + 1}}{(x^2 + 3)^2}\right]$

48. $\ln\sqrt{\dfrac{x^3 + 5}{x^3 - 7}}$

In Exercises 49–54, write *y* as a function of *x*.

49. $\ln y = \ln x + \ln 3$

50. $\ln y = \ln(C) + kx$; C and k are constants.

51. $\ln y = \ln(x - 3) - \ln(y + 2)$

52. $\ln y = \ln x - \ln(x^2y) - 2\ln y$

53. $\ln y = \dfrac{1}{2}\ln(x - 1) + \dfrac{1}{2}\ln(x + 1) - \ln(x^2 + 1)$

54. $\ln(y - 1) = \dfrac{1}{x} + \ln(y)$

In Exercises 55–78, solve each equation.

55. $3^x = 81$

56. $5^{x-1} = 625$

57. $2^{x^2+2x} = 16$

58. $3^{x^2-6x+8} = 1$

59. $3^x = 23$

60. $2^{x-1} = 5.2$

61. $273^x = 19$

62. $27 = 9^x \cdot 3^{x^2}$

63. $3^{2x} = 7^x$

64. $2^{x+4} = 3^{x+1}$

65. $(1.7)^{3x} = 3^{2x-1}$

66. $3(2^{x+5}) = 5(7^{2x-3})$

67. $\log_3(x + 2) - \log_3(x - 1) = 1$

68. $\log_3(x + 12) - \log_3(x + 4) = 2$

69. $\log_2(x + 2) + \log_2(x + 4) = 3$

70. $\log_5(3x + 7) + \log_5(x - 5) = 2$

71. $\log_5(x^2 - 5x + 6) - \log_5(x - 2) = 1$

72. $\log_3(x^2 - x - 6) - \log_3(x - 3) = 1$

73. $\log_6(x - 2) + \log_6(x + 1) = \log_6(x + 4) + \log_6(x - 3)$

74. $\ln(x - 2) - \ln(x + 2) = \ln(x - 1) - \ln(2x + 1)$

75. $2\ln 3x = 3\ln x$

76. $2\log x = \ln e$

77. $2^x - 8 \cdot 2^{-x} - 7 = 0$

78. $3^x - 24 \cdot 3^{-x} = 10$

In Exercises 79–82, solve each inequality.

79. $3(0.2)^x + 5 \le 20$

80. $-4(0.2)^x + 15 < 6$

81. $\log(2x + 7) < 2$

82. $\ln(3x + 5) \le 1$

Applying the Concepts

83. Doubling your money. How much time is required for a \$1000 investment to double in value if interest is earned at the rate of 6.25% compounded annually?

84. Tripling your money. How much time (to the nearest month) is required to triple your money if the interest rate is 100% compounded continuously?

85. Half-life. The half-life of a certain radioactive substance is 20 hours. How long does it take for this substance to fall to 25% of its original value?

86. Plutonium-210. Find the half-life of plutonium-210 assuming that its decay equation is $Q = Q_0 e^{-5 \cdot 10^{-3} t}$, where t is in days.

87. Comparing rates. You have \$7000 to invest for seven years. Which investment will provide the greater return, 5% compounded yearly or 4.75% compounded monthly?

88. Investment growth. How long will it take \$8000 to grow to \$20,000 if the rate of interest is 7% compounded continuously?

89. Population. The formula $P(t) = 33e^{0.003t}$ models the population of Canada, in millions, t years after 2007.

a. Estimate the population of Canada in 2017.
b. According to this model, when will the population of Canada be 60 million?

90. House appreciation. The formula $C(t) = 100 + 25e^{0.03t}$ models the average cost of a house in Sometown, USA, t years after 2000. The cost is expressed in thousands of dollars.

a. Sketch the graph of $y = C(t)$.
b. Estimate the average cost in 2010.
c. According to this model, when will the average cost of a house in Sometown be \$250,000?

91. Drug concentration. An experimental drug was injected into the bloodstream of a rat. The concentration $C(t)$ of the drug (in micrograms per milliliter of blood) after t hours was modeled by the function

$$C(t) = 0.3e^{-0.47t},\ 0.5 \le t \le 10.$$

a. Graph the function $y = C(t)$ for $0.5 \le t \le 10$.
b. When will the concentration of the drug be 0.029 microgram per milliliter? (1 microgram $= 10^{-6}$ gram)

92. X-ray intensity. X-ray technicians are shielded by a wall insulated with lead. The equation $x = \frac{1}{152}\log\left(\frac{I_0}{I}\right)$ measures the thickness x (in centimeters) of the lead insulation required to reduce the initial intensity I_0 of X-rays to the desired intensity I.

a. What thickness of lead is required to reduce the intensity of X-rays to one-tenth their initial intensity?
b. What thickness of lead is required to reduce the intensity of X-rays to $\frac{1}{40}$ their initial intensity?
c. How much is the intensity reduced if the lead insulation is 0.2 millimeters thick?

93. Cooling tea. Chai (tea) is made by adding boiling water (212°F) to the chai mix. Suppose you make chai in a room with the air temperature at 75°F. According to Newton's Law of Cooling, the temperature of the chai t minutes after it is boiled is given by a function of the form $f(t) = 75 + ae^{-kt}$. After one minute, the temperature of the chai falls to 200°F. How long will it take for the chai to be drinkable at 150°F?

94. Spread of influenza. Approximately $P(t) = \frac{6}{1 + 5e^{-0.7t}}$ thousand people caught a new form of influenza within t weeks of its outbreak.

a. Sketch the graph of $y = P(t)$.
b. How many people had the disease initially?
c. How many people contracted the disease within four weeks?
d. If the trend continues, how many people in all will contract the disease?

95. Population density. The population density x miles from the center of a town called Greenville is approximated by the equation $D(x) = 5e^{0.08x}$, in thousands of people per square mile.

a. What is the population density at the center of Greenville?
b. What is the population density 5 miles from the center of Greenville?
c. Approximately how far from the center of Greenville would the density be 15,000 people per square mile?

96. Population. In 2000, the population of the United States was 280 million and the number of vehicles was 200 million. If the population of the United States is growing at the rate of 1% per year while the number of vehicles is growing at the rate of 3%, in what year will there be an average of one vehicle per person?

97. Light intensity. The *Bouguer–Lambert Law* states that the intensity I of sunlight filtering down through water at a depth x (in meters) decreases according to the exponential decay function $I = I_0 e^{-kx}$, where I_0 is the intensity at the surface and $k > 0$ is an absorption constant that depends on the murkiness of the water. Suppose the absorption constant of Carrollwood Lake was experimentally determined to be $k = 0.73$. How much light has been absorbed by the water at a depth of 2 meters?

98. Signal strength. The strength of a TV signal usually fades due to a damping effect of cable lines. If I_0 is the initial strength of the signal, then its strength I at a distance x miles is measured by the formula $I = I_0 e^{-kx}$, where $k > 0$ is a damping constant that depends on the type of wire used. Suppose the damping constant has been measured experimentally to be $k = 0.003$. What percent of the signal is lost at a distance of 10 miles? 20 miles?

99. Walking speed in a city. In 1976, Marc and Helen Bernstein discovered that in a city with population p, the average speed s (in feet per second) that a person walks on main streets can be approximated by the formula

$$s(p) = 0.04 + 0.86 \ln p.$$

a. What is the average walking speed of pedestrians in Tampa (population 470,000)?
b. What is the average walking speed of pedestrians in Bowman, Georgia (population 450)?
c. What is the estimated population of a town in which the estimated average walking speed is 4.6 feet per second?

100. **Drinking and driving.** Just after Eric had his last drink, the alcohol level in his bloodstream was 0.26 (milligram of alcohol per milliliter of blood). After one-half hour, his alcohol level was 0.18. The alcohol level $A(t)$ in a person follows the exponential decay law

$$A(t) = A_0 e^{-kt}, \ (t \text{ in hours}),$$

where $k > 0$ depends on the individual.

a. What is the value of A_0 for Eric?

b. What is the value of k for Eric?

c. If the legal driving limit for alcohol level is 0.08, how long should Eric wait (after his last drink) before he will be able to drive legally?

101. **Bacteria culture.** The mass $m(t)$, in grams, of a bacteria culture grows according to the logistic growth model

$$m(t) = \frac{6}{1 + 5e^{-0.7t}},$$

where t is time measured in days.

a. What is the initial mass of the culture?

b. What happens to the mass in the long run?

c. When will the mass reach 5 grams?

102. **A learning model.** The number of units $n(t)$ produced per day after t days of training is given by

$$n(t) = 60(1 - e^{-kt}),$$

where $k > 0$ is a constant that depends on the individual.

a. Estimate k for Rita, who produced 20 units after one day of training.

b. How many units will Rita produce per day after ten days of training?

c. How many days should Rita train if she is expected to produce 40 units per day?

103. **Music.** A jump of one perfect minor third gives a frequency increase of 20%. A perfect minor third is 300 cents. Find the difference in cents. Is the difference noticeable?

104. **Star brightness.** The magnitude of star A is 3 more than that of the star B. How is their corresponding brightness related?

105. **Acid rain.** In Scotland, the lowest recorded pH level of rainfall was 2.4. What was the concentration of hydrogen ions $[H^+]$ in that rainfall? How much more acidic was this rain than a rain with a pH value of 5.6?

106. **Comparing earthquakes.** The 1997 earthquake in Iran registered 7.5 on the Richter scale, and the 2010 earthquake in Chile registered 8.8 on the Richter scale.

a. Compare the intensities of these earthquakes.

b. Compare the energies released by these quakes.

107. **Comparing two sounds.** How much more intense is a 115 dB sound than a 95 dB sound?

108. **Comparing intensities.** If the intensity of one sound is 5000 times another, what is the difference in the decibel levels of the two sounds?

109. **Sirius-Saturn.** How many times brighter is Sirius than Saturn?

110. **Nuclear bomb.** What magnitude of earthquake releases energy equivalent to that of a 1 megaton nuclear bomb (5×10^{15} joules)?

PRACTICE TEST A

1. Solve the equation $5^{-x} = 125$.

2. Solve the equation $\log_2 x = 5$.

3. State the range of $y = -e^x + 1$ and find the asymptote of its graph.

4. Evaluate $\log_2 \frac{1}{8}$.

5. Solve the exponential equation $\left(\frac{1}{4}\right)^{2-x} = 4$.

6. Evaluate log 0.001 without using a calculator.

7. Rewrite the expression $\ln 3 + 5 \ln x$ in condensed form.

8. Solve the equation $2^{x+1} = 5$.

9. Solve the equation $e^{2x} + e^x - 6 = 0$.

10. Rewrite the expression $\ln \frac{2x^3}{(x+1)^5}$ in expanded logarithmic form.

11. Evaluate $\ln e^{-5}$.

12. Give the equation for the graph obtained by shifting the graph of $y = \ln x$ 3 units up and 1 unit right.

13. Sketch the graph of $y = 3^{x-1} + 2$.

14. State the domain of the function $f(x) = \ln(-x) + 4$.

15. Write $3 \ln x + \ln(x^3 + 2) - \frac{1}{2} \ln(3x^2 + 2)$ in condensed form.

16. Solve the equation $\log x = \log 6 - \log(x - 1)$.

17. Find x if $\log_x 9 = 2$.

18. Suppose \$15,000 is invested in a savings account paying 7% interest per year. Write the formula for the amount in the account after t years if the interest is compounded quarterly.

19. Suppose the number of Hispanic people living in the United States is approximated by $H = 15{,}000e^{0.2t}$, where $t = 0$ represents 1960. According to this model, in which year did the Hispanic population in the United States reach 1.5 million?

20. Compare the intensity of the earthquake in 1960 in Morocco of magnitude 5.8 to that of the 6.2 magnitude earthquake in 1972 in Nicaragua.

PRACTICE TEST B

1. Solve the equation $3^{-x} = 9$.
   a. $\{2\}$ b. $\{-2\}$ c. $\left\{\frac{1}{2}\right\}$ d. $\left\{-\frac{1}{2}\right\}$ e. $\{\ln 2\}$

2. Solve the equation $\log_5 x = 2$.
   a. $\{10\}$ b. $\{25\}$ c. $\left\{\frac{5}{2}\right\}$ d. $\left\{\frac{2}{5}\right\}$ e. $\{2\}$

3. State the range and asymptote of $y = e^{-x} - 1$.
   a. $(0, \infty); y = -1$ b. $(-1, \infty); y = -1$
   c. $(-\infty, \infty); y = -1$ d. $(-1, \infty); y = 1$
   e. $(\infty, 1); y = 1$

4. Evaluate $\log_4 64$.
   a. 16 b. 8 c. 2 d. 3 e. 4

5. Find the solution of the exponential equation $\left(\frac{1}{3}\right)^{1-x} = 3$.
   a. $\left\{-\frac{1}{3}\right\}$ b. $\left\{\frac{1}{3}\right\}$ c. $\{1\}$ d. $\{2\}$

6. Evaluate $\log 0.01$.
   a. -1.99999999 b. -2 c. 2 d. 100

7. Which of the following expressions is equivalent to $\ln 7 + 2 \ln x$?
   a. $\ln (7 + 2x)$ b. $\ln (7x^2)$ c. $\ln (9x)$ d. $\ln (14x)$

8. Solve the equation $2^{x^2} = 3^x$.
   a. $\{0\}$ b. $\left\{\frac{\ln 3}{\ln 2}\right\}$
   c. $\left\{0, \frac{\ln 3}{\ln 2}\right\}$ d. $\{0, \ln 3 - \ln 2\}$

9. Solve the equation $e^{2x} - e^x - 6 = 0$.
   a. $\{-\ln 2\}$ b. $\{-\ln 3\}$ c. $\{\ln 6\}$ d. $\{\ln 3\}$

10. Rewrite the expression $\ln \frac{3x^2}{(x + 1)^{10}}$ in expanded logarithmic form.
    a. $\ln 6x - 10 \ln x + 1$
    b. $2 \ln 3x - 10 \ln (x + 1)$
    c. $2 \ln 3 + 2 \ln x - 10 \ln (x + 1)$
    d. $\ln 3 + 2 \ln x - 10 \ln (x + 1)$

11. Find $\ln e^{3x}$.
    a. 3 b. $3x$ c. $3 + x$ d. x

12. The equation for the graph obtained by shifting the graph of $y = \log_3 x$ 2 units up and 3 units left is
    a. $y = \log_3 (x - 3) + 2$. b. $y = \log_3 (x + 3) - 2$.
    c. $y = \log_3 (x - 3) - 2$. d. $y = \log_3 (x + 3) + 2$.

13. Which of the following graphs is the graph of $y = 5^{x+1} - 4$?

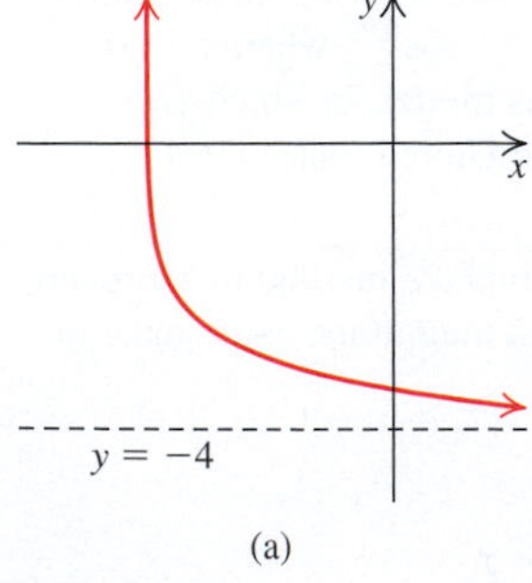

(a)

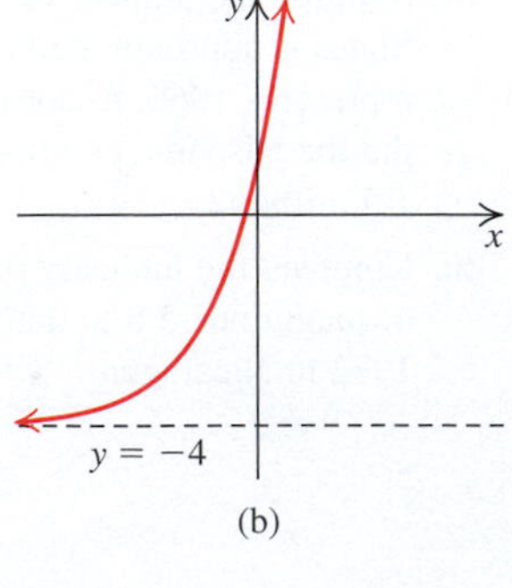

(b)

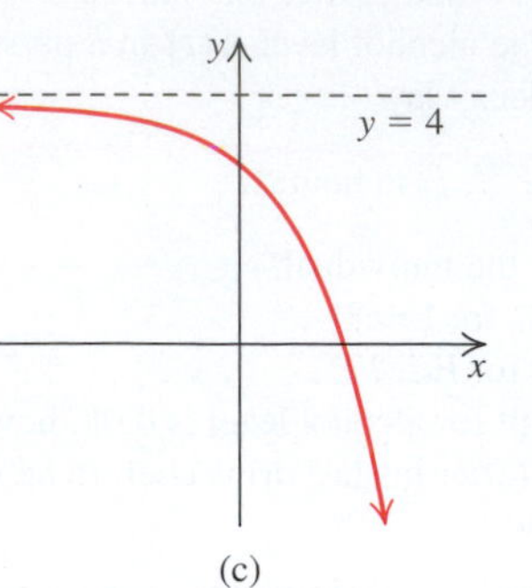

(c)

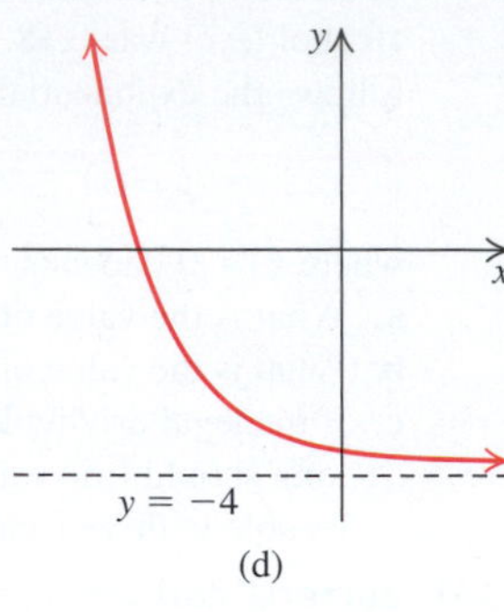

(d)

14. Find the domain of the function $f(x) = \ln (1 - x) + 3$.
    a. $(-\infty, 1)$ b. $(1, \infty)$ c. $(-\infty, -1)$
    d. $(-\infty, 3)$ e. $(3, \infty)$

15. Write $\ln x - 2 \ln (x^2 + 1) + \frac{1}{2} \ln (x^4 + 1)$ in condensed form.
    a. $\ln \frac{x\sqrt{x^4 + 1}}{(x^2 + 1)^2}$ b. $\ln \frac{x}{x^2 + 1}$
    c. $\ln (x - (x^2 + 1)^2 + (x^4 + 1)^{1/2})$
    d. $2 \ln \frac{x(1 + x^4)}{x^2 + 1}$

16. Solve the equation $\log x = \log 12 - \log (x + 1)$.
    a. $\left\{\frac{11}{2}\right\}$ b. $\left\{\frac{13}{2}\right\}$ c. $\{3, -4\}$
    d. $\{3\}$ e. $\left\{\frac{2}{12}\right\}$

17. Find x if $\log_x 16 = 4$.
    a. 4 b. 2 c. 64 d. $\frac{1}{4}$ e. 16

18. Suppose \$12,000 is invested in a savings account paying 10.5% interest per year. Write the formula for the amount in the account after t years assuming that the interest is compounded monthly.
    a. $A = 12{,}000(1.105)^t$ b. $A = 12{,}000(1.525)^{2t}$
    c. $A = 12{,}000(1.2625)^{4t}$ d. $A = 12{,}000(1.00875)^{12t}$

19. The population of a certain city is growing according to the model $P = 10{,}000 \log_5 (t + 5)$, where t is time in years. If $t = 0$ corresponds to the year 2000, what will the population of the city be in 2020?
    a. 30,000 b. 20,000
    c. 50,000 d. 10,000

20. Compare the intensity of the earthquake in 1994 in Northridge, California, of magnitude 6.7 to that of the 7.0 earthquake in 1988 in Armenia.
    a. The Northridge earthquake was about twice as intense.
    b. The Armenia earthquake was about twice as intense.
    c. The Northridge earthquake was about a thousand times as intense.
    d. The Armenia earthquake was about a thousand times as intense.

CUMULATIVE REVIEW EXERCISES CHAPTERS P–4

1. Find the x- and y-intercepts of the graph of the equation $x^2 + (y - 1)^2 = 1$. Check for symmetry with respect to both axes and the origin.
2. Graph the equation $y = x^3$.
3. Find the center and radius of the circle $x^2 + 2x + y^2 - 4y - 20 = 0$. Graph the circle.
4. Write the slope–intercept form of the equation of the line that passes through $(-2, 4)$ and is perpendicular to the line $2x + 3y = 17$.
5. Find the domain of $f(x) = \log \dfrac{x - 2}{x + 3}$.
6. Evaluate the function $f(x) = \begin{cases} 2x - 3, & x < 1 \\ 2x^2 + 1, & x \geq 1 \end{cases}$ at each value specified.
   a. $f(-2)$ b. $f(0)$ c. $f(3)$.
7. Identify the basic function and use transformations to sketch the graph of the function $f(x) = 3\sqrt{x + 1} - 2$.
8. Let $f(x) = 3\sqrt{x}$ and $g(x) = -\dfrac{1}{x}$.
   Find and write the domain of $(f \circ g)(x)$ in interval notation.
9. Determine whether the function has an inverse function; if so, find the inverse function.
   a. $f(x) = 2x^3 + 1$
   b. $f(x) = |x|$
   c. $f(x) = \ln x$.
10. Divide
    a. $\dfrac{2x^3 + 3x + 1}{x^2 + 1}$ using long division.
    b. $\dfrac{2x^4 + 7x^3 - 2x + 3}{x + 3}$ using synthetic division.
11. Find a polynomial function of least degree with integer coefficients that has the given zeros.
    a. $1, 2, -3$
    b. $1, -1, i, -i$
12. Use rules of logarithms to rewrite the following expressions in expanded form.
    a. $\log_2 5x^3$
    b. $\log_a \sqrt[3]{\dfrac{xy^2}{z}}$
    c. $\ln \dfrac{3\sqrt{x}}{5y}$
13. Use rules of logarithms to rewrite the following expressions in condensed form.
    a. $\dfrac{1}{2}(\log x + \log y)$
    b. $3 \ln x - 2 \ln y$
14. Use a calculator to evaluate the expression. Round your answer to three decimal places.
    a. $\log_9 17$
    b. $\log_3 25$
    c. $\log_{1/2} 0.3$
15. Solve each equation.
    a. $\log_6 (x + 3) + \log_6 (x - 2) = 1$
    b. $5^{x^2 - 4x + 5} = 25$
    c. $3.1^{x-1} = 23$
16. Find all possible rational zeros of

    $$f(x) = x^4 - x^3 - 2x^2 - 2x + 4.$$

    Also find upper and lower bounds for the zeros of f.
17. Let $f(x) = (x - 1)^3(x + 2)^2(x - 3)$.
    a. Find the zeros of f along with their multiplicities.
    b. What is the behavior of f near each zero?
    c. What is the end behavior of f?
    d. Sketch the graph of f.
18. Let $f(x) = \dfrac{(x - 1)(x + 2)}{(x - 3)(x + 4)}$.
    a. Find the vertical asymptotes of f (if any).
    b. Find the horizontal asymptotes of f (if any).
    c. Find the intervals over which $f(x) - 1 > 0$ or $f(x) - 1 < 0$.
    d. Sketch the graph of f.

CHAPTER 5

Trigonometric Functions

TOPICS

Angles are vital for activities ranging from taking photographs to designing layouts for city streets. Scientists use angles and basic trigonometry to measure remote distances such as the distance between mountain peaks and the distances between planets. Angles are universally used to specify locations on Earth via longitude and latitude. In this chapter, we begin the study of trigonometry and its many uses.

SECTION 5.1

Angles and Their Measure

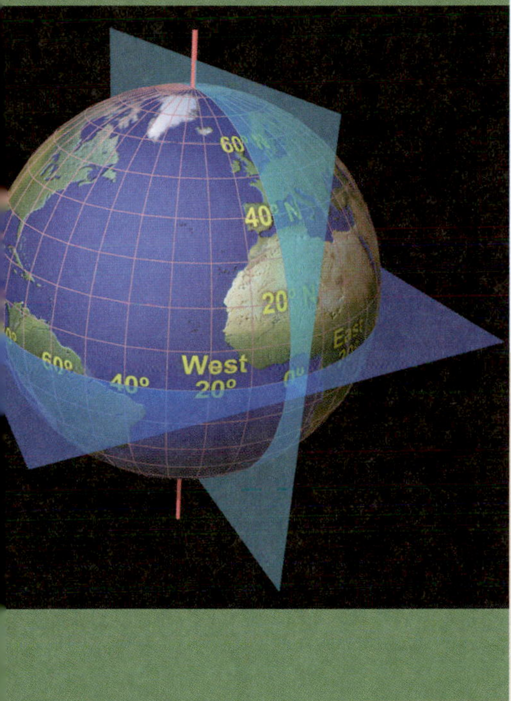

BEFORE STARTING THIS SECTION, REVIEW

1 Circumference and area of a circle (Page 662)

OBJECTIVES

1 Learn vocabulary associated with angles.
2 Use degree and radian measure of an angle.
3 Convert between degree and radian measure.
4 Find complements and supplements.
5 Find the length of an arc of a circle.
6 Find the area of a sector.
7 Compute linear and angular speed.

Latitude and Longitude

Any location on Earth can be described by two numbers, its *latitude* and its *longitude*. To understand how these two location numbers are assigned, we think of Earth as being a perfect sphere.

Lines of latitude are parallel circles of different size around the sphere representing Earth. The longest circle is the equator, with latitude 0, and at the poles, the circles shrink to a point. Lines of longitude (also called *meridians*) are circles of identical size that pass through the North Pole and the South Pole as they circle the globe. Each of these circles crosses the equator. The equator is divided into 360 degrees, and the longitude of a location is the number of degrees assigned to the location where the meridian meets the equator. For historical reasons, the meridian near the old Royal Astronomical Observatory in Greenwich, England, is chosen as 0 longitude. Today this prime meridian is marked with a band of brass that stretches across the yard of the Observatory.

Longitude is measured from the prime meridian, with positive values going east (0 to 180) and negative values going west (0 to -180). Both ± 180-degree longitudes share the same line, in the middle of the Pacific Ocean. At the equator, the distance on Earth's surface for each one degree of latitude or longitude is just over 69 miles.

Every location on Earth is then identified by the meridian and the latitude lines that pass through it. In Example 7, we explain how latitude values are assigned and how they can be used to compute distances between cities having the same longitude.

1 Learn vocabulary associated with angles.

Angles

A **ray** is a part of a line made up of a point, called the **endpoint**, and all of the points on one side of the endpoint. An **angle** is formed by rotating a ray about its endpoint. The angle's **initial side** is the ray's original position, and the angle's **terminal side** is the ray's position after the rotation. The endpoint is called the **vertex** of the angle. A curved arrow drawn near the angle's vertex indicates both the direction and amount of rotation from the initial side to the terminal side. See Figure 5.1.

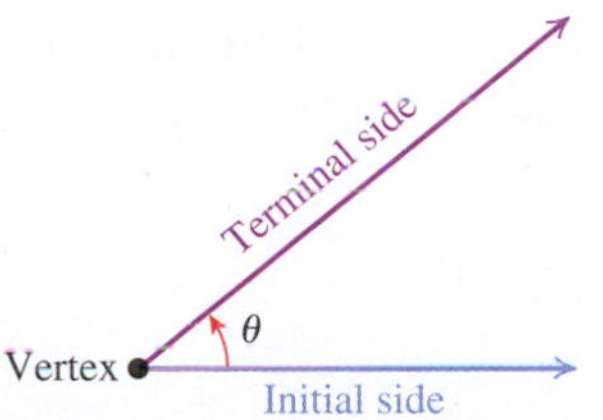

FIGURE 5.1 An angle

The Greek letters α (alpha), β (beta), γ (gamma), and θ (theta) are often used to name angles. If the rotation is counterclockwise, the result is a **positive angle**; if the rotation is clockwise, the result is a **negative angle**. See Figure 5.2(a) and (b). Rotation in both directions is unrestricted. Angles that have the same initial and terminal sides are called **coterminal** angles. In Figure 5.2(c), α and β are different angles but are coterminal.

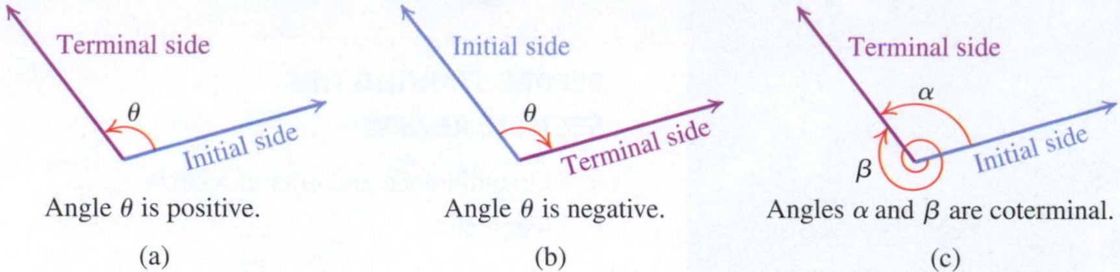

FIGURE 5.2 Positive, negative, and coterminal angles

An angle in a rectangular coordinate system is in **standard position** if its vertex is at the origin and its initial side lies on the positive x-axis. All of the angles in Figure 5.3 are in standard position. An angle in standard position is **quadrantal** if its terminal side lies on a coordinate axis; it is said to **lie in a quadrant** if its terminal side lies in that quadrant. See Figure 5.3.

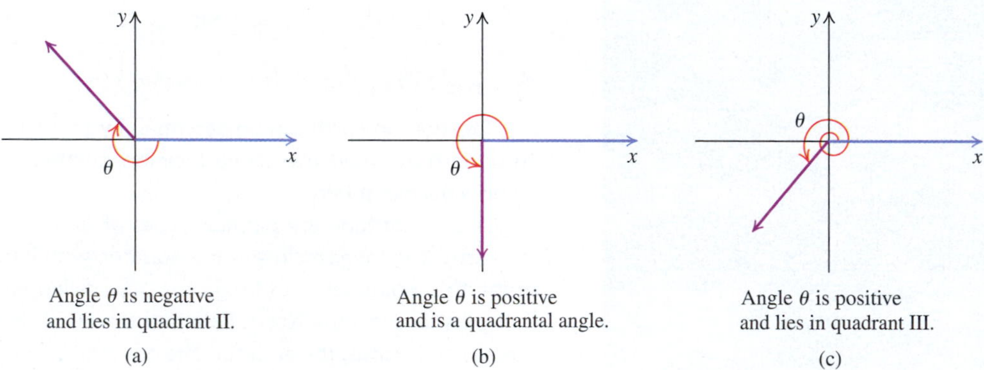

FIGURE 5.3 Angles in standard position

2 Use degree and radian measure of an angle.

Angle Measure

We measure angles by determining the amount of rotation from the initial side to the terminal side, indicated by a curved arrow that also shows direction. Two units of measurement for angles are *degrees* and *radians*.

Degree Measure

A measure of **one degree** (denoted by 1°) is assigned to an angle resulting from a rotation $\frac{1}{360}$ of a complete revolution counterclockwise about the vertex. An angle formed by rotating the initial side counterclockwise one full rotation so that the terminal and initial sides coincide has measure 360 degrees, written as 360°.

Angles are classified according to their measures. As illustrated in Figure 5.4, an **acute angle** has measure between 0° and 90°, a **right angle** has measure 90° (or one-fourth of a revolution), an **obtuse angle** has measure between 90° and 180°, and a **straight angle** has measure 180° (or half of a revolution). Figure 5.4 also shows several angles measured in degrees.

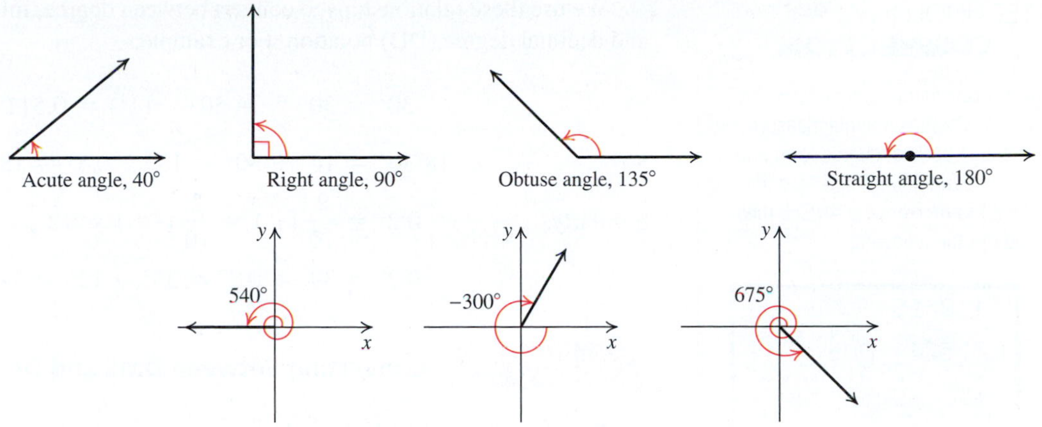

Acute angle, 40° Right angle, 90° Obtuse angle, 135° Straight angle, 180°

540° −300° 675°

FIGURE 5.4 Degree measure of various angles

The measure of an angle has no numerical limit because the terminal side can be rotated without limitation.

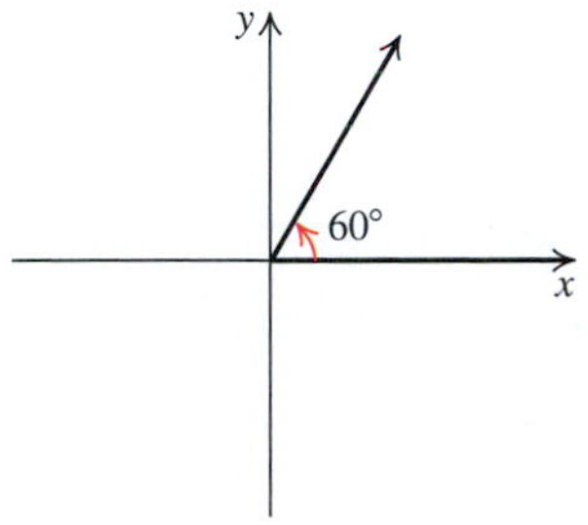

60°

FIGURE 5.5

EXAMPLE 1 Drawing an Angle in Standard Position

Draw each angle in standard position.

a. 60° **b.** 135° **c.** −240° **d.** 405°

Solution

a. Because $60 = \frac{2}{3}(90)$, a 60° angle is $\frac{2}{3}$ of a 90° counterclockwise rotation. See Figure 5.5.

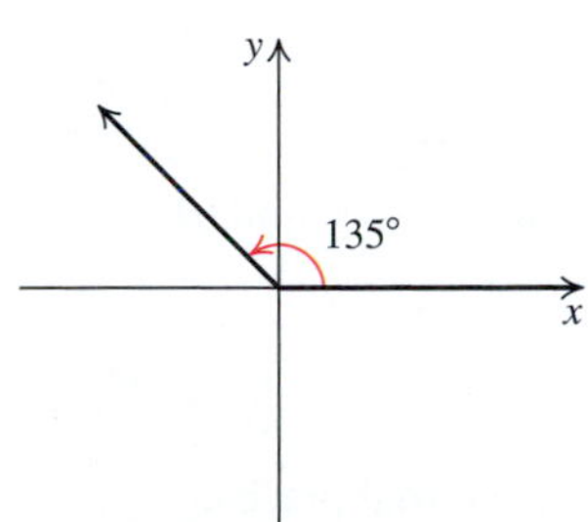

FIGURE 5.6

b. Because $135 = 90 + 45$, a 135° angle is a counterclockwise rotation of 90°, followed by half of a 90° counterclockwise rotation. See Figure 5.6.

c. Because $-240 = -180 + (-60)$, a −240° angle is a clockwise rotation of 180°, followed by a clockwise rotation of 60°. See Figure 5.7.

d. Because $405 = 360 + 45$, a 405° angle is one complete counterclockwise rotation, followed by half of a 90° counterclockwise rotation. See Figure 5.8.

Practice Problem 1 Draw a 225° angle in standard position.

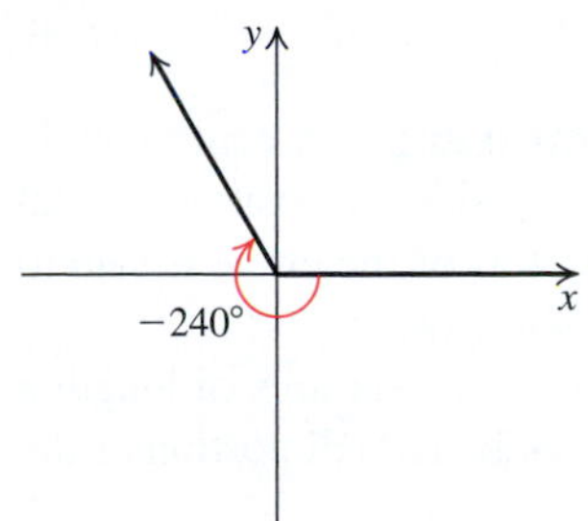

FIGURE 5.7

Today, it is common to divide degrees into fractional parts using *decimal degree notation* such as 30.5°. Traditionally, however, degrees were expressed in terms of *minutes* and *seconds*. One **minute**, denoted 1′, is defined as $\frac{1}{60}(1°)$, and one **second**, denoted 1″, is defined as $\frac{1}{60}(1')$. An angle measuring 27 degrees, 14 minutes, and 39 seconds is written as 27°14′39″. Because $\frac{1}{60} \cdot \frac{1}{60} = \frac{1}{3600}$, we have

$$1'' = \frac{1}{60}(1') = \frac{1}{60}\left[\frac{1}{60}(1°)\right] = \frac{1}{3600}(1°).$$

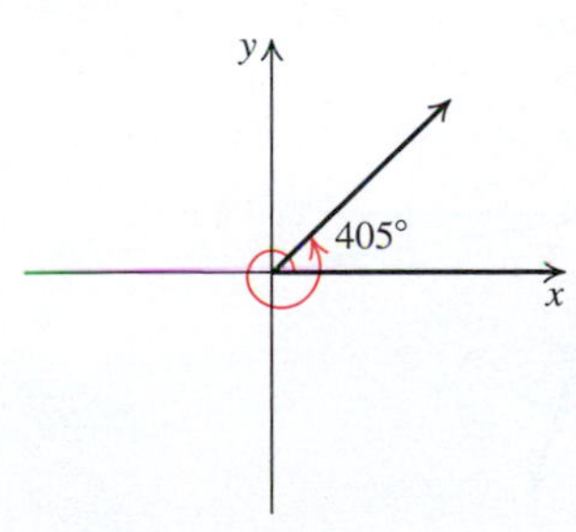

FIGURE 5.8

RELATIONSHIP BETWEEN DEGREES, MINUTES, AND SECONDS

$$1' = \frac{1}{60}(1°) = \left(\frac{1}{60}\right)^{\circ} \qquad 1° = 60' \qquad 1° = 3600''$$

$$1'' = \frac{1}{3600}(1°) = \left(\frac{1}{3600}\right)^{\circ} \qquad 1'' = \frac{1}{60}(1') \qquad 1' = 60''$$

TECHNOLOGY CONNECTION

Many calculators can perform conversions between angle measurements in decimal degree form and in degrees, minutes, and seconds (DMS) form. From the ANGLE menu, you get the following:

```
24°8'15"
               24.1375
67.526°▸DMS
         67°31'33.6"
```

We use these relationships to convert between degree, minute, second (DMS) notation and decimal degree (DD) notation. For example,

$$30' = 30 \cdot 1' = 30 \cdot \frac{1}{60}(1^\circ) = 0.5(1^\circ) = 0.5^\circ.$$

So

$$18^\circ 30' = 18^\circ + 30' = 18^\circ + 0.5^\circ = 18.5^\circ.$$

Similarly,

$$0.2^\circ = \frac{2}{10}(1^\circ) = \frac{2}{10}(60') = 12',$$

$$39.2^\circ = 39^\circ + 0.2^\circ = 39^\circ + 12' = 39^\circ 12'.$$

EXAMPLE 2 Converting Between DMS and DD Notation

a. Convert $24^\circ 8' 15''$ to decimal degree notation, rounded to two decimal places.

b. Convert 67.526° to DMS notation, rounded to the nearest second.

Solution

a.

$24^\circ 8' 15'' = 24^\circ + 8 \cdot 1' + 15 \cdot 1''$	Rewrite.
$= 24^\circ + 8\left(\frac{1}{60}\right)^\circ + 15\left(\frac{1}{3600}\right)^\circ$	$1' = \left(\frac{1}{60}\right)^\circ, 1'' = \left(\frac{1}{3600}\right)^\circ$
$\approx 24.14^\circ$	Use a calculator.

b.

$67.526^\circ = 67^\circ + 0.526 \cdot 1^\circ$	Rewrite.
$= 67^\circ + 0.526(60')$	Replace 1° with $60'$.
$= 67^\circ + 31.56'$	Use a calculator.
$= 67^\circ + 31' + 0.56 \cdot 1'$	Rewrite.
$= 67^\circ + 31' + 0.56(60'')$	Replace $1'$ with $60''$.
$= 67^\circ + 31' + 33.6'' \approx 67^\circ 31' 34''$	Use a calculator and round to the nearest second.

Practice Problem 2

a. Convert $13^\circ 9' 22''$ to decimal degree notation, rounded to two decimal places.

b. Convert 41.275° to DMS notation, rounded to the nearest second.

Radian Measure

An angle whose vertex is at the center of a circle is called a **central angle**. A central angle intercepts the arc of the circle from the initial side to the terminal side. A positive central angle that intercepts an arc of the circle of length equal to the radius of the circle is said to have measure **1 radian**. See Figure 5.9.

Because the circumference of a circle of radius r is $2\pi r \approx 6.28r$, six arcs of length r can be consecutively marked off on a circle of radius r, leaving only a small portion of the circumference uncovered. See Figure 5.10.

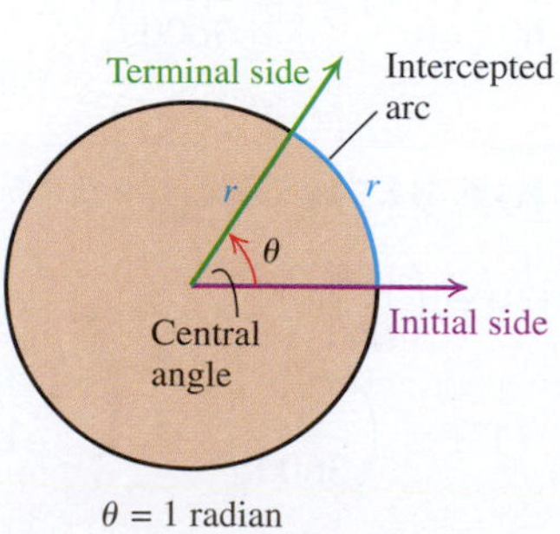

FIGURE 5.9 One Radian

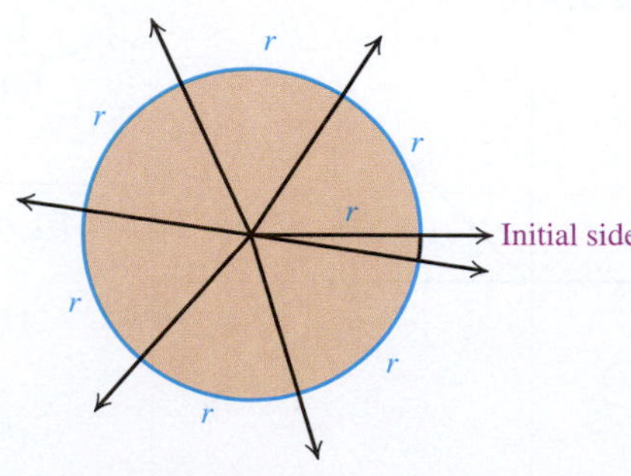

FIGURE 5.10

Each of the positive central angles in Figure 5.10 that intercepts an arc of length r has measure 1 radian; so the measure of one complete revolution is a little more than 6 radians. In this book, we assume (unless otherwise specified) that all central angles are positive. From geometry, we know that the measure of a central angle θ $(0 \le \theta \le 2\pi)$ is proportional to the length, s, of the arc it intercepts. This fact leads to the following relation.

SIDE NOTE

Because both s and r are in units of length, the units "cancel" in

$$\theta = \frac{s \text{ units}}{r \text{ units}} = \frac{s}{r}$$

making radians a unitless measure and consequently a real number.

RADIAN MEASURE OF A CENTRAL ANGLE

The radian measure θ of a central angle that intercepts an arc of length s on a circle of radius r is given by

$$\theta = \frac{s}{r} \text{ radians.}$$

Notice that when $s = r$, we have $\theta = \frac{r}{r} = 1$ radian.

3 Convert between degree and radian measure.

Relationship Between Degrees and Radians

To see the relationship between degrees and radians, consider the central angle in Figure 5.11 with degree measure $\theta = 180°$ in a circle of radius r. The length of the arc, s, that is intercepted by this angle is half the circumference of the circle. So

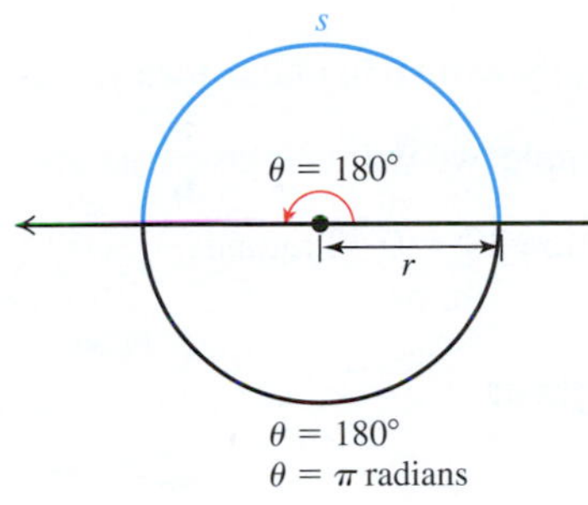

FIGURE 5.11

$$s = \frac{1}{2}\text{(circumference)}$$

$$s = \frac{1}{2}(2\pi r) = \pi r \qquad \text{Circumference} = 2\pi r$$

We can now find the radian measure for $\theta = 180°$.

$$180° = \theta = \frac{s}{r} \text{ radians} \qquad \text{Radian measure of a central angle}$$

$$180° = \frac{\pi r}{r} \text{ radians} \qquad \text{Replace } s \text{ with } \pi r.$$

$$180° = \pi \text{ radians} \qquad \text{Simplify.}$$

CONVERTING BETWEEN DEGREES AND RADIANS

The relationship

$$\mathbf{180° = \pi \text{ radians}}$$

allows us to convert between the degree and the radian measure of an angle.

Degrees to Radians:

$$180° = \pi \text{ radians} \qquad \text{Note that } 180° = (180) \cdot 1°.$$

$$1° = \frac{\pi}{180} \text{ radians} \qquad \text{Divide both sides by 180.}$$

$$\theta° = \frac{\theta\pi}{180} \text{ radians} \qquad \text{Multiply both sides by } \theta.$$

Radians to Degrees:

$$\pi \text{ radians} = 180°$$

$$1 \text{ radian} = \left(\frac{180}{\pi}\right)° \qquad \text{Divide both sides by } \pi.$$

$$\theta \text{ radians} = \left(\frac{180\,\theta}{\pi}\right)° \qquad \text{Multiply both sides by } \theta.$$

TECHNOLOGY CONNECTION

A graphing calculator can convert degrees to radians from the ANGLE menu. The calculator should be set in *radian* mode.

```
180°
        3.141592654
30°
        .5235987756
55°
        .9599310886
```

EXAMPLE 3 Converting from Degrees to Radians

Convert each angle from degrees to radians.

a. 30° **b.** 90° **c.** −225° **d.** 55°

Solution

To convert degrees to radians, multiply degrees by $\frac{\pi}{180°}$.

a. $30° = 30° \cdot \frac{\pi}{180°} = \frac{30\pi}{180} = \frac{\pi}{6}$ radians

b. $90° = 90° \cdot \frac{\pi}{180°} = \frac{90\pi}{180} = \frac{\pi}{2}$ radians

c. $-225° = -225° \cdot \frac{\pi}{180°} = \frac{-225\pi}{180} = -\frac{5\pi}{4}$ radians

d. $55° = 55° \cdot \frac{\pi}{180°} = \frac{55\pi}{180} \approx 0.96$ radians Use a calculator.

Practice Problem 3 Convert −45° to radians.

If an angle is a fraction of a complete revolution, we frequently write it in radian measure as a fractional multiple of π rather than as a decimal. For example, we write 30° as $\frac{\pi}{6}$ radians, which is the exact value, instead of writing the approximation $30° \approx 0.52$ radians.

EXAMPLE 4 Converting from Radians to Degrees

Convert each angle in radians to degrees.

a. $\frac{\pi}{3}$ radians **b.** $-\frac{3\pi}{4}$ radians **c.** 1 radian

Solution

To convert radians to degrees, multiply radians by $\frac{180°}{\pi}$.

a. $\frac{\pi}{3}$ radians $= \frac{\cancel{\pi}}{3} \cdot \frac{180°}{\cancel{\pi}} = \left(\frac{180}{3}\right)^{\circ} = 60°$

b. $-\frac{3\pi}{4}$ radians $= -\frac{3\cancel{\pi}}{4} \cdot \frac{180°}{\cancel{\pi}} = \left(-\frac{3}{4}\right)180° = -135°$

c. 1 radian $= 1 \cdot \frac{180°}{\pi} \approx 57.3°$ Use a calculator.

TECHNOLOGY CONNECTION

A graphing calculator can convert radians to degrees from the ANGLE menu. The calculator should be set in *degree* mode.

```
(π/3)ʳ
                60
(-3π/4)ʳ
              -135
1ʳ
       57.29577951
▮
```

Practice Problem 4 Convert $\frac{3\pi}{2}$ radians to degrees.

Figure 5.12 includes the degree and radian measures of some commonly used angles. With practice, you should be able to make these conversions without using the figure.

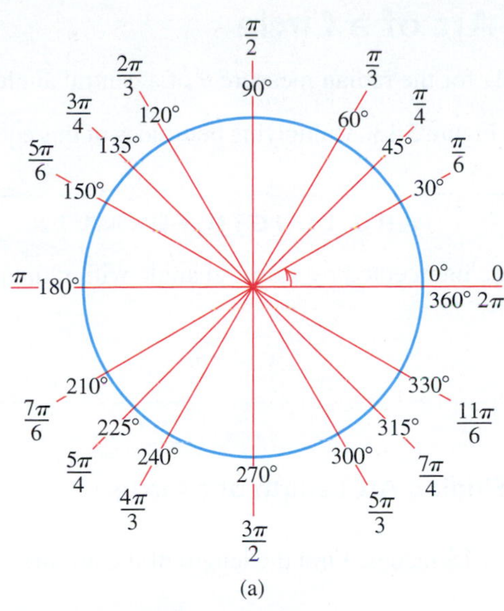

(a)

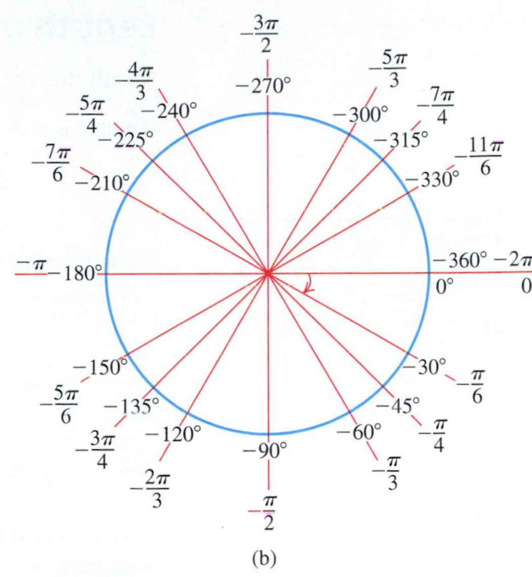

(b)

FIGURE 5.12 Frequently used angles

4 Find complements and supplements.

Complements and Supplements

Two positive angles are **complements** (or **complementary angles**) if the sum of their measures is 90°. So two angles with measures 50° and 40° are complements because $50° + 40° = 90°$. Each angle is the complement of the other. Two positive angles are **supplements** (or **supplementary angles**) if the sum of their measures is 180°. So two angles with measures 120° and 60° are supplements because $120° + 60° = 180°$. Each angle is the supplement of the other.

SIDE NOTE

A positive angle with measure $\geq 90°$ does not have a complement. A positive angle with measure $\geq 180°$ does not have a supplement.

EXAMPLE 5 Finding Complements and Supplements

Find the complement and the supplement of the given angle or explain why the angle has no complement or supplement.

a. 73° **b.** 110°

Solution

a. If θ represents the complement of 73°, then $\theta + 73° = 90°$; so $\theta = 90° - 73° = 17°$. The complement of 73° is 17°.
If α represents the supplement of 73°, then $\alpha + 73° = 180°$; so $\alpha = 180° - 73° = 107°$. The supplement of 73° is 107°.

b. There is no complement of a 110° angle because $110° > 90°$.
If β represents the supplement of 110°, then $\beta = 180° - 110° = 70°$.
The supplement of 110° is 70°.

Practice Problem 5 Find the complement and the supplement of 67° or explain why 67° does not have a complement or a supplement.

With radian measure, two positive angles are complements when the sum of their measures is $\frac{\pi}{2}$ radians and are supplements when that sum is π radians.

5 Find the length of an arc of a circle.

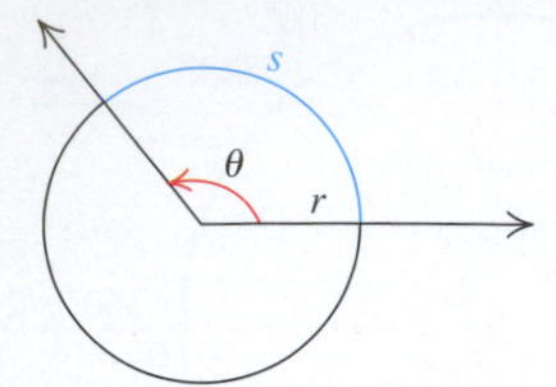

FIGURE 5.13

Length of an Arc of a Circle

Recall that the formula for the radian measure θ of a central angle that intercepts an arc of length s is $\theta = \frac{s}{r}$. See Figure 5.13. Multiplying both sides of this equation by r yields $s = r\theta$.

> **ARC LENGTH FORMULA**
>
> The length s of an arc intercepted by a central angle with radian measure θ in a circle of radius r is given by
>
> $$s = r\theta.$$

EXAMPLE 6 Finding Arc Length of a Circle

A circle has a radius of 18 inches. Find the length of the arc intercepted by a central angle with measure 210°.

Solution

We must first convert the central angle measure from degrees to radians.

$$s = r\theta \qquad \text{Arc length formula, } \theta \text{ in radians}$$

$$s = 18\left(\frac{7\pi}{6}\right) \qquad r = 18;\ \theta = 210° = 210°\left(\frac{\pi}{180°}\right) = \frac{7\pi}{6} \text{ radians}$$

$$= 21\pi \approx 65.97 \text{ inches} \qquad \text{Simplify and use a calculator.}$$

Practice Problem 6 A circle has a radius of 2 meters. Find the length of the arc intercepted by a central angle with measure 225°.

The latitude of L

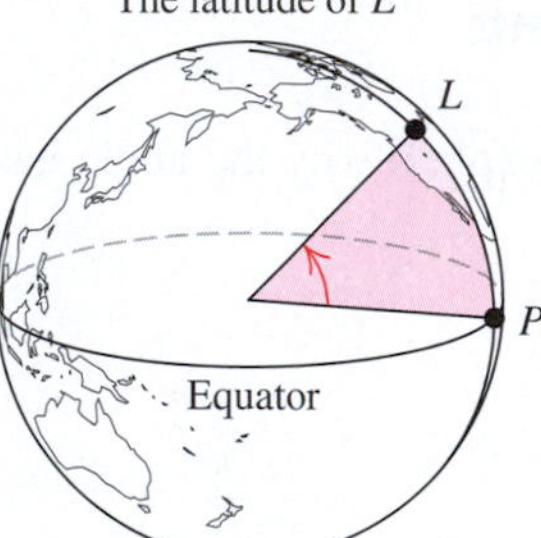

(a)

Distance on a meridian

(b)

FIGURE 5.14 Distance between cities

EXAMPLE 7 Finding the Distance Between Cities

We discussed longitude and latitude in the introduction to this section. We determine the latitude of a location L by first finding the point of intersection, P, of the meridian through L and the equator. The latitude of L is the angle formed by rays drawn from the center of Earth to points L and P, with the ray through P being the initial ray. See Figure 5.14(a).

Billings, Montana, is due north of Grand Junction, Colorado. Find the distance between Billings (latitude 45°48′ N) and Grand Junction (latitude 39° 7′ N). Use 3960 miles as the radius of Earth. The N in 45°48′ N means that the location is north of the equator.

Solution

Because Billings is due north of Grand Junction, the same meridian passes through both cities. The distance between the cities is the length of the arc, s, on this meridian intercepted by the central angle, θ, that is the difference in their latitudes. See Figure 5.14(b).

The measure of angle θ is $45°48' - 39°7' = 6°41'$. To use the arc length formula, we must convert this angle to radians.

$$\theta = 6°41' \approx 6.6833° = 6.6833\left(\frac{\pi}{180}\right) \text{ radian} \approx 0.117 \text{ radian}$$

We use this value of θ and $r = 3960$ miles in the arc length formula:

$$s \approx (3960)(0.117) \text{ miles} \approx 463 \text{ miles} \qquad s = r\theta$$

The distance between Billings and Grand Junction is about 463 miles.

Practice Problem 7 Chicago, Illinois, is due north of Pensacola, Florida. Find the distance between Chicago (latitude 41°51′ N) and Pensacola (latitude 30°25′ N). Use $r = 3960$ miles.

6 Find the area of a sector.

Area of a Sector

A **sector** of a disk is a region bounded by the two sides of a central angle and the intercepted arc. Suppose θ is the radian measure of a central angle of a circle of radius r. See Figure 5.15. To find a formula for the area A of the sector formed by θ, we use the proportion

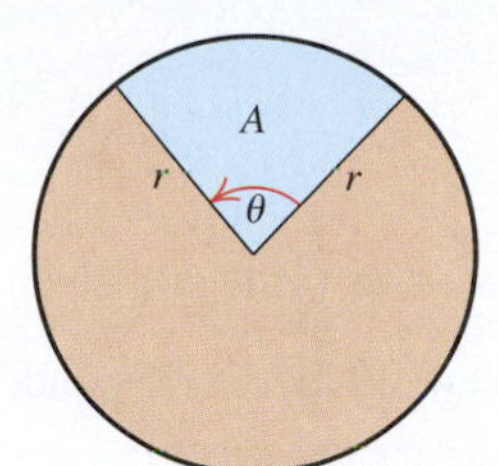

FIGURE 5.15 A sector of a circle

$$\frac{\text{area of a sector}}{\text{area of a circle}} = \frac{\text{length of the intercepted arc}}{\text{circumference of the circle}} \quad \text{Fact from geometry}$$

$$\frac{A}{\pi r^2} = \frac{r\theta}{2\pi r} \quad \text{Area of a circle} = \pi r^2,\ \text{circumference} = 2\pi r,\ \text{Arc length formula } s = r\theta$$

$$A = \pi r^2\left(\frac{r\theta}{2\pi r}\right) \quad \text{Multiply both sides by } \pi r^2.$$

$$= \frac{1}{2}r^2\theta \quad \text{Simplify.}$$

AREA OF A SECTOR

The area A of a sector of a circle of radius r formed by a central angle with radian measure θ is

$$A = \frac{1}{2}r^2\theta.$$

EXAMPLE 8 Finding the Area of a Sector of a Circle

FIGURE 5.16 A slice of pizza

How many square inches of pizza have you eaten (rounded to the nearest square inch) if you eat a sector of an 18-inch-diameter pizza whose edges form a central angle of 30°? See Figure 5.16.

Solution

We must convert 30° to radians to use the sector area formula. The pizza's radius is half its diameter, or 9 inches, and $30° = \frac{\pi}{6}$ radians.

$$A = \frac{1}{2}r^2\theta \quad \text{Area of a sector formula}$$

$$= \frac{1}{2}9^2\left(\frac{\pi}{6}\right) \quad \text{Replace } r \text{ with 9 and } \theta \text{ with } \frac{\pi}{6}.$$

$$\approx 21 \text{ square inches} \quad \text{Use a calculator.}$$

You have eaten about 21 square inches of pizza.

Practice Problem 8 Find the area of a sector of a circle of radius 10 inches formed by an angle of 60°. Round the answer to two decimal places.

7 Compute linear and angular speed.

Linear and Angular Speed

If a skater skates 100 feet in 5 seconds, then her *average speed, v*, is given by

$$v = \frac{\text{distance}}{\text{time}} = \frac{100 \text{ feet}}{5 \text{ seconds}} = 20 \text{ feet per second.}$$

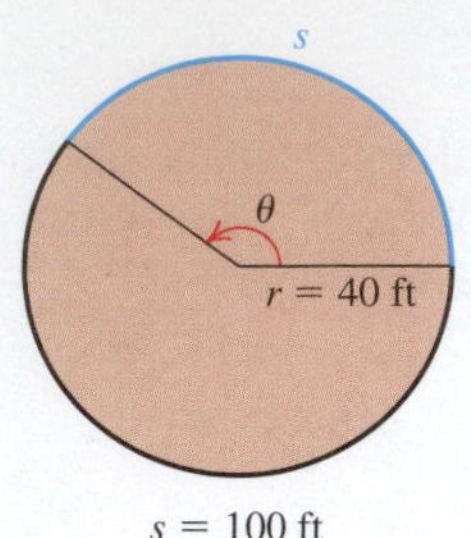

$s = 100$ ft
$\theta = 2.5$ radians

FIGURE 5.17

Suppose now that the skater is skating around the circumference of a circular skating rink with radius 40 feet. See Figure 5.17. In five seconds, the skater travels an arc of length 100 feet and moves through an angle θ, where

$$\theta = \frac{s}{r} = \frac{100}{40} = 2.5 \text{ radians.}$$

We say that the skater is traveling at an (average) *angular speed* ω (omega), given by $\omega = \frac{\theta}{t} = \frac{2.5}{5} = 0.5$ radians per second. We call the average speed, 20 feet per second, the (average) *linear speed* to distinguish it from the (average) angular speed, 0.5 radians per second. Notice that $v = 20 = 40(0.5) = r\omega$. We define these ideas more generally as follows.

LINEAR AND ANGULAR SPEED

Suppose an object travels around a circle of radius r. If the object travels through a central angle of θ radians and an arc of length s, in time t, then

1. $v = \frac{s}{t}$ is the (average) **linear speed** of the object.

2. $\omega = \frac{\theta}{t}$ is the (average) **angular speed** of the object.

Further, because $s = r\theta$, by replacing s with $r\theta$ in **1** and then using **2** to replace $\frac{\theta}{t}$ with ω, we have

3. $v = r\omega$.

EXAMPLE 9 Finding Angular and Linear Speed from Revolutions per Minute

A model plane is attached to a swivel so that it flies in a circular path at the end of a 12-foot wire at the rate of 15 revolutions per minute. Find the angular speed and the linear speed of the plane.

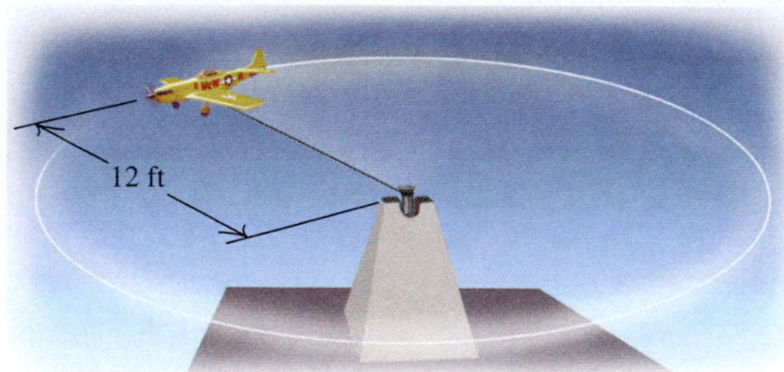

Solution

Because angular speed is measured in radians per unit of time, we first convert revolutions per minute into radians per minute. Recall that

$$1 \text{ revolution} = 2\pi \text{ radians.}$$

Then 15 revolutions $= 15 \cdot 2\pi = 30\pi$ radians. So the angular speed ω is 30π radians per minute. Linear speed is given by

$$\begin{aligned} v &= r\omega \\ &= 12 \cdot 30\pi \text{ feet per minute} && \text{Replace } r \text{ with 12 feet and } \omega \text{ with } 30\pi \text{ radians per minute.} \\ &\approx 1131 \text{ feet per minute} && \text{Use a calculator.} \end{aligned}$$

Practice Problem 9 Find the angular speed and the linear speed of the plane in Example 9 assuming that the wire is 10 feet long and the plane flies at a rate of 18 revolutions per minute.

Answers to Practice Problems

1.

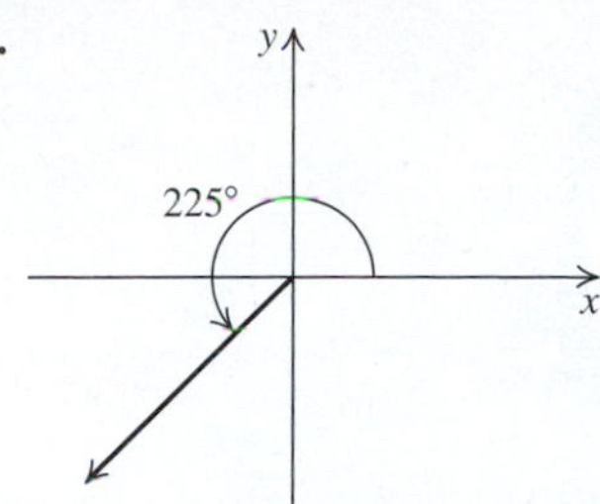

2. a. 13.16°
b. 41°16′30″

3. $-\frac{\pi}{4}$ radians **4.** 270°

5. Complement = 23°; supplement = 113°

6. $\frac{5\pi}{2} \approx 7.85$ meters

7. ≈790 miles

8. $\frac{50\pi}{3} \approx 52.36$ square inches

9. $\omega = 36\pi$ radians per minute; $v = 360\pi \approx 1131$ feet per minute

SECTION 5.1 Exercises

Basic Concepts and Skills

1. A negative angle is formed by rotating the initial side in a(n) ____________ direction.

2. An angle is in standard position if its vertex is at the origin of a coordinate system and its initial side lies on the ____________.

3. One second (1″) is ____________ of a minute.

4. **True or False.** One radian is smaller than one degree.

5. **True or False.** The circumference of a circle with radius 1 meter is approximately 6.28 meters.

6. **True or False.** Every acute angle has a complement and a supplement.

In Exercises 7–14, draw each angle in standard position.

7. 30° **8.** 150°
9. −120° **10.** −330°
11. $\frac{5\pi}{3}$ **12.** $\frac{11\pi}{6}$
13. $-\frac{4\pi}{3}$ **14.** $-\frac{13\pi}{4}$

In Exercises 15–20, convert each angle to decimal degree notation. Round your answers to two decimal places.

15. 70°45′ **16.** 38°38′
17. 23°42′30″ **18.** 45°50′50″
19. −15°42′57″ **20.** −70°18′13″

In Exercises 21–26, convert each angle to DMS notation. Round your answers to the nearest second.

21. 27.32° **22.** 120.64°
23. 13.347° **24.** 110.433°
25. 19.0511° **26.** 82.7272°

In Exercises 27–36, convert each angle from degrees to radians. Express each answer as a multiple of π.

27. 20° **28.** 40° **29.** −180°
30. −210° **31.** 315° **32.** 330°
33. 480° **34.** 450° **35.** −510°
36. −420°

In Exercises 37–46, convert each angle from radians to degrees.

37. $\frac{\pi}{12}$ **38.** $\frac{3\pi}{8}$ **39.** $-\frac{5\pi}{9}$
40. $-\frac{3\pi}{10}$ **41.** $\frac{5\pi}{3}$ **42.** $\frac{11\pi}{6}$
43. $\frac{5\pi}{2}$ **44.** $\frac{17\pi}{6}$ **45.** $-\frac{11\pi}{4}$
46. $-\frac{7\pi}{3}$

In Exercises 47–50, convert each angle from degrees to radians. Round your answers to two decimal places.

47. 12° **48.** 127°
49. −84° **50.** −175°

In Exercises 51–54, convert each angle from radians to degrees. Round your answers to two decimal places.

51. 0.94 **52.** 5
53. −8.21 **54.** −6.28°

In Exercises 55–60, find the complement and the supplement of the given angle or explain why the angle has no complement or supplement.

55. 47° **56.** 75° **57.** 120°
58. 160° **59.** 210° **60.** −50°

In Exercises 61–76, use the following notations: θ = central angle of a circle, r = radius of a circle, s = length of the intercepted arc, v = linear velocity, ω = angular velocity, A = area of the sector of a circle, and t = time.

In each case, find the missing quantity. Find the exact answer.

61. $r = 25$ in, $s = 7$ in, $\theta = ?$

62. $r = 5$ ft, $s = 6$ ft, $\theta = ?$

63. $r = 10.5$ cm, $s = 22$ cm, $\theta = ?$

64. $r = 60$ m, $s = 120$ m, $\theta = ?$

65. $r = 3$ m, $\theta = 25°$, $s = ?$

66. $r = 0.7$ m, $\theta = 357°$, $s = ?$

67. $r = 6.5$ m, $\theta = 12$ radians, $s = ?$

68. $r = 6$ m, $\theta = \frac{\pi}{6}$ radians, $s = ?$

69. $r = 6$ m, $\omega = 10$ rad/min, $v = ?$

70. $r = 3.2$ ft, $\omega = 5$ rad/sec, $v = ?$

71. $v = 20$ ft/sec, $r = 10$ ft, $\omega = ?$

72. $v = 10$ m/min, $r = 6$ m, $\omega = ?$

73. $r = 10$ in, $\theta = 4$ rad, $A = ?$

74. $r = 1.5$ ft, $\theta = 60°$, $A = ?$

75. $A = 20$ ft^2, $\theta = 60°$, $r = ?$

76. $A = 60$ m^2, $\theta = 2$ rad, $r = ?$

Applying the Concepts

77. Wheel ratios. An automobile is pushed so that its wheels turn three-quarters of a revolution. If the tires have a radius of 15 inches, how many inches does the car move?

78. Nautical miles. A nautical mile is the length of an arc of the equator intercepted by a central angle of $1'$. Using 3960 miles for the value of the radius of Earth, express 1 nautical mile in terms of standard (**statute**) miles.

79. Angles on a clock. What is the radian measure of the smaller central angle made by the hands of a clock at 4:00? Express your answer as a rational multiple of π.

80. Angles on a clock. What is the radian measure of the larger central angle made by the hands of a clock at 7:00? Express your answer as a rational multiple of π.

81. Diameter of a pizza. You are told that a slice of pizza whose edges form a 25° angle with an outer crust edge 4 inches long was found in a gym locker. What was the diameter of the original pizza? Round your answer to the nearest inch.

82. Arc of a pendulum. How far does the tip of a 5-foot pendulum travel as it swings through an angle of 30°? Round your answer to two decimal places.

83. Lifting a shark. A shark is suspended by a wire wound around a pulley of radius 4 inches. The shark must be raised 6 inches more to be completely off the ground. Through how many degrees must the pulley be rotated counterclockwise to accomplish the 6-inch lift? Round your answer to the nearest degree.

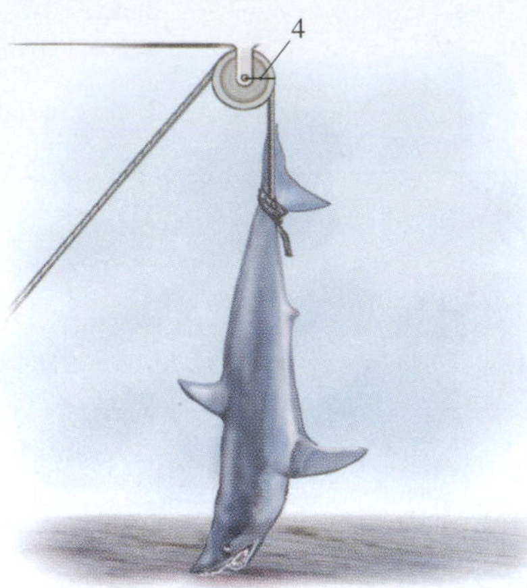

84. Security cameras. A security camera rotates through an angle of 120°. To the nearest foot, what is the arc width of the field of view 40 feet from the camera?

85. Pulley wheels. Two pulleys are connected by a belt so that when one pulley rotates, the linear speeds of the belt and both pulleys are the same. See the figure. The radius of the smaller pulley is 2 inches, and the radius of the larger pulley is 5 inches. A point on the belt travels at a rate of 600 inches per minute.

a. Find the angular speed of the larger pulley.

b. Find the angular speed of the smaller pulley.

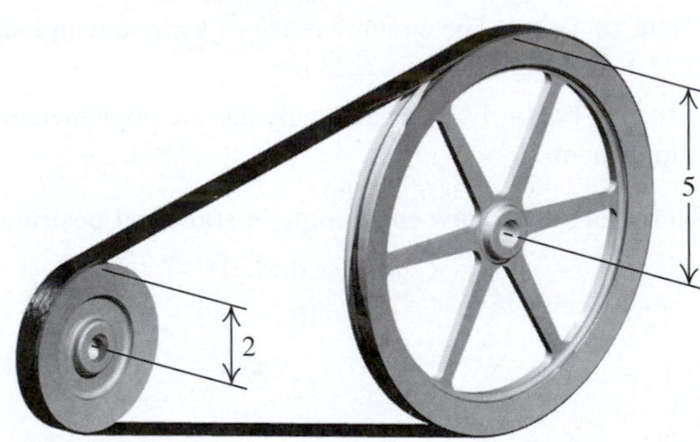

86. Ferris wheel speed. A Ferris wheel in Vienna, Austria, has a diameter of approximately 61 meters. Assume that it takes 90 seconds for the Ferris wheel to make one complete revolution.

a. Find the angular speed of the Ferris wheel. Round your answer to two decimal places.

b. Find the linear speed of the Ferris wheel in meters per second. Round your answer to two decimal places.

87. Bicycle speed. A bicycle's wheels are 24 inches in diameter. If the bike is traveling at a rate of 25 miles per hour, find the angular speed of the wheels. Round your answer to two decimal places.

88. Bicycle speed. A bicycle's wheels are 30 inches in diameter. If the angular speed of the wheels is 11 radians per second, find the speed of the bicycle in inches per second.

89. **Hard disk speed.** Hard disks are circular disks that store data in your computer. Suppose a circular hard disk is 3.75 inches in diameter. If the disk rotates 7200 revolutions per minute, what is the linear speed of a point on the edge of the disk (in inches per minute)?

90. **Linear speed at the equator.** Earth rotates on an axis that goes through both the North and South Poles. It makes one complete revolution in 24 hours. Find the linear speed (in miles per hour) of a location on the equator. (Use 3960 miles as the radius of Earth)

In Exercises 91–96, the latitude of any location on Earth is the angle formed by the two rays drawn from the center of Earth to the location and to the point of intersection of the meridian for the location with the equator. The ray through the location is the initial ray. Use 3960 miles as the radius of Earth.

91. **Distance between cities.** Indianapolis, Indiana, is due north of Montgomery, Alabama. Find the distance between Indianapolis (latitude 39°44′ N) and Montgomery (latitude 32°23′ N).

92. **Distance between cities.** Pittsburgh, Pennsylvania, is due north of Charleston, South Carolina. Find the distance between Pittsburgh (latitude 40°30′ N) and Charleston (latitude 32°54′ N).

93. **Distance between cities.** Amsterdam, Netherlands, is due north of Lyon, France. Find the distance between Amsterdam (latitude 52°23′ N) and Lyon (latitude 45°42′ N).

94. **Distance between cities.** Adana, Turkey (north latitude 36°59′ N), is due north of Jerusalem, Israel (latitude 31°47′ N). Find the distance between Adana and Jerusalem.

95. **Difference in latitudes.** Miles City, Montana, is 440 miles due north of Boulder, Colorado. Find the difference in the latitudes of these two cities.

96. **Difference in latitudes.** Lincoln, Nebraska, is 554 miles due north of Dallas, Texas. Find the difference in the latitudes of these two cities.

Beyond the Basics

97. Use the fact that $\theta + \pi$ and $\theta - \pi$ are coterminal angles to prove that $\frac{5\pi}{3}$ and $-\frac{\pi}{3}$ are coterminal.

98. Show that if α is the complement of θ, then $\alpha + \frac{\theta}{2}$ is the complement of $\frac{\theta}{2}$.

99. Find the smallest positive integer n so that θ and $\theta - \frac{n\pi}{2}$ are coterminal.

100. Find the smallest positive integer n so that θ and $\theta + \frac{n\pi}{3}$ are coterminal.

101. Show that if A is the area of a sector of a circle of radius r formed by a central angle with radian measure θ, then $4A$ is the area of a sector of a circle of radius $2r$ formed by a central angle with radian measure θ.

102. Show that if A is the area of a sector of a circle of radius r formed by a central angle with radian measure θ, then $\frac{1}{2}A$ is the area of a circle of radius r formed by a central angle of radian measure $\frac{1}{2}\theta$.

In Exercises 103–108, use the approximation $\pi \approx \frac{22}{7}$.

103. A racetrack is in the form of a ring with inner circumference 352 meters and outer circumference 396 meters. Find the width of the track.

104. The diameter of a wheel is 84 cm. How many complete revolutions must it take to cover 792 meters?

105. The diameter of a tire is 140 cm. How many revolutions per minute will achieve a speed of 66 kilometers per hour?

106. A car has two wipers that do not overlap. Each wiper has a blade length 14 inches that sweeps through an angle of 110°. Find the total area of the windshield cleaned at each sweep of the blades.

107. An umbrella top is made from a material in the shape of a circle of diameter 3 feet. It has eight equally spaced ribs. Find the area between two consecutive ribs of the umbrella.

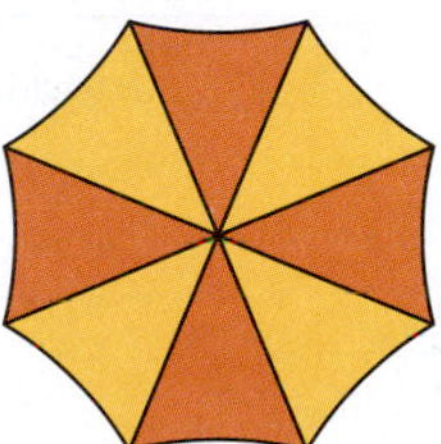

108. A conical drinking cup is made from a circular piece of paper of radius $r = 7$ cm by cutting out a sector of central angle $\frac{\pi}{4}$ and joining the edges OA and OB. Find the volume of the cup. (*Hint:* Volume of a cone $= \frac{1}{3}\pi R^2 h$.)

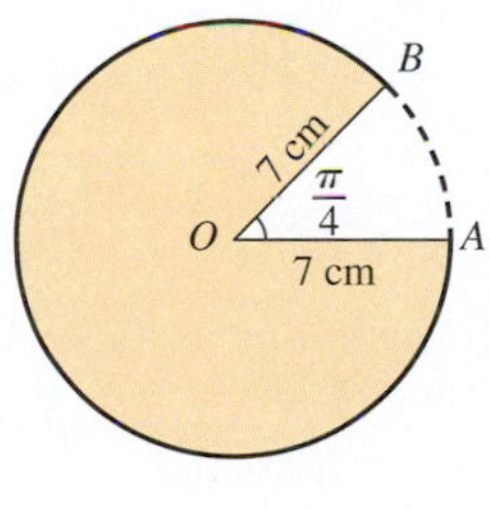

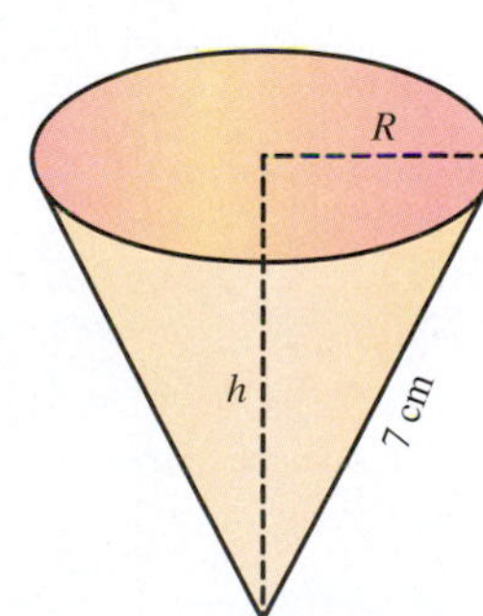

Critical Thinking/Discussion/Writing

109. Arrange the following radian measures from smallest to largest: $\pi, \frac{3\pi}{2}, 3, 4$.

110. Which line of latitude has a smaller radius, the line of latitude through a city of latitude 30°40′13″ or the line of latitude through a city of latitude 40°30′13″.

111. Assume that θ is a central angle in a circle of radius r that intercepts an arc of length s. Show that if θ has radian measure less than $\frac{\pi}{2}$, then the central angles that intercept an arc of length $\frac{\pi r}{2} - s$ and θ are complementary angles.

112. In about 250 B.C., the philosopher Eratosthenes estimated the radius of Earth based on what might appear to be scant information. He knew that the city of Syene (modern Aswan) is directly south of the city of Alexandria and that Syene and Alexandria are (in modern units) about 500 miles apart. He also knew that in Syene at noon on a midsummer day, an upright rod casts no shadow but the noon sun makes a 7°12′ angle with a vertical rod at Alexandria and that then the sun's rays at Syene are effectively parallel to the sun's rays at Alexandria. Explain how Eratosthenes could measure the circumference of Earth by measuring the length of the shadow in Alexandria at noon on the summer solstice when there was no shadow in Syene.

Syene
500 mi
7° 12′
Alexandria
7° 12′

Maintaining Skills

In Exercises 113–118, consider triangle ABC shown in the adjoining figure. Use the Pythagorean theorem to find the third side from the given two sides.

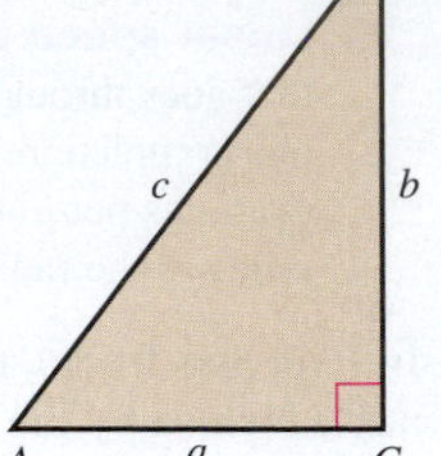

113. $a = 5, b = 12$

114. $a = 12, c = 20$

115. $b = 6, c = 10$

116. $a = 20, b = 21$

117. $b = 3, c = 6$

118. $a = 5, b = 10$

In Exercises 119 and 120, find $\frac{A}{B}$ from the given values of A and B. Then rationalize the denominator of the quotient.

119. $A = \frac{1}{2}, B = \frac{\sqrt{3}}{2}$

120. $A = \frac{\sqrt{3}}{2}, B = \sqrt{\frac{3}{2}}$

SECTION 5.2

Right-Triangle Trigonometry

BEFORE STARTING THIS SECTION, REVIEW

1. Rationalizing the denominator (Section P.6, page 63)
2. Acute angle definition (Section 5.1, page 472)
3. Functions (Section 2.4)

OBJECTIVES

1. Define trigonometric functions of acute angles.
2. Evaluate trigonometric functions of acute angles.
3. Evaluate trigonometric functions for the special angles 30°, 45°, and 60°.
4. Use right-triangle trigonometry in applications.

Mount Kilimanjaro

Measuring Mount Kilimanjaro

Trigonometry developed as a result of attempts to solve practical problems in astronomy, navigation, and land measurement. The word *trigonometry*, coined by Pitiscus in 1594, is derived from two Greek words, *trigonon* (triangle) and *metron* (measure), and means "triangle measurement."

Measuring objects that are generally inaccessible is one of the many successes of trigonometry. In Example 8, we use trigonometry to approximate the height of Mount Kilimanjaro, Tanzania. Mount Kilimanjaro is the highest mountain in Africa.

1 Define trigonometric functions of acute angles.

Trigonometric Ratios and Functions

Consider a right triangle with one of its acute angles labeled θ. In Figure 5.18, the capital letters A, B, and C designate the vertices of the triangle and the lowercase letters a, b, and c, respectively, represent the lengths of the sides opposite these angles.

RECALL

In a right triangle, the side opposite the right angle is called the hypotenuse and the other two sides are called the legs.

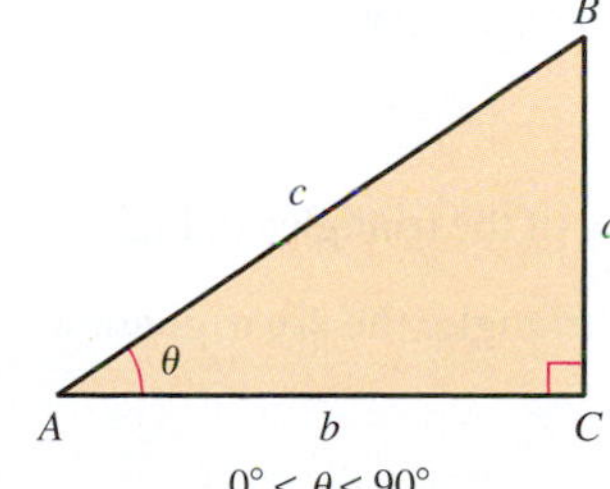

a = length of the side opposite θ
b = length of the side adjacent to θ
c = length of the hypotenuse

FIGURE 5.18 A right triangle

Six ratios can be formed with the lengths of these sides: $\frac{a}{c}, \frac{b}{c}, \frac{a}{b}, \frac{b}{a}, \frac{c}{b}$, and $\frac{c}{a}$. These ratios are used to define the *six trigonometric functions of the acute angle* θ. Later we extend the domains of these functions to include larger sets of angles and then to include sets of real numbers. We use the words *opposite* for the length of the leg opposite to θ, *adjacent* for the length of the leg adjacent to θ, and *hypotenuse* for the length of the hypotenuse. See Figure 5.18.

SIDE **NOTE**

The mnemonic SOHCAHTOA is helpful in remembering that **S**ine is **O**pposite over **H**ypotenuse (SOH), **C**osine is **A**djacent over **H**ypotenuse (CAH), and **T**angent is **O**pposite over **A**djacent (TOA).

TRIGONOMETRIC FUNCTIONS OF AN ANGLE θ IN A RIGHT TRIANGLE

$$\sin\theta = \frac{\text{opposite}}{\text{hypotenuse}} = \frac{a}{c} \qquad \csc\theta = \frac{\text{hypotenuse}}{\text{opposite}} = \frac{c}{a}$$

$$\cos\theta = \frac{\text{adjacent}}{\text{hypotenuse}} = \frac{b}{c} \qquad \sec\theta = \frac{\text{hypotenuse}}{\text{adjacent}} = \frac{c}{b}$$

$$\tan\theta = \frac{\text{opposite}}{\text{adjacent}} = \frac{a}{b} \qquad \cot\theta = \frac{\text{adjacent}}{\text{opposite}} = \frac{b}{a}$$

The full names for these six trigonometric functions are **sine**, **cosine**, **tangent**, **cosecant**, **secant**, and **cotangent**. The customary abbreviations for the values of these functions of an angle θ are **sin θ**, **cos θ**, **tan θ**, **csc θ**, **sec θ**, and **cot θ**, respectively, as shown in the box.

RECALL

Two triangles are *similar* if their corresponding angles are congruent and their corresponding sides are proportional. Two similar triangles have the same shape but not necessarily the same size.

Because any two right triangles having acute angle θ are similar, the ratios of corresponding sides depend only on the angle θ, not on the size of the triangle. Three right triangles with acute angle θ are shown in Figure 5.19.

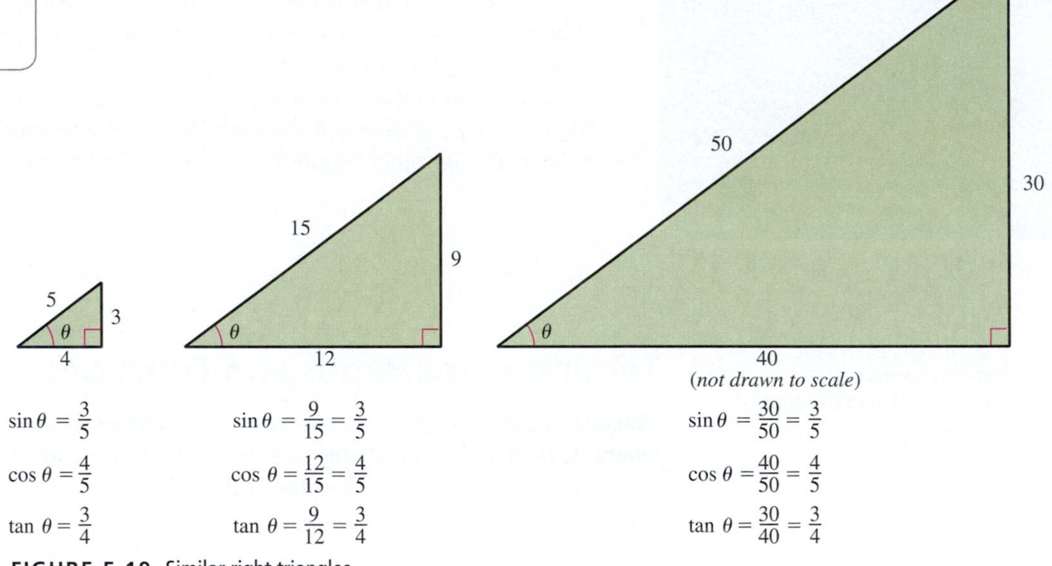

$\sin\theta = \frac{3}{5}$ $\quad$ $\sin\theta = \frac{9}{15} = \frac{3}{5}$ $\quad$ $\sin\theta = \frac{30}{50} = \frac{3}{5}$

$\cos\theta = \frac{4}{5}$ $\quad$ $\cos\theta = \frac{12}{15} = \frac{4}{5}$ $\quad$ $\cos\theta = \frac{40}{50} = \frac{4}{5}$

$\tan\theta = \frac{3}{4}$ $\quad$ $\tan\theta = \frac{9}{12} = \frac{3}{4}$ $\quad$ $\tan\theta = \frac{30}{40} = \frac{3}{4}$

FIGURE 5.19 Similar right triangles

Also, for any of the triangles in Figure 5.19, we have $\csc\theta = \frac{5}{3}$, $\sec\theta = \frac{5}{4}$, and $\cot\theta = \frac{4}{3}$. So for each triangle, the six trigonometric values are identical.

WARNING

We *must* write the variable (such as θ) of each trigonometric function. So $\sin\theta = \frac{1}{2}$ is a correct statement, but $\sin = \frac{1}{2}$ is incorrect.

2 Evaluate trigonometric functions of acute angles.

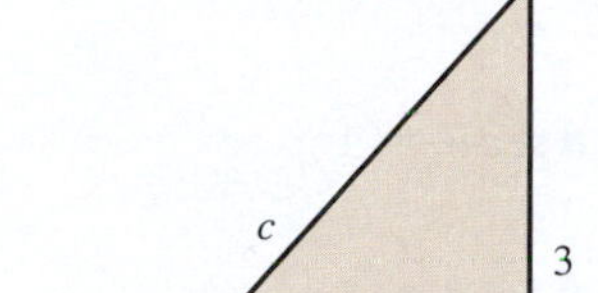

FIGURE 5.20

EXAMPLE 1 **Finding the Values of Trigonometric Functions**

Find the exact values for the six trigonometric functions of the angle θ in Figure 5.20.

Solution

To find the values for the six trigonometric functions of θ, we must first find the value of c, the length of the hypotenuse.

$$\begin{aligned} a^2 + b^2 &= c^2 && \text{Pythagorean theorem} \\ (3)^2 + (\sqrt{7})^2 &= c^2 && \text{Replace } a \text{ with 3 and } b \text{ with } \sqrt{7}. \\ 9 + 7 &= c^2 \\ 16 &= c^2 \\ 4 &= c && c \text{ is a positive number.} \end{aligned}$$

With $c = 4$, $a = 3$, and $b = \sqrt{7}$, we have the following:

$$\sin\theta = \frac{\text{opposite}}{\text{hypotenuse}} = \frac{3}{4} \qquad \csc\theta = \frac{\text{hypotenuse}}{\text{opposite}} = \frac{4}{3}$$

$$\cos\theta = \frac{\text{adjacent}}{\text{hypotenuse}} = \frac{\sqrt{7}}{4} \qquad \sec\theta = \frac{\text{hypotenuse}}{\text{adjacent}} = \frac{4}{\sqrt{7}} = \frac{4\sqrt{7}}{7}$$

$$\tan\theta = \frac{\text{opposite}}{\text{adjacent}} = \frac{3}{\sqrt{7}} = \frac{3\sqrt{7}}{7} \qquad \cot\theta = \frac{\text{adjacent}}{\text{opposite}} = \frac{\sqrt{7}}{3}$$

These are exact values; using a calculator would yield approximate values.

Practice Problem 1 Find the exact values for the six trigonometric functions of the acute angle θ in a right triangle, assuming that the length of the side opposite θ is 4 and the length of the hypotenuse is 5.

Relations between Trigonometric Functions

In the definitions of the trigonometric functions (page 486), we note that the functions in the second column are reciprocals of those in the first column. In other words, for any acute angle θ,

$$\csc\theta = \frac{1}{\sin\theta} \quad \sec\theta = \frac{1}{\cos\theta} \quad \text{and} \quad \cot\theta = \frac{1}{\tan\theta}.$$

We derive the following identity that connects $\sin\theta$, $\cos\theta$, and $\tan\theta$.

$$\begin{aligned} \frac{\sin\theta}{\cos\theta} &= \frac{\frac{a}{c}}{\frac{b}{c}} && \text{Definition of } \sin\theta \text{ and } \cos\theta \\ &= \frac{a}{b} && \text{Simplify.} \\ &= \tan\theta && \text{Definition of } \tan\theta \end{aligned}$$

So for any acute angle θ, we have

$$\tan\theta = \frac{\sin\theta}{\cos\theta} \quad \text{and} \quad \cot = \frac{1}{\tan\theta} = \frac{\cos\theta}{\sin\theta}.$$

We next summarize these identities.

RECIPROCAL AND QUOTIENT IDENTITITES

Reciprocal Identities

$$\csc\theta = \frac{1}{\sin\theta},\qquad \sin\theta = \frac{1}{\csc\theta},\qquad \sin\theta\csc\theta = 1$$

$$\sec\theta = \frac{1}{\cos\theta},\qquad \cos\theta = \frac{1}{\sec\theta},\qquad \cos\theta\sec\theta = 1$$

$$\cot\theta = \frac{1}{\tan\theta},\qquad \tan\theta = \frac{1}{\cot\theta},\qquad \tan\theta\cot\theta = 1$$

Quotient Identities

$$\tan\theta = \frac{\sin\theta}{\cos\theta}\qquad \cot\theta = \frac{\cos\theta}{\sin\theta}$$

EXAMPLE 2 Using Reciprocal and Quotient Identities

Let $\sin\theta = \frac{5}{13}$ and $\cos\theta = \frac{12}{13}$. Find $\tan\theta$, $\csc\theta$, $\sec\theta$, and $\cot\theta$.

Solution

We have

$$\begin{aligned} \tan\theta &= \frac{\sin\theta}{\cos\theta} && \text{Quotient identity} \\ &= \frac{\frac{5}{13}}{\frac{12}{13}} && \text{Given } \sin\theta = \frac{5}{13} \text{ and } \cos\theta = \frac{12}{13} \\ &= \frac{5}{12} && \text{Simplify.} \end{aligned}$$

We now use the reciprocal identities to get

$$\csc\theta = \frac{1}{\sin\theta} = \frac{1}{\frac{5}{13}} = \frac{13}{5},$$

$$\sec\theta = \frac{1}{\cos\theta} = \frac{1}{\frac{12}{13}} = \frac{13}{12}, \text{ and}$$

$$\cot\theta = \frac{1}{\tan\theta} = \frac{1}{\frac{5}{12}} = \frac{12}{5}.$$

Practice Problem 2 Let $\sin\theta = \frac{3}{5}$ and $\cos\theta = \frac{4}{5}$. Find $\tan\theta$, $\csc\theta$, $\sec\theta$, and $\cot\theta$. ■

The next example shows that we can find the values of the remaining five trigonometric functions if we are given the value of one of them.

EXAMPLE 3 **Finding Remaining Trigonometric Function Values from a Given Value**

Find the other five trigonometric function values of θ, given that θ is an angle of a right triangle with $\sin\theta = \dfrac{2}{5}$.

Solution

Because

$$\sin\theta = \frac{2}{5} = \frac{\text{opposite}}{\text{hypotenuse}},$$

we draw a right triangle with hypotenuse of length 5 and the side opposite θ of length 2. See Figure 5.21. Then

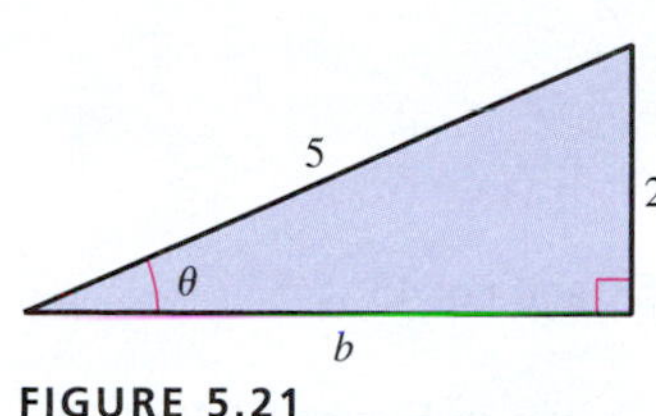

FIGURE 5.21

$5^2 = 2^2 + b^2$	Pythagorean theorem
$b^2 = 5^2 - 2^2$	Subtract 2^2 from both sides; interchange sides.
$b^2 = 21$	Simplify.
$b = \sqrt{21}$	b is a positive number.

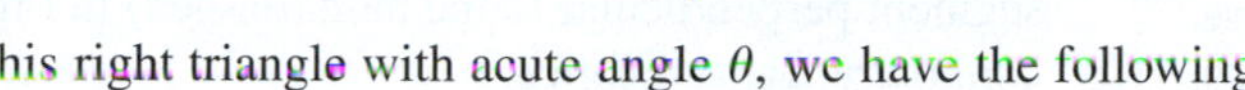

For this right triangle with acute angle θ, we have the following:

$$\begin{aligned} c &= \text{length of the hypotenuse} = 5 \\ a &= \text{length of the opposite side} = 2 \\ b &= \text{length of the adjacent side} = \sqrt{21} \end{aligned}$$

So

$$\sin\theta = \frac{\text{opposite}}{\text{hypotenuse}} = \frac{2}{5} \quad \text{(given)} \qquad \csc\theta = \frac{\text{hypotenuse}}{\text{opposite}} = \frac{5}{2}$$

$$\cos\theta = \frac{\text{adjacent}}{\text{hypotenuse}} = \frac{\sqrt{21}}{5} \qquad \sec\theta = \frac{\text{hypotenuse}}{\text{adjacent}} = \frac{5}{\sqrt{21}} = \frac{5\sqrt{21}}{21}$$

$$\tan\theta = \frac{\text{opposite}}{\text{adjacent}} = \frac{2}{\sqrt{21}} = \frac{2\sqrt{21}}{21} \qquad \cot\theta = \frac{\text{adjacent}}{\text{opposite}} = \frac{\sqrt{21}}{2}$$

Practice Problem 3 Find the other five trigonometric function values of θ given that θ is an angle of a right triangle with $\cos\theta = \dfrac{1}{3}$.

Function Values for Some Special Angles

3 Evaluate trigonometric functions for the special angles 30°, 45°, and 60°.

We can use the Pythagorean theorem to find the exact values of the trigonometric functions for the angles with measure 30°, 45°, and 60° by examining the relationships between the side lengths in two "special" right triangles.

Consider an isosceles right triangle in which the length of each of the two legs of equal length is 1. Because the triangle has two sides of equal length, the two acute angles have equal measure, α. See Figure 5.22.

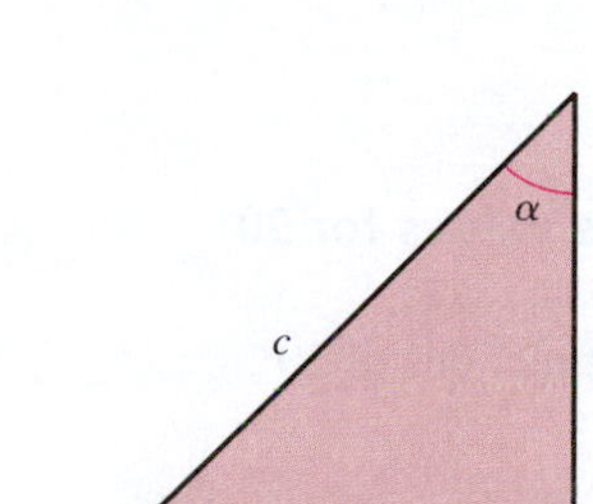

FIGURE 5.22 An isosceles right triangle

Because the sum of the measures of the angles in a triangle is 180°, we have $\alpha + \alpha = 90°$; so $2\alpha = 90°$ and $\alpha = 45°$. As is customary, we let c = length of the hypotenuse; then

$1^2 + 1^2 = c^2$	Pythagorean theorem
$2 = c^2$	
$c = \sqrt{2}$	c is a positive number.

So a right triangle in which both legs have length 1 has a hypotenuse of length $\sqrt{2}$ and contains two 45° angles. See Figure 5.23. We call this an *isosceles right triangle*, or a **45°– 45°– 90° triangle**.

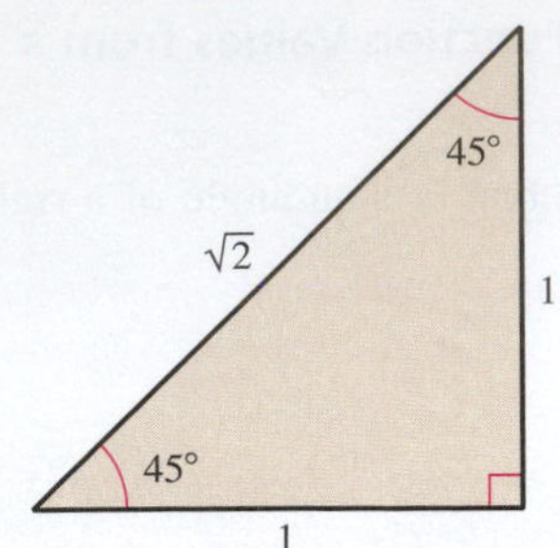

FIGURE 5.23 A 45°– 45°– 90° triangle

EXAMPLE 4 Finding the Trigonometric Functions Values for 45°

Use Figure 5.23 to find sin 45°, cos 45°, and tan 45°.

Solution

Referring to Figure 5.23, we have the following:

$$\sin 45^\circ = \frac{\text{opposite}}{\text{hypotenuse}} = \frac{1}{\sqrt{2}} = \frac{\sqrt{2}}{2}$$

$$\cos 45^\circ = \frac{\text{adjacent}}{\text{hypotenuse}} = \frac{1}{\sqrt{2}} = \frac{\sqrt{2}}{2}$$

$$\tan 45^\circ = \frac{\text{opposite}}{\text{adjacent}} = \frac{1}{1} = 1$$

Practice Problem 4 Use Figure 5.23 to find csc 45°, sec 45°, and cot 45°.

An *equilateral triangle* is a triangle with all three sides of equal length. The measure of each angle of an equilateral triangle is 60°.

Now consider an equilateral triangle with side length 2, as shown in Figure 5.24(a). The segment perpendicular to the base (dashed) in Figure 5.24(b) divides this triangle into two congruent right triangles.

We find the length of the perpendicular segment by applying the Pythagorean Theorem to one of the triangles. Let h denote the length of the perpendicular segment, which is a leg of both congruent right triangles. See Figure 5.24(b).

We can use either triangle to find the exact values of all six trigonometric functions for both 30° and 60° angles. We have

$2^2 = h^2 + 1^2$	Pythagorean theorem applied to Figure 5.24(c)
$h^2 = 2^2 - 1^2$	Subtract 1^2 from both sides and interchange sides.
$h^2 = 3$	Simplify.
$h = \sqrt{3}$	h is positive.

The resulting right triangle whose acute angles measure 30° and 60° is given in Figure 5.24(c). We call this a **30°– 60°– 90° triangle**.

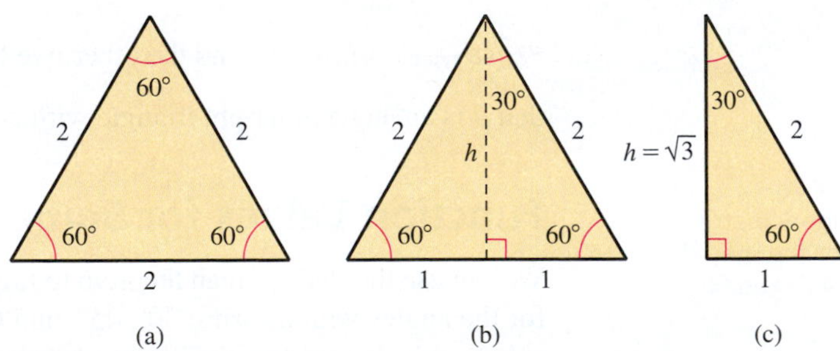

FIGURE 5.24 A 30°– 60°– 90° triangle

EXAMPLE 5 Finding the Trigonometric Functions Values for 30°

Use Figure 5.24(c) to find the values of the six trigonometric functions of 30°.

Solution

Referring to Figure 5.24(c), we have the following:

$$\sin 30^\circ = \frac{\text{opposite}}{\text{hypotenuse}} = \frac{1}{2} \qquad \csc 30^\circ = \frac{\text{hypotenuse}}{\text{opposite}} = \frac{2}{1} = 2$$

$$\cos 30^\circ = \frac{\text{adjacent}}{\text{hypotenuse}} = \frac{\sqrt{3}}{2} \qquad \sec 30^\circ = \frac{\text{hypotenuse}}{\text{adjacent}} = \frac{2}{\sqrt{3}} = \frac{2\sqrt{3}}{3}$$

$$\tan 30^\circ = \frac{\text{opposite}}{\text{adjacent}} = \frac{1}{\sqrt{3}} = \frac{\sqrt{3}}{3} \qquad \cot 30^\circ = \frac{\text{adjacent}}{\text{opposite}} = \frac{\sqrt{3}}{1} = \sqrt{3}$$

Practice Problem 5 Use Figure 5.24(c) to find the values of the six trigonometric functions of 60°.

You should be able to reconstruct the triangles used to find the values for the six trigonometric functions for 30°, 45°, and 60°, as these values occur frequently. Remember that the values depend only on the *acute angle itself,* not on the size of the triangle. We summarize the values obtained so far in Table 5.1 using both degree and radian measures.

TABLE 5.1 Trigonometric Function Values of Some Common Angles

θ (degrees)	θ (radians)	$\sin\theta$	$\cos\theta$	$\tan\theta$	$\csc\theta$	$\sec\theta$	$\cot\theta$
30°	$\frac{\pi}{6}$	$\frac{1}{2}$	$\frac{\sqrt{3}}{2}$	$\frac{\sqrt{3}}{3}$	2	$\frac{2\sqrt{3}}{3}$	$\sqrt{3}$
45°	$\frac{\pi}{4}$	$\frac{\sqrt{2}}{2}$	$\frac{\sqrt{2}}{2}$	1	$\sqrt{2}$	$\sqrt{2}$	1
60°	$\frac{\pi}{3}$	$\frac{\sqrt{3}}{2}$	$\frac{1}{2}$	$\sqrt{3}$	$\frac{2\sqrt{3}}{3}$	2	$\frac{\sqrt{3}}{3}$

TECHNOLOGY CONNECTION

The first screen shows how sin 71° is displayed on a graphing calculator in *degree* mode.

This screen shows how $\tan\frac{5\pi}{7}$ and sec 1.3 are displayed on a graphing calculator in *radian* mode.

Evaluating Trigonometric Functions Using a Calculator

Table 5.1 gives the exact trigonometric function values of 30°, 45°, and 60°. However, we are unable to find the exact trigonometric function values of most angles. In such cases, we settle for an approximate value by using a calculator.

When you are using a calculator to find the values of the trigonometric functions, the first step is to set the *mode* of measurement. If you are working with an angle given in degrees, set the mode to *degrees*; otherwise, set the mode to *radians*.

Your calculator has keys labeled SIN, COS, and TAN for computing the values of the sine, cosine, and tangent functions; but no keys for directly evaluating the cosecant, secant, or cotangent functions. However, these functions are the reciprocal functions of sine, cosine, and tangent functions, respectively. Therefore, to evaluate these functions, take the reciprocal of the appropriate function value.

For example, we use a calculator to find the approximate value of

a. sin 71°. Set the MODE to degrees, $\sin 71° \approx 0.9455185756$

b. $\tan\frac{5\pi}{7}$. Set the MODE to radians, $\tan\frac{5\pi}{7} \approx -1.253960338$

c. sec 1.3. Set the MODE to radians, $\sec 1.3 = \frac{1}{\cos 1.3} \approx 3.738334127$

WARNING

The calculator keys SIN^{-1}, COS^{-1}, and TAN^{-1} do not represent the reciprocal functions for the sine, cosine, and tangent functions, respectively. The functions $\sin^{-1}$, $\cos^{-1}$, and $\tan^{-1}$ are inverse functions and are discussed in Section 5.6.

Complements

Because 30° and 60° are acute angles in the same right triangle, the same six possible ratios of the lengths of the sides of the triangle are used to compute the values of the trigonometric functions for both angles. Notice that $\cos 30° = \sin 60°$, $\sin 30° = \cos 60°$, and $\tan 30° = \cot 60°$. Taking reciprocals of these values extends this relationship to the remaining three trigonometric functions.

The prefix *co-* (in cosine, cotangent, and cosecant) stands for *complement.* The cosine, cotangent, and cosecant functions are called the cofunctions of the sine, tangent, and secant functions, respectively. In general, for any acute angle θ in a right triangle, the other acute angle is its complement, $90° - \theta$. See Figure 5.25.

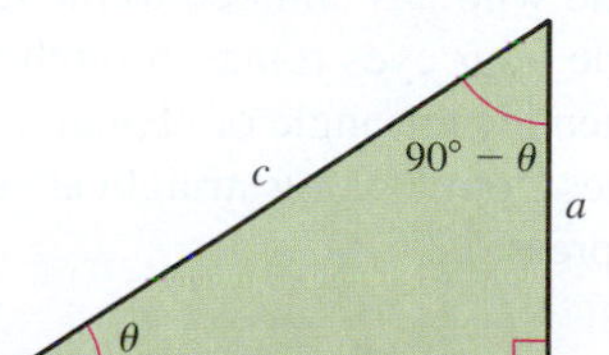

FIGURE 5.25 Side a is opposite θ and adjacent to $90° - \theta$ Side b is adjacent to θ and opposite $90° - \theta$.

SIDE **NOTE**

Here is a scheme for remembering the cofunction identities for acute angles A and B with

$$A + B = 90°, \text{ or } A + B = \frac{\pi}{2},$$

complementary angles

$\sin A = \cos B$

cofunctions

The sine and cosine may be replaced with any other cofunction pairs.

Computations similar to those in Example 5 lead to the relationships, or identities, given next.

COFUNCTION IDENTITIES

The value of any trigonometric function of an angle θ is equal to the cofunction of the complement of θ. This is true whether θ is measured in degrees or in radians.

θ in radians

$$\sin\theta = \cos\left(\frac{\pi}{2} - \theta\right) \qquad \cos\theta = \sin\left(\frac{\pi}{2} - \theta\right)$$

$$\tan\theta = \cot\left(\frac{\pi}{2} - \theta\right) \qquad \cot\theta = \tan\left(\frac{\pi}{2} - \theta\right)$$

$$\sec\theta = \csc\left(\frac{\pi}{2} - \theta\right) \qquad \csc\theta = \sec\left(\frac{\pi}{2} - \theta\right)$$

If θ is measured in degrees, replace $\frac{\pi}{2}$ with 90° in the equations above.

EXAMPLE 6 **Finding Trigonometric Function Values of a Complementary Angle**

a. Given that $\cot 68° \approx 0.4040$, find $\tan 22°$.

b. Given that $\cos 72° \approx 0.3090$, find $\sin 18°$.

Solution

a. Note that $22° = 90° - 68°$. So

$$\tan 22° = \tan(90° - 68°) = \cot 68° \approx 0.4040.$$

b. Here $18° = 90° - 72°$. So

$$\sin 18° = \sin(90° - 72°) = \cos 72° \approx 0.3090.$$

Practice Problem 6

a. Given that $\csc 21° \approx 2.7904$, find $\sec 69°$.

b. Given that $\tan 75° \approx 3.7321$, find $\cot 15°$.

4 Use right-triangle trigonometry in applications.

Applications

Angles that are measured between a line of sight and a horizontal line occur in many applications and are called *angles of elevation* or *angles of depression*. If the line of sight from an observer to an object is *above* the horizontal line, the angle between these two lines is called the **angle of elevation**. If the line of sight is *below* the horizontal line, the angle between the two lines is called the **angle of depression**.

For example, suppose you are taking a picture of a friend who has climbed to the top of the Arc de Triomphe in Paris. See Figure 5.26. The angle your eyes rotate through as your gaze changes from straight ahead to looking at your friend is the angle of elevation.

The angle your friend's eyes rotate through as she changes from looking straight ahead to looking down at you (no pun intended) is the angle of depression.

SIDE **NOTE**

Remember that an angle of elevation or an angle of depression is the angle between the line of sight and a *horizontal* line.

Angle of elevation

FIGURE 5.26 Angles of elevation and depression

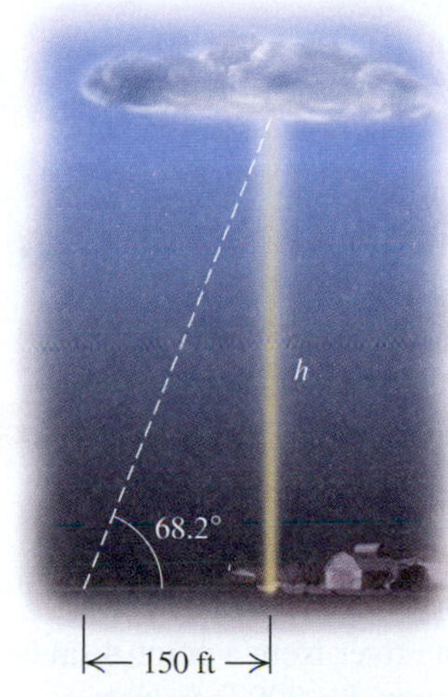

FIGURE 5.27 Height of a cloud

EXAMPLE 7 Finding the Height of a Cloud

To determine the height of a cloud at night, a farmer shines a spotlight straight upward to a spot on the cloud. The angle of elevation to this same spot on the cloud from a point located 150 feet horizontally from the spotlight is 68.2°. Find the height of the cloud.

Solution

In Figure 5.27, h represents the height of the cloud, in feet. We have

$$\tan 68.2° = \frac{h}{150} \quad \text{Definition of tangent}$$

$$h = 150 \tan 68.2° \quad \text{Multiply both sides by 150.}$$

$$h \approx 375 \quad \text{Use a calculator.}$$

The height of the cloud is approximately 375 feet.

Practice Problem 7 From an observation deck 425 feet high on the top of a lighthouse, the angle of depression to a ship at sea is 4.2°. How many miles is the ship from a point at sea level directly below the observation deck?

EXAMPLE 8 Measuring the Height of Mount Kilimanjaro

A surveyor wants to measure the height of Mount Kilimanjaro by using the known height of a nearby mountain. The nearby location is at an altitude of 8720 feet, the distance between that location and Mount Kilimanjaro's peak is 4.9941 miles, and the angle of elevation from the lower location is 23.75°. See Figure 5.28. Use this information to find the approximate height of Mount Kilimanjaro.

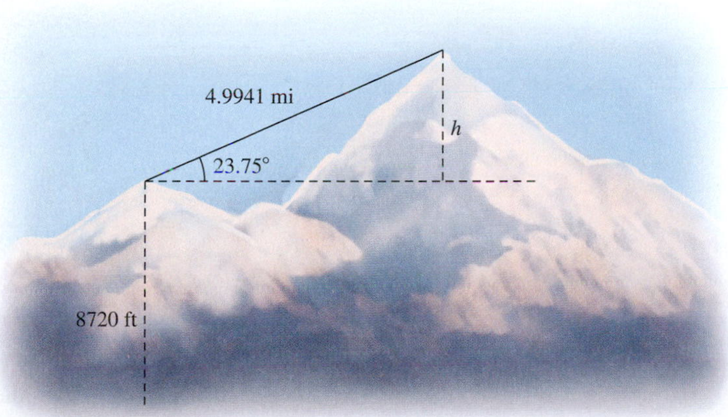

FIGURE 5.28 Height of Mount Kilimanjaro

Solution

The sum of the side length h and the location height of 8720 feet gives the approximate height of Mount Kilimanjaro. Let h be measured in miles. Using the definition of $\sin \theta$, for $\theta = 23.75°$, we have

$$\sin \theta = \frac{\text{opposite}}{\text{hypotenuse}} = \frac{h}{4.9941} \quad \text{From Figure 5.28}$$
$$h = (4.9941) \sin \theta \quad \text{Multiply both sides by 4.9941.}$$
$$= (4.9941) \sin 23.75° \quad \text{Replace } \theta \text{ with } 23.75°.$$
$$h \approx 2.0114 \quad \text{Use a calculator.}$$

Because 1 mile = 5280 feet,

$$2.0114 \text{ miles} = (2.0114)(5280)$$
$$\approx 10{,}620 \text{ feet.}$$

Thus, the height of Mount Kilimanjaro is

$$h + 8720 \approx 10{,}620 + 8720 = 19{,}340 \text{ feet.}$$

Practice Problem 8 The height of a nearby location to Mount McKinley, Alaska, is 12,870 feet, its distance to Mount McKinley's peak is 3.3387 miles, and the angle of elevation θ is 25°. What is the approximate height of Mount McKinley?

EXAMPLE 9 Finding the Width of a River

To find the width of a river, a surveyor sights straight across the river from a point A on her side to a point B on the opposite side. See Figure 5.29. She then walks 200 feet upstream to a point C. The angle θ that the line of sight from point C to point B makes with the riverbank is 58°. How wide is the river, to the nearest foot?

Solution

The points A, B, and C are the vertices of a right triangle with acute angle $\theta = 58°$.

Let w represent the width of the river. From Figure 5.29, we have

$$\frac{w}{200} = \tan 58° \quad \tan \theta = \frac{\text{opposite}}{\text{adjacent}}$$
$$w = 200 \tan 58° \quad \text{Multiply both sides by 200.}$$
$$w \approx 320.07 \text{ feet} \quad \text{Use a calculator.}$$

The river is about 320 feet wide at the point A.

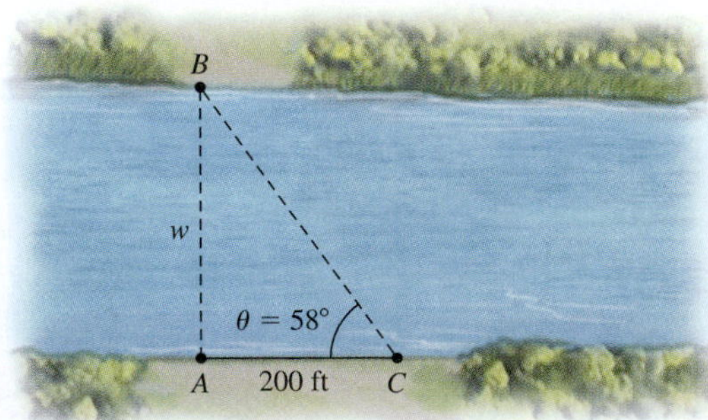

FIGURE 5.29 Width of a river

Practice Problem 9 From a point on the edge of one of two parallel rooftop sides of two building, a point directly opposite is identified on the other rooftop. From a second point 25 feet from the first point along the rooftop edge, a second sighting to the point on the opposite rooftop is made, and the angle θ that this line of sight makes with the rooftop is found to be 60°. How far apart are the two buildings? Round your answer to the nearest foot.

Answers to Practice Problems

1. $\sin\theta = \frac{4}{5}$ $\csc\theta = \frac{5}{4}$

$\cos\theta = \frac{3}{5}$ $\sec\theta = \frac{5}{3}$

$\tan\theta = \frac{4}{3}$ $\cot\theta = \frac{3}{4}$

2. $\tan\theta = \frac{3}{4}, \csc\theta = \frac{5}{3}, \sec\theta = \frac{5}{4}, \cot\theta = \frac{4}{3}$

3. $\sin\theta = \frac{2\sqrt{2}}{3}$ $\csc\theta = \frac{3\sqrt{2}}{4}$

$\cos\theta = \frac{1}{3}$ (given) $\sec\theta = 3$

$\tan\theta = 2\sqrt{2}$ $\cot\theta = \frac{\sqrt{2}}{4}$

4. $\csc 45° = \sqrt{2}, \sec 45° = \sqrt{2}, \cot 45° = 1$

5. $\sin 60° = \frac{\sqrt{3}}{2}, \cos 60° = \frac{1}{2}, \tan 60° = \sqrt{3}$

$\csc 60° = \frac{2\sqrt{3}}{3}, \sec 60° = 2, \cot 60° = \frac{\sqrt{3}}{3}$

6. a. $\sec 69° \approx 2.7904$ **b.** $\cot 15° \approx 3.7321$

7. ≈1.096 miles **8.** ≈20,320 feet **9.** ≈43 feet

SECTION 5.2 Exercises

Basic Concepts and Skills

1. If θ is an acute angle and $\sin\theta = \frac{2\sqrt{2}}{3}$, then $\cos\theta =$ ____________.

2. In a right triangle with sides of lengths 1, 5, and $\sqrt{26}$, the tangent of the angle opposite the side of length 5 is ____________.

3. If θ is an acute angle (measured in degrees) in a right triangle and $\cos\theta = 0.7$, then $\sin(90° - \theta) =$ ____________.

4. **True or False.** If θ is an acute angle in a right triangle and $\tan\theta = \frac{2}{5}$, then the length of the leg opposite θ is always 2.

5. **True or False.** The leg adjacent to angle θ in a right triangle is the leg opposite the angle $90° - \theta$.

6. **True or False.** The length of the hypotenuse of a right triangle having a leg of length 1 must be greater than 1.

In Exercises 7–12, find the exact values for the six trigonometric functions of the angle θ in each figure. Rationalize the denominator where necessary.

7.

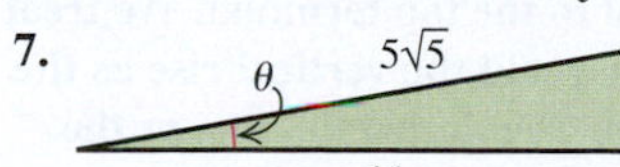

8.

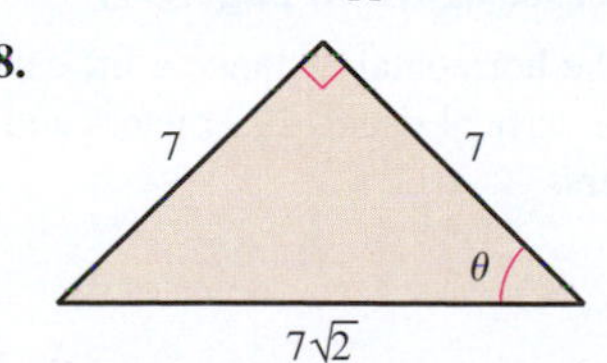

9.

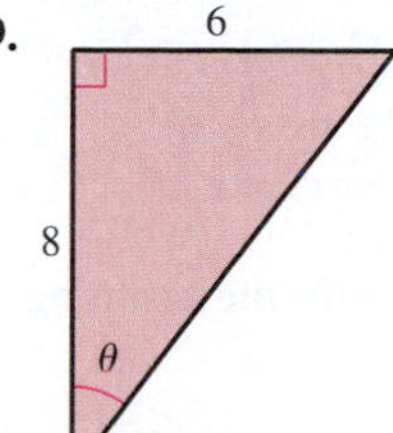

10.

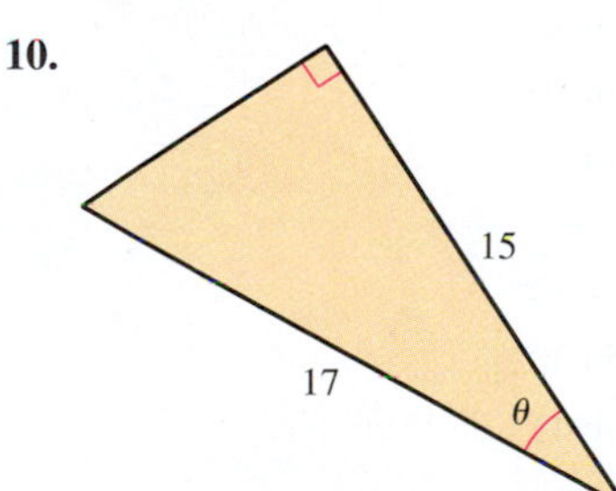

11.

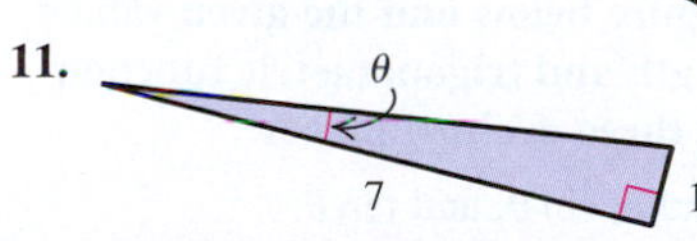

12.

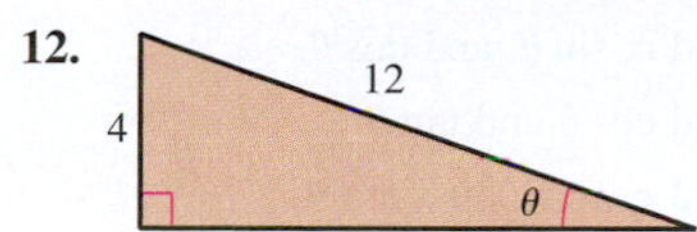

In Exercises 13–18, use reciprocal and quotient identities to find $\tan\theta$, $\csc\theta$, $\sec\theta$, and $\cot\theta$.

13. $\sin\theta = \frac{12}{13}$ and $\cos\theta = \frac{5}{13}$ **14.** $\sin\theta = \frac{8}{17}$ and $\cos\theta = \frac{15}{17}$

15. $\sin\theta = \frac{21}{29}$ and $\cos\theta = \frac{20}{29}$ **16.** $\sin\theta = \frac{40}{41}$ and $\cos\theta = \frac{9}{41}$

17. $\sin\theta = \frac{2}{5}$ and $\cos\theta = \frac{\sqrt{21}}{5}$

18. $\sin\theta = \frac{2\sqrt{6}}{5}$ and $\cos\theta = \frac{1}{5}$

In Exercises 19–24, use each given trigonometric function value of θ to find the five other trigonometric function values of the acute angle θ. Rationalize the denominators where necessary.

19. $\cos\theta = \frac{2}{3}$ **20.** $\sin\theta = \frac{3}{4}$ **21.** $\tan\theta = \frac{5}{3}$

22. $\cot\theta = \frac{6}{11}$ **23.** $\sec\theta = \frac{13}{12}$ **24.** $\csc\theta = 4$

In Exercises 25–30, find the trigonometric function values of each corresponding complementary angle.

25. Given that $\sin 58° \approx 0.8480$, find $\cos 32°$.

26. Given that $\cos 37° \approx 0.7986$, find $\sin 53°$.

27. Given that $\tan 27° \approx 0.5095$, find $\cot 63°$.

28. Given that $\cot 49° \approx 0.8693$, find $\tan 41°$.

29. Given that $\sec 65° \approx 2.3662$, find $\csc 25°$.

30. Given that $\csc 78° \approx 1.0223$, find $\sec 12°$.

In Exercises 31–38, use Table 5.1 and simplify the resulting expression.

31. $\sin 60° \cos 30° + \cos 60° \sin 30°$

32. $\sin 60° \cos 45° + \cos 60° \sin 45°$

33. $\cos 60° \cos 30° + \sin 60° \sin 30°$

34. $\cos 60° \cos 45° - \sin 60° \sin 45°$

35. $\cot 45° + \csc 30°$

36. $\tan 30° \sec 45° + \tan 60° \sec 30°$

37. $\cos 45° \cos 30° - \sin 45° \sin 30°$

38. $\sin 30° \cos 45° + \cos 30° \sin 45°$

In Exercises 39–48, use the figure below and the given values to find each specified side length and trigonometric function value. Round your answer to three decimal places.

39. $a = 8, b = 10$ Find c, $\sin\theta$, and $\tan\theta$.

40. $a = 18, b = 3$ Find c, $\sin\theta$, and $\cos\theta$.

41. $a = 23, b = 7$ Find $\cos\theta$ and $\tan\theta$.

42. $a = 19, b = 27$ Find c, $\cos\theta$, and $\tan\theta$.

43. $\theta = 30°, a = 9$ Find $\sin\theta$, b, and c.

44. $\theta = 30°, a = 5$ Find b and c.

45. $a = 12.5, b = 6.2$ Find c, $\sin\theta$, $\cos\theta$.

46. $a = 4.3, b = 8.1$ Find c, $\cos\theta$, $\tan\theta$.

47. $b = 9.4, c = 14.5$ Find a, $\sin\theta$, $\tan\theta$.

48. $a = 6.2, c = 18.7$ Find b, $\sec\theta$, $\cot\theta$.

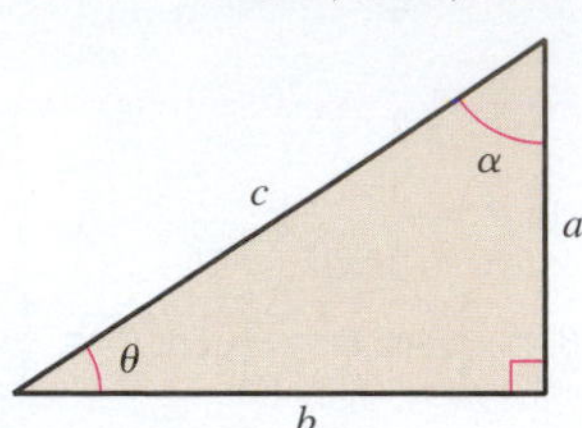

In Exercises 49–54, use the given function value of acute angle θ to find the indicated trigonometric function values.

49. $\sin\theta = \frac{1}{3}$

a. $\csc\theta$ **b.** $\cos\theta$ **c.** $\tan\theta$ **d.** $\cot(90° - \theta)$

50. $\cos\theta = \frac{1}{4}$

a. $\sec\theta$ **b.** $\sin\theta$ **c.** $\tan\theta$ **d.** $\cot(90° - \theta)$

51. $\sec\theta = 5$

a. $\cos\theta$ **b.** $\sin\theta$ **c.** $\tan\theta$ **d.** $\tan(90° - \theta)$

52. $\csc\theta = 2$

a. $\sin\theta$ **b.** $\cos\theta$ **c.** $\tan\theta$ **d.** $\tan(90° - \theta)$

53. $\tan\theta = \frac{5}{2}$

a. $\cot\theta$ **b.** $\sec\theta$ **c.** $\cos\theta$ **d.** $\cos(90° - \theta)$

54. $\cot\theta = \frac{3}{2}$

a. $\tan\theta$ **b.** $\csc\theta$ **c.** $\sin\theta$ **d.** $\sin(90° - \theta)$

Applying the Concepts

55. Positioning a ladder. An 18-foot ladder is resting against a wall of a building in such a way that the top of the ladder is 14 feet above the ground. How far is the foot of the ladder from the base of the building?

56. Positioning a ladder. The ladder from Exercise 55 is placed against a wall so that the foot of the ladder is 7 feet from the base of the wall. How high up the wall will the ladder reach?

In Exercises 57 and 58, use the following definitions: The *vertical rise* of a ski lift is the vertical distance traveled when a lift car goes from the bottom terminal to the top terminal. The *slope length* is the linear distance a lift car travels as it moves from the bottom terminal to the top terminal. We treat the slope length as the hypotenuse and the vertical rise as the side opposite the angle formed by the horizontal line at the top terminal and the lift cables. (See figure on page 497)

57. Ski lift dimensions. Find the horizontal distance a lift car passes over assuming that the vertical rise is 295 meters and the slope length is 1070 meters.

58. **Ski lift dimensions.** Find the vertical rise (to the nearest foot) for a ski lift with slope length 2050 feet assuming that the angle formed by a horizontal line at the top terminal and the lift cable is 16°.

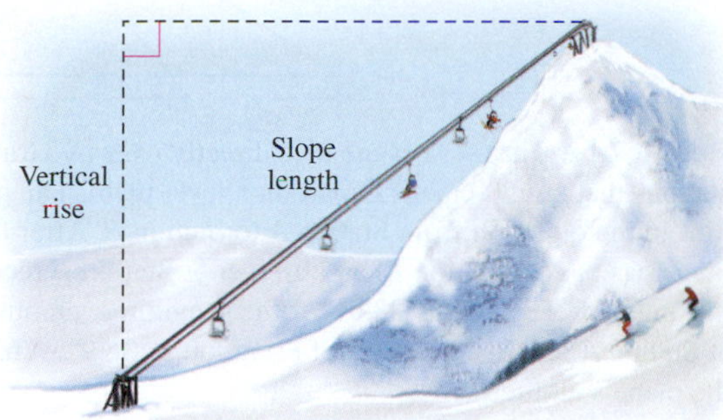

59. **Ski lift height.** If a skier travels 5000 feet up a ski lift whose angle of inclination is 30°, how high above his starting level is he?

60. **Shadow length.** At a time when the sun's rays make a 60° angle with the horizontal, how long is the shadow cast by a 6-foot man standing on flat ground?

61. **Height of a tree.** The angle relative to the horizontal from the top of a tree to a point 12 feet from its base (on flat ground) is 30°. Find the height of the tree.

62. **Height of a cliff.** The angle of elevation from a rowboat moored 75 feet from a cliff is 60°. Find the height of the cliff.

63. **Height of a flagpole.** A flagpole is supported by a 30-foot tension wire attached to the flagpole and to a stake in the ground. If the ground is flat and the angle the wire makes with the ground is 30°, how high up the flagpole is the wire attached?

64. **Flying a kite.** A girl is holding a kite string at a height of 5 feet while flying the kite. How long must the string be for her to raise the kite 200 feet above the ground if the string makes a 30° angle with the ground?

65. **Height of a tree.** Find the height of a pine tree that casts a 93-foot shadow on the ground if the angle of elevation of the sun is 24°35′.

66. **Width of a river.** Sal and his friend are on opposite sides of a river. To find the width of the river, each of them hammers a stake on his bank of the river in such a way that the distance between the two stakes approximates the width of the river. Sal walks 10 feet away from his stake along the river and finds that the line of sight from his new position to his friend's stake across the river and the line of sight to his own stake form a 73°34′ angle. How wide is the river?

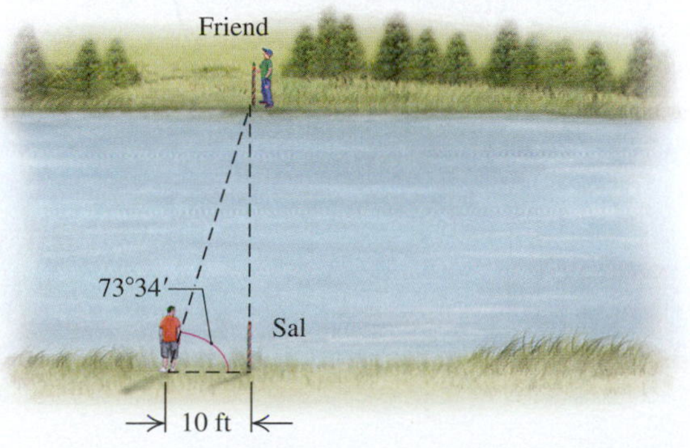

67. **Measuring the Empire State Building.** From a neighboring building's 13th-floor window that is 128 meters above the ground, the angle of elevation of the top of the Empire State Building is 59°20′, and the angle of depression of the base is 40°29′. How far away from the neighboring building and how tall is the Empire State Building?

68. **Sighting a mortar station.** The angle of depression from a helicopter 3000 feet directly above an allied mortar station to an enemy supply depot is 13°. How far is the depot from the mortar station?

69. **Distance between Earth and the moon.** Assuming that the radius of Earth is 3960 miles and angle A (the latitude of A) is 89°3′, find the distance between the center of Earth and the center of the moon.

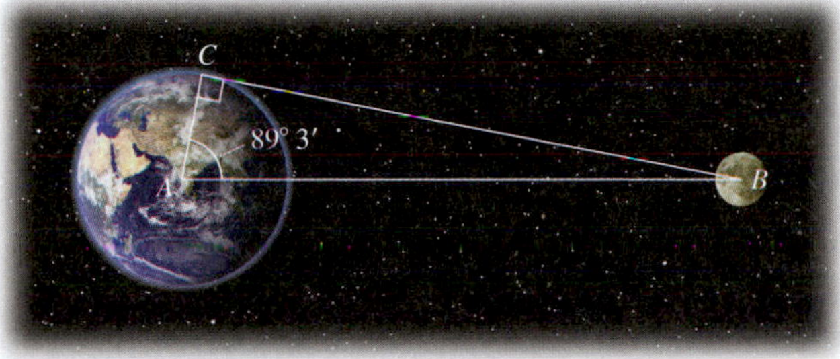

70. **Placing a spotlight.** A spotlight will be set in the ground so that its beam makes a 40° angle with the ground. If the light is to hit a spot 7 feet above the ground on a vertical wall, how far from the wall should the light be placed?

71. **Traffic sign.** A "warning" traffic sign in the shape of a diamond rhombus (rhombus) is 1.8 meters wide. The angle at the top of the sign is 50°. Find the length of each side.

72. **Mooring a boat.** A boat is attached at water level by a taut line to a dock that is 4 feet above the water level. The angle of depression from the dock to the boat is 14°. How long is the rope? Round your answer to the nearest foot.

73. **Wire length.** A horizontal wire between two houses is 100 feet long. The wire is replaced and lowered on one end so that it is now inclined at an angle of 10°. How long is the new wire?

In Exercises 74 and 75, use the figure to find the radius r for the given latitudes (values of θ). Use 3960 miles for the radius of Earth.

74. **Latitude.** Find the radius of the 50th parallel (of latitude).

75. **Latitude.** Find the radius of the 53rd parallel (of latitude).

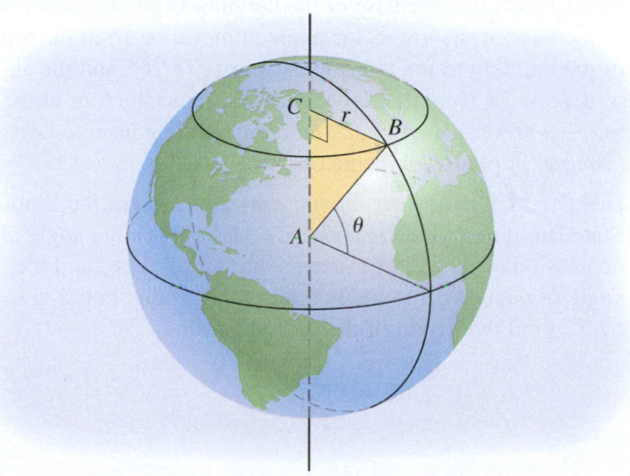

Beyond the Basics

76. In a triangle ABC with right angle at C, if $\tan A = \frac{1}{2}$, find the value of $\sec A + \csc B$.

77. If α is an acute angle and $\sin \alpha = \frac{1}{3}$, find the value of $\cos \alpha \csc \alpha + \tan \alpha \sec \alpha$.

78. If $\sin \theta = \frac{m^2 - n^2}{m^2 + n^2}$, find the value of $\tan \theta + \sec \theta$.

79. If $\tan \theta = \frac{4}{5}$, find the value of $\frac{\cos \theta - \sin \theta}{\cos \theta + \sin \theta}$.

80. Height of an airplane. From a tower 150 feet high with the sun directly overhead, an airplane and its shadow have an angle of elevation of 67.4° and an angle of depression of 8.1°, respectively. Find the height of the airplane.

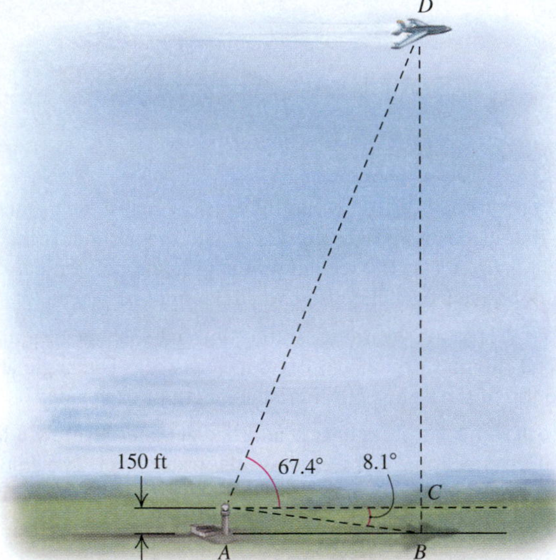

81. Camera angle. Raphael stands 12 feet from a large painting on a wall. He sets his camera 5 feet 6 inches above the floor. If the bottom edge of the painting is 4 feet above the floor and the angle of elevation from the camera to the top of the painting is 50°, find the height of the painting to the nearest tenth of a foot.

82. Height of a building. The WCNC-TV Tower in Dallas, North Carolina, is 600 meters high. Suppose a building is erected such that the base of the building is on the same plane as the base of the tower, the angle of elevation from the top of the building to the top of the tower is 75.24°, and the angle of depression from the top of the building to the foot of the tower is 60.05°. How high would the building have to be? (*Source:* http://en.wikipedia.org)

83. Height of a building. Suppose from the top of the Empire State Building (which is about 1250 feet high), the angle of depression to the top of a second building is 55.8° and the angle of depression to the bottom of the second building is 67.7°. Find the height of the second building.

84. In the figure, show that
$$h = \frac{d}{\cot \alpha - \cot \beta}.$$

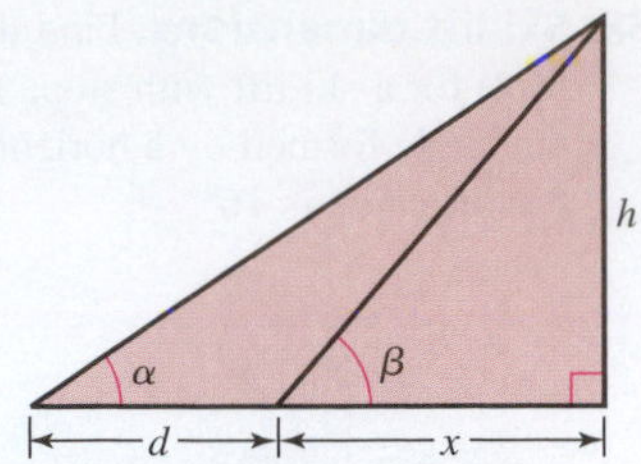

85. Height of a plane. A plane flies directly over two tracking stations that are 500 meters apart on a level plain. The plane flies in the direction from Station A to Station B. After the plane has passed Station B, an observer at Station B records the angle of elevation as 38.9°. At the same time, an observer at Station A records the angle of elevation as 36.9°. What is the plane's altitude?
[*Hint:* Use Exercise 84.]

86. Measuring the Statue of Liberty. While sailing toward the Statue of Liberty, a sailor in a boat observed that at a certain point, the angle of elevation of the tip of the torch was 25°. After sailing another 100 meters toward the statue, the angle of elevation became 41°50′. How tall is the Statue of Liberty?

87. Speed of a plane. A plane flying at an altitude of 12,000 feet finds the angle of depression for a building ahead of the plane to be 10.4°. With the building still straight ahead, two minutes later the angle of depression is 25.6°. Find the speed of the plane relative to the ground (in feet per minute).

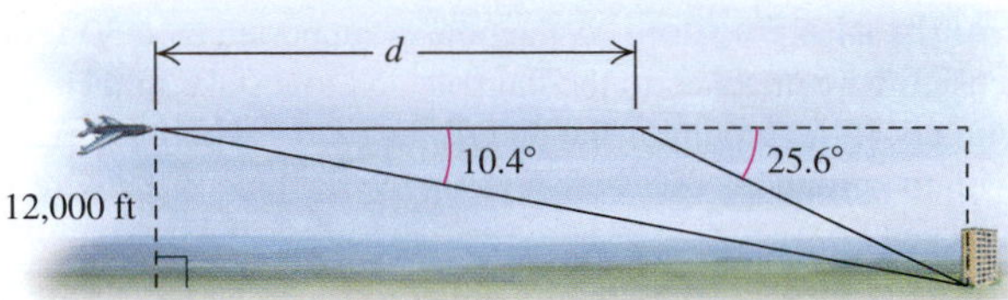

88. In the figure, show that
$$h = \frac{d}{\cot \alpha + \cot \beta}.$$

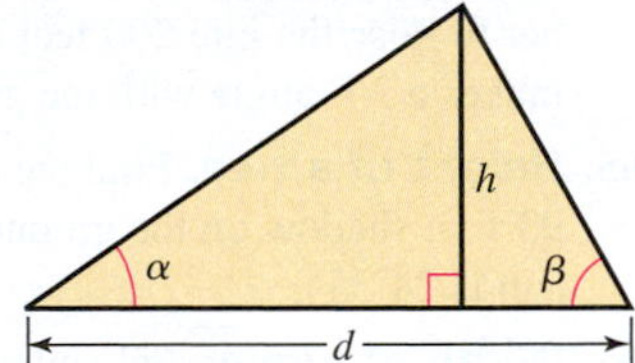

89. Find the area of triangle ABC in the figure.

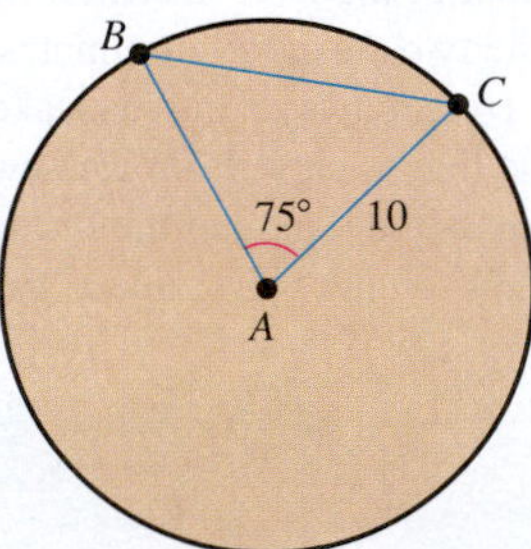

Critical Thinking/Discussion/Writing

90. Draw a right triangle ABC with right angle at C.

(a) Use the definitions of sine and cosine together with the Pythagorean theorem to show that $\sin^2 A + \cos^2 A = 1$.

(b) Use the result in (a) and the cofunction identities (see page 492) to find the value of $\sin A \cos B + \cos A \sin B$.

91. In a right triangle with acute angle θ, what value must the adjacent side have for the length of the opposite side to represent $\tan \theta$?

92. Use the result of Exercise 91 to compare $\tan \alpha$ and $\tan \beta$ assuming that α and β are acute angles and the measure of α is less than the measure of β.

93. All references are to the figure.

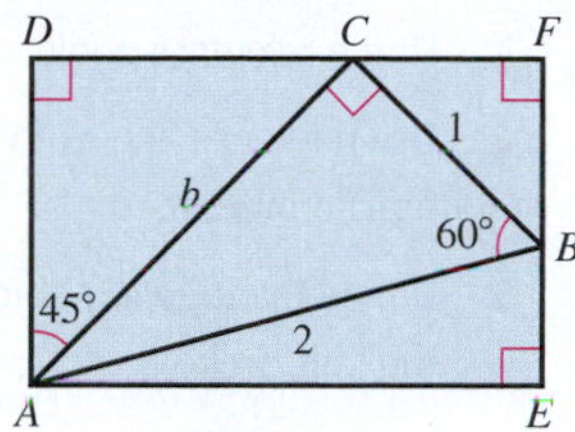

(i) Find b and $\angle CAB$.
(ii) Find AD, DC, and $\angle ACD$.
(iii) Find $\angle BCF$ and $\angle CBF$.
(iv) Find CF and BF.
(v) Find AE and BE.
(vi) Find $\angle BAE$ and $\angle ABE$.
(vii) Find $\sin 15°$ and $\cos 15°$.
(viii) Find $\sin 75°$ and $\cos 75°$.

94. From the foot F, of a mountain, the angle of elevation to its summit is 45°. After a climber has gone 1 kilometer toward the mountain up an incline of 30°, the angle of elevation to the summit is 60°. Find the height of the mountain. (See the figure.)

$\left[\textit{Hint:} \text{ Let } FB = BS = x. \text{ Show } SC = x - \frac{1}{2} \text{ and } DC = x - \frac{\sqrt{3}}{2}.\right]$

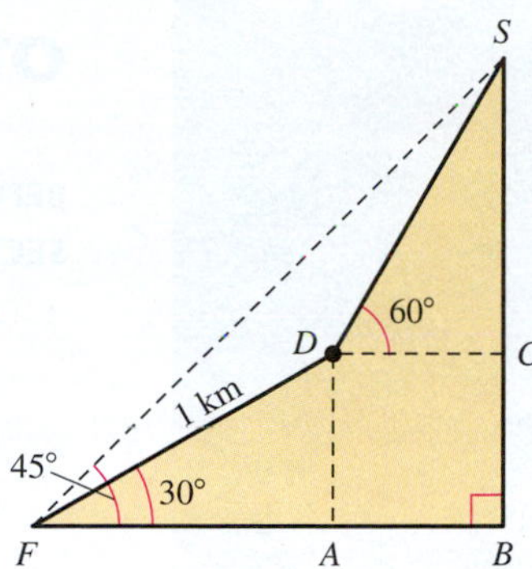

Maintaining Skills

In Exercises 95–98, rationalize the denominator of each expression and simplify.

95. $\dfrac{\sqrt{3} - 1}{\sqrt{3} + 1}$ **96.** $\dfrac{3 + \sqrt{7}}{3 - \sqrt{7}}$

97. $\dfrac{5 + 2\sqrt{3}}{7 + 4\sqrt{3}}$ **98.** $\dfrac{3}{5 - \sqrt{3}} + \dfrac{2}{5 + \sqrt{3}}$

In Exercises 99–102, convert each angle from degrees to radians.

99. 75° **100.** 70° **101.** −510° **102.** 435°

In Exercises 103–106, use the division algorithm to write each number x in the form $x = 360n + r$, where n is an integer and $0 \le r < 360$.

103. $x = 1220$ **104.** $x = 2670$

105. $x = -480$ **106.** $x = -2025$

SECTION 5.3

Trigonometric Functions of Any Angle; The Unit Circle

BEFORE STARTING THIS SECTION, REVIEW

1 Rationalizing a denominator (Section P.6, page 63)
2 Distance between two points (Section 2.1, page 159)
3 The unit circle (Section 2.2, page 172)
4 Trigonometric function values of some common angles (Section 5.2, page 491)

OBJECTIVES

1 Evaluate trigonometric functions of any angle.
2 Determine the signs of the trigonometric functions in each quadrant.
3 Find a reference angle.
4 Use reference angles to find trigonometric function values.
5 Define the trigonometric functions using the unit circle.

The London Eye

The **London Eye**, also known as the **Millennium Wheel**, was constructed in 1999 and is located on the bank of the River Thames in London, England. It is 135 meters high and is the tallest wheel in Europe. The wheel carries 32 sealed and air-conditioned passenger capsules attached to its external circumference. It rotates at a slow rate to complete one revolution in 30 minutes. This slow rotation speed allows passengers to walk on or off the moving capsules at ground level without any need to stop the wheel. Over 3 million people visit the wheel each year, making it the most popular paid tourist attraction in the United Kingdom. In Example 9, we discuss the height of a passenger capsule.

1 Evaluate trigonometric functions of any angle.

Trigonometric Functions of Angles

Consider an angle θ in standard position, as shown in Figure 5.30. Let $P = (x, y)$ be any point (other than the origin) on the terminal ray of θ. The distance between the origin $O = (0, 0)$ and the point P is $r = \sqrt{(x - 0)^2 + (y - 0)^2} = \sqrt{x^2 + y^2}$. Note that $r > 0$.

We define the six trigonometric functions, **sine**, **cosine**, **tangent**, **cosecant**, **secant**, and **cotangent**, for any angle θ by using the same abbreviations we used in the previous section.

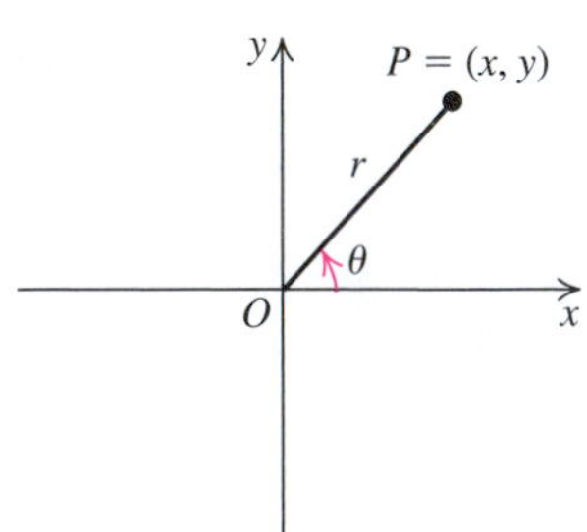

FIGURE 5.30 Angle θ in standard position

The Trigonometric Functions of any Angle θ

Let $P = (x, y)$ be any point on the terminal ray of an angle θ in standard position (other than the origin) and let $r = \sqrt{x^2 + y^2}$. We define the following functions values:

$$\sin\theta = \frac{y}{r} \qquad \csc\theta = \frac{r}{y} \quad (y \neq 0)$$

$$\cos\theta = \frac{x}{r} \qquad \sec\theta = \frac{r}{x} \quad (x \neq 0)$$

$$\tan\theta = \frac{y}{x} \quad (x \neq 0) \qquad \cot\theta = \frac{x}{y} \quad (y \neq 0)$$

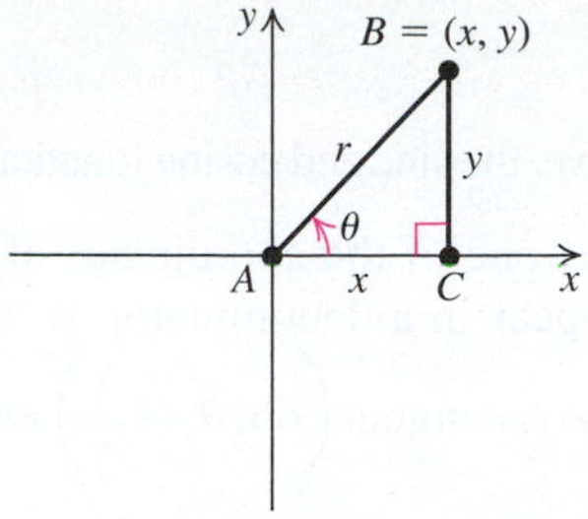

FIGURE 5.31

Now consider a right triangle with acute angle θ and vertices A, B, and C. Place the triangle so that θ is in standard position, as shown in Figure 5.31. If the vertex B has coordinates (x, y), then $r^2 = x^2 + y^2$; so $r = \sqrt{x^2 + y^2}$ and

$$\sin\theta = \frac{\text{opposite}}{\text{hypotenuse}} = \frac{y}{r} \qquad \csc\theta = \frac{\text{hypotenuse}}{\text{opposite}} = \frac{r}{y}$$

$$\cos\theta = \frac{\text{adjacent}}{\text{hypotenuse}} = \frac{x}{r} \qquad \sec\theta = \frac{\text{hypotenuse}}{\text{adjacent}} = \frac{r}{x}$$

$$\tan\theta = \frac{\text{opposite}}{\text{adjacent}} = \frac{y}{x} \qquad \cot\theta = \frac{\text{adjacent}}{\text{opposite}} = \frac{x}{y}$$

Notice that *for acute angles, the definitions from Section 5.2 give the same values for all six trigonometric functions, as do the definitions in this section.* Consequently, the trigonometric function values we computed in Section 5.2, such as those for 30°, 45°, and 60°, are the same values that we would obtain by using the more general definitions given in this section.

If an angle θ is measured in degrees, we use the degree symbol when indicating a trigonometric function value, such as $\cos 60°$ and $\tan(-128)°$. If θ is measured in radians, then no symbol is used when indicating a trigonometric function value such as $\sin\frac{\pi}{4}$, $\sec\pi$, and $\cos(-16)$.

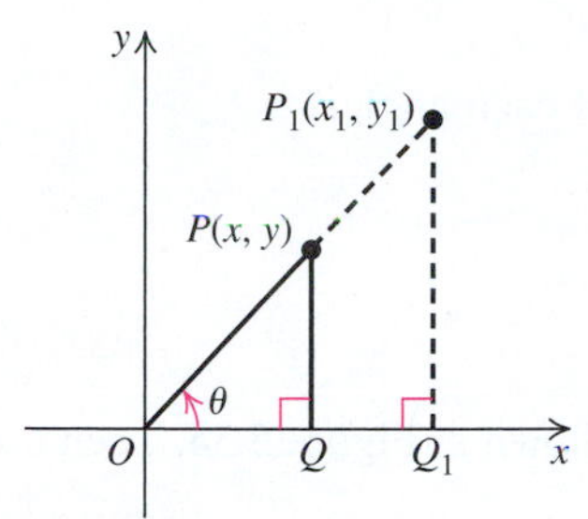

FIGURE 5.32

In Figure 5.32, right triangles POQ and P_1OQ_1 are similar by angle–angle similarity; so the ratios of corresponding sides are equal. Thus, letting $r = \sqrt{x^2 + y^2}$ and $r_1 = \sqrt{x_1^2 + y_1^2}$, we have $\sin\theta = \frac{y}{r} = \frac{y_1}{r_1}$, $\cos\theta = \frac{x}{r} = \frac{x_1}{r_1}$, and so on. This shows that the values of the trigonometric functions do not depend on the choice of the point $P(x, y)$ on the terminal side of θ. Throughout assume that the stated angles are in standard position.

EXAMPLE 1 Finding Trigonometric Function Values

Suppose θ is an angle whose terminal side contains the point $P(-1, 3)$. Find the exact values of the six trigonometric functions of θ.

Solution

Because $x = -1$ and $y = 3$ (see Figure 5.33), we have

$$\begin{aligned} r &= \sqrt{x^2 + y^2} && \text{Definition of } r \\ &= \sqrt{(-1)^2 + 3^2} && \text{Replace } x \text{ with } -1 \text{ and } y \text{ with } 3. \\ &= \sqrt{10} && \text{Simplify.} \end{aligned}$$

Replacing x with -1, y with 3, and r with $\sqrt{10}$ in the definition of the trigonometric functions, we have the following:

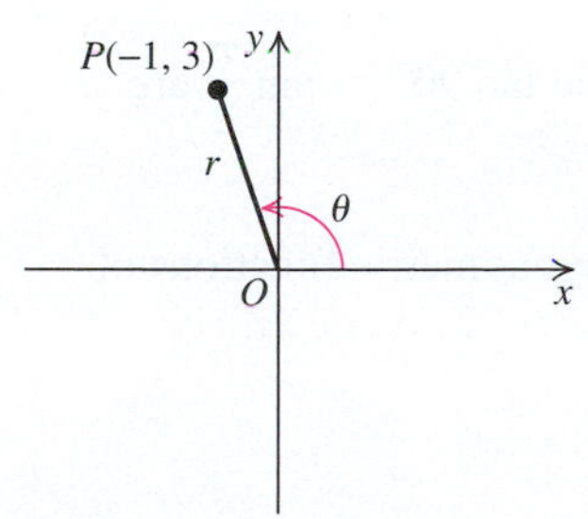

FIGURE 5.33

$$\sin\theta = \frac{y}{r} = \frac{3}{\sqrt{10}} = \frac{3\sqrt{10}}{10} \qquad \csc\theta = \frac{r}{y} = \frac{\sqrt{10}}{3}$$

$$\cos\theta = \frac{x}{r} = \frac{-1}{\sqrt{10}} = -\frac{\sqrt{10}}{10} \qquad \sec\theta = \frac{r}{x} = \frac{\sqrt{10}}{-1} = -\sqrt{10}$$

$$\tan\theta = \frac{y}{x} = \frac{3}{-1} = -3 \qquad \cot\theta = \frac{x}{y} = \frac{-1}{3} = -\frac{1}{3}$$

Practice Problem 1 Suppose θ is an angle whose terminal side contains the point $P(2, -5)$. Find the exact values of the six trigonometric functions of θ.

Quadrantal Angles

Because the denominator r in $\sin\theta = \frac{y}{r}$ and $\cos\theta = \frac{x}{r}$ is positive, the sine and cosine functions are defined for any angle θ. Note that for a quadrantal angle, one of the coordinates of a point on the terminal side is 0; so this coordinate cannot appear in a denominator. If the terminal side is on the x-axis, then the y-coordinate is 0; so the cotangent $\left(\cot\theta = \frac{x}{y}\right)$ and the cosecant $\left(\csc\theta = \frac{r}{y}\right)$ functions are undefined. If the terminal side is on the y-axis, then the x-coordinate is 0 and the tangent $\left(\tan\theta = \frac{y}{x}\right)$ and the secant $\left(\sec\theta = \frac{r}{x}\right)$ functions are undefined.

EXAMPLE 2 Finding the Values of the Trigonometric Functions of Quadrantal Angles

Find the values (if any) of the six trigonometric functions of each angle.

a. $\theta = 0° = 0$ radians **b.** $\theta = 90° = \frac{\pi}{2}$ radians

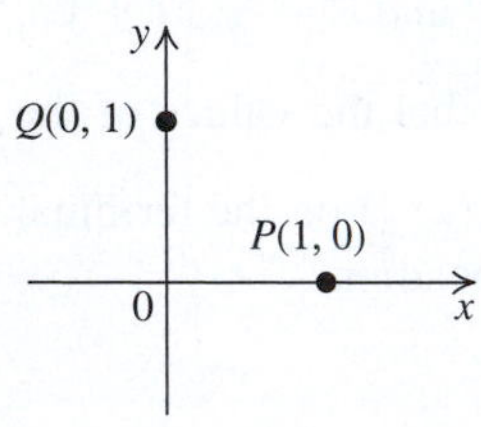

FIGURE 5.34

Solution

First, choose a point on the terminal side of each angle, as shown in Figure 5.34. Then use the definitions of the trigonometric functions.

a. $\theta = 0° = 0$ radians; choose the point $P(1, 0)$, so $x = 1$, $y = 0$, and $r = \sqrt{1^2 + 0^2} = 1$,

$$\sin 0° = \sin 0 = \frac{y}{r} = \frac{0}{1} = 0 \qquad \cos 0° = \cos 0 = \frac{x}{r} = \frac{1}{1} = 1$$

$$\tan 0° = \tan 0 = \frac{y}{x} = \frac{0}{1} = 0 \qquad \sec 0° = \sec 0 = \frac{r}{x} = \frac{1}{1} = 1$$

Because the y-coordinate of P is 0, $\csc 0° = \csc 0$ and $\cot 0° = \cot 0$ are undefined.

b. $\theta = 90° = \frac{\pi}{2}$ radians; choose the point $Q(0, 1)$, so $x = 0$, $y = 1$, and $r = \sqrt{0^2 + 1^2} = 1$,

$$\sin 90° = \sin\frac{\pi}{2} = \frac{y}{r} = \frac{1}{1} = 1 \qquad \cos 90° = \cos\frac{\pi}{2} = \frac{x}{r} = \frac{0}{1} = 0$$

$$\csc 90° = \csc\frac{\pi}{2} = \frac{r}{y} = \frac{1}{1} = 1 \qquad \cot 90° = \cot\frac{\pi}{2} = \frac{x}{y} = \frac{0}{1} = 0$$

Because the x-coordinate of Q is 0, $\sec 90° = \sec\frac{\pi}{2}$ and $\tan 90° = \tan\frac{\pi}{2}$ are undefined.

Practice Problem 2 Find the values (if any) of the six trigonometric functions of each angle.

a. $\theta = 180° = \pi$ **b.** $\theta = 270° = \frac{3\pi}{2}$

Table 5.2 summarizes these facts about quadrantal angles.

TABLE 5.2 Trigonometric Function Values of Quadrantal Angles

θ (degrees)	θ (radians)	$\sin\theta$	$\cos\theta$	$\tan\theta$	$\cot\theta$	$\sec\theta$	$\csc\theta$
$0°$	0	0	1	0	undefined	1	undefined
$90°$	$\frac{\pi}{2}$	1	0	undefined	0	undefined	1
$180°$	π	0	-1	0	undefined	-1	undefined
$270°$	$\frac{3\pi}{2}$	-1	0	undefined	0	undefined	-1
$360°$	2π	0	1	0	undefined	1	undefined

Coterminal Angles

Because the value of each trigonometric function of an angle in standard position is completely determined by the position of the terminal side, the following statements are true:

1. Coterminal angles are assigned identical values by the six trigonometric functions.
2. The *signs* of the values of the trigonometric functions are determined by the quadrant containing the terminal side.
3. For any integer n, θ and $\theta + n360°$ (in degree measure) are coterminal angles and θ and $\theta + 2\pi n$ (in radian measure) are coterminal angles.

   Some frequently used consequences of these observations are given next.

TRIGONOMETRIC FUNCTION VALUES OF COTERMINAL ANGLES

θ in degrees	**θ in radians**
$\sin\theta = \sin(\theta + n360°)$	$\sin\theta = \sin(\theta + 2\pi n)$
$\cos\theta = \cos(\theta + n360°)$	$\cos\theta = \cos(\theta + 2\pi n)$

These equations hold for any integer n.

EXAMPLE 3 Trigonometric Function Values of Coterminal Angles

Find the exact values for the following.

a. $\sin 2580°$ **b.** $\cos\frac{17\pi}{4}$

Solution

a. $2580° = 60° + 2520° = 60° + 7(360°)$, so $\sin 2580° = \sin 60° = \frac{\sqrt{3}}{2}$

b. $\frac{17\pi}{4} = \frac{\pi}{4} + \frac{16\pi}{4} = \frac{\pi}{4} + 2(2\pi)$, so $\cos\frac{17\pi}{4} = \cos\frac{\pi}{4} = \frac{\sqrt{2}}{2}$

Practice Problem 3 Find the exact values for the following

a. $\cos 1830°$ **b.** $\sin\frac{31\pi}{3}$

2 Determine the signs of the trigonometric functions in each quadrant.

Signs of the Trigonometric Functions

Suppose the angle θ is not quadrantal and its terminal side contains the point (x, y). We know that the only value other than x and y used in defining the trigonometric functions, $r = \sqrt{x^2 + y^2}$, is always *positive.* Therefore, the signs of x and y determine the signs of the trigonometric functions.

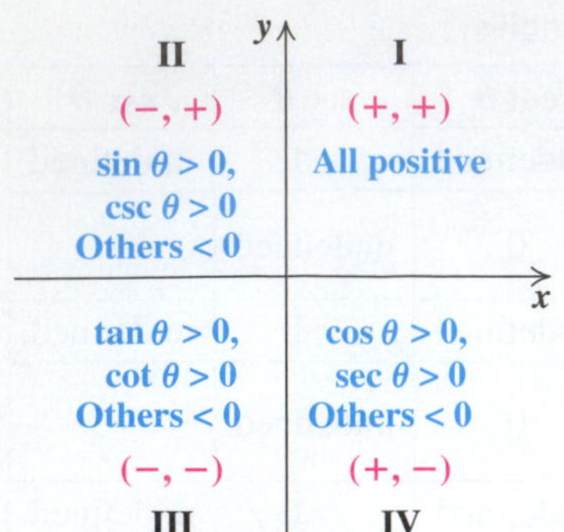

FIGURE 5.35 Signs of the trigonometric functions

SIDE NOTE

The phrase *All Students Take Calculus* can be useful in remembering which trigonometric functions are positive in quadrants I–IV. Use the first letter of each word in the phrase

- **A**ll functions (I)
- **S**ine (II)
- **T**angent (III)
- **C**osine (IV)

and recall that a and $\frac{1}{a}$ have the same sign.

If θ lies in quadrant I, then both x and y are positive; so all six trigonometric function values are positive. However, if θ lies in quadrant II, then x is negative and y is positive; so only $\sin\theta = \frac{y}{r}$ and $\csc\theta = \frac{r}{y}$ are positive. If θ lies in quadrant III, then x and y are both negative; so only $\tan\theta = \frac{y}{x}$ and $\cot\theta = \frac{x}{y}$ are positive. If θ lies in quadrant IV, then x is positive and y is negative; so only $\cos\theta = \frac{x}{r}$ and $\sec\theta = \frac{r}{x}$ are positive. Figure 5.35 summarizes the signs of the trigonometric functions.

EXAMPLE 4 Determining the Quadrant in Which an Angle Lies

If $\tan\theta > 0$ and $\cos\theta < 0$, in which quadrant does θ lie?

Solution

Because $\tan\theta > 0$, θ lies in either quadrant I or quadrant III. However, $\cos\theta > 0$ for θ in quadrant I; so θ must lie in quadrant III.

Practice Problem 4 If $\sin\theta > 0$ and $\cos\theta < 0$, in which quadrant does θ lie?

EXAMPLE 5 Evaluating Trigonometric Functions

Given that $\tan\theta = \frac{3}{2}$ and $\cos\theta < 0$, find the exact values of $\sin\theta$ and $\sec\theta$.

Solution

Because $\tan\theta > 0$ and $\cos\theta < 0$, θ lies in quadrant III. We want to identify a point (x, y) in quadrant III that is on the terminal side of θ. We have

$$\tan\theta = \frac{y}{x} = \frac{3}{2},$$

and because the point (x, y) is in quadrant III, both x and y must be negative. See Figure 5.36. If we choose $x = -2$ and $y = -3$, then

$$\tan\theta = \frac{y}{x} = \frac{-3}{-2} = \frac{3}{2}$$

and $r = \sqrt{x^2 + y^2} = \sqrt{(-2)^2 + (-3)^2} = \sqrt{4 + 9} = \sqrt{13}$.
Using $x = -2$, $y = -3$, and $r = \sqrt{13}$, we can find $\sin\theta$ and $\sec\theta$.

$$\sin\theta = \frac{y}{r} = \frac{-3}{\sqrt{13}} = -\frac{3\sqrt{13}}{13} \qquad \sec\theta = \frac{r}{x} = \frac{\sqrt{13}}{-2} = -\frac{\sqrt{13}}{2}$$

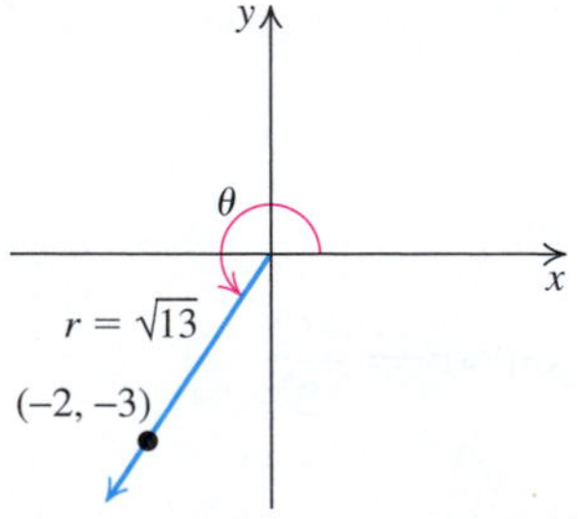

FIGURE 5.36

Practice Problem 5 Given that $\tan\theta = -\frac{4}{5}$ and $\cos\theta > 0$, find the exact values of $\sin\theta$ and $\sec\theta$.

3 Find a reference angle.

Reference Angle

For any angle θ, there is a corresponding acute angle called its *reference angle* whose trigonometric function values are identical to those of θ, except possibly for the sign.

Reference Angle

Let θ be an angle in standard position that is not a quadrantal angle. The **reference angle** for θ is the acute angle θ' ("theta prime") formed by the terminal side of θ and the x**-axis**.

If θ lies in quadrant I, then $\theta = \theta'$. The reference angle θ' for an angle θ $(0° < \theta < 360°)$ in each of the four quadrants is shown in Figure 5.37.

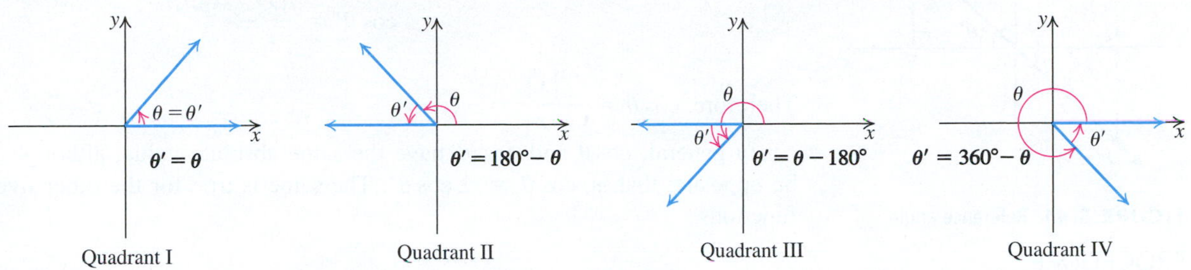

FIGURE 5.37 Reference angles for an angle θ, with $0° < \theta < 360°$

If θ is measured in radians and $0 < \theta < 2\pi$, we have the following:

$$\theta' = \theta \quad \text{for } \theta \text{ in quadrant I}$$
$$\theta' = \pi - \theta \quad \text{for } \theta \text{ in quadrant II}$$
$$\theta' = \theta - \pi \quad \text{for } \theta \text{ in quadrant III}$$
$$\theta' = 2\pi - \theta \quad \text{for } \theta \text{ in quadrant IV}$$

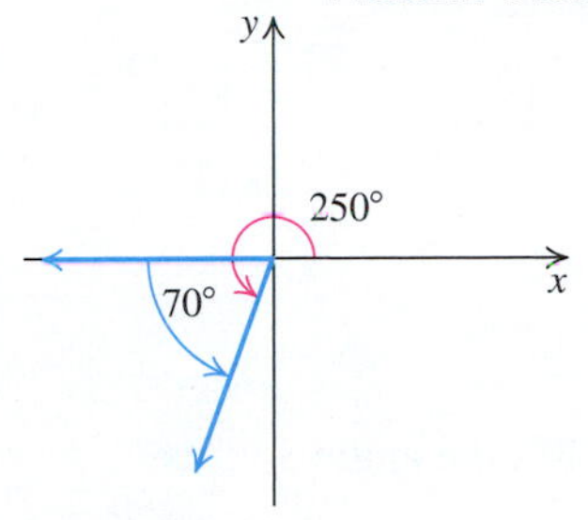

FIGURE 5.38

EXAMPLE 6 Identifying Reference Angles

Find the reference angle θ' for each angle θ.

a. $\theta = 250°$ **b.** $\theta = \frac{3\pi}{5}$ **c.** $\theta = 5.75$

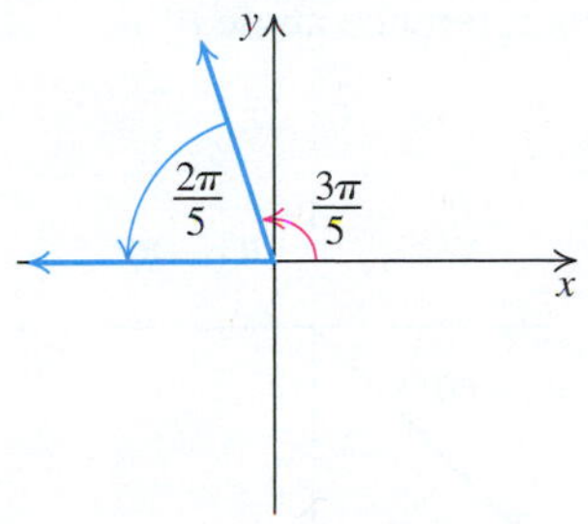

FIGURE 5.39

Solution

a. Because 250° lies in quadrant III, the reference angle is $\theta' = \theta - 180°$. So $\theta' = 250° - 180° = 70°$. See Figure 5.38.

b. Because $\frac{3\pi}{5}$ lies in quadrant II, the reference angle is $\theta' = \pi - \theta$. So $\theta' = \pi - \frac{3\pi}{5} = \frac{5\pi}{5} - \frac{3\pi}{5} = \frac{2\pi}{5}$. See Figure 5.39.

c. Because no degree symbol appears in $\theta = 5.75$, θ has radian measure. Now $\frac{3\pi}{2} \approx 4.71$ and $2\pi \approx 6.28$. So θ lies in quadrant IV and $\theta' = 2\pi - \theta$. So $\theta' = 2\pi - 5.75 \approx 6.28 - 5.75 = 0.53$. See Figure 5.40.

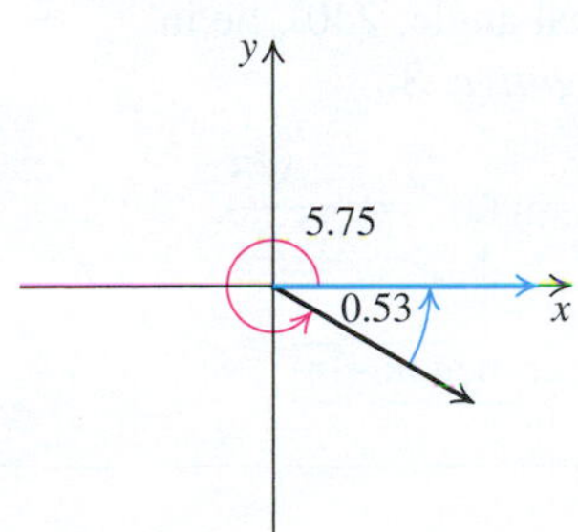

FIGURE 5.40

Practice Problem 6 Find the reference angle θ' for each angle θ.

a. $\theta = 175°$ **b.** $\theta = \frac{5\pi}{3}$ **c.** $\theta = 8.22$

You will learn later that there are convenient methods for finding the values of the trigonometric functions for $-\theta$ from the values of the function for θ. Consequently, we do not need to find reference angles for negative angles.

4 Use reference angles to find trigonometric function values.

Using Reference Angles

A reference angle is used to find the values of trigonometric functions of any angle θ. For example, consider the reference angle θ' of the angle θ in Figure 5.41. Let $P(x, y)$ be a point on the terminal side of θ in quadrant III. Then $Q(|x|, |y|)$ is in quadrant I. From the definition of trigonometric functions, we have

$$\cos\theta = \frac{x}{r} = \frac{-|x|}{r} \qquad \text{Because } x \text{ is negative, } x = -|x|.$$

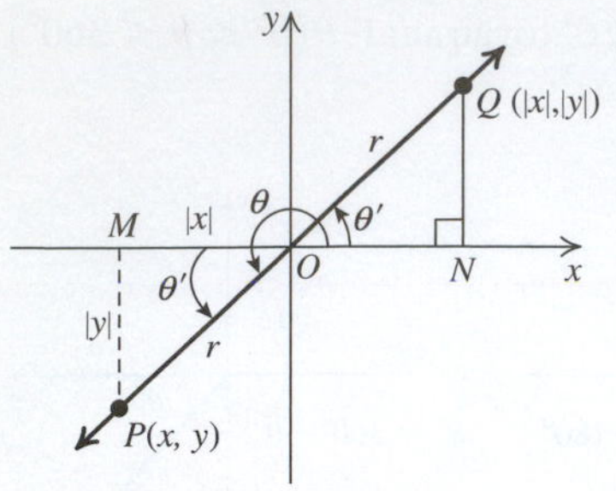

FIGURE 5.41 Reference angle

The right triangles *POM* and *QON* in Figure 5.41 are congruent by Side-Angle-Side. We have

$$\cos\theta' = \frac{|x|}{r}.$$

Therefore, $\cos\theta = -\dfrac{|x|}{r} = -\cos\theta'$.

In general, $\cos\theta$ and $\cos\theta'$ have the same absolute value, although their signs can be opposite; that is, $\cos\theta = \pm\cos\theta'$. The same is true for the other five trigonometric functions.

PROCEDURE **IN ACTION**

EXAMPLE 7 Using Reference Angles to Find Trigonometric Function Values

OBJECTIVE

Find the value of any trigonometric function of a nonquadrantal angle θ.

Step 1 If $\theta > 360°$ or $\theta < 0°$, find a coterminal angle for θ with degree measure between 0° and 360°. Otherwise, go to Step 2.

Step 2 Find the reference angle θ' for the angle resulting from Step 1. Write the trigonometric function of θ'.

Step 3 Choose the correct sign for the trigonometric function value of θ based on the quadrant in which it lies. Write the given trigonometric function of θ in terms of the same trigonometric function of θ' with the appropriate sign.

EXAMPLE

Find sin 1320°.

Because $1320° = 3(360°) + 240°$, the angle 240° is coterminal with the angle 1320°.

Because 240° is in quadrant III, its reference angle θ' is

$$\theta' = 240° - 180° = 60°,$$

and

$$\sin\theta' = \sin 60° = \frac{\sqrt{3}}{2}$$ See page 491.

y, x, θ = 240°, θ′ = 60°

The angle 1320° and its coterminal angle, 240°, lie in quadrant III, where the sine is *negative.* So

$$\sin 1320° = \sin 240° = -\sin 60° = -\frac{\sqrt{3}}{2}.$$

coterminal angles | reference angle

Practice Problem 7 Find cos 1020°.

EXAMPLE 8 Using the Reference Angle to Find Values of Trigonometric Functions

Find the exact value of each expression.

a. tan 330° **b.** $\sec\dfrac{59\pi}{6}$

Solution

a. Step 1 Because 330° is between 0° and 360°, we find its reference angle.

Step 2 Because 330° is in quadrant IV, its reference angle θ' is as follows:

$$\theta' = 360° - 330° = 30°$$

$$\tan\theta' = \tan 30° = \frac{\sqrt{3}}{3}$$ See page 491.

Step 3 In quadrant IV, $\tan \theta$ is negative; so

$$\tan 330° = -\tan 30° = -\frac{\sqrt{3}}{3}.$$

b. Step 1 Because the radian measure of $\frac{59\pi}{6}$ is greater than 2π, its degree measure is greater than 360°.

$$\frac{59\pi}{6} = \frac{11\pi + 48\pi}{6} = \frac{11\pi}{6} + 8\pi;$$

so $\frac{11\pi}{6}$ is an angle between 0 and 2π that is coterminal with $\frac{59\pi}{6}$.

Step 2 Because $\frac{11\pi}{6}$ is in quadrant IV, its reference angle θ' is as follows:

$$\theta' = 2\pi - \frac{11\pi}{6} = \frac{\pi}{6}$$

$$\sec \theta' = \sec \frac{\pi}{6} = \frac{2\sqrt{3}}{3}$$ See page 491.

Step 3 In quadrant IV, $\sec \theta > 0$; so

$$\sec \frac{59\pi}{6} = \sec \frac{11\pi}{6} = \sec \frac{\pi}{6} = \frac{2\sqrt{3}}{3}.$$

Practice Problem 8 Find the exact value of each expression.

a. $\csc 1035°$ **b.** $\cot \frac{17\pi}{6}$

EXAMPLE 9 The Millennium Wheel

The height above the base, h (in meters), of a passenger capsule on the Millennium Wheel is given by $h = 67.8 - 67.2 \cos \theta$, where θ is the angle the capsule arm makes with a vertical ray downward from the wheel's center. See Figure 5.42. How high is the capsule when

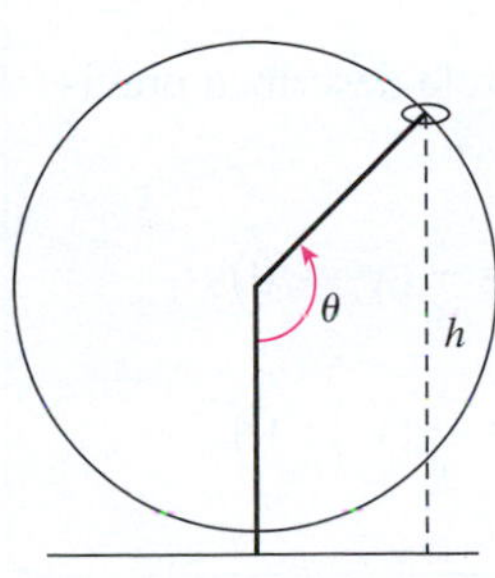

FIGURE 5.42

a. $\theta = 135°$? **b.** $\theta = 300°$? **c.** $\theta = 870°$?

Solution

a.
$$\begin{aligned} h &= 67.8 - 67.2 \cos 135° && \text{Replace } \theta \text{ with } 135°. \\ &= 67.8 - 67.2(\pm\cos 45°) && \text{Because } 135° \text{ is in Q II, } \theta' = 180 - 135 = 45°. \\ &= 67.8 - 67.2(-\cos 45°) && \cos\theta = -\cos\theta' \text{ for } \theta \text{ in Q II} \\ &= 67.8 + 67.2\left(\frac{\sqrt{2}}{2}\right) && \text{Replace } \cos 45° \text{ with } \frac{\sqrt{2}}{2}. \\ &= 67.8 + (33.6)\sqrt{2} \approx 115.32 \text{ meters} \end{aligned}$$

b. $\theta = 300°$ is in Q IV; so $\theta' = 360 - 300 = 60°$ and $\cos \theta = \cos \theta'$.

$$\begin{aligned} h &= 67.8 - 67.2 \cos 300° && \theta = 300° \\ &= 67.8 - 67.2 \cos 60° && \theta' = 60° \text{ and } \cos 300° = \cos 60° \\ &= 67.8 - 67.2\left(\frac{1}{2}\right) && \cos 60° = \frac{1}{2} \\ &= 34.2 \text{ meters} \end{aligned}$$

c. $\theta = 870° = 2(360°) + 150°$; so 870° and 150° are coterminal angles in Q II.

$$\theta' = 180° - 150° = 30° \text{ and } \cos 870° = \cos 150° = -\cos 30° = -\frac{\sqrt{3}}{2}$$

$$\begin{aligned} h &= 67.8 - 67.2 \cos 870° \\ &= 67.8 - 67.2\left(-\frac{\sqrt{3}}{2}\right) \\ &= 67.8 + 33.6\sqrt{3} \approx 126.0 \text{ meters} \end{aligned}$$

Practice Problem 9 In Example 9, find the height of the capsule when $\theta = 600°$.

5 Define the trigonometric functions using the unit circle.

Circular Functions

The relationship $s = r\theta$ between the radian measure of a central angle and the length of the arc it intercepts allows us to define the trigonometric functions so that their domains are sets of numbers rather than angles. In many applications and in advanced mathematics, this is a more useful approach. We begin with a **unit circle**, a circle of radius 1 unit centered at the origin. See Figure 5.43. Recall (Section 2.2) that the equation for this circle is $x^2 + y^2 = 1$.

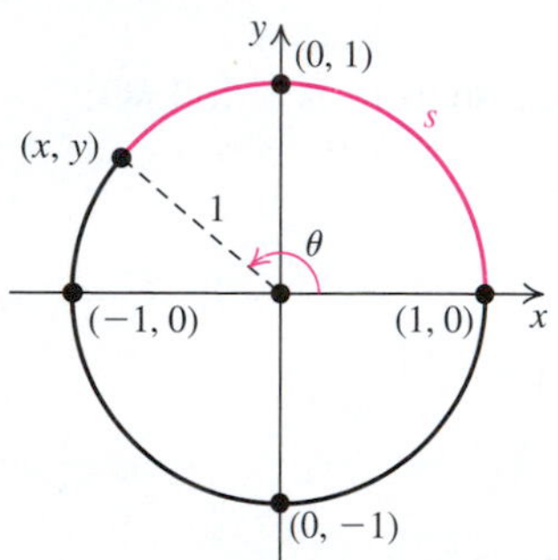

FIGURE 5.43 The unit circle

If s is any positive real number, we measure off an arc of length s (counterclockwise) beginning on the unit circle at $(1, 0)$ and ending at the point (x, y) on the unit circle. Then the point $P(x, y)$ is on the terminal ray of the central angle with radian measure θ that intercepts this arc. See Figure 5.44. Because the distance from the origin to (x, y) is $r = 1$, we have

$$\cos \theta = \frac{x}{r} = \frac{x}{1} = x \quad \text{and} \quad \sin \theta = \frac{y}{r} = \frac{y}{1} = y.$$

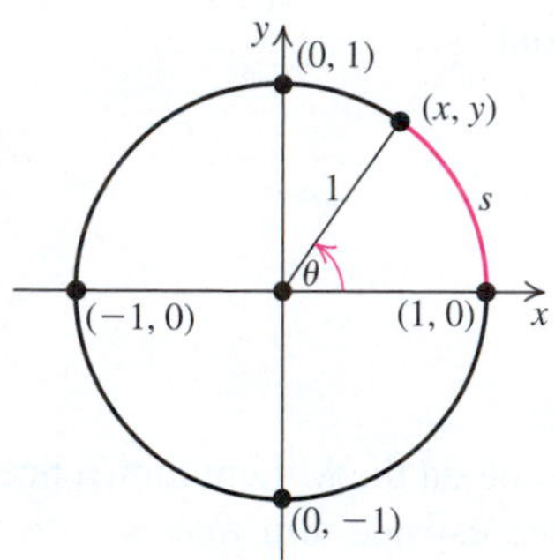

FIGURE 5.44 Circular functions

Because $s = 1 \cdot \theta = \theta$, we can write $\cos s = x$ and $\sin s = y$ and think of these functions as functions of the arc length, s, a real number. If s is a negative real number, we measure off an arc of length $|s|$ in a clockwise direction. Again, $\theta = s$, and we proceed as before. If $s = 0$, we let $(x, y) = (1, 0)$. All six trigonometric functions can now be defined as functions of numbers associated with arc length on a unit circle. When this is done, they are called **circular functions** or **trigonometric functions of real numbers**.

CIRCULAR FUNCTIONS

If s is a real number and $P(x, y)$ is the related point on the unit circle described previously, then

$$\cos s = x \qquad \sin s = y \qquad \tan s = \frac{y}{x}\ (x \neq 0)$$

$$\sec s = \frac{1}{x}\ (x \neq 0) \qquad \csc s = \frac{1}{y}\ (y \neq 0) \qquad \cot s = \frac{x}{y}\ (y \neq 0)$$

A point $P(x, y)$ on the unit circle associated with a real number s has coordinates $(\cos s, \sin s)$ because $x = \cos s$ and $y = \sin s$. See Figure 5.45.

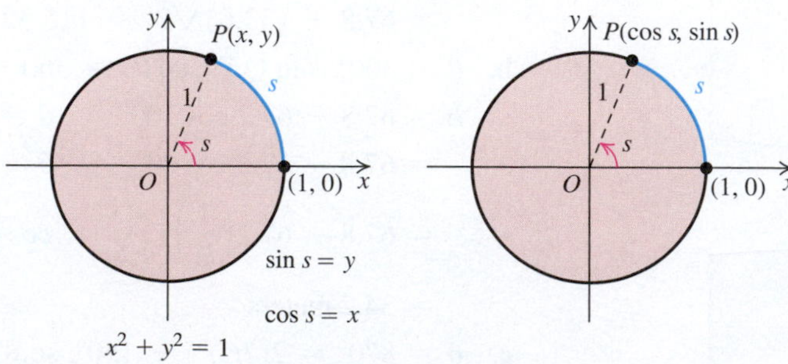

FIGURE 5.45 Unit circle definitions of sin s and cos s

In a unit circle $(r = 1)$, the length of an arc is numerically equal to the measure in radians of the central angle subtended by the arc. Therefore, we view the value of a trigonometric function at a real number s as its value at an angle of $\theta = s$ radians. Thus, we have, for example,

$$\sin s = \sin \theta = \sin\left(\frac{180}{\pi}\theta\right)^\circ.$$

So if s is any real number, we can evaluate the trigonometric function values at s in any of these three ways:

1. Find the endpoint (x, y) of the arc of directed length s units measured from the point $(1, 0)$ on the unit circle and use the circular function definitions.
2. Use the earlier methods for finding the trigonometric function values for a central angle of $s = \theta$ radians.
3. Convert $s = \theta$ from radian measure to degree measure and then find trigonometric function values.

Figure 5.46 shows ordered pairs on the unit circle associated with commonly used angles in degree and radian measure.

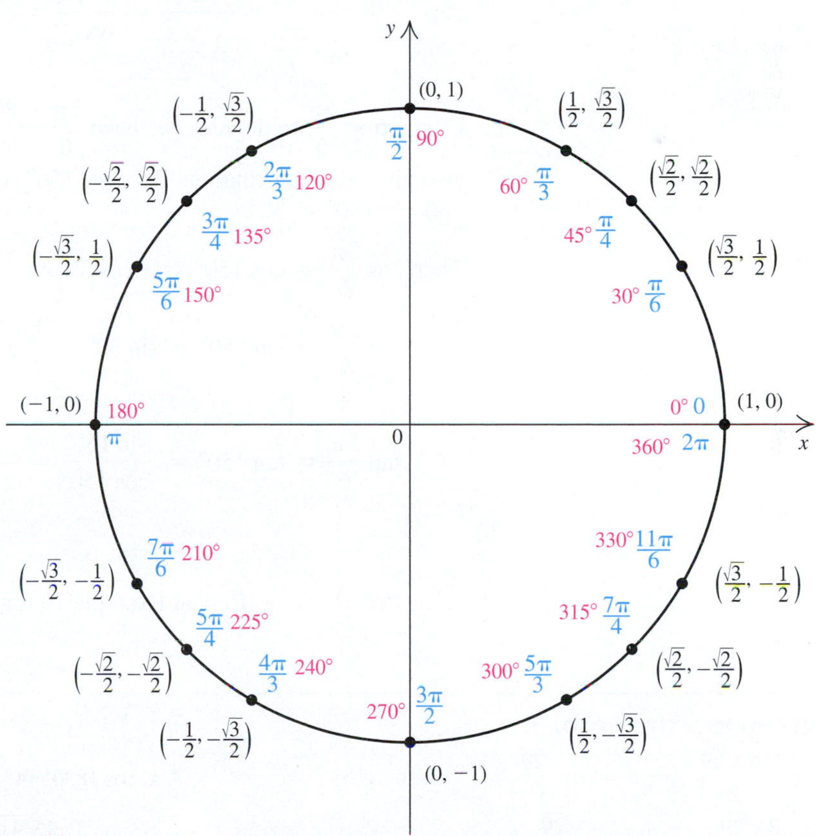

Unit circle: degrees, radians, ordered pairs

FIGURE 5.46

EXAMPLE 10 Finding Exact Circular Function Values in Three Ways

Find $\sin \frac{5\pi}{6}$, $\cos \frac{5\pi}{6}$, and $\tan \frac{5\pi}{6}$ by

a. Using Figure 5.46 to find the endpoint of the arc intercepted by $s = \frac{5\pi}{6}$.

b. Using reference angles and Table 5.1 from Section 5.2.

c. Repeat part **b.** converting $\frac{5\pi}{6}$ to degrees first.

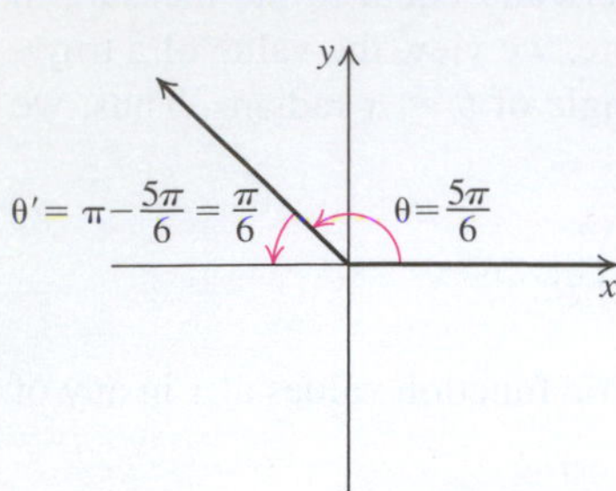

FIGURE 5.47

Solution

a. From Figure 5.46, we see that the endpoint of the arc of length $t = \frac{5\pi}{6}$ is $\left(-\frac{\sqrt{3}}{2}, \frac{1}{2}\right)$.

So $\cos\frac{5\pi}{6} = -\frac{\sqrt{3}}{2}$, $\sin\frac{5\pi}{6} = \frac{1}{2}$, and $\tan\frac{5\pi}{6} = \dfrac{\sin\frac{5\pi}{6}}{\cos\frac{5\pi}{6}} = \dfrac{\frac{1}{2}}{-\frac{\sqrt{3}}{2}} = -\frac{1}{\sqrt{3}} = -\frac{\sqrt{3}}{3}$.

b. Either by comparing $\frac{5\pi}{6}$ to $\frac{6\pi}{6}$ or by referring to Figure 5.46, we see that $t = \theta = \frac{5\pi}{6}$ radians is in Q II. So its cosine is negative and its sine is positive. The reference angle (see Figure 5.47) is

$$\pi - \frac{5\pi}{6} = \frac{6\pi}{6} - \frac{5\pi}{6} = \frac{\pi}{6}.$$

Then $\cos\frac{5\pi}{6} = -\cos\frac{\pi}{6} = -\frac{\sqrt{3}}{2}$, $\sin\frac{5\pi}{6} = \sin\frac{\pi}{6} = \frac{1}{2}$, and

$$\tan\frac{5\pi}{6} = \frac{\sin\frac{5\pi}{6}}{\cos\frac{5\pi}{6}} = \frac{\frac{1}{2}}{-\frac{\sqrt{3}}{2}} = -\frac{1}{\sqrt{3}} = -\frac{\sqrt{3}}{3}.$$

c. Converting $\frac{5\pi}{6}$ to degrees, we have $\frac{5\pi}{6} \cdot \frac{180}{\pi} = 150°$. Notice that Figure 5.46 also gives this information. Because 150° is in Q II, its reference angle is $180° - 150° = 30°$.

$$\text{Then } \cos\frac{5\pi}{6} = \cos 150° = -\cos 30° = -\frac{\sqrt{3}}{2},$$

$$\sin\frac{5\pi}{6} = \sin 150° = \sin 30° = \frac{1}{2}, \text{ and}$$

$$\tan\frac{5\pi}{6} = \tan 150° = \frac{\sin 150°}{\cos 150°} = \frac{\frac{1}{2}}{-\frac{\sqrt{3}}{2}} = -\frac{1}{\sqrt{3}} = -\frac{\sqrt{3}}{3}.$$

Practice Problem 10 Repeat Example 10 for $t = \frac{7\pi}{6}$.

Answers to Practice Problems

1. $\sin\theta = -\frac{5\sqrt{29}}{29}$ $\quad\csc\theta = -\frac{\sqrt{29}}{5}$

$\cos\theta = \frac{2\sqrt{29}}{29}$ $\quad\sec\theta = \frac{\sqrt{29}}{2}$

$\tan\theta = -\frac{5}{2}$ $\quad\cot\theta = -\frac{2}{5}$

2. a. $\sin 180° = \sin\pi = 0$ $\quad\csc 180° = \csc\pi$ is undefined.
$\cos 180° = \cos\pi = -1$ $\quad\sec 180° = \sec\pi = -1$
$\tan 180° = \tan\pi = 0$ $\quad\cot 180° = \cot\pi$ is undefined.

b. $\sin 270° = \sin 3\pi/2 = -1$ $\quad\csc 270° = \csc 3\pi/2 = -1$
$\cos 270° = \cos 3\pi/2 = 0$ $\quad\sec 270° = \sec 3\pi/2$ is undefined.
$\tan 270° = \tan 3\pi/2$ is undefined $\quad\cot 270° = \cot 3\pi/2 = 0$

3. a. $\cos 1830° = \frac{\sqrt{3}}{2}$ **b.** $\sin\frac{31\pi}{3} = \frac{\sqrt{3}}{2}$ **4.** Quadrant II

5. $\sin\theta = -\frac{4\sqrt{41}}{41}$, $\sec\theta = \frac{\sqrt{41}}{5}$ **6. a.** 5° **b.** $\frac{\pi}{3}$

c. $3\pi - 8.22 \approx 1.20$ **7.** $\cos 60° = 0.5$

8. a. $\csc 1035° = -\sqrt{2}$ **b.** $\cot\frac{17\pi}{6} = -\sqrt{3}$ **9.** 101.4 m

10. $\sin t = -\frac{1}{2}$, $\cos t = -\frac{\sqrt{3}}{2}$, $\tan t = \frac{\sqrt{3}}{3}$

SECTION 5.3 Exercises

Basic Concepts and Skills

1. For a point $P(x, y)$ on the terminal side of an angle θ in standard position, we let $r =$ ____________. Then $\sin\theta =$ ____________, $\cos\theta =$ ____________, and $\tan\theta =$ ____________.
2. If $P = (x, y)$ is on the terminal side of a quadrantal angle, then either x or y equals ____________.
3. The reference angle θ' for a nonquadrantal angle θ in standard position is the acute angle formed by the terminal side of θ and the ____________.
4. If θ_1 and θ_2 are coterminal angles, then $\sin\theta_1$ ____________ $\sin\theta_2$.
5. **True or False.** The value of a trigonometric function depends on the choice of the point $P(x, y)$ on the terminal side of θ.
6. **True or False.** In each quadrant, cosine and cosecant are both positive or both negative.

In Exercises 7–22, the terminal side of θ in standard position contains the given point. Find the values of the six trigonometric functions of θ.

7. $(-3, 4)$
8. $(4, -3)$
9. $(5, 12)$
10. $(-12, 5)$
11. $(7, 24)$
12. $(-24, 7)$
13. $(-24, -7)$
14. $(-7, 24)$
15. $(1, 1)$
16. $(-3, -3)$
17. $(\sqrt{2}, \sqrt{2})$
18. $(-3, \sqrt{3})$
19. $(\sqrt{3}, -1)$
20. $(\sqrt{13}, \sqrt{3})$
21. $(5, -2)$
22. $(-3, 5)$

In Exercises 23–44, find the exact value of each quadrantal angle. If any are not defined, write *undefined*.

23. $\sin 450°$
24. $\cos 450°$
25. $\cos(-90°)$
26. $\sin(-90°)$
27. $\tan 450°$
28. $\cot 540°$
29. $\tan(-540°)$
30. $\sec 1080°$
31. $\csc 900°$
32. $\csc 1080°$
33. $\sin(-1530°)$
34. $\cos(-2610°)$
35. $\cos(5\pi)$
36. $\sin(3\pi)$
37. $\tan(4\pi)$
38. $\cot\left(\frac{7\pi}{2}\right)$
39. $\csc\frac{5\pi}{2}$
40. $\sec(7\pi)$
41. $\sin(-2\pi)$
42. $\cos(-5\pi)$
43. $\cos\left(-\frac{3\pi}{2}\right)$
44. $\sin\left(-\frac{\pi}{2}\right)$

In Exercises 45–52, use the given information to find the quadrant in which θ lies.

45. $\sin\theta < 0$ and $\cos\theta < 0$
46. $\sin\theta < 0$ and $\tan\theta > 0$
47. $\sin\theta > 0$ and $\cos\theta < 0$
48. $\tan\theta > 0$ and $\csc\theta < 0$
49. $\cos\theta > 0$ and $\csc\theta < 0$
50. $\cos\theta < 0$ and $\cot\theta > 0$
51. $\sec\theta < 0$ and $\csc\theta > 0$
52. $\sec\theta < 0$ and $\tan\theta > 0$
53. Find the value of x assuming that the point $(x, -5)$ is on the terminal side of θ in quadrant III and $\sin\theta = -\frac{5}{13}$.
54. Repeat Exercise 53 assuming that θ lies in quadrant IV.
55. Find the value of y assuming that the point $(7, y)$ is on the terminal side of θ in quadrant I and $\cos\theta = \frac{7}{25}$.
56. Repeat Exercise 55 assuming that θ lies in quadrant IV.

In Exercises 57–64, find the exact values of the remaining trigonometric functions of θ from the given information.

57. $\cos\theta = -\frac{5}{13}$, θ in quadrant III
58. $\tan\theta = -\frac{3}{4}$, θ in quadrant IV
59. $\cot\theta = -\frac{3}{4}$, θ in quadrant II
60. $\sec\theta = \frac{4}{\sqrt{7}}$, θ in quadrant IV
61. $\sin\theta = \frac{3}{5}$, $\tan\theta < 0$
62. $\cot\theta = \frac{3}{2}$, $\sec\theta > 0$
63. $\sec\theta = 3$, $\sin\theta < 0$
64. $\tan\theta = -2$, $\sin\theta > 0$

In Exercises 65–80, find the exact values of each expression.

65. $\sin 1470°$
66. $\cos 1860°$
67. $\tan 1125°$
68. $\sec 2190°$
69. $\csc(2130°)$
70. $\sec(1410°)$
71. $\tan(690°)$
72. $\cot(2100°)$
73. $\sin\left(\frac{13\pi}{3}\right)$
74. $\cos\left(\frac{19\pi}{3}\right)$
75. $\cos\left(\frac{29\pi}{6}\right)$
76. $\sin\left(\frac{31\pi}{6}\right)$
77. $\tan\left(\frac{37\pi}{4}\right)$
78. $\cot\left(\frac{35\pi}{4}\right)$
79. $\sec\left(\frac{29\pi}{3}\right)$
80. $\csc\left(\frac{43\pi}{6}\right)$

In Exercises 81–86, find the value of sine, cosine, and tangent at each real number s in three ways (see Example 10).

81. $s = \frac{2\pi}{3}$ **82.** $s = \frac{3\pi}{4}$

83. $s = \frac{11\pi}{6}$ **84.** $s = \frac{4\pi}{3}$

85. $s = \frac{5\pi}{4}$ **86.** $s = \frac{5\pi}{3}$

Applying the Concepts

87. A safety chain. A piece of heavy equipment has two 8-foot metal arms that make a V with the horizontal plane and form an angle of 120° at their base. A contractor wants to put a chain from the top of one arm to the top of the other for safety. See the figure. How long must the chain be?

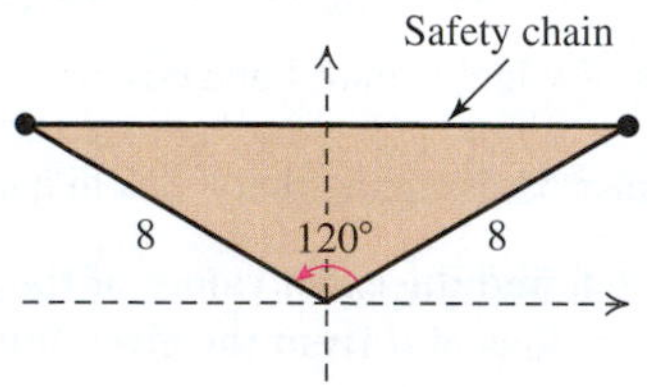

88. Removing a buoy. A crane is situated on the edge of a sheer cliff. It has a 12-foot arm that makes an angle of 120° with the horizontal and drops a hook straight down to a metal buoy that must be pulled from the water. See the figure. How far is the center of the buoy from the cliff?

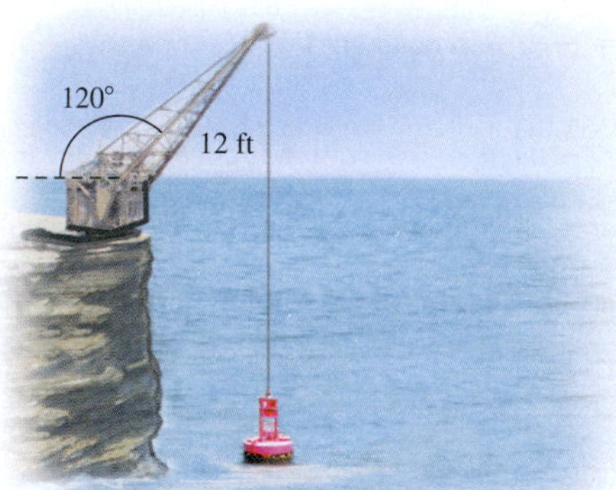

In Exercises 89–96, use the following discussion.

Projectile Motion

Suppose a projectile is launched from the origin into the first quadrant with an initial velocity v_0 feet per second at an angle θ with the horizontal (see figure). If the only force acting on the projectile is the force of gravity (with acceleration, $g_E = 32$ ft/sec²), then the equation of motion is given by the following:

$$y = x\tan\theta - \frac{16\sec^2\theta}{v_0^2}x^2$$

$$\text{Maximum height } H = \frac{1}{64}(v_0\sin\theta)^2$$

$$\text{Time of flight } t = \frac{v_0\sin\theta}{16}$$

$$\text{Horizontal range } R = \frac{v_0^2\sin\theta\cos\theta}{16}$$

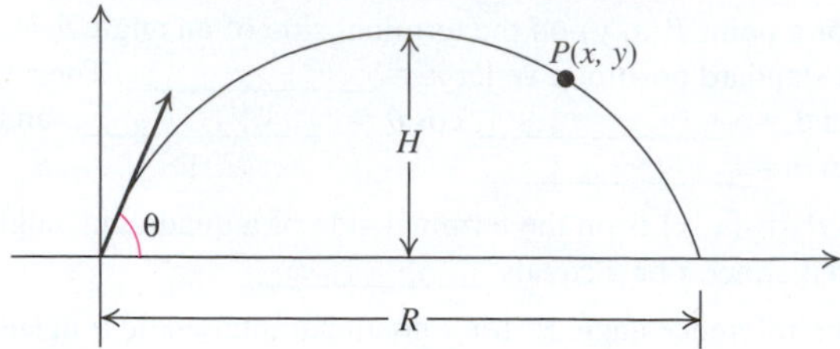

89. Golf. A golf ball is hit with an initial velocity of 44 ft/sec with an angle $\theta = 30°$. Find H, t, and R.

90. Repeat Exercise 89 assuming that $\theta = 45°$.

91. Repeat Exercise 89 assuming that $\theta = 60°$.

92. Repeat Exercise 89 assuming that $\theta = 90°$.

93. Football. A football is kicked with an initial velocity of 80 ft/sec with an angle $\theta = 45°$.

a. Write the equation of motion of the ball.

b. What is the height of the ball at the point 100 feet from the origin?

94. Find the maximum height attained by the football in Exercise 93.

95. Find the time of flight for the ball in Exercise 93.

96. Find the range of the ball in Exercise 93.

97. Light refraction. A fish swimming d feet below the surface is viewed from a line of sight that makes an angle of θ degrees from the vertical. Because the light is refracted by the water, the fish appears to be A feet below the surface, where

$$A = \frac{3d\cos\theta}{\sqrt{7 + 9\cos^2\theta}}.$$

What is the apparent depth of a fish swimming 2 feet below the surface when viewed at an angle of 60° from the vertical?

98. Sound levels. The decibel level, D, of a sound directed eastward is measured 3 yards from the source. If the ray from the sound source through a point 3 yards away makes an angle θ with an eastward ray from the source point, then

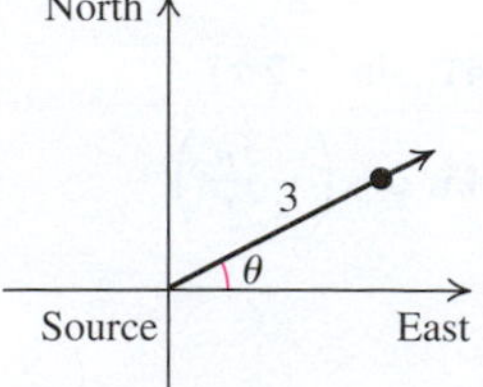

$$D = 25 + 15\cos\theta.$$

Find the decibel level at a point 3 yards from the sound source assuming that

a. The point is due east of the sound source.
b. The point is due west of the sound source.
c. The point is due south of the sound source.

99. **Height of a kite.** A kite is flying at the end of 100 feet of string that is in a taut straight line. The height, h, of the kite is given by $h = 100 \sin \theta$, where θ is the angle the string makes with the ground. Find the kite's height assuming that

a. $\theta = 30°$. b. $\theta = 60°$.

100. **Household voltage.** The voltage in a common household electrical outlet is given by $V = 166 \cos(120\pi t)$, where V is measured in volts ($\pm$ indicates direction) and t in seconds. Find the voltage when t is

a. $\frac{1}{120}$ sec. b. $\frac{1}{60}$ sec. c. $\frac{1}{72}$ sec.

Beyond the Basics

In Exercises 101–104, use the figure to find a triangle congruent to triangle *POM* that justifies the statements in each exercise.

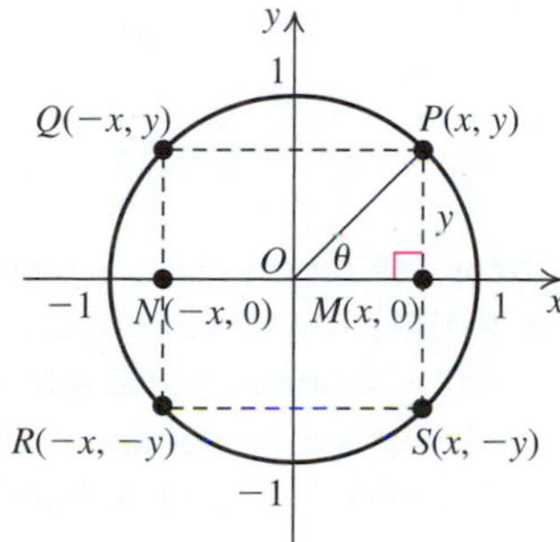

101. $\sin(-\theta) = -\sin\theta$, $\cos(-\theta) = \cos\theta$, and $\tan(-\theta) = -\tan\theta$

102. $\sin(180° - \theta) = \sin\theta$, $\cos(180° - \theta) = -\cos\theta$, and $\tan(180° - \theta) = -\tan\theta$

103. $\sin(180° + \theta) = -\sin\theta$, $\cos(180° + \theta) = -\cos\theta$, and $\tan(180° + \theta) = \tan\theta$

104. $\sin(360° - \theta) = -\sin\theta$, $\cos(360° - \theta) = \cos\theta$, and $\tan(360° - \theta) = -\tan\theta$

In Exercises 105–108, use the results of Exercises 101–104.

105. Find $\sin(-45°)$ and $\tan(-60°)$.

106. Find $\sin 135°$, $\cos 135°$, and $\tan 120°$.

107. Find $\sin 225°$, $\cos 240°$, and $\tan 210°$.

108. Find $\sin 315°$, $\cos 300°$, and $\tan 330°$

109. Show that for any angle θ, $-1 \le \sin\theta \le 1$ and $-1 \le \cos\theta \le 1$.

110. In the figure, show that triangles *POM* and *QON* are congruent.

a. Find coordinates of the point Q.
b. Show that $\sin(\theta + 90°) = \cos\theta$, $\cos(\theta + 90°) = -\sin\theta$, and $\tan(\theta + 90°) = -\cot\theta$.

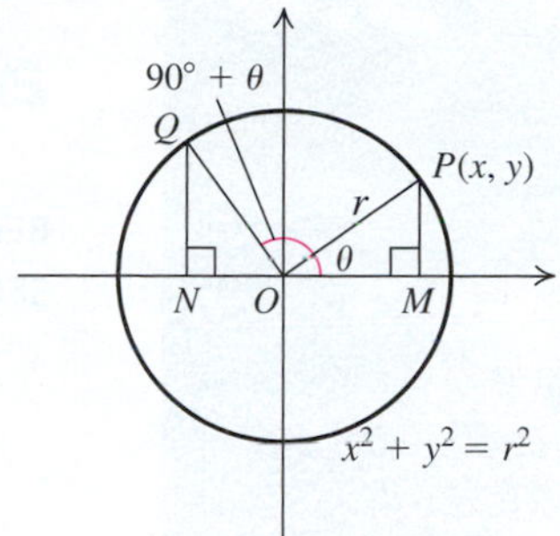

111. In triangle *ABC*, find the value of $\tan(A + B) + \tan C$.

112. What is the smallest positive value for $\sec\theta$?

In Exercises 113–116, without evaluating the functions, explain why each statement is false.

113. $\cos 105° = \cos 60° + \cos 45°$

114. $\sin 260° = 2 \sin 130°$

115. $\tan 123° = \tan 61° + \tan 62°$

116. $\sec 380° = \sec 185° + \sec 195°$

Critical Thinking/Discussion/Writing

117. **True or False.** $\cos(\sin\theta°)° = \sin(\cos\theta°)°$. Explain.

118. **True or False.** $\sec\theta° > \tan\theta°$ for every angle θ in the domain. Explain.

Maintaining Skills

In Exercises 119–122, find the zeros of each function.

119. $f(x) = 3x - 4$

120. $g(x) = 2x + 3$

121. $g(x) = x^2 + 3x - 10$

122. $g(x) = \dfrac{x^2 - 2x - 3}{x - 4}$

In Exercises 123–128, determine whether each function is even, odd, or neither.

123. $f(x) = 2x^3 - 4x$

124. $g(x) = \sqrt{1 - x^2}$

125. $h(x) = x^3 - x^2$

126. $p(x) = \sqrt[3]{x^5}$

127. $g(x) = 3x^4 + 2x^2 - 17$

128. $f(x) = 5x + 3$

SECTION 5.4

Graphs of the Sine and Cosine Functions

BEFORE STARTING THIS SECTION, REVIEW

1. Equation of a circle (Section 2.2, page 171)
2. Transformations of functions (Section 2.7, page 251)
3. Unit circle definitions of the sine and cosine (Section 5.3, page 508)

OBJECTIVES

1. Discuss properties of the sine and cosine functions.
2. Graph the sine and cosine functions.
3. Find the amplitude and period of sinusoidal functions.
4. Find the phase shifts and graph sinusoidal functions of the forms $y = a\sin[b(x - c)]$ and $y = a\cos[b(x - c)]$.

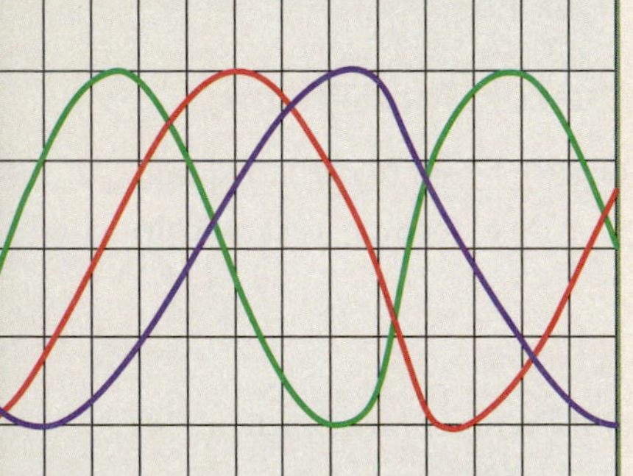

Biorhythms

Biorhythm means "rhythm of life." The theory of biorhythms, a pseudoscience, was developed from the data of several patients in the early twentieth century by Professor Hermann Swoboda (University of Vienna) and Dr. Wilhelm Fliess, a friend of Sigmund Freud. According to the theory, every person has three fundamental biorhythm cycles. All cycles start on a person's day of birth, and each cycle has a fixed period.

1. **Physical cycle** (P) has a period of 23 days. It governs the condition of one's body.
2. **Emotional cycle** (E) has a period of 28 days. It describes one's temperament.
3. **Intellectual cycle** (I) has a period of 33 days. It influences one's thinking capacity.

The periodic functions P, E, and I can each be represented in the form $y = \sin(Bt)$; they are said to determine our biorhythm state on any day, t, since our birth. We experience a "high" or "low" in the corresponding cycle (physical, emotional, or intellectual) on the day if the curve is above or below the midline. We discuss this in Example 5.

1 Discuss properties of the sine and cosine functions.

Properties of Sine and Cosine

We first discuss some properties of the sine and cosine functions that will be useful in sketching the graphs of these functions.

Domain and Range of Sine and Cosine

Each real number t determines a terminal point $P(t) = (x, y)$ on the unit circle. Because $x = \cos t$ and $y = \sin t$, both functions are defined for each real number t. So the domain of both the cosine and sine functions in interval notation is the set of real numbers $(-\infty, \infty)$.

For each real number t, the point $P(t) = (\cos t, \sin t)$ lies on the unit circle $x^2 + y^2 = 1$. Substituting $x = \cos t$ and $y = \sin t$ in the equation, we have the **Pythagorean identity:**

$$(\cos t)^2 + (\sin t)^2 = 1 \text{ for every real number } t.$$

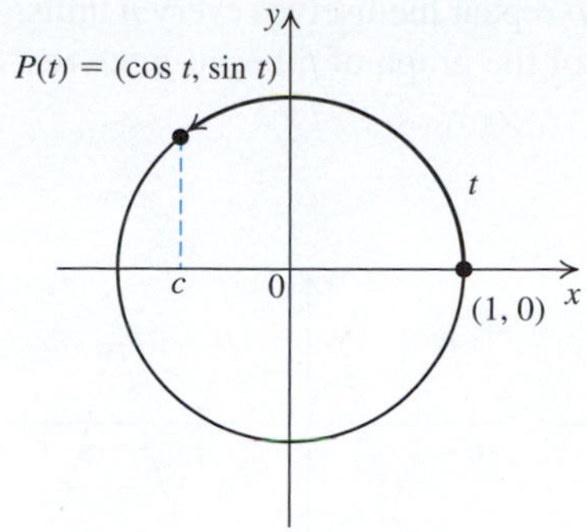

FIGURE 5.48

Because $(\cos t)^2$ and $(\sin t)^2$ are both nonnegative, we have

$$(\cos t)^2 \le 1 \text{ and } (\sin t)^2 \le 1.$$

These equations imply that the values of both cos t and sin t lie between -1 and 1. That is,

$$|\cos t| \le 1 \text{ and } |\sin t| \le 1, \text{ or}$$
$$-1 \le \cos t \le 1 \text{ and } -1 \le \sin t \le 1 \text{ for every number } t.$$

Furthermore, let c be a real number in the interval $[-1, 1]$. Then c is the x-coordinate of some terminal point $P(t) = (\cos t, \sin t)$ on the unit circle. See Figure 5.48. So $c = \cos t$. This shows that the interval $[-1, 1]$ is the range of the cosine function. A similar argument shows that the interval $[-1, 1]$ is also the range of the sine function.

DOMAIN AND RANGE OF SINE AND COSINE

The *domain* of both the sine and cosine functions is $(-\infty, \infty)$.
The *range* of both the sine and cosine functions is $[-1, 1]$.

Zeros of Sine and Cosine Functions

We know that cos $t = 0$ precisely when the terminal point $P(t) = (\cos t, \sin t)$ lies on the y-axis. This happens when $t = \frac{\pi}{2}, -\frac{\pi}{2}, \frac{3\pi}{2}, -\frac{3\pi}{2}, \ldots$; in other words, when t is an odd integer multiple of $\frac{\pi}{2}$. That is, the zeros of cos t are given by

$$t = (2n + 1)\frac{\pi}{2} = \frac{\pi}{2} + n\pi \text{ for any integer } n.$$

Similarly, sin $t = 0$ only if the terminal point $P(t) = (\cos t, \sin t)$ lies on the x-axis. This occurs when $t = 0, \pi, -\pi, 2\pi, -2\pi, \ldots$; that is, when t is an integer multiple of π. So the zeros of sin t are given by

$$t = n\pi \text{ for any integer } n.$$

Even-Odd Properties

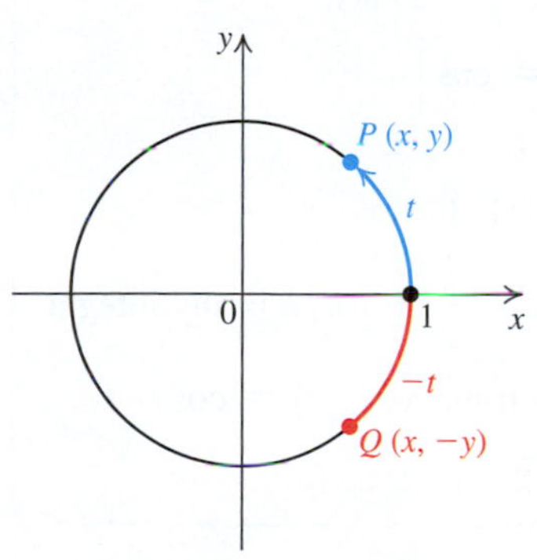

FIGURE 5.49

Let $P(x, y)$ be the terminal point on the unit circle associated with a real number t. Then from Figure 5.49 we see that the point $Q(x, -y)$ is the terminal point associated with the number $-t$. Then

$$\cos(-t) = x = \cos t \quad \text{and}$$
$$\sin(-t) = -y = -\sin t.$$

Because $\sin(-t) = -\sin t$, the sine function is an *odd* function; so its graph is symmetric about the origin. Similarly, because $\cos(-t) = \cos t$, the cosine function is an *even* function; so its graph is symmetric about the y-axis.

See Exercise 101 in Section 5.3.

Periodic Functions

Periodic Functions

A function f is said to be **periodic** if there is a positive number p such that

$$f(x + p) = f(x)$$

for every x in the domain of f.

The smallest value of p (if there is one) for which $f(x + p) = f(x)$ is called the **period** of f. The graph of f over any interval of length p is called one **cycle** of the graph.

We note that the values of a periodic function f of period p repeat themselves every p units. This means that to graph f, we may draw only one cycle of the graph of f, because the rest of the graph is generated by repeated copies of it.

Figure 5.50 shows examples of periodic functions.

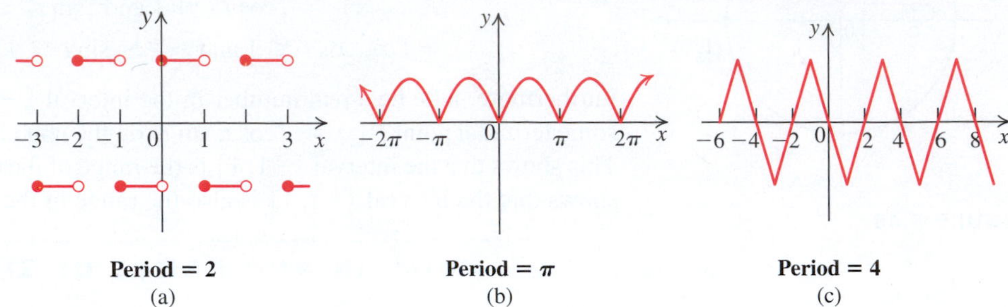

FIGURE 5.50 Periodic functions

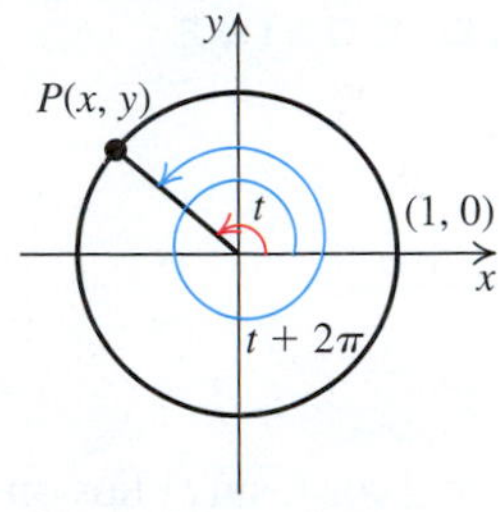

FIGURE 5.51

Let $P(x, y)$ be a point on the unit circle that corresponds to an angle t, measured in radians. If we add 2π to t, we obtain the same point P on the unit circle. See Figure 5.51. Then $\sin(t + 2\pi) = \sin t = y$ and $\cos(t + 2\pi) = \cos t = x$. Therefore, $\sin t$ and $\cos t$ are periodic functions. The period for each of the sine and cosine functions is 2π. See Exercise 86.

PERIOD OF THE SINE AND COSINE FUNCTIONS

For every real number t,

$$\sin(t + 2\pi) = \sin t \quad \text{and} \quad \cos(t + 2\pi) = \cos t.$$

The sine and cosine functions are periodic with **period 2π**.

We summarize the properties of the sine and cosine functions.

PROPERTIES OF SINE AND COSINE

$y = \sin t$	$y = \cos t$
1. Domain: $(-\infty, \infty)$	**1.** Domain: $(-\infty, \infty)$
2. Range: $[-1, 1]$	**2.** Range: $[-1, 1]$
3. Zeros at $t = n\pi$, n is any integer	**3.** Zeros at $t = \frac{\pi}{2} + n\pi$, n is any integer
4. Odd function: $\sin(-t) = -\sin t$	**4.** Even function: $\cos(-t) = \cos t$
5. Period: 2π	**5.** Period: 2π

2 Graph the sine and cosine functions.

Graphs of Sine and Cosine Functions

To graph the sine and cosine functions on an xy-coordinate plane, we use x as the independent variable and y as the dependent variable. So we write $y = \sin x$ and $y = \cos x$ rather than $y = \sin t$ and $y = \cos t$, respectively.

In calculus, it can be shown that the trigonometric functions are continuous on their domains. That is, they are continuous wherever they are defined.

Graph of the Sine Function

To sketch the graph of one cycle of $y = \sin x$ on the interval $[0, 2\pi]$, we use Table 5.3. This table consists of values of $y = \sin x$ for common values of x. In other words, x is a real number that is the radian measure of a common angle. Because the common angles x (in radians) are expressed in multiples of π, we mark off the x-axis in multiples of π. However, the range of $y = \sin x$ is between -1 (minimum value) and 1 (maximum value). We mark off the y-axis in integers.

TABLE 5.3

x	0	$\frac{\pi}{6}$	$\frac{\pi}{4}$	$\frac{\pi}{3}$	$\frac{\pi}{2}$	$\frac{2\pi}{3}$	$\frac{3\pi}{4}$	$\frac{5\pi}{6}$	π	$\frac{7\pi}{6}$	$\frac{5\pi}{4}$	$\frac{4\pi}{3}$	$\frac{3\pi}{2}$	$\frac{5\pi}{3}$	$\frac{7\pi}{4}$	$\frac{11\pi}{6}$	2π
$\sin x$	0	$\frac{1}{2}$	$\frac{1}{\sqrt{2}}$	$\frac{\sqrt{3}}{2}$	1	$\frac{\sqrt{3}}{2}$	$\frac{1}{\sqrt{2}}$	$\frac{1}{2}$	0	$-\frac{1}{2}$	$-\frac{1}{\sqrt{2}}$	$-\frac{\sqrt{3}}{2}$	-1	$-\frac{\sqrt{3}}{2}$	$-\frac{1}{\sqrt{2}}$	$-\frac{1}{2}$	0

Connecting the points $(x, \sin x)$ in Table 5.3 with a smooth curve gives us the graph in Figure 5.52.We have used the points $(x, \sin x)$ where x is a multiple of $\frac{\pi}{4}$ or $\frac{\pi}{6}$. In Figure 5.52 note that for the arc length $\frac{\pi}{6}$, shown on the unit circle, the y-value for the corresponding terminal point $P\left(\frac{\pi}{6}\right)$ is $\frac{1}{2}$; so the point $\left(\frac{\pi}{6}, \frac{1}{2}\right)$ is on the graph of $y = \sin x$.

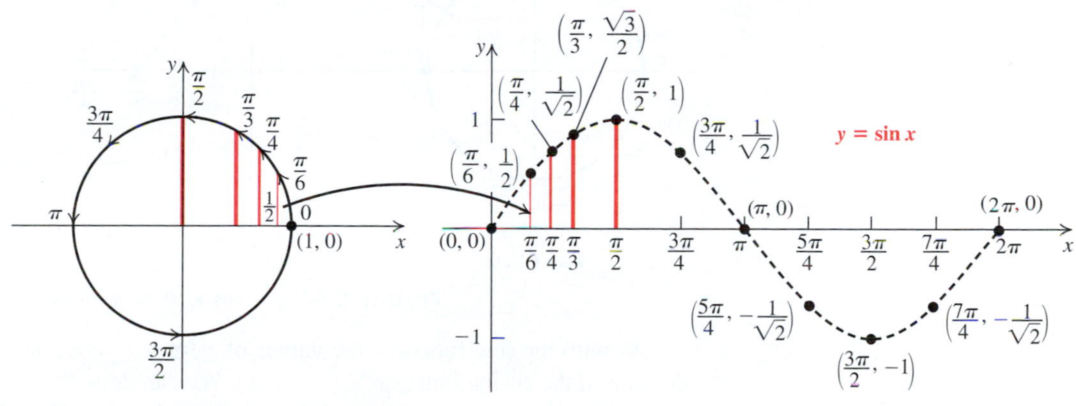

FIGURE 5.52 $y = \sin x, 0 \leq x \leq 2\pi$

Because the sine function is periodic with period 2π, we can sketch the complete graph of $y = \sin x$ by extending the graph in Figure 5.52 indefinitely to the left and to the right in every successive interval of length 2π. See Figure 5.53.

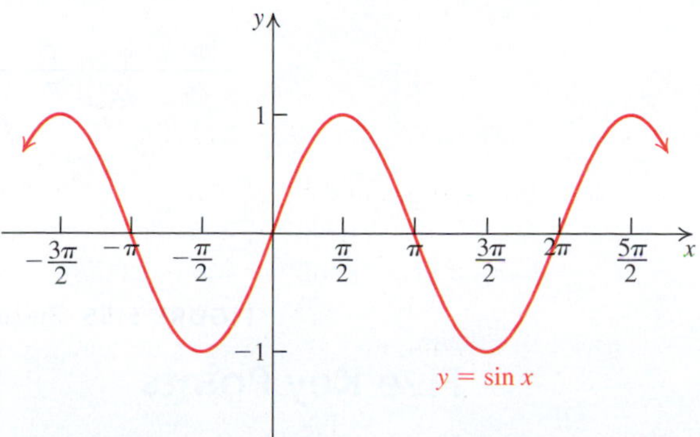

FIGURE 5.53 The sine curve

Graph of the Cosine Function

We can use the same method for graphing $y = \cos x$ that we used for $y = \sin x$.

Table 5.4 gives the values for $y = \cos x$, where x has the same values as in Table 5.3.

TABLE 5.4

x	0	$\frac{\pi}{6}$	$\frac{\pi}{4}$	$\frac{\pi}{3}$	$\frac{\pi}{2}$	$\frac{2\pi}{3}$	$\frac{3\pi}{4}$	$\frac{5\pi}{6}$	π	$\frac{7\pi}{6}$	$\frac{5\pi}{4}$	$\frac{4\pi}{3}$	$\frac{3\pi}{2}$	$\frac{5\pi}{3}$	$\frac{7\pi}{4}$	$\frac{11\pi}{6}$	2π
cos x	1	$\frac{\sqrt{3}}{2}$	$\frac{1}{\sqrt{2}}$	$\frac{1}{2}$	0	$-\frac{1}{2}$	$-\frac{1}{\sqrt{2}}$	$-\frac{\sqrt{3}}{2}$	-1	$-\frac{\sqrt{3}}{2}$	$-\frac{1}{\sqrt{2}}$	$-\frac{1}{2}$	0	$\frac{1}{2}$	$\frac{1}{\sqrt{2}}$	$\frac{\sqrt{3}}{2}$	1

Connecting the points $(x, \cos x)$ in Table 5.4 with a smooth curve gives the graph of $y = \cos x, 0 \le x \le 2\pi$, shown in Figure 5.54. In Figure 5.54 note that for the arc length $\frac{\pi}{6}$, shown on the unit circle, the x-value for the corresponding terminal point $P\left(\frac{\pi}{6}\right)$ is $\frac{\sqrt{3}}{2}$; so the point $\left(\frac{\pi}{6}, \frac{\sqrt{3}}{2}\right)$ is on the graph of $y = \cos x$.

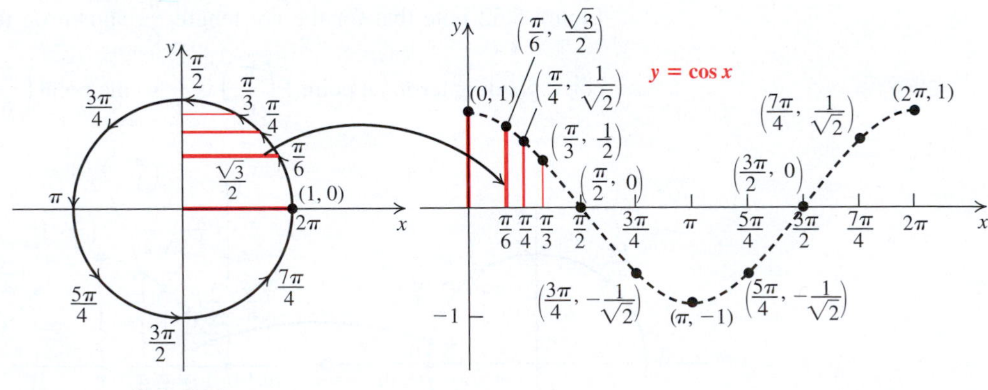

FIGURE 5.54 $y = \cos x, \ 0 \le x \le 2\pi$

As with the sine function, the values of x for $y = \cos x$ are not restricted because the domain of the cosine function is $(-\infty, \infty)$. We can draw the complete graph of $y = \cos x$ by extending the graph in Figure 5.54 indefinitely to the left and right in every successive interval of length 2π. See Figure 5.55.

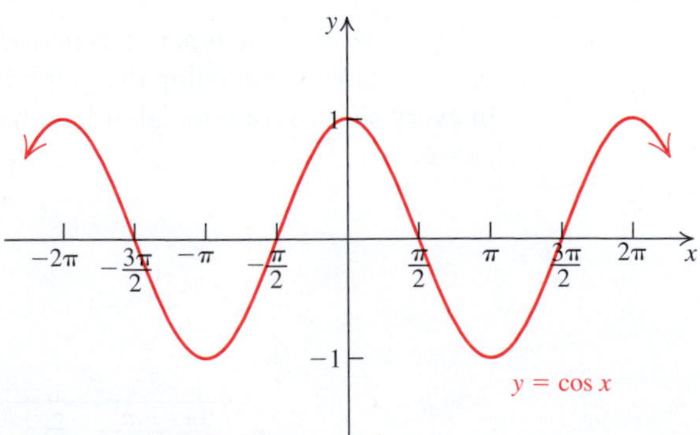

FIGURE 5.55 The cosine curve

Five Key Points

There are five *key points* in every cycle of the graphs of the sine and cosine functions. The key points consist of the x-intercepts, high points, and low points. See Figure 5.56. The key points are very helpful for graphing the sine and cosine functions and their transformations. Note that the x-coordinates of the five key points divide the period into four equal intervals.

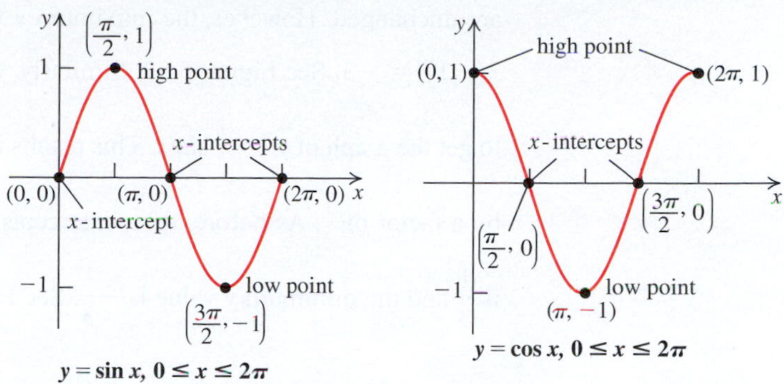

FIGURE 5.56 Key points of the sine and cosine graphs

3 Find the amplitude and period of sinusoidal functions.

Amplitude and Period

The graphs of the sine and cosine functions and their transformations are called **sinusoidal graphs** or **sinusoidal curves**. In this section, we study sinusoidal curves of the form $y = af[b(x - c)] + d$, where f is a sine or cosine function. We first consider transformations of the form $y = af(bx)$.

We start with the effect of the multiplier a in $y = af(bx)$. If the maximum (or minimum) value of $f(x)$ is M and $a > 0$, then the maximum (or minimum) value of $af(x)$is aM. For example, the maximum value of $y = 3 \sin x$ is $3(1) = 3$ and its minimum value is $3(-1) = -3$. If the graph of $y = f(x)$ is "centered" about the x-axis, then the maximum value of $f(x)$ is the *amplitude* of $f(x)$. In general, we have the following definition of the amplitude of a periodic function.

SIDE NOTE

If a function f does not have both a maximum and a minimum value, we say that f has no amplitude or that its amplitude is undefined.

Amplitude

Let f be a periodic function. Suppose M is the maximum value of f and m is the minimum value of f. The amplitude of f is defined by

$$\textbf{Amplitude} = \frac{1}{2}(M - m).$$

Because the maximum value of both functions $y = a \sin x$ and $y = a \cos x$ is $|a|$ and their minimum value is $-|a|$, the amplitude of both functions is $\frac{1}{2}[|a| - (-|a|)] = |a|$.

AMPLITUDES OF SINE AND COSINE

The functions $y = a \sin x$ and $y = a \cos x$ have **amplitude** $= |a|$ and range $= [-|a|, |a|]$.

EXAMPLE 1 **Graphing $y = a \sin x$**

Graph $y = \sin x$, $y = 3 \sin x$, and $y = \frac{1}{3} \sin x$ on the same coordinate system over the interval $[-2\pi, 2\pi]$.

Solution

We begin with the graph of $y = \sin x$ and multiply the y-coordinate of each point (including the key points) on this graph by 3 to get the graph of $y = 3 \sin x$. This results in stretching the graph of $y = \sin x$ vertically by a factor of 3. Because $3(0) = 0$, the x-intercepts are unchanged. However, the maximum y value is $3(1) = 3$ and the minimum y value is $3(-1) = -3$. See Figure 5.57. Similarly, we multiply the y-coordinate on the graph by $\frac{1}{3}$ to get the graph of $y = \frac{1}{3}\sin x$. This results in compressing the graph of $y = \sin x$ vertically by a factor of $\frac{1}{3}$. As before, the x-intercepts are unchanged, but here the maximum y value is $\frac{1}{3}$ and the minimum y value is $-\frac{1}{3}$. See Figure 5.57.

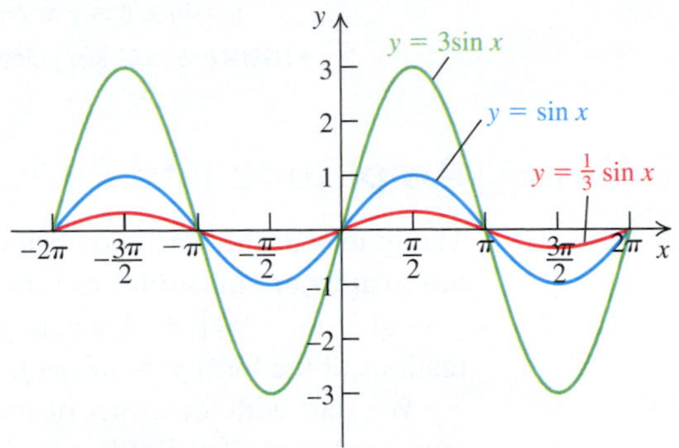

FIGURE 5.57 Varying a in $y = a \sin x$

Practice Problem 1 Graph $y = \sin x$, $y = 5 \sin x$, and $y = \frac{1}{5}\sin x$ on the same coordinate system over the interval $[-2\pi, 2\pi]$.

EXAMPLE 2 Graphing $y = a \cos x$

Graph $y = -2 \cos x$ over the interval $[-2\pi, 2\pi]$. Find the amplitude and range of the function.

Solution

Begin with the graph of $y = \cos x$. Multiply the y-coordinate of each point on that graph by 2 to stretch the graph vertically by a factor of 2 to get the graph of $y = 2 \cos x$. Pay particular attention to the key points. Notice that the x-intercepts are unchanged. Next, reflect the graph of $y = 2 \cos x$ about the x-axis to produce the graph of $y = -2 \cos x$. See Figure 5.58. The amplitude is $|-2| = 2$, the maximum value attained by $y = -2 \cos x$. The minimum function value attained is -2. Therefore, the range of $y = -2 \cos x$ is $[-2, 2]$.

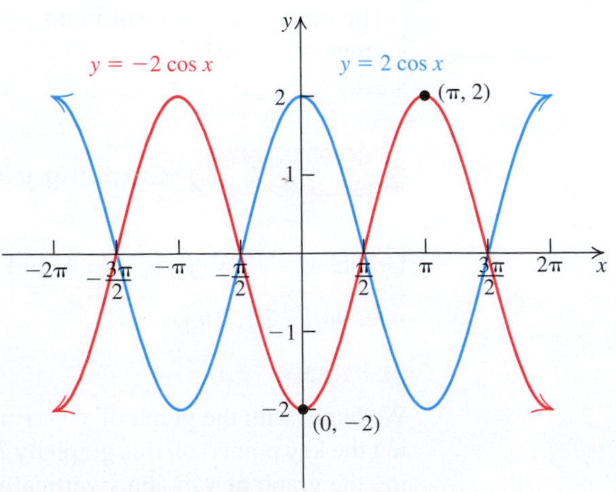

FIGURE 5.58 Graph of $y = -2 \cos x$

Practice Problem 2 Graph $y = 5\cos x$ over the interval $[-2\pi, 2\pi]$. Find the amplitude and range of the function.

Both the sine and cosine functions have period 2π. For $b > 0$, $0 \le bx \le 2\pi$ is equivalent to $0 \le x \le \frac{2\pi}{b}$. Therefore, as x takes on the values from 0 to $\frac{2\pi}{b}$, the expression bx takes on the values from 0 to 2π.

The graphs of $y = \sin bx$ and $y = \cos bx$ have similar shapes to those of $y = \sin x$ and $y = \cos x$, respectively, but they are compressed or expanded horizontally so that they complete one cycle over any interval of length $\frac{2\pi}{b}$. Consequently, the functions $y = \sin bx$ and $y = \cos bx$ both have period $\frac{2\pi}{b}$. When $b < 0$, we can use the facts that $\sin(-x) = -\sin x$ and $\cos(-x) = \cos x$ to rewrite the given function with $b > 0$. For example, $y = \sin(-3x) = -\sin 3x$.

EXAMPLE 3 Graphing $y = \sin bx$

Sketch one cycle of the graphs of $y = \sin x$, $y = \sin 3x$, and $y = \sin \frac{1}{3}x$ on the same coordinate system.

Solution

We begin with the graph of $y = \sin x$ and adjust its period, 2π, by dividing 2π by 3 to find the period $\frac{2\pi}{3}$ of $y = \sin 3x$. We then compress the graph of $y = \sin x$ horizontally so that one cycle of $y = \sin 3x$ is completed on the interval $\left[0, \frac{2\pi}{3}\right]$. Divide the interval $0 \le x \le \frac{2\pi}{3}$ into four equal parts to find the x-coordinates for the key points: $0, \frac{\pi}{6}, \frac{\pi}{3}, \frac{\pi}{2}$ and $\frac{2\pi}{3}$. Because the amplitude is 1, the y values of the key points are unchanged. See Figure 5.59.

Similarly, to find the period of $y = \sin \frac{1}{3}x$, we divide 2π by $\frac{1}{3}$, resulting in $\frac{2\pi}{\frac{1}{3}} = 6\pi$.

Therefore, one cycle for $y = \sin \frac{1}{3}x$ is completed on the interval $[0, 6\pi]$. Divide the interval $0 \le x \le 6\pi$ into four equal parts to get the x-coordinates for the key points: $0, \frac{3\pi}{2}, 3\pi, \frac{9\pi}{2}$, and 6π. Again, the y values of the key points are unchanged. See Figure 5.59.

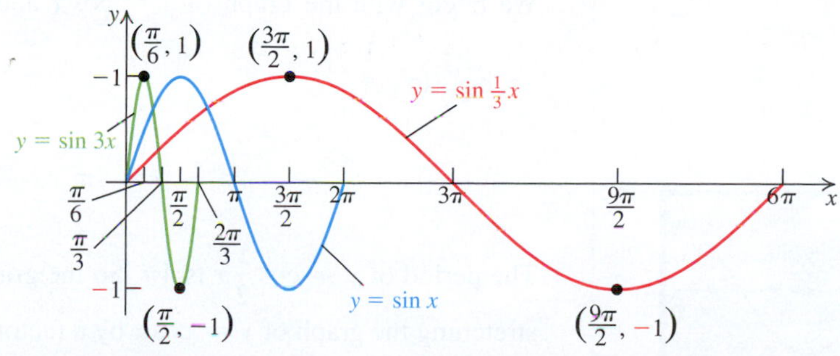

FIGURE 5.59 Horizontally compressing or expanding the graph of $y = \sin x$

Practice Problem 3 Sketch one cycle of the graphs of $y = \sin x$ and $y = \sin \frac{1}{2}x$ on the same coordinate system.

The next box summarizes information used to graph $y = a \sin bx$ and $y = a \cos bx$.

CHANGING THE AMPLITUDE AND PERIOD OF THE SINE AND COSINE FUNCTIONS

The functions $y = a \sin bx$ and $y = a \cos bx$ $(b > 0)$ have amplitude $|a|$ and period $\frac{2\pi}{b}$. If $a > 0$, the graphs of $y = a \sin bx$ and $y = a \cos bx$ are similar to the graphs of $y = \sin x$ and $y = \cos x$, respectively, with three changes.

1. The range is $[-a, a]$.
2. One cycle is completed over the interval $\left[0, \frac{2\pi}{b}\right]$.
3. The x-coordinates for the key points are $0, \frac{\pi}{2b}, \frac{\pi}{b}, \frac{3\pi}{2b}$, and $\frac{2\pi}{b}$. The y-values at each of these numbers is one of $-a$, 0, or a.

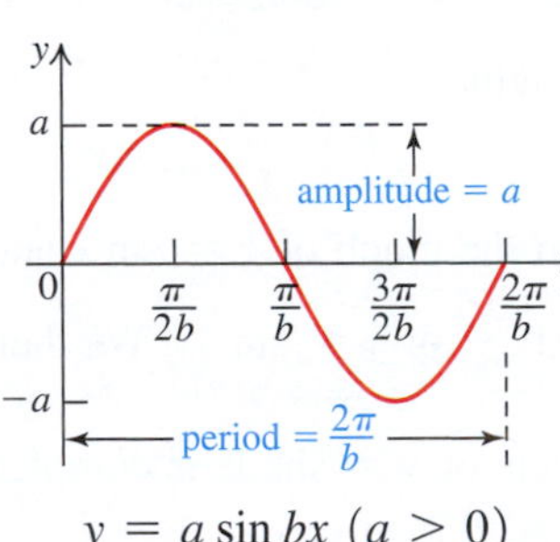

$y = a \sin bx \ (a > 0)$

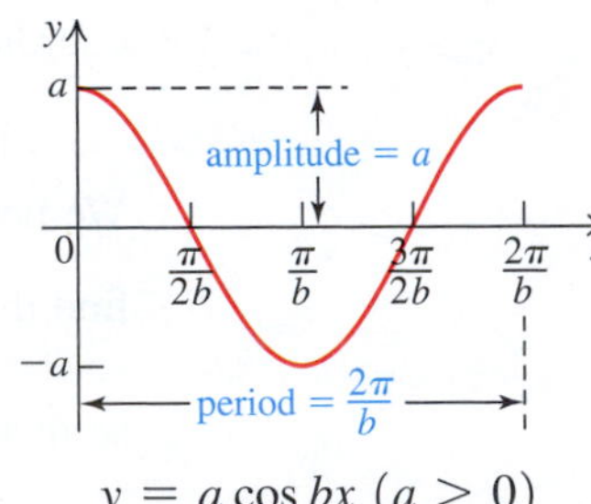

$y = a \cos bx \ (a > 0)$

If $a < 0$, the graphs are the reflections of the graph of $y = |a| \sin bx$ and $y = |a| \cos bx$, respectively, about the x-axis.

If $b < 0$, first use even–odd properties to rewrite the equation and then graph the function. See page 515.

TECHNOLOGY CONNECTION

To graph $y = 3 \cos \frac{1}{2}x$ with a graphing calculator, choose *radian* mode and these WINDOW settings:

$X\min = 0$

$X\max = 4\pi$

$X\text{scl} = \pi/2$

$Y\min = -4$

$Y\max = 4$

$Y\text{scl} = 1$

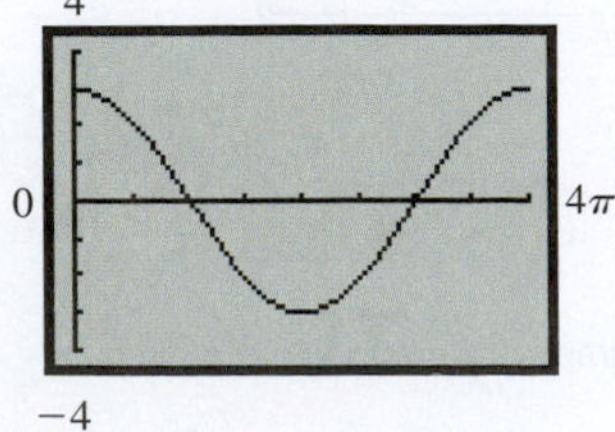

EXAMPLE 4 **Graphing $y = a \cos bx$**

Graph $y = 3 \cos \frac{1}{2}x$ over a one-period interval.

Solution

We begin with the graph of $y = \cos x$ and find the period of $y = \cos \frac{1}{2}x$ by dividing 2π by $\frac{1}{2}$ to get

$$2\pi \div \frac{1}{2} = 4\pi \qquad \text{Replace } b \text{ with } \frac{1}{2} \text{ in } \frac{2\pi}{b}.$$

The period of $y = \cos \frac{1}{2}x$ is 4π. So the graph of $y = \cos \frac{1}{2}x$ is obtained by horizontally stretching the graph of $y = \cos x$ by a factor of 2.

Next, we multiply the y-coordinate of each point on the graph of $y = \cos \frac{1}{2}x$ by 3 to stretch this graph vertically by a factor of 3. To help sketch the graphs, divide the period,

4π, into four equal parts, starting at 0, to find the x-coordinates for the key points. These x-coordinates—$0, \pi, 2\pi, 3\pi$, and 4π—correspond to the highest points, the lowest point, and the x-intercepts of the graph. The key points on the graph are $(0, 3)$, $(\pi, 0)$, $(2\pi, -3)$, $(3\pi, 0)$ and $(4\pi, 3)$. See Figure 5.60. The graph can then be extended indefinitely to the left and right. The amplitude is 3, so the maximum value of y is 3. The range is $[-3, 3]$.

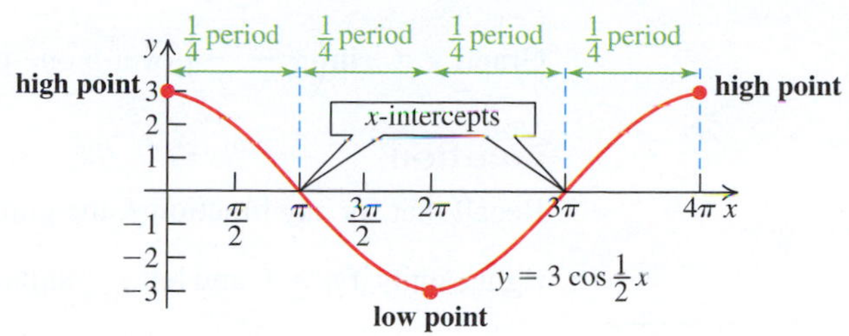

FIGURE 5.60 One cycle of $y = 3\cos\frac{1}{2}x$

Practice Problem 4 Graph $y = 2\cos 2x$ over a one-period interval.

Biorhythm States

The periods for physical (P), emotional (E), and intellectual (I) states are 23, 28, and 33 days, respectively. So your biorhythm states on the tth day since birth are given by

$$P = \sin\left(\frac{2\pi}{23}t\right),\ E = \sin\left(\frac{2\pi}{28}t\right),\ \text{and } I = \sin\left(\frac{2\pi}{33}t\right).$$

To calculate t on a given day, use the formula:

$$t = (\text{your age})(365) + (\text{number of leap years since your birth year}) + (\text{number of days since your last birthday}).$$

EXAMPLE 5 Finding Biorhythm States

Find the biorhythm states of Lisa on May 22, 2012, assuming that she was born on March 13, 1993.

Solution

We first calculate t.

Number of years from March 13, 1993 to March 13, 2012 $= 2012 - 1993 = 19$.

Number of leap years $= 5(1996, 2000, 2004, 2008, 2012)$

Number of days from March 13 to May 22 $= 18(\text{March}) + 30(\text{April}) + 22(\text{May}) = 70$

So $t = (19)(365) + 5 + 70 = 7010$

$$P = \sin\left(\frac{2\pi}{23}t\right) = \sin\left(\frac{2\pi(7010)}{23}\right) \approx -0.98$$

$$E = \sin\left(\frac{2\pi}{28}t\right) = \sin\left(\frac{2\pi(7010)}{28}\right) \approx 0.78$$

$$I = \sin\left(\frac{2\pi}{33}t\right) = \sin\left(\frac{2\pi(7010)}{33}\right) \approx 0.46$$

Substitute $t = 7010$ in each equation; use a calculator in *radian* mode.

Lisa's physical state is way down, her emotional state is excellent, and her intellectual state is good.

Practice Problem 5 Find your biorhythm states today.

4 Find the phase shifts and graph sinusoidal functions of the forms $y = a\sin[b(x - c)]$ and $y = a\cos[b(x - c)]$.

Phase Shift

We now consider sinusoidal graphs with horizontal shifts.

EXAMPLE 6 **Graphing $y = a\sin(x - c)$**

Graph $y = \sin\left(x - \frac{\pi}{2}\right)$ over a one-period interval.

Solution

Recall that for any function f, the graph of $y = f(x - c)$ is the graph of $y = f(x)$ shifted right c units if $c > 0$ and left $|c|$ units if $c < 0$. Comparing the equation $y = \sin\left(x - \frac{\pi}{2}\right)$ with $y = \sin x$, we see that $y = \sin\left(x - \frac{\pi}{2}\right)$ has the form $y = \sin(x - c)$, where $c = \frac{\pi}{2}$. Therefore, the graph of $y = \sin\left(x - \frac{\pi}{2}\right)$ is the graph of $y = \sin x$ shifted right $\frac{\pi}{2}$ units. See Figure 5.61.

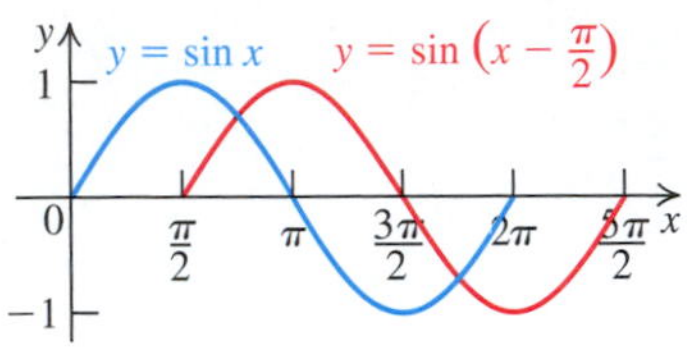

FIGURE 5.61 Graph of $y = \sin\left(x - \frac{\pi}{2}\right)$

Practice Problem 6 Graph $y = \cos\left(x - \frac{\pi}{4}\right)$ over a one-period interval.

Before describing the general procedure for graphing the sinusoidal functions $y = a\sin[b(x - c)]$ and $y = a\cos[b(x - c)]$, with $b > 0$, we find the value of x for which $b(x - c) = 0$. This number is called the **phase shift**. Let

$$b(x - c) = 0$$

$$x - c = 0 \quad \text{Divide both sides by } b.$$

$$x = c \quad \text{Add } c \text{ to both sides.}$$

The phase shift gives the amount by which the graphs of $y = a\sin bx$ or $y = a\cos bx$ are shifted horizontally. These graphs are shifted right if $c > 0$ and left if $c < 0$. To graph a cycle of either function, we start the cycle at $x = c$ and then divide the period into four equal parts. We describe this procedure next.

PROCEDURE **IN ACTION**

EXAMPLE 7 **Graphing $y = a\sin[b(x - c)]$ and $y = a\cos[b(x - c)]$ with $b > 0$**

OBJECTIVE

Graph a function of the form $y = a\sin[b(x - c)]$ or $y = a\cos[b(x - c)]$, with $b > 0$, by finding the amplitude, period, and phase shift.

Step 1 Find the amplitude, period, and phase shift.

amplitude $= |a|$

period $= \frac{2\pi}{b}$

phase shift $= c$

If $c > 0$, shift to the right.

If $c < 0$, shift to the left.

EXAMPLE

Find the amplitude, period, and phase shift and sketch the graph of $y = 3\sin\left[2\left(x - \frac{\pi}{4}\right)\right]$ over a one-period interval.

1. $y = 3\sin\left[2\left(x - \frac{\pi}{4}\right)\right]$

$a = 3$, $b = 2$, $\frac{\pi}{4} = c$

amplitude $= |3| = 3$

period $= \frac{2\pi}{2} = \pi$

phase shift $= \frac{\pi}{4}$ Shift right because $\frac{\pi}{4} > 0$.

(continued)

Step 2 The cycle begins at $x = c$. One complete cycle occurs over the interval $\left[c, c + \frac{2\pi}{b}\right]$.

Step 3 Divide the interval $\left[c, c + \frac{2\pi}{b}\right]$ into four equal parts, each of length $\frac{1}{4}(\text{period}) = \frac{1}{4}\left(\frac{2\pi}{b}\right) = \frac{\pi}{2b}$. This gives the x-coordinates for the five key points:
$c, c + \frac{\pi}{2b}, c + 2\left(\frac{\pi}{2b}\right) = c + \frac{\pi}{b}$,
$c + 3\left(\frac{\pi}{2b}\right)$, and $c + 4\left(\frac{\pi}{2b}\right) = c + \frac{2\pi}{b}$.

Step 4 If $a > 0$, for $y = a \sin[b(x - c)]$, sketch one cycle of the sine curve through the key points
$(c, 0), \left(c + \frac{\pi}{2b}, a\right), \left(c + \frac{\pi}{b}, 0\right)$,
$\left(c + \frac{3\pi}{2b}, -a\right)$, and $\left(c + \frac{2\pi}{b}, 0\right)$.
For $y = a \cos[b(x - c)]$, sketch one cycle of the cosine curve through the key points
$(c, a), \left(c + \frac{\pi}{2b}, 0\right), \left(c + \frac{\pi}{b}, -a\right)$,
$\left(c + \frac{3\pi}{2b}, 0\right)$, and $\left(c + \frac{2\pi}{b}, a\right)$.
If $a < 0$, reflect the graph of
$y = |a| \sin[b(x - c)]$ or
$y = |a| \cos[b(x - c)]$ about the x-axis.

2. Begin the cycle at $x = \frac{\pi}{4}$ and graph one cycle over $\left[\frac{\pi}{4}, \frac{\pi}{4} + \pi\right] = \left[\frac{\pi}{4}, \frac{5\pi}{4}\right]$.

3. Divide the interval $\left[\frac{\pi}{4}, \frac{5\pi}{4}\right]$ into four equal parts each having length $\frac{1}{4}(\text{period}) = \frac{1}{4}(\pi) = \frac{\pi}{4}$.
The x-coordinates of the five key points are
$\frac{\pi}{4}, \frac{\pi}{4} + \frac{\pi}{4} = \frac{\pi}{2}, \frac{\pi}{4} + 2\left(\frac{\pi}{4}\right) = \frac{3\pi}{4}$,
$\frac{\pi}{4} + 3\left(\frac{\pi}{4}\right) = \pi$, and $\frac{\pi}{4} + 4\left(\frac{\pi}{4}\right) = \frac{5\pi}{4}$.

4. Sketch one cycle of the sine curve through the five key points: $\left(\frac{\pi}{4}, 0\right), \left(\frac{\pi}{2}, 3\right), \left(\frac{3\pi}{4}, 0\right), (\pi, -3)$, and $\left(\frac{5\pi}{4}, 0\right)$.

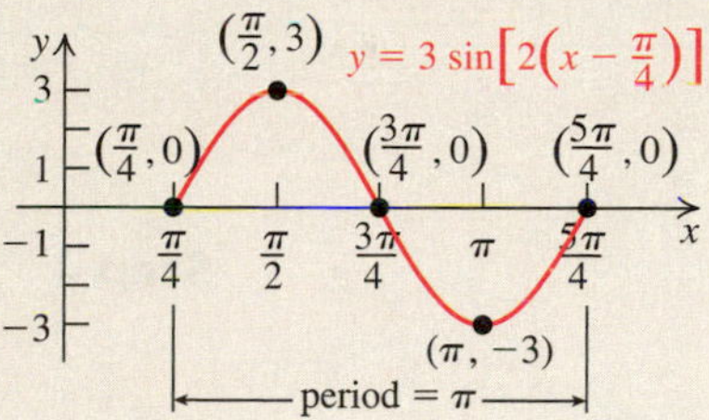

Practice Problem 7 Find the amplitude, period, and phase shift and sketch the graph of the function

$$y = \frac{1}{3}\cos\left[\frac{2}{5}\left(x + \frac{\pi}{6}\right)\right].$$

Graphing when $b < 0$ The equations $\sin(-x) = -\sin x$ and $\cos(-x) = \cos x$ allow us to graph $y = a \sin[b(x - c)]$ and $y = a \cos[b(x - c)]$ when $b < 0$. For example, to graph $y = 3 \sin\left[(-2)\left(x - \frac{\pi}{2}\right)\right]$, we rewrite and graph the equation $y = -3 \sin\left[2\left(x - \frac{\pi}{2}\right)\right]$.

EXAMPLE 8 Graphing $y = a \cos[b(x - c)]$

Graph $y = -\cos\left[2\left(x + \frac{\pi}{2}\right)\right]$ over a one-period interval.

TECHNOLOGY CONNECTION

To graph $y = -\cos\left[2\left(x + \frac{\pi}{2}\right)\right]$ with a graphing calculator, choose *radian* mode and enter these WINDOW settings:

$X\min = -\pi/2$
$X\max = \pi/2$
$X\text{scl} = \pi/4$
$Y\min = -2$
$Y\max = 2$
$Y\text{scl} = 1$

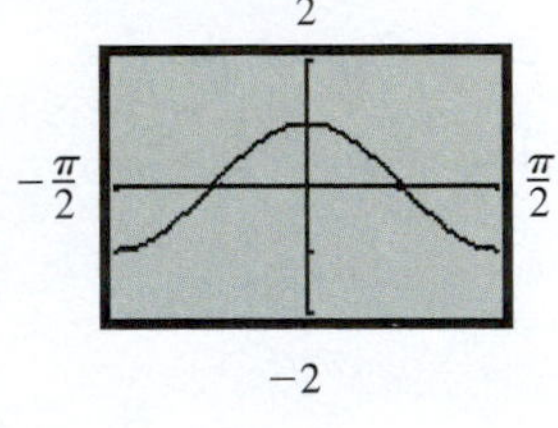

Solution

Use the steps in Example 7 for the procedure for graphing $y = a\cos[b(x - c)]$.

Step 1 Here $y = -\cos\left[2\left(x + \frac{\pi}{2}\right)\right] = (-1)\cos\left[2\left(x - \left(-\frac{\pi}{2}\right)\right)\right]$

$a = -1 \quad b = 2 \quad -\frac{\pi}{2} = c$

$$\text{amplitude} = |a| = |-1| = 1 \qquad \text{period} = \frac{2\pi}{b} = \frac{2\pi}{2} = \pi$$

$$\text{phase shift} = c = -\frac{\pi}{2} \qquad \text{Shift left because } c < 0.$$

Step 2 The cycle begins at $x = -\frac{\pi}{2}$.

One cycle is graphed over the interval $\left[-\frac{\pi}{2}, -\frac{\pi}{2} + \pi\right] = \left[-\frac{\pi}{2}, \frac{\pi}{2}\right]$.

Step 3 $\frac{1}{4}(\text{period}) = \frac{1}{4}(\pi) = \frac{\pi}{4}$

The x-coordinates of the five key points are

starting point $= -\frac{\pi}{2}$, $-\frac{\pi}{2} + \frac{\pi}{4} = -\frac{\pi}{4}$, $-\frac{\pi}{2} + 2\left(\frac{\pi}{4}\right) = 0$,

$-\frac{\pi}{2} + \frac{3\pi}{4} = \frac{\pi}{4}$, and $-\frac{\pi}{2} + 4\left(\frac{\pi}{4}\right) = \frac{\pi}{2}$.

Step 4 Because $a = -1 < 0$, we first graph $y = |-1|\cos\left[2\left(x + \frac{\pi}{2}\right)\right]$ by sketching one cycle, beginning at $\left(-\frac{\pi}{2}, 1\right)$ and going through the points $\left(-\frac{\pi}{4}, 0\right)$, $(0, -1)$, $\left(\frac{\pi}{4}, 0\right)$, and $\left(\frac{\pi}{2}, 1\right)$. We then reflect this graph about the x-axis to obtain the graph of $y = -\cos\left[2\left(x + \frac{\pi}{2}\right)\right]$. See Figure 5.62.

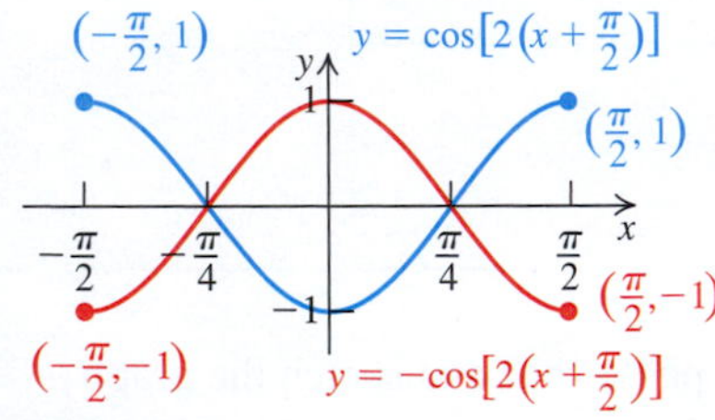

FIGURE 5.62

Practice Problem 8 Graph $y = -2\cos[3(x + \pi)]$ over a one-period interval.

To graph the function of the form $y = a\sin(bx - k)$ or $y = a\cos(bx - k)$, with $b > 0$, we first note that

$$bx - k = b\left(x - \frac{k}{b}\right) \qquad \text{Factor out } b.$$

Consequently, to graph these functions, we rewrite their equations.

$$y = a\sin(bx - k) = a\sin\left[b\left(x - \frac{k}{b}\right)\right]$$

$$y = a\cos(bx - k) = a\cos\left[b\left(x - \frac{k}{b}\right)\right]$$

We can then use the procedures for graphing the functions of the form $y = a\sin[b(x - c)]$ and $y = a\cos[b(x - c)]$ by letting $\frac{k}{b} = c$.

EXAMPLE 9 Finding the Period and Phase Shift

Find the period and the phase shift of each function.

a. $y = 2\cos\left(3x - \dfrac{\pi}{2}\right)$

b. $y = 3\sin(\pi x + 2)$

Solution

a. $y = 2\cos\left(3x - \dfrac{\pi}{2}\right)$ The given function

$= 2\cos\left[3\left(x - \dfrac{\pi}{6}\right)\right]$ Rewrite using the distributive property.

$\uparrow a \quad \uparrow b \quad \uparrow c$ Compare with $y = a\cos[b(x - c)]$.

Period $= \dfrac{2\pi}{b} = \dfrac{2\pi}{3}$ Phase shift $= c = \dfrac{\pi}{6}$

b. $y = 3\sin(\pi x + 2)$ The given function

$y = 3\sin\left[\pi\left(x - \left(-\dfrac{2}{\pi}\right)\right)\right]$ Rewrite using the distributive property.

$\uparrow a \quad \uparrow b \quad \uparrow c$

Period $= \dfrac{2\pi}{b} = \dfrac{2\pi}{\pi} = 2$ Phase shift $= c = -\dfrac{2}{\pi}$

Practice Problem 9 Find the period and the phase shift of each function.

a. $y = 3\sin\left(2x + \dfrac{\pi}{4}\right)$ **b.** $y = \cos(\pi x - 1)$

EXAMPLE 10 Graphing $y = a\sin(bx - k)$

Graph $y = 3\sin\left(2x - \dfrac{\pi}{2}\right)$ over a one-period interval.

Solution

Because $2x - \dfrac{\pi}{2} = 2\left(x - \dfrac{\pi}{4}\right)$, we can write $y = 3\sin\left(2x - \dfrac{\pi}{2}\right) = 3\sin\left[2\left(x - \dfrac{\pi}{4}\right)\right]$. This is the sinusoidal curve we graphed in Example 7 to show the procedure for graphing the functions $y = a\sin[b(x - c)]$ and $y = a\cos[b(x - c)]$. The graph is shown again in Figure 5.63.

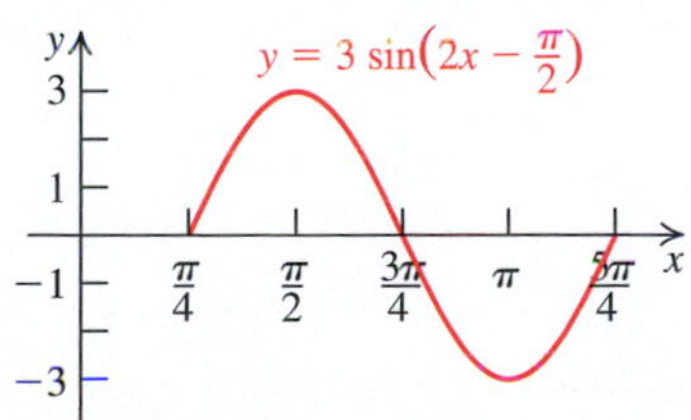

FIGURE 5.63

Practice Problem 10 Graph $y = 2\cos(3x - \pi)$ over a one-period interval.

Vertical Shifts

Recall that the graph of $y = f(x) + d$ results from shifting the graph of $y = f(x)$ vertically d units up if $d > 0$ and $|d|$ units down if $d < 0$. The next example shows this effect on a sinusoidal graph.

EXAMPLE 11 Graphing $y = a\cos[b(x - c)] + d$

Graph $y = \cos\left[2\left(x + \frac{\pi}{2}\right)\right] + 3$ over a one-period interval.

Solution

In Example 8, we graphed the function $y = \cos\left[2\left(x + \frac{\pi}{2}\right)\right]$ in order to graph $y = -\cos\left[2\left(x + \frac{\pi}{2}\right)\right]$. To graph $y = \cos\left[2\left(x + \frac{\pi}{2}\right)\right] + 3$, shift the graph of $y = \cos\left[2\left(x + \frac{\pi}{2}\right)\right]$, up 3 units. See Figure 5.64.

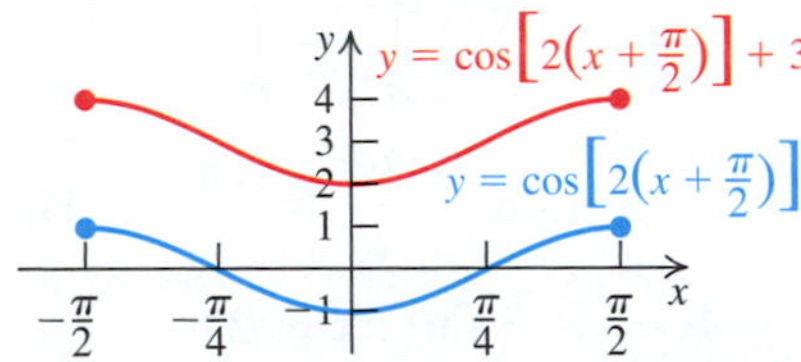

FIGURE 5.64 Graph of $y = \cos\left[2\left(x + \frac{\pi}{2}\right)\right] + 3$

Practice Problem 11 Graph $y = 4\sin\left(3x - \frac{\pi}{3}\right) - 2$ over a one-period interval.

EXAMPLE 12 Modeling the Number of Daylight Hours in Paris

TABLE 5.5

Daylight hours in Paris	
January	8.8
February	10.2
March	11.9
April	13.7
May	15.3
June	16.1
July	15.7
August	14.3
September	12.6
October	10.8
November	9.2
December	8.3

Table 5.5 gives the average number of daylight hours in Paris each month. Let y represent the number of daylight hours in Paris in month x to find a function of the form $y = a\sin[b(x - c)] + d$ that models the hours of daylight throughout the year.

Solution

Plot the values given in Table 5.5, representing the months by the integers 1 through 12 (Jan. = 1, . . . , Dec. = 12). Then sketch a function of the form $y = a\sin[b(x - c)] + d$ that models the points just graphed. See Figure 5.65.

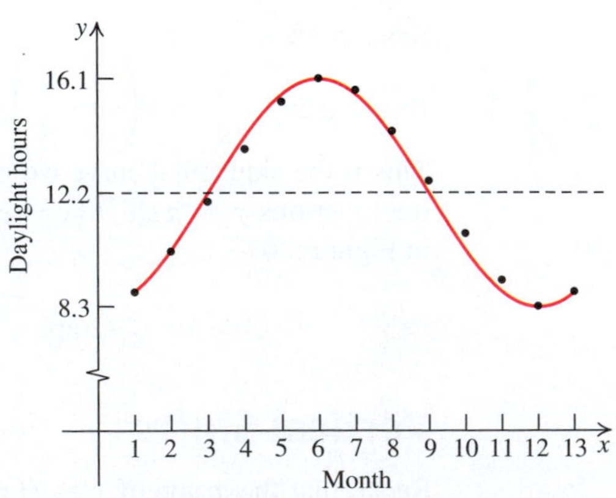

FIGURE 5.65 Hours of daylight in Paris each month

We want to find the values for the constants a, b, c, and d that will produce the graph shown in Figure 5.65. From Table 5.5, the range of the function is $[8.3, 16.1]$; so

$$\text{Amplitude} = a = \frac{\text{highest value} - \text{lowest value}}{2} = \frac{(16.1 - 8.3)}{2} = 3.9.$$

The sine graph has been shifted vertically up by $\frac{1}{2}$(highest value + lowest value), the average of the highest and lowest values. Thus, this average value gives

$$d = \frac{1}{2}(16.1 + 8.3) = 12.2.$$

The weather repeats every 12 months, so the period is 12. Because the period $= \frac{2\pi}{b}$, we write

$$12 = \frac{2\pi}{b} \qquad \text{Replace "period" with 12.}$$

$$b = \frac{2\pi}{12} = \frac{\pi}{6} \qquad \text{Solve for } b.$$

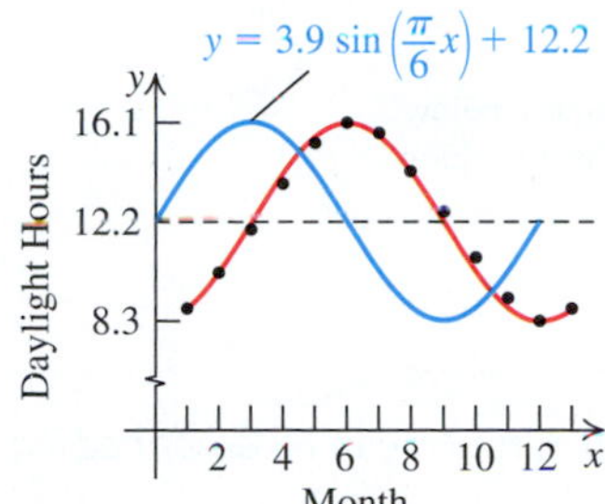

FIGURE 5.65A

So far, we can write $y = 3.9 \sin\left[\left(\frac{\pi}{6}\right)(x - c)\right] + 12.2$. The horizontally "unshifted" graph of $y = 3.9 \sin\left(\frac{\pi}{6}x\right) + 12.2$ is superimposed on Figure 5.65 in Figure 5.65A. This graph must be shifted to the right to fit the plotted data. To determine the phase shift, find the value of c when y has the maximum value. This value is $y = 16.1$ when $x = 6$ (in June), so we now have

$$16.1 = 3.9 \sin\left[\left(\frac{\pi}{6}\right)(6 - c)\right] + 12.2 \qquad \text{Replace } x \text{ with 6 and } y \text{ with 16.1.}$$

$$1 = \sin\left[\left(\frac{\pi}{6}\right)(6 - c)\right] \qquad \text{Subtract 12.2 from both sides and divide by 3.9.}$$

Now $\sin\left[\left(\frac{\pi}{6}\right)(6 - c)\right]$ will first equal 1 when the argument of sine is $\frac{\pi}{2}$. So

$$\frac{\pi}{6}(6 - c) = \frac{\pi}{2} \qquad \sin t = 1 \text{ when } t = \frac{\pi}{2}.$$

$$c = 3 \qquad \text{Solve for } c.$$

The equation describing the hours of daylight in Paris *in month x* is

$$y = 3.9 \sin\left[\frac{\pi}{6}(x - 3)\right] + 12.2.$$

TABLE 5.6

Daylight hours in Fargo	
January	9
February	10.3
March	11.9
April	13.6
May	15.1
June	15.8
July	15.4
August	14.2
September	12.5
October	10.9
November	9.4
December	8.6

Practice Problem 12 Rework Example 12 for the city of Fargo, North Dakota, using the values given in Table 5.6

Simple Harmonic Motion

Trigonometric functions can often describe or model motion caused by vibration, rotation, or oscillation. The motions associated with sound waves, radio waves, alternating electric current, a vibrating guitar string, or the swing of a pendulum are examples of such motion. Another example is the movement of a ball suspended by a spring. If the ball is pulled downward and then released (ignoring the effects of friction and air resistance), the ball will repeatedly move up and down past its rest, or equilibrium, position. See Figure 5.66.

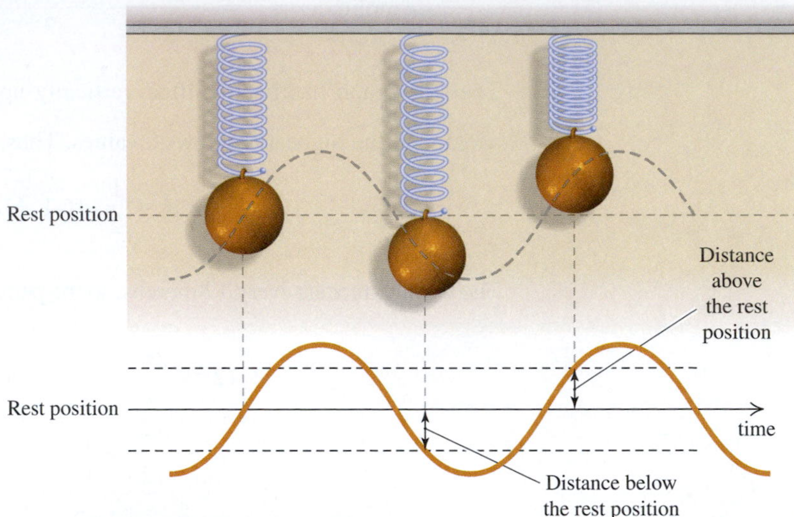

FIGURE 5.66 Simple harmonic motion

Simple Harmonic Motion

An object whose position relative to an equilibrium position at time t can be described by either

$$y = a \sin \omega t \quad \text{or} \quad y = a \cos \omega t \quad (\omega > 0)$$

is in simple harmonic motion.

The *amplitude*, $|a|$, is the maximum distance the object travels from its equilibrium position. The *period* of the motion, $\frac{2\pi}{\omega}$, is the time it takes the object to complete one full cycle. The **frequency** of the motion is $\frac{\omega}{2\pi}$ and gives the number of cycles completed per unit of time.

If the distance above and below the rest position of the ball is graphed as a function of time, a sine or a cosine curve results. The amplitude of the curve is the maximum distance above the rest position the ball travels, and one cycle is completed as the ball travels from its highest position to its lowest position and back to its highest position.

EXAMPLE 13 Simple Harmonic Motion of a Ball Attached to a Spring

Suppose a ball attached to a spring is pulled down 6 inches and released and the resulting simple harmonic motion has a period of eight seconds. Write an equation for the ball's simple harmonic motion.

Solution

We must first choose between an equation of the form $y = a \sin \omega t$ or $y = a \cos \omega t$. We start tracking the motion of the ball when $t = 0$. Note that for $t = 0$,

$$y = a \sin(\omega \cdot 0) = a \cdot \sin 0 = a \cdot 0 = 0$$

and

$$y = a \cos(\omega \cdot 0) = a \cdot \cos 0 = a \cdot 1 = a.$$

If we choose to start tracking the ball's motion when we release it after pulling it down 6 inches, we should choose $a = -6$ and $y = -6 \cos \omega t$. Because we pulled the ball *down* in order to start, a is negative. We now have the form of the equation of motion.

$$y = -6 \cos \omega t \qquad \text{When } t = 0, y = -6.$$

$$\text{period} = \frac{2\pi}{\omega} = 8 \qquad \text{The period is given as eight seconds.}$$

$$\omega = \frac{2\pi}{8} = \frac{\pi}{4} \qquad \text{Solve for } \omega.$$

Replacing ω with $\frac{\pi}{4}$ in $y = -6 \cos \omega t$ yields the equation of the ball's simple harmonic motion:

$$y = -6 \cos \frac{\pi}{4} t$$

Practice Problem 13 Rework Example 13 with a period of three seconds and the ball pulled down 4 inches and released.

Answers to Practice Problems

1.

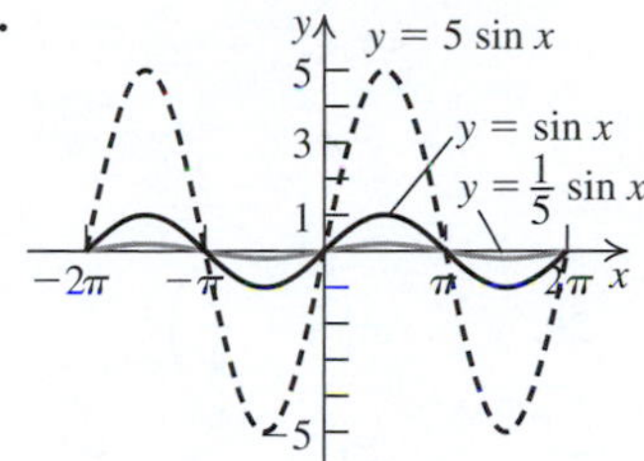

2.

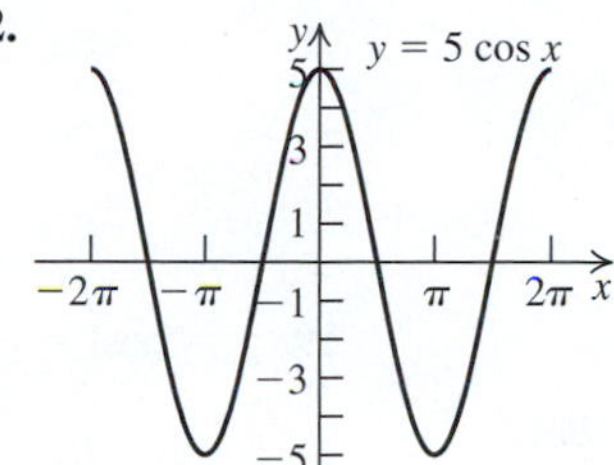

Amplitude = 5; range = [−5, 5]

3.

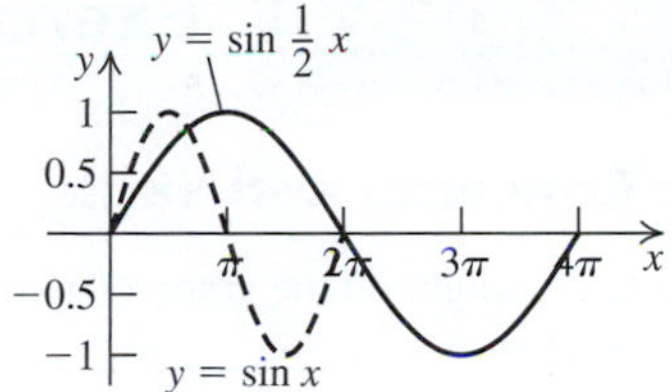

4.

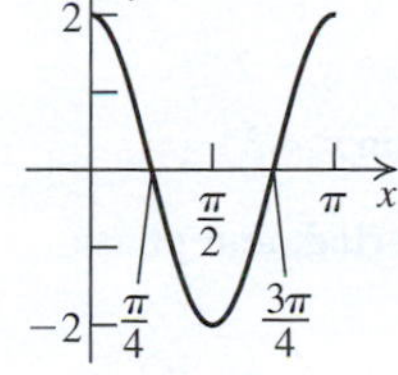

5. Answers will vary.

6.

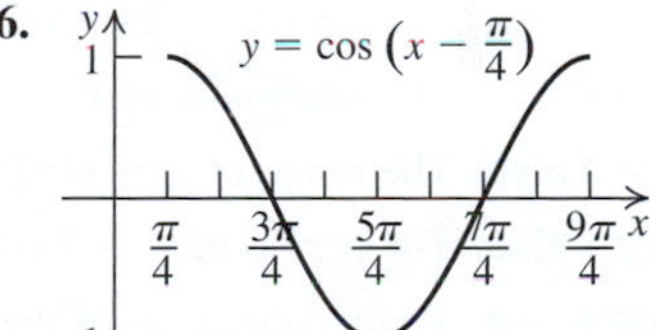

7. Amplitude $= \frac{1}{3}$; period $= 5\pi$; phase shift $= -\frac{\pi}{6}$

$$y = \frac{1}{3} \cos\left[\frac{2}{5}\left(x + \frac{\pi}{6}\right)\right]$$

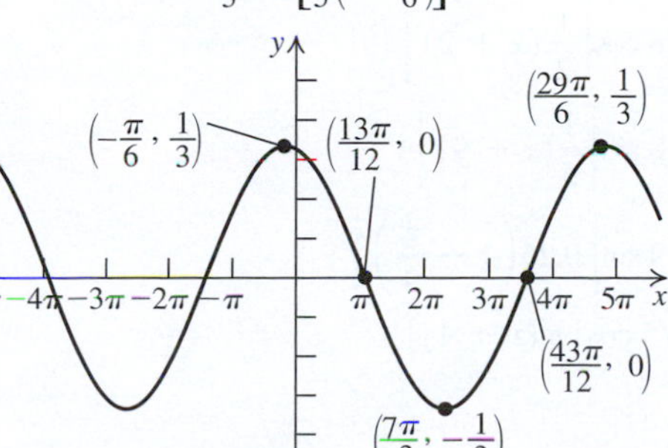

8.

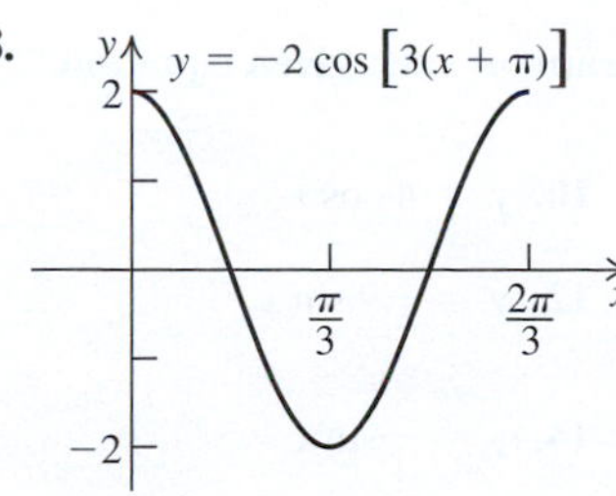

9. a. Period $= \pi$; phase shift $= -\dfrac{\pi}{8}$

b. Period $= 2$; phase shift $= \dfrac{1}{\pi}$

10. $y = 2\cos(3x - \pi)$

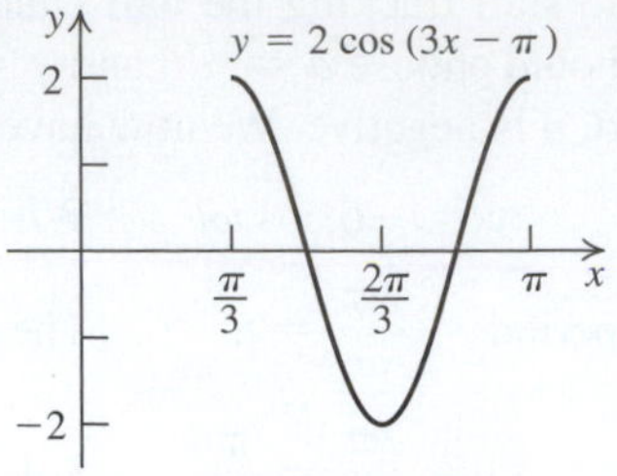

11. $y = 4\sin\left(3x - \frac{\pi}{3}\right) - 2$

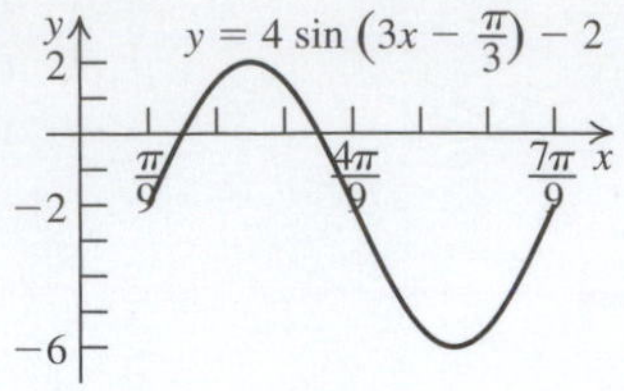

12. $y = 3.6\sin\left[\frac{\pi}{6}(x - 3)\right] + 12.2$

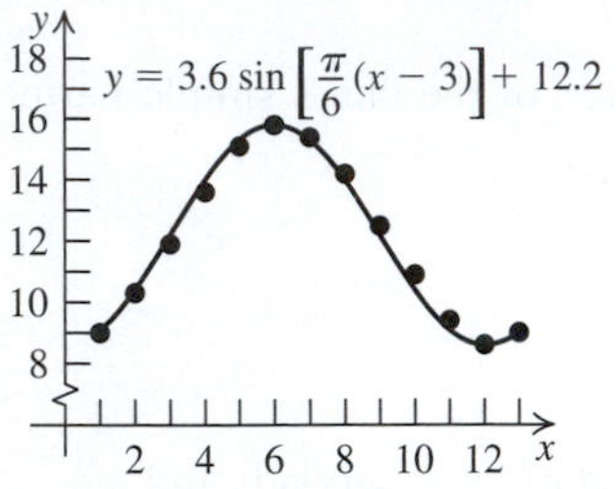

13. $y = -4\cos\dfrac{2\pi}{3}t$

SECTION 5.4 Exercises

Basic Concepts and Skills

1. The lowest point on the graph of $y = \cos x$, $0 \le x \le 2\pi$, occurs when $x =$ ______.

2. The highest point on the graph of $y = \sin x$, $0 \le x \le 2\pi$, occurs when $x =$ ______.

3. The range of the cosine function is ______.

4. The maximum value of $y = -2\sin x$ is ______.

5. True or False. The amplitude of $y = -3\cos x$ is -3.

6. True or False. The range of $y = \sin 3x$ is $[-3, 3]$.

7. True or False. The period of $y = 7\cos 2x$ is π.

8. True or False. The x-coordinate of the first key point for $y = \sin\left(x + \dfrac{\pi}{2}\right)$ is $-\dfrac{\pi}{2}$.

In Exercises 9–28, sketch the graph of each given equation over the interval $[-2\pi, 2\pi]$.

9. $y = 2\sin x$

10. $y = 4\cos x$

11. $y = -\dfrac{1}{2}\sin x$

12. $y = -2\sin x$

13. $y = \dfrac{3}{2}\cos x$

14. $y = \dfrac{5}{4}\sin x$

15. $y = \cos 2x$

16. $y = \sin 4x$

17. $y = \cos\dfrac{2}{3}x$

18. $y = \sin\dfrac{4}{3}x$

19. $y = \cos\left(x + \dfrac{\pi}{2}\right)$

20. $y = \sin\left(x + \dfrac{\pi}{4}\right)$

21. $y = \cos\left(x - \dfrac{\pi}{3}\right)$

22. $y = \sin(x - \pi)$

23. $y = 2\cos\left(x - \dfrac{\pi}{2}\right)$

24. $y = 2\sin\left(x + \dfrac{\pi}{3}\right)$

25. $y = \sin x + 1$

26. $y = \cos x - 2$

27. $y = -\cos x + 1$

28. $y = \sin x - 3$

In Exercises 29–36, find the amplitude, period, and phase shift of each given function.

29. $y = 5\cos(x - \pi)$

30. $y = 3\sin\left(x - \dfrac{\pi}{8}\right)$

31. $y = 7\cos\left[9\left(x + \dfrac{\pi}{6}\right)\right]$

32. $y = 11\sin\left[8\left(x + \dfrac{\pi}{3}\right)\right]$

33. $y = -6\cos\left[\dfrac{1}{2}(x + 2)\right]$

34. $y = -8\sin\left[\dfrac{1}{5}(x + 9)\right]$

35. $y = 0.9\sin\left[0.25\left(x - \dfrac{\pi}{4}\right)\right]$

36. $y = \sqrt{5}\cos[\pi(x + 1)]$

In Exercises 37–44, write an equation of the form $y = a\sin[b(x - c)]$ with the given amplitude, period, and phase shift.

	Amplitude	Period	Phase shift
37.	$\frac{1}{4}$	$\frac{\pi}{8}$	$\frac{\pi}{16}$
38.	2	$\frac{\pi}{6}$	$\frac{\pi}{10}$
39.	6	3π	$-\frac{\pi}{4}$
40.	$\frac{1}{2}$	6π	$-\frac{\pi}{2}$
41.	2.4	6	3
42.	0.8	10	2
43.	$\frac{\pi}{2}$	$\frac{5}{8}$	$-\frac{1}{4}$
44.	$\frac{2}{\pi}$	$\frac{2}{3}$	$-\frac{1}{6}$

In Exercises 45–52, graph each function over a one-period interval.

45. $y = -4\cos\left(x + \frac{\pi}{6}\right)$

46. $y = -3\sin\left(x - \frac{\pi}{6}\right)$

47. $y = \frac{5}{2}\sin\left[2\left(x - \frac{\pi}{4}\right)\right]$

48. $y = \frac{3}{2}\cos\left[2\left(x + \frac{\pi}{3}\right)\right]$

49. $y = -5\cos\left[4\left(x - \frac{\pi}{6}\right)\right]$

50. $y = -3\sin\left[4\left(x + \frac{\pi}{6}\right)\right]$

51. $y = \frac{1}{2}\sin\left[4\left(x + \frac{\pi}{4}\right)\right] + 2$

52. $y = -\frac{1}{2}\cos\left[2\left(x - \frac{\pi}{2}\right)\right] - 3$

In Exercises 53–62, write each function in the form $y = a\sin[b(x - c)]$ or $y = a\cos[b(x - c)]$. Find each period and phase shift.

53. $y = 4\cos\left(2x + \frac{\pi}{3}\right)$

54. $y = 5\sin\left(3x + \frac{\pi}{2}\right)$

55. $y = -\frac{3}{2}\sin(2x - \pi)$

56. $y = -2\cos\left(5x - \frac{\pi}{4}\right)$

57. $y = 3\cos\pi x$

58. $y = \sin\frac{\pi x}{3}$

59. $y = \frac{1}{2}\cos\left(\frac{\pi x}{4} + \frac{\pi}{4}\right)$

60. $y = -\sin\left(\frac{\pi x}{6} + \frac{\pi}{6}\right)$

61. $y = 2\sin(\pi x + 3)$

62. $y = -\cos\left(\pi x - \frac{1}{4}\right)$

Applying the Concepts

63. Biorhythm. Find the biorhythm states of Tina on April 1, 2011, assuming that she was born on July 22, 1990.

64. Biorhythm. Find the biorhythm states of Dave on August 12, 2012, assuming that he was born on March 31, 1991.

65. Pulse and blood pressure. Blood pressure is given by two numbers written as a fraction: $\frac{\text{systolic}}{\text{diastolic}}$. The *systolic* reading is the maximum pressure in an artery, and the *diastolic* reading is the minimum pressure in an artery. As the heart beats, the systolic measurement is taken; when the heart rests, the diastolic measurement is taken. A reading of $\frac{120}{80}$ is considered normal. Your pulse is the number of heartbeats per minute. Suppose Desmond's blood pressure after t minutes is given by

$$p(t) = 20\sin(140\pi t) + 122,$$

where $p(t)$ is the pressure in millimeters of mercury.

a. Find the period and explain how it relates to pulse.
b. Graph the function p over one period.
c. What is Desmond's blood pressure?

66. Electrical circuit voltage. The voltage $V(t)$ in an electrical circuit is given by $V(t) = 4.5\cos(160\pi t)$, where t is measured in seconds.

a. Find the amplitude and the period for $V(t)$.
b. Find the frequency of $V(t)$, that is, the number of cycles completed in one second.
c. Graph $V(t)$ over the interval $\left[0, \frac{1}{40}\right]$.

d. Use a graphing calculator to verify your graph.

67. Voltage of an electrical outlet. The voltage across the terminals of a typical electrical outlet is approximated by $V(t) = 156\sin(120\pi t)$, where t is measured in seconds.

a. Find the amplitude and the period for $V(t)$.
b. Find the frequency of $V(t)$, that is, the number of cycles completed in one second.
c. Graph $V(t)$ over the interval $\left[0, \frac{1}{30}\right]$.

d. Use a graphing calculator to verify your graph.

68. Oscillating ball. Suppose a ball attached to a spring is pulled down 5 inches and released and the resulting simple harmonic motion has a period of ten seconds. Write an equation of the ball's simple harmonic motion.

69. Kangaroo population. The kangaroo population in a certain region is given by the function

$$N(t) = 650 + 150\sin 2t,$$

where the time t is measured in years.

a. What is the largest number of kangaroos present in the region at any time?
b. What is the smallest number of kangaroos present in the region at any time?
c. How much time elapses between occurrences of the maximum and the minimum kangaroo population?

70. Lion and deer populations. The number of deer in a region is given by $D(t) = 450 \sin\left(\frac{3t}{5}\right) + 1200$, where t is in years and the number of lions in that same region is given by $L(t) = 225 \sin\left[\frac{2}{5}(t-2)\right] + 500$, where t is in years.

a. Graph both functions on the same coordinate plane for $0 \le t \le 10$.

b. What can you say about the relationship between these populations based on the graphs obtained in (a)?

c. Use a graphing calculator to verify your graph.

71. Daylight hours in London. The table gives the average number of daylight hours in London each month, where $x = 1$ represents January. Let y represent the number of daylight hours in London in month x. Find a function of the form $y = a \sin[b(x-c)] + d$ that models the hours of daylight throughout the year.

Jan	Feb	Mar	Apr	May	June
8.4	10	11.9	13.9	15.6	16.6
July	Aug	Sept	Oct	Nov	Dec
16.1	14.6	12.6	10.7	8.9	7.9

72. Daylight hours in Washington, D.C. The table gives the average number of daylight hours in Washington, D.C., each month, where $x = 1$ represents January. Let y represent the number of daylight hours in Washington, D.C., in month x. Find a function of the form $y = a \sin[b(x-c)] + d$ that models the hours of daylight throughout the year.

Jan	Feb	Mar	Apr	May	June
9.8	10.8	12	13.2	14.3	14.8
July	Aug	Sept	Oct	Nov	Dec
14.6	13.6	12.4	11.2	10.1	9.5

In Exercises 73 and 74, use the table of average monthly temperatures (where Jan = 1, ..., Dec = 12) to find a function of the form $y = a \sin[b(x-c)] + d$ that models the temperature in the given city. The four seasons of the year match remarkably well with the quarter periods of the sine function. Because the temperatures repeat each year, the period is 12 months; so $b = \frac{2\pi}{12} = \frac{\pi}{6}$. The amplitude $a = \frac{1}{2}$(high temperature − low temperature), and the vertical shift $d = \frac{1}{2}$(high temperature + low temperature). Because the maximum value a of an unshifted sine curve with period 12 occurs at $\frac{1}{4}(12) = 3$ units from the start of the cycle, you can find c by solving the equation c = hottest month −3.

73. Average temperatures. The monthly average temperatures for Augusta, Maine, are given in the table.

Month	1	2	3	4	5	6	7	8	9	10	11	12
°F	19	22	31	43	54	63	70	67	59	48	37	24

Source: www.weatherbase.com

a. Find a function of the form $y = a \sin[b(x-c)] + d$ that models these temperatures.

b. Graph the ordered pairs and the function on the same coordinate system.

c. Compute the function values for January, April, July, and October and compare them with the table values for these months.

d. Use a graphing calculator to verify your graph.

74. Average temperatures. The monthly average temperatures for Nashville, Tennessee, are given in the table.

Month	1	2	3	4	5	6	7	8	9	10	11	12
°F	40	45	53	63	71	79	83	81	74	63	52	44

Source: www.weatherbase.com

a. Find a function of the form $y = a \sin[b(x-c)] + d$ that models these temperatures.

b. Graph the ordered pairs and the function on the same coordinate system.

c. Compute the function values for January, April, July, and October and compare them to the table values for these months.

d. Use a graphing calculator to verify your graph.

Beyond the Basics

In Exercises 75–84, graph each equation over the interval $[0, 2\pi]$.

75. $y = 3 \sin\left(-x + \frac{\pi}{4}\right)$

76. $y = -2 \sin\left(-2x + \frac{\pi}{3}\right)$

77. $y = 2 \cos\left(-3x + \frac{\pi}{2}\right) + 2$

78. $y = -3 \cos\left(-\frac{x}{\pi} + \frac{1}{2}\right) + 3$

79. $y = |\sin x|$

80. $y = |\cos x|$

81. $y = x \sin x$

82. $y = x \cos x$

83. $y = \sqrt{x} \sin x$

84. $y = |x \sin x|$

85. Graph the equations $y = x^2$, $y = -x^2$, and $y = x^2 \sin x$ on the same coordinate plane.

86. We know that $\sin(x + 2\pi) = \sin x$ for all real numbers x. Consequently, if 2π is not the period for the sine function, then $\sin(x + p) = \sin x$ for some real number p, with $0 < p < 2\pi$, and all real numbers x.

a. By evaluating $\sin(x + p) = \sin x$ for $x = 0$, show that if the period is not 2π, then $p = \pi$.

b. Find a value of x for which $\sin(x + \pi) \neq \sin x$ and conclude that the period of the sine function is 2π.

In Exercises 87–94, write an equation for each sinusoidal graph.

87.

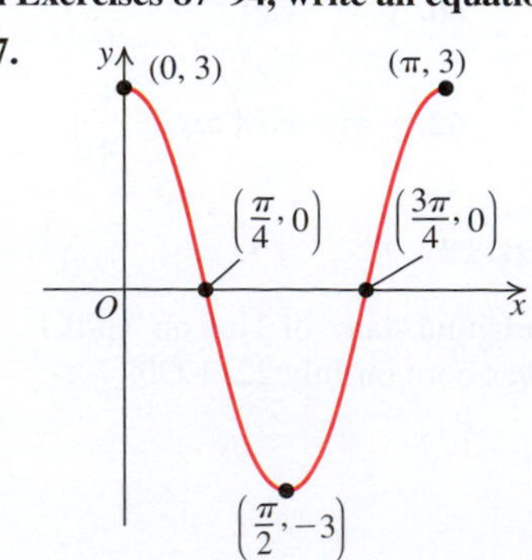

88.

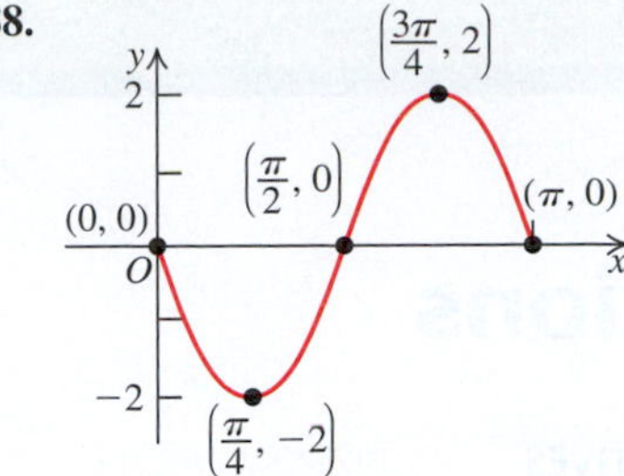

89.

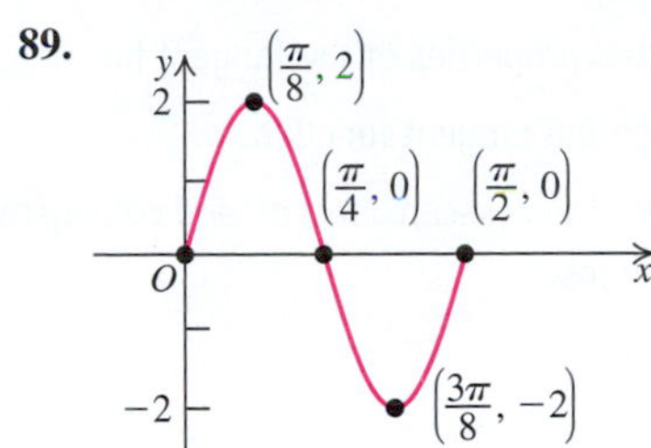

90.

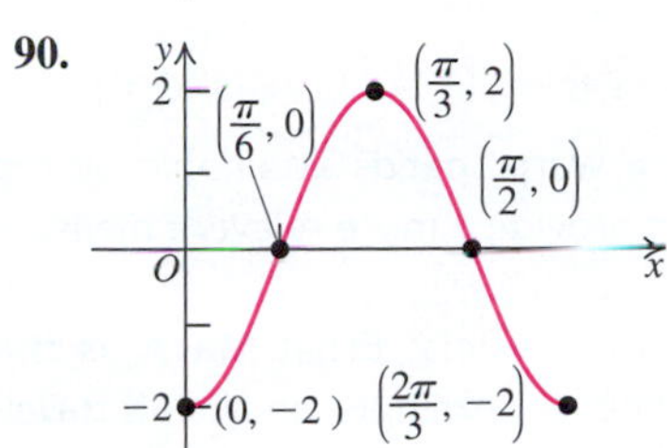

91.

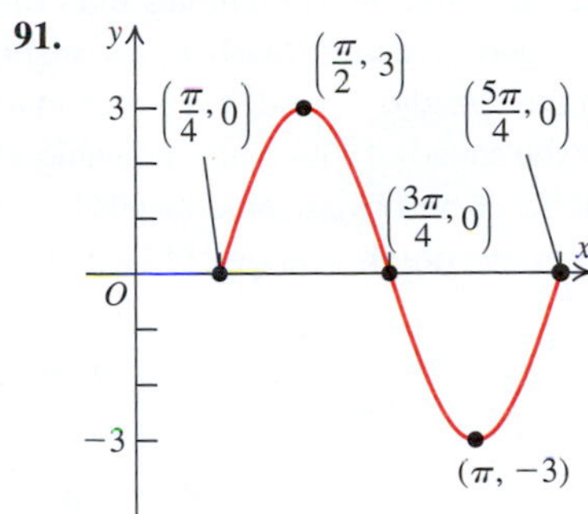

92.

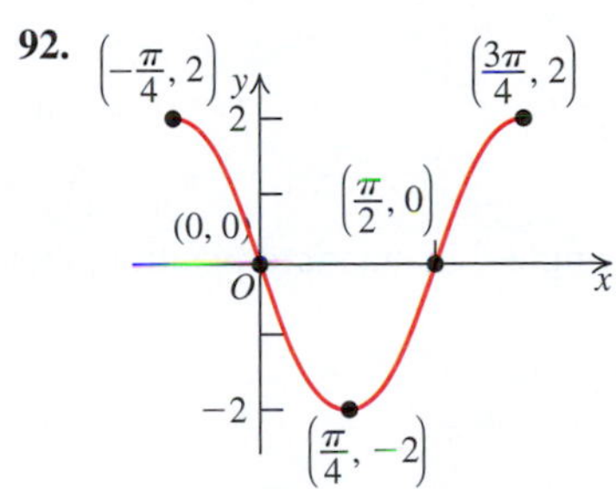

93.

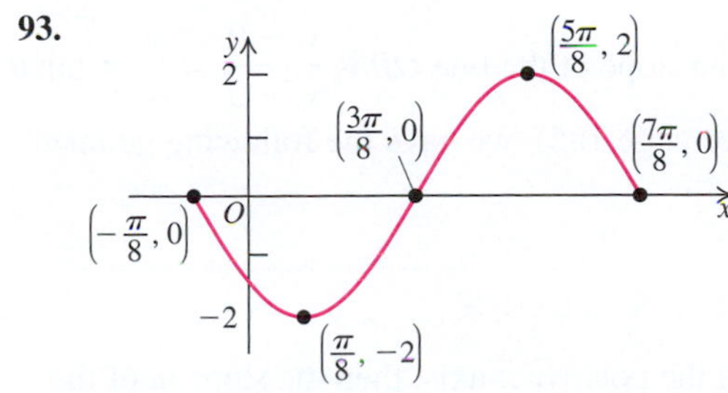

94.

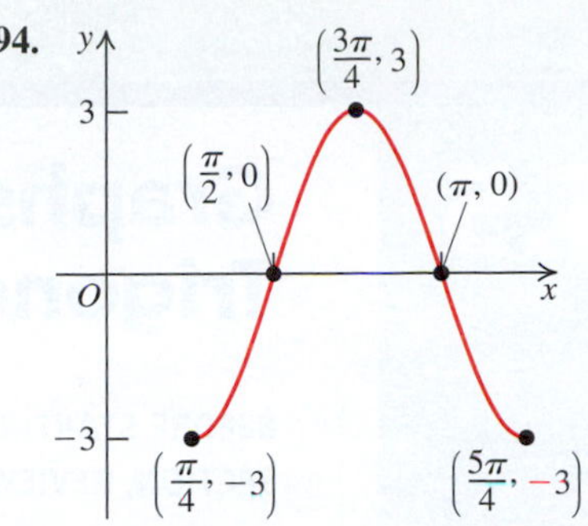

Critical Thinking/Discussion/Writing

95. Suppose a function of the form $y = a\sin[b(x - c)] + d$, with $a > 0$, has period $= 20$.
   a. In a one-period interval, how many units are between the value of x at which the maximum value is attained and the value of x at which the minimum value is attained?
   b. In a one-period interval, how many units are between the value of x at which y has value d and the value of x at which the maximum value is attained?

96. Suppose the minimum value of a function of the form $y = a\cos[b(x - c)] + d$, with $a > 0$, occurs at a value of x that is 5 units from the value of x at which the function has the maximum value. What is the period of the function?

Maintaining Skills

In Exercises 97–100, find the equation of each line in slope–intercept form.

97. Passing through the point $(2, -3)$ with slope 5

98. Passing through the points $(-3, 2)$ and $(1, 4)$

99. Passing through the point $(-3, 4)$ and parallel to the line $2x + 4y = 5$

100. Passing through the point $(3, -1)$ and perpendicular to the line $3x + 6y = 7$

In Exercises 101–104, find the x-intercepts and the vertical asymptotes of the graph of each function.

101. $f(x) = \dfrac{(x+1)(x-2)}{(x+2)(x-3)}$

102. $f(x) = \dfrac{x^2 - 3x - 10}{x^2 - 2x - 3}$

103. $f(x) = \dfrac{x^2 + 6x + 8}{x^2 + 1}$

104. $f(x) = \dfrac{1}{x^2 - 5x - 14}$

In Exercises 105–112, let $f(x) = x^3 - 2x$ and $g(x) = 3x^2 + 1$. State whether each function is even, odd, or neither.

105. $f(x)$

106. $g(x)$

107. $f(x) + g(x)$

108. $f(x)g(x)$

109. $\dfrac{f(x)}{g(x)}$

110. $\dfrac{g(x)}{f(x)}$

111. $\dfrac{1}{f(x)}$

112. $\dfrac{1}{g(x)}$

SECTION 5.5

Graphs of the Other Trigonometric Functions

BEFORE STARTING THIS SECTION, REVIEW

1 Vertical asymptotes (Section 3.6, page 357)
2 Transformations of functions (Section 2.7, page 251)

OBJECTIVES

1 Discuss properties of the tangent function.
2 Graph the tangent function.
3 Graph the cosecant, secant, and cotangent functions.

◆ The U.S.S. Enterprise and Mach Numbers

Star Trek fans probably know quite a bit about the warp speeds attainable by the Starfleet's USS *Enterprise*. Near Earth, Mach numbers provide a more relevant measure of speed.

The Mach number (named after the Austrian physicist Ernst Mach) is the ratio of the speed of the aircraft to the speed of sound. For example, an aircraft traveling at Mach 2 is traveling at twice the speed of sound. An airplane flying at less than the speed of sound is traveling at *subsonic* speed; at the speed of sound (Mach 1), the plane is traveling at *transonic* speed; at speeds greater than the speed of sound, it is traveling at *supersonic* speeds; and at speeds greater than 5 times the speed of sound, it is traveling at *hypersonic* speeds. In Example 5, we see how at supersonic and hypersonic speeds Mach numbers are given as values of the cosecant function and we graph a range of Mach numbers that are attainable by various military aircraft.

1 Discuss properties of the tangent function.

Tangent Function

Recall that if $P(x, y)$ is a point on the terminal side of an angle θ in standard position, then (see Figure 5.67)

$$\tan\theta = \frac{y}{x}.$$

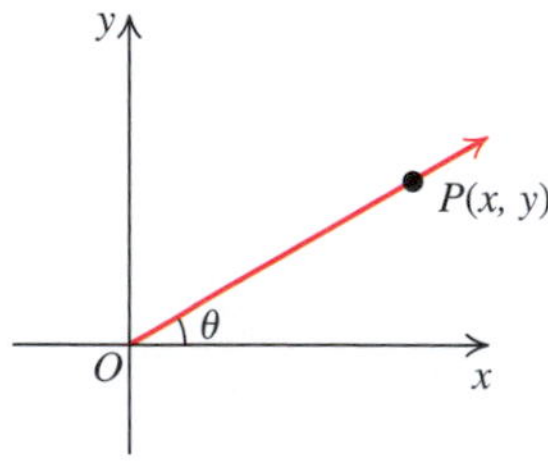

FIGURE 5.67

Because the point $(0, 0)$ is on the line OP, the slope of the line OP is $\frac{y-0}{x-0} = \frac{y}{x} = \tan\theta$.

Because parallel lines have equal slopes (see page 185), we have the following geometric interpretation of the tangent of an angle:

TANGENT AS SLOPE

If a nonvertical line makes an angle θ with the positive x-axis, then the slope m of the line is given by $m = \tan\theta$.

EXAMPLE 1 Equation of a Line

A line ℓ makes an angle $\theta = 60°$ with positive x-axis and passes through the point $P = (7, 5)$. Find the equation of ℓ in slope–intercept form.

Solution

The slope of the line ℓ is $m = \tan 60° = \sqrt{3}$ (page 491). Then

$$y - 5 = \sqrt{3}(x - 7) \qquad \text{Point–slope form: } y - y_1 = m(x - x_1)$$
$$y - 5 = \sqrt{3}x - 7\sqrt{3} \qquad \text{Distributive property}$$
$$y = \sqrt{3}x + (5 - 7\sqrt{3}) \qquad \text{Slope–intercept form: } y = mx + b$$

Practice Problem 1 Repeat Example 1 with $\theta = 30°$ and $P = (2, 3)$.

Domain of tan *x* From the quotient identity (page 488), we have

$$\tan x = \frac{\sin x}{\cos x}.$$

The function $y = \tan x$ is defined for all real numbers x except where $\cos x = 0$. Because $\cos x = 0$ for $x = \frac{\pi}{2} + n\pi$ for any integer n, the domain of $\tan x$ is all real numbers x with $x \neq \frac{\pi}{2} + n\pi$ or $x \neq (2n + 1)\frac{\pi}{2}$. In other words, $\tan x$ is undefined at $x = \pm\frac{\pi}{2}, \pm\frac{3\pi}{2}, \pm\frac{5\pi}{2}, \ldots$. The graph of $y = \tan x$ has a vertical asymptote at $x = (2n + 1)\frac{\pi}{2}$ for every integer n.

Range of tan *x* We note that the tangent of an angle θ in standard position is the slope of the line containing the corresponding terminal side. Because every real number is the slope of some line through the origin, every number is the tangent of some angle in standard position. This means that the range of the tangent function is $(-\infty, \infty)$.

Zeros of tan *x* From $\tan x = \frac{\sin x}{\cos x}$, we have $\tan x = 0$ whenever $\sin x = 0$. Because $\sin x = 0$ for $x = n\pi$ for any integer n, the zeros of $y = \tan x$ are at $x = n\pi$.

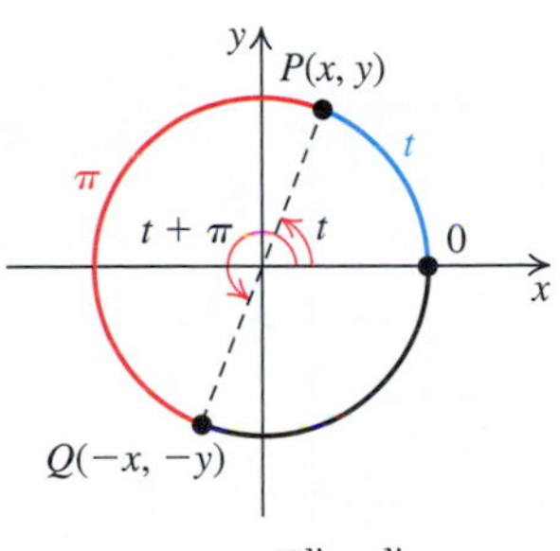

$\tan(t + \pi) = \frac{-y}{-x} = \frac{y}{x} = \tan t$

FIGURE 5.68 $\tan(t + \pi) = \tan t$

Period of tan *x* The tangent function has period π. Figure 5.68 illustrates that $\tan(t + \pi) = \tan t$. Therefore, the values of the tangent function repeat every π units.

Even–Odd Property

We have
$$\tan(-x) = \frac{\sin(-x)}{\cos(-x)} \qquad \text{Quotient identity}$$
$$= \frac{-\sin x}{\cos x} \qquad \text{Sine is odd, and cosine is even.}$$
$$= -\tan x \qquad \text{Quotient identity}$$

So $\tan(-x) = -\tan x$. Therefore, $y = \tan x$ is an odd function and its graph is symmetric about the origin.

2 Graph the tangent function.

Graph of $y = \tan x$

We first summarize the main facts about $y = \tan x$.

MAIN FACTS ABOUT THE TANGENT FUNCTION

1. *Domain*: All real numbers except odd multiples of $\frac{\pi}{2}$
2. *Range*: $(-\infty, \infty)$
3. *Zeros or x-intercepts*: All integer multiples of π or $x = n\pi$
4. *Period*: π
5. *Vertical Asymptotes*: At $x =$ odd integer multiples of $\frac{\pi}{2}$, or $x = (2n + 1)\frac{\pi}{2}$
6. *Symmetry*: Odd function, so the graph is symmetric about the origin.

TECHNOLOGY CONNECTION

To graph the tangent, cotangent, secant, or cosecant function on a graphing calculator, use *dot* mode and appropriate window settings.

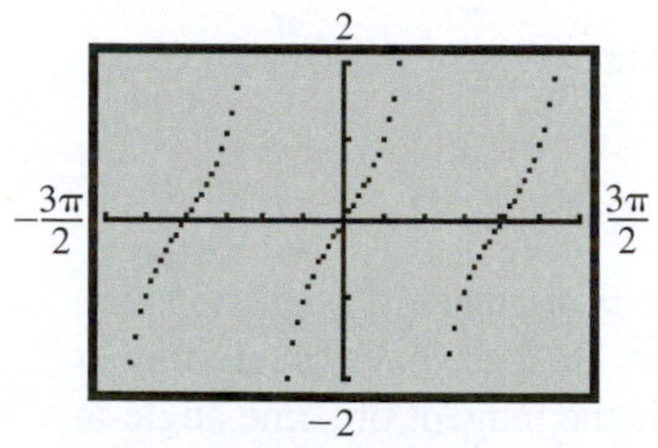

$Y_1 = \tan x$, *dot* mode

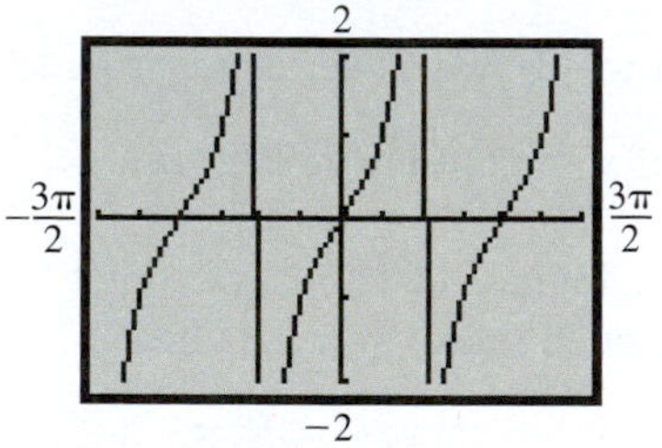

$Y_1 = \tan x$, *connected* mode

To graph $y = \tan x$, we use the approximate values of $\tan x$ for x in the interval $\left[0, \frac{\pi}{2}\right)$. See Table 5.7.

TABLE 5.7

x	0	$\frac{\pi}{6}$	$\frac{\pi}{4}$	$\frac{\pi}{3}$	$\frac{7\pi}{18}$	$\frac{4\pi}{9}$	$\frac{17\pi}{36}$
$y = \tan x$	0	0.6	1	1.7	2.7	5.7	11.4

As x approaches $\frac{\pi}{2}$, the values of $\tan x$ continue to increase; in fact, they increase indefinitely. The decimal value of $\frac{\pi}{2}$ is approximately 1.5707963. When $x = \frac{89\pi}{180} \approx 1.55$, $\tan x > 57$; when $x = 1.57$, $\tan x > 1255$; and when $x = 1.5707$, $\tan x > 10{,}000$. The graph of $y = \tan x$ has a vertical asymptote at $x = \frac{\pi}{2}$. See Figure 5.69(a).

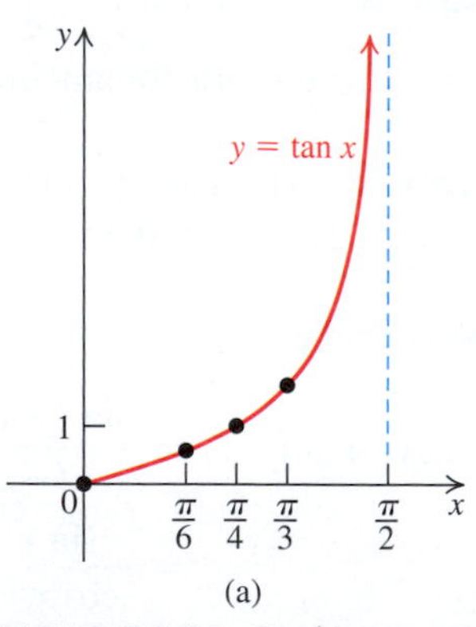

(a)

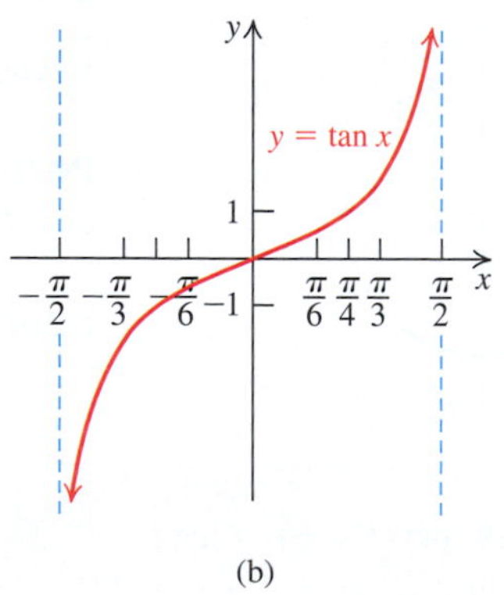

(b)

FIGURE 5.69 Graphing $y = \tan x$

We can extend the graph of $y = \tan x$ to the interval $\left(-\frac{\pi}{2}, \frac{\pi}{2}\right)$ by using the fact that the graph is symmetric about the origin. See Figure 5.69(b).

Because the period of the tangent function is π and we have graphed $y = \tan x$ over an interval of length π, we can draw the complete graph of $y = \tan x$ by repeating the graph in Figure 5.69(b) indefinitely to the left and right, over intervals of length π. Figure 5.70 shows three cycles of the graph of $y = \tan x$.

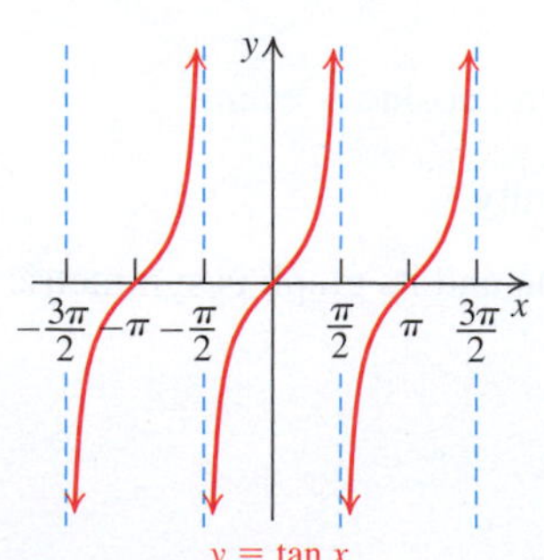

FIGURE 5.70

Graphing $y = a\tan[b(x - c)]$ The procedure for graphing $y = a\tan[b(x - c)]$ is based on the essential features of the graph of $y = \tan x$.

PROCEDURE IN ACTION

EXAMPLE 2 Graphing $y = a \tan(bx - k)$

OBJECTIVE

Graph a function of the form $y = a\tan(bx - k)$, where $b > 0$, by finding the period and phase shift.

Step 1 If necessary, rewrite the equation in the form $y = a\tan[b(x - c)]$, $c = \frac{k}{b}$.

Find the following:

vertical stretch factor $= |a|$

period $= \frac{\pi}{b}$

phase shift $= c$

Step 2 Locate two adjacent vertical asymptotes by solving the following equations for x:

$$b(x - c) = -\frac{\pi}{2} \quad \text{and} \quad b(x - c) = \frac{\pi}{2}.$$

Step 3 Divide the interval on the x-axis between the two vertical asymptotes from Step 2 into four equal parts, each of length $\frac{1}{4}\left(\frac{\pi}{b}\right)$.

Step 4 Evaluate the function at the three x values found in Step 3 that are the division points of the interval.

Step 5 Sketch the vertical asymptotes using the values found in Step 2. Connect the points in Step 4 with a smooth curve in the standard shape of a cycle for the tangent function. Repeat the graph to the left and right over intervals of length $\frac{\pi}{b}$.

EXAMPLE

Graph $y = 3\tan\left(2x - \frac{\pi}{2}\right)$.

1. $y = 3\tan\left[2\left(x - \frac{\pi}{4}\right)\right]$ Rewrite.

$a = 3 \qquad b = 2 \qquad c = \frac{\pi}{4}$

vertical stretch factor $= |a| = |3| = 3$

period $= \frac{\pi}{b} = \frac{\pi}{2}$ phase shift $= c = \frac{\pi}{4}$

2.

$$2\left(x - \frac{\pi}{4}\right) = -\frac{\pi}{2} \qquad 2\left(x - \frac{\pi}{4}\right) = \frac{\pi}{2}$$

$$x - \frac{\pi}{4} = -\frac{\pi}{4} \qquad x - \frac{\pi}{4} = \frac{\pi}{4}$$

$$x = -\frac{\pi}{4} + \frac{\pi}{4} \qquad x = \frac{\pi}{4} + \frac{\pi}{4}$$

$$x = 0 \qquad x = \frac{\pi}{2}$$

3. The interval $\left(0, \frac{\pi}{2}\right)$ has length $\frac{\pi}{2}$, and $\frac{1}{4}\left(\frac{\pi}{2}\right) = \frac{\pi}{8}$.

The division points of the interval $\left(0, \frac{\pi}{2}\right)$ are

$0 + \frac{\pi}{8} = \frac{\pi}{8}$, $0 + 2\left(\frac{\pi}{8}\right) = \frac{\pi}{4}$, and $0 + 3\left(\frac{\pi}{8}\right) = \frac{3\pi}{8}$.

4.

x	$y = 3\tan[2(x - \frac{\pi}{4})]$
$\frac{\pi}{8}$	-3
$\frac{\pi}{4}$	0
$\frac{3\pi}{8}$	3

5.

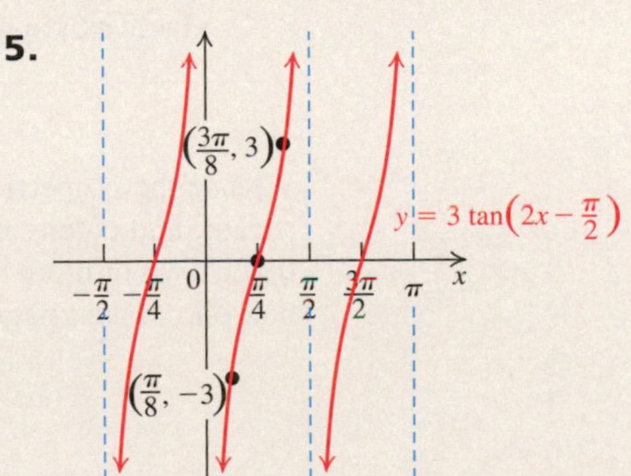

Practice Problem 2 Graph $y = -3\tan\left(x + \frac{\pi}{4}\right)$.

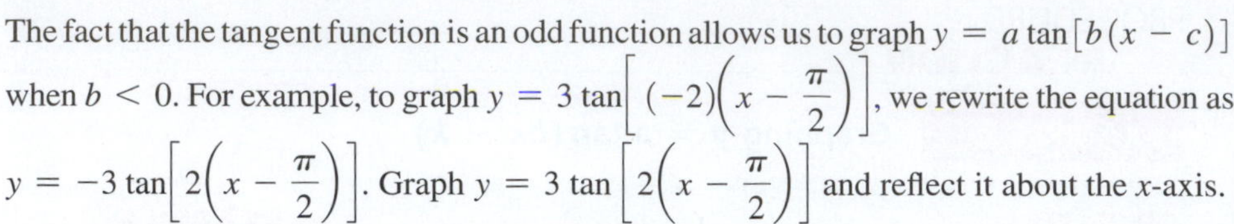

The fact that the tangent function is an odd function allows us to graph $y = a\tan[b(x - c)]$ when $b < 0$. For example, to graph $y = 3\tan\left[(-2)\left(x - \frac{\pi}{2}\right)\right]$, we rewrite the equation as $y = -3\tan\left[2\left(x - \frac{\pi}{2}\right)\right]$. Graph $y = 3\tan\left[2\left(x - \frac{\pi}{2}\right)\right]$ and reflect it about the x-axis.

3 Graph the cosecant, secant, and cotangent functions.

Graphs of the Reciprocal Functions

The cosecant, secant, and cotangent functions are reciprocals of the sine, cosine, and tangent functions, respectively. We first consider some relationships between the properties of any pair of these reciprocal functions. We use these relationships to sketch the graph of the reciprocal function of a trigonometric function $g(x)$.

THE GRAPH OF THE RECIPROCAL OF A TRIGONOMETRIC FUNCTION

Let $f(x)$ be the reciprocal of $g(x)$: $f(x) = \frac{1}{g(x)}$, where g is any trigonometric function.

- **Periodicity** If $g(x)$ has period p, then $f(x)$ also has period p.
- **Zeros** If $g(c) = 0$, then $f(c)$ is undefined. So if c is an x-intercept of the graph of g, then the line $x = c$ is a vertical asymptote of the graph of f. Conversely, if $g(d)$ is undefined, then $f(d) = 0$.
- **Even–Odd**
  a. If $g(x)$ is odd, then $f(x)$ is odd.
  b. If $g(x)$ is even, then $f(x)$ is even.
- **Special Values**
  a. If $g(x_1) = 1$, then $f(x_1) = 1$. Both graphs pass through the point $(x_1, 1)$.
  b. If $g(x_2) = -1$, then $f(x_2) = -1$. Both graphs pass through the point $(x_2, -1)$.
- **Sign**
  a. If $g(x) > 0$ on an interval (a, b), then $f(x) > 0$ on the interval (a, b). Both graphs are above the x-axis on the interval (a, b).
  b. If $g(x) < 0$ on an interval (c, d), then $f(x) < 0$ on the interval (c, d). Both graphs are below the x-axis on the interval (c, d).
- **Increasing–Decreasing**
  a. If $g(x)$ is increasing on an interval (a, b), then $f(x)$ is decreasing on the interval (a, b).
  b. If $g(x)$ is decreasing on an interval (c, d), then $f(x)$ is increasing on the interval (c, d).
- **Magnitude**
  a. If $|g(x)|$ is small, then $|f(x)|$ is large.
  b. If $|g(x)|$ is large, then $|f(x)|$ is small. If c is in the domain of f and $|g(x)|$ is large as x approaches c, then $f(c) = 0$.

We use the properties of the reciprocal functions to sketch the graphs of the cosecant, secant, and cotangent functions from the graphs of the sine, cosine, and tangent functions. Two or more cycles of these graphs along with their reciprocals are shown in the box on the next page.

PROPERTIES OF THE COSECANT, SECANT, AND COTANGENT FUNCTIONS

	$y = \csc x = \dfrac{1}{\sin x}$	$y = \sec x = \dfrac{1}{\cos x}$	$y = \cot x = \dfrac{1}{\tan x}$
For any integer n:			
Domain	All real numbers $x \neq n\pi$	All real numbers $x \neq \dfrac{\pi}{2} + n\pi$	All real numbers $x \neq n\pi$
Range	$(-\infty, -1] \cup [1, \infty)$	$(-\infty, -1] \cup [1, \infty)$	$(-\infty, \infty)$
Period	2π	2π	π
***x*-intercepts**	No x-intercepts	No x-intercepts	$x = \dfrac{\pi}{2} + n\pi$
Even–Odd	Odd	Even	Odd
Vertical Asymptotes	$x = n\pi$	$x = \dfrac{\pi}{2} + n\pi$	$x = n\pi$
Graph	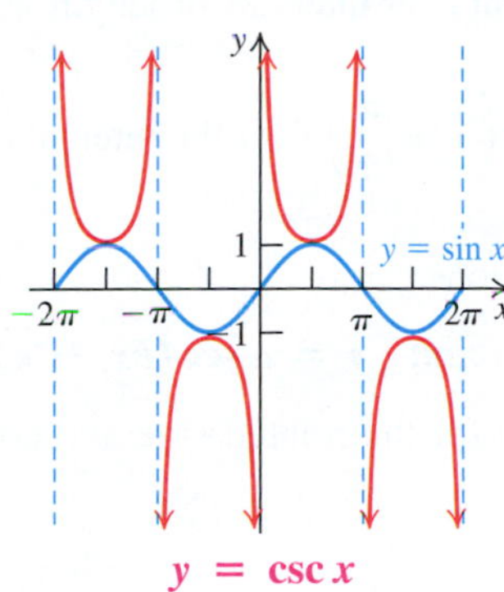 $y = \csc x$	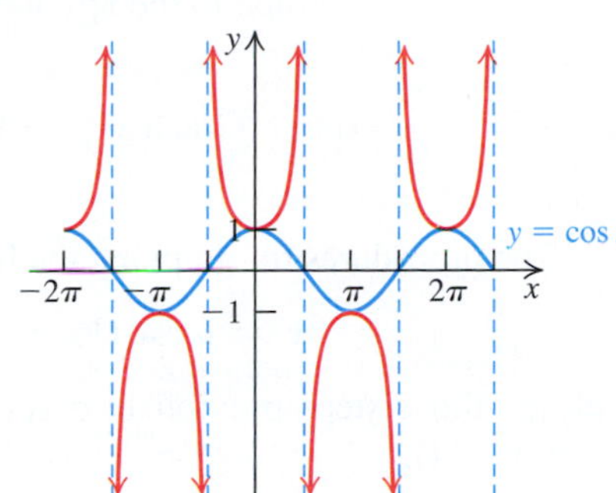 $y = \sec x$	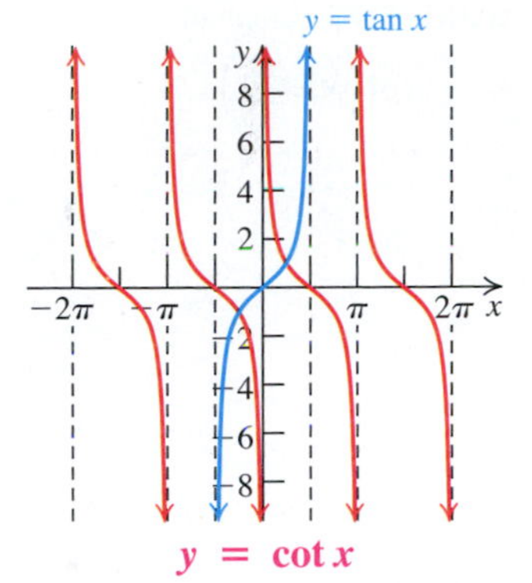 $y = \cot x$

To sketch the graph of $y = a\cot(bx - k)$, we use the steps that are similar to those used in Example 2.

EXAMPLE 3 Graphing $y = a\cot(bx - k)$

Graph $y = -4\cot\left(x - \dfrac{\pi}{2}\right)$ over the interval $[-\pi, 2\pi]$.

Solution

Step 1 For $y = -4\cot\left(x - \dfrac{\pi}{2}\right)$, we have $a = -4$, $b = 1$, and $c = \dfrac{k}{b} = \dfrac{\pi}{2}$. Therefore, the vertical stretch factor $= |-4| = 4$, the period $= \dfrac{\pi}{1} = \pi$, and the phase shift $= \dfrac{\pi}{2}$.

Step 2 Locate two adjacent asymptotes for a cotangent function. Solve the equations:

$$x - \frac{\pi}{2} = 0 \quad \text{and} \quad x - \frac{\pi}{2} = \pi.$$

$$x = \frac{\pi}{2} \qquad\qquad x = \frac{3\pi}{2}$$

Step 3 The interval $\left(\frac{\pi}{2}, \frac{3\pi}{2}\right)$ has length π, the period of the cotangent function. We have $\frac{1}{4}(\pi) = \frac{\pi}{4}$; so the division points of the interval $\left(\frac{\pi}{2}, \frac{3\pi}{2}\right)$ are $\frac{\pi}{2} + \frac{\pi}{4} = \frac{3\pi}{4}, \frac{\pi}{2} + 2\left(\frac{\pi}{4}\right) = \pi$, and $\frac{\pi}{2} + 3\left(\frac{\pi}{4}\right) = \frac{5\pi}{4}$.

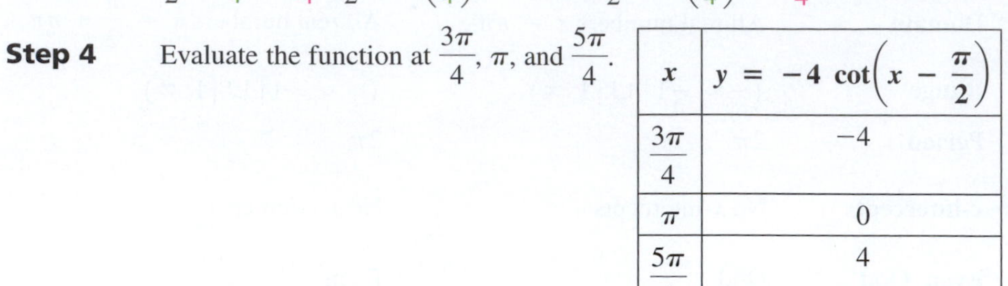

Step 4 Evaluate the function at $\frac{3\pi}{4}$, π, and $\frac{5\pi}{4}$.

x	$y = -4 \cot\left(x - \frac{\pi}{2}\right)$
$\frac{3\pi}{4}$	-4
π	0
$\frac{5\pi}{4}$	4

FIGURE 5.71 Graph of $y = -4 \cot\left(x - \frac{\pi}{2}\right)$

Step 5 Sketch the vertical asymptotes at $x = \frac{\pi}{2}$ and $x = \frac{3\pi}{2}$. Draw the cycle for the cotangent function through the three graph points found in Step 4. Repeat the graph to the left and right over intervals of length π. See Figure 5.71.

Practice Problem 3 Graph $y = -3 \cot\left(x + \frac{\pi}{4}\right)$ over the interval $\left(-\frac{3\pi}{4}, \frac{5\pi}{4}\right)$.

The procedures for graphing the functions

$$y = a \csc(bx - k) \quad \text{and} \quad y = a \sec(bx - k)$$

rely on those steps previously described for the related sine and cosine functions (see Section 5.4).

EXAMPLE 4 Graphing $y = a \csc(bx - k)$

Graph $y = 3 \csc(2x)$ over a two-period interval.

Solution

Because $\csc 2x = \frac{1}{\sin 2x}$, $3 \csc 2x = 3 \sin 2x$ when $\sin 2x = \pm 1$. So we first graph the function $y = 3 \sin 2x$.

We follow the steps for graphing a function of this type given in Section 5.4.

Step 1 $y = 3 \sin 2x$

amplitude $= 3$, period $= \frac{2\pi}{2} = \pi$; no phase shift because

$c = \frac{k}{b} = \frac{0}{2} = 0$

Step 2 The starting point for the cycle is $x = 0$. One cycle is graphed over $[0, \pi]$.

Step 3 $\frac{1}{4}$ (period) $= \frac{1}{4}\pi = \frac{\pi}{4}$

The x-coordinates of the five key points are

$$x = 0, \quad x = \frac{\pi}{4}, \quad x = 2\left(\frac{\pi}{4}\right) = \frac{\pi}{2}, \quad x = 3\left(\frac{\pi}{4}\right) = \frac{3\pi}{4}, \quad x = 4\left(\frac{\pi}{4}\right) = \pi.$$

Step 4 Sketch the graph of $y = 3 \sin 2x$ through the key points $(0, 0)$, $\left(\frac{\pi}{4}, 3\right)$, $\left(\frac{\pi}{2}, 0\right)$, $\left(\frac{3\pi}{4}, -3\right)$, and $(\pi, 0)$. See the blue graph in Figure 5.72.

FIGURE 5.72 Graph of $y = 3 \csc 2x$

Step 5 Use the graph from Step 4 to extend the graph to the interval $\left(-\frac{\pi}{2}, \frac{3\pi}{2}\right)$.

Now use the technique that generated the graph of $y = \csc x$ from that of $y = \sin x$ to graph $y = 3 \csc 2x$. Start at the common graph points $\left(\frac{\pi}{4}, 3\right)$ and $\left(\frac{3\pi}{4}, -3\right)$; then use the reciprocal relationship $y = 3 \csc 2x = 3\left(\frac{1}{\sin 2x}\right)$. See the red graph in Figure 5.72.

Practice Problem 4 Graph $y = 2 \sec 3x$ over a two-period interval.

EXAMPLE 5 Graphing a Range of Mach Numbers

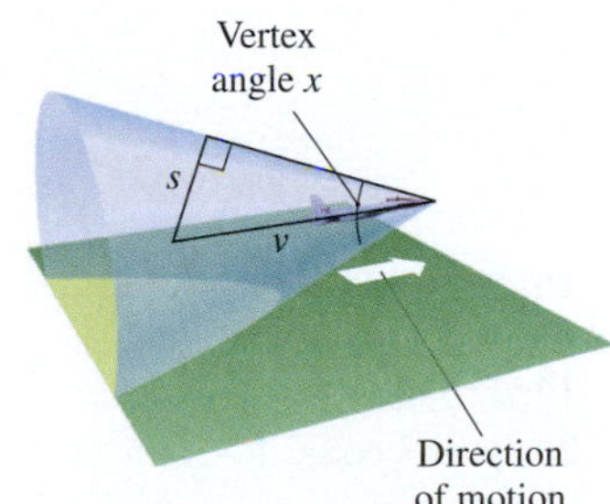

FIGURE 5.73 Sonic cone

When a plane travels at supersonic and hypersonic speeds, small disturbances in the atmosphere are transmitted downstream within a cone. The cone intersects the ground. Figure 5.73 shows the edge of the cone's intersection with the ground. The sound waves strike the edge of the cone at a right angle. The speed of the sound wave is represented by leg s of the right triangle in Figure 5.73. The plane is moving at speed v, which is represented by the hypotenuse of the right triangle in this figure.

The Mach number, M, is given by

$$M = M(x) = \frac{\text{speed of the aircraft}}{\text{speed of sound}} = \frac{v}{s} = \csc\left(\frac{x}{2}\right),$$

where x is the angle at the vertex of the cone. Graph the Mach number function, $M(x)$, as the angle at the vertex of the cone varies. What is the range of Mach numbers associated with the interval $\left[\frac{\pi}{4}, \pi\right)$?

Solution

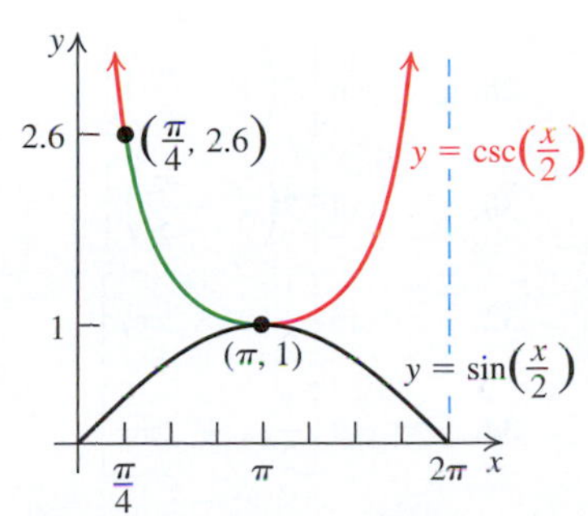

FIGURE 5.74 Mach numbers

Because $\csc \frac{x}{2} = \frac{1}{\sin \frac{x}{2}}$, first graph $y = \sin \frac{x}{2}$. For convenience, we have sketched the graph of $y = \sin \frac{x}{2}$ over the interval $[0, 2\pi]$ in Figure 5.74. The graph of $y = \csc \frac{x}{2}$ is sketched over the interval $(0, 2\pi)$ using the reciprocal relationship between the sine graph and the cosecant graph.

$$\text{For } x = \frac{\pi}{4}, \quad y = \csc \frac{x}{2} = \csc \frac{\pi}{8} \approx 2.6.$$

$$\text{For } x = \pi, \quad y = \csc \frac{x}{2} = \csc \frac{\pi}{2} = 1.$$

The range of Mach numbers associated with the interval $\left[\frac{\pi}{4}, \pi\right)$ is $(1, 2.6]$.

Practice Problem 5 In Example 5, what is the range of Mach numbers associated with the interval $\left[\frac{\pi}{8}, \frac{\pi}{4}\right]$?

Answers to Practice Problems

1. $y = \frac{\sqrt{3}}{3}x + 3 - \frac{2\sqrt{3}}{3}$

2.

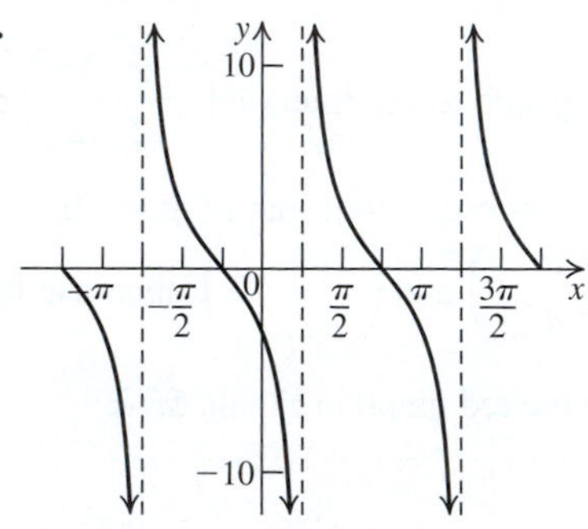

3.

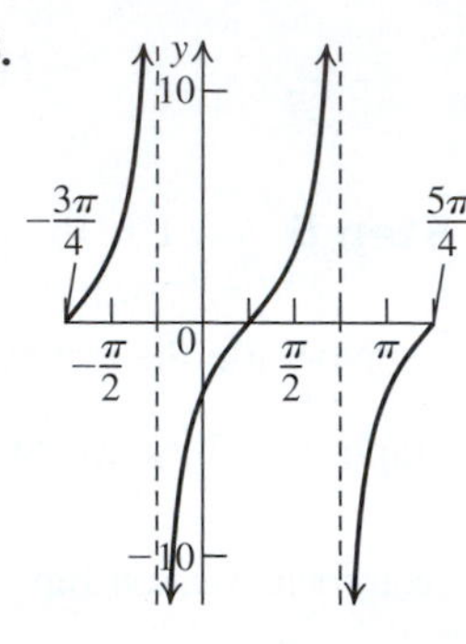

4.

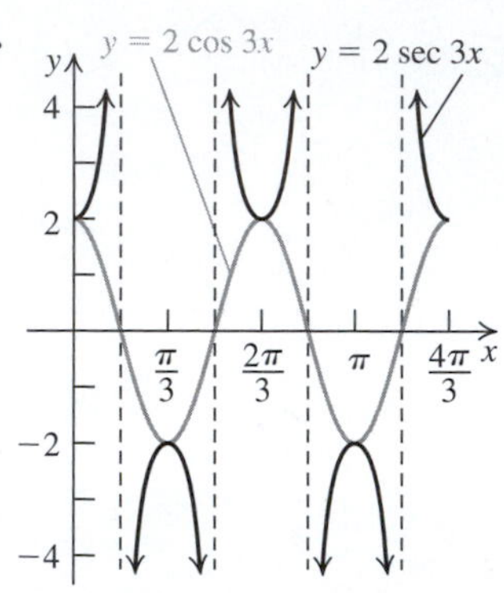

5. $\approx [2.6, 5.1]$

SECTION 5.5 Exercises

Basic Concepts and Skills

1. The tangent function has period ____________.
2. The value of x with $0 \le x \le \pi$ for which the tangent function is undefined is ____________.
3. If $0 \le x \le \pi$ and $\cot x = 0$, then $x =$ ____________.
4. The maximum value of $y = \csc x$ for $\pi \le x \le 2\pi$ is ____________.

In Exercises 5–8, find the slope–intercept form of the equation of each line that passes through the point *P* and makes angle θ with the positive *x*-axis.

5. $P = (-2, 3), \theta = 45°$
6. $P = (3, -1), \theta = 60°$
7. $P = (-3, -2), \theta = 120°$
8. $P = (2, 5), \theta = 135°$

In Exercises 9–26, graph each function over a one-period interval.

9. $y = \tan\left(x - \frac{\pi}{4}\right)$
10. $y = \tan\left(x + \frac{\pi}{4}\right)$
11. $y = \cot\left(x + \frac{\pi}{4}\right)$
12. $y = \cot\left(x - \frac{\pi}{4}\right)$
13. $y = \tan 2x$
14. $y = \tan\frac{x}{2}$
15. $y = \cot\frac{x}{2}$
16. $y = \cot 2x$
17. $y = -\tan x$
18. $y = -\cot x$
19. $y = 3\tan x$
20. $y = 3\cot x$
21. $y = \sec 2x$
22. $y = \sec\frac{x}{2}$
23. $y = \csc 3x$
24. $y = \csc\frac{x}{3}$
25. $y = \sec(x - \pi)$
26. $y = \csc(x - \pi)$

In Exercises 27–44, graph each function over a two-period interval.

27. $y = \tan\left[2\left(x + \frac{\pi}{2}\right)\right]$
28. $y = \tan\left[2\left(x - \frac{\pi}{2}\right)\right]$
29. $y = \cot\left[2\left(x - \frac{\pi}{2}\right)\right]$
30. $y = \cot\left[2\left(x + \frac{\pi}{2}\right)\right]$
31. $y = \tan\left[\frac{1}{2}(x + 2\pi)\right]$
32. $y = \tan\left[\frac{1}{2}(x - 2\pi)\right]$
33. $y = \cot\left[\frac{1}{2}(x - 2\pi)\right]$
34. $y = \cot\left[\frac{1}{2}(x + 2\pi)\right]$

35. $y = \sec\left[4\left(x - \frac{\pi}{4}\right)\right]$ **36.** $y = \sec\left[\frac{1}{2}\left(x - \frac{\pi}{2}\right)\right]$

37. $y = 3\csc\left(x + \frac{\pi}{2}\right)$ **38.** $y = 3\sec\left[2\left(x - \frac{\pi}{6}\right)\right]$

39. $y = \tan\left[\frac{2}{3}\left(x - \frac{\pi}{2}\right)\right]$ **40.** $y = 2\cot\left[2\left(x - \frac{\pi}{6}\right)\right]$

41. $y = -5\tan\left[2\left(x + \frac{\pi}{3}\right)\right]$ **42.** $y = -3\cot\left[\frac{1}{2}\left(x - \frac{\pi}{3}\right)\right]$

43. $y = \frac{1}{3}\cot\left[2(x - \pi)\right]$ **44.** $y = \frac{1}{2}\tan\left[4\left(x - \frac{\pi}{6}\right)\right]$

Applying the Concepts

45. Prison searchlight. A dual-beam rotating light on a movie set is positioned as a spotlight shining on a prison wall. The light is 20 feet from the wall and rotates clockwise. The light shines on point P on the wall when first turned on $(t = 0)$. After t seconds, the distance (in feet) from the beam on the wall to the point P is given by the function

$$d(t) = 20\tan\frac{\pi t}{5}.$$

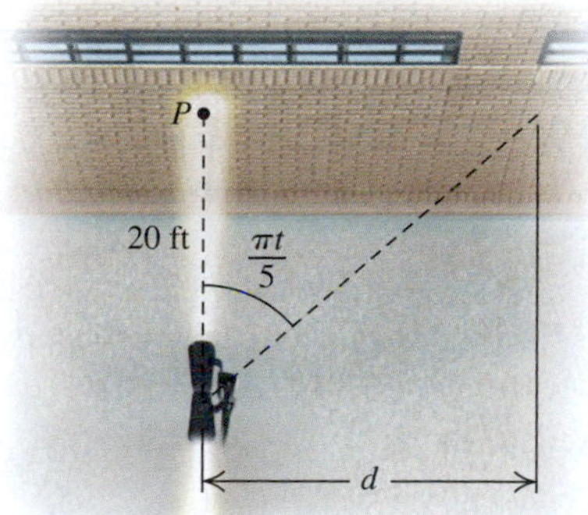

When the light beam is to the right of P, the value of d is positive, and when the beam is to the left of P, the value of d is negative.

a. Graph d over the interval $0 \le t \le 5$.

b. Because $d(t)$ is undefined for $t = 2.5$, where is the rotating light pointing when $t = 2.5$?

46. Prison searchlight. In Exercise 45, find the value for b assuming that the light beam is to sweep the entire wall in 10 seconds and $d(t) = 20\tan bt$.

Modeling with Tangent Function. In Exercises 47 and 48, model the given data by using the function $y = a\tan[b(x - c)]$. Use your equation to find y when $x = -1.5$ and $x = 1.5$. Round your answers to two decimal places.

47.

x	y
−5	undef. $(-\infty)$
−3	−7
−1	0
1	7
3	undef. $(+\infty)$

48.

x	y
−3	undef. $(+\infty)$
−0.5	6
2	0
4.5	−6
7	undef. $(-\infty)$

Modeling with Cotangent Function. In Exercises 49 and 50, model the given data by using the function $y = a\cot[b(x - c)]$.

49.

x	y
1	undef. $(+\infty)$
2	2
3	0
4	−2
5	undef. $(-\infty)$

50.

x	y
2	undef. $(-\infty)$
3.5	−3
5	0
6.5	3
8	undef. $(+\infty)$

51. Modeling with Secant Function. Model the data by using the function $y = a\sec[b(x - c)]$. Use your equation to find y when $x = \frac{1}{3}$.

x	y
−3	2
−2	undef.
−1	−2
0	undef.
1	2

52. Modeling with Cosecant Function. Model the data in Exercise 51 by using the function $y = a\csc[b(x - c)]$.

Beyond the Basics

53. Show that if f is a periodic function with period p, then the reciprocal function $\frac{1}{f}$ is also a periodic function with period p.

54. Show that if f is an odd function, then the reciprocal function $\frac{1}{f}$ is also an odd function.

55. Show that if f is an even function, then the reciprocal function $\frac{1}{f}$ is also an even function.

In Exercises 56–65, graph each function over a two-period interval.

56. $y = 3\tan(-2x)$ **57.** $y = 2\cot(-3x)$

58. $y = -2\sec\left(-\frac{x}{2}\right)$ **59.** $y = -\frac{1}{2}\csc\left(-\frac{x}{5}\right)$

60. $y = \cot(\pi - x)$ **61.** $y = -\tan\left(\frac{\pi}{2} - x\right)$

62. $y = -\sec(\pi - 4x)$

63. $y = 2\sec(\pi - 2x)$

64. $y = 2\tan\left(\frac{\pi}{2} - 2x\right) + 1$

65. $y = 4\cot(\pi - 4x) + 3$

In Exercises 66–69, write an equation for each graph.

66.

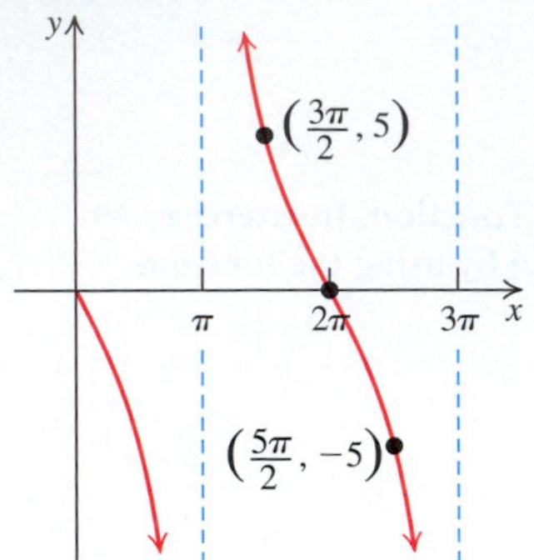

67.

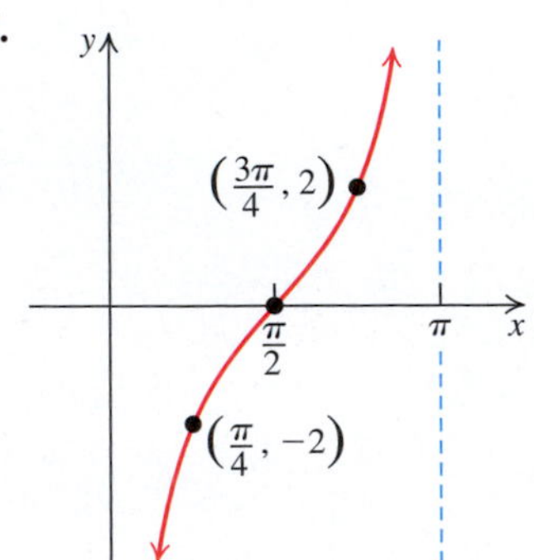

68.

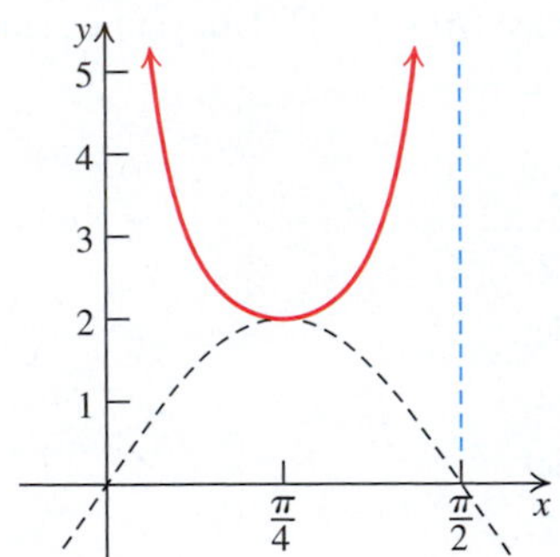

69.

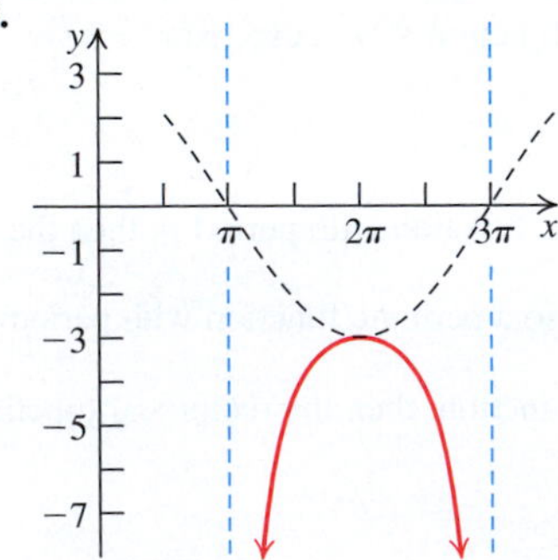

Critical Thinking/Discussion/Writing

70. For what number k, with $-2\pi < k < 0$, is $x = k$ a vertical asymptote for the graph of $y = \cot\left[\frac{1}{2}\left(x - \frac{\pi}{4}\right)\right]$?

71. For what number b does $y = \sec bx$ have period $\frac{\pi}{3}$?

72. Write an equation in the form $y = a\csc[b(x - c)]$ that has the same graph as $y = -2\sec x$.

Maintaining Skills

73. Show that $f(x) = 3x + 7$ is a one-to-one function. Find its inverse.

74. Show that $f(x) = -2x + 5$ is a one-to-one function. Find its inverse.

75. **True or false.** Every increasing function on an interval (a, b) is a one-to-one function.

76. **True or false.** Every decreasing function on an interval (a, b) is a one-to-one function.

77. **True or false.** If f is a one-to-one function, then it has an inverse function.

78. **True or false.** For a one-to-one function, the graph of f^{-1} is the reflection of the graph of f about the line $y = x$.

79. Show that $f(x) = x^2$ is not one-to-one on its domain $(-\infty, \infty)$. Write a subinterval of the domain of f on which f is one-to-one.

80. Is $f(x) = \sin x$ a one-to-one function on the interval $[-\pi, \pi]$? If your answer is no, find a subinterval on which f is one-to-one.

81. Repeat Exercise 80 for $f(x) = \cos x$.

82. Repeat Exercise 80 for $f(x) = \tan x$.

SECTION 5.6

Inverse Trigonometric Functions

BEFORE STARTING THIS SECTION, REVIEW

1 Inverse functions (Section 2.9)

2 Composition of functions (Section 2.8, page 260)

3 Exact values of the trigonometric functions (Section 5.2, page 491)

OBJECTIVES

1 Graph the inverse sine function.

2 Graph the inverse cosine function.

3 Graph the inverse tangent function.

4 Evaluate inverse trigonometric functions.

5 Find exact values of composite functions involving the inverse trigonometric functions.

Retail Theft

In 2005, security cameras at Filene's Basement store in Boston filmed a theft coordinated by a thief and an accomplice. The accomplice distracted the salesperson so that the thief could steal a $16,000 necklace. Retail theft is a major concern for all retail outlets, from Tiffany & Co. to Walmart.

The National Retail Federation, the industry's largest trade group, and the Retail Industry Leaders Association have instituted password-protected national crime databases online. These databases allow retailers to share information about thefts and determine whether they have been a target of individual shoplifters who steal for themselves or a target of organized crime. In addition to participating in the shared databases, many large retailers have their own organized anticrime squads. One estimate puts loss to organized theft at over $30 billion annually. In Example 12, we see how methods in this section can be used in an attempt to prevent loss from theft.

1 Graph the inverse sine function.

The Inverse Sine Function

Recall that a function f has an inverse function if no horizontal line intersects the graph of f in more than one point. Because every horizontal line $y = b$, where $-1 \le b \le 1$, intersects the graph of $y = \sin x$ at more than one point, the sine function fails the horizontal line test; so it is not one-to-one and consequently has no inverse.

The solid portion of the sine graph shown in Figure 5.75 is the graph of $y = \sin x$ for $-\frac{\pi}{2} \le x \le \frac{\pi}{2}$. If we restrict the domain of $y = \sin x$ to the interval $\left[-\frac{\pi}{2}, \frac{\pi}{2}\right]$, the resulting function

$$y = \sin x, -\frac{\pi}{2} \le x \le \frac{\pi}{2}$$

is one-to-one (it passes the horizontal line test); so it has an inverse. Notice that the restricted function takes on all values in the range of $y = \sin x$, which is $[-1, 1]$, and that each of these y-values corresponds to exactly one x-value in the restricted domain $\left[-\frac{\pi}{2}, \frac{\pi}{2}\right]$.

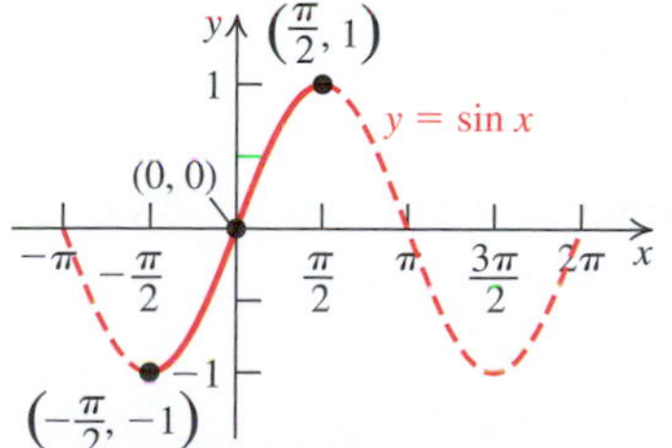

FIGURE 5.75 $y = \sin x$, $-\frac{\pi}{2} \le x \le \frac{\pi}{2}$

The inverse function for $y = \sin x$, $-\frac{\pi}{2} \le x \le \frac{\pi}{2}$, is called the **inverse sine**, or **arcsine**, function and is denoted by $\sin^{-1} x$, or by **arcsin** x.

The range of $y = \sin x$ is $[-1, 1]$; so the domain of $y = \sin^{-1} x$ is $[-1, 1]$. The domain of the restricted sine function is $\left[-\frac{\pi}{2}, \frac{\pi}{2}\right]$; so the range of $y = \sin^{-1} x$ is $\left[-\frac{\pi}{2}, \frac{\pi}{2}\right]$.

Inverse Sine Function

$y = \sin^{-1} x$ means $\sin y = x$, where $-1 \le x \le 1$ and $-\frac{\pi}{2} \le y \le \frac{\pi}{2}$.

Read $y = \sin^{-1} x$ as "y equals inverse sine at x."

RECALL

The graphs of a function and its inverse function are symmetric about the line $y = x$.

We can graph $y = \sin^{-1} x$ by reflecting the graph of $y = \sin x$, for $-\frac{\pi}{2} \le x \le \frac{\pi}{2}$, about the line $y = x$. See Figure 5.76.

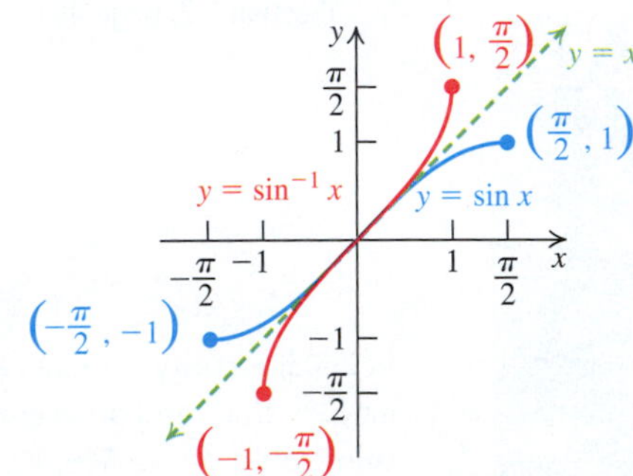

FIGURE 5.76 Graph of $y = \sin^{-1} x$

TABLE 5.8

y	x: $\sin y = x$
$-\frac{\pi}{2}$	-1
$-\frac{\pi}{3}$	$-\frac{\sqrt{3}}{2}$
$-\frac{\pi}{4}$	$-\frac{\sqrt{2}}{2}$
$-\frac{\pi}{6}$	$-\frac{1}{2}$
0	0
$\frac{\pi}{6}$	$\frac{1}{2}$
$\frac{\pi}{4}$	$\frac{\sqrt{2}}{2}$
$\frac{\pi}{3}$	$\frac{\sqrt{3}}{2}$
$\frac{\pi}{2}$	1

Table 5.8 shows some exact values of $y\left(-\frac{\pi}{2} \le y \le \frac{\pi}{2}\right)$ that satisfy the equation $\sin y = x$.

EXAMPLE 1 **Finding the Exact Value for $y = \sin^{-1} x$**

Find the exact values of y.

a. $y = \sin^{-1} \frac{\sqrt{3}}{2}$ **b.** $y = \arcsin\left(-\frac{1}{2}\right)$ **c.** $y = \sin^{-1} 3$

Solution

a. The equation $y = \sin^{-1} \frac{\sqrt{3}}{2}$ means $\sin y = \frac{\sqrt{3}}{2}$ and $-\frac{\pi}{2} \le y \le \frac{\pi}{2}$.

From Table 5.8, $\sin \frac{\pi}{3} = \frac{\sqrt{3}}{2}$ and $-\frac{\pi}{2} \le \frac{\pi}{3} \le \frac{\pi}{2}$; so $y = \sin^{-1} \frac{\sqrt{3}}{2} = \frac{\pi}{3}$.

b. The equation $y = \arcsin\left(-\frac{1}{2}\right)$ means $\sin y = -\frac{1}{2}$ and $-\frac{\pi}{2} \le y \le \frac{\pi}{2}$.

From Table 5.8, $\sin\left(-\frac{\pi}{6}\right) = -\frac{1}{2}$ and $-\frac{\pi}{2} \le -\frac{\pi}{6} \le \frac{\pi}{2}$; so

$y = \arcsin\left(-\frac{1}{2}\right) = -\frac{\pi}{6}$.

c. Because 3 is not in the domain, $[-1, 1]$, of the inverse sine function, $\sin^{-1} 3$ does not exist.

SIDE **NOTE**

You can also read $y = \sin^{-1}(x)$ as "y is the number in the interval $\left[-\frac{\pi}{2}, \frac{\pi}{2}\right]$ whose sine is x."

Practice Problem 1 Find the exact values of y.

a. $y = \sin^{-1}\left(-\frac{\sqrt{3}}{2}\right)$ **b.** $y = \sin^{-1}(-1)$

2 Graph the inverse cosine function.

The Inverse Cosine Function

When we restrict the domain of $y = \cos x$ to the interval $[0, \pi]$, the resulting function, $y = \cos x$ (with $0 \le x \le \pi$), is one-to-one. No horizontal line intersects the graph of $y = \cos x$, with $0 \le x \le \pi$, in more than one point. See Figure 5.77. Consequently, the restricted cosine function has an inverse function.

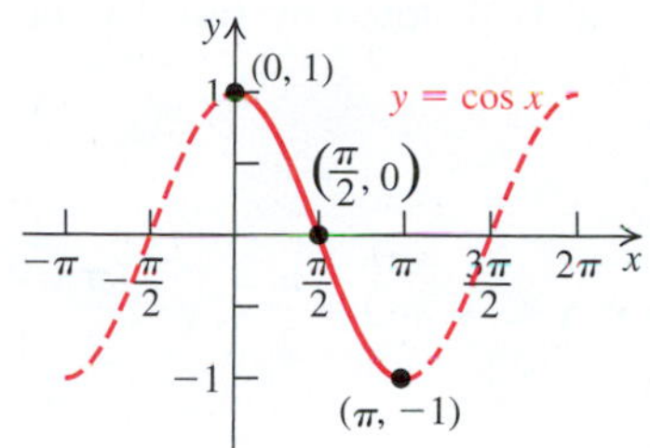

FIGURE 5.77 $y = \cos x$, $0 \le x \le \pi$

The inverse function for $y = \cos x$, $0 \le x \le \pi$, is called the **inverse cosine**, or **arccosine**, function and is denoted by $\cos^{-1} x$, or by $\arccos x$.

Inverse Cosine Function

$y = \cos^{-1} x$ means $\cos y = x$ where $-1 \le x \le 1$ and $0 \le y \le \pi$. Read $y = \cos^{-1} x$ as "y equals inverse cosine at x."

Reflecting the graph of $y = \cos x$, for $0 \le x \le \pi$, about the line $y = x$ produces the graph of $y = \cos^{-1} x$, shown in Figure 5.78.

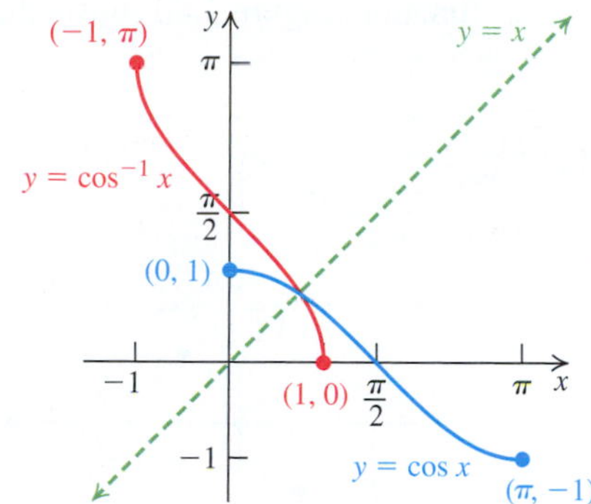

FIGURE 5.78 Graph of $y = \cos^{-1} x$

TABLE 5.9

y	x: $\cos y = x$
0	1
$\frac{\pi}{6}$	$\frac{\sqrt{3}}{2}$
$\frac{\pi}{4}$	$\frac{\sqrt{2}}{2}$
$\frac{\pi}{3}$	$\frac{1}{2}$
$\frac{\pi}{2}$	0
$\frac{2\pi}{3}$	$-\frac{1}{2}$
$\frac{3\pi}{4}$	$-\frac{\sqrt{2}}{2}$
$\frac{5\pi}{6}$	$-\frac{\sqrt{3}}{2}$
π	-1

Table 5.9 shows some exact values of y $(0 \le y \le \pi)$ that satisfy the equation $\cos y = x$.

EXAMPLE 2 Finding the Exact Value for $y = \cos^{-1} x$

Find the exact value of y.

a. $y = \cos^{-1} \frac{\sqrt{2}}{2}$ **b.** $y = \arccos\left(-\frac{1}{2}\right)$

Solution

a. The equation $y = \cos^{-1} \frac{\sqrt{2}}{2}$ means $\cos y = \frac{\sqrt{2}}{2}$ and $0 \le y \le \pi$.

Because $\cos \frac{\pi}{4} = \frac{\sqrt{2}}{2}$ and $0 \le \frac{\pi}{4} \le \pi$, we have $y = \frac{\pi}{4}$.

b. The equation $y = \arccos\left(-\frac{1}{2}\right)$ means $\cos y = -\frac{1}{2}$ and $0 \le y \le \pi$.

Because $\cos \frac{2\pi}{3} = -\frac{1}{2}$ and $0 \le \frac{2\pi}{3} \le \pi$, we have $y = \frac{2\pi}{3}$.

Practice Problem 2 Find the exact value of y.

a. $y = \cos^{-1}\left(-\frac{\sqrt{2}}{2}\right)$ **b.** $y = \arccos\left(\frac{1}{2}\right)$

SIDE NOTE

You can also read $y = \cos^{-1} x$ as "y is the number in the interval $[0, \pi]$ whose cosine is x."

3 Graph the inverse tangent function.

The Inverse Tangent Function

The *inverse tangent function* results from restricting the domain of the tangent function to the interval $\left(-\frac{\pi}{2}, \frac{\pi}{2}\right)$ to obtain a one-to-one function. The inverse of this restricted tangent function is the **inverse tangent**, or **arctangent**, function and is denoted by $\mathbf{tan^{-1}}\,\boldsymbol{x}$, or **arctan** $\boldsymbol{x}$.

Inverse Tangent Function

The equation $y = \tan^{-1} x$ means $\tan y = x$, where $-\infty < x < \infty$ and $-\frac{\pi}{2} < y < \frac{\pi}{2}$.

Read $y = \tan^{-1} x$ as "y equals the inverse tangent at x."

The graph of $y = \tan^{-1} x$ is obtained by reflecting the graph of $y = \tan x$, with $-\frac{\pi}{2} < x < \frac{\pi}{2}$, about the line $y = x$. Figure 5.79 shows the graph of the restricted tangent function. Figure 5.80 shows the graph of $y = \tan^{-1} x$.

TABLE 5.10

y	x: $\tan y = x$
$-\frac{\pi}{3}$	$-\sqrt{3}$
$-\frac{\pi}{4}$	-1
$-\frac{\pi}{6}$	$-\frac{\sqrt{3}}{3}$
0	0
$\frac{\pi}{6}$	$\frac{\sqrt{3}}{3}$
$\frac{\pi}{4}$	1
$\frac{\pi}{3}$	$\sqrt{3}$

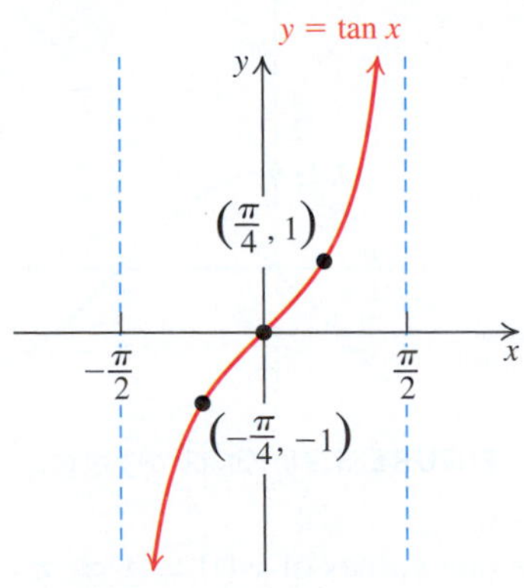

FIGURE 5.79 $y = \tan x, -\frac{\pi}{2} < x < \frac{\pi}{2}$

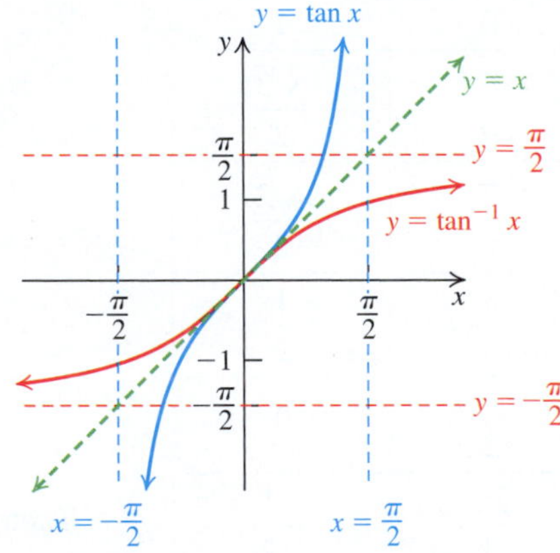

FIGURE 5.80 Graph of $y = \tan^{-1} x$

SIDE NOTE

You can also read $y = \tan^{-1}(x)$ as "y is the number in the interval $\left(-\frac{\pi}{2}, \frac{\pi}{2}\right)$ whose tangent is x."

Table 5.10 shows some exact values of $y\left(-\frac{\pi}{2} < y < \frac{\pi}{2}\right)$ that satisfy the equation $\tan y = x$.

EXAMPLE 3 Finding the Exact Value for $y = \tan^{-1} x$

Find the exact value of y.

a. $y = \tan^{-1} 0$ **b.** $y = \arctan(-\sqrt{3})$

Solution

a. Because $\tan 0 = 0$ and $-\frac{\pi}{2} < 0 < \frac{\pi}{2}$, we have $y = 0$.

b. Because $\tan\left(-\frac{\pi}{3}\right) = -\sqrt{3}$ and $-\frac{\pi}{2} < -\frac{\pi}{3} < \frac{\pi}{2}$, we have $y = -\frac{\pi}{3}$.

Practice Problem 3 Find the exact value of $y = \tan^{-1} \frac{\sqrt{3}}{3}$.

Other Inverse Trigonometric Functions

Sometimes the ranges of the *inverse secant* and *inverse cosecant* functions given in other texts differ from those used in this text. Always check the definitions of the domains of these two functions when they are used outside this course.

Inverse Cotangent, Cosecant, Secant Functions

Inverse cotangent $y = \cot^{-1} x$ means $\cot y = x$, where $-\infty < x < \infty$ and $0 < y < \pi$.

Inverse cosecant $y = \csc^{-1} x$ means $\csc y = x$, where $|x| \ge 1$ and $-\frac{\pi}{2} \le y \le \frac{\pi}{2}, y \ne 0$.

Inverse secant $y = \sec^{-1} x$ means $\sec y = x$, where $|x| \ge 1$ and $0 \le y \le \pi, y \ne \frac{\pi}{2}$.

TECHNOLOGY CONNECTION

The secondary functions on your calculator, labeled SIN^{-1}, COS^{-1}, and TAN^{-1}, are associated with the keys labeled SIN, COS, and TAN, respectively. Consult your manual to learn how to access these secondary functions. The screen shows several values for the trigonometric inverse functions on a calculator set to *radian* mode.

```
cos⁻¹(3/4)
                .723
sin⁻¹(-0.86)
              -1.035
tan⁻¹(-6.25)
              -1.412
```

EXAMPLE 4 Finding the Exact Value for $y = \csc^{-1} x$

Find the exact value for $y = \csc^{-1} 2$.

Solution

Because $\csc \frac{\pi}{6} = 2$ and $-\frac{\pi}{2} \le \frac{\pi}{6} \le \frac{\pi}{2}$, we have $y = \csc^{-1} 2 = \frac{\pi}{6}$.

Practice Problem 4 Find the exact value of $y = \sec^{-1} 2$.

SUMMARY OF MAIN FACTS

Inverse Trigonometric Functions

Inverse Function	Equivalent to	Domain	Range
$y = \sin^{-1} x$	$\sin y = x$	$[-1, 1]$	$\left[-\frac{\pi}{2}, \frac{\pi}{2}\right]$
$y = \cos^{-1} x$	$\cos y = x$	$[-1, 1]$	$[0, \pi]$
$y = \tan^{-1} x$	$\tan y = x$	$(-\infty, \infty)$	$\left(-\frac{\pi}{2}, \frac{\pi}{2}\right)$
$y = \cot^{-1} x$	$\cot y = x$	$(-\infty, \infty)$	$(0, \pi)$
$y = \csc^{-1} x$	$\csc y = x$	$(-\infty, -1] \cup [1, \infty)$	$\left[-\frac{\pi}{2}, 0\right) \cup \left(0, \frac{\pi}{2}\right]$
$y = \sec^{-1} x$	$\sec y = x$	$(-\infty, -1] \cup [1, \infty)$	$\left[0, \frac{\pi}{2}\right) \cup \left(\frac{\pi}{2}, \pi\right]$

4 Evaluate inverse trigonometric functions.

Evaluating Inverse Trigonometric Functions

In Section 5.3, we defined the six trigonometric functions of *real numbers*, and in this section, we have defined the corresponding six inverse trigonometric functions of real numbers. For example,

$$\text{if} \quad \sin \frac{\pi}{4} = \frac{\sqrt{2}}{2}, \quad \text{then} \quad \sin^{-1} \frac{\sqrt{2}}{2} = \frac{\pi}{4}.$$

However, because we also defined the trigonometric functions of *angles* in Section 5.3, it is meaningful when working with angles in degree measure to write a statement such as

$$\sin^{-1}\frac{\sqrt{2}}{2} = 45°.$$

We state some useful identities involving the values of inverse trigonometric functions of $\frac{1}{x}$ and $-x$ in terms of the values of inverse trigonometric functions of x.

INVERSE TRIGONOMETRIC IDENTITIES

For $\frac{1}{x}$	**For $-x$**
$\sin^{-1}\left(\frac{1}{x}\right) = \csc^{-1} x,\ \lvert x\rvert \ge 1$	$\sin^{-1}(-x) = -\sin^{-1} x,\ \lvert x\rvert \le 1$
$\csc^{-1}\left(\frac{1}{x}\right) = \sin^{-1} x,\ \lvert x\rvert \le 1$	$\csc^{-1}(-x) = -\csc^{-1} x,\ \lvert x\rvert \ge 1$
$\cos^{-1}\left(\frac{1}{x}\right) = \sec^{-1} x,\ \lvert x\rvert \ge 1$	$\cos^{-1}(-x) = \pi - \cos^{-1} x,\ \lvert x\rvert \le 1$
$\sec^{-1}\left(\frac{1}{x}\right) = \cos^{-1} x,\ \lvert x\rvert \le 1$	$\sec^{-1}(-x) = \pi - \sec^{-1} x,\ \lvert x\rvert \ge 1$
$\tan^{-1}\left(\frac{1}{x}\right) = \begin{cases} \cot^{-1} x, & x > 0 \\ -\pi + \cot^{-1} x, & x < 0 \end{cases}$	$\tan^{-1}(-x) = -\tan^{-1} x,\ -\infty < x < \infty$
$\cot^{-1}\left(\frac{1}{x}\right) = \begin{cases} \tan^{-1} x, & x > 0 \\ \pi + \tan^{-1} x, & x < 0 \end{cases}$	$\cot^{-1}(-x) = \pi - \cot^{-1} x,\ -\infty < x < \infty$

Note that the equation $\tan^{-1}\left(\frac{1}{x}\right) = \cot^{-1} x$ is true for $x > 0$, because the values of both sides are in the interval $\left(0, \frac{\pi}{2}\right)$. But when $x < 0$, $\tan^{-1}\left(\frac{1}{x}\right)$ is negative and lies in the interval $\left(-\frac{\pi}{2}, 0\right)$ and $\cot^{-1} x$ is positive and lies in the interval $\left(\frac{\pi}{2}, \pi\right)$. So for $x < 0$, the correct equation is $\tan^{-1}\left(\frac{1}{x}\right) = -\pi + \cot^{-1} x$. In other words, for $x < 0$, we have $\cot^{-1} x = \pi + \tan^{-1}\left(\frac{1}{x}\right)$.

EXAMPLE 5 Verifying an Inverse Trigonometric Identity

a. Show that $\cos^{-1}(-x) = \pi - \cos^{-1} x$ for $\lvert x\rvert \le 1$.

b. Use part (a) to evaluate $\sin\left[\cos^{-1}\left(-\frac{1}{2}\right)\right]$.

Solution

a. For $|x| \le 1$, let $\theta = \cos^{-1} x$ with θ in the interval $[0, \pi]$. Then $\cos\theta = x$.

For θ in $[0, \pi]$, $(\pi - \theta)$ is also in $[0, \pi]$ and

$$\cos(\pi - \theta) = -\cos\theta \qquad \text{See page 513.}$$
$$\cos(\pi - \theta) = -x \qquad \text{Replace } \cos\theta \text{ with } x.$$
$$\pi - \theta = \cos^{-1}(-x) \qquad \text{Definition of } \cos^{-1} x$$
$$\pi - \cos^{-1} x = \cos^{-1}(-x) \qquad \text{Replace } \theta \text{ with } \cos^{-1} x.$$

b.
$$\sin\left[\cos^{-1}\left(-\frac{1}{2}\right)\right] = \sin\left[\pi - \cos^{-1}\left(\frac{1}{2}\right)\right] \qquad \text{By part a}$$
$$= \sin\left[\pi - \frac{\pi}{3}\right] \qquad \cos^{-1}\left(\frac{1}{2}\right) = \frac{\pi}{3}$$
$$= \sin\left(\frac{\pi}{3}\right) \qquad \sin(\pi - \theta) = \sin\theta \text{ (page 513).}$$
$$= \frac{\sqrt{3}}{2} \qquad \text{See page 491.}$$

Practice Problem 5 **a.** Show that $\sin^{-1}(-x) = -\sin^{-1} x$ for $|x| \le 1$.

b. Use part **a** to evaluate $\sin\left[\frac{\pi}{3} - \sin^{-1}\left(-\frac{1}{2}\right)\right]$.

Using a Calculator with Inverse Functions When using the inverse trigonometric functions on a calculator to find a real number (or equivalently, an angle measured in radians), make sure you set your calculator to *radian* mode.

When using a calculator to find $\csc^{-1} x$ or $\sec^{-1} x$, find $\sin^{-1}\frac{1}{x}$ and $\cos^{-1}\frac{1}{x}$, respectively. For example, if $\csc^{-1} 5 = \theta$, then $\csc\theta = 5$, or $\frac{1}{\sin\theta} = 5$. So $\sin\theta = \frac{1}{5}$ and $\theta = \sin^{-1}\left(\frac{1}{5}\right)$. However, to find $\cot^{-1} x$, begin by finding $\tan^{-1}\frac{1}{x}$; this gives you a value in the interval $\left(-\frac{\pi}{2}, \frac{\pi}{2}\right)$. If $x > 0$, this is the correct value, but **for $x < 0$, $\cot^{-1}(x) = \pi + \tan^{-1}\frac{1}{x}$,** so that $\cot^{-1}(x)$ is in the interval $\left(\frac{\pi}{2}, \pi\right)$.

When using a calculator to find an unknown angle measure in degrees, make sure you set your calculator to degree measure.

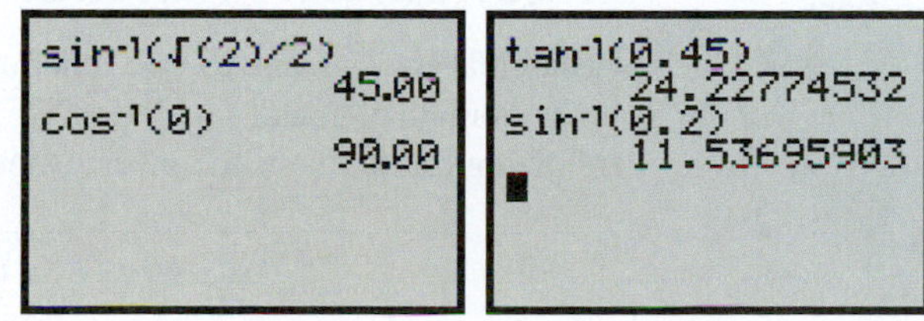

EXAMPLE 6 Using a Calculator to Find the Values of Inverse Functions

Use a calculator to find the value of y in radians rounded to four decimal places.

a. $y = \sin^{-1} 0.75$ **b.** $y = \cot^{-1} 2.8$ **c.** $y = \cot^{-1}(-2.3)$

Solution

Set your calculator to *radian* mode.

a. $y = \sin^{-1} 0.75 \approx 0.8481$

b. $y = \cot^{-1} 2.8 = \tan^{-1}\left(\frac{1}{2.8}\right) \approx 0.3430$

c. $y = \cot^{-1}(-2.3) = \pi + \tan^{-1}\left(-\frac{1}{2.3}\right) \approx 2.7315$

Practice Problem 6 Use a calculator to find the value of y in radians rounded to four decimal places.

a. $y = \cos^{-1} 0.22$ **b.** $y = \csc^{-1} 3.5$ **c.** $y = \cot^{-1}(-4.7)$

EXAMPLE 7 Using a Calculator to Find the Values of Inverse Functions

Use a calculator to find the value of y in degrees rounded to four decimal places.

a. $y = \tan^{-1} 0.99$ **b.** $y = \sec^{-1} 25$ **c.** $y = \cot^{-1}(-1.3)$

Solution

Set your calculator to *degree* mode.

a. $y = \tan^{-1} 0.99 \approx 44.7121°$

b. $y = \sec^{-1} 25 = \cos^{-1} \frac{1}{25} \approx 87.7076°$

c. $y = \cot^{-1}(-1.3) = 180° + \tan^{-1}\left(-\frac{1}{1.3}\right) \approx 142.4314°$

Practice Problem 7 Repeat Example 7 for each expression.

a. $y = \cot^{-1} 0.75$ **b.** $y = \csc^{-1} 13$ **c.** $y = \tan^{-1}(-12)$ 

5 Find exact values of composite functions involving the inverse trigonometric functions.

Composition of Trigonometric and Inverse Trigonometric Functions

Recall that if f is a one-to-one function with inverse f^{-1}, then $f^{-1}[f(x)] = x$ for every x in the domain of f and $f[f^{-1}(x)] = x$ for every x in the domain of f^{-1}. This leads to the following formulas for the inverse sine, cosine, and tangent functions.

INVERSE FUNCTION PROPERTIES

Inverse Sine	Inverse Cosine	Inverse Tangent
$\sin^{-1}(\sin x) = x$, $-\frac{\pi}{2} \le x \le \frac{\pi}{2}$	$\cos^{-1}(\cos x) = x$, $0 \le x \le \pi$	$\tan^{-1}(\tan x) = x$, $-\frac{\pi}{2} < x < \frac{\pi}{2}$
$\sin(\sin^{-1} x) = x$, $-1 \le x \le 1$	$\cos(\cos^{-1} x) = x$, $-1 \le x \le 1$	$\tan(\tan^{-1} x) = x$, $-\infty < x < \infty$

EXAMPLE 8 **Finding the Exact Value of $\sin^{-1}(\sin x)$ and $\cos^{-1}(\cos x)$**

Find the exact value of

a. $\sin^{-1}\left[\sin\left(-\frac{\pi}{8}\right)\right]$. **b.** $\cos^{-1}\left(\cos\frac{5\pi}{4}\right)$.

Solution

a. Because $-\frac{\pi}{2} \le -\frac{\pi}{8} \le \frac{\pi}{2}$, we have $\sin^{-1}\left[\sin\left(-\frac{\pi}{8}\right)\right] = -\frac{\pi}{8}$.

b. We cannot use the formula $\cos^{-1}(\cos x) = x$ for $x = \frac{5\pi}{4}$ because $\frac{5\pi}{4}$ is not in the interval $[0, \pi]$. However, $\cos\frac{5\pi}{4} = \cos\left(2\pi - \frac{5\pi}{4}\right) = \cos\frac{3\pi}{4}$ and $\frac{3\pi}{4}$ is in the interval $[0, \pi]$. Therefore,

$$\cos^{-1}\left(\cos\frac{5\pi}{4}\right) = \cos^{-1}\left(\cos\frac{3\pi}{4}\right) = \frac{3\pi}{4}.$$

Practice Problem 8 Find the exact value of $\sin^{-1}\left(\sin\frac{3\pi}{2}\right)$.

To find the exact values of expressions involving the composition of a trigonometric function and the inverse of a *different* trigonometric function, we use points on the terminal side of an angle in standard position.

EXAMPLE 9 **Finding the Exact Value of a Composite Trigonometric Expression**

Find the exact value of

a. $\cos\left(\tan^{-1}\frac{2}{3}\right)$. **b.** $\sin\left[\cos^{-1}\left(-\frac{1}{4}\right)\right]$.

Solution

a. Let θ represent the radian measure of the angle in the interval $\left(-\frac{\pi}{2}, \frac{\pi}{2}\right)$, with $\tan\theta = \frac{2}{3}$. Then because $\tan\theta$ is positive, θ must be positive. We have

$$\theta = \tan^{-1}\frac{2}{3} \quad \text{and} \quad 0 < \theta < \frac{\pi}{2}.$$

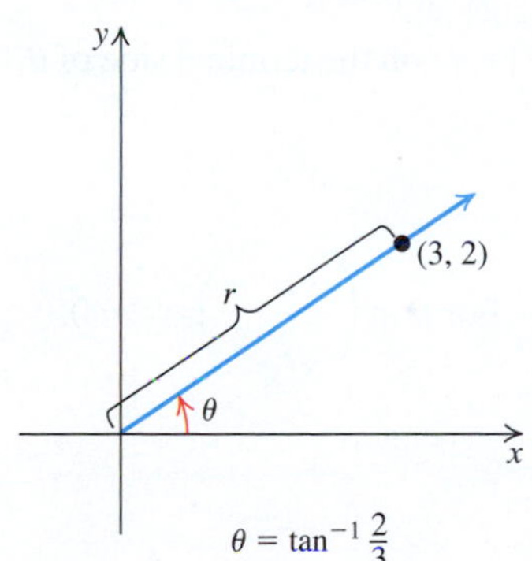

FIGURE 5.81

Figure 5.81 shows θ in standard position. If (x, y) is any point on the terminal side of θ, then $\tan\theta = \frac{y}{x}$.

Consequently, we can choose the point with coordinates $(3, 2)$ to determine the terminal side of θ. Then $x = 3$, $y = 2$ and we have

$$\tan\theta = \frac{2}{3} \quad \text{and} \quad \cos\theta = \frac{x}{r} = \frac{3}{r}, \text{ where}$$

$$r = \sqrt{x^2 + y^2} = \sqrt{3^2 + 2^2} = \sqrt{9 + 4} = \sqrt{13}. \text{ So}$$

$$\cos\left(\tan^{-1}\frac{2}{3}\right) = \cos\theta = \frac{3}{r} = \frac{3}{\sqrt{13}} = \frac{3\sqrt{13}}{13}.$$

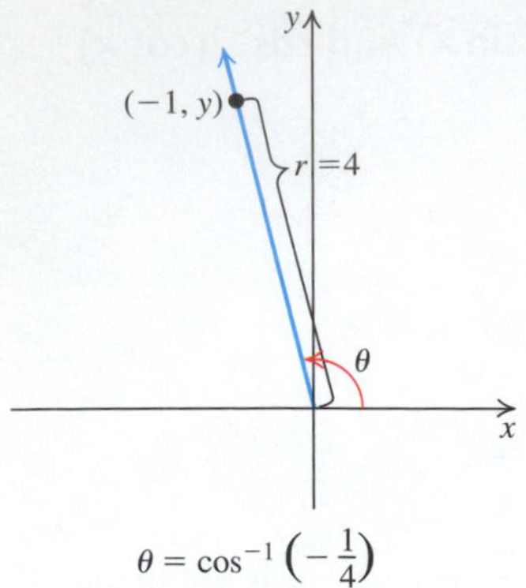

$\theta = \cos^{-1}\left(-\frac{1}{4}\right)$

FIGURE 5.82

b. Let θ represent the radian measure of the angle in $[0, \pi]$, with $\cos\theta = -\frac{1}{4}$. Then because $\cos\theta$ is negative, θ is in quadrant II; so

$$\theta = \cos^{-1}\left(-\frac{1}{4}\right) \quad \text{and} \quad \frac{\pi}{2} < \theta < \pi.$$

Figure 5.82 shows θ in standard position. If (x, y) is a point on the terminal side of θ and r is the distance between (x, y) and the origin, then $\sin\theta = \frac{y}{r}$. We choose the point with coordinates $(-1, y)$, a distance of $r = 4$ units from the origin, on the terminal side of θ. Then

$$\cos\theta = -\frac{1}{4} \quad \text{and} \quad \sin\theta = \frac{y}{4}, \text{ where}$$

$$r = \sqrt{x^2 + y^2} = \sqrt{(-1)^2 + y^2} \quad \text{or} \quad r^2 = 1 + y^2$$

$$4^2 = 1 + y^2 \qquad \text{Replace } r \text{ with 4.}$$

$$15 = y^2 \qquad \text{Simplify.}$$

$$\sqrt{15} = y \qquad y \text{ is positive.}$$

Thus,

$$\sin\left[\cos^{-1}\left(-\frac{1}{4}\right)\right] = \sin\theta = \frac{y}{r} = \frac{\sqrt{15}}{4}.$$

Practice Problem 9 Find the exact value of $\cos\left[\sin^{-1}\left(-\frac{1}{3}\right)\right]$.

EXAMPLE 10 Converting Composite Trigonometric Expressions to Algebraic Expressions

Write an algebraic expression for $y = \tan\left(\sin^{-1}\frac{t}{4}\right)$ for $|t| < 4$.

Solution

Let $\theta = \sin^{-1}\frac{t}{4}$. Then $\sin\theta = \frac{t}{4}$, and θ is in the interval $\left(-\frac{\pi}{2}, \frac{\pi}{2}\right)$. Figure 5.83 shows angle θ in standard position with a point $P(x, t)$ on the terminal side of θ. Then by the Pythagorean theorem, we have

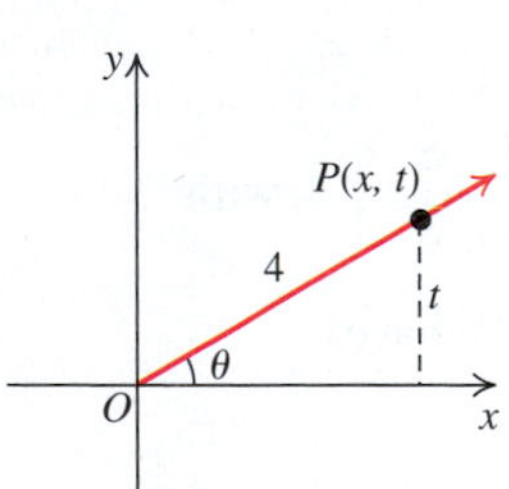

FIGURE 5.83

$$x^2 + t^2 = 4^2$$

$$x^2 = 16 - t^2$$

$$x = \sqrt{16 - t^2} \qquad \text{For } \theta \text{ in } \left(-\frac{\pi}{2}, \frac{\pi}{2}\right), x > 0.$$

The expression $y = \tan\left(\sin^{-1}\frac{t}{4}\right)$ becomes

$$y = \tan\theta \qquad \theta = \sin^{-1}\frac{t}{4}$$

$$y = \frac{t}{x} \qquad \text{Definition of } \tan\theta$$

$$y = \frac{t}{\sqrt{16 - t^2}} \qquad \text{Replace } x \text{ with } \sqrt{16 - t^2}.$$

Practice Problem 10 Write an algebraic expression for

$$y = \tan\left(\cos^{-1}\frac{x}{3}\right) \text{ for } 0 < x < 3.$$

In many applications of calculus, we use trigonometric substitutions to simplify algebraic expressions.

EXAMPLE 11 Using Trigonometry to Simplify Algebraic Expressions

Simplify $y = \dfrac{\sqrt{5 - x^2}}{x}$ using the substitution $x = \sqrt{5} \sin \theta$ for $0 < \theta < \dfrac{\pi}{2}$.

Solution

$$y = \frac{\sqrt{5 - (\sqrt{5} \sin \theta)^2}}{\sqrt{5} \sin \theta} \qquad \text{Substitute for } x.$$

$$= \frac{\sqrt{5 - 5 \sin^2 \theta}}{\sqrt{5} \sin \theta} \qquad (\sqrt{5} \sin \theta)^2 = 5 \sin^2 \theta$$

$$= \frac{\sqrt{5(1 - \sin^2 \theta)}}{\sqrt{5} \sin \theta} \qquad \text{Factor.}$$

$$= \frac{\sqrt{5 \cos^2 \theta}}{\sqrt{5} \sin \theta} \qquad 1 - \sin^2 \theta = \cos^2 \theta$$

$$= \frac{\sqrt{5} \cos \theta}{\sqrt{5} \sin \theta} \qquad \sqrt{5 \cos^2 \theta} = \sqrt{5}\sqrt{\cos^2 \theta} = \sqrt{5} \cos \theta \text{ for } 0 < \theta < \frac{\pi}{2}$$

$$= \cot \theta \qquad \text{Quotient identity}$$

Practice Problem 11 Simplify $y = \dfrac{x}{\sqrt{7 - x^2}}$ using the substitution $x = \sqrt{7} \cos \theta$ for $0 < \theta < \dfrac{\pi}{2}$.

EXAMPLE 12 Finding the Rotation Angle for a Security Camera

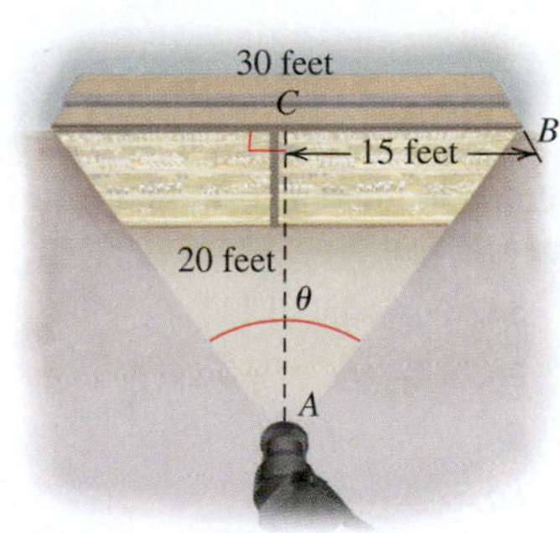

FIGURE 5.84

A security camera is to be installed 20 feet from the center of a jewelry counter. The counter is 30 feet long. What angle, to the nearest degree, should the camera rotate through so that it scans the entire counter? See Figure 5.84.

Solution

The counter center C, the camera A, and a counter end B form a right triangle. The angle at vertex A is $\dfrac{\theta}{2}$, where θ is the angle through which the camera rotates. Note that

$$\tan \frac{\theta}{2} = \frac{15}{20} = \frac{3}{4}$$

$$\frac{\theta}{2} = \tan^{-1} \frac{3}{4} \approx 36.87^\circ \qquad \text{Use a calculator in } \textit{degree} \text{ mode.}$$

$$\theta \approx 73.74^\circ \qquad \text{Multiply both sides by 2.}$$

The camera should rotate through 74° to scan the entire counter.

Practice Problem 12 Rework Example 12 for a counter that is 20 feet long and a camera set 12 feet from the center of the counter.

Answers to Practice Problems

1. a. $-\frac{\pi}{3}$ **b.** $-\frac{\pi}{2}$ **2. a.** $\frac{3\pi}{4}$ **b.** $\frac{\pi}{3}$ **3.** $\frac{\pi}{6}$ **4.** $\frac{\pi}{3}$ **5. b.** 1 **6. a.** 1.3490 **b.** 0.2898 **c.** 2.932 **7. a.** 53.1301° **b.** 4.4117° **c.** −85.2364° **8.** $-\frac{\pi}{2}$ **9.** $\frac{2\sqrt{2}}{3}$ **10.** $\frac{\sqrt{9-x^2}}{x}$ **11.** $\cot\theta$ **12.** 80°

SECTION 5.6 Exercises

Basic Concepts and Skills

1. The domain of $f(x) = \sin^{-1} x$ is ___________.
2. The range of $f(x) = \tan^{-1} x$ is ___________.
3. The exact value of $y = \cos^{-1}\frac{1}{2}$ is ___________.
4. $\sin^{-1}(\sin\pi) =$ ___________.
5. **True or False.** If $-1 \le x \le 0$, then $\sin^{-1} x \le 0$.
6. **True or False.** If $-1 \le x \le 0$, then $\cos^{-1} x \le 0$.
7. **True or False.** The domain of $f(x) = \cos^{-1}x$ is $0 \le x \le \pi$.
8. **True or False.** The value of $\tan^{-1}\left(\frac{1}{\sqrt{3}}\right)$ is $\frac{\pi}{3}$.

In Exercises 9–26, find each exact value of *y* or state that *y* is undefined.

9. $y = \arcsin 0$
10. $y = \cos^{-1} 0$
11. $y = \sin^{-1}\left(-\frac{1}{2}\right)$
12. $y = \arccos\left(-\frac{\sqrt{3}}{2}\right)$
13. $y = \cos^{-1}(-1)$
14. $y = \sin^{-1}\frac{1}{2}$
15. $y = \arccos\frac{\pi}{2}$
16. $y = \sin^{-1}\pi$
17. $y = \tan^{-1}\sqrt{3}$
18. $y = \arctan 1$
19. $y = \tan^{-1}(-1)$
20. $y = \tan^{-1}\left(-\frac{\sqrt{3}}{3}\right)$
21. $y = \cot^{-1}(-1)$
22. $y = \arcsin\left(-\frac{\sqrt{2}}{2}\right)$
23. $y = \arccos(-2)$
24. $y = \sin^{-1}\sqrt{3}$
25. $y = \sec^{-1}(-2)$
26. $y = \csc^{-1}(-2)$

In Exercises 27–36, find each exact value of *y* or state that *y* is undefined.

27. $y = \sin\left(\sin^{-1}\frac{1}{8}\right)$
28. $y = \cos\left(\cos^{-1}\frac{1}{5}\right)$
29. $y = \arctan\left(\tan\frac{\pi}{7}\right)$
30. $y = \arctan\left(\tan\frac{\pi}{4}\right)$
31. $y = \tan(\tan^{-1} 247)$
32. $y = \tan(\tan^{-1} 7)$
33. $y = \arcsin\left(\sin\frac{4\pi}{3}\right)$
34. $y = \arccos\left(\cos\frac{5\pi}{3}\right)$
35. $y = \tan^{-1}\left(\tan\frac{2\pi}{3}\right)$
36. $y = \tan\left(\tan^{-1}\frac{2\pi}{3}\right)$

In Exercise 37–52, use the identities on page 552 to find the exact value of each expression.

37. $\cot^{-1}\left(\frac{1}{\sqrt{3}}\right)$
38. $\sec^{-1}(\sqrt{2})$
39. $\csc^{-1}(2)$
40. $\csc^{-1}\left(\frac{2\sqrt{3}}{3}\right)$
41. $\operatorname{arccot}\left(-\frac{1}{\sqrt{3}}\right)$
42. $\cot^{-1}(-\sqrt{3})$
43. $\arcsin\left(-\frac{\sqrt{3}}{2}\right)$
44. $\csc^{-1}\left(-\frac{2}{\sqrt{3}}\right)$
45. $\cos^{-1}\left(-\frac{1}{2}\right)$
46. $\operatorname{arcsec}\left(-\frac{2}{\sqrt{3}}\right)$
47. $\sin\left[\frac{\pi}{3} - \sin^{-1}\left(-\frac{1}{2}\right)\right]$
48. $\cos\left[\frac{\pi}{6} + \cos^{-1}\left(-\frac{\sqrt{3}}{2}\right)\right]$
49. $\sin\left[\frac{\pi}{2} - \cos^{-1}(-1)\right]$
50. $\tan\left[\frac{\pi}{6} + \cot^{-1}\left(-\frac{1}{\sqrt{3}}\right)\right]$
51. $\sin\left[\tan^{-1}(-\sqrt{3}) + \cos^{-1}\left(-\frac{\sqrt{3}}{2}\right)\right]$
52. $\cos\left[\cot^{-1}(-\sqrt{3}) + \sin^{-1}\left(-\frac{1}{2}\right)\right]$

 **In Exercises 53–62, use a calculator to find each value of *y* in degrees rounded to two decimal places.**

53. $y = \cos^{-1} 0.6$
54. $y = \sin^{-1} 0.23$
55. $y = \sin^{-1}(-0.69)$
56. $y = \cos^{-1}(-0.57)$
57. $y = \sec^{-1}(3.5)$
58. $y = \csc^{-1}(6.8)$
59. $y = \tan^{-1} 14$
60. $y = \tan^{-1} 50$
61. $y = \tan^{-1}(-42.147)$
62. $y = \tan^{-1}(-0.3863)$

In Exercises 63–74, use a sketch to find each exact value of *y*.

63. $y = \cos\left(\sin^{-1}\frac{2}{3}\right)$
64. $y = \sin\left(\cos^{-1}\frac{3}{4}\right)$
65. $y = \sin\left[\cos^{-1}\left(-\frac{4}{5}\right)\right]$
66. $y = \cos\left(\sin^{-1}\frac{3}{5}\right)$
67. $y = \cos\left(\tan^{-1}\frac{5}{2}\right)$
68. $y = \sin\left(\tan^{-1}\frac{13}{5}\right)$

69. $y = \tan\left(\cos^{-1}\frac{4}{5}\right)$

70. $y = \tan\left[\sin^{-1}\left(-\frac{3}{4}\right)\right]$

71. $y = \sin(\tan^{-1} 4)$

72. $y = \cos(\tan^{-1} 3)$

73. $y = \tan(\sec^{-1} 2)$

74. $y = \tan[\csc^{-1}(-2)]$

In Exercises 75–80, write an algebraic expression for each composite trigonometric expression.

75. $y = \sin\left(\tan^{-1}\frac{x}{2}\right);\ x > 0$

76. $y = \sin\left(\cot^{-1}\frac{2x}{3}\right);\ x > 0$

77. $y = \cot\left(\sin^{-1}\frac{3x}{5}\right);\ 0 < x < \frac{5}{3}$

78. $y = \sin\left(\sec^{-1}\frac{x}{3}\right);\ x > 3$

79. $y = \cot\left(\cos^{-1}\frac{x}{2}\right);\ -2 < x < 0$

80. $y = \sin\left(\cot^{-1}\frac{x}{5}\right);\ x < 0$

In Exercises 81–86, simplify the algebraic expression by using the given trigonometric substitution. Assume $0 < \theta < \frac{\pi}{2}$.

81. $y = \frac{x}{\sqrt{9 - x^2}};\ x = 3\cos\theta$

82. $y = \frac{\sqrt{16 - x^2}}{x};\ x = 4\sin\theta$

83. $y = \frac{x}{\sqrt{4 + x^2}};\ x = 2\tan\theta$

84. $y = \frac{\sqrt{9 + x^2}}{x};\ x = 3\cot\theta$

85. $y = \frac{x}{\sqrt{x^2 - 25}};\ x = 5\sec\theta$

86. $y = \frac{\sqrt{x^2 - 36}}{x};\ x = 6\csc\theta$

Applying the Concepts

87. Sprinkler rotation. A sprinkler rotates back and forth through an angle θ, as shown in the figure. At a distance of 5 feet from the sprinkler, the rays that form the sides of angle θ are 6 feet apart. Find θ.

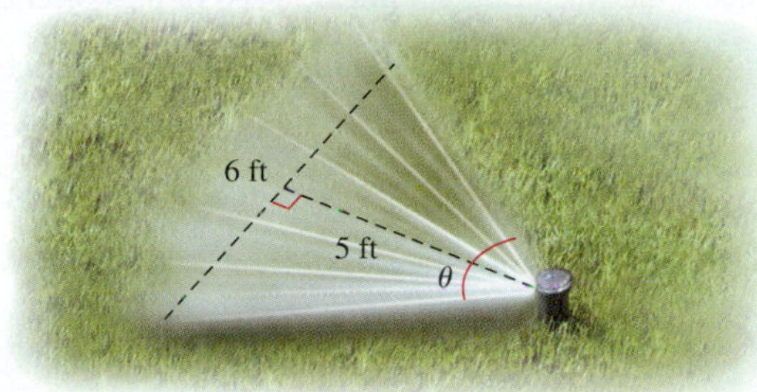

88. Irradiating flowers. A tray of flowers is being irradiated by a beam from a rotating lamp, as shown in the figure. If the tray is 8 feet long and the lamp is 2 feet from the center of the tray, through what angle should the lamp rotate to irradiate the full length of the tray?

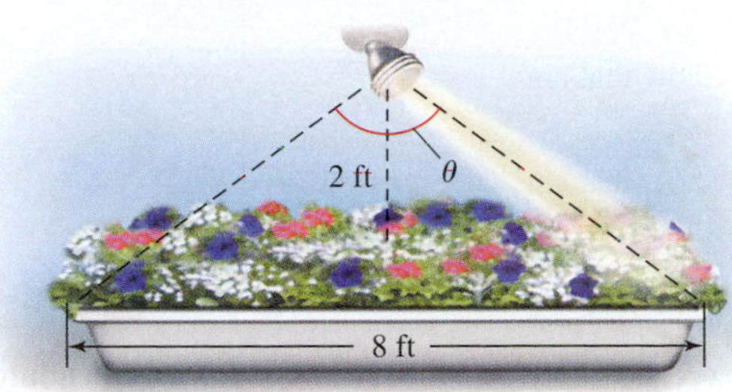

89. Motorcycle racing. A video camera is set up 110 feet away from and at a right angle to a straight quarter-mile racetrack, as shown in the figure. The starting line is to the left, and the finish is to the right. Through what angle must the camera rotate to film the entire race?

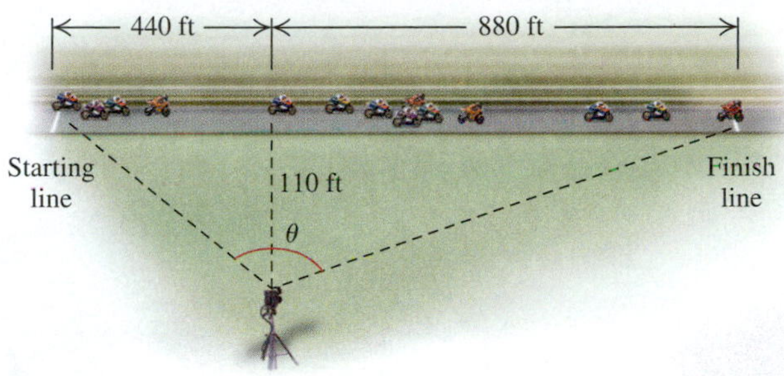

90. Camera's viewing angle. The viewing angle for the 35-millimeter camera is given (in degrees) by $\theta = 2\tan^{-1}\frac{18}{x}$, where x is the focal length of the lens. The focal length on most adjustable cameras is marked in millimeters on the lens mount.

a. Find the viewing angle, in degrees, if the focal length is 50 millimeters.

b. Find the viewing angle, in degrees, if the focal length is 200 millimeters.

Beyond the Basics

In Exercises 91–93, determine whether each function is increasing or decreasing on its domain.

91. $y = \sin^{-1} x$

92. $y = \cos^{-1} x$

93. $y = \tan^{-1} x$

94. Show that

a. $\sec^{-1} x \neq \frac{1}{\cos^{-1} x}$.

b. $\sec^{-1} x = \cos^{-1}\frac{1}{x}$.

95. Graph the function $y = \cot^{-1} x$.

96. Graph the function $y = \csc^{-1} x$.

97. Graph the function $y = \sec^{-1} x$.

98. For what values of x is $\cot^{-1}(\cot x) = x$ true?

99. For what values of x is $\sec^{-1}(\sec x) = x$ true?

100. For what values of x is $\csc^{-1}(\csc x) = x$ true?

Critical Thinking/Discussion/Writing

101. Show that $\sin^{-1} x = \begin{cases} \cos^{-1}(\sqrt{1 - x^2}); & 0 \le x \le 1 \\ -\cos^{-1}(\sqrt{1 - x^2}); & -1 \le x < 0 \end{cases}$

102. Show that $\sin^{-1} x + \cos^{-1} x = \frac{\pi}{2};\ |x| \le 1.$

Maintaining Skills

In Exercises 103–114, state whether the given equation is an identity, a conditional equation, or neither.

(Recall that an equation is an **identity** that is true for all permissible values of the variable and a **conditional equation** is a non-identity that is true for at least one value of the variable.)

103. $5(x + 3) = 8x - 24$

104. $x(20 - 3) - 1 = 10x(2 - 1) + 7x - 1$

105. $3x + 1 = 5(x + 1) - 2x$

106. $x^2 - 4 = 2x^2 + 3x - 2$

107. $\frac{1}{x^2 - 9} = \frac{1}{(x + 3)(x - 3)}$

108. $x^2 - 5x + 6 = 0$

109. $x^2 + 5x = 8x - 2x^2 + 1$

110. $\frac{1}{x^2 - 3x + 2} = \frac{1}{(x - 1)(x - 2)}$

111. $(x - 2)(x + 2) + 5x - 2 = x^2 + 5x - 5$

112. $x(x - 2) + 3 = x(1 - x) + 4$

113. $x^2 + 2x + 9 = (x - 1)(x - 2) + 5x + 7$

114. $3x(x + 2) = x - 1$

In Exercises 115–118, perform the indicated operations and simplify.

115. $\frac{1}{x - 1} - \frac{1}{x + 1}$

116. $\frac{1}{2 - x} + \frac{1}{2 + x}$

117. $\frac{1}{x - 1} - \frac{1}{x^2 - 1}$

118. $\frac{1}{x + 2} - \frac{1}{x + 3}$

In Exercises 119 and 120, rationalize the denominator and simplify.

119. $\frac{1}{\sqrt{2} + 1} + \frac{1}{\sqrt{3} + \sqrt{2}}$

120. $\frac{x}{\sqrt{x + 2} - \sqrt{2}}$

121. If $x^2 + y^2 = 1$, simplify $\frac{x}{1 + y} - \frac{1 - y}{x}$.

122. If $xy = 1$, simplify $\left(x + \frac{1}{y}\right)\left(y + \frac{1}{x}\right)$.

SUMMARY Definitions, Concepts, and Formulas

5.1 Angles and Their Measure

i. An **angle** is formed by rotating a ray about its endpoint.

ii. Angles that have the same initial and terminal sides are **coterminal** angles.

iii. An angle in a rectangular coordinate system is in **standard position** if its vertex is at the origin and its initial side coincides with the positive x-axis. (See page 472 for definitions of positive and negative angles.)

iv. If the terminal side of an angle in standard position lies on the x-axis or the y-axis, the angle is a **quadrantal angle**.

v. Angles can be measured in **degrees**, where $1° = \frac{1}{360}$ of a complete revolution.

vi. An **acute angle** is an angle with measure between 0° and 90°, a **right angle** is an angle with measure 90°, an **obtuse angle** is an angle with measure between 90° and 180°, and a **straight angle** is an angle with measure 180°.

vii. Angles can also be measured in **radians**. The radian measure of a central angle θ that intercepts an arc of length s on a circle of radius r is $\theta = \frac{s}{r}$ radians.

viii. To convert from degrees to radians, multiply degrees by $\frac{\pi}{180°}$. To convert from radians to degrees, multiply radians by $\frac{180°}{\pi}$.

ix. Two positive angles are **complements** if their sum is 90°. Two positive angles are **supplements** if their sum is 180°.

x. The formula for the length s of an arc intercepted by a central angle θ in a circle of radius r is $s = r\theta$.

xi. If an object travels at a constant speed v on a circle of radius r through an angle of θ radians and an arc of length s, in time t, then the (average) **linear speed** of the object is $v = \frac{s}{t}$ and the (average) **angular speed** of the object is $w = \frac{\theta}{t}$. Further, $v = r\omega$.

xii. The area of a sector $= \frac{1}{2}r^2\theta$, where r is the radius of the circle and θ is in radians.

5.2 Right-Triangle Trigonometry

i. The right-triangle definitions for trigonometric functions of an acute angle θ are given by

$$\sin\theta = \frac{\text{opposite}}{\text{hypotenuse}} = \frac{a}{c}; \quad \csc\theta = \frac{\text{hypotenuse}}{\text{opposite}} = \frac{c}{a}$$

$$\cos\theta = \frac{\text{adjacent}}{\text{hypotenuse}} = \frac{b}{c}; \quad \sec\theta = \frac{\text{hypotenuse}}{\text{adjacent}} = \frac{c}{b}$$

$$\tan\theta = \frac{\text{opposite}}{\text{adjacent}} = \frac{a}{b}; \quad \cot\theta = \frac{\text{adjacent}}{\text{opposite}} = \frac{b}{a}$$

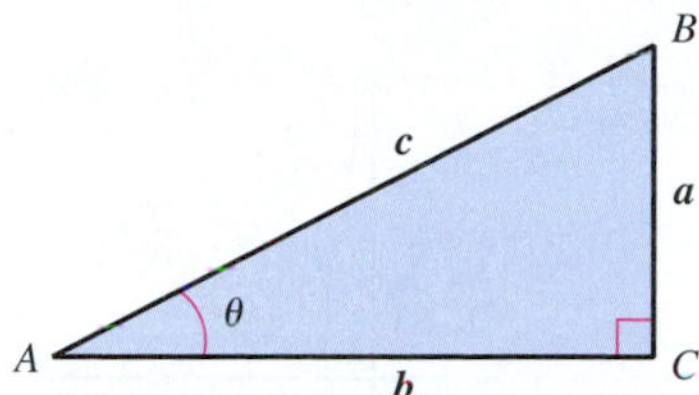

ii. Two acute angles are complements if their sum is $90°$ or $\frac{\pi}{2}$.

The value of any trigonometric function of an acute angle θ is equal to the cofunction of the complement of θ.

iii. To solve a right triangle means to find the measurements of the missing angles and sides.

iv. **Fundamental trigonometric identities**

Reciprocal and Quotient Identities

Reciprocal Identities

$$\sin\theta = \frac{1}{\csc\theta} \quad \cos\theta = \frac{1}{\sec\theta} \quad \tan\theta = \frac{1}{\cot\theta}$$

$$\csc\theta = \frac{1}{\sin\theta} \quad \sec\theta = \frac{1}{\cos\theta} \quad \cot\theta = \frac{1}{\tan\theta}$$

Quotient Identities

$$\tan\theta = \frac{\sin\theta}{\cos\theta} \quad \cot\theta = \frac{\cos\theta}{\sin\theta}$$

Cofunction Identities

$$\sin\theta = \cos(90° - \theta) \quad \cos\theta = \sin(90° - \theta)$$

$$\tan\theta = \cot(90° - \theta) \quad \cot\theta = \tan(90° - \theta)$$

$$\sec\theta = \csc(90° - \theta) \quad \csc\theta = \sec(90° - \theta)$$

If θ is measured in radians, replace $90°$ with $\frac{\pi}{2}$.

Trigonometric Function Values of Some Common Angles

$\theta°$	θ radians	$\sin\theta$	$\cos\theta$	$\tan\theta$	$\csc\theta$	$\sec\theta$	$\cot\theta$
30°	$\frac{\pi}{6}$	$\frac{1}{2}$	$\frac{\sqrt{3}}{2}$	$\frac{\sqrt{3}}{3}$	2	$\frac{2\sqrt{3}}{3}$	$\sqrt{3}$
45°	$\frac{\pi}{4}$	$\frac{\sqrt{2}}{2}$	$\frac{\sqrt{2}}{2}$	1	$\sqrt{2}$	$\sqrt{2}$	1
60°	$\frac{\pi}{3}$	$\frac{\sqrt{3}}{2}$	$\frac{1}{2}$	$\sqrt{3}$	$\frac{2\sqrt{3}}{3}$	2	$\frac{\sqrt{3}}{3}$

5.3 Trigonometric Functions of any Angle; The Unit Circle

i. **Definitions of the Trigonometric Functions of Any Angle θ**

Let $P = (x, y)$ be any point on the terminal ray of an angle θ in standard position (other than the origin) and let $r = \sqrt{x^2 + y^2}$. Then $r > 0$ and

$$\sin\theta = \frac{y}{r}, \cos\theta = \frac{x}{r}, \tan\theta = \frac{y}{x}\ (x \neq 0),$$

$$\csc\theta = \frac{r}{y}\ (y \neq 0), \sec\theta = \frac{r}{x}\ (x \neq 0), \cot\theta = \frac{x}{y}\ (y \neq 0)$$

ii. **Trigonometric function values of coterminal angles**

θ in degrees	θ in radians
$\sin\theta = \sin(\theta + n360°)$	$\sin\theta = \sin(\theta + 2\pi n)$
$\cos\theta = \cos(\theta + n360°)$	$\cos\theta = \cos(\theta + 2\pi n)$

These equations hold for any integer n.

iii. **Signs of the trigonometric functions**

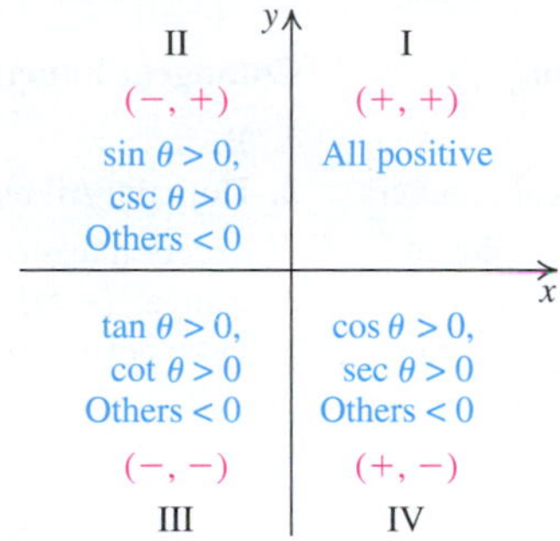

iv. The *reference angle* for θ in standard position is the positive acute angle θ' formed by the terminal side of θ and the x-axis. Reference angles are used to find the values of trigonometric functions for *any* angle θ. See Figure 5.37 on page 504.

v. **The unit circle** If the terminal side of an angle s in standard position intersects the unit circle at the point (x, y), then

$\sin t = y$, $\cos t = x$, $\tan t = \frac{y}{x}$, $\cot t = \frac{x}{y}$, $\sec t = \frac{1}{x}$, and

$\csc t = \frac{1}{y}$. The domain of the sine and cosine functions is $(-\infty, \infty)$, and their range is $[-1, 1]$.

5.4 Graphs of the Sine and Cosine Functions

i. **Sine and cosine function graphs and properties**

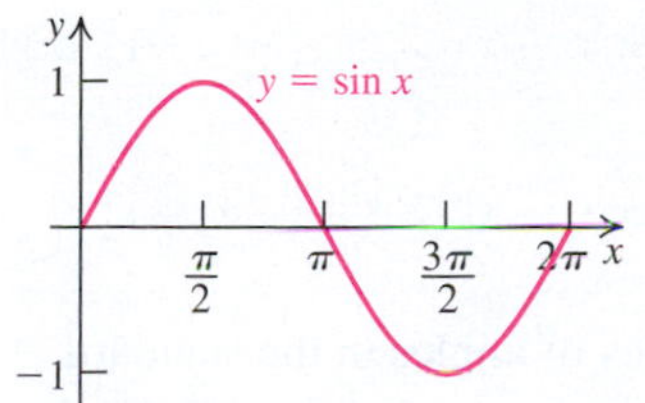

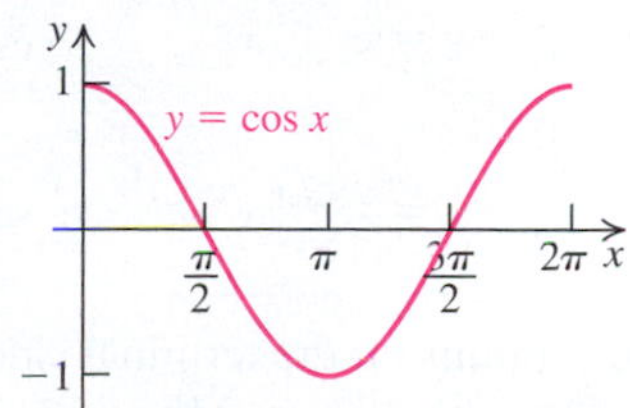

Sine Function

1. Period: 2π
2. Domain: $(-\infty, \infty)$
3. Range: $[-1, 1]$
4. Odd: $\sin(-t) = -\sin t$

Cosine Function

1. Period: 2π
2. Domain: $(-\infty, \infty)$
3. Range: $[-1, 1]$
4. Even: $\cos(-t) = \cos t$

ii. **Pythagorean identity** $(\cos t)^2 + (\sin t)^2 = 1$

iii. The functions $y = a\sin[b(x - c)] + d$ and $y = a\cos[b(x - c)] + d$, $b > 0$, have amplitude $|a|$, period $\frac{2\pi}{b}$, phase shift c, and vertical shift d.

5.5 Graphs of the Other Trigonometric Functions

i. Tangent and cotangent function graphs and properties

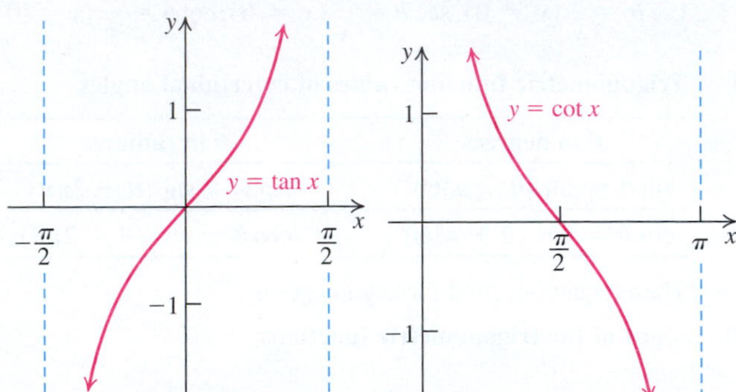

Tangent Function

1. Period: π
2. Domain: All real numbers except odd multiples of $\frac{\pi}{2}$
3. Range: $(-\infty, \infty)$
4. Odd: $\tan(-x) = -\tan x$

Cotangent Function

1. Period: π
2. Domain: All real numbers except integer multiples of π
3. Range: $(-\infty, \infty)$
4. Odd: $\cot(-x) = -\cot x$

ii. Secant and cosecant function graphs and properties

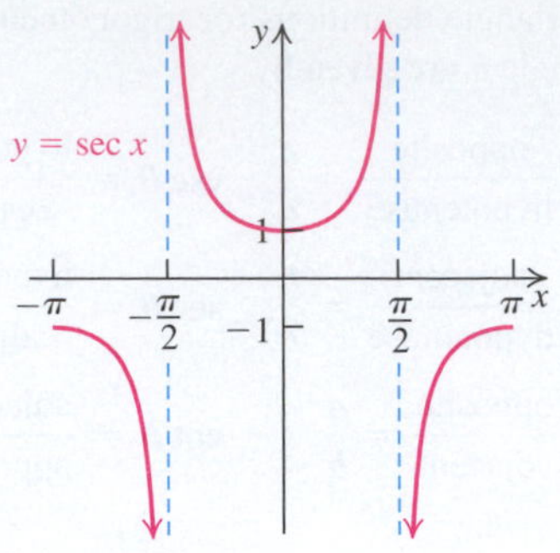

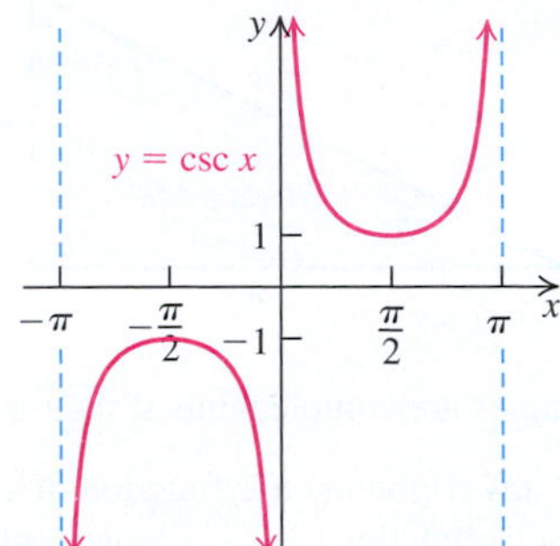

Secant Function

1. Period: 2π
2. Domain: All real numbers except odd multiples of $\frac{\pi}{2}$
3. Range: $(-\infty, -1] \cup [1, \infty)$
4. Even: $\sec(-x) = \sec x$

Cosecant Function

1. Period: 2π
2. Domain: All real numbers except integer multiples of π
3. Range: $(-\infty, -1] \cup [1, \infty)$
4. Odd: $\csc(-x) = -\csc x$

5.6 Inverse Trigonometric Functions

i. **Inverse Trigonometric Functions**

Inverse Function	Equivalent to	Domain	Range
$y = \sin^{-1} x$	$\sin y = x$	$[-1, 1]$	$\left[-\frac{\pi}{2}, \frac{\pi}{2}\right]$
$y = \cos^{-1} x$	$\cos y = x$	$[-1, 1]$	$[0, \pi]$
$y = \tan^{-1} x$	$\tan y = x$	$(-\infty, \infty)$	$\left(-\frac{\pi}{2}, \frac{\pi}{2}\right)$
$y = \cot^{-1} x$	$\cot y = x$	$(-\infty, \infty)$	$(0, \pi)$
$y = \csc^{-1} x$	$\csc y = x$	$(-\infty, -1] \cup [1, \infty)$	$\left[-\frac{\pi}{2}, 0\right) \cup \left(0, \frac{\pi}{2}\right]$
$y = \sec^{-1} x$	$\sec y = x$	$(-\infty, -1] \cup [1, \infty)$	$\left[0, \frac{\pi}{2}\right) \cup \left(\frac{\pi}{2}, \pi\right]$

ii. Points on the terminal sides of angles in the standard position are used to find exact values of the composition of a trigonometric function and a different inverse trigonometric function.

iii. **(a)** $\sin^{-1}(\sin x) = x$ for $-\frac{\pi}{2} \le x \le \frac{\pi}{2}$ and $\sin(\sin^{-1} x) = x$ for $-1 \le x \le 1$

(b) $\cos^{-1}(\cos x) = x$ for $0 \le x \le \pi$ and $\cos(\cos^{-1} x) = x$ for $-1 \le x \le 1$

(c) $\tan^{-1}(\tan x) = x$ for $-\frac{\pi}{2} < x < \frac{\pi}{2}$ and $\tan(\tan^{-1} x) = x$ for $-\infty < x < \infty$

REVIEW EXERCISES

In Exercises 1–4, draw each angle in standard position.

1. $240°$
2. $-150°$
3. $-\frac{2\pi}{3}$
4. $\frac{7\pi}{3}$

In Exercises 5–7, convert each angle from degrees to radians. Express each answer as a multiple of π.

5. $20°$
6. $36°$
7. $-60°$

In Exercises 8–10, convert each angle from radians to degrees.

8. $\frac{7\pi}{10}$
9. $\frac{5\pi}{18}$
10. $-\frac{4\pi}{9}$

In Exercises 11 and 12, find the radian measure of a central angle of a circle of radius *r* that intercepts each arc of length *s*. Round your answers to three decimal places.

11. $r = 15$ inches, $s = 40$ inches
12. $r = 9$ inches, $s = 17$ inches

In Exercises 13 and 14, find the length of the arc on each circle of radius *r* intercepted by a central angle θ. Round your answers to three decimal places.

13. $r = 2$ meters, $\theta = 36°$
14. $r = 0.9$ meter, $\theta = 12°$

In Exercises 15 and 16, find the area of the sector of a circle of radius *r* formed by the central angle θ. Round your answers to three decimal places.

15. $r = 3$ feet, $\theta = 32°$
16. $r = 4$ meters, $\theta = 47°$

In Exercises 17–20, use each given trigonometric function value of θ to find the five other trigonometric function values of an acute angle θ. Rationalize the denominators where necessary.

17. $\cos\theta = \frac{2}{9}$
18. $\sin\theta = \frac{1}{5}$
19. $\tan\theta = \frac{5}{3}$
20. $\cot\theta = \frac{5}{4}$

In Exercises 21–24, a point on the terminal side of an angle θ is given. For each angle, find the exact values of the six trigonometric functions. Rationalize the denominators where necessary.

21. $(2, 8)$
22. $(-3, 7)$
23. $(-\sqrt{5}, 2)$
24. $(-\sqrt{3}, -\sqrt{6})$

In Exercises 25–28, use the given information to find the quadrant in which θ lies.

25. $\sin\theta < 0$ and $\cos\theta > 0$
26. $\sin\theta > 0$ and $\tan\theta > 0$
27. $\sin\theta < 0$ and $\cot\theta > 0$
28. $\cos\theta < 0$ and $\csc\theta > 0$

In Exercises 29–32, find the exact values of the remaining trigonometric functions of θ from the given information.

29. $\cos\theta = -\frac{4}{5}$, θ in quadrant III
30. $\tan\theta = -\frac{5}{12}$, θ in quadrant IV
31. $\sin\theta = \frac{3}{5}$, θ in quadrant II
32. $\csc\theta = -\frac{5}{4}$, θ in quadrant III

In Exercises 33–36, find the exact value of each trigonometric function by using reference angles. Do not use a calculator.

33. $\cos 150°$
34. $\sin(-300°)$
35. $\tan 390°$
36. $\cot -405°$

In Exercises 37–40, sketch the graph of each given equation over the interval $[-2\pi, 2\pi]$.

37. $y = -\frac{3}{2}\cos x$
38. $y = \frac{5}{2}\sin x$
39. $y = 3\sin\left(x - \frac{\pi}{3}\right)$
40. $y = 4\cos\left(x + \frac{3\pi}{2}\right)$

In Exercises 41–44, find the amplitude or vertical stretch factor, period, and phase shift of each given function.

41. $y = 14\sin\left(2x + \frac{\pi}{7}\right)$
42. $y = 21\cos\left(8x + \frac{\pi}{9}\right)$
43. $y = 6\tan\left(2x + \frac{\pi}{5}\right)$
44. $y = -11\cot\left(6x + \frac{\pi}{12}\right)$

In Exercises 45–50, graph each function over a two-period interval.

45. $y = -2\cos\left(x + \frac{2\pi}{3}\right)$
46. $y = 3\cos(x - \pi)$
47. $y = 5\tan\left[2\left(x - \frac{\pi}{4}\right)\right]$
48. $y = -\cot\left[2\left(x + \frac{\pi}{4}\right)\right]$
49. $y = \frac{1}{2}\sec\frac{x}{2}$
50. $y = -3\csc\frac{x}{2}$

In Exercises 51–58, find the exact value of *y* or state that *y* is undefined.

51. $y = \cos^{-1}\left(\cos\frac{5\pi}{8}\right)$
52. $y = \sin^{-1}\left(\sin\frac{7\pi}{6}\right)$
53. $y = \tan^{-1}\left[\tan\left(-\frac{2\pi}{3}\right)\right]$
54. $y = \sin\left(\cos^{-1}\frac{1}{2}\right)$
55. $y = \cos\left(\sin^{-1}\frac{\sqrt{2}}{2}\right)$
56. $y = \cos\left(\tan^{-1}\frac{3}{4}\right)$
57. $y = \tan\left[\cos^{-1}\left(-\frac{1}{2}\right)\right]$
58. $y = \tan\left[\sin^{-1}\left(-\frac{\sqrt{3}}{2}\right)\right]$

59. A tractor rolls backward so that its wheels turn a quarter of a revolution. If the tires have a radius of 28 inches, how many inches has the tractor moved?

60. What is the radian measure of the smaller central angle made by the hands of a clock at 5:00? Express your answer as a rational multiple of π.

61. New Orleans, Louisiana, is due south of Dubuque, Iowa. Find the distance between New Orleans (north latitude 29°59′ N) and Dubuque (north latitude 42°31′ N). Use 3960 miles as the value of the radius of Earth.

62. A horse on the outside row of a merry-go-round is 20 feet from the center, and its linear speed is 314 feet per minute. Find its angular speed.

63. A point on the rim of a pottery wheel with a 28-inch diameter has a linear speed of 21 inches per second. Find its angular speed.

64. A NASA spacecraft has a circular orbit around Mars about 400 km above its surface. Assuming that it takes two hours to complete one orbit and the radius of Mars is 6780 km, find the linear speed of the spacecraft.

65. A geostationary satellite has a circular orbit 36,000 km above Earth's surface. It takes 24 hours for the satellite to complete its orbit, appearing to be stationary above a fixed location on Earth. Assuming that the radius of Earth is 6400 km, find the linear speed of the satellite.

66. A surveyor wants to measure the width of a river. She stands at a point A, with point B opposite her on the other side of the river. From A, she walks 320 feet along the bank to a point C. The line CA makes an angle of 40° with the line CB. What is the width of the river?

67. The table gives the average number of daylight hours in Santa Fe, New Mexico, each month. Let y represent the number of daylight hours in Santa Fe in month x and find a function of the form $y = a \sin b(x - c) + d$ that models the hours of daylight throughout the year, where $x = 1$ represents January.

Jan	Feb	Mar	Apr	May	June
10.1	9.9	12.0	12.7	14.1	14.1
July	Aug	Sept	Oct	Nov	Dec
14.3	13.5	12.0	11.3	10.0	9.8

68. Suppose a ball attached to a spring is pulled down 11 inches and released and the resulting simple harmonic motion has a period of five seconds. Write an equation of the ball's simple harmonic motion.

PRACTICE TEST A

1. Convert 140° to radians.

2. Convert $\frac{7\pi}{5}$ to degrees.

3. If $(2, -5)$ is a point on the terminal side of an angle θ, find $\cos\theta$.

4. Find the radian measure of a central angle of a circle of radius 10 centimeters that intercepts an arc of length 15 centimeters.

5. Find the area of a sector of a circle of radius 15 centimeters that intersects an arc of length 10 centimeters.

6. Assuming that $\cos\theta = \frac{2}{7}$ for an acute angle θ, find $\sin\theta$.

7. Assuming that $\tan\theta < 0$ and $\csc\theta > 0$, find the quadrant in which θ lies.

8. Assuming that $\cot\theta = -\frac{5}{12}$ and θ is in quadrant IV, find $\sec\theta$.

9. What is the reference angle for 217°?

10. Assuming that $\cos\frac{3\pi}{7} = 0.223$, find $\sin\frac{\pi}{14}$.

11. From a seat 65 meters high on a Ferris wheel, the angle of depression to a hot dog stand is 42°. How many meters, to the nearest meter, is the hot dog stand from a point directly below the seat?

12. Give the amplitude and range for $y = -7 \sin x$.

13. Find the amplitude, period, and phase shift for $y = 19 \sin[12(x + 3\pi)]$.

14. Suppose a ball attached to a spring is pulled down 7 inches and released and the resulting simple harmonic motion has a period of four seconds. Write an equation for the ball's simple harmonic motion.

15. Graph $y = -\cos\frac{x}{2}$ over a one-period interval.

16. Graph $y = \sin[2(x - \pi)]$ over a one-period interval.

17. Graph $y = \tan\left(x + \frac{\pi}{4}\right)$ over a one-period interval.

18. Find the exact value of $y = \sin^{-1}\left(-\frac{\sqrt{3}}{2}\right)$.

19. Find the exact value of $y = \cos^{-1}\left[\cos\frac{4\pi}{3}\right]$.

20. Find the exact value of $y = \tan\left[\sin^{-1}\left(-\frac{1}{3}\right)\right]$.

PRACTICE TEST B

1. Convert 160° to radians.

a. $\frac{7\pi}{9}$ **b.** $\frac{8\pi}{9}$ **c.** $\frac{4\pi}{3}$ **d.** $\frac{7\pi}{3}$

2. Convert $\frac{13\pi}{5}$ to degrees.

a. 36° **b.** 72° **c.** 108° **d.** 468°

3. Assuming that $(-3, 1)$ is a point on the terminal side of an angle θ, find $\cos\theta$.

a. $-\frac{3\sqrt{10}}{10}$ **b.** $-\frac{\sqrt{10}}{10}$ **c.** $\frac{\sqrt{10}}{10}$ **d.** $\frac{3\sqrt{10}}{10}$

4. Find the radian measure of a central angle of a circle of radius 8 that intercepts an arc of length 4.

a. $\frac{1}{2}$ **b.** 2 **c.** 32 **d.** 128

5. Find the area of a sector of a circle of radius 6 centimeters that intercepts an arc of length 3 centimeters.

a. $\frac{9}{2}\text{ cm}^2$ **b.** $\frac{27}{2}\text{ cm}^2$ **c.** 6 cm^2 **d.** 9 cm^2

6. Assuming that $\cos\theta = \frac{3}{4}$ for an acute angle θ, find $\sin\theta$.

a. $\frac{3}{\sqrt{7}}$ **b.** $\frac{\sqrt{7}}{4}$ **c.** $\frac{\sqrt{7}}{3}$ **d.** $\frac{4}{3}$

7. Assuming that $\sin\theta < 0$ and $\sec\theta > 0$, find the quadrant in which θ lies.

a. I **b.** II **c.** III **d.** IV

8. Assuming that $\tan\theta = \frac{12}{5}$ and θ is in quadrant III, find $\csc\theta$.

a. $-\frac{5}{12}$ **b.** $-\frac{13}{12}$ **c.** $\frac{5}{13}$ **d.** $\frac{5}{12}$

9. What is the reference angle for 640°?

a. 10° **b.** 60° **c.** 80° **d.** 280°

10. In which quadrant is it true that $\cos\theta > 0$ and $\csc\theta > 0$?

a. I **b.** II **c.** III **d.** IV

11. A point on the rim of a wheel with a radius of 50 centimeters has an angular speed of 0.8 radian per second. What is the linear speed of the point?

a. 4 centimeters per second
b. 6.25 centimeters per second
c. 40 centimeters per second
d. 62.5 centimeters per second

12. If $y = a\cos bx$ has an amplitude of $\frac{1}{2}$ and a period of 4π, which are possible values of a and b?

a. $a = \frac{1}{2}, b = \frac{1}{2}$ **b.** $a = 2, b = \frac{1}{2}$
c. $a = \frac{1}{2}, b = 2$ **d.** $a = 2, b = 2$

13. Which of the following statements is *false*?

a. $\sin(-x) = -\sin x$ **b.** $\cos(-x) = -\sin x$
c. $\tan(-x) = -\tan x$ **d.** $\cot(-x) = -\cot x$

14. The equation $y = -5\cos\frac{\pi}{4}t$ describes the harmonic of a ball attached to a spring. What is the period of the motion?

a. $\frac{\pi}{4}$ **b.** -5 **c.** 5 **d.** 8

15. Find the amplitude a, period b, and phase shift c for $y = -7\cos\left[\frac{1}{6}\left(x - \frac{\pi}{12}\right)\right]$.

a. $a = 7, b = \frac{1}{6}\pi, c = \frac{\pi}{12}$
b. $a = 7, b = \frac{1}{6}\pi, c = -\frac{\pi}{12}$
c. $a = 7, b = 12\pi, c = -\frac{\pi}{12}$
d. $a = 7, b = 12\pi, c = \frac{\pi}{12}$

16. This is the graph of which function?

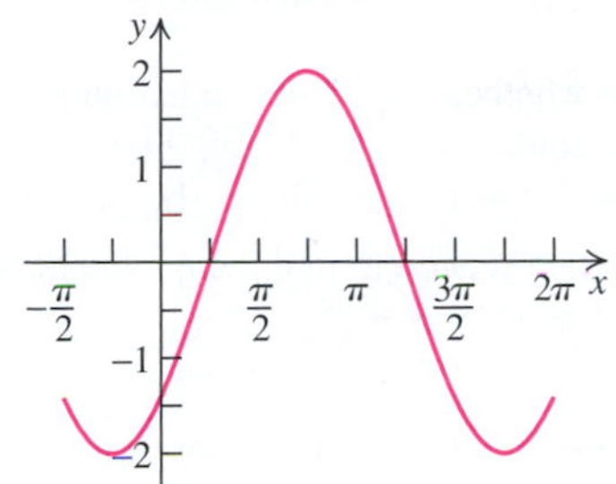

a. $y = -2\sin\left(x + \frac{\pi}{2}\right)$ **b.** $y = 2\sin\left(x + \frac{\pi}{2}\right)$
c. $y = -2\cos\left(x + \frac{\pi}{4}\right)$ **d.** $y = 2\cos\left(x + \frac{\pi}{4}\right)$

17. Find the exact value of $y = \cos^{-1}\left(\cos\frac{11\pi}{3}\right)$.

a. $\frac{11\pi}{3}$ **b.** $\frac{5\pi}{3}$ **c.** $\frac{\pi}{3}$ **d.** $-\frac{\pi}{3}$

18. Find the exact value of $y = \cos\left[\sin^{-1}\left(-\frac{4}{5}\right)\right]$.

a. $\frac{3}{5}$ **b.** $-\frac{3}{5}$ **c.** $\frac{3}{4}$ **d.** $-\frac{4}{3}$

19. Find the exact value of $y = \cos^{-1}\left(\cos\frac{7\pi}{4}\right)$.

a. $\frac{2\pi}{3}$ **b.** $\frac{-2\pi}{3}$ **c.** $\frac{7\pi}{4}$ **d.** $\frac{\pi}{4}$

20. Find the exact value of $y = \cos\left[\tan^{-1}\left(\frac{5}{3}\right)\right]$.

a. $\frac{3}{5}$ **b.** $\frac{5}{34}$ **c.** $\frac{5\sqrt{34}}{34}$ **d.** $\frac{3\sqrt{34}}{34}$

CUMULATIVE REVIEW EXERCISES CHAPTERS P–5

1. Solve the equation $3x^2 - 30x + 50 = -24$.
2. Find all real number solutions of the equation $(\sqrt{t} + 1)^2 + 2(\sqrt{t} + 1) = 3$.
3. Solve the inequality $\frac{4 - x}{2x - 4} > 0$ and write the answer in interval notation.
4. Write the slope–intercept form of the equation of the line containing the points $(3, -1)$ and $(-6, 5)$.
5. Find the domain of the function $f(x) = \ln\left(\frac{x}{2 - x}\right)$.
6. Identify the basic function and then use transformations to sketch the graph of the function $f(x) = 2(x - 3)^2 - 5$.
7. Use transformations to sketch the graph of the function $f(x) = 4e^x + 1$.
8. Find and write the domain of $(f \circ g)(x)$ in interval notation assuming that $f(x) = \ln x$ and $g(x) = -\frac{1}{x}$.
9. Determine whether each function has an inverse function. If so, find an equation for $f^{-1}(x)$.
   a. $f(x) = 3x + 7$ **b.** $f(x) = |x|$
10. Use synthetic division to find the quotient and remainder.
   $$\frac{2x^4 + 4x^3 + x^2 + x - 2}{x + 2}$$
11. Find the horizontal and vertical asymptotes of the graph of $f(x) = \frac{10 - 3x^2}{x^2 + 4x - 5}$.
12. Use rules of logarithms to rewrite each expression in the expanded form.
   a. $\log \sqrt[3]{\frac{x^2 y}{z}}$ **b.** $\ln\left(\frac{7x^4}{5\sqrt{y}}\right)$
13. Sketch the graph of $f(x) = \begin{cases} -\ln x & \text{if } x > 0 \\ \ln|x| & \text{if } x < 0 \end{cases}$.
14. Solve the equation $\log(x^2 + 9x) = 1$.
15. Find all possible rational zeros of the function $f(x) = 2x^4 + 5x^3 + x^2 + 10x - 6$.
16. Use the given trigonometric function value of θ to find the five other trigonometric function values of an acute angle θ. Rationalize the denominators where necessary.
   a. $\cos\theta = \frac{3}{7}$ **b.** $\sin\theta = \frac{2}{11}$
17. Find the exact values of the remaining trigonometric functions of θ from the given information.
   a. $\cos\theta = -\frac{4}{5}$, θ in quadrant II
   b. $\cot\theta = \frac{5}{12}$, θ in quadrant III
18. Sketch the graph of each equation over the interval $[-2\pi, 2\pi]$.
   a. $y = -\frac{1}{2}\cos(x + \pi)$
   b. $y = -5\sin\left(x + \frac{5\pi}{3}\right)$
19. Find the exact value of y or state that y is undefined.
   a. $y = \cos\left[\sin^{-1}\left(-\frac{1}{5}\right)\right]$
   b. $y = \sin\left(\tan^{-1}\frac{7}{2}\right)$
20. Find the exact value of $y = \cos^{-1}\left(\cos\frac{5\pi}{3}\right)$.

CHAPTER 6

Trigonometric Identities and Equations

TOPICS

Such varied phenomena as sonic booms from aircraft, musical tones, hours of daylight in a given location, current in an electric circuit, tides, and rainfall can be described using trigonometric functions. Solutions of trigonometric equations often provide answers to questions about the conditions under which specific characteristics of these phenomena occur. In this chapter, we study techniques for solving trigonometric equations.

SECTION 6.1

Verifying Identities

BEFORE STARTING THIS SECTION, REVIEW

1. Signs of the trigonometric functions (Section 5.3, page 504)
2. LCD and adding fractions (Section P.5, page 50)
3. Factoring and special products (Section P.4, page 40)
4. Reciprocal, quotient, and Pythagorean identities (Sections 5.2 and 5.4)

OBJECTIVES

1. Use fundamental trigonometric identities to evaluate trigonometric functions.
2. Simplify a complicated trigonometric expression.
3. Prove that a given equation is not an identity.
4. Verify a trigonometric identity.

Hipparchus of Rhodes (190–120 B.C.) Hipparchus was born in Nicaea (now called Iznik) in Bithynia (northwest Turkey) but spent much of his life in Rhodes (Greece). Many historians consider him the founder of trigonometry. He introduced trigonometric functions in the form of a chord table—the ancestor of the sine table found in ancient Indian astronomical works. With his chord table, Hipparchus could solve the height–distance problems of plane trigonometry. In his time, there was no Greek term for trigonometry because it was not counted as a branch of mathematics. Rather, trigonometry was just an aid to solve problems in astronomy.

Visualizing Trigonometric Functions

Trigonometric functions are commonly defined as the ratios of two sides corresponding to an angle in a right triangle. The Greek mathematician Hipparchus was the first person to relate trigonometric functions to a circle. His work allows us to visualize the trigonometric functions of an acute angle θ as the length of various segments associated with a unit circle. See Figure 6.1. The figure shows several similar triangles. For example, triangles PCT and OCP are similar; so

$$\frac{CT}{PC} = \frac{PC}{OC} \quad \text{Sides are proportional.}$$

From Figure 6.1, we use the definitions of trigonometric functions to get

$$PC = \sin\theta, \quad OC = \cos\theta, \quad \text{and} \quad CT = OT - OC = \sec\theta - \cos\theta.$$

Substituting these values into the equation $\frac{CT}{PC} = \frac{PC}{OC}$, we obtain

$$\frac{\sec\theta - \cos\theta}{\sin\theta} = \frac{\sin\theta}{\cos\theta}.$$

In Example 6, we verify that this equation is an identity. Recall that an *identity* is an equation that is true for all numbers for which both sides are defined.

In Section 5.3, the six trigonometric functions were defined in terms of the coordinates of a point $P(x, y)$ on a unit circle. Consequently, these functions are related to each other and any expression containing these functions can be written in several equivalent ways.

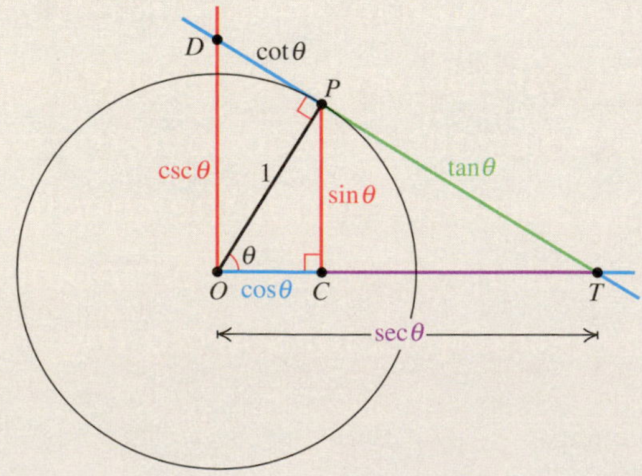

FIGURE 6.1 The unit circle and trigonometric functions

Fundamental Trigonometric Identities

In Chapter 5, we used the definitions of the trigonometric functions to establish reciprocal, quotient, and even–odd identities. We also introduced the identity

$$\sin^2 x + \cos^2 x = 1.$$

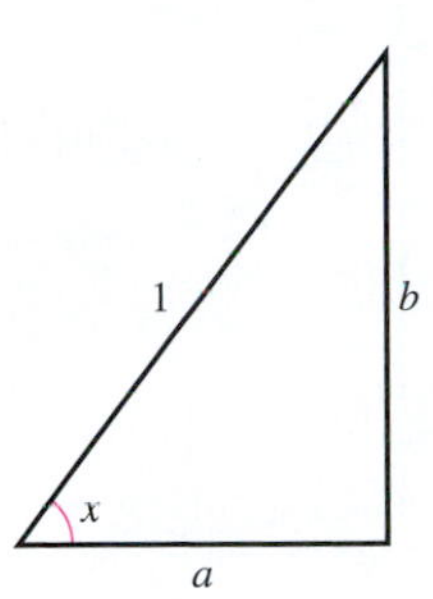

$a^2 + b^2 = 1$

$\cos x = \frac{a}{1} = a$

$\sin x = \frac{b}{1} = b$

$\sin^2 x + \cos^2 x = 1$

FIGURE 6.2

This is one of the **Pythagorean identities**. The name comes from its connection to the Pythagorean theorem. (See Figure 6.2.) From

$$\sin^2 x + \cos^2 x = 1,$$

$$\frac{\sin^2 x}{\sin^2 x} + \frac{\cos^2 x}{\sin^2 x} = \frac{1}{\sin^2 x} \quad \text{Divide both sides by } \sin^2 x.$$

$$1 + \cot^2 x = \csc^2 x \quad \frac{a^2}{b^2} = \left(\frac{a}{b}\right)^2\text{; simplify.}$$

This is also a Pythagorean identity. Next, from

$$\sin^2 x + \cos^2 x = 1,$$

$$\frac{\sin^2 x}{\cos^2 x} + \frac{\cos^2 x}{\cos^2 x} = \frac{1}{\cos^2 x} \quad \text{Divide both sides by } \cos^2 x.$$

$$\tan^2 x + 1 = \sec^2 x \quad \frac{a^2}{b^2} = \left(\frac{a}{b}\right)^2\text{; simplify.}$$

These three Pythagorean identities, together with the reciprocal, quotient, and even–odd identities, are called the **fundamental trigonometric identities**.

FUNDAMENTAL TRIGONOMETRIC IDENTITIES

1. Reciprocal Identities

$$\csc x = \frac{1}{\sin x} \qquad \sec x = \frac{1}{\cos x} \qquad \cot x = \frac{1}{\tan x}$$

$$\sin x = \frac{1}{\csc x} \qquad \cos x = \frac{1}{\sec x} \qquad \tan x = \frac{1}{\cot x}$$

2. Quotient Identities

$$\tan x = \frac{\sin x}{\cos x} \qquad \cot x = \frac{\cos x}{\sin x}$$

3. Pythagorean Identities

$$\sin^2 x + \cos^2 x = 1 \qquad 1 + \tan^2 x = \sec^2 x \qquad 1 + \cot^2 x = \csc^2 x$$

4. Even–Odd Identities

$$\sin(-x) = -\sin x \qquad \cos(-x) = \cos x \qquad \tan(-x) = -\tan x$$

$$\csc(-x) = -\csc x \qquad \sec(-x) = \sec x \qquad \cot(-x) = -\cot x$$

1 Use fundamental trigonometric identities to evaluate trigonometric functions.

Evaluating Trigonometric Functions Using the Fundamental Identities

It is frequently helpful to express the Pythagorean identities in equivalent forms.

Pythagorean Identity	Equivalent Forms
$\sin^2 x + \cos^2 x = 1$	$\begin{cases} \sin^2 x = 1 - \cos^2 x \\ \cos^2 x = 1 - \sin^2 x \end{cases}$
$1 + \cot^2 x = \csc^2 x$	$\begin{cases} 1 = \csc^2 x - \cot^2 x \\ \cot^2 x = \csc^2 x - 1 \end{cases}$
$1 + \tan^2 x = \sec^2 x$	$\begin{cases} 1 = \sec^2 x - \tan^2 x \\ \tan^2 x = \sec^2 x - 1 \end{cases}$

RECALL

Remember that the terminal side of an angle determines the quadrant in which the angle lies.

EXAMPLE 1 Using Pythagorean and Reciprocal Identities

If $\cos\theta = -\frac{5}{13}$ and $\pi < \theta < \frac{3\pi}{2}$, find $\tan\theta$.

Solution

Because $\cos\theta = -\frac{5}{13}$, the reciprocal identity says that $\sec\theta = -\frac{13}{5}$.

$$\tan^2\theta = \sec^2\theta - 1 \qquad \text{Equivalent form of } 1 + \tan^2\theta = \sec^2\theta,$$

$$\tan\theta = \sqrt{\sec^2\theta - 1} \qquad \text{Because } \theta \text{ is in quadrant III, } \tan\theta \text{ is positive.}$$

$$= \sqrt{\left(-\frac{13}{5}\right)^2 - 1} \qquad \text{Replace } \sec\theta \text{ with } -\frac{13}{5}.$$

$$= \sqrt{\frac{169}{25} - 1} = \sqrt{\frac{144}{25}} = \frac{12}{5}$$

Practice Problem 1 If $\sin\theta = \frac{4}{5}$ and $\frac{\pi}{2} < \theta < \pi$, find $\cot\theta$.

The next example shows how fundamental trigonometric identities are used to evaluate the remaining trigonometric functions from one specified trigonometric function value and the quadrant in which the angle lies.

EXAMPLE 2 Using the Fundamental Trigonometric Identities

If $\cot x = \frac{3}{4}$ and $\pi < x < \frac{3\pi}{2}$, find the values of the remaining trigonometric functions.

Solution

When given $\cot x$, we can find $\csc x$ by using the identity:

$$\csc^2 x = 1 + \cot^2 x \qquad \text{Pythagorean identity}$$

$$= 1 + \left(\frac{3}{4}\right)^2 \qquad \text{Replace } \cot x \text{ with } \frac{3}{4}.$$

$$\csc^2 x = 1 + \frac{9}{16} = \frac{25}{16} \qquad \text{Simplify.}$$

$$\csc x = -\frac{5}{4} \qquad \text{Square root property: Because } x \text{ is in quadrant III, } \csc x \text{ is negative.}$$

$$\sin x = -\frac{4}{5} \qquad \sin x = \frac{1}{\csc x}$$

Multiplying both sides of the quotient identity $\frac{\cos x}{\sin x} = \cot x$ by $\sin x$, we have

$$\begin{aligned} \cos x &= \cot x \sin x \\ &= \left(\frac{3}{4}\right)\left(-\frac{4}{5}\right) && \text{Substitute values of } \cot x \text{ and } \sin x. \\ &= -\frac{3}{5} && \text{Simplify.} \end{aligned}$$

We have the following values of the trigonometric functions:

$$\sin x = -\frac{4}{5} \qquad \cos x = -\frac{3}{5} \qquad \tan x = \frac{4}{3}$$

$$\csc x = -\frac{5}{4} \qquad \sec x = -\frac{5}{3} \qquad \cot x = \frac{3}{4} \qquad \text{Use the reciprocal identities.}$$

Practice Problem 2 If $\tan x = -\frac{1}{2}$ and $\frac{\pi}{2} < x < \pi$, find the values of the remaining trigonometric functions.

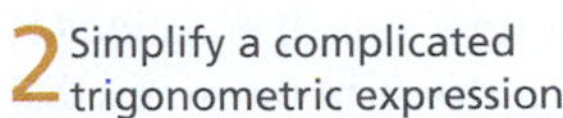

2 Simplify a complicated trigonometric expression.

Simplifying a Trigonometric Expression

Recall that when simplifying algebraic expressions, we can use properties of real numbers, factoring techniques, and special product formulas. We can also rationalize the denominator or find the least common denominator. To simplify trigonometric expressions, we use these same techniques together with the fundamental trigonometric identities.

EXAMPLE 3 Expressing All Trigonometric Functions in Terms of Sines and Cosines to Simplify an Expression

Rewrite the given expression in terms of sines and cosines and then simplify the resulting expression.

$$\cot x + \tan x + \csc x \sec x$$

Solution

$$\begin{aligned} &\cot x + \tan x + \csc x \sec x \\ &= \frac{\cos x}{\sin x} + \frac{\sin x}{\cos x} + \frac{1}{\sin x} \cdot \frac{1}{\cos x} && \text{Quotient and Reciprocal identities} \\ &= \frac{\cos^2 x}{\sin x \cos x} + \frac{\sin^2 x}{\sin x \cos x} + \frac{1}{\sin x \cos x} && \text{Write each fraction with common denominator } \sin x \cos x. \\ &= \frac{\cos^2 x + \sin^2 x + 1}{\sin x \cos x} && \text{Add numerators.} \\ &= \frac{1 + 1}{\sin x \cos x} = \frac{2}{\sin x \cos x} && \sin^2 x + \cos^2 x = 1 \end{aligned}$$

Practice Problem 3 Rewrite the expression in terms of sines and cosines and then simplify the resulting expression.

$$\frac{\tan x}{\sec x + 1} + \frac{\tan x}{\sec x - 1}$$

3 Prove that a given equation is not an identity.

Trigonometric Equations and Identities

To verify that an equation is a trigonometric identity, you must *verify* that both sides of the equation are equal for all values of the variable for which both sides are defined.

Consider the following equation:

$$\cos x = \sqrt{1 - \sin^2 x} \tag{1}$$

Equation (1) is true for all values of x in the interval $\left[-\frac{\pi}{2}, \frac{\pi}{2}\right]$. See Figure 6.3(a).

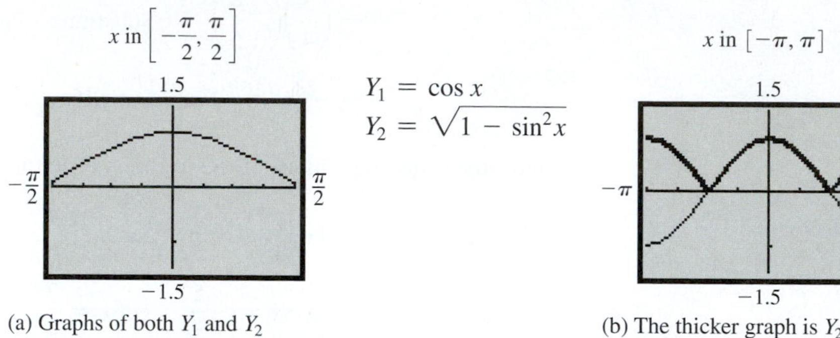

(a) Graphs of both Y_1 and Y_2

(b) The thicker graph is Y_2

FIGURE 6.3

Because the domain of the sine and cosine functions is the interval $(-\infty, \infty)$ and the range of the sine function is $[-1, 1]$, both sides of equation (1) are defined for all real numbers. But for any value of x in the interval $\left(\frac{\pi}{2}, \pi\right]$, the left side of equation (1) has a negative value (because $\cos x$ is negative in quadrant II) and the right side of equation (1) has a positive value. Therefore, equation (1) is *not* an identity. Figure 6.3(b) illustrates that a graphing calculator can help verify that a given equation is *not* an identity. Figure 6.3(a), however, shows that a graphing calculator cannot prove that a given equation *is* an identity because you might happen to use a viewing window where the graphs coincide.

EXAMPLE 4 Proving That an Equation Is Not an Identity

Prove that the following equation is not an identity.

$$(\sin x - \cos x)^2 = \sin^2 x - \cos^2 x$$

Solution

SIDE NOTE

A **counterexample** is an example that disproves a statement.

To prove that the equation is not an identity, we look for a counterexample. That is, we find at least one value of x that results in both sides being defined but not equal.

Let $x = 0$. We know that $\sin 0 = 0$ and $\cos 0 = 1$.

The left side $= (\sin x - \cos x)^2 = (0 - 1)^2 = 1$ Replace x with 0.

The right side $= \sin^2 x - \cos^2 x = (0)^2 - (1)^2 = -1$ Replace x with 0.

For $x = 0$, the two sides of the equation are not equal; so it is not an identity.

Practice Problem 4 Prove that the equation $\cos x = 1 - \sin x$ is not an identity. 

4 Verify a trigonometric identity.

Process of Verifying Trigonometric Identities

Verifying a trigonometric identity differs from solving an equation. In verifying an identity, we are given an equation and want to show that it is true for *all* values of the variable for which both sides of the equation are defined. We will use the following approach.

TECHNOLOGY CONNECTION

In Example 5, we use a graphing calculator to graph $Y_1 = \dfrac{\csc^2 x - 1}{\cot x}$ and $Y_2 = \cot x$. Although not a definitive method of proof, the graphs appear identical; so the equation in Example 5 *appears* to be an identity.

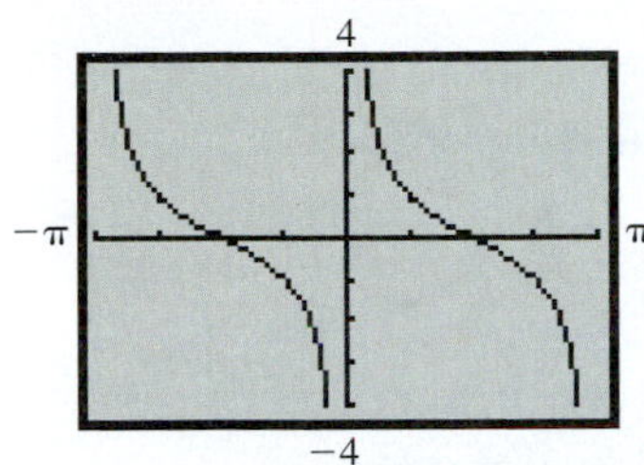

Graph of Y_1

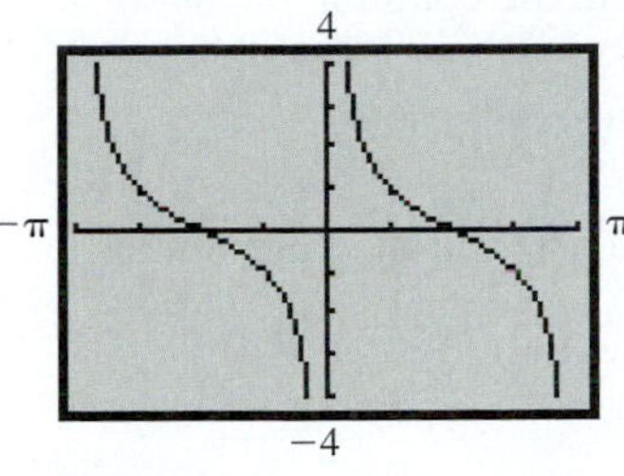

Graph of Y_2

VERIFYING TRIGONOMETRIC IDENTITIES

To verify that an equation is an identity, transform one side of the equation into the other side by a sequence of steps, each of which produces an identity. The steps involved can be algebraic manipulations or can use fundamental identities. Note that in verifying an identity, we *do not* just perform the same operation on both sides of the equation; see Exercise 125.

Methods of Verifying Trigonometric Identities

Verifying trigonometric identities requires practice and experience; there is no set procedure. In the following examples, we suggest five guidelines for verifying trigonometric identities, with the first two guidelines being more widely used.

1. Start with the more complicated side and transform it to the simpler side.

EXAMPLE 5 Verifying an Identity

Verify the identity:

$$\frac{\csc^2 x - 1}{\cot x} = \cot x$$

Solution

We start with the left side of the equation because it appears more complicated than the right side.

$$\frac{\csc^2 x - 1}{\cot x} = \frac{\cot^2 x}{\cot x} \qquad \csc^2 x = 1 + \cot^2 x, \text{ so } \csc^2 x - 1 = \cot^2 x$$

$$= \cot x \qquad \text{Divide out the common factor, } \cot x.$$

We have shown that the left side of the equation is equal to the right side, which verifies the identity.

Practice Problem 5 Verify the identity:

$$\frac{1 - \sin^2 x}{\cos x} = \cos x$$

2. Stay focused on the final expression.
While working on one side of the equation, stay focused on your goal of converting it to the form on the other side. This often helps in deciding what your next step should be.

EXAMPLE 6 Verifying an Identity

Verify the following identity proposed in the introduction to this section:

$$\frac{\sec\theta - \cos\theta}{\sin\theta} = \frac{\sin\theta}{\cos\theta}$$

TECHNOLOGY CONNECTION

In Example 6, we use the "Table" feature of a graphing calculator in *radian* mode to create a table of values for $Y_1 = \dfrac{\sec x - \cos x}{\sin x}$ and $Y_2 = \dfrac{\sin x}{\cos x}$ for different values of x.

X	Y1	Y2
-3	-.288	-.288
-2.5	-1.865	-1.865
-2	-10.5	-10.5
-1.5	-398.7	-398.7
-1	-5.765	-5.765
-.5	-1.245	-1.245
0	0	ERROR

X=0

X	Y1	Y2
0	0	ERROR
.5	1.245	1.245
1	5.7649	5.7649
1.5	398.7	398.7
2	10.501	10.501
2.5	1.8649	1.8649
3	.28797	.28797

X=0

Note that Y_2 equals 0 when $x = 0$ but that Y_1 is undefined when $x = 0$. While not a definitive method of proof, the table shows that the equation in Example 6 *appears* to be an identity because the table values are identical for values of x for which both Y_1 and Y_2 are defined.

Solution

We start with the more complicated left side; note the other side involves only $\sin\theta$ and $\cos\theta$.

$$\frac{\sec\theta - \cos\theta}{\sin\theta} = \frac{\frac{1}{\cos\theta} - \cos\theta}{\sin\theta} \qquad \sec\theta = \frac{1}{\cos\theta}$$

$$= \frac{\left(\frac{1}{\cos\theta} - \cos\theta\right)\cos\theta}{\sin\theta\cos\theta} \qquad \text{Multiply numerator and denominator by } \cos\theta.$$

$$= \frac{\frac{1}{\cos\theta}\cdot\cos\theta - \cos^2\theta}{\sin\theta\cos\theta} \qquad \text{Distributive property}$$

$$= \frac{1 - \cos^2\theta}{\sin\theta\cos\theta} \qquad \text{Simplify.}$$

$$= \frac{\sin^2\theta}{\sin\theta\cos\theta} \qquad \sin^2\theta + \cos^2\theta = 1, \text{ so } \sin^2\theta = 1 - \cos^2\theta$$

$$= \frac{\sin\theta}{\cos\theta} \qquad \text{Divide out the common factor } \sin\theta.$$

The left side is identical to the right side; the given equation is an identity.

Practice Problem 6 In Figure 6.1 on page 568, triangles OPD and TCP are similar. So $\dfrac{OD}{OP} = \dfrac{PT}{CT}$ leads to

$$\frac{\csc\theta}{1} = \frac{\tan\theta}{\sec\theta - \cos\theta}.$$

Verify that this equation is an identity.

3. Convert to sines and cosines.
It may be helpful to rewrite all trigonometric functions in the equation in terms of sines and cosines and then simplify.

EXAMPLE 7 Verifying by Rewriting with Sines and Cosines

Verify the identity $\cot^4 x + \cot^2 x = \cot^2 x \csc^2 x$.

Solution

We start with the more complicated left side.

$$\cot^4 x + \cot^2 x = \frac{\cos^4 x}{\sin^4 x} + \frac{\cos^2 x}{\sin^2 x} \qquad \cot x = \frac{\cos x}{\sin x}$$

$$= \frac{\cos^4 x}{\sin^4 x} + \frac{\cos^2 x}{\sin^2 x}\cdot\frac{\sin^2 x}{\sin^2 x} \qquad \text{The LCD is } \sin^4 x.$$

$$= \frac{\cos^4 x + \cos^2 x \sin^2 x}{\sin^4 x} \qquad \text{Add rational expressions.}$$

$$= \frac{\cos^2 x(\cos^2 x + \sin^2 x)}{\sin^4 x} \qquad \text{Factor out } \cos^2 x.$$

$$= \frac{\cos^2 x(1)}{\sin^4 x} \qquad \cos^2 x + \sin^2 x = 1$$

$$= \frac{\cos^2 x}{\sin^2 x} \cdot \frac{1}{\sin^2 x} \qquad \text{Factor to obtain the form on the right side.}$$

$$= \cot^2 x \csc^2 x \qquad \cot x = \frac{\cos x}{\sin x}, \csc x = \frac{1}{\sin x}$$

Because the left side is identical to the right side, the given equation is an identity.

Practice Problem 7 Verify the identity $\tan^4 x + \tan^2 x = \tan^2 x \sec^2 x$.

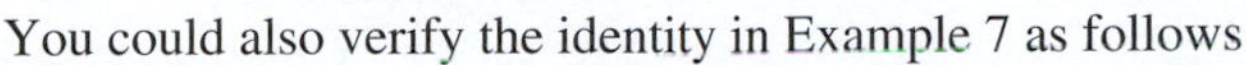

You could also verify the identity in Example 7 as follows:

$$\begin{aligned} \cot^4 x + \cot^2 x &= \cot^2 x(\cot^2 x + 1) && \text{Distributive property} \\ &= \cot^2 x \csc^2 x && \text{Pythagorean identity} \end{aligned}$$

This approach shows that rewriting the expression using only sines and cosines is not always the quickest way to verify an identity. It is a useful approach when you are stuck, however.

4. Work on both sides.
Sometimes it is helpful to work separately on both sides of the equation. To verify the identity $P(x) = Q(x)$, for example, we transform the left side $P(x)$ into $R(x)$ by using algebraic manipulations and known identities. Then $P(x) = R(x)$ is an identity. Next, we transform the right side, $Q(x)$, into $R(x)$ so that the equation $Q(x) = R(x)$ is an identity. It then follows that $P(x) = Q(x)$ is an identity. See the next example.

EXAMPLE 8 Verifying an Identity by Transforming Both Sides Separately

Verify the identity:

$$\frac{1}{1 - \sin x} - \frac{1}{1 + \sin x} = \frac{\tan^2 x + \sec^2 x + 1}{\csc x}$$

Solution

We start with the left side of the equation.

$$\frac{1}{1 - \sin x} - \frac{1}{1 + \sin x} \qquad \text{Begin with the left side.}$$

$$= \frac{1}{1 - \sin x} \cdot \frac{1 + \sin x}{1 + \sin x} - \frac{1}{1 + \sin x} \cdot \frac{1 - \sin x}{1 - \sin x} \qquad \text{Rewrite using the LCD, } (1 - \sin x)(1 + \sin x).$$

$$= \frac{(1 + \sin x) - (1 - \sin x)}{(1 - \sin x)(1 + \sin x)} \qquad \text{Subtract fractions.}$$

$$= \frac{1 + \sin x - 1 + \sin x}{1 - \sin^2 x} \qquad \text{Rewrite the numerator; multiply in the denominator.}$$

$$= \frac{2 \sin x}{\cos^2 x} \qquad \text{Simplify; } 1 - \sin^2 x = \cos^2 x$$

We now try to convert the right side to the form $\frac{2\sin x}{\cos^2 x}$.

$$\begin{aligned}
&\frac{\tan^2 x + \sec^2 x + 1}{\csc x} \\
&= \frac{(\tan^2 x + 1) + \sec^2 x}{\csc x} && \text{Regroup terms.} \\
&= \frac{\sec^2 x + \sec^2 x}{\csc x} && \tan^2 x + 1 = \sec^2 x \\
&= \frac{2\sec^2 x}{\csc x} && \text{Simplify.} \\
&= 2\left(\frac{1}{\csc x}\right)\sec^2 x && \text{Rewrite.} \\
&= 2\sin x\left(\frac{1}{\cos^2 x}\right) && \text{Reciprocal identities} \\
&= \frac{2\sin x}{\cos^2 x} && \text{Multiply.}
\end{aligned}$$

From the original equation, we see that

Left-Hand Side — Third Expression — Right-Hand Side

$$\frac{1}{1-\sin x} - \frac{1}{1+\sin x} = \frac{2\sin x}{\cos^2 x} = \frac{\tan^2 x + \sec^2 x + 1}{\csc x}.$$

Because both sides of the original equation are equal to $\frac{2\sin x}{\cos^2 x}$, the identity is verified.

Practice Problem 8 Verify the identity:

$$\tan\theta + \sec\theta = \frac{\csc\theta + 1}{\cot\theta}$$

5. Use conjugates.

It may be helpful to multiply both the numerator and denominator of a fraction by the same factor. Recall that the sum $a + b$ and the difference $a - b$ are the *conjugates* of each other and that $(a + b)(a - b) = a^2 - b^2$.

SIDE NOTE

You cannot verify a trigonometric identity that involves rational expressions by using cross multiplication because you do NOT know the expressions are equal. Equality is what you are trying to verify.

EXAMPLE 9 Verifying an Identity by Using a Conjugate

Verify the identity:

$$\frac{\cos x}{1+\sin x} = \frac{1-\sin x}{\cos x}$$

Solution

We start with the left side of the equation.

$$\frac{\cos x}{1 + \sin x} = \frac{\cos x(1 - \sin x)}{(1 + \sin x)(1 - \sin x)}$$ Multiply the numerator and the denominator by $1 - \sin x$, the conjugate of $1 + \sin x$.

$$= \frac{\cos x(1 - \sin x)}{1 - \sin^2 x}$$ $(a + b)(a - b) = a^2 - b^2$

$$= \frac{\cos x(1 - \sin x)}{\cos^2 x}$$ $1 - \sin^2 x = \cos^2 x$

$$= \frac{1 - \sin x}{\cos x}$$ Remove the common factor $\cos x$.

Because the left side is identical to the right side, the given equation is an identity.

Practice Problem 9 Verify the identity:

$$\frac{\tan x}{\sec x + 1} = \frac{\sec x - 1}{\tan x}$$

SUMMARY OF MAIN FACTS

Guidelines for Verifying Trigonometric Identities

Algebra Operations

Review the procedure for combining fractions by finding the least common denominator.

Fundamental Trigonometric Identities

Review the fundamental trigonometric identities summarized on page 569. Look for an opportunity to apply the fundamental trigonometric identities when working on either side of the identity to be verified. Become thoroughly familiar with alternative forms of fundamental identities. For example, $\sin^2 x = 1 - \cos^2 x$ and $\sec^2 x - \tan^2 x = 1$ are alternative forms of the fundamental identities $\sin^2 x + \cos^2 x = 1$ and $\sec^2 x = 1 + \tan^2 x$, respectively.

1. **Start with the more complicated side and transform it to the simpler side.**
   It is generally helpful to start with the more complicated side of an identity and simplify it until it becomes identical to the other side. See Example 5.
2. **Stay focused on the final expression.**
   While working on one side of the identity, stay focused on your goal of converting it to the form on the other side. See Example 6.

   The following three techniques are *sometimes* helpful.
3. **Option: Convert to sines and cosines.**
   Rewrite one side of the identity in terms of sines and cosines. See Example 7.
4. **Option: Work on both sides.**
   Transform each side separately to the same equivalent expression. See Example 8.
5. **Option: Use conjugates.**
   In expressions containing $1 + \sin x$, $1 - \sin x$, $1 + \cos x$, $1 - \cos x$, $\sec x + \tan x$, and so on, multiply the numerator and the denominator by its conjugate and then use the appropriate Pythagorean identity. See Example 9.

In calculus, it is sometimes helpful to convert algebraic expressions to trigonometric ones. This technique is called *trigonometric substitution* and is illustrated in the next example.

EXAMPLE 10 Trigonometric Substitution

Replace x with $\cos\theta$ in the expression $\sqrt{1-x^2}$ and simplify. Assume $0 \leqslant \theta \leqslant \pi/2$.

Solution

$$\begin{aligned} \sqrt{1-x^2} &= \sqrt{1-\cos^2\theta} && \text{Replace } x \text{ with } \cos\theta. \\ &= \sqrt{\sin^2\theta} && \text{Pythagorean identity} \\ &= \sin\theta && \sin\theta \geqslant 0 \text{ because } 0 \leqslant \theta \leqslant \frac{\pi}{2} \end{aligned}$$

Practice Problem 10 Replace x with $3\sin\theta$ in the expression $\sqrt{9-x^2}$ and simplify. Assume $0 \le \theta \le \dfrac{\pi}{2}$.

Answers to Practice Problems

1. $-\dfrac{3}{4}$ **2.** $\sin x = \dfrac{\sqrt{5}}{5}$, $\cos x = -\dfrac{2\sqrt{5}}{5}$, $\sec x = -\dfrac{\sqrt{5}}{2}$, $\csc x = \sqrt{5}$, $\cot x = -2$ **3.** $\dfrac{2}{\sin x}$ **4.** At $x = \dfrac{3\pi}{2}$, $0 \neq 2$. **10.** $3\cos\theta$

SECTION 6.1 Exercises

Basic Concepts and Skills

1. An equation that is true for all values of the variable in its domain is called a(n) ____________.
2. To show that an equation is not an identity, we must find at least ____________ value(s) of the variable that results in both sides being defined but ____________.
3. $\sin^2 x +$ ____________ $= 1$; $1 +$ ____________ $= \sec^2 x$; $\csc^2 x - \cot^2 x =$ ____________.
4. **True or False.** $\tan^2 x \cot x = \tan x$ is not an identity because $\tan^2 x \cot x$ is not defined for $x = 0$ but $\tan x$ is defined for $x = 0$.
5. **True or False.** The value of $x = \dfrac{3\pi}{2}$ can be used to show that $\sin x = \sqrt{1-\cos^2 x}$ is not an identity.
6. **True or False.** If $\cos\theta < 0$ and $\pi < \theta < \dfrac{3\pi}{2}$, then $\tan\theta < 0$.
7. If $\sin\theta = -\dfrac{12}{13}$ and $\pi < \theta < \dfrac{3\pi}{2}$, find $\cos\theta$.
8. If $\sec\theta = \dfrac{5}{4}$ and $\dfrac{3\pi}{2} < \theta < 2\pi$, find $\tan\theta$.
9. If $\cot\theta = \dfrac{1}{2}$ and $\pi < \theta < \dfrac{3\pi}{2}$, find $\csc\theta$.
10. If $\cos\theta = -\dfrac{1}{3}$ and $\dfrac{\pi}{2} < \theta < \pi$, find $\tan\theta$.
11. If $\sin x = -\dfrac{2}{3}$ and $\pi < x < \dfrac{3\pi}{2}$, find $\cot x$.
12. If $\tan x = \dfrac{1}{2}$ and $\pi < x < \dfrac{3\pi}{2}$, find $\cos x$.

In Exercises 13–24, use the fundamental identities and the given information to find the exact values of the remaining trigonometric functions of x.

13. $\cos x = -\dfrac{3}{5}$ and $\sin x = \dfrac{4}{5}$
14. $\cos x = \dfrac{3}{5}$ and $\sin x = -\dfrac{4}{5}$
15. $\sin x = \dfrac{1}{\sqrt{3}}$ and $\cos x = \sqrt{\dfrac{2}{3}}$
16. $\sin x = \dfrac{1}{\sqrt{3}}$ and $\cos x = -\sqrt{\dfrac{2}{3}}$

17. $\tan x = \frac{1}{2}$ and $\sec x = -\frac{\sqrt{5}}{2}$

18. $\tan x = \frac{1}{2}$ and $\sec x = \frac{\sqrt{5}}{2}$

19. $\csc x = 3$ and $\cot x = 2\sqrt{2}$

20. $\csc x = 3$ and $\cot x = -2\sqrt{2}$

21. $\cot x = \frac{12}{5}$ and $\sin x = -\frac{5}{13}$

22. $\cot x = -\frac{12}{5}$ and $\sin x = -\frac{5}{13}$

23. $\sec x = 3$ and $\frac{3\pi}{2} < x < 2\pi$

24. $\tan x = -2$ and $\frac{\pi}{2} < x < \pi$

In Exercises 25–30, for each expression in Column I, find the expression in Column II that results in an identity when the expressions are equated.

I	II
25. $\dfrac{\cos^2 x + \sin^2 x}{\sin x \cos x}$	**a.** $\cos x$
26. $\dfrac{1}{1 + \tan^2 x}$	**b.** $(1 + \cos x)^2$
27. $-\cot x \sin(-x)$	**c.** $\dfrac{1}{\sin x \cos x}$
28. $1 + 2\cos x + \cos^2 x$	**d.** $\sin^4 x$
29. $1 - 2\cos^2 x + \cos^4 x$	**e.** 1
30. $\sin^2 x(1 + \cot^2 x)$	**f.** $\cos^2 x$

In Exercises 31–40, use the fundamental identities and appropriate algebraic operations to simplify each expression.

31. $(1 + \tan x)(1 - \tan x) + \sec^2 x$

32. $(\sec x - 1)(\sec x + 1) - \tan^2 x$

33. $(\sec x + \tan x)(\sec x - \tan x)$

34. $\dfrac{\sec^2 x - 4}{\sec x - 2}$

35. $\csc^4 x - \cot^4 x$

36. $\sin x \cos x(\tan x + \cot x)$

37. $\dfrac{\sec x \csc x(\sin x + \cos x)}{\sec x + \csc x}$

38. $\dfrac{1}{\csc x + 1} - \dfrac{1}{\csc x - 1}$

39. $\dfrac{\tan^2 x - 2\tan x - 3}{\tan x + 1}$

40. $\dfrac{\tan^2 x + \sec x - 1}{\sec x - 1}$

In Exercises 41–46, prove that the given equation is not an identity by finding a value of x for which the two sides have different values. The answers may vary.

41. $\sin x = 1 - \cos x$

42. $\tan x = \sec x - 1$

43. $\cos x = \sqrt{1 - \sin^2 x}$

44. $\sec x = \sqrt{1 + \tan^2 x}$

45. $\sin^2 x = (1 - \cos x)^2$

46. $\cot^2 x = (\csc x + 1)^2$

In Exercises 47–52, use a graphing calculator to determine whether the equation could be an identity by graphing each side of the equation on the same screen. Then prove or disprove the statement algebraically.

47. $\cos^2 x - \sin^2 x = 2\cos^2 x - 1$

48. $(\sin x + \cos x)^2 - 1 = 2\sin x \cos x$

49. $\sin x + \cos x = 1$

50. $\dfrac{\sin 2x}{2} = \sin x$

51. $\dfrac{\sin^2 x}{1 + \cos x} = 1 - \cos x$

52. $\sin x = \cos x \tan x$

In Exercises 53–58, use the "Table" feature of a graphing calculator in *radian* mode to calculate each side of the equation for different values of x. Determine whether the equation could be an identity. Then prove or disprove the statement algebraically.

53. $\sin x \cot x = \cos x$

54. $\sec x \cot x = \csc x$

55. $\dfrac{\tan x - 1}{\tan x + 1} = \dfrac{1 - \cot x}{1 + \cot x}$

56. $\sin^2 x \sec^2 x + 1 = \sec^2 x$

57. $\tan 2x = 2\tan x$

58. $(1 - \sin x)^2 = \cos x$

In Exercises 59–94, verify each identity.

59. $\sin x \tan x + \cos x = \sec x$

60. $\cos x \cot x + \sin x = \csc x$

61. $\dfrac{1 - 4\cos^2 x}{1 - 2\cos x} = 1 + 2\cos x$

62. $\dfrac{9 - 16\sin^2 x}{3 + 4\sin x} = 3 - 4\sin x$

63. $(\cos x - \sin x)(\cos x + \sin x) = 1 - 2\sin^2 x$

64. $(\sin x - \cos x)(\sin x + \cos x) = 1 - 2\cos^2 x$

65. $\sin^2 x \cot^2 x + \sin^2 x = 1$

66. $\tan^2 x - \sin^2 x = \sin^4 x \sec^2 x$

67. $\sin^3 x - \cos^3 x = (\sin x - \cos x)(1 + \sin x \cos x)$

68. $\sin^3 x + \cos^3 x = (\sin x + \cos x)(1 - \sin x \cos x)$

69. $\cos^4 x - \sin^4 x = 1 - 2\sin^2 x$

70. $\cos^4 x - \sin^4 x = 2\cos^2 x - 1$

71. $\dfrac{1}{1 - \sin x} + \dfrac{1}{1 + \sin x} = 2\sec^2 x$

72. $\dfrac{1}{1 - \cos x} + \dfrac{1}{1 + \cos x} = 2\csc^2 x$

73. $\dfrac{1}{\csc x - 1} - \dfrac{1}{\csc x + 1} = 2\tan^2 x$

74. $\dfrac{1}{\sec x - 1} + \dfrac{1}{\sec x + 1} = 2\sec x \cot^2 x$

75. $\sec^2 x + \csc^2 x = \sec^2 x \csc^2 x$

76. $\cot^2 x + \tan^2 x = \sec^2 x \csc^2 x - 2$

77. $\dfrac{1}{\sec x - \tan x} + \dfrac{1}{\sec x + \tan x} = \dfrac{2}{\cos x}$

78. $\dfrac{1}{\csc x + \cot x} + \dfrac{1}{\csc x - \cot x} = \dfrac{2}{\sin x}$

79. $\dfrac{\sin x}{1 + \cos x} = \csc x - \cot x$

80. $\dfrac{\sin x}{1 - \cos x} = \csc x + \cot x$

81. $(\sin x + \cos x)^2 = 1 + 2 \sin x \cos x$

82. $(\sin x - \cos x)^2 = 1 - 2 \sin x \cos x$

83. $(1 + \tan x)^2 = \sec^2 x + 2 \tan x$

84. $(1 - \cot x)^2 = \csc^2 x - 2 \cot x$

85. $\dfrac{\tan x \sin x}{\tan x + \sin x} = \dfrac{\tan x - \sin x}{\tan x \sin x}$

86. $\dfrac{\cot x \cos x}{\cot x + \cos x} = \dfrac{\cot x - \cos x}{\cot x \cos x}$

87. $(\tan x + \cot x)^2 = \sec^2 x + \csc^2 x$

88. $(1 + \cot^2 x)(1 + \tan^2 x) = \dfrac{1}{\sin^2 x \cos^2 x}$

89. $\dfrac{\sin^2 x - \cos^2 x}{\sec^2 x - \csc^2 x} = \sin^2 x \cos^2 x$

90. $\left(\tan x + \dfrac{1}{\cot x}\right)\left(\cot x + \dfrac{1}{\tan x}\right) = 4$

91. $\dfrac{\tan x}{1 + \sec x} + \dfrac{1 + \sec x}{\tan x} = 2 \csc x$

92. $\dfrac{\cot x}{1 + \csc x} + \dfrac{1 + \csc x}{\cot x} = 2 \sec x$

93. $\dfrac{\sin x + \tan x}{\cos x + 1} = \tan x$

94. $\dfrac{\sin x}{1 + \tan x} = \dfrac{\cos x}{1 + \cot x}$

In Exercises 95–100, make the indicated trigonometric substitution in the given algebraic expression and simplify. Assume $0 < \theta < \dfrac{\pi}{2}$.

95. $\sqrt{1 + x^2}, x = \tan \theta$

96. $\sqrt{4 - x^2}, x = 2 \cos \theta$

97. $\sqrt{x^2 - 1}, x = \sec \theta$

98. $\sqrt{x^2 - 4}, x = 2 \sec \theta$

99. $\dfrac{x}{\sqrt{1 - x^2}}, x = \cos \theta$

100. $\dfrac{x}{\sqrt{x^2 - 1}}, x = \sec \theta$

Applying the Concepts

101. Length of a ladder. A ladder x feet long makes an angle θ with the horizontal and reaches a height of 20 feet. Then $x = \dfrac{20}{\sin \theta}$. Use a reciprocal identity to rewrite this formula.

102. Distance from a building. From a distance of x feet to a 60-foot-high building, the angle of elevation is θ degrees. Then $x = \dfrac{60}{\tan \theta}$. Use a reciprocal identity to rewrite this formula.

103. Intersecting lines. If two nonvertical and nonperpendicular lines with slopes m_1 and m_2 $(m_1 > m_2)$ intersect, then the acute angle θ between the lines satisfies the equation $m_1 \cos \theta - \sin \theta = m_2 \cos \theta + m_1 m_2 \sin \theta$. Rewrite this equation in terms of a single trigonometric function.

104. Tower height. The angles to the top of a tower measured from two locations d units apart are α and β as shown in the figure. The height h of the tower satisfies the equation $h = \dfrac{d \sin \alpha \sin \beta}{\cos \alpha \sin \beta - \sin \alpha \cos \beta}$. Rewrite this equation in terms of cotangents.

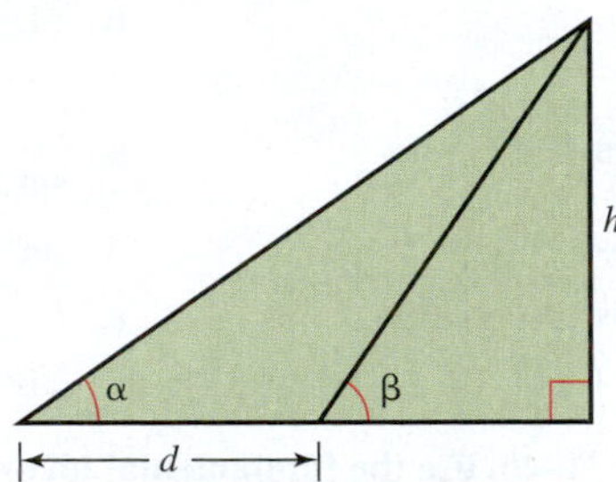

Beyond the Basics

In Exercises 105–109, verify each identity.

105. $\dfrac{1 - \sin x}{1 - \sec x} - \dfrac{1 + \sin x}{1 + \sec x} = 2 \cot x(\cos x - \csc x)$

106. $\dfrac{\sec x + \tan x}{\csc x + \cot x} - \dfrac{\sec x - \tan x}{\csc x - \cot x} = 2(\sec x - \csc x)$

107. $(1 - \tan x)^2 + (1 - \cot x)^2 = (\sec x - \csc x)^2$

108. $\sec^6 x - \tan^6 x = 1 + 3 \tan^2 x + 3 \tan^4 x$

109. $\dfrac{\cot x + \csc x - 1}{\cot x + \csc x + 1} = \dfrac{1 - \sin x}{\cos x}$

110. $\dfrac{\tan x + \sec x - 1}{\tan x - \sec x + 1} = \dfrac{1 + \sin x}{\cos x}$

In Exercises 111–120, simplify each expression. Answer may vary.

111. $(\sin u \cos v + \cos u \sin v)^2 + (\cos u \cos v - \sin u \sin v)^2$

112. $(\sin u \cos v - \cos u \sin v)^2 + (\cos u \cos v + \sin u \sin v)^2$

113. $3(\sin^4 x + \cos^4 x) - 2(\sin^6 x + \cos^6 x)$

114. $\sin^6 x + \cos^6 x - 3\cos^4 x$

115. $\dfrac{1 - \cos\theta}{1 + \cos\theta} + \dfrac{1 + \cos\theta}{1 - \cos\theta}$

116. $\dfrac{\cos^3\theta + \sin^3\theta}{\cos\theta + \sin\theta} + \dfrac{\cos^3\theta - \sin^3\theta}{\cos\theta - \sin\theta}$

117. $\dfrac{\sec^2\theta + 2\tan^2\theta}{1 + 3\tan^2\theta}$

118. $\dfrac{\csc^2\alpha + \sec^2\alpha}{\csc^2\alpha - \sec^2\alpha} - \dfrac{1 + \tan^2\alpha}{1 - \tan^2\alpha}$

119. $\cos^2 x(3 - 4\cos^2 x)^2 + \sin^2 x(3 - 4\sin^2 x)^2$

120. $(\sin\theta + \csc\theta)^2 + (\cos\theta + \sec\theta)^2 - \cot^2\theta$

121. If $x = r\cos u\cos v$, $y = r\cos u\sin v$, and $z = r\sin u$, verify that $x^2 + y^2 + z^2 = r^2$.

122. If $\tan x + \cot x = 2$, show that

a. $\tan^2 x + \cot^2 x = 2$.

b. $\tan^3 x + \cot^3 x = 2$.

123. If $\sec x + \cos x = 2$, show that

a. $\sec^2 x + \cos^2 x = \sec^4 x + \cos^4 x = 2$

b. $\sec^3 x + \cos^3 x = 2$.

124. If $\sin x + \sin^2 x = 1$, show that $\cos^2 x + \cos^4 x = 1$.

Critical Thinking/Discussion/Writing

125. Consider the equation $x = -x$, which is not an identity. If you square both sides, you get $x^2 = (-x)^2$, which is an identity. This shows that performing the same operation on both sides of an equation (a nonidentity) may lead to an identity.

Consider the equation $\sqrt{\sin^2 x} = \sin x$.

a. Find a value of x that shows that this equation is not an identity.

b. Show that squaring both sides of this equation results in an identity.

126. If x and y are real numbers, explain why $\sec\theta$ cannot be equal to $\dfrac{xy}{x^2 + y^2}$.

127. Find values of t for which $\sin\theta = \dfrac{1 + t^2}{1 - t^2}$ is possible.

128. Find values of a and b for which $\cos^2\theta = \dfrac{a^2 + b^2}{2ab}$ is possible.

129. Find values of a and b for which $\csc^2\theta = \dfrac{2ab}{a^2 + b^2}$ is possible.

130. Prove that the following inequalities are true for $0 < \theta < \dfrac{\pi}{2}$.

a. $\sin^2\theta + \csc^2\theta \ge 2$

b. $\cos^2\theta + \sec^2\theta \ge 2$

c. $\sec^2\theta + \csc^2\theta \ge 4$

Maintaining Skills

In Exercises 131 and 132, expand each expression.

131. $(2x - y)^2$ **132.** $(3x - 5y)^2$

In Exercises 133–135, find the exact value.

133. $\sin\left(-\dfrac{\pi}{6}\right)$ **134.** $\cos\left(-\dfrac{\pi}{6}\right)$ **135.** $\tan\left(-\dfrac{\pi}{6}\right)$

In Exercises 136–138, identify the Pythagorean identity for which each expression is an equivalent form.

136. $\cos^2 x = 1 - \sin^2 x$

137. $\sec^2 x - \tan^2 x = 1$

138. $\cot^2 x = \csc^2 x - 1$

139. Find $\sin t$, given that $\cos t = -\dfrac{\sqrt{3}}{4}$ and $\sin t > 0$.

140. Find $\cos t$, given that $\tan t = \dfrac{4}{3}$ and t is in quadrant III.

SECTION 6.2

Sum and Difference Formulas

BEFORE STARTING THIS SECTION, REVIEW

1. Distance formula (Section 2.1, page 159)
2. Fundamental trigonometric identities (Section 6.1, page 569)
3. Trigonometric functions of common angles (Section 5.2, page 491)

OBJECTIVES

1. Use the sum and difference formulas for cosine.
2. Use the cofunction identities.
3. Use the sum and difference formulas for sine.
4. Use the sum and difference formulas for tangent.

◆ Pure Tones in Music

A vibrating part of an instrument transmits its vibrations to the air, and they travel away as sound waves. The number of complete cycles per second traveled by a wave is called its *frequency* and is denoted by f. The unit of frequency, cycles per second, is also called hertz (abbreviated Hz) after Gustav Hertz, the discoverer of radio waves. The note A in the octave above middle C has a frequency of 440 Hz, or 440 cycles per second. Therefore, the time between waves is $\frac{1}{440}$ second. We define this time between waves as the period $T: T = \frac{1}{f}$. You hear the sound produced by note A when the vibration from the source travels to your ear and puts pressure on your eardrum.

A *pure tone* is a tone in which the vibration is a simple harmonic of just one frequency. The simple harmonic of a pure tone with frequency f can be described by the equation

$$y = a\sin(2\pi ft),$$

where $|a|$ is the amplitude (indicating loudness), t is the time in seconds, and y is the pressure of a pure tone on an eardrum (in pounds per square foot). In Example 10, we consider the pressure due to two pure tones of the same frequency.

1 Use the sum and difference formulas for cosine.

Sum and Difference Formulas for Cosine

In this section, we develop identities involving the sum and difference of two variables, u and v. The variables represent any two real numbers, or angles (in radian or degree measure).

SUM AND DIFFERENCE FORMULAS FOR COSINE

$$\cos(u + v) = \cos u\cos v - \sin u\sin v$$
$$\cos(u - v) = \cos u\cos v + \sin u\sin v$$

To prove the second formula, we assume that $0 < v < u < 2\pi$, although this identity is true for all real numbers u and v. Figure 6.4(a) shows points P and Q on the unit circle on the terminal sides of angles u and v. The definition of the circular functions tells us that $P = (\cos u, \sin u)$ and $Q = (\cos v, \sin v)$. In Figure 6.4(b), we rotated angle $(u - v)$ to standard position and labeled point B on the unit circle and the terminal side of the angle $(u - v)$.

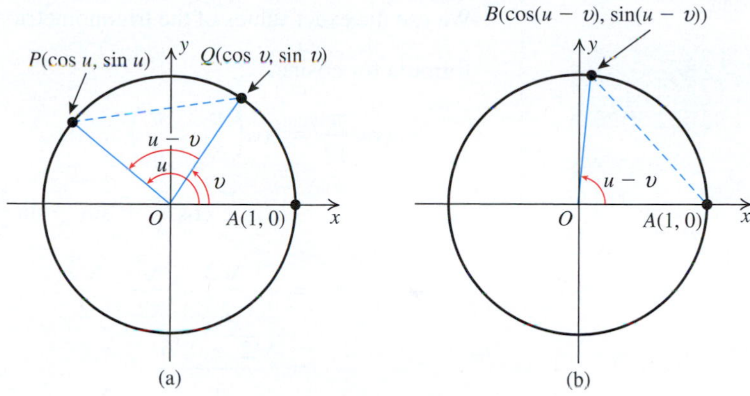

FIGURE 6.4

RECALL

The distance $d(P, Q)$ between two points $P(x_1, y_1)$ and $Q(x_2, y_2)$ is given by

$d(P, Q) = \sqrt{(x_2 - x_1)^2 + (y_2 - y_1)^2}$.

We know that triangle POQ in Figure 6.4(a) is congruent to triangle BOA in Figure 6.4(b) by SAS (side-angle-side) congruence; so

$$d(P, Q) = d(A, B)$$
$$[d(P, Q)]^2 = [d(A, B)]^2 \quad (1) \qquad \text{Square both sides.}$$

Now use the distance formula to simplify the left side of equation (1).

$$\begin{aligned}
[d(P, Q)]^2 &= (\cos u - \cos v)^2 + (\sin u - \sin v)^2 && \text{Distance formula}\\
&= \cos^2 u - 2\cos u \cos v + \cos^2 v && (a - b)^2 = a^2 - 2ab + b^2\\
&\quad + \sin^2 u - 2 \sin u \sin v + \sin^2 v &&\\
&= (\cos^2 u + \sin^2 u) + (\cos^2 v + \sin^2 v) && \text{Combine terms.}\\
&\quad - 2(\cos u \cos v + \sin u \sin v) &&\\
&= 1 + 1 - 2(\cos u \cos v + \sin u \sin v) && \text{Pythagorean identity}\\
[d(P, Q)]^2 &= 2 - 2(\cos u \cos v + \sin u \sin v) && \text{Simplify.}
\end{aligned}$$

Now simplify the right side of equation (1).

$$\begin{aligned}
[d(A, B)]^2 &= [\cos(u - v) - 1]^2 + [\sin(u - v) - 0]^2 && \text{Distance formula}\\
&= \cos^2(u - v) - 2\cos(u - v) + 1 + \sin^2(u - v) && \text{Expand binomial.}\\
&= [\cos^2(u - v) + \sin^2(u - v)] + 1 - 2\cos(u - v) && \text{Combine terms.}\\
&= 1 + 1 - 2\cos(u - v) && \text{Pythagorean identity}\\
[d(A, B)]^2 &= 2 - 2\cos(u - v) && \text{Simplify.}
\end{aligned}$$

Because $[d(A, B)]^2 = [d(P, Q)]^2$,

$$\begin{aligned}
2 - 2\cos(u - v) &= 2 - 2(\cos u \cos v + \sin u \sin v) && \text{Substitute in each side.}\\
\mathbf{\cos(u - v)} &= \mathbf{\cos u \cos v + \sin u \sin v} && \text{Simplify.}
\end{aligned}$$

We have proved the formula for the cosine of the difference of two angles or two real numbers.

> **TECHNOLOGY CONNECTION**
>
> Although a calculator will give only a decimal approximation for irrational numbers, you can use one to check your work. Remember to use *radian* mode.
>
> 
>

EXAMPLE 1 **Using the Difference Formula for Cosine**

Find the exact value of $\cos\frac{\pi}{12}$ by using $\frac{\pi}{12} = \frac{\pi}{3} - \frac{\pi}{4}$.

Solution

We use the exact values of the trigonometric functions of $\frac{\pi}{3}$ and $\frac{\pi}{4}$ as well as the difference formula for cosine.

$$\cos\frac{\pi}{12} = \cos\left(\frac{\pi}{3} - \frac{\pi}{4}\right) \qquad \frac{\pi}{3} - \frac{\pi}{4} = \frac{4\pi}{12} - \frac{3\pi}{12} = \frac{\pi}{12}$$

$$= \cos\frac{\pi}{3}\cos\frac{\pi}{4} + \sin\frac{\pi}{3}\sin\frac{\pi}{4} \qquad \text{Formula for } \cos(u - v)$$

$$= \frac{1}{2}\cdot\frac{\sqrt{2}}{2} + \frac{\sqrt{3}}{2}\cdot\frac{\sqrt{2}}{2} \qquad \text{Use exact values. See page 491.}$$

$$= \frac{\sqrt{2} + \sqrt{6}}{4} \qquad \text{Multiply and add.}$$

Practice Problem 1 Find the exact value of $\cos 15°$ by using $15° = 45° - 30°$.

> **RECALL**
>
> The cosine function is even: $\cos(-x) = \cos(x)$.
>
> The sine function is odd: $\sin(-x) = -\sin(x)$.

The difference formula for $\cos(u - v)$ is true for all real numbers and angles u and v. We can use this formula and the even–odd identities to prove the formula for $\cos(u + v)$.

$$\cos(u + v) = \cos[u - (-v)] \qquad u + v = u - (-v)$$

$$= \cos u \cos(-v) + \sin u \sin(-v) \qquad \text{Difference formula for cosine}$$

$$= \cos u \cos v + \sin u(-\sin v) \qquad \cos(-v) = \cos v;\ \sin(-v) = -\sin v$$

$$\mathbf{\cos(u + v) = \cos u \cos v - \sin u \sin v} \qquad \text{Simplify.}$$

We have proved the formula for the cosine of the sum of two angles or two real numbers.

EXAMPLE 2 **Using the Sum Formula for Cosine**

Find the exact value of $\cos 75°$ by using $75° = 45° + 30°$.

Solution

$$\cos 75° = \cos(45° + 30°) \qquad 75° = 45° + 30°$$

$$= \cos 45° \cos 30° - \sin 45° \sin 30° \qquad \text{Sum formula for cosine}$$

$$= \frac{\sqrt{2}}{2}\cdot\frac{\sqrt{3}}{2} - \frac{\sqrt{2}}{2}\cdot\frac{1}{2} \qquad \text{Use exact values.}$$

$$= \frac{\sqrt{6}}{4} - \frac{\sqrt{2}}{4} \qquad \text{Multiply.}$$

$$= \frac{\sqrt{6} - \sqrt{2}}{4} \qquad \text{Simplify.}$$

Practice Problem 2 Find the exact value of $\cos\frac{7\pi}{12}$ by using $\frac{7\pi}{12} = \frac{\pi}{3} + \frac{\pi}{4}$.

2 Use the cofunction identities.

Cofunction Identities

Two trigonometric functions f and g are called **cofunctions** if

$$f\left(\frac{\pi}{2} - x\right) = g(x) \quad \text{and} \quad g\left(\frac{\pi}{2} - x\right) = f(x).$$

We derive two *cofunction identities.*

$$\cos(u - v) = \cos u \cos v + \sin u \sin v \qquad \text{Difference formula for cosine}$$

$$\cos\left(\frac{\pi}{2} - v\right) = \cos\frac{\pi}{2}\cos v + \sin\frac{\pi}{2}\sin v \qquad \text{Replace } u \text{ with } \frac{\pi}{2}.$$

$$= 0 \cdot \cos v + 1 \cdot \sin v = \sin v \qquad \cos\frac{\pi}{2} = 0;\ \sin\frac{\pi}{2} = 1$$

The result is a cofunction identity that holds for any real number v or angle v in radian measure:

$$\mathbf{\cos\left(\frac{\pi}{2} - v\right) = \sin v}$$

If we replace v with $\frac{\pi}{2} - v$ in the identity $\cos\left(\frac{\pi}{2} - v\right) = \sin v$, we have

$$\cos\left[\frac{\pi}{2} - \left(\frac{\pi}{2} - v\right)\right] = \sin\left(\frac{\pi}{2} - v\right) \qquad \text{Replace } v \text{ with } \frac{\pi}{2} - v.$$

$$\cos v = \sin\left(\frac{\pi}{2} - v\right) \qquad \text{Simplify.}$$

So $$\mathbf{\sin\left(\frac{\pi}{2} - v\right) = \cos v.}$$

SIDE NOTE

For an acute angle v, these cofunction identities were established in Section 5.2.

BASIC COFUNCTION IDENTITIES

If v is any real number or angle measured in radians, then

$$\cos\left(\frac{\pi}{2} - v\right) = \sin v$$

$$\sin\left(\frac{\pi}{2} - v\right) = \cos v$$

If angle v is measured in degrees, then replace $\frac{\pi}{2}$ with 90° in these identities.

EXAMPLE 3 Using Cofunction Identities

Prove that for any real number x, $\tan\left(\frac{\pi}{2} - x\right) = \cot x$.

Solution

$$\tan\left(\frac{\pi}{2} - x\right) = \frac{\sin\left(\frac{\pi}{2} - x\right)}{\cos\left(\frac{\pi}{2} - x\right)} \qquad \text{Quotient identity}$$

$$= \frac{\cos x}{\sin x} \qquad \text{Use cofunction identities.}$$

$$= \cot x \qquad \text{Quotient identity}$$

Practice Problem 3 Prove that for any real number x, $\sec\left(\frac{\pi}{2} - x\right) = \csc x$.

3 Use the sum and difference formulas for sine.

Sum and Difference Formulas for Sine

To prove the difference formula for the sine function, we start with a cofunction identity.

$$\begin{aligned}
\sin(u - v) &= \cos\left[\frac{\pi}{2} - (u - v)\right] && \text{Cofunction identity} \\
&= \cos\left[\left(\frac{\pi}{2} - u\right) + v\right] && \frac{\pi}{2} - (u - v) = \left(\frac{\pi}{2} - u\right) + v \\
&= \cos\left(\frac{\pi}{2} - u\right)\cos v - \sin\left(\frac{\pi}{2} - u\right)\sin v && \text{Sum formula for cosine} \\
&= \sin u \cos v - \cos u \sin v && \text{Cofunction identities}
\end{aligned}$$

We have the difference formula for sine:

$$\sin(u - v) = \sin u \cos v - \cos u \sin v,$$

which holds for all real numbers u and v. If we replace v with $-v$, we derive the sum formula for sine.

$$\begin{aligned}
\sin(u + v) &= \sin[u - (-v)] && u + v = u - (-v) \\
&= \sin u \cos(-v) - \cos u \sin(-v) && \text{Difference formula for sine} \\
&= \sin u \cos v - \cos u(-\sin v) && \cos(-v) = \cos v;\ \sin(-v) = -\sin v \\
&= \sin u \cos v + \cos u \sin v && \text{Simplify.}
\end{aligned}$$

This proves the sum formula for sine:

$$\sin(u + v) = \sin u \cos v + \cos u \sin v$$

SUM AND DIFFERENCE FORMULAS FOR SINE

$$\sin(u + v) = \sin u \cos v + \cos u \sin v$$
$$\sin(u - v) = \sin u \cos v - \cos u \sin v$$

WARNING

In general,

$$\sin(u \pm v) \neq \sin u \pm \sin v$$
$$\cos(u \pm v) \neq \cos u \pm \cos v$$

EXAMPLE 4 Using the Sum and Difference Formulas for Sine

Prove the identity $\sin(\pi - x) = \sin x$.

Solution

$$\begin{aligned}
\sin(\pi - x) &= \sin \pi \cos x - \cos \pi \sin x && \text{Difference formula for sine} \\
&= (0)\cos x - (-1)\sin x && \sin \pi = 0;\ \cos \pi = -1 \\
&= \sin x && \text{Simplify.}
\end{aligned}$$

Practice Problem 4 Prove the identity $\sin(\pi + x) = -\sin x$.

EXAMPLE 5 **Using the Sum and Difference Formulas for Sine**

Find the exact value of $\sin 63° \cos 27° + \cos 63° \sin 27°$ without using a calculator.

Solution

The given expression is the right side of the sum formula for sine:

$$\sin(u + v) = \sin u \cos v + \cos u \sin v$$

$$\sin 63° \cos 27° + \cos 63° \sin 27° = \sin(63° + 27°) \qquad u = 63°, v = 27°$$

$$= \sin 90° = 1$$

Practice Problem 5 Find the exact value of $\sin 43° \cos 13° - \cos 43° \sin 13°$ without using a calculator.

EXAMPLE 6 **Finding the Exact Value of a Sum**

Let $\sin u = -\frac{3}{5}$ and $\cos v = \frac{12}{13}$, with $\pi < u < \frac{3\pi}{2}$ and $\frac{3\pi}{2} < v < 2\pi$. Find the exact value of $\sin(u + v)$.

Solution

Given $\sin u = -\frac{3}{5}$ with u in quadrant III, we find the exact value of $\cos u$.

$$\cos^2 u = 1 - \sin^2 u \qquad \text{Pythagorean identity}$$

$$\cos u = -\sqrt{1 - \sin^2 u} \qquad \text{In quadrant III, } \cos u \text{ is negative.}$$

$$= -\sqrt{1 - \left(-\frac{3}{5}\right)^2} \qquad \text{Replace } \sin u \text{ with } -\frac{3}{5}.$$

$$= -\sqrt{1 - \frac{9}{25}} \qquad \left(-\frac{3}{5}\right)^2 = \frac{9}{25}$$

$$= -\sqrt{\frac{16}{25}} \qquad 1 - \frac{9}{25} = \frac{25}{25} - \frac{9}{25} = \frac{16}{25}$$

$$\cos u = -\frac{4}{5}$$

Similarly, given $\cos v = \frac{12}{13}$ with v in quadrant IV, we find the exact value of $\sin v$.

$$\sin v = -\sqrt{1 - \cos^2 v} \qquad \text{In quadrant IV, } \sin v \text{ is negative.}$$

$$= -\sqrt{1 - \left(\frac{12}{13}\right)^2} \qquad \text{Replace } \cos v \text{ with } \frac{12}{13}.$$

$$= -\sqrt{\frac{25}{169}} \qquad 1 - \left(\frac{12}{13}\right)^2 = 1 - \frac{144}{169} = \frac{169 - 144}{169}$$

$$\sin v = -\frac{5}{13}$$

$$\sin(u + v) = \sin u \cos v + \cos u \sin v \qquad \text{Sum formula for sine}$$

$$= \left(-\frac{3}{5}\right)\left(\frac{12}{13}\right) + \left(-\frac{4}{5}\right)\left(-\frac{5}{13}\right) \qquad \text{Replace values of } \sin u, \sin v, \cos u, \text{ and } \cos v.$$

$$= -\frac{36}{65} + \frac{20}{65} = -\frac{16}{65} \qquad \text{Multiply and add.}$$

The exact value of $\sin(u + v)$ is $-\frac{16}{65}$.

SIDE NOTE

Because

$$\sec(u \pm v) = \frac{1}{\cos(u \pm v)}$$

and

$$\csc(u \pm v) = \frac{1}{\sin(u \pm v)}$$

the sum and difference formulas for the sine and cosine can be used to find secant and cosecant values as well.

Practice Problem 6 For the angles u and v in Example 6, find the exact value of $\cos(u + v)$.

EXAMPLE 7 **Verifying an Identity by Using a Sum or Difference Formula**

Verify the identity $\dfrac{\sin(x - y)}{\cos x \cos y} = \tan x - \tan y$.

Solution

We start with the more complicated left side.

$$\frac{\sin(x - y)}{\cos x \cos y} = \frac{\sin x \cos y - \cos x \sin y}{\cos x \cos y} \qquad \text{Formula for } \sin(u - v)$$

$$= \frac{\sin x \cos y}{\cos x \cos y} - \frac{\cos x \sin y}{\cos x \cos y} \qquad \frac{a - b}{c} = \frac{a}{c} - \frac{b}{c}$$

$$= \frac{\sin x}{\cos x} - \frac{\sin y}{\cos y} \qquad \text{Remove common factors.}$$

$$= \tan x - \tan y \qquad \text{Quotient identity}$$

Because the left side is identical to the right side, the given equation is an identity.

Practice Problem 7 Verify the following identity: $\dfrac{\cos(x + y)}{\sin x \sin y} = \cot x \cot y - 1$

Reduction Formula

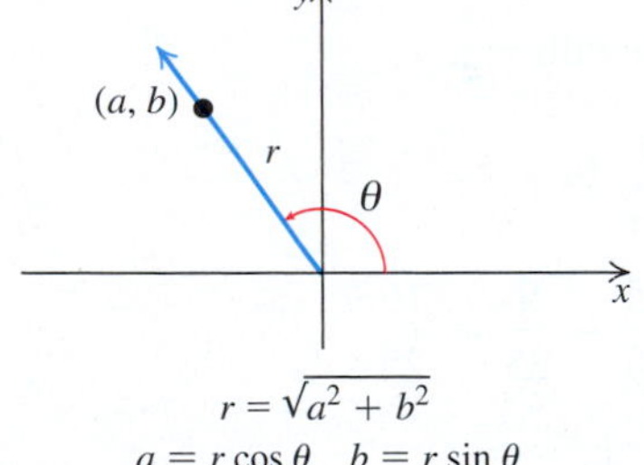

$r = \sqrt{a^2 + b^2}$

$a = r\cos\theta \quad b = r\sin\theta$

FIGURE 6.5

Figure 6.5 suggests that a point (a, b) is on the terminal side of an angle θ in standard position if and only if

$$\cos\theta = \frac{a}{\sqrt{a^2 + b^2}} \quad \text{and} \quad \sin\theta = \frac{b}{\sqrt{a^2 + b^2}}.$$

Using these values for $\cos\theta$ and $\sin\theta$ in the sum formula for the sine, we obtain

$$\sin(x + \theta) = \sin x \cos\theta + \cos x \sin\theta \qquad \text{Sum formula for sine}$$

$$\sin(x + \theta) = \sin x \frac{a}{\sqrt{a^2 + b^2}} + \cos x \frac{b}{\sqrt{a^2 + b^2}}$$

Multiplying both sides of this equation by $\sqrt{a^2 + b^2}$, we have the reduction formula:

$$\sqrt{a^2 + b^2}\sin(x + \theta) = a\sin x + b\cos x$$

REDUCTION FORMULA

If (a, b) is any point on the terminal side of an angle θ (radians) in standard position, then

$$a\sin x + b\cos x = \sqrt{a^2 + b^2}\sin(x + \theta)$$

for any real number x, where $\cos\theta = \dfrac{a}{\sqrt{a^2 + b^2}}$ and $\sin\theta = \dfrac{b}{\sqrt{a^2 + b^2}}$.

EXAMPLE 8 Using the Reduction Formula

Find an angle θ, in radians, and a real number A such that

$$\sin x - \sqrt{3}\cos x = A\sin(x + \theta).$$

Solution

Note that by the reduction formula,

$$\begin{aligned}\sin x - \sqrt{3}\cos x &= a\sin x + b\cos x \qquad a = 1 \text{ and } b = -\sqrt{3}\\ &= A\sin(x + \theta),\end{aligned}$$

where $A = \sqrt{a^2 + b^2} = \sqrt{1^2 + (-\sqrt{3})^2} = \sqrt{4} = 2$ and θ is any angle in standard position that has the point $(a, b) = (1, -\sqrt{3})$ on its terminal side. One such angle is $\theta = \tan^{-1}\left(\frac{b}{a}\right) = \tan^{-1}(-\sqrt{3}) = -\frac{\pi}{3}$. Then with $A = 2$ and $\theta = -\frac{\pi}{3}$, we have

$$\sin x - \sqrt{3}\cos x = 2\sin\left[x + \left(-\frac{\pi}{3}\right)\right] = 2\sin\left(x - \frac{\pi}{3}\right).$$

Practice Problem 8 Find an angle θ, in radians, and a real number A such that

$$\sin x + \sqrt{3}\cos x = A\sin(x + \theta).$$

EXAMPLE 9 Using the Reduction Formula to Sketch a Graph

Sketch two cycles of the graph of the equation $y = \sin x - \sqrt{3}\cos x$.

Solution

Example 8 shows that $\sin x - \sqrt{3}\cos x = 2\sin\left(x - \frac{\pi}{3}\right)$. Therefore, the graphs of $y = \sin x - \sqrt{3}\cos x$ and $y = 2\sin\left(x - \frac{\pi}{3}\right)$ are the same. The graph of $y = 2\sin\left(x - \frac{\pi}{3}\right)$ is a sine wave with amplitude 2, period 2π, and phase shift $\frac{\pi}{3}$ units to the right. Two cycles of the graph of $y = 2\sin\left(x - \frac{\pi}{3}\right) = \sin x - \sqrt{3}\cos x$ are shown in Figure 6.6.

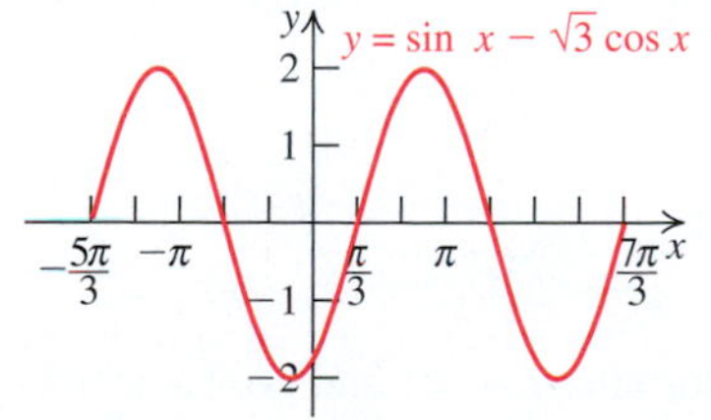

FIGURE 6.6

Practice Problem 9 Sketch two cycles of the graph of $y = \sin x + \sqrt{3}\cos x$.

EXAMPLE 10 Combining Two Pure Tones

Suppose the pressure exerted by two pure tones (in pounds per square foot) after t seconds is given by

$$y_1 = 0.3\sin(800\pi t) \quad \text{and} \quad y_2 = 0.4\cos(800\pi t).$$

Find the amplitude, period, frequency, and phase shift for the total pressure $y = y_1 + y_2$.

Solution

We have $y = 0.3\sin(800\pi t) + 0.4\cos(800\pi t)$. We use the reduction formula with $a = 0.3$ and $b = 0.4$ to rewrite this equation in the form

$$y = A\sin(800\pi t + \theta) = A\sin\left[800\pi\left(t + \frac{\theta}{800\pi}\right)\right],$$

where

$$A = \sqrt{a^2 + b^2} = \sqrt{(0.3)^2 + (0.4)^2} = 0.5$$

and $(0.3, 0.4)$ is a point on the terminal side of θ. One such angle is $\theta = \tan^{-1}\left(\frac{0.4}{0.3}\right)$. For the total pressure,

$$y = A \sin\left[800\pi\left(t + \frac{\theta}{800\pi}\right)\right]$$

$$= 0.5 \sin\left[800\pi\left(t + \frac{\theta}{800\pi}\right)\right] \qquad A = 0.5$$

$$\text{amplitude} = 0.5$$

$$\text{period} = \frac{2\pi}{800\pi} = \frac{1}{400}$$

$$\text{frequency} = 400 \qquad \text{Frequency} = \frac{1}{\text{Period}}$$

$$\text{phase shift} = -\frac{\theta}{800\pi} \approx -0.00037$$

Use a calculator in *radian* mode with $\theta = \tan^{-1}\left(\frac{0.4}{0.3}\right)$.

Practice Problem 10 Find the amplitude, period, frequency, and phase shift of the total pressure due to the pressure from the two pure tones

$$y_1 = 0.1 \sin(400t) \quad \text{and} \quad y_2 = 0.2 \cos(400t).$$

4 Use the sum and difference formulas for tangent.

Sum and Difference Formulas for Tangent

We use the quotient identity $\tan x = \frac{\sin x}{\cos x}$ and the formulas for $\sin(u - v)$ and $\cos(u - v)$ to derive a difference formula for tangent.

$$\tan(u - v) = \frac{\sin(u - v)}{\cos(u - v)} \qquad \text{Quotient identity}$$

$$= \frac{\sin u \cos v - \cos u \sin v}{\cos u \cos v + \sin u \sin v} \qquad \text{Formulas for } \sin(u - v) \text{ and } \cos(u - v)$$

$$= \frac{\frac{\sin u \cos v}{\cos u \cos v} - \frac{\cos u \sin v}{\cos u \cos v}}{\frac{\cos u \cos v}{\cos u \cos v} + \frac{\sin u \sin v}{\cos u \cos v}} \qquad \text{Divide numerator and denominator by } \cos u \cos v.$$

$$= \frac{\frac{\sin u}{\cos u} - \frac{\sin v}{\cos v}}{1 + \frac{\sin u \sin v}{\cos u \cos v}} \qquad \text{Simplify.}$$

$$= \frac{\tan u - \tan v}{1 + \tan u \tan v} \qquad \text{Quotient identity}$$

We have derived the difference formula for the tangent:

$$\mathbf{\tan(u - v) = \frac{\tan u - \tan v}{1 + \tan u \tan v}}$$

Replacing v with $-v$ in the difference formula, we have

$$\tan(u + v) = \tan[u - (-v)] \qquad u + v = u - (-v)$$
$$= \frac{\tan u - \tan(-v)}{1 + \tan u \tan(-v)} \qquad \text{Formula for } \tan(u - v)$$
$$= \frac{\tan u - (-\tan v)}{1 + \tan u\,(-\tan v)} \qquad \tan(-v) = -\tan v$$
$$= \frac{\tan u + \tan v}{1 - \tan u \tan v} \qquad \text{Simplify.}$$

We now have the sum formula for tangent:

$$\mathbf{\tan(u + v) = \frac{\tan u + \tan v}{1 - \tan u \tan v}}$$

EXAMPLE 11 Verifying an Identity

Verify the identity $\tan(\pi - x) = -\tan x$.

Solution

Apply the difference formula for the tangent to $\tan(\pi - x)$.

$$\tan(\pi - x) = \frac{\tan \pi - \tan x}{1 + \tan \pi \tan x} \qquad \text{Replace } u \text{ with } \pi \text{ and } v \text{ with } x \text{ in the formula for } \tan(u - v).$$
$$= \frac{0 - \tan x}{1 + 0 \cdot \tan x} \qquad \tan \pi = 0$$
$$= -\tan x \qquad \text{Simplify.}$$

Therefore, the given equation is an identity.

Practice Problem 11 Verify the identity $\tan(\pi + x) = \tan x$.

The important identities involving the sum, difference, and cofunctions of two numbers or angles are summarized next.

SUMMARY OF MAIN FACTS

Sum and Difference Formula

$$\cos(u - v) = \cos u \cos v + \sin u \sin v \qquad \cos(u + v) = \cos u \cos v - \sin u \sin v$$
$$\sin(u - v) = \sin u \cos v - \cos u \sin v \qquad \sin(u + v) = \sin u \cos v + \cos u \sin v$$
$$\tan(u - v) = \frac{\tan u - \tan v}{1 + \tan u \tan v} \qquad \tan(u + v) = \frac{\tan u + \tan v}{1 - \tan u \tan v}$$

Cofunction Identities

$$\sin\left(\frac{\pi}{2} - x\right) = \cos x \qquad \cos\left(\frac{\pi}{2} - x\right) = \sin x \qquad \tan\left(\frac{\pi}{2} - x\right) = \cot x$$
$$\csc\left(\frac{\pi}{2} - x\right) = \sec x \qquad \sec\left(\frac{\pi}{2} - x\right) = \csc x \qquad \cot\left(\frac{\pi}{2} - x\right) = \tan x$$

Reduction Formula

If (a, b) is any point on the terminal side of an angle θ (radians) in standard position, then for any real number x,

$$a \sin x + b \cos x = \sqrt{a^2 + b^2}\,\sin(x + \theta),$$

where $\cos\theta = \dfrac{a}{\sqrt{a^2 + b^2}}$ and $\sin\theta = \dfrac{b}{\sqrt{a^2 + b^2}}$.

Answers to Practice Problems

1. $\frac{\sqrt{6}+\sqrt{2}}{4}$ 2. $\frac{\sqrt{2}-\sqrt{6}}{4}$ 5. $\frac{1}{2}$ 6. $-\frac{63}{65}$ 8. $A = 2, \theta = \frac{\pi}{3}$

9.

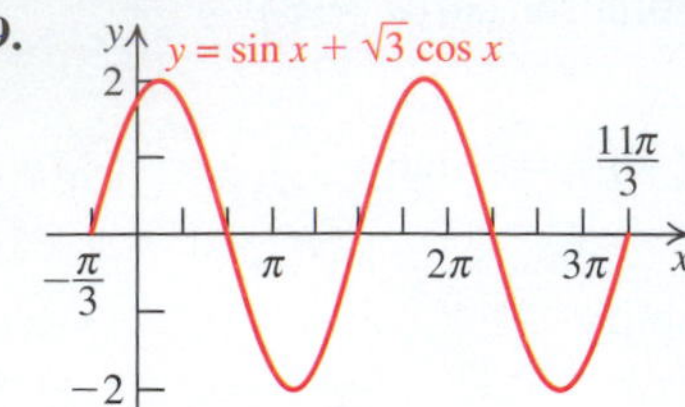

10. Amplitude $= 0.1\sqrt{5}$; period $= \frac{\pi}{200}$; frequency $= \frac{200}{\pi}$; phase shift ≈ -0.0028

SECTION 6.2 Exercises

Basic Concepts and Skills

1. $\sin(A + B) =$ ____________.
2. $\cos A \cos B - \sin A \sin B =$ ____________.
3. $\tan(A + B) =$ ____________.
4. **True or False.** $\cos\left[\frac{\pi}{2} + x\right] = \sin x$
5. **True or False.** $\sin\left[\frac{\pi}{2} + x\right] = \cos x$
6. **True or False.** $\tan(x + y) = \tan x + \tan y$

In Exercises 7–26, find the exact value of each expression.

7. $\sin(45° + 30°)$
8. $\sin(45° - 30°)$
9. $\sin(60° - 45°)$
10. $\sin(60° + 45°)$
11. $\sin(-105°)$
12. $\cos 285°$
13. $\tan 225°$
14. $\tan(-165°)$
15. $\sin\left(\frac{\pi}{6} + \frac{\pi}{4}\right)$
16. $\cos\left(\frac{\pi}{3} - \frac{\pi}{4}\right)$
17. $\tan\left(\frac{\pi}{4} - \frac{\pi}{6}\right)$
18. $\cot\left(\frac{\pi}{3} - \frac{\pi}{4}\right)$
19. $\sec\left(\frac{\pi}{3} + \frac{\pi}{4}\right)$
20. $\csc\left(\frac{\pi}{4} - \frac{\pi}{3}\right)$
21. $\cos\frac{-5\pi}{12}$
22. $\sin\frac{7\pi}{12}$
23. $\tan\frac{19\pi}{12}$
24. $\sec\frac{\pi}{12}$
25. $\tan\frac{17\pi}{12}$
26. $\csc\frac{11\pi}{12}$

In Exercises 27–40, verify each identity.

27. $\sin\left(x + \frac{\pi}{2}\right) = \cos x$
28. $\cos\left(x + \frac{\pi}{2}\right) = -\sin x$
29. $\sin\left(x - \frac{\pi}{2}\right) = -\cos x$
30. $\cos\left(x - \frac{\pi}{2}\right) = \sin x$
31. $\tan\left(x + \frac{\pi}{2}\right) = -\cot x$
32. $\tan\left(x - \frac{\pi}{2}\right) = -\cot x$
33. $\csc(x + \pi) = -\csc x$
34. $\sec(x + \pi) = -\sec x$
35. $\cos\left(x + \frac{3\pi}{2}\right) = \sin x$
36. $\cos\left(x - \frac{3\pi}{2}\right) = -\sin x$
37. $\tan\left(x - \frac{3\pi}{2}\right) = -\cot x$
38. $\tan\left(x + \frac{3\pi}{2}\right) = -\cot x$
39. $\cot(3\pi - x) = -\cot x$
40. $\csc\left(\frac{5\pi}{2} - x\right) = \sec x$

In Exercises 41–50, find the exact value of each expression without using a calculator.

41. $\sin 56° \cos 34° + \cos 56° \sin 34°$
42. $\cos 57° \cos 33° - \sin 57° \sin 33°$
43. $\cos 331° \cos 61° + \sin 331° \sin 61°$
44. $\cos 110° \sin 70° + \sin 110° \cos 70°$
45. $\frac{\tan 129° - \tan 84°}{1 + \tan 129° \tan 84°}$
46. $\frac{\tan 28° + \tan 17°}{1 - \tan 28° \tan 17°}$
47. $\sin\frac{7\pi}{12}\cos\frac{3\pi}{12} - \cos\frac{7\pi}{12}\sin\frac{3\pi}{12}$
48. $\cos\frac{5\pi}{12}\cos\frac{\pi}{12} - \sin\frac{5\pi}{12}\sin\frac{\pi}{12}$
49. $\dfrac{\tan\frac{5\pi}{12} - \tan\frac{2\pi}{12}}{1 + \tan\frac{5\pi}{12}\tan\frac{2\pi}{12}}$
50. $\dfrac{\tan\frac{5\pi}{12} + \tan\frac{7\pi}{12}}{1 - \tan\frac{5\pi}{12}\tan\frac{7\pi}{12}}$

In Exercises 51–56, find the exact value of each expression, given that $\tan u = \frac{3}{4}$, with u in quadrant III, and $\sin v = \frac{5}{13}$, with v in quadrant II.

51. $\sin(u - v)$
52. $\sin(u + v)$
53. $\cos(u + v)$
54. $\cos(u - v)$
55. $\tan(u + v)$
56. $\tan(u - v)$

In Exercises 57–62, find the exact value of each expression, given that $\cos\alpha = -\frac{2}{5}$, with α in quadrant II, and $\sin\beta = -\frac{3}{7}$, with β in quadrant IV.

57. $\sin(\alpha - \beta)$

58. $\cos(\alpha - \beta)$

59. $\csc(\alpha + \beta)$

60. $\sec(\alpha + \beta)$

61. $\cot(\alpha - \beta)$

62. $\cot(\alpha + \beta)$

In Exercises 63–70, verify each identity.

63. $\dfrac{\sin(x + y)}{\cos x \cos y} = \tan x + \tan y$

64. $\dfrac{\sin(x + y)}{\sin x \sin y} = \cot x + \cot y$

65. $\dfrac{\cos(x + y)}{\cos x \cos y} = 1 - \tan x \tan y$

66. $\dfrac{\cos(x - y)}{\sin x \sin y} = \cot x \cot y + 1$

67. $\dfrac{\cos(x - y)}{\sin(x + y)} = \dfrac{1 + \cot x \cot y}{\cot x + \cot y}$

68. $\dfrac{\cos(x + y)}{\sin(x - y)} = \dfrac{1 - \cot x \cot y}{\cot x - \cot y}$

69. $\dfrac{\sin(x - y)}{\sin(x + y)} = \dfrac{\cot y - \cot x}{\cot y + \cot x}$

70. $\dfrac{\cos(x + y)}{\cos(x - y)} = \dfrac{\cot x \cot y - 1}{\cot x \cot y + 1}$

In Exercises 71–78, rewrite each equation in the form $y = A\sin(x + \theta)$; graph the resulting equation. If θ is not a common angle, approximate its value to the nearest thousandth of the radian.

71. $y = \sin x + \cos x$

72. $y = \sin x - \cos x$

73. $y = 3\sin x + 4\cos x$

74. $y = 4\sin x - 3\cos x$

75. $y = 5\sin x - 12\cos x$

76. $y = 12\sin x + 5\cos x$

77. $y = \cos x - 3\sin x$

78. $y = 3\cos x - 4\sin x$

Applying the Concepts

79. Analytic geometry. Suppose two nonvertical lines l_1 and l_2 lie in a plane. The slope of l_1 is m_1, and the slope of l_2 is m_2. Note that $m_1 = \tan\theta_1$ and $m_2 = \tan\theta_2$. Show that the tangent of the smallest nonnegative angle α between the lines l_1 and l_2 is given by $\tan\alpha = \left|\dfrac{m_2 - m_1}{1 + m_1 m_2}\right|$.

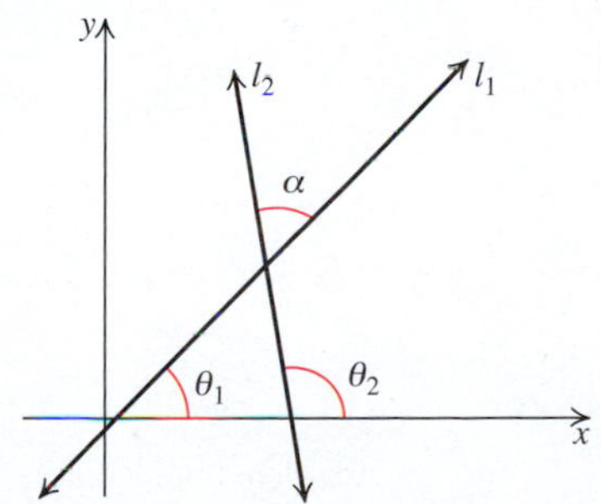

In Exercises 80–82, use Exercise 79 to find the smallest nonnegative angle in radians between the lines l_1 and l_2.

80. Angle between lines. $l_1: y = x + 5$, $l_2: y = 4x + 2$

81. Angle between lines. $l_1: y = 2x + 5$, $l_2: 6x - 3y = 21$

82. Angle between lines. $l_1: y = 2x - 4$, $l_2: x + 2y = 7$

83. Combining two pure tones. The pressure exerted by two pure tones in pounds per square foot is given by the equations $y_1 = 0.4\sin(400t)$ and $y_2 = 0.3\cos(400t)$. Find the amplitude and the phase shift for the total pressure $y = y_1 + y_2$.

84. Combining two pure tones. Repeat Exercise 83 for $y_1 = 0.05\sin(600t)$ and $y_2 = 0.12\cos(600t)$.

85. Simple harmonic motion. A simple harmonic motion is described by the equation $x = 0.12\sin 2t + 0.5\cos 2t$, where x is the distance from the equilibrium position and t is the time in seconds.

a. Find the amplitude of the motion.

b. Find the frequency and the phase shift.

86. Simple harmonic motion. Repeat Exercise 85 assuming that the motion is described by the equation

$$x = \frac{1}{2}\sin 3t + \frac{1}{3}\cos 3t.$$

Beyond the Basics

87. Difference quotient. Let $f(x) = \sin x$. Show that
$$\frac{f(x + h) - f(x)}{h} = \sin x\left(\frac{\cos h - 1}{h}\right) + \cos x\left(\frac{\sin h}{h}\right).$$

88. Difference quotient. Let $f(x) = \cos x$. Show that
$$\frac{f(x + h) - f(x)}{h} = \cos x\left(\frac{\cos h - 1}{h}\right) - \sin x\left(\frac{\sin h}{h}\right).$$

In Exercises 89–100, verify each identity.

89. $\cos u\cos(u + v) + \sin u\sin(u + v) = \cos v$

90. $\sin(x + y)\cos y - \cos(x + y)\sin y = \sin x$

91. $\sin 5x\cos 3x - \cos 5x\sin 3x = \sin 2x$

92. $\sin(2x - y)\cos y + \cos(2x - y)\sin y = \sin 2x$

93. $\sin\left(\dfrac{\pi}{2} + x - y\right) = \cos x\cos y + \sin x\sin y$

94. $\cos\left(\dfrac{\pi}{2} + x - y\right) = \cos x\sin y - \sin x\cos y$

95. $\sin(x + y)\sin(x - y) = \sin^2 x - \sin^2 y$

96. $\cos(\alpha + \beta)\cos(\alpha - \beta) = \cos^2\alpha - \sin^2\beta$

97. $\sin^2\left(\dfrac{\pi}{4} + \dfrac{x}{2}\right) - \sin^2\left(\dfrac{\pi}{4} - \dfrac{x}{2}\right) = \sin x$

98. $\cos^2\left(\dfrac{\pi}{4} + \dfrac{x}{2}\right) - \sin^2\left(\dfrac{\pi}{4} - \dfrac{x}{2}\right) = 0$

99. $\dfrac{\sin(\alpha - \beta)}{\sin\alpha\sin\beta} + \dfrac{\sin(\beta - \gamma)}{\sin\beta\sin\gamma} + \dfrac{\sin(\gamma - \alpha)}{\sin\gamma\sin\alpha} = 0$

100. $\dfrac{\sin(\alpha - \beta)}{\cos\alpha\cos\beta} + \dfrac{\sin(\beta - \gamma)}{\cos\beta\cos\gamma} + \dfrac{\sin(\gamma - \alpha)}{\cos\gamma\cos\alpha} = 0$

In Exercises 101–106, solve each equation in the interval $[0, 2\pi)$ by first converting the left side of the equation to the form $A \sin(x + \theta)$ (see page 588).

101. $\sin x + \cos x = 1$

102. $\sin x - \cos x = 1$

103. $\sin x + \sqrt{3}\cos x = -1$

104. $\sqrt{3}\sin x - \cos x = 1$

105. $\sin x - \sqrt{3}\cos x = 2$

106. $5\sin x + 12\cos x = 14$

In Exercises 107–116, find the exact value of each expression.

107. $\cos^2 15° - \cos^2 30° + \cos^2 45° - \cos^2 60° + \cos^2 75°$

108. $\cos^2\left(\frac{\pi}{8}\right) + \cos^2\left(\frac{3\pi}{8}\right) + \cos^2\left(\frac{5\pi}{8}\right) + \cos^2\left(\frac{7\pi}{8}\right)$

109. $\cos 60° + \cos 80° + \cos 100°$

110. $\sin 30° - \sin 70° + \sin 110°$

111. $\sin\left[\tan^{-1}\left(-\frac{3}{4}\right) + \cos^{-1}\left(\frac{4}{5}\right)\right]$

112. $\cos\left[\sin^{-1}\left(-\frac{3}{5}\right) + \cos^{-1}\left(\frac{3}{5}\right)\right]$

113. $\sin\left[\sin^{-1}\left(\frac{3}{5}\right) - \cos^{-1}\left(\frac{4}{5}\right)\right]$

114. $\cos\left[\sin^{-1}\left(\frac{3}{5}\right) + \tan^{-1}\left(-\frac{4}{3}\right)\right]$

115. $\tan\left[\cos^{-1}\left(\frac{4}{5}\right) + \tan^{-1}\left(\frac{2}{3}\right)\right]$

116. $\tan\left[\sin^{-1}\left(-\frac{3}{5}\right) + \cos^{-1}\left(\frac{4}{5}\right)\right]$

117. Show that $\tan 70° - \tan 20° = 2\tan 50°$.

118. Show that $\sin(\theta + n\pi) = (-1)^n \sin\theta$.

119. Find the exact value of $\tan 1° \tan 2° \tan 3° \cdots \tan 88° \tan 89°$.

In Exercises 120–122, assume that x and y are positive numbers in the domain of the inverse function considered. Prove each identity.

120. $\sin^{-1}x \pm \sin^{-1}y = \sin^{-1}(x\sqrt{1 - y^2} \pm y\sqrt{1 - x^2})$
[*Hint:* Let $u = \sin^{-1}x$ and $v = \sin^{-1}y$ and consider $\sin[(u \pm v)]$.]

121. $\cos^{-1}x + \cos^{-1}y = \cos^{-1}(xy - \sqrt{1 - x^2}\sqrt{1 - y^2})$

122. $\sin^{-1}x + \cos^{-1}x = \frac{\pi}{2}$

Critical Thinking/Discussion/Writing

123. Find values of u and v such that $\cos(u + v) \neq \cos u + \cos v$.

124. Find values of u and v such that $\cos(u - v) \neq \cos u - \cos v$.

125. Explain why we cannot use the formula for $\tan(u - v)$ to verify that $\tan\left(\frac{\pi}{2} - x\right) = \cot x$.

126. Find a formula for $\cot(x - y)$ in terms of $\cot x$ and $\cot y$.

127. Find values of u and v such that $\cos(u + v) = \cos u \cos v$.

Maintaining Skills

In Exercises 128 and 129, find the exact value of each expression.

128. $\sin\theta$, if $\cos\theta = \frac{4}{5}$ and θ is in quadrant IV

129. $\tan\theta$, if $\sin\theta = -\frac{1}{\sqrt{17}}$ and θ is in quadrant III

In Exercises 130–133, determine whether the expression is positive or negative if θ is in the given quadrant.

130. $\sin\theta\cos\theta$, quadrant III

131. $\tan\theta\csc\theta$, quadrant IV

132. $\frac{\cot\theta\tan\theta}{\sin\theta}$, quadrant III

133. $\frac{\cos\theta\sin\theta}{\sec\theta}$, quadrant III

In Exercises 134–137, find the exact value of each expression.

134. $\theta = \frac{\pi}{6}$, find $\sin 2\theta$

135. $\theta = -\frac{13\pi}{12}$, find $\tan 2\theta$

136. $\theta = -\frac{\pi}{3}$, find $\sin\frac{\theta}{2}$

137. $\theta = \frac{39\pi}{2}$, find $\cos\frac{\theta}{2}$

SECTION 6.3

Double-Angle and Half-Angle Formulas

BEFORE STARTING THIS SECTION, REVIEW

1. Sum formulas for sine, cosine, and tangent functions (Section 6.2, page 591)
2. Pythagorean identities (Section 6.1, page 569)
3. Trigonometric functions of common angles (Section 5.2, page 491)

OBJECTIVES

1. Use double-angle formulas.
2. Use power-reducing formulas.
3. Use half-angle formulas.

◆ Cost of Using an Electric Blanket

Three basic properties of electricity—*voltage, current,* and *power*—together with the rate charged by your electric company, are used to find the cost of using an electric device.

Voltage (*V*) is measured in *volts*. It is the force that pushes electricity through a wire. *Current* (*I*), measured in *amperes* (amps), measures how much electricity is moving through the device per second. *Power* (*P*), measured in *watts,* gives the energy consumed per second by an electric device and is defined by the equation

$$P = VI \qquad \text{Power} = (\text{Voltage})(\text{Current})$$

Your electric company bills you by the kilowatt-hour. When you turn on an electric device that consumes 1000 watts for one hour, it consumes 1 kilowatt-hour. Suppose your electric company charges 8¢ per kilowatt-hour. An electric blanket might use 250 watts (depending on the setting). If you turn it on for ten hours, it will consume $250 \times 10 = 2500 = 2.5$ kilowatt-hours. This will cost you $(2.5)(8) = 20$¢.

In Example 5, we use trigonometric identities to compute the *wattage rating* of an electric blanket.

1 Use double-angle formulas.

Double-Angle Formulas

Suppose an angle measures x (radians or degrees); then $2x$ is double the measure of x. *Double-angle formulas* express trigonometric functions of $2x$ in terms of functions of x. We begin with the sum formulas for the sine, cosine, and tangent functions.

$$\sin(u + v) = \sin u \cos v + \cos u \sin v$$
$$\cos(u + v) = \cos u \cos v - \sin u \sin v$$
$$\tan(u + v) = \frac{\tan u + \tan v}{1 - \tan u \tan v}$$

To find the double-angle formula for the sine, replace u and v with x in the sum formula for sine.

$$\sin(u + v) = \sin u \cos v + \cos u \sin v \quad \text{Sum formula for sine}$$
$$\sin(x + x) = \sin x \cos x + \cos x \sin x \quad \text{Replace both } u \text{ and } v \text{ with } x.$$
$$\sin 2x = 2 \sin x \cos x \quad \text{Simplify.}$$

The identity $\mathbf{\sin 2x = 2 \sin x \cos x}$

is the double-angle formula for the sine function. We derive double-angle formulas for the cosine and tangent functions by replacing both u and v with x in their sum formulas to obtain:

$$\cos 2x = \cos^2 x - \sin^2 x \qquad \tan 2x = \frac{2 \tan x}{1 - \tan^2 x}$$

To derive two other useful forms for $\cos 2x$, first replace $\cos^2 x$ with $1 - \sin^2 x$ and then replace $\sin^2 x$ with $1 - \cos^2 x$ in the formula for $\cos 2x$.

$$\begin{aligned} \cos 2x &= \cos^2 x - \sin^2 x \\ &= (1 - \sin^2 x) - \sin^2 x \\ \cos 2x &= 1 - 2\sin^2 x \end{aligned} \qquad \begin{aligned} \cos 2x &= \cos^2 x - \sin^2 x \\ &= \cos^2 x - (1 - \cos^2 x) \\ \cos 2x &= 2\cos^2 x - 1 \end{aligned}$$

DOUBLE-ANGLE FORMULAS

$$\sin 2x = 2 \sin x \cos x \qquad \cos 2x = \cos^2 x - \sin^2 x$$
$$\tan 2x = \frac{2 \tan x}{1 - \tan^2 x} \qquad \cos 2x = 1 - 2\sin^2 x$$
$$\cos 2x = 2\cos^2 x - 1$$

EXAMPLE 1 Using Double-Angle Formulas

If $\cos \theta = -\frac{3}{5}$ and θ is in quadrant II, find the exact value of each expression.

a. $\sin 2\theta$ **b.** $\cos 2\theta$ **c.** $\tan 2\theta$

Solution

We first find $\sin \theta$ and $\tan \theta$.

$$\sin \theta = \sqrt{1 - \cos^2 \theta} \quad \theta \text{ is in quadrant II.}$$
$$= \sqrt{1 - \left(-\frac{3}{5}\right)^2} \quad \cos \theta = -\frac{3}{5}$$
$$= \frac{4}{5} \quad \text{Simplify.}$$
$$\tan \theta = \frac{\sin \theta}{\cos \theta} = \frac{4/5}{-3/5} = -\frac{4}{3} \quad \cos \theta = -\frac{3}{5} \text{ is given.}$$

a. $\sin 2\theta = 2 \sin \theta \cos \theta$ Double-angle formula for sine

$= 2\left(\frac{4}{5}\right)\left(-\frac{3}{5}\right)$ Replace $\sin \theta$ with $\frac{4}{5}$ and $\cos \theta$ with $-\frac{3}{5}$.

$= -\frac{24}{25}$ Simplify.

b. $\cos 2\theta = \cos^2\theta - \sin^2\theta$ Double-angle formula for cosines

$= \left(-\frac{3}{5}\right)^2 - \left(\frac{4}{5}\right)^2$ Replace $\cos\theta$ with $-\frac{3}{5}$ and $\sin\theta$ with $\frac{4}{5}$.

$= \frac{9}{25} - \frac{16}{25} = -\frac{7}{25}$ Simplify.

c. $\tan 2\theta = \frac{2\tan\theta}{1 - \tan^2\theta}$ Double-angle formula for tangent

$= \frac{2\left(-\frac{4}{3}\right)}{1 - \left(-\frac{4}{3}\right)^2}$ Replace $\tan\theta$ with $-\frac{4}{3}$.

$= \frac{-\frac{8}{3}}{1 - \frac{16}{9}} = \frac{24}{7}$ Simplify.

You can also find $\tan 2\theta$ by using parts **a** and **b**:

$$\tan 2\theta = \frac{\sin 2\theta}{\cos 2\theta} = \frac{-\frac{24}{25}}{-\frac{7}{25}} = \frac{24}{7}$$

Practice Problem 1 If $\sin x = \frac{12}{13}$ and $\frac{\pi}{2} < x < \pi$, find the exact value of each expression.

a. $\sin 2x$ **b.** $\cos 2x$ **c.** $\tan 2x$

EXAMPLE 2 Using Double-Angle Formulas

Find the exact value of each expression.

a. $1 - 2\sin^2\left(\frac{\pi}{12}\right)$ **b.** $\frac{2\tan 22.5^\circ}{1 - \tan^2 22.5^\circ}$

Solution

a. The given expression is the right side of the following formula, where $\theta = \frac{\pi}{12}$.

$\cos 2\theta = 1 - 2\sin^2\theta$ A double-angle formula for cosine

$1 - 2\sin^2\left(\frac{\pi}{12}\right) = \cos\left[2\left(\frac{\pi}{12}\right)\right]$ Replace θ with $\frac{\pi}{2}$; interchange sides.

$= \cos\frac{\pi}{6} = \frac{\sqrt{3}}{2}$

b. The given expression is the right side of the following formula, where $\theta = 22.5^\circ$.

$\tan 2\theta = \frac{2\tan\theta}{1 - \tan^2\theta}$ Double-angle formula for tangent

$\frac{2\tan(22.5^\circ)}{1 - \tan^2(22.5^\circ)} = \tan[2(22.5^\circ)]$ Replace θ with 22.5°; interchange sides.

$= \tan 45^\circ = 1$

Practice Problem 2 Find the exact value of each expression.

a. $2\cos^2\left(\frac{\pi}{12}\right) - 1$ **b.** $\cos^2 22.5° - \sin^2 22.5°$

EXAMPLE 3 Finding a Triple-Angle Formula for Sine

Verify the identity $\sin 3x = 3\sin x - 4\sin^3 x$.

Solution

We begin by writing $3x$ as $2x + x$ and use the sum formula for the sine and the double-angle formulas to verify this identity.

$$\begin{aligned}
\sin 3x &= \sin(2x + x) \\
&= \sin 2x \cos x + \cos 2x \sin x && \text{Sum formula for sine} \\
&= (2\sin x\cos x)\cos x + (1 - 2\sin^2 x)\sin x && \text{Replace } \sin 2x \text{ with } 2\sin x\cos x \text{ and } \cos 2x \text{ with } 1 - 2\sin^2 x. \\
&= 2\sin x\cos^2 x + \sin x - 2\sin^3 x && \text{Multiply; distributive property} \\
&= 2\sin x(1 - \sin^2 x) + \sin x - 2\sin^3 x && \cos^2 x = 1 - \sin^2 x \\
&= 2\sin x - 2\sin^3 x + \sin x - 2\sin^3 x && \text{Distributive property} \\
&= 3\sin x - 4\sin^3 x && \text{Simplify.}
\end{aligned}$$

We have verified the identity by showing that the left and right sides of the equation are equal.

Practice Problem 3 Verify the identity $\cos 3x = 4\cos^3 x - 3\cos x$.

We can use the double-angle formulas to express trigonometric functions of 4θ, 6θ, and 8θ in terms of 2θ, 3θ, and 4θ, respectively. For example, the identities

$$\cos 4\theta = 2\cos^2 2\theta - 1, \quad \sin 6\theta = 2\sin 3\theta\cos 3\theta, \quad \text{and} \quad \tan 8\theta = \frac{2\tan 4\theta}{1 - \tan^2 4\theta}$$

can be directly derived from the appropriate double-angle formulas.

2 Use power-reducing formulas.

Power-Reducing Formulas

The purpose of *power-reducing formulas* is to express $\sin^2 x$, $\cos^2 x$, and $\tan^2 x$ in terms of trigonometric functions with powers not greater than 1. These formulas are useful in calculus.

POWER-REDUCING FORMULAS

$$\sin^2 x = \frac{1 - \cos 2x}{2} \qquad \cos^2 x = \frac{1 + \cos 2x}{2} \qquad \tan^2 x = \frac{1 - \cos 2x}{1 + \cos 2x}$$

We can derive the first two power-reducing formulas by using the appropriate formula for $\cos 2x$.

$$\begin{aligned}
\cos 2x &= 1 - 2\sin^2 x && \text{Double-angle formula for } \cos 2x \text{ in terms of sine} \\
2\sin^2 x &= 1 - \cos 2x && \text{Add } 2\sin^2 x - \cos 2x \text{ to both sides and simplify.} \\
\sin^2 x &= \frac{1 - \cos 2x}{2} && \text{Divide both sides by 2.}
\end{aligned}$$

Similarly, we can derive the second formula.

$$\cos 2x = 2\cos^2 x - 1 \quad \text{Double-angle formula for } \cos 2x \text{ in terms of cosine}$$
$$1 + \cos 2x = 2\cos^2 x \quad \text{Add 1 to both sides.}$$
$$\frac{1 + \cos 2x}{2} = \cos^2 x \quad \text{Divide both sides by 2.}$$

We begin with the quotient identity to prove the third formula.

$$\tan^2 x = \frac{\sin^2 x}{\cos^2 x} \quad \text{Quotient identity}$$
$$= \frac{\frac{1 - \cos 2x}{2}}{\frac{1 + \cos 2x}{2}} \quad \text{Power-reducing formulas for } \sin^2 x \text{ and } \cos^2 x$$
$$= \frac{1 - \cos 2x}{1 + \cos 2x} \quad \text{Multiply numerator and denominator by 2. Simplify.}$$

EXAMPLE 4 **Using Power-Reducing Formulas**

Write an equivalent expression for $\cos^4 x$ that contains only first powers of cosines of multiple angles.

Solution

We use the power-reducing formulas repeatedly.

$$\cos^4 x = (\cos^2 x)^2 \quad a^4 = (a^2)^2$$
$$= \left(\frac{1 + \cos 2x}{2}\right)^2 \quad \text{Power-reducing formula}$$
$$= \frac{1}{4}(1 + 2\cos 2x + \cos^2 2x) \quad \text{Expand the binomial.}$$
$$= \frac{1}{4}\left(1 + 2\cos 2x + \frac{1 + \cos 4x}{2}\right) \quad \text{Power-reducing formula for } \cos^2 x\text{; replace } x \text{ with } 2x.$$
$$= \frac{1}{4}\left(1 + 2\cos 2x + \frac{1}{2} + \frac{1}{2}\cos 4x\right) \quad \frac{a + b}{c} = \frac{a}{c} + \frac{b}{c}$$
$$= \frac{1}{4} + \frac{2}{4}\cos 2x + \frac{1}{8} + \frac{1}{8}\cos 4x \quad \text{Distributive property}$$
$$= \frac{3}{8} + \frac{1}{2}\cos 2x + \frac{1}{8}\cos 4x \quad \text{Simplify.}$$

Practice Problem 4 Write an equivalent expression for $\sin^4 x$ that contains only first powers of cosines of multiple angles.

Alternating Current

Alternating current (AC) is the electric current that reverses direction, usually many times per second. Most electrical generators produce alternating current. The *wattage rating* of an electric device is $\frac{1}{\sqrt{2}} \approx 0.7071$ times the maximum wattage of the device.

EXAMPLE 5 **Finding the Wattage Rating of an Electric Blanket**

The voltage V of a household current is given by $V = 170 \sin(120\pi t)$ volts, where t is in seconds. Suppose the amount of current passing through an electric blanket is $I = 0.1 \sin(120\pi t)$ amps, where t is in seconds. Find the wattage rating of the blanket.

Solution

We have $P = VI$ — Power = voltage × current; watts = volts × amps

$$P = [170 \sin(120\pi t)][0.1 \sin(120\pi t)]$$ Substitute the given values of V and I.

$$= 17 \sin^2(120\pi t)$$ Multiply.

$$= 17\left[\frac{1 - \cos(240\pi t)}{2}\right]$$ Power-reducing formula

$$P = 8.5 - 8.5 \cos(240\pi t)$$ Simplify.

The maximum value of P is 17 watts when $\cos(240\pi t) = -1$, so the wattage rating for this blanket is $\frac{17}{\sqrt{2}} \approx 12$ watts.

Practice Problem 5 In Example 5, find the wattage rating of a lightbulb for which $I = 0.83 \sin(120\pi t)$ amps.

3 Use half-angle formulas.

Half-Angle Formulas

If an angle measures θ, then $\frac{\theta}{2}$ is half the measure of θ. Half-angle formulas express trigonometric functions of $\frac{\theta}{2}$ in terms of functions of θ. To derive the half-angle formulas, we replace x with $\frac{\theta}{2}$ in the power-reducing formulas and then take the square root of both sides. For example,

$$\cos^2 x = \frac{1 + \cos 2x}{2}$$ Power-reducing formula for cosines

$$\cos^2 \frac{\theta}{2} = \frac{1 + \cos\left[2\left(\frac{\theta}{2}\right)\right]}{2}$$ Replace x with $\frac{\theta}{2}$.

$$\cos^2 \frac{\theta}{2} = \frac{1 + \cos\theta}{2}$$ Simplify.

$$\cos \frac{\theta}{2} = \pm\sqrt{\frac{1 + \cos\theta}{2}}$$ Square root property

We call the last equation a *half-angle formula* for cosine. The sign + or − depends on the quadrant in which $\frac{\theta}{2}$ lies. We can derive half-angle formulas for sine and tangent in a similar manner.

HALF-ANGLE FORMULAS

$$\sin\frac{\theta}{2} = \pm\sqrt{\frac{1-\cos\theta}{2}} \qquad \cos\frac{\theta}{2} = \pm\sqrt{\frac{1+\cos\theta}{2}} \qquad \tan\frac{\theta}{2} = \pm\sqrt{\frac{1-\cos\theta}{1+\cos\theta}}$$

The sign $+$ or $-$ depends on the quadrant in which $\frac{\theta}{2}$ lies. Alternate formulas for $\tan\frac{\theta}{2}$ are:

$$\tan\frac{\theta}{2} = \frac{\sin\theta}{1+\cos\theta} \qquad \tan\frac{\theta}{2} = \frac{1-\cos\theta}{\sin\theta}$$

(See Exercises 67 and 68.)

EXAMPLE 6 Using Half-Angle Formulas

Use a half-angle formula to find the exact value of $\cos 157.5°$.

Solution

Because $157.5° = \frac{315°}{2}$, we use the half-angle formula for $\cos\frac{\theta}{2}$ with $\theta = 315°$.

Because $\frac{\theta}{2} = 157.5°$ lies in quadrant II, $\cos\frac{\theta}{2}$ is negative.

$$\cos 157.5° = \cos\frac{315°}{2}$$

$$= -\sqrt{\frac{1+\cos 315°}{2}} \qquad \text{Half-angle formula for cosine}$$

$$= -\sqrt{\frac{1+\cos 45°}{2}} \qquad \cos 315° = \cos(360° - 45°) = \cos 45°$$

$$= -\sqrt{\frac{\left(1+\frac{\sqrt{2}}{2}\right)\cdot 2}{2\cdot 2}} \qquad \text{Replace } \cos 45° \text{ with } \frac{\sqrt{2}}{2}\text{. Multiply numerator and denominator by 2.}$$

$$= -\sqrt{\frac{2+\sqrt{2}}{2\cdot 2}} = -\frac{\sqrt{2+\sqrt{2}}}{2} \qquad \text{Simplify.}$$

Practice Problem 6 Find the exact value of $\sin 112.5°$.

EXAMPLE 7 Finding the Exact Value

Given that $\sin\theta = -\frac{5}{13}$, $\pi < \theta < \frac{3\pi}{2}$, find the exact value of each expression.

a. $\sin\frac{\theta}{2}$ **b.** $\tan\frac{\theta}{2}$

Solution

Because θ lies in quadrant III, $\cos\theta$ is negative; so

$$\cos\theta = -\sqrt{1-\sin^2\theta} \qquad \text{Pythagorean identity}$$

$$\cos\theta = -\sqrt{1-\left(-\frac{5}{13}\right)^2} \qquad \text{Replace } \sin\theta \text{ with } -\frac{5}{13}.$$

$$= -\sqrt{1-\frac{25}{169}} = -\sqrt{\frac{169-25}{169}} = -\frac{12}{13} \qquad \text{Simplify.}$$

a. Because $\pi < \theta < \frac{3\pi}{2}, \frac{\theta}{2}$ lies in quadrant II $\left(\text{if } \pi < \theta < \frac{3\pi}{2}, \text{ then } \frac{\pi}{2} < \frac{\theta}{2} < \frac{3\pi}{4}\right)$.

The half-angle formula gives

$$\sin\frac{\theta}{2} = \sqrt{\frac{1-\cos\theta}{2}} \qquad \text{In quadrant II, } \sin\frac{\theta}{2} \text{ is positive.}$$

$$= \sqrt{\frac{1-\left(-\frac{12}{13}\right)}{2}} \qquad \text{Replace } \cos\theta \text{ with } -\frac{12}{13}.$$

$$= \sqrt{\frac{\frac{25}{13}}{2}} = \frac{5}{\sqrt{26}} = \frac{5\sqrt{26}}{26} \qquad \text{Simplify.}$$

b. From the half-angle formula, we have

$$\tan\frac{\theta}{2} = -\sqrt{\frac{1-\cos\theta}{1+\cos\theta}} \qquad \text{In quadrant II, } \tan\frac{\theta}{2} \text{ is negative.}$$

$$= -\sqrt{\frac{1-\left(-\frac{12}{13}\right)}{1+\left(-\frac{12}{13}\right)}} \qquad \cos\theta = -\frac{12}{13}$$

$$= -\sqrt{\frac{\frac{25}{13}}{\frac{1}{13}}} = -5 \qquad \text{Simplify.}$$

Practice Problem 7 For θ in Example 7, find $\cos\frac{\theta}{2}$.

EXAMPLE 8 Verifying an Identity Containing Half-Angles

Verify the identity $\sin x \cos\frac{x}{2} = \sin\frac{x}{2}(1 + \cos x)$.

Solution

The left side contains $\sin x$, and the right side contains $\sin\frac{x}{2}$. We replace x with $\frac{x}{2}$ in the double-angle identity for sine.

$$\sin 2x = 2\sin x\cos x \qquad \text{Double-angle identity for sine}$$

$$\sin x = 2\sin\frac{x}{2}\cos\frac{x}{2} \qquad \text{Replace } x \text{ with } \frac{x}{2}; \sin\left[2\left(\frac{x}{2}\right)\right] = \sin x.$$

Begin with the left side of the original equation and use this expression for $\sin x$.

$$\sin x\cos\frac{x}{2} = 2\sin\frac{x}{2}\cos\frac{x}{2}\cos\frac{x}{2} \qquad \text{Replace } \sin x \text{ with } 2\sin\frac{x}{2}\cos\frac{x}{2}.$$

$$= 2\sin\frac{x}{2}\cos^2\frac{x}{2}$$

$$= 2\left(\sin\frac{x}{2}\right)\left(\frac{1+\cos\left[2\left(\frac{x}{2}\right)\right]}{2}\right) \qquad \text{Power-reducing formula for cosine}$$

$$= \sin\frac{x}{2}(1+\cos x) \qquad \cos\left[2\left(\frac{x}{2}\right)\right] = \cos x; \text{ simplify.}$$

Because the left side is identical to the right side of the equation, the identity is verified.

Practice Problem 8 Verify the identity $\sin\frac{x}{2}\sin x = \cos\frac{x}{2}(1 - \cos x)$.

SUMMARY OF MAIN FACTS

Double-Angle Formulas

$$\sin 2x = 2\sin x\cos x \qquad \cos 2x = \cos^2 x - \sin^2 x$$

$$\tan 2x = \frac{2\tan x}{1 - \tan^2 x} \qquad \cos 2x = 1 - 2\sin^2 x$$

$$\cos 2x = 2\cos^2 x - 1$$

Power-Reducing Formulas

$$\sin^2 x = \frac{1 - \cos 2x}{2} \qquad \cos^2 x = \frac{1 + \cos 2x}{2} \qquad \tan^2 x = \frac{1 - \cos 2x}{1 + \cos 2x}$$

Half-Angle Formulas

$$\sin\frac{\theta}{2} = \pm\sqrt{\frac{1 - \cos\theta}{2}} \qquad \cos\frac{\theta}{2} = \pm\sqrt{\frac{1 + \cos\theta}{2}} \qquad \tan\frac{\theta}{2} = \pm\sqrt{\frac{1 - \cos\theta}{1 + \cos\theta}}$$

$$= \frac{\sin\theta}{1 + \cos\theta}$$

$$= \frac{1 - \cos\theta}{\sin\theta}$$

The sign + or − depends on the quadrant in which $\frac{\theta}{2}$ lies.

Answers to Practice Problems

1. a. $-\frac{120}{169}$ **b.** $-\frac{119}{169}$ **c.** $\frac{120}{119}$ **2. a.** $\frac{\sqrt{3}}{2}$ **b.** $\frac{\sqrt{2}}{2}$ **6.** $\frac{\sqrt{2 + \sqrt{2}}}{2}$ **7.** $-\frac{\sqrt{26}}{26}$

4. $\frac{3}{8} - \frac{1}{2}\cos 2x + \frac{1}{8}\cos 4x$ **5.** ≈ 99.8 watts

SECTION 6.3 Exercises

Basic Concepts and Skills

1. The double-angle identity for $\sin 2x$ is $\sin 2x =$ __________.

2. In the double-angle identity $\cos 2x = \cos^2 x - \sin^2 x$, replace $\cos^2 x$ with $1 - \sin^2 x$ to obtain a double-angle identity $\cos 2x =$ __________ in terms of $\sin^2 x$. Solve this identity for $\sin^2 x$ to obtain the power-reducing identity $\sin^2 x =$ __________.

3. The identity for $\cos 2x$ in terms of $\cos^2 x$ is $\cos 2x =$ __________. Solve this identity for $\cos^2 x$ to obtain the power-reducing identity $\cos^2 x =$ __________.

4. **True or False.** $\tan 2x = \frac{1 - \cos 2x}{1 + \cos 2x}$

5. **True or False.** $\frac{1}{2}\tan 2x = \tan x$

6. **True or False.** $\cos\frac{\theta}{2} = -\sqrt{\frac{1 + \cos\theta}{2}}, \pi < \theta < 2\pi$

In Exercises 7–12, use the given information about the angle θ to find the exact value of

a. $\sin 2\theta$. **b.** $\cos 2\theta$. **c.** $\tan 2\theta$.

7. $\sin\theta = \frac{3}{5}$, θ in quadrant II

8. $\cos\theta = -\frac{5}{13}$, θ in quadrant III

9. $\tan\theta = 4, \sin\theta < 0$

10. $\sec\theta = -\sqrt{3}, \sin\theta > 0$

11. $\tan\theta = -2, \frac{\pi}{2} < \theta < \pi$

12. $\cot\theta = -7, \frac{3\pi}{2} < \theta < 2\pi$

In Exercises 13–22, use a double-angle identity to find the exact value of each expression.

13. $1 - 2\sin^2 75°$

14. $\frac{2\tan 75°}{1 - \tan^2 75°}$

15. $2\cos^2 105° - 1$

16. $1 - 2\sin^2 165°$

17. $\frac{2\tan 165°}{1 - \tan^2 165°}$

18. $2\cos^2 165° - 1$

19. $1 - 2\sin^2\frac{\pi}{8}$

20. $2\cos^2\left(-\frac{\pi}{8}\right) - 1$

21. $\frac{2\tan\left(-\frac{5\pi}{12}\right)}{1 - \tan^2\left(-\frac{5\pi}{12}\right)}$

22. $1 - 2\sin^2\left(-\frac{7\pi}{12}\right)$

In Exercises 23 and 24, verify each "quadruple-angle" identity.

23. $\sin 4\theta = \cos\theta(4\sin\theta - 8\sin^3\theta)$

24. $\cos 4\theta = 8\cos^4\theta - 8\cos^2\theta + 1$

In Exercises 25–32, verify each identity.

25. $\cos^4 x - \sin^4 x = \cos 2x$

26. $1 + \cos 2x + 2\sin^2 x = 2$

27. $\frac{1 + \sin 2x}{\cos 2x} = \frac{\cos x + \sin x}{\cos x - \sin x}$

28. $\frac{\cos 2x}{\sin 2x} + \frac{\sin x}{\cos x} = \csc 2x$

29. $\frac{\sin 2x}{\sin x} - \frac{\cos 2x}{\cos x} = \sec x$

30. $\frac{\cos 3x}{\sin 3x} + \frac{\sin x}{\cos x} = \frac{\cos 2x}{\sin 3x\cos x}$

31. $\tan 2x + \tan x = \frac{\sin 3x}{\cos 2x\cos x}$

32. $\tan 2x - \tan x = \tan x\sec 2x$

In Exercises 33–42, use the power-reducing identities to rewrite each expression to one that contains a single trigonometric function of power 1.

33. $4\sin^2 x\cos^2 x$

34. $\sin^2 x\cos^2 x$

35. $4\sin x\cos x(1 - 2\sin^2 x)$

36. $4\sin x\cos x(2\cos^2 x - 1)$

37. $2\sin 3x\cos 3x(2\cos^2 3x - 1)$

38. $\sin 8x(1 - 2\sin^2 4x)$

39. $\sin\frac{x}{2}\cos\frac{x}{2}\left(1 - 2\sin^2\frac{x}{2}\right)$

40. $\sin x\left(2\cos^2\frac{x}{2} - 1\right)$

41. $8\sin^4\frac{x}{2}$

42. $8\cos^4\frac{x}{2}$

In Exercises 43–54, use half-angle identities to find the exact value of each expression.

43. $\sin\frac{\pi}{12}$

44. $\sin\frac{\pi}{8}$

45. $\cos\frac{\pi}{8}$

46. $\tan\frac{\pi}{8}$

47. $\sin\left(-\frac{3\pi}{8}\right)$

48. $\cos\left(-\frac{3\pi}{8}\right)$

49. $\tan\left(\frac{7\pi}{8}\right)$

50. $\sec\left(-\frac{7\pi}{8}\right)$

51. $\tan 112.5°$

52. $\cos 112.5°$

53. $\sin(-75°)$

54. $\tan(-105°)$

In Exercises 55–62, use the information about the angle θ to find the exact value of

a. $\sin\frac{\theta}{2}$. **b.** $\cos\frac{\theta}{2}$. **c.** $\tan\frac{\theta}{2}$.

55. $\sin\theta = \frac{4}{5}, \frac{\pi}{2} < \theta < \pi$

56. $\cos\theta = -\frac{12}{13}, \pi < \theta < \frac{3\pi}{2}$

57. $\tan\theta = -\frac{2}{3}, \frac{\pi}{2} < \theta < \pi$

58. $\cot\theta = \frac{3}{4}, \pi < \theta < \frac{3\pi}{2}$

59. $\sin\theta = \frac{1}{5}, \cos\theta < 0$

60. $\cos\theta = \frac{2}{3}, \sin\theta < 0$

61. $\sec\theta = \sqrt{5}, \sin\theta > 0$

62. $\csc\theta = \sqrt{7}, \tan\theta < 0$

In Exercises 63–72, verify each identity.

63. $\left(\sin\frac{t}{2} + \cos\frac{t}{2}\right)^2 = 1 + \sin t$

64. $\left(\sin\frac{t}{2} - \cos\frac{t}{2}\right)^2 = 1 - \sin t$

65. $2\cos^2\frac{x}{2} = \frac{\sin^2 x}{1 - \cos x}$

66. $2\sin^2\frac{x}{2} = \frac{\sin^2 x}{1 + \cos x}$

67. $\tan\frac{x}{2} = \frac{\sin x}{1 + \cos x}$

68. $\tan\frac{x}{2} = \frac{1 - \cos x}{\sin x}$

69. $\sin^2\frac{x}{2} + \cos x = \cos^2\frac{x}{2}$

70. $\cos^2 x + \cos x = 2\cos^2\frac{x}{2} - \sin^2 x$

71. $\sin x = \frac{2\tan\frac{x}{2}}{1 + \tan^2\frac{x}{2}}$

72. $\cos x = \frac{1 - \tan^2\frac{x}{2}}{1 + \tan^2\frac{x}{2}}$

Applying the Concepts

In Exercises 73–78, assume that the voltage of the current is given by $V = 170 \sin(120\pi t)$, where t is in seconds. Find the wattage rating $\left(\frac{\text{maximum wattage}}{\sqrt{2}}\right)$ of each electric device assuming that the current flowing through the device is I amps and the power in watts is given by $P = VI$.

73. **Lightbulb.** $I = 0.832 \sin(120\pi t)$
74. **Microwave oven.** $I = 7.487 \sin(120\pi t)$
75. **Toaster.** $I = 9.983 \sin(120\pi t)$
76. **Refrigerator.** $I = 6.655 \sin(120\pi t)$
77. **Vacuum cleaner.** $I = 4.991 \sin(120\pi t)$
78. **Television.** $I = 2.917 \sin(120\pi t)$
79. **Throwing a football.** A quarterback throws a ball with an initial velocity of v_0 feet per second at an angle θ with the horizontal. The horizontal distance x in feet the ball is thrown is modeled by the equation $x = \frac{v_0^2}{16} \sin\theta\cos\theta$. For a fixed v_0, use a double-angle identity to find the angle that produces the maximum distance x.

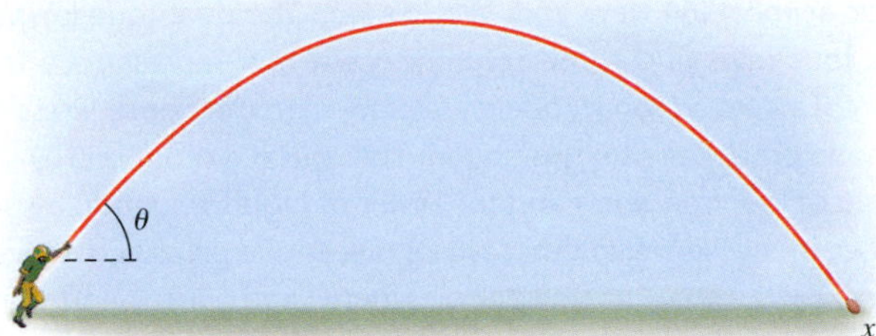

Beyond the Basics

In Exercises 80–90, verify each identity.

80. $\tan 3x = \dfrac{3\tan x - \tan^3 x}{1 - 3\tan^2 x}$

81. $\dfrac{\sin 3x + \cos 3x}{\cos x - \sin x} = 1 + 2\sin 2x$

82. $\dfrac{\sin x - \cos x}{\sin x + \cos x} - \dfrac{\sin x + \cos x}{\sin x - \cos x} = 2\tan 2x$

83. $\dfrac{2}{\tan x + \cot x} = \sin 2x$

84. $\dfrac{2\sin x}{\cos 3x} = \tan 3x - \tan x$

85. $\cot x - \tan x = 2\cot 2x$

86. $\dfrac{1 - \tan^2\left(\frac{\pi}{4} - x\right)}{1 + \tan^2\left(\frac{\pi}{4} - x\right)} = \sin 2x$

87. $\dfrac{1 + \sin 2x - \cos 2x}{1 + \sin 2x + \cos 2x} = \tan x$

88. $\cos\theta\cos 2\theta\cos 4\theta = \dfrac{\sin 8\theta}{8\sin\theta}$

89. $\sin^2\left(\frac{\pi}{8} + \frac{x}{2}\right) - \sin^2\left(\frac{\pi}{8} - \frac{x}{2}\right) = \frac{1}{\sqrt{2}}\sin x$

90. $\sqrt{2 + \sqrt{2 + 2\cos 4x}} = 2\cos x,\ 0 < x < \frac{\pi}{4}$

Critical Thinking/Discussion/Writing

91. Use the figure to find the exact value of each expression.
   a. $\sin\theta$ b. $\cos\theta$
   c. $\sin 2\theta$ d. $\cos 2\theta$
   e. $\tan 2\theta$ f. $\sin\frac{\theta}{2}$
   g. $\cos\frac{\theta}{2}$ h. $\tan\frac{\theta}{2}$

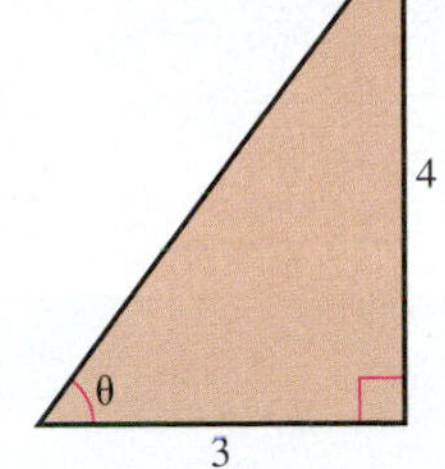

Maintaining Skills

In Exercises 92 and 93, use the cofunction identities to find an exact value for each expression.

92. $\sec\left(\frac{\pi}{2} - \theta\right)$, if $\sin\theta = \sqrt{6} - \sqrt{3}$

93. $\csc\theta$, if $\cos\left(\frac{\pi}{2} - \theta\right) = \sqrt{5} - \sqrt{2}$

In Exercises 94–97, use the even–odd identities to find an exact value for each expression. (Other identities may also be required.)

94. $\sin\theta$, if $\sin(-\theta) = 0.76$
95. $\cos\theta$, if $\cos(-\theta) = 0.87$
96. $\sin\theta\cos^2\theta$, if $\sin(-\theta) = \frac{1}{3}$
97. $\tan\theta\sec^2\theta$, if $\tan(-\theta) = -2$

In Exercises 98–101, use the sum and difference formulas to find an exact value for each expression.

98. $\cos\left(\frac{\pi}{2} + \frac{\pi}{6}\right)$

99. $\sin\left(\frac{\pi}{6} - \frac{3\pi}{4}\right)$

100. $\sin\frac{2\pi}{9}\cos\frac{8\pi}{9} - \cos\frac{2\pi}{9}\sin\frac{8\pi}{9}$

101. $\cos\frac{\pi}{2}\cos\frac{\pi}{6} + \sin\frac{\pi}{2}\sin\frac{\pi}{6}$

SECTION 6.4

Product-to-Sum and Sum-to-Product Formulas

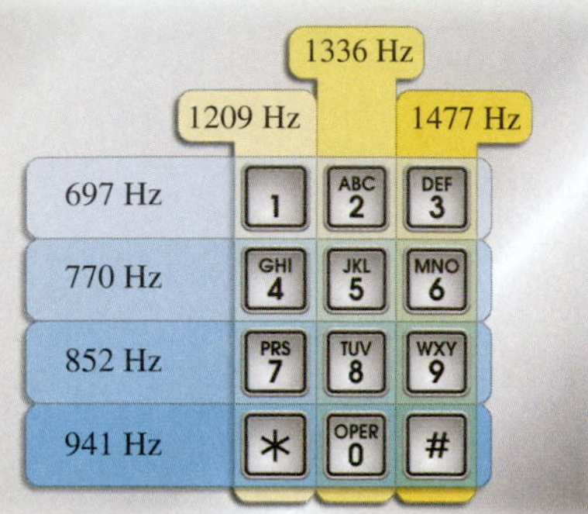

BEFORE STARTING THIS SECTION, REVIEW

1. Sum and difference formulas for sines and cosines (Section 6.2, page 591)
2. Fundamental identities (Section 6.1, page 569)

OBJECTIVES

1. Derive product-to-sum formulas.
2. Derive sum-to-product formulas.
3. Verify trigonometric identities involving multiple angles.

◆ Touch-Tone Phones

Dual-tone multi-frequency (DTMF), also known as Touch-Tone, tone dialing, or push-button dialing, is a method for instructing a telephone switching network to dial a telephone number. The system was developed by Bell Labs in the late 1950s, and Touch-Tone phones were introduced to the public at the 1964 New York World's Fair. The dual-tone keypad on a Touch-Tone phone has four rows and three columns. Each row represents a *low* frequency, and each column represents a *high* frequency, as shown in the figure. Pressing a button on a Touch-Tone phone produces a unique sound. The sound is produced by the combination of two tones (one of low frequency and the other of high frequency) associated with that button, as shown in the figure. In other words, the sound produced is given by $y = y_1 + y_2$, with $y_1 = \sin(2\pi lt)$ and $y_2 = \sin(2\pi ht)$, where l and h are the button's low and high frequencies (in hertz = Hz), respectively.

For example, if you press a single button, such as "1," it will send a sinusoidal tone of two frequencies 697 Hz and 1209 Hz. The sound produced by pressing "1" is represented by the equation

$$y = \sin[2\pi(697)t] + \sin[2\pi(1209)t]$$

or

$$y = \sin(1394\pi t) + \sin(2418\pi t).$$

The two tones are the reason for calling such phones *dual-tone multi-frequency phones*. In Example 6, we analyze the sound produced by the sum of two sine waves. (*Source:* http://en.wikipedia.org)

1 Derive product-to-sum formulas.

Product-to-Sum Formulas

The following formulas allow us to rewrite products of sines or cosines as sums or differences.

PRODUCT-TO-SUM FORMULAS

$$\cos x \cos y = \frac{1}{2}[\cos(x - y) + \cos(x + y)]$$

$$\sin x \sin y = \frac{1}{2}[\cos(x - y) - \cos(x + y)]$$

$$\sin x \cos y = \frac{1}{2}[\sin(x + y) + \sin(x - y)]$$

$$\cos x \sin y = \frac{1}{2}[\sin(x + y) - \sin(x - y)]$$

These formulas are fairly easy to prove. For example, to prove the first formula

$$\cos x \cos y = \frac{1}{2}[\cos(x - y) + \cos(x + y)],$$

we write the difference and sum formulas for cosine and add.

$$\cos(x - y) = \cos x \cos y + \sin x \sin y$$
$$\underline{\cos(x + y) = \cos x \cos y - \sin x \sin y}$$

$$\cos(x - y) + \cos(x + y) = 2\cos x \cos y + 0 \quad \text{Add.}$$

$$2\cos x \cos y = \cos(x - y) + \cos(x + y) \quad \text{Interchange sides.}$$

$$\mathbf{\cos x \cos y = \frac{1}{2}[\cos(x - y) + \cos(x + y)]} \quad \text{Multiply both sides by } \frac{1}{2}.$$

Similarly, subtracting $\cos(x + y)$ from $\cos(x - y)$ gives the second formula:

$$\mathbf{\sin x \sin y = \frac{1}{2}[\cos(x - y) - \cos(x + y)]}$$

We can prove the third and fourth formulas in the box on page 606, by using the sum and difference formulas for sine.

EXAMPLE 1 Using a Product-to-Sum Formula

Write $\sin 5\theta \cos 3\theta$ as the sum or difference of two trigonometric functions.

Solution

Use the formula $\sin x \cos y = \frac{1}{2}[\sin(x + y) + \sin(x - y)]$.

$$\sin 5\theta \cos 3\theta = \frac{1}{2}[\sin(5\theta + 3\theta) + \sin(5\theta - 3\theta)] \quad \text{Replace } x \text{ with } 5\theta \text{ and } y \text{ with } 3\theta.$$

$$= \frac{1}{2}[\sin 8\theta + \sin 2\theta] \quad \text{Simplify.}$$

$$= \frac{1}{2}\sin 8\theta + \frac{1}{2}\sin 2\theta \quad \text{Distributive property}$$

Practice Problem 1 Write $\cos 3x \cos x$ as the sum or difference of two trigonometric functions.

EXAMPLE 2 Using a Product-to-Sum Formula

Find the exact value of $\sin 75° \sin 15°$.

Solution

Use the formula $\sin x \sin y = \frac{1}{2}[\cos(x - y) - \cos(x + y)]$.

$$\sin 75° \sin 15° = \frac{1}{2}[\cos(75° - 15°) - \cos(75° + 15°)] \quad \text{Replace } x \text{ with } 75° \text{ and } y \text{ with } 15°.$$

$$= \frac{1}{2}[\cos 60° - \cos 90°] \quad \text{Simplify.}$$

$$= \frac{1}{2}\left[\frac{1}{2} - 0\right] = \frac{1}{4} \quad \cos 60° = \frac{1}{2};\ \cos 90° = 0$$

Practice Problem 2 Find the exact value of $\sin 15° \cos 75°$.

2 Derive sum-to-product formulas.

Sum-to-Product Formulas

The following formulas allow us to do just the opposite—rewrite sums or differences of sines or cosines as products.

SUM-TO-PRODUCT FORMULAS

$$\cos x + \cos y = 2\cos\left(\frac{x+y}{2}\right)\cos\left(\frac{x-y}{2}\right)$$

$$\cos x - \cos y = -2\sin\left(\frac{x+y}{2}\right)\sin\left(\frac{x-y}{2}\right)$$

$$\sin x + \sin y = 2\sin\left(\frac{x+y}{2}\right)\cos\left(\frac{x-y}{2}\right)$$

$$\sin x - \sin y = 2\sin\left(\frac{x-y}{2}\right)\cos\left(\frac{x+y}{2}\right)$$

We prove these formulas by using the product-to-sum formulas. For example, to prove the first formula, we start with the right side of the equation and use the product-to-sum formulas.

$$2\cos\left(\frac{x+y}{2}\right)\cos\left(\frac{x-y}{2}\right)$$

$$= 2\cdot\frac{1}{2}\left[\cos\left(\frac{x+y}{2}-\frac{x-y}{2}\right)+\cos\left(\frac{x+y}{2}+\frac{x-y}{2}\right)\right]$$ Use the formula for the product of cosines.

$$= \cos\left(\frac{x+y-x+y}{2}\right)+\cos\left(\frac{x+y+x-y}{2}\right)$$ Simplify.

$$= \cos\left(\frac{2y}{2}\right)+\cos\left(\frac{2x}{2}\right)$$ Simplify.

$$= \cos y + \cos x$$ Simplify.

You can verify the other three formulas in the box in a similar way. See Exercise 87.

EXAMPLE 3 Using Sum-to-Product Formulas

Write each expression as a product of two trigonometric functions and simplify where possible.

a. $\sin 4\theta - \sin 6\theta$ **b.** $\cos 65° + \cos 55°$

Solution

a. Use the formula $\sin x - \sin y = 2\sin\left(\frac{x-y}{2}\right)\cos\left(\frac{x+y}{2}\right)$.

$$\sin 4\theta - \sin 6\theta = 2\sin\left(\frac{4\theta-6\theta}{2}\right)\cos\left(\frac{4\theta+6\theta}{2}\right)$$ Replace x with 4θ and y with 6θ.

$$= 2\sin(-\theta)\cos(5\theta)$$ Simplify.

$$= -2\sin\theta\cos 5\theta$$ $\sin(-\theta) = -\sin\theta$

b. Use the formula $\cos x + \cos y = 2\cos\left(\frac{x+y}{2}\right)\cos\left(\frac{x-y}{2}\right)$.

$$\cos 65° + \cos 55° = 2\cos\left(\frac{65° + 55°}{2}\right)\cos\left(\frac{65° - 55°}{2}\right) \quad \text{Replace } x \text{ with } 65° \text{ and } y \text{ with } 55°.$$

$$= 2\cos 60° \cos 5° \quad \text{Simplify.}$$

$$= 2\left(\frac{1}{2}\right)\cos 5° = \cos 5° \quad \cos 60° = \frac{1}{2}$$

Practice Problem 3 Write each expression as a product of two trigonometric functions and simplify where possible.

a. $\cos 2x - \cos 4x$ **b.** $\sin 43° + \sin 17°$

In the next example, we show how to express a sum of a sine and a cosine as a product by first using a cofunction identity.

EXAMPLE 4 Expressing a Sum of a Sine and a Cosine as a Product

Write $\sin 5\theta + \cos 3\theta$ as a product of two trigonometric functions.

Solution

Use the formula

$$\cos x + \cos y = 2\cos\left(\frac{x+y}{2}\right)\cos\left(\frac{x-y}{2}\right).$$

$$\sin 5\theta + \cos 3\theta = \cos\left(\frac{\pi}{2} - 5\theta\right) + \cos 3\theta \quad \text{Cofunction identity}$$

$$= 2\cos\left(\frac{\frac{\pi}{2} - 5\theta + 3\theta}{2}\right)\cos\left(\frac{\frac{\pi}{2} - 5\theta - 3\theta}{2}\right) \quad \text{Sum-to-product formula for cosines}$$

$$= 2\cos\left(\frac{\pi}{4} - \theta\right)\cos\left(\frac{\pi}{4} - 4\theta\right) \quad \text{Simplify.}$$

SIDE NOTE

In Example 4, you could also write $\sin 5\theta + \cos 3\theta = \sin 5\theta + \sin\left(\frac{\pi}{2} - 3\theta\right)$ and then use the appropriate sum-to-product formula.

Practice Problem 4 Write $\sin 3x - \cos x$ as a product of two trigonometric functions.

3 Verify trigonometric identities involving multiple angles.

Verify Trigonometric Identities

The next example illustrates the use of sum-to-product formulas to verify identities.

EXAMPLE 5 Verifying an Identity

Verify the identity $\dfrac{\sin 5\theta + \sin 9\theta}{\cos 5\theta - \cos 9\theta} = \cot 2\theta$.

Solution

$$\frac{\sin 5\theta + \sin 9\theta}{\cos 5\theta - \cos 9\theta} = \frac{2\sin\frac{5\theta + 9\theta}{2}\cos\frac{5\theta - 9\theta}{2}}{-2\sin\frac{5\theta + 9\theta}{2}\sin\frac{5\theta - 9\theta}{2}}$$ Use the sum-to-product formulas for $\sin x + \sin y$ and $\cos x - \cos y$.

$$= \frac{2\sin 7\theta\cos(-2\theta)}{-2\sin 7\theta\sin(-2\theta)}$$ Simplify.

$$= \frac{\cos(-2\theta)}{-\sin(-2\theta)}$$ Remove the common factors.

$$= \frac{\cos 2\theta}{\sin 2\theta}$$ $\cos(-x) = \cos x$ and $\sin(-x) = -\sin x$

$$= \cot 2\theta$$ Quotient identity

Practice Problem 5 Verify the identity $\dfrac{\cos 5x - \cos x}{\sin x - \sin 5x} = \tan 3x$.

Analyzing Touch-Tone Phones

When a sound of frequency f_1 is combined with a sound of frequency f_2, the resulting sound has frequency $\dfrac{f_1 + f_2}{2}$. In a Touch-Tone phone, when you press a button with a low frequency l and a high frequency h, the sound produced is modeled by

$$y = \sin(2\pi lt) + \sin(2\pi ht)$$

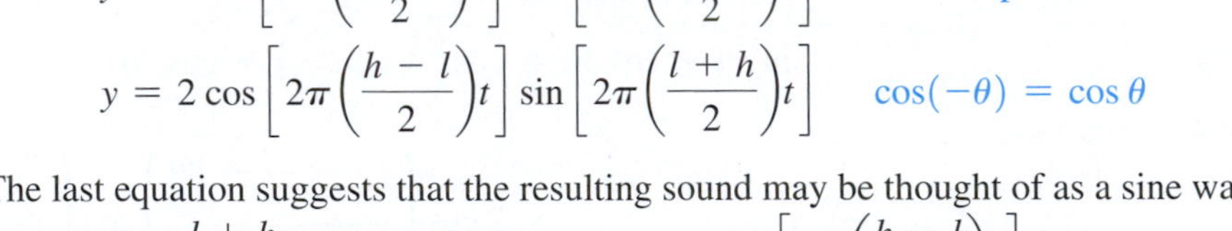

$$y = 2\sin\left[2\pi\left(\frac{l + h}{2}\right)t\right]\cos\left[2\pi\left(\frac{l - h}{2}\right)t\right]$$ Sum-to-product formula

$$y = 2\cos\left[2\pi\left(\frac{h - l}{2}\right)t\right]\sin\left[2\pi\left(\frac{l + h}{2}\right)t\right]$$ $\cos(-\theta) = \cos\theta$

The last equation suggests that the resulting sound may be thought of as a sine wave with frequency of $\dfrac{l + h}{2}$ and variable amplitude of $2\cos\left[2\pi\left(\dfrac{h - l}{2}\right)t\right]$.

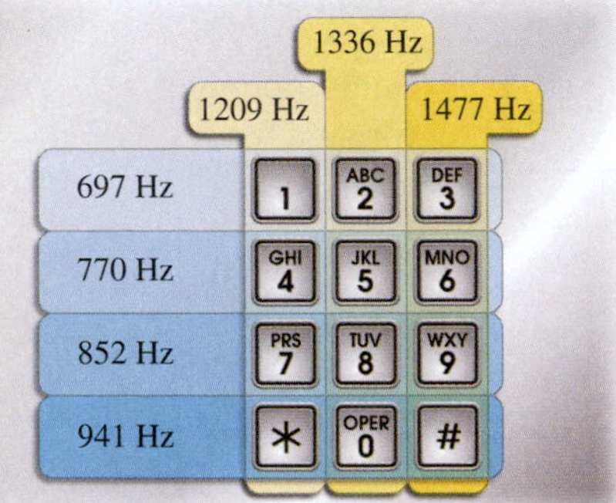

FIGURE 6.7

EXAMPLE 6 **Touch-Tone Phones**

a. Write an expression that models the tone of the sound produced by pressing the "8" button on your Touch-Tone phone. See Figure 6.7.

b. Rewrite the expression from part **a** as a product of two trigonometric functions.

c. Write the frequency and the variable amplitude of the sound from part **a**.

Solution

a. Pressing the "8" key produces the sound given by

$$y = \sin[2\pi(852)t] + \sin[2\pi(1336)t]$$ $l = 852, h = 1336$

b. $y = 2\sin\left[2\pi\left(\dfrac{852 + 1336}{2}\right)t\right]\cos\left[2\pi\left(\dfrac{852 - 1336}{2}\right)t\right]$ Sum-to-product formula

$$= 2\cos[2\pi(242)t]\sin[2\pi(1094)t]$$ $\cos(-\theta) = \cos\theta$

$$= 2\cos(484\pi t)\sin[2\pi(1094)t]$$

c. The frequency is 1094 Hz, and the variable amplitude is $2\cos(484\pi t)$.

Practice Problem 6 Repeat Example 6 assuming that you press the "1" button on your Touch-Tone phone.

TECHNOLOGY **CONNECTION**

Graph of $y = 2\cos(484\pi t)\sin[2\pi(1094)t]$

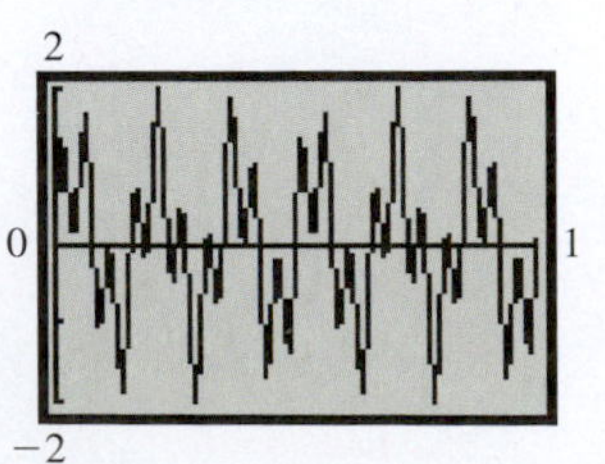

SUMMARY OF MAIN FACTS

Product-to-Sum Formulas

$$\cos x \cos y = \frac{1}{2}[\cos(x - y) + \cos(x + y)]$$

$$\sin x \sin y = \frac{1}{2}[\cos(x - y) - \cos(x + y)]$$

$$\sin x \cos y = \frac{1}{2}[\sin(x + y) + \sin(x - y)]$$

$$\cos x \sin y = \frac{1}{2}[\sin(x + y) - \sin(x - y)]$$

Sum-to-Product Formulas

$$\cos x + \cos y = 2\cos\left(\frac{x+y}{2}\right)\cos\left(\frac{x-y}{2}\right)$$

$$\cos x - \cos y = -2\sin\left(\frac{x+y}{2}\right)\sin\left(\frac{x-y}{2}\right)$$

$$\sin x + \sin y = 2\sin\left(\frac{x+y}{2}\right)\cos\left(\frac{x-y}{2}\right)$$

$$\sin x - \sin y = 2\sin\left(\frac{x-y}{2}\right)\cos\left(\frac{x+y}{2}\right)$$

Answers to Practice Problems

1. $\frac{1}{2}\cos 2x + \frac{1}{2}\cos 4x$ **2.** $\frac{2 - \sqrt{3}}{4}$

3. a. $2\sin 3x \sin x$ **b.** $\cos 13°$ **4.** $2\sin\left(2x - \frac{\pi}{4}\right)\cos\left(x + \frac{\pi}{4}\right)$

6. a. $y = \sin(2\pi \cdot 697t) + \sin(2\pi \cdot 1209t)$
b. $y = 2\cos(512\pi t)\sin(1906\pi t)$
c. $f = 953$ Hz, $A = 2\cos(512\pi t)$

SECTION 6.4 Exercises

Basic Concepts and Skills

1. We can rewrite the product of two sines as a difference of two cosines by using the formula $\sin x \sin y =$ ______________.
2. We can rewrite the product of a sine and a cosine as the sum of two sines by using the formula $\sin x \cos y =$ ______________.
3. We can rewrite the sum of two cosines as a product of two cosines by using the formula $\cos x + \cos y =$ ______________.
4. **True or False.** $\cos x - \cos y = 2\sin\left(\frac{x+y}{2}\right)\sin\left(\frac{y-x}{2}\right)$.
5. **True or False.** $\sin x + \sin y = \sin(x + y)$.
6. **True or False.** $\sin(2x + 1) - \sin(2x - 1) = 2\sin(1)\cos 2x$.

In Exercises 7–22, use the product-to-sum formulas to rewrite each expression as the sum or difference of two functions. Simplify where possible.

7. $\sin x \cos x$
8. $\cos x \cos x$
9. $\sin x \sin x$
10. $\cos x \sin x$
11. $\sin 25° \cos 5°$
12. $\sin 40° \sin 20°$
13. $\cos 140° \cos 20°$
14. $\cos 70° \sin 20°$
15. $\sin\frac{7\pi}{12}\sin\frac{\pi}{12}$
16. $\sin\frac{3\pi}{8}\cos\frac{\pi}{8}$
17. $\cos\frac{5\pi}{8}\sin\frac{\pi}{8}$
18. $\cos\frac{5\pi}{3}\cos\frac{\pi}{3}$
19. $\sin 5\theta \cos\theta$
20. $\cos 3\theta \sin 2\theta$
21. $\cos 4x \cos 3x$
22. $\sin 5x \sin 2x$

In Exercises 23–30, find the exact value of each expression.

23. $\sin 37.5° \sin 7.5°$

24. $\cos 52.5° \cos 7.5°$

25. $\sin 67.5° \cos 22.5°$

26. $\cos 105° \sin 75°$

27. $\sin \frac{5\pi}{24} \cos \frac{\pi}{24}$

28. $\sin \frac{7\pi}{12} \sin \frac{\pi}{12}$

29. $\cos \frac{13\pi}{24} \cos \frac{5\pi}{24}$

30. $\cos \frac{7\pi}{24} \sin \frac{\pi}{24}$

In Exercises 31–50, use sum-to-product formulas to rewrite each expression as a product. Simplify where possible.

31. $\cos 40° - \cos 20°$

32. $\sin 22° + \sin 8°$

33. $\sin 32° - \sin 16°$

34. $\cos 47° + \cos 13°$

35. $\sin \frac{\pi}{5} + \sin \frac{2\pi}{5}$

36. $\cos \frac{\pi}{12} + \cos \frac{\pi}{3}$

37. $\cos \frac{1}{2} + \cos \frac{1}{3}$

38. $\sin \frac{2}{3} - \sin \frac{1}{4}$

39. $\cos 3x + \cos 5x$

40. $\sin 5x - \sin 3x$

41. $\sin 7x + \sin(-x)$

42. $\cos 7x - \cos 3x$

43. $\sin x + \cos x$

44. $\cos x - \sin x$

45. $\sin 2x - \cos 2x$

46. $\cos 3x + \sin 3x$

47. $\sin 3x + \cos 5x$

48. $\sin 5x - \cos x$

49. $a(\sin x + \cos x)$

50. $a(\sin bx + \cos bx)$

In Exercises 51–60, verify each identity.

51. $\dfrac{\sin x + \sin 3x}{\cos x + \cos 3x} = \tan 2x$

52. $\dfrac{\sin 2x + \sin 4x}{\cos 2x + \cos 4x} = \tan 3x$

53. $\dfrac{\cos 3x - \cos 7x}{\sin 7x + \sin 3x} = \tan 2x$

54. $\dfrac{\cos 12x - \cos 4x}{\sin 4x - \sin 12x} = \tan 8x$

55. $\dfrac{\cos 2x + \cos 2y}{\cos 2x - \cos 2y} = \cot (y - x) \cot (y + x)$

56. $\dfrac{\sin 2x + \sin 2y}{\sin 2x - \sin 2y} = \dfrac{\tan(x + y)}{\tan(x - y)}$

57. $\sin x + \sin 2x + \sin 3x = \sin 2x(1 + 2\cos x)$

58. $\cos x + \cos 2x + \cos 3x = \cos 2x(1 + 2\cos x)$

59. $\sin 2x + \sin 4x + \sin 6x = 4 \cos x \cos 2x \sin 3x$

60. $\cos x + \cos 3x + \cos 5x + \cos 7x = 4 \cos x \cos 2x \cos 4x$

Applying the Concepts

61. Touch-Tone phone.

a. Write an expression that models the tone by pressing the "7" button on your Touch-Tone phone. See Figure 6.7, page 610.

b. Rewrite the expression from part (a) as a product of two functions.

c. Write the frequency and the variable amplitude of the sound from part (a).

62. Touch-Tone phone. Repeat Exercise 61 assuming that you press the "#" button on your Touch-Tone phone.

63. Beats in music. When two musical tones of slightly different frequencies are played, the sound you hear will fluctuate in volume according to the difference in their frequencies. These are called *beat frequencies.* (Piano tuners use the beat frequency to adjust piano wire until it is at the same frequency as the tuning fork.) Let $y_1 = 0.05 \cos(112\pi t)$ and $y_2 = 0.05 \cos(120\pi t)$ model two tones.

a. Write $y = y_1 + y_2$ as the product of two functions.

b. How many beats are there?

64. Beats in music. Repeat Exercise 63 for the two tones represented by $y_1 = 0.04 \sin(110\pi t)$ and $y_2 = 0.04 \sin(114\pi t)$.

Beyond the Basics

In Exercises 65 and 66, sketch the graph of *f*. What is the amplitude and period?

65. $f(x) = \cos(2x + 1) + \cos(2x - 1)$

66. $f(x) = \sin(3x + 1) + \sin(3x - 1)$

In Exercises 67–76, verify each identity.

67. $\cos 40° + \cos 50° + \cos 70° + \cos 80° = \cos 10° + \cos 20°$

68. $\sin 10° + \sin 20° + \sin 40° + \sin 50° = 2 \cos 5° \cos 15°$

69. $\cos 4\theta \cos \theta - \cos 6\theta \cos 9\theta = \sin 5\theta \sin 10\theta$

70. $\cos 38° \cos 46° - \sin 14° \sin 22° = \frac{1}{2} \cos 24°$

71. $\sin 25° \sin 35° - \sin 25° \sin 85° - \sin 35° \sin 85° = -\frac{3}{4}$

72. $\cos 20° \cos 40° \cos 60° \cos 80° = \frac{1}{16}$

73. $\dfrac{\sin 3x \cos 5x - \sin x \cos 7x}{\sin x \sin 7x + \cos 3x \cos 5x} = \tan 2x$

74. $\dfrac{\sin 11x \sin x + \sin 7x \sin 3x}{\cos 11x \sin x + \cos 7x \sin 3x} = \tan 8x$

75. $\dfrac{\cos x + \cos 3x + \cos 5x + \cos 7x}{\sin x + \sin 3x + \sin 5x + \sin 7x} = \cot 4x$

76. $\dfrac{\sin 3x + \sin 5x + \sin 7x + \sin 9x}{\cos 3x + \cos 5x + \cos 7x + \cos 9x} = \tan 6x$

77. Verify the identity $\cos 7x = 2 \cos 6x \cos x - 2 \cos 4x \cos x + 2 \cos 2x \cos x - \cos x$.

78. a. Verify the identity

$$4 \sin \theta \sin\left(\frac{\pi}{3} + \theta\right) \sin\left(\frac{\pi}{3} - \theta\right) = \sin 3\theta.$$

[*Hint:* Recall that $\sin 3\theta = 3 \sin \theta - 4 \sin^3 \theta$.]

b. Use part (a) to show that

$$\sin 20° \sin 40° \sin 60° \sin 80° = \frac{3}{16}.$$

79. Verify the identity

$$\frac{\sin(x + 3y) + \sin(3x + y)}{\sin 2x + \sin 2y} = 2 \cos(x + y).$$

80. Verify the identity $\sin^3 x \sin 3x + \cos^3 x \cos 3x = \cos^3 2x$.

81. Verify the following sum-to-product formulas.

a. $\cos x - \cos y = -2 \sin\left(\frac{x + y}{2}\right) \sin\left(\frac{x - y}{2}\right)$

b. $\sin x + \sin y = 2 \sin\left(\frac{x + y}{2}\right) \cos\left(\frac{x - y}{2}\right)$

c. $\sin x - \sin y = 2 \sin\left(\frac{x - y}{2}\right) \cos\left(\frac{x + y}{2}\right)$

Critical Thinking/Discussion/Writing

82. Supply the reason for each step in verifying the identity

$\sin\frac{\pi}{14}\sin\frac{3\pi}{14}\sin\frac{5\pi}{14} = \frac{1}{8}.$

$$\sin\frac{\pi}{14}\sin\frac{3\pi}{14}\sin\frac{5\pi}{14} = \cos\frac{3\pi}{7}\cos\frac{2\pi}{7}\cos\frac{\pi}{7}$$

$$= \frac{8\sin\frac{\pi}{7}\cos\frac{\pi}{7}\cos\frac{2\pi}{7}\cos\frac{3\pi}{7}}{8\sin\frac{\pi}{7}}$$

$$= \frac{2\sin\frac{4\pi}{7}\cos\frac{3\pi}{7}}{8\sin\frac{\pi}{7}}$$

$$= \frac{\sin\pi + \sin\frac{\pi}{7}}{8\sin\frac{\pi}{7}}$$

$$= \frac{1}{8}$$

Assume that $A + B + C = 180°$. Verify each identity.

83. $\sin 2A + \sin 2B + \sin 2C = 4\sin A \sin B \sin C$

84. $\cos 2A + \cos 2B + \cos 2C = -1 - 4\cos A \cos B \cos C$

85. Explain how to obtain the identity $\cos^2 x = \frac{1 + \cos 2x}{2}$ using the product-to-sum formula for $\cos x \cos y$.

Maintaining Skills

In Exercises 86–91, identify each equation as an identity, conditional, or inconsistent equation.

86. $\sin x = \frac{1}{2}$

87. $\cos x = 3$

88. $\sin x = \cos x$

89. $\cos^2 x = 1 - \sin^2 x$

90. $\sin 2x = 2\sin x \cos x$

91. $\tan x = \sqrt{\sec^2 x - 1}$

In Exercises 92–95, find the reference angle for each angle.

92. $\frac{5\pi}{3}$

93. $\frac{7\pi}{8}$

94. $-\frac{3\pi}{7}$

95. $-\frac{17\pi}{4}$

96. Describe all angles coterminal with $-\frac{\pi}{7}$.

97. Find a third quadrant angle with reference angle $\frac{\pi}{5}$.

SECTION 6.5

Trigonometric Equations I

BEFORE STARTING THIS SECTION, REVIEW

1 Reference angle (Section 5.3, page 504)

2 Periods of trigonometric functions (Sections 5.4 and 5.5)

3 Factoring techniques (Section P.4)

4 Solving linear and quadratic equations (Sections 1.1 and 1.2)

OBJECTIVES

1 Solve trigonometric equations of the form $a \sin(x - c) = k$, $a \cos(x - c) = k$, and $a \tan(x - c) = k$.

2 Solve trigonometric equations by using the zero-product property.

3 Solve trigonometric equations that contain more than one trigonometric function.

4 Solve trigonometric equations by squaring both sides.

Projectile Motion

The angle that a projectile makes with a horizontal plane on which it rests at takeoff and its initial velocity determine the rest of its flight. The time it takes for the projectile to hit the ground after reaching its maximum height is the same as if it had been dropped straight down from that height. The horizontal motion has no effect on the vertical motion. Suppose we ignore air resistance and measure time in seconds and distance in feet. Then the equation $h = v_0 t \sin\theta - 16t^2$ gives the height, h, of the projectile after t seconds, where θ is the initial angle the projectile makes with the ground and v_0 is the projectile's initial velocity. The horizontal distance, d, the projectile travels in t seconds is $d = tv_0 \cos\theta$. In Example 4, we use these equations to investigate the flight of a projectile.

Trigonometric Equations

Recall that a *trigonometric equation* is a conditional equation that contains only trigonometric functions of a variable. An identity is an equation that is true for all values in the domain of the variable. We verified trigonometric identities in Sections 6.1–6.4. Now we work with trigonometric equations that are satisfied by some but not all values of the variable. For example, $\sin x = \frac{1}{2}$ is satisfied by $x = \frac{\pi}{6}$ but not by $x = 0$.

Solving a trigonometric equation means finding its *solution set.* Some trigonometric equations do not have a solution. For example, the equation $\sin x = 2$ has no solution because $-1 \le \sin x \le 1$ for all real numbers x. However, if a trigonometric equation has a solution, it has infinitely many solutions because the trigonometric functions are periodic. Also note that we cannot find an exact solution for every equation. In such cases, we approximate the solution by using a calculator.

In this section, we study three types of equations:

1. Equations of the form

$$a \sin(x - c) = k, \qquad a \cos(x - c) = k, \qquad a \tan(x - c) = k,$$

where a, c, and k are constants.

2. Equations that can be solved by factoring using the zero-product property.
3. Equations that are equivalent to a polynomial equation in one trigonometric function.

1 Solve trigonometric equations of the form $a\sin(x - c) = k$, $a\cos(x - c) = k$, and $a\tan(x - c) = k$.

Trigonometric Equations of the Form $a\sin(x - c) = k$, $a\cos(x - c) = k$, and $a\tan(x - c) = k$

We begin with some simple equations.

EXAMPLE 1 **Solving a Trigonometric Equation**

Find all solutions of each equation. Express all solutions in radians.

a. $\sin x = \dfrac{\sqrt{2}}{2}$ **b.** $\cos\theta = -\dfrac{\sqrt{3}}{2}$ **c.** $\tan x = -\sqrt{3}$

Solution

FIGURE 6.8 $\sin x = \dfrac{\sqrt{2}}{2}$

a. We first find the solutions of $\sin x = \dfrac{\sqrt{2}}{2}$ in the interval $[0, 2\pi)$. We know that $\sin\dfrac{\pi}{4} = \dfrac{\sqrt{2}}{2}$ and that $\sin x$ is positive only in quadrants I and II. See Figure 6.8. The first and second quadrant angles with reference angle $\dfrac{\pi}{4}$ are

$$x = \frac{\pi}{4} \quad \text{and} \quad x = \pi - \frac{\pi}{4} = \frac{3\pi}{4}.$$

So $x = \dfrac{\pi}{4}$ and $x = \dfrac{3\pi}{4}$ are the only solutions in the interval $[0, 2\pi)$. Because the sine function is periodic with period 2π, adding any integer multiple of 2π to either of these values still makes the sine of the resulting value $\dfrac{\sqrt{2}}{2}$. All solutions of the equation $\sin x = \dfrac{\sqrt{2}}{2}$ are therefore given by

$$x = \frac{\pi}{4} + 2n\pi \quad \text{or} \quad x = \frac{3\pi}{4} + 2n\pi, \text{ for any integer } n.$$

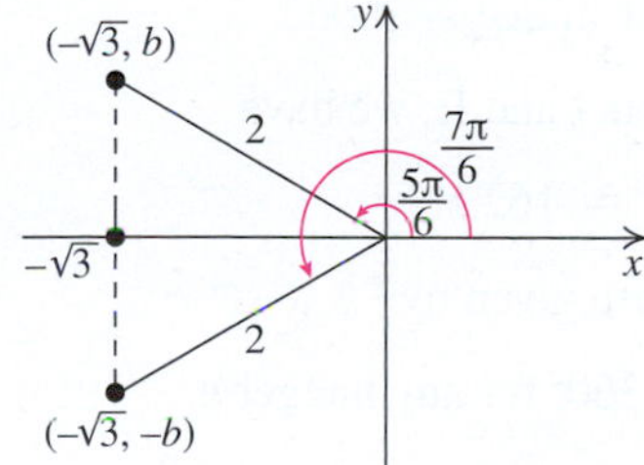

FIGURE 6.9 $\cos\theta = -\dfrac{\sqrt{3}}{2}$

b. As in part **a**, we first find all solutions in the interval $[0, 2\pi)$. We know that $\cos\dfrac{\pi}{6} = \dfrac{\sqrt{3}}{2}$ and that $\cos\theta$ is negative only in quadrants II and III. See Figure 6.9. The second and third quadrant angles having reference angle $\dfrac{\pi}{6}$ are

$$\theta = \pi - \frac{\pi}{6} = \frac{5\pi}{6} \quad \text{and} \quad \theta = \pi + \frac{\pi}{6} = \frac{7\pi}{6}.$$

So $\theta = \dfrac{5\pi}{6}$ and $\theta = \dfrac{7\pi}{6}$ are the only solutions in the interval $[0, 2\pi)$. The period of the cosine function is 2π, so all solutions are given by

$$\theta = \frac{5\pi}{6} + 2n\pi \quad \text{or} \quad \theta = \frac{7\pi}{6} + 2n\pi, \text{ for any integer } n.$$

c. Because $\tan x$ is a periodic function with period π, we first find all solutions of $\tan x = -\sqrt{3}$ in the interval $[0, \pi)$. We know that $\tan\dfrac{\pi}{3} = \sqrt{3}$ and that $\tan x$ is

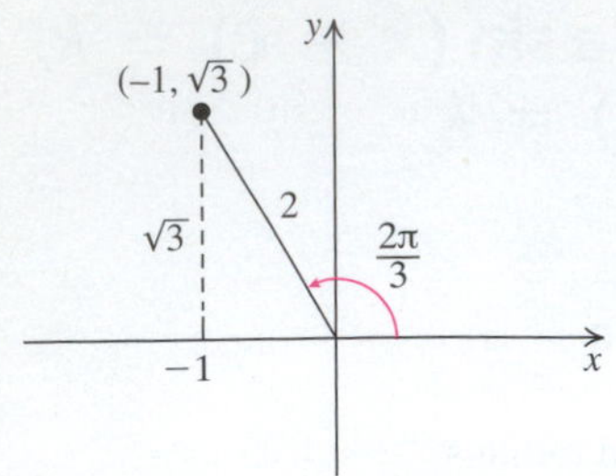

FIGURE 6.10 $\tan \frac{2\pi}{3} = -\sqrt{3}$

negative in quadrant II. See Figure 6.10. The second quadrant angle with reference angle $\frac{\pi}{3}$ is

$$x = \pi - \frac{\pi}{3} = \frac{2\pi}{3}.$$

So $x = \frac{2\pi}{3}$ is the only solution in the interval $[0, \pi)$. Because the period of $\tan x$ is π, all solutions of $\tan x = -\sqrt{3}$ are given by

$$x = \frac{2\pi}{3} + n\pi, \text{ for any integer } n.$$

Practice Problem 1 Find all solutions of each equation. Express solutions in radians.

a. $\sin x = 1$ **b.** $\cos x = 1$ **c.** $\tan x = 1$

In Example 1, we were able to find the exact solutions of trigonometric equations because the angles involved were common angles. In the next example, we find approximate solutions of trigonometric equations involving other angles.

EXAMPLE 2 Finding Approximate Solutions

a. Find all solutions of $\sin x = 0.3$. Round the solutions to the nearest tenth of a degree.

b. Find all solutions of $\cot x = -3.5$ in the interval $[0, 2\pi)$. Round the solutions to four decimal places.

RECALL

If x' denotes the reference angle for x, then $\sin x = \pm \sin x'$, $\cos x = \pm \cos x'$, and $\tan x = \pm \tan x'$. Similar identities hold for $\sec x$, $\csc x$, and $\cot x$.

Solution

a. We use the function $\sin^{-1}$ to find the reference angle x' for any angle x with $\sin x = 0.3$.

$$\sin x' = |\sin x| = 0.3 \quad \text{Given equation}$$
$$x' = \sin^{-1}(0.3) \quad 0° < x' < 90°$$
$$x' \approx 17.5° \quad \text{Use a calculator in } \textit{degree} \text{ mode.}$$

Because the sine function has positive values in quadrants I and II, we have

$$x \approx 17.5° \quad \text{or} \quad x \approx 180° - 17.5° = 162.5°.$$

The period of the sine function is 360°, so all solutions are given by

$$x \approx 17.5° + n \cdot 360° \quad \text{or} \quad x \approx 162.5° + n \cdot 360° \text{ for any integer } n.$$

b. Because $\cot x = -3.5 = -\frac{7}{2}$, we have

$$\tan x = -\frac{2}{7} \quad \text{Reciprocal identity}$$

First, we find the reference angle x' for an angle x with $\tan x = -\frac{2}{7}$.

$$\tan x' = |\tan x| = \left|-\frac{2}{7}\right| = \frac{2}{7}$$
$$x' = \tan^{-1}\left(\frac{2}{7}\right) \quad 0 < x' < \frac{\pi}{2}$$
$$x' \approx 0.2783 \quad \text{Use a calculator in } \textit{radian} \text{ mode.}$$

Because the cotangent function is negative in quadrants II and IV, we have

$$x \approx \pi - 0.2783 \approx 2.8633 \quad \text{and} \quad x \approx 2\pi - 0.2783 \approx 6.0049.$$

These are the only solutions in the interval $[0, 2\pi)$.

Practice Problem 2

a. Find all solutions of $\sec x = 1.5$. Round the solutions to the nearest tenth of a degree.

b. Find all solutions of $\tan x = -2$ in the interval $[0, 2\pi)$. Round the solutions to four decimal places.

RECALL

The period of $y = \tan x$ and $y = \cot x$ is π. The period of the other four trigonometric functions is 2π.

In most of our examples, we find the solutions in the interval $[0, 2\pi)$ for equations involving sine, cosine, secant, and cosecant functions. To find all solutions of such equations, add $2n\pi$ to these solutions. Similarly, to find all solutions of an equation involving tangent or cotangent function, add $n\pi$ to each solution found in $[0, \pi)$. When an equation has only a finite number of solutions, we write the solutions using set notation.

EXAMPLE 3 Solving a Linear Trigonometric Equation

Find all solutions in the interval $[0, 2\pi)$ of the equation

$$2 \sin\left(x - \frac{\pi}{4}\right) + 1 = 2.$$

Solution

Replace $x - \frac{\pi}{4}$ with θ in the given equation.

$$2 \sin \theta + 1 = 2$$

$$\sin \theta = \frac{1}{2} \qquad \text{Solve for } \sin \theta.$$

The reference angle for θ is $\frac{\pi}{6}$ because $\sin \frac{\pi}{6} = \frac{1}{2}$. See page 491 or recall from memory.

Because $\sin \theta$ is positive only in quadrants I and II, the solutions of $\sin \theta = \frac{1}{2}$ in $[0, 2\pi)$ are

$$\theta = \frac{\pi}{6} \quad \text{or} \quad \theta = \pi - \frac{\pi}{6} = \frac{5\pi}{6}.$$

Since $\theta = x - \frac{\pi}{4}$, $\theta = \frac{\pi}{6}$ or $\theta = \frac{5\pi}{6}$ gives

$x - \frac{\pi}{4} = \frac{\pi}{6}$	$x - \frac{\pi}{4} = \frac{5\pi}{6}$	Replace θ with $x - \frac{\pi}{4}$.
$x = \frac{\pi}{6} + \frac{\pi}{4}$	$x = \frac{5\pi}{6} + \frac{\pi}{4}$	Add $\frac{\pi}{4}$ to both sides.
$x = \frac{2\pi}{12} + \frac{3\pi}{12} = \frac{5\pi}{12}$	$x = \frac{10\pi}{12} + \frac{3\pi}{12} = \frac{13\pi}{12}$	Simplify.

The solution set in the interval $[0, 2\pi)$ is $\left\{\frac{5\pi}{12}, \frac{13\pi}{12}\right\}$.

Practice Problem 3 Find all solutions in the interval $[0°, 360°)$ of the equation $2 \sec (x - 30)° + 1 = \sec (x - 30)° + 3$.

EXAMPLE 4 Finding the Angle from Which a Bullet Was Fired

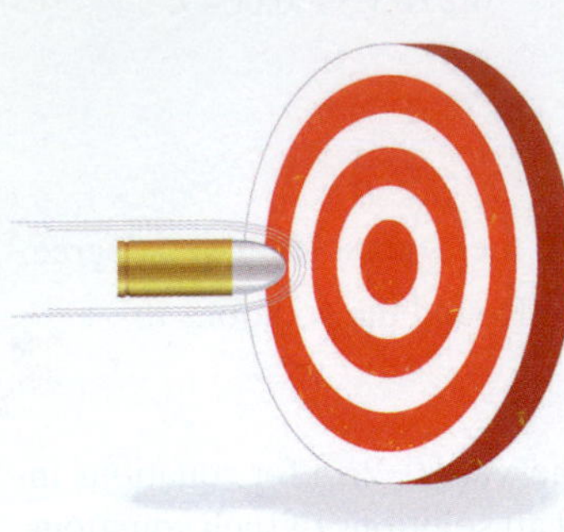

A bullet hits a target 1500 feet away 0.9 second after being fired. If the bullet's velocity is 3280 ft/sec, from what angle to the ground was the bullet fired? Give your answer to the nearest degree.

Solution

We use the equation $d = tv_0 \cos\theta$ (from the section introduction) for the horizontal distance, d, that the projectile travels in t seconds, where θ is the initial angle the projectile makes with the ground and v_0 is the projectile's initial velocity. Because $d = tv_0 \cos\theta$,

$$1500 = 0.9(3280)\cos\theta \quad \text{Replace } d \text{ with 1500, } t \text{ with 0.9, and } v_0 \text{ with 3280.}$$

$$\cos\theta = \frac{1500}{0.9(3280)} \quad \text{Solve for } \cos\theta\text{: interchange sides.}$$

$$\theta = \cos^{-1}\left(\frac{1500}{0.9(3280)}\right) \quad \theta \text{ is in Q I.}$$

$$\theta \approx 59° \quad \text{Use a calculator.}$$

Practice Problem 4 Rework Example 4 assuming that the target is 2500 feet away and the bullet takes 1 second to reach the target.

2 Solve trigonometric equations by using the zero-product property.

Trigonometric Equations and the Zero-Product Property

EXAMPLE 5 Solving an Equation by Using the Zero-Product Property

Find all solutions of the equation $(2 \sin x - 1)(7 \tan x + 2) = 0$ in the interval $[0, 2\pi)$. Round the solutions to four decimal places.

Solution

$$(2\sin x - 1)(7\tan x + 2) = 0 \quad \text{Given equation}$$

$$2\sin x - 1 = 0 \quad \text{or} \quad 7\tan x + 2 = 0 \quad \text{Zero-product property}$$

$$\sin x = \frac{1}{2} \qquad \tan x = -\frac{2}{7} \quad \text{Solve for } \sin x \text{ and } \tan x.$$

$$x = \frac{\pi}{6} \quad \text{or} \quad x = \frac{5\pi}{6}$$

See Example 3.

$$x \approx \pi - 0.2783 \quad \text{or} \quad x \approx 2\pi - 0.2783$$

$$x \approx 2.8633 \quad \text{or} \quad x \approx 6.0049$$

See Example 2(b).

So the solution set of the given equation is

$$\left\{\frac{\pi}{6}, \frac{5\pi}{6}, 2.8633, 6.0049\right\}.$$

Practice Problem 5 Find all solutions of the equation

$$(\sin x - 1)(\sqrt{3}\tan x + 1) = 0$$

in the interval $[0, 2\pi)$.

TECHNOLOGY CONNECTION

You can confirm that the equation in Example 6 has the two solutions we found in $[0, 2\pi)$ by graphing $Y_1 = 2\sin^2 x - 5\sin x + 2$ and inspecting the x-intercepts using TRACE.

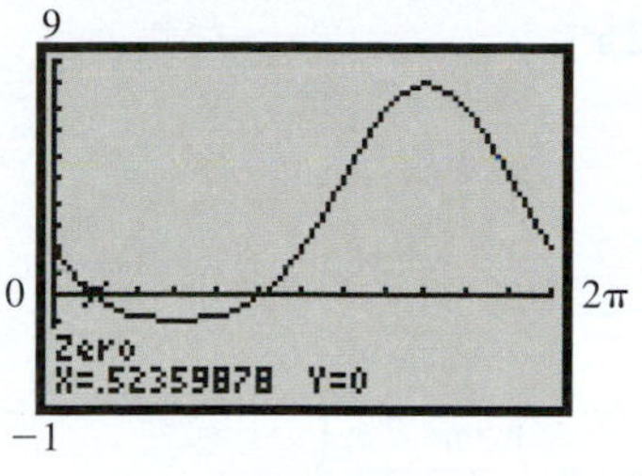

EXAMPLE 6 Solving a Quadratic Trigonometric Equation

Find all solutions of the equation $2\sin^2\theta - 5\sin\theta + 2 = 0$. Express the solutions in radians.

Solution

The equation is quadratic in $\sin\theta$. We use the zero-product property to solve for $\sin\theta$ and solve the resulting equation for θ.

$$2\sin^2\theta - 5\sin\theta + 2 = 0 \qquad 2x^2 - 5x + 2 = 0 \text{ if } x = \sin\theta$$

$$(2\sin\theta - 1)(\sin\theta - 2) = 0 \qquad (2x - 1)(x - 2) = 0\text{: factor.}$$

$$2\sin\theta - 1 = 0 \quad \text{or} \quad \sin\theta - 2 = 0 \qquad \text{Zero-product property}$$

$$\sin\theta = \frac{1}{2} \quad \Bigg| \quad \sin\theta = 2 \qquad \text{Solve for } \sin\theta.$$

$$\theta = \frac{\pi}{6} \text{ or } \theta = \frac{5\pi}{6} \quad \Bigg| \quad \text{No solution} \qquad -1 \le \sin\theta \le 1$$

So $\theta = \frac{\pi}{6}$ and $\theta = \frac{5\pi}{6}$ are the only solutions in the interval $[0, 2\pi)$. Because $\sin\theta$ has period 2π, all possible solutions are

$$\theta = \frac{\pi}{6} + 2n\pi \quad \text{or} \quad \theta = \frac{5\pi}{6} + 2n\pi, \text{ for any integer } n.$$

Practice Problem 6 Find all solutions of the equation $2\cos^2\theta - \cos\theta - 1 = 0$. Express the solutions in radians.

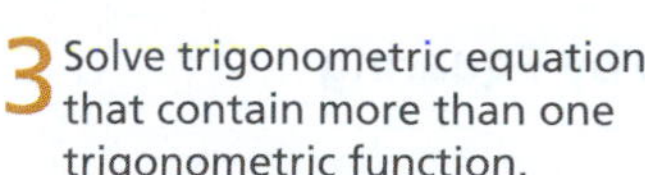

3 Solve trigonometric equations that contain more than one trigonometric function.

Equations with More Than One Trigonometric Function

When a trigonometric equation contains more than one trigonometric function, sometimes we can use trigonometric identities to convert the equation into an equation containing only one trigonometric function.

EXAMPLE 7 Solving a Trigonometric Equation Using Identities

Find all solutions of the equation $2\sin^2\theta + \sqrt{3}\cos\theta + 1 = 0$ in the interval $[0, 2\pi)$.

Solution

Because the equation contains sine and cosine, we use the Pythagorean identity $\sin^2\theta + \cos^2\theta = 1$ to convert the equation into one containing only cosine.

$$2\sin^2\theta + \sqrt{3}\cos\theta + 1 = 0 \qquad \text{Given equation}$$

$$2(1 - \cos^2\theta) + \sqrt{3}\cos\theta + 1 = 0 \qquad \text{Replace } \sin^2\theta \text{ with } 1 - \cos^2\theta.$$

$$2 - 2\cos^2\theta + \sqrt{3}\cos\theta + 1 = 0 \qquad \text{Distributive property}$$

$$3 - 2\cos^2\theta + \sqrt{3}\cos\theta = 0 \qquad \text{Combine terms.}$$

$$2\cos^2\theta - \sqrt{3}\cos\theta - 3 = 0 \qquad \text{Multiply both sides by } -1.$$

Solve the last equation for $\cos\theta$.

$$\cos\theta = \frac{-(-\sqrt{3}) \pm \sqrt{(-\sqrt{3})^2 - 4(2)(-3)}}{2(2)} \qquad \text{Use } a = 2, b = -\sqrt{3}, \text{ and } c = -3 \text{ in the quadratic formula.}$$

$$= \frac{\sqrt{3} \pm \sqrt{3 + 24}}{4} \qquad \text{Simplify.}$$

$$= \frac{\sqrt{3} \pm \sqrt{27}}{4} = \frac{\sqrt{3} \pm 3\sqrt{3}}{4} \qquad \sqrt{27} = \sqrt{9 \cdot 3} = \sqrt{9}\sqrt{3} = 3\sqrt{3}$$

So

$$\cos\theta = \frac{\sqrt{3} + 3\sqrt{3}}{4} = \frac{4\sqrt{3}}{4} = \sqrt{3}$$

$\cos\theta = \sqrt{3} > 1$

Because $-1 \le \cos\theta \le 1$, the equation $\cos\theta = \sqrt{3}$ has no solution.

or

$$\cos\theta = \frac{\sqrt{3} - 3\sqrt{3}}{4} = \frac{(-2)\sqrt{3}}{4} = -\frac{\sqrt{3}}{2}$$

The equation $\cos\theta = -\frac{\sqrt{3}}{2}$ has two solutions in the interval $[0, 2\pi)$:

$$\theta = \pi - \frac{\pi}{6} = \frac{5\pi}{6} \text{ and}$$

$$\theta = \pi + \frac{\pi}{6} = \frac{7\pi}{6}$$

The solution set for the given equations in the interval $[0, 2\pi)$ is $\left\{\frac{5\pi}{6}, \frac{7\pi}{6}\right\}$.

SIDE NOTE

The Pythagorean identities

$\sin^2 x + \cos^2 x = 1$

$\tan^2 x + 1 = \sec^2 x$

$\cot^2 x + 1 = \csc^2 x$

can sometimes be used to rewrite a trigonometric equation after both sides are squared.

Practice Problem 7 Find all solutions of the equation $2\cos^2\theta + 3\sin\theta - 3 = 0$ in the interval $[0, 2\pi)$.

4 Solve trigonometric equations by squaring both sides.

Extraneous Solutions

Next, we solve a trigonometric equation by squaring both sides of the equation and then using an identity. Recall that squaring both sides of an equation may result in extraneous solutions. Therefore, you must check all possible solutions.

EXAMPLE 8 **Solving a Trigonometric Equation by Squaring**

Find all solutions of the equation $\sqrt{3}\cos x = \sin x + 1$ in the interval $[0, 2\pi)$.

Solution

$(\sqrt{3}\cos x)^2 = (\sin x + 1)^2$	Square both sides of the equation.
$3\cos^2 x = \sin^2 x + 2\sin x + 1$	Expand the binomial.
$3(1 - \sin^2 x) = \sin^2 x + 2\sin x + 1$	Pythagorean identity
$3 - 3\sin^2 x = \sin^2 x + 2\sin x + 1$	Distributive property
$-4\sin^2 x - 2\sin x + 2 = 0$	Collect like terms.
$2\sin^2 x + \sin x - 1 = 0$	Divide both sides by -2.
$(2\sin x - 1)(\sin x + 1) = 0$	Factor.
$2\sin x - 1 = 0$ or $\sin x + 1 = 0$	Zero-product property
$\sin x = \frac{1}{2}$	$\sin x = -1$	Solve for $\sin x$.
$x = \frac{\pi}{6}$ or $x = \pi - \frac{\pi}{6} = \frac{5\pi}{6}$	$x = \frac{3\pi}{2}$	Solve for x.

The possible solutions are $\frac{\pi}{6}, \frac{5\pi}{6}$, and $\frac{3\pi}{2}$.

TECHNOLOGY CONNECTION

You can confirm that the equation in Example 8 has two solutions in $[0, 2\pi)$ by graphing $Y_1 = \sqrt{3}\cos x$ and the right side $Y_2 = \sin x + 1$. Find the points of intersection using INTERSECT.

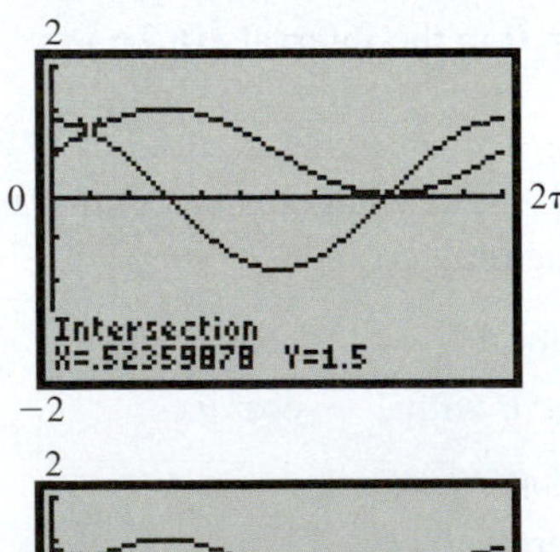

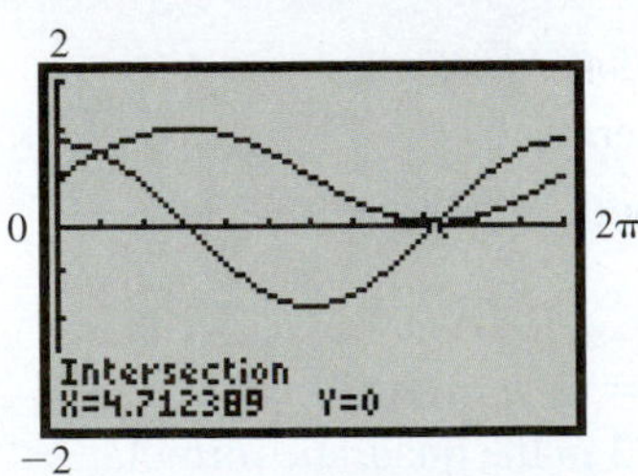

Check:

$x = \frac{\pi}{6}$	$x = \frac{5\pi}{6}$	$x = \frac{3\pi}{2}$
$\sqrt{3}\cos\frac{\pi}{6} \stackrel{?}{=} \sin\frac{\pi}{6} + 1$	$\sqrt{3}\cos\frac{5\pi}{6} \stackrel{?}{=} \sin\frac{5\pi}{6} + 1$	$\sqrt{3}\cos\frac{3\pi}{2} \stackrel{?}{=} \sin\frac{3\pi}{2} + 1$
$\sqrt{3}\left(\frac{\sqrt{3}}{2}\right) \stackrel{?}{=} \frac{1}{2} + 1$	$\sqrt{3}\left(-\frac{\sqrt{3}}{2}\right) \stackrel{?}{=} \frac{1}{2} + 1$	$\sqrt{3}(0) \stackrel{?}{=} -1 + 1$
$\frac{3}{2} \stackrel{?}{=} \frac{3}{2}$ ✓	$-\frac{3}{2} \stackrel{?}{=} \frac{3}{2}$ ×	$0 \stackrel{?}{=} 0$ ✓
Yes	No	Yes

The solution set of the equation in the interval $[0, 2\pi)$ is $\left\{\frac{\pi}{6}, \frac{3\pi}{2}\right\}$.

Practice Problem 8 Find all solutions of the equation $\sqrt{3}\cot\theta + 1 = \sqrt{3}\csc\theta$ in the interval $[0, 2\pi)$.

Answers to Practice Problems

1. a. $x = \frac{\pi}{2} + 2n\pi$ **b.** $x = 2n\pi$ **c.** $x = \frac{\pi}{4} + n\pi$
2. a. $x = 48.2° + n \cdot 360°, x = 311.8° + n \cdot 360°$
b. $x \approx 2.0344, x \approx 5.1760$ **3.** $x = 90°, x = 330°$
4. 40° **5.** $\left\{\frac{\pi}{2}, \frac{5\pi}{6}, \frac{11\pi}{6}\right\}$
6. $\theta = 2n\pi, \theta = \frac{2\pi}{3} + 2n\pi, \theta = \frac{4\pi}{3} + 2n\pi$
7. $\theta = \frac{\pi}{6}, \theta = \frac{\pi}{2}, \theta = \frac{5\pi}{6}$ **8.** $\left\{\frac{\pi}{3}\right\}$

SECTION 6.5 Exercises

Basic Concepts and Skills

1. The equation $\sin x = \frac{1}{2}$ has ____________ solution(s) in $[0, 2\pi)$.
2. All solutions of $\sin x = \frac{1}{2}$ are given by ____________ and ____________ for any integer n.
3. The equation $\cos x = 1$ has ____________ solution(s) in $[0, 2\pi)$.
4. All solutions of $\tan x = 1$ are given by ____________ for any integer n.
5. **True or False.** The equation $\sec x = \frac{1}{2}$ has two solutions in $[0, 2\pi)$.
6. **True or False.** All solutions of $\sin x = 0$ are given by $x = n\pi$ for any integer n.

In Exercises 7–16, find all solutions of each equation. Express the solutions in radians.

7. $\cos x = 0$
8. $\sin x = 0$
9. $\tan x = -1$
10. $\cot x = -1$
11. $\cos x = \frac{\sqrt{2}}{2}$
12. $\sin x = \frac{\sqrt{3}}{2}$
13. $\cot x = \sqrt{3}$
14. $\tan x = -\frac{\sqrt{3}}{3}$
15. $\cos x = -\frac{1}{2}$
16. $\sin x = -\frac{\sqrt{3}}{2}$

In Exercises 17–26, find all solutions of each equation. Express the solutions in degrees.

17. $\tan x = \frac{\sqrt{3}}{3}$

18. $\cot x = 1$

19. $\sin x = -\frac{1}{2}$

20. $\cos x = \frac{1}{2}$

21. $\csc x = 1$

22. $\sec x = -1$

23. $\sqrt{3}\csc x - 2 = 0$

24. $\sqrt{3}\sec x + 2 = 0$

25. $2\sec x - 4 = 0$

26. $2\csc x + 4 = 0$

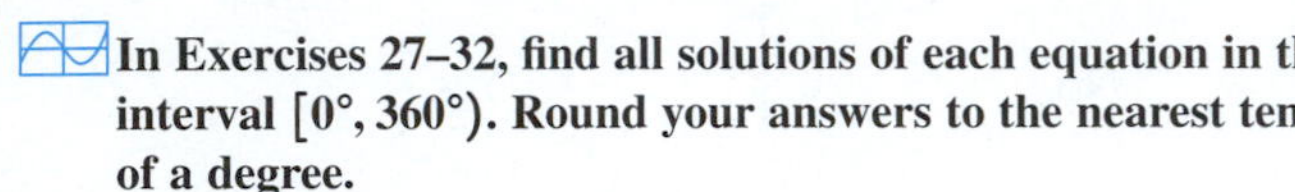

In Exercises 27–32, find all solutions of each equation in the interval $[0°, 360°)$. Round your answers to the nearest tenth of a degree.

27. $\sin\theta = 0.4$

28. $\cos\theta = 0.6$

29. $\sec\theta = 7.2$

30. $\csc\theta = -4.5$

31. $\tan(\theta - 30°) = -5$

32. $\cot(\theta + 30°) = 6$

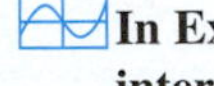

In Exercises 33–38, find all solutions of each equation in the interval $[0, 2\pi)$. Round the solutions to four decimal places.

33. $\csc x = -3$

34. $3\sin x - 1 = 0$

35. $3\tan x + 4 = 0$

36. $2\sec x - 7 = 0$

37. $2\csc x + 5 = 0$

38. $\cos x = 0.1106$

In Exercises 39–46, find all solutions of each equation in the interval $[0, 2\pi)$.

39. $\sin\left(x + \frac{\pi}{4}\right) = \frac{1}{2}$

40. $2\cos\left(x - \frac{\pi}{4}\right) + 1 = 0$

41. $\sec\left(x - \frac{\pi}{8}\right) + 2 = 0$

42. $\csc\left(x + \frac{\pi}{8}\right) - 2 = 0$

43. $\sqrt{3}\tan\left(x - \frac{\pi}{6}\right) - 1 = 0$

44. $\cot\left(x + \frac{\pi}{6}\right) + 1 = 0$

45. $2\sin\left(x - \frac{\pi}{3}\right) + 1 = 0$

46. $2\cos\left(x + \frac{\pi}{3}\right) + \sqrt{2} = 0$

In Exercises 47–54, find all solutions of each equation in the interval $[0, 2\pi)$.

47. $(\sin x + 1)(\tan x - 1) = 0$

48. $(2\cos x + 1)(\sqrt{3}\tan x - 1) = 0$

49. $(\csc x - 2)(\cot x + 1) = 0$

50. $(\sqrt{3}\sec x - 2)(\sqrt{3}\cot x + 1) = 0$

51. $(\tan x + 1)(2\sin x - 1) = 0$

52. $(2\sin x - \sqrt{3})(2\cos x - 1) = 0$

53. $(\sqrt{2}\sec x - 2)(2\sin x + 1) = 0$

54. $(\cot x - 1)(\sqrt{2}\csc x + 2) = 0$

In Exercises 55–62, find all solutions of each equation in the interval $[0, 2\pi)$.

55. $4\sin^2 x = 1$

56. $4\cos^2 x = 1$

57. $\tan^2 x = 1$

58. $\sec^2 x = 2$

59. $3\csc^2 x = 4$

60. $3\cot^2 x = 1$

61. $2\sin^2\theta - \sin\theta - 1 = 0$

62. $2\cos^2\theta - 5\cos\theta + 2 = 0$

In Exercises 63–72, use trigonometric identities to solve each equation in the interval $[0, 2\pi)$.

63. $\sin x = \cos x$

64. $\sqrt{3}\sin x + \cos x = 0$

65. $3\sin^2 x = \cos^2 x$

66. $3\cos^2 x = \sin^2 x$

67. $\cos^2 x - \sin^2 x = 1$

68. $2\sin^2 x + \cos x - 1 = 0$

69. $2\cos^2 x - 3\sin x - 3 = 0$

70. $2\sin^2 x - \cos x - 1 = 0$

71. $\sqrt{3}\sec^2 x - 2\tan x - 2\sqrt{3} = 0$

72. $\csc^2 x - (\sqrt{3} + 1)\cot x + (\sqrt{3} - 1) = 0$

In Exercises 73–76, solve each trigonometric equation in the interval $[0, 2\pi)$ by first squaring both sides.

73. $\sqrt{3}\sin x = 1 + \cos x$

74. $\tan x + 1 = \sec x$

75. $\sqrt{3}\tan\theta + 1 = \sqrt{3}\sec\theta$

76. $\sqrt{3}\cot\theta + 1 = \sqrt{3}\csc\theta$

Applying the Concepts

77. Angle of elevation. Find the angle of elevation of the sun assuming that the shadow of a tree 24 feet high is $8\sqrt{3}$ feet in length.

78. Leaning Tower of Pisa. The Leaning Tower of Pisa is 179 feet tall, and when the sun is directly overhead, its shadow measures 16.5 feet. At what angle is the tower inclined from the vertical?

79. Returning spacecraft. When a spacecraft reenters the atmosphere, the angle of reentry from the horizontal must be between 5.1° and 7.1°. Under 5.1°, the craft would "skip" back into space, and over 7.1°, the acceleration forces would be too high and the craft would crash. On reentry, a spacecraft descends 60 kilometers vertically while traveling 605 kilometers. Find the angle of reentry off the horizontal to the nearest tenth of a degree.

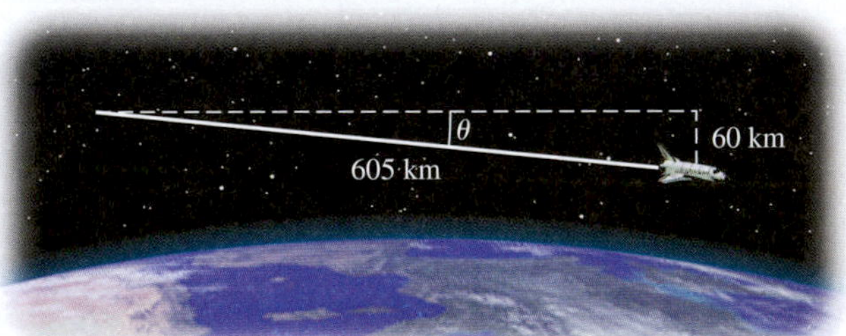

80. A flare is fired from the deck of an anchored ship at an initial velocity of 78 feet per second. After three seconds, the flare is directly over a lifeboat stalled 112 feet from the boat. At what angle to the deck was the flare fired? Give your answer to the nearest degree.

81. Find the angle that a golf ball's path makes with the ground, assuming that its initial velocity is 150 feet per second and after four seconds it is 25 feet above the ground. Give your answer to the nearest degree.

82. Find the angle that a baseball's path makes with the ground, assuming that its initial velocity is 140 feet per second and after two seconds it is 40 feet higher than when it was initially hit. Give your answer to the nearest degree.

Beyond the Basics

In Exercises 83–88, find all solutions of each equation in the interval $[0, 2\pi)$.

83. $\sin x \cos x - \frac{\sqrt{3}}{2}\sin x + \frac{1}{2}\cos x - \frac{\sqrt{3}}{4} = 0$

[*Hint:* Factor by grouping.]

84. $2\sin x \tan x + 2\sin x + \tan x + 1 = 0$

[*Hint:* Factor by grouping.]

85. $5\sin^2\theta + \cos^2\theta - \sec^2\theta = 0$

86. $2\sin\theta = \cos\theta$

87. $2\cos x = 1 - \sin x$

88. $\cos^4 x - \cos^2 x + 1 = 0$

Critical Thinking/Discussion/Writing

In Exercises 89–92, find all solutions of each equation in the interval $[0, 2\pi)$.

89. $\tan^2 x = 4$

90. $3\cos^2 x + \cos x = 0$

91. $\cos x - \sec x + 1 = 0$

92. $3\csc x + 2\cot^2 x = 5$

93. Solve for x in $[0, 2\pi)$: $\sin x \cos x = \frac{1}{2}$.

[*Hint:* Multiply both sides by 2 and use a double-angle formula (Section 6.3)]

94. Follow the hint given for Exercise 93 to show that the equation $\sin x \cos x = 1$ has no real-number solutions.

Maintaining Skills

In Exercises 95–98, find all solutions of the given equation in the interval $[0, 2\pi)$.

95. $\sin x = -\frac{1}{2}$

96. $\tan x - 1 = 0$

97. $2\cos x = \sqrt{3}$

98. $4\sin^2 x = 1$

In Exercises 99–102, solve each equation.

99. $x = \cos^{-1}\left(-\frac{\sqrt{2}}{2}\right)$

100. $x = \sin^{-1}\left(\frac{1}{2}\right)$

101. $x = \tan^{-1}(1)$

102. $x = \tan^{-1}(-\sqrt{3})$

103. Use a sum-to-product formula to express $\cos 3x + \cos 5x$ as a product.

104. If $\sin^{-1} x = \frac{\pi}{5}$, then $\cos\left(\frac{\pi}{2} - \frac{\pi}{5}\right) =$ ____________.

SECTION 6.6

Trigonometric Equations II

BEFORE STARTING THIS SECTION, REVIEW

1. Reference angle (Section 5.3, page 504)
2. Periods of trigonometric functions (Sections 5.4 and 5.5)
3. Factoring techniques (Section P.4)
4. Solving linear and quadratic equations (Sections 1.1 and 1.2)

OBJECTIVES

1. Solve trigonometric equations involving multiple angles.
2. Solve equations using sum-to-product identities.
3. Solve equations containing inverse trigonometric functions.

The Phases of the Moon

The moon orbits the earth and completes one revolution in about 29.5 days (a lunar month). During this period, the angle between the sun, the earth, and the moon changes. See Figure 6.11. This causes different amounts of the illuminated moon to face the earth; consequently, the shape of the moon appears to change. The shape varies from a **new moon**—the phase of the moon when the moon is not visible from the earth (this happens when the moon is between the sun and the earth)—to a **half moon**—the phase of the moon when the moon looks like half of a circular disk—to a **full moon**—the phase of the moon when the moon looks like a full circular disk in the sky—back to a half moon and then to a new moon. The portion of the moon visible from the earth on any given day is called the **phase of the moon**. In Example 3, we express the phases of the moon by a sinusoidal equation.

DO YOU KNOW?

The expression *once in a blue moon* means "not often," but what exactly is a blue moon? The popular definition says that a second full moon in a single calendar month is called a *blue moon*. On average, a blue moon occurs once every two and one-half years.

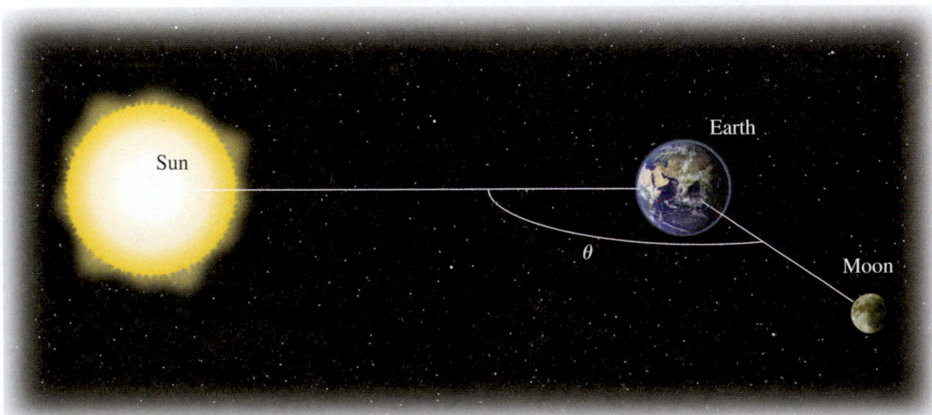

FIGURE 6.11 Moon phases

In this section, we solve equations involving trigonometric functions of multiple angles and those containing inverse trigonometric functions.

1 Solve trigonometric equations involving multiple angles.

Equations Involving Multiple Angles

If x is the measure of an angle, then for any real number k, the number kx is a **multiple angle** of x. Equations such as

$$\sin 3x = \frac{1}{2}, \qquad \cos \frac{1}{2}x = 0.7, \qquad \text{and} \qquad \sin 2x + \sin 4x = 0$$

involve multiple angles. We use the same techniques to solve the equations that we used in Section 6.5. We need to be careful when finding all solutions of such equations; the variable itself, such as x, must belong to the specified interval (if any), not its multiple (see Example 1).

EXAMPLE 1 Solving a Trigonometric Equation Containing Multiple Angles

Find all solutions of the equation $\cos 3x = \frac{1}{2}$ in the interval $[0, 2\pi)$.

Solution

Recall that $\cos \theta = \frac{1}{2}$ for the acute angle $\theta = \frac{\pi}{3}$. Because $\cos \theta$ is positive in quadrants I and IV, we also have $\theta = 2\pi - \frac{\pi}{3} = \frac{5\pi}{3}$. The period of the cosine function is 2π; so replacing θ with $3x$, all solutions of the equation $\cos 3x = \frac{1}{2}$ are given by

$$3x = \frac{\pi}{3} + 2n\pi \quad \text{or} \quad 3x = \frac{5\pi}{3} + 2n\pi \qquad \text{For any integer } n$$

$$x = \frac{\pi}{9} + \frac{2n\pi}{3} \quad \text{or} \quad x = \frac{5\pi}{9} + \frac{2n\pi}{3} \qquad \text{Divide both sides by 3.}$$

To find solutions in the interval $[0, 2\pi)$, try the following:

$n = -1$	$x = \frac{\pi}{9} - \frac{2\pi}{3} = -\frac{5\pi}{9}$ ✗	$x = \frac{5\pi}{9} - \frac{2\pi}{3} = -\frac{\pi}{9}$ ✗
$n = 0$	$x = \frac{\pi}{9}$ ✓	$x = \frac{5\pi}{9}$ ✓
$n = 1$	$x = \frac{\pi}{9} + \frac{2\pi}{3} = \frac{7\pi}{9}$ ✓	$x = \frac{5\pi}{9} + \frac{2\pi}{3} = \frac{11\pi}{9}$ ✓
$n = 2$	$x = \frac{\pi}{9} + \frac{4\pi}{3} = \frac{13\pi}{9}$ ✓	$x = \frac{5\pi}{9} + \frac{4\pi}{3} = \frac{17\pi}{9}$ ✓
$n = 3$	$x = \frac{\pi}{9} + 2\pi = \frac{19\pi}{9}$ ✗	$x = \frac{5\pi}{9} + 2\pi = \frac{23\pi}{9}$ ✗

The values resulting from $n = -1$ are too small and those resulting from $n = 3$ are too large to be in the interval $[0, 2\pi)$. The solution set corresponding to $n = 0$, 1, and 2 is

$$\left\{\frac{\pi}{9}, \frac{5\pi}{9}, \frac{7\pi}{9}, \frac{11\pi}{9}, \frac{13\pi}{9}, \frac{17\pi}{9}\right\}.$$

TECHNOLOGY CONNECTION

On a graphing calculator, you can see that the graphs of $Y_1 = \cos 3x$ and $Y_2 = \frac{1}{2}$ intersect in six points, which correspond to the six solutions in the interval $[0, 2\pi)$.

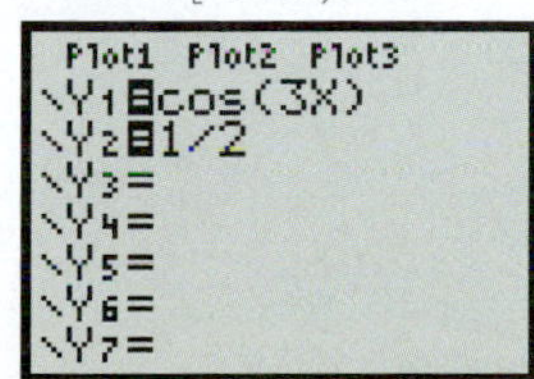

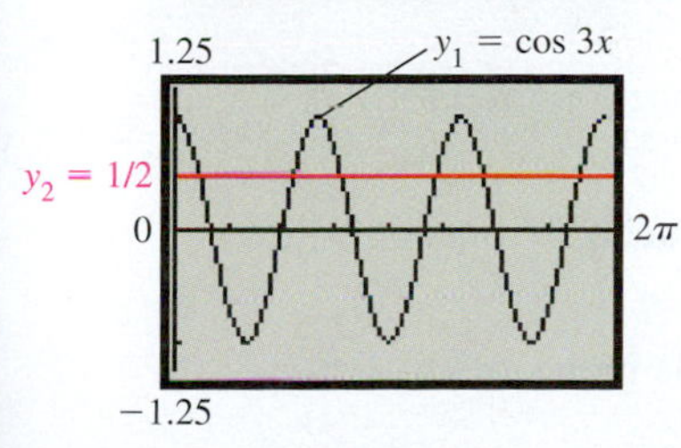

Practice Problem 1 Find all solutions of the equation $\sin 2x = \frac{1}{2}$ in the interval $[0, 2\pi)$.

TECHNOLOGY CONNECTION

The graphs of $Y_1 = \sin \frac{x}{2}$ and $Y_2 = \frac{-\sqrt{3}}{2}$ confirm that the equation in Example 2 has no solution.

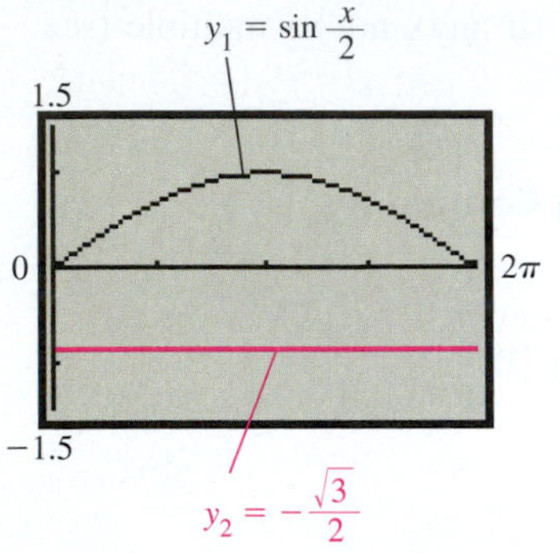

EXAMPLE 2 Solving a Trigonometric Equation Containing a Multiple Angle

Find all solutions of the equation $\sin \frac{x}{2} = -\frac{\sqrt{3}}{2}$ in the interval $[0, 2\pi)$.

Solution

We know that $\sin \frac{\pi}{3} = \frac{\sqrt{3}}{2}$. Because $\sin \theta$ is negative only in quadrants III and IV, the solutions for $\sin \theta = -\frac{\sqrt{3}}{2}$ are

$$\theta = \pi + \frac{\pi}{3} = \frac{4\pi}{3} \quad \text{and} \quad \theta = 2\pi - \frac{\pi}{3} = \frac{5\pi}{3}.$$

Because the sine function has period 2π, all solutions of $\sin \frac{x}{2} = -\frac{\sqrt{3}}{2}$ are given by

$$\frac{x}{2} = \frac{4\pi}{3} + 2n\pi \quad \text{or} \quad \frac{x}{2} = \frac{5\pi}{3} + 2n\pi \qquad \text{For any integer } n$$

$$x = \frac{8\pi}{3} + 4n\pi \quad \text{or} \quad x = \frac{10\pi}{3} + 4n\pi \qquad \text{Multiply both sides by 2.}$$

For any integer n, the values of x in both expressions lie outside the interval $[0, 2\pi)$. For example, for $n = 0$, $x = \frac{8\pi}{3} > 2\pi$ and for $n = -1$, $x = \frac{8\pi}{3} - 4\pi = -\frac{2\pi}{3} < 0$. The equation $\sin \frac{x}{2} = -\frac{\sqrt{3}}{2}$, therefore, has no solution in the interval $[0, 2\pi)$; so the solution set is $\varnothing$.

Practice Problem 2 Find all solutions of the equation $\tan \frac{x}{2} = \frac{\sqrt{3}}{3}$ in the interval $[0, 2\pi)$.

EXAMPLE 3 The Phases of the Moon

The moon orbits Earth in 29.5 days. The portion F of the moon visible from Earth x days after the new moon is given by the equation:

$$F = 0.5 \sin\left[\frac{\pi}{14.75}\left(x - \frac{14.75}{2}\right)\right] + 0.5.$$

Find x when 75% of the moon is visible from Earth.

Solution

We are asked to find x when $F = 75\% = 0.75$. We solve the equation.

$$0.75 = 0.5 \sin\left[\frac{\pi}{14.75}\left(x - \frac{14.75}{2}\right)\right] + 0.5 \qquad \text{Replace } F \text{ with 0.75.}$$

$$0.25 = 0.5 \sin\left[\frac{\pi}{14.75}\left(x - \frac{14.75}{2}\right)\right] \qquad \text{Subtract 0.5 from both sides.}$$

$$\frac{1}{2} = \sin\left[\frac{\pi}{14.75}\left(x - \frac{14.75}{2}\right)\right] \qquad \text{Divide both sides by 0.5.}$$

Let $\theta = \frac{\pi}{14.75}\left(x - \frac{14.75}{2}\right)$ be a solution of the equation $\sin \theta = \frac{1}{2}$.

We know that $\sin \frac{\pi}{6} = \frac{1}{2}$. Because $\sin \theta$ is positive only in quadrants I and II, we have

$$\frac{\pi}{14.75}\left(x - \frac{14.75}{2}\right) = \frac{\pi}{6} \quad \text{or} \quad \frac{\pi}{14.75}\left(x - \frac{14.75}{2}\right) = \frac{5\pi}{6}$$

$$x - \frac{14.75}{2} = \frac{14.75}{6} \quad \Bigg| \quad x - \frac{14.75}{2} = \frac{5(14.75)}{6} \qquad \text{Multiply both sides by } 14.75/\pi.$$

$$x = \frac{14.75}{2} + \frac{14.75}{6} \quad \Bigg| \quad x = \frac{14.75}{2} + \frac{5(14.75)}{6} \qquad \text{Solve for } x.$$

$$x \approx 10 \quad \Bigg| \quad x \approx 20 \qquad \text{Use a calculator and round to the nearest day.}$$

So 75% of the moon is visible 10 days and 20 days after the new moon.

Practice Problem 3 In Example 3, find x when 30% of the moon is visible from Earth.

EXAMPLE 4 Solving a Trigonometric Equation

Find all solutions of $\tan 2\theta = \sqrt{3}$ for $0° \le \theta < 360°$.

Solution

We have $\tan x = \sqrt{3}$ for the acute angle $x = 60°$. Replacing x with 2θ in the equation $\tan x = \sqrt{3}$, we have all solutions of the equation $\tan 2\theta = \sqrt{3}$ given by

$$2\theta = 60° + 180°(n) \qquad \text{For any integer } n\text{; the period of the tangent function is } \pi = 180°.$$

$$\theta = 30° + 90°(n) \qquad \text{Divide both sides by 2.}$$

For $n = 0, 1, 2,$ and 3, we have all values of θ between 0° and 360° that satisfy the equation $\tan 2\theta = \sqrt{3}$.

$$\begin{aligned} n = 0, \quad \theta &= 30° + 90°(0) = 30° \\ n = 1, \quad \theta &= 30° + 90°(1) = 120° \\ n = 2, \quad \theta &= 30° + 90°(2) = 210° \\ n = 3, \quad \theta &= 30° + 90°(3) = 300° \end{aligned}$$

The solution set is $\{30°, 120°, 210°, 300°\}$.

Practice Problem 4 Find all solutions of $\cot 3\theta = 1$ for $0 \le \theta < 360°$.

EXAMPLE 5 Solving an Equation by Squaring

Solve $\sin 2x + \cos 2x = \sqrt{2}$ over the interval $[0, 2\pi)$.

Solution

$$\sin 2x + \cos 2x = \sqrt{2} \qquad \text{Original equation}$$

$$(\sin 2x + \cos 2x)^2 = (\sqrt{2})^2 \qquad \text{Square both sides.}$$

$$\sin^2 2x + \cos^2 2x + 2 \sin 2x \cos 2x = 2 \qquad (a + b)^2 = a^2 + b^2 + 2ab$$

$$1 + \sin 4x = 2 \qquad \sin^2 A + \cos^2 A = 1;\ 2 \sin A \cos A = \sin 2A$$

$$\sin 4x = 1 \qquad \text{Solve for } \sin 4x.$$

We have $\sin\theta = 1$ for $\theta = \frac{\pi}{2}$. Replacing θ with $4x$ in the equation $\sin\theta = 1$, we have all solutions of the equation $\sin 4x = 1$ given by

$$4x = \frac{\pi}{2} + 2n\pi \quad \text{For any integer } n\text{; the period of the sine function is } 2\pi.$$

$$x = \frac{\pi}{8} + \frac{n\pi}{2} \quad \text{Divide by 4; simplify.}$$

For $n = 0, 1, 2$, and 3, we have all possible values of x between 0 and 2π that satisfy the equation $\sin 4x = 1$.

$$n = 0 \quad x = \frac{\pi}{8}$$

$$n = 1 \quad x = \frac{\pi}{8} + \frac{\pi}{2} = \frac{5\pi}{8}$$

$$n = 2 \quad x = \frac{\pi}{8} + \pi = \frac{9\pi}{8}$$

$$n = 3 \quad x = \frac{\pi}{8} + \frac{3\pi}{2} = \frac{13\pi}{8}$$

The possible solutions of the original equation are $\frac{\pi}{8}, \frac{5\pi}{8}, \frac{9\pi}{8}$, and $\frac{13\pi}{8}$. Because squaring both sides of an equation may result in extraneous solutions, we must check all possible solutions.

You can verify that the solution set of the equation is $\left\{\frac{\pi}{8}, \frac{9\pi}{8}\right\}$.

Practice Problem 5 Solve $\sin 3x + \cos 3x = \sqrt{\frac{3}{2}}$ over the interval $[0, 2\pi)$.

2 Solve equations using sum-to-product identities.

Using Sum-to-Product Identities

The next example illustrates a procedure for solving a trigonometric equation of the form $\sin ax \pm \sin bx = 0$ or $\cos ax \pm \cos bx = 0$. To solve such equations, review the sum-to-product identities on page 611.

EXAMPLE 6 Solving an Equation Using a Sum-to-Product Identity

Solve $\cos 2x + \cos 3x = 0$ over the interval $[0, 2\pi)$.

Solution

$$\cos 2x + \cos 3x = 0 \quad \text{Original equation}$$

$$2\cos\frac{5x}{2}\cos\frac{x}{2} = 0 \quad \cos u + \cos v = 2\cos\frac{u+v}{2}\cos\frac{u-v}{2}$$

So

$$\cos\frac{5x}{2} = 0 \quad \text{or} \quad \cos\frac{x}{2} = 0 \quad \text{Zero-product property}$$

We know that if $\cos\theta = 0$, then for any integer n,

$$\theta = \frac{\pi}{2} + 2n\pi \quad \text{or} \quad \theta = \frac{3\pi}{2} + 2n\pi$$

We replace θ with $\frac{5x}{2}$ and $\frac{x}{2}$ and find the values of x that lie in the interval $[0, 2\pi)$.

	$\cos\frac{5x}{2} = 0$		$\cos\frac{x}{2} = 0$	
	$\frac{5x}{2} = \frac{\pi}{2} + 2n\pi$	$\frac{5x}{2} = \frac{3\pi}{2} + 2n\pi$	$\frac{x}{2} = \frac{\pi}{2} + 2n\pi$	$\frac{x}{2} = \frac{3\pi}{2} + 2n\pi$
	$x = \frac{\pi}{5} + \frac{4n\pi}{5}$	$x = \frac{3\pi}{5} + \frac{4n\pi}{5}$	$x = \pi + 4n\pi$	$x = 3\pi + 4n\pi$
$n = 0$	$x = \frac{\pi}{5}$ ✓	$x = \frac{3\pi}{5}$ ✓	$x = \pi$ ✓	$x = 3\pi$ ×
$n = 1$	$x = \pi$ ✓	$x = \frac{7\pi}{5}$ ✓	$x = 5\pi$ ×	
$n = 2$	$x = \frac{9\pi}{5}$ ✓	$x = \frac{11\pi}{5}$ ×		
$n = 3$	$x = \frac{13\pi}{5}$ ×			

The solution set of the given equation is $\left\{\frac{\pi}{5}, \frac{3\pi}{5}, \pi, \frac{7\pi}{5}, \frac{9\pi}{5}\right\}$.

Practice Problem 6 Solve $\sin 2x + \sin 3x = 0$ over the interval $[0, 2\pi)$.

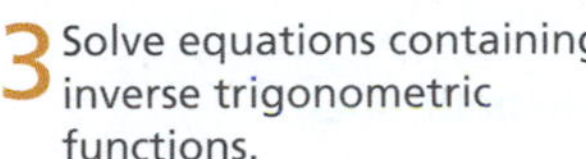

3 Solve equations containing inverse trigonometric functions.

Equations Containing Inverse Functions

We now consider equations containing inverse trigonometric functions.

EXAMPLE 7 Solving an Equation Containing an Inverse Tangent Function

Solve $\frac{\pi}{6} + 2\tan^{-1}(x - 1) = \frac{\pi}{2}$.

Solution

$$\frac{\pi}{6} + 2\tan^{-1}(x - 1) = \frac{\pi}{2} \qquad \text{Original equation}$$

$$2\tan^{-1}(x - 1) = \frac{\pi}{2} - \frac{\pi}{6} = \frac{\pi}{3} \qquad \text{Subtract } \frac{\pi}{6} \text{ and simplify.}$$

$$\tan^{-1}(x - 1) = \frac{\pi}{6} \qquad \text{Divide both sides by 2.}$$

$$x - 1 = \tan\frac{\pi}{6} \qquad \text{Definition of inverse function}$$

$$x = 1 + \tan\frac{\pi}{6} \qquad \text{Solve for } x$$

$$x = 1 + \frac{\sqrt{3}}{3} = \frac{3 + \sqrt{3}}{3} \qquad \tan\frac{\pi}{6} = \frac{\sqrt{3}}{3}$$

Practice Problem 7 Solve $\frac{\pi}{4} + 3\sin^{-1}(x + 1) = \frac{5\pi}{4}$.

The equation $\sin^{-1}\left(\frac{1}{2}\right) = x$ has only one solution, $x = \frac{\pi}{6}$. However, taking the sine of both sides gives $\sin\left(\sin^{-1}\frac{1}{2}\right) = \sin x$, or $\frac{1}{2} = \sin x$. The last equation has infinitely many solutions. As with squaring both sides of an equation, taking the sine (or any other trigonometric function) of both sides of an equation may result in *extraneous solutions*. In this case, you must check all solutions.

EXAMPLE 8 Solving an Equation Containing Inverse Functions

RECALL

If $y = \sin^{-1} x$, then $\sin y = x$ and $-\frac{\pi}{2} \le y \le \frac{\pi}{2}$. If $t = \cos^{-1} x$, then $\cos t = x$ and $0 \le t \le \pi$.

a. Verify the identity: $\sin^{-1} x + \cos^{-1} x = \frac{\pi}{2}$, x in $[-1, 1]$.

b. Use part a to solve: $\sin^{-1} x - \cos^{-1} x = \frac{\pi}{4}$, x in $[-1, 1]$.

Solution

a. Let $\sin^{-1} x = \theta$. Then $-\frac{\pi}{2} \le \theta \le \frac{\pi}{2}$, so $\frac{\pi}{2} \ge -\theta \ge -\frac{\pi}{2}$ and $0 \le \frac{\pi}{2} - \theta \le \pi$.

$$x = \sin\theta \qquad \text{From } \sin^{-1} x = \theta.$$

$$x = \cos\left(\frac{\pi}{2} - \theta\right) \qquad \text{Cofunction identity}$$

$$\cos^{-1} x = \frac{\pi}{2} - \theta \qquad \text{Definition of } \cos^{-1} x.$$

$$\theta + \cos^{-1} x = \frac{\pi}{2} \qquad \text{Add } \theta \text{ to both sides.}$$

$$\sin^{-1} x + \cos^{-1} x = \frac{\pi}{2} \qquad \text{Replace } \theta \text{ with } \sin^{-1} x.$$

b.

$$\sin^{-1} x - \cos^{-1} x = \frac{\pi}{4} \qquad \text{Given equation}$$

$$\sin^{-1} x - \left(\frac{\pi}{2} - \sin^{-1} x\right) = \frac{\pi}{4} \qquad \text{From part } \mathbf{a}.\ \cos^{-1} x = \frac{\pi}{2} - \sin^{-1} x.$$

$$2\sin^{-1} x = \frac{\pi}{4} + \frac{\pi}{2} = \frac{3\pi}{4} \qquad \text{Simplify.}$$

$$x = \sin\frac{3\pi}{8} \qquad \text{Solve for } x.$$

$$x = \sqrt{\frac{1 - \cos\left(\frac{3\pi}{4}\right)}{2}} \qquad \text{Half-angle formula}$$

$$x = \sqrt{\frac{1 + \frac{\sqrt{2}}{2}}{2}} \qquad \cos\frac{3\pi}{4} = -\frac{\sqrt{2}}{2}$$

$$x = \frac{\sqrt{2 + \sqrt{2}}}{2} \qquad \text{Simplify}$$

The solution set is $\left\{\frac{\sqrt{2 + \sqrt{2}}}{2}\right\}$.

Practice Problem 8

a. Verify the identity: $\tan^{-1} x + \cot^{-1} x = \frac{\pi}{2}$, x in $(-\infty, \infty)$.

b. Use part a. to solve: $\tan^{-1} x - \cot^{-1} x = \frac{\pi}{4}$.

Answers to Practice Problems

1. $\left\{\frac{\pi}{12}, \frac{5\pi}{12}, \frac{13\pi}{12}, \frac{17\pi}{12}\right\}$ **2.** $\left\{\frac{\pi}{3}\right\}$ **3.** After about 5 days and 24 days **4.** $\{15°, 75°, 135°, 195°, 255°, 315°\}$ **5.** $\left\{\frac{\pi}{36}, \frac{5\pi}{36}, \frac{25\pi}{36}, \frac{29\pi}{36}, \frac{49\pi}{36}, \frac{53\pi}{36}\right\}$ **6.** $\left\{0, \frac{2\pi}{5}, \frac{4\pi}{5}, \pi, \frac{6\pi}{5}, \frac{8\pi}{5}\right\}$ **7.** $\left\{\frac{\sqrt{3}-2}{2}\right\}$ **8. b.** $\left\{\sqrt{\frac{2+\sqrt{2}}{2-\sqrt{2}}}\right\}$

SECTION 6.6 Exercises

Basic Concepts and Skills

1. If $\sin 2x_1 = \sin 2x_2$ and $0 < x_1 < x_2 < \frac{\pi}{2}$, then $x_2 =$ ________.

2. If $\cos 2x_1 = \cos 2x_2$ and $0 < x_1 < x_2 < \pi$, then $x_2 =$ ________.

3. If $\tan 2x_1 = \tan 2x_2$ and $0 < x_1 < x_2 < \pi$, then $x_2 =$ ________.

4. True or False. $\sin(2\sin^{-1} x) = 2x\sqrt{1 - x^2}$ if $-1 \le x \le 1$.

5. True or False. $\frac{\sin 2x}{2} = \sin x$

6. True or False. $\frac{2\tan\frac{x}{2}}{x} = \tan 1$

In Exercises 7–26, find all solutions of each equation in the interval $[0, 2\pi)$.

7. $\cos 2x = \frac{1}{2}$ **8.** $\cos 2x = 0$

9. $\csc 2x = \frac{1}{2}$ **10.** $\sec 2x = \frac{\sqrt{3}}{2}$

11. $\tan 2x = \frac{\sqrt{3}}{3}$ **12.** $\cot 2x = \frac{\sqrt{3}}{3}$

13. $\sin 3x = \frac{1}{2}$ **14.** $\cos 3x = \frac{\sqrt{3}}{2}$

15. $\cos\frac{x}{2} = \frac{1}{2}$ **16.** $\csc\frac{x}{2} = 2$

17. $\tan\frac{x}{3} = 1$ **18.** $\cot\frac{x}{3} = \sqrt{3}$

19. $2\sin 3x = \sqrt{2}$ **20.** $2\cos 3x = \sqrt{2}$

21. $2\cos(2x + 1) = 1$ **22.** $2\sin(2x - 1) = \sqrt{3}$

23. $2\sin(4x - 1) = 1$ **24.** $2\cos(4x + 1) = \sqrt{3}$

25. $2\cos\left(\frac{3x}{2} - 1\right) = \sqrt{3}$ **26.** $2\sin\left(\frac{3x}{2} + 1\right) = 1$

In Exercises 27–38, solve each equation for θ in the interval $[0°, 360°)$. If necessary, write your answer to the nearest tenth of a degree.

27. $3\cos 2\theta = 1$ **28.** $4\sin 2\theta = 1$

29. $2\cos 3\theta + 1 = 2$ **30.** $2\sin 3\theta - 1 = -2$

31. $3\sin 3\theta + 1 = 0$ **32.** $4\cos 3\theta + 1 = 0$

33. $2\sin\frac{\theta}{2} + 1 = 0$ **34.** $2\cos\frac{\theta}{2} + \sqrt{3} = 0$

35. $2\sec 2\theta + 3 = 7$ **36.** $2\csc 2\theta + 1 = 0$

37. $3\csc\frac{\theta}{2} - 1 = 5$ **38.** $5\sec\frac{\theta}{2} - 3 = 7$

In Exercises 39–46, solve each equation by using sum-to-product identities in the interval $[0, 2\pi)$.

39. $\sin 2x + \sin x = 0$ **40.** $\cos 2x + \cos x = 0$

41. $\cos 3x - \cos x = 0$ **42.** $\sin 3x - \sin x = 0$

43. $\cos 3x + \cos 5x = 0$ **44.** $\cos 3x - \cos 5x = 0$

45. $\sin 3x - \sin 5x = 0$ **46.** $\sin 3x + \sin 5x = 0$

In Exercises 47–54, solve each equation for x.

47. $\sin^{-1} x - \cos^{-1} x = \frac{\pi}{6}$ **48.** $\sin^{-1} x + \cos^{-1} x = \frac{\pi}{6}$

49. $\sin^{-1} x + \cos^{-1} x = \frac{3\pi}{4}$ **50.** $\sin^{-1} x - \cos^{-1} x = \frac{\pi}{3}$

51. $\sin^{-1} x - \cos^{-1} x = -\frac{\pi}{6}$ **52.** $\sin^{-1} x + \cos^{-1} x = -\frac{3\pi}{4}$

53. $\sin^{-1} x - \tan^{-1} \frac{2}{3} = \frac{\pi}{4}$ **54.** $\sin^{-1} x - \cos^{-1} \frac{2}{3} = \frac{\pi}{6}$

Applying the Concepts

55. Electric current. The electric current I (in amperes) produced by an alternator in time t (in seconds) is given by $I = 60 \sin(120\pi t)$. Find the smallest possible positive value of t (rounded to four decimal places) such that

a. $I = 30$ amperes.
b. $I = -20$ amperes.

56. Simple harmonic motion. A simple harmonic motion is described by the equation $x = 6 \sin\left(\frac{\pi}{2} t\right)$, where x is the displacement in feet and t is time in seconds. Find positive values of t for which the displacement is 3 feet.

57. Simple harmonic motion. Repeat Exercise 56 assuming that the simple harmonic motion is described by the equation $x = 6 \cos\left(\frac{\pi}{2} t\right)$.

58. Overcoat sales. The monthly sales of overcoats at the menswear store Suit Yourself in Winterland are approximated by the equation

$$S(x) = 500 + 500 \sin\left[\frac{\pi}{4}(x - 2)\right],$$

where x is the month, with $x = 1$ for January. Find the months (after appropriate rounding) in which the number of overcoats sold is

a. 0.
b. 1000.
c. 500.

59. Number of tourists. The weekly number of tourists visiting a tropical island is approximated by the equation

$$y = 20 + 10 \sin\left[\frac{\pi}{26}(x - 14)\right],$$

where y is the number of visitors (in thousands) in the xth week of the year, starting with $x = 1$ for the first week in January. Find the week (after appropriate rounding) in which the number of tourists to the island is

a. 30,000.
b. 25,000.
c. 15,000.

60. Average monthly temperature. The average monthly sea surface temperature T (in degrees Fahrenheit) on Paradise Island is approximated by the equation

$$T = 10.5 \sin\left[\frac{\pi}{6}(x - 5)\right] + 73.5,$$

where x is the month, with $x = 1$ representing January. Find the months when the average sea surface temperature is

a. 78.75° F.
b. 84° F.

61. Average monthly snowfall. The average monthly snowfall y (in inches) in Rochester, New York, in the xth month can be approximated by

$$y = 12 + 12 \sin\left[\frac{\pi}{4}(x - 2)\right], 1 \le x \le 8,$$

where October = 1, November = 2, . . . May = 8. Find the months (after appropriate rounding) when the average snowfall in Rochester is

a. 24 inches.
b. 18 inches.

62. Watching TV. A 3-foot-high LCD television is mounted on a wall with its base 1.5 feet above the level of the observer's eye. Let θ be the angle of vision subtended by the TV when you sit x feet from the wall.

a. Show that $\theta = \tan^{-1} \frac{9}{2x} - \tan^{-1} \frac{3}{2x}$.

b. Find the maximum value of x (to the nearest tenth foot) assuming that $\tan \theta = \frac{1}{4}$.

Beyond the Basics

In Exercises 63–80, find all solutions of each equation in the interval $[0, 2\pi)$.

63. $\sin^4 2x = 1$ **64.** $\cos^4 2x = 1$

65. $4 \cos^2 \frac{x}{2} = 3$ **66.** $4 \sin^2 \frac{x}{2} = 1$

67. $\tan^2 2x = 1$ **68.** $\cot^2 \frac{x}{2} = 1$

69. $3 \sec^2 \frac{x}{2} = 12$

70. $5 \csc^2 \frac{x}{2} - 11 = 9$

71. $(\tan 2x - 1)(\sin x + 1) = 0$

72. $(\sqrt{3} \cot 2x - 1)(2 \cos x - 1) = 0$

73. $(1 - 2 \sin 2x - 1)(\sqrt{3} + 2 \cos x) = 0$

74. $(\sqrt{3} \tan 2x + 1)(2 \cos x + 1) = 0$

75. $\sin 2x \cos 2x + \frac{\sqrt{3}}{2} \sin 2x + \frac{1}{2} \cos 2x + \frac{\sqrt{3}}{4} = 0$

76. $2 \sin 3x \tan 2x - 2 \sin 3x + \tan 2x - 1 = 0$

77. $\sin 2x + \sin 4x = 2 \sin 3x$

78. $\cos x + \cos 5x = 2 \cos 2x$

79. $\sin x + \sin 3x = \cos x + \cos 3x$

80. $\sin 2x + \sin 4x = \cos 2x + \cos 4x$

In Exercises 81–84,

a. Write the left side of the equation in the form $A \sin(Bx - C)$.

b. Solve for x in the interval $[0, 2\pi)$.

81. $3 \sin 2x + 4 \cos 2x = 0$

82. $3 \cos 2x - 4 \sin 2x = \frac{5}{2}$

83. $5 \sin 3x - 12 \cos 3x = \frac{13\sqrt{3}}{2}$

84. $12 \sin 3x + 5 \cos 3x = \frac{13}{2}$

Critical Thinking/Discussion/Writing

In Exercises 85 and 86, solve each equation for x.

85. $\cos(2 \sin^{-1} x) = \frac{1}{2}$

86. $\tan(\tan^{-1} x + \tan^{-1} 1) = 5$

Maintaining Skills

In Exercises 87 and 88, rewrite each equation to obtain an equivalent equation having the variable in the numerator.

87. $\frac{3}{x} = \frac{5}{\pi}$

88. $\frac{\sin 12°}{4} = \frac{\sqrt{2}}{2x}$

In Exercises 89 and 90, solve each equation for x.

89. $\frac{12}{x} = \frac{3}{4}$

90. $\frac{7}{9} = \frac{2}{3x}$

In Exercises 91–94, give the exact value for each expression.

91. $\sin 60°$

92. $\sin 45°$

93. $\sin(180° - 30°)$

94. $\sin(180° - 45°)$

In Exercises 95 and 96, answer *True* or *False*.

95. If A, B, and C represent the angles in a triangle and $C = 90°$, then $A + B = 90°$.

96. If a, b, and c, represent the lengths of the sides in a triangle, then $a^2 + b^2 = c$.

SUMMARY Definitions, Concepts, and Formulas

6.1 Verifying Identities

Reciprocal Identities

$$\csc x = \frac{1}{\sin x} \qquad \sec x = \frac{1}{\cos x} \qquad \cot x = \frac{1}{\tan x}$$

Quotient Identities

$$\tan x = \frac{\sin x}{\cos x} \qquad \cot x = \frac{\cos x}{\sin x}$$

Pythagorean Identities

$$\sin^2 x + \cos^2 x = 1$$
$$1 + \tan^2 x = \sec^2 x$$
$$1 + \cot^2 x = \csc^2 x$$

Even–Odd Identities

$$\sin(-x) = -\sin x \quad \cos(-x) = \cos x \quad \tan(-x) = -\tan x$$
$$\csc(-x) = -\csc x \quad \sec(-x) = \sec x \quad \cot(-x) = -\cot x$$

See page 577 for *guidelines for verifying trigonometric identities.*

6.2 Sum and Difference Formulas

Sum and Difference Formulas

$$\cos(u - v) = \cos u \cos v + \sin u \sin v$$
$$\cos(u + v) = \cos u \cos v - \sin u \sin v$$
$$\sin(u - v) = \sin u \cos v - \cos u \sin v$$
$$\sin(u + v) = \sin u \cos v + \cos u \sin v$$
$$\tan(u - v) = \frac{\tan u - \tan v}{1 + \tan u \tan v}$$
$$\tan(u + v) = \frac{\tan u + \tan v}{1 - \tan u \tan v}$$

Cofunction Identities

$$\sin\left(\frac{\pi}{2} - x\right) = \cos x \qquad \cos\left(\frac{\pi}{2} - x\right) = \sin x$$
$$\tan\left(\frac{\pi}{2} - x\right) = \cot x \qquad \csc\left(\frac{\pi}{2} - x\right) = \sec x$$
$$\sec\left(\frac{\pi}{2} - x\right) = \csc x \qquad \cot\left(\frac{\pi}{2} - x\right) = \tan x$$

Reduction Formula

If (a, b) is any point on the terminal side of an angle θ in standard position, then $a \sin x + b \cos x = \sqrt{a^2 + b^2} \sin(x + \theta)$ for any real number x, where

$$\cos\theta = \frac{a}{\sqrt{a^2 + b^2}} \quad \text{and} \quad \sin\theta = \frac{b}{\sqrt{a^2 + b^2}}.$$

6.3 Double-Angle and Half-Angle Formulas

Double-Angle Formulas

$$\sin 2x = 2 \sin x \cos x$$
$$\cos 2x = \cos^2 x - \sin^2 x$$
$$\cos 2x = 1 - 2 \sin^2 x$$
$$\cos 2x = 2 \cos^2 x - 1$$
$$\tan 2x = \frac{2 \tan x}{1 - \tan^2 x}$$

Power-Reducing Formulas

$$\sin^2 x = \frac{1 - \cos 2x}{2}$$
$$\cos^2 x = \frac{1 + \cos 2x}{2}$$
$$\tan^2 x = \frac{1 - \cos 2x}{1 + \cos 2x}$$

Half-Angle Formulas

$$\sin \frac{\theta}{2} = \pm \sqrt{\frac{1 - \cos \theta}{2}}$$
$$\cos \frac{\theta}{2} = \pm \sqrt{\frac{1 + \cos \theta}{2}}$$
$$\tan \frac{\theta}{2} = \pm \sqrt{\frac{1 - \cos \theta}{1 + \cos \theta}} = \frac{\sin \theta}{1 + \cos \theta} = \frac{1 - \cos \theta}{\sin \theta}$$

where the sign + or − depends on the quadrant in which $\frac{\theta}{2}$ lies.

6.4 Product-to-Sum and Sum-to-Product Formulas

Product-to-Sum Formulas

$$\cos x \cos y = \frac{1}{2}[\cos(x - y) + \cos(x + y)]$$
$$\sin x \sin y = \frac{1}{2}[\cos(x - y) - \cos(x + y)]$$
$$\sin x \cos y = \frac{1}{2}[\sin(x + y) + \sin(x - y)]$$
$$\cos x \sin y = \frac{1}{2}[\sin(x + y) - \sin(x - y)]$$

Sum-to-Product Formulas

$$\cos x + \cos y = 2 \cos\left(\frac{x + y}{2}\right) \cos\left(\frac{x - y}{2}\right)$$
$$\cos x - \cos y = -2 \sin\left(\frac{x + y}{2}\right) \sin\left(\frac{x - y}{2}\right)$$
$$\sin x + \sin y = 2 \sin\left(\frac{x + y}{2}\right) \cos\left(\frac{x - y}{2}\right)$$
$$\sin x - \sin y = 2 \sin\left(\frac{x - y}{2}\right) \cos\left(\frac{x + y}{2}\right)$$

6.5 Trigonometric Equations I

Trigonometric Equations I A trigonometric equation is an equation that contains a trigonometric function with a variable. Values of the variable that satisfy a trigonometric equation are its *solutions*. Solving a trigonometric equation means finding its *solution set*.

Algebraic techniques are used to solve equations that are in linear, quadratic, or factorable form. Sometimes trigonometric identities are used to solve trigonometric equations.

In this section, we solve three types of equations.

1. Equation of the form

   $a \sin(x - c) = k$, $a \cos(x - c) = k$, and $a \tan(x - c) = k$

2. Equations that can be solved by factoring using the zero-product property
3. Equations that are equivalent to a polynomial equation in one trigonometric function

6.6 Trigonometric Equations II

In this section, we solve trigonometric equations involving multiple angles. We use sum-to-product and other identities to solve trigonometric equations, including those containing inverse trigonometric functions.

REVIEW EXERCISES

Basic Concepts and Skills

In Exercises 1–4, use the information given to find the exact value of the remaining trigonometric functions of θ.

1. $\sin \theta = -\frac{2}{3}$ and $\cos \theta < 0$

2. $\tan \theta = -\frac{1}{2}$ and $\csc \theta > 0$

3. $\sec \theta = 3$ and $\tan \theta < 0$

4. $\csc \theta = 5$ and $\cot \theta < 0$

In Exercises 5–20, verify each identity.

5. $(\sin x + \cos x)^2 + (\sin x - \cos x)^2 = 2$

6. $(1 - \tan x)^2 + (1 + \tan x)^2 = 2 \sec^2 x$

7. $\frac{1 - \tan^2 \theta}{1 + \tan^2 \theta} = \cos^2 \theta - \sin^2 \theta$

8. $\frac{\sin x + \tan x}{\csc x + \cot x} = \sin^2 x \sec x$

9. $\dfrac{\sin\theta}{1+\cos\theta}+\dfrac{\sin\theta}{1-\cos\theta}=2\csc\theta$

10. $\tan^2 x\sin^2 x=\tan^2 x-\sin^2 x$

11. $\dfrac{\sin\theta}{1-\cot\theta}+\dfrac{\cos\theta}{1-\tan\theta}=\cos\theta+\sin\theta$

12. $\dfrac{\tan\theta-\sin\theta}{\sin^3\theta}=\dfrac{\sec\theta}{1+\cos\theta}$

13. $\dfrac{\tan x}{\sec x-1}+\dfrac{\tan x}{\sec x+1}=2\csc x$

14. $\dfrac{1}{\csc x-\cot x}-\dfrac{1}{\cot x+\csc x}=2\cot x$

15. $\dfrac{1+\sin\theta}{1-\sin\theta}=(\sec\theta+\tan\theta)^2$

16. $\dfrac{1-\cos\theta}{1+\cos\theta}=(\csc\theta-\cot\theta)^2$

17. $\dfrac{\sec x-\tan x}{\sec x+\tan x}=(\sec x-\tan x)^2$

18. $\dfrac{\csc x+\cot x}{\csc x-\cot x}=(\csc x+\cot x)^2$

19. $\dfrac{\sin x-\cos x+1}{\sin x+\cos x-1}=\dfrac{\sin x+1}{\cos x}$
[*Hint:* Multiply the numerator and the denominator of the left side by $\sin x+(1-\cos x)$.]

20. $\sin^2 x\cot x=\dfrac{2\cos x}{\sin x+\csc x+\cos^2 x\csc x}$

In Exercises 21–28, find the exact value of each expression.

21. $\cos 15°$

22. $\sin 105°$

23. $\csc 75°$

24. $\tan 75°$

25. $\sin 41°\cos 49°+\cos 41°\sin 49°$

26. $\cos 50°\cos 10°-\sin 50°\sin 10°$

27. $\dfrac{\tan 69°+\tan 66°}{1-\tan 69°\tan 66°}$

28. $2\cos 75°\cos 15°$

In Exercises 29–32, let $\sin u=\frac{4}{5}$, $\cos v=\frac{5}{13}$, for $0<u\le\frac{\pi}{2}$, $0<v\le\frac{\pi}{2}$. Find the value of each expression.

29. $\sin(u-v)$

30. $\cos(u+v)$

31. $\cos(u-v)$

32. $\tan(u-v)$

In Exercises 33–50, verify each identity.

33. $\sin(x-y)\cos y+\cos(x-y)\sin y=\sin x$

34. $\cos(x-y)\cos y-\sin(x-y)\sin y=\cos x$

35. $\dfrac{\sin(u+v)}{\sin(u-v)}=\dfrac{\tan u+\tan v}{\tan u-\tan v}$

36. $\dfrac{\tan(x+y)}{\cot(x-y)}=\dfrac{\tan^2 x-\tan^2 y}{1-\tan^2 x\tan^2 y}$

37. $\dfrac{\sin 4x}{\sin 2x}-\dfrac{\cos 4x}{\cos 2x}=\sec 2x$

38. $\dfrac{\sin 3x}{\sin x}-\dfrac{\cos 3x}{\cos x}=2$

39. $\sin\left(x-\dfrac{\pi}{6}\right)+\cos\left(x+\dfrac{\pi}{3}\right)=0$

40. $\cos 2x=\cos^4 x-\sin^4 x$

41. $1+\tan\theta\tan 2\theta=\sec 2\theta$

42. $\dfrac{\sin\theta+\sin 2\theta}{1+\cos\theta+\cos 2\theta}=\tan\theta$

43. $\dfrac{\sin\theta+\sin 2\theta}{\cos\theta+\cos 2\theta}=\tan\dfrac{3\theta}{2}$

44. $\dfrac{\sin\theta-\sin 2\theta}{\cos\theta-\cos 2\theta}=-\cot\dfrac{3\theta}{2}$

45. $\dfrac{\sin 5x-\sin 3x}{\sin 5x+\sin 3x}=\dfrac{\tan x}{\tan 4x}$

46. $\dfrac{\cos 5x-\cos 3x}{\cos 5x+\cos 3x}=-\tan x\tan 4x$

47. $\dfrac{\tan 3x+\tan x}{\tan 3x-\tan x}=2\cos 2x$

48. $\dfrac{\sin 4x}{2(1+\cos 4x)}=\dfrac{\tan x}{1-\tan^2 x}$

49. $\dfrac{\sin 3x+\sin 5x+\sin 7x+\sin 9x}{\cos 3x+\cos 5x+\cos 7x+\cos 9x}=\tan 6x$

50. $\dfrac{\cos x+\cos 3x+\cos 5x+\cos 7x}{\sin x+\sin 3x+\sin 5x+\sin 7x}=\cot 4x$

In Exercises 51 and 52, sketch the graph of each equation by first converting it to the form $y=A\sin(x+\theta)$.

51. $y=\sqrt{3}\sin x+\cos x$

52. $y=\sin x+\sqrt{3}\cos x$

In Exercises 53 and 54, assume that $A+B+C=180°$. Verify each identity.

53. $\sin 2A+\sin 2B-\sin 2C=4\cos A\cos B\sin C$

54. $\tan A+\tan B+\tan C=\tan A\tan B\tan C$

In Exercises 55–60, find all solutions of each equation in the interval $[0, 2\pi)$. Use a graphing calculator to verify your solutions.

55. $2\cos^2 x-1=0$

56. $3\tan^2 x-1=0$

57. $2\cos^2 x-\cos x-1=0$

58. $2\sin^2 x-5\sin x-3=0$

59. $2\sin 3x-1=0$

60. $2\cos(2x-1)-1=0$

In Exercises 61–66, solve each equation for θ in the interval $[0°, 360°)$. If necessary, write your answers to the nearest tenth of a degree. Use a graphing calculator to verify your solutions.

61. $3\sin\theta-1=0$

62. $4\cos\theta+1=0$

63. $2\sqrt{3}\cos 2\theta-3=0$

64. $(1-2\sin 2\theta)(\sqrt{3}+2\cos 2\theta)=0$

65. $(\sqrt{3}\tan\theta+1)(2\cos\theta+1)=0$

66. $\sqrt{3}\cos\theta=\sin\theta+1$

Applying the Concepts

67. A searchlight beam from a Coast Guard boat shines on a lake at an angle θ to the surface of the water. Viewed by a swimmer underwater, the beam appears to make an angle α with the surface of the water, as shown in the figure. Under these circumstances, $\cos\theta = \frac{4}{3}\cos\alpha$. Find α if $\cos\theta = \frac{2}{3}$.

68. The rabbit population of an area is described by $N(t) = 2400 + 500\sin\pi t$, where $N(t)$ gives the number of rabbits in the area after t years. When will the rabbit population first reach 1900?

69. The voltage from an alternating current generator is given by $V(t) = 160\cos 120\pi t$, where t is measured in seconds. Find the time t when the voltage first equals 80.

70. A jeweler wants to make a gold pendant in the shape of an isosceles triangle. Express the area of the pendant in terms of $\frac{\theta}{2}$ and x, where θ is the angle included by two sides of equal length x.

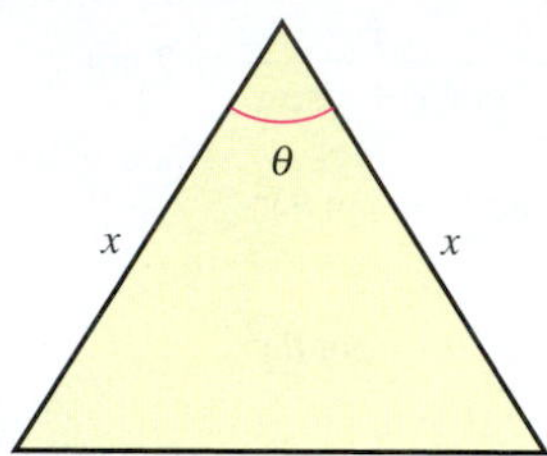

71. Assuming that the area of the pendant in Exercise 70 is 1 square inch, find the angle included between the two sides of equal length when each measures 2 inches.

PRACTICE TEST A

1. Assuming that $\sin\theta = \frac{3}{5}$ and $\cos\theta < 0$, find $\tan\theta$.

2. Assuming that $\tan x = \frac{2}{3}$ and $\csc x < 0$, find $\cos x$.

In Problems 3–8, verify each identity.

3. $\dfrac{1 - \sin^2 x}{\sin^2 x} = \cot^2 x$

4. $2\sin x\cos x = (\sin x + \cos x + 1)(\sin x + \cos x - 1)$

5. $\sin x \sin\left(\frac{\pi}{2} - x\right) = \dfrac{\sin 2x}{2}$

6. $\dfrac{\sin 2x}{2(\cos x + \sin x)} = \dfrac{\sin x}{1 + \tan x}$

7. $\dfrac{\sin 2x + \sin 4x}{\cos 2x + \cos 4x} = \tan 3x$

8. $\cos(x + y)\cos(x - y) = \cos^2 x + \cos^2 y - 1$

In Problems 9 and 10, find all solutions of each equation.

9. $\tan(-x) = 1$

10. $\sin 4x = \frac{1}{2}$

In Problems 11 and 12, find all solutions of each equation in the interval $[0, 2\pi)$.

11. $\cos\frac{x}{3} = \frac{\sqrt{2}}{2}$

12. $\sin 2x + \cos x = 0$

In Problems 13–15, find the exact value of each expression.

13. $\sin 56° + \cos 146°$

14. $\cos 48°\cos 12° - \sin 48°\sin 12°$

15. $\sin\frac{5\pi}{12}$

16. If $\sin\theta = \frac{4}{5}$, find the exact value of $\cos 2\theta$.

17. If $\tan\theta = \frac{3}{4}$, find the exact value of $\sin 2\theta$.

18. Determine whether the function $y = \cos\left(\frac{\pi}{2} - x\right) + \tan x$ is odd, even, or neither.

19. Determine whether $\sin x + \sin y = \sin(x + y)$ is an identity.

20. A golf ball is hit on a level fairway with an initial velocity of $v_0 = 128$ ft/sec on an initial angle of flight θ (in degrees). The range (in feet) of the ball is given by the expression $\dfrac{v_0^2\sin 2\theta}{32}$. Find the value(s) of θ if the range is 512 feet.

PRACTICE TEST B

1. If $\sin\theta = -\frac{12}{13}$ and $\pi < \theta < \frac{3\pi}{2}$, find $\sec\theta$.

   a. $-\frac{13}{5}$ b. $-\frac{12}{5}$

   c. $-\frac{5}{12}$ d. $\frac{13}{5}$

2. Which expression is equal to $(\sin x + \cos x)^2 + (\sin x - \cos x)^2$?

   a. 1 b. 2

   c. $\sin x \cos x$ d. $4 \sin x \cos x$

3. Which expression is equal to $\frac{\sin x}{1 + \cos x} + \frac{\sin x}{1 - \cos x}$?

   a. $2 \sin x$ b. $2 \cos x$

   c. $2 \sec x$ d. $2 \csc x$

4. Which expression is equal to $\frac{\tan x + \tan y}{\cot x + \cot y}$?

   a. $\cot x \cot y$ b. $\tan x \tan y$

   c. $\sec x \csc y$ d. $\sin x \cos y$

5. Which expression is equal to $\frac{\sin x + \sin y}{\cos x + \cos y} + \frac{\cos x - \cos y}{\sin x - \sin y}$?

   a. 0 b. 1

   c. $\tan x \tan y$ d. $\tan x + \tan y$

6. Which expression is equal to $\frac{\sin 2x}{1 + \cos 2x}$?

   a. $\cot x$ b. $\tan x$

   c. $\tan 2x$ d. $\cot 2x$

7. Which expression is equal to $\frac{1 - \cos 2x}{1 + \cos 2x}$?

   a. $\sin^2 x$ b. $\cos^2 x$

   c. $\tan^2 x$ d. $\cot^2 x$

8. Which expression is equal to $\frac{\sin 3x - \sin x}{\cos x - \cos 3x}$?

   a. $\tan x$ b. $\tan 2x$

   c. $\cot x$ d. $\cot 2x$

In Problems 9–11, find all solutions of each equation.

9. $\cot(-x) = -1$

   a. $x = \frac{\pi}{4} + 2\pi n$, for any integer n

   b. $x = \frac{3\pi}{4} + 2\pi n$, for any integer n

   c. $x = \frac{\pi}{4} + \pi n$, for any integer n

   d. $x = \frac{3\pi}{4} + \pi n$, for any integer n

10. $2 \cos 4x = -1$

   a. $x = \frac{2\pi}{3} + 2\pi n$ and $x = \frac{4\pi}{3} + 2\pi n$, for any integer n

   b. $x = \frac{\pi}{3} + \frac{n\pi}{2}$ and $x = \frac{5\pi}{3} + \frac{n\pi}{2}$, for any integer n

   c. $x = \frac{\pi}{6} + \frac{n\pi}{2}$ and $x = \frac{\pi}{3} + \frac{n\pi}{2}$, for any integer n

   d. $x = \frac{\pi}{12} + 2\pi n$ and $x = \frac{5\pi}{12} + 2\pi n$, for any integer n

11. $\sin\frac{x}{3} = \frac{\sqrt{2}}{2}$

   a. $x = \frac{\pi}{4} + 2\pi n$ and $x = \frac{3\pi}{4} + 2\pi n$, for any integer n

   b. $x = \frac{\pi}{4} + 6n\pi$ and $x = \frac{\pi}{12} + 6n\pi$, for any integer n

   c. $x = \frac{3\pi}{4} + 6\pi n$ and $x = \frac{9\pi}{4} + 6\pi n$, for any integer n

   d. $x = \frac{3\pi}{4} + 2\pi n$ and $x = \frac{5\pi}{4} + 2\pi n$, for any integer n

12. Find all solutions of $\cos x - \sin 2x = 0$ for the interval $[0, 2\pi)$.

   a. $\left\{\frac{\pi}{6}, \frac{\pi}{2}, \frac{5\pi}{6}, \frac{3\pi}{2}\right\}$ b. $\left\{\frac{\pi}{6}, \frac{\pi}{3}, \frac{5\pi}{6}, \frac{7\pi}{6}\right\}$

   c. $\left\{\frac{\pi}{2}, \frac{5\pi}{6}, \pi, \frac{3\pi}{2}\right\}$ d. $\left\{\frac{\pi}{6}, \frac{\pi}{2}, \frac{5\pi}{6}, \frac{11\pi}{6}\right\}$

In Problems 13 and 14, find the exact value of each expression.

13. $\sin 71° + \cos 161°$

   a. 1.892 b. −1.892

   c. 1 d. 0

14. $\sin 53° \cos 37° + \cos 53° \sin 37°$

   a. $\frac{1}{2}$ b. $\frac{\sqrt{3}}{2}$

   c. 0 d. 1

15. Assuming that $\sin\theta = \frac{2}{3}$, find the exact value of $\cos 2\theta$.

   a. $\frac{1}{3}$ b. $\frac{1}{9}$

   c. $\frac{14}{9}$ d. $\frac{\sqrt{5}}{3}$

16. Assuming that $\tan\theta = \frac{4}{3}$, find the exact value of $\sin 2\theta$.

   a. $\frac{24}{25}$ b. $\frac{4}{5}$

   c. $\frac{3}{5}$ d. $\frac{8}{5}$

17. Which of the following best describes the symmetry of the graph of $y = \sin\left(\frac{\pi}{2} - x\right) + \cot x$?

   a. Symmetric with respect to the origin

   b. Symmetric with respect to the x-axis

   c. Symmetric with respect to the y-axis

   d. Not symmetric to either axis or the origin

18. Which pair of values results in a false statement for $\tan(x + y) = \tan x + \tan y$?

   a. $x = 0, y = \pi$ b. $x = \frac{\pi}{4}, y = -\frac{\pi}{4}$

   c. $x = \frac{\pi}{4}, y = \frac{5\pi}{4}$ d. $x = \frac{\pi}{4}, y = \frac{3\pi}{4}$

19. Identify all solutions of $2\cos^2 x = \frac{3}{2}$ in the interval $[0, 2\pi)$.

a. $\left\{\frac{\pi}{6}\right\}$ **b.** $\left\{\frac{\pi}{6}, \frac{5\pi}{6}\right\}$

c. $\left\{\frac{\pi}{6}, \frac{11\pi}{6}\right\}$ **d.** $\left\{\frac{\pi}{6}, \frac{5\pi}{6}, \frac{7\pi}{6}, \frac{11\pi}{6}\right\}$

20. A golf ball is hit on a level fairway with an initial velocity of $v_0 = 128$ ft/sec and an initial angle of flight θ (in degrees). The range (in feet) of the ball is given by the expression $\frac{v_0^2 \sin 2\theta}{32}$. Find the value(s) of θ if the range is $256\sqrt{3}$ feet.

a. $45°$ **b.** $30°, 60°$

c. $45°, 135°$ **d.** $150°, 175°$

CUMULATIVE REVIEW EXERCISES CHAPTERS P–6

In Exercises 1–6, solve each equation or inequality.

1. $x^2 + x - 1 = 0$

2. $\log_2(x + 1) + \log_2(x - 1) = 1$

3. $5^{-x} = 9$

4. $\sin 2x - \cos x = 0 \ (0 \le x < 2\pi)$

5. $x^3 - 4x > 0$

6. $\frac{2x - 3}{x + 2} - 1 < 0$

7. Assuming that $f(x) = 2x^2 - x + 3$, find $\frac{f(x + h) - f(x)}{h}$.

8. Assuming that $f(x) = \frac{x + 2}{2x - 1}$, find $f^{-1}(x)$.

In Exercises 9–16, sketch the graph of each function $f(x)$ by using transformations on the given basic function.

9. $f(x) = x^2 - 4x - 7$; use $y = x^2$.

10. $f(x) = 2\sqrt{x - 3} + 1$; use $y = \sqrt{x}$.

11. $f(x) = 2^{x-1} - 2$; use $y = 2^x$.

12. $f(x) = 2\ln(x + 1)$; use $y = \ln x$.

13. $f(x) = \frac{1}{2}\sin 2x$; use $y = \sin x$.

14. $f(x) = -2\cos(3x - 1)$; use $y = \cos x$.

15. $f(x) = -3\csc x$; use $y = 3\sin x$.

16. $f(x) = 2\tan 4x$; use $y = \tan x$.

17. Simplify $\frac{2\csc^2 x - 5\cot x - 5}{2\cot x + 1}$.

18. Use an identity to find the exact value of $\sec 15°$.

19. Verify the identity $\sin\left(x - \frac{3\pi}{2}\right) = \cos x$.

20. The angles to the top of a tower from two points on the ground at distances 800 and 3200 feet are complementary angles. Find the height of the tower.

CHAPTER 7

Applications of Trigonometric Functions

TOPICS

Applications of trigonometry permeate virtually every field, from architecture and bionics to medicine and physics. In this chapter, we introduce vectors and polar coordinates and investigate the many applications of trigonometry to real-world problems.

SECTION 7.1

The Law of Sines

Mount Everest

In 1852, an unsung hero Radhanath Sikdar managed to *calculate* the height of Peak XV, an icy peak in the Himalayas. The highest mountain in the world—later named Mount Everest—stood at 29,002 feet. Sikdar's calculations used triangulations as well as the phenomenon called refraction—the bending of light rays by the density of Earth's atmosphere. Like George Everest, Sikdar may have never seen Mount Everest.

BEFORE STARTING THIS SECTION, REVIEW

1. Trigonometric functions of angles (Section 5.3, page 500)
2. Inverse trigonometric functions (Section 5.6, page 551)

OBJECTIVES

1. Learn vocabulary and conventions for solving triangles.
2. Derive the Law of Sines.
3. Solve AAS and ASA triangles by using the Law of Sines.
4. Solve for possible triangles in the ambiguous SSA case.

◆ The Great Trigonometric Survey of India

Triangulation is the process of finding the coordinates and the distance to a point by using the Law of Sines. Triangulation was used in surveying the Indian subcontinent. The survey, which was called "one of the most stupendous works in the whole history of science," was begun by William Lambton, a British army officer, in 1802. It started in the south of India and extended north to Nepal, a distance of approximately 1600 miles. Lasting several decades and employing thousands of workers, the survey was named the Great Trigonometric Survey (GTS). In 1818, George Everest (1790–1866), a Welsh surveyor and geographer, was appointed assistant to Lambton. Everest, who succeeded Lambton in 1823 and was later promoted as surveyor general of India, completed the GTS. Mount Everest was surveyed and named after Everest by his successor Andrew Waugh. In Example 2, we use the Law of Sines to estimate the height of a mountain.

1 Learn vocabulary and conventions for solving triangles.

Solving Oblique Triangles

The process of finding the unknown side lengths and angle measures in a triangle is called *solving a triangle*. In Section 5.2, you learned techniques of solving right triangles. We now discuss triangles that are not necessarily right triangles. A triangle with no right angle is called an **oblique triangle**. From geometry, we know that two triangles with any of the following sets of equal parts will always be **congruent**. In other words, the following sets of given parts determine a unique triangle:

ASA: Two angles and the included side

AAS: Two angles and a nonincluded side

SAS: Two sides and the included angle

SSS: All three sides

RECALL

Two triangles are *congruent* if corresponding sides are congruent (equal in length) and corresponding angles are congruent (equal in measure). An *included side* is the side between two angles, and an *included angle* is the angle between two sides.

Although **SSA** (two sides and a nonincluded angle) does not guarantee a unique triangle, there can be, at most, two triangles with the given parts. On the other hand, **AAA** (or **AA**) guarantees only *similar* triangles, not necessarily congruent ones; so there are infinitely many triangles with the same three angle measures.

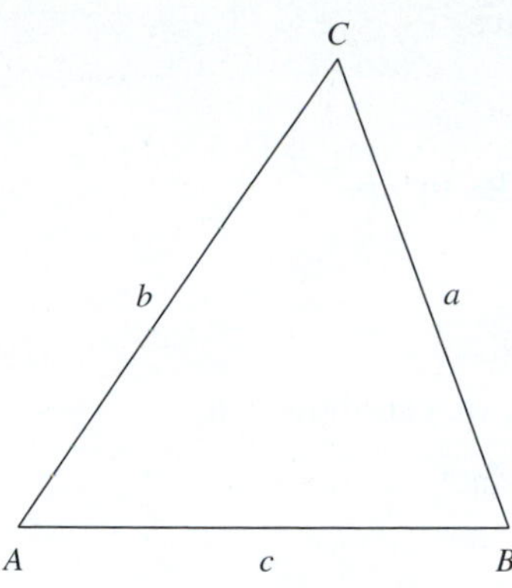

FIGURE 7.1 Labeling a triangle

For any triangle ABC, we let each letter, such as A, stand for both the vertex A and the measure of the angle at A. As usual, the angles of a triangle ABC are labeled A, B, and C and the lengths of their opposite sides are labeled, a, b, and c, respectively. See Figure 7.1.

To solve an oblique triangle, given at least one side and two other measures, we consider four possible cases:

Case 1. Two angles and a side are known (**AAS** and **ASA** triangles).

Case 2. Two sides and an angle opposite one of the given sides are known (**SSA** triangles).

Case 3. Two sides and their included angle are known (**SAS** triangles).

Case 4. All three sides are known (**SSS** triangles).

We use the Law of Sines to analyze triangles in Cases 1 and 2. In the next section, we use the *Law of Cosines* to solve triangles in Cases 3 and 4.

2 Derive the Law of Sines.

The Law of Sines

The Law of Sines states that in a triangle ABC with sides of length a, b, and c,

$$\frac{\sin A}{a} = \frac{\sin B}{b} = \frac{\sin C}{c}.$$

We can derive the Law of Sines by using the right triangle definition of the sine of an acute angle θ:

$$\sin\theta = \frac{\text{opposite leg}}{\text{hypotenuse}}$$

RECALL

An *altitude* of a triangle is a segment drawn from any vertex of the triangle perpendicular to the opposite side, or to an extension of the opposite side.

We begin with an oblique triangle ABC. See Figures 7.2 and 7.3. Draw altitude CD from the vertex C to the side AB (or its extension, as in Figure 7.3). Let h be the length of segment CD.

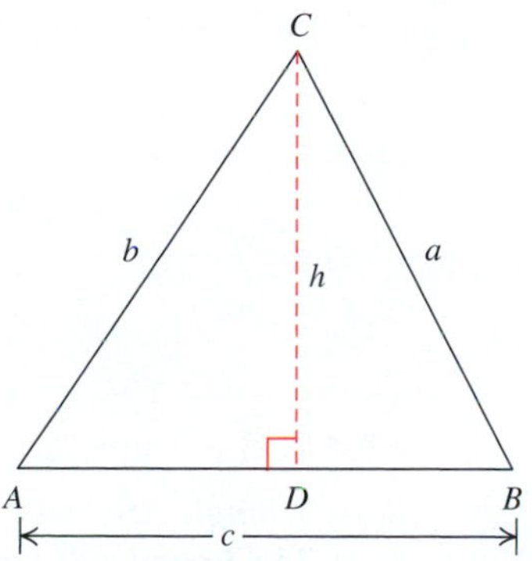

FIGURE 7.2 Altitude CD with acute $\angle B$

In right triangle CDA,

$$\frac{h}{b} = \sin A \quad \text{or}$$

$$h = b \sin A.$$

In right triangle CDB,

$$\frac{h}{a} = \sin B \quad \text{or}$$

$$h = a \sin B.$$

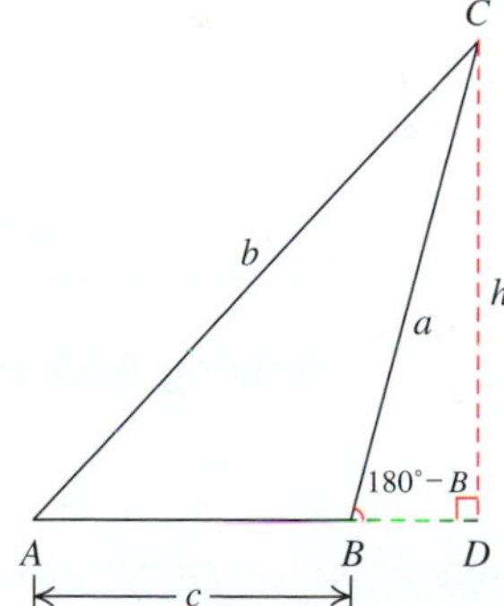

FIGURE 7.3 Altitude CD with obtuse $\angle B$

In right triangle CDA,

$$\frac{h}{b} = \sin A \quad \text{or}$$

$$h = b \sin A.$$

In right triangle CDB,

$$\frac{h}{a} = \sin(180° - B) = \sin B \quad \text{or}$$

$$h = a \sin B.$$

Because in each figure $h = b \sin A$ and $h = a \sin B$, we have

$$b \sin A = a \sin B$$
$$\frac{b \sin A}{ab} = \frac{a \sin B}{ab} \quad \text{Divide both sides by } ab.$$
$$\frac{\sin A}{a} = \frac{\sin B}{b} \quad \text{Simplify.}$$

Similarly, by drawing altitudes from the other two vertices, we can show that

$$\frac{\sin B}{b} = \frac{\sin C}{c} \quad \text{and} \quad \frac{\sin C}{c} = \frac{\sin A}{a}.$$

We have proved the Law of Sines.

THE LAW OF SINES

In any triangle ABC, with sides of length a, b, and c,

$$\frac{\sin A}{a} = \frac{\sin B}{b}, \quad \frac{\sin B}{b} = \frac{\sin C}{c}, \quad \text{and} \quad \frac{\sin C}{c} = \frac{\sin A}{a}.$$

We can rewrite these relations in compact notation:

$$\frac{\sin A}{a} = \frac{\sin B}{b} = \frac{\sin C}{c}, \quad \text{or equivalently,} \quad \frac{a}{\sin A} = \frac{b}{\sin B} = \frac{c}{\sin C}$$

3 Solve AAS and ASA triangles by using the Law of Sines.

Solving AAS and ASA Triangles

We solve AAS and ASA triangles of Case 1, where two angles and a side are given. Note that if two angles of a triangle are given, then you also know the third angle (because the sum of the angles of a triangle is 180°). So in Case 1, all three angles and one side of a triangle are known.

PROCEDURE IN ACTION

EXAMPLE 1 Solving AAS and ASA Triangles

OBJECTIVE

Solve a triangle given two angles and a side.

Step 1 Find the third angle. Find the measure of the third angle by subtracting the measures of the known angles from 180°.

Step 2 Make a chart. Make a chart of the six parts of the triangle, including the known and the unknown parts. Sketch the triangle.

EXAMPLE

Solve triangle ABC with $A = 62°$, $c = 14$ feet, and $B = 74°$. Round side lengths to the nearest tenth.

1. $C = 180° - A - B$
$= 180° - 62° - 74°$
$= 44°$

2.

$A = 62°$	$a = ?$
$B = 74°$	$b = ?$
$C = 44°$	$c = 14$

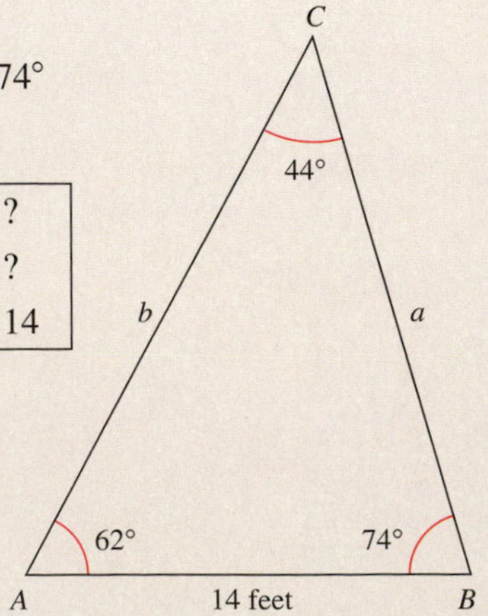

(continued)

Step 3 Apply the Law of Sines. Select two ratios from the Law of Sines in which three of the four quantities are known. Solve for the fourth quantity. Use the form of the Law of Sines in which the unknown quantity is in the numerator.

3.
$$\frac{b}{\sin B} = \frac{c}{\sin C} \qquad \frac{a}{\sin A} = \frac{c}{\sin C}$$
$$\frac{b}{\sin 74°} = \frac{14}{\sin 44°} \qquad \frac{a}{\sin 62°} = \frac{14}{\sin 44°}$$
$$b = \frac{14 \sin 74°}{\sin 44°} \qquad a = \frac{14 \sin 62°}{\sin 44°}$$
$$b \approx 19.4 \text{ feet} \qquad a \approx 17.8 \text{ feet}$$

Step 4 Show the solution. Show the solution by completing the chart.

4.

$A = 62°$	$a \approx 17.8$ feet
$B = 74°$	$b \approx 19.4$ feet
$C = 44°$	$c = 14$

Practice Problem 1 Solve triangle ABC with $A = 70°$, $B = 65°$, and $a = 16$ inches. Round side lengths to the nearest tenth.

EXAMPLE 2 Height of a Mountain

From a point on a level plain at the foot of a mountain, a surveyor finds the angle of elevation of the peak of the mountain to be 20°. She walks 3465 meters closer (on a direct line between the first point and the base of the mountain) and finds the angle of elevation to be 23°. Estimate the height of the mountain to the nearest meter.

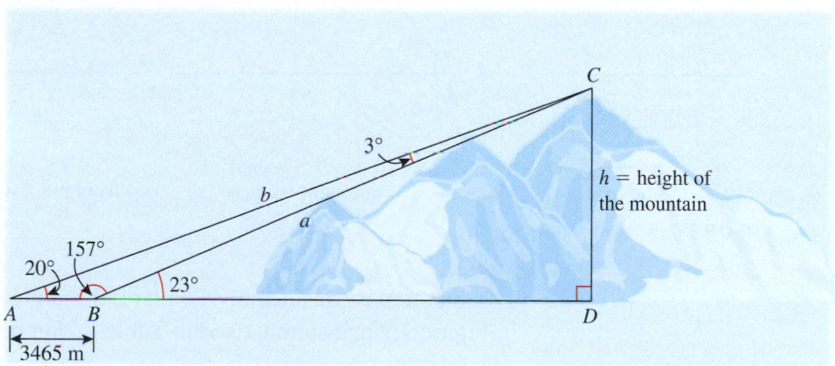

FIGURE 7.4 Height of a mountain

Solution

In Figure 7.4, consider triangle ABC.

$$\angle ABC = 180° - 23° = 157° \qquad \angle ABC + \angle CBD = 180°$$

Find the third angle in triangle ABC.

$$C = 180° - 20° - 157° = 3°$$

Apply the Law of Sines.

$$\frac{a}{\sin 20°} = \frac{3465}{\sin 3°}$$
$$a = \frac{3465 \sin 20°}{\sin 3°} \qquad \text{Multiply both sides by } \sin 20°.$$
$$\approx 22{,}644 \text{ meters} \qquad \text{Use a calculator.}$$

Now in triangle BCD, $\sin 23° = \frac{h}{a}$. So $h = a \sin 23° \approx 22{,}644 \sin 23° \approx 8848$ meters. The mountain is approximately 8848 meters high.

SIDE NOTE

When using your calculator to solve triangles, make sure it is in *degree* mode.

Practice Problem 2 From a point on a level plain at the foot of a mountain, the angle of elevation of the peak is 40°. If you move 2500 feet farther (on a direct line extending from the first point and the base of the mountain), the angle of elevation of the peak is 35°. How high above the plain is the peak? Round your answer to the nearest foot.

4 Solve for possible triangles in the ambiguous SSA case.

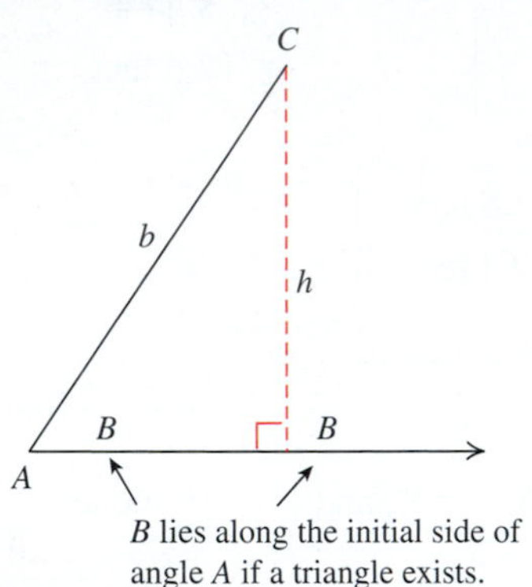

FIGURE 7.5 Locating *B*

Solving SSA Triangles—the Ambiguous Case

In Case 1, where we know two angles and a side, we obtain a unique triangle. However, if we know the lengths of two sides and the measure of the angle opposite one of these sides, then we could have a result that is (1) not a triangle, (2) exactly one triangle, or (3) two different triangles. For this reason, Case 2 is called the **ambiguous case**.

Suppose we want to draw triangle ABC with given measures of A, a, and b. We draw angle A with a horizontal initial side, and we place the vertex C on the terminal side of angle A so that the length of segment AC is b. We need to locate vertex B on the initial side of angle A so that the measure of the segment CB is a. See Figure 7.5.

The number of possible triangles depends on the length h of the altitude from C to the initial side of angle A. Because $\sin A = \frac{h}{b}$, we have $h = b \sin A$. Various possibilities for forming triangle ACB are illustrated in Figure 7.6, where A is an acute angle.

We summarize these results in Table 7.1 on page 645.

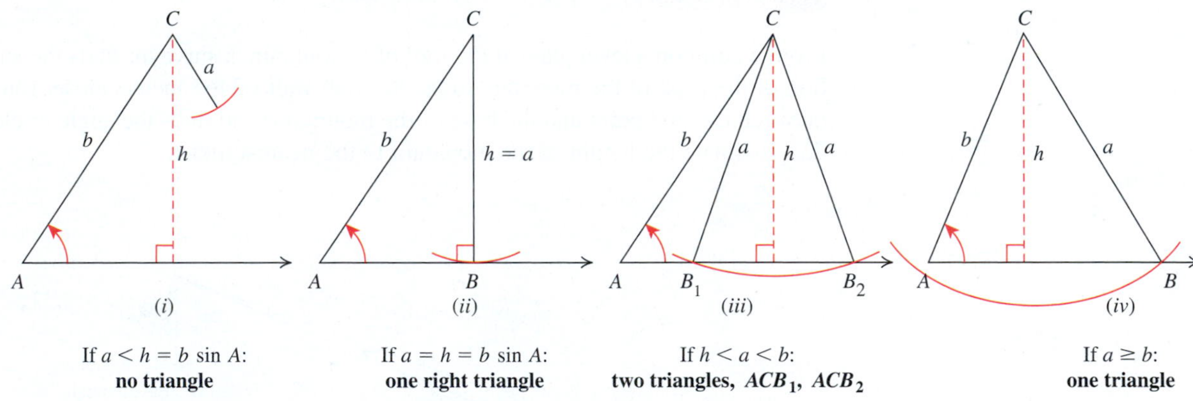

FIGURE 7.6 Acute angle *A*

The situation is considerably simpler if the angle A is obtuse. The possibilities are illustrated in Figure 7.7 and summarized in Table 7.2 on page 645.

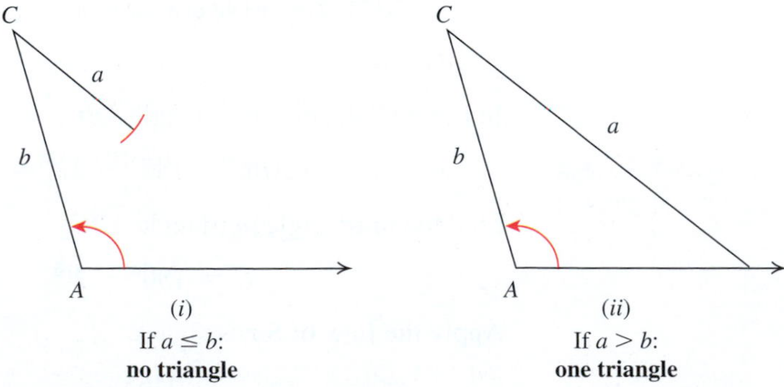

FIGURE 7.7 Obtuse angle *A*

In general in an SSA triangle, we are given:

(i) the known angle θ,
(ii) opposite side (side opposite to angle θ), and
(iii) adjacent side (side adjacent to angle θ).

Then $\quad$ **altitude = (adjacent side) sin θ.**

We summarize the situations from Figures 7.6 and 7.7 in Tables 7.1 and 7.2, respectively

TABLE 7.1 If θ is an acute angle

Case	Condition on Opposite Side	Number of Triangles
1	opposite side $<$ altitude	None
2	opposite side $=$ altitude	One right triangle
3	altitude $<$ opposite side $<$ adjacent side	Two triangles
4	opposite side $\geq$ adjacent side	One triangle

TABLE 7.2 If θ is an obtuse angle

Case	Condition on Opposite Side	Number of Triangles
1	opposite side $\leq$ adjacent side	None
2	opposite side $>$ adjacent side	One triangle

EXAMPLE 3 Finding the Number of Triangles

How many triangles can be drawn with $A = 42°$, $b = 2.4$, and $a = 2.1$?

Solution

$A = 42°$ is an acute angle, and opposite side $= a = 2.1 < 2.4 = b =$ adjacent side. We calculate

$$\text{altitude} = b \sin A = 2.4 \sin(42°) \quad \text{Replace } b \text{ with 2.4 and } A \text{ with } 42°.$$
$$\approx 1.6 \quad \text{Use a calculator.}$$

We have: altitude $<$ opposite side $<$ adjacent side. So by Case 3 of Table 7.1, two triangles can be drawn with the given measurements.

Practice Problem 3 How many triangles can be drawn with $A = 35°$, $b = 6.5$, and $a = 3.5$?

EXAMPLE 4 Finding the Number of Triangles

How many triangles can be drawn with $A = 98°$, $b = 4.8$, and $a = 5.1$?

Solution

$A = 98°$ is an obtuse angle, and opposite side $= a = 5.1 > 4.8 = b =$ adjacent side. So by Case 2 of Table 7.2, exactly one triangle can be drawn with the given measurements.

Practice Problem 4 How many triangles can be drawn with $A = 105°$, $b = 12.5$ and $a = 11.2$?

We strongly suggest that you begin with a tentative sketch of the triangle ABC.

PROCEDURE
IN ACTION

EXAMPLE 5 Solving the SSA Triangles

OBJECTIVE

Solve a triangle if two sides and an angle θ opposite one of them is given.

EXAMPLE

Solve triangle ABC with $B = 32°$, $b = 100$ feet, and $c = 150$ feet. Round each answer to the nearest tenth.

Step 1 Make a chart of the six parts. Identify θ, opposite and adjacent sides.

1. $\theta = B$, opposite side $= b$, adjacent side $= c$

$A = ?$	$a = ?$
$B = 32°$	$b = 100$
$C = ?$	$c = 150$

Three quantities are known.

Step 2 Count the solutions.
There may be none, one, or two solutions.
If the known angle
(i) θ is acute—use Table 7.1
(ii) θ is obtuse—use Table 7.2
(iii) $\theta = 90°$—use procedures in Section 5.2.
If there are no solutions, you are done. Otherwise, continue with Step 3.

2. Here $\theta = B = 32°$ is an acute angle, so we use Table 7.1.

$$\begin{aligned}\text{Altitude} &= (\text{adjacent side}) \sin B \\ &= 150 \sin 32° && \text{Substitute values.} \\ &\approx 79.488 && \text{Use a calculator.}\end{aligned}$$

We have
altitude < opposite side < adjacent side, so there will be two solutions.

Step 3 Apply the Law of Sines to the two ratios in which two sides and one angle are known, see the chart in Step 1. Solve for the sine of the unknown angle. Use the form of the Law of Sines in which the unknown angle is in the numerator.

3. $\dfrac{\sin C}{c} = \dfrac{\sin B}{b}$ The Law of Sines

$\sin C = \dfrac{c \sin B}{b}$ Multiply both sides by c.

$= \dfrac{(150) \sin 32°}{100}$ Substitute values.

$\sin C \approx 0.7949$ Use a calculator.

Step 4 Find the second angle(s)
If you determined in step 2 that:
(i) there are two solutions, find two angles; one acute and one obtuse.
(ii) there is one solution, find one acute angle.

4. Because there are two solutions, we find two angles:

$$C_1 \approx \sin^{-1}(0.7949) \approx 52.6°$$
$$C_2 \approx 180° - 52.6° = 127.4°$$

Step 5 Find the third angle of the triangle(s).

5. $\angle BAC_1 \approx 180° - 32° - 52.6° = 95.4°$
$\angle BAC_2 \approx 180° - 32° - 127.4° = 20.6°$

Step 6 Use the Law of Sines to find the remaining side(s).

6.

$\dfrac{a_1}{\sin \angle BAC_1} = \dfrac{b}{\sin B}$	$\dfrac{a_2}{\sin \angle BAC_2} = \dfrac{b}{\sin B}$
$a_1 = \dfrac{b \sin \angle BAC_1}{\sin B}$	$a_2 = \dfrac{b \sin \angle BAC_2}{\sin B}$
$a_1 = \dfrac{100 \sin 95.4°}{\sin 32°}$	$a_2 = \dfrac{(100) \sin 20.6°}{\sin 32°}$
$a_1 \approx 187.9$ feet	$a_2 \approx 66.4$ feet

Step 7 Show the solution(s).

7.

$\triangle BAC_1$

$\angle BAC_1 \approx 95.4°$	$a_1 \approx 187.9$
$B = 32°$	$b = 100$
$C_1 \approx 52.6°$	$c = 150$

$\triangle BAC_2$

$\angle BAC_2 \approx 20.6°$	$a_2 \approx 66.4$
$B = 32°$	$b = 100$
$C_2 \approx 127.4°$	$c = 150$

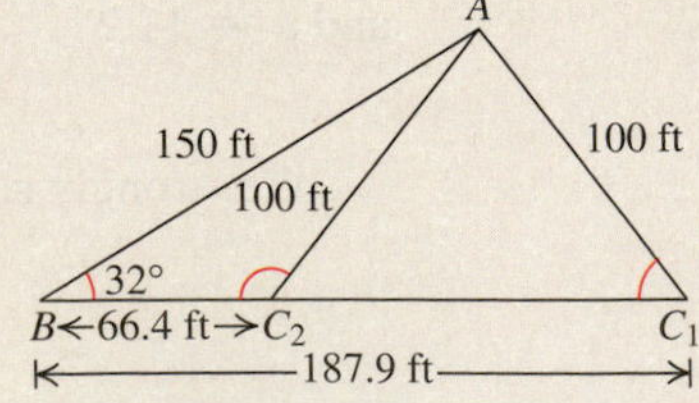

Practice Problem 5 Solve triangle ABC with $C = 35°$, $b = 15$ feet, and $c = 12$ feet. Round each answer to the nearest tenth.

EXAMPLE 6 Solving an SSA Triangle (No Solution)

Solve triangle ABC with $A = 50°$, $a = 8$ inches, and $b = 15$ inches.

Solution

Step 1 **Make a chart.**

$A = 50°$	$a = 8$
$B = ?$	$b = 15$
$C = ?$	$c = ?$

Three quantities are known.

Step 2 **Count the solutions.** Because $\theta = A = 50°$ is acute, we use Table 7.1. The opposite side $a = 8$ is not greater than the adjacent side $b = 15$, so we need to compute the altitude.

$$\begin{aligned} \text{altitude} &= (\text{adjacent side}) \sin A \\ &= 15 \sin(50°) && \text{Replace: adjacent side with 15, } A = 50° \\ &\approx 11.491 && \text{Use a calculator.} \end{aligned}$$

We have opposite side $a = 8 < 11.491 =$ altitude, so there is no triangle.
Alternatively, from the Law of Sines:

$$\begin{aligned} \frac{\sin B}{b} &= \frac{\sin A}{a} \quad \text{or} \\ \sin B &= \frac{b \sin A}{a} && \text{Multiply both sides by } b. \\ &= \frac{15 \sin 50°}{8} && \text{Substitute values.} \\ &\approx 1.44 && \text{Use a calculator.} \end{aligned}$$

Because $\sin B$ is never greater than 1, we conclude that there is no triangle with the given measurements.

Practice Problem 6 Solve triangle ABC with $A = 65°$, $a = 16$ meters, and $b = 30$ meters.

EXAMPLE 7 Solving an SSA Triangle (One Solution)

Solve triangle ABC with $C = 40°$, $c = 20$ meters, and $a = 15$ meters. Round each answer to the nearest tenth of a meter.

Solution

Step 1 **Make a chart.**

$A = ?$	$a = 15$
$B = ?$	$b = ?$
$C = 40°$	$c = 20$

Three quantities are known.

Step 2 **Find the number of triangles.** Because $\theta = C = 40°$ is acute, we use Table 7.1. The opposite side $c = 20$ is greater than the adjacent side $a = 15$, so there is one triangle.

Step 3 **Apply the Law of Sines.**

$$\frac{\sin A}{a} = \frac{\sin C}{c} \quad \text{The Law of Sines}$$

$$\sin A = \frac{a \sin C}{c} \quad \text{Multiply both sides by } a.$$

$$= \frac{(15) \sin 40°}{20} \quad \text{Substitute values.}$$

$$\sin A \approx 0.4821 \quad \text{Use a calculator.}$$

Step 4 **Find the second angle(s).** Because there is one solution, we find one angle.

$$A = \sin^{-1}(0.4821) \approx 28.8°$$

Step 5 The third angle at B has measure $\approx 180° - 40° - 28.8° = 111.2°$.

Step 6 **Find the remaining side length.**

$$\frac{b}{\sin B} = \frac{c}{\sin C} \quad \text{The Law of Sines}$$

$$b = \frac{c \sin B}{\sin C} \quad \text{Multiply both sides by } \sin B.$$

$$= \frac{(20) \sin(111.2°)}{\sin 40°} \quad \text{Substitute values.}$$

$$b \approx 29.0 \text{ m} \quad \text{Use a calculator.}$$

Step 7 **Show the solution.** See Figure 7.8.

$A \approx 28.8°$	$a = 15$ meters
$B \approx 111.2°$	$b \approx 29.0$ meters
$C = 40°$	$c = 20$ meters

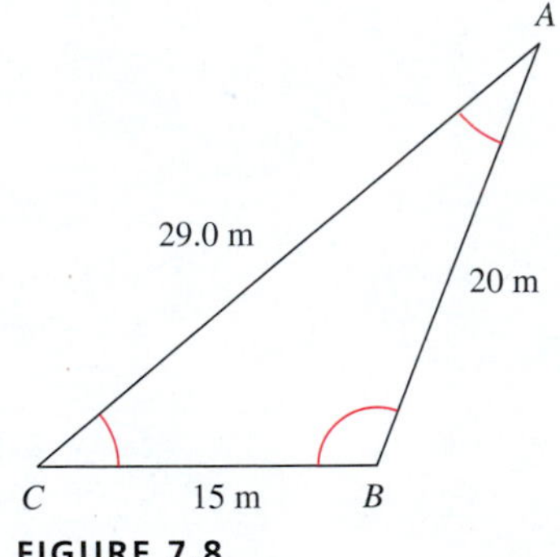

FIGURE 7.8

Practice Problem 7 Solve triangle ABC with $C = 60°$, $c = 50$ feet, and $a = 30$ feet.

Bearings

In navigation and surveying, directions are usually given by using *bearings*. A **bearing** is the measure of an acute angle from due north or due south. The bearing N 40° E means 40° to the east of due north, whereas S 30° W means 30° to the west of due south. See Figure 7.9.

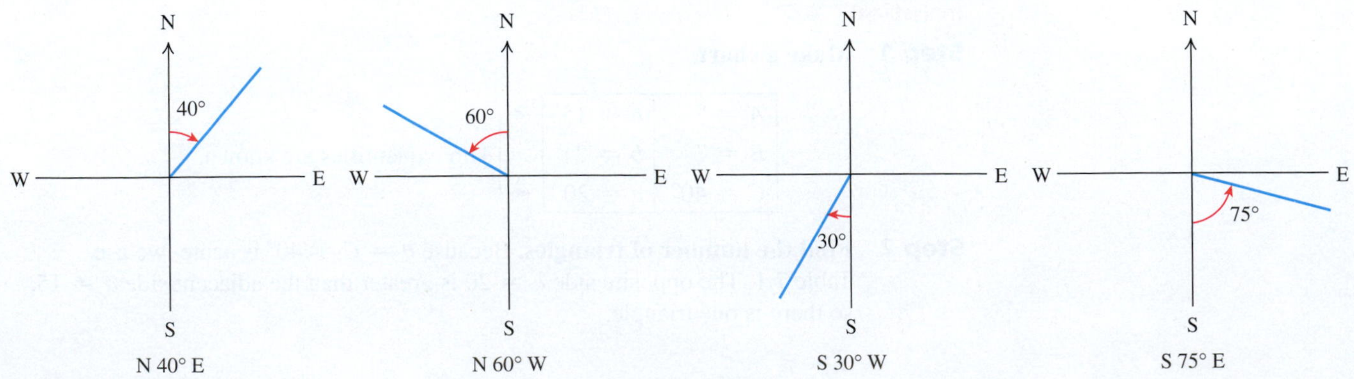

FIGURE 7.9 Bearings

EXAMPLE 8 Navigation Using Bearings

A ship sailing due west at 20 miles per hour records the bearing of an oil rig at N 55.4° W. An hour and a half later the bearing of the same rig is N 66.8° E.

a. How far is the ship from the oil rig the second time?

b. How close did the ship pass to the oil rig?

Solution

a. In an hour and a half, the ship travels $(1.5)(20) = 30$ miles due west. The oil rig (C), the starting point for the ship (A), and the position of the ship after an hour and a half (B) form the vertices of the triangle ABC shown in Figure 7.10.

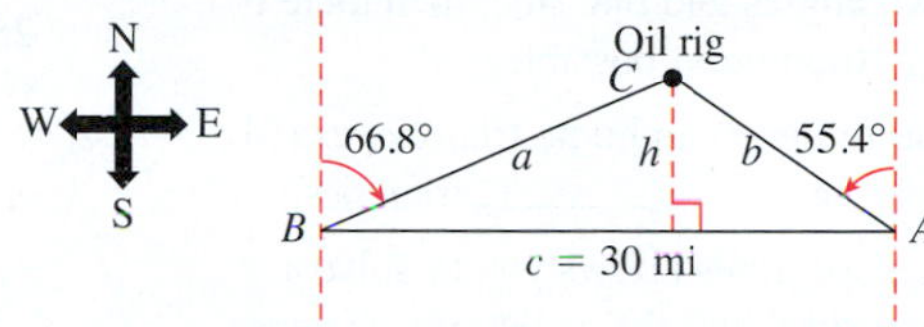

FIGURE 7.10 Navigation

Then $A = 90° - 55.4° = 34.6°$ and $B = 90° - 66.8° = 23.2°$. Therefore, $C = 180° - 34.6° - 23.2° = 122.2°$.

$$\frac{a}{\sin A} = \frac{c}{\sin C}$$ The Law of Sines

$$a = \frac{c \sin A}{\sin C}$$ Multiply both sides by $\sin A$.

$$= \frac{30 \sin 34.6°}{\sin 122.2°} \approx 20 \text{ miles}$$ Substitute values and use a calculator.

The ship is approximately 20 miles from the oil rig when the second bearing is taken.

b. The shortest distance between the ship and the oil rig is the length of the segment h in triangle ABC.

$$\sin B = \frac{h}{a}$$ Right-triangle definition of the sine

$$h = a \sin B$$ Multiply both sides by a.

$$h \approx 20 \sin 23.2°$$ Substitute values.

$$\approx 7.9$$ Use a calculator.

The ship passes within 7.9 miles of the oil rig.

Practice Problem 8 Rework Example 8 assuming that the second bearing is taken after two hours and is N 60.4° E and the ship is traveling due west at 25 miles per hour.

Answers to Practice Problems

1. $C = 45°, b \approx 15.4$ in, $c \approx 12.0$ in **2.** 10,576 ft **3.** None **4.** None **5.** Solution 1: $A \approx 99.2°, B \approx 45.8°, a \approx 20.7$ ft; Solution 2: $A \approx 10.8°, B \approx 134.2°, a \approx 3.9$ ft **6.** No triangle exists. **7.** $A \approx 31.3°, B \approx 88.7°, b \approx 57.7$ ft **8. a.** ≈ 31.5 mi **b.** ≈ 15.6 mi

SECTION 7.1 Exercises

Basic Concepts and Skills

1. If you know two angles of a triangle, then you can determine the third angle because the sum of all three angles is ______________ degrees.

2. The Law of Sines states that if a, b, and c are the sides opposite angles A, B, and C, then $\frac{a}{\sin A} =$ ______________ $=$ ______________.

3. If you are given any two angles and one side, then there is exactly ______________ triangle(s) possible.

4. For given data a, A, and b, there can be no triangle, exactly ______________ triangle, or ______________ triangles.

5. **True or False.** The Law of Sines allows you to solve a triangle if you know two sides and the angle opposite one of them.

6. **True or False.** A triangle is uniquely determined if any two sides and one angle are given.

In Exercises 7–12, determine the number of triangles that can be drawn with the given data.

7. $a = 40, b = 70, A = 30°$
8. $a = 24, b = 32, A = 45°$
9. $b = 15, c = 19, B = 60°$
10. $b = 75, c = 85, B = 135°$
11. $a = 50, b = 70, B = 120°$
12. $c = 85, a = 45, C = 150°$

In Exercises 13–20, find the exact value requested in a triangle ABC

13. Given $a = 2$, $b = 3$, and $\sin A = \frac{2}{3}$, find B.
14. Given $b = \sqrt{6}$, $c = 2$, and $B = 60°$, find C.
15. Given $b = \sqrt{3}$, $c = \sqrt{2}$, and $B = \frac{1}{2}(A + C)$, find A.
16. Given $a = 4$, $b = \sqrt{8}$, and $A = \frac{1}{3}(B + C)$, find B.
17. Given $B = 45°$, $C = 105°$, and $a = 2$, find b.
18. Given $A = 45°$, $B = 60°$, and $a = 10$, find b.
19. Given $B = 45°$, $C = 75°$, and $b = \sqrt{6}$, find a.
20. Given $A = 105°$, $C = 45°$, and $b = \sqrt{12}$, find c.

In Exercises 21–28, solve each triangle. Round each answer to the nearest tenth.

21.

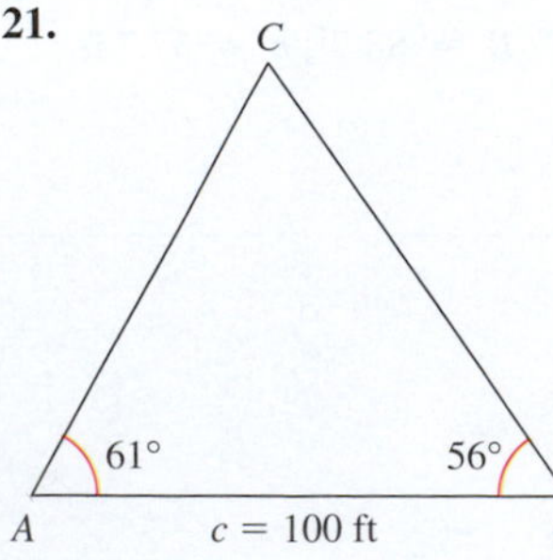

22.

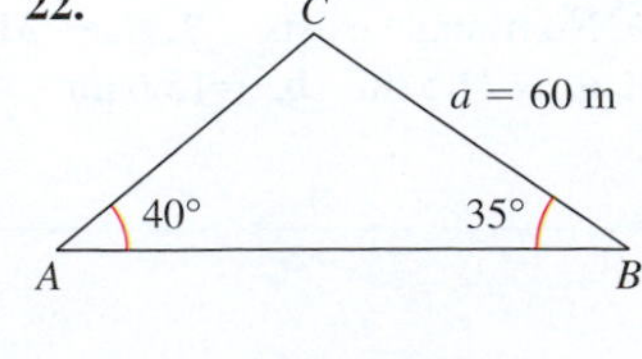

23.

C
43°
110°
A
c = 47 m
B

24.

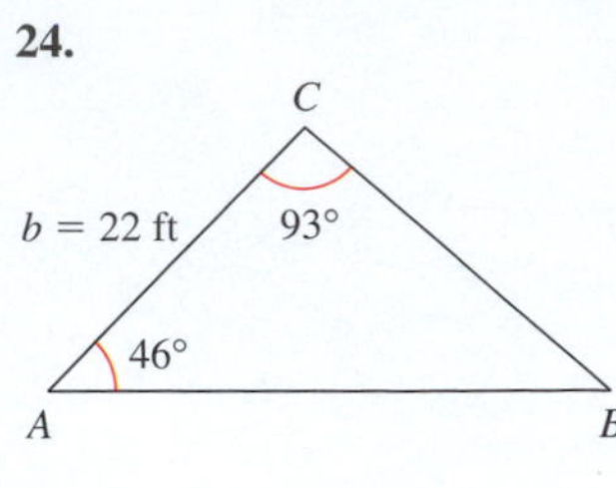

25.

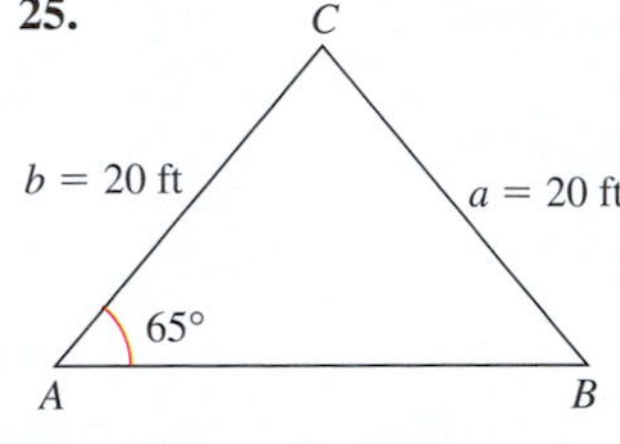

26.

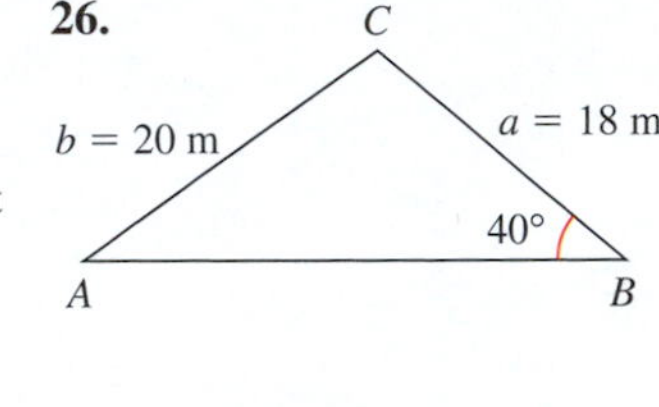

27.

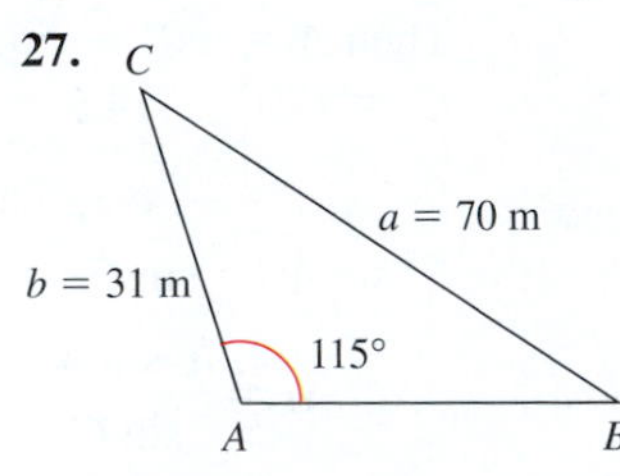

28.

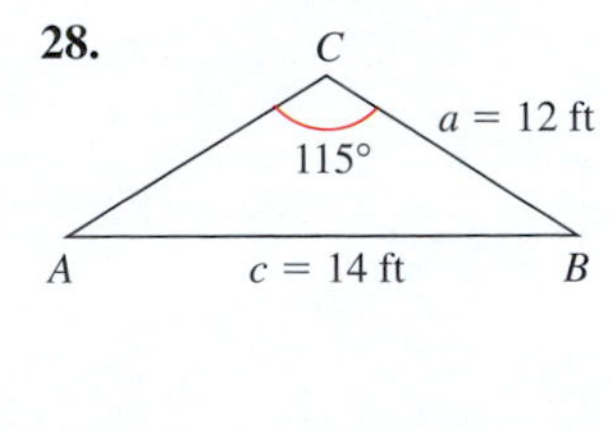

In Exercises 29–40, solve triangle ABC. Round each answer to the nearest tenth.

29. $A = 40°, B = 35°, a = 100$ meters
30. $A = 80°, B = 20°, a = 100$ meters
31. $A = 46°, C = 55°, a = 75$ centimeters
32. $A = 35°, C = 98°, a = 75$ centimeters
33. $A = 35°, C = 47°, c = 60$ feet
34. $A = 44°, C = 76°, c = 40$ feet
35. $B = 43°, C = 67°, b = 40$ inches
36. $B = 95°, C = 35°, b = 100$ inches
37. $B = 110°, C = 46°, c = 23.5$ feet
38. $B = 67°, C = 63°, c = 16.8$ feet
39. $A = 35.7°, B = 45.8°, c = 30$ meters
40. $A = 64.5°, B = 54.3°, c = 40$ meters

In Exercises 41–60, solve each SSA triangle. Indicate whether the given measurements result in no triangle, one triangle, or two triangles. Solve each resulting triangle. Round each answer to the nearest tenth.

41. $A = 40°, a = 23, b = 20$
42. $A = 36°, a = 30, b = 24$
43. $A = 30°, a = 25, b = 50$
44. $A = 60°, a = 20\sqrt{3}, b = 40$
45. $A = 40°, a = 10, b = 20$
46. $A = 62°, a = 30, b = 40$
47. $A = 95°, a = 18, b = 21$
48. $A = 110°, a = 37, b = 41$
49. $A = 100°, a = 40, b = 34$

50. $A = 105°, a = 70, b = 30$

51. $B = 50°, b = 22, c = 40$

52. $B = 64°, b = 45, c = 60$

53. $B = 46°, b = 35, c = 40$

54. $B = 32°, b = 50, c = 60$

55. $B = 97°, b = 27, c = 30$

56. $B = 110°, b = 19, c = 21$

57. $A = 42°, a = 55, c = 62$

58. $A = 34°, a = 6, c = 8$

59. $C = 40°, a = 3.3, c = 2.1$

60. $C = 62°, a = 50, c = 100$

Applying the Concepts

61. Finding distance. Angela wants to find the distance from point A to her friend Carmen's house at point C on the other side of the river. She knows the distance from A to Betty's house at B is 540 feet. See the figure. The measurement of angles A and B are 57° and 46°, respectively. Calculate the distance from A to C. Round to the nearest foot.

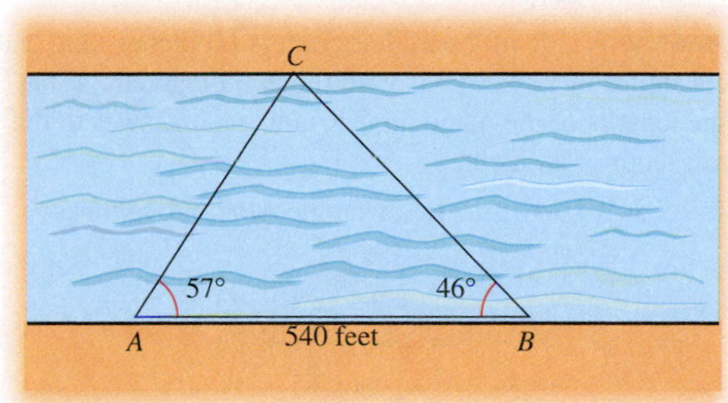

62. Finding distance. In Exercise 61, find the width of the river assuming that the houses are on the (very straight) banks of the river.

63. Target. A laser beam with an angle of elevation of 42° is reflected by a target and is received 1200 yards from the point of origin. Assume that the trajectory of the beam forms (approximately) an isosceles triangle.

a. Find the total distance the beam travels. Round to the nearest yard.

b. What is the height of the target? Round to the nearest yard.

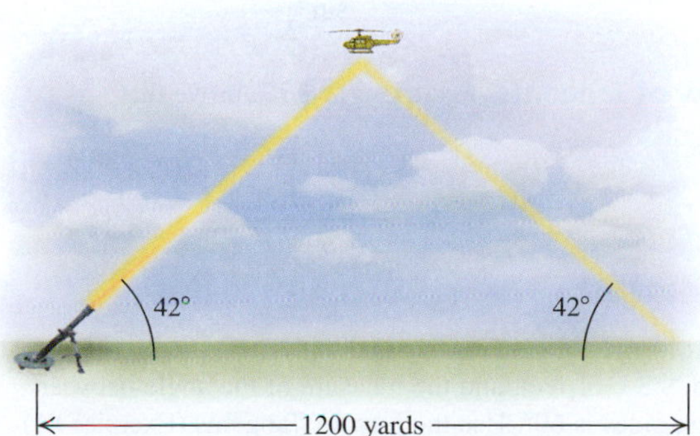

64. Height of a flagpole. Two surveyors stand 200 feet apart with a flagpole between them. Suppose the "transit" at each location is 5 feet high. The transit measures the angles of elevation of the top of the flagpole at the two locations to be 30° and 25°, respectively. Find the height of the flagpole. Round to the nearest foot.

65. Radio beacon. A ship sailing due east at the rate of 16 miles per hour records the bearing of a radio beacon at N 36.5° E. Two hours later the bearing of the same beacon is N 55.7° W. Round each answer to the nearest tenth of a mile.

a. How far is the ship from the beacon the second time?

b. How close to the beacon did the ship pass?

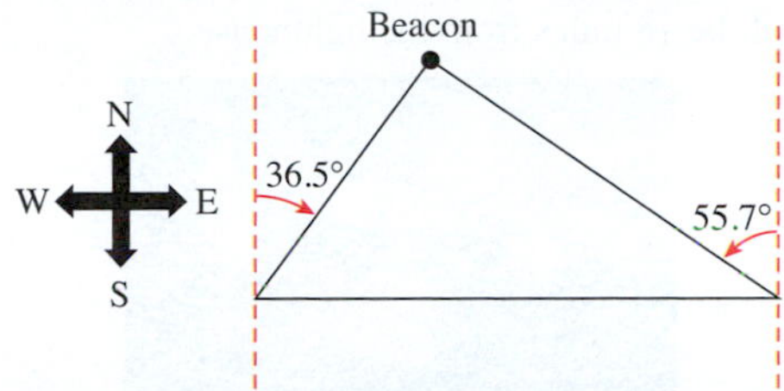

66. Lighthouse. A boat sailing due north at the rate of 14 miles per hour records the bearing of a lighthouse as N 8.4° E. Two hours later the bearing of the same lighthouse is N 30.9° E. Round the answers to the nearest tenth of a mile.

a. How far is the ship from the lighthouse the second time?

b. If the boat follows the same course and speed, how close to the lighthouse will it approach?

67. Geostationary satellite. A camel rider was traveling due west at night in the desert. At midnight, his angle of elevation to a satellite was 89°. After traveling for 20 miles, his angle of elevation to the satellite was 89.05°. The satellite is at a constant height above the ground. How high is the satellite? Round the answer to the nearest mile.

68. Height of a tower. A flagpole 20 feet tall is placed on top of a tower. At a point on the ground, a surveyor measures the angle of elevation of the bottom of the flagpole to be 69.6° and that of the top of the flagpole to be 70.9°. What is the height of the tower to the nearest foot?

69. Distance between planets. The angle subtended at Earth by lines joining Venus and the Sun is 31°. If Venus is 67 million miles from the Sun and Earth is 93 million miles from the Sun, what is the distance (to the nearest million) between Earth and Venus? [*Hint:* There are two possible answers.]

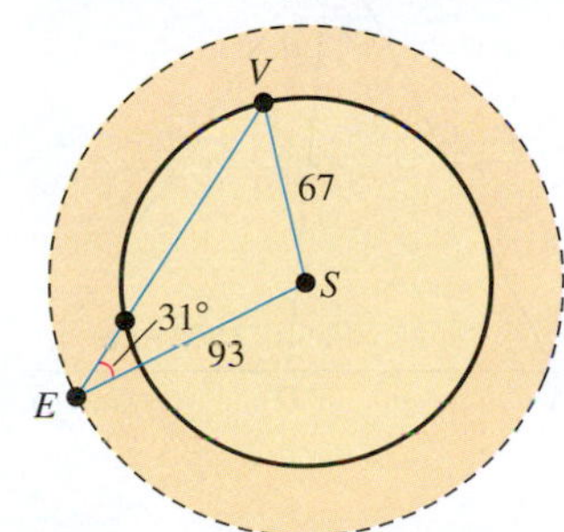

70. Distance between planets. The angle subtended at Mars by the lines joining Mars to Earth and the Sun is 23°. If Mars is 128 million miles from the Sun and Earth is 93 million miles from the Sun, what is the distance (to the nearest million) between Earth and Mars?

71. Navigation. A point on an island is located 24 miles southwest of a dock. A ship leaves the dock at 1 P.M. traveling west at 12 mph. At what time(s) to the nearest minute is the ship 20 miles from the point?

72. Navigation. A lighthouse is 23 miles N 55° E of a dock. A ship leaves the dock at 3 P.M. and sails due east at a speed of 15 mph. Find the time(s) to the nearest minute when the ship will be 18 miles from the lighthouse.

73. Geometry. The sides of a parallelogram are 15 m and 11 m, and the longer diagonal makes an angle of 18° with the longer side. Find the length of the longer diagonal.

74. Geometry. In the trapezoid shown, find the length x to the nearest tenth.

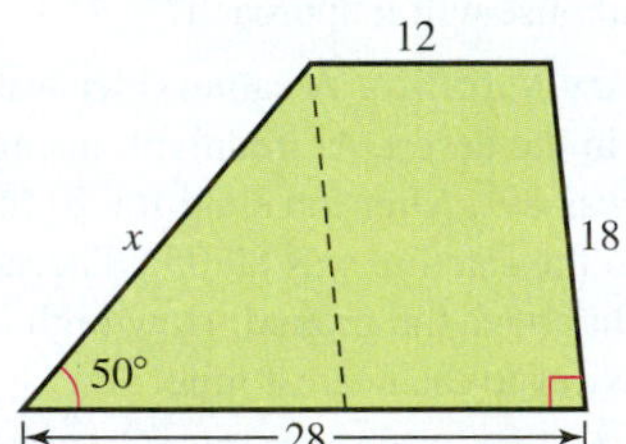

Beyond the Basics

75. Suppose CD bisects angle C of triangle ABC. Show that $\frac{AD}{DB} = \frac{AC}{CB}$. See the figure.
[*Hint:* Apply the Law of Sines to triangles ACD and DCB.]

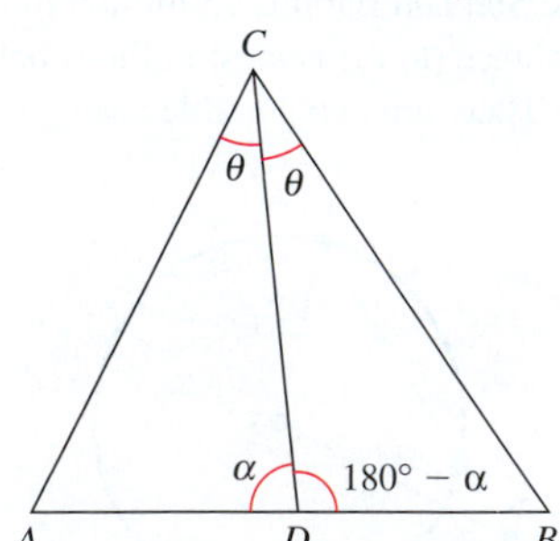

76. Distance. A flagpole 15 feet high stands on a building 75 feet high. To an observer at a height of 90 feet, the building and the flagpole intercept equal angles. Find the distance of the observer from the top of the flagpole. [*Hint:* Use Exercise 75.]

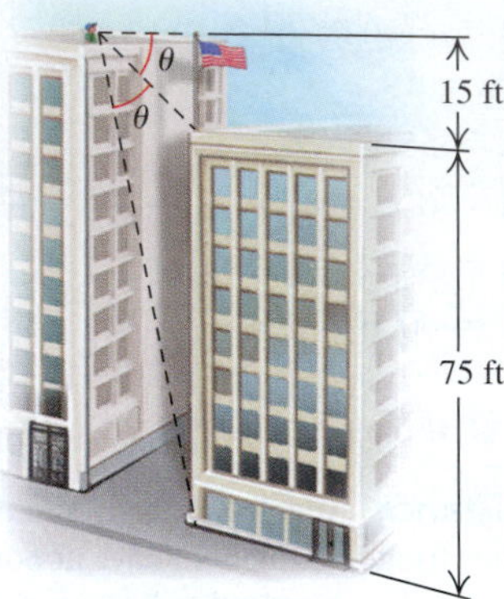

77. Geometry. In a triangle ABC, $a \cos A = b \cos B$. Show that the triangle is either isosceles or right angled.
[*Hint*: Show $\sin 2A = \sin 2B$. Then $2A = 2B$ or $2A = 180° - 2B$.]

78. Geometry. A side of a parallelogram is 60 inches long and makes angles of 28° and 42° with the two diagonals. What are the lengths of the diagonals? Round each answer to the nearest tenth of an inch.

79. Mollweide's formula. In a triangle ABC, prove that

$$\frac{b - c}{a} = \frac{\sin \frac{B - C}{2}}{\cos \frac{A}{2}}.$$

[*Hint:* Let $\frac{a}{\sin A} = \frac{b}{\sin B} = \frac{c}{\sin C} = k$. Write an expression for $\frac{b - c}{a}$ in terms of sines and then use Sum-to-Product and Half-Angle formulas. Note that because the formula contains all six parts of a triangle, it can be used as a check for the solutions of a triangle.]

80. Mollweide's formula. In a triangle ABC, prove that

$$\frac{b + c}{a} = \frac{\cos \frac{B - C}{2}}{\sin \frac{A}{2}}.$$

81. Law of Tangents. In a triangle ABC, prove that

$$\frac{b - c}{b + c} = \frac{\tan\left(\frac{B - C}{2}\right)}{\tan\left(\frac{B + C}{2}\right)}.$$

82. Angle measure. Two sides of a triangle are $\sqrt{3} + 1$ feet and $\sqrt{3} - 1$ feet, and the measure of the angle between these sides is 60°. Use the Law of Tangents (Exercise 81) to find the measure of the difference of the remaining angles.

Critical Thinking/Discussion/Writing

In Exercises 83 and 84, give the most specific description you can for a triangle satisfying the specified conditions. Explain your reasoning.

83. In a triangle ABC, $a \sin A = b \sin B$.

84. In a triangle ABC, $a \cos A = b \cos B$.

85. Use the Law of Sines to show that any isosceles triangle has two equal angles.

86. Use the Law of Sines to show that any triangle with two equal angles is isosceles.

Maintaining Skills

In Exercises 87–102, write the exact value of each expression without using a calculator.

87. $\cos 30°$

88. $\cos 120°$

89. $\cos 135°$

90. $\cos 150°$

91. $\cos^{-1}\left(\frac{1}{2}\right)$

92. $\cos^{-1}(2)$

93. $\cos^{-1}\left(-\frac{1}{2}\right)$

94. $\cos^{-1}\left(-\frac{\sqrt{3}}{2}\right)$

95. If $a = 2$, $b = 3$, and $\cos C = \frac{1}{6}$, find $\sqrt{a^2 + b^2 - 2ab \cos C}$.

96. If $b = 4$, $c = 5$, and $\cos A = \frac{1}{10}$, find $\sqrt{b^2 + c^2 - 2bc \cos A}$.

97. If $c = 3$, $a = 1$, and $B = 60°$, find $\sqrt{c^2 + a^2 - 2ca \cos B}$.

98. If $a = \sqrt{8}$, $b = 4$, and $C = 45°$, find $\sqrt{a^2 + b^2 - 2ab \cos C}$.

99. If $a = 2$, $b = 3$, and $c = 4$, find $\frac{a^2 + b^2 - c^2}{2ab}$.

100. If $a = 5$, $b = 4$, and $c = 3$, find $\frac{b^2 + c^2 - a^2}{2bc}$.

101. If $a = 1$, $b = 1$, and $c = \sqrt{3}$, find $\cos^{-1}\left(\frac{b^2 + c^2 - a^2}{2bc}\right)$.

102. If $a = 3$, $b = 5$, and $c = 7$, find $\cos^{-1}\left(\frac{a^2 + b^2 - c^2}{2ab}\right)$.

SECTION 7.2

The Law of Cosines

BEFORE STARTING THIS SECTION, REVIEW

1. Distance formula (Section 2.1, page 159)
2. Inverse trigonometric functions (Section 5.6, page 551)

OBJECTIVES

1. Derive the Law of Cosines.
2. Use the Law of Cosines to solve SAS triangles.
3. Use the Law of Cosines to solve SSS triangles.

◆ Generalizing the Pythagorean Theorem

Suppose a Boeing 747 jumbo jet is flying over Disney World in Orlando, Florida, at 552 miles per hour and is heading due south to Brazil. Twenty minutes later an F-16 fighter jet heading due east passes over Disney World at a speed of 1250 miles per hour. By using the Pythagorean theorem, we can easily calculate the distance *d* between the two planes *t* hours after the F-16 passes over Disney World. We have

$$d = \sqrt{(1250t)^2 + \left[\frac{1}{3}(552) + 552t\right]^2} \qquad \text{20 minutes} = \frac{1}{3}\text{ hour}$$

Suppose all the other facts in the problem are unchanged except that the F-16 has a bearing of N 37° E. Now we can no longer apply the Pythagorean theorem. In Example 2, we use the Law of Cosines, described in this section, to solve this problem.

1 Derive the Law of Cosines.

The Law of Cosines

The Pythagorean theorem states that the relationship $c^2 = a^2 + b^2$ holds in a right triangle *ABC*, where *c* represents the length of the hypotenuse. This relationship is not true if the triangle is not a right triangle. The **Law of Cosines** is a generalization of the Pythagorean theorem that is true in any triangle. This law will be used to solve triangles in which two sides and the included angle are known (SAS triangles), as well as those triangles in which all three sides are known (SSS triangles).

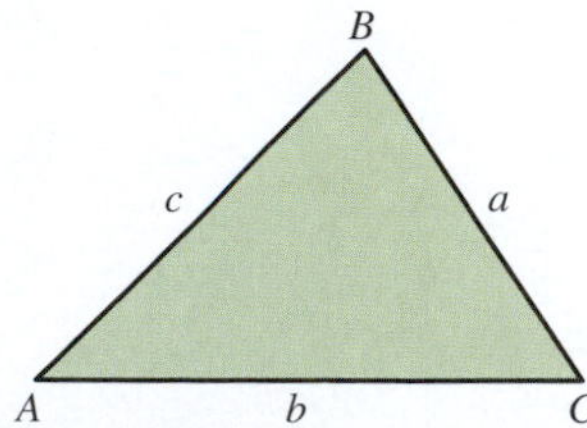

FIGURE 7.11

THE LAW OF COSINES

In triangle *ABC* with sides of lengths *a*, *b*, and *c* (as in Figure 7.11),

$$a^2 = b^2 + c^2 - 2bc \cos A,$$
$$b^2 = c^2 + a^2 - 2ca \cos B,$$
$$c^2 = a^2 + b^2 - 2ab \cos C.$$

In words, the square of any side of a triangle is equal to the sum of the squares of the length of the other two sides, less twice the product of the lengths of the other sides and the cosine of their included angle.

Derivation of the Law of Cosines

To derive the Law of Cosines, place triangle ABC in a rectangular coordinate system with the vertex A at the origin and the side c along the positive x-axis. See Figure 7.12.

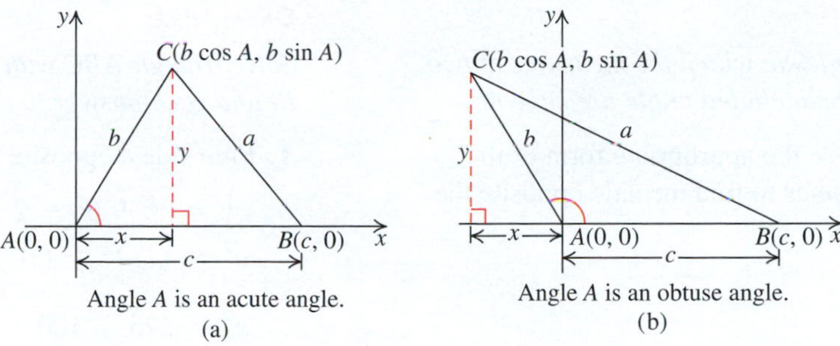

FIGURE 7.12 Two cases for an angle A

Label the coordinates of the vertices as shown in Figure 7.12. In Figures 7.12(a) and 7.12(b), the point $C(x, y)$ on the terminal side of angle A has coordinates $(b\cos A, b\sin A)$, which are found using the cosine and sine definitions.

$$\cos A = \frac{x}{b} \qquad \sin A = \frac{y}{b}$$

$$x = b\cos A \qquad y = b\sin A$$

The point B has coordinates $(c, 0)$. Applying the distance formula to the line segment joining the points $B(c, 0)$ and $C(b\cos A, b\sin A)$, we have

$$\begin{aligned}
a &= d(C, B) \\
a^2 &= [d(C, B)]^2 && \text{Square both sides.} \\
a^2 &= (b\cos A - c)^2 + (b\sin A - 0)^2 && \text{Distance formula} \\
a^2 &= b^2\cos^2 A - 2bc\cos A + c^2 + b^2\sin^2 A && \text{Expand binomial.} \\
a^2 &= b^2(\sin^2 A + \cos^2 A) + c^2 - 2bc\cos A && \text{Regroup terms.} \\
a^2 &= b^2 + c^2 - 2bc\cos A && \sin^2 A + \cos^2 A = 1
\end{aligned}$$

The last equation is one of the forms of the Law of Cosines. Similarly, by placing the vertex B and then the vertex C at the origin, we obtain the other two forms. Notice that the Law of Cosines becomes the Pythagorean theorem if the included angle is 90° because $\cos 90° = 0$.

2 Use the Law of Cosines to solve SAS triangles.

Solving SAS Triangles

We solve SAS triangles (Case 3 of Section 7.1) using the Law of Cosines.

PROCEDURE IN ACTION

EXAMPLE 1 Solving the SAS Triangles

OBJECTIVE

Solve triangles in which the measures of two sides and the included angle are known.

EXAMPLE

Solve triangle ABC with $a = 15$ inches, $b = 10$ inches, and $C = 60°$. Round each answer to the nearest tenth.

Step 1 Use the appropriate form of the Law of Cosines to find the side opposite the given angle.

1. Find side c opposite angle C.

$$c^2 = a^2 + b^2 - 2ab \cos C \qquad \text{The Law of Cosines}$$
$$c^2 = (15)^2 + (10)^2 - 2(15)(10)\cos 60° \qquad \text{Substitute values.}$$
$$c^2 = 225 + 100 - 2(15)(10)\left(\frac{1}{2}\right) \qquad \cos 60° = \frac{1}{2}$$
$$c^2 = 175 \qquad \text{Simplify.}$$
$$c = \sqrt{175} \approx 13.2 \qquad \text{Use a calculator.}$$

Step 2 Use the **Law of Sines** to find the angle opposite the shorter of the two given sides. Note that this angle is always an acute angle.

Alternatively, you can use Step 1 and the Law of Cosines again to find another angle. Here

$$\cos B = \frac{c^2 + a^2 - b^2}{2ca} = \frac{175 + (15)^2 - (10)^2}{2(\sqrt{175})(15)}$$
$$\approx 0.75593$$

So, $B = \cos^{-1}(0.75593) \approx 40.9°$.

2. Find angle B.

$$\frac{\sin B}{b} = \frac{\sin C}{c} \qquad \text{The Law of Sines}$$
$$\sin B = \frac{b \sin C}{c} \qquad \text{Multiply both sides by } b.$$
$$\sin B = \frac{10 \sin 60°}{\sqrt{175}} \qquad \text{Substitute values.}$$
$$B = \sin^{-1}\left(\frac{10 \sin 60°}{\sqrt{175}}\right) \approx 40.9° \qquad 0° < B \le 90°$$

Step 3 Use the angle sum formula to find the third angle.

3. $A \approx 180° - 60° - 40.9° \approx 79.1° \qquad A + B + C = 180°$

Step 4 Write the solution.

4.

$A \approx 79.1°$	$a = 15$ inches
$B \approx 40.9°$	$b = 10$ inches
$C = 60°$	$c \approx 13.2$ inches

Practice Problem 1 Solve triangle ABC with $c = 25$ inches, $a = 15$ inches, and $B = 60°$. Round each answer to the nearest tenth.

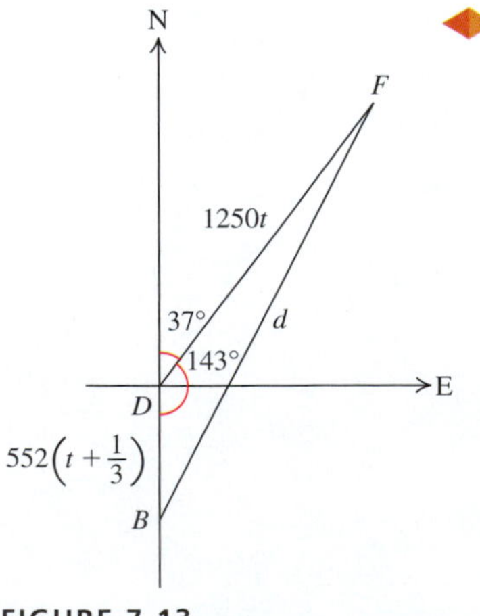

FIGURE 7.13

EXAMPLE 2 Using the Law of Cosines

A Boeing 747 is flying over Disney World headed due south at 552 miles per hour. Twenty minutes later an F-16 passes over Disney World with a bearing of N 37° E at a speed of 1250 miles per hour. Find the distance between the two planes three hours after the F-16 passes over Disney World. Round the answer to the nearest tenth.

Solution

Suppose the F-16 has been traveling for t hours after passing over Disney World. Then because the Boeing 747 had a head start of 20 minutes $= \frac{1}{3}$ hour, the Boeing 747 has been traveling $\left(t + \frac{1}{3}\right)$ hours due south. The distance d between the two planes is shown in Figure 7.13. Using the Law of Cosines in triangle FDB, we have

$$d^2 = (1250t)^2 + \left[552\left(t + \frac{1}{3}\right)\right]^2 - 2(1250t)\cdot 552\left(t + \frac{1}{3}\right)\cos 143°$$

$$d^2 \approx 28{,}469{,}270.04 \quad \text{Substitute } t = 3\text{; use a calculator.}$$

$$d \approx 5335.7 \text{ miles} \quad \text{Use a calculator.}$$

Practice Problem 2 Repeat Example 2 assuming that the F-16 is traveling at a speed of 1375 miles per hour with bearing of N 75° E and the Boeing 747 is traveling with a bearing of S 12° W at 550 miles per hour.

3 Use the Law of Cosines to solve SSS triangles.

Solving SSS Triangles

We solve SSS triangles (Case 4 of Section 7.1).

PROCEDURE IN ACTION

EXAMPLE 3 Solving SSS Triangles

OBJECTIVE

Solve triangles in which the measures of the three sides are known.

EXAMPLE

Solve triangle ABC with $a = 3.1$ feet, $b = 5.4$ feet, and $c = 7.2$ feet. Round answers to the nearest tenth.

Step 1 Use the Law of Cosines to find the angle opposite the longest side.

1. Because c is the longest side, we first find angle C.

$$c^2 = a^2 + b^2 - 2ab\cos C \quad \text{The Law of Cosines}$$

$$2ab\cos C = a^2 + b^2 - c^2 \quad \text{Add } 2ab\cos C - c^2 \text{ to both sides.}$$

$$\cos C = \frac{a^2 + b^2 - c^2}{2ab} \quad \text{Solve for } \cos C.$$

$$\cos C = \frac{(3.1)^2 + (5.4)^2 - (7.2)^2}{2(3.1)(5.4)} \quad \text{Substitute values.}$$

$$\cos C \approx -0.39 \quad \text{Use a calculator.}$$

$$C \approx \cos^{-1}(-0.39) \approx 113.0° \quad 0° < C < 180°$$

Step 2 Use the Law of Sines to find either of the two remaining acute angles.

Alternatively, you can use the Law of Cosines again to find another angle.

Here

$$\cos B = \frac{c^2 + a^2 - b^2}{2ca} = \frac{(7.2)^2 + (3.1)^2 - (5.4)^2}{2(7.2)(3.1)} \approx 0.7233.$$

So $B \approx \cos^{-1}(0.7233) \approx 43.7°$.

2. Find angle B.

$$\frac{\sin B}{b} = \frac{\sin C}{c} \quad \text{The Law of Sines}$$

$$\sin B = \frac{b\sin C}{c} \quad \text{Multiply both sides by } b.$$

$$B = \sin^{-1}\left(\frac{b\sin C}{c}\right) \quad 0° < B < 90°$$

$$B = \sin^{-1}\left(\frac{5.4\sin 113.0°}{7.2}\right) \quad \text{Substitute values.}$$

$$B \approx 43.7° \quad \text{Use a calculator.}$$

Step 3 Use the angle sum formula to find the third angle.

3. $A \approx 180° - 43.7° - 113.0°$ $\quad A + B + C = 180°$

$A \approx 23.3°$ $\quad$ Simplify.

Step 4 Write the solution.

4.

$A \approx 23.3°$	$a = 3.1$ feet
$B = 43.7°$	$b = 5.4$ feet
$C \approx 113.0°$	$c = 7.2$ feet

Practice Problem 3 Solve triangle ABC with $a = 4.5$, $b = 6.7$, and $c = 5.3$. Round each answer to the nearest tenth.

EXAMPLE 4 **Solving an SSS Triangle**

Solve triangle ABC with $a = 2$ meters, $b = 9$ meters, and $c = 5$ meters. Round each answer to the nearest tenth.

Solution

Step 1 We first find B, the angle opposite the longest side.

$$b^2 = c^2 + a^2 - 2ca\cos B \quad \text{The Law of Cosines}$$

$$\cos B = \frac{c^2 + a^2 - b^2}{2ca} \quad \text{Solve for } \cos B.$$

$$\cos B = \frac{5^2 + 2^2 - 9^2}{2(5)(2)} \quad \text{Substitute values.}$$

$$\cos B = -2.6 \quad \text{Simplify.}$$

RECALL

The triangle inequality states that in any triangle, the sum of the lengths of any two sides of a triangle is greater than the length of the third side.

Because the range of the cosine function is $[-1, 1]$, there is no angle B for which $\cos B = -2.6$. This means that a triangle with the given information cannot exist. Because $2 + 5 < 9$, you can also use the triangle inequality from geometry to see that there is no such triangle.

Practice Problem 4 Solve triangle ABC with $a = 2$ inches, $b = 3$ inches, and $c = 6$ inches.

Answers to Practice Problems

1. $b \approx 21.8, A \approx 36.6°, C \approx 83.4°$ **2.** 5219.5 miles **3.** $A \approx 41.9°, B \approx 85.9°, C \approx 52.1°$ **4.** No triangle exists.

SECTION 7.2 Exercises

Basic Concepts and Skills

1. One form of the Law of Cosines is $c^2 = a^2 + b^2 - 2ab\cos($____________$)$.

2. If we take the angle in the Law of Cosines to be 90°, then we get the ____________ theorem.

3. Triangles with SAS given are solved by the Law of Cosines, as are triangles with ____________ sides given.

4. When one angle is found by the Law of Cosines, the other can be found with the Law of ____________.

5. **True or False.** The Law of Cosines is used to solve triangles when two angles and a side are given.

6. **True or False.** For a triangle with SSS given, we can find angle A by the formula

$$A = \cos^{-1}\left(\frac{b^2 + c^2 - a^2}{2bc}\right).$$

In Exercises 7–14, find the exact value requested in each triangle *ABC*.

7. If $b = 4$, $c = 6$, and $\cos A = \frac{1}{16}$, find a.

8. If $a = 13$, $b = 4$, and $\cos C = -\frac{5}{13}$, find c.

9. If $a = 5$, $b = 3$, and $C = 60°$, find c.

10. If $a = 6$, $c = 8$, and $B = 120°$, find b.

11. If $a = 7$, $b = 6$, and $c = 5$, find $\cos A$.

12. If $a = 7$, $b = 6$, and $c = 5$, find $\cos C$.

13. If $a = 4$, $b = 5$, and $c = \sqrt{21}$, find C.

14. If $a = 7$, $b = \sqrt{109}$, and $c = 5$, find B.

In Exercises 15–18, solve each triangle. Round each answer to the nearest tenth.

15.

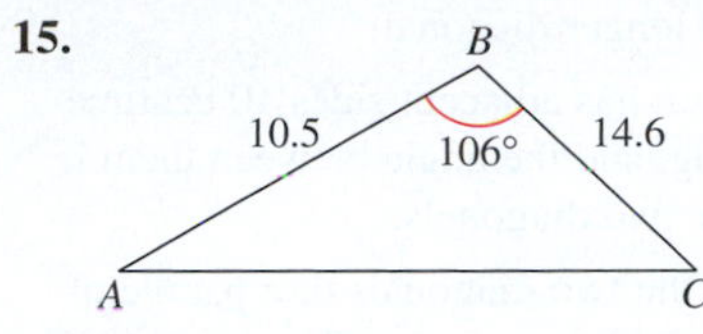

16.

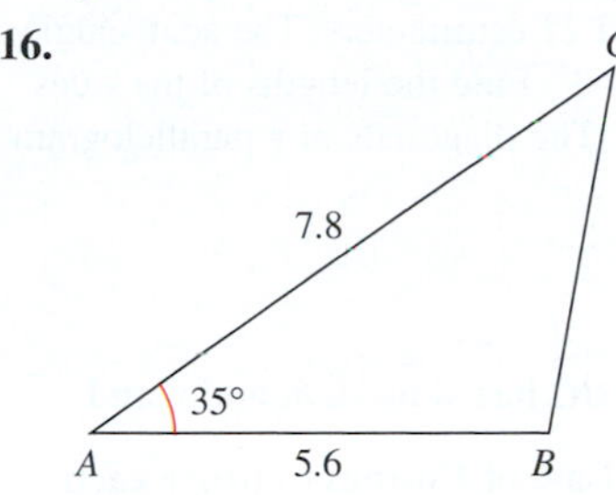

17.

C
30
18
A
15
B

18.

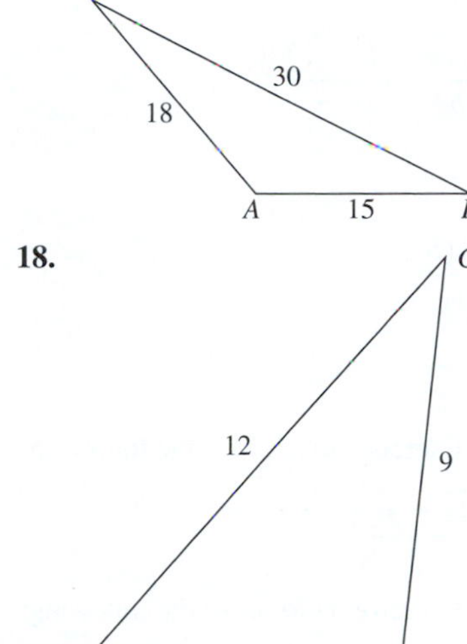

In Exercises 19–34, solve each triangle *ABC*. Round each answer to the nearest tenth. All sides are measured in feet.

19. $a = 15, b = 9, C = 120°$

20. $a = 14, b = 10, C = 75°$

21. $b = 10, c = 12, A = 62°$

22. $b = 11, c = 16, A = 110°$

23. $c = 12, a = 15, b = 11$

24. $c = 16, a = 11, b = 13$

25. $a = 9, b = 13, c = 18$

26. $a = 14, b = 6, c = 10$

27. $a = 2.5, b = 3.7, c = 5.4$

28. $a = 4.2, b = 2.9, c = 3.6$

29. $b = 3.2, c = 4.3, A = 97.7°$

30. $b = 5.4, c = 3.6, A = 79.2°$

31. $c = 4.9, a = 3.9, B = 68.3°$

32. $c = 7.8, a = 9.8, B = 95.6°$

33. $a = 2.3, b = 2.8, c = 3.7$

34. $a = 5.3, b = 2.9, c = 4.6$

Applying the Concepts

In Exercises 35–48, round each answer to the nearest tenth.

35. Chord length. Find the length of the chord intercepted by a central angle of 42° in a circle of radius 8 feet.

36. Central angle. Find the measure of the central angle of a circle of radius 6 feet that intercepted a chord of length 3.5 feet.

37. Tunnel length. Engineers must bore a straight tunnel through the base of a mountain. They select a reference point *C* on the plain surrounding the mountain. The distance from *C* to portal *A* (the starting tunnel point) and portal *B* (the ending tunnel point) is 2352 yards and 1763 yards, respectively. The measure of $\angle ACB$ is 41°. How long is the tunnel?

38. Pond length. A surveyor needs to find the length of a pond (see figure), but does so indirectly. She selects a point *A* on one side of the pond and measures distances to the points *B* and *C* at opposite ends of the pond to be 537 yards and 823 yards, respectively. The measure of $\angle BAC$ is 130°. See figure. How long is the pond?

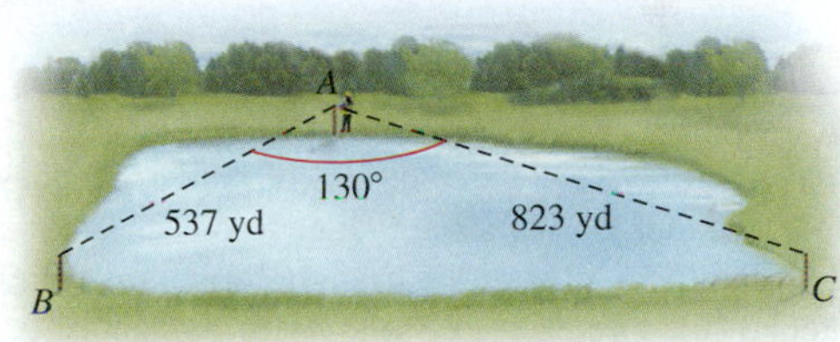

39. Roof truss. A roof truss is made in the shape of an inverted V. The lengths of the two edges are 11 feet and 23 feet. The edges meet at the peak, making a 65° angle. Find

a. The width of the truss.

b. The height of the peak.

40. Solar panels. A roof of a house addition is being built to accept solar energy panels as shown in the figure. Find

a. The length of the edge *AC* of the truss.

b. The measure of $\angle BAC$.

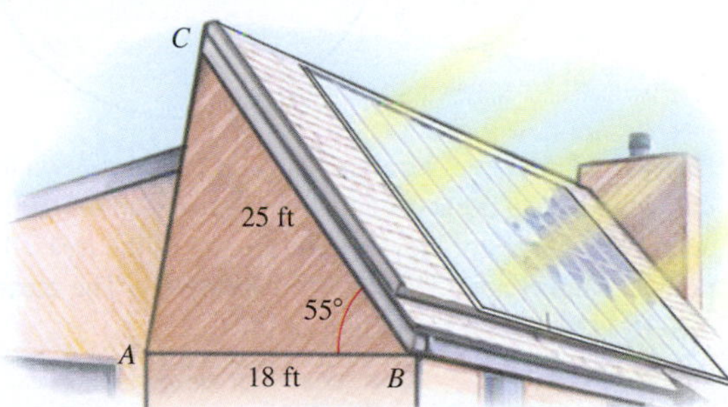

41. Hikers. Two hikers, Sonia and Tony, leave the same point at the same time. Sonia walks due east at the rate of 3 miles per hour, and Tony walks 45° northeast at the rate of 4.3 miles per hour. How far apart are the hikers after three hours?

42. Distance between ships. Two ships leave the same port—Ship A at 1.00 P.M. and ship B at 3:30 P.M. Ship A sails on a bearing of S 37° E at 18 miles per hour, and B sails on a bearing of N 28° E at 20 miles per hour. How far apart are the ships at 8.00 P.M.?

43. Navigation. A ship is traveling due north. At two different points A and B, the navigator of the ship sites a lighthouse at the point C, as shown in the accompanying figure.

a. Determine the distance from B to C.

b. How much farther due north must the ship travel to reach the point closest to the lighthouse?

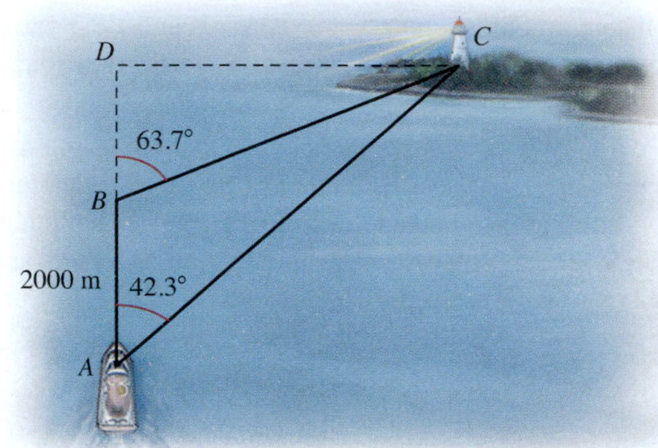

44. Tangential circles. Three circles of radii 1.2 inches, 2.2 inches, and 3.1 inches are tangent to each other externally. Find the angles of the triangle formed by joining the centers of the circles.

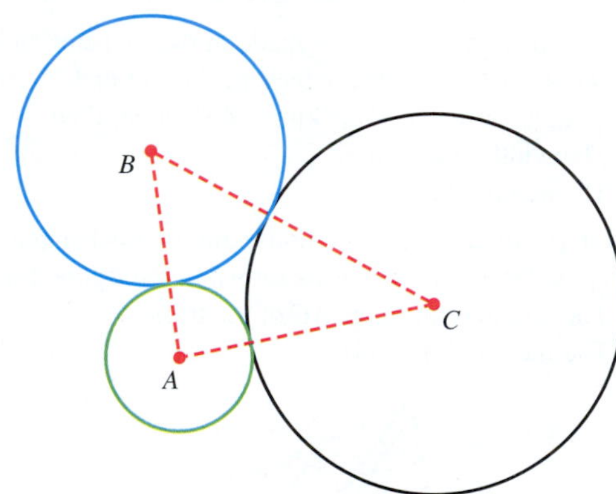

45. Height of a tree. A tree is planted at a point O on horizontal ground. Two points A and B on the ground are 100 feet apart. The angles of elevation of the top of the tree T from the points A and B are 45° and 30°, respectively. The measure of $\angle AOB$ is 60°. Find the height of the tree.

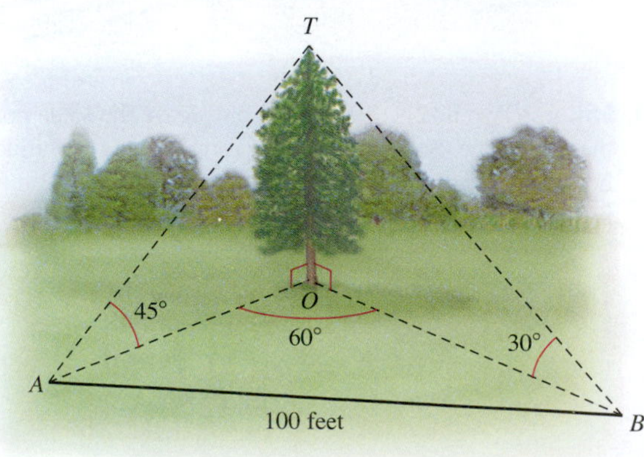

46. Geometry. A parallelogram has adjacent sides 8 centimeters and 13 centimeters. If the shorter diagonal is 11 centimeters long, find the length of the longer diagonal.

47. Geometry. A parallelogram has adjacent sides 10 centimeters and 15 centimeters long, and the angle between them is 40°. Find the lengths of the two diagonals.

48. Geometry. The length of the two diagonals of a parallelogram are 16 centimeters and 22 centimeters. The acute angle between these diagonals is 63°. Find the lengths of the sides of the parallelogram. [*Hint:* The diagonals of a parallelogram bisect each other.]

Beyond the Basics

In Exercises 49–55, triangle ABC has sides a, b, and c and $s = \frac{1}{2}(a + b + c)$. Use the Law of Cosines to prove each identity.

49. $1 - \cos A = \dfrac{(a - b + c)(a + b - c)}{2bc}$

50. $1 - \cos A = \dfrac{2(s - b)(s - c)}{bc}$

51. $1 + \cos A = \dfrac{(b + c + a)(b + c - a)}{2bc}$

52. $1 + \cos A = \dfrac{2s(s - a)}{bc}$

53. Use half-angle formula and Exercise 50 to prove the following:

$$\sin \frac{A}{2} = \sqrt{\frac{(s - b)(s - c)}{bc}}$$

54. Use half-angle formula and Exercise 52 to prove the following:

$$\cos \frac{A}{2} = \sqrt{\frac{s(s - a)}{bc}}$$

55. Use Exercises 53 and 54 to prove that

$$\sin A = \frac{2}{bc}\sqrt{s(s - a)(s - b)(s - c)}.$$

56. Use the Law of Cosines to show that for a triangle with sides of lengths a, b, and c (with $a \le b$),

$$b - a < c < b + a.$$

Critical Thinking/Discussion/Writing

57. Solve triangle ABC with vertices $A(-2, 1)$, $B(5, 3)$, and $C(3, 6)$.

58. Solve triangle ABC with vertices $A(-3, -5)$, $B(6, 10)$, and $C(3, -2)$.

You can use the Law of Cosines to solve triangle ABC in the ambiguous case of Section 7.1. For example, to solve triangle ABC with $B = 150°, b = 10$, and $c = 6$, use the Law of Cosines to write $b^2 = a^2 + c^2 - 2ac \cos B$.

Substitute values of b, c, and B in this equation to obtain a quadratic equation in a. Find the roots of this equation and interpret your results. Again use the Law of Cosines to find A. Then find $C = 180° - A - B$.

In Exercises 59–61, use the technique described above to solve each triangle ABC.

59. $B = 150°, b = 10$, and $c = 6$

60. $A = 30°, a = 6$, and $b = 10$

61. $A = 60°, a = 12$, and $c = 15$

62. Find values of b, c, and A in the form of the Law of Cosines $a^2 = b^2 + c^2 - 2bc \cos A$ so that $a^2 = b^2 + c^2 - 1$.

63. Explain why there are an unlimited number of choices for b and c in Exercise 62.

64. In the equation $a^2 = b^2 + c^2 - 2bc \cos A$, a form of the Law of Cosines, discuss the conditions under which a^2 is less than $b^2 + c^2$.

Maintaining Skills

In Exercises 65–70, use $a = 6, b = 4, c = 3, \alpha = 30°, \beta = 45°$, and $\theta = 60°$ to find the exact value of each expression without using a calculator.

65. $b \sin \alpha$

66. $c \sin \beta$

67. $\frac{1}{2}ab \sin \theta$

68. $\frac{1}{2}bc \sin \alpha$

69. $\frac{1}{2}ca \sin \beta$

70. $\frac{1}{2}ab \sin \alpha$

In Exercises 71–78, for the given values of a, b, and c, find the exact value of

(a) $s = \dfrac{a + b + c}{2}$.

(b) $K = \sqrt{s(s - a)(s - b)(s - c)}$.

71. $a = 3, \quad b = 4, \quad c = 5$

72. $a = 5, \quad b = 12, \quad c = 13$

73. $a = 6, \quad b = 6, \quad c = 6$

74. $a = 15, \quad b = 11, \quad c = 6$

75. $a = 18, \quad b = 10, \quad c = 14$

76. $a = 12, \quad b = 17, \quad c = 25$

77. $a = 10, \quad b = 13, \quad c = 13$

78. $a = 26, \quad b = 28, \quad c = 30$

SECTION 7.3

Areas of Polygons Using Trigonometry

BEFORE STARTING THIS SECTION, REVIEW

1. Congruent triangles (Section 7.1, page 640)
2. Solving SSA triangles (Section 7.1, page 644)

OBJECTIVES

1. Use geometry formulas.
2. Find the area of SAS triangles.
3. Find the area of AAS and ASA triangles.
4. Find the area of SSS triangles.
5. Learn vocabulary associated with polygons.

◆ The Bermuda Triangle

The Bermuda Triangle, or Devil's Triangle, is the expanse of the Atlantic Ocean between Florida, Bermuda, and Puerto Rico. This mysterious stretch of sea has an unusually high occurrence of disappearing ships and planes.

For example, on Halloween in 1991, pilot John Verdi and his copilot were flying a Grumman Cougar jet over the triangle. They radioed the nearest tower to get permission to increase their altitude. The tower agreed and watched as the jet began the ascent and then disappeared off the radar. The jet didn't fly out of range of the radar, didn't descend, and didn't radio a mayday (distress call). It just vanished, and the plane and crew were never recovered. Popular culture has attributed such disappearances to extraterrestrial forces. However, documented evidence indicates that the nature of disappearances in the region is similar to that in any region of the ocean. In Exercise 43, you are asked to find the area of the Bermuda Triangle.

1 Use geometry formulas.

Geometry Formulas

Although the area of a rectangle is just the product of its length and width, the general concept of area (and the related concepts of perimeter and volume) is developed with the use of calculus. Fortunately, the formulas for calculating these quantities (common plane regions and solids in space) require only algebra. Figure 7.14 lists some useful formulas from geometry.

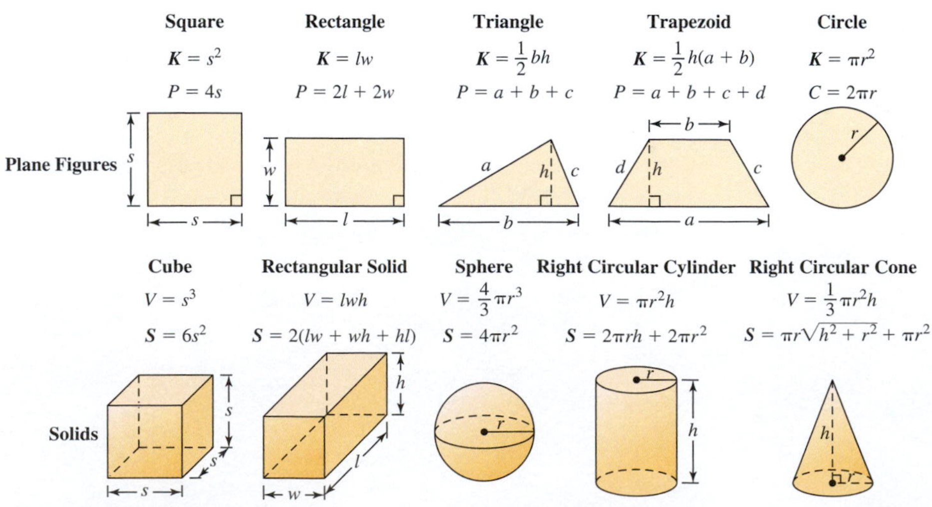

FIGURE 7.14 Geometry formulas

EXAMPLE 1 Finding Perimeter and Area

Figure 7.15 shows a rectangle with a semicircle on the top. Find

a. Its perimeter. **b.** The shaded area enclosed by the figure.

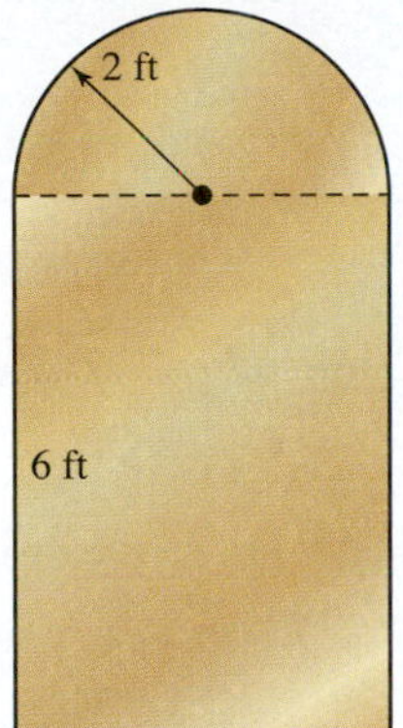

FIGURE 7.15

Solution

a. The figure consists of three sides of a rectangle and a semicircle. Because the radius shown in the figure is 2 ft, its diameter is $2(2) = 4$ ft. So the sum of the lengths of the three sides of the rectangle is $6 + 4 + 6 = 16$ ft. The circumference (perimeter) of the semicircle is $\frac{1}{2}(\text{circumference}) = \frac{1}{2}(2\pi r) = \frac{1}{2}(2\pi \cdot 2) = 2\pi$ ft.

The perimeter of the figure $= 16 \text{ ft} + 2\pi \text{ ft}$

$= (16 + 2\pi) \text{ ft}$ Exact answer

$\approx 22.2832 \text{ ft}$ Use a calculator.

b. Area enclosed by the figure

$= \text{Area of the rectangle} + \text{area of the semicircle}$

$= l \cdot w + \frac{1}{2}(\pi r^2)$

$= (4)(6) + \frac{1}{2}[\pi(2)^2]$ $l = 4, w = 6, r = 2$

$= (24 + 2\pi) \text{ ft}^2$ Simplify. Exact answer

$\approx 30.2832 \text{ ft}^2$ Use a calculator.

Practice Problem 1 One side of a rectangular plot is 48 feet, and its diagonal is 50 feet. Find

a. Its perimeter. **b.** Its area.

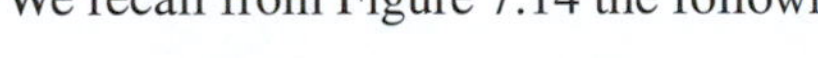

We recall from Figure 7.14 the following formula.

Area of a Triangle

The area K of a triangle is

$$K = \frac{1}{2}(\text{base})(\text{height})$$

or

$$K = \frac{1}{2}bh,$$

where b is the base and h is the height (the length of the altitude to the base).

If a base, b, and the height, h, of a triangle are given, then we use this formula to calculate the area of the triangle. In this section, we discover alternative formulas that can be used if other sets of data are given instead.

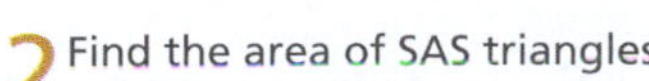

2 Find the area of SAS triangles.

Area of SAS Triangles

We first derive a formula for the area of a triangle in which two sides and the included angle are known.

Let ABC be a triangle with known sides b and c and included angle $A = \theta$. Let h be the altitude from the vertex B to the side b.

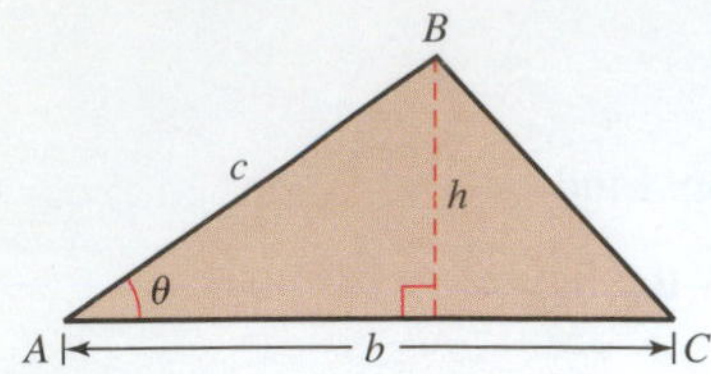

FIGURE 7.16 Acute angle A

If $A = \theta$ is an acute angle in a triangle (see Figure 7.16), then $\sin A = \frac{h}{c}$; so $h = c \sin A$. The area of this triangle is given by

$$K = \frac{1}{2}bh = \frac{1}{2}bc \sin A \qquad \text{Replace } h \text{ with } c \sin A.$$

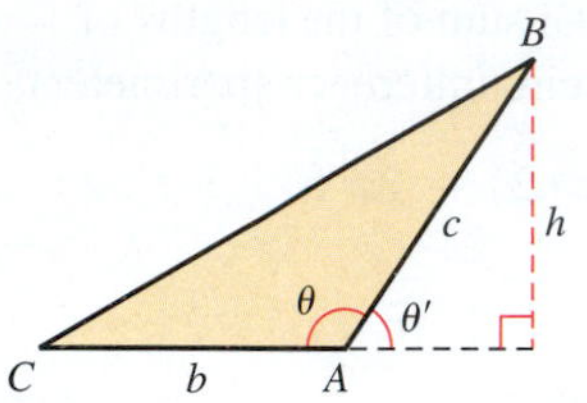

FIGURE 7.17 Obtuse angle A

The angle $A = \theta$ shown in Figure 7.17 is an obtuse angle. Here $\theta' = 180° - \theta$ and $\sin \theta = \sin \theta' = \frac{h}{c}$.

So

$$h = c \sin \theta' = c \sin \theta = c \sin A$$

$$K = \frac{1}{2}bh = \frac{1}{2}bc \sin A \qquad \text{Replace } h \text{ with } c \sin A.$$

In both cases (A is acute or obtuse), the area K of the triangle is $\frac{1}{2}bc \sin A$. By dropping altitudes from A and C, we have the following result.

Area of an SAS Triangle

The area K of a triangle ABC with sides a, b, and c is

$$K = \frac{1}{2}bc \sin A \qquad K = \frac{1}{2}ca \sin B \qquad K = \frac{1}{2}ab \sin C.$$

In words: The area K of a triangle is one-half the product of two of its sides and the sine of the included angle.

EXAMPLE 2 Finding the Area of a Triangle

Find the area of the triangle ABC in Figure 7.18.

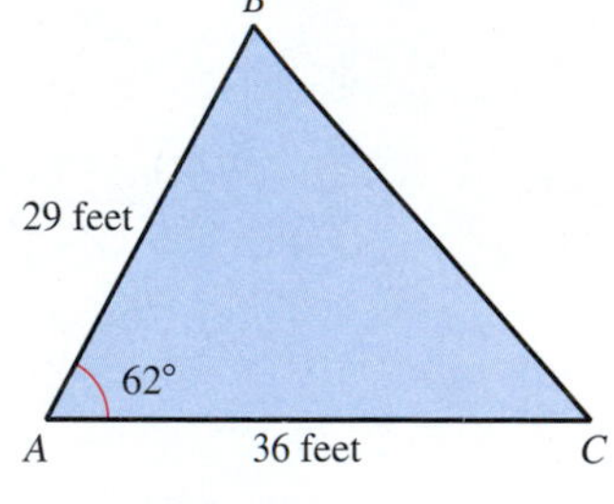

FIGURE 7.18

Solution

We are given the angle of measure 62° between the sides of lengths 36 feet and 29 feet. The area K of the triangle ABC is given by

$$K = \frac{1}{2}bc \sin \theta \qquad \text{Area formula}$$

$$= \frac{1}{2}(36)(29) \sin 62° \qquad b = 36, c = 29, \theta = 62°$$

$$\approx 460.9 \text{ square feet} \qquad \text{Use a calculator.}$$

Practice Problem 2 Find the area of triangle ABC assuming that the lengths of the sides AC and BC are 27 feet and 38 feet, respectively, and the measure of the angle between these sides is 47°.

EXAMPLE 3 The Ambiguous Case

A triangle ABC has area $K = 20 \text{ ft}^2$. The lengths of two of its sides are $a = 10$ ft and $b = 8$ ft. Find the measure of the angle between these sides.

Solution

Let θ be the angle between the sides of lengths a and b. Then

$$\frac{1}{2}ab\sin\theta = K \qquad \text{Area formula}$$

$$\frac{1}{2}(10)(8)\sin\theta = 20 \qquad a = 10, b = 8, K = 20$$

$$\sin\theta = \frac{1}{2} \qquad \text{Solve for } \sin\theta \text{ and simplify.}$$

Then $\theta = \sin^{-1}\left(\frac{1}{2}\right) = 30°$ is an angle between the sides of lengths a and b. However, because

$$\sin(180° - \theta) = \sin\theta, \qquad \text{See page 513, Exercise 102.}$$

we have

$$\sin 150° = \sin(180° - 30°) = \sin 30° = \frac{1}{2}.$$

So the angle between the sides of lengths a and b is also 150°. Figure 7.19 shows two possible triangles with the given information.

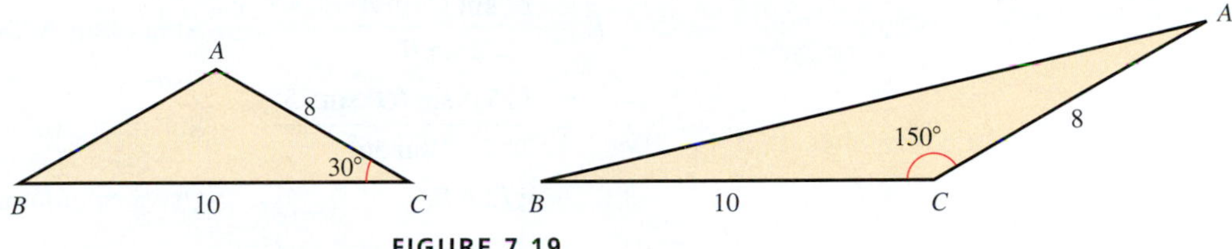

FIGURE 7.19

Practice Problem 3 Repeat Example 3 with $K = 6$, $a = 4$, and $b = 3$.

3 Find the area of AAS and ASA triangles.

Area of AAS and ASA Triangles

Suppose we are given two angles A and C and the side a of triangle ABC. We find the third angle B by the angle sum formula. So

$$B = 180° - A - C.$$

We find b by the Law of Sines.

$$\frac{b}{\sin B} = \frac{a}{\sin A}$$

$$b = \frac{a\sin B}{\sin A} \qquad \text{Multiply both sides by } \sin B.$$

The area, K, of triangle ABC is

$$K = \frac{1}{2}ab\sin C \qquad \text{SAS formula}$$

$$= \frac{1}{2}a\left(\frac{a\sin B}{\sin A}\right)\sin C \qquad \text{Replace } b \text{ with } \frac{a\sin B}{\sin A}.$$

$$= \frac{a^2\sin B\sin C}{2\sin A}.$$

We obtain similar results if any two angles and the side b or the side c is given.

Area of AAS and ASA Triangles

The area K of an AAS triangle ABC with given side a, b, or c is

$$K = \frac{a^2 \sin B \sin C}{2 \sin A}, \quad K = \frac{b^2 \sin C \sin A}{2 \sin B}, \quad \text{or} \quad K = \frac{c^2 \sin A \sin B}{2 \sin C}.$$

EXAMPLE 4 **Area of an AAS Triangle**

Find the area of triangle ABC with

$$B = 36°, C = 69°, \text{and } b = 15 \text{ ft}.$$

Solution

First, we find the third angle of the triangle. We have

$$\begin{aligned} A &= 180° - B - C && \text{Angle sum formula} \\ &= 180° - 36° - 69° && \text{Substitute for } B \text{ and } C. \\ &= 75°. \end{aligned}$$

Because we are given side b, we use the area formula for an AAS triangle that contains side b. Then

$$\begin{aligned} K &= \frac{b^2 \sin C \sin A}{2 \sin B} && \text{Area of an AAS triangle} \\ &= \frac{(15)^2 \sin 69° \sin 75°}{2 \sin 36°} && \text{Substitute values for } b, C, A, \text{ and } B. \\ &\approx 172.6 \text{ ft}^2 && \text{Use a calculator.} \end{aligned}$$

Practice Problem 4 Find the area of triangle ABC with $A = 63°$, $B = 74°$, and $c = 18$ in.

4 Find the area of SSS triangles.

Area of SSS Triangles

The Law of Cosines can be used to derive a formula for the area of a triangle if the lengths of the three sides (SSS triangles) are known. The formula is called **Heron's formula**.

HERON'S FORMULA FOR SSS TRIANGLES

The area K of an SSS triangle with sides of lengths a, b, and c is given by

$$K = \sqrt{s(s - a)(s - b)(s - c)},$$

where $s = \frac{1}{2}(a + b + c)$ is the **semiperimeter**.

The derivation of Heron's formula is given in Exercise 55.

EXAMPLE 5 **Using Heron's Formula**

Find the area of triangle ABC with $a = 29$ inches, $b = 25$ inches, and $c = 40$ inches. Round your answer to the nearest tenth.

Solution

We first find s.

$$s = \frac{a + b + c}{2} = \frac{29 + 25 + 40}{2} = 47$$

$$\begin{aligned} \text{Area of the triangle} &= \sqrt{s(s-a)(s-b)(s-c)} && \text{Heron's formula} \\ &= \sqrt{47(47-29)(47-25)(47-40)} && \text{Substitute values.} \\ &\approx 360.9 \text{ square inches} && \text{Use a calculator.} \end{aligned}$$

Practice Problem 5 Find the area of triangle ABC with $a = 11$ meters, $b = 17$ meters, and $c = 20$ meters. Round your answer to the nearest tenth.

EXAMPLE 6 Using Heron's Formula

A triangular swimming pool has side lengths 23 feet, 17 feet, and 26 feet. How many gallons of water will fill the pool to a depth of 5 feet? Round your answer to the nearest whole number.

Solution

To calculate the volume of water, we first calculate the area of the triangular surface.

We have $a = 23$, $b = 17$, and $c = 26$. So $s = \frac{1}{2}(a + b + c) = 33$.

By Heron's formula, the area K of the triangular surface is

$$\begin{aligned} K &= \sqrt{s(s-a)(s-b)(s-c)} \\ &= \sqrt{33(33-23)(33-17)(33-26)} && \text{Substitute values for } a, b, c, \text{ and } s. \\ &\approx 192.2498 \text{ square feet.} \end{aligned}$$

$$\begin{aligned} \text{The volume of water} &= \text{surface area} \times \text{depth} \\ &\approx 192.2498 \times 5 \\ &\approx 961.25 \text{ cubic feet} \end{aligned}$$

One cubic foot contains approximately 7.5 gallons of water. So $961.25 \times 7.5 \approx 7209$ gallons of water will fill the pool.

Practice Problem 6 Repeat Example 6, assuming that the swimming pool has side lengths of 25 feet, 30 feet, and 33 feet and the depth of the pool is 5.5 feet.

5 Learn vocabulary associated with polygons.

Polygon Names

Sides	Name
n	n-gon
3	Triangle
4	Quadrilateral
5	Pentagon
6	Hexagon
7	Heptagon
8	Octagon
10	Decagon
12	Dodecagon

Polygons

A closed plane figure bounded by at least three line segments (called *sides*) is called a **polygon**. Generally accepted names of some polygons with n sides are given in the margin. A point where two adjacent sides of a polygon meet is called the **vertex** (plural *vertices*) of a polygon. An angle formed inside the polygon by two adjacent sides is called an **interior angle**. A line joining any two nonconsecutive vertices is called a **diagonal** of the polygon. If all the sides and all the interior angles of a polygon are equal, it is called a **regular polygon**.

Here are some facts about polygons with n sides:

1. The **number of diagonals** in a polygon $= \frac{1}{2}n(n - 3)$.
2. The **number of triangles** formed by drawing all the diagonals from one vertex $= (n - 2)$.
3. The **sum of the interior angles** $= 180(n - 2)$ degrees.
One way to find the area of a polygon with n sides is to decompose it into $(n - 2)$ triangles, then calculate the sum of the areas of the triangles. In Exercises 47–54, we discuss the area and the perimeter of some regular polygons.

Answers to Practice Problems

1. a. 124 ft **b.** 672 ft^2 **2.** 375.2 ft^2 **3.** 90° **4.** 203.4 in^2 **5.** 93.5 m^2 **6.** 14,801 gal

SECTION 7.3 Exercises

Basic Concepts and Skills

1. The area K of a triangle with base b and height h is $K =$ ____________.

2. An *SAS* triangle is a triangle in which two sides and the ____________ are known.

3. The area of an *SAS* triangle ABC with sides a and c is $K =$ ____________.

4. Heron's formula states that the area of an *SSS* is $K =$ ____________, where $s =$ ____________.

In Exercises 5–10, find the exact value of the area of each *SAS* triangle ABC.

5. $A = 30°, b = 6, c = 5$

6. $B = 60°, a = 12, c = 7$

7. $C = 120°, a = 8, b = 5$

8. $A = 135°, b = 4, c = 6$

9. $B = 150°, a = 12, c = 9$

10. $C = 45°, a = \sqrt{8}, b = 5$

In Exercises 11–18, find the area of each SAS triangle ABC. Round your answers to the nearest tenth.

11. $A = 57°, b = 30$ in., $c = 52$ in.

12. $B = 110°, a = 20$ cm, $c = 27$ cm

13. $C = 46°, a = 15$ km, $b = 22$ km

14. $A = 146.7°, b = 16.7$ ft, $c = 18$ ft

15. $B = 38.6°, a = 12$ mm, $c = 16.7$ mm

16. $B = 112.5°, a = 151.6$ ft, $c = 221.8$ ft

17. $C = 107.3°, a = 271$ ft, $b = 194.3$ ft

18. $A = 131.8°, b = 15.7$ mm, $c = 18.2$ mm

In Exercises 19–26, find the area of each AAS or ASA triangle ABC. Round your answers to the nearest tenth.

19. $a = 16$ ft, $B = 57°, C = 49°$

20. $a = 12$ ft, $A = 73°, C = 64°$

21. $b = 15.3$ yd, $A = 64°, B = 38°$

22. $b = 10$ m, $B = 53.4°, C = 65.6°$

23. $c = 16.3$ cm, $B = 55°, C = 37.5°$

24. $c = 20.5$ ft, $A = 64°, B = 84.2°$

25. $a = 65.4$ ft, $A = 62°15', B = 44°30'$

26. $b = 24.3$ m, $B = 56°18', C = 37°36'$

In Exercises 27–34, find the area of each SSS triangle ABC. Round your answers to the nearest tenth.

27. $a = 2, b = 3, c = 4$

28. $a = 50, b = 100, c = 130$

29. $a = 50, b = 50, c = 75$

30. $a = 100, b = 100, c = 125$

31. $a = 7.5, b = 4.5, c = 6.0$

32. $a = 8.5, b = 5.1, c = 4.5$

33. $a = 3.7, b = 5.1, c = 4.2$

34. $a = 9.8, b = 5.7, c = 6.5$

In Exercises 35–38, find an angle θ between the sides a and b of a triangle ABC with given area K.

35. $K = 12, a = 6, b = 8$

36. $K = 6, a = 4, b = 2\sqrt{3}$

37. $K = 30, a = 12, b = 5$

38. $K = 15, a = 5\sqrt{2}, b = 6$

In Exercises 39–42, use the following figure of a parallelogram (not drawn to scale).

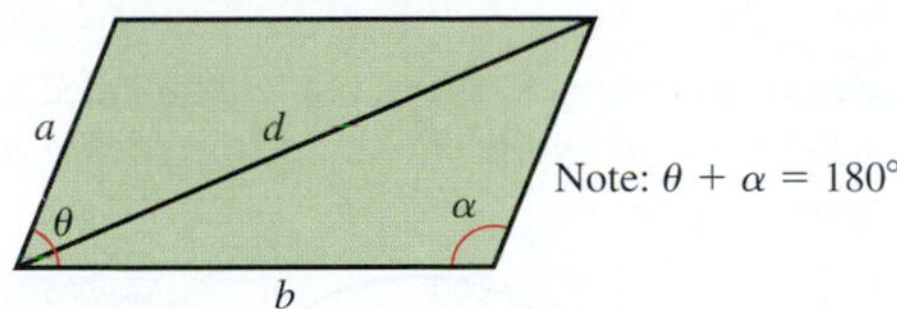

39. Show that the area of the parallelogram is $ab \sin \theta = ab \sin \alpha$.

40. Find the area of the parallelogram if $a = 8$ cm, $b = 13$ cm, and $\theta = 60°$.

41. Find the area of the parallelogram if $a = 12$ cm, $b = 16$ cm, and $d = 22$ cm.

42. If the area of parallelogram is 10 cm^2 with $b = 4$ cm and $\alpha = \frac{5\pi}{6}$, find a.

Applying the Concepts

43. Area of the Bermuda Triangle. The angle formed by the lines of sight from Fort Lauderdale, Florida, to Bermuda and to San Juan, Puerto Rico, is approximately 67°. The distance from Fort Lauderdale to Bermuda is 1026 miles, and the distance from Fort Lauderdale to San Juan is 1046 miles. Find the area of the Bermuda Triangle, which is formed with those cities as vertices.

44. Landscaping. A triangular region between the three streets shown in the figure will be landscaped for \$30 a square foot. How much will the landscaping cost?

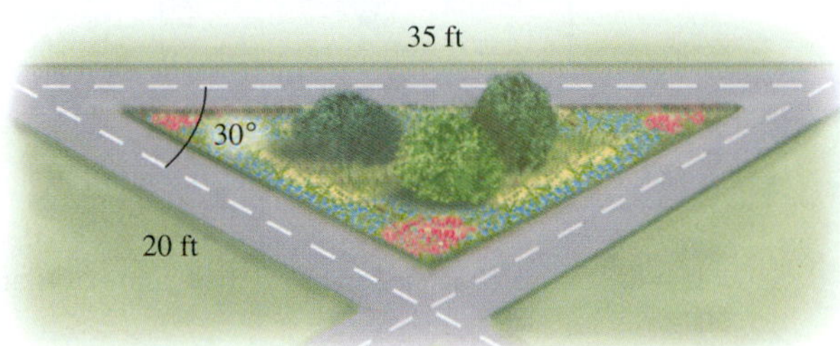

45. Real estate. You have inherited a commercial triangular-shaped lot of side lengths 400 feet, 250 feet, and 274 feet. The neighboring property is selling for \$1 million dollars per acre. How much is your lot worth? (*Hint:* 1 acre = 43,560 square feet)

46. Swimming pool. Find the number of gallons of water in a triangular swimming pool with sides 11 feet, 16 feet, and 19 feet and a depth of 5 feet. Round your answer to the nearest whole number. (Recall: Approximately 7.5 gallons of water are in 1 cubic foot.)

Beyond the Basics

Exercises 47–54 deal with a regular polygon. For $n \geq 3$, a regular n-gon is an n-sided polygon, all of whose sides are equal in length and all of whose angles are equal in measure.

Inscribed polygon. A regular n-gon inscribed in a circle of radius r has n vertices that are equally spaced on the circle. The adjoining figure is an inscribed octagon ($n = 8$) on a circle of radius r. Use this figure for Exercises 47–50.

47. a. Write the coordinates of the vertices of the inscribed octagon with one of the vertices at $A = (r, 0)$.

b. Find the length of the side AB.

c. Use part (b) to find the *perimeter* of the octagon.

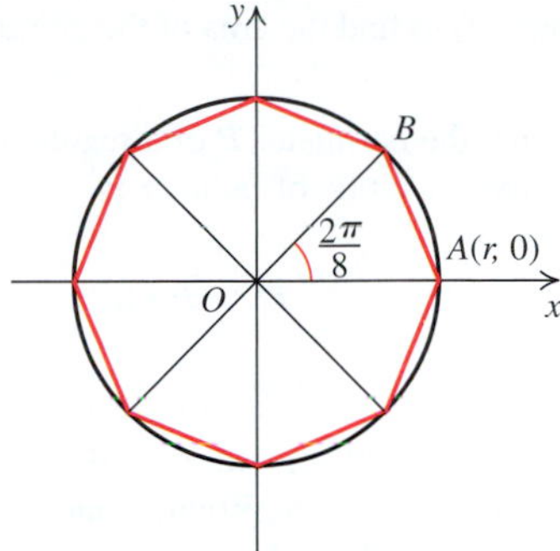

48. a. In the figure, find the area of triangle AOB.

b. Use part (a) to find the area of the inscribed octagon.

49. a. Show that the *perimeter* $\boldsymbol{P}$ of a regular n-gon inscribed in a circle of radius r is

$$\boldsymbol{P} = nr\sqrt{2 - 2\cos\frac{2\pi}{n}}.$$

b. What is the circumference (perimeter) of a circle of radius $r = 5$?

c. Calculate $\boldsymbol{P}$ in part (a) with $r = 5$ and $n = 4$, 10, 50, and 100. Round your answer to four decimal places. What do you observe?

50. a. Show that the area $\boldsymbol{A}$ of a regular n-gon inscribed in a circle of radius r is

$$\boldsymbol{A} = \frac{nr^2}{2}\sin\frac{2\pi}{n}.$$

b. What is the area of a disc of radius $r = 5$?

c. Calculate $\boldsymbol{A}$ from part (a) with $r = 5$ and $n = 4$, 10, 50, and 100. Round your answer to four decimal places. What do you observe?

Circumscribed polygon. A regular circumscribed n-gon about a circle of radius r has n vertices, and its sides are tangent to n equally spaced points on the circumference of the circle. The adjoining figure is a circumscribed hexagon ($n = 6$) about a circle of radius r. Use this figure for Exercises 51–52.

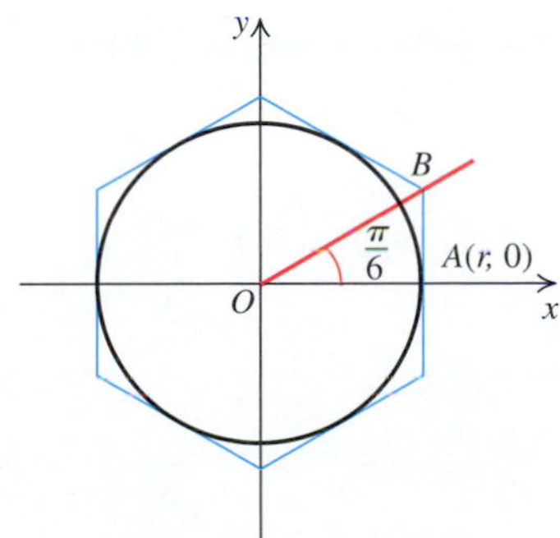

51. a. In the figure, show that $B = \left(r, r\tan\frac{\pi}{6}\right)$ and conclude that the length of each side of the hexagon is $2r\tan\frac{\pi}{6}$.

b. Use part (a) to find the perimeter of the hexagon.

52. a. Use the figure to show that the area of the triangle AOB is $\frac{1}{2}r^2\tan\frac{\pi}{6}$.

b. Use part (a) to find the area of the circumscribed hexagon.

53. a. Show that the perimeter P of a regular circumscribed n-gon about a circle of radius r is

$$P = 2nr\tan\frac{\pi}{n}.$$

b. What is the circumference of a circle of radius $r = 10$?

c. Calculate P in part (a) with $r = 10$ and $n = 4, 10, 50,$ and 100. Round your answer to four decimal places. What do you observe?

54. a. Show that the area A of a regular n-gon circumscribed about a circle of radius r is

$$A = nr^2\tan\frac{\pi}{n}.$$

b. What is the area of a disc of radius $r = 10$?

c. Calculate A by using the formula in part (a) with $r = 10$ and $n = 4, 10, 50,$ and 100. Round your answer to four decimal places. What do you observe?

55. Derive Heron's formula by justifying each step.

In a triangle ABC,

$$K = \frac{1}{2}ab\sin C$$

$$4K^2 = a^2b^2(1 - \cos^2 C)$$

$$4K^2 = a^2b^2\left[1 - \left(\frac{a^2 + b^2 - c^2}{2ab}\right)^2\right]$$

$$16K^2 = 4a^2b^2 - (a^2 + b^2 - c^2)^2$$

$$16K^2 = [2ab + (a^2 + b^2 - c^2)][2ab - (a^2 + b^2 - c^2)]$$

$$16K^2 = [(a + b)^2 - c^2][c^2 - (a - b)^2]$$

$$K^2 = \left(\frac{a + b + c}{2}\right)\left(\frac{a + b - c}{2}\right)\left(\frac{c + a - b}{2}\right)\left(\frac{c + b - a}{2}\right)$$

$$K = \sqrt{s(s - a)(s - b)(s - c)}, \quad s = \frac{1}{2}(a + b + c)$$

56. In a triangle ABC, show that

$$\sin A = \frac{2}{bc}\sqrt{s(s - a)(s - b)(s - c)}.$$

57. Use Heron's formula to show that the area of an equilateral triangle whose sides have length a is $K = \frac{\sqrt{3}}{4}a^2$.

58. A **segment** of a circle is a part of a circle bounded by a chord and the arc cut by the chord. See the shaded region in the figure.

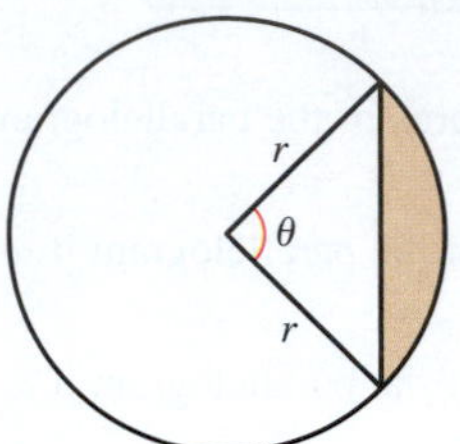

a. Show that the area of the segment of a circle of radius r subtended (formed) by a central angle of θ radians is $\frac{1}{2}r^2(\theta - \sin\theta)$.

b. Find the area of the segment subtended by a central angle of 60° in a circle of radius 6 inches.

59. Circumscribed circle. Perpendicular bisectors of the sides of a triangle meet at a point O, and the distance from O to each of the vertices is the same number (say, R). The circle with center at O and radius R is called the **circumscribed circle** of the triangle. See the figure.

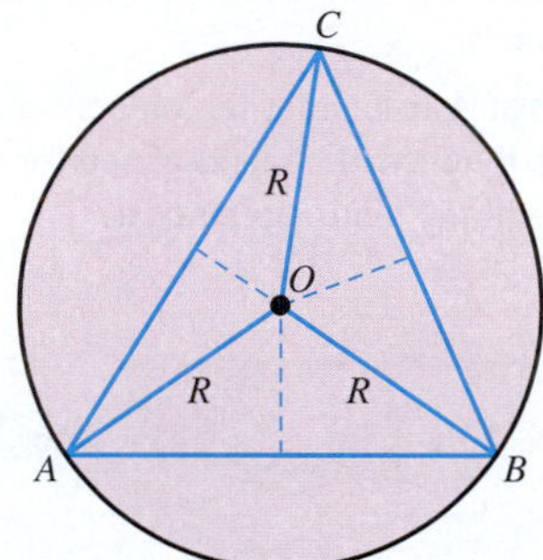

Show that the radius R of the circumscribed circle of triangle ABC is given by

$$R = \frac{a}{2\sin A} = \frac{b}{2\sin B} = \frac{c}{2\sin C}.$$

(*Hint:* Use a theorem of geometry: $\angle BOC = 2\angle A$.)

60. A triangle ABC with given angles A, B, and C is inscribed in a circle of diameter 1. Find a, b, and c.

61. Use Exercises 56 and 59 to show that

$$R = \frac{abc}{4K},$$

where K is the area of the triangle ABC.

62. A triangle ABC has $a = 18$, $b = 24$, and $c = 30$. Find K and R. [*Hint*: Use Exercise 61.]

63. Inscribed circle. The bisectors of the angles of a triangle meet at a point I, and the perpendicular distance from I to each of the sides is the same number (say, r). The circle with center I and radius r is called the **inscribed circle** of the triangle. See the figure.

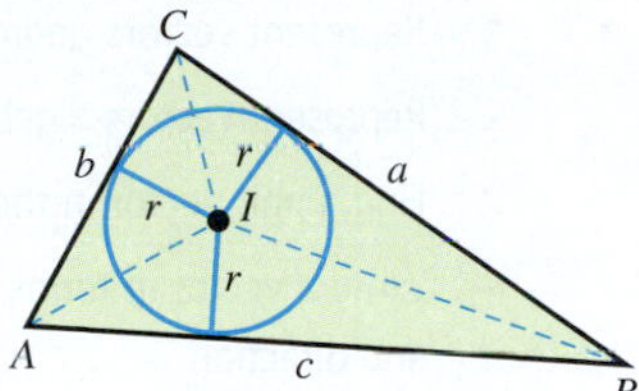

Show that the radius r of the inscribed circle of triangle ABC is given by

$$r = \frac{\sqrt{s(s-a)(s-b)(s-c)}}{s}.$$

(*Hint:* Show that $rs = K$.)

64. From Exercises 61 and 63, conclude that

$$rR = \frac{abc}{2(a+b+c)}.$$

65. In triangle ABC, let p_1, p_2, and p_3 be the lengths of the perpendiculars from the vertices A, B, and C to the sides a, b, and c, respectively. Show that

$$\frac{1}{p_1} + \frac{1}{p_2} + \frac{1}{p_3} = \sqrt{\frac{s}{(s-a)(s-b)(s-c)}}.$$

$$\left(\textit{Hint}: \frac{1}{2}ap_1 = \frac{1}{2}bp_2 = \frac{1}{2}cp_3 = K\right)$$

66. From Exercises 61 and 64, conclude that

$$\frac{1}{p_1} + \frac{1}{p_2} + \frac{1}{p_3} = \frac{1}{r}.$$

Critical Thinking/Discussion/Writing

67. Show that the area, K, of a triangle ABC is

$$K = \frac{a^2 \sin 2B + b^2 \sin 2A}{4} = \frac{b^2 \sin 2C + c^2 \sin 2B}{4} = \frac{c^2 \sin 2A + a^2 \sin 2C}{4}.$$

68. Let r and R be the radii of the inscribed circle and the circumscribed circle, respectively, of triangle ABC.

Show that $r = 4R \sin\frac{A}{2} \sin\frac{B}{2} \sin\frac{C}{2}$.

Maintaining Skills

In Exercises 69 and 70, write each expression in the form $ax + by$ and compute $\sqrt{a^2 + b^2}$.

69. $2(-x + 3y) + 3(x - y)$

70. $3(2x - 3y) - 2(x - 2y)$

71. Write the exact value of

a. $\sin\frac{\pi}{6}$ **b.** $\cos\frac{\pi}{4}$ **c.** $\tan\frac{\pi}{3}$

72. Write the exact value of

a. $\sin^{-1}\left(-\frac{1}{2}\right)$

b. $\cos^{-1}\left(-\frac{1}{2}\right)$

c. $\tan^{-1}(-\sqrt{3})$

73. Find angle θ in quadrant III whose reference angle $\theta' = \tan^{-1}(\sqrt{3})$.

74. Find angle θ in quadrant II whose reference angle $\theta' = \left|\tan^{-1}\left(-\frac{1}{\sqrt{3}}\right)\right|$.

In Exercises 75–78, let l_1 be the line passing through the points S and T and l_2 the line passing through the points O and P. In each case, find

a. $d(S, T)$ and $d(O, P)$.

b. slope m_1 of l_1 and m_2 of l_2.

c. Are l_1 and l_2 parallel, perpendicular, or neither?

75. $S = (1, 3), T = (6, 15);\quad O = (0, 0), P = (5, 12)$

76. $S = (2, -3), T = (-1, -7);$
$O = (0, 0), P = (-3, 4)$

77. $S = (-1, 2), T = (-2, 1);$
$O = (0, 0), P = (1, -1)$

78. $S = (-1, 1), T = (3, -2);$
$O = (0, 0), P = (4, -3)$

SECTION 7.4

Vectors

BEFORE STARTING THIS SECTION, REVIEW

1 Trigonometric functions of angles (Section 5.3, page 500)

2 Inverse trigonometric functions (Section 5.6, page 551)

OBJECTIVES

1 Represent vectors geometrically.

2 Represent vectors algebraically.

3 Find a unit vector in the direction of **v**.

4 Write a vector in terms of its magnitude and direction.

5 Use vectors in applications.

The Meaning of Force

A **force** is a push or pull resulting from an object's *interaction* with another object. When the interacting objects are physically in contact with each other, the resulting force is called a *contact force*. Examples of contact forces include friction, tension, air resistance, and applied forces. A noncontact force is called an *action-at-a-distance* force. Examples of such forces include gravitational, electrical, and magnetic forces. The standard units of measurement for the magnitude of a force are pounds (lb) in the English system and newtons (N) in the metric system. The conversion factor between the systems is

$$1 \text{ pound of force} = 4.45 \text{ newtons}.$$

In Examples 8–10, we discuss the *resultant* of two or more forces acting on an object.

SIDE NOTE

The *magnitude* of a vector is similar to the *absolute value* of a real number. Like absolute value, the magnitude of a vector cannot be negative.

Vectors

Many physical quantities such as length, area, volume, mass, and temperature are completely described by their magnitudes in appropriate units. Such quantities are called **scalar quantities**. Other physical quantities such as velocity, acceleration, and force are completely described only if *both* a magnitude (size) and a direction are specified. For example, the movement of wind is usually described by its speed (magnitude) and its direction, say, 15 miles per hour southwest. The wind speed and wind direction together form a **vector quantity** called the *wind velocity*.

1 Represent vectors geometrically.

Geometric Vectors

We can represent a **vector** geometrically by a directed line segment with an arrowhead. The arrow specifies the direction of the vector, and its length describes the magnitude. The tail of the arrow is the vector's **initial point**, and the tip of the arrow is its **terminal point**. We denote vectors by lowercase boldface type, such as **a**, **b**, **i**, **j**, **u**, **v**, and **w**. When discussing vectors, we refer to real numbers as **scalars**. Scalars will be denoted by lowercase italic type, such as a, b, x, y, and z.

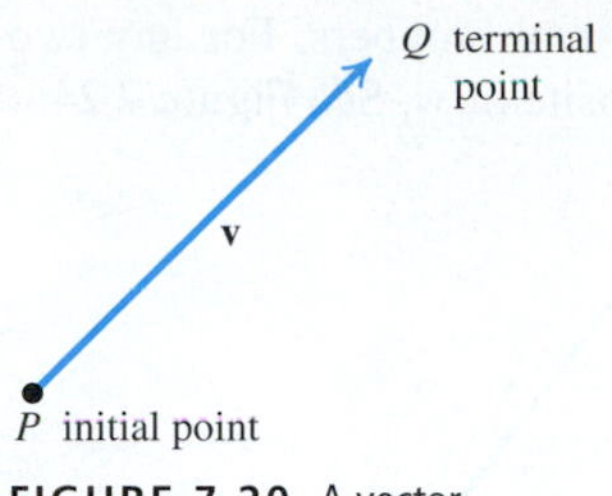

FIGURE 7.20 A vector

As in Figure 7.20, if the initial point of a vector **v** is P and the terminal point is Q, we write the vector **v** as:

$$\mathbf{v} = \overrightarrow{PQ}.$$

The **magnitude** (or **norm**) of a vector $\mathbf{v} = \overrightarrow{PQ}$, denoted by $\|\mathbf{v}\|$, or $\|\overrightarrow{PQ}\|$, is the length of the vector **v** and is a scalar quantity.

Equivalent Vectors

Two vectors having the same length and same direction are called **equivalent vectors**. Because a vector is determined by its length and direction only, equivalent vectors are regarded as **equal** even though they may be located in different positions. If **v** and **w** are equivalent, we write $\mathbf{v} = \mathbf{w}$. See Figure 7.21.

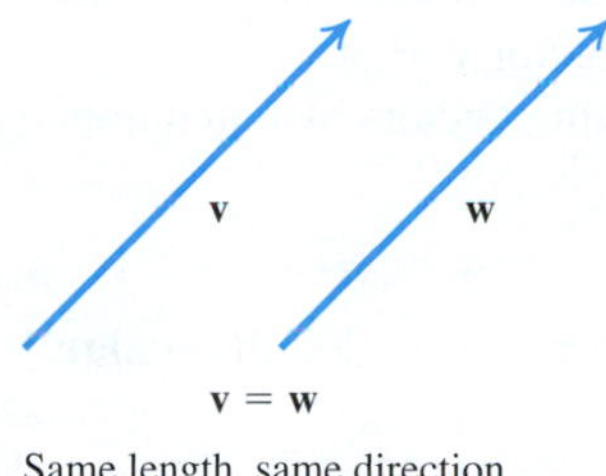

$\mathbf{v} = \mathbf{w}$

Same length, same direction

(a)

$\mathbf{v} \neq \mathbf{u}$

Different length, same direction

(b)

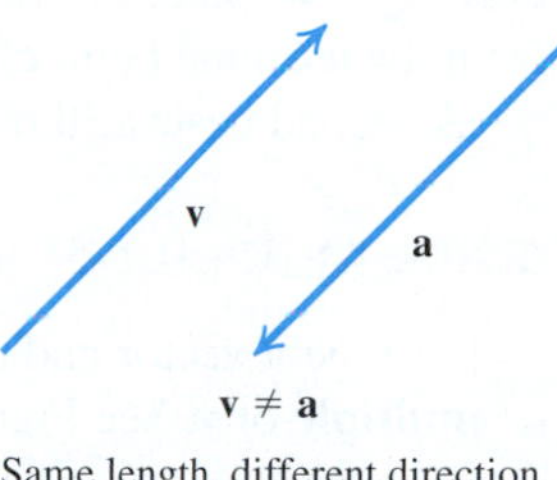

$\mathbf{v} \neq \mathbf{a}$

Same length, different direction

(c)

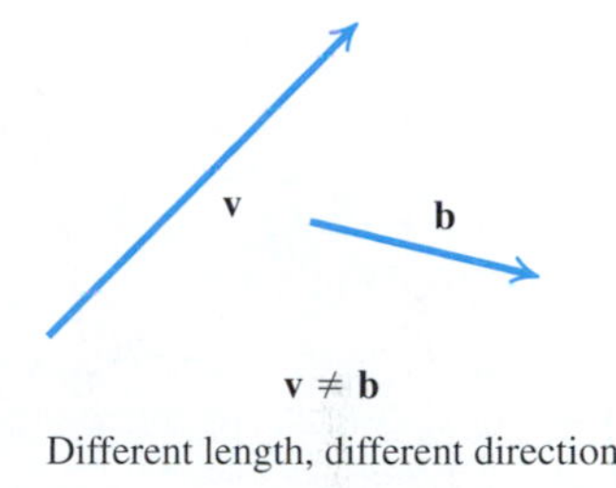

$\mathbf{v} \neq \mathbf{b}$

Different length, different direction

(d)

FIGURE 7.21 Equivalent vectors have the same length and direction

The vector of length zero is called the **zero vector** and is denoted by **0**. The zero vector has zero magnitude and arbitrary direction. If vectors **v** and **a**, as in Figure 7.21(c), have the same length and opposite direction, then **a** is the **opposite vector** of **v** and we write $\mathbf{a} = -\mathbf{v}$.

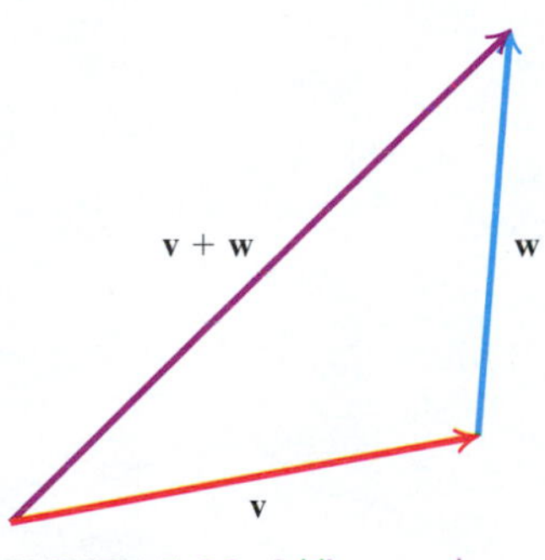

FIGURE 7.22 Adding **v** and **w**

Adding Vectors

We add two vectors **v** and **w**, as shown in Figure 7.22.

Geometric Vector Addition

Let **v** and **w** be any two vectors. Position the terminal point of **v** so that it coincides with the initial point of **w**. The *sum* $\mathbf{v} + \mathbf{w}$ is the **resultant vector** whose initial point coincides with the initial point of **v**, and its terminal point coincides with the terminal point of **w**.

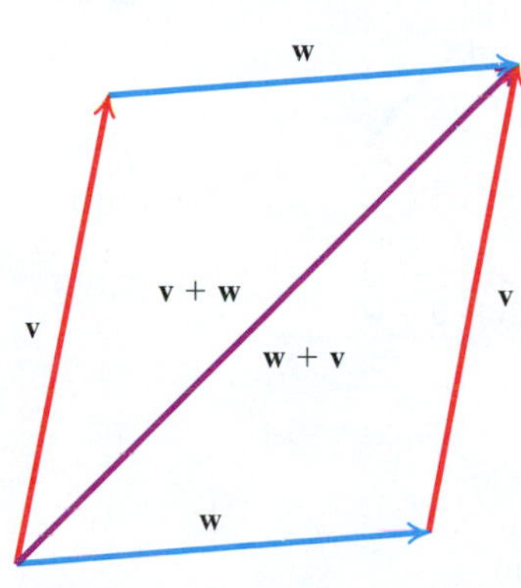

FIGURE 7.23 $\mathbf{v}+\mathbf{w}=\mathbf{w}+\mathbf{v}$

In Figure 7.23, we have constructed two sums, $\mathbf{v} + \mathbf{w}$ and $\mathbf{w} + \mathbf{v}$. It shows that

$$\boxed{\mathbf{v} + \mathbf{w} = \mathbf{w} + \mathbf{v}}$$

and that the sum coincides with the diagonal of the parallelogram determined by **v** and **w** when **v** and **w** have the same initial point.

Vector subtraction is defined just like the subtraction of real numbers. For any two vectors **v** and **w**, $\mathbf{v} - \mathbf{w} = \mathbf{v} + (-\mathbf{w})$, where $-\mathbf{w}$ is the opposite of **w**. See Figure 7.24.

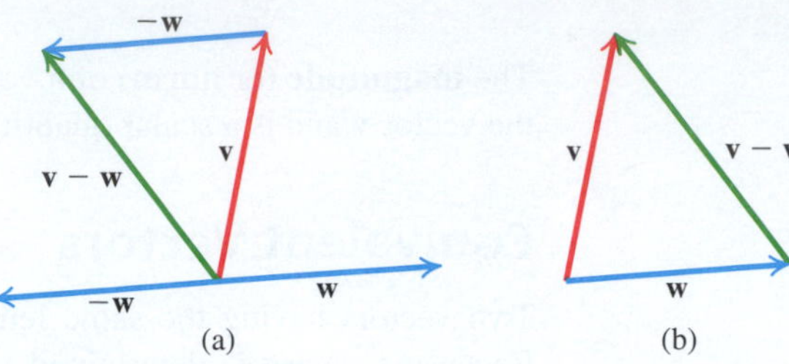

FIGURE 7.24 Vector **v − w**

Figure 7.24(b) shows that you can construct the difference vector $\mathbf{v} - \mathbf{w}$ without first drawing $-\mathbf{w}$: place the vectors **v** and **w** so that their initial points coincide. Then the vector from the terminal point of **w** to the terminal point of **v** is the vector $\mathbf{v} - \mathbf{w}$.

A second basic arithmetic operation for vectors is multiplying vectors by real numbers.

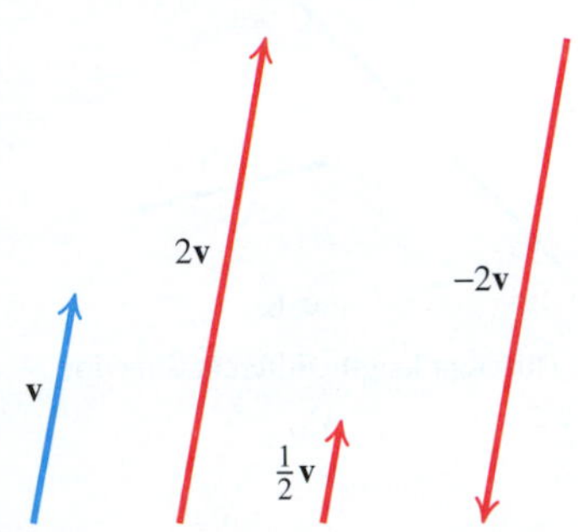

FIGURE 7.25 Scalar multiples of **v**

Scalar Multiples of Vectors

Let **v** be a vector and c a scalar (a real number). The vector $c\mathbf{v}$ is called the **scalar multiple** of **v**. See Figure 7.25.

If $c > 0$, $c\mathbf{v}$ has the same direction as **v** and magnitude $c\|\mathbf{v}\|$.
If $c < 0$, $c\mathbf{v}$ has the opposite direction of **v** and magnitude $|c|\|\mathbf{v}\|$.
If $c = 0$, $c\mathbf{v} = 0\mathbf{v} = \mathbf{0}$.

EXAMPLE 1 Geometric Vectors

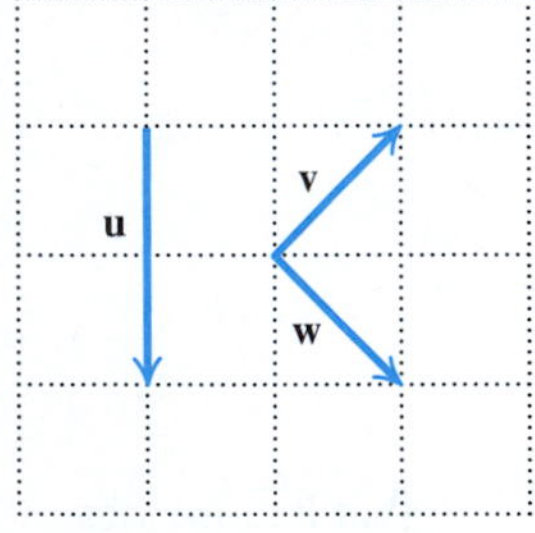

FIGURE 7.26

Use the vectors **u**, **v**, and **w** in Figure 7.26 to graph each vector.

a. $\mathbf{u} - 2\mathbf{w}$ **b.** $2\mathbf{v} - \mathbf{u} + \mathbf{w}$

Solution

The graphs are shown in Figure 7.27.

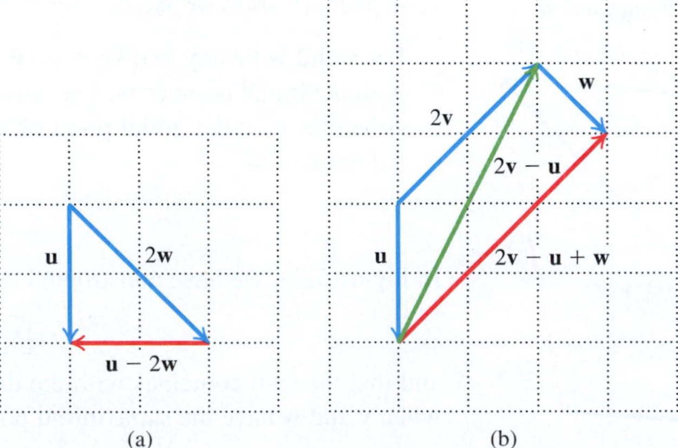

FIGURE 7.27

Practice Problem 1 Use the vectors **u**, **v**, and **w** of Figure 7.26 to graph each vector.

a. $2\mathbf{w} + \mathbf{v}$ **b.** $2\mathbf{w} + \mathbf{v} - \mathbf{u}$

2 Represent vectors algebraically.

Algebraic Vectors

We now consider vectors in the Cartesian coordinate plane. Because the location of the initial point of a vector is not relevant, we typically draw vectors with their initial point at the origin. Such a vector is called a **position vector**. Note that the terminal point of a position vector will completely determine the vector, so specifying the terminal point will specify the vector. For the position vector **v** (see Figure 7.28) with initial point at the origin O and terminal point at $P(v_1, v_2)$, we denote the vector by

$$\mathbf{v} = \overrightarrow{OP} = \langle v_1, v_2 \rangle.$$

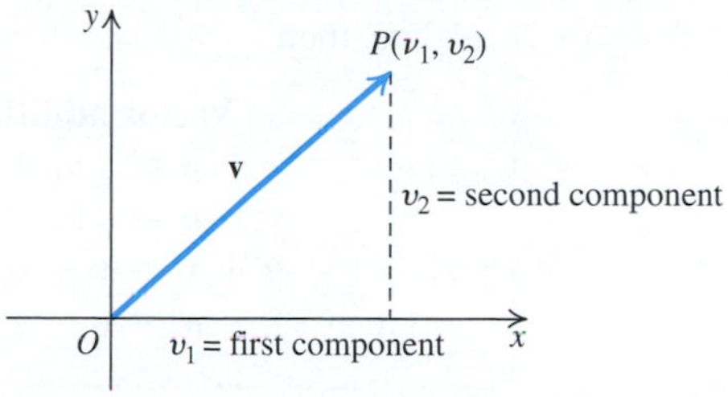

FIGURE 7.28 A position vector

We call v_1 and v_2 the **components** of the vector **v**; v_1 is the **first component**, and v_2 is the **second component**. Note the difference between the notations for the *point* (v_1, v_2) and the *position vector* $\langle v_1, v_2 \rangle$. The magnitude of the position vector $\mathbf{v} = \langle v_1, v_2 \rangle$ follows directly from the Pythagorean theorem. We have $\|\mathbf{v}\| = \sqrt{v_1^2 + v_2^2}$.

If equivalent vectors **v** and **w** are located so that their initial points are at the origin, then their terminal points must coincide (because the equivalent vectors have the same length and same direction). Thus, for the vectors

$$\mathbf{v} = \langle v_1, v_2 \rangle \quad \text{and} \quad \mathbf{w} = \langle w_1, w_2 \rangle,$$
$$\mathbf{v} = \mathbf{w} \quad \text{if and only if} \quad v_1 = w_1 \quad \text{and} \quad v_2 = w_2.$$

We can use congruent triangles to show that any vector **v** with initial point $P(x_1, y_1)$ and terminal point $Q(x_2, y_2)$ is equivalent to the position vector $\mathbf{w} = \langle x_2 - x_1, y_2 - y_1 \rangle$. Any vector in the Cartesian plane can therefore be represented by a position vector. See Figure 7.29.

SIDE NOTE

Notice that the point (v_1, v_2) has neither magnitude nor direction, whereas the position vector $\langle v_1, v_2 \rangle$ has both.

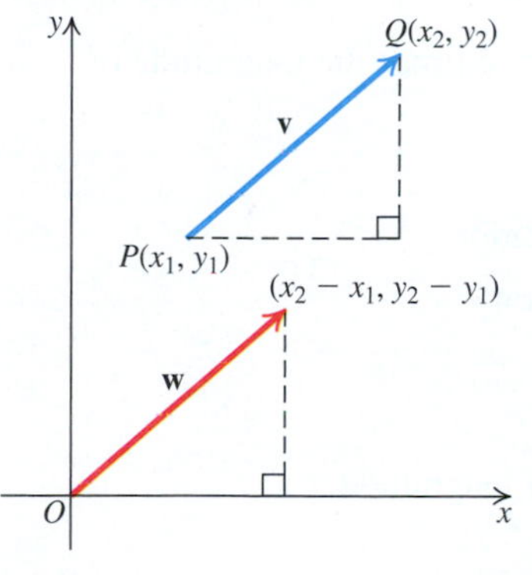

FIGURE 7.29 Two equivalent vectors

REPRESENTING A VECTOR AS A POSITION VECTOR

The vector $\overrightarrow{PQ}$ with initial point $P(x_1, y_1)$ and terminal point $Q(x_2, y_2)$ is equal to the position vector

$$\mathbf{w} = \langle x_2 - x_1, y_2 - y_1 \rangle.$$

EXAMPLE 2 Representing a Vector in the Cartesian Plane

Let **v** be the vector with initial point $P(4, -2)$ and terminal point $Q(-1, 3)$. Write **v** as a position vector.

Solution

Because **v** has initial point $P(4, -2)$, $x_1 = 4$ and $y_1 = -2$.
Because **v** has terminal point $Q(-1, 3)$, $x_2 = -1$ and $y_2 = 3$.

So

$\mathbf{v} = \langle x_2 - x_1, y_2 - y_1 \rangle$ Position vector representation of $\overrightarrow{PQ}$

$\mathbf{v} = \langle -1 - 4, 3 - (-2) \rangle$ Substitute values.

$\mathbf{v} = \langle -5, 5 \rangle$ Simplify.

Practice Problem 2 Let $\mathbf{w}$ be the vector with initial point $P(-2, 7)$ and terminal point $Q(1, -3)$. Write $\mathbf{w}$ as a position vector.

The zero vector is $\mathbf{0} = \langle 0, 0 \rangle$, and the opposite of the vector $\mathbf{v} = \langle v_1, v_2 \rangle$ is $-\mathbf{v} = \langle -v_1, -v_2 \rangle$. The following properties describe the arithmetic operations on vectors, using components.

ARITHMETIC OPERATIONS AND PROPERTIES OF VECTORS

If $\mathbf{u} = \langle u_1, u_2 \rangle$, $\mathbf{v} = \langle v_1, v_2 \rangle$, and $\mathbf{w} = \langle w_1, w_2 \rangle$ are vectors and c and d are any scalars, then

Vector addition	Scalar multiplication
$\mathbf{v} + \mathbf{u} = \langle v_1 + u_1, v_2 + u_2 \rangle$	$c\mathbf{v} = \langle cv_1, cv_2 \rangle$
$\mathbf{v} - \mathbf{u} = \langle v_1 - u_1, v_2 - u_2 \rangle$	$(c + d)\mathbf{v} = c\mathbf{v} + d\mathbf{v}$
$\mathbf{u} + \mathbf{v} = \mathbf{v} + \mathbf{u}$	$c(\mathbf{v} + \mathbf{w}) = c\mathbf{v} + c\mathbf{w}$
$(\mathbf{u} + \mathbf{v}) + \mathbf{w} = \mathbf{u} + (\mathbf{v} + \mathbf{w})$	$c(d\mathbf{v}) = (cd)\mathbf{v}$

EXAMPLE 3 Operations on Vectors

Let $\mathbf{v} = \langle 2, 3 \rangle$ and $\mathbf{w} = \langle -4, 1 \rangle$. Find each expression.

a. $\mathbf{v} + \mathbf{w}$ **b.** $-2\mathbf{v}$ **c.** $2\mathbf{v} - \mathbf{w}$ **d.** $\|2\mathbf{v} - \mathbf{w}\|$

Solution

a. $\mathbf{v} + \mathbf{w} = \langle 2, 3 \rangle + \langle -4, 1 \rangle = \langle 2 - 4, 3 + 1 \rangle = \langle -2, 4 \rangle$

b. $-2\mathbf{v} = -2\langle 2, 3 \rangle = \langle -2 \cdot 2, -2 \cdot 3 \rangle = \langle -4, -6 \rangle$

Note that $-2\mathbf{v}$ is a vector whose magnitude is $|-2| = 2$ times the magnitude of $\mathbf{v}$ and whose direction is opposite that of $\mathbf{v}$.

c.
$$\begin{aligned} 2\mathbf{v} - \mathbf{w} &= 2\langle 2, 3 \rangle - \langle -4, 1 \rangle \\ &= \langle 4, 6 \rangle - \langle -4, 1 \rangle && \text{Scalar multiplication} \\ &= \langle 4 - (-4), 6 - 1 \rangle && \text{Vector subtraction} \\ &= \langle 8, 5 \rangle && \text{Simplify.} \end{aligned}$$

d.
$$\begin{aligned} \|2\mathbf{v} - \mathbf{w}\| &= \|\langle 8, 5 \rangle\| && \text{From part } \mathbf{c} \\ &= \sqrt{8^2 + 5^2} && \text{Definition of magnitude} \\ &= \sqrt{64 + 25} = \sqrt{89} && \text{Simplify.} \end{aligned}$$

Practice Problem 3 Let $\mathbf{v} = \langle -1, 2 \rangle$ and $\mathbf{w} = \langle 2, -3 \rangle$. Find each expression.

a. $\mathbf{v} + \mathbf{w}$ **b.** $3\mathbf{w}$ **c.** $3\mathbf{w} - 2\mathbf{v}$ **d.** $\|3\mathbf{w} - 2\mathbf{v}\|$

3 Find a unit vector in the direction of v.

Unit Vectors

If $c > 0$ and $\mathbf{v}$ is any vector, then the vector $c\mathbf{v}$ has the same direction as $\mathbf{v}$ and its magnitude is given by

$$\|c\mathbf{v}\| = c\|\mathbf{v}\|.$$

If $\mathbf{v}$ is a nonzero vector and $c = \frac{1}{\|\mathbf{v}\|}$, then $\left\| \frac{1}{\|\mathbf{v}\|}\mathbf{v} \right\| = \frac{1}{\|\mathbf{v}\|}\|\mathbf{v}\| = 1$. Consequently, $\frac{1}{\|\mathbf{v}\|}\mathbf{v}$ is a vector of magnitude 1 in the same direction as $\mathbf{v}$. A vector of magnitude 1 is called a **unit vector**.

EXAMPLE 4 Finding a Unit Vector

Let $\mathbf{v} = \langle 3, -4 \rangle$.

a. Find a unit vector $\mathbf{u}$ in the direction of $\mathbf{v}$.

b. Find a vector of magnitude 15 in the direction of $\mathbf{v}$.

Solution

a. First, find the magnitude of $\mathbf{v} = \langle 3, -4 \rangle$.

$$\|\mathbf{v}\| = \sqrt{(3)^2 + (-4)^2} \quad \text{Formula for magnitude}$$
$$\|\mathbf{v}\| = \sqrt{25} = 5 \quad \text{Simplify.}$$

Now let

$$\mathbf{u} = \frac{1}{\|\mathbf{v}\|}\mathbf{v} \quad \text{Multiply } \mathbf{v} \text{ by the scalar } \frac{1}{\|\mathbf{v}\|}.$$
$$\mathbf{u} = \frac{1}{5}\langle 3, -4 \rangle \quad \text{Substitute values.}$$
$$\mathbf{u} = \left\langle \frac{3}{5}, -\frac{4}{5} \right\rangle \quad \text{Scalar multiplication}$$

Check that $\|\mathbf{u}\| = 1$:

$$\|\mathbf{u}\| = \sqrt{\left(\frac{3}{5}\right)^2 + \left(-\frac{4}{5}\right)^2} = \sqrt{\frac{9}{25} + \frac{16}{25}} = \sqrt{\frac{25}{25}} = 1$$

Therefore, $\mathbf{u} = \left\langle \frac{3}{5}, -\frac{4}{5} \right\rangle$ is a unit vector in the same direction as $\mathbf{v}$. (Note that $\mathbf{u}$ is a scalar multiple of $\mathbf{v}$.)

b. The required vector $= 15\,\mathbf{u} = 15\left\langle \frac{3}{5}, -\frac{4}{5} \right\rangle = \langle 9, -12 \rangle$.

Practice Problem 4 Find a unit vector in the direction of $\mathbf{v} = \langle -12, 5 \rangle$.

Vectors in i, j Form

In a Cartesian coordinate plane, two important unit vectors lie along the positive coordinate axes:

$$\mathbf{i} = \langle 1, 0 \rangle \quad \text{and} \quad \mathbf{j} = \langle 0, 1 \rangle$$

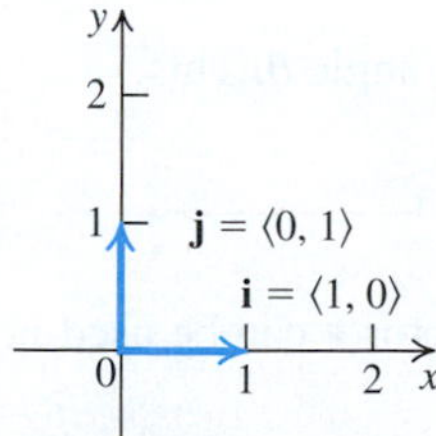

FIGURE 7.30 Standard unit vectors

The unit vectors $\mathbf{i}$ and $\mathbf{j}$ are called **standard unit vectors**. See Figure 7.30.

Every vector $\mathbf{v} = \langle v_1, v_2 \rangle$ can be expressed in terms of $\mathbf{i}$ and $\mathbf{j}$ as follows:

$$\mathbf{v} = \langle v_1, v_2 \rangle = \langle v_1, 0 \rangle + \langle 0, v_2 \rangle \quad \text{Vector addition}$$
$$= v_1\langle 1, 0 \rangle + v_2\langle 0, 1 \rangle \quad \text{Scalar multiplication}$$
$$= v_1\mathbf{i} + v_2\mathbf{j} \quad \text{Replace } \langle 1, 0 \rangle \text{ with } \mathbf{i} \text{ and } \langle 0, 1 \rangle \text{ with } \mathbf{j}.$$

The scalars v_1 and v_2 are called the *horizontal* and *vertical components of* $\mathbf{v}$, respectively. A vector $\mathbf{v} = \langle v_1, v_2 \rangle$ from $(0, 0)$ to (v_1, v_2) can therefore be represented in the form

$$\mathbf{v} = v_1\mathbf{i} + v_2\mathbf{j}$$

with

$$\|\mathbf{v}\| = \sqrt{v_1^2 + v_2^2}.$$

EXAMPLE 5 Vectors Involving i and j

Find each expression for $\mathbf{u} = 4\mathbf{i} + 7\mathbf{j}$ and $\mathbf{v} = 2\mathbf{i} + 5\mathbf{j}$.

a. $\mathbf{u} - 3\mathbf{v}$ **b.** $\|\mathbf{u} - 3\mathbf{v}\|$

Solution

a.
$$\begin{aligned} \mathbf{u} - 3\mathbf{v} &= (4\mathbf{i} + 7\mathbf{j}) - 3(2\mathbf{i} + 5\mathbf{j}) \\ &= 4\mathbf{i} + 7\mathbf{j} - 6\mathbf{i} - 15\mathbf{j} && \text{Scalar multiplication} \\ &= (4 - 6)\mathbf{i} + (7 - 15)\mathbf{j} && \text{Group terms.} \\ &= -2\mathbf{i} - 8\mathbf{j} && \text{Simplify.} \end{aligned}$$

b.
$$\begin{aligned} \|\mathbf{u} - 3\mathbf{v}\| &= \|-2\mathbf{i} - 8\mathbf{j}\| \\ &= \sqrt{(-2)^2 + (-8)^2} && \text{Definition of magnitude} \\ &= \sqrt{68} = 2\sqrt{17} && \text{Simplify.} \end{aligned}$$

Practice Problem 5 Find each expression for $\mathbf{u} = -3\mathbf{i} + 2\mathbf{j}$ and $\mathbf{v} = \mathbf{i} + 4\mathbf{j}$.

a. $3\mathbf{u} + 2\mathbf{v}$ **b.** $\|3\mathbf{u} + 2\mathbf{v}\|$

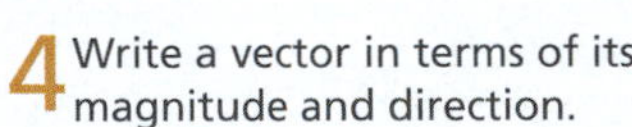

Vector in Terms of Magnitude and Direction

Let $\mathbf{v} = \langle v_1, v_2 \rangle$ be a position vector and suppose θ is the smallest nonnegative angle that $\mathbf{v}$ makes with the positive x-axis. The angle θ is called the **direction angle** of $\mathbf{v}$. See Figure 7.31.

The length (or magnitude) of the vector $\overrightarrow{OP}$ is $\|\mathbf{v}\|$; so

$$\frac{v_1}{\|\mathbf{v}\|} = \cos\theta \quad \text{and} \quad \frac{v_2}{\|\mathbf{v}\|} = \sin\theta \qquad \text{Definitions of cosine and sine}$$

$$v_1 = \|\mathbf{v}\|\cos\theta \qquad v_2 = \|\mathbf{v}\|\sin\theta \qquad \text{Multiply both sides by } \|\mathbf{v}\|.$$

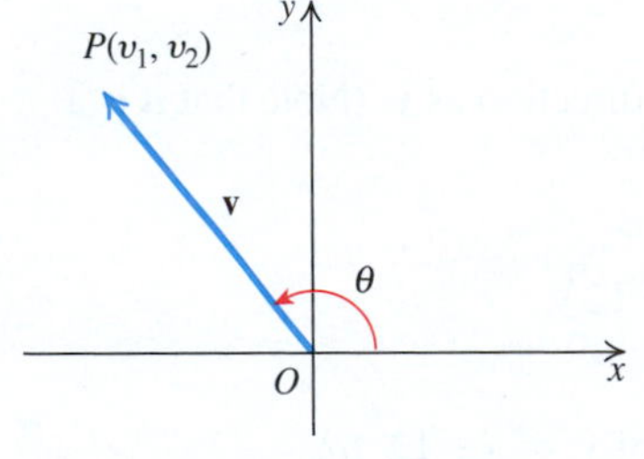

FIGURE 7.31 $\mathbf{v} = \|\mathbf{v}\|(\cos\theta\mathbf{i} + \sin\theta\mathbf{j})$

Hence,

$$\begin{aligned} \mathbf{v} &= v_1\mathbf{i} + v_2\mathbf{j} && \text{Write } \mathbf{v} \text{ in terms of } \mathbf{i} \text{ and } \mathbf{j}. \\ \mathbf{v} &= \|\mathbf{v}\|\cos\theta\,\mathbf{i} + \|\mathbf{v}\|\sin\theta\,\mathbf{j} && \text{Replace values of } v_1 \text{ and } v_2. \\ \mathbf{v} &= \|\mathbf{v}\|(\cos\theta\,\mathbf{i} + \sin\theta\,\mathbf{j}) && \text{Factor out } \|\mathbf{v}\|. \end{aligned}$$

Trigonometric Form of a Vector

The formula

$$\mathbf{v} = \|\mathbf{v}\|(\cos\theta\,\mathbf{i} + \sin\theta\,\mathbf{j})$$

expresses a vector $\mathbf{v}$ in terms of its magnitude $\|\mathbf{v}\|$ and its direction angle θ. This representation is called the **trigonometric form** of the vector $\mathbf{v}$.

Note that any angle coterminal with the direction angle θ of a vector $\mathbf{v}$ can be used in place of θ in the formula $\mathbf{v} = \|\mathbf{v}\|(\cos\theta\,\mathbf{i} + \sin\theta\,\mathbf{j})$.

EXAMPLE 6 Writing a Vector with Given Length and Direction Angle

Write the vector of magnitude 3 whose direction angle is $\frac{\pi}{3}$.

Solution

If **v** is the required vector, then

$$\mathbf{v} = \|\mathbf{v}\|(\cos\theta\,\mathbf{i} + \sin\theta\,\mathbf{j}) \quad \text{Write } \mathbf{v} \text{ in terms of its magnitude and direction.}$$

$$\mathbf{v} = 3\left(\cos\frac{\pi}{3}\mathbf{i} + \sin\frac{\pi}{3}\mathbf{j}\right) \quad \text{Substitute } \|\mathbf{v}\| = 3 \text{ and } \theta = \frac{\pi}{3}.$$

$$\mathbf{v} = 3\left(\frac{1}{2}\mathbf{i} + \frac{\sqrt{3}}{2}\mathbf{j}\right) \quad \cos\frac{\pi}{3} = \frac{1}{2},\ \sin\frac{\pi}{3} = \frac{\sqrt{3}}{2}$$

$$\mathbf{v} = \frac{3}{2}\mathbf{i} + \frac{3\sqrt{3}}{2}\mathbf{j} \quad \text{Distributive property}$$

Practice Problem 6 Write the vector of magnitude 2 that makes an angle of $\frac{11\pi}{6}$ with the positive x-axis.

We can find the direction angle θ of a position vector $\mathbf{v} = \langle v_1, v_2 \rangle$ by using the equation $\tan\theta = \frac{v_2}{v_1}$ and the quadrant in which the point (v_1, v_2) lies.

EXAMPLE 7 Finding the Direction Angle of a Vector

Find the direction angle θ of the vector $\mathbf{v} = -4\mathbf{i} + 3\mathbf{j}$. Round to the nearest hundredth of a degree.

Solution

$$\mathbf{v} = -4\mathbf{i} + 3\mathbf{j}$$

$$= \langle -4, 3 \rangle \quad \text{Write } \mathbf{v} \text{ as a position vector.}$$

$$\tan\theta = \frac{3}{-4} = -\frac{3}{4}$$

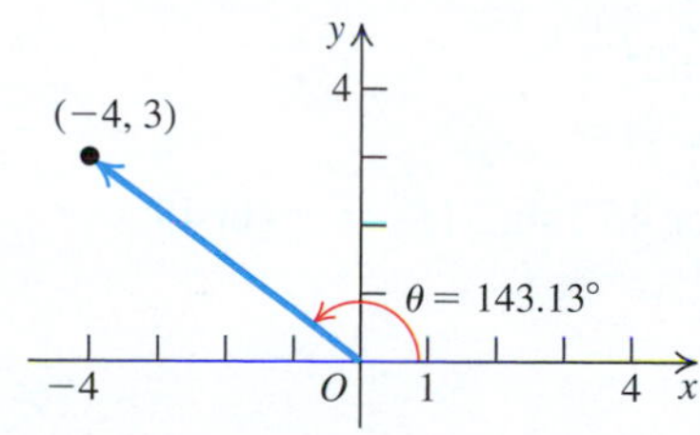

FIGURE 7.32 The direction angle of $\mathbf{v} = -4\mathbf{i} + 3\mathbf{j}$

The reference angle θ' is given by

$$\theta' = \left|\tan^{-1}\left(-\frac{3}{4}\right)\right| \approx |-36.87°| = 36.87° \quad \text{Use a calculator.}$$

Because the point $(-4, 3)$ lies in quadrant II, we have

$$\theta = 180° - \theta' \approx 180° - 36.87° = 143.13°.$$

The direction angle of **v** is 143.13°. See Figure 7.32.

Practice Problem 7 Find the direction angle of $\mathbf{v} = 2\mathbf{i} - 3\mathbf{j}$.

5 Use vectors in applications.

Applications of Vectors

It is an experimental fact that forces behave like vectors. This means that if a system of forces acts on a particle, that particle will move as though it were acted on by a single force equal to the vector sum of the forces. This single force is called the **resultant** of the system of forces. If the particle does not move, the resultant is zero. We say that the particle is in **equilibrium**.

EXAMPLE 8 Finding the Resultant

Find the magnitude and bearing of the resultant **R** of two forces $\mathbf{F}_1$ and $\mathbf{F}_2$, where $\mathbf{F}_1$ is a 50 lb force acting northward and $\mathbf{F}_2$ is a 40 lb force acting eastward.

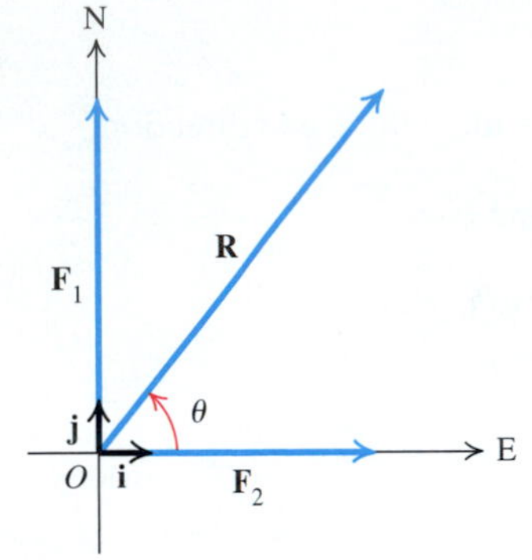

Solution

We set up the coordinate system with the y-axis pointing northward. Then $\mathbf{F}_1 = 50\mathbf{j}$ and $\mathbf{F}_2 = 40\mathbf{i}$. See Figure 7.33.

$$\mathbf{R} = \mathbf{F}_1 + \mathbf{F}_2 = \mathbf{F}_2 + \mathbf{F}_1 = 40\mathbf{i} + 50\mathbf{j}$$

$$\|\mathbf{R}\| = \sqrt{(40)^2 + (50)^2} \approx 64.0 \qquad \text{Use a calculator.}$$

$$\tan\theta = \frac{50}{40} \qquad \text{Direction angle}$$

$$\theta = \tan^{-1}\left(\frac{50}{40}\right) \approx 51.3° \qquad \text{Use a calculator.}$$

FIGURE 7.33 The resultant **R**

The angle between **R** and the y-axis (north direction) is $90° - 51.3° = 38.7°$. Therefore, **R** is a force of approximately 64.0 lb in the direction N 38.7° E.

Practice Problem 8 Repeat Example 8 assuming that $\mathbf{F}_1$ is a 40 lb force acting northward and $\mathbf{F}_2$ is a 30 lb force acting westward.

EXAMPLE 9 Using Vectors in Air Navigation

An F-15 fighter jet is flying over Mount Rushmore at an airspeed (speed in still air) of 800 miles per hour on a bearing of N 30° E. The velocity of wind is 40 miles per hour in the direction of S 45° E. Find the actual speed and direction (relative to the ground) of the plane. Round each answer to the nearest tenth.

Solution

Set up a coordinate system with north along the positive y-axis. See Figure 7.34.

Let **v** be the air velocity of the plane,
w be the wind velocity, and
r be the resultant ground velocity of the plane.

Writing each vector in terms of its magnitude and the direction angle θ that the vector makes with the positive x-axis, we have

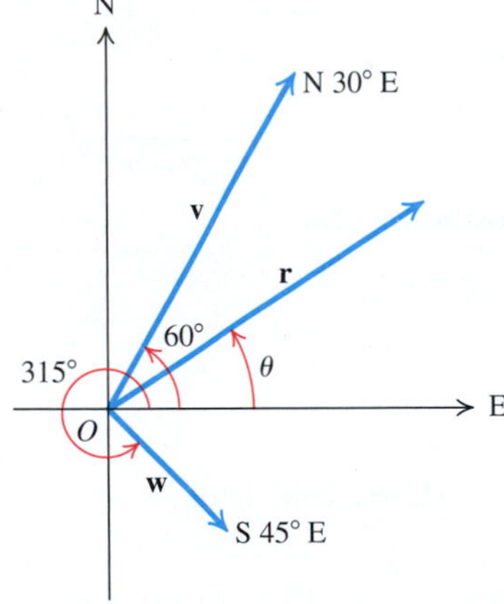

Vectors are not drawn to scale.

FIGURE 7.34 Velocity vectors

$$\mathbf{v} = 800(\cos 60°\mathbf{i} + \sin 60°\mathbf{j}) \qquad 90° - 30° = 60°$$

$$\mathbf{w} = 40[\cos 315°\mathbf{i} + \sin 315°\mathbf{j}] \qquad 270° + 45° = 315°$$

$$= 40[\cos 45°\mathbf{i} - \sin 45°\mathbf{j}] \qquad \cos 315° = \cos 45°;\ \sin 315° = -\sin 45°$$

The resultant **r** is given by

$$\mathbf{r} = \mathbf{v} + \mathbf{w}$$

$$\mathbf{r} = 800(\cos 60°\mathbf{i} + \sin 60°\mathbf{j}) + 40(\cos 45°\mathbf{i} - \sin 45°\mathbf{j}) \qquad \text{Substitute values of } \mathbf{v} \text{ and } \mathbf{w}.$$

$$\mathbf{r} = (800\cos 60° + 40\cos 45°)\mathbf{i} + (800\sin 60° - 40\sin 45°)\mathbf{j} \qquad \text{Add vectors.}$$

$$\|\mathbf{r}\| = \sqrt{(800\cos 60° + 40\cos 45°)^2 + (800\sin 60° - 40\sin 45°)^2} \qquad \|r_1\mathbf{i} + r_2\mathbf{j}\| = \sqrt{r_1^2 + r_2^2}$$

$$\|\mathbf{r}\| \approx 790.6 \qquad \text{Use a calculator.}$$

The ground speed of the F-15 is approximately 790.6 miles per hour. To find the actual direction (bearing) of the plane, we first find the direction angle θ of **r**.

$$\theta = \tan^{-1}\left(\frac{800 \sin 60° - 40 \sin 45°}{800 \cos 60° + 40 \cos 45°}\right) \qquad \text{For } \mathbf{r} = r_1\mathbf{i} + r_2\mathbf{j},\ \theta = \tan^{-1}\left(\frac{r_2}{r_1}\right).$$

$$\approx 57.2° \qquad \text{Use a calculator.}$$

The angle between **r** and the y-axis (direction north) is

$$90° - 57.2° = 32.8°.$$

The bearing of the F-15 is approximately N 32.8° E.

Practice Problem 9 Repeat Example 9 assuming that the bearing of the plane is N 60° W and the direction of the wind is S 30° W.

EXAMPLE 10 Obtaining an Equilibrium

What single force must be added to the system of forces **A**, **B**, **C**, and **D** shown in Figure 7.35 to obtain a system that is in equilibrium?

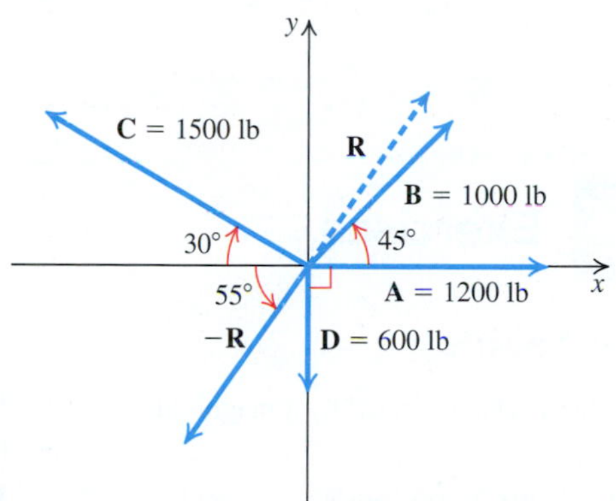

FIGURE 7.35 A system in equilibrium

Solution

We first find the resultant **R** of the given system. The simplest procedure is to find and add the corresponding horizontal components and the vertical components of each force.

The horizontal component r_1 of **R** is given by

$$\begin{aligned} r_1 &= a_1 + b_1 + c_1 + d_1 \\ &= 1200 \cos 0° + 1000 \cos 45° + 1500 \cos 150° + 600 \cos 270° \\ &\approx 608.07 \end{aligned}$$

RECALL

For a vector **v** with direction angle θ, the horizontal component $v_1 = \|\mathbf{v}\| \cos \theta$ and vertical component $v_2 = \|\mathbf{v}\| \sin \theta$.

Similarly, the vertical component r_2 of **R** is given by

$$\begin{aligned} r_2 &= a_2 + b_2 + c_2 + d_2 \\ &= 1200 \sin 0° + 1000 \sin 45° + 1500 \sin 150° + 600 \sin 270° \\ &\approx 857.11 \end{aligned}$$

$$\begin{aligned} \mathbf{R} &= r_1\mathbf{i} + r_2\mathbf{j} \\ &= 608.07\,\mathbf{i} + 857.11\,\mathbf{j} \end{aligned}$$

$$\|\mathbf{R}\| = \sqrt{(608.07)^2 + (857.11)^2} \approx 1051 \text{ lb}$$

$$\theta = \tan^{-1}\left(\frac{r_2}{r_1}\right) = \tan^{-1}\left(\frac{857.11}{608.07}\right)$$

$$\approx 55.0° \qquad \text{Use a calculator.}$$

The force that must be added to obtain equilibrium is the negative of **R**. This is a force of 1051 lb, and $-\mathbf{R}$ makes an angle of $180° + 55° = 235°$ with the positive x-axis. See Figure 7.35.

Practice Problem 10 Find the resultant of the following system of forces acting simultaneously at a point: $\mathbf{u} = 200$ lb in the direction N 40° E, $\mathbf{v} = 300$ lb in the direction N 70° W, and $\mathbf{w} = 400$ lb in the direction S 20° E.

Answers to Practice Problems

1. a.

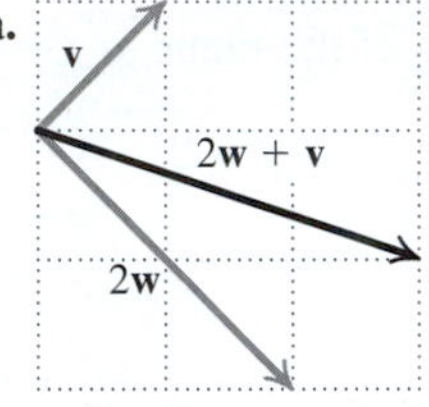

b.

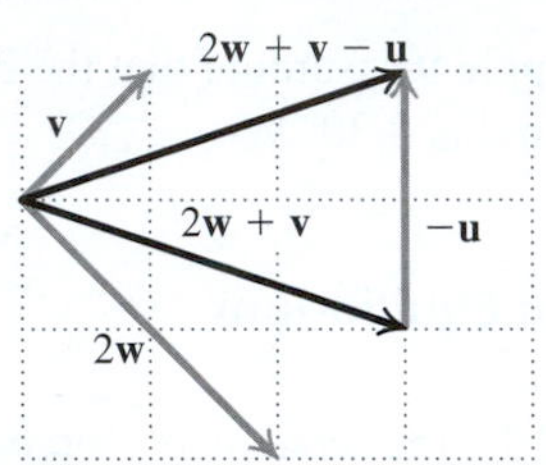

2. $\langle 3, -10 \rangle$ **3. a.** $\langle 1, -1 \rangle$ **b.** $\langle 6, -9 \rangle$ **c.** $\langle 8, -13 \rangle$ **d.** $\sqrt{233}$

4. $\left\langle -\frac{12}{13}, \frac{5}{13} \right\rangle$ **5. a.** $-7\mathbf{i} + 14\mathbf{j}$ **b.** $7\sqrt{5}$ **6.** $\sqrt{3}\mathbf{i} - \mathbf{j}$

7. $\tan^{-1}\left(-\frac{3}{2}\right) + 360° \approx 303.69°$ **8. R** is a force of 50 pounds in the direction N 36.9° W. **9.** The ground speed is approximately 801.0 miles per hour, and the bearing is approximately N 62.9° W. **10.** The magnitude of the resultant is approximately 121.2 lb, and the resultant is in the direction S 7.8° W.

SECTION 7.4 Exercises

Basic Concepts and Skills

1. A vector is a quantity that is characterized by a magnitude and a(n) ____________.
2. The resultant of **v** and **w** is the vector sum $\mathbf{v} + \mathbf{w}$ and is represented by the diagonal of the ____________ with adjacent sides **v** and **w**.
3. If $\mathbf{v} = \langle a, b \rangle$ and $a > 0$, then $\|\mathbf{v}\| =$ ____________, and its direction angle $\theta = \tan^{-1}$(____________).
4. If **v** is a nonzero vector, then the unit vector **u** in the direction of **v** is given by $\mathbf{u} = \frac{1}{\|\mathbf{v}\|}$(____________).
5. **True or False.** The zero vector **0** is the only vector with no direction specified.
6. **True or False.** The vector $-\mathbf{v}$ has the same magnitude as **v** but the opposite direction.

In Exercises 7–14, use the vectors u, v, and w in the accompanying figure to graph each vector.

7. $\mathbf{u} + \mathbf{v}$
8. $3\mathbf{v}$
9. $2\mathbf{u} - \mathbf{w}$
10. $\mathbf{w} - 2\mathbf{v}$
11. $\mathbf{u} + \mathbf{v} + \mathbf{w}$
12. $2\mathbf{u} - \mathbf{w} + \mathbf{v}$
13. $\mathbf{w} - 2\mathbf{v} + \mathbf{u}$
14. $2\mathbf{v} + 3\mathbf{w} - \mathbf{u}$

In Exercises 15–22, the vector v has initial point P and terminal point Q. Write v as a position vector.

15. $P(3, 6), Q(2, 9)$
16. $P(6, -4), Q(1, 1)$
17. $P(-5, -2), Q(-3, -4)$
18. $P(0, 0), Q(-3, -6)$
19. $P(-1, 4), Q(2, -3)$
20. $P(3.5, 2.7), Q(-1.5, 1.3)$
21. $P\left(\frac{1}{2}, \frac{3}{4}\right), Q\left(-\frac{1}{2}, -\frac{7}{4}\right)$
22. $P\left(-\frac{2}{3}, \frac{4}{9}\right), Q\left(\frac{1}{3}, -\frac{2}{3}\right)$

In Exercises 23–26, determine whether the vectors $\overrightarrow{AB}$ and $\overrightarrow{CD}$ are equivalent. [*Hint:* Write $\overrightarrow{AB}$ and $\overrightarrow{CD}$ as position vectors.]

23. $A(1, 0), B(3, 4), C(-1, 2), D(1, 6)$
24. $A(-1, 2), B(3, -2), C(2, 5), D(6, 1)$
25. $A(2, -1), B(3, 5), C(-1, 3), D(-2, -3)$
26. $A(5, 7), B(6, 3), C(-2, 1), D(-3, 5)$

In Exercises 27–34, let $\mathbf{v} = \langle -1, 2 \rangle$ and $\mathbf{w} = \langle 3, -2 \rangle$. Find each expression.

27. $\|\mathbf{v}\|$
28. $\|\mathbf{w}\|$
29. $\mathbf{v} - \mathbf{w}$
30. $\mathbf{v} + \mathbf{w}$
31. $2\mathbf{v} - 3\mathbf{w}$
32. $2\mathbf{w} - 3\mathbf{v}$
33. $\|2\mathbf{v} - 3\mathbf{w}\|$
34. $\|2\mathbf{w} - 3\mathbf{v}\|$

In Exercises 35–40, find a unit vector u in the direction of the given vector.

35. $\langle 1, -1 \rangle$
36. $\langle 1, 3 \rangle$
37. $\langle -4, 3 \rangle$
38. $\langle 5, -12 \rangle$
39. $\langle \sqrt{2}, \sqrt{2} \rangle$
40. $\langle \sqrt{5}, -2 \rangle$

In Exercises 41–46, find each expression for $\mathbf{u} = 2\mathbf{i} - 5\mathbf{j}$ and $\mathbf{v} = -3\mathbf{i} - 2\mathbf{j}$.

41. $\mathbf{u} + \mathbf{v}$
42. $\mathbf{u} - \mathbf{v}$
43. $2\mathbf{u} - 3\mathbf{v}$
44. $2\mathbf{v} + 3\mathbf{u}$
45. $\|2\mathbf{u} - 3\mathbf{v}\|$
46. $\|2\mathbf{v} + 3\mathbf{u}\|$

In Exercises 47–54, write the vector v in the form $v_1\mathbf{i} + v_2\mathbf{j}$, given $\|\mathbf{v}\|$ and the angle θ that v makes with the positive *x*-axis.

47. $\|\mathbf{v}\| = 2, \theta = 30°$
48. $\|\mathbf{v}\| = 5, \theta = 45°$
49. $\|\mathbf{v}\| = 4, \theta = 120°$
50. $\|\mathbf{v}\| = 3, \theta = 150°$
51. $\|\mathbf{v}\| = 3, \theta = \frac{5\pi}{3}$
52. $\|\mathbf{v}\| = 4, \theta = \frac{11\pi}{6}$
53. $\|\mathbf{v}\| = 7, \theta = -\frac{\pi}{3}$
54. $\|\mathbf{v}\| = 8, \theta = \frac{3\pi}{4}$

In Exercises 55–62, find the magnitude and the direction angle of the vector v.

55. $\mathbf{v} = 10(\cos 60°\mathbf{i} + \sin 60°\mathbf{j})$
56. $\mathbf{v} = -4(\cos 30°\mathbf{i} + \sin 30°\mathbf{j})$
57. $\mathbf{v} = -3(\cos 30°\mathbf{i} - \sin 30°\mathbf{j})$
58. $\mathbf{v} = 2(\cos 300°\mathbf{i} - \sin 300°\mathbf{j})$
59. $\mathbf{v} = 5\mathbf{i} + 12\mathbf{j}$
60. $\mathbf{v} = 12\mathbf{i} - 5\mathbf{j}$
61. $\mathbf{v} = -4\mathbf{i} - 3\mathbf{j}$
62. $\mathbf{v} = -5\mathbf{i} + 12\mathbf{j}$

Applying the Concepts

In Exercises 63–74 write each answer to the nearest tenth of a unit.

63. **Wind and vectors.** A wind is blowing 25 miles per hour in the direction N 67° E. Express the wind velocity **v** in the form $a\mathbf{i} + b\mathbf{j}$.

64. **Wind and vectors.** The velocity **v** of a wind is given by $\mathbf{v} = 5\mathbf{i} - 12\mathbf{j}$, where the vectors **i** and **j** represent 1-mile-per-hour winds blowing east and north, respectively. Find the speed (magnitude of **v**) and the direction of the wind.

65. **Resultant force.** Find the magnitude and bearing of the resultant **R** of two forces $\mathbf{F}_1$ and $\mathbf{F}_2$, where $\mathbf{F}_1$ is a force of 25 pounds acting due south and $\mathbf{F}_2$ is a force of 32 pounds acting due west.

66. **Resultant force.** Repeat Exercise 65 assuming that $\mathbf{F}_1$ is doubled.

67. **Finding components.** A force of 80 pounds acts in the direction N 65° E. Find its east and north components.

68. **Finding components.** Repeat Exercise 67 assuming that a force of 60 pounds acts in the direction S 32° W.

69. **Resultant force.** Two forces of 300 pounds and 400 pounds act at the origin and make angles of 30° and 60°, respectively, with the positive *x*-axis. Find the resultant.

70. **Resultant force.** Repeat Exercise 69 assuming that the forces have the same magnitude but that the angle each force makes with the positive *x*-axis is doubled.

71. **Air navigation.** A plane flies at an airspeed (speed in still air) of 500 miles per hour on a bearing of N 35° E. An east wind (wind from east to west) is blowing at 30 miles per hour. Find the plane's ground speed and direction.

72. **Air navigation.** A plane flies at an airspeed of 550 miles per hour on a straight course from an airfield *F*. A 30-mile-per-hour wind is blowing from the west. What must be the plane's bearing so that after one hour of flying time, the plane is due north of *F*?

73. **River navigation.** A river flowing southward has a current of 4 miles per hour. A motorboat in still water maintains a speed of 15 miles per hour. The motorboat starts at the west shore on a bearing of N 40° E. What is the actual speed and direction of the boat?

74. **River crossing.** A fisherwoman on the west bank of the river of Exercise 73 sees several people fishing at a spot *P* directly east of her. If she owns the boat of Exercise 73, what direction should she take to land on the spot *P*?

In Exercises 75 and 76, find a single force that must be added to the system of forces shown in each figure to obtain a system that is in equilibrium.

75.

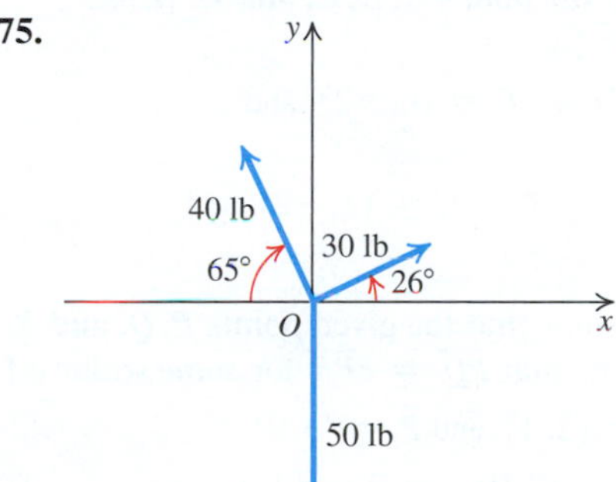

76.

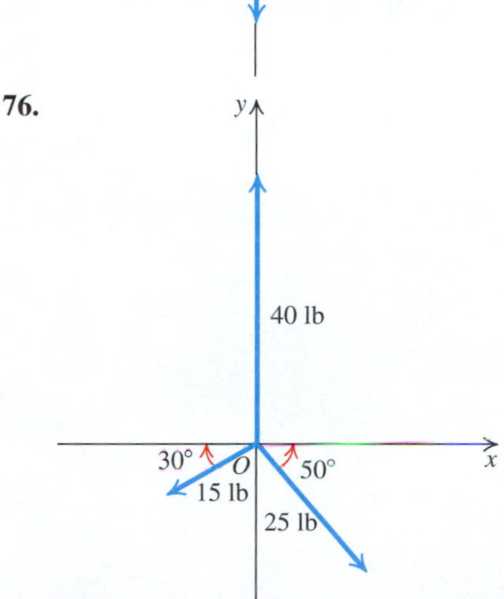

Beyond the Basics

In Exercises 77 and 78 calculate the magnitude of the forces $\mathbf{F}_1$ and $\mathbf{F}_2$ assuming that the system is in equilibrium.

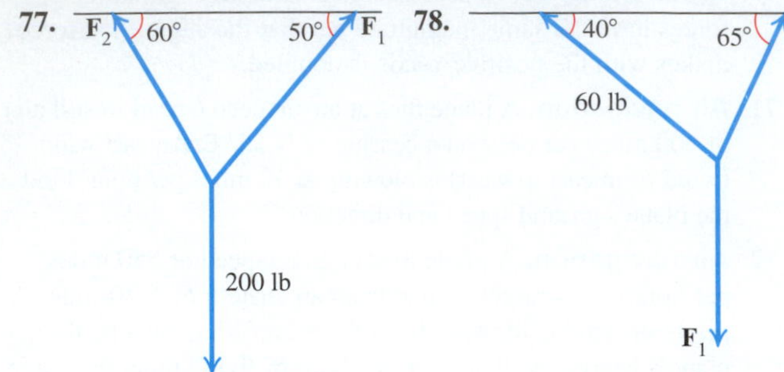

79. Find the terminal point of $\mathbf{v} = 4\mathbf{i} - 3\mathbf{j}$ assuming that the initial point is $(-2, 1)$.

80. Find the initial point of $\mathbf{v} = \langle -3, 5 \rangle$ assuming that the terminal point is $(4, 0)$.

81. Let $\mathbf{u} = \langle -1, 2 \rangle$ and $\mathbf{v} = \langle 3, 5 \rangle$. Find a vector $\mathbf{x} = \langle x_1, x_2 \rangle$ that satisfies the equation $2\mathbf{u} - \mathbf{x} = 2\mathbf{x} + 3\mathbf{v}$.

82. Let $P(3, 5)$ and $Q(7, -4)$ be two points in a coordinate plane. Use vectors to find the coordinates of the point on the line segment joining P and Q that is $\frac{3}{4}$ of the way from P to Q.

In Exercises 83 and 84, show that *ABCD* is a parallelogram, given the coordinates of the points *A*, *B*, *C*, and *D*. [*Hint:* Show that $\overrightarrow{AB} = \pm\overrightarrow{CD}$.]

83. $A = (2, 3)$, $B = (1, 4)$, $C = (0, -2)$, and $D = (1, -3)$

84. $A = (-2, -1)$, $B = (3, 0)$, $C = (1, -2)$, and $D = (-4, -3)$

In Exercise 85 and 86, show that the given points *P*, *Q*, and *R* are collinear. [*Hint:* Show that $\overrightarrow{PQ} = c\overrightarrow{PR}$ for some scalar *c*.]

85. $P = (0, -3)$, $Q = (2, 1)$, and $R = (3, 3)$

86. $P = (-2, -6)$, $Q = (2, 0)$, and $R = (4, 3)$

87. Use vectors to find the lengths of the diagonals of the parallelogram determined by $\mathbf{i} + 2\mathbf{j}$ and $\mathbf{i} - \mathbf{j}$.

88. Use vectors to find the fourth vertex of a parallelogram, with three vertices at $(0, 0)$, $(2, 3)$, and $(6, 4)$. [*Hint:* There is more than one answer.]

Critical Thinking/Discussion/Writing

Let $A = (1, 2)$, $B = (4, 3)$, and $C = (6, 1)$ be points in the plane. Find $P = (x, y)$ so that each of the following is true.

89. $\overrightarrow{AB} = \overrightarrow{CP}$

90. $\overrightarrow{PA} = \overrightarrow{CB}$

91. $\overrightarrow{AP} = \overrightarrow{CB}$

92. $\overrightarrow{PA} = \overrightarrow{BC}$

93. Explain why two equivalent vectors with the same initial point must have the same terminal point.

94. Give an example of a physical environment in which two equivalent vectors have a different effect in that environment.

Maintaining Skills

95. **True or False.** If $\cos\theta > 0$, then θ is an acute angle.

96. **True or False.** If $\cos\theta = 0$, then $\theta = \frac{\pi}{2}$.

97. Find $\cos^{-1}(-1)$, $\cos^{-1}(0)$, and $\cos^{-1}(1)$.

98. Show that the lines $2x + 3y = 6$ and $3x - 2y = 12$ are perpendicular to each other.

In Exercises 99–102, let $\mathbf{v} = \langle 1, -2 \rangle$ and $\mathbf{w} = \langle 3, 2 \rangle$.

99. Compute $\|\mathbf{v}\|$.

100. Compute $\|\mathbf{w}\|$.

101. Compute $\dfrac{(1)(3) + (-2)(2)}{\|\mathbf{v}\|\|\mathbf{w}\|}$.

102. Verify that $\|\mathbf{v} + \mathbf{w}\|^2 = \|\mathbf{v}\|^2 + \|\mathbf{w}\|^2 + 2[(1)(3) + (-2)(2)]$.

SECTION 7.5

The Dot Product

BEFORE STARTING THIS SECTION, REVIEW

1 The Law of Sines (Section 7.1, page 642)
2 The Law of Cosines (Section 7.2, page 654)

OBJECTIVES

1 Define the dot product of two vectors.
2 Find the angle between two vectors.
3 Define orthogonal vectors.
4 Find the projection of a vector onto another vector.
5 Decompose a vector into two orthogonal vectors.
6 Use the definition of work.

How to Measure Work

If you move an object a distance of d units by applying a force F in the direction of motion, then physicists define the work W done on the object by the force F to be

$$W = Fd$$

$$\text{Work} = (\text{force})(\text{distance}).$$

If the force is measured in pounds and the distance is measured in feet, then the unit of work is *foot-pounds.* Unfortunately, it is not always possible to exert a force in the direction you would like. For example, if you are pulling a child's wagon, you exert a force in a direction dictated by the position of the hand you use to pull the wagon. In this section, we give a precise definition of work done by a constant force, and in Example 10, we compute the work done in one such instance.

1 Define the dot product of two vectors.

The Dot Product

We know that scalar multiplication and vector addition and subtraction produce vectors. A new operation, the *dot product* of two vectors, produces a *scalar.*

The Dot Product

For two vectors $\mathbf{v} = \langle v_1, v_2 \rangle$ and $\mathbf{w} = \langle w_1, w_2 \rangle$, the **dot product** of $\mathbf{v}$ and $\mathbf{w}$, denoted $\mathbf{v} \cdot \mathbf{w}$, is defined by

$$\mathbf{v} \cdot \mathbf{w} = \langle v_1, v_2 \rangle \cdot \langle w_1, w_2 \rangle = v_1 w_1 + v_2 w_2.$$

EXAMPLE 1 Finding the Dot Product

Find the dot product $\mathbf{v} \cdot \mathbf{w}$.

a. $\mathbf{v} = \langle 2, -3 \rangle$ and $\mathbf{w} = \langle 3, 4 \rangle$

b. $\mathbf{v} = -3\mathbf{i} + 5\mathbf{j}$ and $\mathbf{w} = 2\mathbf{i} + 3\mathbf{j}$

Solution

a. $\mathbf{v}\cdot\mathbf{w} = \langle 2, -3\rangle \cdot \langle 3, 4\rangle$

$= (2)(3) + (-3)(4) = -6$ — Definition of dot product

b. $\mathbf{v}\cdot\mathbf{w} = (-3\mathbf{i} + 5\mathbf{j})\cdot(2\mathbf{i} + 3\mathbf{j})$

$= \langle -3, 5\rangle \cdot \langle 2, 3\rangle$ — Rewrite as position vectors.

$= (-3)(2) + (5)(3) = 9$ — Definition of dot product

Practice Problem 1 Find the dot product $\mathbf{v}\cdot\mathbf{w}$.

a. $\mathbf{v} = \langle 1, 2\rangle$ and $\mathbf{w} = \langle -2, 5\rangle$

b. $\mathbf{v} = 4\mathbf{i} - 3\mathbf{j}$ and $\mathbf{w} = -3\mathbf{i} - 4\mathbf{j}$

2 Find the angle between two vectors.

The Angle Between Two Vectors

We use the Law of Cosines to find another formula for the dot product. The new formula will be used to find the angle between any two vectors.

Let θ be the angle between the two vectors $\mathbf{v} = \langle v_1, v_2\rangle$ and $\mathbf{w} = \langle w_1, w_2\rangle$. See Figure 7.36. Apply the Law of Cosines to the triangle OPQ in Figure 7.36.

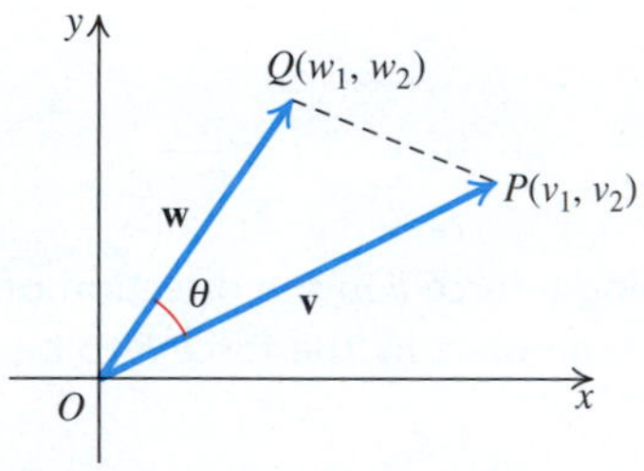

FIGURE 7.36 Angle between two vectors

$[d(P, Q)]^2 = \|\mathbf{v}\|^2 + \|\mathbf{w}\|^2 - 2\|\mathbf{v}\|\,\|\mathbf{w}\|\cos\theta$ — The Law of Cosines

$(v_1 - w_1)^2 + (v_2 - w_2)^2$

$= (v_1^2 + v_2^2) + (w_1^2 + w_2^2) - 2\|\mathbf{v}\|\,\|\mathbf{w}\|\cos\theta$ — Use the distance formula and square of the magnitudes of **v** and **w**.

$v_1^2 - 2v_1w_1 + w_1^2 + v_2^2 - 2v_2w_2 + w_2^2$

$= v_1^2 + v_2^2 + w_1^2 + w_2^2 - 2\|\mathbf{v}\|\,\|\mathbf{w}\|\cos\theta$ — $(A - B)^2 = A^2 - 2AB + B^2$

$-2v_1w_1 - 2v_2w_2 = -2\|\mathbf{v}\|\,\|\mathbf{w}\|\cos\theta$ — Subtract $v_1^2 + w_1^2 + v_2^2 + w_2^2$ from both sides.

$v_1w_1 + v_2w_2 = \|\mathbf{v}\|\,\|\mathbf{w}\|\cos\theta$ — Divide both sides by -2.

$\mathbf{v}\cdot\mathbf{w} = \|\mathbf{v}\|\,\|\mathbf{w}\|\cos\theta$ — Definition of $\mathbf{v}\cdot\mathbf{w}$

The last equation provides us with another form for the dot product $\mathbf{v}\cdot\mathbf{w}$. Solving the last equation for $\cos\theta$ gives us a formula for finding the angle between two vectors.

THE DOT PRODUCT AND THE ANGLE BETWEEN TWO VECTORS

If θ $(0° \le \theta \le 180°)$ is the angle between two nonzero vectors **v** and **w**, then

$$\mathbf{v}\cdot\mathbf{w} = \|\mathbf{v}\|\|\mathbf{w}\|\cos\theta \quad \text{or} \quad \cos\theta = \frac{\mathbf{v}\cdot\mathbf{w}}{\|\mathbf{v}\|\|\mathbf{w}\|}.$$

EXAMPLE 2 **Finding the Dot Product**

If **v** and **w** are two vectors of magnitudes 4 and 7, respectively, and the angle between them is 30°, find $\mathbf{v}\cdot\mathbf{w}$. Round the answer to the nearest tenth.

Solution

$\mathbf{v}\cdot\mathbf{w} = \|\mathbf{v}\|\,\|\mathbf{w}\|\cos\theta$ — Formula for $\mathbf{v}\cdot\mathbf{w}$

$= (4)(7)\cos 30°$ — Substitute given values.

$= \frac{28\sqrt{3}}{2} = 14\sqrt{3}$ — Exact answer

≈ 24.2 — Use a calculator.

Practice Problem 2 Repeat Example 2 assuming that the angle between **v** and **w** is 60°.

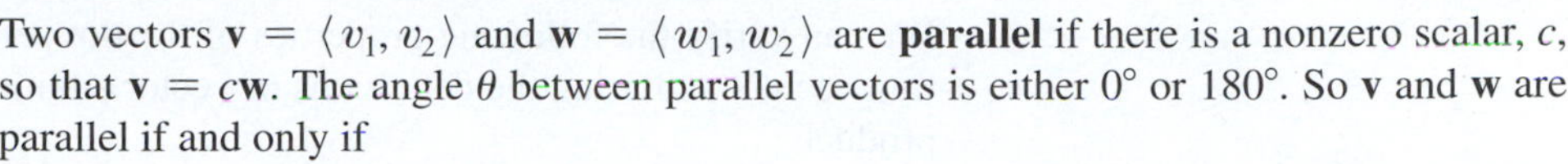

Two vectors $\mathbf{v} = \langle v_1, v_2 \rangle$ and $\mathbf{w} = \langle w_1, w_2 \rangle$ are **parallel** if there is a nonzero scalar, c, so that $\mathbf{v} = c\mathbf{w}$. The angle θ between parallel vectors is either 0° or 180°. So **v** and **w** are parallel if and only if

(i) $\mathbf{v} \cdot \mathbf{w} = \pm \|\mathbf{v}\| \|\mathbf{w}\|$ or

(ii) $\dfrac{v_1}{w_1} = \dfrac{v_2}{w_2}$.

EXAMPLE 3 Deciding Whether Two Vectors Are Parallel

Determine whether the vectors $\mathbf{v} = 2\mathbf{i} - 3\mathbf{j}$ and $\mathbf{w} = -4\mathbf{i} + 6\mathbf{j}$ are parallel.

Solution

Because $\mathbf{w} = -2(2\mathbf{i} - 3\mathbf{j}) = -2\mathbf{v}$, the vectors **v** and **w** are parallel. Note that

$$\cos\theta = \frac{\mathbf{v} \cdot \mathbf{w}}{\|\mathbf{v}\| \|\mathbf{w}\|} = \frac{-8 - 18}{\sqrt{13}\sqrt{52}} = \frac{-26}{\sqrt{13}(2\sqrt{13})} = \frac{-26}{26} = -1,$$

so that $\theta = 180°$ is the angle between **v** and **w**. The vectors **v** and **w** are parallel.

Alternate Solution $\quad \dfrac{2}{-4} \stackrel{?}{=} \dfrac{-3}{6} \quad$ check $\dfrac{v_1}{w_1} = \dfrac{v_2}{w_2}$

or $-\dfrac{1}{2} \stackrel{?}{=} -\dfrac{1}{2} \quad$ YES!

The vectors **v** and **w** are parallel.

Practice Problem 3 Determine whether $\mathbf{v} = -\mathbf{i} - 2\mathbf{j}$ and $\mathbf{w} = 2\mathbf{i} + \mathbf{j}$ are parallel.

Because the angle θ between two nonzero vectors **v** and **w** satisfies $\cos\theta = \dfrac{\mathbf{v} \cdot \mathbf{w}}{\|\mathbf{v}\| \|\mathbf{w}\|}$ and $0° \le \theta \le 180°$, we have $\theta = \cos^{-1}\left(\dfrac{\mathbf{v} \cdot \mathbf{w}}{\|\mathbf{v}\| \|\mathbf{w}\|}\right)$.

EXAMPLE 4 Finding the Angle Between Two Vectors

Find the angle θ (in degrees) between the vectors $\mathbf{v} = 2\mathbf{i} + 3\mathbf{j}$ and $\mathbf{w} = -3\mathbf{i} + 4\mathbf{j}$. Round the answer to the nearest tenth of a degree.

Solution

$$\cos\theta = \frac{\mathbf{v} \cdot \mathbf{w}}{\|\mathbf{v}\| \|\mathbf{w}\|}$$ Formula for the cosine of the angle between **v** and **w**

$$\cos\theta = \frac{(2\mathbf{i} + 3\mathbf{j}) \cdot (-3\mathbf{i} + 4\mathbf{j})}{\|2\mathbf{i} + 3\mathbf{j}\| \|-3\mathbf{i} + 4\mathbf{j}\|}$$ Substitute expressions for **v** and **w**.

$$\cos\theta = \frac{(2)(-3) + (3)(4)}{\sqrt{2^2 + 3^2}\sqrt{(-3)^2 + 4^2}}$$ Use formulas for the dot product and the magnitude.

$$\cos\theta = \frac{6}{5\sqrt{13}}$$ Simplify.

$$\theta = \cos^{-1}\left(\frac{6}{5\sqrt{13}}\right)$$ $0° \le \theta \le 180°$

$$\theta \approx 70.6°$$ Use a calculator.

Practice Problem 4 Find the angle (in degrees) between the vectors $\mathbf{v} = 2\mathbf{i} - 3\mathbf{j}$ and $\mathbf{w} = 3\mathbf{i} + 5\mathbf{j}$. Round the answer to the nearest tenth of a degree.

We can verify the following properties of the dot product by writing the vectors as position vectors and using the definitions of vector addition, scalar multiplication, and the dot product.

Properties of the Dot Product

If $\mathbf{u}$, $\mathbf{v}$, and $\mathbf{w}$ are vectors and c is a scalar, then

1. $\mathbf{u} \cdot \mathbf{v} = \mathbf{v} \cdot \mathbf{u}$.
2. $\mathbf{u} \cdot (\mathbf{v} + \mathbf{w}) = \mathbf{u} \cdot \mathbf{v} + \mathbf{u} \cdot \mathbf{w}$.
3. $\mathbf{0} \cdot \mathbf{v} = 0$.
4. $\mathbf{v} \cdot \mathbf{v} = \|\mathbf{v}\|^2$.
5. $(c\mathbf{u}) \cdot \mathbf{v} = c(\mathbf{u} \cdot \mathbf{v}) = \mathbf{u} \cdot (c\mathbf{v})$.

For the vectors $\mathbf{i} = \langle 1, 0 \rangle$ and $\mathbf{j} = \langle 0, 1 \rangle$, we have

$$\mathbf{i} \cdot \mathbf{i} = \|\mathbf{i}\|^2 = 1, \quad \mathbf{j} \cdot \mathbf{j} = \|\mathbf{j}\|^2 = 1, \quad \text{and} \quad \mathbf{i} \cdot \mathbf{j} = \mathbf{j} \cdot \mathbf{i} = 0.$$

EXAMPLE 5 Finding the Magnitude of the Sum of Two Vectors

Let $\mathbf{v}$ and $\mathbf{w}$ be two vectors in the plane of magnitudes 12 and 7, respectively. The angle between $\mathbf{v}$ and $\mathbf{w}$ is 65°. Find the magnitude of $\mathbf{v} + 2\mathbf{w}$ to the nearest tenth.

Solution

$$\begin{aligned}
\|\mathbf{v} + 2\mathbf{w}\|^2 &= (\mathbf{v} + 2\mathbf{w}) \cdot (\mathbf{v} + 2\mathbf{w}) && \text{Property 4} \\
&= (\mathbf{v} + 2\mathbf{w}) \cdot \mathbf{v} + (\mathbf{v} + 2\mathbf{w}) \cdot (2\mathbf{w}) && \text{Property 2} \\
&= \mathbf{v} \cdot \mathbf{v} + 2(\mathbf{w} \cdot \mathbf{v}) + 2(\mathbf{v} \cdot \mathbf{w}) + 4(\mathbf{w} \cdot \mathbf{w}) && \text{Properties 1 and 5} \\
&= \|\mathbf{v}\|^2 + 4(\mathbf{v} \cdot \mathbf{w}) + 4\|\mathbf{w}\|^2 && \text{Properties 1 and 4} \\
&= \|\mathbf{v}\|^2 + 4\|\mathbf{v}\|\|\mathbf{w}\| \cos\theta + 4\|\mathbf{w}\|^2 && \mathbf{v} \cdot \mathbf{w} = \|\mathbf{v}\|\|\mathbf{w}\| \cos\theta \\
&= (12)^2 + 4(12)(7)\cos 65^\circ + 4(7)^2 && \text{Substitute values.} \\
\|\mathbf{v} + 2\mathbf{w}\|^2 &\approx 481.9997 && \text{Use a calculator.} \\
\|\mathbf{v} + 2\mathbf{w}\| &\approx 21.95 \approx 22.0 && \text{Take the positive square root.}
\end{aligned}$$

Practice Problem 5 Repeat Example 5 assuming that $\|\mathbf{v}\| = 20$ and $\|\mathbf{w}\| = 12$ and the angle between $\mathbf{v}$ and $\mathbf{w}$ is 54°.

The procedure in Example 5 can be used to prove the following identities.

For any two vectors $\mathbf{v}$ and $\mathbf{w}$,

(i) $\|\mathbf{v} + \mathbf{w}\|^2 = \|\mathbf{v}\|^2 + \|\mathbf{w}\|^2 + 2(\mathbf{v} \cdot \mathbf{w})$.
(ii) $\|\mathbf{v} - \mathbf{w}\|^2 = \|\mathbf{v}\|^2 + \|\mathbf{w}\|^2 - 2(\mathbf{v} \cdot \mathbf{w})$.
(iii) $(\mathbf{v} + \mathbf{w}) \cdot (\mathbf{v} - \mathbf{w}) = \|\mathbf{v}\|^2 - \|\mathbf{w}\|^2$.

3 Define orthogonal vectors.

Orthogonal Vectors

We know that if $\mathbf{v}$ and $\mathbf{w}$ are two nonzero vectors and $\theta\,(0° \le \theta \le 180°)$ is the measure of the angle between them, then

$$\mathbf{v} \cdot \mathbf{w} = \|\mathbf{v}\|\|\mathbf{w}\| \cos \theta \qquad \text{Form of the dot product}$$

From this equation, it follows that

(i) $\mathbf{v} \cdot \mathbf{w} > 0$ if and only if θ is zero or an acute angle,

(ii) $\mathbf{v} \cdot \mathbf{w} < 0$ if and only if θ is a straight or an obtuse angle, and

(iii) $\mathbf{v} \cdot \mathbf{w} = 0$ if and only if $\cos \theta = 0$.

Because $0° \le \theta \le 180°$, we conclude that

$$\mathbf{v} \cdot \mathbf{w} = 0 \text{ if and only if } \theta = 90°.$$

Orthogonal Vectors

Two vectors $\mathbf{v}$ and $\mathbf{w}$ are **orthogonal** (perpendicular) if and only if $\mathbf{v} \cdot \mathbf{w} = 0$.

Because $\mathbf{0} \cdot \mathbf{w} = 0$ for any vector $\mathbf{w}$, it follows from the definition that the zero vector is orthogonal to every vector.

EXAMPLE 6 Deciding Whether Vectors Are Orthogonal

Determine whether $\mathbf{v} = 2\mathbf{i} - 3\mathbf{j}$ and $\mathbf{w} = 3\mathbf{i} + 2\mathbf{j}$ are orthogonal.

Solution

Because $\mathbf{v} \cdot \mathbf{w} = (2\mathbf{i} - 3\mathbf{j}) \cdot (3\mathbf{i} + 2\mathbf{j}) = \langle 2, -3 \rangle \cdot \langle 3, 2 \rangle = 6 - 6 = 0$, $\mathbf{v}$ and $\mathbf{w}$ are orthogonal.

Practice Problem 6 Determine whether $\mathbf{v} = 4\mathbf{i} + 8\mathbf{j}$ and $\mathbf{w} = 2\mathbf{i} - \mathbf{j}$ are orthogonal.

EXAMPLE 7 Orthogonal Vectors

Find the scalar c so that the vectors $\mathbf{v} = 3\mathbf{i} + 2\mathbf{j}$ and $\mathbf{w} = 4\mathbf{i} + c\mathbf{j}$ are orthogonal.

Solution

The vectors $\mathbf{v} = 3\mathbf{i} + 2\mathbf{j} = \langle 3, 2 \rangle$ and $\mathbf{w} = 4\mathbf{i} + c\mathbf{j} = \langle 4, c \rangle$ are orthogonal if and only if

$$\begin{aligned} \mathbf{v} \cdot \mathbf{w} &= 0 && \text{Definition of orthogonality} \\ \langle 3, 2 \rangle \cdot \langle 4, c \rangle &= 0 && \\ (3)(4) + (2)(c) &= 0 && \text{Definition of dot product} \\ 12 + 2c &= 0 && \text{Simplify.} \\ c &= -6 && \text{Solve for } c. \end{aligned}$$

Practice Problem 7 Find the scalar k so that the vectors $\mathbf{v} = \langle k, -2 \rangle$ and $\mathbf{w} = \langle 4, 6 \rangle$ are orthogonal.

4 Find the projection of a vector onto another vector.

Projection of a Vector

A geometric interpretation of the dot product is obtained by considering the *projection* of a vector onto another vector.

Let $\overrightarrow{OP}$ and $\overrightarrow{OQ}$ be representations of the nonzero vectors **v** and **w**, respectively, The projections of $\overrightarrow{OP}$ in the direction of $\overrightarrow{OQ}$ is the directed line segment $\overrightarrow{OR}$, where R is the foot of the perpendicular from P to the line containing $\overrightarrow{OQ}$. See Figure 7.37. Then the vector represented by $\overrightarrow{OR}$ is called the **vector projection of v onto w** and is denoted by $\text{proj}_{\mathbf{w}}\mathbf{v}$.

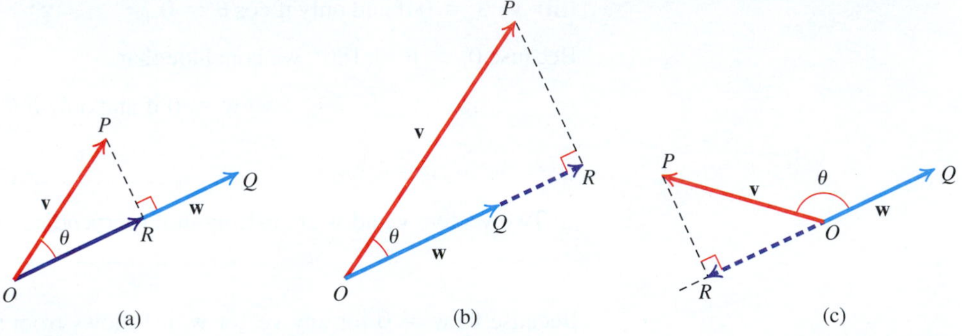

FIGURE 7.37 Vector projection of **v** onto **w**

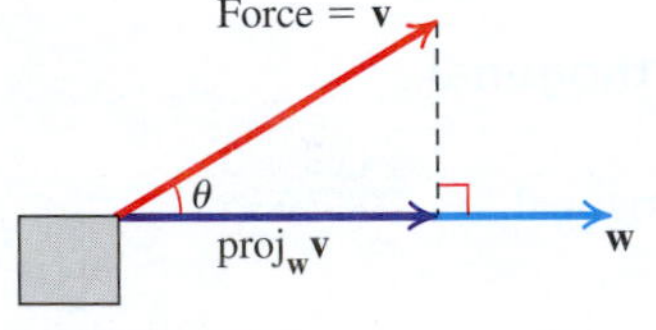

FIGURE 7.38

In Figure 7.38, if a force **v** is used to pull a box, then the effective force that moves the box in the direction of **w** is the vector projection of **v** onto **w**.

Let $\theta\,(0° \le \theta \le 180°)$ be the measure of the angle between nonzero vectors **v** and **w**. The number $\|\mathbf{v}\|\cos\theta$ is called the **scalar projection of v onto w**. We can express this number in terms of the dot product.

$$\begin{aligned}\text{Scalar projection of } \mathbf{v} \text{ onto } \mathbf{w} &= \|\mathbf{v}\|\cos\theta \\ &= \frac{\|\mathbf{v}\|\,\|\mathbf{w}\|\cos\theta}{\|\mathbf{w}\|} && \text{Multiply numerator and denominator by } \|\mathbf{w}\|. \\ &= \frac{\mathbf{v}\cdot\mathbf{w}}{\|\mathbf{w}\|} && \mathbf{v}\cdot\mathbf{w} = \|\mathbf{v}\|\,\|\mathbf{w}\|\cos\theta\end{aligned}$$

If θ is acute, then $\text{proj}_{\mathbf{w}}\mathbf{v}$ has magnitude $\|\mathbf{v}\|\cos\theta$ and direction $\dfrac{\mathbf{w}}{\|\mathbf{w}\|}$.

If θ is obtuse, then $\cos\theta < 0$ and $\text{proj}_{\mathbf{w}}\mathbf{v}$ has magnitude $-\|\mathbf{v}\|\cos\theta$ and direction $-\dfrac{\mathbf{w}}{\|\mathbf{w}\|}$.

See Figure 7.39.

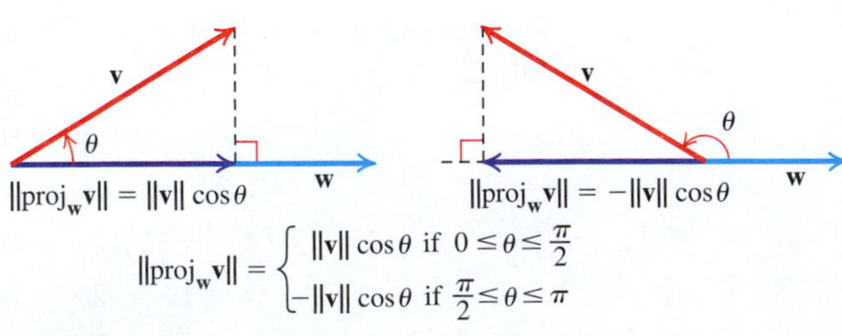

FIGURE 7.39 Magnitude and direction of $\text{proj}_{\mathbf{w}}\mathbf{v}$

SIDE **NOTE**

Any nonzero vector can be written as

$$\mathbf{R} = \|\mathbf{R}\|\mathbf{u},$$

where $\mathbf{u} = \dfrac{\mathbf{R}}{\|\mathbf{R}\|}$ is the unit vector in the direction of $\mathbf{R}$.

Because $(-\|\mathbf{v}\| \cos\theta)\left(-\dfrac{\mathbf{w}}{\|\mathbf{w}\|}\right) = (\|\mathbf{v}\| \cos\theta)\left(\dfrac{\mathbf{w}}{\|\mathbf{w}\|}\right)$, in both cases ($\theta$ acute or obtuse), we have

$$\text{proj}_{\mathbf{w}}\mathbf{v} = (\|\mathbf{v}\| \cos\theta)\left(\frac{\mathbf{w}}{\|\mathbf{w}\|}\right) \quad \text{See side note.}$$

$$= \left(\frac{\mathbf{v}\cdot\mathbf{w}}{\|\mathbf{w}\|}\right)\frac{\mathbf{w}}{\|\mathbf{w}\|} \quad \|\mathbf{v}\| \cos\theta = \frac{\mathbf{v}\cdot\mathbf{w}}{\|\mathbf{w}\|}$$

$$= \frac{\mathbf{v}\cdot\mathbf{w}}{\|\mathbf{w}\|^2}\mathbf{w}.$$

VECTOR AND SCALAR PROJECTIONS OF v ONTO w

Let **v** and **w** be two nonzero vectors and let $\theta\,(0° \le \theta \le 180°)$ be the angle between them.

The **vector projection of v onto w** is

$$\text{proj}_{\mathbf{w}}\mathbf{v} = \left(\frac{\mathbf{v}\cdot\mathbf{w}}{\|\mathbf{w}\|^2}\right)\mathbf{w}.$$

The **scalar projection of v onto w** is

$$\|\mathbf{v}\|\cos\theta = \frac{\mathbf{v}\cdot\mathbf{w}}{\|\mathbf{w}\|}.$$

EXAMPLE 8 Finding the Vector Projection of v onto w

Given $\mathbf{v} = \langle -2, 3\rangle$ and $\mathbf{w} = \langle 3, -4\rangle$, find the vector and scalar projections of **v** onto **w**.

Solution

$$\|\mathbf{v}\| = \sqrt{(-2)^2 + 3^2} = \sqrt{13} \text{ and } \|\mathbf{w}\| = \sqrt{3^2 + (-4)^2} = \sqrt{25} = 5$$

$$\mathbf{v}\cdot\mathbf{w} = \mathbf{w}\cdot\mathbf{v} = \langle -2, 3\rangle \cdot \langle 3, -4\rangle$$

$$= (-2)(3) + 3(-4) = -6 - 12 = -18$$

a. $\text{proj}_{\mathbf{w}}\mathbf{v} = \dfrac{\mathbf{v}\cdot\mathbf{w}}{\|\mathbf{w}\|^2}\mathbf{w}$ Vector projection of **v** onto **w**

$= -\dfrac{18}{25}\langle 3, -4\rangle$ Substitute values.

$= \left\langle -\dfrac{54}{25}, \dfrac{72}{25}\right\rangle$ Scalar multiplication

b. Scalar projection of **v** onto **w** $= \dfrac{\mathbf{v}\cdot\mathbf{w}}{\|\mathbf{w}\|}$

$= -\dfrac{18}{5}$ $\mathbf{v}\cdot\mathbf{w} = -18,\ \|\mathbf{w}\| = 5$

Practice Problem 8 Given $\mathbf{v} = \langle 1, -2\rangle$ and $\mathbf{w} = \langle 2, 5\rangle$, find the vector and scalar projections of **v** onto **w**.

5 Decompose a vector into two orthogonal vectors.

Decomposition of a Vector

Given two vectors $\mathbf{v}$ and $\mathbf{w}$, adding these vectors produces the resultant vector $\mathbf{v} + \mathbf{w}$. In many applications, it is necessary to *decompose* a vector into two orthogonal vectors. We use the vector $\text{proj}_{\mathbf{w}}\mathbf{v}$ to write $\mathbf{v}$ as the sum of two orthogonal vectors.

Decomposition of a Vector

Let $\mathbf{v}$ and $\mathbf{w}$ be two nonzero vectors. The vector $\mathbf{v}$ can be written in the form

$$\mathbf{v} = \mathbf{v}_1 + \mathbf{v}_2,$$

where $\mathbf{v}_1$ is parallel to $\mathbf{w}$ and $\mathbf{v}_2$ is orthogonal to $\mathbf{w}$. We let

$$\mathbf{v}_1 = \text{proj}_{\mathbf{w}}\mathbf{v} = \left(\frac{\mathbf{v}\cdot\mathbf{w}}{\|\mathbf{w}\|^2}\right)\mathbf{w} \text{ and } \mathbf{v}_2 = \mathbf{v} - \mathbf{v}_1.$$

The expression $\mathbf{v} = \mathbf{v}_1 + \mathbf{v}_2$ is called the **decomposition of v with respect to w.** The vectors $\mathbf{v}_1$ and $\mathbf{v}_2$ are called **components of v.**

Because $\mathbf{v}_1$ is a scalar multiple of $\mathbf{w}$, $\mathbf{v}_1$ is parallel to $\mathbf{w}$. In Exercise 84, we ask you to prove that $\mathbf{v}_2$ is orthogonal to $\mathbf{w}$.

PROCEDURE IN ACTION

EXAMPLE 9 Decomposing a Vector into Components

OBJECTIVE

Express $\mathbf{v} = \mathbf{v}_1 + \mathbf{v}_2$, where $\mathbf{v}_1$ is parallel to $\mathbf{w}$ and $\mathbf{v}_2$ is orthogonal to $\mathbf{w}$.

Step 1 Compute.

$$\mathbf{v}\cdot\mathbf{w} \text{ and } \|\mathbf{w}\|^2$$

Step 2 Compute.

$$\mathbf{v}_1 = \text{proj}_{\mathbf{w}}\mathbf{v} = \left(\frac{\mathbf{v}\cdot\mathbf{w}}{\|\mathbf{w}\|^2}\right)\mathbf{w}$$

Step 3 Compute.

$$\mathbf{v}_2 = \mathbf{v} - \mathbf{v}_1$$

Check that $\mathbf{v}_2$ is orthogonal to $\mathbf{w}$ by showing $\mathbf{v}_2\cdot\mathbf{w} = 0$.

Step 4 Write the decomposition.

$$\mathbf{v} = \mathbf{v}_1 + \mathbf{v}_2$$

EXAMPLE

Find the decomposition $\mathbf{v} = \mathbf{v}_1 + \mathbf{v}_2$ of $\mathbf{v} = \langle -2, 10\rangle$ with respect to $\mathbf{w} = \langle 3, -2\rangle$.

1. $\mathbf{v}\cdot\mathbf{w} = \langle -2, 10\rangle\cdot\langle 3, -2\rangle$

$= (-2)(3) + (10)(-2) = -26$

$\|\mathbf{w}\|^2 = (3)^2 + (-2)^2 = 9 + 4 = 13$

2. $\mathbf{v}_1 = \left(\frac{\mathbf{v}\cdot\mathbf{w}}{\|\mathbf{w}\|^2}\right)\mathbf{w} = \frac{-26}{13}\langle 3, -2\rangle$ From Step 1

$= -2\langle 3, -2\rangle = \langle -6, 4\rangle$ Simplify.

3. $\mathbf{v}_2 = \mathbf{v} - \mathbf{v}_1 = \langle -2, 10\rangle - \langle -6, 4\rangle$

$= \langle -2 - (-6), 10 - 4\rangle$

$= \langle 4, 6\rangle$

Check

$\mathbf{v}_2\cdot\mathbf{w} = \langle 4, 6\rangle\cdot\langle 3, -2\rangle = (4)(3) + (6)(-2) = 0$

4. $\mathbf{v} = \langle -2, 10\rangle = \underbrace{\mathbf{v}_1}_{} + \underbrace{\mathbf{v}_2}_{}$

$= \underbrace{\langle -6, 4\rangle}_{\text{Parallel to } \mathbf{w}} + \underbrace{\langle 4, 6\rangle}_{\text{Orthogonal to } \mathbf{w}}$

Practice Problem 9 Repeat Example 9 with $\mathbf{v} = \langle 5, 1\rangle$ and $\mathbf{w} = \langle 4, 4\rangle$.

7 Use the definition of work.

Work

Work

The work W done by a constant force $\mathbf{F}$ in moving an object from a point P to a point Q is defined by

$$W = \mathbf{F} \cdot \overrightarrow{PQ} = \|\mathbf{F}\| \|\overrightarrow{PQ}\| \cos\theta,$$

where θ is the angle between $\mathbf{F}$ and $\overrightarrow{PQ}$.

EXAMPLE 10 Computing Work

A child pulls a wagon along level ground, with a force of 40 pounds along the handle on the wagon that makes an angle of 42° with the horizontal. How much work has she done by pulling the wagon 150 feet?

Solution

The work done is as follows:

$W = \mathbf{F} \cdot \overrightarrow{PQ}$	Definition of work
$= \|\mathbf{F}\| \|\overrightarrow{PQ}\| \cos\theta$	$\mathbf{u} \cdot \mathbf{v} = \|\mathbf{u}\| \|\mathbf{v}\| \cos\theta$
$= (40)(150) \cos 42°$	Substitute given values.
≈ 4458.87 foot-pounds	Use a calculator.

Practice Problem 10 In Example 10, compute the work done if the handle on the wagon makes an angle of 60° with the horizontal.

Answers to Practice Problems

1. a. 8 **b.** 0 **2.** 14 **3.** Not parallel
4. 115.3° **5.** ≈39 **6.** $\mathbf{v}$ and $\mathbf{w}$ are orthogonal.
7. $k = 3$ **8. a.** $\left(-\frac{16}{29}, -\frac{40}{29}\right)$ **b.** $-\frac{8\sqrt{29}}{29}$
9. $\langle 3, 3 \rangle + \langle 2, -2 \rangle$
10. 3000 foot-pounds

SECTION 7.5 Exercises

Basic Concepts and Skills

1. The dot product of $\mathbf{v} = \langle a_1, a_2 \rangle$ and $\mathbf{w} = \langle b_1, b_2 \rangle$ is defined as $\mathbf{v} \cdot \mathbf{w} =$ ______________.
2. If θ is the angle between the vectors $\mathbf{v}$ and $\mathbf{w}$, then $\mathbf{v} \cdot \mathbf{w} =$ ______________.
3. If $\mathbf{v}$ and $\mathbf{w}$ are orthogonal, then $\mathbf{v} \cdot \mathbf{w} =$ ______________.
4. If $\mathbf{v}$ and $\mathbf{w}$ are nonzero vectors with $\mathbf{v} \cdot \mathbf{w} = 0$, then $\mathbf{v}$ and $\mathbf{w}$ are ______________.
5. **True or False.** If $\mathbf{v} \cdot \mathbf{w} < 0$, then the angle between $\mathbf{v}$ and $\mathbf{w}$ is an obtuse angle.
6. **True or False.** To find a vector perpendicular to a given vector, switch components and change the sign of one of them.

In Exercises 7–14, find u · v.

7. $\mathbf{u} = \langle 1, -2 \rangle, \mathbf{v} = \langle 3, 5 \rangle$
8. $\mathbf{u} = \langle 1, 3 \rangle, \mathbf{v} = \langle -3, 1 \rangle$
9. $\mathbf{u} = \langle 2, -6 \rangle, \mathbf{v} = \langle 3, 1 \rangle$
10. $\mathbf{u} = \langle -1, -2 \rangle, \mathbf{v} = \langle -3, -4 \rangle$
11. $\mathbf{u} = \mathbf{i} - 3\mathbf{j}, \mathbf{v} = 8\mathbf{i} - 2\mathbf{j}$
12. $\mathbf{u} = 6\mathbf{i} - \mathbf{j}, \mathbf{v} = 2\mathbf{i} + 7\mathbf{j}$
13. $\mathbf{u} = 4\mathbf{i} - 2\mathbf{j}, \mathbf{v} = 3\mathbf{j}$
14. $\mathbf{u} = -2\mathbf{i}, \mathbf{v} = 5\mathbf{j}$

In Exercises 15–20, find u · v, where θ is the angle between the vectors u and v. Round answers to the nearest tenth.

15. $\|\mathbf{u}\| = 2, \|\mathbf{v}\| = 5, \theta = \frac{\pi}{6}$
16. $\|\mathbf{u}\| = 3, \|\mathbf{v}\| = 4, \theta = \frac{\pi}{3}$
17. $\|\mathbf{u}\| = 5, \|\mathbf{v}\| = 4, \theta = 65°$

18. $\|\mathbf{u}\| = 5, \|\mathbf{v}\| = 3, \theta = 78°$

19. $\|\mathbf{u}\| = 3, \|\mathbf{v}\| = 7, \theta = 120°$

20. $\|\mathbf{u}\| = 6, \|\mathbf{v}\| = 4, \theta = 150°$

In Exercises 21–30, find the angle between the vectors v and w.

21. $\|\mathbf{v}\| = 2, \|\mathbf{w}\| = \sqrt{3}$, and $\mathbf{v} \cdot \mathbf{w} = \sqrt{3}$

22. $\|\mathbf{v}\| = 1, \|\mathbf{w}\| = 4$ and $\mathbf{v} \cdot \mathbf{w} = -\sqrt{8}$

23. $\mathbf{v} = -2\mathbf{i} - 3\mathbf{j}, \mathbf{w} = \mathbf{i} + \mathbf{j}$

24. $\mathbf{v} = \mathbf{i} - \mathbf{j}, \mathbf{w} = 3\mathbf{i} - 4\mathbf{j}$

25. $\mathbf{v} = 2\mathbf{i} + 5\mathbf{j}, \mathbf{w} = -5\mathbf{i} + 2\mathbf{j}$

26. $\mathbf{v} = 3\mathbf{i} - 7\mathbf{j}, \mathbf{w} = 7\mathbf{i} + 3\mathbf{j}$

27. $\mathbf{v} = \mathbf{i} + \mathbf{j}, \mathbf{w} = 3\mathbf{i} + 3\mathbf{j}$

28. $\mathbf{v} = -2\mathbf{i} + 3\mathbf{j}, \mathbf{w} = -4\mathbf{i} + 6\mathbf{j}$

29. $\mathbf{v} = 2\mathbf{i} - 3\mathbf{j}, \mathbf{w} = -4\mathbf{i} + 6\mathbf{j}$

30. $\mathbf{v} = -3\mathbf{i} + 4\mathbf{j}, \mathbf{w} = 6\mathbf{i} - 8\mathbf{j}$

In Exercises 31–38, let v and w be two vectors in the plane of magnitudes 4 and 5, respectively. The angle between v and w is 60°. Find each expression.

31. $\|\mathbf{v} + \mathbf{w}\|$

32. $\|\mathbf{v} - \mathbf{w}\|$

33. $\|2\mathbf{v} + \mathbf{w}\|$

34. $\|2\mathbf{v} - \mathbf{w}\|$

35. Angle between $\mathbf{v} + \mathbf{w}$ and $\mathbf{w}$

36. Angle between $\mathbf{v} - \mathbf{w}$ and $\mathbf{v}$

37. Angle between $2\mathbf{v} + \mathbf{w}$ and $\mathbf{v}$

38. Angle between $2\mathbf{v} - \mathbf{w}$ and $\mathbf{w}$

In Exercises 39–44, determine whether the vectors v and w are orthogonal.

39. $\mathbf{v} = 2\mathbf{i} + 3\mathbf{j}, \mathbf{w} = -3\mathbf{i} + 2\mathbf{j}$

40. $\mathbf{v} = -4\mathbf{i} - 5\mathbf{j}, \mathbf{w} = 5\mathbf{i} - 4\mathbf{j}$

41. $\mathbf{v} = \langle 2, 7 \rangle, \mathbf{w} = \langle -7, -2 \rangle$

42. $\mathbf{v} = \langle -3, 4 \rangle, \mathbf{w} = \langle 4, -3 \rangle$

43. $\mathbf{v} = \langle 12, 6 \rangle, \mathbf{w} = \langle 2, -4 \rangle$

44. $\mathbf{v} = \langle -9, 6 \rangle, \mathbf{w} = \langle 4, 6 \rangle$

In Exercises 45–50, determine whether the vectors v and w are parallel.

45. $\mathbf{v} = 2\mathbf{i} + 3\mathbf{j}, \mathbf{w} = \frac{2}{3}\mathbf{i} + \mathbf{j}$

46. $\mathbf{v} = 5\mathbf{i} - 2\mathbf{j}, \mathbf{w} = -10\mathbf{i} + 4\mathbf{j}$

47. $\mathbf{v} = \langle 3, 5 \rangle, \mathbf{w} = \langle -6, 10 \rangle$

48. $\mathbf{v} = \langle -4, -2 \rangle, \mathbf{w} = \left\langle -1, \frac{1}{2} \right\rangle$

49. $\mathbf{v} = \langle 6, 3 \rangle, \mathbf{w} = \langle 2, 1 \rangle$

50. $\mathbf{v} = \langle 2, 5 \rangle, \mathbf{w} = \langle -4, -10 \rangle$

In Exercises 51–54, determine the scalar *c* so that the vectors v and w are (a) orthogonal and (b) parallel.

51. $\mathbf{v} = \langle 2, 3 \rangle, \mathbf{w} = \langle 3, c \rangle$

52. $\mathbf{v} = \langle 1, -2 \rangle, \mathbf{w} = \langle c, 4 \rangle$

53. $\mathbf{v} = \langle -4, 3 \rangle, \mathbf{w} = \langle c, -6 \rangle$

54. $\mathbf{v} = \langle -1, -3 \rangle, \mathbf{w} = \langle -2, c \rangle$

In Exercises 55–62, find $\text{proj}_{\mathbf{w}}\mathbf{v}$.

55. $\mathbf{v} = \langle 3, 5 \rangle, \mathbf{w} = \langle 4, 1 \rangle$

56. $\mathbf{v} = \langle 3, 5 \rangle, \mathbf{w} = \langle -4, 2 \rangle$

57. $\mathbf{v} = \langle 0, 2 \rangle, \mathbf{w} = \langle 3, 4 \rangle$

58. $\mathbf{v} = \langle 2, 0 \rangle, \mathbf{w} = \langle 4, -3 \rangle$

59. $\mathbf{v} = 5\mathbf{i} - 3\mathbf{j}, \mathbf{w} = \mathbf{i}$

60. $\mathbf{v} = 2\mathbf{i} + 7\mathbf{j}, \mathbf{w} = \mathbf{j}$

61. $\mathbf{v} = \langle a, b \rangle, \mathbf{w} = \mathbf{i} + \mathbf{j}$

62. $\mathbf{v} = \langle a, b \rangle, \mathbf{w} = \mathbf{i} - \mathbf{j}$

In Exercises 63–70, decompose v with respect to w.

63. $\mathbf{v} = \langle 2, 3 \rangle, \mathbf{w} = \mathbf{i}$

64. $\mathbf{v} = \langle 4, -3 \rangle, \mathbf{w} = \mathbf{j}$

65. $\mathbf{v} = \mathbf{i}, \mathbf{w} = \langle 1, 1 \rangle$

66. $\mathbf{v} = \mathbf{j}, \mathbf{w} = \langle 4, -3 \rangle$

67. $\mathbf{v} = \langle 4, 2 \rangle, \mathbf{w} = \langle -2, 4 \rangle$

68. $\mathbf{v} = \langle 5, 2 \rangle, \mathbf{w} = \langle 3, 5 \rangle$

69. $\mathbf{v} = \langle 4, -1 \rangle, \mathbf{w} = \langle 3, 4 \rangle$

70. $\mathbf{v} = \langle -3, 5 \rangle, \mathbf{w} = \langle 3, -4 \rangle$

Applying the Concepts

71. Finding resultant force. Two forces of 6 and 8 pounds act at an angle of 30° to each other. Find the resultant force and the angle of the resultant force to each of the two forces.

72. Finding resultant force. Repeat Exercise 71 assuming that the two forces act at an angle of 60° to each other.

73. Work. A force $\mathbf{F} = 64\mathbf{i} - 20\mathbf{j}$ (in pounds) is applied to an object. The resulting movement of the object is represented by the vector $\mathbf{d} = 10\mathbf{i} + 4\mathbf{j}$ (in feet). Find the work done by the force.

74. Work. A car is pushed 150 feet on a level street by a force of 60 pounds at an angle of 18° with the horizontal. Find the work done.

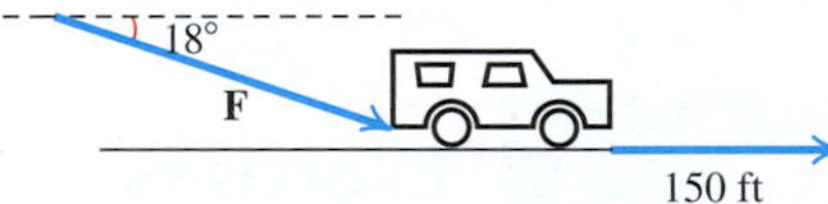

75. Inclined plane. A car is being towed up an inclined plane making an angle of 18° with the horizontal as shown in the figure. Find the force **F** needed to make the component of **F** parallel to the ground equal to 5250 pounds.

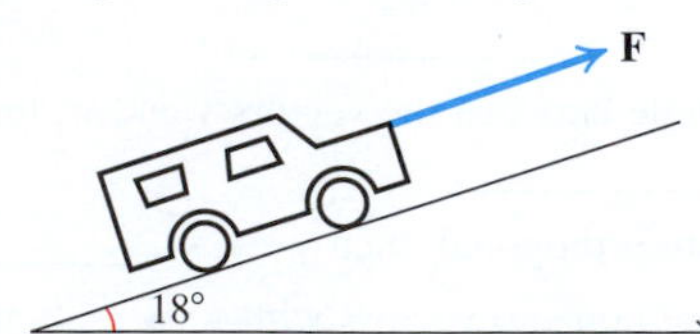

76. Inclined plane. In Exercise 75, find the work done by the force **F** to tow the car up 10 feet along the inclined plane.

77. Work. A vector $\mathbf{F}$ represents a force that has a magnitude of 10 pounds, and $\frac{2\pi}{3}$ is the angle for its direction. Find the work done by the force in moving an object from the origin to the point $(-6, 3)$. Distance is measured in feet.

78. Work. Two forces $\mathbf{F}_1 = 3\mathbf{i} - 2\mathbf{j}$ and $\mathbf{F}_2 = -4\mathbf{i} + 3\mathbf{j}$ act on an object and move it along a straight line from the point $(8, 4)$ to the point $(3, 5)$. Find the work done by the two forces acting together. The magnitudes of $\mathbf{F}_1$ and $\mathbf{F}_2$ are measured in pounds, and distance is measured in feet.

Beyond the Basics

79. Orthogonality on a circle. In the figure, AB is the diameter of a circle with center $O(0, 0)$ and radius a. Show that $\overrightarrow{CA}$ and $\overrightarrow{CB}$ are perpendicular. [*Hint:* $x^2 + y^2 = a^2$.]

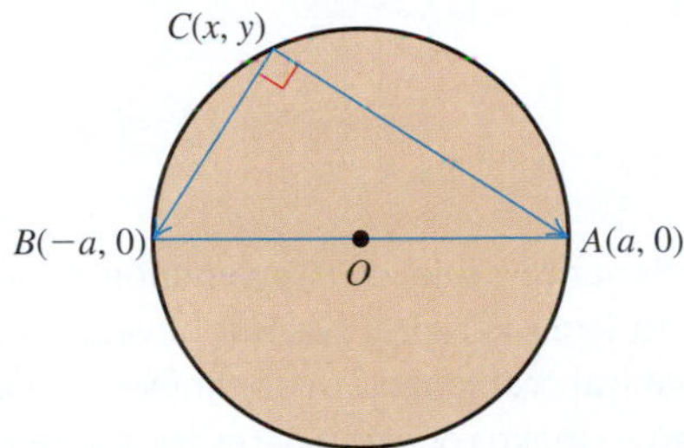

80. Cauchy-Schwarz inequality. If $\mathbf{v}$ and $\mathbf{w}$ are two vectors, then $|\mathbf{v} \cdot \mathbf{w}| \le \|\mathbf{v}\|\|\mathbf{w}\|$. When is $|\mathbf{v} \cdot \mathbf{w}| = \|\mathbf{v}\|\|\mathbf{w}\|$? [*Hint:* $\mathbf{v} \cdot \mathbf{w} = \|\mathbf{v}\|\|\mathbf{w}\| \cos\theta$.]

81. Verify that for any vectors $\mathbf{u}$, $\mathbf{v}$, and $\mathbf{w}$, $\mathbf{u} \cdot (\mathbf{v} + \mathbf{w}) = \mathbf{u} \cdot \mathbf{v} + \mathbf{u} \cdot \mathbf{w}$.

82. Prove that $(\mathbf{v} + \mathbf{w}) \cdot (\mathbf{v} + \mathbf{w}) = \|\mathbf{v}\|^2 + 2(\mathbf{v} \cdot \mathbf{w}) + \|\mathbf{w}\|^2$.

83. Triangle inequality. Use Exercises 80 and 82 to prove that $\|\mathbf{v} + \mathbf{w}\| \le \|\mathbf{v}\| + \|\mathbf{w}\|$.

84. In the decomposition $\mathbf{v} = \mathbf{v}_1 + \mathbf{v}_2$ (see page 692) with respect to $\mathbf{w}$, show that $\mathbf{v}_2 = \mathbf{v} - \text{proj}_{\mathbf{w}}\mathbf{v}$ is orthogonal to $\mathbf{w}$.

85. Show that the vectors $\mathbf{v} = \langle 2, -5 \rangle$ and $\mathbf{w} = \langle 5, 2 \rangle$ are perpendicular to each other.

86. Let $\mathbf{v} = 3\mathbf{i} + 7\mathbf{j}$. Find a vector perpendicular to $\mathbf{v}$.

87. Orthogonal unit vectors. Find the unit vectors orthogonal to $\mathbf{v} = 3\mathbf{i} - 4\mathbf{j}$.

88. Orthogonal unit vectors. Let $\mathbf{u}_1$ and $\mathbf{u}_2$ be orthogonal unit vectors and $\mathbf{v} = x\mathbf{u}_1 + y\mathbf{u}_2$. Find

a. $\mathbf{v} \cdot \mathbf{u}_1$ and **b.** $\mathbf{v} \cdot \mathbf{u}_2$.

89. Right triangle. Use the dot product to show that ABC is a right triangle and identify the right angle assuming that $A = (-2, 1)$, $B = (3, 1)$, and $C = (-1, -1)$.

90. Parallelogram law. Show that for any two vectors $\mathbf{u}$ and $\mathbf{v}$,

$$\|\mathbf{u} + \mathbf{v}\|^2 + \|\mathbf{u} - \mathbf{v}\|^2 = 2\|\mathbf{u}\|^2 + 2\|\mathbf{v}\|^2.$$

[*Hint:* $\|\mathbf{a}\|^2 = \mathbf{a} \cdot \mathbf{a}$.]

91. Polarization identity. Show that for any two vectors $\mathbf{u}$ and $\mathbf{v}$,

$$\|\mathbf{u} + \mathbf{v}\|^2 - \|\mathbf{u} - \mathbf{v}\|^2 = 4(\mathbf{u} \cdot \mathbf{v}).$$

92. Show that $\mathbf{u}$ and $\mathbf{v}$ are orthogonal if and only if $\|\mathbf{u} + \mathbf{v}\| = \|\mathbf{u} - \mathbf{v}\|$.

93. Show that the distance d from a point $P(x_0, y_0)$ to the line l: $ax + by + c = 0$ is

$$d = \frac{|ax_0 + by_0 + c|}{\sqrt{a^2 + b^2}}.$$

[*Hint:* **i.** Choose two points $Q(x_1, y_1)$ and $R(x_2, y_2)$ on l.
ii. Show that $\mathbf{n} = \langle a, b \rangle$ is perpendicular to l by showing that $\mathbf{n} \cdot \overrightarrow{QR} = 0$.
iii. Show that $\|\text{proj}_{\mathbf{n}}\overrightarrow{QP}\| = d$.]

Critical Thinking/Discussion/Writing

94. True or False. If $\mathbf{u} \cdot \mathbf{v} = \mathbf{u} \cdot \mathbf{w}$ and $\mathbf{u} \neq 0$, then $\mathbf{v} = \mathbf{w}$. Give reasons for your answer.

95. Explain why $\frac{\mathbf{v} \cdot \mathbf{w}}{\|\mathbf{v}\|\|\mathbf{w}\|}$ is a number between -1 and 1 for any two nonzero vectors $\mathbf{v}$ and $\mathbf{w}$.

96. Is the angle between $-\mathbf{v}$ and $-\mathbf{w}$ the same as the angle between $\mathbf{v}$ and $\mathbf{w}$ for any two nonzero vectors $\mathbf{v}$ and $\mathbf{w}$? Explain your answer.

Maintaining Skills

In Exercises 97–104, describe the graph of each equation. Find the x- and y-intercepts of each graph.

97. $x = -1$

98. $y = 2$

99. $y = x^2 - x - 6$

100. $x = y^2 + 4y + 3$

101. $x^2 + (y - 1)^2 = 4$

102. $(x + 1)^2 + y^2 = 9$

103. $x^2 + y^2 - 2x + 4y + 2 = 0$

104. $x^2 + y^2 + 6x - 4y - 12 = 0$

In Exercises 105–108, test each equation for symmetry about the x-axis, the y-axis, and the origin.

105. $x^3 + y^2 = 5$

106. $x^3 + y^3 - 3xy^2 = 0$

107. $2x^2 - 3y^3 = 7$

108. $2x^2 + 3y^2 = 6$

SECTION 7.6 Polar Coordinates

BEFORE STARTING THIS SECTION, REVIEW

1. Coordinate plane (Section 2.1, page 156)
2. Degree and radian measure of angles (Section 5.1, pages 472 and 475)
3. Trigonometric functions of angles (Section 5.3, page 500)

OBJECTIVES

1. Plot points using polar coordinates.
2. Convert points between polar and rectangular forms.
3. Convert equations between rectangular and polar forms.
4. Graph polar equations.

◆ Bionics

On large construction sites and maintenance projects, it is common to see mechanical systems that lift, dig, and scoop just like biological systems having arms, legs, and hands. The technique of solving mechanical problems with our knowledge of biological systems is called **bionics**. Bionics is used to make robotic devices for the remote-control handling of radioactive materials and the machines that are sent to the moon or to Mars to scoop soil samples, bring them into the space vehicle, perform analysis on them, and radio the results back to Earth. In Example 4, we use polar coordinates to find the reach of a robotic arm.

Up to now, we have used a rectangular coordinate system to identify points in the plane. In this section, we introduce another coordinate system, called a **polar coordinate system**, that allows us to describe many curves using rather simple equations. The polar coordinate system is believed to have been introduced by Jakob Bernoulli (1654–1705).

1 Plot points using polar coordinates.

Polar Coordinates

In a rectangular coordinate system, a point in the plane is described in terms of the horizontal and vertical directed distances from the origin. In a polar coordinate system, we draw a ray (usually, horizontal directed to the right), called the **polar axis**, in the plane; its endpoint is called the **pole**. A point P in the plane is described by an ordered pair of numbers (r, θ), the **polar coordinates** of P.

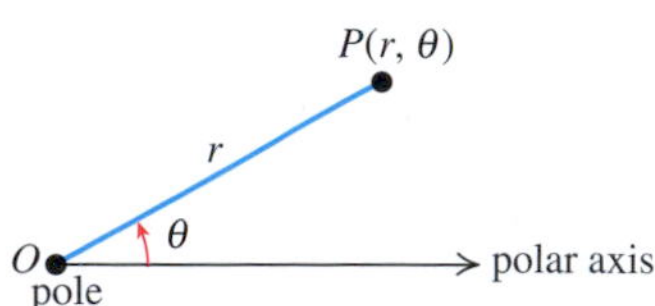

FIGURE 7.40 Polar coordinates (r, θ) of a point

Figure 7.40 shows the point $P(r, \theta)$ in the polar coordinate system, where r is the "directed distance" from the pole O to the point P and θ is a directed angle from the polar axis to the line segment OP (see Figure 7.43). The angle θ may be measured in degrees or radians. As usual, positive angles are measured counterclockwise from the polar axis and negative angles are measured clockwise from the polar axis.

To locate a point with polar coordinates $(5, 40°)$, we draw an angle $\theta = 40°$ counterclockwise from the polar axis. We then measure 5 units along the terminal side of θ. In Figure 7.41, we have plotted the points with polar coordinates $(5, 40°)$, $(3, 220°)$, $(4, -30°)$, and $(2, 110°)$.

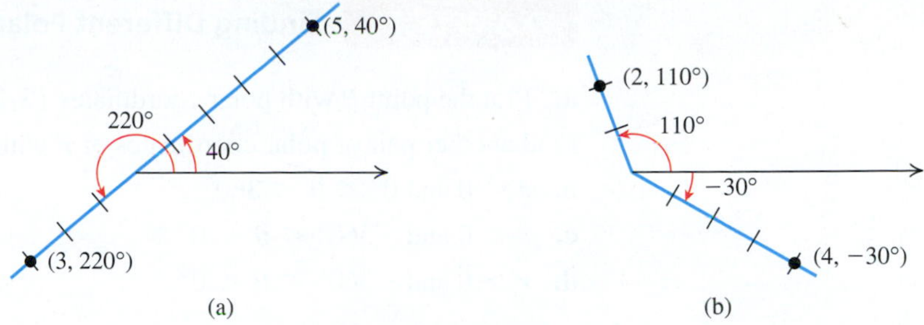

FIGURE 7.41 Plotting points using polar coordinates

Unlike rectangular coordinates, the polar coordinates of a point are not unique. For example, the polar coordinates $(3, 60°)$, $(3, 420°)$, and $(3, -300°)$ represent the same point. See Figure 7.42. In general, if a point P has polar coordinates (r, θ), then for any integer n,

$$\underbrace{(r, \theta + n \cdot 360°)}_{\theta \text{ in degrees}} \quad \text{or} \quad \underbrace{(r, \theta + 2n\pi)}_{\theta \text{ in radians}}$$

are also polar coordinates of P.

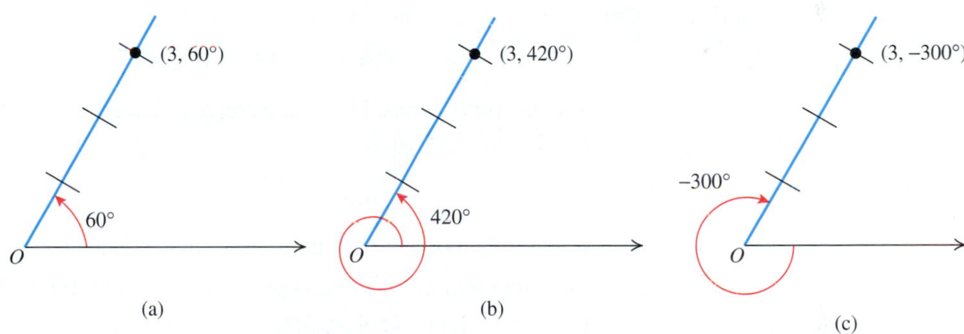

FIGURE 7.42 Polar coordinates of a point are not unique

SIDE **NOTE**

When plotting a point in polar coordinates, measure the angle θ first. Then (i) for $r > 0$, measure a distance of r units from the pole along the terminal ray of θ and (ii) for $r < 0$, plot the point $(|r|, \theta + \pi)$.

When equations are graphed in polar coordinates, it is convenient to allow r to be negative. That's why we defined r as *a directed distance*. If r is negative, then instead of the point being on the terminal side of the angle, it is $|r|$ units on the extension of the terminal side, which is the ray from the pole that extends in the direction opposite the terminal side of θ. In other words, to graph a point (r, θ) for $r < 0$, measure $|r|$ units along the terminal side of $\theta + \pi$ (or $\theta + 180°$ if θ is in degrees). For example, the point $P(-4, -30°)$ shown in Figure 7.43 is the same point as $(|-4|, -30° + 180°) = (4, 150°)$. Some other polar coordinates for the same point P are $(-4, 330°)$, $(4, -210°)$, and $(4, 510°)$.

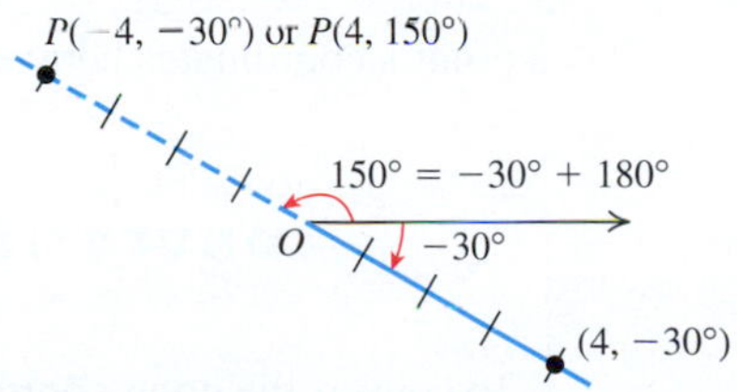

FIGURE 7.43 Negative r

EXAMPLE 1 Finding Different Polar Coordinates

a. Plot the point P with polar coordinates $(3, 225°)$.

Find another pair of polar coordinates of P with the following properties.

b. $r < 0$ and $0° < \theta < 360°$

c. $r < 0$ and $-360° < \theta < 0°$

d. $r > 0$ and $-360° < \theta < 0°$

Solution

a. The point $P(3, 225°)$ is plotted by first drawing $\theta = 225°$ counterclockwise from the polar axis. Because $r = 3 > 0$, the point P is on the terminal side of the angle, 3 units from the pole. See Figure 7.44(a).

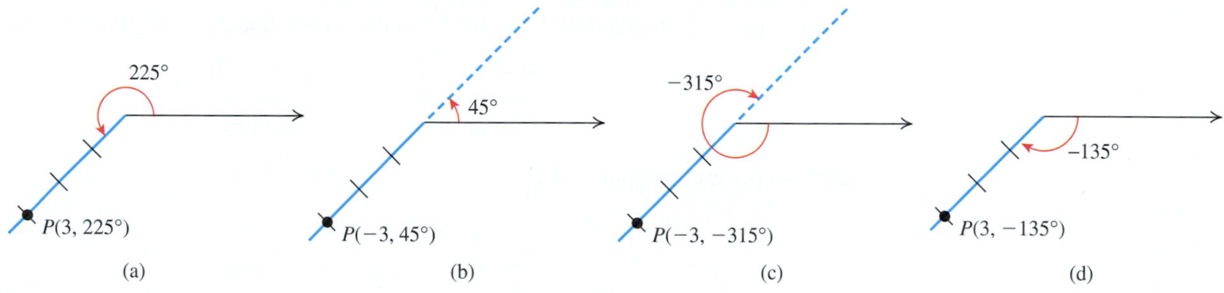

FIGURE 7.44 Plotting points

The answers to parts, **b, c,** and **d** are, respectively, $(-3, 45°)$, $(-3, -315°)$, and $(3, -135°)$. See Figure 7.44(b)–(d).

Practice Problem 1

a. Plot the point P with polar coordinates $(2, -150°)$.

Find another pair of polar coordinates of P with the following properties.

b. $r < 0$ and $0° < \theta < 360°$

c. $r > 0$ and $0° < \theta < 360°$

d. $r < 0$ and $-360° < \theta < 0°$

2 Convert points between polar and rectangular forms.

Converting Between Polar and Rectangular Forms

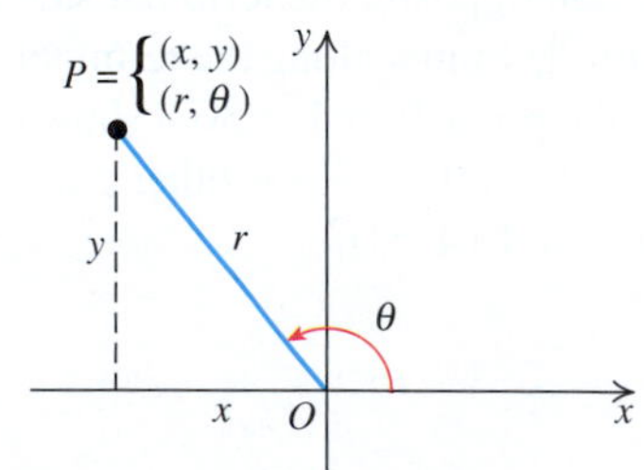

FIGURE 7.45 $P = (x, y)$ or $P = (r, \theta)$

Frequently, we need to use polar and rectangular coordinates in the same problem. To do this, we let the positive x-axis of the rectangular coordinate system serve as the polar axis and the origin serve as the pole. Each point P then has polar coordinates (r, θ) and rectangular coordinates (x, y), as shown in Figure 7.45.

From Figure 7.45, we see the following relationships: $x^2 + y^2 = r^2$, $\sin\theta = \frac{y}{r}$, $\cos\theta = \frac{x}{r}$, and $\tan\theta = \frac{y}{x}$. These relationships hold whether $r > 0$ or $r < 0$, and they are used to convert a point's coordinates between rectangular and polar forms.

CONVERTING FROM POLAR TO RECTANGULAR COORDINATES

To convert the polar coordinates (r, θ) of a point to its rectangular coordinates (x, y), use the equations

$$x = r\cos\theta \qquad \text{and} \qquad y = r\sin\theta.$$

TECHNOLOGY CONNECTION

A graphing calculator can be used to check conversions from polar to rectangular coordinates. Choose the appropriate options from the ANGLE menu. The *x*- and *y*-coordinates must be calculated separately. The screen shown here is from a calculator set in *degree* mode.

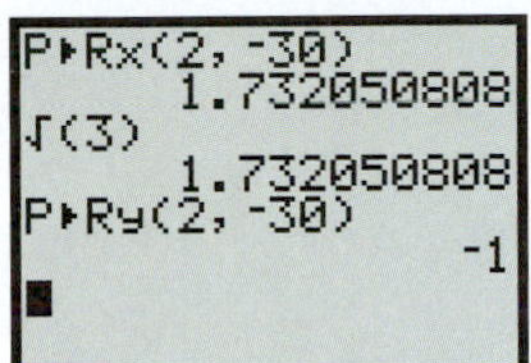

EXAMPLE 2 Converting from Polar to Rectangular Coordinates

Convert the polar coordinates of each point to its rectangular coordinates.

a. $(2, -30°)$ **b.** $\left(-4, \frac{\pi}{3}\right)$

Solution

a. $x = r\cos\theta$ $\quad y = r\sin\theta$ Conversion equations

$x = 2\cos(-30°)$ $\quad y = 2\sin(-30°)$ Replace r with 2 and θ with $-30°$.

$x = 2\cos 30°$ $\quad y = -2\sin 30°$ $\cos(-\theta) = \cos\theta$, $\sin(-\theta) = -\sin\theta$

$x = 2\left(\frac{\sqrt{3}}{2}\right) = \sqrt{3}$ $\quad y = -2\left(\frac{1}{2}\right) = -1$ $\cos 30° = \frac{\sqrt{3}}{2}$, $\sin 30° = \frac{1}{2}$

The rectangular coordinates of $(2, -30°)$ are $(\sqrt{3}, -1)$.

b. $x = r\cos\theta$ $\quad y = r\sin\theta$ Conversion equations

$x = -4\cos\frac{\pi}{3}$ $\quad y = -4\sin\frac{\pi}{3}$ Replace r with -4 and θ with $\frac{\pi}{3}$.

$x = -4\left(\frac{1}{2}\right) = -2$ $\quad y = -4\left(\frac{\sqrt{3}}{2}\right) = -2\sqrt{3}$ $\cos\frac{\pi}{3} = \frac{1}{2}$, $\sin\frac{\pi}{3} = \frac{\sqrt{3}}{2}$

The rectangular coordinates of $\left(-4, \frac{\pi}{3}\right)$ are $(-2, -2\sqrt{3})$.

Practice Problem 2 Convert the polar coordinates of each point to its rectangular coordinates.

a. $(-3, 60°)$ **b.** $\left(2, -\frac{\pi}{4}\right)$

When converting from rectangular coordinates (x, y) of a point to its polar coordinates (r, θ), remember that infinitely many representations are possible. We will find the polar coordinates (r, θ) of the point (x, y) that have $r > 0$ and $0° \le \theta < 360°$ (or $0 \le \theta < 2\pi$).

CONVERTING FROM RECTANGULAR TO POLAR COORDINATES

To convert the rectangular coordinates (x, y) of a point to its polar coordinates (r, θ),

1. Find the quadrant in which the given point (x, y) lies.
2. Use $r = \sqrt{x^2 + y^2}$ to find r.
3. Find θ by using $\tan\theta = \frac{y}{x}$ and choose θ so that it lies in the same quadrant as the point (x, y).

TECHNOLOGY CONNECTION

A graphing calculator can also be used to check conversions from rectangular to polar coordinates. Choose the appropriate options on the ANGLE menu. The r and θ coordinates must be calculated separately. The screen shown here is from a calculator set in *radian* mode

```
R▸Pr(-2,2√(3))
                4
R▸Pθ(-2,2√(3))
      2.094395102
2π/3
      2.094395102
```

EXAMPLE 3 Converting from Rectangular to Polar Coordinates

Find polar coordinates (r, θ) of the point P with $r > 0$ and $0 \le \theta < 2\pi$, whose rectangular coordinates are $(x, y) = (-2, 2\sqrt{3})$.

Solution

1. The point $P(-2, 2\sqrt{3})$ lies in quadrant II with $x = -2$ and $y = 2\sqrt{3}$.
2. $r = \sqrt{x^2 + y^2}$ Formula for computing r
   $= \sqrt{(-2)^2 + (2\sqrt{3})^2}$ Substitute values of x and y
   $= \sqrt{4 + 12} = \sqrt{16} = 4$ Simplify
3. $\tan\theta = \dfrac{y}{x}$ Definition of $\tan\theta$
   $\tan\theta = \dfrac{2\sqrt{3}}{-2} = -\sqrt{3}$ Substitute values and simplify.

Because $\tan\theta$ is negative in quadrants II and IV, the solutions of equation $\tan\theta = -\sqrt{3}$ are: $\theta = \pi - \dfrac{\pi}{3} = \dfrac{2\pi}{3}$ and $\theta = 2\pi - \dfrac{\pi}{3} = \dfrac{5\pi}{3}$. Because $P = (-2, 2\sqrt{3})$ lies in quadrant II, we choose $\theta = \dfrac{2\pi}{3}$. So the required polar coordinates are $\left(4, \dfrac{2\pi}{3}\right)$.

Practice Problem 3 Find the polar coordinates of the point P with $r > 0$ and $0 \le \theta < 2\pi$ whose rectangular coordinates are $(x, y) = (-1, -1)$.

EXAMPLE 4 Positioning a Robotic Hand

For the robotic arm of Figure 7.46, find the polar coordinates of the hand relative to the shoulder. Round all answers to the nearest tenth.

Solution

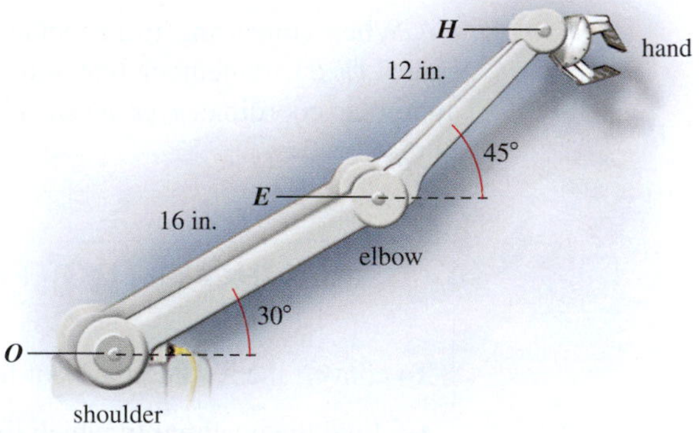

FIGURE 7.46 Robotic arm

We need to find the polar coordinates (r, θ) of the hand H, with the shoulder O as the pole. Because segment $\overline{OE}$ measures 16 inches and makes an angle of 30° with the horizontal, we can write the vector $\overrightarrow{OE}$ in terms of its components:

$$\overrightarrow{OE} = \langle 16\cos 30°, 16\sin 30° \rangle$$

Similarly,

$$\overrightarrow{EH} = \langle 12\cos 45°, 12\sin 45° \rangle.$$

We know that

$$\overrightarrow{OH} = \overrightarrow{OE} + \overrightarrow{EH}$$ Resultant vector

$$= \langle 16 \cos 30°, 16 \sin 30° \rangle + \langle 12 \cos 45°, 12 \sin 45° \rangle$$

$$= \langle 16 \cos 30° + 12 \cos 45°, 16 \sin 30° + 12 \sin 45° \rangle$$ Vector addition

So the rectangular coordinates (x, y) of H are

$$x = 16 \cos 30° + 12 \cos 45°, \quad y = 16 \sin 30° + 12 \sin 45°.$$

We now convert the coordinates (x, y) of H to polar coordinates (r, θ).

$$r = \sqrt{x^2 + y^2}$$ Formula for r

$$= \sqrt{(16 \cos 30° + 12 \cos 45°)^2 + (16 \sin 30° + 12 \sin 45°)^2}$$ Substitute values for x and y.

$$\approx 27.8 \text{ inches}$$ Use a calculator.

$$\theta = \theta' = \tan^{-1}\left(\frac{y}{x}\right)$$ Definition of inverse tangent

$$= \tan^{-1}\left(\frac{16 \sin 30° + 12 \sin 45°}{16 \cos 30° + 12 \cos 45°}\right) \approx 36.4°$$ Substitute values and use a calculator.

The polar coordinates of H relative to the pole O are $(r, \theta) = (27.8, 36.4°)$.

Practice Problem 4 Repeat Example 4 with $\overline{OE} = 15$ inches and $\overline{EH} = 10$ inches.

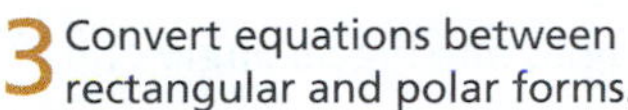
3 Convert equations between rectangular and polar forms.

Converting Equations Between Rectangular and Polar Forms

An equation that has the rectangular coordinates x and y as variables is called a **rectangular** (or **Cartesian**) **equation**. Similarly, an equation where the polar coordinates r and θ are the variables is called a **polar equation**. Some examples of polar equations are

$$r = \sin \theta, \quad r = 1 + \cos \theta, \quad \text{and} \quad r = \theta.$$

To convert a rectangular equation to a polar equation, we simply replace x with $r \cos \theta$ and y with $r \sin \theta$ and then simplify where possible.

EXAMPLE 5 **Converting an Equation from Rectangular to Polar Form**

Convert the equation $x^2 + y^2 - 3x + 4 = 0$ to polar form.

Solution

$$x^2 + y^2 - 3x + 4 = 0$$ Given equation

$$(r \cos \theta)^2 + (r \sin \theta)^2 - 3(r \cos \theta) + 4 = 0$$ $x = r \cos \theta, y = r \sin \theta$

$$r^2 \cos^2 \theta + r^2 \sin^2 \theta - 3r \cos \theta + 4 = 0$$ $(ab)^2 = a^2b^2$

$$r^2(\cos^2 \theta + \sin^2 \theta) - 3r \cos \theta + 4 = 0$$ Factor out r^2 from the first two terms.

$$r^2 - 3r \cos \theta + 4 = 0$$ $\sin^2 \theta + \cos^2 \theta = 1$

The equation $r^2 - 3r \cos \theta + 4 = 0$ is a polar form of the given rectangular equation.

Practice Problem 5 Convert the equation $x^2 + y^2 - 2y + 3 = 0$ to polar form.

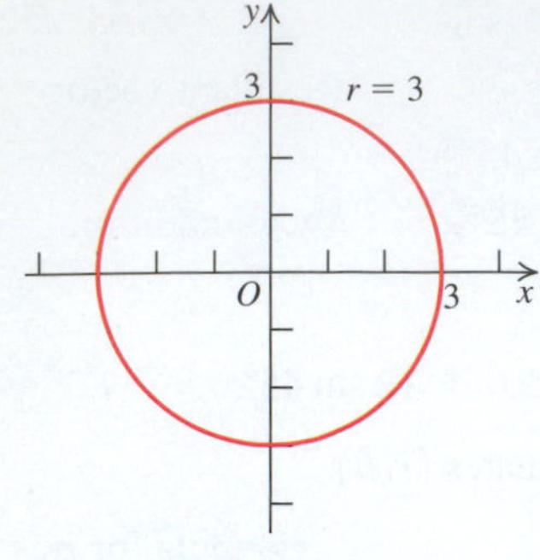

FIGURE 7.47 $x^2 + y^2 = 9$ or $r = 3$

Converting an equation from polar to rectangular form will frequently require some ingenuity to use the substitutions:

$$r^2 = x^2 + y^2, \quad r\cos\theta = x, \quad r\sin\theta = y, \quad \text{and} \quad \tan\theta = \frac{y}{x}.$$

EXAMPLE 6 **Converting an Equation from Polar to Rectangular Form**

Convert each polar equation to a rectangular equation and identify its graph.

a. $r = 3$ **b.** $\theta = 45°$ **c.** $r = \csc\theta$ **d.** $r = 2\cos\theta$

Solution

a.

$r = 3$	Given polar equation
$r^2 = 9$	Square both sides.
$x^2 + y^2 = 9$	Replace r^2 with $x^2 + y^2$.

The graph of $r = 3$ (or $x^2 + y^2 = 3^2$) is a circle with center at the pole and radius = 3 units. See Figure 7.47.

b.

$\theta = 45°$	Given polar equation
$\tan\theta = \tan 45°$	If $\alpha = \beta$, then $\tan\alpha = \tan\beta$.
$\frac{y}{x} = 1$	$\tan\theta = \frac{y}{x}$ and $\tan 45° = 1$
$y = x$	Multiply both sides by x.

FIGURE 7.48 $y = x$ or $\theta = 45°$

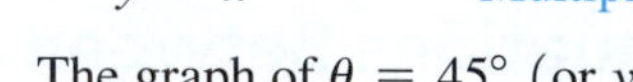

The graph of $\theta = 45°$ (or $y = x$) is a line through the origin, making an angle of 45° with the positive x-axis. See Figure 7.48.

c.

$r = \csc\theta$	Given polar equation
$r = \frac{1}{\sin\theta}$	Reciprocal identity
$r\sin\theta = 1$	Multiply both sides by $\sin\theta$.
$y = 1$	Replace $r\sin\theta$ with y.

The graph of $r = \csc\theta$ (or $y = 1$) is a horizontal line. See Figure 7.49.

FIGURE 7.49 $y = 1$ or $r = \csc\theta$

d.

$r = 2\cos\theta$	Given polar equation
$r^2 = 2r\cos\theta$	Multiply both sides by r.
$x^2 + y^2 = 2x$	Replace r^2 with $x^2 + y^2$ and $r\cos\theta$ with x.
$x^2 - 2x + y^2 = 0$	Subtract $2x$ from both sides and simplify.
$(x^2 - 2x + 1) + y^2 = 1$	Add 1 to both sides to complete the square.
$(x - 1)^2 + y^2 = 1$	Factor perfect square trinomial.

The graph of $r = 2\cos\theta$ (or $(x - 1)^2 + y^2 = 1$) is a circle with center at $(1, 0)$ and radius 1 unit. See Figure 7.50.

Practice Problem 6 Convert each polar equation to a rectangular equation and identify its graph.

a. $r = -5$ **b.** $\theta = -\frac{\pi}{3}$ **c.** $r = \sec\theta$ **d.** $r = 2\sin\theta$

FIGURE 7.50 $(x - 1)^2 + y^2 = 1$ or $r = 2\cos\theta$

4 Graph polar equations.

The Graph of a Polar Equation

Many important polar equations do not convert "nicely" to rectangular equations. To graph such equations, we plot points in polar coordinates. The graph of a polar equation

$$r = f(\theta)$$

is the set of all points $P(r, \theta)$ that have at least one polar coordinate representation that satisfies the equation. We make a table of several ordered pair solutions (r, θ) of the equation, plot these points, and join them with a smooth curve.

EXAMPLE 7 Sketching a Graph by Plotting Points

Sketch the graph of the polar equation $r = \theta$ $(\theta \geq 0)$ by plotting points.

Solution

We choose values of θ that are integer multiples of $\frac{\pi}{2}$. Because $r = \theta$, the polar coordinates of points on the curve are as follows:

$$(0, 0), \left(\frac{\pi}{2}, \frac{\pi}{2}\right), (\pi, \pi), \left(\frac{3\pi}{2}, \frac{3\pi}{2}\right), (2\pi, 2\pi), \left(\frac{5\pi}{2}, \frac{5\pi}{2}\right), \ldots$$

Plot these points on a polar grid and join them in order of increasing θ to obtain the graph of $r = \theta$ shown in Figure 7.51. This curve is called the **spiral of Archimedes**.

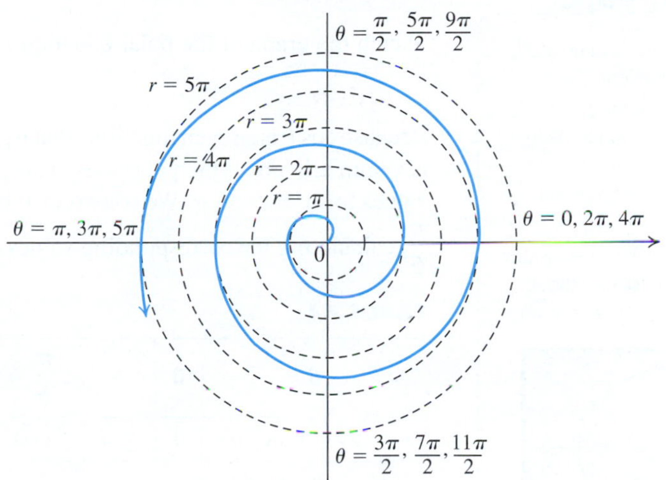

FIGURE 7.51 Spiral of Archimedes $r = \theta$

Practice Problem 7 Sketch the graph of the polar equation $r = 2\theta$ $(\theta \leq 0)$ by plotting points.

As with Cartesian equations, it is helpful to determine symmetry in the graph of a polar equation. If the graph of a polar equation has symmetry, then we can sketch its graph by plotting fewer points.

TESTS FOR SYMMETRY IN POLAR COORDINATES

The graph of a polar equation has the symmetry described below if an equivalent equation results when making the replacements shown.

Symmetry with respect to the polar axis	Symmetry with respect to the line $\theta = \frac{\pi}{2}$	Symmetry with respect to the pole
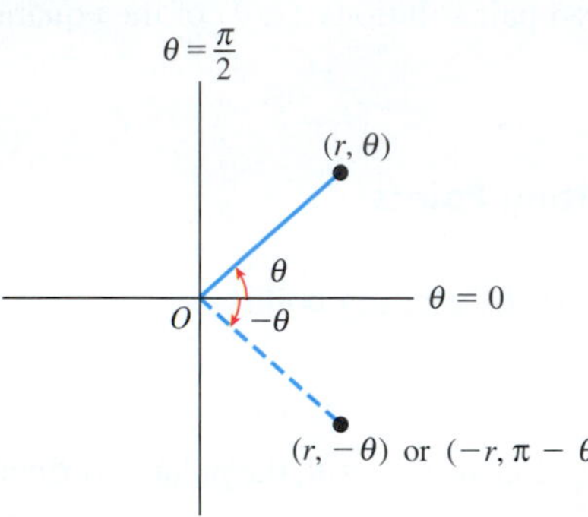	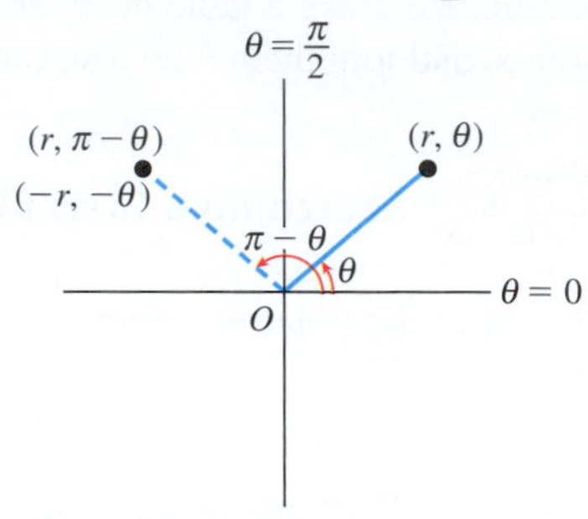	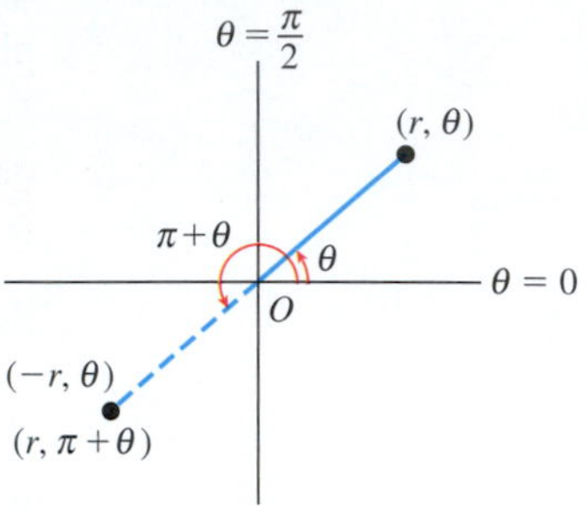
Replace (r, θ) with $(r, -\theta)$ or $(-r, \pi - \theta)$.	Replace (r, θ) with $(r, \pi - \theta)$ or $(-r, -\theta)$.	Replace (r, θ) with $(r, \pi + \theta)$ or $(-r, \theta)$.

Note that if a polar equation has two of these symmetries, then it must have the third symmetry.

TECHNOLOGY  **CONNECTION**

A graphing calculator can be used to graph polar equations. The calculator must be set in *polar* (Pol) mode. Either *radian* or *degree* mode may be used, but make sure your WINDOW settings are appropriate for your choice. See your owner's manual for details on polar graphing. The screen shows the graph of $r = 2(1 + \cos\theta)$.

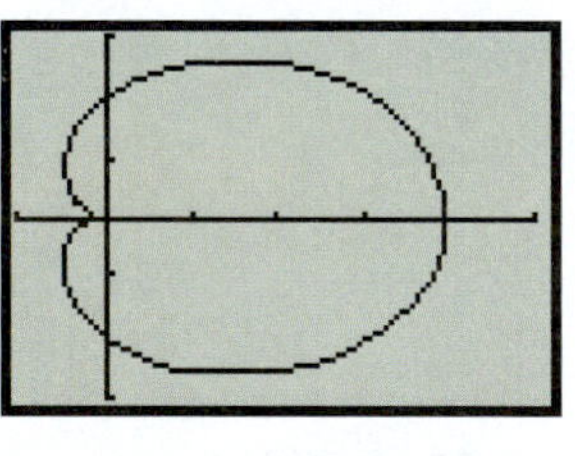

EXAMPLE 8 **Sketching the Graph of a Polar Equation**

Sketch the graph of the polar equation $r = 2(1 + \cos\theta)$.

Solution

Because $\cos\theta$ is an even function (that is, $\cos(-\theta) = \cos\theta$), the graph of $r = 2(1 + \cos\theta)$ is symmetric about the polar axis. Thus, to graph this equation, it is sufficient to compute values for $0 \le \theta \le \pi$. We construct Table 7.3 by substituting values for θ at increments of $\frac{\pi}{6}$, calculating the corresponding values of r.

TABLE 7.3

θ	0	$\frac{\pi}{6}$	$\frac{\pi}{3}$	$\frac{\pi}{2}$	$\frac{2\pi}{3}$	$\frac{5\pi}{6}$	π
$r = 2(1 + \cos\theta)$	4	$2 + \sqrt{3} \approx 3.73$	3	2	1	$2 - \sqrt{3} \approx 0.27$	0

We plot the points (r, θ) and draw a smooth curve through them. Because the graph is symmetric about the polar axis, we also reflect the curve about the polar axis.

The graph of the equation $r = 2(1 + \cos\theta)$ is shown in Figure 7.52. This type of curve is called a *cardioid* because it resembles a heart.

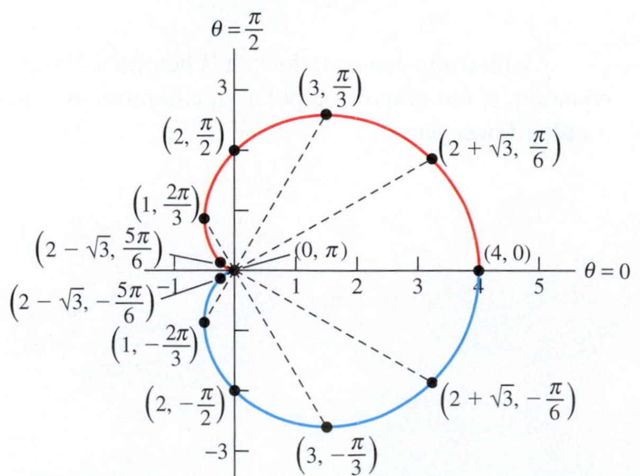

FIGURE 7.52 $r = 2(1 + \cos\theta)$

Practice Problem 8 Sketch the graph of the polar equation $r = 2(1 + \sin\theta)$.

Limaçons

The graph of an equation of the form

$$r = a + b\cos\theta \quad \text{or} \quad r = a + b\sin\theta$$

is called a limaçon of Pascal.

If $|b| > |a|$, the limaçon has an interior loop. See Figure 7.53.

If $|b| = |a|$, the limaçon is a cardioid. See Figure 7.52.

If $|b| < |a| < 2|b|$, the limaçon is dimpled. See Figure 7.54.

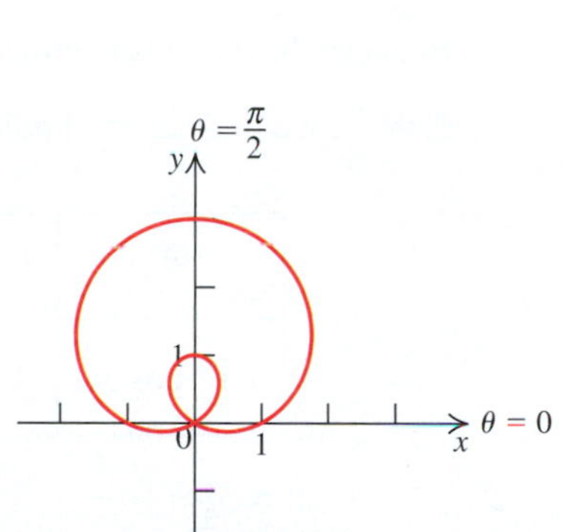

FIGURE 7.53 Graph of $r = 1 + 2\sin\theta$

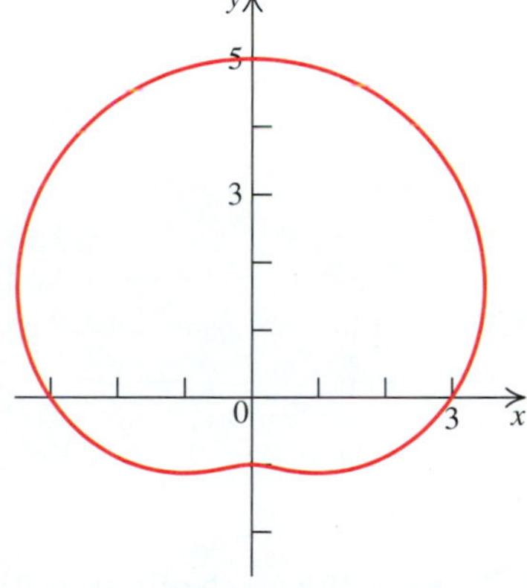

FIGURE 7.54 Graph of $r = 3 + 2\sin\theta$

In Figures 7.53 and 7.54, note that even though both graphs are symmetric about the line $\theta = \frac{\pi}{2}$ (y-axis), the test "replace (r, θ) with $(-r, -\theta)$" fails. So if a test for polar coordinate symmetry fails, the best we can say is "no conclusion." In other words, the tests for polar coordinate symmetry are *sufficient* but *not necessary* for showing symmetry.

PROCEDURE **IN ACTION**

EXAMPLE 9 **Graphing a Polar Equation**

OBJECTIVE

Sketch the graph of a polar equation $r = f(\theta)$, where f is a periodic function.

Step 1 Test for symmetry.

EXAMPLE

Sketch the graph of $r = \cos 2\theta$.

1. Replace θ with $-\theta$,

$$\cos(-2\theta) = \cos 2\theta \qquad (1)$$

Replace θ with $\pi - \theta$,

$$\cos[2(\pi - \theta)] = \cos(2\pi - 2\theta) = \cos 2\theta \qquad (2)$$

From (1) and (2), we conclude that the graph is symmetric about the polar axis $(\theta = 0)$ and about the line $\theta = \frac{\pi}{2}$. So, it is also symmetric about the pole.

(continued)

Step 2 Analyze the behavior of r, symmetries from Step 1, and then find selected points.

2. To get a sense of how r behaves for different values of θ, we make the following table:

As θ goes from:	r varies from:
0 to $\frac{\pi}{4}$	1 to 0
$\frac{\pi}{4}$ to $\frac{\pi}{2}$	0 to -1
$\frac{\pi}{2}$ to $\frac{3\pi}{4}$	-1 to 0
$\frac{3\pi}{4}$ to π	0 to 1

Because the curve is symmetric about $\theta = 0$ and $\theta = \frac{\pi}{2}$, we find selected points for $0 \le \theta \le \frac{\pi}{2}$:

θ	0	$\frac{\pi}{6}$	$\frac{\pi}{4}$	$\frac{\pi}{3}$	$\frac{\pi}{2}$
$r = \cos 2\theta$	1	$\frac{1}{2}$	0	$-\frac{1}{2}$	-1

Step 3 Sketch the graph of $r = f(\theta)$ using the points (r, θ) found in Step 2.

3.

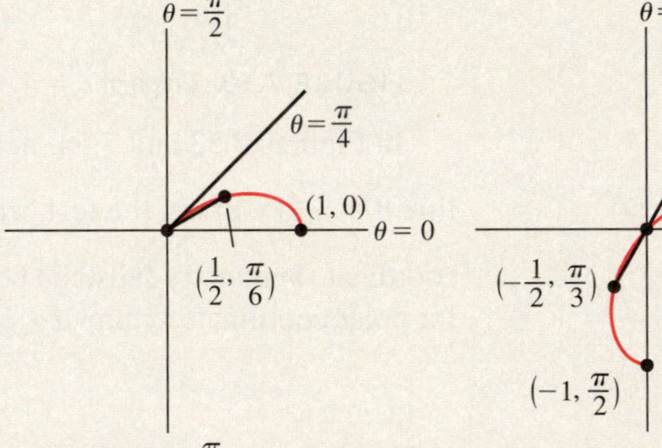

(r is negative for $\frac{\pi}{4} < \theta \le \frac{\pi}{2}$.)

Step 4 Use symmetries to complete the graph.

4.

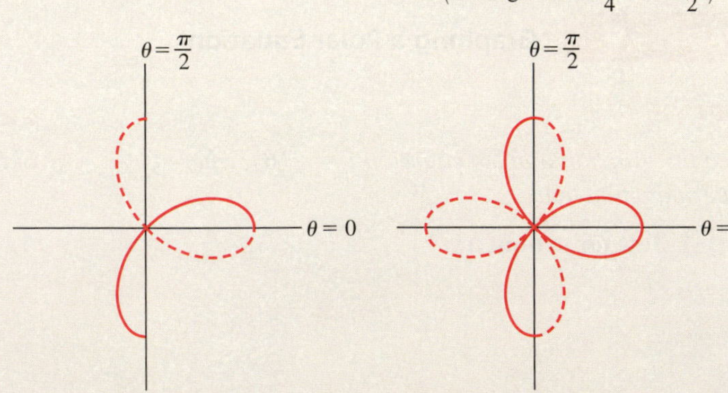

Using symmetry about the polar axis   Using symmetry about the line $\theta = \frac{\pi}{2}$

The curve of $r = \cos 2\theta$ is called a *four-leafed rose.*

Practice Problem 9 Sketch the graph of $r = \sin 2\theta$.

The following graphs have simple polar equations.

LIMAÇONS

The graphs of $r = a \pm b\cos\theta$, $r = a \pm b\sin\theta$ $(a > 0, b > 0)$ are called **limaçons** (French for "snail").

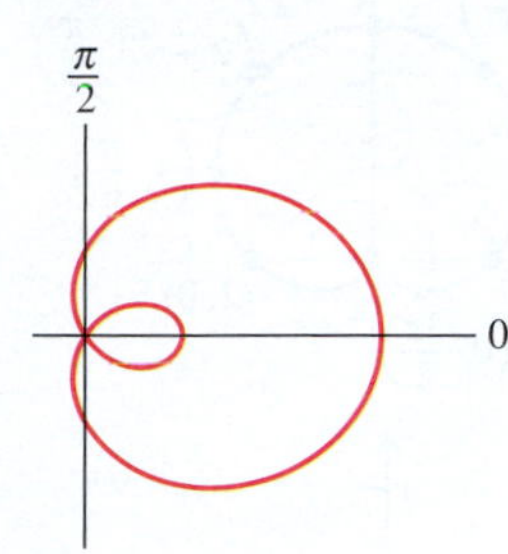

Inner loop

if $\frac{a}{b} < 1$

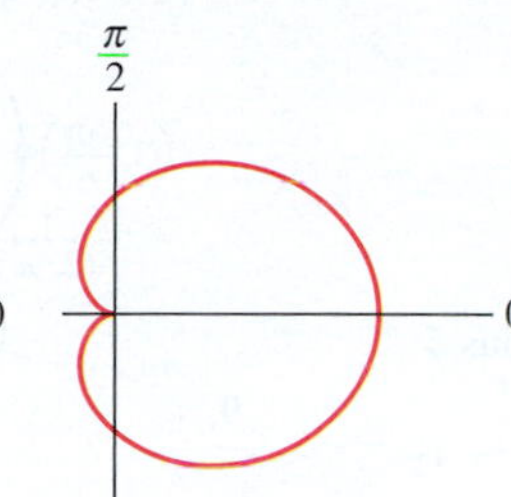

Cardioid

if $\frac{a}{b} = 1$

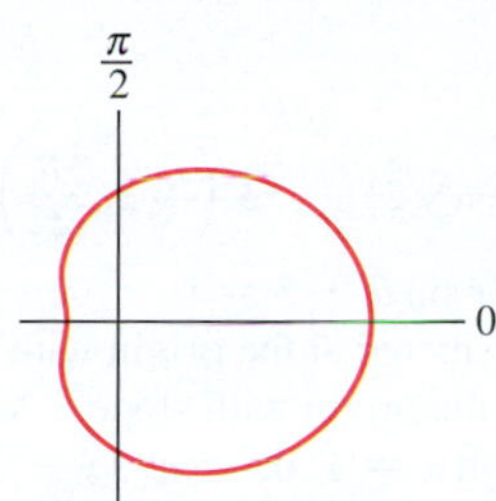

Dimpled

if $1 < \frac{a}{b} < 2$

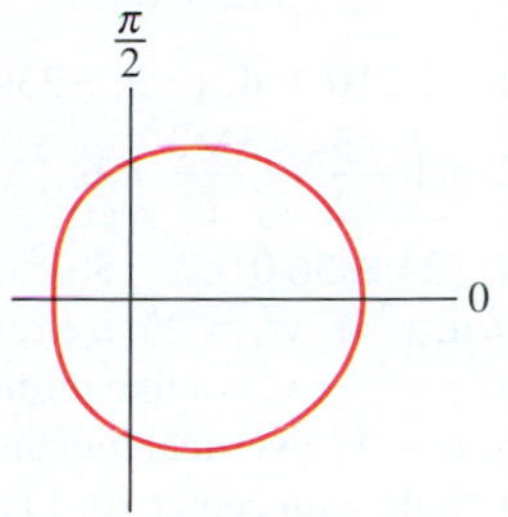

Not Dimpled

if $\frac{a}{b} \geq 2$

ROSE CURVES

The graphs of $r = a\cos n\theta$, $r = a\sin n\theta$ $(a > 0)$ are called **rose curves**. If n is odd, the rose has n petals. If n is even, the rose has $2n$ petals.

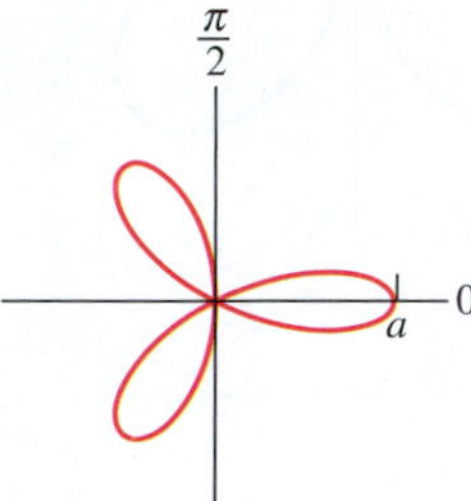

$r = a\cos 3\theta$

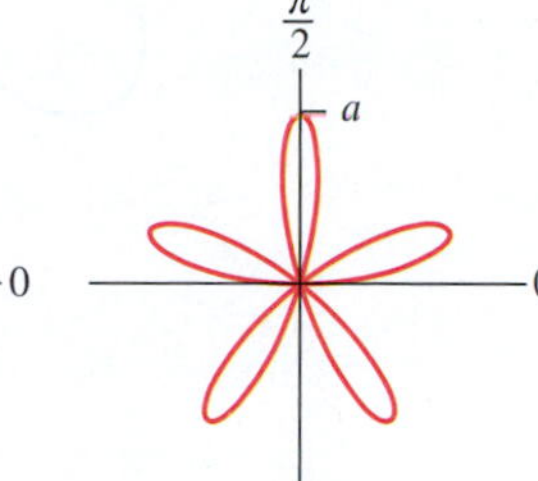

$r = a\sin 5\theta$

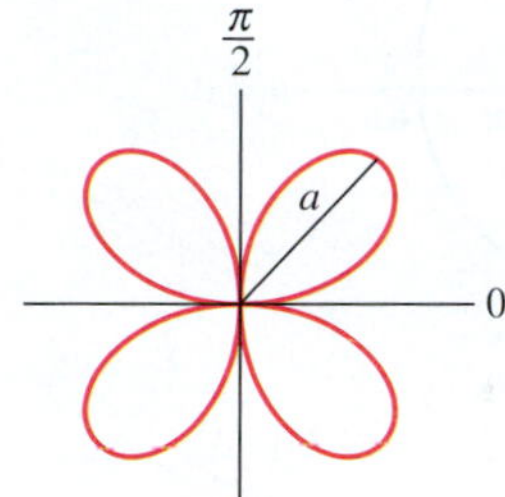

$r = a\sin 2\theta$

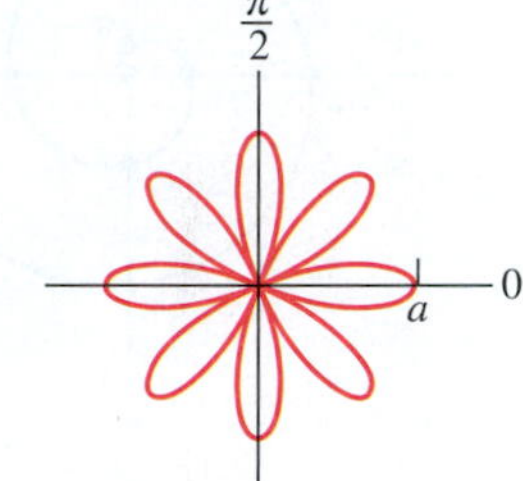

$r = a\cos 4\theta$

CIRCLES AND LEMNISCATES

$(a > 0)$

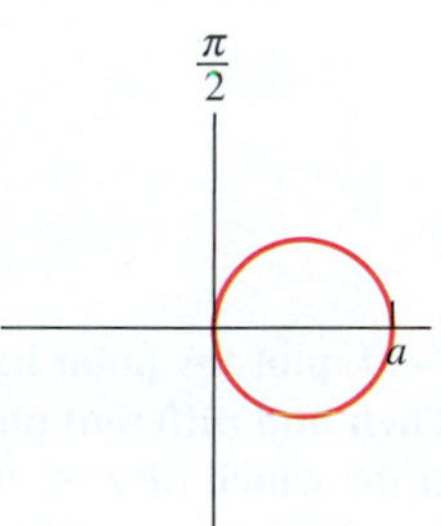

$r = a\cos\theta$
Circle

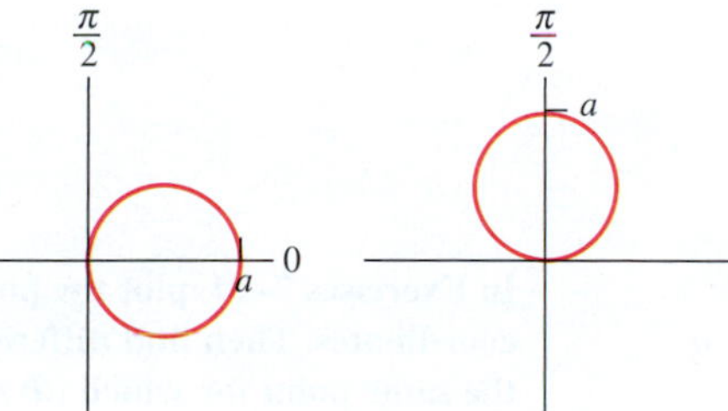

$r = a\sin\theta$
Circle

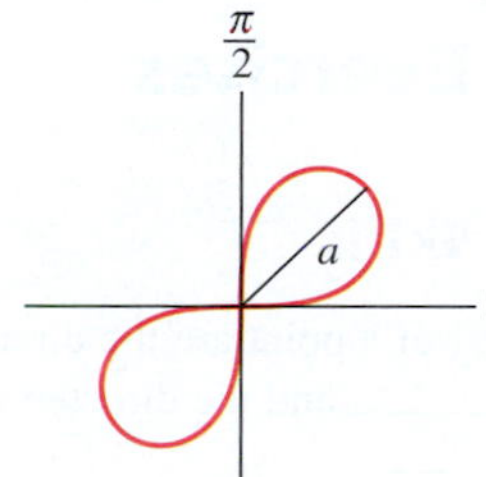

$r^2 = a^2\sin 2\theta$
Lemniscate

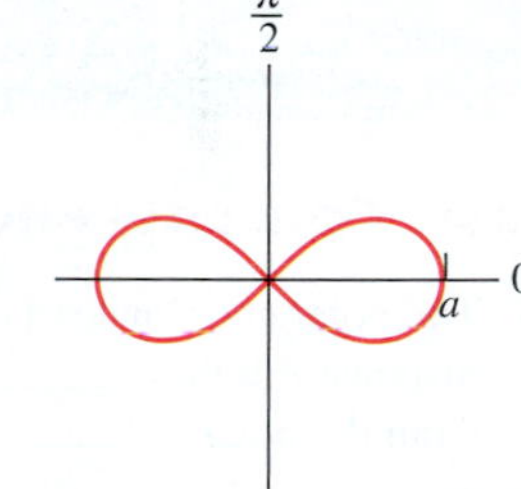

$r^2 = a^2\cos 2\theta$
Lemniscate

SPIRALS

$(a > 0)$

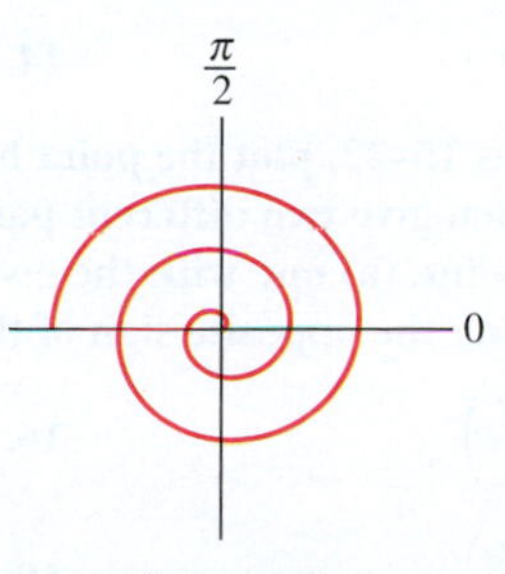

$r = a\theta, \theta \geq 0$

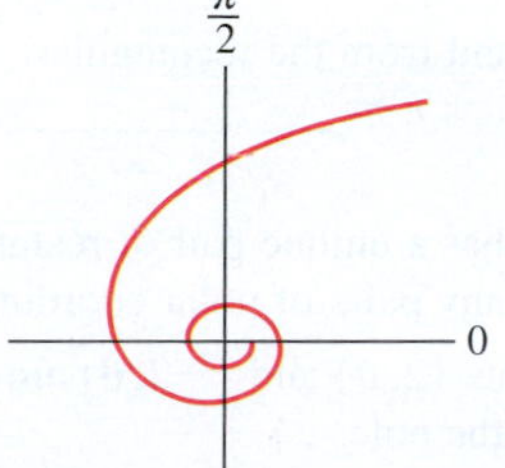

$r = a^{k\theta}, \theta \leq 0, 0 < a < 1, k > 0$

Answers to Practice Problems

1. a. $P(2, -150°)$ **b.** $(-2, 30°)$

c. $(2, 210°)$ **d.** $(-2, -330°)$

2. a. $\left(-\frac{3}{2}, -\frac{3\sqrt{3}}{2}\right)$ **b.** $(\sqrt{2}, -\sqrt{2})$ **3.** $\left(\sqrt{2}, \frac{5\pi}{4}\right)$

4. $(24.8, 36.0°)$ **5.** $r^2 - 2r\sin\theta + 3 = 0$

6. a. $x^2 + y^2 = 25$; a circle centered at the origin with radius 5

b. $y = -x\sqrt{3}$; a line through the origin with slope $-\sqrt{3}$

c. $x = 1$; a vertical line through $x = 1$ **d.** $x^2 + (y - 1)^2 = 1$; a circle with center $(0, 1)$ and radius 1

7.

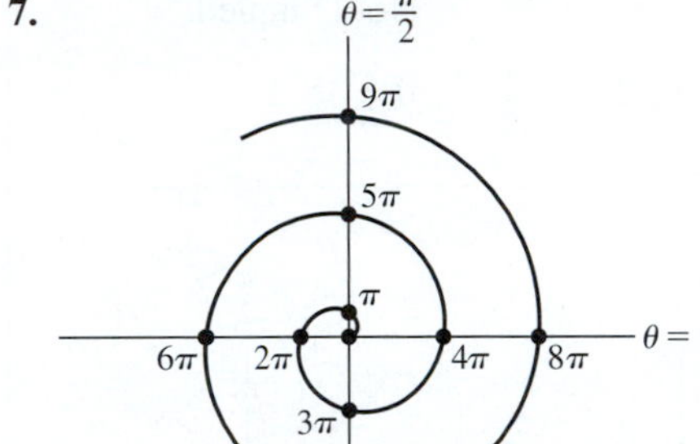

8.

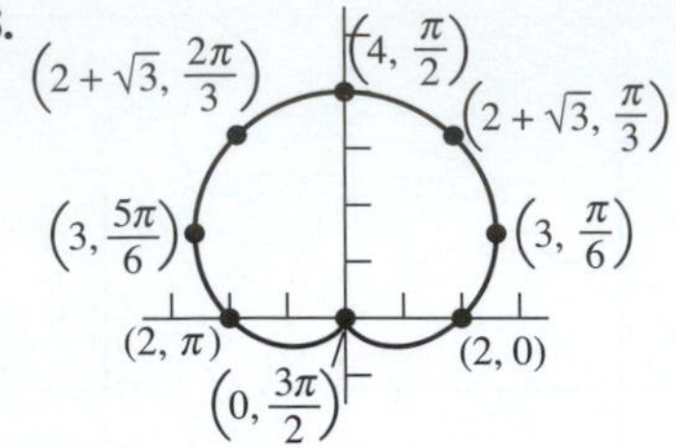

9.

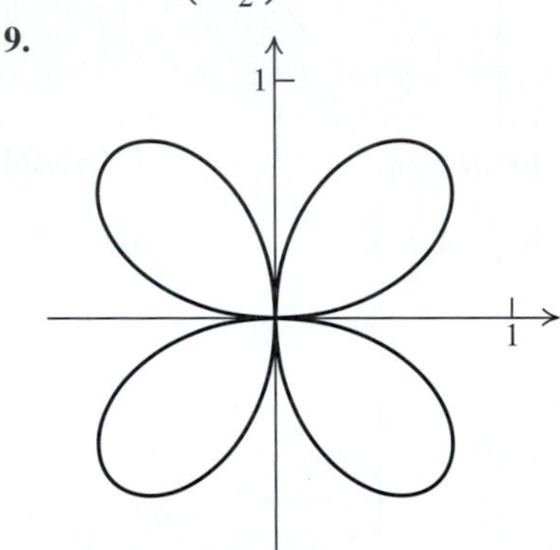

SECTION 7.6 Exercises

Basic Concepts and Skills

1. The polar coordinates (r, θ) of a point are the directed distance r to the ____________ and the directed angle θ from the polar ____________.
2. The positive value of r corresponds to a distance r on the terminal side of θ, and a negative value corresponds to a distance $|r|$ in the ____________ direction.
3. The substitutions $x =$ ____________ and $y =$ ____________ transform a rectangular equation into a polar equation for the same curve.
4. Polar coordinates are found from the rectangular coordinates by the formulas $r =$ ____________ and $\theta =$ ____________.
5. **True or False.** A point has a unique pair of rectangular coordinates, but it has many pairs of polar coordinates.
6. **True or False.** The points $(2, \theta)$ and $(-2, \theta)$ are on opposite sides of the line through the pole.

In Exercises 7–14, plot the point having the given polar coordinates. Then find different polar coordinates (r, θ) for the same point for which (a) $r < 0$ and $0° \le \theta < 360°$; (b) $r < 0$ and $-360° < \theta < 0°$; (c) $r > 0$ and $-360° < \theta < 0°$.

7. $(4, 45°)$
8. $(5, 150°)$
9. $(3, 90°)$
10. $(2, 240°)$
11. $(3, 60°)$
12. $(4, 210°)$
13. $(6, 300°)$
14. $(2, 270°)$

In Exercises 15–22, plot the point having the given polar coordinates. Then give two different pairs of polar coordinates of the same point, (a) one with the given value of r and (b) one with r having the opposite sign of the given value of r.

15. $\left(2, -\frac{\pi}{3}\right)$
16. $\left(\sqrt{2}, -\frac{\pi}{4}\right)$
17. $\left(-3, \frac{\pi}{6}\right)$
18. $\left(-2, \frac{3\pi}{4}\right)$

19. $\left(-2, -\frac{\pi}{6}\right)$

20. $(-2, -\pi)$

21. $\left(4, \frac{7\pi}{6}\right)$

22. $(2, 3)$

In Exercises 23–30, convert the given polar coordinates of each point to rectangular coordinates.

23. $(3, 60°)$

24. $(-2, -30°)$

25. $(5, -60°)$

26. $(-3, 90°)$

27. $(3, \pi)$

28. $\left(\sqrt{2}, -\frac{\pi}{4}\right)$

29. $\left(-2, -\frac{5\pi}{6}\right)$

30. $\left(-1, \frac{7\pi}{6}\right)$

In Exercises 31–38, convert the rectangular coordinates of each point to polar coordinates (r, θ) with $r > 0$ and $0 \le \theta < 2\pi$.

31. $(1, -1)$

32. $(-\sqrt{3}, 1)$

33. $(3, 3)$

34. $(-4, 0)$

35. $(3, -3)$

36. $(2\sqrt{3}, 2)$

37. $(-1, \sqrt{3})$

38. $(-2, -2\sqrt{3})$

In Exercises 39–46, convert each rectangular equation to polar form.

39. $x^2 + y^2 = 16$

40. $x + y = 1$

41. $y^2 = 4x$

42. $x^3 = 3y^2$

43. $y^2 = 6y - x^2$

44. $x^2 - y^2 = 1$

45. $xy = 1$

46. $x^2 + y^2 - 4x + 6y = 12$

In Exercises 47–54, convert each polar equation to rectangular form. Identify each curve.

47. $r = 2$

48. $r = -3$

49. $\theta = \frac{3\pi}{4}$

50. $\theta = -\frac{\pi}{6}$

51. $r = 4\cos\theta$

52. $r = 4\sin\theta$

53. $r = -2\sin\theta$

54. $r = -3\cos\theta$

In Exercises 55–62, sketch the graph of each polar equation by transforming it to rectangular coordinates.

55. $r = -2$

56. $r\cos\theta = -1$

57. $r\sin\theta = 2$

58. $r = \sin\theta$

59. $r = 2\sec\theta$

60. $r + 3\cos\theta = 0$

61. $r = \dfrac{1}{\cos\theta + \sin\theta}$

62. $r = \dfrac{6}{2\cos\theta + 3\sin\theta}$

In Exercises 63–78, sketch the graph of each polar equation.

63. $r = \sin\theta$

64. $r = \cos\theta$

65. $r = 1 - \cos\theta$

66. $r = 1 + \sin\theta$

67. $r = \cos 3\theta$

68. $r = \sin 3\theta$

69. $r = \sin 4\theta$

70. $r = \cos 4\theta$

71. $r = 1 - 2\sin\theta$

72. $r = 2 - 4\sin\theta$

73. $r = 2 + 4\cos\theta$

74. $r = 2 - 4\cos\theta$

75. $r = 3 - 2\sin\theta$

76. $r = 5 + 3\sin\theta$

77. $r = 4 + 3\cos\theta$

78. $r = 5 + 2\cos\theta$

Applying the Concepts

In Exercises 79–82, determine the position of the hand relative to the shoulder by considering the robotic arm illustrated in the figure. Round all answers to the nearest tenth.

79. $\alpha = 45°, \beta = 30°$

80. $\alpha = -30°, \beta = 60°$

81. $\alpha = -70°, \beta = 0°$

82. $\alpha = 47°, \beta = 17°$

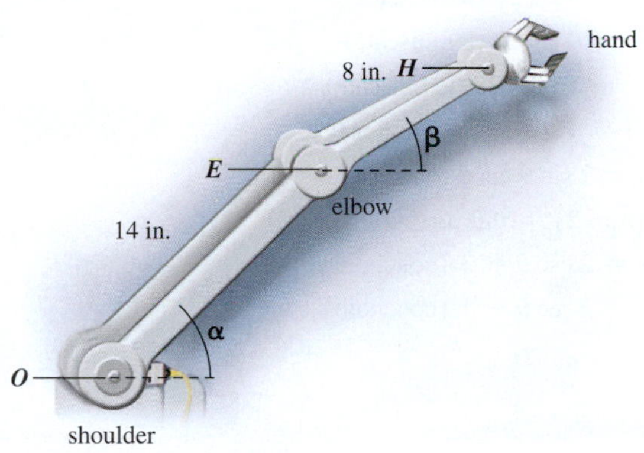

Beyond the Basics

In Exercises 83 and 84, convert each rectangular equation to polar form. Assume that $x \ge 0, y \ge 0$ and that $\sqrt{x^2 + y^2} < \frac{\pi}{2}$.

83. $y = x\tan\left(\sqrt{x^2 + y^2}\right)$

84. $y = x\tan\left(\ln\sqrt{x^2 + y^2}\right)$

In Exercises 85–90, convert each polar equation to rectangular form.

85. $r(1 - \sin\theta) = 3$

86. $r(1 + \cos\theta) = 2$

87. $r\left(1 + \frac{1}{2}\cos\theta\right) = 1$

88. $r\left(1 + \frac{3}{4}\sin\theta\right) = 3$

89. $r(1 - 3\cos\theta) = 5$

90. $r(1 - 2\sin\theta) = 4$

91. Prove that the distance between the points $P(r_1, \theta_1)$ and $Q(r_2, \theta_2)$ is given by

$$d(P, Q) = \sqrt{r_1^2 + r_2^2 - 2r_1r_2\cos(\theta_1 - \theta_2)}.$$

92. Prove that the equation $r = a\sin\theta + b\cos\theta$ represents a circle. Find its center and radius.

93. Prove that the area K of the triangle whose polar coordinates are $(0, 0)$, (r_1, θ_1), and (r_2, θ_2) is given by $K = \frac{1}{2} r_1 r_2 \sin(\theta_2 - \theta_1)$. [Assume that $0 \le \theta_1 < \theta_2 \le \pi$ and $r_1 > 0, r_2 > 0$.]

94. Show that the polar equation of a circle with center (a, θ_0) and radius a is given by $r = 2a \cos(\theta - \theta_0)$. Discuss the cases $\theta_0 = 0$ and $\theta_0 = \frac{\pi}{2}$.

In Exercises 95–104, use a graphing calculator to graph the polar equation. Identify any horizontal or vertical asymptotes that appear on the screen but are not part of the graph.

95. $r = e^{\theta}$ (logarithmic spiral)

96. $r = e^{\frac{\theta}{3}}$ (logarithmic spiral)

97. $r^2 = 4 \sin 2\theta$ (lemniscate)

98. $r^2 = 9 \sin 2\theta$ (lemniscate)

99. $r^2 = 16 \cos 2\theta$ (lemniscate)

100. $r^2 = 4 \cos 2\theta$ (lemniscate)

101. $r = 2 \sin \theta \tan \theta$ (cissoid)

102. $(r - 2)^2 = 8\theta$ (parabolic spiral)

103. $r = 2 \csc \theta + 3$ (conchoid)

104. $r = 2 \sec \theta - 1$ (conchoid)

Critical Thinking/Discussion/Writing

Determine whether each statement is true or false. Explain your reasoning.

105. The points (r, θ) and $(-r, \theta)$ are symmetric with respect to the y-axis.

106. The points (r, θ) and $(-r, -\theta)$ are symmetric with respect to the y-axis.

107. The points (r, θ) and $(r, -\theta)$ are symmetric with respect to the x-axis.

108. The polar coordinates $(r, -\theta)$ and $(-r, \pi - \theta)$ represent the same point in the plane.

Maintaining Skills

In Exercises 109–120, write each complex number in the form $a + bi$.

109. $(2 - 3i) + (-5 + i)$

110. $(5 + 3i) - (4 - 2i)$

111. $i(2 + 5i)$

112. $-i(5 - 3i)$

113. $(2 + i)(3 - i)$

114. $(4 - 3i)(2 - 5i)$

115. $(2 - 3i)^2$

116. $(1 + 2i)^3$

117. $\frac{1}{3 - 4i}$

118. $\frac{2i}{2 + 3i}$

119. $\frac{2 - 3i}{3 + 4i}$

120. $\frac{1 + i}{2 - i}$

In Exercises 121–124, write the trigonometric form (see page 678) for each vector.

121. $\mathbf{i} + \mathbf{j}$

122. $2\mathbf{i} - 2\mathbf{j}$

123. $-\sqrt{3}\mathbf{i} + \mathbf{j}$

124. $-\mathbf{i} - \sqrt{3}\mathbf{j}$

SECTION 7.7

Polar Form of Complex Numbers; DeMoivre's Theorem

BEFORE STARTING THIS SECTION, REVIEW

1 Addition of complex numbers (Section 1.3, page 106)

2 Conversion between polar and rectangular coordinates (Section 7.6, pages 698–699)

OBJECTIVES

1 Represent complex numbers geometrically.

2 Find the absolute value of a complex number.

3 Write a complex number in polar form.

4 Find products and quotients of complex numbers in polar form.

5 Use DeMoivre's theorem to find powers of a complex number.

6 Use DeMoivre's theorem to find the *n*th roots of a complex number.

Alternating Current Circuits

In the early days of the study of alternate current (AC) circuits, scientists concluded that AC circuits were somehow different from the battery-powered direct current (DC) circuits. However, both types of circuits obey the same physical and mathematical laws. The breakthrough in understanding the AC circuits came in 1893 by Charles Steinmetz. He explained that in AC circuits, the voltage, current, and resistance (called *impedance* in AC circuits) do not act like scalars, but alternate in direction and possess frequency and phase shift. He advocated the use of the polar form of complex numbers to represent the magnitude, frequency, and phase shift for the AC circuit quantities: voltage, current, and impedance. Steinmetz eventually became known as "the wizard who generated electricity from the square root of minus one." In Example 6, we use complex numbers to compute the total impedance in an AC circuit.

1 Represent complex numbers geometrically.

Geometric Representation of Complex Numbers

Because each complex number $a + bi$ determines a unique ordered pair (a, b) of real numbers, we can represent the set of complex numbers geometrically by points in a rectangular coordinate system. Specifically, we can represent the complex number $a + bi$ by the point (a, b) in a rectangular coordinate system. We call the plane in this system the **complex plane**. The x-axis is also called the **real axis** because the real part of a complex number is plotted along the x-axis. Similarly, the y-axis is also called the **imaginary axis**. See Figure 7.55. We can think of a complex number $a + bi$ as a position vector with initial point $(0, 0)$ and terminal point (a, b).

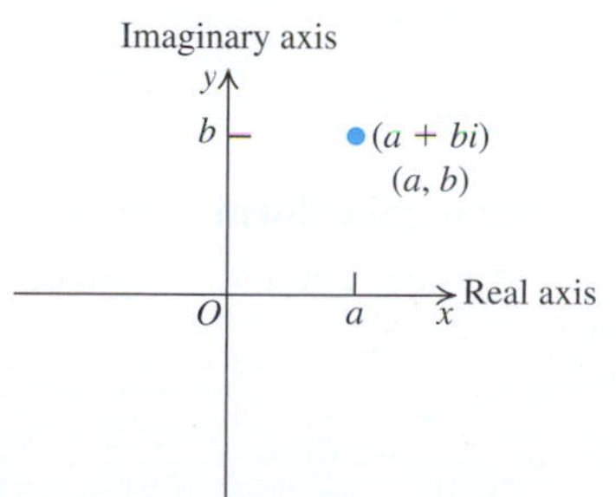

FIGURE 7.55 Complex plane

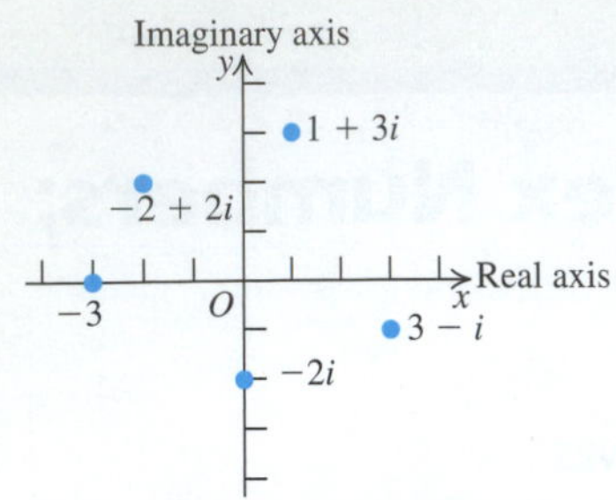

FIGURE 7.56 Plotting complex numbers

EXAMPLE 1 Plotting Complex Numbers

Plot each complex number in the complex plane.

$$1 + 3i, \quad -2 + 2i, \quad -3, \quad -2i, \quad \text{and} \quad 3 - i$$

Solution

Figure 7.56 shows the points representing the complex numbers $1 + 3i$, $-2 + 2i$, -3, $-2i$, and $3 - i$.

Practice Problem 1 Plot each complex number in the complex plane.

$$2 + 3i, \quad -3 + 2i, \quad 4, \quad -2 - 3i, \quad \text{and} \quad 3 - 2i$$

2 Find the absolute value of a complex number.

The Absolute Value of a Complex Number

Let (a, b) represent the complex number $z = a + bi$ in a rectangular coordinate system. Then the distance from the origin O to the point (a, b) is $\sqrt{a^2 + b^2}$. See Figure 7.57. The number $\sqrt{a^2 + b^2}$ is called the *absolute value* (or **magnitude** or **modulus**) of the complex number $z = a + bi$ and is denoted by $|z|$.

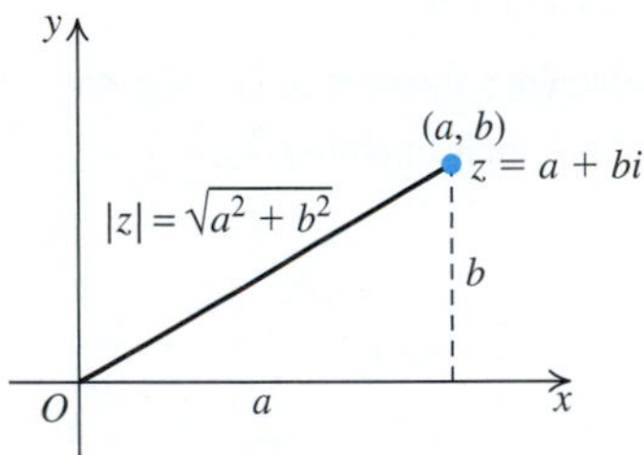

FIGURE 7.57 Absolute value of z

Absolute Value of a Complex Number

The absolute value of a complex number $z = a + bi$:

$$|z| = |a + bi| = \sqrt{a^2 + b^2}$$

EXAMPLE 2 Finding the Absolute Value of a Complex Number

Find the absolute value of each complex number.

a. $4 + 3i$ **b.** $2 - 3i$ **c.** $-4 + i$ **d.** $-2 - 2i$ **e.** $-3i$

Solution

In each case, we use the formula $|a + bi| = \sqrt{a^2 + b^2}$ and simplify.

a. $|4 + 3i| = \sqrt{4^2 + 3^2} = \sqrt{25} = 5$

b. $|2 - 3i| = \sqrt{2^2 + (-3)^2} = \sqrt{4 + 9} = \sqrt{13}$

c. $|-4 + i| = |-4 + 1 \cdot i| = \sqrt{(-4)^2 + (1)^2} = \sqrt{16 + 1} = \sqrt{17}$

d. $|-2 - 2i| = \sqrt{(-2)^2 + (-2)^2} = \sqrt{4 + 4} = \sqrt{4 \cdot 2} = 2\sqrt{2}$

e. $|-3i| = |0 - 3i| = \sqrt{0^2 + (-3)^2} = \sqrt{0 + 9} = 3$

Practice Problem 2 Find the absolute value of each complex number.

a. $-5 + 12i$ **b.** -7 **c.** i **d.** $a - bi$

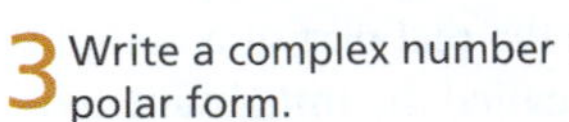
3 Write a complex number in polar form.

Polar Form of a Complex Number

A complex number z written as $z = a + bi$ is said to be in **rectangular form**. The point (a, b) has polar coordinates (r, θ), where $r = \sqrt{a^2 + b^2}$, $a = r\cos\theta$, and $b = r\sin\theta$.

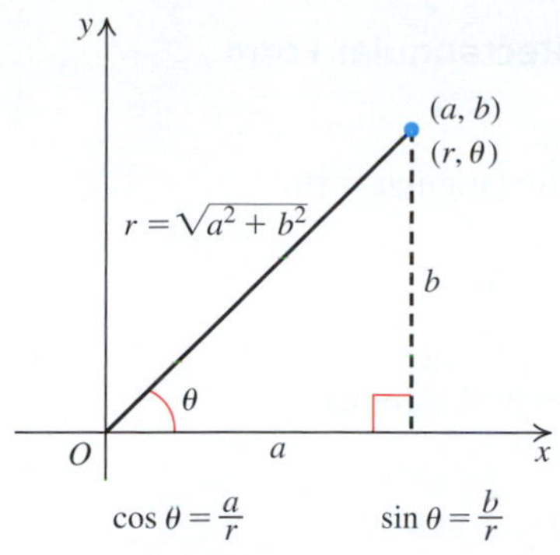

FIGURE 7.58 Polar form of $z = a + bi$

The complex number $z = a + bi$ (see Figure 7.58) can therefore be written in the form

$$z = r\cos\theta + (r\sin\theta)i = r(\cos\theta + i\sin\theta).$$

We call $r(\cos\theta + i\sin\theta)$ the **polar form** or the **trigonometric form** of z.

POLAR FORM OF A COMPLEX NUMBER

The complex number $z = a + bi$ can be written in **polar form**

$$z = r(\cos\theta + i\sin\theta),$$

where $a = r\cos\theta$, $b = r\sin\theta$, $r = \sqrt{a^2 + b^2}$, and $\tan\theta = \dfrac{b}{a}$.

When a nonzero complex number is written in polar form, the positive number r is the **modulus** or **absolute value** of z; the angle θ is called the **argument** of z (written $\theta = \arg z$).

Note that the angle θ in the polar representation of z is not unique because for any integer n,

$$r(\cos\theta + i\sin\theta) = r[\cos(\theta + n\cdot 360°) + i\sin(\theta + n\cdot 360°)].$$

Two complex numbers in polar form are therefore equal if and only if their *moduli* (plural of *modulus*) are equal and their arguments differ by an integer multiple of 360° (or an integer multiple of 2π).

PROCEDURE **IN ACTION**

EXAMPLE 3 Writing a Complex Number in Polar Form

OBJECTIVE

Write a complex number $z = a + bi$ in polar form.

EXAMPLE

Write $z = \sqrt{3} - i$ in polar form. Express the argument θ in degrees, $0° \le \theta < 360°$.

Step 1 Find r. Identify a and b. Use the formula $r = \sqrt{a^2 + b^2}$.

1. $z = \sqrt{3} - i$, $a = \sqrt{3}$, $b = -1$

$r = \sqrt{(\sqrt{3})^2 + (-1)^2} = \sqrt{3 + 1} = 2$

Step 2 Find θ. Use $\tan\theta = \dfrac{b}{a}$ to find the possible values of θ. Choose θ in the quadrant in which (a, b) lies. In general, using a calculator, a value of θ is given by

$$\theta = \begin{cases} \tan^{-1}\dfrac{b}{a} & \text{if } a > 0 \\ 180° + \tan^{-1}\dfrac{b}{a} & \text{if } a < 0 \end{cases}.$$

2. $\tan\theta = \dfrac{b}{a} = \dfrac{-1}{\sqrt{3}} = -\dfrac{1}{\sqrt{3}}$. We know that $\tan 30° = \dfrac{1}{\sqrt{3}}$ and that $\tan\theta$ is negative in quadrants II and IV; so either $\theta = 180° - 30° = 150°$ or $\theta = 360° - 30° = 330°$. Because $(a, b) = (\sqrt{3}, -1)$ lies in quadrant IV, we have $\theta = 330°$. See the figure.

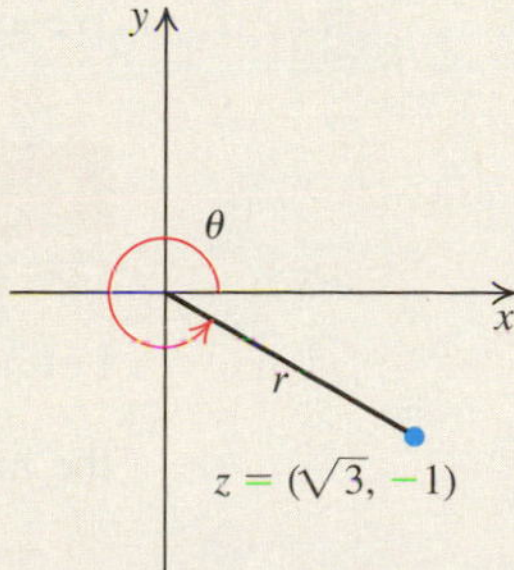

Step 3 Write in polar form. From Steps 1 and 2, write the polar form $z = r(\cos\theta + i\sin\theta)$.

3. $z = 2(\cos 330° + i\sin 330°)$ $\quad r = 2, \theta = 330°$

Practice Problem 3 Write $z = -1 - i$ in polar form, letting $0 \le \theta < 2\pi$.

EXAMPLE 4 Writing a Complex Number in Rectangular Form

Write the complex number $z = 2\left(\cos\frac{\pi}{6} + i\sin\frac{\pi}{6}\right)$ in rectangular form.

Solution

$$\begin{aligned} z &= 2\left(\cos\frac{\pi}{6} + i\sin\frac{\pi}{6}\right) && \text{Given complex number} \\ &= 2\cos\frac{\pi}{6} + 2i\sin\frac{\pi}{6} && \text{Distributive property} \\ &= 2\left(\frac{\sqrt{3}}{2}\right) + 2i\left(\frac{1}{2}\right) && \cos\frac{\pi}{6} = \frac{\sqrt{3}}{2}, \sin\frac{\pi}{6} = \frac{1}{2} \\ &= \sqrt{3} + i && \text{Simplify.} \end{aligned}$$

The rectangular form of $z = 2\left(\cos\frac{\pi}{6} + i\sin\frac{\pi}{6}\right)$ is $\sqrt{3} + i$.

Practice Problem 4 Write $z = 4\left(\cos\frac{5\pi}{3} + i\sin\frac{5\pi}{3}\right)$ in rectangular form.

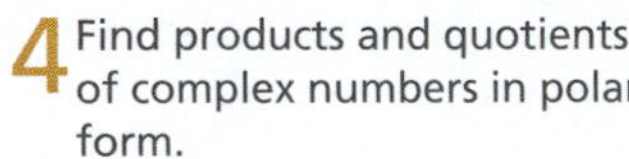
4 Find products and quotients of complex numbers in polar form.

Product and Quotient in Polar Form

The polar representation of complex numbers leads to an interesting interpretation of the product and quotient of two complex numbers.

PRODUCT AND QUOTIENT RULES FOR TWO COMPLEX NUMBERS IN POLAR FORM

Let $z_1 = r_1(\cos\theta_1 + i\sin\theta_1)$ and $z_2 = r_2(\cos\theta_2 + i\sin\theta_2)$ be two complex numbers in polar form. Then

$$z_1z_2 = r_1r_2[\cos(\theta_1 + \theta_2) + i\sin(\theta_1 + \theta_2)] \qquad \text{Product rule}$$

and

$$\frac{z_1}{z_2} = \frac{r_1}{r_2}[\cos(\theta_1 - \theta_2) + i\sin(\theta_1 - \theta_2)], z_2 \neq 0 \qquad \text{Quotient rule}$$

In words: To multiply two complex numbers in polar form, we multiply their moduli and add their arguments; to divide two complex numbers, we divide their moduli and subtract their arguments.

We ask you to prove these rules in Exercises 95 and 96.

EXAMPLE 5 Finding the Product and Quotient of Two Complex Numbers

Let $z_1 = 3(\cos 65° + i\sin 65°)$ and $z_2 = 4(\cos 15° + i\sin 15°)$. Find z_1z_2 and $\frac{z_1}{z_2}$. Leave the answers in polar form.

Solution

$$\begin{aligned}
z_1 z_2 &= 3(\cos 65° + i \sin 65°) \cdot 4(\cos 15° + i \sin 15°) && \text{Product of given numbers} \\
&= 3 \cdot 4[\cos(65° + 15°) + i \sin(65° + 15°)] && \text{Multiply moduli and add arguments.} \\
&= 12(\cos 80° + i \sin 80°) && \text{Simplify.} \\
\frac{z_1}{z_2} &= \frac{3(\cos 65° + i \sin 65°)}{4(\cos 15° + i \sin 15°)} && \text{Quotient of given numbers} \\
&= \frac{3}{4}[\cos(65° - 15°) + i \sin(65° - 15°)] && \text{Divide moduli and subtract arguments.} \\
&= \frac{3}{4}(\cos 50° + i \sin 50°) && \text{Simplify.}
\end{aligned}$$

Practice Problem 5 Let $z_1 = 5(\cos 75° + i \sin 75°)$ and $z_2 = 2(\cos 60° + i \sin 60°)$. Find $z_1 z_2$ and $\frac{z_1}{z_2}$. Leave the answers in polar form.

EXAMPLE 6 Using Complex Numbers in AC Circuits

In a parallel circuit, the total impedance Z_t is given by

$$Z_t = \frac{Z_1 Z_2}{Z_1 + Z_2}.$$

Find Z_t, if

$$Z_1 = 9(\cos 90° + i \sin 90°) \text{ and } Z_2 = 4[\cos(-60°) + i \sin(-60°)].$$

Solution

We first calculate $Z_1 Z_2$ and $Z_1 + Z_2$.

$$\begin{aligned}
Z_1 Z_2 &= 9(\cos 90° + i \sin 90°) \cdot 4[\cos(-60°) + i \sin(-60°)] && \text{Substitute values for } Z_1 \text{ and } Z_2. \\
&= 9 \cdot 4[\cos(90° - 60°) + i \sin(90° - 60°)] && \text{Multiply moduli and add arguments.} \\
&= 36(\cos 30° + i \sin 30°) && \text{Simplify.}
\end{aligned}$$

To add complex numbers in polar form, convert to rectangular form, perform addition, and then convert back to polar form.

$$\begin{aligned}
Z_1 &= 9 \cos 90° + 9i \sin 90° = 9i && \cos 90° = 0, \sin 90° = 1 \\
Z_2 &= 4[\cos(-60°) + i \sin(-60°)] \\
&= 4 \cos 60° - 4i \sin 60° && \cos(-\theta) = \cos\theta, \sin(-\theta) = -\sin\theta \\
&= 4\left(\frac{1}{2}\right) - 4i\left(\frac{\sqrt{3}}{2}\right) && \cos 60° = \frac{1}{2}, \sin 60° = \frac{\sqrt{3}}{2} \\
&= 2 - 2\sqrt{3}i && \text{Simplify.} \\
Z_1 + Z_2 &= 9i + 2 - 2\sqrt{3}i \\
Z_1 + Z_2 &= 2 + (9 - 2\sqrt{3})i && \text{Regroup.}
\end{aligned}$$

To write $Z_1 + Z_2$ in polar form, find the values of r and θ.

$$\begin{aligned}
r &= \sqrt{a^2 + b^2} \\
&= \sqrt{2^2 + (9 - 2\sqrt{3})^2} && \text{Replace } a \text{ with 2 and } b \text{ with } 9 - 2\sqrt{3}. \\
&\approx 5.89 && \text{Use a calculator.} \\
\tan\theta &= \frac{b}{a} = \frac{9 - 2\sqrt{3}}{2} \approx 2.768 \\
\theta &\approx \tan^{-1}(2.768) \approx 70° && \text{Use a calculator.}
\end{aligned}$$

The polar form of $Z_1 + Z_2$ is approximately $5.89(\cos 70° + i \sin 70°)$. Thus,

$$Z_t = \frac{Z_1Z_2}{Z_1 + Z_2} \approx \frac{36(\cos 30° + i \sin 30°)}{5.89(\cos 70° + i \sin 70°)}$$ Substitute values for Z_1Z_2 and $Z_1 + Z_2$.

$$= \frac{36}{5.89}[\cos(30° - 70°) + i \sin(30° - 70°)]$$ Divide moduli and subtract.

$$= \frac{36}{5.89}[\cos(-40°) + i \sin(-40°)]$$ Simplify.

$$= \frac{36}{5.89}(\cos 40° - i \sin 40°)$$ $\cos(-\theta) = \cos\theta$, $\sin(-\theta) = -\sin\theta$

$$\approx 6.11(\cos 40° - i \sin 40°)$$ $\frac{36}{5.89} \approx 6.11$

Practice Problem 6 Repeat Example 6 with

$$Z_1 = 4(\cos 45° + i \sin 45°) \quad \text{and} \quad Z_2 = 6(\cos 0° + i \sin 0°).$$

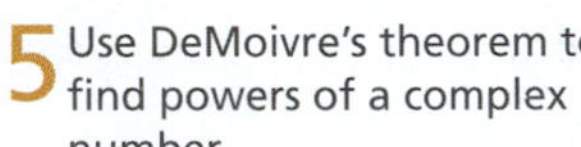
5 Use DeMoivre's theorem to find powers of a complex number.

Powers of Complex Numbers in Polar Form

Let $z = r(\cos\theta + i \sin\theta)$ be a complex number in polar form. Then

$$z^2 = z \cdot z$$
$$= r(\cos\theta + i \sin\theta) \cdot r(\cos\theta + i \sin\theta)$$ Form product $z \cdot z$.
$$= (r \cdot r)[\cos(\theta + \theta) + i \sin(\theta + \theta)]$$ Multiply moduli and add arguments.
$$= r^2(\cos 2\theta + i \sin 2\theta)$$ Simplify.
$$z^3 = z^2 \cdot z$$
$$= r^2(\cos 2\theta + i \sin 2\theta) \cdot r(\cos\theta + i \sin\theta)$$ Polar form of z^2 and z
$$= (r^2 \cdot r)[\cos(2\theta + \theta) + i \sin(2\theta + \theta)]$$ Multiply moduli and add arguments.
$$= r^3(\cos 3\theta + i \sin 3\theta).$$ Simplify.

In a similar way, you can show that

$$z^4 = r^4(\cos 4\theta + i \sin 4\theta).$$

DeMoivre's theorem (whose proof is omitted) follows this pattern.

DEMOIVRE'S THEOREM

Let $z = r(\cos\theta + i \sin\theta)$ be a complex number in polar form. Then for any integer n,

$$z^n = r^n(\cos n\theta + i \sin n\theta).$$

EXAMPLE 7 Finding the Power of a Complex Number

Let $z = 1 + i$. Use DeMoivre's theorem to find each power of z. Write answers in rectangular form.

a. z^{16} **b.** z^{-10}

Solution

We first convert $z = 1 + i$ to polar form. We find r and θ.

$$r = \sqrt{a^2 + b^2} = \sqrt{1^2 + 1^2} = \sqrt{2}$$

TECHNOLOGY CONNECTION

A graphing calculator can be used to find powers of complex numbers or to check your work. The calculator must be set in $a + bi$ mode.

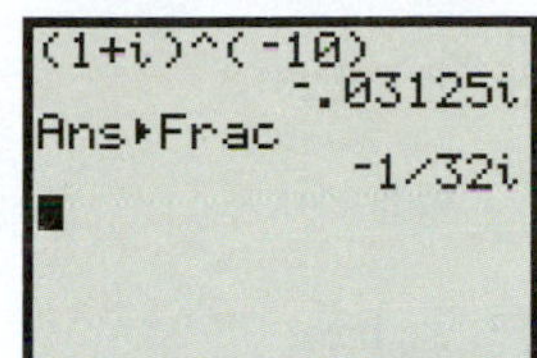

and

$$\tan\theta = \frac{b}{a} = \frac{1}{1} = 1, \text{ so } \theta = \frac{\pi}{4}$$

a. $z = \sqrt{2}\left(\cos\frac{\pi}{4} + i\sin\frac{\pi}{4}\right)$ Polar form of $z = 1 + i$

$z^{16} = \left[\sqrt{2}\left(\cos\frac{\pi}{4} + i\sin\frac{\pi}{4}\right)\right]^{16}$ Raise both sides to the 16th power.

$z^{16} = (\sqrt{2})^{16}\left[\cos\left(16\cdot\frac{\pi}{4}\right) + i\sin\left(16\cdot\frac{\pi}{4}\right)\right]$ DeMoivre's theorem

$= 2^8[\cos(4\pi) + i\sin(4\pi)]$ $(\sqrt{2})^{16} = (2^{\frac{1}{2}})^{16} = 2^{\frac{1}{2}\cdot 16} = 2^8$

$= 256(1 + i\cdot 0) = 256$ $\cos(4\pi) = 1, \sin(4\pi) = 0$

b. $z = \sqrt{2}\left(\cos\frac{\pi}{4} + i\sin\frac{\pi}{4}\right)$ Polar form of $z = 1 + i$

$z^{-10} = \left[\sqrt{2}\left(\cos\frac{\pi}{4} + i\sin\frac{\pi}{4}\right)\right]^{-10}$ Raise both sides to the -10th power.

$z^{-10} = (\sqrt{2})^{-10}\left[\cos\left(-10\cdot\frac{\pi}{4}\right) + i\sin\left(-10\cdot\frac{\pi}{4}\right)\right]$ DeMoivre's theorem

$= \frac{1}{32}\left[\cos\left(-\frac{5\pi}{2}\right) + i\sin\left(-\frac{5\pi}{2}\right)\right]$ $(\sqrt{2})^{-10} = (2^{1/2})^{-10} = 2^{-5} = \frac{1}{32}$

$= \frac{1}{32}[0 + i(-1)] = -\frac{1}{32}i$ $\cos\left(-\frac{5\pi}{2}\right) = 0, \sin\left(-\frac{5\pi}{2}\right) = -1$

Practice Problem 7 Let $z = -1 + i$. Use DeMoivre's theorem to find each power of z. Write the answers in rectangular form.

a. z^8 **b.** z^{-12}

6 Use DeMoivre's theorem to find the nth roots of a complex number.

Roots of Complex Numbers

Let z and w be two complex numbers and let n be a positive integer. The complex number z is called an ***n*th root of *w*** if

$$z^n = w.$$

We can use DeMoivre's theorem to find the nth roots of a complex number.

DEMOIVRE'S *n*th ROOTS THEOREM

The nth roots of a complex number $w = r(\cos\theta + i\sin\theta)$, where $r > 0$ and θ is in degrees, are given by

$$z_k = r^{1/n}\left[\cos\left(\frac{\theta + 360°k}{n}\right) + i\sin\left(\frac{\theta + 360°k}{n}\right)\right], \text{ for } k = 0, 1, 2, \ldots, n - 1.$$

If θ is in radians, replace 360° with 2π in z_k.

The result says that there are exactly n nth roots of a nonzero complex number.

EXAMPLE 8 Finding the Roots of a Complex Number

Find the three cube roots of $1 + i$ in polar form, with the argument in degrees.

Solution

In Example 7, we showed that

$$1 + i = \sqrt{2}\left(\cos\frac{\pi}{4} + i\sin\frac{\pi}{4}\right) \qquad \text{Polar form of } 1 + i$$

$$1 + i = \sqrt{2}(\cos 45° + i\sin 45°). \qquad \text{Replace } \frac{\pi}{4} \text{ with } 45°.$$

RECALL

$\sqrt{2} = 2^{1/2}$, so
$(\sqrt{2})^{1/3} = (2^{1/2})^{1/3} = 2^{1/6}$.

From DeMoivre's theorem for finding complex roots, with $n = 3$, we have

$$z_k = (\sqrt{2})^{1/3}\left[\cos\left(\frac{45° + 360°k}{3}\right) + i\sin\left(\frac{45° + 360°k}{3}\right)\right], \quad k = 0, 1, 2.$$

By substituting $k = 0$, 1, and 2 in the expression for z_k and simplifying, we find the three cube roots.

$$z_0 = 2^{1/6}\left[\cos\left(\frac{45° + 360°\cdot 0}{3}\right) + i\sin\left(\frac{45° + 360°\cdot 0}{3}\right)\right] \qquad k = 0$$

$$= 2^{1/6}(\cos 15° + i\sin 15°) \qquad \text{Simplify.}$$

$$z_1 = 2^{1/6}\left[\cos\left(\frac{45° + 360°\cdot 1}{3}\right) + i\sin\left(\frac{45° + 360°\cdot 1}{3}\right)\right] \qquad k = 1$$

$$= 2^{1/6}(\cos 135° + i\sin 135°) \qquad \text{Simplify.}$$

$$z_2 = 2^{1/6}\left[\cos\left(\frac{45° + 360°\cdot 2}{3}\right) + i\sin\left(\frac{45° + 360°\cdot 2}{3}\right)\right] \qquad k = 2$$

$$= 2^{1/6}(\cos 255° + i\sin 255°) \qquad \text{Simplify.}$$

The complex numbers z_0, z_1, and z_2 are the three cube roots of the complex number $1 + i$.

TECHNOLOGY CONNECTION

A graphing calculator can be used to check complex roots. The calculator must be set in $a + bi$ mode. The screen shown here is from a calculator also set in *degree* mode.

Practice Problem 8 Find the three cube roots of $-1 + i$ in polar form, with the argument in degrees.

The solutions of the so-called *cyclotomic* equation $z^n = 1$ are called the ***n*th roots of unity**. The name refers to the close association of this equation with the regular n-gons inscribed in the unit circle.

EXAMPLE 9 Finding *n*th Roots of Unity

Find the complex sixth roots of unity. Write the roots in polar form and represent them geometrically.

Solution

The polar form for $1 = 1 + 0i$ is

$$1 = 1(\cos 0° + i\sin 0°).$$

From DeMoivre's theorem for finding complex roots with $n = 6$, we have

$$z_k = (1)^{1/6}\left[\cos\left(\frac{0° + 360°\cdot k}{6}\right) + i\sin\left(\frac{0° + 360°\cdot k}{6}\right)\right] \qquad k = 0, 1, 2, 3, 4, 5$$

$$z_0 = 1(\cos 0° + i\sin 0°) \qquad \text{Let } k = 0 \text{ in } z_k \text{ and simplify.}$$

$$z_1 = 1(\cos 60° + i\sin 60°) \qquad \text{Let } k = 1 \text{ in } z_k \text{ and simplify.}$$

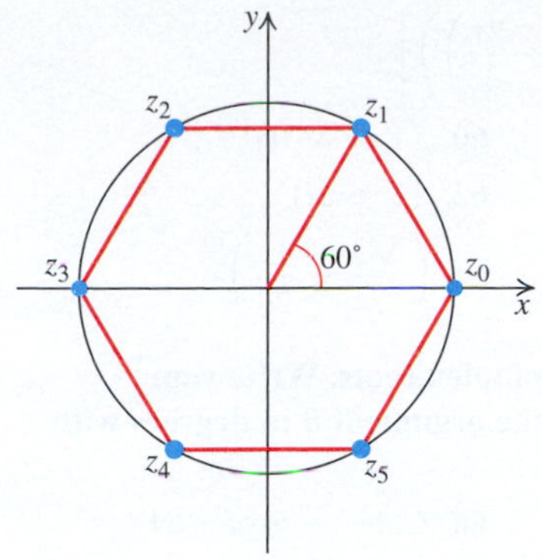

$z_2 = 1(\cos 120° + i \sin 120°)$ Let $k = 2$ in z_k and simplify.
$z_3 = 1(\cos 180° + i \sin 180°)$ Let $k = 3$ in z_k and simplify.
$z_4 = 1(\cos 240° + i \sin 240°)$ Let $k = 4$ in z_k and simplify.
$z_5 = 1(\cos 300° + i \sin 300°)$ Let $k = 5$ in z_k and simplify.

Geometrically, the sixth roots of unity form a regular hexagon and are equally spaced at 60° intervals on a unit circle. See Figure 7.59.

FIGURE 7.59 Sixth roots of unity

Practice Problem 9 Find the complex fourth roots of unity. Write the roots in polar form.

Answers to Practice Problems

1.

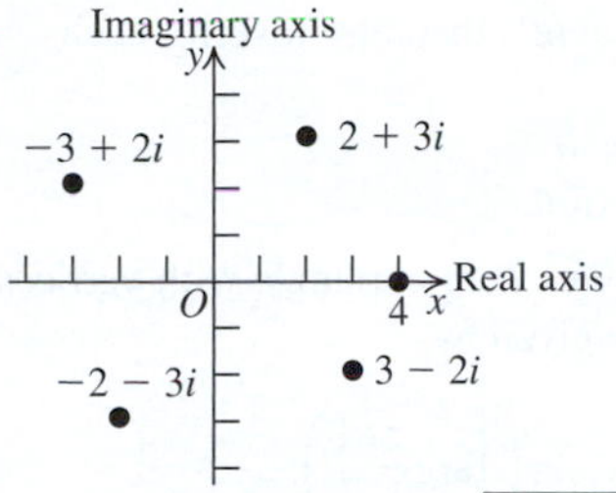

2. a. 13 **b.** 7 **c.** 1 **d.** $\sqrt{a^2 + b^2}$

3. $z = \sqrt{2}\left(\cos \frac{5\pi}{4} + i \sin \frac{5\pi}{4}\right)$ **4.** $2 - 2\sqrt{3}i$

5. $z_1 z_2 = 10(\cos 135° + i \sin 135°)$; $\frac{z_1}{z_2} = \frac{5}{2}(\cos 15° + i \sin 15°)$

6. $2.6(\cos 27.2° + i \sin 27.2°)$ **7. a.** 16 **b.** $-\frac{1}{64}$

8. $2^{\frac{1}{6}}(\cos 45° + i \sin 45°)$, $2^{\frac{1}{6}}(\cos 165° + i \sin 165°)$, $2^{\frac{1}{6}}(\cos 285° + i \sin 285°)$ **9.** $\cos 0° + i \sin 0°$, $\cos 90° + i \sin 90°$, $\cos 180° + i \sin 180°$, $\cos 270° + i \sin 270°$

SECTION 7.7 Exercises

Basic Concepts and Skills

1. If $z = a + bi$, $a > 0$, then $|z| =$ ____________ and $\arg z = \theta =$ ____________.
2. If $|z| = r$ and $\arg z = \theta$, then the polar form of z is $z =$ ____________.
3. To multiply two complex numbers in polar form, multiply their ____________ and ____________ their arguments.
4. DeMoivre's theorem states that $[r(\cos \theta + i \sin \theta)]^n =$ ____________.
5. **True or False.** If $z = r(\cos \theta + i \sin \theta)$, then $\frac{1}{z} = z^{-1} = \frac{1}{r}(\cos \theta - i \sin \theta)$.
6. **True or False.** For $n \geq 2$, the n complex nth roots of 1 are equally spaced points on the unit circle.

In Exercises 7–14, plot each complex number and find its absolute value.

7. $z = 2$
8. $z = -3$
9. $z = -4i$
10. $z = 2i$
11. $z = 3 + 4i$
12. $z = -1 + 2i$
13. $z = -2 - 3i$
14. $z = 2 - 5i$

In Exercises 15–26, write each complex number in polar form. Express the argument θ in degrees, with $0 \leq \theta < 360°$.

15. $1 + \sqrt{3}i$
16. $-1 + \sqrt{2}i$
17. $-1 + i$
18. $1 - i$
19. i
20. $-i$
21. 1
22. -1
23. $3 - 3i$
24. $4\sqrt{3} + 4i$
25. $2 - 2\sqrt{3}i$
26. $2 + 3i$

In Exercises 27–38, write each complex number in rectangular form.

27. $2(\cos 60° + i \sin 60°)$
28. $4(\cos 120° + i \sin 120°)$
29. $3(\cos \pi + i \sin \pi)$
30. $5\left(\cos \frac{\pi}{2} + i \sin \frac{\pi}{2}\right)$
31. $5(\cos 240° + i \sin 240°)$
32. $2(\cos 300° + i \sin 300°)$
33. $8(\cos 0° + i \sin 0°)$
34. $2(\cos(-90°)° + i \sin(-90°))$
35. $6\left(\cos \frac{5\pi}{6} + i \sin \frac{5\pi}{6}\right)$
36. $4\left(\cos \frac{3\pi}{4} + i \sin \frac{3\pi}{4}\right)$
37. $3\left(\cos\left(-\frac{\pi}{3}\right) + i \sin\left(-\frac{\pi}{3}\right)\right)$
38. $5\left(\cos\left(-\frac{7\pi}{6}\right) + i \sin\left(-\frac{7\pi}{6}\right)\right)$

In Exercises 39–50, find z_1z_2 and $\frac{z_1}{z_2}$. Write each answer in polar form.

39. $z_1 = 4(\cos 75° + i \sin 75°), z_2 = 2(\cos 15° + i \sin 15°)$
40. $z_1 = 6(\cos 90° + i \sin 90°), z_2 = 2(\cos 45° + i \sin 45°)$
41. $z_1 = 5(\cos 240° + i \sin 240°),$
$z_2 = 2(\cos 60° + i \sin 60°)$
42. $z_1 = 10(\cos 135° + i \sin 135°),$
$z_2 = 4(\cos 225° + i \sin 225°)$
43. $z_1 = 3(\cos 40° + i \sin 40°), z_2 = 5(\cos 20° + i \sin 20°)$
44. $z_1 = 5(\cos 65° + i \sin 65°), z_2 = 2(\cos 25° + i \sin 25°)$
45. $z_1 = 1 + i, z_2 = 1 - i$
46. $z_1 = 1 + \sqrt{3}i, z_2 = 1 + i$
47. $z_1 = -\sqrt{3} + i, z_2 = 2 + 2i$
48. $z_1 = 3 + 3i, z_2 = 2 - 2\sqrt{3}i$
49. $z_1 = 2\sqrt{3} + 2i, z_2 = \sqrt{3} - i$
50. $z_1 = 4\sqrt{3} + 4i, z_2 = 3 - 3i$

In Exercises 51–64, use DeMoivre's theorem to find the indicated power. Write the answers in rectangular form.

51. $\left[2\left(\cos \frac{\pi}{3} + i \sin \frac{\pi}{3}\right)\right]^{12}$
52. $\left[\sqrt{2}\left(\cos \frac{\pi}{4} + i \sin \frac{\pi}{4}\right)\right]^{20}$
53. $\left[2\left(\cos\left(-\frac{3\pi}{4}\right) + i \sin\left(-\frac{3\pi}{4}\right)\right)\right]^{6}$
54. $\left(\cos \frac{\pi}{27} + i \sin \frac{\pi}{27}\right)^{-9}$
55. $\left[2\left(\cos\left(\frac{3\pi}{4}\right) + i \sin\left(\frac{3\pi}{4}\right)\right)\right]^{-6}$
56. $\left[3\left(\cos\left(-\frac{5\pi}{6}\right) + i \sin\left(-\frac{5\pi}{6}\right)\right)\right]^{16}$
57. $\left[2\left(\cos\left(-\frac{\pi}{4}\right) + i \sin\left(-\frac{\pi}{4}\right)\right)\right]^{-10}$
58. $\left[\frac{1}{2}\left(\cos\left(-\frac{3\pi}{4}\right) + i \sin\left(-\frac{3\pi}{4}\right)\right)\right]^{-6}$
59. $(1 - i)^{12}$
60. $(1 - \sqrt{3}i)^{10}$
61. i^{25}
62. $(3 + 3i)^8$
63. $\left(\frac{1}{2} - \frac{\sqrt{3}}{2}i\right)^{-8}$
64. $\left(\frac{\sqrt{3}}{2} + \frac{1}{2}i\right)^{10}$

In Exercises 65–72, find all complex roots. Write your answers in polar form, with the argument θ in degrees with $0 \le \theta < 360°$.

65. Cube roots of 64
66. Cube roots of -64
67. Square roots of i
68. Fourth roots of $-i$
69. Sixth roots of -1
70. Eighth roots of unity
71. Square roots of $1 - \sqrt{3}i$
72. Cube roots of $4 + 4\sqrt{3}i$

Applying the Concepts

73. **Trigonometry.** Use DeMoivre's theorem to verify each identity.
   a. $\sin 3\theta = 3 \sin \theta - 4 \sin^3 \theta$
   b. $\cos 3\theta = 4 \cos^3 \theta - 3 \cos \theta$
74. **Geometry.** Show that the area K of a triangle with vertices at the origin O, z_1, and z_2 is given by
$$K = \frac{1}{2}|z_1||z_2|\sin\left(\left|\arg \frac{z_1}{z_2}\right|\right).$$
75. **AC circuits.** Use Ohm's law, $I = \frac{V}{Z}$, to find the current I, given voltage $V = 120(\cos 60° + i \sin 60°)$ and impedance $Z = 8 \cos(30° + i \sin 30°)$.
76. **AC circuits.** Use Ohm's law from Exercise 75 to find the voltage V, given $I = 6(\cos 40° + i \sin 40°)$ and $Z = 16 \cos(110° + i \sin 110°)$.

In Exercises 77 and 78, use the formula $Z_t = \frac{Z_1Z_2}{Z_1 + Z_2}$ to find the total impedance in a parallel circuit for the given values of Z_1 and Z_2.

77. $Z_1 = 16(\cos 180° + i \sin 180°)$ and
$Z_2 = 2(\cos 150° + i \sin 150°)$
78. $Z_1 = 12(\cos 270° + i \sin 270°)$ and
$Z_2 = 3(\cos 60° + i \sin 60°)$

Beyond the Basics

In Exercises 79–84, use DeMoivre's theorem to compute each power in polar form, with θ in degrees, with $0 \le \theta < 360°$.

79. $(1 - \sqrt{3}i)^{10}(-2 + 2i)^{-6}$
80. $\left(\frac{1}{2} + \frac{\sqrt{3}}{2}i\right)^8(2 - 2i)^{-6}$
81. $\left(\sin \frac{\pi}{6} + i \cos \frac{\pi}{6}\right)^{10}$
82. $\left(\sin \frac{2\pi}{3} + i \cos \frac{2\pi}{3}\right)^{6}$
83. $\left(\sin \frac{5\pi}{3} + i \cos \frac{5\pi}{3}\right)^{-8}$
84. $\left(\sin \frac{7\pi}{6} + i \cos \frac{7\pi}{6}\right)^{-10}$

In Exercises 85–88, find all complex solutions of each equation.

85. $z^4 = 1$

86. $z^8 = -1$

87. $z^3 = 1 + i$

88. $z^7 = (1 - i)^2$

89. Let $z = \cos\theta + i\sin\theta, z \neq 0$. Prove that $\frac{1}{z} = \cos\theta - i\sin\theta$.

In Exercises 90–93, let $z = \cos\theta + i\sin\theta, z \neq 0$. Use Exercise 89 and DeMoivre's theorem to prove each identity.

90. $z + \frac{1}{z} = 2\cos\theta$

91. $z - \frac{1}{z} = 2i\sin\theta$

92. $z^n + \frac{1}{z^n} = 2\cos n\theta$

93. $z^n - \frac{1}{z^n} = 2i\sin n\theta$

94. Prove that

$$(1 + \cos\theta + i\sin\theta)^n = \left(2\cos\frac{\theta}{2}\right)^n\left(\cos\frac{n\theta}{2} + i\sin\frac{n\theta}{2}\right).$$

95. Prove the product rule for two complex numbers in polar form.

96. Prove the quotient rule for two complex numbers in polar form.

97. Find the product $(3 + 2i)(5 + i)$ using FOIL. Then deduce the identity: $\tan^{-1}\left(\frac{2}{3}\right) + \tan^{-1}\left(\frac{1}{5}\right) = \frac{\pi}{4}$. [*Hint:* $\arg(z_1z_2) = \arg z_1 + \arg z_2$.]

98. Find the product $(p + q + i)(p^2 + pq + 1 + iq)$; then deduce the identity:

$$\tan^{-1}\left(\frac{1}{p+q}\right) + \tan^{-1}\left(\frac{q}{p^2 + pq + 1}\right) = \tan^{-1}\left(\frac{1}{p}\right).$$

99. Expand $(2 + 3i)^4$; then deduce the identity:

$$4\tan^{-1}\left(\frac{3}{2}\right) - \tan^{-1}\left(\frac{120}{119}\right) = \pi.$$

100. Find the product $(5 + i)^4(-239 + i)$; then deduce the identity: $4\tan^{-1}\left(\frac{1}{5}\right) - \tan^{-1}\left(\frac{1}{239}\right) = \frac{\pi}{4}$.

101. Find the three solutions of the equation $x^3 + x^2 + x + 1 = 0$. [*Hint:* Multiply both sides by $(x - 1)$, solve the new equation, and reject the root $x = 1$.]

102. Find the four solutions of the equation $x^4 + x^3 + x^2 + x + 1 = 0$.

Critical Thinking/Discussion/Writing

103. Let n be a positive integer. State whether each of the following is true or false.

a. $\left(\cos\frac{\pi}{6} + i\sin\frac{\pi}{6}\right)^n = \cos\frac{n\pi}{6} + i\sin\frac{n\pi}{6}$

b. $\left(\sin\frac{\pi}{3} + i\cos\frac{\pi}{3}\right)^n = \sin\frac{n\pi}{3} + i\cos\frac{n\pi}{3}$

c. $\left[\cos\left(-\frac{\pi}{3}\right) + i\sin\left(-\frac{\pi}{3}\right)\right]^n = \cos\frac{n\pi}{3} - i\sin\frac{n\pi}{3}$

d. $\left(\cos\frac{\pi}{6} + i\sin\frac{\pi}{3}\right)^n = \cos\frac{n\pi}{6} + i\sin\frac{n\pi}{3}$

e. $(\cos\theta + i\sin\theta)^n = \cos n\theta + i\sin n\theta$

f. $(\cos\theta - i\sin\theta)^n = \cos n\theta - i\sin n\theta$

g. $(\cos\theta + i\sin\theta)^{-n} = \cos n\theta - i\sin n\theta$

h. $(\cos\theta - i\sin\theta)^{-n} = \cos n\theta + i\sin n\theta$

i. $(\sin\theta + i\cos\theta)^n = \sin n\theta + i\cos n\theta$

j. $(\sin\theta + i\cos\theta)^n = \cos\left[n\left(\frac{\pi}{2} - \theta\right)\right] + i\sin\left[n\left(\frac{\pi}{2} - \theta\right)\right]$

Maintaining Skills

In Exercises 104–107, find the value of a variable that satisfies the given condition.

104. Find x if $3x + 2y = 7$ and $y = \frac{1}{2}$.

105. Find x if $-2x + 9y = 5$ and $y = 3$.

106. Find y if $-4x + 7y = 7$ and $x = 0$.

107. Find y if $23x - 14y = -5$ and $x = 1$.

108. Find the slope of the line with equation $10x - 2y = 28$.

109. Find the slope of the line with equation $15x + 5y = 2$.

110. Find an equation of the line through the point $(5, -1)$ parallel to the line with equation $6x - 3y = 7$.

111. Find an equation of the line through the point $(3, 3)$ parallel to the line with equation $x + 2y = 1$.

112. Find values for a and b so that the equation $ax + by = 3$ has the same graph as the equation $2x - 4y = 12$.

113. Find values for a and b so that the equation $ax + by = 2$ has the same graph as the equation $4x - y = -1$.

SUMMARY Definitions, Concepts, and Formulas

7.1 The Law of Sines

i. In a triangle ABC, with sides of length a, b, and c,
$\frac{\sin A}{a} = \frac{\sin B}{b} = \frac{\sin C}{c}$, or equivalently,
$\frac{a}{\sin A} = \frac{b}{\sin B} = \frac{c}{\sin C}$.

ii. The Law of Sines is used to solve AAS, ASA, and SSA triangles. The SSA case is called the *ambiguous case* because the information provided may lead to two triangles, one triangle, or no triangle.

7.2 The Law of Cosines

i. In a triangle ABC with sides of lengths a, b, and c,

$$a^2 = b^2 + c^2 - 2\,bc\cos A,$$
$$b^2 = c^2 + a^2 - 2\,ca\cos B,$$
$$c^2 = a^2 + b^2 - 2\,ab\cos C.$$

ii. The Law of Cosines is used to (a) solve *SAS* and *SSS* triangles, and (b) Prove Heron's formula.

7.3 Areas of Polygons Using Trigonometry

i. **Area of an SAS triangle**
The area K of a triangle ABC with sides a, b, and c is

$$K = \frac{1}{2}bc\sin A = \frac{1}{2}ca\sin B = \frac{1}{2}ab\sin C.$$

Area K is one-half the product of two of its sides and the sine of the included angle.

ii. **Area of AAS and ASA triangles**
The area K of a triangle ABC with sides a, b, and c is

$$K = \frac{a^2\sin B\sin C}{2\sin A} = \frac{b^2\sin C\sin A}{2\sin B} = \frac{c^2\sin A\sin B}{2\sin C}.$$

iii. **Area of SSS triangles**
Heron's formula The area K of a triangle with sides of lengths a, b, and c is given by

$$K = \sqrt{s(s-a)(s-b)(s-c)},$$

where $s = \frac{1}{2}(a + b + c)$ is the *semiperimeter*.

7.4 Vectors

i. A vector $\mathbf{v}$ is a quantity that has both magnitude and direction. A vector is geometrically represented by an arrow.

ii. The magnitude of the scalar product $c\mathbf{v}$ is $|c|$ times the magnitude of $\mathbf{v}$. If $c > 0$, $c\mathbf{v}$ has the direction of $\mathbf{v}$. If $c < 0$, $c\mathbf{v}$ has the opposite direction of $\mathbf{v}$. The zero vector $\mathbf{0}$ has magnitude 0 but arbitrary direction.

iii. To add the vectors $\mathbf{v}$ and $\mathbf{w}$ geometrically, place $\mathbf{w}$ so that its initial point coincides with the terminal point of $\mathbf{v}$ and draw a vector from the initial point of $\mathbf{v}$ to the terminal point of $\mathbf{w}$.

iv. **Algebraic vectors.** A vector $\mathbf{v}$ with initial point at the origin and terminal point (v_1, v_2) is written as $\mathbf{v} = \langle v_1, v_2\rangle$; v_1 and v_2 are the horizontal and vertical components of $\mathbf{v}$, respectively.

- The magnitude of $\mathbf{v} = \|\mathbf{v}\| = \sqrt{v_1^2 + v_2^2}$. The direction angle of $\mathbf{v}$ is the angle θ that $\mathbf{v}$ makes with the positive x-axis. We have $\cos\theta = \frac{v_1}{\|\mathbf{v}\|}$ and $\sin\theta = \frac{v_2}{\|\mathbf{v}\|}$, $0° \le \theta < 360°$.
- The horizontal and vertical components of vector $\mathbf{v}$ with magnitude r and direction angle θ are given by

$$v_1 = r\cos\theta,\ v_2 = r\sin\theta.$$

v. **Vector operations.** For vectors $\mathbf{v} = \langle v_1, v_2\rangle = v_1\mathbf{i} + v_2\mathbf{j}$ and $\mathbf{w} = \langle w_1, w_2\rangle = w_1\mathbf{i} + w_2\mathbf{j}$, the following operations apply:

- **Vector addition:** $\mathbf{v} + \mathbf{w} = \langle v_1 + w_1, v_2 + w_2\rangle$
- **Scalar multiplication:** $c\mathbf{v} = \langle cv_1, cv_2\rangle$

7.5 The Dot Product

i. For two vectors $\mathbf{v} = \langle v_1, v_2\rangle$ and $\mathbf{w} = \langle w_1, w_2\rangle$, the dot product

$$\mathbf{v}\cdot\mathbf{w} = v_1w_1 + v_2w_2 = \|\mathbf{v}\|\,\|\mathbf{w}\|\cos\theta,$$

where θ $(0° \le \theta \le 180°)$ is the angle between $\mathbf{v}$ and $\mathbf{w}$.

ii. Vectors $\mathbf{v}$ and $\mathbf{w}$ are
1. Orthogonal (perpendicular) if $\mathbf{v}\cdot\mathbf{w} = 0$.
2. Parallel if $\mathbf{v}\cdot\mathbf{w} = \pm\|\mathbf{v}\|\,\|\mathbf{w}\|$ or $\frac{v_1}{w_1} = \frac{v_2}{w_2}$.

iii. Vector projection of $\mathbf{v}$ onto $\mathbf{w}$ is given by $\text{proj}_{\mathbf{w}}\mathbf{v} = \frac{\mathbf{v}\cdot\mathbf{w}}{\|\mathbf{w}\|^2}\mathbf{w}$.

The scalar projection of $\mathbf{v}$ onto $\mathbf{w}$ is $\frac{\mathbf{v}\cdot\mathbf{w}}{\|\mathbf{w}\|}$.

iv. A nonzero vector $\mathbf{v}$ can be written as $\mathbf{v} = \mathbf{v}_1 + \mathbf{v}_2$, where $\mathbf{v}_1 = \text{proj}_{\mathbf{w}}\mathbf{v}$ is parallel to $\mathbf{w}$ and $\mathbf{v}_2 = \mathbf{v} - \mathbf{v}_1$ is orthogonal to $\mathbf{w}$.

v. The **work** W done by a constant force $\mathbf{F}$ in moving an object from a point P to a point Q is $W = \mathbf{F}\cdot\overrightarrow{PQ}$.

7.6 Polar Coordinates

i. The ordered pair (r, θ) represents a point P in the plane that is a directed distance of r units from the origin (pole) O, where θ is the directed angle between the polar axis (positive x-axis) and the line segment OP. The ordered pair (r, θ) gives the polar coordinates of P.

ii. The polar coordinates of a point $P(r, \theta)$ are not unique. In fact, if n is any integer, then

$$(r, \theta),\ (r, \theta + 2n\pi),\text{ and }(-r, \theta + \pi + 2n\pi)$$

are all polar coordinates of the same point.

iii. To convert a point from polar coordinates (r, θ) to rectangular coordinates, use the equations $x = r\cos\theta$ and $y = \sin\theta$.

iv. To convert a point from rectangular coordinates (x, y) to polar coordinates (r, θ), use the procedure on page 699.

v. A polar equation is an equation whose variables are r and θ.

7.7 Polar Form of Complex Numbers; DeMoivre's Theorem

i. **Geometric representation.** A complex number $z = a + bi$ can be represented as an ordered pair (a, b) in the complex plane.

ii. **Modulus.** $|z| = |a + bi| = \sqrt{a^2 + b^2}$, $|z|$ is called the **modulus** of z.

iii. **Polar form.** The polar form of $z = a + bi$ is $z = r(\cos\theta + i\sin\theta)$, where $r = \sqrt{a^2 + b^2}$, $a = r\cos\theta$, $b = r\sin\theta$, and $\tan\theta = \frac{b}{a}$. The number r is the modulus of z, and θ is called the argument of z (written arg z).

iv. **Product and quotient rules.** Let $z_1 = r_1(\cos\theta_1 + i\sin\theta_1)$ and $z_2 = r_2(\cos\theta_2 + i\sin\theta_2)$. Then

$$z_1z_2 = r_1r_2[\cos(\theta_1 + \theta_2) + i\sin(\theta_1 + \theta_2)]$$

and

$$\frac{z_1}{z_2} = \frac{r_1}{r_2}[\cos(\theta_1 - \theta_2) + i\sin(\theta_1 - \theta_2)], \quad z_2 \neq 0.$$

v. **DeMoivre's theorem.** Let $z = r(\cos\theta + i\sin\theta)$ be a complex number in polar form. Then for any integer n,

$$[r(\cos\theta + i\sin\theta)]^n = r^n(\cos n\theta + i\sin n\theta).$$

vi. **Roots of complex numbers.** The n nth roots of a complex number $r(\cos\theta + i\sin\theta)$ are given by

$$z_k = \sqrt[n]{r}\left[\cos\left(\frac{\theta + 2\pi k}{n}\right) + i\sin\left(\frac{\theta + 2\pi k}{n}\right)\right] \text{ or}$$

$$z_k = \sqrt[n]{r}\left[\cos\left(\frac{\theta + 360°k}{n}\right) + i\sin\left(\frac{\theta + 360°k}{n}\right)\right],$$

where $k = 0, 1, 2, \ldots, n - 1$.

REVIEW EXERCISES

Basic Concepts and Skills

In Exercises 1–10, solve right triangle *ABC* with right angle at *C*. Round each answer to the nearest tenth.

1. $A = 30°, a = 6$
2. $B = 35°, b = 5$
3. $A = 37°, b = 4$
4. $B = 43°, a = 10$
5. $A = 40°, c = 12$
6. $B = 50°, c = 8$
7. $a = 3, b = 5$
8. $a = 5, b = 10$
9. $a = 4, c = 6$
10. $b = 3, c = 7$

In Exercises 11–18, use the Law of Sines to solve each triangle *ABC*. Round each answer to the nearest tenth.

11. $A = 40°, B = 35°, c = 100$
12. $B = 30°, C = 80°, a = 100$
13. $A = 45°, a = 25, b = 75$
14. $B = 36°, a = 12.5, b = 8.7$
15. $A = 48.5°, C = 57.3°, b = 47.3$
16. $A = 67°, a = 100, c = 125$
17. $A = 65.2°, a = 21.3, b = 19$
18. $C = 53°, a = 140, c = 115$
19. In triangle *ABC*, let $c = 20$ and $B = 60°$. Find a value of b such that C has **(i)** two possible values, **(ii)** exactly one value, **(iii)** no value.
20. Repeat Exercise 19 for $c = 20$ and $B = 150°$.

In Exercises 21–28, use the Law of Cosines to solve each triangle *ABC*. Round each answer to the nearest tenth.

21. $a = 60, b = 90, c = 125$
22. $a = 15, b = 9, C = 120°$
23. $a = 40, c = 38, B = 80°$
24. $a = 10, b = 20, c = 22$
25. $a = 2.6, b = 3.7, c = 4.8$
26. $a = 15, c = 26, B = 115°$
27. $a = 12, b = 7, C = 130°$
28. $b = 75, c = 100, A = 80°$

In Exercises 29–32, find the area of each triangle *ABC* with the given information. Round each answer to the nearest square unit.

29. $a = 5$ meters, $b = 7$ meters, $c = 10$ meters
30. $a = 2.4$ meters, $b = 3.4$ meters, $c = 4.4$ meters
31. $A = 65°, b = 6$ feet, $c = 4$ feet
32. $A = 115°, b = 20$ inches, $a = 30$ inches

In Exercises 33 and 34, let v be the vector with initial point *P* and terminal point *Q*. Write v as a position vector in terms of i and j.

33. $P(3, 5), Q(2, 7)$
34. $P(-1, 3), Q(5, -4)$

In Exercises 35–38, let $\mathbf{v} = \langle -2, 3\rangle$ and $\mathbf{w} = \langle 5, -6\rangle$. Find each expression.

35. $5\mathbf{v}$
36. $2\mathbf{v} + \mathbf{w}$
37. $3\mathbf{v} - 2\mathbf{w}$
38. $\|\mathbf{v} + \mathbf{w}\|$

In Exercises 39–42, find the unit vector u in the direction of the vector v.

39. $\mathbf{v} = \mathbf{i} + \mathbf{j}$
40. $\mathbf{v} = 2\mathbf{i} - 7\mathbf{j}$
41. $\mathbf{v} = \langle 3, -5\rangle$
42. $\mathbf{v} = \langle -5, -2\rangle$

In Exercises 43–46, write v in terms of i and j for the given magnitude $\|\mathbf{v}\|$ and direction angle θ.

43. $\|\mathbf{v}\| = 6, \theta = 30°$
44. $\|\mathbf{v}\| = 20, \theta = 120°$
45. $\|\mathbf{v}\| = 12, \theta = 225°$
46. $\|\mathbf{v}\| = 10, \theta = -30°$

In Exercises 47–50, find the dot product $\mathbf{v} \cdot \mathbf{w}$ and $\text{proj}_\mathbf{w}\mathbf{v}$.

47. $\mathbf{v} = \langle 2, -3\rangle, \mathbf{w} = \langle 3, 4\rangle$
48. $\mathbf{v} = \langle -1, -2\rangle, \mathbf{w} = \langle 4, -1\rangle$
49. $\mathbf{v} = 2\mathbf{i} - 5\mathbf{j}, \mathbf{w} = 5\mathbf{i} + 2\mathbf{j}$
50. $\mathbf{v} = 2\mathbf{i} + \mathbf{j}, \mathbf{w} = 2\mathbf{i} - \mathbf{j}$

In Exercises 51–54, use the dot product to find the angle θ $(0 \leq \theta \leq \pi)$ between the vectors v and w. Round your answers to the nearest tenth of a degree.

51. $\mathbf{v} = 2\mathbf{i} + 3\mathbf{j}, \mathbf{w} = -\mathbf{i} + 2\mathbf{j}$
52. $\mathbf{v} = \mathbf{i} + 4\mathbf{j}, \mathbf{w} = -4\mathbf{i} - \mathbf{j}$
53. $\mathbf{v} = \langle 1, 1\rangle, \mathbf{w} = \langle -3, 2\rangle$
54. $\mathbf{v} = \langle 1, 5\rangle, \mathbf{w} = \langle 3, -1\rangle$

In Exercises 55–58, plot each point in polar coordinates and find its rectangular coordinates.

55. $(24, 30°)$
56. $(12, -60°)$
57. $\left(-2, -\frac{\pi}{4}\right)$
58. $\left(-3, -\frac{3\pi}{4}\right)$

In Exercises 59–62, convert the rectangular coordinates of each point to polar coordinates (r, θ), with $r > 0$ and $0 \le \theta < 2\pi$.

59. $(-2, 2)$
60. $(\sqrt{3}, 1)$
61. $(2\sqrt{3}, -2)$
62. $(-2, -2\sqrt{3})$

In Exercises 63–66, convert each rectangular equation to a polar equation.

63. $3x + 2y = 12$
64. $x^2 + y^2 = 36$
65. $x^2 + y^2 = 8x$
66. $x^2 + y^2 = 6y$

In Exercises 67–72, convert each polar equation to a rectangular equation.

67. $r = -3$
68. $\theta = \frac{5\pi}{6}$
69. $r = 3 \csc \theta$
70. $r = 2 \sec \theta$
71. $r = 1 - 2 \sin \theta$
72. $r = 3 \cos \theta$

In Exercises 73–76, write each complex number in polar form. Express the arguments in radians.

73. $-3i$
74. $-1 + i$
75. $5\sqrt{3} - 5i$
76. $-2 - 2\sqrt{3}i$

In Exercises 77–80, write each complex number in rectangular form.

77. $2(\cos 45^\circ + i \sin 45^\circ)$
78. $3(\cos 240^\circ + i \sin 240^\circ)$
79. $6\left(\cos \frac{3\pi}{4} + i \sin \frac{3\pi}{4}\right)$
80. $4\left(\cos \frac{7\pi}{6} + i \sin \frac{7\pi}{6}\right)$

In Exercises 81–84, find $z_1 z_2$ and $\frac{z_1}{z_2}$. Leave your answers in polar form.

81. $z_1 = 3(\cos 25^\circ + \sin 25^\circ), z_2 = 2(\cos 10^\circ + i \sin 10^\circ)$
82. $z_1 = 4(\cos 300^\circ + i \sin 300^\circ), z_2 = 2(\cos 20^\circ + i \sin 20^\circ)$
83. $z_1 = 2\left(\cos \frac{5\pi}{6} + i \sin \frac{5\pi}{6}\right), z_2 = 3\left(\cos \frac{\pi}{3} + i \sin \frac{\pi}{3}\right)$
84. $z_1 = 5\left(\cos \frac{4\pi}{3} + i \sin \frac{4\pi}{3}\right), z_2 = 15\left(\cos \frac{\pi}{3} + i \sin \frac{\pi}{3}\right)$

In Exercises 85–88, use DeMoivre's theorem to find the indicated power. Write your answers in polar form.

85. $[3(\cos 40^\circ + i \sin 40^\circ)]^3$
86. $\left[4\left(\cos \frac{\pi}{6} + i \sin \frac{\pi}{6}\right)\right]^6$
87. $(2 - 2\sqrt{3}i)^6$
88. $(2 - 2i)^7$

In Exercises 89–92, find all complex roots. Write your answers in polar form, with the arguments in degrees.

89. Cube roots of -125
90. Fourth roots of $-16i$
91. Fifth roots of $-1 + \sqrt{3}i$
92. Sixth roots of $1 - i$

Applying the Concepts

93. **Width of a river.** A surveyor wants to measure the width of a river. She stands at a point A, with point B opposite her on the other side of the river. From A, she walks 320 feet along the bank to a point C. The line CA makes an angle of 40° with the line CB. What is the width of the river?

94. **Angle of elevation of a hill.** A tower 120 feet tall is located at the top of a hill. At a point 460 feet down the hill, the angle between the surface of the hill and the line of sight to the top of the tower is 12°. Find the angle of elevation of the hill to a horizontal plane.

95. **Distance between cars.** Two cars start from a point A along two straight roads. The angle between the two roads is 72°. The speeds of the two cars are 55 miles per hour and 65 miles per hour. How far apart are the two cars after 80 minutes? Round to the nearest tenth of a mile.

96. **Triangular plot.** A triangular plot of land has sides of lengths 310 feet, 415 feet, and 175 feet. Find the largest angle between the sides.

97. **Area.** Find the area of the triangular plot in Exercise 96. Round your answer to the nearest square foot.

98. **Resultant force.** Two forces of magnitudes 16 and 20 pounds are acting on an object. The bearings of the forces are N 75° E and S 20° E, respectively. Find the magnitude and the direction of the resultant force, to the nearest tenth.

99. **Work.** A force of magnitude 30 pounds at an angle of 48° is used to pull a wagon 60 feet. Find the work done.

100. **AC current.** The total impedance in an AC circuit is given by $Z = \frac{Z_1 Z_2}{Z_1 + Z_2}$. Find Z if $Z_1 = 20\left(\cos \frac{\pi}{6} + i \sin \frac{\pi}{6}\right)$ and $Z_2 = 10\left(\cos \frac{2\pi}{3} + i \sin \frac{2\pi}{3}\right)$.

PRACTICE TEST A

1. In Problems 1–4, solve triangle ABC. Round each answer to the nearest tenth.

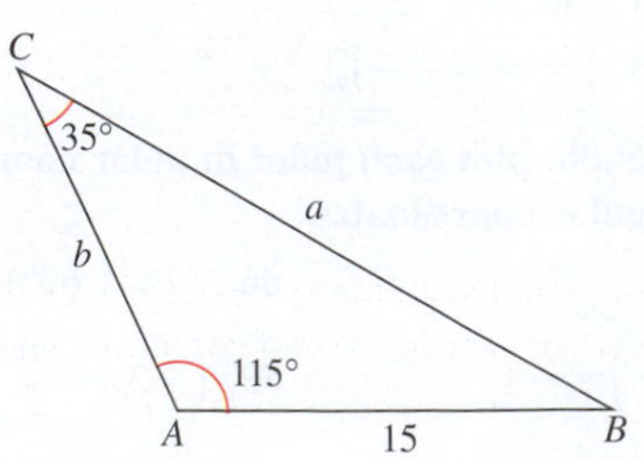

2. $A = 42^\circ, B = 37^\circ, a = 50$ meters.

3.

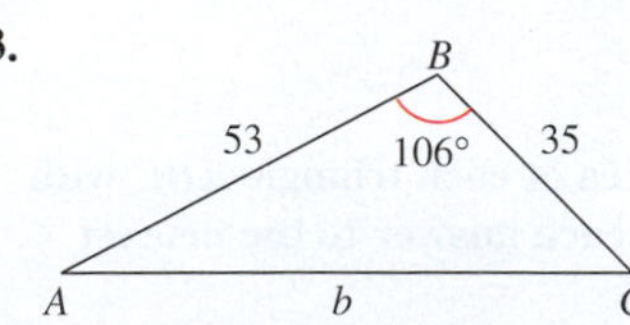

4. $a = 30$ feet, $b = 20$ feet, $c = 25$ feet.

5. Solve right triangle ABC if $a = 5.6$ and $b = 4.1$. Round each answer to the nearest tenth.

6. Find the measure of the central angle of a circle of radius 8 feet that intercepts a chord of length 4.5 feet. Round to the nearest tenth of a degree.

7. A surveyor, starting from point A, walks 580 feet in the direction N 70.0° E. From that point, she walks 725 feet in the direction N 35.0° W. How far, to the nearest foot, is she from her starting point?

8. The vector **v** has initial point $P(3, 5)$ and terminal point $Q(2, -7)$. Write **v** as a position vector.

9. If $\mathbf{v} = \langle -2, 3 \rangle$ and $\mathbf{w} = \langle 1, 5 \rangle$, find $\mathbf{v} - \mathbf{w}$.

10. If $\mathbf{v} = -2\mathbf{i} + 3\mathbf{j}$ and $\mathbf{w} = \mathbf{i} + 5\mathbf{j}$, find $2\mathbf{v} - 3\mathbf{w}$.

11. Let $\|\mathbf{v}\| = 3$ and suppose **v** makes an angle of $\theta = -30°$ with the positive x-axis. Write the vector **v** in the form $v_1\mathbf{i} + v_2\mathbf{j}$.

12. Find the dot product $\mathbf{v} \cdot \mathbf{w}$ if $\mathbf{v} = 4\mathbf{i} + 3\mathbf{j}$ and $\mathbf{w} = -\mathbf{i} + 7\mathbf{j}$.

13. Find the angle $\theta\,(0 \le \theta \le 180°)$ between the vectors $\mathbf{v} = 3\mathbf{i} - 4\mathbf{j}$ and $\mathbf{w} = -2\mathbf{i} + 5\mathbf{j}$. Round the answer to the nearest tenth of a degree.

14. Convert $(-2, -45°)$ to rectangular coordinates.

15. Convert $(-\sqrt{3}, -1)$ to polar coordinates with $r > 0$ and $0 \le \theta < 2\pi$.

16. Convert the polar equation $r = -3$ to rectangular form. Identify the curve having this equation.

17. Write $3\left(\cos\left(\frac{2\pi}{3}\right) + i \sin\left(\frac{2\pi}{3}\right)\right)$ in rectangular form.

18. Write $\frac{z_1}{z_2}$ in polar form, if $z_1 = 1 - \sqrt{3}i$, $z_2 = -1 + i$, with the argument given in degrees.

19. Use DeMoivre's theorem to find $\left[\sqrt{5}\left(\cos\left(\frac{\pi}{4}\right) + i \sin\left(\frac{\pi}{4}\right)\right)\right]^{-6}$. Write the answer in rectangular form.

20. Find all fourth roots of $1 + i$. Express the argument in degrees.

PRACTICE TEST B

1. In triangle ABC, which is closest to the correct value for b?
   a. $b = 13.4$
   b. $b = 20.8$
   c. $b = 20.2$
   d. $b = 0.05$

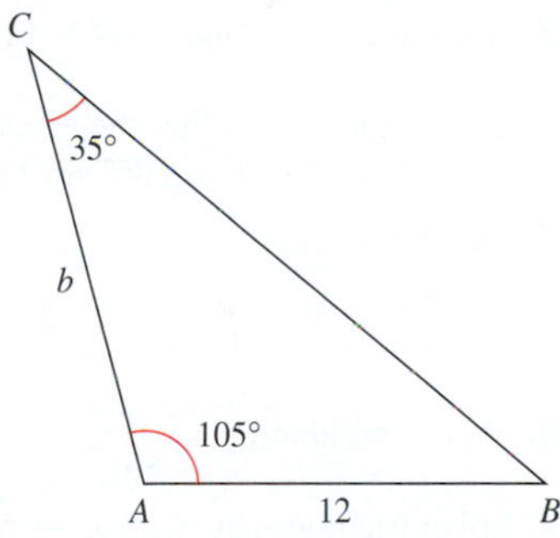

2. In triangle ABC, if $A = 27°$, $B = 50°$, and $a = 25$ meters, which is closest to the correct value for c?
   a. $c = 0.03$ meter b. $c = 42.2$ meters
   c. $c = 53.7$ meters d. $c = 0.02$ meter

3. In a triangle ABC, which statement is true for the measurements $A = 25°$, $a = 25$ feet, $b = 30$ feet? (The angle measurements in choices (c) and (d) are given to the nearest tenth.)
   a. No triangle has these measurements.
   b. One triangle has these measurements.
   c. Two triangles, with $c_1 = 48.7$ and $c_2 = 5.6$, have these measurements (in feet).
   d. Two triangles, with $c_1 = 46$ and $c_2 = 6.4$, have these measurements (in feet).

4. In triangle ABC, which is closest to the correct value for b?
   a. $b = 35.8$
   b. $b = 57.2$
   c. $b = 18.7$
   d. $b = 48.4$

B
21 110° 37
A b C

5. A 3-foot shrub is 4.5 feet away from a 12-foot streetlight. What length shadow is cast by the shrub?
   a. 1.5 feet b. 2 feet c. 2.67 feet d. 4 feet

6. In triangle ABC, with $a = 12$ feet, $b = 13$ feet, and $c = 7$ feet, which is closest to the correct value for A?
   a. 24° b. 66° c. 81.8° d. 32.2°

7. Find the measure of the central angle of a circle of radius 9 feet that intercepts a chord of length 3 feet. Round to the nearest tenth of a degree.
   a. 70.8° b. 63.6° c. 26.4° d. 19.2°

8. A hunter leaves a straight east–west road at a point A, walking 2 miles at a bearing N 40° W to a point B. From point B, he walks at a bearing of S 20° E and arrives back on the road at a point C. How far, to the nearest tenth of a mile, is point B from point C?
   a. 4.5 miles b. 2.9 miles
   c. 1.6 miles d. 5.3 miles

9. The vector **v** has initial point $P(-2, 5)$ and terminal point $Q(3, -1)$. Write **v** as a position vector.
   a. $\langle 5, -6 \rangle$ b. $\langle 1, -6 \rangle$ c. $\langle 1, -4 \rangle$ d. $\langle 5, -4 \rangle$

10. If $\mathbf{v} = \langle 1, -4 \rangle$ and $\mathbf{w} = \langle 3, -2 \rangle$, find $\mathbf{v} - \mathbf{w}$.
   a. $\langle 4, -2 \rangle$ b. $\langle 2, 2 \rangle$ c. $\langle -2, -2 \rangle$ d. $\langle 4, -6 \rangle$

11. Let $\mathbf{v} = 7\mathbf{i} + 3\mathbf{j}$ and $\mathbf{w} = -2\mathbf{i} + \mathbf{j}$. Find $2\mathbf{v} - 3\mathbf{w}$.
   a. $20\mathbf{i} + 3\mathbf{j}$ b. $20\mathbf{i} - 3\mathbf{j}$ c. $8\mathbf{i} - 6\mathbf{j}$ d. $8\mathbf{i} + 6\mathbf{j}$

12. Let $\|\mathbf{v}\| = 6$ and suppose **v** makes an angle of $\theta = -60°$ with the positive x-axis. Write the vector **v** in the form $v_1\mathbf{i} + v_2\mathbf{j}$.
   a. $3\sqrt{3}\,\mathbf{i} - 3\mathbf{j}$ b. $3\sqrt{3}\,\mathbf{i} + 3\mathbf{j}$
   c. $3\mathbf{i} + 3\sqrt{3}\mathbf{j}$ d. $3\,\mathbf{i} - 3\sqrt{3}\mathbf{j}$

13. Find the dot product $\mathbf{v} \cdot \mathbf{w}$ if $\mathbf{v} = -2\mathbf{i} - 5\mathbf{j}$ and $\mathbf{w} = 9\mathbf{i} - 4\mathbf{j}$.
   a. -38 b. 38 c. $2\mathbf{i} + 20\mathbf{j}$ d. 2

14. Find the angle between the vectors $\mathbf{v} = 7\mathbf{i} - \mathbf{j}$ and $\mathbf{w} = \mathbf{i} + \sqrt{2}\mathbf{j}$. Round the answer to the nearest degree.
   a. 84° b. 63° c. 32° d. 17°

15. Convert $(-4, -30°)$ to rectangular coordinates.

a. $(-2\sqrt{3}, -2)$ **b.** $(2, -2\sqrt{3})$
c. $(-2\sqrt{3}, 2)$ **d.** $(2\sqrt{3}, -2)$

16. Convert $(3, -\sqrt{3})$ to polar coordinates, with $r > 0$ and $0 \le \theta < 360°$.

a. $(12, 330°)$ **b.** $(2\sqrt{3}, 330°)$
c. $(2\sqrt{3}, 150°)$ **d.** $(\sqrt{3}, 150°)$

17. Which polar equation has a circle for its graph?

a. $r = -4\cos\theta + 1$ **b.** $r = -2$
c. $r\sin\theta = 2$ **d.** $r\cos\theta = 1$

18. Write $5\left(\cos\left(\frac{7\pi}{3}\right) + i\sin\left(\frac{7\pi}{3}\right)\right)$ in rectangular form.

a. $5 + 5\sqrt{3}\,\mathbf{i}$ **b.** $5 - 5\sqrt{3}\,\mathbf{i}$
c. $\frac{5}{2} + \frac{5\sqrt{3}}{2}\mathbf{i}$ **d.** $\frac{5}{2} - \frac{5\sqrt{3}}{2}\mathbf{i}$

19. Write $\frac{z_1}{z_2}$ in polar form, if $z_1 = i$ and $z_2 = 1 - i$.

a. $1(\cos 135° + i\sin 135°)$
b. $\sqrt{2}(\cos 135° + i\sin 135°)$
c. $\frac{\sqrt{2}}{2}(\cos 135° + i\sin 135°)$
d. $\frac{\sqrt{2}}{2}(\cos 315° + i\sin 315°)$

20. Use DeMoivre's theorem to find $\left[\sqrt{2}\left(\cos\left(\frac{\pi}{12}\right) + i\sin\left(\frac{\pi}{12}\right)\right)\right]^4$. Write the answer in rectangular form.

a. $-2 + 2\sqrt{3}i$ **b.** $\frac{1}{2} + \frac{\sqrt{3}}{2}i$
c. $2 + 2\sqrt{3}i$ **d.** $\frac{1}{2} - \frac{\sqrt{3}}{2}i$

CUMULATIVE REVIEW EXERCISES CHAPTERS P–7

1. If $f(x) = \sqrt{x}$ and $g(x) = \frac{1}{x-2}$, find the domain of $\frac{g}{f}$.

2. If $f(x) = 2x + 7$, find $f^{-1}(x)$.

3. Sketch the graph of $y = -2x^2 + 8x + 10$.

4. Solve the inequality $x^2 - 6x + 8 > 0$.

In Exercises 5 and 6, solve each equation for all solutions in the interval $[0, 2\pi)$. Give your answers in radians.

5. $\sin\theta\cos\theta = -\frac{\sqrt{3}}{4}$ **6.** $4\sin^2\theta - 1 = 0$

In Exercises 7 and 8, find all complex roots. Write your answers in polar form, with the argument in degrees.

7. Fourth roots of $\sqrt{3} + i$

8. Sixth roots of -64

9. Find the equation of the line in slope–intercept form that passes through the point $(-1, 5)$ and is perpendicular to the line $2x + 3y + 6 = 0$.

10. Write the following in condensed form:

$$2\log x + \frac{1}{2}\log(y + 1) - \log(3x + 1).$$

In Exercises 11–14, use transformations to sketch the graph of each function.

11. $f(x) = 2\sqrt{x - 1} - 3$ **12.** $f(x) = -|x + 1| + 2$

13. $f(x) = -2\cos(3x + \pi)$ **14.** $f(x) = 2\sec 3x$

15. If $\cos\theta = -\frac{5}{13}$ and $\sin\theta > 0$, find $\tan\theta$.

16. Use an identity to find the exact value of $\cos 65°\cos 35° + \sin 65°\sin 35°$.

17. Verify the identity

$$\frac{1}{1 - \cos x} + \frac{1}{1 + \cos x} = 2(1 + \cot^2 x).$$

18. Verify the identity $\frac{1 + \cos 2x}{\sin 2x} = \cot x$.

19. Solve triangle ABC with $A = 65°$, $B = 46°$, $c = 60$ meters. Round your answers to the nearest tenth if necessary.

20. A ladder leaning against a vertical wall makes an angle of 47° with the ground. The foot of the ladder is 2.3 meters from the wall. Find the length of the ladder. Round to the nearest tenth.

CHAPTER 8

Systems of Equations and Inequalities

TOPICS

Systems of equations are frequently the tools used to describe relationships among variables representing different interrelated quantities. These same systems of equations are used to predict results associated with changes in the quantities.

The quantities involved can come from the study of medicine, taxes, elections, business, agriculture, city planning, or any discipline in which multiple quantities interact.

SECTION 8.1

Systems of Linear Equations in Two Variables

Al-Khwarizmi (780–850)
Al-Khwarizmi came to Baghdad from the town of Khwarizm, which is now called Khiva and is part of Uzbekistan. The term *algorithm* is a corruption of the name al-khwarizmi. Originally, the word *algorism* was used for the rules for performing arithmetic by using decimal notation. *Algorism* evolved into the word *algorithm* by the eighteenth century. The word *algorithm* now means a finite set of precise instructions for performing a computation or for solving a problem. (The picture shown here is a Soviet postage stamp from 1983 depicting Al-Khwarizmi.)

BEFORE STARTING THIS SECTION, REVIEW

1. Graphs of equations (Section 2.2, page 165)
2. Graphs of linear equations (Section 2.3, page 184)

OBJECTIVES

1. Verify a solution to a system of equations.
2. Solve a system of equations by the graphical method.
3. Solve a system of equations by the substitution method.
4. Solve a system of equations by the elimination method.
5. Solve applied problems by solving systems of equations.

Algebra–Iraq Connection

The most important contributions of medieval Islamic mathematicians lie in the area of algebra. One of the greatest Islamic scholars was Muhammad ibn Musa al-Khwarizmi (780–850). Al-Khwarizmi was one of the first scholars in the House of Wisdom, an academy of scientists, established by caliph al-Mamun in the city of Baghdad in what is now Iraq. Jews, Christians, and Muslims worked together in scholarly pursuits in the academy during this period. The eminent scholar al-Khwarizmi wrote several books on astronomy and mathematics. Western Europeans first learned about algebra from his books. The word *algebra* comes from the Arabic *al-jabr*, part of the title of his book *Kitab al-jabr wal-muqabala*.

This book was translated into Latin and was a widely used text. The Arabic word *al-jabr* means "restoration," as in restoring broken parts. (At one time, it was not unusual to see the sign "Algebrista y Sangrador," meaning "bone setter and blood letter," at the entrance of a Spanish barber's shop. The sign informed customers of the barber's side business.) Al-Khwarizmi used the term in the mathematical sense of removing a negative quantity on one side of an equation and restoring it as a positive quantity on the other side.

1 Verify a solution to a system of equations.

System of Equations

A set of equations with common variables is called a **system of equations**. If each equation in a system of equations is linear, then the set of equations is called a **system of linear equations** or a **linear system of equations**. However, if at least one equation in a system of equations is nonlinear, then the set of equations is called a **system of nonlinear equations**. In this section, we solve systems of two linear equations in two variables, such as

$$\begin{cases} 2x - y = 5 \\ x + 2y = 5. \end{cases}$$

Notice that we designate a system of equations by using a left brace. A system of equations is sometimes referred to as a set of *simultaneous equations*. A **solution of a system**

of equations in two variables x and y is an ordered pair of numbers (a, b) such that when x is replaced with a and y is replaced with b, the resulting equations are true. The **solution set of a system of equations** is the set of all solutions of the system.

EXAMPLE 1 **Verifying a Solution**

Check whether each ordered pair is a solution of the system of linear equations:

$$\begin{cases} 2x - y = 5 & (1) \\ x + 2y = 5 & (2) \end{cases}$$

a. $(3, 1)$ **b.** $(4, 3)$

Solution

a. Check (3, 1)

Equation (1)	Equation (2)
$2x - y = 5$	$x + 2y = 5$
$2(3) - 1 \stackrel{?}{=} 5$	$3 + 2(1) \stackrel{?}{=} 5$
$5 = 5$ ✓	$5 = 5$ ✓

Because $(3, 1)$ satisfies *both* equations, it is a solution of the system.

b. Check (4, 3)

Equation (1)	Equation (2)
$2x - y = 5$	$x + 2y = 5$
$2(4) - 3 \stackrel{?}{=} 5$	$4 + 2(3) \stackrel{?}{=} 5$
$5 = 5$ ✓	$10 \stackrel{?}{=} 5$ ✗

Because $(4, 3)$ does not satisfy *both* equations, it is not a solution of the system.

Practice Problem 1 Check whether each ordered pair is a solution of the system:

$$\begin{cases} x + y = 4 & (1) \\ 3x - y = 0 & (2) \end{cases}$$

a. $(2, 2)$ **b.** $(1, 3)$

In this section, you will learn three methods of solving a system of two linear equations in two variables: the *graphical method*, the *substitution method*, and the *elimination method*.

2 Solve a system of equations by the graphical method.

Graphical Method

Recall that the graph of a linear equation

$$ax + by = c \qquad (a \text{ and } b \text{ not both zero})$$

is a line. Consider the following system of linear equations:

$$\begin{cases} a_1x + b_1y = c_1 \\ a_2x + b_2y = c_2 \end{cases}$$

A solution of this system is a point (an ordered pair) that satisfies both equations. Therefore, to find or estimate the solution set of this system, we graph both equations on the same coordinate axes and find the coordinates of any points of intersection. This procedure is called the **graphical method**.

EXAMPLE 2 Solving a System by the Graphical Method

Use the graphical method to solve the system of equations.

$$\begin{cases} 2x - y = 4 & (1) \\ 2x + 3y = 12 & (2) \end{cases}$$

Solution

Step 1 **Graph both equations on the same coordinate axes.**

(i) We first graph equation (1) by finding its intercepts.

a. To find the y-intercept, set $x = 0$ and solve for y. We have $2(0) - y = 4$, or $y = -4$; so the y-intercept is -4.

b. To find the x-intercept, set $y = 0$ and solve for x. We have $2x - 0 = 4$, or $x = 2$; so the x-intercept is 2.

The points $(0, -4)$ and $(2, 0)$ are on the graph of equation (1), which is sketched in red in Figure 8.1.

(ii) We now graph equation (2) using intercepts.

Here the x-intercept is 6 and the y-intercept is 4; so joining the points $(0, 4)$ and $(6, 0)$ produces the line sketched in blue in Figure 8.1.

FIGURE 8.1 Graphical solution

Step 2 **Find the point(s) of intersection of the two graphs.** We observe in Figure 8.1 that the point of intersection of the two graphs is $(3, 2)$.

Step 3 **Check your solution(s).** Replace x with 3 and y with 2 in equations (1) and (2).

Equation (1)	Equation (2)
$2x - y = 4$	$2x + 3y = 12$
$2(3) - 2 \stackrel{?}{=} 4$	$2(3) + 3(2) \stackrel{?}{=} 12$
$4 = 4$ ✓	$12 = 12$ ✓

Step 4 **Write the solution set for the system.**
The solution set is $\{(3, 2)\}$.

Practice Problem 2 Solve the system of equations graphically.

$$\begin{cases} x + y = 2 & (1) \\ 4x + y = -1 & (2) \end{cases}$$

TECHNOLOGY CONNECTION

In Example 2, solve each equation for y. Graph

$y_1 = 2x - 4$ and

$y_2 = -\frac{2}{3}x + 4$

on the same screen.

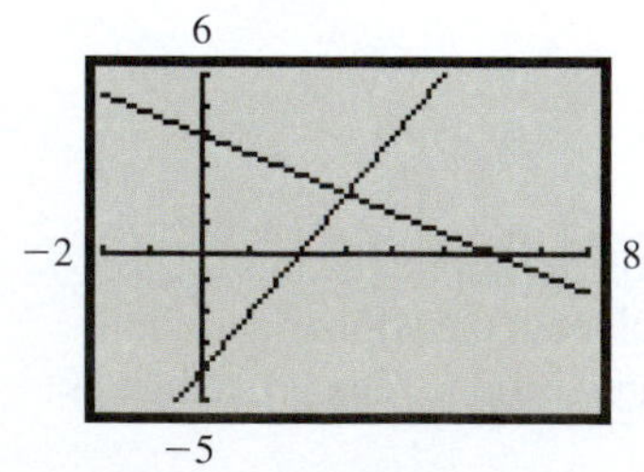

Use the Intersect feature on your calculator to find the point of intersection.

If a system of equations has at least one solution (as in Example 2), the system is **consistent**. A system of equations with no solution is **inconsistent**, and its solution set is the empty set, $\varnothing$.

A system of two linear equations in two variables must have one of the following types of solution sets:

1. One solution, also called **unique solution** (the lines intersect; see Figure 8.2(a)): the system is consistent, and the equations in the system are said to be **independent**.
2. No solution (the lines are parallel; see Figure 8.2(b)): the system is inconsistent.
3. Infinitely many solutions (the lines coincide; see Figure 8.2(c)): the system is consistent, and the equations in the system are said to be **dependent**.

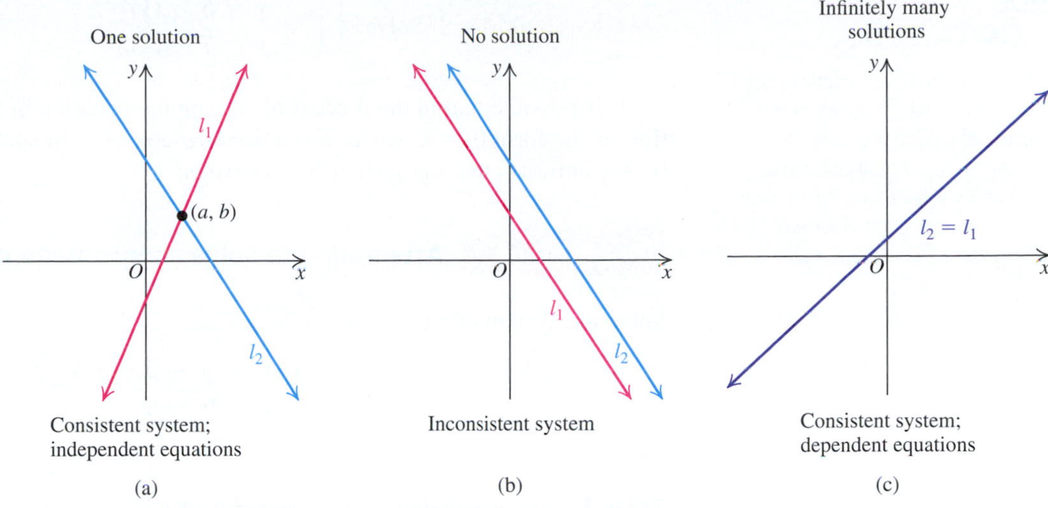

FIGURE 8.2 Possible solution sets of a system of two linear equations

3 Solve a system of equations by the substitution method.

Substitution Method

Another way to solve a linear system of equations is with the **substitution method**.

PROCEDURE IN ACTION

EXAMPLE 3 The Substitution Method

OBJECTIVE

Reduce the solution of the system to the solution of one equation in one variable by substitution.

EXAMPLE

Solve the system:

$$\begin{cases} 2x - 5y = 3 & (1) \\ y - 2x = 9 & (2) \end{cases}$$

Step 1 Solve for one variable. Choose one of the equations and express one of its variables in terms of the other variable.

1. We choose equation (2) and express y in terms of x.

$y = 2x + 9 \quad (3)$ — Solve equation (2) for y.

Step 2 Substitute. Substitute the expression found in Step 1 into the other equation to obtain an equation in one variable.

2.

$2x - 5y = 3$ — Equation (1)

$2x - 5(2x + 9) = 3$ — Substitute $2x + 9$ for y.

$2x - 10x - 45 = 3$ — Distributive property

Step 3 Solve the equation obtained in Step 2.

3.

$-8x - 45 = 3$ — Combine like terms.

$-8x = 3 + 45$ — Add 45 to both sides.

$x = -6$ — Solve for x.

Step 4 Back-substitute. Substitute the value(s) you found in Step 3 back into the expression you found in Step 1. The result is the solution(s).

4. $y = 2x + 9$ — Equation (3) from Step 1

$y = 2(-6) + 9 = -3$ — Substitute $x = -6$.

Because $x = -6$ and $y = -3$, the solution set is $\{(-6, -3)\}$.

Step 5 Check. Check your answer(s) in the original equations.

5. Check: $x = -6$ and $y = -3$.

Equation (1)	Equation (2)
$2(-6) - 5(-3) \stackrel{?}{=} 3$	$-3 - 2(-6) \stackrel{?}{=} 9$
$-12 + 15 = 3$ ✓	$-3 + 12 = 9$ ✓

SIDE NOTE

It is easy to make arithmetic errors in the many steps involved in the process of solving a system of equations. You should check your solution by substituting it into each equation in the original system.

Practice Problem 3 Solve: $\begin{cases} x - y = 5 & (1) \\ 2x + y = 7 & (2) \end{cases}$

It is possible that in the process of solving the equation in Step 3, you obtain an equation of the form $0 = k$, where k is a *nonzero* constant. In such cases, the false statement $0 = k$ indicates that the system is *inconsistent*.

EXAMPLE 4 **Attempting to Solve an Inconsistent System of Equations**

Solve the system of equations.

$$\begin{cases} x + y = 3 & (1) \\ 2x + 2y = 9 & (2) \end{cases}$$

Solution

Step 1 Solve equation (1) for y in terms of x.

$$y = 3 - x \qquad \text{Add } -x \text{ to both sides.}$$

Step 2 Substitute this expression into equation (2).

$$2x + 2y = 9 \qquad \text{Equation (2)}$$
$$2x + 2(3 - x) = 9 \qquad \text{Replace } y \text{ with } 3 - x \text{ (from Step 1).}$$

Step 3 Solve for x.

$$2x + 6 - 2x = 9 \qquad \text{Distributive property}$$
$$\underbrace{0 = 3}_{\text{False}} \qquad \text{Subtract 6 from both sides and simplify.}$$

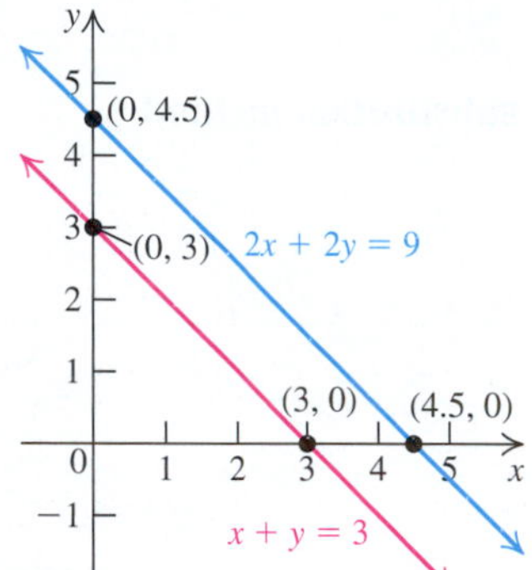

FIGURE 8.3 Inconsistent system

Because equation $0 = 3$ is false, the system is inconsistent. Figure 8.3 shows that the graphs of the two equations in this system are parallel lines. Because the lines do not intersect, the system has no solution. The solution set is $\varnothing$.

Practice Problem 4 Solve the system of equations.

$$\begin{cases} x - 3y = 1 & (1) \\ -2x + 6y = 3 & (2) \end{cases}$$

In Step 3, it is also possible to end with an equation of the form $k = k$. In such cases, the equations are *dependent* and the system has infinitely many solutions.

EXAMPLE 5 **Solving a Dependent System**

Solve the system of equations.

$$\begin{cases} 4x + 2y = 12 & (1) \\ -2x - y = -6 & (2) \end{cases}$$

Solution

Step 1 Solve equation (2) for y in terms of x.

$$-2x - y = -6 \qquad \text{Equation (2)}$$
$$-y = -6 + 2x \qquad \text{Add } 2x \text{ to both sides.}$$
$$y = 6 - 2x \qquad \text{Multiply both sides by } -1.$$

Step 2 Substitute $(6 - 2x)$ for y in equation (1).

$$4x + 2y = 12 \qquad \text{Equation (1)}$$
$$4x + 2(6 - 2x) = 12 \qquad \text{Replace } y \text{ with } 6 - 2x \text{ (from Step 1).}$$

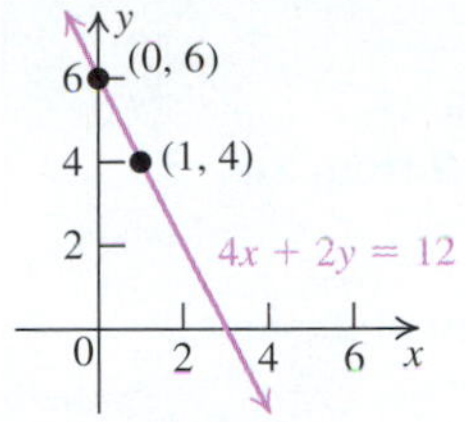

FIGURE 8.4 Dependent equations

Step 3 Solve for x.

$$4x + 12 - 4x = 12 \quad \text{Distributive property}$$
$$\underbrace{0 = 0}_{\text{True}} \quad \text{Subtract 12 from both sides and simplify.}$$

The equation $0 = 0$ is true for *every* value of x. Thus, *any* value of x can be used in the equation $y = 6 - 2x$ for back-substitution.

The solutions of the system are of the form $(x, 6 - 2x)$, and the solution set is

$$\{(x, 6 - 2x)\}.$$

In other words, the solution set consists of the infinite number of ordered pairs (x, y) lying on the line with equation $4x + 2y = 12$, as shown in Figure 8.4. You can find particular solutions by replacing x with any real number. For example, if we let $x = 0$ in $(x, 6 - 2x)$, we find that $(0, 6 - 2(0)) = (0, 6)$ is a solution of the system. Similarly, letting $x = 1$, we find that $(1, 4)$ is a solution.

DO YOU **KNOW?**

We use the informal notation

$$\{(x, 6 - 2x)\}$$

rather than

$$\{(x, 6 - 2x) \mid x \text{ is a real number}\}.$$

Practice Problem 5 Solve the system of equations.

$$\begin{cases} -2x + y = -3 & (1) \\ 4x - 2y = 6 & (2) \end{cases}$$

4 Solve a system of equations by the elimination method.

Elimination Method

The **elimination method** of solving a system of equations is also called the **addition method**.

PROCEDURE **IN ACTION**

EXAMPLE 6 The Elimination Method

OBJECTIVE

Solve a system of two linear equations by first eliminating one variable.

EXAMPLE

Solve the system:

$$\begin{cases} 2x + 3y = 21 & (1) \\ 3x - 4y = 23 & (2) \end{cases}$$

Step 1 Adjust the coefficients. If necessary, multiply both equations by appropriate numbers to get two new equations in which the coefficients of the variable to be eliminated are opposites.

1. We arbitrarily select y as the variable to be eliminated.

$$\begin{cases} 8x + 12y = 84 & \text{Multiply equation (1) by 4.} \\ 9x \underbrace{- 12y}_{} = 69 & \text{Multiply equation (2) by 3.} \end{cases}$$

Coefficients are opposites of each other.

Step 2 Add the equations. Add the resulting equations to get an equation in one variable.

2.
$$\begin{aligned} 8x + 12y &= 84 \\ 9x - 12y &= 69 \\ \hline 17x \qquad &= 153 \end{aligned}$$

Adding the two equations from Step 1 eliminates the y-terms.

Step 3 Solve the resulting equation.

3. $17x = 153$ Equation from Step 2

$x = \frac{153}{17}$ Divide both sides by 17.

$x = 9$ Simplify.

(continued)

Step 4 Back-substitute the value you found into one of the original equations to solve for the other variable.

4.
$$\begin{aligned} 2x + 3y &= 21 && \text{Equation (1)} \\ 2(9) + 3y &= 21 && \text{Replace } x \text{ with 9 from Step 3.} \\ 18 + 3y &= 21 && \text{Simplify.} \\ 3y &= 3 && \text{Subtract 18 from both sides.} \\ y &= 1 && \text{Solve for } y. \end{aligned}$$

Step 5 Write the solution set from Steps 3 and 4.

5. The solution set is $\{(9, 1)\}$.

Step 6 Check your solution(s) in the original equations (1) and (2).

6. Check $x = 9$ and $y = 1$.

Equation (1)	Equation (2)
$2(9) + 3(1) \stackrel{?}{=} 21$	$3(9) - 4(1) = 23$
$18 + 3 = 21$ ✓	$27 - 4 = 23$ ✓

Practice Problem 6 Solve the system.

$$\begin{cases} 3x + 2y = 3 & (1) \\ 9x - 4y = 4 & (2) \end{cases}$$

As in the substitution method, if you add the equations in Step 2 and the resulting equation becomes $0 = k$, where $k \neq 0$, then the system is inconsistent: it has no solutions. As an illustration, if you solve the system of equations in Example 4 by the elimination method, you will obtain the equation $0 = 3$. You now conclude that the system is inconsistent. Similarly, in Step 2, if you obtain $0x + 0y = 0$ (or $0 = 0$), then the equations are dependent. See Example 5.

5 Solve applied problems by solving systems of equations.

Applications

We can deduce from Section 2.4 that as the price of a product increases, demand for it decreases and as the price increases, the supply of the product also increases. The **equilibrium point** is the ordered pair (x, p) such that the number of units, x, and the price per unit, p, satisfy *both* the demand and supply equations.

SIDE NOTE

The graph of a system of equations allows you to use geometric intuition to understand algebra. If the geometric and algebraic representations of a solution do not agree, you must go back and find the error(s).

EXAMPLE 7 Finding the Equilibrium Point

Find the equilibrium point if the supply and demand functions for a new brand of digital video recorder (DVR) are given by the system

$$\begin{aligned} p &= 60 + 0.0012x \quad (1) && \text{Supply equation} \\ p &= 80 - 0.0008x \quad (2) && \text{Demand equation} \end{aligned}$$

where p is the price in dollars and x is the number of units.

Solution

We substitute the value of p from equation (1) into equation (2) and solve the resulting equation.

$$\begin{aligned} p &= 80 - 0.0008x && \text{Demand equation (2)} \\ 60 + 0.0012x &= 80 - 0.0008x && \text{Replace } p \text{ with } 60 + 0.0012x \text{ in (1).} \\ 0.002x &= 20 && \text{Collect like terms and simplify.} \\ x &= \frac{20}{0.002} = 10{,}000 && \text{Solve for } x. \end{aligned}$$

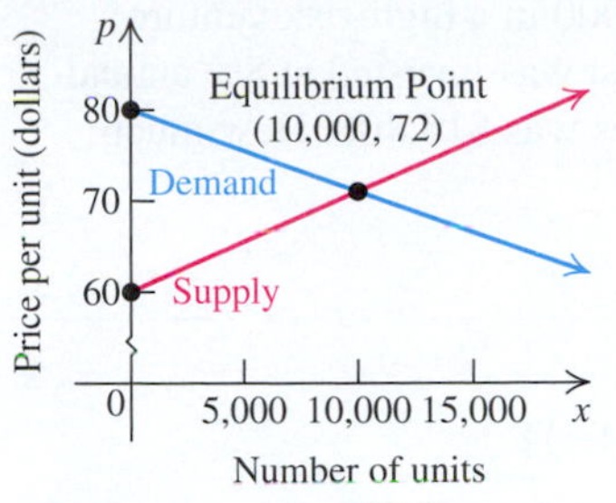

FIGURE 8.5 Price of DVRs

The equilibrium point occurs when the supply and demand for DVRs is 10,000 units. To find the price p, we back-substitute $x = 10,000$ into either of the original equations (1) or (2).

$$p = 60 + 0.0012x \quad \text{Equation (1)}$$
$$= 60 + 0.0012(10,000) \quad \text{Replace } x \text{ with } 10,000.$$
$$= 72 \quad \text{Simplify.}$$

The equilibrium point is (10,000, 72). See Figure 8.5. This means that at a price of \$72 per unit, the supply and demand for DVRs is 10,000 units.

Check. Verify that the ordered pair (10,000, 72) satisfies both equations (1) and (2).

Practice Problem 7 Find the equilibrium point.

$$\begin{cases} p = 20 + 0.002x & (1) \quad \text{Supply equation} \\ p = 77 - 0.008x & (2) \quad \text{Demand equation} \end{cases}$$

In general, to solve an applied problem involving two unknowns, we need to set up two equations using two variables that represent the two unknowns. The general rule is that **we need two equations to find two unknowns.**

EXAMPLE 8 Analyzing Investments

Last year Mrs. Rogers invested \$40,000. Part was invested at 6% interest rate per year, and the rest was invested in a risky venture at 10% per year. Combined income for the year totaled \$3,120. How much did she invest at each rate?

Solution

Two amounts are to be determined, so we will represent them using two variables. Let

$$x = \text{amount invested at } 6\%$$
$$y = \text{amount invested at } 10\%.$$

Because 6% of x is $0.06x$ and 10% of y is $0.10y$, we have

$$0.06x = \text{income from the } 6\% \text{ investment}$$
$$0.10y = \text{income from the } 10\% \text{ investment.}$$

We now write our two equations in two variables:

$$x + y = 40,000 \quad (1) \quad \text{Total amount invested}$$
$$0.06x + 0.10y = 3,120 \quad (2) \quad \text{Total income from both investments}$$

Let's use the elimination method to solve these equations.

$$-6x - 6y = -240,000 \quad \text{Multiply equation (1) by } -6.$$
$$6x + 10y = 312,000 \quad \text{Multiply equation (2) by 100.}$$
$$4y = 72,000 \quad \text{Add}$$
$$y = \frac{72,000}{4} = 18,000 \quad \text{Solve for } y.$$

Back-substitute $y = 18,000$ in equation (1) to get

$$x + 18,000 = 40,000$$
$$x = 40,000 - 18,000 \quad \text{Solve for } x.$$
$$= 22,000.$$

Mrs. Rogers invested \$22,000 at 6% and \$18,000 at 10%.

Check. 6% of \$22,000 is $0.06 \times 22,000 = 1320$, and 10% of \$18,000 is $0.10 \times 18,000 = 1800$.

Finally, \$22,000 + \$18,000 = \$40,000, and \$1320 + \$1800 = \$3120, as given.

Practice Problem 8 A speculator invested part of \$150,000 in a high-risk venture and received a return of 12% at the end of the year. The rest was invested at 8% annual return. The combined annual income from the two ventures was \$15,400. How much was invested at each rate?

Answers to Practice Problems

1. a. No **b.** Yes **2.** $\{(-1, 3)\}$ **3.** $\{(4, -1)\}$ **4.** $\varnothing$ **5.** $\{(x, 2x - 3)\}$ **6.** $\left\{\left(\frac{2}{3}, \frac{1}{2}\right)\right\}$ **7.** $\{(5700, 31.4)\}$ **8.** \$85,000 at 12%, \$65,000 at 8%

SECTION 8.1 Exercises

Basic Concepts and Skills

1. The ordered pair (a, b) is a(n) ____________ of a system of equations in x and y provided that when x is replaced with a and y is replaced with b, the resulting equations are true.
2. The two nongraphical methods for solving a system of equations are the ____________ and ____________ methods.
3. If in the process of solving a system of equations you get an equation of the form $0 = k$, where k is not zero, then the system is ____________.
4. If in the process of solving a system of equations you get an equation of the form $0 = 0$, then the system has ____________.
5. **True or False.** A system consisting of two identical equations has no solution.
6. **True or False.** If in the process of solving a system of two equations in x and y you get the equation $5 = 5$, then the system has exactly one solution.

In Exercises 7–12, determine which ordered pairs are solutions of each system of equations.

7. $\begin{cases} 2x + 3y = 3 \\ 3x - 4y = 13 \end{cases}$ $(1, -3), (3, -1), (6, 3), \left(5, \frac{1}{2}\right)$

8. $\begin{cases} x + 2y = 6 \\ 3x + 6y = 18 \end{cases}$ $(2, 2), (-2, 4), (0, 3), (1, 2)$

9. $\begin{cases} 5x - 2y = 7 \\ -10x + 4y = 11 \end{cases}$ $\left(\frac{5}{4}, 1\right), \left(0, \frac{11}{4}\right), (1, -1), (3, 4)$

10. $\begin{cases} x - 2y = -5 \\ 3x - y = 5 \end{cases}$ $(1, 3), (-5, 0), (3, 4), (3, -4)$

11. $\begin{cases} x + y = 1 \\ \frac{1}{2}x + \frac{1}{3}y = 2 \end{cases}$ $(0, 1), (1, 0), \left(\frac{2}{3}, \frac{3}{2}\right), (10, -9)$

12. $\begin{cases} \frac{2}{x} + \frac{3}{y} = 2 \\ \frac{6}{x} + \frac{18}{y} = 9 \end{cases}$ $(3, 2), (2, 3), (4, 3), (3, 4)$

In Exercises 13–22, estimate the solution(s) (if any) of each system by the graphical method. Check your solution(s). For any dependent equations, write your answer with x being arbitrary.

13. $\begin{cases} x + y = 3 \\ x - y = 1 \end{cases}$

14. $\begin{cases} x + y = 10 \\ x - y = 2 \end{cases}$

15. $\begin{cases} x + 2y = 6 \\ 2x + y = 6 \end{cases}$

16. $\begin{cases} 2x - y = 4 \\ x - y = 3 \end{cases}$

17. $\begin{cases} 3x - y = -9 \\ y = 3x + 6 \end{cases}$

18. $\begin{cases} 5x + 2y = 10 \\ y = -\frac{5}{2}x - 5 \end{cases}$

19. $\begin{cases} x + y = 7 \\ y = 2x \end{cases}$

20. $\begin{cases} y - x = 2 \\ y + x = 9 \end{cases}$

21. $\begin{cases} 3x + y = 12 \\ y = -3x + 12 \end{cases}$

22. $\begin{cases} 2x + 3y = 6 \\ 6y = -4x + 12 \end{cases}$

In Exercises 23–36, determine whether each system is *consistent* or *inconsistent*. If the system is consistent, determine whether the equations are dependent or independent. Do not solve the system.

23. $\begin{cases} y = -2x + 3 \\ y = 3x + 5 \end{cases}$

24. $\begin{cases} 3x + y = 5 \\ 2x + y = 4 \end{cases}$

25. $\begin{cases} 2x + 3y = 5 \\ 3x + 2y = 7 \end{cases}$

26. $\begin{cases} 2x - 4y = 5 \\ 3x + 5y = -6 \end{cases}$

27. $\begin{cases} 3x + 5y = 7 \\ 6x + 10y = 14 \end{cases}$

28. $\begin{cases} 3x - y = 2 \\ 9x - 3y = 6 \end{cases}$

29. $\begin{cases} x + 2y = -5 \\ 2x - y = 4 \end{cases}$

30. $\begin{cases} x + 2y = -2 \\ 2x - 3y = 5 \end{cases}$

31. $\begin{cases} 2x - 3y = 5 \\ 6x - 9y = 10 \end{cases}$

32. $\begin{cases} 3x + y = 2 \\ 15x + 5y = 15 \end{cases}$

33. $\begin{cases} -3x + 4y = 5 \\ \frac{9}{2}x - 6y = \frac{15}{2} \end{cases}$

34. $\begin{cases} 6x + 5y = 11 \\ 9x + \frac{15}{2}y = 21 \end{cases}$

35. $\begin{cases} 7x - 2y = 3 \\ 11x - \frac{3}{2}y = 8 \end{cases}$

36. $\begin{cases} 4x + 7y = 10 \\ 10x + \frac{35}{2}y = 25 \end{cases}$

In Exercises 37–46, solve each system of equations by the substitution method. Check your solutions. For any dependent equations, write your answer in the ordered pair form given in Example 5.

37. $\begin{cases} y = 2x + 1 \\ 5x + 2y = 9 \end{cases}$ **38.** $\begin{cases} x = 3y - 1 \\ 2x - 3y = 7 \end{cases}$

39. $\begin{cases} 3x - y = 5 \\ x + y = 7 \end{cases}$ **40.** $\begin{cases} 2x + y = 2 \\ 3x - y = -7 \end{cases}$

41. $\begin{cases} 2x - y = 5 \\ -4x + 2y = 7 \end{cases}$ **42.** $\begin{cases} 3x + 2y = 5 \\ -9x - 6y = 15 \end{cases}$

43. $\begin{cases} \frac{2}{3}x + y = 3 \\ 3x + 2y = 1 \end{cases}$ **44.** $\begin{cases} x - 2y = 3 \\ 4x + 6y = 3 \end{cases}$

45. $\begin{cases} x - 2y = 5 \\ -3x + 6y = -15 \end{cases}$ **46.** $\begin{cases} x + y = 3 \\ 2x + 2y = 6 \end{cases}$

In Exercises 47–56, solve each system of equations by the elimination method. Check your solutions. For any dependent equations, write your answer as in Example 5.

47. $\begin{cases} x - y = 1 \\ x + y = 5 \end{cases}$ **48.** $\begin{cases} 2x - 3y = 5 \\ 3x + 2y = 14 \end{cases}$

49. $\begin{cases} x + y = 0 \\ 2x + 3y = 3 \end{cases}$ **50.** $\begin{cases} x + y = 3 \\ 3x + y = 1 \end{cases}$

51. $\begin{cases} 5x - y = 5 \\ 3x + 2y = -10 \end{cases}$ **52.** $\begin{cases} 3x - 2y = 1 \\ -7x + 3y = 1 \end{cases}$

53. $\begin{cases} x - y = 2 \\ -2x + 2y = 5 \end{cases}$ **54.** $\begin{cases} x + y = 5 \\ 2x + 2y = -10 \end{cases}$

55. $\begin{cases} 4x + 6y = 12 \\ 2x + 3y = 6 \end{cases}$ **56.** $\begin{cases} 4x + 7y = -3 \\ -8x - 14y = 6 \end{cases}$

In Exercises 57–74, use any method to solve each system of equations. For any dependent equations, write your answer as in Example 5.

57. $\begin{cases} 2x + y = 9 \\ 2x - 3y = 5 \end{cases}$ **58.** $\begin{cases} x + 2y = 10 \\ x - 2y = -6 \end{cases}$

59. $\begin{cases} 2x + 5y = 2 \\ x + 3y = 2 \end{cases}$ **60.** $\begin{cases} 4x - y = 6 \\ 3x - 4y = 11 \end{cases}$

61. $\begin{cases} 2x + 3y = 7 \\ 3x + y = 7 \end{cases}$ **62.** $\begin{cases} x = 3y + 4 \\ x = 5y + 10 \end{cases}$

63. $\begin{cases} 2x + 3y = 9 \\ 3x + 2y = 11 \end{cases}$ **64.** $\begin{cases} 3x - 4y = 0 \\ y = \dfrac{2x + 1}{3} \end{cases}$

65. $\begin{cases} \dfrac{x}{4} + \dfrac{y}{6} = 1 \\ x + 2(x - y) = 7 \end{cases}$ **66.** $\begin{cases} \dfrac{x}{3} + \dfrac{y}{5} = 12 \\ x - y = 4 \end{cases}$

67. $\begin{cases} 3x = 2(x + y) \\ 3x - 5y = 2 \end{cases}$ **68.** $\begin{cases} y + x + 2 = 0 \\ y + 2x + 1 = 0 \end{cases}$

69. $\begin{cases} 0.2x + 0.7y = 1.5 \\ 0.4x - 0.3y = 1.3 \end{cases}$ **70.** $\begin{cases} 0.6x + y = -1 \\ x - 0.5y = 7 \end{cases}$

71. $\begin{cases} \dfrac{x}{2} + \dfrac{y}{3} = 1 \\ \dfrac{3x}{4} + \dfrac{y}{2} = 1 \end{cases}$ **72.** $\begin{cases} \dfrac{x}{3} - \dfrac{y}{5} = 2 \\ 6y - 10x = 25 \end{cases}$

73. $\begin{cases} \dfrac{x}{3} - \dfrac{y}{2} = 1 \\ \dfrac{3y}{8} - \dfrac{x}{4} = -\dfrac{3}{4} \end{cases}$ **74.** $\begin{cases} \dfrac{x}{5} - \dfrac{y}{2} = 1 \\ 15y - 6x = -30 \end{cases}$

In Exercises 75–82, let $u = \frac{1}{x}$ and $v = \frac{1}{y}$. Solve for u and v; then solve for x and y.

75. $\begin{cases} \dfrac{2}{x} + \dfrac{5}{y} = -5 \\ \dfrac{3}{x} - \dfrac{2}{y} = -17 \end{cases}$ **76.** $\begin{cases} \dfrac{2}{x} + \dfrac{1}{y} = 3 \\ \dfrac{4}{x} - \dfrac{2}{y} = 0 \end{cases}$

77. $\begin{cases} \dfrac{3}{x} + \dfrac{1}{y} = 4 \\ \dfrac{6}{x} - \dfrac{1}{y} = 2 \end{cases}$ **78.** $\begin{cases} \dfrac{6}{x} + \dfrac{3}{y} = 0 \\ \dfrac{4}{x} + \dfrac{9}{y} = -1 \end{cases}$

79. $\begin{cases} \dfrac{5}{x} + \dfrac{10}{y} = 3 \\ \dfrac{2}{x} - \dfrac{12}{y} = -2 \end{cases}$ **80.** $\begin{cases} \dfrac{3}{x} + \dfrac{4}{y} = 1 \\ \dfrac{6}{x} + \dfrac{4}{y} = 3 \end{cases}$

81. $\begin{cases} \dfrac{2}{x} + \dfrac{1}{y} = 4 \\ x + 2y = 6xy \end{cases}$ **82.** $\begin{cases} \dfrac{1}{x} - \dfrac{2}{y} = 3 \\ 2x - y = 5xy \end{cases}$

Applying the Concepts

In Exercises 83–86, the demand and supply functions of a product are given. In each case, p represents the price in dollars per unit and x represents the number of units in hundreds. Find the equilibrium point.

83. $\begin{cases} 2p + x = 140 & \text{Demand equation} \\ 12p - x = 280 & \text{Supply equation} \end{cases}$

84. $\begin{cases} 7p + x = 150 & \text{Demand equation} \\ 10p - x = 20 & \text{Supply equation} \end{cases}$

85. $\begin{cases} 2p + x = 25 & \text{Demand equation} \\ x - p = 13 & \text{Supply equation} \end{cases}$

86. $\begin{cases} p + 2x = 96 & \text{Demand equation} \\ p - x = 39 & \text{Supply equation} \end{cases}$

In Exercises 87–102, use a system of equations to solve each problem.

87. Diameter of a pizza. The sum of the diameters of the largest and smallest pizzas sold at the Monster Pizza Shop is 29 inches. The difference in their diameters is 13 inches. Find the diameters of the largest and smallest pizzas.

88. Calories in hamburgers. The sum of the number of calories in a hamburger from Boston Burger and a hamburger from Carmen's Broiler is 1130. The difference in the number of calories in the hamburgers is 40. If the Boston Burger has the larger number of calories, how many calories are in each restaurant's hamburger?

89. Trash composition. Paper and plastic together account for 48% (by weight) of the total trash collected. If the weight of paper trash collected is five times the weight of plastic trash, what percent of the total trash collected is paper and what percent is plastic?

90. **Cost of food and clothing.** The average monthly combined cost of food and clothing for the Martínez family is \$1000. If they spend four times as much for food as they do for clothes, what is the average monthly cost of each?

91. **Mardi Gras parade "throws."** Levon paid 40¢ for each string of beads and 30¢ for each doubloon (a special coin) he bought to throw from his Mardi Gras float. He paid a total of \$265 for the two items. His doubloons and beads combined to a total of 770 items. How many doubloons and how many strings of beads did Levon buy?

92. **Halloween candy.** Janet bought 135 pieces of candy to give away on Halloween. She bought two kinds of candy, paying 24¢ apiece for one kind and 18¢ apiece for the other. If she spent \$26.70 for the candy, how many pieces of each kind did she buy?

In Exercises 93 and 94, use the information in the following table.

Food Description McDonald's	Fat (gms)	Carb (gms)	Protein (gms)
Breakfast Burrito	20	21	13
Egg McMuffin	12	27	17

Source: www.shapefit.com/mcdonalds.html

Suppose you ate breakfast at McDonald's during one workweek (Monday–Friday).

93. **McDonald's breakfast.** If you consumed 123 grams of carbohydrates and 77 grams of protein, how many of each breakfast item did you eat during the workweek?

94. **McDonald's breakfast.** If you consumed 68 grams of fat and 129 grams of carbohydrates, how many of each breakfast item did you eat during the workweek?

95. **Investment.** Last year Mrs. García invested \$50,000. She put part of the money in a real estate venture that paid 7.5% for the year and the rest in a small-business venture that returned 12% for the year. The combined income from the two investments for the year totaled \$5190. How much did she invest at each rate?

96. **Investment.** Mr. Sharma invested a total of \$30,000 in two ventures for a year. The annual return from one of them was 8%, and the other paid 10.5% for the year. He received a total income of \$2550 from both investments. How much did he invest at each rate?

97. **Tutoring other students.** A student earns twice as much per hour for tutoring as she does working at McDougal's. If her average wage is \$11.25 per hour, how much does she earn per hour at each job?

98. **Plumber's earnings.** A plumber earns \$15 per hour more than her apprentice. For a 40-hour week, their combined earnings are \$2200. What is the hourly rate for each?

99. **Mixture problem.** An herb that sells for \$5.50 per pound is mixed with tea that sells for \$3.20 per pound to produce a 100-pound mix that is worth \$3.66 per pound. How many pounds of each ingredient does the mix contain?

100. **Mixture problem.** A chemist had a solution of 60% acid. She added some distilled water, reducing the acid concentration to 40%. She then added 1 more liter of water to further reduce the acid concentration to 30%. How much of the 30% solution did she then have?

101. **Airplane speed.** With the help of a tail wind, a plane travels 3000 kilometers in five hours. The return trip against the wind requires six hours. Assume that the direction and the wind speed are constant. Find both the speed of the plane in still air and the wind speed. [*Hint:* Remember that the wind helps the plane in one direction and hinders it in the other.]

102. **Motorboat speed.** A motorboat travels upstream a distance of 12 miles in two hours. If the current had been twice as strong, the trip would have taken three hours. How long should it take for the return trip downstream?

103. **Break-even analysis.** A local publishing company publishes *City Magazine*. The production and setup costs are \$30,000, and the cost of producing each magazine is \$2. Each magazine sells for \$3.50. Assume that x magazines are published and sold.

a. Write the total cost function $y = C(x)$ and the revenue function $y = R(x)$.

b. Graph both functions from part (a) on the same coordinate plane.

c. How many magazines must be sold to break even?

104. **Break-even analysis.** An electronic manufacturer plans to make digital portable music players (MP4s). Fixed costs will be \$750,000, and it will cost \$50 to manufacture each MP4, which will be sold for \$125 to the retailers. Assume that x MP4s are manufactured and sold.

a. Write the total cost function $y = C(x)$ and the revenue function $y = R(x)$.

b. Graph both functions from part (a) on the same coordinate plane.

c. How many MP4s must be sold to break even?

105. **Making a job decision.** Shanaysha has job offers from department stores A and B to sell major appliances. Store A offers her a fixed salary of \$400 per week. Store B offers her \$150 per week plus a commission of 4% of her weekly sales. How much should Shanaysha's weekly sales be for the offer from Store B to be better than that from Store A?

106. **Making a job decision.** Sheena has job offers from companies A and B to sell insurance. Company A offers her \$25,000 per year plus 2% of her yearly sales premiums. Company B offers her \$30,000 per year plus 1% of her yearly sales premiums. How much must Sheena earn in yearly sales premiums for the offer from Company B to be the better offer?

Beyond the Basics

In Exercises 107–110, solve each system for x and y.

107. $\begin{cases} 2\log_3 x + 3\log_3 y = 8 \\ 3\log_3 x - \log_3 y = 1 \end{cases}$

108. $\begin{cases} 3\log_2(x + y) - 5\log_2(x - y) = 2 \\ 2\log_2(x + y) + 3\log_2(x - y) = 14 \end{cases}$

109. $\begin{cases} 3e^x - 4e^y = 4 \\ 2e^x + 5e^y = 18 \end{cases}$ **110.** $\begin{cases} e^{x+2y} - 3e^{x-y} = 16 \\ 3e^{x+2y} - 12e^{x-y} = 46 \end{cases}$

111. For what value of the constant c is the following system consistent?

$$\begin{cases} x + 2y = 7 \\ 3x + 5y = 11 \\ cx + 3y = 4 \end{cases}$$

[*Hint:* Solve the first two equations for x and y and then substitute the solution into the third equation.]

112. Find the value of c for which the following system of equations is inconsistent.

$$\begin{cases} 2x + cy = 11 \\ 5x - 7y = 5 \end{cases}$$

113. If $(x - 2)$ is a factor of both $f(x) = x^3 - 4x^2 + ax + b$ and $g(x) = x^3 - ax^2 + bx + 8$, find the values of a and b.

114. Let l_1: $a_1x + b_1y + c_1 = 0$ and l_2: $a_2x + b_2y + c_2 = 0$ be two nonparallel lines. Then for any constant k, show that the line $(a_1x + b_1y + c_1) + k(a_2x + b_2y + c_2) = 0$ is an equation of the line passing through the point of intersection of l_1 and l_2.

In Exercises 115–117, write your answer in slope–intercept form. [*Hint:* Use Exercise 114.]

115. Find an equation of the line that passes through the point of intersection of the lines with equations $2x + y = 3$ and $x - 3y = 12$ and that is parallel to the line with equation $3x + 2y = 8$.

116. Find an equation of the line that passes through the point of intersection of the lines with equations $5x + 2y = 7$ and $6x - 5y = 38$ and that passes through the point $(1, 3)$.

117. Find an equation of the line that passes through the point of intersection of the lines with equations $2x + 5y + 7 = 0$ and $13x - 10y + 3 = 0$ and that is perpendicular to the line with equation $7x + 13y = 8$.

Critical Thinking/Discussion/Writing

118. Consider the system of equations

$$\begin{cases} a_1x + b_1y = c_1 \\ a_2x + b_2y = c_2 \end{cases}$$

where $a_1, b_1, c_1, a_2, b_2,$ and c_2 are constants. Find a relationship (an equation) among these constants such that the system of equations has

a. Only one solution. Find the solution in terms of the constants.
b. No solution.
c. Infinitely many solutions.

119. Let $3x + 4y - 12 = 0$ be the equation of a line l_1 and let $P(5, 8)$ be a point.

a. Find an equation of a line l_2 passing through the point P and perpendicular to l_1.
b. Find the point Q of the intersection of the two lines l_1 and l_2.
c. Find the distance $d(P, Q)$. This is the distance from the point P to the line l_1.

120. Use the procedure outlined in Exercise 119 to show that the (perpendicular) distance from a point $P(x_1, y_1)$ to the line $ax + by + c = 0$ is given by $\dfrac{|ax_1 + by_1 + c|}{\sqrt{a^2 + b^2}}$.

121. Use the formula from Exercise 120 to find the distance from the given point P to the given line l.

a. $P(2, 3)$, l: $x + y - 7 = 0$
b. $P(-2, 5)$, l: $2x - y + 3 = 0$
c. $P(3, 4)$, l: $5x - 2y - 7 = 0$
d. $P(0, 0)$, l: $ax + by + c = 0$

122. Find the reflection of the point $P = (2, 1)$ about the line l: $2x + 3y = 20$. [Let $Q = (a, b)$ be the reflection point; then you get two equations in a and b. (*i*) The midpoint of $\overline{PQ}$ lies on l, (*ii*) l is perpendicular to the line containing the segment $\overline{PQ}$.]

123. Find the reflection of the point $P = (2, 1)$ about the line l: $3x + y = 12$.

Maintaining Skills

In Exercises 124–130, find the requested variable.

124. Find x if $3x - (2x - 1) - (4x + 3) = -5$.

125. Find y if $y + 2z = 7$ and $z = 2$.

126. Find x if $x - 3y = 5$ and $y = -3$.

127. Find x if $2x + 3y + 5z = 21$, $y = 1$, and $z = 2$.

128. Find y if $3x - 5y + z = 2$, $x = -2$, and $z = 3$.

129. Find x if $2x + 3y - 2z = 5$, $y - z = 4$, and $z = -3$.

130. Find x if $3x + 2y + z = 2$, $y + 2z = 1$, and $z = 2$.

SECTION 8.2

Systems of Linear Equations in Three Variables

BEFORE STARTING THIS SECTION, REVIEW

1 Solving a system of equations in two variables (Section 8.1)

OBJECTIVES

1 Solve a system of linear equations in three variables.

2 Classify systems as consistent and inconsistent.

3 Solve nonsquare systems.

4 Interpret linear systems in three variables geometrically.

5 Use linear systems in applications.

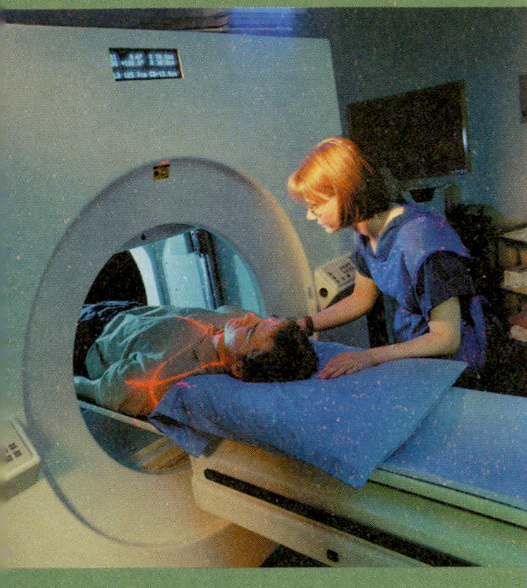

CAT Scans

Computerized axial tomography imaging is more commonly known as a "CAT scan" or "CT scan." Tomography is from the Greek word *tomos*, meaning "slice or section," and *graphia*, meaning "describing." The CAT scan machine was invented in 1972 by British engineer Godfrey Hounsfield and, independently, by physicist Allan Cormack of Tufts University in Massachusetts. Hounsfield never attended a university, but he started experimenting with electrical and mechanical devices as a boy. He shared a Nobel Prize in 1979 with Cormack for inventing the scanner technology and was honored with knighthood in England for his contributions to medicine and science. CAT scanners use X-rays measured outside the patient's body, together with algebraic computations, to construct pictures of the patient's internal body organs, including the brain. The scanner does not "take" any pictures; instead, it *computes* pictures. The idea behind the CAT scanner is simple. When an X-ray passes through an object, it loses intensity. The denser the object, the more intensity the X-ray loses. Because tumors are denser than healthy tissues, an unexpected loss of intensity may indicate the presence of a tumor. To obtain sufficient information, a large number of properly arranged X-rays is needed. Imagine a cross section of body tissue with a superimposed mathematical grid. Each region bounded by the grid lines is called a *grid cell*. In an actual CAT scan, each grid cell is 2 square millimeters and contains 10,000 or more biological cells. The grid shown in Figure 8.6 would require many X-rays. Actual CAT scanners work with grids that have 512 × 512 grid cells and require hundreds of thousands of X-ray beams. In Example 6, we discuss a simple situation in which X-rays pass through three grid cells.

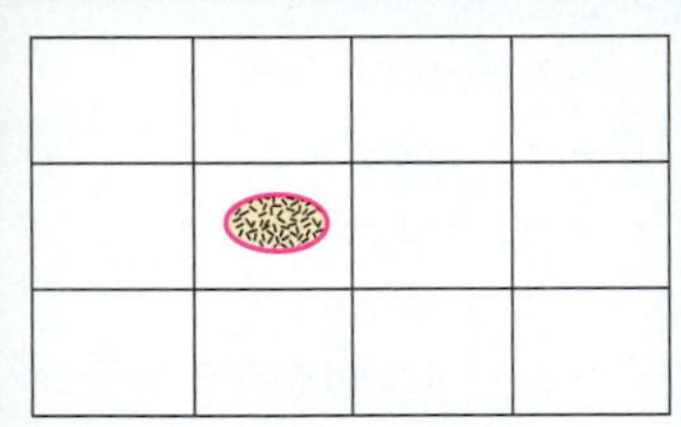

Cross section with tumor

FIGURE 8.6 Cross section with tumor

1 Solve a system of linear equations in three variables.

Systems of Linear Equations

A **linear equation in the variables** $x_1, x_2, \ldots, x_n$ is an equation that can be written in the form

$$a_1x_1 + a_2x_2 + \cdots + a_nx_n = b,$$

where b and the *coefficients* $a_1, a_2, \ldots, a_n$ are real numbers. The subscript n can be any positive integer. When n is small—say, between 2 and 5 (as in most textbook examples and exercises)—we normally use x, y, z, u, and v instead of x_1, x_2, x_3, x_4, and x_5. However, in real-life applications, n might be 100 or 1000 or even greater.

A **system of linear equations** (or a **linear system**) in three variables is a collection of two or more linear equations involving the same three variables. For example,

$$\begin{cases} x + 3y + z = 0 \\ 2x - y + z = 5 \\ 3x - 3y + 2z = 10 \end{cases}$$

is a system of three linear equations in the three variables x, y, and z. An ordered triple (a, b, c) is a **solution** of a system of three linear equations in three variables x, y, and z if each equation in the system is a true statement when a, b, and c are substituted for x, y, and z, respectively.

EXAMPLE 1 Verifying a Solution

Determine whether the ordered triple $(2, -1, 3)$ is a solution of the following linear system.

$$\begin{cases} 5x + 3y - 2z = 1 & (1) \\ x - y + z = 6 & (2) \\ 2x + 2y - z = -1 & (3) \end{cases}$$

Solution

We replace x with 2, y with -1, and z with 3 in all three equations.

Equation (1)	Equation (2)	Equation (3)
$5x + 3y - 2z = 1$	$x - y + z = 6$	$2x + 2y - z = -1$
$5(2) + 3(-1) - 2(3) \stackrel{?}{=} 1$	$2 - (-1) + 3 \stackrel{?}{=} 6$	$2(2) + 2(-1) - (3) \stackrel{?}{=} -1$
$10 - 3 - 6 \stackrel{?}{=} 1$	$2 + 1 + 3 \stackrel{?}{=} 6$	$4 - 2 - 3 \stackrel{?}{=} -1$
$1 = 1$ ✓	$6 = 6$ ✓	$-1 = -1$ ✓

Because the ordered triple $(2, -1, 3)$ satisfies all three equations, it is a solution of the system.

Practice Problem 1 Determine whether the ordered triple $(2, -3, 2)$ is a solution of the following linear system.

$$\begin{cases} x + y + z = 1 \\ 3x + 4y + z = -4 \\ 2x + y + 2z = 5 \end{cases}$$

We will now use a procedure called **Gaussian elimination** to convert a system of three equations in the variables x, y, and z into an equivalent system in *triangular form* with leading coefficient 1, as shown below.

$$\begin{cases} x + ay + bz = p & (1) \\ \quad\;\; y + cz = q & (2) \\ \quad\quad\quad\; z = z_0 & (3) \end{cases}$$

We can then solve the system by back-substitution. To get a system in triangular form, we use the following three operations to obtain equivalent systems (systems with the same solution set as the original system).

OPERATIONS THAT PRODUCE EQUIVALENT SYSTEMS

1. Interchange the position of any two equations.
2. Multiply any equation by a nonzero constant.
3. Add a multiple of one equation to another.

PROCEDURE IN ACTION

EXAMPLE 2 The Gaussian Elimination Method

OBJECTIVE

Solve a system of three equations in three variables by first converting the system to the form

$$\begin{cases} x + ay + bz = p \\ \quad\;\; y + cz = q \\ \qquad\quad\; z = z_0 \end{cases}$$

EXAMPLE

Solve the system of equations.

$$\begin{cases} \qquad 3y - \;\; z = 5 & (1) \\ 2x + \;\; y + \;\; z = 9 & (2) \\ \;\; x + 2y + 2z = 3 & (3) \end{cases}$$

Step 1 Rearrange the equations, if necessary, to obtain an x-term with a nonzero coefficient in the first equation. Then multiply the first equation by the reciprocal of the coefficient of the x-term to get 1 as a leading coefficient.

1. Interchange equations (1) and (3).

$$\begin{cases} \;\; x + 2y + 2z = 3 & (3) \\ 2x + \;\; y + \;\; z = 9 & (2) \\ \qquad 3y - \;\; z = 5 & (1) \end{cases}$$

The first equation (equation (3)) already has a leading coefficient of 1.

Step 2 By adding appropriate multiples of the first equation, eliminate any x-terms from the second and third equations. Then multiply the resulting second equation by the reciprocal of the coefficient of the y-term to get 1 as a leading coefficient.

2. To eliminate x from equation (2), add -2 times equation (3) to equation (2).

$$\begin{array}{rll} -2x - 4y - 4z = -6 & & -2(x + 2y + 3z = 3) \\ 2x + y + z = 9 & (2) & \\ \hline -3y - 3z = 3 & (4) & \text{Add.} \end{array}$$

$$\begin{cases} x + 2y + 2z = 3 & (3) \\ \qquad y + \;\; z = -1 & (5) & \text{Multiply equation (4) by } -\frac{1}{3}. \\ \qquad 3y - \;\; z = 5 & (1) \end{cases}$$

Step 3 Add an appropriate multiple of the second equation from Step 2 to eliminate any y-term from the third equation. Then solve the resulting equation for z to obtain an equivalent system in triangular form.

3. To eliminate y from equation (1), add -3 times equation (5) to equation (1).

$$\begin{array}{rll} -3y - 3z = 3 & & -3(y + z = -1) \\ 3y - z = 5 & (1) & \\ \hline -4z = 8 & & \text{Add.} \\ z = -2 & & \text{Solve for } z. \end{array}$$

Equivalent system in triangular form:

$$\begin{array}{rl} x + 2y + 2z = 3 & (3) \\ y + z = -1 & (5) \\ z = -2 & \end{array}$$

Step 4 Back-substitute the value of z from Step 3 into one of the equations containing only y and z in Step 3 and solve for y.

4.

$$\begin{array}{rl} y + z = -1 & \text{Equation (5)} \\ y + (-2) = -1 & \text{Substitute } z = -2. \\ y = 1 & \text{Solve for } y. \end{array}$$

(continued)

Step 5 Back-substitute the values of y and z from Steps 3 and 4 in any equation containing x, y, and z and solve for x.

5.
$$x + 2y + 2z = 3 \quad \text{Equation (3) from Step 1}$$
$$x + 2(1) + 2(-2) = 3 \quad \text{Substitute } y = 1, z = -2.$$
$$x = 5 \quad \text{Solve for } x.$$

Step 6 Write the solution set.

6. The solution set is $\{(5, 1, -2)\}$.

Step 7 Check your answer in the original equations.

7. **Check:** Verify the solution by substituting $x = 5$, $y = 1$, and $z = -2$ in each of the original equations.

$$3(1) - (-2) = 5 ✓ \quad (1)$$
$$2(5) + (1) + (-2) = 9 ✓ \quad (2)$$
$$5 + 2(1) + 2(-2) = 3 ✓ \quad (3)$$

Practice Problem 2 Solve the system.

$$\begin{cases} 2x + 5y \phantom{{}+2z} = 1 \\ x - 3y + 2z = 1 \\ -x + 2y + z = 7 \end{cases}$$

2 Classify systems as consistent and inconsistent.

Number of Solutions of a Linear System

As in the case of two variables, a system of linear equations in three variables may be **consistent** (having at least one solution) or **inconsistent** (having no solution). The equations in a consistent system may be **dependent** (the system has infinitely many solutions) or **independent** (the system has one solution).

INCONSISTENT SYSTEM

If in the process of converting a linear system to triangular form an equation of the form $0 = k$ occurs, where $k \neq 0$, then the system has no solution and is *inconsistent*.

EXAMPLE 3 Attempting to Solve a Linear System with No Solution

Solve the system of equations.

$$\begin{cases} x - y + 2z = 5 & (1) \\ 2x + y + z = 7 & (2) \\ 3x - 2y + 5z = 20 & (3) \end{cases}$$

Solution

Steps 1–2 To eliminate x from equation (2), add -2 times equation (1) to equation (2).

$$\begin{array}{rll} -2x + 2y - 4z = -10 & & -2(x - y + 2z = 5) \\ 2x + y + z = 7 & (2) & \\ \hline 3y - 3z = -3 & (4) & \text{Add.} \end{array}$$

Next, we eliminate x from equation (3) using equations (1) and (3).

$$\begin{array}{rll} -3x + 3y - 6z = -15 & & -3(x - y + 2z = 5) \\ 3x - 2y + 5z = 20 & (3) & \\ \hline y - z = 5 & (5) & \text{Add.} \end{array}$$

We now have the equivalent system:

$$\begin{cases} x - y + 2z = 5 & (1) \\ 3y - 3z = -3 & (4) \\ y - z = 5 & (5) \end{cases}$$

SIDE NOTE

An alternate method for Step 3 is to solve the 2×2 system of equations (4) and (5), then back-substitute into equation (1) to solve for the third variable.

Step 3 Multiply equation (4) by $\frac{1}{3}$ to obtain

$$y - z = -1 \quad (6)$$

We can eliminate y in equation (5) using equation (6).

$$\begin{array}{rll} -y + z = 1 & & -1(y - z = -1) \\ y - z = 5 & (5) & \\ \hline 0 = 6 & (7) & \text{Add.} \end{array}$$

We now have the system in triangular form:

$$\begin{cases} x - y + 2z = 5 & (1) \\ y - z = -1 & (6) \\ \underbrace{0 = 6}_{\text{False}} & (7) \end{cases}$$

This system is equivalent to the original system. Because equation (7) is false, we conclude that the solution set of the system is $\varnothing$ and the system is inconsistent.

Practice Problem 3 Solve the system of equations.

$$\begin{cases} 2x + 2y + 2z = 12 \\ -3x + y - 11z = -6 \\ 2x + y + 4z = -8 \end{cases}$$

DEPENDENT EQUATIONS

If in the process of converting a linear system to triangular form

(i) an equation of the form $0 = k (k \neq 0)$ *does not* occur but

(ii) an equation of the form $0 = 0$ *does* occur, then the system of equations has infinitely many solutions and the equations are *dependent*.

EXAMPLE 4 Solving a System with Infinitely Many Solutions

Solve the system of equations.

$$\begin{cases} x - y + z = 7 & (1) \\ 3x + 2y - 12z = 11 & (2) \\ 4x + y - 11z = 18 & (3) \end{cases}$$

Solution

Steps 1–2 Eliminate x using equations (1) and (2).

$$\begin{array}{rll} -3x + 3y - 3z = -21 & & -3(x - y + z = 7) \\ 3x + 2y - 12z = 11 & (2) & \\ \hline 5y - 15z = -10 & (4) & \text{Add.} \end{array}$$

Eliminate x using equations (1) and (3).

$$\begin{array}{rlr} -4x + 4y - 4z = -28 & & -4(x - y + z = 7) \\ 4x + y - 11z = 18 & (3) & \\ \hline 5y - 15z = -10 & (5) & \text{Add.} \end{array}$$

We now have the following equivalent system:

$$\begin{cases} x - y + z = 7 & (1) \\ 5y - 15z = -10 & (4) \\ 5y - 15z = -10 & (5) \end{cases}$$

Step 3 We eliminate y using equations (4) and (5).

$$\begin{array}{rlr} -5y + 15z = 10 & & -1(5y - 15z = -10) \\ 5y - 15z = -10 & (5) & \\ \hline 0 = 0 & (6) & \text{Add.} \end{array}$$

We finally have the system in triangular form:

$$\begin{cases} x - y + z = 7 & (1) \\ 5y - 15z = -10 & (4) \\ \underbrace{0 = 0}_{\text{True}} & (6) \end{cases}$$

The equation $0 = 0$ is equivalent to $0z = 0$, which is true for every value of z. Solving equation (4) for y, we have $y = 3z - 2$. Substituting $y = 3z - 2$ into equation (1) and solving for x, we obtain

$$\begin{array}{ll} x = y - z + 7 & \text{Solve equation (1) for } x. \\ x = (3z - 2) - z + 7 & \text{Substitute } y = 3z - 2. \\ x = 2z + 5 & \text{Simplify.} \end{array}$$

Thus, for every real number z, the triple $(x, y, z) = (2z + 5, 3z - 2, z)$ is a solution of the system. For example, when $z = 0$, the triple $(5, -2, 0)$ is a solution. When $z = 1$, the triple $(7, 1, 1)$ is a solution. Therefore, the given system has infinitely many solutions, one for each real number z, and the equations are dependent. The solution set for the system is $\{(2z + 5, 3z - 2, z)\}$.

Practice Problem 4 Solve the system of equations.

$$\begin{cases} x + y + z = 5 \\ -4x - y - 8z = -29 \\ 2x + 5y - 2z = 1 \end{cases}$$

3 Solve nonsquare systems.

Nonsquare Systems

Sometimes in a linear system, the number of equations is not the same as the number of variables. Such systems are called **nonsquare linear systems**.

EXAMPLE 5 **Solving a Nonsquare Linear System**

Solve the system of linear equations.

$$\begin{cases} x + 5y + 3z = 7 & (1) \\ 2x + 11y - 4z = 6 & (2) \end{cases}$$

Solution

Steps 1–2 We eliminate x using equations (1) and (2).

$$\begin{array}{rlr} -2x - 10y - 6z = -14 & & -2(x + 5y + 3z = 7) \\ 2x + 11y - 4z = 6 & (2) & \\ \hline y - 10z = -8 & (3) & \text{Add.} \end{array}$$

We obtain the equivalent system:

$$\begin{cases} x + 5y + 3z = 7 & (1) \\ \quad y - 10z = -8 & (3) \end{cases}$$

Steps 3–4 Because there is no third equation, Steps 3 and 4 are not needed.

Step 5 Solve equation (3) for y in terms of z to obtain

$$y = 10z - 8.$$

Step 6 Back-substitute $y = 10z - 8$ into equation (1) and solve for x.

$$\begin{aligned} x + 5y + 3z &= 7 && \text{Equation (1)} \\ x + 5(10z - 8) + 3z &= 7 && \text{Replace } y \text{ with } 10z - 8. \\ x &= -53z + 47 && \text{Solve for } x. \end{aligned}$$

Steps 5 and 6 mean that z determines both the x and y values. The system has infinitely many solutions, one for each real number z. The solution set is $\{(-53z + 47, 10z - 8, z)\}$.

Practice Problem 5 Solve the system of linear equations.

$$\begin{cases} x + 3y + 2z = 4 \\ 2x + 7y - z = 5 \end{cases}$$

4 Interpret linear systems in three variables geometrically.

Geometric Interpretation

The graph of a linear equation in three variables, such as $ax + by + cz = d$ (where a, b, and c are not all zero), is a plane in three-dimensional space. Figure 8.7 shows possible situations for a system of three linear equations in three variables.

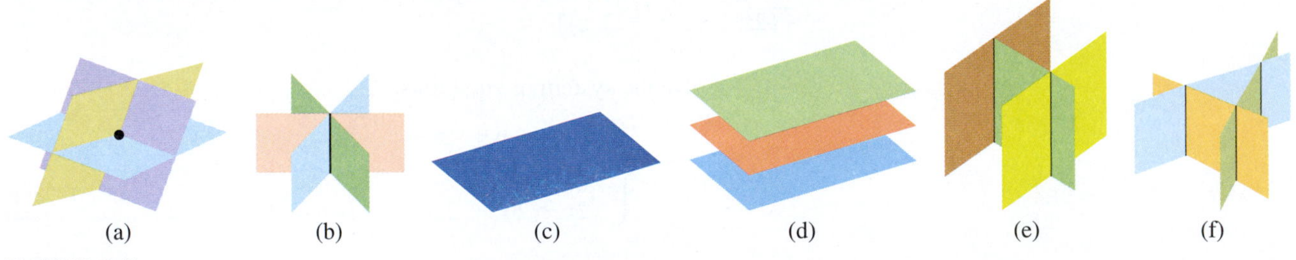

FIGURE 8.7

a. Three planes intersect in a single point. The system has only one solution.

b. Three planes intersect in one line. The system has infinitely many solutions.

c. Three planes coincide with each other. The system has infinitely many solutions.

d. There are three parallel planes. The system has no solution.

e. Two parallel planes are intersected by a third plane. The system has no solution.

f. Three planes have no point in common. The system has no solution.

5 Use linear systems in applications.

An Application to CAT Scans

CAT scanners report X-ray data in units called *linear attenuation units* (LAUs), also called *Hounsfield numbers*. The data are reported in such a way that if an X-ray passes through consecutive grid cells, the total weakening of the beam is the sum of the individual decreases. Table 8.1 gives the range of LAUs that determine whether the grid cell contains healthy tissue, tumorous tissue, a bone, or a metal.

TABLE 8.1 CAT Scanner Ranges

Type of Tissue	LAU Values
Healthy tissue	0.1625–0.2977
Tumorous tissue	0.2679–0.3930
Bone	0.38757–0.5108
Metal	1.54–∞

Reprinted with permission from *Mathematics Teacher,* May 1996. © 1996 by the National Council of Teachers of Mathematics.

EXAMPLE 6 A CAT Scan with Three Grid Cells

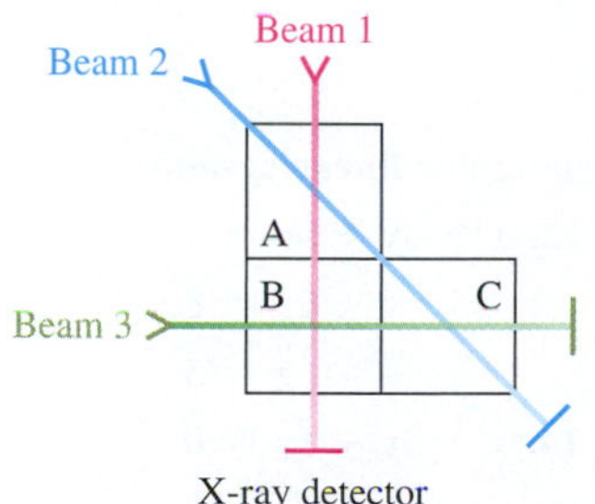

Three grid cells, three X-rays

FIGURE 8.8 Reprinted with permission from *Mathematics Teacher,* May 1996. © 1996 by the National Council of Teachers of Mathematics

Let A, B, and C be three grid cells as shown in Figure 8.8. A CAT scanner reports the following data for a patient named Monica.

(i) Beam 1 is weakened by 0.80 unit as it passes through grid cells A and B.
(ii) Beam 2 is weakened by 0.55 unit as it passes through grid cells A and C.
(iii) Beam 3 is weakened by 0.65 unit as it passes through grid cells B and C.

Use Table 8.1 to determine which grid cells contain each type of tissue listed.

Solution

Suppose grid cell A weakens the beam by x units, grid cell B weakens it by y units, and grid cell C weakens it by z units. Then we have the system:

$$\begin{cases} x + y = 0.80 & (1) \quad \text{(Beam 1)} \\ x + z = 0.55 & (2) \quad \text{(Beam 2)} \\ y + z = 0.65 & (3) \quad \text{(Beam 3)} \end{cases}$$

To solve this system of equations, we use the elimination procedure. Subtracting equation (1) from equation (2), we obtain the equivalent system:

$$\begin{cases} x + y = 0.80 & (1) \\ -y + z = -0.25 & (4) \quad \text{Subtract equation (1) from equation (2).} \\ y + z = 0.65 & (3) \end{cases}$$

Add equation (4) to equation (3) to get the system:

$$\begin{cases} x + y = 0.80 & (1) \\ -y + z = -0.25 & (4) \\ 2z = 0.40 & (5) \quad \text{Add equation (4) to equation (3).} \end{cases}$$

Multiply equation (5) by $\frac{1}{2}$ to obtain $z = 0.20$. Then back-substitute $z = 0.20$ in equation (4) to get $-y + 0.20 = -0.25$, or $y = 0.45$. Next, back-substitute $y = 0.45$ in equation (1) to get $x + 0.45 = 0.80$. Solving for x gives $x = 0.35$. Now referring to Table 8.1, we conclude that cell A contains tumorous tissue (because $x = 0.35$), cell B contains a bone (because $y = 0.45$), and cell C contains healthy tissue (because $z = 0.20$).

Practice Problem 6 In Example 6, suppose **(i)** Beam 1 is weakened by 0.65 unit as it passes through grid cells A and B, **(ii)** Beam 2 is weakened by 0.70 unit as it passes through grid cells A and C, and **(iii)** Beam 3 is weakened by 0.55 unit as it passes through grid cells B and C. Use Table 8.1 to determine which grid cells contain each type of tissue listed.

Answers to Practice Problems

1. Yes:
$$\begin{aligned} (2) + (-3) + (2) &= 1 \\ 3(2) + 4(-3) + (2) &= -4 \\ 2(2) + (-3) + 2(2) &= 5 \end{aligned}$$

2. $\{(-2, 1, 3)\}$ **3.** $\varnothing$ **4.** $\left\{\left(8 - \frac{7}{3}z, \frac{4}{3}z - 3, z\right)\right\}$

5. $\{(13 - 17z, -3 + 5z, z)\}$ **6.** Cell A contains bone (because $x = 0.4$), cell B contains healthy tissue (because $y = 0.25$), and cell C contains tumorous tissue (because $z = 0.3$).

SECTION 8.2 Exercises

Basic Concepts and Skills

1. Systems of equations that have the same solution set are called ____________ systems.
2. Adding a(n) ____________ multiple of one equation to another equation in the system produces an equivalent system.
3. If any of the equations in a system has no solution, then the system is ____________.
4. If the number of equations is different from the number of variables in a linear system, then the system is called a(n) ____________ system.
5. **True or False.** The graph of the equation $x - 2y + z = 1$ is a plane in three-dimensional space.
6. **True or False.** If $\{(2z, 1 - z, z + 3)\}$ is the solution set of a linear system in three variables, then $(0, 1, 3)$ is a solution of this system.

In Exercises 7–10, determine whether the given ordered triple (x, y, z) is a solution of the given linear system.

7. $\begin{cases} 2x - 2y - 3z = 1 \\ 3y + 2z = -1 \\ y + z = 0 \end{cases}$ $(1, -1, 1)$

8. $\begin{cases} 3x - 2y + z = 2 \\ y + z = 5 \\ x + 3z = 8 \end{cases}$ $(2, 3, 2)$

9. $\begin{cases} x + 3y - 2z = 0 \\ 2x - y + 4z = 5 \\ x - 11y + 14z = 0 \end{cases}$ $(-10, 8, 7)$

10. $\begin{cases} x - 4y + 7z = 14 \\ 3x + 8y - 2z = 13 \\ 7x - 8y + 26z = 5 \end{cases}$ $(4, 1, 2)$

In Exercises 11–14, solve each triangular linear system.

11. $\begin{aligned} x + y + z &= 4 \\ y - 2z &= 4 \\ z &= -1 \end{aligned}$

12. $\begin{aligned} x + 3y + z &= 3 \\ y + 2z &= 8 \\ z &= 5 \end{aligned}$

13. $\begin{aligned} x - 5y + 3z &= -1 \\ y - 2z &= -6 \\ z &= 4 \end{aligned}$

14. $\begin{aligned} x + 3y + 5z &= 0 \\ y + 7z &= 2 \\ z &= \frac{1}{2} \end{aligned}$

15. In the system in Exercise 7, interchange equations (2) and (3). Write the new equivalent system. In the new system, eliminate y from the last equation. Convert the system to triangular form.
16. In Exercise 8, interchange equations (1) and (3). In the new equivalent system, eliminate x from the last equation. Convert the system to triangular form.
17. Convert the system of Exercise 9 to triangular form.
18. Convert the system of Exercise 10 to triangular form.

In Exercises 19–22, convert the system to triangular form and then use back-substitution to solve the system.

19. $\begin{cases} x - y + z = 6 \\ 2y + 3z = 5 \\ 2z = 6 \end{cases}$

20. $\begin{cases} 4x + 5y + 2z = -3 \\ 3y - z = 14 \\ -3z = 15 \end{cases}$

21. $\begin{cases} 4x + 4y + 4z = 7 \\ 3x - 8y = 14 \\ 4z = -1 \end{cases}$

22. $\begin{cases} 5x + 10y + 10z = 0 \\ 2y + 3z = -0.6 \\ 4z = 1.6 \end{cases}$

In Exercises 23–44, find the solution set of each linear system. Identify inconsistent systems and dependent equations. For dependent equations, write your answer as in Example 4.

23. $\begin{cases} x + y + z = 6 \\ x - y + z = 2 \\ 2x + y - z = 1 \end{cases}$

24. $\begin{cases} x + y + z = 6 \\ 2x + 3y - z = 5 \\ 3x - 2y + 3z = 8 \end{cases}$

25. $\begin{cases} 2x + 3y + z = 9 \\ x + 2y + 3z = 6 \\ 3x + y + 2z = 8 \end{cases}$

26. $\begin{cases} 4x + 2y + 3z = 6 \\ x + 2y + 2z = 1 \\ 2x - y + z = -1 \end{cases}$

27. $\begin{cases} x + y - 2z = 5 \\ 2x - y - z = -4 \\ x - 2y + z = -2 \end{cases}$

28. $\begin{cases} x + y + 4z = 6 \\ x + 2y - 2z = 8 \\ 7x + 10y + 10z = 60 \end{cases}$

29. $\begin{cases} 2x + 3y + 2z = 7 \\ x + 3y - z = -2 \\ x - y + 2z = 8 \end{cases}$

30. $\begin{cases} x - y + 2z = 3 \\ 2x + 2y + z = 3 \\ x + y + 3z = 4 \end{cases}$

31. $\begin{cases} 4x - 2y + z = 5 \\ 2x + y - 2z = 4 \\ x + 3y - 2z = 6 \end{cases}$

32. $\begin{cases} x - 3y + 2z = 9 \\ 2x + 4y - 3z = -9 \\ 3x - 2y + 5z = 12 \end{cases}$

33. $\begin{cases} 2x + y + z = 6 \\ x + y - z = 1 \\ x + y + 2z = 4 \end{cases}$

34. $\begin{cases} 2x + y - 3z = 7 \\ x - y - 2z = 4 \\ 3x + 3y + 2z = 4 \end{cases}$

35. $\begin{cases} x - y - z = 3 \\ x + 9y + z = 3 \\ 2x + 3y - z = 6 \end{cases}$

36. $\begin{cases} x + y - z = 2 \\ 3x - y - z = 10 \\ 3x + y - 2z = 8 \end{cases}$

37. $\begin{cases} x + y = 0 \\ y + 2z = -4 \\ y + z = 4 - x \end{cases}$

38. $\begin{cases} 2x + 4y + 3z = 6 \\ x + 2z = -1 \\ x - 2y + z = -5 \end{cases}$

39. $\begin{cases} x + y + z = 6 \\ x + 2y + 3z = 14 \\ x + 4y + 7z = 30 \end{cases}$

40. $\begin{cases} x - y + z = 3 \\ 2x + y - z = 2 \\ x + 2y - 2z = -1 \end{cases}$

41. $\begin{cases} 3x - 2z = 11 \\ 2x + y = 8 \\ 2y + 3z = 1 \end{cases}$

42. $\begin{cases} 2x + y = 4 \\ x + 2z = 3 \\ 3y - z = 5 \end{cases}$

43. $\begin{cases} 2x + 6y + 11 = 0 \\ 6y - 18z + 1 = 0 \end{cases}$

44. $\begin{cases} 3x + 5y - 15 = 0 \\ 6x + 20y - 6z = 11 \end{cases}$

Applying the Concepts

In Exercises 45–52, use a system of equations to solve each problem.

45. Investment. Miguel invested \$20,000 in three different funds that paid 4%, 5%, and 6% for the year. The total income for the year from the three funds was \$1060. The income from the 6% fund was twice the income from the 5% fund. What amount was invested in each fund?

46. Number problem. The sum of the digits in a three-digit number is 14. The sum of the hundreds digit and the units digit is equal to the tens digit. If the hundreds digit and the units digit are interchanged, the number is increased by 297. What is the original number?

47. Election campaign. Alex, Becky, and Courtney volunteered to stuff 741 envelopes with newsletters for Senator Douglas's reelection campaign. Alex could assemble 124 per hour; Becky, 118 per hour; and Courtney, 132 per hour. They worked a total of 6 hours. The sum of the number of hours that Becky and Courtney spent was twice what Alex spent. How long did each of them work?

48. Age of students. A college algebra class of 38 students at Central State College was made up of people who were 18, 19, and 20 years of age. The average of their ages was 18.5 years. How many of each age were in the class if the number of 18-year-olds was eight more than the combined number of 19- and 20-year-olds?

49. Coins in a machine. A vending machine's coin box contains nickels, dimes, and quarters. The total number of coins in the box is 300. The number of dimes is three times the number of nickels and quarters together. If the box contains \$30.05, find the number of nickels, dimes, and quarters that it contains.

50. Sports. The Wildcats scored 46 points in a football game. Twice the number of points resulting from the sum of field goals and extra points equals two more than the number of points from touchdowns. Five times the number of points scored by field goals equals twice the number of points from touchdowns. Find the number of points resulting from touchdowns, field goals, and extra points. [A touchdown = 6 points, a field goal = 3 points, and an extra point = 1 point.]

51. Weekly wage. Amy worked 53 hours one week and was paid at three different rates. She earned \$7.40 per hour for her normal daytime work, \$9.20 per hour for night work, and \$11.75 per hour for holiday work. If her total gross wages for the week were \$452.20 and the number of regular daytime hours she worked exceeded the combined hours of night and holiday work by 9 hours, how many hours of each category of work did Amy perform?

52. Components of a product. A manufacturer buys three components—*A*, *B*, and *C*—for use in making a toaster. She used as many units of *A* as she did of *B* and *C* combined. The cost of *A*, *B*, and *C* is \$4, \$5, and \$6 per unit, respectively. If she purchased 100 units of these components for a total of \$480, how many units of each did she purchase?

In Exercises 53–56, use Table 8.1 and Figure 8.8 on page 747 to determine which grid cells of the patients contain healthy tissue, tumorous tissue, bone, or metal.

Beam decrease, in LAUs

Patient	Beam 1	Beam 2	Beam 3
53. Vicky	0.54	0.40	0.52
54. Yolanda	0.65	0.80	0.75
55. Nina	0.51	0.49	0.44
56. Srinivasan	0.44	2.21	2.23

Beyond the Basics

In Exercises 57–60, write a linear equation of the form $x + by + cz = d$ that is satisfied by all three of the given ordered triples.

57. $(1, 0, 0), (0, 1, 0), (0, 0, 1)$

58. $\left(\frac{1}{3}, 0, 0\right), (0, 4, 3), (1, 2, 2)$

59. $(3, -4, 0), \left(0, \frac{1}{4}, \frac{1}{2}\right), (1, 1, -4)$

60. $(0, 1, -10), \left(\frac{1}{8}, 0, \frac{1}{4}\right), \left(1, \frac{1}{3}, -2\right)$

In Exercises 61–64, find an equation of the parabola of the form $y = ax^2 + bx + c$ that passes through the three given points.

61. $(0, 1), (-1, 0), (1, 4)$

62. $(0, 2), (-1, 30), (2, 6)$

63. $(1, 2), (-1, 4), (2, 4)$

64. $(0, 3), (-1, 4), (1, 6)$

In Exercises 65–68, find an equation of the circle of the form $x^2 + y^2 + ax + by + c = 0$ that passes through the three given points.

65. $(0, 4), (2\sqrt{2}, 2\sqrt{2}), (-4, 0)$

66. $(0, 3), (0, -1), (\sqrt{3}, 2)$

67. $(1, 2), (6, -3), (4, 1)$

68. $(5, 6), (-1, 6), (3, 2)$

In Exercises 69 and 70, solve the system of equations.

69. $$\begin{cases} \frac{1}{x} + \frac{3}{y} - \frac{1}{z} = 5 \\ \frac{2}{x} + \frac{4}{y} + \frac{6}{z} = 4 \\ \frac{2}{x} + \frac{3}{y} + \frac{1}{z} = 3 \end{cases}$$

70. $$\begin{cases} \frac{1}{x} + \frac{2}{y} + \frac{3}{z} = 8 \\ \frac{2}{x} + \frac{5}{y} + \frac{9}{z} = 16 \\ \frac{3}{x} - \frac{4}{y} - \frac{5}{z} = 32 \end{cases}$$

$\left[\textit{Hint: } \text{Let } \frac{1}{x} = u, \frac{1}{y} = v, \text{ and } \frac{1}{z} = w. \text{ Solve for } u, v, \text{ and } w.\right]$

In Exercises 71 and 72, find the value of c for which the given system has a unique solution.

71. $$\begin{cases} x + 2y - 5z - 9 = 0 \\ 3x - y + 2z - 14 = 0 \\ 2x + 3y - z - 3 = 0 \\ cx - 5y + z + 3 = 0 \end{cases}$$

72. $$\begin{cases} x + 5y + z = 42 \\ 3x + y - 3z = c \\ x - y + 3z = 4 \\ x + y - z = 0 \end{cases}$$

73. Find a quadratic function of the form $y = ax^2 + bx + c$ whose graph passes through the points $(-1, -1)$, $(0, 5)$, and $(2, 5)$.

74. Use Table 8.1 to investigate four grid cells, arranged in a square, as shown in the given figure. Figures (a) and (b) reprinted with permission from *Mathematics Teacher,* May 1996. © 1996 by National Council of Teachers of Mathematics.

a. For Figure (a), the following data were observed.

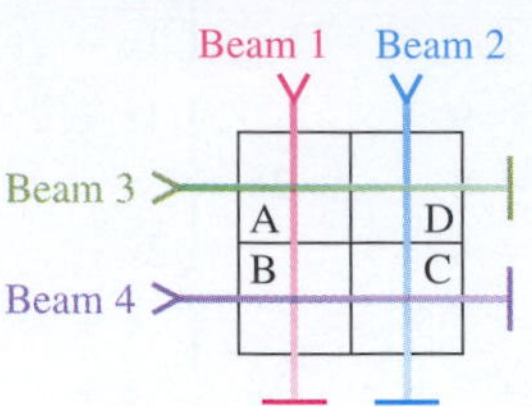

Four grid cells and four X-rays

(a)

1. Beam 1 decreased by 0.60 unit.
2. Beam 2 decreased by 0.75 unit.
3. Beam 3 decreased by 0.65 unit.
4. Beam 4 decreased by 0.70 unit.

Is there sufficient information in Figure (a) to determine which grid cells may contain tumorous tissue?

b. From two additional X-rays, as shown in Figure (b), the following data were observed.

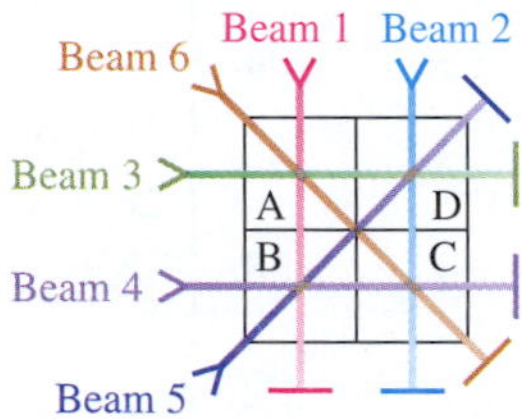

Four grid cells and six X-rays

(b)

5. Beam 5 decreased by 0.85 unit.
6. Beam 6 decreased by 0.50 unit.

Show that the six X-rays represented in the two figures are sufficient to locate tumors in this situation but that, in fact, it is not necessary to use all six rays.

Critical Thinking/Discussion/Writing

75. Write a linear system of three equations in the variables x, y, and z that has $\{(1, -1, 2)\}$ as its solution set.

76. Write a linear system of three equations in the variables x, y, and z such that

a. The system is inconsistent.

b. The system has infinitely many solutions.

Maintaining Skills

In Exercises 77–80, solve each equation for the specified variable.

77. $\dfrac{1}{2 \cdot 3} = \dfrac{1}{2} + \dfrac{B}{3}$, for B

78. $\dfrac{2}{5 \cdot 7} = \dfrac{A}{5} - \dfrac{1}{7}$, for A

79. $\dfrac{1}{n \cdot (n + 1)} = \dfrac{A}{n} - \dfrac{1}{n + 1}$, for A

80. $\dfrac{4}{n \cdot (n + 2)} = \dfrac{2}{n} + \dfrac{B}{n + 2}$, for B

81. Find the values of A and B for which the equation

$$2x + 3 = A(x + 3) + B(x - 1)$$

is an identity (true for all values of x). [*Hint:* Evaluate both sides at $x = 1$ and $x = -3$]

82. Find the values of A and B for which the equation

$$3x + 2 = A(x - 2) + B(x + 2)$$

is an identity

In Exercises 83–88, factor each expression.

83. $x^2 + 5x + 6$

84. $x^2 - 3x - 10$

85. $2x^2 + 5x - 3$

86. $3x^2 + x - 2$

87. $4x^2 - 9$

88. $x^3 - 8$

SECTION 8.3

Partial–Fraction Decomposition

Georg Simon Ohm (1787–1854) Georg Ohm was born in Erlangen, Germany. His father, Johann, a mechanic and a self-educated man, was interested in philosophy and mathematics and gave his children an excellent education in physics, chemistry, mathematics, and philosophy through his own teachings. The Ohm's law named in Georg Ohm's honor appeared in his famous pamphlet *Die galvanische Kette, mathematisch bearbeitet* (1827). Ohm's work was not appreciated in Germany, where it was first published. Eventually, it was recognized, first by the British Royal Society, which awarded Ohm the Copley Medal in 1841. In 1849, Ohm became a curator of the Bavarian Academy's science museum and began to lecture at the University of Munich. Finally, in 1852 (two years before his death), Ohm achieved his lifelong ambition of becoming chair of physics at the University of Munich.

BEFORE STARTING THIS SECTION, REVIEW

1. Division of polynomials (Section 3.3)
2. Factoring polynomials (Section P.4)
3. Irreducible polynomials (Section P.4, page 39)
4. Rational expressions (Section P.5, page 47)

OBJECTIVES

1. Become familiar with partial-fraction decomposition.
2. Decompose $\dfrac{P(x)}{Q(x)}$ when $Q(x)$ has only distinct linear factors.
3. Decompose $\dfrac{P(x)}{Q(x)}$ when $Q(x)$ has repeated linear factors.
4. Decompose $\dfrac{P(x)}{Q(x)}$ when $Q(x)$ has distinct irreducible quadratic factors.
5. Decompose $\dfrac{P(x)}{Q(x)}$ when $Q(x)$ has repeated irreducible quadratic factors.

Ohm's Law

If a voltage is applied across a resistor (such as a wire), a current will flow. Ohm's law states that in a circuit at constant temperature,

$$I = \frac{V}{R}$$

where V is the voltage (measured in volts), I is the current (measured in amperes), and R is the resistance (measured in ohms). A **parallel circuit** consists of two or more resistors connected as shown in Figure 8.9. In the figure, the current I from the source divides into two currents, I_1 and I_2, with I_1 going through one resistor and I_2 going through the other, so that $I = I_1 + I_2$. Suppose we want to find the total resistance R in the circuit due to the two resistors R_1 and R_2 connected in parallel. Ohm's law can be used to show that in the parallel circuit in Figure 8.9, $\frac{1}{R} = \frac{1}{R_1} + \frac{1}{R_2}$. This result is called the *parallel property of resistors.* In Example 7, we examine the parallel property of resistors by using partial fractions.

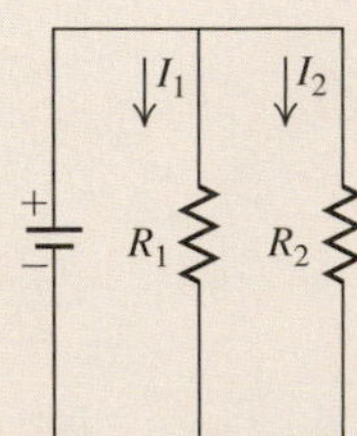

FIGURE 8.9 Parallel resistors

1 Become familiar with partial–fraction decomposition.

Partial Fractions

Recall that to add two rational expressions such as $\frac{2}{x+3}$ and $\frac{3}{x-1}$, required you to find the least common denominator, rewrite expressions with the same denominator, and then: add the corresponding numerators. This procedure gives

$$\frac{2}{x+3}+\frac{3}{x-1}=\frac{2(x-1)}{(x+3)(x-1)}+\frac{3(x+3)}{(x-1)(x+3)}=\frac{5x+7}{(x+3)(x-1)}$$

In some applications of algebra and more advanced mathematics, you need a *reverse* procedure for *splitting* a fraction such as $\frac{5x+7}{(x-1)(x+3)}$ into the simpler fractions to obtain:

$$\frac{5x+7}{(x+3)(x-1)}=\frac{2}{x+3}+\frac{3}{x-1}$$

Each of the two fractions on the right is called a **partial fraction**. Their sum is called the **partial-fraction decomposition** of the rational expression on the left.

A rational expression $\frac{P(x)}{Q(x)}$ is called **improper** if the degree $P(x) \geq$ degree $Q(x)$ and is called **proper** if the degree $P(x) <$ degree $Q(x)$. Examples of improper rational expressions are

$$\frac{x^3}{x^2-1},\quad \frac{(x+2)(x-1)}{(x-2)(x+3)},\quad \text{and} \quad \frac{x^2+x+1}{2x+3}.$$

Using long division, we can express an improper rational fraction $\frac{P(x)}{Q(x)}$ as the sum of a polynomial and a proper rational fraction:

$$\underset{\substack{\uparrow\\ \text{Improper}\\ \text{Fraction}}}{\frac{P(x)}{Q(x)}} = \underset{\substack{\uparrow\\ \text{Polynomial}}}{S(x)} + \underset{\substack{\uparrow\\ \text{Proper}\\ \text{Fraction}}}{\frac{R(x)}{Q(x)}}$$

RECALL

Any polynomial $Q(x)$ with real coefficients can be factored so that each factor is linear or is an irreducible quadratic factor.

We therefore restrict our discussion of the decomposition of $\frac{P(x)}{Q(x)}$ into partial fractions to cases involving proper rational fractions. We also assume that $P(x)$ and $Q(x)$ have no common factor.

The problem of decomposing a proper fraction $\frac{P(x)}{Q(x)}$ into partial fractions depends on the type of factors in the denominator $Q(x)$. In this section, we consider four cases for $Q(x)$.

2 Decompose $\frac{P(x)}{Q(x)}$ when $Q(x)$ has only distinct linear factors.

$Q(x)$ Has Only Distinct Linear Factors

CASE 1: THE DENOMINATOR IS THE PRODUCT OF DISTINCT (NONREPEATED) LINEAR FACTORS

Suppose $Q(x)$ can be factored as

$$Q(x) = c(x - a_1)(x - a_2)\cdots(x - a_n),$$

with no factor repeated. The partial-fraction decomposition of $\frac{P(x)}{Q(x)}$ is of the form

$$\frac{P(x)}{Q(x)} = \frac{A_1}{x - a_1} + \frac{A_2}{x - a_2} + \cdots + \frac{A_n}{x - a_n},$$

where $A_1, A_2, \ldots, A_n$ are constants to be determined.

We can find the constants $A_1, A_2, \ldots, A_n$ by using the following procedure. Note that if the number of constants is small, we use the letters $A, B, C, \ldots,$ instead of $A_1, A_2, A_3, \ldots$.

PROCEDURE **IN ACTION**

EXAMPLE 1 **Partial-Fraction Decomposition**

OBJECTIVE	EXAMPLE
Find the partial-fraction decomposition of a rational expression.	*Find the partial-fraction decomposition of* $\frac{3x + 26}{(x - 3)(x + 4)}$.
Step 1 Write the form of the partial-fraction decomposition with the unknown constants $A, B, C, \ldots$ in the numerators of the decomposition.	**1.** $\frac{3x + 26}{(x - 3)(x + 4)} = \frac{A}{x - 3} + \frac{B}{x + 4}$
Step 2 Multiply both sides of the equation in Step 1 by the original denominator. Use the distributive property and eliminate common factors. Simplify.	**2.** Multiply both sides by $(x - 3)(x + 4)$ and simplify to obtain $3x + 26 = (x + 4)A + (x - 3)B.$
Step 3 Write both sides of the equation in Step 2 in descending powers of x and equate the coefficients of like powers of x.	**3.** $3x + 26 = (A + B)x + (4A - 3B)$ $\begin{cases} 3 = A + B \\ 26 = 4A - 3B \end{cases}$ Equate coefficients of x. Equate constant coefficients.
Step 4 Solve the linear system resulting from Step 3 for the constants $A, B, C, \ldots$.	**4.** Solving the system of equations in Step 3, we obtain $A = 5$ and $B = -2$.
Step 5 Substitute the values you found for $A, B, C, \ldots$ into the equation in Step 1 and write the partial-fraction decomposition.	**5.** $\frac{3x + 26}{(x - 3)(x + 4)} = \frac{5}{x - 3} + \frac{-2}{x + 4}$ Replace A with 5 and B with -2 in Step 1. $\frac{3x + 26}{(x - 3)(x + 4)} = \frac{5}{x - 3} - \frac{2}{x + 4}$

Practice Problem 1 Find the partial-fraction decomposition of $\frac{2x - 7}{(x + 1)(x - 2)}$.

EXAMPLE 2 **Finding the Partial-Fraction Decomposition When the Denominator Has Only Distinct Linear Factors**

Find the partial-fraction decomposition of the expression.

$$\frac{10x - 4}{x^3 - 4x}$$

Solution

First, factor the denominator.

$$\begin{aligned} x^3 - 4x &= x(x^2 - 4) && \text{Distributive property} \\ &= x(x - 2)(x + 2) && x^2 - 4 = (x - 2)(x + 2) \end{aligned}$$

Each of the factors x, $x - 2$, and $x + 2$ becomes a denominator for a partial fraction.

Step 1 The partial-fraction decomposition is given by

$$\frac{10x - 4}{x(x - 2)(x + 2)} = \frac{A}{x} + \frac{B}{x - 2} + \frac{C}{x + 2}$$ Use A, B, and C instead of A_1, A_2, and A_3.

Step 2 Multiply both sides by the common denominator $x(x - 2)(x + 2)$.

Distribute

$$\cancel{x(x-2)(x+2)}\left[\frac{10x - 4}{\cancel{x(x-2)(x+2)}}\right] = x(x - 2)(x + 2)\left[\frac{A}{x} + \frac{B}{x - 2} + \frac{C}{x + 2}\right]$$

$$\begin{aligned} 10x - 4 &= A(x - 2)(x + 2) + Bx(x + 2) + Cx(x - 2) \quad (1) && \text{Distributive property; simplify.} \\ &= A(x^2 - 4) + B(x^2 + 2x) + C(x^2 - 2x) && \text{Multiply.} \\ &= Ax^2 - 4A + Bx^2 + 2Bx + Cx^2 - 2Cx && \text{Distributive property} \\ &= (A + B + C)x^2 + (2B - 2C)x - 4A. && \text{Combine like terms.} \end{aligned}$$

Step 3 Now use the fact that *two equal polynomials have equal corresponding coefficients*. Writing $10x - 4 = 0x^2 + 10x - 4$, we have

$$0x^2 + 10x - 4 = (A + B + C)x^2 + (2B - 2C)x - 4A.$$

Equating corresponding coefficients leads to the system of equations.

$$\begin{cases} A + B + C = 0 & \text{Equate coefficients of } x^2. \\ 2B - 2C = 10 & \text{Equate coefficients of } x. \\ -4A = -4 & \text{Equate constant coefficients.} \end{cases}$$

Step 4 Solve the system of equations in Step 3 to obtain $A = 1$, $B = 2$, and $C = -3$. See Exercise 57.

Step 5 The partial-fraction decomposition is

$$\frac{10x - 4}{x^3 - 4x} = \frac{1}{x} + \frac{2}{x - 2} + \frac{-3}{x + 2} = \frac{1}{x} + \frac{2}{x - 2} - \frac{3}{x + 2}.$$

Alternative Solution of Example 2

In Example 2, to find the constants A, B, and C, we *equated coefficients* of like powers of x and solved the resulting system. An alternative (and sometimes quicker) method is to *substitute* well-chosen values for x in equation (1) found in Step 2:

$$10x - 4 = A(x - 2)(x + 2) + Bx(x + 2) + Cx(x - 2) \quad (1)$$

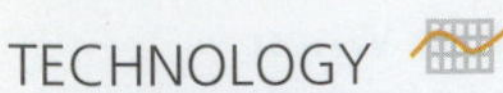

TECHNOLOGY CONNECTION

You can reinforce your partial-fraction connection graphically: The graph of

$$Y_1 = \frac{10x - 4}{x^3 - 4x}$$

appears to be identical to the graph of

$$Y_2 = \frac{1}{x} + \frac{2}{x-2} - \frac{3}{x+2}.$$

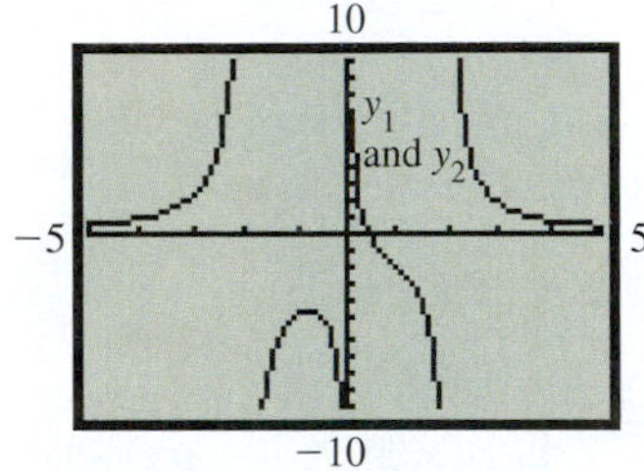

The graphs of Y_1 and Y_2 are the same.

Substitute $x = 2$ in equation (1) to cause the terms containing A and C to be 0.

$$10(2) - 4 = A(2-2)(2+2) + B(2)(2+2) + C(2)(2-2)$$
$$16 = 8B \quad \text{Simplify.}$$
$$2 = B \quad \text{Solve for } B.$$

Now substitute $x = -2$ in equation (1) to get

$$10(-2) - 4 = A(-2-2)(-2+2) + B(-2)(-2+2) + C(-2)(-2-2)$$
$$-24 = 8C \quad \text{Simplify.}$$
$$-3 = C \quad \text{Solve for } C.$$

Finally, substitute $x = 0$ in equation (1) to get

$$10(0) - 4 = A(0-2)(0+2) + B(0)(0+2) + C(0)(0-2)$$
$$-4 = -4A \quad \text{Simplify.}$$
$$1 = A \quad \text{Solve for } A.$$

Because we found the same values for A, B, and C, the partial-fraction decomposition will also be the same.

Practice Problem 2 Find the partial-fraction decomposition of

$$\frac{3x^2 + 4x + 3}{x^3 - x}.$$

$Q(x)$ Has Repeated Linear Factors

3 Decompose $\frac{P(x)}{Q(x)}$ when $Q(x)$ has repeated linear factors.

CASE 2: THE DENOMINATOR HAS A REPEATED LINEAR FACTOR

Let $(x - a)^m$ be the linear factor $(x - a)$ that is repeated m times in $Q(x)$. Then the portion of the partial-fraction decomposition of $\frac{P(x)}{Q(x)}$ that corresponds to the factor $(x - a)^m$ is

$$\frac{A_1}{x-a} + \frac{A_2}{(x-a)^2} + \cdots + \frac{A_m}{(x-a)^m},$$

where $A_1, A_2, \ldots, A_m$ are constants.

EXAMPLE 3 **Finding the Partial-Fraction Decomposition When the Denominator Has Repeated Linear Factors**

Find the partial-fraction decomposition of: $\dfrac{x+4}{(x+3)(x-1)^2}$

Solution

Step 1 The linear factor $(x - 1)$ is repeated twice, and the factor $(x + 3)$ is nonrepeating. So the partial-fraction decomposition has the form

$$\frac{x+4}{(x+3)(x-1)^2} = \frac{A}{x+3} + \frac{B}{x-1} + \frac{C}{(x-1)^2}.$$

Step 2–4 Multiply both sides of the equation in Step 1 by the original denominator, $(x + 3)(x - 1)^2$; then use the distributive property on the right side and simplify to get

$$x + 4 = A(x-1)^2 + B(x+3)(x-1) + C(x+3).$$

Substitute $x = 1$ in the last equation to obtain

$$1 + 4 = A(1 - 1)^2 + B(1 + 3)(1 - 1) + C(1 + 3).$$

This simplifies to $5 = 4C$, or $C = \frac{5}{4}$.

To find A, we substitute $x = -3$ to get

$$-3 + 4 = A(-3 - 1)^2 + B(-3 + 3)(-3 - 1) + C(-3 + 3).$$

This simplifies to $1 = 16A$, or $A = \frac{1}{16}$.

To find B, we replace x with any convenient number, say, 0. We have

$$0 + 4 = A(0 - 1)^2 + B(0 + 3)(0 - 1) + C(0 + 3)$$
$$4 = A - 3B + 3C \quad \text{Simplify.}$$

Now replace A with $\frac{1}{16}$ and C with $\frac{5}{4}$ in the last equation.

$$4 = \frac{1}{16} - 3B + 3\left(\frac{5}{4}\right)$$
$$64 = 1 - 48B + 60 \quad \text{Multiply both sides by 16.}$$
$$B = -\frac{1}{16} \quad \text{Solve for } B.$$

We have $A = \frac{1}{16}$, $B = -\frac{1}{16}$, and $C = \frac{5}{4}$.

Step 5 Substitute $A = \frac{1}{16}$, $B = -\frac{1}{16}$, and $C = \frac{5}{4}$ in the decomposition in Step 1 to get

$$\frac{x + 4}{(x + 3)(x - 1)^2} = \frac{\frac{1}{16}}{x + 3} + \frac{-\frac{1}{16}}{x - 1} + \frac{\frac{5}{4}}{(x - 1)^2},$$

or

$$\frac{x + 4}{(x + 3)(x - 1)^2} = \frac{1}{16(x + 3)} - \frac{1}{16(x - 1)} + \frac{5}{4(x - 1)^2}.$$

Practice Problem 3 Find the partial-fraction decomposition of

$$\frac{x + 5}{x(x - 1)^2}.$$

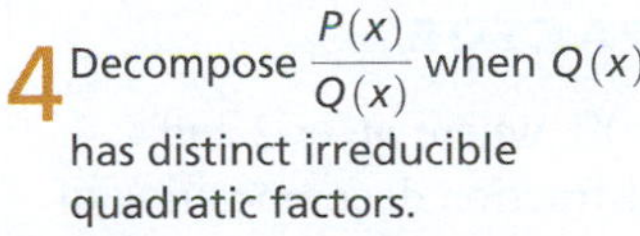
4 Decompose $\frac{P(x)}{Q(x)}$ when $Q(x)$ has distinct irreducible quadratic factors.

$Q(x)$ Has Distinct Irreducible Quadratic Factors

CASE 3: THE DENOMINATOR HAS A NONREPEATED IRREDUCIBLE QUADRATIC FACTOR

Suppose $ax^2 + bx + c$ is an irreducible quadratic factor of $Q(x)$. Then the portion of the partial-fraction decomposition of $\frac{P(x)}{Q(x)}$ that corresponds to $ax^2 + bx + c$ has the form

$$\frac{Ax + B}{ax^2 + bx + c}.$$

RECALL

The expression $ax^2 + bx + c$ is irreducible if $b^2 - 4ac < 0$. In that case, $ax^2 + bx + c = 0$ has no real solutions.

EXAMPLE 4 **Finding the Partial-Fraction Decomposition When the Denominator Has a Distinct Irreducible Quadratic Factor**

Find the partial-fraction decomposition of $\dfrac{3x^2 - 8x + 1}{(x - 4)(x^2 + 1)}$.

Solution

Step 1 The factor $x - 4$ is linear, and $x^2 + 1$ is irreducible; so the partial-fraction decomposition has the form

$$\frac{3x^2 - 8x + 1}{(x - 4)(x^2 + 1)} = \frac{A}{x - 4} + \frac{Bx + C}{x^2 + 1}.$$

Step 2 Multiply both sides of the decomposition in Step 1 by the original denominator, $(x - 4)(x^2 + 1)$, and simplify to get

$$3x^2 - 8x + 1 = A(x^2 + 1) + (Bx + C)(x - 4).$$

Substitute $x = 4$ to find $17 = 17A$, or $A = 1$.

Step 3 Collect like terms and write both sides of the equation in Step 2 in descending powers of x.

$$3x^2 - 8x + 1 = (A + B)x^2 + (-4B + C)x + (A - 4C)$$

Equate corresponding coefficients to get

$$\begin{cases} A + B = 3 & (1) \quad \text{Equate the coefficients of } x^2. \\ -4B + C = -8 & (2) \quad \text{Equate the coefficients of } x. \\ A - 4C = 1 & (3). \quad \text{Equate the constant coefficients.} \end{cases}$$

Step 4 Substitute $A = 1$ from Step 2 in equation (1) to get $1 + B = 3$, or $B = 2$. Substitute $A = 1$ in equation (3) to get $1 - 4C = 1$, or $C = 0$.

Step 5 Substitute $A = 1$, $B = 2$, and $C = 0$ into the decomposition in Step 1 to get

$$\frac{3x^2 - 8x + 1}{(x - 4)(x^2 + 1)} = \frac{1}{x - 4} + \frac{2x}{x^2 + 1}.$$

Practice Problem 4 Find the partial-fraction decomposition of

$$\frac{3x^2 + 5x - 2}{x(x^2 + 2)}.$$

5 Decompose $\dfrac{P(x)}{Q(x)}$ when $Q(x)$ has repeated irreducible quadratic factors.

$Q(x)$ Has Repeated Irreducible Quadratic Factors

CASE 4: THE DENOMINATOR HAS A REPEATED IRREDUCIBLE QUADRATIC FACTOR

Suppose the denominator $Q(x)$ has a factor $(ax^2 + bx + c)^m$, where $m \geq 2$ and $ax^2 + bx + c$ is irreducible. Then the portion of the partial-fraction decomposition of $\dfrac{P(x)}{Q(x)}$ that corresponds to the factor $(ax^2 + bx + c)^m$ has the form

$$\frac{A_1x + B_1}{ax^2 + bx + c} + \frac{A_2x + B_2}{(ax^2 + bx + c)^2} + \cdots + \frac{A_mx + B_m}{(ax^2 + bx + c)^m}.$$

EXAMPLE 5 **Finding the Partial-Fraction Decomposition When the Denominator Has a Repeated Irreducible Quadratic Factor**

Find the partial-fraction decomposition of $\dfrac{2x^4 - x^3 + 13x^2 - 2x + 13}{(x - 1)(x^2 + 4)^2}$.

Solution

Step 1 The denominator has a nonrepeating linear factor $(x - 1)$ and an irreducible quadratic factor $(x^2 + 4)$ that appears twice. The decomposition therefore has the form

$$\frac{2x^4 - x^3 + 13x^2 - 2x + 13}{(x - 1)(x^2 + 4)^2} = \frac{A}{x - 1} + \frac{Bx + C}{x^2 + 4} + \frac{Dx + E}{(x^2 + 4)^2}.$$

Step 2 Multiply both sides of the decomposition in Step 1 by the original denominator $(x - 1)(x^2 + 4)^2$, use the distributive property, and eliminate common factors:

$$2x^4 - x^3 + 13x^2 - 2x + 13 = A(x^2 + 4)^2 + (Bx + C)(x - 1)(x^2 + 4) + (Dx + E)(x - 1).$$

Substitute $x = 1$ and simplify to obtain $25 = 25A$, or $A = 1$.

Steps 3–4 Multiply the factors on the right side of the equation in Step 2 and collect like terms to obtain

$$2x^4 - x^3 + 13x^2 - 2x + 13 = (A + B)x^4 + (-B + C)x^3 + (8A + 4B - C + D)x^2 + (-4B + 4C - D + E)x + (16A - 4C - E).$$

Equate corresponding coefficients to get the system:

$$\begin{cases} A + B = 2 & (1) & \text{Coefficient of } x^4 \\ -B + C = -1 & (2) & \text{Coefficient of } x^3 \\ 8A + 4B - C + D = 13 & (3) & \text{Coefficient of } x^2 \\ -4B + 4C - D + E = -2 & (4) & \text{Coefficient of } x \\ 16A - 4C - E = 13 & (5) & \text{Constant coefficient} \end{cases}$$

Now back-substitute $A = 1$ from Step 2 in equation (1) to get $B = 1$. Again, back-substitute $B = 1$ in equation (2) to get $C = 0$.

Back-substitute $A = 1$ and $C = 0$ in equation (5) to obtain $E = 3$. Again, back-substitute $A = 1, B = 1$, and $C = 0$ in equation (3) to get $D = 1$. Hence, $A = 1, B = 1, C = 0, D = 1$, and $E = 3$.

Step 5 Substitute $A = 1, B = 1, C = 0, D = 1$, and $E = 3$ in the equation in Step 1 to obtain the partial-fraction decomposition.

$$\frac{2x^4 - x^3 + 13x^2 - 2x + 13}{(x - 1)(x^2 + 4)^2} = \frac{1}{x - 1} + \frac{x}{x^2 + 4} + \frac{x + 3}{(x^2 + 4)^2}$$

Practice Problem 5 Find the partial-fraction decomposition of

$$\frac{x^2 + 3x + 1}{(x^2 + 1)^2}.$$

EXAMPLE 6 **Learning a Clever Way to Add**

Express the sum

$$\frac{1}{2 \cdot 3} + \frac{1}{3 \cdot 4} + \frac{1}{4 \cdot 5} + \cdots + \frac{1}{2009 \cdot 2010}$$

as a fraction of whole numbers in lowest terms.

Solution

Each term in the sum is of the form $\frac{1}{k(k+1)}$. From the partial-fraction decomposition, we have

$$\frac{1}{k(k+1)} = \frac{1}{k} - \frac{1}{k+1}.$$

Therefore, by using this decomposition for each term, we get

$$\frac{1}{2\cdot 3} + \frac{1}{3\cdot 4} + \frac{1}{4\cdot 5} + \cdots + \frac{1}{2009\cdot 2010}$$
$$= \left(\frac{1}{2} - \frac{1}{3}\right) + \left(\frac{1}{3} - \frac{1}{4}\right) + \left(\frac{1}{4} - \frac{1}{5}\right) + \cdots + \left(\frac{1}{2009} - \frac{1}{2010}\right).$$

After removing parentheses, we see that most terms cancel in pairs $\left(\text{such as } -\frac{1}{3} \text{ and } \frac{1}{3}\right)$, leaving only the first and the last term. The original sum is therefore referred to as a *telescoping sum* that "collapses" to

$$\frac{1}{2} - \frac{1}{2010} = \frac{1005 - 1}{2010} = \frac{1004}{2010} = \frac{502}{1005}.$$

Practice Problem 6 Express the sum $\frac{1}{4\cdot 5} + \frac{1}{5\cdot 6} + \frac{1}{6\cdot 7} + \cdots + \frac{1}{3111\cdot 3112}$ as a fraction of whole numbers in lowest terms.

EXAMPLE 7 Calculating the Resistance in a Circuit

The total resistance R due to two resistors connected in parallel is given by

$$R = \frac{x(x+5)}{3x+10}.$$

Use partial-fraction decomposition to write $\frac{1}{R}$ in terms of partial fractions. What does each term represent?

Solution

From $R = \frac{x(x+5)}{3x+10}$, we get the following:

$$\frac{1}{R} = \frac{3x+10}{x(x+5)} \qquad \text{Take the reciprocal of both sides.}$$

$$\frac{3x+10}{x(x+5)} = \frac{A}{x} + \frac{B}{x+5} \qquad \text{Form of partial-fraction decomposition}$$

$$3x + 10 = A(x+5) + Bx \qquad \text{Multiply both sides by } x(x+5) \text{ and simplify.}$$

RECALL

The total resistance R in a circuit due to two resistors R_1 and R_2 connected in parallel satisfies the equation

$$\frac{1}{R} = \frac{1}{R_1} + \frac{1}{R_2}.$$

Substitute $x = 0$ in the last equation to obtain $10 = 5A$, or $A = 2$. Now substitute $x = -5$ in the same equation to get $-5 = -5B$, or $B = 1$. Thus, we have

$$\frac{1}{R} = \frac{3x+10}{x(x+5)} = \frac{2}{x} + \frac{1}{x+5}$$

$$\frac{1}{R} = \frac{1}{\frac{x}{2}} + \frac{1}{x+5} \qquad \text{Rewrite the right side.}$$

The last equation says that if two resistors with resistances $R_1 = \frac{x}{2}$ and $R_2 = x + 5$ are connected in parallel, they will produce a total resistance R, where $R = \frac{x(x+5)}{3x+10}$.

Practice Problem 7 Repeat Example 7 assuming that $R = \frac{(x+1)(x+2)}{4x+7}$.

Answers to Practice Problems

1. $\frac{3}{x+1} - \frac{1}{x-2}$
2. $\frac{5}{x-1} - \frac{3}{x} + \frac{1}{x+1}$
3. $\frac{5}{x} + \frac{6}{(x-1)^2} - \frac{5}{x-1}$
4. $\frac{-1}{x} + \frac{4x+5}{x^2+2}$
5. $\frac{1}{x^2+1} + \frac{3x}{(x^2+1)^2}$
6. $\frac{777}{3112}$
7. $\frac{1}{R} = \frac{3}{x+1} + \frac{1}{x+2} = \frac{1}{\frac{x+1}{3}} + \frac{1}{x+2}$; $R_1 = \frac{x+1}{3}$, $R_2 = x + 2$

SECTION 8.3 Exercises

Basic Concepts and Skills

1. In a rational expression, if the degree of the numerator, $P(x)$, is less than the degree of the denominator, $Q(x)$, then the expression is ______________.
2. The numerators in the partial-fraction decomposition of a rational expression are constants if the denominator, $Q(x)$, can be factored into ______________.
3. In a rational expression, if $(x - 8)$ is a linear factor that is repeated three times in the denominator, then the portion of the expression's partial-fraction decomposition that corresponds to $(x - 8)^3$ has ______________ terms.
4. **True or False.** The form of the partial-fraction decomposition of a rational expression depends only on its denominator.
5. **True or False.** Both $(x - 1)^2$ and $(x + 2)^2$ are irreducible quadratic factors of the polynomial $Q(x) = (x - 1)^2(x + 2)^2$.
6. **True or False.** If the denominator of a rational expression is $Q(x) = x^3 - 9x$, then x, $x - 3$, and $x + 3$ become denominators in its partial-fraction decomposition.

In Exercises 7–16, write the form of the partial-fraction decomposition of each rational expression. You do not need to solve for the constants.

7. $\frac{1}{(x-1)(x+2)}$
8. $\frac{x}{(x+1)(x-3)}$
9. $\frac{1}{x^2+7x+6}$
10. $\frac{3}{x^2-6x+8}$
11. $\frac{2}{x^3-x^2}$
12. $\frac{x-1}{(x+2)^2(x-3)}$
13. $\frac{x^2-3x+3}{(x+1)(x^2-x+1)}$
14. $\frac{2x+3}{(x-1)^2(x^2+x+1)}$
15. $\frac{3x-4}{(x^2+1)^2}$
16. $\frac{x-2}{(2x+3)^2(x^2+3)^2}$

In Exercises 17–48, find the partial-fraction decomposition of each rational expression.

17. $\frac{2x+1}{(x+1)(x+2)}$
18. $\frac{7}{(x-2)(x+5)}$
19. $\frac{1}{x^2+4x+3}$
20. $\frac{x}{x^2+5x+6}$
21. $\frac{2}{x^2+2x}$
22. $\frac{x+9}{x^2-9}$
23. $\frac{x}{(x+1)(x+2)(x+3)}$
24. $\frac{x^2}{(x-1)(x^2+5x+4)}$
25. $\frac{x-1}{(x+1)^2}$
26. $\frac{3x+2}{x^2(x+1)}$
27. $\frac{2x^2+x}{(x+1)^3}$
28. $\frac{x}{(x-1)(x+1)}$
29. $\frac{-x^2+3x+1}{x^3+2x^2+x}$
30. $\frac{5x^2-8x+2}{x^3-2x^2+x}$
31. $\frac{1}{x^2(x+1)^2}$
32. $\frac{2}{(x-1)^2(x+3)^2}$
33. $\frac{1}{(x^2-1)^2}$
34. $\frac{3}{(x^2-4)^2}$
35. $\frac{x-1}{(2x-3)^2}$
36. $\frac{6x+1}{(3x+1)^2}$
37. $\frac{6x+7}{4x^2+12x+9}$
38. $\frac{2x+3}{9x^2+30x+25}$
39. $\frac{x-3}{x^3+x^2}$
40. $\frac{1}{x^3+x}$
41. $\frac{x^2+2x+4}{x^3+x^2}$
42. $\frac{x^2+2x-1}{(x-1)(x^2+1)}$

43. $\frac{x}{x^4 - 1}$

44. $\frac{3x^3 - 5x^2 + 12x + 4}{x^4 - 16}$

45. $\frac{1}{x(x^2 + 1)^2}$

46. $\frac{x^2}{(x^2 + 2)^2}$

47. $\frac{2x^2 + 3x}{(x^2 + 1)(x^2 + 2)}$

48. $\frac{x^2 - 2x}{(x^2 + 9)(x^2 + x + 7)}$

Applying the Concepts

In Exercises 49–52, express each sum as a fraction of whole numbers in lowest terms.

49. $\frac{1}{1 \cdot 2} + \frac{1}{2 \cdot 3} + \frac{1}{3 \cdot 4} + \cdots + \frac{1}{n(n + 1)}$

50. $\frac{2}{1 \cdot 3} + \frac{2}{2 \cdot 4} + \frac{2}{3 \cdot 5} + \cdots + \frac{2}{100 \cdot 102}$

51. $\frac{2}{1 \cdot 3} + \frac{2}{3 \cdot 5} + \frac{2}{5 \cdot 7} + \cdots + \frac{2}{(2n - 1)(2n + 1)}$

52. $\frac{2}{1 \cdot 2 \cdot 3} + \frac{2}{2 \cdot 3 \cdot 4} + \frac{2}{3 \cdot 4 \cdot 5} + \cdots + \frac{2}{100 \cdot 101 \cdot 102}$

[*Hint*: Write

$$\frac{2}{k(k + 1)(k + 2)} = \frac{1}{k} - \frac{2}{k + 1} + \frac{1}{k + 2}$$
$$= \frac{1}{k} - \frac{1}{k + 1} - \frac{1}{k + 1} + \frac{1}{k + 2}.$$]

In Exercises 53–56, the total resistance R in a circuit is given. Use partial-fraction decomposition to write $\frac{1}{R}$ in terms of partial fractions. Interpret each term in the partial fraction.

53. Circuit resistance. $R = \frac{(x + 1)(x + 3)}{2x + 4}$

54. Circuit resistance. $R = \frac{(x + 2)(x + 4)}{7x + 20}$

55. Circuit resistance. $R = \frac{R_1 R_2 R_3}{R_1 R_2 + R_2 R_3 + R_3 R_1}$

56. Circuit resistance. $R = \frac{x(x + 2)(x + 4)}{3x^2 + 12x + 8}$

Beyond the Basics

57. Solve the system of equations in Example 2 to verify the values of the constants A, B, and C.

In Exercises 58–66, find the partial-fraction decomposition of each rational expression.

58. $\frac{1}{x^3 + 1}$

59. $\frac{4x}{(x^2 - 1)^2}$

60. $\frac{2x + 3}{(x^2 + 1)(x + 1)}$

61. $\frac{x + 1}{(x^2 + 1)(x - 1)^2}$

62. $\frac{x}{(x^2 + 1)^2(x - 1)}$

63. $\frac{x^3}{(x + 1)^2(x + 2)^2}$

64. $\frac{2x^3 + 2x^2 - 1}{(x^2 + x + 1)^2}$

65. $\frac{15x}{x^3 - 27}$

66. $\frac{2x^2 - 9x + 10}{x^3 + 8}$

67. Find the partial-fraction decomposition of

$$\frac{4x}{x^4 + 4}.$$

[*Hint*: $x^4 + 4 = x^4 + 4 + 4x^2 - 4x^2 = (x^2 + 2)^2 - (2x)^2$
$= (x^2 - 2x + 2)(x^2 + 2x + 2)$.]

68. Find the partial-fraction decomposition of

$$\frac{x^2 - 8x + 18}{(x - 5)^3}.$$

In Exercises 69–72, find the partial-fraction decomposition by first using long division.

69. $\frac{x^2 + 4x + 5}{x^2 + 3x + 2}$

70. $\frac{(x - 1)(x + 2)}{(x + 3)(x - 4)}$

71. $\frac{2x^4 + x^3 + 2x^2 - 2x - 1}{(x - 1)(x^2 + 1)}$

72. $\frac{x^4}{(x - 1)(x - 2)(x - 3)}$

Critical Thinking/Discussion/Writing

73. In the partial-fraction decomposition of an expression, we obtain a system of equations by comparing coefficients of the like terms. Justify the validity of this step.

74. Let $f(x) = \frac{x + 8}{x^2 + x - 2} = g(x) + h(x)$, where $g(x)$ and $h(x)$ are the terms in the partial-fraction decomposition of $f(x)$.

a. Sketch the graphs of f, g, and h.
b. Compare the graphs of f and g. What do you observe?
c. Compare the graphs of f and h. What do you observe?

Maintaining Skills

In Exercises 75–78, solve each equation for x.

75. $x^2 + 10x + 21 = 0$

76. $x^2 - 2x - 24 = 0$

77. $2x^2 + x - 10 = 0$

78. $6x^2 + x - 2 = 0$

In Exercises 79–82, solve each system of equations by the substitution method.

79. $\begin{cases} x + y = 3 \\ 3x + 2y = 7 \end{cases}$

80. $\begin{cases} 3x + y = -3 \\ 4x + 3y = 1 \end{cases}$

81. $\begin{cases} x - 2y = 1 \\ 2x + 3y = 16 \end{cases}$

82. $\begin{cases} x - 2y = 10 \\ \frac{1}{2}x + \frac{1}{3}y = 1 \end{cases}$

In Exercises 83–86, solve each system of equations by the elimination method.

83. $\begin{cases} 2x - y = 4 \\ 3x + 2y = 13 \end{cases}$

84. $\begin{cases} 2x + 3y = -1 \\ 3x - 2y = 5 \end{cases}$

85. $\begin{cases} 2x + 5y = 1 \\ 3x - 2y = -8 \end{cases}$

86. $\begin{cases} 2x - 3y = 6 \\ -3x + 2y = 1 \end{cases}$

SECTION 8.4

Systems of Nonlinear Equations

BEFORE STARTING THIS SECTION, REVIEW

1 Solve systems of linear equations (Sections 8.1 and 8.2)

OBJECTIVES

1 Solve systems of nonlinear equations by substitution.

2 Solve systems of nonlinear equations by elimination.

3 Solve applied problems using nonlinear systems.

New York Stock Exchange

◆ Investing in the Stock Market

A *stock* is a certificate that shows that you own a small fraction of a corporation. When you own stock in a company, you are referred to as a *shareholder*. When you buy stock, you are paying for a small percentage of everything the company owns, including a slice of its profits (or losses).

Suppose you inherit some money and want to invest \$1000 in the stock market. Where do you invest your money? (This is a serious question. You may have to do a great deal of research on the companies that interest you, or you may have to get help from a broker.)

Suppose you get a tip from a friend who heard it from his brother-in-law (a taxi driver who overheard his customer's conversation) that Microsoft is coming out with a new version of Microsoft Windows that is going to double the company's business. Where do you buy Microsoft stock? You go to a broker. A brokerage house (such as Fidelity Investments) is a supermarket of stocks. You can tell any broker that you have \$1000 and want to buy as much Microsoft stock as you can. The broker might reply, "A share of Microsoft at this time [the price may fluctuate every second] costs \$31 [the actual closing price on June 12, 2013, was \$34.40], and I am going to charge you \$55 for my services. So you can buy $\frac{1000 - 55}{31} \approx 30.48$, or 30 shares of Microsoft." You then give the broker the money, and you become part owner of Microsoft. (Usually, the broker does not give you the paper stock certificate, but rather transfers ownership to you.)

You can make a profit on your investment if you sell your stock for more than the total amount you paid for it. You can also make money while you own certain stocks by receiving dividends. A *dividend* on a share of stock is a return on your investment and represents the portion of the company's profits allocated to that share. The dividend can be in the form of cash or additional shares of the company. In Example 3, we discuss such a scenario.

RECALL

An equation of the form

$$ax + by = c$$

is a linear equation in the variables x and y.

Systems of Nonlinear Equations

In Sections 8.1 and 8.2, we discussed systems of *linear* equations. We now consider systems of *nonlinear* equations involving two variables: In these systems, at least one equation is nonlinear.

To determine whether an ordered pair is a solution of a nonlinear system, you check whether it is a solution of each equation in the system. Thus, $(2, 1)$ is a solution of the system

$$\begin{cases} x^2 + 3y = 7 & (1) \\ 5x^2 - 4y^2 = 16 & (2) \end{cases}$$

because $2^2 + 3(1) = 4 + 3 = 7$ and $5(2)^2 - 4(1)^2 = 20 - 4 = 16$ are both true statements.

1 Solve systems of nonlinear equations by substitution.

Solving Systems of Nonlinear Equations by Substitution

We follow the same steps used in the substitution method for solving a system of linear equations in Section 8.1. See page 731. This method is useful when one of the equations in the system is linear in one of the variables.

EXAMPLE 1 **Using Substitution to Solve a Nonlinear System**

Solve the system of equations by the substitution method.

$$\begin{cases} 4x + y = -3 & (1) \\ -x^2 + y = 1 & (2) \end{cases}$$

Solution

Because both equations are linear in y, we solve either equation for y.

Step 1 **Solve for one variable.** In equation (2), express y in terms of x.

$$y = x^2 + 1 \quad (3) \qquad \text{Solve equation (2) for } y.$$

Step 2 **Substitute.** Substitute $x^2 + 1$ for y in equation (1).

$$\begin{aligned} 4x + y &= -3 && \text{Equation (1)} \\ 4x + (x^2 + 1) &= -3 && \text{Replace } y \text{ with } (x^2 + 1). \\ 4x + x^2 + 4 &= 0 && \text{Add 3 to both sides.} \\ x^2 + 4x + 4 &= 0 && \text{Rewrite in descending powers of } x. \end{aligned}$$

Step 3 **Solve** the resulting equation from Step 2.

$$\begin{aligned} (x + 2)(x + 2) &= 0 && \text{Factor.} \\ x + 2 &= 0 && \text{Zero-product property} \\ x &= -2 && \text{Solve for } x. \end{aligned}$$

Step 4 **Back-substitute.** Substitute $x = -2$ in equation (3) to find the corresponding y-value.

$$\begin{aligned} y &= x^2 + 1 && \text{Equation (3)} \\ y &= (-2)^2 + 1 && \text{Replace } x \text{ with } -2. \\ y &= 5 && \text{Simplify.} \end{aligned}$$

Step 5 From Steps 3 and 4, we have $x = -2$ and $y = 5$. The apparent solution set of the system is $\{(-2, 5)\}$.

Step 6 **Check:** Replace x with -2 and y with 5 in both equations.

Equation (1)	Equation (2)	
$4x + y = -3$	$-x^2 + y = 1$	
$4(-2) + 5 \stackrel{?}{=} -3$	$-(-2)^2 + 5 \stackrel{?}{=} 1$	Replace x with -2 and y with 5
$-8 + 5 = -3$ ✓	$-4 + 5 = 1$ ✓	

The graphs of the line $4x + y = -3$ and the parabola $y = x^2 + 1$ confirm that the solution set is $\{(-2, 5)\}$. See Figure 8.10.

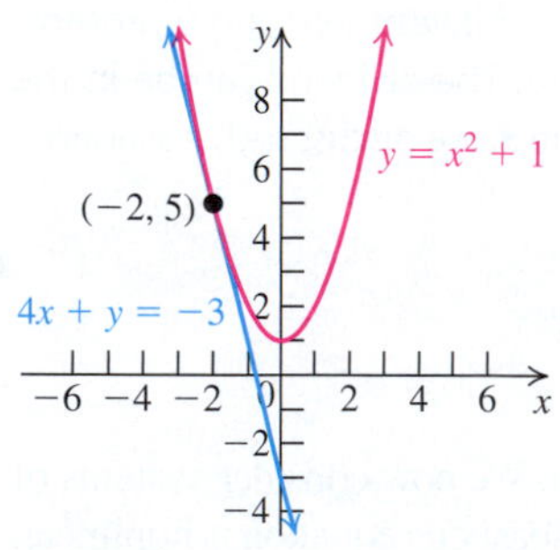

FIGURE 8.10

Practice Problem 1 Solve the system of equations by the substitution method.

$$\begin{cases} x^2 + y = 2 & (1) \\ 2x + y = 3 & (2) \end{cases}$$

2 Solve systems of nonlinear equations by elimination.

Solving Systems of Nonlinear Equations by Elimination

We follow the same steps outlined for solving systems of linear equations by the elimination method. See page 733. This method works well when both equations of the system are of the same degree in the same variable.

TECHNOLOGY CONNECTION

To use a graphing calculator to graph the system in Example 2, solve $x^2 + y^2 = 25$ for y and enter the two resulting equations.

$$y_1 = \sqrt{25 - x^2}$$
$$y_2 = -\sqrt{25 - x^2}$$

Then enter

$$y_3 = x^2 - 5$$

and use the ZSquare feature.

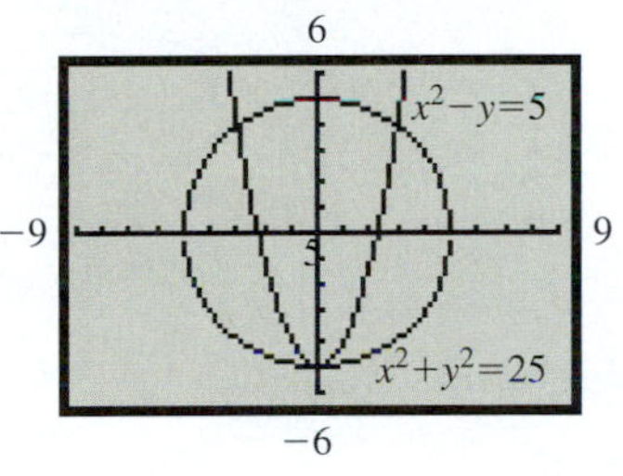

EXAMPLE 2 Using Elimination to Solve a Nonlinear System

Solve the system of equations by the elimination method.

$$\begin{cases} x^2 + y^2 = 25 & (1) \\ x^2 - y = 5 & (2) \end{cases}$$

Solution

Step 1 **Adjust the coefficients.** Because x has the same power in both equations, we choose the variable x for elimination. We multiply equation (2) by -1 and obtain the equivalent system.

$$\begin{cases} x^2 + y^2 = 25 & (1) \\ -x^2 + y = -5 & (3) \quad \text{Multiply equation (2) by } -1. \end{cases}$$

Step 2

$$y^2 + y = 20 \quad (4) \quad \text{Add equations (1) and (3).}$$

Step 3 Solve the equation found in Step 2: $y^2 + y = 20$.

$$y^2 + y - 20 = 0 \quad \text{Subtract 20 from both sides.}$$
$$(y + 5)(y - 4) = 0 \quad \text{Factor.}$$
$$y + 5 = 0 \quad \text{or} \quad y - 4 = 0 \quad \text{Zero-product property}$$
$$y = -5 \quad \text{or} \quad y = 4 \quad \text{Solve for } y.$$

Step 4 **Back-substitute** the values in one of the original equations to solve for the other variable.

(i) Substitute $y = -5$ in equation (2) and solve for x.

$$x^2 - y = 5 \quad \text{Equation (2)}$$
$$x^2 - (-5) = 5 \quad \text{Replace } y \text{ with } -5.$$
$$x^2 = 0 \quad \text{Simplify.}$$
$$x = 0 \quad \text{Solve for } x.$$

So $(0, -5)$ is a solution of the system.

(ii) Substitute $y = 4$ in equation (2) and solve for x.

$$x^2 - y = 5 \quad \text{Equation (2)}$$
$$x^2 - 4 = 5 \quad \text{Replace } y \text{ with 4.}$$
$$x^2 = 9 \quad \text{Add 4 to both sides.}$$
$$x = \pm 3 \quad \text{Square root property}$$

So $(3, 4)$ and $(-3, 4)$ are solutions of the system.
From **(i)** and **(ii)**, the apparent solution set for the system is $\{(0, -5), (3, 4), (-3, 4)\}$.

Step 5 Check:

	$(0, -5)$	$(3, 4)$	$(-3, 4)$
$x^2 + y^2 = 25$	$0^2 + (-5)^2 \stackrel{?}{=} 25$ $25 = 25$ ✓	$3^2 + 4^2 \stackrel{?}{=} 25$ $9 + 16 = 25$ $25 = 25$ ✓	$(-3)^2 + 4^2 \stackrel{?}{=} 25$ $9 + 16 \stackrel{?}{=} 25$ $25 = 25$ ✓
$x^2 - y = 5$	$0^2 - (-5) \stackrel{?}{=} 5$ $5 = 5$ ✓	$3^2 - 4 \stackrel{?}{=} 5$ $9 - 4 = 5$ ✓	$(-3)^2 - 4 \stackrel{?}{=} 5$ $9 - 4 = 5$ ✓

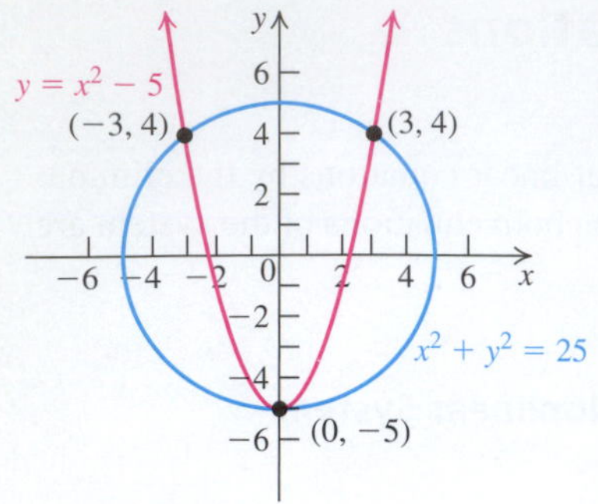

FIGURE 8.11

The graphs of the circle $x^2 + y^2 = 25$ and the parabola $y = x^2 - 5$ sketched in Figure 8.11 confirm the three solutions.

Practice Problem 2 Solve the system of equations.

$$\begin{cases} x^2 + 2y^2 = 34 \\ x^2 - y^2 = 7 \end{cases}$$

Applications

In the next example, we solve an investment problem that involves a nonlinear system of equations.

3 Solve applied problems using nonlinear systems.

EXAMPLE 3 Investing in Stocks

Danielle bought 240 shares of Alpha Airlines stock at \$40 per share and paid \$100 in broker's fees. She kept these shares for three years, during which time she received a number of additional shares of the stock as dividends. At the end of three years, her dividends were worth \$2400 and she sold all of her stock and made a profit of \$7100 (after paying \$100 in initial broker's commission). How many shares of the stock did she receive as dividends? What was the selling price of each share of the stock?

Solution

Let x = number of shares she received as dividends
p = selling price per share

Then

$$xp = 2400 \quad (1) \qquad \text{The dividend total is \$2400.}$$

The number of shares she sold at p dollars per share is $240 + x$.

$$\text{Revenue} = (240 + x)p$$
$$\text{Cost} = (240)(40) + 100 = 9700$$

number of shares bought (240); price she paid per share (40); commission to purchase (100)

and

$$\begin{aligned} \text{Revenue} - \text{Cost} &= \text{Profit} \\ (240 + x)p - 9700 &= 7100 \\ (240 + x)p &= 16{,}800 && \text{Add 9700 to both sides.} \\ 240p + xp &= 16{,}800 \quad (2) && \text{Distributive property} \end{aligned}$$

We solve the system of equations:

$$\begin{cases} xp = 2400 & (1) \quad \text{Revenue from the dividends} \\ 240p + xp = 16{,}800 & (2) \quad \text{Revenue} - \text{Cost} = \text{Profit} \end{cases}$$

Substitute $xp = 2400$ from equation (1) into equation (2).

$$\begin{aligned} 240p + xp &= 16{,}800 \quad (2) \\ 240p + 2400 &= 16{,}800 \quad (3) && \text{Substitute 2400 for } xp \text{ in (2).} \\ 240p &= 14{,}400 && \text{Subtract 2400 from both sides.} \\ p &= 60 && \text{Solve for } p. \end{aligned}$$

Back-substitute $p = 60$ in equation (1).

$$xp = 2400 \quad \text{Equation (1)}$$
$$x(60) = 2400 \quad \text{Replace } p \text{ with } 60.$$
$$x = \frac{2400}{60} = 40 \quad \text{Solve for } x.$$

Danielle received $x = 40$ shares as dividends and sold her stock at $p = \$60$ per share.

Practice Problem 3 Rework Example 3 assuming that Danielle's dividends were worth \$1950 after three years and her profit was \$7850.

Answers to Practice Problems

1. $\{(1, 1)\}$ **2.** $\{(4, 3), (-4, 3), (4, -3), (-4, -3)\}$ **3.** Danielle received $x = 30$ shares as dividends and sold her stock at $p = \$65$ per share.

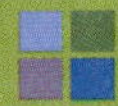

SECTION 8.4 Exercises

Basic Concepts and Skills

1. In a system of nonlinear equations, ____________ equation must be nonlinear.

2. Both the ____________ and ____________ methods can also be used to solve systems of nonlinear equations.

3. The solutions of the system $\begin{cases} ax + by = c & (1) \\ (x - h)^2 + (y - k)^2 = r^2 & (2) \end{cases}$ represent the points of ____________ of the graphs of equations (1) and (2).

4. The system of equations in Example 1 has exactly one solution because the graph of equation (1) is a(n) ____________ line to the graph of equation (2).

5. True or False. The system of equations $\begin{cases} y = ax^2 \\ (x - h)^2 + (y - k)^2 = r^2 \end{cases}$ may have 0, 1, 2, 3, or 4 real solutions.

6. True or False. If $4x + 3y = 8$ is one of the equations in a system of equations, then the system is linear.

In Exercises 7–14, determine which of the given ordered pairs are solutions of each system of equations.

7. $\begin{cases} 2x + 3y = 3 \\ x - y^2 = 2 \end{cases}$ $(1, -3), (3, -1), (6, 3), \left(5, \frac{1}{2}\right)$

8. $\begin{cases} x + 2y = 6 \\ y = x^2 \end{cases}$ $(2, 2), (-2, 4), (0, 3), (1, 2)$

9. $\begin{cases} 5x - 2y = 7 \\ x^2 + y^2 = 2 \end{cases}$ $\left(\frac{5}{4}, 1\right), \left(0, \frac{11}{4}\right), (1, -1), (3, 4)$

10. $\begin{cases} x - 2y = -5 \\ x^2 + y^2 = 25 \end{cases}$ $(1, 3), (-5, 0), (3, 4), (3, -4)$

11. $\begin{cases} 4x^2 + 5y^2 = 180 \\ x^2 - y^2 = 9 \end{cases}$ $(5, 4), (-5, 4), (3, 0), (-5, -4)$

12. $\begin{cases} x^2 - y^2 = 3 \\ x^2 - 4x + y^2 = -3 \end{cases}$ $(2, 1), (2, -1), (-2, 1), (-2, -1)$

13. $\begin{cases} y = e^{x-1} \\ y = 2x - 1 \end{cases}$ $(1, 1), (0, -1), (2, e), (-1, -3)$

14. $\begin{cases} y = \ln(x + 1) \\ y = x \end{cases}$ $(1, 1), (0, 0), (-1, -1), (e - 1, 1)$

In Exercises 15–30, solve each system of equations by the substitution method. Check your solutions.

15. $\begin{cases} y = x^2 \\ y = x + 2 \end{cases}$

16. $\begin{cases} y = x^2 \\ x + y = 6 \end{cases}$

17. $\begin{cases} x^2 - y = 6 \\ x - y = 0 \end{cases}$

18. $\begin{cases} x^2 - y = 6 \\ 5x - y = 0 \end{cases}$

19. $\begin{cases} x^2 + y^2 = 9 \\ x = 3 \end{cases}$

20. $\begin{cases} x^2 + y^2 = 9 \\ y = 3 \end{cases}$

21. $\begin{cases} x^2 + y^2 = 5 \\ x - y = -3 \end{cases}$

22. $\begin{cases} x^2 + y^2 = 13 \\ 2x - 3y = 0 \end{cases}$

23. $\begin{cases} x^2 - 4x + y^2 = -2 \\ x - y = 2 \end{cases}$

24. $\begin{cases} x^2 - 8y + y^2 = -6 \\ 2x - y = 1 \end{cases}$

25. $\begin{cases} x - y = -2 \\ xy = 3 \end{cases}$

26. $\begin{cases} x - 2y = 4 \\ xy = 6 \end{cases}$

27. $\begin{cases} 4x^2 + y^2 = 25 \\ x + y = 5 \end{cases}$

28. $\begin{cases} x^2 + 4y^2 = 16 \\ x + 2y = 4 \end{cases}$

29. $\begin{cases} x^2 - y^2 = 24 \\ 5x - 7y = 0 \end{cases}$

30. $\begin{cases} x^2 - 7y^2 = 9 \\ x - y = 3 \end{cases}$

In Exercises 31–40, solve each system of equations (only real solutions) by the elimination method. Check your solutions.

31. $\begin{cases} x^2 + y^2 = 20 \\ x^2 - y^2 = 12 \end{cases}$

32. $\begin{cases} x^2 + 8y^2 = 9 \\ 3x^2 + y^2 = 4 \end{cases}$

33. $\begin{cases} x^2 + 2y^2 = 12 \\ 7y^2 - 5x^2 = 8 \end{cases}$

34. $\begin{cases} 3x^2 + 2y^2 = 77 \\ x^2 - 6y^2 = 19 \end{cases}$

35. $\begin{cases} x^2 - y = 2 \\ 2x - y = 4 \end{cases}$

36. $\begin{cases} x^2 + 3y = 0 \\ x - y = -12 \end{cases}$

37. $\begin{cases} x^2 + y^2 = 5 \\ 3x^2 - 2y^2 = -5 \end{cases}$

38. $\begin{cases} x^2 + 4y^2 = 5 \\ 9x^2 - y^2 = 8 \end{cases}$

39. $\begin{cases} x^2 + y^2 + 2x = 9 \\ x^2 + 4y^2 + 3x = 14 \end{cases}$

40. $\begin{cases} x^2 + y^2 = 9 \\ x^2 + y^2 - 18x = 0 \end{cases}$

In Exercises 41–54, use any method to solve each system of equations.

41. $\begin{cases} x + y = 8 \\ xy = 15 \end{cases}$

42. $\begin{cases} 2x + y = 8 \\ x^2 - y^2 = 5 \end{cases}$

43. $\begin{cases} x^2 + y^2 = 2 \\ 3x^2 + 3y^2 = 9 \end{cases}$

44. $\begin{cases} x^2 + y^2 = 5 \\ 3x^2 - y^2 = -11 \end{cases}$

45. $\begin{cases} y^2 = 4x + 4 \\ y = 2x - 2 \end{cases}$

46. $\begin{cases} xy = 125 \\ y = x^2 \end{cases}$

47. $\begin{cases} x^2 + 4y^2 = 25 \\ x - 2y + 1 = 0 \end{cases}$

48. $\begin{cases} y = x^2 - 5x + 4 \\ 3x + y = 3 \end{cases}$

49. $\begin{cases} x^2 - 3y^2 = 1 \\ x^2 + 4y^2 = 8 \end{cases}$

50. $\begin{cases} 4x^2 - y^2 = 12 \\ 4y^2 - x^2 = 12 \end{cases}$

51. $\begin{cases} x^2 - xy + 5x = 4 \\ 2x^2 - 3xy + 10x = -2 \end{cases}$

52. $\begin{cases} x^2 - xy + x = -4 \\ 3x^2 - 2xy - 2x = 4 \end{cases}$

53. $\begin{cases} x^2 + y^2 - 8x = -8 \\ x^2 - 4y^2 + 6x = 0 \end{cases}$

54. $\begin{cases} 4x^2 + y^2 - 9y = -4 \\ 4x^2 - y^2 - 3y = 0 \end{cases}$

Applying the Concepts

Exercises 55 and 56 show the demand and supply functions of a product. In each case, *p* represents the price per unit and *x* represents the number of units, in hundreds. Determine the price that gives market equilibrium and the number of units that are demanded and supplied at that price.

55. $\begin{cases} p + 2x^2 = 96 \\ p - 13x = 39 \end{cases}$

56. $\begin{cases} p^2 + 6p + 3x = 75 \\ -p + x = 13 \end{cases}$

In Exercises 57–64, use systems of equations to solve the problem.

57. **Numbers.** Find two positive numbers whose sum is 24 and whose product is 143.

58. **Numbers.** Find two positive numbers whose difference is 13 and whose product is 114.

59. **Commercial land.** A commercial parcel of land is in the form of a trapezoid with two right angles, as shown in the figure. The oblique side is 100 meters in length, and two sides that meet at the vertex of one of the right angles are equal. The perimeter of the land is 360 meters. Find
 a. The lengths of the remaining sides.
 b. The area of the land.

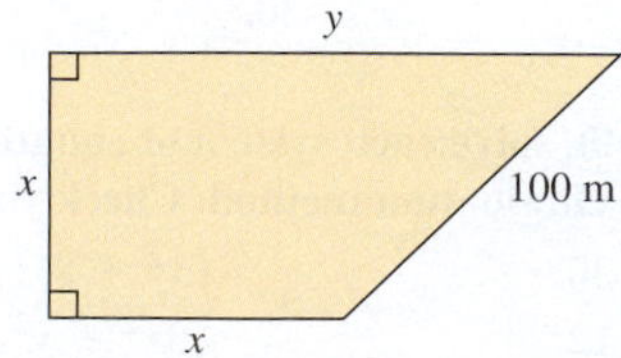

60. **Dividing a pasture.** The area of a rectangular pasture is 8750 square feet. The pasture is divided into three smaller pastures by two fences parallel to the shorter sides. The widths of two of the larger pastures are the same, and the width of the third is one-half that of the others. Find the possible dimensions of the original pasture if the perimeter of the smaller of the subdivisions is 190 feet.

61. **Chartering a fishing boat.** A group of students of a fraternity chartered a fishing boat that would cost \$960. Before the date of the fishing trip, eight more students joined the group and agreed to pay their share of the cost of the charter. The cost per student was reduced by \$6. Find the original number of students in the group and the original cost per student.

62. **Carpet sale.** A salesperson sold a square carpet and a rectangular carpet whose length was 6 feet more than its width. The combined area of the two carpets was 540 square feet. The price of the square carpet was \$12 per square yard, and the price of the rectangular carpet was \$10 per square yard. Assuming that the rectangular piece was \$50 more than the square piece, find the dimensions of each carpet.

63. **Buying and selling stocks.** Flat Fee Brokerage, Inc., charges \$100 commission for every transaction (purchase or sale) you make. Anju bought a block of stock containing over 400 shares that cost her \$10,000, including the commission. After one year, she received a stock dividend of 30 shares. She then sold all of her stock for \$3 more per share than it cost and made a profit of \$1900 (after paying the commission). How many shares did she buy, and what was the original price of each share?

64. **Trading stocks.** Repeat Exercise 63 with the following details: Anju does not receive dividends and decides to keep 100 shares of the stock. She sells the remaining stock for \$10 more per share than it cost and makes a profit of \$3900 after paying the commission.

Beyond the Basics

In Exercises 65–68, the given circle intersects the given line at the points *A* and *B*. Find the coordinates of *A* and *B* and compute the distance $d(A, B)$.

65. Circle: $x^2 + y^2 - 7x + 5y + 6 = 0$; line: x-axis

66. The circle of Exercise 65 and the y-axis

67. Circle: $x^2 + y^2 + 2x - 4y - 5 = 0$; line: $x - y + 1 = 0$

68. Circle: $x^2 + y^2 - 6x - 8y - 50 = 0$; line: $2x + y - 5 = 0$

69. Show that the line with equation $x + 2y + 6 = 0$ is a tangent line to the circle with equation $x^2 + y^2 - 2x + 2y - 3 = 0$. (A line that intersects the circle at exactly one point is a **tangent line** to the circle.)

70. Repeat Exercise 69 for the line with equation $3x - 4y = 0$ and the circle with equation $x^2 + y^2 - 2x - 4y + 4 = 0$.

In Exercises 71–74, find the value(s) of c so that the graphs of the equations in the system are tangent.

71. $\begin{cases} x + y = 2 \\ x^2 + y^2 = c^2 \end{cases}$ **72.** $\begin{cases} cx - y = 5 \\ x^2 + y^2 = 16 \end{cases}$

73. $\begin{cases} 4x^2 + y^2 = 25 \\ 8x + 3y = c \end{cases}$ **74.** $\begin{cases} x^2 + 4y^2 = 9 \\ cx + 2y = -4 \end{cases}$

In Exercises 75–78, solve by letting $u = \dfrac{1}{x}$ and $v = \dfrac{1}{y}$.

75. $\begin{cases} \dfrac{1}{x} - \dfrac{1}{y} = 5 \\ \dfrac{1}{xy} - 24 = 0 \end{cases}$ **76.** $\begin{cases} \dfrac{1}{x} + \dfrac{1}{y} = 3 \\ \dfrac{1}{x^2} - \dfrac{1}{y^2} = -3 \end{cases}$

77. $\begin{cases} \dfrac{5}{x^2} - \dfrac{3}{y^2} = 2 \\ \dfrac{6}{x^2} + \dfrac{1}{y^2} = 7 \end{cases}$ **78.** $\begin{cases} \dfrac{3}{x^2} - \dfrac{2}{y^2} = 1 \\ \dfrac{6}{x^2} + \dfrac{5}{y^2} = 11 \end{cases}$

In Exercises 79–84, find square roots of each complex number w. [$z = a + bi$ is a square root of w if $z^2 = w$.]

79. $w = -5 + 12i$ **80.** $w = -16 - 30i$

81. $w = 7 - 24i$ **82.** $w = 3 - 4i$

83. $w = 13 + 8\sqrt{3}i$ **84.** $w = -1 - 4\sqrt{5}i$

 **In Exercises 85–90, solve each system of equations and verify the solutions graphically.**

85. $\begin{cases} y = 3^x + 4 \\ y = 3^{2x} - 2 \end{cases}$ **86.** $\begin{cases} y = \ln x + 1 \\ y = \ln x - 2 \end{cases}$

87. $\begin{cases} y = 2^x + 3 \\ y = 2^{2x} + 1 \end{cases}$ **88.** $\begin{cases} y = 2^x + 3 \\ y = 4^x + 1 \end{cases}$

89. $\begin{cases} x = 3^y \\ 3^{2y} = 3x - 2 \end{cases}$ **90.** $\begin{cases} x = 3^y \\ x^2 = 3^{y+1} - 2 \end{cases}$

Critical Thinking/Discussion/Writing

91. Consider a system of two equations in two variables.

$$\begin{cases} y = mx + b \\ y = Ax^3 + Bx^2 + Cx + D,\ A \neq 0 \end{cases}$$

Explain whether it is possible for this system to have

a. No real solutions.
b. One real solution.
c. Two real solutions.
d. Three real solutions.
e. Four real solutions.
f. More than four real solutions.

92. Repeat Exercise 91 for the system.

$$\begin{cases} y = mx + b \\ y = Ax^4 + Bx^3 + Cx^2 + Dx + E,\ A \neq 0 \end{cases}$$

Maintaining Skills

In Exercises 93–100, sketch the graph of each equation.

93. $2x + y = 0$ **94.** $2y - x = 0$

95. $3x + 4y = 12$ **96.** $4x - 3y = 24$

97. $y - 2x^2 = 0$ **98.** $y - x^2 = 3$

99. $y + x^2 - 2 = 0$ **100.** $x^2 + y^2 = 16$

In Exercises 101–104, determine which of the given points are on the graph of the equation.

	Equation	Points
101.	$5x + 3y = 0$	$(0, 0)$, $(-3, 5)$, $(1, -1)$, $(-1, 1)$
102.	$3x + 4y = 12$	$(0, 0)$, $(4, 0)$, $(0, 3)$, $(3, 4)$
103.	$y = x^2 + 3$	$(0, 0)$, $(1, 4)$, $(-1, 4)$, $(2, 5)$
104.	$y + x^2 = 2$	$(0, 0)$, $(0, 2)$, $(2, -2)$, $(1, 1)$

SECTION 8.5

Systems of Inequalities

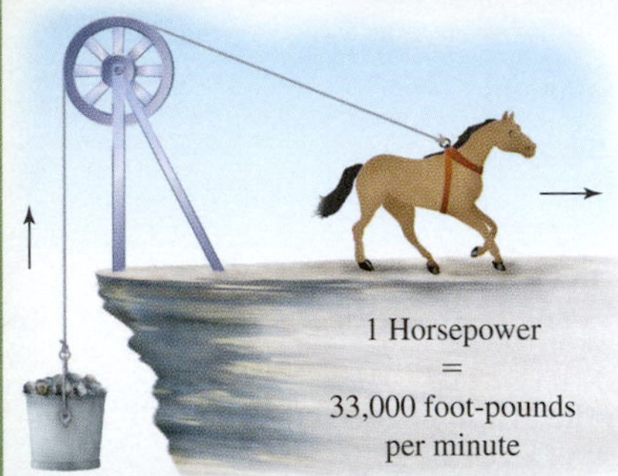

BEFORE STARTING THIS SECTION, REVIEW

1. Intersection of sets (Section P.1, page 6)
2. Graphing a line (Section 2.3, page 184)
3. Solving systems of equations (Sections 8.1, 8.2, and 8.4)

OBJECTIVES

1. Graph a linear inequality in two variables.
2. Graph systems of linear inequalities in two variables.
3. Graph a nonlinear inequality in two variables.
4. Graph systems of nonlinear inequalities in two variables.

◆ What is Horsepower?

The definition of horsepower was originated by the inventor James Watt (1736–1819). To demonstrate the power (work capability) of the steam engine that he invented and patented in 1783, he rigged his favorite horse with a rope and some pulleys and showed that his horse could raise a 3300-pound weight 10 feet in the air in one minute. Watt called this amount of work 1 horsepower (hp). So he defined

$$\begin{aligned} 1 \text{ hp} &= 33{,}000 \text{ foot-pounds per minute } (3300 \times 10 = 33{,}000) \\ &= 550 \text{ foot-pounds per second} \left(\frac{33{,}000}{60} = 550\right). \end{aligned}$$

Because horses were one of the main energy sources during that time, the public easily understood and accepted the horsepower measure. Watt labeled his steam engines in terms of equivalent horsepower. A 10 hp engine could do the work of 10 horses, or could lift 5500 pounds up 1 foot in one second. In the usual definition of power $= \dfrac{(\text{force})(\text{distance})}{\text{time}}$, the horse provides the force. In Exercise 83, we discuss engines with different amounts of horsepower.

1 Graph a linear inequality in two variables.

Graph of a Linear Inequality in Two Variables

The statements $x + y > 4$, $2x + 3y < 7$, $y \ge x$, and $x + y \le 9$ are examples of linear inequalities in the variables x and y. As with equations, a **solution of an inequality in two variables** x and y is an ordered pair (a, b) that results in a true statement when x is replaced with a and y is replaced with b in the inequality. For example, the ordered pair $(3, 2)$ is a solution of the inequality $4x + 5y > 11$ because $4(3) + 5(2) = 22 > 11$ is a true statement. The ordered pair $(1, 1)$ is *not* a solution of the inequality $4x + 5y > 11$ because $4(1) + 5(1) = 9 > 11$ is a false statement. The set of all solutions of an inequality is called the **solution set of the inequality**. The **graph of an inequality in two variables** is the graph of the solution set of the inequality.

EXAMPLE 1 Graphing a Linear Inequality

Graph the inequality $2x + y > 6$.

Solution

First, graph the *equation* $2x + y = 6$. See Figure 8.12(a). Notice that if we solve this equation for y, we have $y = -2x + 6$. Any point $P(a, b)$ whose y-coordinate, b, is *greater* than $6 - 2a$ must be above this line. Therefore, the graph of $y > -2x + 6$, which is equivalent to $2x + y > 6$, consists of all points (x, y) in the plane that lie above the line $2x + y = 6$. The graph of the inequality $2x + y > 6$ is shaded in Figure 8.12(b).

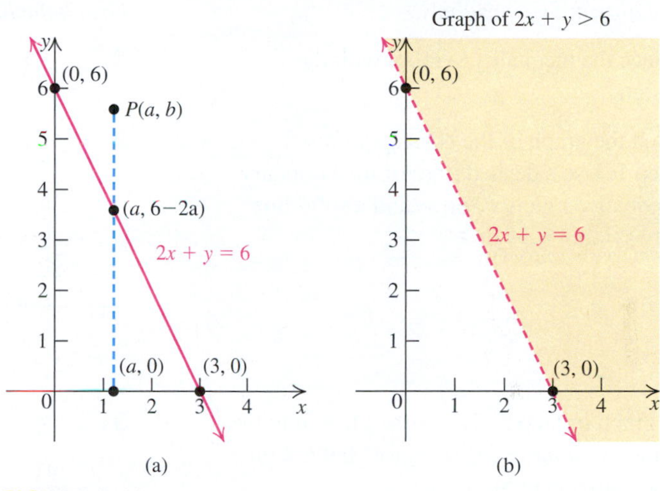

FIGURE 8.12

Practice Problem 1 Graph the inequality $3x + y < 6$.

In Figure 8.12(b), the graph of the equation $2x + y = 6$ is drawn as a dashed line to indicate that points on the line are *not* included in the solution set of the inequality $2x + y > 6$. In Figure 8.13(a), however, the graph of the inequality $2x + y \geq 6$ shows $2x + y = 6$ as a solid line. The solid line indicates that the points on the line *are* included in the solution set. Note that the region below the line $2x + y = 6$ in Figure 8.13(b) is the graph of the inequality $2x + y < 6$.

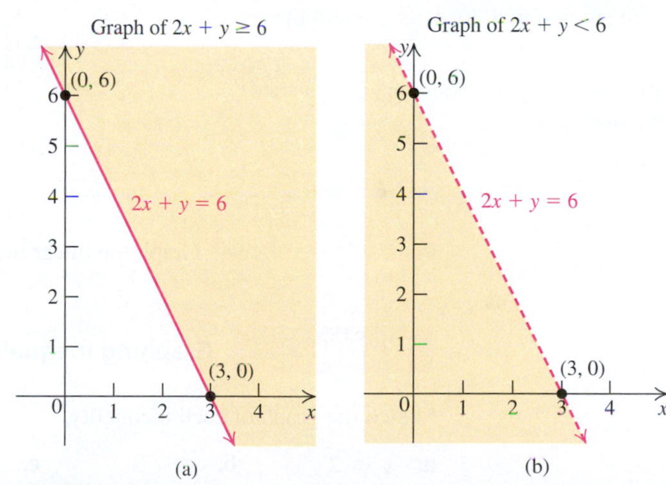

FIGURE 8.13

In general, the graph of a linear inequality's *corresponding equation* (found by changing the inequality symbol to the equality sign) is a line that separates the plane into two regions, each called a *half plane*. The line itself is called the **boundary** of each region. *All* of the points in one region satisfy the inequality, and *none* of the points in the other region are solutions. You can therefore test one point not on the boundary, called a **test point**, to determine which region represents the solution set.

PROCEDURE IN ACTION

EXAMPLE 2 Graphing a Linear Inequality in Two Variables

OBJECTIVE

Graph an inequality in two variables.

Step 1 Replace the inequality symbol with the equality $(=)$ sign.

Step 2 Sketch the graph of the *corresponding equation* in Step 1. Use a dashed line for the boundary if the given inequality sign is $<$ or $>$ and a solid line if the inequality symbol is $\le$ or $\ge$.

Step 3 The graph in Step 2 divides the plane into two regions. Select a *test point* in either region, but not on the graph of the equation in Step 1.

Step 4 **(i)** If the coordinates of the test point satisfy the inequality, then so do all of the points in that region. Shade that region.

(ii) If the test point's coordinates do not satisfy the inequality, shade the region that does not contain the test point.

The shaded region (including the boundary if it is solid) is the graph of the inequality.

EXAMPLE

Graph the linear inequality $y \ge -2x + 4$.

1. $y = -2x + 4$

2.

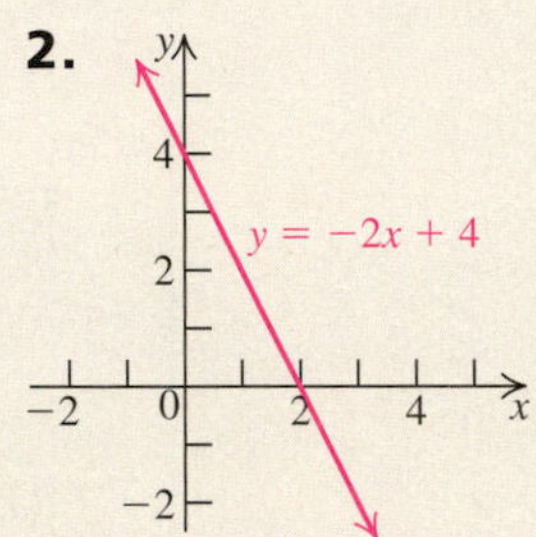

The graph of the equation $y = -2x + 4$ is shown in the figure.

The line drawn is solid because the given inequality sign is $\ge$.

3.

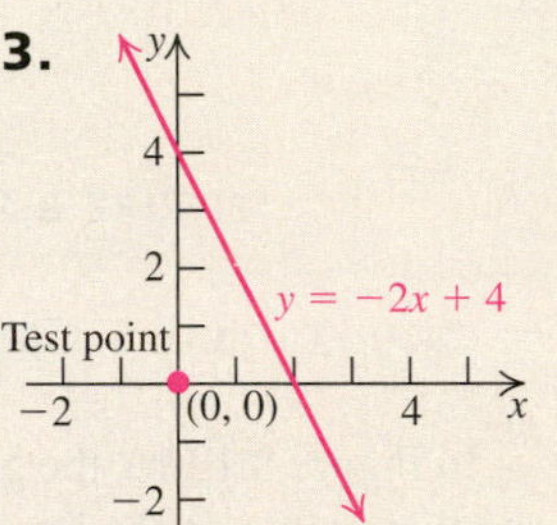

Select $(0, 0)$ as a test point.

4.

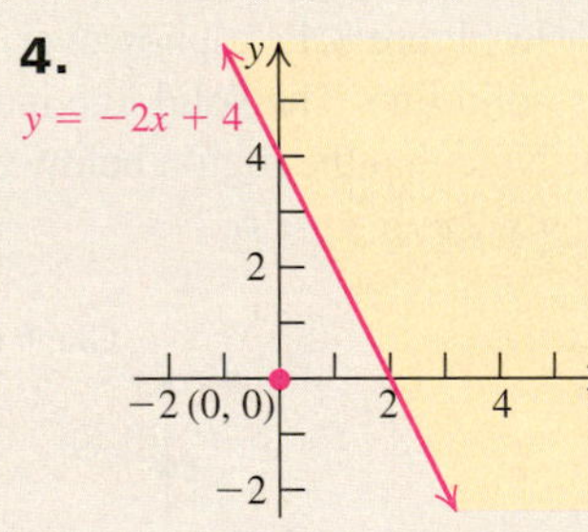

Because $0 \ge -2(0) + 4$ is false, the coordinates of the test point $(0, 0)$ do *not* satisfy the inequality $y \ge -2x + 4$.

The graph of the solution set of $y \ge -2x + 4$ is on the side of the line that does not contain $(0, 0)$. It is shaded in the figure.

Practice Problem 2 Graph the linear inequality $y \le -2x + 4$.

EXAMPLE 3 Graphing Inequalities

Sketch the graph of each inequality.

a. $x \ge 2$ **b.** $y < 3$ **c.** $x + y < 4$

Solution

Steps	Inequality		
1. Change the inequality to an equality.	**a.** $x \geq 2$ $x = 2$	**b.** $y < 3$ $y = 3$	**c.** $x + y < 4$ $x + y = 4$

2. Graph the equation in Step 1 with a dashed line ($<$ or $>$) or a solid line ($\leq$ or $\geq$).

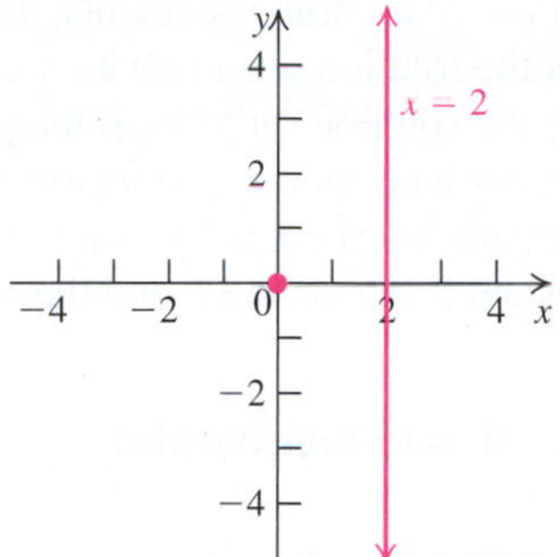

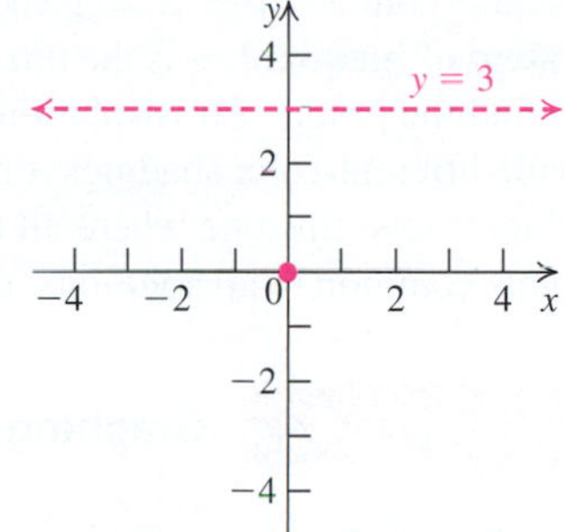

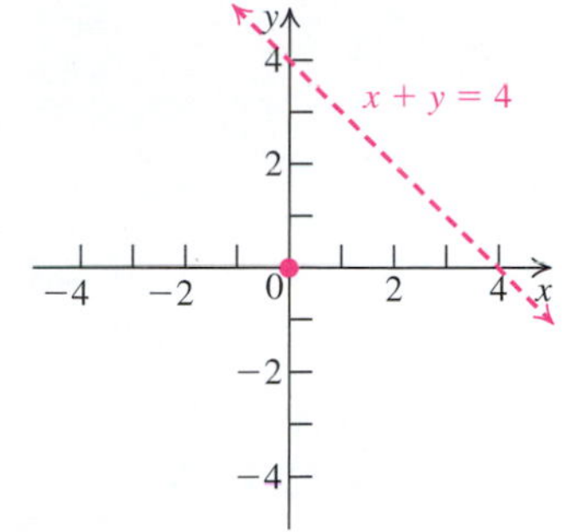

3. Select a test point.

- **a.** Test $(0, 0)$ in $x \geq 2$; $0 \geq 2$ is a false statement. The region not containing $(0, 0)$, together with the vertical line, is the solution set.
- **b.** Test $(0, 0)$ in $y < 3$; $0 < 3$ is a true statement. The region containing $(0, 0)$ is the solution set.
- **c.** Test $(0, 0)$ in $x + y < 4$; $0 + 0 < 4$ is a true statement. The region containing $(0, 0)$ is the solution set.

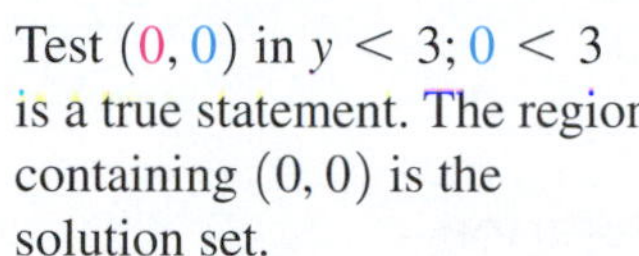

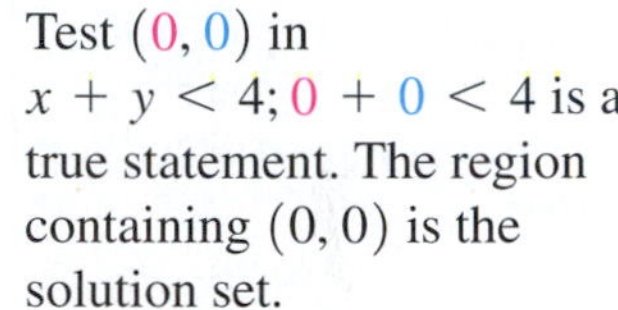

4. Shade the solution set.

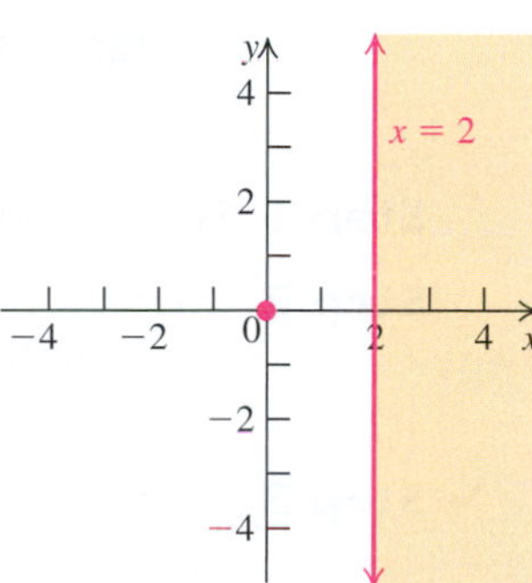

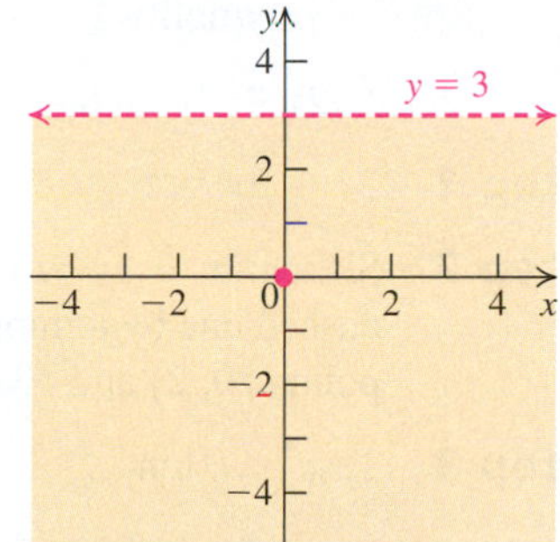

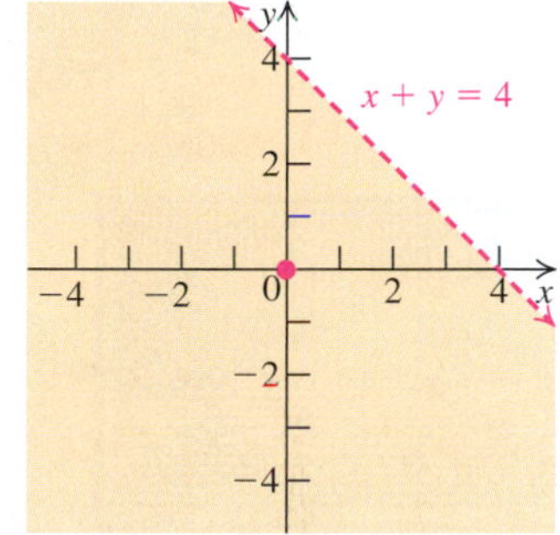

Practice Problem 3 Graph the inequality $x + y > 9$.

TECHNOLOGY CONNECTION

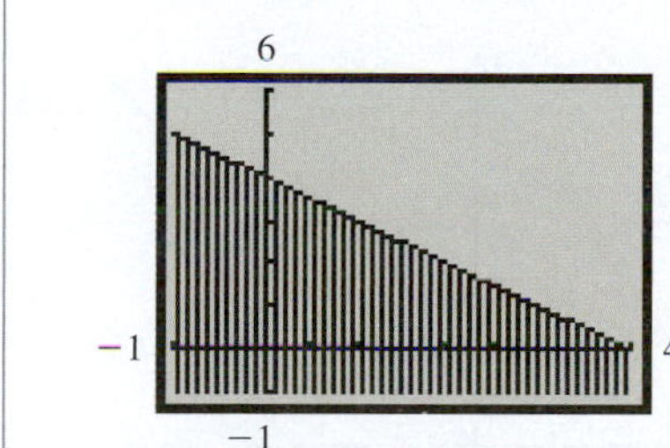

To shade a graph on a graphing calculator, first solve the corresponding equation for y. Enter the equation using the [Y =] key. On some calculators, you can move the cursor to the symbol to the left of Y_1. Press [ENTER] to cycle through the shading options: ◣ shades below the line, and ◥ shades above it.

2 Graph systems of linear inequalities in two variables.

Systems of Linear Inequalities in Two Variables

An ordered pair (a, b) is a **solution of a system of inequalities** involving two variables if it is a solution of each of its inequalities. For example, $(0, 1)$ is a solution of the system

$$\begin{cases} 2x + y \le 1 \\ x - 3y < 0 \end{cases}$$

because both $2(0) + 1 \le 1$ and $0 - 3(1) < 0$ are true statements. The **solution set of a system of inequalities** is the *intersection* of the solution sets of all inequalities in the system. To find the graph of this solution set, graph the solution set of each inequality in the system (with different-color shadings) on the same coordinate plane. The region common to all of the solution sets, (the one where all the shaded parts overlap), is the solution set. If there are no points common to all solutions, then the system is inconsistent, and the answer is $\varnothing$.

TECHNOLOGY CONNECTION

A graphing calculator uses overlapping crosshatch shading to display the solution set of a system of inequalities. The solution set for the system

$$\begin{cases} 2x + 3y > 6 \\ y - x \le 0 \end{cases}$$

is shown here.

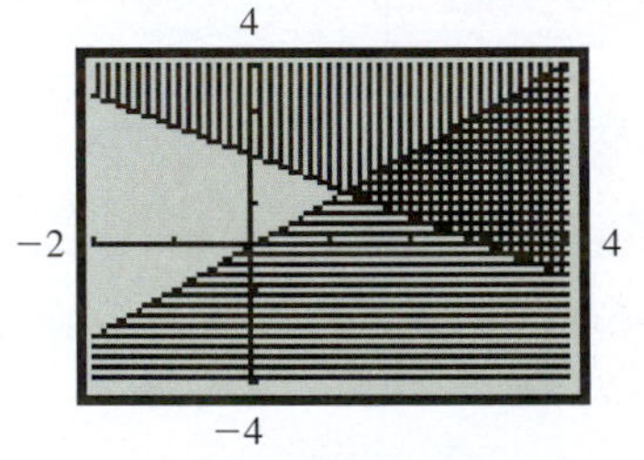

EXAMPLE 4 Graphing a System of Two Inequalities

Graph the solution set of the system of inequalities.

$$\begin{cases} 2x + 3y > 6 & (1) \\ y - x \le 0 & (2) \end{cases}$$

Solution

Graph each inequality separately in the same coordinate plane.

Inequality 1

$2x + 3y > 6$

Step 1 $2x + 3y = 6$

Step 2 Sketch $2x + 3y = 6$ as a dashed line by joining the points $(0, 2)$ and $(3, 0)$.

Step 3 Test $(0, 0)$ in

$$2x + 3y > 6$$
$$2 \cdot 0 + 3 \cdot 0 \overset{?}{>} 6$$
$$0 \overset{?}{>} 6 \quad \text{False}$$

Step 4 Shade as in Figure 8.14(a).

Inequality 2

$y - x \le 0$

Step 1 $y - x = 0$

Step 2 Sketch $y - x = 0$ as a solid line by joining the points $(0, 0)$ and $(1, 1)$.

Step 3 Test $(1, 0)$ in

$$y - x \le 0$$
$$0 - 1 \overset{?}{\le} 0$$
$$-1 \overset{?}{\le} 0 \quad \text{True}$$

Step 4 Shade as in Figure 8.14(b).

The graph of the solution set of the system of inequalities (1) and (2) (the region where the shadings overlap) is shown shaded in purple in Figure 8.14(c).

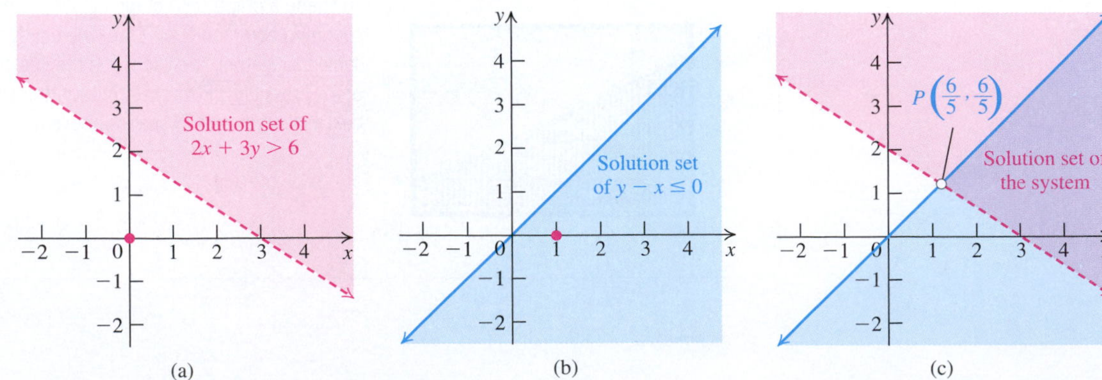

FIGURE 8.14 Graph of a system of two inequalities

Practice Problem 4 Graph the solution set of the system of inequalities.

$$\begin{cases} 3x - y > 3 \\ x - y \geq 1 \end{cases}$$

The point $P\left(\frac{6}{5}, \frac{6}{5}\right)$ in Figure 8.14(c) is the point of intersection of the two lines $2x + 3y = 6$ and $y - x = 0$. Such a point is called a **corner point**, or **vertex**, of the solution set. It is found by solving the system of equations $\begin{cases} 2x + 3y = 6 \\ y - x = 0 \end{cases}$.

In this case, however, the point P is not in the solution set of the given system because the ordered pair $\left(\frac{6}{5}, \frac{6}{5}\right)$ satisfies only inequality (2), but not inequality (1), of Example 4. We show P as an open circle in Figure 8.14(c).

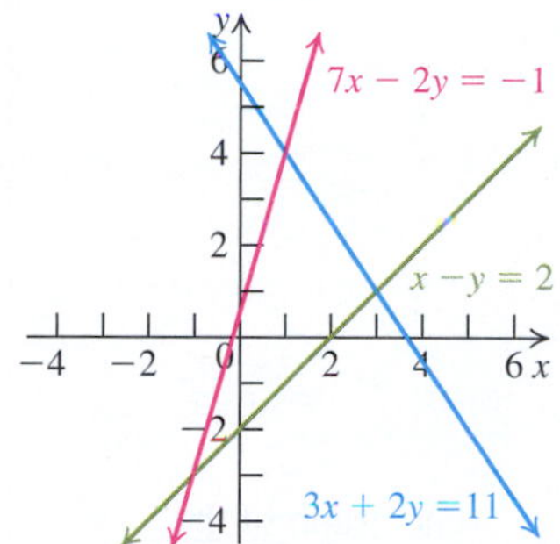

FIGURE 8.15

EXAMPLE 5 Solving a System of Three Linear Inequalities

Sketch the graph and label the vertices of the solution set of the system of linear inequalities.

$$\begin{cases} 3x + 2y \leq 11 & (1) \\ x - y \leq 2 & (2) \\ 7x - 2y \geq -1 & (3) \end{cases}$$

Solution

First, on the same coordinate plane, sketch the graphs of the three linear equations that correspond to the three inequalities. Because either $\leq$ or $\geq$ occurs in all three inequalities, all of the equations are graphed as *solid* lines. See Figure 8.15.

Notice that $(0, 0)$ satisfies each of the inequalities but none of the equations. You can use $(0, 0)$ as the test point for each inequality.

Make the following conclusions:

(i) Because $(0, 0)$ lies below the line $3x + 2y = 11$, the region on or below the line $3x + 2y = 11$ is in the solution set of inequality (1).

(ii) Because $(0, 0)$ is above the line $x - y = 2$, the region on or above the line $x - y = 2$ is in the solution set of inequality (2).

(iii) Finally, because $(0, 0)$ lies below the line $7x - 2y = -1$, the region on or below the line $7x - 2y = -1$ belongs to the solution set of inequality (3).

SIDE NOTE

In working with linear inequalities, you must use the test point to graph the solution set. *Do not* assume that just because the symbol $\leq$ or $<$ appears in the inequality, the graph of the inequality is below the line.

The region common to the regions in **(i)**, **(ii)**, and **(iii)** (including the boundaries) is the solution set of the given system of inequalities. The solution set is the shaded region in Figure 8.16, including the sides of the triangle.

Figure 8.16 also shows the vertices of the solution set. These vertices are obtained by solving each pair of equations in the system. Because all vertices are solutions of the given system, they are shown as (solid) points.

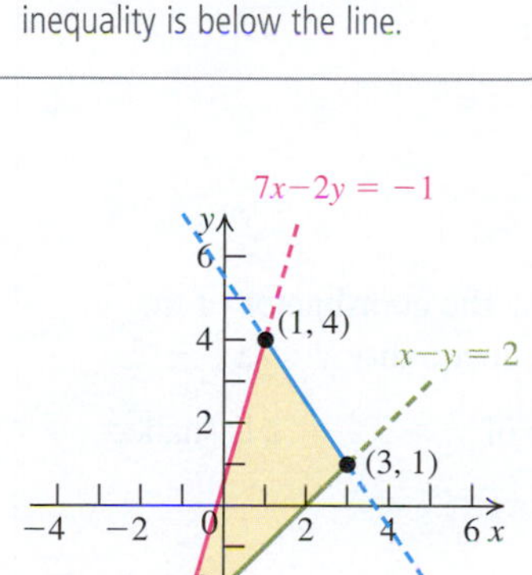

FIGURE 8.16 Graph of the solution set

a. To find the vertex $(3, 1)$, solve the system

$$\begin{cases} 3x + 2y = 11 & (1) \\ x - y = 2 & (2) \end{cases}$$

Solve equation (2) for x to obtain $x = 2 + y$. Substitute $x = 2 + y$ in equation (1).

$$3(2 + y) + 2y = 11$$
$$6 + 3y + 2y = 11 \quad \text{Distributive property}$$
$$y = 1 \quad \text{Solve for } y.$$

Back-substitute $y = 1$ in equation (2) to find $x = 3$. The vertex is the point $(3, 1)$.

b. Solve the system of equations $\begin{cases} x - y = 2 \\ 7x - 2y = -1 \end{cases}$ by the substitution method to find the vertex $(-1, -3)$.

c. Solve the system of equations $\begin{cases} 3x + 2y = 11 \\ 7x - 2y = -1 \end{cases}$ by the elimination method to find the vertex $(1, 4)$.

Practice Problem 5 Sketch the graph (and label the corner points) of the solution set of the system of inequalities.

$$\begin{cases} 2x + 3y < 16 \\ 4x + 2y \geq 16 \\ 2x - \ \ y \leq 8 \end{cases}$$

3 Graph a nonlinear inequality in two variables.

Nonlinear Inequality

We follow the same steps that we used in solving a linear inequality.

PROCEDURE IN ACTION

EXAMPLE 6 Graphing a Nonlinear Inequality in Two Variables

OBJECTIVE

Graph a nonlinear inequality in two variables.

Step 1 Replace the inequality symbol with an equal $(=)$ sign.

Step 2 Sketch the graph of the *corresponding equation* in Step 1. Use a dashed curve if the given inequality sign is $<$ or $>$ and a solid curve if the inequality symbol is $\leq$ or $\geq$.

Step 3 The graph in Step 2 divides the plane into two regions. Select a *test point* in either region, but not on the graph.

Step 4 **(i)** If the coordinates of the test point satisfy the inequality, then so do all of the points in that region. Shade that region.

(ii) If the test point's coordinates do not satisfy the inequality, shade the region that does not contain the test point.

The shaded region (including the boundary if it is a solid curve) is the graph of the inequality.

EXAMPLE

Graph $y > x^2 - 2$.

1. $y = x^2 - 2$

2.

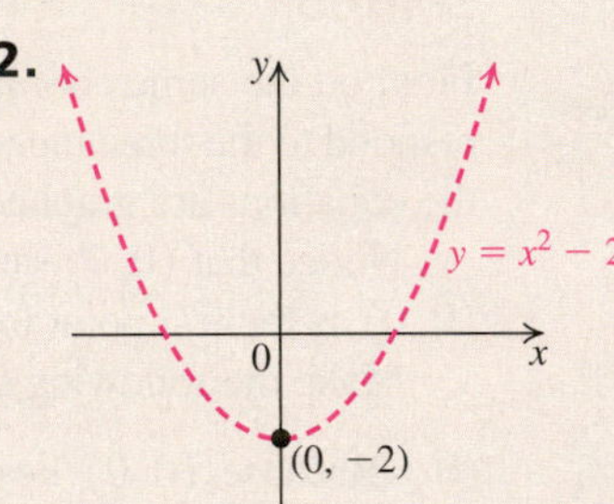

The graph of the parabola with vertex $(0, -2)$ is sketched as a dashed curve in the figure.

3.

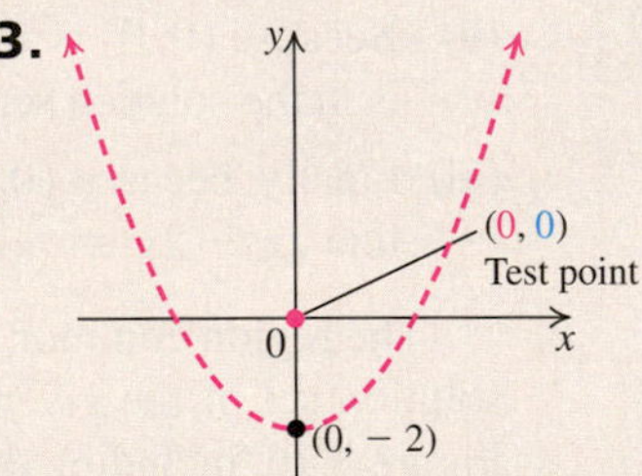

4. Because $0 > 0^2 - 2 = -2$, the coordinates of the test point, $(0, 0)$, satisfy the inequality $y > x^2 - 2$.

The graph of the solution set of $y > x^2 - 2$ is shaded.

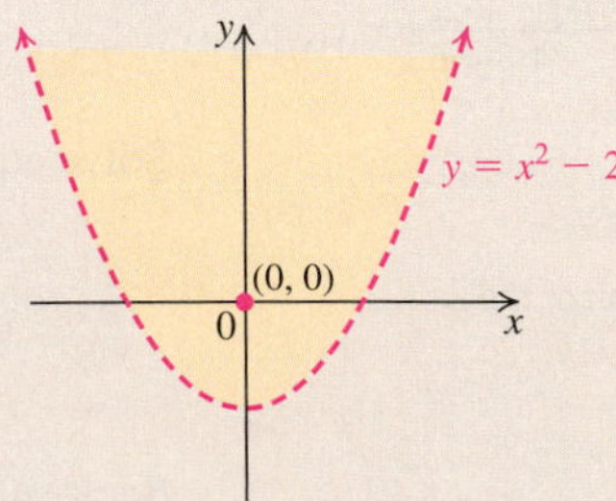

Practice Problem 6 Graph $y \leq x^2 - 2$.

4 Graph systems of nonlinear inequalities in two variables.

Nonlinear Systems

The procedure for solving a nonlinear system of inequalities is identical to the procedure for solving a linear system of inequalities.

EXAMPLE 7 **Solving a Nonlinear System of Inequalities**

Graph the solution set of the system of inequalities.

$$\begin{cases} y \le 4 - x^2 & (1) \\ y \ge \dfrac{3}{2}x - 3 & (2) \\ y \ge -6x - 3 & (3) \end{cases}$$

Solution

Graph each inequality separately in the same coordinate plane. Because $(0, 0)$ is not a solution of any of the corresponding equations, use $(0, 0)$ as a test point for each inequality.

Inequality (1)	Inequality (2)	Inequality (3)
$y \le 4 - x^2$	$y \ge \frac{3}{2}x - 3$	$y \ge -6x - 3$
Step 1 $y = 4 - x^2$	$y = \frac{3}{2}x - 3$	$y = -6x - 3$
Step 2 The graph of $y = 4 - x^2$ is a parabola opening down with vertex $(0, 4)$. Sketch it as a solid curve.	The graph of $y = \frac{3}{2}x - 3$ is a line through the points $(0, -3)$ and $(2, 0)$. Sketch a solid line.	The graph of $y = -6x - 3$ is a line through the points $(0, -3)$ and $\left(-\frac{1}{2}, 0\right)$. Sketch a solid line.
Step 3 Testing $(0, 0)$ in $y \le 4 - x^2$ gives $0 \le 4$, a true statement. The solution set is below the parabola.	Testing $(0, 0)$ in $y \ge \frac{3}{2}x - 3$ gives $0 \ge -3$, a true statement. The solution set is above the line.	Testing $(0, 0)$ in $y \ge -6x - 3$ gives $0 \ge -3$, a true statement. The solution set is above the line.
Step 4 Shade the region as in Figure 8.17(a).	Shade the region as in Figure 8.17(b).	Shade the region as in Figure 8.17(c).

The region common to all three graphs is the graph of the solution set of the given system of inequalities. See Figure 8.17(d).

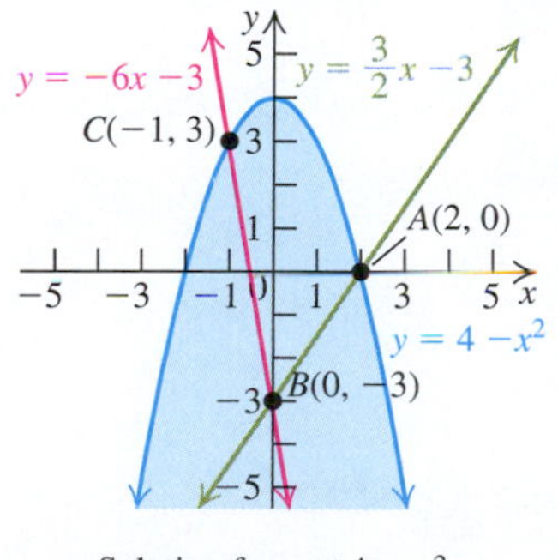

Solution for $y \le 4 - x^2$

(a)

Solution for $y \ge \frac{3}{2}x - 3$

(b)

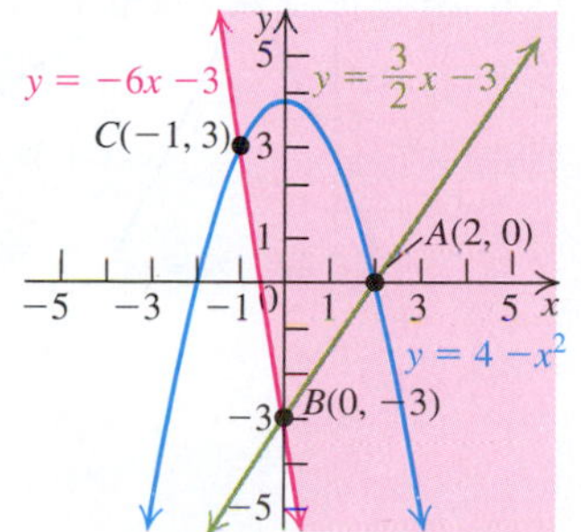

Solution for $y \ge -6x - 3$

(c)

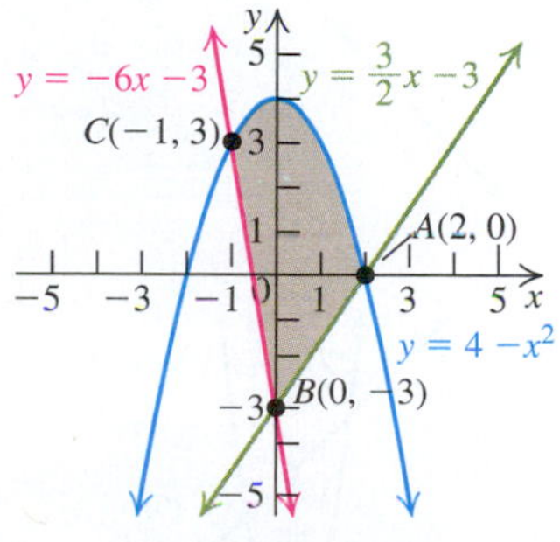

Solution of the system

(d)

FIGURE 8.17

Use the substitution method to locate the points of intersection A, B, and C of the corresponding equations. See Figure 8.17(d). Solve all pairs of corresponding equations.

The point $A(2, 0)$ is the solution of the system

$$\begin{cases} y = 4 - x^2 & \text{Boundary of inequality (1)} \\ y = \frac{3}{2}x - 3 & \text{Boundary of inequality (2)} \end{cases}$$

The point $B(0, -3)$ is the solution of the system

$$\begin{cases} y = \frac{3}{2}x - 3 & \text{Boundary of inequality (2)} \\ y = -6x - 3 & \text{Boundary of inequality (3)} \end{cases}$$

Finally, the point $C(-1, 3)$ is the solution of the system

$$\begin{cases} y = -6x - 3 & \text{Boundary of inequality (1)} \\ y = 4 - x^2 & \text{Boundary of inequality (2)} \end{cases}$$

Practice Problem 7 Graph the solution set of the system of inequalities.

$$\begin{cases} y \geq x^2 + 1 \\ y \leq -x + 13 \\ y < 4x + 13 \end{cases}$$

Answers to Practice Problems

1.

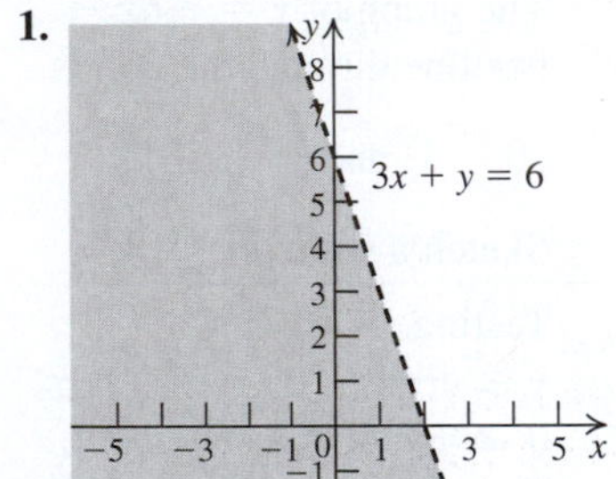

2.

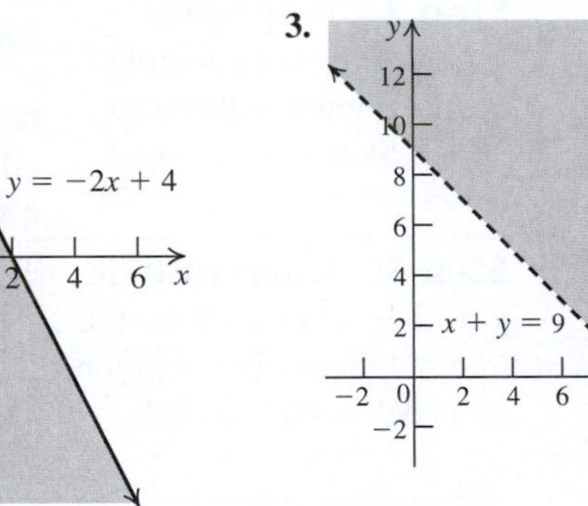

3.

$x + y = 9$

4.

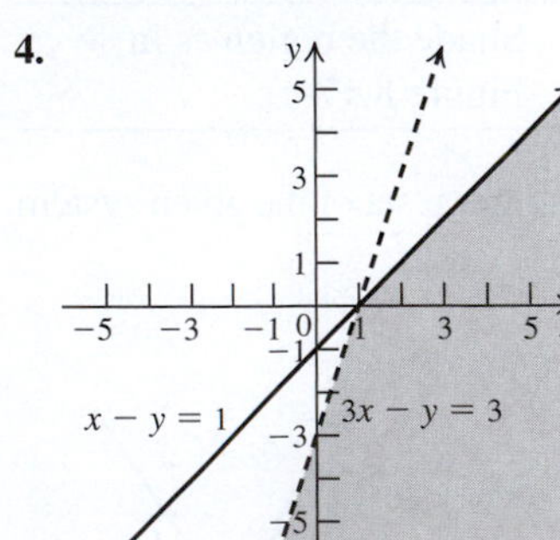

5. $(4, 0)$, $(5, 2)$, and $(2, 4)$

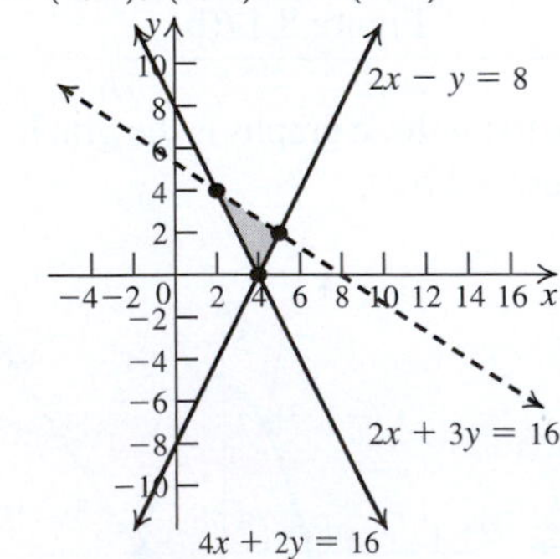

6.

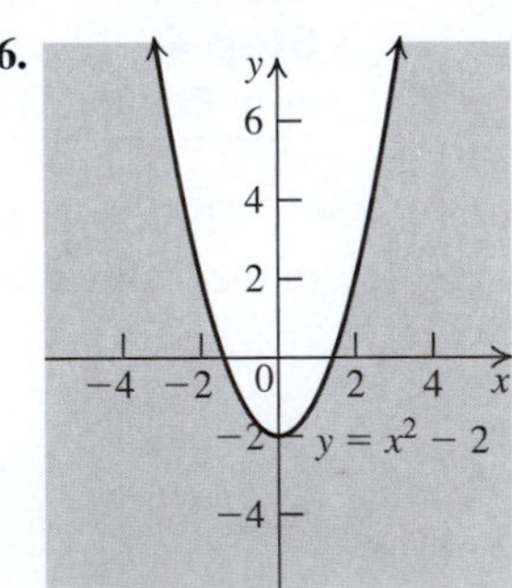

7.

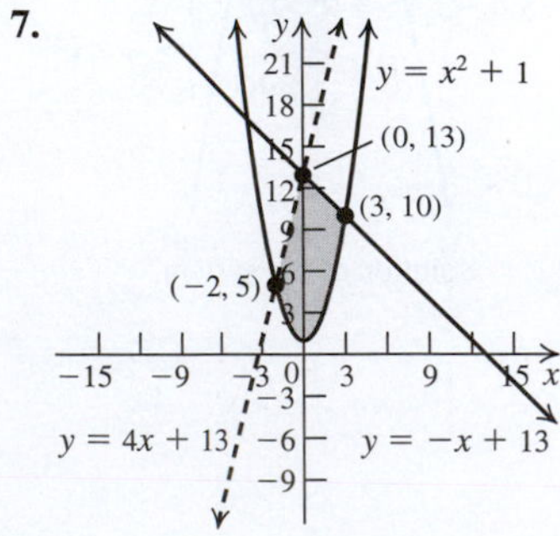

SECTION 8.5 Exercises

Basic Concepts and Skills

1. In the graph of $x - 3y > 1$, the corresponding equation $x - 3y = 1$ is graphed as a(n) ____________ line.
2. If a test point from one of the two regions determined by an inequality's corresponding equation satisfies the inequality, then ____________ in that region satisfy the inequality.
3. In a system of inequalities containing both $2x + y > 5$ and $2x - y \le 3$, the point of intersection of the lines $2x + y = 5$ and $2x - y = 3$ is ____________.
4. In a nonlinear system of equations or inequalities, ____________ equation or inequality must be nonlinear.
5. **True or False.** There are always infinitely many solutions of an inequality $ax + by < c$, where a, b, and c are real numbers and $a \ne 0$.
6. **True or False.** If $x^2 + 2y < 5$ is one of the inequalities in a system of inequalities, then the point $(2, 1)$ cannot be used as a test point in graphing the system.

In Exercises 7–22, graph each inequality.

7. $x \ge 0$
8. $y \ge 0$
9. $x > -1$
10. $y > 2$
11. $x \ge 2$
12. $y \le 3$
13. $y - x < 0$
14. $y + 2x \le 0$
15. $x + 2y < 6$
16. $3x + 2y < 12$
17. $2x + 3y \ge 12$
18. $2x + 5y \ge 10$
19. $x - 2y \le 4$
20. $3x - 4y \le 12$
21. $3x + 5y < 15$
22. $5x + 7y < 35$

In Exercises 23–28, determine which ordered pairs are solutions of each system of inequalities.

23. $\begin{cases} x + y < 2 \\ 2x + y \ge 6 \end{cases}$
a. $(0, 0)$ b. $(-4, -1)$ c. $(3, 0)$ d. $(0, 3)$

24. $\begin{cases} x - y < 2 \\ 3x + 4y \ge 12 \end{cases}$
a. $(0, 0)$ b. $\left(\frac{16}{5}, \frac{3}{5}\right)$ c. $(4, 0)$ d. $\left(\frac{1}{2}, \frac{1}{2}\right)$

25. $\begin{cases} y < x + 2 \\ x + y \le 4 \end{cases}$
a. $(0, 0)$ b. $(1, 0)$ c. $(0, 1)$ d. $(1, 1)$

26. $\begin{cases} y \le 2 - x \\ x + y \le 1 \end{cases}$
a. $(0, 0)$ b. $(0, 1)$ c. $(1, 2)$ d. $(-1, -1)$

27. $\begin{cases} 3x - 4y \le 12 \\ x + y \le 4 \\ 5x - 2y \ge 6 \end{cases}$
a. $(0, 0)$ b. $(2, 0)$ c. $(3, 1)$ d. $(2, 2)$

28. $\begin{cases} 2 + y \le 3 \\ x - 2y \le 3 \\ 5x + 2y \ge 3 \end{cases}$
a. $(0, 0)$ b. $(1, 0)$ c. $(1, 2)$ d. $(2, 1)$

In Exercises 29–44, graph the solution set of each system of linear inequalities and label the vertices (if any) of the solution set.

29. $\begin{cases} x + y \le 1 \\ x - y \le -1 \end{cases}$

30. $\begin{cases} x + y \ge 1 \\ y - x \ge 1 \end{cases}$

31. $\begin{cases} 3x + 5y \le 15 \\ 2x + 2y \le 8 \end{cases}$

32. $\begin{cases} -2x + y \le 2 \\ 3x + 2y \le 4 \end{cases}$

33. $\begin{cases} 2x + 3y \le 6 \\ 4x + 6y \ge 24 \end{cases}$

34. $\begin{cases} x + 2y \ge 6 \\ 2x + 4y \le 4 \end{cases}$

35. $\begin{cases} 3x + y \le 8 \\ 4x + 2y \ge 4 \end{cases}$

36. $\begin{cases} x + 2y \le 12 \\ 2x + 4y \ge 8 \end{cases}$

37. $\begin{cases} x \ge 0 \\ y \ge 0 \\ x + y \le 1 \end{cases}$

38. $\begin{cases} x \ge 0 \\ y \ge 0 \\ x + y > 2 \end{cases}$

39. $\begin{cases} x \ge 0 \\ y \ge 0 \\ 2x + 3y \ge 6 \end{cases}$

40. $\begin{cases} x \ge 0 \\ y \ge 0 \\ 3x + 4y \le 12 \end{cases}$

41. $\begin{cases} x + y \le 1 \\ x - y \ge 1 \\ 2x + y \le 1 \end{cases}$

42. $\begin{cases} x + y \le 1 \\ x - y \ge 1 \\ 2x + y \ge 1 \end{cases}$

43. $\begin{cases} x + y \le 1 \\ x - y < 2 \\ -x + y \le 3 \\ x + y \ge -4 \end{cases}$

44. $\begin{cases} x + y \le -1 \\ 2x - y > -2 \\ -2x + y \ge -3 \\ 2x + 2y \ge -4 \end{cases}$

In Exercises 45–50, use the following figure to indicate the regions (A–G) that correspond to the graph of the given system of inequalities.

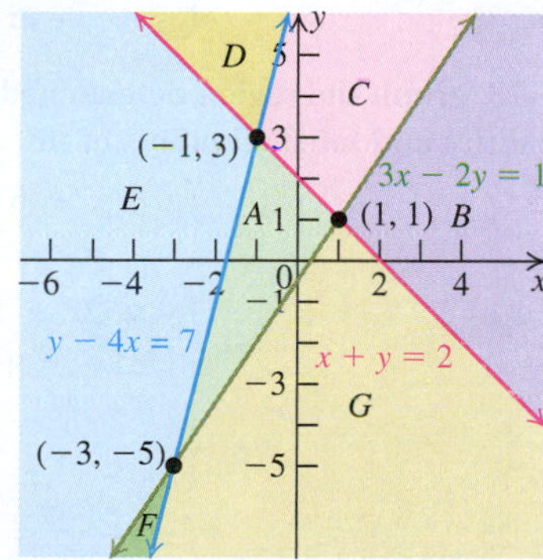

45. $\begin{cases} x + y \le 2 \\ 3x - 2y \le 1 \\ y - 4x \le 7 \end{cases}$

46. $\begin{cases} x + y \ge 2 \\ 3x - 2y \le 1 \\ y - 4x \le 7 \end{cases}$

47. $\begin{cases} x + y \ge 2 \\ 3x - 2y \ge 1 \\ y - 4x \le 7 \end{cases}$

48. $\begin{cases} x + y \ge 2 \\ 3x - 2y \le 1 \\ y - 4x \ge 7 \end{cases}$

49. $\begin{cases} x + y \le 2 \\ 3x - 2y \le 1 \\ y - 4x \ge 7 \end{cases}$

50. $\begin{cases} x + y \le 2 \\ 3x - 2y \ge 1 \\ y - 4x \ge 7 \end{cases}$

In Exercises 51–64, use the procedure for graphing a nonlinear inequality in two variables to graph each inequality.

51. $y \le x^2 + 2$

52. $y < x^2 - 3$

53. $y > x^2 - 1$

54. $y \ge x^2 + 1$

55. $y \le (x - 1)^2 + 3$

56. $y > -(x + 1)^2 + 4$

57. $x^2 + y^2 > 4$

58. $x^2 + y^2 \le 9$

59. $(x - 1)^2 + (y - 2)^2 \le 9$

60. $(x + 2)^2 + (y - 1)^2 > 4$

61. $y < 2^x$

62. $y \ge e^x$

63. $y < \log x$

64. $y \ge \ln x$

For Exercises 65–70, use the following figure to indicate the region or regions that correspond to the graph of each system of inequalities.

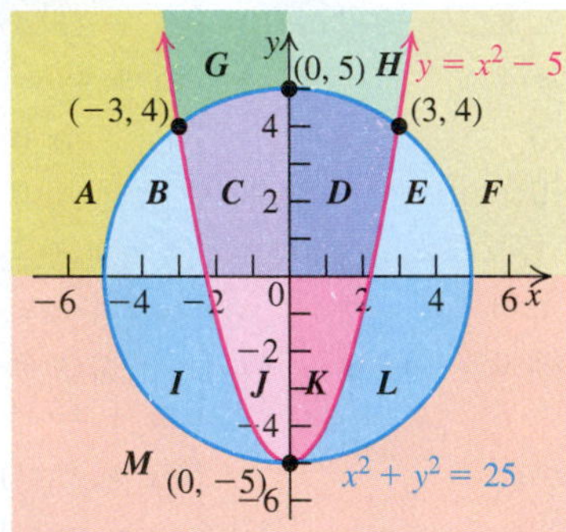

65. $\begin{cases} y \ge x^2 - 5 \\ x^2 + y^2 \le 25 \\ x \ge 0 \end{cases}$

66. $\begin{cases} y \ge x^2 - 5 \\ x^2 + y^2 \ge 25 \\ x \ge 0 \end{cases}$

67. $\begin{cases} y \le x^2 - 5 \\ x^2 + y^2 \le 25 \\ x \ge 0 \end{cases}$

68. $\begin{cases} y \le x^2 - 5 \\ x^2 + y^2 \ge 25 \\ y \ge 0 \end{cases}$

69. $\begin{cases} y \le x^2 - 5 \\ x^2 + y^2 \le 25 \\ y \ge 0 \end{cases}$

70. $\begin{cases} y \ge x^2 - 5 \\ x^2 + y^2 \le 25 \\ y \ge 0 \end{cases}$

In Exercises 71–78, graph the region determined by each system of inequalities and label all points of intersection.

71. $\begin{cases} x + y \le 6 \\ y \ge x^2 \end{cases}$

72. $\begin{cases} x + y \ge 6 \\ y \ge x^2 \end{cases}$

73. $\begin{cases} y \ge 2x + 1 \\ y \le -x^2 + 2 \end{cases}$

74. $\begin{cases} y \le 2x + 1 \\ y \le -x^2 + 2 \end{cases}$

75. $\begin{cases} y \ge x \\ x^2 + y^2 \le 1 \end{cases}$

76. $\begin{cases} y \le x \\ x^2 + y^2 \le 1 \end{cases}$

77. $\begin{cases} x^2 + y^2 \le 25 \\ x^2 + y \le 5 \end{cases}$

78. $\begin{cases} x^2 + y^2 \le 25 \\ x^2 + y \ge 5 \end{cases}$

Applying the Concepts

79. Grocer's shelf space. Your local grocery store has shelf space for, at most, 40 cases of Coke and Sprite. The grocer wants at least ten cases of Coke and five cases of Sprite. Graph the possible stock arrangements and find the vertices.

80. DVD players. An electronic store sells two types, A and B, of DVD players. Each player of type A and type B costs \$60 and \$80, respectively. The store wants at least 20 players in stock and does not want its investment in the players to exceed \$6000. Graph the possible inventory of the players and find the vertices.

81. Home builder. A builder has 43 units of material and 25 units of labor available to build one- and/or two-story houses within one year. Assume that a two-story house requires seven units of material and four units of labor and a one-story house requires five units of material and three units of labor. Graph the possible inventory of houses and find the vertices.

82. Financial planning. Sabrina has \$100,000 to invest in stocks and bonds. The annual rate of return for stocks and bonds is 4% and 7%, respectively. She wants at least \$5600 income from her investments. She also wants to invest at least \$60,000 in bonds. Graph her possible investment portfolio and find the vertices.

83. Engine manufacturing. A company manufactures 3 hp and 5 hp engines. The time (in hours) required to assemble, test, and package each type of engine is shown in the table below.

	Process		
Engine	**Assemble**	**Test**	**Package**
3 hp	3	2	0.5
5 hp	4.5	1	0.75

The maximum times (in hours) available for assembling, testing, and packaging are 360 hours, 200 hours, and 60 hours, respectively. Graph the possible processing activities and find the vertices.

84. Pet food. Vege Pet Food manufactures two types of dog food: Vegies and Yummies. Each bag of Vegies contains 3 pounds of vegetables and 2 pounds of cereal; each bag of Yummies contains 1.5 pounds of vegetables and 3.5 pounds of cereal. The total pounds of vegetables and cereal available per week are 20,000 pounds and 27,000 pounds, respectively. Graph the possible number of bags of each type of food and find the vertices.

Beyond the Basics

In Exercises 85–92, write a system of linear inequalities that has the given graph.

85.

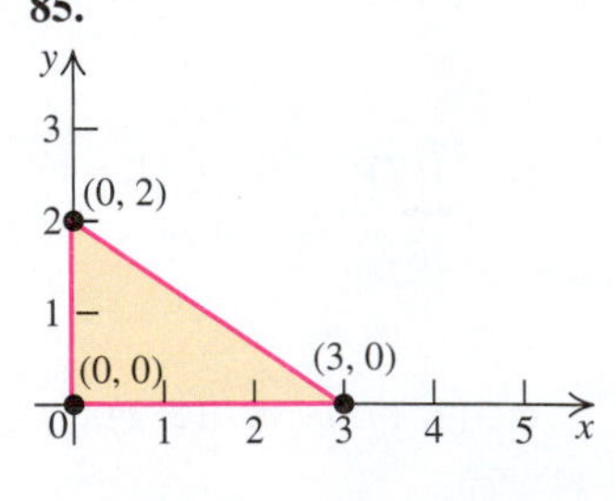

86.

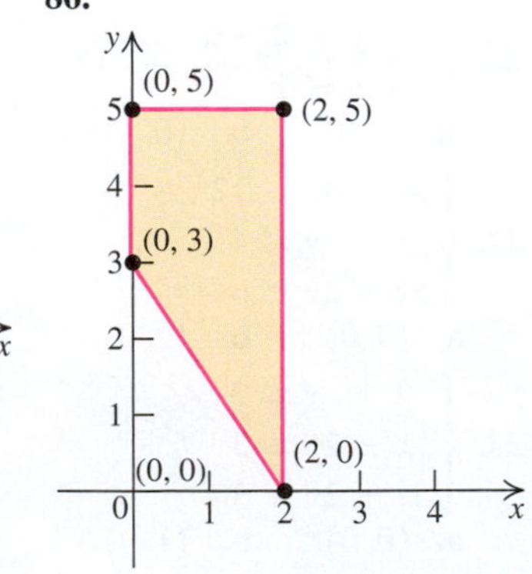

87. **88.**

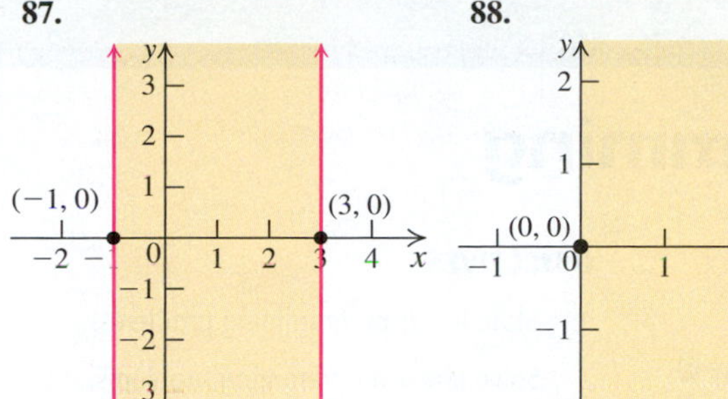

89.

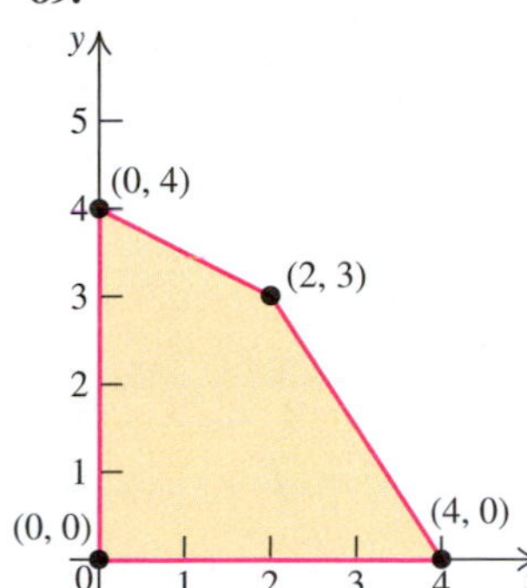

90.

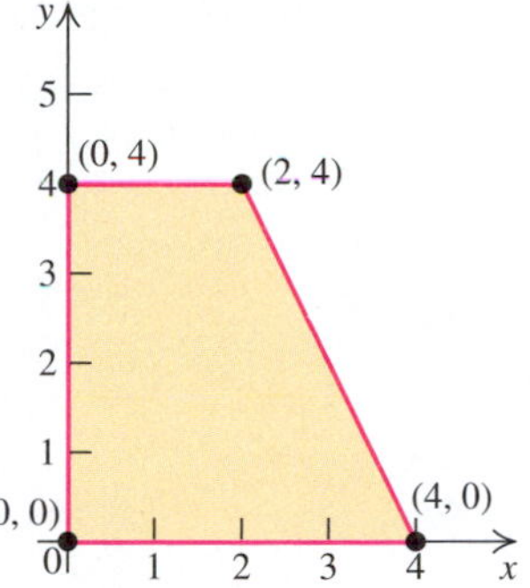

91.

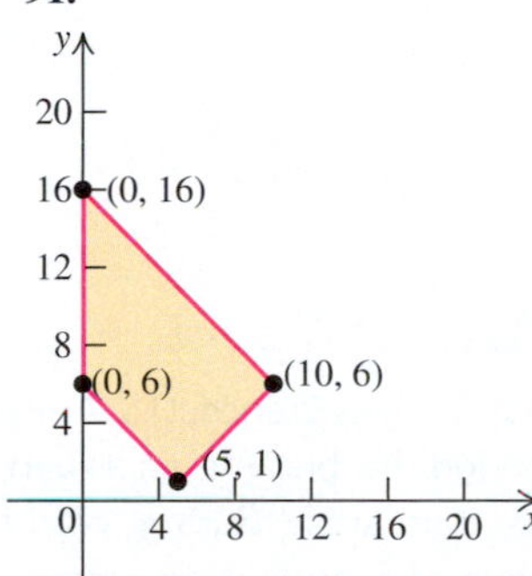

92.

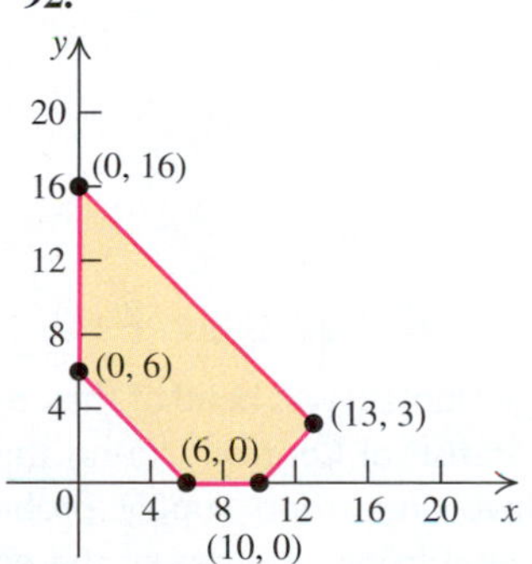

In Exercises 93–102, graph each inequality.

93. $xy \geq 1$

94. $xy < 3$

95. $y < 2^x$

96. $y \leq 2^{x-1}$

97. $y \geq 3^{x-2}$

98. $y > -3^x$

99. $y \geq \ln x$

100. $y < -\ln x$

101. $y > \ln(x - 1)$

102. $y \leq -\ln(x + 1)$

In Exercises 103 and 104, graph the solution set of each inequality.

103. $1 \leq x^2 + y^2 \leq 9$

104. $x^2 \leq y \leq 4x^2$

Critical Thinking/Discussion/Writing

In Exercises 105–106, graph the solution set of each inequality.

105. $|x| + |y| < 1$

106. $|x + 2y| < 4$

Maintaining Skills

In Exercises 107–110, find the point of intersection of the two lines.

107. $x + 2y = 40$ and $3x + y = 30$

108. $3x + y = 30$ and $4x + 3y = 60$

109. $x - 2y = 2$ and $3x + 2y = 12$

110. $3x + 2y = 12$ and $-3x + 2y = 3$

In Exercises 111–112, find the vertices of the triangle whose sides lie along the given lines.

111. $y - x = 1$, $2y + x = 8$, and $y + 2x = 13$

112. $y + x = 3$, $y - 3x = -5$, and $5y - 3x = 23$

113. Consider the following system of linear inequalities:

$$\begin{cases} 3x + 4y \leq 12 \\ 4x + 3y \leq 12 \\ x \geq 0 \\ y \geq 0 \end{cases}$$

a. Graph and shade the solution set of the system of inequalities.

b. List the coordinates of the corner points of the region in part (a).

SECTION 8.6

Linear Programming

George Dantzig 1914–2005
George Dantzig received numerous honors and awards for his work. Dantzig attributed his success in mathematics to his father, who encouraged him to develop his analytic power by making him solve thousands of geometry problems while George was still in high school.

During his first year of graduate studies, Dantzig arrived late in one of Professor Neyman's classes at the University of California at Berkeley. On the blackboard were two problems that Dantzig assumed had been assigned for homework. He submitted his homework a little late. He was surprised to learn that these were the two famous unsolved problems in statistics. The solutions of these problems became part of Dantzig's dissertation for his PhD degree from Berkeley.

BEFORE STARTING THIS SECTION, REVIEW

1 Intersection of sets (Section P.1, page 6)

2 Graphing a line (Section 2.3, page 184)

3 Solving systems of linear inequalities (Section 8.5, page 774)

OBJECTIVES

1 State linear programming problems.

2 Solve linear programming problems.

3 Use linear programming in applications.

◆ The Success of Linear Programming

From 1941 to 1946, George Dantzig was head of the Combat Analysis Branch, U.S. Army Air Corps Headquarters Statistical Control. During this period, he became an expert on planning methods involving troop supply problems that arose during World War II. With a war on several fronts, the size of the problem of coordinating supplies was daunting. This problem was known as *programming*, a military term which at that time referred to plans or schedules for training, logistical supply, or the deployment of forces. Dantzig mechanized the planning process by introducing *linear programming*, where *programming* has the military meaning.

In recent years, linear programming has been applied to problems in almost all areas of human life. A quick check in the library at a university may uncover entire books on linear programming applied to business, agriculture, city planning, and many other areas. Even feedlot operators use linear programming to decide how much and what kinds of food they should give their livestock each day. In Example 3 and in the exercises, we consider several linear programming problems involving two variables.

1 State linear programming problems.

Linear Programming

Recall that if a function f with domain $[a, b]$ has a largest and a smallest value, the largest value is called the *maximum value* and the smallest value is called the *minimum value*. The process of finding the maximum or minimum value of a quantity is called **optimization**. For example, for the function

$$f(x) = x^2, \quad -2 \le x \le 3,$$

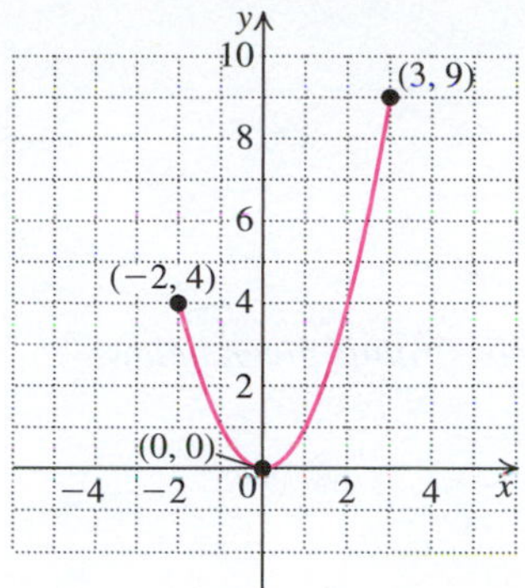

FIGURE 8.18 Optimization

the maximum value of f occurs at $x = 3$ and is given by $f(3) = 3^2 = 9$, while the minimum value occurs at $x = 0$ and is given by $f(0) = 0^2 = 0$. See Figure 8.18. This is an example of an optimization problem involving one variable.

Now we will investigate optimization problems involving two variables, say, x and y. Assume the following conditions:

1. The quantity f to be maximized or minimized can be written as a linear expression in x and y; that is,

$$f = ax + by, \text{ where } a \neq 0, b \neq 0 \text{ are constants.}$$

2. The domain of the variables x and y is restricted to a region S that is determined by (is a solution set of) a system of linear inequalities.

A problem that satisfies conditions (1) and (2) is called a **linear programming problem**.

The inequalities that determine the region S are called **constraints**, the region S is called the **set of feasible solutions**, and $f = ax + by$ is called the **objective function**. A point in S at which f reaches its maximum (or minimum) value, together with the value of f at that point, is called an **optimal solution**.

Consider the following linear programming problem:

Find values of x and y that will make $f = 2x + 3y$ as large as possible (in other words, that will result in a maximum value for f), where x and y are restricted by the following constraints:

$$\begin{cases} 3x + 5y \leq 15 \\ x \geq 0 \\ y \geq 0 \end{cases}$$

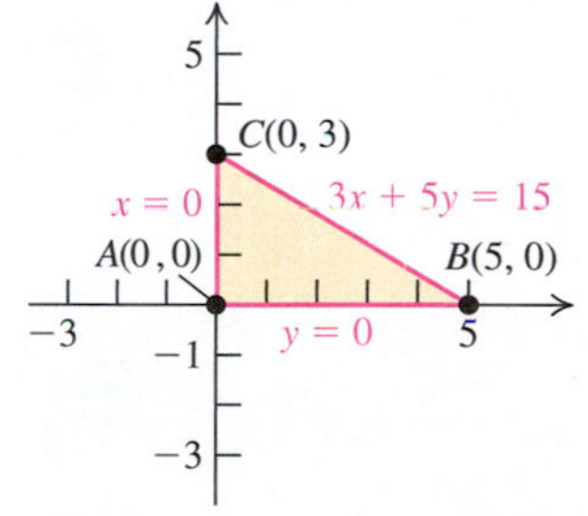

FIGURE 8.19 Feasible solutions

To solve the problem, first sketch the graph of the permissible values of x and y; that is, find the set of feasible solutions determined by the constraints of the problem. Using the method explained in Section 8.5, you find that the set of feasible solutions is the shaded region shown in Figure 8.19.

Now you have to find the values of $f = 2x + 3y$ at each of the points in the shaded region of Figure 8.19 and then see where f takes on the maximum value. The shaded region, however, has infinitely many points. Because you cannot evaluate f at each point, you need another method to solve the problem.

Fortunately, it can be shown that if there is an optimal solution, it must occur at one of the vertices (corner points) of the feasible solution set. This means that when there is an optimal solution, *we can find the maximum (or minimum) value of f by computing the value of f at each vertex and then picking the largest (or smallest) value that results.*

SOLUTION OF A LINEAR PROGRAMMING PROBLEM

1. If a linear programming problem has a solution, it must occur at one of the vertices of the feasible solution set.
2. A linear programming problem may have many solutions, but at least one of them occurs at a vertex of the feasible solution set.
3. In any case, the value of the objective function is unique.

2 Solve linear programming problems.

Solving Linear Programming Problems

We next describe a procedure for solving linear programming problems that have a solution.

PROCEDURE IN ACTION

EXAMPLE 1 Solving a Linear Programming Problem

OBJECTIVE

Solve a linear programming problem.

EXAMPLE

Maximize $f = 2x + 3y$ subject to the constraints given below.

Step 1 Write an expression for the quantity to be maximized or minimized. This expression is the objective function.

1. Maximize the objective function

$$f = 2x + 3y$$

Step 2 Write all constraints as linear inequalities.

2. subject to the constraints

$$\begin{cases} 3x + 5y \le 15 \\ x \ge 0 \\ y \ge 0. \end{cases}$$

Step 3 Graph the solution set of the constraint inequalities. This set is the set of feasible solutions.

3. See Figure 8.19 for the solution set of the constraints.

Step 4 Find all vertices of the solution set in Step 3 by solving all pairs of equations corresponding to the constraint inequalities.

4. Find

$A(0, 0)$ by solving $\begin{cases} x = 0 \\ y = 0 \end{cases}$

$B(5, 0)$ by solving $\begin{cases} y = 0 \\ 3x + 5y = 15 \end{cases}$

and

$C(0, 3)$ by solving $\begin{cases} x = 0 \\ 3x + 5y = 15. \end{cases}$

Step 5 Find the values of the objective function at each of the vertices in Step 4.

5. At $A(0, 0), f = 2(0) + 3(0) = 0.$
At $B(5, 0), f = 2(5) + 3(0) = 10.$
At $C(0, 3), f = 2(0) + 3(3) = 9.$

Step 6 The largest of the values (if any) in Step 5 is the maximum value of the objective function, and the smallest of the values (if any) in Step 5 is the minimum value of the objective function.

6. At $B(5, 0)$, the objective function f has a maximum value of 10. Although you are not asked for the minimum value, it is 0 at $A(0, 0)$.

Practice Problem 1 Maximize $f = 4x + 5y$ subject to the constraints

$$\begin{cases} 5x + 7y \le 35 \\ x \ge 0 \\ y \ge 0\ . \end{cases}$$

We stated in the box on page 783 that the optimal value of the objective function f is unique, but the solution (the point (x, y) at which f attains this optimal value) may not be unique. We next solve a linear programming problem that has infinitely many solutions.

EXAMPLE 2 Solving a Linear Programming Problem Having a Nonunique Solution

Minimize $f = 60x + 40y$ subject to the constraints $x + y \leq 8$, $x + 2y \geq 4$, $3x + 2y \geq 6$, $x \geq 0$, $y \geq 0$.

Solution

Step 1 The objective function is $f = 60x + 40y$.

Step 2 We want to find the minimum value of f that can occur at the points (x, y) that satisfy the constraints

$$\begin{cases} x + y \leq 8 \\ x + 2y \geq 4 \\ 3x + 2y \geq 6 \\ x \geq 0 \\ y \geq 0. \end{cases}$$

Step 3 The graph of the solution of the inequalities in Step 2 is shaded in Figure 8.20.

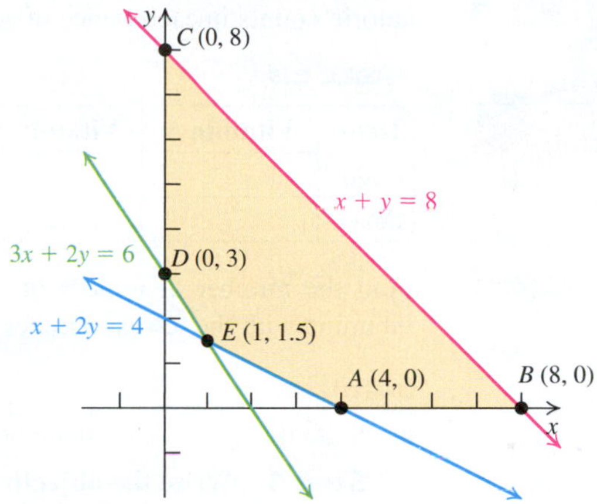

FIGURE 8.20 Feasible solutions

Step 4 Solving the pairs of equations corresponding to the constraint inequalities, we get the coordinates of the vertices $A = (4, 0)$, $B = (8, 0)$, $C = (0, 8)$, $D = (0, 3)$, and $E = (1, 1.5)$.

Step 5 Table 8.2 lists the vertices and the corresponding values of the objective function.

TABLE 8.2

Vertex (x, y)	Value of the Objective Function $f = 60x + 40y$
$A = (4, 0)$	$f = (60)(4) + (40)(0) = 240$
$B = (8, 0)$	$f = (60)(8) + (40)(0) = 480$
$C = (0, 8)$	$f = (60)(0) + (40)(8) = 320$
$D = (0, 3)$	$f = (60)(0) + (40)(3) = \boxed{120}$
$E = (1, 1.5)$	$f = (60)(1) + (40)(1.5) = \boxed{120}$

Step 6 We observe from Table 8.2 that the minimum value of f is 120 and that it occurs at two vertices, D and E. In fact, the unique minimum value occurs at every point on the line segment DE. So the linear programming problem has infinitely many solutions, each of which satisfies $3x + 2y = 6$.

Practice Problem 2 Repeat Example 2 with $f = \frac{1}{2}x + y$.

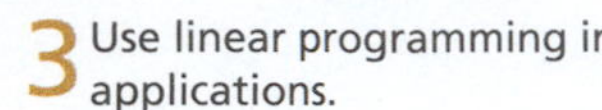
3 Use linear programming in applications.

Applications

The next example shows how linear programming is used to minimize the number of calories in a diet.

EXAMPLE 3 Nutrition; Minimizing Calories

Peter Griffin wants to go on a crash diet and needs your help in designing a lunch menu. The menu is to include two items: soup and salad. The vitamin units (milligrams) and calorie counts in each ounce of soup and salad are given in Table 8.3.

TABLE 8.3

Item	Vitamin A	Vitamin C	Calories
Soup	1	3	50
Salad	1	2	40

The menu must provide at least
10 units of vitamin A.
24 units of vitamin C.

Find the number of ounces of each item in the menu needed to provide the required vitamins with the fewest number of calories.

Solution

a. State the problem mathematically.

Step 1 **Write the objective function.** Let the lunch menu contain x ounces of soup and y ounces of salad. You need to minimize the total number of calories in the menu. Because the soup has 50 calories per ounce (see Table 8.3), x ounces of soup has $50x$ calories. Likewise, the salad has $40y$ calories. The number f of calories in the two items is therefore given by the objective function

$$f = 50x + 40y.$$

Step 2 **Write the constraints.** Because each ounce of soup provides 1 unit of vitamin A and each ounce of salad provides 1 unit of vitamin A, the two items together provide $1 \cdot x + 1 \cdot y = x + y$ units of vitamin A. The lunch must have at least 10 units of vitamin A; this means that

$$x + y \geq 10.$$

Similarly, the two items provide $3x + 2y$ units of vitamin C. The menu must provide at least 24 units of vitamin C, so

$$3x + 2y \geq 24.$$

The fact that x and y cannot be negative means that $x \geq 0$ and $y \geq 0$. You now have four constraints.

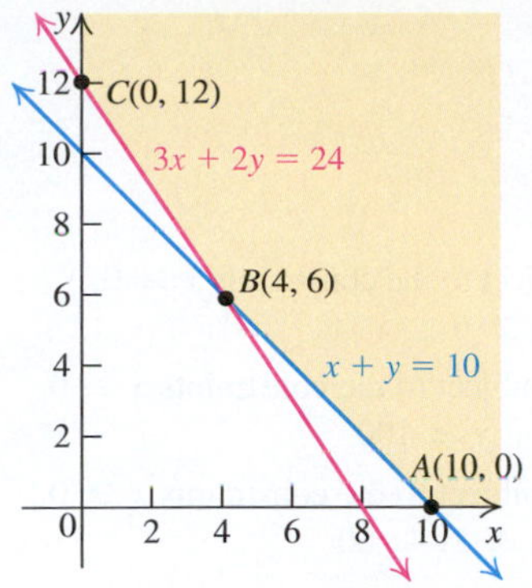

FIGURE 8.21 Feasible solutions

Summarize your information. Find x and y such that the value of

$$f = 50x + 40y \quad \text{Objective function from Step 1}$$

is a *minimum*, with the restrictions

$$\begin{cases} x + \ \ y \geq 10 \\ 3x + 2y \geq 24 \\ \qquad x \geq 0 \\ \qquad y \geq 0. \end{cases} \quad \text{Constraints from Step 2}$$

b. Solve the linear programming problem.

Step 3 **Graph the set of feasible solutions.** The set of feasible solutions is shaded in Figure 8.21. This set is bounded by the lines whose equations are

$$x + y = 10, 3x + 2y = 24, x = 0, \text{and } y = 0.$$

Step 4 **Find the vertices.** The vertices of the set of feasible solutions are

$$A(10, 0), \text{ found by solving } \begin{cases} \qquad y = 0 \\ x + y = 10 \end{cases}$$

$$B(4, 6), \text{ found by solving } \begin{cases} x + \ \ y = 10 \\ 3x + 2y = 24 \end{cases}$$

$$C(0, 12), \text{ found by solving } \begin{cases} x \qquad = 0 \\ 3x + 2y = 24. \end{cases}$$

Step 5 **Find the value of *f* at the vertices.** The value of f at each of the vertices is given in Table 8.4.

TABLE 8.4

Vertex (x, y)	Value of $f = 50x + 40y$
$(10, 0)$	$50(10) + 40(0) = 500$
$(4, 6)$	$50(4) + 40(6) = 440$
$(0, 12)$	$50(0) + 40(12) = 480$

Step 6 **Find the maximum or minimum value of *f*.** From Table 8.4, the smallest value of f is 440, which occurs when $x = 4$ and $y = 6$.

c. State the conclusion.

The lunch menu for Peter Griffin should contain 4 ounces of soup and 6 ounces of salad. His intake of 440 calories will be as small as possible under the given constraints.

Practice Problem 3 How does the answer to Example 3 change if 1 ounce of soup contains 30 calories and 1 ounce of salad contains 60 calories? 

Notice in Example 3 that the value of the objective function $f = 50x + 40y$ can be made as large as you want by choosing x or y (or both) to be very large. So f has a minimum value of 440 but no maximum value.

Answers to Practice Problems

1. The objective function f has a maximum of 28 at the point $(7, 0)$. **2.** The minimum value is 2 at $(4, 0)$ and $(1, 1.5)$. **3.** The minimum number of calories is 300. The lunch then consists of 10 ounces of soup and no salad.

SECTION 8.6 Exercises

Basic Concepts and Skills

1. The process of finding the maximum or minimum value of a quantity is called ____________.

2. In a linear programming problem, the linear expression f that is to be maximized or minimized is called a(n) ____________.

3. The inequalities that determine the region S in a linear programming problem are called ____________, and S is called the set of ____________ solutions.

4. In linear programming, the expression for the objective function is ____________, and all of the constraint inequalities are ____________.

5. **True or False.** In a linear programming problem, the optimal value of the objective function is unique.

6. **True or False.** In a linear programming problem, the optimal value of the objective function occurs at a unique point.

In Exercises 7–12, find the maximum and the minimum value of each objective function *f*. The graph of the set of feasible solutions is as follows:

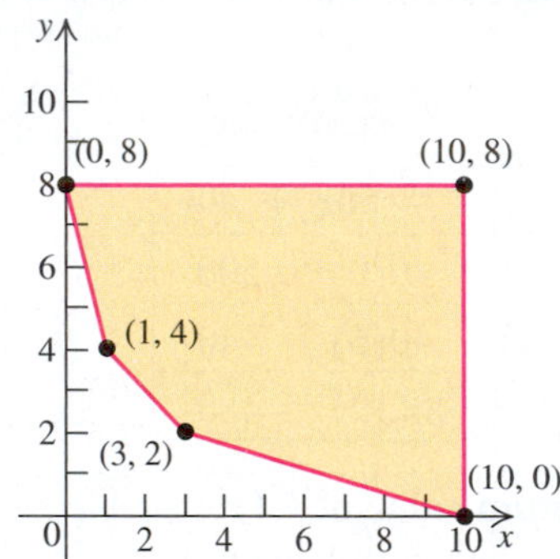

7. $f = x + y$
8. $f = 2x + y$
9. $f = x + 2y$
10. $f = 2x + 5y$
11. $f = 5x + 2y$
12. $f = 8x + 5y$

In Exercises 13–24, you are given an objective function *f* and a set of constraints. Find the set of feasible solutions determined by the given constraints and then maximize or minimize the objective function as directed.

13. Maximize $f = 9x + 13y$, subject to the constraints $x \geq 0, y \geq 0, 5x + 8y \leq 40, 3x + y \leq 12$

14. Maximize $f = 7x + 6y$, subject to the constraints $x \geq 0, y \geq 0, 2x + 3y \leq 13, x + y \leq 5$

15. Maximize $f = 5x + 7y$, subject to the constraints $x \geq 0$, $y \geq 0, x + y \leq 70, x + 2y \leq 100, 2x + y \leq 120$

16. Maximize $f = 2x + y$, subject to the constraints $x \geq 0$, $y \geq 0, x + y \geq 5, 2x + 3y \leq 21, 4x + 3y \leq 24$

17. Minimize $f = x + 4y$, subject to the constraints $x \geq 0$, $y \geq 0, x + 3y \geq 3, 2x + y \geq 2$

18. Minimize $f = 5x + 2y$, subject to the constraints $x \geq 0$, $y \geq 0, 5x + y \geq 10, x + y \geq 6$

19. Minimize $f = 13x + 15y$, subject to the constraints $x \geq 0$, $y \geq 0, 3x + 4y \geq 360, 2x + y \geq 100$

20. Minimize $f = 40x + 37y$, subject to the constraints $x \geq 0$, $y \geq 0, 10x + 3y \geq 180, 2x + 3y \geq 60$

21. Maximize $f = 15x + 7y$, subject to the constraints $3x + 8y \leq 24, 2x + 2y \leq 8, 5x + 3y \leq 18, x \geq 0$, $y \geq 0$.

22. Maximize $f = 17x + 10y$, subject to the constraints $2x + 3y \geq 12, 5y - 6x \leq 6, x \leq 4, x \geq 0, y \geq 2$.

23. Minimize $f = 8x + 16y$, subject to the constraints $2x + y \geq 40, x + 2y \geq 50, x + y \geq 35, x \geq 0, y \geq 0$.

24. Minimize $f = 12x + 12y$, subject to the constraints $2x + 3y \geq 4, 3x + y \leq 6, x \geq 0, y \geq 0$.

Applying the Concepts

25. **Agriculture: crop planning.** A farmer owns 240 acres of cropland. He can use his land to produce corn and soybeans. A total of 320 labor hours is available to him during the production period of both of these commodities. Each acre for corn production requires two hours of labor, and each acre for soybean production requires one hour of labor. Assuming that the profit per acre in corn production is \$50 and that in soybean production is \$40, find the number of acres that he should allocate to the production of corn and soybeans so as to maximize his profit.

26. Repeat Exercise 25 assuming that the profit per acre in corn production is \$30 and that in soybean production is \$40.

27. **Manufacturing: production scheduling.** Two machines (I and II) each produce two grades of plywood: Grade A and Grade B. In one hour of operation, machine I produces 20 units of Grade A and 10 units of Grade B plywood. Also in one hour of operation, machine II produces 30 units of Grade A and 40 units of Grade B plywood. The machines are required to meet a production schedule of at least 1400 units of Grade A and 1200 units of Grade B plywood. Assuming that the cost of operating machine I is \$50 per hour and the cost of operating machine II is \$80 per hour, find the number of hours each machine should be operated so as to minimize the cost.

28. Repeat Exercise 27 assuming that the cost of operating machine I is \$70 per hour and the cost of operating machine II is \$90 per hour.

29. **Advertising.** The Sundial Cheese Company wants to allocate its \$6000 advertising budget to two local media—television and newspaper—so that the exposure (number of viewers) to its advertisements is maximized. There are 60,000 viewers exposed to each minute of television time and 20,000 readers exposed to each page of newspaper advertising. Each minute

of television time costs $1000, and each page of newspaper advertising costs $500. How should the company allocate its budget if it has to buy at least one minute of television time and at least two pages of newspaper advertising?

30. **Advertising.** The Fantastic Bread Company allocates a maximum of $24,000 to advertise its product to television and newspaper. Each hour of television costs $4000, and each page of newspaper advertising costs $3000. The exposure from each hour of television time is assumed to be 120,000, and the exposure from each page of newspaper advertising is 80,000. Furthermore, the board of directors requires at least two hours of television time and one page of newspaper advertising. How should the advertising budget be divided to maximize exposure to the advertisements?

31. **Agriculture.** A Florida citrus company has 480 acres of land for growing oranges and grapefruit. Profits per acre are $40 for oranges and $30 for grapefruit. The total labor hours available during the production are 800. Each acre for oranges uses two hours of labor, and each acre for grapefruit uses one hour of labor. If the fixed costs are $3000, what is the maximum profit?

32. **Agriculture.** The Florida Juice Company makes two types of fruit punch—Fruity and Tangy—by blending orange juice and apple juice into a mixture. The fruit punch is sold in 5-gallon bottles. A bottle of Fruity earns a profit of $3, and a bottle of Tangy earns a $2 profit. A bottle of Fruity requires 3 gallons of orange juice and 2 gallons of apple juice, and a bottle of Tangy requires 4 gallons of orange juice and 1 gallon of apple juice. If 200 gallons of orange juice and 120 gallons of apple juice are available, find the maximum profit for the company.

33. **Manufacturing.** The Fantastic Furniture Company manufactures rectangular and circular tables. Each rectangular table requires one hour of assembling and one hour of finishing, and each circular table requires one hour of assembling and two hours of finishing. The company's 20 assemblers and 30 finishers each work 40 hours per week. Each rectangular table brings a profit of $3, and each circular table brings a profit of $4. If all of the tables manufactured are sold, how many of each type should be made to maximize profits?

34. **Manufacturing.** A company produces a portable global positioning system (GPS) and a portable digital video disc (DVD) player, each of which requires three stages of processing. The length of time for processing each unit is given in the following table.

Stage	GPS (hr/unit)	DVD (hr/unit)	Maximum Process Capacity (hr/day)
I	12	12	840
II	3	6	300
III	8	4	480
Profit per unit	$6	$4	

How many of each product should the company produce per day to maximize profit?

35. **Building houses.** A contractor has 100 units of concrete, 160 units of wood, and 400 units of glass. He builds terraced houses and cottages. A terraced house requires 1 unit of concrete, 2 units of wood, and 2 units of glass; a cottage requires 1 unit of concrete, 1 unit of wood, and 5 units of glass. A terraced house sells for $40,000, and a cottage sells for $45,000. How many houses of each type should the contractor build to obtain the maximum amount of money?

36. **Hospitality.** Mrs. Adams owns a motel consisting of 300 single rooms and an attached restaurant that serves breakfast. She knows that 30% of the male guests and 50% of the female guests will eat in the restaurant. Suppose she makes a profit of $18.50 per day from every guest who eats in her restaurant and $15 per day from every guest who does not eat in her restaurant. Assuming that Mrs. Adams never has more than 125 female guests and never more than 220 male guests, find the number of male and female guests that would provide the maximum profit.

37. **Hospitality.** In Exercise 36, find Mrs. Adams's maximum profit during a season in which she always has at least 100 guests, assuming that she makes a profit of $15 per day from every guest who eats in her restaurant and she suffers a loss of $2 per day from every guest who does not eat in her restaurant.

38. **Transportation.** Major Motors, Inc., must produce at least 5000 luxury cars and 12,000 medium-priced cars. It must produce, at most, 30,000 units of compact cars. The company owns one factory in Michigan and one in North Carolina. The Michigan factory produces 20, 40, and 60 units of luxury, medium-priced, and compact cars, respectively, per day, and the North Carolina factory produces 10, 30, and 20 luxury, medium-priced, and compact cars, respectively, per day. If the Michigan factory costs $60,000 per day to operate and the North Carolina factory costs $40,000 per day to operate, find the number of days each factory should operate to minimize the cost and meet the requirements.

39. **Nutrition.** Elisa Epstein is buying two types of frozen meals: a meal of enchiladas with vegetables and a vegetable loaf meal. She wants to guarantee that she gets a minimum of 60 grams of carbohydrates, 40 grams of proteins, and 35 grams of fats. The enchilada meal contains 5, 3, and 5 grams of carbohydrates, proteins, and fats, respectively, per kilogram, and the vegetable loaf contains 2, 2, and 1 gram of carbohydrates, proteins, and fats, respectively, per kilogram. If the enchilada meal costs $3.50 per kilogram and the vegetable loaf costs $2.25 per kilogram, how many kilograms of each should Elisa buy to minimize the cost and still meet the minimum requirement?

40. **Temporary help.** The personnel director of the main post office plans to hire extra helpers during the holiday season. Because office space is limited, the number of temporary workers cannot exceed ten. She hires workers sent by Super Temps and by Ready Aid. From her past experience, she knows that a typical worker sent by Super Temps can handle 300 letters and 80 packages per day and that a typical worker from Ready Aid can handle 600 letters and 40 packages per day. The post office expects that the additional daily volume of letters and packages will be at least 3900 and 560, respectively. The daily wages for a typical worker from Super Temps and Ready Aid are $96 and $86, respectively. How many workers from each agency should be hired to keep the daily wages to a minimum?

Beyond the Basics

41. Maximize $f = 8x + 7y$, subject to the constraints $x + 2y \geq 10$, $8y - 7x \leq 40$, $x + y \leq 20$, $4x - y \leq 40$, $x \geq 0$, $y \geq 0$.

42. Minimize $f = 3x + 2y$, subject to the constraints $6y - 5x \leq 80$, $7x - 5y \leq 80$, $5x + 8y \geq 115$, $10x - y \leq 60$, $0 \leq x \leq 20$, $y \geq 0$.

43. Minimize $f = 5x + 6y$, subject to the constraints $y \leq x + 5$, $7x + 2y \leq 28$, $0 \leq x \leq 3$, $0 \leq y \leq 6$.

44. Maximize $f = 7x + 3y$, subject to the constraints $x + y \leq 100$, $x + y \geq 50$, $2x + 5y \geq 150$, $2x + y \geq 80$, $x \geq 0$, $y \geq 0$.

45. Suppose an objective function $f = ax + by$ has the same value M at two different points $P(x_1, y_1)$ and $Q(x_2, y_2)$. Prove that f has the value M at every point of the line segment PQ.

46. Suppose the feasible solution set of a linear programming problem is a quadrilateral. Is it possible that the optimal solution also occurs at an interior point of the quadrilateral?

Critical Thinking/Discussion/Writing

47. Consider this problem: Maximize $f = 2x + 3y$, subject to the constraints $y - x \leq 6$, $x \geq 0$, $0 \leq y \leq 8$.

a. Explain why the region S of feasible solutions is unbounded.

b. Show that the problem has no solution.

48. Diet problem. The objective of a diet problem is to have certain quantities of food on the menu that meet nutritional requirements at a minimum cost. Suppose the consideration is limited to milk, chicken, and vegetables and to vitamins A, B, and C. The number of milligrams of each of these vitamins contained within a unit of each food is given below.

Vitamin	Liters of Milk	Pounds of Chicken	Pounds of Vegetable	Minimum Daily Requirement
A	1	2	5	1 mg
B	10	5	4	50 mg
C	6	20	7	10 mg
Cost	\$3.20	\$2.50	\$1.30	

Write a linear programming formulation of this problem if a daily diet consists of x liters of milk, y pounds of chicken, and z pounds of vegetables.

Maintaining Skills

In Exercises 49–52, solve each system of linear equations by the Gaussian elimination method (See page 742.)

49. $\begin{cases} x - 2y = 4 \\ -3x + 5y = -7 \end{cases}$

50. $\begin{cases} 3x + 5y = 6 \\ 2x - y = 17 \end{cases}$

51. $\begin{cases} x - 4y - z = 11 \\ 2x - 5y + 2z = 39 \\ -3x + 2y + z = 1 \end{cases}$

52. $\begin{cases} x + 3y - 2z = 5 \\ 2x + y + 4z = 8 \\ 6x + y - 3z = 5 \end{cases}$

53. Consider the following rectangular array of numbers:

$$\begin{bmatrix} 2 & 4 & -6 & 10 \\ 3 & -2 & 4 & 12 \end{bmatrix} \begin{matrix} \leftarrow \text{row 1} \\ \leftarrow \text{row 2} \end{matrix}$$

Write the new rectangular array after multiplying each number in row 1 by $\frac{1}{2}$.

54. In the rectangular array obtained in Exercise 53, add -3 times each number in row 1 to the corresponding number in row 2.

55. In the rectangular array obtained in Exercise 54, multiply each number in row 2 by $-\frac{1}{8}$.

56. In the rectangular array obtained in Exercise 55, add -2 times each number in row 2 to the corresponding number in row 1.

SUMMARY Definitions, Concepts, and Formulas

8.1 Systems of Linear Equations in Two Variables

A **system of equations** is a set of equations with common variables.

A **solution** of a system of equations in two variables x and y is an ordered pair of numbers (a, b) that satisfies all equations in the system. The solution set of a system of two linear equations of the form $ax + by = c$ is the point(s) of intersection of the graphs of the equations.

Systems with no solution are called **inconsistent**, and systems with at least one solution are **consistent**. A system of two linear equations in two variables may have one solution (**independent equations**), no solution (**inconsistent system**), or infinitely many solutions (**dependent equations**).

Three methods of solving a system of equations are the graphical method, the substitution method, and the elimination or addition method.

8.2 Systems of Linear Equations in Three Variables

A **linear equation** in n variables $x_1, x_2, \ldots, x_n$ is an equation that can be written in the form

$$a_1x_1 + a_2x_2 + \cdots + a_nx_n = b,$$

where b and the coefficients $a_1, a_2, \ldots, a_n$ are real numbers.

A system of linear equations can be solved by transforming it into an *equivalent system* that is easier to solve.

The following operations produce equivalent systems:

1. Interchange the position of any two equations.
2. Multiply any equation by a nonzero constant.
3. Add a multiple of one equation to another.

The **Gaussian elimination method** is a procedure for converting a system of linear equations into an equivalent system in *triangular form*.

A linear system may have only one solution, infinitely many solutions, or no solution.

8.3 Partial-Fraction Decomposition

In partial-fraction decomposition, we undo the addition of rational expressions. A rational expression $\dfrac{P(x)}{Q(x)}$ is called a proper fraction if degree $P(x) <$ degree $Q(x)$.

There are four cases to consider in decomposing $\dfrac{P(x)}{Q(x)}$ into partial fractions:

1. $Q(x)$ has only distinct linear factors. See page 754.
2. $Q(x)$ has repeated linear factors. See page 756.
3. $Q(x)$ has distinct irreducible quadratic factors. See page 757.
4. $Q(x)$ has repeated irreducible quadratic factors. See page 758.

8.4 Systems of Nonlinear Equations

A system of equations in which at least one equation is nonlinear is called a **nonlinear system of equations**. The method of substitution, elimination, or graphing is sometimes used to solve a nonlinear system.

8.5 Systems of Inequalities

A statement of the form $ax + by < c$ is a linear inequality in the variables x and y. The symbol $<$ may be replaced with $\le, >,$ or $\ge$. A procedure for graphing a linear inequality in two variables is described on page 772.

The graph of the solution set of a system of inequalities is obtained by graphing each inequality of the system in the same coordinate plane and then finding the region that is common to every graph in the system.

8.6 Linear Programming

In a linear programming model, a quantity f (to be maximized or minimized) that can be expressed in the form $f = ax + by$ is called an **objective function** of the variables x and y. The **constraints** are the restrictions placed on the variables x and y that can be expressed as a system of linear inequalities.

A procedure for solving a linear programming problem is given on page 784.

REVIEW EXERCISES

Basic Concepts and Skills

In Exercises 1–18, solve each system of equations by using the method of your choice. Identify systems with no solution and systems with infinitely many solutions.

1. $\begin{cases} 3x - y = -5 \\ x + 2y = 3 \end{cases}$

2. $\begin{cases} x + 3y + 6 = 0 \\ y = 4x - 2 \end{cases}$

3. $\begin{cases} 2x + 4y = 3 \\ 3x + 6y = 10 \end{cases}$

4. $\begin{cases} x - y = 2 \\ 2x - 2y = 9 \end{cases}$

5. $\begin{cases} 3x - y = 3 \\ \frac{1}{2}x + \frac{1}{3}y = 2 \end{cases}$

6. $\begin{cases} 0.02y - 0.03x = -0.04 \\ 1.5x - 2y = 3 \end{cases}$

7. $\begin{cases} x + 3y + z = 0 \\ 2x - y + z = 5 \\ 3x - 3y + 2z = 10 \end{cases}$

8. $\begin{cases} 2x + y = 11 \\ 3y - z = 5 \\ x + 2z = 1 \end{cases}$

9. $\begin{cases} x + y = 1 \\ 3y + 2z = 0 \\ 2x - 3z = 7 \end{cases}$

10. $\begin{cases} 2x - 3y + z = 2 \\ x - 3y + 2z = -1 \\ 2x + 3y + 2z = 3 \end{cases}$

11. $\begin{cases} x + y + z = 1 \\ x + 5y + 5z = -1 \\ 3x - y - z = 4 \end{cases}$

12. $\begin{cases} x + 3y - 2z = -4 \\ 2x + 6y - 4z = 3 \\ x + y + z = 1 \end{cases}$

13. $\begin{cases} x + 4y + 3z = 1 \\ 2x + 5y + 4z = 4 \end{cases}$

14. $\begin{cases} 3x - y + 2z = 9 \\ x - 2y + 3z = 2 \end{cases}$

15. $\begin{cases} 3x - y = 2 \\ x + 2y = 9 \\ 3x + y = 10 \end{cases}$

16. $\begin{cases} x - 2y = 1 \\ 3x + 4y = 11 \\ 2x + 2y = 7 \end{cases}$

17. $\begin{cases} x + y + z = 3 \\ 2x - y + z = 4 \\ x + 4y + 2z = 5 \end{cases}$

18. $\begin{cases} x - y + z = 4 \\ x + 2y + 3z = 2 \\ 2x + y + 4z = 6 \end{cases}$

In Exercises 19–24, graph each system of inequalities.

19. $\begin{cases} x + y \le 1 \\ x - y \le 1 \\ x \ge 0 \end{cases}$

20. $\begin{cases} 2x + 3y \le 6 \\ 4x - 3y \le 12 \\ x \ge 0 \end{cases}$

21. $\begin{cases} 4x + y \le 7 \\ 2x + 5y \ge -1 \\ x - 2y \ge -5 \end{cases}$

22. $\begin{cases} 7x - 2y + 6 \le 0 \\ x + y + 14 \ge 0 \\ 2x - 3y + 9 \ge 0 \end{cases}$

23. $\begin{cases} x + y - 3 \le 0 \\ 3x - y - 9 \le 0 \\ y + 3 \ge 0 \\ 7x + 4y + 23 \ge 0 \end{cases}$

24. $\begin{cases} x + 5y - 11 \le 0 \\ 5x + y - 7 \le 0 \\ x - 5y - 17 \le 0 \\ 7x + y + 25 \ge 0 \end{cases}$

In Exercises 25–28, solve each linear programming problem.

25. Maximize $z = 2x + 3y$, subject to the constraints $x \ge 0, y \ge 0, 3x + 7y \le 21$.

26. Minimize $z = -5x + 2y$, subject to the constraints $x \ge -2$, $y \le 1, x - 2y + 2 \le 0$.

27. Minimize $z = x + 3y$, subject to the constraints $2x - 3y + 2 \ge 0, 4x + y - 10 \le 0, x - 3y - 9 \le 0$, $3x + y + 3 \ge 0$.

28. Maximize $z = 2x + 5y$, subject to the constraints $x + 3y - 3 \le 0, 3x - y - 9 \le 0, x + 4y + 10 \ge 0$, $3x - 2y + 2 \ge 0$.

In Exercises 29–34, solve each nonlinear system of equations.

29. $\begin{cases} x + 3y = 1 \\ x^2 - 3x = 7y + 3 \end{cases}$

30. $\begin{cases} 4x - y = 3 \\ y^2 - 2y = x - 2 \end{cases}$

31. $\begin{cases} x - y = 4 \\ 5x^2 + y^2 = 24 \end{cases}$

32. $\begin{cases} x - 2y = 7 \\ 2x^2 + 3y^2 = 29 \end{cases}$

33. $\begin{cases} xy = 2 \\ x^2 + 2y^2 = 9 \end{cases}$

34. $\begin{cases} xy = 3 \\ 2x^2 + 3y^2 = 21 \end{cases}$

In Exercises 35–40, graph each nonlinear system of inequalities.

35. $\begin{cases} x - y \le 1 \\ x^2 + y^2 \le 13 \end{cases}$

36. $\begin{cases} x - y \ge 1 \\ x^2 + y^2 \le 13 \end{cases}$

37. $\begin{cases} x - y \le 1 \\ x^2 + y^2 \le 13 \\ x \ge 0 \\ y \ge 0 \end{cases}$

38. $\begin{cases} x - y \ge 1 \\ x^2 + y^2 \ge 13 \\ x \le 4 \\ y \ge 0 \end{cases}$

39. $\begin{cases} y \le 3x \\ x^2 + y^2 \le 25 \\ x \ge 0 \\ y \ge 0 \end{cases}$

40. $\begin{cases} y \ge 3x \\ x^2 + y^2 \le 25 \\ x \ge 0 \\ y \ge 0 \end{cases}$

In Exercises 41–46, find the partial-fraction decomposition of each rational expression.

41. $\dfrac{x + 4}{x^2 + 5x + 6}$

42. $\dfrac{x + 14}{x^2 + 3x - 4}$

43. $\dfrac{3x^2 + x + 1}{x(x - 1)^2}$

44. $\dfrac{3x^2 + 2x + 3}{(x^2 - 1)(x + 1)}$

45. $\dfrac{x^2 + 2x + 3}{(x^2 + 4)^2}$

46. $\dfrac{2x}{x^4 - 1}$

Applying the Concepts

47. Investment. A speculator invested part of $15,000 in a high-risk venture and received a return of 12% at the end of the year. The rest of the $15,000 was invested at 4% annual interest. The combined annual income from the two sources was $1300. How much was invested at each rate?

48. Agriculture. A farmer earns a profit of $525 per acre of tomatoes and $475 per acre of soybeans. His soybean acreage is 5 acres more than twice his tomato acreage. If his total profit from the two crops is $24,500, how many acres of tomatoes and how many acres of soybeans does he have?

49. Geometry. The area of a rectangle is 63 square feet, and its perimeter is 33 feet. What are the dimensions of the rectangle?

50. Numbers. Twice the sum of the reciprocals of two numbers is 13, and the product of the numbers is $\frac{1}{9}$. Find the numbers.

51. Geometry. The hypotenuse of a right triangle is 17. If one leg of the triangle is increased by 1 and the other leg is increased by 4, the hypotenuse becomes 20. Find the sides of the triangle.

52. Paper route. Chris covers her paper route, which is 21 miles long, by 7:30 A.M. each day. If her average rate of travel were 1 mile faster each hour, she would cover the route by 7 A.M. What time does she start in the morning?

53. Agriculture. A rectangular pasture with an area of 6400 square meters is divided into three smaller pastures by two fences parallel to the shorter sides. The width of two of the smaller pastures is the same, and the width of the third is twice that of the others. Find the dimensions of the original pasture if the perimeter of the larger of the subdivisions is 240 m.

54. Leasing. Budget Rentals leases its compact cars for $23.00 per day plus $0.17 per mile. Dollar Rentals leases the same car for $24.00 per day plus $0.22 per mile.

a. Find the cost functions describing the daily cost of leasing from each company.

b. Graph the two functions in part (a) on the same coordinate axes.

c. From which company should you lease the car if you plan to drive **(i)** 50 miles per day; **(ii)** 60 miles per day; **(iii)** 70 miles per day?

55. Break-even analysis. Auto-Sprinkler Corp. manufactures seven-day, 24-hour variable timers for lawn sprinklers. The corporation has a monthly fixed cost of $60,000 and a production cost of $12 for each timer manufactured. Each timer sells for $20.

a. Write the cost function and the revenue function for selling x timers per month.

b. Graph the two functions in part (a) on the same coordinate plane and find the break-even point graphically.

c. Find the break-even point algebraically.

d. How many timers should be sold in a month to realize a profit (before taxes) of 15% of the cost?

56. Equilibrium quantity and price. The demand equation for a product is $p = \dfrac{4000}{x}$, where p is the price per unit and x is the number of units of the product. The supply equation for the product is $p = \dfrac{x}{20} + 10$.

a. Find the equilibrium quantity.

b. Find the equilibrium price.

c. Graph the supply and demand functions on the same coordinate plane and label the equilibrium point.

57. **Apartment lease.** Alisha and Sunita signed a lease on an apartment for nine months and split the rent evenly. At the end of six months, Alisha got married and moved out. She paid the landlady an amount equal to the difference between double-occupancy rental and single-occupancy rental for the remaining three months, and Sunita paid the single-occupancy rate for the same three months. If the nine-month rental cost Alisha \$2250 and Sunita \$3150, what were the single and double monthly rates?

58. **Seating arrangement.** The 600 graduating seniors at Central State College are seated in rows, each of which contains the same number of chairs, and every chair is occupied. If five more chairs were in each row, everyone could be seated in four fewer rows. How many chairs are in each row?

59. **Passing a final exam.** Twenty-six students in a college algebra class took a final exam on which the passing score was 70. The mean score of those who passed was 78, and the mean score of those who failed was 26. The mean of all scores was 72. How many students failed the exam?

60. **Finding numbers.** The average of a and b is 2.5, the average of b and c is 3.8, and the average of a and c is 3.1. Find the numbers a, b, and c.

61. **May–December match.** Steve and his wife, Janet, have the same birthday. When Steve was as old as Janet is now, he was twice as old as Janet was then. When Janet becomes as old as Steve is now, the sum of their ages will be 119. How old are Steve and Janet now?

62. **Percentage increase.** Let x and y be positive real numbers. Increasing x by y percent gives 46, and increasing y by x percent gives 21. Find x and y.

63. **Criminals rob a bank.** Three criminals—Butch, Sundance, and Billy—robbed a bank and divided the loot in the following manner: Because Butch planned the job and drove the getaway car, he got 75% as much as Sundance and Billy put together. Because Sundance was an expert safecracker, he got \$500 more than 50% of Billy's and Butch's cut put together. Billy got \$1000 less than three times the difference of Butch and Sundance's cut. How much did they steal, and what was each criminal's take?

64. **Curve fitting.** Find all of the curves of the form $y = ax^2 + bx + c$ that contain the points $(0, 1)$, $(1, 0)$, and $(-1, 6)$.

65. **Using exponents.** Given that $2^x = 8^y$ and $9^y = 3^{x-2}$, find x and y.

66. **Area of an octagon.** A square of side length 8 has its corners cut off to make it a regular octagon. Find the area of the octagon. (See the figure.)

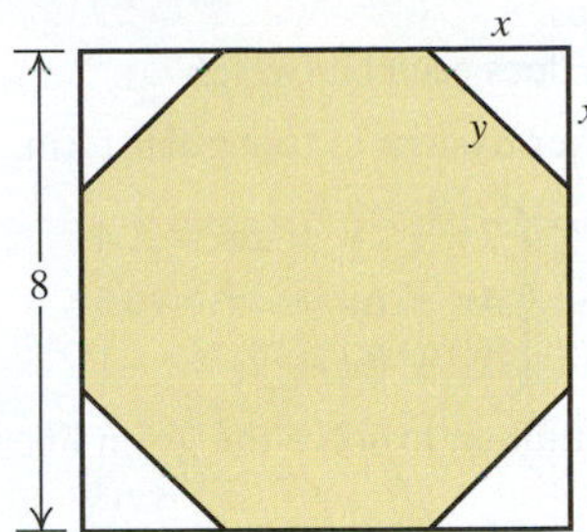

67. **Building houses.** A builder has 42 units of material and 32 units of labor available during a given period. She builds two-story houses or one-story houses or some of both. Suppose she makes a profit of \$10,000 on each two-story house and \$4000 on each one-story house. A two-story house requires 7 units of material and 1 unit of labor, whereas a one-story house requires 1 unit of material and 2 units of labor. How many houses of each type should she build to make a maximum profit? What is the maximum profit?

68. **Minimizing cost.** An animal food is to be a mixture of two products X and Y. The content and cost of 1 pound of each product is given in the following table.

Product	Protein Grams	Fat Grams	Carbohydrates Grams	Cost
X	180	2	240	\$0.75
Y	36	8	200	\$0.56

How much of each product should be used to minimize the cost if each bag must contain at least 612 grams of protein; at least 22 grams of fat; and, at most, 1880 grams of carbohydrates?

69. **Lobster prices.** The Captain's Catch classifies its lobsters as small, medium, and large. A package of one large, two medium, and four small lobsters sells for \$344; a package of two large and three small lobsters sells for \$255; and a package of two large, two medium, and five small lobsters sells for \$449. What is the individual price of each size lobster?

70. **Ancient mathematics.** The *Nine Chapters in the Mathematical Art*, a Chinese work written about 250 B.C. contains the following problem: Three sheafs of good crop, two sheaf of mediocre crop, and one sheaf of bad crop are sold for 39 dou. Two sheafs of good, three mediocre, and one bad are sold for 34 dou. And one good, two mediocre, and three bad are sold for 26 dou. What price is received for a sheaf of each of good crop, mediocre crop, and bad crop?

PRACTICE TEST A

In Problems 1–7, solve each system of equations.

1. $\begin{cases} 2x - y = 4 \\ 2x + y = 4 \end{cases}$

2. $\begin{cases} x + 2y = 8 \\ 3x + 6y = 24 \end{cases}$

3. $\begin{cases} -2x + y = 4 \\ 4x - 2y = 4 \end{cases}$

4. $\begin{cases} 3x + 3y = -15 \\ 2x - 2y = -10 \end{cases}$

5. $\begin{cases} \dfrac{5x}{3} + \dfrac{y}{2} = 14 \\ \dfrac{2x}{3} - \dfrac{y}{8} = 3 \end{cases}$

6. $\begin{cases} y = x^2 \\ 3x - y + 4 = 0 \end{cases}$

7. $\begin{cases} x - 3y = -4 \\ 2x^2 + 3x - 3y = 8 \end{cases}$

8. Two gold bars together weigh a total of 485 pounds. One bar weighs 15 pounds more than the other.
   a. Write a system of equations that describes these relationships.
   b. How much does each bar weigh?

9. Convert the given system to triangular form.

$$\begin{cases} 2x + y + 2z = 4 \\ 4x + 6y + z = 15 \\ 2x + 2y + 7z = -1 \end{cases}$$

10. Use back-substitution to solve the given linear system.

$$\begin{cases} x - 6y + 3z = -2 \\ 9y - 5z = 2 \\ 2z = 10 \end{cases}$$

In Exercises 11–13, solve the system of equations.

11. $\begin{cases} x + y + z = 8 \\ 2x - 2y + 2z = 4 \\ x + y - z = 12 \end{cases}$

12. $\begin{cases} x + z = -1 \\ 3y + 2z = 5 \\ 3x - 3y + z = -8 \end{cases}$

13. $\begin{cases} 2x - y + z = 2 \\ x + y - z = -1 \\ x - 5y + 5z = 7 \end{cases}$

14. A vending machine's coin box contains nickels, dimes, and quarters. The total number of coins in the box is 300. The number of dimes is three times the number of nickels and quarters together. Assuming that \$30.65 is in the box, find the number of nickels, dimes, and quarters that it contains.

For Problems 15 and 16, write the form of the partial-fraction decomposition of the given rational expression. You do not need to solve for the constants.

15. $\dfrac{2x}{(x-5)(x+1)}$

16. $\dfrac{-5x^2 + x - 8}{(x-2)(x^2+1)^2}$

17. Find the partial-fraction decomposition of the rational expression

$$\frac{x+3}{(x+4)^2(x-7)}.$$

18. Graph the inequality $3x + y < 6$.

19. Graph the solution set of the system of inequalities.

$$\begin{cases} y - x^2 \le 3 \\ y - x > 0 \end{cases}$$

20. Maximize $z = 2x + y$, subject to the constraints $x \ge 0, y \ge 0, x + 3y \le 3$.

PRACTICE TEST B

In Problems 1–7, solve each system of equations.

1. $\begin{cases} 2x - 3y = 16 \\ x - y = 7 \end{cases}$
   a. $\{(5, -2)\}$ b. $\{(8, 0)\}$
   c. $\{(0, -7)\}$ d. $\{(4, -3)\}$

2. $\begin{cases} 6x - 9y = -2 \\ 3x - 5y = -6 \end{cases}$
   a. $\{(-3, 0)\}$ b. $\left\{\left(0, \frac{5}{6}\right)\right\}$
   c. $\left\{\left(\frac{44}{3}, 10\right)\right\}$ d. $\{(1, 1)\}$

3. $\begin{cases} 2x - 5y = 9 \\ 4x - 10y = 18 \end{cases}$
   a. $\left\{\left(0, \frac{9}{2}\right)\right\}$ b. $\left\{\left(\frac{5y}{2} + \frac{9}{2}, y\right)\right\}$
   c. $\left\{\left(x, \frac{5x}{2} + \frac{9}{2}\right)\right\}$ d. $\varnothing$

4. $\begin{cases} 3x + 5y = 1 \\ -6x - 10y = 2 \end{cases}$
   a. $\left\{\left(0, -\frac{1}{5}\right)\right\}$ b. $\{(2, -1)\}$
   c. $\varnothing$ d. $\left\{\left(-\frac{5y}{3} + \frac{1}{3}, y\right)\right\}$

5. $\begin{cases} \dfrac{x}{5} + \dfrac{2y}{5} = 1 \\ \dfrac{x}{4} - \dfrac{y}{3} = -\dfrac{5}{12} \end{cases}$
   a. $\{(-15, 10)\}$ b. $\{(1, 2)\}$
   c. $\{(-10, -15)\}$ d. $\{(1, 0)\}$

6. $\begin{cases} x - 3y = \dfrac{1}{2} \\ -2x + 6y = -1 \end{cases}$
   a. $\left\{\left(\frac{1}{2}, 0\right)\right\}$ b. $\left\{\left(2, \frac{1}{2}\right)\right\}$
   c. $\varnothing$ d. $\left\{\left(3y + \frac{1}{2}, y\right)\right\}$

7. $\begin{cases} 3x + 4y = 12 \\ 3x^2 + 16y^2 = 48 \end{cases}$
   a. $\{(0, 4), (3, 0)\}$ b. $\{(0, -4)\}$
   c. $\{(0, -4), (3, 0)\}$ d. $\left\{(4, 0), \left(2, \frac{3}{2}\right)\right\}$

8. Student tickets for a dance cost \$2, and nonstudent tickets cost \$5. Three hundred tickets were sold, and the total ticket receipts were \$975. How many of each type of ticket were sold?

a. 150 student tickets, 150 nonstudent tickets
b. 200 student tickets, 100 nonstudent tickets
c. 210 student tickets, 90 nonstudent tickets
d. 175 student tickets, 125 nonstudent tickets

9. Convert the given system to triangular form.

$$\begin{cases} x + 3y + 3z = 4 \\ 2x + 5y + 4z = 5 \\ x + 2y + 2z = 6 \end{cases}$$

a. $x + 3y + 3z = 4$, $y + 2z = 8$, $z = 2$
b. $x + 3y + 3z = 4$, $2y + z = 1$, $z = 5$
c. $x + 3y + 3z = 4$, $y + 2z = 3$, $z = 5$
d. $x + 3y + 3z = 4$, $2y + z = -2$, $z = 5$

10. Use back-substitution to solve the linear system.

$$\begin{cases} 2x + y + z = 0 \\ 10y - 2z = 4 \\ 3z = 9 \end{cases}$$

a. $\{(-2, 1, 3)\}$
b. $\{(0, -3, 3)\}$
c. $\{(0, 2, 8)\}$
d. $\{(-1, 1, 1)\}$

11. Solve the given system of equations or state that the system is inconsistent.

$$\begin{cases} 2x + 13y + 6z = 1 \\ 3x + 10y + 11z = 15 \\ 2x + 10y + 8z = 8 \end{cases}$$

a. $\{(3, -1, -4)\}$
b. $\{(1, -1, 2)\}$
c. $\varnothing$
d. $(1, 6, 8)\}$

12. Solve the given system of equations or state that the system is inconsistent.

$$\begin{cases} 4x - y + 7z = -2 \\ 2x + y + 11z = 13 \\ 3x - y + 4z = -3 \end{cases}$$

a. $\{(-3z + 1, -5z + 11, z)\}$
b. $\{(-14, -14, 5)\}$
c. $\left\{\left(-\frac{19}{5}, 3, \frac{8}{5}\right)\right\}$
d. $\varnothing$

13. Solve the given system of equations or state that the system is inconsistent.

$$\begin{cases} x - 2y + 3z = 4 \\ 2x - y + z = 1 \\ x + y - 2z = -3 \end{cases}$$

a. $\{(1, 6, 5)\}$
b. $\{(0, 1, 2)\}$
c. $\{(x, 5x + 1, 3x + 2)\}$
d. $\varnothing$

14. Forty-six students will go to France, Italy, or Spain for six weeks during the summer. The number of students going to France or Italy is four more than the number going to Spain. The number of students going to France is two less than the number going to Spain. How many students are going to each country?

a. France: 23, Italy: 21, Spain: 2
b. France: 19, Italy: 6, Spain: 21
c. France: 21, Italy: 6, Spain: 19
d. France: 2, Italy: 21, Spain: 23

For Problems 15 and 16, write the form of the partial-fraction decomposition of the given rational functions. You do not need to solve for the constants.

15. $\dfrac{x}{(x + 2)(x - 7)}$

a. $\dfrac{A}{x + 2} + \dfrac{Bx}{x - 7}$
b. $\dfrac{Ax}{x + 2} + \dfrac{Bx}{x - 7}$
c. $\dfrac{A}{x + 2} + \dfrac{B}{x - 7}$
d. $\dfrac{A}{x + 2} + \dfrac{Bx + C}{x - 7}$

16. $\dfrac{7 - x}{(x - 3)(x + 5)^2}$

a. $\dfrac{A}{x - 3} + \dfrac{B}{x + 5} + \dfrac{C}{(x + 5)^2}$
b. $\dfrac{A}{x - 3} + \dfrac{B}{x + 5} + \dfrac{Cx + D}{(x + 5)^2}$
c. $\dfrac{A}{x + 3} + \dfrac{B}{(x + 5)^2}$
d. $\dfrac{A}{x + 3} + \dfrac{Bx + C}{(x + 5)^2}$

17. Find the partial-fraction decomposition of the rational expression

$$\frac{x^2 + 15x + 18}{x^3 - 9x}.$$

a. $\dfrac{26}{x - 9} - \dfrac{2}{x}$
b. $\dfrac{4}{x + 3} - \dfrac{2}{x} - \dfrac{1}{x - 3}$
c. $\dfrac{4}{x - 3} - \dfrac{2}{x} - \dfrac{1}{x + 3}$
d. $\dfrac{3x - 15}{x^2 - 9} - \dfrac{2}{x}$

18. Which of the graphs at the right is the graph of $x + 3y > 3$?

19. Which of the graphs at the right is the graph of the solution set of the system of inequalities

$$\begin{cases} y - x^2 + 4 \ge 0 \\ 3x - y \le 0 \end{cases}?$$

20. Maximize $z = 3x + 21y$, subject to the constraints $x \ge 0, y \ge 0, 2x + y \le 8, 2x + 3y \le 16$.

a. 84
b. 112
c. $\dfrac{16}{3}$
d. 12

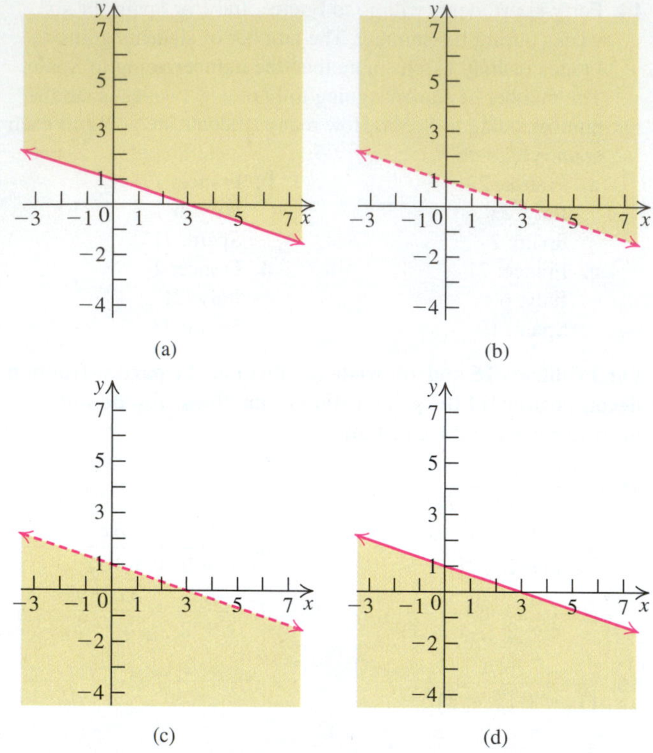

Figure for Exercise 18

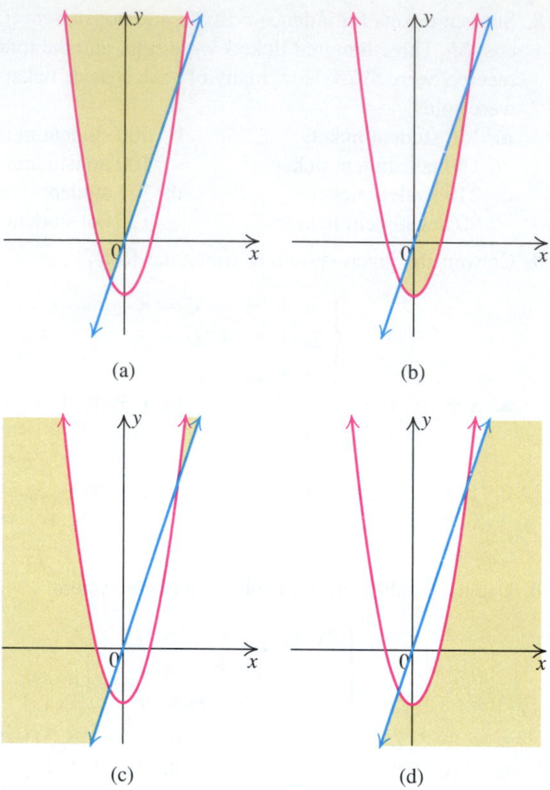

Figure for Exercise 19

CUMULATIVE REVIEW EXERCISES CHAPTERS P–8

In Exercises 1–8, solve each equation or inequality.

1. $\dfrac{1}{x-1} + \dfrac{4}{x-4} = \dfrac{5}{x-5}$
2. $2\left(x + \dfrac{1}{x}\right)^2 - 7\left(x + \dfrac{1}{x}\right) + 5 = 0$
3. $\sqrt{3x - 5} = x - 3$
4. $\dfrac{x-1}{x+3} \le 0$
5. $x^2 - 9x + 20 > 0$
6. $2^{x-1} = 5$
7. $\log_x 16 = 4$
8. $\log(x - 3) + \log(x - 1) = \log(2x - 5)$

In Exercises 9–12, use transformations to graph each function.

9. $f(x) = |x + 1| - 2$
10. $f(x) = -(x - 2)^2 + 3$
11. $f(x) = 2^{x-1} - 3$
12. $f(x) = \sqrt{x - 1} + 3$
13. Let $f(x) = 2x - 2$.
    a. Find $f^{-1}(x)$.
    b. Graph f and f^{-1} on the same coordinate plane.
14. Let $f(x) = x^3 - x^2 + x - 6$.
    a. List all possible rational zeros of f.
    b. Use synthetic division to test the possible rational zeros and find a real zero.
    c. Use the zero from part (b) to find all of the (real or complex) zeros of $f(x)$.
15. Expand and simplify: $\log_3(9x^4)$
16. A radioactive substance decays so that the amount present in t years is given by $A = A_0e^{-kt}$, where A_0 is the initial amount. Find the half-life of the substance if
    a. $k = 0.05$
    b. $k = 0.0002$
17. If f is a one-to-one function and $f(2) = 9$, find $f^{-1}(9)$.
18. Find the time required to double your money if it is compounded continuously at 7.5%.

In Exercises 19 and 20, solve the system of equations.

19. $\begin{cases} 5x - 2y + 25 = 0 \\ 4y - 3x - 29 = 0 \end{cases}$
20. $\begin{cases} 2x - y + z = 3 \\ x + 3y - 2z = 11 \\ 3x - 2y + z = 4 \end{cases}$

CHAPTER 9

Matrices and Determinants

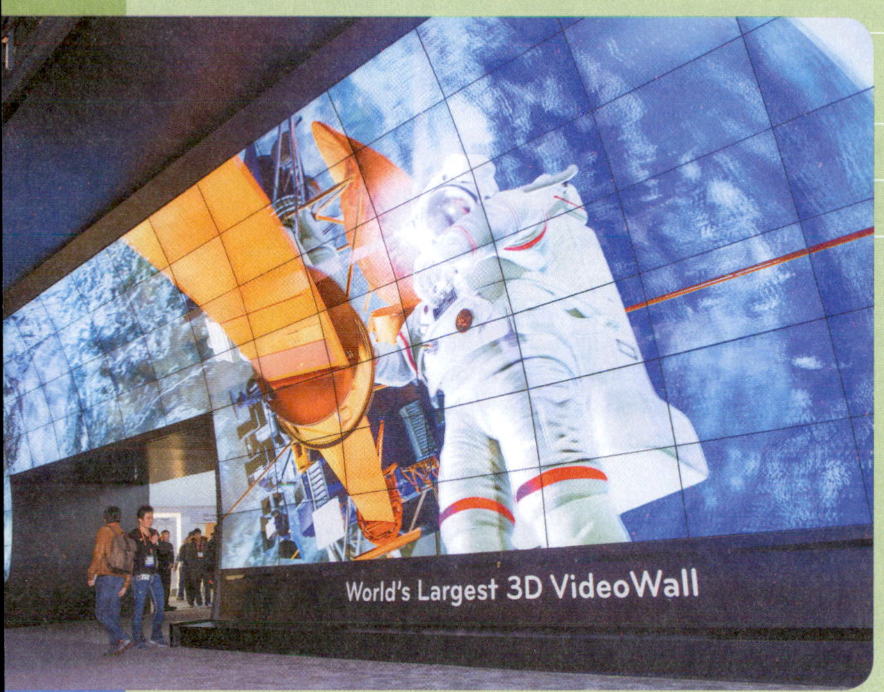

TOPICS

The linear programming problems introduced in the previous chapter involving quantities from the study of such disciplines as medicine, traffic control, structural design, and city planning often involve an extremely large number of variables. To solve such problems successfully requires the use of powerful computers and new techinques. In this chapter, we study matrices and several methods of solving linear systems that are easily implemented on computers.

SECTION 9.1

Matrices and Systems of Equations

Wassily Leontief 1906–1999
Leontief studied philosophy, sociology, and economics at the University of Leningrad and earned the degree of Learned Economist in 1925. He continued his studies at the University of Berlin, where he earned his PhD degree. In 1973, the Nobel Prize in Economic Sciences was awarded to Leontief for his work on input–output analysis of the U.S. economy.

BEFORE STARTING THIS SECTION, REVIEW

1 Systems of equations (Sections 8.1 and 8.2)
2 Equivalent systems (Section 8.2, page 742)
3 Gaussian elimination (Section 8.2, page 741)

OBJECTIVES

1 Define a matrix.
2 Use matrices to solve a system of linear equations.
3 Use Gaussian elimination to solve a system.
4 Use Gauss–Jordan elimination to solve a system.

◆ Leontief Input–Output Model

In 1949, the Russian-born Harvard Professor Wassily Leontief opened the door to a new era in mathematical modeling in economics. He divided the U.S. economy into 500 "sectors," such as the coal, automotive, a communications industries and agriculture. For each sector, he wrote a linear equation that described how that sector distributed its output to other sectors of the economy. In 1949, the largest computer was the Mark II, and it could not handle the resulting system of 500 equations in 500 variables. Leontief then combined some of the sectors and reduced the system to 42 equations in 42 variables to solve the problem. As computers have become more powerful, scientists and engineers can now work on problems far more complex than they ever dreamed of a few decades ago. In Example 10, we explore Leontief's input–output model applied to a simple economy.

1 Define a matrix.

Definition of a Matrix

Suppose Great Builders Incorporated (GBI), builds ranch, colonial, and modern houses. GBI wants to compare the cost of labor and the cost of material involved in building each type of house during one year. These data could be represented as in Table 9.1.

TABLE 9.1

Type of House	Ranch	Colonial	Modern
Cost of labor (in millions)	12	13	14
Cost of material (in millions)	9	11	10

Or it could be represented more compactly by the following array:

$$\begin{bmatrix} 12 & 13 & 14 \\ 9 & 11 & 10 \end{bmatrix}$$

This rectangular array of numbers is called a *matrix* (plural, *matrices*). This matrix has two rows (the types of costs) and three columns (the types of houses).

Matrix

A **matrix** is a rectangular array of numbers.

$$A = \begin{bmatrix} a_{11} & a_{12} & \cdots & a_{1n} \\ a_{21} & a_{22} & \cdots & a_{2n} \\ \vdots & \vdots & & \vdots \\ a_{m1} & a_{m2} & \cdots & a_{mn} \end{bmatrix} \begin{matrix} \leftarrow \text{Row } 1 \\ \leftarrow \text{Row } 2 \\ \\ \leftarrow \text{Row } m \end{matrix}$$

Column 1 ↑, Column 2 ↑, Column n ↑

If a matrix A has m rows and n columns, then we say that A is of **order m by n** (written $m \times n$).

The **entry** or **element** in the ith row and jth column is a real number and is denoted by the *double-subscript* notation a_{ij}. We call a_{ij} the **(i, j)th entry**. For example, a_{24} is the entry in the second row and fourth column of the matrix A. In general, we will use capital letters $A, B, \ldots$ to denote matrices and the corresponding lowercase letter $a_{ij}, b_{ij}, \ldots$ for their entries.

If A has n rows and n columns, then A is called a **square matrix of order n**. The entries $a_{11}, a_{22}, \ldots, a_{nn}$ form the **main diagonal** of A. A $1 \times n$ matrix is called a **row matrix**, and an $n \times 1$ matrix is called a **column matrix**.

EXAMPLE 1 Determining the Order of Matrices

Determine the order of each matrix. Identify square, row, and column matrices. Identify entries in the main diagonal of each square matrix.

a. $A = [3]$ **b.** $B = [3 \quad 5 \quad -7]$ **c.** $C = \begin{bmatrix} 0 & 1 \\ -3 & 4 \end{bmatrix}$ **d.** $D = \begin{bmatrix} 1 & 2 & 3 \\ 4 & 5 & 6 \\ 7 & 8 & 9 \end{bmatrix}$

Solution

a. Matrix A with one row and one column is a 1×1 matrix. A is a square matrix of order 1. In A, $a_{11} = 3$ is the main diagonal. A is also a column and a row matrix.

b. Matrix B with one row and three columns is a 1×3 matrix. B is a row matrix.

c. Matrix C, a 2×2 matrix, is a square matrix of order 2. In C, the entries $c_{11} = 0$ and $c_{22} = 4$ form the main diagonal.

d. Matrix D, a 3×3 matrix, is a square matrix of order 3. In D, the entries $d_{11} = 1$, $d_{22} = 5$, and $d_{33} = 9$ form the main diagonal.

Practice Problem 1 Determine the order of each matrix.

a. $\begin{bmatrix} -1 & 3 \\ 7 & 4 \\ 0 & 0 \end{bmatrix}$ **b.** $[3 \quad -8]$

2 Use matrices to solve a system of linear equations.

Using Matrices to Solve Linear Systems

To solve a system of linear equations by the elimination method, the particular symbols used for the variables do not matter; only the coefficients and the constants are important. We can display the constants and coefficients of a system in a matrix called the **augmented**

matrix of the system. The matrix containing only the coefficients of the variables is the **coefficient matrix**. Consider the following system.

$$\begin{cases} x - y - z = 1 \\ 2x - 3y + z = 10 \\ x + y - 2z = 0 \end{cases}$$

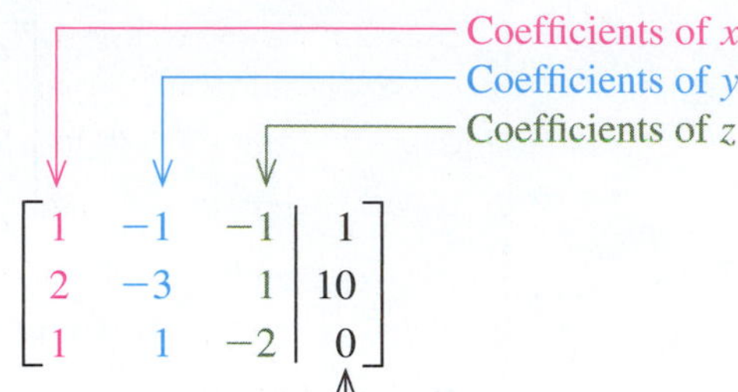

Augmented Matrix: $\left[\begin{array}{ccc|c} 1 & -1 & -1 & 1 \\ 2 & -3 & 1 & 10 \\ 1 & 1 & -2 & 0 \end{array}\right]$

Coefficient Matrix: $\begin{bmatrix} 1 & -1 & -1 \\ 2 & -3 & 1 \\ 1 & 1 & -2 \end{bmatrix}$

Notice that the numbers in the first column of the augmented matrix are the coefficients of x, those in the second column are the coefficients of y, and those in the third column are the coefficients of z. The constants on the right side of the equations in the system are found in the fourth column. The vertical line in the augmented matrix is to remind you of the equal signs in the equations. If a variable does not appear in one of the equations, represent it with a zero in the matrix.

EXAMPLE 2 Writing the Augmented Matrix of a Linear System

Write the augmented matrix of the linear system.

$$\begin{cases} 2x + 3z = 1 \\ 2z + y = 5 \\ -4x + 5y = 7 \end{cases}$$

Solution

First, write the system with the variables lined in columns and insert zeros as coefficients of any missing variables.

$$\begin{cases} 2x + 0y + 3z = 1 \\ 0x + 1y + 2z = 5 \\ -4x + 5y + 0z = 7 \end{cases}$$

The augmented matrix of the given system is as follows:

$$\left[\begin{array}{ccc|c} 2 & 0 & 3 & 1 \\ 0 & 1 & 2 & 5 \\ -4 & 5 & 0 & 7 \end{array}\right]$$

Practice Problem 2 Write the augmented matrix of the linear system.

$$\begin{cases} 3y - z = 8 \\ x + 4y = 14 \\ -2y + 9z = 0 \end{cases}$$

TECHNOLOGY CONNECTION

Many graphing calculators let you first name a matrix, specify the size, and then insert the entries for the matrix. No vertical line separates the column of constants, but this does not affect any of the matrix operations. The augmented matrix from Example 2 has been named *A*.

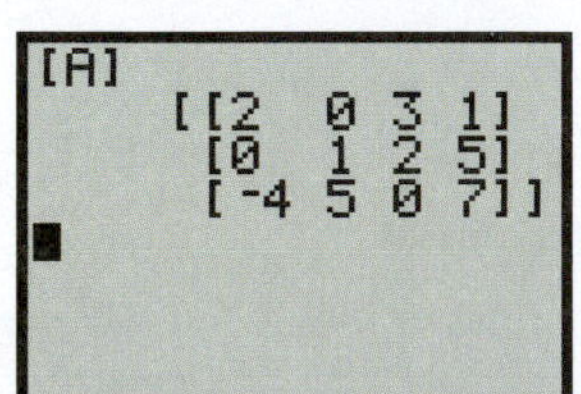

We can reverse this process and write a linear system from a given augmented matrix. For example, the augmented matrix

$$\left[\begin{array}{ccc|c} 9 & 2 & -1 & 6 \\ 3 & 4 & 7 & -8 \\ 0 & 1 & 5 & 11 \end{array}\right] \text{ corresponds to the system of equations } \begin{cases} 9x + 2y - z = 6 \\ 3x + 4y + 7z = -8. \\ y + 5z = 11 \end{cases}$$

The basic strategy for solving a system of equations is to *replace the given system with an equivalent system* (one with the same solution set) *that is easier to solve.* In the previous chapter, we used three basic operations to solve a linear system:

1. Interchange two equations.
2. Multiply all the terms in an equation by a nonzero constant.
3. Add a multiple of one equation to another equation. In other words, replace one equation with the sum of itself and a multiple of another equation.

In matrix terminology, the three corresponding operations are called the **elementary row operations**. Two matrices are **row equivalent** if one can be obtained from the other by a sequence of elementary row operations, given next. The symbol R_i represents the ith row of a matrix.

ELEMENTARY ROW OPERATIONS

Row Operation	In Symbols	Description
1. Interchange two rows.	$R_i \leftrightarrow R_j$	Interchange the ith and jth rows.
2. Multiply a row by a nonzero constant.	cR_j	Multiply the jth row by c, $c \neq 0$.
3. Add a multiple of one row to another row.	$cR_i + R_j \rightarrow R_j$	Replace the jth row by adding c times the ith row to it.

EXAMPLE 3 Applying Elementary Row Operations

Perform the indicated row operations (**a**), (**b**), and (**c**) in order on the following matrix:

$$A = \begin{bmatrix} 3 & -2 & -3 \\ 2 & 4 & 14 \end{bmatrix}$$

a. $R_1 \leftrightarrow R_2$, **b.** $\frac{1}{2}R_1$, **c.** $-3R_1 + R_2 \rightarrow R_2$

Solution

a. $A = \begin{bmatrix} 3 & -2 & -3 \\ 2 & 4 & 14 \end{bmatrix} \xrightarrow{R_1 \leftrightarrow R_2} \begin{bmatrix} 2 & 4 & 14 \\ 3 & -2 & -3 \end{bmatrix} = B$ Interchange the 1st and 2nd rows.

b. $B = \begin{bmatrix} 2 & 4 & 14 \\ 3 & -2 & -3 \end{bmatrix} \xrightarrow{\frac{1}{2}R_1} \begin{bmatrix} 1 & 2 & 7 \\ 3 & -2 & -3 \end{bmatrix} = C$ Multiply the 1st row by $\frac{1}{2}$.

c. $C = \begin{bmatrix} 1 & 2 & 7 \\ 3 & -2 & -3 \end{bmatrix} \xrightarrow{-3R_1 + R_2 \rightarrow R_2} \begin{bmatrix} 1 & 2 & 7 \\ 0 & -8 & -24 \end{bmatrix}$ Add -3 times the 1st row to the 2nd row.

Practice Problem 3 Perform the row operations of Example 3 in order on the following matrix:

$$A = \begin{bmatrix} 3 & 4 & 5 \\ 2 & 4 & 6 \end{bmatrix}$$

In the next example, we solve a system of equations both with and without matrix notation and place the results side by side for comparison. We solve the system by converting it into triangular form by first using elimination and then back-substitution.

TECHNOLOGY CONNECTION

Many graphing calculators provide all three row operations. Consider the matrix *A*.

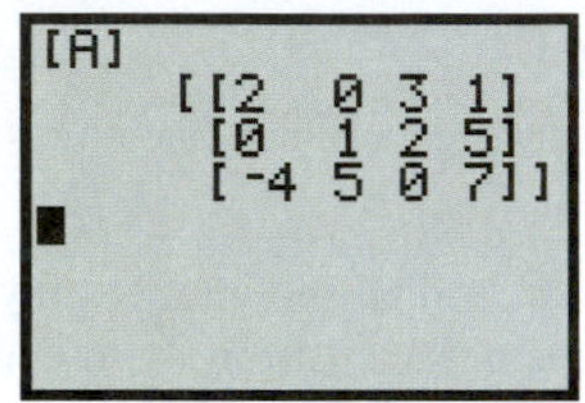

Here are some examples of row operations on the matrix A.

Swap rows 1 and 3.

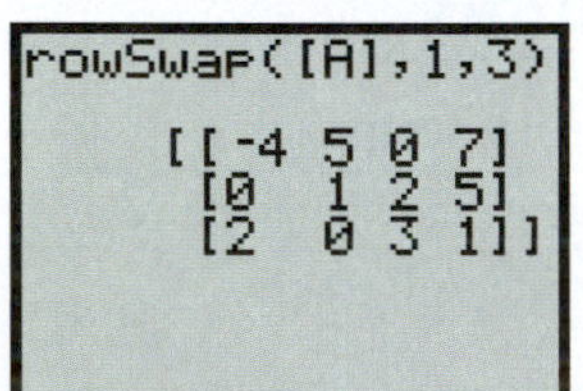

Multiply row 1 by 4.

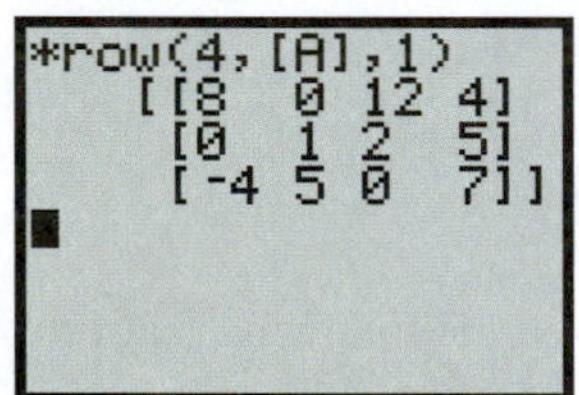

Add 3 times row 1 to row 2.

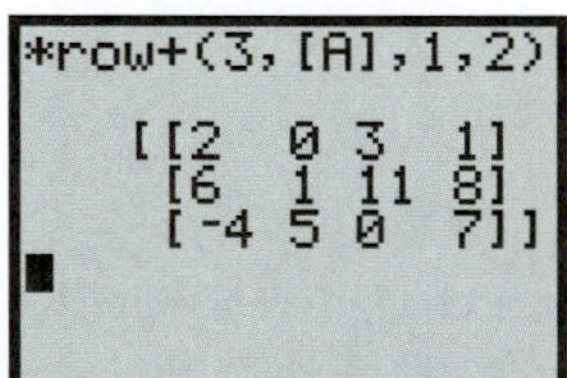

EXAMPLE 4 Comparing Linear Systems and Matrices

Solve the system of linear equations:

$$\begin{cases} x - y - z = 1 & (1) \\ 2x - 3y + z = 10 & (2) \\ x + y - 2z = 0 & (3) \end{cases}$$

Solution

Linear System

$$\begin{cases} x - y - z = 1 & (1) \\ 2x - 3y + z = 10 & (2) \\ x + y - 2z = 0 & (3) \end{cases}$$

Augmented Matrix

$$A = \left[\begin{array}{rrr|r} 1 & -1 & -1 & 1 \\ 2 & -3 & 1 & 10 \\ 1 & 1 & -2 & 0 \end{array}\right]$$

Add -2 times equation (1) to equation (2). Add -1 times equation (1) to equation (3).

Use the first row to produce zeros at the (2, 1) and (3, 1) positions.

$$\begin{cases} x - y - z = 1 & (1) \\ -y + 3z = 8 & (4) \\ 2y - z = -1 & (5) \end{cases}$$

$$\xrightarrow[(-1)R_1 + R_3 \to R_3]{-2R_1 + R_2 \to R_2} \left[\begin{array}{rrr|r} 1 & -1 & -1 & 1 \\ 0 & -1 & 3 & 8 \\ 0 & 2 & -1 & -1 \end{array}\right]$$

Multiply equation (4) by -1.

Produce a 1 at the (2, 2) position.

$$\begin{cases} x - y - z = 1 & (1) \\ y - 3z = -8 & (6) \\ 2y - z = -1 & (5) \end{cases}$$

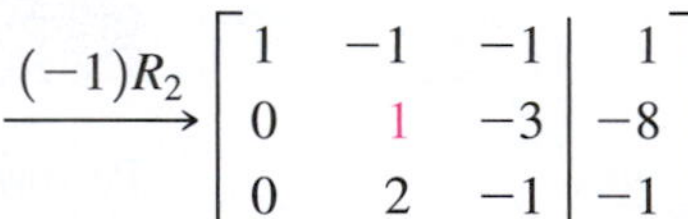

$$\xrightarrow{(-1)R_2} \left[\begin{array}{rrr|r} 1 & -1 & -1 & 1 \\ 0 & 1 & -3 & -8 \\ 0 & 2 & -1 & -1 \end{array}\right]$$

Add -2 times equation (6) to equation (5).

Use the second row to produce a zero at the (3, 2) position.

$$\begin{cases} x - y - z = 1 & (1) \\ y - 3z = -8 & (6) \\ 5z = 15 & (7) \end{cases}$$

$$\xrightarrow{-2R_2 + R_3 \to R_3} \left[\begin{array}{rrr|r} 1 & -1 & -1 & 1 \\ 0 & 1 & -3 & -8 \\ 0 & 0 & 5 & 15 \end{array}\right]$$

Equation (7), $5z = 15$, which corresponds to row 3 of the final matrix, gives the value $z = 3$. We find y by back-substitution.

$y - 3z = -8$	Equation (6); final matrix row 2
$y - 3(3) = -8$	Replace z with 3.
$y = 1$	Solve for y.

We find x by back-substitution.

$x - y - z = 1$	Equation (1); final matrix row 1
$x - 1 - 3 = 1$	Replace y with 1 and z with 3.
$x = 5$	Solve for x.

The solution of the system is $x = 5, y = 1$, and $z = 3$. You should check this solution by substituting it into the original system of equations. The solution set is $\{(5, 1, 3)\}$.

Practice Problem 4 Solve the system of linear equations:

$$\begin{cases} x - 6y + 3z = -2 \\ 3x + 3y - 2z = -2 \\ 2x - 3y + z = -2 \end{cases}$$

The last matrix in the solution of Example 4 is in *row-echelon form,* which is defined next. In the definition, a **nonzero row** means a row that contains at least one nonzero entry; a **leading entry** of a row is the leftmost nonzero entry in a nonzero row.

ROW-ECHELON FORM AND REDUCED ROW-ECHELON FORM

An $m \times n$ matrix is in **row-echelon form** if it has the following three properties:

1. The leading entry of each nonzero row is 1.
2. The leading entry in any row is to the right of the leading entry in the row above it.
3. All nonzero rows are above the rows consisting entirely of zeros.

A matrix in row-echelon form having the following property is in **reduced row-echelon form:** Each leading 1 is the only nonzero entry in its column.

Property 2 says that the leading entries form an echelon (steplike) pattern that moves down and to the right.

The following matrices are in row-echelon form; a starred entry may be any value, including zero.

$$\begin{bmatrix} 1 & * & * \\ 0 & 1 & * \\ 0 & 0 & 1 \end{bmatrix} \quad \begin{bmatrix} 1 & * & * & * \\ 0 & 1 & * & * \\ 0 & 0 & 1 & * \end{bmatrix} \quad \begin{bmatrix} 1 & * & * & * \\ 0 & 0 & 1 & * \\ 0 & 0 & 0 & 1 \\ 0 & 0 & 0 & 0 \end{bmatrix}$$

The following matrices are in *reduced* row-echelon form because the entries above and below each leading 1 are zero. (Again, the starred entries may be any value, including zero.)

$$\begin{bmatrix} 0 & 1 & 0 & 0 \\ 0 & 0 & 1 & 0 \\ 0 & 0 & 0 & 1 \end{bmatrix} \quad \begin{bmatrix} 1 & 0 & 0 & * \\ 0 & 1 & 0 & * \\ 0 & 0 & 1 & * \end{bmatrix} \quad \begin{bmatrix} 1 & * & 0 & 0 \\ 0 & 0 & 1 & 0 \\ 0 & 0 & 0 & 1 \\ 0 & 0 & 0 & 0 \end{bmatrix}$$

An augmented matrix may be transformed into several different row-echelon form matrices by using different sequences of row operations. However, its *reduced* row-echelon form is *unique*. See Exercise 91.

3 Use Gaussian elimination to solve a system.

Gaussian Elimination

The method of solving a system of linear equations by transforming the augmented matrix of the system into row-echelon form and then using back-substitution to find the solution set is known as **Gaussian elimination**. See Section 8.2.

PROCEDURE IN ACTION

EXAMPLE 5 **Solving Linear Systems by Using Gaussian Elimination**

OBJECTIVE

Solve a system of linear equations by Gaussian elimination.

Step 1 Write the augmented matrix of the system.

Step 2 Use elementary row operations to transform the augmented matrix into row-echelon form.

Step 3 Write the system of linear equations that corresponds to the last matrix in Step 2.

Step 4 Use the system of equations from Step 3, together with back-substitution, to find the system's solution set.

EXAMPLE

Solve the system of equations.

$$\begin{cases} x + 2y = -2 \\ 4x + 3y = 7 \end{cases}$$

1. $A = \left[\begin{array}{cc|c} 1 & 2 & -2 \\ 4 & 3 & 7 \end{array}\right]$ Augmented matrix

2. $\xrightarrow{-4R_1 + R_2 \to R_2} \left[\begin{array}{cc|c} 1 & 2 & -2 \\ 0 & -5 & 15 \end{array}\right]$ This operation produces a 0 in the $(2, 1)$ position.

$\xrightarrow{-\frac{1}{5}R_2} \left[\begin{array}{cc|c} 1 & 2 & -2 \\ 0 & 1 & -3 \end{array}\right]$ This operation produces a 1 in the $(2, 2)$ position.

3. $\begin{cases} x + 2y = -2 & (1) \\ \quad\;\; y = -3 & (2) \end{cases}$

4.

$x + 2y = -2$ Equation (1)

$x + 2(-3) = -2$ Replace y with -3.

$x = 4$ Solve for x.

You should check the solution set, $\{(4, -3)\}$, by substituting $x = 4$ and $y = -3$ into the original system of equations.

Practice Problem 5 Solve by Gaussian elimination.

$$\begin{cases} x - 2y = 1 \\ 2x + 3y = 16 \end{cases}$$

EXAMPLE 6 **Solving a System by Using Gaussian Elimination**

Solve by Gaussian elimination.

$$\begin{cases} 2x + y + z = 6 \\ -3x - 4y + 2z = 4 \\ x + y - z = -2 \end{cases}$$

Solution

Step 1 $A = \left[\begin{array}{ccc|c} 2 & 1 & 1 & 6 \\ -3 & -4 & 2 & 4 \\ 1 & 1 & -1 & -2 \end{array}\right]$ The augmented matrix of the system

Step 2 $\xrightarrow{R_1 \leftrightarrow R_3} \left[\begin{array}{ccc|c} 1 & 1 & -1 & -2 \\ -3 & -4 & 2 & 4 \\ 2 & 1 & 1 & 6 \end{array}\right]$

$\xrightarrow[-2R_1 + R_3 \to R_3]{3R_1 + R_2 \to R_2} \left[\begin{array}{ccc|c} 1 & 1 & -1 & -2 \\ 0 & -1 & -1 & -2 \\ 0 & -1 & 3 & 10 \end{array}\right]$

SIDE NOTE

Note that when we perform a row operation such as

$3R_1 + R_2 \to R_2$ on a matrix,

1. the entries in row 1 are *unchanged* and
2. we multiply each entry of row 1 by 3 and add this product to the corresponding entries of row 2 to obtain a *new row 2*.

$$\xrightarrow{(-1)R_2}\left[\begin{array}{ccc|c} 1 & 1 & -1 & -2 \\ 0 & 1 & 1 & 2 \\ 0 & -1 & 3 & 10 \end{array}\right]$$

$$\xrightarrow{R_2 + R_3 \rightarrow R_3}\left[\begin{array}{ccc|c} 1 & 1 & -1 & -2 \\ 0 & 1 & 1 & 2 \\ 0 & 0 & 4 & 12 \end{array}\right]$$

$$\xrightarrow{\frac{1}{4}R_3}\left[\begin{array}{ccc|c} 1 & 1 & -1 & -2 \\ 0 & 1 & 1 & 2 \\ 0 & 0 & 1 & 3 \end{array}\right]$$

The last matrix is in row-echelon form.

Step 3 The system of equations corresponding to the last matrix in Step 2 is

$$\begin{cases} x + y - z = -2 & (1) \\ \quad y + z = 2 & (2) \\ \qquad z = 3 & (3) \end{cases}$$

Step 4 Equation (3) in Step 3 gives the value $z = 3$. Back-substitute $z = 3$ in equation (2).

$$\begin{aligned} y + z &= 2 && \text{Equation (2)} \\ y + 3 &= 2 && \text{Replace } z \text{ with 3.} \\ y &= -1 && \text{Solve for } y. \end{aligned}$$

Now back-substitute $z = 3$ and $y = -1$ in equation (1).

$$\begin{aligned} x + y - z &= -2 && \text{Equation (1)} \\ x - 1 - 3 &= -2 && \\ x &= 2 && \text{Solve for } x. \end{aligned}$$

You should check the solution set, $\{(2, -1, 3)\}$, by substituting these values into the original system of equations.

Practice Problem 6 Solve the system of equations:

$$\begin{cases} 2x + y - z = 7 \\ x - 3y - 3z = 4 \\ 4x + y + z = 3 \end{cases}$$

EXAMPLE 7 Attempting to Solve a System with No Solution

Solve the system of equations by Gaussian elimination:

$$\begin{cases} \quad y + 5z = -4 \\ x + 4y + 3z = -2 \\ 2x + 7y + z = 8 \end{cases}$$

Solution

Step 1 $A = \left[\begin{array}{ccc|c} 0 & 1 & 5 & -4 \\ 1 & 4 & 3 & -2 \\ 2 & 7 & 1 & 8 \end{array}\right]$. The augmented matrix of the system

Step 2 $\xrightarrow{R_1 \leftrightarrow R_2}\left[\begin{array}{ccc|c} 1 & 4 & 3 & -2 \\ 0 & 1 & 5 & -4 \\ 2 & 7 & 1 & 8 \end{array}\right]$

$$\xrightarrow{-2R_1 + R_3 \to R_3} \left[\begin{array}{rrr|r} 1 & 4 & 3 & -2 \\ 0 & 1 & 5 & -4 \\ 0 & -1 & -5 & 12 \end{array}\right]$$ Row 2 already begins with a 0.

$$\xrightarrow{R_2 + R_3 \to R_3} \left[\begin{array}{rrr|r} 1 & 4 & 3 & -2 \\ 0 & 1 & 5 & -4 \\ 0 & 0 & 0 & 8 \end{array}\right]$$

$$\xrightarrow{\frac{1}{8}R_3} \left[\begin{array}{rrr|r} 1 & 4 & 3 & -2 \\ 0 & 1 & 5 & -4 \\ 0 & 0 & 0 & 1 \end{array}\right]$$ The matrix is now in row-echelon form.

Step 3 The system of equations corresponding to the last matrix in Step 2 is

$$\begin{cases} x + 4y + 3z = -2 \\ \quad\;\; y + 5z = -4 \\ \qquad\quad\;\; 0 = 1 \end{cases}$$ A false statement

RECALL

A system is called *inconsistent* if it has no solution.

Because the third equation $0 = 1$ is a false statement, we conclude that this system is inconsistent. Because this system is equivalent to the original system, the original system is also inconsistent.

Step 4 The solution set for the system is $\varnothing$.

Practice Problem 7 Solve the following system of equations by first transforming the augmented matrix into row-echelon form.

$$\begin{cases} 6x + 8y - 14z = 3 \\ 3x + 4y - 7z = 12 \\ 6x + 3y + z = 0 \end{cases}$$

4 Use Gauss–Jordan elimination to solve a system.

Gauss–Jordan Elimination

If we continue the Gaussian elimination procedure until we have a *reduced* row-echelon form, the procedure is called **Gauss–Jordan elimination**.

TECHNOLOGY CONNECTION

Many graphing calculators can produce the unique reduced row-echelon form of an augmented matrix. Consider the augmented matrix *A* in Example 4.

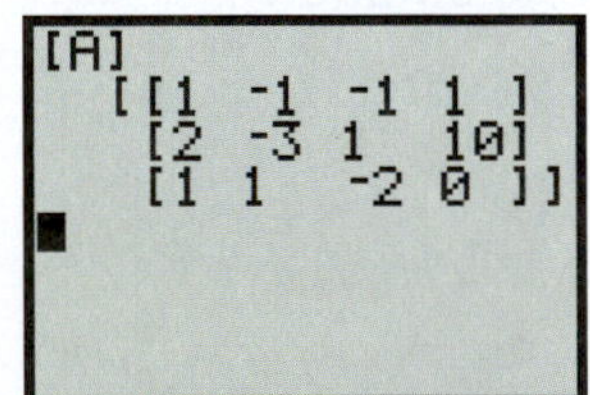

We use the "rref" (reduced row-echelon form) option to obtain the reduced row-echelon form.

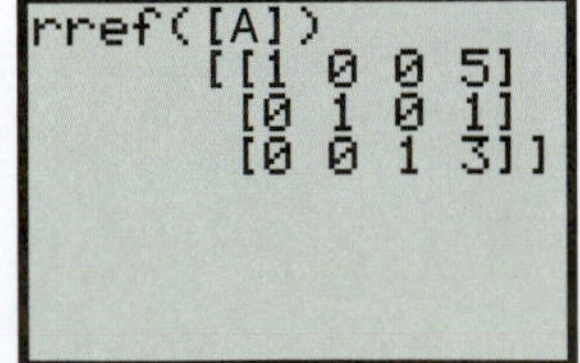

EXAMPLE 8 **Solving a System of Equations by Gauss–Jordan Elimination**

Solve the system given in Example 4 by Gauss–Jordan elimination.

$$\begin{cases} x - y - z = 1 \\ 2x - 3y + z = 10 \\ x + y - 2z = 0 \end{cases}$$ The given system

Solution

$$\left[\begin{array}{rrr|r} 1 & -1 & -1 & 1 \\ 0 & 1 & -3 & -8 \\ 0 & 0 & 5 & 15 \end{array}\right]$$ The final augmented matrix of the system in Example 4

$$\xrightarrow{\frac{1}{5}R_3} \left[\begin{array}{rrr|r} 1 & -1 & -1 & 1 \\ 0 & 1 & -3 & -8 \\ 0 & 0 & 1 & 3 \end{array}\right]$$ The matrix is now in row-echelon form.

$$\xrightarrow{R_2 + R_1 \to R_1} \left[\begin{array}{rrr|r} 1 & 0 & -4 & -7 \\ 0 & 1 & -3 & -8 \\ 0 & 0 & 1 & 3 \end{array}\right]$$

$$\xrightarrow[3R_3 + R_2 \to R_2]{4R_3 + R_1 \to R_1} \left[\begin{array}{rrr|r} 1 & 0 & 0 & 5 \\ 0 & 1 & 0 & 1 \\ 0 & 0 & 1 & 3 \end{array}\right]$$

We now have an equivalent matrix in reduced row-echelon form. The corresponding system of equations for the last augmented matrix is as follows:

$$\begin{cases} x = 5 \\ y = 1 \\ z = 3 \end{cases}$$

The solution set is therefore $\{(5, 1, 3)\}$, as in Example 4.

Practice Problem 8 Solve the system by Gauss–Jordan elimination:

$$\begin{cases} 2x - 3y - 2z = 0 \\ x + y - 2z = 7 \\ 3x - 5y - 5z = 3 \end{cases}$$

EXAMPLE 9 Solving a System with Infinitely Many Solutions

Solve the system of equations:

$$\begin{cases} x + 2y + 5z = 4 \\ y + 4z = 4 \\ 2x + 4y + 10z = 8 \end{cases}$$

Solution

The augmented matrix A of the system is given below. We want to find the equivalent system in reduced row-echelon form.

Because $a_{11} = 1$ and $a_{21} = 0$, we need a zero at the $(3, 1)$ position.

$$A = \left[\begin{array}{ccc|c} 1 & 2 & 5 & 4 \\ 0 & 1 & 4 & 4 \\ 2 & 4 & 10 & 8 \end{array}\right] \xrightarrow{-2R_1 + R_3 \to R_3} \left[\begin{array}{ccc|c} 1 & 2 & 5 & 4 \\ 0 & 1 & 4 & 4 \\ 0 & 0 & 0 & 0 \end{array}\right]$$

Next, we need a zero at the $(1, 2)$ position.

$$\xrightarrow{-2R_2 + R_1 \to R_1} \left[\begin{array}{ccc|c} 1 & 0 & -3 & -4 \\ 0 & 1 & 4 & 4 \\ 0 & 0 & 0 & 0 \end{array}\right]$$

The matrix is now in reduced row-echelon form.

The equivalent system is

$$\begin{aligned} x - 3z &= -4 \\ y + 4z &= 4. \end{aligned}$$

Solving for x and y in terms of z, we obtain

$$\begin{aligned} x &= 3z - 4 \\ y &= -4z + 4. \end{aligned}$$

Each real number z results in a solution with $y = -4z + 4$ and $x = 3z - 4$, giving infinitely many solutions of the form $(3z - 4, -4z + 4, z)$. The solution set is $\{(3z - 4, -4z + 4, z)\}$.

Practice Problem 9 Solve the system of equations:

$$\begin{cases} x + z = -1 \\ 3y + 2z = 5 \\ 3x - 3y + z = -8 \end{cases}$$

EXAMPLE 10 Leontief Input–Output Model

Consider an economy that has steel, coal, and transportation industries. There are two types of demands (measured in dollars) on the production of each industry: interindustry demand and external consumer demand. The outputs and requirements of the three industries are shown in Figure 9.1. For example, \$1.00 of transportation output requires \$0.10 from steel and \$0.01 from coal.

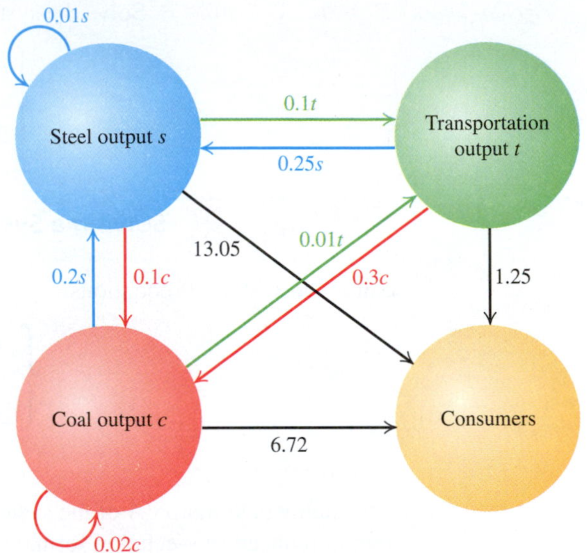

FIGURE 9.1 Output diagram of an economy

a. Write a system of equations that expresses the outputs of the three industries. Assume that all quantities are given in millions of dollars.

b. Verify that $s = 15$, $c = 10$, and $t = 8$ will meet both interindustry and consumer demand.

Solution

a. To satisfy both consumer and interindustry demand, we obtain the following system of outputs. (s denotes total steel output; c, total coal output; t, total transportation output.)

$$\begin{cases} s = 0.01s + 0.1t + 0.1c + 13.05 & \text{Distribution of steel output} \\ c = 0.2s + 0.01t + 0.02c + 6.72 & \text{Distribution of coal output} \\ t = 0.25s + 0.3c + 1.25 & \text{Distribution of transportation output} \end{cases}$$

You can rewrite this system as follows:

$$\begin{cases} 0.99s - 0.1c - 0.1t = 13.05 & (1) \\ -0.2s + 0.98c - 0.01t = 6.72 & (2) \\ -0.25s - 0.3c + t = 1.25 & (3) \end{cases}$$

b. To verify that the given numbers (obtained by solving the system above using matrices) satisfy these equations, we substitute $s = 15$, $c = 10$, and $t = 8$ into each equation.

$$(0.99)(15) - (0.1)(10) - (0.1)(8) = 13.05 \checkmark \quad (1)$$
$$(-0.20)(15) + (0.98)(10) - (0.01)(8) = 6.72 \checkmark \quad (2)$$
$$(-0.25)(15) - (0.30)(10) + 8 = 1.25 \checkmark \quad (3)$$

In other words, output levels of $s = 15$, $c = 10$, and $t = 8$ million dollars will meet interindustry and consumer demand.

Practice Problem 10 In Example 10, assume that consumer demand (in millions of dollars) is 12.46 for steel, 3 for coal, and 2.7 for transportation.

a. Write a system of equations to express the outputs for the three industries.

b. Verify that $s = 14, c = 6, t = 8$ will meet both interindustry and consumer demand.

Answers to Practice Problems

1. a. 3×2 **b.** 1×2 **2.** $\left[\begin{array}{ccc|c} 0 & 3 & -1 & 8 \\ 1 & 4 & 0 & 14 \\ 0 & -2 & 9 & 0 \end{array}\right]$

3. $B = \begin{bmatrix} 2 & 4 & 6 \\ 3 & 4 & 5 \end{bmatrix} \xrightarrow{\frac{1}{2}R_1} \begin{bmatrix} 1 & 2 & 3 \\ 3 & 4 & 5 \end{bmatrix} \xrightarrow{-3R_1 + R_2 \to R_2}$

$C = \begin{bmatrix} 1 & 2 & 3 \\ 0 & -2 & -4 \end{bmatrix}$ **4.** $\left\{\left(-1, -\frac{1}{3}, -1\right)\right\}$ **5.** $\{(5, 2)\}$

6. $\{(1, 2, -3)\}$ **7.** $\varnothing$ **8.** $\{(1, 2, -2)\}$

9. $\left\{\left(-z - 1, \frac{5}{3} - \frac{2}{3}z, z\right)\right\}$

10. a. $\begin{cases} s = 0.01s + 0.1c + 0.1t + 12.46 \\ c = 0.2s + 0.02c + 0.01t + 3 \\ t = 0.25s + 0.3c + 2.7 \end{cases}$

b. $(0.99)(14) - (0.1)(6) - (0.1)(8) = 12.46$ ✓
$(-0.20)(14) + (0.98)(6) - (0.01)(8) = 3$ ✓
$(-0.25)(14) - (0.30)(6) + 8 = 2.7$ ✓

SECTION 9.1 Exercises

Basic Concepts and Skills

1. A matrix is any rectangular array of ____________.
2. The array of coefficients and constants in a linear system is called the ____________ of the system.
3. Two matrices are row-equivalent if one can be obtained from the other by a sequence of ____________.
4. If a matrix is in row-echelon form and each leading entry 1 is the only nonzero entry in its column, then the matrix is in ____________ form.
5. **True or False.** If A is a 3×4 matrix, then each row of A has three entries.
6. **True or False.** Every $m \times n$ $(n \geq 2)$ matrix is the matrix of some linear system.
7. **True or False.** The augmented matrix for the system of equations

$$\begin{cases} 2x + 3y = 5 \\ 3y - 4z = -1 \\ x + 2z = 3 \end{cases} \quad \text{is} \quad \left[\begin{array}{cc|c} 2 & 3 & 5 \\ 3 & -4 & -1 \\ 1 & 2 & 3 \end{array}\right].$$

8. **True or False.** The augmented matrix for the system of equations

$$\begin{cases} 2x - y = 1 \\ 3x + y = 9 \\ 5x - 2y = 4 \end{cases} \quad \text{is} \quad \left[\begin{array}{cc|c} 2x & -y & 1 \\ 3x & y & 9 \\ 5x & -2y & 4 \end{array}\right].$$

In Exercises 9–14, determine the order of each matrix.

9. $[7]$ **10.** $[1 \;\; 3 \;\; 5 \;\; 9]$

11. $\begin{bmatrix} 3 & 4.3 & 7.5 & 8 \\ 2 & -1 & -3 & 0 \end{bmatrix}$ **12.** $\begin{bmatrix} -1 & 2 & 3 \\ 4 & -5 & 7 \\ 1 & 0 & 0 \end{bmatrix}$

13. $\begin{bmatrix} 4 & 5 & 1 \\ 0 & 0 & 0 \end{bmatrix}$ **14.** $\begin{bmatrix} -5 & 2 & 3 & 7 \\ 2.5 & e & -\pi & \frac{1}{2} \\ -e & \pi & 0 & 1 \end{bmatrix}$

15. Let $A = \begin{bmatrix} 1 & 2 & 3 & 4 \\ 5 & 6 & 7 & 8 \\ 9 & 10 & 11 & 12 \end{bmatrix}$. Identify the entries a_{13}, a_{31}, a_{33}, and a_{34}.

16. In matrix A from Exercise 15, identify the (i, j)th location a_{ij} of each entry.

a. 7 **b.** 10
c. 4 **d.** 12

17. Is the following array a matrix? Why or why not?

$$\begin{bmatrix} 2 & 0 & 4 \\ 1 & -3 & 5 \\ 0 & 0 & \end{bmatrix}$$

18. Write the matrix A with the following entries:
$a_{13} = 5, a_{22} = 2, a_{11} = 6, a_{12} = -5, a_{23} = 4$, and $a_{21} = 7$

In Exercises 19–24, write the augmented matrix for each system of linear equations.

19. $\begin{cases} 2x + 4y = 2 \\ x - 3y = 1 \end{cases}$ **20.** $\begin{cases} x + 2y = 7 \\ 3x + 5y = 11 \end{cases}$

21. $\begin{cases} 5x - 11 = 7y \\ 17y - 19 = 13x \end{cases}$ **22.** $\begin{cases} 2y - 10 = 3x \\ 5x + 7 = y \end{cases}$

23. $\begin{cases} -x + 2y + 3z = 8 \\ 2x - 3y + 9z = 16 \\ 4x - 5y - 6z = 32 \end{cases}$ **24.** $\begin{cases} x - y = 2 \\ 2x + 3z = -5 \\ y - 2z = 7 \end{cases}$

In Exercises 25–28, write the system of linear equations represented by each augmented matrix. Use x, y, and z as the variables.

25. $\left[\begin{array}{ccc|c} 1 & 2 & -3 & 4 \\ -2 & -3 & 1 & 5 \\ 3 & -3 & 2 & 7 \end{array}\right]$

26. $\left[\begin{array}{ccc|c} -1 & 2 & 3 & 6 \\ 2 & 3 & 1 & 2 \\ 4 & 3 & 2 & 1 \end{array}\right]$

27. $\left[\begin{array}{ccc|c} 1 & -1 & 1 & 2 \\ 2 & 1 & -3 & 6 \end{array}\right]$

28. $\left[\begin{array}{ccc|c} 1 & 1 & 1 & 2 \\ 1 & -1 & -1 & 4 \\ 2 & 3 & 1 & 6 \\ -1 & 1 & -1 & 8 \end{array}\right]$

In Exercises 29–32, perform the indicated elementary row operations in the stated order.

29. $\begin{bmatrix} 2 & 3 & 5 \\ 1 & 2 & 3 \end{bmatrix}$; (i) $R_1 \leftrightarrow R_2$, (ii) $-2R_1 + R_2 \rightarrow R_2$, (iii) $-R_2$

30. $\begin{bmatrix} 2 & 4 & 2 \\ 1 & 5 & 7 \end{bmatrix}$; (i) $\frac{1}{2}R_1$, (ii) $(-1)\,R_1 + R_2 \rightarrow R_2$, (iii) $\frac{1}{3}R_2$

31. $\begin{bmatrix} 1 & 2 & 3 & 4 \\ 0 & 4 & -3 & 11 \\ 0 & 1 & 5 & -3 \end{bmatrix}$; (i) $R_2 \leftrightarrow R_3$, (ii) $-4R_2 + R_3 \rightarrow R_3$, (iii) $-\frac{1}{23}R_3$

32. $\begin{bmatrix} 1 & \frac{3}{2} & -2 & \frac{1}{2} \\ 0 & 2 & 3 & 4 \\ -2 & -1 & 7 & 3 \end{bmatrix}$; (i) $2R_1 + R_3 \rightarrow R_3$, (ii) $\frac{1}{2}R_2 + R_3 \rightarrow R_3$

In Exercises 33–36, identify the elementary row operation used and supply the missing entries in each row-equivalent matrix.

33. $\begin{bmatrix} 4 & 5 & -7 \\ 5 & 4 & -2 \end{bmatrix} \rightarrow \begin{bmatrix} 1 & ? & -\frac{7}{4} \\ 5 & 4 & -2 \end{bmatrix} \rightarrow \begin{bmatrix} 1 & ? & -\frac{7}{4} \\ 0 & -\frac{9}{4} & ? \end{bmatrix} \rightarrow \begin{bmatrix} 1 & ? & -\frac{7}{4} \\ 0 & 1 & ? \end{bmatrix}$

34. $\begin{bmatrix} 2 & 6 & -8 \\ 3 & -1 & 2 \end{bmatrix} \rightarrow \begin{bmatrix} 1 & 3 & -? \\ 3 & -1 & 2 \end{bmatrix} \rightarrow \begin{bmatrix} 1 & 3 & ? \\ 0 & -? & 14 \end{bmatrix} \rightarrow \begin{bmatrix} 1 & 3 & -? \\ 0 & 1 & ? \end{bmatrix}$

35. $\begin{bmatrix} 1 & 4 & 3 & 1 \\ 0 & -3 & -2 & 0 \\ 0 & 7 & 5 & -3 \end{bmatrix} \rightarrow \begin{bmatrix} 1 & 4 & 3 & 1 \\ 0 & 1 & ? & 0 \\ 0 & 7 & 5 & -3 \end{bmatrix} \rightarrow \begin{bmatrix} 1 & 4 & 3 & 1 \\ 0 & 1 & ? & 0 \\ 0 & 0 & ? & -3 \end{bmatrix} \rightarrow \begin{bmatrix} 1 & 4 & 3 & 1 \\ 0 & 1 & ? & 0 \\ 0 & 0 & 1 & ? \end{bmatrix}$

36. $\begin{bmatrix} 1 & 1 & 1 & 3 \\ 2 & 3 & 3 & 8 \\ 1 & -3 & -2 & 5 \end{bmatrix} \rightarrow \begin{bmatrix} 1 & 1 & 1 & 3 \\ 0 & 1 & ? & 2 \\ 1 & -3 & -2 & 5 \end{bmatrix} \rightarrow \begin{bmatrix} 1 & 1 & 1 & 3 \\ 0 & 1 & ? & 2 \\ 0 & -4 & ? & 2 \end{bmatrix} \rightarrow \begin{bmatrix} 1 & 1 & 1 & 3 \\ 0 & 1 & ? & 2 \\ 0 & 0 & ? & 10 \end{bmatrix}$

In Exercises 37–44, determine whether each matrix is in row-echelon form. If your answer is no, explain why. If your answer is yes, is the matrix in reduced row-echelon form?

37. $\begin{bmatrix} 0 & 1 & 2 \\ 1 & 0 & 3 \end{bmatrix}$

38. $\begin{bmatrix} 1 & 2 & 5 \\ 0 & 3 & 6 \end{bmatrix}$

39. $\begin{bmatrix} 1 & 0 & 0 & 2 \\ 0 & 1 & 0 & 3 \\ 0 & 0 & 1 & 4 \end{bmatrix}$

40. $\begin{bmatrix} 0 & 0 & 1 & 2 \\ 0 & 1 & 0 & 1 \\ 1 & 0 & 0 & 3 \end{bmatrix}$

41. $\begin{bmatrix} 1 & 2 & 0 & 2 \\ 0 & 0 & 1 & 5 \end{bmatrix}$

42. $\begin{bmatrix} 0 & 1 & -1 & -2 & 0 & 2 & 0 \\ 0 & 0 & 0 & 0 & 1 & 2 & 0 \end{bmatrix}$

43. $\begin{bmatrix} 1 & 0 & 0 & -2 \\ 0 & 1 & 0 & 3 \\ 0 & 0 & 1 & 2 \\ 0 & 0 & 0 & 0 \end{bmatrix}$

44. $\begin{bmatrix} 1 & 0 & 0 & 3 \\ 0 & 1 & 0 & 4 \\ 0 & 0 & 0 & 5 \\ 0 & 0 & 1 & 3 \end{bmatrix}$

In Exercises 45–54, the augmented matrix of a system of equations has been transformed to an equivalent matrix in row-echelon form or reduced row-echelon form. Using x, y, z, and w as variables, write the system of equations corresponding to the matrix. If the system is consistent, solve it.

45. $\left[\begin{array}{cc|c} 1 & 2 & 1 \\ 0 & 1 & -2 \end{array}\right]$

46. $\left[\begin{array}{cc|c} 1 & 0 & 2 \\ 0 & 1 & 3 \end{array}\right]$

47. $\left[\begin{array}{ccc|c} 1 & 4 & 2 & 2 \\ 0 & 0 & 1 & 3 \end{array}\right]$

48. $\left[\begin{array}{ccc|c} 1 & 2 & 3 & 4 \\ 0 & 0 & 0 & 1 \end{array}\right]$

49. $\left[\begin{array}{ccc|c} 1 & 2 & 3 & 2 \\ 0 & 1 & -2 & 4 \\ 0 & 0 & 1 & -1 \end{array}\right]$

50. $\left[\begin{array}{ccc|c} 1 & 0 & 2 & 12 \\ 0 & 1 & 3 & 12 \\ 0 & 0 & 1 & 5 \end{array}\right]$

51. $\left[\begin{array}{cccc|c} 1 & 0 & 0 & 0 & 2 \\ 0 & 1 & 0 & 0 & -5 \\ 0 & 0 & 1 & 2 & 3 \\ 0 & 0 & 0 & 0 & 0 \end{array}\right]$

52. $\left[\begin{array}{cccc|c} 1 & 0 & 0 & 0 & 4 \\ 0 & 1 & 0 & 0 & 3 \\ 0 & 0 & 0 & 1 & 2 \end{array}\right]$

53. $\left[\begin{array}{cccc|c} 1 & 0 & 0 & 0 & -5 \\ 0 & 1 & 0 & 0 & 4 \\ 0 & 0 & 1 & 2 & 3 \\ 0 & 0 & 0 & 1 & 0 \end{array}\right]$

54. $\left[\begin{array}{cccc|c} 1 & 0 & 0 & 0 & 3 \\ 0 & 1 & 0 & 0 & 2 \\ 0 & 0 & 1 & 0 & 0 \\ 0 & 0 & 0 & 0 & 1 \end{array}\right]$

In Exercises 55–68, solve each system of equations by Gaussian elimination.

55. $\begin{cases} x - 2y = 11 \\ 2x - y = 13 \end{cases}$

56. $\begin{cases} 3x - 2y = 4 \\ 4x - 3y = 5 \end{cases}$

57. $\begin{cases} 2x - 3y = 3 \\ 4x - y = 11 \end{cases}$

58. $\begin{cases} 3x + 2y = 1 \\ 6x + 4y = 3 \end{cases}$

59. $\begin{cases} 3x - 5y = 4 \\ 4x - 15y = 13 \end{cases}$

60. $\begin{cases} -2x + 4y = 1 \\ 3x - 5y = -9 \end{cases}$

61. $\begin{cases} x - y = 1 \\ 2x + y = 5 \\ 3x - 4y = 2 \end{cases}$

62. $\begin{cases} y = 2x + 1 \\ 3x + 2y + 1.5 = 0 \\ 4x - 2y + 2 = 0 \end{cases}$

63. $\begin{cases} x + y + z = 6 \\ x - y + z = 2 \\ 2x + y - z = 1 \end{cases}$

64. $\begin{cases} 2x + 4y + z = 5 \\ x + y + z = 6 \\ 2x + 3y + z = 6 \end{cases}$

65. $\begin{cases} 2x + 3y - z = 9 \\ x + y + z = 9 \\ 3x - y - z = -1 \end{cases}$

66. $\begin{cases} x + y + 2z = 4 \\ 2x - y + 3z = 9 \\ 3x - y - z = 2 \end{cases}$

67. $\begin{cases} 3x + 2y + 4z = 19 \\ 2x - y + z = 3 \\ 6x + 7y - z = 17 \end{cases}$

68. $\begin{cases} 4x + 3y + z = 8 \\ 2x + y + 4z = -4 \\ 3x + z = 1 \end{cases}$

In Exercises 69–74, solve each system of equations by Gauss–Jordan elimination.

69. $\begin{cases} x - y = 1 \\ x - z = -1 \\ 2x + y - z = 3 \end{cases}$

70. $\begin{cases} 4x + 5z = 7 \\ y - 6z = 8 \\ 3x + 4y = 9 \end{cases}$

71. $\begin{cases} x + y - z = 4 \\ x + 3y + 5z = 10 \\ 3x + 5y + 3z = 18 \end{cases}$

72. $\begin{cases} x + y + z = -5 \\ 2x - y - z = -4 \\ y + z = -2 \end{cases}$

73. $\begin{cases} x + 2y - z = 6 \\ 3x + y + 2z = 3 \\ 2x + 5y + 3z = 9 \end{cases}$

74. $\begin{cases} 2x + 4y - z = 9 \\ x + 3y - 3z = 4 \\ 3x + y + 2z = 7 \end{cases}$

In Exercises 75 and 76, the reduced row-echelon forms of the augmented matrices of three systems of equations are given. How many solutions does each system have?

75. a. $\left[\begin{array}{cc|c} 1 & 0 & 2 \\ 0 & 1 & 3 \end{array}\right]$ b. $\left[\begin{array}{ccc|c} 1 & 2 & 3 & 8 \\ 0 & 0 & 1 & 2 \\ 0 & 0 & 0 & 0 \end{array}\right]$

c. $\left[\begin{array}{cc|c} 1 & 0 & 2 \\ 0 & 0 & 3 \end{array}\right]$

76. a. $\left[\begin{array}{ccc|c} 1 & 0 & 0 & -1 \\ 0 & 1 & 0 & 2 \\ 0 & 0 & 1 & 3 \end{array}\right]$ b. $\left[\begin{array}{ccc|c} 1 & 0 & 0 & 4 \\ 0 & 1 & 2 & 5 \\ 0 & 0 & 0 & 6 \end{array}\right]$

c. $\left[\begin{array}{ccc|c} 1 & 1 & 2 & 3 \\ 0 & 0 & 0 & 0 \\ 0 & 0 & 0 & 0 \end{array}\right]$

In Exercises 77 and 78, determine whether the statement is true or false and justify your answer.

77. If A is a 5×7 matrix, then each column of A has seven entries.

78. If a matrix A is in reduced row-echelon form, then at least one of the entries in each column must be a 1.

Applying the Concepts

In Exercises 79–82, Leontief's input–output model of a simplified economy is described. See Example 10. For each exercise, do the following.

a. Set up (without solving) a linear system whose solution will represent the required production schedule.

b. Write an augmented matrix for the economy.

c. Solve the system of equations by using part (b) to find the production schedule that will meet interindustry and consumer demand.

79. **Two-sector economy.** Consider an economy with only two industries: A and B. Suppose industry B needs \$0.10 worth of A's product for each \$1.00 of output B produces and that industry A needs \$0.20 worth of B's product for each \$1.00 of output A produces. Consumer demand for A's product is \$1000, and consumer demand for B's product is \$780.

80. **A company with two branches.** A company has two interacting branches: A and B. Branch A consumes \$0.50 of its own output and \$0.20 of B's output for every \$1.00 it produces. Branch B consumes \$0.60 of A's output and \$0.30 of its own output for \$1.00 of output. Consumer demand for A's product is \$50,000, and consumer demand for B's product is \$40,000.

81. **Three-sector economy.** An economy has three sectors: labor, transportation, and food industries. Suppose the demand on \$1.00 in labor is \$0.40 for transportation and \$0.20 for food; the demand on \$1.00 in transportation is \$0.50 for labor and \$0.30 for transportation; and the demand on \$1.00 in food production is \$0.50 for labor, \$0.05 for transportation, and \$0.35 for food. Outside consumer demand for the current production period is \$10,000 for labor, \$20,000 for transportation, and \$10,000 for food.

82. **Three-sector economy.** Consider an economy with three industries A, B, and C, with outputs a, b, and c, respectively. Demand on the three industries is shown in the figure.

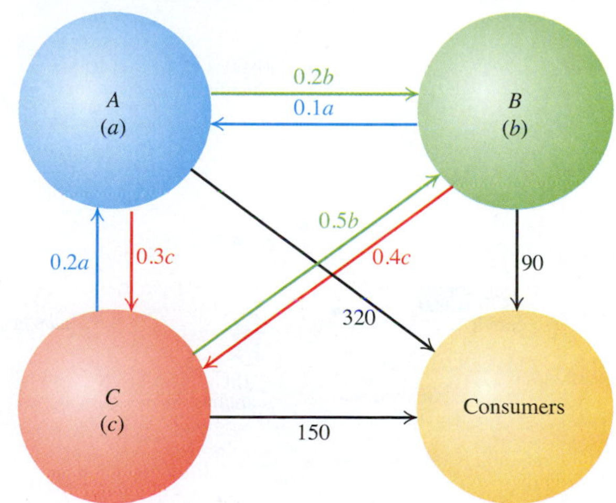

83. Heat transfer. In a study of heat transfer in a grid of wires, the temperature at an exterior node is maintained at a constant value (in °F) as shown in the accompanying figure. When the grid is in thermal equilibrium, the temperature at an interior node is the average of the temperatures at the four adjacent nodes. For instance, $T_1 = \dfrac{0 + 0 + 300 + T_2}{4}$, or $4T_1 - T_2 = 300$. Find the temperatures T_1, T_2, and T_3 when the grid is in thermal equilibrium.

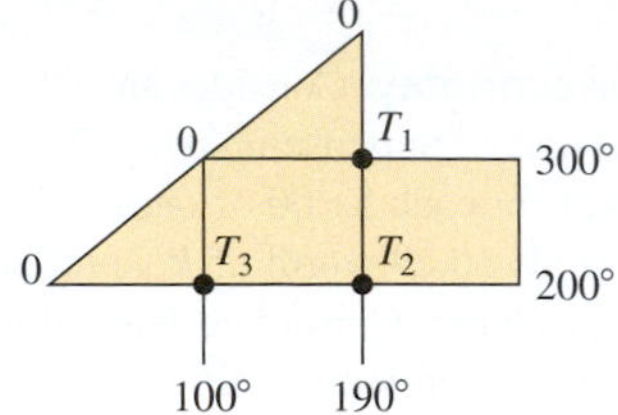

84. Heat transfer. Repeat Exercise 83 for the following grid of wires.

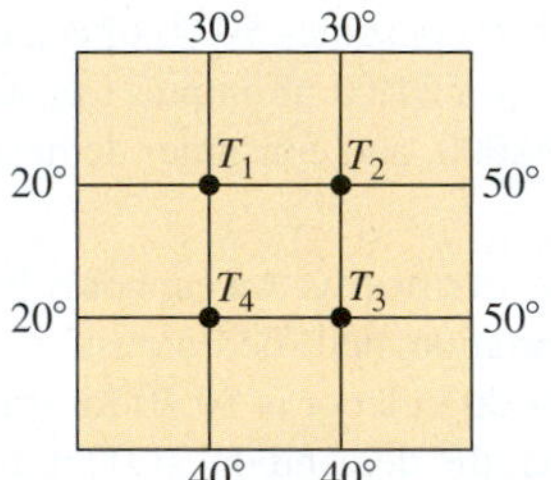

85. Traffic flow. The sketch shows the intersections of certain one-way streets. The traffic volume during one hour was observed. To keep the traffic moving, traffic controllers use the formula that the number of cars per hour entering an intersection equals the number of cars per hour exiting the intersection.

a. Set up a system of equations that keeps traffic moving.

b. Solve the system of equations in part (a).

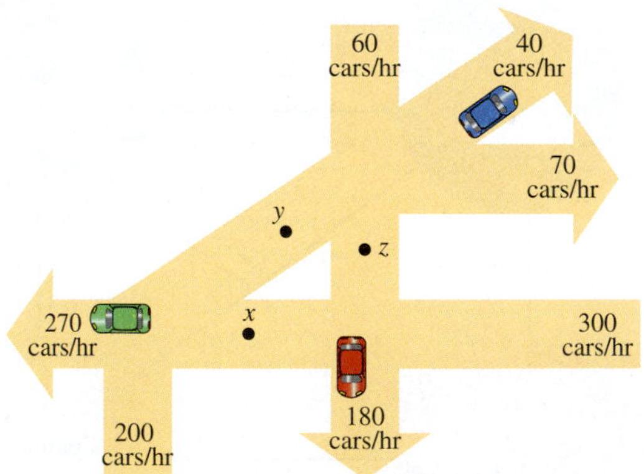

86. Traffic flow. Repeat Exercise 85 for the following pattern of one-way streets.

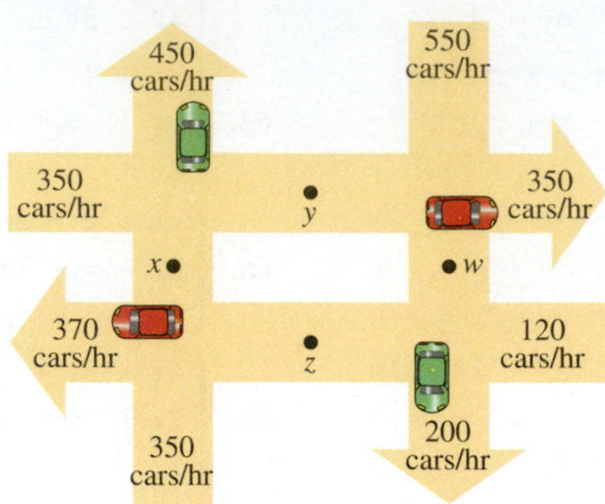

87. Parabola. Find the equation of the parabola $y = ax^2 + bx + c$ that passes through the points $(-1, 9)$, $(1, 3)$, and $(2, 6)$.

88. Modeling. A football was punted. Its height h above the ground at time t is given by the following table:

Time t (seconds)	Height h (feet)
$t = 0.5$	31
$t = 1$	51
$t = 2$	67

a. Use the system of equations to find the equation of the parabola of the form $h = at^2 + bt + c$.

b. What is the hang time (the time it takes the ball to touch the ground) of the punt?

c. What is the maximum height of the ball?

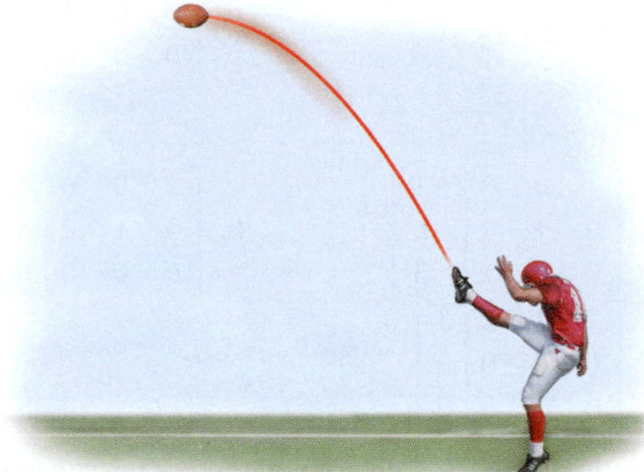

Beyond the Basics

In Exercises 89 and 90, solve each system of equations by Gauss–Jordan elimination.

89. $\begin{cases} x + y + z + w = 0 \\ x + 3y + 2z + 4w = 0 \\ 2x + z - w = 0 \end{cases}$

90. $\begin{cases} x - y + z - w = 4 \\ x + 2y + z + w = 2 \\ 2x + 3y + 4z + 5w = 5 \\ 3x + 4y + 2z - w = 8 \end{cases}$

91. Let $A = \begin{bmatrix} 1 & 2 & -3 & 1 \\ 1 & 0 & -3 & 2 \\ 0 & 1 & 1 & 0 \\ 2 & 3 & 0 & -2 \end{bmatrix}$.

a. Find two different matrices B and C in row-echelon form that are row-equivalent to A.

b. Show that the reduced row-echelon form of the matrices B and C of part (a) produce the same matrix.

92. Consider the following system of linear equations:

$$\begin{cases} x + y + z = 6 \\ x - y + z = 2 \\ 2x + y - z = 1 \end{cases}$$

a. Find two different matrices B and C in row-echelon form that yield the same solution set.

b. Show that the reduced row-echelon form of the matrices B and C of part (a) produces the same matrix.

93. a. Use the methods of this section to solve a general 2×2 system:

$$\begin{cases} ax + by = m \\ cx + dy = n \end{cases}$$

b. Under what conditions on the coefficients does the system have **(i)** a unique solution, **(ii)** no solution, and **(iii)** infinitely many solutions?

94. Construct three different augmented matrices for linear systems whose solution set is $x = 1, y = 0, z = -2$. There are many different correct answers.

In Exercises 95 and 96, solve each system of equations by using row operations.

95. $\begin{cases} \log x + \log y + \log z = 6 \\ 3 \log x - \log y + 3 \log z = 10 \\ 5 \log x + 5 \log y - 4 \log z = 3 \end{cases}$

[*Hint:* Let $u = \log x$, $v = \log y$, and $w = \log z$.]

96. $\begin{cases} 4 \cdot 2^x - 3 \cdot 3^y + 5^z = 1 \\ 2^x + 4 \cdot 3^y - 2 \cdot 5^z = 10 \\ 2 \cdot 2^x - 2 \cdot 3^y + 3 \cdot 5^z = 4 \end{cases}$

[*Hint:* Let $u = 2^x$, $v = 3^y$, and $w = 5^z$.]

97. Find the cubic function $y = ax^3 + bx^2 + cx + d$ whose graph passes through the points $(1, 5)$, $(-1, 1)$, $(2, 7)$, and $(-2, 11)$.

98. Find the cubic function $y = ax^3 + bx^2 + cx + d$ whose graph passes through the points $(1, 8)$, $(-1, 2)$, $(2, 8)$, and $(-2, 20)$.

Critical Thinking/Discussion/Writing

99. Find the form of all 2×1 matrices in reduced row-echelon form.

100. Find the form of all 2×2 matrices in reduced row-echelon form.

101. Suppose a matrix A is transformed to a matrix B by an elementary row operation. Is there an elementary row operation that transforms B to A? Explain.

102. Are the following statements true or false? Justify your answers. Suppose a matrix A is in reduced row-echelon form.

a. If we delete a row of A, then the remaining matrix is in reduced row-echelon form.

b. Repeat part (a), replacing *row* with *column*.

Maintaining Skills

In Exercises 103–106, solve each equation.

103. $3(x - 1) = 5 - x$

104. $5(x - 2) = 3(x - 3) + 13$

105. $-3(x + 4) + 2 = 8 - x$

106. $6x - 3(5x + 2) = 4 - 5x$

In Exercises 107–110, solve each system.

107. $\begin{cases} 2x - y = 5 \\ x + 2y = 25 \end{cases}$

108. $\begin{cases} x - 3y = 1 \\ 2x + y = -5 \end{cases}$

109. $\begin{cases} x + 3y = 6 \\ 2x + 6y = 8 \end{cases}$

110. $\begin{cases} 2x - y = 3 \\ 6x - 3y = 9 \end{cases}$

In Exercises 111–114, identify each statement as True or False.

111. If $x(3t^2 + \pi\sqrt{t^4 + t^2 + 5}) = 0$ where t is a real number, then $x = 0$.

112. If x, y, and z are any real numbers, then $xy + z = x + yz$.

113. If x, y, and z are real numbers, then $xy(z + 5) = xyz + 5xy$.

114. If $(x^4 + x^2)(y - 3) = 1$ and $y = \frac{10}{3}$, then $x^4 + x^2 = 3$.

SECTION 9.2

Matrix Algebra

BEFORE STARTING THIS SECTION, REVIEW

1 Solving systems of equations (Sections 8.1 and 8.2)

OBJECTIVES

1 Define equality of two matrices.

2 Define matrix addition and scalar multiplication.

3 Define matrix multiplication.

4 Apply matrix multiplication to computer graphics.

Computer Graphics

The following fields illustrate the widespread and growing applications of computer graphics:

1. The *automobile industry*. Both computer-aided design (CAD) and computer-aided manufacturing (CAM) have revolutionized the automotive industry. Many months before a new car is built, engineers design and construct a *mathematical car*—a wire-frame model that exists only in computer memory and on graphics display terminals. The mathematical model organizes and influences each step of the design and manufacture of the car.
2. The *entertainment industry*. Computer graphics are used by the entertainment industry with spectacular success. They were used in *AVATAR* and other action movies, such as the latest James Bond thriller, for special effects, and they are used in the design and creation of arcade games.
3. *Military*. From the beginning of the widespread use of computers (in the 1960s), computer graphics have been used by the military to prepare for or defend against war. Modern weaponry depends heavily on computer-generated images.

Engineers use matrix algebra techniques on graphic images to change their orientation of scale, to zoom in on them, or to switch between two- and three-dimensional views. In Example 10 and in the exercises, we examine how basic matrix algebra can be used to manipulate graphical images.

1 Define equality of two matrices.

Equality of Matrices

In Section 9.1, we defined a matrix as a rectangular array of numbers written within brackets. In this and the next sections, you will learn about the operations with and uses of matrices. A matrix may be represented in any of the following ways:

1. A matrix may be denoted by capital letters: A, B, C, and so on.
2. A matrix may be denoted by its representative (i, j)th entry: $[a_{ij}]$, $[b_{ij}]$, $[c_{ij}]$, and so on.
3. A matrix may be written as a rectangular array of numbers:

$$A = [a_{ij}] = \begin{bmatrix} a_{11} & a_{12} & \cdots & a_{1n} \\ a_{21} & a_{22} & \cdots & a_{2n} \\ \vdots & \vdots & \cdots & \vdots \\ a_{m1} & a_{m2} & \cdots & a_{mn} \end{bmatrix}$$

James Joseph Sylvester (1814–1897)

In 1850, Sylvester coined the term **matrix** to denote a rectangular array of numbers. Sylvester, who was born into a Jewish family in London and studied for several years at Cambridge, was not permitted to take his degree there for religious reasons. Therefore, he received his degree from Trinity College, Dublin. In 1871, he accepted the chair of mathematics at the newly opened Johns Hopkins University in Baltimore. While at Hopkins, Sylvester founded the *American Journal of Mathematics* and helped develop a tradition of graduate education in mathematics in the United States.

We now examine some *algebraic* properties of matrices.

Equality of Matrices

Two matrices $A = [a_{ij}]$ and $B = [b_{ij}]$ are said to be **equal**, written $A = B$, if

1. A and B have the same order $m \times n$ (that is, A and B have the same number m of rows and same number n of columns).
2. $a_{ij} = b_{ij}$ for all i and j; that is, each (i, j)th entry of A is equal to the corresponding (i, j)th entry of B.

EXAMPLE 1 **Determining Equality of Matrices**

Find x and y such that

$$\begin{bmatrix} x + y & 2 \\ 3 & x - y \end{bmatrix} = \begin{bmatrix} 4 & 2 \\ 3 & 1 \end{bmatrix}.$$

Solution

Because equal matrices must have identical entries in corresponding positions, we have the following system of equations:

$$\begin{cases} x + y = 4 & (1) \\ x - y = 1 & (2) \end{cases}$$

Solve this system: Add the two equations to obtain $x = \frac{5}{2}$. Substitute $x = \frac{5}{2}$ in equation $x - y = 1$ and solve for y to obtain $y = \frac{3}{2}$. So $x = \frac{5}{2}$ and $y = \frac{3}{2}$.

Practice Problem 1 Find x and y such that

$$\begin{bmatrix} 1 & 2x - y \\ x + y & 5 \end{bmatrix} = \begin{bmatrix} 1 & 1 \\ 2 & 5 \end{bmatrix}.$$

2 Define matrix addition and scalar multiplication.

Matrix Addition and Scalar Multiplication

Matrix addition differs somewhat from real number addition. You can add any two real numbers, but you can add two matrices only if they have the same order.

Matrix Addition

If $A = [a_{ij}]$ and $B = [b_{ij}]$ are two $m \times n$ matrices, their **sum** $A + B$ is the $m \times n$ matrix defined by

$$A + B = [a_{ij} + b_{ij}],$$

for all i and j.

The definition of matrix addition says that

(i) we can add two matrices of the same order simply by adding their corresponding entries, but

(ii) the sum of two matrices of different orders is not defined.

TECHNOLOGY CONNECTION

To find the sum of two matrices with a graphing calculator, name the matrices and then use the regular addition key with the names of the matrices.

Consider the matrices A and B in Example 2 as follows:

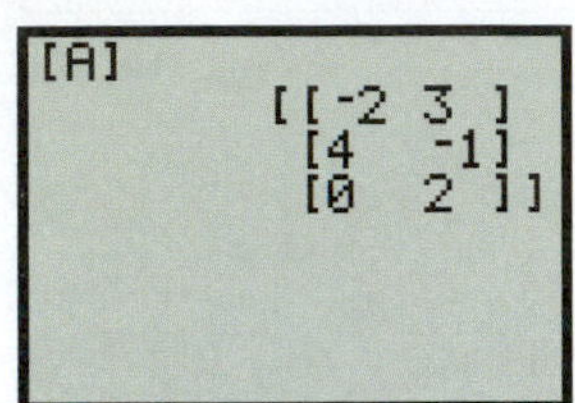

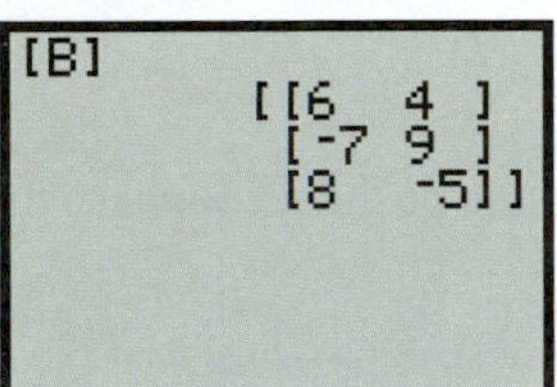

The sum is

```
[A]+[B]
      [[4  7 ]
       [-3 8 ]
       [8  -3]]
```

EXAMPLE 2 **Adding Matrices**

Let $A = \begin{bmatrix} -2 & 3 \\ 4 & -1 \\ 0 & 2 \end{bmatrix}$, $B = \begin{bmatrix} 6 & 4 \\ -7 & 9 \\ 8 & -5 \end{bmatrix}$, and $C = \begin{bmatrix} 1 & 2 \\ 3 & 4 \end{bmatrix}$. Find each sum if possible.

a. $A + B$ **b.** $A + C$

Solution

a. Because A and B have the same order, $A + B$ is defined. We have

$$A + B = \begin{bmatrix} -2 & 3 \\ 4 & -1 \\ 0 & 2 \end{bmatrix} + \begin{bmatrix} 6 & 4 \\ -7 & 9 \\ 8 & -5 \end{bmatrix} = \begin{bmatrix} -2+6 & 3+4 \\ 4-7 & -1+9 \\ 0+8 & 2-5 \end{bmatrix} = \begin{bmatrix} 4 & 7 \\ -3 & 8 \\ 8 & -3 \end{bmatrix}$$

b. $A + C$ is not defined because A and C do not have the same order.

Practice Problem 2 Let $A = \begin{bmatrix} 2 & -1 & 4 \\ 5 & 0 & 9 \end{bmatrix}$ and $B = \begin{bmatrix} -8 & 2 & 9 \\ 7 & 3 & 6 \end{bmatrix}$. Find $A + B$.

You can also multiply any matrix A by any real number c. The real number c is called a **scalar**, and the multiplication process is called *scalar multiplication*.

Scalar Multiplication

Let $A = [a_{ij}]$ be an $m \times n$ matrix and let c be a real number. Then the **scalar product** of A and c is denoted by cA and is defined by

$$cA = [ca_{ij}].$$

In other words, to multiply a matrix A by a real number c, you multiply each entry of A by c. When $c = -1$, we frequently write $-A$ instead of $(-1)A$. The matrix $-A$ is called the **negative** of A. Further, the *difference* of two matrices, $A - B$, is defined to be the sum $A + (-B)$.

Matrix Subtraction

If A and B are two $m \times n$ matrices, then their **difference** is defined by

$$A - B = A + (-B).$$

Subtraction $A - B$ is performed by subtracting the corresponding entries of B from those of A.

EXAMPLE 3 **Performing Scalar Multiplication and Matrix Subtraction**

Let $A = \begin{bmatrix} 1 & 2 & 0 \\ -1 & 3 & 1 \\ 2 & -1 & 4 \end{bmatrix}$ and $B = \begin{bmatrix} 2 & 1 & 3 \\ 1 & 0 & -2 \\ -3 & 4 & 5 \end{bmatrix}$. Find the following.

a. $3A$ **b.** $2B$ **c.** $3A - 2B$

TECHNOLOGY CONNECTION

To find the difference of two matrices with a graphing calculator, name the matrices and then use the regular subtraction key with the names of the matrices. Similarly, to multiply a matrix by a scalar, use the name of the matrix and the regular multiplication key.

Consider the matrices A and B in Example 3 as follows:

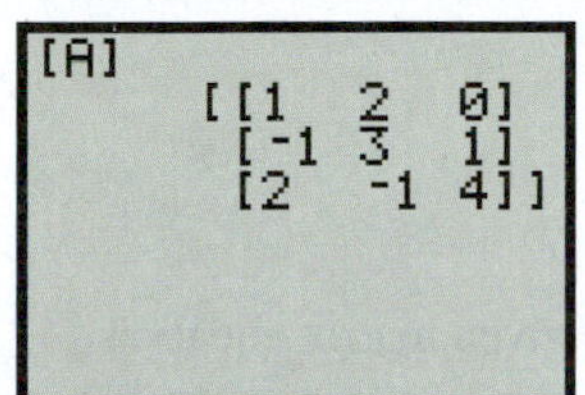


```
[B]
     [[2  1 3 ]
      [1  0 -2]
      [-3 4 5 ]]
```

We find $3A - 2B$:

```
3*[A]-2*[B]
   [[-1 4    -6]
    [-5 9     7]
    [12 -11  2 ]]
```

Solution

a. $3A = 3\begin{bmatrix} 1 & 2 & 0 \\ -1 & 3 & 1 \\ 2 & -1 & 4 \end{bmatrix} = \begin{bmatrix} 3(1) & 3(2) & 3(0) \\ 3(-1) & 3(3) & 3(1) \\ 3(2) & 3(-1) & 3(4) \end{bmatrix} = \begin{bmatrix} 3 & 6 & 0 \\ -3 & 9 & 3 \\ 6 & -3 & 12 \end{bmatrix}$

b. $2B = 2\begin{bmatrix} 2 & 1 & 3 \\ 1 & 0 & -2 \\ -3 & 4 & 5 \end{bmatrix} = \begin{bmatrix} 2(2) & 2(1) & 2(3) \\ 2(1) & 2(0) & 2(-2) \\ 2(-3) & 2(4) & 2(5) \end{bmatrix} = \begin{bmatrix} 4 & 2 & 6 \\ 2 & 0 & -4 \\ -6 & 8 & 10 \end{bmatrix}$

c. $3A - 2B = 3A + (-1)2B$

$$= \begin{bmatrix} 3 & 6 & 0 \\ -3 & 9 & 3 \\ 6 & -3 & 12 \end{bmatrix} + \begin{bmatrix} -4 & -2 & -6 \\ -2 & 0 & 4 \\ 6 & -8 & -10 \end{bmatrix}$$

Substitute from parts **a** and **b**, changing the sign of each entry in **b**.

$$= \begin{bmatrix} 3+(-4) & 6+(-2) & 0+(-6) \\ -3+(-2) & 9+0 & 3+4 \\ 6+6 & -3+(-8) & 12+(-10) \end{bmatrix} = \begin{bmatrix} -1 & 4 & -6 \\ -5 & 9 & 7 \\ 12 & -11 & 2 \end{bmatrix}$$

Practice Problem 3 Let $A = \begin{bmatrix} 7 & -4 \\ 3 & 6 \\ 0 & -2 \end{bmatrix}$ and $B = \begin{bmatrix} 1 & -3 \\ 2 & 2 \\ 5 & 8 \end{bmatrix}$. Find $2A - 3B$.

An $m \times n$ matrix whose entries are all equal to 0 is called the $m \times n$ **zero matrix** and is denoted by $\mathbf{0}_{mn}$. When m and n are understood from the context, we simply write $\mathbf{0}$. Because the entries of a matrix are real numbers, many of the algebraic properties of matrix addition and scalar multiplication are similar to the familiar properties of real numbers.

MATRIX ADDITION AND SCALAR MULTIPLICATION PROPERTIES

Let A, B, and C be $m \times n$ matrices and c and d be scalars.

1. $A + B = B + A$	Commutative property of addition
2. $A + (B + C) = (A + B) + C$	Associative property of addition
3. $A + \mathbf{0} = \mathbf{0} + A = A$	Additive identity property
4. $A + (-A) = (-A) + A = \mathbf{0}$	Additive inverse property
5. $(cd)A = c(dA)$	Associative property of scalar multiplication
6. $1A = A$	Scalar identity property
7. $c(A + B) = cA + cB$	Distributive property
8. $(c + d)A = cA + dA$	Distributive property

In the next example, we use these properties to solve an equation involving matrices. Such an equation is called a **matrix equation**.

EXAMPLE 4 Solving a Matrix Equation

Solve the matrix equation $3A + 2X = 4B$ for X, where

$$A = \begin{bmatrix} 2 & 0 \\ 4 & 6 \end{bmatrix} \text{ and } B = \begin{bmatrix} 1 & 3 \\ 5 & 2 \end{bmatrix}.$$

Solution

$$\begin{aligned} 3A + 2X &= 4B \\ 2X &= 4B - 3A \\ X &= \frac{1}{2}(4B - 3A) \\ &= \frac{1}{2}\left(4\begin{bmatrix} 1 & 3 \\ 5 & 2 \end{bmatrix} - 3\begin{bmatrix} 2 & 0 \\ 4 & 6 \end{bmatrix}\right) \\ &= \frac{1}{2}\left(\begin{bmatrix} 4 & 12 \\ 20 & 8 \end{bmatrix} - \begin{bmatrix} 6 & 0 \\ 12 & 18 \end{bmatrix}\right) \\ &= \frac{1}{2}\begin{bmatrix} -2 & 12 \\ 8 & -10 \end{bmatrix} \\ &= \begin{bmatrix} -1 & 6 \\ 4 & -5 \end{bmatrix} \end{aligned}$$

You should check that the matrix $X = \begin{bmatrix} -1 & 6 \\ 4 & -5 \end{bmatrix}$ satisfies the given matrix equation.

Practice Problem 4 Solve the matrix equation $5A + 3X = 2B$ for X, where

$$A = \begin{bmatrix} 1 & -1 \\ 3 & 5 \end{bmatrix} \text{ and } B = \begin{bmatrix} 2 & 7 \\ -3 & -5 \end{bmatrix}.$$

3 Define matrix multiplication.

Matrix Multiplication

You know that the sum of two matrices is defined only if both matrices are of the same order. In the case of multiplication, however, the general rule is that if A is an $m \times p$ matrix and B is a $p \times n$ matrix, then the *product* AB is defined; otherwise, it is not defined.

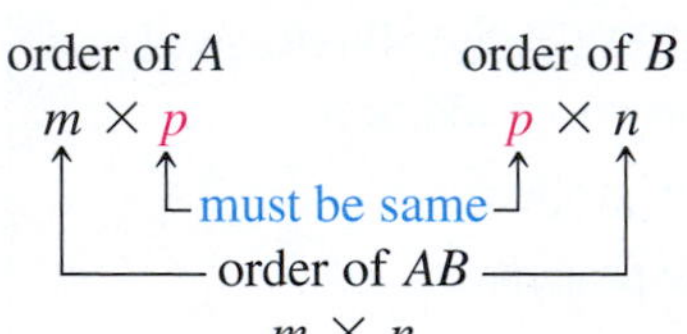

RULE FOR DEFINING THE PRODUCT *AB*

To define the product AB of two matrices A and B, the number of columns of A must be equal to the number of rows of B. If A is an $m \times p$ matrix and B is a $p \times n$ matrix, then the product AB is an $m \times n$ matrix.

EXAMPLE 5 **Determining the Order of the Product**

For each pair of matrices A and B, state whether the product matrix AB is defined. If it is, find the order of AB.

a. $A = \begin{bmatrix} 1 & 3 & 5 \end{bmatrix}, B = \begin{bmatrix} 2 \\ 3 \\ 7 \end{bmatrix}$ **b.** $A = \begin{bmatrix} 2 & 3 \\ -1 & 4 \\ 5 & 6 \end{bmatrix}, B = \begin{bmatrix} 2 & 1 & 3 & 5 \\ 4 & -1 & -2 & 6 \end{bmatrix}$

c. $A = \begin{bmatrix} 1 \\ -1 \\ 2 \end{bmatrix}, B = \begin{bmatrix} 2 & 1 & 4 \\ 0 & 2 & 2 \end{bmatrix}$

Solution

a. Yes, the product AB is defined. A is a 1×3 matrix, and B is a 3×1 matrix; so the number of columns of A equals the number of rows of B. The product AB is a 1×1 matrix that has a single entry.

b. Yes, AB is defined. A is a 3×2 matrix, and B is a 2×4 matrix; so the number of columns of A equals the number of rows of B. The product AB is a 3×4 matrix.

c. No, AB is not defined. A is a 3×1 matrix, and B is a 2×3 matrix; so the number of columns of A does not equal the number of rows of B.

Practice Problem 5 State whether the product matrix AB is defined. If AB is defined, find the order of the product matrix AB.

$$A = \begin{bmatrix} 1 & 4 & 7 \\ 0 & 2 & -2 \end{bmatrix}, \quad B = \begin{bmatrix} 4 \\ 3 \\ -1 \end{bmatrix}$$

Before you learn how to multiply two matrices A and B, consider the next example.

EXAMPLE 6 Calculating the Total Revenue from Car Sales

A car dealership sold 10 sports cars (S), 30 compacts (C), and 45 mini-compacts (M). The average price per car of type S, C, and M was 42, 24, and 17 (in thousands of dollars), respectively. Find the total revenue received by the dealership from the sale of these cars.

Solution

You can represent the number of cars of each type sold by a 1×3 matrix:

$$N = [S \quad C \quad M] = [10 \quad 30 \quad 45]$$

Using the price of each type of car, you can construct a 3×1 price matrix:

$$P = \begin{bmatrix} 42 \\ 24 \\ 17 \end{bmatrix}$$

Then the total revenue R received by the dealership from the sale of these cars last month is given by $R = 10(42) + 30(24) + 45(17) = 1905$ (or \$1,905,000).

We define the right side of this equation to be the product NP of the two matrices N and P. In matrix notation,

$$NP = [10 \quad 30 \quad 45] \begin{bmatrix} 42 \\ 24 \\ 17 \end{bmatrix} = 10(42) + 30(24) + 45(17) = 1905.$$

Practice Problem 6 In Example 6, suppose the average price per car of types S, C, and M is 41, 26, and 19 (in thousands of dollars), respectively. Find the total revenue received by the dealership from the sale of these cars.

Example 6 leads to the following definition.

PRODUCT OF $1 \times n$ AND $n \times 1$ MATRICES

Suppose A is a $1 \times n$ matrix and B is an $n \times 1$ matrix:

$$A = [a_1 \quad a_2 \quad a_3 \cdots a_n] \text{ and } B = \begin{bmatrix} b_1 \\ b_2 \\ \vdots \\ b_n \end{bmatrix}.$$

We define the product AB by

$$AB = [a_1 b_1 + a_2 b_2 + \cdots + a_n b_n].$$

We make the following observations about this definition.

1. Because A is a $1 \times n$ matrix and B is an $n \times 1$ matrix, the product AB is defined and is a 1×1 matrix.
2. To find the product AB, multiply the leftmost entry of A and the top entry of B; then move to the right in A and down in B while multiplying corresponding (first, second, third, and so on) entries together; finally, add all of the resulting products.

EXAMPLE 7 Finding the Product AB

Find AB, where $A = \begin{bmatrix} 2 & 0 & 1 & -2 \end{bmatrix}$ and $B = \begin{bmatrix} 3 \\ 1 \\ -1 \\ 4 \end{bmatrix}$.

Solution

Because A has order 1×4 and B has order 4×1, AB is defined. The product rule gives

$$AB = [2(3) + 0(1) + 1(-1) + (-2)(4)] = [-3].$$

The product of the 1×4 matrix A and the 4×1 matrix B is therefore the 1×1 matrix $[-3]$.

Practice Problem 7 Find AB:

$$A = \begin{bmatrix} 3 & -1 & 2 & 7 \end{bmatrix} \quad \text{and} \quad B = \begin{bmatrix} -2 \\ 0 \\ 1 \\ 5 \end{bmatrix}$$

In Example 7, you may think of the answer -3 as the 1×1 matrix $[-3]$ or simply as the number -3. We will have occasion to make use of both points of view.

The rule for multiplying any two matrices can be reduced to the special case of multiplying a $1 \times p$ matrix by a $p \times 1$ matrix repeatedly.

Matrix Multiplication

Let $A = [a_{ij}]$ be an $m \times p$ matrix and $B = [b_{ij}]$ be a $p \times n$ matrix. Then the **product** AB is the $m \times n$ matrix $C = [c_{ij}]$, where the entry c_{ij} of C is obtained by matrix multiplying the ith row of A by the jth column of B.

The definition of the product AB says that

$$c_{ij} = a_{i1}b_{1j} + a_{i2}b_{2j} + \cdots + a_{ip}b_{pj}.$$

Written in full, the matrix product $AB = C$ in the definition is as follows:

$$\begin{bmatrix} a_{11} & a_{12} & \cdots & a_{1p} \\ a_{21} & a_{22} & \cdots & a_{2p} \\ \vdots & \vdots & & \vdots \\ a_{i1} & a_{i2} & \cdots & a_{ip} \\ \vdots & \vdots & & \vdots \\ a_{m1} & a_{m2} & \cdots & a_{mp} \end{bmatrix} \begin{bmatrix} b_{11} & b_{12} & \cdots & b_{1j} & \cdots & b_{1n} \\ b_{21} & b_{22} & \cdots & b_{2j} & \cdots & b_{2n} \\ \vdots & \vdots & \cdots & \vdots & \cdots & \vdots \\ b_{p1} & b_{p2} & \cdots & b_{pj} & \cdots & b_{pn} \end{bmatrix} = \begin{bmatrix} c_{11} & c_{12} & \cdots & c_{1j} & \cdots & c_{1n} \\ \vdots & \vdots & \cdots & \vdots & & \vdots \\ c_{i1} & c_{i2} & \cdots & c_{ij} & \cdots & c_{in} \\ \vdots & \vdots & \vdots & \vdots & & \vdots \\ c_{m1} & c_{m2} & \cdots & c_{mj} & \cdots & c_{mn} \end{bmatrix}$$

TECHNOLOGY CONNECTION

To find the product of two matrices using a graphing calculator, name the matrices and then use the regular multiplication key with the names of the matrices.

Consider the matrices A and B in Example 8.

We find the product AB.

```
[A]*[B]
        [[5 2 7]
         [0 8 8]]
```

EXAMPLE 8 **Finding the Product of Two Matrices**

Find the products AB and BA, if possible.

$$A = \begin{bmatrix} 1 & 2 \\ -1 & 3 \end{bmatrix} \quad \text{and} \quad B = \begin{bmatrix} 3 & -2 & 1 \\ 1 & 2 & 3 \end{bmatrix}$$

Solution

Because A is of order 2×2 and the order of B is 2×3, the product AB is defined and has order 2×3.

Let $AB = C = [c_{ij}]$. By the definition of the product AB, each entry c_{ij} of C is found by multiplying the ith row of A by the jth column of B. For example c_{11} is found by multiplying the first row of A by the first column of B:

$$\begin{bmatrix} 1 & 2 \\ * & * \end{bmatrix}\begin{bmatrix} 3 & * & * \\ 1 & * & * \end{bmatrix} = \begin{bmatrix} 1(3) + 2(1) & * & * \\ * & * & * \end{bmatrix}$$

So

$$AB = \begin{bmatrix} 1 & 2 \\ -1 & 3 \end{bmatrix}\begin{bmatrix} 3 & -2 & 1 \\ 1 & 2 & 3 \end{bmatrix}$$

1st row of A times 1st column of B | 1st row of A times 2nd column of B | 1st row of A times 3rd column of B

$$AB = \begin{bmatrix} 1(3) + 2(1) & 1(-2) + 2(2) & 1(1) + 2(3) \\ -1(3) + 3(1) & -1(-2) + 3(2) & -1(1) + 3(3) \end{bmatrix} = \begin{bmatrix} 5 & 2 & 7 \\ 0 & 8 & 8 \end{bmatrix}$$

2nd row of A times 1st column of B | 2nd row of A times 2nd column of B | 2nd row of A times 3rd column of B

The product BA is not defined because B is of order 2×3 and A is of order 2×2; that is, the number of columns of B is not equal to the number of rows of A.

Practice Problem 8 Find the products AB and BA if possible.

$$A = \begin{bmatrix} 5 & 0 \\ 2 & -1 \end{bmatrix} \quad \text{and} \quad B = \begin{bmatrix} 8 & 1 \\ -2 & 6 \\ 0 & 4 \end{bmatrix}$$

In Example 8, it is obvious that $AB \neq BA$ because BA is undefined. Even if both AB and BA are defined, AB may not equal BA, as the next example illustrates.

EXAMPLE 9 **Finding the Product of Two Matrices**

Find the products AB and BA.

$$A = \begin{bmatrix} 2 & 3 \\ -1 & 0 \end{bmatrix} \quad \text{and} \quad B = \begin{bmatrix} 3 & 4 \\ 5 & 1 \end{bmatrix}$$

Solution

$$AB = \begin{bmatrix} 2 & 3 \\ -1 & 0 \end{bmatrix}\begin{bmatrix} 3 & 4 \\ 5 & 1 \end{bmatrix} = \begin{bmatrix} 2(3) + 3(5) & 2(4) + 3(1) \\ -1(3) + 0(5) & -1(4) + 0(1) \end{bmatrix} = \begin{bmatrix} 21 & 11 \\ -3 & -4 \end{bmatrix}$$

$$BA = \begin{bmatrix} 3 & 4 \\ 5 & 1 \end{bmatrix}\begin{bmatrix} 2 & 3 \\ -1 & 0 \end{bmatrix} = \begin{bmatrix} 3(2) + 4(-1) & 3(3) + 4(0) \\ 5(2) + 1(-1) & 5(3) + 1(0) \end{bmatrix} = \begin{bmatrix} 2 & 9 \\ 9 & 15 \end{bmatrix}$$

Practice Problem 9 Find the products AB and BA.

$$A = \begin{bmatrix} 7 & 1 \\ 0 & 3 \end{bmatrix} \text{ and } B = \begin{bmatrix} 2 & -1 \\ 4 & 4 \end{bmatrix}$$

In Example 9, observe that $AB \neq BA$. Thus, in general, *matrix multiplication is not commutative*. Matrix multiplication, however, does share many of the properties of the multiplication of real numbers.

Properties of Matrix Multiplication

Let A, B, and C be matrices and let c be a scalar. Assume that each product and sum is defined. Then

1. $(AB)C = A(BC)$ — Associative property of multiplication

2. **(i)** $A(B + C) = AB + AC$ — Distributive property

(ii) $(A + B)C = AC + BC$ — Distributive property

3. $c(AB) = (cA)B = A(cB)$ — Associative property of scalar multiplication

In the exercises, you will be asked to verify these properties and compare them to similar properties of real numbers. Also, integral powers of square matrices are defined exactly as for the real numbers. We write A^2 to mean AA, A^3 to mean AAA, and so on. We define A^0 by

$$A^0 = I_n = \begin{bmatrix} 1 & 0 & 0 & \cdots & 0 \\ 0 & 1 & 0 & \cdots & 0 \\ 0 & 0 & 1 & \cdots & 0 \\ \vdots & \vdots & \vdots & \vdots & \\ 0 & 0 & 0 & \cdots & 1 \end{bmatrix}.$$

The matrix I_n is called the **identity matrix**.

When the order of the matrix is clear from the context, we can drop the subscript n and write I for I_n. For any square matrix A, we have

$$IA = A \text{ and } AI = A$$

So in matrix multiplication, the matrix I plays a role similar to that of the number 1 in the multiplication of real numbers.

4 Apply matrix multiplication to computer graphics.

Computer Graphics

Letters used for labels on a computer screen are very simple two-dimensional graphics. The next example illustrates how a letter (or a figure) is transformed when the coordinates of the points on the figure are all multiplied by the same matrix.

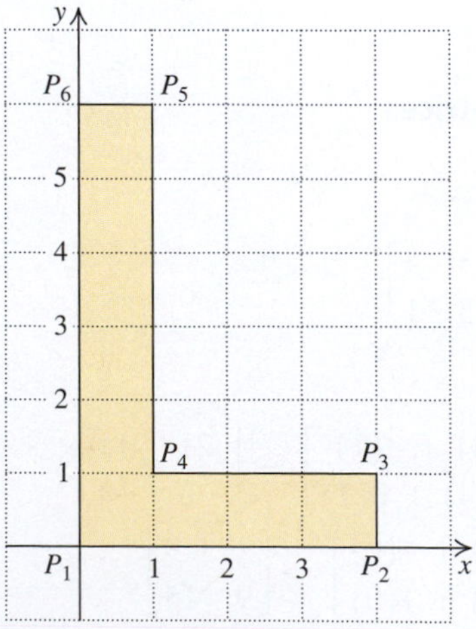

FIGURE 9.2

EXAMPLE 10 **Transforming a Letter**

The capital letter L in Figure 9.2 is determined by six points (or *vertices*) P_1 through P_6. The coordinates of the six points can be stored in a *data matrix D*. We draw line segments connecting these vertices in order.

$$\begin{array}{r} \\ x\text{-coordinate} \\ y\text{-coordinate} \end{array} \begin{array}{c} \begin{matrix} P_1 & P_2 & P_3 & P_4 & P_5 & P_6 \end{matrix} \\ \begin{bmatrix} 0 & 4 & 4 & 1 & 1 & 0 \\ 0 & 0 & 1 & 1 & 6 & 6 \end{bmatrix} \end{array} = D$$

If $A = \begin{bmatrix} 1 & 0.25 \\ 0 & 1 \end{bmatrix}$, compute AD and graph the figure it represents.

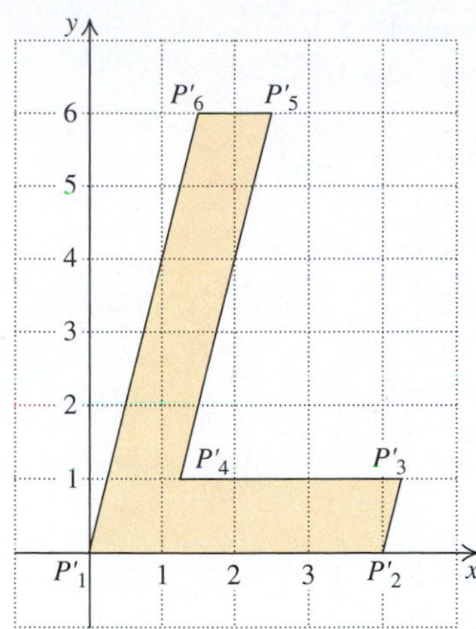

FIGURE 9.3

Solution

The columns of the product matrix AD represent the transformed vertices of the letter L.

$$AD = \begin{bmatrix} 1 & 0.25 \\ 0 & 1 \end{bmatrix}\begin{bmatrix} 0 & 4 & 4 & 1 & 1 & 0 \\ 0 & 0 & 1 & 1 & 6 & 6 \end{bmatrix}$$

$$= \begin{bmatrix} 1\cdot 0 + 0.25\cdot 0 & 1\cdot 4 + 0.25\cdot 0 & 1\cdot 4 + 0.25\cdot 1 & 1\cdot 1 + 0.25\cdot 1 & 1\cdot 1 + 0.25\cdot 6 & 1\cdot 0 + 0.25\cdot 6 \\ 0\cdot 0 + 1\cdot 0 & 0\cdot 4 + 1\cdot 0 & 0\cdot 4 + 1\cdot 1 & 0\cdot 1 + 1\cdot 1 & 0\cdot 1 + 1\cdot 6 & 0\cdot 0 + 1\cdot 6 \end{bmatrix}$$

$$\begin{matrix} P'_1 & P'_2 & P'_3 & P'_4 & P'_5 & P'_6 \end{matrix}$$
$$= \begin{bmatrix} 0 & 4 & 4.25 & 1.25 & 2.5 & 1.5 \\ 0 & 0 & 1 & 1 & 6 & 6 \end{bmatrix}$$

Figure 9.3 shows the transformed vertices and the transformed figure "L" formed by these vertices.

Practice Problem 10 In Example 10, use the matrix $A = \begin{bmatrix} 0 & 1 \\ 1 & 0.25 \end{bmatrix}$. Graph the figure represented by the matrix AD.

Answers to Practice Problems

1. $x = 1, y = 1$ **2.** $\begin{bmatrix} -6 & 1 & 13 \\ 12 & 3 & 15 \end{bmatrix}$

3. $\begin{bmatrix} 11 & 1 \\ 0 & 6 \\ -15 & -28 \end{bmatrix}$ **4.** $X = \begin{bmatrix} -\frac{1}{3} & \frac{19}{3} \\ -7 & -\frac{35}{3} \end{bmatrix}$

5. Yes, the product AB is defined with order 2×1. **6.** \$2045 thousand **7.** $[31]$ **8.** The product AB is not defined. $BA = \begin{bmatrix} 42 & -1 \\ 2 & -6 \\ 8 & -4 \end{bmatrix}$ **9.** $AB = \begin{bmatrix} 18 & -3 \\ 12 & 12 \end{bmatrix}$, $BA = \begin{bmatrix} 14 & -1 \\ 28 & 16 \end{bmatrix}$

10.

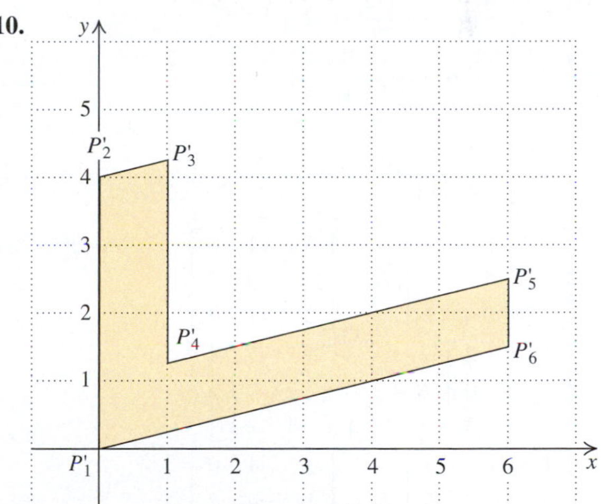

SECTION 9.2 Exercises

Basic Concepts and Skills

1. Two $m \times n$ matrices $A = [a_{ij}]$ and $B = [b_{ij}]$ are equal if ____________ for all i and j.
2. Let $A = [a_{ij}]$, $B = [b_{ij}]$, and $C = [c_{ij}]$. If $A + B = C$, then $c_{ij} =$ ____________ for all i and for all j.
3. The product of a $1 \times n$ matrix A and an $n \times 1$ matrix B is a ____________ matrix.
4. If A is an $m \times n$ matrix and B is an $n \times p$ matrix, then AB is defined and is a(n) ____________ matrix.
5. **True or False.** If AB and BA are defined, then $AB = BA$.
6. **True or False.** Any two square matrices can be multiplied.

In Exercises 7–14, find the values of all variables.

7. $\begin{bmatrix} 2 \\ -3 \end{bmatrix} = \begin{bmatrix} x \\ u \end{bmatrix}$ **8.** $\begin{bmatrix} -\frac{y}{2} \\ x \end{bmatrix} = \begin{bmatrix} 4 \\ 3 \end{bmatrix}$

9. $\begin{bmatrix} 2 & x \\ y & -3 \end{bmatrix} = \begin{bmatrix} 2 & -1 \\ 3 & -3 \end{bmatrix}$

10. $\begin{bmatrix} 2 - x & 1 \\ -2 & 3 + y \end{bmatrix} = \begin{bmatrix} 3 & 1 \\ -2 & -3 \end{bmatrix}$

11. $\begin{bmatrix} 2x - 3y & -4 \\ 5 & 3x + y \end{bmatrix} = \begin{bmatrix} 1 & -4 \\ 5 & 7 \end{bmatrix}$

12. $\begin{bmatrix} 3x + y & -17 \\ 19 & 2 \end{bmatrix} = \begin{bmatrix} -\frac{1}{2} & -17 \\ 19 & 2x + 3y \end{bmatrix}$

13. $\begin{bmatrix} x - y & 1 & 2 \\ 4 & 3x - 2y & 3 \\ 5 & 6 & 5x - 10y \end{bmatrix} = \begin{bmatrix} -1 & 1 & 2 \\ 4 & -1 & 3 \\ 5 & 6 & 6 \end{bmatrix}$

14. $\begin{bmatrix} x + y & 2 & 3 \\ 2 & x - y & 4 \\ 3 & 4 & 2x + 3y \end{bmatrix} = \begin{bmatrix} -1 & 2 & 3 \\ 2 & 5 & 4 \\ 3 & 4 & -5 \end{bmatrix}$

In Exercises 15–22, find each of the following if possible.

a. $A + B$ **b.** $A - B$
c. $-3A$ **d.** $3A - 2B$
e. $(A + B)^2$ **f.** $A^2 - B^2$

15. $A = \begin{bmatrix} 1 & 2 \\ 3 & 4 \end{bmatrix}, B = \begin{bmatrix} -1 & 0 \\ 2 & -3 \end{bmatrix}$

16. $A = \begin{bmatrix} \frac{1}{3} & 1 \\ -1 & 2 \end{bmatrix}, B = \begin{bmatrix} 1 & 0 \\ 2 & -\frac{1}{2} \end{bmatrix}$

17. $A = \begin{bmatrix} 2 & 3 \\ -4 & 5 \end{bmatrix}, B = \begin{bmatrix} 1 & 0 & 2 \\ 3 & 1 & 4 \end{bmatrix}$

18. $A = \begin{bmatrix} 1 & 2 & 3 \\ -1 & -3 & 4 \end{bmatrix}, B = \begin{bmatrix} 1 & 0 \\ 2 & 3 \\ 1 & 4 \end{bmatrix}$

19. $A = \begin{bmatrix} 4 & 0 & -1 \\ -2 & 5 & 2 \\ 0 & 0 & 1 \end{bmatrix}, B = \begin{bmatrix} 3 & 1 & 0 \\ 1 & -4 & 2 \\ 2 & 1 & 3 \end{bmatrix}$

20. $A = \begin{bmatrix} 1 & 0 & 2 \\ 2 & 1 & 0 \\ 0 & 1 & 3 \end{bmatrix}, B = \begin{bmatrix} 3 & -1 & 2 \\ 0 & 2 & 1 \\ 1 & 4 & -1 \end{bmatrix}$

21. $A = \begin{bmatrix} 1 & 2 & -3 \\ 3 & 4 & 5 \\ 2 & -1 & 0 \end{bmatrix}, B = \begin{bmatrix} 3 & 1 & 0 \\ 1 & -4 & 2 \\ 2 & 1 & 3 \end{bmatrix}$

22. $A = \begin{bmatrix} 1 & 0 & 2 \\ 2 & 1 & 0 \\ 0 & 1 & 3 \end{bmatrix}, B = \begin{bmatrix} 3 & -1 & 2 \\ 0 & 2 & 1 \\ 1 & 4 & -1 \end{bmatrix}$

In Exercises 23–30, solve each matrix equation for X, where

$$A = \begin{bmatrix} 2 & 3 & -1 \\ 1 & -2 & 4 \end{bmatrix} \text{ and } B = \begin{bmatrix} -2 & 1 & 0 \\ 2 & 3 & 4 \end{bmatrix}.$$

23. $A + X = B$ **24.** $B + X = A$
25. $2X - A = B$ **26.** $2X - B = A$
27. $2X + 3A = B$ **28.** $3X + 2A = B$
29. $2A + 3B + 4X = 0$ **30.** $3X - 2A + 5B = 0$

In Exercises 31–40, find each product if possible.

a. AB **b.** BA

31. $A = \begin{bmatrix} 1 & 2 \\ 3 & 4 \end{bmatrix}, B = \begin{bmatrix} -2 & 1 \\ 3 & 5 \end{bmatrix}$

32. $A = \begin{bmatrix} 3 & 2 \\ 1 & 5 \\ 0 & 1 \end{bmatrix}, B = \begin{bmatrix} 1 & 3 & -2 \\ 2 & 5 & 0 \end{bmatrix}$

33. $A = \begin{bmatrix} 2 & -1 & 0 \\ -3 & 1 & 2 \end{bmatrix}, B = \begin{bmatrix} 1 & 5 \\ -2 & 3 \\ 4 & 0 \end{bmatrix}$

34. $A = \begin{bmatrix} 1 & 2 & 3 \\ 4 & 5 & 6 \end{bmatrix}, B = \begin{bmatrix} -1 & -3 & -6 \\ 2 & 5 & 7 \end{bmatrix}$

35. $A = \begin{bmatrix} 2 & 3 & 5 \end{bmatrix}, B = \begin{bmatrix} 1 \\ -2 \\ 4 \end{bmatrix}$

36. $A = \begin{bmatrix} 1 \\ 2 \\ -1 \end{bmatrix}, B = \begin{bmatrix} -3 & 0 & 2 \end{bmatrix}$

37. $A = \begin{bmatrix} 1 & 2 & 3 \end{bmatrix}, B = \begin{bmatrix} 1 & 2 & -1 \\ 0 & 3 & 1 \\ 2 & 0 & -3 \end{bmatrix}$

38. $A = \begin{bmatrix} 4 & -6 & 2 \\ 2 & 3 & 0 \\ 1 & 2 & -3 \end{bmatrix}, B = \begin{bmatrix} -2 & 0 & 1 \end{bmatrix}$

39. $A = \begin{bmatrix} 2 & 0 & 1 \\ 1 & 4 & 2 \\ 3 & -1 & 0 \end{bmatrix}, B = \begin{bmatrix} 3 & 1 & 0 \\ -1 & 2 & 0 \\ 4 & 5 & 2 \end{bmatrix}$

40. $A = \begin{bmatrix} 3 & -1 & 2 \\ 0 & 4 & -3 \\ 1 & -2 & 2 \end{bmatrix}, B = \begin{bmatrix} 2 & -5 & 0 \\ -1 & 2 & -1 \\ 3 & 0 & 2 \end{bmatrix}$

41. Let $A = \begin{bmatrix} 3 & 1 & 2 \\ 0 & 4 & 3 \\ 1 & -2 & 2 \end{bmatrix}$ and $B = \begin{bmatrix} 2 & 5 & 0 \\ 1 & 2 & -1 \\ 3 & 0 & 2 \end{bmatrix}$.
Verify that $AB \neq BA$.

42. Let $A = \begin{bmatrix} 5 & 2 & 4 \\ -6 & -3 & 10 \\ -2 & 6 & 5 \end{bmatrix}$ and $B = \begin{bmatrix} 4 & 1 & 2 \\ -3 & 0 & 5 \\ -1 & 3 & 4 \end{bmatrix}$.
Verify that $AB = BA$.

In Exercises 43–46, let

$$A = \begin{bmatrix} 1 & 2 \\ 3 & 4 \end{bmatrix}, B = \begin{bmatrix} 2 & -3 \\ 3 & 5 \end{bmatrix}, \text{ and } C = \begin{bmatrix} 0 & 1 \\ 2 & 4 \end{bmatrix}.$$

43. Verify the associative property of multiplication: $(AB)C = A(BC)$.

44. Verify the distributive property: $A(B + C) = AB + AC$.

45. Verify the distributive property: $(A + B)C = AC + BC$.

46. For any real number c, verify that $c(AB) = (cA)(B) = A(cB)$.

Applying the Concepts

47. Cost matrix. The Build-Rite Co. is building an apartment complex. The cost of purchasing and transporting specific amounts of steel, glass, and wood (in appropriate units) from two different locations is given by the following matrices:

$$A = \begin{bmatrix} 7 & 3 & 18 \\ 4 & 1 & 3 \end{bmatrix} \begin{matrix} \text{Cost of material} \\ \text{Transportation cost} \end{matrix}$$
(columns: Steel, Glass, Wood)

$$B = \begin{bmatrix} 6 & 2 & 20 \\ 3 & 1 & 4 \end{bmatrix} \begin{matrix} \text{Cost of material} \\ \text{Transportation cost} \end{matrix}$$

Find the matrix representing the total cost of material and transportation for steel, glass, and wood from both locations.

48. **Cost matrix.** Repeat Exercise 47 if the order from location A is tripled and the order from location B is doubled.

49. **Stock purchase.** Ms. Goodbyer plans to buy 100 shares of computer stock, 300 shares of oil stock, and 400 shares of automobile stock. The computer stock is selling for \$60 a share, oil stock is selling for \$38 a share, and automobile stock is selling for \$17 a share. Use matrix multiplication to calculate the total cost of Ms. Goodbyer's purchases.

50. **Product export.** The International Export Corporation received an export order for three of its products, say, A, B, and C. The export order is (in thousands) for 50 of product A, 75 of product B, and 150 of product C. The material cost, labor, and profit (each in appropriate units) required to produce each unit of the product are given in the following table:

	Material	Labor	Profit
Product A	20	2	20
Product B	15	3	25
Product C	25	1	15

Use matrix multiplication to compute the total material cost, total labor, and total profit if the entire export order is filled.

51. **Executive compensation.** The Multinational Oil Corporation pays its top executives a salary, a cash bonus, and shares of its stock annually. In 2012, the chairman of the board received \$2.5 million in salary, a \$1.5 million bonus, and 50,000 shares of stock; the president of the company received one-half the compensation of the chairman; and each of the four vice presidents was paid \$100,000 in salary, a \$150,000 bonus, and 5000 shares of stock.
   a. Express payments to these executives in salary, bonus, and stock as a 3×3 matrix.
   b. Express the number of executives of each rank as a column matrix.
   c. Use matrix multiplication to compute the total amount the company paid to each executive in each category in 2012.

52. **Calories and protein.** The Browns (B) and Newgards (N) are neighboring families. The Brown family has two men, three women, and one child, and the Newgard family has one man, one woman, and two children. Both families are weight watchers. They have the same dietician, who recommends the following daily allowance of calories and proteins:

Calories
male adults: 2400 calories
female adults: 1900 calories
children: 1800 calories
Protein (in grams)
male adults: 55
female adults: 45
children: 33

Represent the preceding information in matrix form. Use matrix multiplication to compute the total requirements of calories and proteins for each of the two families.

In Exercise 53–56, a matrix A is given. Graph the figure represented by the matrix AD, where D is the matrix representation of the letter L of Example 10.

53. $A = \begin{bmatrix} 1 & 0 \\ 0 & -1 \end{bmatrix}$ **54.** $A = \begin{bmatrix} -1 & 0 \\ 0 & 1 \end{bmatrix}$

55. $A = \begin{bmatrix} 1 & 0 \\ 0.25 & 1 \end{bmatrix}$ **56.** $A = \begin{bmatrix} 1 & -0.25 \\ 0 & 1 \end{bmatrix}$

Beyond the Basics

57. Let $A = \begin{bmatrix} 0 & 3 \\ 0 & 0 \end{bmatrix}$, $B = \begin{bmatrix} 2 & 1 \\ 3 & 0 \end{bmatrix}$, and $C = \begin{bmatrix} 5 & 4 \\ 3 & 0 \end{bmatrix}$.
Verify that $AB = AC$. This example shows that $AB = AC$ does not, in general, imply that $B = C$.

58. Let $A = \begin{bmatrix} 2 & -3 & -5 \\ -1 & 4 & 5 \\ 1 & -3 & -4 \end{bmatrix}$ and $B = \begin{bmatrix} -1 & 3 & 5 \\ 1 & -3 & -5 \\ -1 & 3 & 5 \end{bmatrix}$.
Verify that $AB = \mathbf{0}$. This example shows that $AB = \mathbf{0}$ does not imply that $A = \mathbf{0}$ or $B = \mathbf{0}$.

59. Select appropriate 2×2 matrices A and B to show that, in general, $(A + B)^2 \neq A^2 + 2AB + B^2$.

60. Repeat Exercise 59 to show that, in general, $A^2 - B^2 \neq (A - B)(A + B)$.

61. Let $A = \begin{bmatrix} 1 & -2 & 3 \\ 2 & -4 & 1 \\ 3 & -5 & 2 \end{bmatrix}$. Show that

$$3A^2 - 2A + I = \begin{bmatrix} 17 & -23 & 15 \\ -13 & 30 & 10 \\ -9 & 22 & 21 \end{bmatrix},$$

where $I = \begin{bmatrix} 1 & 0 & 0 \\ 0 & 1 & 0 \\ 0 & 0 & 1 \end{bmatrix}$.

62. Let $A = \begin{bmatrix} 1 & 3 & 2 \\ 2 & 0 & 3 \\ 1 & -1 & 1 \end{bmatrix}$. Show that

$$A^3 - 3A^2 + A - I = \begin{bmatrix} -3 & 14 & -3 \\ 5 & -2 & 8 \\ 5 & -7 & 6 \end{bmatrix},$$

where I is given in Exercise 61.

63. If possible, find a matrix B such that $\begin{bmatrix} 2 & 3 \\ 1 & 2 \end{bmatrix} B = \begin{bmatrix} 1 & 0 \\ 0 & 1 \end{bmatrix}$.

64. If possible find a matrix B such that

$$\begin{bmatrix} 1 & -2 \\ -2 & 4 \end{bmatrix} B = \begin{bmatrix} 1 & 0 \\ 0 & 1 \end{bmatrix}.$$

65. Find x and y if

$$\left(4\begin{bmatrix} 2 & -1 & 3 \\ 1 & 0 & 2 \end{bmatrix} - \begin{bmatrix} 1 & 2 & -1 \\ 2 & -3 & 4 \end{bmatrix}\right)\begin{bmatrix} 2 \\ -1 \\ 1 \end{bmatrix} = \begin{bmatrix} x \\ y \end{bmatrix}.$$

66. Find x, y, and z if

$$\begin{bmatrix} 3 & 2 & -1 \\ 4 & 9 & 2 \\ 5 & 0 & -2 \end{bmatrix}\begin{bmatrix} x \\ y \\ z \end{bmatrix} = \begin{bmatrix} 0 \\ 7 \\ 2 \end{bmatrix}.$$

67. $A = \begin{bmatrix} 0 & 1 \\ 0 & 0 \end{bmatrix}, B = \begin{bmatrix} 1 & 0 \\ 0 & 0 \end{bmatrix}$

Show that $AB = \mathbf{0}$ but $BA = \begin{bmatrix} 0 & 1 \\ 0 & 0 \end{bmatrix} \neq \mathbf{0}$.

68. Find matrices A and B such that

$$A + B = \begin{bmatrix} 5 & 2 \\ 0 & 9 \end{bmatrix} \text{ and } A - B = \begin{bmatrix} 3 & 6 \\ 0 & -1 \end{bmatrix}.$$

69. Find matrices A and B such that

$$2A - B = \begin{bmatrix} 6 & -6 & 0 \\ -4 & 2 & 1 \end{bmatrix} \text{ and } A + B = \begin{bmatrix} 3 & 0 & 3 \\ -2 & 1 & -4 \end{bmatrix}.$$

70. Let $A = \begin{bmatrix} 2 & -3 & -5 \\ -1 & 4 & 5 \\ 1 & -3 & -4 \end{bmatrix}, B = \begin{bmatrix} -1 & 3 & 5 \\ 1 & -3 & -5 \\ -1 & 3 & 5 \end{bmatrix}$.

Show that $AB = \mathbf{0}$ and $BA = \mathbf{0}$.

71. Let $A = \begin{bmatrix} 2 & -3 & -5 \\ -1 & 4 & 5 \\ 1 & -3 & -4 \end{bmatrix}, B = \begin{bmatrix} 2 & -2 & -4 \\ -1 & 3 & 4 \\ 1 & -2 & -3 \end{bmatrix}$.

Show that **(i)** $AB = A$, **(ii)** $BA = B$, **(iii)** $A^2 = A$.

72. Let $A = \begin{bmatrix} 2 & 3 \\ -1 & 2 \end{bmatrix}$.

a. Show that $A^2 = 4A - 7I$.

b. Use part (a) to show that $A^4 = 8A - 63I$.

c. Use part (b) to show that $A^4 = \begin{bmatrix} -47 & 24 \\ -8 & -47 \end{bmatrix}$.

73. Let $A = \begin{bmatrix} 4 & -1 & -4 \\ 3 & 0 & -4 \\ 3 & -1 & -3 \end{bmatrix}$. Show that $A^2 = I$.

Critical Thinking/Discussion/Writing

74. Let A be a $3 \times n$ matrix and B be a $5 \times m$ matrix.

a. Under what conditions is AB defined? What is the order of AB when this product is defined?

b. Repeat part (a) for BA.

75. Let $A = \begin{bmatrix} 3 & -1 & 0 \\ -2 & 3 & 1 \end{bmatrix}, B = \begin{bmatrix} 2 \\ 3 \\ 4 \end{bmatrix}$, and $C = \begin{bmatrix} -1 & 2 \end{bmatrix}$.

In which order should all three matrices be multiplied to produce a number?

76. Market share. The Asma Corporation (A), Bronkial Brothers (B), and Coufmore Company (C) simultaneously decide to introduce a new cigarette at a time when each company has one-third of the market. During the year, the following occurs:

(i) A retains 40% of its customers and loses 30% to B and 30% to C.

(ii) B retains 30% of its customers and loses 60% to A and 10% to C.

(iii) C retains 30% of its customers and loses 60% to A and 10% to B.

The foregoing information can be represented by the following *transition matrix*:

$$P = \begin{matrix} & A & B & C \\ A & 0.4 & 0.3 & 0.3 \\ B & 0.6 & 0.3 & 0.1 \\ C & 0.6 & 0.1 & 0.3 \end{matrix}$$

Write the initial market share as $X = \begin{bmatrix} \frac{1}{3} & \frac{1}{3} & \frac{1}{3} \end{bmatrix}$.

a. Compute XP^2 and interpret your result.

b. Compute XP^3, XP^4, and XP^5; observe the pattern for the long run. Describe the long-term market share for each company.

Maintaining Skills

In Exercises 77 and 78, rewrite each expression without exponents.

77. $\left(\frac{1}{2}\right)^{-1}$

78. $\left(\frac{5}{12}\right)^{-1}$

In Exercises 79–84, solve each equation.

79. $x^{-1} = \frac{1}{8}$

80. $\left(\frac{x}{35}\right)^{-1} = 7$

81. $\frac{1}{4}x - \frac{7}{12} = \frac{11}{12} - \frac{5}{4}x$

82. $\frac{2}{3}x - \frac{3}{5} = \frac{4}{3}x + \frac{2}{15}$

83. $\frac{2}{4x - 1} = \frac{3}{4x + 1}$

84. $\frac{2}{x + 1} = \frac{1}{x - 1}$

In Exercises 85–88, solve each system.

85. $\begin{cases} 3x + 5y = 2 \\ 6x + 10y = 4 \end{cases}$

86. $\begin{cases} -9x + 2y = 8 \\ 5x + 8y = 32 \end{cases}$

87. $\begin{cases} 3x + 2y = 6 \\ 6x + 4y = -13 \end{cases}$

88. $\begin{cases} 3x - 4y = 13 \\ 2x + 5y = 1 \end{cases}$

SECTION 9.3

The Matrix Inverse

BEFORE STARTING THIS SECTION, REVIEW

1. Matrix multiplication (Section 9.2, page 818)
2. Equality of matrices (Section 9.2, page 815)
3. Gauss–Jordan elimination procedure (Section 9.1, page 806)
4. Leontieff input–output model (Section 9.1, page 808)

OBJECTIVES

1. Verify the multiplicative inverse of a matrix.
2. Find the inverse of a matrix.
3. Find the inverse of a 2×2 matrix.
4. Use matrix inverses to solve systems of linear equations.
5. Use matrix inverses in applied problems.

Coding and Decoding Messages

Imagine you are a secret agent who is looking for spies in different parts of the world. Periodically, you communicate with your headquarters via your laptop or by satellite. You expect that your communications will be intercepted. So before you broadcast any message, you must write it using a secret code. A message written with a secret code is called a **cryptogram**.

First, you might replace each letter in the message with a number and send the message as a sequence of numbers. The problem with this approach is that it is rather easy to crack such a code. For example, if letters such as e are always represented by the same symbol, then a code breaker (hacker) can guess which symbol represents each letter and eventually translate your message. In Example 8, we show you a better way to encode and decode a message using matrix multiplication.

1 Verify the multiplicative inverse of a matrix.

The Multiplicative Inverse of a Matrix

In arithmetic, you know that if a is a nonzero real number, then a has a unique reciprocal, denoted by $\frac{1}{a}$, or a^{-1}. The number a^{-1} is called the **multiplicative inverse** of a, and

$$aa^{-1} = a^{-1}a = 1.$$

Similarly, certain square matrices have *multiplicative inverses,* with an identity matrix acting as "1."

The Inverse of a Matrix

Let A be an $n \times n$ matrix and let I be the $n \times n$ identity matrix. If there is an $n \times n$ matrix B such that

$$AB = I \quad \text{and} \quad BA = I,$$

then B is called the **inverse** of A and we write $B = A^{-1}$ (read "A inverse").

Notice that if B is the inverse of A, then A is the inverse of B.

Arthur Cayley (1821–1895)

English mathematicians of the first third of the 19th century attempted to extend the basic ideas of algebra and to determine exactly how much one can generalize the properties of integers to other types of quantities. Although Sylvester (see page 799) coined the term *matrix*, it was Arthur Cayley who developed the algebra of matrices. Cayley studied mathematics at Trinity College, Cambridge, but because there was no suitable teaching job available, he decided to become a lawyer. Although he became skilled in legal work, he regarded the law as just a way to support himself. During his legal career, he was able to write more than 300 mathematical papers. In 1863, Cambridge University established a new post in mathematics and offered it to Cayley. He accepted the job eagerly even though it paid less money than he made as a lawyer.

Recall that in general (even for square matrices), $AB \neq BA$. However, if A and B are square matrices with $AB = I$, then it can be shown that $BA = I$. Therefore, to determine whether A and B are inverses, you need only check that either $AB = I$ or $BA = I$.

EXAMPLE 1 Verifying the Inverse of a Matrix

Show that B is the inverse of A.

$$A = \begin{bmatrix} 1 & 3 \\ 2 & 7 \end{bmatrix} \text{ and } B = \begin{bmatrix} 7 & -3 \\ -2 & 1 \end{bmatrix}$$

Solution

You need to verify that $AB = I$ or $BA = I$.

$$AB = \begin{bmatrix} 1 & 3 \\ 2 & 7 \end{bmatrix}\begin{bmatrix} 7 & -3 \\ -2 & 1 \end{bmatrix} = \begin{bmatrix} 1(7) + 3(-2) & 1(-3) + 3(1) \\ 2(7) + 7(-2) & 2(-3) + 7(1) \end{bmatrix} = \begin{bmatrix} 1 & 0 \\ 0 & 1 \end{bmatrix} = I$$

Because $AB = I$, it follows that $B = A^{-1}$.

Practice Problem 1 Show that B is the inverse of A.

$$A = \begin{bmatrix} 3 & 2 \\ 2 & 1 \end{bmatrix} \quad \text{and} \quad B = \begin{bmatrix} -1 & 2 \\ 2 & -3 \end{bmatrix}$$

In working with real numbers, you know that if $a = 0$, then a^{-1} does not exist. So it should not come as a surprise that if A is a square *zero* matrix, then A^{-1} does not exist. Surprisingly, there are also nonzero matrices that do not have inverses.

EXAMPLE 2 Proving That a Particular Nonzero Matrix Has No Inverse

Show that the matrix A does not have an inverse.

$$A = \begin{bmatrix} 2 & 0 \\ 0 & 0 \end{bmatrix}$$

Solution

Suppose A has an inverse B, where

$$B = \begin{bmatrix} x & y \\ z & w \end{bmatrix}.$$

Then you have

$$\begin{bmatrix} 2 & 0 \\ 0 & 0 \end{bmatrix}\begin{bmatrix} x & y \\ z & w \end{bmatrix} = \begin{bmatrix} 1 & 0 \\ 0 & 1 \end{bmatrix} \qquad AB = I$$

$$\begin{bmatrix} 2x & 2y \\ 0 & 0 \end{bmatrix} = \begin{bmatrix} 1 & 0 \\ 0 & 1 \end{bmatrix} \qquad \text{Multiply the two matrices on the left side.}$$

Because these two matrices are equal, you must have

$$0 = 1 \qquad \text{Equate the entries in the } (2, 2) \text{ position.}$$

Because $0 = 1$ is a false statement, the matrix A does not have an inverse.

Practice Problem 2 Show that the matrix $A = \begin{bmatrix} 3 & 1 \\ 3 & 1 \end{bmatrix}$ does not have an inverse.

2 Find the inverse of a matrix.

Finding the Inverse of a Matrix

It can be shown that if a square matrix A has an inverse, then the inverse is unique. This means that a square matrix cannot have more than one inverse. If a matrix A has an inverse, then A is called **invertible** or **nonsingular**. A square matrix that does not have an inverse is called a **singular** matrix. The following discussion derives a technique for finding the inverse of an invertible matrix.

Suppose we want to find the inverse of the matrix $A = \begin{bmatrix} 1 & 2 \\ 3 & 5 \end{bmatrix}$. We first let $B = \begin{bmatrix} x & y \\ z & w \end{bmatrix}$ be the inverse of the matrix A. Then

$$AB = I \quad \text{Definition of inverse}$$

$$\begin{bmatrix} 1 & 2 \\ 3 & 5 \end{bmatrix}\begin{bmatrix} x & y \\ z & w \end{bmatrix} = \begin{bmatrix} 1 & 0 \\ 0 & 1 \end{bmatrix} \quad \text{Substitute for } A, B, \text{ and } I.$$

$$\begin{bmatrix} x + 2z & y + 2w \\ 3x + 5z & 3y + 5w \end{bmatrix} = \begin{bmatrix} 1 & 0 \\ 0 & 1 \end{bmatrix} \quad \text{Multiply matrices } A \text{ and } B.$$

Equating corresponding entries in the last equation yields two systems of linear equations.

$$\begin{cases} x + 2z = 1 \\ 3x + 5z = 0 \end{cases} \quad \text{Equate the entries in the } (1, 1) \text{ and } (2, 1) \text{ positions.}$$

$$\begin{cases} y + 2w = 0 \\ 3y + 5w = 1 \end{cases} \quad \text{Equate the entries in the } (1, 2) \text{ and } (2, 2) \text{ positions.}$$

Let's solve the two systems by Gauss–Jordan elimination.

System of Equations	Matrix Form
$\begin{cases} x + 2z = 1 \\ 3x + 5z = 0 \end{cases}$	$\left[\begin{array}{cc\|c} 1 & 2 & 1 \\ 3 & 5 & 0 \end{array}\right]$
$\begin{cases} y + 2w = 0 \\ 3y + 5w = 1 \end{cases}$	$\left[\begin{array}{cc\|c} 1 & 2 & 0 \\ 3 & 5 & 1 \end{array}\right]$

TECHNOLOGY CONNECTION

To use a graphing calculator to find the inverse of an invertible matrix, name the matrix and then use the regular reciprocal key with the name of the matrix.

```
[A]
        [[1 2]
         [3 5]]
```

```
[A]-1
      [[-5 2 ]
       [3  -1]]
```

Because the coefficient matrix for both systems of equations is the original matrix

$$A = \begin{bmatrix} 1 & 2 \\ 3 & 5 \end{bmatrix},$$

the calculations involved in the Gauss–Jordan elimination on the coefficient matrix A will be the same for both systems. Therefore, you can solve both systems at once by considering the "doubly augmented" matrix

$$C = \left[\begin{array}{cc|cc} 1 & 2 & 1 & 0 \\ 3 & 5 & 0 & 1 \end{array}\right].$$

Note that the matrix C is formed by adjoining the identity matrix I to matrix A:

$$C = [A \mid I]$$

Now apply Gauss–Jordan elimination to the matrix C.

$$C = \left[\begin{array}{cc|cc} 1 & 2 & 1 & 0 \\ 3 & 5 & 0 & 1 \end{array}\right] \xrightarrow{-3R_1 + R_2 \to R_2} \left[\begin{array}{cc|cc} 1 & 2 & 1 & 0 \\ 0 & -1 & -3 & 1 \end{array}\right] \xrightarrow{(-1)R_2} \left[\begin{array}{cc|cc} 1 & 2 & 1 & 0 \\ 0 & 1 & 3 & -1 \end{array}\right]$$

$$\xrightarrow{-2R_2 + R_1 \to R_1} \left[\begin{array}{cc|cc} 1 & 0 & -5 & 2 \\ 0 & 1 & 3 & -1 \end{array}\right]$$

Converting the last matrix back into two systems of equations, we have

$$\left[\begin{array}{cc|c} 1 & 0 & -5 \\ 0 & 1 & 3 \end{array}\right] \text{ represents } \begin{cases} 1\cdot x + 0\cdot z = -5 \\ 0\cdot x + 1\cdot z = 3 \end{cases} \quad \text{or} \quad \begin{cases} x = -5 \\ z = 3 \end{cases}$$

Similarly,

$$\left[\begin{array}{cc|c} 1 & 0 & 2 \\ 0 & 1 & -1 \end{array}\right] \text{ represents } \begin{cases} 1\cdot y + 0\cdot w = 2 \\ 0\cdot y + 1\cdot w = -1 \end{cases} \quad \text{or} \quad \begin{cases} y = 2 \\ w = -1 \end{cases}$$

We conclude that $x = -5$, $y = 2$, $z = 3$, and $w = -1$.

Because $B = \begin{bmatrix} x & y \\ z & w \end{bmatrix}$, it follows that

$$B = \begin{bmatrix} -5 & 2 \\ 3 & -1 \end{bmatrix} = A^{-1}.$$

You should check that $AB = I$.

We have shown that to find the inverse of an invertible matrix A, we transform the matrix $[A \mid I]$ by a sequence of elementary row operations into $[I \mid B]$, where $B = A^{-1}$.

PROCEDURE FOR FINDING THE INVERSE OF A MATRIX

Let A be an $n \times n$ matrix.

1. Form the $n \times 2n$ augmented matrix $[A \mid I]$, where I is the $n \times n$ identity matrix.
2. If there is a sequence of elementary row operations that transforms A into I, then this same sequence of row operations will transform $[A \mid I]$ into $[I \mid B]$, where $B = A^{-1}$.
3. Check your work by showing that $AA^{-1} = I$.

If it is not possible to transform A into I by row operations, then A does not have an inverse. (This occurs if, at any step in the process, you obtain a matrix $[C \mid D]$ in which C has a row of zeros.)

EXAMPLE 3 Finding the Inverse of a Matrix

Find the inverse (if it exists) of the matrix $A = \begin{bmatrix} 1 & 2 & 3 \\ 2 & 5 & 7 \\ 2 & 4 & 6 \end{bmatrix}$.

Solution

Step 1 Start by adjoining the identity matrix $I = \begin{bmatrix} 1 & 0 & 0 \\ 0 & 1 & 0 \\ 0 & 0 & 1 \end{bmatrix}$ to the matrix A to form the augmented matrix $[A \mid I]$.

$$[A \mid I] = \left[\begin{array}{ccc|ccc} 1 & 2 & 3 & 1 & 0 & 0 \\ 2 & 5 & 7 & 0 & 1 & 0 \\ 2 & 4 & 6 & 0 & 0 & 1 \end{array}\right]$$

Step 2 Perform row operations on $[A \mid I]$ to transform it to a matrix of the form $[I \mid B]$.

$$\left[\begin{array}{ccc|ccc} 1 & 2 & 3 & 1 & 0 & 0 \\ 2 & 5 & 7 & 0 & 1 & 0 \\ 2 & 4 & 6 & 0 & 0 & 1 \end{array}\right] \xrightarrow[-2R_1 + R_3 \to R_3]{-2R_1 + R_2 \to R_2} \left[\begin{array}{ccc|ccc} 1 & 2 & 3 & 1 & 0 & 0 \\ 0 & 1 & 1 & -2 & 1 & 0 \\ 0 & 0 & 0 & -2 & 0 & 1 \end{array}\right] = [C \mid D]$$

Because the third row of matrix C consists of zeros, it is impossible to transform A into I by row operations. Hence, A is *not* invertible.

Practice Problem 3 Find the inverse (if it exists) of the matrix $A = \begin{bmatrix} 1 & 4 & -2 \\ -1 & 1 & 2 \\ 3 & 7 & -6 \end{bmatrix}$.

EXAMPLE 4 **Finding the Inverse of a Matrix**

Find the inverse (if it exists) of the matrix $A = \begin{bmatrix} 1 & 1 & 0 \\ 0 & 3 & 1 \\ 2 & 3 & 3 \end{bmatrix}$.

Solution

Step 1 Start with the matrix $[A \mid I]$:

$$[A \mid I] = \left[\begin{array}{ccc|ccc} 1 & 1 & 0 & 1 & 0 & 0 \\ 0 & 3 & 1 & 0 & 1 & 0 \\ 2 & 3 & 3 & 0 & 0 & 1 \end{array}\right] \qquad I = I_3$$

Step 2 Use row operations on $[A \mid I]$. Then $[A \mid I] \rightarrow$

$$\xrightarrow[-2R_1 + R_3 \rightarrow R_3]{\frac{1}{3}R_2 \rightarrow R_2} \left[\begin{array}{ccc|ccc} 1 & 1 & 0 & 1 & 0 & 0 \\ 0 & 1 & \frac{1}{3} & 0 & \frac{1}{3} & 0 \\ 0 & 1 & 3 & -2 & 0 & 1 \end{array}\right] \xrightarrow{(-1)R_2 + R_3 \rightarrow R_3} \left[\begin{array}{ccc|ccc} 1 & 1 & 0 & 1 & 0 & 0 \\ 0 & 1 & \frac{1}{3} & 0 & \frac{1}{3} & 0 \\ 0 & 0 & \frac{8}{3} & -2 & -\frac{1}{3} & 1 \end{array}\right]$$

$$\xrightarrow{\frac{3}{8}R_3} \left[\begin{array}{ccc|ccc} 1 & 1 & 0 & 1 & 0 & 0 \\ 0 & 1 & \frac{1}{3} & 0 & \frac{1}{3} & 0 \\ 0 & 0 & 1 & -\frac{6}{8} & -\frac{1}{8} & \frac{3}{8} \end{array}\right] \xrightarrow{-\frac{1}{3}R_3 + R_2 \rightarrow R_2} \left[\begin{array}{ccc|ccc} 1 & 1 & 0 & 1 & 0 & 0 \\ 0 & 1 & 0 & \frac{2}{8} & \frac{3}{8} & -\frac{1}{8} \\ 0 & 0 & 1 & -\frac{6}{8} & -\frac{1}{8} & \frac{3}{8} \end{array}\right]$$

$$\xrightarrow{(-1)R_2 + R_1 \rightarrow R_1} \left[\begin{array}{ccc|ccc} 1 & 0 & 0 & \frac{6}{8} & -\frac{3}{8} & \frac{1}{8} \\ 0 & 1 & 0 & \frac{2}{8} & \frac{3}{8} & -\frac{1}{8} \\ 0 & 0 & 1 & -\frac{6}{8} & -\frac{1}{8} & \frac{3}{8} \end{array}\right].$$

Hence, A is invertible and

$$A^{-1} = \begin{bmatrix} \frac{6}{8} & -\frac{3}{8} & \frac{1}{8} \\ \frac{2}{8} & \frac{3}{8} & -\frac{1}{8} \\ -\frac{6}{8} & -\frac{1}{8} & \frac{3}{8} \end{bmatrix} = \frac{1}{8}\begin{bmatrix} 6 & -3 & 1 \\ 2 & 3 & -1 \\ -6 & -1 & 3 \end{bmatrix}.$$

Step 3 You should verify that $AA^{-1} = I$.

Practice Problem 4 Find the inverse (if it exists) of the matrix $A = \begin{bmatrix} 1 & 2 & 3 \\ -2 & 3 & 1 \\ 4 & 5 & -2 \end{bmatrix}$.

3 Find the inverse of a 2×2 matrix.

A Rule for Finding the Inverse of a 2 × 2 Matrix

You can quickly determine whether any 2×2 matrix is invertible and, if so, what its inverse is.

A Rule for Finding the Inverse of a 2 × 2 Matrix

The matrix

$$A = \begin{bmatrix} a & b \\ c & d \end{bmatrix}$$

is invertible if and only if $ad - bc \neq 0$. Moreover, if $ad - bc \neq 0$, then

$$A^{-1} = \frac{1}{ad - bc}\begin{bmatrix} d & -b \\ -c & a \end{bmatrix}.$$

If $ad - bc = 0$, the matrix A does not have an inverse.

EXAMPLE 5 Finding the Inverse of a 2 × 2 Matrix

Find the inverse (if it exists) of each matrix.

a. $A = \begin{bmatrix} 5 & 2 \\ 4 & 3 \end{bmatrix}$ **b.** $B = \begin{bmatrix} 4 & 6 \\ 2 & 3 \end{bmatrix}$

Solution

a. For the matrix A, $a = 5, b = 2, c = 4$, and $d = 3$.
Here $ad - bc = (5)(3) - (2)(4) = 15 - 8 = 7 \neq 0$; so A is invertible and

$$A^{-1} = \frac{1}{7}\begin{bmatrix} 3 & -2 \\ -4 & 5 \end{bmatrix}$$ Substitute for a, b, c, and d in $\frac{1}{ad - bc}\begin{bmatrix} d & -b \\ -c & a \end{bmatrix}$.

$$= \begin{bmatrix} \frac{3}{7} & -\frac{2}{7} \\ -\frac{4}{7} & \frac{5}{7} \end{bmatrix}$$ Scalar multiplication

You should verify that $AA^{-1} = I$.

b. For the matrix B, $a = 4, b = 6, c = 2$, and $d = 3$; so

$$ad - bc = (4)(3) - (6)(2) = 12 - 12 = 0.$$

Therefore, the matrix B does not have an inverse.

Practice Problem 5 Find the inverse (if it exists) of each matrix.

a. $A = \begin{bmatrix} 8 & 2 \\ 4 & 1 \end{bmatrix}$ **b.** $B = \begin{bmatrix} 8 & -2 \\ 3 & 1 \end{bmatrix}$

4 Use matrix inverses to solve systems of linear equations.

Solving Systems of Linear Equations by Using Matrix Inverses

Matrix multiplication can be used to write a system of linear equations in matrix form.

System of Equations	Matrix Form
$\begin{cases} 3x - 2y = 4 \\ 4x - 3y = 5 \end{cases}$	$\begin{bmatrix} 3 & -2 \\ 4 & -3 \end{bmatrix}\begin{bmatrix} x \\ y \end{bmatrix} = \begin{bmatrix} 4 \\ 5 \end{bmatrix}$
$\begin{cases} 2x_1 + 4x_2 - x_3 = 9 \\ 3x_1 + x_2 + 2x_3 = 7 \\ x_1 + 3x_2 - 3x_3 = 4 \end{cases}$	$\begin{bmatrix} 2 & 4 & -1 \\ 3 & 1 & 2 \\ 1 & 3 & -3 \end{bmatrix}\begin{bmatrix} x_1 \\ x_2 \\ x_3 \end{bmatrix} = \begin{bmatrix} 9 \\ 7 \\ 4 \end{bmatrix}$

Solving a system of linear equations amounts to solving a corresponding matrix equation of the form

$$AX = B,$$

where A is the coefficient matrix of the system, X is the column matrix containing the variables, and B is the column matrix of the system's constants.

If A is a square matrix *and* A is invertible, then the matrix equation $AX = B$ has a unique solution, $X = A^{-1}B$, as the following derivation shows.

$$AX = B \quad \text{Given equation}$$

$$A^{-1}(AX) = A^{-1}B \quad \text{Be careful to multiply by } A^{-1} \text{ on the left on both sides of the equation because multiplication of matrices is not commutative.}$$

$$(A^{-1}A)X = A^{-1}B \quad \text{Associative property}$$

$$IX = A^{-1}B \quad A^{-1}A = I$$

$$X = A^{-1}B \quad I \text{ is the identity matrix, and } IX = X.$$

EXAMPLE 6 **Solving a Linear System by Using an Inverse Matrix**

Use a matrix inverse to solve the linear system: $\begin{cases} x + y = 4 \\ 3y + z = 7 \\ 2x + 3y + 3z = 21 \end{cases}$

Solution

Write the linear system in matrix form.

$$\underbrace{\begin{bmatrix} 1 & 1 & 0 \\ 0 & 3 & 1 \\ 2 & 3 & 3 \end{bmatrix}}_{A}\underbrace{\begin{bmatrix} x \\ y \\ z \end{bmatrix}}_{X} = \underbrace{\begin{bmatrix} 4 \\ 7 \\ 21 \end{bmatrix}}_{B}$$

Use zeros for coefficients of missing variables.

We need to solve the matrix equation $AX = B$ for X. Because the matrix A is invertible (see Example 4), the system has a unique solution $X = A^{-1}B$. Use the following:

$$A^{-1} = \frac{1}{8}\begin{bmatrix} 6 & -3 & 1 \\ 2 & 3 & -1 \\ -6 & -1 & 3 \end{bmatrix}$$ Computed in Example 4

$$X = A^{-1}B$$

$$= \frac{1}{8}\begin{bmatrix} 6 & -3 & 1 \\ 2 & 3 & -1 \\ -6 & -1 & 3 \end{bmatrix}\begin{bmatrix} 4 \\ 7 \\ 21 \end{bmatrix}$$ Substitute for A^{-1} and B.

$$= \frac{1}{8}\begin{bmatrix} 6(4) - 3(7) + 1(21) \\ 2(4) + 3(7) - 1(21) \\ -6(4) - 1(7) + 3(21) \end{bmatrix}$$ Matrix multiplication

$$= \frac{1}{8}\begin{bmatrix} 24 \\ 8 \\ 32 \end{bmatrix} = \begin{bmatrix} 3 \\ 1 \\ 4 \end{bmatrix}$$ Scalar multiplication

The solution set is $\{(3, 1, 4)\}$, which you can check in the original system.

Practice Problem 6 Use a matrix inverse to solve the linear system.

$$\begin{cases} 3x + 2y + 3z = 9 \\ 3x + y = 12 \\ x + z = 6 \end{cases}$$

5 Use matrix inverses in applied problems.

Applications of Matrix Inverses

The Leontief Input–Output Model We can use the inverse of a matrix to analyze input–output models that we studied in Section 9.1.

Suppose a simplified economy depends on two products: energy (E) and food (F). To produce 1 unit of E requires $\frac{1}{4}$ unit of E and $\frac{1}{2}$ unit of F. To produce 1 unit of F requires $\frac{1}{3}$ unit of E and $\frac{1}{4}$ unit of F. Then the interindustry consumption is given by the following matrix.

Input

$$A = \begin{matrix} & E & F & \\ & \begin{bmatrix} \frac{1}{4} & \frac{1}{2} \\ \frac{1}{3} & \frac{1}{4} \end{bmatrix} & & \begin{matrix} E \\ F \end{matrix} \end{matrix}$$ **Output**

The matrix A is the **interindustry technology input–output matrix**, or simply the **technology matrix**, of the system. If the system is producing x_1 units of energy and x_2 units of food, then the column matrix $X = \begin{bmatrix} x_1 \\ x_2 \end{bmatrix}$ is called the **gross production matrix**.

Consequently,

$$AX = \begin{bmatrix} \frac{1}{4} & \frac{1}{2} \\ \frac{1}{3} & \frac{1}{4} \end{bmatrix}\begin{bmatrix} x_1 \\ x_2 \end{bmatrix} = \begin{bmatrix} \frac{1}{4}x_1 + \frac{1}{2}x_2 \\ \frac{1}{3}x_1 + \frac{1}{4}x_2 \end{bmatrix} \begin{matrix} \leftarrow \text{units consumed by } E \\ \leftarrow \text{units consumed by } F \end{matrix}$$

represents the interindustry consumptions. If the column matrix $D = \begin{bmatrix} d_1 \\ d_2 \end{bmatrix}$ represents consumer demand, then

$$\begin{aligned}
D &= X - AX && \text{Gross production minus interindustry consumption} \\
&= IX - AX && I \text{ is the identity matrix.} \\
D &= (I - A)X && \text{Distributive property} \\
(I - A)^{-1}D &= (I - A)^{-1}(I - A)X && \text{Multiply both sides by } (I - A)^{-1} \text{ if } I - A \text{ is invertible.} \\
(I - A)^{-1}D &= IX && (I - A)^{-1}(I - A) = I \\
&= X && I \text{ is the identity matrix.}
\end{aligned}$$

So if $(I - A)$ is invertible, then the gross production matrix is $X = (I - A)^{-1}D$.

EXAMPLE 7 Using the Leontief Input–Output Model

In the preceding discussion, suppose the consumer demand for energy is 1000 units and for food is 3000 units. Find the level of production (X) that will meet interindustry and consumer demand.

Solution

For the matrix A, we have

$$I - A = \begin{bmatrix} 1 & 0 \\ 0 & 1 \end{bmatrix} - \begin{bmatrix} \frac{1}{4} & \frac{1}{2} \\ \frac{1}{3} & \frac{1}{4} \end{bmatrix} = \begin{bmatrix} \frac{3}{4} & -\frac{1}{2} \\ -\frac{1}{3} & \frac{3}{4} \end{bmatrix}.$$

Use the formula for the inverse of a 2×2 matrix.

RECALL

If $ad - bc \neq 0$, then for $A = \begin{bmatrix} a & b \\ c & d \end{bmatrix}$, we have

$$A^{-1} = \frac{1}{ad - bc}\begin{bmatrix} d & -b \\ -c & a \end{bmatrix}.$$

$$(I - A)^{-1} = \frac{1}{\frac{3}{4} \cdot \frac{3}{4} - \left(-\frac{1}{2}\right)\left(-\frac{1}{3}\right)} \begin{bmatrix} \frac{3}{4} & \frac{1}{2} \\ \frac{1}{3} & \frac{3}{4} \end{bmatrix}$$

$$= \frac{48}{19}\begin{bmatrix} \frac{3}{4} & \frac{1}{2} \\ \frac{1}{3} & \frac{3}{4} \end{bmatrix} = \frac{1}{19}\begin{bmatrix} 36 & 24 \\ 16 & 36 \end{bmatrix} \qquad \text{Simplify.}$$

The matrix $D = \begin{bmatrix} 1000 \\ 3000 \end{bmatrix}$. Therefore, the gross production matrix X is

$$X = (I - A)^{-1}D = \frac{1}{19}\begin{bmatrix} 36 & 24 \\ 16 & 36 \end{bmatrix}\begin{bmatrix} 1000 \\ 3000 \end{bmatrix} \qquad \text{Substitute for } (I - A)^{-1} \text{ and } D.$$

$$= \frac{1}{19}\begin{bmatrix} 108{,}000 \\ 124{,}000 \end{bmatrix} = \begin{bmatrix} \frac{108{,}000}{19} \\ \frac{124{,}000}{19} \end{bmatrix}.$$

To meet the consumer demand for 1000 units of energy and 3000 units of food, the energy produced must be $\frac{108{,}000}{19}$ units and food production must be $\frac{124{,}000}{19}$ units.

Practice Problem 7 In Example 7, find the level of production (X) that will meet both the interindustry and consumer demand if consumer demand for energy is 800 units and consumer demand for food is 3400 units.

Cryptography

To encode or decode a message, first associate each letter in the message with a number, in the obvious manner. The blank corresponds to 0.

blank	A	B	C	D	E	F	G	H	I	J	K	L	M
0	1	2	3	4	5	6	7	8	9	10	11	12	13
	N	O	P	Q	R	S	T	U	V	W	X	Y	Z
	14	15	16	17	18	19	20	21	22	23	24	25	26

For example, the message

JOHNSON IS IN DANGER

becomes

10	15	8	14	19	15	14	0	9	19	0	9	14	0	4	1	14	7	5	18
J	O	H	N	S	O	N		I	S		I	N		D	A	N	G	E	R

This message is easy to decode, so we will make the coding more complicated. First, partition these 20 numbers into groups of three, enclosed in brackets, and insert zeros at the end of the last bracket if necessary.

[10 15 8] [14 19 15] [14 0 9] [19 0 9] [14 0 4] [1 14 7] [5 18 0]
J O H N S O N I S I N D A N G E R

Let's write these seven groups of three numbers as the seven columns of a 3×7 matrix.

$$M = \begin{bmatrix} 10 & 14 & 14 & 19 & 14 & 1 & 5 \\ 15 & 19 & 0 & 0 & 0 & 14 & 18 \\ 8 & 15 & 9 & 9 & 4 & 7 & 0 \end{bmatrix}$$

Now choose any invertible 3×3 matrix A, say,

$$A = \begin{bmatrix} 1 & 2 & 3 \\ 1 & 3 & 3 \\ 1 & 2 & 4 \end{bmatrix}.$$

We can use the matrix A to convert the message into code and then use A^{-1} to decode the message, as in the next example. The matrix A is called the **coding matrix**.

EXAMPLE 8 Encoding and Decoding a Message

Encode and decode the message

JOHNSON IS IN DANGER

by using the matrix A given above.

Solution

Step 1 Express the message numerically and partition the numbers into groups of three.

[10 15 8] [14 19 15] [14 0 9] [19 0 9] [14 0 4] [1 14 7] [5 18 0]

Step 2 Write each group of three numbers in Step 1 as a column of a matrix M.

$$M = \begin{bmatrix} 10 & 14 & 14 & 19 & 14 & 1 & 5 \\ 15 & 19 & 0 & 0 & 0 & 14 & 18 \\ 8 & 15 & 9 & 9 & 4 & 7 & 0 \end{bmatrix}$$

Step 3 Select an invertible 3×3 matrix A, such as

$$A = \begin{bmatrix} 1 & 2 & 3 \\ 1 & 3 & 3 \\ 1 & 2 & 4 \end{bmatrix}.$$

Step 4 Multiply the matrices in Steps 2 and 3 to form AM.

$$AM = \begin{bmatrix} 1 & 2 & 3 \\ 1 & 3 & 3 \\ 1 & 2 & 4 \end{bmatrix} \begin{bmatrix} 10 & 14 & 14 & 19 & 14 & 1 & 5 \\ 15 & 19 & 0 & 0 & 0 & 14 & 18 \\ 8 & 15 & 9 & 9 & 4 & 7 & 0 \end{bmatrix}$$

$$= \begin{bmatrix} 64 & 97 & 41 & 46 & 26 & 50 & 41 \\ 79 & 116 & 41 & 46 & 26 & 64 & 59 \\ 72 & 112 & 50 & 55 & 30 & 57 & 41 \end{bmatrix} \quad \text{Matrix multiplication}$$

The coded message sent will be the numbers from column 1 of AM, followed by the numbers from column 2, and so on.

64 79 72 97 116 112 41 41 50 46 46 55 26 26 30 50 64 57 41 59 41

Step 5 To decode the message in Step 4, write the numbers received in the message in groups of three and as columns of the matrix AM. Find the **decoding matrix** A^{-1} of the coding matrix A. You can verify that

$$A^{-1} = \begin{bmatrix} 6 & -2 & -3 \\ -1 & 1 & 0 \\ -1 & 0 & 1 \end{bmatrix}.$$

Step 6 Compute the message matrix M.

$$M = A^{-1}(AM) = \begin{bmatrix} 6 & -2 & -3 \\ -1 & 1 & 0 \\ -1 & 0 & 1 \end{bmatrix} \begin{bmatrix} 64 & 97 & 41 & 46 & 26 & 50 & 41 \\ 79 & 116 & 41 & 46 & 26 & 64 & 59 \\ 72 & 112 & 50 & 55 & 30 & 57 & 41 \end{bmatrix}$$

$$= \begin{bmatrix} 10 & 14 & 14 & 19 & 14 & 1 & 5 \\ 15 & 19 & 0 & 0 & 0 & 14 & 18 \\ 8 & 15 & 9 & 9 & 4 & 7 & 0 \end{bmatrix} \quad \text{Matrix multiplication}$$

Step 7 Finally, convert the numerical message from the matrix M in Step 6 back into English.

JOHNSON IS IN DANGER

Practice Problem 8 Rework Example 8, replacing the message JOHNSON IS IN DANGER with the message JACK IS NOW SAFE.

Answers to Practice Problems

1. $\begin{bmatrix} 3 & 2 \\ 2 & 1 \end{bmatrix} \begin{bmatrix} -1 & 2 \\ 2 & -3 \end{bmatrix} = \begin{bmatrix} 1 & 0 \\ 0 & 1 \end{bmatrix}$

2. $\begin{bmatrix} 3x + z & 3y + w \\ 3x + z & 3y + w \end{bmatrix} \neq \begin{bmatrix} 1 & 0 \\ 0 & 1 \end{bmatrix}$
because this implies that $1 = 3x + z = 0$, which is false.

3. The inverse of matrix A does not exist.

4. $A^{-1} = \begin{bmatrix} \frac{1}{7} & -\frac{19}{77} & \frac{1}{11} \\ 0 & \frac{2}{11} & \frac{1}{11} \\ \frac{2}{7} & -\frac{3}{77} & -\frac{1}{11} \end{bmatrix}$

5. a. The inverse of matrix A does not exist;

b. $B^{-1} = \begin{bmatrix} \frac{1}{14} & \frac{1}{7} \\ -\frac{3}{14} & \frac{4}{7} \end{bmatrix}$

6. $\left\{\left(\frac{11}{2}, -\frac{9}{2}, \frac{1}{2}\right)\right\}$

7. $X = \begin{bmatrix} \frac{110{,}400}{19} \\ \frac{135{,}200}{19} \end{bmatrix}$

8. $AM = \begin{bmatrix} 1 & 2 & 3 \\ 1 & 3 & 3 \\ 1 & 2 & 4 \end{bmatrix} \begin{bmatrix} 10 & 11 & 19 & 15 & 19 & 5 \\ 1 & 0 & 0 & 23 & 1 & 0 \\ 3 & 9 & 14 & 0 & 6 & 0 \end{bmatrix}$

$$= \begin{bmatrix} 21 & 38 & 61 & 61 & 39 & 5 \\ 22 & 38 & 61 & 84 & 40 & 5 \\ 24 & 47 & 75 & 61 & 45 & 5 \end{bmatrix}$$

SECTION 9.3 Exercises

Basic Concepts and Skills

1. For $n \times n$ matrices A and B, if $AB = I = BA$, then B is called the ____________ of A.
2. An $n \times n$ matrix A is invertible if there is a matrix B such that ____________ $= I$.
3. To find the inverse of an invertible matrix A, we transform $[A \mid I]$ by a sequence of row operations into $[I \mid B]$, where $B =$ ____________.
4. $A = \begin{bmatrix} a & b \\ c & d \end{bmatrix}$ is invertible if and only if ____________.
5. **True or False.** Every square matrix is invertible.
6. **True or False.** $A = \begin{bmatrix} 8 & 16 \\ 4 & 8 \end{bmatrix}$ is invertible.

In Exercises 7–16, determine whether B is the inverse of A by computing AB and BA.

7. $A = \begin{bmatrix} 1 & 2 \\ 1 & 3 \end{bmatrix}, B = \begin{bmatrix} 3 & -2 \\ -1 & 1 \end{bmatrix}$
8. $A = \begin{bmatrix} 3 & 2 \\ 4 & 3 \end{bmatrix}, B = \begin{bmatrix} 3 & -2 \\ -4 & 3 \end{bmatrix}$
9. $A = \begin{bmatrix} 3 & 2 \\ 1 & 4 \end{bmatrix}, B = \begin{bmatrix} \frac{2}{5} & -\frac{1}{5} \\ -\frac{1}{10} & \frac{3}{10} \end{bmatrix}$
10. $A = \begin{bmatrix} 2 & -3 \\ 4 & -3 \end{bmatrix}, B = \begin{bmatrix} -\frac{1}{2} & \frac{1}{2} \\ -\frac{2}{3} & \frac{1}{3} \end{bmatrix}$
11. $A = \begin{bmatrix} 1 & 0 & -1 \\ -1 & 1 & 1 \end{bmatrix}, B = \begin{bmatrix} 2 & 1 \\ 1 & 1 \\ 1 & 1 \end{bmatrix}$
12. $A = \begin{bmatrix} -2 & 1 & 3 \\ 0 & -1 & 1 \\ 1 & 2 & 0 \end{bmatrix}, B = \frac{1}{8}\begin{bmatrix} -2 & 6 & 4 \\ 1 & -3 & 2 \\ 1 & 5 & 2 \end{bmatrix}$
13. $A = \begin{bmatrix} 1 & -2 & 1 \\ -8 & 6 & -2 \\ 5 & -3 & 1 \end{bmatrix}, B = \frac{1}{2}\begin{bmatrix} 0 & 1 & 2 \\ 1 & 2 & 3 \\ 3 & 1 & 1 \end{bmatrix}$
14. $A = \begin{bmatrix} 2 & 3 & 1 \\ 1 & 2 & 3 \\ 3 & 1 & 2 \end{bmatrix}, B = \frac{1}{18}\begin{bmatrix} 1 & -5 & 7 \\ 7 & 1 & -5 \\ -5 & 7 & 1 \end{bmatrix}$
15. $A = \begin{bmatrix} 1 & 1 & 1 \\ 1 & 2 & 3 \\ -1 & 1 & -1 \end{bmatrix}, B = \frac{1}{4}\begin{bmatrix} 5 & -2 & -1 \\ 2 & 0 & 2 \\ -3 & 2 & -1 \end{bmatrix}$
16. $A = \begin{bmatrix} 1 & -1 & 0 \\ 0 & 1 & -1 \\ 1 & 0 & 1 \end{bmatrix}, B = \frac{1}{2}\begin{bmatrix} 1 & 1 & 1 \\ -1 & 1 & 1 \\ -1 & -1 & 1 \end{bmatrix}$

In Exercises 17–26, find A^{-1} (if it exists) by forming the matrix $[A \mid I]$ and using row operation to obtain $[I \mid B]$, where $B = A^{-1}$.

17. $A = \begin{bmatrix} 2 & 0 \\ 1 & 3 \end{bmatrix}$
18. $A = \begin{bmatrix} 4 & 3 \\ 1 & 0 \end{bmatrix}$
19. $A = \begin{bmatrix} 2 & 4 \\ 3 & 6 \end{bmatrix}$
20. $A = \begin{bmatrix} 9 & 6 \\ 6 & 4 \end{bmatrix}$
21. $A = \begin{bmatrix} 1 & 6 & 4 \\ 0 & 2 & 3 \\ 0 & 1 & 2 \end{bmatrix}$
22. $A = \begin{bmatrix} 2 & 0 & -1 \\ 5 & 1 & 0 \\ 0 & 1 & 3 \end{bmatrix}$
23. $A = \begin{bmatrix} 2 & 3 & 1 \\ 2 & 4 & 1 \\ 3 & 7 & 2 \end{bmatrix}$
24. $A = \begin{bmatrix} 1 & 1 & 5 \\ 4 & 3 & -5 \\ 1 & 1 & 0 \end{bmatrix}$
25. $A = \begin{bmatrix} 3 & -3 & 4 \\ 2 & -3 & 4 \\ 0 & -1 & 1 \end{bmatrix}$
26. $A = \begin{bmatrix} 1 & 2 & -2 \\ -1 & 3 & 0 \\ 0 & -2 & 1 \end{bmatrix}$

In Exercises 27–32, find the inverse (if it exists) of $A = \begin{bmatrix} a & b \\ c & d \end{bmatrix}$ by using the formula $A^{-1} = \frac{1}{ad - bc}\begin{bmatrix} d & -b \\ -c & a \end{bmatrix}$. Check that $A^{-1}A = I$.

27. $A = \begin{bmatrix} 1 & 0 \\ 3 & 2 \end{bmatrix}$
28. $A = \begin{bmatrix} 3 & 4 \\ 5 & 6 \end{bmatrix}$
29. $A = \begin{bmatrix} 2 & -3 \\ -3 & 5 \end{bmatrix}$
30. $A = \begin{bmatrix} 3 & -4 \\ 2 & -2 \end{bmatrix}$
31. $A = \begin{bmatrix} a & -b \\ b & -a \end{bmatrix}, a^2 \neq b^2$
32. $A = \begin{bmatrix} 2 & -1 \\ 1 & -1 \end{bmatrix}$

In Exercises 33–36, write the given linear system of equations as a matrix equation $AX = B$.

33. $\begin{cases} 2x + 3y = -9 \\ x - 3y = 13 \end{cases}$
34. $\begin{cases} 5x - 4y = 7 \\ 4x - 3y = 5 \end{cases}$
35. $\begin{cases} 3x + 2y + z = 8 \\ 2x + y + 3z = 7 \\ x + 3y + 2z = 9 \end{cases}$
36. $\begin{cases} x + 3y + z = 4 \\ x - 5y + 2z = 7 \\ 3x + y - 4z = -9 \end{cases}$

In Exercises 37–40, write each matrix equation as a linear system of equations.

37. $\begin{bmatrix} 1 & -2 \\ 2 & 1 \end{bmatrix}\begin{bmatrix} x \\ y \end{bmatrix} = \begin{bmatrix} 0 \\ 5 \end{bmatrix}$
38. $\begin{bmatrix} 2 & 3 \\ 3 & -1 \end{bmatrix}\begin{bmatrix} x_1 \\ x_2 \end{bmatrix} = \begin{bmatrix} 0 \\ 11 \end{bmatrix}$
39. $\begin{bmatrix} 2 & 3 & 1 \\ 5 & 7 & -1 \\ 4 & 3 & 0 \end{bmatrix}\begin{bmatrix} x_1 \\ x_2 \\ x_3 \end{bmatrix} = \begin{bmatrix} -1 \\ 5 \\ 5 \end{bmatrix}$
40. $\begin{bmatrix} 3 & -2 & 3 \\ 5 & 0 & 4 \\ 2 & 7 & 0 \end{bmatrix}\begin{bmatrix} r \\ s \\ t \end{bmatrix} = \begin{bmatrix} 4 \\ 3 \\ -8 \end{bmatrix}$

41. Let $A = \begin{bmatrix} 1 & 2 & 5 \\ 2 & 3 & 8 \\ -1 & 1 & 2 \end{bmatrix}$. Show that $A^{-1} = \begin{bmatrix} 2 & -1 & -1 \\ 12 & -7 & -2 \\ -5 & 3 & 1 \end{bmatrix}$.

In Exercises 42–44, use the result of Exercise 41 to solve the system of equations.

42. $\begin{cases} x + 2y + 5z = 4 \\ 2x + 3y + 8z = 6 \\ -x + y + 2z = 3 \end{cases}$

43. $\begin{cases} x_1 + 2x_2 + 5x_3 = 1 \\ 2x_1 + 3x_2 + 8x_3 = 3 \\ -x_1 + x_2 + 2x_3 = -3 \end{cases}$

44. $\begin{cases} x + 2y + 5z = -4 \\ 2x + 3y + 8z = -6 \\ -x + y + 2z = -\dfrac{5}{2} \end{cases}$

45. a. Find the inverse of the matrix $A = \begin{bmatrix} 1 & 1 & 1 \\ 1 & 2 & 3 \\ 1 & 4 & 9 \end{bmatrix}$.

b. Use the result of part (a) to solve the linear system.

$$\begin{cases} x + y + z = 6 \\ x + 2y + 3z = 14 \\ x + 4y + 9z = 36 \end{cases}$$

46. a. Find the inverse of the matrix $A = \begin{bmatrix} 2 & 4 & -1 \\ 3 & 1 & 2 \\ 1 & 3 & -3 \end{bmatrix}$.

b. Use the result of part (a) to solve the linear system.

$$\begin{cases} 2x + 4y - z = 9 \\ 3x + y + 2z = 7 \\ x + 3y - 3z = 4 \end{cases}$$

Applying the Concepts

In Exercises 47–52, solve each linear system by using an inverse matrix.

47. $\begin{cases} 3x + 7y = 11 \\ -5x + 4y = 13 \end{cases}$

48. $\begin{cases} x - 7y = 3 \\ 2x + 3y = 23 \end{cases}$

49. $\begin{cases} x + y + 2z = 7 \\ x - y - 3z = -6 \\ 2x + 3y + z = 4 \end{cases}$

50. $\begin{cases} x + y + z = 6 \\ 2x - 3y + 3z = 5 \\ 3x - 2y - z = -4 \end{cases}$

51. $\begin{cases} 2x + 2y + 3z = 7 \\ 5x + 3y + 5z = 3 \\ 3x + 5y + z = -5 \end{cases}$

52. $\begin{cases} 9x + 7y + 4z = 12 \\ 6x + 5y + 4z = 5 \\ 4x + 3y + z = 7 \end{cases}$

In Exercises 53–56, use the method presented in this section to solve the problem.

53. Investment. At the beginning of 2012, Liz decided to invest \$90,000, splitting it three different ways. Part of the money she invested in a treasury bill that yields 3% annual interest, she invested another part in bonds with an annual yield 7%, and the rest she invested in a mutual fund that estimated an annual interest rate of 8%. Liz expected her net income for 2012 to be \$5920. Unfortunately, the mutual fund lost 11% of its assets during the year so that Liz received only \$980 income from all three investments combined. How much money did Liz put in each investment?

54. Sandwich sales. The Fresh Sandwich specializes in chicken, fish, and ham sandwiches. In April, May, and June 2012, they sold the following quantities.

	Sales Volume in Hundreds		
	April	**May**	**June**
Chicken	8	7	4
Fish	11	8	6
Ham	6	7	3
Monthly revenue, in hundreds, from the sales of these sandwiches	\$164	\$141	\$86

Find the selling price of each of the three types of sandwiches.

55. Euler's formula. A plane graph connects points called vertices with nonintersecting lines called edges and divides the plane into regions. The accompanying figure has seven vertices, nine edges, and four regions. Euler's theorem states that in a plane graph,

$$V - E + R = 2,$$

where $V =$ the number of vertices,
$E =$ the number of edges, and
$R =$ the number of regions.

In a certain plane graph, twice the number of edges is three times the number of vertices and twice the number of regions is one less than the number of edges. Find the number of vertices, edges, and regions for this graph. Sketch a plane graph that satisfies the conditions of this exercise.

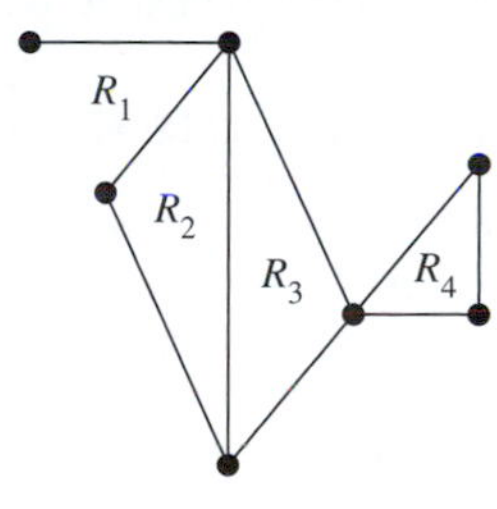

56. Projectile. On a planet, the height h of a projectile fired up from an initial height h_0 with initial vertical velocity of v_0 ft/sec after t seconds is given by $h = \frac{1}{2}at^2 + v_0 t + h_0$, where a is the gravitational constant. The following measurements were recorded.

Time t in seconds	1	2	3
Height h in feet	100	92	76

a. Find the initial height.
b. Find the height of the projectile after 2.5 seconds.

In Exercises 57–60, technology matrix A, representing inter-industry demand, and matrix D, representing consumer demand for the sectors in a two- or three-sector economy, are given. Use the matrix equation $X = (I - A)^{-1}D$ to find the production level X that will satisfy both demands.

57. Two-sector economy. In a two-sector economy,

$$A = \begin{bmatrix} 0.1 & 0.4 \\ 0.5 & 0.2 \end{bmatrix}, D = \begin{bmatrix} 50 \\ 30 \end{bmatrix}.$$

58. Two-sector economy. In a two-sector economy,

$$A = \begin{bmatrix} 0.2 & 0.5 \\ 0.1 & 0.6 \end{bmatrix}, D = \begin{bmatrix} 11 \\ 18 \end{bmatrix}.$$

59. Three-sector economy. In a three-sector economy,

$$A = \begin{bmatrix} 0.2 & 0.2 & 0 \\ 0.1 & 0.1 & 0.3 \\ 0.1 & 0 & 0.2 \end{bmatrix}, D = \begin{bmatrix} 400 \\ 600 \\ 800 \end{bmatrix}.$$

60. Three-sector economy. In a three-sector economy,

$$A = \begin{bmatrix} 0.5 & 0.4 & 0.2 \\ 0.2 & 0.3 & 0.1 \\ 0.1 & 0.1 & 0.3 \end{bmatrix}, D = \begin{bmatrix} 500 \\ 300 \\ 200 \end{bmatrix}.$$

In Exercises 61 and 62, use a 2×2 invertible coding matrix A of your choice. Remember to partition the numerical message into groups of two. (a) Write a cryptogram for each message. (b) Check by decoding.

61. Coding and decoding. CANNOT FIND COLOMBO.

62. Coding and decoding. FOUND HEAD OF TERRORIST NETWORK.

In Exercises 63 and 64, use a 3×3 invertible coding matrix A of your choice.

63. Code and then decode the message of Exercise 61.

64. Code and then decode the message of Exercise 62.

Beyond the Basics

65. Suppose A is an invertible square matrix.
  a. Is A^{-1} invertible? Why or why not?
  b. If your answer to part (a) is yes, then what is $(A^{-1})^{-1}$?

66. Let $A^{-1} = \begin{bmatrix} 2 & 1 \\ 3 & -4 \end{bmatrix}$. Find A.

67. Let A be a square matrix. Show that if A^2 is invertible, then A must be invertible. [*Hint:* $A^2B = A(AB)$.]

68. Show that if A is invertible, then A^2 is invertible.

69. Show that if A and B are $n \times n$ invertible matrices, then AB is invertible and $(AB)^{-1} = B^{-1}A^{-1}$. (Note the order.)

70. Verify the result of Exercise 69 for

$$A = \begin{bmatrix} 1 & 2 \\ 3 & 4 \end{bmatrix}, B = \begin{bmatrix} -1 & 0 \\ 3 & -2 \end{bmatrix}.$$

71. Let A, B, and C be $n \times n$ matrices such that $AB = AC$. Show that if A is invertible, then $B = C$.

72. Is the result of Exercise 71 true if A is not invertible?

73. Let A and B be $n \times n$ matrices such that $AB = A$. Show that if A is invertible, then $B = I$.

74. Give an example to show that the result of Exercise 73 is not true if A is not invertible.

75. Let $A = \begin{bmatrix} 3 & 4 \\ 2 & 3 \end{bmatrix}$.
  a. Show that the matrix A satisfies the equation

$$A^2 - 6A + I = 0.$$

  b. Use the equation in part (a) to show that $A^{-1} = 6I - A$.
  c. Use the equation in part (b) to find A^{-1}.

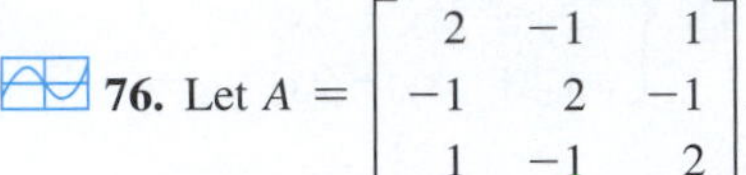

76. Let $A = \begin{bmatrix} 2 & -1 & 1 \\ -1 & 2 & -1 \\ 1 & -1 & 2 \end{bmatrix}$.
  a. Show that A satisfies the equation $A^3 - 6A^2 + 9A - 4I = 0$.
  b. Use the equation in part (a) to show that

$$A^{-1} = \frac{1}{4}(A^2 - 6A + 9I).$$

  [*Hint:* Consider the product $\frac{1}{4}(A^2 - 6A + 9I)A$.]
  c. Use part (b) to find A^{-1}.

In Exercises 77 and 78, use your graphing utility to find the inverse of the coefficient matrix. Then use the inverse to solve each linear system.

77. $\begin{cases} x - y - z - w = -4 \\ x + y + z - w = 2 \\ 2x + y + z - w = 3 \\ x - y + z - w = -2 \end{cases}$

78. $\begin{cases} x + y + z - w = 1 \\ x - y - z + w = 1 \\ x - y + z + 3w = 1 \\ x + y + z + w = 3 \end{cases}$

79. Suppose A and B are 2×2 matrices and $B^{-1} = \begin{bmatrix} 1 & 2 \\ 3 & 4 \end{bmatrix}$.
  Find A
  a. If $AB = \begin{bmatrix} -1 & 3 \\ 0 & 2 \end{bmatrix}$.
  b. If $BA = \begin{bmatrix} -1 & 3 \\ 0 & 2 \end{bmatrix}$.
  c. If $BAB^{-1} = \begin{bmatrix} 1 & 1 \\ 1 & 2 \end{bmatrix}$.
  d. If $B^{-1}AB = \begin{bmatrix} 1 & 1 \\ 1 & 2 \end{bmatrix}$.

80. **a.** Is the matrix $\begin{bmatrix} x & -1 \\ 2 & x + 2 \end{bmatrix}$ invertible for all real numbers x?
  b. Is the matrix $\begin{bmatrix} x - 1 & -1 \\ 3 & x - 5 \end{bmatrix}$ invertible for all real numbers x?

Critical Thinking/Discussion/Writing

81. True or False. Justify your answers.
  a. If A and B are matrices for which $AB = BA$, then the formula $A^2B = BA^2$ must hold.
  b. If A and B are invertible matrices, then $A + B$ must be invertible.

82. Give an example of a nonzero 2×2 matrix A such that $A^2 = \mathbf{0}$. Show that the matrix $I + A$ is invertible and that $(I + A)^{-1} = I - A$.

83. Is it true that every nonsingular matrix is a square matrix? Why or why not?

Maintaining Skills

In Exercises 84–87, state whether each expression is equal to 1 or −1.

84. $(-1)^0$

85. $(-1)^{10}$

86. $(-1)^{17}$

87. $(-1)^{6+9}$

In Exercises 88–91, perform the indicated operation.

88. $[-1, 4]\begin{bmatrix} 2 \\ 3 \end{bmatrix}$

89. $\begin{bmatrix} -1 \\ -3 \end{bmatrix}[5, 0]$

90. $3\begin{bmatrix} 7 & 1 \\ 5 & -2 \end{bmatrix}$

91. $(-1)\begin{bmatrix} -6 & 7 \\ 4 & -1 \end{bmatrix}$

In Exercises 92–95, solve each system of equations.

92. $\begin{cases} 2x - 3y = 16 \\ x - y = 7 \end{cases}$

93. $\begin{cases} 16x - 9y = -5 \\ 10x + 18y = -11 \end{cases}$

94. $\begin{cases} x - y + 5z = -6 \\ 3x + 3y - z = 10 \\ x + 3y + 2z = 5 \end{cases}$

95. $\begin{cases} 2x - y + 2z = 3 \\ 2x + 2y - z = 0 \\ -x + 2y + 2z = -12 \end{cases}$

SECTION 9.4

Determinants and Cramer's Rule

BEFORE STARTING THIS SECTION, REVIEW

1. Definition of a matrix (Section 9.1, page 799)
2. Systems of equations (Section 8.1, page 728)

OBJECTIVES

1. Calculate the determinant of a 2×2 matrix.
2. Find minors and cofactors.
3. Evaluate the determinant of an $n \times n$ matrix.
4. Apply Cramer's Rule.

Seki Kōwa (1642–1708)
Seki Kōwa has been called "the Japanese Newton." He was born in the same year as Newton. Like Newton, he acquired much of his knowledge by self-study and was a brilliant problem solver. But the greatest similarity between the two men lies in the claim that Seki Kowa was the creator of *yenri* (calculus). Thus, Seki Kōwa probably invented the calculus in the East just as Newton and Leibniz invented the calculus in the West.

A Little History of Determinants

It appears that determinants were first investigated in 1683 by the outstanding Japanese mathematician Seki Kōwa in connection with systems of linear equations. Determinants were independently investigated in 1693 by the German mathematician Gottfried Leibniz (1646–1716). The French mathematician Alexandre-Théophile Vandermonde (1735–1796) was the first to give a coherent and systematic exposition of the theory of determinants. Determinants were used extensively in the 19th century by eminent mathematicians such as Cauchy, Jacobi, and Kronecker. Today determinants are of little numerical value in the large-scale matrix computations that occur so often. Nevertheless, a knowledge of determinants is useful in some applications of matrices.

1 Calculate the determinant of a 2×2 matrix.

The Determinant of a 2×2 Matrix

Associated with each square matrix A is a number called the *determinant* of A.

SIDE NOTE

A determinant can be thought of as a function whose domain is the set of all square matrices and whose range is the set of real numbers.

Determinant of a 2×2 Matrix

The **determinant** of a 2×2 matrix

$$A = \begin{bmatrix} a & b \\ c & d \end{bmatrix},$$

denoted by $\det(A)$, $|A|$, or $\begin{vmatrix} a & b \\ c & d \end{vmatrix}$, is called the **2 by 2 determinant** of A and is defined by

$$\det(A) = ad - bc.$$

To remember this definition, note that you form products along diagonal elements as shown by the arrows.

$$\begin{bmatrix} a & b \\ c & d \end{bmatrix}$$
$$-bc \quad +ad$$

You attach a plus sign if you descend from left to right and a minus sign if you descend from right to left. Then you form the sum $ad - bc$.

EXAMPLE 1 Calculating the Determinant of a 2×2 Matrix

Evaluate each determinant.

a. $\begin{vmatrix} 3 & -4 \\ 1 & 5 \end{vmatrix}$ **b.** $\begin{vmatrix} -2 & 3 \\ 4 & -5 \end{vmatrix}$

Solution

a. $\begin{vmatrix} 3 & -4 \\ 1 & 5 \end{vmatrix} = (3)(5) - (-4)(1) = 15 - (-4) = 15 + 4 = 19$

b. $\begin{vmatrix} -2 & 3 \\ 4 & -5 \end{vmatrix} = (-2)(-5) - (3)(4) = 10 - 12 = -2$

Practice Problem 1 Evaluate each determinant.

a. $\begin{vmatrix} 5 & 1 \\ 3 & -7 \end{vmatrix}$ **b.** $\begin{vmatrix} 2 & -9 \\ -4 & 18 \end{vmatrix}$

WARNING

Note carefully the notations used for matrices and determinants. Brackets, [ ], denote a matrix, the vertical bars, | |, denote the determinant of a matrix. These bars are *not* the same as the absolute value bars. Remember that *the determinant of a matrix is a number that may be positive, negative, or zero.*

2 Find minors and cofactors.

Minors and Cofactors

To define the determinants of square matrices of order 3 or higher, we need some additional terminology.

Minors and Cofactors in an $n \times n$ Matrix

Let A be an $n \times n$ square matrix. The **minor** M_{ij} of the element a_{ij} is the determinant of the $(n-1) \times (n-1)$ matrix obtained by deleting the ith row and the jth column of A. The **cofactor** A_{ij} of the entry a_{ij} is defined by

$$A_{ij} = (-1)^{i+j} M_{ij}.$$

SIDE NOTE

Signs of cofactors

$$\begin{bmatrix} + & - & + & - & \cdots \\ - & + & - & + & \cdots \\ + & - & + & - & \cdots \\ - & + & - & + & \cdots \\ \vdots & \vdots & \vdots & \vdots & \end{bmatrix}$$

Note how the positions where $i + j$ is even alternate with those where $i + j$ is odd.

When $i + j$ is an even integer, $(-1)^{i+j} = 1$. When $i + j$ is an odd integer, $(-1)^{i+j} = -1$. Therefore,

$$A_{ij} = \begin{cases} M_{ij} \text{ if } i + j \text{ is an even integer} \\ -M_{ij} \text{ if } i + j \text{ is an odd integer.} \end{cases}$$

Notice that to compute a minor in an $n \times n$ matrix, you must know how to compute the determinant of an $(n-1) \times (n-1)$ matrix. At this point in the text, because you have learned the definition of the determinant of a 2×2 matrix, you can find the minors and cofactors of entries in a 3×3 matrix.

EXAMPLE 2 **Finding Minors and Cofactors**

For the matrix $A = \begin{bmatrix} 2 & 3 & 4 \\ 5 & -3 & -6 \\ 0 & 1 & 7 \end{bmatrix}$, find the following.

a. The minors M_{11}, M_{23}, and M_{32}

b. The cofactors A_{11}, A_{23}, and A_{32}

Solution

a. (i) To find M_{11}, delete the first row and first column of the matrix A.

$$\begin{bmatrix} 2 & 3 & 4 \\ 5 & -3 & -6 \\ 0 & 1 & 7 \end{bmatrix}$$

Now find the determinant of the resulting matrix.

$$M_{11} = \begin{vmatrix} -3 & -6 \\ 1 & 7 \end{vmatrix} = (-3)(7) - (-6)(1) = -21 + 6 = -15$$

(ii) To find the minor M_{23}, delete the second row and third column of the matrix A.

$$\begin{bmatrix} 2 & 3 & 4 \\ 5 & -3 & -6 \\ 0 & 1 & 7 \end{bmatrix}$$

Now find the determinant of the resulting matrix.

$$M_{23} = \begin{vmatrix} 2 & 3 \\ 0 & 1 \end{vmatrix} = (2)(1) - (3)(0) = 2$$

(iii) To find M_{32}, delete the third row and second column of A.

$$\begin{bmatrix} 2 & 3 & 4 \\ 5 & -3 & -6 \\ 0 & 1 & 7 \end{bmatrix}$$

So $M_{32} = \begin{vmatrix} 2 & 4 \\ 5 & -6 \end{vmatrix} = (2)(-6) - (4)(5) = -12 - 20 = -32.$

b. To find the cofactors, use the formula $A_{ij} = (-1)^{i+j} M_{ij}$.

$A_{11} = (-1)^{1+1} M_{11} = (1)(-15) = -15$ $\quad (-1)^{1+1} = (-1)^2 = 1; M_{11} = -15$

$A_{23} = (-1)^{2+3} M_{23} = (-1)(2) = -2$ $\quad (-1)^{2+3} = (-1)^5 = -1; M_{23} = 2$

$A_{32} = (-1)^{3+2} M_{32} = (-1)(-32) = 32$ $\quad (-1)^{3+2} = (-1)^5 = -1; M_{32} = -32$

Practice Problem 2 For the matrix $A = \begin{bmatrix} 3 & -1 & 2 \\ 4 & 5 & 6 \\ 7 & 1 & 2 \end{bmatrix}$, find the following.

a. The minors M_{11}, M_{23}, and M_{32}

b. The cofactors A_{11}, A_{23}, and A_{32}

3 Evaluate the determinant of an $n \times n$ matrix.

The Determinant of an $n \times n$ Matrix

There are several ways of defining the determinant of a square matrix. Here we give an *inductive definition*. This means that the definition of the determinant of a matrix of order n uses the determinants of the matrices of order $(n - 1)$.

n by *n* Determinant

Let A be a square matrix of order $n \geq 3$. The **determinant** of A is the sum of the entries in any row of A (or column of A) multiplied by their respective cofactors.

Applying the preceding definition to find the determinant of a matrix is called **expanding by cofactors**. Expanding by the cofactors of the ith row gives

$$\det(A) = a_{i1}A_{i1} + a_{i2}A_{i2} + \cdots + a_{in}A_{in}.$$

Similarly, expanding by the cofactors of the jth column yields

$$\det(A) = a_{1j}A_{1j} + a_{2j}A_{2j} + \cdots + a_{nj}A_{nj}.$$

We usually say "expanding by the ith row (column)" instead of "expanding by the cofactors of the ith row (column)."

TECHNOLOGY CONNECTION

Use the Det option on a graphing calculator to find the determinant of a matrix.

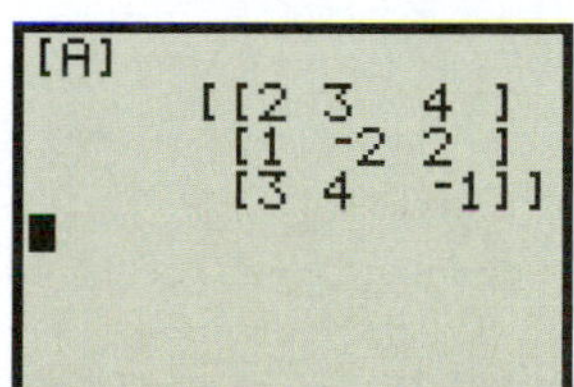

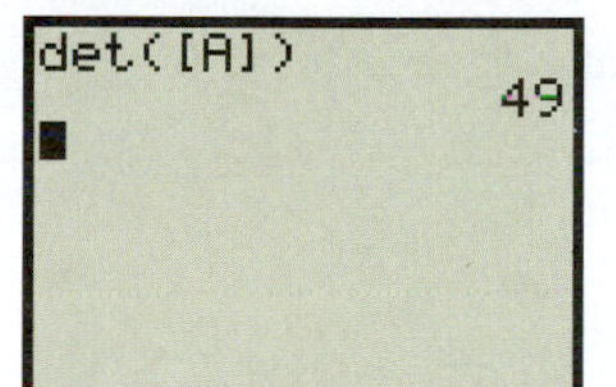

EXAMPLE 3 Evaluating a 3 by 3 Determinant

Find the determinant of $A = \begin{bmatrix} 2 & 3 & 4 \\ 1 & -2 & 2 \\ 3 & 4 & -1 \end{bmatrix}$.

Solution

Expanding by the first row, we have $|A| = a_{11}A_{11} + a_{12}A_{12} + a_{13}A_{13}$ where

$$A_{11} = (-1)^{1+1}M_{11} = (1)\begin{vmatrix} -2 & 2 \\ 4 & -1 \end{vmatrix}$$

$$A_{12} = (-1)^{1+2}M_{12} = (-1)\begin{vmatrix} 1 & 2 \\ 3 & -1 \end{vmatrix}$$

$$A_{13} = (-1)^{1+3}M_{13} = (1)\begin{vmatrix} 1 & -2 \\ 3 & 4 \end{vmatrix}.$$

Thus,

$$\begin{vmatrix} 2 & 3 & 4 \\ 1 & -2 & 2 \\ 3 & 4 & -1 \end{vmatrix} = 2(1)\begin{vmatrix} -2 & 2 \\ 4 & -1 \end{vmatrix} + 3(-1)\begin{vmatrix} 1 & 2 \\ 3 & -1 \end{vmatrix} + 4(1)\begin{vmatrix} 1 & -2 \\ 3 & 4 \end{vmatrix}$$

$$= 2(2 - 8) - 3(-1 - 6) + 4(4 + 6)$$ Evaluate determinants of order 2.

$$= -12 + 21 + 40 = 49$$ Simplify.

Practice Problem 3 Find the determinant of $A = \begin{bmatrix} 2 & -3 & 7 \\ -2 & -1 & 9 \\ 0 & 2 & -9 \end{bmatrix}$.

SIDE **NOTE**

In the definition, we note that the value $\det(A)$ *does not* depend on the choice of a row or a column. The proof of this fact is outside the scope of this book.

The definition says that we can expand the determinant by any row or column. If we expand by the third column of A in Example 3, we have

$$\begin{vmatrix} 2 & 3 & 4 \\ 1 & -2 & 2 \\ 3 & 4 & -1 \end{vmatrix} = 4(1)\begin{vmatrix} 1 & -2 \\ 3 & 4 \end{vmatrix} + 2(-1)\begin{vmatrix} 2 & 3 \\ 3 & 4 \end{vmatrix} + (-1)(1)\begin{vmatrix} 2 & 3 \\ 1 & -2 \end{vmatrix}$$

$$= 4(4 + 6) - 2(8 - 9) - 1(-4 - 3)$$

$$= 40 + 2 + 7 = 49.$$

Some properties of determinants are discussed in Exercises 70–73. They are useful in evaluating determinants (see Exercise 74).

4 Apply Cramer's Rule.

Cramer's Rule

You have already learned several methods for solving a system of linear equations. Another method called **Cramer's Rule** uses determinants. To see a derivation of the rule, consider a system of two linear equations in two variables.

$$\begin{cases} a_1x + b_1y = c_1 & (1) \\ a_2x + b_2y = c_2 & (2) \end{cases}$$

We can solve this system by using the elimination method. Let's first eliminate y and solve for x.

$$\begin{aligned} a_1b_2x + b_1b_2y &= c_1b_2 && \text{Multiply equation (1) by } b_2 \\ -a_2b_1x - b_1b_2y &= -c_2b_1 && \text{Multiply equation (2) by } -b_1. \\ \hline (a_1b_2 - a_2b_1)x &= c_1b_2 - c_2b_1 && \text{Add to eliminate } y. \\ x &= \frac{c_1b_2 - c_2b_1}{a_1b_2 - a_2b_1} && \text{Divide both sides by } a_1b_2 - a_2b_1 \text{ if } a_1b_2 - a_2b_1 \neq 0. \end{aligned}$$

Similarly, by eliminating x and solving for y, we have

$$y = \frac{a_1c_2 - a_2c_1}{a_1b_2 - a_2b_1} \qquad \text{if } a_1b_2 - a_2b_1 \neq 0$$

The solution of the system of equations (1) and (2) is therefore given by

$$x = \frac{c_1b_2 - c_2b_1}{a_1b_2 - a_2b_1}, \quad y = \frac{a_1c_2 - a_2c_1}{a_1b_2 - a_2b_1} \quad \text{provided that } a_1b_2 - a_2b_1 \neq 0.$$

You can write the solution in the form of determinants:

$$x = \frac{\begin{vmatrix} c_1 & b_1 \\ c_2 & b_2 \end{vmatrix}}{\begin{vmatrix} a_1 & b_1 \\ a_2 & b_2 \end{vmatrix}}, \quad y = \frac{\begin{vmatrix} a_1 & c_1 \\ a_2 & c_2 \end{vmatrix}}{\begin{vmatrix} a_1 & b_1 \\ a_2 & b_2 \end{vmatrix}} \quad \text{provided that } D = \begin{vmatrix} a_1 & b_1 \\ a_2 & b_2 \end{vmatrix} = a_1b_2 - a_2b_1 \neq 0.$$

Note that the denominator of both the x- and y-values in the solution is the determinant D of the coefficient matrix of the linear system.

Suppose we write $D_x = \begin{vmatrix} c_1 & b_1 \\ c_2 & b_2 \end{vmatrix}$ and $D_y = \begin{vmatrix} a_1 & c_1 \\ a_2 & c_2 \end{vmatrix}$, which are obtained by replacing the column containing the coefficients of x in D and the column containing the coefficients of y in D with the column of constants, respectively. Then the solution just found can be written as

$$x = \frac{D_x}{D}, \quad y = \frac{D_y}{D} \quad \text{provided that } D \neq 0.$$

This representation of the solution is Cramer's Rule for solving a system of linear equations.

Gabriel Cramer (1704–1752)

Gabriel Cramer was one of three sons of a medical doctor in Geneva, Switzerland. Gabriel moved rapidly through his education in Geneva, and in 1722, while he was still 18 years old, he was awarded a PhD degree. Two years later he accepted the chair of mathematics at the Académie de Clavin in Geneva.

Cramer taught and traveled extensively and yet found time to produce articles and books covering a wide range of subjects. Although Cramer was not the first person to state the rule, it is universally known as "Cramer's Rule."

CRAMER'S RULE FOR SOLVING TWO EQUATIONS IN TWO VARIABLES

The system

$$\begin{cases} a_1x + b_1y = c_1 \\ a_2x + b_2y = c_2 \end{cases}$$

of two equations in two variables has a unique solution (x, y) given by

$$x = \frac{D_x}{D} \quad \text{and} \quad y = \frac{D_y}{D}$$

provided that $D \neq 0$, where

$$D = \begin{vmatrix} a_1 & b_1 \\ a_2 & b_2 \end{vmatrix}, \quad D_x = \begin{vmatrix} c_1 & b_1 \\ c_2 & b_2 \end{vmatrix}, \quad \text{and} \quad D_y = \begin{vmatrix} a_1 & c_1 \\ a_2 & c_2 \end{vmatrix}.$$

EXAMPLE 4 **Using Cramer's Rule**

Use Cramer's Rule to solve the system: $\begin{cases} 5x + 4y = 1 \\ 2x + 3y = 6 \end{cases}$

Solution

Step 1 Form the determinant D of the coefficient matrix.

$$D = \begin{vmatrix} 5 & 4 \\ 2 & 3 \end{vmatrix} = 15 - 8 = 7$$

Because $D = 7 \neq 0$, the system has a unique solution.

Step 2 Replace the column of coefficients of x in D with the constant terms to get

$$D_x = \begin{vmatrix} 1 & 4 \\ 6 & 3 \end{vmatrix} = 3 - 24 = -21.$$

Step 3 Replace the column of coefficients of y in D with the constant terms to get

$$D_y = \begin{vmatrix} 5 & 1 \\ 2 & 6 \end{vmatrix} = 30 - 2 = 28.$$

Step 4 By Cramer's Rule,

$$x = \frac{D_x}{D} = \frac{-21}{7} = -3 \quad \text{and} \quad y = \frac{D_y}{D} = \frac{28}{7} = 4.$$

The solution is $x = -3$ and $y = 4$, or $\{(-3, 4)\}$.

Check: You should check the solution in the original system.

Practice Problem 4 Use Cramer's Rule to solve the following system:

$$\begin{cases} 2x + 3y = 7 \\ 5x + 9y = 4 \end{cases}$$

Cramer's Rule, as used in Example 4, can be generalized to a system of equations with more than two variables. We state the rule for three equations in three variables.

CRAMER'S RULE FOR SOLVING THREE EQUATIONS IN THREE VARIABLES

The system

$$\begin{cases} a_1x + b_1y + c_1z = k_1 \\ a_2x + b_2y + c_2z = k_2 \\ a_3x + b_3y + c_3z = k_3 \end{cases}$$

of three equations in three variables has a unique solution (x, y, z) given by

$$x = \frac{D_x}{D}, \quad y = \frac{D_y}{D}, \quad \text{and} \quad z = \frac{D_z}{D}$$

provided that $D \neq 0$, where

$$D = \begin{vmatrix} a_1 & b_1 & c_1 \\ a_2 & b_2 & c_2 \\ a_3 & b_3 & c_3 \end{vmatrix}, \quad D_x = \begin{vmatrix} k_1 & b_1 & c_1 \\ k_2 & b_2 & c_2 \\ k_3 & b_3 & c_3 \end{vmatrix}, \quad D_y = \begin{vmatrix} a_1 & k_1 & c_1 \\ a_2 & k_2 & c_2 \\ a_3 & k_3 & c_3 \end{vmatrix}, \quad \text{and} \quad D_z = \begin{vmatrix} a_1 & b_1 & k_1 \\ a_2 & b_2 & k_2 \\ a_3 & b_3 & k_3 \end{vmatrix}.$$

SIDE NOTE

When you apply Cramer's Rule, make sure that all linear equations are written in the form

$$ax + by + cz = k.$$

EXAMPLE 5 Using Cramer's Rule

Use Cramer's Rule to solve the system of equations.

$$\begin{cases} 7x + y + z - 1 = 0 \\ 5x + 3z - 2y = 4 \\ 4x - z + 3y = 0 \end{cases}$$

Solution

First, rewrite the system of equations so that the terms on the left side of the equal sign are in the proper order and only the constant terms appear on the right side.

$$\begin{cases} 7x + y + z = 1 \\ 5x - 2y + 3z = 4 \\ 4x + 3y - z = 0 \end{cases}$$

Step 1 $D = \begin{vmatrix} 7 & 1 & 1 \\ 5 & -2 & 3 \\ 4 & 3 & -1 \end{vmatrix}$ — D is the determinant of the coefficient matrix.

$= 7\begin{vmatrix} -2 & 3 \\ 3 & -1 \end{vmatrix} - 5\begin{vmatrix} 1 & 1 \\ 3 & -1 \end{vmatrix} + 4\begin{vmatrix} 1 & 1 \\ -2 & 3 \end{vmatrix}$ — Expand by column 1.

$= 7(2 - 9) - 5(-1 - 3) + 4(3 + 2)$ — Evaluate determinants of order 2.

$= 7(-7) - 5(-4) + 4(5) = -9$ — Simplify.

Because $D \neq 0$, the system has a unique solution.

Step 2 $D_x = \begin{vmatrix} 1 & 1 & 1 \\ 4 & -2 & 3 \\ 0 & 3 & -1 \end{vmatrix}$ — Replace the x-coefficients in the first column in D with the constants.

$= 1\begin{vmatrix} -2 & 3 \\ 3 & -1 \end{vmatrix} - 4\begin{vmatrix} 1 & 1 \\ 3 & -1 \end{vmatrix} + 0\begin{vmatrix} 1 & 1 \\ -2 & 3 \end{vmatrix}$ — Expand by column 1.

$= 1(2 - 9) - 4(-1 - 3)$ — Evaluate determinants of order 2.

$= 1(-7) - 4(-4) = 9$ — Simplify.

Step 3 $D_y = \begin{vmatrix} 7 & 1 & 1 \\ 5 & 4 & 3 \\ 4 & 0 & -1 \end{vmatrix}$ Replace column 2 in D with the constants.

$= -1\begin{vmatrix} 5 & 3 \\ 4 & -1 \end{vmatrix} + 4\begin{vmatrix} 7 & 1 \\ 4 & -1 \end{vmatrix} - 0\begin{vmatrix} 7 & 1 \\ 5 & 3 \end{vmatrix}$ Expand by column 2 because it contains a zero.

$= -1(-5 - 12) + 4(-7 - 4)$ Evaluate the determinants.

$= 17 - 44 = -27$ Simplify.

Step 4 $D_z = \begin{vmatrix} 7 & 1 & 1 \\ 5 & -2 & 4 \\ 4 & 3 & 0 \end{vmatrix}$ Replace column 3 in D with the constants.

$= 1\begin{vmatrix} 5 & -2 \\ 4 & 3 \end{vmatrix} - 4\begin{vmatrix} 7 & 1 \\ 4 & 3 \end{vmatrix} + 0\begin{vmatrix} 7 & 1 \\ 5 & -2 \end{vmatrix}$ Expand by column 3.

$= (15 + 8) - 4(21 - 4)$ Evaluate the determinants.

$= 23 - 4(17) = -45$ Simplify.

Step 5 Cramer's Rule gives the following values:

$x = \frac{D_x}{D} = \frac{9}{-9} = -1$ From Steps 1 and 2

$y = \frac{D_y}{D} = \frac{-27}{-9} = 3$ From Steps 1 and 3

$z = \frac{D_z}{D} = \frac{-45}{-9} = 5$ From Steps 1 and 4

The solution set is therefore $\{(-1, 3, 5)\}$.

Check: You should check the solution.

Practice Problem 5 Use Cramer's Rule to solve the system.

$$\begin{cases} 3x + 2y + z = 4 \\ 4x + 3y + z = 5 \\ 5x + y + z = 9 \end{cases}$$

WARNING

It is important to note that Cramer's Rule does not apply if $D = 0$. In this case, the system is either consistent and has dependent equations (has infinitely many solutions) or inconsistent (has no solution). When $D = 0$, use a different method for solving systems of equations to find the solution set.

Answers to Practice Problems

1. a. -38 **b.** 0 **2. a.** $M_{11} = 4, M_{23} = 10, M_{32} = 10$ **b.** $A_{11} = 4, A_{23} = -10, A_{32} = -10$ **3.** 8 **4.** $\{(17, -9)\}$ **5.** $\{(2, -1, 0)\}$

SECTION 9.4 Exercises

Basic Concepts and Skills

1. The determinant of $\begin{bmatrix} a & b \\ c & d \end{bmatrix}$ is ____________.

2. The minor of an element is the determinant you get by deleting the ____________ and the ____________ containing that element.

3. To expand an $n \times n$ determinant, you multiply each element of some row (or column) by its ____________ and add the result.

4. A system of n linear equations in n variables has a unique solution provided the determinant of the ____________ is not zero.

5. **True or False.** A 2 by 2 determinant is always positive.

6. **True or False.** To expand an $n \times n$ determinant, you multiply each element of some row by its minor and add the result.

In Exercises 7–16, evaluate each determinant.

7. $\begin{vmatrix} 2 & 3 \\ 4 & 5 \end{vmatrix}$

8. $\begin{vmatrix} 3 & -5 \\ 1 & 4 \end{vmatrix}$

9. $\begin{vmatrix} 4 & -2 \\ 3 & -3 \end{vmatrix}$

10. $\begin{vmatrix} -2 & \frac{1}{2} \\ 3 & 1 \end{vmatrix}$

11. $\begin{vmatrix} -1 & -3 \\ -4 & -5 \end{vmatrix}$

12. $\begin{vmatrix} \frac{1}{2} & \frac{1}{3} \\ \frac{1}{4} & \frac{1}{6} \end{vmatrix}$

13. $\begin{vmatrix} \frac{3}{8} & \frac{1}{2} \\ -\frac{1}{9} & 5 \end{vmatrix}$

14. $\begin{vmatrix} -\frac{1}{2} & \frac{1}{3} \\ \frac{1}{4} & -\frac{1}{3} \end{vmatrix}$

15. $\begin{vmatrix} \sqrt{a} & \sqrt{b} \\ \sqrt{b} & \sqrt{a} \end{vmatrix}$

16. $\begin{vmatrix} \sqrt{3} & 1 \\ -1 & \sqrt{3} \end{vmatrix}$

For Exercises 17–20, consider the matrix

$$A = \begin{bmatrix} 2 & -3 & 4 \\ 1 & -1 & 2 \\ 0 & 1 & 2 \end{bmatrix}.$$

17. Find the minors.
   a. M_{21} **b.** M_{23} **c.** M_{32}

18. Find the cofactors.
   a. A_{21} **b.** A_{23} **c.** A_{32}

19. Find the minors.
   a. M_{11} **b.** M_{22} **c.** M_{31}

20. Find the cofactors.
   a. A_{11} **b.** A_{22} **c.** A_{31}

In Exercises 21–36, evaluate each determinant.

21. $\begin{vmatrix} 1 & 0 & -1 \\ 0 & 2 & 2 \\ -1 & 0 & 0 \end{vmatrix}$

22. $\begin{vmatrix} 2 & 0 & \frac{1}{2} \\ 1 & 0 & 2 \\ 4 & 0 & -5 \end{vmatrix}$

23. $\begin{vmatrix} 1 & 2 & 3 \\ 0 & 3 & 4 \\ 0 & 0 & 4 \end{vmatrix}$

24. $\begin{vmatrix} 2 & 3 & 4 \\ 0 & -4 & 6 \\ 0 & 0 & -5 \end{vmatrix}$

25. $\begin{vmatrix} 1 & 0 & 0 \\ 2 & 0 & 0 \\ 3 & 4 & 5 \end{vmatrix}$

26. $\begin{vmatrix} -1 & 0 & 0 \\ 3 & 2 & 0 \\ 4 & 5 & -3 \end{vmatrix}$

27. $\begin{vmatrix} 1 & 6 & 0 \\ 2 & 5 & 3 \\ 3 & 4 & 0 \end{vmatrix}$

28. $\begin{vmatrix} 1 & 0 & 2 \\ 0 & 5 & 7 \\ 3 & 1 & 0 \end{vmatrix}$

29. $\begin{vmatrix} 3 & 4 & 1 \\ 1 & 4 & 3 \\ 4 & 3 & 1 \end{vmatrix}$

30. $\begin{vmatrix} 3 & 1 & -2 \\ 4 & 0 & -4 \\ 2 & -1 & -3 \end{vmatrix}$

31. $\begin{vmatrix} 0 & 1 & 6 \\ 1 & 0 & 4 \\ 8 & 3 & 1 \end{vmatrix}$

32. $\begin{vmatrix} 0 & 2 & 3 \\ 1 & 0 & 1 \\ 3 & 2 & 0 \end{vmatrix}$

33. $\begin{vmatrix} a & b & 0 \\ 0 & a & b \\ b & 0 & a \end{vmatrix}$

34. $\begin{vmatrix} 0 & z & y \\ z & 0 & x \\ y & x & 0 \end{vmatrix}$

35. $\begin{vmatrix} a & b & c \\ c & a & b \\ b & c & a \end{vmatrix}$

36. $\begin{vmatrix} u & v & w \\ w & v & u \\ u & v & w \end{vmatrix}$

In Exercises 37–46, use Cramer's Rule (if applicable) to solve each system of equations.

37. $\begin{cases} x + y = 8 \\ x - y = -2 \end{cases}$

38. $\begin{cases} 4x + 3y = -1 \\ 2x - 5y = 9 \end{cases}$

39. $\begin{cases} 5x + 3y = 11 \\ 2x + y = 4 \end{cases}$

40. $\begin{cases} 2x - 7y = 13 \\ 5x + 6y = 9 \end{cases}$

41. $\begin{cases} 2x + 9y = 4 \\ 3x - 2y = 6 \end{cases}$

42. $\begin{cases} 5x + 3y = 1 \\ 2x - 5y = -12 \end{cases}$

43. $\begin{cases} 2x - 3y = 4 \\ 4x - 6y = 8 \end{cases}$

44. $\begin{cases} 3x + y = 2 \\ 6x + 2y = 4 \end{cases}$

45. $\begin{cases} \frac{2}{x} + \frac{3}{y} = 2 \\ \frac{5}{x} + \frac{8}{y} = \frac{31}{6} \end{cases}$

46. $\begin{cases} \frac{3}{x} - \frac{6}{y} = 2 \\ \frac{4}{x} + \frac{7}{y} = -3 \end{cases}$

$\left[\textit{Hint: } \text{For Exercises 45 and 46, let } u = \frac{1}{x} \text{ and } v = \frac{1}{y}.\right]$

In Exercises 47–56, solve each system of equations by means of Cramer's Rule. You may use technology to compute the determinants involved.

47. $\begin{cases} x - 2y + z = -1 \\ 3x + y - z = 4 \\ y + z = 1 \end{cases}$

48. $\begin{cases} x + 3y = 4 \\ y + 3z = 7 \\ z + 4x = 6 \end{cases}$

49. $\begin{cases} x + y - z = -3 \\ 2x + 3y + z = 2 \\ 2y + z = 1 \end{cases}$

50. $\begin{cases} 3x + y + z = 10 \\ x + y - z = 0 \\ 5x - 9y = 1 \end{cases}$

51. $\begin{cases} x + y + z = 3 \\ 2x - 3y + 5z = 4 \\ x + 2y - 4z = -1 \end{cases}$

52. $\begin{cases} x + 2y - z = 3 \\ 3x - y + z = 8 \\ x + y + z = 0 \end{cases}$

53. $\begin{cases} 2x - 3y + 5z = 11 \\ 3x + 5y - 2z = 7 \\ x + 2y - 3z = -4 \end{cases}$

54. $\begin{cases} x - 3y - 1 = 0 \\ 2x - y - 4z = 2 \\ y + 2z - 4 = 0 \end{cases}$

55. $\begin{cases} 5x + 2y + z = 12 \\ 2x + y + 3z = 13 \\ 3x + 2y + 4z = 19 \end{cases}$

56. $\begin{cases} 2x + y + z = 7 \\ 3x - y - z = -2 \\ x + 2y - 3z = -4 \end{cases}$

Applying the Concepts

It is known that the *area of a triangle* with vertices (x_1, y_1), (x_2, y_2), and (x_3, y_3) is equal to the absolute value of D, where

$$D = \frac{1}{2}\begin{vmatrix} x_1 & y_1 & 1 \\ x_2 & y_2 & 1 \\ x_3 & y_3 & 1 \end{vmatrix}.$$

In Exercises 57–60, use determinants to find the area of each triangle.

57. **Area of a triangle.** The triangle with vertices $(1, 2)$, $(-3, 4)$, and $(4, 6)$.

58. **Area of a triangle.** The triangle with vertices $(3, 1)$, $(4, 2)$, and $(5, 4)$.

59. **Area of a triangle.** The triangle with vertices $(-2, 1)$, $(-3, -5)$, and $(2, 4)$.

60. **Area of a triangle.** The triangle with vertices $(-1, 2)$ and $(2, 4)$ and the origin.

Collinearity. Three points $A(x_1, y_1)$, $B(x_1, y_2)$, and $C(x_3, y_3)$ are *collinear* (lie on the same line) if and only if the area of the triangle ABC is 0. In terms of determinants, three points (x_1, y_1), (x_2, y_2), and (x_3, y_3) are collinear if and only if

$$\begin{vmatrix} x_1 & y_1 & 1 \\ x_2 & y_2 & 1 \\ x_3 & y_3 & 1 \end{vmatrix} = 0.$$

In Exercises 61–64, use determinants to find whether the three given points are collinear.

61. $(0, 3)$, $(-1, 1)$, and $(2, 7)$

62. $(2, 0)$, $\left(1, \frac{1}{2}\right)$, and $(4, -1)$

63. $(0, -4)$, $(3, -2)$, and $(1, -4)$

64. $\left(0, -\frac{1}{4}\right)$, $(1, -1)$, and $(2, -2)$

Equation of a line. Three points (x, y), (x_1, y_1), and (x_2, y_2) all lie on the same line if and only if

$$D = \begin{vmatrix} x & y & 1 \\ x_1 & y_1 & 1 \\ x_2 & y_2 & 1 \end{vmatrix} = 0.$$

In other words, a point $P(x, y)$ lies on the line passing through the points (x_1, y_1) and (x_2, y_2) if and only if the determinant $D = 0$. Therefore, the equation

$$\begin{vmatrix} x & y & 1 \\ x_1 & y_1 & 1 \\ x_2 & y_2 & 1 \end{vmatrix} = 0$$

is an equation of the line passing through the points (x_1, y_1) and (x_2, y_2).

In Exercises 65–68, use determinants to write an equation of the line passing through the given points. Then evaluate the determinant and write the equation of the line in slope–intercept form.

65. $(-1, -1)$, $(1, 3)$

66. $(0, 4)$, $(1, 1)$

67. $\left(0, \frac{1}{3}\right)$, $(1, 1)$

68. $\left(1, -\frac{1}{3}\right)$, $(2, 0)$

Beyond the Basics

In Exercises 69–73, we list some properties of determinants. These properties are true for $n \times n$ matrices. Verify the properties for $n = 2$ or $n = 3$.

69. If each entry in any row or each entry in any column of a matrix A is zero, then $|A| = 0$. Verify each of the following.

a. $\begin{vmatrix} 0 & 0 \\ 2 & 5 \end{vmatrix} = 0$

b. $\begin{vmatrix} 1 & 0 \\ 3 & 0 \end{vmatrix} = 0$

c. $\begin{vmatrix} 1 & -2 & 3 \\ 0 & 0 & 0 \\ 4 & 5 & -7 \end{vmatrix} = 0$

d. $\begin{vmatrix} 4 & 5 & 0 \\ 6 & -7 & 0 \\ 8 & 15 & 0 \end{vmatrix} = 0$

70. If a matrix B is obtained from matrix A by interchanging two rows (or columns), then $|B| = -|A|$. Verify each of the following.

a. $\begin{vmatrix} 2 & 3 \\ 4 & 5 \end{vmatrix} = -\begin{vmatrix} 4 & 5 \\ 2 & 3 \end{vmatrix}$

b. $\begin{vmatrix} -5 & 3 \\ 2 & -4 \end{vmatrix} = -\begin{vmatrix} 3 & -5 \\ -4 & 2 \end{vmatrix}$

c. $\begin{vmatrix} 1 & 3 & 5 \\ 0 & 1 & 2 \\ 3 & -1 & 4 \end{vmatrix} = -\begin{vmatrix} 1 & 3 & 5 \\ 3 & -1 & 4 \\ 0 & 1 & 2 \end{vmatrix}$

71. If two rows (or columns) of a matrix A have corresponding entries that are equal, then $|A| = 0$. Verify the following.

a. $\begin{vmatrix} 2 & 3 & 5 \\ -1 & 4 & 8 \\ 2 & 3 & 5 \end{vmatrix} = 0$

b. $\begin{vmatrix} 3 & 4 & 3 \\ 1 & 2 & 1 \\ -1 & 6 & -1 \end{vmatrix} = 0$

72. If a matrix B is obtained from a matrix A by multiplying every entry of one row (or one column) by a real number c, then $|B| = c|A|$. Verify each of the following.

a. $\begin{vmatrix} -1 & 2 & 3 \\ 1\cdot 5 & 4\cdot 5 & 8\cdot 5 \\ 2 & 3 & 6 \end{vmatrix} = 5\begin{vmatrix} -1 & 2 & 3 \\ 1 & 4 & 8 \\ 2 & 3 & 6 \end{vmatrix}$

b. $\begin{vmatrix} 1 & 2 & 3 \\ -1 & 5 & 6 \\ 0 & 2 & 9 \end{vmatrix} = 3\begin{vmatrix} 1 & 2 & 1 \\ -1 & 5 & 2 \\ 0 & 2 & 3 \end{vmatrix}$

73. If a matrix B is obtained from a matrix A by replacing any row (or column) of A by the sum of that row (or column) and c times another row (or column), then $|B| = |A|$.

Verify that the following two determinants are equal:

$$\begin{vmatrix} 2 & -3 & 4 \\ -4 & 7 & -8 \\ 5 & -1 & 3 \end{vmatrix} \xrightarrow{2R_1 + R_2 \rightarrow R_2} \begin{vmatrix} 2 & -3 & 4 \\ 0 & 1 & 0 \\ 5 & -1 & 3 \end{vmatrix}$$

74. Evaluate the determinant

$$D = \begin{vmatrix} 1 & 2 & 3 & 3 \\ 4 & 5 & 2 & 1 \\ 1 & 2 & 5 & 3 \\ 2 & 4 & 3 & 7 \end{vmatrix}$$

by following these steps:

(i) $(-1)R_1 + R_3 \rightarrow R_3$

(ii) Expand by the cofactors of row 3.

(iii) In **(ii)**, you will have a determinant of order 3. Evaluate it to obtain the value of D.

In Exercises 75–78, use the properties of determinants in Exercises 69–73 to obtain three zeros in one row (or column) and then evaluate the given determinant by evaluating the resulting determinant of order 3.

75. $\begin{vmatrix} 2 & 0 & -3 & 4 \\ 0 & 1 & 0 & 5 \\ 5 & 0 & -9 & 8 \\ 1 & 2 & 0 & 7 \end{vmatrix}$ **76.** $\begin{vmatrix} -3 & 1 & 2 & 3 \\ 0 & -2 & 4 & 7 \\ -9 & 3 & 5 & 2 \\ 2 & -4 & 3 & -12 \end{vmatrix}$

77. $\begin{vmatrix} 5 & 7 & 1 & 2 \\ 6 & 8 & 9 & 3 \\ 24 & 22 & 6 & 10 \\ 21 & 17 & 7 & 10 \end{vmatrix}$ **78.** $\begin{vmatrix} 1 & 2 & 3 & 0 \\ 2 & -3 & 1 & 0 \\ -1 & -3 & 4 & 5 \\ 1 & 3 & 7 & 4 \end{vmatrix}$

In Exercises 79–86, solve each equation for *x*.

79. $\begin{vmatrix} 3 & 2 \\ 6 & x \end{vmatrix} = 0$ **80.** $\begin{vmatrix} 3 & 2 \\ x & 12 \end{vmatrix} = 0$

81. $\begin{vmatrix} x & -2 \\ 1 & x-1 \end{vmatrix} = 0$ **82.** $\begin{vmatrix} 2x+7 & 4 \\ 1 & x \end{vmatrix} = 0$

83. $\begin{vmatrix} 1 & -3 & 1 \\ 4 & 7 & x \\ 0 & 2 & 2 \end{vmatrix} = 0$ **84.** $\begin{vmatrix} x & x+1 & x+2 \\ 2 & 3 & -1 \\ 3 & -2 & 4 \end{vmatrix} = 0$

85. $\begin{vmatrix} x & 0 & 1 \\ 0 & x & 0 \\ 1 & 0 & x \end{vmatrix} = 0$ **86.** $\begin{vmatrix} x & 2 & 3 \\ x & x & 1 \\ 2 & 0 & 1 \end{vmatrix} = -8$

In Exercises 87 and 88, use the formula on page 851 for the area of a triangle.

87. If the area of the triangle formed by the points $(-4, 2)$, $(0, k)$, and $(-2, k)$ is 28 square units, find k.

88. If the area of the triangle formed by the points $(2, 3)$, $(1, 2)$, and $(-2, k)$ is 3 square units, find k.

Critical Thinking/Discussion/Writing

89. Solve the system by using Cramer's Rule.

$$\begin{cases} \dfrac{1}{x-2} + \dfrac{3}{y+1} = 13 \\ \dfrac{4}{x-2} - \dfrac{5}{y+1} = 1 \end{cases}$$

[*Hint:* Do not try to simplify the equations by clearing the denominators because the resulting equations will not be linear. Instead, let $u = \dfrac{1}{x-2}$ and $v = \dfrac{1}{y+1}$. Solve the system for u and v and then use this solution to solve for x and y.]

In Exercises 90 and 91, use Cramer's Rule to solve each system of equations.

90. $\begin{cases} \dfrac{6}{x+1} + \dfrac{4}{y-1} = 7 \\ \dfrac{8}{x+1} + \dfrac{5}{y-1} = 9 \end{cases}$

91. $\begin{cases} \dfrac{1}{3^x} + \dfrac{4}{4^y} = 25 \\ \dfrac{2}{3^x} - \dfrac{1}{4^y} = 14 \end{cases}$

92. Evaluate.

a. $\begin{vmatrix} \log_2 3 & \log_8 3 \\ \log_3 4 & \log_3 4 \end{vmatrix}$ **b.** $\begin{vmatrix} \log_3 512 & \log_4 3 \\ \log_3 8 & \log_4 9 \end{vmatrix}$

93. Find the value of k for which the following system has no solution.

$$\begin{cases} kx + 3y - z = 1 \\ x + 2y + z = 2 \\ -kx + y + 2z = -1 \end{cases}$$

94. Find the values of a and b for which the system of equations

$$\begin{cases} 2x + ay + 6z = 8 \\ x + 2y + bz = 5 \\ x + y + 3z = 4 \end{cases}$$

has

(i) A unique solution.
(ii) Infinitely many solutions.
(iii) No solution.

Maintaining Skills

In Exercises 95 and 98, solve each equation for *y*.

95. $x^2 = 8y$

96. $(x+3)^2 = 4(y-1)$

97. $(x-2)^2 = 6(y-3)$

98. $(x-5)^2 = -2(y+7)$

In Exercises 99–102, find the quadratic function, $y = f(x)$, with the given properties.

99. Having form $y = ax^2$, and passing through the point $(-2, -2)$

100. Having form $y = ax^2 + bx + c$, vertex $(0, 0)$, and passing through the point $(2, -20)$

101. Having form $y = ax^2 + bx + c$, vertex $(-3, -8)$, and passing through the point $(0, 10)$

102. Having form $y = ax^2 + bx + c$, x-intercepts 1 and -3, and y-intercept 6

In Exercises 103 and 104, determine whether the given quadratic function
a. Opens up or down.
b. Find its vertex.
c. Find the axis of symmetry.
d. Find the *x*-intercepts, if any.
e. Find the *y*-intercept.

103. $f(x) = 2x^2 + 4x - 6$ **104.** $f(x) = -2x^2 + 12x - 10$

SUMMARY Definitions, Concepts, and Formulas

9.1 Matrices and Systems of Equations

i. **Matrix.** An $m \times n$ matrix is a rectangular array of numbers with m rows and n columns.

ii. **Augmented matrix.** The augmented matrix of a system of linear equations is a matrix whose entries are the coefficients of the variables in the system together with the constants.

iii. **Elementary row operations.** For an augmented matrix of a system of linear equations, the following row operations will result in the matrix of an equivalent system:
1. Interchange any two rows.
2. Multiply any row by a nonzero constant.
3. Add a multiple of one row to another row.

iv. **Row-echelon form and reduced row-echelon form.** The elementary row operations are used to transform an augmented matrix to row-echelon form or reduced row-echelon form. See page 803.

v. **Gaussian elimination** is the method of transforming the augmented matrix to row-echelon form and then using back-substitution to find the solution set of a system of linear equations.

vi. **Gauss–Jordan elimination.** If you continue the Gaussian elimination procedure until a reduced row-echelon form is obtained, the procedure is called Gauss–Jordan elimination.

9.2 Matrix Algebra

i. **Equality of matrices.** Two matrices $A = [a_{ij}]$ and $B = [b_{ij}]$ are equal if **(i)** A and B have the same order and **(ii)** all corresponding entries are equal.

ii. **Matrix addition.** The matrices A and B can be added or subtracted if they have the same order. Their sum or difference can be obtained by adding or subtracting the corresponding entries.

iii. **Scalar multiplication.** The product of a matrix by a real number (scalar) is obtained by multiplying every entry of the matrix by the real number.

iv. **Matrix multiplication.** The product AB of two matrices is defined only if the number of columns of A is equal to the number of rows of B. If A is an $m \times p$ matrix and B is a $p \times n$ matrix, then the product AB is an $m \times n$ matrix.

The (i, j)th entry of AB is the sum of the products of the corresponding entries in the ith row of A and the jth column of B.

v. **The identity matrix.** The $n \times n$ identity matrix denoted by I_n or I is a matrix that has 1s on the main diagonal and 0s elsewhere. If A is an $n \times n$ matrix, then $AI = IA = A$.

9.3 The Matrix Inverse

i. **Inverse of a matrix.** Let A be an $n \times n$ matrix and I be the $n \times n$ identity matrix. If there is an $n \times n$ matrix B such that $AB = BA = I$, then A is **invertible** and B is called the **inverse** of A. We write $B = A^{-1}$.

ii. **Finding the inverse.** A procedure for finding A^{-1} (if it exists) is given on page 830.

iii. **Inverse of a 2 × 2 matrix.** The matrix $A = \begin{bmatrix} a & b \\ c & d \end{bmatrix}$ has an inverse if and only if $ad - bc \neq 0$. Moreover, if $ad - bc \neq 0$, then

$$A^{-1} = \frac{1}{ad - bc}\begin{bmatrix} d & -b \\ -c & a \end{bmatrix}.$$

iv. If A is invertible, then the solution of the matrix equation $AX = B$ is $X = A^{-1}B$.

9.4 Determinants and Cramer's Rule

i. **Determinant.** The determinant of a square matrix A is a real number denoted by $\det(A)$ or $|A|$.

ii. The determinant of a 2×2 matrix $A = \begin{bmatrix} a & b \\ c & d \end{bmatrix}$ is defined by $\det(A) = \begin{vmatrix} a & b \\ c & d \end{vmatrix} = ad - bc$.

iii. The determinant of a matrix of order $n \geq 3$ can be found by expanding by cofactors. See page 845.

iv. **Cramer's Rule** gives formulas for solving a square system of linear equations for which the determinant of the coefficient matrix is nonzero. See page 847 for solving a system of two equations in two variables and page 848 for solving a system of three equations in three variables.

REVIEW EXERCISES

Basic Concepts and Skills

In Exercises 1–4, determine the order of each matrix.

1. $\begin{bmatrix} -1 & 2 & 3 & 4 \end{bmatrix}$

2. $\begin{bmatrix} 5 \end{bmatrix}$

3. $\begin{bmatrix} 2 & 3 \\ -1 & 2 \\ 0 & 1 \end{bmatrix}$

4. $\begin{bmatrix} 1 & -1 & 2 & -4 \\ 5 & 4 & 3 & 2 \end{bmatrix}$

5. Let A be the matrix of Exercise 4. Identify the entries a_{12}, a_{14}, a_{23}, and a_{21}.

6. Write the 2×3 matrix A with entries $a_{22} = 3, a_{21} = 4, a_{12} = 2, a_{13} = 5, a_{11} = -1$, and $a_{23} = 1$.

In Exercises 7 and 8, write the augmented matrix for each system of linear equations.

7. $\begin{cases} 2x - 3y = 7 \\ 3x + y = 6 \end{cases}$

8. $\begin{cases} x + y - z = 6 \\ 2x - 3y - 2z = 2 \\ 5x - 3y + z = 8 \end{cases}$

In Exercises 9 and 10, convert each matrix to row-echelon form (Answers may vary).

9. $\begin{bmatrix} 0 & 1 & 2 & 1 \\ 2 & 0 & 3 & 4 \\ 1 & -2 & 1 & 7 \end{bmatrix}$

10. $\begin{bmatrix} 3 & -1 & 2 & 12 \\ 1 & 1 & -1 & 2 \\ 1 & 2 & 1 & 7 \end{bmatrix}$

In Exercises 11 and 12, convert each matrix to reduced row-echelon form.

11. $\begin{bmatrix} 3 & 1 & 3 & 1 \\ 2 & 1 & 1 & 1 \\ 1 & -1 & -1 & 0 \end{bmatrix}$

12. $\begin{bmatrix} 3 & 4 & -4 & 2 \\ 2 & 1 & -2 & 1 \\ 1 & 1 & 2 & 1 \end{bmatrix}$

In Exercises 13–16, solve each system of equations by Gaussian elimination.

13. $\begin{cases} x + y - z = 0 \\ 2x + y - 2z = 3 \\ 3x - 2y + 3z = 9 \end{cases}$

14. $\begin{cases} x - y + 2z = 1 \\ x + 3y - z = 6 \\ 2x + y - 3z = 1 \end{cases}$

15. $\begin{cases} x - 2y + 3z = -2 \\ 2x - 3y + z = 9 \\ 3x - y + 2z = 5 \end{cases}$

16. $\begin{cases} 2x + y + z = 7 \\ x + 2y + z = 3 \\ x + 2y + 2z = 6 \end{cases}$

In Exercises 17–20, solve each system of equations by Gauss–Jordan elimination.

17. $\begin{cases} x - 2y - 2z = 11 \\ 3x + 4y - z = -2 \\ 4x + 5y + 7z = 7 \end{cases}$

18. $\begin{cases} x - y - 9z = 1 \\ 3x + 2y - z = 2 \\ 4x + 3y + 3z = 0 \end{cases}$

19. $\begin{cases} 2x - y + 3z = 4 \\ x + 3y + 3z = -2 \\ 3x + 2y - 6z = 6 \end{cases}$

20. $\begin{cases} x - y - 9z = 1 \\ 3x + 2y - z = 2 \\ 4x + 3y + 3z = 0 \end{cases}$

21. Find x and y if $\begin{bmatrix} x - y & 0 \\ 1 & x + y \end{bmatrix} = \begin{bmatrix} 1 & 0 \\ 1 & 3 \end{bmatrix}$.

22. Find x and y if
$$\begin{bmatrix} 2x + 3y & -1 & -2 \\ 0 & 1 & x - y \\ 2 & 3x + 4y & 5 \end{bmatrix} = \begin{bmatrix} 4 & -1 & -2 \\ 0 & 1 & -3 \\ 2 & 5 & 5 \end{bmatrix}.$$

23. Let $A = \begin{bmatrix} 1 & 2 \\ -3 & 4 \end{bmatrix}$ and $B = \begin{bmatrix} 2 & -3 \\ -5 & 6 \end{bmatrix}$. Find each of the following.
    a. $A + B$ b. $A - B$ c. $2A$
    d. $-3B$ e. $2A - 3B$

24. Let $A = \begin{bmatrix} 2 & 0 & -1 \\ 1 & 2 & 2 \\ -2 & 0 & 3 \end{bmatrix}$ and $B = \begin{bmatrix} 0 & 1 & -1 \\ 1 & -1 & 0 \\ -1 & 0 & 1 \end{bmatrix}$.
    Find each of the following.
    a. $A + B$ b. $A - B$ c. $2A + 3B$

25. For the matrices A and B of Exercise 23, solve the matrix equation $3A + 2B - 3X = \mathbf{0}$ for X.

26. For the matrices A and B of Exercise 24, solve the matrix equation $A - 2X + 2B = \mathbf{0}$ for X.

In Exercises 27–30, find the products, if possible.
a. AB **b.** BA

27. $A = \begin{bmatrix} 0 & 1 \\ 2 & 3 \end{bmatrix}$, $B = \begin{bmatrix} -1 & -1 \\ -3 & 4 \end{bmatrix}$

28. $A = \begin{bmatrix} 0 & 1 & 2 \\ -1 & 0 & 1 \end{bmatrix}$, $B = \begin{bmatrix} 1 & 2 \\ 3 & -1 \\ 0 & 4 \end{bmatrix}$

29. $A = \begin{bmatrix} 1 & 2 & -1 \end{bmatrix}$, $B = \begin{bmatrix} 2 \\ 3 \\ 1 \end{bmatrix}$

30. $A = \begin{bmatrix} 1 & 0 \\ 2 & -1 \\ 3 & -2 \end{bmatrix}$, $B = \begin{bmatrix} 1 & 2 & 3 & 4 \\ 2 & -1 & 2 & 3 \end{bmatrix}$

31. Find a matrix $A = \begin{bmatrix} x & y \\ z & w \end{bmatrix}$ such that
$$\begin{bmatrix} 1 & 2 \\ 3 & -1 \end{bmatrix} A = \begin{bmatrix} 5 & 6 \\ 1 & -3 \end{bmatrix}.$$

32. Find a matrix $B = \begin{bmatrix} x & y \\ z & w \end{bmatrix}$ such that
$$B \begin{bmatrix} 1 & 2 \\ 3 & -1 \end{bmatrix} = \begin{bmatrix} 5 & 6 \\ 1 & -3 \end{bmatrix}.$$
Compare your answers for Exercises 31 and 32.

33. Find a matrix $A = \begin{bmatrix} x & y \\ z & w \end{bmatrix}$ such that
$$\begin{bmatrix} 5 & 3 \\ 4 & 2 \end{bmatrix} A = \begin{bmatrix} 1 & 0 \\ 0 & 1 \end{bmatrix}.$$

34. Find a matrix $A = \begin{bmatrix} x & y \\ z & w \end{bmatrix}$ such that
$$A \begin{bmatrix} 5 & 3 \\ 4 & 2 \end{bmatrix} = \begin{bmatrix} 1 & 0 \\ 0 & 1 \end{bmatrix}.$$
Compare your answers for Exercises 33 and 34.

35. Let $A = \begin{bmatrix} 0 & 1 \\ -1 & 3 \end{bmatrix}$.
    a. Find A^{-1}. (Use the formula on page 832.)
    b. Find $(A^2)^{-1}$.
    c. Find $(A^{-1})^2$.
    d. Use your answers in parts (b) and (c) to show that $(A^{-1})^2$ is the inverse of A^2.

36. Repeat Exercise 35 for $A = \begin{bmatrix} 3 & 4 \\ 2 & 3 \end{bmatrix}$.

In Exercises 37–40, show that B is the inverse of A.

37. $A = \begin{bmatrix} 7 & 6 \\ 6 & 5 \end{bmatrix}$, $B = \begin{bmatrix} -5 & 6 \\ 6 & -7 \end{bmatrix}$

38. $A = \begin{bmatrix} -1 & 3 \\ 1 & -4 \end{bmatrix}$, $B = \begin{bmatrix} -4 & -3 \\ -1 & -1 \end{bmatrix}$

39. $A = \begin{bmatrix} 2 & -1 & 0 \\ 1 & 0 & 4 \\ 1 & -1 & 1 \end{bmatrix}$, $B = \frac{1}{5}\begin{bmatrix} 4 & 1 & -4 \\ 3 & 2 & -8 \\ -1 & 1 & 1 \end{bmatrix}$

40. $A = \begin{bmatrix} 2 & 0 & 1 \\ 0 & -1 & 2 \\ 1 & 0 & 1 \end{bmatrix}$, $B = \begin{bmatrix} 1 & 0 & -1 \\ -2 & -1 & 4 \\ -1 & 0 & 2 \end{bmatrix}$

In Exercises 41–44, find the inverse (if it exists) of each matrix.

41. $\begin{bmatrix} 3 & 1 \\ 2 & 4 \end{bmatrix}$

42. $\begin{bmatrix} 3 & -4 \\ 1 & 2 \end{bmatrix}$

43. $\begin{bmatrix} 1 & 2 & -2 \\ -1 & 3 & 0 \\ 0 & -2 & 1 \end{bmatrix}$

44. $\begin{bmatrix} 1 & 2 & 1 \\ 2 & 2 & 4 \\ 0 & 0 & 3 \end{bmatrix}$

In Exercises 45–50, use an inverse matrix (if possible) to solve each system of linear equations.

45. $\begin{cases} x + 3y = 7 \\ 2x + 5y = 4 \end{cases}$

46. $\begin{cases} 3x + 5y = 4 \\ 2x + 4y = 5 \end{cases}$

47. $\begin{cases} x + 3y + 3z = 3 \\ x + 4y + 3z = 5 \\ x + 3y + 4z = 6 \end{cases}$

48. $\begin{cases} x + 2y + 3z = 6 \\ 2x + 4y + 5z = 8 \\ 3x + 5y + 6z = 10 \end{cases}$

49. $\begin{cases} x - y + z = 3 \\ 4x + 2y = 5 \\ 7x - y - z = 6 \end{cases}$

50. $\begin{cases} x + 2y + 4z = 7 \\ 4x + 3y - 2z = 6 \\ x - 3z = 4 \end{cases}$

In Exercises 51–54, find the determinant of each matrix.

51. $\begin{bmatrix} 2 & -5 \\ 3 & 4 \end{bmatrix}$

52. $\begin{bmatrix} -1 & 4 \\ 11 & -3 \end{bmatrix}$

53. $\begin{bmatrix} 12 & 21 \\ 4 & 7 \end{bmatrix}$

54. $\begin{bmatrix} -7 & 9 \\ -4 & 5 \end{bmatrix}$

In Exercises 55 and 56, find (a) the minors M_{12}, M_{23}, and M_{22} and (b) the cofactors A_{12}, A_{23}, and A_{22} of the given matrix A.

55. $A = \begin{bmatrix} 4 & -1 & 2 \\ -2 & -3 & 5 \\ 0 & 2 & -4 \end{bmatrix}$

56. $A = \begin{bmatrix} 1 & 2 & 3 \\ 4 & 5 & 6 \\ 7 & 8 & 9 \end{bmatrix}$

In Exercises 57 and 58, find the determinant of each matrix A by expanding (a) by the second row and (b) by the third column.

57. $A = \begin{bmatrix} 1 & -2 & 3 \\ 4 & -1 & -2 \\ -2 & 1 & 5 \end{bmatrix}$

58. $A = \begin{bmatrix} 1 & 4 & 3 \\ 6 & 8 & 10 \\ 2 & 5 & 4 \end{bmatrix}$

In Exercises 59 and 60, find the determinant of each matrix A.

59. $A = \begin{bmatrix} 1 & 2 & 0 \\ 0 & 1 & 2 \\ 1 & 0 & 2 \end{bmatrix}$

60. $A = \begin{bmatrix} 2 & 3 & 4 \\ 2 & 5 & 9 \\ 2 & 7 & 16 \end{bmatrix}$

In Exercises 61–64, use Cramer's Rule to solve each system of equations.

61. $\begin{cases} 5x + 3y = 11 \\ 2x + y = 4 \end{cases}$

62. $\begin{cases} 2x - 7y - 13 = 0 \\ 5x + 6y - 9 = 0 \end{cases}$

63. $\begin{cases} 3x + y - z = 14 \\ x + 3y - z = 16 \\ x + y - 3z = -10 \end{cases}$

64. $\begin{cases} 2x + 3y - 2z = 0 \\ 5y - 3x + 4z = 9 \\ 3x + 7y - 6z = 4 \end{cases}$

In Exercises 65 and 66, solve each equation for x.

65. $\begin{vmatrix} 1 & 2 & 4 \\ -3 & 5 & 7 \\ 1 & x & 4 \end{vmatrix} = 0$

66. $\begin{vmatrix} 1 & -1 & 2 \\ 0 & x & 1 \\ 3 & 2 & x - 1 \end{vmatrix} = 14$

67. Show that the equation of the line passing through the points $(2, 3)$ and $(1, -4)$ is equivalent to the equation

$$\begin{vmatrix} x & y & 1 \\ 2 & 3 & 1 \\ 1 & -4 & 1 \end{vmatrix} = 0.$$ (See page 851.)

68. Use a determinant to find the area of the triangle with vertices $A(1, 1)$, $B(4, 3)$, and $C(2, 7)$. (See page 851.)

Applying the Concepts

69. Maximizing profit. A company is considering which of three methods of production it should use in producing the three products A, B, and C. The number of units of each product produced by each method is shown in this matrix:

$$\begin{matrix} & A & B & C \\ \textit{Method } 1 & 4 & 8 & 2 \\ \textit{Method } 2 & 5 & 7 & 1 \\ \textit{Method } 3 & 5 & 4 & 8 \end{matrix}$$

The profit per unit of the products A, B, and C is \$10, \$4, and \$6, respectively. Use matrix multiplication to find which method maximizes total profit for the company.

70. Nutrition. Consider two families: Ardestanis (A) and Barkley (B). Family A consists of two adult males, three adult females, and one child. Family B consists of one adult male, one adult female, and two children. Their mutual dietician considers the age and weight of each member of the two families and recommends daily allowances for calories—adult males 2200, adult females 1700, and children 1500—and for protein—adult males 50 grams, adult females 40 grams, and children 30 grams.

Represent the information in the preceding paragraph by matrices. Use matrix multiplication to calculate the total requirements of calories and proteins for each family.

In Exercises 71–76, each problem produces a system of linear equations in two or three variables. Use the method of your choice from this chapter to solve the system of equations.

71. Airplane speed. An airplane traveled 1680 miles in 3 hours with the wind. It would have taken 3.5 hours to make the same trip against the wind. Find the speed of the plane and the velocity of the wind assuming that both are constant.

72. Walking speed. In 5 hours, Nertha walks 2 miles farther than Kristina walks in 4 hours; then in 12 hours, Kristina walks 10 miles farther than Nertha walks in 10 hours. How many miles per hour does each woman walk?

73. Friends. Andrew, Bonnie, and Chauncie have \$320 between them. Andrew has twice as much as Chauncie, and Bonnie and Chauncie together have \$20 less than Andrew. How much money does each person have?

74. Ratio. Steve and Nicole entered a double-decker bus in London on which there were $\frac{5}{8}$ as many men as women. After they boarded the bus, there were $\frac{7}{11}$ as many men as women on the bus. How many people were on the bus before they entered it?

75. Hospital workers. Metropolitan Hospital employs three types of caregivers: registered nurses, licensed practical nurses, and nurse's aides. Each registered nurse is paid \$75 per hour, each licensed practical nurse is paid \$20 per hour, and each nurse's aide is paid \$30 per hour. The hospital has budgeted \$4850 per hour for the three categories of caregivers. The hospital requires that each caregiver spend some time in a class learning about advanced medical practices each week: registered nurses, 3 hours; licensed practical nurses,

5 hours; and nurse's aides, 4 hours. The advanced classes can accommodate a maximum of 530 person-hours each week. The hospital needs a total of 130 registered nurses, licensed practical nurses, and nurse's aides to meet the daily needs of the patients. Assume that the allowable payroll is met, the training classes are full, and the patients' needs are met. How many registered nurses, licensed practical nurses, and nurse's aides should the hospital employ?

76. Sales tax. A drugstore sells three categories of items:

(i) Medicine, which is not taxed
(ii) Nonmedical items, which are taxed at the rate of 8%
(iii) Beer and cigarettes, which are taxed at the "sin tax" rate of 20%

Last Friday, the total sales (excluding taxes) of all three items were \$18,500. The total tax collected was \$1020. The sales of medicines exceeded the combined sales of the other items by \$3500. Find the sales in each category.

PRACTICE TEST A

1. Give the order of the matrix and the value of the designated entry of the matrix.

$$\begin{bmatrix} 3 & -1 & -3 & 0 \\ 2 & 4 & 1 & 7 \\ 5 & 0 & 3 & 2 \\ 8 & 10 & 9 & 6 \\ -4 & 0 & 0 & 3 \end{bmatrix} \quad a_{43}$$

2. Write an augmented matrix for the system of equations.

$$\begin{cases} 7x - 3y + 9z = 5 \\ -2x + 4y + 3z = -12 \\ 8x - 5y + z = -9 \end{cases}$$

3. Write the system of linear equations represented by the augmented matrix

$$\left[\begin{array}{ccc|c} 4 & 0 & -1 & -3 \\ 1 & 3 & 0 & 9 \\ 2 & 7 & 5 & 8 \end{array}\right].$$

Use x, y, and z for the variables.

4. Find values for the variables so that the matrices

$$\begin{bmatrix} x-5 & y+3 \\ z & 9 \end{bmatrix} = \begin{bmatrix} 5 & 13 \\ 0 & 9 \end{bmatrix}$$

are equal.

In Problems 5 and 6, solve the system of equations by using matrices.

5. $\begin{cases} x + 2y + z = 6 \\ x + y - z = 7 \\ 2x - y + 2z = -3 \end{cases}$

6. $\begin{cases} 2x + y - 4z = 6 \\ -x + 3y - z = -2 \\ 2x - 6y + 2z = 4 \end{cases}$

7. Let $A = \begin{bmatrix} 5 & -2 \\ 4 & 0 \\ 7 & 6 \end{bmatrix}$ and $B = \begin{bmatrix} -3 & -1 \\ 0 & -8 \\ -4 & 6 \end{bmatrix}$. Find $A - B$.

In Problems 8 and 9, find the product AB or state that AB is not defined.

8. $A = \begin{bmatrix} 3 & -7 & 2 \end{bmatrix}$, $B = \begin{bmatrix} 0 \\ 1 \\ 4 \end{bmatrix}$

9. $A = \begin{bmatrix} -4 & 1 & 2 \\ -9 & 0 & 3 \end{bmatrix}$, $B = \begin{bmatrix} 2 & 8 \\ 7 & -5 \end{bmatrix}$

For Problems 10–13, let

$$A = \begin{bmatrix} -3 & 1 & 0 \\ 5 & 7 & 2 \end{bmatrix}, \quad B = \begin{bmatrix} -1 & 4 \\ 8 & 2 \end{bmatrix}, \quad \text{and} \quad C = \begin{bmatrix} 1 & 5 \\ 0 & 4 \end{bmatrix}.$$

10. Find $2A$.

11. Find $A + BA$.

12. Find C^2.

13. Find C^{-1}.

14. Find the inverse of the matrix

$$A = \begin{bmatrix} 2 & 1 & 3 \\ 1 & 2 & -1 \\ 3 & 1 & 5 \end{bmatrix}.$$

15. Write the linear system

$$\begin{cases} 5x + 2y = 32 \\ 3x + y = 18 \end{cases}$$

as a matrix equation in the form $AX = B$, where A is the coefficient matrix and B is the constant matrix.

16. Write the matrix equation

$$\begin{bmatrix} 12 & -3 \\ -2 & 7 \end{bmatrix}\begin{bmatrix} x \\ y \end{bmatrix} = \begin{bmatrix} 5 \\ -9 \end{bmatrix}$$

as a system of linear equations without matrices.

17. Solve the matrix equation $A - 5X = 2B$ for X, where

$$A = \begin{bmatrix} 1 & 5 & -2 \\ 4 & -2 & 7 \end{bmatrix} \text{ and } B = \begin{bmatrix} 2 & 5 & -11 \\ 18 & 8 & 11 \end{bmatrix}.$$

In Problems 18 and 19, evaluate the determinant.

18. $\begin{vmatrix} \frac{1}{2} & -\frac{1}{4} \\ \frac{1}{2} & \frac{3}{4} \end{vmatrix}$

19. $\begin{vmatrix} 1 & 3 & 5 \\ 2 & 0 & 10 \\ -3 & 1 & -15 \end{vmatrix}$

20. Use Cramer's Rule to write the solution of the system

$$\begin{cases} 2x - y + z = 3 \\ x + y + z = 6 \\ 4x + 3y - 2z = 4 \end{cases}$$

in the determinant form. (You do not need to find the solution.)

PRACTICE TEST B

1. Give the order of the matrix and the value of the designated entries of the matrix.

$$A = \begin{bmatrix} 0 & 3 & 0 & -10 & 6 \\ 14 & -5 & 7 & -7 & 8 \\ -15 & 9 & -6 & 13 & -2 \\ 0.4 & 11 & 4 & -3 & 0 \end{bmatrix} \quad a_{34}$$

a. Order: 5×4; $a_{34} = 6$
b. Order: 4×5; $a_{34} = 11$
c. Order: 20; $a_{34} = -4$
d. Order: 4×5; $a_{34} = 13$

2. Write an augmented matrix for the following system of equations.

$$\begin{cases} 8x + 4y + 5z = 10 \\ -2x + 3y + 6z = -9 \\ -2x + 8y + 6z = 4 \end{cases}$$

a. $\begin{bmatrix} 8 & 4 & 5 \\ -2 & 3 & 6 \\ -2 & 8 & 6 \end{bmatrix}$ **b.** $\left[\begin{array}{ccc|c} 10 & 5 & 4 & 8 \\ -9 & 6 & 3 & -2 \\ 4 & 6 & 8 & -2 \end{array}\right]$

c. $\left[\begin{array}{ccc|c} 8 & -2 & -2 & 10 \\ 4 & 3 & 8 & -9 \\ 5 & 6 & 6 & 4 \end{array}\right]$ **d.** $\left[\begin{array}{ccc|c} 8 & 4 & 5 & 10 \\ -2 & 3 & 6 & -9 \\ -2 & 8 & 6 & 4 \end{array}\right]$

3. Write the system of linear equations represented by the following augmented matrix.

$$\left[\begin{array}{ccc|c} -1 & 7 & 0 & -6 \\ 2 & 0 & 6 & 5 \\ 0 & 8 & -8 & 4 \end{array}\right]$$

Use x, y, and z for the variables.

a. $\begin{cases} x + 7y + z = -6 \\ 2x + 6z = 5 \\ 8y + 8z = 4 \end{cases}$ **b.** $\begin{cases} -x + 7y + z = -6 \\ 2x + 6y = 5 \\ 8x - 8y = 4 \end{cases}$

c. $\begin{cases} -x + 7y + z = -6 \\ 2x + y + 6z = 5 \\ x + 8y - 8z = 4 \end{cases}$ **d.** $\begin{cases} -x + 7y = -6 \\ 2x + 6z = 5 \\ 8y - 8z = 4 \end{cases}$

4. Find values for the variables so that the matrices

$$\begin{bmatrix} x + 3 & y + 4 \\ 7 & 3 \end{bmatrix} \text{ and } \begin{bmatrix} 9 & 2 \\ 7 & z \end{bmatrix}$$

are equal.

a. $x = -6; y = 2; z = 3$ **b.** $x = 6; y = 9; z = 3$
c. $x = 6; y = -2; z = 3$ **d.** $x = 9; y = 2; z = 3$

5. Find the value of z in the following system.

$$\begin{cases} 2x + y = 15 \\ 2y + z = 25 \\ 2z + x = 26 \end{cases}$$

a. 4 **b.** 9 **c.** 11 **d.** 12

6. Find the value of $4x + 3y + z$ in the following system.

$$\begin{cases} 2x + y = 17 \\ y + 2z = 15 \\ x + z = 9 \end{cases}$$

a. 41 **b.** 43 **c.** 55 **d.** 45

7. Let $A = \begin{bmatrix} -1 & 4 \\ 0 & 4 \\ 8 & -4 \end{bmatrix}$ and $B = \begin{bmatrix} 7 & 2 \\ 17 & 4 \\ 2 & 2 \end{bmatrix}$. Find $A - B$.

a. $\begin{bmatrix} 1 & 2 \\ 7 & 0 \\ 6 & -2 \end{bmatrix}$ **b.** $\begin{bmatrix} 3 & -3 \\ 7 & 0 \\ -6 & 6 \end{bmatrix}$

c. $\begin{bmatrix} -8 & 2 \\ -17 & 0 \\ 6 & -6 \end{bmatrix}$ **d.** $\begin{bmatrix} 1 & 5 \\ 7 & 8 \\ 10 & 0 \end{bmatrix}$

In Problems 8 and 9, find the product AB if possible.

8. $A = \begin{bmatrix} -8 & 2 & 9 \end{bmatrix}$, $B = \begin{bmatrix} 3 \\ 0 \\ -3 \end{bmatrix}$

a. $\begin{bmatrix} -24 & 0 & -27 \end{bmatrix}$ **b.** $\begin{bmatrix} -51 \end{bmatrix}$

c. $\begin{bmatrix} 210 \end{bmatrix}$ **d.** $\begin{bmatrix} -24 \\ 0 \\ -27 \end{bmatrix}$

9. $A = \begin{bmatrix} 3 & -2 & 1 \\ 0 & 4 & -2 \end{bmatrix}$, $B = \begin{bmatrix} 5 & 0 \\ -2 & 1 \end{bmatrix}$

a. AB is not defined. **b.** $\begin{bmatrix} 15 & -10 & 5 \\ -6 & 8 & -4 \end{bmatrix}$

c. $\begin{bmatrix} 15 & -6 \\ -10 & 8 \\ 5 & -4 \end{bmatrix}$ **d.** $\begin{bmatrix} 15 & 0 \\ 0 & 4 \end{bmatrix}$

For Problems 10–13, let

$A = \begin{bmatrix} 2 & 1 & -3 \\ -5 & 2 & 1 \end{bmatrix}$, $B = \begin{bmatrix} -3 & 7 \\ 2 & 4 \end{bmatrix}$, and $C = \begin{bmatrix} 5 & 4 \\ 1 & 0 \end{bmatrix}$.

10. Find $2A$.

a. $\begin{bmatrix} 1 & \frac{1}{2} & -\frac{3}{2} \\ -\frac{5}{2} & 1 & \frac{1}{2} \end{bmatrix}$ **b.** $\begin{bmatrix} 4 & 2 & -6 \\ -10 & 4 & 2 \end{bmatrix}$

c. $\begin{bmatrix} 4 & 3 & -1 \\ -3 & 4 & 3 \end{bmatrix}$ **d.** $\begin{bmatrix} 0 & -1 & -5 \\ -7 & 0 & -1 \end{bmatrix}$

11. Find $A + BA$.

a. $\begin{bmatrix} 2 & -4 & 10 \\ -5 & 4 & 5 \end{bmatrix}$ **b.** $\begin{bmatrix} 4 & -2 & 6 \\ -10 & 4 & 2 \end{bmatrix}$

c. $\begin{bmatrix} -39 & 12 & 13 \\ -21 & 12 & -1 \end{bmatrix}$ **d.** $\begin{bmatrix} -41 & 17 & -2 \\ -16 & 6 & 10 \end{bmatrix}$

12. Find C^2.

a. $\begin{bmatrix} 10 & 8 \\ 2 & 0 \end{bmatrix}$ **b.** $\begin{bmatrix} 41 & 5 \\ 5 & 1 \end{bmatrix}$

c. $\begin{bmatrix} 29 & 20 \\ 5 & 4 \end{bmatrix}$ **d.** $\begin{bmatrix} 0 & 4 \\ 1 & 5 \end{bmatrix}$

13. Find C^{-1}.

a. $\begin{bmatrix} -\frac{5}{4} & 1 \\ \frac{1}{4} & 0 \end{bmatrix}$ b. $\begin{bmatrix} 0 & 1 \\ \frac{1}{4} & -\frac{5}{4} \end{bmatrix}$

c. $\begin{bmatrix} \frac{1}{4} & -\frac{5}{4} \\ 0 & 1 \end{bmatrix}$ d. $\begin{bmatrix} 0 & -1 \\ -\frac{1}{4} & -\frac{5}{4} \end{bmatrix}$

14. Find the inverse of the following matrix.

$$A = \begin{bmatrix} 1 & 0 & 0 \\ 2 & 1 & 0 \\ 3 & -4 & 1 \end{bmatrix}$$

a. $\begin{bmatrix} 1 & 0 & 0 \\ -2 & 1 & 0 \\ 3 & 4 & 1 \end{bmatrix}$ b. $\begin{bmatrix} 1 & 0 & 0 \\ -2 & 1 & 0 \\ -8 & 3 & 1 \end{bmatrix}$

c. $\begin{bmatrix} 1 & -2 & -11 \\ 0 & 1 & 4 \\ 0 & 0 & 1 \end{bmatrix}$ d. $\begin{bmatrix} 1 & 0 & 0 \\ -2 & 1 & 0 \\ -11 & 4 & 1 \end{bmatrix}$

15. Write the linear system as a matrix equation

$$\begin{cases} 4x + 2y = 15 \\ -2x + 4y = 9 \end{cases}$$

in the form $AX = B$, where A is the coefficient matrix and B is the constant matrix.

a. $\begin{bmatrix} 4 & 2 \\ -2 & 4 \end{bmatrix}\begin{bmatrix} x \\ y \end{bmatrix} = \begin{bmatrix} 15 \\ 9 \end{bmatrix}$ b. $\begin{bmatrix} 4 & 2 \\ 4 & -2 \end{bmatrix}\begin{bmatrix} x \\ y \end{bmatrix} = \begin{bmatrix} 15 \\ 9 \end{bmatrix}$

c. $\begin{bmatrix} 15 & 2 \\ 9 & -2 \end{bmatrix}\begin{bmatrix} x \\ y \end{bmatrix} = \begin{bmatrix} 4 \\ 4 \end{bmatrix}$ d. $\begin{bmatrix} 4 & -2 \\ 2 & 4 \end{bmatrix}\begin{bmatrix} x \\ y \end{bmatrix} = \begin{bmatrix} 15 \\ 9 \end{bmatrix}$

16. Write the matrix equation

$$\begin{bmatrix} -3 & 9 \\ -8 & -4 \end{bmatrix}\begin{bmatrix} x \\ y \end{bmatrix} = \begin{bmatrix} -4 \\ -5 \end{bmatrix}$$

as a system of linear equations.

a. $\begin{cases} -3x + 9y = -4 \\ -8x - 4y = -5 \end{cases}$ b. $\begin{cases} -3x + 9y = -4 \\ -4x - 8y = -5 \end{cases}$

c. $\begin{cases} 9x - 3y = -4 \\ -8x - 4y = -5 \end{cases}$ d. $\begin{cases} -3x + 9y = 4 \\ -8x - 4y = 5 \end{cases}$

17. Let $A = \begin{bmatrix} -2 & -3 & 1 \\ -5 & 3 & -2 \end{bmatrix}$ and $B = \begin{bmatrix} -1 & -2 & -1 \\ 1 & 0 & 1 \end{bmatrix}$. Solve the matrix equation $A - 3X = -5B$ for X.

a. $X = \begin{bmatrix} -\frac{5}{2} & -\frac{9}{2} & -1 \\ -1 & \frac{3}{2} & \frac{1}{2} \end{bmatrix}$

b. $X = \begin{bmatrix} -\frac{7}{3} & -\frac{13}{3} & -\frac{4}{3} \\ 0 & 1 & 1 \end{bmatrix}$

c. $X = \begin{bmatrix} -1 & \frac{3}{2} & -1 \\ -\frac{5}{2} & -\frac{9}{2} & -1 \end{bmatrix}$

d. $X = \begin{bmatrix} -\frac{1}{3} & \frac{1}{3} & \frac{8}{3} \\ -\frac{20}{3} & 3 & -\frac{11}{3} \end{bmatrix}$

In Problems 18 and 19, evaluate the determinant.

18. $\begin{vmatrix} -8 & 5 \\ -4 & -1 \end{vmatrix}$

a. -28 b. -44 c. -12 d. 28

19. $\begin{vmatrix} 2 & 3 & -2 \\ 3 & 0 & -3 \\ -3 & 0 & -5 \end{vmatrix}$

a. -72 b. 18 c. 72 d. -18

20. Solve the system of equations

$$\begin{cases} x + y + z = -6 \\ x - y + 3z = -22 \\ 2x + y + z = -10 \end{cases}$$

by using Cramer's Rule.

a. $\varnothing$ b. $\{(-4, 3, -5)\}$

c. $\{(-5, -4, 3)\}$ d. $\{(-5, 3, -4)\}$

CUMULATIVE REVIEW EXERCISES CHAPTERS P–9

1. Assuming that the distance between the points $(2, -3)$ and $(-1, y)$ is 5 units, find y.

In Problems 2–8, solve each equation or inequality.

2. $\dfrac{4}{x - 1} - \dfrac{3}{x + 2} = \dfrac{18}{(x + 2)(x - 1)}$

3. $|2x - 5| = 3$

4. $4x^2 = 8x - 13$

5. $\left(\dfrac{3x - 1}{x + 5}\right)^2 - 3\left(\dfrac{3x - 1}{x + 5}\right) - 28 = 0$

6. $\log_2|x| + \log_2|x + 6| = 4$

7. $\dfrac{x + 2}{2x - 1} > 0$

8. $x^2 - 7x + 6 \le 0$

9. Write all possible rational zeros of

$$f(x) = 4x^3 + 8x^2 - 11x + 3.$$

10. Show that $\frac{1}{2}$ is a zero of multiplicity 2 of the function f in Exercise 9.

11. The horsepower required to propel a ship varies as the cube of the ship's speed. If the horsepower required for a speed of 15 miles per hour is 10,125, find the horsepower required for a speed of 20 miles per hour.

12. A city covers an area of 400 square miles. At present, only 2% of its area is reserved for parks. In the unincorporated area outside the city limits, 20% of the area could be developed into parks. How many square miles of the incorporated area had to be annexed so that the city could develop 12% of its area as parks?

In Problems 13 and 14, solve the system of equations.

13. $\begin{cases} 3x + y = 2 \\ 4x + 5y = -1 \end{cases}$

14. $\begin{cases} 2x + y = 5 \\ y^2 - 2y = -3x + 5 \end{cases}$

15. Use transformations to sketch the graph of

$$f(x) = 3|x + 1| + 2.$$

16. Find the inverse of the matrix $\begin{bmatrix} 1 & 2 & -2 \\ -1 & 3 & 0 \\ 0 & -2 & 1 \end{bmatrix}$.

17. Use the result in Exercise 16 to solve the system of equations.

$$\begin{cases} x + 2y - 2z = 5 \\ -x + 3y \quad\quad = 2 \\ \quad - 2y + z = -3 \end{cases}$$

18. Assuming that $f(x) = x^2 + 3x - 1$ and $g(x) = x + 2$, find **(a)** $F(x) = (f \circ g)(x)$ and **(b)** $F(4)$.

19. Let $f(x) = \dfrac{x}{x + 4}$. Find $f^{-1}(x)$.

20. In Exercise 19, find the domain and range of *f*.

CHAPTER 10

Conic Sections

Greek mathematicians were fascinated by certain curves, called conic sections, now used to describe natural phenomena such as planetary orbits. Modern applications also include the construction of lenses, whispering galleries, navigation systems, and even medical devices. In this chapter, we investigate the conic sections and their many uses.

SECTION 10.1

Conic Sections: Overview

Most of the sections in this chapter focus on the plane curves called **conics** or **conic sections**. As the name implies, these curves are the sections of a cone (similar to an ice cream cone) formed when a plane intersects the cone. We also discuss parametric equations of conics and other curves in the plane.

Apollonius of Perga (c. 262 B.C.–190 B.C.)
The mathematician Apollonius was born in Perga, in southern Asia Minor (today known as Murtina in Turkey). He went to Alexandria to study with the successors of Euclid. He became famous in ancient times for his work on astronomy and his eight books on conics. He was able to discover and prove hundreds of beautiful and difficult theorems without modern algebraic symbolism. He is credited with introducing the terms *ellipse*, *hyperbola*, and *parabola*.

Euclid defined a cone as a surface generated by rotating a right triangle about one of its legs. However, the following description of a cone given by Apollonius is more appropriate.

Right Circular Cone

Draw a circle on a plane (say, horizontal). Draw a line l, called the **axis**, passing through the center of the circle and perpendicular to the plane. Choose a point V above the plane on this line. The surface consisting of all lines that simultaneously pass through the point V and the circle is called a **right circular cone** with **vertex** V. The vertex V separates the surface into two parts called *nappes of the cone*. See Figure 10.1.

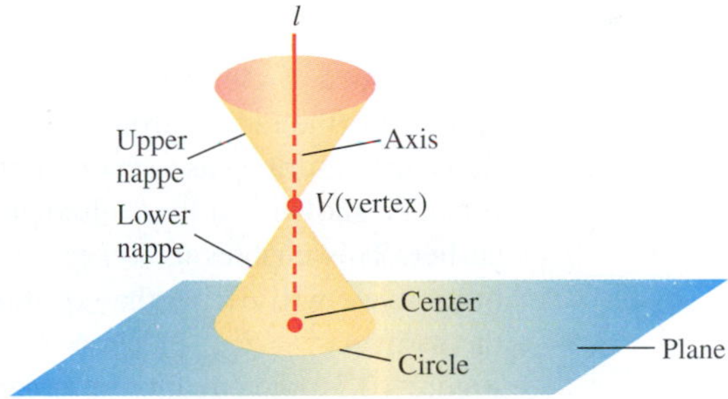

FIGURE 10.1 Parts of a cone

If we slice the cone with a plane not passing through the vertex, the intersections of the slicing plane and the cone are **conic sections**. If the slicing plane is horizontal (parallel to the first plane), the intersection is a **circle** (Figure 10.2(a)). If the slicing plane is inclined slightly from the horizontal, the intersection is an oval-shaped curve called an **ellipse**. As the angle of the slicing plane increases, more elongated ellipses are formed (Figure 10.2(b)). If the angle of the slicing plane increases still further so that the slicing plane is parallel to one of the lines generating the cone, the intersection is called a **parabola** (Figure 10. 2(c)). If the angle of the slicing plane increases yet further so that the slicing plane intersects both nappes of the cone, the resulting intersection is called a **hyperbola** (Figure 10. 2(d)).

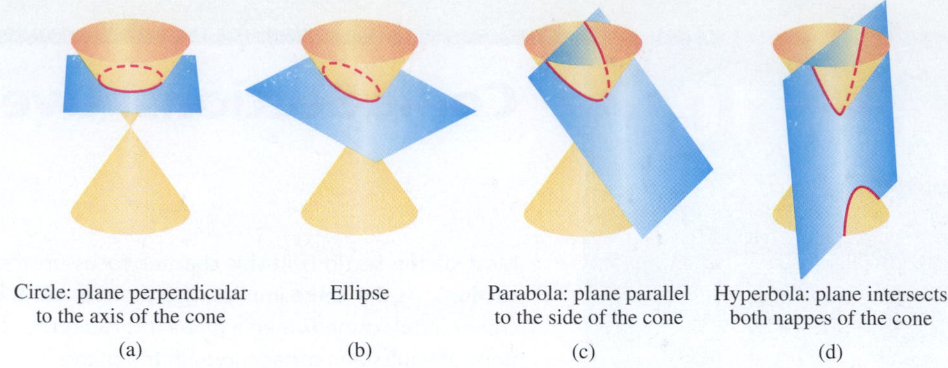

FIGURE 10.2 Conic sections

Intersections of the slicing plane through the vertex results in points and lines called **degenerate conic sections**. See Figure 10.3.

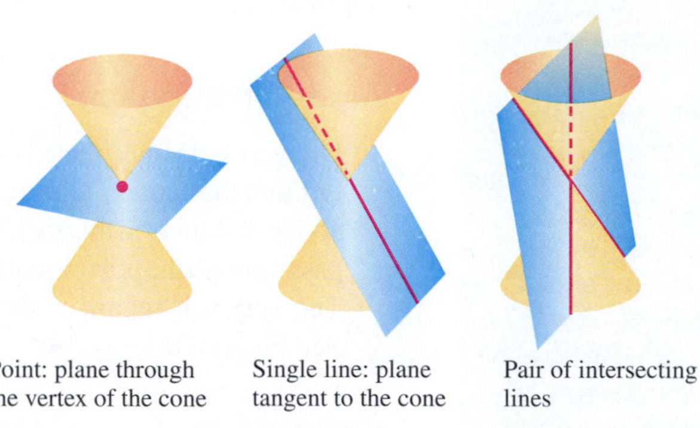

FIGURE 10.3 Degenerate conic sections

Just as a circle (see page 171) is defined as the set of points in the plane at a fixed distance r (the radius) from a fixed point (the center), we can define each of the conic sections just described as a set of points in the plane that satisfies certain geometric conditions. It can be shown that these alternative definitions give the same family of curves described earlier. This enables us to keep our work in a two-dimensional setting. Using these definitions, we will derive the equations of the conic sections and show that an equation of the form $Ax^2 + Cy^2 + Dx + Ey + F = 0$ is (except in degenerate cases represented in Figure 10.3.) the equation of a parabola, an ellipse, or a hyperbola. (A circle is a special case of an ellipse.)

We study these conic sections because of their practical applications. For example, a satellite dish, a flashlight lens, and a telescope lens have a parabolic shape; the planets travel in elliptical orbits, and comets travel in orbits that are either elliptical or hyperbolic. A comet with an elliptical orbit can be viewed from Earth more than once, while those with a hyperbolic orbit can be viewed only once!

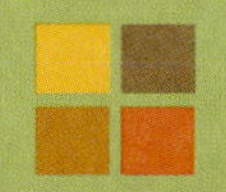

SECTION 10.2

The Parabola

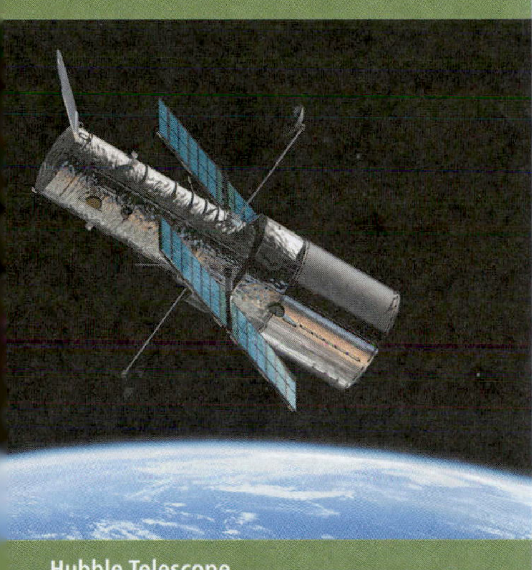

Hubble Telescope

BEFORE STARTING THIS SECTION, REVIEW

1. Distance formula (Section 2.1, page 159)
2. Midpoint formula (Section 2.1, page 161)
3. Completing the square (Section 3.1, page 295)
4. Line of symmetry (Section 2.2, page 168)
5. Slopes of perpendicular lines (Section 2.3, page 185)

OBJECTIVES

1. Define a parabola geometrically.
2. Find an equation of a parabola.
3. Translate a parabola.
4. Use the reflecting property of parabolas.

Edwin Hubble (1889–1953)
Edwin Hubble is renowned for discovering that there are other galaxies in the universe beyond the Milky Way. Hubble then wanted to classify the galaxies according to their content, distance, shape, and brightness patterns. He made the momentous discovery that the galaxies were moving away from each other at a rate proportional to the distance between them (Hubble's Law). This led to the calculation of the point where the expansion began and to confirmation of the Big Bang theory of the origin of the universe. Recent estimates place the age of the universe at about 14 billion years.

The Hubble Space Telescope

In 1918, the 2.5-meter (100-inch) Hooker Telescope was installed at Mount Wilson Observatory in Pasadena, California. With this telescope, astronomer Edwin Hubble measured the distances and velocities of the galaxies. His work led to today's concept of an expanding universe. In 1977, the National Aeronautics and Space Administration (NASA) named its largest, most complex, and most capable orbiting telescope in honor of Edwin Hubble.

On April 25, 1990, the Hubble Space Telescope was deployed into orbit from the space shuttle *Discovery*. However, due to a spherical aberration in the telescope's parabolic mirror, the Hubble had "blurred vision." (See Example 4.) In 1993, astronauts from the space shuttle *Endeavor* installed replacement instruments and supplemental optics to restore the telescope to full optimal performance.

Not since Galileo turned his telescope toward the heavens in 1610 has any event so changed our understanding of the universe as did the deployment of the Hubble Space Telescope. The first instrument of any kind that was specifically designed for routine servicing by spacewalking astronauts, the Hubble telescope brought stunning pictures from the ends of the universe to the general public.

1 Define a parabola geometrically.

Geometric Definition of a Parabola

In Section 3.1, we named the graph of a quadratic function $y = ax^2 + bx + c, a \neq 0$, a parabola. In this section, we give a *geometric* definition of a parabola. Then we show that this definition leads to the graph of a quadratic function.

RECALL

The distance from a point to a line is defined as the length of the perpendicular line segment from the point to the line.

Parabola

Let l be a line and F a point not on the line l on the plane determined by F and l. See Figure 10.4(a). Then the set of all points P in the plane that are the same distance from F as they are from the line l is called a **parabola**. See Figure 10.4(b). Thus, a parabola is the set of points P for which $d(F, P) = d(P, l)$ where $d(P, l)$ denotes the distance between the point P and the line l.

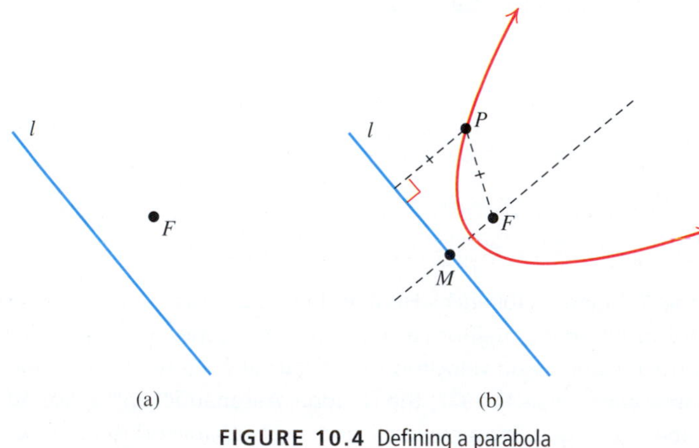

FIGURE 10.4 Defining a parabola

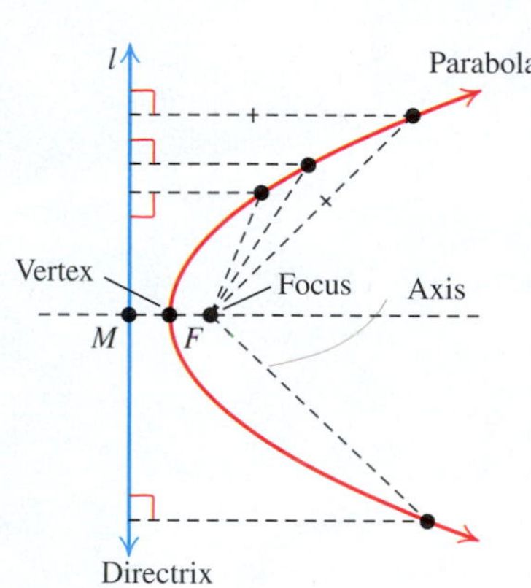

FIGURE 10.5

The line l is the **directrix** of the parabola, and the point F is the **focus**. The line through the focus *perpendicular* to the directrix is called the **axis**, or the **axis of symmetry**, of the parabola. The point at which the axis intersects the parabola is called the **vertex**. See Figure 10.5. Note that the vertex of a parabola is the midpoint of the segment $\overline{FM}$.

2 Find an equation of a parabola.

Equation of a Parabola

A coordinate system allows us to transform the geometric description of a parabola into an algebraic *equation of the parabola*. This equation is particularly simple if we place the vertex V at the origin and the focus on one of the coordinate axes. Suppose for a moment that the focus F is on the positive x-axis. Then F has coordinates of the form $(p, 0)$ with $p > 0$, as shown in Figure 10.6.

The vertex V lies on the parabola with focus F and directrix l; so we have the following:

$$\begin{aligned} d(V, l) &= d(V, F) \qquad \text{By the definition of parabola} \\ &= p \end{aligned}$$

The vertex V at the origin is located midway between the focus and the directrix, so the equation of the directrix is $x = -p$.

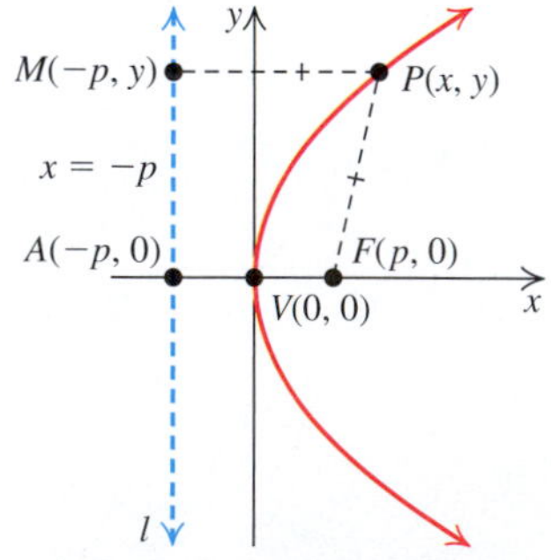

FIGURE 10.6 Deriving the equation of a parabola

Suppose a point $P(x, y)$ lies on the parabola shown in Figure 10.6. Then the distance from the point P to the directrix l is the distance between the points $P(x, y)$ and $M(-p, y)$. So we have

$$d(P, F) = d(P, M).$$

Using the distance formula on each side gives

$$\sqrt{(x-p)^2+y^2} = \sqrt{(x+p)^2}$$

$$(x-p)^2+y^2 = (x+p)^2 \qquad \text{Square both sides.}$$

$$x^2-2px+p^2+y^2 = x^2+2px+p^2 \qquad \text{Expand squares.}$$

$$y^2 = 4px \qquad \text{Simplify.}$$

The equation $y^2 = 4px$ is called the **standard equation of a parabola with vertex (0, 0) and focus (*p*, 0).** Similarly, if the focus of a parabola is placed on the negative x-axis, we obtain the equation $y^2 = -4px$ as the standard equation of a parabola with vertex $(0, 0)$ and focus $(-p, 0)$.

By interchanging the roles of x and y, we find that $x^2 = 4py$ is the standard equation of a parabola with vertex $(0, 0)$ and focus $(0, p)$. Similarly, the equation $x^2 = -4py$ is the standard equation of a parabola with vertex $(0, 0)$ and focus $(0, -p)$.

SUMMARY OF MAIN FACTS

Main facts about a parabola with $p > 0$

Standard Equation	$y^2 = 4px$	$y^2 = -4px$	$x^2 = 4py$	$x^2 = -4py$
Vertex	$(0, 0)$	$(0, 0)$	$(0, 0)$	$(0, 0)$
Description	Opens right	Opens left	Opens up	Opens down
Axis of Symmetry	$y = 0$ (x-axis)	$y = 0$ (x-axis)	$x = 0$ (y-axis)	$x = 0$ (y-axis)
Focus	$(p, 0)$	$(-p, 0)$	$(0, p)$	$(0, -p)$
Directrix	$x = -p$	$x = p$	$y = -p$	$y = p$

Graph

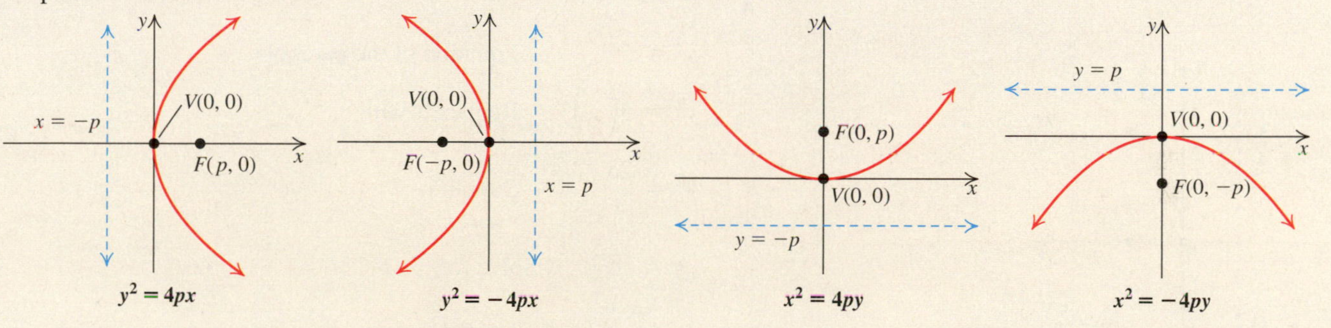

EXAMPLE 1 Graphing a Parabola

Graph each parabola and specify the vertex, focus, directrix, and axis.

a. $x^2 = -8y$ **b.** $y^2 = 5x$

Solution

a. The equation $x^2 = -8y$ has the standard form $x^2 = -4py$, so

$$-4p = -8 \qquad \text{Equate coefficients of } y.$$

$$p = 2 \qquad \text{Divide both sides by } -4.$$

The parabola opens down. The vertex is the origin, and the focus is $(0, -2)$. The directrix is the horizontal line $y = 2$; the axis of the parabola is the y-axis. You should plot some additional points to ensure the accuracy of your sketch. Because the focus is $(0, -2)$, substitute $y = -2$ in the equation $x^2 = -8y$ of the parabola to obtain

$$x^2 = (-8)(-2) \qquad \text{Replace } y \text{ with } -2 \text{ in } x^2 = -8y.$$

$$x^2 = 16 \qquad \text{Simplify.}$$

$$x = \pm 4 \qquad \text{Solve for } x.$$

TECHNOLOGY CONNECTION

You can use a graphing calculator to graph parabolas.

To graph $x^2 = 4py$, just graph $Y_1 = \frac{1}{4p}x^2$.

For example, when $x^2 = -8y$, you have $4p = -8$; so you graph $Y_1 = -\frac{1}{8}x^2$.

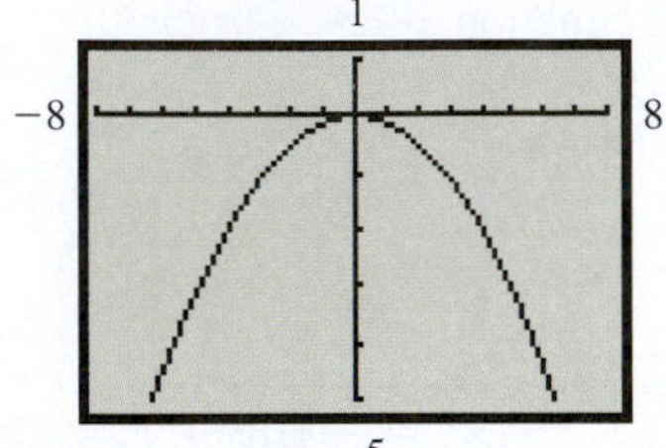

Thus, the points $(4, -2)$ and $(-4, -2)$ are two symmetric points on the parabola to the right and left of the focus. The graph of this parabola is sketched in Figure 10.7.

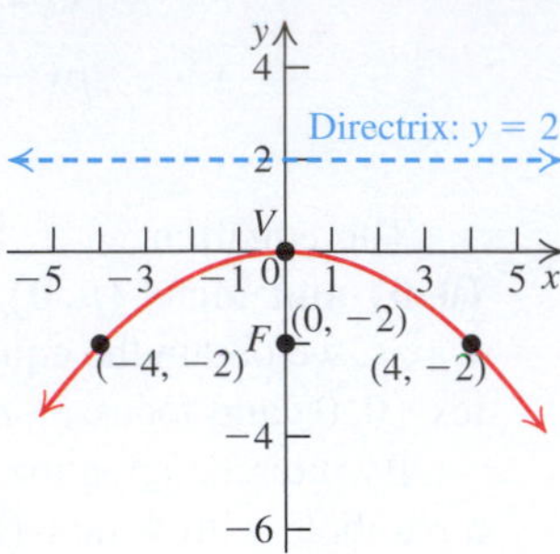

FIGURE 10.7 Graph of $x^2 = -8y$

b. The equation $y^2 = 5x$ has the standard form $y^2 = 4px$.

$$4p = 5 \quad \text{Equate coefficients of } x.$$

$$p = \frac{5}{4} \quad \text{Divide both sides by 4.}$$

The parabola opens to the right. The vertex is the origin, and the focus is $\left(\frac{5}{4}, 0\right)$. The directrix is the vertical line $x = -p = -\frac{5}{4}$; the axis of the parabola is the x-axis. To find two symmetric points on the parabola that are above and below the focus, substitute $x = \frac{5}{4}$ in the equation of the parabola.

$$y^2 = 5x \quad \text{Equation of the parabola}$$

$$y^2 = 5\left(\frac{5}{4}\right) \quad \text{Replace } x \text{ with } \frac{5}{4}.$$

$$y^2 = \frac{25}{4} \quad \text{Simplify.}$$

$$y = \pm\frac{5}{2} \quad \text{Solve for } y.$$

Plot the two additional points $\left(\frac{5}{4}, \frac{5}{2}\right)$ and $\left(\frac{5}{4}, -\frac{5}{2}\right)$ on the parabola. The graph of $y^2 = 5x$ is sketched in Figure 10.8.

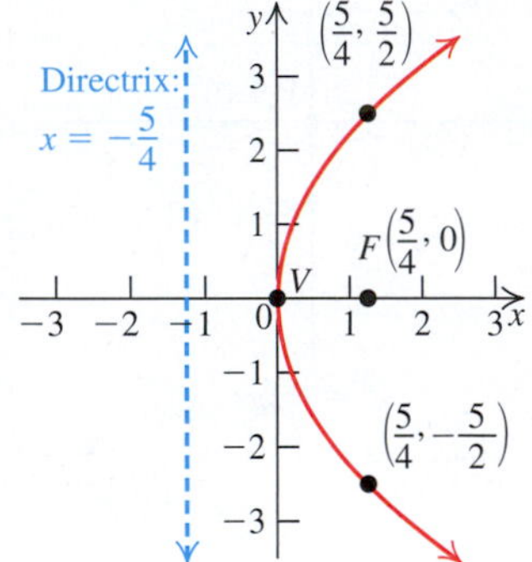

FIGURE 10.8 Graph of $y^2 = 5x$

Practice Problem 1 Graph the parabola and specify the vertex, focus, directrix, and axis.

a. $x^2 = 12y$ **b.** $y^2 = -6x$ 

The line segment joining the focus and the two points symmetric about the axis on a parabola is called the *latus rectum* of the parabola. Intuitively, the length of the latus rectum determines the "size of the opening" of the parabola.

TECHNOLOGY CONNECTION

To graph $y^2 = 4px$, solve the equation for y to obtain

$$y = \pm\sqrt{4px}.$$

Then graph $Y_1 = \sqrt{4px}$ and $Y_2 = -\sqrt{4px}$. For example, to graph $y^2 = 5x$, you graph

$$Y_1 = \sqrt{5x},\ Y_2 = -\sqrt{5x}.$$

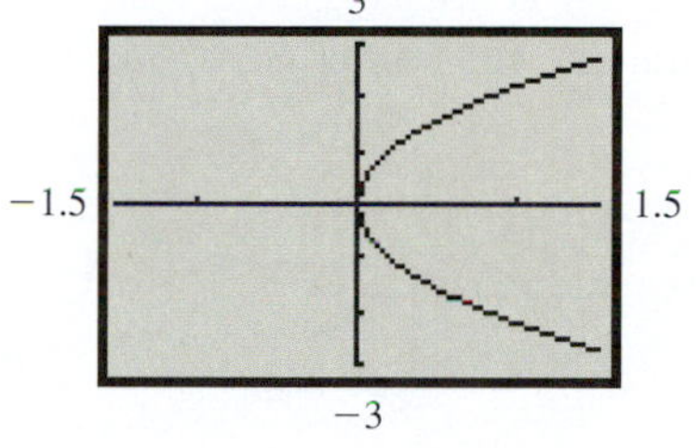

Latus Rectum

The line segment that passes through the focus of a parabola, is perpendicular to the axis, and has endpoints on the parabola is called the **latus rectum of the parabola**. Figure 10.9 shows that the length of the latus rectum for each of the graphs $y^2 = \pm 4px$ and $x^2 = \pm 4py$ for $p > 0$ is $4p$.

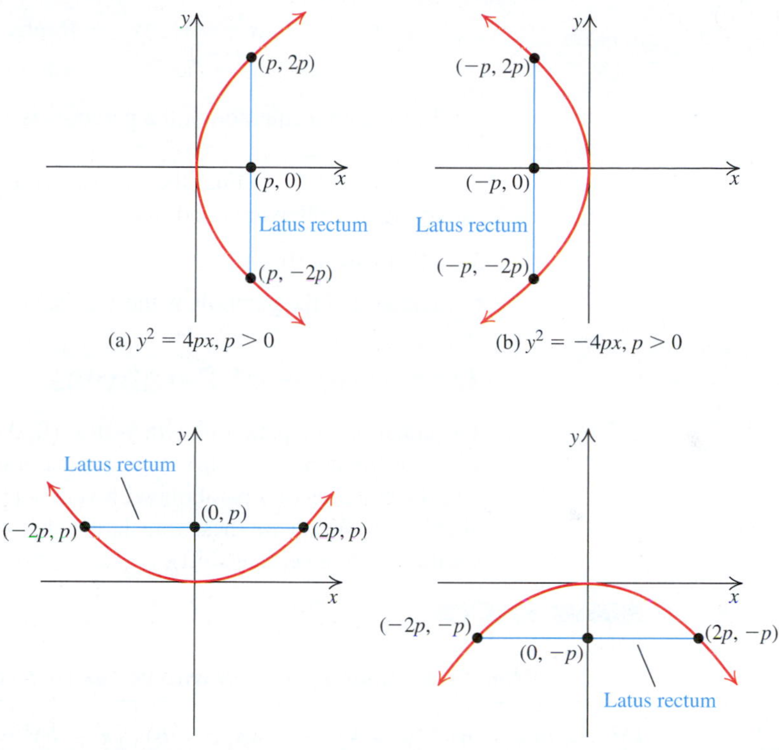

FIGURE 10.9 The latus rectum

EXAMPLE 2 Finding the Equation of a Parabola

Find the standard equation of each parabola with vertex $(0, 0)$ satisfying the given description.

a. The focus is $(-3, 0)$.

b. The axis of the parabola is the y-axis, and the graph passes through the point $(-4, 2)$.

Solution

a. The vertex $(0, 0)$ of the parabola and its focus $(-3, 0)$ are on the x-axis, so the parabola opens to the left. Its equation must be of the form $y^2 = -4px$ with $p = 3$.

$$
\begin{aligned}
y^2 &= -4px && \text{Form of the equation}\\
y^2 &= -4(3)x && \text{Replace } p \text{ with } 3.\\
y^2 &= -12x && \text{Simplify.}
\end{aligned}
$$

The equation of the parabola is $y^2 = -12x$.

b. Because the vertex of the parabola is $(0, 0)$, the axis is the y-axis, and the point $(-4, 2)$ is above the x-axis, the parabola opens up and the equation of the parabola is of the form

$$x^2 = 4py.$$

This equation is satisfied by $x = -4$ and $y = 2$ because the parabola passes through the point $(-4, 2)$.

$$(-4)^2 = 4p(2) \quad \text{Replace } x \text{ with } -4 \text{ and } y \text{ with } 2.$$
$$16 = 8p \quad \text{Simplify.}$$
$$2 = p \quad \text{Divide both sides by 8.}$$

Therefore, the equation of the parabola is

$$x^2 = 4(2)y \quad \text{Replace } p \text{ with 2 in } x^2 = 4py.$$
$$x^2 = 8y \quad \text{Simplify.}$$

The required equation of the parabola is $x^2 = 8y$.

Practice Problem 2 Find the standard equation of each parabola with vertex $(0, 0)$ satisfying the following conditions.

a. The focus is $(0, 2)$.

b. The axis of the parabola is the x-axis, and the graph passes through $(1, 2)$.

Translations of Parabolas

The equation of a parabola with vertex $(0, 0)$ and axis on a coordinate axis is of the form $x^2 = \pm 4py$ or $y^2 = \pm 4px$. You can use translations of the graphs of these equations to find the equation of a parabola with vertex (h, k) and axis of symmetry parallel to a coordinate axis. We obtain these translations by replacing x with $x - h$ (a horizontal shift) and y with $y - k$ (a vertical shift).

SUMMARY OF MAIN FACTS

Main facts about a parabola with vertex (h, k) and $p > 0$

Standard Equation	$(y - k)^2 = 4p(x - h)$	$(y - k)^2 = -4p(x - h)$	$(x - h)^2 = 4p(y - k)$	$(x - h)^2 = -4p(y - k)$
Equation of axis	$y = k$	$y = k$	$x = h$	$x = h$
Description	Opens right	Opens left	Opens up	Opens down
Vertex	(h, k)	(h, k)	(h, k)	(h, k)
Focus	$(h + p, k)$	$(h - p, k)$	$(h, k + p)$	$(h, k - p)$
Directrix	$x = h - p$	$x = h + p$	$y = k - p$	$y = k + p$

Graph

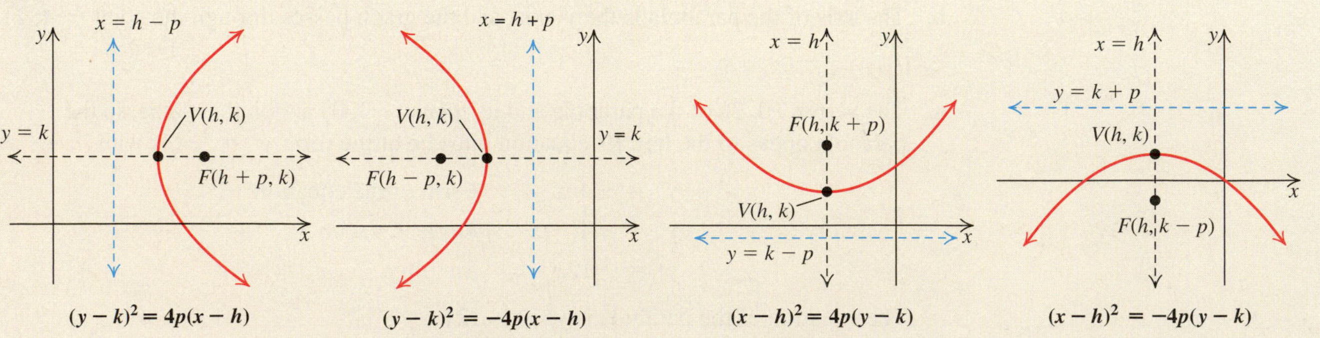

Expanding the standard equations of a parabola and collecting like terms results in an equation of the form

$$Ax^2 + Cy^2 + Dx + Ey + F = 0$$

with $A = 0$ or $C = 0$ but not both. Conversely, except in degenerate cases, any equation of this form can be put into one of the standard forms of a parabola by completing the square on the x- or y-terms.

EXAMPLE 3 Graphing a Parabola

Find the vertex, focus, and directrix of the parabola $2y^2 - 8y - x + 7 = 0$. Sketch the graph of the parabola.

Solution

First, complete the square on y. Then you can compare the resulting equation with one of the standard forms of the equation of a parabola.

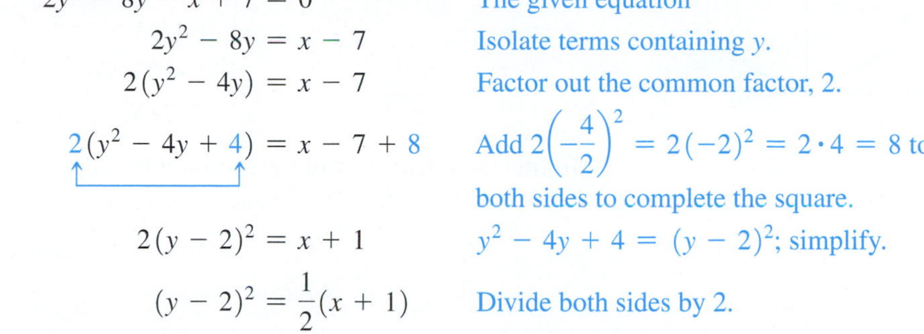

$$2y^2 - 8y - x + 7 = 0 \qquad \text{The given equation}$$

$$2y^2 - 8y = x - 7 \qquad \text{Isolate terms containing } y.$$

$$2(y^2 - 4y) = x - 7 \qquad \text{Factor out the common factor, 2.}$$

$$2(y^2 - 4y + 4) = x - 7 + 8 \qquad \text{Add } 2\left(-\frac{4}{2}\right)^2 = 2(-2)^2 = 2 \cdot 4 = 8 \text{ to both sides to complete the square.}$$

$$2(y - 2)^2 = x + 1 \qquad y^2 - 4y + 4 = (y - 2)^2\text{; simplify.}$$

$$(y - 2)^2 = \frac{1}{2}(x + 1) \qquad \text{Divide both sides by 2.}$$

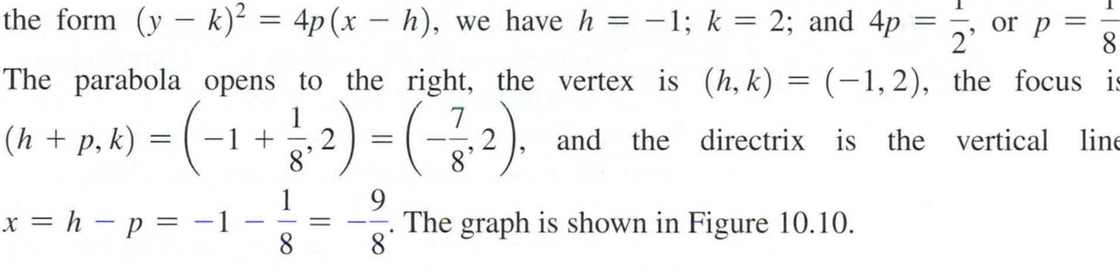

The last equation is in one of the standard forms of the parabola. Comparing it with the form $(y - k)^2 = 4p(x - h)$, we have $h = -1$; $k = 2$; and $4p = \frac{1}{2}$, or $p = \frac{1}{8}$. The parabola opens to the right, the vertex is $(h, k) = (-1, 2)$, the focus is $(h + p, k) = \left(-1 + \frac{1}{8}, 2\right) = \left(-\frac{7}{8}, 2\right)$, and the directrix is the vertical line $x = h - p = -1 - \frac{1}{8} = -\frac{9}{8}$. The graph is shown in Figure 10.10.

FIGURE 10.10 Converting $2y^2 - 8y - x + 7 = 0$ to standard form

Practice Problem 3 Find the vertex, focus, and directrix of the parabola

$$2x^2 - 8x - y + 7 = 0.$$

Sketch the graph of the parabola.

4 Use the reflecting property of parabolas.

Reflecting Property of Parabolas

A property of parabolas that is useful in applications is the **reflecting property**: if a reflecting surface has parabolic cross sections with a common focus, then all light rays entering the surface parallel to the axis will be reflected through the focus. See Figure 10.11(a). This property is used in reflecting telescopes and satellite antennas because the light rays or radio waves bouncing off a parabolic surface are reflected to the focus, where they are collected and amplified.

Conversely, if a light source is located at the focus of a parabolic reflector, the reflected rays will form a beam parallel to the axis. See Figure 10.11(b). This principle is used in flashlights, searchlights, and other such devices.

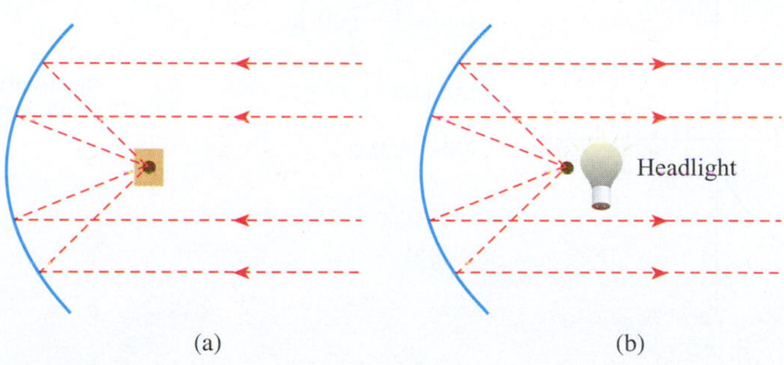

FIGURE 10.11

EXAMPLE 4 Calculating Some Properties of the Hubble Space Telescope

The parabolic mirror used in the Hubble Space Telescope has a diameter of 94.5 inches. See Figure 10.12. Find the equation of the parabola if its focus is 2304 inches from the vertex. What is the thickness of the mirror at the edges?

FIGURE 10.12 Hubble Space Telescope

Solution

Position the parabola so that its vertex is at the origin and its focus is on the positive y-axis. The equation of the parabola is of the form

$$x^2 = 4py$$
$$x^2 = 4(2304)y \qquad p = 2304$$
$$x^2 = 9216y \qquad \text{Simplify.}$$

To find the thickness y of the mirror at the edge, substitute $x = 47.25$ (half the diameter) in the equation $x^2 = 9216y$ and solve for y.

$$(47.25)^2 = 9216y \qquad \text{Replace } x \text{ with } 47.25.$$
$$\frac{(47.25)^2}{9216} = y \qquad \text{Divide both sides by 9216.}$$
$$0.2422485 \approx y \qquad \text{Use a calculator.}$$

Thus, the thickness of the mirror at the edges is approximately 0.2422485 inch.

The Hubble telescope initially had "blurred vision" because the mirror was ground at the edges to a thickness of 0.24224 inch. In other words, it had been ground 0.000008 inch, or two-millionths of a meter, smaller than it should have been. What a difference a rounding error can make! The problem was fixed in 1993.

Practice Problem 4 A small imitation of the Hubble's parabolic mirror has a diameter of 3 inches. Find the equation of the parabola if its focus is 7.3 inches from the vertex. What is the thickness of the mirror at its edges, correct to four decimal places?

Answers to Practice Problems

1. a.

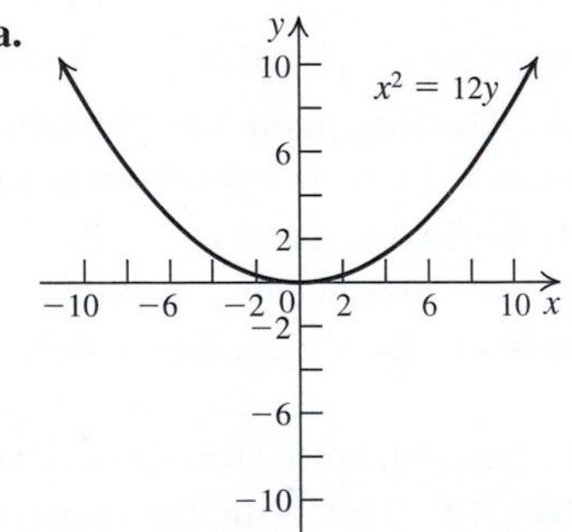

Vertex: $(0, 0)$; focus: $(0, 3)$; directix: $y = -3$; axis: y-axis

b.

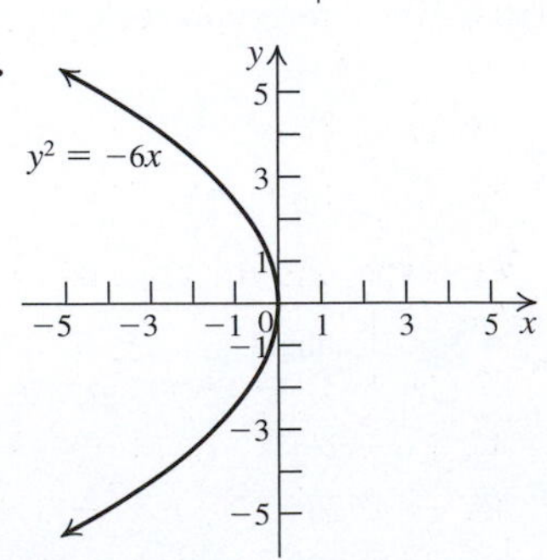

Vertex: $(0, 0)$; focus: $\left(-\frac{3}{2}, 0\right)$; directix: $x = \frac{3}{2}$; axis: x-axis

2. a. $x^2 = 8y$ **b.** $y^2 = 4x$

3. Vertex: $(2, -1)$; focus: $\left(2, -\frac{7}{8}\right)$; directrix: $y = -\frac{9}{8}$

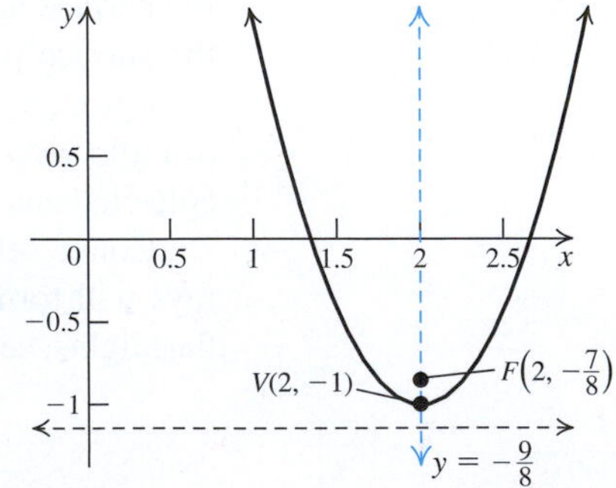

4. Equation: $x^2 = 29.2y$. The thickness of the mirror at the edges is 0.077055 in.

SECTION 10.2 Exercises

Basic Concepts and Skills

1. A parabola is the set of all points P in the plane that are equidistant from a fixed line called the ______________ and a fixed point not on the line called the ______________.
2. The line segment through the focus perpendicular to the directrix with endpoints on the parabola is called the ______________ of the parabola.
3. The point at which the axis intersects the parabola is called the ______________ of the parabola.
4. The graph of $y + 2 = 3(x - 4)^2$ is obtained by shifting the graph of $y = 3x^2$ to the right ______________ units and shifting ______________ 2 units.
5. **True or False.** The vertex of a parabola is midway between its focus and its directrix.
6. **True or False.** A parabola is symmetric about its axis.

In Exercises 7–14, find the focus and directrix of the parabola with the given equation. Then match each equation to one of the graphs labeled (a) through (h).

7. $x^2 = 2y$	**8.** $9x^2 = 4y$
9. $16x^2 = -9y$	**10.** $x^2 = -2y$
11. $y^2 = 2x$	**12.** $9y^2 = 16x$
13. $9y^2 = -16x$	**14.** $y^2 = -2x$

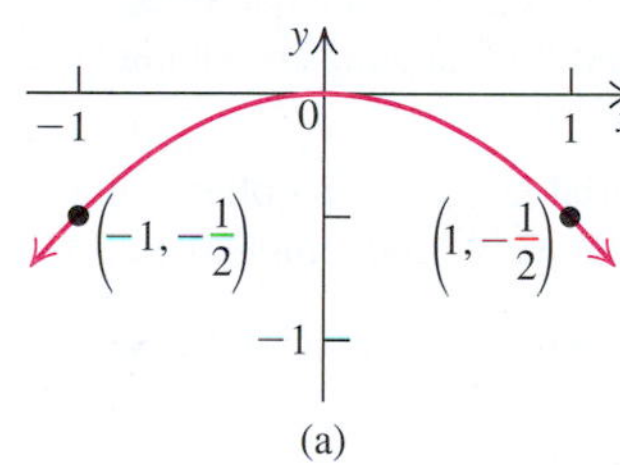

(a)

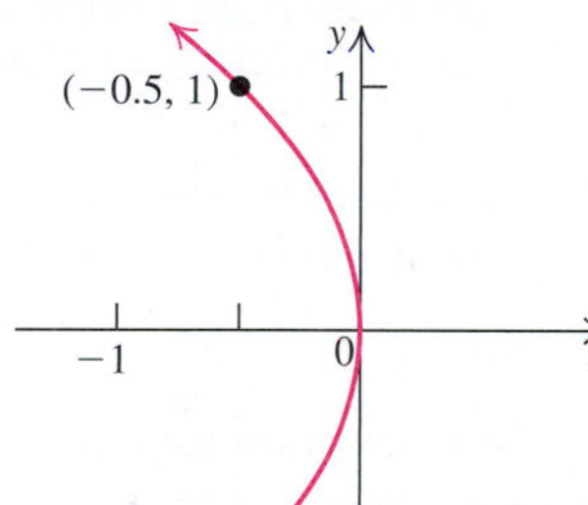

(b)

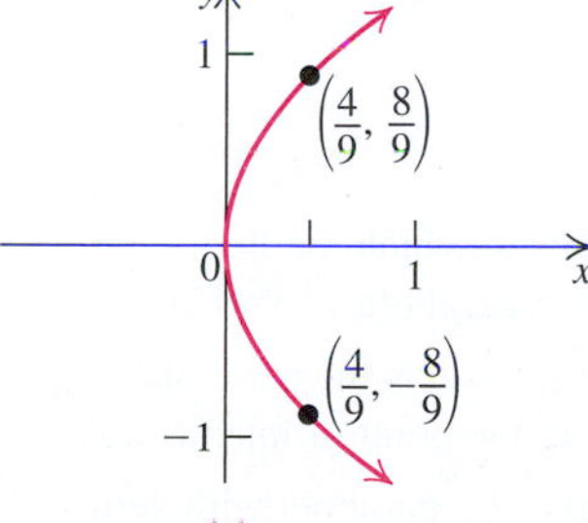

(c)

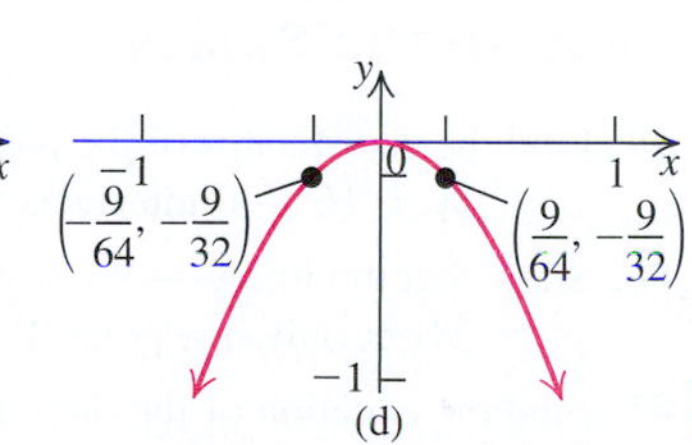

(d)

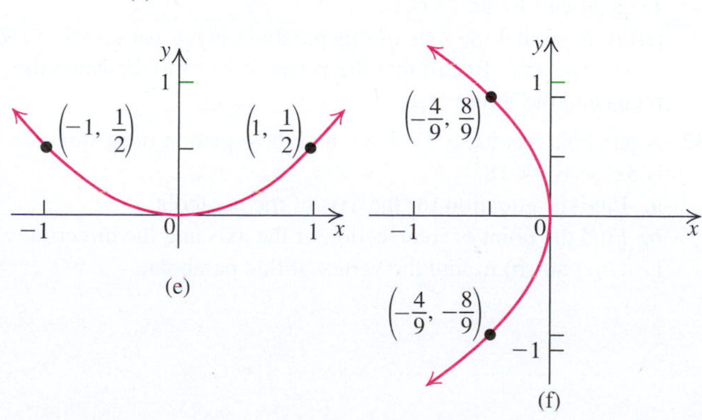

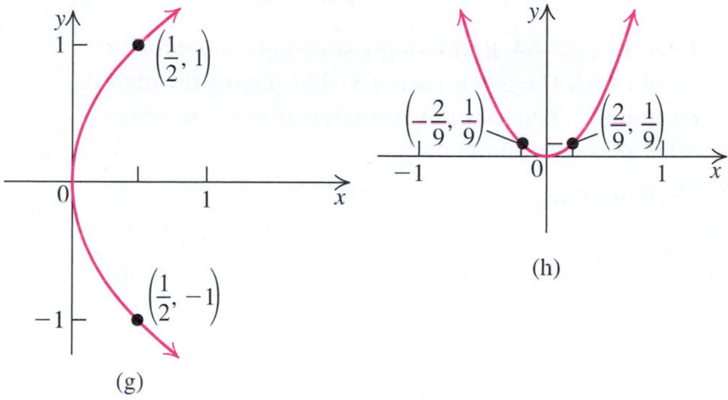

In Exercises 15–20, use a graphing device to graph each parabola.

15. $x^2 - 2y = 0$	**16.** $x^2 = 6y$
17. $y^2 = -3x$	**18.** $y^2 + 5x = 0$
19. $3x + 2y^2 = 0$	**20.** $y^2 - 5x = 0$

In Exercises 21–36, find the standard equation of the parabola that satisfies the given conditions. Also find the length of the latus rectum of each parabola.

21. Focus: $(0, 2)$; directrix: $y = 4$
22. Focus: $(0, 4)$; directrix: $y = -2$
23. Focus: $(-2, 0)$; directrix: $x = 3$
24. Focus: $(-1, 0)$; directrix: $x = -2$
25. Vertex: $(1, 1)$; directrix: $x = 3$
26. Vertex: $(1, 1)$; directrix: $y = 2$
27. Vertex: $(1, 1)$; directrix: $y = -3$
28. Vertex: $(1, 1)$; directrix: $x = -2$
29. Vertex: $(1, 0)$; focus: $(3, 0)$
30. Vertex: $(0, 1)$; focus: $(0, 2)$
31. Vertex: $(0, 1)$; focus: $(0, -2)$
32. Vertex: $(-1, 0)$; focus: $(-3, 0)$
33. Vertex: $(2, 3)$; directrix: $x = 4$
34. Vertex: $(2, 3)$; directrix: $y = 4$
35. Vertex: $(2, 3)$; directrix: $y = 1$
36. Vertex: $(2, 3)$; directrix: $x = 1$

In Exercises 37–40, use a graphing device to graph each parabola.

37. $(x + 1) = 4(y - 1)^2$	**38.** $(x - 2) = -4(y - 1)^2$
39. $(x - 1)^2 = -9(y + 1)$	**40.** $(x + 2)^2 = 6(y + 1)$

In Exercises 41–56, find the vertex, focus, and directrix of each parabola. Graph the equation.

41. $(y - 1)^2 = 2(x + 1)$	**42.** $(y + 2)^2 = 8(x - 3)$
43. $(x + 2)^2 = 3(y - 2)$	**44.** $(x - 3)^2 = 4(y + 1)$
45. $(y + 1)^2 = -6(x - 2)$	**46.** $(y - 2)^2 = -12(x + 3)$
47. $(x - 1)^2 = -10(y - 3)$	**48.** $(x + 2)^2 = -8(y + 3)$

49. $y = x^2 + 2x + 2$

50. $y = 3x^2 + 6x + 2$

51. $2y^2 + 4y - 2x + 1 = 0$

52. $3y^2 - 6y + x - 1 = 0$

53. $y + x^2 + x + \frac{5}{4} = 0$

54. $x = 8y^2 + 8y$

55. $x = 24y - 3y^2 - 2$

56. $y = 16x - 2x^2 + 3$

In Exercises 57–64, find two possible equations for the parabola with the given vertex *V* that passes through the given point *P*. The axis of symmetry of each parabola is parallel to a coordinate axis.

57. $V(0, 0), P(1, 2)$

58. $V(0, 0), P(-3, 2)$

59. $V(0, 1), P(2, 3)$

60. $V(1, 2), P(2, 1)$

61. $V(-2, 1), P(-3, 0)$

62. $V(1, -1), P(0, 0)$

63. $V(-1, 1), P(0, 2)$

64. $V(2, 3), P(3, -1)$

Applying the Concepts

65. **Satellite dish.** A parabolic satellite dish is 40 inches across and 20 inches deep (from the vertex to the plane of the rim). How far from the vertex must the signal-receiving (receptor) unit be located to ensure that it is at the focus of the parabolic cross sections?

66. Repeat Exercise 65 assuming that the dish is 8 feet across and 3 feet deep.

67. **Solar heating.** A parabolic reflector mirror is to be used for the solar heating of water. The mirror is 18 feet across and 6 feet deep. Where should the heating element be placed to heat the water most rapidly?

68. **Flashlights.** A parabolic flashlight reflector is to be 4 inches across and 2 inches deep. Where should the lightbulb be placed?

69. **Flashlight.** The shape of a parabolic flashlight reflector can be described by the equation $x = 4y^2$. Where should the lightbulb be placed?

70. Repeat Exercise 69 assuming that the parabolic flashlight reflector has the shape described by the equation $x = \frac{1}{2}y^2$.

71. **Satellite dish.** The shape of a parabolic satellite dish can be described by the equation $y = 4x^2$. Where should the microphone be placed?

72. Repeat Exercise 71 assuming that the shape of the dish can be described by the equation $y = x^2$.

73. **Suspension bridge.** The ends of a suspension bridge cable are fastened to two towers that are 800 feet apart and 120 feet high. Assume that the cable hangs in the shape of a parabola and touches the roadbed midway between the towers. Find the length of the straight wire needed to suspend the roadbed from the cable at a distance of 250 feet from the tower.

74. **Suspension bridge.** Repeat Exercise 73 assuming that the suspension bridge cable is at a height of 8 feet above the midway between the towers.

75. **Kicking a football.** A football is kicked from the ground 70 yards along a parabolic path. The maximum height of the ball is 30 yards. Find the height of the ball 5 yards before it hits the ground.

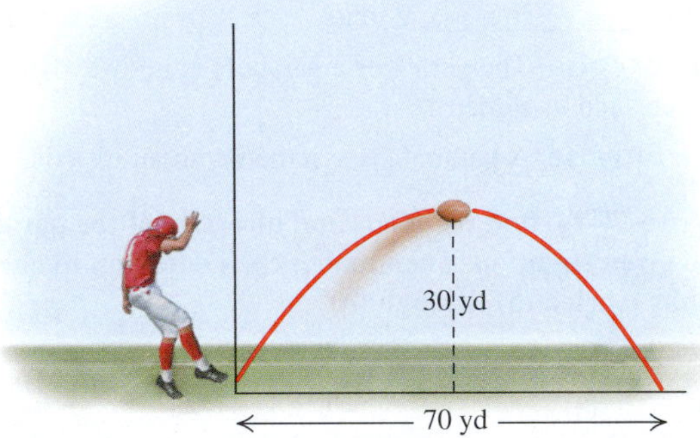

76. **Parabolic arch bridge.** A parabolic arch of a bridge has a 60-foot base and a height of 24 feet. Find the height of the arch at distances of 5, 10, and 20 feet from the center of the base.

77. **Variable cost.** The average variable cost y, in dollars, of a monthly output of x tons of a company producing a metal is given by $y = \frac{1}{10}x^2 - 3x + 50$. Find the output and cost at the vertex of the parabola.

78. Repeat Exercise 77 assuming that the variable cost is given by $y = \frac{1}{8}x^2 - 2x + 60$.

Beyond the Basics

79. Find the coordinates of the points at which the line $2x - 3y + 16 = 0$ intersects the parabola $y^2 = 16x$.

80. Show that the line $y = 2x + 3$ intersects the parabola $y^2 = 24x$ at only one point. Find the point of intersection.

81. Find the equation of the directrix of a parabola with vertex $(-2, 2)$ and focus $(-6, 6)$.
[*Hint*: Note that the axis of this parabola is *not* parallel to a coordinate axis. Recall that the vertex is midway between the focus and the directrix.]

82. A parabola has focus $(-4, 5)$, and the equation of its directrix is $3x - 4y = 18$.
   a. Find the equation for the axis of the parabola.
   b. Find the point of intersection of the axis and the directrix.
   c. Use part (b) to find the vertex of this parabola.

83. Find an equation of two possible parabolas whose latus rectum is the line segment joining the points $(3, 5)$ and $(3, -3)$.

84. Repeat Exercise 83 assuming that the latus rectum is the line segment joining the points $(-5, 1)$ and $(3, 1)$.

85. Find an equation of the parabola with axis parallel to the y-axis, passing through the points $(0, 5)$, $(1, 4)$, and $(2, 7)$.

86. Find an equation of the parabola with axis parallel to the x-axis, passing through the points $(-1, -1)$, $(3, 1)$, and $(4, 0)$.

87. Find the vertex, focus, directrix, and axis of the parabola

$$x^2 - 8x + 2y + 4 = 0.$$

88. Repeat Exercise 87 for the parabola

$$Ax^2 + Dx + Ey + F = 0, \quad A \neq 0, E \neq 0.$$

89. Find the vertex, focus, directrix, and axis of the parabola

$$y^2 + 3x - 6y + 15 = 0.$$

90. Repeat Exercise 89 for the parabola

$$Cy^2 + Dx + Ey + F = 0, \quad C \neq 0, D \neq 0.$$

In Exercises 91–94, use the following definition: A *tangent line* to a parabola is a line that is not parallel to the axis of the parabola and that intersects the parabola at exactly one point.

91. Find an equation of the tangent line at the point $(1, 3)$ on the parabola $y = 3x^2$ by using the following steps.

Step 1 Let m be the slope of the tangent line. Then the equation of the tangent line is $y - 3 = m(x - 1)$. Solve this equation for y.

Step 2 Substitute the value of y from Step 1 in the equation $y = 3x^2$. You will obtain a quadratic equation in x.

Step 3 For the line in Step 1 to be the tangent line to the parabola, the roots of the quadratic equation in Step 2 must be equal. Set discriminant = 0 and solve for m.

Step 4 Use the value of m from Step 3 to write the equation of the tangent line to the parabola $y = 3x^2$.

92. Show that an equation of the tangent line at the point (x_1, y_1) on the parabola $y = 4ax^2$ is $y = 8ax_1x - y_1$. [*Hint*: Follow the steps in the Exercise 91 and note that $y_1 = 4ax_1^2$.]

93. Show that an equation of the tangent line at the point (x_1, y_1) on the parabola $x = 4ay^2$ is $y = \dfrac{1}{8ay_1}x + \dfrac{y_1}{2}$.

94. Reflection property of a parabola. Let $P(x_1 y_1)$ be a point on the parabola $y = 4ax^2$ with focus $F = (0, a)$ and the directrix $y = -a$. Let AP be the tangent line to the parabola. See the figure.

a. Use Exercise 92 to show that $d(F, A) = a + y_1$. Then use the definition of the parabola to show that $d(P, F) = a + y_1$. Conclude that $\angle\beta = \angle\gamma$.

This will suggest the following construction of the tangent line: Let B be a point on the axis of the parabola such that BP is perpendicular to the axis. Let V be the vertex of the parabola. Find the point A on the axis of the parabola such that $d(B, V) = d(A, V)$. Then the line AP is the tangent line to the parabola at the point P.

b. Use geometry to show that the angle between the y-axis and the line AP equals the angle between the lines AP and the line $x = x_1$ (which is parallel to the axis of parabola).

This demonstrates the reflection property of a parabola, which says that the tangent line to a parabola at a point P makes equal angles with

1. The line through P and the focus.
2. The axis of the parabola.

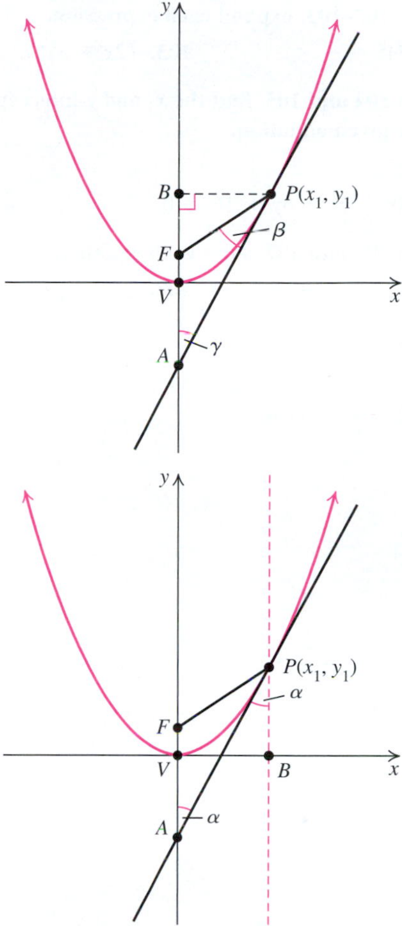

Critical Thinking/Discussion/Writing

95. a. Is a parabola always the graph of a function? Why or why not?

b. Find all possible values of the slope of the directrix of a parabola such that the parabola is always the graph of a function.

96. The equation $y^2 = 4kx$ describes a family of curves for each value of k. Sketch the graphs of the members of this family for $k = \pm 1, \pm 2, \pm 3$. What is the common characteristic of this family?

97. True or False. The graph of $x = ay^2 + by + c$ $(a \neq 0)$ is a parabola whose axis is parallel to the y-axis.

98. It can be proved that the distance d from a point (x_1, y_1) to a line $ax + by + c = 0$ is given by

$$d = \frac{|ax_1 + by_1 + c|}{\sqrt{a^2 + b^2}}$$

Use this fact and the definition of a parabola to find an equation of the parabola whose focus is $(4, 5)$ and whose directrix is the line $3x + 4y = 7$. Find an equation for the axis of this parabola.

99. Given that the vertex of a parabola is $(6, -3)$ and the directrix is the line $3x - 5y + 1 = 0$, find the coordinates of the focus.

Maintaining Skills

In Exercises 100 and 101, find the distance between the given points.

100. $(-1, 3)$ and $(6, 2)$ **101.** $(-1, 2)$ and $(-1, -5)$

In Exercises 102–103, expand each expression.

102. $(x - 5)^2$ **103.** $(2x + 3)^2$

In Exercises 104 and 105, find the *x*- and *y*-intercepts of the graph of the given equation.

104. $x^2 + x - 2y = 2$

105. $x^2 + 8x + y^2 - 6y + 16 = 0$

In Exercises 106 and 107, solve each equation.

106. $\sqrt{6x + 7} = x + 2$ **107.** $2 + \sqrt{3x - 5} = x - 1$

In Exercises 108 and 109, rewrite each expression in the form $a(x - b)^2 + c$ by completing the square.

108. $x^2 - 6x + 7$

109. $4x^2 - 8x + 14$

In Exercises 110 and 111, use the horizontal and vertical shifts of the parabola with the given equation to find an equation of a parabola with the specified vertex.

110. $y = 4x^2$, vertex $(1, -2)$

111. $y = (x + 2)^2 + 3$, vertex $(-1, 2)$

SECTION 10.3

The Ellipse

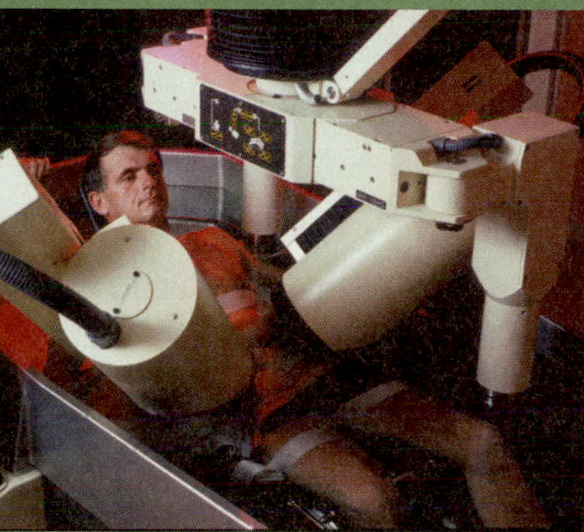

Lithotripter

BEFORE STARTING THIS SECTION, REVIEW

1. Distance formula (Section 2.1, page 159)
2. Completing the square (Section 3.1, page 295)
3. Midpoint formula (Section 2.1, page 161)
4. Transformations of graphs (Section 2.7, page 251)
5. Symmetry (Section 2.2, page 168)

OBJECTIVES

1. Define an ellipse.
2. Find the equation of an ellipse.
3. Translate ellipses.
4. Use ellipses in applications.

What Is Lithotripsy?

Lithotripsy is a combination of two words: *litho* and *tripsy*. In Greek, the word *lith* denotes a stone and the word *tripsy* means "crushing." The complete medical term for lithotripsy is *extracorporeal shock-wave lithotripsy* (ESWL). Lithotripsy is a medical procedure in which a kidney stone is crushed into small sandlike pieces with the help of ultrasound high-energy shock waves, without breaking the skin.

The lithotripsy machine is called a *lithotripter*. The technology for the lithotripter was developed in Germany. The first successful treatment of a patient by a lithotripter was in February 1980 at Munich University. In the beginning, the patient had to lie in an elliptical tank filled with water and only small kidney stones were crushed. But as the medical research advanced, newer machines with better technology were manufactured. Now there is no need to keep the patient unconscious with an empty stomach, nor is the tub of water required.

In Example 5, you will see how the *reflecting property* of an ellipse is used in lithotripsy.

1 Define an ellipse.

Definition of Ellipse

Ellipse

An **ellipse** is the set of all points in the plane, the sum of whose distances from two fixed points is a constant. The fixed points are called the **foci** (the plural of *focus*) of the ellipse.

You can draw an ellipse with the use of thumbtacks, a fixed length of string, and a pencil. Stick a thumbtack at each of the two points (the foci) and tie one end of the string to each tack. Place a pencil inside the loop of the string and pull it taut. Move the pencil, keeping the string taut at all times. The pencil traces an ellipse, as shown in Figure 10.13.

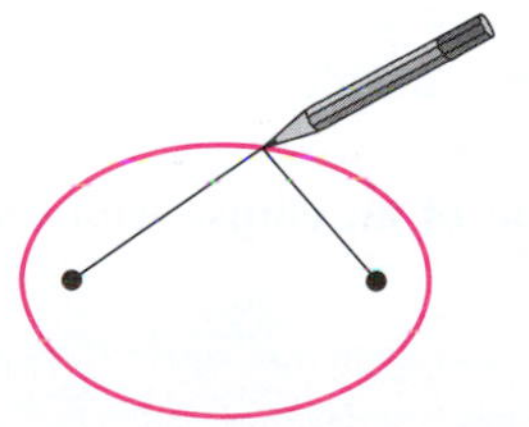

FIGURE 10.13 Drawing an ellipse

The points of intersection of the ellipse with the line through the foci are called the **vertices** (plural of *vertex*). The line segment connecting the vertices is the **major axis** of the ellipse. The midpoint of the major axis is the **center** of the ellipse. The line segment that is perpendicular to the major axis at the center and with endpoints on the ellipse is called the **minor axis**. See Figure 10.14.

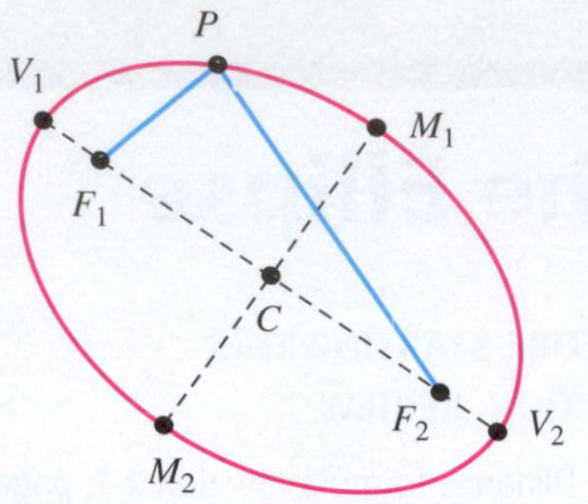

Points F_1 and F_2 are the foci.
Points V_1 and V_2 are the vertices.
Point C is the center.
Segment $\overline{V_1 V_2}$ is the major axis.
Segment $\overline{M_1 M_2}$ is the minor axis.
The sum of the distances $PF_1 + PF_2$ from any point P on the ellipse to the foci is constant.

FIGURE 10.14

2 Find the equation of an ellipse.

Equation of an Ellipse

To find a simple equation of an ellipse, we place the ellipse's major axis along one of the coordinate axes and the center of the ellipse at the origin. We will begin with the case in which the major axis lies along the x-axis. Let $F_1(-c, 0)$ and $F_2(c, 0)$ be the coordinates of the foci. See Figure 10.15. Note that the center at the origin is the midpoint of the line segment having the foci as endpoints. To simplify the equation, it is customary to let $2a$ be the constant distance referred to in the definition. Let $P(x, y)$ be a point on the ellipse. Then

$$d(P, F_1) + d(P, F_2) = 2a \quad \text{Definition of ellipse}$$
$$\sqrt{(x+c)^2 + y^2} + \sqrt{(x-c)^2 + y^2} = 2a \quad \text{Distance formula}$$
$$\sqrt{(x+c)^2 + y^2} = 2a - \sqrt{(x-c)^2 + y^2} \quad \text{Isolate a radical.}$$
$$(x+c)^2 + y^2 = \left(2a - \sqrt{(x-c)^2 + y^2}\right)^2 \quad \text{Square both sides.}$$
$$x^2 + 2cx + c^2 + y^2 = 4a^2 - 4a\sqrt{(x-c)^2 + y^2} + (x-c)^2 + y^2 \quad \text{Expand squares.}$$
$$x^2 + 2cx + c^2 + y^2 = 4a^2 - 4a\sqrt{(x-c)^2 + y^2} + x^2 - 2cx + c^2 + y^2 \quad \text{Expand squares.}$$
$$4cx - 4a^2 = -4a\sqrt{(x-c)^2 + y^2} \quad \text{Isolate the radical and simplify.}$$
$$cx - a^2 = -a\sqrt{(x-c)^2 + y^2} \quad \text{Divide both sides by 4.}$$
$$(cx - a^2)^2 = a^2[(x-c)^2 + y^2] \quad \text{Square both sides.}$$
$$c^2x^2 - 2a^2cx + a^4 = a^2(x^2 - 2cx + c^2 + y^2) \quad \text{Expand squares.}$$
$$(c^2 - a^2)x^2 - a^2y^2 = a^2c^2 - a^4 \quad \text{Simplify and rearrange.}$$
$$(a^2 - c^2)x^2 + a^2y^2 = a^2(a^2 - c^2) \quad (1) \quad \text{Multiply both sides by } -1 \text{ and factor } a^4 - a^2c^2.$$

Now consider triangle F_1PF_2 in Figure 10.15. The *triangle inequality* from geometry states that the sum of the lengths of any two sides of a triangle is greater than the length of the third side; so for this triangle, we have

$$d(P, F_1) + d(P, F_2) > d(F_1, F_2)$$
$$2a > 2c \quad 2a = d(P, F_1) + d(P, F_2);\ d(F_1, F_2) = 2c$$
$$a > c \quad \text{Divide both sides by 2.}$$
$$a^2 > c^2 \quad \text{Square both sides.}$$

y
$P(x, y)$
C
$V_2(a, 0)$
$V_1(-a, 0)$ $F_1(-c, 0)$ $F_2(c, 0)$ x

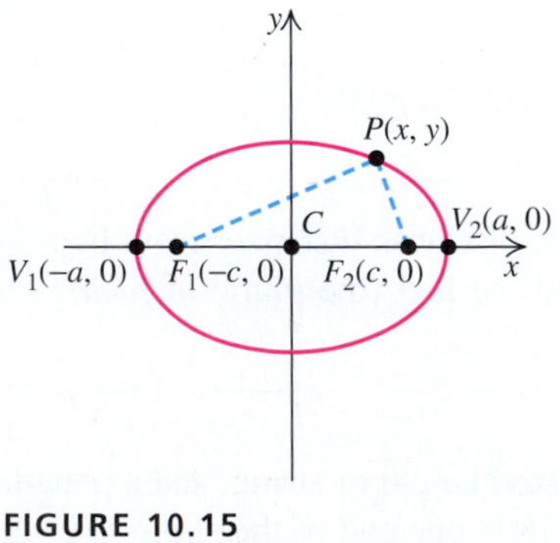

FIGURE 10.15

Therefore, $a^2 - c^2 > 0$. Letting $b^2 = a^2 - c^2$, we obtain

$$(a^2 - c^2)x^2 + a^2y^2 = a^2(a^2 - c^2) \quad \text{Equation (1)}$$
$$b^2x^2 + a^2y^2 = a^2b^2 \quad \text{Replace } a^2 - c^2 \text{ with } b^2.$$
$$\frac{x^2}{a^2} + \frac{y^2}{b^2} = 1 \quad (2) \quad \text{Divide both sides by } a^2b^2.$$

The last equation is the **standard form of the equation of an ellipse** with center $(0, 0)$ and foci $(-c, 0)$ and $(c, 0)$, where $b^2 = a^2 - c^2$.

To find the x-intercepts for the graph of this ellipse, set $y = 0$ in equation (2) to get $\frac{x^2}{a^2} = 1$. The solutions are $x = \pm a$, so the x-intercepts of the graph are $-a$ and a.

Therefore, the ellipse's vertices are $(-a, 0)$ and $(a, 0)$. The distance between the vertices $V_1(-a, 0)$ and $V_2(a, 0)$ is $2a$, so the length of the major axis is $2a$.

The y-intercepts of the ellipse (found by setting $x = 0$ in equation (2)) are $-b$ and b, so the endpoints of the minor axis are $(0, -b)$ and $(0, b)$ Consequently, the length of the minor axis is $2b$. See the left-hand figure in the table below.

Similarly, by reversing the roles of x and y, an equation of the ellipse with center $(0, 0)$ and foci $(0, -c)$ and $(0, c)$ on the y-axis is given by

$$\frac{x^2}{b^2} + \frac{y^2}{a^2} = 1, \text{ where } b^2 = a^2 - c^2 \qquad (3)$$

In equation (3), the major axis, of length $2a$, is along the y-axis and the minor axis, of length $2b$, is along the x-axis. See the right-hand figure in the table below.

If the major axis of an ellipse is along or parallel to the x-axis, the ellipse is called a **horizontal ellipse**, an ellipse with major axis along or parallel to the y-axis is called a **vertical ellipse**.

We summarize the facts about horizontal and vertical ellipses with center $(0, 0)$. By the equation of a major or minor axis, we mean the equation of the line on which that axis lies.

SUMMARY OF MAIN FACTS

Main facts about an ellipse with center (0, 0)

Standard Equation	$\frac{x^2}{a^2} + \frac{y^2}{b^2} = 1; a > b > 0$ **(Horizontal ellipse)**	$\frac{x^2}{b^2} + \frac{y^2}{a^2} = 1; a > b > 0$ **(Vertical ellipse)**
Relationship between a, b, and c	$b^2 = a^2 - c^2$	$b^2 = a^2 - c^2$
Major axis along	x-axis	y-axis
Length of major axis	$2a$	$2a$
Minor axis along	y-axis	x-axis
Length of minor axis	$2b$	$2b$
Vertices	$(\pm a, 0)$	$(0, \pm a)$
Foci	$(\pm c, 0)$	$(0, \pm c)$
Endpoints of minor axis	$(0, \pm b)$	$(\pm b, 0)$
Symmetry	The graph is symmetric about the x-axis, y-axis, and origin.	The graph is symmetric about the x-axis, y-axis, and origin.
Graph	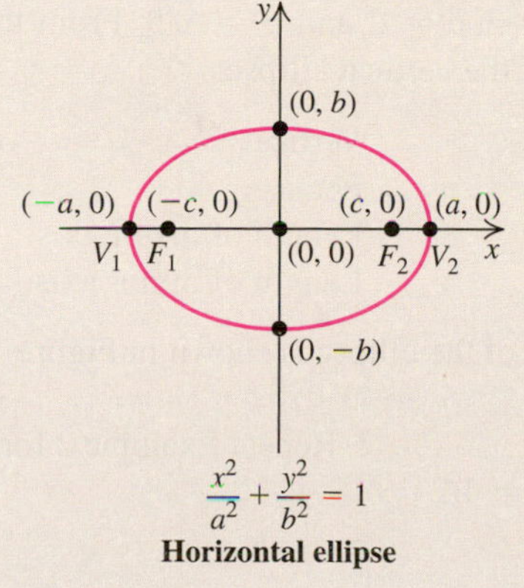 $\frac{x^2}{a^2} + \frac{y^2}{b^2} = 1$ **Horizontal ellipse**	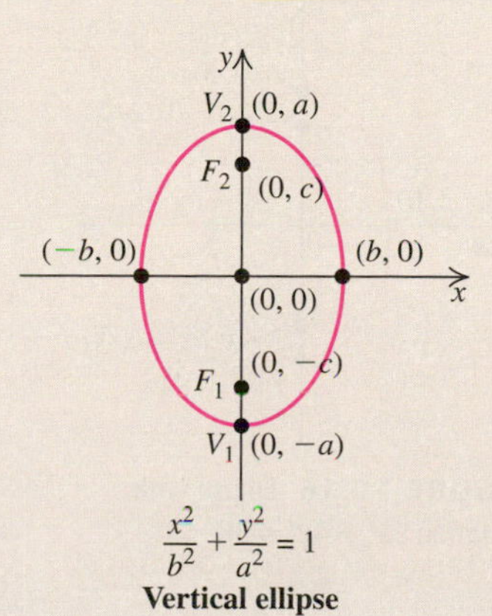 $\frac{x^2}{b^2} + \frac{y^2}{a^2} = 1$ **Vertical ellipse**

Notice that if $a = b$, then $a^2 = b^2$ and equations (2) and (3) of an ellipse are equivalent to $x^2 + y^2 = a^2$; this is the equation for a circle with center at (0, 0) and radius a. Consequently, a circle is a special case of an ellipse.

EXAMPLE 1 Finding an Equation of an Ellipse

Find the standard form of the equation of the ellipse that has vertex $(5, 0)$ and foci $(\pm 4, 0)$.

Solution

Because the foci are $(-4, 0)$ and $(4, 0)$, the major axis is on the x-axis. We know that $c = 4$ and $a = 5$. Then find b^2:

$$b^2 = a^2 - c^2 \qquad \text{Relationship between } a, b, \text{ and } c$$
$$b^2 = (5)^2 - (4)^2 = 9 \qquad \text{Replace } c \text{ with 4 and } a \text{ with 5.}$$

Substituting 25 for a^2 and 9 for b^2 in the standard form for a horizontal ellipse, we get

$$\frac{x^2}{25} + \frac{y^2}{9} = 1.$$

Practice Problem 1 Find the standard form of the equation of the ellipse that has vertex $(0, 10)$ and foci $(0, -8)$ and $(0, 8)$.

TECHNOLOGY CONNECTIONS

To graph the ellipse $\frac{x^2}{4} + \frac{y^2}{9} = 1$, solve the equation for y:

$$\frac{y^2}{9} = 1 - \frac{x^2}{4}$$

or $y^2 = 9\left(1 - \frac{x^2}{4}\right)$

so that $y = \pm 3\sqrt{1 - \frac{x^2}{4}}$.

Then graph both functions

$Y_1 = 3\sqrt{1 - \frac{x^2}{4}}$, and

$Y_2 = -3\sqrt{1 - \frac{x^2}{4}}$.

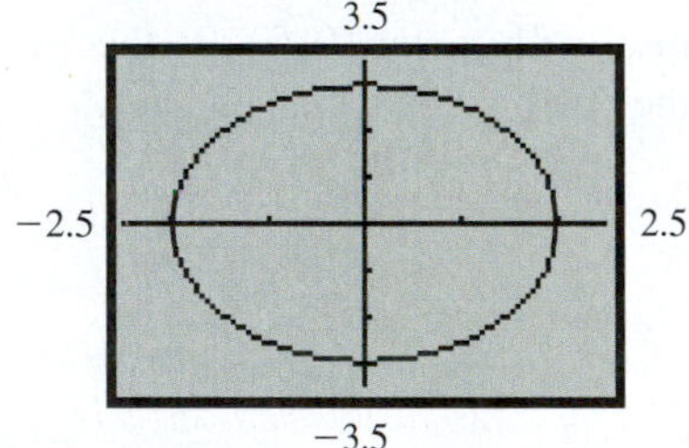

EXAMPLE 2 Graphing an Ellipse

Graph the ellipse whose equation is $9x^2 + 4y^2 = 36$. Find the foci of the ellipse and the lengths of its axes.

Solution

First, write the equation in standard form.

$$9x^2 + 4y^2 = 36 \qquad \text{Given equation}$$
$$\frac{9x^2}{36} + \frac{4y^2}{36} = 1 \qquad \text{Divide both sides by 36.}$$
$$\frac{x^2}{4} + \frac{y^2}{9} = 1 \qquad \text{Simplify.}$$

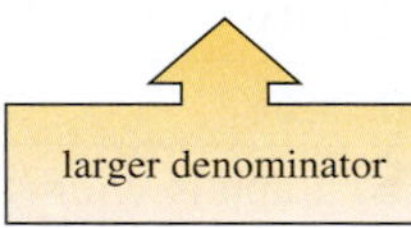

Because the denominator in the y^2-term is larger than the denominator in the x^2-term, the ellipse is a vertical ellipse. Here $a^2 = 9$ and $b^2 = 4$, so $c^2 = a^2 - b^2 = 9 - 4 = 5$. Thus, $a = 3, b = 2$ and $c = \sqrt{5}$. From the table on page 877, we know the following features of the vertical ellipse:

Vertices:	$(0, \pm a) = (0, \pm 3)$
Foci:	$(0, \pm c) = (0, \pm\sqrt{5})$
Length of major axis:	$2a = 2(3) = 6$
Length of minor axis:	$2b = 2(2) = 4$

The graph of the ellipse is shown in Figure 10.16.

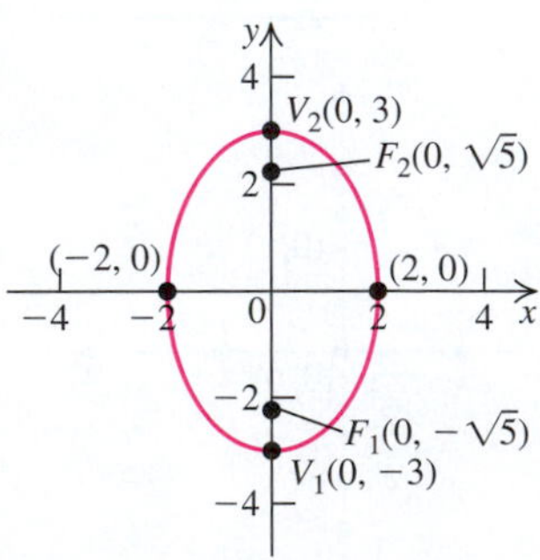

FIGURE 10.16 Ellipse with equation $9x^2 + 4y^2 = 36$

Practice Problem 2 Repeat Example 2 for the ellipse whose equation is $4x^2 + y^2 = 16$.

3 Translate ellipses.

Translations of Ellipses

The graph of the equation

$$\frac{x^2}{a^2} + \frac{y^2}{b^2} = 1$$

is an ellipse with center $(0, 0)$ and major axis along a coordinate axis. As in the case of a parabola, horizontal and vertical shifts can be used to obtain the graph of an ellipse whose equation is

$$\frac{(x-h)^2}{a^2} + \frac{(y-k)^2}{b^2} = 1.$$

The center of such an ellipse is (h, k), and its axes are parallel to the coordinate axes.

SUMMARY OF MAIN FACTS

Main facts about horizontal and vertical ellipses with center (h, k)

Standard Equation	$\frac{(x-h)^2}{a^2} + \frac{(y-k)^2}{b^2} = 1;$ $a > b > 0$ **(Horizontal ellipse)**	$\frac{(x-h)^2}{b^2} + \frac{(y-k)^2}{a^2} = 1;$ $a > b > 0$ **(Vertical ellipse)**
Center	(h, k)	(h, k)
Major axis along the line	$y = k$	$x = h$
Length of major axis	$2a$	$2a$
Minor axis along the line	$x = h$	$y = k$
Length of minor axis	$2b$	$2b$
Vertices	$(h + a, k), (h - a, k)$	$(h, k + a), (h, k - a)$
Endpoints of minor axis	$(h, k - b), (h, k + b)$	$(h - b, k), (h + b, k),$
Foci	$(h + c, k), (h - c, k)$	$(h, k + c), (h, k - c)$
Relations between a, b, and c	$c^2 = a^2 - b^2$	$c^2 = a^2 - b^2$
Symmetry	The graph is symmetric about the lines $x = h$ and $y = k$.	The graph is symmetric about the lines $x = h$ and $y = k$.
Graph	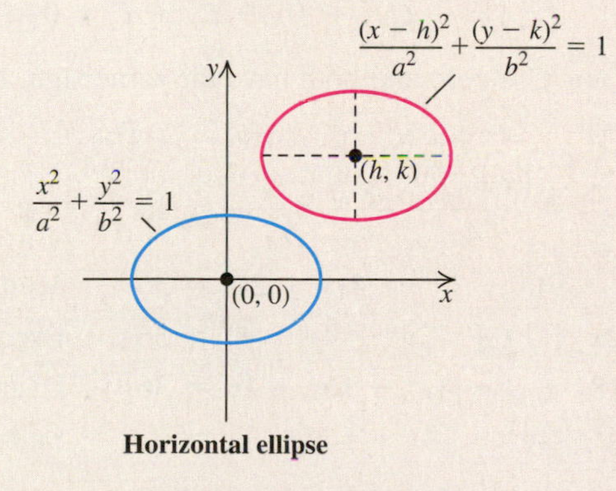 Horizontal ellipse	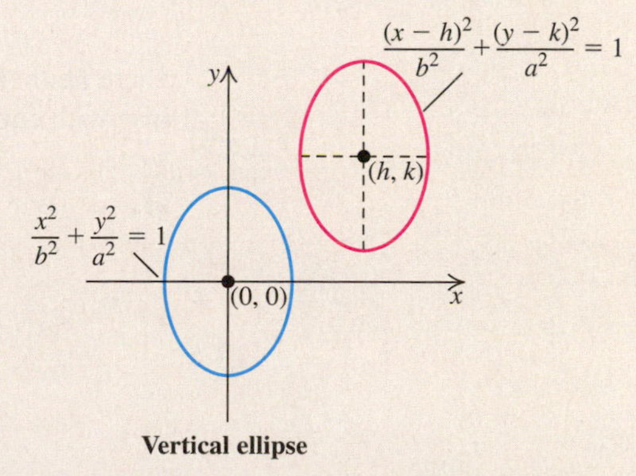 Vertical ellipse

RECALL

The midpoint of the segment joining (x_1, y_1) and (x_2, y_2) is

$$\left(\frac{x_1 + x_2}{2}, \frac{y_1 + y_2}{2}\right).$$

EXAMPLE 3 Finding the Equation of an Ellipse

Find an equation of the ellipse that has foci $(-3, 2)$ and $(5, 2)$ and that has a major axis of length 10.

Solution

Because the foci $(-3, 2)$ and $(5, 2)$ lie on the horizontal line $y = 2$, the ellipse is a horizontal ellipse. The center of the ellipse is the midpoint of the line segment joining the foci. We have $k = 2$ and using the midpoint formula, we obtain

$$h = \frac{-3 + 5}{2} = 1.$$

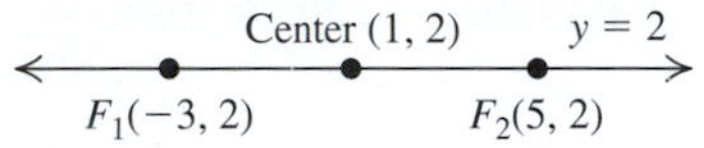

Center (1, 2) $y = 2$

$F_1(-3, 2)$ $F_2(5, 2)$

FIGURE 10.17 Finding the center given the foci

The center of the ellipse is $(1, 2)$. See Figure 10.17. Because the length of the major axis is 10, the vertices must be at a distance $a = 5$ units from the center.

In Figure 10.17, the foci are 4 units from the center. Thus, $c = 4$. Now find b^2:

$$b^2 = a^2 - c^2 = (5)^2 - (4)^2 = 9$$

Because the major axis is horizontal, the standard form of the equation of the ellipse is

$$\frac{(x - h)^2}{a^2} + \frac{(y - k)^2}{b^2} = 1 \qquad a > b > 0$$

$$\frac{(x - 1)^2}{25} + \frac{(y - 2)^2}{9} = 1 \qquad \text{Replace } h \text{ with 1, } k \text{ with 2, } a^2 \text{ with 25, and } b^2 \text{ with 9.}$$

For this horizontal ellipse with center $(1, 2)$ we have $a = 5$, $b = 3$, and $c = 4$.

Vertices:

$$(h \pm a, k) = (1 \pm 5, 2) = (-4, 2) \text{ and } (6, 2)$$

Endpoints of Minor Axis:

$$(h, k \pm b) = (1, 2 \pm 3) = (1, -1) \text{ and } (1, 5)$$

Using the center and these four points, we sketch the graph shown in Figure 10.18.

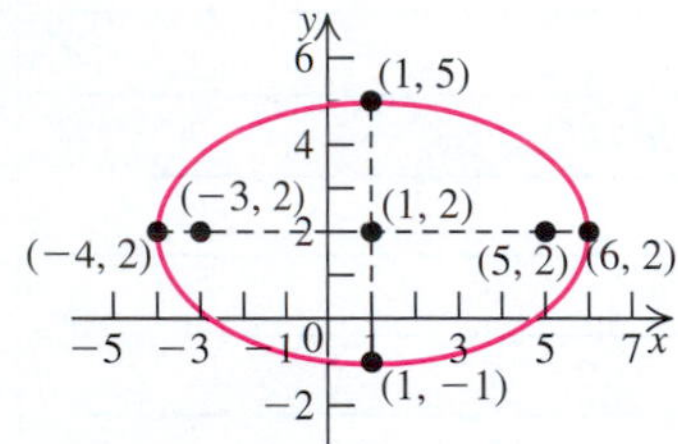

FIGURE 10.18 Ellipse with foci (−3, 2) and (5, 2), and major axis 10.

Practice Problem 3 Find an equation of the ellipse that has foci at $(2, -3)$ and $(2, 5)$ and has a major axis of length 10.

Both standard forms of the equations of an ellipse with center (h, k) can be written in the form:

$$Ax^2 + Cy^2 + Dx + Ey + F = 0 \qquad (4)$$

where neither A nor C is zero and both have the same sign. For example, consider the ellipse with equation

$$\frac{(x - 1)^2}{4} + \frac{(y + 2)^2}{9} = 1,$$

$$9(x - 1)^2 + 4(y + 2)^2 = 36 \qquad \text{Multiply both sides by 36.}$$

$$9(x^2 - 2x + 1) + 4(y^2 + 4y + 4) = 36 \qquad \text{Expand squares.}$$

$$9x^2 - 18x + 9 + 4y^2 + 16y + 16 = 36 \qquad \text{Distributive property}$$

$$9x^2 + 4y^2 - 18x + 16y - 11 = 0 \qquad \text{Simplify.}$$

Comparing this equation with equation (4), we have

$$A = 9, C = 4, D = -18, E = 16, \text{ and } F = -11.$$

Conversely, equation (4) can be written in a standard form for an ellipse by completing the squares in both of the variables x and y.

EXAMPLE 4 **Converting to Standard Form**

Find the center, vertices, and foci of the ellipse with equation

$$3x^2 + 4y^2 + 12x - 8y - 32 = 0.$$

Solution

Complete squares on x and y.

$$3x^2 + 4y^2 + 12x - 8y - 32 = 0 \quad \text{Original equation}$$
$$(3x^2 + 12x) + (4y^2 - 8y) = 32 \quad \text{Group like terms.}$$
$$3(x^2 + 4x) + 4(y^2 - 2y) = 32 \quad \text{Factor out 3 and 4.}$$
$$3(x^2 + 4x + 4) + 4(y^2 - 2y + 1) = 32 + 12 + 4 \quad \text{Complete squares on } x \text{ and } y\text{: add } 3 \cdot 4 \text{ and } 4 \cdot 1 \text{ to each side.}$$
$$3(x + 2)^2 + 4(y - 1)^2 = 48 \quad \text{Factor and simplify.}$$
$$\frac{(x + 2)^2}{16} + \frac{(y - 1)^2}{12} = 1 \quad \text{Divide both sides by 48. (16 is the larger denominator.)}$$

The last equation is the standard form for a horizontal ellipse with center $(-2, 1)$, $a^2 = 16$, $b^2 = 12$, and $c^2 = a^2 - b^2 = 16 - 12 = 4$. Thus, $a = 4$, $b = \sqrt{12} = 2\sqrt{3}$, and $c = 2$.

From the table on page 879, we have the following information:
The length of the major axis is $2a = 8$.
The length of the minor axis is $2b = 4\sqrt{3}$.

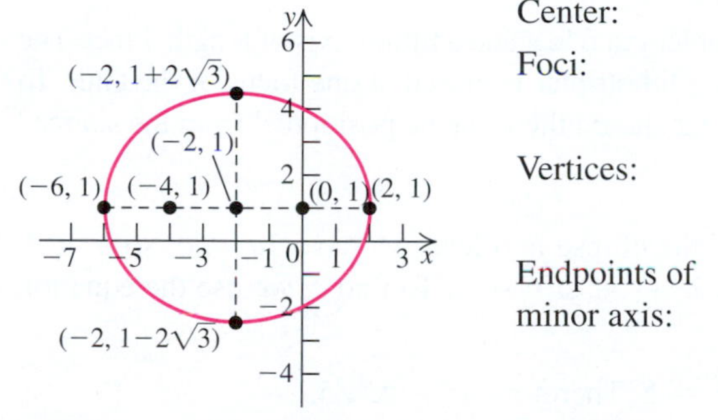

FIGURE 10.19 Ellipse with equation $3x^2 + 4y^2 + 12x - 8y - 32 = 0$

Center: $(h, k) = (-2, 1)$

Foci: $(h \pm c, k) = (-2 \pm 2, 1) = (-4, 1)$ and $(0, 1)$

Vertices: $(h \pm a, k) = (-2 \pm 4, 1) = (-6, 1)$ and $(2, 1)$

Endpoints of minor axis: $(h, k \pm b) = (-2, 1 \pm 2\sqrt{3}) = (-2, 1 + 2\sqrt{3})$ and $(-2, 1 - 2\sqrt{3}) \approx (-2, 4.46)$ and $(-2, -2.46)$

The graph of the ellipse is shown in Figure 10.19.

Practice Problem 4 Find the center, vertices, and foci of the ellipse with equation

$$x^2 + 4y^2 - 6x + 8y - 29 = 0.$$

A *circle* is considered an ellipse in which the two foci coincide at the center. Conversely, if the two foci of an ellipse coincide, then $2c = 0$ (remember, $2c$ is the distance between the two foci of an ellipse); so $c = 0$ Again, from the relationship $c^2 = a^2 - b^2 = 0$ we have $a = b$. Thus, the equation for the ellipse becomes the equation for a circle of radius $r = a$ having the same center as that of the ellipse.

4 Use ellipses in applications.

Applications

Ellipses have many applications. We mention just a few.

1. The orbits of the planets are ellipses with the sun at one focus. This fact alone is sufficient to explain the overwhelming importance of ellipses since the 17th century, when Kepler and Newton did their monumental work in astronomy.
2. Newton also reasoned that comets move in elliptical orbits about the sun. His friend Edmund Halley used this information to predict that a certain comet (now called Halley's comet) reappears about every 77 years. Halley saw the comet in 1682 and

Edmund Halley (1656–1742)

Edmund Halley was a friend of Isaac Newton. Although Halley is known primarily for calculating the orbit of Halley's comet, he was also a biologist, a geologist, a sea captain, a spy, and the author of the first actuarial mortality tables.

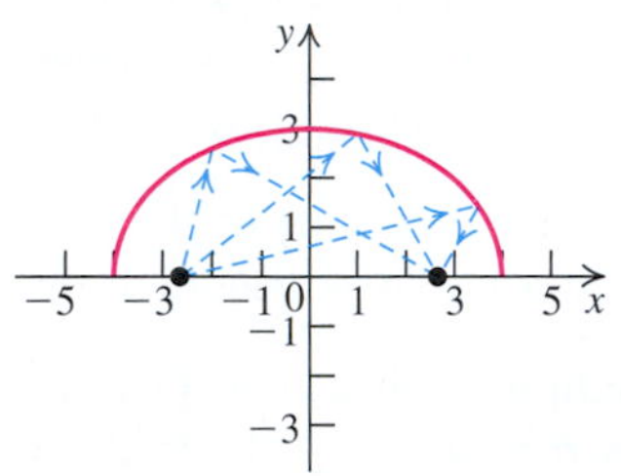

FIGURE 10.20 Reflective property

correctly predicted its return in 1759. The last sighting of the comet was in 1986, and it is due to return in 2061. A recent study shows that Halley's comet has made approximately 2000 cycles so far and has about the same number to go before the sun erodes it away completely. See Exercise 73.

3. We can calculate the distance that a planet travels in one orbit around the sun.
4. **The reflecting property of ellipses.** The *reflecting property* for an ellipse says that a ray of light originating at one focus will be reflected to the other focus. See Figure 10.20. Sound waves also follow such paths. This property is used in the construction of "whispering galleries," such as the gallery at St. Paul's Cathedral in London. Such rooms have ceilings whose cross sections are elliptical with common foci. As a result, sounds emanating from one focus are reflected by the ceiling to the other focus. Thus, a whisper at one focus may not be audible at all at a nearby place, but may nevertheless be clearly heard far off at the other focus.

Example 5 illustrates how the *reflecting property* of an ellipse is used in lithotripsy. Recall from the introduction to this section that a *lithotripter* is a machine designed to crush kidney stones into small sandlike pieces with the help of high-energy shock waves. To focus these shock waves accurately, a lithotripter uses the reflective property of ellipses. A patient is carefully positioned so that the kidney stone is at one focus of the ellipsoid (a three-dimensional ellipse) and the shock-producing source is placed at the other focus. The high-energy shock waves generated at this focus are concentrated on the kidney stone at the other focus, thus pulverizing it without harming the patient.

EXAMPLE 5 Lithotripsy

An elliptical water tank has a major axis of length 6 feet and a minor axis of length 4 feet. The source of high-energy shock waves from a lithotripter is placed at one focus of the tank. To smash the kidney stone of a patient, how far should the stone be positioned from the source?

Solution

Because the length of the major axis of the ellipse is 6 feet, we have $2a = 6$; so $a = 3$. Similarly, the minor axis of 4 feet gives $2b = 4$, or $b = 2$. To find c, we use the equation $c^2 = a^2 - b^2$. We have

$$c^2 = (3)^2 - (2)^2 = 5. \text{ Therefore, } c = \pm\sqrt{5}.$$

If we position the center of the ellipse at $(0, 0)$ and the major axis along the x-axis, then the foci of the ellipse are $(-\sqrt{5}, 0)$ and $(\sqrt{5}, 0)$. The distance between these foci is $2\sqrt{5} \approx 4.472$ feet. The kidney stone should be positioned 4.472 feet from the source of the shock waves.

Practice Problem 5 In Example 5, if the tank has a major axis of 8 feet and a minor axis of 4 feet, how far should the stone be positioned from the source?

Answers to Practice Problems

1. $\dfrac{x^2}{36} + \dfrac{y^2}{100} = 1$

2.

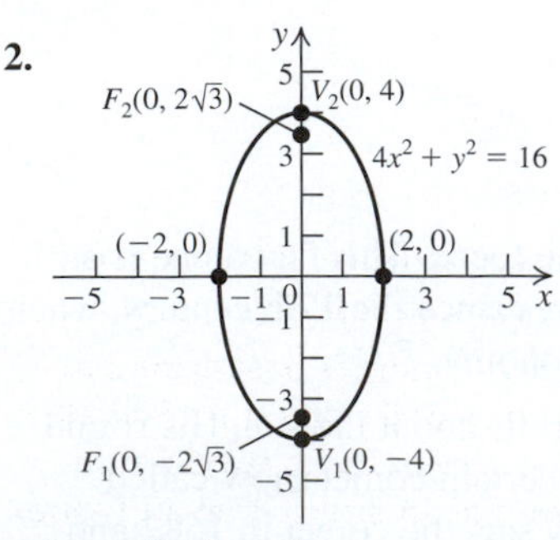

$2a = 8; 2b = 4$

3. $\dfrac{(x-2)^2}{9} + \dfrac{(y-1)^2}{25} = 1$

4. Center: $(3, -1)$; vertices: $(3 + \sqrt{42}, -1)$ and $(3 - \sqrt{42}, -1)$; foci: $\left(3 + \dfrac{3\sqrt{14}}{2}, -1\right)$ and $\left(3 - \dfrac{3\sqrt{14}}{2}, -1\right)$

5. $4\sqrt{3} \approx 6.9282$ ft

SECTION 10.3 Exercises

Basic Concepts and Skills

1. An ellipse is the set of all points in the plane, the ____________ of whose distances from two fixed points is a constant.
2. The points of intersection of the ellipse with the line through the foci are called ____________ of the ellipse.
3. The standard equation of an ellipse with center $(0, 0)$ vertices $(\pm a, 0)$ and foci $(\pm c, 0)$ is $\frac{x^2}{a^2} + \frac{y^2}{b^2} = 1$, where $b^2 =$ ____________.
4. In the equation $\frac{x^2}{m^2} + \frac{y^2}{n^2} = 1$
   (i) if $m^2 > n^2$, the graph is a(n) ____________ ellipse;
   (ii) if $m^2 < n^2$, the graph is a(n) ____________ ellipse;
   (iii) if $m^2 = n^2$, the graph is a(n) ____________.
5. **True or False.** The ellipse $\frac{x^2}{a^2} + \frac{y^2}{b^2} = 1$ lies inside the box formed by the four lines $x = \pm a, y = \pm b$.
6. **True or False.** The graph of $Ax^2 + Cy^2 + Dx + Ey + F = 0$ is a degenerate conic or an ellipse if $AC > 0$.

In Exercises 7–26, find the vertices and foci for each ellipse. Graph each equation.

7. $\frac{x^2}{16} + \frac{y^2}{4} = 1$
8. $\frac{x^2}{4} + \frac{y^2}{16} = 1$
9. $\frac{x^2}{9} + y^2 = 1$
10. $x^2 + \frac{y^2}{4} = 1$
11. $\frac{x^2}{25} + \frac{y^2}{16} = 1$
12. $\frac{x^2}{25} + \frac{y^2}{9} = 1$
13. $\frac{x^2}{16} + \frac{y^2}{36} = 1$
14. $\frac{x^2}{9} + \frac{y^2}{16} = 1$
15. $x^2 + y^2 = 4$
16. $x^2 + y^2 = 16$
17. $x^2 + 4y^2 = 4$
18. $9x^2 + y^2 = 9$
19. $9x^2 + 4y^2 = 36$
20. $4x^2 + 25y^2 = 100$
21. $3x^2 + 4y^2 = 12$
22. $4x^2 + 5y^2 = 20$
23. $5x^2 - 10 = -2y^2$
24. $3y^2 = 21 - 7x^2$
25. $2x^2 + 3y^2 = 7$
26. $3x^2 + 4y^2 = 11$

In Exercises 27–40, find the standard form of the equation for the ellipse satisfying the given conditions. Graph the equation.

27. Foci: $(\pm 1, 0)$; vertex $(3, 0)$
28. Foci: $(\pm 3, 0)$; vertex $(5, 0)$
29. Foci: $(0, \pm 2)$; vertex $(0, 4)$
30. Foci: $(0, \pm 3)$; vertex $(0, -6)$
31. Foci: $(\pm 4, 0)$; y-intercepts: ± 3
32. Foci: $(\pm 3, 0)$; y-intercepts: ± 4
33. Foci: $(0, \pm 2)$ x-intercepts: ± 4
34. Foci: $(0, \pm 3)$; x-intercepts: ± 5
35. Center: $(0, 0)$; vertical major axis of length 10; minor axis of length 6
36. Center: $(0, 0)$; vertical major axis of length 8; minor axis of length 4
37. Center: $(0, 0)$; vertices: $(\pm 6, 0)$; $c = 3$
38. Center: $(0, 0)$; vertices: $(0, \pm 5)$; $c = 3$
39. Center: $(0, 0)$; foci: $(0, \pm 2)$; $b = 3$
40. Center: $(0, 0)$; foci: $(\pm 3, 0)$; $b = 2$

In Exercises 41–56, find the center, foci, and vertices of each ellipse. Graph each equation.

41. $\frac{(x-1)^2}{4} + \frac{(y-1)^2}{9} = 1$
42. $\frac{(x-1)^2}{4} + \frac{(y-1)^2}{16} = 1$
43. $\frac{x^2}{16} + \frac{(y+3)^2}{4} = 1$
44. $\frac{(x+2)^2}{4} + \frac{y^2}{9} = 1$
45. $3(x-1)^2 + 4(y+2)^2 = 12$
46. $9(x+1)^2 + 4(y-2)^2 = 36$
47. $4(x+3)^2 + 5(y-1)^2 = 20$
48. $25(x-2)^2 + 4(y+5)^2 = 100$
49. $5x^2 + 9y^2 + 10x - 36y - 4 = 0$
50. $9x^2 + 4y^2 + 72x - 24y + 144 = 0$
51. $9x^2 + 5y^2 + 36x - 40y + 71 = 0$
52. $3x^2 + 4y^2 + 12x - 16y - 32 = 0$
53. $x^2 + 2y^2 - 2x + 4y + 1 = 0$
54. $2x^2 + 2y^2 - 12x - 8y + 3 = 0$
55. $2x^2 + 9y^2 - 4x + 18y + 12 = 0$
56. $3x^2 + 2y^2 + 12x - 4y + 15 = 0$

In Exercises 57–60, use a graphing utility to sketch the graph of each ellipse.

57. $\frac{x^2}{3} + \frac{y^2}{5} = 1$
58. $\frac{x^2}{6} + \frac{y^2}{4} = 1$
59. $x^2 + 3y^2 = 21$
60. $5x^2 + y^2 = 10$

Applying the Concepts

61. **Semielliptical arch.** A portion of an arch in the shape of a *semiellipse* (half an ellipse) is 50 feet wide at the base and has a maximum height of 20 feet. Find the height of the arch at a distance of 10 feet from the center. See the figure.

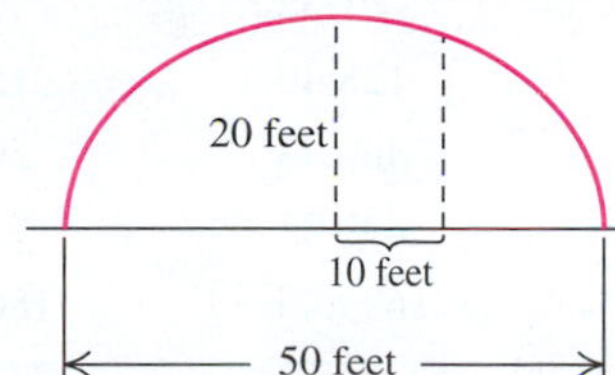

62. In Exercise 61, find the distance between the two points at the base where the height of the arch is 10 feet.
63. **Semielliptical arch bridge.** A bridge is built in the shape of a semielliptical arch. The span of the bridge is 150 meters, and its maximum height is 45 meters. There are two vertical supports, each at a distance of 25 meters from the central position. Find their heights.

64. Elliptical billiard table. Some billiard tables (similar to pool tables, but without pockets) are manufactured in the shape of an ellipse. The foci of such tables are plainly marked for the convenience of the players. An elliptical billiard table is 6 feet long and 4 feet wide. Find the location of the foci.

65. "Ellipti-pool." In Exercise 64, suppose there is a pocket at one of the foci. A pool shark places the ball at the table and bets that he will be able to make the ball fall into the pocket 10 out of 10 times. Explain why you should or should not accept the bet.

66. Whispering gallery. The vertical cross section of the dome of Statuary Hall in Washington, D.C., is semielliptical. The hall is 96 feet long and 23 feet high. It is said that John C. Calhoun used the whispering-gallery phenomenon to eavesdrop on his adversaries. Where should Calhoun and his foes have been standing for the maximum whispering effect?

67. Whispering gallery. The vertical cross section of the dome of the Mormon Tabernacle in Salt Lake City, Utah, is semielliptical in shape. The cross section is 250 feet long with a maximum height of 80 feet. Where should two people stand to maximize the whispering effect?

In Exercises 68–71, use the fact (discovered by Johannes Kepler in 1609) that the planets move in elliptical orbits with the Sun at one of the foci. Astronomers have measured the *perihelion* (the smallest distance from the planet to the Sun) and *aphelion* (the largest distance from the planet to the Sun) for each of the planets. The distances given in Table 10.1 are in millions of miles.

TABLE 10.1

Planet	Perihelion	Aphelion
Mercury	28.56	43.88
Venus	66.74	67.68
Earth	91.38	94.54
Mars	128.49	154.83
Jupiter	460.43	506.87
Saturn	837.05	936.37
Uranus	1699.45	1866.59
Neptune	2771.72	2816.42

68. Mars. Find an equation of Mars's orbit about the Sun.

[*Hint*: Set up a coordinate system with the sun at one focus and the major axis lying on the x-axis. See the figure. Calculate a from the equation $2a$ = aphelion + perihelion. Calculate c from the equation $c = a -$ perihelion. Calculate b^2 from the equation $b^2 = a^2 - c^2$. Write the equation $\frac{x^2}{a^2} + \frac{y^2}{b^2} = 1.$]

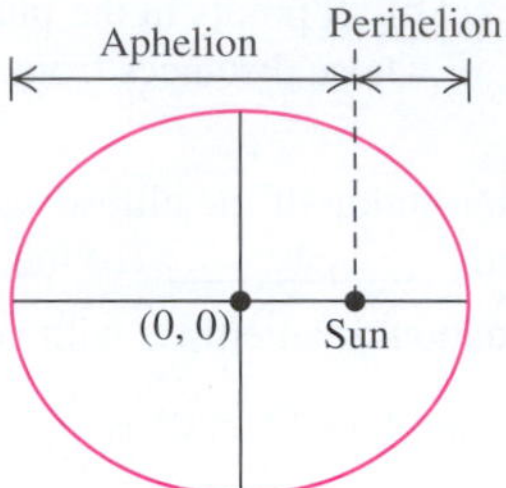

69. Earth. Write an equation for the orbit of Earth about the Sun.

70. Mercury. Write an equation for the orbit of the planet Mercury about the Sun.

71. Saturn. Write an equation for the orbit of the planet Saturn about the Sun.

72. Moon. The moon orbits Earth in an elliptical path with Earth at one focus. The major and minor axes of the orbit have lengths of 768,806 kilometers and 767,746 kilometers, respectively. Find the perihelion and aphelion of the orbit.

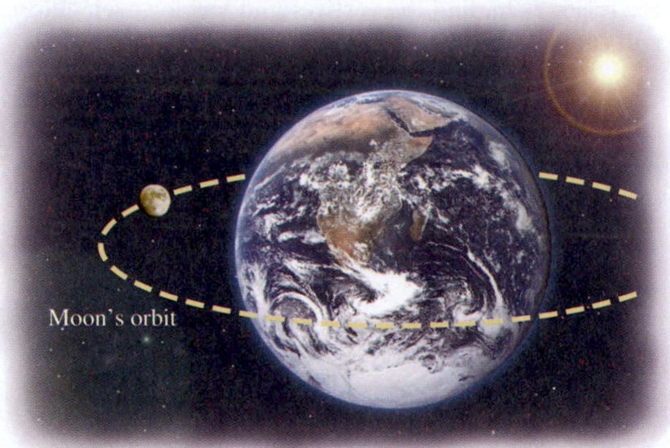

73. Halley's comet. In 1682, Edmund Halley (1656–1742) identified the orbit of a certain comet (now called Halley's comet) as an elliptical orbit about the Sun, with the sun at one focus. He calculated the length of its major axis to be approximately 5.39×10^9 kilometers and its minor axis to be approximately 1.36×10^9 kilometers. Find the perihelion and aphelion of the orbit of Halley's comet.

74. Orbit of a satellite. Assume that Earth is a sphere with a radius of 3960 miles. A satellite has an elliptical orbit around Earth with the center of Earth as one of its foci. The maximum and minimum heights of the satellite from the surface of Earth are 164 miles and 110 miles, respectively. Find an equation for the orbit of the satellite.

Beyond the Basics

75. Find an equation of the ellipse in standard form with center $(0, 0)$ and with major axis of length 10 passing through the point $\left(-3, \frac{16}{5}\right)$.

76. Find an equation of the ellipse in standard form with center $(0, 0)$ and with major axis of length 6 passing through the point $\left(1, \frac{3\sqrt{5}}{2}\right)$.

77. Find the lengths of the major and minor axes of the ellipse

$$\frac{x^2}{b^2} + \frac{y^2}{a^2} = 1,$$

given that the ellipse passes through the points $(2, 1)$ and $(1, -3)$.

78. Find an equation of the ellipse with center $(2, 1)$ and passing through the points $(10, 1)$ and $(6, 2)$.

79. The **eccentricity** of an ellipse, denoted by e, is defined as $e = \frac{\text{distance between the foci}}{\text{distance between the vertices}} = \frac{2c}{2a} = \frac{c}{a}$. What happens when $e = 0$?

In Exercises 80–84, use the definition of *e* given in Exercise 79 to determine the eccentricity of each ellipse. (This *e* has nothing to do with the number *e*, discussed in Chapter 3, that is the base for natural logarithms.)

80. $\frac{x^2}{16} + \frac{y^2}{9} = 1$

81. $20x^2 + 36y^2 = 720$

82. $x^2 + 4y^2 = 1$

83. $\frac{(x+1)^2}{25} + \frac{(y-2)^2}{9} = 1$

84. $x^2 + 2y^2 - 2x + 4y + 1 = 0$

85. An ellipse has its major axis along the x-axis and its minor axis along the y-axis. Its eccentricity is $\frac{1}{2}$, and the distance between the foci is 4. Find an equation of the ellipse.

86. Find an equation of the ellipse whose major axis has endpoints $(-3, 0)$ and $(3, 0)$ and that has eccentricity $\frac{1}{3}$.

87. A point $P(x, y)$ moves so that its distance from the point $(4, 0)$ is always one-half its distance from the line $x = 16$ Show that the equation of the path of P is an ellipse of eccentricity $\frac{1}{2}$.

88. A point, $P(x, y)$ moves so that its distance from the point $(0, 2)$ is always one-third its distance from the line $y = 10$ Show that the equation of the path of P is an ellipse of eccentricity $\frac{1}{3}$.

89. The **latus rectum of an ellipse** is a line segment that passes through a focus and that is perpendicular to the major axis and has endpoints on the ellipse. Show that the length of the latus rectum of an ellipse is $\frac{2b^2}{a}$, where b is half the length of the minor axis and a is half the length of the major axis.

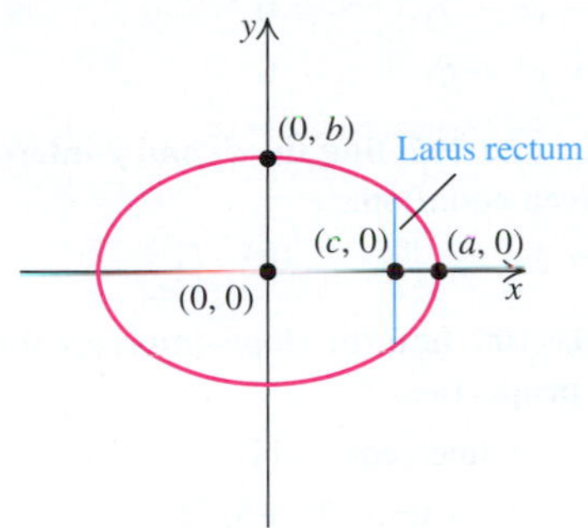

90. Find an equation of the ellipse with center $(0, 0)$, major axis along the x-axis, eccentricity $\frac{1}{\sqrt{2}}$ and latus rectum of length 3 units.

In Exercises 91–94, find all points of intersection of the given curves by solving systems of nonlinear equations. Sketch the graphs of the curves and show the points of intersection.

91. $\begin{cases} x + 3y = -2 \\ 4x^2 + 3y^2 = 7 \end{cases}$

92. $\begin{cases} x^2 = 2y \\ 2x^2 + y^2 = 12 \end{cases}$

93. $\begin{cases} x^2 + y^2 = 20 \\ 9x^2 + y^2 = 36 \end{cases}$

94. $\begin{cases} 9x^2 + 16y^2 = 36 \\ 18x^2 + 5y^2 = 45 \end{cases}$

In Exercises 95 and 96, use the following definition: A *tangent line* to an ellipse is a line that intersects the ellipse at exactly one point.

95. Find the equation of the tangent line to the ellipse with equation $\frac{x^2}{a^2} + \frac{y^2}{b^2} = 1$, with $a = \frac{5}{2}$ and $b = 5$ at the point $(2, 3)$
[*Hint*: See the steps for Exercise 91 in Section 10.2 and (in Step 3) write the discriminant as a perfect square.]

96. Show that the equation of the tangent line to the ellipse with equation $\frac{x^2}{a^2} + \frac{y^2}{b^2} = 1$, at the point (x_1, y_1), is
$y = -\frac{b^2 x_1}{a^2 y_1} x + \frac{b^2 x_1^2}{b^2 y_1^2} + y_1$.
[*Hint*: See the steps for Exercise 91 in Section 10.2 and show that $(a^2 y_1 m + b^2 x_1)^2 = 0$.]

Critical Thinking/Discussion/Writing

97. Find the number of circles with a radius of 3 units that touch both of the coordinate axes. Write the equation of each circle.

98. Find the number of ellipses with a major axis and a minor axis of lengths 6 units and 4 units, respectively, that touch both axes. Write the equation of each ellipse.

Maintaining Skills

In Exercises 99 and 100, find the center and radius of each circle whose equation is given.

99. $x^2 + 6x + y^2 - 2y - 5 = 0$

100. $x^2 + 4x + y^2 + 6y + 5 = 0$

In Exercises 101 and 102, find the *x*- and *y*-intercepts of the graph of the given equation.

101. $x^2 + y^2 = 25$

102. $(x - 2)^2 + (y + 3)^2 = 8$

In Exercises 103–106, find the slope–intercept form of the line with the given properties.

103. Slope $= -4$, y-intercept $= 11$

104. Slope $= 0$, passes through $(-5, 7)$

105. Passes through $(-3, 7)$ and $(-1, -5)$

106. Passes through $(2, -4)$ parallel to the line with equation $6x + 2y = 7$

In Exercises 107 and 108, find the asymptotes of the graph of the given function.

107. $f(x) = \dfrac{2x^2 - 3x + 1}{x - 2}$

108. $f(x) = \dfrac{x^2 - 4x - 5}{x - 3}$

SECTION 10.4 The Hyperbola

Alfred Lee Loomis (1887–1975)
Alfred Loomis was an American lawyer, investment banker, physicist, and patron of scientific research. Besides inventing LORAN, he made significant contributions to the ground-based technology for landing airplanes by instruments. President Roosevelt recognized the value of Loomis's work and described him as second perhaps only to Churchill as the civilian most responsible for the Allied victory in World War II.

BEFORE STARTING THIS SECTION, REVIEW

1. Distance formula (Section 2.1, page 159)
2. Midpoint formula (Section 2.1, page 161)
3. Completing the square (Section 3.1, page 295)
4. Oblique asymptotes (Section 3.6, page 366)
5. Transformations of graphs (Section 2.7, page 251)
6. Symmetry (Section 2.2, page 168)

OBJECTIVES

1. Define a hyperbola.
2. Find the asymptotes of a hyperbola.
3. Graph a hyperbola.
4. Translate hyperbolas.
5. Use hyperbolas in applications.

What Is Loran?

LORAN stands for LOng-RAnge Navigation. The federal government runs the LORAN system, which uses land-based radio navigation transmitters to provide users with information on position and timing. During World War II, Alfred Lee Loomis was selected to chair the National Defense Research Committee. Much of his work focused on the problem of creating a light system for plane-carried radar. As a result of this effort, he invented LORAN, the long-range navigation system whose offshoot, LORAN-C, remains in widespread use. The navigational method provided by LORAN is based on the principle of determining the points of intersection of two hyperbolas to fix the two-dimensional position of the receiver. In Example 8, we illustrate how this is done.

The LORAN-C system is now being supplemented by the global positioning system (GPS), which works on the same principle as LORAN-C. The GPS system consists of 24 satellites that orbit 11,000 miles above Earth. The GPS was created by the U.S. Department of Defense to provide precise navigation information for targeting weapons systems anywhere in the world. The GPS is available to civilian users worldwide in a degraded mode. It is now used in aviation, navigation, and automobiles. The system can provide your exact position on Earth anywhere, anytime.

The GPS and LORAN-C work together, giving a highly accurate, duplicate navigation system for the Defense Department.

1 Define a hyperbola.

P
F_1
foci
F_2

FIGURE 10.21

Definition of Hyperbola

Hyperbola

A **hyperbola** is the set of all points in the plane, the positive difference of whose distances from two fixed points is constant. The fixed points are called the **foci** of the hyperbola. See Figure 10.21.

Figure 10.22 shows a hyperbola in *standard position*, with foci $F_1(-c, 0)$ and $F_2(c, 0)$ on the x-axis at equal distances from the origin. The two parts of the hyperbola are called **branches**.

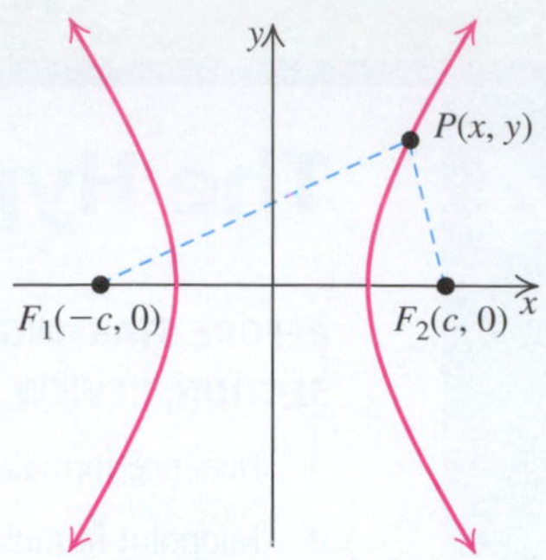

FIGURE 10.22 $|PF_1 - PF_2|$ is a constant for any P on the hyperbola

As in the discussion of ellipses in Section 10.3 (see page 876), we take $2a$ as the positive constant referred to in the definition.

A point $P(x, y)$ lies on a hyperbola if and only if

$$|d(P, F_1) - d(P, F_2)| = 2a \quad \text{Definition of hyperbola}$$

$$\sqrt{(x + c)^2 + y^2} - \sqrt{(x - c)^2 + y^2} = \pm 2a \quad \text{Distance formula}$$

As in the case of an ellipse, we eliminate radicals and simplify (see Exercise 93) to obtain the equation.

$$(c^2 - a^2)x^2 - a^2y^2 = a^2(c^2 - a^2) \quad (1)$$

$$b^2x^2 - a^2y^2 = a^2b^2 \quad \text{Replace } (c^2 - a^2) \text{ with } b^2.$$

$$\frac{x^2}{a^2} - \frac{y^2}{b^2} = 1 \quad (2) \quad \text{Divide both sides by } a^2b^2.$$

Equation (2) is called the **standard form of the equation of a hyperbola** with center $(0, 0)$. The x-intercepts of the graph of equation (2) are $-a$ and a. The points corresponding to these x-intercepts are the **vertices** of the hyperbola. The distance between the vertices $V_1(-a, 0)$ and $V_2(a, 0)$ is $2a$. The line segment joining the two vertices is called the **transverse axis**. The midpoint of the transverse axis is the **center** of the hyperbola. The *center* is also the midpoint of the line segment joining the foci. See Figure 10.23. The line segment joining the points $(0, -b)$ and $(0, b)$ is called the **conjugate axis**.

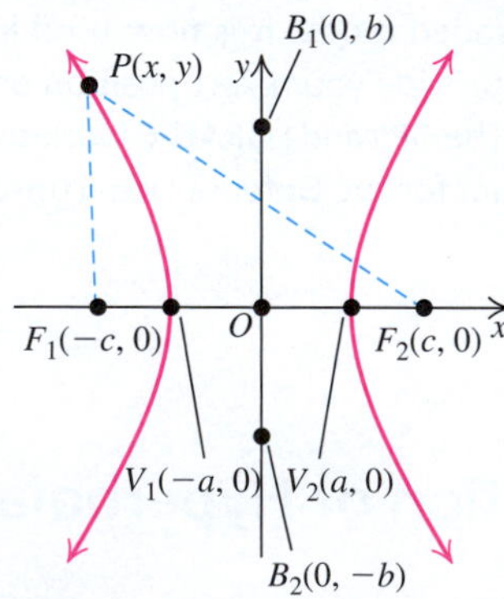

Points F_1 and F_2 are foci.
Points V_1 and V_2 are vertices.
Point O is the center.
Segment $\overline{V_1V_2}$ is the transverse axis.
Segment $\overline{B_1B_2}$ is the conjugate axis.
The difference of the distances $|PF_1 - PF_2|$ from any point P on the hyperbola is constant.

FIGURE 10.23 A hyperbola

Similarly, an equation of a hyperbola with center $(0, 0)$ and foci $(0, -c)$ and $(0, c)$ on the y-axis is given by:

$$\frac{y^2}{a^2} - \frac{x^2}{b^2} = 1, \text{ where } b^2 = c^2 - a^2 \quad (3)$$

Here the vertices are $(0, -a)$ and $(0, a)$. The transverse axis of length $2a$ of the graph of equation (3) lies on the y-axis, and its conjugate axis is the segment joining the points $(-b, 0)$ and $(0, b)$ of length $2b$ that lies on the x-axis.

SUMMARY OF **MAIN FACTS**

Main facts about hyperbolas centered at (0, 0)

Standard Equation	$\frac{x^2}{a^2} - \frac{y^2}{b^2} = 1; a > 0, b > 0$	$\frac{y^2}{a^2} - \frac{x^2}{b^2} = 1; a > 0, b > 0$
Transverse axis along	$y = 0$ (x-axis)	$x = 0$ (y-axis)
Length of transverse axis	$2a$	$2a$
Conjugate axis along	$x = 0$ (y-axis)	$y = 0$ (x-axis)
Length of conjugate axis	$2b$	$2b$
Vertices	$(\pm a, 0)$	$(0, \pm a)$
Endpoints of conjugate axis	$(0, \pm b)$	$(\pm b, 0)$
Foci	$(\pm c, 0)$, where $c^2 = a^2 + b^2$	$(0, \pm c)$, where $c^2 = a^2 + b^2$
Description	Hyperbola has a *left branch* and a *right branch*. (Hyperbola opens left and right.)	Hyperbola has an *upper branch* and a *lower branch*. (Hyperbola opens up and down.)
Graph	Horizontal transverse axis	Vertical transverse axis

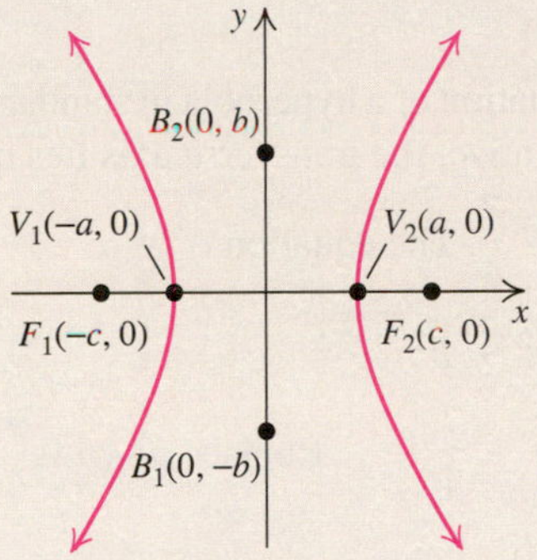

Horizontal transverse axis

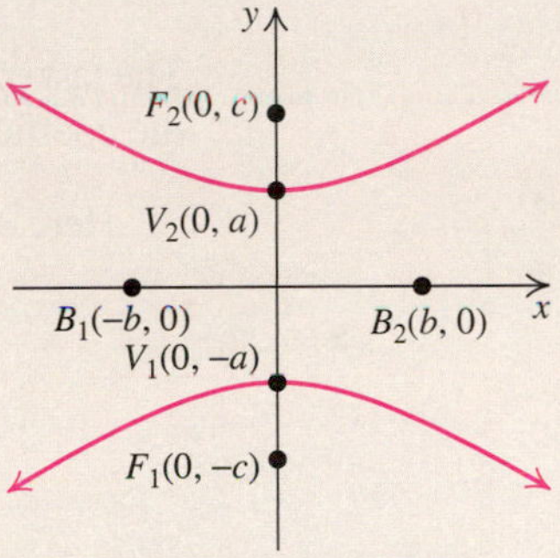

Vertical transverse axis

EXAMPLE 1 Determining the Orientation of a Hyperbola

Does the hyperbola

$$\frac{x^2}{4} - \frac{y^2}{10} = 1$$

have its transverse axis on the x-axis or y-axis?

Solution

The orientation (left–right branches or up–down branches) of a hyperbola is determined *by noting where the minus sign occurs in the standard equation*. From the table above, we see that the transverse axis is on the x-axis when the minus sign precedes the y^2-term and is on the y-axis when the minus sign precedes the x^2-term. In the equation in this example, the minus sign precedes the y^2-term; so the transverse axis is on the x-axis. See Figure 10.24.

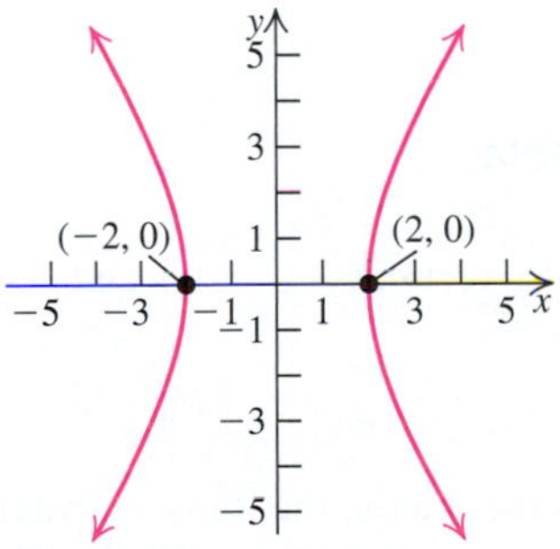

FIGURE 10.24 $\frac{x^2}{4} - \frac{y^2}{10} = 1$

Practice Problem 1 Does the hyperbola $\frac{y^2}{8} - \frac{x^2}{5} = 1$ have its transverse axis on the x-axis or the y-axis?

EXAMPLE 2 **Finding the Vertices and Foci from the Equation of a Hyperbola**

Find the vertices and foci for the hyperbola

$$28y^2 - 36x^2 = 63.$$

Solution

First, convert the equation to the standard form.

$$28y^2 - 36x^2 = 63 \quad \text{Given equation}$$

$$\frac{28y^2}{63} - \frac{36x^2}{63} = 1 \quad \text{Divide both sides by 63.}$$

$$\frac{4y^2}{9} - \frac{4x^2}{7} = 1 \quad \text{Simplify.}$$

$$\frac{y^2}{\frac{9}{4}} - \frac{x^2}{\frac{7}{4}} = 1 \quad \text{Divide the numerator and denominator of each term by 4.}$$

The last equation is the equation of a hyperbola in standard form with center $(0, 0)$. Because the coefficient of x^2 is negative, the transverse axis lies on the y-axis.

Here $a^2 = \frac{9}{4}$ and $b^2 = \frac{7}{4}$. The equation $c^2 = a^2 + b^2$ gives the value of c^2.

$$c^2 = a^2 + b^2$$

$$= \frac{9}{4} + \frac{7}{4} \quad \text{Replace } a^2 \text{ with } \frac{9}{4} \text{ and } b^2 \text{ with } \frac{7}{4}.$$

$$= \frac{16}{4} = 4 \quad \text{Simplify.}$$

Now $a^2 = \frac{9}{4}$ gives $a = \frac{3}{2}$ and $c^2 = 4$ gives $c = 2$ because we require $a > 0$ and $c > 0$.

The vertices of the hyperbola are $(0, -a) = \left(0, -\frac{3}{2}\right)$ and $(0, a) = \left(0, \frac{3}{2}\right)$. The foci of the hyperbola are $(0, -c) = (0, -2)$ and $(0, c) = (0, 2)$.

Practice Problem 2 Find the vertices and foci for the hyperbola

$$x^2 - 4y^2 = 8.$$

EXAMPLE 3 **Finding the Equation of a Hyperbola**

Find the standard form of the equation of a hyperbola with vertices $(\pm 4, 0)$ and foci $(\pm 5, 0)$.

Solution

Because the foci of the hyperbola $(-5, 0)$ and $(5, 0)$ are on the x-axis, the transverse axis lies on the x-axis. The center of the hyperbola is midway between the foci, at $(0, 0)$. The standard form of such a hyperbola is

$$\frac{x^2}{a^2} - \frac{y^2}{b^2} = 1.$$

We need to find a^2 and b^2.

The distance a between the center $(0, 0)$ to either vertex, $(-4, 0)$ or $(4, 0)$ is 4; so $a = 4$ and $a^2 = 16$. The distance c between the center $(0, 0)$ to either focus, $(-5, 0)$ or $(5, 0)$, is 5; so $c = 5$ and $c^2 = 25$. Now use the equation $b^2 = c^2 - a^2$ to find

$b^2 = c^2 - a^2 = 25 - 16 = 9$. Substitute $a^2 = 16$ and $b^2 = 9$ in $\frac{x^2}{a^2} - \frac{y^2}{b^2} = 1$ to get the standard form of the equation of the hyperbola

$$\frac{x^2}{16} - \frac{y^2}{9} = 1.$$

Practice Problem 3 Find the standard form of the equation of a hyperbola with vertices at $(0, \pm 4)$ and foci at $(0, \pm 6)$.

2 Find the asymptotes of a hyperbola.

The Asymptotes of a Hyperbola

Recall from Section 3.6 that a line $y = mx + b$ is a horizontal asymptote (if $m = 0$) or an oblique (slant) asymptote (if $m \neq 0$) of the graph of $y = f(x)$ if the difference $f(x) - (mx + b)$ approaches 0 as $x \to \infty$ or as $x \to -\infty$. When horizontal or oblique asymptotes exist, they provide us with information about the end behavior of the graph of f. This is especially helpful in graphing a hyperbola.

To find the asymptotes of the hyperbola with equation $\frac{x^2}{a^2} - \frac{y^2}{b^2} = 1$, we first solve this equation for y.

$$\frac{x^2}{a^2} - \frac{y^2}{b^2} = 1 \quad \text{Given equation of hyperbola}$$

$$-\frac{y^2}{b^2} = -\frac{x^2}{a^2} + 1 \quad \text{Subtract } \frac{x^2}{a^2} \text{ from both sides.}$$

$$\frac{y^2}{b^2} = \frac{x^2}{a^2} - 1 \quad \text{Multiply both sides by } -1.$$

$$y^2 = b^2\left(\frac{x^2}{a^2} - 1\right) \quad \text{Multiply both sides by } b^2.$$

$$y^2 = b^2\left(\frac{x^2}{a^2} - \frac{x^2}{a^2} \cdot \frac{a^2}{x^2}\right) \quad \text{Write } 1 = \frac{x^2}{a^2} \cdot \frac{a^2}{x^2}.$$

$$y^2 = \frac{b^2x^2}{a^2}\left(1 - \frac{a^2}{x^2}\right) \quad \text{Factor out } \frac{x^2}{a^2}.$$

$$y = \pm\frac{bx}{a}\sqrt{1 - \frac{a^2}{x^2}} \quad \text{Square root property}$$

As $x \to \infty$ or as $x \to -\infty$, the quantity $\frac{a^2}{x^2}$ approaches 0. Thus, $\sqrt{1 - \frac{a^2}{x^2}}$ approaches 1 and the value of y approaches $\pm\frac{bx}{a}$. Therefore, the lines $y = \frac{b}{a}x$ and $y = -\frac{b}{a}x$ are the oblique asymptotes for the graph of the hyperbola $\frac{x^2}{a^2} - \frac{y^2}{b^2} = 1$. Similarly, you can show that the lines $y = \frac{a}{b}x$ and $y = -\frac{a}{b}x$ are the oblique asymptotes for the graph of the hyperbola $\frac{y^2}{a^2} - \frac{x^2}{b^2} = 1$.

SIDE NOTE

A convenient device can be used to obtain equations of the asymptotes of a hyperbola in standard form. Substitute 0 for the 1 on the right side of the equation of the hyperbola and then solve for y in terms of x. For example, for the hyperbola $\frac{x^2}{a^2} - \frac{y^2}{b^2} = 1$, replace the 1 on the right side with 0 to obtain $\frac{x^2}{a^2} - \frac{y^2}{b^2} = 0$.

Solve this equation for y.

$$\frac{y^2}{b^2} = \frac{x^2}{a^2}$$

$$y^2 = \frac{b^2}{a^2}x^2$$

The lines $y = \pm\frac{b}{a}x$ are the slant asymptotes of the hyperbola.

THE ASYMPTOTES OF A HYPERBOLA WITH CENTER (0, 0)

1. The graph of the hyperbola $\dfrac{x^2}{a^2} - \dfrac{y^2}{b^2} = 1$ has transverse axis along the x-axis and has two asymptotes:

$$y = \frac{b}{a}x \quad \text{and} \quad y = -\frac{b}{a}x$$

2. The graph of the hyperbola $\dfrac{y^2}{a^2} - \dfrac{x^2}{b^2} = 1$ has transverse axis along the y-axis and has two asymptotes:

$$y = \frac{a}{b}x \quad \text{and} \quad y = -\frac{a}{b}x$$

EXAMPLE 4 Finding the Asymptotes of a Hyperbola

Determine the asymptotes of each hyperbola.

a. $\dfrac{x^2}{4} - \dfrac{y^2}{9} = 1$ **b.** $\dfrac{y^2}{9} - \dfrac{x^2}{16} = 1$

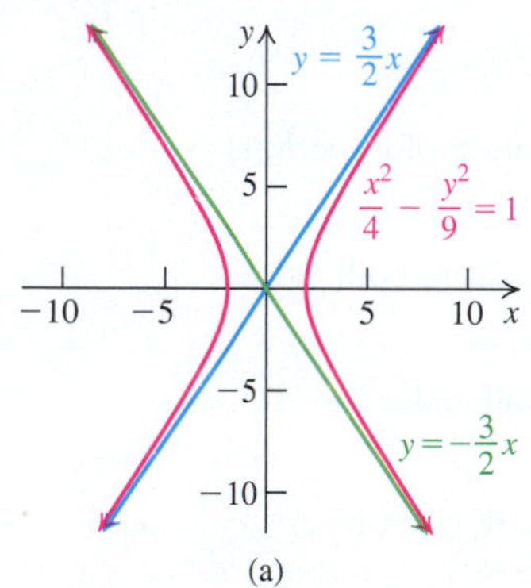

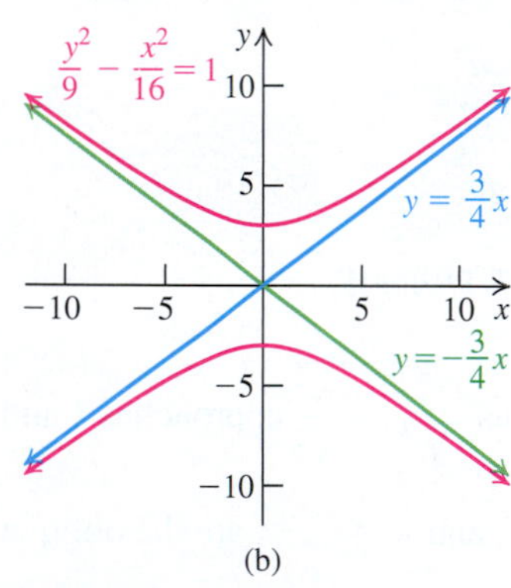

FIGURE 10.25 Asymptotes

Solution

a. The hyperbola $\dfrac{x^2}{4} - \dfrac{y^2}{9} = 1$ is of the form $\dfrac{x^2}{a^2} - \dfrac{y^2}{b^2} = 1$; so $a = 2$ and $b = 3$. Substituting these values into $y = \dfrac{b}{a}x$ and $y = -\dfrac{b}{a}x$, we get the asymptotes

$$y = \frac{3}{2}x \text{ and } y = -\frac{3}{2}x.$$

b. The hyperbola $\dfrac{y^2}{9} - \dfrac{x^2}{16} = 1$ has the form $\dfrac{y^2}{a^2} - \dfrac{x^2}{b^2} = 1$; so $a = 3$ and $b = 4$. Substituting these values into $y = \dfrac{a}{b}x$ and $y = -\dfrac{a}{b}x$, we get the asymptotes

$$y = \frac{3}{4}x \text{ and } y = -\frac{3}{4}x.$$

The hyperbola of parts **a** and **b** and their asymptotes are shown in Figure 10.25(a) and Figure 10.25(b), respectively.

Practice Problem 4 Determine the asymptotes of the hyperbola $\dfrac{y^2}{4} - \dfrac{x^2}{9} = 1$.

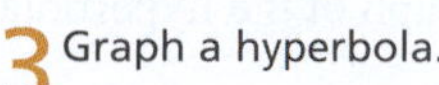
3 Graph a hyperbola.

Graphing a Hyperbola with Center (0,0)

Consider the hyperbola with equation

$$\frac{x^2}{a^2} - \frac{y^2}{b^2} = 1.$$

The vertices of this hyperbola are $(-a, 0)$ and $(a, 0)$. The endpoints of the conjugate axis are $(0, -b)$ and $(0, b)$. The rectangle with vertices (a, b), $(-a, b)$, $(-a, -b)$, and $(a, -b)$ is called the **fundamental rectangle** of the hyperbola. See Figure 10.26 on page 893 (Step **3**). The diagonals of the fundamental rectangle have slopes $\dfrac{b}{a}$ and $-\dfrac{b}{a}$. Thus, the extensions of these diagonals are the asymptotes of the hyperbola.

PROCEDURE
IN ACTION

EXAMPLE 5 Graphing a Hyperbola Centered at (0, 0)

OBJECTIVE

Sketch the graph of a hyperbola in the form

(i) $Ax^2 - By^2 = C$ *or* (ii) $Ay^2 - Bx^2 = C$

$A, B, C > 0.$

EXAMPLE

Sketch the graph of

a. $16x^2 - 9y^2 = 144.$ **b.** $25y^2 - 4x^2 = 100.$

Step 1 Write the equation in standard form. Determine transverse axis and the orientation of the hyperbola.

1. $16x^2 - 9y^2 = 144$

$$\frac{16x^2}{144} - \frac{9y^2}{144} = \frac{144}{144}$$

$$\frac{x^2}{9} - \frac{y^2}{16} = 1$$

Minus sign precedes y^2-term, and transverse axis is along the x-axis; hyperbola opens left and right.

$25y^2 - 4x^2 = 100$

$$\frac{25y^2}{100} - \frac{4x^2}{100} = \frac{100}{100}$$

$$\frac{y^2}{4} - \frac{x^2}{25} = 1$$

Minus sign precedes x^2-term, and transverse axis is along the y-axis; hyperbola opens up and down.

Step 2 Locate vertices and the endpoints of the conjugate axis.

2. Because $a^2 = 9, a = 3.$ Also, $b^2 = 16$; so $b = 4$. The vertices are $(\pm 3, 0)$, and the endpoints of the conjugate axis are $(0, \pm 4)$.

Because $a^2 = 4, a = 2$. Also, $b^2 = 25$; so $b = 5$. The vertices are $(0, \pm 2)$, and the endpoints of the conjugate axis are $(\pm 5, 0)$.

Step 3 Lightly sketch the fundamental rectangle by drawing dashed lines parallel to the coordinate axes through the points in Step 2.

3. Draw dashed lines $x = 3, x = -3, y = 4$, and $y = -4$ to form the fundamental rectangle with vertices $(3, 4), (-3, 4), (-3, -4)$, and $(3, -4)$.

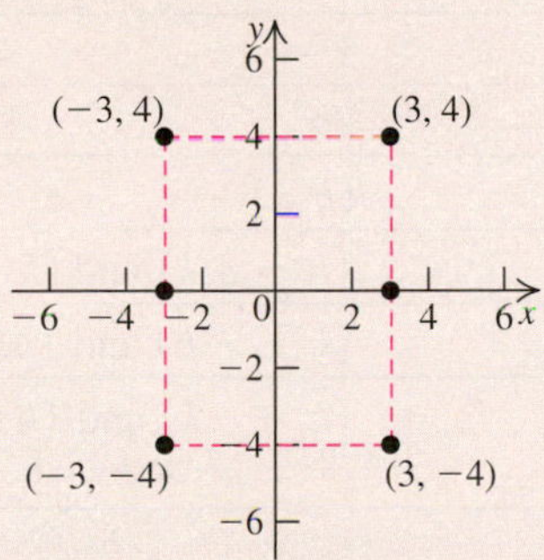

Draw dashed lines $y = 2, y = -2, x = 5$, and $x = -5$ to form the fundamental rectangle with vertices $(5, 2), (-5, 2), (-5, -2)$, and $(5, -2)$.

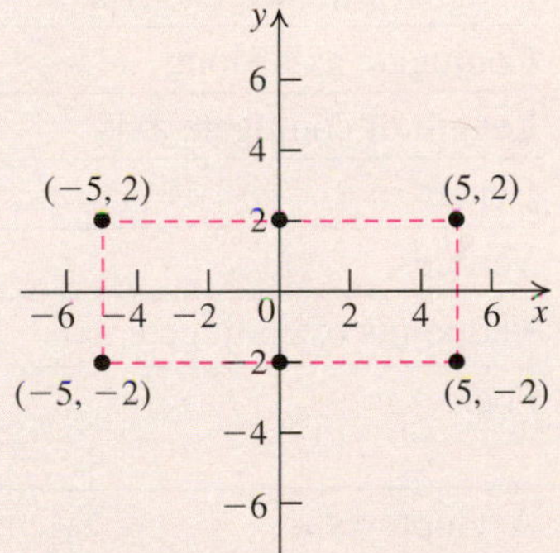

Step 4 Sketch the asymptotes. Extend the diagonals of the fundamental rectangle. These are the asymptotes.

4.

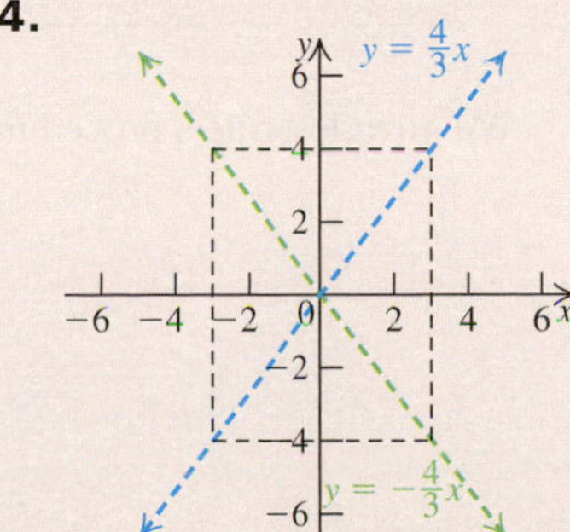

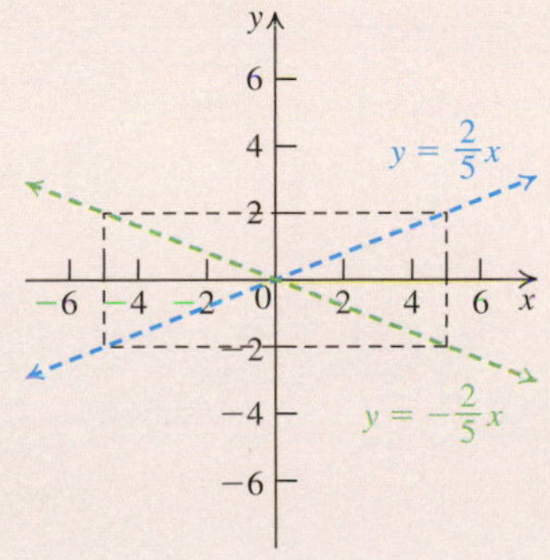

(continued)

Step 5 Sketch the graph. Draw both branches of the hyperbola through the vertices, approaching the asymptotes. See Figures 10.26 and 10.27. The foci are located on the transverse axis, c units from the center, where $c^2 = a^2 + b^2$.

5.

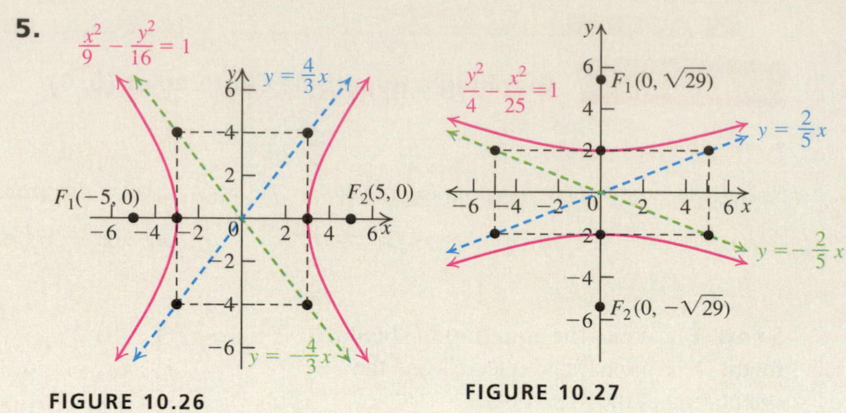

FIGURE 10.26

FIGURE 10.27

Practice Problem 5 Sketch the graph of the equations.

a. $25x^2 - 4y^2 = 100$ **b.** $9y^2 - x^2 = 1$

4 Translate hyperbolas.

Translations of Hyperbolas

We can use horizontal and vertical shifts to find the standard form of the equations of hyperbolas centered at (h, k).

SUMMARY OF MAIN FACTS

Main Properties of Hyperbolas Centered at (h, k)

Standard Equation	$\frac{(x-h)^2}{a^2} - \frac{(y-k)^2}{b^2} = 1$; $a > 0, b > 0$	$\frac{(y-k)^2}{a^2} - \frac{(x-h)^2}{b^2} = 1$; $a > 0, b > 0$
Transverse axis along	$y = k$	$x = h$
Length of transverse axis	$2a$	$2a$
Conjugate axis along	$x = h$	$y = k$
Length of conjugate axis	$2b$	$2b$
Center	(h, k)	(h, k)
Vertices	$(h - a, k)$ and $(h + a, k)$	$(h, k - a)$ and $(h, k + a)$
Endpoints of conjugate axis	$(h, k - b)$ and $(h, k + b)$	$(h - b, k)$ and $(h + b, k)$
Foci	$(h - c, k)$ and $(h + c, k)$; $c^2 = a^2 + b^2$	$(h, k - c)$ and $(h, k + c)$; $c^2 = a^2 + b^2$
Asymptotes	$y - k = \pm\frac{b}{a}(x - h)$	$y - k = \pm\frac{a}{b}(x - h)$

We now describe a procedure for sketching the graph of a hyperbola centered at (h, k).

PROCEDURE
IN ACTION

EXAMPLE 6 Graphing a Hyperbola Centered at (h, k)

OBJECTIVE

Sketch the graph of either

a. $\dfrac{(x-h)^2}{a^2} - \dfrac{(y-k)^2}{b^2} = 1$ *or*

b. $\dfrac{(y-k)^2}{a^2} - \dfrac{(x-h)^2}{b^2} = 1.$

EXAMPLE

Sketch the graph of each equation.

a. $\dfrac{(x-1)^2}{4} - \dfrac{(y+2)^2}{9} = 1.$ **b.** $\dfrac{(y+2)^2}{9} - \dfrac{(x-1)^2}{4} = 1.$

Step 1 Plot the center (h, k) **and draw horizontal and vertical dashed lines through the center.**

1.

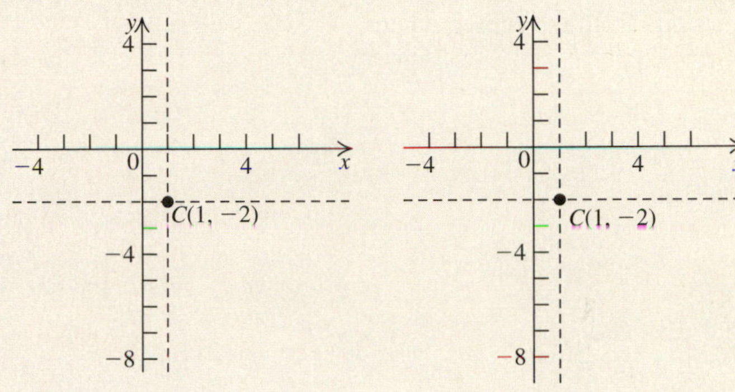

Step 2 Locate the vertices and the endpoints of the conjugate axis. Lightly sketch the fundamental rectangle, with sides parallel to the coordinate axes, through these points.

2.

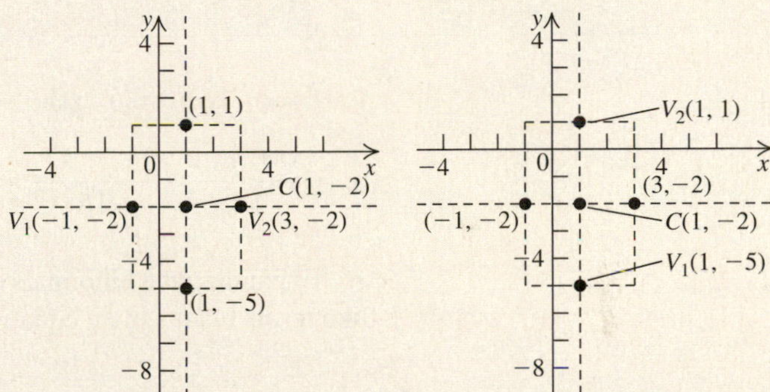

Step 3 Sketch dashed lines through opposite vertices of the fundamental rectangle. These are the asymptotes.

3.

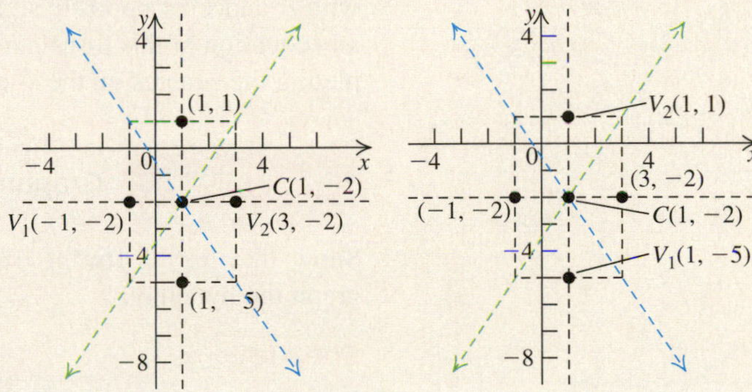

(continued)

Step 4 **Draw both branches** of the hyperbola through the vertices, approaching the asymptotes.

Step 5 **Locate the foci** on the transverse axis, c units from the center, where $c^2 = a^2 + b^2$.

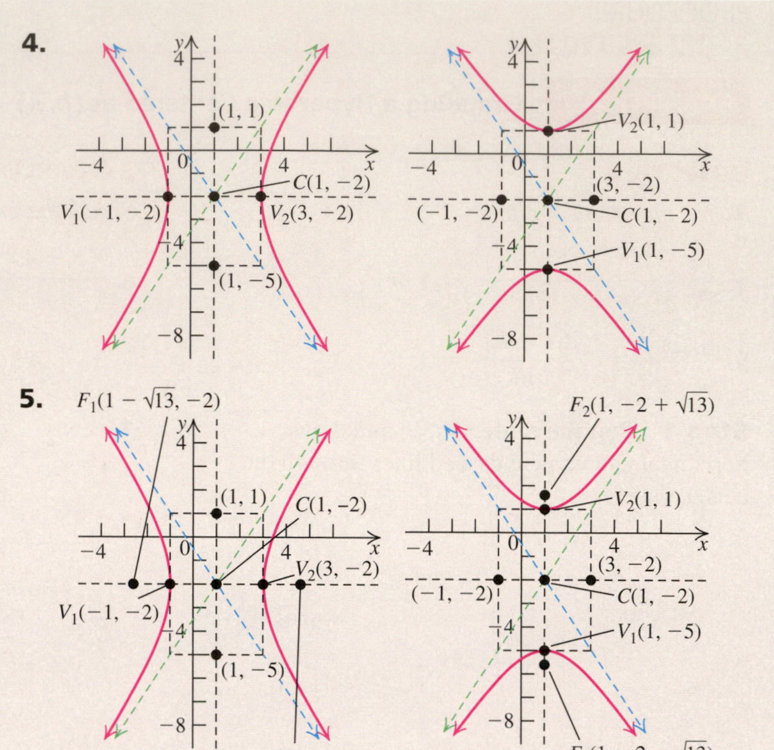

Practice Problem 6 Sketch the graph of each equation.

a. $\dfrac{(x+1)^2}{4} - \dfrac{(y-1)^2}{16} = 1$ **b.** $\dfrac{(y-1)^2}{4} - \dfrac{(x+1)^2}{9} = 1$

Expanding the binomials in the equation of a hyperbola in standard form and collecting like terms results in an equation of the form

$$Ax^2 + Cy^2 + Dx + Ey + F = 0,$$

with A and C of opposite sign (that is, $AC < 0$.) Conversely, except in degenerate cases, any equation of this form can be put into one of the standard forms of a hyperbola by completing the squares on the x- and y-terms.

EXAMPLE 7 Graphing a Hyperbola

Show that $9x^2 - 16y^2 + 18x + 64y - 199 = 0$ is an equation of a hyperbola; then graph the hyperbola.

Solution

Because we plan on completing squares, first group the x- and y-terms on the left side and the constants on the right side.

$9x^2 - 16y^2 + 18x + 64y - 199 = 0$	Given equation
$(9x^2 + 18x) + (-16y^2 + 64y) = 199$	Group terms in each variable.
$9(x^2 + 2x) - 16(y^2 - 4y) = 199$	Factor out 9 and −16.
$9(x^2 + 2x + 1) - 16(y^2 - 4y + 4) = 199 + 9 - 64$	Complete the squares: add $9 \cdot 1$ and $-16 \cdot 4$.

$$9(x + 1)^2 - 16(y - 2)^2 = 144 \quad \text{Factor and simplify.}$$

$$\frac{9(x + 1)^2}{144} - \frac{16(y - 2)^2}{144} = 1 \quad \text{Divide both sides by 144 to obtain 1 on the right side.}$$

$$\frac{(x + 1)^2}{16} - \frac{(y - 2)^2}{9} = 1 \quad \text{Simplify.}$$

The last equation is the standard form of the equation of a hyperbola with center $(-1, 2)$. Now we sketch the graph of this hyperbola.

Steps 1–2 **Locate the vertices.** For this hyperbola with center $(-1, 2)$, $a^2 = 16$ and $b^2 = 9$. Therefore, $a = 4$ and $b = 3$. From the table on page 894, with $h = -1$ and $k = 2$, the vertices are $(h - a, k) = (-1 - 4, 2) = (-5, 2)$ and $(h + a, k) = (-1 + 4, 2) = (3, 2)$.

Draw the fundamental rectangle. The vertices of the fundamental rectangle are $(3, -1)$, $(3, 5)$, $(-5, 5)$, and $(5, -1)$.

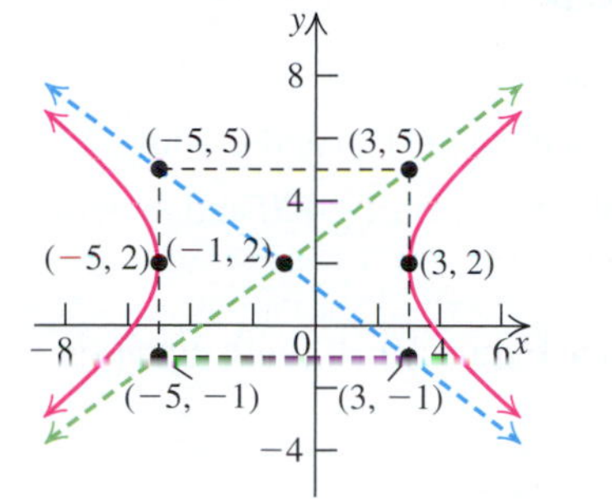

Steps 3 **Sketch the asymptotes.** Extend the diagonals of the fundamental rectangle to sketch the asymptotes:

$$y - 2 = \frac{3}{4}(x + 1) \text{ and } y - 2 = -\frac{3}{4}(x + 1)$$

Steps 4 **Sketch the graph.** Draw two branches of the hyperbola opening to the left and right, starting from the vertices $(-5, 2)$ and $(3, 2)$ and approaching the asymptotes. See Figure 10.28.

FIGURE 10.28 The hyperbola $9x^2 - 16y^2 + 18x + 64y - 199 = 0$

Practice Problem 7 Show that $x^2 - 4y^2 - 2x + 16y - 20 = 0$ is an equation of a hyperbola and graph the hyperbola.

5 Use hyperbolas in applications.

Applications

Hyperbolas have many applications. We list a few of them here.

1. Comets that do not move in elliptical orbits around the sun almost always move in a hyperbolic orbit. (In theory, they also can move in parabolic orbits.)
2. Boyle's Law states that if a perfect gas is kept at a constant temperature, then its pressure P and volume V are related by the equations $PV = c$, where c is a constant. The graph of this equation is a hyperbola. In Exercise 94(b), you are asked to explore the case where $c = 8$. In this case, the transverse axis is not parallel to a coordinate axis.
3. The hyperbola has the reflecting property that a ray of light from a source at one focus of a hyperbolic mirror (a mirror with hyperbolic cross sections) is reflected along the line through the other focus.

   The reflecting properties of the parabola and hyperbola are combined into one design for a reflecting telescope. See Figure 10.29. The parallel rays from a star are finally focused at the eyepiece at F_2.
4. The definition of a hyperbola forms the basis of several important navigational systems.

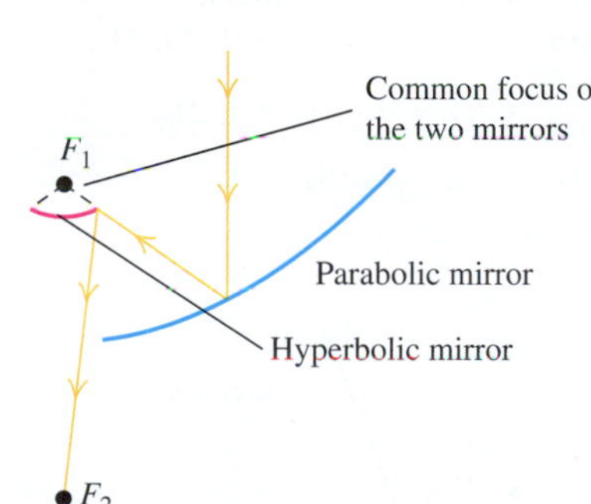

FIGURE 10.29

How Does LORAN Work?

Suppose two stations A and B (several miles apart) transmit synchronized radio signals. The LORAN receiver in your ship measures the difference in reception times of these synchronized signals. The radio signals travel at a speed of 186,000 miles per second. Using this information, you can determine the difference $2a$ in the distance of your ship's receiver from the two transmitters. By the definition of a hyperbola, this information places your ship somewhere on a hyperbola with foci A and B. With two pairs of transmitters, the position of your ship can be determined at a point of intersection of the two hyperbolas. See Exercises 87 and 88.

EXAMPLE 8 Using LORAN

LORAN navigational transmitters A and B are located at $(-130, 0)$ and $(130, 0)$, respectively. A receiver P on a fishing boat somewhere in the first quadrant listens to the pair (A, B) of transmissions and computes the difference of the distances from the boat to A and B as 240 miles. Find the equation of the hyperbola on which P is located.

Solution

With reference to the transmitters A and B, P is located on a hyperbola with $2a = 240$, or $a = 120$, and the hyperbola has foci $(-130, 0)$ and $(130, 0)$.

$$\frac{x^2}{a^2} - \frac{y^2}{b^2} = 1 \quad \text{Standard from of the equation of a hyperbola}$$

$$a^2 = (120)^2 = 14{,}400 \quad \text{Replace } a \text{ with 120.}$$

$$b^2 = c^2 - a^2 = (130)^2 - (120)^2 = 2500 \quad \text{Replace } c \text{ with 130 and } a \text{ with 120.}$$

$$\frac{x^2}{14{,}400} - \frac{y^2}{2500} = 1 \quad \text{Replace } a^2 \text{ with 14,400 and } b^2 \text{ with 2500.}$$

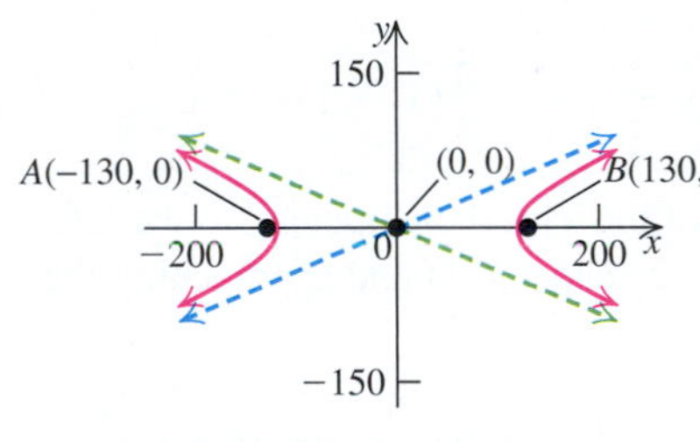

FIGURE 10.30

Therefore, the location of the ship lies on a hyperbola (see Figure 10.30) with equation

$$\frac{x^2}{14{,}400} - \frac{y^2}{2500} = 1.$$

Practice Problem 8 In Example 8, the transmitters A and B are located at $(-150, 0)$ and $(150, 0)$, respectively, and the difference of the distances from the boat to A and B is 260 miles. Find the equation of the hyperbola on which P is located.

Answers to Practice Problems

1. y-axis **2.** Vertices: $(2\sqrt{2}, 0)$ and $(-2\sqrt{2}, 0)$; foci: $(\sqrt{10}, 0)$ and $(-\sqrt{10}, 0)$ **3.** $\frac{y^2}{16} - \frac{x^2}{20} = 1$ **4.** $y = \frac{2}{3}x$ and $y = -\frac{2}{3}x$

5. a. The standard form of the equation is $\frac{x^2}{4} - \frac{y^2}{25} = 1$. Vertices: $(2, 0)$ and $(-2, 0)$; Foci: $(\pm\sqrt{29}, 0)$; endpoints of the conjugate axis: $(0, 5)$ and $(0, -5)$; asymptotes: $y = \frac{5}{2}x$ and $y = -\frac{5}{2}x$

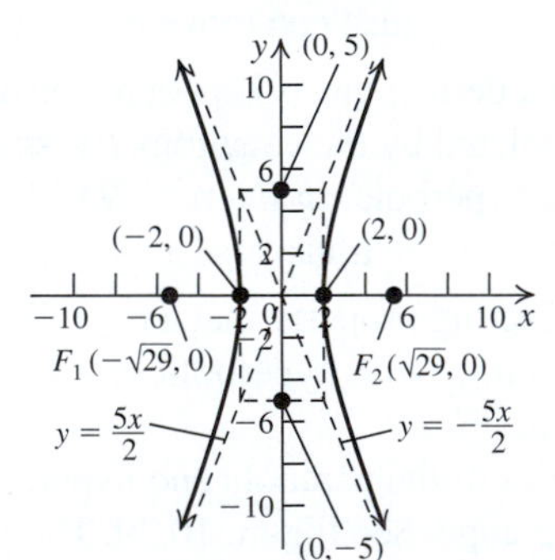

b. The standard form of the equation is $\frac{y^2}{\frac{1}{9}} - \frac{x^2}{1} = 1$. Vertices: $\left(0, \frac{1}{3}\right)$ and $\left(0, -\frac{1}{3}\right)$; Foci: $\left(0, \pm\frac{\sqrt{10}}{3}\right)$; endpoints of the conjugate axis: $(1, 0)$ and $(-1, 0)$; asymptotes: $y = \frac{1}{3}x$ and $y = -\frac{1}{3}x$

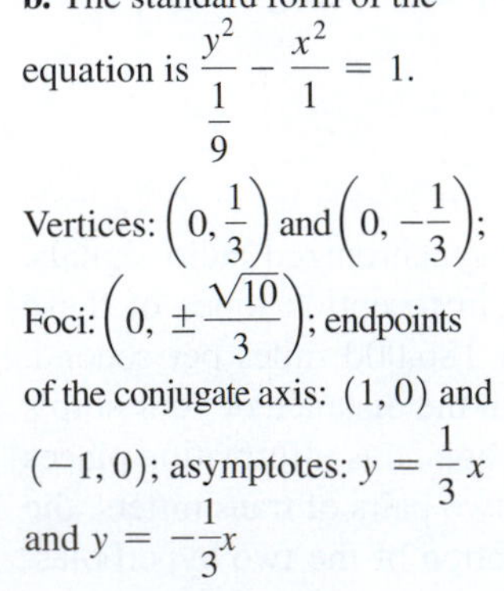

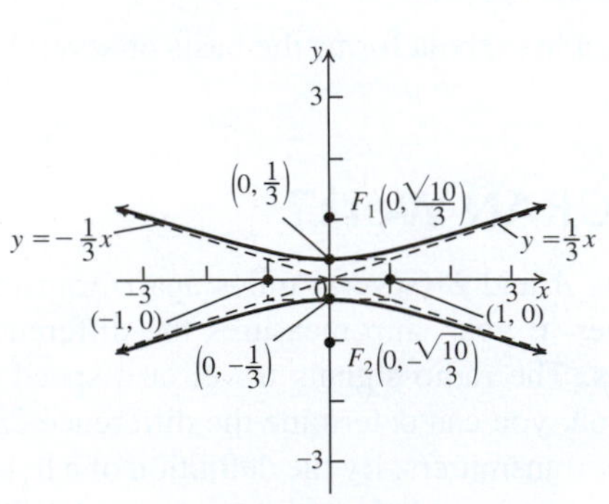

6. a. $\frac{(x+1)^2}{4} - \frac{(y-1)^2}{16} = 1 \Rightarrow a = 2, b = 4$, and center $(-1, 1)$. The vertices are $(-1 - 2, 1) = (-3, 1)$ and $(-1 + 2, 1) = (1, 1)$. The endpoints of the conjugate axes are $(-1, 1 - 4) = (-1, -3)$ and $(-1, 1 + 4) = (-1, 5)$. The asymptotes are $y - k = \pm\frac{b}{a}(x - h) \Rightarrow y - 1 = \pm 2(x + 1)$. $c^2 = a^2 + b^2 \Rightarrow c^2 = 4 + 16 \Rightarrow c = \pm\sqrt{20}$ The foci are $(-\sqrt{20} - 1, 1)$ and $(\sqrt{20} - 1, 1)$.

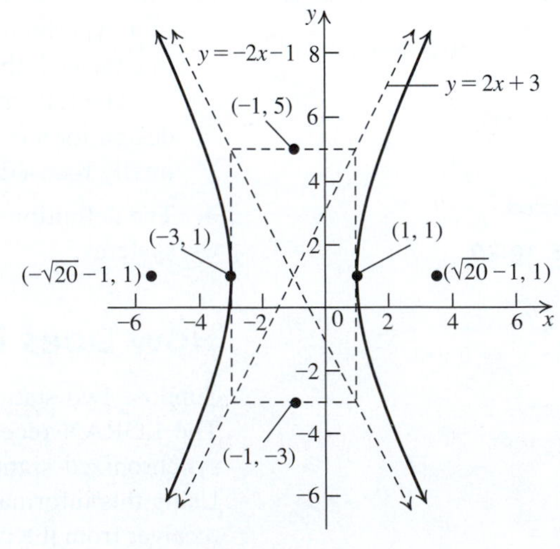

b. $\frac{(y-1)^2}{4} - \frac{(x+1)^2}{9} = 1 \Rightarrow a = 2, b = 3$, and center $(-1, 1)$. The vertices are $(-1, 1-2) = (-1, -1)$ and $(-1, 1+2) = (-1, 3)$. The endpoints of the conjugate axes are $(-1-3, 1) = (-4, 1)$ and $(-1+3, 1) = (2, 1)$ The asymptotes are $y - k = \pm\frac{a}{b}(x - h) \Rightarrow y - 1 = \pm\frac{2}{3}(x+1)$. $c^2 = a^2 + b^2 \Rightarrow c^2 = 4 + 9 \Rightarrow c = \pm\sqrt{13}$ The foci are $(-1, 1+\sqrt{13})$ and $(-1, 1-\sqrt{13})$.

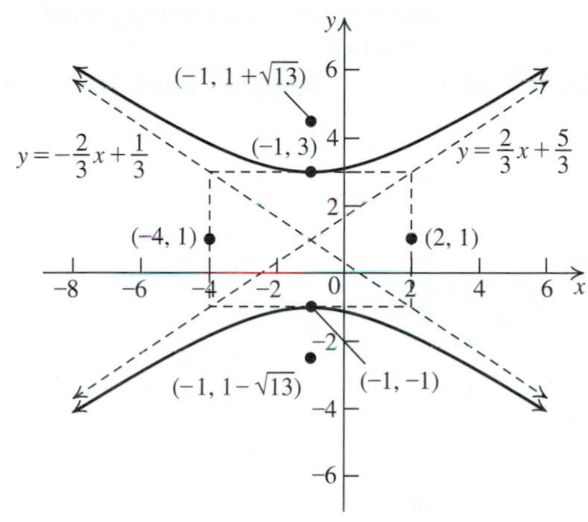

7. $\frac{(x-1)^2}{5} - \frac{(y-2)^2}{\frac{5}{4}} = 1$. Center: $(1, 2)$; vertices: $(1+\sqrt{5}, 2)$ and $(1-\sqrt{5}, 2)$; endpoints of the conjugate axis: $\left(1, 2+\frac{\sqrt{5}}{2}\right)$ and $\left(1, 2-\frac{\sqrt{5}}{2}\right)$; asymptotes: $y - 2 = \pm\frac{1}{2}(x-1)$; foci: $\left(\frac{7}{2}, 2\right)$ and $\left(-\frac{3}{2}, 2\right)$

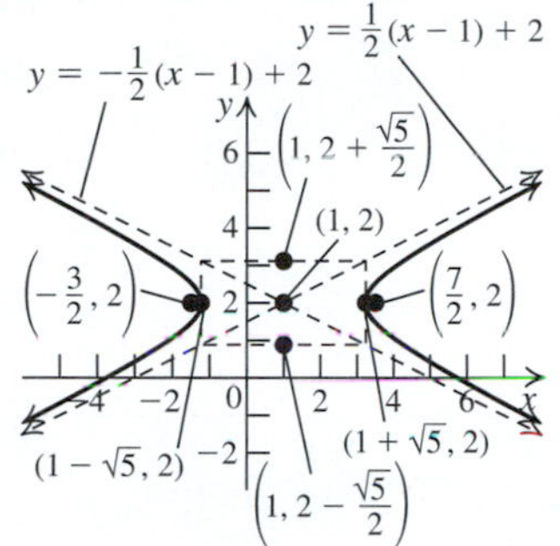

8. $\frac{x^2}{16{,}900} - \frac{y^2}{5{,}600} = 1$

SECTION 10.4 Exercises

Basic Concepts and Skills

1. A hyperbola is a set of all points in the plane the absolute value of the ______________ of whose distances from two fixed points is constant.
2. The line segment joining the two vertices of a hyperbola is called the ______________ axis.
3. The standard equation of the hyperbola with center $(0, 0)$, vertices $(\pm a, 0)$, and foci $(\pm c, 0)$ is
$$\frac{x^2}{a^2} - \frac{y^2}{b^2} = 1, \text{ where } b^2 = \underline{\qquad\qquad}.$$
4. For the hyperbola $\frac{y^2}{a^2} - \frac{x^2}{b^2} = 1$, the vertices are ______________ and ______________ and the foci are $(0, \pm c)$, where $c^2 =$ ______________.
5. **True or False.** The graph of $\frac{x^2}{a^2} - \frac{y^2}{b^2} = -1$ is a hyperbola.
6. **True or False.** The graph of $Ax^2 + Cy^2 + Dx + Ey + F = 0$ is a degenerate conic or a hyperbola if $AC < 0$.

In Exercises 7–14, the equation of a hyperbola is given. Match each equation with its graph shown in (a)–(h).

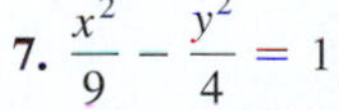

7. $\frac{x^2}{9} - \frac{y^2}{4} = 1$
8. $\frac{x^2}{4} - \frac{y^2}{25} = 1$
9. $\frac{y^2}{25} - \frac{x^2}{4} = 1$
10. $\frac{y^2}{4} - \frac{x^2}{9} = 1$
11. $4x^2 - 25y^2 = 100$
12. $16x^2 - 49y^2 = 196$

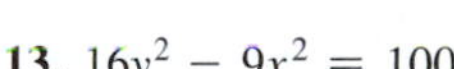

13. $16y^2 - 9x^2 = 100$
14. $9y^2 - 16x^2 = 144$

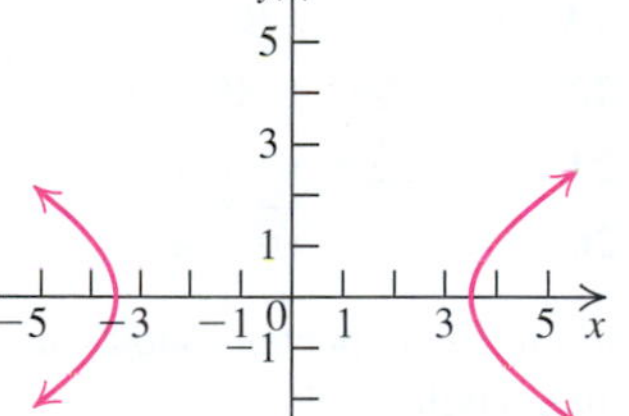

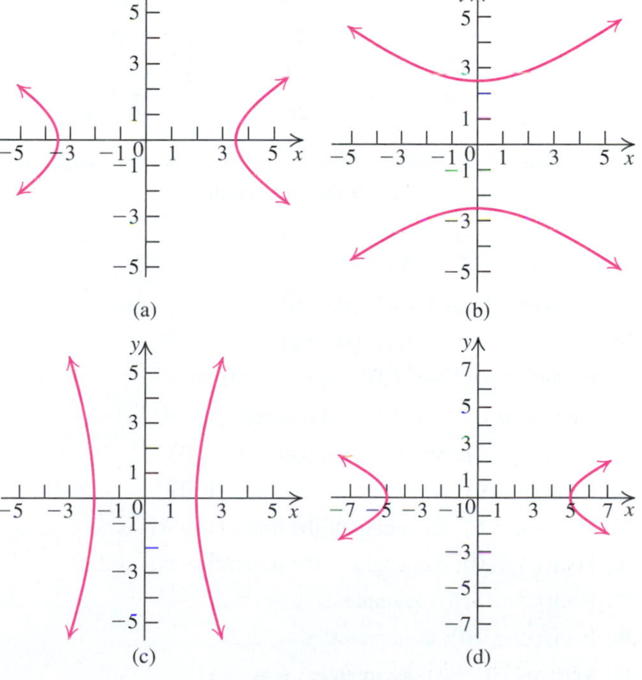

(a) (b) (c) (d)

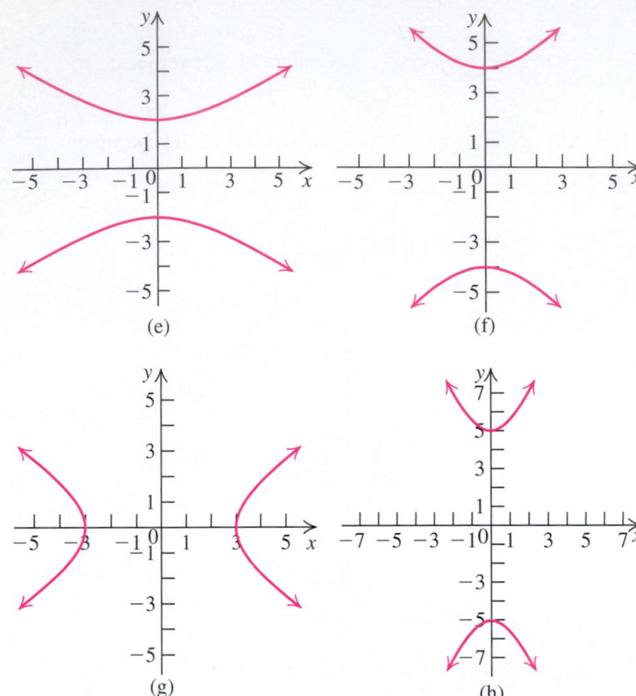

In Exercises 15–26, the equation of a hyperbola is given.

a. Find and plot the vertices, the foci, and the transverse axis of the hyperbola.
b. State how the hyperbola opens.
c. Find and plot the vertices of the fundamental rectangle.
d. Sketch the asymptotes and write their equations.
e. Graph the hyperbola by using the vertices and the asymptotes.

15. $x^2 - \frac{y^2}{4} = 1$

16. $y^2 - \frac{x^2}{4} = 1$

17. $x^2 - y^2 = -1$

18. $y^2 - x^2 = -1$

19. $9y^2 - x^2 = 36$

20. $9x^2 - y^2 = 36$

21. $4x^2 - 9y^2 - 36 = 0$

22. $4x^2 - 9y^2 + 36 = 0$

23. $y = \pm\sqrt{4x^2 + 1}$

24. $y = \pm 2\sqrt{x^2 + 1}$

25. $y = \pm\sqrt{9x^2 - 1}$

26. $y = \pm 3\sqrt{x^2 - 4}$

In Exercises 27–40, find an equation of the hyperbola satisfying the given conditions. Graph the hyperbola.

27. Vertices: $(\pm 2, 0)$; foci: $(\pm 3, 0)$

28. Vertices: $(\pm 3, 0)$; foci: $(\pm 5, 0)$

29. Vertices: $(0, \pm 4)$; foci: $(0, \pm 6)$

30. Vertices: $(0, \pm 5)$; foci: $(0, \pm 8)$

31. Center: $(0, 0)$; vertex: $(0, 2)$; focus: $(0, 5)$

32. Center: $(0, 0)$; vertex: $(0, -1)$; focus: $(0, -4)$

33. Center: $(0, 0)$; vertex: $(1, 0)$; focus: $(-5, 0)$

34. Center: $(0, 0)$; vertex: $(-3, 0)$; focus: $(6, 0)$

35. Foci: $(0, \pm 5)$; the length of the transverse axis is 6.

36. Foci: $(\pm 2, 0)$; the length of the transverse axis is 2.

37. Foci: $(\pm\sqrt{5}, 0)$; asymptotes: $y = \pm 2x$

38. Foci: $(0, \pm 10)$; asymptotes: $y = \pm 3x$

39. Vertices: $(0, \pm 4)$; asymptotes: $y = \pm x$

40. Vertices: $(\pm 3, 0)$; asymptotes: $y = \pm 2x$

In Exercises 41–48, use a graphing device to graph each hyperbola.

41. $\frac{x^2}{2} - \frac{y^2}{4} = 1$

42. $\frac{x^2}{6} - \frac{y^2}{2} = 1$

43. $2y^2 - x^2 = 6$

44. $y^2 - 3x^2 = 12$

45. $(x + 1)^2 - 2y^2 = 4$

46. $3x^2 - (y - 2)^2 = 6$

47. $2x^2 - y^2 - 4x - 2y - 7 = 0$

48. $4y^2 - 3x^2 - 16y - 6x + 1 = 0$

In Exercises 49–68, the equation of a hyperbola is given.

a. Find and plot the center, vertices, transverse axis, and asymptotes of the hyperbola.
b. Use the vertices and the asymptotes to graph the hyperbola.

49. $\frac{(x - 1)^2}{9} - \frac{(y + 1)^2}{16} = 1$

50. $\frac{(y - 1)^2}{16} - \frac{(x + 1)^2}{4} = 1$

51. $\frac{(x + 2)^2}{25} - \frac{y^2}{49} = 1$

52. $\frac{(x + 1)^2}{9} - \frac{(y + 2)^2}{36} = 1$

53. $\frac{(x + 4)^2}{25} - \frac{(y + 3)^2}{49} = 1$

54. $\frac{(x - 3)^2}{9} - \frac{(y + 1)^2}{9} = 1$

55. $4x^2 - (y + 1)^2 = 25$

56. $9(x - 1)^2 - y^2 = 144$

57. $(y + 1)^2 - 9(x - 2)^2 = 25$

58. $6(x - 4)^2 - 3(y + 3)^2 = 4$

59. $x^2 - y^2 + 6x = 36$

60. $x^2 + 4x - 4y^2 = 12$

61. $x^2 - 4y^2 - 4x = 0$

62. $x^2 - 2y^2 - 8y = 12$

63. $2x^2 - y^2 + 12x - 8y + 3 = 0$

64. $y^2 - 9x^2 - 4y - 30x = 33$

65. $3x^2 - 18x - 2y^2 - 8y + 1 = 0$

66. $4y^2 - 9x^2 - 8y - 36x = 68$

67. $y^2 + 2\sqrt{2} - x^2 + 2\sqrt{2}x = 1$

68. $x^2 + x = \frac{y^2 - 1}{4}$

In Exercises 69–78, identify the conic section represented by each equation and sketch the graph.

69. $x^2 - 6x + 12y + 33 = 0$

70. $9y^2 = x^2 - 4x$

71. $y^2 - 9x^2 = -1$

72. $4x^2 + 9y^2 - 8x + 36y + 4 = 0$

73. $x^2 + y^2 - 4x + 8y = 16$

74. $2y^2 + 3x - 4y - 7 = 0$

75. $2x^2 - 4x + 3y + 8 = 0$

76. $y^2 - 9x^2 + 18x - 4y = 14$

77. $4x^2 + 9y^2 + 8x - 54y + 49 = 0$

78. $x^2 - 4y^2 + 6x + 12y = 0$

Applying the Concepts

79. Sound of an explosion. Points A and B are 1000 meters apart, and it is determined from the sound of an explosion heard at these points at different times that the location of the explosion is 600 meters closer to A than to B. Show that the location of the explosion is restricted to points on a hyperbola and find the equation of the hyperbola. [*Hint:* Let the coordinates of A be $(-500, 0)$ and those of B be $(500, 0)$.]

80. Sound of an explosion. Points A and B are 2 miles apart. The sound of an explosion at A was heard three seconds before it was heard at B. Assume that the sound travels at 1100 feet/second. Show that the location of the explosion is restricted to a hyperbola and find the equation of the hyperbola. [*Hint:* Recall that 1 mile = 5280 feet.]

81. Thunder and lightning. Nicole and Juan, who are 8000 feet apart, hear thunder. Nicole hears the thunder three seconds before Juan does. Find the equation of the hyperbola whose points are locations where the lightning might have struck. Assume that the sound travels at 1100 feet/second.

82. Locating source of thunder. In Exercise 81, Valerie is at a location midway between Nicole and Juan, and she hears the thunder two seconds after Juan does. Determine the location where the lightning strikes in relation to the three people involved.

83. LORAN. Two LORAN stations, A and B, are situated 300 kilometers apart along a straight coastline. Simultaneous radio signals are sent from each station to a ship. The ship receives the signal from A 0.0005 second before the signal from B. Assume that the radio signals travel 300,000 kilometers per second. Find the equation of the hyperbola on which the ship is located.

84. LORAN. In Exercise 83, assuming that the ship follows the graph of the hyperbola, determine the location of the ship when it reaches the coastline.

85. Target practice. A gun at G and a target at T are 1600 feet apart. The muzzle velocity of the bullet is 2000 feet per second. A person at P hears the crack of the gun and the thud of the bullet at the same time. Show that the possible locations of the person are on a hyperbola. Find the equation of the hyperbola with G at $(-800, 0)$ and T at $(800, 0)$. Assume that the speed of sound is 1100 feet/second. [*Hint:* Show that $|d(P, G) - (P, T)|$ is a constant.]

86. LORAN. Two LORAN stations, A and B, lie on an east–west line with A 250 miles west of B. A plane is flying west on a line 50 miles north of the line AB. Radio signals are sent (traveling at 980 feet per microsecond) simultaneously from A and B, and the one sent from B arrives at the plane 500 microseconds before the one from A. Where is the plane?

87. Using LORAN. LORAN navigational transmitters A, B, C, and D are located at $(-100, 0)$, $(100, 0)$, $(0, -150)$, and $(0, 150)$, respectively. A navigator P on a ship somewhere in the second quadrant listens to the pair (A, B) of transmitters and computes the difference of the distances from the ship to A and B as 120 miles. Similarly, by listening to the pair (C, D) of transmitters, navigator P computes the difference of the distances from the ship to C and D as 80 miles. Use a calculator to find the coordinates of the ship.

88. Using LORAN. LORAN navigational transmitters A, B, and C are located at $(200, 0)$, $(-200, 0)$, and $(200, 1000)$, respectively. A navigator on a fishing boat listens to the pair (A, B) of transmitters and finds that the difference of the distances from the boat to A and B is 300 miles. She also finds that the difference of the distances from the boat to A and C is 400 miles. Use a calculator to find the possible locations of the boat.

Beyond the Basics

89. Write the standard form of the equation of the hyperbola for which the difference of the distances from any point on the hyperbola to $(-5, 0)$ and $(5, 0)$ is equal to (a) 2 units; (b) 4 units; (c) 8 units.

90. Sketch the graphs of all three hyperbolas in Exercise 89 on the same coordinate plane.

91. Write the standard form of the equation of the hyperbola for which the difference of the distances from any point on the hyperbola to the points F_1 and F_2 is 2 units, where

a. F_1 is $(0, 6)$ and F_2 is $(0, -6)$.
b. F_1 is $(0, 4)$ and F_2 $(0, -4)$.
c. F_1 is $(0, 3)$ and F_2 $(0, -3)$.

92. Sketch the graphs of all three hyperbolas in Exercise 91 on the same coordinate plane.

93. Rewrite equation

$$\sqrt{(x + c)^2 + y^2} - \sqrt{(x - c)^2 + y^2} = \pm 2a$$

from page 888 as

$$\sqrt{(x + c)^2 + y^2} = \pm 2a + \sqrt{(x - c)^2 + y^2}.$$

Square both sides to obtain

$$(x + c)^2 + y^2 = 4a^2 \pm 4a\sqrt{(x - c)^2 + y^2} + (x - c)^2 + y^2.$$

Simplify and isolate the radical; then square both sides again to show that the equation can be simplified to

$$(c^2 - a^2)x^2 - a^2y^2 = a^2(c^2 - a^2).$$

94. a. Let $p > 0$. Prove that the graph of the equation $xy = \frac{p^2}{2}$ is a hyperbola by showing that this equation is the equation of the hyperbola with foci (p, p) and $(-p, -p)$ and with difference $2p$ in distance.

[*Hint:* Show that the equation

$$\sqrt{(x + p)^2 + (y + p)^2} - \sqrt{(x - p)^2 + (y - p)^2} = 2p$$

simplifies to $xy = \frac{p^2}{2}$.]

b. Use part (a) to sketch the graph of the hyperbola $xy = 8$. Find the coordinates of the foci.

95. A hyperbola for which the lengths of the transverse and conjugate axes are equal is called an **equilateral hyperbola**. Show that the asymptotes of an equilateral hyperbola are perpendicular to one another.

96. The **latus rectum of a hyperbola** is a line segment that passes through a focus, is perpendicular to the transverse axis, and has endpoints on the hyperbola. Show that the length of the latus rectum of the hyperbola

$$\frac{x^2}{a^2} - \frac{y^2}{b^2} = 1 \text{ is } \frac{2b^2}{a}.$$

97. The ***eccentricity* of a hyperbola**, denoted by e, is defined by

$$e = \frac{\text{Distance between the foci}}{\text{Distance between the vertices}} = \frac{2c}{2a} = \frac{c}{a}.$$

Show that for every hyperbola, $e > 1$. What happens when $e = 1$?

In Exercises 98–102, find the eccentricity and the length of the latus rectum of each hyperbola.

98. $\frac{x^2}{16} - \frac{y^2}{9} = 1$ **99.** $36x^2 - 25y^2 = 900$

100. $x^2 - y^2 = 49$

101. $8(x - 1)^2 - (y + 2)^2 = 2$

102. $5x^2 - 4y^2 - 10x - 8y - 19 = 0$

103. For the hyperbola $\frac{x^2}{a^2} - \frac{y^2}{b^2} = 1$ with eccentricity e, show that $b^2 = a^2(e^2 - 1)$.

104. Find an equation of a hyperbola assuming that its foci are $(\pm 6, 0)$ and the length of its latus rectum is 10.

105. A point $P(x, y)$ moves in the plane so that its distance from the point $(3, 0)$ is always twice its distance from the line $x = -1$. Show that the equation of the path of P is a hyperbola of eccentricity 2.

106. A point $P(x, y)$ moves in the plane so that its distance from the point $(0, -3)$ is always three times its distance from the line $y = 1$. Show that the equation of the path of P is a hyperbola of eccentricity 3.

In Exercises 107–112, find all points of intersection of the given curves. Make a sketch of the curves that shows the points of intersection.

107. $y - 2x - 20 = 0$ and $y^2 - 4x^2 = 36$

108. $x - 2y^2 = 0$ and $x^2 = 8y^2 + 5$

109. $x^2 + y^2 = 15$ and $x^2 - y^2 = 1$

110. $x^2 + 9y^2 = 9$ and $4x^2 - 25y^2 = 36$

111. $9x^2 + 4y^2 = 72$ and $x^2 - 9y^2 = -77$

112. $3y^2 - 7x^2 = 5$ and $9x^2 - 2y^2 = 1$

In Exercises 113 and 114, use the following definition. A *tangent line* to a hyperbola is a line that is not parallel to either asymptote of the hyperbola and that intersects the hyperbola at only one point.

113. Find the equation of the tangent line to the hyperbola with equation $\frac{x^2}{a^2} - \frac{y^2}{b^2} = 1$, with $a = \sqrt{\frac{5}{3}}$ and $b = \sqrt{\frac{5}{7}}$ at the point $(2, 1)$.
[*Hint:* See the steps for Exercise 91 in Section 10.2 and (in Step 3) write the discriminant as a perfect square.]

114. Show that the equation of the tangent line to the hyperbola with equation $\frac{x^2}{a^2} - \frac{y^2}{b^2} = 1$, at the point (x_1, y_1), is

$$y = \frac{b^2x_1}{a^2y_1}x - \frac{b^2x_1^2}{a^2y_1^2} + y_1.$$

[*Hint:* See guidelines for Exercise 92 in Section 10.2 and show that $(a^2y_1m - b^2x_1)^2 = 0$.]

Critical Thinking/Writing/Discussion

115. Explain why each of the following represents a degenerate conic section. Where possible, sketch the graph.

a. $4x^2 - 9y^2 = 0$
b. $x^2 + 4y^2 + 6 = 0$
c. $8x^2 + 5y^2 = 0$
d. $x^2 + y^2 - 4x + 8y = -20$
e. $y^2 - 2x^2 = 0$

116. Discuss the graph of a quadratic equation in x and y of the form

$$Ax^2 + Cy^2 + Dx + Ey + F = 0, \quad AC \neq 0.$$

Rewrite that equation in the form

$$A\left(x^2 + \frac{D}{A}x + \frac{D^2}{4A^2}\right) + C\left(y^2 + \frac{E}{C}y + \frac{E^2}{4C^2}\right)$$
$$= \frac{D^2}{4A} + \frac{E^2}{4C} - F.$$

Include in your discussion the types of graphs you will obtain in each of the following cases:

i. $A > 0, C > 0$ **ii.** $A > 0, C < 0$
iii. $A < 0, C > 0$ **iv.** $A < 0, C < 0$

In each case, discuss how the classification is affected by the sign of

$$\frac{D^2}{4A} + \frac{E^2}{4C} - F.$$

117. Explain why the hyperbolas with equations $x^2 - y^2 = \pm 1$ have the same asymptotes. Sketch the graphs on the same coordinate system.

Maintaining Skills

In Exercises 118–121, find the value of each function for the given number.

118. $f(x) = \frac{1}{1 + x}; x = 3$ **119.** $g(x) = 2x + 1; x = 10$

120. $h(x) = (-1)^3 3^{x-1}; x = 5$

121. $f(x) = \frac{16x}{x^2 + 12}; x = 6$

In Exercises 122–125, find the values of the given expression for the specified integer, *n*.

122. $(-1)^n; n = 17$ **123.** $(-1)^n \frac{1}{n + 1}; n = 8$

124. $(-1)^{n+1}2^n; n = 5$

125. $(-1)^{n-2}(2n + 1) + (-1)^n; n = 7$

In Exercises 126 and 127, specify the integer that is the result of the computation, where *a* represents a nonzero real number.

126. $\frac{3 \cdot \pi \cdot e^2 \cdot 7 \cdot 11 \cdot 19 \cdot a}{a \cdot 3 \cdot \pi \cdot e^2 \cdot 19}$

127. $\frac{a \cdot 2 \cdot 3 \cdot 4 \cdot 5 \cdot 6 \cdot 7 \cdot 8 \cdot 9 \cdot 10}{3 \cdot 5 \cdot 7 \cdot 8 \cdot 9 \cdot 10 \cdot a}$

In Exercises 128 and 129, specify whether the given statement is true or false.

128. $c(1 - 5a + 3b) = c - 5ac + 3bc$

129. $3a - 2b + 6c - 5d = (3a + 6c) - (2b + 5d)$

SUMMARY Definitions, Concepts, and Formulas

10.1 Conic Sections: Overview

Conic sections are the curves formed when a plane intersects the surface of a right circular cone. Except in some degenerate cases, the curves formed are the circle, the ellipse, the parabola, and the hyperbola.

10.2 The Parabola

i. **Definition.** A **parabola** is the set of all points in the plane that are the same distance from a fixed line and a fixed point not on the line. The fixed line is called the **directrix**, and the fixed point is called the **focus**.

ii. **Axis or axis of symmetry.** This is the line that passes through the focus and is perpendicular to the directrix.

iii. **Vertex.** This is the point at which the parabola intersects its axis. The vertex is halfway between the focus and the directrix.

iv. **Standard forms of equations of parabolas with vertex at (0,0) and $p > 0$ are**

$y^2 = \pm 4px$ and $x^2 = \pm 4py$. (See the table on page 865.)

v. The **latus rectum** of a parabola is the line segment that passes through the focus, is perpendicular to the axis of the parabola, and has endpoints on the parabola.

vi. **Standard forms of equations of parabolas with vertex (h, k) and $p > 0$** are $(y - k)^2 = \pm 4p(x - h)$ and $(x - h)^2 = \pm 4p(y - k)$. (See the table on page 868.)

10.3 The Ellipse

i. **Definition.** An **ellipse** is the set of all points in the plane, the sum of whose distances from two fixed points is a constant. The fixed points are called the **foci** of the ellipse.

ii. **Vertices.** These are the two points where the line through the foci intersects the ellipse.

iii. **Major axis.** This is the line segment joining the two vertices of the ellipse.

iv. **Center.** This is the midpoint of the line segment joining the foci. It is also the midpoint of the major axis.

v. **Minor axis.** This is the line segment that passes through the center of the ellipse, is perpendicular to the major axis, and has endpoints on the ellipse.

vi. **Standard forms of equations of ellipses with center $(0, 0)$, $a > b > 0$, $c < a$,** and $b^2 = a^2 - c^2$ are

$$\frac{x^2}{a^2} + \frac{y^2}{b^2} = 1, \text{ with foci } (\pm c, 0) \text{ and vertices } (\pm a, 0),$$

and

$$\frac{x^2}{b^2} + \frac{y^2}{a^2} = 1, \text{ with foci } (0, \pm c) \text{ and vertices } (0, \pm a).$$

See the table on page 877.

vii. **Standard forms of equations of ellipses with center (h, k), $a > b > 0$,** and $b^2 = a^2 - c^2$ are

$$\frac{(x-h)^2}{a^2} + \frac{(y-k)^2}{b^2} = 1 \text{ and } \frac{(x-h)^2}{b^2} + \frac{(y-k)^2}{a^2} = 1.$$

See the table on page 879.

10.4 The Hyperbola

i. **Definition.** A **hyperbola** is the set of all points in the plane, the difference of whose distances from two fixed points is a constant. The fixed points are called the **foci** of the hyperbola.

ii. **Vertices.** These are the two points where the line through the foci intersects the hyperbola.

iii. **Transverse axis.** This is the line segment joining the two vertices of the hyperbola.

iv. **Center.** This is the midpoint of the line segment joining the foci. It is also the midpoint of the transverse axis.

v. **Standard forms of equations of hyperbolas with center $(0, 0)$, $c > a$,** and $b^2 = c^2 - a^2$ are

$$\frac{x^2}{a^2} - \frac{y^2}{b^2} = 1, \text{ with foci } (\pm c, 0) \text{ and vertices } (\pm a, 0)$$

and

$$\frac{y^2}{a^2} - \frac{x^2}{b^2} = 1, \text{ with foci } (0, \pm c) \text{ and vertices } (0, \pm a).$$

See the table on page 889.

vi. **Conjugate axis.** This is a line segment of length $2b$ that passes through the center of the hyperbola and is perpendicularly bisected by the transverse axis.

vii. **Asymptotes.** The asymptotes of the hyperbola $\frac{x^2}{a^2} - \frac{y^2}{b^2} = 1$ are the two lines $y = \pm\frac{b}{a}x$; the asymptotes of the hyperbola $\frac{y^2}{a^2} - \frac{x^2}{b^2} = 1$ are the lines $y = \pm\frac{a}{b}x$.

viii. **Graphing.** A procedure for graphing a hyperbola centered at $(0, 0)$ is given on page 893.

ix. **Standard forms of equations of hyperbolas centered at (h, k)** are

$$\frac{(x-h)^2}{a^2} - \frac{(y-k)^2}{b^2} = 1 \text{ and } \frac{(y-k)^2}{a^2} - \frac{(x-h)^2}{b^2} = 1.$$

See the table on page 894. Also see page 895 for a procedure for graphing such hyperbolas.

REVIEW EXERCISES

In Exercises 1–8, sketch the graph of each parabola. Determine the vertex, focus, axis, and directrix of each parabola.

1. $y^2 = -6x$ **2.** $y^2 = 12x$

3. $x^2 = 7y$ **4.** $x^2 = -3y$

5. $(x - 2)^2 = -(y + 3)$ **6.** $(y + 1)^2 = 5(x + 2)$

7. $y^2 = -4y + 2x + 1$ **8.** $-x^2 + 2x + y = 0$

In Exercises 9–12, find an equation of the parabola satisfying the given conditions.

9. Vertex: $(0, 0)$ focus: $(-3, 0)$

10. Vertex: $(0, 0)$ focus: $(0, 4)$

11. Focus: $(0, 4)$ directrix: $y = -4$

12. Focus: $(-3, 0)$; directrix: $x = 3$

In Exercises 13–20, sketch the graph of each ellipse. Determine the foci, vertices, and endpoints of the minor axis of the ellipse.

13. $\dfrac{x^2}{25} + \dfrac{y^2}{4} = 1$ **14.** $\dfrac{x^2}{9} + \dfrac{y^2}{36} = 1$

15. $4x^2 + y^2 = 4$ **16.** $16x^2 + y^2 = 64$

17. $16(x + 1)^2 + 9(y + 4)^2 = 144$

18. $4(x - 1)^2 + 3(y + 2)^2 = 12$

19. $x^2 + 9y^2 + 2x - 18y + 1 = 0$

20. $4x^2 + y^2 + 8x - 10y + 13 = 0$

In Exercises 21–24, find an equation of the ellipse satisfying the given conditions.

21. Vertices: $(\pm 4, 0)$; endpoints of minor axis: $(0, \pm 2)$

22. Vertices: $(0, \pm 6)$; endpoints of minor axis: $(\pm 2, 0)$

23. Length of major axis: 20; foci: $(\pm 5, 0)$

24. Length of minor axis: 16; foci: $(0, \pm 6)$

In Exercises 25–32, sketch the graph of each hyperbola. Determine the vertices, the foci, and the asymptotes of each hyperbola.

25. $\dfrac{y^2}{16} - \dfrac{x^2}{4} = 1$ **26.** $\dfrac{x^2}{16} - \dfrac{y^2}{9} = 1$

27. $8x^2 - y^2 = 8$ **28.** $4y^2 - 4x^2 = 1$

29. $\dfrac{(x + 2)^2}{9} - \dfrac{(y - 3)^2}{4} = 1$

30. $\dfrac{(y + 1)^2}{6} - \dfrac{(x - 2)^2}{8} = 1$

31. $4y^2 - x^2 + 40y - 4x + 60 = 0$

32. $4x^2 - 9y^2 + 16x - 54y - 29 = 0$

In Exercises 33–36, find an equation of the hyperbola satisfying the given condition.

33. Vertices: $(\pm 1, 0)$; foci: $(\pm 2, 0)$

34. Vertices: $(0, \pm 2)$; foci: $(0, \pm 4)$

35. Vertices: $(\pm 2, 0)$; asymptotes: $y = \pm 3x$

36. Vertices: $(0, \pm 3)$; asymptotes: $y = \pm x$

In Exercises 37–48, identify each equation as representing a circle, a parabola, an ellipse, or a hyperbola. Sketch the graph of the conic.

37. $5x^2 - 4y^2 = 20$ **38.** $x^2 - x + y = 1$

39. $3x^2 + 4y^2 + 8y - 12x - 6 = 0$

40. $2y - x^2 = 0$ **41.** $x^2 + y^2 + 2x - 3 = 0$

42. $2x + y^2 = 0$ **43.** $y^2 = x^2 + 3$

44. $x^2 = 10 - 3y^2$

45. $3x^2 + 3y^2 - 6x + 12y + 5 = 0$

46. $y + 2x - x^2 + 2y^2 = 0$ **47.** $9x^2 + 8y^2 = 36$

48. $2y^2 + 4y = 3x^2 - 6x + 9$

49. Find an equation of the hyperbola whose foci are the vertices of the ellipse $4x^2 + 9y^2 = 36$ and whose vertices are the foci of this ellipse.

50. Find an equation of the ellipse whose foci are the vertices of the hyperbola $9x^2 - 16y^2 = 144$ and whose vertices are the foci of this hyperbola.

In Exercises 51–54, find all points of intersection of the given curves and make a sketch.

51. $x^2 - 4y^2 = 36$ and $x - 2y - 20 = 0$

52. $y^2 - 8x^2 = 5$ and $y - 2x^2 = 0$

53. $3x^2 - 7y^2 = 5$ and $9y^2 - 2x^2 = 1$

54. $x^2 - y^2 = 1$ and $x^2 + y^2 = 7$

55. Parabolic curve. Water flowing from the end of a horizontal pipe 20 feet above the ground describes a parabolic curve whose vertex is at the end of the pipe. If at a point 6 feet below the line of the pipe the flow of water has curved outward 8 feet beyond a vertical line through the end of the pipe, how far beyond this vertical line will the water strike the ground?

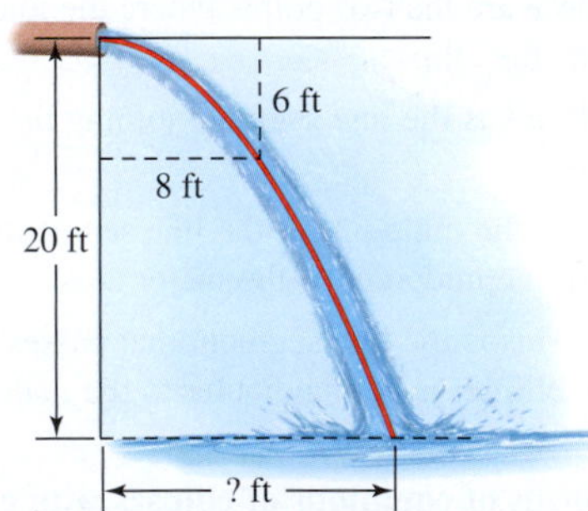

56. Parabolic arch. A parabolic arch has a height of 20 meters and a width of 36 meters at the base. Assuming that the vertex of the parabola is at the top of the arch, find the height of the arch at a distance of 9 meters from the center of the base.

57. Football. A football is 12 inches long, and a cross section containing a seam is an ellipse with a minor axis of length 7 inches. Suppose every cross section of the ball formed by a plane perpendicular to the major axis of the ellipse is a circle. Find the circumference of such a circular cross section located 2 inches from an end of the ball.

58. Production cost. A company assembles computers at two locations A and B that are 1000 miles apart. The per unit cost of production at location A is \$20 less than at location B. Assume that the route of delivery of the computers is along a straight line and that the delivery cost is 25¢ per unit per mile. Find the equation of the curve at any point the computers can be supplied from either location at the same cost. [*Hint:* Take A at $(-500, 0)$ and B at $(500, 0)$.]

PRACTICE TEST A

1. Find the standard form of the parabola with focus $(0, 12)$ and directrix with equation $x = -12$.
2. Convert the equation $y^2 - 2y + 8x + 25 = 0$ to the standard form for a parabola by completing the square.
3. Find the focus and directrix of the parabola with equation $x^2 = -9y$.
4. Graph the parabola with equation $y^2 = -5x$.
5. Find the vertex, focus, and directrix of the parabola with equation $(x + 2)^2 = -8(y - 1)$.
6. The base of a water slide is parabolic in shape and is 6 feet wide and 2 feet deep. Find the height of the slide 1 foot from its center.
7. Find an equation of the parabola with vertex $(3, -1)$ and directrix $x = -3$.

In Problems 8–10, find the standard form of the equation of the ellipse satisfying the given conditions.

8. Foci: $(0, -2), (0, 2)$ vertices: $(0, -4), (0, 4)$
9. Foci: $(-2, 0)(2, 0)$; y-intercepts: $-5, 5$
10. Major axis horizontal with length 18; minor axis of length 4; center $(0, 0)$
11. Graph the ellipse with equation $9x^2 + y^2 = 9$.
12. Find the vertices and foci for the hyperbola with equation $49y^2 - x^2 = 49$.
13. Graph the hyperbola with equation $\frac{x^2}{49} - \frac{y^2}{4} = 1$.
14. Find the standard form of the equation of the hyperbola with foci $(0, -\sqrt{45}), (0, \sqrt{45})$ and vertices $(0, -6), (0, 6)$.

In Problems 15 and 16, convert the equation to the standard form for a hyperbola by completing the squares on x and y.

15. $y^2 - x^2 + 2x = 2$
16. $x^2 - 2x - 4y^2 - 16y = 19$

In Problems 17–20, identify the conic section and sketch its graph.

17. $x^2 - 9y^2 = -16$
18. $x^2 - x + 4y + \frac{1}{4} = 0$
19. $x^2 + y^2 + 2x - 6y = 3$
20. $25x^2 + 4y^2 = 100$

PRACTICE TEST B

1. Find the standard form of the equation of the parabola with focus $(-10, 0)$ and directrix $x = 10$.
   a. $y^2 = -10x$ **b.** $y^2 = -40x$
   c. $x^2 = -40y$ **d.** $y^2 = 40x$
2. Convert the equation $y^2 - 4y - 5x + 24 = 0$ to the standard form for a parabola by completing the square.
   a. $(y + 2)^2 = 5(x - 4)$ **b.** $(y - 2)^2 = 5(x - 4)$
   c. $(y + 2)^2 = -5(x - 4)$ **d.** $(y - 2)^2 = 5(x + 4)$
3. Find the focus and directrix of the parabola with equation $x = 7y^2$.
   a. focus: $\left(\frac{1}{28}, 0\right)$; directrix: $x = -\frac{1}{28}$
   b. focus: $\left(0, \frac{1}{28}\right)$; directrix: $y = -\frac{1}{28}$
   c. focus: $\left(\frac{1}{28}, 0\right)$; directrix: $x = \frac{1}{28}$
   d. focus: $\left(\frac{1}{7}, 0\right)$ directrix: $x = -\frac{1}{7}$

4. Graph the parabola with equation $y^2 = -7x$.

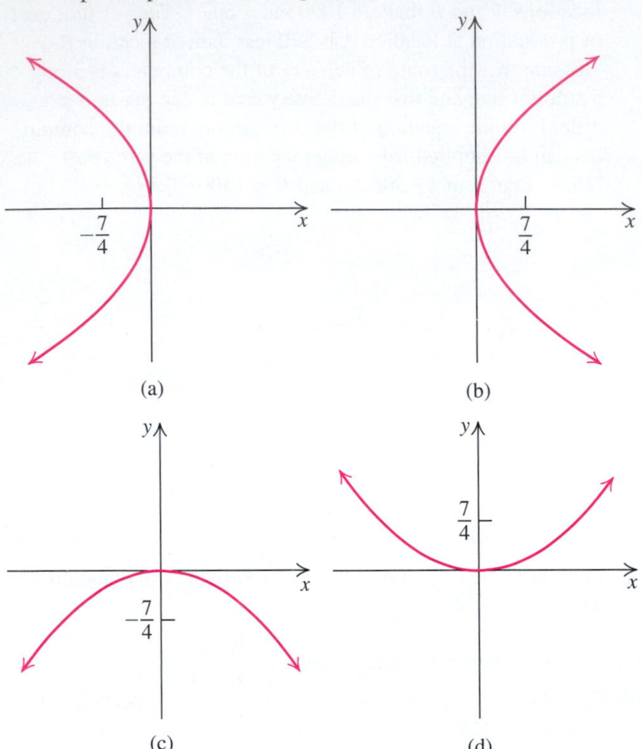

5. Find the vertex, focus, and directrix of the parabola with equation $(x - 3)^2 = 12(y - 1)$.
   a. vertex: $(3, 1)$; focus: $(3, 4)$; directrix: $y = -2$
   b. vertex: $(3, -1)$; focus: $(-3, 2)$; directrix: $y = -4$
   c. vertex: $(3, 1)$; focus: $(3, -2)$; directrix: $x = 4$
   d. vertex: $(1, 3)$; focus: $(1, 6)$; directrix: $y = 0$

6. The parabolic arch of a bridge has a 180-foot base and a height of 25 feet. Find the height of the arch 45 feet from the center of the base.
   a. 12.5 ft b. 6.25 ft
   c. 16.7 ft d. 18.75 ft

7. Find an equation of the parabola with vertex $(-2, 1)$ and directrix $x = 2$.
   a. $(y - 1)^2 = -16(x + 2)$
   b. $(y - 1)^2 = 16(x + 2)$
   c. $(y + 2)^2 = -16(x - 1)$
   d. $(y + 2)^2 = 16(x - 1)$

In Problems 8–10, find the standard form of the equation of the ellipse satisfying the given conditions.

8. Foci: $(-3, 0)$, $(3, 0)$; vertices: $(-5, 0)$, $(5, 0)$
   a. $\frac{x^2}{9} + \frac{y^2}{16} = 1$ b. $\frac{x^2}{25} + \frac{y^2}{16} = 1$
   c. $\frac{x^2}{9} + \frac{y^2}{25} = 1$ d. $\frac{x^2}{16} + \frac{y^2}{25} = 1$

9. Foci: $(0, -3)$, $(0, 3)$; y-intercepts: $-7, 7$
   a. $\frac{x^2}{49} + \frac{y^2}{40} = 1$ b. $\frac{x^2}{9} + \frac{y^2}{49} = 1$
   c. $\frac{x^2}{9} + \frac{y^2}{40} = 1$ d. $\frac{x^2}{40} + \frac{y^2}{49} = 1$

10. Major axis vertical with length 16; length of minor axis = 8; center: $(0, 0)$
   a. $\frac{x^2}{16} + \frac{y^2}{64} = 1$ b. $\frac{x^2}{64} + \frac{y^2}{256} = 1$
   c. $\frac{x^2}{8} + \frac{y^2}{64} = 1$ d. $\frac{x^2}{64} + \frac{y^2}{16} = 1$

11. Graph the ellipse with equation $9(x - 1)^2 + 4(y - 2)^2 = 36$.

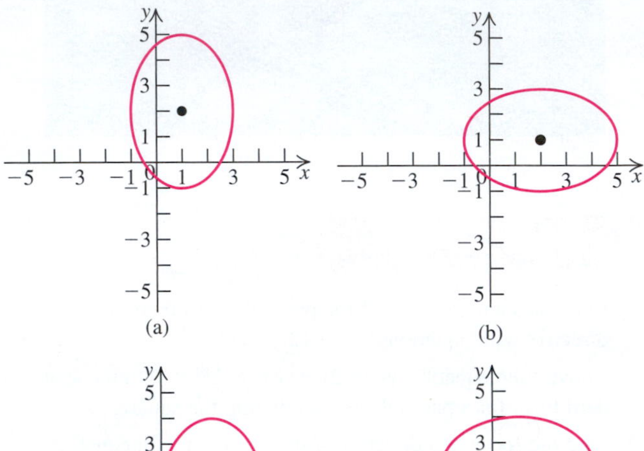

12. Find the vertices and foci of the hyperbola with equation $\frac{x^2}{121} - \frac{y^2}{4} = 1$.
   a. vertices: $(-11, 0), (11, 0)$; foci: $(-2, 0)$, $(2, 0)$
   b. vertices: $(0, -1), (0, 11)$; foci: $(-5\sqrt{5}, 0)(5\sqrt{5}, 0)$
   c. vertices: $(-11, 0), (11, 0)$; foci: $(-5\sqrt{5}, 0)(5\sqrt{5}, 0)$
   d. vertices: $(-2, 0), (2, 0)$; foci: $(-5\sqrt{5}, 0)(5\sqrt{5}, 0)$

13. Graph the hyperbola with equation $\frac{x^2}{25} - \frac{y^2}{4} = 1$.

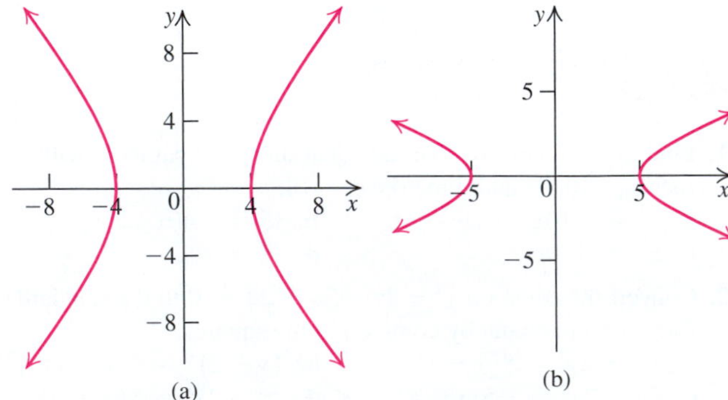

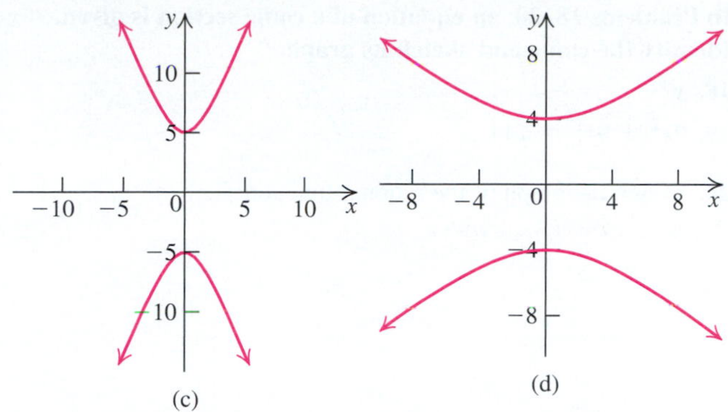

(c) (d)

14. Find the standard form of the equation of the hyperbola with foci $(0, -10)$, $(0, 10)$ and vertices $(0, -5)$, $(0, 5)$.

a. $\dfrac{y^2}{25} - \dfrac{x^2}{100} = 1$ **b.** $\dfrac{y^2}{25} - \dfrac{x^2}{75} = 1$

c. $\dfrac{x^2}{25} - \dfrac{y^2}{100} = 1$ **d.** $\dfrac{x^2}{25} - \dfrac{y^2}{75} = 1$

In Problems 15 and 16, convert the equation to the standard form for a hyperbola by completing the square on *x* and *y*.

15. $y^2 - 4x^2 - 2y - 16x - 19 = 0$

a. $\dfrac{(y-2)^2}{4} - (x+4)^2 = 1$

b. $\dfrac{(x-1)^2}{4} - (y+2)^2 = 1$

c. $\dfrac{(y-1)^2}{4} - (x+2)^2 = 1$

d. $(x+2)^2 - \dfrac{(y-1)^2}{4} = 1$

16. $4y^2 - 9x^2 - 16y - 36x - 56 = 0$

a. $\dfrac{(y-2)^2}{4} - \dfrac{(x+2)^2}{9} = 1$

b. $\dfrac{(x-2)^2}{4} - \dfrac{(y+2)^2}{9} = 1$

c. $\dfrac{(y-2)^2}{9} - \dfrac{(x+2)^2}{4} = 1$

d. $\dfrac{(y+2)^2}{9} - \dfrac{(x-2)^2}{4} = 1$

In Problems 17–20, identify the conic section.

17. $3x^2 + 2y^2 - 6x + 4y - 1 = 0$

a. parabola **b.** circle

c. ellipse **d.** hyperbola

18. $x^2 + 2x + 4y + 5 = 0$

a. parabola **b.** circle

c. ellipse **d.** hyperbola

19. $4x^2 - y^2 + 16x + 2y + 11 = 0$

a. parabola **b.** circle

c. ellipse **d.** hyperbola

20. $5y^2 - 3x^2 + 20y - 6x + 2 = 0$

a. parabola **b.** circle

c. ellipse **d.** hyperbola

CUMULATIVE REVIEW EXERCISES CHAPTERS P–10

1. Let $f(x) = x^2 - 3x + 2$. Find $\dfrac{f(x+h) - f(x)}{h}$.

2. Sketch the graph of $f(x) = \sqrt{x+1} - 2$.

3. Let $f(x) = 2x - 3$. Find the inverse function f^{-1}. Verify that $f(f^{-1}(x)) = x$.

4. Sketch the graph of $f(x) = \left(\dfrac{1}{2}\right)^{x+2}$.

5. Solve the equation $\log_5(x-1) + \log_5(x-2) = 3\log_5\sqrt[3]{6}$.

6. Express

$$\log_a \sqrt[3]{x\sqrt{yz}}$$

in terms of logarithms of x, y, and z.

7. Solve the inequality $\dfrac{x}{x-2} \geq 1$.

8. The current I in an electric circuit is given by

$$I = \frac{V}{R}(1-e^{-0.3t}).$$

Use natural logarithms to solve for t.

In Problems 9–12, solve each system of equations.

9. $\begin{cases} 1.4x - 0.5y = 1.3 \\ 0.4x + 1.1y = 4.1 \end{cases}$

10. $\begin{cases} 2x + y - 4z = 3 \\ x - 2y + 3z = 4 \\ -3x + 4y - z = -2 \end{cases}$

11. $\begin{cases} y = 2 - \log x \\ y - \log(x+3) = 1 \end{cases}$

12. $\begin{cases} y = x^2 - 1 \\ 3x^2 + 8y^2 = 8 \end{cases}$

13. Find the determinant of the matrix $\begin{bmatrix} 1 & 4 & 7 \\ 2 & 5 & 8 \\ 3 & 6 & 9 \end{bmatrix}$.

14. Use Cramer's Rule to solve the system of equations.

$$\begin{cases} 2x - 3y = -4 \\ 5x + 7y = 1 \end{cases}$$

15. Find the inverse of the matrix $A = \begin{bmatrix} 3 & -2 \\ -5 & 4 \end{bmatrix}$.

16. Find an equation of the line that passes through the point of intersection of the lines $x + 2y - 3 = 0$ and $3x + 4y - 5 = 0$ and that is perpendicular to the line $x - 3y + 5 = 0$. Write your answer in slope–intercept form.

17. Solve the equation $2x^4 - 5x^2 + 3 = 0$.

In Problems 18–20, an equation of a conic section is given. Identify the conic and sketch its graph.

18. $x^2 - y^2 = -4$

19. $9x^2 + 9y^2 = 144$

20. Sketch the graph of the rational function $f(x) = \dfrac{x}{x^2 - 16}$.

CHAPTER 11

Further Topics in Algebra

TOPICS

Probability was studied in ancient Eastern civilizations, including those in Babylonia and Egypt. Probability is most famously used in gambling, and Egyptians even used objects made from the heel bones of animals, called *tali*, as a type of dice. Today, probability is used not only in gambling, but also in polling, insurance, business, the life sciences, and many other areas. In this chapter, we learn to understand and compute the probability of an event.

SECTION 11.1

Sequences and Series

BEFORE STARTING THIS SECTION, REVIEW

1. Algebraic expressions (Section P.1, page 13)
2. Finding functional values (Section 2.4, page 198)

OBJECTIVES

1. Use sequence notation and find specific and general terms in a sequence.
2. Use factorial notation.
3. Use summation notation to write partial sums of a series.

◆ The Family Tree of the Honeybee

The bee that most of us know best is the honeybee, an insect that lives in a colony and has an unusual family tree. Perhaps the most surprising fact about honeybees is that not all of them have two parents. This phenomenon begins with a special female in the colony called the *queen*. Many other female bees, called *worker bees*, live in the colony, but unlike the queen bee, they produce no eggs. Male bees do no work and are produced from the queen's unfertilized eggs. The female bees are produced as a result of the queen bee mating with a male bee. Consequently, all female bees have two parents—a male and a female—whereas male bees have just one parent—a female. In this section, we study sequences, and in Example 4, we see how a famous sequence accurately counts a honeybee's ancestors.

1 Use sequence notation and find specific and general terms in a sequence.

Sequences

The word *sequence* is used in mathematics in much the same way it is used in ordinary English. If someone saw a *sequence* of bad movies, you know that the person could list the first bad movie he or she saw, the second bad movie, and so on.

Sequence

An **infinite sequence** is a function whose domain is the set of positive integers. The function values, written as

$$a_1, a_2, a_3, a_4, \ldots, a_n, \ldots,$$

are called the **terms** of the sequence. The ***n*th term**, a_n, is called the **general term** of the sequence.

If the domain of a function consists of only the first n positive integers, the sequence is called a **finite sequence**.

In computer science, finite sequences of the form $a_1, a_2, \ldots, a_n$ are called *strings*. Sometimes for convenience, we include 0 in the domain of the function that defines a sequence, and we write the sequence terms as $a_0, a_1, a_2, a_3, \ldots$. The nth term of such a sequence is a_{n-1}.

DO YOU **KNOW?**

Notice in Example 1 that converting $a_n = 5n - 1$ to standard function notation gives us $f(n) = 5n - 1$, a linear function.

TECHNOLOGY **CONNECTION**

A graphing calculator can write the terms of a sequence and graph the sequence. You must first select *sequence* mode.

The first four terms of the sequence $a_n = \dfrac{1}{n+1}$ and the graph of this sequence are shown here. Use the right arrow key ▶ to scroll to the right to display the sequence values corresponding to $n = 3$ and $n = 4$.

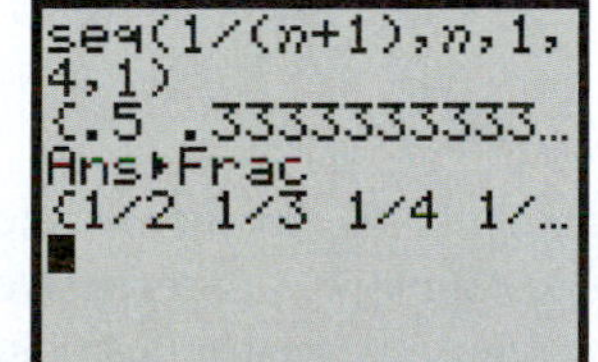

Now press GRAPH.

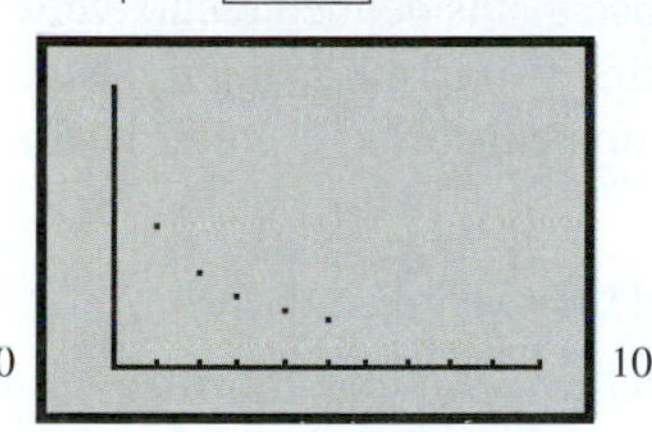

[0, 10, 1] by [0, 1, 1]

We can use subscripts on variables other than a to represent the terms of a sequence. In Example 1(c), we use the variable b.

EXAMPLE 1 Writing the Terms of a Sequence from the General Term

Write the first four terms of each sequence.

a. $a_n = 5n - 1$ **b.** $a_n = \dfrac{1}{n+1}$ **c.** $b_n = (-1)^{n+1}\left(\dfrac{1}{n}\right)$

Solution

The first four terms of each sequence are found by replacing n with the integers 1, 2, 3, and 4 in the equation defining a_n.

a.
$a_n = 5n - 1$ General term of the sequence
$a_1 = 5(1) - 1 = 4$ 1st term: Replace n with 1.
$a_2 = 5(2) - 1 = 9$ 2nd term: Replace n with 2.
$a_3 = 5(3) - 1 = 14$ 3rd term: Replace n with 3.
$a_4 = 5(4) - 1 = 19$ 4th term: Replace n with 4.

The first four terms of the sequence are 4, 9, 14, and 19.

b.
$$a_n = \frac{1}{n+1}$$
$$a_1 = \frac{1}{1+1} = \frac{1}{2}$$
$$a_2 = \frac{1}{2+1} = \frac{1}{3}$$
$$a_3 = \frac{1}{3+1} = \frac{1}{4}$$
$$a_4 = \frac{1}{4+1} = \frac{1}{5}$$

The first four terms are $\frac{1}{2}, \frac{1}{3}, \frac{1}{4}$, and $\frac{1}{5}$.

c.
$$b_n = (-1)^{n+1}\left(\frac{1}{n}\right)$$
$$b_1 = (-1)^{1+1}\left(\frac{1}{1}\right) = (-1)^2(1) = 1$$
$$b_2 = (-1)^{2+1}\left(\frac{1}{2}\right) = (-1)^3\left(\frac{1}{2}\right) = -\frac{1}{2}$$
$$b_3 = (-1)^{3+1}\left(\frac{1}{3}\right) = (-1)^4\left(\frac{1}{3}\right) = \frac{1}{3}$$
$$b_4 = (-1)^{4+1}\left(\frac{1}{4}\right) = (-1)^5\left(\frac{1}{4}\right) = -\frac{1}{4}$$

The first four terms are $1, -\frac{1}{2}, \frac{1}{3}$, and $-\frac{1}{4}$.

Practice Problem 1 Write the first four terms of the sequence with general term $a_n = -2^n$.

Frequently, the first few terms of a sequence exhibit a pattern that *suggests* a natural choice for the general term of the sequence.

EXAMPLE 2 Finding a General Term of a Sequence from a Pattern

Write the general term a_n for a sequence whose first five terms are given.

a. $1, 4, 9, 16, 25, \ldots$ **b.** $0, \frac{1}{2}, -\frac{2}{3}, \frac{3}{4}, -\frac{4}{5}, \ldots$

Solution

Write the position number of the term above each term of the sequence and look for a pattern that connects the term to the position number of the term.

a.

n:	1,	2,	3,	4,	5,	. . . ,	n
term:	1,	4,	9,	16,	25,	. . . ,	a_n

Apparent pattern: Here $1 = 1^2, 4 = 2^2, 9 = 3^2, 16 = 4^2$, and $25 = 5^2$. Each term is the square of the position number of that term. This suggests $a_n = n^2$.

b.

$$\begin{array}{lcccccccc} n: & 1, & 2, & 3, & 4, & 5, & \ldots, & n \\ \text{term:} & 0, & \frac{1}{2}, & -\frac{2}{3}, & \frac{3}{4}, & -\frac{4}{5}, & \ldots, & a_n \end{array}$$

Apparent pattern: When the terms alternate in sign and $n = 1$, we use the factor $(-1)^n$ if we want to begin with the factor -1 and we use the factor $(-1)^{n+1}$ if we want to begin with the factor 1. Notice that each term can be written as a quotient with denominator equal to the position number and numerator equal to one less than the position number, suggesting the general term $a_n = (-1)^n \dfrac{n-1}{n}$.

Practice Problem 2 Write a general term for a sequence whose first five terms are $0, -\frac{1}{2}, \frac{2}{3}, -\frac{3}{4}, \frac{4}{5}, \ldots$.

WARNING

In Example 2, we found the general term of a sequence by finding a pattern in the first few terms of the sequence. However, when only a finite number of successive terms are given without a rule that defines the general term, a *unique* general term cannot be found.

To indicate why this is so, consider the two sequences:

$$a_n = n^2, \text{ and}$$
$$b_n = n^2 + (n-1)(n-2)(n-3)(n-4).$$

The first four terms of these two sequences are identical: 1, 4, 9, 16. However, the fifth term is different: $a_5 = 25$, but $b_5 = 49$. Thus, either a_n or b_n could correctly describe a sequence whose first four terms are 1, 4, 9, and 16. So you can *never* find a unique general term from a finite number of terms in a sequence.

Leonardo of Pisa (Fibonacci)

(1175–1250)

Leonardo, often known today by the name Fibonacci (son of Bonaccio) was born around 1175. He traveled extensively through North Africa and the Mediterranean, probably on business with his father. At each location, he met with Islamic scholars and absorbed the mathematical knowledge of the Islamic world. After returning to Pisa around 1200, he started writing books on mathematics. He was one of the earliest European writers on algebra. In his most famous book *Liber abaci*, or *Book of Calculations*, he introduced the Western world to the rules of computing with the new Hindu–Arabic numerals.

Recursive Formulas

Up to this point, we have expressed the general term as a function of the position of the term in the sequence. A second approach is to define a sequence *recursively*. A **recursive formula** requires that one or more of the first few terms of the sequence be specified and all other terms be defined in relation to previously defined terms.

For example, the **Fibonacci sequence** is a famous sequence that is defined recursively and occurs often in nature. In this sequence, we specify the first two terms as $a_0 = 1$ and $a_1 = 1$; each subsequent term is the sum of the two terms immediately preceding it. So we have

$$\begin{array}{ll} a_0 = 1, a_1 = 1, a_2 = a_0 + a_1 = 1 + 1 = 2 & \text{Add the two given terms.} \\ a_1 = 1, a_2 = 2, a_3 = a_1 + a_2 = 1 + 2 = 3 & \text{Add the two previous terms.} \\ a_2 = 2, a_3 = 3, a_4 = a_2 + a_3 = 2 + 3 = 5 & \text{Add the two previous terms.} \\ a_3 = 3, a_4 = 5, a_5 = a_3 + a_4 = 3 + 5 = 8 & \text{Add the two previous terms.} \end{array}$$

The first six terms of the sequence are then

$$1, 1, 2, 3, 5, 8.$$

This sequence can also be defined in subscript notation:

$$a_0 = 1, a_1 = 1, a_k = a_{k-2} + a_{k-1}, \text{ for all } k \geq 2$$

Replacing k with 2, 3, 4, and 5, respectively, in the equation $a_k = a_{k-2} + a_{k-1}$ will produce the values $a_2 = 2, a_3 = 3, a_4 = 5$, and $a_5 = 8$.

EXAMPLE 3 Finding Terms of a Recursively Defined Sequence

Write the first five terms of the recursively defined sequence

$$a_1 = 4, \; a_{n+1} = 2a_n - 9.$$

Solution

We are given the first term of the sequence: $a_1 = 4$.

$a_2 = 2a_1 - 9$	Replace n with 1 in $a_{n+1} = 2a_n - 9$.
$a_2 = 2(4) - 9 = -1$	Replace a_1 with 4.
$a_3 = 2a_2 - 9$	Replace n with 2 in $a_{n+1} = 2a_n - 9$.
$a_3 = 2(-1) - 9 = -11$	Replace a_2 with -1.
$a_4 = 2a_3 - 9$	Replace n with 3 in $a_{n+1} = 2a_n - 9$.
$a_4 = 2(-11) - 9 = -31$	Replace a_3 with -11.
$a_5 = 2a_4 - 9$	Replace n with 4 in $a_{n+1} = 2a_n - 9$.
$a_5 = 2(-31) - 9 = -71$	Replace a_4 with -31.

Thus, the first five terms of the sequence are

$$4, -1, -11, -31, -71.$$

Practice Problem 3 Write the first five terms of the recursively defined sequence $a_1 = -3, a_{n+1} = 2a_n + 5$.

EXAMPLE 4 Diagramming the Family Tree of a Honeybee

Find a sequence that accurately counts the ancestors of a male honeybee. The first term of the sequence should give the number of parents of a male honeybee, the second term the number of grandparents, the third term the number of great-grandparents, and so on.

Solution

Recall from the introduction to this section that female honeybees have two parents—a male and a female—but that male honeybees have just one parent—a female.

First, we consider the family tree of a male honeybee. A male bee has one parent: a female. Because his mother had two parents, he has two grandparents. Because his grandmother had two parents and his grandfather had only one parent, he has three great-grandparents.

Now to count the bee's ancestors, we count backward using Figure 11.1. You should recognize the first six terms of the Fibonacci sequence: 1, 1, 2, 3, 5, 8. This sequence is defined by $a_0 = 1, a_1 = 1$, and $a_k = a_{k-2} + a_{k-1}$, for all $k \geq 2$.

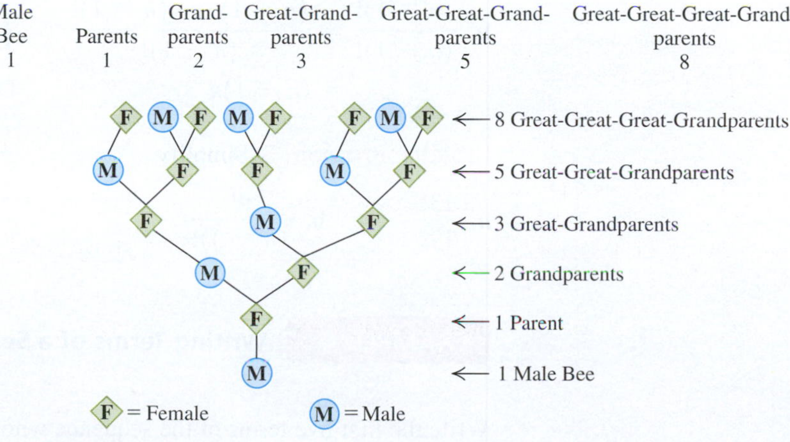

FIGURE 11.1

Practice Problem 4 Find a sequence that accurately counts the ancestors of a female honeybee.

2 Use factorial notation.

Factorial Notation

Special types of products, called factorials, appear in some very important sequences.

DO YOU **KNOW?**

The symbol $n!$ was first introduced by Christian Kramp of Strasbourg in 1808 as a convenience for the printer. Alternative symbols $\underline{|n}$ and $\pi(n)$ were used until late in the 19th century.

A typical calculator will compute approximate values of $n!$ up to $n = 69$.

$$69! \approx 1.7112 \times 10^{98}$$

Factorial

For any positive integer n, ***n factorial*** (written as $n!$) is defined as

$$n! = n \cdot (n-1) \cdots 4 \cdot 3 \cdot 2 \cdot 1.$$

As a special case, ***zero factorial*** $(0!)$ is defined by

$$0! = 1.$$

Here are the first eight values of $n!$

$$0! = 1 \qquad 4! = 4 \cdot 3 \cdot 2 \cdot 1 = 24$$
$$1! = 1 \qquad 5! = 5 \cdot 4 \cdot 3 \cdot 2 \cdot 1 = 120$$
$$2! = 2 \cdot 1 = 2 \qquad 6! = 6 \cdot 5 \cdot 4 \cdot 3 \cdot 2 \cdot 1 = 720$$
$$3! = 3 \cdot 2 \cdot 1 = 6 \qquad 7! = 7 \cdot 6 \cdot 5 \cdot 4 \cdot 3 \cdot 2 \cdot 1 = 5040$$

The values of $n!$ get large very quickly. For example, $10! = 3{,}628{,}800$ and $12! = 479{,}001{,}600$.

In simplifying some expressions involving factorials, it is helpful to note that

$$\boxed{n! = n \cdot (n-1)!}$$

TECHNOLOGY **CONNECTION**

EXAMPLE 5 **Simplifying a Factorial Expression**

Simplify.

a. $\dfrac{16!}{14!}$ **b.** $\dfrac{(n+1)!}{(n-1)!}$

Solution

a. $\dfrac{16!}{14!} = \dfrac{16 \cdot 15 \cdot 14!}{14!}$ $\quad 16! = 16 \cdot 15! = 16 \cdot 15 \cdot 14!$

$= 16 \cdot 15 = 240$ $\quad$ Divide out the factor 14!

b. $\dfrac{(n+1)!}{(n-1)!} = \dfrac{(n+1) \cdot n \cdot (n-1)!}{(n-1)!}$ $\quad (n+1)! = (n+1) \cdot n! = (n+1) \cdot n \cdot (n-1)!$

$= (n+1)n$ $\quad$ Divide out the common factor $(n-1)!$

Practice Problem 5 Simplify.

a. $\dfrac{13!}{12!}$ **b.** $\dfrac{n!}{(n-3)!}$

EXAMPLE 6 **Writing Terms of a Sequence Involving Factorials**

Write the first five terms of the sequence whose general term is $a_n = \dfrac{(-1)^{n+1}}{n!}$.

Solution

Replace n in the formula for the general term with each positive integer from 1 through 5.

$$a_1 = \frac{(-1)^{1+1}}{1!} = \frac{(-1)^2}{1} = 1 \qquad a_4 = \frac{(-1)^{4+1}}{4!} = \frac{(-1)^5}{24} = -\frac{1}{24}$$

$$a_2 = \frac{(-1)^{2+1}}{2!} = \frac{(-1)^3}{2} = -\frac{1}{2} \qquad a_5 = \frac{(-1)^{5+1}}{5!} = \frac{(-1)^6}{120} = \frac{1}{120}$$

$$a_3 = \frac{(-1)^{3+1}}{3!} = \frac{(-1)^4}{6} = \frac{1}{6}$$

Using these five terms, we could write this sequence as

$$1, -\frac{1}{2}, \frac{1}{6}, -\frac{1}{24}, \frac{1}{120}, \ldots .$$

Practice Problem 6 Write the first five terms of the sequence whose general term is

$$a_n = \frac{(-1)^n 2^n}{n!}.$$

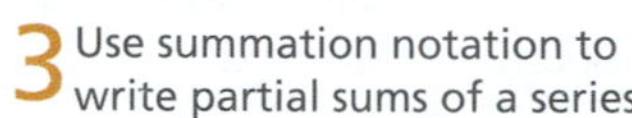

3 Use summation notation to write partial sums of a series.

Summation Notation

If we know the general term of a sequence, we can represent a *sum* of terms of the sequence by using *summation* (or *sigma*) *notation*. In this notation, the Greek letter Σ (capital sigma) indicates that we are to *add* the given terms. The letter Σ corresponds to the English letter S, the first letter of the word *sum*.

SUMMATION NOTATION

The sum of the first $n(\geq 1)$ terms of a sequence $a_1, a_2, a_3, \ldots, a_n, \ldots$ is denoted by

$$\sum_{i=1}^{n} a_i = a_1 + a_2 + a_3 + \cdots + a_n.$$

The letter i in the summation notation is called the **index of summation**, n is called the **upper limit**, and 1 is called the **lower limit** of the summation.

In summation notation, the index need not start at $i = 1$. Also, any letter may be used in place of the index i. In general, we have

$$\sum_{j=k}^{n} a_j = a_k + a_{k+1} + \cdots + a_n, \quad n \geq k \geq 0.$$

TECHNOLOGY CONNECTION

A graphing calculator in *sequence* mode can sum terms of a series.

The viewing window shows $\sum_{j=4}^{7}(2j^2 - 1)$. Note that the calculator replaces the variable j with n.

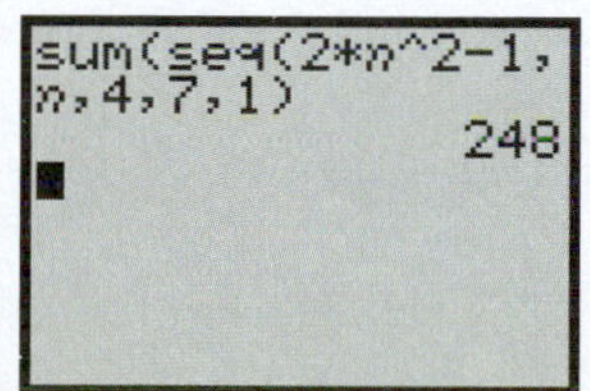

EXAMPLE 7 Evaluating Sums Given in Summation Notation

Find each sum.

a. $\sum_{i=1}^{9} i$ **b.** $\sum_{j=4}^{7}(2j^2 - 1)$ **c.** $\sum_{k=0}^{4} \frac{2^k}{k!}$

Solution

a. Replace i with successive integers from 1 to 9 inclusive; then add.

$$\sum_{i=1}^{9} i = 1 + 2 + 3 + 4 + 5 + 6 + 7 + 8 + 9 = 45$$

b. The index of summation is j. Evaluate $(2j^2 - 1)$ for all consecutive integers from 4 through 7 inclusive; then add.

$$\sum_{j=4}^{7}(2j^2 - 1) = [2(4)^2 - 1] + [2(5)^2 - 1] + [2(6)^2 - 1] + [2(7)^2 - 1]$$
$$= 31 + 49 + 71 + 97 = 248$$

c. The index of summation is k. Evaluate $\frac{2^k}{k!}$ for consecutive integers 0 through 4 inclusive; then add.

$$\sum_{k=0}^{4}\frac{2^k}{k!} = \frac{2^0}{0!} + \frac{2^1}{1!} + \frac{2^2}{2!} + \frac{2^3}{3!} + \frac{2^4}{4!}$$
$$= \frac{1}{1} + \frac{2}{1} + \frac{4}{2} + \frac{8}{6} + \frac{16}{24} = 1 + 2 + 2 + \frac{4}{3} + \frac{2}{3} = 7$$

Practice Problem 7 Find the following sum: $\sum_{k=0}^{3}(-1)^k k!$

The familiar properties of real numbers, such as the distributive, commutative, and associative properties, can be used to prove the summation properties listed next.

Summation Properties

Let a_k and b_k represent the general terms of two sequences and let c represent any real number. Then

1. $\sum_{k=1}^{n} c = cn$
2. $\sum_{k=1}^{n} ca_k = c\sum_{k=1}^{n} a_k$
3. $\sum_{k=1}^{n}(a_k + b_k) = \sum_{k=1}^{n} a_k + \sum_{k=1}^{n} b_k$
4. $\sum_{k=1}^{n}(a_k - b_k) = \sum_{k=1}^{n} a_k - \sum_{k=1}^{n} b_k$
5. $\sum_{k=1}^{n} a_k = \sum_{k=1}^{j} a_k + \sum_{k=j+1}^{n} a_k$, for $1 \le j < n$

Series

We all know how to add numbers, but do you think it is possible to add infinitely many numbers? It is sometimes possible to add infinitely many numbers, although a complete explanation of this kind of addition must be put off until a calculus course. For now, we need some definitions.

Series

Let $a_1, a_2, a_3, \ldots, a_k, \ldots$ be an infinite sequence. Then

1. The sum of all terms of the sequence is called a **series** and is denoted by

$$a_1 + a_2 + a_3 + \cdots = \sum_{i=1}^{\infty} a_i.$$

2. The sum of the first n terms of the sequence

$$a_1 + a_2 + a_3 + \cdots + a_n = \sum_{i=1}^{n} a_i$$

is called the ***n*th partial sum** of the series.

EXAMPLE 8 **Writing a Sum in Summation Notation**

Write each sum in summation notation.

a. $3 + 5 + 7 + \cdots + 21$ **b.** $\frac{1}{4} + \frac{1}{9} + \cdots + \frac{1}{49}$

Solution

a. The finite series $3 + 5 + 7 + \cdots + 21$ is a sum of consecutive odd numbers from 3 to 21. Each of these numbers can be expressed in the form $2k + 1$, starting with $k = 1$ and ending with $k = 10$. With k as the index of summation, 1 as the lower limit, and 10 as the upper limit, we write

$$3 + 5 + 7 + \cdots + 21 = \sum_{k=1}^{10} (2k + 1).$$

b. The finite series $\frac{1}{4} + \frac{1}{9} + \cdots + \frac{1}{49}$ is a sum of fractions, each of which has numerator 1 and denominator k^2, starting with $k = 2$ and ending with $k = 7$. We write

$$\frac{1}{4} + \frac{1}{9} + \cdots + \frac{1}{49} = \sum_{k=2}^{7} \frac{1}{k^2}.$$

Practice Problem 8 Write the following in summation notation: $2 - 4 + 6 - 8 + 10 - 12 + 14$

Answers to Practice Problems

1. $a_1 = -2, a_2 = -4, a_3 = -8, a_4 = -16$ **2.** $a_n = (-1)^{n+1}\left(1 - \frac{1}{n}\right)$ **3.** $a_1 = -3, a_2 = -1, a_3 = 3, a_4 = 11, a_5 = 27$ **4.** $a_1 = 2, a_2 = 3$, and $a_k = a_{k-2} + a_{k-1}$, for all $k \geq 3$ **5. a.** 13 **b.** $n(n - 1)(n - 2)$ **6.** $a_1 = -2, a_2 = 2, a_3 = -\frac{4}{3}, a_4 = \frac{2}{3}, a_5 = -\frac{4}{15}$ **7.** -4 **8.** $\sum_{k=1}^{7}(-1)^{k+1}(2k)$

SECTION 11.1 Exercises

Basic Concepts and Skills

1. An infinite sequence is a function whose domain is the set of ______________.

2. If a sequence is defined by $a_n = 2n - \frac{10}{n}$, then $a_5 =$ ______________

3. By definition, $0! =$ ______________.

4. Evaluate $\sum_{k=0}^{5} \frac{2k + 1}{2}$.

5. True or False. If $a_1 = 3$ and $a_{n+1} = a_n^2$, then $a_3 = 27$.

6. True or False. $\frac{(2n)!}{n!} = 2$.

In Exercises 7–26, write the first four terms of each sequence.

7. $a_n = 3n - 2$

8. $a_n = 2n + 1$

9. $a_n = 1 - \frac{1}{n}$

10. $a_n = 1 + \frac{1}{n}$

11. $a_n = -n^2$

12. $a_n = n^3$

13. $a_n = \frac{2n}{n + 1}$

14. $a_n = \frac{3n}{n^2 + 1}$

15. $a_n = (-1)^{n+1}$

16. $a_n = (-3)^{n-1}$

17. $a_n = 3 - \frac{1}{2^n}$

18. $a_n = \left(\frac{3}{2}\right)^n$

19. $a_n = 0.6$

20. $a_n = -0.4$

21. $a_n = \frac{(-1)^n}{n!}$

22. $a_n = \frac{n}{n!}$

23. $a_n = (-1)^n 3^{-n}$

24. $a_n = (-1)^n 3^{n-1}$

25. $a_n = \frac{e^n}{2n}$

26. $a_n = \frac{2^n}{e^n}$

In Exercises 27–40, write a general term a_n for each sequence. Assume that n begins with 1.

27. 1, 4, 7, 10, . . .

28. 7, 9, 11, 13, . . .

29. $\frac{1}{2}, \frac{1}{3}, \frac{1}{4}, \frac{1}{5}, \ldots$

30. $\frac{2}{3}, \frac{3}{4}, \frac{4}{5}, \frac{5}{6}, \ldots$

31. 2, −2, 2, −2, . . .

32. −3, 6, −9, 12, . . .

33. $\frac{1}{2}, \frac{3}{4}, \frac{9}{8}, \frac{27}{16}, \ldots$

34. $-\frac{1}{2}, \frac{1}{4}, -\frac{1}{8}, \frac{1}{16}, \ldots$

35. $1 \cdot 2, 2 \cdot 3, 3 \cdot 4, 4 \cdot 5, \ldots$

36. $-\frac{1}{1 \cdot 2}, \frac{1}{2 \cdot 3}, -\frac{1}{3 \cdot 4}, \frac{1}{4 \cdot 5}, \ldots$

37. $2 + \frac{1}{2}, 2 - \frac{1}{3}, 2 + \frac{1}{4}, 2 - \frac{1}{5}, \ldots$

38. $1 + \frac{1}{2}, 1 + \frac{1}{3}, 1 + \frac{1}{4}, 1 + \frac{1}{5}, \ldots$

39. $\frac{3^2}{2}, \frac{3^3}{3}, \frac{3^4}{4}, \frac{3^5}{5}, \ldots$

40. $\frac{e}{2}, \frac{e^2}{4}, \frac{e^3}{8}, \frac{e^4}{16}, \ldots$

In Exercises 41–50, write the first five terms of each recursively defined sequence.

41. $a_1 = 2, a_{n+1} = a_n + 3$

42. $a_1 = 5, a_{n+1} = a_n - 1$

43. $a_1 = 3, a_{n+1} = 2a_n$

44. $a_1 = 1, a_{n+1} = \frac{1}{2}a_n$

45. $a_1 = 7, a_{n+1} = -2a_n + 3$

46. $a_1 = -4, a_{n+1} = -3a_n - 5$

47. $a_1 = 2, a_{n+1} = \frac{1}{a_n}$

48. $a_1 = -1, a_{n+1} = \frac{-1}{a_n}$

49. $a_1 = 25, a_{n+1} = \frac{(-1)^n}{5a_n}$

50. $a_1 = 12, a_{n+1} = \frac{(-1)^n}{3a_n}$

In Exercises 51–56, use a graphing calculator to (a) find the first ten terms of the sequence and (b) graph the first ten terms of the sequence.

51. $a_n = 3n^2 - 1$

52. $a_n = 4 - \frac{3}{n}$

53. $a_n = n\left(1 - \frac{1}{n}\right)$

54. $a_n = n^3 - n^2$

55. $a_n = 1 - \frac{1}{a_{n-1}}, a_1 = \frac{1}{2}$

56. $a_n = a_{n-1}^2, a_1 = 1$

In Exercises 57–64, simplify the factorial expression.

57. $\frac{3!}{5!}$

58. $\frac{8!}{10!}$

59. $\frac{12!}{11!}$

60. $\frac{20!}{18!}$

61. $\frac{n!}{(n+1)!}$

62. $\frac{(n-1)!}{(n-2)!}$

63. $\frac{(2n+1)!}{(2n)!}$

64. $\frac{(2n+1)!}{(2n-1)!}$

In Exercises 65–76, find each sum.

65. $\sum_{k=1}^{7} 5$

66. $\sum_{j=1}^{4} 12$

67. $\sum_{j=0}^{5} j^2$

68. $\sum_{k=0}^{4} k^3$

69. $\sum_{i=1}^{5} (2i - 1)$

70. $\sum_{k=0}^{6} (1 - 3k)$

71. $\sum_{j=3}^{7} \frac{j+1}{j}$

72. $\sum_{k=3}^{8} \frac{1}{k+1}$

73. $\sum_{i=2}^{6} (-1)^i 3^{i-1}$

74. $\sum_{k=2}^{4} (-1)^{k+1} k$

75. $\sum_{k=4}^{7} (2 - k^2)$

76. $\sum_{j=4}^{9} (j^3 + 1)$

In Exercises 77–84, write each sum in summation notation.

77. $1 + 3 + 5 + 7 + \cdots + 101$

78. $2 + 4 + 6 + 8 + \cdots + 102$

79. $\frac{1}{5(1)} + \frac{1}{5(2)} + \frac{1}{5(3)} + \frac{1}{5(4)} + \cdots + \frac{1}{5(11)}$

80. $\frac{2}{1 \cdot 2} + \frac{2}{2 \cdot 3} + \frac{2}{3 \cdot 4} + \frac{2}{4 \cdot 5} + \cdots + \frac{2}{9 \cdot 10}$

81. $1 - \frac{1}{2} + \frac{1}{3} - \frac{1}{4} + \cdots - \frac{1}{50}$

82. $1 - 2 + 3 - 4 + \cdots + (-50)$

83. $\frac{1}{2} + \frac{2}{3} + \frac{3}{4} + \frac{4}{5} + \cdots + \frac{10}{11}$

84. $\frac{2}{1} + \frac{2^2}{2} + \frac{2^3}{3} + \frac{2^4}{4} + \cdots + \frac{2^{10}}{10}$

In Exercises 85–90, use a graphing calculator to find each sum.

85. $\sum_{i=1}^{10} 12i^2$

86. $\sum_{i=1}^{50} (2i + 7)$

87. $\sum_{k=5}^{30} \frac{7}{1 - k^2}$

88. $\sum_{k=10}^{25} \frac{(k+1)^2}{k}$

89. $\sum_{j=1}^{100} (-1)^j j$

90. $\sum_{j=8}^{50} \left(7 + \frac{(-1)^j}{j}\right)$

Applying the Concepts

91. **Free fall.** In the absence of friction, a freely falling body will fall about 16 feet the first second, 48 feet the next second, 80 feet the third second, 112 feet the fourth second, and so on. How far has it fallen during
a. The seventh second?
b. The nth second?

92. **Bacterial growth.** A colony of 1000 bacteria doubles in size every hour. How many bacteria will there be in
a. 2 hours?
b. 5 hours?
c. n hours?

93. **Workplace fines.** A contractor who quits work on a house was told she would be fined \$50 if she did not resume work on Monday, \$75 if she failed to resume work on Tuesday, \$100 if she did not resume work on Wednesday, and so on (including weekends). How much will her fine be on the ninth day she fails to show up for work?

94. **Appreciating value.** A painting valued at \$30,000 is expected to appreciate \$1280 the first year, \$1240 the second year, \$1200 the third year, and so on. How much will the painting appreciate in
a. The seventh year?
b. The tenth year?

95. **Cell phone use.** At the end of the first six months after a company began providing cell phones to its sales force, it averaged 600 cell minutes per month. For the next four years, its monthly cell phone use doubled every six months. How many cell minutes per month were being used at the end of three years?

96. **Motorcycle acceleration.** A motorcycle travels 10 yards the first second and then increases its speed by 20 yards per second in each succeeding second. At this rate, how far will the motorcycle travel during
a. The eighth second?
b. The nth second?

97. **Compound interest.** Suppose \$10,000 is deposited into an account that earns 6% interest compounded semiannually. The balance in the account after n compounding periods is given by the sequence

$$A_n = 10{,}000\left(1 + \frac{0.06}{2}\right)^n, n = 1, 2, 3, \ldots.$$

a. Find the first six terms of this sequence.
b. Find the balance in the account after 8 years.

98. **Compound interest.** Suppose \$100 is deposited at the beginning of a year into an account that earns 8% interest compounded quarterly. The balance in the account after n compounding periods is

$$A_n = 100\left(1 + \frac{0.08}{4}\right)^n, n = 1, 2, 3, \ldots.$$

a. Find the first six terms of this sequence.
b. Find the balance in the account after ten years.

99. **Real estate value.** Analysts estimate that a \$100,000 condominium will increase 5% in value each year for the next seven years. Find the value of the condominium in each of those years. Write a formula for a sequence whose first seven terms give these values.

100. **Diminishing savings.** Laura withdraws 10% of her \$50,000 savings each year. Find the amount remaining in her account for each of the next ten years. Write a formula for a sequence whose first ten terms give these values.

Beyond the Basics

101. Find a formula for the nth term of the sequence defined recursively by $a_1 = \sqrt{2}, a_{n+1} = \sqrt{2a_n}$. [*Hint*: Write each of the first five terms as a power of 2.]

102. The triangular tiles used in the figures shown have white interiors and gold edges. A sequence of figures is obtained by adding one triangle to the previous figure.

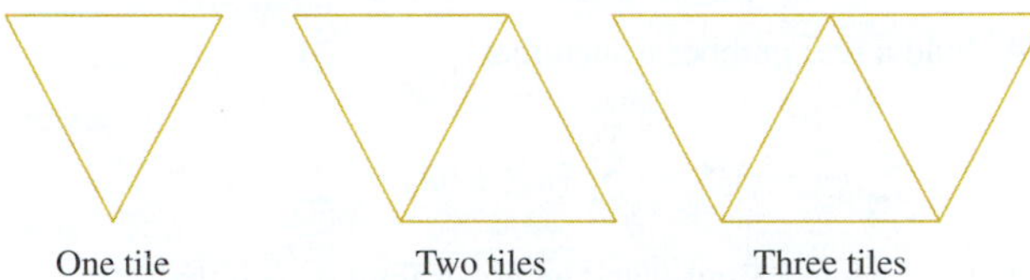

One tile Two tiles Three tiles

a. Write a recursive sequence whose nth term, a_n, gives the number of gold edges in the nth figure.
b. Write a recursive sequence whose nth term, b_n, gives the number of gold edges that lie on the perimeter of the nth figure.

103. The figure shows a sequence of successively smaller squares formed by connecting the midpoints of the sides of the preceding larger square. The area of the largest square is 1.
a. Write the first five terms of the sequence whose general term, a_n, gives the area of the nth square.
b. Write the general term, a_n, of the sequence.

104. A sequence of concentric circles is designed so that the radius of the first circle is 1 and each successive circle has a radius that is twice the length of the radius of the preceding circle. Write a recursive sequence whose nth term, a_n, gives the area of the nth circle.

105. Find the first ten terms of the sequence defined recursively by $a_1 = a_2 = 1, a_n = a_{a_{n-1}} + a_{n-a_{n-1}}$. In 1989, a prize of \$1000 was offered to whoever first discovered (for this sequence) a value of n for which $\left|\left|\frac{a_i}{i}\right| - \left|\frac{1}{2}\right|\right| < \frac{1}{20}$ for all $i > n$.

A mathematician named Colin L. Mallows of AT&T/Bell Laboratories found that $n = 1489$ and claimed the prize. (*Source:* http://el.media.mit.edu/logo-foundation/pubs/papers/easy_as_11223.html)

106. A sequence of numbers $a_0, a_1, a_2, a_3, \ldots$ satisfies the equation

$$a_n^2 = (-1)^n a_{n-1} + a_{n+1}.$$

If $a_0 = 1$ and $a_1 = 3$, find a_3.

107. Find a sequence with general term a_m such that

$$\sum_{n=1}^{20} n^2 = \sum_{m=0}^{19} a_m.$$

108. Find a real number c such that

$$\sum_{n=1}^{50} (n - 3)^2 = \sum_{n=1}^{50} (n^2 - 6n) + \sum_{n=1}^{50} c.$$

109. Find a lower limit p and an upper limit q such that

$$\sum_{n=0}^{10} n^3 = \sum_{m=p}^{q} (m - 2)^3.$$

110. Give an example of two sequences a_k and b_k such that

$$\sum_{k=1}^{20} a_k b_k \neq \left(\sum_{k=1}^{20} a_k\right)\left(\sum_{k=1}^{20} b_k\right).$$

111. Show that $7! - 2(5!) = 40(5!)$.

112. Show that $\frac{10!}{6!\,4!} + \frac{10!}{5!\,5!} = \frac{11!}{6!\,5!}$.

113. If $\frac{1}{5!} + \frac{1}{6!} = \frac{x}{7!}$, find x.

114. If $(n + 1)! = 12(n - 1)!$, find n.

Critical Thinking/Discussion/Writing

115. Find the largest integer value for the upper limit k if

$$\sum_{n=1}^{k} (n^2 + n)$$

$$= \sum_{n=1}^{k} [(n - 1)(n - 2)(n - 3)(n - 4)(n - 5) + n^2 + n].$$

116. **The Ulam conjecture.** Define the sequence a_n recursively by

$$a_n = \begin{cases} \frac{a_{n-1}}{2} & \text{if } a_{n-1} \text{ is even} \\ 3a_{n-1} + 1 & \text{if } a_{n-1} \text{ is odd.} \end{cases}$$

The Ulam conjecture says that if your first term a_0 is any positive integer, the sequence a_n will eventually reach the integer 1. Verify the conjecture for $a_0 = 13$.

117. Show that $\frac{(2n)!}{n!} = [1 \cdot 3 \cdot 5 \cdot \cdots \cdot (2n - 1)]2^n$.

118. Explain why $(n! + 1)$ is not divisible by any integer between 2 and n.

Maintaining Skills

For each sequence in Exercises 119–122, find $a_2 - a_1, a_3 - a_2, a_4 - a_3$, and $a_5 - a_4$.

119. 1, 5, 9, 13, 17

120. 2, 5.2, 8.4, 11.6, 14.8

121. 6, 1, −4, −9, −14

122. 5, 1.5, −2, −5.5, −9

In Exercises 123–126, the first term a_1 and a number d are given. Write the next four terms of the sequence defined by $a_2 = a_1 + d, a_3 = a_2 + d, a_4 = a_3 + d$, and $a_5 = a_4 + d$.

123. $a_1 = 3, d = 4$

124. $a_1 = 4, d = 2.5$

125. $a_1 = 7, d = -3$

126. $a_1 = 8, d = -4.5$

In Exercises 127–130, the nth term a_n is given. Find $a_2 - a_1, a_3 - a_2, a_4 - a_3$, and $a_5 - a_4$.

127. $a_n = 3n + 5$

128. $a_n = 5n - 7$

129. $a_n = -2n - 6$

130. $a_n = -3n + 8$

131. Let $a_n = 2n - 5$. Find a_{n+1} and a_{n-1}.

132. Let $a_n = -3n + 4$. Find a_{n+1} and a_{n-1}.

SECTION 11.2

Arithmetic Sequences; Partial Sums

BEFORE STARTING THIS SECTION, REVIEW

1. The general term of a sequence (Section 11.1, page 910)
2. Partial sums (Section 11.1, page 916)

OBJECTIVES

1. Identify an arithmetic sequence and find its common difference.
2. Find the sum of the first n terms of an arithmetic sequence.

Falling Space Junk

One April more than 300 kilograms (700 pounds) of "space junk" crashed to the ground in South Africa. The material was eventually identified as part of a Delta 2 rocket used to launch a global positioning satellite. Once an object begins a free fall toward Earth, it falls (in the absence of friction) 16 feet in the first second, 48 feet in the next second, 80 feet in the third second, and so on. The number of feet traveled in succeeding seconds is the sequence

$$16, 48, 80, 112, \ldots.$$

This is an example of an *arithmetic sequence*. We investigate such a sequence in Example 5.

1 Identify an arithmetic sequence and find its common difference.

Arithmetic Sequence

When the difference $(a_{n+1} - a_n)$ between any two consecutive terms of a sequence is always the same number, the sequence is called an *arithmetic sequence*. In other words, the sequence $a_1, a_2, a_3, a_4, \ldots$ is an arithmetic sequence if $a_2 - a_1 = a_3 - a_2 = a_4 - a_3 = \cdots$.

Arithmetic Sequence

The sequence

$$a_1, a_2, a_3, a_4, \ldots, a_n, \ldots$$

is an **arithmetic sequence**, or an **arithmetic progression**, if there is a number d such that each term in the sequence except the first is obtained from the preceding term by adding d to it. The number d is called the **common difference** of the arithmetic sequence. We have

$$d = a_{n+1} - a_n, \quad n \geq 1.$$

EXAMPLE 1 Finding the Common Difference

Find the common difference of each arithmetic sequence.

a. $3, 23, 43, 63, 83, \ldots$ **b.** $29, 19, 9, -1, -11, \ldots$

Solution

a. The common difference is $23 - 3 = 20$ (or $43 - 23$, $63 - 43$, or $83 - 63$).

b. The common difference is $19 - 29 = -10$ (or $9 - 19$, $-1 - 9$, or $-11 - (-1)$).

Practice Problem 1 Find the common difference of the arithmetic sequence

$$3, -2, -7, -12, -17, \ldots.$$

An arithmetic sequence can be completely specified by giving the first term a_1 and the common difference d. Let's see why. Suppose we are told that the sequence $a_1, a_2, a_3, a_4, \ldots$ is an arithmetic sequence and that $a_1 = 5$ and $d = 4$. Then

$$a_2 - a_1 = 4, \quad \text{so} \quad a_2 = a_1 + 4 = 5 + 4 = 9;$$
$$a_3 - a_2 = 4, \quad \text{so} \quad a_3 = a_2 + 4 = 9 + 4 = 13;$$
$$a_4 - a_3 = 4, \quad \text{so} \quad a_4 = a_3 + 4 = 13 + 4 = 17; \text{ and so on.}$$

Rewriting $d = a_{n+1} - a_n$ as $a_{n+1} = a_n + d$, we have a recursive definition of an arithmetic sequence.

Recursive Definition of an Arithmetic Sequence

An arithmetic sequence $a_1, a_2, a_3, a_4, \ldots, a_n, \ldots$ can be defined recursively. The recursive formula

$$a_{n+1} = a_n + d \text{ for } n \geq 1$$

defines an arithmetic sequence with first term a_1 and common difference d.

Consider an arithmetic sequence with first term a_1 and common difference d. Using the recursive definition $a_{n+1} = a_n + d$, for $n \geq 1$, we write

$a_1 = a_1$	The first term is a_1.
$a_2 = a_1 + d$	Replace n with 1 in $a_{n+1} = a_n + d$.
$a_3 = a_2 + d = (a_1 + d) + d = a_1 + 2d$	Replace a_2 with $a_1 + d$; simplify.
$a_4 = a_3 + d = (a_1 + 2d) + d = a_1 + 3d$	Replace a_3 with $a_1 + 2d$; simplify.
$a_5 = a_4 + d = (a_1 + 3d) + d = a_1 + 4d$	Replace a_4 with $a_1 + 3d$; simplify.
$\vdots$	
$a_n = a_{n-1} + d$	
$= [a_1 + (n - 2)d] + d$	Replace a_{n-1} with $a_1 + (n - 2)d$.
$= a_1 + (n - 1)d$	Simplify.

Thus, we can find the general term a_n from the values a_1 and d.

*n*TH TERM OF AN ARITHMETIC SEQUENCE

If a sequence $a_1, a_2, a_3, \ldots$ is an arithmetic sequence, then its ***n*th** term, a_n, is given by

$$a_n = a_1 + (n - 1)d,$$

where a_1 is the first term and d is the common difference.

EXAMPLE 2 **Finding the *n*th Term of an Arithmetic Sequence**

Find an expression for the nth term of the arithmetic sequence

$$3, 7, 11, 15, \ldots$$

Solution

Because $a_1 = 3$ and $d = 7 - 3 = 4$, we have

$$\begin{aligned} a_n &= a_1 + (n-1)d && \text{Formula for the } n\text{th term} \\ &= 3 + (n-1)4 && \text{Replace } a_1 \text{ with 3 and } d \text{ with 4.} \\ &= 3 + 4n - 4 = 4n - 1 && \text{Simplify.} \end{aligned}$$

Note that in the expression $a_n = 4n - 1$, we get the sequence 3, 7, 11, 15, . . . by substituting $n = 1, 2, 3, 4, \ldots$.

Practice Problem 2 Find an expression for the nth term of the arithmetic sequence

$$-3, 1, 5, 9, 13, 17, \ldots.$$

EXAMPLE 3 Finding the Common Difference of an Arithmetic Sequence

Find the common difference d and the nth term a_n of the arithmetic sequence whose 5th term is 15 and whose 20th term is 45.

Solution

$$\begin{aligned} a_n &= a_1 + (n-1)d && \text{Formula for the } n\text{th term} \\ 45 &= a_1 + (20-1)d && \text{Replace } n \text{ with 20; } a_n = a_{20} = 45 \\ (1) \quad 45 &= a_1 + 19d. && \text{Simplify.} \end{aligned}$$

Also,

$$\begin{aligned} 15 &= a_1 + (5-1)d && \text{Replace } n \text{ with 5; } a_n = a_5 = 15 \\ (2) \quad 15 &= a_1 + 4d. && \text{Simplify.} \end{aligned}$$

Solving the system of equations

$$\begin{cases} a_1 + 19d = 45 & (1) \\ a_1 + 4d = 15 & (2) \end{cases}$$

gives $d = 2$ and $a_1 = 7$. We substitute $a_1 = 7$ and $d = 2$ in the formula for the nth term.

$$\begin{aligned} a_n &= a_1 + (n-1)d && \text{Formula for the } n\text{th term} \\ a_n &= 7 + (n-1)2 && \text{Replace } a_1 \text{ with 7 and } d \text{ with 2.} \\ &= 7 + 2n - 2 = 2n + 5 && \text{Simplify.} \end{aligned}$$

The nth term of this sequence is given by

$$a_n = 2n + 5, \quad n \geq 1.$$

Practice Problem 3 Find the common difference d and the nth term a_n of the arithmetic sequence whose 4th term is 41 and whose 15th term is 8.

2 Find the sum of the first n terms of an arithmetic sequence.

Sum of a Finite Arithmetic Sequence

We can use the following formula in many applications involving the sum of a finite arithmetic sequence.

$$1 + 2 + 3 + \cdots + n = \frac{n(n+1)}{2}$$

Karl Friedrich Gauss (1777–1855)

Karl Friedrich Gauss discovered this formula in his arithmetic class at the age of 8. One day, the story goes, the teacher became so incensed with the class that he assigned the students the task of adding up all of the numbers from 1 to 100. As Gauss's classmates dutifully began to add, Gauss walked up to the teacher and presented the answer, 5050. The story goes that the teacher was neither impressed nor amused. Here are Gauss's calculations:

$$\begin{aligned} S &= 1 + 2 + 3 + \cdots + 100 \\ S &= \underline{100 + 99 + 98 + \cdots + 1} \\ 2S &= \underbrace{101 + 101 + 101 + \cdots + 101}_{100 \text{ terms}} \\ 2S &= 100 \times 101 \\ S &= \frac{100 \times 101}{2} = 5050 \end{aligned}$$

It is said that Karl Friedrich Gauss (1777–1855) discovered this formula when he realized that any sum of numbers added in reverse order produces the same sum. Therefore, if S denotes the sum of the first n natural numbers, then

$$\begin{aligned} S &= 1 + 2 + 3 + \cdots + n \\ S &= n + (n-1) + (n-2) + \cdots + 1 \\ 2S &= \underbrace{(n+1) + (n+1) + (n+1) + \cdots + (n+1)}_{n \text{ terms}} && \text{Add } S + S = 2S. \\ 2S &= n(n+1) \\ S &= \frac{n(n+1)}{2} && \text{Divide both sides by 2.} \end{aligned}$$

For example, if $n = 100$, we have

$$\begin{aligned} S = 1 + 2 + 3 + \cdots + 100 &= \frac{100(100+1)}{2} \\ &= \frac{100(101)}{2} = 5050. \end{aligned}$$

We can use this method to calculate the sum S_n of the first n terms of any arithmetic sequence.

First, we write the sum S_n of the first n terms:

$$S_n = a_1 + (a_1 + d) + (a_1 + 2d) + (a_1 + 3d) + \cdots + a_n$$

We can also write S_n (with the terms in reverse order) by starting with a_n and *subtracting* the common difference d:

$$S_n = a_n + (a_n - d) + (a_n - 2d) + (a_n - 3d) + \cdots + a_1$$

Adding the two equations for S_n, we find that the d's in the sums "drop out" (for example, $(a_1 + d) + (a_n - d) = a_1 + a_n$) and we have

$$\begin{aligned} 2S_n &= \underbrace{(a_1 + a_n) + (a_1 + a_n) + \cdots + (a_1 + a_n)}_{n \text{ terms}} \\ 2S_n &= n(a_1 + a_n) \\ S_n &= n\left(\frac{a_1 + a_n}{2}\right) && \text{Divide both sides by 2.} \end{aligned}$$

SUM OF THE FIRST n TERMS OF AN ARITHMETIC SEQUENCE

Let $a_1, a_2, a_3, \ldots a_n$ be the first n terms of an arithmetic sequence with common difference d. The sum S_n of these n terms is given by

$$S_n = n\left(\frac{a_1 + a_n}{2}\right)$$

or

$$S_n = n\left(\frac{2a_1 + (n-1)d}{2}\right) \qquad a_n = a_1 + (n-1)d$$

EXAMPLE 4 **Finding the Sum of Terms of a Finite Arithmetic Sequence**

Consider the arithmetic sequence of numbers 1, 4, 7, . . . 25. Find

a. The number of terms in the sequence.

b. The sum $1 + 4 + 7 + \cdots + 25$.

Solution

a. In this arithmetic sequence, $a_1 = 1$ and $d = 3$. Let's find the number of terms.

$$\begin{aligned} a_n &= a_1 + (n-1)d && \text{Formula for } n\text{th term} \\ 25 &= 1 + (n-1)3 && \text{Replace } a_n \text{ with 25, } a_1 \text{ with 1, and } d \text{ with 3.} \\ 24 &= (n-1)3 && \text{Subtract 1 from both sides.} \\ 8 &= n - 1 && \text{Divide both sides by 3.} \\ n &= 9 && \text{Solve for } n. \end{aligned}$$

So there are 9 terms in the sequence.

b.
$$\begin{aligned} S_n &= n\left(\frac{a_1 + a_n}{2}\right) && \text{Formula for } S_n \\ S_9 &= 9\left(\frac{1 + 25}{2}\right) && \text{Replace } n \text{ with 9, } a_1 \text{ with 1, and } a_n (= a_9) \text{ with 25.} \\ &= 9(13) = 117 && \text{Simplify.} \end{aligned}$$

So $1 + 4 + 7 + \cdots + 25 = 117$.

Practice Problem 4 Find the sum $\frac{2}{3} + \frac{5}{6} + 1 + \frac{7}{6} + \frac{4}{3} + \frac{3}{2} + \frac{5}{3} + \frac{11}{6} + 2 + \frac{13}{6}$.

EXAMPLE 5 Calculating the Distance Traveled by a Freely Falling Object

In the introduction to this section, we described the arithmetic sequence 16, 48, 80, 112, . . . whose terms gave the number of feet that freely falling space junk falls in successive seconds. For this sequence, find the following:

a. The common difference d

b. The nth term a_n

c. The distance a piece of space junk travels in 10 seconds

Solution

a. The common difference $d = a_2 - a_1 = 48 - 16 = 32$.

b.
$$\begin{aligned} a_n &= a_1 + (n-1)d && \text{Formula for the } n\text{th term} \\ &= 16 + (n-1)32 && \text{Replace } a_1 \text{ with 16 and } d \text{ with 32.} \\ &= 32n - 16 && \text{Simplify.} \end{aligned}$$

c. The sum of the first ten terms of this sequence gives the distance the piece travels in ten seconds. We find a_{10} from the formula:

$$\begin{aligned} a_n &= 32n - 16 && \text{From part } \mathbf{b} \\ a_{10} &= 32(10) - 16 && \text{Replace } n \text{ with 10.} \\ &= 304 \\ S_n &= n\left(\frac{a_1 + a_n}{2}\right) && \text{Formula for } S_n \\ S_{10} &= 10\left(\frac{16 + 304}{2}\right) && \text{Replace } n \text{ with 10, } a_1 \text{ with 16, and } a_n (= a_{10}) \text{ with 304.} \\ &= 1600 && \text{Simplify.} \end{aligned}$$

The piece falls 1600 feet in ten seconds.

Practice Problem 5 In Example 5, find the distance the piece travels during the 11th through 15th seconds.

Answers to Practice Problems

1. -5 **2.** $a_n = 4n - 7$ **3.** $d = -3, a_n = 53 - 3n$ **4.** $\frac{85}{6}$ **5.** 2000 ft

SECTION 11.2 Exercises

Basic Concepts and Skills

1. If 5 is the common difference of an arithmetic sequence with general term a_n, then $a_{17} - a_{16} =$ ____________.

2. If 5 is the common difference of an arithmetic sequence with a general term a_n and $a_{21} = 7$, then $a_{22} =$ ____________.

3. If 14 is the term immediately following the sequence term 17 in an arithmetic sequence, then the common difference is ____________.

4. If $a_1 = 2$ and $a_{11} = 22$ for an arithmetic sequence, then the common difference, d, is ____________.

5. **True or False.** The common difference of an arithmetic sequence is always positive.

6. **True or False.** The sum of the first n terms of an arithmetic sequence always equals n times the average of its first and nth terms.

In Exercises 7–20, determine whether each sequence is arithmetic. For those that are, find the first term a_1 and the common difference d.

7. $1, 2, 3, 4, 5, \ldots.$

8. $1, 3, 5, 7, 9, \ldots$

9. $2, 5, 8, 11, 14, \ldots$

10. $10, 7, 4, 1, -2, \ldots$

11. $1, \frac{1}{2}, 0, -\frac{1}{2}, -\frac{1}{4}, \ldots$

12. $2, 4, 8, 16, 32, \ldots$

13. $1, -1, 2, -2, 3, \ldots$

14. $-\frac{1}{4}, \frac{1}{4}, \frac{3}{4}, \frac{5}{4}, \frac{7}{4}, \ldots$

15. $0.6, 0.2, -0.2, -0.6, -1, \ldots$

16. $2.3, 2.7, 3.1, 3.5, 3.9, \ldots$

17. $a_n = 2n + 6$

18. $a_n = 1 - 5n$

19. $a_n = 1 - n^2$

20. $a_n = 2n^2 - 3$

In Exercises 21–30, find an expression for the nth term of the arithmetic sequence.

21. $5, 8, 11, 14, 17, \ldots$

22. $4, 7, 10, 13, 16, \ldots$

23. $11, 6, 1, -4, -9, \ldots$

24. $9, 5, 1, -3, -7, \ldots$

25. $\frac{1}{2}, \frac{1}{4}, 0, -\frac{1}{4}, -\frac{1}{2}, \ldots$

26. $\frac{2}{3}, \frac{5}{6}, 1, \frac{7}{6}, \frac{4}{3}, \ldots$

27. $-\frac{3}{5}, -1, -\frac{7}{5}, -\frac{9}{5}, -\frac{11}{5}, \ldots$

28. $\frac{1}{2}, 2, \frac{7}{2}, 5, \frac{13}{2}, \ldots$

29. $e, 3 + e, 6 + e, 9 + e, 12 + e, \ldots$

30. $2\pi, 2(\pi + 2), 2(\pi + 4), 2(\pi + 6), 2(\pi + 8) \ldots$

In Exercises 31–36, find the common difference d and the nth term a_n of the arithmetic sequence with the specified terms.

31. 4th term 21; 10th term 60

32. 3rd term 15; 21st term 87

33. 7th term 8; 15th term -8

34. 5th term 12; 18th term -1

35. 3rd term 7; 23rd term 17

36. 11th term -1; 31st term 5

In Exercises 37–46, find the sum of each arithmetic sequence.

37. $1 + 2 + 3 + \cdots + 50$

38. $2 + 4 + 6 + \cdots + 102$

39. $1 + 3 + 5 + \cdots + 99$

40. $5 + 10 + 15 + \cdots + 200$

41. $3 + 6 + 9 + \cdots + 300$

42. $4 + 7 + 10 + \cdots + 301$

43. $2 - 1 - 4 - \cdots - 34$

44. $-3 - 8 - 13 - \cdots - 48$

45. $\frac{1}{3} + 1 + \frac{5}{3} + \cdots + 7$

46. $\frac{3}{5} + 2 + \frac{17}{5} + \cdots + \frac{101}{5}$

In Exercises 47–52, find the sum of the first n terms of the given arithmetic sequence.

47. $2, 7, 12, \ldots; n = 50$

48. $8, 10, 12, \ldots; n = 40$

49. $-15, -11, -7, \ldots; n = 20$

50. $-20, -13, -6, \ldots; n = 25$

51. $3.5, 3.7, 3.9, \ldots; n = 100$

52. $-7, -6.5, -6, \ldots; n = 80$

In Exercises 53–58, find n for the given value of a_n in each arithmetic sequence.

53. $a_n = 75; 1, 3, 5, \ldots$

54. $a_n = 120; 2, 4, 6, \ldots$

55. $a_n = 95; -1, 3, 7, \ldots$

56. $a_n = 83; -5, -3, -1, \ldots$

57. $a_n = 50\sqrt{3}; 2\sqrt{3}, 4\sqrt{3}, 6\sqrt{3}, \ldots$

58. $a_n = 73\pi; 3\pi, 5\pi, 7\pi, \ldots$

Applying the Concepts

In Exercises 59–67, assume that the indicated sequence is arithmetic.

59. Orange picking. When Eric started work as an orange picker, he picked 10 oranges in the first minute, 12 in the second minute, 14 in the third minute, and so on. How many oranges did Eric pick in the first half hour?

60. Exercise. Walking up a steep hill, Jan walks 60 feet in the first minute, 57 feet in the second minute, 54 feet in the third minute, and so on.

a. How far will Jan walk in the *n*th minute?

b. How far will Jan walk in the first 15 minutes?

61. Contest winner. A contest winner will receive money each month for three years. The winner receives \$50 the first month, \$75 the second month, \$100 the third month, and so on. How much money will the winner have collected after 30 months?

62. Salary. Darren took a 12-month temporary job with a monthly salary that increased a fixed amount each month. He can't recall the starting salary, but does remember that he was paid \$820 at the end of the third month and \$910 for his last month's work. How much was Darren's total pay for the entire 12 months?

63. Hourly wage. Antonio's new weekend job started at an hourly wage of \$12.75. He is guaranteed a raise of 25¢ an hour every three months for the next four years. What will Antonio's hourly wage be at the end of four years?

64. Competing job offers. Denzel is considering offers from two companies. A marketing company pays \$32,500 the first year and guarantees a raise of \$1300 each year; an exporting company pays \$36,000 the first year, with a guaranteed raise of \$400 each year. Over a five-year period, which company will pay more? How much more?

65. Theater seating. A theater has 25 rows of seats. The first row has 20 seats, the second row has 22 seats, the third row has 24 seats, and so on. How many seats are in the theater?

66. Bricks in a driveway. A brick driveway has 50 rows of bricks. The first row has 16 bricks, and the fiftieth row has 65 bricks. How many bricks does the driveway contain?

67. Stacked boxes. A stack of boxes has 17 rows. The bottom row has 40 boxes, and the top row has 8 boxes. How many boxes does the stack contain?

68. Elena wanted to teach her young daughter Sophia about saving. She gave Sophia 10¢ the first day and an additional 5¢ per day on each subsequent day. After a while, Sophia counted her money and found that she had saved a total of \$3.25. How many days had she been saving?

69. Only 3 customers showed up for the opening day of a flea market. However, 9 came the second day, and an additional 6 customers came on each subsequent day. After several days, there was a total of 192 customers. How many days had the flea market been open?

70. Two cars, A and B, start together in the same direction from the same place. Car A goes with a uniform speed of 60 mph. Car B goes at a speed of 50 mph in the first hour and increases its speed by 5 mph each succeeding hour. After how many hours will car B overtake car A?

Beyond the Basics

In Exercises 71 and 72, show that the given sequence a_n is an arithmetic sequence. Find the common difference *d*.

71. $a_n = \log\left(\frac{a^n}{b^{n-1}}\right)$

72. $a_n = \log(ab^n)$

73. If the sum S_n of the first *n* terms of a sequence a_n is $3n^2 + 4n$, show that the sequence a_n is an arithmetic sequence [Hint: $a_n = S_n - S_{n-1}$; then show $a_{n+1} - a_n$ is a constant *d*.]

74. Prove that a sequence a_n is an arithmetic sequence if and only if the sum S_n of its first *n* terms is of the form $An^2 + Bn$.

75. Find the sum of all natural numbers between 45 and 100 that are divisible by 3.

76. Find the sum of all natural numbers between 26 and 120 that are divisible by 7.

77. A sequence is **harmonic** if the reciprocals of the terms of the sequence form an arithmetic sequence. Is the sequence $\frac{1}{2}, \frac{3}{5}, \frac{3}{4}, 1, \frac{3}{2}, 3 \ldots$ harmonic? Explain your reasoning.

78. Consider the harmonic sequence $\frac{2}{5}, \frac{2}{7}, \frac{2}{9}, \ldots$.

a. Find the fourth term.

b. Find the *n*th term.

79. Show that if the numbers *a*, *m*, *b* are consecutive terms in an arithmetic sequence, then $m = \frac{a + b}{2}$. We call *m* the **arithmetic mean** of *a* and *b*.

80. If the numbers $a, m_1, m_2, \ldots, m_k, b$ form an arithmetic sequence, we say that the numbers $m_1, m_2, \ldots, m_k$ are ***k*-arithmetic means** between *a* and *b*. Insert *k*-arithmetic means between 1 and 30 so that the sum of the resulting series is 465.

Critical Thinking/Discussion/Writing

81. Find the *n*th term of an arithmetic sequence whose first term is 10 and whose 21st term is 0.

82. If a_1 and *d* are the first term and common difference, respectively, of an arithmetic sequence, find the first term and difference of an arithmetic sequence whose terms are the negatives of the terms in the given sequence.

83. How many terms are in the sum $\sum_{i=22}^{100} 17i^3$?

84. The sum of the first n counting numbers is 12,403. Find the value of n.

Maintaining Skills

For each sequence in Exercises 85–88, find $\frac{a_2}{a_1}, \frac{a_3}{a_2}, \frac{a_4}{a_3}$, and $\frac{a_5}{a_4}$.

85. 3, 6, 12, 24, 48

86. $2, \frac{4}{3}, \frac{8}{9}, \frac{16}{27}, \frac{32}{81}$

87. 4, −12, 36, −108, 324

88. $3, -\frac{3}{2}, \frac{3}{4}, -\frac{3}{8}, \frac{3}{16}$

In Exercises 89–92, the first term a_1 and a number r are given. Write the next four terms of the sequence defined by $a_2 = a_1r, a_3 = a_2r, a_4 = a_3r$, and $a_5 = a_4r$.

89. $a_1 = 2, r = 3$

90. $a_1 = 3, r = \frac{2}{5}$

91. $a_1 = 1, r = -2$

92. $a_1 = 1, r = -\frac{3}{2}$

In Exercises 93–96, the nth term a_n of a sequence is given. Find $\frac{a_2}{a_1}, \frac{a_3}{a_2}, \frac{a_4}{a_3}$, and $\frac{a_5}{a_4}$.

93. $a_n = 5 \cdot 2^n$

94. $a_n = 4 \cdot 3^n$

95. $a_n = 2(-3)^n$

96. $a_n = 5(-2)^{-n}$

97. Let $a_n = -3 \cdot 2^n$. Find a_{n+1} and a_{n-1}.

98. Let $a_n = 5(-3)^{-n}$. Find a_{n+1} and a_{n-1}.

SECTION **11.3**

Geometric Sequences and Series

BEFORE STARTING THIS SECTION, REVIEW

1. The general term of a sequence (Section 11.1, page 910)
2. Partial sums (Section 11.1, page 916)
3. Compound interest (Section 4.1, page 400)

OBJECTIVES

1. Identify a geometric sequence and find its common ratio.
2. Find the sum of a finite geometric sequence.
3. Solve annuity problems.
4. Find the sum of an infinite geometric sequence.

Spreading the Wealth

Cities across the nation compete to host a Super Bowl. Cities across the world compete to host the Olympic Games. What is the incentive? The *multiplier effect* is often listed among the benefits for hosting one of these events. The effect refers to the phenomenon that occurs when money spent directly on an event has a ripple effect on the economy that is many times the amount spent directly on the event.

Suppose, for example, that \$10 million is spent directly by tourists for hotels, meals, tickets, cabs, and so on. Some portion of that amount will be respent by its recipients on groceries, clothes, gas, and so on.

This process is repeated over and over, affecting the economy much more than the initial \$10 million spent. Under certain assumptions, the spending just described results in an *infinite geometric series*, whose sum is called the *multiplier*. Example 9 illustrates this effect.

1 Identify a geometric sequence and find its common ratio.

Geometric Sequence

When the *ratio* $\left(\frac{a_{n+1}}{a_n}\right)$ of any two consecutive terms of a sequence is always the same number, the sequence is called a *geometric sequence*. In other words, the sequence $a_1, a_2, a_3, a_4, \ldots$ is a geometric sequence if

$$\frac{a_2}{a_1} = \frac{a_3}{a_2} = \frac{a_4}{a_3} = \cdots.$$

Geometric Sequence

The sequence

$$a_1, a_2, a_3, a_4, \ldots, a_n, \ldots$$

is a **geometric sequence**, or a **geometric progression**, if there is a number r such that each term except the first in the sequence is obtained by multiplying the previous term by r. The number r is called the **common ratio** of the geometric sequence. We have

$$\frac{a_{n+1}}{a_n} = r, \quad n \geq 1.$$

EXAMPLE 1 **Finding the Common Ratio**

Find the common ratio for each geometric sequence.

a. 2, 10, 50, 250, 1250, . . . **b.** $-162, -54, -18, -6, -2, \ldots$

Solution

a. The common ratio is $\frac{10}{2} = 5\left(\text{or } \frac{50}{10}, \frac{250}{50}, \text{or } \frac{1250}{250}\right)$.

b. The common ratio is $\frac{-54}{-162} = \frac{1}{3}\left(\text{or } \frac{-18}{-54}, \frac{-6}{-18}, \text{or } \frac{-2}{-6}\right)$.

Practice Problem 1 Find the common ratio for the geometric sequence 6, 18, 54, 162, 486, ...

A geometric sequence can be completely specified by giving the first term a_1 and the common ratio r. Suppose, for example, that we are told that the sequence $a_1, a_2, a_3, a_4, \ldots$ is geometric and that $a_1 = 3$ and $r = 4$. Then we can find a_2:

$$\frac{a_2}{a_1} = 4, \text{ so } a_2 = 4a_1 = 4 \cdot 3 = 12$$

Now we continue:

$$\frac{a_3}{a_2} = 4, \text{ so } a_3 = 4a_2 = 4 \cdot 12 = 48;$$

$$\frac{a_4}{a_3} = 4, \text{ so } a_4 = 4a_3 = 4 \cdot 48 = 192; \text{ and so on.}$$

By rewriting $\frac{a_{n+1}}{a_n} = r$ as $a_{n+1} = ra_n$, we have a recursive definition of a geometric sequence.

Recursive Definition of a Geometric Sequence

A geometric sequence $a_1, a_2, a_3, a_4, \ldots, a_n, \ldots$ can be defined recursively. The recursive formula

$$a_{n+1} = ra_n, \quad n \geq 1$$

defines a geometric sequence with the first term a_1 and the common ratio r.

EXAMPLE 2 **Determining Whether a Sequence Is Geometric**

Determine whether each sequence is geometric. If it is, find the first term and the common ratio.

a. $a_n = 4^n$ **b.** $b_n = \frac{3}{2^n}$ **c.** $c_n = 1 - 5^n$

Solution

a. The sequence $4, 16, 64, 256, \ldots, 4^n, \ldots$ appears to be geometric, with $a_1 = 4^1 = 4$ and $r = 4$. To verify this, we check that $\frac{a_{n+1}}{a_n} = 4$ for all $n \geq 1$.

$a_{n+1} = 4^{n+1}$ Replace n with $n + 1$ in the general term $a_n = 4^n$.

$\frac{a_{n+1}}{a_n} = \frac{4^{n+1}}{4^n} = 4$ $\frac{4^{n+1}}{4^n} = \frac{4^n 4}{4^n} = 4$

Consequently, $\frac{a_{n+1}}{a_n} = 4$ for all $n \geq 1$, and the sequence is geometric.

b. The sequence $\frac{3}{2}, \frac{3}{4}, \frac{3}{8}, \frac{3}{16}, \ldots$ appears to be geometric, with $b_1 = \frac{3}{2}$ and $r = \frac{1}{2}$.

To verify this, we check that $\frac{b_{n+1}}{b_n} = \frac{1}{2}$ for all $n \geq 1$.

$$b_{n+1} = \frac{3}{2^{n+1}} \qquad \text{Replace } n \text{ with } n + 1 \text{ in the general term } b_n = \frac{3}{2^n}.$$

$$\frac{b_{n+1}}{b_n} = \frac{3}{2^{n+1}} \div \frac{3}{2^n} \qquad \frac{b_{n+1}}{b_n} = b_{n+1} \div b_n$$

$$= \frac{3}{2^{n+1}} \cdot \frac{2^n}{3} = \frac{1}{2}$$

Consequently, $\frac{b_{n+1}}{b_n} = \frac{1}{2}$ for all $n \geq 1$, and the sequence is geometric.

c. The sequence $-4, -24, -124, -624, \ldots, 1 - 5^n, \ldots$ is *not* a geometric sequence because not all pairs of consecutive terms have identical quotients, $\frac{a_{n+1}}{a_n}$. The quotient of the first two terms is $\frac{-24}{-4} = 6$, but the quotient of the third and second terms is $\frac{-124}{-24} = \frac{31}{6}$.

Practice Problem 2 Determine whether the sequence $a_n = \left(\frac{3}{2}\right)^n$ is geometric. If so, find the first term and the common ratio.

From the equation $a_{n+1} = a_n r$ used in the recursive definition of a geometric sequence, we see that each term after the first is obtained by multiplying the preceding term by the common ratio r.

THE GENERAL TERM OF A GEOMETRIC SEQUENCE

Every geometric sequence can be written in the form

$$a_1, a_1 r, a_1 r^2, a_1 r^3, \ldots\ a_1 r^{n-1}, \ldots,$$

where r is the common ratio. Because $a_1 = a_1(1) = a_1 r^0$, the **nth term of the geometric sequence** is

$$a_n = a_1 r^{n-1}, \text{ for } n \geq 1.$$

EXAMPLE 3 Finding Terms in a Geometric Sequence

For the geometric sequence $1, 3, 9, 27, \ldots,$ find each of the following:

a. a_1 **b.** r **c.** a_n

Solution

a. The first term of the sequence is given: $a_1 = 1$.

b. The sequence is geometric, so we can take the ratio of any two consecutive terms, $\frac{a_{n+1}}{a_n}$. The ratio of the first two terms gives us

$$r = \frac{3}{1} = 3.$$

c. $a_n = a_1 r^{n-1}$ — Formula for the nth term

$= (1)(3^{n-1})$ — Replace a_1 with 1 and r with 3.

$= 3^{n-1}$

Practice Problem 3 For the geometric sequence $2, \frac{6}{5}, \frac{18}{25}, \frac{54}{125}, \ldots$, find the following:

a. a_1 **b.** r **c.** a_n

Notice in Example 3 that converting $a_n = 3^{n-1}$ to standard function notation gives us $f(n) = 3^{n-1}$, an exponential function. A geometric sequence with first term a_1 and common ratio r can be viewed as an exponential function $f(n) = a_1 r^{n-1}$ with the set of natural numbers as its domain.

EXAMPLE 4 Finding a Particular Term in a Geometric Sequence

Find the 23rd term of a geometric sequence whose first term is 10 and whose common ratio is 1.2.

Solution

$a_n = a_1 r^{n-1}$ — Formula for the nth term

$a_{23} = 10(1.2)^{23-1}$ — Replace n with 23, a_1 with 10, and r with 1.2.

$= 10(1.2)^{22}$ — Exact value

≈ 552.06 — Use a calculator.

Practice Problem 4 Find the 18th term of a geometric sequence whose first term is 7 and whose common ratio is 1.5.

EXAMPLE 5 Finding the Number of Terms of a Geometric Sequence

Find the number of terms of the geometric sequence $4, 2, 1, \frac{1}{2}, \frac{1}{4}, \ldots, \frac{1}{64}$.

Solution

The given geometric sequence has first term $a_1 = 4$ and common ratio $r = \frac{2}{4} = \frac{1}{2}$.

Let the nth term be $\frac{1}{64}$. Then

$a_n = \frac{1}{64}$

$a_1 r^{n-1} = \frac{1}{64}$ — $a_n = a_1 r^{n-1}$

$4\left(\frac{1}{2}\right)^{n-1} = \frac{1}{64}$ — $a_1 = 4, r = \frac{1}{2}$

$\left(\frac{1}{2}\right)^{n-1} = \frac{1}{4 \cdot 64} = \frac{1}{256}$ — Divide both sides by 4.

$\left(\frac{1}{2}\right)^{n-1} = \left(\frac{1}{2}\right)^{8}$ — $\frac{1}{256} = \frac{1}{2^8} = \left(\frac{1}{2}\right)^8$

$n - 1 = 8$ — $a^m = a^n$ implies $m = n$.

$n = 9$ — Solve for n.

There are 9 terms in the given sequence.

Practice Problem 5 Find the number of terms of the geometric sequence $\frac{1}{9}, \frac{1}{3}, 1, 3, \ldots, 243$.

2 Find the sum of a finite geometric sequence.

Finding the Sum of a Finite Geometric Sequence

We can find a formula for the sum S_n of the first n terms in a geometric sequence. (We have already found a similar formula for arithmetic sequences.)

$$S_n = a_1 + a_1 r + a_1 r^2 + a_1 r^3 + \cdots + a_1 r^{n-1}$$ Definition of S_n

$$rS_n = a_1 r + a_1 r^2 + a_1 r^3 + a_1 r^4 + \cdots + a_1 r^n$$ Multiply both sides by r.

$$S_n - rS_n = a_1 - a_1 r^n$$ Subtract the second equation from the first.

$$(1 - r)S_n = a_1(1 - r^n)$$ Factor both sides.

$$S_n = \frac{a_1(1 - r^n)}{1 - r}, r \neq 1$$ Divide both sides by $1 - r$.

SUM OF THE TERMS OF A FINITE GEOMETRIC SEQUENCE

Let $a_1, a_2, a_3, \ldots, a_n$ be the first n terms of a geometric sequence with first term a_1, nth term a_n, and common ratio r. The sum S_n of these terms is

$$S_n = \sum_{i=1}^{n} a_1 r^{i-1} = \frac{a_1(1 - r^n)}{1 - r}, r \neq 1.$$

A geometric sequence with $r = 1$ is a sequence in which all terms are identical; that is, $a_n = a_1(1)^{n-1} = a_1$. Consequently, the sum of the first n terms, S_n, is

$$\underbrace{a_1 + a_1 + \cdots + a_1}_{n \text{ terms}} = na_1.$$

TECHNOLOGY CONNECTION

The following screen shows the result of Example 6a obtained using a graphing calculator.

EXAMPLE 6 **Finding the Sum of Terms of a Finite Geometric Sequence**

Find each sum.

a. $\sum_{i=1}^{15} 5(0.7)^{i-1}$ **b.** $\sum_{i=1}^{15} 5(0.7)^{i}$

Solution

a. We can evaluate $\sum_{i=1}^{15} 5(0.7)^{i-1}$ by using the formula for $S_n = \sum_{i=1}^{n} a_1 r^{i-1} = \frac{a_1(1 - r^n)}{1 - r}$, where $a_1 = 5$, $r = 0.7$, and $n = 15$.

$$S_{15} = \sum_{i=1}^{15} a_1 r^{i-1} = 5\left[\frac{1 - (0.7)^{15}}{1 - 0.7}\right]$$ Replace a_1 with 5, $r = 0.7$, and $n = 15$.

$$\approx 16.5875$$ Use a calculator.

b. $$\sum_{i=1}^{15} 5(0.7)^{i} = (0.7)\sum_{i=1}^{15} 5(0.7)^{i-1}$$ Factor out 0.7 from each term.

$$\approx (0.7)(16.5875)$$ From part **a**

$$= 11.6113$$

Practice Problem 6 Find the sum $\sum_{i=1}^{17} 3(0.4)^{i}$.

3 Solve annuity problems.

Annuities

One of the most important applications of the sum of a finite geometric sequence is the computation of the value of an *annuity*. An **annuity** is a sequence of equal periodic payments. When a fixed rate of compound interest applies to all payments, the sum of all of the payments made plus all interest can be found by using the formula for the sum of a finite geometric sequence. The **value of an annuity** (also called the **future value of an annuity**) is the sum of all payments and interest. To develop a formula for the value of an annuity, we suppose $\$P$ is deposited at the end of each year at an annual interest rate i compounded annually.

The compound interest formula $A = P(1 + i)^t$ gives the total value after t years when $\$P$ earns an annual interest rate i (in decimal form) compounded once a year. At the end of the first year, the initial payment of $\$P$ is made and the annuity's value is $\$P$. At the end of the second year, $\$P$ is again deposited. At this time, the first deposit has earned interest during the second year. The value of the annuity after two years is

$$P \quad + \quad P(1 + i).$$

Payment made at the end of year 2 (↑ P); First payment plus interest earned for one year (↑ $P(1+i)$)

The value of the annuity after three years is

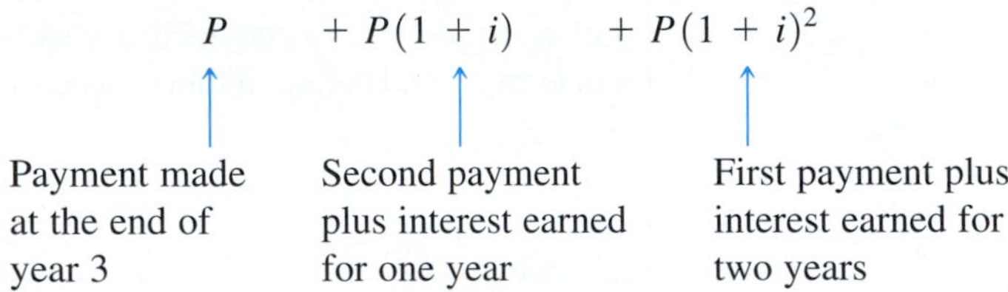

The value of the annuity after t years is

$$P + P(1 + i) + P(1 + i)^2 + P(1 + i)^3 + \cdots + P(1 + i)^{t-1}.$$

Payment made at the end of year t (↑ P); First payment plus interest earned for $t - 1$ year (↑ $P(1+i)^{t-1}$)

The value of the annuity after t years is the sum of a geometric sequence with first term $a_1 = P$ and common ratio $r = 1 + i$. We compute this sum A as follows:

$$A = S_n = \frac{a_1(1 - r^n)}{1 - r} \qquad \text{Formula for } S_n$$

$$A = \frac{P[1 - (1 + i)^t]}{1 - (1 + i)} \qquad \text{Replace } n \text{ with } t,\ a_1 \text{ with } P, \text{ and } r \text{ with } 1 + i.$$

$$= \frac{P[1 - (1 + i)^t]}{-i} \qquad \text{Simplify.}$$

$$= \frac{P[(1 + i)^t - 1]}{i} \qquad \text{Multiply numerator and denominator by } -1.$$

RECALL

The sum S_n of the first n terms of a geometric sequence with first term a_1 and common ratio r is given by the formula

$$S_n = \frac{a_1(1 - r^n)}{1 - r}.$$

If interest is compounded n times per year and equal payments are made at the end of each compounding period, we adjust the formula as we did in Section 3.1 to account for the more frequent compounding.

VALUE OF AN ANNUITY

Let P represent the payment in dollars made at the end of each of n compounding periods per year and let i be the annual interest rate. Then the value A of the annuity after t years is

$$A = P\left[\frac{\left(1 + \frac{i}{n}\right)^{nt} - 1}{\frac{i}{n}}\right].$$

EXAMPLE 7 Finding the Value of an Annuity

An individual retirement account (IRA) is a common way to save money to provide funds after retirement. Suppose you make payments of \$1200 into an IRA at the end of each year at an annual interest rate of 4.5% per year, compounded annually. What is the value of this annuity after 35 years?

Solution

Each annuity payment is $P = \$1200$. The annual interest rate is 4.5% (so $i = 0.045$), and the number of years is $t = 35$. Because interest is compounded annually, $n = 1$. The value of the annuity is

$$A = 1200\left[\frac{\left(1 + \frac{0.045}{1}\right)^{(1)35} - 1}{\frac{0.045}{1}}\right]$$

$$= \$97{,}795.94 \qquad \text{Use a calculator.}$$

The value of the IRA after 35 years is \$97,795.94.

Practice Problem 7 If in Example 7 the end-of-year payments are \$1500, the annual interest rate remains 4.5%, compounded annually, and the payments are made for 30 years, what is the value of the annuity?

4 Find the sum of an infinite geometric sequence.

Infinite Geometric Series

The sum S_n of the first n terms of a geometric series is given by the formula $S_n = \frac{a_1(1 - r^n)}{1 - r}$. If this finite sum S_n approaches a number S as $n \to \infty$ (that is, as n gets larger and larger), we say that S is the **sum of the infinite geometric series**, and we write

$$S = \sum_{i=1}^{\infty} a_1 r^{i-1}.$$

If r is any real number with $-1 < r < 1$ (equivalently, $|r| < 1$), then the value of r^n, like the value of $\left(\frac{1}{2}\right)^n$, gets closer and closer to 0 as n gets larger and larger. We indicate that the values of r^n approach 0 by writing

$$\lim_{n\to\infty} r^n = 0 \text{ if } |r| < 1.$$

The expression $\lim_{n\to\infty} r^n$ is read "the limit, as n approaches infinity, of r to the n" or "the limit, as n approaches infinity, of r to the nth power."

When $|r| < 1$,

$$S_n = \frac{a_1(1 - r^n)}{1 - r} \to \frac{a_1(1 - 0)}{1 - r} = \frac{a_1}{1 - r} \text{ as } n \to \infty.$$

SUM OF THE TERMS OF AN INFINITE GEOMETRIC SEQUENCE

If $|r| < 1$, the infinite sum

$$a_1 + a_1 r + a_1 r^2 + a_1 r^3 + \cdots + a_1 r^{n-1} + \cdots$$

is given by

$$S = \sum_{i=1}^{\infty} a_1 r^{i-1} = \frac{a_1}{1 - r}.$$

When $|r| \geq 1$, the infinite geometric series does not have a sum. This is because the value of r^n does not approach 0 as $n \to \infty$. For example, if $r = 2$, we have

$$2^1 = 2, 2^2 = 4, 2^3 = 8, 2^4 = 16, 2^5 = 32, 2^6 = 64, \ldots.$$

EXAMPLE 8 Finding the Sum of an Infinite Geometric Series

Find the sum $2 + \frac{3}{2} + \frac{9}{8} + \frac{27}{32} + \cdots$.

Solution

The first term is $a_1 = 2$, and the common ratio is

$$r = \frac{3/2}{2} = \frac{3}{4}.$$

Because $|r| = \frac{3}{4} < 1$, we can use the formula for the sum of an infinite geometric series.

$$\begin{aligned} S &= \frac{a_1}{1 - r} && \text{Formula for } S \\ &= \frac{2}{1 - \frac{3}{4}} && \text{Replace } a_1 \text{ with 2 and } r \text{ with } \tfrac{3}{4}. \\ &= 8 && \text{Simplify.} \end{aligned}$$

Practice Problem 8 Find the sum $3 + 2 + \frac{4}{3} + \frac{8}{9} + \cdots$.

EXAMPLE 9 Calculating the Multiplier Effect

The host city for the Super Bowl expects that tourists will spend \$10,000,000. Assume that 80% of this money is spent again in the city, then 80% of this second round of spending is spent again, and so on. In the introduction to this section, we said that such a spending pattern results in a geometric series whose sum is called the *multiplier*. Find this series and its sum.

Solution

We start our series with the \$10,000,000 brought into the city and add the subsequent amounts spent.

$$10{,}000{,}000 + 10{,}000{,}000(0.80) + 10{,}000{,}000(0.80)^2 + 10{,}000{,}000(0.80)^3 + \cdots$$

↑	↑	↑	↑
original 10,000,000	80% of 10,000,000	80% of previous amount	80% of previous amount

Using the formula $\sum_{i=1}^{\infty} ar^{i-1} = \frac{a}{1 - r}$ for the sum of an infinite geometric series, we have

$$\sum_{i=1}^{\infty} (10{,}000{,}000)(0.80)^{i-1} = \frac{\$10{,}000{,}000}{1 - 0.80} = \$50{,}000{,}000.$$

This amount should shed some light on why there is so much competition to serve as the host city for a major sports event.

Practice Problem 9 Find the series and the sum that results in Example 9 assuming that 85%, rather than 80%, occurs in the repeated spending.

Answers to Practice Problems

1. 3 **2.** It is geometric; $a_1 = \frac{3}{2}, r = \frac{3}{2}$ **3. a.** $a_1 = 2$ **b.** $r = \frac{3}{5}$ **c.** $a_n = 2\left(\frac{3}{5}\right)^{n-1}$ **4.** $7(1.5)^{17} \approx 6896.8288$ **5.** 8 **6.** ≈ 2 **7.** \$91,510.60 **8.** 9 **9.** $\sum_{n=0}^{\infty}(10{,}000{,}000)(0.85)^n \approx \$66{,}666{,}666.67$

SECTION 11.3 Exercises

Basic Concepts and Skills

1. If 5 is the common ratio of a geometric sequence with general term a_n, then $\frac{a_{63}}{a_{62}} =$ ________.
2. If 5 is the common ratio of a geometric sequence with general term a_n and $a_{15} = -2$, then $a_{16} =$ ________.
3. If 24 is the term immediately following the sequence term 8 in a geometric sequence, then the common ratio is ________.
4. An infinite geometric series $a_1 + a_1 r + a_1 r^2 + \cdots$ does not have a sum if ________ ≥ 1.
5. **True or False.** If the sum of an infinite geometric series having the common ratio $r = \frac{1}{3}$ is $S = 15$, then $a_1 = 10$.
6. **True or False.** The sum of an infinite geometric series can never be negative.

In Exercises 7–26, determine whether each sequence is geometric. If it is, find the first term and the common ratio.

7. 3, 6, 12, 24, ...
8. 2, 4, 8, 16, . . .
9. 1, 5, 10, 20, . . .
10. 1, 1, 3, 3, 9, 9, . . .
11. 1, −3, 9, −27, . . .
12. −1, 2, −4, 8, . . .
13. 7, −7, 7, −7, . . .
14. 1, −2, 4, −8, . . .
15. $9, 3, 1, \frac{1}{3}, \ldots$
16. $5, 2, \frac{4}{5}, \frac{8}{25}, \ldots$
17. $a_n = \left(-\frac{1}{2}\right)^n$
18. $a_n = 5\left(\frac{2}{3}\right)^n$
19. $a_n = 2^{n-1}$
20. $a_n = -(1.06)^{n-1}$
21. $a_n = 7n^2 + 1$
22. $a_n = 1 - (2n)^2$
23. $a_n = 3^{-n}$
24. $a_n = 50(0.1)^{-n}$
25. $a_n = 5^{n/2}$
26. $a_n = 7^{\sqrt{n}}$

In Exercises 27–34, find the first term a_1, the common ratio r, and the nth term a_n for each geometric sequence.

27. 2, 10, 50, 250, . . .
28. −3, −6, −12, −24, . . .
29. $5, \frac{10}{3}, \frac{20}{9}, \frac{40}{27}, \ldots$
30. $1, \sqrt{3}, 3, 3\sqrt{3}, \ldots$
31. 0.2, −0.6, 1.8, −5.4, . . .
32. 1.3, −0.26, 0.052, −0.0104, . . .
33. $\pi^4, \pi^6, \pi^8, \pi^{10}, \ldots$
34. $e^2, 1, e^{-2}, e^{-4}, \ldots$

In Exercises 35–44, find the indicated term of each geometric sequence.

35. a_7 when $a_1 = 5$ and $r = 2$
36. a_7 when $a_1 = 8$ and $r = 3$
37. a_{10} when $a_1 = 3$ and $r = -2$
38. a_{10} when $a_1 = 7$ and $r = -2$
39. a_6 when $a_1 = \frac{1}{16}$ and $r = 3$
40. a_6 when $a_1 = \frac{1}{81}$ and $r = 3$
41. a_9 when $a_1 = -1$ and $r = \frac{5}{2}$
42. a_9 when $a_1 = -4$ and $r = \frac{3}{4}$
43. a_{20} when $a_1 = 500$ and $r = -\frac{1}{2}$
44. a_{20} when $a_1 = 1000$ and $r = -\frac{1}{10}$

In Exercises 45–50, find the number of terms of the given geometric sequence.

45. 5, 10, 20, . . . , 5120
46. 2, 6, 18, . . . , 4374
47. $16, 8, 4, \ldots, \frac{1}{32}$
48. $27, 9, 3, \ldots, \frac{1}{27}$
49. $18, -12, 8, \ldots, \frac{512}{729}$
50. $25, -15, 9, \ldots, \frac{729}{625}$

In Exercises 51–56, find the sum S_n of the first n terms of each geometric sequence.

51. $\frac{1}{10}, \frac{1}{2}, \frac{5}{2}, \frac{25}{2}, \ldots; n = 10$
52. $6, 2, \frac{2}{3}, \frac{2}{9}, \ldots; n = 10$
53. $\frac{1}{25}, -\frac{1}{5}, 1, -5, \ldots; n = 12$
54. $-10, \frac{1}{10}, \frac{-1}{1000}, \frac{1}{100{,}000}, \ldots; n = 12$
55. $5, \frac{5}{4}, \frac{5}{4^2}, \frac{5}{4^3}, \ldots; n = 8$
56. $2, \frac{2}{5}, \frac{2}{5^2}, \frac{2}{5^3}, \ldots; n = 8$

In Exercises 57–64, find each sum.

57. $\sum_{i=1}^{5}\left(\frac{1}{2}\right)^{i-1}$

58. $\sum_{i=1}^{5}\left(\frac{1}{5}\right)^{i-1}$

59. $\sum_{i=1}^{8}3\left(\frac{2}{3}\right)^{i-1}$

60. $\sum_{i=1}^{8}2\left(\frac{3}{5}\right)^{i-1}$

61. $\sum_{i=3}^{10}\frac{2^{i-1}}{4}$

62. $\sum_{i=3}^{10}\frac{5^{i-1}}{2}$

63. $\sum_{i=1}^{20}\left(-\frac{3}{5}\right)\left(-\frac{5}{2}\right)^{i-1}$

64. $\sum_{i=1}^{20}\left(-\frac{1}{4}\right)(3^{2-i})$

In Exercises 65–70, use a calculator to find each sum. Round your answers to three decimal places.

65. $\frac{1}{5}+\frac{1}{10}+\frac{1}{20}+\cdots+\frac{1}{320}$

66. $2+6+18+\cdots+1458$

67. $\sum_{n=1}^{10}3^{n-2}$

68. $\sum_{n=1}^{15}\left(\frac{1}{7}\right)^{n-1}$

69. $\sum_{n=1}^{10}(-1)^{n-1}3^{2-n}$

70. $\sum_{n=1}^{20}(-4)^{3-n}$

In Exercises 71–80, find each sum.

71. $\frac{1}{3}+\frac{1}{9}+\frac{1}{27}+\frac{1}{81}+\cdots$

72. $\frac{5}{2}+\frac{5}{4}+\frac{5}{8}+\frac{5}{16}+\cdots$

73. $-\frac{1}{2}+\frac{1}{4}-\frac{1}{8}+\frac{1}{16}-\cdots$

74. $-\frac{3}{2}+\frac{3}{4}-\frac{3}{8}+\frac{3}{16}+\cdots$

75. $8-2+\frac{1}{2}-\frac{1}{8}+\cdots$

76. $1-\frac{3}{5}+\frac{9}{25}-\frac{27}{125}+\cdots$

77. $\sum_{n=0}^{\infty}5\left(\frac{1}{3}\right)^{n}$

78. $\sum_{n=0}^{\infty}3\left(\frac{1}{4}\right)^{n}$

79. $\sum_{n=0}^{\infty}\left(-\frac{1}{4}\right)^{n}$

80. $\sum_{n=0}^{\infty}\left(-\frac{1}{3}\right)^{n}$

Applying the Concepts

81. **Population growth.** The population in a small town is increasing at the rate of 3% per year. If the present population is 20,000, what will the population be at the end of five years?

82. **Growth of a zoo collection.** The butterfly collection at a local zoo started with only 25 butterflies. If the number of butterflies doubled each month, how many butterflies were in the collection after 12 months?

83. **Savings growth.** Ramón deposits $100 on the last day of each month into a savings account that pays 6% annually, compounded monthly. What is the balance in the account after 36 compounding periods?

84. **Savings growth.** Kat deposits $300 semiannually into a savings account for her son. If the account pays 8% annually, compounded semiannually, what is the balance in the account after 20 compounding periods?

85. **Ancestors.** Every person has two parents, four grandparents, eight great-grandparents, and so on. Find the number of ancestors a person has in the tenth generation back.

86. **Bacterial growth.** A colony of bacteria doubles in number every day. If there are 1000 bacteria now, how many will there be
   a. On the 7th day?
   b. On the nth day?

87. **Investment.** An actuarial firm budgets a new position at $36,000 for the first year and a 5% raise each year thereafter. Find the total compensation a successful applicant for the job would receive during the first 20 years of employment.

88. **Investment.** A realtor offers two choices of payment for a small condominium:
   a. Pay $4000 per month for the next 25 months.
   b. Pay 1¢ the first month, 2¢ the second month, 4¢ the third month, and so on for 25 months.
   Which option, a or b, is the better choice for the buyer? Explain your reasoning.

89. **Pendulum motion.** The distance traveled by any point on a certain pendulum is 20% less than in the preceding swing. If the length traveled by the pendulum bob during the first swing is 56.25 centimeters, find the total distance the bob has traveled at the end of the fifth swing.

90. **Depreciating a tractor.** Every year a tractor loses 20% of the value it had at the beginning of the year. If the tractor now has a value of $120,000, what will its value be in five years?

91. **Rebounding ball.** A particular ball always rebounds $\frac{3}{5}$ the distance it falls. If the ball is dropped from a height of 5 meters, how far will it travel before coming to a stop?

92. **Rebounding ball.** Repeat Exercise 91 assuming that the ball is dropped from a height of 9 meters.

Beyond the Basics

93. The accompanying figure shows the first six squares in an infinite sequence of successively smaller squares formed by connecting the midpoints of the sides of the preceding larger square. The area of the largest square is 1. Find the total area of the indicated sections that are shaded as the number of squares increases to infinity.

94. The accompanying figure shows the first three equilateral triangles in an infinite sequence of successively smaller equilateral triangles formed by connecting the midpoints of the sides of the preceding larger triangle. The largest

triangle has sides of length 4. Find the total perimeter of all of the triangles.

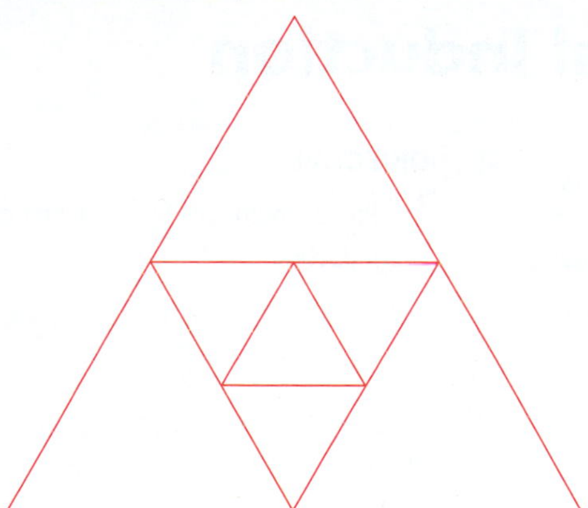

95. Prove that if $a_1, a_2, a_3, \ldots$ is a geometric sequence of positive terms, then the sequence $\ln a_1, \ln a_2, \ln a_3, \ldots$ is an arithmetic sequence.

96. Prove that if $a_1, a_2, a_3, \ldots$ is a geometric sequence, then $a_1^2, a_2^2, a_3^2, \ldots$ is also a geometric sequence.

97. Show that if $a_1, a_2, a_3, \ldots$ is a geometric sequence with common ratio r, then $\frac{1}{a_1}, \frac{1}{a_2}, \frac{1}{a_3}, \ldots$ is also a geometric sequence. Also find its common ratio.

98. Show that if c is any nonzero real number, then $c^2, c, 1, \frac{1}{c}, \ldots$ is a geometric sequence.

99. Show that if x is any nonzero real number, then $x, 2, \frac{4}{x}, \frac{8}{x^2}, \ldots$ is a geometric sequence.

100. Show that if a and x are any two nonzero real numbers, then $\frac{x}{a}, -1, \frac{a}{x}, -\frac{a^2}{x^2}, \ldots$ is a geometric sequence.

101. Show that if a, x, and y are any three nonzero real numbers, then $\frac{a}{x}, -\frac{a}{xy}, \frac{a}{xy^2}, \frac{a}{xy^3}, \ldots$ is a geometric sequence.

102. Show that if $a_1, a_2, a_3, \ldots$ is an arithmetic sequence with common difference d, then $2^{a_1}, 2^{a_2}, 2^{a_3}, \ldots$ is a geometric sequence. Also find its common ratio.

Critical Thinking/Discussion/Writing

103. If a_n denotes the nth term of an arithmetic sequence with $a_1 = 10$ and $d = 2.7$ and if b_n denotes the nth term of a geometric sequence with $b_1 = 10$ and $r = 2$, which number is larger, a_{1001} or b_{1001}?

104. If the sum $\frac{a_1(1 - r^n)}{1 - r}$ of the first n terms of a geometric sequence is 1023, find a_1 when $n = 10$ and $r = \frac{1}{2}$.

105. Find n and k so that the sum $5 + 5 \cdot 2 + 5 \cdot 2^2 + \cdots + 5 \cdot 2^{15}$ can be written as $\sum_{i=0}^{n} 5 \cdot 2^i$ or as $\sum_{i=1}^{k} 5 \cdot 2^{i-1}$.

106. The sum of three consecutive terms of a geometric sequence is 35, and their product is 1000. Find the numbers. [Hint: Let $\frac{a}{r}$, a, and ar be the three terms of the geometric sequence.]

107. The sum of three consecutive terms of an arithmetic sequence is 15. If 1, 4, and 19 are respectively added to the three terms, the resulting numbers form three consecutive terms of a geometric sequence. Find the numbers.

108. The product of the first three terms of a geometric sequence is 1000. If 6 is added to the second term and 7 is added to the third term, the resulting numbers form the three consecutive terms of an arithmeti sequences. Find the numbers.

Maintaining Skills

In Exercises 109–112, a statement P_n involving positive integers n is given. Write the statements P_3 and P_4. State whether P_3 and P_4 are *true* or *false*.

109. P_n: $n(n + 1)$ is even.

110. P_n: $n^3 + n$ is divisible by 3.

111. P_n: $1 + 2 + 3 + \cdots + n = \frac{n(n + 1)}{2}$

112. P_n: $2^n > 3n$

In Exercises 113–116, a statement P_n involving a positive integer n is given. Write the statement P_{n+1}.

113. P_n: $n^2 + n$ is divisible by 2.

114. P_n: $2^{3n} - 1$ is divisible by 7.

115. P_n: $3^n > 5n$

116. P_n: $1 + 4 + 7 + \cdots + (3n - 2) = \frac{1}{2}n(3n - 1)$

117. Show that $\frac{k(k + 1)}{2} + (k + 1) = \frac{(k + 1)(k + 2)}{2}$.

118. Show that $\frac{1}{6}k(k + 1)(2k + 1) + (k + 1)^2 = \frac{1}{6}(k + 1)(k + 2)(2k + 3)$.

SECTION 11.4

Mathematical Induction

BEFORE STARTING THIS SECTION, REVIEW

1. Natural numbers (Section P.1, page 3)
2. Inequalities (Section 1.5, page 129)
3. Exponents (Section P.2, page 18)
4. Series (Section 11.1, page 916)

OBJECTIVE

1. Prove statements by mathematical induction.

Jumping to Conclusions; Good Induction Versus Bad Induction

A scientist had two large jars before him on the laboratory table. The jar on his left contained a hundred fleas; the jar on his right was empty. The scientist carefully lifted a flea from the jar on the left; placed the flea on the table between the two jars; stepped back; and said in a loud voice, "Jump!" The flea jumped and was put into the jar on the right. A second flea was carefully lifted from the jar on the left and placed on the table between the two jars. Again, the scientist stepped back and said in a loud voice, "Jump!" The flea jumped and was put into the jar on the right. In the same manner, the scientist treated each of the hundred fleas in the jar on the left, and each flea jumped as ordered. The two jars were then interchanged, and the experiment was continued with a slight difference. This time the scientist carefully lifted a flea from the jar on the left; *yanked off its hind legs*; placed the flea on the table between the jars; stepped back; and said in a loud voice, "Jump!" The flea did not jump and was put into the jar on the right. A second flea was carefully lifted from the jar on the left, its hind legs yanked off, and then placed on the table between the two jars. Again, the scientist stepped back and said in a loud voice, "Jump!" The flea did not jump and was put into the jar on the right. In this manner, the scientist treated each of the hundred fleas in the jar on the left, and in no case did a flea jump when ordered. So the scientist recorded the following inductive statement in his notebook: "A flea, if its hind legs are yanked off, cannot hear." (*Source:* Modified from Howard Eves, *Mathematical Circles.*)

This, of course, is an example of truly bad reasoning. The principle of mathematical induction, demonstrated in Example 2, is the gold standard for reasoning in proving that statements about natural numbers are true.

1 Prove statements by Mathematical Induction.

Mathematical Induction

Suppose your boss wants you to verify that each of a thousand bags contains two specific gold coins. There is only one way to be absolutely sure: You must check every bag. That would take a very long time, but you can eventually finish the job. However, consider the problem of verifying that a statement about the natural numbers such as

$$1^2 + 2^2 + 3^2 + \cdots + n^2 = \frac{n(n+1)(2n+1)}{6}$$

is true for all natural numbers n. Let's first test the statement for several values of n by comparing the value of $1^2 + 2^2 + 3^2 + \cdots + n^2$ in the second column of Table 11.1 with the value of $\frac{n(n+1)(2n+1)}{6}$ the third column.

TABLE 11.1

n	$1^2 + 2^2 + 3^2 + \cdots + n^2$	$\dfrac{n(n + 1)(2n + 1)}{6}$	Equal?
1	$1^2 = 1$	$\dfrac{1(1 + 1)[2(1) + 1]}{6} = 1$	Yes
2	$1^2 + 2^2 = 5$	$\dfrac{2(2 + 1)[2(2) + 1]}{6} = 5$	Yes
3	$1^2 + 2^2 + 3^2 = 14$	$\dfrac{3(3 + 1)[2(3) + 1]}{6} = 14$	Yes
4	$1^2 + 2^2 + 3^2 + 4^2 = 30$	$\dfrac{4(4 + 1)[2(4) + 1]}{6} = 30$	Yes
5	$1^2 + 2^2 + 3^2 + 4^2 + 5^2 = 55$	$\dfrac{5(5 + 1)[2(5) + 1]}{6} = 55$	Yes

We see that the statement is correct for the natural numbers 1, 2, 3, 4, and 5; furthermore, you can continue to verify that the statement is correct for any particular natural number n you try. But is it *always* correct? Clearly, we can't check every natural number!

The principle of *mathematical induction* provides a method for proving that statements about natural numbers are true for *all* natural numbers. The principle is based on the fact that after any natural number n, there is a next-larger natural number $n + 1$ and that any specified natural number can be reached by a finite number of such steps, starting from the natural number 1.

THE PRINCIPLE OF MATHEMATICAL INDUCTION

Let P_n be a statement that involves the natural number n with the following two properties:

1. P_1 is true (the statement is true for the natural number 1).

2. If P_k is a true statement, then P_{k+1} is a true statement.

Then the statement P_n is true for every natural number n.

FIGURE 11.2

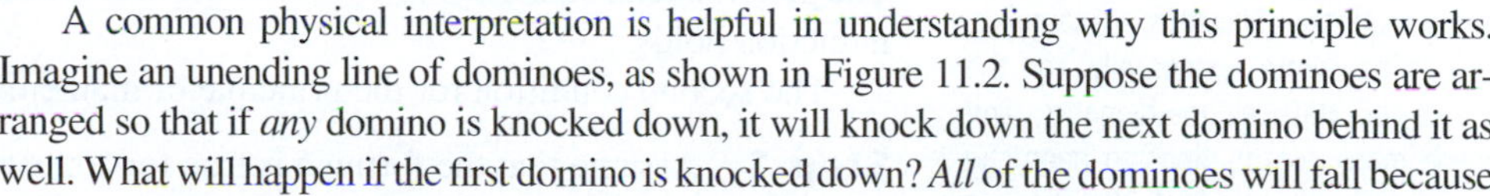

A common physical interpretation is helpful in understanding why this principle works. Imagine an unending line of dominoes, as shown in Figure 11.2. Suppose the dominoes are arranged so that if *any* domino is knocked down, it will knock down the next domino behind it as well. What will happen if the first domino is knocked down? *All* of the dominoes will fall because

1. knocking down the first domino causes the second domino to be knocked down;

2. when the second domino is knocked down, it knocks down the third domino, and so on.

Determining the Statement P_{k+1} from the Statement P_k

To successfully use the Principle of Mathematical Induction, you must be able to determine the statement P_{k+1} from a given statement P_k. Suppose the given statement is

$$P_k\colon k \geq 1;$$

then we have

$$P_{k+1}\colon k + 1 \geq 1. \qquad \text{Replace } k \text{ with } k + 1 \text{ in } P_k.$$

It is important to recognize that because P_k says "k is greater than or equal to 1," the statement P_{k+1} says, "$k + 1$ is greater than or equal to 1." That is, P_{k+1} asserts the same property for $k + 1$ that P_k asserts for k.

EXAMPLE 1 Determining P_{k+1} from P_k

Find the statement P_{k+1} from the given statement P_k.

a. $P_k: k < 2^k$

b. $P_k: S_k = 3k^2 + 9$

c. $P_k: 1 + 2 + 2^2 + 2^3 + \cdots + 2^{k-1} = 2^k - 1$

d. $P_k: 3 + 6 + 9 + 12 + \cdots + 3k = \dfrac{3k(k + 1)}{2}$

Solution

a. $P_k: k < 2^k$, $P_{k+1}: k + 1 < 2^{k+1}$ — Replace k with $k + 1$.

b. $P_k: S_k = 3k^2 + 9$, $P_{k+1}: S_{k+1} = 3(k + 1)^2 + 9$ — Replace k with $k + 1$.

c. $P_k: 1 + 2 + 2^2 + 2^3 + \cdots + 2^{k-1} = 2^k - 1$,

$P_{k+1}: 1 + 2 + 2^2 + 2^3 + \cdots + 2^{(k+1)-1} = 2^{k+1} - 1$ — Replace k with $k + 1$.

d. $P_k: 3 + 6 + 9 + 12 + \cdots + 3k = \dfrac{3k(k + 1)}{2}$,

$P_{k+1}: 3 + 6 + 9 + 12 + \cdots + 3(k + 1) = \dfrac{3(k + 1)[(k + 1) + 1]}{2}$ — Replace k with $k + 1$.

Practice Problem 1 Find P_{k+1} for P_k: $(k + 3)^2 > k^2 + 9$.

EXAMPLE 2 Using Mathematical Induction

Use mathematical induction to prove that for all natural numbers n,

$$2 + 4 + 6 + \cdots + 2n = n(n + 1).$$

Solution

First, we verify that this statement is true for $n = 1$.

Check: $2(1) = 1(1 + 1)$ — Replace n with 1 in the original statement.

$2 = 2.$ ✓

The given statement is true for $n = 1$, so the first condition for the principle of mathematical induction holds.

The second condition for the principle of mathematical induction requires two steps.

Step 1 Assume that the formula is true for some unspecified natural number k:

$$P_k: 2 + 4 + 6 + \cdots + 2k = k(k + 1) \quad \text{Assumed true}$$

Step 2 On the basis of the assumption that P_k is true, show that P_{k+1} is true, which is the same as showing that

$$P_{k+1}: 2 + 4 + 6 + \cdots + 2(k + 1) = (k + 1)[(k + 1) + 1] \quad \text{Replace } k \text{ with } k + 1 \text{ in } P_k.$$

Begin by using P_k, the statement assumed to be true. Add $2(k + 1)$ to both sides of P_k in Step 1, which again results in a true statement.

Francesco Maurolico (1494–1575)

The first known use of mathematical induction is in the work of the 16th-century mathematician Francesco Maurolico. Maurolico wrote extensively on the works of classical mathematics and made many contributions to geometry and optics. In his book *Arithmeticorum Libri Duo*, Maurolico presented a variety of properties of the integers, together with proofs of these properties. To prove some of the properties, he devised the method of mathematical induction. His first use of mathematical induction in this book was to prove that the sum of the first n odd positive integers equals n^2.

$2 + 4 + 6 + \cdots + 2k = k(k + 1)$	Assumed true
$2 + 4 + 6 + \cdots + 2k + 2(k + 1) = k(k + 1) + 2(k + 1)$	Add $2(k + 1)$ to both sides
$2 + 4 + 6 + \cdots + 2k + 2(k + 1) = (k + 1)(k + 2)$	Factor out the common factor $(k + 1)$.
$2 + 4 + 6 + \cdots + 2(k + 1) = (k + 1)[(k + 1) + 1]$	Rewrite $(k + 2)$ as $(k + 1) + 1$.

It is not *necessary* to write $2k$ here because it is understood that $2k$ is the even integer that precedes $2(k + 1)$.

This last equation says that P_{k+1} is true if P_k is assumed to be true. Therefore, by the principle of mathematical induction, the statement

$$2 + 4 + 6 + \cdots + 2n = n(n + 1)$$

is true for every natural number n.

Practice Problem 2 Use mathematical induction to prove that for all natural numbers n,

$$1 + 2 + 3 + \cdots + n = \frac{n(n + 1)}{2}.$$

EXAMPLE 3 Using Mathematical Induction

Use mathematical induction to prove that

$$2^n > n$$

for all natural numbers n.

Solution

First, we show that the given statement is true for $n = 1$.

$2^1 > 1$ Replace n with 1 in the original statement.

Thus, the inequality is true for $n = 1$.

Next, we assume that for some unspecified natural number k,

$$P_k\text{: } 2^k > k \quad \text{is true.}$$

Then we must use P_k to prove that P_{k+1} is true; that is,

$$P_{k+1}\text{: } 2^{k+1} > k + 1.$$

Now by the product rule of exponents, $2^{k+1} = 2^k \cdot 2^1 = 2^k \cdot 2$; so we can get some information about 2^{k+1} by multiplying both sides of P_k: $2^k > k$ by 2.

$2^k > k$	P_k is assumed true.
$2^k \cdot 2 > 2 \cdot k$	Multiply both sides by 2.
$2^{k+1} = 2^k \cdot 2 > 2k = k + k \geq k + 1$	Because $k \geq 1$, $k + k \geq k + 1$.

Thus, $2^{k+1} > k + 1$ is true.

By the principle of mathematical induction, the statement

$$2^n > n$$

is true for every natural number n.

Practice Problem 3 Use mathematical induction to prove that $3^n > 2n$ for all natural numbers n.

Answer to Practice Problem

1. $P_{k+1}: (k+4)^2 > (k+1)^2 + 9$

SECTION 11.4 Exercises

Basic Concepts and Skills

1. Mathematical induction can only be used to prove statements about the ______________.
2. The first step in a mathematical induction proof that a statement P_n is true for all natural numbers n is that ______________.
3. The second step in a mathematical induction proof is to assume that P_k is true for a natural number k and then show that ______________.
4. **True or False.** If P_n is a statement about natural numbers and there is exactly one natural number, r, for which P_r is false, then P_n is false.
5. **True or False.** The statement $e^n \geq n$ cannot be proved by mathematical induction because e is not a natural number.
6. **True or False.** The first step in proving that $2n \neq n$ for all natural numbers n is to note that $2 \neq 1$.

In Exercises 7–10, find P_{k+1} from the given statement P_k.

7. P_k: $(k+1)^2 - 2k = k^2 + 1$
8. P_k: $(1+k)(1-k) = 1 - k^2$
9. P_k: $2^k > 5k$
10. P_{k+1}: $1 + 3 + 5 + \cdots + (2k-1) = k^2$

In Exercises 11–34, use mathematical induction to prove that each statement is true for all natural numbers n.

11. $2 + 4 + 6 + \cdots + 2n = n^2 + n$
12. $1 + 3 + 5 + \cdots + (2n-1) = n^2$
13. $4 + 8 + 12 + \cdots + 4n = 2n(n+1)$
14. $3 + 6 + 9 + \cdots + 3n = \dfrac{3n(n+1)}{2}$
15. $1 + 5 + 9 + \cdots + (4n-3) = n(2n-1)$
16. $3 + 8 + 13 + \cdots + (5n-2) = \dfrac{n(5n+1)}{2}$
17. $3 + 9 + 27 + \cdots + 3^n = \dfrac{3(3^n - 1)}{2}$
18. $5 + 25 + 125 + \cdots + 5^n = \dfrac{5(5^n - 1)}{4}$
19. $\dfrac{1}{1\cdot 2} + \dfrac{1}{2\cdot 3} + \dfrac{1}{3\cdot 4} + \cdots + \dfrac{1}{n(n+1)} = \dfrac{n}{n+1}$
20. $\dfrac{1}{2\cdot 4} + \dfrac{1}{4\cdot 6} + \dfrac{1}{6\cdot 8} + \cdots + \dfrac{1}{2n(2n+2)} = \dfrac{n}{4(n+1)}$
21. $2 \leq 2^n$
22. $2n + 1 \leq 3^n$
23. $n(n+2) < (n+1)^2$
24. $n \leq n^2$
25. $\dfrac{n!}{n} = (n-1)!$
26. $\dfrac{n!}{(n+1)!} = \dfrac{1}{n+1}$
27. $1^2 + 2^2 + 3^2 + \cdots + n^2 = \dfrac{n(n+1)(2n+1)}{6}$
28. $1^3 + 2^3 + 3^3 + \cdots + n^3 = \dfrac{n^2(n+1)^2}{4}$
29. 2 is a factor of $n^2 + n$.
30. 2 is a factor of $n^3 + 5n$.
31. 6 is a factor of $n(n+1)(n+2)$.
32. 3 is a factor of $n(n+1)(n-1)$.
33. $(ab)^n = a^n b^n$
34. $\left(\dfrac{a}{b}\right)^n = \dfrac{a^n}{b^n}$

Applying the Concepts

35. **Counting hugs.** At a family reunion of n people $(n \geq 2)$. each person hugs everyone else. Use mathematical induction to show that the number of hugs is $\dfrac{n^2 - n}{2}$.
36. **Winning prizes.** In a contest in which n prizes are possible, a contestant can win any number of the prizes (from 0 to n). Use mathematical induction to show that there are 2^n possible outcomes for the sets of prizes the contestant might win. [*Hint:* If $n = 1$, there are $2(=2^1)$ outcomes: winning 0 prizes or winning the one prize.]
37. **Koch's snowflake.** A geometric shape called the Koch snowflake can be formed by starting with an equilateral triangle, as in the figure. Assume that each side of the triangle has length 1. On the middle part of each side, construct an equilateral triangle with sides of length $\dfrac{1}{3}$ and erase the side of the smaller triangle that is on the side of the original triangle. Repeat this procedure for each of the smaller triangles. This process produces a figure that resembles a snowflake.
    a. Find a formula for the number of sides of the nth figure. Use mathematical induction to prove that this formula is correct.
    b. Find a formula for the perimeter of the nth figure. Use mathematical induction to prove that this formula is correct.

38. **Sierpinski's triangle.** A geometric figure known as Sierpinski's triangle is constructed by starting with an equilateral triangle as in the figure. Assume that each side of the triangle has length 1. Inside the original triangle, draw a second triangle by connecting the midpoints of the sides of the original triangle. This divides the original triangle into four identical equilateral triangles. Repeat the process by constructing equilateral triangles inside each of the three smaller blue triangles, again using the midpoints of the sides of the triangle as vertices. Continuing this process with each group of smaller triangles produces Sierpinski's triangle. Coloring each set of smaller triangles helps in following the construction. (Be sure to count blue as well as white triangles.)

a. When a process is repeated over and over, each repetition is called an **iteration.** How many triangles will the fourth iteration have?

b. How many triangles will the fifth iteration have?

c. Find a formula for the number of triangles in the nth iteration. Use mathematical induction to prove that this formula is correct.

39. **Towers of Hanoi.** Three pegs are attached vertically to a horizontal board, as shown in the figure. One peg has n rings stacked on it, each ring smaller than the one below it. In a game known as the Tower of Hanoi puzzle, all of the rings must be moved to a different peg, with only one ring moved at a time, and no ring can be moved on top of a smaller ring. Determine the least number of moves that will accomplish this transfer. Use mathematical induction to prove that your answer is correct.

40. **Exam answer sheets.** An exam in which each question is answered with either "True" or "False" has n questions.

a. If every question is answered, how many answer sheets are possible

(i) If $n = 1$?

(ii) If $n = 2$?

b. Find a formula for the number of possible answer sheets for an exam with n questions (for any natural number n) and use mathematical induction to prove that the formula is correct.

Beyond the Basics

In Exercises 41–49, use mathematical induction to prove each statement for all natural numbers n.

41. 5 is a factor of $8^n - 3^n$.

42. 24 is a factor of $5^{2n} - 1$.

43. 64 is a factor of $3^{2n+2} - 8n - 9$.

44. 64 is a factor of $9^n - 8n - 1$.

45. 3 is a factor of $2^{2n+1} + 1$.

46. 5 is a factor of $2^{4n} - 1$.

47. $a - b$ is a factor of $a^n - b^n$.
[*Hint:* $a^{k+1} - b^{k+1} = a(a^k - b^k) + b^k(a - b)$.]

48. If $a \neq 1$, then $1 + a + a^2 + \cdots + a^{n-1} = \dfrac{a^n - 1}{a - 1}$.

49. $\displaystyle\sum_{k=1}^{n+1} \frac{1}{n + k} \leq \frac{5}{6}$

Critical Thinking/Discussion/Writing

50. Consider the statement "2 is a factor of $4n - 1$ for all natural numbers n." Then P_k: 2 is a factor of $4k - 1$.

Assume that P_k is true so that $4k - 1 = 2m$ for some integer m. Then

$$\begin{aligned} P_{k+1} &= 4(k + 1) - 1 \\ &= 4k + 4 - 1 \\ &= (4k - 1) + 4 \\ &= 2m + 4 \\ &= 2(m + 2) \end{aligned}$$

Thus, if P_k is true, then P_{k+1} is also true. However, 2 is never a factor of $4n - 1$ because $4n - 1$ is always an odd integer. Explain why this "proof" is not valid.

Extended principle of mathematical induction. **This principle states that if a statement P_n about natural numbers satisfies the two conditions**

(1) P_m is true for some natural number m,

(2) P_k is true implies that P_{k+1} is also true,
Then P_n is true for all natural numbers n, with $n \geq m$.

In Exercises 51 and 52, use the extended principle of mathematical induction to prove the statement.

51. Prove that $2^n > n^2$ for all natural numbers n, $n > 4$.

52. Prove that $n! > n^3$ for all natural numbers n, $n \geq 6$.

Maintaining Skills

In Exercises 53–56, verify each product by using the product rules.

53. $(x + y)^2 = (x + y)(x + y) = x^2 + 2xy + y^2$

54. $(x + y)^3 = (x + y)(x + y)^2 = x^3 + 3x^2y + 3xy^2 + y^3$

55. $(x + y)^4 = (x + y)(x + y)^3$
$= x^4 + 4x^3y + 6x^2y^2 + 4xy^3 + y^4$

56. $(x + y)^5 = (x + y)(x + y)^4$
$= x^5 + 5x^4y + 10x^3y^2 + 10x^2y^3 + 5xy^4 + y^5$

57. In Exercises 53–56, what is the pattern on the exponents of x?

58. In Exercises 53–56, what is the pattern on the exponents of y?

From the pattern of expansions in Exercises 53–56, if we expand $(x + y)^n$ for a positive integer n, what will be your answer to each of the following?

59. The sum of the exponents on x and y in each term

60. The number of terms in the expansion

SECTION 11.5

The Binomial Theorem

BEFORE STARTING THIS SECTION, REVIEW

1. Special products (Section P.3, page 34)
2. Factorials (Section 11.1, page 914)
3. Summation notation (Section 11.1, page 915)

OBJECTIVES

1. Use Pascal's Triangle to compute binomial coefficients.
2. Use Pascal's Triangle to expand a binomial power.
3. Use the Binomial Theorem to expand a binomial power.
4. Find the coefficient of a term in a binomial expansion.

Blaise Pascal (1623–1662)

Pascaline

Blaise Pascal

Blaise Pascal was a French mathematician. An evident genius, he was 14 when he began to accompany his father to gatherings of mathematicians that were arranged by Father Marin Mersenne. At the age of 16, Pascal presented his own results at one of Mersenne's meetings.

Pascal's desire to help his father with his work collecting taxes led him to invent the first digital calculator, called the Pascaline. Pascal was also interested in atmospheric pressure, and in 1648, he observed that the pressure of the atmosphere decreased with height and that a vacuum existed above the atmosphere.

Pascal's intense interest in mathematics led him to produce important results related to conic sections and to engage in correspondence with Fermat in which he formulated the foundations for the *theory of probability*. Pascal died a painful death from cancer at the age of 39.

In this section, we study *Pascal's Triangle*. This triangle of numbers contains the important *binomial coefficients*. See page 947. Pascal's work was influential in Newton's discovery of the general *Binomial Theorem* for fractional and negative powers.

Binomial Expansions Recall that a polynomial that has exactly two terms is called a *binomial*. In the current section, we study a method for expanding $(x + y)^n$ for any positive integer n.

First, consider these binomial expansions:

$$(x + y)^1 = x + y$$
$$(x + y)^2 = x^2 + 2xy + y^2$$
$$(x + y)^3 = x^3 + 3x^2y + 3xy^2 + y^3$$
$$(x + y)^4 = x^4 + 4x^3y + 6x^2y^2 + 4xy^3 + y^4$$
$$(x + y)^5 = x^5 + 5x^4y + 10x^3y^2 + 10x^2y^3 + 5xy^4 + y^5$$

The expansions of $(x + y)^2$ and $(x + y)^3$ should be familiar to you from Section P.3; the last two are left for you to verify (see Exercises 55 and 56 in Section 11.4). The key point is that each is a product of the previous binomial and $(x + y)$. For example, $(x + y)^4 = (x + y)^3(x + y)$.

The patterns for expansions of $(x + y)^n$ (with $n = 1, 2, 3, 4, 5$) suggest the following:

1. The expansion of $(x + y)^n$ has $n + 1$ terms.
2. The sum of the exponents on x and y in each term equals n.
3. The exponent on x starts at $n\,(x^n = x^n \cdot y^0)$ in the first term and decreases by 1 for each term until it is 0 in the last term $(x^0 \cdot y^n = y^n)$.
4. The exponent on y starts at $0\,(x^n = x^n \cdot y^0)$ in the first term and increases by 1 for each term until it is n in the last term $(x^0 \cdot y^n = y^n)$.
5. The variables x and y have symmetrical roles. That is, replacing x with y and y with x in the expansion of $(x + y)^n$ yields the same terms, just in the reverse order.

You may also have noticed that the coefficients of the first and last terms are both 1 and the coefficients of the second and the next-to-last terms are equal. In general, the coefficients of

$$x^{n-j}y^j \quad \text{and} \quad x^j y^{n-j}$$

are equal for $j = 0, 1, 2, \dots, n$.

The coefficients in the expansion of $(x + y)^n$ are called the **binomial coefficients**.

1 Use Pascal's Triangle to compute binomial coefficients.

Pascal's Triangle

As early as A.D. 1100, the Chinese scholar Chia Hsien had discovered the secret of the binomial coefficients that was later rediscovered by the French philosopher and mathematician Blaise Pascal (1623–1662). To understand Chia Hsien's and Pascal's construction of binomial coefficients, let's first look at the coefficients in these binomial expansions:

$$\begin{aligned}
(x + y)^0 &= 1\\
(x + y)^1 &= 1x + 1y\\
(x + y)^2 &= 1x^2 + 2xy + 1y^2\\
(x + y)^3 &= 1x^3 + 3x^2y + 3xy^2 + 1y^3\\
(x + y)^4 &= 1x^4 + 4x^3y + 6x^2y^2 + 4xy^3 + 1y^4\\
(x + y)^5 &= 1x^5 + 5x^4y + 10x^3y^2 + 10x^2y^3 + 5xy^4 + 1y^5
\end{aligned}$$

If we remove the variables and the plus signs and list only the coefficients, we get a triangle of numbers composed of the binomial coefficients. This triangle of numbers is known as *Pascal's Triangle*.

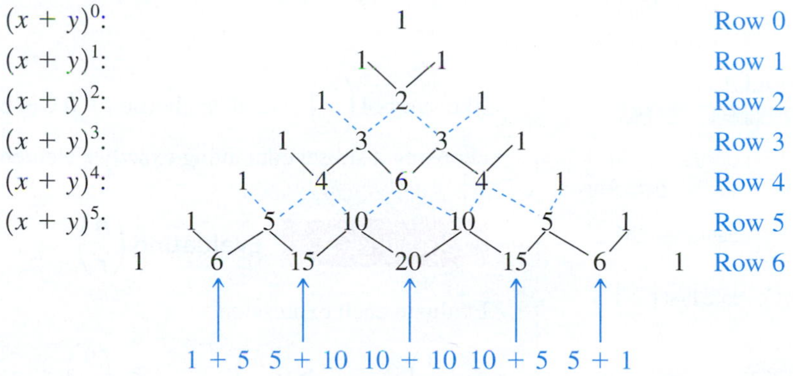

$(x + y)^0$:	1	Row 0
$(x + y)^1$:	1 1	Row 1
$(x + y)^2$:	1 2 1	Row 2
$(x + y)^3$:	1 3 3 1	Row 3
$(x + y)^4$:	1 4 6 4 1	Row 4
$(x + y)^5$:	1 5 10 10 5 1	Row 5
	1 6 15 20 15 6 1	Row 6

1 + 5 5 + 10 10 + 10 10 + 5 5 + 1

Note the symmetry in Pascal's Triangle. If the triangle were folded vertically down the middle, the numbers on each side of the crease would match. To create a new bottom row in the triangle, put the number 1 in the first and last places of the new row and add two neighboring entries in the previous row.

The top row is called the *zeroth row* because it corresponds to the binomial expansion of $(x + y)^0$. The next row is called the *first row* because it corresponds to the binomial expansion of $(x + y)^1$. All of the rows are named so that the nth *row* corresponds to the coefficients of $(x + y)^n$.

2 Use Pascal's Triangle to expand a binomial power.

EXAMPLE 1 **Using Pascal's Triangle to Expand a Binomial Power**

Expand $(4y - 2x)^5$.

Solution

From the fifth row of Pascal's Triangle, we see that the binomial coefficients are

$$1, 5, 10, 10, 5, 1.$$

We must make some changes in the expansion

$$(x + y)^5 = x^5 + 5x^4y + 10x^3y^2 + 10x^2y^3 + 5xy^4 + y^5$$

to get the expansion for $(4y - 2x)^5$:

1. Replace x with $4y$.
2. Replace y with $-2x$.

$$\begin{aligned}(4y - 2x)^5 &= [4y + (-2x)]^5 = (4y)^5 + 5(4y)^4(-2x)\\ &\quad + 10(4y)^3(-2x)^2 + 10(4y)^2(-2x)^3 + 5(4y)(-2x)^4 + (-2x)^5\\ &= 1024y^5 - 2560y^4x + 2560y^3x^2 - 1280y^2x^3 + 320yx^4 - 32x^5\end{aligned}$$

We note that expanding a *difference* results in alternating signs between terms.

Practice Problem 1 Expand $(3y - x)^6$.

3 Use the Binomial Theorem to expand a Binomial power.

The Binomial Theorem

The coefficients in a binomial expansion can be computed by using ratios of certain factorials. We first introduce the symbol $\binom{n}{r}$.

The Symbol $\binom{n}{r}$

If r and n are integers with $0 \le r \le n$, then we define

$$\binom{n}{r} = \frac{n!}{r!(n-r)!}, \text{ Note that } \binom{n}{0} = 1, \text{ and } \binom{n}{n} = 1, \text{ becuase } 0! = 1.$$

TECHNOLOGY CONNECTION

Most graphing calculators can compute the binomial coefficients $\binom{n}{r}$. They frequently use the function *nCr*. Note that $\binom{5}{3}$ is shown as 5 nCr 3 on the viewing screen.

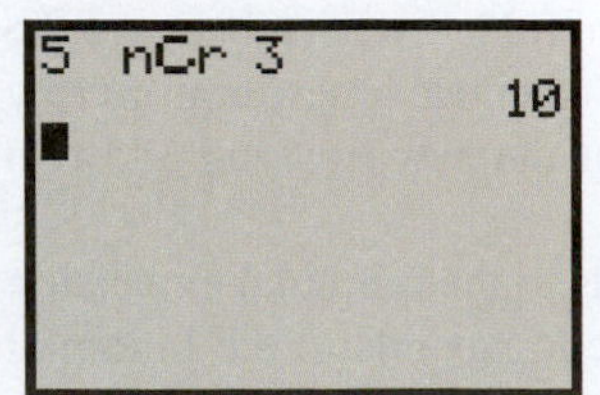

The symbol $\binom{n}{r}$ is read "n choose r." It can be shown that $\binom{n}{r}$ is the number of ways of choosing a subset containing exactly r elements from a set with n elements.

EXAMPLE 2 **Evaluating $\binom{n}{r}$**

Evaluate each expression.

a. $\binom{4}{1}$ **b.** $\binom{5}{3}$ **c.** $\binom{9}{0}$ **d.** $\binom{35}{35}$

Solution

a. $\binom{4}{1} = \frac{4!}{1!(4-1)!} = \frac{4!}{1!3!} = \frac{4\cdot\cancel{3}\cdot\cancel{2}\cdot1}{1(\cancel{3}\cdot\cancel{2}\cdot1)} = \frac{4}{1} = 4$

b. $\binom{5}{3} = \frac{5!}{3!(5-3!)} = \frac{5!}{3!2!} = \frac{5\cdot4\cdot\cancel{3!}}{\cancel{3!}2!} = \frac{5\cdot4}{2} = 5\cdot2 = 10$

c. $\binom{9}{0} = \frac{9!}{0!(9-0)!} = \frac{9!}{0!\,9!} = \frac{1}{1} = 1$ Recall that $0! = 1$.

d. $\binom{35}{35} = \frac{35!}{35!(35-35)!} = \frac{35!}{35!\,0!} = \frac{1}{1} = 1$

Practice Problem 2 Evaluate each expression.

a. $\binom{6}{2}$ **b.** $\binom{12}{9}$

The numbers $\binom{n}{r}$ show up as the coefficients in the expansion of a binomial power. For example, $(x+y)^4$ can be written as either

$$(x+y)^4 = x^4 + 4x^3y + 6x^2y^2 + 4xy^3 + y^4$$

or

$$(x+y)^4 = \binom{4}{0}x^4 + \binom{4}{1}x^3y + \binom{4}{2}x^2y^2 + \binom{4}{3}xy^3 + \binom{4}{4}y^4.$$

This result is the *Binomial Theorem* (for $n = 4$), which can be proved by mathematical induction. It provides an efficient method for expanding a binomial power. (See Exercise 64.) The Binomial Theorem can be used to expand a binomial power directly, without reference to Pascal's Triangle. This technique is particularly useful in expanding large powers of a binomial. For example, the expansion of $(x+y)^{20}$ would require that you produce 20 rows of the Pascal Triangle.

THE BINOMIAL THEOREM

If n is a natural number, then the binomial expansion of $(x+y)^n$ is given by

$$(x+y)^n = \binom{n}{0}x^n + \binom{n}{1}x^{n-1}y + \binom{n}{2}x^{n-2}y^2 + \cdots + \binom{n}{r}x^{n-r}y^r + \cdots + \binom{n}{n}y^n$$
$$= \sum_{r=0}^{n}\binom{n}{r}x^{n-r}y^r.$$

The coefficient of $x^{n-r}y^r$ is $\binom{n}{r} = \frac{n!}{r!(n-r)!}$.

EXAMPLE 3 Expanding a Binomial Power by Using the Binomial Theorem

Find the binomial expansion of $(x-3y)^4$.

Solution

$$\begin{aligned}(x-3y)^4 &= [x+(-3y)]^4\\
&= \binom{4}{0}x^4 + \binom{4}{1}x^3(-3y) + \binom{4}{2}x^2(-3y)^2 + \binom{4}{3}x(-3y)^3 + \binom{4}{4}(-3y)^4\\
&= \frac{4!}{0!\,4!}x^4 + \frac{4!}{1!\,3!}x^3(-3y) + \frac{4!}{2!\,2!}x^2(-3y)^2 + \frac{4!}{3!\,1!}x(-3y)^3 + \frac{4!}{4!\,0!}(-3y)^4\\
&= x^4 + \frac{4\cdot 3!}{1!\,3!}x^3(-3y) + \frac{4\cdot 3\cdot 2!}{2!\,2!}x^2(-3y)^2 + \frac{4\cdot 3!}{3!\,1!}x(-3y)^3 + (-3y)^4\\
&= x^4 + 4x^3(-3y) + \frac{4\cdot 3}{2!}x^2(9y^2) + 4x(-27y^3) + (81y^4)\\
&= x^4 - 12x^3y + 54x^2y^2 - 108xy^3 + 81y^4\end{aligned}$$

Practice Problem 3 Find the binomial expansion of $(3x-y)^4$.

4 Find the coefficient of a term in a binomial expansion.

Binomial Coefficients

The Binomial coefficients $\binom{n}{r}$ are useful for finding a particular term in a binomial expansion.

EXAMPLE 4 Finding a Particular Coefficient in a Binomial Expansion

Find the coefficient of x^9y^3 in the expansion of $(x + y)^{12}$.

Solution

The form of the expansion is

$$(x + y)^{12} = \binom{12}{0}x^{12} + \binom{12}{1}x^{11}y + \binom{12}{2}x^{10}y^2 + \binom{12}{3}x^9y^3 + \cdots + \binom{12}{11}xy^{11} + \binom{12}{12}y^{12}.$$

From the fourth term, $\binom{12}{3}x^9y^3$, the coefficient of x^9y^3 is

$$\binom{12}{3} = \frac{12!}{3!(12 - 3!)} = \frac{12!}{3!\,9!} = \frac{12 \cdot 11 \cdot 10 \cdot \cancel{9!}}{3!\,\cancel{9!}} = 220.$$

Practice Problem 4 Find the coefficient of x^3y^9 in the expansion of $(x + y)^{12}$.

The method illustrated in Example 4 allows us to find any particular term in a binomial expansion without writing out the complete expansion.

PARTICULAR TERM IN A BINOMIAL EXPANSION

The term containing the factor x^r in the expansion of $(x + y)^n$ is

$$\binom{n}{n - r}x^ry^{n-r}.$$

EXAMPLE 5 Finding a Particular Term in a Binomial Expansion

Find the term containing x^{10} in the expansion of $(x + 2a)^{15}$.

Solution

We begin with the formula for the term containing the factor x^r.

$$\binom{n}{n - r}x^ry^{n-r} = \binom{15}{15 - 10}x^{10}(2a)^{15-10}$$ Replace n with 15, r with 10, and y with $2a$.

$$= \binom{15}{5}x^{10}(2a)^5$$ Simplify.

$$= \frac{15!}{5!(15 - 5)!}x^{10}2^5a^5$$ Recall that $\binom{n}{r} = \frac{n!}{r!(n - r)!}$

$$= \frac{15!}{5!\,10!} \cdot 32x^{10}a^5$$ $2^5 = 32$

$$= \frac{15 \cdot 14 \cdot 13 \cdot 12 \cdot 11 \cdot \cancel{10!}}{5! \cdot \cancel{10!}} \cdot 32x^{10}a^5$$ Use a calculator.

$$= 96{,}096x^{10}a^5$$

Practice Problem 5 Find the term containing x^3 in the expansion of $(x + 2a)^{15}$.

Answers to Practice Problems

1. $(3y - x)^6 = 729y^6 - 1458y^5x + 1215y^4x^2 - 540y^3x^3 + 135y^2x^4 - 18yx^5 + x^6$

2. a. $\binom{6}{2} = 15$ **b.** $\binom{12}{9} = 220$

3. $(3x - y)^4 = 81x^4 - 108x^3y + 54x^2y^2 - 12xy^3 + y^4$

4. 220 **5.** $1{,}863{,}680x^3a^{12}$

SECTION 11.5 Exercises

Basic Concepts and Skills

1. The expansion of $(x + y)^n$ has ____________ terms.
2. In the expansion of $(x + y)^5$, the coefficient of x^2y^3 is $\binom{n}{r}$ when $n =$ ____________ and $r =$ ____________.
3. Expanding a difference such as $(2x - y)^{10}$ results in ____________ between terms.
4. For any positive integer n, $\binom{n}{n} =$ ____________.
5. **True or False.** Every coefficient in the expansion of $(x + y)^n$, for $n > 1$, appears exactly twice.
6. **True or False.** If $n > 3$, then $2\binom{n}{2} = n$.

In Exercises 7–16, evaluate each expression.

7. $\frac{6!}{3!}$ **8.** $\frac{11!}{9!}$

9. $\frac{12!}{11!}$ **10.** $\frac{3!}{0!}$

11. $\binom{6}{4}$ **12.** $\binom{6}{3}$

13. $\binom{9}{0}$ **14.** $\binom{12}{0}$

15. $\binom{7}{1}$ **16.** $\binom{7}{3}$

In Exercises 17–24, use Pascal's Triangle to expand each binomial.

17. $(x + 2)^4$ **18.** $(x + 3)^4$

19. $(x - 2)^5$ **20.** $(3 - x)^5$

21. $(2 - 3x)^3$ **22.** $(3 - 2x)^3$

23. $(2x + 3y)^4$ **24.** $(2x + 5y)^4$

In Exercises 25–46, use either the Binomial Theorem or Pascal's Triangle to expand each binomial.

25. $(x + 1)^4$ **26.** $(x + 2)^4$

27. $(x - 1)^5$ **28.** $(1 - x)^5$

29. $(y - 3)^3$ **30.** $(2 - y)^5$

31. $(x + y)^6$ **32.** $(x - y)^6$

33. $(1 + 3y)^5$ **34.** $(2x + 1)^5$

35. $(2x + 1)^4$ **36.** $(3x - 1)^4$

37. $(x - 2y)^3$ **38.** $(2x - y)^3$

39. $(2x + y)^4$ **40.** $(3x - 2y)^4$

41. $\left(\frac{x}{2} + 2\right)^7$ **42.** $\left(2 - \frac{x}{2}\right)^7$

43. $\left(a^2 - \frac{1}{3}\right)^4$ **44.** $\left(\frac{1}{2} - a^2\right)^4$

45. $\left(\frac{1}{x} + y\right)^3$ **46.** $\left(x + \frac{2}{y}\right)^3$

In Exercises 47–54, find the specified term.

47. $(x + y)^{10}$; term containing x^7
48. $(x + y)^{10}$; term containing y^7
49. $(x - 2)^{12}$; term containing x^3
50. $(2 - x)^{12}$; term containing x^3
51. $(2x + 3y)^8$; term containing x^6
52. $(2x + 3y)^8$; term containing y^6
53. $(5x - 2y)^{11}$; term containing y^9
54. $(7x - y)^{15}$; term containing x^9
55. Find the value of $(1.2)^5$ by writing it in the form $(1 + 0.2)^5$ and applying the Binomial Theorem.
56. Use the method in Exercise 55 to evaluate each expression.
    a. $(2.9)^4$ **b.** $(10.4)^3$

Beyond the Basics

57. Find the middle term in the expansion of $\left(\sqrt{x} - \frac{2}{x^2}\right)^{10}$.
58. Find the middle term in the expansion of $\left(\sqrt{x} + \frac{1}{x^2}\right)^{10}$.
59. Find the middle term in the expansion of $(1 - x^2y^{-3})^{12}$.
60. Show that the middle term in the expansion of $(1 + x)^{2n}$ is
    $$\frac{1 \cdot 3 \cdot 5 \cdot \cdots \cdot (2n - 1)}{n!} 2^n x^n.$$
61. Prove that $\binom{n}{0} + \binom{n}{1} + \binom{n}{2} + \cdots + \binom{n}{n} = 2^n$.
    [*Hint*: Expand $(1 + 1)^n = 2^n$.]

62. Prove that $\binom{n}{0} - \binom{n}{1} + \binom{n}{2} - \binom{n}{3} + \cdots + (-1)^n\binom{n}{n} = 0.$

[*Hint:* Expand $(1 - 1)^n = 0$.]

63. Prove that $\binom{k}{j} + \binom{k}{j-1} = \binom{k+1}{j}$.

64. Prove the Binomial Theorem by using the Principle of Mathematical Induction.

[*Hint:*

$$(x + y)^{k+1} = (x + y)(x + y)^k = (x + y)\sum_{j=0}^{k}\binom{k}{j}x^{k-j}y^j$$

$$= \sum_{j=0}^{k}\binom{k}{j}x^{k+1-j}y^j + \sum_{j=0}^{k}\binom{k}{j}x^{k-j}y^{j+1}.$$

Use Exercise 63 to show that

$$(x + y)^{k+1} = \sum_{j=0}^{k+1}\binom{k+1}{j}x^{k+1-j}y^j.\ ]$$

In Exercises 65–67, find the value of the expression without expanding any term.

65. $(2x - 1)^4 + 4(2x - 1)^3(3 - 2x) + 6(2x - 1)^2(3 - 2x)^2 + 4(2x - 1)(3 - 2x)^3 + (3 - 2x)^4$

66. $(x + 1)^4 - 4(x + 1)^3(x - 1) + 6(x + 1)^2(x - 1)^2 - 4(x + 1)(x - 1)^3 + (x - 1)^4$

67. $(3x - 1)^5 + 5(3x - 1)^4(1 - 2x) + 10(3x - 1)^3(1 - 2x)^2 + 10(3x - 1)^2(1 - 2x)^3 + 5(3x - 1)(1 - 2x)^4 + (1 - 2x)^5$

68. Find the constant term in each expansion.

a. $\left(x^2 - \frac{1}{x}\right)^9$ **b.** $\left(\sqrt{x} - \frac{2}{x^2}\right)^{10}$

69. If the constant term in the expansion of $\left(kx - \frac{1}{x^2}\right)^6$ is 240, find k.

70. If the constant term in the expansion of $\left(x^3 + \frac{k}{x^8}\right)^{11}$ is 1320, find k.

71. Show that there is no constant term in the expansion of $\left(2x^2 - \frac{1}{4x}\right)^{11}$.

Critical Thinking/Discussion/Writing

72. Use the binomial expansion of $(x + y)^6$, with $x = 1$ and $y = 1$, to show that

$$2^6 = \binom{6}{0} + \binom{6}{1} + \binom{6}{2} + \binom{6}{3} + \binom{6}{4} + \binom{6}{5} + \binom{6}{6}.$$

73. Use the binomial expansion of $(x + y)^{10}$ to show that

$$\binom{10}{0} - \binom{10}{1} + \binom{10}{2} - \binom{10}{3} + \binom{10}{4} - \binom{10}{5} + \binom{10}{6} - \binom{10}{7} + \binom{10}{8} - \binom{10}{9} + \binom{10}{10} = 0.$$

74. Show that if $x \geq 0$ and $y \geq 0$, then $(x + y) \geq \sqrt{x^2 + y^2}$.

75. Use the binomial expansion of $(x + y)^n$ to show that if $x > 0$ and $y > 0$, then $(x + y)^n > x^n + y^n$.

76. Use the binomial expansion of $(1 + x)^n$ to show that if $x > 0$, then $(1 + x)^n > 1 + nx$.

Maintaining Skills

In Exercises 77–80, write each expression using factorial notation.

77. $5 \cdot 6 \cdot 7 \cdot 8$

78. $10 \cdot 9 \cdot 8 \cdot 7 \cdot 6$

79. $2 \cdot 4 \cdot 6 \cdot 8 \cdot 10 \cdot 12$

80. $3 \cdot 6 \cdot 9 \cdot 12 \cdot 15$

81. Compute $\frac{12!}{10!}$.

82. Compute $\frac{n!}{(n - r)!}$. for $n = 100$ and $r = 2$.

83. Compute $\frac{8!}{5!3!}$.

84. Compute $\frac{n!}{(n - r)!r!}$ for $n = 10$ and $r = 3$.

SECTION 11.6

Counting Principles

BEFORE STARTING THIS SECTION, REVIEW

1. Exponents (Section P.2, page 18)
2. Factorials (Section 11.1, page 914)

OBJECTIVES

1. Use the Fundamental Counting Principle.
2. Use the formula for permutations.
3. Use the formula for combinations.
4. Use the formula for distinguishable permutations.

◆ The Mystery of Social Security Numbers

Social Security numbers may seem to be assigned randomly, but they are not. Although the details are complicated, here is a brief introduction: First, the format for all Social Security numbers is NNN-NN-NNNN, where N must be one of the numbers 0, 1, 2, 3, 4, 5, 6, 7, 8, 9. U.S. citizens, permanent residents, and certain temporary residents are issued these numbers. For working people, the numbers help track tax contributions and Social Security benefits.

The first three digits in a Social Security number make up an *area number* assigned to a geographic location. The area number indicates the state in which the card was issued or the ZIP Code in the mailing address provided in the original application.

The middle two digits constitute a *group number* and are designed simply to break the Social Security number into convenient blocks for issuance. The exact procedure is not very enlightening.

The last four digits are *serial numbers* and are assigned in each group from 0001 through 9999.

Various restrictions limit the numbers that can be used. For example, numbers with all zeros in any of the digit blocks (000-xx-xxxx, xxx-00-xxxx, or xxx-xx-0000) are never issued. Even if there were no restrictions, at some point, there might be *more people than available numbers*. How many numbers can be issued in the Social Security format with no restrictions? The answer is 1,000,000,000 (1 billion). (The U.S. population in January 2013 was about 315,000,000 (315 million).)

In Example 3, you learn how to calculate the total number of possible Social Security numbers.

1 Use the Fundamental Counting Principle.

Fundamental Counting Principle

You may think that you already know all there is to know about counting. However, mathematicians have developed a variety of techniques that allow you to determine the number of items that are of interest to you without making the effort of listing and counting all of them. We start with an example that easily lists the items to be counted, but that also demonstrates a general counting procedure.

EXAMPLE 1 **Counting Possible Car Selections**

Suppose you are trying to decide between buying a sport-utility vehicle (SUV) and a four-door sedan. The SUV is available in black, red, and silver. The sedan is available in black, blue, and green. In how many ways can you choose a type of car and its color?

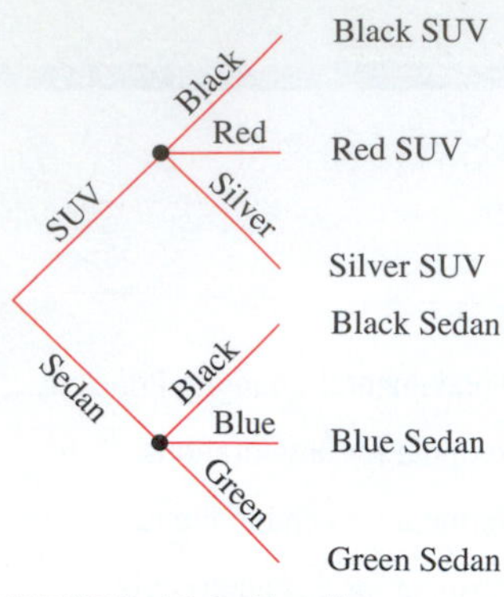

FIGURE 11.3 Tree diagram

Solution

We can represent the possible solutions by using a *tree diagram*, as shown in Figure 11.3. There are two initial choices: an SUV and a sedan. For each of these two choices, there are three additional choices of color; so a total of $2 \cdot 3 = 6$ car selections are possible.

Practice Problem 1 Construct a tree diagram and determine the number of ways you can choose to take one of three courses—English, French, or Math—in the morning, afternoon, or evening.

The car selection process in Example 1 involves making two choices. First, choose a type of car; second, choose a color. It is important in this process that the same *number* of choices be available for the second choice regardless of the result of the first choice. The exact rules for counting by this method are given next.

> **FUNDAMENTAL COUNTING PRINCIPLE**
>
> If a first choice can be made in p different ways, a second choice can be made in q different ways, a third choice can be made in r different ways, and so on, then the sequence of choices can be made in $p \cdot q \cdot r \ldots$ different ways.

EXAMPLE 2 Counting Music Programs

A disc jockey wants to open a program by picking a song from among 40 songs by one artist and close the program by picking a song from among 32 songs by a second artist. How many different choices are possible?

Solution

Each choice of an opening and ending song sequence can be obtained as follows:

First, choose an opening song in 40 ways.

Second, choose the song to end the program in 32 ways.

Because no matter which song is chosen to open the program, all 32 songs are available to end the program, the counting principle applies, and there are a total of

$$40 \cdot 32 = 1280$$

possible choices for the opening and ending sequences.

Practice Problem 2 Reba gets to choose both dinner and a movie for a Saturday night date. She can choose from any of seven restaurants and five movies. How many different choices are possible?

WARNING

Be careful when using the Fundamental Counting Principle. Suppose you are trying to decide between buying an SUV available only in black, red, or silver and a convertible available only in black or red. In how many ways can you choose a type of car and its color? The car and color can be selected by making two choices: type (SUV or convertible) followed by color. But in this situation, the number of ways you can make the second choice depends on the first choice. You can choose from *three* colors for an SUV, but only *two* colors for a convertible. The Fundamental Counting Principle *does not apply*. Fortunately, there are so few different choices here that you can easily list all five: black SUV, red SUV, silver SUV, black convertible, and red convertible.

EXAMPLE 3 Counting Possible Social Security Numbers

Social Security numbers have the format NNN-NN-NNNN, where each N must be one of the integers 0, 1, 2, 3, 4, 5, 6, 7, 8, and 9. Assuming that there are no other restrictions, how many such numbers are possible?

Solution

There are nine positions in the format

NNN-NN-NNNN.

Because the number of choices available at each position is not affected by previous choices, the Fundamental Counting Principle can be used. Replace each of the letters N with a blank to be filled in with one of the ten digits 0, 1, 2, 3, 4, 5, 6, 7, 8, and 9.

We have

$$\underline{10}\cdot\underline{10}\cdot\underline{10}\cdot\underline{10}\cdot\underline{10}\cdot\underline{10}\cdot\underline{10}\cdot\underline{10}\cdot\underline{10} = 10^9.$$

Thus, there are $10^9 = 1{,}000{,}000{,}000$ possible Social Security numbers.

Practice Problem 3 In Example 3, suppose we fix the area number for the Social Security number to be 433. How many social security numbers can be assigned to this area—that is, numbers having the form 433-NN-NNNN?

2 Use the formula for permutations.

Permutations

Permutation

A **permutation** is an arrangement of n distinct objects in a fixed order in which no object is used more than once. The specific order is important: Each different ordering of the same objects is a different permutation.

EXAMPLE 4 Counting Possible Arrangements for a Group Photograph

How many different ways can five people be arranged in a row for a group photograph?

Solution

To count the number of ways we can arrange five people in a row for a group photograph, we can assign numbers to each of the five photograph positions.

1	2	3	4	5

Then we proceed as follows:

First, choosing any of the five people for Position 1 gives 5 ways.
Second, choosing one of the four remaining people for Position 2 gives 4 ways.
Third, choosing one of the three remaining people for Position 3 gives 3 ways.
Fourth, choosing one of the remaining two people for Position 4 gives 2 ways.
Fifth, choosing the last person remaining for Position 5 gives 1 way.
Using the Fundamental Counting Principle, we see that there are

$$5\cdot 4\cdot 3\cdot 2\cdot 1 = 5!, \text{ or } 120,$$

possible arrangements.

Practice Problem 4 How many different ways can seven books be arranged on a shelf?

This technique gives a general formula for counting permutations of n distinct objects.

NUMBER OF PERMUTATIONS OF n OBJECTS

The number of permutations of n distinct objects is

$$n! = n(n-1)\cdot \cdots \cdot 4\cdot 3\cdot 2\cdot 1.$$

That is, n distinct objects can be arranged in $n!$ different ways.

Sometimes only some, but not all, of the available objects are to be arranged in a specific order, again with no object being used more than once. You may want to choose r objects from among n available objects. This type of ordering is called a **permutation of *n* objects taken *r* at a time**.

EXAMPLE 5 Counting Prizewinners

Three prizes (for first, second, and third place) are to be given out in a dog show having 27 contestants. In how many different ways can dogs come in first, second, and third?

Solution

Consider how we might list the prize winners:

Any dog might win first prize; there are 27 possibilities.

Any remaining dog might win second prize; there are 26 possibilities.

Any remaining dog might win third prize; there are 25 possibilities. We can use the Fundamental Counting Principle: there are

$$\underset{\substack{\uparrow\\ \text{1st}\\ \text{prize}}}{27} \cdot \underset{\substack{\uparrow\\ \text{2nd}\\ \text{prize}}}{26} \cdot \underset{\substack{\uparrow\\ \text{3rd}\\ \text{prize}}}{25} = 17{,}550$$

distinct ways that the dogs can come in first, second, and third. This is an example of permutations of 27 objects taken 3 at a time.

Practice Problem 5 There are nine different rides at a state fair. A group has to decide which four rides they will go on and in what order. How many possibilities are there?

PERMUTATIONS OF n OBJECTS TAKEN r AT A TIME

The number of permutations of n distinct objects taken r at a time is denoted by $P(n, r)$, where

$$\begin{aligned} P(n,r) &= n(n-1)(n-2)\ldots\ (n-r+1) \\ &= \frac{n!}{(n-r)!}. \end{aligned}$$

EXAMPLE 6 Using the Permutations Formula

Use the formula for $P(n, r)$ to evaluate each expression.

a. $P(7, 3)$ **b.** $P(6, 0)$

TECHNOLOGY CONNECTION

Most graphing calculators can compute $P(n, r)$. They frequently use the symbol *nPr*. Note that $P(7, 3)$ is shown as 7 *nPr* 3 on the viewing screen.

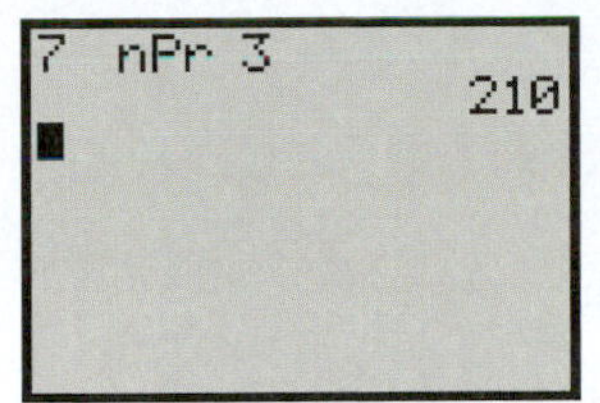

Solution

a. $P(7, 3) = \frac{7!}{(7-3!)}$ Replace n with 7 and r with 3 in $P(n, r)$.

$= \frac{7!}{4!} = \frac{7 \cdot 6 \cdot 5 \cdot 4!}{4!} = 7 \cdot 6 \cdot 5 = 210$

b. $P(6, 0) = \frac{6!}{(6-0)!}$ Replace n with 6 and r with 0 in $P(n, r)$.

$= \frac{6!}{6!} = 1$

Practice Problem 6 Evaluate.

a. $P(9, 2)$ **b.** $P(n, 0)$

Let's apply the permutations formula to the problem of counting prizewinners in Example 5. The number of permutations of 27 dogs taken 3 at a time is

$$P(27, 3) = \frac{27!}{(27-3)!}$$

$$= \frac{27!}{24!} = \frac{27 \cdot 26 \cdot 25 \cdot 24!}{24!} = 27 \cdot 26 \cdot 25 = 17{,}550.$$

This is the same answer we found in Example 5.

3 Use the formula for combinations.

Combinations

Order is not always important in selecting objects. For example, if someone is bringing six movies on a vacation, the order in which the movies were chosen is unimportant.

A Combination of *n* Distinct Objects Taken *r* at a Time

When r objects are chosen from n distinct objects without regard to order, we call the set of r objects a **combination of n objects taken r at a time**. The symbol $C(n, r)$ denotes the total number of combinations of n objects taken r at a time.

EXAMPLE 7 Listing Combinations

List all of the combinations of the four mascot names "bears," "bulls," "lions," and "tigers," taken two at a time. Find $C(4, 2)$.

Solution

If "bears" is one of the mascots chosen, the possibilities are {bears, bulls}, {bears, lions}, and {bears, tigers}.

If "bulls" is one of the mascots chosen, the remaining possibilities (excluding {bears, bulls}) are {bulls, lions} and {bulls, tigers}.

The final choice of two mascot names is {lions, tigers}.

The combinations of the four mascot names taken two at a time are {bears, bulls}, {bears, lions}, {bears, tigers}, {bulls, lions}, {bulls, tigers}, and {lions, tigers}.

Because there are six such combinations, we have $C(4, 2) = 6$.

Practice Problem 7 List all of the combinations of the four mascot names "bears," "bulls," "lions," and "tigers," taken three at a time. Find $C(4, 3)$.

To find a formula for $C(n, r)$, we begin by comparing $C(n, r)$ with $P(n, r)$, the number of permutations of n objects taken r at a time. Suppose, for example, we have seven distinct objects—A, B, C, D, E, F, and G—and we pick three of them—A, D, and E (a combination of seven objects taken three at a time). This one combination *ADE* gives $3! = 6$ permutations of seven objects taken three at a time. Here are all of the possible arrangements of the letters *A*, *D*, and *E*: *ADE*, *AED*, *DAE*, *DEA*, *EAD*, and *EDA*. Because one combination generates six permutations, there are six $(3!)$ times as many permutations of seven objects taken three at a time as there are combinations of seven objects taken three at a time. Thus, 3! (number of combinations) = (number of permutations).

$$3!\,C(7,3) = P(7,3)$$

$$C(7,3) = \frac{P(7,3)}{3!}$$

$$= \frac{\frac{7!}{(7-3)!}}{3!} \quad \text{Replace } P(7,3) \text{ with } \frac{7!}{(7-3)!}.$$

$$= \frac{7!}{(7-3)!\,3!} = \frac{7!}{4!\,3!} = \frac{7\cdot 6\cdot 5\cdot \cancel{4!}}{\cancel{4!}\,3!} = \frac{7\cdot 6\cdot 5}{3\cdot 2\cdot 1} = 35$$

A similar process leads to a more general formula.

RECALL

Notice that the value of $C(n, r)$ is the same as that of $\binom{n}{r}$.

THE NUMBER OF COMBINATIONS OF n DISTINCT OBJECTS TAKEN r AT A TIME

The number of combinations of n distinct objects taken r at a time is

$$C(n, r) = \frac{n!}{(n-r)!\,r!}.$$

EXAMPLE 8 Using the Combinations Formula

Use the formula $C(n, r)$ to evaluate each expression.

a. $C(7, 5)$ **b.** $C(8, 0)$ **c.** $C(4, 4)$

Solution

a. $C(7,5) = \frac{7!}{(7-5)!\,5!}$ Replace n with 7 and r with 5 in $C(n, r)$.

$$= \frac{7!}{2!\,5!} = \frac{7\cdot \overset{3}{\cancel{6}}\cdot \cancel{5!}}{2\cdot \cancel{1}\cdot \cancel{5!}} = 21$$

b. $C(8,0) = \frac{8!}{(8-0)!\,0!}$ Substitute $n = 8$ and $r = 0$ in $C(n, r)$.

$$= \frac{8!}{8!\,0!} = 1$$

c. $C(4,4) = \frac{4!}{(4-4)!\,4!}$ Substitute $n = 4$ and $r = 4$ in $C(n, r)$.

$$= \frac{4!}{0!\,4!} = 1$$

TECHNOLOGY CONNECTION

Most graphing calculators can compute $C(n, r)$. Note that $C(7, 5)$ is shown as 7 nCr 5 on the viewing screen.

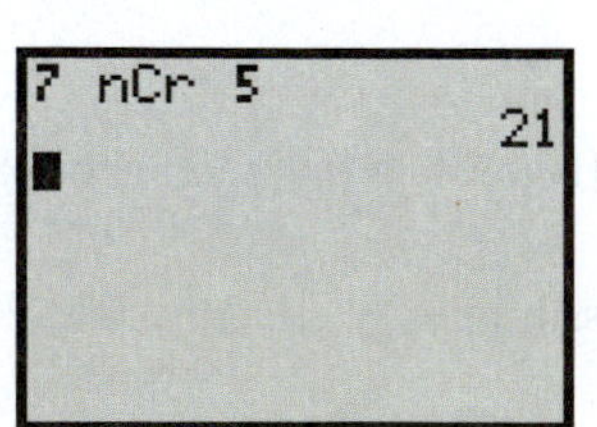

Practice Problem 8 Evaluate.

a. $C(9, 2)$ **b.** $C(n, 0)$ **c.** $C(n, n)$

EXAMPLE 9 **Choosing Pizza Toppings**

How many different ways can five pizza toppings be chosen from the following choices: pepperoni, onions, mushrooms, green peppers, olives, tomatoes, mozzarella, and anchovies?

Solution

Eight toppings are listed, so the question is, in how many ways can we pick five things from a collection of eight things? That is, we need to find the number of combinations of eight objects taken five at a time, or $C(8, 5)$.

Use the formula $C(n, r) = \dfrac{n!}{(n-r)!\,r!}$.

$$C(8, 5) = \frac{8!}{(8-5)!\,5!} \quad \text{Replace } n \text{ with 8 and } r \text{ with 5.}$$

$$= \frac{8!}{3!\,5!} = \frac{8 \cdot 7 \cdot 6}{3 \cdot 2 \cdot 1} = 56.$$

There are 56 ways that five of the eight pizza toppings can be chosen.

Practice Problem 9 A box of assorted chocolates contains 12 different kinds of chocolates. In how many ways can three chocolates be chosen?

4 Use the formula for distinguishable permutations.

Distinguishable Permutations

Suppose you were to stand in a line with identical twins. Although the three of you can arrange yourselves in $3! = 6$ different ways, someone who could not tell the twins apart could distinguish only three different arrangements. These three arrangements are called *distinguishable permutations*. To better understand the idea of distinguishable permutations, suppose we have three tiles with the numbers 1, 2, and 3 printed on the fronts and the letters *M O M* printed on the backs, respectively.

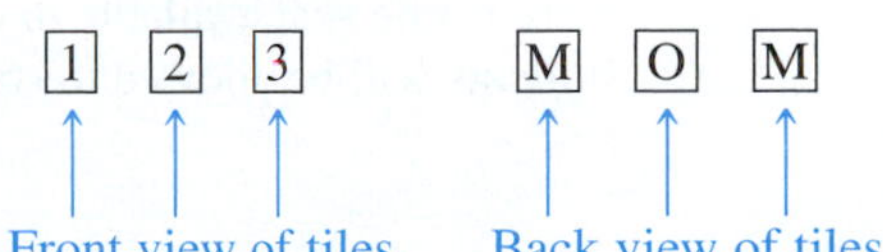

There are $3! = 6$ permutations of the numbers 1, 2, 3.

Now consider the corresponding arrangements of the letters *M, O,* and *M* on the back view of the tiles as the front view of the tiles are arranged to show each of the six permutations of 1, 2, 3. (See margin.)

Front View of Tiles	Back View of Tiles
1 2 3	M O M
1 3 2	M M O
2 1 3	O M M
2 3 1	O M M
3 1 2	M M O
3 2 1	M O M

There are only three distinguishable arrangements that appear when the tiles are viewed from the back: MOM, MMO, and OMM.

The front view shows the permutations of three distinct objects, but the back view shows the permutations of three objects of which one is of one kind (the letter *O*) and two are of a second kind (the letter *M*).

DISTINGUISHABLE PERMUTATIONS

The number of distinguishable permutations of n objects of which n_1 are of one kind, n_2 are of a second kind, . . . , and n_k are of a kth kind is

$$\frac{n!}{n_1! \cdot n_2! \cdot \cdots \cdot n_k!},$$

where $n_1 + n_2 + \cdots + n_k = n$.

EXAMPLE 10 Distributing Gifts

In how many ways can nine gifts be distributed among three children if each child receives three gifts?

Solution

If we number the gifts 1 through 9 and use A, B, and C to represent the children, we can visualize any distribution of gifts as an arrangement of the letters A, B, and C (each letter being used three times) above the numbers 1 through 9. Here is one possible arrangement:

$$\begin{matrix} B & A & A & C & C & B & A & C & B \\ 1 & 2 & 3 & 4 & 5 & 6 & 7 & 8 & 9 \end{matrix}$$

In this arrangement, child A gets gifts 2, 3, and 7; child B gets gifts 1, 6, and 9; and child C gets gifts 4, 5, and 8.

To count all of the possible distributions of gifts, we count the number of permutations of nine objects, of which three are of one kind, three are of a second kind, and three are of a third kind:

$$\frac{9}{3!\,3!\,3!} = 1680$$

There are 1680 ways the nine gifts can be distributed among the three children.

Practice Problem 10 In how many ways can six counselors be sent in pairs to three different locations? 

Deciding Whether to Use Permutations, Combinations, or the Fundamental Counting Principle

Suppose 23 students show up late for a private screening of a new movie. Suppose also that there is one seat available in each of the first, second, eighth, and tenth rows and that five students will be allowed to stand in the back of the theater.

SUMMARY OF MAIN FACTS

Permutations If order is important, use permutations.	In how many ways can 4 of the 23 students be assigned the available seats in rows 1, 2, 8, and 10? ***Answer:*** 23 (Row 1) · 22 (Row 2) · 21 (Row 8) · 20 (Row 10) · = 212,520 ways
Combinations If order is not important, use combinations.	In how many ways can 5 of the 19 students *not given seats* be chosen to stand in the back? ***Answer:*** $C(19, 5) = 11{,}628$ ways
Fundamental Counting Principle Whenever you are making consecutive choices and the number of choices at each stage is not affected by the way earlier choices were made, you can use the Fundamental Counting Principle.	In how many ways can students be assigned to the available seats and five students be chosen to stand in the back? ***Answer:*** $(212{,}520 \times 11{,}628)$ ways ↑ Assign the four seats. ↑ Pick five students. $= 2{,}471{,}182{,}560$ ways

Notice that

- Because the Fundamental Counting Principle is used to count permutations, you can always use that principle instead of the formula for permutations.
- There are frequently equally good, although different, ways to use counting techniques to solve a problem. For example, we could first choose 5 students to stand in the back of the theater in $C(23, 5)$ ways and then assign the seats in rows 1, 2, 8, and 10 to the 18 remaining students in $18 \cdot 17 \cdot 16 \cdot 15$ ways. The total number of ways students can be assigned to stand in the back and assigned to the available seats is then $C(23, 5) \cdot 18 \cdot 17 \cdot 16 \cdot 15$. We verify that

$$C(23, 5) \cdot 18 \cdot 17 \cdot 16 \cdot 15 = 33{,}649 \times 73{,}440 = 2{,}471{,}182{,}560 \text{ ways.}$$

Answers to Practice Problems

1.

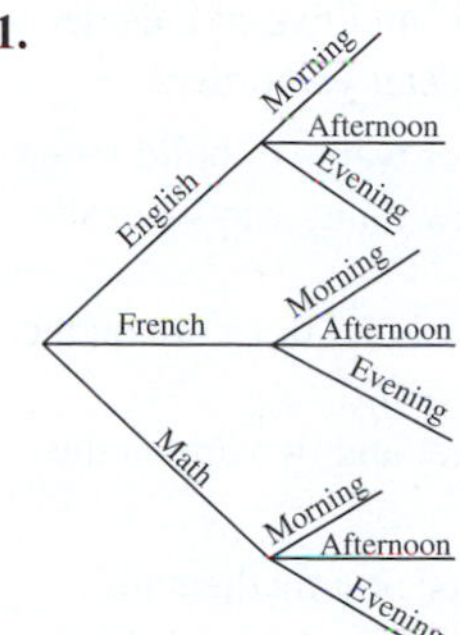

There are $(3)(3) = 9$ ways to choose a course.

2. 35 **3.** 1,000,000 **4.** 5040 **5.** 3024 **6. a.** 72 **b.** 1
7. {bears, bulls, lions}, {bears, bulls, tigers}, {bears, lions, tigers}, {bulls, lions, tigers}; $C(4, 3) = 4$
8. a. 36 **b.** 1 **c.** 1 **9.** 220 **10.** 90

SECTION 11.6 Exercises

Basic Concepts and Skills

1. Any arrangement of n distinct objects in a fixed order in which no object is used more than once is called a(n) ____________.
2. The number of permutations of five distinct objects is ____________.
3. When r objects are chosen from n distinct objects, the set of r objects is called a(n) ____________ of n objects taken r at a time.
4. The number of distinguishable permutations of 10 objects of which 3 are of one kind and 7 are of a second kind is ____________.
5. **True or False.** When order is important in counting objects, we use permutations.
6. **True or False.** There are more permutations of n objects taken r at a time than there are combinations of n objects taken r at a time.

In Exercises 7–14, use the formula for $P(n, r)$ to evaluate each expression.

7. $P(6, 1)$ **8.** $P(7, 3)$
9. $P(8, 2)$ **10.** $P(10, 6)$
11. $P(9, 9)$ **12.** $P(5, 5)$
13. $P(7, 0)$ **14.** $P(4, 0)$

In Exercises 15–22, use the formula for $C(n, r)$ to evaluate each expression.

15. $C(8, 3)$ **16.** $C(7, 2)$
17. $C(9, 4)$ **18.** $C(10, 5)$
19. $C(5, 5)$ **20.** $C(6, 6)$
21. $C(3, 0)$ **22.** $C(5, 0)$

In Exercises 23–26, use the Fundamental Counting Principle to solve each problem.

23. How many different two-letter codes can be made from the 26 capital letters of the alphabet if (a) repeated letters are allowed; (b) repeated letters are not allowed.
24. How many possible answer sheets are there for a ten-question true–false exam if no answer is left blank?
25. In how many ways can a president and vice president be chosen from an organization with 50 members?
26. How many four-digit numbers can be formed with the digits 0, 1, 2, 3, 4, 5, 6, 7, 8, and 9 if the first digit cannot be 0?

In Exercises 27–30, use the formula for counting permutations to solve each problem.

27. In how many different ways can the members of a family of four be seated in a row of four chairs?

28. Ten teams are entered in a bowling tournament. In how many ways can first, second, and third prizes be awarded?

29. Frederico has five shirts, three pairs of pants, and four ties that are appropriate for an interview. If he wears one of each type of apparel, how many different outfits can he wear?

30. A choreographer has to arrange eight pieces for a dance program. In how many ways can this be done?

In Exercises 31–34, use the formula for counting combinations to solve each problem.

31. A pizza menu offers pepperoni, peppers, mushrooms, sausage, and meatballs as extra toppings that can be added to your pizza. How many different pizzas can you order with two additional toppings?

32. In how many ways can 2 student representatives be chosen from a class of 15 students?

33. Students are allowed to choose 9 out of 11 problems to work for credit on an exam. How many ways can this be done?

34. You have to choose 3 out of 18 potential teammates to help you with a group assignment. How many ways can this be done?

In Exercises 35–40, solve the problem by any appropriate counting method.

35. How many ways can Ashley choose 4 out of 11 different canned goods to donate to charity?

36. How many ways can a computer store hire a salesclerk and a technician from seven applicants for the clerk position and four applicants for the technician position?

37. How many ways can six people be seated in a row with eight seats, leaving two seats vacant?

38. Seven students arrive to rent two canoes and three kayaks. How many ways can two students be selected to rent the canoes and three to rent the kayaks?

39. Dakota is trying to decide in which order to visit five prospective colleges. How many ways can this be done?

40. How many ways can the first 4 numbers be called from the 75 possible bingo numbers?

Applying the Concepts

41. Area codes. How many three-digit area codes for telephones are possible if
- **a.** The first digit cannot be a 0 or 1 and the second digit must be a 0 or 1?
- **b.** There are no restrictions on the second digit? (Previous restrictions were abandoned in 1995.)

42. Radio station call letters. Radio stations in the United States have call letters that begin with either *K* or *W* (for example, WXYZ in Detroit). Some have a total of three letters, and others have four. How many different call-letter selections are possible?

43. Fraternity letters. The Greek alphabet contains 24 letters. How many three-letter fraternity names can be made with Greek letters
- **a.** If no repetitions are allowed?
- **b.** If repetitions are allowed?

44. Codes. How many three-letter codes can be made from the first ten letters of the alphabet if
- **a.** No letters are repeated?
- **b.** Letters may be repeated?

45. Supreme Court decisions. In how many ways can all of the nine justices of the U.S. Supreme Court reach a majority decision? (The U.S. Supreme Court has nine justices who are appointed for life.)

46. Committee selection. A club consists of ten members. How many ways can a committee of four be selected?

47. Wardrobe choices. Al's wardrobe consists of eight pairs of slacks, eight shirts, and eight pairs of shoes. He does not want to wear the same outfit on any two days of the year. Will his current wardrobe allow him to do this?

48. Menu selection. A French restaurant offers two choices of soup, three of salad, eight of an entree, and five of a dessert. How many different complete dinners can you order?

49. Housing development. A developer wants to build seven houses, each of a different design. How many ways can she arrange these homes on a street if
- **a.** Four lots are on one side of the street and three are on the opposite side?
- **b.** Five lots are on one side of the street and two are on the opposite side?

50. Designing exercise sets. The author of a mathematics textbook has seven problems for an exercise set, and she is trying to arrange them in increasing order of difficulty. How many ways can she select the problems if no two of them are equally difficult? How many ways can the author fail?

51. Letter selection. From the letters of the word *CHARITY*, groups of four letters are formed (without regard to order). How many of them will contain
- **a.** Both *A* and *R*?
- **b.** *A* but not *R*?
- **c.** Neither *A* nor *R*?

52. Committee selection. How many ways can a committee of 4 students and 2 professors be formed from 20 students and 8 professors?

53. Inventory orders. A manufacturer makes shirts in five different colors, seven different neck sizes, and three different sleeve lengths. How many shirts should a department store order to have one of each type?

54. International book codes. All recent books are identified by their International Standard Book Number (ISBN), a ten-digit code assigned by the publisher. A typical ISBN is 0-321-75526-2. The first digit, 0, represents the language of the book (English); the second block of digits, 321, represents the publishing company (Pearson); the third block of digits, 75526, is the number assigned by the publishing company to that book; and the final digit, 2, is the check digit, used to detect the most commonly made errors when ISBNs are copied. How many ISBNs are possible? (Remember that the previous nine digits determine the last digit.)

55. Political polling. A senator sends out questionnaires asking his constituents to rank ten issues of concern to them, such as crime, education, and taxes, in order of importance. Replies are filed in folders, and two or more replies are put in the same folder if and only if they give the same ranking to all ten issues. Find the maximum number of folders that might be needed.

56. **Parcel delivery.** A UPS driver must deliver 12 parcels to 12 different addresses. In how many orders can the driver make these deliveries?

57. **Circus visit.** A father with seven children takes four of them at a time to a circus as often as he can, without taking the same four children more than once. What is the maximum number of times each child will go to the circus, and how many times will the father go?

58. **Arranging books.** From six history books, four biology books, and five economics books, how many ways can a person select three history books, two biology books, and four economics books and arrange them on a shelf?

59. **Bingo.** Each of four columns on a bingo card contains 5 different numbers chosen in any order from 15. Column "B" chooses from 1 to 15, column "I" from 16 to 30, . . . , and column "O" from 61 to 75. Only four numbers are under column "N" because of the free space. How many different cards are possible?

60. **Football conferences.** The Grid-Iron Football League in Sardonia is divided into two conferences—East and West—each having six members. Each team in the league must play each team in its conference twice and each team in the other conference once during a season. What is the total number of games played in the league?

61. **Married couples prohibited.** How many ways can a committee of four people be selected from five married couples if no committee is to include husband-and-wife pairs?

62. **Social club committees.** A social club contains seven women and four men. The club wants to select a committee of three to represent it at a state convention. How many of the possible committees contain at least one man?

63. **Coin gifts.** Among Corey's collection of early American coins, he has a half-dollar, a quarter, a nickel, and a penny. He wants to give his younger sister some or all of these four coins. How many different sets of coins can Cory give?

64. **Meal possibilities.** The university cafeteria offers two choices of soups, three of salad, five of main dishes, and six of desserts. How many different four-course meals can you choose?

In Exercises 65–68, how many distinguishable ways can the letters of each word be arranged?

65. TAIWAN
66. AMERICA
67. SUCCESS
68. ARRANGE

Beyond the Basics

69. A company that supplies temporary help has a contract for 20 weeks to provide three workers per week to an insurance firm. The agreement specifies that in no two weeks can the same three workers be sent to work at the firm. How many different workers are necessary?

70. An FM radio station wants to start its evening programming with five songs every day for 21 days without using the same five songs on any two days. What is the smallest number of songs that can be used to accomplish this result?

71. In geometry, any two points determine a line. How many lines are determined by 30 given points, no 3 of which lie on the same line?

72. In geometry, any 3 noncollinear points determine a triangle. How many triangles can be determined by 21 given points, no 3 of which are collinear?

In Exercises 73–80, solve each equation.

73. $\binom{m+1}{m-1} = 3!$

74. $6\binom{n-1}{2} = \binom{n+1}{4}$

75. $2\binom{n-1}{2} = \binom{n}{3}$

76. $\frac{4}{3}\binom{k}{2} = \binom{k+1}{3}$

77. $\binom{n}{0} + \binom{n}{1} + \binom{n}{2} + \cdots + \binom{n}{n} = 64$

78. $\binom{k}{1} + \binom{k}{2} + \binom{k}{3} + \cdots + \binom{k}{k-1} = 126$

79. $\binom{m}{4} = \binom{m}{5}$

80. $\binom{n}{7} = \binom{n}{3}$

Critical Thinking/Discussion/Writing

81. How many terms of the form $x^n y^m$ are possible if n and m can be any integer such that $1 \le n \le 5$ and $1 \le m \le 5$?

82. Use the Fundamental Counting Principle to determine the number of terms in the product

$$(a + b)(x^2 + 2xy + y^2).$$

Explain your reasoning.
[*Hint:* Each term in the product can be obtained by multiplying one term chosen from each factor.]

Maintaining Skills

In Exercises, 83–86 the notation $n(E)$ means the number of elements in the set E.

83. If $E = \{2, 5, 7\}$, find $n(E)$.

84. If $E = \{-3, -1\}$, find $n(E)$.

85. If $E = \{0\}$, find $n(E)$.

86. If $E = \varnothing$, find $n(E)$.

In Exercises 87–90, let $A = \{1, 2, 3, 5\}$ and $B = \{3, 4, 5, 6, 7\}$.

87. Find $n(A)$ and $n(B)$.

88. Find $A \cup B$ and $n(A \cup B)$.

89. Find $A \cap B$ and $n(A \cap B)$.

90. Verify that $n(A \cup B) = n(A) + n(B) - n(A \cap B)$.

In Exercise 91–94, let $A = \{a, b, c\}$ and $B = \{2, 4, 6, 8\}$.

91. Find $n(A)$ and $n(B)$.

92. Find $A \cup B$ and $n(A \cup B)$.

93. Find $A \cap B$ and $n(A \cap B)$.

94. Verify that $n(A \cup B) = n(A) + n(B) - n(A \cap B)$.

SECTION 11.7

Probability

BEFORE STARTING THIS SECTION, REVIEW

1. Set notation (Section P.1, page 3)
2. Fundamental Counting Principle (Section 11.6, page 954)
3. Permutations (Section 11.6, page 955)
4. Combinations (Section 11.6, page 957)

OBJECTIVES

1. Find the probability of an event.
2. Use the Additive Rule for finding probabilities.
3. Find the probability of mutually exclusive events.
4. Find the probability of the complement of an event.
5. Find experimental probabilities.

◆ Double Lottery Winner

Many people dream of winning a lottery, but only the most optimistic dream of winning it twice. When Evelyn Marie Adams won the New Jersey state lottery for the second time, the *New York Times* (February 14, 1986) claimed that the chances of one person winning the lottery twice were about 1 in 17 trillion. Two weeks later a letter from 2 statisticians appeared in the *Times* challenging this claim. Although they agreed that any particular person had very little chance of winning a lottery twice, they said that it was almost certain that someone in the United States would win twice. Further, they said that the odds were even for another double lottery win to occur within seven years. In less than two years, Robert Humphries won his second Pennsylvania lottery prize.

Evaluating how likely an event is to occur is the realm of probability, which we study in this section. In Example 4, we determine the probability of winning a lottery in which 6 numbers are randomly selected from the numbers 1 through 53.

1 Find the probability of an event.

The Probability of an Event

Any process that terminates in one or more outcomes (results) is an **experiment**. For example, if we ask people what time of the day they typically eat their first meal or if we pull balls from a bag of colored balls, we have performed an experiment. The outcomes of the first experiment are the times of day given as responses to our question. The outcomes of the second experiment are the colors of the balls drawn from the bag.

The set of all possible outcomes of a given experiment is called the **sample space** of the experiment. Suppose the experiment is tossing a coin. Then we could denote the outcome "heads" by H; denote the outcome "tails" by T; and write the sample space, S, in set notation as

$$S = \{H, T\}.$$

A single die is a cube, each face of which contains one, two, three, four, five, or six dots and no two faces contain the same number of dots. A roll of the die is an experiment, the

outcome of which is the number of dots showing on the upper face, and the sample space S can be written in set notation as

$$S = \{1, 2, 3, 4, 5, 6\}.$$

An event is any set of possible outcomes; that is, an **event** is any subset of a sample space. For the die-rolling experiment, one possible event is "The number showing is even." In set notation, we write this event as $E = \{2, 4, 6\}$. The event "The number showing is a 6" is written in set notation as $A = \{6\}$.

Outcomes of an experiment are said to be **equally likely** if no outcome should result more often than any other outcome when the experiment is run repeatedly. In both the coin-tossing and die-rolling experiments, all of the outcomes are equally likely.

Probability of an Event

If all of the outcomes in a sample space S are equally likely, then the **probability** of an event E, denoted $P(E)$, is the ratio of the number of outcomes in E, denoted $n(E)$, to the total number of outcomes in S, denoted $n(S)$. That is,

$$P(E) = \frac{n(E)}{n(S)}.$$

Because E is a subset of S, $0 \le n(E) \le n(S)$. Dividing by $n(S)$ yields $0 \le \frac{n(E)}{n(S)} \le 1$. So the probability $P(E)$ of an event E is a number between 0 and 1 inclusive.

EXAMPLE 1 Finding the Probability of an Event

A single die is rolled. Write each event in set notation and find the probability of the event.

a. E_1: The number 2 is showing.

b. E_2: The number showing is odd.

c. E_3: The number showing is less than 5.

d. E_4: The number showing is greater than or equal to 1.

e. E_5: The number showing is 0.

Solution

The sample space for the experiment of rolling a die is $S = \{1, 2, 3, 4, 5, 6\}$, so $n(S) = 6$. For any event E, $P(E) = \frac{n(E)}{n(S)}$.

a. $E_1 = \{2\}$, so $n(E_1) = 1$. $P(E_1) = \frac{n(E_1)}{n(S)} = \frac{1}{6}$

b. $E_2 = \{1, 3, 5\}$, so $n(E_2) = 3$. $P(E_2) = \frac{n(E_2)}{n(S)} = \frac{3}{6} = \frac{1}{2}$

c. $E_3 = \{1, 2, 3, 4\}$, so $n(E_3) = 4$. $P(E_3) = \frac{n(E_3)}{n(S)} = \frac{4}{6} = \frac{2}{3}$

d. $E_4 = \{1, 2, 3, 4, 5, 6\}$, so $n(E_4) = 6$. $P(E_4) = \frac{n(E_4)}{n(S)} = \frac{6}{6} = 1$

e. $E_5 = \varnothing$, so $n(E_s) = 0$. $P(E_5) = \frac{n(E_5)}{n(S)} = \frac{0}{6} = 0$

Practice Problem 1 A single die is rolled. Write each event in set notation and give the probability of each event.

a. E_1: The number showing is odd.

b. E_2: The number showing is greater than 4.

Notice in Example 1**d** that $E_4 = S$; so the event E_4 is certain to occur on every roll of the die. Every **certain event** has probability 1. On the other hand, in Example 1**e**, the event $E_5 = \varnothing$; so E_5 never occurs, and $P(E_5) = 0$. An **impossible event** always has probability 0.

EXAMPLE 2 Tossing a Coin

What is the probability of getting at least one tail when a coin is tossed twice?

Solution

Tossing a coin once results in two equally likely outcomes: heads (H) and tails (T). Tossing the coin twice results in two equally likely outcomes for the first toss and, regardless of the first outcome, two equally likely outcomes for the second toss. By the Fundamental Counting Principle, there are $2 \cdot 2$, or 4, outcomes when we toss the coin twice. We can represent these outcomes as (first toss, second toss) pairs:

$$S = \{(H, H), (H, T), (T, H), (T, T)\}.$$

Because three of these outcomes, $E = \{(H, T), (T, H), (T, T)\}$, result in at least one tail we have

$$P(E) = \frac{n(E)}{n(S)} = \frac{3}{4}.$$

Practice Problem 2 What is the probability of getting exactly one head when a coin is tossed twice?

Equally likely outcomes occur in most games of chance that involve tossing coins, rolling dice, or drawing cards. Lotteries and games involving spinning wheels also are assumed to generate equally likely outcomes.

EXAMPLE 3 Rolling a Pair of Dice

What is the probability of getting a sum of 8 when a pair of dice is rolled?

Solution

There are six equally likely outcomes for each die; so by the Fundamental Counting Principle, there are $6 \cdot 6$ or 36, equally likely outcomes when a pair of dice is rolled. It is useful to think of having two dice of different colors—say, red and green—in distinguishing the 36 outcomes. In this way, 2 on the red die and 6 on the green die is easily seen to be a different physical outcome from 6 on the red die and 2 on the green die, even though the same sum results.

We can represent the 36 outcomes as (red die, green die) pairs:

$$\begin{aligned} S = \{&(1, 1), (1, 2), (1, 3), (1, 4), (1, 5), (1, 6),\\ &(2, 1), (2, 2), (2, 3), (2, 4), (2, 5), (2, 6),\\ &(3, 1), (3, 2), (3, 3), (3, 4), (3, 5), (3, 6),\\ &(4, 1), (4, 2), (4, 3), (4, 4), (4, 5), (4, 6),\\ &(5, 1), (5, 2), (5, 3), (5, 4), (5, 5), (5, 6),\\ &(6, 1), (6, 2), (6, 3), (6, 4), (6, 5), (6, 6)\}\end{aligned}$$

The highlighted ordered pairs in the sample space are the outcomes that have a sum of 8. In set notation, the event "The sum of the numbers on the dice is 8" is

$$E = \{(6, 2), (5, 3), (4, 4), (3, 5), (2, 6)\}.$$

Because $n(E) = 5$ and $n(S) = 36$,

$$P(E) = \frac{n(E)}{n(S)} = \frac{5}{36}.$$

So the probability of getting a sum of 8 is $\frac{5}{36}$.

Practice Problem 3 What is the probability of getting a sum of 7 when a pair of dice is rolled?

EXAMPLE 4 **Winning the Florida Lottery**

Table tennis balls numbered 1 through 53 are mixed together in a turning drum or basket. Six of the balls are chosen in a random way. To win the lottery, a player must have selected all six of the numbers chosen. The order in which the numbers are chosen does not matter. What is the probability of matching all six numbers? (*Source:* www.fllottery.com)

Solution

Because the order in which the numbers are selected is not important, the sample space S consists of all sets of 6 numbers that can be selected from 53 numbers. Therefore, $n(S) = C(53, 6) = 22{,}957{,}480$. Because only 1 set of 6 numbers matches the 6 numbers drawn, the probability of the event "guessed all 6 numbers" is

$$\begin{aligned} P(\text{winning the lottery}) &= P(\text{guessed all 6 numbers}) \\ &= \frac{1}{22{,}957{,}480} \approx 0.00000004. \end{aligned}$$

Whether you say your chances of winning are 1 in 22,957,480 or about 0.00000004, you definitely know they are not very good!

Practice Problem 4 A state lottery requires a winner to correctly select 6 out of 50 numbers. The order of the numbers does not matter. What is the probability of matching all six numbers?

2 Use the Additive Rule for finding probabilities.

The Additive Rule

Consider the experiment of rolling one die. The sample space is $S = \{1, 2, 3, 4, 5, 6\}$, and $n(S) = 6$.

Let E be the event "The number rolled is greater than 4" and let F be the event "The number rolled is even." Then

$$E = \{5, 6\} \quad \text{and} \quad F = \{2, 4, 6\}.$$

The event "The number rolled is greater than 4 *and* even" is $E \cap F$.

$$E \cap F = \{6\}$$

Because $n(E) = 2$, $n(F) = 3$, $n(E \cap F) = 1$, and $n(S) = 6$, we have

$$P(E) = \frac{2}{6}, P(F) = \frac{3}{6}, P(E \cap F) = \frac{1}{6}.$$

The event "The number rolled is either greater than 4 *or* even" is $E \cup F$.

$$E \cup F = \{2, 4, 5, 6\}$$

Note that 6 is in *both* E and F, so it is counted once when $n(E)$ is counted and a second time when $n(F)$ is counted. However, 6 is counted only once when $n(E \cup F)$ is counted. We have $n(E \cup F) = n(E) + n(F) - n(E \cap F)$. That is, because 6 was counted twice in adding $n(E) + n(F)$, we must subtract $n(E \cap F)$ to get an accurate count for $E \cup F$.

$$n(E \cup F) = n(E) + n(F) - n(E \cap F)$$

$$\frac{n(E \cup F)}{n(S)} = \frac{n(E)}{n(S)} + \frac{n(F)}{n(S)} - \frac{n(E \cap F)}{n(S)} \quad \text{Divide both sides by } n(S).$$

$$P(E \cup F) = P(E) + P(F) - P(E \cap F) \quad \text{Definition of probability}$$

This example illustrates the *Additive Rule.*

THE ADDITIVE RULE

If E and F are events in a sample space, then

$$P(E \cup F) = P(E) + P(F) - P(E \cap F).$$

3 Find the probability of mutually exclusive events.

Mutually Exclusive Events

Two events are *mutually exclusive* if it is impossible for both to occur simultaneously. That is, E and F are **mutually exclusive** events if $E \cap F$ contains no outcomes. In this case, $P(E \cap F) = 0$ and the Additive Rule is somewhat simplified.

MUTUALLY EXCLUSIVE EVENTS

If E and F are any events in a sample space and $E \cap F = \varnothing$, then

$$P(E \cup F) = P(E) + P(F).$$

A standard deck of 52 playing cards uses four different designs called suits:

clubs,  ; diamonds,  ; hearts,  ; and spades, 

Each suit has 13 cards: an ace, a king, a queen, and a jack, and cards numbered 2 through 10. The diamond and heart symbols are red, and the club and spade symbols are black. For the purposes of this section, we assume that all decks of cards are standard and well shuffled, so that all cards drawn or dealt from a deck are equally likely.

EXAMPLE 5 Drawing a Card from a Deck

A card is drawn from a deck. What is the probability that the card is

a. A king?
b. A spade?
c. A king or a spade?
d. A heart or a spade?

Solution

The sample space S is the 52-card deck; so $n(S) = 52$.

a. There are four kings, one in each suit; so

$$P(\text{king}) = \frac{4}{52} = \frac{1}{13}.$$

b. There are 13 spades in the deck; so

$$P(\text{spade}) = \frac{13}{52} = \frac{1}{4}.$$

c. By the Additive Rule, we have

$$P(\text{king or spade}) = P(\text{king}) + P(\text{spade}) - P(\text{king and spade}).$$

Because there is only one card that is both a king and a spade, $P(\text{king and spade}) = \frac{1}{52}$. From parts **a** and **b** in this example,

$$\begin{aligned} P(\text{king or spade}) &= \frac{4}{52} + \frac{13}{52} - \frac{1}{52} \\ &= \frac{16}{52} = \frac{4}{13}. \end{aligned}$$

d. Because no card is both a heart and a spade, the events "Draw a heart" and "Draw a spade" are mutually exclusive. Thus,

$$\begin{aligned} P(\text{heart or spade}) &= P(\text{heart}) + P(\text{spade}) \\ &= \frac{13}{52} + \frac{13}{52} = \frac{1}{2}. \end{aligned}$$

Practice Problem 5 What is the probability that a card pulled from a deck is a jack or a king?

4 Find the probability of the complement of an event.

The Complement of an Event

The **complement of an event** E is the set of all outcomes in the sample space that are *not* in E. The complement of E is denoted by E'. Because every outcome in the sample space is in E or in E', the event $E \cup E'$ is a certain event and $P(E \cup E') = 1$. Also, E and E' are mutually exclusive events ($E \cap E' = \phi$, and $P(E \cap E') = 0$), so $P(E \cup E') = P(E) + P(E')$. Thus, $P(E) + P(E') = 1$ and $P(E') = 1 - P(E)$.

PROBABILITY OF THE COMPLEMENT

If E is any event and E' is its complement, then

$$P(E') = 1 - P(E).$$

EXAMPLE 6 Finding the Probability of the Complement

Brandy has applied to receive funding from her company to attend a conference. She knows that her name, along with the names of nine other deserving candidates, was put in a box from which two names will be drawn. What is the probability that Brandy will be one of the two selected?

Solution

It is easier to determine the probability that Brandy will *not* be selected. Because there are nine candidates other than Brandy, there are $C(9, 2)$ ways of selecting two employees not including Brandy. There are $C(10, 2)$ ways of selecting two employees from all ten candidates. Hence, the probability that Brandy is not selected is

$$\begin{aligned} P(\text{Brandy is not selected}) &= \frac{C(9, 2)}{C(10, 2)} \\ &= \frac{9!}{2!\,7!} \cdot \frac{2!\,8!}{10!} = \frac{8}{10} = \frac{4}{5}. \end{aligned}$$

The probability that Brandy is selected is

$$P(\text{Brandy is selected}) = 1 - P(\text{Brandy is not selected})$$
$$= 1 - \frac{4}{5} = \frac{1}{5}.$$

Practice Problem 6 In Example 6, what is the probability of Brandy being selected if three names are drawn from the box?

5 Find experimental probabilities.

Experimental Probabilities

The probabilities we have seen up to this point have been determined by some form of reasoning, not data. Data support the decision to assign the value $\frac{1}{2}$ as the probability of a head resulting from one toss of a coin, but the assignment is not *based* on data. This is also true for the probabilities associated with drawing a card, rolling a pair of dice, drawing a ball from an urn, and so on. Probability assigned this way is called **theoretical probability**.

In contrast, we can assign probability values based on observation and data. If an experiment is performed n times (where n is large) and a particular outcome occurs k times, then the number $\frac{k}{n}$ is the value for the probability of that outcome. Probability determined this way is called **experimental probability**. A statement such as "The probability that a 10-year-old child will live to be 95 years old is $\frac{3}{100{,}000}$" refers to experimental probability.

RELATIVE FREQUENCY PRINCIPLE

If an experiment is performed n times and an event E occurs k times, then we say that the experimental (or statistical) probability of the event, $P(E)$, is

$$P(E) = \frac{k}{n}.$$

EXAMPLE 7 **Using Data to Estimate Probabilities**

Table 11.2 shows the gender distribution of students at four California city colleges.

TABLE 11.2

College	Male	Female
Los Angeles	6,995	8,179
San Diego	12,251	14,914
San Francisco	16,857	2,000
Sacramento	9,040	11,838

Source: www.areaconnect.com

Find the probability that a student selected *at random* is:

a. A female.

b. Attending Los Angeles City College.

c. A female attending Los Angeles City College.

Round your answers to the nearest hundredth.

Solution

The term *randomly selected* means that every possible selection is equally likely. In our example, this means that each student is equally likely to be selected. With random selection, the percentage of students in a given category gives the probability that a student in that category will be chosen.

a. The total number of students is 82,074. The total number of female students is 36,931. Thus, $P(\text{female}) = \dfrac{36{,}931}{82{,}074} \approx 0.45$.

b. The total number of students is 82,074. The total number of students attending Los Angeles City College is 15,174. Hence,

$$P(\text{attends Los Angeles City College}) = \frac{15{,}174}{82{,}074} \approx 0.18.$$

c. There are 8179 female students attending Los Angeles City College. Consequently,

$$P(\text{female and attending Los Angeles City College}) = \frac{8179}{82{,}074} \approx 0.10.$$

Practice Problem 7 In Example 7, find the probability that a student selected at random is a male or attends Sacramento College.

Answers to Practice Problems

1. a. $E_1 = \{1, 3, 5\}, P(E_1) = \frac{1}{2}$ **b.** $E_2 = \{5, 6\}, P(E_2) = \frac{1}{3}$ **2.** $\frac{1}{2}$ **3.** $\frac{1}{6}$ **4.** $\frac{1}{15{,}890{,}700}$ **5.** $\frac{2}{13}$ **6.** $\frac{3}{10}$ **7.** 0.69

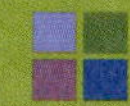

SECTION 11.7 Exercises

Basic Concepts and Skills

1. If no outcome of an experiment results more often than any other outcome, the outcomes are said to be ____________.
2. A *certain* event has probability ____________, and an *impossible* event has probability ____________.
3. If it is impossible for two events to occur simultaneously, then the events are said to be ____________.
4. If the probability that an event occurs is 0.7, then the probability that the event does not occur is ____________.
5. **True or False.** The probability of any event is a number between 0 and 1.
6. **True or False.** If an experiment is performed 100 times and an event E occurs 65 times, then the *experimental probability* of the event is 0.65.

In Exercises 7–12, write the sample space for each experiment.

7. Students are asked to rank Wendy's, McDonald's, and Burger King in order of preference.
8. Students are asked to choose either a hamburger or a cheeseburger and then pick Wendy's, McDonalds, or Burger King as the restaurant from which they would prefer to order it.
9. Two albums are selected from among *Thriller* (Michael Jackson), *The Wall* (Pink Floyd), *Eagles: Their Greatest Hits* (Eagles), and *Led Zeppelin IV* (Led Zeppelin).
10. Three types of potato chips are selected from among regular salted, regular unsalted, barbecue, cheddar cheese, and sour cream and onion.
11. Two blanks in a questionnaire are filled in, the first identifying race as white, African American, Native American, Asian, other, or multiracial, and the second indicating gender as male or female
12. A six-sided die is chosen from a box containing a white, a red, and a green die. Then the die is tossed.

In Exercises 13–16, find the probability of the given event if a year is selected at random.

13. Thanksgiving falls on a Sunday.
14. Thanksgiving falls on a Thursday.
15. Christmas in the United States falls on December 25.
16. Christmas falls on January 1.

In Exercises 17–26, classify each statement as an example of *theoretical probability* or *experimental probability.*

17. The probability that Ken will go to church next Sunday is 0.85.

18. The probability of exactly two heads when a fair coin is tossed twice is $\frac{1}{4}$.

19. The probability of drawing a heart from a deck of cards is 0.25.

20. The probability that an American adult will watch a major network's evening newscast is 0.17.

21. The probability that a person driving during the Memorial Day weekend will die in an automobile accident is $\frac{1}{58{,}000}$.

22. The probability that a student will get a D or an F in a college statistics course is 0.21.

23. The probability of drawing an ace from a deck of cards is $\frac{1}{13}$.

24. The probability that a person will be struck by lightning this year is $\frac{1}{727{,}000}$.

25. The probability that a plane will take off on time from JFK International airport is 0.78.

26. The probability that a college student cheats on an exam is 0.32.

27. Match each given sentence with an appropriate probability.

An event that	has probability
is very likely to happen	0
will surely happen	0.5
is a rare event	0.001
equally likely to happen or not	1
will never happen	0.999

28. A coin is tossed three times. Rearrange the order of the following events from the most probable event (first) to the least probable event (last).
{exactly two heads}, {the sure event},
{at least one head}, {at least two heads}.

In Exercises 29–34, a die is rolled. Find the probability of each event.

29. The number showing is a 1 or a 6.

30. The number 5 is showing.

31. The number showing is greater than 4.

32. The number showing is a multiple of 3.

33. The number showing is even.

34. The number showing is less than 1.

In Exercises 35–40, a card is drawn randomly from a standard 52-card deck. Find the probability of each event.

35. A queen is drawn.

36. An ace is drawn.

37. A heart is drawn.

38. An ace of hearts or a king of hearts is drawn.

39. A face card (jack, queen, or king) is drawn.

40. A heart that is not a face card is drawn.

In Exercises 41–46, a digit is chosen randomly from the digits 0 through 9; then 3 is added to the digit and the sum recorded. Find the probability of each event.

41. The sum is 11.

42. The sum is 0.

43. The sum is less than 4.

44. The sum is greater than 8.

45. The sum is even.

46. The sum is odd.

Applying the Concepts

47. A blind date. The probability that Tony will like his blind date is 0.3. What is the probability that he will not like his blind date?

48. Drawing cards. The probability of drawing a spade from a standard 52-card deck is 0.25. What is the probability of drawing a card that is not a spade?

49. Gender and the Air Force. In 2012, about 20% of members of the U.S. Air Force were women. What is the probability that an Air Force member chosen at random is a woman? (*Source:* www.af.mil/news)

50. Movie attendance. If 85% of people between the ages of 18 and 24 go to a movie at least once a month, what is the probability that a randomly chosen person in this age group will go to a movie this month?

51. Gender and the Marines. In 2012, about 6% of members of the U.S. Marines were women. What is the probability that a Marine chosen at random is a man? (*Source:* DMDC-June 2005, TFDW-June 2005)

52. Movie attendance. If 20% of people over 65 go to a movie at least once a month, what is the probability that a randomly chosen person over 65 will *not* go to a movie this month?

53. Raffles. One hundred twenty-five tickets were sold at a raffle. If you have ten tickets, what is the probability of your winning the (only) prize? If there are two prizes, what is the probability of your winning both?

54. Committee selection. To select a committee with two members, two people will be selected at random from a group of four men and four women. What is the probability that a man and a woman will be selected?

55. Gender of children. A couple has two children. Assuming that each child is equally likely to be a boy or a girl, find the probability of having

a. Two boys.
b. Two girls.
c. One boy and one girl.

56. Gender of children. A couple has three children. Assuming that each child is equally likely to be a boy or a girl, find the probability that the couple will have

a. All girls.
b. At least two girls.
c. At most two girls.
d. Exactly two girls.

57. Car ownership. The following table shows the number of cars owned by each of the 4440 families in a small town.

No. of families	37	1256	2370	526	131	120
No. of cars owned	0	1	2	3	4	5

A family is selected at random. Find the probability that this family owns

a. Exactly two cars.
b. At most two cars.
c. Exactly six cars.
d. At most six cars.
e. At least one car.

58. Mammograms. According to the American Cancer Society, 199 of 200 mammograms (for breast cancer screening) turn out to be normal. Find the probability that a mammogram of a woman chosen at random is not normal.

59. Lactose intolerance. Lactose intolerance (the inability to digest sugars found in dairy products) affects about 20% of non-Hispanic white Americans and about 75% of African, Asian, and Native Americans. What is the probability that a non-Hispanic white American has no lactose intolerance? Answer the same question for a person in the second group.

60. Guessing on an exam. A student has just taken a ten-question true–false test. If he must guess to answer each of the ten questions, what is the probability that

a. He scores 100% on the test.
b. He scores 90% on the test.
c. He scores at least 90% on the test.

61. Committee selection. A five-person committee is choosing a subcommittee of two of its members. Each member of the committee is a different age. The subcommittee members are chosen at random.

a. What is the probability that the subcommittee will consist of the two oldest members?
b. What is the probability that the subcommittee will consist of the oldest and the youngest members?

62. Children in U.S. In 2012, the total number of children ages 0–17 in U.S. was 73.7 million. The age distribution (in millions) is given in the following table.

Age	Number of Children
0–5	24.1
6–11	24.5
12–17	25.1

Source: www.childstats.gov

If a child from this age group is chosen at random, what is the probability that he/she is at least 6 years old?

63. A test for depression. Suppose that 10% of the U.S. population suffers from depression. A pharmaceutical company is believed to have developed a test with the following results: The test works for 90% of the people who are tested and are actually depressed (as diagnosed through other acceptable methods). In other words, if a patient is actually depressed, there is a probability of 0.9 that this patient will be diagnosed as a depressed patient by the pharmaceutical company. The test also works for 85% of people who are tested and not depressed. Suppose the company tests 1000 people for depression. Complete the following table.

Number of people who are	Depressed according to the test	Normal according to the test
actually depressed		
actually normal		

64. Playing marbles. Natasha and Deshawn are playing a game with marbles. Natasha has one marble (call it N), and Deshawn has two (call them M_1 and M_2). They roll their marbles toward a stake in the ground. The person whose marble is closest to the stake wins. Assume that measurements are accurate enough to eliminate ties. List all of the sample points. Use the notation M_1NM_2 to denote the sample point when M_1 is closest to the stake and M_2 is farthest from the stake. If the children are equally skillful (when the sample points are all equally likely), find the probability that Deshawn wins.

65. Gender among siblings. If a family with four children is chosen at random, are the children more likely to be three of one gender and one of the other gender than two boys and two girls? (Assume that each child in a family is equally likely to be a boy or a girl.)

66. Hospital treatments. Hospital records show that 11% of all patients admitted are admitted for surgical treatment, 15% are admitted for obstetrics, and 3% are admitted for both obstetrics and surgery. What is the probability that a new patient admitted is an obstetrics patient or a surgery patient or both?

67. Pain relief medicine. A pharmaceutical company testing a new over-the-counter pain relief medicine gave the new medicine, their old medicine, and a placebo to different groups of patients. (A placebo is a sugar pill with no medicinal ingredients. However, it sometimes works, supposedly for psychological reasons.) The following table shows the results of the tests.

	Amount of Pain Relief		
Medicine	**Some**	**Complete**	**None**
Placebo	12	4	34
Old	25	25	10
New	18	40	12
Total	55	69	56

Find the probability that a randomly chosen person from the group of 180 people receiving one of the three types of medicines

a. Received the placebo or had no pain relief.
b. Received the new medicine or had complete pain relief.

Beyond the Basics

68. A famous problem called the *birthday problem* asks for the probability that at least two people in a group of people have the same birthday (the same day and the same month but not necessarily the same year). The best way to approach this problem is to consider the complementary event that no two people have the same birthday. Assume that all birthdays are equally likely.

a. Find the probability that in a group of two people, both have the same birthday.

b. Find the probability that in a group of three people, at least two have the same birthday.

c. Find the probability that in a group of 50 people, at least two have the same birthday.

69. What is the probability that a number between 1 and 1,000,000 contains the digit 3? [*Hint:* Consider the complementary event.]

70. A business executive types four letters to her clients and then types four envelopes addressed to the clients. What is the probability that if the four letters are randomly placed in envelopes, they are placed in correct envelopes?

71. Juan gave his telephone number to Maggie. She remembers that the first three digits are 407 and the remaining four digits consist of two 3s, one 6, and one 8. She is not certain about the order of the digits, however. Maggie dials 407-3638. What is the probability that Maggie dialed the correct number?

Critical Thinking/Discussion/Writing

72. The March 2005 population survey by the U.S. Census Bureau found that among the 187 million people aged 25 or older, 59,840,000 reported high school graduation as the highest level of education attained. Find the probability that a person chosen at random from the 187 million in the survey would report his or her highest level of education as high school graduation.

73. Find the probability of randomly choosing a dark caramel from a box that contains only light and dark caramels and has twice as many light as dark caramels.

74. If a single die is rolled two times, what is the probability that the second roll will result in a number larger than the first roll?

SUMMARY Definitions, Concepts, and Formulas

11.1 Sequences and Series

i. An infinite sequence is a function whose domain is the set of positive integers. The function values $a_1, a_2, a_3, \dots, a_n, \dots$ are called the **terms** of the sequence. The term a_n is the ***n*th term**, or **general term**, of the sequence.

ii. ***Recursive formula***
A sequence may be defined recursively, with the nth term of the sequence defined in relation to previous terms.

iii. ***Factorial notation***

$$n! = n(n-1)\cdots 3\cdot 2\cdot 1 \text{ and } 0! = 1$$

iv. ***Summation notation***

$$\sum_{i=1}^{n} a_i = a_1 + a_2 + a_3 + \cdots + a_n$$

v. ***Series***
Given an infinite sequence $a_1, a_2, a_3, \dots, a_n, \dots$, the sum $\sum_{i=1}^{n} a_i = a_1 + a_2 + a_3 + \cdots$ is called a **series** and the sum $\sum_{i=1}^{n} a_i$ is called the ***n*th partial sum** of the series.

11.2 Arithmetic Sequences; Partial Sums

i. A sequence is an **arithmetic sequence** if each term after the first differs from the preceding term by a constant. The constant difference d between the consecutive terms is called the **common difference**. We have $d = a_n - a_{n-1}$ for all $n \ge 2$.

ii. The ***n*th term** a_n of the arithmetic sequence $a_1, a_2, a_3, \dots, a_n, \dots$ is given by $a_n = a_1 + (n-1)d$, $n \ge 1$ where d is the common difference.

iii. The **sum** S_n of the first n terms of the arithmetic sequence $a_1, a_2, \dots, a_n, \dots$ is given by the formula

$$S_n = n\left(\frac{a_1 + a_n}{2}\right) = n\left(\frac{2a_1 + (n-1)d}{2}\right).$$

11.3 Geometric Sequences and Series

i. A sequence is a **geometric sequence** if each term after the first is a constant multiple of the preceding term. The constant ratio r between the consecutive terms is called the **common ratio**. We have $\frac{a_{n+1}}{a_n} = r, n \ge 1$.

ii. **The *n*th term a_n of the geometric sequence** $a_1, a_2, \dots, a_n, \dots$ is given by $a_n = a_1 r^{n-1}$, $n \ge 1$, where r is the common ratio.

iii. The **sum** S_n of the first n terms of the geometric sequence $a_1, a_2, \dots, a_n, \dots$ with common ratio $r \ne 1$ is given by the formula

$$S_n = \frac{a_1(1 - r^n)}{1 - r}.$$

The **sum S of an infinite geometric sequence** is given by

$$S = \sum_{i=1}^{\infty} a_1 r^{i-1} = \frac{a_1}{1-r} \text{ if } |r| < 1.$$

iv. An **annuity** is a sequence of equal payments made at equal time intervals. Suppose \$$P$ is the payment made at the end of each of n compounding periods per year and i is the annual interest rate. Then the value A of the annuity after t years is

$$A = P\frac{\left(1 + \frac{i}{n}\right)^{nt} - 1}{\frac{i}{n}}.$$

v. If $|r| < 1$, the infinite geometric series $a_1 + a_1 r + a_1 r^2 + \cdots + a_1 r^{n-1} + \cdots$ has the sum $S = \frac{a_1}{1 - r}$. When $|r| \geq 1$, the series does not have a finite sum.

11.4 Mathematical Induction

The statement P_n is true for all positive integers n if the following properties hold:

i. P_1 is true.

ii. If P_k is a true statement, then P_{k+1} is a true statement.

11.5 The Binomial Theorem

i. Binomial coefficient: $\binom{n}{j} = \frac{n!}{j!(n - j)!}, 0 \leq j \leq n$

$$\binom{n}{0} = 1 = \binom{n}{n}$$

ii. Binomial theorem:

$$(x + y)^n = \binom{n}{0}x^n + \binom{n}{1}x^{n-1}y + \binom{n}{2}x^{n-2}y^2 + \cdots + \binom{n}{j}x^{n-j}y^j + \cdots + \binom{n}{n}y^n$$

$$= \sum_{j=0}^{n}\binom{n}{j}x^{n-j}y^j$$

iii. The term containing the factor x^j in the expansion of $(x + y)^n$ is

$$\binom{n}{n - j}x^j y^{n-j}.$$

11.6 Counting Principles

i. **Fundamental Counting Principle:** If a first choice can be made in p different ways, a second choice in q different ways, a third choice in r different ways, and so on, then the sequence of choices can be made in $p \cdot q \cdot r \ldots$ different ways.

ii. A **permutation** is an arrangement of n distinct objects in a fixed order in which no object is used more than once. The order in an arrangement is important.

iii. **Permutation formula:** The number of permutations of n distinct objects taken r at a time is

$$P(n, r) = n(n - 1)(n - 2) \ldots (n - r + 1) = \frac{n!}{(n - r)!}.$$

iv. **Combination formula:** When r objects are chosen from n distinct objects without regard to order, we call the set of r objects a **combination** of n objects taken r at a time. The number of combinations of n distinct objects taken r at a time is denoted by $C(n, r)$, where

$$C(n, r) = \frac{n!}{(n - r)!r!}.$$

v. **Distinguishable permutations:** The number of permutations of n objects of which n_1 are of one kind n_2 are of a second kind, . . . , and n_k are of a kth kind is

$$\frac{n!}{n_1!n_2! \ldots n_k!},$$

where $n_1 + n_2 + \cdots + n_k = n$.

11.7 Probability

i. **Experiment:** Any process that terminates in one or more **outcomes** (results) is an experiment.

ii. **Sample space:** The set of all possible outcomes of an experiment is called the sample space of the experiment.

iii. **Event:** An event is any subset of a sample space.

iv. **Equally likely events:** Outcomes of an experiment are said to be equally likely if no outcome should result more often than any other outcome when the experiment is performed repeatedly.

v. **Probability:** If all of the outcomes in a sample space S are equally likely, the probability of an event E, denoted $P(E)$, is defined by

$$P(E) = \frac{n(E)}{n(S)},$$

where $n(E)$ is the number of outcomes in E and $n(S)$ is the total number of outcomes in S.

a. **The Additive Rule:** If E and F are events in a sample space S, then

$$P(E \text{ or } F) = P(E \cup F) = P(E) + P(F) - P(E \cap F).$$

b. If $E \cap F = \varnothing$, E and F are called **mutually exclusive** events. For mutually exclusive events E and F, $P(E \cup F) = P(E) + P(F)$.

c. $P(E') = P(\text{not } E) = 1 - P(E)$; the event E' is called the **complement** of the event E.

REVIEW EXERCISES

In Exercises 1–4, write the first five terms of each sequence.

1. $a_n = 2n - 3$

2. $a_n = \frac{n(n - 2)}{2}$

3. $a_n = \frac{n}{2n + 1}$

4. $a_n = (-2)^{n-1}$

In Exercises 5 and 6, write a general term a_n for each sequence.

5. 30, 28, 26, 24, . . .

6. −1, 2, −4, 8, . . .

In Exercises 7–10, simplify each expression.

7. $\frac{9!}{8!}$ **8.** $\frac{10!}{7!}$

9. $\frac{(n+1)!}{n!}$ **10.** $\frac{n!}{(n-2)!}$

In Exercises 11–14, find each sum.

11. $\sum_{k=1}^{4} k^3$ **12.** $\sum_{j=1}^{5} \frac{1}{2j}$

13. $\sum_{k=1}^{7} \frac{k+1}{k}$ **14.** $\sum_{k=1}^{5} (-1)^k 3^{k+1}$

In Exercises 15 and 16, write each sum in summation notation.

15. $1 + \frac{1}{2} + \frac{1}{3} + \frac{1}{4} + \cdots + \frac{1}{50}$

16. $2 - 4 + 8 - 16 + 32$

In Exercises 17–20, determine whether each sequence is arithmetic. If a sequence is arithmetic, find the first term a_1 and the common difference d.

17. $11, 6, 1, -4, \ldots$ **18.** $\frac{2}{3}, \frac{5}{6}, 1, \frac{7}{6}, \ldots$

19. $a_n = n^2 + 4$ **20.** $a_n = 3n - 5$

In Exercises 21–24, find an expression for the *n*th term of each arithmetic sequence.

21. $3, 6, 9, 12, 15, \ldots$ **22.** $5, 9, 13, 17, 21, \ldots$

23. $x, x + 1, x + 2, x + 3, x + 4, \ldots$

24. $3x, 5x, 7x, 9x, 11x, \ldots$

In Exercises 25 and 26, find the common difference d and the *n*th term a_n for each arithmetic sequence.

25. 3rd term: 7; 8th term: 17

26. 5th term: -16; 20th term: -46

In Exercises 27 and 28, find each sum.

27. $7 + 9 + 11 + 13 + \cdots + 37$

28. $\frac{1}{4} + \frac{1}{2} + \frac{3}{4} + 1 + \cdots + 15$

In Exercises 29 and 30, find the sum of the first *n* terms of each arithmetic sequence.

29. $3, 8, 13, \ldots; n = 40$ **30.** $-6, -5.5, -5, \ldots; n = 60$

In Exercises 31–34, determine whether each sequence is geometric. If a sequence is geometric, find the first term a_1 and the common ratio r.

31. $4, -8, 16, -32, \ldots$ **32.** $\frac{1}{5}, \frac{1}{10}, \frac{1}{20}, \frac{1}{40}, \ldots$

33. $\frac{1}{3}, 1, \frac{5}{3}, 9, \ldots$ **34.** $a_n = 2^{-n}$

In Exercises 35 and 36, for each geometric sequence, find the first term a_1, the common ratio r, and the *n*th term a_n.

35. $16, -4, 1, -\frac{1}{4}, \ldots$ **36.** $-\frac{5}{6}, -\frac{1}{3}, -\frac{2}{15}, -\frac{4}{75}, \ldots$

In Exercises 37 and 38, find the indicated term of each geometric sequence.

37. a_{10} when $a_1 = 2, r = 3$ **38.** a_{12} when $a_1 = -2, r = \frac{3}{2}$

In Exercises 39 and 40, find the sum S_n of the first *n* terms of each geometric sequence.

39. $\frac{1}{10}, \frac{1}{5}, \frac{2}{5}, \frac{4}{5}, \ldots; n = 12$ **40.** $2, -1, \frac{1}{2}, -\frac{1}{4}, \ldots; n = 10$

In Exercises 41–44, find each sum.

41. $\frac{1}{2} + \frac{1}{6} + \frac{1}{18} + \frac{1}{54} + \cdots$ **42.** $-5 - 2 - \frac{4}{5} - \frac{8}{25} - \ldots$

43. $\sum_{i=1}^{\infty} \left(\frac{3}{5}\right)^i$ **44.** $\sum_{i=1}^{\infty} 7\left(-\frac{1}{4}\right)^i$

In Exercises 45 and 46, use mathematical induction to prove that each statement is true for all natural numbers *n*.

45. $\sum_{k=1}^{n} 2^k = 2^{n+1} - 2$

46. $\sum_{k=1}^{n} k(k+1) = \frac{n(n+1)(n+2)}{3}$

In Exercises 47 and 48, evaluate each binomial coefficient.

47. $\binom{12}{7}$ **48.** $\binom{11}{0}$

In Exercises 49 and 50, find each binomial expansion.

49. $(x - 3)^4$ **50.** $\left(\frac{x}{2} + 2\right)^6$

In Exercises 51 and 52, find the specified term.

51. $(x + 2)^{12}$; term containing x^5

52. $(x + 2y)^{12}$; term containing x^7

In Exercises 53–64, use any method to solve the problem.

53. How many different numbers can be written with the digits 1, 2, 3, and 4 if no digit may be repeated?

54. How many ways can horses finish in first, second, and third place in a ten-horse race?

55. How many ways can seven people line up at a ticket window?

56. In a building with seven entrances, how many ways can you enter the building and leave by a different entrance?

57. How many ways can five movies be listed on a display (vertically, one movie per line)?

58. How many different amounts of money can you make with a nickel, a dime, a quarter, and a half-dollar?

59. How many ways can three candies be taken from a box of ten candies?

60. How many ways can 2 face cards be drawn from a standard 52-card deck?

61. How many ways can 2 shirts and 3 pairs of pants be chosen from 12 shirts and 8 pairs of pants?

62. How many doubles teams can be formed from eight tennis players?

63. In how many distinguishable ways can the letters of the word R E I T E R A T E be arranged?

64. A bookshelf has room for four books. How many different arrangements can be made on the shelf from six available books?

In Exercises 65–74, find the probability requested.

65. Four silver dollars, all with different dates, are randomly arranged on a horizontal display. What is the probability that the two with the most recent dates will be next to each other?

66. A jar contains five white, three black, and two green stones. What is the probability of drawing a black stone on a single draw?

67. The word R A N D O M has been spelled in a game of Scrabble. If two letters are chosen from this word at random, what is the probability that they are both consonants?

68. The names of all 30 party guests are placed in a box, and one name is drawn to award a door prize. If you and your date are among the guests, what is the probability that one of you will win the prize?

69. If two people are chosen at random from a group consisting of five men and three women, what is the probability that one man and one woman are chosen?

70. A company determines that 32% of the e-mail it receives is junk mail. What is the probability that a randomly chosen e-mail at this company is not junk mail?

71. If 2 cards are drawn from a standard 52-card deck, what is the probability that both are clubs?

72. A battery manufacturer inspects 500 batteries and finds that 30 are defective. What is the probability that a randomly selected battery is not defective?

73. A ball is taken at random from a pool table containing balls numbered 1 through 9. What is the probability of obtaining (a) ball number 7? (b) an even-numbered ball? (c) an odd-numbered ball?

74. A clothes dryer load contains two shirts, four pairs of socks, and three nightgowns. You pull one item randomly from the dryer.
   a. What is the probability that it is a sock?
   b. What is the probability that it is a nightgown?

PRACTICE TEST A

In Problems 1 and 2, write the first five terms of each sequence. State whether the sequence is arithmetic or geometric.

1. $a_n = 3(5 - 4n)$

2. $a_n = -3(2^n)$

3. Write the first five terms of the sequence defined by $a_1 = -2$ and $a_n = 3a_{n-1} + 5$ for $n \geq 2$.

4. Evaluate $\dfrac{(n-1)!}{n!}$.

5. Find a_7 for the arithmetic sequence with $a_1 = -3$ and $d = 4$.

6. Find a_8 for the geometric sequence with $a_1 = 13$ and $r = -\dfrac{1}{2}$.

In Problems 7–9, find each sum.

7. $\displaystyle\sum_{k=1}^{20} (3k - 2)$

8. $\displaystyle\sum_{k=1}^{5} \left(\frac{3}{4}\right)(2^{-k})$

9. $\displaystyle\sum_{k=1}^{\infty} 18\left(\frac{1}{100}\right)^k$; write the answer as a fraction in lowest terms.

10. Evaluate $\dbinom{13}{0}$

11. Expand $(1 - 2x)^4$.

12. Find the term containing x^1 in the expansion of $(2x + 1)^4$.

13. In how many ways can you answer every question on a true–false test containing ten questions?

In Problems 14 and 15, evaluate each expression.

14. $P(9, 2)$

15. $C(7, 5)$

16. A firm needs to fill two positions—one in accounting and one in human resources. How many ways can these positions be filled if there are six applicants for the accounting position and ten applicants for the human resources position?

17. A state is considering using license plates consisting of two letters followed by a four-digit number. How many distinct license plate numbers can be formed?

18. A box contains five red balls and seven black balls. What is the probability that a ball drawn at random will be red?

19. What is the probability that a 4 does *not* come up on one roll of a die?

20. A box contains seven quarters, six dimes, and five nickels. If two coins are drawn at random, what is the probability that both are quarters?

PRACTICE TEST B

In Problems 1 and 2, write the first five terms of each sequence. State whether the sequence is arithmetic or geometric.

1. $a_n = 4(2n - 3)$
a. $-1, 1, 3, 5, 7$; arithmetic
b. $-1, 1, 3, 5, 7$; geometric
c. $-4, 4, 12, 20, 28$; arithmetic
d. $-4, 4, 12, 20, 28$; geometric

2. $a_n = 2(4^n)$
a. $2, 8; 32, 128, 512$; arithmetic
b. $2, 8, 32, 128, 512$; geometric
c. $8, 32, 128, 512, 2048$; arithmetic
d. $8, 32, 128, 512, 2048$; geometric

3. Write the first five terms of the sequence defined by $a_1 = 5$ and $a_n = 2a_{n-1} + 4$, for $n \geq 2$.
a. $5, 14, 32, 68, 140$ **b.** $5, 14, 19, 24, 29$
c. $5, 9, 13, 17, 21$ **d.** $5, 8, 10, 12, 14$

4. Evaluate $\dfrac{(n + 2)!}{n + 2}$.
a. $2!$ **b.** 1
c. $(n + 1)!$ **d.** $n + 1$

5. Find a_8 for the arithmetic sequence with $a_1 = -6$ and $d = 3$.
a. 15 **b.** -271
c. 30 **d.** 18

6. Find a_{10} for the geometric sequence with $a_1 = 247$ and $r = \dfrac{1}{3}$.
a. 250 **b.** $\dfrac{247}{19,683}$
c. $\dfrac{247}{59,049}$ **d.** $\dfrac{247}{177,147}$

In Problems 7 and 8, find each sum.

7. $\sum_{k=1}^{45} (4k - 7)$
a. 4185 **b.** 3735
c. 4027.5 **d.** 3825

8. $\sum_{k=1}^{5} \left(\dfrac{4}{3}\right)(2^k)$
a. $\dfrac{248}{3}$ **b.** $\dfrac{251}{3}$
c. $\dfrac{242}{3}$ **d.** $\dfrac{287}{3}$

9. Evaluate $\sum_{k=1}^{\infty} 8(-0.3)^{k-1}$.
a. 6 **b.** $\dfrac{80}{13}$
c. $-\dfrac{80}{13}$ **d.** $-\dfrac{83}{20}$

10. Evaluate $\dbinom{12}{11}$.
a. $1.\overline{09}$ **b.** 12
c. 1 **d.** $11!$

11. Expand $(3x - 1)^4$.
a. $81x^4 - 108x^3 + 54x^2 - 12x + 1$
b. $-81x^4 - 108x^3 - 54x^2 + 12x - 1$
c. $-81x^4 + 108x^3 + 54x^2 + 12x + 1$
d. $81x^4 - 108x^2 + 54x - 12$

12. Find the coefficient of x in the expansion of $(2x + 3)^3$.
a. 108 **b.** 9
c. 36 **d.** 54

13. Karen has four necklaces, ten pairs of earrings, and three bracelets. How many ways can she choose a necklace, a pair of earrings, and a bracelet?
a. 40 **b.** 17
c. 120 **d.** 240

In Problems 14 and 15, evaluate each expression.

14. $P(7, 3)$
a. 210 **b.** 840
c. 1680 **d.** 5040

15. $C(10, 7)$
a. 3 **b.** 720
c. 620,400 **d.** 120

16. How many two-digit numbers can be formed from the digits 0, 1, 2, 3, 4, 5, 6, 7, 8, and 9? No digit can be used more than once.
a. 45 **b.** 81
c. 3,628,880 **d.** 100

17. You have selected eight CDs for your dad, but have only enough money to buy five of them. In how many ways can you select the five if you decide that a particular CD is a "must buy"?
a. 56 **b.** 70
c. 1680 **d.** 35

18. What is the probability of rolling a 5 on one roll of a die?
a. $\dfrac{1}{5}$ **b.** 4
c. $\dfrac{1}{6}$ **d.** 0.05

19. What is the probability of getting a sum greater than 9 when a pair of dice is rolled?
a. $\dfrac{1}{6}$ **b.** $\dfrac{1}{4}$
c. $\dfrac{1}{12}$ **d.** $\dfrac{1}{18}$

20. What is the probability that a card drawn at random from a standard 52-card deck is *not* a spade?
a. $\dfrac{2}{5}$ **b.** $\dfrac{1}{4}$
c. $\dfrac{4}{13}$ **d.** $\dfrac{3}{4}$

CUMULATIVE REVIEW EXERCISES CHAPTERS P–11

In Problems 1–9, solve each equation or inequality.

1. $|3x - 8| = |x|$

2. $x^2(x^2 - 5) = -4$

3. $\log_2(3x - 5) + \log_2 x = 1$

4. $e^{2x} - e^x - 2 = 0$

5. $\dfrac{6}{x + 2} = \dfrac{4}{x}$

6. $2(x - 8)^{-1} = (x - 2)^{-1}$

7. $\sqrt{x - 3} = \sqrt{x} - 1$

8. $x^2 + 4x \geq 0$

9. $18 \leq x^2 + 6x$

In Problems 10–11, solve each system.

10. $\begin{cases} 5x + 4y = 6 \\ 4x - 3y = 11 \end{cases}$

11. $\begin{cases} x - 3y + 6z = -8 \\ 5x - 6y - 2z = 7 \\ 3x - 2y - 10z = 11 \end{cases}$

12. Graph the function $y = \cos\frac{2}{3}x$ over one period.

13. Use identities to find the exact value of $\sin\left(-\frac{7\pi}{12}\right)$.

14. Write the complex number $2 + 2\sqrt{3}i$ in polar form.

15. Use transformations to sketch the graph of $f(x) = \sqrt{x - 3}$.

16. Find the domain of $f(x) = \dfrac{x + 3}{x(x + 1)}$.

17. Write $7 \ln x - \ln(x - 5)$ in condensed form.

18. If $A = \begin{bmatrix} 1 & 0 & -2 \\ 4 & 1 & 0 \\ 1 & 1 & 7 \end{bmatrix}$, find A^{-1}.

19. Four students are pairing up to ride two tandem bicycles, each of which accommodates two riders, one in back and one in front. How many ways can this be done if we count arrangements as different if either the front or back rider is different?

20. A room has 6 people who are 25 years or older and12 people whose age is between 17 and 24. Half the members of each group have gotten speeding tickets at some time. What is the probability that a person chosen at random from the room will be a person who is 25 years or older or has gotten a speeding ticket at some time?

Answers

CHAPTER P

Section P.1

Answers to Practice Problems:

1. $\frac{710}{333}$ **2. a.** true **b.** true **c.** false
3. $A \cap B = \{0\}, A \cup B = \{-4, -3, -2, -1, 0, 1, 2, 3, 4\}$
4. a. $(-\infty, \infty)$ **b.** $[2, 5)$ **5. a.** 10 **b.** 1 **c.** 1
6. 9 **7. a.** 5 **b.** 1 **c.** -33 **d.** 1 **8. a.** $\frac{17}{8}$ **b.** $\frac{34}{15}$ **c.** $\frac{3}{2}$ **d.** $\frac{2}{3}$
9. a. $\frac{10}{3}$ **b.** $\frac{22}{3}$ **10.** 17°C

Basic Concepts and Skills:

1. zero **2.** rational number; real number **3.** left **4.** irrational
5. true **6.** false **7.** $0.\overline{3}$ repeating **8.** $0.\overline{6}$ repeating
9. -0.8 terminating **10.** -0.25 terminating **11.** $0.\overline{27}$ repeating
12. $0.\overline{3}$ repeating **13.** $3.1\overline{6}$ repeating **14.** $2.7\overline{3}$ repeating **15.** $\frac{15}{4}$
16. $\frac{47}{20}$ **17.** $-\frac{16}{3}$ **18.** $\frac{29}{3}$ **19.** $\frac{211}{99}$ **20.** $\frac{320}{99}$ **21.** $\frac{2239}{495}$
22. $\frac{14{,}093}{9900}$ **23.** Rational **24.** Rational **25.** Rational
26. Rational **27.** Rational **28.** Rational **29.** Irrational
30. Irrational **31.** $3 > -2$ **32.** $-3 < -2$ **33.** $\frac{1}{2} \geq \frac{1}{2}$
34. $x < x + 1$ **35.** $5 \leq 2x$ **36.** $x - 1 > 2$ **37.** $-x > 0$
38. $x < 0$ **39.** $2x + 7 \leq 14$ **40.** $2x + 3 \leq 5$ **41.** $=$
42. $<$ **43.** $<$ **44.** $=$ **45.** $\{-4, -3, -2, 1, 2, 3, 4\}$
46. $\{0, 2, 4\}$ **47.** $\{-4, -2, 0, 2\}$ **48.** $\{-4, 3, 2, -1, 0, 1, 2, 3, 4\}$
49. $\{-4, -3, -2, 0, 2, 4\}$ **50.** $\{-3, 0, 2, 4\}$
51. $\{-4, -3, -2, 0, 2\}$ **52.** $\{-4, -3, -2, -1, 0, 1, 2, 3, 4\}$
53. $(-2, 5), [1, 3)$ **54.** $[1, 7], (3, 5)$ **55.** $(-6, 10), \varnothing$
56. $(-\infty, \infty), \varnothing$ **57.** $(-\infty, \infty), [2, 5)$ **58.** $(-2, \infty), (0, \infty)$
59. $(-\infty, \infty), [-1, 3) \cup [5, 7]$ **60.** $(-\infty, 2] \cup (6, \infty), [-3, 0]$
61. 4 **62.** -17 **63.** $\frac{5}{7}$ **64.** $\frac{3}{5}$ **65.** $5 - \sqrt{2}$ **66.** $5 - \sqrt{2}$
67. $2 - \sqrt{3}$ **68.** $\pi - 3$ **69.** 1 **70.** -1 **71.** 2 **72.** 3
73. $|3 - 8| = 5$ **74.** $|2 - 14| = 12$ **75.** $|-6 - 9| = 15$
76. $|-12 - 3| = 15$ **77.** $|-20 - (-6)| = 14$
78. $|-14 - (-1)| = 13$ **79.** $\left|\frac{22}{7} - \left(-\frac{4}{7}\right)\right| = \frac{26}{7}$
80. $\left|\frac{16}{5} - \left(-\frac{3}{5}\right)\right| = \frac{19}{5}$

81. $-3 < x \leq 1$

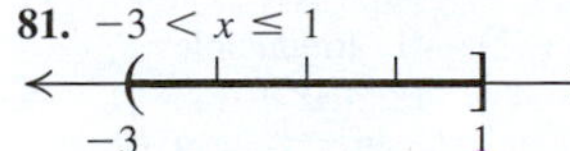

82. $-6 \leq x < -2$

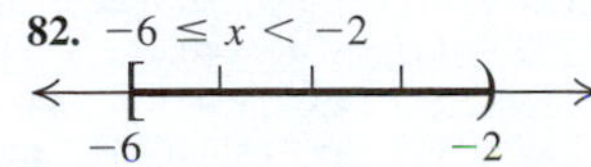

83. $x \geq -3$

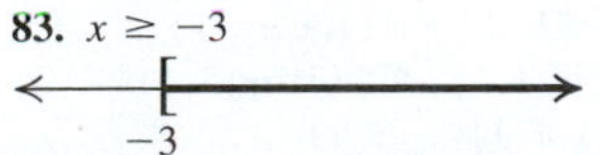

84. $x \geq 0$

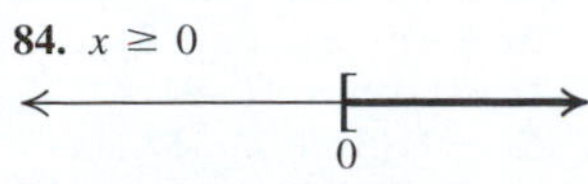

85. $x \leq 5$

86. $x \leq -1$

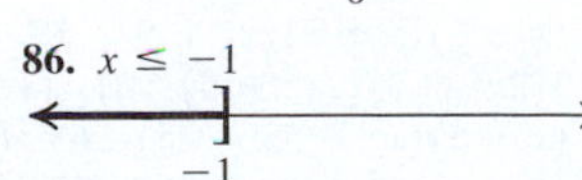

87. $-\frac{3}{4} < x < \frac{9}{4}$ **88.** $-3 < x < -\frac{1}{2}$

89. $4x + 4$ **90.** $-6 + 3x$ **91.** $5x - 5y + 5$ **92.** $6x + 10 - 2y$
93. $-5; \frac{1}{5}$ **94.** $\frac{2}{3}; -\frac{3}{2}$ **95.** 0; No reciprocal **96.** $-1.7; \frac{10}{17}$
97. Additive inverse **98.** Additive inverse **99.** Multiplicative identity
100. Multiplicative identity **101.** Multiplicative associative
102. Multiplicative associative **103.** Multiplicative inverse
104. Multiplicative inverse **105.** Additive identity
106. Additive identity **107.** Additive associative
108. Additive, commutative, and associative **109.** $\frac{29}{15}$ **110.** $\frac{29}{20}$
111. $\frac{67}{35}$ **112.** $\frac{59}{12}$ **113.** $\frac{17}{15}$ **114.** $\frac{34}{45}$ **115.** $-\frac{11}{40}$ **116.** $\frac{27}{40}$
117. $-\frac{8}{99}$ **118.** $\frac{1}{24}$ **119.** $-\frac{1}{10}$ **120.** $\frac{1}{12}$ **121.** $\frac{2}{9}$ **122.** $\frac{2}{3}$ **123.** $\frac{3}{2}$
124. $\frac{1}{3}$ **125.** $\frac{2}{3}$ **126.** $\frac{9}{14}$ **127.** 1 **128.** $\frac{11}{2}$ **129.** $\frac{1}{15}$ **130.** $\frac{11}{6}$
131. 11 **132.** -21 **133.** -1 **134.** 56 **135.** -6 **136.** $\frac{43}{3}$
137. -13 **138.** $-\frac{77}{5}$ **139.** $\frac{62}{15}$ **140.** $-\frac{104}{75}$ **141.** $\frac{3x + xy}{3y}$
142. $(x + 2)(x + 3) = x^2 + 5x + 6$ **143.** $5(x + 3) = 5x + 15$
144. $(25x)(4x) = 100x^2$ **145.** $x - (3y + 2) = x - 3y - 2$
146. $2x - (4y - 5) = 2x - 4y + 5$ **147.** $\frac{x + y}{x} = 1 + \frac{y}{x}$
148. $\frac{x + y}{x + z} = \frac{x}{x + z} + \frac{y}{x + z}$ **149.** $(x + 1)(y + 1) = xy + x + y + 1$
150. $\frac{\frac{1}{x}}{\frac{1}{y}} = \frac{y}{x}$

Applying the Concepts:

153. a. People who own either an MP3 or DVD player
b. People who own both an MP3 and DVD player
154. a. $A = \{$2011 Lexus LS 460$\}$
b. $B = \{$2011 Lexus LS 460, 2011 Lincoln Town Car$\}$
c. $C = \{$2011 Lexus LS 460, 2011 Lincoln Town Car, Audi A4$\}$
d. $A \cap B = \{$2011 Lexus LS 460$\}$
e. $B \cap C = \{$2011 Lexus LS 460, 2011 Lincoln Town Car$\}$
f. $A \cup B = \{$2011 Lexus LS 460, 2011 Lincoln Town Car$\}$
g. $A \cup C = \{$2011 Lexus LS 460, 2011 Lincoln Town Car, 2011 Audi A4$\}$
155. $119.5 \leq x \leq 134.5$ **156.** $30 \leq x \leq 107$

157. a. $|124 - 120| = 4$ **b.** $|137 - 120| = 17$
c. $|114 - 120| = 6$ **158. a.** $|15 - 14| = 1$ **b.** $|15 - 17.5| = 2.5$
c. $|15 - 9| = 6$ **159.** 950 g **160.** Two orders

Section P.2

Answers to Practice Problems:

1. a. 8 **b.** $9a^2$ **c.** $\frac{1}{16}$ **2. a.** $\frac{1}{2}$ **b.** 1 **c.** $\frac{4}{9}$ **3. a.** $3x^9$ **b.** 16
4. a. 81 **b.** 125 **c.** $\frac{2x^7}{3}$ **5. a.** 1 **b.** 1 **c.** $\frac{1}{x^8}$ **d.** x^{10}
6. a. $\frac{2}{x}$ **b.** $\frac{25}{x^2}$ **c.** x^3y^6 **d.** $\frac{x^6}{y^3}$ **7. a.** $\frac{1}{9}$ **b.** $\frac{49}{100}$
8. a. $\frac{1}{4x^8}$ **b.** $-\frac{1}{xy^3}$ **9.** 7.32×10^5 **10.** \$89 per person

Basic Concepts and Skills:

1. exponent **2.** 3 **3.** 16 **4.** 5^3a^3 **5.** false **6.** false
7. Exponent: 3; base: 17 **8.** Exponent: 2; base: 10
9. Exponent: 0; base: 9 **10.** Exponent: 0; base: -2
11. Exponent: 5; base: -5 **12.** Exponent: 2; base: -99
13. Exponent: 3; base: 10 **14.** Exponent: 7; base: -3
15. Exponent: 2; base: a **16.** Exponent: 3; base: $-b$ **17.** 6 **18.** 81
19. 1 **20.** 1 **21.** 64 **22.** 729 **23.** $\frac{1}{81}$ **24.** $\frac{1}{49}$ **25.** $\frac{1}{15{,}625}$
26. $\frac{1}{125}$ **27.** 16 **28.** 7 **29.** 2 **30.** 0 **31.** $\frac{2}{9}$ **32.** $\frac{2}{25}$ **33.** $-\frac{1}{8}$
34. $-\frac{1}{9}$ **35.** $\frac{1}{9}$ **36.** $-\frac{1}{8}$ **37.** 2 **38.** $\frac{1}{9}$ **39.** 1 **40.** 81 **41.** 6
42. 100 **43.** $-\frac{2}{25}$ **44.** $-\frac{3}{49}$ **45.** $\frac{49}{121}$ **46.** $\frac{25}{169}$ **47.** x^4
48. $\frac{1}{x}$ **49.** $\frac{y}{x}$ **50.** $\frac{x^2}{y^2}$ **51.** $\frac{1}{xy^2}$ **52.** $\frac{1}{x^3y^2}$ **53.** $\frac{1}{x^{12}}$ **54.** $\frac{1}{x^{10}}$
55. x^{33} **56.** x^{48} **57.** $-3x^5y^5$ **58.** $-8x^6y^6$ **59.** $\frac{4x^2}{y^2}$ **60.** $\frac{6y^3}{x^3}$
61. $\frac{3x^5}{y^5}$ **62.** $-\frac{5y^6}{x^6}$ **63.** $\frac{1}{x^4}$ **64.** $\frac{1}{x^{10}}$ **65.** $\frac{8y^3}{x^3}$ **66.** $\frac{625y^4}{x^8}$
67. $-243x^5y^5$ **68.** $-8x^3y^3$ **69.** $\frac{25}{9x^2}$ **70.** $\frac{81}{625y^4}$ **71.** $\frac{64}{x^9y^{15}}$
72. $\frac{243x^{10}}{y^{10}}$ **73.** $\frac{x^5}{y^4}$ **74.** $\frac{x^3}{y^4}$ **75.** $\frac{3x}{y^2}$ **76.** $\frac{5y}{x^2}$ **77.** $\frac{1}{x}$ **78.** $\frac{1}{b^{16}}$
79. $-\frac{1}{16x^5y^2}$ **80.** $\frac{y^{12}}{x^2}$ **81.** $\frac{16}{x^{15}y^{10}z^5}$ **82.** $\frac{4z^3}{x^3}$ **83.** $\frac{5b^4}{a^6}$
84. $-\frac{9a^3}{8bc^2}$ **85.** $\frac{x^3z^{15}}{y^3}$ **86.** $\frac{y^3}{x^6z^7}$ **87.** 1.25×10^2 **88.** 2.47×10^2
89. 8.5×10^5 **90.** 2.05×10^5 **91.** 7×10^{-3} **92.** 1.9×10^{-3}
93. 2.75×10^{-6} **94.** 3.8×10^{-6}

Applying the Concepts:

95. 1080 ft^3 **96.** 25 in^3 **97. a.** $(2x)(2x) = 4x^2 = 2^2A$
b. $(3x)(3x) = 9x^2 = 3^2A$ **98. a.** $\pi\left(\frac{2d}{2}\right)^2 = 4\pi\left(\frac{d}{2}\right)^2 = 2^2A$
b. $\pi\left(\frac{3d}{2}\right)^2 = 9\pi\left(\frac{d}{2}\right)^2 = 3^2A$ **99.** $F = 1562.5$ **100.** $F = 17{,}662.5$
101. 1.27×10^4 (km); 3480; 1.39×10^6 (km); 134,000; 4.8×10^3 (km) **102.** 3.76×10^{22} atoms **103.** 6.02×10^{20} atoms
104. 2.66×10^{-20} kg **105.** 1.67×10^{-21} kg **106.** 3.8×10^8 m
107. 5.98×10^{24} kg

Section P.3

Answers to Practice Problems:

1. 889 ft **2.** $5x^3 + 5x^2 + 2x - 4$ **3.** $5x^4 - 5x^3 - x^2 - x + 12$
4. $-8x^5 - 4x^4 + 10x^3$ **5.** $-10x^4 + x^3 - 33x^2 - 14x$
6. a. $4x^2 + 27x - 7$ **b.** $6x^2 - 19x + 10$ **7.** $9x^2 + 12x + 4$
8. $1 - 4x^2$ **9.** $x^2y + x^3$

Basic Concepts and Skills:

1. -3; 7 **2.** standard **3.** 4 **4.** 7 **5.** true **6.** true
7. Yes; $x^2 + 2x + 1$ **8.** No **9.** No
10. Yes; $x^7 + 3x^5 + 3x^4 - 2x + 1$ **11.** No **12.** No **13.** Yes
14. Yes **15.** Degree: 1; terms: $7x$, 3 **16.** Degree: 2; terms: $-3x^2$, 7
17. Degree: 4; terms: $-x^4, x^2, 2x, -9$ **18.** Degree: 7; terms: $9x^7, 2x^3, x, -21$ **19.** $2x^2 - 3x - 1$ **20.** $2x^3 - x^2 - 2x - 2$
21. $x^3 - x^2 + 5x - 8$ **22.** $-2x^3 - 3x^2 + 9x - 6$
23. $-10x^4 - 6x^3 + 12x^2 - 7x + 17$ **24.** $13x^2 - 8x + 14$
25. $-24x^2 - 14x - 14$ **26.** $-2x^2 - 30x + 2$ **27.** $2y^3 + y^2 - 2y - 1$
28. $6y^2 + 6y + 1$ **29.** $12x^2 + 18x$ **30.** $21x^2 - 28x$
31. $x^3 + 3x^2 + 4x + 2$ **32.** $2x^3 - 13x^2 + 16x - 5$
33. $3x^3 - 5x^2 + 5x - 2$ **34.** $2x^3 - 5x^2 + 5x + 4$ **35.** $x^2 + 3x + 2$
36. $x^2 + 5x + 6$ **37.** $9x^2 + 9x + 2$ **38.** $2x^2 + 11x + 15$
39. $-4x^2 - 7x + 15$ **40.** $-2x^2 + 11x - 5$ **41.** $6x^2 - 7x + 2$
42. $5x^2 - 8x + 3$ **43.** $4x^2 + 4ax - 15a^2$ **44.** $5x^2 + 23ax - 10a^2$
45. $4x + 4$ **46.** $-6x + 9$ **47.** $9x^2 + 27x + 27$ **48.** $-6x^2 + 12x - 8$
49. $16x^2 + 8x + 1$ **50.** $9x^2 + 12x + 4$ **51.** $27x^3 + 27x^2 + 9x + 1$
52. $8x^3 + 36x^2 + 54x + 27$ **53.** $-4x^2 + 25$ **54.** $-16x^2 + 9$
55. $x^2 + \frac{3}{2}x + \frac{9}{16}$ **56.** $x^2 + \frac{4}{5}x + \frac{4}{25}$ **57.** $2x^3 - 9x^2 + 19x - 15$
58. $x^3 - 6x^2 + 5x + 6$ **59.** $y^3 + 1$ **60.** $y^3 + 64$ **61.** $x^3 - 216$
62. $x^3 - 1$ **63.** $3x^2 + 11xy + 10y^2$ **64.** $14x^2 + 11xy + 2y^2$
65. $6x^2 + 11xy - 7y^2$ **66.** $2x^2 - xy - 15y^2$ **67.** $x^4 - 2x^2y^2 + y^4$
68. $16x^4 - 8x^2y^2 + y^4$ **69.** $x^3 - 3x^2y + 4y^3$ **70.** $x^3 + 3x^2y - 4y^3$
71. $x^4 - 4x^3y + 16xy^3 - 16y^4$ **72.** $16x^4 + 16x^3y - 4xy^3 - y^4$
73. 10 **74.** 13 **75. a.** 7 **b.** 47 **76. a.** 6 **b.** 34 **77.** 72 **78.** 58

Applying the Concepts:

79. In 2012, it was \$6.02. **80.** In 2012, it was \$6.33 billion.
81. \$250 **82.** \$100 **83.** 500 ft **84.** 84 ft **85. a.** $-x + 22.50$
b. $-10x^2 + 195x + 675$ **86. a.** $10n + 250$ **b.** $-20n^2 + 12{,}500$

Section P.4

Answers to Practice Problems:

1. a. $2x^3(3x^2 + 7)$ **b.** $7x^2(x^3 + 3x^2 + 5)$
2. a. $(x + 4)(x + 2)$ **b.** $(x - 5)(x + 2)$
3. a. $(x + 2)^2$ **b.** $(3x - 1)^2$ **4. a.** $(x + 4)(x - 4)$
b. $(2x - 5)(2x + 5)$ **5.** $(x^2 + 9)(x - 3)(x + 3)$
6. a. $(x - 5)(x^2 + 5x + 25)$ **b.** $(3x + 2)(9x^2 - 6x + 4)$
7. \$16.224 million **8. a.** $(x + 3)(x^2 + 1)$
b. $(7x - 5)(2x + 1)(2x - 1)$ **c.** $(x + y + 1)(x - y - 1)$
9. $(x^2 + x + 3)(x^2 - x + 3)$ **10. a.** $(5x + 1)(x + 2)$
b. $(4x - 3)(x + 1)$

Basic Concepts and Skills:

1. factors **2.** greatest common monomial **3.** $10x^2$ **4.** irreducible
5. true **6.** true **7.** $8(x - 3)$ **8.** $5(x + 5)$ **9.** $-6x(x - 2)$
10. $-3(x^2 - 7)$ **11.** $7x^2(1 + 2x)$ **12.** $9x^3(1 - 2x)$
13. $x^2(x^2 + 2x + 1)$ **14.** $x^2(x^2 - 5x + 7)$ **15.** $x^2(3x - 1)$
16. $2x^2(x + 1)$ **17.** $4ax^2(2x + 1)$ **18.** $ax(x^3 - 2x + 1)$
19. $(x + 3)(x^2 + 1)$ **20.** $(x + 5)(x^2 + 1)$ **21.** $(x - 5)(x^2 + 1)$
22. $(x - 7)(x^2 + 1)$ **23.** $(3x + 2)(2x^2 + 1)$ **24.** $(x + 2)(3x^2 + 1)$
25. $x^2(3x^2 + 1)(4x^3 + 1)$ **26.** $x^2(x^2 + 1)(3x^3 + 1)$
27. $(x + 3)(x + 4)$ **28.** $(x + 3)(x + 5)$ **29.** $(x - 2)(x - 4)$
30. $(x - 2)(x - 7)$ **31.** $(x + 1)(x - 4)$ **32.** $(x + 1)(x - 6)$
33. Irreducible **34.** Irreducible **35.** $(2x + 9)(x - 4)$
36. $(2x + 9)(x - 3)$ **37.** $(2x + 3)(3x + 4)$ **38.** $(2x - 3)(4x + 1)$
39. $(3x + 1)(x - 4)$ **40.** $(x + 1)(5x + 2)$ **41.** Irreducible
42. Irreducible **43.** $(x + 3)^2$ **44.** $(x + 4)^2$ **45.** $(3x + 1)^2$
46. $(6x + 1)^2$ **47.** $(5x - 2)^2$ **48.** $(8x + 2)^2$ **49.** $(7x + 3)^2$
50. $(3x + 4)^2$ **51.** $(x - 8)(x + 8)$ **52.** $(x - 11)(x + 11)$
53. $(2x - 1)(2x + 1)$ **54.** $(3x - 1)(3x + 1)$ **55.** $(4x - 3)(4x + 3)$
56. $(5x - 7)(5x + 7)$ **57.** $(x - 1)(x + 1)(x^2 + 1)$
58. $(x - 3)(x + 3)(x^2 + 9)$ **59.** $5(2x^2 - 1)(2x^2 + 1)$
60. $3(2x^2 - 5)(2x^2 + 5)$ **61.** $(x + 4)(x^2 - 4x + 16)$
62. $(x + 5)(x^2 - 5x + 25)$ **63.** $(x - 3)(x^2 + 3x + 9)$
64. $(x - 6)(x^2 + 6x + 36)$ **65.** $(2 - x)(4 + 2x + x^2)$
66. $(3 - x)(9 + 3x + x^2)$ **67.** $(2x - 3)(4x^2 + 6x + 9)$
68. $(2x - 5)(4x^2 + 10x + 25)$ **69.** $5(2x + 1)(4x^2 - 2x + 1)$
70. $7(x + 2)(x^2 - 2x + 4)$ **71.** $(x^2 + 3x + 5)(x^2 - 3x - 5)$

72. $(x^2 + x + 1)(x^2 - x + 1)$ **73.** $(x^2 + x + 8)(x^2 - x + 8)$
74. $(x^2 + x - 3)(x^2 - x - 3)$ **75.** $(x^2 + 3x + 4)(x^2 - 3x + 4)$
76. $(x^2 + 2x - 6)(x^2 - 2x - 6)$ **77.** $(x^2 + 2x + 2)(x^2 - 2x + 2)$
78. $(x^2 + 4x + 8)(x^2 - 4x + 8)$ **79.** $(1 - 4x)(1 + 4x)$
80. $(2 - 5x)(2 + 5x)$ **81.** $(x - 3)^2$ **82.** $(x - 4)^2$ **83.** $(2x + 1)^2$
84. $(4x + 1)^2$ **85.** $2(x + 1)(x - 5)$ **86.** $5(x + 2)(x - 4)$
87. $(2x - 5)(x + 4)$ **88.** $(2x + 5)(x - 6)$ **89.** Irreducible
90. Irreducible **91.** $3x^3(x + 2)^2$ **92.** $2x^3(x + 4)^2$
93. $(3x + 1)(3x - 1)$ **94.** $(4x + 5)(4x - 5)$ **95.** $(4x + 3)^2$
96. $(2x + 5)^2$ **97.** Irreducible **98.** Irreducible
99. $x(5x + 2)(9x - 2)$ **100.** $x(5x - 1)(5x + 9)$
101. $a(x + a)(x - 8a)$ **102.** $a(x + 2a)(x - 12a)$
103. $(x + 4 + 4a)(x + 4 - 4a)$ **104.** $(x + 3 + 6a)(x + 3 - 6a)$
105. $3x^3(x + 2y)^2$ **106.** $4x^3(x + 3y)^2$
107. $(x + 2 + 5a)(x + 2 - 5a)$ **108.** $(x + 2 + 4a)(x + 2 - 4a)$
109. $2x^4(3x + y)^2$ **110.** $3x^4(2x + y)^2$

Applying the Concepts:

111. $x(8 - x)$ **112.** $x(21 - x)$ **113.** $4x(18 - x)(8 - x)$
114. $4(12 - x)(12 + x)$ **115.** $\pi(2 - x)(2 + x)$
116. $8\pi(x - 3)(x + 3)$ **117.** $2x(1400 - x)$
118. $\frac{1}{4}x(96 - \pi x)$

Section P.5

Answers to Practice Problems:

1. 0.32 **2. a.** $\frac{2x}{3}$ **b.** $\frac{x - 2}{x + 2}$ **3.** $\frac{x(x - 3)}{14}$ **4. a.** $\frac{7}{x - 6}$
b. $\frac{1}{x + 4}$ **5. a.** $\frac{x(5x - 4)}{(x + 2)(x - 5)}$ **b.** $\frac{x(3x + 23)}{(x + 3)(x - 4)}$
6. a. $3x^2(x + 2)^2(x - 2)^2$ **b.** $(x - 5)(x + 5)(x - 1)$
7. a. $\frac{x^2 + 2x + 8}{(x + 2)(x - 2)^2}$ **b.** $\frac{45 - 7x}{3(x - 5)^2}$ **8.** $\frac{1}{x - 5}$ **9.** $\frac{25x}{15x - 12}$

Basic Concepts and Skills:

1. each denominator **2.** factor **3.** $x(x - 2)(x + 1)$
4. complex rational expression **5.** false **6.** true
7. $\frac{2}{x + 1}, x \neq -1$ **8.** $\frac{3}{x - 2}, x \neq 2$ **9.** $\frac{3}{x - 1}, x \neq -1, x \neq 1$
10. $\frac{5}{2 + x}, x \neq -2, x \neq 2$ **11.** $-\frac{2}{x + 3}, x \neq -3, x \neq 3$
12. $\frac{3}{x - 5}, x \neq -5, x \neq 5$ **13.** $-1, x \neq \frac{1}{2}$ **14.** $-1, x \neq \frac{2}{5}$
15. $\frac{x - 3}{4}, x \neq 3$ **16.** $\frac{x - 5}{3}, x \neq 5$ **17.** $\frac{7x}{x + 1}, x \neq -1$
18. $\frac{4x}{x + 3}, x \neq -3$ **19.** $\frac{x - 10}{x + 7}, x \neq -7, x \neq 1$
20. $\frac{x + 5}{x - 4}, x \neq 3, x \neq 4$ **21.** $\frac{x + 2}{x - 2}, x \neq -\frac{1}{3}, x \neq 0, x \neq 2$
22. $\frac{1}{x - 4}, x \neq -\frac{1}{3}, x \neq 0, x \neq 4$ **23.** 1 **24.** $\frac{1}{3}$ **25.** $\frac{x + 3}{2(x - 3)}$
26. $\frac{(5x + 3)(x + 2)}{4}$ **27.** $\frac{x - 1}{x}$ **28.** 5 **29.** $\frac{x - 1}{x + 3}$
30. $\frac{(x - 2)(x + 4)}{(x + 5)(x + 1)}$ **31.** -1 **32.** -1 **33.** $\frac{3}{8}$ **34.** $\frac{9}{80}$
35. $\frac{5x(x - 3)}{2}$ **36.** $\frac{(x + 1)^2}{21}$ **37.** $\frac{3(x + 3)}{x + 4}$ **38.** $\frac{x + 4}{x + 1}$
39. $\frac{x - 3}{x^2 - 2x + 4}$ **40.** $\frac{x - 1}{x - 4}$ **41.** $\frac{x + 3}{5}$ **42.** $\frac{7 - x}{4}$ **43.** $\frac{x + 4}{2x + 1}$
44. $\frac{3x}{7x - 3}$ **45.** $\frac{1}{x + 1}$ **46.** $\frac{x + 9}{3x + 2}$ **47.** $\frac{2(x - 2)}{x - 3}$ **48.** $-\frac{x - 4}{x - 1}$
49. $\frac{7x}{x^2 + 1}$ **50.** $\frac{2x}{(x - 1)^2}$ **51.** $\frac{4x}{x - 3}$ **52.** $\frac{3x}{(x + 5)^2}$ **53.** $\frac{1}{x + 2}$
54. $\frac{5}{x + 1}$ **55.** $\frac{2(1 - 3x)}{(2x - 1)(2x + 1)}$ **56.** $\frac{1 - 8x}{(4x - 1)(4x + 1)}$
57. $-\frac{x^3 + 2x^2 + 4x - 8}{x(x - 2)(x + 2)}$ **58.** $\frac{2x^3 + 6x^2 - x - 1}{x(x - 1)(x + 1)}$ **59.** $12(x - 2)$
60. $21(1 + 3x)$ **61.** $(2x - 1)(2x + 1)^2$ **62.** $(3x + 1)(3x - 1)^2$
63. $(x - 1)(x + 1)(x + 2)$ **64.** $(x - 3)(x + 3)(x + 2)$
65. $(x - 1)(x - 4)(x + 2)$ **66.** $(x - 3)(x + 1)(x + 2)$
67. $\frac{7x + 15}{(x - 3)(x + 3)}$ **68.** $\frac{x(3x + 4)}{(x - 1)(x + 1)}$ **69.** $\frac{x(4 - x)}{(x - 2)(x + 2)}$
70. $-\frac{2x^2 + 6x + 5}{(x - 4)(x + 4)}$ **71.** $\frac{2x + 3}{(x - 2)(x + 5)}$ **72.** $\frac{2x^2 + x + 5}{(x - 2)(x + 1)^2}$
73. $\frac{2(9x^2 - 5x + 1)}{(3x + 1)(3x - 1)^2}$ **74.** $\frac{2(4x^2 + 3x + 1)}{(2x - 1)(2x + 1)^2}$
75. $\frac{9(x - 3)}{(x + 4)(x + 5)(x - 5)}$ **76.** $-\frac{7}{(x + 4)(x - 3)}$
77. $\frac{2x - 3}{(2 + x)(2 - x)}$ **78.** $\frac{5(x + 8)}{(5 - x)(5 + x)}$ **79.** $\frac{16a^2}{(x - 5a)(x - 3a)}$
80. $\frac{5x^2}{(2x - a)(3x + a)}$ **81.** $-\frac{h}{x(x + h)}$ **82.** $-\frac{h(2x + h)}{x^2(x + h)^2}$ **83.** $\frac{2x}{3}$
84. $-3x^2$ **85.** $\frac{1}{x - 1}$ **86.** $\frac{1}{(1 - x)(1 + x)}$ **87.** $\frac{1 - x}{1 + x}$
88. $\frac{2(x - 1)}{x + 2}$ **89.** $-x$ **90.** $-x$ **91.** $\frac{x(2x - 1)}{2x + 1}$ **92.** $\frac{x(2x + 3)}{2x + 1}$
93. $-\frac{1}{x(x + h)}$ **94.** $-\frac{2x + h}{x^2(x + h)^2}$ **95.** $\frac{x}{a}, x \neq -a$ **96.** $\frac{2x}{x^2 + a^2}$

Applying the Concepts

97. 10 cm **98.** 1 ft **99. a.** $\frac{3}{100 + x}$ **b.** 2%
100. a. $\frac{1500}{200{,}000 + x}$ **b.** 0.5% **101. a.** $\$\frac{0.1\pi x^3 + 2.4}{x}$ **b.** \$2.46
102. a. $\$\frac{0.4x^3 + 3.6}{x}$ **b.** \$3.30

Section P.6

Answers to Practice Problems:

1. a. 12 **b.** $\frac{1}{7}$ **c.** $\frac{1}{4}$ **2. a.** $2\sqrt{5}$ **b.** $4\sqrt{3}$ **c.** $2|x|\sqrt{3}$
d. $\frac{2y\sqrt{5y}}{3\sqrt{3}|x|}$ **3.** ≈ 67 mph **4. a.** -2 **b.** 2 **c.** 3
d. Not a real number **5. a.** $2\sqrt[3]{9}$ **b.** $2\sqrt[4]{3a^2}$ **c.** a^2 **6. a.** $13\sqrt{3}$
b. 0 **7.** $\sqrt[10]{972}$ **8. a.** $\frac{7\sqrt{2}}{4}$ **b.** $\frac{\sqrt[3]{6}}{2}$ **9. a.** $\sqrt{7} + \sqrt{5}$
b. $\sqrt{x} - \sqrt{2}$ **10. a.** $\frac{1}{2}$ **b.** 5 **c.** -2 **11. a.** $5\sqrt[3]{5}$ **b.** -216
c. $\frac{1}{1024}$ **d.** Not a real number **12. a.** $12x^{7/10}$ **b.** $\frac{5}{x^{7/12}}$ **c.** $\frac{1}{x^{2/15}}$
13. a. $\sqrt[3]{x^2}$ **b.** 5 **c.** x^2 **14.** $\frac{(2x + 3)(x + 3)^{1/2}}{x + 3}$

Basic Concepts and Skills:

1. two **2.** numerator; denominator; $\sqrt{5}$. **3.** like radicals **4.** $\sqrt[3]{7}$
5. false **6.** true **7.** 8 **8.** 10 **9.** 4 **10.** 5 **11.** -3 **12.** -6
13. $-\frac{1}{2}$ **14.** $\frac{3}{4}$ **15.** 3 **16.** -6 **17.** Not a real number
18. Not a real number **19.** 1 **20.** -1 **21.** -7 **22.** 4 **23.** $4\sqrt{2}$
24. $5\sqrt{5}$ **25.** $3x\sqrt{2}$ **26.** $3x\sqrt{3}$ **27.** $3x\sqrt{x}$ **28.** $2x\sqrt{2x}$
29. $3x\sqrt{2}$ **30.** $5x\sqrt{2}$ **31.** $3x\sqrt{5x}$ **32.** $6x\sqrt{x}$ **33.** $-2x$
34. $4x$ **35.** $-x^2$ **36.** $-3x^2$ **37.** $\frac{\sqrt{10}}{8}$ **38.** $\frac{\sqrt{6}}{10}$ **39.** $\frac{2}{x}$
40. $\frac{\sqrt[3]{7}}{2x}$ **41.** $\frac{\sqrt[4]{2x^3}}{x^2}$ **42.** $-\frac{\sqrt[4]{5x^3}}{2x^2}$ **43.** x^2 **44.** $\frac{2x^2}{5y^3}$ **45.** $\frac{3x^3}{y^2}$
46. x^3y **47.** $7\sqrt{3}$ **48.** $6\sqrt{2}$ **49.** $9\sqrt{5}$ **50.** $6\sqrt{7}$ **51.** $3\sqrt{2x}$
52. $5\sqrt{5x}$ **53.** $-\sqrt[3]{3}$ **54.** $5\sqrt[3]{2}$ **55.** $2\sqrt[3]{3x}$ **56.** $10\sqrt[3]{2x}$
57. $\sqrt{2x}(x^2 + 3x - 20)$ **58.** $\sqrt{2x}(10x^2 + 7x - 18)$
59. $(2x - y)^2\sqrt{3xy}$ **60.** $\sqrt{2xy}(2x + 4y - 3x^2)$ **61.** $\frac{2\sqrt{3}}{3}$

62. $2\sqrt{5}$ **63.** $\frac{7\sqrt{15}}{15}$ **64.** $-\frac{3\sqrt{2}}{4}$ **65.** $\frac{\sqrt{2}-x}{2-x^2}$ **66.** $\frac{\sqrt{5}-2x}{5-4x^2}$
67. $3(2+\sqrt{3})$ **68.** $5(1+\sqrt{2})$ **69.** $\sqrt{3}-\sqrt{2}$
70. $3(\sqrt{5}+\sqrt{3})$ **71.** $\frac{7-2\sqrt{10}}{3}$ **72.** $\frac{5+\sqrt{21}}{2}$
73. $\frac{2x+h-2\sqrt{x(x+h)}}{h}$ **74.** $\frac{a^2x+6a\sqrt{x}+9}{a^2x-9}$ **75.** $\frac{1}{\sqrt{4+h}+2}$
76. $\frac{1}{\sqrt{y+9}+3}$ **77.** $\frac{1}{2+\sqrt{4-x}}$ **78.** $\frac{1}{\sqrt{x+2}+\sqrt{2}}$
79. $\frac{1}{\sqrt{x}+2}$ **80.** $\frac{1}{(x+5)(\sqrt{x}+\sqrt{5})}$ **81.** $\frac{2x}{\sqrt{x^2+4x}+x}$
82. $\frac{x}{\sqrt{x^2+2x+3}+\sqrt{x^2+3}}$ **83.** 5 **84.** 12 **85.** −2
86. −3 **87.** 4 **88.** 64 **89.** $-\frac{1}{125}$ **90.** $-\frac{1}{27}$ **91.** $\frac{125}{27}$ **92.** 9
93. $x^{9/10}$ **94.** $5x^{19/15}$ **95.** $x^{1/10}$ **96.** $x^{11/12}$ **97.** $4x^4$ **98.** $4x^{3/2}$
99. $\frac{1}{9x^4y^2}$ **100.** $\frac{1}{64x^6y^9}$ **101.** $5x^{5/4}$ **102.** $5x^{11/6}$ **103.** $\frac{x^3}{y^8}$
104. $\frac{x^{5/6}}{3y}$ **105.** $3^{1/2}$ **106.** $5^{1/2}$ **107.** x^3 **108.** x^4 **109.** x^2y^3
110. $2xy^4$ **111.** 3 **112.** 7 **113.** $x^{5/3}$ **114.** $2x$ **115.** $\sqrt[15]{864}$
116. $\sqrt[12]{648}$ **117.** $x(\sqrt[12]{x^5})$ **118.** $x(\sqrt[35]{x^4})$ **119.** $m\sqrt[9]{250m^8n^7}$
120. $3xy(\sqrt[6]{3xy})$ **121.** $xy(\sqrt[30]{x^9y^{13}})$ **122.** $ab\sqrt[30]{3^67^5a^4b^{12}}$
123. $\frac{2}{3}x^{1/3}(7x-6)$ **124.** $\frac{1}{3}x^{1/3}(49x+4)$
125. $2(3x+1)^{1/3}(7x-1)$ **126.** $7(3x-1)^{1/3}(x+1)$
127. $(x^2+1)^{1/2}(14x^3-5x^2+8x-2)$
128. $(x^2+2)^{1/2}(5x^3-8x^2+4x-4)$

Applying the Concepts:
133. 40 cm **134.** 8% **135.** 2 cm/sec **136.** 21 m/sec
137. 23 amp **138.** 1.74×10^6 m **139.** 0.46 lb/in.^{2} **140.** 218 lb

Review Exercises

1. a. $\{4\}$ **b.** $\{0,4\}$ **c.** $\{-5,0,4\}$ **d.** $\left\{-5,0,0.2,0.\overline{31},\frac{1}{2},4\right\}$
e. $\{\sqrt{7},\sqrt{12}\}$ **f.** $\left\{-5,0,0.2,0.\overline{31},\frac{1}{2},\sqrt{7},\sqrt{12},4\right\}$ **2.** Additive commutative **3.** Distributive **4.** Multiplicative commutative
5. Multiplicative identity
6. $-2\le x<3$

−2 3

7. $x\le 1$

1

8. $[1,4)$

1 4

9. $(0,\infty)$

0

10. −6 **11.** 1 **12.** 12 **13.** $\sqrt{15}-1$ **14.** −1 **15.** −64
16. 2 **17.** −29 **18.** 64 **19.** −5 **20.** $\frac{1}{2}$ **21.** 16 **22.** 9
23. $\frac{216}{125}$ **24.** $\frac{1}{729}$ **25.** 250 **26.** 288 **27.** 0.7 **28.** 10 **29.** 5
30. 9 **31.** −5 **32.** 9 **33.** $\frac{1}{4}$ **34.** 10 **35.** −1 **36.** 64 **37.** 2
38. 48 **39.** $-\frac{5}{4}$ **40.** 0 **41.** $\frac{1}{4}$ **42.** x^3 **43.** x^{10} **44.** $\frac{y^2}{x^3}$ **45.** $\frac{x^6}{y^2}$
46. $\frac{x^8}{y^4}$ **47.** $\frac{64}{xy^2}$ **48.** $\frac{x^2}{343y}$ **49.** $\frac{y^3}{16x^2}$ **50.** $10x^{17/12}$ **51.** $21x^{7/4}$
52. $4x^{5/12}$ **53.** $2x^5$ **54.** $2x^2$ **55.** $x^{10}y^2$ **56.** $22\sqrt{3}$ **57.** $\frac{8\sqrt{11}}{11}$
58. $6x\sqrt{5}$ **59.** $\sqrt{6}$ **60.** $25\sqrt[3]{2}$ **61.** $2x\sqrt{3}$ **62.** $5x\sqrt{3}$ **63.** $5x$
64. $\sqrt{2x}$ **65.** $14\sqrt[3]{5}$ **66.** $-y\sqrt[3]{7x}$ **67.** $\frac{7\sqrt{3}}{3}$ **68.** $\frac{36+4\sqrt{6}}{75}$
69. $-2+\sqrt{3}$ **70.** $-\frac{49+18\sqrt{5}}{71}$ **71.** 2.3051×10^{13}
72. 0.0001595 **73.** $4x^3-12x^2+9x-6$ **74.** $5x^3+9x^2-11x+6$
75. $9x^4+11x^3-12x^2+14$ **76.** $15x^4+7x^3+11x^2+2$
77. $x^2-15x+36$ **78.** $x^2-14x+49$ **79.** $x^{10}-4$ **80.** $16x^2-9$
81. $6x^2-7x-55$ **82.** $3x^3-24$ **83.** $(x-5)(x+2)$
84. $(x+1)(x+9)$ **85.** $(6x-11)(4x+1)$ **86.** $3(5x+6)(x+1)$
87. $(x+5)(x+11)$ **88.** $(x-1)(x^2+2)$ **89.** $(x-1)(x^3+7)$
90. $(3x+4)^2$ **91.** $(5x+4)(2x+3)$ **92.** $(4x+3)(2x+3)$
93. $(4x-3)(3x+4)$ **94.** $x^2(x-2)(x+2)$ **95.** $(2x-7)(2x+7)$
96. $(4x-9)(4x+9)$ **97.** $(x+6)^2$ **98.** Irreducible **99.** $(8x+3)^2$
100. $(2x-1)(4x^2+2x+1)$ **101.** $(2x+3)(4x^2-6x+9)$
102. $7(x-1)(x^2+x+1)$ **103.** $(x-4)(x+4)(x+5)$
104. $(x-3)(x+3)(x+6)$ **105.** $\frac{14}{x-9}$
106. $\frac{2(4x-5)}{(x-1)(x+1)(x-2)}$ **107.** $\frac{22-5x}{(x+2)(x+6)(x+8)}$
108. $\frac{4x^2+3x+33}{(x-1)(x-6)(x+4)}$ **109.** $-\frac{4x}{(x-1)(x+1)}$ **110.** 2
111. $\frac{2x+3}{2x+1}$ **112.** $\frac{x-2}{x+3}$ **113.** $\frac{x(x-2)}{2(2x+3)}$ **114.** $\frac{3x-2}{x-1}$
115. $\frac{(1-x)(1+x+x^2)}{(1+x)(1-x+x^2)}$ **116.** $x(x-1)(x+2)$ **117.** −1
118. $\frac{x+4}{x+1}$

Applying the Concepts:
119. 29 **120.** 13 in. **121.** $A=96, P=48$ **122.** 140 **123.** 48 ft^2
124. 1.05 g **125.** 74.46 mi **126.** $\frac{20{,}000}{1{,}000{,}000+x}$ **127.** 1.5 ft
128. 1521 ft

Practice Test

1. 4 **2.** $100-\sqrt{2}$ **3.** −8 **4.** $(-\infty,-9)\cup(-9,3)\cup(3,\infty)$
5. $-27x^3y^3$ **6.** $-2x^2$ **7.** $2\sqrt{3x}$ **8.** $-\frac{1}{64}$ **9.** $\frac{5x^{5/2}}{y^{1/2}}$
10. $-\frac{5+5\sqrt{3}}{2}$ **11.** $19x^2-18x+11$ **12.** $5x^2-11x+2$
13. $x^4+6x^2y^2+9y^4$ **14.** $(x-5)(x+5)\pi$ **15.** $(x-3)(x-2)$
16. $(3x+2)^2$ **17.** $(2x-3)(4x^2+6x+9)$ **18.** $\frac{-9}{(x+2)}$
19. $\frac{1}{(3x+1)(3x-1)}$ **20.** 37

CHAPTER 1

Section 1.1

Answers to Practice Problems:
1. a. $(-\infty,\infty)$ **b.** $(-\infty,2)\cup(2,\infty)$ **c.** $[1,\infty)$ **2.** $\left\{-\frac{3}{5}\right\}$ **3.** $\varnothing$
4. $(-\infty,\infty)$ **5.** 10°C **6.** $w=\frac{P-2\ell}{2}$ **7.** Width: 3 m; length: 11 m
8. \$3750 in bonds and \$11,250 in stocks **9.** \$1500 **10.** 395 m
11. 34.3 ft

Basic Concepts and Skills:
1. defined **3.** equivalent **5.** false **7. a.** No **b.** Yes **9. a.** Yes
b. No **11. a.** Yes **b.** Yes **13.** $(-\infty,-2)\cup(-2,1)\cup(1,\infty)$
15. $(-\infty,3)\cup(3,4)\cup(4,\infty)$ **17.** No, $x=0$ **19.** Yes
21. $\{3\}$ **23.** $\{-2\}$ **25.** $\{7\}$ **27.** $\{-1\}$ **29.** $\{6\}$ **31.** $\{12\}$
33. $\{1\}$ **35.** $\{-11\}$ **37.** $\left\{\frac{23}{6}\right\}$ **39.** $\left\{\frac{7}{8}\right\}$ **41.** $\{2\}$ **43.** $\{28\}$
45. $\left\{\frac{4}{5}\right\}$ **47.** $r=\frac{d}{t}$ **49.** $r=\frac{C}{2\pi}$ **51.** $R=\frac{E}{I}$ **53.** $h=\frac{2A}{a+b}$
55. $u=\frac{fv}{v-f}$ **57.** $m=\frac{y-b}{x}$ **59.** $0.065x$ **61.** $\$22{,}000-x$

Applying the Concepts:
63. 13 ft **65.** Length: 25 ft; width: 15 ft **67.** 57 cm **69.** 2 m
71. 19 ft **73.** \$62,000 and \$85,000 **75.** July: 530 tickets; August: 583 tickets **77.** Younger son: \$45,000; older son: \$180,000
79. a. 81 **b.** 78 **81.** \$4000 in a tax shelter and \$3000 in a bank
83. \$2500 **85.** 35,000 shaving sets **87.** 2.5 hr **89.** 30 mph
91. 65 mph and 70 mph **93.** 1500 cameras **95.** 7 and 21
97. 14 and 28

Beyond the Basics:
99. a. No, because the solution sets are $\{0, 1\}$ and $\{1\}$, respectively.
b. No, because the solution sets are $\{-3, 3\}$ and $\{3\}$, respectively.
c. No, because the solution sets are $\{0, 1\}$ and $\{0\}$, respectively.
d. No, because the solution sets are $\{\varnothing\}$ and $\{2\}$, respectively.
101. 50 mph **103.** 25 liters **105.** 72 yr **107.** 26 min
109. a. 112 mi **b.** $\frac{448}{3} = 149.33$ mph

Critical Thinking/Discussion/Writing:
111. (c)

Maintaining Skills:
113. $2\sqrt{2}$ **115.** $2\sqrt{3}$ **117.** $1 + 3\sqrt{2}$ **119.** $3 - \sqrt{3}$
121. $x(x + 1)$ **123.** $(x - 2)(x + 2)$ **125.** $(x + 2)^2$
127. $(x - 1)(x - 7)$ **129.** $(3x + 1)(2x - 1)$ **131.** $(5x + 2)(1 - x)$

Section 1.2

Answers to Practice Problems:
1. $\{-21, -4\}$ **2.** $\left\{0, \frac{5}{2}\right\}$ **3.** $\{3\}$ **4.** $\{-2 - \sqrt{5}, -2 + \sqrt{5}\}$
5. $\{3 - \sqrt{2}, 3 + \sqrt{2}\}$ **6.** $\left\{3 - \frac{\sqrt{11}}{2}, 3 + \frac{\sqrt{11}}{2}\right\}$ **7.** $\left\{-\frac{1}{2}, \frac{2}{3}\right\}$
8. 25.48 ft by 127.41 ft **9.** $18 + 18\sqrt{5} \approx 58.25$ ft

Basic Concepts and Skills:
1. quadratic **3.** $\pm\sqrt{5}$ **5.** true **7.** Yes **9.** Yes **11.** Yes **13.** No
15. $k = 2$ **17.** $\{0, 5\}$ **19.** $\{-7, 2\}$ **21.** $\{-1, 6\}$ **23.** $\{-4, 4\}$
25. $\{-2, 2\}$ **27.** $\{-3, 5\}$ **29.** 4 **31.** 9 **33.** $\frac{49}{4}$ **35.** $\frac{1}{36}$
37. $\frac{a^2}{4}$ **39.** $\{-1 - \sqrt{6}, -1 + \sqrt{6}\}$ **41.** $\left\{\frac{3 - \sqrt{13}}{2}, \frac{3 + \sqrt{13}}{2}\right\}$
43. $\left\{-3, \frac{3}{2}\right\}$ **45.** $\{-1 - \sqrt{5}, -1 + \sqrt{5}\}$ **47.** $\left\{-\frac{1}{2}, \frac{5}{3}\right\}$
49. $\left\{\frac{1 - \sqrt{22}}{3}, \frac{1 + \sqrt{22}}{3}\right\}$ **51.** $\left\{-3, \frac{1}{2}\right\}$ **53.** $\left\{2, -\frac{2}{3}\right\}$
55. $\{6 - \sqrt{33}, 6 + \sqrt{33}\}$ **57.** $\left\{-2, -\frac{2}{3}\right\}$ **59.** $\left\{-1, -\frac{2}{3}\right\}$
61. $\left\{-2, -\frac{2}{5}\right\}$ **63.** $\left\{-\frac{7 + \sqrt{13}}{6}, -\frac{7 - \sqrt{13}}{6}\right\}$ **65.** $\left\{-3, \frac{5}{2}\right\}$
67. $\left\{-\frac{2}{3}, \frac{3}{2}\right\}$ **69.** $\{-9 - \sqrt{14}, -9 + \sqrt{14}\}$ **71.** $\left\{\frac{1}{6}, \frac{7}{3}\right\}$
73. $\left\{-\frac{\sqrt{10}}{2}, \frac{\sqrt{10}}{2}\right\}$ **75.** $\left\{-\frac{25}{2}, 15\right\}$ **77.** 23.82 in. **79.** 5.23 cm

Applying the Concepts:
81. 60 ft by 180 ft **83.** 7 and 21 **85.** 23, 34 **87.** 20 cm by 25 cm
89. 4 in. and 12 in. **91.** 2 in. **93.** 19.8 in. by 19.8 in.
95. After 6 hours **97.** 6 ft **99. a.** 160 ft **b.** After 1 sec and after 6 sec
c. 7 sec **101. a.** 1.59 sec and 4.41 sec **b.** 9.24 sec
103. a. 0.45 sec and 15.55 sec **b.** 17.70 sec

Beyond the Basics:
105. $-2\sqrt{3}$ or $2\sqrt{3}$ **107.** 0 or 8 **109.** $-\frac{1}{2}$
113. 2 **115.** $ax^2 + bx + c = ax^2 - a(r + s)x + ars$
$= a(x^2 - rx - sx + rs)$
$= a[x(x - r) - s(x - r)] = a(x - r)(x - s)$
117. a. $x^2 - x - 12 = 0$ **b.** $x^2 - 10x + 25 = 0$
c. $x^2 - 6x + 7 = 0$ **123.** $\{1, 9\}$ **125.** b

Maintaining Skills:
129. -27 **131.** $5 - 2\sqrt{6}$ **133.** $2 + \sqrt{3}$ **135.** $\frac{3 - \sqrt{3}}{2}$
137. $4x - 3$ **139.** $3x - 9$ **141.** $6x^2 - 7x - 20$ **143.** $25x^2 - 4$
145. $\frac{-5 \pm \sqrt{5}}{2}$ **147.** $\frac{5 \pm \sqrt{17}}{4}$

Section 1.3

Answers to Practice Problems:
1. a. Real $= -1$, imaginary $= 2$ **b.** Real $= -\frac{1}{3}$, imaginary $= -6$
c. Real $= 8$, imaginary $= 0$ **2.** $a = -2; b = 1$ **3. a.** $4 - 2i$
b. $-1 + 4i$ **c.** $-2 + 5i$ **4. a.** $26 + 2i$ **b.** $-15 - 21i$
5. a. $5 - 12i$ **b.** $16 + 14\sqrt{2}i$ **6. a.** 37 **b.** 4
7. a. $1 + i$ **b.** $-\frac{15}{41} - \frac{12}{41}i$ **8.** $\frac{23}{10} + \frac{11}{10}i$ **9. a.** $\left\{-\frac{3}{2}i, \frac{3}{2}i\right\}$
b. $\{2 - 3i, 2 + 3i\}$ **10. a.** One real **b.** Two unequal real
c. Two nonreal complex

Basic Concepts and Skills:
1. $\sqrt{-1}; -1$ **3.** $i\sqrt{b}$ **5.** true **7.** $x = 3; y = 2$
9. $x = 2; y = -4$ **11.** $8 + 3i$ **13.** $-1 - 6i$ **15.** $-5 - 5i$
17. $15 + 6i$ **19.** $-8 + 12i$ **21.** $-3 + 15i$ **23.** $20 + 8i$
25. $3 + 11i$ **27.** 13 **29.** $24 + 7i$ **31.** $-141 - 24\sqrt{3}i$
33. $26 - 2i$ **35.** $\bar{z} = 2 + 3i, z\bar{z} = 13$ **37.** $\bar{z} = \frac{1}{2} + 2i, z\bar{z} = \frac{17}{4}$
39. $\bar{z} = \sqrt{2} + 3i, z\bar{z} = 11$ **41.** $5i$ **43.** $-\frac{1}{2} + \frac{1}{2}i$ **45.** $1 + 2i$
47. $\frac{5}{2} + \frac{1}{2}i$ **49.** $\frac{43}{65} - \frac{6}{65}i$ **51.** 2 **53.** $-\frac{19}{13} + \frac{4}{13}i$
55. $\{-2i, 2i\}$ **57.** $\{1 - i, 1 + i\}$ **59.** $\{5 - i\sqrt{3}, 5 + i\sqrt{3}\}$
61. $\left\{\frac{4}{7} - \frac{\sqrt{5}}{7}i, \frac{4}{7} + \frac{\sqrt{5}}{7}i\right\}$ **63.** $\left\{-\frac{5}{3}i, \frac{5}{3}i\right\}$

Applying the Concepts:
65. $9 + i$ **67.** $Z = \frac{595}{74} + \frac{315}{74}i$ **69.** $V = 45 + 11i$
71. $I = \frac{17}{15} + \frac{4}{15}i$

Beyond the Basics:
73. i **75.** i **77.** 6 **79.** $5i$ **81.** $-4i$ **89.** $x = 1; y = 4$
91. $x = 1; y = 0$ **93.** $-\frac{3}{5} + \frac{4}{5}i$

Critical Thinking/Discussion/Writing:
97. $n = 4$

Maintaining Skills:
99. $(x - 1)(x - 3)(x + 3)$ **101.** $x(x - 1)$ **103.** $(x - 3)(x + 3)$
105. $(x - 3)(x - 2)(x + 1)$ **107.** 9 **109.** 32 **111.** $3 + x^2$
113. $3x^2 + 7$

Section 1.4

Answers to Practice Problems:
1. $\{-2, -1, 1\}$ **2.** $\{-2, 0, 2\}$ **3.** $\{-2, 2, 5\}$ **4.** $\{-10, 1\}$ **5.** $\varnothing$
6. $\{0, 3\}$ **7.** $\{10\}$ **8.** $\{9\}$ **9. a.** $\{3\}$ **b.** $\left\{-\frac{9}{2}, \frac{7}{2}\right\}$
10. $\{1, 216\}$ **11.** $\left\{\frac{1}{3}, 1\right\}$ **12.** $0.6c$ or 60% of the speed of light

Basic Concepts and Skills:
1. extraneous **3.** 16 **5.** false **7.** $\{0, 2\}$ **9.** $\{0, 1\}$ **11.** $\{0\}$
13. $\{-1, 0, 1\}$ **15.** $\{0, 3\}$ **17.** $\left\{\frac{1}{4}, 2\right\}$ **19.** $\{1 - \sqrt{2}, 1 + \sqrt{2}\}$
21. $\left\{\frac{7}{2} - \frac{\sqrt{53}}{2}, \frac{7}{2} + \frac{\sqrt{53}}{2}\right\}$ **23.** $\{-3, 5\}$ **25.** $\left\{\frac{1}{26}, 5\right\}$ **27.** $\varnothing$
29. $\varnothing$ **31.** $\{-1\}$ **33.** $\{-2, 7\}$ **35.** $\{3\}$ **37.** $\varnothing$ **39.** $\{-2\}$

41. $\{3\}$ **43.** $\{6\}$ **45.** $\{-2, 1\}$ **47.** $\{3, 7\}$ **49.** $\{0, 8\}$
51. $\{9\}$ **53.** $\{-2\}$ **55.** $\{3, 7\}$ **57.** $\{13\}$ **59.** $\left\{-\frac{24}{5}, 6\right\}$
61. $\{4, 9\}$ **63.** $\left\{\frac{1}{4}, 49\right\}$ **65.** $\left\{-\frac{1}{6}, \frac{1}{7}\right\}$ **67.** $\{8, 64\}$ **69.** $\left\{\frac{1}{16}\right\}$
71. $\{-3, -2, 2, 3\}$ **73.** $\left\{-\frac{\sqrt{2}}{2}, \frac{\sqrt{2}}{2}, -i, i\right\}$ **75.** $\{-2, 2\}$
77. $\left\{0, \frac{1}{3}\right\}$ **79.** $\{0\}$ **81.** $\{-2\sqrt{2}, -\sqrt{3}, \sqrt{3}, 2\sqrt{2}\}$
83. $\{-1, 1, 2, 4\}$

Applying the Concepts:
85. $\frac{6}{8}$ **87. a.** 20 shares **b.** \$90 **89.** 230 ft **91.** 2 mph **93.** 40 min
95. 120,000 ft^2 **97.** $CE = 6$ mi or $CE = \frac{3162}{205} \approx 15.4$ mi

Beyond the Basics:
99. $\{-1, 2 - \sqrt{7}, 3, 2 + \sqrt{7}\}$ **101.** $\{2\}$ **103.** $\{-7, -2\}$
105. $\left\{-3, -\frac{5}{3}\right\}$ **107.** $\left\{\frac{1}{8}, 8\right\}$ **109.** $\{\pm 2\sqrt{2}, \pm 3\sqrt{3}\}$ **111.** $\{36\}$
113. $\{-5b, 2b\}$ **115.** $\left\{\frac{2a}{3}, a\right\}$

Critical Thinking/Discussion/Writing:
117. $x = \frac{1 + \sqrt{5}}{2}$ **119.** $x = \frac{1}{2} + \frac{\sqrt{4n + 1}}{2}$

Maintaining Skills:
121. $\{2\}$ **123.** $\{-11\}$ **125.** $\{0\}$ **127.** $\{-\sqrt{17}, \sqrt{17}\}$
129. $\left\{\frac{1}{5}\right\}$ **131.** $\{-3, 2\}$ **133.** $\frac{6x + 5}{2x}$ **135.** $\frac{-x^2 + 4x + 11}{(x + 1)(x + 3)}$

Section 1.5

Answers to Practice Problems:

1. a. $(2, \infty)$
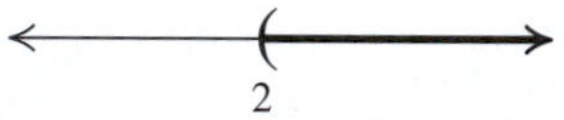

b. $(-\infty, 5]$

2. 2.5 hours **3. a.** $(-\infty, \infty)$ **b.** $\varnothing$ **4.** $(-\infty, 2] \cup [4, \infty)$

5. $(-1, 3]$
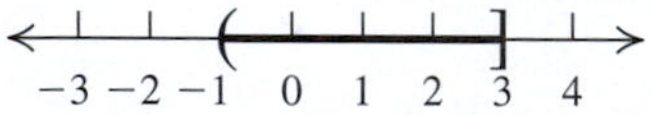

6. $\left[-1, \frac{3}{2}\right)$
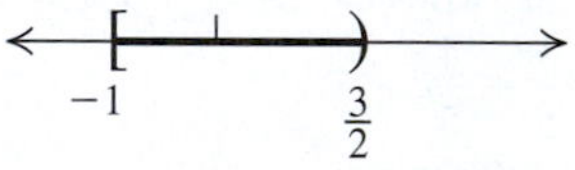

7. $a = -4, b = 11$

8. 59°F to 77°F **9.** $(-1, 4)$ **10.** $[-6, 1)$

Basic Concepts and Skills:
1. $<$ **3.** $\geq$ **5.** $<$ **7.** $\geq$ **9.** $<$

11. $(-2, 5)$
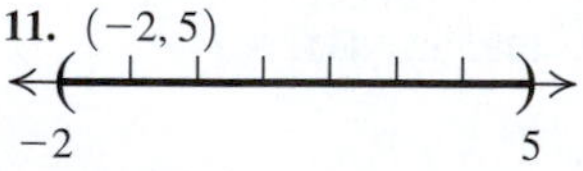

13. $(0. 4]$

15. $[-1, \infty)$
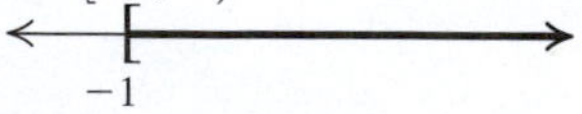

17. $(-\infty, -2]$
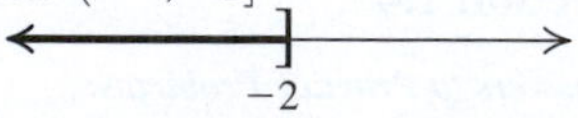

19. $(-\infty, 3)$
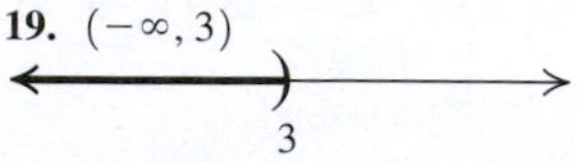

21. $[-3, \infty)$
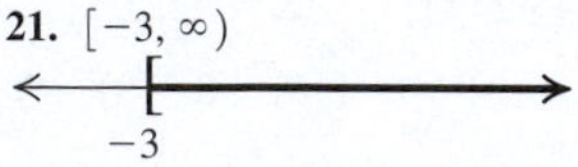

23. $(-\infty, 2)$
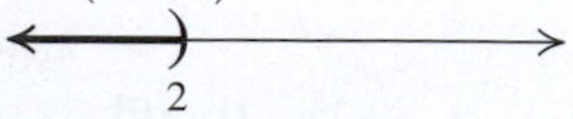

25. $(-\infty, -4)$
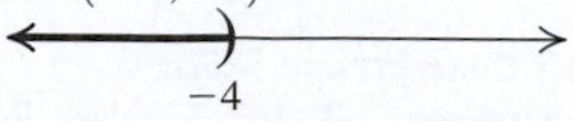

27. $(-\infty, -1)$
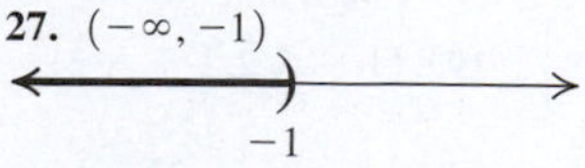

29. $(-\infty, 3]$
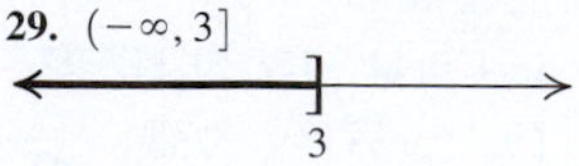

31. $(2, \infty)$
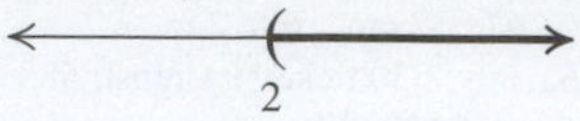

33. $\varnothing$
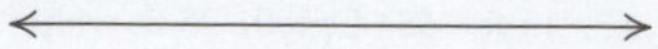

35. $\left(-\infty, \frac{3}{2}\right]$
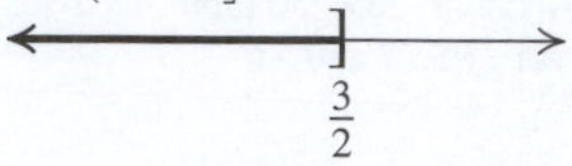

37. $(-\infty, \infty)$

39. $(-\infty, 3]$
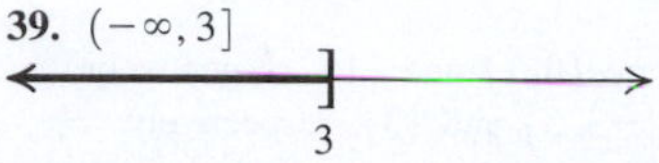

41. $[2, \infty)$ **43.** $[-18, \infty)$

45. $\varnothing$ **47.** $[15, \infty)$ **49.** $(-\infty, \infty)$ **51.** $(-\infty, -2) \cup (2, \infty)$
53. $\left(-\infty, \frac{11}{2}\right]$ **55.** $(-\infty, -2] \cup (6, \infty)$ **57.** $(-\infty, \infty)$ **59.** $[-2, 5]$
61. $[-2, 5)$ **63.** $\varnothing$ **65.** $(-\infty, 5]$ **67.** $(-2, -1)$ **69.** $[-2, 4)$
71. $[-6, 1]$ **73.** $(-3, 3]$ **75.** $\left(-1, \frac{3}{2}\right]$ **77.** $\varnothing$ **79.** $a = 5, b = 8$
81. $a = 1, b = 3$ **83.** $a = -1, b = 19$ **85.** $[-6, 2]$
87. $\left(-\infty, -\frac{3}{2}\right] \cup \left[\frac{1}{3}, \infty\right)$ **89.** $[-3, -1] \cup [1, \infty)$
91. $(-2, 0) \cup (6, \infty)$ **93.** $\varnothing$ **95.** $\{0\} \cup [1, \infty)$
97. $(-\infty, -1] \cup [1, \infty)$ **99.** $(-\infty, -2)$ **101.** $(-2, 5)$ **103.** $(-4, 0)$
105. $\left(-\infty, -\frac{5}{2}\right] \cup (-2, \infty)$ **107.** $(-2, 0) \cup (2, \infty)$
109. $(-\infty, -5) \cup [-1, 2] \cup (3, \infty)$ **111.** $(-2, -1] \cup [1, 2)$
113. $\left(\frac{2}{3}, 3\right]$ **115.** $\left(-\frac{1}{2}, \frac{3}{4}\right]$ **117.** $\left(-3, -\frac{1}{3}\right] \cup (3, \infty)$
119. $\left(-1, \frac{1}{7}\right] \cup (3, \infty)$

Applying the Concepts:
121. \$2012.50 to \$2100 **123.** 7.5 gal to 12 gal **125.** At least 15 qt
127. At least 8 **129.** 32°F to 100°F
131. More than 133 cards but not more than 180

Beyond the Basics:
133. $(-\infty, -4) \cup (4, \infty)$ **135.** $(-\infty, 0) \cup (4, \infty)$ **137.** $\varnothing$
139. $(-\infty, -12] \cup [12, \infty)$ **141.** At most, 4000 radios

Critical Thinking/Discussion/Writing:
143. a. $(x + 4)(x - 5) < 0$ **b.** $(x + 2)(x - 6) \leq 0$ **c.** $x^2 \geq 0$
d. $x^2 < 0$ **e.** $(x - 3)^2 \leq 0$ **f.** $(x - 2)^2 > 0$

Maintaining Skills:
145. 3 **147.** 2 **149.** 0 **151.** 7 **153.** 3.4 **155.** $|x + 2| = 5$
157. $|x - 4| \leq 2$ **159.** $|x - 5| \leq 3$

Section 1.6

Answers to Practice Problems:
1. a. $\{2\}$ **b.** $\{-1, 2\}$ **2.** $\left\{\frac{1}{2}\right\}$ **3.** $\left\{\frac{6}{5}, 2\right\}$

4. $[-3, 1]$
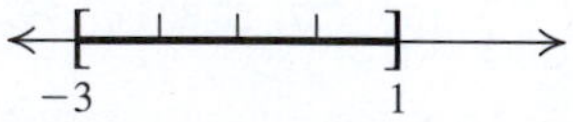

5. $[270, 420]$

6. $\left(-\infty, -\frac{9}{2}\right] \cup \left[\frac{3}{2}, \infty\right)$
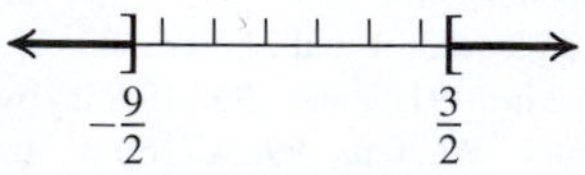

7. a. $(-\infty, \infty)$ **b.** $\varnothing$ **8.** $(-\infty, -6) \cup \left(-\frac{14}{5}, \infty\right)$

Basic Concepts and Skills:
1. $\{-a, a\}$ **3.** $(-\infty, -a] \cup [a, \infty)$ **5.** true **7.** $\{-3, 3\}$
9. $\{-3, 3\}$ **11.** $\{-5, -1\}$ **13.** $\{-1, 7\}$ **15.** $\left\{-\frac{7}{6}, \frac{11}{6}\right\}$
17. $\{-2, -1\}$ **19.** $\{-6, 6\}$ **21.** $\{-20, 4\}$ **23.** $\{-1, 2\}$ **25.** $\varnothing$
27. $\varnothing$ **29.** $\{-2, 2\}$ **31.** $\{-1, 2\}$ **33.** $\left\{-\frac{1}{3}, 1\right\}$ **35.** $\{-4\}$

37. $\left\{-3, -\frac{5}{9}\right\}$ **39.** $\{0, 2\}$ **41.** $\left\{\frac{5}{4}, \frac{3}{2}\right\}$ **43.** $\varnothing$ **45.** $(-4, 4)$
47. $(-\infty, -4) \cup (4, \infty)$ **49.** $(-4, 2)$ **51.** $(-\infty, \infty)$ **53.** $\left(-\frac{1}{2}, \frac{7}{2}\right)$
55. $(-\infty, 1) \cup (4, \infty)$ **57.** $\left[-\frac{23}{3}, 5\right]$ **59.** $\varnothing$ **61.** $\left(-\frac{1}{2}, \infty\right)$
63. $(-\infty, -6] \cup [0, \infty)$ **65.** $[-5, -2) \cup (-2, -1]$
67. $\left(\frac{1}{2}, 1\right) \cup \left(1, \frac{5}{2}\right)$ **69.** $\left(-\infty, \frac{11}{5}\right] \cup [3, \infty)$

Applying the Concepts:
71. The temperatures in Tampa during December are between 55°F and 95°F. **73.** $|x - 700| \le 50$ **75.** $|x - 120| \le 6.75$
77. Between \$5040 and \$6480 **79.** Between \$18.90 and \$19.50

Beyond the Basics:
81. a. $\{3\}$ **b.** $\{-4, -2\}$ **c.** $\{-1, 2, 3, 6\}$ **d.** $\{-4, -3, 0, 1\}$
83. $\left\{\frac{1}{3}, 3\right\}$ **85.** $\{-6, 6\}$ **91.** $|x - 4| < 3$ **93.** $|x - 4| \le 6$
95. $|x - 7| > 4$ **97.** $|2x - 5| \ge 15$ **99.** $|2x - a - b| < b - a$
101. $|2x - a - b| > b - a$ **103.** $|x - 39| \le 31$
105. $[-1, 1] \cup [3, 5]$ **107.** $(-\infty, -1] \cup [7, \infty)$ **109.** $\left(-2, -\frac{1}{2}\right]$

Critical Thinking/Discussion/Writing:
111. $(-\infty, 2] \cup [4, \infty)$

Maintaining Skills:
113. 3.5 **115.** −5 **117.** 5 **119.** $\sqrt{13}$ **121.** 4, $(x + 2)^2$
123. $\frac{25}{4}, \left(x - \frac{5}{2}\right)^2$ **125.** $\frac{9}{16}, \left(x + \frac{3}{4}\right)^2$

Review Exercises

1. $\{3\}$ **3.** $\{2\}$ **5.** $(-\infty, \infty)$ **7.** $\{1\}$ **9.** $\{1\}$ **11.** $\{11\}$
13. $\left\{-4, -\frac{1}{3}\right\}$ **15.** $\left\{-\frac{3}{2}\right\}$ **17.** $\left\{\frac{1}{5}, 3\right\}$ **19.** $g = \frac{p - k}{t}$
21. $B = \frac{T}{T - 2}$ **23.** $\{0, 7\}$ **25.** $\{-2, 5\}$ **27.** $\{-6, 1\}$
29. $\{-5, 1\}$ **31.** $\left\{-1, \frac{2}{3}\right\}$ **33.** $\left\{\frac{3}{2} - \frac{\sqrt{13}}{2}, \frac{3}{2} + \frac{\sqrt{13}}{2}\right\}$
35. $\left\{-1, \frac{1}{2}\right\}$ **37.** $\{2 - 2\sqrt{3}, 2 + 2\sqrt{3}\}$ **39.** $\left\{\frac{1}{4} - \frac{\sqrt{17}}{4}, \frac{1}{4} + \frac{\sqrt{17}}{4}\right\}$
41. $\left\{\frac{1}{2} - \frac{\sqrt{3}}{2}i, \frac{1}{2} + \frac{\sqrt{3}}{2}i\right\}$ **43.** $\{3 - 2i, 3 + 2i\}$
45. $\left\{1 - \frac{3}{2}i, 1 + \frac{3}{2}i\right\}$ **47.** $D = 49$, two real, unequal **49.** $D = -16$, two unequal complex roots **51.** $\{-4, 4\}$ **53.** $\left\{\frac{1}{2}\right\}$ **55.** $\{9\}$
57. $\{16\}$ **59.** $\left\{-\frac{10}{7}, -\frac{2}{7}\right\}$ **61.** $\{-64, 1\}$ **63.** $\{0\}$
65. $\left\{-3, -\frac{1}{2}, \frac{1}{2}, 3\right\}$ **67.** $\{0\}$ **69.** $\left\{-\frac{3}{10}, 6\right\}$ **71.** $\{-3y, y\}$
73. $\{y - 1, y - 2\}$ **75.** $(-\infty, -2)$ **77.** $\left(-\infty, \frac{17}{3}\right]$ **79.** $\left[-\frac{6}{7}, \infty\right)$
81. $\left(\frac{7}{6}, \infty\right)$ **83.** $(-\infty, 21]$ **85.** $\left(-\infty, \frac{41}{10}\right)$ **87.** $(-\infty, 1) \cup (3, \infty)$
89. $(2, 3)$ **91.** $[-2, 3)$ **93.** $[-47, 3)$ **95.** $(-\infty, -3] \cup [2, \infty)$
97. $(-\infty, -5) \cup [-3, -2) \cup [1, \infty)$ **99.** $\left[-3, \frac{5}{3}\right]$
101. $(-\infty, 1) \cup (3, \infty)$ **103.** $(-\infty, -1] \cup [9, \infty)$
105. $\left(-\infty, -\frac{7}{2}\right] \cup \left[-\frac{5}{4}, \infty\right)$ **107.** $(-\infty, -2) \cup \left(-2, -\frac{1}{4}\right]$

Applying the Concepts:
109. $\frac{11}{\pi}$ cm **111.** 3 m **113.** 13 cm **115.** 350 cm
117. 35°, 35°, 110° **119.** \$12,000 at 6%; \$18,000 at 8%
121. 8 liters of the $4\frac{1}{2}$% and 2 liters of the 12% **123.** 8 people
125. 220 **127.** 36 **129.** 40

Practice Test A

1. $\{2\}$ **2.** $\{1\}$ **3.** $\left\{-\frac{2\sqrt{30}}{5}, \frac{2\sqrt{30}}{5}\right\}$ **4.** 6 cm by 9 cm
5. $\{-9, -4\}$ **6.** \$8200 **7.** 1 in. **8.** $\left\{-4, -\frac{5}{2}\right\}$
9. $\frac{1}{9}, \left(x + \frac{1}{3}\right)^2$ **10.** $\left\{\frac{5}{6} - \frac{\sqrt{37}}{6}, \frac{5}{6} + \frac{\sqrt{37}}{6}\right\}$ **11.** 9 in.
12. $\{-6 - \sqrt{3}, -6 + \sqrt{3}\}$ **13.** $\{-5, 0, 5\}$ **14.** $\{4\}$
15. $\{-12, -6\}$ **16.** $[90, \infty)$ **17.** $\left(-\frac{1}{2}, \frac{7}{2}\right)$
18. $(-\infty, -2) \cup (5, \infty)$ **19.** $(-\infty, 2]$ **20.** $\left[\frac{2}{5}, 2\right]$

Practice Test B

1. d **2.** d **3.** a **4.** c **5.** b **6.** a **7.** d **8.** d **9.** b **10.** a **11.** b
12. d **13.** a **14.** a **15.** d **16.** d **17.** b **18.** c **19.** a **20.** a

CHAPTER 2

Section 2.1

Answers to Practice Problems:
1.

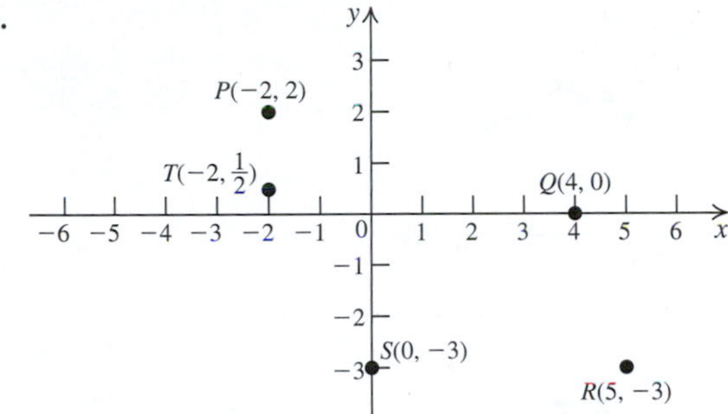

2. (2005, 17.9), (2006, 17.8), (2007, 17.8) (2008, 17.6), (2009, 17.6), (2010, 16.3), (2011, 16.0), and (2012, 18.0)

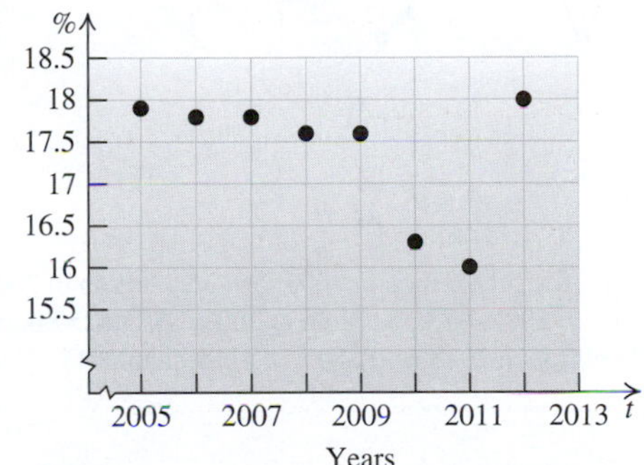

3. $\sqrt{2} \approx 1.4$ **4.** Yes **5.** $60\sqrt{2} \approx 84.85$ feet **6.** $\left(\frac{11}{2}, -\frac{3}{2}\right)$

Basic Concepts and Skills:
1. second **3.** $\sqrt{(x_2 - x_1)^2 + (y_2 - y_1)^2}$ **5.** true
7.

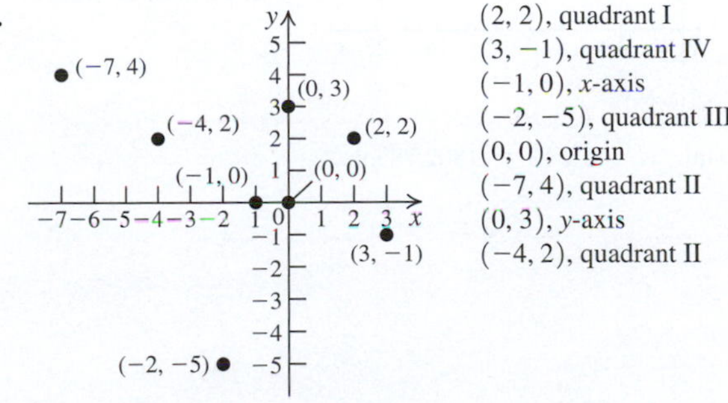

(2, 2), quadrant I
(3, −1), quadrant IV
(−1, 0), x-axis
(−2, −5), quadrant III
(0, 0), origin
(−7, 4), quadrant II
(0, 3), y-axis
(−4, 2), quadrant II

9. a. On the y-axis **b.** Vertical line intersecting the x-axis at -1
11. a. $y > 0$ **b.** $y < 0$ **c.** $x < 0$ **d.** $x > 0$ **13. a.** 4 **b.** $(2, 3)$
15. a. $\sqrt{13}$ **b.** $(0.5, -4)$ **17. a.** $4\sqrt{5}$ **b.** $(1, -2.5)$
19. a. 1 **b.** $(\sqrt{2}, 4.5)$ **21. a.** $\sqrt{2}|t - k|$ **b.** $\left(\frac{t+k}{2}, \frac{t+k}{2}\right)$
23. Yes **25.** Yes **27.** Yes **29.** No **31.** $(-3, 2), (-2, 4), (-1, 6)$
33. Isosceles **35.** Scalene **37.** Scalene **39.** Equilateral
41. $d(P, R) = d(Q, S) = 17\sqrt{2}$ **43.** -2 or 6 **45.** $\left(-\frac{8}{7}, 0\right)$

Applying the Concepts:

47.

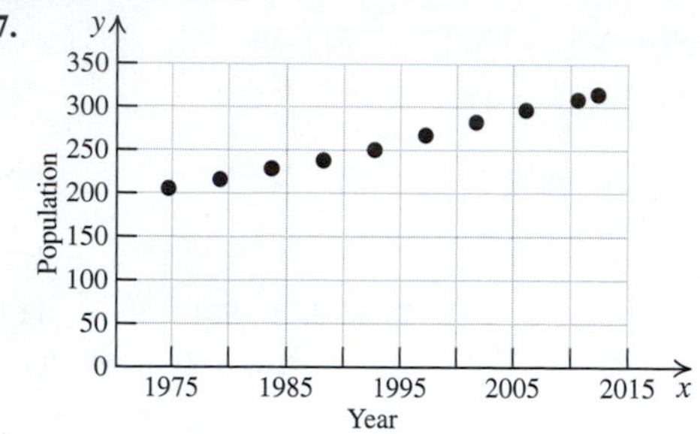

49.

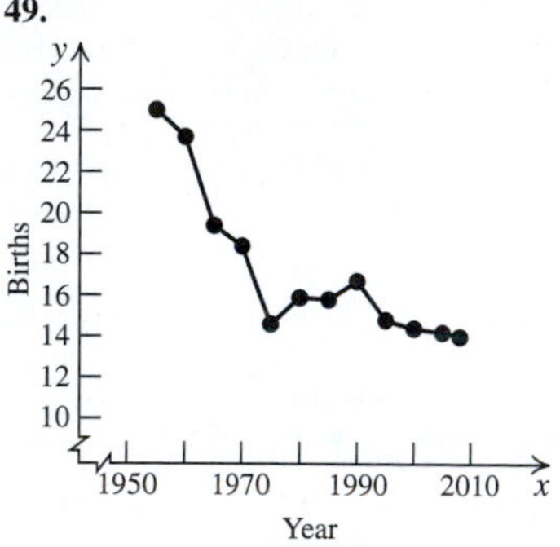

51.

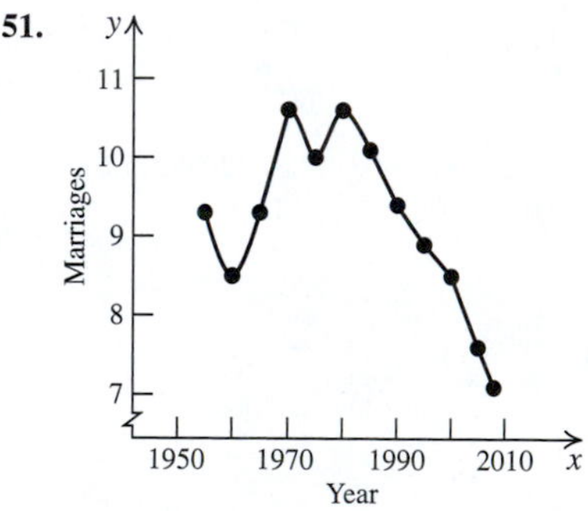

53. 15,771 **55.** \$595.15 billion in 2005, \$642.25 billion in 2006, \$689.375 billion in 2007, \$736.5 billion in 2008, \$783.625 billion in 2009, \$830.75 billion in 2010, and \$877.875 billion in 2011

57. a.

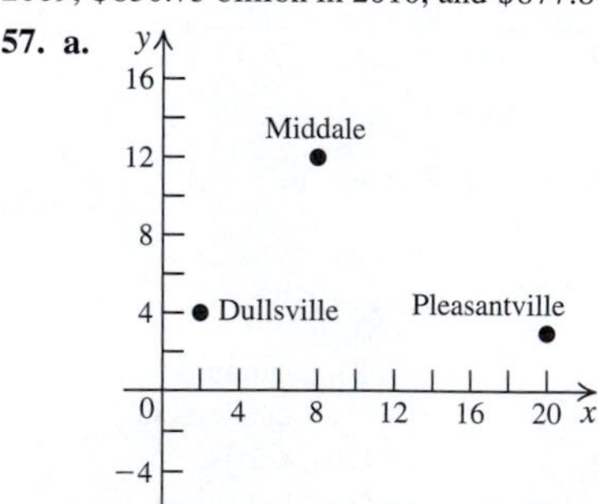

b. 2500 mi **c.** $500\sqrt{13} \approx 1802.78$ mi

Beyond the Basics:

65. $\frac{\sqrt{2}}{2}c$ **67. c.** $\left(\frac{2}{3}, \frac{5}{3}\right)$ and $\left(\frac{7}{3}, \frac{4}{3}\right)$

Critical Thinking/Discussion/Writing:

69. a. y-axis **b.** x-axis **71. a.** Quadrants I and III **b.** Quadrants II and IV **73. a.** Right half-plane **b.** Upper half-plane

Maintaining Skills:

75. a. $\frac{1}{2}$ **b.** 1 **77. a.** 0 **b.** 0 **79.** 9 **81.** $\frac{9}{4}$ **83.** $\frac{a^2}{4}$

Section 2.2

Answers to Practice Problems:

1.

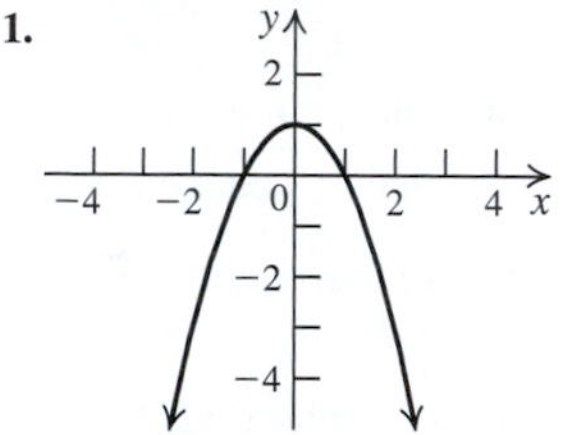

2. x-intercepts: $-2, \frac{1}{2}$; y-intercept: -2 **3.** Symmetric about the y-axis **4.** Not symmetric about the x-axis, symmetric about the y-axis, not symmetric about the origin

5. a.

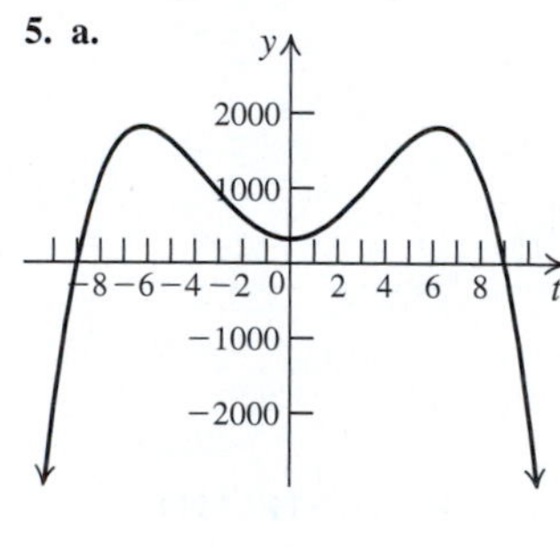

b.

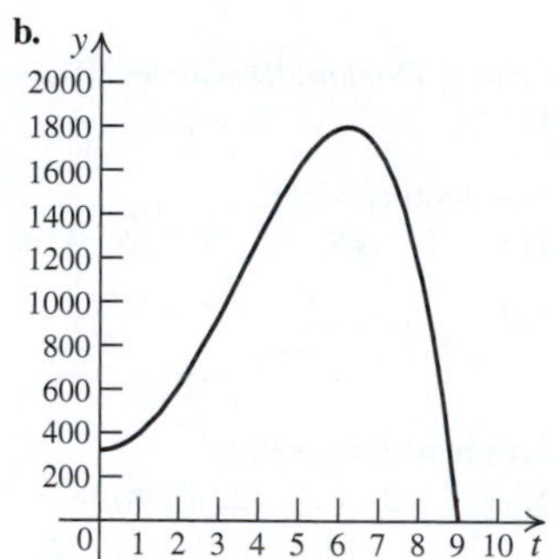

c. After 9 years **6.** $(x - 3)^2 + (y + 6)^2 = 100$

7.

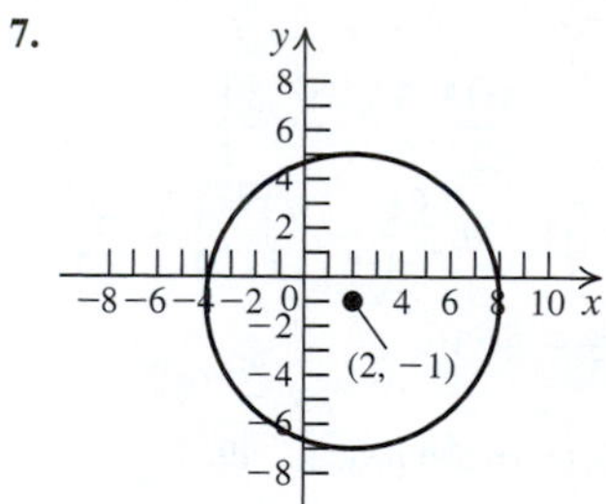

8. $(x + 2)^2 + (y - 3)^2 = 25$: $C = (-2, 3), r = 5$

Basic Concepts and Skills:

1. that satisfy the equation **3.** y- **5.** false
7. On the graph: $(-3, -4), (1, 0), (4, 3)$; not on the graph: $(2, 3)$
9. On the graph: $(3, 2), (0, 1), (8, 3)$; not on the graph: $(8, -3)$
11. On the graph: $(1, 0), (2, \sqrt{3}), (2, -\sqrt{3})$; not on the graph: $(0, -1)$
13. a. x-intercepts -3 and 3; no y-intercepts **b.** Symmetric about x-axis, y-axis, and origin **15. a.** x-intercepts -3 and 3; y-intercepts -2 and 2 **b.** Symmetric about x-axis, y-axis, and origin **17. a.** x-intercept 0; y-intercept 0 **b.** Symmetric about x-axis **19. a.** x-intercepts -3 and 3; y-intercept 2 **b.** Symmetric about y-axis **21. a.** No x-intercept; y-intercept 1 **b.** No Symmetries **23. a.** x-intercepts 0 and 3; y-intercept 0 **b.** Symmetric about x-axis

25. (−3, −2), (−2, −1), (−1, 0), (0, 1), (1, 2), (2, 3), (3, 4)

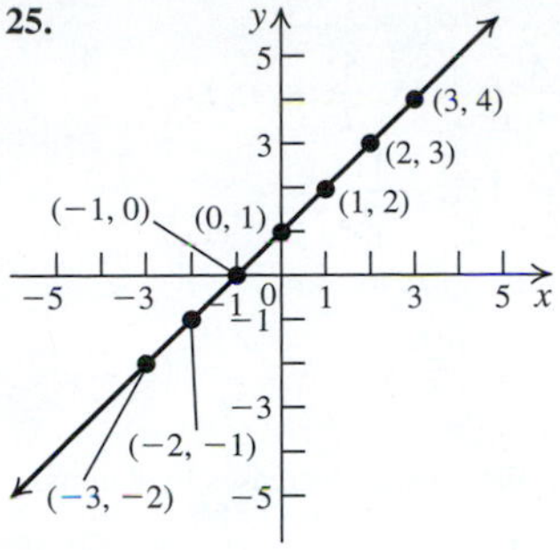

27. (−3, −6), (−2, −4), (−1, −2), (0, 0), (1, 2), (2, 4), (3, 6)

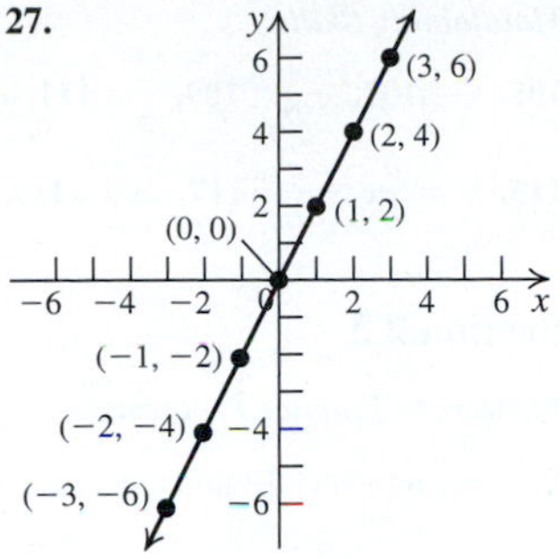

29. (−3, 9), (−2, 4), (−1, 1), (0, 0), (1, 1), (2, 4), (3, 9)

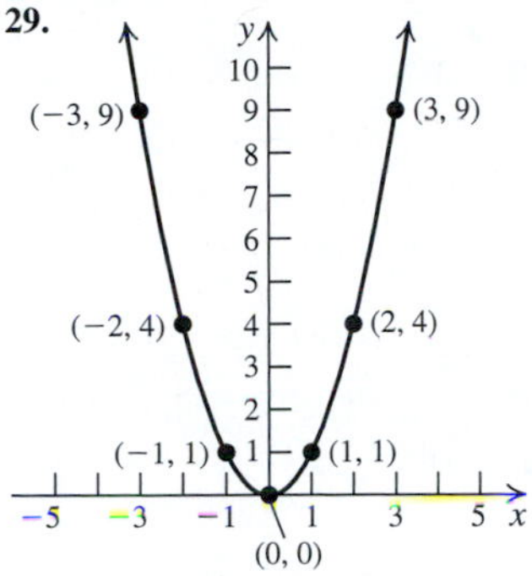

31. (−3, 3), (−2, 2), (−1, 1), (0, 0), (1, 1), (2, 2), (3, 3)

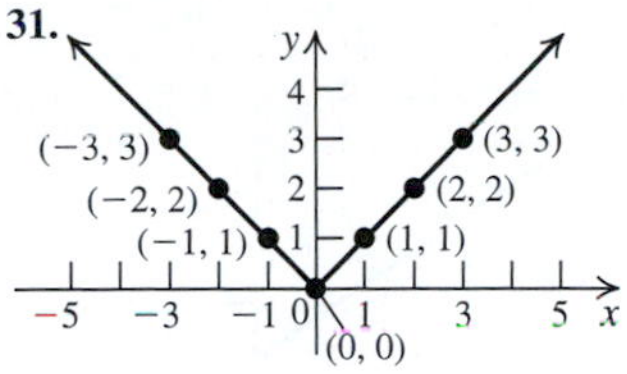

33. (−3, 4), (−2, 3), (−1, 2), (0, 1), (1, 2), (2, 3), (3, 4)

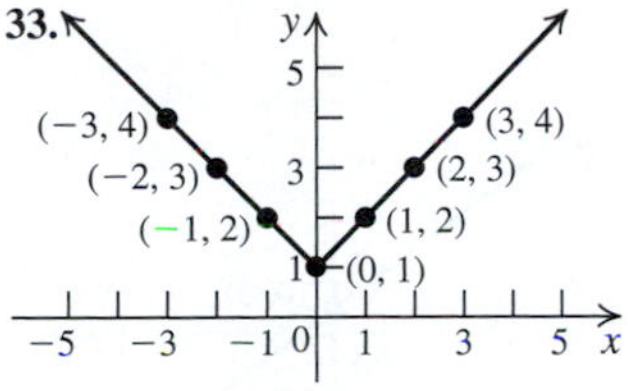

35. (−3, −5), (−2, 0), (−1, 3), (0, 4), (1, 3), (2, 0), (3, −5)

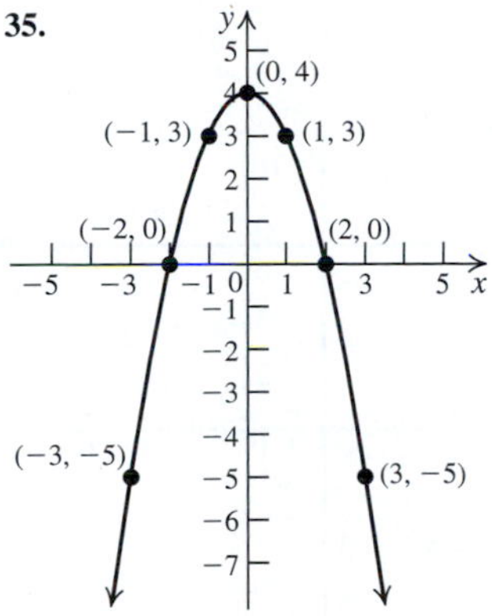

37. (−3, 0), (−2, 2.24), (−1, 2.83), (0, 3), (1, 2.83), (2, 2.24), (3, 0)

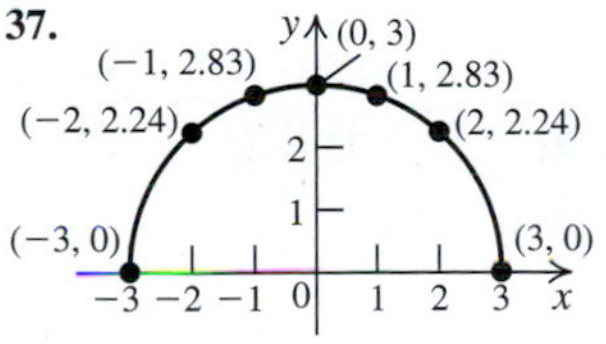

39. (−3, −27), (−2, −8), (−1, −1), (0, 0), (1, 1), (2, 8), (3, 27)

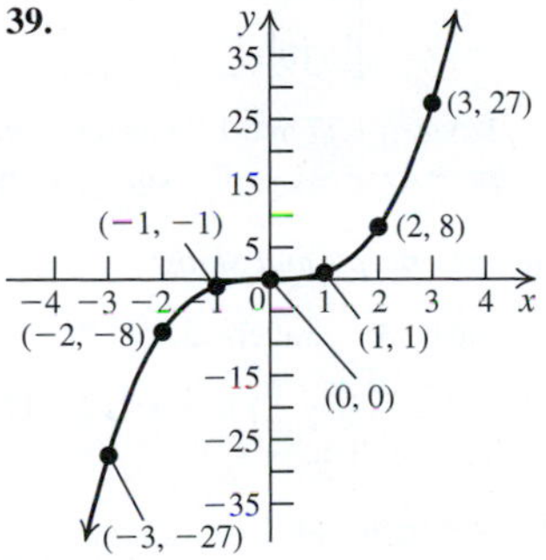

41. (−3, −1.44), (−2, −1.26), (−1, −1), (0, 0), (1, 1), (2, 1.26), (3, 1.44)

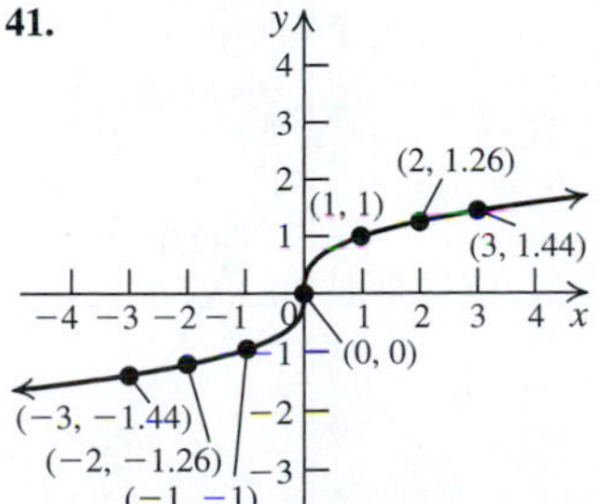

43. (0, 0), (1, 1), (2, 2), (3, 3), (1, −1), (2, −2), (3, −3)

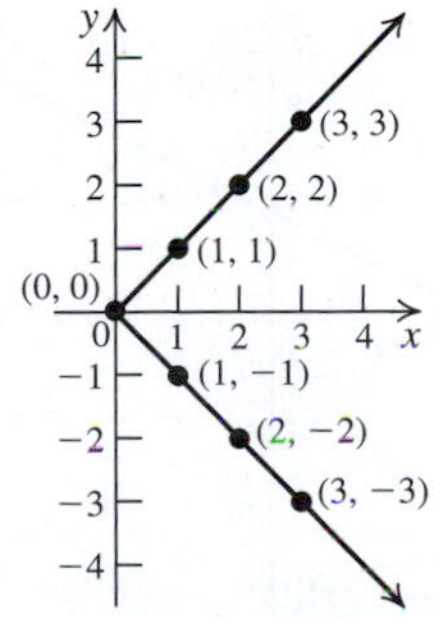

45.

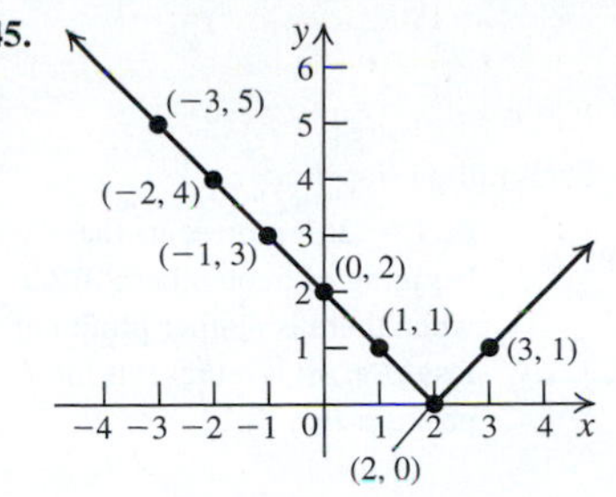

47. x-intercept: 4; y-intercept: 3 **49.** x-intercepts: 2, 4; y-intercept: 8 **51.** x-intercepts: −2, 2; y-intercepts: −2, 2 **53.** x-intercepts: −1, 1; no y-intercept **55.** No x-intercepts; y-intercept 1 **57.** Not symmetric with respect to the x-axis, symmetric with respect to the y-axis, not symmetric with respect to the origin **59.** Not symmetric with respect to the x-axis, not symmetric with respect to the y-axis, symmetric with respect to the origin **61.** Not symmetric with respect to the x-axis, symmetric with respect to the y-axis, not symmetric with respect to the origin **63.** Not symmetric with respect to the x-axis, not symmetric with respect to the y-axis, symmetric with respect to the origin **65.** Not symmetric with respect to the x-axis, not symmetric with respect to the y-axis, symmetric with respect to the origin **67.** Center: (2, 3); radius: 6

69. Center: (−2, −3); radius: $\sqrt{11}$

71. $x^2 + (y - 1)^2 = 4$ **73.** $(x + 1)^2 + (y - 2)^2 = 2$

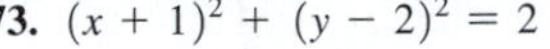

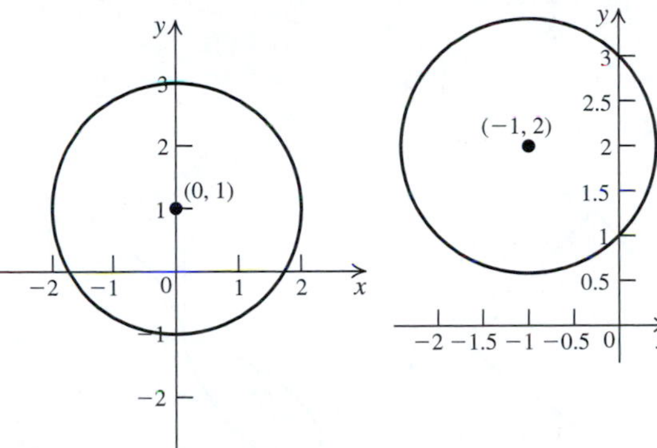

75. $(x - 3)^2 + (y + 4)^2 = 97$ **77.** $(x - 1)^2 + (y - 2)^2 = 4$

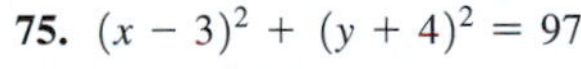

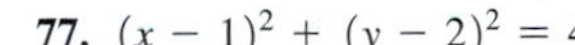

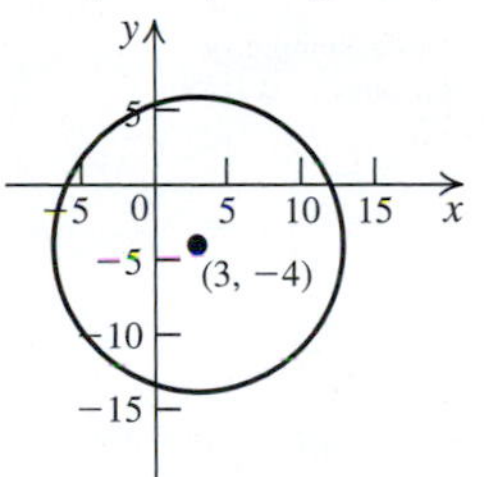

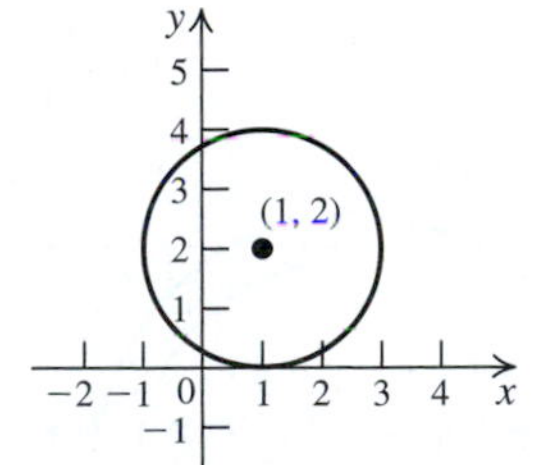

79. $(x - 2)^2 + (y - 5)^2 = 26$

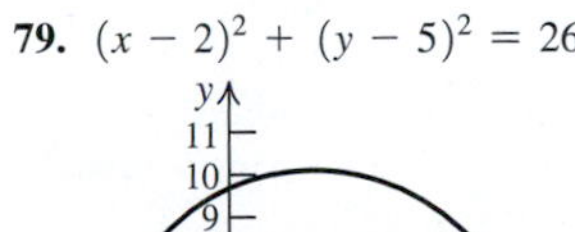

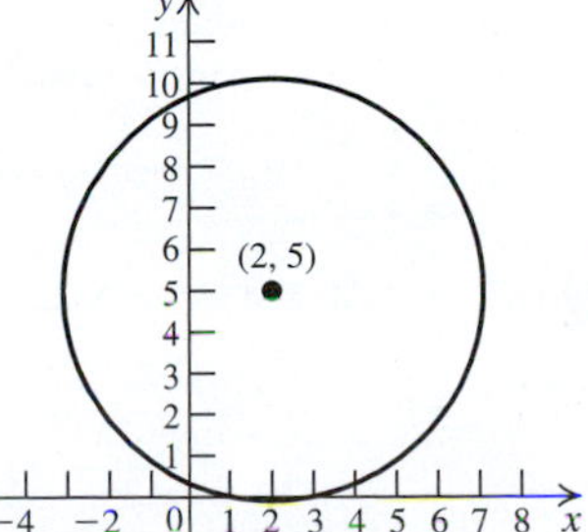

81. a. Circle; center: (1, 1); radius: $\sqrt{6}$
b. $(1 - \sqrt{5}, 0), (1 + \sqrt{5}, 0), (0, 1 - \sqrt{5}), (0, 1 + \sqrt{5})$

83. a. Circle; center: (0, −1); radius: 1 **b.** (0, 0), (0, −2)

85. a. Circle; center: $\left(\frac{1}{2}, 0\right)$; radius: $\frac{1}{2}$ **b.** (1, 0), (0, 0)

Applying the Concepts:

87. $|x| = |y|$ **89.** $x = \dfrac{y^2}{4} + 1$

91. a. \$12 million (profit) **b.** $-$\$5.5 million (loss)

c.

d. $t = 2$; it represents the begining of September 2012, when there is neither profit nor loss. **e.** 8; it represents the profit in July 2012.

93. a. After 0 second: 320 ft; after 1 second: 432 ft; after 2 seconds: 512 ft; after 3 seconds: 560 ft; after 4 seconds: 576 ft; after 5 seconds: 560 ft; after 6 seconds: 512 ft

b.

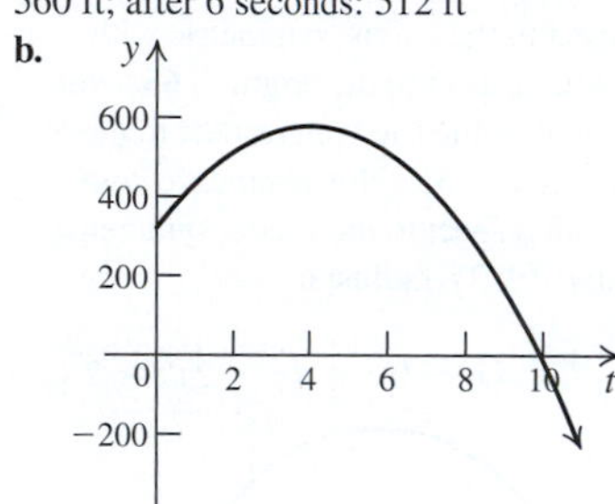

c. $0 \le t \le 10$ **d.** t-intercept: 10; it represents the time when the object hits the ground. y-intercept: 320; it shows the height of the building.

Beyond the Basics:

95.

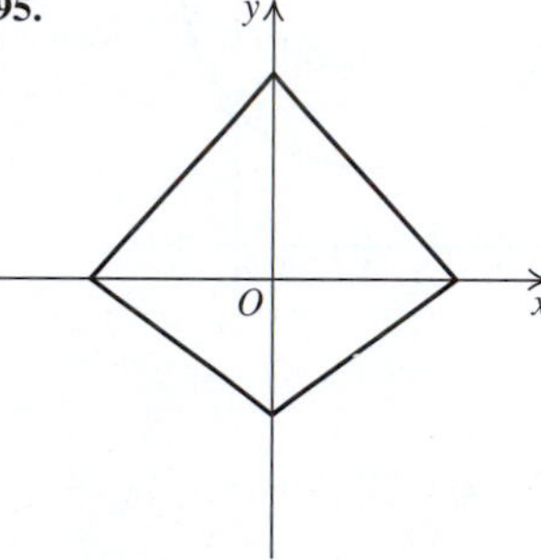

97.

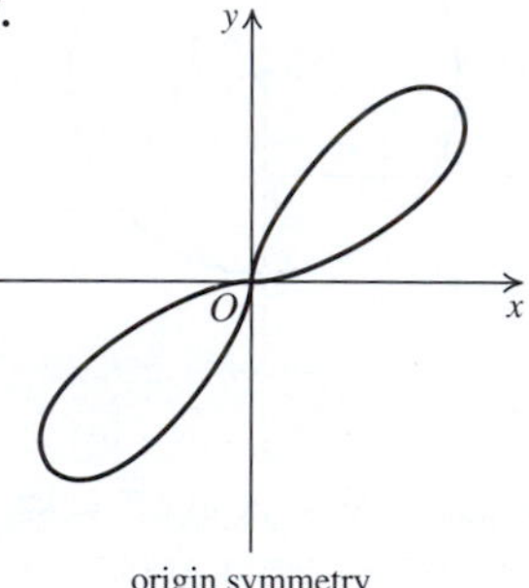

origin symmetry

99. Area: 11π

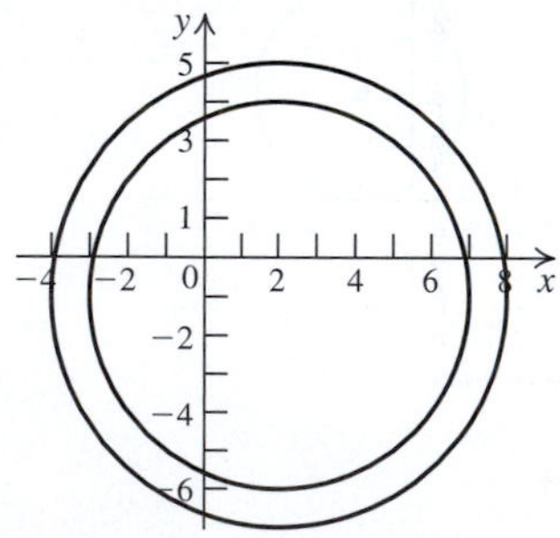

Critical Thinking/Discussion/Writing:

101. The graph is the union of the graphs of $y = \sqrt{2x}$ and $y = -\sqrt{2x}$.

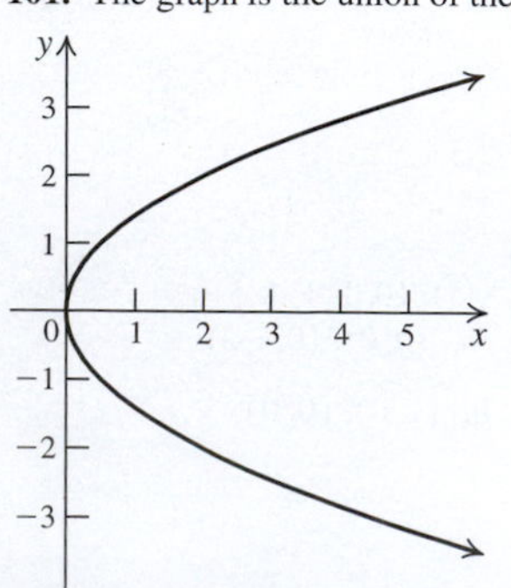

103. c. $(0.4, 0)$ and $(2.6, 0)$

Maintaining Skills:

105. $\dfrac{1}{2}$ **107.** $-\dfrac{1}{2}$ **109.** $\dfrac{2}{5}$ **111.** $\dfrac{A}{1 + rt}$ **113.** $\dfrac{5}{2}x - 15$

115. $y = \dfrac{a}{b}x - \dfrac{c}{b}$ **117.** $-\dfrac{1}{2}$ **119.** $\dfrac{3}{2}$ **121.** $-\dfrac{11}{2}$

Section 2.3

Answers to Practice Problems:

1. $-\dfrac{8}{13}$; for every 13 units increase in x, the y-value decreases by 8 units.

2. $y + 3 = -\dfrac{2}{3}(x + 2)$; $y = -\dfrac{2}{3}x - \dfrac{13}{3}$ **3.** $y + 4 = 5(x + 3)$; $y = 5x + 11$ **4.** $y + 3 = 2x$; $y = 2x - 3$

5.

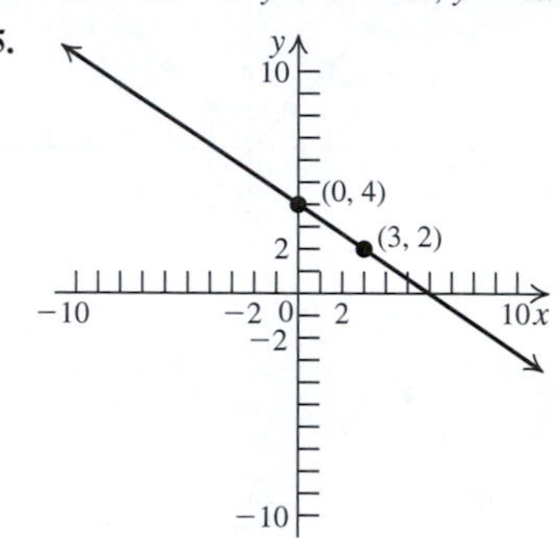

6. **7.**

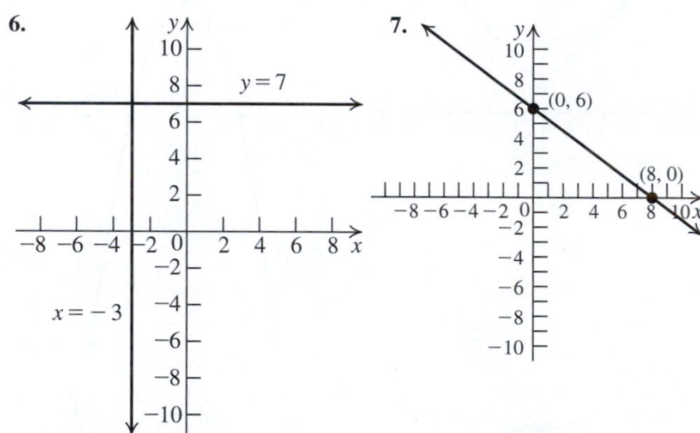

8. Between 1.77 and 1.79 meters **9. a.** $4x - 3y + 23 = 0$ **b.** $5x - 4y - 31 = 0$ **10.** 10.22 million

Basic Concepts and Skills:

1. slope **3.** parallel **5.** 5 **7.** false **9.** $\dfrac{4}{3}$, rising **11.** 0, horizontal **13.** -2.2, falling **15.** 4, rising **17. a.** ℓ_3 **b.** ℓ_2 **c.** ℓ_4 **d.** ℓ_1 **19. a.** $y = 0$ **b.** $x = 0$

21. $y = \dfrac{1}{2}x + 4$ **23.** $y = -\dfrac{3}{2}x + 4$

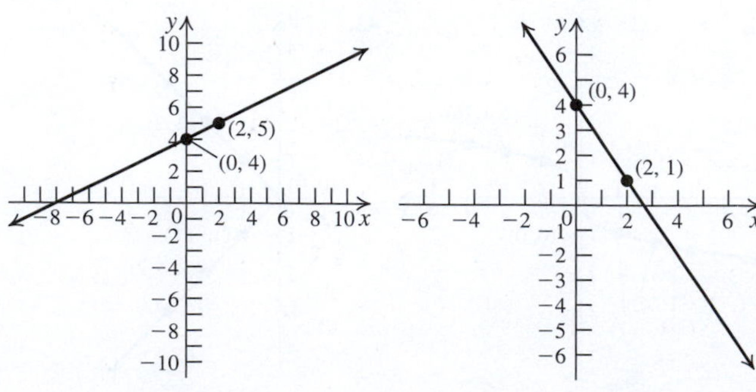

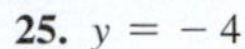

25. $y = -4$

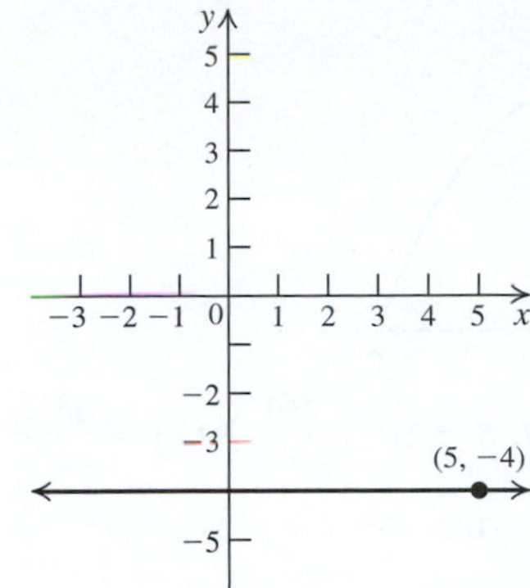

27. $y = -x + 1$ **29.** $y = 3$ **31.** $y = \frac{2}{3}x + \frac{1}{3}$

33. $y = -\frac{7}{2}x + 2$ **35.** $x = 5$ **37.** $y = 0$ **39.** $y = 14$

41. $y = -\frac{2}{3}x - 4$ **43.** $y = \frac{4}{3}x + 4$ **45.** $y = 7$ **47.** $y = -5$

49. a. Parallel **b.** Neither **c.** Perpendicular

51. $m = -\frac{1}{2}$; y-intercept $= 2$; x-intercept $= 4$ **53.** $m = \frac{3}{2}$; y-intercept $= 3$; x-intercept $= -2$

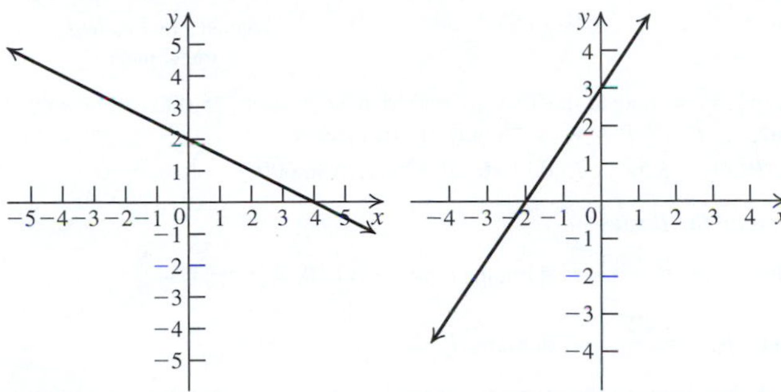

55. m is undefined; no y-intercept; x-intercept $= 5$ **57.** m is undefined; y-intercept $=$ y-axis; x-intercept $= 0$

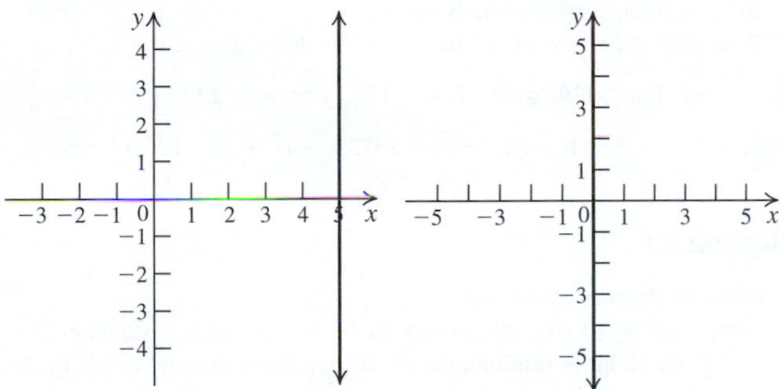

61. $\frac{x}{3} + \frac{y}{2} = 1$; x-intercept $= 3$ y-intercept $= 2$ **63.** $y = -x - 2$

65. $y = x + 2$; because the coordinates of the point $(-1, 1)$ satisfy this equation, $(-1, 1)$ also lies on this line. **67.** $y = -\frac{1}{2}x + \frac{37}{4}$

69. Parallel **71.** Neither **73.** Perpendicular **75.** Neither

77. a. $y = 3x - 9$ **b.** $y = 2x + 4$ **79.** $y = -x + 2$

81. $y = -3x - 2$ **83.** $y = -\frac{1}{6}x + 4$ **85.** $y = \frac{3}{2}x + \frac{13}{4}$

Applying the Concepts:

87. $\frac{1}{10}$ **89. a.** x: time in weeks; y: amount of money after x weeks in the account; $y = 7x + 130$ **b.** Slope: weekly deposit in the account; y-intercept: initial deposit

91. a. x: number of hours worked; y: amount of money earned per week;

$$y = \begin{cases} 11x & x \leq 40 \\ 16.5x - 220 & x > 40 \end{cases}$$

b. Slope: hourly wage; y-intercept: salary for 0 hours of work

93. a. $x =$ amount of rupees; $y =$ amount of dollars equal to x rupees; $y = \frac{1}{53.87}x$, or $y = 0.0186x$ **b.** Slope: the number of dollars per rupee; when $x = 0$, $y = 0$ **95. a.** Initial value **b.** $1000 **c.** The value of the machine decreased $800 per year. **d.** $y = -0.8x + 9$ **e.** The slope gives the machine's yearly depreciation, $800 or 0.8 Thous.

97. $y = 5x + 40{,}000$ **99. a.** $y = 266.8t + 2425$

b.

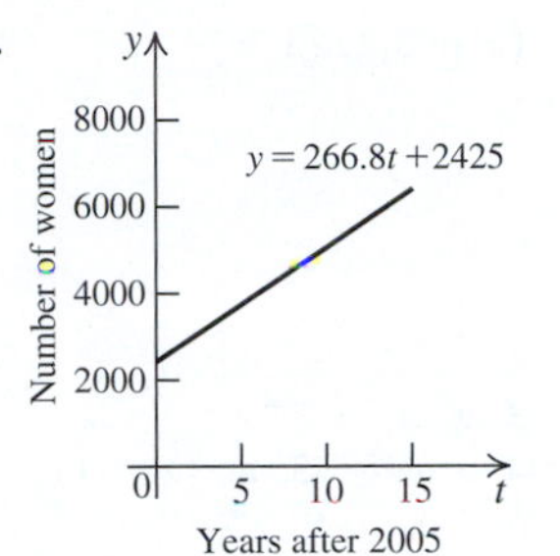

c. 3225 **d.** 5627

101. a. $y = 23.1x + 117.8$

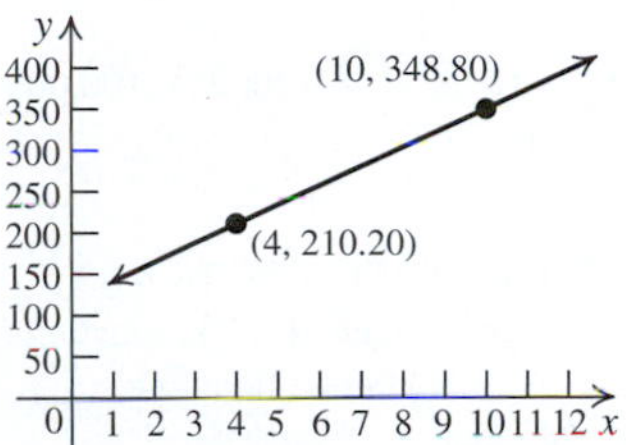

b. Slope: cost of producing one modem; y-intercept: fixed overhead cost **c.** 395 **103.** 9.98% **105. a.** $y = -2x + 12.4$

b.

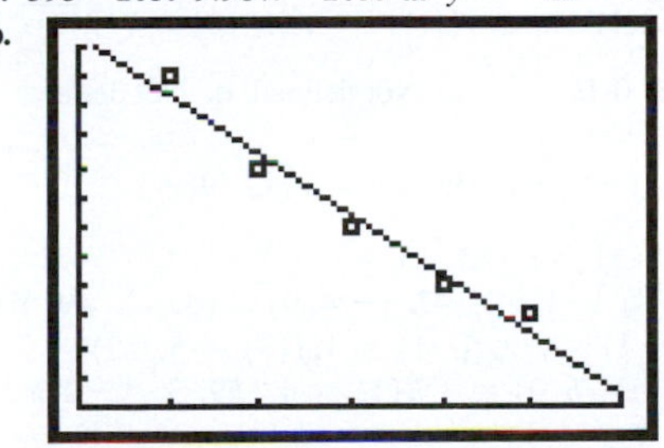

c. 0 papers

Beyond the Basics:

109. $c = -1$ **115.** $(9.5, 12)$ or $(-5.5, -8)$

117. $x^2 + y^2 + x - 2y - 41 = 0$ **119.** $y = \frac{4}{3}x - \frac{25}{3}$

121. Lines with slope 2 and different y-intercepts

123. Lines passing through the point $(1, 0)$ with different slopes

Critical Thinking/Discussion/Writing:

125. a. This is a family of lines parallel to the line $y = -2x$, and they all have slope equal to -2.

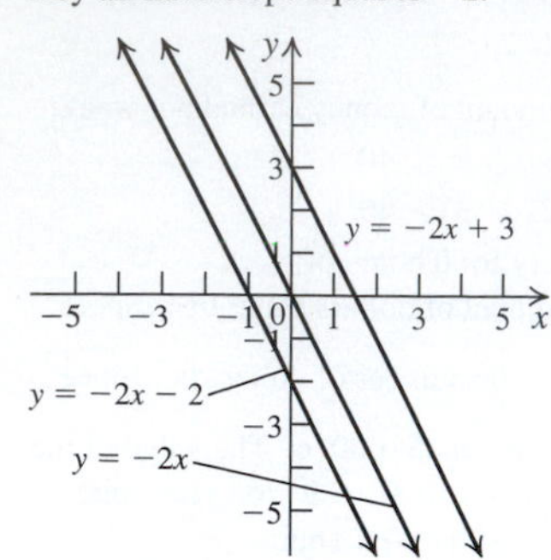

b. This is a family of all lines that pass through the point $(0, -4)$.

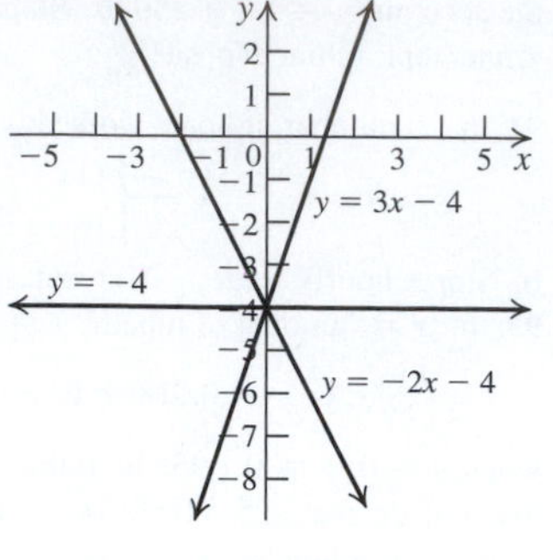

Maintaining Skills:

127. $\{-1, 2\}$ **129.** $\{-3, -2\}$ **131.** $(-\infty, 2] \cup [3, \infty)$

133. $-\dfrac{1}{x(x+h)}$ **135.** $\dfrac{1}{\sqrt{5}+\sqrt{2}}$

Section 2.4

Answers to Practice Problems:

1. a. Yes **b.** No **2. a.** No **b.** Yes **3. a.** 0 **b.** -7 **c.** $-2x^2 - 4hx + 5x - 2h^2 + 5h$ **4.** 44 sq. units **5. a.** $(-\infty, 1)$ **b.** $(-\infty, -2] \cup (3, \infty)$ **6. a.** No **b.** Yes **c.** $[0, 9]$ **7.** No **8. a.** Yes **b.** $-3, -1$ **c.** -5 **d.** $-5, 1$ **9.** Domain: $(-3, \infty)$; range: $(-2, 2] \cup \{3\}$ **10.** Range: $[6, 12)$; $C(11) = 11.994$ **11.** If c = cost per foot on land, then $C = c\left[1.3\sqrt{(500)^2 + x^2} + 1200 - x\right]$. **12. a.** $C = 12002 + 100{,}000$ **b.** $R = 2500x$ **c.** $P = 1300x - 100{,}000$ **d.** 77

Basic Concepts and Skills:

1. independent **3.** -14 **5.** $y = 0$ **7.** true **9.** Domain: $\{a, b, c\}$; range: $\{d, e\}$; function **11.** Domain; $\{a, b, c\}$; range: $\{1, 2\}$; function **13.** Domain: $\{0, 3, 8\}$: range: $\{-3, -2, -1, 1, 2\}$; not a function **15.** Yes **17.** Yes **19.** No **21.** Yes **23.** Yes **25.** No **27.** Yes **29.** $f(0) = 1, g(0)$ is not defined, $h(0) = \sqrt{2}, f(a) = a^2 - 3a + 1, f(-x) = x^2 + 3x + 1$ **31.** $f(-1) = 5, g(-1)$ is not defined, $h(-1) = \sqrt{3}, h(c) = \sqrt{2-c}$, $h(-x) = \sqrt{2+x}$ **33. a.** 0 **b.** $\dfrac{2\sqrt{3}}{3}$ **c.** Not defined **d.** Not defined **e.** $\dfrac{-2x}{\sqrt{4-x^2}}$ **35.** 9 **37.** $(-\infty, \infty)$ **39.** $(-\infty, 9) \cup (9, \infty)$ **41.** $(-\infty, -1) \cup (-1, 1) \cup (1, \infty)$ **43.** $[3, \infty)$ **45.** $(-\infty, -2) \cup (-2, -1) \cup (-1, \infty)$ **47.** $(-\infty, 0) \cup (0, \infty)$ **49.** Yes **51.** Yes **53.** No **55.** $f(-4) = -2, f(-1) = 1, f(3) = 5, f(5) = 7$ **57.** $h(-2) = -5, h(-1) = 4, h(0) = 3, h(1) = 4$ **59.** $x = -2$ or 3 **61. a.** No **b.** $x = -1 \pm \sqrt{3}$ **c.** 5 **d.** $x = -1 \pm \frac{1}{2}\sqrt{14}$ **63.** Domain: $[-3, 2]$; range: $[-3, 3]$ **65.** Domain: $[-4, \infty)$; range: $[-2, 3]$ **67.** Domain: $[-3, \infty)$; range: $\{-3\} \cup [-1, 4]$ **69.** Domain: $(-\infty, -4] \cup [-2, 2] \cup [4, \infty)$; range: $[-2, 2] \cup \{3\}$ **71.** $[-9, \infty)$ **73.** $-3, 4, 7, 9$ **75.** $f(-7) = 4, f(1) = 5, f(5) = 2$ **77.** $\{-3.75, -2.14, 3\} \cup [12, \infty)$ **79.** $[-9, \infty)$ **81.** $g(-4) = -1, g(1) = 3, g(3) = 4$ **83.** $[-9, -5)$

Applying the Concepts:

85. Yes, because there is only one high temperature every day. **87.** No, because Nevada and North Carolina both start with *N*. **89.** $A(x) = x^2$; $A(4) = 16$; $A(4)$ represents the area of a square tile with side of length 4. **91.** It is a function. $S(x) = 6x^2$; $S(3) = 54$ **93. a.** $[0, 8]$ **b.** $h(2) = 192, h(4) = 256, h(6) = 192$ **c.** 8 sec

d.

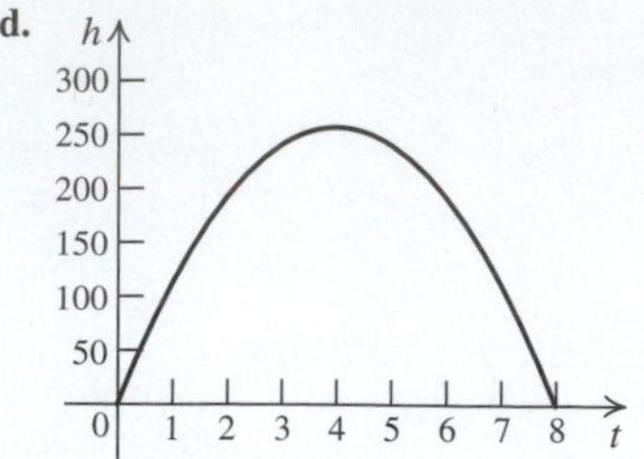

95. $P = 28x - x^2$ **97.** $S = 2x^2 + \dfrac{256}{x}$ **99.** $A = \dfrac{x^2}{4\pi} + \dfrac{1}{16}(20 - x)^2$ **101.** $C = 2x^2 + \dfrac{6912}{x}$ **103.** $d = [(x-2)^2 + (x^3 - 3x + 5)^2]^{1/2}$

105. a. $p(5) = 1150$; if 5000 TVs can be sold, the price per TV is \$1150. $p(15) = 900$; if 15,000 TVs can be sold, the price per TV is \$900. $p(30) = 525$; if 30,000 TVs can be sold, the price per TV is \$525.

b.

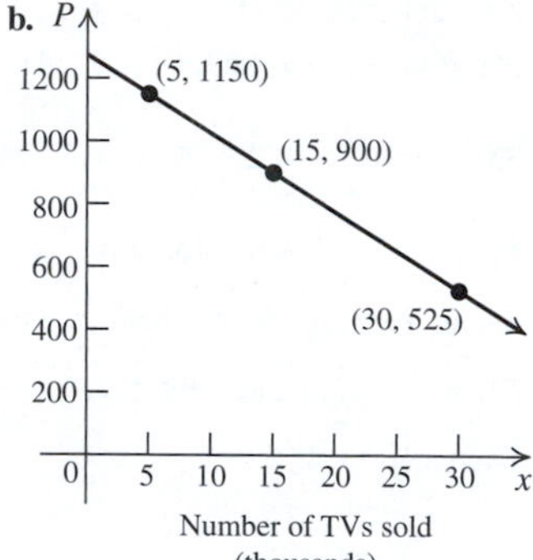

c. $p(25) = 650$; if the TVs are priced at \$650 each, 25,000 can be sold. **107. a.** $C(x) = 5.5x + 75{,}000$ **b.** $R(x) = 9x$ **c.** $P(x) = 3.5x - 75{,}000$ **d.** 21,429 **e.** \$86,000

Beyond the Basics:

109. $f(x) = \dfrac{4x+2}{x}$; domain: $(-\infty, 0) \cup (0, \infty)$; $f(4) = \dfrac{9}{2}$

111. $f(x) = \dfrac{2-x}{x^2+1}$; domain: $(-\infty, \infty)$; $f(4) = -\dfrac{2}{17}$

113. No, because they have different domains **115.** No, because g is not defined at -1 **117.** Yes **119.** 3 **121.** $a = -9, b = 6$

Critical Thinking/Discussion/Writing:

127. a. $ax^2 + bx + c = 0$ **b.** $y = c$ **c.** $b^2 - 4ac < 0$ **d.** Not possible **129. a.** 9 **b.** 8 **131.** $y = -x$ **133.** $y = \dfrac{1}{2}x + \dfrac{7}{2}$ **135. a.** $2x^2 + 3x$ **b.** $-2x^2 + 3x$ **137. a.** $-x^3 + 2x$ **b.** $-x^3 + 2x$

Section 2.5

Answers to Practice Problems:

1. Increasing on $(2, \infty)$, decreasing on $(-\infty, -3)$, and constant on $(-3, 2)$ **2.** Relative minimum at $(0, 0)$; relative maximum at $(1, 1)$ **3.** Mrs. Osborn's windpipe should be contracted to a radius of 7.33 mm for maximizing the airflow velocity.

4. $f(-x) = -(-x)^2 = -x^2 = f(x)$; therefore, f is even.

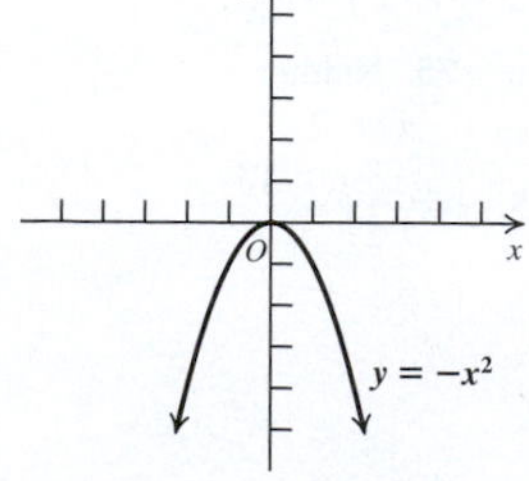

5. $f(-x) = -(-x^3) = x^3 = -f(x)$; therefore, f is odd.

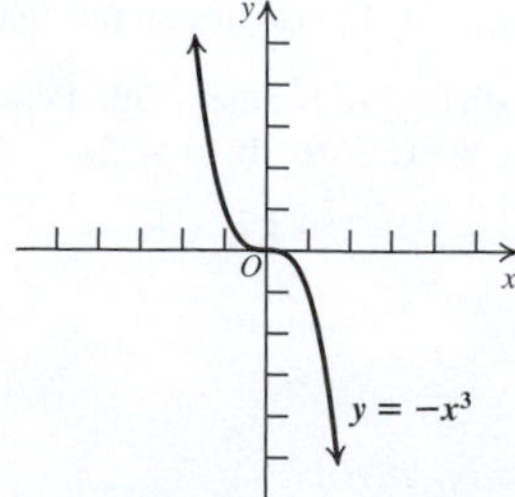

6. a. Even **b.** Odd **c.** Neither **7.** -6 **8.** -1 **9.** $-2x + 1 - h$

Basic Concepts and Skills:

1. > **3.** $f(-x) = f(x)$ for all x in the domain of f **5.** true **7.** Increasing on $(-\infty, \infty)$ **9.** Increasing on $(\infty, 2)$, decreasing on $(2, \infty)$ **11.** Increasing on $(-\infty, -2)$ and $(2, \infty)$, constant on $(-2, 2)$ **13.** Increasing on $(-\infty, -3)$ and $\left(-\frac{1}{2}, 2\right)$, decreasing on $\left(-3, -\frac{1}{2}\right)$ and $(2, \infty)$ **15.** No relative extrema **17.** $(2, 10)$ is a relative maximum point and a turning point. **19.** Any point $(x, 2)$ is a relative maximum and a relative minimum point on the interval $(-2, 2)$; none of these points are turning points. **21.** $(-3, 4)$ and $(2, 5)$ are relative maxima points and turning points; $\left(-\frac{1}{2}, -2\right)$ is a relative minimum and a turning point. **23.** Odd **25.** Neither **27.** Odd **29.** Even **31.** Neither **33.** Even **35.** Odd **37.** Neither **39.** Even **41.** Odd **43.** Even **45.** Neither **47. a.** Domain: $(-\infty, \infty)$; range: $(-\infty, 3]$ **b.** x-intercepts: $-3, 3$; y-intercept: 3 **c.** Increasing on $(-\infty, 0)$; decreasing on $(0, \infty)$ **d.** Relative maximum at $(0, 3)$ **e.** Even **49. a.** Domain: $(-3, 4)$; range: $[-2, 2]$ **b.** x-intercept: 1; y-intercept: -1 **c.** Constant on $(-3, -1)$ and $(3, 4)$; increasing on $(-1, 3)$ **d.** Any point $(x, -2)$ is a relative maximum and a relative minimum on $(-3, -1)$, and $(-1, -2)$ is a relative minimum. Any point $(x, 2)$ is a relative maximum and a relative minimum on $(3, 4)$, and $(3, 2)$ is a relative maximum. **e.** Neither **51. a.** Domain: $(-2, 4)$; range: $[-2, 3]$ **b.** x- and y-intercept: 0 **c.** Decreasing on $(-2, -1)$ and $(3, 4)$; increasing on $(-1, 3)$ **d.** Relative maximum at $(3, 3)$; relative minimum at $(-1, -2)$ **e.** Neither **53. a.** Domain: $(-\infty, \infty)$; range: $(0, \infty)$ **b.** x-intercepts: none; y-intercept: 1 **c.** Increasing on $(-\infty, \infty)$ **d.** None **e.** Neither **55.** -2 **57.** 3 **59.** -2 **61.** -2 **63.** 7 **65.** $-\frac{1}{12}$ **67.** 2 **69.** $-x - 1$ **71.** $3x + 7$ **73.** $-\frac{4}{x}$ **75.** 1 **77.** -2 **79.** m **81.** $2x + h$ **83.** $4x + 2h + 3$ **85.** 0 **87.** $-\frac{1}{x(x + h)}$

Applying the Concepts:

89. $[0, \infty)$; the particle's motion is tracked indefinitely from time $t = 0$. **91.** The graph is above the t-axis on the intervals $(0, 9)$ and $(21, 24)$; the particle is moving forward between 0 and 9 sec and again between 21 and 24 sec. **93.** The particle is moving with increasing speed forward on $(0, 3)$, $(5, 6)$, and $(21, 23)$ and moving backward on $(11, 15)$. **95.** Maximum speed is attained between times $t = 15$ and $t = 16$. **97.** The particle is moving forward with increasing velocity. **99. b.** $(0, 6)$ **c.** $[0, 128]$ **d.** When $x = 2$, V has a maximum value of 128. **101. a.** $C(x) = 210x + 10{,}500$ **b.** \$21,000 **c.** \$420 **d.** 100 **103. a.** 4 ft from the origin **b.** 28 ft **c.** 6 ft/sec **d.** 10 ft/sec

Beyond the Basics:

107. $(-1, 5)$; no **109.** $\frac{1}{\sqrt{x + h} + \sqrt{x}}$ **111.** $\frac{1}{\sqrt{x(x + h)}(\sqrt{x} + \sqrt{x + h})}$

Critical Thinking/Discussion/Writing:

113. f has a relative maximum (minimum) at $x = a$ if there is an interval $[a, x_1)$ with $a < x_1 < b$ such that $f(a) \ge f(x)$ $(f(a) \le f(x))$ for every x in the interval. **115. a.** $f(x) = x$ on the interval $[-1, 1]$

b. $f(x) = \begin{cases} x & \text{if } 0 \le x < 1 \\ 0 & \text{if } x = 1 \end{cases}$ **c.** $f(x) = \begin{cases} x & \text{if } 0 < x \le 1 \\ 1 & \text{if } x = 0 \end{cases}$

d. $f(x) = \begin{cases} 0 & \text{if } x = 0 \text{ or } x = 1 \\ 1 & \text{if } 0 < x < 1 \text{ and } x \text{ is rational} \\ -1 & \text{if } 0 < x < 1 \text{ and } x \text{ is irrational} \end{cases}$

Maintaining Skills:

117. $y = -3x + 20$ **119.** (*i*) $2\sqrt{2}$, (*ii*) 8, (*iii*) $-8i$

Section 2.6

Answers to Practice Problems:

1. $g(x) = 2x + 6$ **2.** 15.524 m **3.** Domain: $(-\infty, 0]$; range: $[0, \infty)$

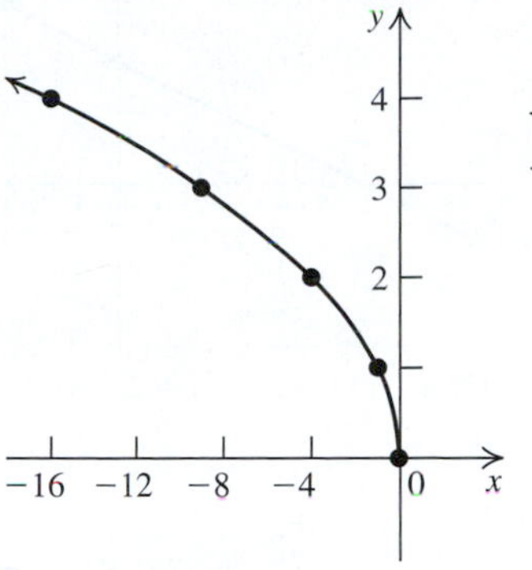

4. Domain: $(-\infty, \infty]$; range: $[-\infty, \infty)$

5. $f(-2) = 4, f(3) = 6$

6. a. $f(x) = \begin{cases} 50 + 4(x - 55), 56 \le x < 75 \\ 200 + 5(x - 75), x \ge 75 \end{cases}$ **b.** \$70 **c.** \$275

7. $f(x) = -3x, x \le -1$ $f(x) = 2x, x > -1$

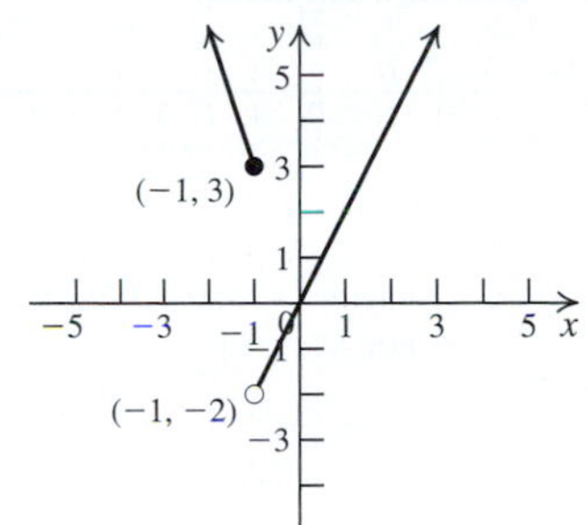

8. $f(x) = \begin{cases} -x - 4 & \text{if} & x \le -2 \\ \frac{4}{5}x - \frac{2}{5} & \text{if} & -2 < x < 3 \\ -2x + 8 & \text{if} & x \ge 3 \end{cases}$

9. $[\![-3.4]\!] = -4$; $[\![4.7]\!] = 4$

Basic Concepts and Skills:

1. horizontal **3.** 3

5. $f(x) = x + 1$ **7.** $f(x) = 2x + 3$

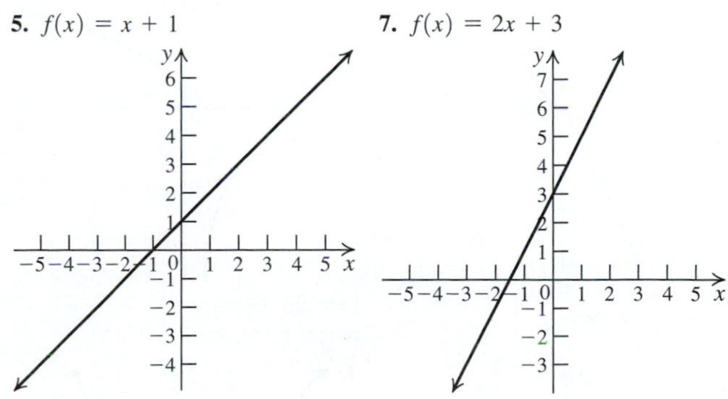

9. $f(x) = -3x + 4$

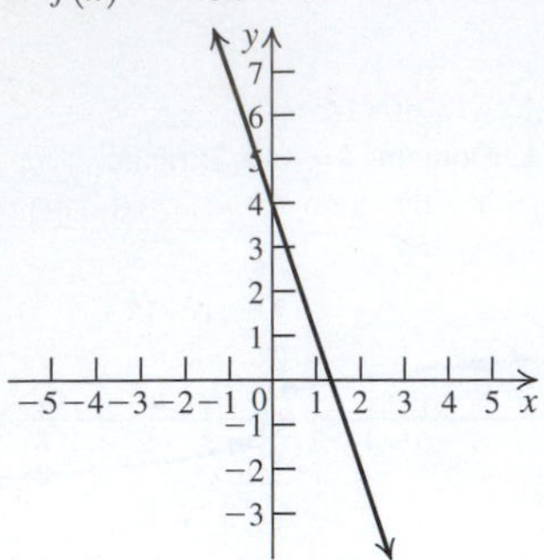

11. $f(x) = \frac{1}{2}x + 3$

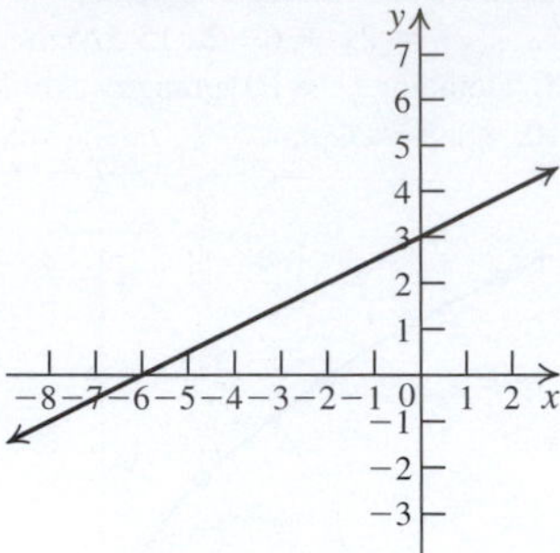

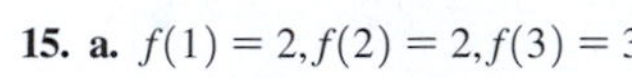

13. $f(x) = -\frac{2}{3}x - 1$

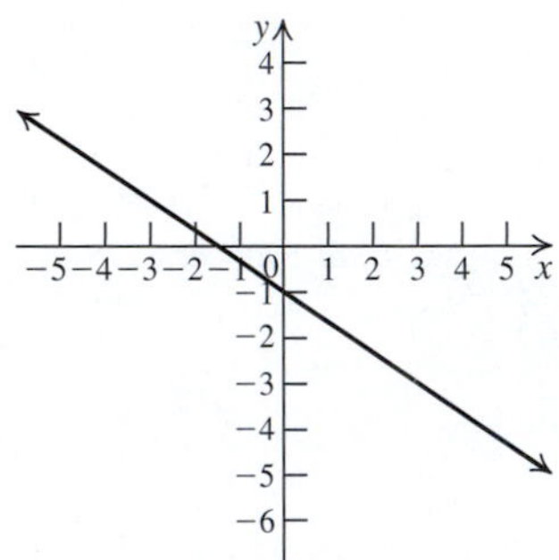

15. a. $f(1) = 2, f(2) = 2, f(3) = 3$

b.

17. a. $f(-15) = -1, f(12) = 1$

b.

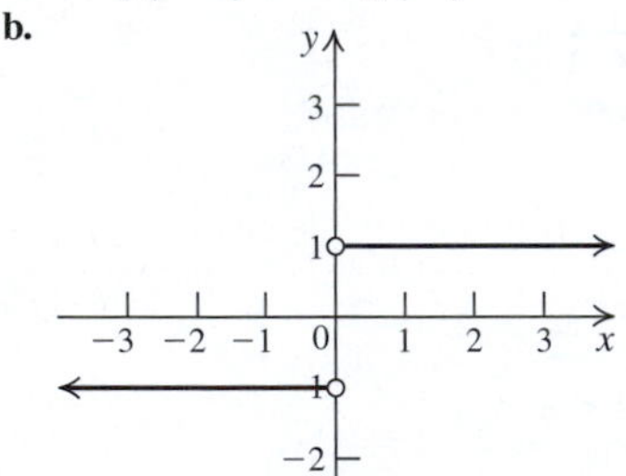

c. Domain: $(-\infty, 0) \cup (0, \infty)$; range: $\{-1, 1\}$

19.

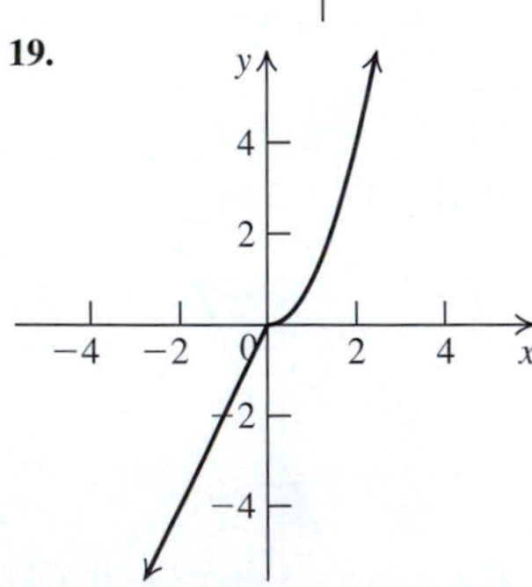

Domain: $(-\infty, \infty)$; range: $(-\infty, \infty)$

21.

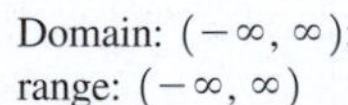

Domain: $(-\infty, \infty)$; range: $(-\infty, \infty)$

23.

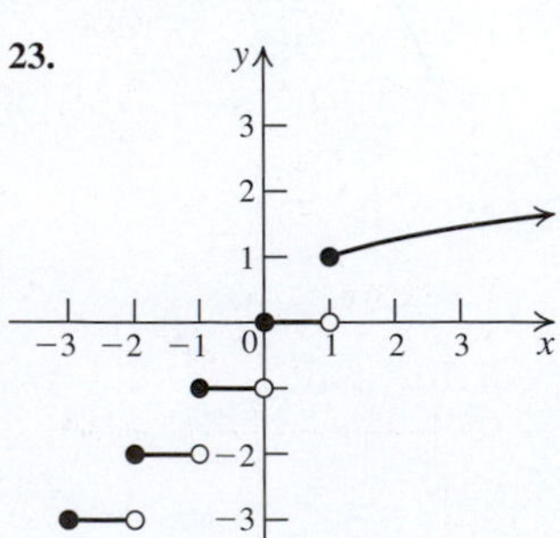

Domain: $(-\infty, \infty)$; range: $\{\dots, -3, -2, -1, 0\} \cup [1, \infty)$

25.

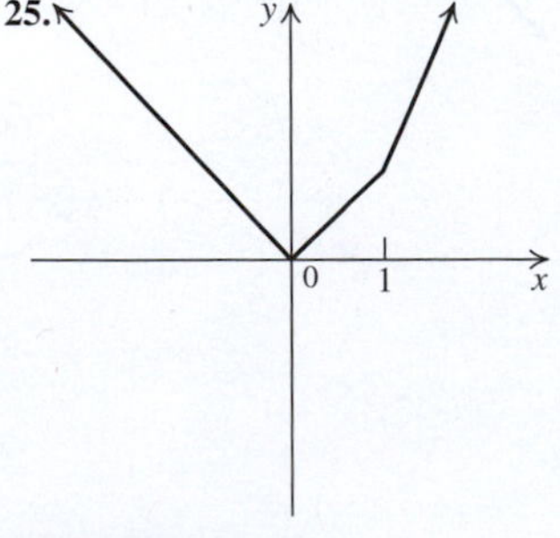

Domain: $(-\infty, \infty)$; range: $[0, \infty)$

27.

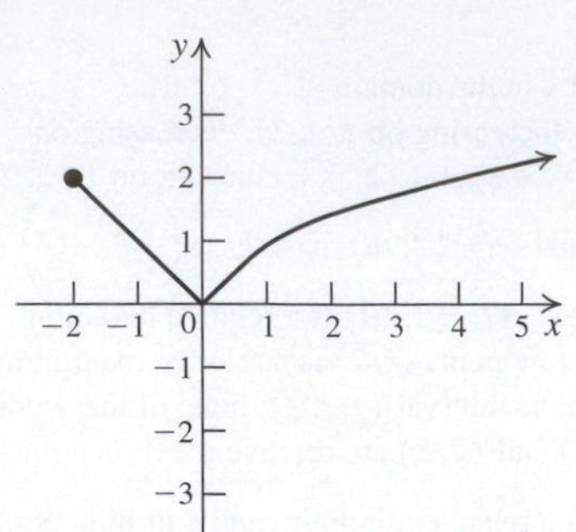

Domain: $[-2, \infty)$; range: $[0, \infty)$

29. **31.**

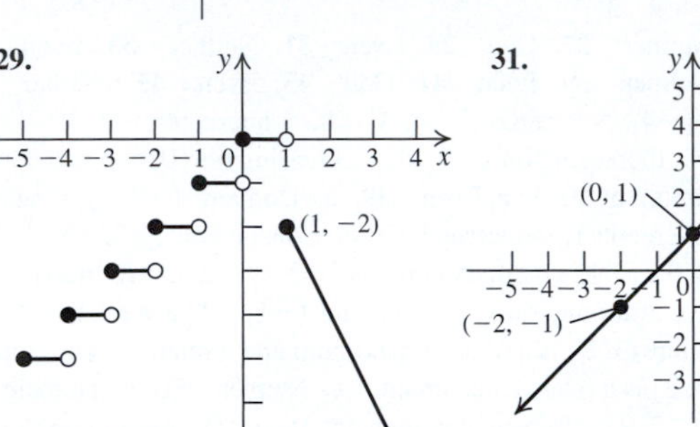

29. Domain: $(-\infty, 4]$; range: $\{\dots, -3, -2, -1, 0\} \cup [-8, -2]$

31. Domain: $(-\infty, \infty)$; range: $(-\infty, 2)$

33. $f(x) = \begin{cases} x + 1 & \text{if } x < 2 \\ -3x + 9 & \text{if } x \geq 2 \end{cases}$

35. $f(x) = \begin{cases} 2x + 6 & \text{if } x < -2 \\ 4 & \text{if } -2 \leq x < 2 \\ -x + 3 & \text{if } x \geq 2 \end{cases}$

Applying the Concepts:

37. a. $f(x) = \frac{1}{33.81}x$; domain: $[0, \infty)$; range: $[0, \infty)$ **b.** 0.0887; it means that 3 oz $=$ 0.0887 liters. **c.** 0.3549 **39. a.** No d-intercept because $d \geq 0$; y-intercept: 1. It means that the pressure at sea level $(d = 0)$ is 1 atm. **b.** $P(0) = 1, P(10) \approx 1.3, P(33) = 2, P(100) \approx 4.03$ **c.** 132 ft

41. a. $C = 50x + 6000$

b. The y-intercept is the fixed overhead cost.

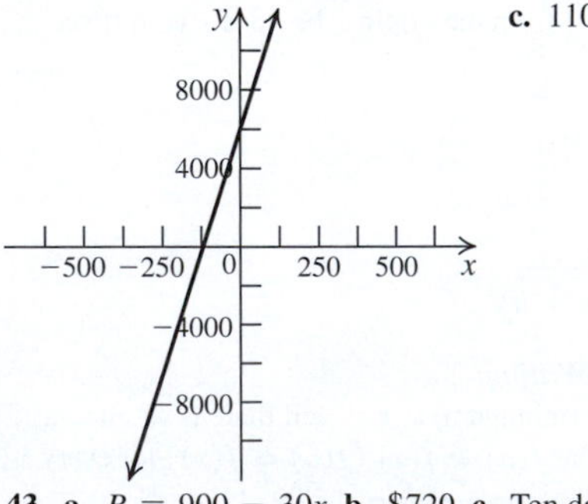

c. 110

43. a. $R = 900 - 30x$ **b.** \$720 **c.** Ten days after the first of the month

45. 70 **47. a.** $y = \frac{2}{27}(x - 150) + 30$ **b.** 45 **c.** 690 mg/m^3

49. a.

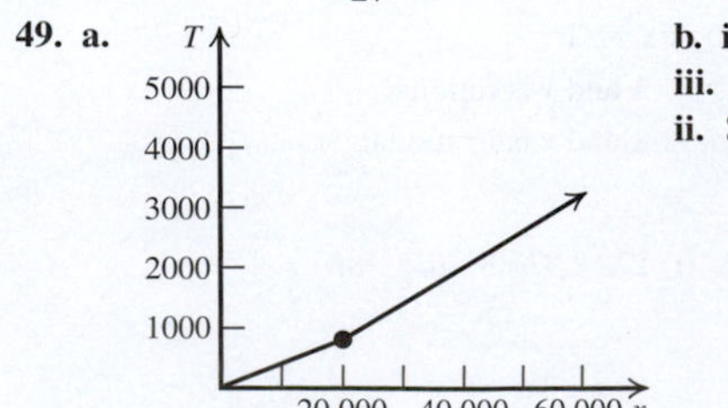

b. i. \$480 **ii.** \$800 **iii.** \$2600 **c. i.** \$15,000 **ii.** \$26,667 **iii.** \$45,000

Beyond the Basics:

51. a. i. -1 ii. -1 iii. -1 **b.** $x = \frac{1}{2}$ **c.**

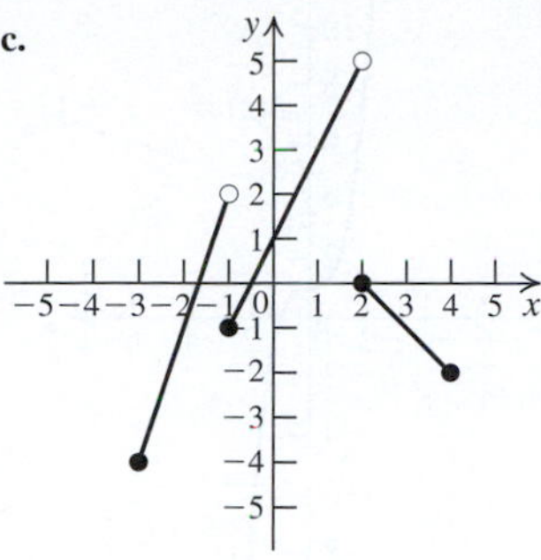

53. a. Domain: $(-\infty, \infty)$; range $[0, 1)$ **b.** Increasing on $(n, n + 1)$ for every integer n **c.** Neither **55.** 2

Critical Thinking/Discussion/Writing:

57. a. $C(x) = 17(f(x) - 1) + 44$

b.

Cost ($); Weight (oz)

c. Domain: $(0, \infty)$; range: $\{17n + 44 : n \text{ a nonnegative integer}\}$

59. a. $C(x) = \begin{cases} 150, & \text{if } x \leq 100 \\ -0.2[\![100 - x]\!] + 150, & \text{if } x > 100 \end{cases}$

b.

Cost; Miles driven

c. $(299, 300]$

Maintaining Skills:

61. $f(x) + 3$ **63.** $f(-x)$

65.

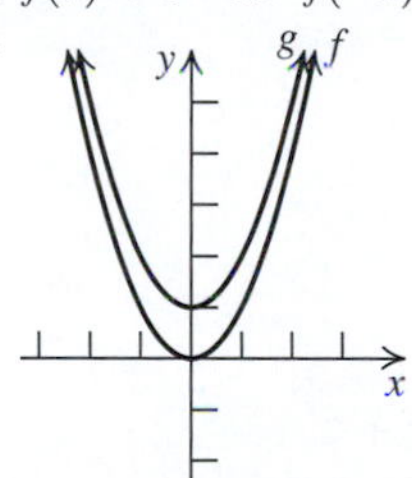

67.

Section 2.7

Answers to Practice Problems:

1. The graph of g is the graph of f shifted 1 unit up; the graph of h is the graph of f shifted 2 units down.

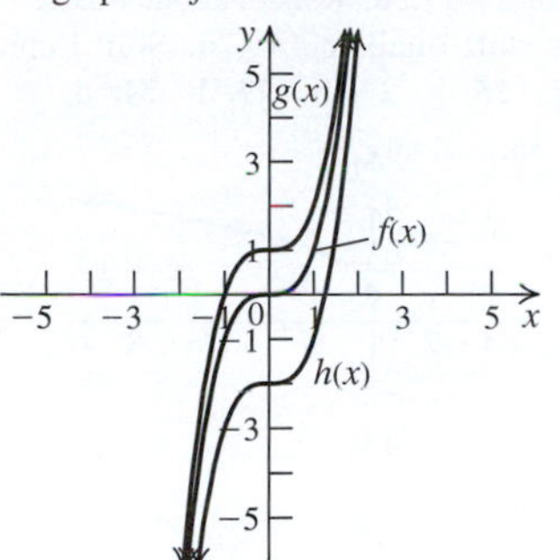

2. The graph of g is the graph of f shifted 1 unit to the right; the graph of h is the graph of f shifted 2 units to the left.

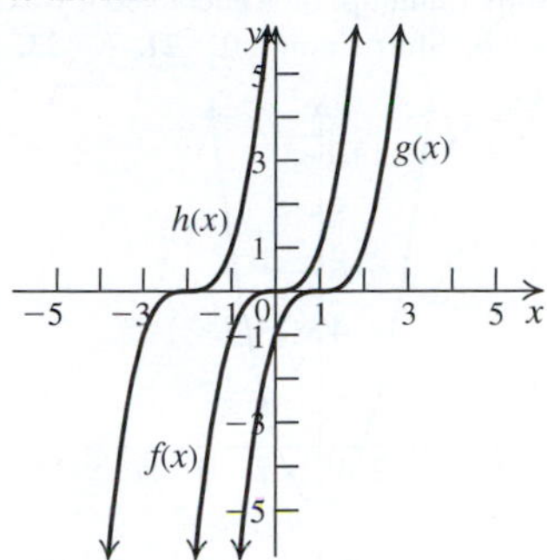

3.

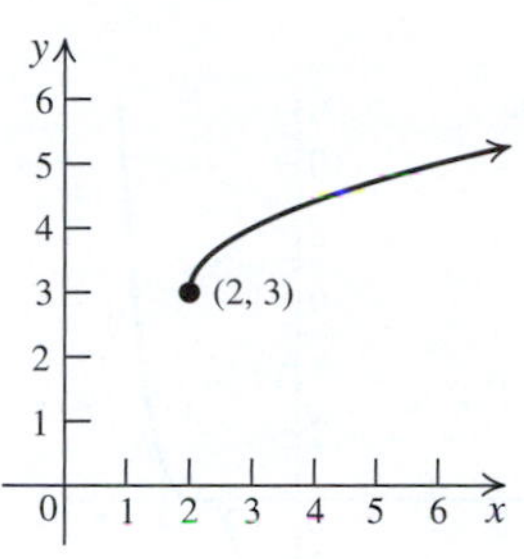

4. Shift the graph of $y = x^2$ to the right 1 unit, reflect the resulting graph about the x-axis, and shift it up 2 units.

5.

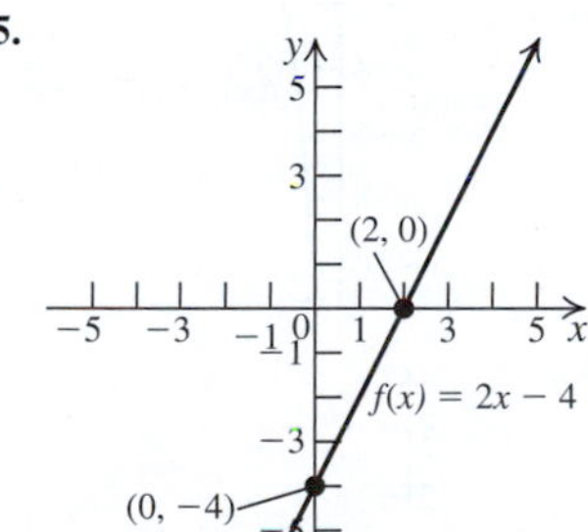

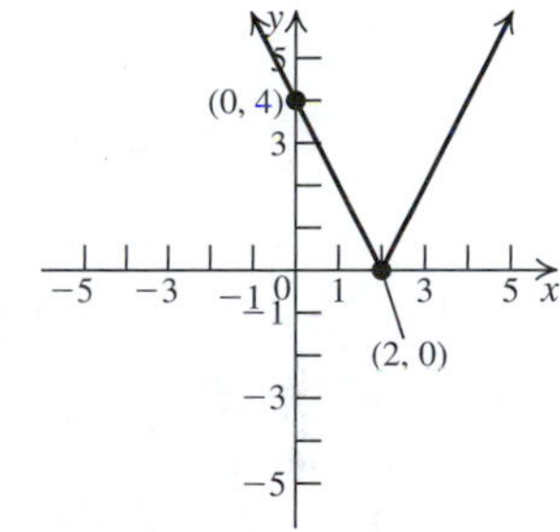

6. The graph of g is the graph of f vertically stretched by multiplying each of its y-coordinates by 2.

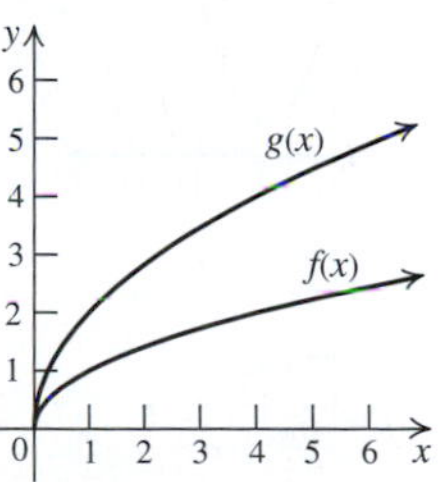

7. a.

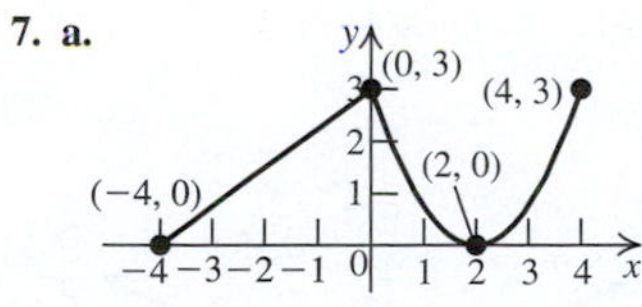

b.

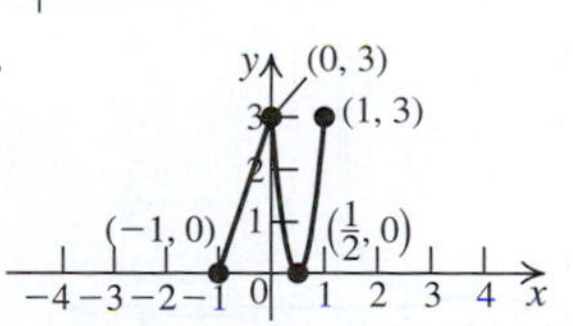

8.

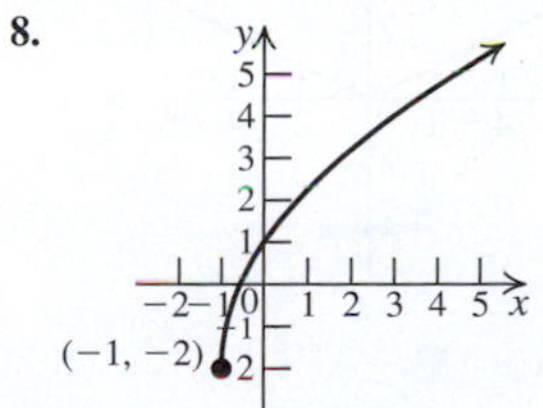

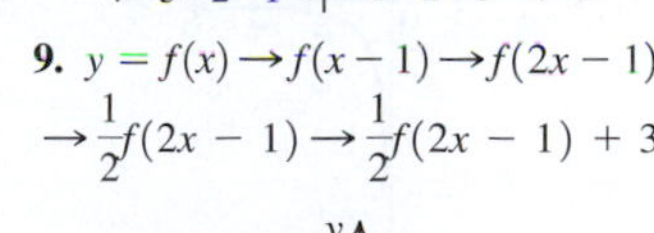

9. $y = f(x) \rightarrow f(x - 1) \rightarrow f(2x - 1) \rightarrow \frac{1}{2}f(2x - 1) \rightarrow \frac{1}{2}f(2x - 1) + 3$

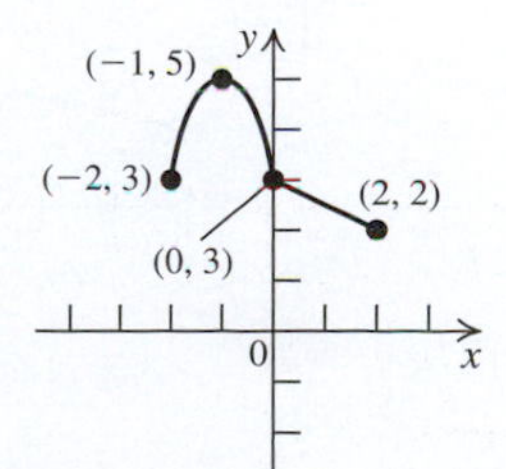

Basic Concepts and Skills:

1. down **3.** >1 **5.** false **7. a.** Shift 2 units up **b.** Shift 1 units down **9. a.** Shift 1 units left **b.** Shift 2 units right **11. a.** Shift 1 unit left and 2 units down **b.** Shift 1 unit right and 3 units up **13. a.** Reflect about x-axis **b.** Reflect about y-axis **15. a.** Stretch vertically by a factor of 2 **b.** Compress horizontally by a factor of 2 **17. a.** Reflect about x-axis, shift 1 unit up **b.** Reflect about y-axis, shift 1 unit up **19. a.** Shift 1 unit up **b.** Shift 1 unit left **21.** e **23.** g **25.** i **27.** b **29.** l **31.** d

33. $f(x) = x^2 - 2$

35.

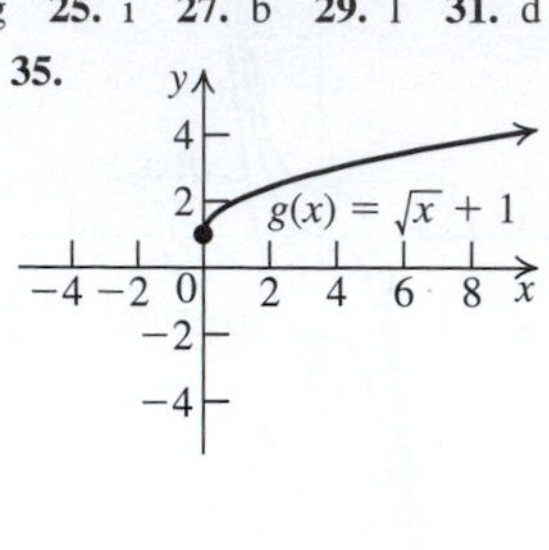

37.

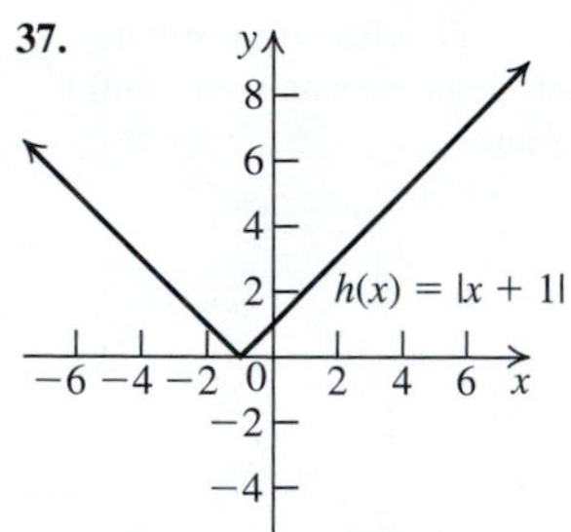

39.

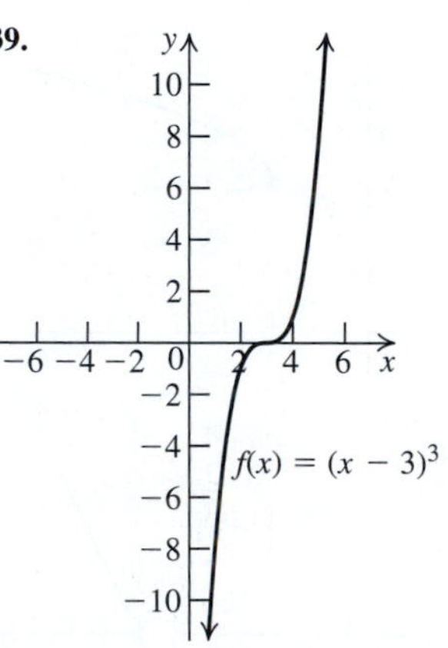

41. $g(x) = (x - 2)^2 + 1$

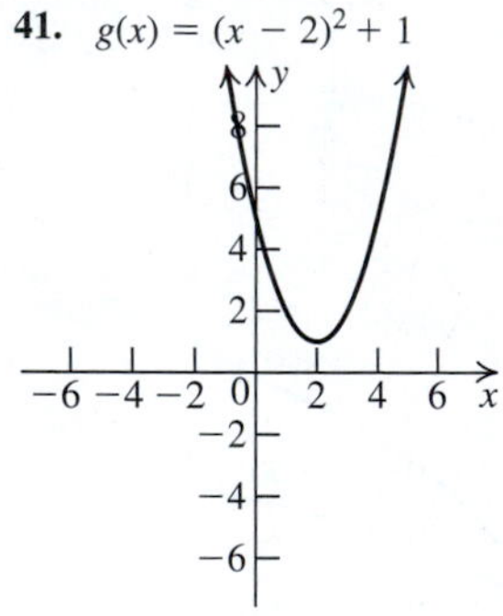

43.

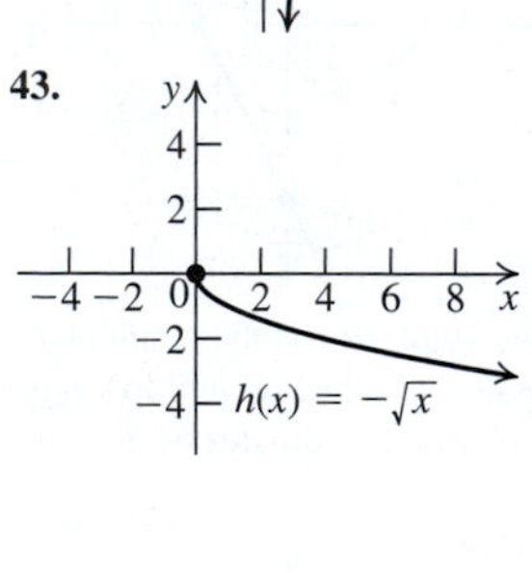

45.

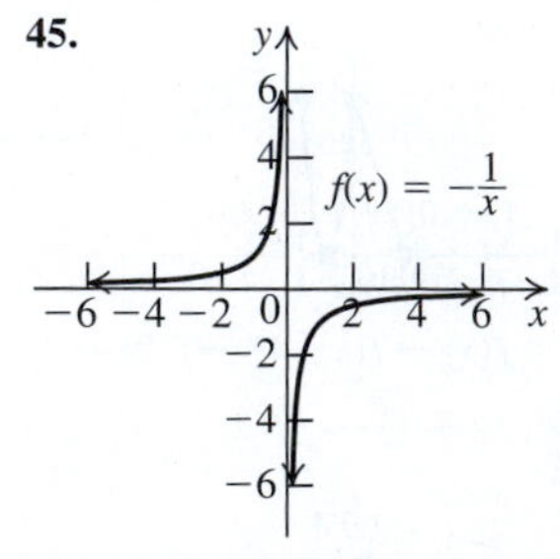

47.

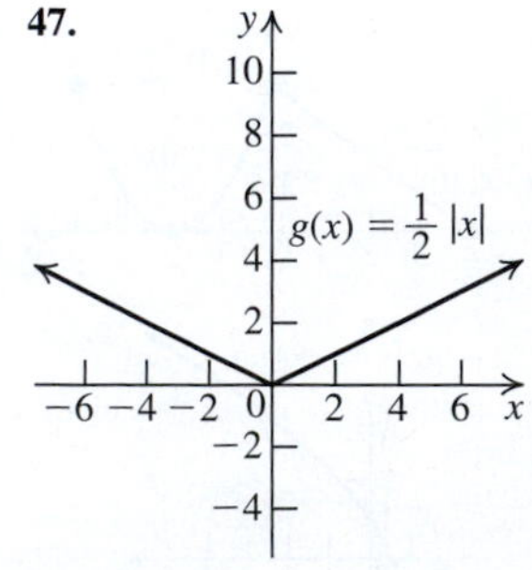

49.

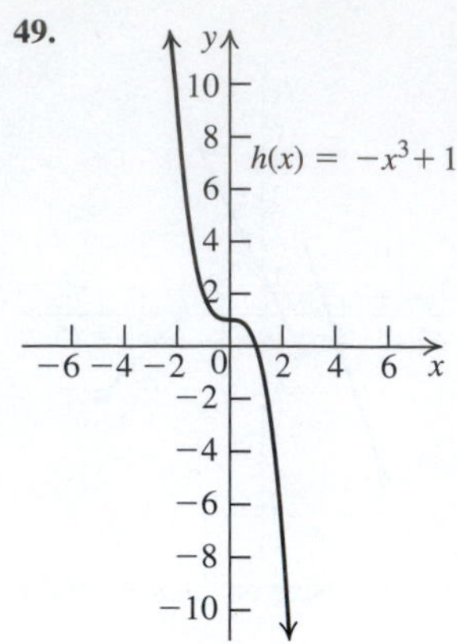

51.

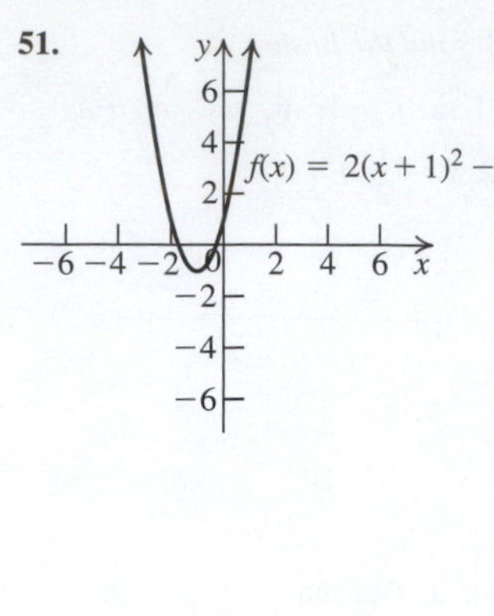

53.

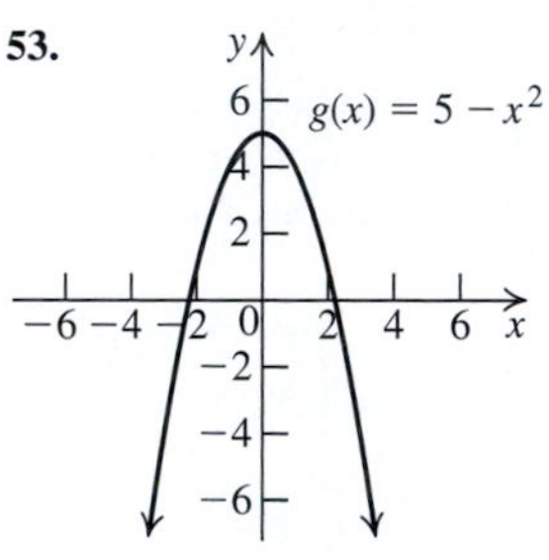

55.

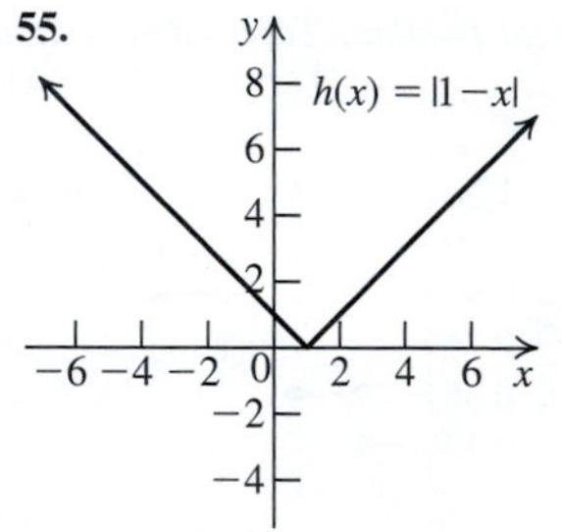

57.

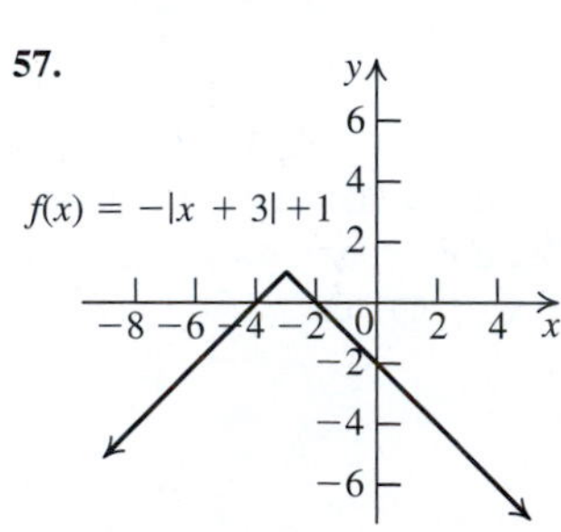

59.

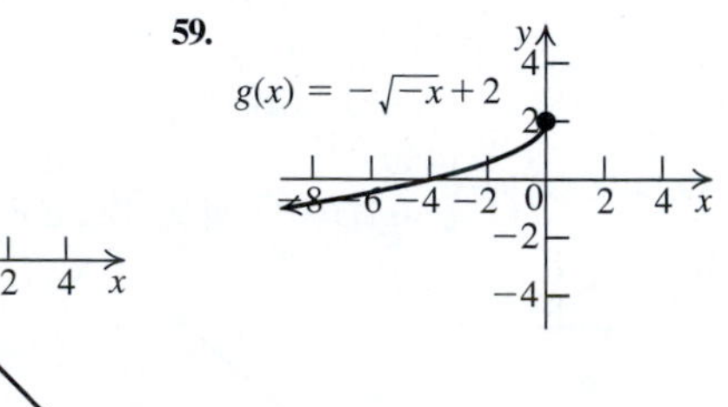

61. $h(x) = 2[[x + 1]]$

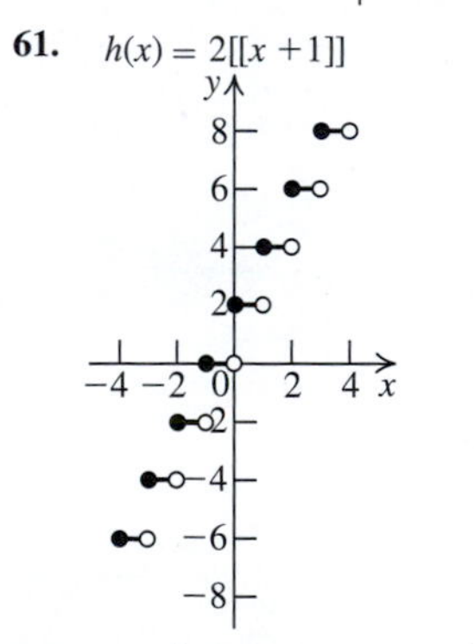

63. $y = x^3 + 2$ **65.** $y = -|x|$ **67.** $y = (x - 3)^2 + 2$

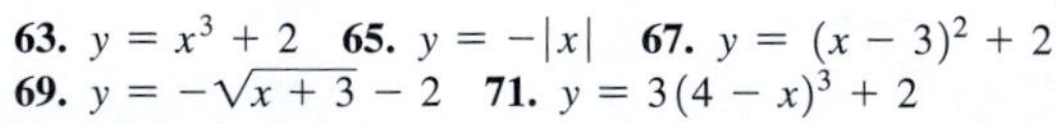

69. $y = -\sqrt{x + 3} - 2$ **71.** $y = 3(4 - x)^3 + 2$

73. $y = -|2x - 4| - 3$

75.

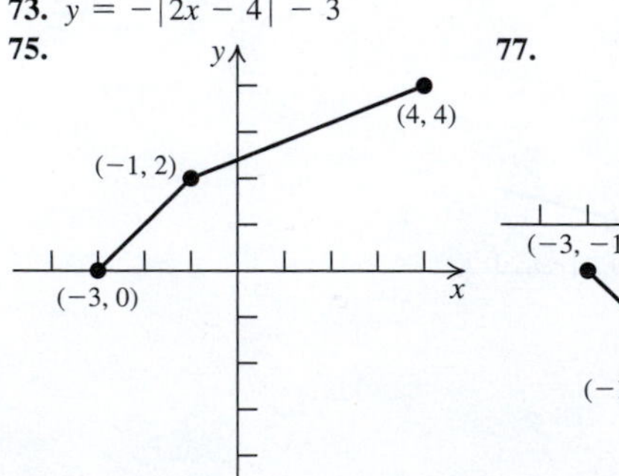

77.

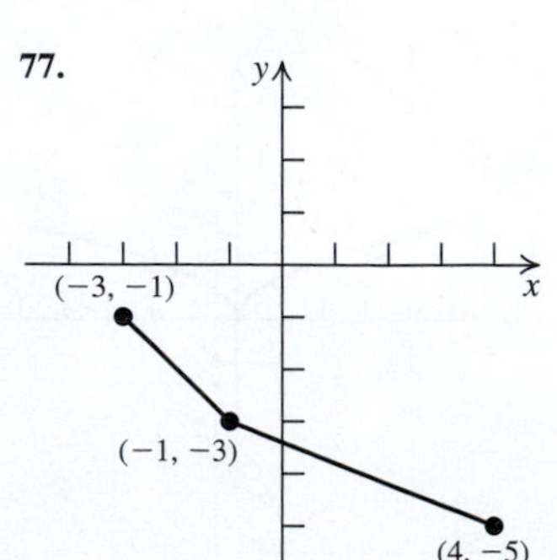

79.

$\left(-\frac{3}{2}, 1\right)$, $\left(-\frac{1}{2}, 3\right)$, (2, 5)

81.

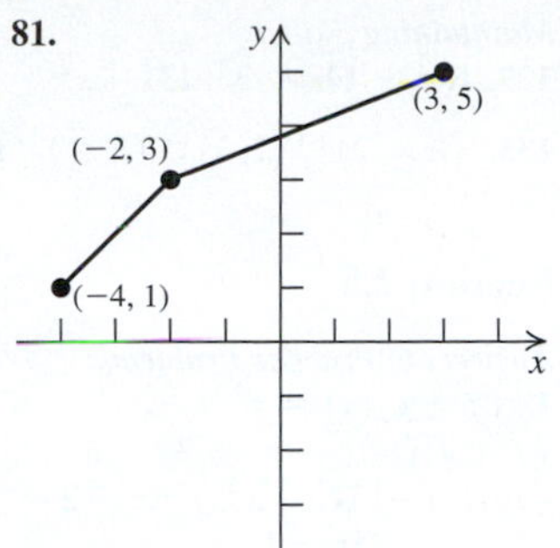

83.

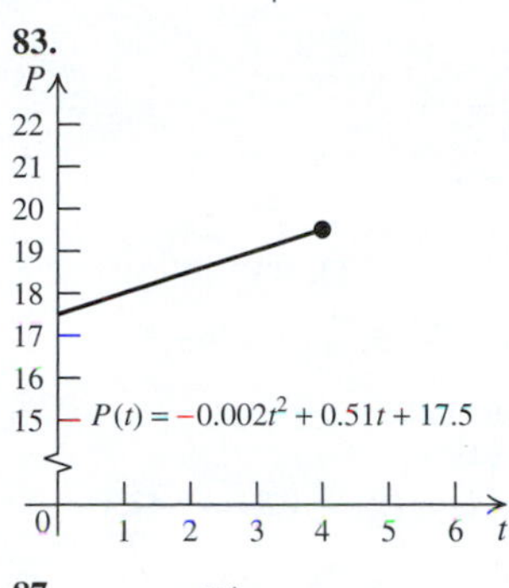

85.

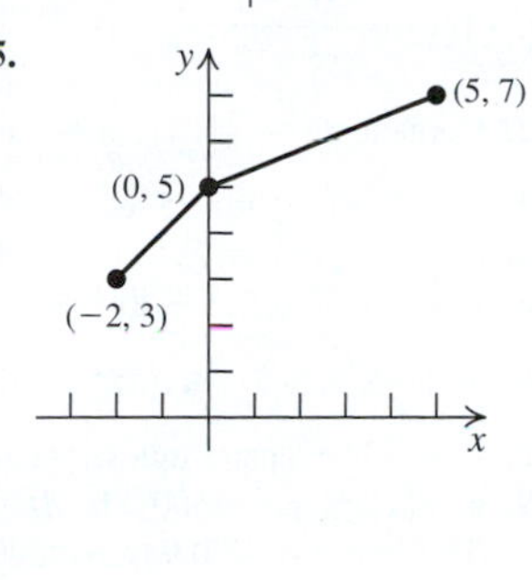

87.

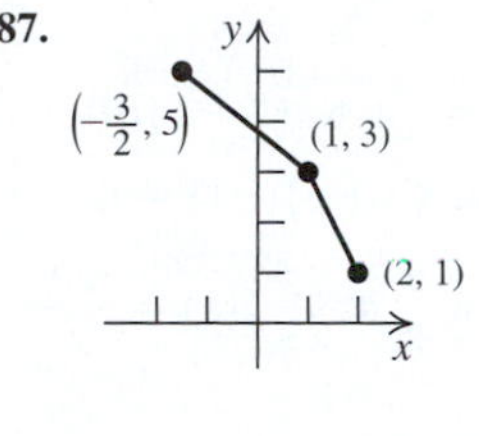

89. a.

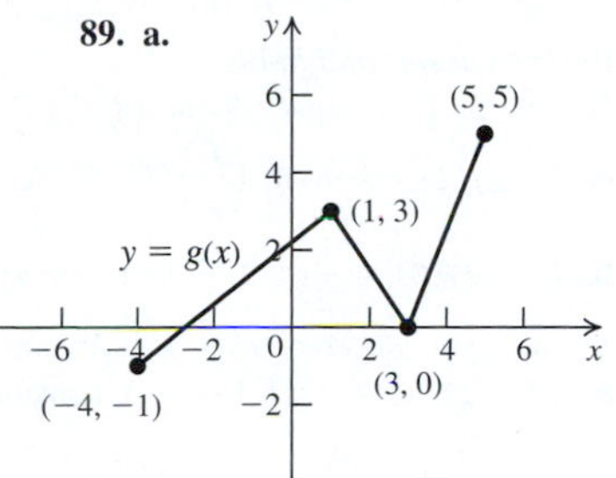

b.

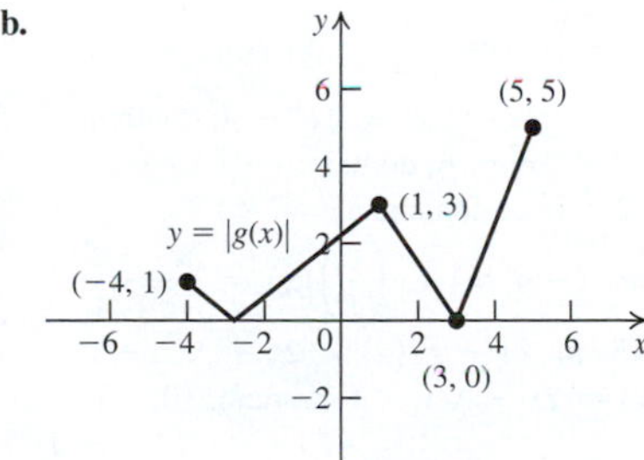

91. a.

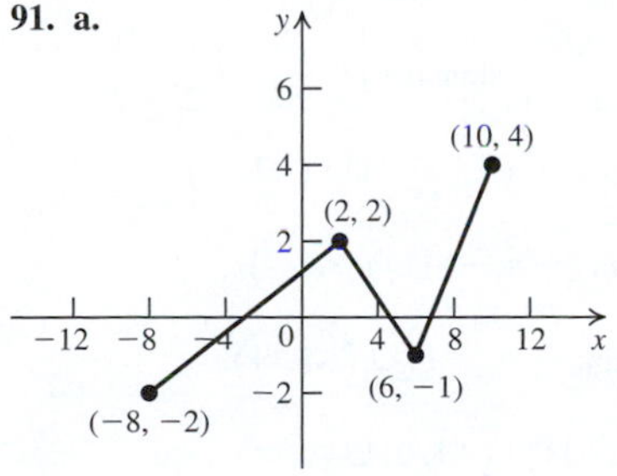

b.

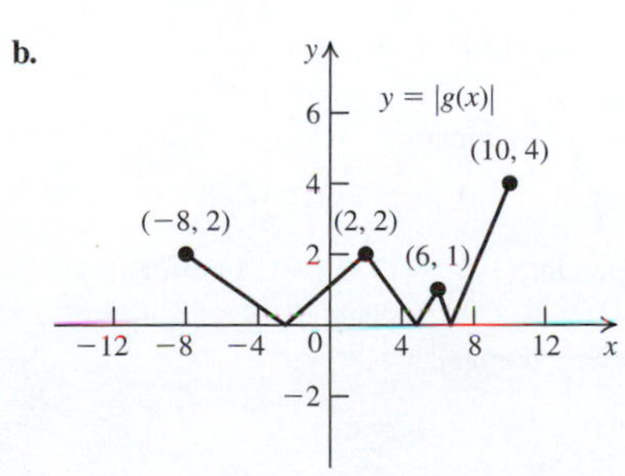

93. a. **b.**

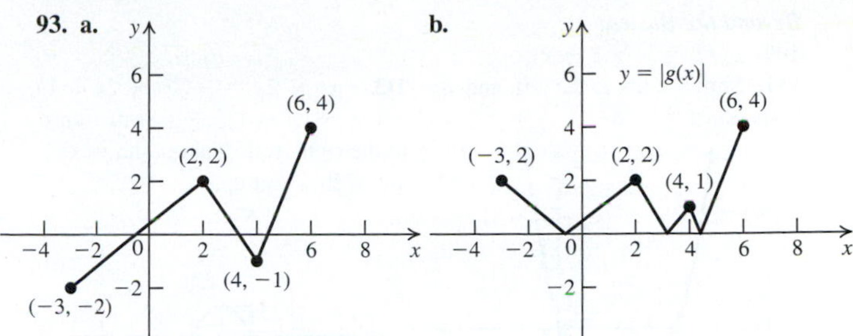

95. a. **b.**

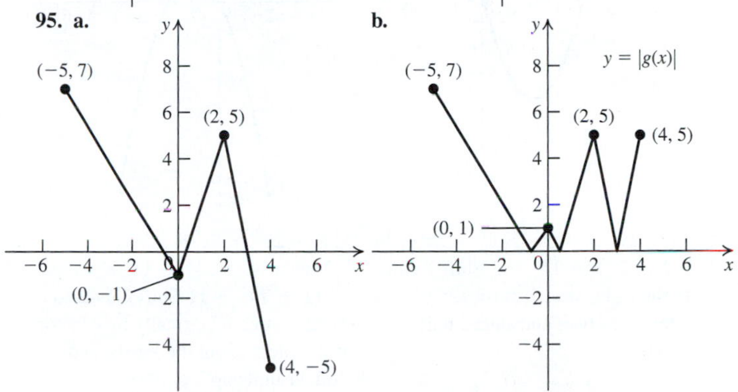

97. a. **b.**

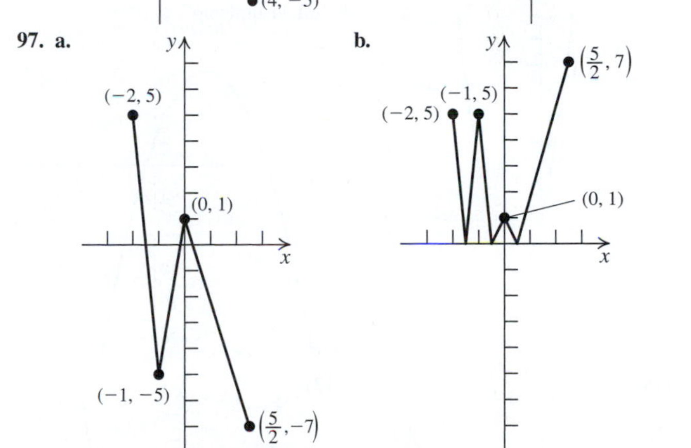

Applying the Concepts:

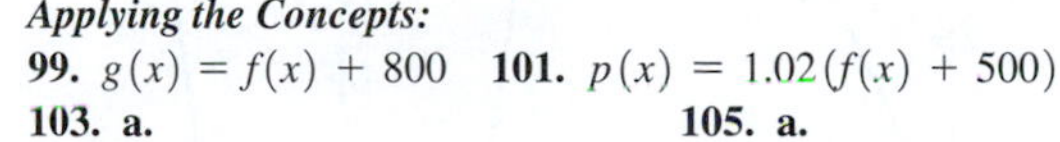

99. $g(x) = f(x) + 800$ **101.** $p(x) = 1.02(f(x) + 500)$

103. a.

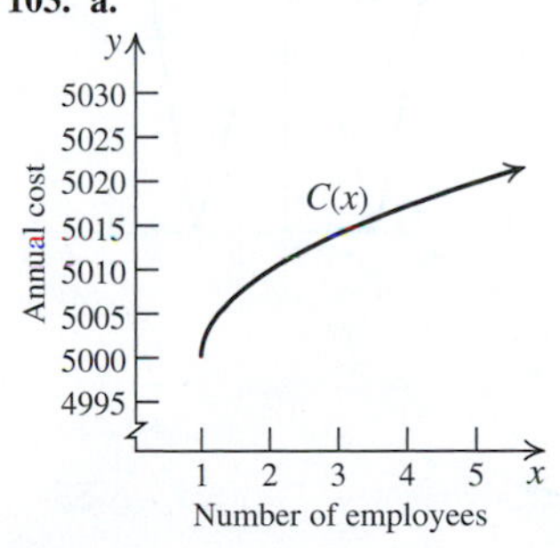

b. \$5199.75

105. a.

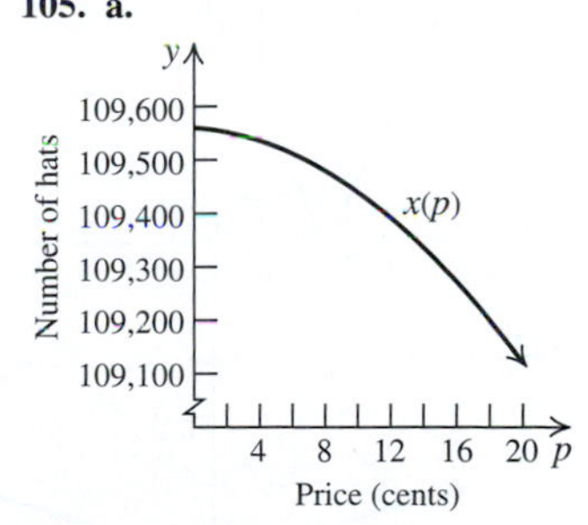

b. \$2 **c.** \$3.30

107. The first coordinate gives the month. The second coordinate gives the hours of daylight.

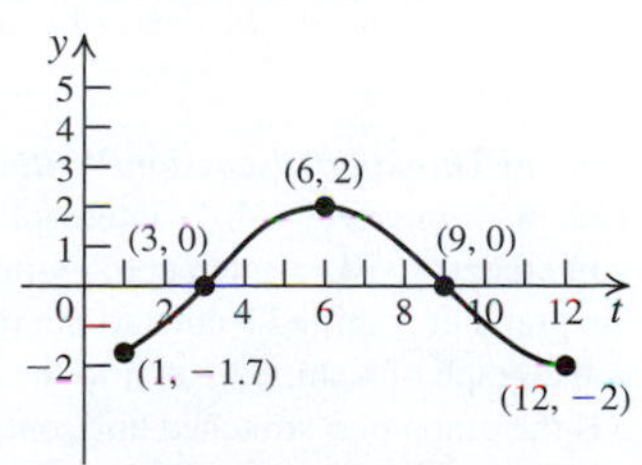

Beyond the Basics:

109. $g(x) = -\sqrt{1-x}$

111. Shift 2 units to the left, and 4 units down.

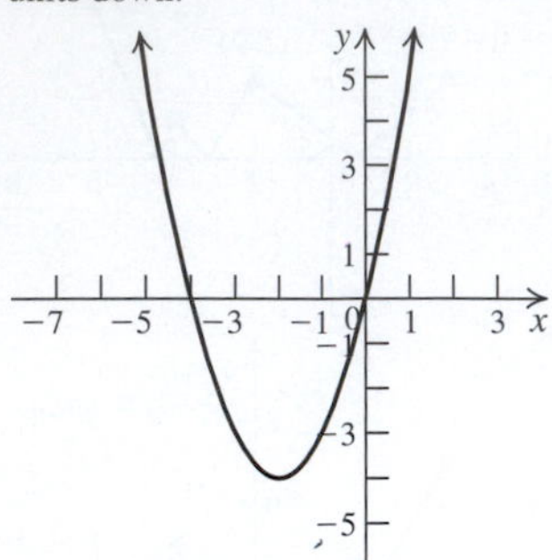

113. $-x^2 + 2x = -(x^2 - 2x + 1) + 1 = -(x-1)^2 + 1$ Shift 1 unit to the right, reflect about the x-axis, and shift 1 unit up.

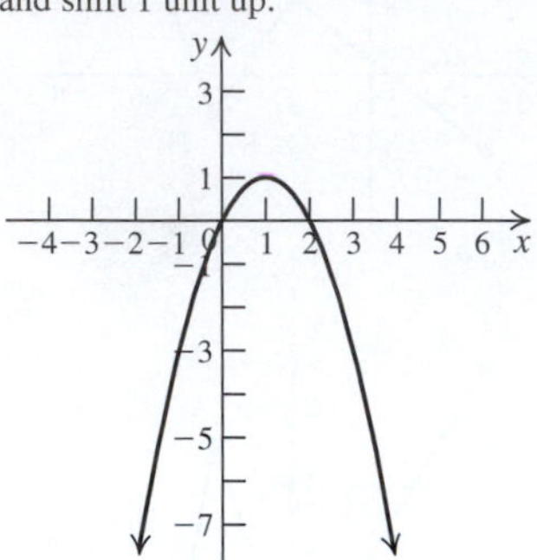

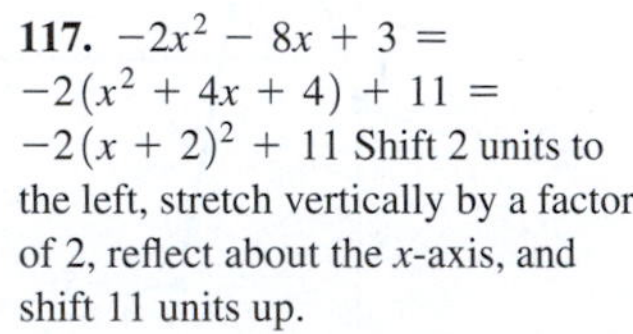

115. $2x^2 - 4x = 2(x^2 - 2x + 1) - 2 = 2(x-1)^2 - 2$ Shift 1 unit to the right, stretch vertically by a factor of two, and shift 2 units down.

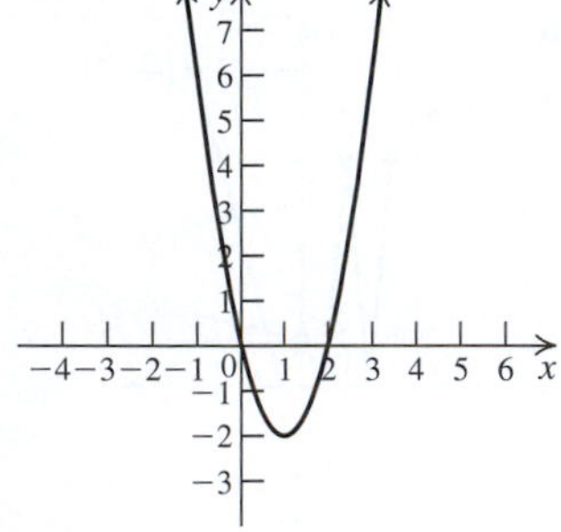

117. $-2x^2 - 8x + 3 = -2(x^2 + 4x + 4) + 11 = -2(x+2)^2 + 11$ Shift 2 units to the left, stretch vertically by a factor of 2, reflect about the x-axis, and shift 11 units up.

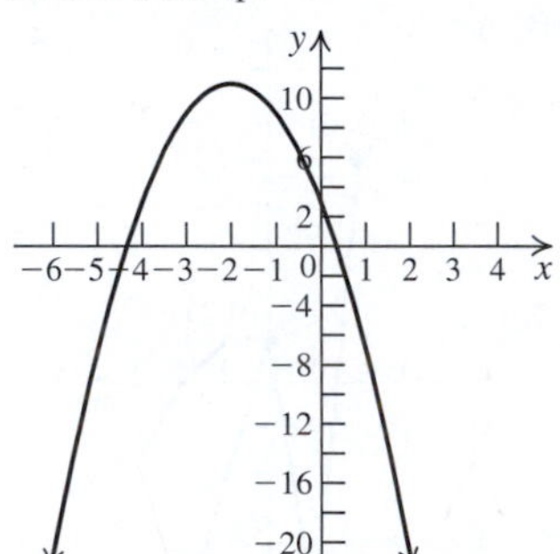

119.

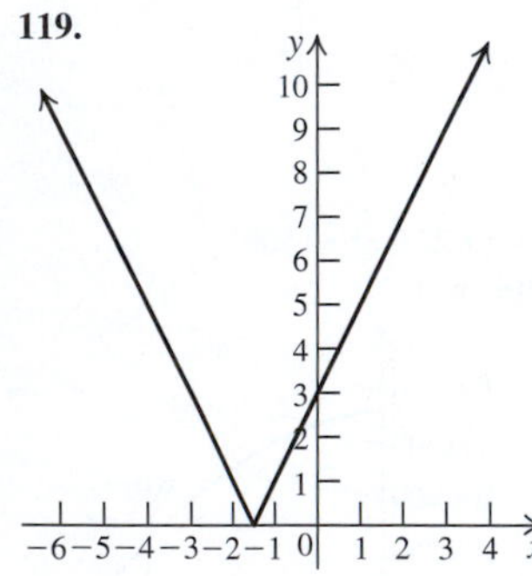

121.

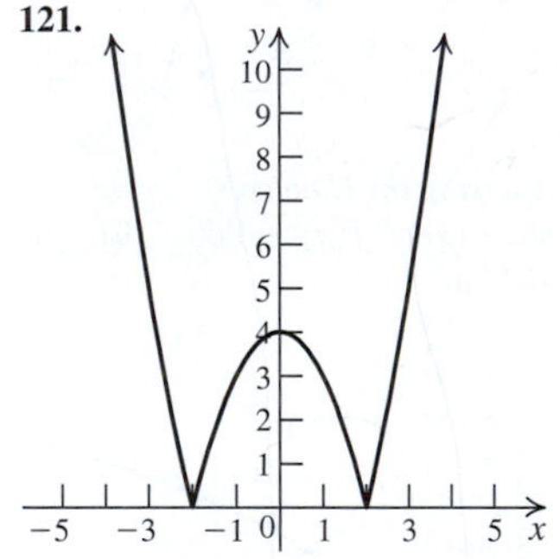

123.

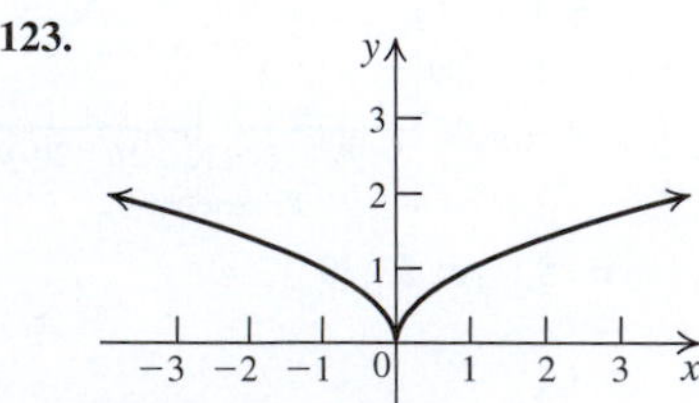

Critical Thinking/Discussion/Writing:

125. a. x-intercept 1 **b.** x-intercept -2; y-intercept 15 **c.** x-intercept -2; y-intercept -3 **d.** x-intercept 2; y-intercept 3 **127. a.** The graph of g is the graph of h shifted 3 units to the right and 3 units up. **b.** The graph of g is the graph of h shifted 1 unit to the left and 1 unit down. **c.** The graph of g is the graph of h stretched horizontally and vertically by a factor of 2. **d.** The graph of g is the graph of h stretched horizontally by a factor of 3, reflected about the y-axis, stretched vertically by a factor of 3, and reflected about the x-axis.

Maintaining Skills:

129. $6x^2 + 14x + 3$ **131.** $2x^2 + 15x - 3$ **133.** $x^3 - 8$ **135.** $(-\infty, 2) \cup (2, 3) \cup (3, \infty)$ **137.** $\left[\frac{3}{2}, \infty\right)$

Section 2.8

Answers to Practice Problems:

1. $(f+g)(x) = x^2 + 3x + 1$
$(f-g)(x) = -x^2 + 3x - 3$
$(fg)(x) = 3x^3 - x^2 + 6x - 2$
$\left(\frac{f}{g}\right)(x) = \frac{3x-1}{x^2+2}$

2. Domain: $fg = [1, 3]; \frac{f}{g} = [1, 3); \frac{g}{f} = (1, 3]$

3. a. $(f \circ g)(0) = -5$ **b.** $(g \circ f)(0) = 1$

4. a. $(g \circ f)(x) = 2x^2 - 8x + 9$ **b.** $(f \circ g)(x) = 1 - 2x^2$ **c.** $(g \circ g)(x) = 8x^4 + 8x^2 + 3$ **5.** $(-\infty, 1] \cup (3, \infty)$

6. a. $[-\sqrt{3}, \sqrt{3}]$ **b.** $[1, 5]$ **7.** $f(x) = \frac{1}{x}$ **8. a.** $A = 9\pi t^2$ **b.** $A = 324\pi$ square miles; ≈ 1018 square miles

9. a. $r(x) = x - 4500$ **b.** $d(x) = 0.94x$ **c. (i)** $(r \circ d)(x) = 0.94x - 4500$ **(ii)** $(d \circ r)(x) = 0.94x - 4230$ **d.** $(d \circ r)(x) - (r \circ d)(x) = 270$

Basic Concepts and Skills:

1. -8 **3.** 1 **5.** false **7. a.** $(f+g)(-1) = -1$ **b.** $(f-g)(0) = 0$ **c.** $(f \cdot g)(2) = -8$ **d.** $\left(\frac{f}{g}\right)(1) = -2$ **9. a.** $(f+g)(-1) = 0$ **b.** $(f-g)(0) = \frac{1}{2}(\sqrt{2} - 2)$ **c.** $(f \cdot g)(2) = \frac{5}{2}$ **d.** $\left(\frac{f}{g}\right)(1) = \frac{\sqrt{3}}{9}$

11. a. $(f+g)(x) = x^2 + x - 3$; domain: $(-\infty, \infty)$ **b.** $(f-g)(x) = -x^2 + x - 3$; domain: $(-\infty, \infty)$ **c.** $(f \cdot g)(x) = x^3 - 3x^2$; domain: $(-\infty, \infty)$ **d.** $\left(\frac{f}{g}\right)(x) = \frac{x-3}{x^2}$; domain: $(-\infty, 0) \cup (0, \infty)$ **e.** $\left(\frac{g}{f}\right)(x) = \frac{x^2}{x-3}$; domain: $(-\infty, 3) \cup (3, \infty)$ **13. a.** $(f+g)(x) = x^3 + 2x^2 + 4$; domain: $(-\infty, \infty)$ **b.** $(f-g)(x) = x^3 - 2x^2 - 6$; domain: $(-\infty, \infty)$ **c.** $(f \cdot g)(x) = 2x^5 + 5x^3 - 2x^2 - 5$; domain: $(-\infty, \infty)$ **d.** $\left(\frac{f}{g}\right)(x) = \frac{x^3-1}{2x^2+5}$; domain: $(-\infty, \infty)$ **e.** $\left(\frac{g}{f}\right)(x) = \frac{2x^2+5}{x^3-1}$; domain: $(-\infty, 1) \cup (1, \infty)$ **15. a.** $(f+g)(x) = 2x + \sqrt{x} - 1$; domain: $[0, \infty)$ **b.** $(f-g)(x) = 2x - \sqrt{x} - 1$; domain: $[0, \infty)$ **c.** $(f \cdot g)(x) = 2x\sqrt{x} - \sqrt{x}$; domain: $[0, \infty)$ **d.** $\left(\frac{f}{g}\right)(x) = \frac{2x-1}{\sqrt{x}}$; domain: $(0, \infty)$ **e.** $\left(\frac{g}{f}\right)(x) = \frac{\sqrt{x}}{2x-1}$; domain: $\left[0, \frac{1}{2}\right) \cup \left(\frac{1}{2}, \infty\right)$

17. a. $(f+g)(x) = \frac{2+x}{x+1}$; domain: $(-\infty, -1) \cup (-1, \infty)$

b. $(f-g)(x) = \frac{2-x}{x+1}$; domain: $(-\infty, -1) \cup (-1, \infty)$

c. $(f \cdot g)(x) = \frac{2x}{(x+1)^2}$; domain: $(-\infty, -1) \cup (-1, \infty)$

d. $\left(\frac{f}{g}\right)(x) = \frac{2}{x}$; domain: $(-\infty, -1) \cup (-1, 0) \cup (0, \infty)$

e. $\left(\frac{g}{f}\right)(x) = \frac{x}{2}$; domain: $(-\infty, -1) \cup (-1, \infty)$

19. a. $(f+g)(x) = \frac{x^3 - x^2 + 2x}{x^2 - 1}$; domain: $(-\infty, -1) \cup (-1, 1) \cup (1, \infty)$

b. $(f-g)(x) = \frac{x^2 - 2x}{x - 1}$; domain: $(-\infty, -1) \cup (-1, 1) \cup (1, \infty)$

c. $(f \cdot g)(x) = \frac{2x^3}{x^3 + x^2 - x - 1}$; domain: $(-\infty, -1) \cup (-1, 1) \cup (1, \infty)$

d. $\left(\frac{f}{g}\right)(x) = \frac{x^2 - x}{2}$; domain: $(-\infty, -1) \cup (-1, 0) \cup (0, 1) \cup (1, \infty)$
e. $\left(\frac{g}{f}\right)(x) = \frac{2}{x^2 - x}$; domain: $(-\infty, -1) \cup (-1, 0) \cup (0, 1) \cup (1, \infty)$
21. a. $(f + g)(x) = \frac{4x^2 - 5x - 28}{x^3 - 3x^2 - 16x + 48}$;
domain: $(-\infty, -4) \cup (-4, 3) \cup (3, 4) \cup (4, \infty)$
b. $(f - g)(x) = -\frac{7}{(x - 3)(x + 4)}$;
domain: $(-\infty, -4) \cup (-4, 3) \cup (3, 4) \cup (4, \infty)$
c. $(f \cdot g)(x) = \frac{4x^2 - 14x}{x^4 - 7x^3 - 4x^2 + 112x - 192}$;
domain: $(-\infty, -4) \cup (-4, 3) \cup (3, 4) \cup (4, \infty)$
d. $\left(\frac{f}{g}\right)(x) = \frac{2x^2 - 6x}{2x^2 + x - 28}$;
domain: $(-\infty, -4) \cup (-4, 3) \cup \left(3, \frac{7}{2}\right) \cup \left(\frac{7}{2}, 4\right) \cup (4, \infty)$
e. $\left(\frac{g}{f}\right)(x) = \frac{2x^2 + x - 28}{2x^2 - 6x}$;
domain: $(-\infty, -4) \cup (-4, 0) \cup (0, 3) \cup (3, 4) \cup (4, \infty)$
23. a. $[1, 5]$ **b.** $[1, 5)$ **25. a.** $[-2, 3]$ **b.** $[-2, 3)$
27. $g(f(x)) = 2x^2 + 1$; $g(f(2)) = 9$; $g(f(-3)) = 19$ **29.** 11
31. 31 **33.** $8a^2 + 8a - 1$ **35.** 7 **37.** $\frac{2x}{x + 1}$; $(-\infty, -1) \cup (-1, 0) \cup (0, \infty)$
39. $\sqrt{-3x - 1}$; $\left(-\infty, -\frac{1}{3}\right]$ **41.** $|x^2 - 1|$; $(-\infty, \infty)$
43. a. $(f \circ g)(x) = 2x + 5$; domain: $(-\infty, \infty)$
b. $(g \circ f)(x) = 2x + 1$; domain: $(-\infty, \infty)$ **c.** $(f \circ f)(x) = 4x - 9$; domain: $(-\infty, \infty)$ **d.** $(g \circ g)(x) = x + 8$; domain: $(-\infty, \infty)$
45. a. $(f \circ g)(x) = -2x^2 - 1$; domain: $(-\infty, \infty)$
b. $(g \circ f)(x) = 4x^2 - 4x + 2$; domain: $(-\infty, \infty)$
c. $(f \circ f)(x) = 4x - 1$; domain: $(-\infty, \infty)$
d. $(g \circ g)(x) = x^4 + 2x^2 + 2$; domain: $(-\infty, \infty)$
47. a. $(f \circ g)(x) = x$; domain: $[0, \infty)$ **b.** $(g \circ f)(x) = |x|$; domain: $(-\infty, \infty)$ **c.** $(f \circ f)(x) = x^4$; domain: $(-\infty, \infty)$ **d.** $(g \circ g)(x) = \sqrt[4]{x}$; domain: $[0, \infty)$ **49. a.** $(f \circ g)(x) = -\frac{x^2}{x^2 - 2}$; domain: $(-\infty, -\sqrt{2}) \cup (-\sqrt{2}, 0) \cup (0, \sqrt{2}) \cup (\sqrt{2}, \infty)$
b. $(g \circ f)(x) = (2x - 1)^2$; domain: $\left(-\infty, \frac{1}{2}\right) \cup \left(\frac{1}{2}, \infty\right)$
c. $(f \circ f)(x) = -\frac{2x - 1}{2x - 3}$; domain: $\left(-\infty, \frac{1}{2}\right) \cup \left(\frac{1}{2}, \frac{3}{2}\right) \cup \left(\frac{3}{2}, \infty\right)$
d. $(g \circ g)(x) = x^4$; domain: $(-\infty, 0) \cup (0, \infty)$
51. a. $(f \circ g)(x) = \sqrt{\sqrt{4 - x} - 1}$; domain: $(-\infty, 3]$
b. $(g \circ f)(x) = \sqrt{4 - \sqrt{x - 1}}$; domain: $[1, 17]$
c. $(f \circ f)(x) = \sqrt{\sqrt{x - 1} - 1}$; domain: $[2, \infty)$
d. $(g \circ g)(x) = \sqrt{4 - \sqrt{4 - x}}$; domain: $[-12, 4]$
53. a. $(f \circ g)(x) = \sqrt{3 - x^2}$; domain $= [-\sqrt{3}, \sqrt{3}]$
b. $(g \circ f)(x) = \sqrt{5 - x^2}$; domain $= [-\sqrt{5}, -1] \cup [1, \sqrt{5}]$
c. $(f \circ f)(x) = \sqrt{x^2 - 2}$; domain $= [-\infty, \sqrt{2}] \cup [\sqrt{2}, \infty)$
d. $(g \circ g)(x) = |x|$; domain $= [-2, 2]$ **55. a.** $(f \circ g)(x) = \frac{2}{1 + x}$; domain: $(-\infty, -1) \cup (-1, 1) \cup (1, \infty)$ **b.** $(g \circ f)(x) = -2x - 1$; domain: $(-\infty, 0) \cup (0, \infty)$ **c.** $(f \circ f)(x) = \frac{2x + 1}{x + 1}$; domain: $(-\infty, -1) \cup (-1, 0) \cup (0, \infty)$ **d.** $(g \circ g)(x) = -\frac{1}{x}$; domain: $(-\infty, 0) \cup (0, 1) \cup (1, \infty)$ **57.** $f(x) = \sqrt{x}$; $g(x) = x + 2$
59. $f(x) = x^{10}$; $g(x) = x^2 - 3$ **61.** $f(x) = \frac{1}{x}$; $g(x) = 3x - 5$
63. $f(x) = \sqrt[3]{x}$; $g(x) = x^2 - 7$ **65.** $f(x) = \frac{1}{|x|}$; $g(x) = x^3 - 1$

Applying the Concepts:
67. a. Cost function **b.** Revenue function **c.** Selling price of x shirts after sale tax **d.** Profit function **69. a.** $P(x) = 20x - 350$
b. $P(20) = 50$; profit made after 20 radios were sold **c.** 43
d. $(R \circ x)(C) = 5C - 1{,}750$. The function represents the revenue in terms of the cost C. **71. a.** $f(x) = 0.7x$ **b.** $g(x) = x - 5$
c. $(g \circ f)(x) = 0.7x - 5$ **d.** $(f \circ g)(x) = 0.7(x - 5)$ **e.** \$1.50
73. a. $f(x) = 1.1x$; $g(x) = x + 8$ **b.** $(f \circ g)(x) = 1.1x + 8.8$; a final test score, which originally was x, after adding 8 points and then increasing the score by 10% **c.** $(g \circ f)(x) = 1.1x + 8$; a final test score, which originally was x, after increasing the score by 10% and then adding 8 points **d.** 85.8 and 85.0 **e.** No **f. (i)** 73.82 **(ii)** 74.55
75. a. $f(x) = \pi x^2$ **b.** $g(x) = \pi(x + 30)^2$ **c.** The area between the fountain and the fence **d.** \$16,052 **77. a.** $(f \circ g)(t) = \pi(2t + 1)^2$
b. $A(t) = \pi(2t + 1)^2$ **c.** They are the same.

Beyond the Basics:
79. a. $f + g = \{(-2, 3), (1, 0), (3, 2)\}$
b. $fg = \{(-2, 0), (1, -4), (3, 0)\}$ **c.** $\frac{f}{g} = \{(1, -1), (3, 0)\}$
d. $f \circ g = \{(0, 1), (1, 3), (3, 1)\}$
81. c. $h(x) = \frac{h(x) + h(-x)}{2} + \frac{h(x) - h(-x)}{2}$
83. $(-\infty, -2) \cup [-1, 1] \cup (2, \infty)$

Critical Thinking/Discussion/Writing:
85. a. Domain: $(-\infty, 0) \cup [1, \infty)$ **b.** Domain: $[0, 2]$ **c.** Domain: $[1, 2]$ **d.** Domain: $[1, 2)$ **87. a.** Even function **b.** Odd function **c.** Neither **d.** Even function **e.** Even function **f.** Odd function

Maintaining Skills:
89. a. Yes **b.** $S = \{(2, -3), (1, -1), (3, 1), (1, 2)\}$ S is not a function.
91. $y = \frac{x - 3}{2}$ **93.** $x = -\sqrt{4 - y^2}$

Section 2.9

Answers to Practice Problems:
1. Not one-to-one; the horizontal line $y = 1$ intersects the graph at two different points.

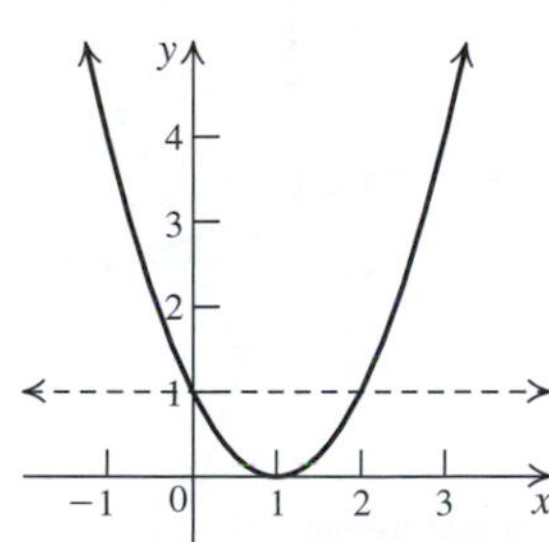

2. a. −3 **b.** 4
3. $(f \circ g)(x) = 3\left(\frac{x + 1}{3}\right) - 1 = x + 1 - 1 = x$ and $(g \circ f)(x) = \frac{3x - 1 + 1}{3} = \frac{3x}{3} = x$; therefore, f and g are inverses of each other.
4.

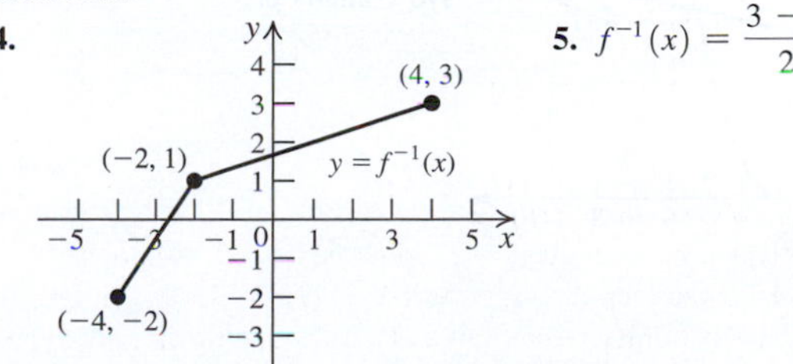

5. $f^{-1}(x) = \frac{3 - x}{2}$

6. $f^{-1}(x) = \frac{3x}{1 - x}, x \neq 1$ **7.** Domain: $(-\infty, -3) \cup (-3, \infty)$; range: $(-\infty, 1) \cup (1, \infty)$ **8.** $G^{-1}(x) = -\sqrt{x + 1}$ **9.** 3597 ft

Basic Concepts and Skills:

1. one-to-one **3.** $\frac{1}{3}x$ **5.** true **7.** One-to-one **9.** Not one-to-one
11. Not one-to-one **13.** One-to-one **15.** 2 **17.** −1 **19.** *a*
21. 337 **23.** −1580 **25. a.** 3 **b.** 3 **c.** 19 **d.** 5 **27. a.** 2 **b.** 1
c. 269

35. **37.**

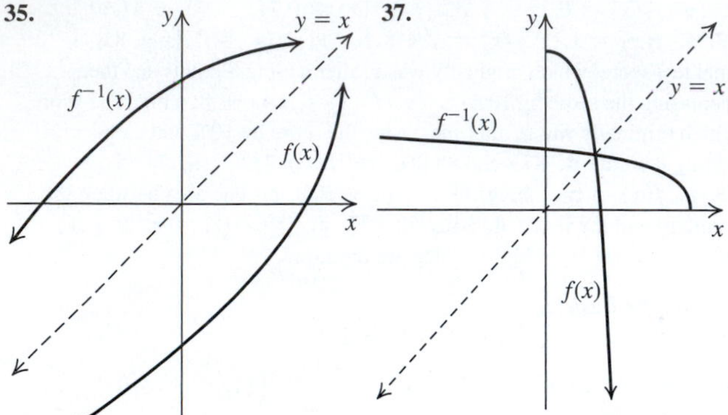

39.

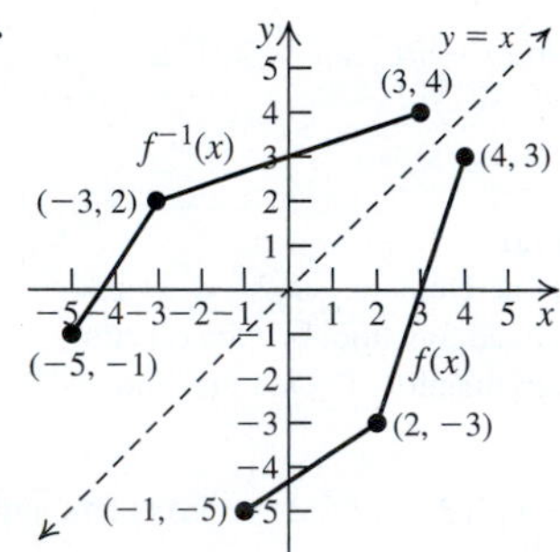

41. a. One-to-one **b.** $f^{-1}(x) = 5 - \frac{1}{3}x$

c.

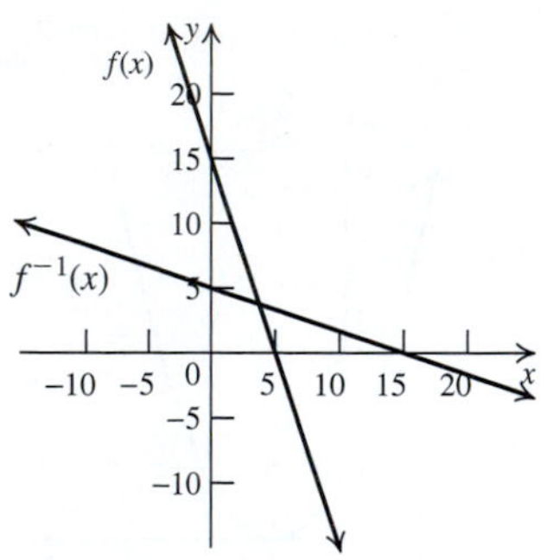

d. Domain f: $(-\infty, \infty)$
x-intercept: 5
y-intercept: 15
Domain f^{-1}: $(-\infty, \infty)$
x-intercept: 15
y-intercept: 5

43. Not one-to-one

45. a. One-to-one **b.** $f^{-1}(x) = (x - 3)^2$

c.

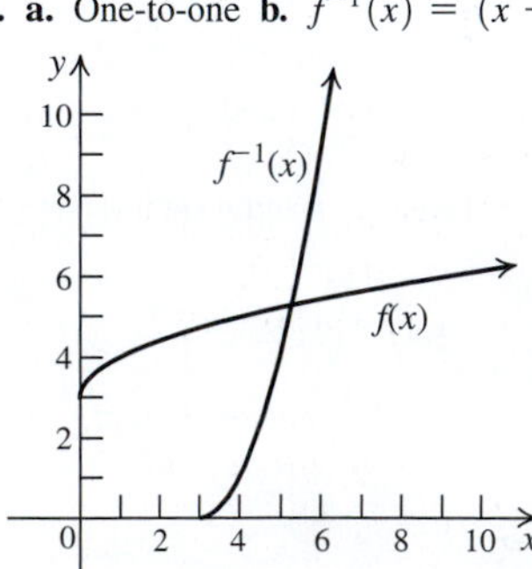

d. Domain f: $[0, \infty)$
No x-intercept
y-intercept: 3
Domain f^{-1}: $[3, \infty)$
x-intercept: 3
No y-intercept

47. a. One-to-one **b.** $g^{-1}(x) = (x^3 - 1)$

c.

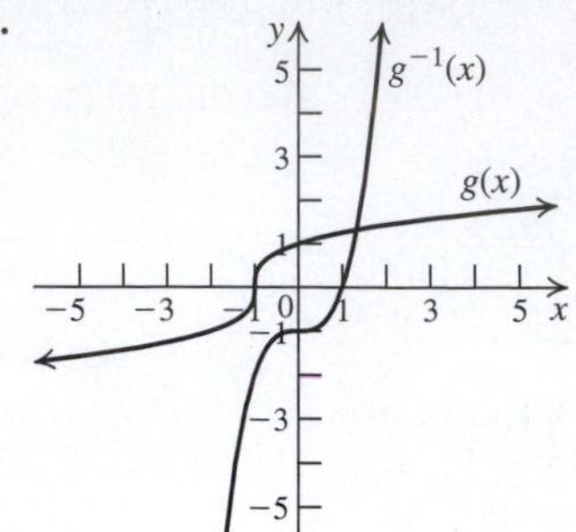

d. Domain g: $(-\infty, \infty)$
x-intercept: −1
y-intercept: 1
Domain g^{-1}: $(-\infty, \infty)$
x-intercept: 1
y-intercept: −1

49. a. One-to-one **b.** $f^{-1}(x) = \frac{x + 1}{x}$

c.

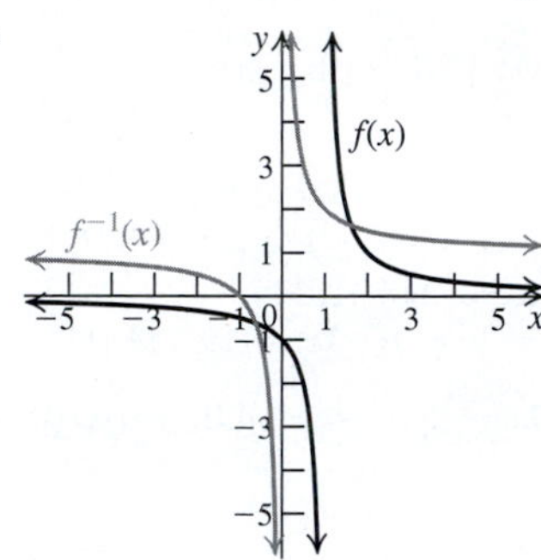

d. Domain f: $(-\infty, 1) \cup (1, \infty)$
No x-intercept
y-intercept: −1
Domain f^{-1}: $(-\infty, 0) \cup (0, \infty)$
x-intercept: −1
No y-intercept

51. a. One-to-one **b.** $f^{-1}(x) = x^2 - 4x + 3$

c.

$f^{-1}(x)$
$f(x)$
(−1, 2)
(2, −1)

d. Domain f: $[-1, \infty)$.
y-intercept: 3
No x-intercept
Domain f^{-1}: $[2, \infty)$
x-intercept is 3.
No y-intercept

53. Domain: $(-\infty, -2) \cup (-2, \infty)$; range: $(-\infty, 1) \cup (1, \infty)$

55. $f^{-1}(x) = \frac{2x + 1}{x - 1}$ Domain: $(-\infty, 2) \cup (2, \infty)$;
range: $(-\infty, 1) \cup (1, \infty)$ **57.** $f^{-1}(x) = \frac{1 - x}{x + 2}$
Domain: $(-\infty, -1) \cup (-1, \infty)$; range: $(-\infty, -2) \cup (-2, \infty)$

59. $y = \sqrt{-x}, x \le 0$

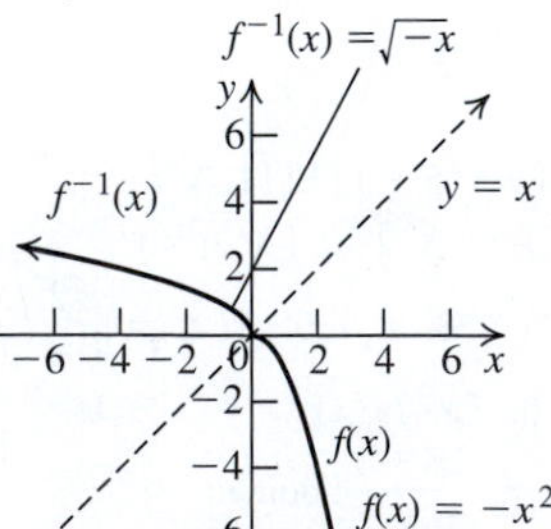

61. $y = x, x \ge 0$

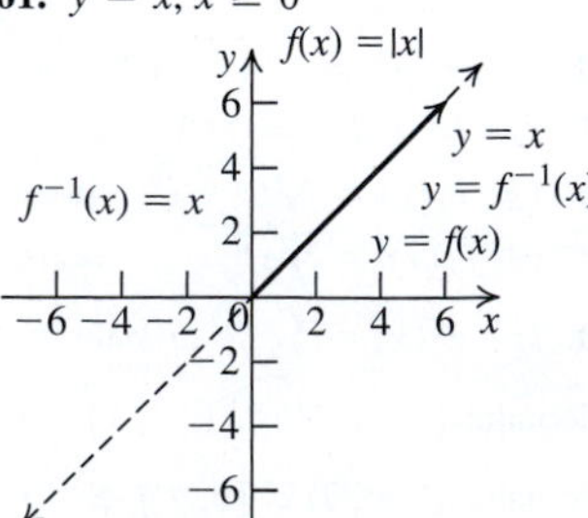

63. $y = -\sqrt{x-1}, x \geq 1$

$f(x) = x^2 + 1$

$f^{-1}(x) = -\sqrt{x-1}$

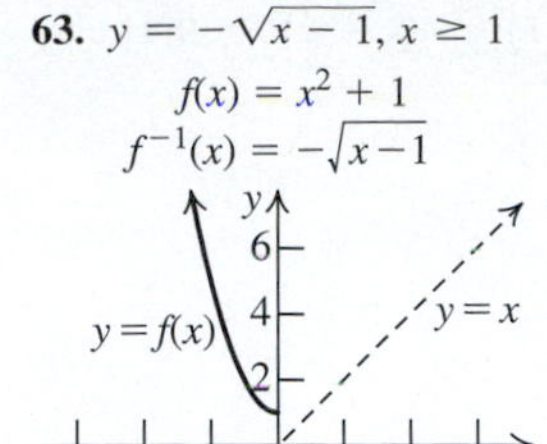

65. $y = -\sqrt{2-x}, x \leq 2$

$f(x) = -x^2 + 2$

$f^{-1}(x) = -\sqrt{2-x}$

Applying the Concepts:

67. a. $C(K) = K - 273$; it represents the Celsius temperature corresponding to a given Kelvin temperature. **b.** 27°C **c.** 295 K

69. a. $F(K(C)) = \frac{9}{5}C + 32$ **b.** $C(K(F)) = \frac{5}{9}F = \frac{160}{9}$

71. a. Dollars to euros: $f(x) = 0.75x$ (x is the number of dollars; $f(x)$ is the number of euros); euros to dollars: $g(x) = 1.25x$ (x is the number of euros; $g(x)$ is the number of dollars) **c.** She loses money.

73. a. $w = \begin{cases} 4 + 0.05x, & \text{if } x > 60 \\ 7, & \text{if } x \leq 60 \end{cases}$

b. It does not have an inverse because it is constant on $(0, 60)$ and hence is not one-to-one. **c.** Restriction of the domain: $[60, \infty)$ **75. a.** $x = \frac{1}{64}V^2$; it gives the height of the water in terms of the velocity. **b. (i)** 14.0625 ft **(ii)** 6.25 ft **77. a.** The balance due after x months

b. $f^{-1}(x) = 60 - \frac{1}{600}x$; it shows the number of months that have passed from the first payment until the balance due is $\$x$. **c.** $23.333 \approx 24$ months

Beyond the Basics:

81. a.

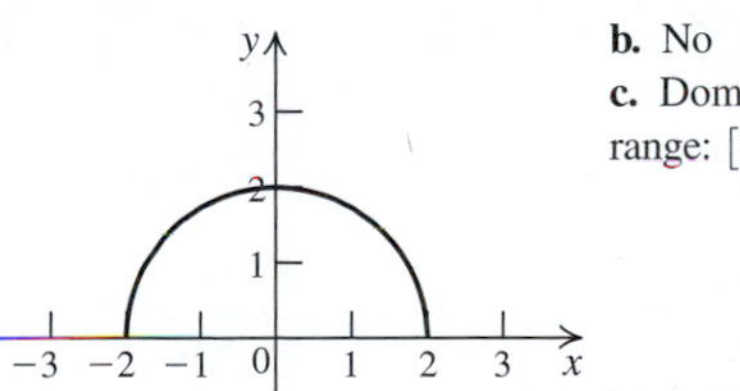

b. No
c. Domain: $[-2, 2]$; range: $[0, 2]$

83. a.

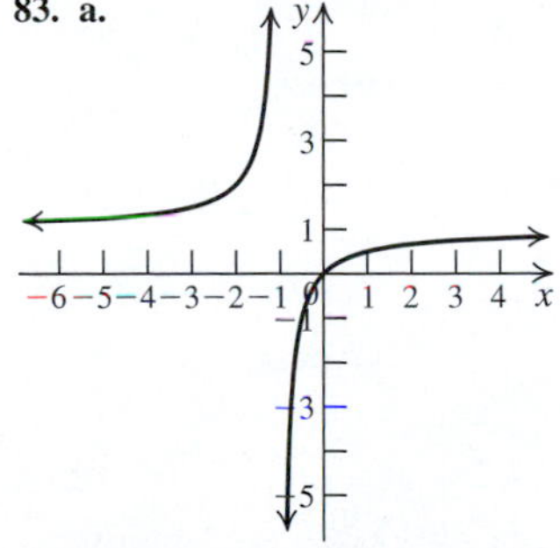

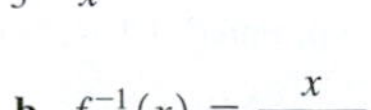

b. $f^{-1}(x) = \frac{x}{1-x}$

c. Domain: $(-\infty, -1) \cup (-1, \infty)$; range: $(-\infty, 1) \cup (1, \infty)$

f satisfies the horizontal-line test.

85. a. The midpoint is $M(5, 5)$. Because its coordinates satisfy the equation $y = x$, it lies on the line $y = x$. **b.** The slope of the line segment $\overline{PQ}$ is -1, and the slope of the line $y = x$ is 1. Because $(-1)(1) = -1$, the two lines are perpendicular.

87. a. The graph of g is the graph of f shifted 1 unit to the right and 2 units up. **b.** $g^{-1}(x) = \sqrt[3]{x-2} + 1$ **c.**

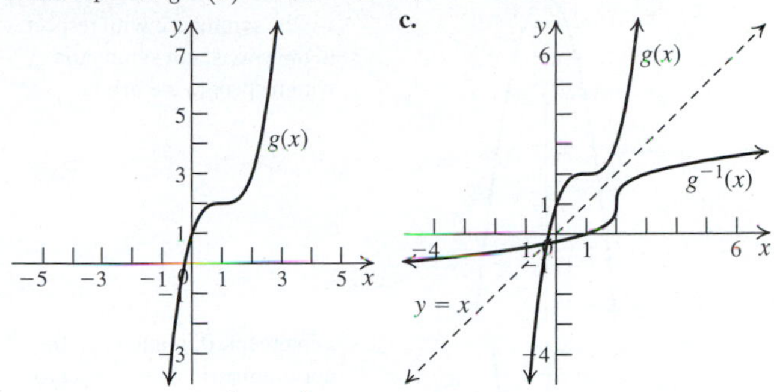

89. a. (i) $f^{-1}(x) = \frac{1}{2}x - \frac{3}{2}$ **(ii)** $g^{-1}(x) = \sqrt[3]{x+1}$

(iii) $(f \circ g)(x) = 2x^3 + 1$ **(iv)** $(g \circ f)(x) = 8x^3 + 36x^2 + 54x + 26$

(v) $(f \circ g)^{-1}(x) = \sqrt[3]{\frac{x-1}{2}}$ **(vi)** $(g \circ f)^{-1}(x) = \frac{1}{2}\sqrt[3]{x+1} - \frac{3}{2}$

(vii) $(f^{-1} \circ g^{-1})(x) = \frac{1}{2}\sqrt[3]{x+1} - \frac{3}{2}$ **(viii)** $(g^{-1} \circ f^{-1})(x) = \sqrt[3]{\frac{x-1}{2}}$

b. (i) $(f \circ g)^{-1}(x) = \sqrt[3]{\frac{x-1}{2}} = (g^{-1} \circ f^{-1})(x)$

(ii) $(g \circ f)^{-1}(x) = \frac{1}{2}\sqrt[3]{x+1} - \frac{3}{2} = (f^{-1} \circ g^{-1})(x)$

Critical Thinking/Discussion/Writing:

91. No. $f(x) = x^3 - x$ is odd, but it does not have an inverse because $f(0) = f(1)$. So it is not one-to-one. **93.** Yes, because increasing and decreasing functions are one-to-one

Maintaining Skills:

95. $x = 3, x = 4$ **97.** $x = -1, x = 3$

99.

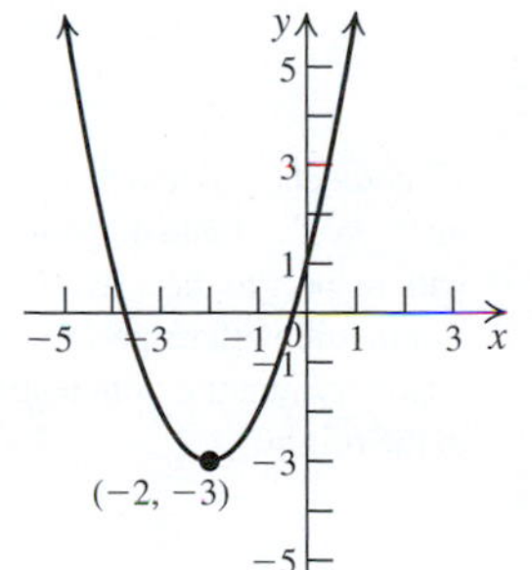

Review Exercises

1. false **3.** true **5.** false **7.** true **9. a.** $2\sqrt{5}$ **b.** $(1, 4)$ **c.** $\frac{1}{2}$ **11. a.** $5\sqrt{2}$ **b.** $\left(\frac{13}{2}, -\frac{11}{2}\right)$ **c.** -1 **13. a.** $\sqrt{34}$ **b.** $\left(\frac{7}{2}, -\frac{9}{2}\right)$ **c.** $\frac{5}{3}$ **15.** $AC^2 + BC^2 = AB^2$ $(34 + 34 = 68)$ **17.** $(4, 5)$ **19.** $\left(\frac{31}{18}, 0\right)$

21. Not symmetric with respect to the x-axis, symmetric with respect to the y-axis, not symmetric with respect to the origin **23.** Symmetric with respect to the x-axis, not symmetric with respect to the y-axis, not symmetric with respect to the origin

25.

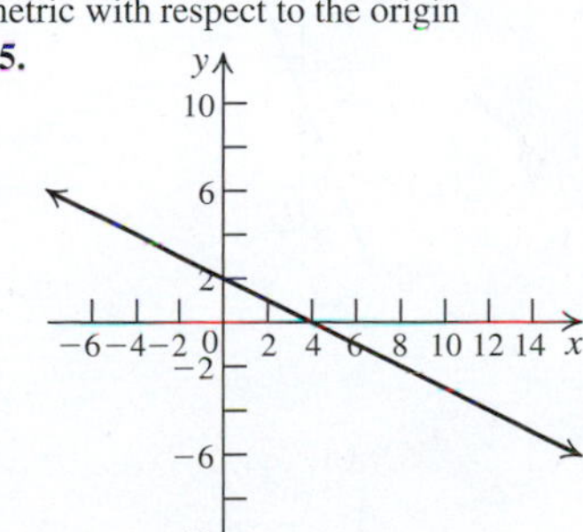

x-intercept: 4; y-intercept: 2; not symmetric with respect to the x-axis, not symmetric with respect to the y-axis, not symmetric with respect to the origin

27. 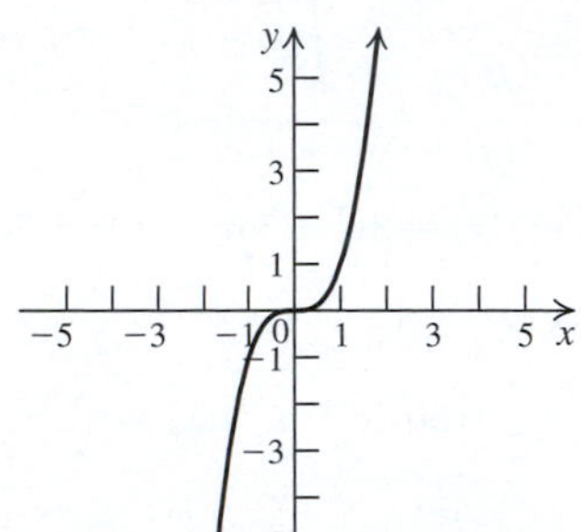
x-intercept: 0; y-intercept: 0; not symmetric with respect to the x-axis, symmetric with respect to the y-axis, not symmetric with respect to the origin

29. 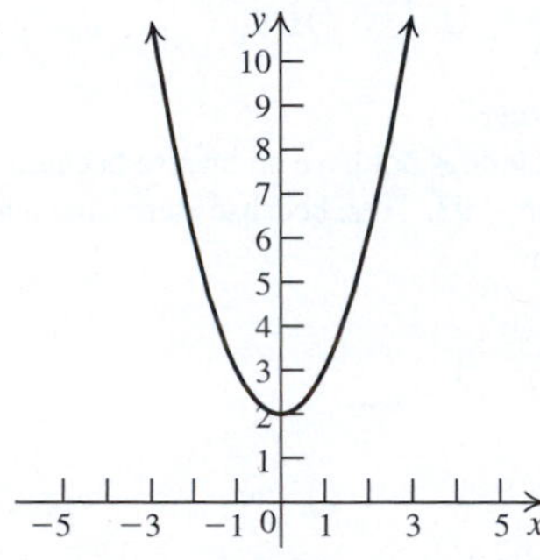
x-intercept: 0; y-intercept: 0; not symmetric with respect to the x-axis, not symmetric with respect to the y-axis, symmetric with respect to the origin

31. 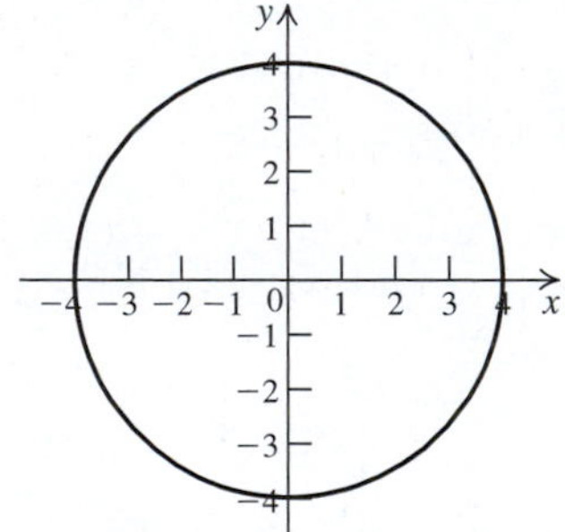
No x-intercept; y-intercept: 2; not symmetric with respect to the x-axis, symmetric with respect to the y-axis, not symmetric with respect to the origin

33.
x-intercepts: -4 and 4; y-intercepts: -4 and 4; symmetric with respect to the x-axis, symmetric with respect to the y-axis, symmetric with respect to the origin

35. $(x - 2)^2 + (y + 3)^2 = 25$ **37.** $(x + 2)^2 + (y + 5)^2 = 4$

39. $\frac{x}{2} - \frac{y}{5} = 1 \Rightarrow y = \frac{5}{2}x - 5$;

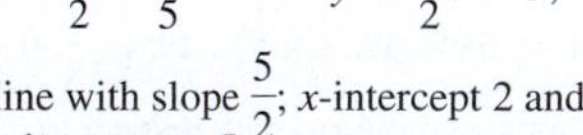

line with slope $\frac{5}{2}$; x-intercept 2 and y-intercept -5

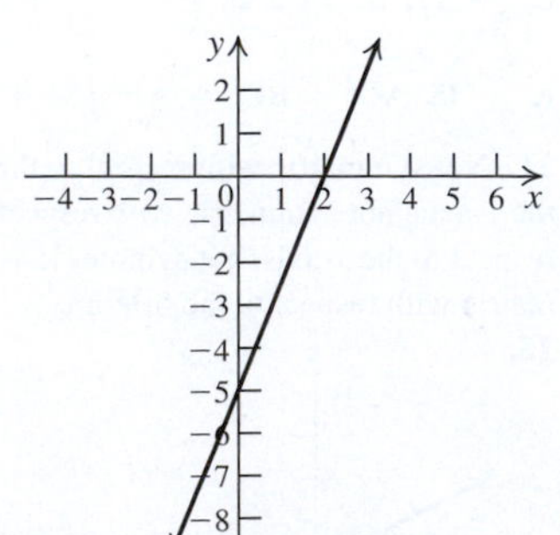

41. $x^2 + y^2 - 2x + 4y - 4 = 0 \Rightarrow (x - 1)^2 + (y + 2)^2 = 9$; circle centered at $(1, -2)$ and with radius 3

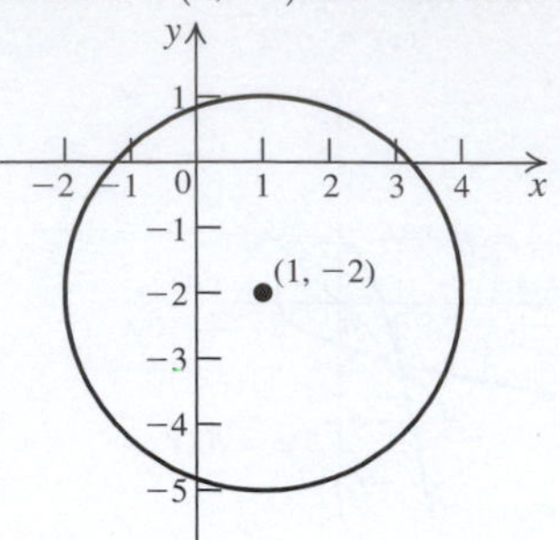

43. $y = -2x + 4$ **45.** $y = -2x + 5$ **47. a.** Parallel **b.** Neither **c.** Perpendicular **d.** Neither

49. 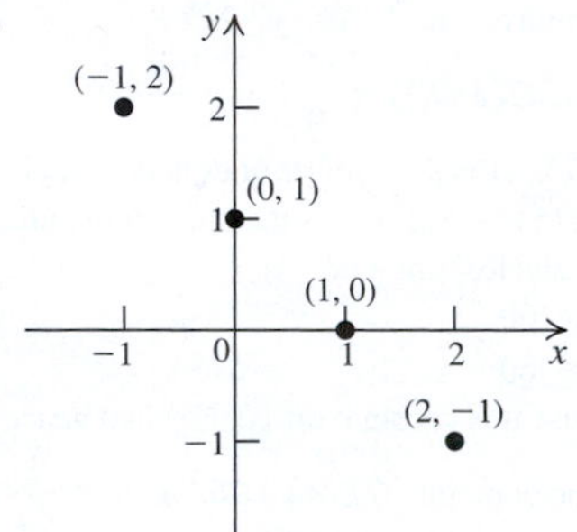

Domain: $\{-1, 0, 1, 2\}$; range: $\{-1, 0, 1, 2\}$; function

51. 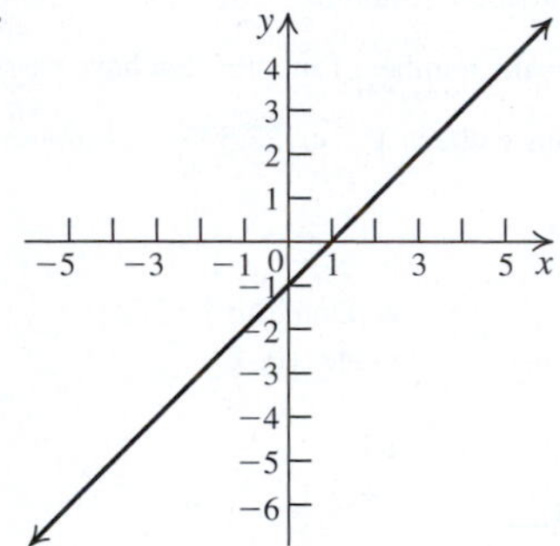

Domain: $(-\infty, \infty)$; range: $(-\infty, \infty)$; function

53. 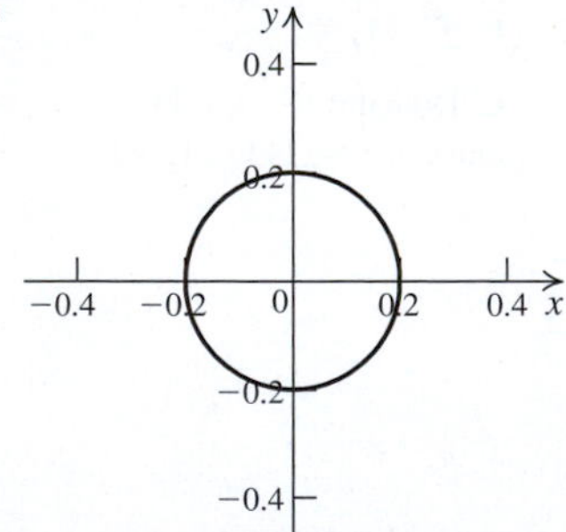

Domain: $[-0.2, 0.2]$; range: $[-0.2, 0.2]$; not a function

55. 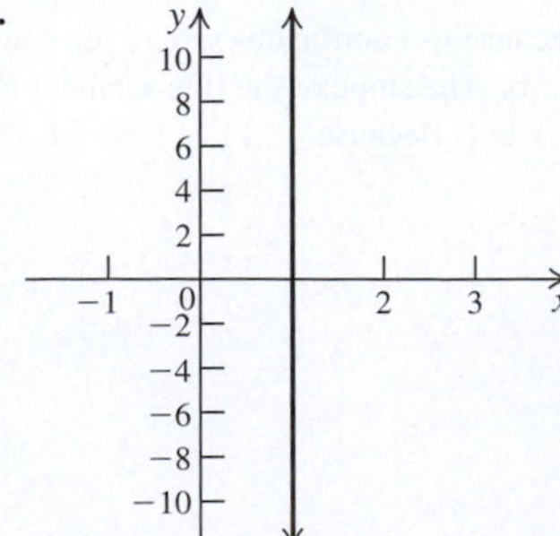

Domain: $\{1\}$; range: $(-\infty, \infty)$; not a function

57.

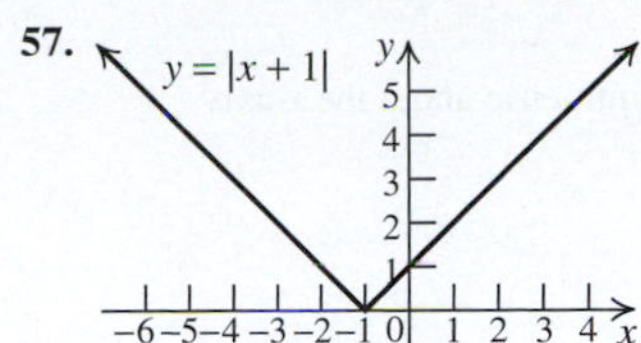

Domain: $(-\infty, \infty)$; range: $[0, \infty)$; function
59. -5 **61.** $x = 1$ **63.** 3 **65.** -10 **67.** 22 **69.** $3x^2 - 5$
71. $9x + 4$ **73.** $3a + 3h + 1$ **75.** 3
77.

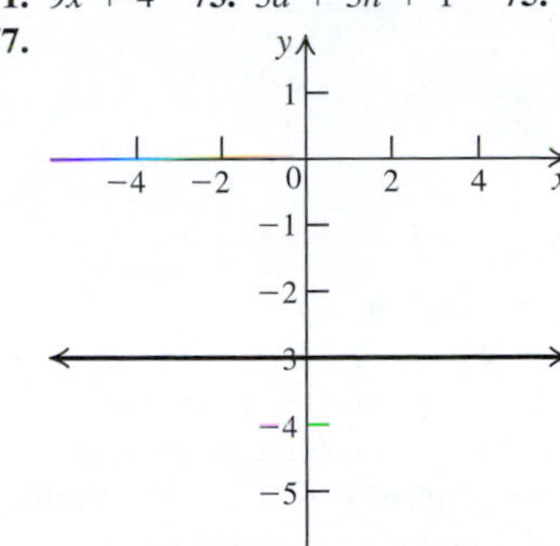

Domain: $(-\infty, \infty)$; range: $\{-3\}$; constant on $(-\infty, \infty)$
79.

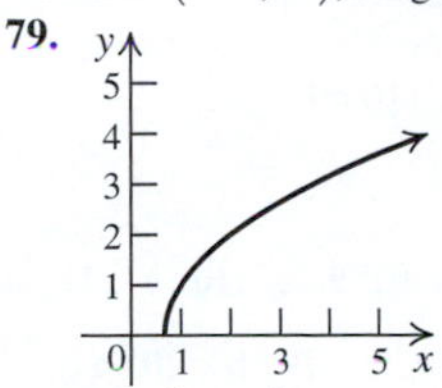

Domain: $\left[\frac{2}{3}, \infty\right)$; range: $[0, \infty)$; increasing on $\left(\frac{2}{3}, \infty\right)$

81.

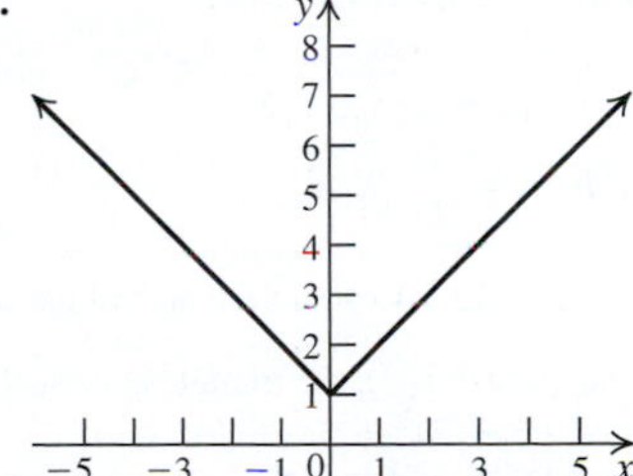

Domain: $(-\infty, \infty)$; range: $[1, \infty)$; decreasing on $(-\infty, 0)$, increasing on $(0, \infty)$
83. The graph of g is the graph of f shifted 1 unit to the left.

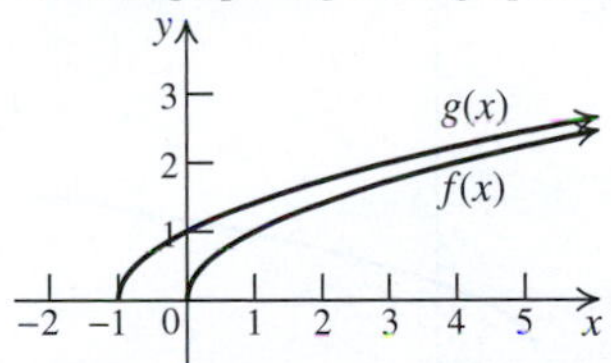

85. The graph of g is the graph of f shifted 2 units to the right and reflected in the x-axis.

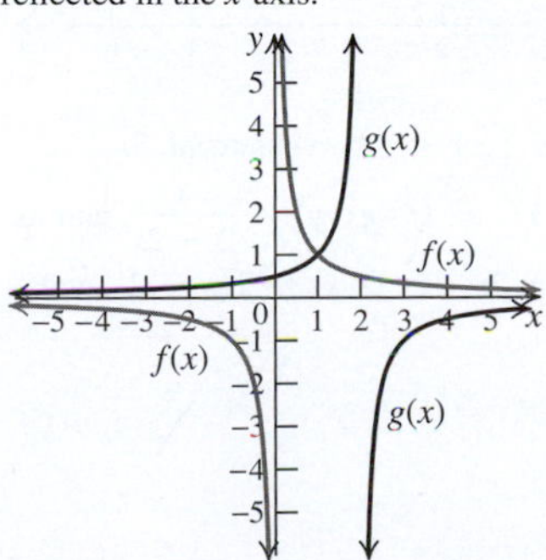

87. Even; not symmetric with respect to the x-axis, symmetric with respect to the y-axis, not symmetric with respect to the origin
89. Even; not symmetric with respect to the x-axis, symmetric with respect to the y-axis, not symmetric with respect to the origin
91. Neither even nor odd; not symmetric with respect to the x-axis, not symmetric with respect to the y-axis, not symmetric with respect to the origin **93.** $f(x) = (g \circ h)(x)$, where $g(x) = \sqrt{x}$ and $h(x) = x^2 - 4$
95. $h(x) = (f \circ g)(x)$, where $f(x) = \sqrt{x}$ and $g(x) = \dfrac{x - 3}{2x + 5}$
97. One-to-one; $f^{-1}(x) = x - 2$

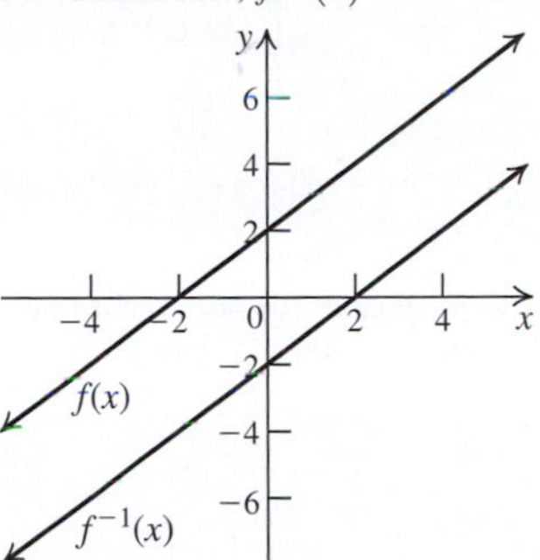

99. One-to-one; $f^{-1}(x) = x^3 + 2$

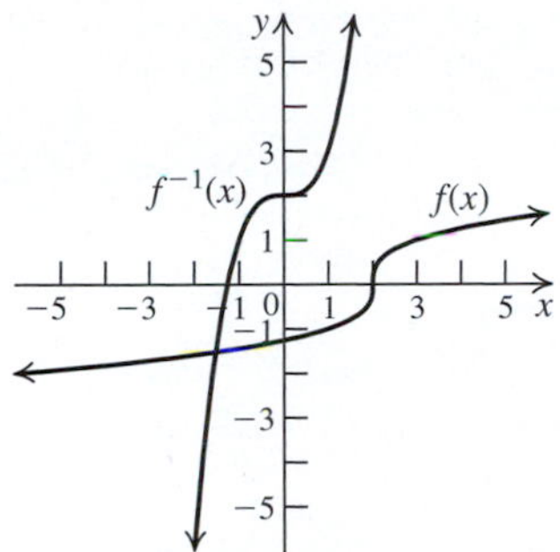

101. $f^{-1}(x) = \dfrac{2x + 1}{1 - x}$; domain: $(-\infty, -2) \cup (-2, \infty)$; range: $(-\infty, 1) \cup (1, \infty)$ **103. a.** $f(x) = \begin{cases} 3x + 6, & \text{if } -3 \le x \le -2 \\ \frac{1}{2}x + 1, & \text{if } -2 < x < 0 \\ x + 1, & \text{if } 0 \le x \le 3 \end{cases}$
b. Domain: $[-3, 3]$; range: $[-3, 4]$ **c.** x-intercept: -2; y-intercept: 1
d. **e.**

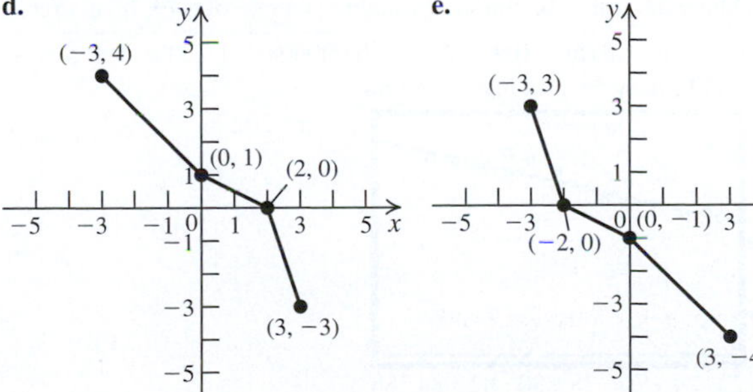

f. **g.**

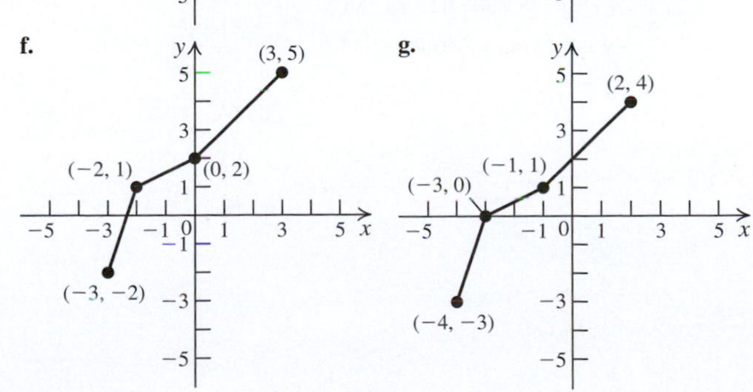

h.

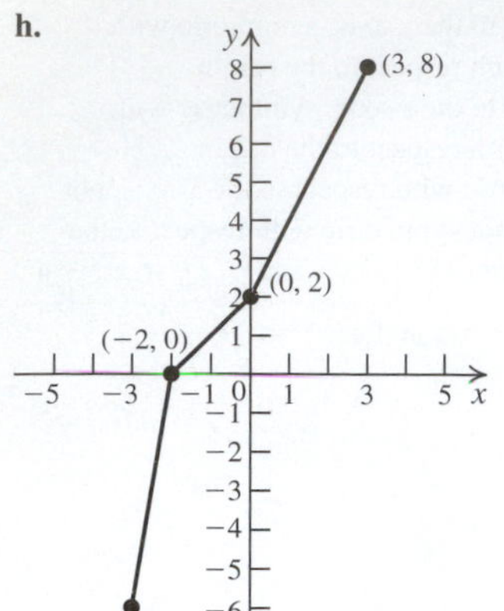

i.

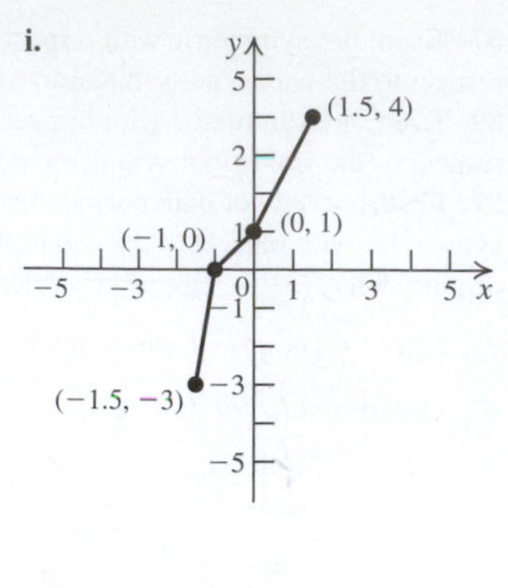

j.

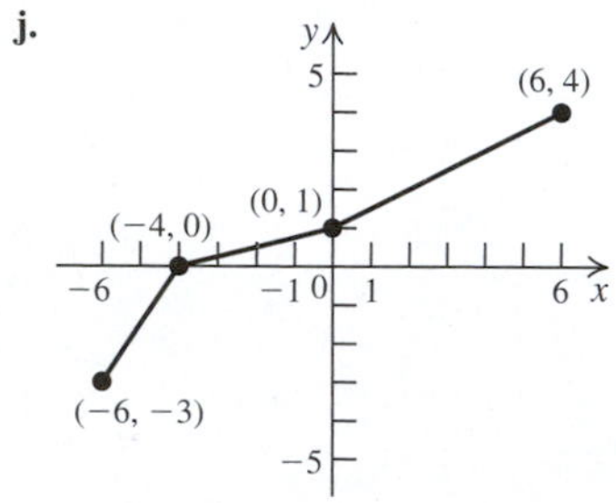

k. It satisfies the horizontal-line test.

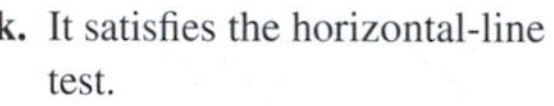

l.

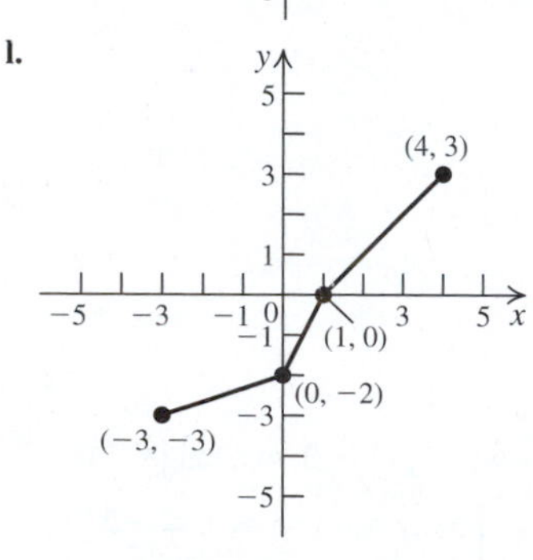

Applying the Concepts:

105. a. $C = 0.875w - 22{,}125$ **b.** Slope: the cost of disposing 1 pound of waste; x-intercept: the amount of waste that can be disposed with no cost; y-intercept: the fixed cost **c.** \$510,750 **d.** 1,168, 142.86 lb **107. a.** She won \$98. **b.** She was winning at a rate of \$49/h. **c.** Yes, after 20 hours **d.** \$5/hr **109. a.** $0.5\sqrt{0.000004t^4 + 0.004t^2 + 5}$ **b.** 1.13 **111. a.** $y = 7.4474x - 363.88$

b.

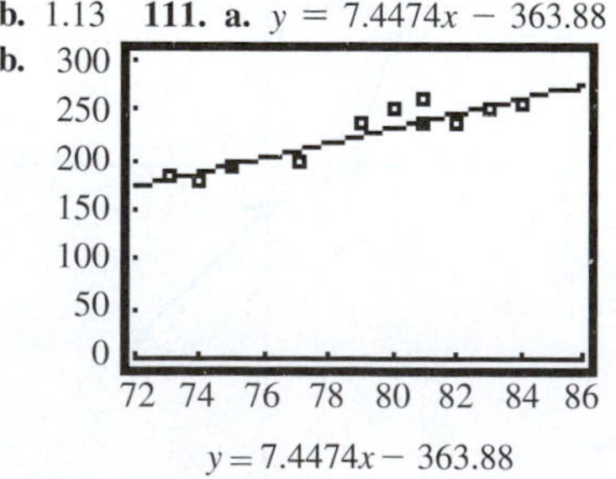

c. 202 lb

Practice Test A

1. $(x + 3)^2 + (y - 4)^2 = 2$ **2.** Symmetric about the x-axis **3.** x-intercept: $-1, 0, 3$; y-intercept: 0

4.

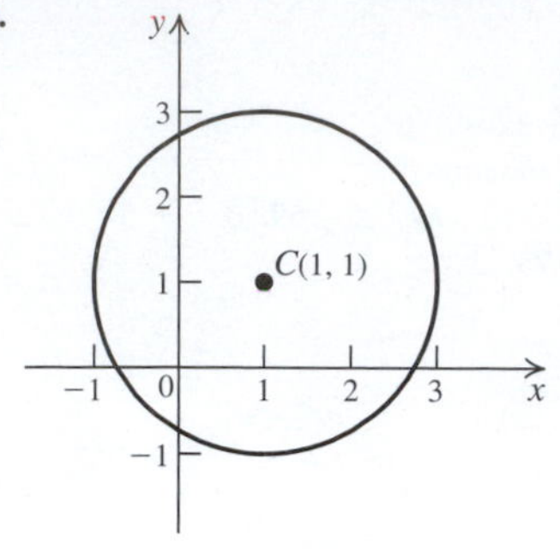

5. $y = -x + 9$ **6.** $y = 4x - 9$ **7.** -36 **8.** -1 **9.** $x^4 - 4x^3 + 2x^2 + 4x$ **10. a.** $f(-1) = -3$ **b.** $f(0) = -2$ **c.** $f(1) = -1$ **11.** $[0, 1)$ **12.** 2 **13.** Even **14.** Increasing on $(-\infty, 0) \cup (2, \infty)$, decreasing on $(0, 2)$ **15.** Shift the graph of $y = \sqrt{x}$ 3 units to the right, stretch the graph of $y = \sqrt{x - 3}$ vertically by a factor of 2, and shift the graph of $y = 2\sqrt{x - 3}$ 4 units up. **16.** 2.5 sec **17.** 2 **18.** $f^{-1}(x) = \dfrac{x}{x - 2}$ **19.** $A(x) = 100x + 1000$ **20. a.** \$87.50 **b.** 110 mi

Practice Test B

1. d **2.** b **3.** d **4.** d **5.** c **6.** c **7.** d **8.** b **9.** a **10.** c **11.** a **12.** d **13.** a **14.** a **15.** b **16.** d **17.** c **18.** c **19.** b **20.** a

Cumulative Review Exercises (Chapters P–2)

1. a. $\dfrac{y^2}{x^3}$ **b.** $\dfrac{1}{x + y}$ **3. a.** $3\sqrt{3}$ **b.** $\dfrac{2x}{(x + 1)(x + 2)}$ **5. a.** $x = 4$ **b.** No solution **7. a.** $\dfrac{1}{4} \pm \dfrac{\sqrt{23}}{4}i$ **b.** $x = \dfrac{3}{2}$ **9. a.** $x = \dfrac{7 + \sqrt{17}}{8}$ **b.** $x = 0; \dfrac{8}{9}$ **11. a.** $(0, 4)$ **b.** $[-3, -2]$ **13.** Because the length of the side BC is equal to the length of the side AC (which is 5), the triangle is isosceles. **15.** All of the points satisfy the equation $(x - 2)^2 + (y + 1)^2 = 5^2$, which is a circle with center $(2, -1)$ and radius 5. **17.** $y = -3x + 5$ **19.** $y = -0.5x$ **21.** $y = -\dfrac{1}{4}x + 4$ **23.** $x = 2$

25. **27.**

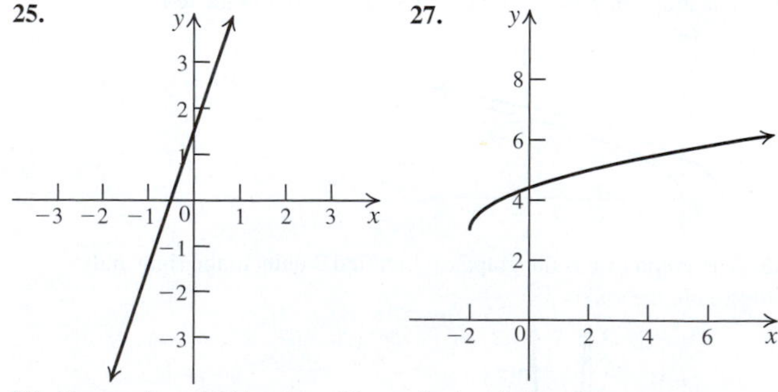

29. She purchased 60 books. **31. a.** $[-1, \infty)$ **b.** x-intercept: 8; y-intercept: -2 **c.** -3 **d.** $(8, \infty)$ **33. a.** $(f \circ g)(x) = \dfrac{x}{2 - 2x}$, and its domain is $(-\infty, 0) \cup (0, 1) \cup (1, \infty)$. **b.** $(g \circ f)(x) = 2(x - 2)$, and its domain is $(-\infty, 2) \cup (2, \infty)$.

CHAPTER 3

Section 3.1

Answers to Practice Problems:

1. $y = 3(x - 1)^2 - 5$; f has minimum value -5.

2. The graph is a parabola. $a = -2, h = -1, k = 3$. It opens downward. Vertex $(-1, 3)$. It has maximum value 3; x-intercepts: $\pm\sqrt{\frac{3}{2}} - 1$; y-intercept: 1

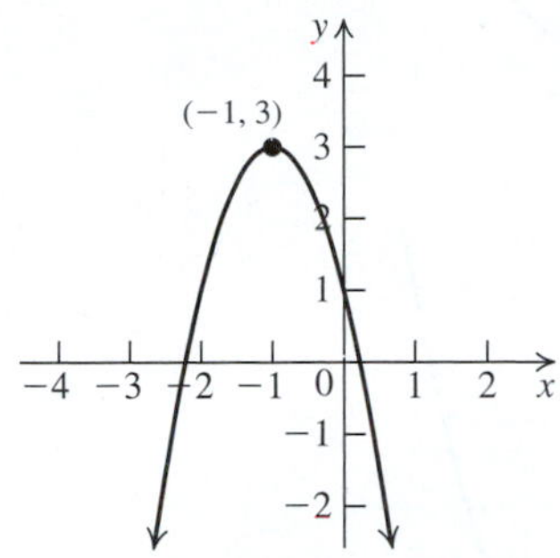

3. The graph is a parabola. It opens upward. Vertex $\left(\frac{1}{2}, \frac{-27}{4}\right)$; It has a minimum value $-\frac{27}{4}$; x-intercepts: $-1, 2$; y-intercept: -6

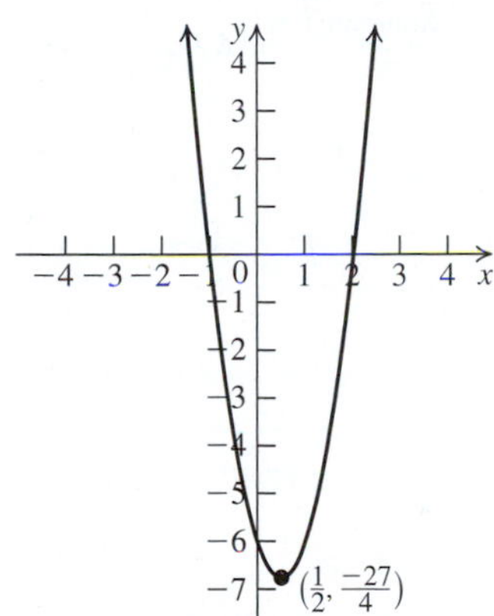

4. $\left[\frac{3 - 2\sqrt{3}}{3}, \frac{3 + 2\sqrt{3}}{3}\right]$

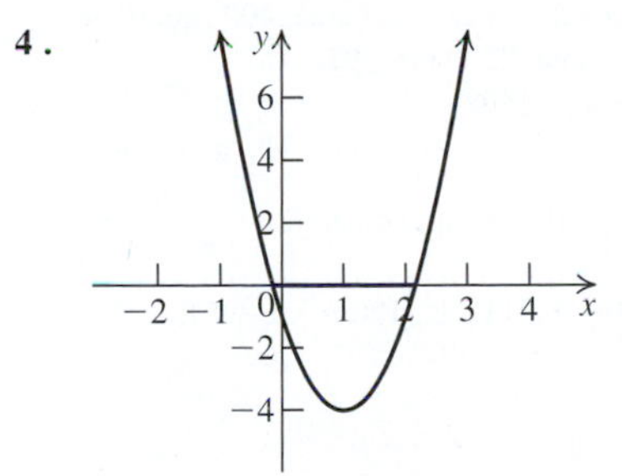

5. $a < 0, b < 0, c > 0$ **6. a.** $h(t) = -16t^2 + 96t + 100$ (feet) **b.** 3 seconds **7.** Maximum area is 62,500 sq ft. The playground is square, 250 feet on each side.

Basic Concepts and Skills:

1. vertex **3.** true **5.** $\frac{1}{2}$ **7.** true **9.** f **11.** a **13.** h **15.** g **17.** $y = -2x^2$ **19.** $y = 5x^2$ **21.** $y = 2x^2$ **23.** $y = -(x + 3)^2$ **25.** $y = 2(x - 2)^2 + 5$ **27.** $y = \frac{11}{49}(x - 2)^2 - 3$ **29.** $y = -12\left(x - \frac{1}{2}\right)^2 + \frac{1}{2}$ **31.** $y = \frac{3}{4}(x + 2)^2$ **33.** $y = \frac{3}{4}(x - 3)^2 - 1$ **35.** $a > 0$ $b < 0$ $c < 0$ **37.** $a > 0$ $b < 0$ $c > 0$ **39.** $a < 0$ $b < 0$ $c < 0$

41.

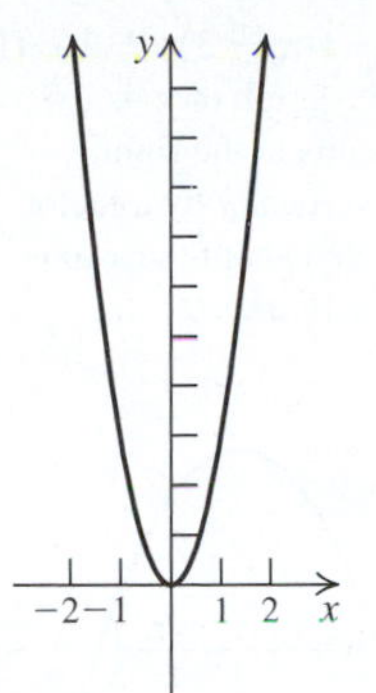

The graph is the graph of $y = x^2$ stretched vertically by a factor of 3.

43.

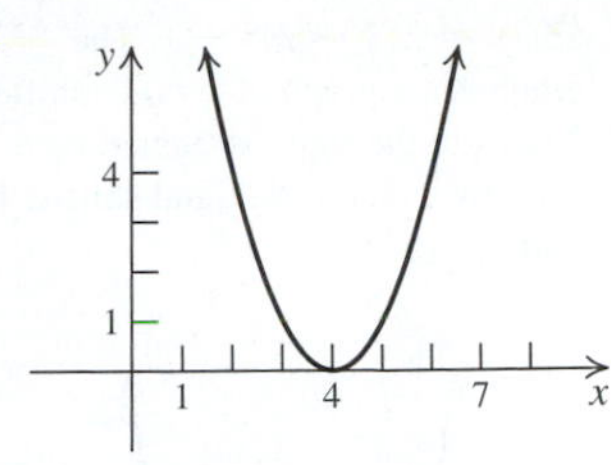

The graph is the graph of $y = x^2$ shifted 4 units to the right.

45.

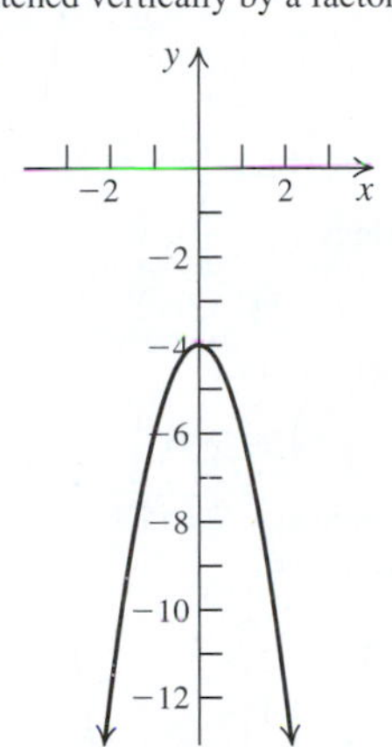

The graph is the graph of $y = x^2$ stretched vertically by a factor of 2, reflected about the x-axis, and shifted down 4 units.

47.

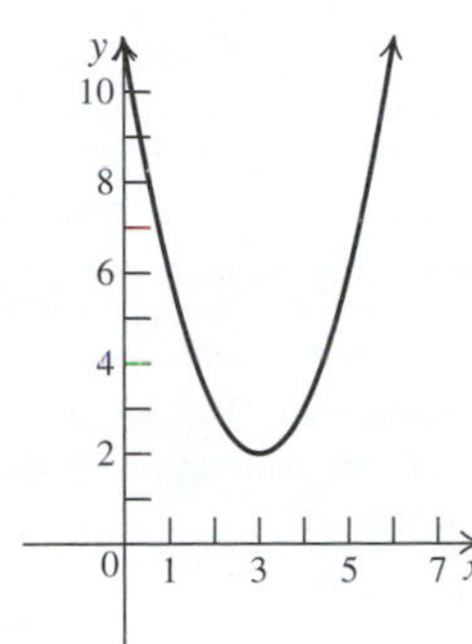

The graph is the graph of $y = x^2$ shifted 3 units right and 2 units up.

49. The graph is the graph of $y = x^2$ shifted 2 units right, stretched vertically by a factor of 3, reflected about the x-axis, and shifted 4 units up.

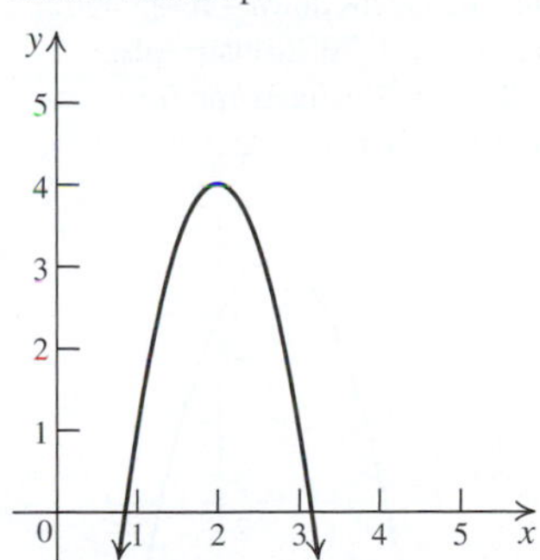

51. $y = (x + 2)^2 - 4$. The graph is the graph of $y = x^2$ shifted 2 units to the left and 4 units down. Vertex: $(-2, -4)$; axis: $x = -2$; x-intercepts: -4 and 0; y-intercept: 0

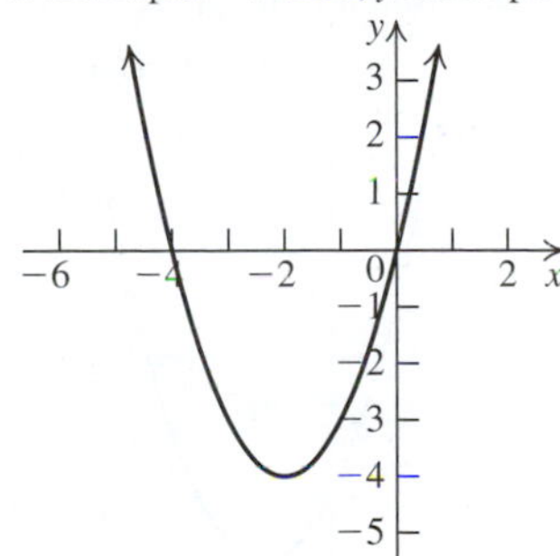

53. $y = -(x - 3)^2 - 1$. The graph is the graph of $y = x^2$ shifted 3 units to the right, reflected about the x-axis, and shifted 1 unit down. Vertex: $(3, -1)$; axis: $x = 3$; no x-intercept; y-intercept: -10

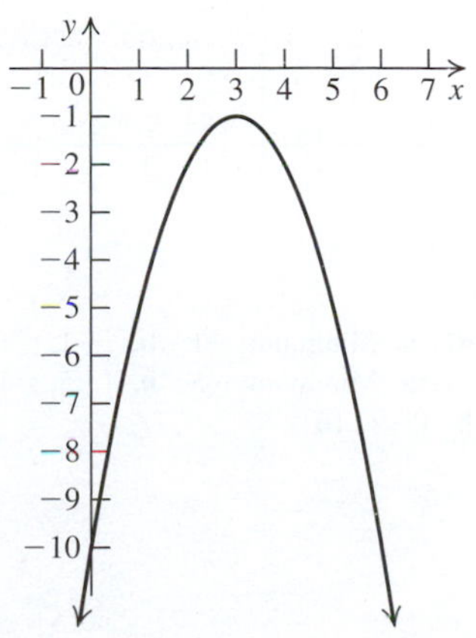

55. $y = 2(x - 2)^2 + 1$. The graph is the graph of $y = x^2$ shifted 2 units to the right, stretched vertically by a factor of 2, and shifted 1 unit up.

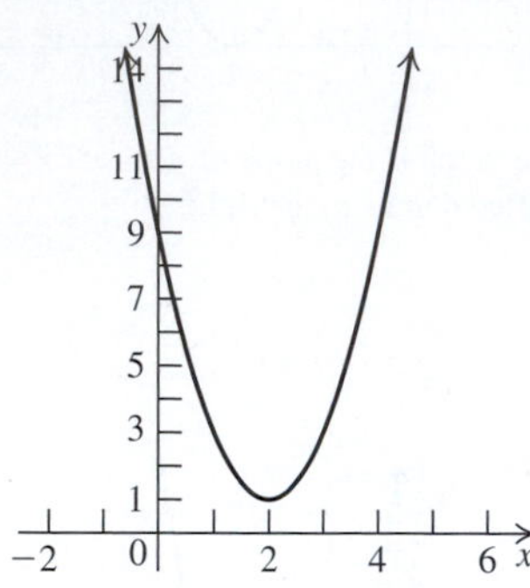

Vertex: $(2, 1)$; axis: $x = 2$; no x-intercept; y-intercept: 9

57. $y = -3(x - 3)^2 + 16$. The graph is the graph of $y = x^2$ shifted 3 units to the right, stretched vertically by a factor of 3, reflected about the x-axis, and shifted 16 units up.

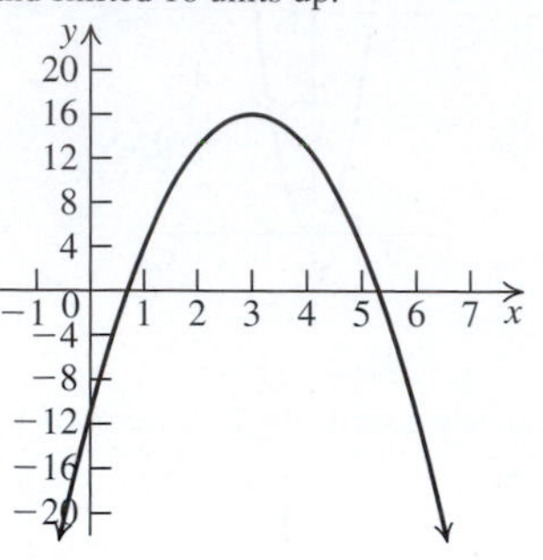

Vertex: $(3, 16)$; axis: $x = 3$; x-intercepts: $3 \pm \frac{4}{3}\sqrt{3}$; y-intercept: -11

59. a. Opens up **b.** $(4, -1)$ **c.** $x = 4$ **d.** x-intercepts: 3 and 5; y-intercept: 15
e.

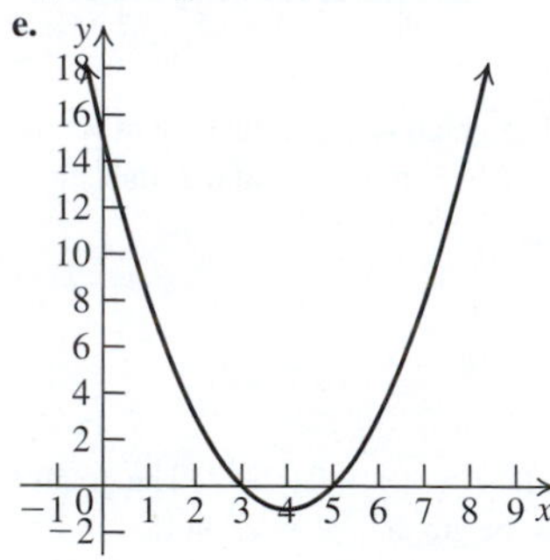

61. a. Opens up **b.** $\left(\frac{1}{2}, \frac{-25}{4}\right)$ **c.** $x = \frac{1}{2}$ **d.** x-intercepts: -2 and 3; y-intercept: -6
e.

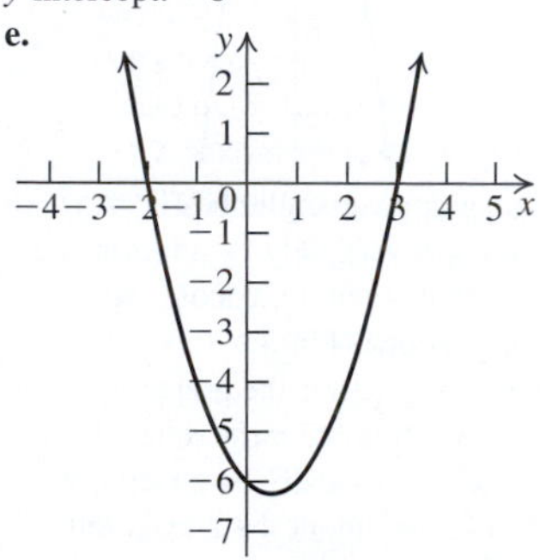

63. a. Opens up **b.** $(1, 3)$ **c.** $x = 1$ **d.** x-intercepts: none; y-intercept: 4
e.

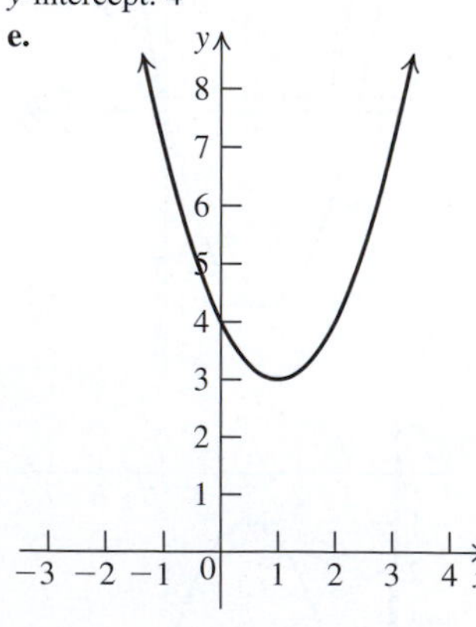

65. a. Opens down **b.** $(-1, 7)$ **c.** $x = -1$ **d.** x-intercepts: $-1 \pm \sqrt{7}$; y-intercept: 6
e.

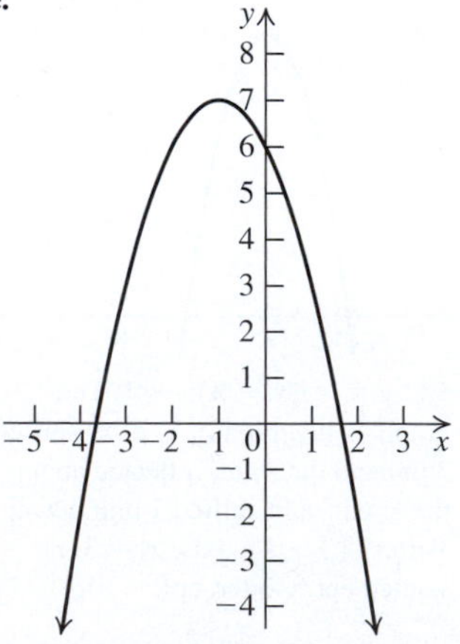

67. a. Minimum: -1 **b.** $[-1, \infty)$ **69. a.** Maximum: 0 **b.** $(-\infty, 0]$
71. a. Minimum: -5 **b.** $[-5, \infty)$ **73. a.** Maximum: 16 **b.** $(-\infty, 16]$

75.

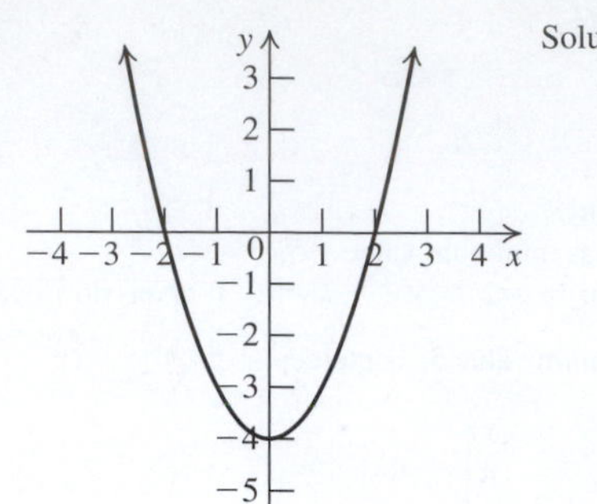

Solution: $[-2, 2]$

77.

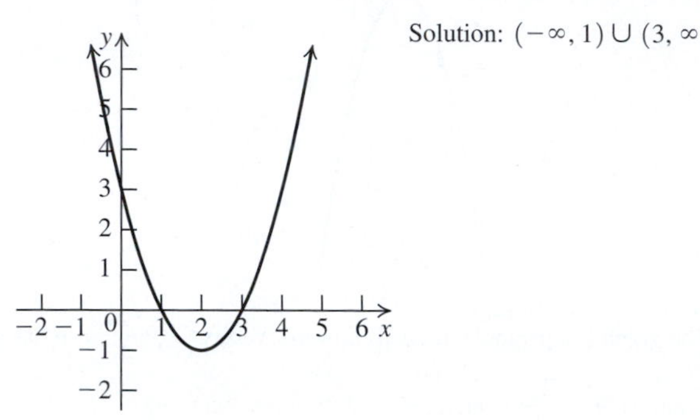

Solution: $(-\infty, 1) \cup (3, \infty)$

79.

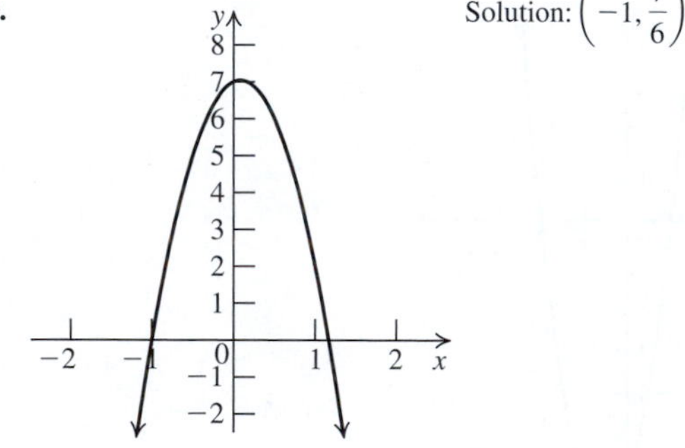

Solution: $\left(-1, \frac{7}{6}\right)$

Applying the Concepts:

81. a. ≈ 550 ft **b.** ≈ 140 ft **c.** 547 ft; 139 ft **83.** 19
85. 25 **87.** Square with 20-unit sides; maximum area: 400 square units
89. Dimensions: 75 m × 100 m; area: 7500 m^2 **91.** 38
93. 28 students; maximum revenue: \$1568
95. a. $h(t) = -0.8t^2 + 8t + 5$ (meters) **b.** 5 sec **97. a.** 64 ft
b. After 4 sec **99.** Dimensions of the rectangle: height $= \frac{18}{\pi + 4}$ ft; width $= \frac{36}{\pi + 4}$ ft **101. a.** $(59, 36.81)$ **b.** 120 ft **c.** 37 ft
d. 9 ft **e.** 113.6 ft

Beyond the Basics:

103. $f(x) = 3x^2 + 12x + 15$ **105.** $f(x) = 6x^2 - 12x + 4$
107. $f(x) = -2x^2 + 12x + 14$ **109.** Opening up: $x^2 - 4x - 12$; opening down: $-x^2 + 4x + 12$ **111.** Opening up: $x^2 + 8x + 7$; opening down: $-x^2 - 8x - 7$ **113.** $a = 2$

Maintaining Skills:

121. -1 **123.** $-\frac{1}{16}$ **125.** 1 **127.** $\frac{9x^2}{4}$ **129.** $(2x + 3)(2x - 3)$
131. $(3x + 4)(5x - 3)$ **133.** $(x - 2)(x - 1)(x + 2)$ **135.** $(x^2 + 3)(x + 4)$

Section 3.2

Answers to Practice Problems:

1. a. Not a polynomial function **b.** A polynomial function; the degree is 7, the leading term is $2x^7$, and the leading coefficient is 2.

2. $P(x) = 4x^3 + 2x^2 + 5x - 17 = x^3\left(4 + \frac{2}{x} + \frac{5}{x^2} - \frac{17}{x^3}\right)$

When $|x|$ is large, the terms $\frac{2}{x}, \frac{5}{x^2}$, and $-\frac{17}{x^3}$ are close to 0. Therefore, $P(x) \approx x^3(4 + 0 + 0 - 0) = 4x^3$. **3.** $y \to -\infty$ as $x \to -\infty$ and $y \to -\infty$ as $x \to \infty$ **4.** $\frac{3}{2}$ **5.** $f(1) = -7, f(2) = 4$. Because $f(1)$ and $f(2)$ have opposite signs, by this Intermediate Value Theorem, f has a real zero between 1 and 2. **6.** $-5, -1$, and 3 **7.** -5 and -3 with multiplicity 1 and 1 with multiplicity 2 **8.** 3

9.

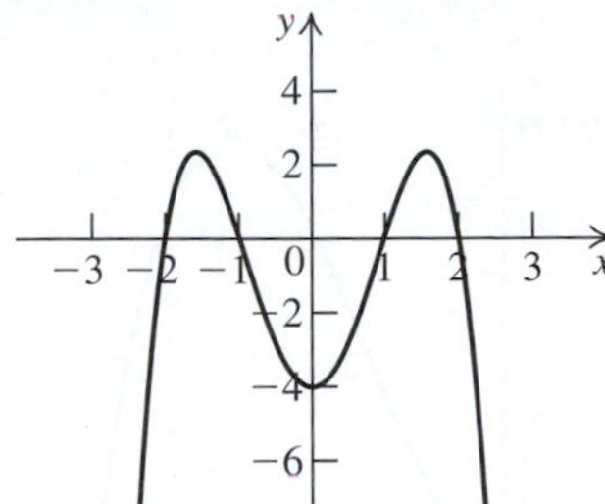

10. 207.378 liters

Basic Concepts and Skills:

1. 5; $2x^5$; 2; -6 **3.** end behavior **5.** zero **7.** $n - 1$
9. Polynomial function; degree: 5; leading term: $2x^5$; leading coefficient: 2
11. Polynomial function; degree: 3; leading term: $\frac{2}{3}x^3$; leading coefficient: $\frac{2}{3}$ **13.** Polynomial function; degree: 4; leading term: πx^4; leading coefficient: π **15.** Presence of $|x|$ **17.** Domain not $(-\infty, \infty)$ **19.** Presence of $\sqrt{x}$ **21.** Domain not $(-\infty, \infty)$ **23.** Graph not continuous **25.** Graph not continuous **27.** Not a function **29.** c **31.** a **33.** d **35.** Zeros: $x = -5$, multiplicity: 1, the graph crosses the x-axis; $x = 1$, multiplicity: 1, the graph crosses the x-axis; maximum number of turning points: 1 **37.** Zeros: $x = -1$, multiplicity: 2, the graph touches but does not cross the x-axis; $x = 1$, multiplicity: 3, the graph crosses the x-axis; maximum number of turning points: 4

39. Zeros: $x = -\frac{2}{3}$, multiplicity: 1, the graph crosses the x-axis; $x = \frac{1}{2}$, multiplicity: 2, the graph touches but does not cross the x-axis; maximum number of turning points: 2 **41.** Zeros: $x = 0$, multiplicity: 2, the graph touches but does not cross the x-axis; $x = 3$, multiplicity: 2, the graph touches but does not cross the x-axis; maximum number of turning points: 3 **43.** 2.09 **45.** 2.28 **47. a.** $y \to \infty$ as $x \to -\infty$ and $y \to \infty$ as $x \to \infty$ **b.** No zeros **c.** The graph is above the x-axis on $(-\infty, \infty)$. **d.** 3 **e.** y-axis symmetry **f.** 1 **g.**

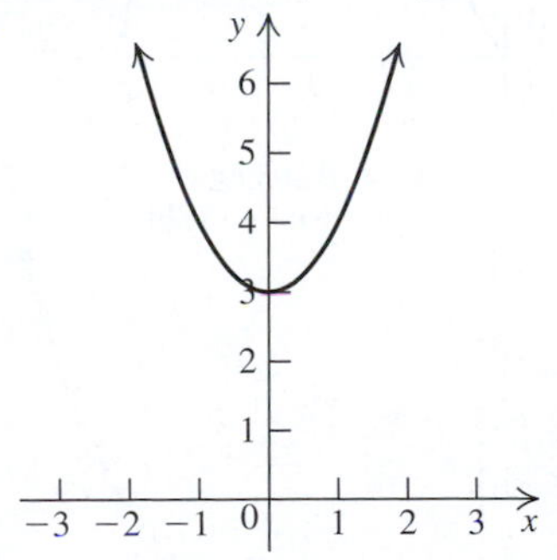

49. a. $y \to \infty$ as $x \to -\infty$ and $y \to \infty$ as $x \to \infty$ **b.** Zeros: $x = -7$, the graph crosses the x-axis; $x = 3$, the graph crosses the x-axis. **c.** The graph is above the x-axis on $(-\infty, -7)$ and $(3, \infty)$ and below the x-axis on $(-7, 3)$. **d.** -21 **e.** symmetric about $x = -2$ **f.** 1

g.

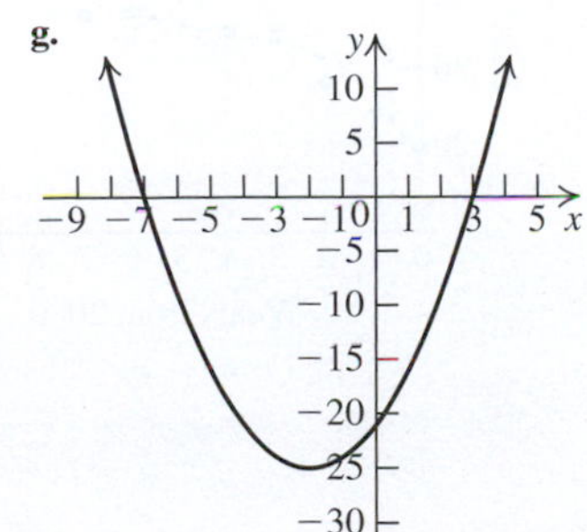

51. a. $y \to \infty$ as $x \to -\infty$ and $y \to -\infty$ as $x \to \infty$ **b.** Zeros: $x = -1$, the graph crosses the x-axis; $x = 0$, the graph touches but does not cross the x-axis.
c. The graph is above the x-axis on $(-\infty, -1)$ and below the x-axis on $(-1, 0) \cup (0, \infty)$. **d.** 0 **e.** no symmetries **f.** 2

g.

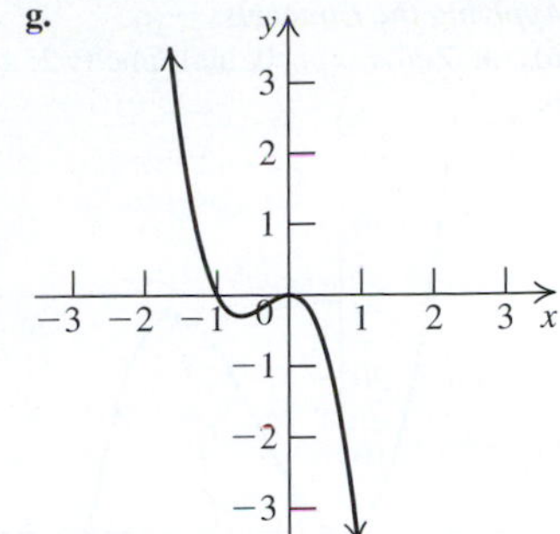

53. a. $y \to \infty$ as $x \to -\infty$ and $y \to \infty$ as $x \to \infty$
b. Zeros: $x = 0$, the graph touches but does not cross the x-axis; $x = 1$, the graph touches but does not cross the x-axis.
c. The graph is above the x-axis on $(-\infty, 0) \cup (0, 1) \cup (1, \infty)$.
d. 0 **e.** no symmetries **f.** 3

g.

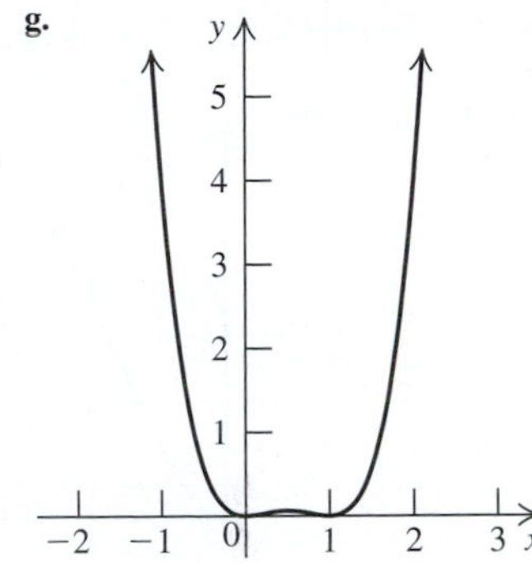

55. a. $y \to \infty$ as $x \to -\infty$ and $y \to \infty$ as $x \to \infty$ **b.** Zeros: $x = -3$, the graph crosses the x-axis; $x = 1$, the graph touches but does not cross the x-axis; $x = 4$, the graph crosses the x-axis. **c.** The graph is above the x-axis on $(-\infty, -3)$ and $(4, \infty)$ and below the x-axis on $(-3, 1) \cup (1, 4)$. **d.** -12
e. no symmetries **f.** 3

g.

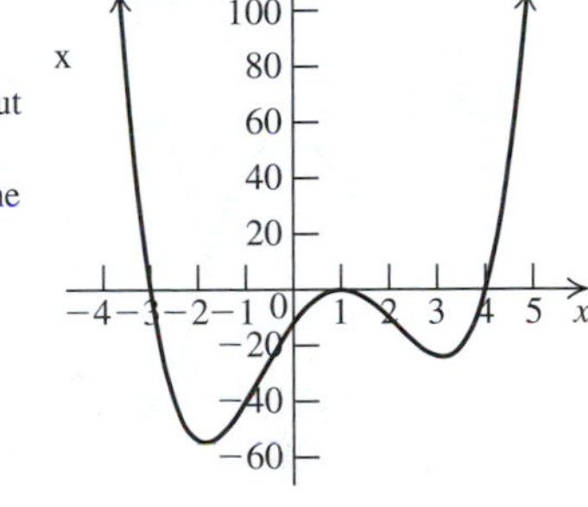

57. a. $y \to \infty$ as $x \to -\infty$ and $y \to -\infty$ as $x \to \infty$ **b.** Zeros: $x = -1$, the graph touches but does not cross the x-axis; $x = 0$, the graph touches but does not cross the x-axis; $x = 1$, the graph crosses the x-axis. **c.** The graph is above the x-axis on $(-\infty, -1) \cup (-1, 0) \cup (0, 1)$ and below the x-axis on $(1, \infty)$. **d.** 0 **e.** no symmetries **f.** 4

g.

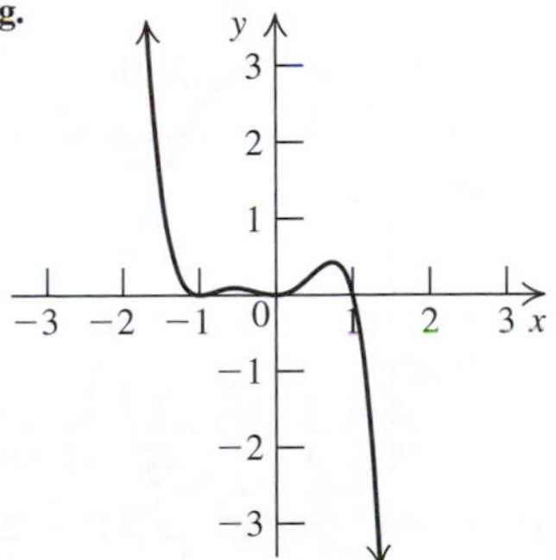

59. a. $y \to \infty$ as $x \to -\infty$ and $y \to \infty$ as $x \to \infty$ **b.** Zeros: $x = -2$, the graph crosses the x-axis; $x = -1$, the graph crosses the x-axis; $x = 0$, the graph crosses the x-axis; $x = 1$, the graph crosses the x-axis. **c.** The graph is above the x-axis on $(-\infty, -2) \cup (-1, 0) \cup (1, \infty)$ and below the x-axis on $(-2, -1) \cup (0, 1)$. **d.** 0 **e.** no symmetries **f.** 3

g.

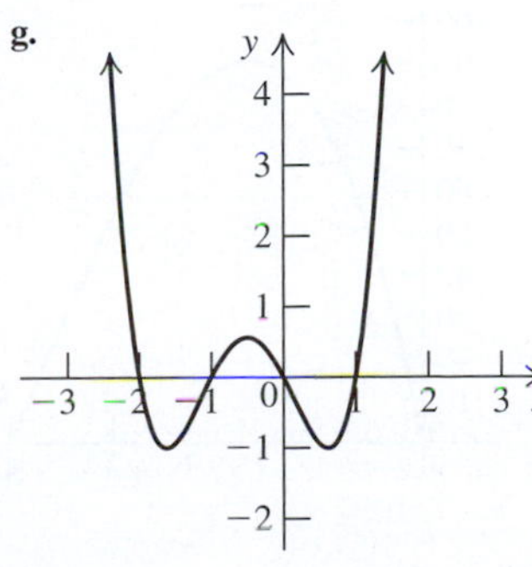

Applying the Concepts:

61. a. Zeros: $x = 0$, multiplicity 2; $x = 4$, multiplicity 1

b.

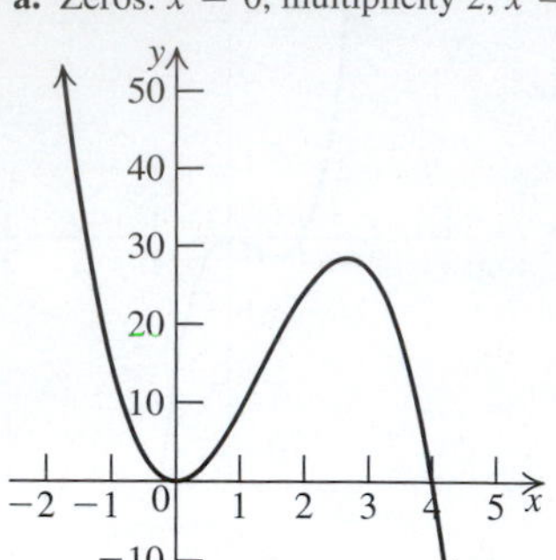

c. 2 **d.** Domain: $[0, 4]$. The portion between the x-intercepts constitutes the graph of $R(x)$.

63. a. $[0, 1]$ **b.**

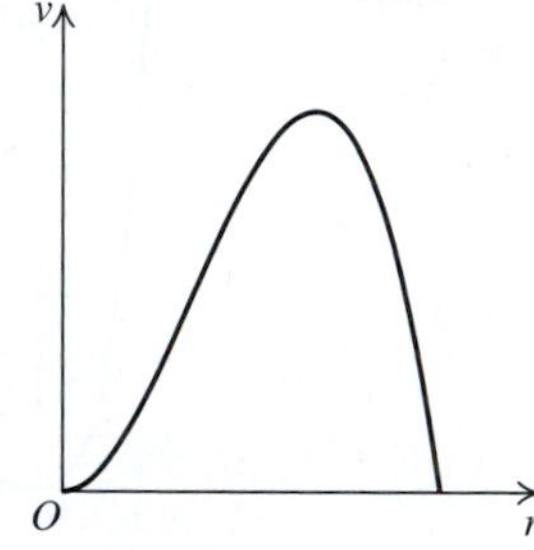

65. a. $R(x) = 27x - \frac{1}{90{,}000}x^3$ **b.** $[0, 900\sqrt{3}]$

c.

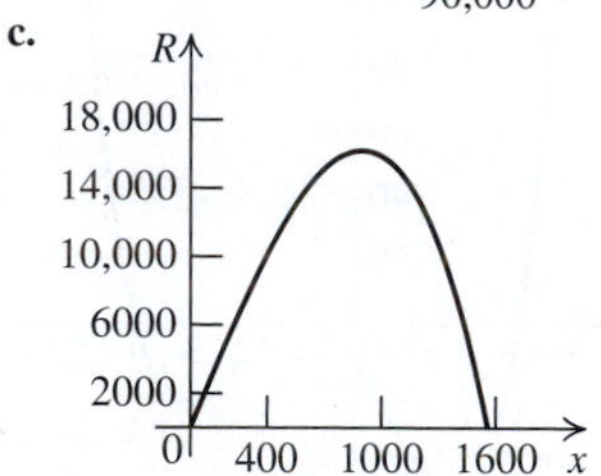

67. a. $N(x) = (x + 12)(400 - 2x^2)$ **b.** $[0, 10\sqrt{2}]$

c.

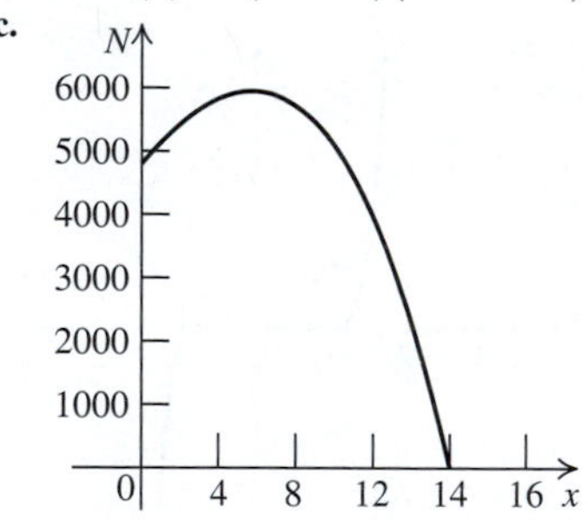

69. a. $V(x) = x(8 - 2x)(15 - 2x)$

b.

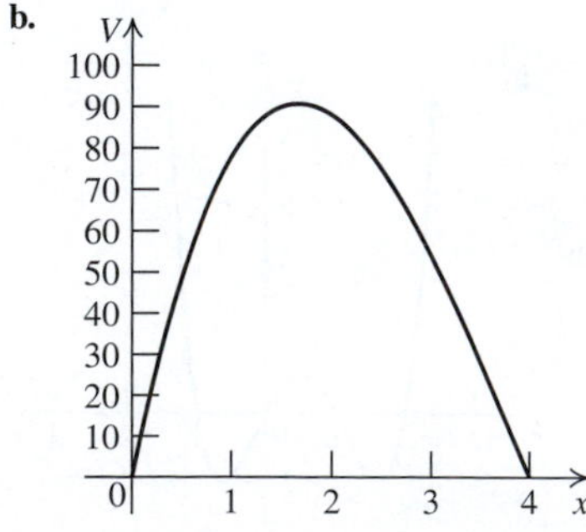

71. a. $V(x) = x^2(108 - 4x)$. **b.**

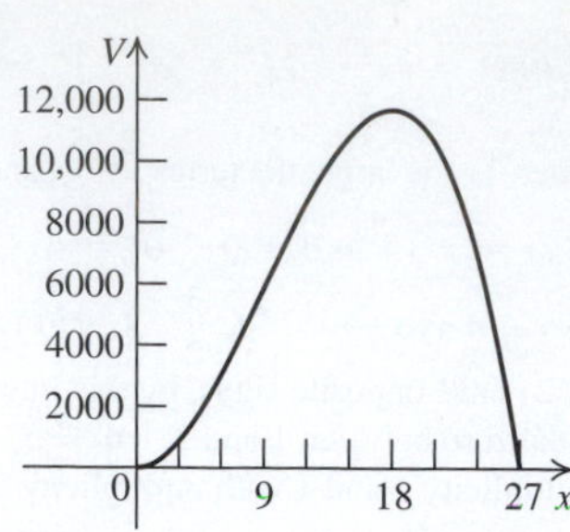

73. a. $V(x) = x^2(62 - 2x)$ **b.**

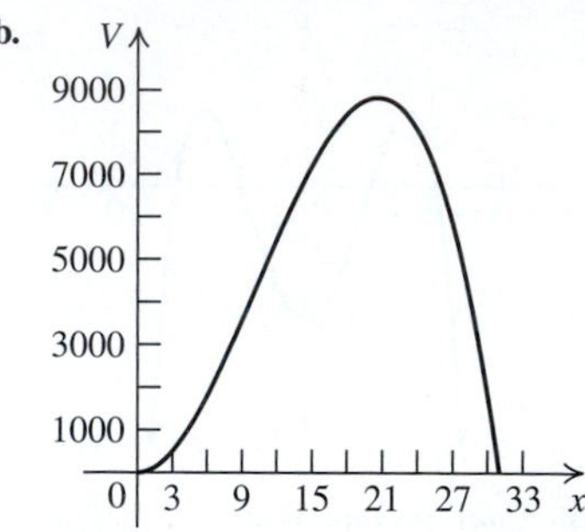

75. a. $V(x) = 6x^2(15 - x)$ **b.**

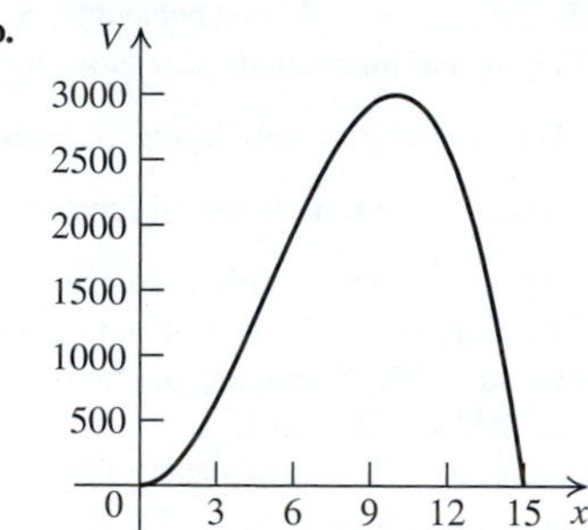

77.

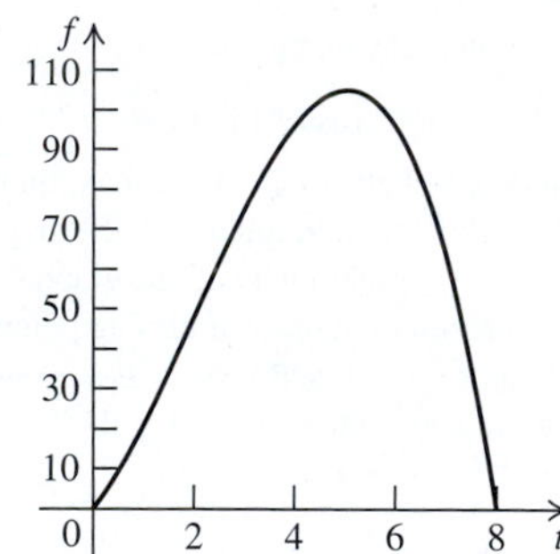

79.

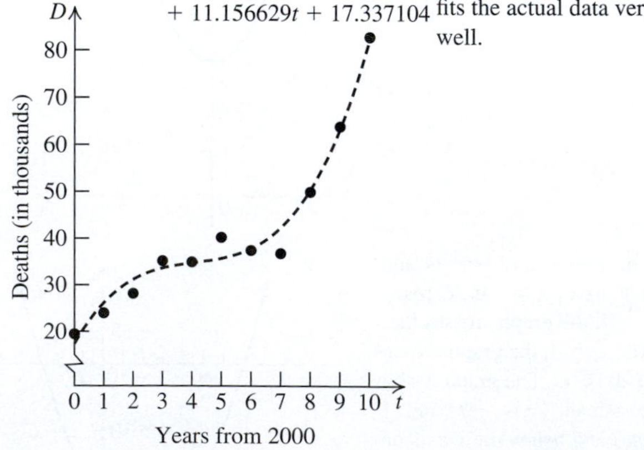

The model function $D(t)$ fits the actual data very well.

Beyond the Basics:

81. The graph of f is the graph of $y = x^4$ shifted 1 unit to the right.

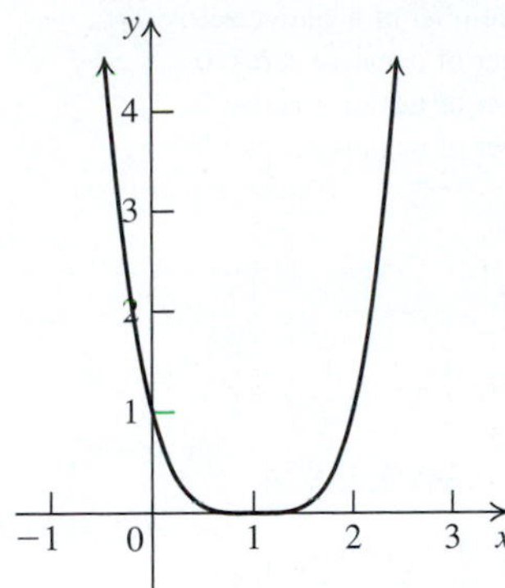

Zero: $x = 1$, multiplicity: 4

83. The graph of f is the graph of $y = x^4$ shifted 2 units up.

No zeros

85. The graph of f is the graph of $y = x^4$ shifted 1 unit to the right and reflected about the x-axis.

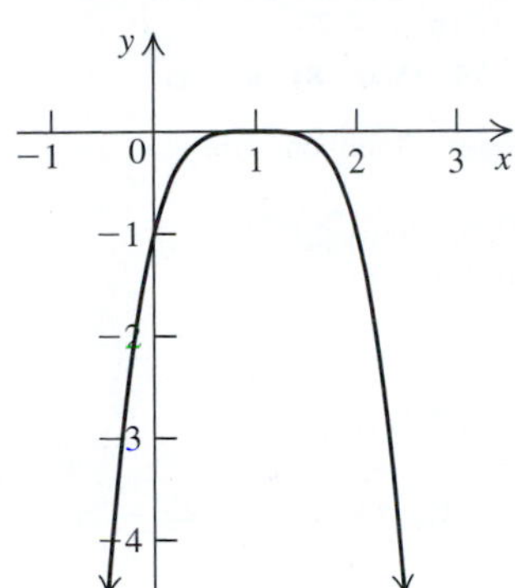

Zero: $x = 1$, multiplicity: 4

87. The graph of f is the graph of $y = x^5$ shifted 1 unit up.

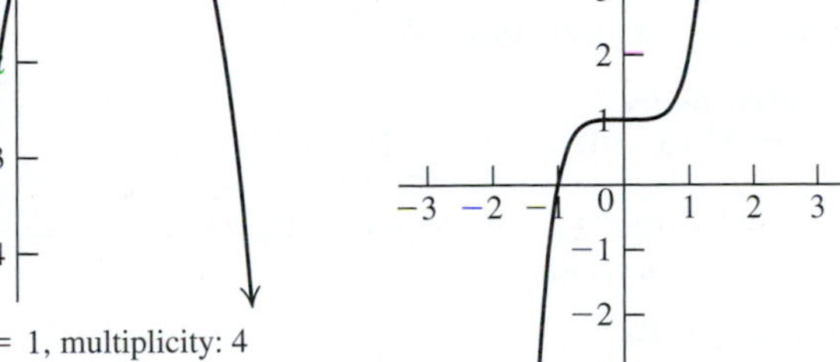

Zero: $x = -1$, multiplicity: 1

89. The graph of f is the graph of $y = x^5$ shifted 1 unit to the left, compressed by a factor of $\frac{1}{4}$, reflected about the x-axis, and shifted 8 units up.

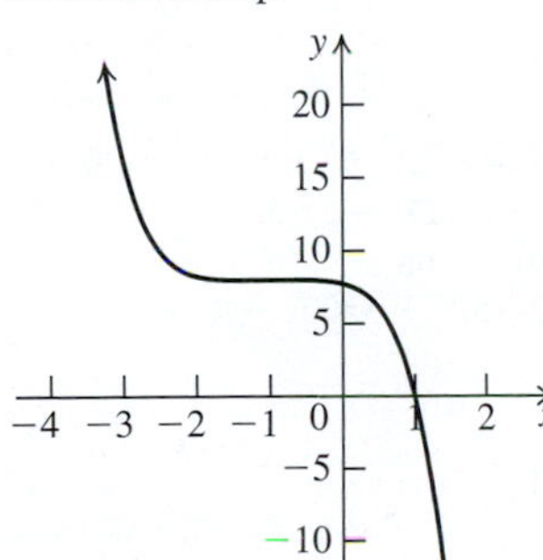

Zero: $x = 1$, multiplicity: 1

93. Maximum vertical distance: ≈ 0.25.It occurs at $x \approx 0.71$.

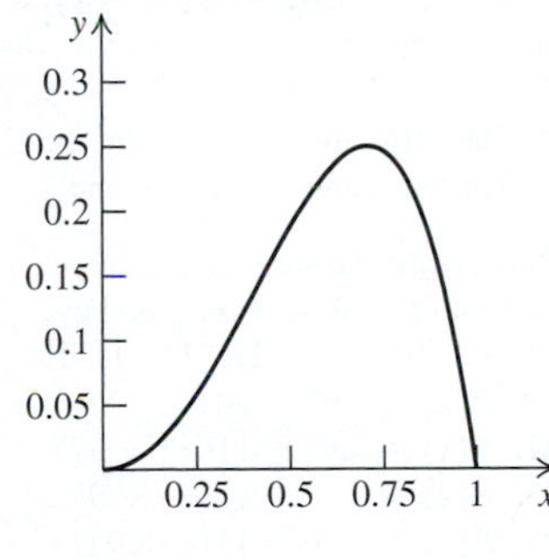

95. $f(x) = x^2(x + 1)$ **97.** $f(x) = 1 - x^4$ **99.** Degree 5 because the graph has 5 x-intercepts and four turning points **101.** Degree 6 because the graph has five turning points

Critical Thinking/Discussion/Writing:

103. No, because the domain of any polynomial function is $(-\infty, \infty)$, which includes the point $x = 0$. **105.** Not possible; this is inconsistent with its end behavior. **107.** $P \circ Q$ is a polynomial of degree $m \cdot n$

Maintaining Skills:

109. $\frac{1}{x^2}$ **111.** $x^3 - 3$ **113. a.** $7 = 1 \cdot 7$ **b.** $7 = 7 \cdot 1$ **c.** $7 = (-1)(-7)$ **d.** $7 = (-7)(-1)$

115. $\pm 1, \pm 2, \pm 3, \pm 4, \pm 5, \pm 6, \pm 8, \pm 10, \pm 12, \pm 15, \pm 20, \pm 24, \pm 30, \pm 40, \pm 60, \pm 120$ **117.** $g(3) = 70$ **119.** $f(-12) = 0$ **121.** $(2x + 1)(1 - x)$ **123.** $(x - 9)(3x + 1)(2x - 3)$

Section 3.3

Answers to Practice Problems:

1. Quotient: $3x + 4$; remainder $= -2x + 3$

2. $x^2 + x + 3 + \frac{2x - 3}{x^2 - x + 3}$ **3.** $2x^2 - x - 3 + \frac{-4}{x - 3}$

4. $2x^2 - 5x - 3 + \frac{2}{x - 3}$ **5.** 4 **6.** 32 **7.** $\left\{-2, -\frac{2}{3}, 3\right\}$ **8.** 18

Basic Concepts and Skills:

1. $x^4 - 2x^3 + 5x^2 - 2x + 1$; $x^2 - 2x + 3$; $x^2 + 2$; $2x - 5$ **3.** $F(a)$ **5.** true **7.** Quotient: $3x - 2$; remainder: 0 **9.** Quotient: $3x^3 - 3x^2 - 3x + 6$; remainder: -13 **11.** Quotient: $2x - 1$; remainder: 0 **13.** Quotient: $z^2 + 2z + 1$; remainder: 0 **15.** Quotient: $x^2 - 7$; remainder: -5 **17.** Quotient: $x^2 + 2x - 11$; remainder: 12 **19.** Quotient: $x^3 - x^2 + 4$; remainder: 13

21. Quotient: $2x^2 + 5x - \frac{1}{2}$; remainder: $\frac{3}{4}$

23. Quotient: $2x^2 - 6x + 6$; remainder: -1
25. Quotient: $x^4 + 2x^3 - 5x^2 - 3x - 2$; remainder: -3
27. Quotient: $x^4 - x^3 + x^2 - x + 1$; remainder: 0
29. Quotient: $x^5 - x^4 + 3x^3 - 4x^2 + 4x - 4$; remainder: 9

31. a. 5 **b.** 3 **c.** $\frac{15}{8}$ **d.** 1301 **33. a.** -17 **b.** -27 **c.** -56 **d.** 24 **43.** $k = -1$ **45.** $k = -7$ **47.** Remainder: 1 **49.** Remainder: 6

Applying the Concepts:

51. $2x^2 + 1$ **53. a.** $R(x) = -2x^2 + 200x - 4800$ **b.** $R(x) = -2x^2 + 200x - 1800$ **c.** The maximum weekly revenue is \$3200 if the phone is priced at \$50. **55.** 2008 **57.** 2006

Beyond the Basics:

61. a. n is an odd integer. **b.** n is an even integer.
c. There is no such n. **d.** n is any positive integer.
63. a. Remainder: 2 **b.** Remainder: -2
69. For example, $g(x) = x^4 + 1$

Maintaining Skills:

71. $x^3 + 2x + 3; 0$ **73.** $-x^5 + 2x; 3$ **75.** $\pm 1, \pm 11$ **77.** $\pm 1, \pm 3, \pm 17, \pm 51$ **79.** degree $= 2$; leading coefficient $= 1$; constant term $= -1$ **81.** degree $= 10$; leading coefficient $= -7$; constant term $= 4$

Section 3.4

Answers to Practice Problems:

1. $\{-2\}$ **2.** $\frac{1}{2}, 2 + \sqrt{3}, 2 - \sqrt{3}$ **3.** Number of positive zeros: 1; number of negative zeros: 0 or 2 **4.** Upper bound: 1; lower bound: -3

5. $-\sqrt{2}, \frac{2}{3}, \sqrt{2}, 3$ **6.**

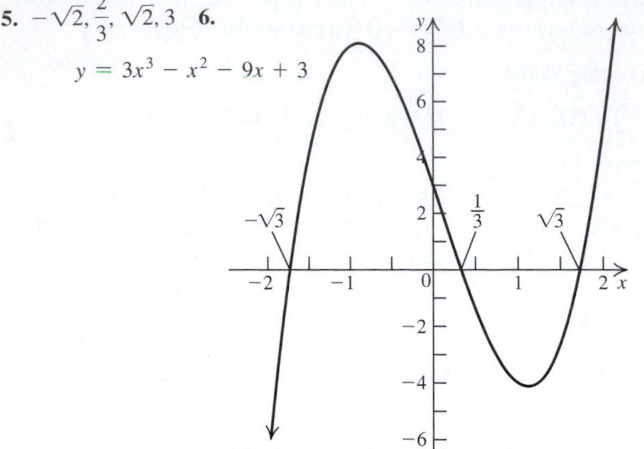

Basic Concepts and Skills:

1. a_0; a_n **3.** variations; even integer **5.** false

7. $\left\{\pm\frac{1}{3}, \pm 1, \pm\frac{5}{3}, \pm 5\right\}$

9. $\left\{\pm\frac{1}{4}, \pm\frac{1}{2}, \pm\frac{3}{4}, \pm 1, \pm\frac{3}{2}, \pm 2, \pm 3, \pm 6\right\}$

11. $\{-2, 1, 2\}$ **13.** $\{-1\}$ **15.** $\left\{-3, \frac{1}{2}, 2\right\}$ **17.** $\left\{\frac{2}{3}\right\}$ **19.** $\left\{-\frac{1}{3}\right\}$

21. $\{-1, 2\}$ **23.** $\{-3, -1, 1, 4\}$ **25.** No rational zeros

27.

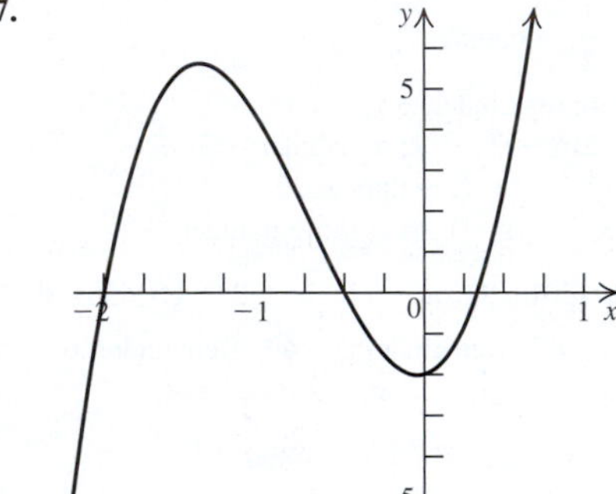

29.

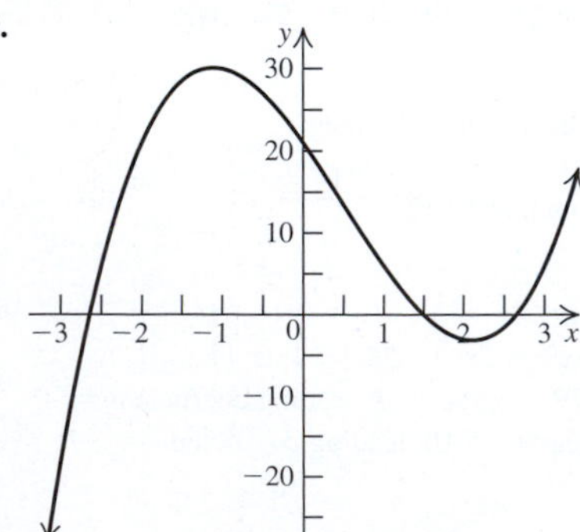

31.

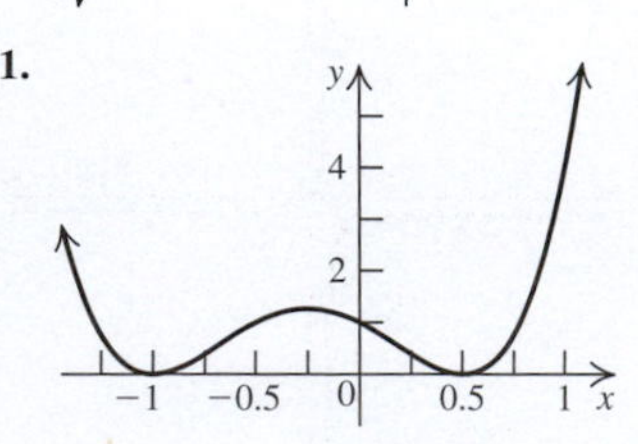

33. Number of positive zeros: 0 or 2; number of negative zeros: 1
35. Number of positive zeros: 0 or 2; number of negative zeros: 1
37. Number of positive zeros: 1 or 3; number of negative zeros: 0 or 2
39. Number of positive zeros: 1 or 3; number of negative zeros: 1
41. Number of positive zeros: 0; number of negative zeros: 0
43. Number of positive zeros: 0; number of negative zeros: 0
45. Number of positive zeros: 1; number of negative zeros: 0
47. Upper bound: 1; lower bound: -1
49. Upper bound: 1; lower bound: -3
51. Upper bound: 31; lower bound: -31
53. Upper bound: 4 lower bound: -4

55. $\{1, -3 + \sqrt{11}, -3 - \sqrt{11}\}$ **57.** $\left\{-\sqrt{3}, \frac{1}{2}, \sqrt{3}\right\}$

59. $\left\{\frac{1}{2}, 2 - \sqrt{3}, 2 + \sqrt{3}\right\}$ **61.** $\{-2, -\sqrt{3}, 1, \sqrt{3}\}$

63. $\left\{-1, 1, \frac{3 + \sqrt{5}}{2}, \frac{3 - \sqrt{5}}{2}\right\}$ **65.** $\left\{-\sqrt{3}, \frac{1}{2}, \sqrt{3}, 2\right\}$

67. $\left\{-\sqrt{2}, -\frac{1}{3}, \frac{1}{2}, \sqrt{2}\right\}$ **69.** $\{-\sqrt{3}, -1, 1, \sqrt{3}, 2\}$

71. $\left\{-2, -\sqrt{3}, \frac{1}{2}, 1, \sqrt{3}\right\}$ **73.** $\left\{-\frac{1}{2}, 1, 2, 3\right\}$

Applying the Concepts:

75. Length: 51; width: 6 **77.** $x = 4$ **79.** 1500 **81. a.** 5 yr **b.** $y = -\frac{1}{18}x(x - 3)(x - 8)$ **c.** export 1.7 million barrels

83. a. $y = \frac{1}{54}x(x - 6)(x - 8)$ **b.** \$592.59 profit **c.** 2004, 2007 **d.** \$898.46 in 2006

Beyond the Basics:

89. $y = x^3 - x$ **91.** $y = 6x^3 - 23x^2 + 15x + 14$
93. $y = x^3 - 5x^2 + 5x + 3$ **95.** $y = x^4 - 8x^3 + 17x^2 - 2x - 14$
97. $y = x^4 - 10x^2 + 1$ **99.** $y = x^4 - 12x^3 + 40x^2 - 24x - 36$
103. a. False **b.** False

Maintaining Skills:

105. $-32i$ **107.** $-4x^2 - 1$ **109.** $25x^2 + 4$
111. $x^2 - 2x + 5$ **113.** $x^2 - 4x + 13$

Section 3.5

Answers to Practice Problems:

1. a. $P(x) = 3(x + 2)(x - 1)(x - 1 - i)(x - 1 + i)$
b. $P(x) = 3x^4 - 3x^3 - 6x^2 + 18x - 12$
2. $-3, -3, 2 + 3i, 2 + 3i, 2 - 3i, 2 - 3i, i, -i$
3. The zeros are 1, 2, $2i$, and $-2i$.
4. The zeros are 2, 4, $1 + i$, and $1 - i$.

Beyond the Basics:

1. complex **3.** $a - bi$ **5.** false **7.** $x = \pm 5i$ **9.** $x = -2 \pm 3i$
11. $\{2, 3i, -3i\}$ **13.** $3 - i$ **15.** $-5 - i$ **17.** $-i, -3i$
19. $P(x) = 2x^4 - 20x^3 + 70x^2 - 180x + 468$
21. $P(x) = 7x^5 - 119x^4 + 777x^3 - 2415x^2 + 3500x - 1750$
23. $P(x) = x(x + 1)(x^2 + 9)$
25. $P(x) = x(x - 1)(x + 2)(x^2 - 6x + 10)$
27. $1, 4 + i, 4 - i$ **29.** $-\frac{4}{3}, 1 + 3i, 1 - 3i$
31. 1 (with multiplicity 2), $\frac{3}{2} + \frac{3}{2}i, \frac{3}{2} - \frac{3}{2}i$ **33.** $-3, 1, 3 + i, 3 - i$
35. $\frac{1}{2}, 2, 3, i, -i$ **37.** $f(x) = \frac{1}{3}(x^2 + 9)$
39. $f(x) = \frac{1}{2}(2 - x)(x^2 + 1)(x^2 + 4)$

Beyond the Basics:

41. There are three cube roots: 1, $-\frac{1}{2} + \frac{\sqrt{3}}{2}i$, and $-\frac{1}{2} - \frac{\sqrt{3}}{2}i$
43. $P(x) = (x - 2)(x + i)$ **45.** $P(x) = (x - 4)(x + 1)(x - i)$

47. $f(x) = -4(x - 2)(x^2 - 2x + 5)$
$y \to \infty$ as $x \to -\infty$, $y \to -\infty$ as $x \to \infty$
49. $f(x) = -2(x^2 - 1)(x^2 - 6x + 10)$
$y \to -\infty$ as $x \to -\infty$, $y \to -\infty$ as $x \to \infty$

Maintaining Skills:

53. $\{-4, -1 + i, -1 - i\}$ **55.** $\left\{1, \frac{3}{2}\right\}$
57. $\{-3, 0, 4\}$ **59.** $Q(x) = x + 1; R(x) = -2$
61. $Q(x) = 4x + 3; R(x) = -4x^2 - 3x - 5$ **63.** -8 **65.** 11

Section 3.6

Answers to Practice Problems:

1. Domain: $(-\infty, -1) \cup (-1, 5) \cup (5, \infty)$
2. a. Domain: $(-\infty, 2) \cup (2, \infty)$; range: $(-\infty, 0) \cup (0, \infty)$; Vertical asymptote: $x = 2$; horizontal asymptote: $y = 0$

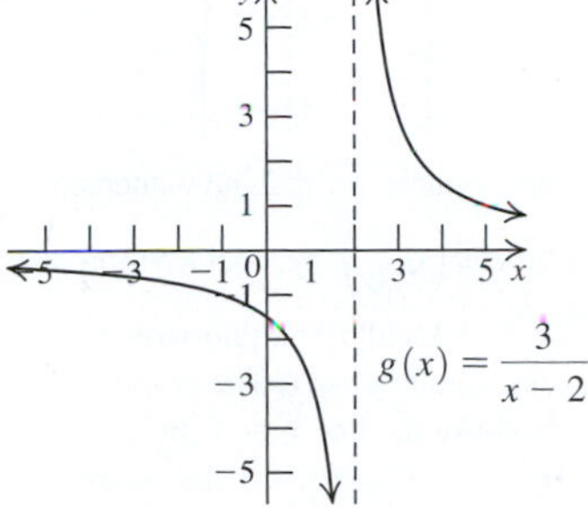

b. Domain: $(-\infty, -1) \cup (-1, \infty)$; range: $(-\infty, 2) \cup (2, \infty)$; vertical asymptote: $x = -1$; Horizontal asymptote: $y = 2$

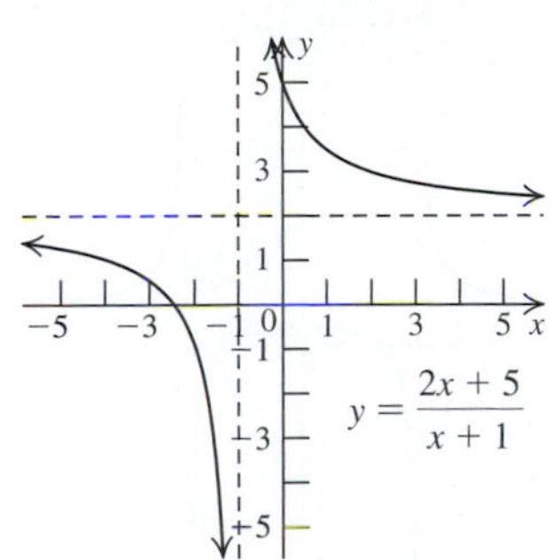

3. $x = -5$ and $x = 2$ **4.** $x = -3$ **5. a.** $y = \frac{2}{3}$
b. No horizontal asymptote **c.** $y = 0$

6. x-intercept: 0; y-intercept: 0; vertical asymptotes: $x = -1$ and $x = 1$; horizontal asymptote: x-axis. The graph is above the x-axis on $(-1, 0) \cup (1, \infty)$ and below the x-axis on $(-\infty, -1) \cup (0, 1)$.

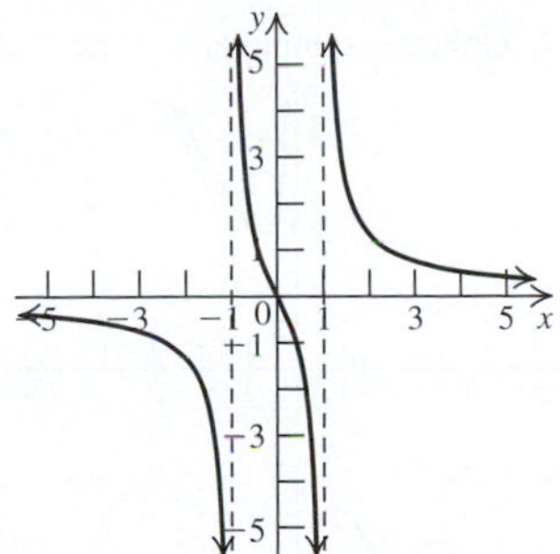

7. x-intercepts: $\pm\frac{\sqrt{2}}{2}$; y-intercept: $\frac{1}{3}$; vertical asymptotes: $x = -\frac{3}{2}$ and $x = 1$; horizontal asymptote: $y = 1$; The graph is above the line $y = 1$ on $\left(-\infty, -\frac{3}{2}\right) \cup (1, \infty)$ and below it on $\left(-\frac{3}{2}, 1\right)$. The graph crosses the horizontal at $(2, 1)$.

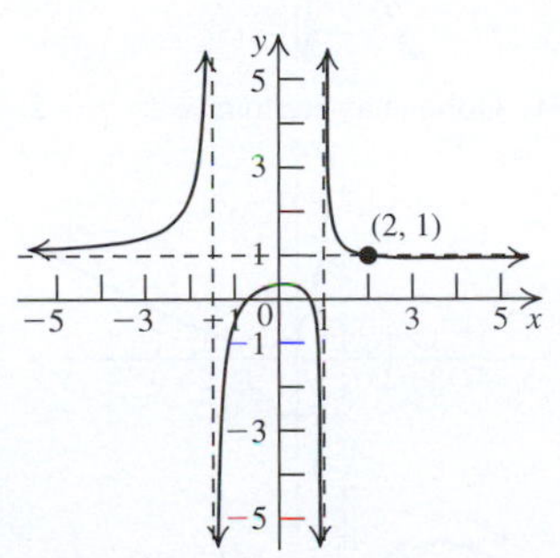

8. No x-intercept; y-intercept: $\frac{1}{2}$; no vertical asymptote; horizontal asymptote: $y = 1$. The graph is below the line $y = 1$ on $(-\infty, \infty)$.

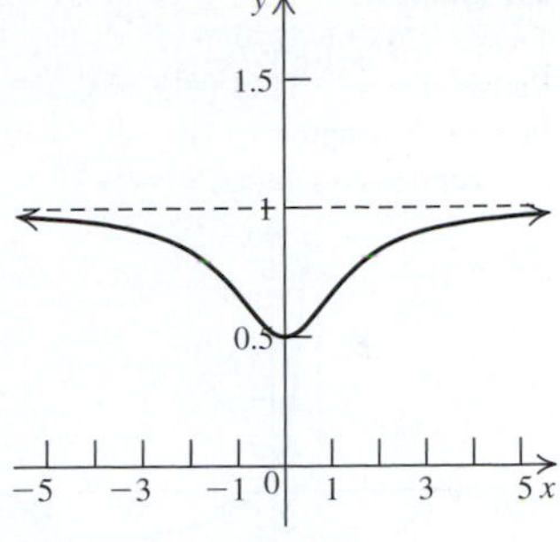

9. No x-intercept; y-intercept: -2; vertical asymptote: $x = 1$; no horizontal asymptote; $y = x + 1$ is an oblique asymptote. The graph is above the line $y = x + 1$ on $(1, \infty)$ and below it on $(-\infty, 1)$.

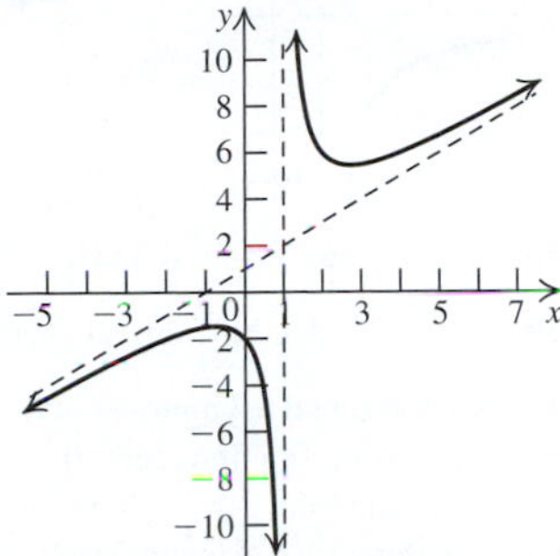

10. a. $R(10) = \$30$ billion
$R(20) = \$40$ billion
$R(30) = \$42$ billion
$R(40) = \$40$ billion
$R(50) = \$35.7$ billion
$R(60) = \$30$ billion

b.

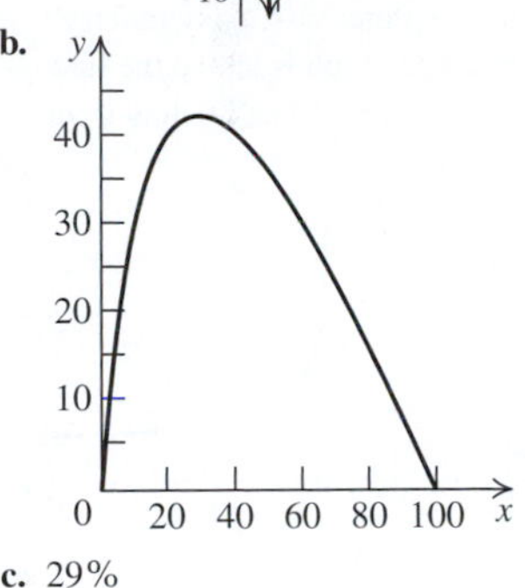

c. 29%

Basic Concepts and Skills:

1. $\frac{N(x)}{D(x)}$, where $N(x)$ and $D(x)$ are polynomials and $D(x)$ is not the zero polynomial **3.** $\infty; -\infty$ **5.** false **7.** $(-\infty, -4) \cup (-4, \infty)$
9. $(-\infty, \infty)$ **11.** $(-\infty, -2) \cup (-2, 3) \cup (3, \infty)$
13. $(-\infty, 2) \cup (2, 4) \cup (4, \infty)$ **15.** ∞ **17.** ∞ **19.** 1
21. $(-\infty, -2) \cup (-2, 1) \cup (1, \infty)$ **23.** $x = -2; x = 1$
25. Domain: $(-\infty, 4) \cup (4, \infty)$
Range: $(-\infty, 0) \cup (0, \infty)$
Vertical Asymptote: $x = 4$
Horizontal Asymptote: $y = 0$

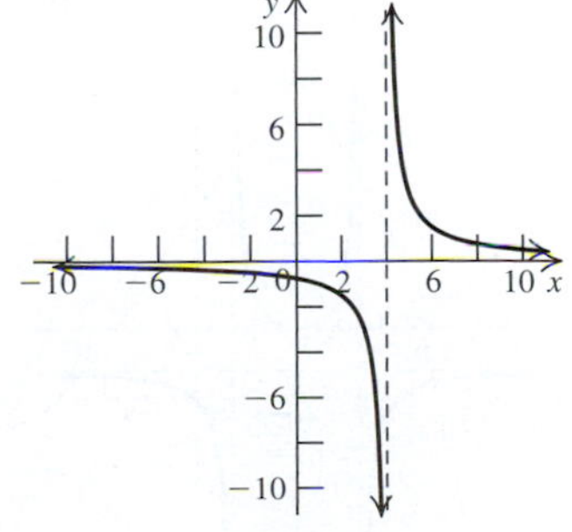

27. Domain: $\left(-\infty, -\frac{1}{3}\right) \cup \left(-\frac{1}{3}, \infty\right)$

Range: $\left(-\infty, -\frac{1}{3}\right) \cup \left(-\frac{1}{3}, \infty\right)$

Vertical Asymptote: $x = -\frac{1}{3}$

Horizontal Asymptote: $y = -\frac{1}{3}$

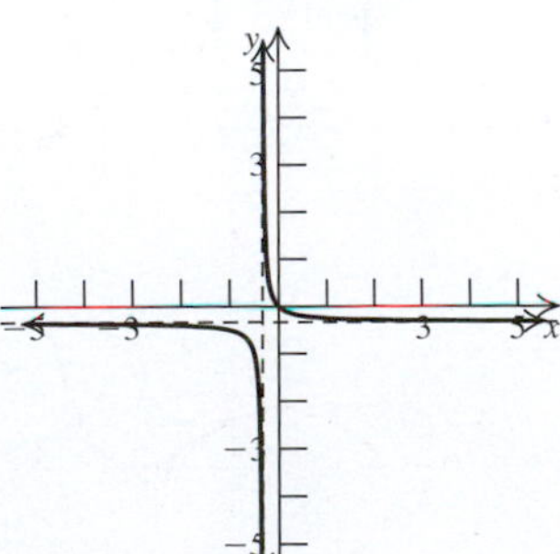

29. Domain: $(-\infty, -2) \cup (-2, \infty)$
Range: $(-\infty, -3) \cup (-3, \infty)$
Vertical Asymptote: $x = -2$
Horizontal Asymptote: $y = -3$

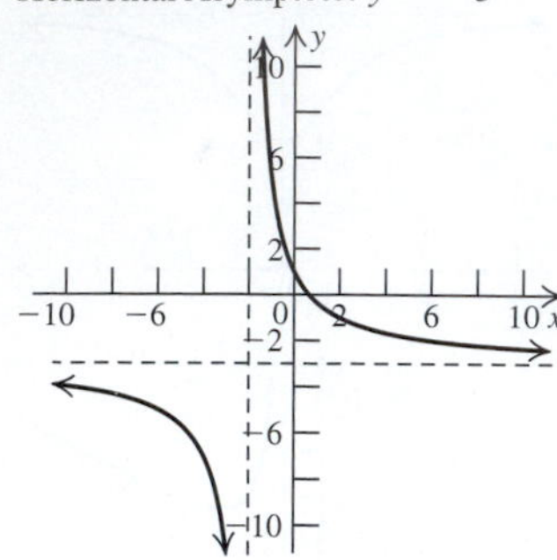

31. Domain: $(-\infty, 4) \cup (4, \infty)$
Range: $(-\infty, 5) \cup (5, \infty)$
Vertical Asymptote: $x = 4$
Horizontal Asymptote: $y = 5$

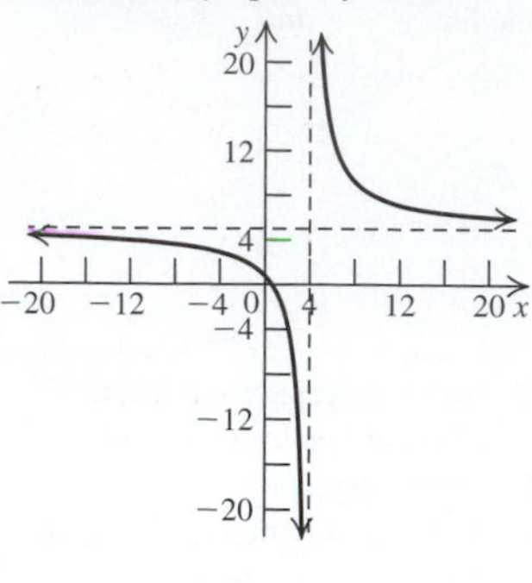

33. $x = 1$ **35.** $x = -4$ and $x = 3$ **37.** $x = -3$ and $x = 2$

39. $x = -3$ **41.** No vertical asymptote **43.** $y = 0$ **45.** $y = \frac{2}{3}$

47. No horizontal asymptote **49.** $y = 0$ **51.** d **53.** e **55.** a

57. x-intercept: 0; y-intercept: 0; vertical asymptote: $x = 3$; horizontal asymptote: $y = 2$; symmetry: none. The graph is above the line $y = 2$ on $(3, \infty)$ and below it on $(-\infty, 3)$.

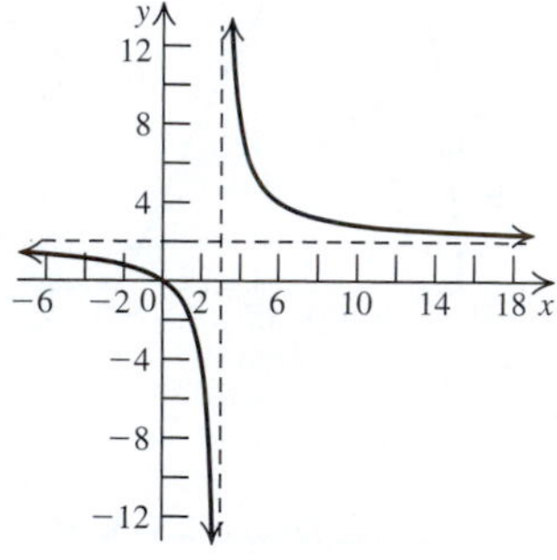

59. x-intercept: 0; y-intercept: 0; vertical asymptotes: $x = -2$ and $x = 2$; horizontal asymptote: x-axis. The graph is above the x-axis on $(-2, 0) \cup (2, \infty)$ and below the x-axis on $(-\infty, -2) \cup (0, 2)$.

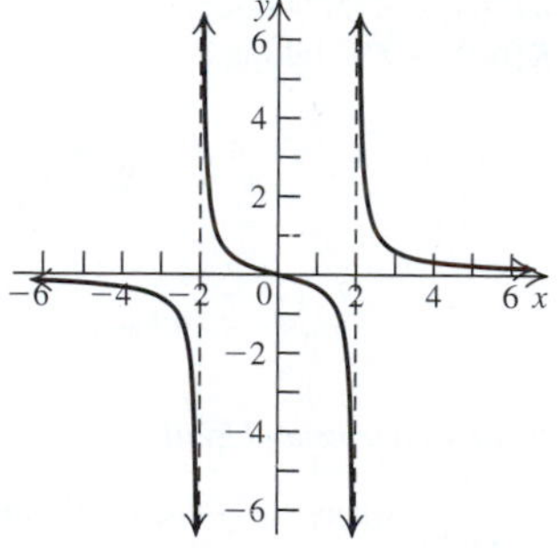

61. x-intercept: 0; y-intercept: 0; vertical asymptotes: $x = -3$ and $x = 3$; horizontal asymptote: $y = -2$. The graph is above the line $y = -2$ on $(-3, 3)$ and below it on $(-\infty, -3) \cup (3, \infty)$.

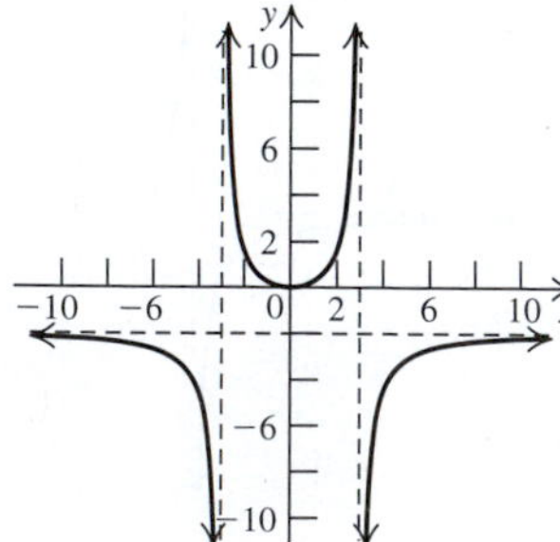

63. No x-intercept; y-intercept: -1; vertical asymptotes: $x = -\sqrt{2}$ and $x = \sqrt{2}$; horizontal asymptote: x-axis. The graph is above the x-axis on $(-\infty, -\sqrt{2}) \cup (\sqrt{2}, \infty)$ and below the x-axis on $(-\sqrt{2}, \sqrt{2})$.

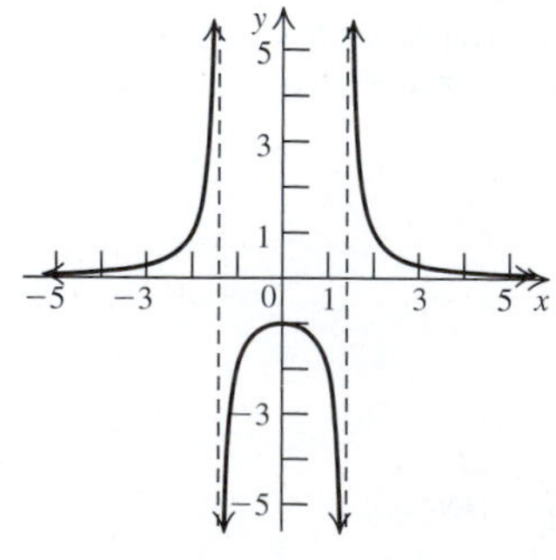

65. x-intercept: -1; y-intercept: $-\frac{1}{6}$; vertical asymptotes: $x = -3$ and $x = 2$; horizontal asymptote: x-axis. The graph is above the x-axis on $(-3, -1) \cup (2, \infty)$ and below the x-axis on $(-\infty, -3) \cup (-1, 2)$.

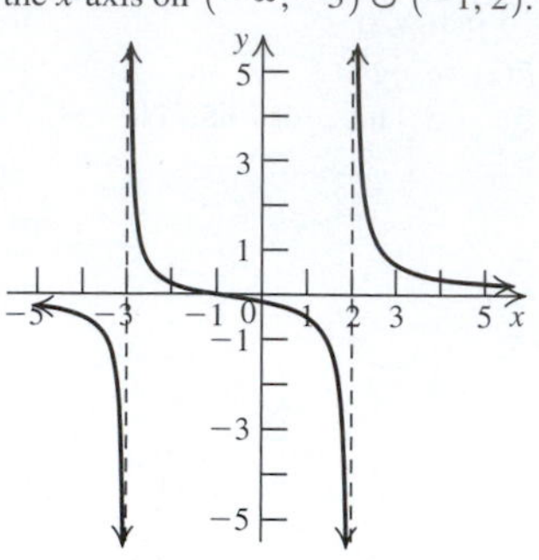
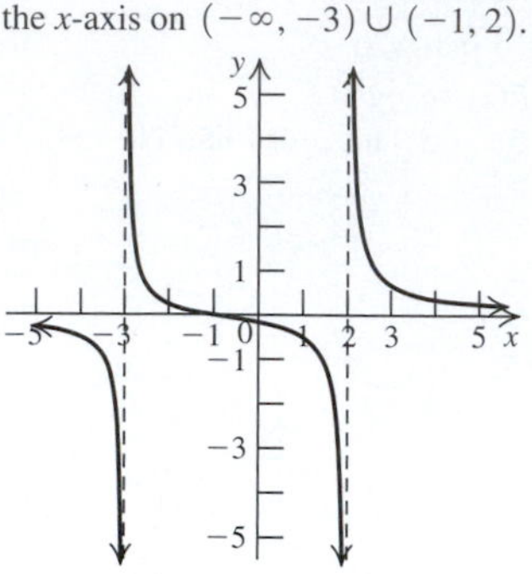

67. x-intercept: 0; y-intercept: 0; no vertical asymptote; horizontal asymptote: $y = 1$. The graph is above the x-axis on $(-\infty, 0) \cup (0, \infty)$ and below the line $y = 1$ on $(-\infty, \infty)$.

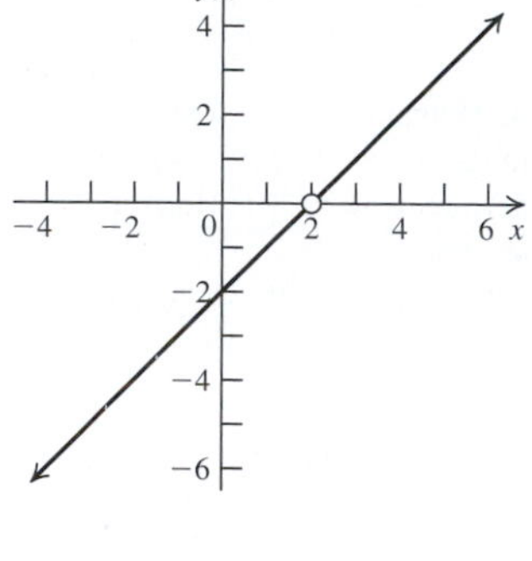

69. x-intercept: ± 2; no y-intercept: a hole at $\left(0, \frac{4}{9}\right)$; vertical asymptotes: $x = -3$ and $x = 3$; horizontal asymptote: $y = 1$. The graph is above the line $y = 1$ on $(-\infty, -3) \cup (3, \infty)$ and below it on $(-3, 0) \cup (0, 3)$.

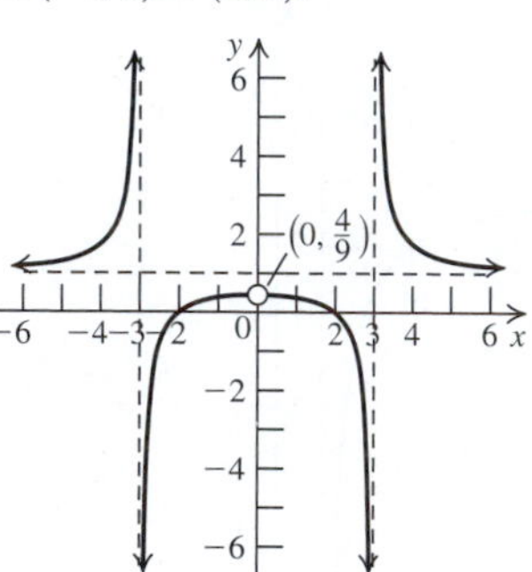

71. No x-intercept; y-intercept: -2; a hole at $(2, 0)$; no vertical asymptote; no horizontal asymptote. The graph is above the x-axis on $(2, \infty)$ and below the x-axis on $(-\infty, 2)$.

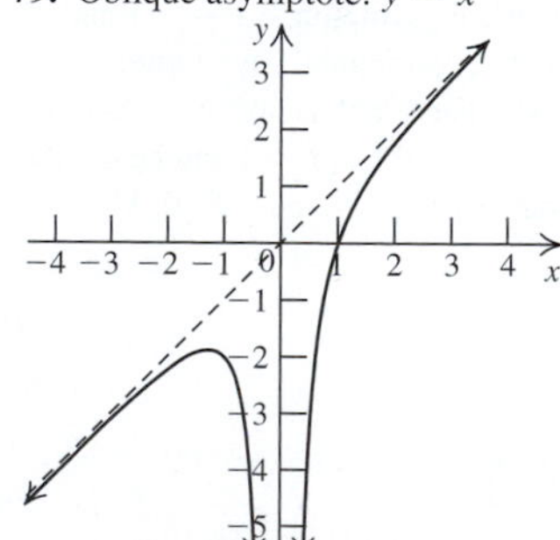

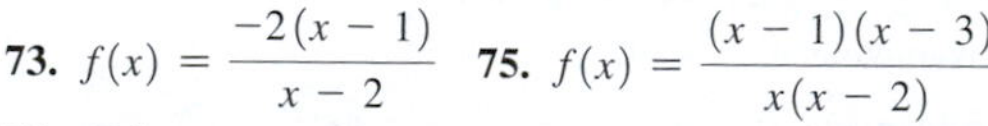

73. $f(x) = \dfrac{-2(x - 1)}{x - 2}$ **75.** $f(x) = \dfrac{(x - 1)(x - 3)}{x(x - 2)}$

77. Oblique asymptote: $y = 2x$

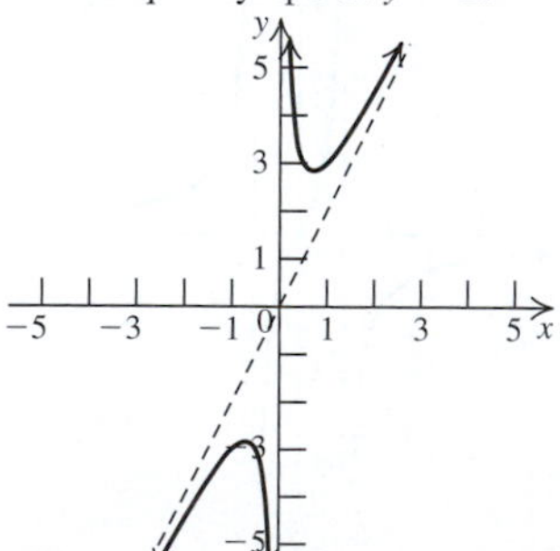

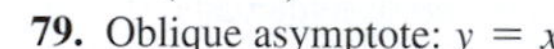

79. Oblique asymptote: $y = x$

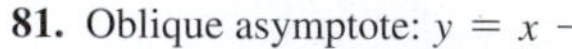

81. Oblique asymptote: $y = x - 2$

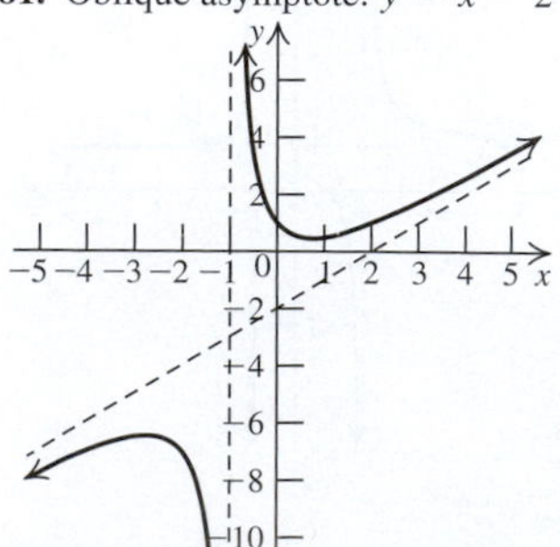

83. Oblique asymptote: $y = x - 2$

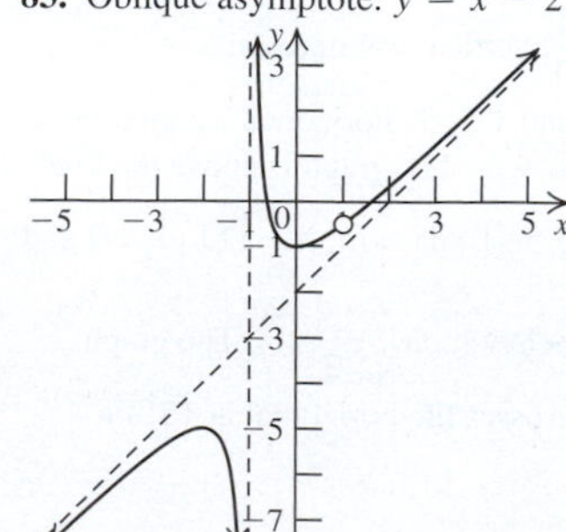

Applying the Concepts:

85. c. $\overline{C}(100) = 20.5$, $\overline{C}(500) = 4.5$, $\overline{C}(1000) = 2.5$. These show the average cost of producing 100, 500, and 1000 trinkets, respectively. **d.** Horizontal asymptote: $y = 0.5$. It means the average cost approaches the daily fixed cost of producing each trinket if the number of trinkets produced approaches ∞.

87. a.

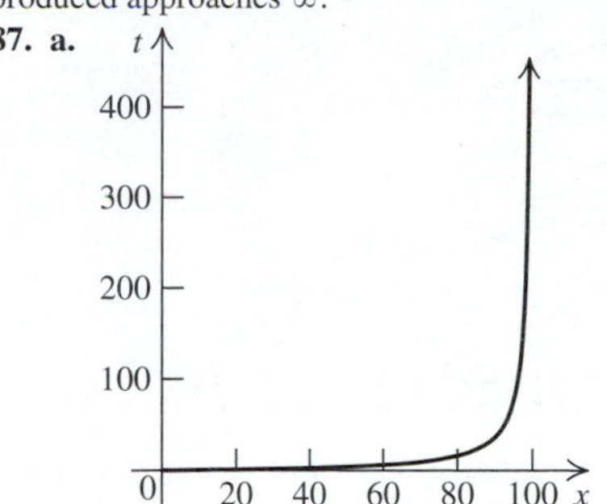

b. about 4 min; about 12 min; about 76 min; 397 min; **c. (i)** ∞ **(ii)** Not applicable; the domain is $x < 100$.

89. a. \$3.02 billion

b.

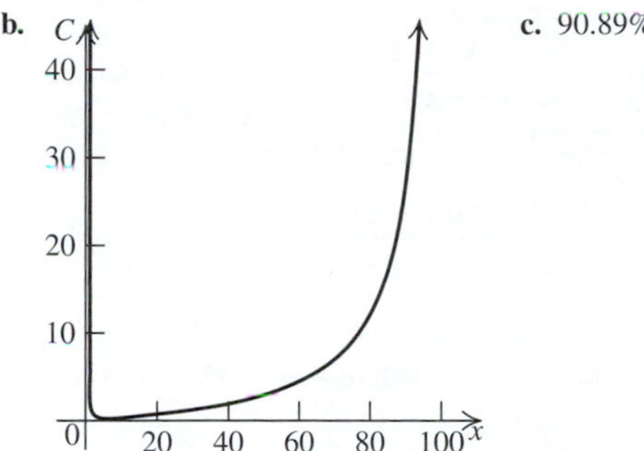

c. 90.89%

91. a. 16,000 **b.** The population will stabilize at 4000.

93. a. $f(x) = (10x + 200{,}000)/(x - 2500)$ **b.** \$40 **c.** More than 25,000 **d.** $x = 2500$ (free books); $y = 10$ (cost per book)

Beyond the Basics:

95. Stretch the graph of $y = \dfrac{1}{x}$ vertically by a factor of 2 and reflect the graph about the x-axis.

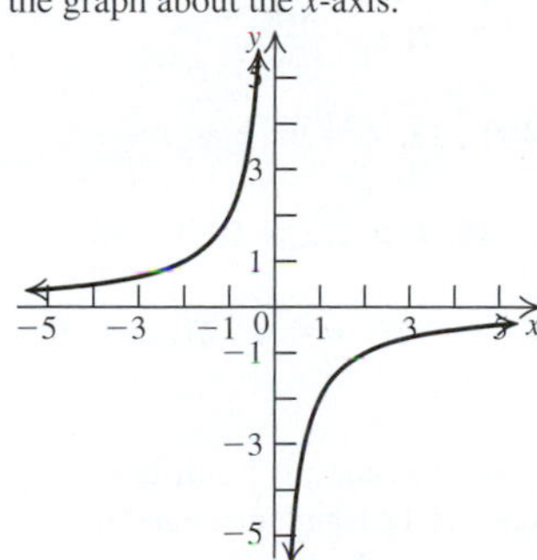

97. Shift the graph of $y = \dfrac{1}{x^2}$ two units to the right.

99. Shift the graph of $y = \dfrac{1}{x^2}$ one unit to the right and two units down.

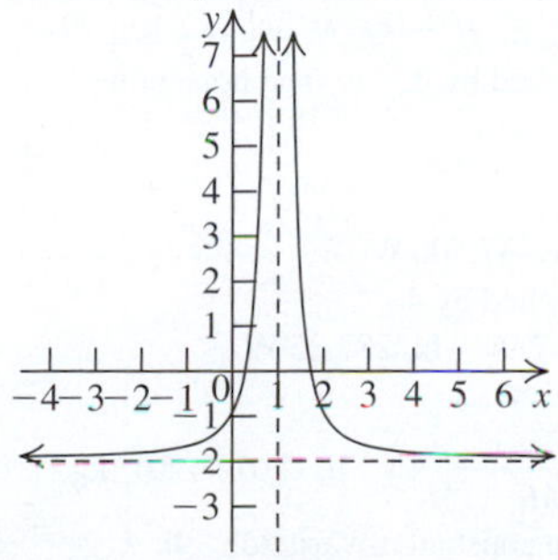

101. Shift the graph of $y = \dfrac{1}{x^2}$ six units to the left.

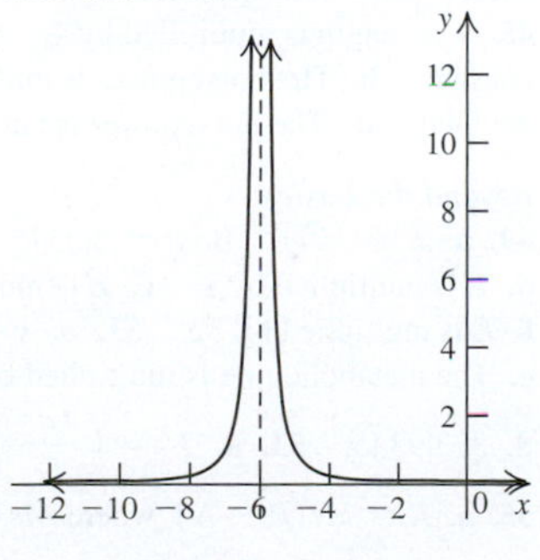

103. Shift the graph of $y = \dfrac{1}{x^2}$ one unit to the right and one unit up.

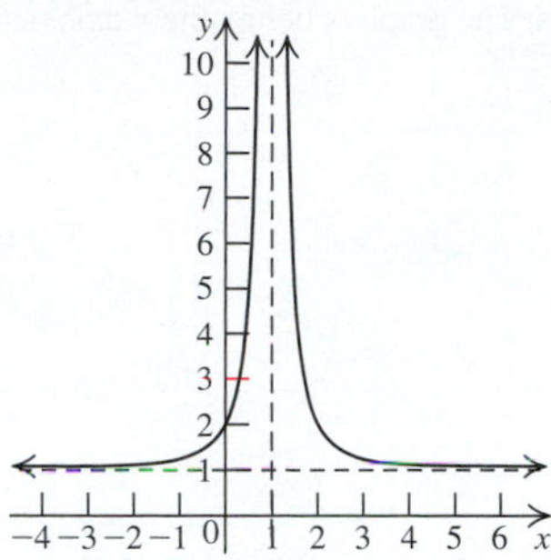

105. a. $f(x) \to 0$ as $x \to -\infty$, $f(x) \to 0$ as $x \to \infty$, $f(x) \to -\infty$ as $x \to 0^-$, and $f(x) \to \infty$ as $x \to 0^+$; no x-intercept; no y-intercept; vertical asymptote: y-axis; horizontal asymptote: x-axis. The graph is above the x-axis on $(0, \infty)$ and below the x-axis on $(-\infty, 0)$.

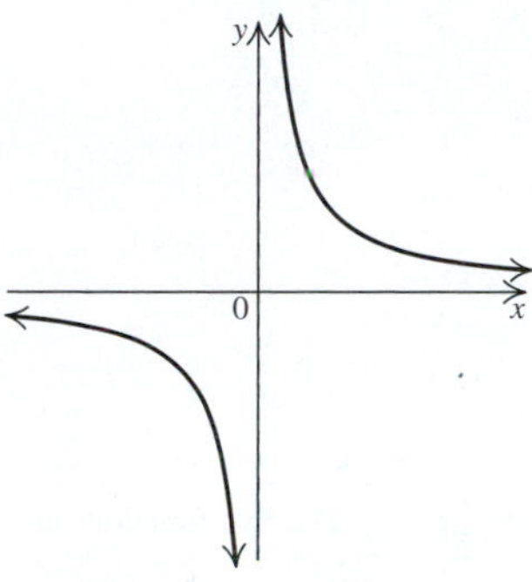

b. $f(x) \to 0$ as $x \to -\infty$, $f(x) \to 0$ as $x \to \infty$, $f(x) \to \infty$ as $x \to 0^-$, and $f(x) \to -\infty$ as $x \to 0^+$; no x-intercept; no y-intercept; vertical asymptote: y-axis; horizontal asymptote: x-axis. The graph is above the x-axis on $(-\infty, 0)$ and below the x-axis on $(0, \infty)$.

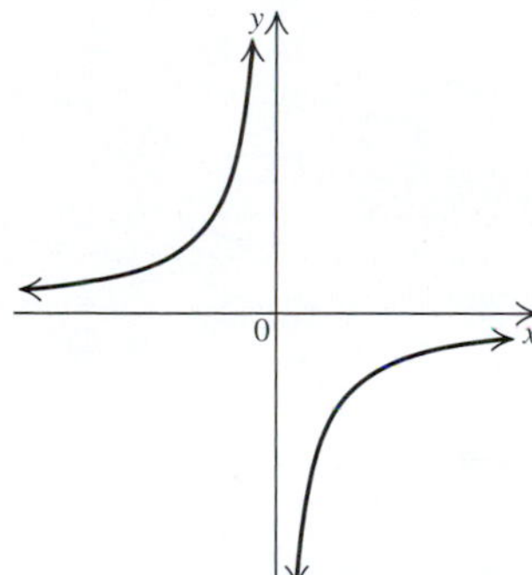

c. $f(x) \to 0$ as $x \to \infty$, $f(x) \to 0$ as $x \to -\infty$, $f(x) \to \infty$ as $x \to 0^-$, and $f(x) \to \infty$ as $x \to 0^+$; no x-intercept; no y-intercept; vertical asymptote: y-axis; horizontal asymptote: x-axis. The graph is above the x-axis on $(-\infty, 0) \cup (0, \infty)$.

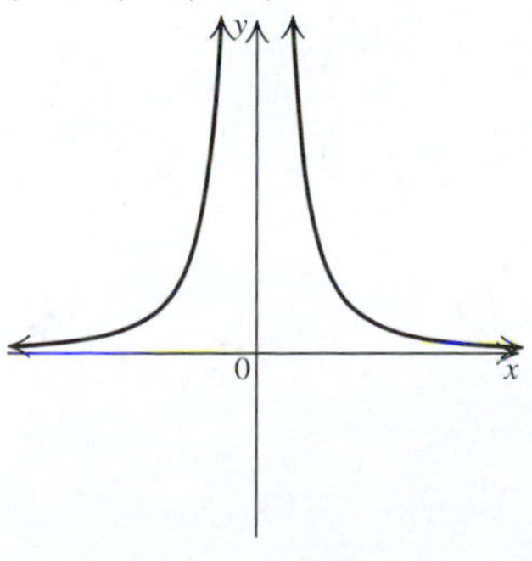

d. $f(x) \to 0$ as $x \to -\infty$, $f(x) \to 0$ as $x \to \infty$, $f(x) \to -\infty$ as $x \to 0^-$, and $f(x) \to -\infty$ as $x \to 0^+$; no x-intercept; no y-intercept; vertical asymptote: y-axis; horizontal asymptote: x-axis. The graph is below the x-axis on $(-\infty, 0) \cup (0, \infty)$.

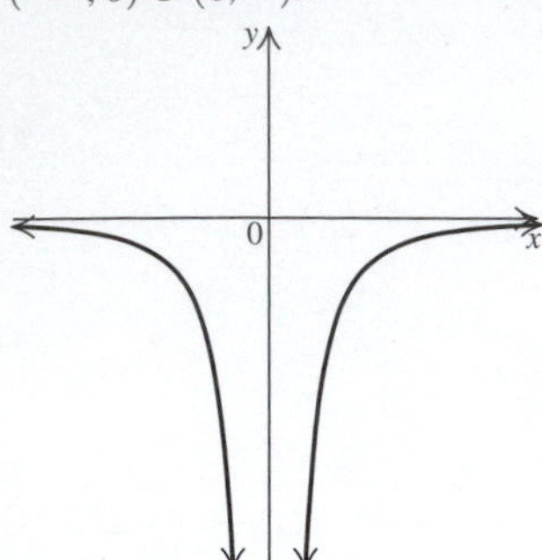

107.

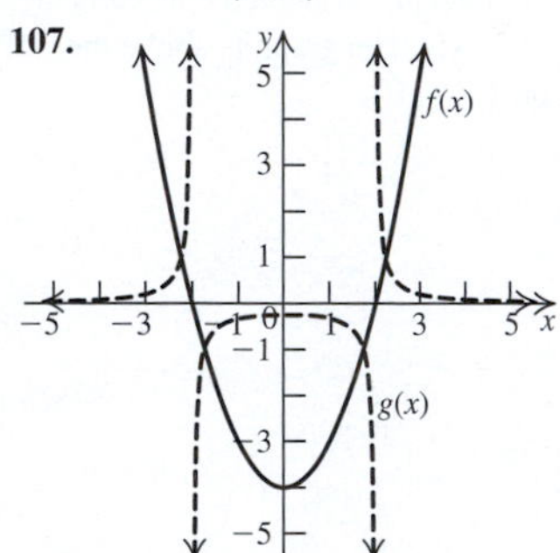

109. $[f(x)]^{-1} = \dfrac{1}{2x+3}$ and $f^{-1}(x) = \dfrac{1}{2}x - \dfrac{3}{2}$. The two functions are different.

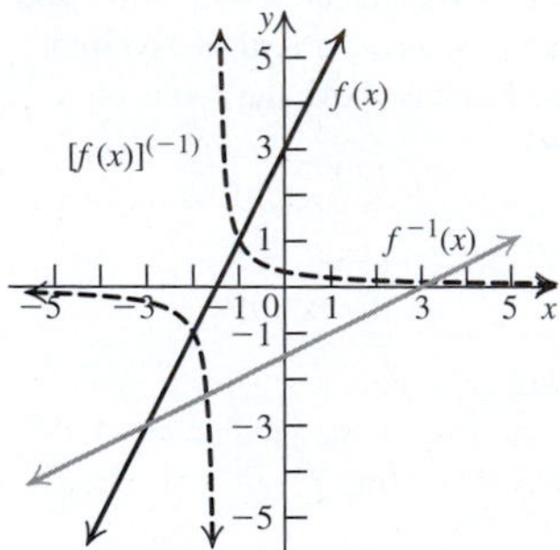

111. $g(x)$ has the oblique asymptote $y = 2x + 3$; $y \to -\infty$ as $x \to -\infty$, and $y \to \infty$ as $x \to \infty$.

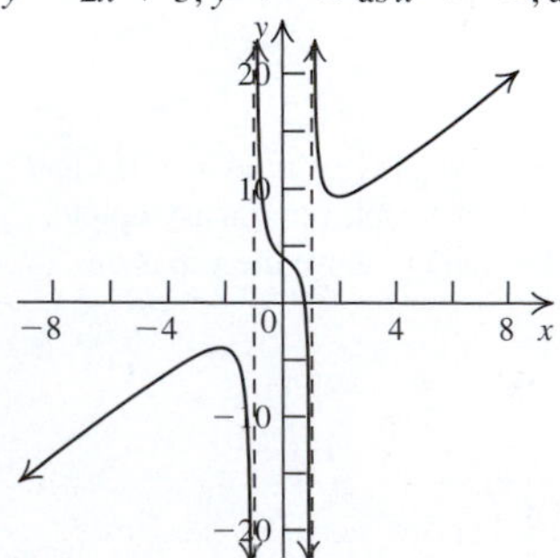

113. Point of intersection: $\left(-\dfrac{1}{3}, 1\right)$; horizontal asymptote $y = 1$

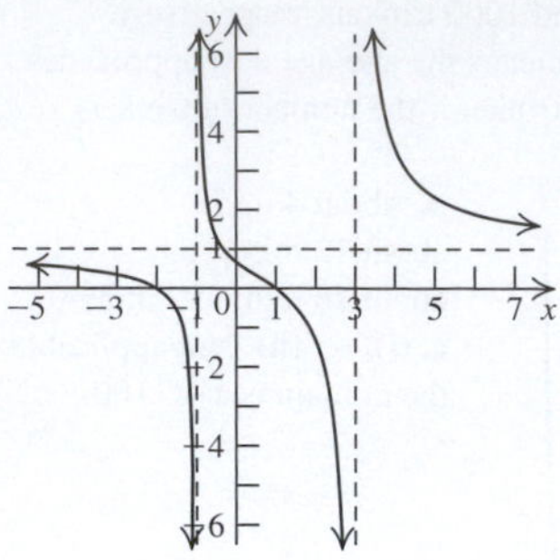

115. $f(x) = \dfrac{2-x}{x-3}$ **117.** $f(x) = \dfrac{-x^2+1}{x}$
119. $f(x) = (x+4)/(x-2)$ **121.** $f(x) = (4x^2-1)/(x-1)^2$
123. $f(x) = (3x^2 - x + 1)/(x - 1)$

Critical Thinking/Discussion/Writing:

125. a. $f(x) = \dfrac{1}{x}$ **b.** $f(x) = \dfrac{x^2+2}{x^2+x-2}$
c. $f(x) = \dfrac{x^3 + 2x^2 - 7x - 1}{x^3 + x^2 - 6x + 5}$
127. $R(x) = \dfrac{(x+1)x^3 + (x-2)(x-5)}{x^3}$

Maintaining Skills:

129. a. $y = -\dfrac{2}{3}x - \dfrac{1}{3}$ **b.** $y = -1$ **131.** $c = 4$ **133.** $k = 3; y = \dfrac{3}{4}$

Section 3.7

Answers to Practice Problems:

1. $y = 24$ **2.** 275 **3.** $y = 300$ **4.** $A = 20$ **5.** $x = \dfrac{9}{4}$
6. $\approx 3.7 \text{ m/sec}^2$

Basic Concepts and Skills:

1. $P = kT$ **3.** $y = k\sqrt{x}$ **5.** $V = kx^3y^4$ **7.** $z = \dfrac{kxu}{v^2}$
9. $k = \dfrac{1}{2}, x = 14$ **11.** $k = 16, s = 400$ **13.** $k = 33, r = 99$
15. $k = 8, B = \dfrac{1}{8}$ **17.** $k = 7, y = 4$ **19.** $k = 2, z = 50$
21. $k = \dfrac{1}{17}, P = \dfrac{324}{17}$ **23.** $k = 54, x = 4$ **25.** $y = 6$ **27.** $y = 200$

Applying the Concepts:

29. $y = Hx$, where y is the speed of the galaxies and x is the distance between them **31. a.** $y = 30.5x$, where y is the length measured in centimeters and x is the length measured in feet. **b. i.** 244 cm **ii.** 162.67 cm **c. i.** 1.87 ft **ii.** 4.07 ft **33.** 35 g **35.** $\frac{3}{4}$ sec **37.** 60 lb/in^2 **39. a.** 18.97 lb **b.** 201.01 lb **41.** 1.63 m/s^2 **43. a.** 1280 candlepower **b.** 8.94 ft from the source **45.** The length is multiplied by 4. **47. a.** $H = kR^2N$, where k is a constant **b.** The horsepower is multiplied by 4. **c.** The horsepower is doubled. **d.** The horsepower is cut in half.

Beyond the Basics:

49. a. $E = kl^2v^3$ **b.** $k = 0.0375$ **c.** 37,500 W **d.** E is multiplied by 8. **e.** E is multiplied by 4. **f.** E is multiplied by 32. **51. a.** $k = 2.94$ **b.** 287.25 W **c.** The metabolic rate is multiplied by 2.83. **d.** 373.93 kg **53. a.** $T^2 = \left(\dfrac{4\pi^2}{G}\right)\left(\dfrac{r^3}{M_1 + M_2}\right)$ **b.** 2.01×10^{30}kg **55. a.** $R = kN(P - N)$, where k is the constant of variation **b.** $k = \dfrac{5}{10^6}$ **c.** 125 people per day **d.** 2764 or 7236

Critical Thinking/Discussion/Writing:

57. a. \$480 **b.** \$15,920 **c.** The value of the original diamond is \$112,500. The value of a diamond whose weight is twice that of the original is \$450,000. **59.** 1:2

Maintaining Skills:

61. 1 **63.** $\frac{1}{2}$ **65.** 16 **67.** $5x$ **69.** 2^{x^2-x} **71.** $m = \frac{y-b}{x}$ **73.** $C = \frac{A}{B} - 1$ **75.** $B = A \cdot 10^n$

Review Exercises

Basic Skills and Concepts:

1.
i. Opens up
ii. Vertex: $(1, 2)$
iii. Axis: $x = 1$
iv. No x-intercept
v. y-intercept: 3
vi. Decreasing on $(-\infty, 1)$, increasing on $(1, \infty)$

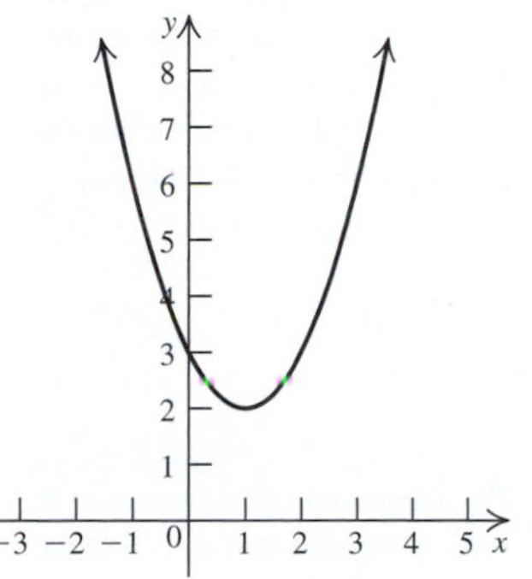

3. i. Opens down
ii. Vertex: $(3, 4)$
iii. Axis: $x = 3$
iv. x-intercepts: $3 \pm \sqrt{2}$
v. y-intercept: -14
vi. Increasing on $(-\infty, 3)$, decreasing on $(3, \infty)$

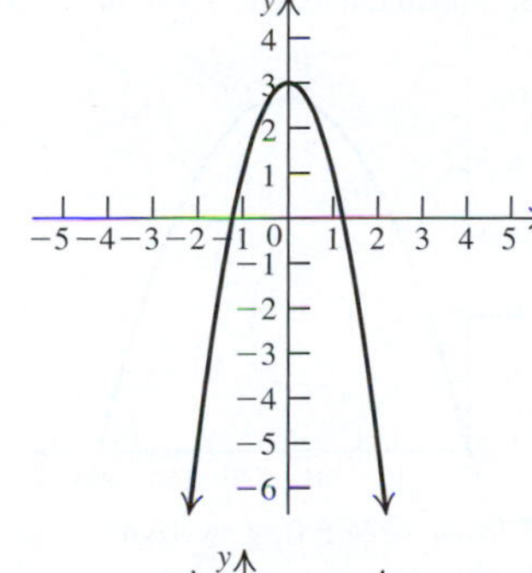

5. i. Opens down
ii. Vertex: $(0, 3)$
iii. Axis: y-axis
iv. x-intercepts: $\pm\frac{\sqrt{6}}{2}$
v. y-intercept: 3
vi. Increasing on $(-\infty, 0)$, decreasing on $(0, \infty)$

7. i. Opens up
ii. Vertex: $(1, 1)$
iii. Axis: $x = 1$
iv. No x-intercept
v. y-intercept: 3
vi. Decreasing on $(-\infty, 1)$, increasing on $(1, \infty)$

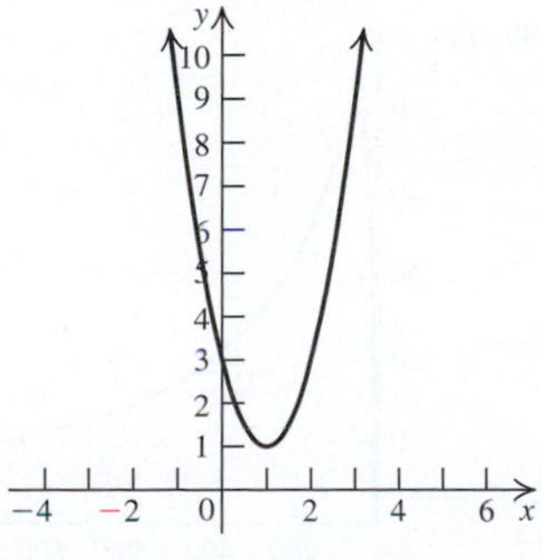

9. i. Opens up
ii. Vertex: $\left(\frac{1}{3}, \frac{2}{3}\right)$
iii. Axis: $x = \frac{1}{3}$
iv. No x-intercept
v. y-intercept: 1
vi. Decreasing on $\left(-\infty, \frac{1}{3}\right)$, increasing on $\left(\frac{1}{3}, \infty\right)$

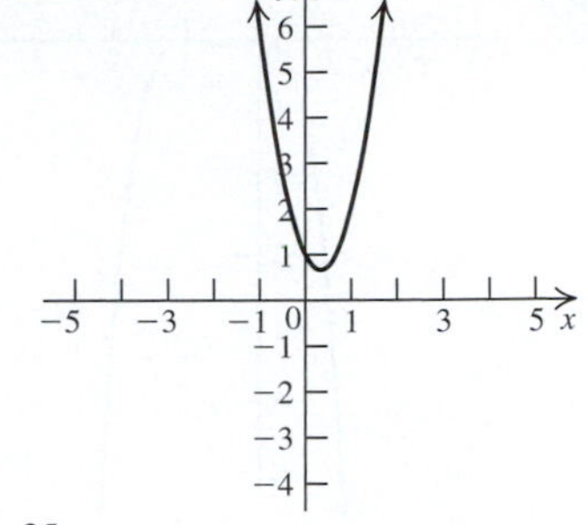

11. Minimum: -1 **13.** Maximum: $\frac{25}{8}$

15. Shift the graph of $y = x^3$ one unit to the left and two units down.

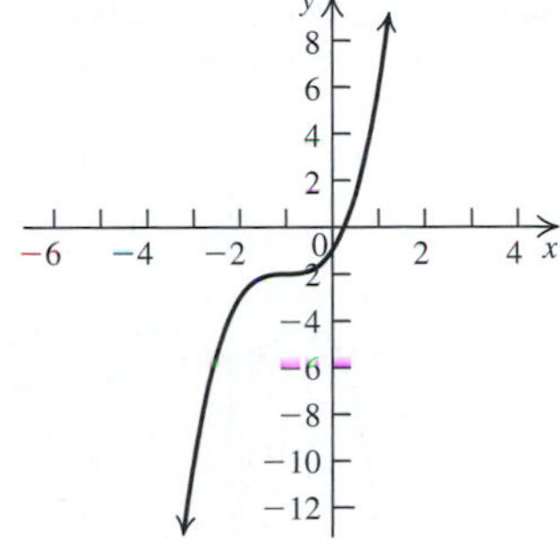

17. Shift the graph of $y = x^3$ one unit right, reflect the resulting graph in the x-axis, and shift it one unit up.

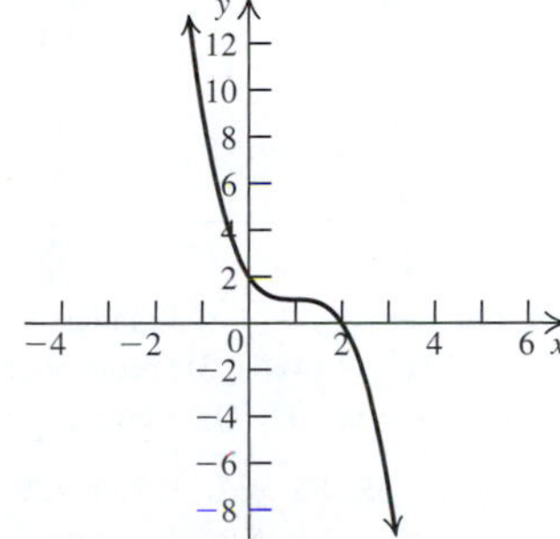

19. (i) $f(x) \to -\infty$ as $x \to -\infty$ and $f(x) \to \infty$ as $x \to \infty$ **(ii)** Zeros: $x = -2$, multiplicity: 1, the graph crosses the x-axis; $x = 0$, multiplicity: 1, the graph crosses the x-axis; $x = 1$, multiplicity: 1, the graph crosses the x-axis. **(iii)** x-intercepts: $-2, 0, 1$; y-intercept: 0 **(iv)** The graph is above the x-axis on $(-2, 0) \cup (1, \infty)$ and below the x-axis on $(-\infty, -2) \cup (0, 1)$.
(v)

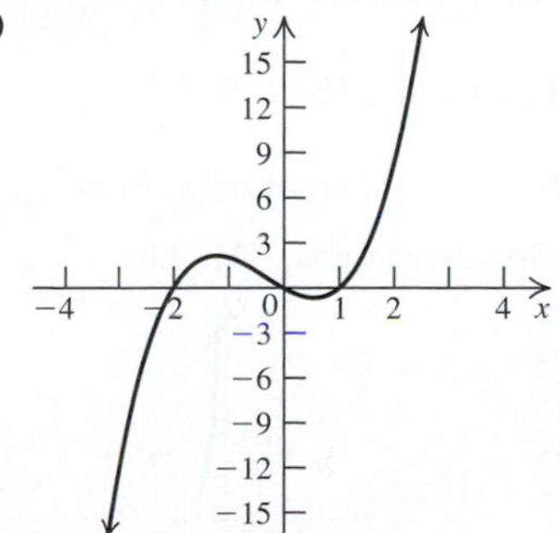

21. (i) $f(x) \to -\infty$ as $x \to -\infty$ and $f(x) \to -\infty$ as $x \to \infty$ **(ii)** Zeros: $x = 0$, multiplicity: 2, the graph touches but does not cross the x-axis; $x = 1$, multiplicity: 2, the graph touches but does not cross the x-axis. **(iii)** x-intercepts: 0, 1; y-intercept: 0 **(iv)** The graph is below the x-axis on $(-\infty, 0) \cup (0, 1) \cup (1, \infty)$.

(v)

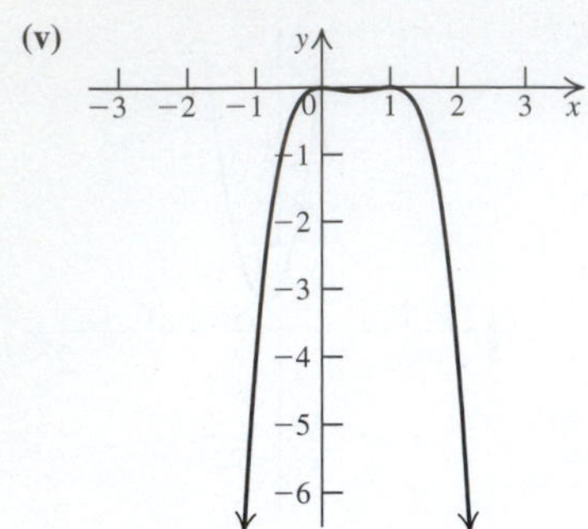

23. **(i)** $f(x) \to -\infty$ as $x \to -\infty$ and $f(x) \to -\infty$ as $x \to \infty$ **(ii)** Zeros: $x = -1$, multiplicity: 1, the graph crosses the x-axis; $x = 0$, multiplicity: 2, the graph touches but does not cross the x-axis; $x = 1$ multiplicity: 1, the graph crosses the x-axis. **(iii)** x-intercepts: $-1, 0, 1$; y-intercept: 0 **(iv)** The graph is above the x-axis on $(-1, 0) \cup (0, 1)$ and below the x-axis on $(-\infty, -1) \cup (1, \infty)$.
(v)

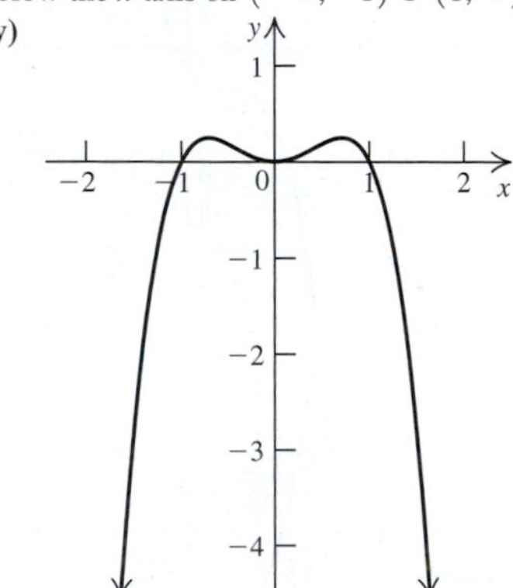

25. Quotient: $2x + 3$; remainder: -7 **27.** Quotient: $8x^3 - 12x^2 + 14x - 21$; remainder: 186 **29.** Quotient: $x^2 + 3x - 3$; remainder: -6 **31.** Quotient: $2x^3 - 5x^2 + 10x - 17$; remainder: 182

33. -11 **35.** 88 **37.** $1, 2, 4$ **39.** $-3, -2, \frac{1}{3}$

41. $-6, -3, -2, -1, 1, 2, 3, 6$ **43.** $-2, -\frac{1}{5}, 0$ **45.** $-3, -2, 2$

47. $-2, 3 \pm \sqrt{2}$ **49.** $\frac{1}{2}, 1 \pm \sqrt{3}$ **51.** $-3, 1, 2$ **53.** $-1, 3, 3i, -3i$

55. $2i, -2i, -1 + 2i, -1 - 2i$ **57.** $\{-2, 1, 2\}$ **59.** $\left\{-1, -\frac{1}{2}, \frac{3}{2}\right\}$

61. $\{2, 3 \pm i\sqrt{2}\}$ **63.** $\left\{-\frac{1}{3}, 1 \pm i\sqrt{5}\right\}$ **65.** $\{-1, 2, i, -i\}$

67. $\left\{-2, \frac{1}{2}, 1 \pm i\sqrt{2}\right\}$ **69.** The only possible rational roots are $-2, -1, 1$, and 2. None of them satisfy the equation. **71.** 1.88

73. x-intercept: -1; no y-intercept; vertical asymptote: y-axis; horizontal asymptote: $y = 1$; symmetry: none. The graph is above the x-axis on $(-\infty, -1) \cup (0, \infty)$ and below the x-axis on $(-1, 0)$.

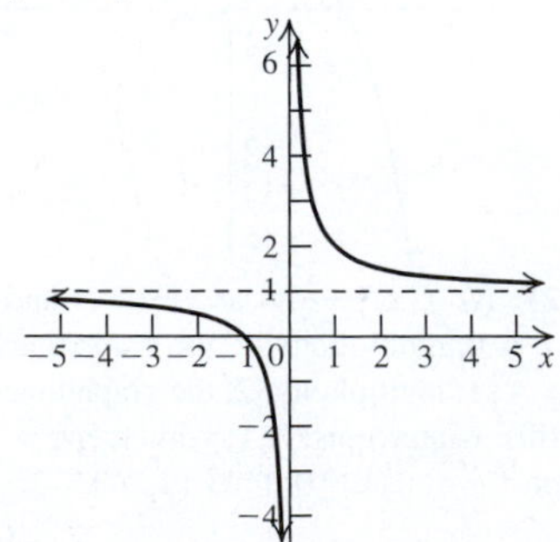

75. x-intercept: 0; y-intercept: 0; vertical asymptotes: $x = -1$ and $x = 1$; horizontal asymptote: x-axis. The graph is above the x-axis on $(-1, 0) \cup (1, \infty)$ and below the x-axis on $(-\infty, -1) \cup (0, 1)$.

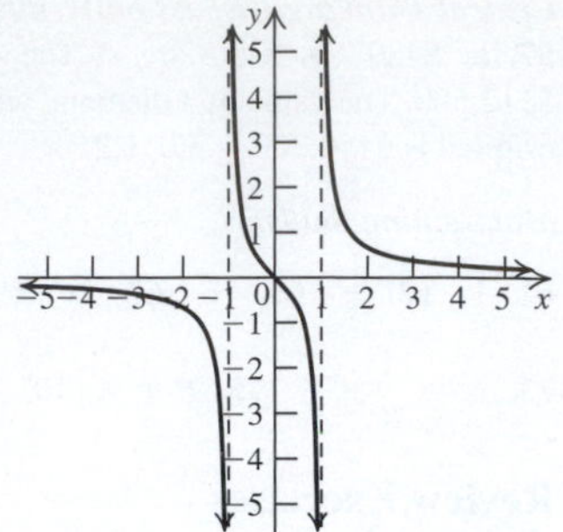

77. x-intercept: 0; y-intercept: 0; vertical asymptotes: $x = -3$ and $x = 3$; no horizontal asymptote. The graph is above the x-axis on $(-3, 0) \cup (3, \infty)$ and below the x-axis on $(-\infty, -3) \cup (0, 3)$.

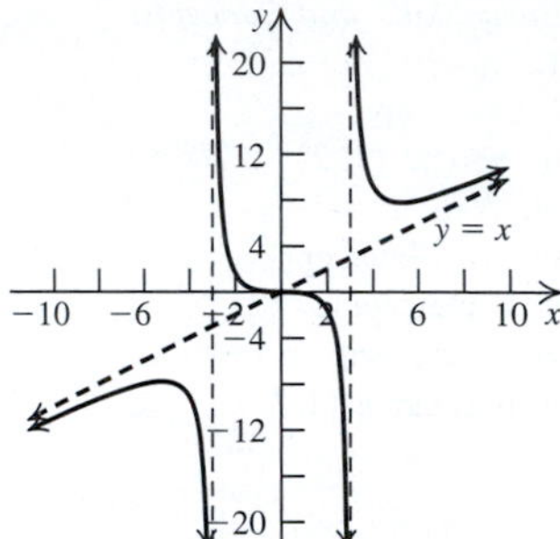

79. x-intercept: 0; y-intercept: 0; vertical asymptotes: $x = -2$ and $x = 2$; no horizontal asymptote. The graph is above the x-axis on $(-\infty, -2) \cup (2, \infty)$ and below the x-axis on $(-2, 0) \cup (0, 2)$.

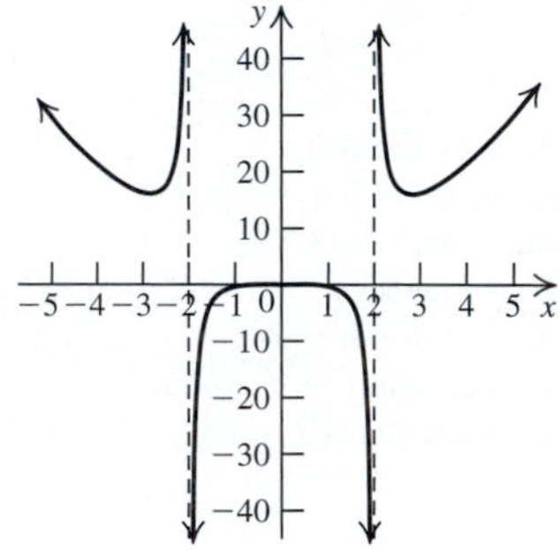

Applying the Concepts:

81. $y = 15$ **83.** $s = 45$

85. Maximum height: 1000; the missile hits the ground at $x = 200$.

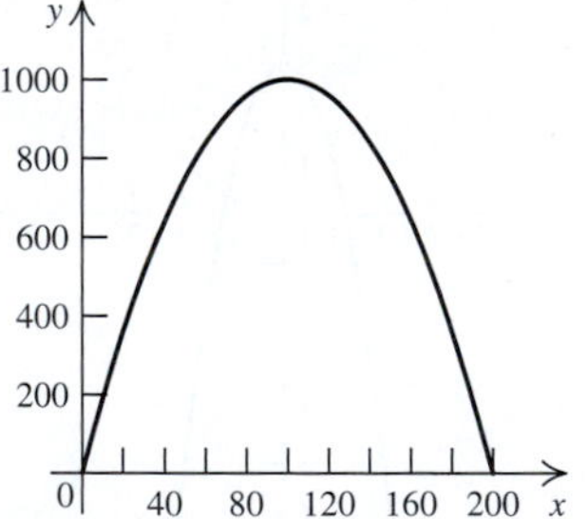

87. $x = \approx 24.5$ ft, $y \approx 16.3$ ft

89. **a.** **b.** $x = 50$

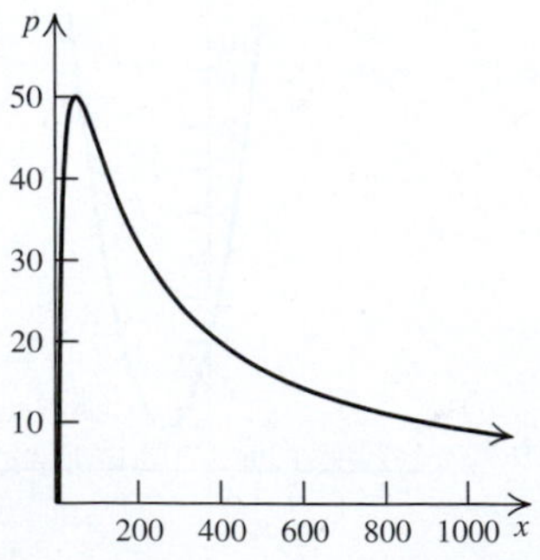

91. a. $P(x) = -\frac{3}{1000}x^2 + 20.1x - 150$ **b.** 3350

c. $\overline{C}(x) = \frac{3}{1000}x + 3.9 + \frac{150}{x}$

93. \$245 **95.** $6\sqrt{2}$ in **97. a.** 36 amp **b.** 150 Ω

99. a. $k = \frac{1}{200{,}000}$ **b.** 500 people every day **c.** 10,00 or 19,000

Practice Test A

1. x-intercepts: $3 \pm \sqrt{7}$

2.

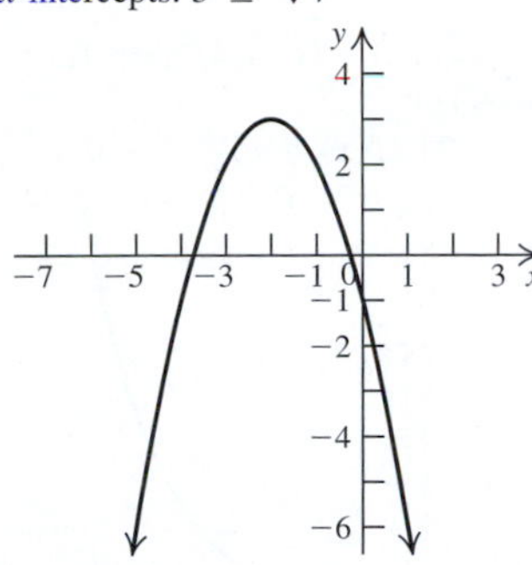

3. $(1, 10)$ **4.** $(-\infty, -4) \cup (-4, 1) \cup (1, \infty)$

5. Quotient: $x^2 - 4x + 3$; remainder: 0

6.

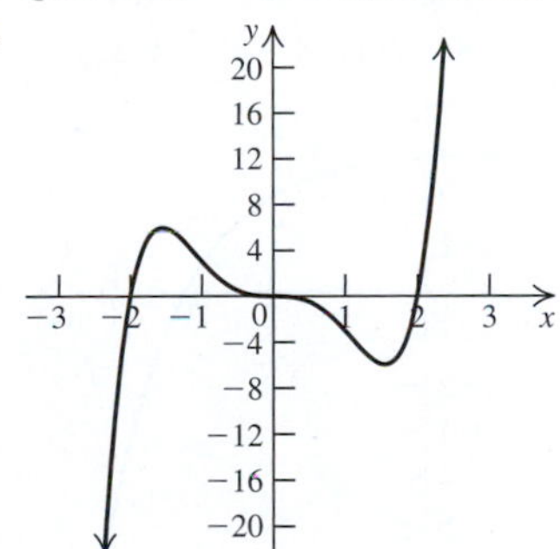

7. $-2, 1, 2$ **8.** Quotient: $-3x^2 + 5x + 1$ **9.** $P(-2) = -53$

10. $-2, 2, 5$ **11.** $0, -\frac{1}{2} + \frac{\sqrt{61}}{2}, -\frac{1}{2} - \frac{\sqrt{61}}{2}$

12. $-9, -\frac{9}{2}, -3, -\frac{3}{2}, -1, -\frac{1}{2}, \frac{1}{2}, 1, \frac{3}{2}, 3, \frac{9}{2}, 9$

13. $f(x) \to -\infty$ as $x \to -\infty$ and $f(x) \to \infty$ as $x \to \infty$

14. $x = -2$, multiplicity: 3; $x = 2$, multiplicity: 1

15. Number of positive zeros: 0 or 2; number of negative zeros: 1

16. Horizontal asymptote: $y = 2$; vertical asymptotes: $x = -4$ and $x = 5$

17. $y = \frac{kx}{t^2}$, where $k \neq 0$ is a constant **18.** $y = 4$ **19.** $x = 15$

20. $V(x) = x(8 - 2x)(17 - 2x)$

Practice Test B

1. b **2.** d **3.** a **4.** b **5.** d **6.** c **7.** c **8.** b **9.** c **10.** a **11.** a **12.** d **13.** c **14.** b **15.** c **16.** c **17.** d **18.** b **19.** b **20.** d

Cumulative Review Exercises (Chapters P–2)

1. $\sqrt{13}$

3. x-intercepts: 4, -2 y-intercept: -8

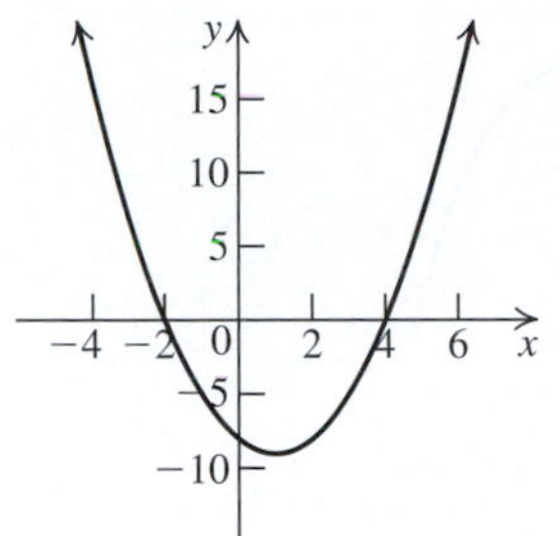

5. Slope is $\frac{1}{2}$; x-intercept: -6; y-intercept: 3

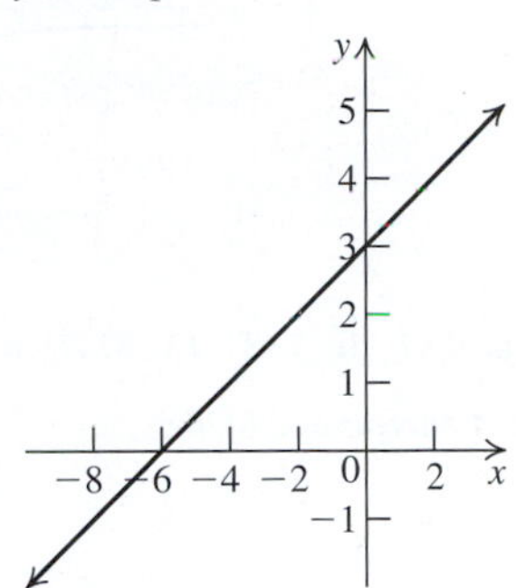

7. Center: $(-1, 2)$; radius: 3 **9.** $y = -\frac{2}{3}x + \frac{11}{3}$ **11.** Domain $(-\infty, 2)$.

13. a. $f(g(x)) = \sqrt{x^2 + 1}$ **b.** $g(f(x)) = x + 1$ **c.** $f(f(x)) = \sqrt[4]{x}$

d. $g(g(x)) = x^4 + 2x^2 + 2$ **15.** $f^{-1}(x) = \frac{1}{2}x + \frac{3}{2}$

17. $-6, -3, -2, -\frac{3}{2}, -1, -\frac{1}{2}, \frac{1}{2}, 1, \frac{3}{2}, 2, 3, 6$

19.

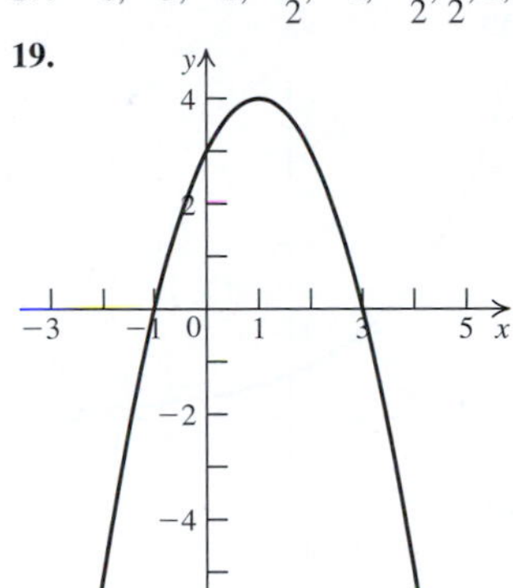

21.

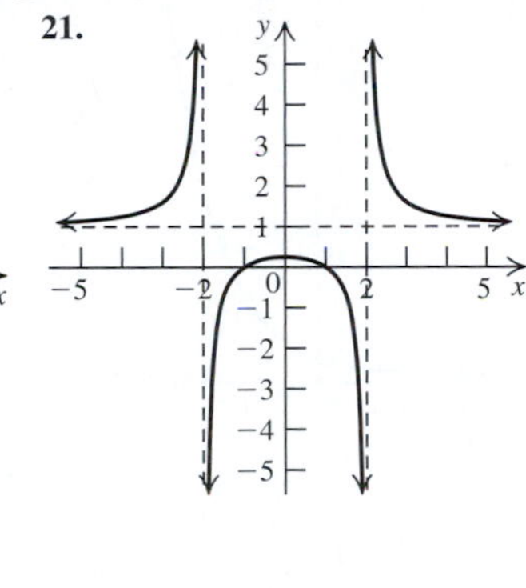

23. 9 **25.** $x = 50$

CHAPTER 4

Section 4.1

Answers to Practice Problems:

1. $f(2) = \frac{1}{16}$; $f(0) = 1$; $f(-1) = 4$; $f\left(\frac{5}{2}\right) = \frac{1}{32}$; $f\left(-\frac{3}{2}\right) = 8$

2. a. $3^{3\sqrt{2}} = 27^{\sqrt{2}}$ **b.** a^4

3.

4.

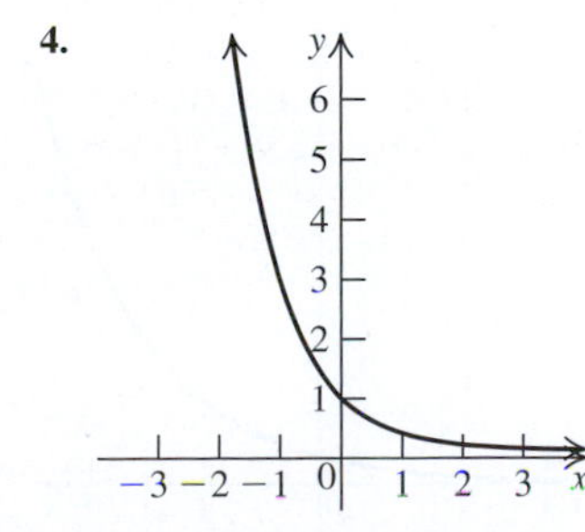

5. a. 7^x **b.** $4\left(\frac{1}{2}\right)^x$ **6.** \$11,500 **7. a.** $A = \$11{,}485.03$

b. $I = \$3{,}485.03$ **8. i.** \$5325.00 **ii.** \$5330.28 **iii.** \$5333.01 **iv.** \$5334.86 **v.** \$5335.76 **9.** 10.023% **10.** \$14,764.48

11. $y = -e^{x-1} - 2$

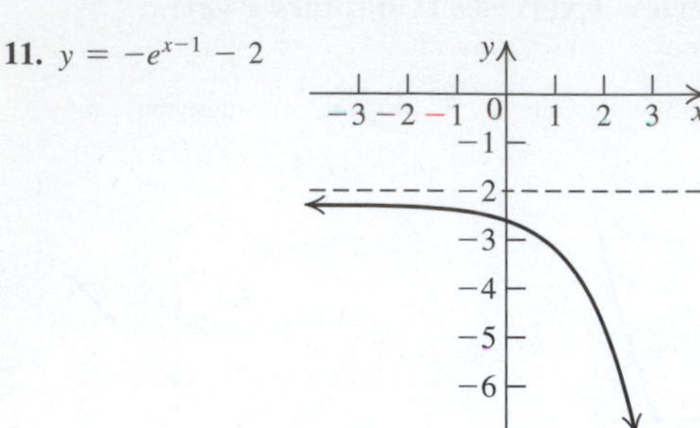

12. a. 9.82 **b.** 5.18 **13.** \$7471.10

Basic Concepts and Skills:

1. $(-\infty, \infty)$; $(0, \infty)$ **3.** x-axis **5.** $P\left(1 + \frac{r}{n}\right)^{nt}$ **7.** false

9. true **11.** No, the base is not a constant. **13.** Yes, the base is $\frac{1}{2}$.

15. No, the base is not a constant. **17.** No, the base is not a positive constant. **19.** $\frac{1}{5}$ **21.** 0.0892, 11.2116 **23.** $\frac{4}{9}, \frac{729}{64}$ **25.** $3^{2\sqrt{2}}$ **27.** 2^{π}

29. $3^{\sqrt{6}}$ **31.** a^6 **33. a.** 4^x **b.** 3^x **35. a.** $\frac{1}{5} \cdot 5^x$ **b.** $5 \cdot \left(\frac{1}{5}\right)^x$

37.

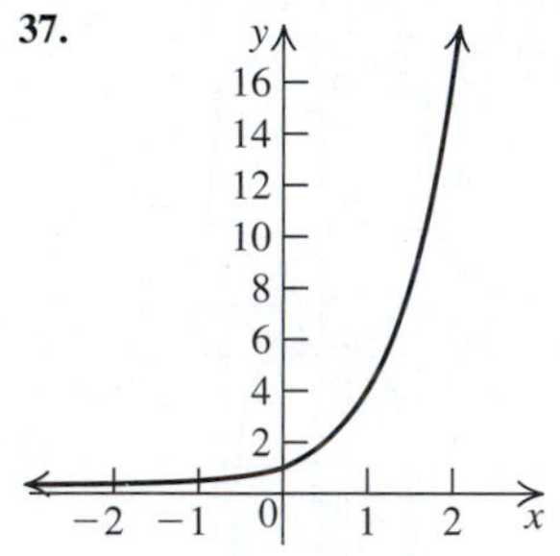

39.

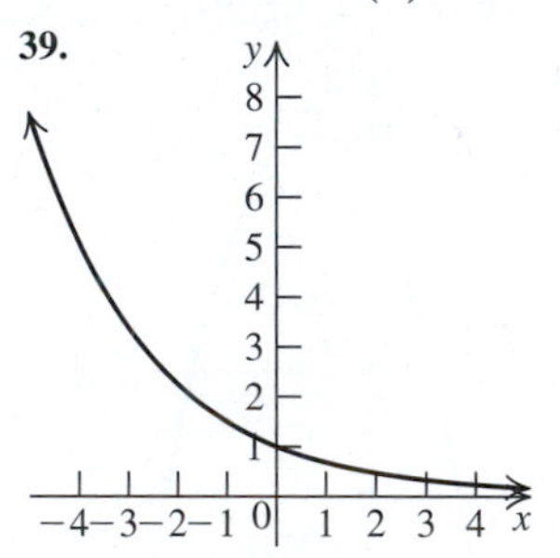

41.

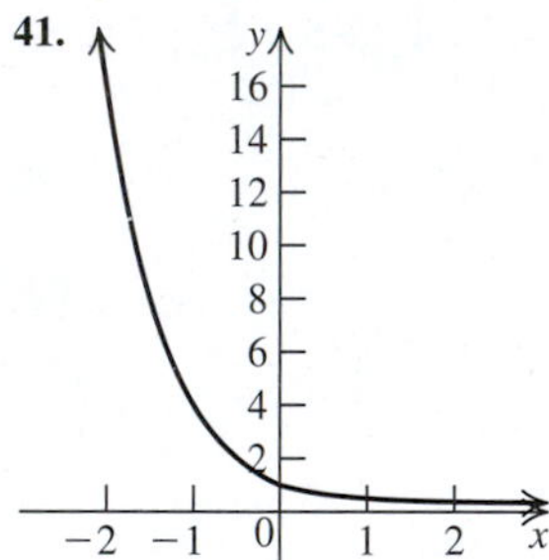

43.

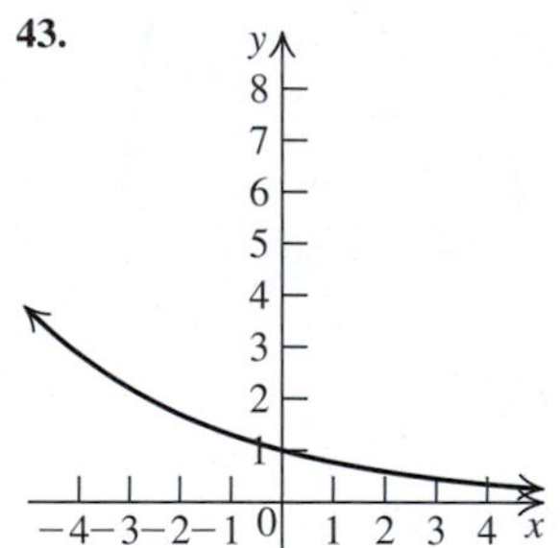

45. c **47.** a

49. Domain: $(-\infty, \infty)$; range: $(0, \infty)$; horizontal asymptote: $y = 0$

51. Domain: $(-\infty, \infty)$; range: $(0, \infty)$; horizontal asymptote: $y = 0$

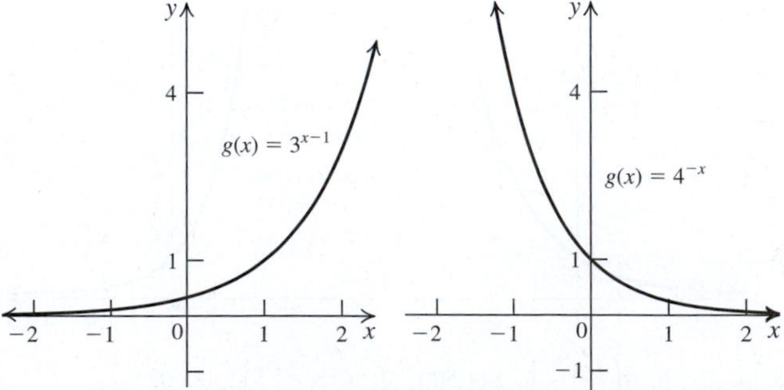

53. Domain: $(-\infty, \infty)$; range: $(-\infty, 4)$; horizontal asymptote: $y = 4$

55. Domain: $(-\infty, \infty)$; range: $(-\infty, 3)$; horizontal asymptote: $y = 3$

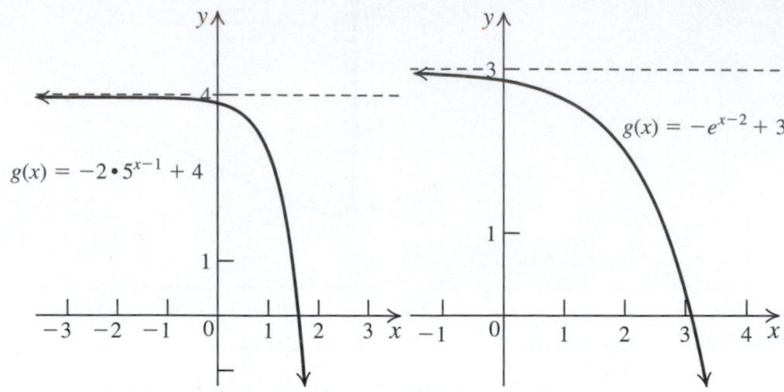

57. $f(x) = (1.5)^x + 2$; $f(2) = 4.25$ **59.** $f(x) = \left(\frac{1}{3}\right)^x - 2$; $f(2) = -\frac{17}{9}$ **61.** $y = 2^{x+2} + 5$ **63.** $y = 2\left(\frac{1}{2}\right)^x - 5$ **65.** \$2500

67. \$5764.69 **69. a.** \$7936.21 **b.** \$4436.21 **71. a.** \$12,365.41 **b.** \$4865.41 **73.** \$4631.93 **75.** \$4493.68

77. $f(x) = e^{-x}; y = 0$

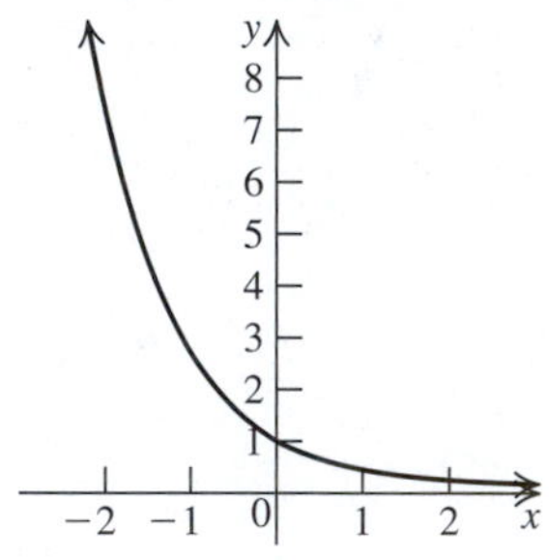

79. $f(x) = e^{x-2}; y = 0$

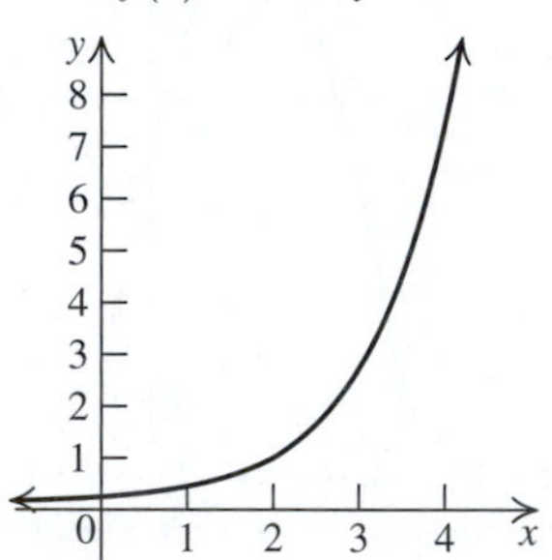

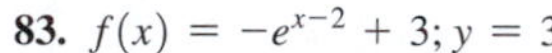

81. $f(x) = 1 + e^x; y = 1$

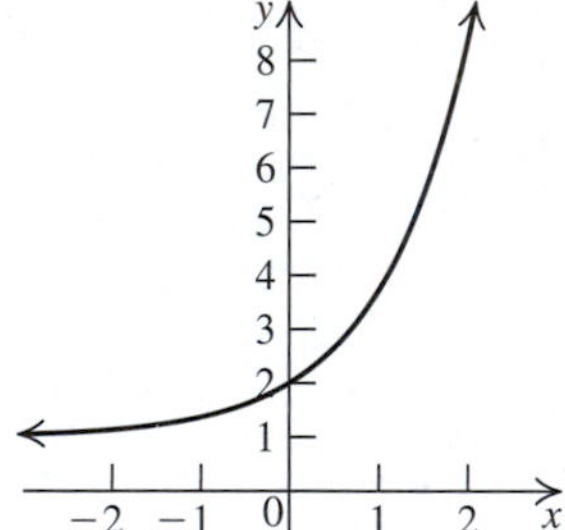

83. $f(x) = -e^{x-2} + 3; y = 3$

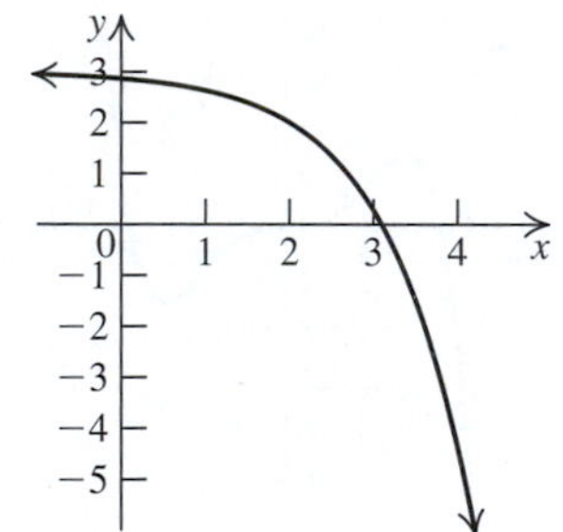

Applying the Concepts:

85. a. 176.6°C; 148.1°C **b.** 5 hours **c.** 25°C **87.** \$220,262 **89.** \$35,496.43 **91.** 16.81% **93.** 0.1357 mm^2 **95.** 3.9%

97. a. $0.015 \times 2^{30} \approx 16{,}106{,}127$ cm
b. $0.015 \times 2^{40} \approx 16{,}492{,}674{,}420$ cm
c. $0.015 \times 2^{50} \approx 1.689 \times 10^{13}$ cm

Beyond the Basics:

101. $S_5 = 2.716666$, $S_{10} = 2.7182818$, $S_{15} = 2.718281828$
105. $y = e^x \rightarrow y = e^{x-1} \rightarrow y = e^{2x-1} \rightarrow y = 3e^{2x-1}$
107. $y = e^x \rightarrow y = e^{2+x} \rightarrow y = e^{2+3x} \rightarrow y = 5e^{2+3x} \rightarrow y = 5e^{2-3x} \rightarrow y = 5e^{2-3x} + 4$

Maintaining Skills:

115. 1 **117.** −2 **119.** 49 **121.** $3\sqrt{2}$

123. $f^{-1}(x) = \frac{x-4}{3}$ Domain and range of f^{-1} is $(-\infty, \infty)$.

125. $f^{-1}(x) = x^2, x \geq 0$ Domain and range of f^{-1} is $[0, \infty)$.

127. $f^{-1}(x) = 1 + \frac{1}{x}$ Domain of $f^{-1} = (-\infty, 0) \cup (0, \infty)$; range of $f^{-1} = (-\infty, 1) \cup (1, \infty)$.

Section 4.2

Answers to Practice Problems:

1. a. $\log_2 1024 = 10$ **b.** $\log_9\left(\frac{1}{3}\right) = -\frac{1}{2}$ **c.** $\log_a p = q$
2. a. $2^6 = 64$ **b.** $v^w = u$ **3. a.** 2 **b.** $-1/2$ **c.** -5
4. a. 0 **b.** 5 **c.** 5 **5.** $(-\infty, 1)$
6.

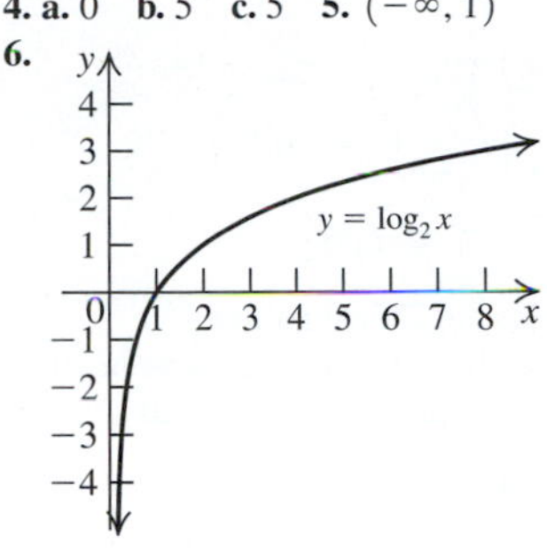

7. Shift the graph of $y = \log_2 x$ 3 units right and reflect the graph about the x-axis.

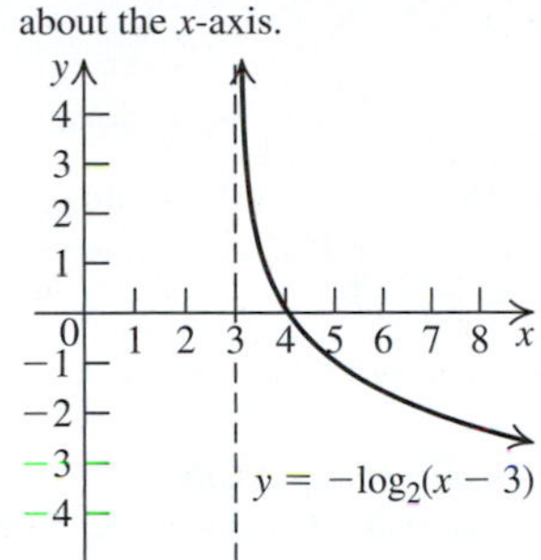

8. $P_2 = \log 3 - \log 2 \approx 0.176$. About 17.6% of the data is expected to have 2 as the first digit. **9. a.** -1 **b.** ≈ 0.693 **10. a.** ≈ 17 years **b.** 21.97% **11.** ≈ 5.4 min

Basic Concepts and Skills:

1. $(0, \infty)$; $(-\infty, \infty)$ **3.** common; natural **5.** false **7.** $\log_5 25 = 2$
9. $\log_{1/16} 4 = -\frac{1}{2}$ **11.** $\log_{10} 1 = 0$ **13.** $\log_{10} 0.1 = -1$
15. $\log_a 5 = 2$ **17.** $\log_a (13/2) = 3$ **19.** $2^5 = 32$ **21.** $10^2 = 100$
23. $10^0 = 1$ **25.** $10^{-2} = 0.01$ **27.** $8^{1/3} = 2$ **29.** $e^x = 2$ **31.** 3
33. 4 **35.** -3 **37.** $\frac{3}{2}$ **39.** $\frac{1}{4}$ **41.** 0 **43.** 1 **45.** 7 **47.** 5 **49.** 4
51. 8 **53.** $(-1, \infty)$ **55.** $(1, \infty)$ **57.** $(2, \infty)$ **59.** $(1, 2)$
61. a. f **b.** a **c.** d **d.** b **e.** e **f.** c

63. Domain = $(-3, \infty)$
Range = $(-\infty, \infty)$
Asymptote: $x = -3$

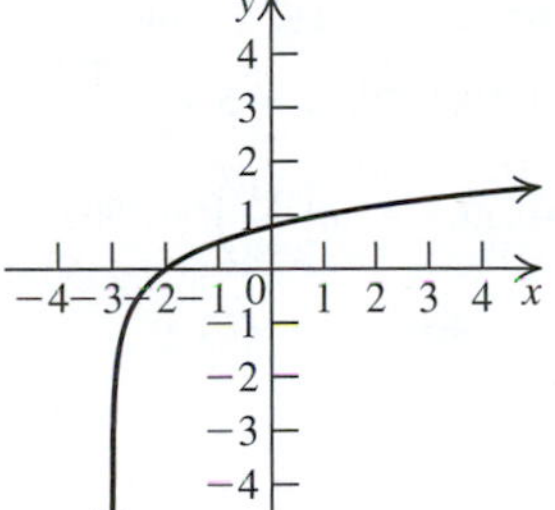

65. Domain = $(0, \infty)$
Range = $(-\infty, \infty)$
Asymptote: $x = 0$

67. Domain = $(-\infty, 0)$
Range = $(-\infty, \infty)$
Asymptote: $x = 0$

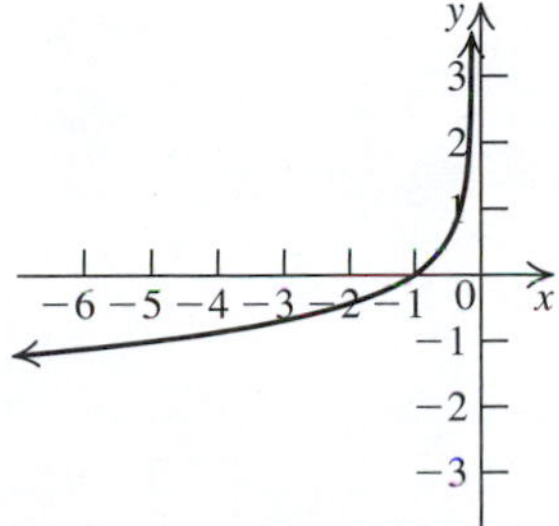

69. Domain = $(0, \infty)$
Range = $[0, \infty)$
Asymptote: $x = 0$

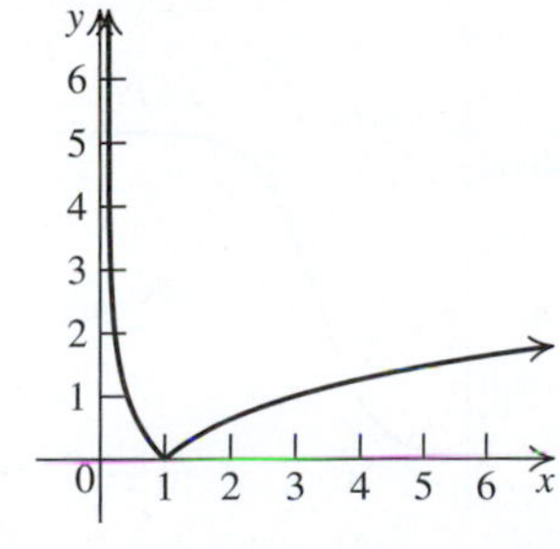

71. Domain = $(1, \infty)$
Range = $(-\infty, \infty)$
Asymptote: $x = 1$

73. Domain = $(-\infty, 3)$
Range = $(-\infty, \infty)$
Asymptote: $x = 3$

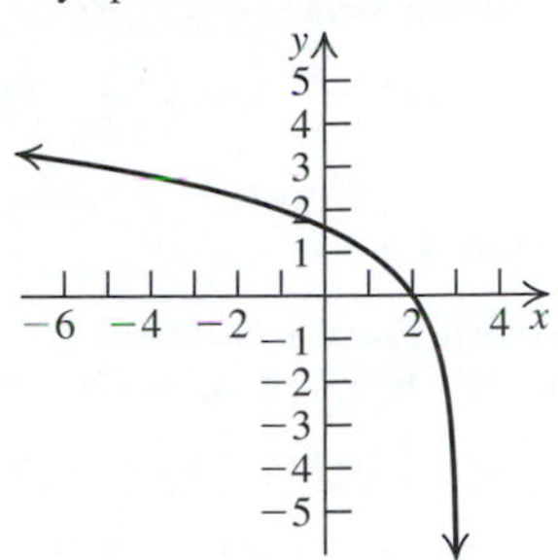

75. Domain = $(-\infty, 3)$
Range = $(-\infty, \infty)$
Asymptote: $x = 3$

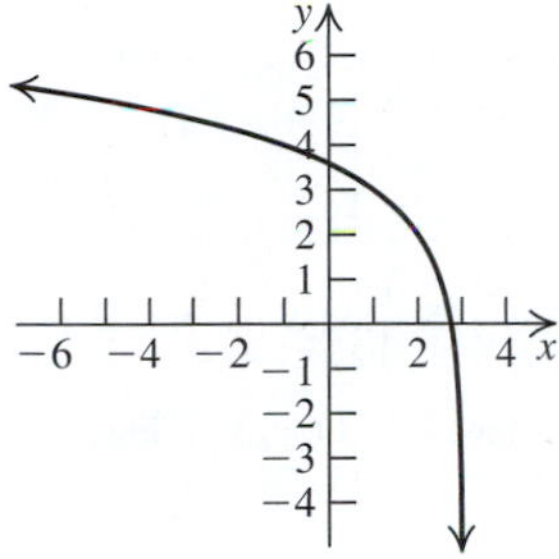

77. Domain = $(-\infty, 0) \cup (0, \infty)$
Range = $(-\infty, \infty)$
Asymptote: $x = 0$

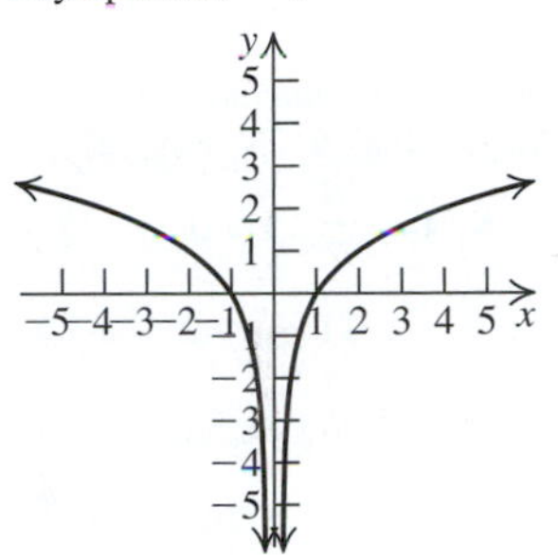

79. 1 **81.** 2 **83.** 4 **85.** $x = 100$ **87.** $x = e$

89.

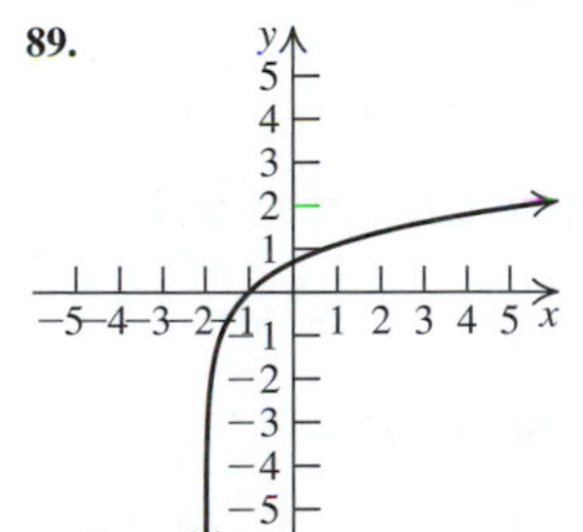

91.

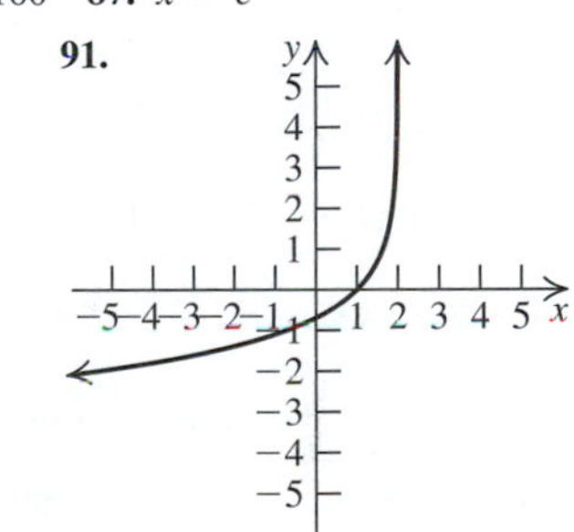

93.

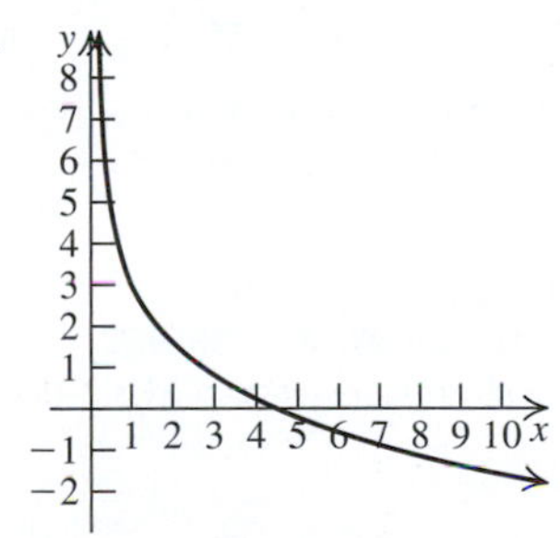

Applying the Concepts:

95. 8.66 yr **97.** 11.55% **99. a.** ≈ 280.8 million **b.** 0.93% **c.** 2039 **101. a. i.** 22.66°F **ii.** 20.13°F **iii.** 20°F **b.** 5.5 min **103.** 17.6 min **105. a.** 0.27% **b.** 26.85 months **107. a.** $P \approx 1500e^{0.231t}$ **b.** 7557 **c.** 10 yr **109. a.** $\log(4/3) \approx 0.125 = 12.5\%$ **b.** The sum is 1. One of the digits 1 … 9 will appear at the first place.

Beyond the Basics:

111. $(-\infty, -1) \cup (2, \infty)$ **113. a.** $(1, \infty)$ **b.** $(2, \infty)$ **c.** $(2, \infty)$ **d.** $(11, \infty)$ **115.** $y = \log x \to y = \log(x - 2) \to$
$y = \log\left(\frac{1}{2}x - 2\right) \to y = 3\log\left(\frac{1}{2}x - 2\right) \to$
$y = -3\log\left(\frac{1}{2}x - 2\right) \to y = -3\log\left(\frac{1}{2}x - 2\right) + 4$
117. a. \$24,659.70 **b.** 4.05%

Critical Thinking/Discussion/Writing:

119. 1 **121.** 16 **123. a.** Yes **b.** The increasing property

Maintaining Skills:

125. a^9 **127.** a^4 **129.** $\left(\frac{2}{3}\right)^4$ **131.** 3.807×10^{13}

Section 4.3

Answers to Practice Problems:

1. a. -1 **b.** 13 **2. a.** $\ln(2x - 1) - \ln(x + 4)$
b. $\log 2 + \frac{1}{2}\log x + \frac{1}{2}\log y - \frac{1}{2}\log z$ **3.** $\log\sqrt{x^2 - 1}$
4. $2.215088 \times 10^{1343}$ **5.** $\frac{\log 15}{\log 3} = \frac{\ln 15}{\ln 3} \approx 2.46497$
6. a. $y = 5 - \frac{2}{\log 3}\log x$ **b.** $y = 5 - 2\log_3 x$ **7.** ≈ 25 years
8. $\approx 65.34\%$

Basic Concepts and Skills:

1. $\log_a M$; $\log_a N$ **3.** $r\log_a M$ **5.** false **7.** 0.78 **9.** 0.7 **11.** -1.7
13. 7.3 **15.** $\frac{16}{3}$ **17.** 0.56 **19.** $\ln(x) + \ln(x - 1)$
21. $\frac{1}{2}\log_a x + 3\log_a y$ **23.** $\frac{1}{3}\log_a x - \frac{1}{3}\log_a y$ **25.** $\frac{1}{4}\log_2 x + \frac{1}{2}\log_2 y - \frac{3}{4}$
27. $\frac{1}{2}\log(x^2 + 1) - \log(x + 3)$ **29.** $2\log_b x + 3\log_b y + \log_b z$
31. $\ln x + \frac{1}{2}\ln(x - 1) - \ln(x^2 + 2)$
33. $2\ln(x + 1) - \ln(x - 3) - \frac{1}{2}\ln(x + 4)$
35. $\ln(x + 1) + \frac{1}{2}\ln(x^2 + 2) - \frac{1}{2}\ln(x^2 + 5)$
37. $3\ln x + 4\ln(3x + 1) - \frac{1}{2}\ln(x^2 + 1) + 5\ln(x + 2) - 2\ln(x - 3)$
39. $\log_2(7x)$ **41.** $\log\left(\frac{z\sqrt{x}}{y}\right)$ **43.** $\log_2(zy^2)^{1/5}$ **45.** $\ln(xy^2z^3)$
47. $\ln\left(\frac{x^2}{\sqrt{x^2 + 1}}\right)$ **49.** $1.4036 \cdot 10^{217}$ **51.** $9.3762 \cdot 10^{1897}$
53. 234^{567} **55.** 362 **57.** 2.322 **59.** -1.585 **61.** 1.760 **63.** 3.6
65. $\frac{1}{2}$ **67.** 1 **69.** 18 **71.** 2 **73.** $y = 2 - \log x$ **75.** $y = 2 - \ln x$
77. $y = 1 + 3\log_5 x$ **79.** $y = -1 + 5\log_2 x$ **81.** ≈ 10.7 yrs
83. ≈ 10.4 hrs

Applying the Concepts:

85. 1.2775% **87.** 1.0498% **89.** 12.53 yr **91.** 12.969 g
93. 12.60 yr **95.** $k = 0.1204$ **97.** 13.7 g **99.** $P = \$859.72$; interest $\approx$ \$86,332,80 **101.** He can afford a mortgage of \$110,545.60

Beyond the Basics:

107.

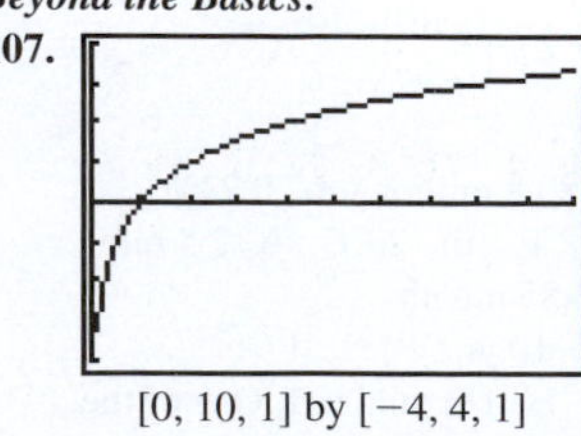

[0, 10, 1] by [−4, 4, 1]

109. a. 0 **b.** 0 **c.** 2 **d.** 1 **111.** (8, 10) **113. a.** False **b.** True **c.** False **d.** False **e.** True **f.** True **g.** False **h.** True **i.** False **j.** True

Critical Thinking/Discussion/Writing:

117. Because $\log\frac{1}{2}$ is a negative number, $3 < 4$ implies that $3\log\frac{1}{2} > 4\log\frac{1}{2}$. **119.** 12978189

Maintaining Skills:

121. 11 **123.** 2 **125.** $t^2 - t + 1 = 0$ **127.** $t^2 + 4 = 0$ **129.** -2
131. $-4, 1$ **133.** $(-\infty, 7)$ **135.** $(2, \infty)$

Section 4.4

Answers to Practice Problems:

1. a. $x = 5$ **b.** $x = \frac{2}{3}$ **2.** $x = \frac{\ln\left(\frac{11}{7}\right)}{\ln 3} - 1 \approx -0.59$
3. $x = \frac{\ln 3}{2\ln 2 - \ln 3} \approx 3.82$ **4.** $x = \ln 5 \approx 1.609$
5. a. United States: 343.61 million; Pakistan: 255.96 million
b. Sometime in 2022 (after 11.69 yr) **c.** In 23.68 yrs (sometime in 2033)
6. $x = e^{3/2}$ **7.** $x = 9$ **8.** $\varnothing$ **9.** The average growth rate was approximately 1.74%. **10.** $x < -2$ **11.** $x < \frac{1 - e^2}{3}$

Basic Concepts and Skills:

1. exponential **3.** logistic **5.** false **7.** $x = 4$ **9.** $x = \frac{5}{3}$
11. $x = \pm\frac{7}{2}$ **13.** $\varnothing$ **15.** $x = 1$ **17.** $x = \frac{1}{2}$ **19.** $x = \pm 9$
21. $x = 10{,}000$ **23.** $x = \frac{\ln 3}{\ln 2} = 1.585$
25. $x = \left(\frac{\ln 15 - 3\ln 2}{2\ln 2}\right) = 0.453$ **27.** $x = \frac{\ln 17 - \ln 5}{\ln 2} = 1.766$
29. $x = \frac{\ln 10 - \ln 3 + \ln 4}{2\ln 4} = 0.934$ **31.** $x = \frac{\ln 5}{\ln 5 + \ln 2} = 0.699$
33. $x = \frac{\ln 2 - 6\ln 3}{4\ln 3 + \ln 2} = -1.159$ **35.** $x = \frac{\ln 2 - \ln 3 - \ln 5}{\ln 5 - \ln 3} = -3.944$
37. $t = \frac{\ln 2}{\ln(1.065)} = 11.007$ **39.** $x = \frac{\ln 7}{\ln 2} = 2.807$
41. $x = \frac{\ln 2}{\ln 3}, \frac{\ln 4}{\ln 3} = 0.631, 1.262$ **43.** $x = \frac{\ln 4}{\ln 3} = 1.262$
45. $x = \frac{\ln 5 - \ln 3}{2\ln 3} = 0.232$ **47.** $x = \frac{\ln 2}{\ln 3} \approx 0.631$
49. $x = \frac{\ln 18 - \ln 7}{\ln 3} \approx 0.860$ **51.** $\varnothing$ **53.** $x = -\frac{49}{20}$ **55.** $x = 3, -2$
57. $x = 5, 2$ **59.** $x = 8$ **61.** $x = 1$ **63.** $x = 3$ **65.** $x = 4$
67. $\varnothing$ **69.** $x = \frac{2}{3}, \frac{5}{2}$ **71.** $a = 30, k = 2$; 500 **73.** $a = 5, k = \frac{1}{2}$; 31
75. $a = 2, k = \ln\left(\frac{17}{2}\right)$; 0.068 **77.** $a = -2, k = \ln\left(\frac{11}{18}\right)$; -7.902
79. $x \geq \frac{\ln 2}{\ln 0.3}$ **81.** $x \leq 0$ **83.** $(-3, 17)$ **85.** $x \geq e + 5$
87. $\left(\frac{7}{3}, 5\right)$

Applying the Concepts:

89. a. 10.087 yr; **b.** 9.870 yr; **c.** 9.821 yr; **d.** 9.797 yr; **e.** 9.796 yr
91. $r \approx 8.66\%$; \$30,837.52 **93. a.** 2.134 lm **b.** 19.08 ft
95. a. 18,542 **b.** 5 wk **97. a.** 20 **b.** 4490
c.

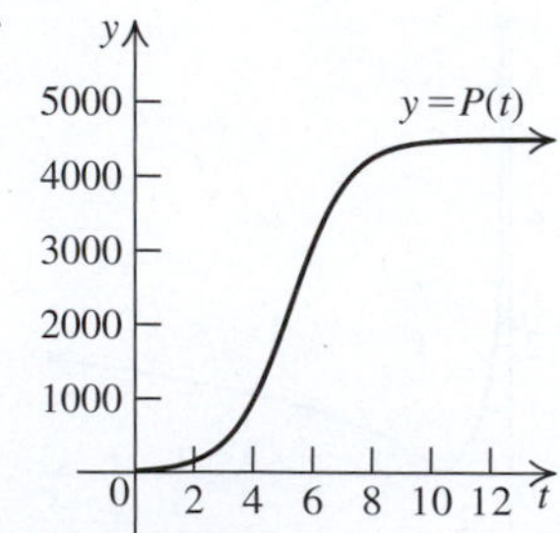

Beyond the Basics:

99. $t = \frac{1}{k}\ln\left(\frac{P}{M - P}\right)$ **103.** $x = 1, 10$ **105.** $\frac{1}{9}, 3$ **107.** $x = 2$
109. $\varnothing$ **111.** $f^{-1}(x) = \log_3(x - 5)$ **113.** $f^{-1}(x) = \log_4\frac{x - 7}{3}$

115. $f^{-1}(x) = 2^{x-1} + 1$ **117.** $f^{-1}(x) = \frac{1 + e^{2x}}{1 - e^{2x}}$ **119.** 130
121. 209 **123.** 380 **125.** 1

Critical Thinking/Discussion/Writing:
127. $t = \frac{\ln a}{k}$

Maintaining Skills:
129. $\log_9 81 = 2$ **131.** $\log 3 = x$ **133.** $64 = 2^6$ **135.** $A = 2 \cdot 10^3$
137. $x = \frac{1}{125}$ **139.** $x = 10$ **141.** $x = 4$ **143.** $x = \frac{\ln 3 + 3\ln 5}{2\ln 5 - \ln 3}$

Section 4.5

Answers to Practice Problems:
1. a. 5.57 **b.** $3.39 \cdot 10^{-9}$ **2.** ≈ 2512 **3.** $3{,}162{,}278 I_0$ **4.** $I_{Moz} \approx 63 I_{Cal}$
5. 178 times **6.** 73 dB **7.** 100 times more **8.** $6.3 \cdot 10^{-8}$ W/m^2
9. ≈ 494 hertz **10.** $\approx 400 - 386 = 14$ (cents) **11. a.** $10^{0.8} \approx 6.3$ times brighter **b.** ≈ 4.1598

Basic Concepts and Skills:
1. acidic **3.** $\log\left(\frac{I}{I_0}\right)$ **5.** $10\log\left(\frac{I}{I_0}\right)$ **7.** $2.5\log\left(\frac{b_1}{b_2}\right)$ **9.** true
11. pH = 8, base **13.** pH $\approx$ 4.64, acid
15. $[\text{H}^+] = 10^{-6}$, $[\text{OH}^-] = 10^{-8}$
17. $[\text{H}^+] = 3.16 \cdot 10^{-10}$ $[\text{OH}^-] = 3.16 \cdot 10^{-5}$ **19. a.** $10^5 I_0$
b. $7.94 \cdot 10^{11}$ **21. a.** $6.3 \cdot 10^7 I_0$ **b.** $1.26 \cdot 10^{16}$ **23. a.** 6; **b.** $10^6 I_0$
25. a. 5.1 **b.** $1.26 \cdot 10^5 I_0$ **27.** 40 **29.** 55.4 **31.** 10^{-4}
33. $2.95 \cdot 10^{-6}$ **35.** A#466 Hz, C523 Hz **37.** ≈ 18 cents; noticeable
39. A is $\approx 2.5 \times 10^6$ times brighter than B.
41. B is ≈ 6.31 times brighter than A.

Applying the Concepts:
43. 7.4; basic **45. a.** $7.1 \cdot 10^{-4}$ **b.** $6.3 \cdot 10^{-8}$ **c.** $1.7 \cdot 10^{-8}$ **d.** 10^{-3}
47. $1.6 \cdot 10^{-4}$; about 160 times more **49.** $\text{pH}_A = \text{pH}_B - 2$
51. pH decreased by $\log 50 \approx 1.7$; more acidic
53. a. $6.3 \cdot 10^7 I_0$ **b.** $1.259 \cdot 10^{16}$ **55. a.** $I_A = 10 I_B$ **b.** $E_A = 31.6 E_B$
57. $\log 150 \approx 2.18$ **59.** ≈ 77 dB **61.** $\approx 3.16 \cdot 10^6$ times as intense
63. 160 dB **65.** $I_2 \approx 1.26 I_1$. **67.** ≈ 4 cents more **69.** $\approx 3.98 \cdot 10^5$
71. $\approx 1.58 \cdot 10^9$ **73.** ≈ -5.4

Beyond the Basics:
75. a. 26 dB; 18 dB; 12 dB; 6 dB **b.** 200 ft
c. $a = 10(4 + \log 4) \approx 46$; $b = -20$ **d.** $r = \frac{200}{10^{L/20}}$
77. 5×10^{15} vs 2.5×10^{16}; earthquake is 5 times more powerful.
79. $[875, 885]$ Hz **81.** -2.2

Maintaining Skills:
83. $\frac{3}{2}$; -6 **85.** $-e^{-1.2}$; $e^{2.2}$ **87.** (0, 2); 3 **89.** (3, −1); $\sqrt{7}$ **91.** (1, 0); 2
93. (1, −2); 3

Review Exercises

Basic Concepts and Skills:
1. False **3.** False **5.** True **7.** False **9.** True **11.** h **13.** f
15. d **17.** a

19. Domain = $(-\infty, \infty)$
Range = $(0, \infty)$ Asymptote: $y = 0$

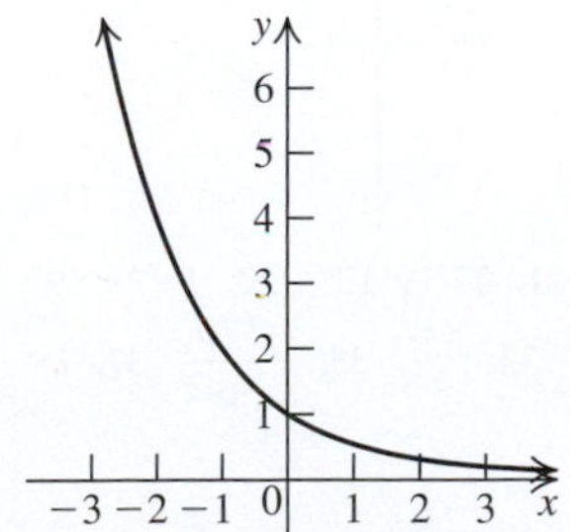

21. Domain = $(-\infty, \infty)$
Range = $(3, \infty)$ Asymptote: $y = 3$

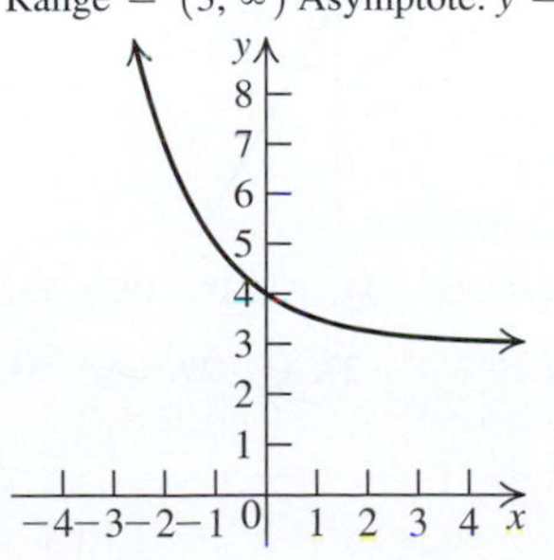

23. Domain = $(-\infty, \infty)$
Range = $(0, 1]$ Asymptote: $y = 0$

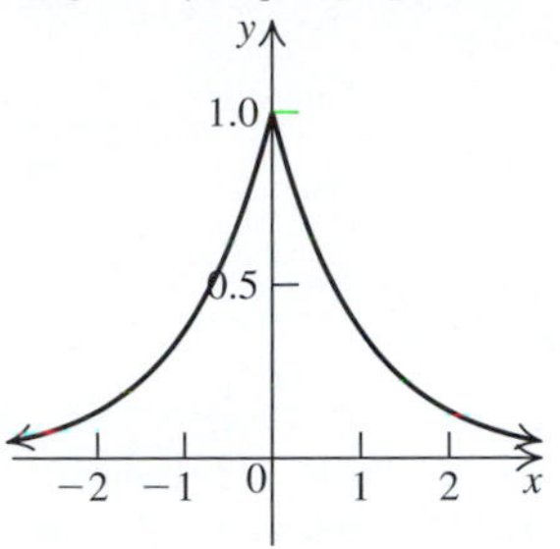

25. Domain = $(-\infty, 0)$
Range = $(-\infty, \infty)$
Asymptote: $x = 0$

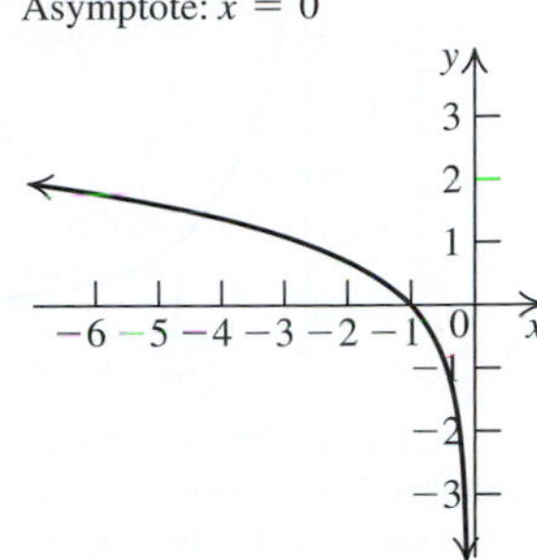

27. Domain = $(1, \infty)$
Range = $(-\infty, \infty)$
Asymptote: $x = 1$

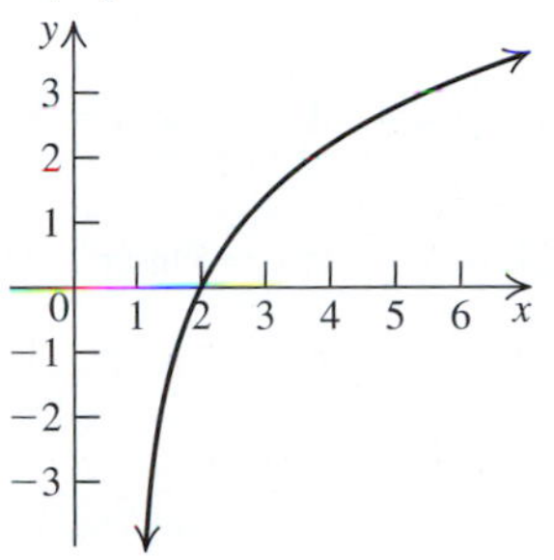

29. Domain = $(-\infty, 0)$
Range = $(-\infty, \infty)$
Asymptote: $x = 0$

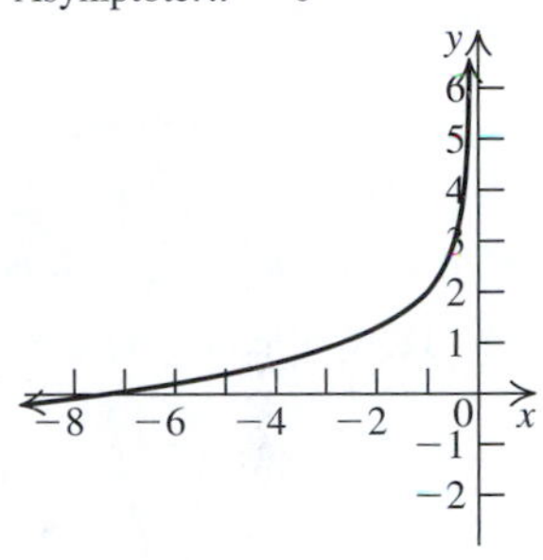

31. a. x-intercept = $\ln(2/3)$,
y-intercept = 1
b. $\lim_{x\to\infty} f(x) = 3$ $\lim_{x\to-\infty} f(x) = -\infty$

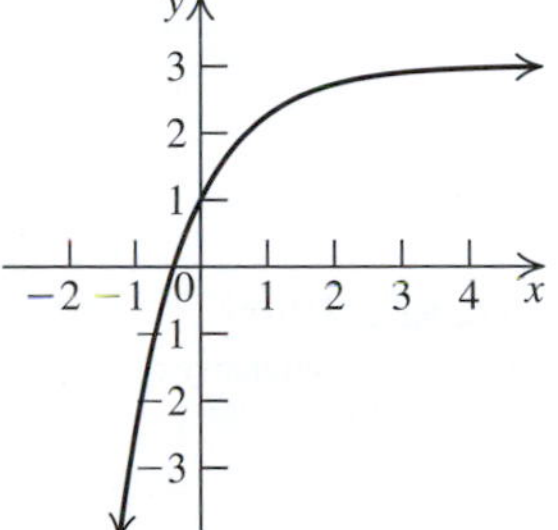

33. a. x-intercept: none,
y-intercept = 1
b. as $x \to \infty$, $y \to 0$
as $x \to -\infty$, $y \to 0$

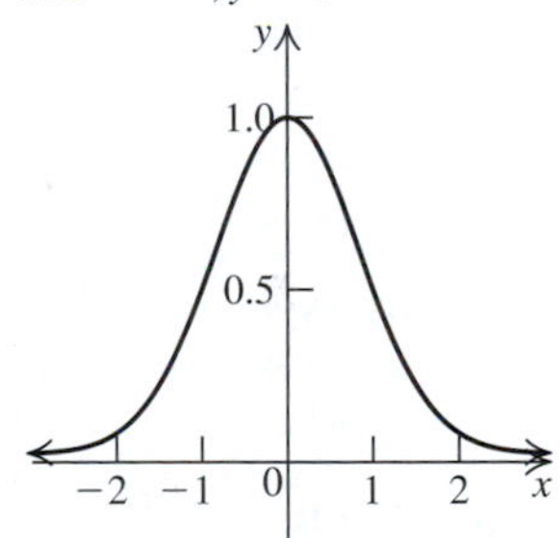

35. $a = 10, k = 2, f(2) = 160$
37. $a = 3, k = -0.2824, f(4) = 0.38898$ **39.** $y = 3 \cdot 2^x$
41. $y = \log_5(x - 1)$ **43. a.** $y = -2^{(x-1)} - 3$ **b.** $y = -2^{(x-1)} + 3$
45. $\ln x + 2\ln y + 3\ln z$ **47.** $\ln x + \frac{1}{2}\ln(x^2 + 1) - 2\ln(x^2 + 3)$
49. $y = 3x$ **51.** $y = -1 + \sqrt{x - 2}$ **53.** $y = \frac{\sqrt{x^2 - 1}}{x^2 + 1}$ **55.** 4
57. $-1 \pm \sqrt{5}$ **59.** $\frac{\ln 23}{\ln 3} \approx 2.854$ **61.** $\frac{\ln 19}{\ln 273} \approx 0.525$ **63.** 0
65. $\frac{-\ln 3}{\ln\left(\frac{(1.7)^3}{9}\right)} \approx 1.815$ **67.** $\frac{5}{2}$ **69.** 0 **71.** 8 **73.** 5 **75.** 9
77. 3 **79.** $x \geq -1$ **81.** $\left(-\frac{7}{2}, \frac{93}{3}\right)$

Applying the Concepts:
83. 11.4 yr **85.** 40 hr **87.** At 5% compounded yearly, A = \$9849.70. At 4.75% compounded monthly, A = \$9754.74. The 5% investment will provide the greater return. **89. a.** 34 million **b.** In 2206 (after 199.3 yr)

91. a.

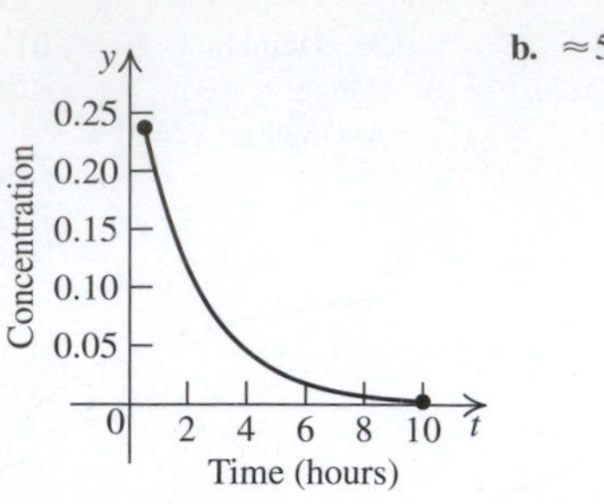

b. ≈ 5 hr

93. $t = 6.57$ min **95. a.** 5000 people/square mile **b.** 7459 people/square mile **c.** ≈ 13.73 miles **97.** $I = 0.2322I_0$ **99. a.** 11.27 ft/sec **b.** 5.29 ft/sec **c.** 201 **101. a.** 1 g **b.** Remains stationary at 6 g **c.** 4.6 days **103.** 16 cents; yes, it is noticeable. **105.** $4.0 \cdot 10^{-3}$; 1585 times more **107.** 100 times more intense **109.** ≈ 9.7

Practice Test A

1. $x = -3$ **2.** $x = 32$

3. Range $= (-\infty, 1)$; horizontal asymptote: $y = 1$ **4.** -3 **5.** $x = 3$

6. -3 **7.** $\ln 3x^5$ **8.** $\dfrac{\ln\left(\frac{5}{2}\right)}{\ln 2}$ **9.** $\ln 2$ **10.** $\ln 2 + 3\ln x - 5\ln(x+1)$

11. -5 **12.** $y = \ln(x-1) + 3$

13.

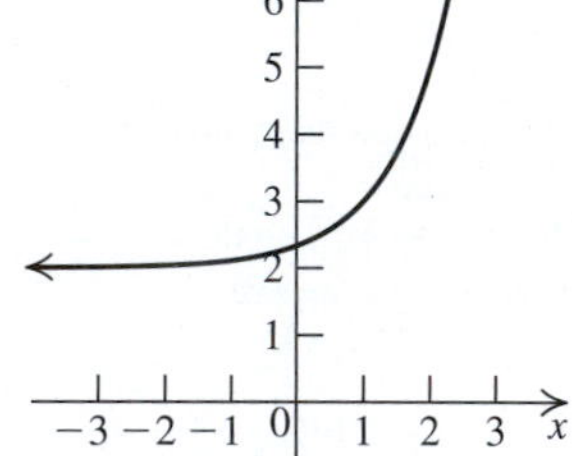

14. $(-\infty, 0)$ **15.** $\ln \dfrac{x^3(x^3+2)}{\sqrt{3x^2+2}}$ **16.** $x = 3$ **17.** $x = 3$

18. $A(t) = 15{,}000(1.0175)^{4t}$ **19.** In 1983 (after 23.03 yr)

20. The Nicaragua earthquake was about 2.5 times the intensity of the Morocco earthquake.

Practice Test B

1. b **2.** b **3.** b **4.** d **5.** d **6.** b **7.** b **8.** c **9.** d **10.** d **11.** b **12.** d **13.** b **14.** a **15.** a **16.** d **17.** b **18.** d **19.** b **20.** b

Cumulative Review Exercises (Chapters P–4)

1. x-intercept: 0; y-intercepts: 0, 2; symmetric about y-axis

3. Center: $(-1, 2)$; radius: 5

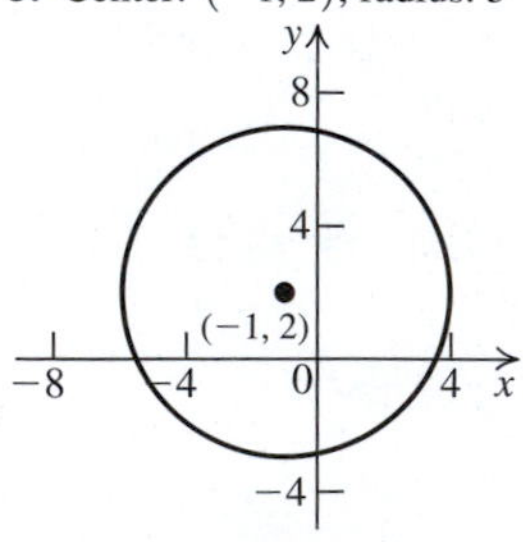

5. $(-\infty, -3) \cup (2, \infty)$

7. Start with $y = \sqrt{x}$, shift 1 unit left, stretch vertically by a factor of 3, and shift 2 units down.

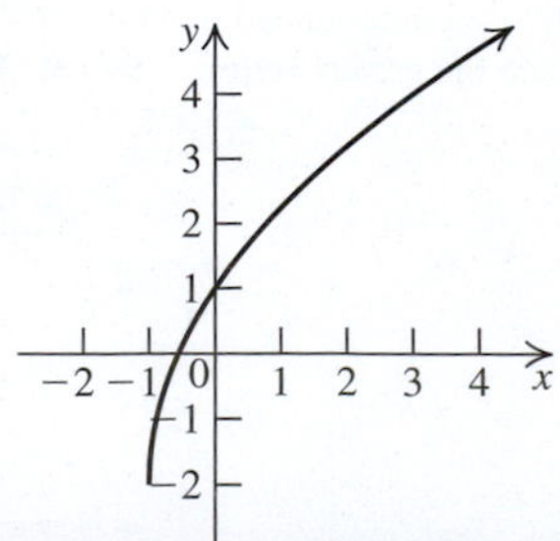

9. a. Yes, $f^{-1}(x) = \left(\dfrac{x-1}{2}\right)^{1/3}$ **b.** No **c.** Yes, $f^{-1}(x) = e^x$

11. a. $(x-1)(x-2)(x+3) = x^3 - 7x + 6$ **b.** $x^4 - 1$

13. a. $\log\sqrt{xy}$ **b.** $\ln\dfrac{x^3}{y^2}$ **15. a.** $\{3\}$ **b.** $\{1, 3\}$ **c.** 3.77

17. a. 1, 1, 1, −2, −2, 3 **b.** At $x = 1$ and $x = 3$, the graph crosses the x-axis. At $x = -2$, the graph touches but does not cross the x-axis. **c.** $f(x) \to \infty$ as $x \to \infty$; $f(x) \to \infty$ as $x \to -\infty$

d.

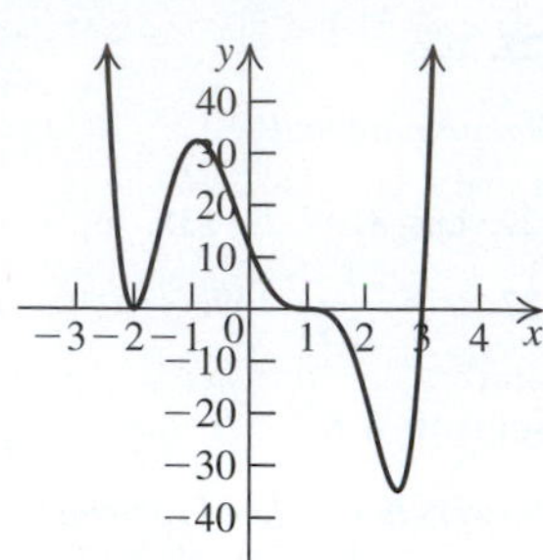

CHAPTER 5

Section 5.1

Answers to Practice Problems:

1.

225°

2. a. 13.16° **b.** 41°16′30″ **3.** $-\dfrac{\pi}{4}$ radians **4.** 270°

5. Complement = 23°; supplement = 113° **6.** $\dfrac{5\pi}{2} \approx 7.85$ meters

7. ≈ 790 miles **8.** $\dfrac{50\pi}{3} \approx 52.36$ square inches

9. $\omega = 36\pi$ radians per minute; $v = 360\pi \approx 1131$ feet per minute

Basic Concepts and Skills:

1. clockwise **3.** one-sixtieth **5.** true

7.

30°

9.

−120°

11. **13.**

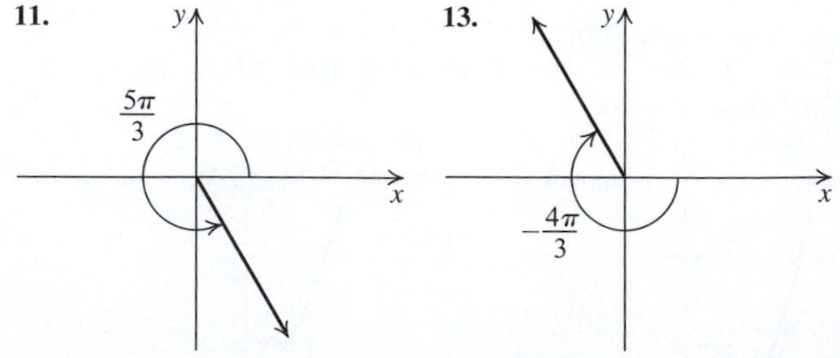

15. 70.75° **17.** 23.71° **19.** −15.72° **21.** 27°19′12″ **23.** 13°20′49″

25. 19°3′4″ **27.** $\dfrac{\pi}{9}$ **29.** $-\pi$ **31.** $\dfrac{7\pi}{4}$ **33.** $\dfrac{8\pi}{3}$ **35.** $-\dfrac{17\pi}{6}$ **37.** 15°

39. $-100°$ **41.** $300°$ **43.** $450°$ **45.** $-495°$ **47.** 0.21 radian **49.** -1.47 radians **51.** $53.86°$ **53.** $-470.40°$ **55.** Complement: 43°; supplement: 133° **57.** Complement: none because the measure of the angle is greater than 90°; supplement: 60° **59.** Complement: none because the measure of the angle is greater than 90°; supplement: none because the measure of the angle is greater than 180°

61. $7/25 \approx 0.28$ radian **63.** $\frac{44}{21} \approx 2.095$ radians **65.** $\frac{5\pi}{12} \approx 1.309$ m

67. 78 m **69.** 60 m/min **71.** 2 rad/sec **73.** 200 in^2

75. $\sqrt{\frac{120}{\pi}} \approx 6.180$ ft

Applying the Concepts:

77. 70.69 in **79.** $\frac{2\pi}{3}$ radians **81.** 18 in **83.** 86° **85. a.** 120 radians per minute **b.** 300 radians per minute **87.** 132,000 radians per hour **89.** 84,823 inches per minute **91.** ≈ 508 mi **93.** ≈ 462 mi **95.** $\approx 6.366°$

Beyond the Basics:

99. 4 **103.** 7 m **105.** 250 **107.** $\frac{99}{112} \approx 0.884\ \text{ft}^2$

Critical Thinking/Discussion/Writing:

109. $3, \pi, 4, \frac{3\pi}{2}$

Maintaining Skills:

113. $c = 13$ **115.** $a = 8$ **117.** $a = 3\sqrt{3}$ **119.** $\frac{\sqrt{3}}{3}$

Section 5.2

Answers to Practice Problems:

1. $\sin\theta = \frac{4}{5}$ $\csc\theta = \frac{5}{4}$

$\cos\theta = \frac{3}{5}$ $\sec\theta = \frac{5}{3}$

$\tan\theta = \frac{4}{3}$ $\cot\theta = \frac{3}{4}$

2. $\tan\theta = \frac{3}{4}$, $\csc\theta = \frac{5}{3}$, $\sec\theta = \frac{5}{4}$, $\cot\theta = \frac{4}{3}$

3. $\sin\theta = \frac{2\sqrt{2}}{3}$ $\csc\theta = \frac{3\sqrt{2}}{4}$

$\cos\theta = \frac{1}{3}$ (given) $\sec\theta = 3$

$\tan\theta = 2\sqrt{2}$ $\cot\theta = \frac{\sqrt{2}}{4}$

4. $\csc 45° = \sqrt{2}$, $\sec 45° = \sqrt{2}$, $\cot 45° = 1$

5. $\sin 60° = \frac{\sqrt{3}}{2}$, $\cos 60° = \frac{1}{2}$, $\tan 60° = \sqrt{3}$

$\csc 60° = \frac{2\sqrt{3}}{3}$, $\sec 60° = 2$, $\cot 60° = \frac{\sqrt{3}}{3}$ **6. a.** $\sec 69° \approx 2.7904$ **b.** $\cot 15° \approx 3.7321$ **7.** ≈ 1.096 miles **8.** $\approx 20{,}320$ feet **9.** ≈ 43 feet

Basic Concepts and Skills:

1. $\frac{1}{3}$ **3.** 0.7 **5.** true

7. $\sin\theta = \frac{2\sqrt{5}}{25}$ $\csc\theta = \frac{5\sqrt{5}}{2}$

$\cos\theta = \frac{11\sqrt{5}}{25}$ $\sec\theta = \frac{5\sqrt{5}}{11}$

$\tan\theta = \frac{2}{11}$ $\cot\theta = \frac{11}{2}$

9. $\sin\theta = \frac{3}{5}$ $\csc\theta = \frac{5}{3}$

$\cos\theta = \frac{4}{5}$ $\sec\theta = \frac{5}{4}$

$\tan\theta = \frac{3}{4}$ $\cot\theta = \frac{4}{3}$

11. $\sin\theta = \frac{\sqrt{2}}{10}$ $\csc\theta = 5\sqrt{2}$

$\cos\theta = \frac{7\sqrt{2}}{10}$ $\sec\theta = \frac{5\sqrt{2}}{7}$

$\tan\theta = \frac{1}{7}$ $\cot\theta = 7$

13. $\tan\theta = \frac{12}{5}$, $\csc\theta = \frac{13}{12}$, $\sec\theta = \frac{13}{5}$, $\cot\theta = \frac{5}{12}$

15. $\tan\theta = \frac{21}{20}$, $\csc\theta = \frac{29}{21}$, $\sec\theta = \frac{29}{20}$, $\cot\theta = \frac{20}{21}$

17. $\tan\theta = \frac{2\sqrt{21}}{21}$, $\csc\theta = \frac{5}{2}$, $\sec\theta = \frac{5\sqrt{21}}{21}$, $\cot\theta = \frac{\sqrt{21}}{2}$

19. $\sin\theta = \frac{\sqrt{5}}{3}$ $\csc\theta = \frac{3\sqrt{5}}{5}$

$\cos\theta = \frac{2}{3}$ $\sec\theta = \frac{3}{2}$

$\tan\theta = \frac{\sqrt{5}}{2}$ $\cot\theta = \frac{2\sqrt{5}}{5}$

21. $\sin\theta = \frac{5\sqrt{34}}{34}$ $\csc\theta = \frac{\sqrt{34}}{5}$

$\cos\theta = \frac{3\sqrt{34}}{34}$ $\sec\theta = \frac{\sqrt{34}}{3}$

$\tan\theta = \frac{5}{3}$ $\cot\theta = \frac{3}{5}$

23. $\sin\theta = \frac{5}{13}$ $\csc\theta = \frac{13}{5}$

$\cos\theta = \frac{12}{13}$ $\sec\theta = \frac{13}{12}$

$\tan\theta = \frac{5}{12}$ $\cot\theta = \frac{12}{5}$ **25.** 0.8480 **27.** 0.5095 **29.** 2.3662

31. 1 **33.** $\frac{\sqrt{3}}{2}$ **35.** 3 **37.** $\frac{\sqrt{6}-\sqrt{2}}{4}$

39. $c \approx 12.806$, $\sin\theta \approx 0.625$, $\tan\theta = 0.8$ **41.** $c \approx 24.042$; $\cos\theta \approx 0.291$, $\tan\theta \approx 3.286$ **43.** $\sin\theta = 0.5$, $b \approx 15.588$, $c = 18$ **45.** $c \approx 13.953$, $\sin\theta \approx 0.896$, $\cos\theta \approx 0.444$ **47.** $a \approx 11.040$, $\sin\theta \approx 0.761$, $\tan\theta \approx 1.174$

49. a. 3 **b.** $\frac{2\sqrt{2}}{3}$ **c.** $\frac{\sqrt{2}}{4}$ **d.** $\frac{\sqrt{2}}{4}$ **51. a.** $\frac{1}{5}$ **b.** $\frac{2\sqrt{6}}{5}$ **c.** $2\sqrt{6}$ **d.** $\frac{\sqrt{6}}{12}$ **53. a.** $\frac{2}{5}$ **b.** $\frac{\sqrt{29}}{2}$ **c.** $\frac{2\sqrt{29}}{29}$ **d.** $\frac{5\sqrt{29}}{29}$

Applying the Concepts:

55. ≈ 11.3 ft **57.** ≈ 1029 m **59.** 2500 ft **61.** $4\sqrt{3} \approx 6.9$ ft **63.** 15 ft **65.** ≈ 43 ft **67.** The Empire State Building is approximately 150 m away and is approximately 381 m tall. **69.** $\approx 238{,}844$ mi **71.** 2.1 m **73.** 101.5 ft **75.** ≈ 2383 miles

Beyond the Basics:

77. $\frac{16\sqrt{2}+3}{8}$ **79.** $\frac{1}{9}$ **81.** ≈ 15.8 ft **83.** ≈ 496 ft **85.** ≈ 5402 m **87.** $\approx 20{,}168$ feet per minute **89.** ≈ 48 square units

Critical Thinking/Discussion/Writing:

91. 1 **93. (i)** $\sqrt{3}$; 30° **(ii)** $\frac{\sqrt{6}}{2}$; $\frac{\sqrt{6}}{2}$; 45° **(iii)** 45°; 45° **(iv)** $\frac{\sqrt{2}}{2}$; $\frac{\sqrt{2}}{2}$ **(v)** $\frac{\sqrt{6}+\sqrt{2}}{2}$; $\frac{\sqrt{6}-\sqrt{2}}{2}$ **(vi)** 15°; 75°

(vii) $\sin 15° = \frac{\sqrt{6}-\sqrt{2}}{4}$ $\cos 15° = \frac{\sqrt{6}+\sqrt{2}}{4}$

(viii) $\sin 75° \approx \frac{\sqrt{6}+\sqrt{2}}{4}$; $\cos 75° \approx \frac{\sqrt{6}-\sqrt{2}}{4}$

Maintaining Skills:

95. $2 - \sqrt{3}$ **97.** $11 - 6\sqrt{3}$ **99.** $\frac{5\pi}{12}$ **101.** $\frac{-17}{6}\pi$

103. $360 \cdot 3 + 140$ **105.** $360 \cdot (-2) + 240$

Section 5.3

Answers to Practice Problems:

1. $\sin\theta = -\frac{5\sqrt{29}}{29}$ $\csc\theta = -\frac{\sqrt{29}}{5}$ $\cos\theta = \frac{2\sqrt{29}}{29}$ $\sec\theta = \frac{\sqrt{29}}{2}$

$\tan\theta = -\frac{5}{2}$ $\cot\theta = -\frac{2}{5}$

2. a. $\sin 180° = \sin\pi = 0$ $\csc 180° = \csc\pi$ is undefined.
$\cos 180° = \cos\pi = -1$ $\sec 180° = \sec\pi = -1$
$\tan 180° = \tan\pi = 0$ $\cot 180° = \cot\pi$ is undefined.
b. $\sin 270° = \sin 3\pi/2 = -1$ $\csc 270° = \csc 3\pi/2 = -1$
$\cos 270° = \cos 3\pi/2 = 0$ $\sec 270° = \sec 3\pi/2$ is undefined.
$\tan 270° = \tan 3\pi/2$ is undefined $\cot 270° = \cot 3\pi/2 = 0$

3. a. $\cos 1830° = \frac{\sqrt{3}}{2}$ **b.** $\sin\frac{31\pi}{3} = \frac{\sqrt{3}}{2}$ **4.** Quadrant II

5. $\sin\theta = -\frac{4\sqrt{41}}{41}$, $\sec\theta = \frac{\sqrt{41}}{5}$ **6. a.** 5° **b.** $\frac{\pi}{3}$

c. $3\pi - 8.22 \approx 1.20$ **7.** $\cos 60° \approx 0.5$ **8. a.** $\csc 1035° = -\sqrt{2}$

b. $\cot\frac{17\pi}{6} = -\sqrt{3}$ **9.** 101.4 m

10. $\sin t = -\frac{1}{2}$, $\cos t = -\frac{\sqrt{3}}{2}$, $\tan t = \frac{\sqrt{3}}{3}$

Basic Concepts and Skills:

1. $\sqrt{x^2 + y^2}$; $\frac{y}{r}$; $\frac{x}{r}$; $\frac{y}{x}$ **3.** x-axis **5.** false

7. $\sin\theta = \frac{4}{5}$ $\csc\theta = \frac{5}{4}$
$\cos\theta = -\frac{3}{5}$ $\sec\theta = -\frac{5}{3}$
$\tan\theta = -\frac{4}{3}$ $\cot\theta = -\frac{3}{4}$

9. $\sin\theta = \frac{12}{13}$ $\csc\theta = \frac{13}{12}$
$\cos\theta = \frac{5}{13}$ $\sec\theta = \frac{13}{5}$
$\tan\theta = \frac{12}{5}$ $\cot\theta = \frac{5}{12}$

11. $\sin\theta = \frac{24}{25}$ $\csc\theta = \frac{25}{24}$
$\cos\theta = \frac{7}{25}$ $\sec\theta = \frac{25}{7}$
$\tan\theta = \frac{24}{7}$ $\cot\theta = \frac{7}{24}$

13. $\sin\theta = -\frac{7}{25}$ $\csc\theta = -\frac{25}{7}$
$\cos\theta = -\frac{24}{25}$ $\sec\theta = -\frac{25}{24}$
$\tan\theta = \frac{7}{24}$ $\cot\theta = \frac{24}{7}$

15. $\sin\theta = \frac{1}{\sqrt{2}} = \frac{\sqrt{2}}{2}$ $\csc\theta = \frac{\sqrt{2}}{1} = \sqrt{2}$
$\cos\theta = \frac{1}{\sqrt{2}} = \frac{\sqrt{2}}{2}$ $\sec\theta = \frac{\sqrt{2}}{1} = \sqrt{2}$
$\tan\theta = \frac{1}{1} = 1$ $\cot\theta = \frac{1}{1} = 1$

17. $\sin\theta = \frac{\sqrt{2}}{2}$ $\csc\theta = \frac{2}{\sqrt{2}} = \sqrt{2}$
$\cos\theta = \frac{\sqrt{2}}{2}$ $\sec\theta = \frac{2}{\sqrt{2}} = \sqrt{2}$
$\tan\theta = \frac{\sqrt{2}}{\sqrt{2}} = 1$ $\cot\theta = \frac{\sqrt{2}}{\sqrt{2}} = 1$

19. $\sin\theta = -\frac{1}{2}$ $\csc\theta = -\frac{2}{1} = -2$
$\cos\theta = \frac{\sqrt{3}}{2}$ $\sec\theta = \frac{2}{\sqrt{3}} = \frac{2\sqrt{3}}{3}$
$\tan\theta = -\frac{1}{\sqrt{3}} = -\frac{\sqrt{3}}{3}$ $\cot\theta = -\frac{\sqrt{3}}{1} = -\sqrt{3}$

21. $\sin\theta = -\frac{2}{\sqrt{29}} = -\frac{2\sqrt{29}}{29}$ $\csc\theta = -\frac{\sqrt{29}}{2}$
$\cos\theta = \frac{5}{\sqrt{29}} = \frac{5\sqrt{29}}{29}$ $\sec\theta = \frac{\sqrt{29}}{2}$
$\tan\theta = -\frac{2}{5}$ $\cot\theta = -\frac{5}{2}$

23. 1 **25.** 0 **27.** Undefined **29.** 0 **31.** Undefined **33.** -1 **35.** -1 **37.** 0 **39.** 1 **41.** 0 **43.** 0 **45.** Quadrant III **47.** Quadrant II **49.** Quadrant IV **51.** Quadrant II **53.** -12 **55.** 24

57. $\sin\theta = -\frac{12}{13}$ $\csc\theta = -\frac{13}{12}$
$\cos\theta = -\frac{5}{13}$ $\sec\theta = -\frac{13}{5}$
$\tan\theta = \frac{12}{5}$ $\cot\theta = \frac{5}{12}$

59. $\sin\theta = \frac{4}{5}$ $\csc\theta = \frac{5}{4}$
$\cos\theta = -\frac{3}{5}$ $\sec\theta = -\frac{5}{3}$
$\tan\theta = -\frac{4}{3}$ $\cot\theta = -\frac{3}{4}$

61. $\sin\theta = \frac{3}{5}$ $\csc\theta = \frac{5}{3}$ $\cos\theta = -\frac{4}{5}$ $\sec\theta = -\frac{5}{4}$
$\tan\theta = -\frac{3}{4}$ $\cot\theta = -\frac{4}{3}$

63. $\sin\theta = -\frac{2\sqrt{2}}{3}$ $\csc\theta = -\frac{3\sqrt{2}}{4}$ $\cos\theta = \frac{1}{3}$ $\sec\theta = 3$
$\tan\theta = -2\sqrt{2}$ $\cot\theta = -\frac{\sqrt{2}}{4}$ **65.** $\frac{1}{2}$ **67.** 1 **69.** -2

71. $-\frac{\sqrt{3}}{3}$ **73.** $\frac{\sqrt{3}}{2}$ **75.** $-\frac{\sqrt{3}}{2}$ **77.** 1 **79.** 2

81. $\sin s = \frac{\sqrt{3}}{2}$, $\cos s = -\frac{1}{2}$, $\tan s = -\sqrt{3}$

83. $\sin s = -\frac{1}{2}$, $\cos s = -\frac{\sqrt{3}}{2}$, $\tan s = -\frac{\sqrt{3}}{3}$

85. $\sin s = -\frac{\sqrt{2}}{2}$, $\cos s = -\frac{\sqrt{2}}{2}$, $\tan s = 1$

Applying the Concepts:

87. $8\sqrt{3} \approx 13.86$ ft **89.** $H = 7.5625$ ft; $t = 1.375$ sec; $R = 52.39$ ft **91.** $H = 22.6875$ ft; $t = 2.38$ sec; $R = 52.39$ ft **93. a.** $y = x - \frac{x^2}{200}$. **b.** 50 ft **95.** 3.54 sec **97.** 0.99 ft **99. a.** 50 ft **b.** $50\sqrt{3} \approx 86.6$ ft

Beyond the Basics:

101. Triangle *SOM* **103.** Triangle *RON* **105.** $-\frac{\sqrt{2}}{2}$; $-\sqrt{3}$

107. $-\frac{\sqrt{2}}{2}$; $-\frac{1}{2}$; $\frac{\sqrt{3}}{3}$ **111.** 0 **113.** The value on the left is negative, but the value on the right is positive. **115.** The value on the left is negative, but the value on the right is positive.

Critical Thinking/Discussion/Writing:

117. False. The two sides are unequal for $\theta = 90°$.

Maintaining Skills:

119. $\frac{4}{3}$ **121.** $-5, 2$ **123.** Odd **125.** Neither **127.** Even

Section 5.4

Answers to Practice Problems:

1. **2.**

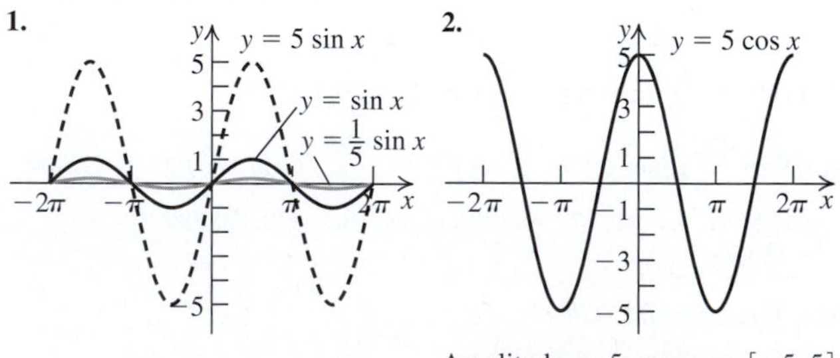

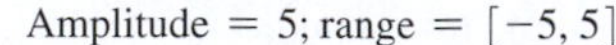
Amplitude = 5; range = [−5, 5]

3. **4.**

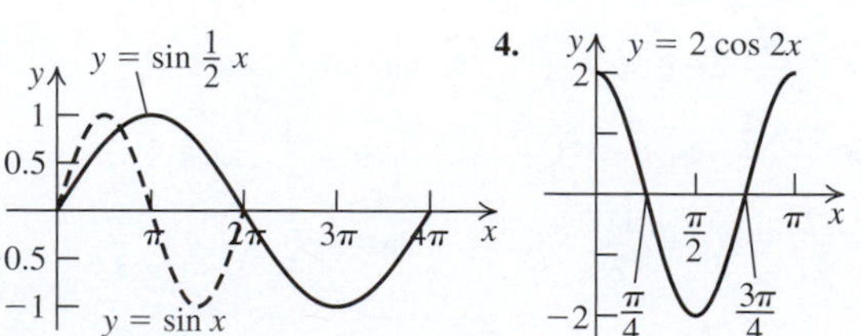

5. Answers will vary.

6.

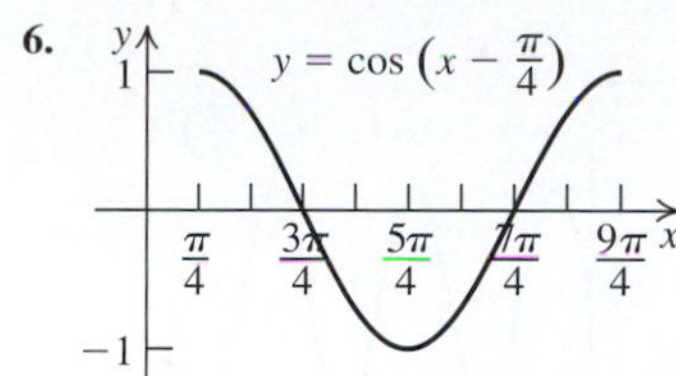

7. Amplitude $= \frac{1}{3}$; period $= 5\pi$; phase shift $= -\frac{\pi}{6}$

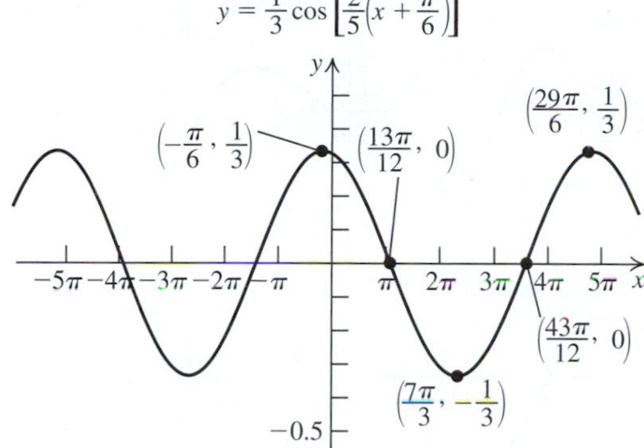

8.

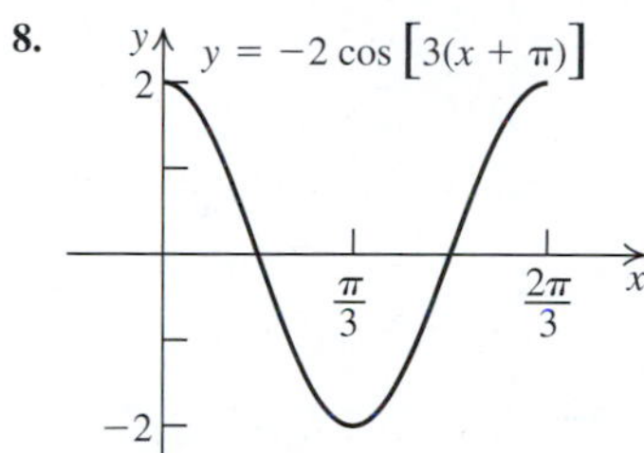

9. a. Period $= \pi$; phase shift $= -\frac{\pi}{8}$ **b.** Period $= 2$; phase shift $= \frac{1}{\pi}$

10.

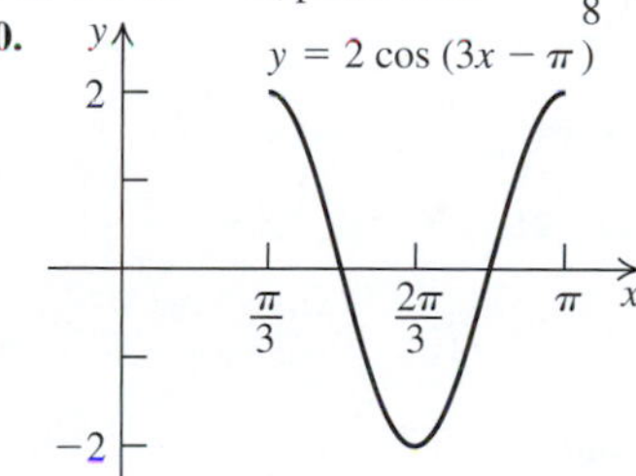

11.

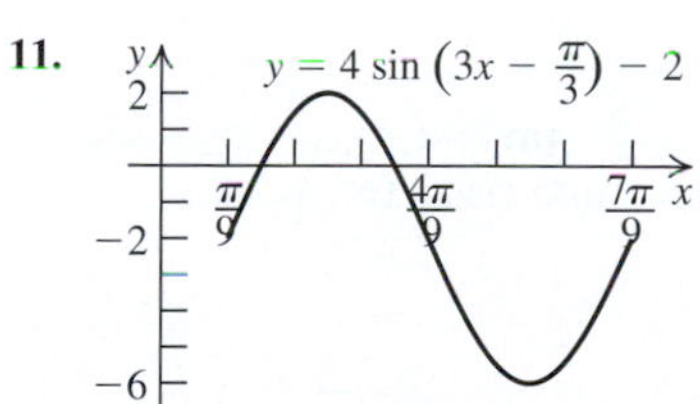

12.

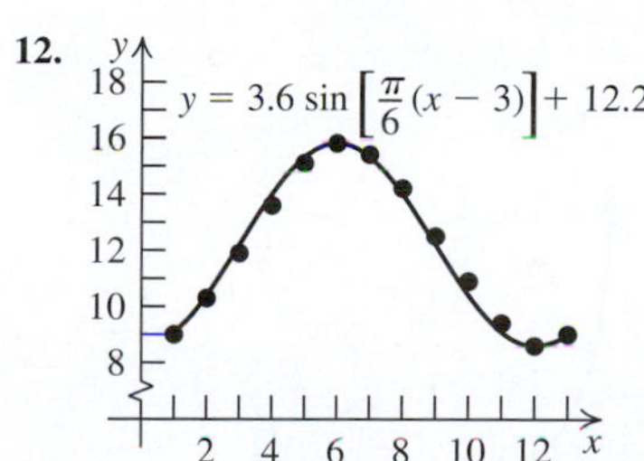

13. $y = -4\cos\frac{2\pi}{3}t$

Basic Concepts and Skills:

1. π **3.** $[-1, 1]$ **5.** false **7.** true

9.

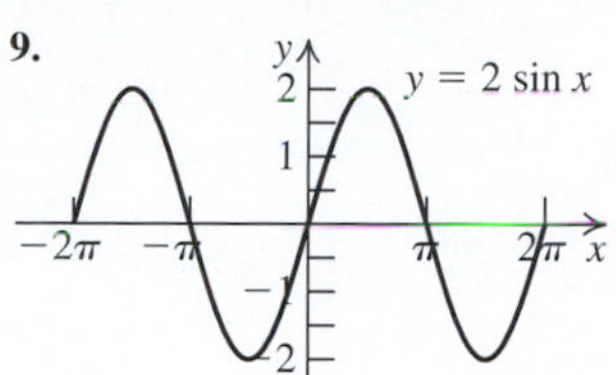

11.

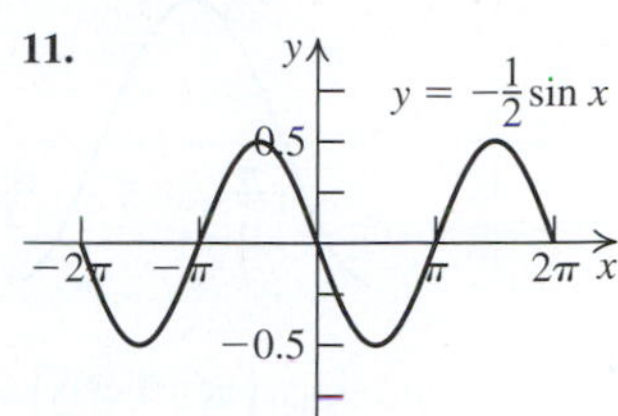

13.

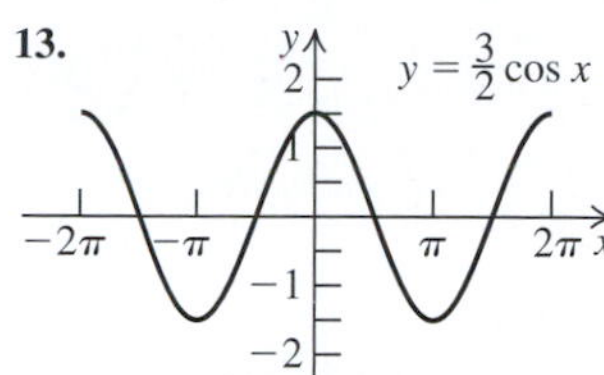

15.

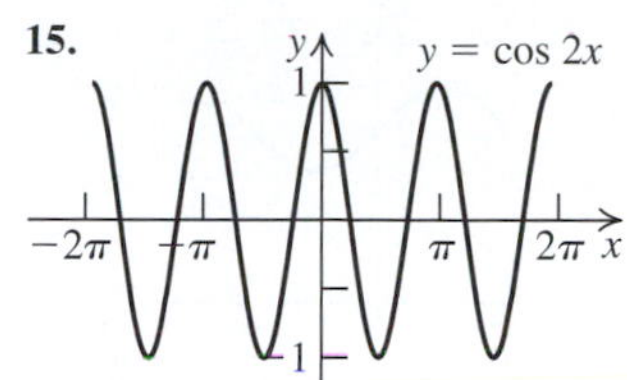

17.

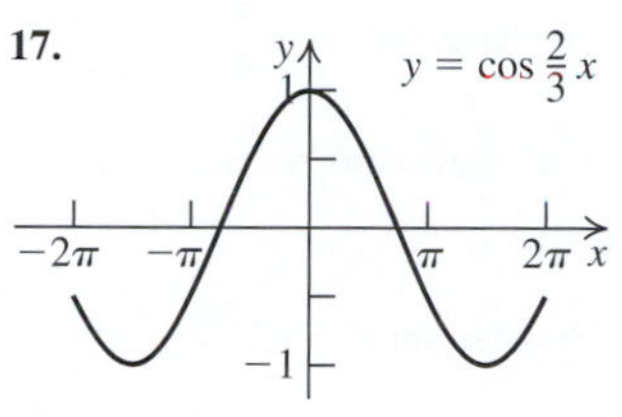

19.

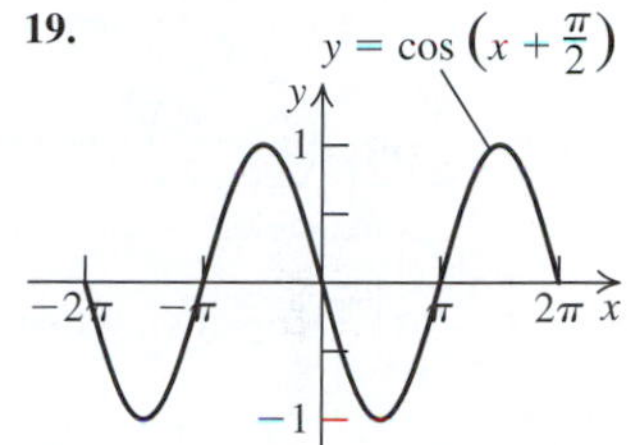

21.

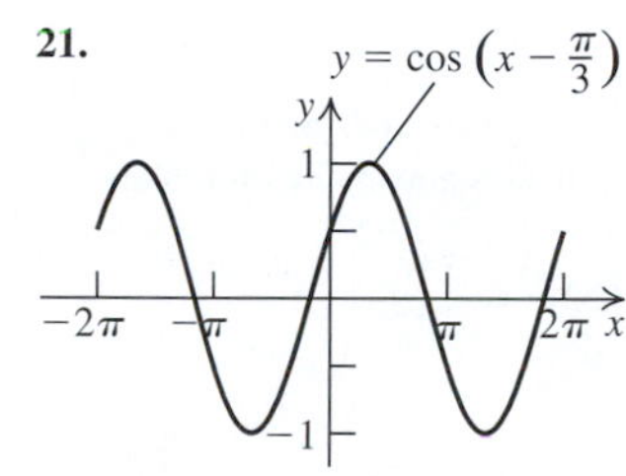

23.

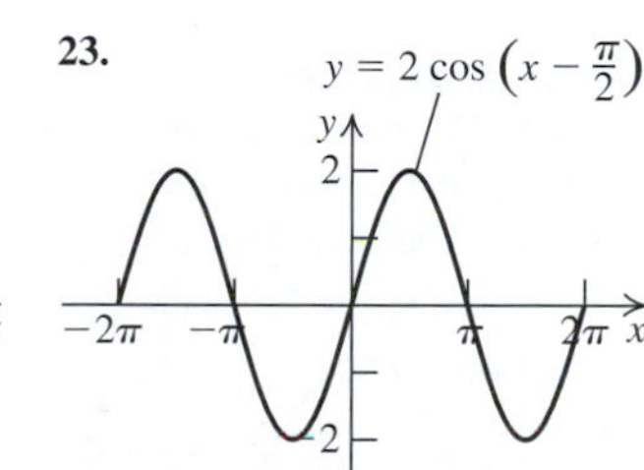

25.

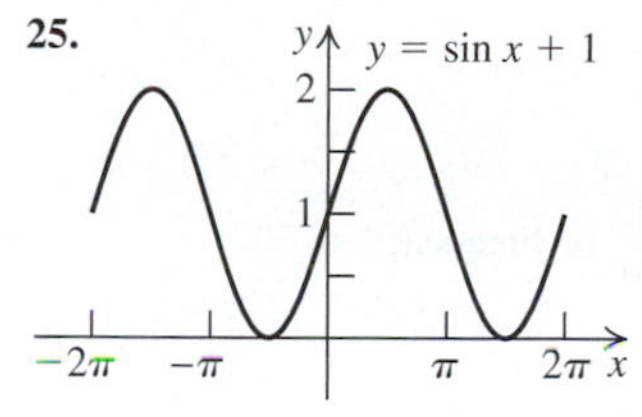

27.

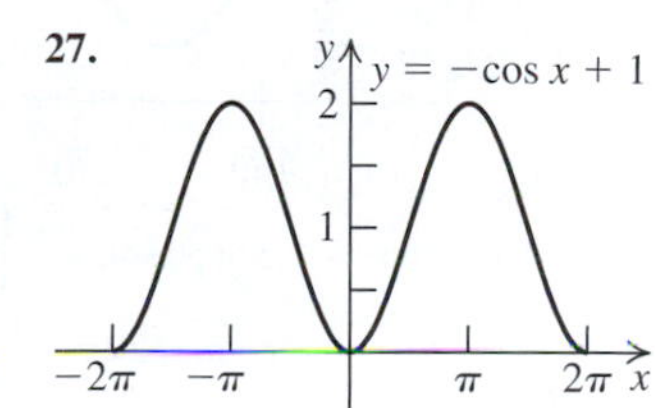

29. Amplitude $= 5$; period $= 2\pi$; phase shift $= \pi$

31. Amplitude $= 7$; period $= \frac{2\pi}{9}$; phase shift $= -\frac{\pi}{6}$

33. Amplitude $= 6$; period $= 4\pi$; phase shift $= -2$

35. Amplitude $= 0.9$; period $= 8\pi$; phase shift $= \frac{\pi}{4}$

37. $y = \frac{1}{4}\sin\left[16\left(x - \frac{\pi}{16}\right)\right]$ **39.** $y = 6\sin\left[\frac{2}{3}\left(x + \frac{\pi}{4}\right)\right]$

41. $y = 2.4\sin\left[\frac{\pi}{3}(x - 3)\right]$ **43.** $y = \frac{\pi}{2}\sin\left[\frac{16\pi}{5}\left(x + \frac{1}{4}\right)\right]$

45.

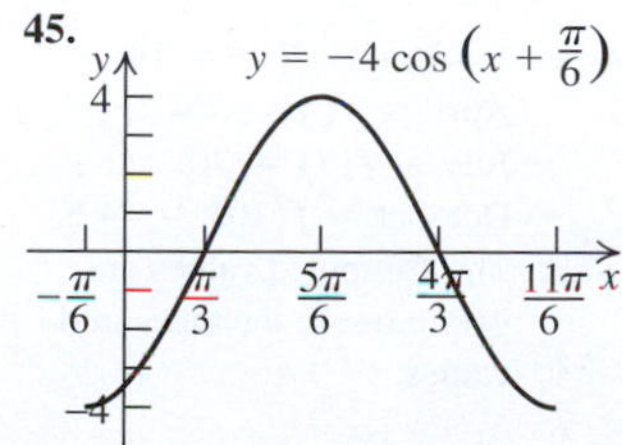

47.

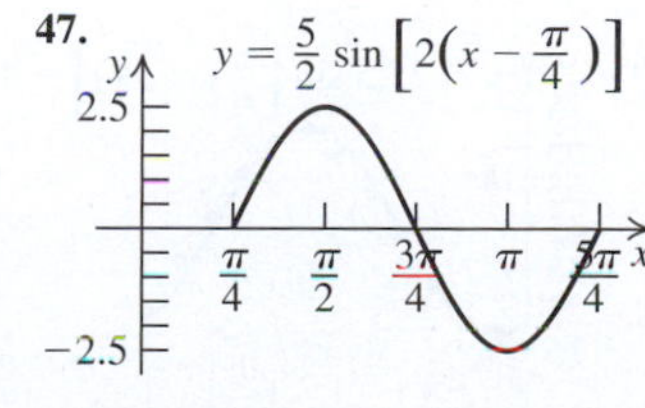

49.

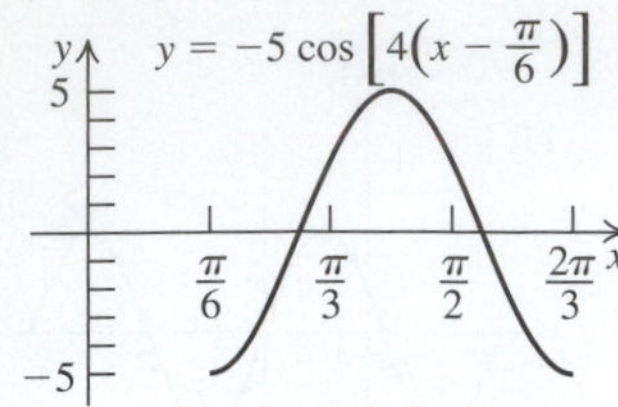

51.

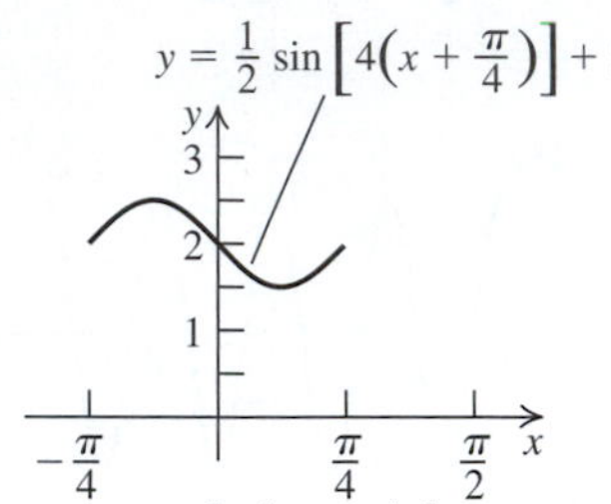

53. $y = 4\cos\left[2\left(x + \frac{\pi}{6}\right)\right]$; period $= \pi$; phase shift $= -\frac{\pi}{6}$

55. $y = -\frac{3}{2}\sin\left[2\left(x - \frac{\pi}{2}\right)\right]$; period $= \pi$; phase shift $= \frac{\pi}{2}$

57. $y = 3\cos \pi x$; period $= 2$; phase shift $= 0$

59. $y = \frac{1}{2}\cos\left[\frac{\pi}{4}(x + 1)\right]$; period $= 8$; phase shift $= -1$

61. $y = 2\sin\left[\pi\left(x + \frac{3}{\pi}\right)\right]$; period $= 2$; phase shift $= -\frac{3}{\pi}$

Applying the Concepts:

63. $t = 7558$; $P \approx -0.63$; $E \approx -0.43$; $I \approx 0.19$ **65. a.** Period $= \frac{1}{70}$. The pulse is the frequency of the function; it says how many times the heart beats in one minute.

b.

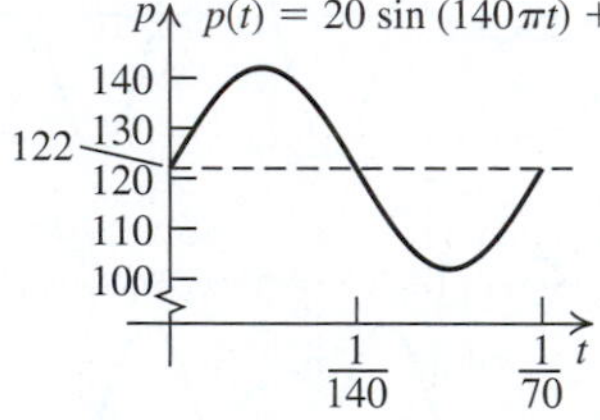

c. $\frac{142}{102}$

67. a. Amplitude $= 156$; period $= \frac{1}{60}$ **b.** Frequency $= 60$

c. $V(t) = 156\sin(120\pi t)$

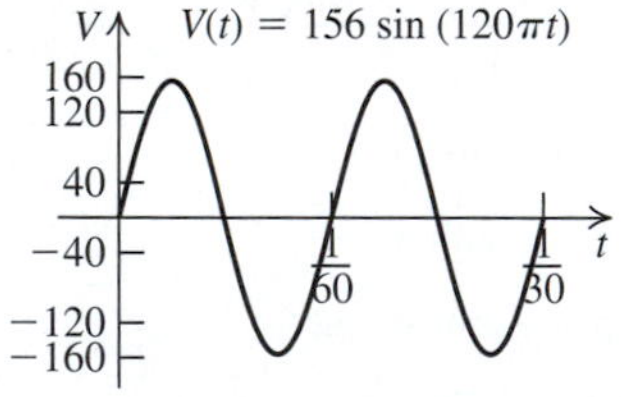

69. a. 800 kangaroos **b.** 500 kangaroos **c.** $\frac{\pi}{2}$ yrs ≈ 1.57 yr

71. $y = 4.35\sin\left[\frac{\pi}{6}(x - 3)\right] + 12.25$

73. a. $y = 25.5\sin\left[\frac{\pi}{6}(x - 4)\right] + 44.5$

b. $y = 25.5\sin\left[\frac{\pi}{6}(x - 4)\right] + 44.5$

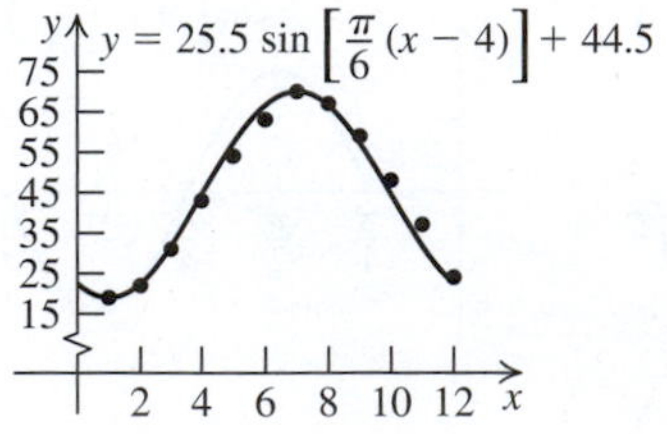

c. January $= f(1) = 19$; April $= f(4) = 44.5$; July $= f(7) = 70$; October $= f(10) = 44.5$. The computed values are very close to the measured values.

Beyond the Basics:

75.

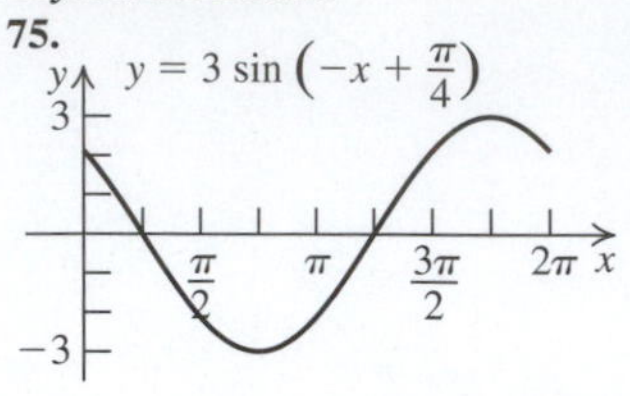

77.

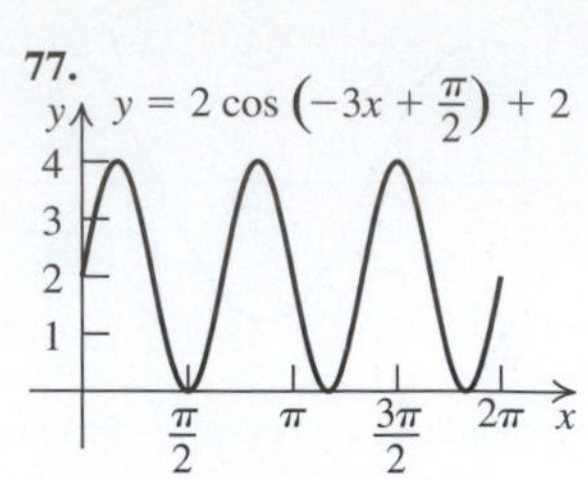

79. $y = |\sin x|$

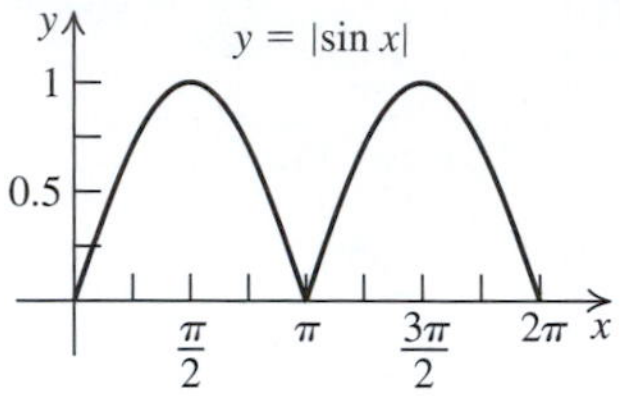

81.

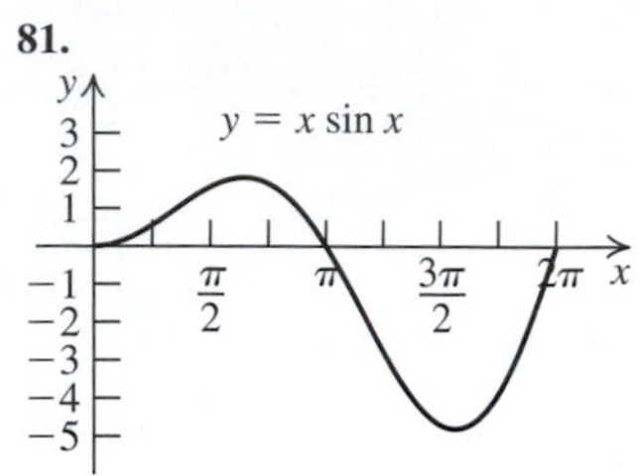

83.

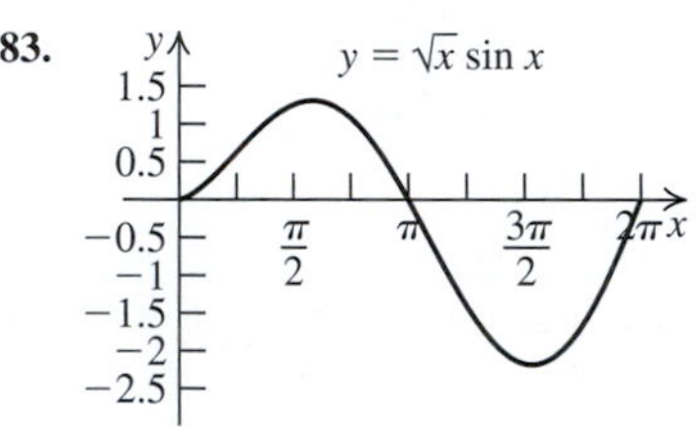

85.

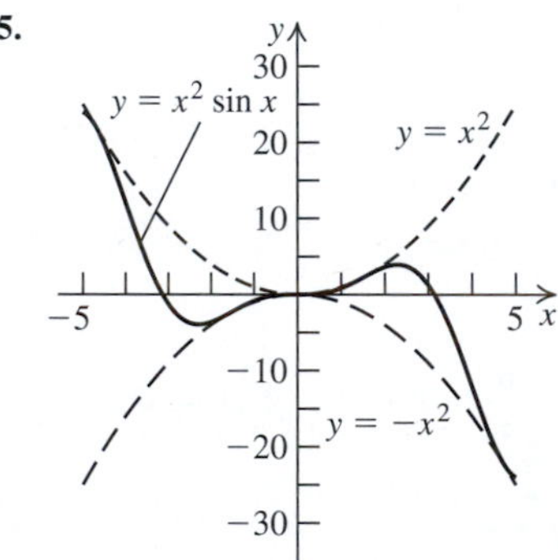

87. $y = 3\cos 2x$ **89.** $y = 2\sin 4x$ **91.** $y = 3\sin\left[2\left(x - \frac{\pi}{4}\right)\right]$

93. $y = -2\sin\left[2\left(x + \frac{\pi}{8}\right)\right]$

Critical Thinking/Discussion/Writing:

95. a. 10 units **b.** 5 units

Maintaining Skills:

97. $y = 5x - 13$ **99.** $y = -\frac{1}{2}x + \frac{5}{2}$ **101.** $-1, 2; x = -2, x - 3$ **103.** $-2, -4$; no vertical asymptotes **105.** Odd **107.** Neither **109.** Odd **111.** Odd

Section 5.5

Answers to Practice Problems

1. $y = \frac{\sqrt{3}}{3}x + 3 - \frac{2\sqrt{3}}{3}$

2.

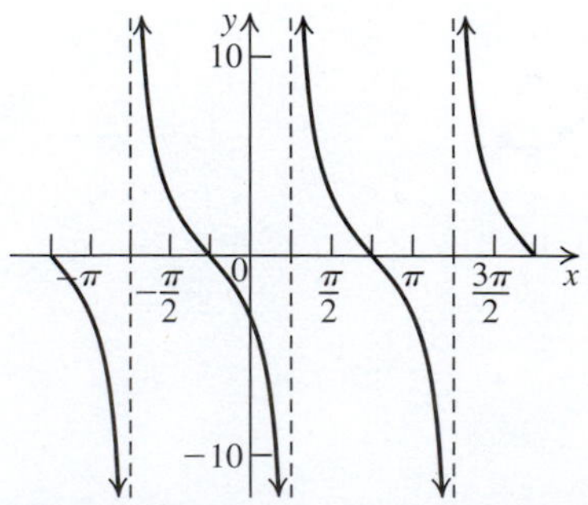

3.
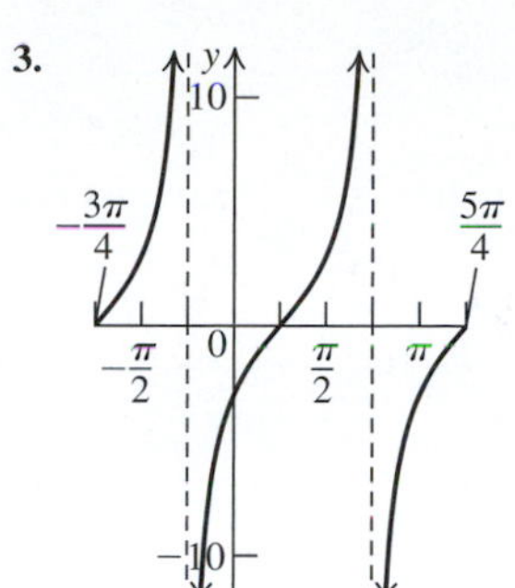

4.
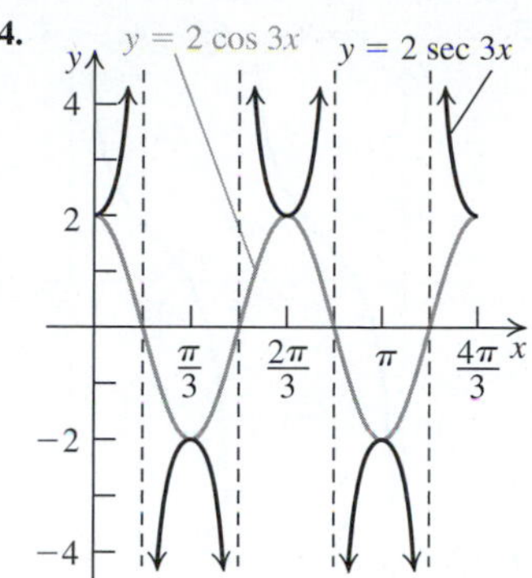

5. $\approx [2.6, 5.1]$

Basic Concepts and Skills:

1. π 3. $\frac{\pi}{2}$ 5. $y = x + 5$ 7. $y = -\sqrt{3}x - (3\sqrt{3} + 2)$

9.
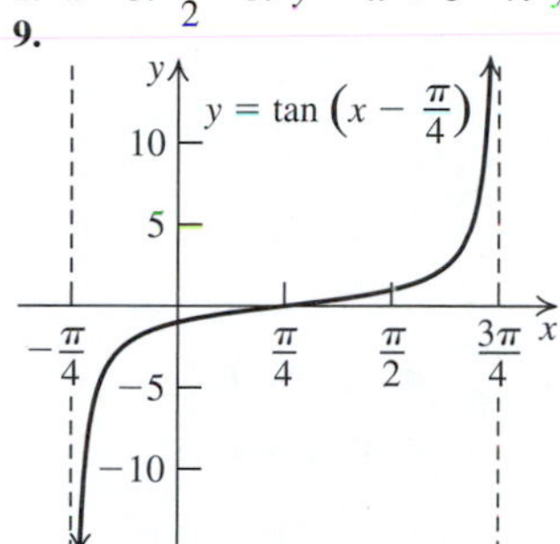

11.
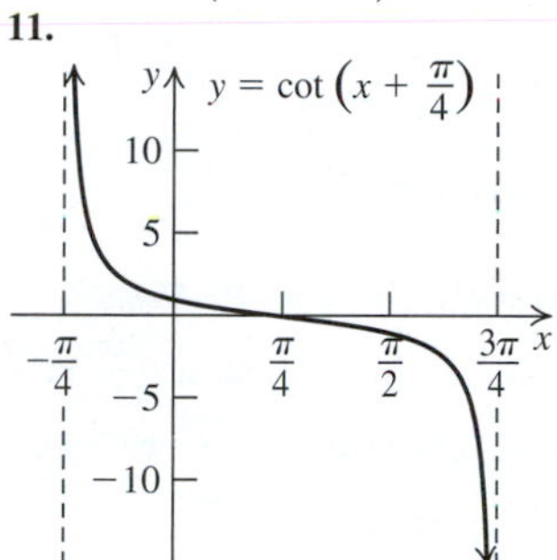

13.
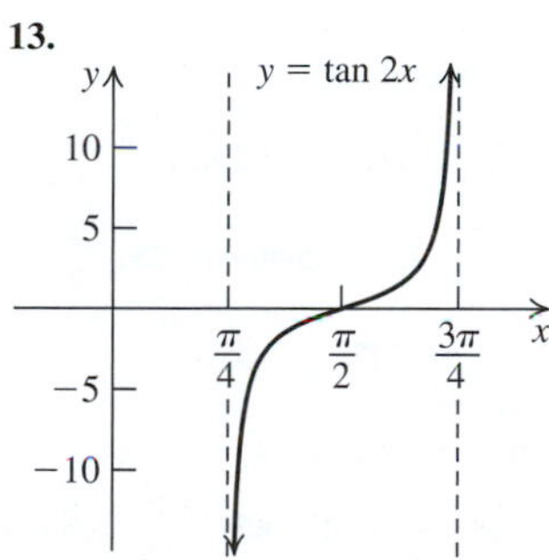

15.
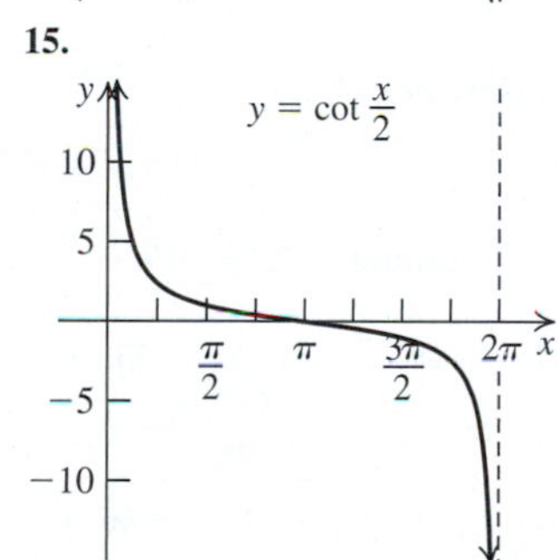

17.
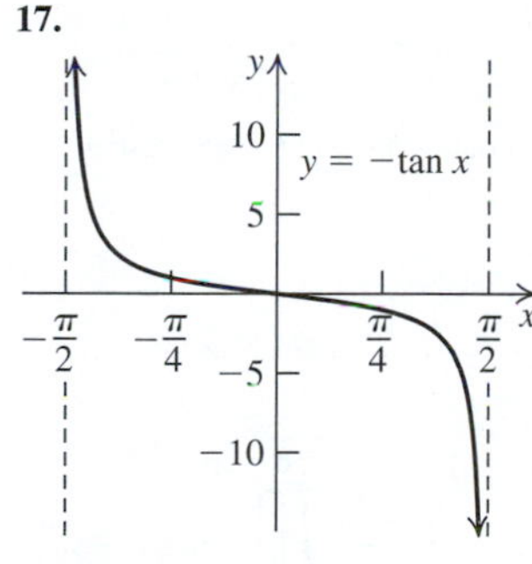

19.
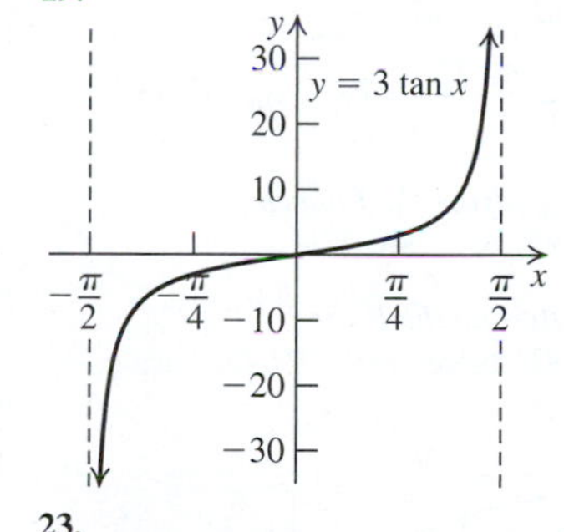

21.
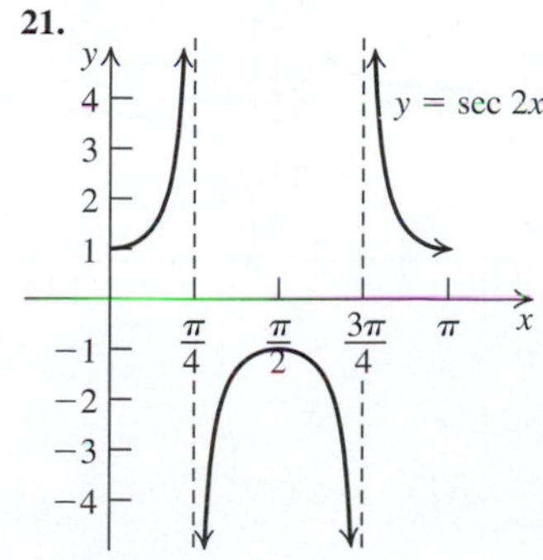

23.
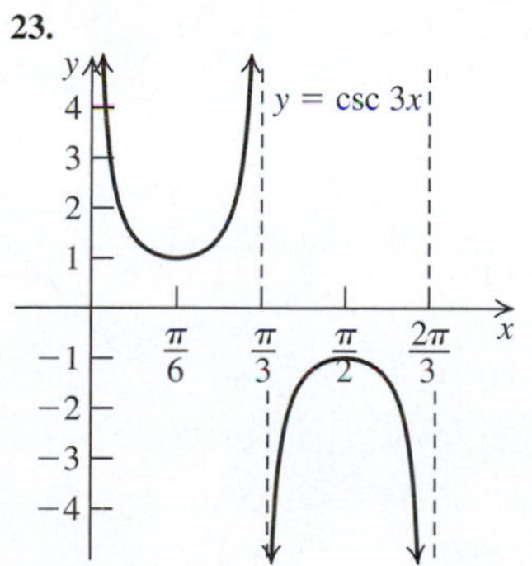

25.
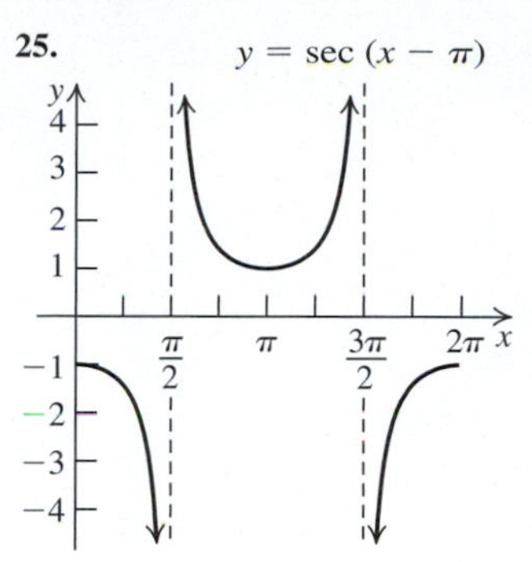

27.
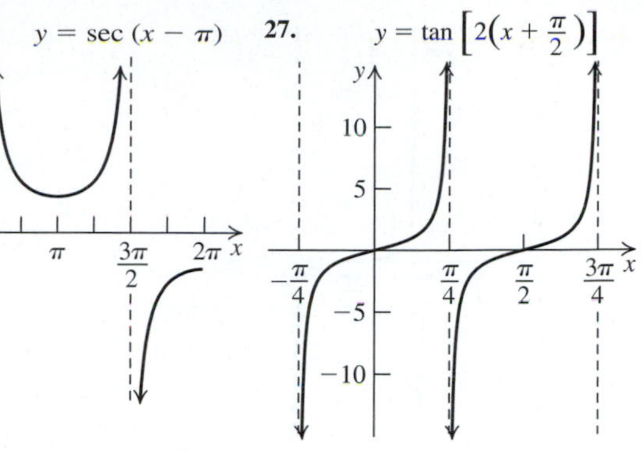

29.
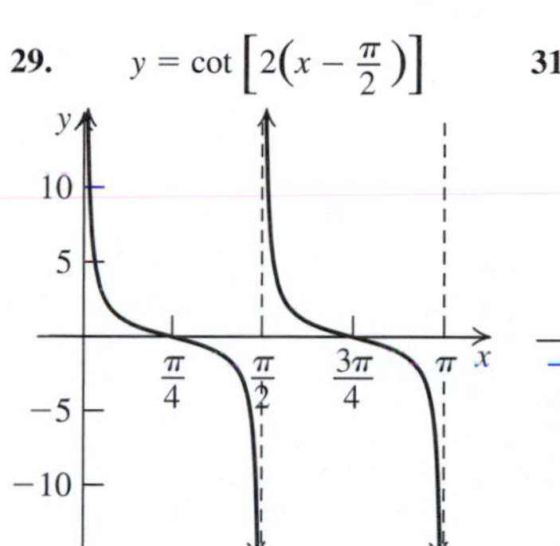

31.
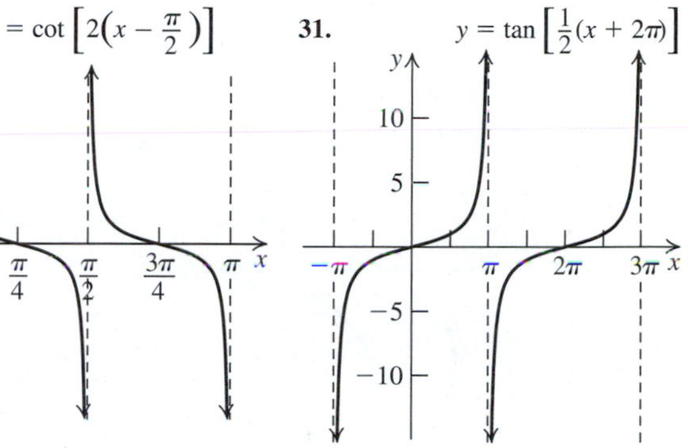

33.
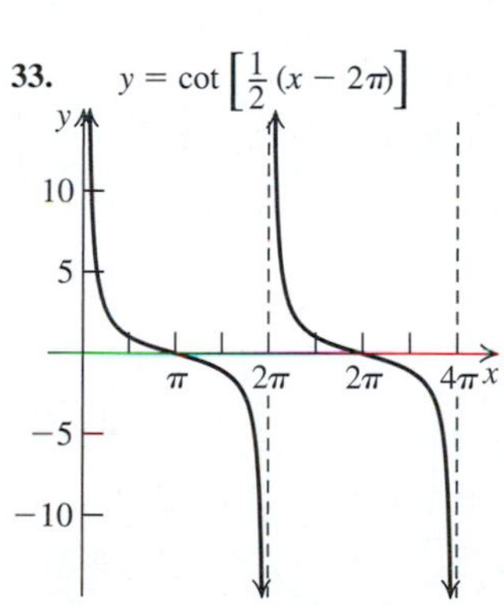

35.
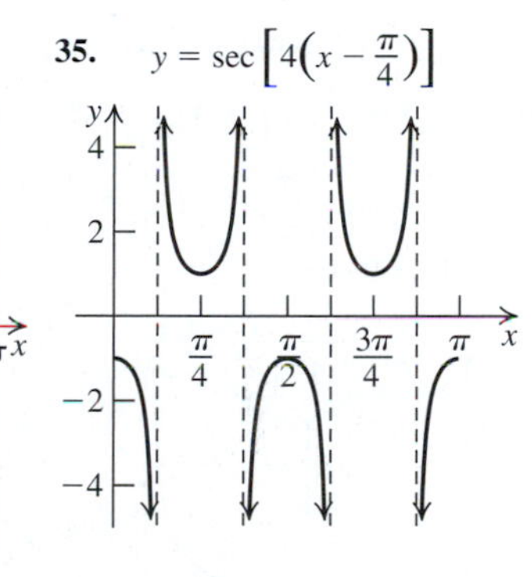

37.
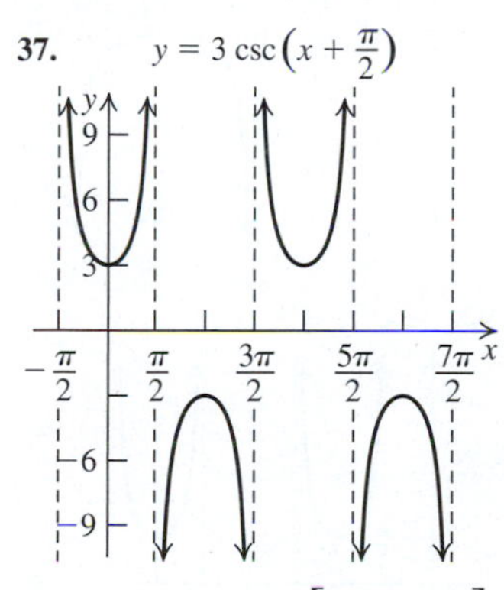

39.
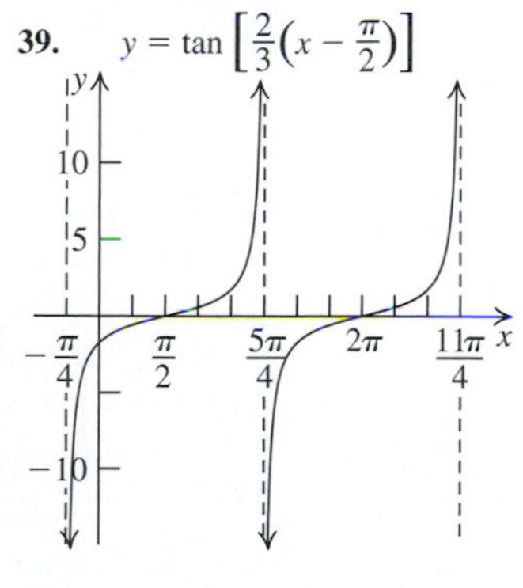

41.
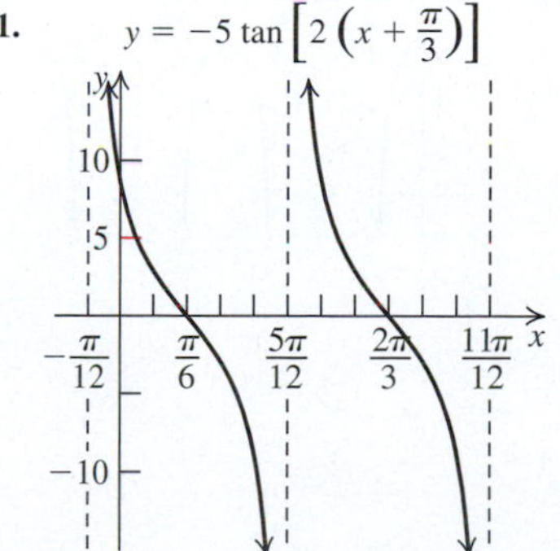

43. $y = \frac{1}{3}\cot\left[2(x - \pi)\right]$

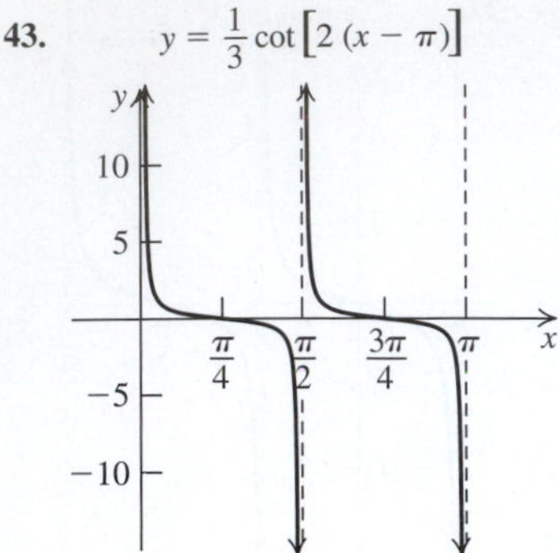

Applying the Concepts:

45. a.

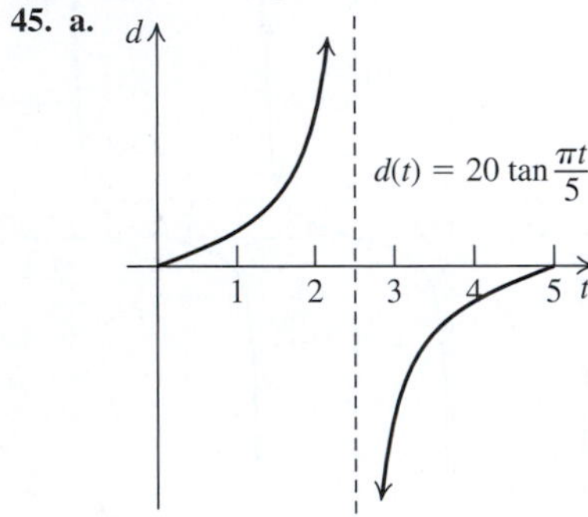

b. The light is pointing parallel to the wall.

47. $y = 7 \tan\left[\frac{\pi}{8}(x + 1)\right]$; −1.39, 10.48 49. $y = 2 \cot\left[\frac{\pi}{4}(x - 1)\right]$

51. $y = -2 \sec\left[\frac{\pi}{2}(x + 1)\right]$; 4

Beyond the Basics:

57. $y = 2\cot(-3x)$

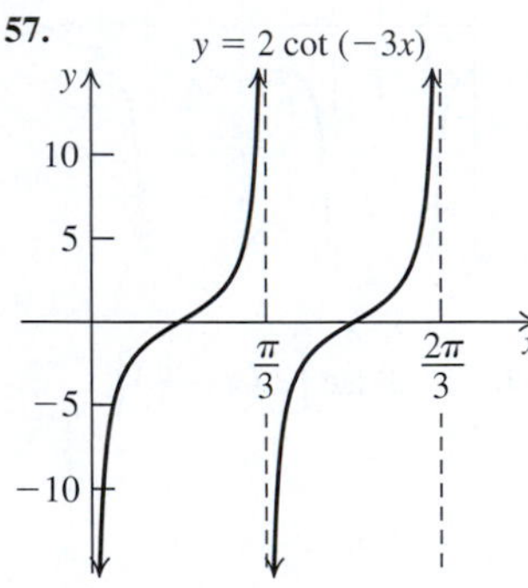

59. $y = -\frac{1}{2}\csc\left(-\frac{x}{5}\right)$

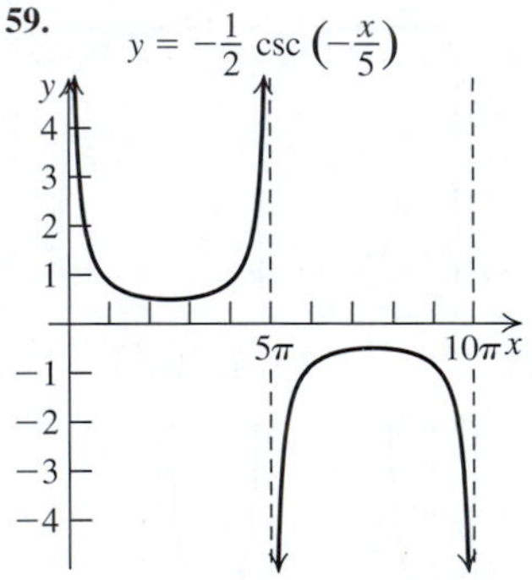

61. $y = -\tan\left(\frac{\pi}{2} - x\right)$

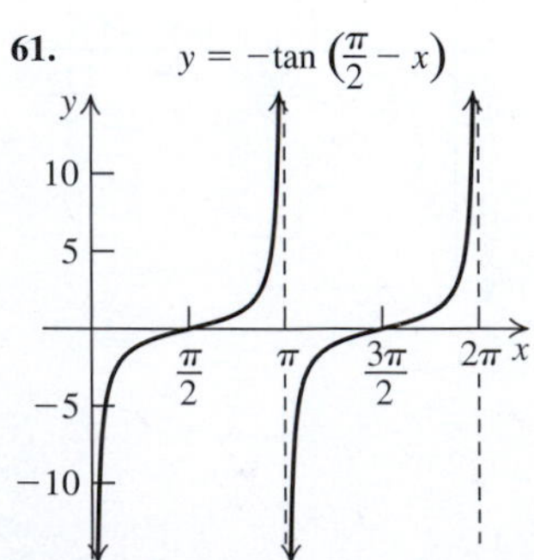

63. $y = 2\sec(\pi - 2x)$

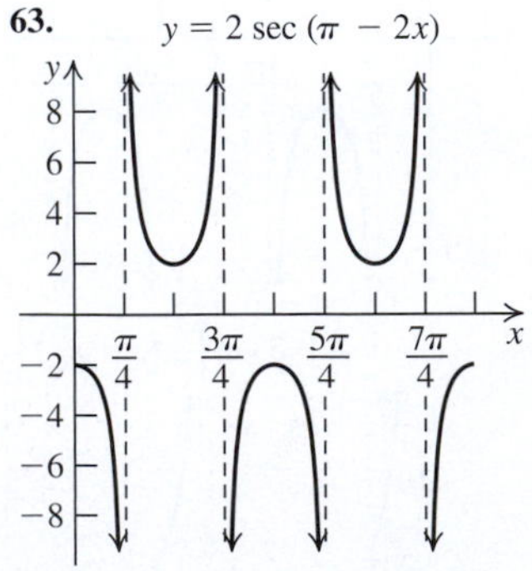

65. $y = 4\cot(\pi - 4x) + 3$

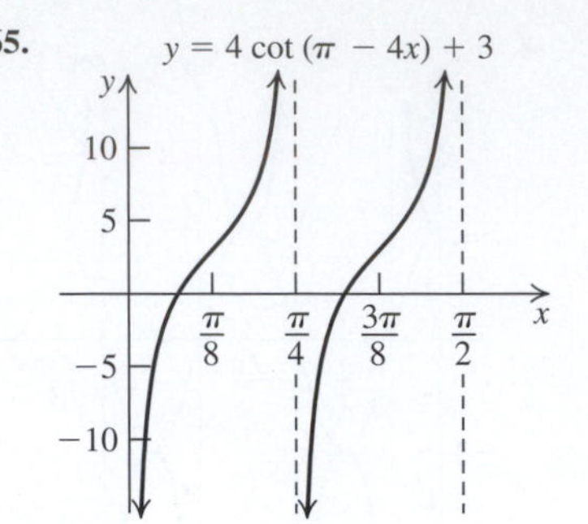

67. $y = 2\tan\left(x + \frac{\pi}{2}\right)$ 69. $y = -3\sec\left(\frac{x}{2} - \pi\right)$

Critical Thinking/Discussion/Writing:

71. 6

Maintaining Skills:

73. $f^{-1}(x) = \frac{x - 7}{3}$ 75. True 77. True 79. $[0, \infty)$ 81. No; $[0, \pi]$

Section 5.6

Answers to Practice Problems:

1. a. $-\frac{\pi}{3}$ b. $-\frac{\pi}{2}$ 2. a. $\frac{3\pi}{4}$ b. $\frac{\pi}{3}$ 3. $\frac{\pi}{6}$ 4. $\frac{\pi}{3}$ 5. b. 1
6. a. 1.3490 b. 0.2898 c. 2.932 7. a. 53.1301° b. 4.4117°
c. −85.2364° 8. $-\frac{\pi}{2}$ 9. $\frac{2\sqrt{2}}{3}$ 10. $\frac{\sqrt{9 - x^2}}{x}$ 11. $\cot\theta$
12. 80°

Basic Concepts and Skills:

1. $[-1, 1]$ 3. $\frac{\pi}{3}$ 5. true 7. false 9. 0 11. $-\frac{\pi}{6}$ 13. π
15. Undefined 17. $\frac{\pi}{3}$ 19. $-\frac{\pi}{4}$ 21. $\frac{3\pi}{4}$ 23. Undefined 25. $\frac{2\pi}{3}$
27. $\frac{1}{8}$ 29. $\frac{\pi}{7}$ 31. 247 33. $-\frac{\pi}{3}$ 35. $-\frac{\pi}{3}$ 37. $\frac{\pi}{3}$ 39. $\frac{\pi}{6}$
41. $\frac{2\pi}{3}$ 43. $-\frac{\pi}{3}$ 45. $\frac{2\pi}{3}$ 47. 1 49. −1 51. 1 53. 53.13°
55. −43.63° 57. 73.40° 59. 85.91° 61. −88.64° 63. $\frac{\sqrt{5}}{3}$
65. $\frac{3}{5}$ 67. $\frac{2\sqrt{29}}{29}$ 69. $\frac{3}{4}$ 71. $\frac{4\sqrt{17}}{17}$ 73. $\sqrt{3}$ 75. $\frac{x}{\sqrt{4 + x^2}}$
77. $\frac{\sqrt{25 - 9x^2}}{3x}$ 79. $\frac{x}{\sqrt{4 - x^2}}$ 81. $\cot\theta$ 83. $\sin\theta$ 85. $\csc\theta$

Applying the Concepts:

87. 62° 89. 159°

Beyond the Basics:

91. Increasing 93. Increasing

95. 97.

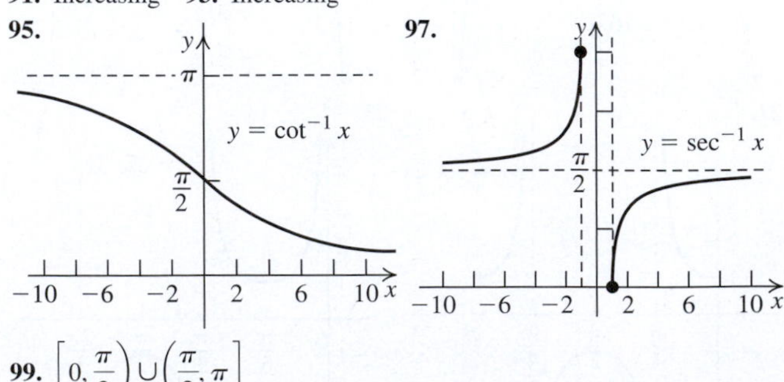

99. $\left[0, \frac{\pi}{2}\right) \cup \left(\frac{\pi}{2}, \pi\right]$

Maintaining Skills:

103. Conditional equation **105.** Neither **107.** Identity

109. Conditional equation **111.** Neither **113.** Identity **115.** $\frac{2}{x^2 - 1}$

117. $\frac{x}{x^2 - 1}$ **119.** $\sqrt{3} - 1$ **121.** 0

Review Exercises

1. **3.**

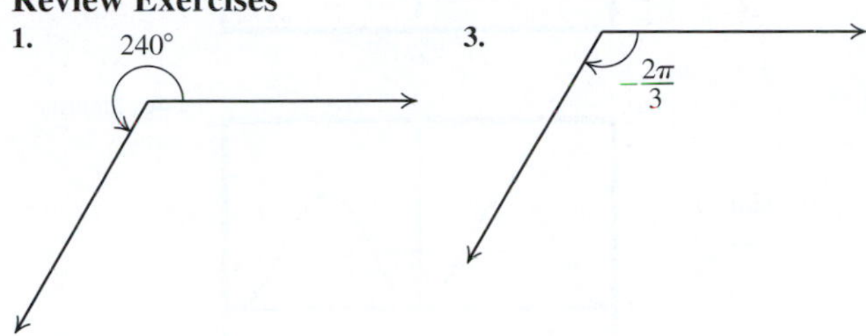

5. $\frac{\pi}{9}$ **7.** $-\frac{\pi}{3}$ **9.** 50° **11.** 2.667 **13.** 1.257 m **15.** 2.513 ft^2

17. $\sin\theta = \frac{\sqrt{77}}{9}$, $\tan\theta = \frac{\sqrt{77}}{2}$, $\cot\theta = \frac{2\sqrt{77}}{77}$, $\sec\theta = \frac{9}{2}$, $\csc\theta = \frac{9\sqrt{77}}{77}$

19. $\sin\theta = \frac{5\sqrt{34}}{34}$, $\cos\theta = \frac{3\sqrt{34}}{34}$, $\cot\theta = \frac{3}{5}$, $\sec\theta = \frac{\sqrt{34}}{3}$, $\csc\theta = \frac{\sqrt{34}}{5}$

21. $\sin\theta = \frac{4\sqrt{17}}{17}$, $\cos\theta = \frac{\sqrt{17}}{17}$, $\tan\theta = 4$, $\cot\theta = \frac{1}{4}$, $\sec\theta = \sqrt{17}$, $\csc\theta = \frac{\sqrt{17}}{4}$ **23.** $\sin\theta = \frac{2}{3}$, $\cos\theta = -\frac{\sqrt{5}}{3}$, $\tan\theta = -\frac{2\sqrt{5}}{5}$, $\cot\theta = -\frac{\sqrt{5}}{2}$, $\sec\theta = -\frac{3\sqrt{5}}{5}$, $\csc\theta = \frac{3}{2}$ **25.** IV **27.** III

29. $\sin\theta = -\frac{3}{5}$, $\tan\theta = \frac{3}{4}$, $\cot\theta = \frac{4}{3}$, $\sec\theta = -\frac{5}{4}$, $\csc\theta = -\frac{5}{3}$

31. $\cos\theta = -\frac{4}{5}$, $\tan\theta = -\frac{3}{4}$, $\cot\theta = -\frac{4}{3}$, $\sec\theta = -\frac{5}{4}$, $\csc\theta = \frac{5}{3}$

33. $-\frac{\sqrt{3}}{2}$ **35.** $\frac{\sqrt{3}}{3}$

37.

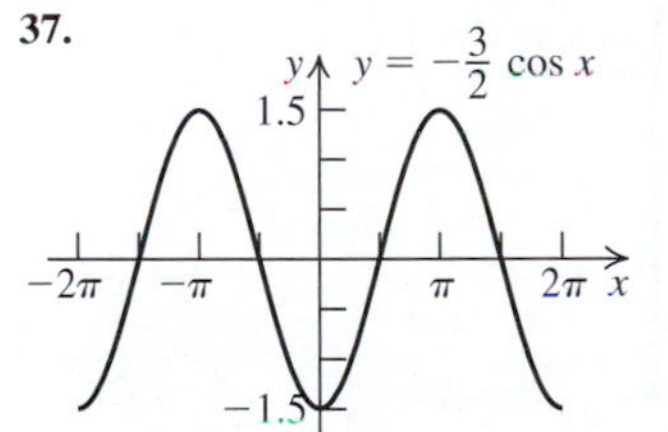

39.

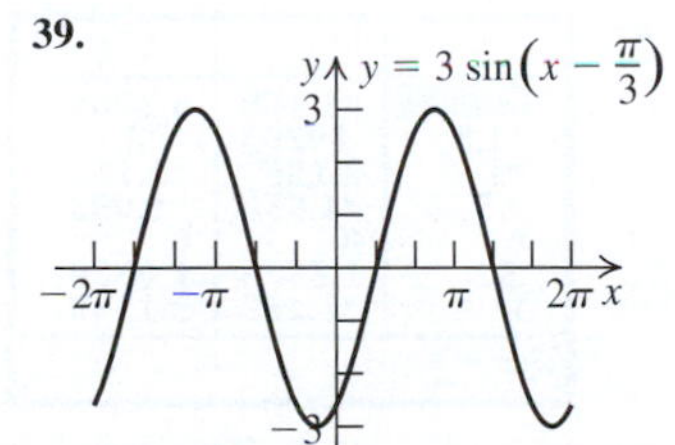

41. Amplitude = 14; period = π; phase shift = $-\frac{\pi}{14}$

43. Vertical stretch = 6; period = $\frac{\pi}{2}$; phase shift = $-\frac{\pi}{10}$

45.

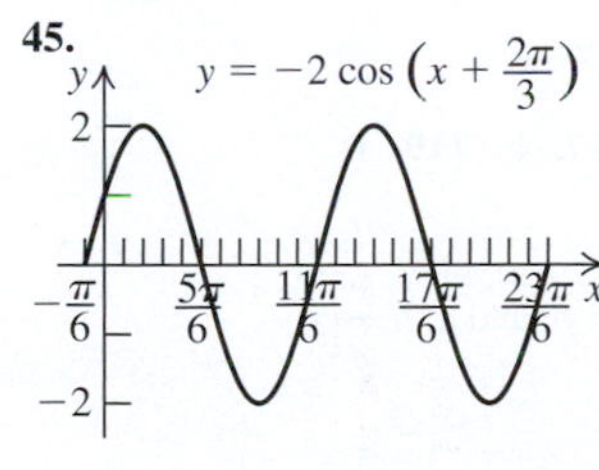

47.

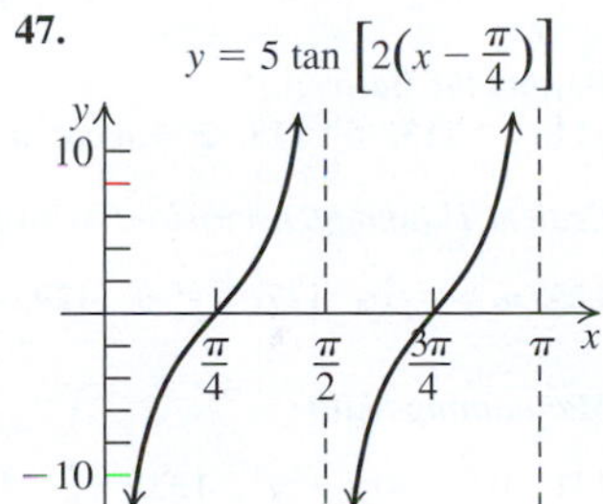

49.

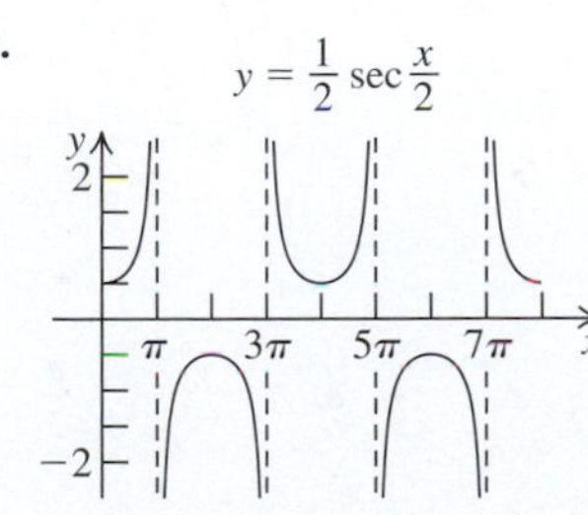

51. $\frac{5\pi}{8}$ **53.** $\frac{\pi}{3}$ **55.** $\frac{\sqrt{2}}{2}$ **57.** $-\sqrt{3}$ **59.** ≈ 44 in **61.** ≈ 866 mi

63. $\frac{3}{2}$ radians/sec **65.** 11,100 km/hr

67. $y = 2.25\sin\left[\frac{\pi}{6}(x - 4)\right] + 12.05$

Practice Test A

1. 2.4435 **2.** 252° **3.** $\frac{2\sqrt{29}}{29}$ **4.** 1.5 **5.** 75 cm^2 **6.** $\frac{3\sqrt{5}}{7}$

7. Quadrant II **8.** $\frac{13}{5}$ **9.** 37° **10.** 0.223 **11.** 72 m

12. Amplitude = 7; range = [−7, 7]

13. Amplitude = 19; period = $\frac{\pi}{6}$; phase shift = −3π

14. $y = -7\cos\left(\frac{\pi}{2}t\right)$

15.

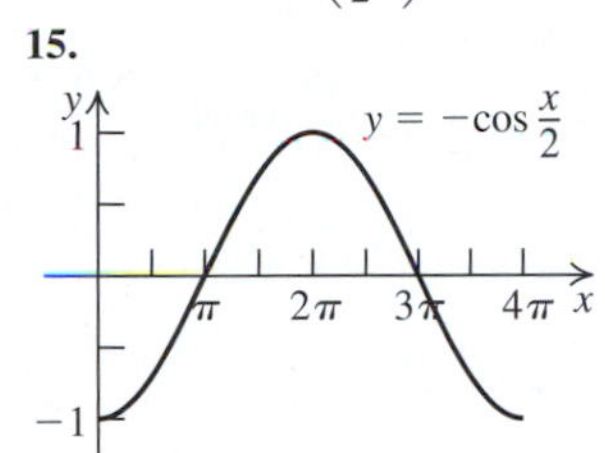

16.

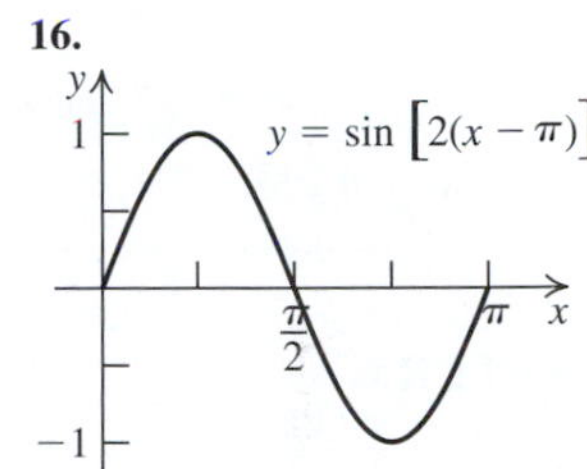

17.

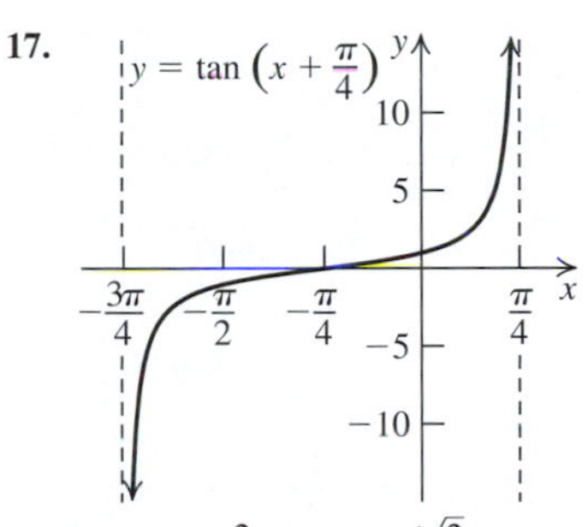

18. $-\frac{\pi}{3}$ **19.** $\frac{2\pi}{3}$ **20.** $-\frac{\sqrt{2}}{4}$

Practice Test B

1. b **2.** d **3.** a **4.** a **5.** d **6.** b **7.** d **8.** b **9.** c **10.** a **11.** c **12.** a **13.** b **14.** d **15.** d **16.** c **17.** c **18.** a **19.** d **20.** d

Cumulative Review Exercises (Chapters P–5)

1. $5 \pm \frac{\sqrt{3}}{3}$ **3.** (2, 4) **5.** (0, 2)

7.

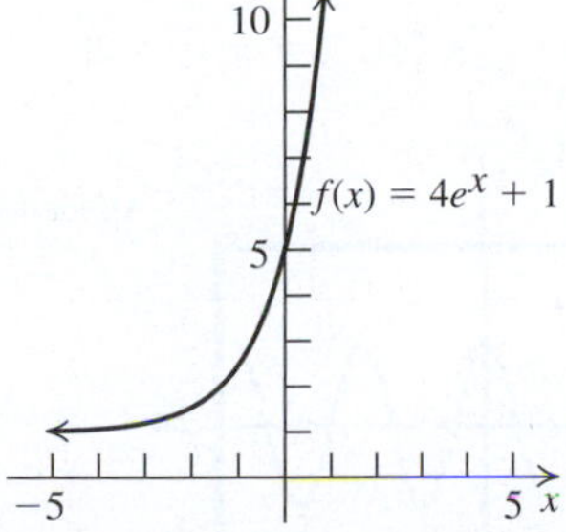

y
10
$f(x) = 4e^x + 1$
5
−5
5 x

9. a. $f^{-1}(x) = \frac{x}{3} - \frac{7}{3}$ **b.** The function does not have an inverse.

11. $y = -3, x = -5, x = 1$

13.

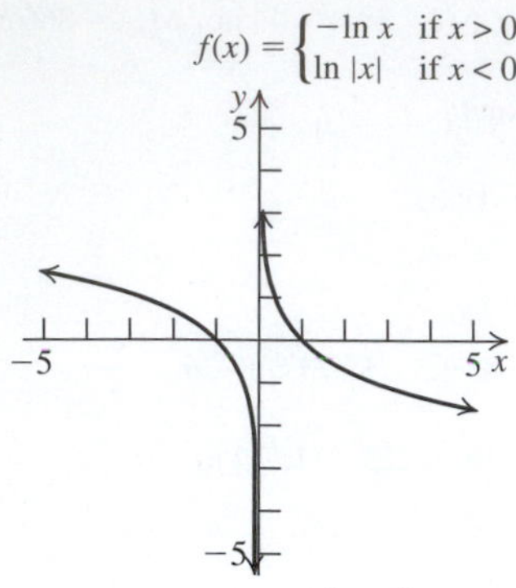

15. $\pm 1, \pm 2, \pm 3, \pm 6, \pm\frac{1}{2}, \pm\frac{3}{2}$

17. a. $\sin\theta = \frac{3}{5}, \tan\theta = -\frac{3}{4}, \cot\theta = -\frac{4}{3}, \sec\theta = -\frac{5}{4}, \csc\theta = \frac{5}{3}$

b. $\sin\theta = -\frac{12}{13}, \cos\theta = -\frac{5}{13}, \tan\theta = \frac{12}{5}, \sec\theta = -\frac{13}{5}, \csc\theta = -\frac{13}{12}$

19. a. $\frac{2\sqrt{6}}{5}$ **b.** $\frac{7\sqrt{53}}{53}$

CHAPTER 6

Section 6.1

Answers to Practice Problems:

1. $-\frac{3}{4}$ **2.** $\sin x = \frac{\sqrt{5}}{5}, \cos x = -\frac{2\sqrt{5}}{5}, \sec x = -\frac{\sqrt{5}}{2}, \csc x = \sqrt{5},$ $\cot x = -2$ **3.** $\frac{2}{\sin x}$ **4.** At $x = \frac{3\pi}{2}, 0 \neq 2.$ **10.** $3\cos\theta$

Basic Concepts and Skills:

1. identity **3.** $\cos^2 x; \tan^2 x;$ 1 **5.** true **7.** $-\frac{5}{13}$ **9.** $-\frac{\sqrt{5}}{2}$ **11.** $\frac{\sqrt{5}}{2}$

13. $\tan x = -\frac{4}{3}, \cot x = -\frac{3}{4}, \sec x = -\frac{5}{3}, \csc x = \frac{5}{4}$

15. $\tan x = \frac{\sqrt{2}}{2}, \cot x = \sqrt{2}, \sec x = \frac{\sqrt{6}}{2}, \csc x = \sqrt{3}$

17. $\sin x = -\frac{\sqrt{5}}{5}, \cos x = -\frac{2\sqrt{5}}{5}, \cot x = 2, \csc = -\sqrt{5}$

19. $\sin x = \frac{1}{3}, \cos x = \frac{2\sqrt{2}}{3}, \tan x = \frac{\sqrt{2}}{4}, \sec x = \frac{3\sqrt{2}}{4}$

21. $\cos x = -\frac{12}{13}, \tan x = \frac{5}{12}, \sec x = -\frac{13}{12}, \csc x = -\frac{13}{5}$

23. $\sin x = -\frac{2\sqrt{2}}{3}, \cos x = \frac{1}{3}, \tan x = -2\sqrt{2},$ $\cot x = -\frac{\sqrt{2}}{4}, \csc x = -\frac{3\sqrt{2}}{4}$ **25.** c **27.** a **29.** d **31.** 2 **33.** 1

35. $\csc^2 x + \cot^2 x$ **37.** 1 **39.** $\tan x - 3$ **41.** $x = \pi; 0 \neq 2$

43. $x = \pi; -1 \neq 1$ **45.** $x = \frac{\pi}{3}; \frac{3}{4} \neq \frac{1}{4}$

47. An identity

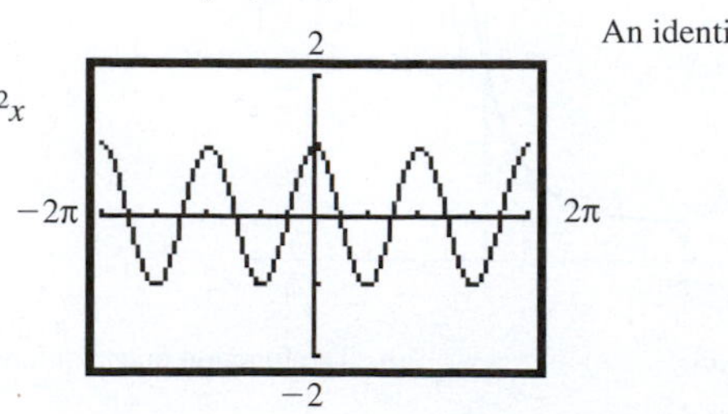

49. Not an identity

$y = \sin(x) + \cos(x)$
$y = 1$
2
−2π
2π
−2

51. An identity

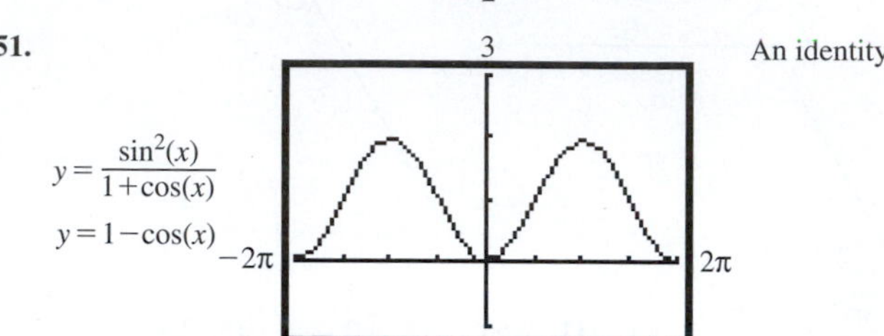

53. An identity

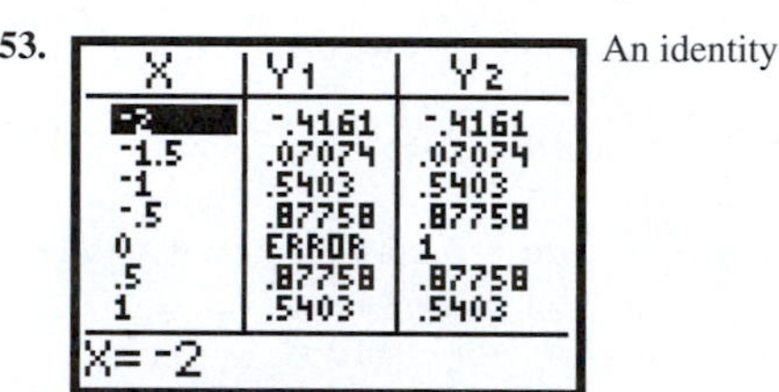

X	Y1	Y2
-2	-.4161	-.4161
-1.5	.07074	.07074
-1	.5403	.5403
-.5	.87758	.87758
0	ERROR	1
.5	.87758	.87758
1	.5403	.5403

X= -2

55. An identity

X	Y1	Y2
-2	.37206	.37206
-1.5	1.1527	1.1527
-1	4.588	4.588
-.5	-3.408	-3.408
0	-1	ERROR
.5	-.2934	-.2934
1	.21796	.21796

X= -2

57. Not an identity

X	Y1	Y2
-2	-1.158	4.3701
-1.5	.14255	-28.2
-1	2.185	-3.115
-.5	-1.557	-1.093
0	0	0
.5	1.5574	1.0926
1	-2.185	3.1148

X= -2

95. $\sec\theta$ **97.** $\tan\theta$ **99.** $\cot\theta$

Applying the Concepts:

101. $20\csc\theta$ **103.** $\tan\theta = \frac{m_1 - m_2}{1 + m_1 m_2}$

Beyond the Basics:

111. 1 **113.** 1 **115.** $2 + 4\cot^2\theta$ **117.** 1 **119.** 1

Critical Thinking/Discussion/Writing:

125. a. $x = \frac{\pi}{2}$ **127.** $t = 0$ **129.** $a = b$, and $a, b \neq 0$

Maintaining Skills:

131. $4x^2 - 4xy + y^2$ **133.** $-\frac{1}{2}$ **135.** $\frac{\sqrt{3}}{3}$ **137.** $\tan^2 x + 1 = \sec^2 x$

139. $\sin t = \frac{\sqrt{13}}{4}$

Section 6.2

Answers to Practice Problems:

1. $\frac{\sqrt{6} + \sqrt{2}}{4}$ **2.** $\frac{\sqrt{2} - \sqrt{6}}{4}$ **5.** $\frac{1}{2}$ **6.** $-\frac{63}{65}$ **8.** $A = 2, \theta = \frac{\pi}{3}$

9. $y = \sin x + \sqrt{3}\cos x$

10. Amplitude $= 0.1\sqrt{5}$; period $= \dfrac{\pi}{200}$; frequency $= \dfrac{200}{\pi}$; phase shift ≈ -0.0028

Basic Concepts and Skills:

1. $\sin A\cos B + \cos A\sin B$ **3.** $\dfrac{\tan A + \tan B}{1 - \tan A\tan B}$ **5.** true **7.** $\dfrac{\sqrt{6}+\sqrt{2}}{4}$ **9.** $\dfrac{\sqrt{6}-\sqrt{2}}{4}$ **11.** $-\dfrac{\sqrt{6}+\sqrt{2}}{4}$ **13.** 1 **15.** $\dfrac{\sqrt{6}+\sqrt{2}}{4}$ **17.** $2-\sqrt{3}$ **19.** $-\sqrt{6}-\sqrt{2}$ **21.** $\dfrac{\sqrt{6}-\sqrt{2}}{4}$ **23.** $-2-\sqrt{3}$ **25.** $2+\sqrt{3}$ **41.** 1 **43.** 0 **45.** 1 **47.** $\dfrac{\sqrt{3}}{2}$ **49.** 1 **51.** $\dfrac{56}{65}$ **53.** $\dfrac{63}{65}$ **55.** $\dfrac{16}{63}$ **57.** $\dfrac{2\sqrt{210}-6}{35}$ **59.** $\dfrac{35\sqrt{210}-105}{402}$ **61.** $\dfrac{-4\sqrt{10}-3\sqrt{21}}{2\sqrt{210}-6}$

71.

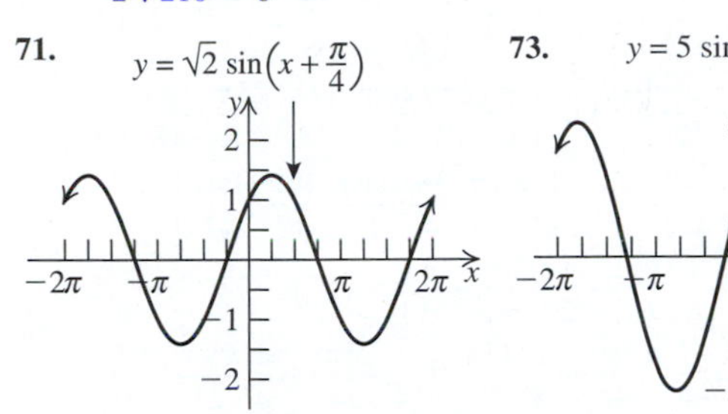

73.

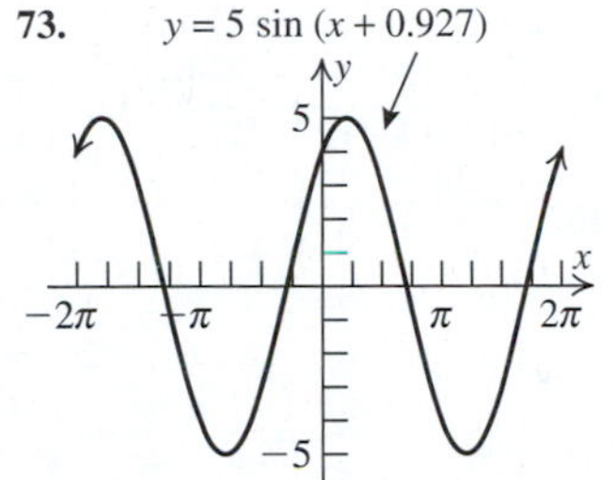

75.

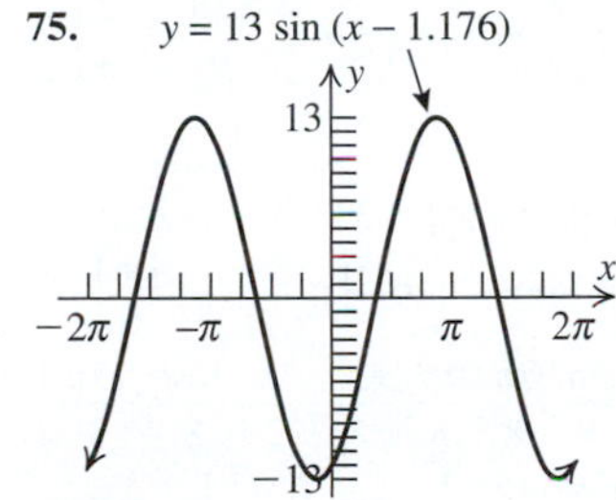

77.

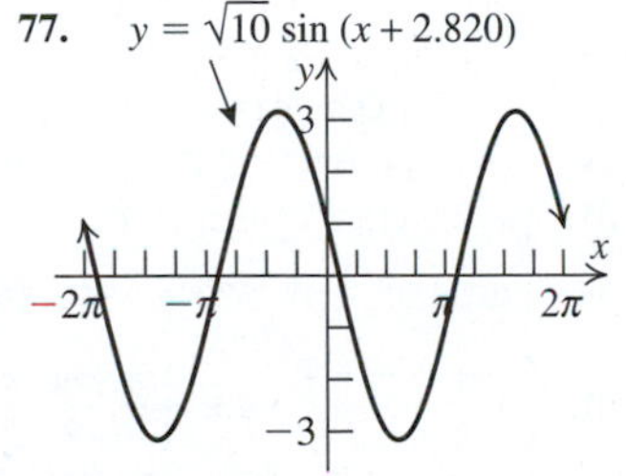

Applying the Concepts:

81. 0 **83.** $A = 0.5; \theta \approx -0.00161$ **85. a.** $A \approx 0.5142$ **b.** Frequency $= \dfrac{1}{\pi}$; phase shift ≈ -0.6676

Beyond the Basics:

101. $0, \dfrac{\pi}{2}$ **103.** $\dfrac{5\pi}{6}, \dfrac{3\pi}{2}$ **105.** $\dfrac{5\pi}{6}$ **107.** $\dfrac{1}{2}$ **109.** $\dfrac{1}{2}$ **111.** 0 **113.** 0 **115.** $\dfrac{17}{6}$ **119.** 1

Critical Thinking/Discussion/Writing:

123. $u = 0, v = 0$ **127.** $u =$ any real number, $v = k\pi$, k any integer

Maintaining Skills:

129. $\tan\theta = \dfrac{1}{4}$ **131.** Positive **133.** Negative **135.** $-\dfrac{\sqrt{3}}{3}$ **137.** $\dfrac{\sqrt{2}}{2}$

Section 6.3

Answers to Practice Problems:

1. a. $-\dfrac{120}{169}$ **b.** $-\dfrac{119}{169}$ **c.** $\dfrac{120}{119}$ **2. a.** $\dfrac{\sqrt{3}}{2}$ **b.** $\dfrac{\sqrt{2}}{2}$ **4.** $\dfrac{3}{8}-\dfrac{1}{2}\cos 2x+\dfrac{1}{8}\cos 4x$ **5.** ≈ 99.8 watts **6.** $\dfrac{\sqrt{2+\sqrt{2}}}{2}$ **7.** $-\dfrac{\sqrt{26}}{26}$

Basic Concepts and Skills:

1. $2\sin x\cos x$ **3.** $2\cos^2 x - 1$; $\dfrac{1+\cos 2x}{2}$ **5.** false **7. a.** $-\dfrac{24}{25}$ **b.** $\dfrac{7}{25}$ **c.** $-\dfrac{24}{7}$ **9. a.** $\dfrac{8}{17}$ **b.** $-\dfrac{15}{17}$ **c.** $-\dfrac{8}{15}$ **11. a.** $-\dfrac{4}{5}$ **b.** $-\dfrac{3}{5}$ **c.** $\dfrac{4}{3}$ **13.** $-\dfrac{\sqrt{3}}{2}$ **15.** $-\dfrac{\sqrt{3}}{2}$ **17.** $-\dfrac{\sqrt{3}}{3}$ **19.** $\dfrac{\sqrt{2}}{2}$ **21.** $\dfrac{\sqrt{3}}{3}$ **33.** $\dfrac{1-\cos 4x}{2}$ **35.** $\sin 4x$ **37.** $\dfrac{\sin 12x}{2}$ **39.** $\dfrac{\sin 2x}{4}$ **41.** $\cos 2x - 4\cos x + 3$ **43.** $\dfrac{\sqrt{2-\sqrt{3}}}{2}$ **45.** $\dfrac{\sqrt{2+\sqrt{2}}}{2}$ **47.** $-\dfrac{\sqrt{2+\sqrt{2}}}{2}$ **49.** $-\dfrac{\sqrt{2-\sqrt{2}}}{\sqrt{2+\sqrt{2}}}$ **51.** $-\sqrt{2}-1$ **53.** $-\dfrac{\sqrt{2+\sqrt{3}}}{2}$ **55. a.** $\dfrac{2\sqrt{5}}{5}$ **b.** $\dfrac{\sqrt{5}}{5}$ **c.** 2 **57. a.** $\sqrt{\dfrac{13+3\sqrt{13}}{26}}$ **b.** $\sqrt{\dfrac{13-3\sqrt{13}}{26}}$ **c.** $\sqrt{\dfrac{13+3\sqrt{13}}{13-3\sqrt{13}}}$ **59. a.** $\sqrt{\dfrac{5+2\sqrt{6}}{10}}$ **b.** $\sqrt{\dfrac{5-2\sqrt{6}}{10}}$ **c.** $\sqrt{\dfrac{5+2\sqrt{6}}{5-2\sqrt{6}}}$ **61. a.** $\sqrt{\dfrac{5-\sqrt{5}}{10}}$ **b.** $\sqrt{\dfrac{5+\sqrt{5}}{10}}$ **c.** $\sqrt{\dfrac{5-\sqrt{5}}{5+\sqrt{5}}}$

Applying the Concepts:

73. 100 watts **75.** 1200 watts **77.** 600 watts **79.** $\dfrac{\pi}{4}$

Critical Thinking/Discussion/Writing:

91. a. $\dfrac{4}{5}$ **b.** $\dfrac{3}{5}$ **c.** $\dfrac{24}{25}$ **d.** $-\dfrac{7}{25}$ **e.** $-\dfrac{24}{7}$ **f.** $\dfrac{\sqrt{5}}{5}$ **g.** $\dfrac{2\sqrt{5}}{5}$ **h.** $\dfrac{1}{2}$

Maintaining Skills:

93. $\dfrac{1}{\sqrt{5}-\sqrt{2}}$ **95.** 0.87 **97.** 10 **99.** $-\dfrac{\sqrt{2}+\sqrt{6}}{4}$ **101.** $\dfrac{1}{2}$

Section 6.4

Answers to Practice Problems:

1. $\dfrac{1}{2}\cos 2x + \dfrac{1}{2}\cos 4x$ **2.** $\dfrac{2-\sqrt{3}}{4}$ **3. a.** $2\sin 3x\sin x$ **b.** $\cos 13°$ **4.** $2\sin\left(2x-\dfrac{\pi}{4}\right)\cos\left(x+\dfrac{\pi}{4}\right)$ **6. a.** $y = \sin(2\pi\cdot 697t) + \sin(2\pi\cdot 1209t)$ **b.** $y = 2\cos(512\pi t)\sin(1906\pi t)$ **c.** $f = 953$ Hz, $A = 2\cos(512\pi t)$

Basic Concepts and Skills:

1. $\dfrac{1}{2}[\cos(x-y) - \cos(x+y)]$ **3.** $2\cos\left(\dfrac{x+y}{2}\right)\cos\left(\dfrac{x-y}{2}\right)$ **5.** false **7.** $\dfrac{1}{2}\sin 2x$ **9.** $\dfrac{1}{2}-\dfrac{1}{2}\cos 2x$ **11.** $\dfrac{1}{4}+\dfrac{1}{2}\sin 20°$ **13.** $-\dfrac{1}{4}-\dfrac{1}{2}\cos 20°$ **15.** $\dfrac{1}{4}$ **17.** $\dfrac{\sqrt{2}}{4}-\dfrac{1}{2}$ **19.** $\dfrac{1}{2}\sin 6\theta + \dfrac{1}{2}\sin 4\theta$ **21.** $\dfrac{1}{2}\cos x + \dfrac{1}{2}\cos 7x$ **23.** $\dfrac{\sqrt{3}}{4}-\dfrac{\sqrt{2}}{4}$ **25.** $\dfrac{1}{2}+\dfrac{\sqrt{2}}{4}$ **27.** $\dfrac{1}{4}+\dfrac{\sqrt{2}}{4}$ **29.** $\dfrac{1}{4}-\dfrac{\sqrt{2}}{4}$ **31.** $-\sin 10°$ **33.** $2\sin 8°\cos 24°$

35. $2\sin\frac{3\pi}{10}\cos\frac{\pi}{10}$ **37.** $2\cos\frac{5}{12}\cos\frac{1}{12}$ **39.** $2\cos 4x\cos x$

41. $2\sin 3x\cos 4x$ **43.** $\sqrt{2}\cos\left(x-\frac{\pi}{4}\right)$ **45.** $\sqrt{2}\sin\left(2x-\frac{\pi}{4}\right)$

47. $2\sin\left(\frac{\pi}{4}-x\right)\cos\left(4x-\frac{\pi}{4}\right)$ **49.** $a\sqrt{2}\cos\left(x-\frac{\pi}{4}\right)$

Applying the Concepts:

61. a. $y=\sin[2\pi(852)t]+\sin[2\pi(1209)t]$
b. $y=2\cos(357\pi t)\sin[2\pi(1030.5)t]$
c. $f=1030.5$ Hz, $A=2\cos(357\pi t)$
63. a. $y=0.1\cos(116\pi t)\cos(4\pi t)$ **b.** $f=4$ beats per unit time

Beyond the Basics:

65. The amplitude is $2\cos 1\approx 1.081$. The period is π.

$y=\cos(2x+1)+\cos(2x-1)$

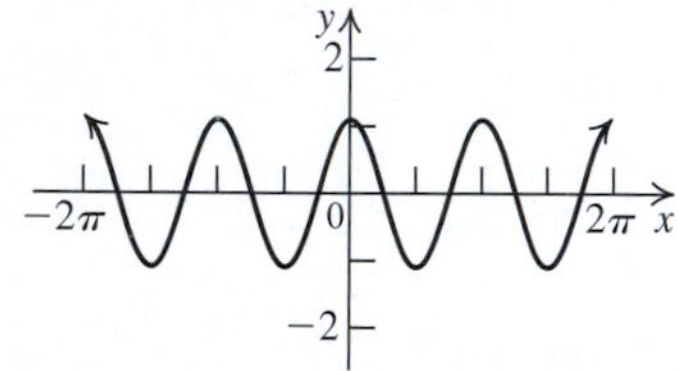

Maintaining Skills:

87. Inconsistent **89.** Identity **91.** Conditional **93.** $\frac{\pi}{8}$ **95.** $\frac{\pi}{4}$
97. $\frac{6\pi}{5}$

Section 6.5

Answers to Practice Problems:

1. a. $x=\frac{\pi}{2}+2n\pi$ **b.** $x=2n\pi$ **c.** $x=\frac{\pi}{4}+n\pi$
2. a. $x=48.2°+n\cdot 360°, x=311.8°+n\cdot 360°$
b. $x\approx 2.0345, x\approx 5.1761$ **3.** $x=90°, x=330°$ **4.** $40°$
5. $\left\{\frac{\pi}{2},\frac{5\pi}{6},\frac{11\pi}{6}\right\}$ **6.** $\theta=2n\pi, \theta=\frac{2\pi}{3}+2n\pi, \theta=\frac{4\pi}{3}+2n\pi$
7. $\theta=\frac{\pi}{6}, \theta=\frac{\pi}{2}, \theta=\frac{5\pi}{6}$ **8.** $\left\{\frac{\pi}{3}\right\}$

Basic Concepts and Skills:

1. two **3.** one **5.** false **7.** $\frac{\pi}{2}+2n\pi, \frac{3\pi}{2}+2n\pi$ **9.** $\frac{3\pi}{4}+n\pi$
11. $\frac{\pi}{4}+2n\pi, \frac{7\pi}{4}+2n\pi$ **13.** $\frac{\pi}{6}+n\pi$ **15.** $\frac{2\pi}{3}+2n\pi, \frac{4\pi}{3}+2n\pi$
17. $30°+180°n$ **19.** $210°+360°n, 330°+360°n$ **21.** $90°+360°n$
23. $60°+360°n, 120°+360°n$ **25.** $60°+360°n, 300°+360°n$
27. $\{23.6°, 156.4°\}$ **29.** $\{82.0°, 278.0°\}$ **31.** $\{131.3°, 311.3°\}$
33. $\{3.4814, 5.9433\}$ **35.** $\{2.2143, 5.3559\}$ **37.** $\{3.5531, 5.8717\}$
39. $\left\{\frac{7\pi}{12},\frac{23\pi}{12}\right\}$ **41.** $\left\{\frac{19\pi}{24},\frac{35\pi}{24}\right\}$ **43.** $\left\{\frac{\pi}{3},\frac{4\pi}{3}\right\}$ **45.** $\left\{\frac{\pi}{6},\frac{3\pi}{2}\right\}$
47. $\left\{\frac{\pi}{4},\frac{5\pi}{4}\right\}$ **49.** $\left\{\frac{\pi}{6},\frac{3\pi}{4},\frac{5\pi}{6},\frac{7\pi}{4}\right\}$ **51.** $\left\{\frac{\pi}{6},\frac{3\pi}{4},\frac{5\pi}{6},\frac{7\pi}{4}\right\}$
53. $\left\{\frac{\pi}{4},\frac{7\pi}{6},\frac{7\pi}{4},\frac{11\pi}{6}\right\}$ **55.** $\left\{\frac{\pi}{6},\frac{5\pi}{6},\frac{7\pi}{6},\frac{11\pi}{6}\right\}$ **57.** $\left\{\frac{\pi}{4},\frac{3\pi}{4},\frac{5\pi}{4},\frac{7\pi}{4}\right\}$
59. $\left\{\frac{\pi}{3},\frac{2\pi}{3},\frac{4\pi}{3},\frac{5\pi}{3}\right\}$ **61.** $\left\{\frac{\pi}{2},\frac{7\pi}{6},\frac{11\pi}{6}\right\}$ **63.** $\left\{\frac{\pi}{4},\frac{5\pi}{4}\right\}$
65. $\left\{\frac{\pi}{6},\frac{5\pi}{6},\frac{7\pi}{6},\frac{11\pi}{6}\right\}$ **67.** $0, \pi$ **69.** $\left\{\frac{7\pi}{6},\frac{3\pi}{2},\frac{11\pi}{6}\right\}$
71. $\left\{\frac{\pi}{3},\frac{5\pi}{6},\frac{4\pi}{3},\frac{11\pi}{6}\right\}$ **73.** $\left\{\frac{\pi}{3},\pi\right\}$ **75.** $\left\{\frac{\pi}{6}\right\}$

Applying the Concepts:

77. $60°$ **79.** $\approx 5.7°$ **81.** $28°$

Beyond the Basics:

83. $\left\{\frac{\pi}{6},\frac{7\pi}{6},\frac{11\pi}{6}\right\}$ **85.** $\left\{0,\frac{\pi}{3},\frac{2\pi}{3},\pi,\frac{4\pi}{3},\frac{5\pi}{3}\right\}$ **87.** $\left\{\frac{\pi}{2}, 5.6397\right\}$

Critical Thinking/Discussion/Writing:

89. $\{1.107, 2.034, 4.249, 5.176\}$ **91.** $\{0.905, 5.379\}$
93. $\left\{\frac{\pi}{4},\frac{5\pi}{4}\right\}$ **95.** $\left\{\frac{7\pi}{6},\frac{11\pi}{6}\right\}$ **97.** $\left\{\frac{\pi}{6},\frac{11\pi}{6}\right\}$ **99.** $\left\{\frac{3\pi}{4}\right\}$
101. $\left\{\frac{\pi}{4}\right\}$ **103.** $2\cos 4x\cos x$

Section 6.6

Answers to Practice Problems:

1. $\left\{\frac{\pi}{12},\frac{5\pi}{12},\frac{13\pi}{12},\frac{17\pi}{12}\right\}$ **2.** $\left\{\frac{\pi}{3}\right\}$ **3.** After about 5 days and 24 days
4. $\{15°, 75°, 135°, 195°, 255°, 315°\}$ **5.** $\left\{\frac{\pi}{36},\frac{5\pi}{36},\frac{25\pi}{36},\frac{29\pi}{36},\frac{49\pi}{36},\frac{53\pi}{36}\right\}$
6. $\left\{0,\frac{2\pi}{5},\frac{4\pi}{5},\pi,\frac{6\pi}{5},\frac{8\pi}{5}\right\}$ **7.** $\left\{\frac{\sqrt{3}-2}{2}\right\}$ **8. b.** $\left\{\frac{\sqrt{2+\sqrt{2}}}{2}\right\}$

Basic Concepts and Skills:

1. $\frac{\pi}{2}+x_1$ **3.** $\frac{\pi}{2}-x_1$ **5.** false **7.** $\left\{\frac{\pi}{6},\frac{5\pi}{6},\frac{7\pi}{6},\frac{11\pi}{6}\right\}$ **9.** $\varnothing$
11. $\left\{\frac{\pi}{12},\frac{7\pi}{12},\frac{13\pi}{12},\frac{19\pi}{12}\right\}$ **13.** $\left\{\frac{\pi}{18},\frac{5\pi}{18},\frac{13\pi}{18},\frac{17\pi}{18},\frac{25\pi}{18},\frac{29\pi}{18}\right\}$
15. $\left\{\frac{2\pi}{3}\right\}$ **17.** $\left\{\frac{3\pi}{4}\right\}$ **19.** $\left\{\frac{\pi}{12},\frac{\pi}{4},\frac{3\pi}{4},\frac{11\pi}{12},\frac{17\pi}{12},\frac{19\pi}{12}\right\}$
21. $\left\{\frac{\pi}{6}-\frac{1}{2},\frac{5\pi}{6}-\frac{1}{2},\frac{7\pi}{6}-\frac{1}{2},\frac{11\pi}{6}-\frac{1}{2}\right\}$
23. $\left\{\frac{\pi}{24}+\frac{1}{4},\frac{5\pi}{24}+\frac{1}{4},\frac{13\pi}{24}+\frac{1}{4},\frac{17\pi}{24}+\frac{1}{4},\frac{25\pi}{24}+\frac{1}{4},\frac{29\pi}{24}+\frac{1}{4},\right.$
$\left.\frac{37\pi}{24}+\frac{1}{4},\frac{41\pi}{24}+\frac{1}{4}\right\}$ **25.** $\left\{-\frac{\pi}{9}+\frac{2}{3},\frac{\pi}{9}+\frac{2}{3},\frac{11\pi}{9}+\frac{2}{3},\frac{13\pi}{9}+\frac{2}{3}\right\}$
27. $\{35.3°, 144.7°, 215.3°, 324.7°\}$
29. $\{20°, 100°, 140°, 220°, 260°, 340°\}$
31. $\{66.5°, 113.5°, 186.5°, 233.5°, 306.5°, 353.5°\}$ **33.** $\varnothing$
35. $\{30°, 150°, 210°, 330°\}$ **37.** $\{60°, 300°\}$ **39.** $\left\{0,\frac{2\pi}{3},\pi,\frac{4\pi}{3}\right\}$
41. $\left\{0,\frac{\pi}{2},\pi,\frac{3\pi}{2}\right\}$ **43.** $\left\{\frac{\pi}{8},\frac{3\pi}{8},\frac{\pi}{2},\frac{5\pi}{8},\frac{7\pi}{8},\frac{9\pi}{8},\frac{11\pi}{8},\frac{3\pi}{2},\frac{13\pi}{8},\frac{15\pi}{8}\right\}$
45. $\left\{0,\frac{\pi}{8},\frac{3\pi}{8},\frac{5\pi}{8},\frac{7\pi}{8},\pi,\frac{9\pi}{8},\frac{11\pi}{8},\frac{13\pi}{8},\frac{15\pi}{8}\right\}$ **47.** $\left\{\frac{\sqrt{3}}{2}\right\}$ **49.** $\varnothing$
51. $\left\{\frac{1}{2}\right\}$ **53.** $\left\{\frac{5\sqrt{26}}{26}\right\}$

Applying the Concepts:

55. a. ≈ 0.0014 sec **b.** ≈ 0.0092 sec
57. $\frac{2}{3}+4n$ sec, $\frac{10}{3}+4n$ sec, n an integer **59. a.** 27 **b.** 18, 36
c. 10, 44 **61. a.** January **b.** February, December

Beyond the Basics:

63. $\left\{\frac{\pi}{4},\frac{3\pi}{4},\frac{5\pi}{4},\frac{7\pi}{4}\right\}$ **65.** $\left\{\frac{\pi}{3},\frac{5\pi}{3}\right\}$
67. $\left\{\frac{\pi}{8},\frac{3\pi}{8},\frac{5\pi}{8},\frac{7\pi}{8},\frac{9\pi}{8},\frac{11\pi}{8},\frac{13\pi}{8},\frac{15\pi}{8}\right\}$ **69.** $\left\{\frac{2\pi}{3},\frac{4\pi}{3}\right\}$
71. $\left\{\frac{\pi}{8},\frac{5\pi}{8},\frac{9\pi}{8},\frac{3\pi}{2},\frac{13\pi}{8}\right\}$ **73.** $\left\{0,\frac{\pi}{2},\frac{5\pi}{6},\pi,\frac{7\pi}{6},\frac{3\pi}{2}\right\}$
75. $\left\{\frac{5\pi}{12},\frac{7\pi}{12},\frac{11\pi}{12},\frac{17\pi}{12},\frac{19\pi}{12},\frac{23\pi}{12}\right\}$ **77.** $\left\{0,\frac{\pi}{3},\frac{2\pi}{3},\pi,\frac{4\pi}{3},\frac{5\pi}{3}\right\}$
79. $\left\{\frac{\pi}{8},\frac{\pi}{2},\frac{5\pi}{8},\frac{9\pi}{8},\frac{3\pi}{2},\frac{13\pi}{8}\right\}$ **81. a.** $5\sin\left[2x+\tan^{-1}\left(\frac{4}{3}\right)\right]$
b. $\{1.107, 2.678, 4.249, 5.820\}$ **83. a.** $13\sin\left[3x+\tan^{-1}\left(-\frac{12}{5}\right)\right]$
b. $\{0.741, 1.090, 2.835, 3.185, 4.930, 5.279\}$

Critical Thinking/Discussion/Writing:

85. $\left\{-\frac{1}{2}, \frac{1}{2}\right\}$

Maintaining Skills:

87. $\frac{x}{3} = \frac{\pi}{5}$ **89.** $\{16\}$ **91.** $\frac{\sqrt{3}}{2}$ **93.** $\frac{1}{2}$ **95.** True

Review Exercises

1. $\cos\theta = -\frac{\sqrt{5}}{3}, \tan\theta = \frac{2\sqrt{5}}{5}, \cot\theta = \frac{\sqrt{5}}{2},$
$\sec\theta = -\frac{3\sqrt{5}}{5}, \csc\theta = -\frac{3}{2}$

3. $\sin\theta = -\frac{2\sqrt{2}}{3}, \cos\theta = \frac{1}{3}, \tan\theta = -2\sqrt{2},$
$\cot\theta = -\frac{\sqrt{2}}{4}, \csc\theta = -\frac{3\sqrt{2}}{4}$ **21.** $\frac{\sqrt{2}+\sqrt{3}}{2}$ **23.** $\sqrt{6} - \sqrt{2}$

25. 1 **27.** -1 **29.** $-\frac{16}{65}$ **31.** $\frac{63}{65}$

51.

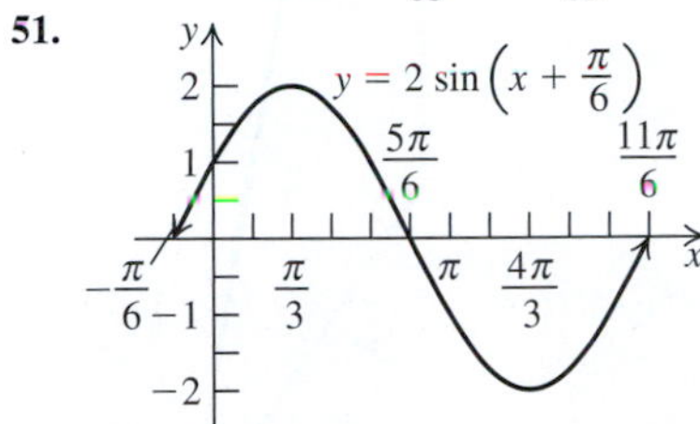

55. $\left\{\frac{\pi}{4}, \frac{3\pi}{4}, \frac{5\pi}{4}, \frac{7\pi}{4}\right\}$ **57.** $\left\{0, \frac{2\pi}{3}, \frac{4\pi}{3}\right\}$

59. $\left\{\frac{\pi}{18}, \frac{5\pi}{18}, \frac{13\pi}{18}, \frac{17\pi}{18}, \frac{25\pi}{18}, \frac{29\pi}{18}\right\}$ **61.** $\{19.5°, 160.5°\}$

63. $\{15°, 165°, 195°, 345°\}$ **65.** $\{120°, 150°, 240°, 330°\}$

Applying the Concepts:

67. 60° **69.** $\frac{1}{360}$ sec **71.** 30°

Practice Test A

1. $-\frac{3}{4}$ **2.** $-\frac{3\sqrt{13}}{13}$ **9.** $-\frac{\pi}{4} + n\pi$ **10.** $\frac{\pi}{24} + n\frac{\pi}{2}, \frac{5\pi}{24} + n\frac{\pi}{2}$

11. $\left\{\frac{3\pi}{4}\right\}$ **12.** $\left\{\frac{\pi}{2}, \frac{7\pi}{6}, \frac{3\pi}{2}, \frac{11\pi}{6}\right\}$ **13.** 0 **14.** $\frac{1}{2}$ **15.** $\frac{\sqrt{6}+\sqrt{2}}{4}$

16. $-\frac{7}{25}$ **17.** $\frac{24}{25}$ **18.** Odd **19.** No, let $x = y = \frac{\pi}{2}$. **20.** 45°

Practice Test B

1. a **2.** b **3.** d **4.** b **5.** a **6.** b **7.** c **8.** d **9.** c **10.** c **11.** c **12.** a **13.** d **14.** d **15.** b **16.** a **17.** d **18.** c **19.** d **20.** b

Cumulative Review Exercises (Chapters P–6)

1. $x = -\frac{1}{2} \pm \frac{\sqrt{5}}{2}$ **3.** $x = -\frac{\log 9}{\log 5}$ **5.** $(-2, 0) \cup (2, \infty)$

7. $4x - 1 + 2h$

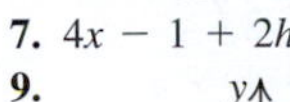

9. **11.**

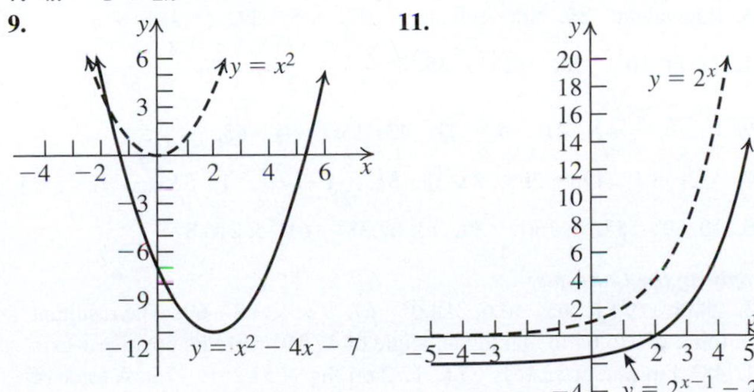

13. **15.**

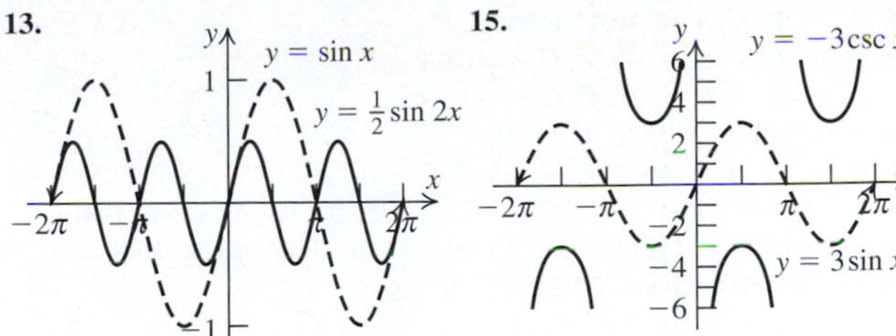

17. $\cot x - 3$

CHAPTER 7

Section 7.1

Answers to Practice Problems:

1. $C = 45°, b \approx 15.4$ in, $c \approx 12.0$ in **2.** 10,576 ft **3.** None **4.** None **5.** Solution 1: $A \approx 99.2°, B \approx 45.8°, a \approx 20.7$ ft; Solution 2: $A \approx 10.8°, B \approx 134.2°, a \approx 3.9$ ft **6.** No triangle exists. **7.** $A \approx 31.3°, B \approx 88.7°, b \approx 57.7$ ft **8. a.** ≈ 31.5 mi **b.** ≈ 15.6 mi

Basic Concepts and Skills:

1. 180 **3.** one **5.** true **7.** Two **9.** None **11.** One **13.** 90° **15.** 75° **17.** $2\sqrt{2}$ **19.** 3 **21.** $C = 63°, a \approx 98.2$ ft, $b \approx 93.0$ ft
23. $B = 27°, a \approx 64.8$ m, $b \approx 31.3$ m
25. $B = 65°, C = 50°, c \approx 16.9$ ft
27. $B \approx 23.7°, C \approx 41.3°, c \approx 51.0$ m
29. $C = 105°, b \approx 89.2$ m, $c \approx 150.3$ m
31. $B = 79°, b \approx 102.3$ cm, $c \approx 85.4$ cm
33. $B = 98°, a \approx 47.1$ ft, $b \approx 81.2$ ft
35. $A = 70°, a \approx 55.1$ in, $c \approx 54.0$ in
37. $A = 24°, a \approx 13.3$ ft, $b \approx 30.7$ ft
39. $C = 98.5°, a \approx 17.7$ m, $b \approx 21.7$ m
41. $B \approx 34°, C \approx 106°, c \approx 34.4$
43. $C = 60°, B = 90°, c = 25\sqrt{3}$
45. No triangle exists. **47.** No triangle exists.
49. $B \approx 56.8°, C \approx 23.2°, c \approx 16.0$ **51.** No triangle exists.
53. Solution 1: $C \approx 55.3°, A \approx 78.7°, a \approx 47.7$
Solution 2: $C \approx 124.7°, A \approx 9.3°, a \approx 7.9$ **55.** No triangle exists.
57. Solution 1: $C \approx 49.0°, B \approx 89.0°, b \approx 82.2$
Solution 2: $C \approx 131.0°, B \approx 7.0°, b \approx 10.0$ **59.** No triangle exists.

Applying the Concepts:

61. ≈ 399 ft **63. a.** 1615 yd **b.** 540 yd **65. a.** 25.7 mi **b.** 14.5 mi **67.** 22,912 mi **69.** 33 or 127 million miles **71.** 1:32 P.M.; 3:18 P.M. **73.** 24.2 m

Critical Thinking/Discussion/Writing:

83. An isosceles triangle with $\angle A = \angle B$

Maintaining Skills:

87. $\frac{\sqrt{3}}{2}$ **89.** $-\frac{\sqrt{2}}{2}$ **91.** 60° **93.** 120° **95.** $\sqrt{11}$ **97.** $\sqrt{7}$ **99.** $-\frac{1}{4}$ **101.** 30°

Section 7.2

Answers to Practice Problems:

1. $b \approx 21.8, A \approx 36.6°, C \approx 83.4°$ **2.** 5219.5 miles **3.** $A \approx 41.9°, B \approx 85.9°, C \approx 52.1°$ **4.** No triangle exists.

Basic Concepts and Skills:

1. C **3.** three **5.** false **7.** 7 **9.** $\sqrt{19}$ **11.** $\frac{1}{5}$ **13.** 60°
15. $b \approx 20.2, A \approx 44°, C \approx 30°$
17. $A \approx 130.5°, B \approx 27.1°, C \approx 22.4°$
19. $c = 21, A \approx 38.2°, B \approx 21.8°$
21. $a \approx 11.5, B \approx 50.4°, C \approx 67.6°$

23. $A \approx 81.3°, B \approx 46.4°, C \approx 52.3°$
25. $A \approx 28.3°, B \approx 43.3°, C \approx 108.4°$
27. $A \approx 23.7°, B \approx 36.4°, C \approx 119.9°$
29. $a \approx 5.7, B \approx 33.8°, C \approx 48.5°$
31. $b \approx 5.0, A \approx 46.3°, C \approx 65.4°$
33. $A \approx 38.4°, B \approx 49.1°, C \approx 92.5°$ **35.** 5.7 ft **37.** 1543.1 yd
39. a. 20.9 ft **b.** 11.0 ft **41.** 9.1 mi **43. a.** 3689 m **b.** 1634.5 m
45. 66.4 ft **47.** 9.8 cm, 23.6 cm

Critical Thinking/Discussion/Writing:

57. $a = \sqrt{13}, b = 5\sqrt{2}, c = \sqrt{53}, A \approx 29.1°, B \approx 72.2°, C \approx 78.7°$
59. $a \approx 4.3, A \approx 12.5°, C \approx 17.5°$ **61.** No triangle can be formed. **63.** We need $2bc \cos A = 1$, for example $A = 60°$, $b > 0$ and $c = \dfrac{1}{b}$

Maintaining Skills:

65. 2 **67.** $6\sqrt{3}$ **69.** $\dfrac{9\sqrt{2}}{2}$ **71. a.** 6 **b.** 6 **73. a.** 9 **b.** $9\sqrt{3}$
75. a. 21 **b.** $21\sqrt{11}$ **77. a.** 18 **b.** 60

Section 7.3

Answers to Practice Problems:

1. a. 124 ft **b.** 672 ft^2 **2.** 375.2 ft^2 **3.** 90° **4.** 203.4 in^2
5. 93.5 m^2 **6.** 14,801 gal

Basic Concepts and Skills:

1. $\dfrac{1}{2}bh$ **3.** $\dfrac{1}{2}ac \sin B$ **5.** 7.5 **7.** $10\sqrt{3}$ **9.** 27 **11.** 654.2 sq in
13. 118.7 sq km **15.** 62.5 sq mm **17.** 25,136.6 sq ft **19.** 84.3 sq ft
21. 167.1 sq yd **23.** 178.6 sq cm **25.** 1621.9 sq ft **27.** 2.9
29. 1240.2 **31.** 13.5 **33.** 7.7 **35.** 30° or 150° **37.** 90°
41. 187.35 cm^2 **43.** 493,941 sq mi **45.** $\approx$\$775,668
47. a. $(r, 0), \left(\dfrac{r\sqrt{2}}{2}, \dfrac{r\sqrt{2}}{2}\right), (0, r), \left(-\dfrac{r\sqrt{2}}{2}, \dfrac{r\sqrt{2}}{2}\right), (-r, 0),$
$\left(-\dfrac{r\sqrt{2}}{2}, -\dfrac{r\sqrt{2}}{2}\right), (0, -r), \left(\dfrac{r\sqrt{2}}{2}, -\dfrac{r\sqrt{2}}{2}\right)$ **b.** $r\sqrt{2 - \sqrt{2}}$
c. $8r\sqrt{2 - \sqrt{2}}$ **49. b.** $10\pi \approx 31.4159$ **c.** 28.2843, 30.9017, 31.3953, 31.4108, resp. Perimeters approach the circumference of the circle.
51. b. $4r\sqrt{3}$ **53. b.** $20\pi \approx 62.8319$ **c.** 80.0, 64.9839, 62.9147, 62.8525, resp. Perimeters approach the circumference of the circle.

Maintaining Skills:

69. $x + 3y, \sqrt{10}$ **71. a.** $\dfrac{1}{2}$ **b.** $\dfrac{\sqrt{2}}{2}$ **c.** $\sqrt{3}$ **73.** $\theta = \dfrac{4\pi}{3}$
75. a. $d(S, T) = d(O, P) = 13$ **b.** $m_1 = m_2 = \dfrac{12}{5}$ **c.** Parallel
77. a. $d(S, T) = d(O, P) = \sqrt{2}$ **b.** $m_1 = 1, m_2 = -1$
c. Perpendicular

Section 7.4

Answers to Practice Problems:

1. a.

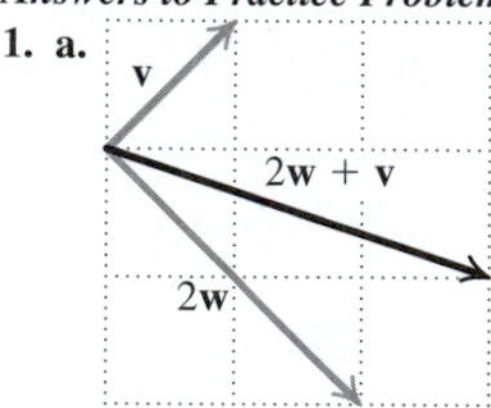

b.

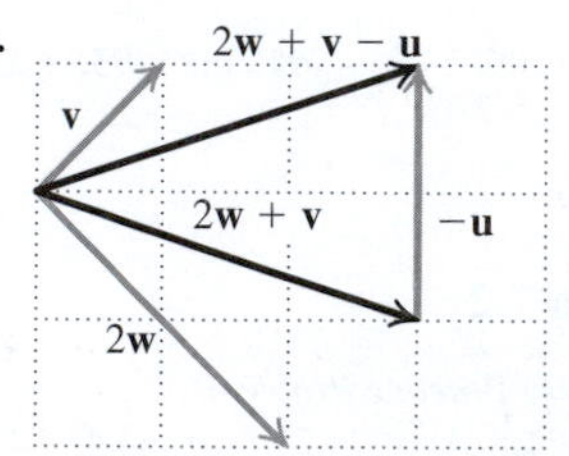

2. $\langle 3, -10\rangle$ **3. a.** $\langle 1, -1\rangle$ **b.** $\langle 6, -9\rangle$ **c.** $\langle 8, -13\rangle$ **d.** $\sqrt{233}$
4. $\left\langle -\dfrac{12}{13}, \dfrac{5}{13}\right\rangle$ **5. a.** $-7\mathbf{i} + 14\mathbf{j}$ **b.** $7\sqrt{5}$ **6.** $\sqrt{3}\mathbf{i} - \mathbf{j}$
7. $\tan^{-1}\left(-\dfrac{3}{2}\right) + 360° \approx 303.69°$ **8. R** is a force of 50 pounds in the direction N 36.9° W. **9.** The ground speed is approximately 801.0 miles per hour, and the bearing is approximately N 62.9° W. **10.** The magnitude of the resultant is approximately 121.2 lb, and the resultant is in the direction S 7.8° W.

Basic Concepts and Skills:

1. direction **3.** $\sqrt{a^2 + b^2}, \dfrac{b}{a}$ **5.** true

7.

u + v
u
v

9.

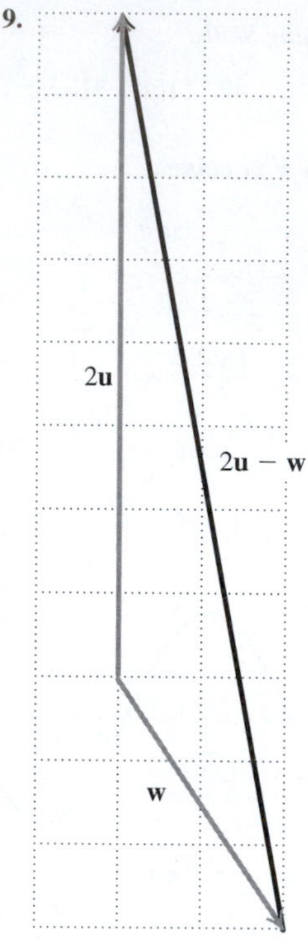

11. **13.**

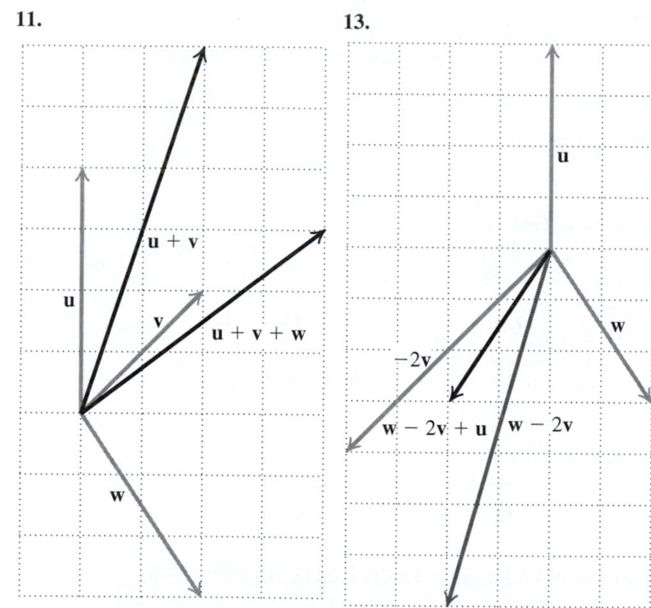

15. $\langle -1, 3\rangle$ **17.** $\langle 2, -2\rangle$ **19.** $\langle 3, -7\rangle$ **21.** $\left\langle -1, -\dfrac{5}{2}\right\rangle$
23. Equivalent **25.** Not equivalent **27.** $\sqrt{5}$ **29.** $\langle -4, 4\rangle$
31. $\langle -11, 10\rangle$ **33.** $\sqrt{221}$ **35.** $\left\langle \dfrac{\sqrt{2}}{2}, -\dfrac{\sqrt{2}}{2}\right\rangle$ **37.** $\left\langle -\dfrac{4}{5}, \dfrac{3}{5}\right\rangle$
39. $\left\langle \dfrac{\sqrt{2}}{2}, \dfrac{\sqrt{2}}{2}\right\rangle$ **41.** $-\mathbf{i} - 7\mathbf{j}$ **43.** $13\mathbf{i} - 4\mathbf{j}$ **45.** $\sqrt{185}$
47. $\sqrt{3}\mathbf{i} + \mathbf{j}$ **49.** $-2\mathbf{i} + 2\sqrt{3}\mathbf{j}$ **51.** $\dfrac{3}{2}\mathbf{i} - \dfrac{3\sqrt{3}}{2}\mathbf{j}$ **53.** $\dfrac{7}{2}\mathbf{i} - \dfrac{7\sqrt{3}}{2}\mathbf{j}$
55. 10, 60° **57.** 3, 150° **59.** 13, 67.38° **61.** 5, 216.87°

Applying the Concepts:

63. $23.0\mathbf{i} + 9.8\mathbf{j}$ **65.** 40.6, 218.0° **67.** 72.5, 33.8 **69.** The resultant is a force of 676.64 lb making an angle of 47.19° with the positive x-axis. **71.** 483.4 mi/hr, N 32.1° E **73.** 12.2 mi/hr, N 52.2° E **75.** A force of magnitude 10.08 lb making an angle of 176.61° with the positive x-axis

Beyond the Basics:
77. $\|\mathbf{F}_1\| = 106.42$ lb, $\|\mathbf{F}_2\| = 136.81$ lb **79.** $(2, -2)$
81. $\left\langle -\frac{11}{3}, -\frac{11}{3} \right\rangle$ **83.** $\overrightarrow{AB} = -\overrightarrow{CD}$ **85.** $\overrightarrow{PQ} = \frac{2}{3}\overrightarrow{PR}$
87. $3, \sqrt{5}$ **89.** $(9, 2)$ **91.** $(-1, 4)$ **95.** false **97.** $\pi, \frac{\pi}{2}, 0$
99. $\sqrt{5}$ **101.** $-\frac{1}{\sqrt{65}}$

Section 7.5

Answers to Practice Problems:
1. a. 8 **b.** 0 **2.** 14 **3.** Not parallel **4.** 115.3° **5.** ≈ 39 **6.** **v** and **w** are orthogonal. **7.** $k = 3$ **8. a.** $\left(-\frac{16}{29}, -\frac{40}{29}\right)$ **b.** $-\frac{8\sqrt{29}}{29}$
9. $\langle 3, 3 \rangle + \langle 2, -2 \rangle$ **10.** 3000 foot-pounds

Basic Concepts and Skills:
1. $a_1b_1 + a_2b_2$ **3.** 0 **5.** true **7.** -7 **9.** 0 **11.** 14
13. -6 **15.** $5\sqrt{3} \approx 8.7$ **17.** 8.5 **19.** -10.5 **21.** 60°
23. $\cos^{-1}\left(-\frac{5}{\sqrt{26}}\right) \approx 168.7°$ **25.** 90° **27.** 0° **29.** 180°
31. $\sqrt{61}$ **33.** $\sqrt{129}$ **35.** $\cos^{-1}\left(\frac{7}{\sqrt{61}}\right)$ **37.** $\cos^{-1}\left(\frac{21}{2\sqrt{129}}\right)$
39. Orthogonal **41.** Not orthogonal **43.** Orthogonal
45. Parallel **47.** Not parallel **49.** Parallel
51. a. $c = -2$, **b.** $c = \frac{9}{2}$ **53. a.** $c = -\frac{9}{2}$, **b.** $c = 8$
55. $\langle 4, 1 \rangle$ **57.** $\left\langle \frac{24}{25}, \frac{32}{25} \right\rangle$ **59.** $\langle 5, 0 \rangle$ **61.** $\left\langle \frac{a}{2} + \frac{b}{2}, \frac{a}{2} + \frac{b}{2} \right\rangle$
63. $\langle 2, 0 \rangle + \langle 0, 3 \rangle$ **65.** $\left\langle \frac{1}{2}, \frac{1}{2} \right\rangle + \left\langle \frac{1}{2}, -\frac{1}{2} \right\rangle$
67. $\langle 0, 0 \rangle + \langle 4, 2 \rangle$ **69.** $\left\langle \frac{24}{25}, \frac{32}{25} \right\rangle + \left\langle \frac{76}{25}, -\frac{57}{25} \right\rangle$

Applying the Concepts:
71. 13.5 lb, 12.8°, 17.2° **73.** 560 foot-pounds **75.** 5520.2 lb
77. 56 foot-pounds

Beyond the Basics:
87. $\pm\left\langle \frac{4}{5}, \frac{3}{5} \right\rangle$

Maintaining Skills:
97. Vertical line; x-intercept: -1, no y-intercept **99.** Parabola; x-intercepts: $-2, 3$, y-intercept: -6 **101.** Circle; x-intercepts: $\pm\sqrt{3}$, y-intercepts: $-1, 3$ **103.** Circle; no x-intercepts, y-intercepts: $-2 \pm \sqrt{2}$
105. Only the x-axis **107.** Only the y-axis

Section 7.6

Answers to Practice Problems:
1. a. $P(2, -150°)$ **b.** $(-2, 30°)$
c. $(2, 210°)$ **d.** $(-2, -330°)$
2. a. $\left(-\frac{3}{2}, -\frac{3\sqrt{3}}{2}\right)$ **b.** $(\sqrt{2}, -\sqrt{2})$ **3.** $\left(\sqrt{2}, \frac{5\pi}{4}\right)$
4. $(24.8, 36.0°)$ **5.** $r^2 - 2r\sin\theta + 3 = 0$
6. a. $x^2 + y^2 = 25$; a circle centered at the origin with radius 5
b. $y = -x\sqrt{3}$; a line through the origin with slope $-\sqrt{3}$
c. $x = 1$; a vertical line through $x = 1$ **d.** $x^2 + (y - 1)^2 = 1$; a circle with center $(0, 1)$ and radius 1

7.
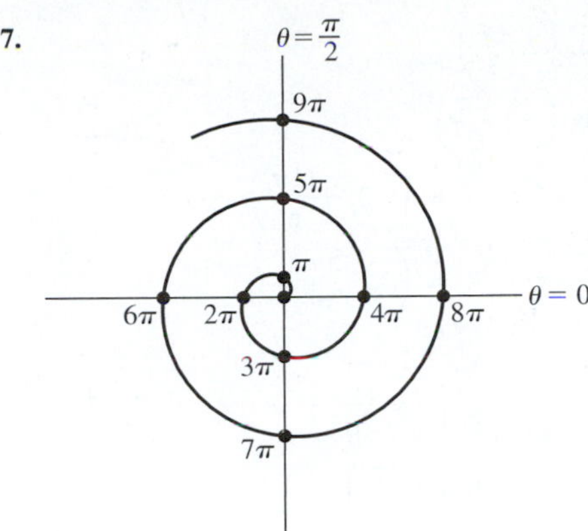

8.
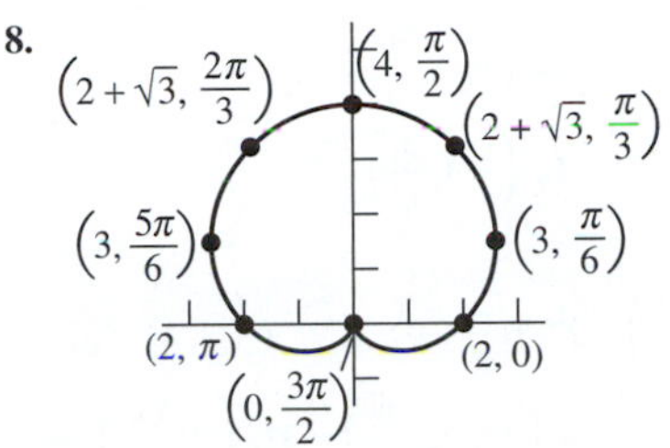

9.
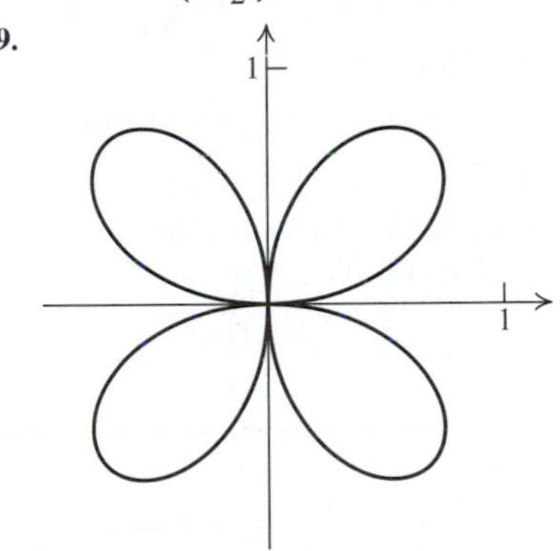

Basic Concepts and Skills:
1. origin, axis **3.** $r\cos\theta$, $r\sin\theta$ **5.** true

7.
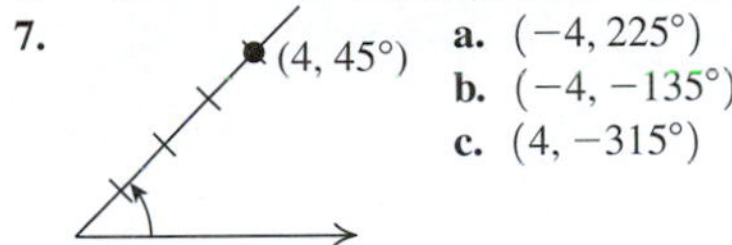

a. $(-4, 225°)$ **b.** $(-4, -135°)$ **c.** $(4, -315°)$

9.
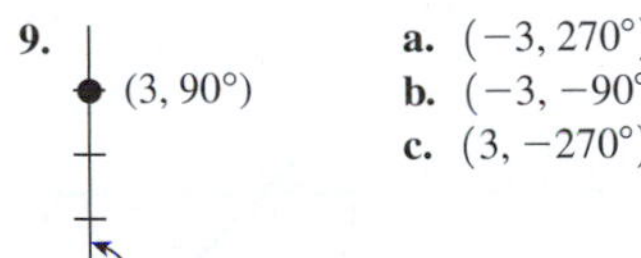

a. $(-3, 270°)$ **b.** $(-3, -90°)$ **c.** $(3, -270°)$

11.
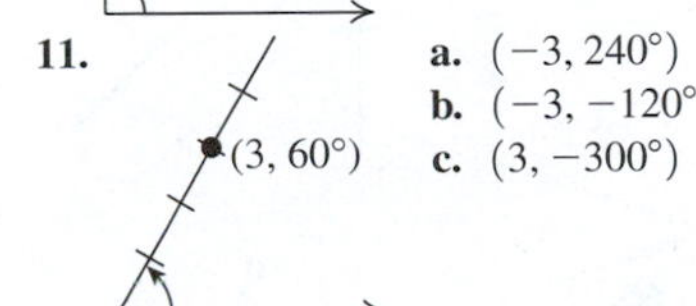

a. $(-3, 240°)$ **b.** $(-3, -120°)$ **c.** $(3, -300°)$

13.
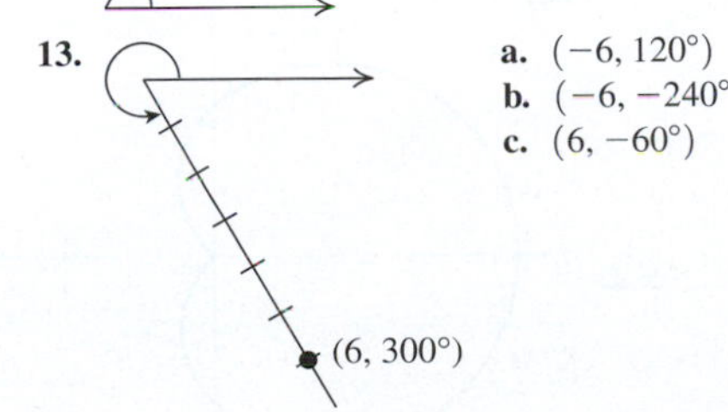

a. $(-6, 120°)$ **b.** $(-6, -240°)$ **c.** $(6, -60°)$

15. 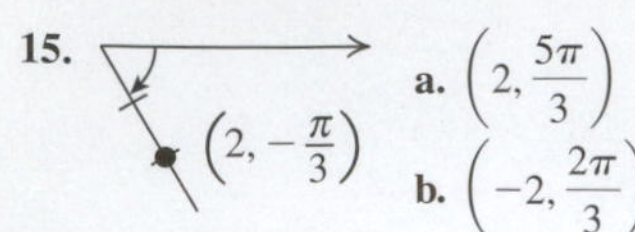

a. $\left(2, \frac{5\pi}{3}\right)$ **b.** $\left(-2, \frac{2\pi}{3}\right)$

17. **a.** $\left(-3, -\frac{11\pi}{6}\right)$ **b.** $\left(3, -\frac{5\pi}{6}\right)$ $\left(-3, \frac{\pi}{6}\right)$

19. 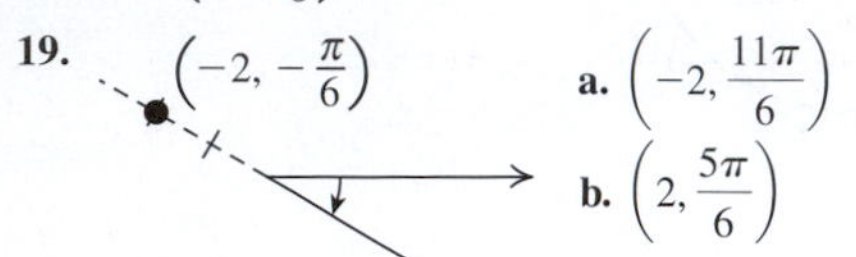

a. $\left(-2, \frac{11\pi}{6}\right)$ **b.** $\left(2, \frac{5\pi}{6}\right)$

21. **a.** $\left(4, -\frac{5\pi}{6}\right)$ **b.** $\left(-4, \frac{\pi}{6}\right)$ $\left(4, \frac{7\pi}{6}\right)$

23. $\left(\frac{3}{2}, \frac{3\sqrt{3}}{2}\right)$ **25.** $\left(\frac{5}{2}, -\frac{5\sqrt{3}}{2}\right)$ **27.** $(-3, 0)$ **29.** $(\sqrt{3}, 1)$
31. $\left(\sqrt{2}, \frac{7\pi}{4}\right)$ **33.** $\left(3\sqrt{2}, \frac{\pi}{4}\right)$ **35.** $\left(3\sqrt{2}, \frac{7\pi}{4}\right)$ **37.** $\left(2, \frac{2\pi}{3}\right)$
39. $r = 4$ **41.** $r = 4 \cot\theta \csc\theta$ **43.** $r = 6\sin\theta$
45. $r^2 = \frac{1}{\sin\theta\cos\theta}$ **47.** $x^2 + y^2 = 4$; circle centered at $(0, 0)$ with radius 2 **49.** $y = -x$; a line with slope -1 and y-intercept 0
51. $(x - 2)^2 + y^2 = 4$; a circle centered at $(2, 0)$ with radius 2
53. $x^2 + (y + 1)^2 = 1$; a circle centered at $(0, -1)$ with radius 1

55. **57.**
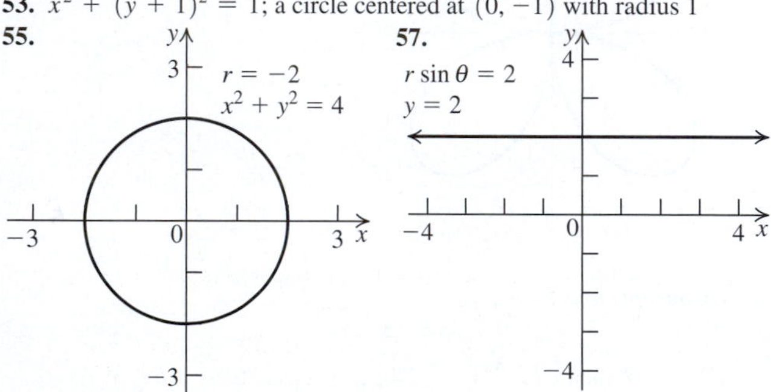

59. **61.**
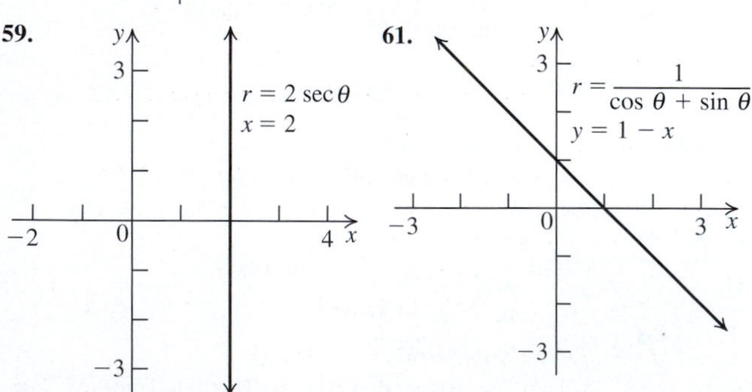

63. **65.**
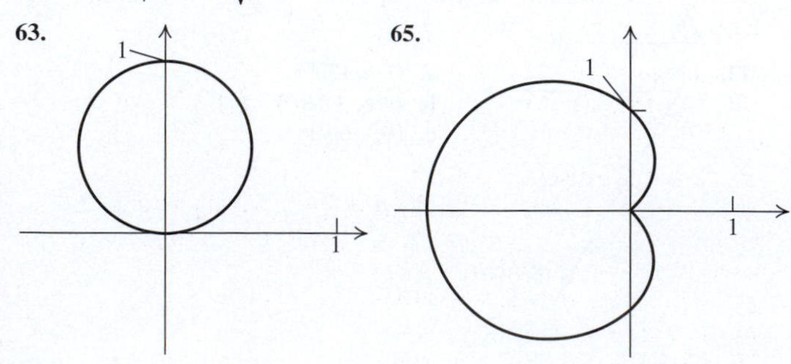

67.
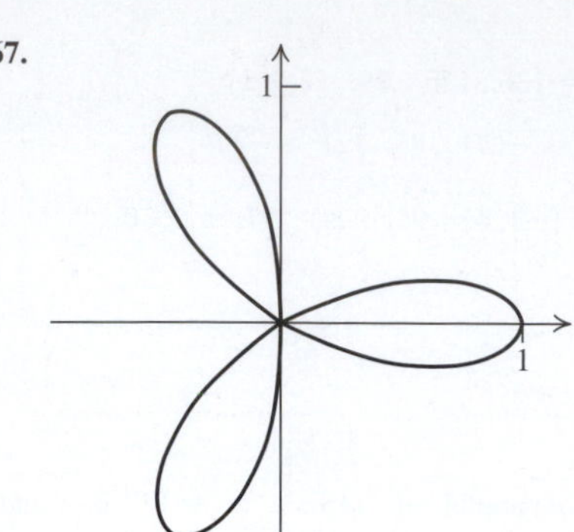

69.
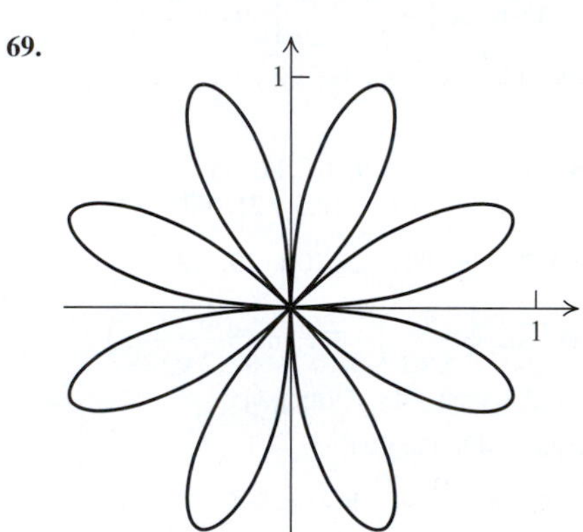

71. **73.**
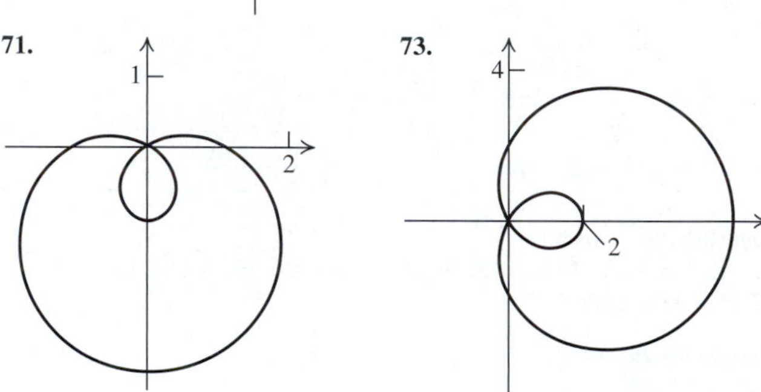

75. **77.**
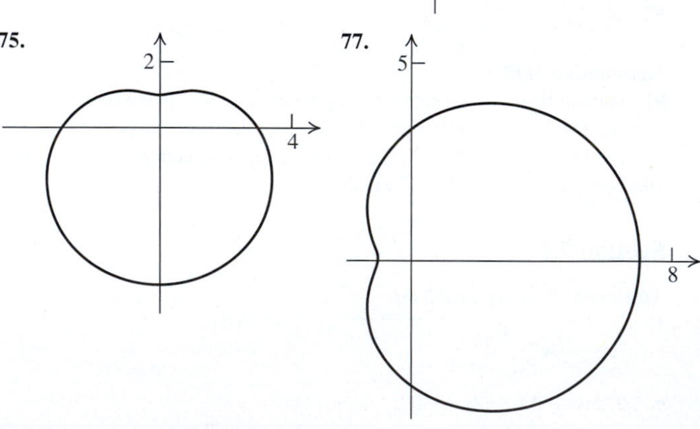

79. (21.8, 39.6°) **81.** (18.3, −45.8°)

Beyond the Basics:

83. $r = \theta$ **85.** $y = \frac{x^2}{6} - \frac{3}{2}$ **87.** $y^2 = 1 - x - \frac{3x^2}{4}$
89. $y^2 - 8x^2 - 30x - 25 = 0$

Applying the Concepts:

95. 5, −40, 10, 40, −140

97.

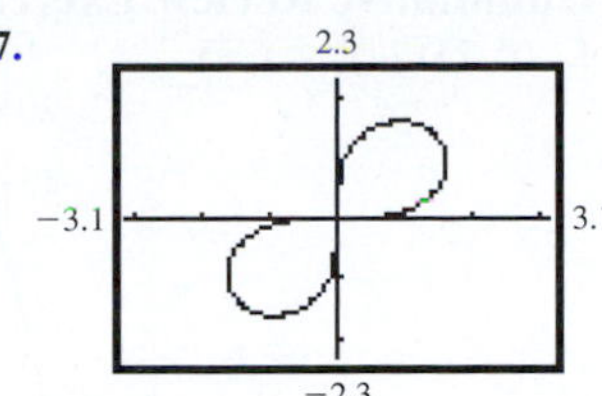

99.

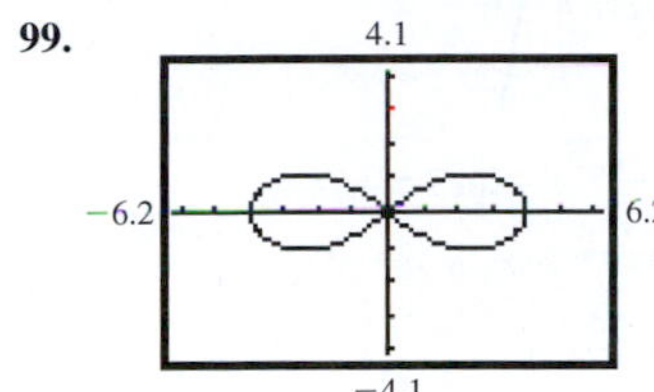

101.

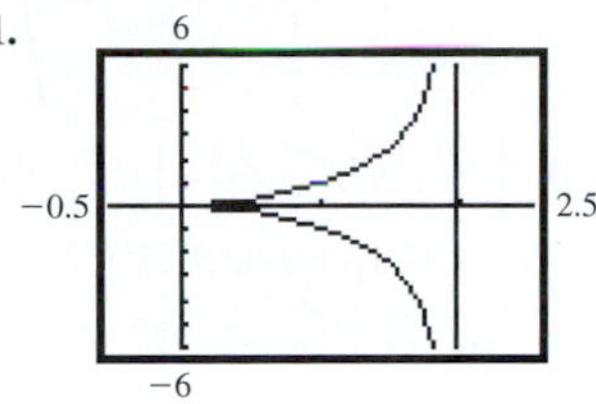

103.

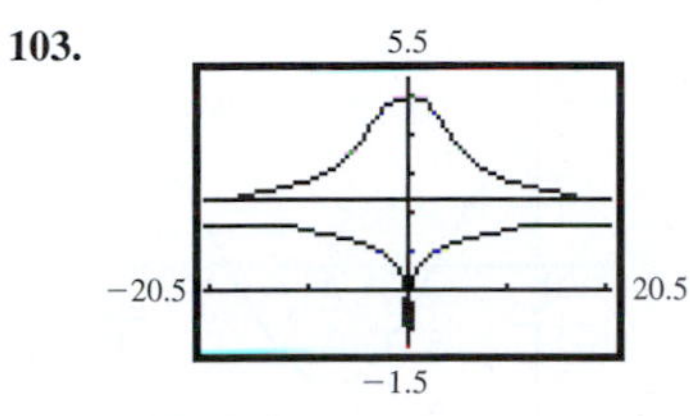

Critical Thinking/Discussion/Writing:

105. False **107.** True

Maintaining Skills:

109. $-3 - 2i$ **111.** $-5 + 2i$ **113.** $7 + i$ **115.** $-5 - 12i$

117. $\frac{3}{25} + \frac{4}{25}i$ **119.** $-\frac{6}{25} - \frac{17}{25}i$ **121.** $\sqrt{2}\left(\cos\frac{\pi}{4}\mathbf{i} + \sin\frac{\pi}{4}\mathbf{j}\right)$

123. $2\left(\cos\frac{5\pi}{6}\mathbf{i} + \sin\frac{5\pi}{6}\mathbf{j}\right)$

Section 7.7

Answers to Practice Problems:

1. Imaginary axis, y; $-3 + 2i$; $2 + 3i$; Real axis, x; O; 4; $-2 - 3i$; $3 - 2i$

2. a. 13 **b.** 7 **c.** 1 **d.** $\sqrt{a^2 + b^2}$

3. $z = \sqrt{2}\left(\cos\frac{5\pi}{4} + i\sin\frac{5\pi}{4}\right)$ **4.** $2 - 2\sqrt{3}i$

5. $z_1z_2 = 10(\cos 135° + i\sin 135°)$; $\frac{z_1}{z_2} = \frac{5}{2}(\cos 15° + i\sin 15°)$

6. $2.6(\cos 27.2° + i\sin 27.2°)$ **7. a.** 16 **b.** $-\frac{1}{64}$

8. $2^{\frac{1}{6}}(\cos 45° + i\sin 45°)$, $2^{\frac{1}{6}}(\cos 165° + i\sin 165°)$, $2^{\frac{1}{6}}(\cos 285° + i\sin 285°)$ **9.** $\cos 0° + i\sin 0°$, $\cos 90° + i\sin 90°$, $\cos 180° + i\sin 180°$, $\cos 270° + i\sin 270°$

Basic Concepts and Skills:

1. $\sqrt{a^2 + b^2}$, $\tan^{-1}\left(\frac{b}{a}\right)$ **3.** moduli, add **5.** true **7.** 2 **9.** 4 **11.** 5 **13.** $\sqrt{13}$

7–14.

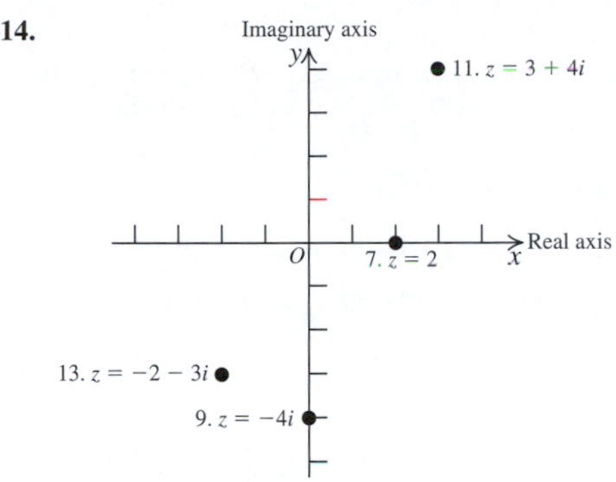

15. $2(\cos 60° + i\sin 60°)$ **17.** $\sqrt{2}(\cos 135° + i\sin 135°)$

19. $\cos 90° + i\sin 90°$ **21.** $\cos 0° + i\sin 0°$

23. $3\sqrt{2}(\cos 315° + i\sin 315°)$

25. $4(\cos 300° + i\sin 300°)$ **27.** $1 + \sqrt{3}i$ **29.** -3

31. $-\frac{5}{2} - \frac{5\sqrt{3}}{2}i$ **33.** 8 **35.** $-3\sqrt{3} + 3i$ **37.** $\frac{3}{2} - \frac{3\sqrt{3}}{2}i$

39. $z_1z_2 = 8(\cos 90° + i\sin 90°)$; $\frac{z_1}{z_2} = 2(\cos 60° + i\sin 60°)$

41. $z_1z_2 = 10(\cos 300° + i\sin 300°)$; $\frac{z_1}{z_2} = \frac{5}{2}(\cos 180° + i\sin 180°)$

43. $z_1z_2 = 15(\cos 60° + i\sin 60°)$; $\frac{z_1}{z_2} = \frac{3}{5}(\cos 20° + i\sin 20°)$

45. $z_1z_2 = 2(\cos 0° + i\sin 0°)$; $\frac{z_1}{z_2} = \cos 90° + i\sin 90°$

47. $z_1z_2 = 4\sqrt{2}(\cos 195° + i\sin 195°)$; $\frac{z_1}{z_2} = \frac{\sqrt{2}}{2}(\cos 105° + i\sin 105°)$

49. $z_1z_2 = 8$; $\frac{z_1}{z_2} = 2(\cos 60° + i\sin 60°)$ **51.** 2^{12} **53.** $-64i$

55. $-\frac{1}{64}i$ **57.** $\frac{1}{2^{10}}i$ **59.** -64 **61.** i **63.** $-\frac{1}{2} + \frac{\sqrt{3}}{2}i$

65. $4(\cos 0° + i\sin 0°)$, $4(\cos 120° + i\sin 120°)$, $4(\cos 240° + i\sin 240°)$

67. $\cos 45° + i\sin 45°$, $\cos 225° + i\sin 225°$

69. $\cos 30° + i\sin 30°$, $\cos 90° + i\sin 90°$, $\cos 150° + i\sin 150°$, $\cos 210° + i\sin 210°$, $\cos 270° + i\sin 270°$, $\cos 330° + i\sin 330°$

71. $\sqrt{2}(\cos 150° + i\sin 150°)$, $\sqrt{2}(\cos 330° + i\sin 330°)$

Applying the Concepts:

75. $15(\cos 30° + i\sin 30°)$ **77.** $1.8(\cos 153.2° + i\sin 153.2°)$

Beyond the Basics:

79. $2(\cos 30° + i\sin 30°)$ **81.** $\cos 240° + i\sin 240°$

83. $\cos 240° + i\sin 240°$ **85.** $\cos 0° + i\sin 0°$, $\cos 90° + i\sin 90°$, $\cos 180° + i\sin 180°$, $\cos 270° + i\sin 270°$

87. $2^{\frac{1}{6}}(\cos 15° + i\sin 15°)$, $2^{\frac{1}{6}}(\cos 135° + i\sin 135°)$, $2^{\frac{1}{6}}(\cos 255° + i\sin 255°)$ **101.** $\{-1, i, -i\}$

Critical Thinking/Discussion/Writing:

103. a. True **b.** False **c.** True **d.** False **e.** True **f.** True **g.** True **h.** True **i.** False **j.** True

Maintaining Skills:

105. 11 **107.** 2 **109.** -3 **111.** $x + 2y = 9$ **113.** $a = -8, b = 2$

Review Exercises

1. $B = 60°, b \approx 10.4, c = 12$ **3.** $B = 53°, a \approx 3.0, c \approx 5.0$

5. $B = 50°, a \approx 7.7, b \approx 9.2$ **7.** $A \approx 31.0°, B \approx 59.0°, c \approx 5.8$

9. $A \approx 41.8°, B \approx 48.2°, b \approx 4.5$

11. $C = 105°, a \approx 66.5, b \approx 59.4$

13. No triangle exists

15. $B = 74.2°, a \approx 36.8, c \approx 41.4$

17. $B \approx 54.1°, C \approx 60.7°, c \approx 20.5$
19. Answers will vary. Sample answers: **i.** 18, **ii.** $10\sqrt{3}$, **iii.** 10
21. $A \approx 26.6°, B \approx 42.1°, C \approx 111.3°$
23. $A \approx 51.7°, C \approx 48.3°, b \approx 50.2$
25. $A \approx 32.5°, B \approx 49.8°, C \approx 97.7°$
27. $A \approx 32.0°, B \approx 18.0°, c \approx 17.3$
29. 16 square meters **31.** 11 square feet **33.** $-\mathbf{i} + 2\mathbf{j}$ **35.** $\langle -10, 15 \rangle$
37. $\langle -16, 21 \rangle$ **39.** $\frac{\sqrt{2}}{2}\mathbf{i} + \frac{\sqrt{2}}{2}\mathbf{j}$ **41.** $\left\langle \frac{3\sqrt{34}}{34}, -\frac{5\sqrt{34}}{34} \right\rangle$
43. $3\sqrt{3}\mathbf{i} + 3\mathbf{j}$ **45.** $-6\sqrt{2}\mathbf{i} - 6\sqrt{2}\mathbf{j}$
47. -6; $\text{proj}_{\mathbf{w}}\mathbf{v} = \left\langle -\frac{18}{25}, -\frac{24}{25} \right\rangle$ **49.** 0; $\text{proj}_{\mathbf{w}}\mathbf{v} = \langle 0, 0 \rangle$
51. 60.3° **53.** 101.3° **55.** (24, 30°) $(12\sqrt{3}, 12)$

57. $\left(-2, -\frac{\pi}{4}\right)$ $(-\sqrt{2}, \sqrt{2})$ **59.** $\left(2\sqrt{2}, \frac{3\pi}{4}\right)$

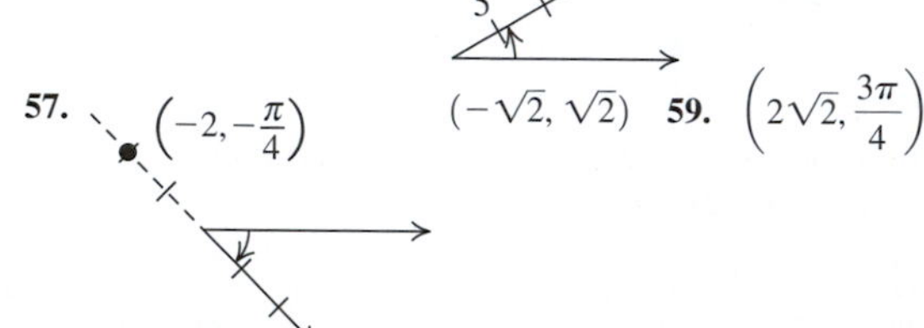

61. $\left(4, \frac{11\pi}{6}\right)$ **63.** $r = \frac{12}{3\cos\theta + 2\sin\theta}$ **65.** $r = 8\cos\theta$
67. $x^2 + y^2 = 9$ **69.** $y = 3$ **71.** $(x^2 + y^2 + 2y)^2 = x^2 + y^2$
73. $3\left(\cos\frac{3\pi}{2} + i\sin\frac{3\pi}{2}\right)$ **75.** $10\left(\cos\frac{11\pi}{6} + i\sin\frac{11\pi}{6}\right)$
77. $\sqrt{2} + \sqrt{2}i$ **79.** $-3\sqrt{2} + 3\sqrt{2}i$
81. $z_1z_2 = 6(\cos 35° + i\sin 35°)$; $\frac{z_1}{z_2} = \frac{3}{2}(\cos 15° + i\sin 15°)$
83. $z_1z_2 = 6\left(\cos\frac{7\pi}{6} + i\sin\frac{7\pi}{6}\right)$; $\frac{z_1}{z_2} = \frac{2}{3}\left(\cos\frac{\pi}{2} + i\sin\frac{\pi}{2}\right)$
85. $27(\cos 120° + i\sin 120°)$ **87.** $4096(\cos 0 + i\sin 0)$
89. $5(\cos 60° + i\sin 60°)$, $5(\cos 180° + i\sin 180°)$, $5(\cos 300° + i\sin 300°)$ **91.** $2^{\frac{1}{5}}(\cos 24° + i\sin 24°)$, $2^{\frac{1}{5}}(\cos 96° + i\sin 96°)$, $2^{\frac{1}{5}}(\cos 168° + i\sin 168°)$, $2^{\frac{1}{5}}(\cos 240° + i\sin 240°)$, $2^{\frac{1}{5}}(\cos 312° + i\sin 312°)$ **93.** 268.5 ft
95. 94.7 mi **97.** 24,625 square feet **99.** 1204.4 foot-pounds

Practice Test A

1. $B = 30°, a \approx 23.7, b \approx 13.1$
2. $C = 101°, b \approx 45.0$ m, $c \approx 73.4$ m
3. $A \approx 28.2°, C \approx 45.8°, b \approx 71.1$
4. $A \approx 82.8°, B \approx 41.4°, C \approx 55.8°$
5. $c \approx 6.9, A \approx 53.8°, B \approx 36.2°$
6. 32.7° **7.** 803 ft **8.** $\langle -1, -12 \rangle$ **9.** $\langle -3, -2 \rangle$ **10.** $-7\mathbf{i} - 9\mathbf{j}$
11. $\frac{3\sqrt{3}}{2}\mathbf{i} - \frac{3}{2}\mathbf{j}$ **12.** 17 **13.** 164.9° **14.** $(-\sqrt{2}, \sqrt{2})$
15. $\left(2, \frac{7\pi}{6}\right)$ **16.** $x^2 + y^2 = 9$; circle with center $(0, 0)$ and radius 3
17. $-\frac{3}{2} + \frac{3\sqrt{3}}{2}i$ **18.** $\sqrt{2}(\cos 165° + i\sin 165°)$ **19.** $\frac{1}{125}i$
20. $2^{\frac{1}{8}}(\cos 11.25° + i\sin 11.25°)$, $2^{\frac{1}{8}}(\cos 101.25° + i\sin 101.25°)$, $2^{\frac{1}{8}}(\cos 191.25° + i\sin 191.25°)$, $2^{\frac{1}{8}}(\cos 281.25° + i\sin 281.25°)$

Practice Test B

1. a **2.** c **3.** c **4.** d **5.** a **6.** b **7.** d **8.** c **9.** a **10.** c **11.** a
12. d **13.** d **14.** b **15.** c **16.** b **17.** b **18.** c **19.** c **20.** c

Cumulative Review Exercises (Chapters P–7)

1. $(0, 2) \cup (2, \infty)$ **3.**

5. $\left\{\frac{2\pi}{3}, \frac{5\pi}{6}, \frac{5\pi}{3}, \frac{11\pi}{6}\right\}$ **7.** $2^{\frac{1}{4}}(\cos 7.5° + i\sin 7.5°)$, $2^{\frac{1}{4}}(\cos 97.5° + i\sin 97.5°)$, $2^{\frac{1}{4}}(\cos 187.5° + i\sin 187.5°)$, $2^{\frac{1}{4}}(\cos 277.5° + i\sin 227.5°)$ **9.** $y = \frac{3}{2}x + \frac{13}{2}$

11. **13.**

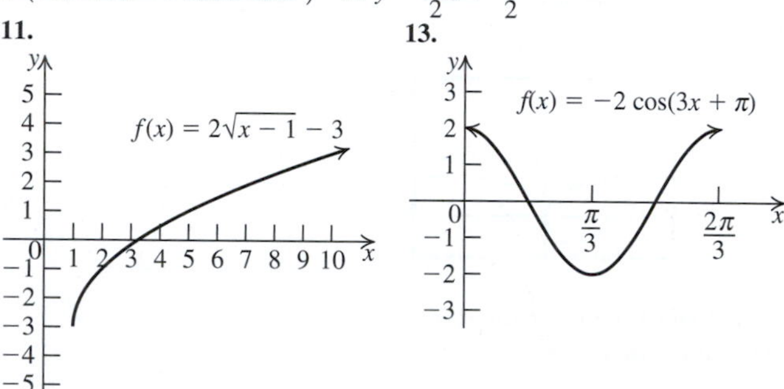

15. $-\frac{12}{5}$ **19.** $C = 69°, a \approx 58.2$ m, $b \approx 46.2$ m

CHAPTER 8

Section 8.1

Answers to Practice Problems:

1. a. No **b.** Yes **2.** $\{(-1, 3)\}$ **3.** $\{(4, -1)\}$ **4.** $\varnothing$
5. $\{(x, 2x - 3)\}$ **6.** $\left\{\left(\frac{2}{3}, \frac{1}{2}\right)\right\}$ **7.** $\{(5700, 31.4)\}$
8. \$85,000 at 12%, \$65,000 at 8%

Basic Concepts and Skills:

1. solution **3.** inconsistent **5.** false **7.** $(3, -1)$
9. None of them **11.** $(10, -9)$
13. $\{(2, 1)\}$ **15.** $\{(2, 2)\}$

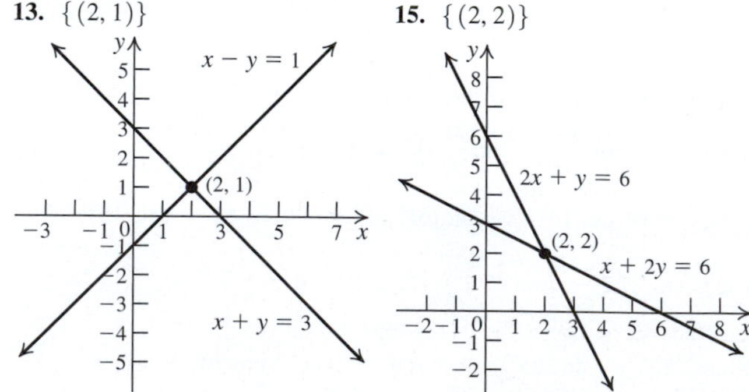

17. Inconsistent

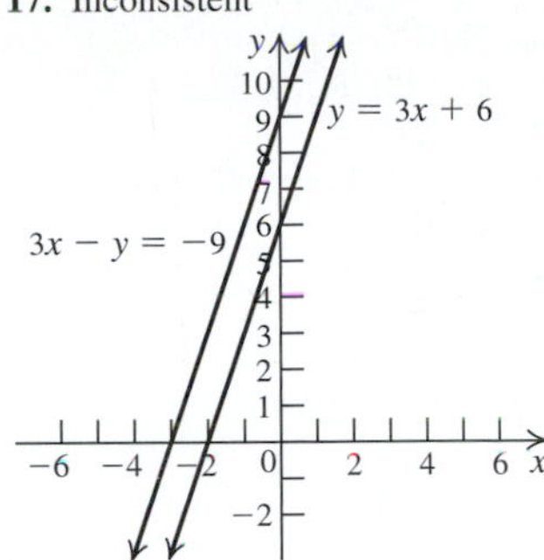

19. $\left\{\left(\frac{7}{3}, \frac{14}{3}\right)\right\}$

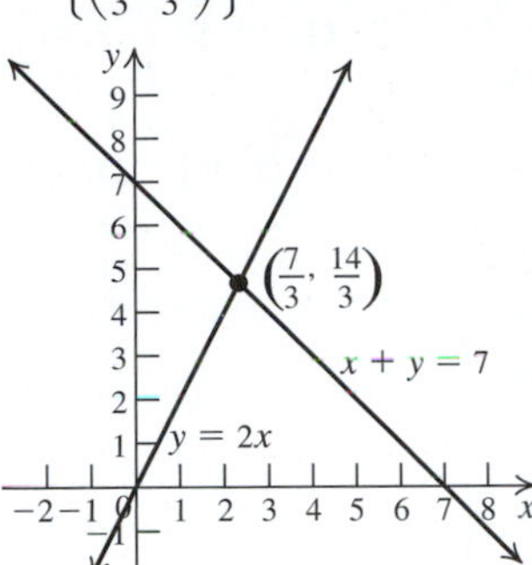

21. $\{(x, 12 - 3x)\}$

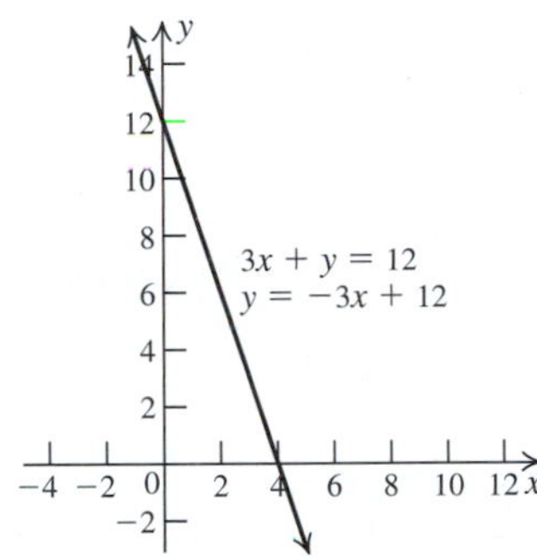

23. Consistent; Independent **25.** Consistent; Independent **27.** Consistent; Dependent **29.** Consistent; Independent **31.** Inconsistent **33.** Inconsistent **35.** Consistent; Independent **37.** $\left\{\left(\frac{7}{9}, \frac{23}{9}\right)\right\}$ **39.** $\{(3, 4)\}$ **41.** $\varnothing$ **43.** $\{(-3, 5)\}$ **45.** $\left\{\left(x, \frac{1}{2}(x-5)\right)\right\}$ **47.** $\{(3, 2)\}$ **49.** $\{(-3, 3)\}$ **51.** $\{(0, -5)\}$ **53.** $\varnothing$ **55.** $\left\{\left(x, 2 - \frac{2}{3}x\right)\right\}$ **57.** $\{(4, 1)\}$ **59.** $\{(-4, 2)\}$ **61.** $\{(2, 1)\}$ **63.** $\{(3, 1)\}$ **65.** $\left\{\left(\frac{19}{6}, \frac{5}{4}\right)\right\}$ **67.** $\{(4, 2)\}$ **69.** $\{(4, 1)\}$ **71.** $\varnothing$ **73.** $\left(x, \frac{2}{3}(x-3)\right)$ **75.** $\left\{\left(-\frac{1}{5}, 1\right)\right\}$ **77.** $\left\{\left(\frac{3}{2}, \frac{1}{2}\right)\right\}$ **79.** $\{(5, 5)\}$ **81.** $\varnothing$

Applying the Concepts:
83. $(80, 30)$ **85.** $(17, 4)$ **87.** The diameter of the largest pizza is 21 inches, and the diameter of the smallest pizza is 8 inches. **89.** 8% of the total trash collected is plastic, and 40% is paper. **91.** 340 beads and 430 doubloons **93.** 3 Egg McMuffins and 2 Breakfast Burritos **95.** \$18,000 at 7.5% and \$32,000 at 12% **97.** \$7.50 per hour at McDougal's and \$15 per hour for tutoring **99.** 20 pounds of herb and 80 pounds of tea **101.** The speed of the plane is 550 km/hr, and the wind speed is 50 km/hr. **103. a.** $C(x) = 30{,}000 + 2x$; $R(x) = 3.5x$

b.

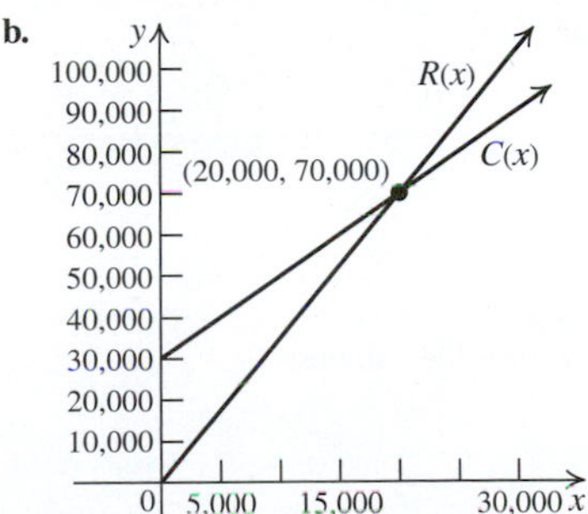

c. 20,000 magazines

105. Her weekly sales should be more than \$6250. **107.** $(3, 9)$ **109.** $(\ln 4, \ln 2)$ **111.** $c = 2$ **113.** $a = 4, b = 0$

115. $y = -\frac{3}{2}x + \frac{3}{2}$ **117.** $y = \frac{13}{7}x + \frac{6}{7}$ **119. a.** $3y - 4x = 4$ **b.** $\left(\frac{4}{5}, \frac{12}{5}\right)$ **c.** 7 **121. a.** $\sqrt{2}$ **b.** $\frac{6\sqrt{5}}{5}$ **c.** 0 **d.** $\frac{|c|}{\sqrt{a^2 + b^2}}$ **123.** $(5, 2)$

Maintaining Skills:
125. 3 **127.** 4 **129.** −2

Section 8.2

Answers to Practice Problems:
1. Yes: $(2) + (-3) + (2) = 1$
$3(2) + 4(-3) + (2) = -4$
$2(2) + (-3) + 2(2) = 5$
2. $\{(-2, 1, 3)\}$ **3.** $\varnothing$
4. $\left\{\left(8 - \frac{7}{3}z, \frac{4}{3}z - 3, z\right)\right\}$ **5.** $\{(13 - 17z, -3 + 5z, z)\}$
6. Cell *A* contains bone (because $x = 0.4$), cell *B* contains healthy tissue (because $y = 0.25$), and cell *C* contains tumorous tissue (because $z = 0.3$).

Basic Concepts and Skills:
1. equivalent **3.** inconsistent **5.** true **7.** Yes **9.** No **11.** $\{(3, 2, -1)\}$ **13.** $\{(-3, 2, 4)\}$ **15.** $\begin{cases} x - y - \frac{3}{2}z = \frac{1}{2} \\ y + z = 0 \\ z = 1 \end{cases}$
17. $\begin{cases} x + 3y - 2z = 0 \\ y - \frac{8}{7}z = -\frac{5}{7} \\ 0 = k, k \neq 0 \end{cases}$ **19.** $\{(1, -2, 3)\}$
21. $\left\{\left(\frac{30}{11}, -\frac{8}{11}, -\frac{1}{4}\right)\right\}$ **23.** $\{(1, 2, 3)\}$ **25.** $\left\{\left(\frac{35}{18}, \frac{29}{18}, \frac{5}{18}\right)\right\}$ **27.** Inconsistent **29.** $\{(3, -1, 2)\}$ **31.** $\{(2, 2, 1)\}$ **33.** $\{(3, -1, 1)\}$ **35.** $\left\{\left(3 + \frac{4}{5}z, -\frac{1}{5}z, z\right)\right\}$ **37.** $\{(12, -12, 4)\}$ **39.** Dependent: $\{(z - 2, 8 - 2z, z)\}$ **41.** $\{(3, 2, -1)\}$ **43.** Dependent; $\left\{\left(-5 - 9z, 3z - \frac{1}{6}, z\right)\right\}$ **45.** \$4000 invested at 4%, \$6000 invested at 5%, and \$10,000 invested at 6% **47.** Alex worked 2 hours, Becky worked 2.5 hours, and Courtney worked 1.5 hours. **49.** 56 nickels, 225 dimes, and 19 quarters **51.** 31 hours for normal daytime work, 14 hours at night, and 8 hours on a holiday

Beyond the Basics:
57. $x + y + z = 1$ **59.** $x + \frac{2}{3}y + \frac{1}{3}z = \frac{1}{3}$ **61.** $y = x^2 + 2x + 1$ **63.** $y = x^2 - x + 2$ **65.** $x^2 + y^2 - 16 = 0$ **67.** $x^2 + y^2 - 2x + 6y - 15 = 0$ **69.** $\left\{\left(-\frac{9}{14}, \frac{9}{19}, -\frac{9}{2}\right)\right\}$ **71.** $c = -\frac{16}{5}$ **73.** $y = -2x^2 + 4x + 5$

Applying the Concepts:
47. Alex worked 2 hours, Becky worked 2.5 hours, and Courtney worked 1.5 hours. **53.** Cell *A* contains healthy tissue (because $x = 0.21$), cell *B* contains tumorous tissue (because $y = 0.33$), and cell *C* contains healthy tissue (because $z = 0.19$). **55.** Cells *A*, *B*, and *C* contain healthy tissue (because $x = 0.28$, $y = 0.23$, and $z = 0.21$).

Critical Thinking/Discussion/Writing:
75. $\begin{cases} x + y - z = -2 \\ 2x - y + 3z = 9 \\ x + y + z = 2 \end{cases}$ Answers may vary.

Maintaining Skills:
77. −1 **79.** 1 **81.** $A = \frac{5}{4}$ $B = \frac{3}{4}$ **83.** $(x + 2)(x + 3)$ **85.** $(2x - 1)(x + 3)$ **87.** $(2x - 3)(2x + 3)$

Section 8.3

Answers to Practice Problems:

1. $\frac{3}{x+1} - \frac{1}{x-2}$ **2.** $\frac{5}{x-1} - \frac{3}{x} + \frac{1}{x+1}$

3. $\frac{5}{x} + \frac{6}{(x-1)^2} - \frac{5}{x-1}$ **4.** $\frac{-1}{x} + \frac{4x+5}{x^2+2}$

5. $\frac{1}{x^2+1} + \frac{3x}{(x^2+1)^2}$ **6.** $\frac{777}{3112}$

7. $\frac{1}{R} = \frac{3}{x+1} + \frac{1}{x+2} = \frac{1}{\frac{x+1}{3}} + \frac{1}{x+2}$;

$R_1 = \frac{x+1}{3}, R_2 = x+2$

Basic Concepts and Skills:

1. proper **3.** three **5.** false **7.** $\frac{A}{x-1} + \frac{B}{x+2}$

9. $\frac{A}{x+6} + \frac{B}{x+1}$ **11.** $\frac{A}{x^2} + \frac{B}{x} + \frac{C}{x-1}$

13. $\frac{A}{x+1} + \frac{Bx+C}{x^2-x+1}$ **15.** $\frac{Ax+B}{x^2+1} + \frac{Cx+D}{(x^2+1)^2}$

17. $\frac{3}{x+2} - \frac{1}{x+1}$ **19.** $-\frac{1}{2(x+3)} + \frac{1}{2(x+1)}$

21. $\frac{1}{x} - \frac{1}{x+2}$ **23.** $\frac{2}{x+2} - \frac{3}{2(x+3)} - \frac{1}{2(x+1)}$

25. $\frac{1}{x+1} - \frac{2}{(x+1)^2}$ **27.** $\frac{2}{x+1} - \frac{3}{(x+1)^2} + \frac{1}{(x+1)^3}$

29. $\frac{1}{x} + \frac{3}{(x+1)^2} - \frac{2}{x+1}$ **31.** $-\frac{2}{x} + \frac{1}{x^2} + \frac{1}{(x+1)^2} + \frac{2}{x+1}$

33. $\frac{1}{4(x-1)^2} + \frac{1}{4(x+1)^2} + \frac{1}{4(x+1)} - \frac{1}{4(x-1)}$

35. $\frac{1}{2(2x-3)} + \frac{1}{2(2x-3)^2}$ **37.** $-\frac{2}{(2x+3)^2} + \frac{3}{2x+3}$

39. $\frac{4}{x} - \frac{3}{x^2} - \frac{4}{x+1}$ **41.** $\frac{3}{x+1} - \frac{2}{x} + \frac{4}{x^2}$

43. $-\frac{x}{2(x^2+1)} + \frac{1}{4(x+1)} + \frac{1}{4(x-1)}$

45. $-\frac{x}{x^2+1} + \frac{1}{x} - \frac{x}{(x^2+1)^2}$ **47.** $\frac{4-3x}{x^2+2} + \frac{3x-2}{x^2+1}$

Applying the Concepts:

49. $\frac{n}{n+1}$ **51.** $\frac{2n}{2n+1}$ **53.** $\frac{1}{R} = \frac{1}{x+1} + \frac{1}{x+3}$

55. $\frac{1}{R} = \frac{1}{R_1} + \frac{1}{R_2} + \frac{1}{R_3}$

Beyond the Basics:

59. $\frac{1}{(x-1)^2} - \frac{1}{(x+1)^2}$ **61.** $-\frac{1}{2(x-1)} + \frac{x-1}{2(x^2+1)} + \frac{1}{(x-1)^2}$

63. $-\frac{4}{x+2} - \frac{8}{(x+2)^2} - \frac{1}{(x+1)^2} + \frac{5}{x+1}$

65. $\frac{5}{3(x-3)} + \frac{-5x+15}{3(x^2+3x+9)}$ **67.** $\frac{1}{x^2-2x+2} - \frac{1}{x^2+2x+2}$

69. $1 + \frac{2}{x+1} - \frac{1}{x+2}$ **71.** $2x + 3 + \frac{2x-1}{x^2+1} + \frac{1}{x-1}$

Maintaining Skills:

75. $-7, -3$ **77.** $-\frac{5}{2}, 2$ **79.** $(1, 2)$ **81.** $(5, 2)$

83. $(3, 2)$ **85.** $(-2, 1)$

Section 8.4

Answers to Practice Problems:

1. $\{(1, 1)\}$ **2.** $\{(4, 3), (-4, 3), (4, -3), (-4, -3)\}$

3. Danielle received $x = 30$ shares as dividends and sold her stock at $p = \$65$ per share.

Basic Concepts and Skills:

1. at least one **3.** intersection **5.** true **7.** $(3, -1)$

9. $(1, -1)$ **11.** $(5, 4), (-5, 4), (-5, -4)$ **13.** $(1, 1)$

15. $\{(-1, 1), (2, 4)\}$ **17.** $\{(3, 3), (-2, -2)\}$

19. $\{(3, 0)\}$ **21.** $\{(-2, 1), (-1, 2)\}$ **23.** $\{(1, -1), (3, 1)\}$

25. $\{(-3, -1), (1, 3)\}$ **27.** $\{(2, 3), (0, 5)\}$

29. $\{(7, 5), (-7, -5)\}$ **31.** $\{(4, 2), (-4, 2), (4, -2), (-4, -2)\}$

33. $\{(2, 2), (-2, 2), (2, -2), (-2, -2)\}$

35. $\varnothing$ **37.** $\{(1, 2), (-1, 2), (1, -2), (-1, -2)\}$

39. $\left\{(2, 1), (2, -1), \left(-\frac{11}{3}, -\frac{\sqrt{26}}{3}\right), \left(-\frac{11}{3}, -\frac{\sqrt{26}}{3}\right)\right\}$

41. $\{(5, 3), (3, 5)\}$ **43.** $\varnothing$ **45.** $\{(0, -2), (3, 4)\}$

47. $\left\{(3, 2), \left(-4, -\frac{3}{2}\right)\right\}$ **49.** $\{(-2, 1), (2, 1), (-2, -1), (2, -1)\}$

51. $\left\{(2, 5), \left(-7, -\frac{10}{7}\right)\right\}$ **53.** $\left\{(2, -2), (2, 2), \left(\frac{16}{5}, \frac{2\sqrt{46}}{5}\right), \left(\frac{16}{5}, -\frac{2\sqrt{46}}{5}\right)\right\}$

Applying the Concepts:

55. $(3, 78)$ **57.** 11 and 13 **59. a.** $x = 60$ m; $y = 140$ m **b.** 6000 m^2 **61.** 32 people; \$30 **63.** 450 original shares; purchased at \$22 per share

Beyond the Basics:

65. $A(1, 0), B(6, 0), d(A, B) = 5$

67. $A(-2, -1), B(2, 3), d(A, B) = 4\sqrt{2}$ **69.** $(0, -3)$

71. $c = \pm\sqrt{2}$ **73.** $c = \pm 25$ **75.** $\left\{\left(-\frac{1}{3}, -\frac{1}{8}\right), \left(\frac{1}{8}, \frac{1}{3}\right)\right\}$

77. $\{(-1, 1), (1, 1), (-1, -1), (1, -1)\}$ **79.** $\pm(2 + 3i)$

81. $\pm(4 - 3i)$ **83.** $\pm(4 + \sqrt{3}i)$

85. $\{(1, 7)\}$ **87.** $\{(1, 5)\}$

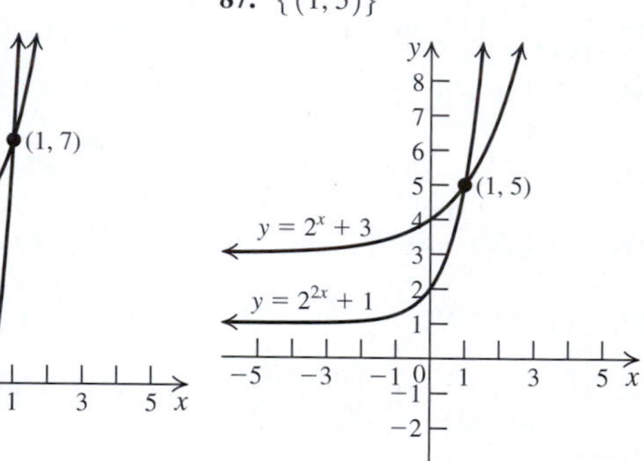

89. 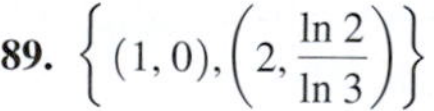 $\left\{(1, 0), \left(2, \frac{\ln 2}{\ln 3}\right)\right\}$

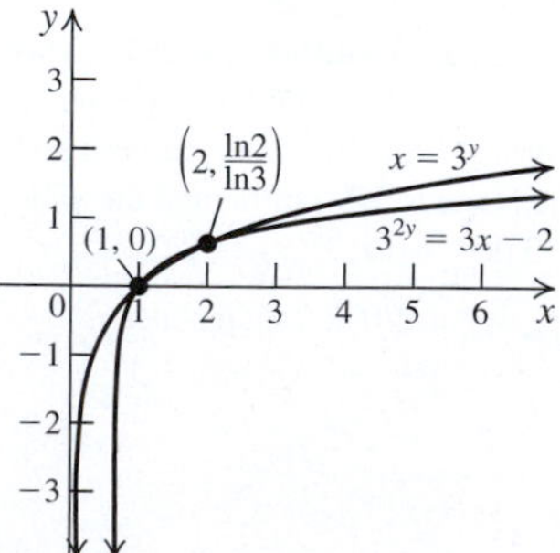

Critical Thinking/Discussion/Writing:

91. a. Not possible **b.** Possible **c.** Possible **d.** Possible **e.** Not possible **f.** Not possible

93.

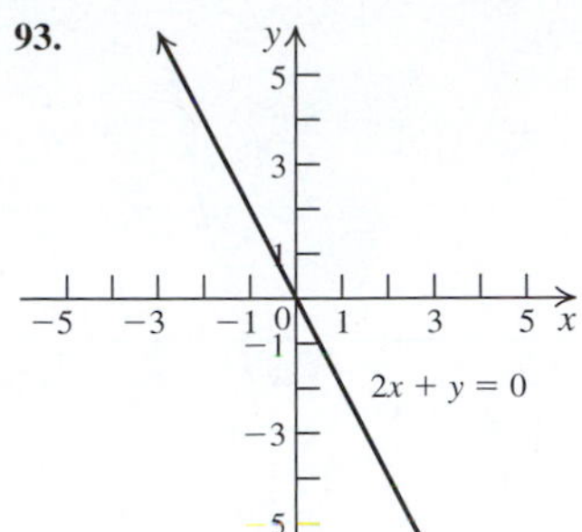

95.

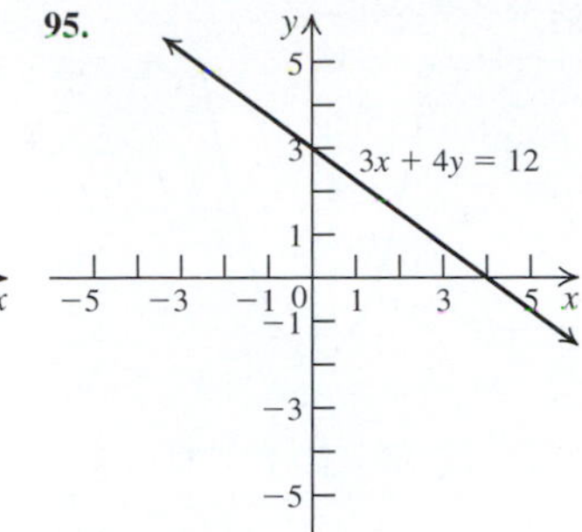

97.

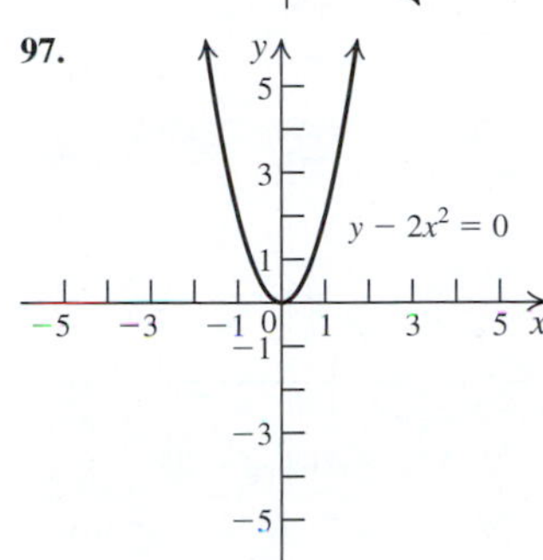

99.

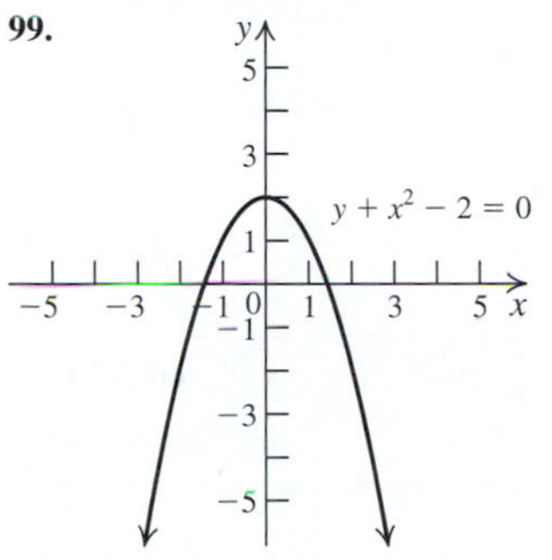

Maintaining Skills:

101. $(0, 0), (-3, 5)$ **103.** $(1, 4), (-1, 4)$

Section 8.5

Answers to Practice Problems:

1.

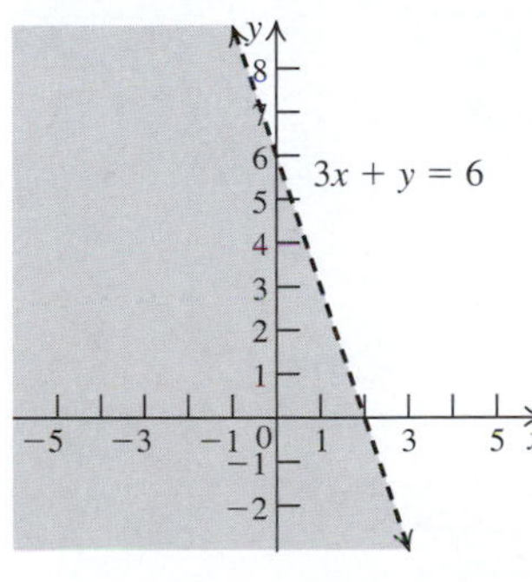

2.

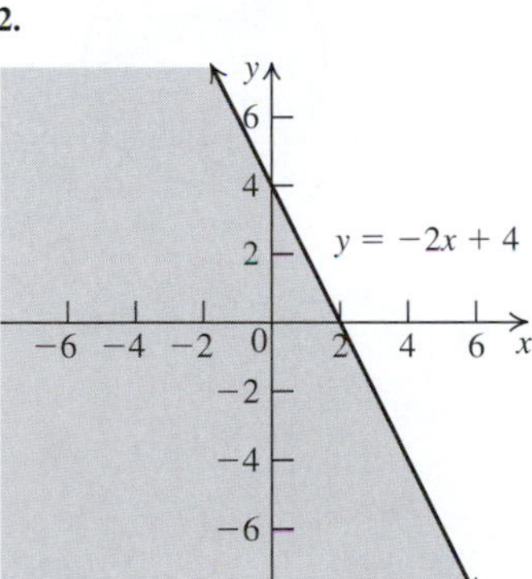

3.

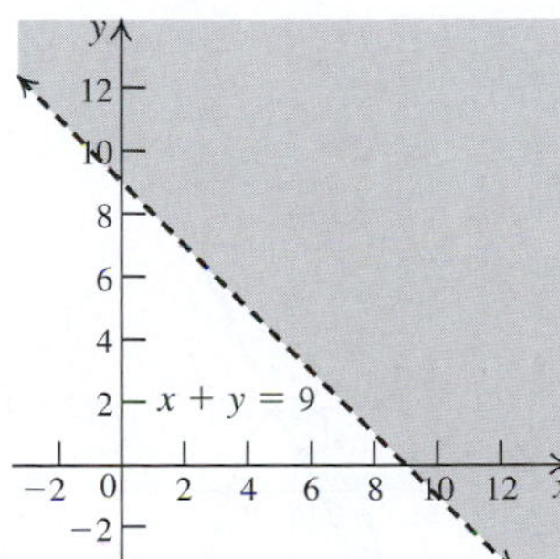

4.

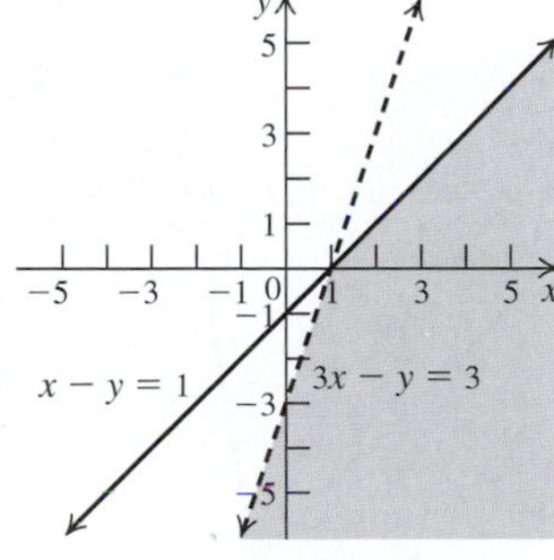

5. $(4, 0), (5, 2)$, and $(2, 4)$

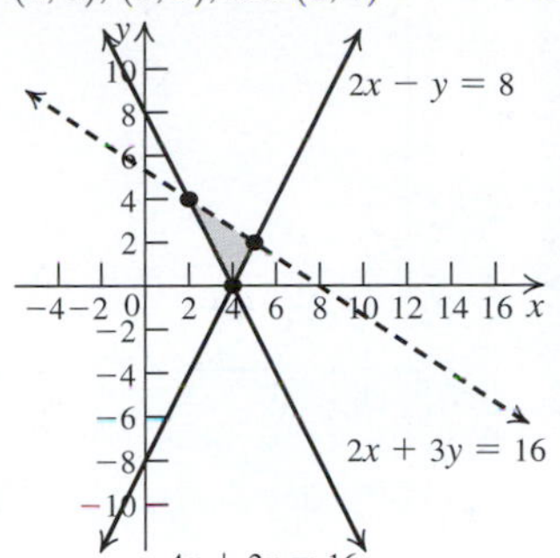

6.

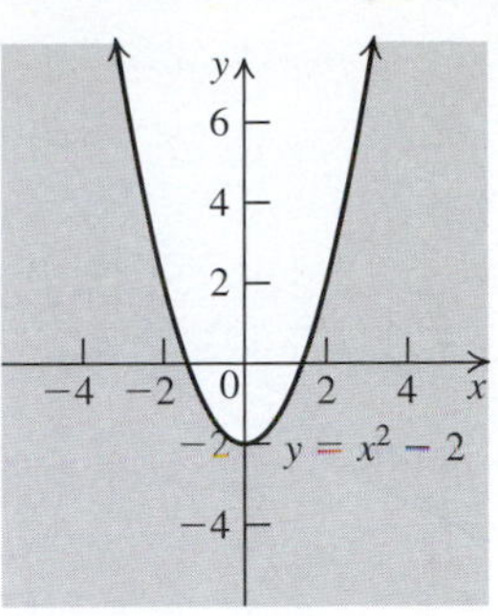

7.

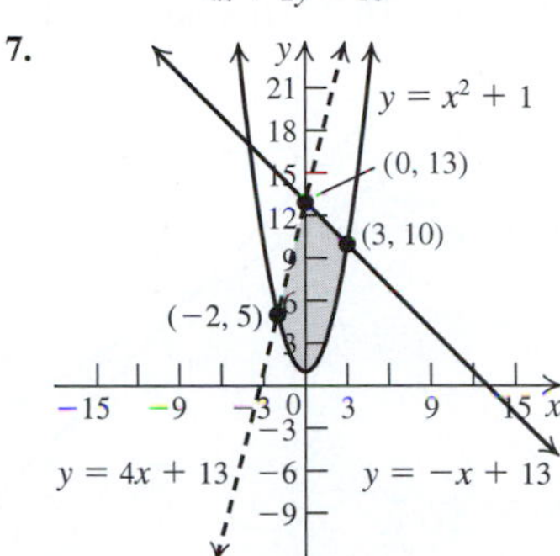

Basic Concepts and Skills:

1. dashed **3.** not a solution of the system **5.** true

7.

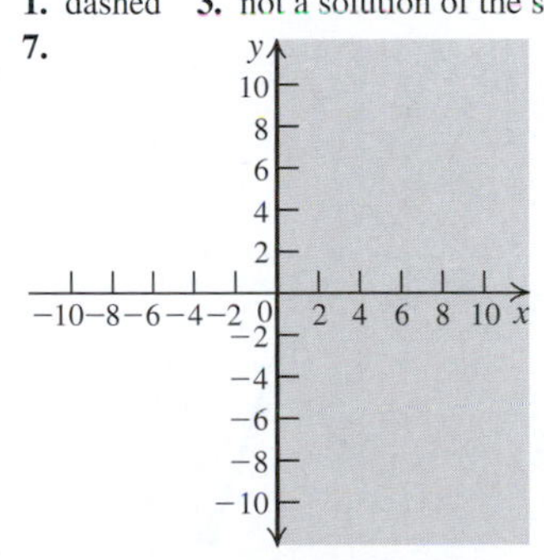

9.

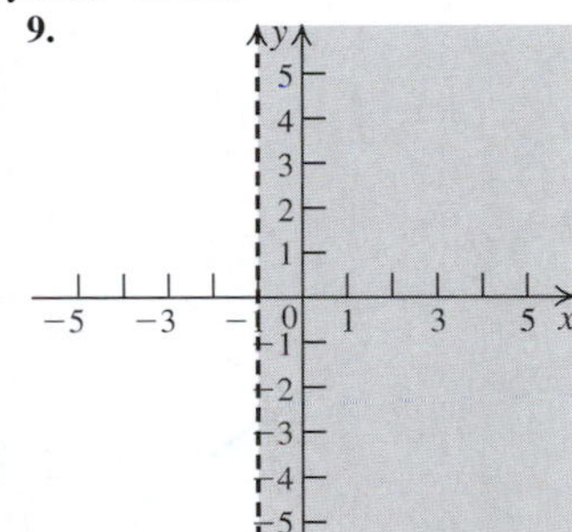

11.

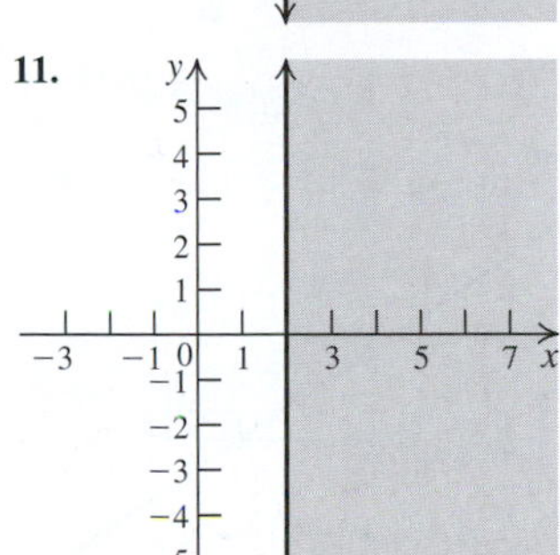

13.

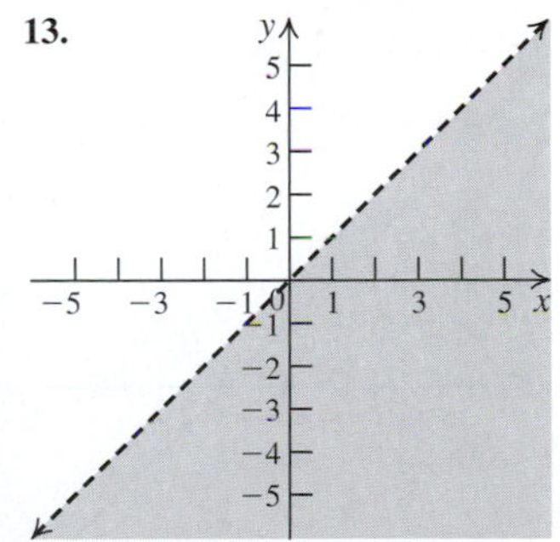

15.

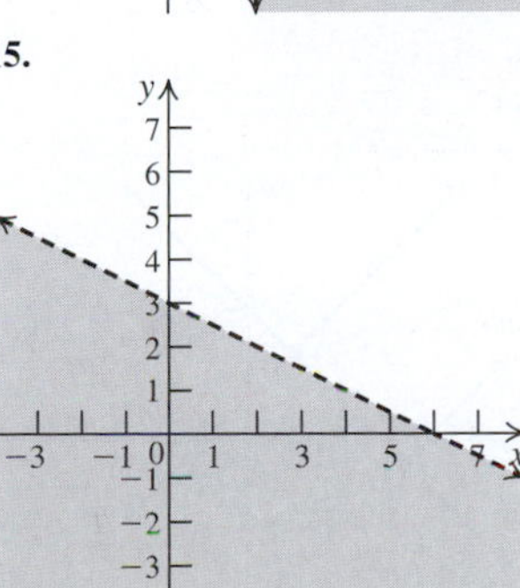

17.

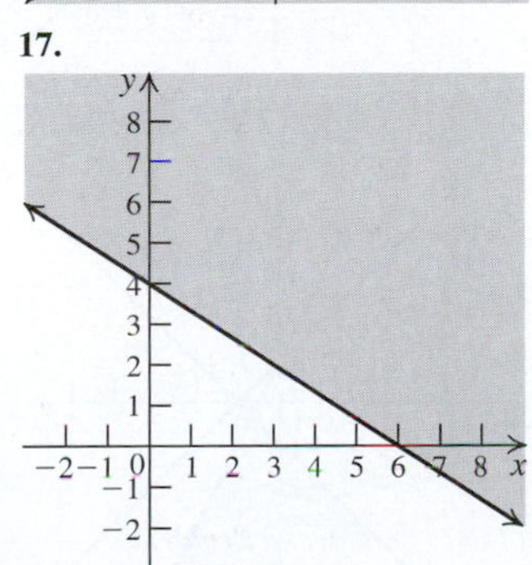

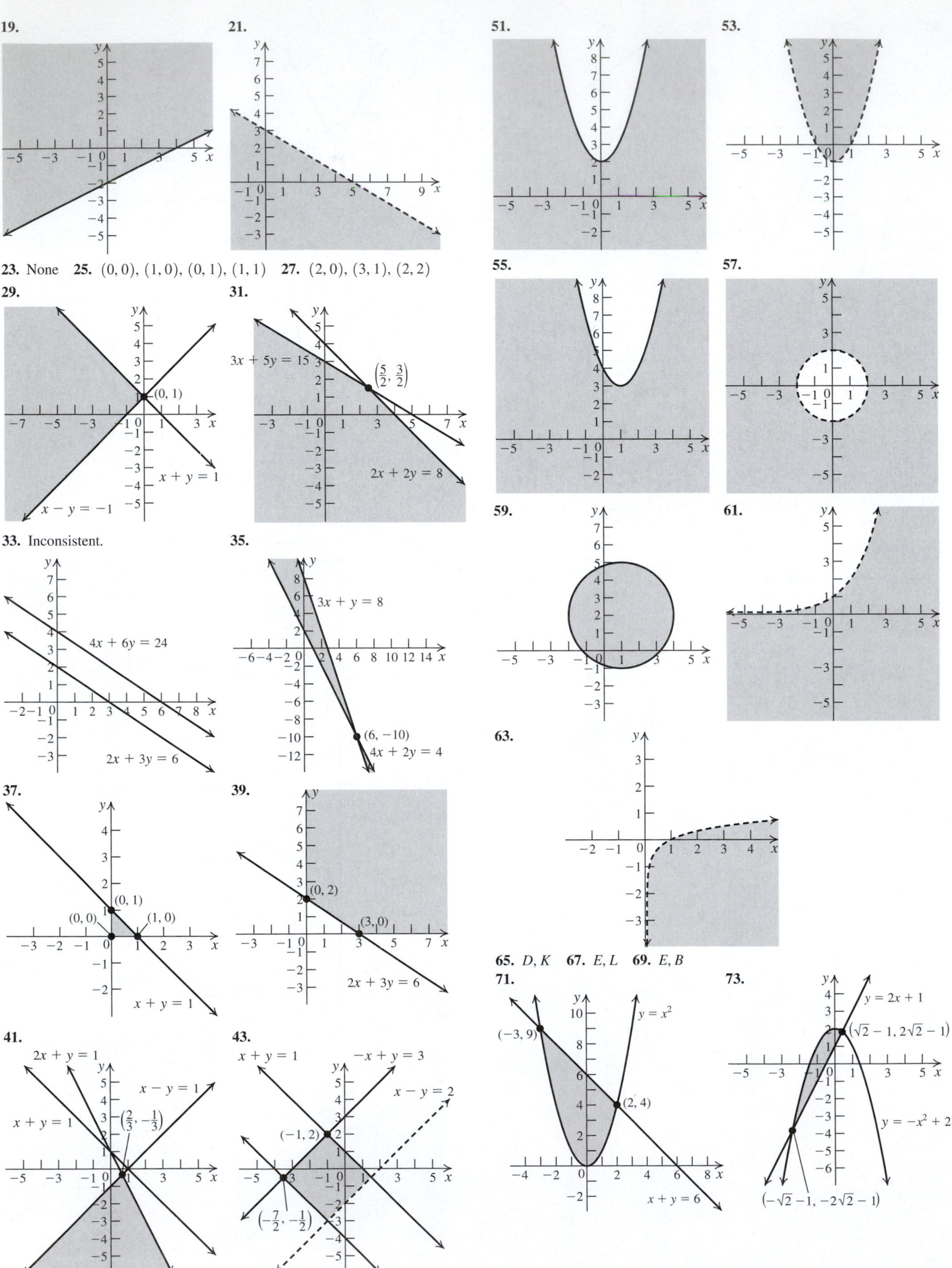

23. None **25.** (0, 0), (1, 0), (0, 1), (1, 1) **27.** (2, 0), (3, 1), (2, 2)

33. Inconsistent.

65. *D*, *K* **67.** *E*, *L* **69.** *E*, *B*

45. *A* **47.** *B* **49.** *E*

75. **77.**

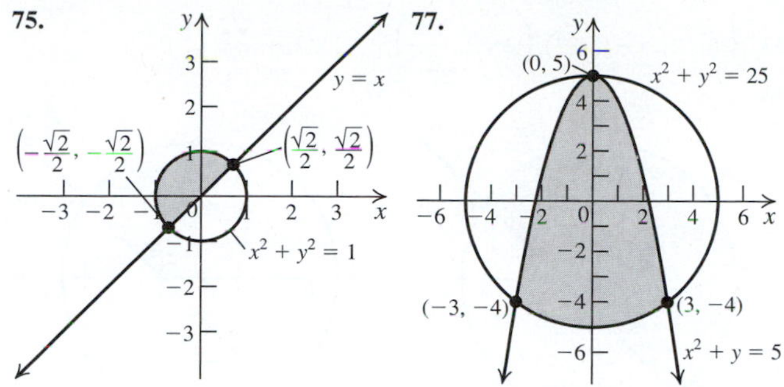

Applying the Concepts:

79.

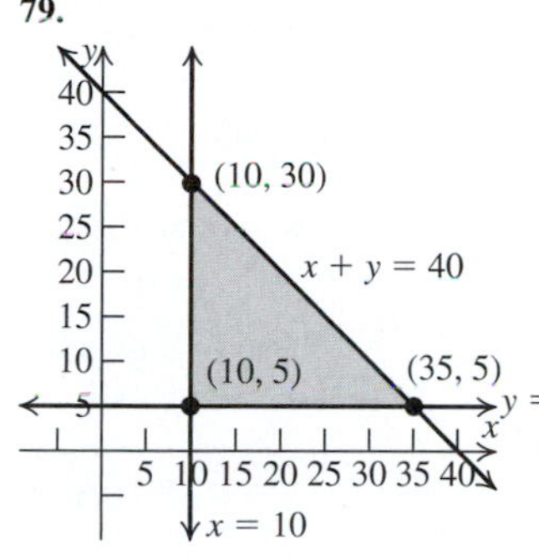

81.

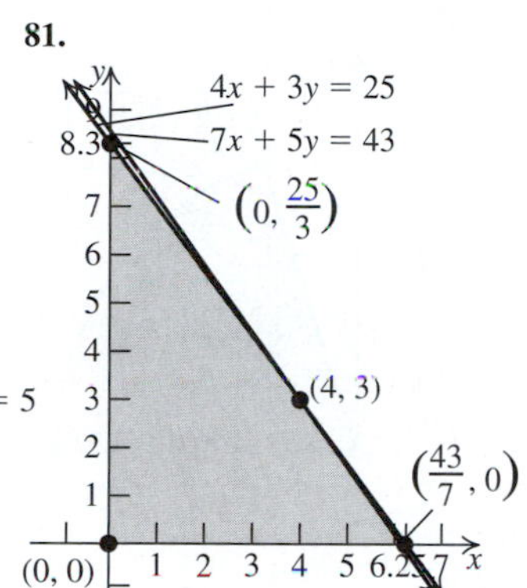

83.

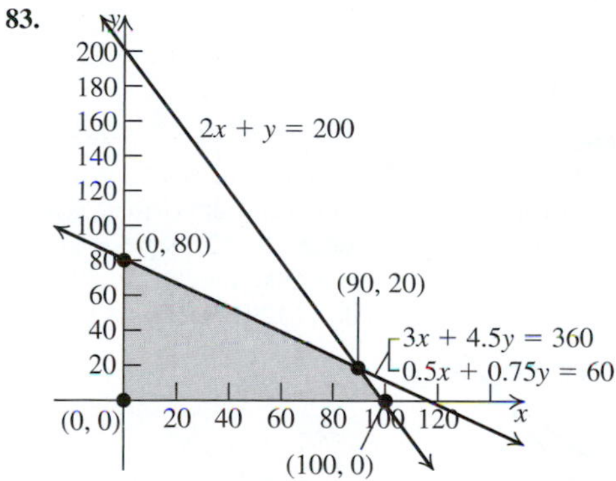

Beyond the Basics:

85. $\begin{cases} x \geq 0 \\ y \geq 0 \\ y \leq \frac{-2}{3}x + 2 \end{cases}$ **87.** $\begin{cases} x \geq -1 \\ x \leq 3 \end{cases}$ **89.** $\begin{cases} x \geq 0 \\ y \geq 0 \\ y \leq -\frac{1}{2}x + 4 \\ y \leq -\frac{3}{2}x + 6 \end{cases}$

91. $\begin{cases} x \geq 0 \\ y \geq x - 4 \\ y \geq -x + 6 \\ y \leq -x + 16 \end{cases}$

93. **95.**

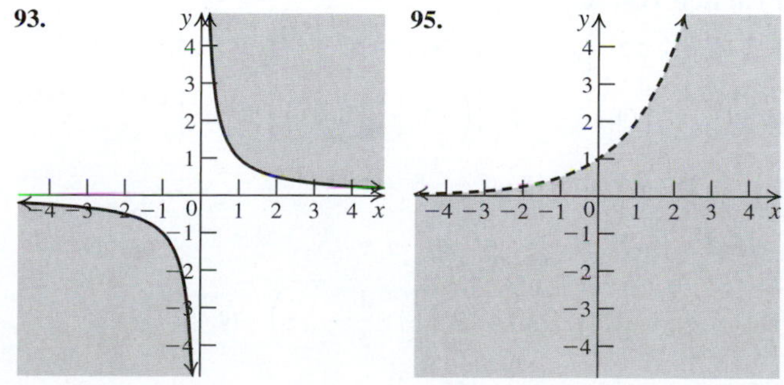

97. **99.**

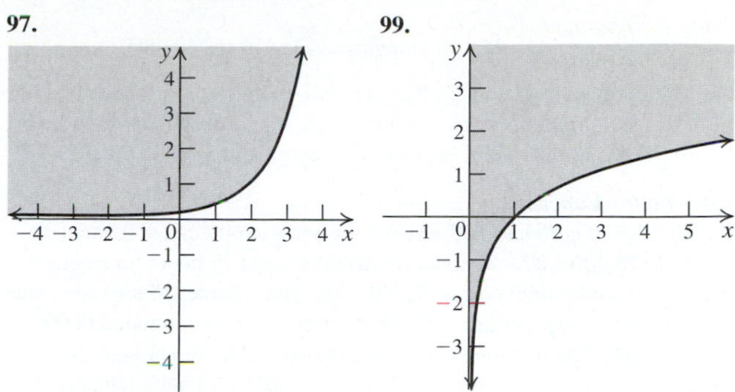

101. **103.**

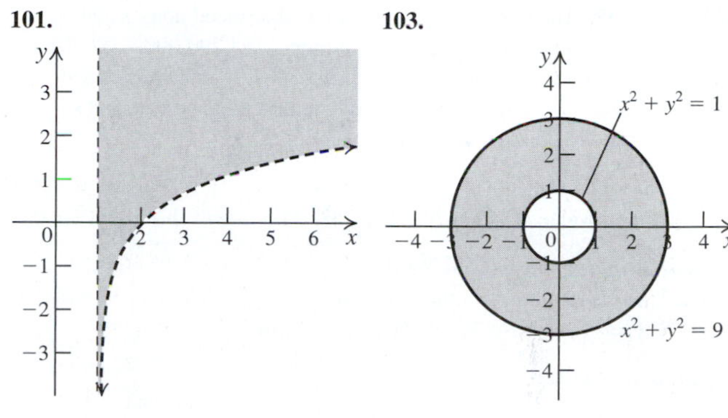

Critical Thinking/Discussion/Writing:

105.

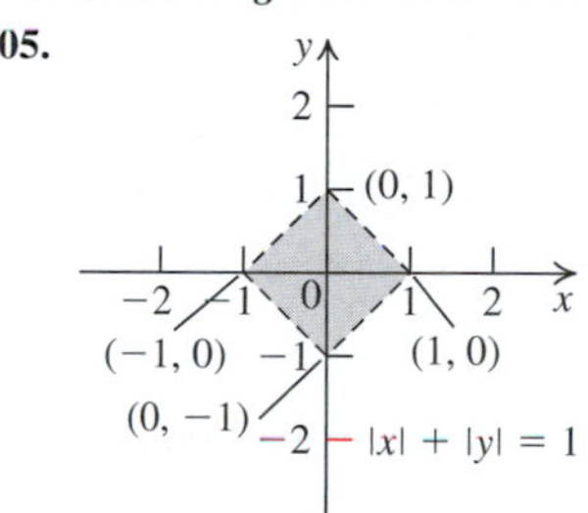

Maintaining Skills:

107. (4, 18) **109.** $\left(\frac{7}{2}, \frac{3}{4}\right)$ **111.** (2, 3), (4, 5), (6, 1)

113. a. and b.

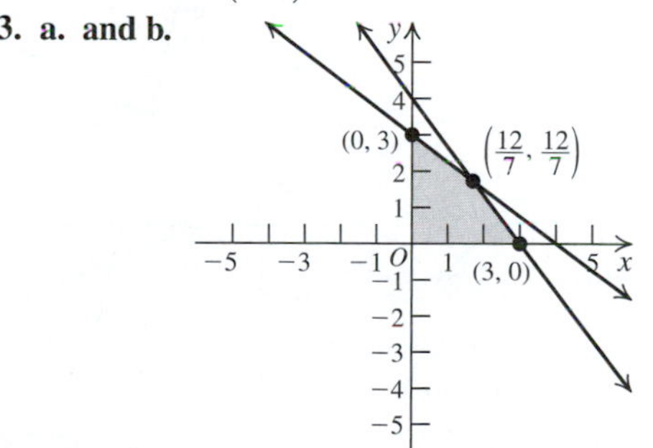

Section 8.6

Answers to Practice Problems:

1. The objective function f has a maximum of 28 at the point $(7, 0)$. **2.** The minimum value is 2 at $(4, 0)$ and $(1, 1.5)$. **3.** The minimum number of calories is 300. The lunch then consists of 10 ounces of soup and no salad.

Basic Concepts and Skills:

1. optimization **3.** constraints, feasible **5.** true **7.** Max 18, Min 5 **9.** Max 26, Min 7 **11.** Max 66, Min 13 **13.** Maximum value is $\frac{1284}{19}$ at $\left(\frac{56}{19}, \frac{60}{19}\right)$. **15.** Maximum value is 410 at $(40, 30)$. **17.** Minimum value is 3 at $(3, 0)$. **19.** Minimum value is 1364 at $(8, 84)$. **21.** Maximum value is 54 at $(3.6, 0)$. **23.** Minimum value is 400 for all (x, y) on the line segment between $(20, 15)$ and $(50, 0)$.

Applying the Concepts:
25. 80 acres of corn and 160 acres of soybeans to obtain a maximum profit of \$10,400 **27.** 40 hours for machine I and 20 hours for machine II to obtain a minimum cost of \$3600 **29.** Five minutes of television and two pages of newspaper advertisement to obtain a maximum of 340,000 viewers **31.** The maximum profit is \$14,600. **33.** 400 of both the rectangular and circular table should be made to have a maximum profit of \$2800. **35.** The contractor should build 33 terraced houses and 66 cottages to obtain a maximum amount of money: \$4,290,000. **37.** Mrs. Adams's maximum profit is \$1355 when she has 175 male guests and 125 female guests. **39.** Elisa should buy ten enchilada meals and five vegetable loafs to obtain a minimum cost of \$46.25.

Beyond the Basics:
41. Maximum value is 152 at $(12, 8)$. **43.** Minimum value is 0 at $(0, 0)$.

Critical Thinking/Discussion/Writing
47. a. There is no upper bound on the values for x. **b.** For $x_1 > x_0 \geq 0$ and $0 \leq y \leq 8$, we have $2x_1 + 3y > 2x_0 + 3y$.

Maintaining Skills:
49. $(-6, -5)$ **51.** $(-1, -5, 8)$ **53.** $\begin{bmatrix} 1 & 2 & -3 & 5 \\ 3 & -2 & 4 & 12 \end{bmatrix}$

55. $\begin{bmatrix} 1 & 2 & -3 & 5 \\ 0 & 1 & -\frac{13}{8} & \frac{3}{8} \end{bmatrix}$

Review Exercises

Basic Concepts and Skills:
1. $\{(-1, 2)\}$ **3.** $\varnothing$ **5.** $\{(2, 3)\}$ **7.** $\{(1, -1, 2)\}$ **9.** $\{(-1, 2, -3)\}$ **11.** $\varnothing$ **13.** $\{x, 2x - 8, 11 - 3x\}$ **15.** $\varnothing$

17. $\left(\frac{7}{3} - \frac{2}{3}z, \frac{2}{3} - \frac{1}{3}z, z\right)$

19.

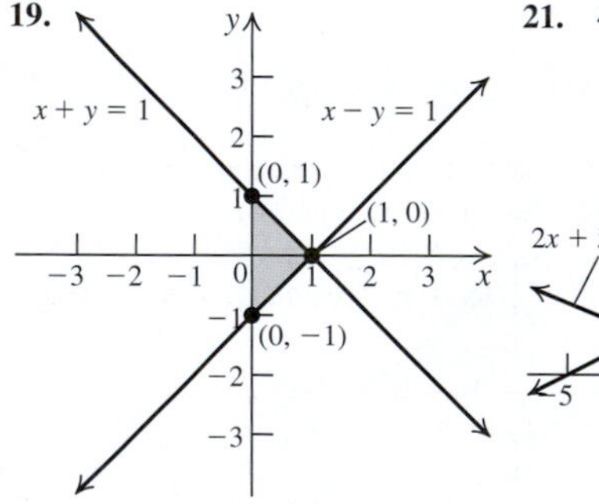

21.

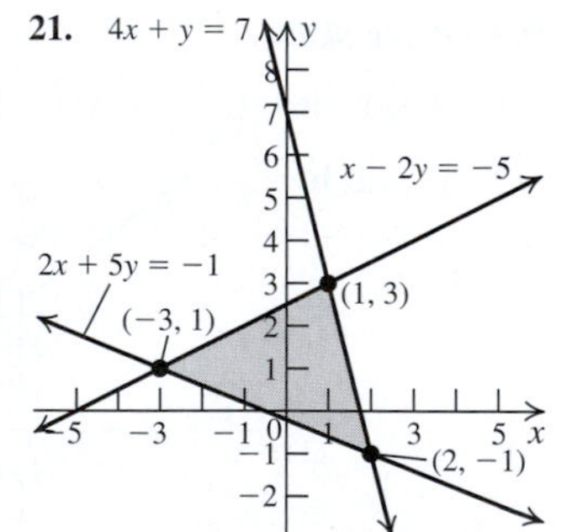

23.

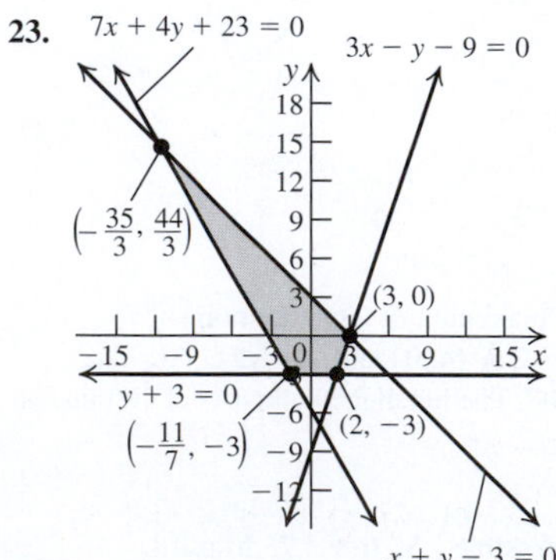

25. Max of 14 at $x = 7, y = 0$ **27.** Min of -9 at $x = 0, y = -3$

29. $\left\{(-2, 1), \left(\frac{8}{3}, -\frac{5}{9}\right)\right\}$ **31.** $\left\{(2, -2), \left(-\frac{2}{3}, -\frac{14}{3}\right)\right\}$

33. $\left\{(1, 2), (-1, -2), \left(2\sqrt{2}, \frac{\sqrt{2}}{2}\right), \left(-2\sqrt{2}, -\frac{\sqrt{2}}{2}\right)\right\}$

35. **37.**

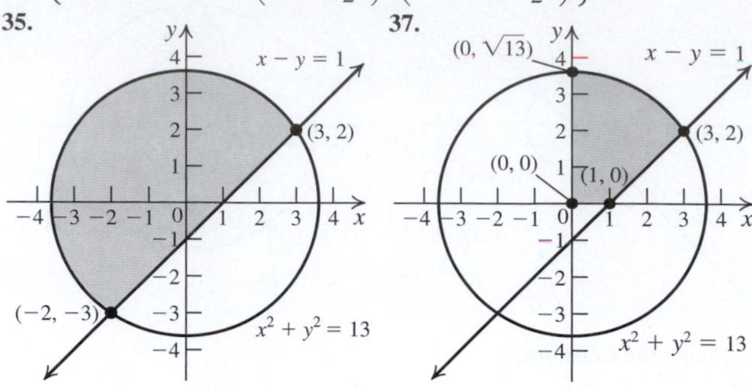

39.

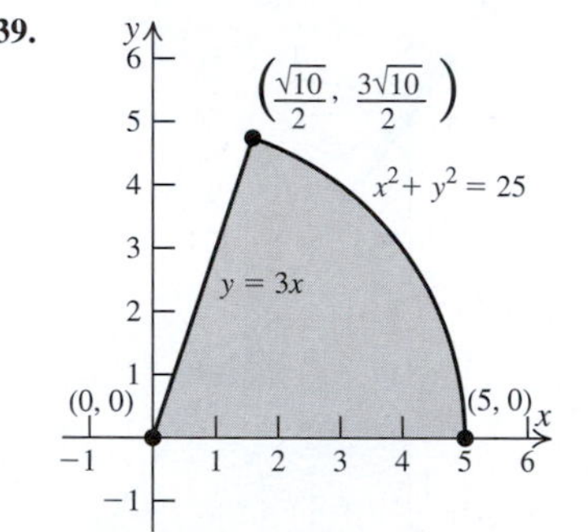

41. $\frac{2}{x + 2} - \frac{1}{x + 3}$ **43.** $\frac{2}{x - 1} + \frac{1}{x} + \frac{5}{(x - 1)^2}$

45. $\frac{-1 + 2x}{(x^2 + 4)^2} + \frac{1}{x^2 + 4}$

Applying the Concepts:
47. \$8750 was invested at 12%, and \$6250 was invested at 4%. **49.** 10.5 ft by 6 ft **51.** 15 and 8 **53.** The original width is 40 m, and the original length is 160 m. **55. a.** $C(x) = 60{,}000 + 12x, R(x) = 20x$
b.

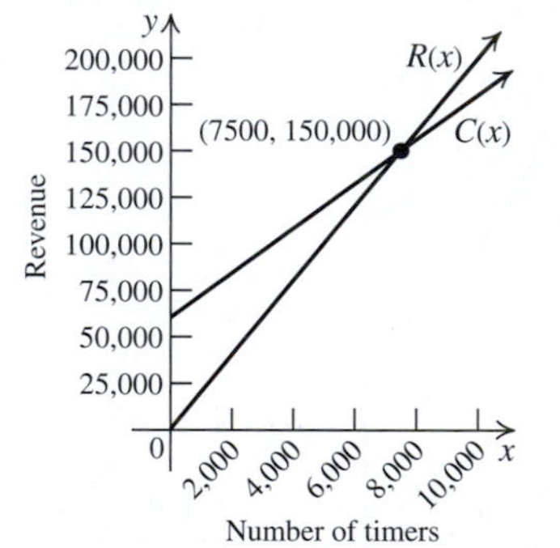

c. $(7500, 150{,}000)$ **d.** 11,129 timers

57. \$450 for the single and \$600 for the double **59.** 3 **61.** Janet is 34, and Steve is 51. **63.** They stole \$35,000, and \$15,000 went to Butch, \$12,000 went to Sundance, and \$8000 went to Billy. **65.** $x = 6, y = 2$ **67.** She should build 4 two-story houses and 14 one-story houses to obtain a maximum profit of \$96,000. **69.** large: \$60; medium: \$52; small: \$45

Practice Test A

1. $\{(2, 0)\}$ **2.** $\left\{x, 4 - \frac{x}{2}\right\}$ **3.** $\varnothing$ **4.** $\{(-5, 0)\}$ **5.** $\{(6, 8)\}$
6. $\{(-1, 1), (4, 16)\}$ **7.** $\left\{\left(-3, \frac{1}{3}\right), (2, 2)\right\}$ **8. a.** $\begin{cases} x + y = 485 \\ x - y = 15 \end{cases}$
b. $x = 250, y = 235$ **9.** $x + \frac{1}{2}y + z = 2$ $\quad y - \frac{3}{4}z = \frac{7}{4}$ $\quad z = \frac{-27}{23}$ **10.** $\{(1, 3, 5)\}$ **11.** $(7, 3, -2)$
12. $\left(x, \frac{2x}{3} + \frac{7}{3}, -x - 1\right)$ **13.** $\left(\frac{1}{3}, -\frac{4}{3} + z, z\right)$ **14.** 53 nickels, 225 dimes, and 22 quarters **15.** $\frac{A}{(x - 5)} + \frac{B}{(x + 1)}$

16. $\dfrac{A}{x-2}+\dfrac{Bx+C}{x^2+1}+\dfrac{Dx+E}{(x^2+1)^2}$

17. $-\dfrac{10}{121(x+4)}+\dfrac{1}{11(x+4)^2}+\dfrac{10}{121(x-7)}$

18.

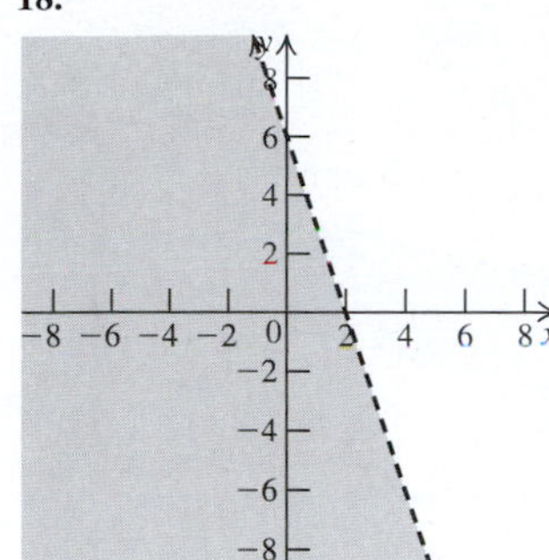

19.

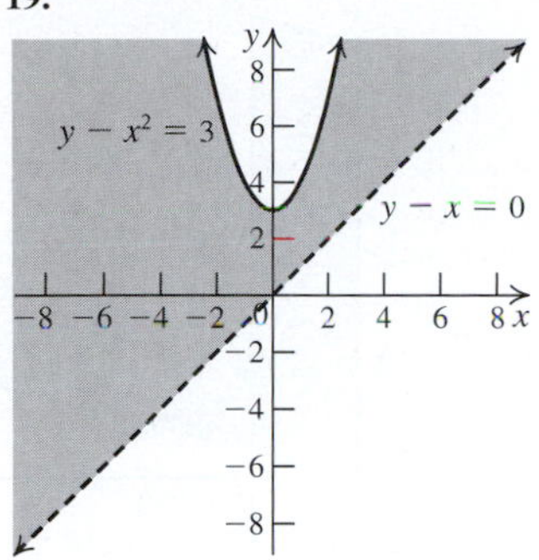

20. Maximum value is 6 at $(3, 0)$.

Practice Test B

1. a **2.** c **3.** b **4.** c **5.** b **6.** d **7.** d **8.** d **9.** c **10.** a **11.** b **12.** d **13.** c **14.** b **15.** c **16.** a **17.** c **18.** b **19.** a **20.** b

Cumulative Review Exercises (Chapters P–8)

1. $\left\{\frac{5}{2}\right\}$ **3.** $\{7\}$ **5.** $(-\infty, 4) \cup (5, \infty)$ **7.** $\{2\}$

9. **11.**

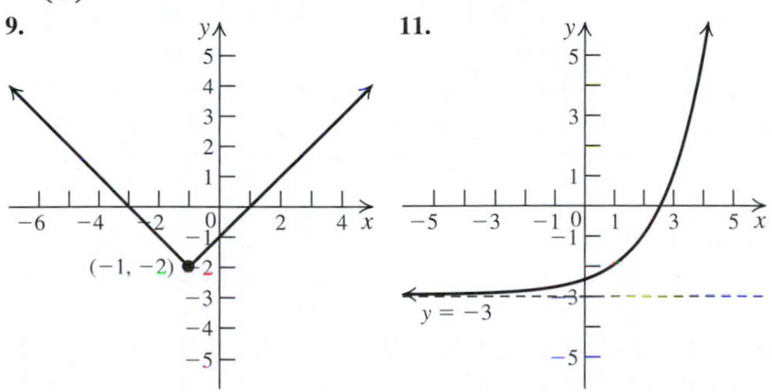

13. a. $f^{-1}(x) = \dfrac{x+2}{2}$ **b.**

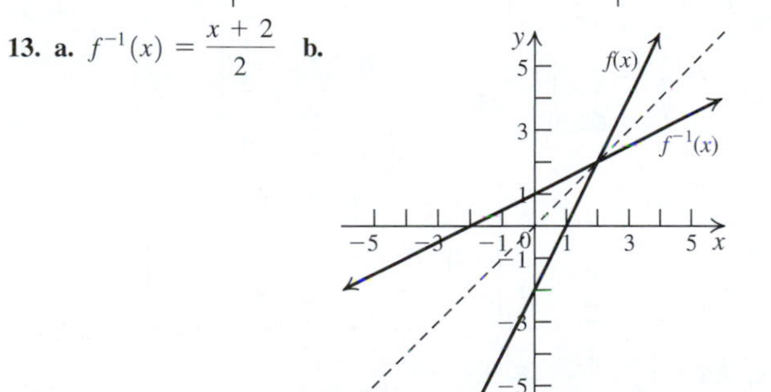

15. $2 + 4\log_3(x)$ **17.** 2 **19.** $\{(-3, 5)\}$

CHAPTER 9

Section 9.1

Answers to Practice Problems:

1. a. 3×2 **b.** 1×2 **2.** $\left[\begin{array}{ccc|c} 0 & 3 & -1 & 8 \\ 1 & 4 & 0 & 14 \\ 0 & -2 & 9 & 0 \end{array}\right]$

3. $B = \begin{bmatrix} 2 & 4 & 6 \\ 3 & 4 & 5 \end{bmatrix} \xrightarrow{\frac{1}{2}R_1} \begin{bmatrix} 1 & 2 & 3 \\ 3 & 4 & 5 \end{bmatrix} \xrightarrow{-3R_1 + R_2 \to R_2}$
$C = \begin{bmatrix} 1 & 2 & 3 \\ 0 & -2 & -4 \end{bmatrix}$ **4.** $\left\{\left(-1, -\frac{1}{3}, -1\right)\right\}$ **5.** $\{(5, 2)\}$

6. $\{(1, 2, -3)\}$ **7.** $\varnothing$ **8.** $\{(1, 2, -2)\}$

9. $\left\{\left(-z-1, \frac{5}{3}-\frac{2}{3}z, z\right)\right\}$

10. a. $\begin{cases} s = 0.01s + 0.1c + 0.1t + 12.46 \\ c = 0.2s + 0.02c + 0.01t + 3 \\ t = 0.25s + 0.3c + 2.7 \end{cases}$

b. $(0.99)(14) - (0.1)(6) - (0.1)(8) = 12.46$ ✓
$(-0.20)(14) + (0.98)(6) - (0.01)(8) = 3$ ✓
$(-0.25)(14) - (0.30)(6) + 8 = 2.7$ ✓

Basic Concepts and Skills:

1. numbers **3.** row operations **5.** false **7.** false **9.** 1×1 **11.** 2×4 **13.** 2×3 **15.** $a_{13} = 3, a_{31} = 9, a_{33} = 11, a_{34} = 12$

17. No, it is not a rectangular array of numbers. **19.** $\left[\begin{array}{cc|c} 2 & 4 & 2 \\ 1 & -3 & 1 \end{array}\right]$

21. $\left[\begin{array}{cc|c} 5 & -7 & 11 \\ -13 & 17 & 19 \end{array}\right]$ **23.** $\left[\begin{array}{ccc|c} -1 & 2 & 3 & 8 \\ 2 & -3 & 9 & 16 \\ 4 & -5 & -6 & 32 \end{array}\right]$

25. $\begin{cases} x + 2y - 3z = 4 \\ -2x - 3y + z = 5 \\ 3x - 3y + 2z = 7 \end{cases}$ **27.** $\begin{cases} x - y + z = 2 \\ 2x + y - 3z = 6 \end{cases}$ **29.** $\begin{bmatrix} 1 & 2 & 3 \\ 0 & 1 & 1 \end{bmatrix}$

31. $\begin{bmatrix} 1 & 2 & 3 & 4 \\ 0 & 1 & 5 & -3 \\ 0 & 0 & 1 & -1 \end{bmatrix}$ **33.** $\begin{bmatrix} 4 & 5 & -7 \\ 5 & 4 & -2 \end{bmatrix} \xrightarrow{\frac{1}{4}R_1} \begin{bmatrix} 1 & \frac{5}{4} & -\frac{7}{4} \\ 5 & 4 & -2 \end{bmatrix}$

$\xrightarrow{-5R_1 + R_2 \to R_2} \begin{bmatrix} 1 & \frac{5}{4} & -\frac{7}{4} \\ 0 & -\frac{9}{4} & \frac{27}{4} \end{bmatrix} \xrightarrow{-\frac{4}{9}R_2} \begin{bmatrix} 1 & \frac{5}{4} & -\frac{7}{4} \\ 0 & 1 & -3 \end{bmatrix}$

35. $\begin{bmatrix} 1 & 4 & 3 & 1 \\ 0 & -3 & -2 & 0 \\ 0 & 7 & 5 & -3 \end{bmatrix} \xrightarrow{-\frac{1}{3}R_2} \begin{bmatrix} 1 & 4 & 3 & 1 \\ 0 & 1 & \frac{2}{3} & 0 \\ 0 & 7 & 5 & -3 \end{bmatrix}$

$\xrightarrow{-7R_2 + R_3 \to R_3} \begin{bmatrix} 1 & 4 & 3 & 1 \\ 0 & 1 & \frac{2}{3} & 0 \\ 0 & 0 & \frac{1}{3} & -3 \end{bmatrix} \xrightarrow{3R_3} \begin{bmatrix} 1 & 4 & 3 & 1 \\ 0 & 1 & \frac{2}{3} & 0 \\ 0 & 0 & 1 & -9 \end{bmatrix}$

37. No, because Property 2 is not satisfied **39.** The matrix is in reduced row-echelon form because Properties 1–4 are satisfied **41.** The matrix is in reduced row-echelon form because Properties 1–4 are satisfied. **43.** The matrix is in reduced row-echelon form because Properties 1–4 are satisfied.

45. $(5, -2)$; $\begin{cases} x + 2y = 1 \\ y = -2 \end{cases}$

47. $(-4 - 4y, y, 3)$; $\begin{cases} x + 4y + 2z = 2 \\ z = 3 \end{cases}$

49. $(1, 2, -1)$; $\begin{cases} x + 2y + 3z = 2 \\ y - 2z = 4 \\ z = -1 \end{cases}$

51. $(2, -5, -2w + 3, w)$; $\begin{cases} x = 2 \\ y = -5 \\ z + 2w = 3 \end{cases}$

53. $(-5, 4, 3, 0)$; $\begin{cases} x = -5 \\ y = 4 \\ z + 2w = 3 \\ w = 0 \end{cases}$ **55.** $\{(5, -3)\}$

57. $\{(3, 1)\}$ **59.** $\left\{\left(-\frac{1}{5}, -\frac{23}{25}\right)\right\}$ **61.** $\{(2, 1)\}$ **63.** $\{(1, 2, 3)\}$

65. $\{(2, 3, 4)\}$ **67.** $\{(1, 2, 3)\}$ **69.** $\left\{\left(\frac{5}{2}, \frac{3}{2}, \frac{7}{2}\right)\right\}$

71. $\{(1 + 4z, 3 - 3z, z)\}$ **73.** $\{(1, 2, -1)\}$

75. **a.** One solution **b.** Infinitely many solutions **c.** No solution
77. False. Each row of A has seven entries.

Applying the Concepts:

79. **a.** $\begin{cases} a = 0.1b + 1000 \\ b = 0.2a + 780 \end{cases}$ **b.** $\left[\begin{array}{cc|c} 1 & -0.1 & 1000 \\ -0.2 & 1 & 780 \end{array}\right]$

c. $a = 1100, b = 1000$ **81.** **a.** $\begin{aligned} l &= 0.4t + 0.2f + 10{,}000 \\ t &= 0.5l + 0.3t + 20{,}000 \\ f &= 0.5l + 0.05t + 0.35f + 10{,}000 \end{aligned}$

b. $\left[\begin{array}{ccc|c} 1 & -0.4 & -0.2 & 10{,}000 \\ -0.5 & 0.7 & 0 & 20{,}000 \\ -0.5 & -0.05 & 0.65 & 10{,}000 \end{array}\right]$ **c.** food, \$55,000; transportation, \$61,000; labor, \$45,400 **83.** $T_1 = 110, T_2 = 140, T_3 = 60$

85. **a.** $\begin{cases} x - y = 70 \\ -x + z = -120 \\ y - z = 50 \end{cases}$ **b.** $\{(120 + z, 50 + z, z)\} : 0 \le z \le 130$

87. $y = 2x^2 - 3x + 4$

Beyond the Basics:

89. $\{(y + 2w, y, -2y - 3w, w)\}$

91. **a.** $B = \begin{bmatrix} 1 & \frac{3}{2} & 0 & -1 \\ 0 & 1 & 0 & \frac{1}{2} \\ 0 & 0 & 1 & \frac{1}{2} \\ 0 & 0 & 0 & 1 \end{bmatrix}, C = \begin{bmatrix} 1 & 2 & -3 & 1 \\ 0 & 1 & 0 & -\frac{1}{2} \\ 0 & 0 & 1 & \frac{1}{2} \\ 0 & 0 & 0 & 1 \end{bmatrix}$

b. The reduced row-echelon form of the matrices B and C is

$\begin{bmatrix} 1 & 0 & 0 & 0 \\ 0 & 1 & 0 & 0 \\ 0 & 0 & 1 & 0 \\ 0 & 0 & 0 & 1 \end{bmatrix}$. **93.** **a.** $\left\{\left(\frac{dm - nb}{ad - bc}, \frac{an - cm}{ad - bc}\right)\right\}$

b. **(i)** $cb \ne ad$ **(ii)** $cb = ad$ and $\frac{m}{b} \ne \frac{n}{d}$ **(iii)** $cb = ad$ and $\frac{m}{b} = \frac{n}{d}$

95. $\{(10, 100, 1000)\}$ **97.** $y = -x^3 + 2x^2 + 3x + 1$

99. $\begin{bmatrix} 0 \\ 0 \end{bmatrix}, \begin{bmatrix} 1 \\ 0 \end{bmatrix}$ **101.** Yes, the inverse of the operation used to transform A to B

Maintaining Skills:

103. $\{2\}$ **105.** $\{-9\}$ **107.** $\{(7, 9)\}$ **109.** $\varnothing$
111. true **113.** true

Section 9.2

Answers to Practice Problems:

1. $x = 1, y = 1$ **2.** $\begin{bmatrix} -6 & 1 & 13 \\ 12 & 3 & 15 \end{bmatrix}$ **3.** $\begin{bmatrix} 11 & 1 \\ 0 & 6 \\ -15 & -28 \end{bmatrix}$

4. $X = \begin{bmatrix} -\frac{1}{3} & \frac{19}{3} \\ -7 & -\frac{35}{3} \end{bmatrix}$ **5.** Yes, the product AB is defined with order 2×1.

6. \$2045 thousand **7.** $[31]$ **8.** The product AB is not defined. $BA = \begin{bmatrix} 42 & -1 \\ 2 & -6 \\ 8 & -4 \end{bmatrix}$ **9.** $AB = \begin{bmatrix} 18 & -3 \\ 12 & 12 \end{bmatrix}, BA = \begin{bmatrix} 14 & -1 \\ 28 & 16 \end{bmatrix}$

10.

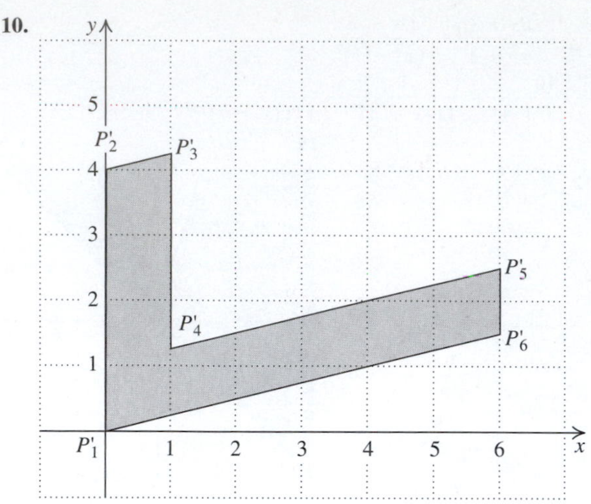

Basic Concepts and Skills:

1. $a_{ij} = b_{ij}$ **3.** 1×1 **5.** false **7.** $x = 2, u = -3$

9. $x = -1, y = 3$ **11.** $x = 2, y = 1$ **13.** $\varnothing$ **15.** **a.** $\begin{bmatrix} 0 & 2 \\ 5 & 1 \end{bmatrix}$

b. $\begin{bmatrix} 2 & 2 \\ 1 & 7 \end{bmatrix}$ **c.** $\begin{bmatrix} -3 & -6 \\ -9 & -12 \end{bmatrix}$ **d.** $\begin{bmatrix} 5 & 6 \\ 5 & 18 \end{bmatrix}$ **e.** $\begin{bmatrix} 10 & 2 \\ 5 & 11 \end{bmatrix}$

f. $\begin{bmatrix} 6 & 10 \\ 23 & 13 \end{bmatrix}$ **17.** a, b, d, e, and f are not defined. **c.** $\begin{bmatrix} -6 & -9 \\ 12 & -15 \end{bmatrix}$

19. **a.** $\begin{bmatrix} 7 & 1 & -1 \\ -1 & 1 & 4 \\ 2 & 1 & 4 \end{bmatrix}$ **b.** $\begin{bmatrix} 1 & -1 & -1 \\ -3 & 9 & 0 \\ -2 & -1 & -2 \end{bmatrix}$ **c.** $\begin{bmatrix} -12 & 0 & 3 \\ 6 & -15 & -6 \\ 0 & 0 & -3 \end{bmatrix}$

d. $\begin{bmatrix} 6 & -2 & -3 \\ -8 & 23 & 2 \\ -4 & -2 & -3 \end{bmatrix}$ **e.** $\begin{bmatrix} 46 & 7 & -7 \\ 0 & 4 & 21 \\ 21 & 7 & 18 \end{bmatrix}$ **f.** $\begin{bmatrix} 6 & 1 & -7 \\ -21 & 6 & 16 \\ -13 & -1 & -10 \end{bmatrix}$

21. **a.** $\begin{bmatrix} 4 & 3 & -3 \\ 4 & 0 & 7 \\ 4 & 0 & 3 \end{bmatrix}$ **b.** $\begin{bmatrix} -2 & 1 & -3 \\ 2 & 8 & 3 \\ 0 & -2 & -3 \end{bmatrix}$ **c.** $\begin{bmatrix} -3 & -6 & 9 \\ -9 & -12 & -15 \\ -6 & 3 & 0 \end{bmatrix}$

d. $\begin{bmatrix} -3 & 4 & -9 \\ 7 & 20 & 11 \\ 2 & -5 & -6 \end{bmatrix}$ **e.** $\begin{bmatrix} 16 & 12 & 0 \\ 44 & 12 & 9 \\ 28 & 12 & -3 \end{bmatrix}$ **f.** $\begin{bmatrix} -9 & 14 & 5 \\ 22 & -2 & 13 \\ -14 & -1 & -22 \end{bmatrix}$

23. $X = \begin{bmatrix} -4 & -2 & 1 \\ 1 & 5 & 0 \end{bmatrix}$ **25.** $X = \begin{bmatrix} 0 & 2 & -\frac{1}{2} \\ \frac{3}{2} & \frac{1}{2} & 4 \end{bmatrix}$

27. $X = \begin{bmatrix} -4 & -4 & \frac{3}{2} \\ -\frac{1}{2} & \frac{9}{2} & -4 \end{bmatrix}$ **29.** $X = \begin{bmatrix} \frac{1}{2} & -\frac{9}{4} & \frac{1}{2} \\ -2 & -\frac{5}{4} & -5 \end{bmatrix}$

31. **a.** $AB = \begin{bmatrix} 4 & 11 \\ 6 & 23 \end{bmatrix}$ **b.** $BA = \begin{bmatrix} 1 & 0 \\ 18 & 26 \end{bmatrix}$

33. **a.** $AB = \begin{bmatrix} 4 & 7 \\ 3 & -12 \end{bmatrix}$ **b.** $BA = \begin{bmatrix} -13 & 4 & 10 \\ -13 & 5 & 6 \\ 8 & -4 & 0 \end{bmatrix}$

35. **a.** $AB = [16]$ **b.** $BA = \begin{bmatrix} 2 & 3 & 5 \\ -4 & -6 & -10 \\ 8 & 12 & 20 \end{bmatrix}$

37. **a.** $AB = [7 \quad 8 \quad -8]$ **b.** The product BA is not defined.

39. **a.** $AB = \begin{bmatrix} 10 & 7 & 2 \\ 7 & 19 & 4 \\ 10 & 1 & 0 \end{bmatrix}$ **b.** $BA = \begin{bmatrix} 7 & 4 & 5 \\ 0 & 8 & 3 \\ 19 & 18 & 14 \end{bmatrix}$

41. $AB = \begin{bmatrix} 13 & 17 & 3 \\ 13 & 8 & 2 \\ 6 & 1 & 6 \end{bmatrix} \ne \begin{bmatrix} 6 & 22 & 19 \\ 2 & 11 & 6 \\ 11 & -1 & 10 \end{bmatrix} = BA$

Applying the Concepts:

47.
$$\begin{array}{l} \\ C = \end{array}\begin{array}{c} \text{Steel} \quad \text{Glass} \quad \text{Wood} \\ \begin{bmatrix} 13 & 5 & 38 \\ 7 & 2 & 7 \end{bmatrix} \end{array}\begin{array}{l} \\ \text{Cost of material} \\ \text{Transportation cost} \end{array}$$

49. \$24,200

51. a.
$$\begin{array}{l} \\ \text{Salary} \\ \text{Bonus} \\ \text{Stock} \end{array}\begin{array}{c} \text{Chairman} \quad \text{President} \quad \text{Vice president} \\ \begin{bmatrix} 2{,}500{,}000 & 1{,}250{,}000 & 100{,}000 \\ 1{,}500{,}000 & 750{,}000 & 150{,}000 \\ 50{,}000 & 25{,}000 & 5000 \end{bmatrix} \end{array}$$

b. $\begin{bmatrix} 1 \\ 1 \\ 4 \end{bmatrix}\begin{array}{l} \text{Chairman} \\ \text{President} \\ \text{Vice president} \end{array}$ **c.** $\begin{bmatrix} 4{,}150{,}000 \\ 2{,}850{,}000 \\ 95{,}000 \end{bmatrix}\begin{array}{l} \text{Total salary} \\ \text{Total bonuses} \\ \text{Total stocks} \end{array}$

53. $AD = \begin{bmatrix} 0 & 4 & 4 & 1 & 1 & 0 \\ 0 & 0 & -1 & -1 & -6 & -6 \end{bmatrix}$

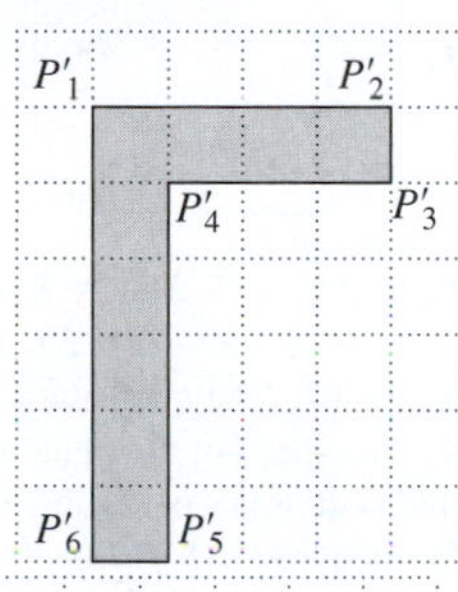

55. $AD = \begin{bmatrix} 0 & 4 & 4 & 1 & 1 & 0 \\ 0 & 1 & 2 & 1.25 & 6.25 & 6 \end{bmatrix}$

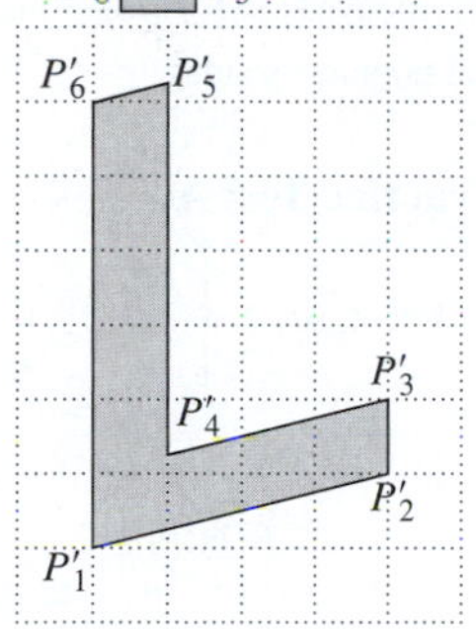

57. $AB = \begin{bmatrix} 9 & 0 \\ 0 & 0 \end{bmatrix} = AC$

Beyond the Basics:

59. Let $A = \begin{bmatrix} 1 & 2 \\ 3 & 4 \end{bmatrix}$ and $B = \begin{bmatrix} -1 & 0 \\ 2 & -3 \end{bmatrix}$. Then
$(A + B)^2 = \begin{bmatrix} 10 & 2 \\ 5 & 11 \end{bmatrix} \neq \begin{bmatrix} 14 & -2 \\ 17 & 7 \end{bmatrix} = A^2 + 2AB + B^2.$

63. $B = \begin{bmatrix} 2 & -3 \\ -1 & 2 \end{bmatrix}$ **65.** $x = 33, y = 5$

69. $A = \begin{bmatrix} 3 & -2 & 1 \\ -2 & 1 & -1 \end{bmatrix}, B = \begin{bmatrix} 0 & 2 & 2 \\ 0 & 0 & -3 \end{bmatrix}$

Critical Thinking/Discussion/Writing:

75. $(CA)B$

Maintaining Skills:

77. 2 **79.** $\{8\}$ **81.** $\{1\}$ **83.** $\left\{\frac{5}{4}\right\}$ **85.** $\left\{\left(x, \frac{2 - 3x}{5}\right)\right\}$ **87.** $\varnothing$

Section 9.3

Answers to Practice Problems:

1. $\begin{bmatrix} 3 & 2 \\ 2 & 1 \end{bmatrix}\begin{bmatrix} -1 & 2 \\ 2 & -3 \end{bmatrix} = \begin{bmatrix} 1 & 0 \\ 0 & 1 \end{bmatrix}$ **2.** $\begin{bmatrix} 3x + z & 3y + w \\ 3x + z & 3y + w \end{bmatrix} \neq \begin{bmatrix} 1 & 0 \\ 0 & 1 \end{bmatrix}$
because this implies that $1 = 3x + z = 0$, which is false.

3. The inverse of matrix A does not exist. **4.** $A^{-1} = \begin{bmatrix} \frac{1}{7} & -\frac{19}{77} & \frac{1}{11} \\ 0 & \frac{2}{11} & \frac{1}{11} \\ \frac{2}{7} & -\frac{3}{77} & -\frac{1}{11} \end{bmatrix}$

5. a. The inverse of matrix A does not exist; **b.** $B^{-1} = \begin{bmatrix} \frac{1}{14} & \frac{1}{7} \\ -\frac{3}{14} & \frac{4}{7} \end{bmatrix}$

6. $\left\{\left(\frac{11}{2}, -\frac{9}{2}, \frac{1}{2}\right)\right\}$ **7.** $X = \begin{bmatrix} \frac{110{,}400}{19} \\ \frac{135{,}200}{19} \end{bmatrix}$

8. $AM = \begin{bmatrix} 1 & 2 & 3 \\ 1 & 3 & 3 \\ 1 & 2 & 4 \end{bmatrix}\begin{bmatrix} 10 & 11 & 19 & 15 & 19 & 5 \\ 1 & 0 & 0 & 23 & 1 & 0 \\ 3 & 9 & 14 & 0 & 6 & 0 \end{bmatrix}$
$= \begin{bmatrix} 21 & 38 & 61 & 61 & 39 & 5 \\ 22 & 38 & 61 & 84 & 40 & 5 \\ 24 & 47 & 75 & 61 & 45 & 5 \end{bmatrix}$

Basic Concepts and Skills:

1. inverse **3.** A^{-1} **5.** false **7.** Yes **9.** Yes **11.** No **13.** No

15. Yes **17.** $A^{-1} = \begin{bmatrix} \frac{1}{2} & 0 \\ -\frac{1}{6} & \frac{1}{3} \end{bmatrix}$ **19.** Does not exist

21. $A^{-1} = \begin{bmatrix} 1 & -8 & 10 \\ 0 & 2 & -3 \\ 0 & -1 & 2 \end{bmatrix}$ **25.** $\begin{bmatrix} 1 & -1 & 0 \\ -2 & 3 & -4 \\ -2 & 3 & -3 \end{bmatrix}$

27. $A^{-1} = \begin{bmatrix} 1 & 0 \\ -\frac{3}{2} & \frac{1}{2} \end{bmatrix}$ **29.** $A^{-1} = \begin{bmatrix} 5 & 3 \\ 3 & 2 \end{bmatrix}$

31. $A^{-1} = \frac{1}{b^2 - a^2}\begin{bmatrix} -a & b \\ -b & a \end{bmatrix}$, where $a^2 \neq b^2$

33. $\begin{bmatrix} 2 & 3 \\ 1 & -3 \end{bmatrix}\begin{bmatrix} x \\ y \end{bmatrix} = \begin{bmatrix} -9 \\ 13 \end{bmatrix}$ **35.** $\begin{bmatrix} 3 & 2 & 1 \\ 2 & 1 & 3 \\ 1 & 3 & 2 \end{bmatrix}\begin{bmatrix} x \\ y \\ z \end{bmatrix} = \begin{bmatrix} 8 \\ 7 \\ 9 \end{bmatrix}$

37. $\begin{cases} x - 2y = 0 \\ 2x + y = 5 \end{cases}$ **39.** $\begin{cases} 2x_1 + 3x_2 + x_3 = -1 \\ 5x_1 + 7x_2 - x_3 = 5 \\ 4x_1 + 3x_2 = 5 \end{cases}$

41. $\begin{bmatrix} 1 & 2 & 5 \\ 2 & 3 & 8 \\ -1 & 1 & 2 \end{bmatrix}\begin{bmatrix} 2 & -1 & -1 \\ 12 & -7 & -2 \\ -5 & 3 & 1 \end{bmatrix} = \begin{bmatrix} 1 & 0 & 0 \\ 0 & 1 & 0 \\ 0 & 0 & 1 \end{bmatrix}$

43. $\{(2, -3, 1)\}$ **45. a.** $A^{-1} = \begin{bmatrix} 3 & -\frac{5}{2} & \frac{1}{2} \\ -3 & 4 & -1 \\ 1 & -\frac{3}{2} & \frac{1}{2} \end{bmatrix}$ **b.** $\{(1, 2, 3)\}$

Applying the Concepts:

47. $\{(-1, 2)\}$ **49.** $\{(2, -1, 3)\}$ **51.** $\{(-5, 1, 5)\}$
53. Treasury: \$16,000; bonds: \$48,000; fund: \$26,000

55. $V = 10, E = 15, R = 7$

57. $X = \begin{bmatrix} 100 \\ 100 \end{bmatrix}$ **59.** $X = \begin{bmatrix} \frac{216{,}000}{277} \\ \frac{310{,}000}{277} \\ \frac{304{,}000}{277} \end{bmatrix}$

Beyond the Basics:

65. a. Yes. By the definition of inverse, if $AB = I$, then $BA = I$ and A and B are inverses. **b.** $(A^{-1})^{-1} = A$ **73.** If A is invertible, then $I = A^{-1}A = A^{-1}(AB) = (A^{-1}A)B = IB = B$.

75. c. $A^{-1} = \begin{bmatrix} 3 & -4 \\ -2 & 3 \end{bmatrix}$ **77.** $\{(1, 2, 1, 2)\}$ **79. a.** $A = \begin{bmatrix} 8 & 10 \\ 6 & 8 \end{bmatrix}$

b. $A = \begin{bmatrix} -1 & 7 \\ -3 & 17 \end{bmatrix}$ **c.** $A = \begin{bmatrix} \frac{3}{2} & \frac{1}{2} \\ \frac{5}{2} & \frac{3}{2} \end{bmatrix}$ **d.** $A = \begin{bmatrix} -1 & -2 \\ \frac{5}{2} & 4 \end{bmatrix}$

Critical Thinking/Discussion/Writing:

81. a. True. $A^2B = AAB = ABA = BAA = BA^2$ **b.** False. Let $A = I$ and $B = -I$. **83.** True

Maintaining Skills:

85. 1 **87.** -1 **89.** $\begin{bmatrix} -5 & 0 \\ -15 & 0 \end{bmatrix}$ **91.** $(-1)\begin{bmatrix} 6 & -7 \\ -4 & 1 \end{bmatrix}$

93. $\left\{\left(-\frac{1}{2}, -\frac{1}{3}\right)\right\}$ **95.** $\{(2, -3, -2)\}$

Section 9.4

Answers to Practice Problems:

1. a. -38 **b.** 0 **2. a.** $M_{11} = 4, M_{23} = 10, M_{32} = 10$ **b.** $A_{11} = 4, A_{23} = -10, A_{32} = -10$ **3.** 8 **4.** $\{(17, -9)\}$ **5.** $\{(2, -1, 0)\}$

Basic Concepts and Skills:

1. $ad - bc$ **3.** cofactor **5.** false **7.** -2 **9.** -6 **11.** -7

13. $\frac{139}{72}$ **15.** $a - b$ **17. a.** -10 **b.** 2 **c.** 0 **19. a.** -4 **b.** 4 **c.** -2 **21.** -2 **23.** 12 **25.** 0 **27.** 42 **29.** 16 **31.** 49 **33.** $a^3 + b^3$ **35.** $a^3 + b^3 + c^3 - 3abc$ **37.** $\{(3, 5)\}$ **39.** $\{(1, 2)\}$ **41.** $\{(2, 0)\}$ **43.** $\left\{\left(\frac{3}{2}y + 2, y\right)\right\}$ **45.** $\{(2, 3)\}$ **47.** $\{(1, 1, 0)\}$ **49.** $\{(1, -1, 3)\}$ **51.** $\{(1, 1, 1)\}$ **53.** $\{(1, 2, 3)\}$ **55.** $\{(1, 2, 3)\}$

Applying the Concepts:

57. 11 **59.** 10.5 **61.** Yes **63.** No **65.** $y = 2x + 1$ **67.** $y = \frac{2}{3}x + \frac{1}{3}$

Beyond the Basics:

75. 21 **77.** 476 **79.** $\{4\}$ **81.** $\left\{\frac{1}{2} \pm \frac{\sqrt{7}}{2}i\right\}$ **83.** $\{23\}$ **85.** $\{-1, 0, 1\}$ **87.** 30

Critical Thinking/Discussion/Writing:

89. $\left\{\left(\frac{9}{4}, -\frac{2}{3}\right)\right\}$ **91.** $\{(-2, -1)\}$ **93.** $k = -\frac{7}{2}$ **95.** $y = \frac{1}{8}x^2$ **97.** $y = \frac{1}{6}x^2 - \frac{2}{3}x + \frac{11}{3}$ **99.** $y = -\frac{1}{2}x^2$ **101.** $y = 2x^2 + 12x + 10$ **103. a.** Opens up **b.** $(-1, -8)$ **c.** $x = -1$ **d.** -3 and 1 **e.** -6

Review Exercises

1. 1×4 **3.** 3×2 **5.** $a_{12} = -1, a_{14} = -4, a_{23} = 3, a_{21} = 5$

7. $\left[\begin{array}{cc|c} 2 & -3 & 7 \\ 3 & 1 & 6 \end{array}\right]$ **9.** $\begin{bmatrix} 1 & -2 & 1 & 7 \\ 0 & 1 & 2 & 1 \\ 0 & 0 & 1 & 2 \end{bmatrix}$ **11.** $\begin{bmatrix} 1 & 0 & 0 & \frac{1}{3} \\ 0 & 1 & 0 & \frac{1}{2} \\ 0 & 0 & 1 & -\frac{1}{6} \end{bmatrix}$

13. $\{(2, -3, -1)\}$ **15.** $\{(3, -2, -3)\}$ **17.** $\{(5, -4, 1)\}$ **19.** $\left\{\left(2, -1, -\frac{1}{3}\right)\right\}$ **21.** $\{(2, 1)\}$ **23. a.** $\begin{bmatrix} 3 & -1 \\ -8 & 10 \end{bmatrix}$

b. $\begin{bmatrix} -1 & 5 \\ 2 & -2 \end{bmatrix}$ **c.** $\begin{bmatrix} 2 & 4 \\ -6 & 8 \end{bmatrix}$ **d.** $\begin{bmatrix} -6 & 9 \\ 15 & -18 \end{bmatrix}$ **e.** $\begin{bmatrix} -4 & 13 \\ 9 & -10 \end{bmatrix}$

25. $X = \begin{bmatrix} \frac{7}{3} & 0 \\ -\frac{19}{3} & 8 \end{bmatrix}$ **27.** $AB = \begin{bmatrix} -3 & 4 \\ -11 & 10 \end{bmatrix}, BA = \begin{bmatrix} -2 & -4 \\ 8 & 9 \end{bmatrix}$

29. $AB = [7], BA = \begin{bmatrix} 2 & 4 & -2 \\ 3 & 6 & -3 \\ 1 & 2 & -1 \end{bmatrix}$ **31.** $\begin{bmatrix} 1 & 0 \\ 2 & 3 \end{bmatrix}$ **33.** $\begin{bmatrix} -1 & \frac{3}{2} \\ 2 & -\frac{5}{2} \end{bmatrix}$

35. a. $\begin{bmatrix} 3 & -1 \\ 1 & 0 \end{bmatrix}$ **b.** $\begin{bmatrix} 8 & -3 \\ 3 & -1 \end{bmatrix}$ **c.** $\begin{bmatrix} 8 & -3 \\ 3 & -1 \end{bmatrix}$

41. $\begin{bmatrix} \frac{2}{5} & -\frac{1}{10} \\ -\frac{1}{5} & \frac{3}{10} \end{bmatrix}$ **43.** $\begin{bmatrix} 3 & 2 & 6 \\ 1 & 1 & 2 \\ 2 & 2 & 5 \end{bmatrix}$ **45.** $\{(-23, 10)\}$

47. $\{(-12, 2, 3)\}$ **49.** $\left\{\left(\frac{7}{6}, \frac{1}{6}, 2\right)\right\}$ **51.** 23 **53.** 0

55. a. $M_{12} = 8, M_{23} = 8, M_{22} = -16$ **b.** $A_{12} = -8, A_{23} = -8, A_{22} = -16$ **57.** 35 **59.** 6 **61.** $\{(1, 2)\}$ **63.** $\{(5, 6, 7)\}$ **65.** 2 **67.** They both satisfy the equation $y = 7x - 11$ **69.** Method 3 **71.** The speed of the plane is 520 miles per hour, and the velocity of the wind is 40 miles per hour. **73.** Andrew has \$170, Bonnie has \$65, and Chauncie has \$85. **75.** 30 registered nurses, 40 licensed practical nurses, and 60 nurse's aides

Practice Test A

1. $5 \times 4; 9$ **2.** $\left[\begin{array}{ccc|c} 7 & -3 & 9 & 5 \\ -2 & 4 & 3 & -12 \\ 8 & -5 & 1 & -9 \end{array}\right]$ **3.** $\begin{cases} 4x - \quad\quad z = -3 \\ x + 3y \quad\quad = 9 \\ 2x + 7y + 5z = 8 \end{cases}$

4. $(10, 10, 0)$ **5.** $\{(2, 3, -2)\}$ **6.** $\left\{\left(\frac{20}{7} + \frac{11}{7}z, \frac{2}{7} + \frac{6}{7}z, z\right)\right\}$

7. $A - B = \begin{bmatrix} 8 & -1 \\ 4 & 8 \\ 11 & 0 \end{bmatrix}$ **8.** $AB = [1]$ **9.** The product AB is not defined.

10. $\begin{bmatrix} -6 & 2 & 0 \\ 10 & 14 & 4 \end{bmatrix}$ **11.** $\begin{bmatrix} 20 & 28 & 8 \\ -9 & 29 & 6 \end{bmatrix}$ **12.** $\begin{bmatrix} 1 & 25 \\ 0 & 16 \end{bmatrix}$

13. $\begin{bmatrix} 1 & -\frac{5}{4} \\ 0 & \frac{1}{4} \end{bmatrix}$ **14.** $\begin{bmatrix} -11 & 2 & 7 \\ 8 & -1 & -5 \\ 5 & -1 & -3 \end{bmatrix}$ **15.** $\begin{bmatrix} 5 & 2 \\ 3 & 1 \end{bmatrix}\begin{bmatrix} x \\ y \end{bmatrix} = \begin{bmatrix} 32 \\ 18 \end{bmatrix}$

16. $\begin{cases} 12x - 3y = 5 \\ -2x + 7y = -9 \end{cases}$ **17.** $X = \begin{bmatrix} -\frac{3}{5} & -1 & 4 \\ -\frac{32}{5} & -\frac{18}{5} & -3 \end{bmatrix}$

18. $\frac{1}{2}$ **19.** 0 **20.** $x = \dfrac{\begin{vmatrix} 3 & -1 & 1 \\ 6 & 1 & 1 \\ 4 & 3 & -2 \end{vmatrix}}{\begin{vmatrix} 2 & -1 & 1 \\ 1 & 1 & 1 \\ 4 & 3 & -2 \end{vmatrix}}, y = \dfrac{\begin{vmatrix} 2 & 3 & 1 \\ 1 & 6 & 1 \\ 4 & 4 & -2 \end{vmatrix}}{\begin{vmatrix} 2 & -1 & 1 \\ 1 & 1 & 1 \\ 4 & 3 & -2 \end{vmatrix}},$

$z = \dfrac{\begin{vmatrix} 2 & -1 & 3 \\ 1 & 1 & 6 \\ 4 & 3 & 4 \end{vmatrix}}{\begin{vmatrix} 2 & -1 & 1 \\ 1 & 1 & 1 \\ 4 & 3 & -2 \end{vmatrix}}$

Practice Test B

1. d **2.** d **3.** d **4.** c **5.** c **6.** d **7.** c **8.** b **9.** a **10.** b **11.** c **12.** c **13.** b **14.** d **15.** a **16.** a **17.** b **18.** d **19.** c **20.** b

Cumulative Review Exercises (Chapters P–9)

1. $-7, 1$ **3.** $4, 1$ **5.** $-9, \frac{-19}{7}$ **7.** $x < -2$ or $x > \frac{1}{2}$

9. $\pm 1, \pm\frac{1}{2}, \pm\frac{1}{4}, \pm 3, \pm\frac{3}{2}, \pm\frac{3}{4}$ **11.** 24,000 **13.** $\{(1, -1)\}$

15.

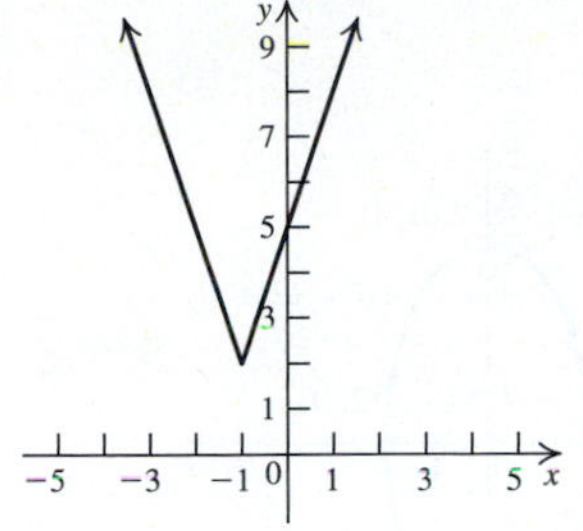

17. $\{(1, 1, -1)\}$ **19.** $f^{-1}(x) = -\frac{4x}{x-1}$

CHAPTER 10

Section 10.2

Answers to Practice Problems:

1. a.

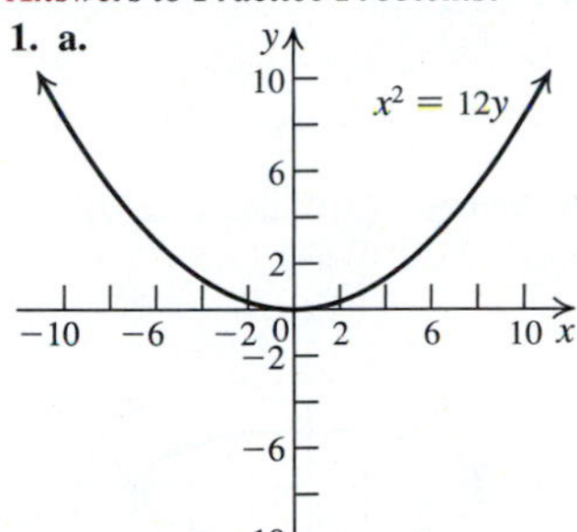

Vertex: $(0, 0)$; focus: $(0, 3)$; directrix: $y = -3$; axis: y-axis

b.

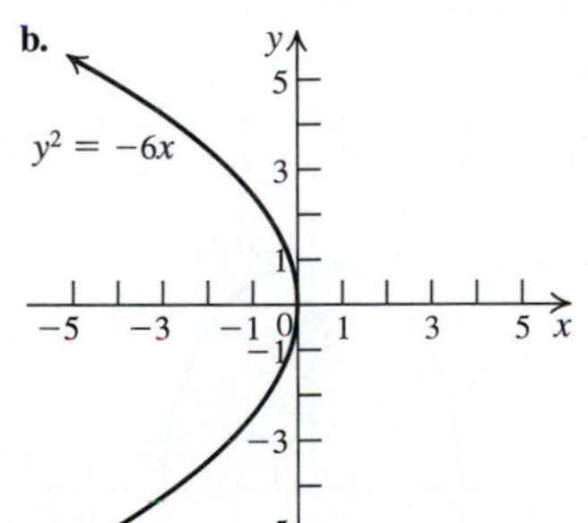

Vertex: $(0, 0)$; focus: $\left(-\frac{3}{2}, 0\right)$; directrix: $x = \frac{3}{2}$; axis: x-axis

2. a. $x^2 = 8y$ **b.** $y^2 = 4x$ **3.** Vertex: $(2, -1)$; focus: $\left(2, -\frac{7}{8}\right)$; directrix: $y = -\frac{9}{8}$

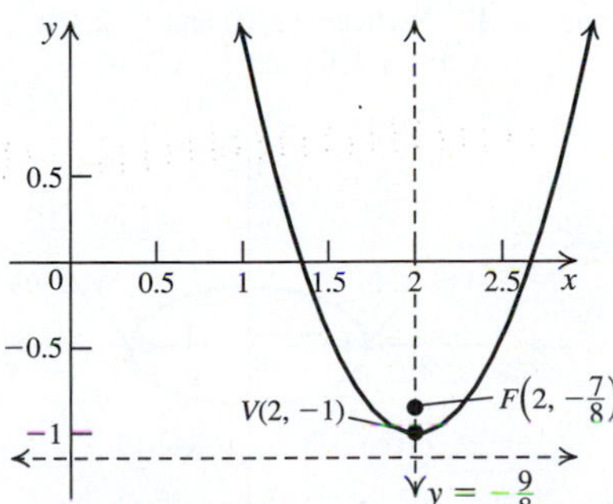

4. Equation: $x^2 = 29.2y$. The thickness of the mirror at the edges is 0.077055 in.

Basic Concepts and Skills:

1. directrix, focus **3.** vertex **5.** true

7. Focus: $\left(0, \frac{1}{2}\right)$ directrix: $y = -\frac{1}{2}$; graph: e

9. Focus: $\left(0, -\frac{9}{64}\right)$ directrix: $y = \frac{9}{64}$; graph: d

11. Focus: $\left(\frac{1}{2}, 0\right)$ directrix: $x = -\frac{1}{2}$; graph: g

13. Focus: $\left(-\frac{4}{9}, 0\right)$ directrix: $x = \frac{4}{9}$; graph: f

15.

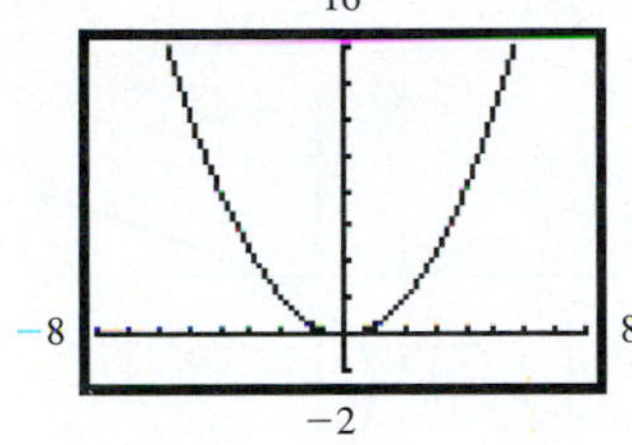

17.

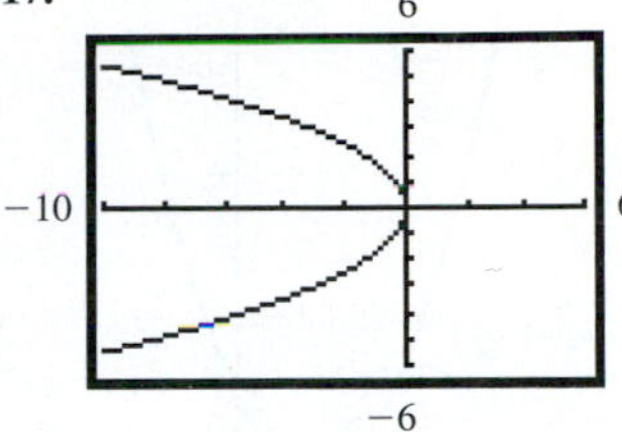

19.

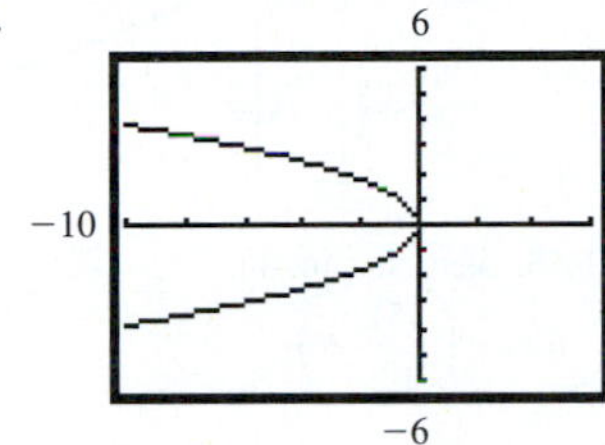

21. $x^2 = -4(y-3); 4$
23. $y^2 = -10\left(x - \frac{1}{2}\right); 10$
25. $(y-1)^2 = -8(x-1); 8$
27. $(x-1)^2 = 16(y-1); 16$
29. $y^2 = 8(x-1); 8$
31. $x^2 = -12(y-1); 12$
33. $(y-3)^2 = -8(x-2); 8$
35. $(x-2)^2 = 8(y-3); 8$

37.

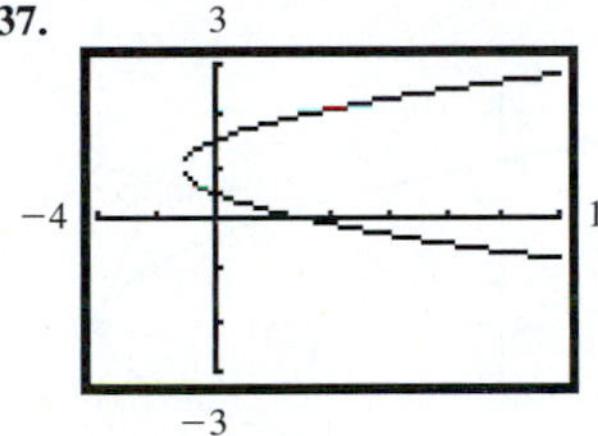

39.

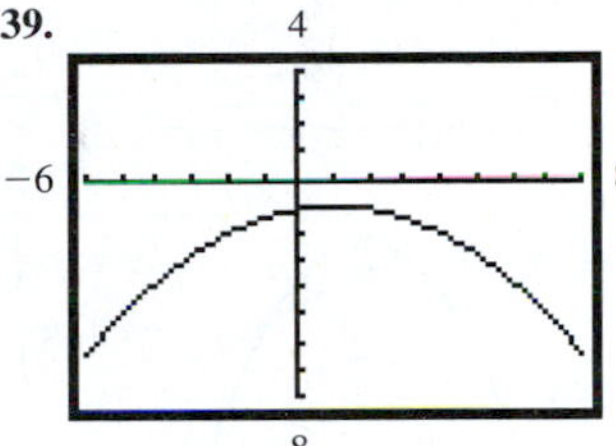

41. Vertex: $(-1, 1)$; focus: $\left(-\frac{1}{2}, 1\right)$; directrix: $x = -\frac{3}{2}$

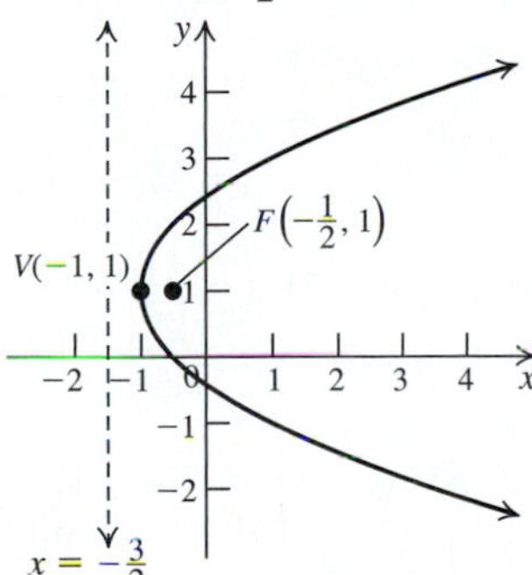

43. Vertex $(-2, 2)$; focus: $\left(-2, \frac{11}{4}\right)$; directrix: $y = \frac{5}{4}$

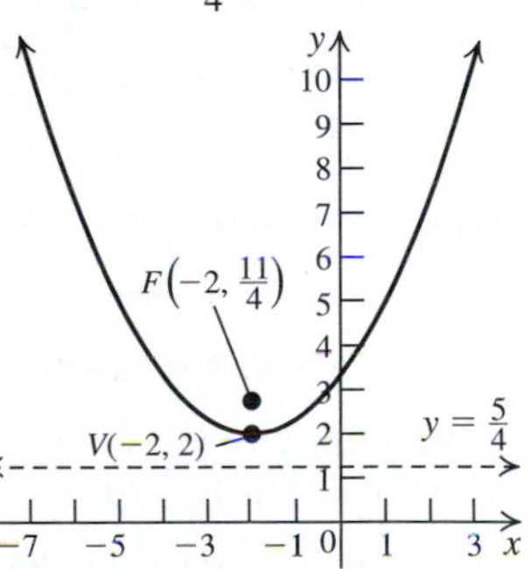

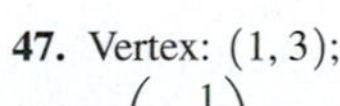

45. Vertex: $(2, -1)$; focus: $\left(\frac{1}{2}, -1\right)$; directrix: $x = \frac{7}{2}$

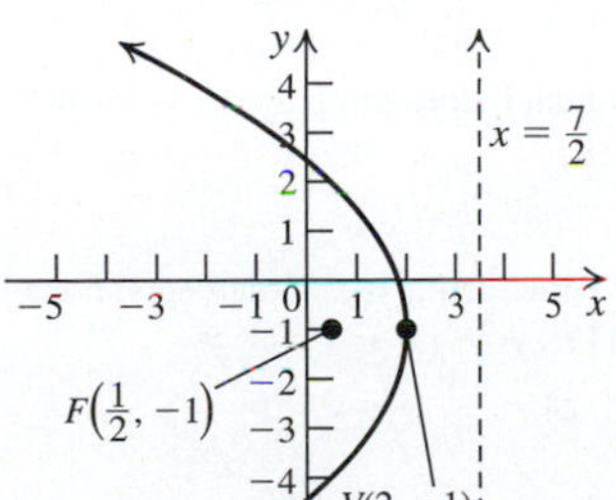

47. Vertex: $(1, 3)$; focus: $\left(1, \frac{1}{2}\right)$; directrix: $y = \frac{11}{2}$

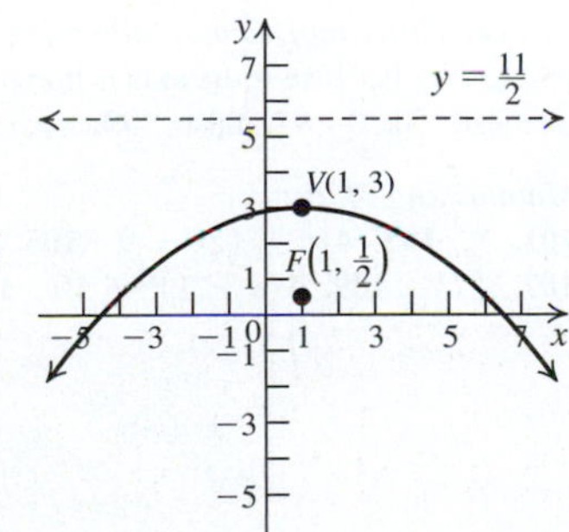

49. Vertex: $(-1, 1)$; focus: $\left(-1, \frac{5}{4}\right)$; directrix: $y = \frac{3}{4}$

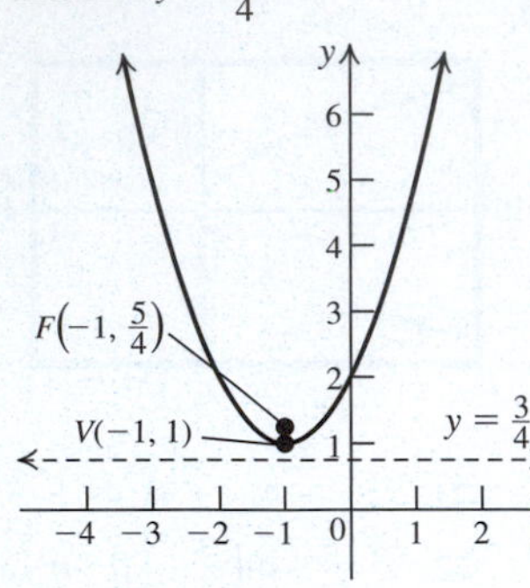

51. Vertex: $\left(-\frac{1}{2}, -1\right)$; focus: $\left(-\frac{1}{4}, -1\right)$; directrix: $x = -\frac{3}{4}$

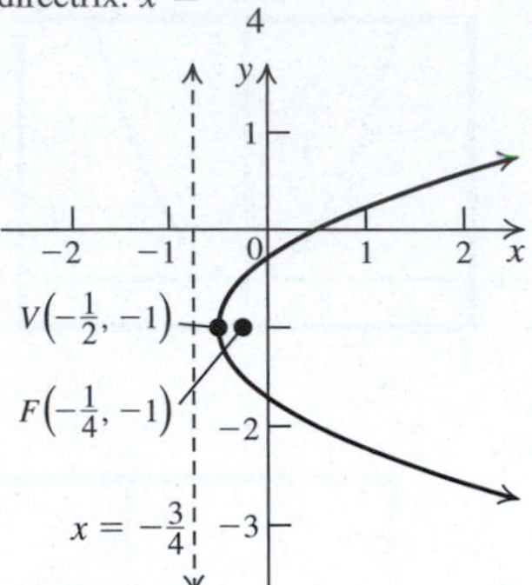

53. Vertex: $\left(-\frac{1}{2}, -1\right)$; focus: $\left(-\frac{1}{2}, -\frac{5}{4}\right)$; directrix: $y = -\frac{3}{4}$

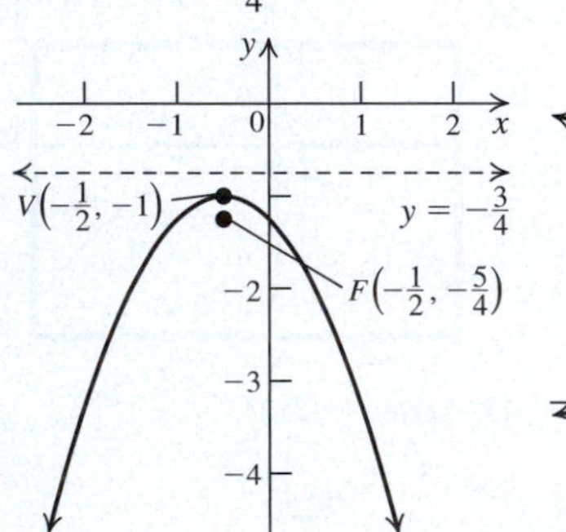

55. Vertex: $(46, 4)$; focus: $\left(\frac{551}{12}, 4\right)$; directrix: $x = \frac{553}{12}$

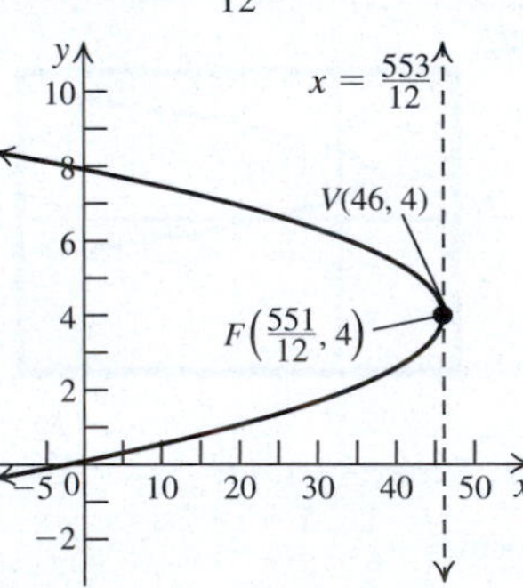

57. $y^2 = 4x$ and $x^2 = \frac{1}{2}y$ **59.** $(y - 1)^2 = 2x$ and $x^2 = 2(y - 1)$

61. $(y - 1)^2 = -(x + 2)$ and $(x + 2)^2 = -(y - 1)$

63. $(y - 1)^2 = x + 1$ and $(x + 1)^2 = y - 1$

Applying the Concepts:

65. 5 in **67.** $3\frac{3}{8}$ ft from the vertex **69.** $\left(\frac{1}{16}, 0\right)$ **71.** $\left(0, \frac{1}{16}\right)$

73. 16.875 ft **75.** 7.96 yd **77.** Output: 15 tons; cost: \$27.50

Beyond the Basics:

79. $(4, 8)$ and $(16, 16)$ **81.** $y = x - 4$ **83.** $(y - 1)^2 = 8(x - 1)$ and $(y - 1)^2 = -8(x - 5)$ **85.** $\left(x - \frac{3}{4}\right)^2 = \frac{1}{2}\left(y - \frac{31}{8}\right)$

87. Vertex: $(4, 6)$; focus: $\left(4, \frac{11}{2}\right)$; directrix: $y = \frac{13}{2}$; axis: $x = 4$

89. Vertex: $(-2, 3)$; focus: $\left(-\frac{11}{4}, 3\right)$; directrix: $x = -\frac{5}{4}$; axis: $y = 3$

91. $y = 6x - 3$

Critical Thinking/Discussion/Writing:

95. a. No, because if the axis is horizontal, it does not pass the vertical line test. **b.** 0 **97.** false **99.** $(9, -8)$

Maintaining Skills:

101. 7 **103.** $4x^2 + 12x + 9$ **105.** x-intercept: -4; y-intercepts: none **107.** $\{7\}$ **109.** $4(x - 1)^2 + 10$ **111.** $y = (x + 1)^2 + 2$

Section 10.3

Answers to Practice Problems:

1. $\frac{x^2}{36} + \frac{y^2}{100} = 1$ **2.**

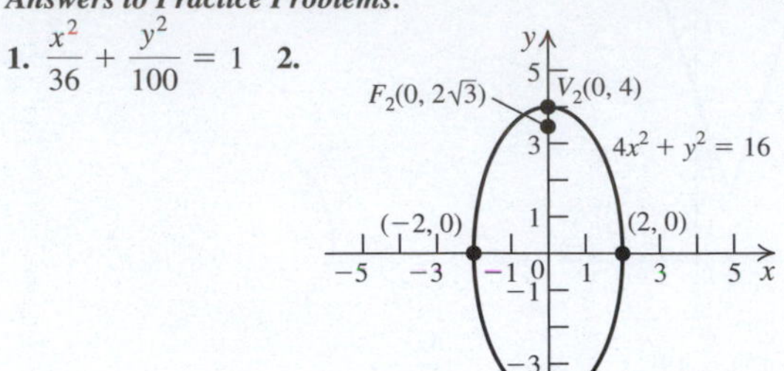

3. $\frac{(x - 2)^2}{9} + \frac{(y - 1)^2}{25} = 1$ **4.** Center: $(3, -1)$; vertices: $(3 + \sqrt{42}, -1)$ and $(3 - \sqrt{42}, -1)$; foci: $\left(3 + \frac{3\sqrt{14}}{2}, -1\right)$ and $\left(3 - \frac{3\sqrt{14}}{2}, -1\right)$ **5.** $4\sqrt{3} \approx 6.9282$ ft

Basic Concepts and Skills:

1. sum **3.** $a^2 - c^2$ **5.** true

7. Vertices: $(4, 0)$ and $(-4, 0)$; focus: $(2\sqrt{3}, 0)$ and $(-2\sqrt{3}, 0)$

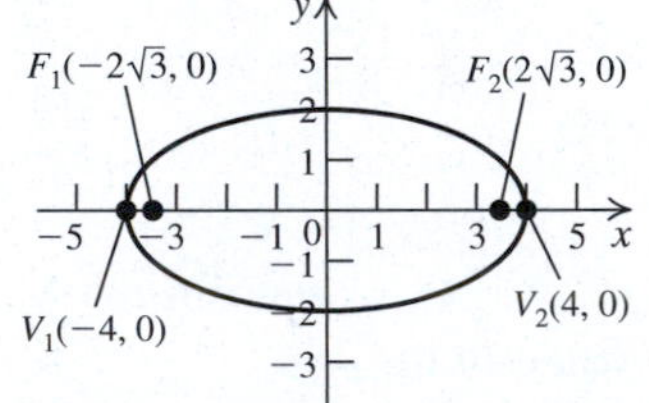

9. Vertices: $(3, 0)$ and $(-3, 0)$; foci: $(2\sqrt{2}, 0)$ and $(-2\sqrt{2}, 0)$

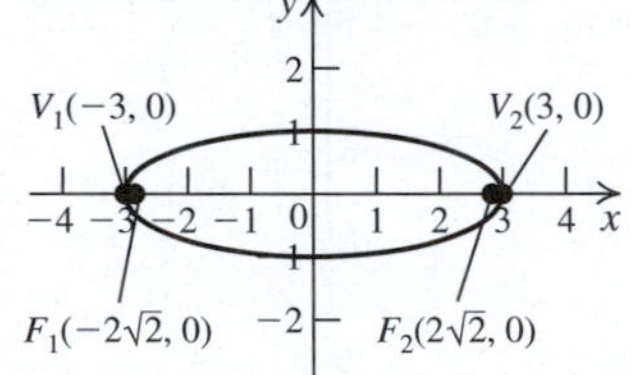

11. Vertices: $(5, 0)$ and $(-5, 0)$; foci: $(3, 0)$ and $(-3, 0)$

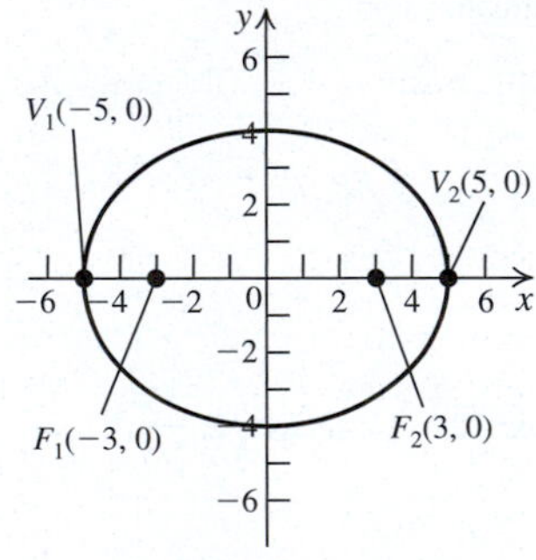

13. Vertices: $(0, 6)$ and $(0, -6)$; foci: $(0, 2\sqrt{5})$ and $(0, -2\sqrt{5})$

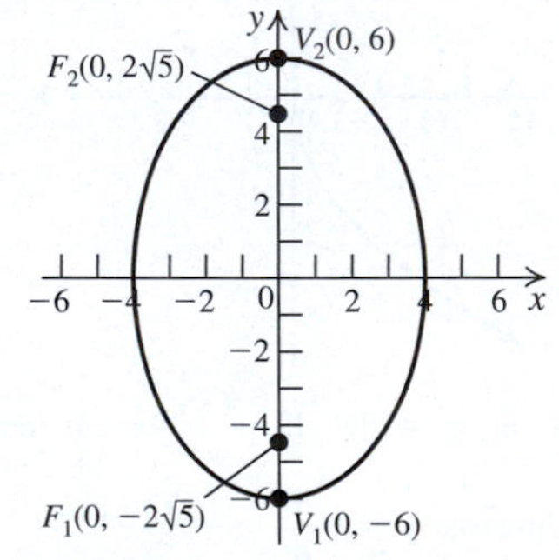

15. Circle centered at the origin with radius 2

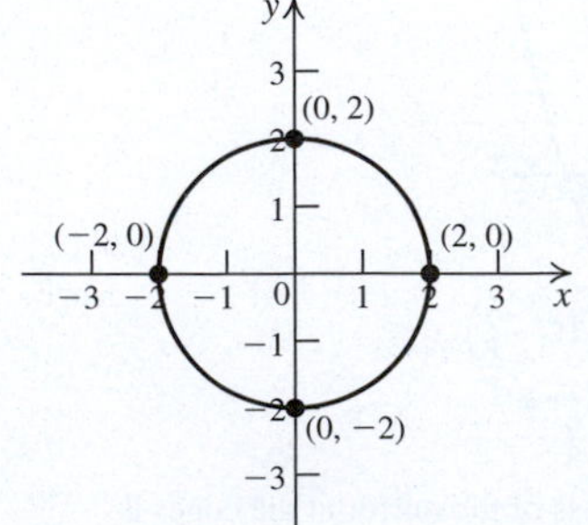

17. Vertices: $(2, 0)$ and $(-2, 0)$; foci: $(\sqrt{3}, 0)$ and $(-\sqrt{3}, 0)$

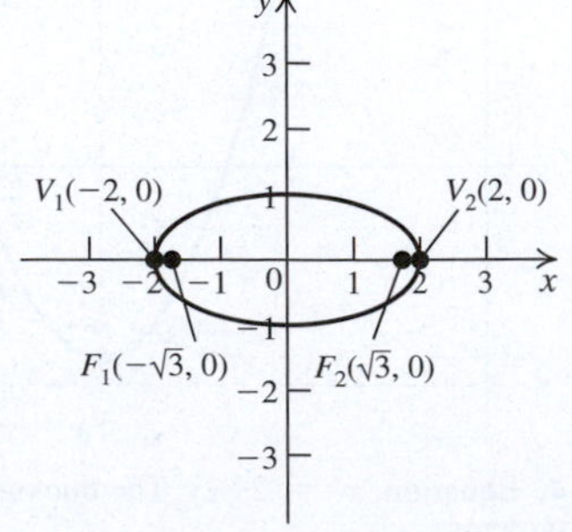

19. Vertices: $(0, 3)$ and $(0, -3)$; foci: $(0, \sqrt{5})$ and $(0, -\sqrt{5})$

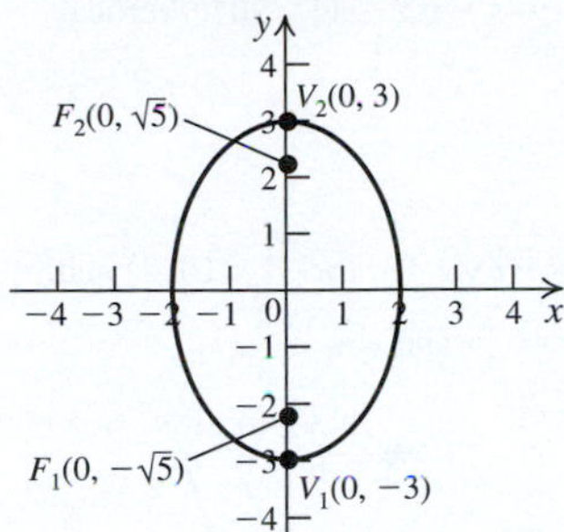

21. Vertices: $(2, 0)$ and $(-2, 0)$; foci: $(1, 0)$ and $(-1, 0)$

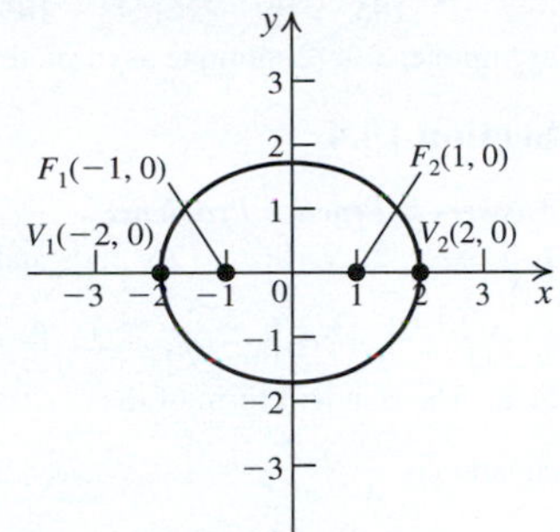

23. Vertices: $(0, \sqrt{5})$ and $(0, -\sqrt{5})$; foci: $(0, \sqrt{3})$ and $(0, -\sqrt{3})$

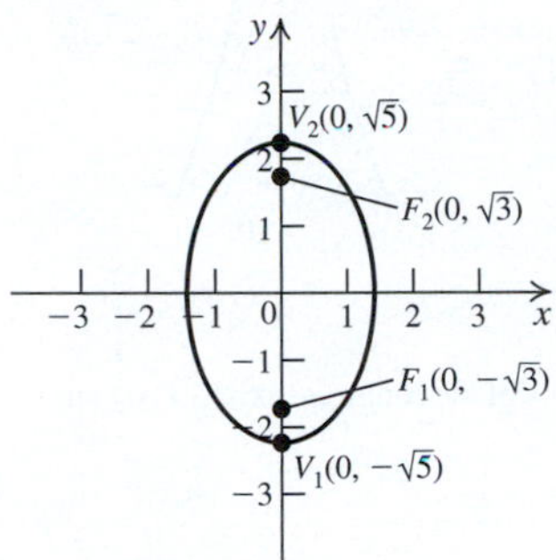

25. Vertices: $\left(\frac{\sqrt{14}}{2}, 0\right)$ and

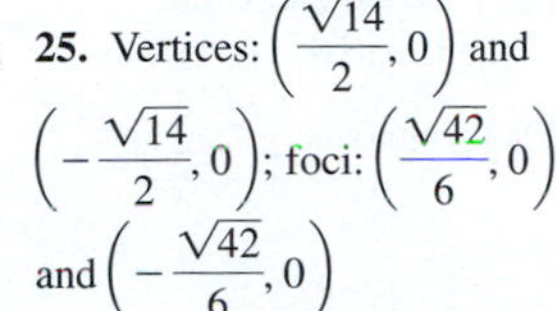

$\left(-\frac{\sqrt{14}}{2}, 0\right)$; foci: $\left(\frac{\sqrt{42}}{6}, 0\right)$ and $\left(-\frac{\sqrt{42}}{6}, 0\right)$

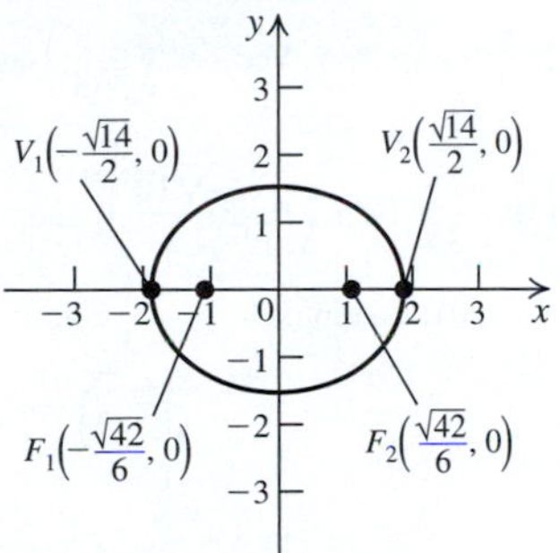

27. $\dfrac{x^2}{9} + \dfrac{y^2}{8} = 1$

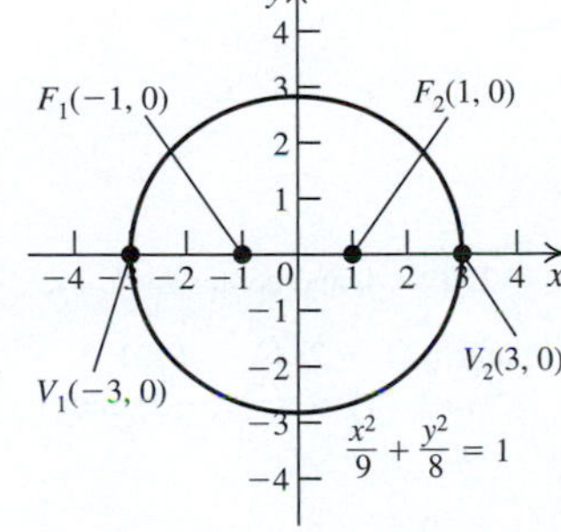

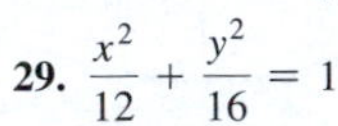

29. $\dfrac{x^2}{12} + \dfrac{y^2}{16} = 1$

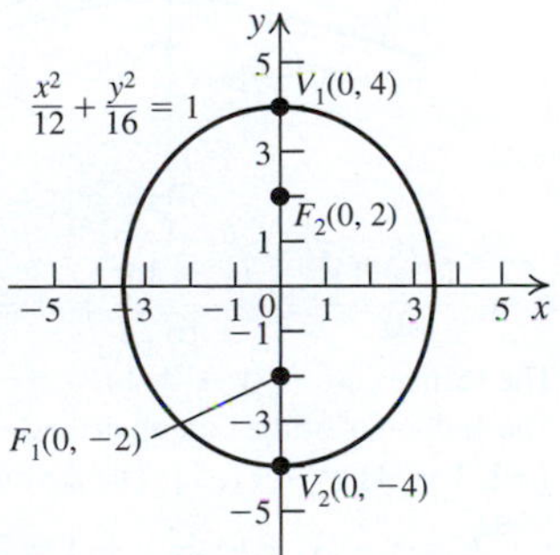

31. $\dfrac{x^2}{25} + \dfrac{y^2}{9} = 1$

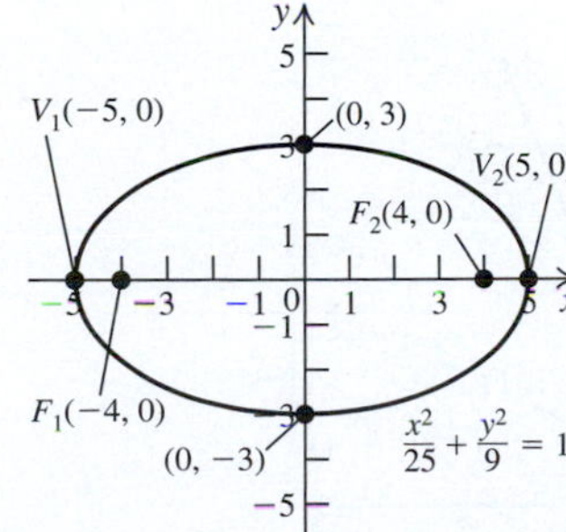

33. $\dfrac{x^2}{16} + \dfrac{y^2}{20} = 1$

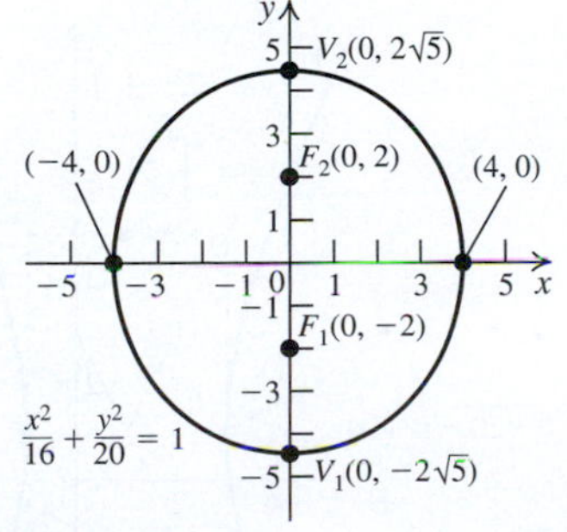

35. $\dfrac{x^2}{9} + \dfrac{y^2}{25} = 1$

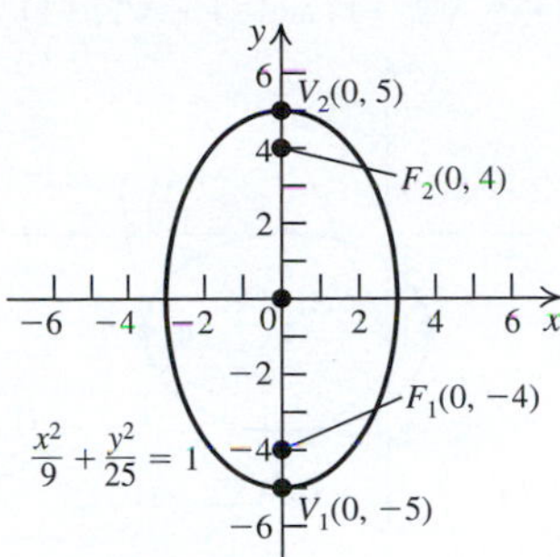

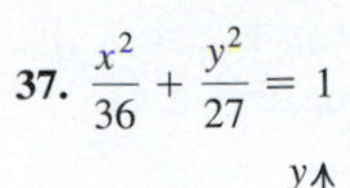

37. $\dfrac{x^2}{36} + \dfrac{y^2}{27} = 1$

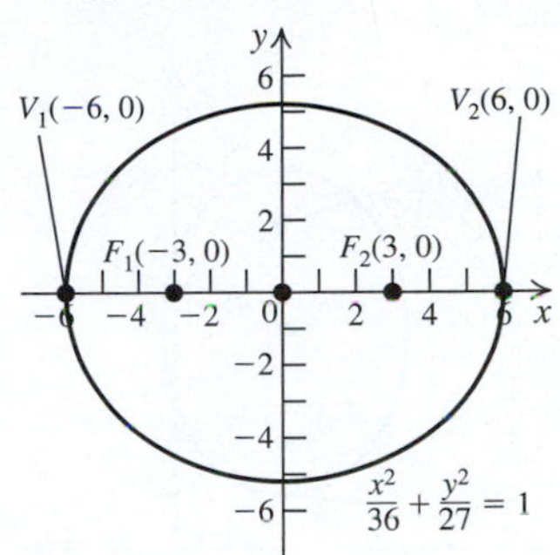

39. $\dfrac{x^2}{9} + \dfrac{y^2}{13} = 1$

$V_2(0, \sqrt{13})$, $F_2(0, 2)$, $F_1(0, -2)$, $V_1(0, -\sqrt{13})$, $\frac{x^2}{9} + \frac{y^2}{13} = 1$

41. Center: $(1, 1)$; foci:

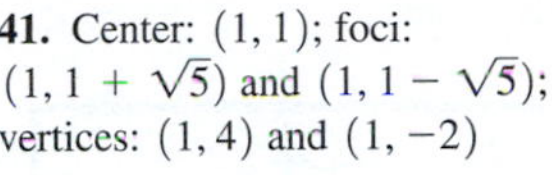

$(1, 1 + \sqrt{5})$ and $(1, 1 - \sqrt{5})$; vertices: $(1, 4)$ and $(1, -2)$

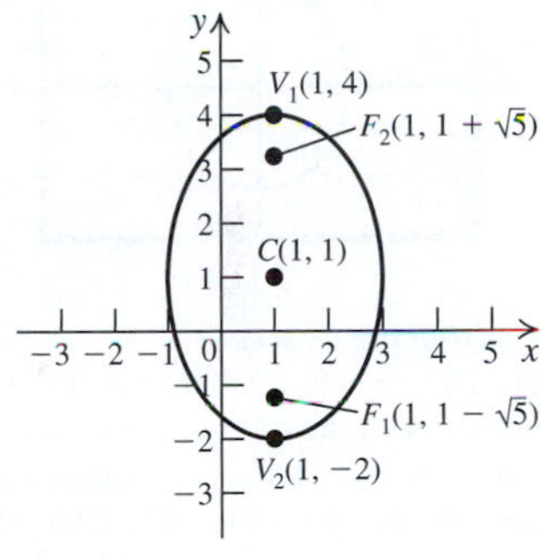

43. Center: $(0, -3)$; foci: $(2\sqrt{3}, -3)$ and $(-2\sqrt{3}, -3)$; vertices: $(4, -3)$ and $(-4, -3)$

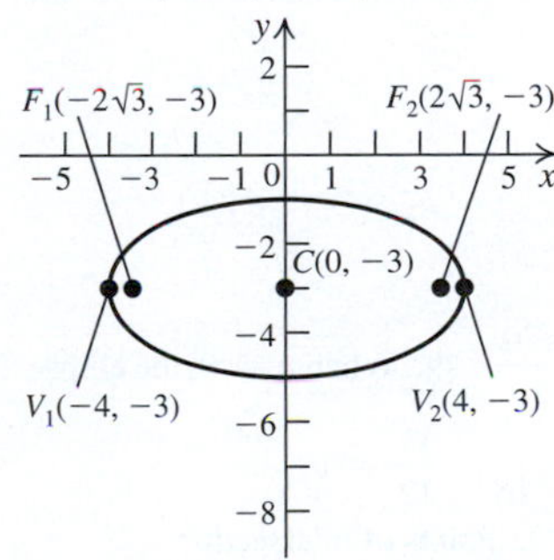

45. Center: $(1, -2)$; foci: $(2, -2)$ and $(0, -2)$; vertices: $(3, -2)$ and $(-1, -2)$

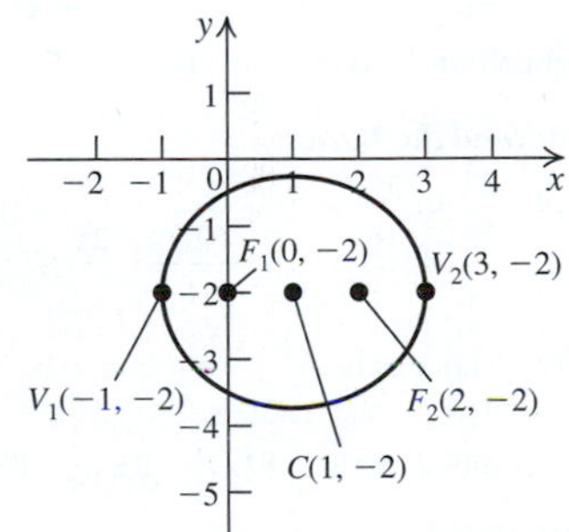

47. Center: $(-3, 1)$; foci: $(-2, 1)$ and $(-4, 1)$; vertices: $(-3 + \sqrt{5}, 1)$ and $(-3 - \sqrt{5}, 1)$

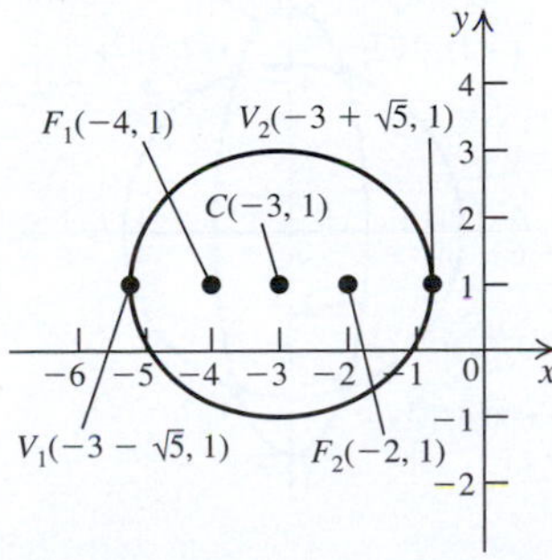

49. Center: $(-1, 2)$ foci: $(1, 2)$ and $(-3, 2)$; vertices: $(2, 2)$ and $(-4, 2)$

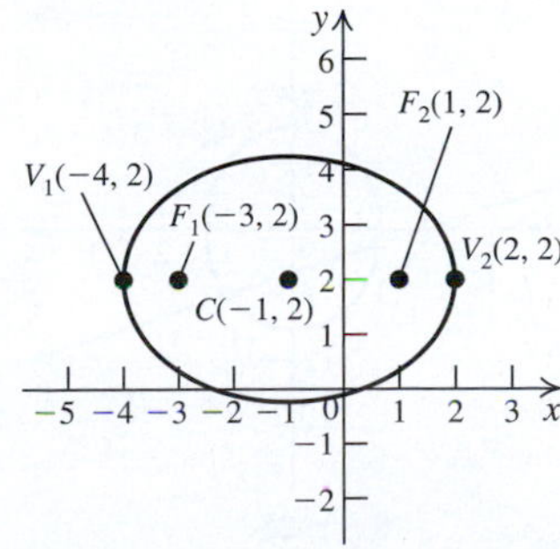

51. Center: $(-2, 4)$; foci: $(-2, 6)$ and $(-2, 2)$; vertices: $(-2, 7)$ and $(-2, 1)$

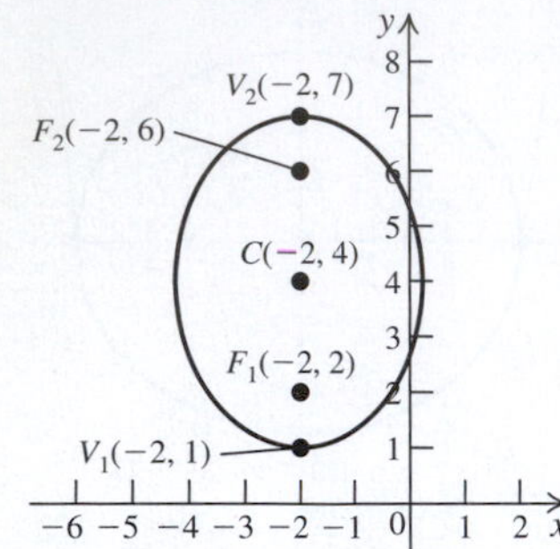

53. Center $(1, -1)$; foci: $(2, -1)$ and $(0, -1)$; vertices: $(1 + \sqrt{2}, -1)$ and $(1 - \sqrt{2}, -1)$

55. No graph

57.

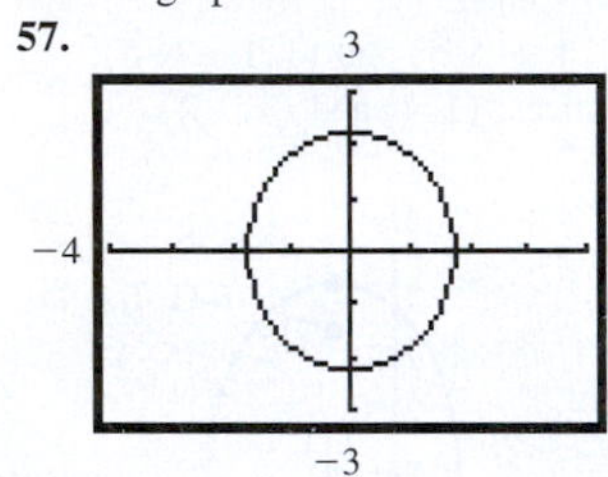

59.

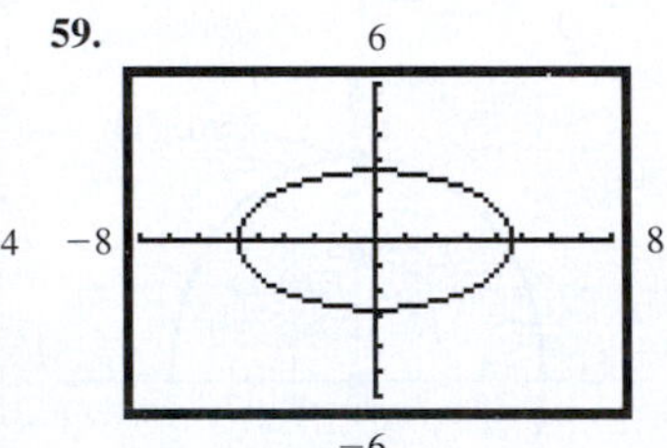

Applying the Concepts:

61. 18.3 ft **63.** 42.4 m **65.** The bet should not be accepted. The pool shark can hit the ball from any point straight into the pocket. or he can shoot it through the other focus and it will fall into the pocket because of the reflecting property. **67.** 29 ft from the walls on the opposite sides, along the major axis. **69.** $\frac{x^2}{8,641.5616} + \frac{y^2}{8,639.0652} = 1$

71. $\frac{x^2}{786,254.6241} + \frac{y^2}{783,788.5085} = 1$ **73.** Perihelion: 8.72×10^7 km; aphelion: 5.3028×10^9 km

Beyond the Basics:

75. $\frac{x^2}{25} + \frac{y^2}{16} = 1$ or $\frac{x^2}{\frac{625}{41}} + \frac{y^2}{25} = 1$

77. Major axis: $\frac{2\sqrt{105}}{3}$; minor axis: $\frac{\sqrt{70}}{2}$ **79.** When $e = 0$, the ellipse becomes a circle. **81.** $\frac{2}{3}$ **83.** $\frac{4}{5}$ **85.** $\frac{x^2}{16} + \frac{y^2}{12} = 1$

91. Points of intersection: $\left(-\frac{17}{13}, -\frac{3}{13}\right)$ and $(1, -1)$

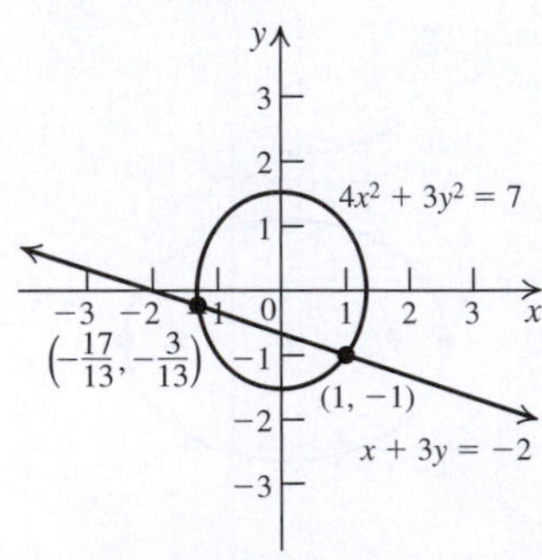

93. Points of intersection: $(\sqrt{2}, 3\sqrt{2})$, $(\sqrt{2}, -3\sqrt{2})$, $(-\sqrt{2}, 3\sqrt{2})$, $(-\sqrt{2}, -3\sqrt{2})$

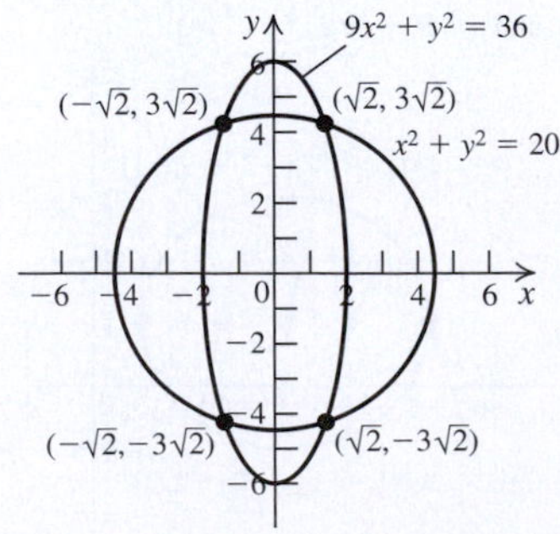

95. $y = -\frac{8}{3}x + \frac{25}{3}$

Critical Thinking/Discussion/Writing:

97. Four circles: $(x - 3)^2 + (y - 3)^2 = 9$, $(x + 3)^2 + (y - 3)^2 = 9$, $(x + 3)^2 + (y + 3)^2 = 9$, $(x - 3)^2 + (y + 3)^2 = 9$

Maintaining Skills:

99. Center: $(-3, 1)$; radius: $\sqrt{15}$ **101.** x-intercepts: none; y-intercepts: $-5, -1$ **103.** $y = -4x + 11$ **105.** $y = -6x - 11$ **107.** Vertical asymptote: $x = 2$; oblique asymptote: $y = 2x + 1$

Section 10.4

Answers to Practice Problems:

1. y-axis **2.** Vertices: $(2\sqrt{2}, 0)$ and $(-2\sqrt{2}, 0)$; foci: $(\sqrt{10}, 0)$ and $(-\sqrt{10}, 0)$ **3.** $\frac{y^2}{16} - \frac{x^2}{20} = 1$ **4.** $y = \frac{2}{3}x$ and $y = -\frac{2}{3}x$

5. a. The standard form of the equation is $\frac{x^2}{4} - \frac{y^2}{25} = 1$. Vertices: $(2, 0)$ and $(-2, 0)$; Foci: $(\pm\sqrt{29}, 0)$; endpoints of the conjugate axis: $(0, 5)$ and $(0, -5)$; asymptotes: $y = \frac{5}{2}x$ and $y = -\frac{5}{2}x$

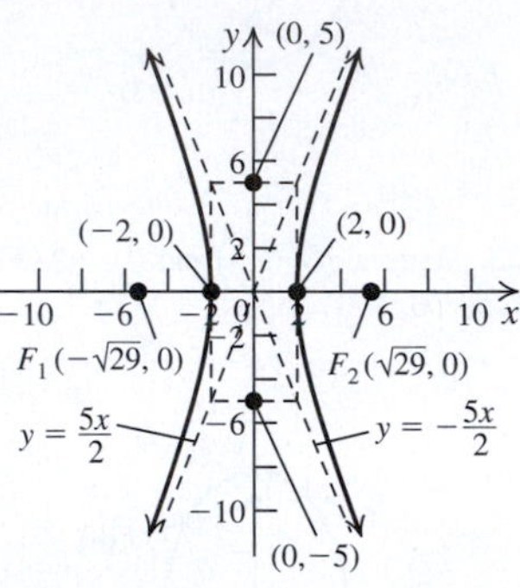

b. The standard form of the equation is $\frac{y^2}{\frac{1}{9}} - \frac{x^2}{1} = 1$. Vertices: $\left(0, \frac{1}{3}\right)$ and $\left(0, -\frac{1}{3}\right)$; Foci: $\left(0, \pm\frac{\sqrt{10}}{3}\right)$; endpoints of the conjugate axis: $(1, 0)$ and $(-1, 0)$; asymptotes: $y = \frac{1}{3}x$ and $y = -\frac{1}{3}x$

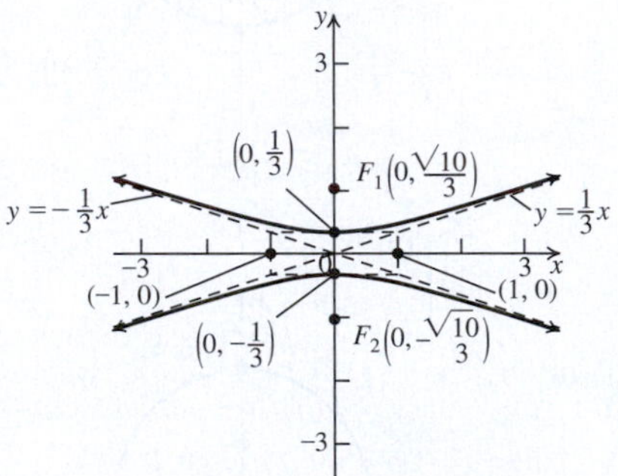

6. a. $\frac{(x + 1)^2}{4} - \frac{(y - 1)^2}{16} = 1 \Rightarrow a = 2, b = 4$, and center $(-1, 1)$. The vertices are $(-1 - 2, 1) = (-3, 1)$ and $(-1 + 2, 1) = (1, 1)$. The endpoints of the conjugate axes are $(-1, 1 - 4) = (-1, -3)$ and $(-1, 1 + 4) = (-1, 5)$. The asymptotes are $y - k = \pm\frac{b}{a}(x - h) \Rightarrow y - 1 = \pm 2(x + 1)$.

$c^2 = a^2 + b^2 \Rightarrow c^2 = 4 + 16 \Rightarrow c = \pm\sqrt{20}$

The foci are $(-\sqrt{20} - 1, 1)$ and $(\sqrt{20} - 1, 1)$.

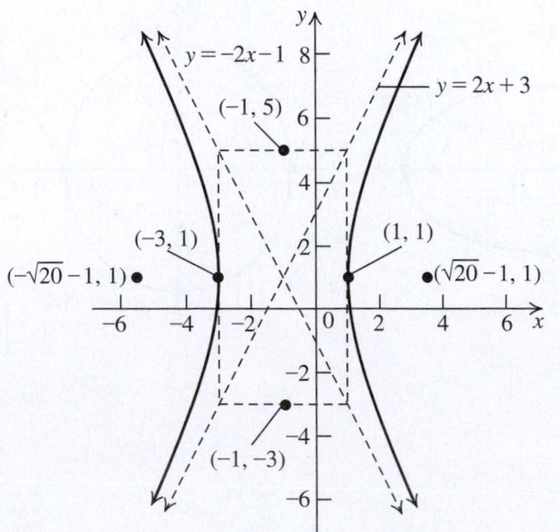

b. $\dfrac{(y-1)^2}{4} - \dfrac{(x+1)^2}{9} = 1 \Rightarrow a = 2, b = 3$, and center $(-1, 1)$.
The vertices are $(-1, 1-2) = (-1, -1)$ and $(-1, 1+2) = (-1, 3)$. The endpoints of the conjugate axes are $(-1-3, 1) = (-4, 1)$ and $(-1+3, 1) = (2, 1)$ The asymptotes are
$y - k = \pm\dfrac{a}{b}(x-h) \Rightarrow y - 1 = \pm\dfrac{2}{3}(x+1)$.
$c^2 = a^2 + b^2 \Rightarrow c^2 = 4 + 9 \Rightarrow c = \pm\sqrt{13}$ The foci are $(-1, 1+\sqrt{13})$ and $(-1, 1-\sqrt{13})$.

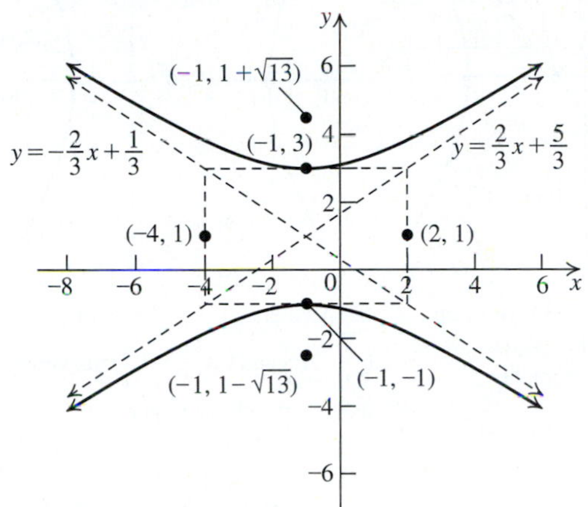

7. $\dfrac{(x-1)^2}{5} - \dfrac{(y-2)^2}{\frac{5}{4}} = 1$.

Center: $(1, 2)$; vertices: $(1+\sqrt{5}, 2)$ and $(1-\sqrt{5}, 2)$; endpoints of the conjugate axis: $\left(1, 2+\dfrac{\sqrt{5}}{2}\right)$ and $\left(1, 2-\dfrac{\sqrt{5}}{2}\right)$; asymptotes: $y - 2 = \pm\dfrac{1}{2}(x-1)$; foci: $\left(\dfrac{7}{2}, 2\right)$ and $\left(-\dfrac{3}{2}, 2\right)$

8. $\dfrac{x^2}{16{,}900} - \dfrac{y^2}{5{,}600} = 1$

Basic Concepts and Skills:
1. difference **3.** $c^2 - a^2$ **5.** true **7.** g **9.** h **11.** d **13.** b

15. Vertices: $(1, 0)$ and $(-1, 0)$; foci: $(\sqrt{5}, 0)$ and $(-\sqrt{5}, 0)$; transverse axis: x-axis; the hyperbola opens left and right; vertices of the fundamental rectangle: $(1, 2), (-1, 2), (-1, -2), (1, -2)$; asymptotes: $y = \pm 2x$

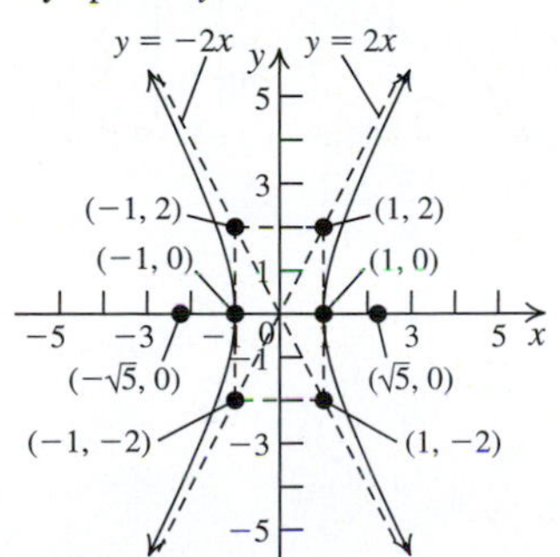

17. Vertices: $(0, 1)$ and $(0, -1)$; foci: $(0, \sqrt{2})$ and $(0, -\sqrt{2})$; transverse axis: y-axis; the hyperbola opens up and down; vertices of the fundamental rectangle: $(1, 1), (-1, 1), (-1, -1), (1, -1)$; asymptotes: $y = \pm x$

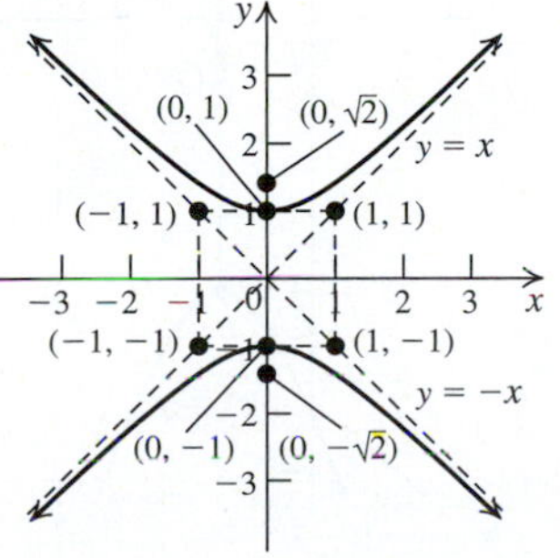

19. Vertices: $(0, 2)$ and $(0, -2)$; foci: $(0, 2\sqrt{10})$ and $(0, -2\sqrt{10})$; transverse axis: y-axis; the hyperbola opens up and down; vertices of the fundamental rectangle: $(6, 2), (-6, 2), (-6, -2), (6, -2)$; asymptotes: $y = \pm\dfrac{1}{3}x$

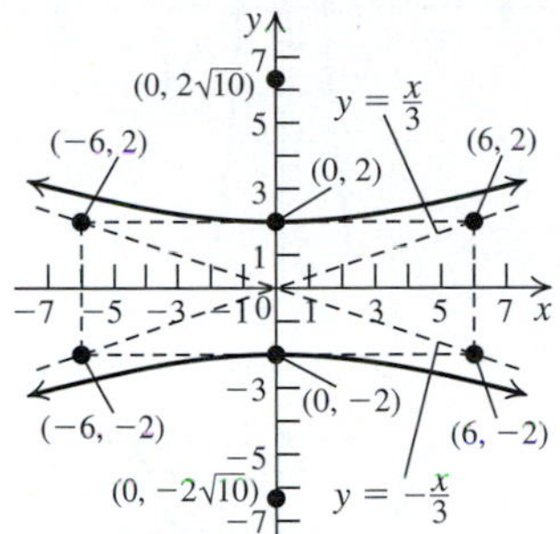

21. Vertices: $(3, 0)$ and $(-3, 0)$; foci: $(\sqrt{13}, 0)$ and $(-\sqrt{13}, 0)$; transverse axis: x-axis; the hyperbola opens left and right; vertices of the fundamental rectangle: $(3, 2), (-3, 2), (-3, -2), (3, -2)$; asymptotes: $y = \pm\dfrac{2}{3}x$

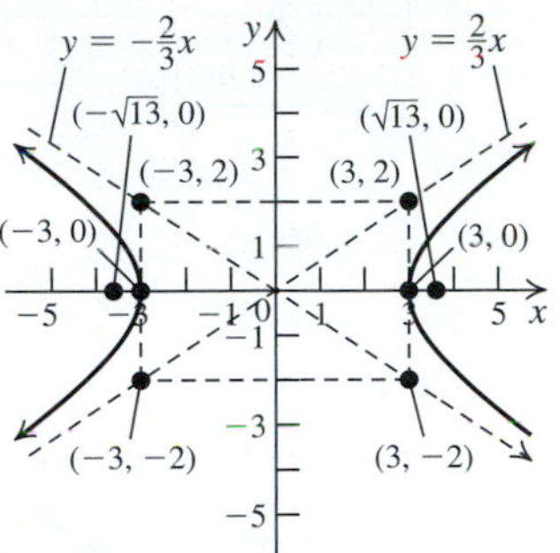

23. Vertices: $(0, 1)$ and $(0, -1)$; foci: $\left(0, \dfrac{\sqrt{5}}{2}\right)$ and $\left(0, -\dfrac{\sqrt{5}}{2}\right)$; transverse axis: y-axis; the hyperbola opens up and down; vertices of the fundamental rectangle: $\left(\dfrac{1}{2}, 1\right), \left(-\dfrac{1}{2}, 1\right), \left(-\dfrac{1}{2}, -1\right), \left(\dfrac{1}{2}, -1\right)$; asymptotes: $y = \pm 2x$

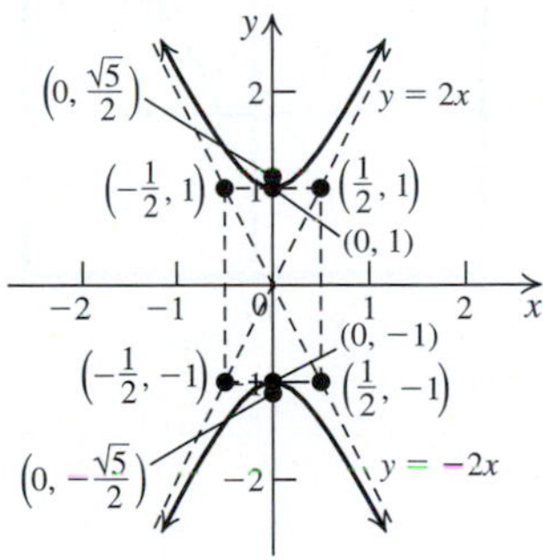

25. Vertices: $\left(\dfrac{1}{3}, 0\right)$ and $\left(-\dfrac{1}{3}, 0\right)$; foci: $\left(\dfrac{\sqrt{10}}{3}, 0\right)$ and $\left(-\dfrac{\sqrt{10}}{3}, 0\right)$; transverse axis: x-axis; the hyperbola opens left and right; vertices of the fundamental rectangle: $\left(\dfrac{1}{3}, 1\right), \left(-\dfrac{1}{3}, 1\right), \left(-\dfrac{1}{3}, -1\right), \left(\dfrac{1}{3}, -1\right)$; asymptotes: $y = \pm 3x$

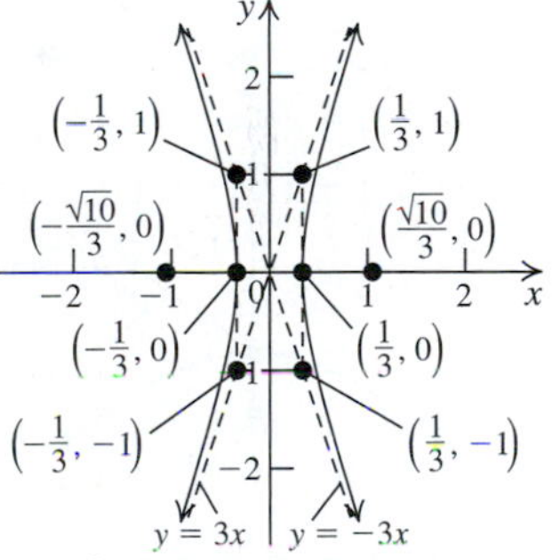

27. $\dfrac{x^2}{4} - \dfrac{y^2}{5} = 1$

29. 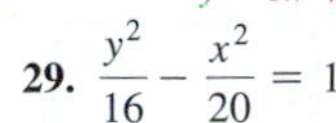 $\dfrac{y^2}{16} - \dfrac{x^2}{20} = 1$

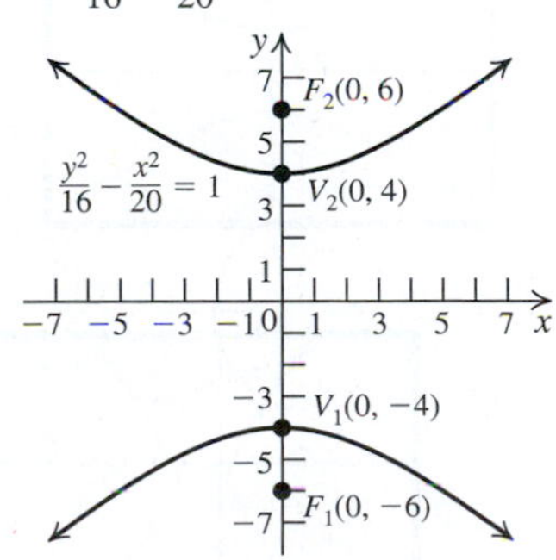

31. $\dfrac{y^2}{4} - \dfrac{x^2}{21} = 1$

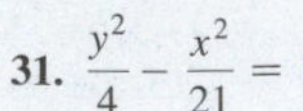

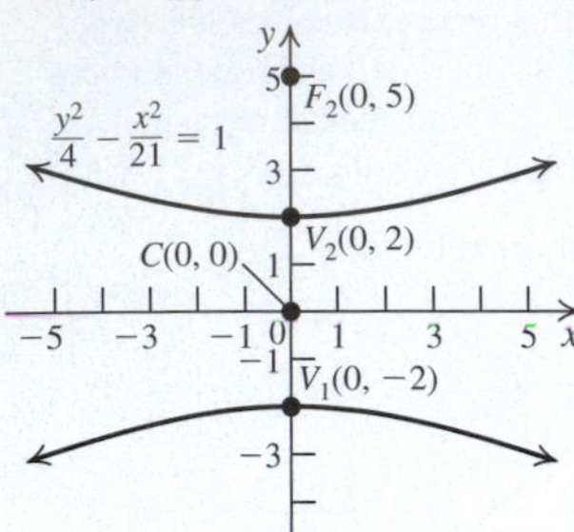

33. $x^2 - \dfrac{y^2}{24} = 1$

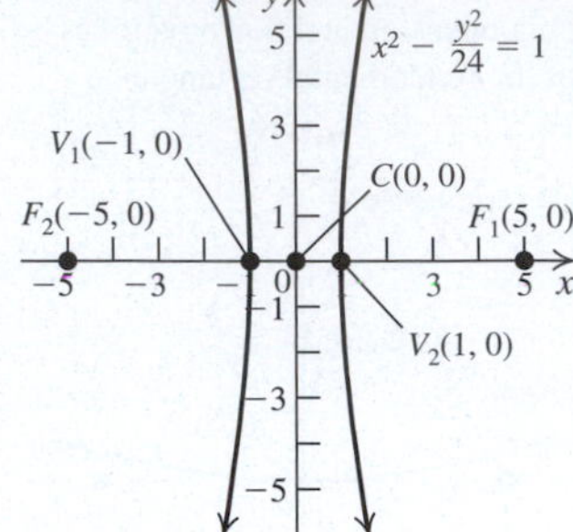

35. $\dfrac{y^2}{9} - \dfrac{x^2}{16} = 1$

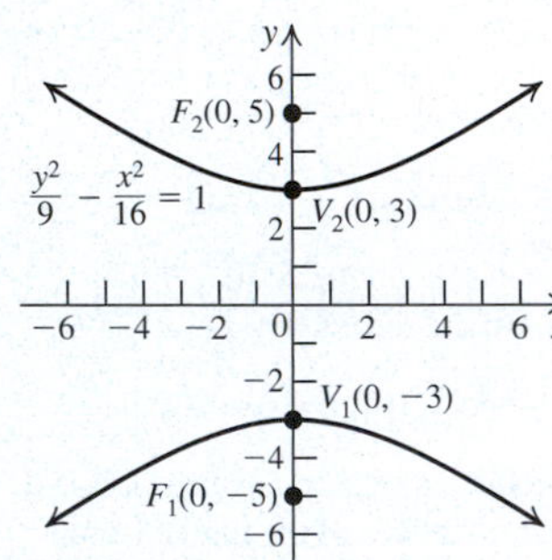

37. $x^2 - \dfrac{y^2}{4} = 1$

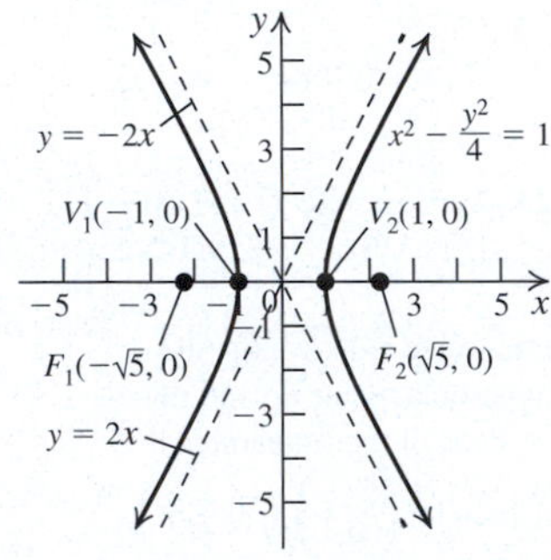

39. $\dfrac{y^2}{16} - \dfrac{x^2}{16} = 1$

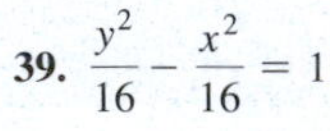

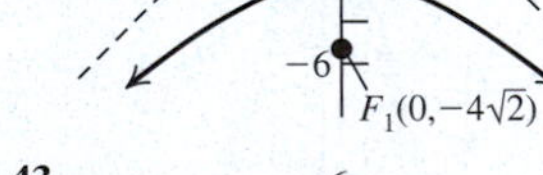

41.

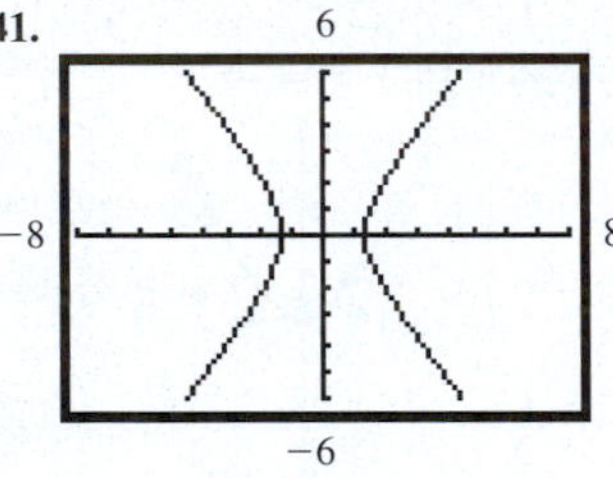

43.

45.

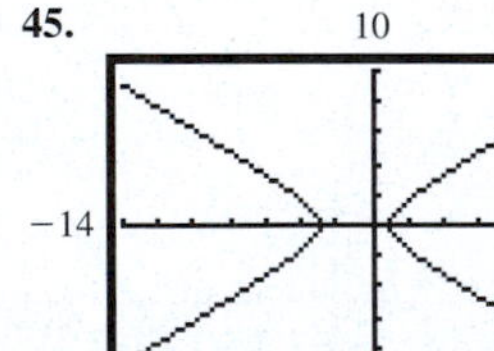

47.

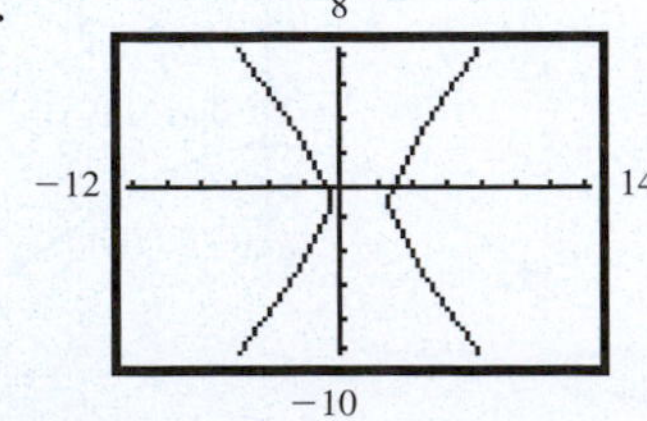

49. Center: $(1, -1)$; vertices: $(4, -1)$ and $(-2, -1)$; transverse axis: $y = -1$; asymptotes: $y + 1 = \pm\dfrac{4}{3}(x + 1)$

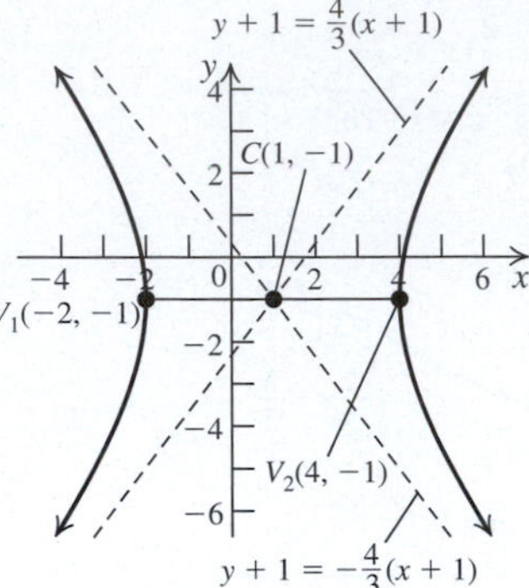

51. Center: $(-2, 0)$; vertices: $(3, 0)$ and $(-7, 0)$; transverse axis: x-axis; asymptotes: $y = \pm\dfrac{7}{5}(x + 2)$

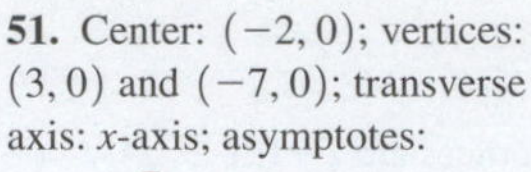

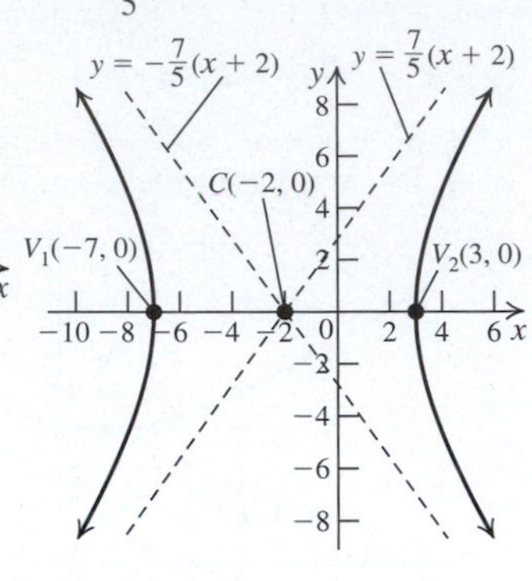

53. Center: $(-4, -3)$; vertices: $(1, -3)$ and $(-9, -3)$; transverse axis: $y = -3$; asymptotes: $y + 3 = \pm\dfrac{7}{5}(x + 4)$

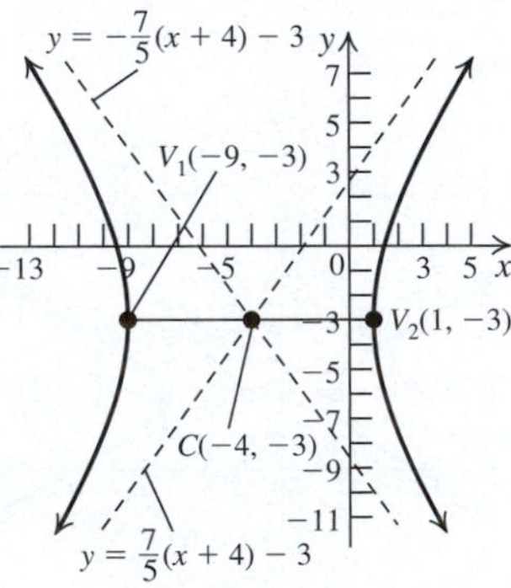

55. Center: $(0, -1)$; vertices: $\left(\dfrac{5}{2}, -1\right)$ and $\left(-\dfrac{5}{2}, -1\right)$; transverse axis: $y = -1$; asymptotes: $y + 1 = \pm 2x$

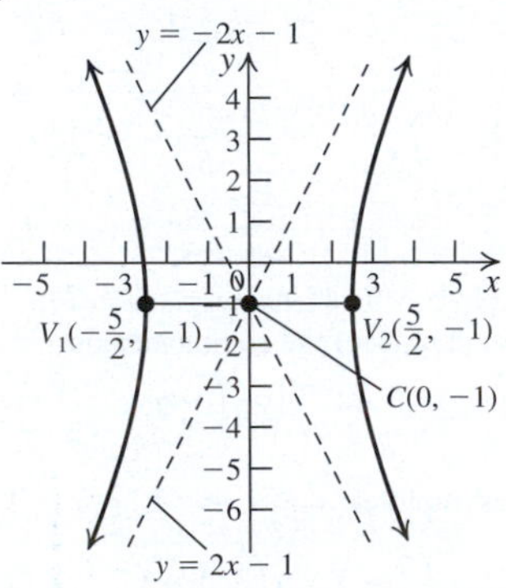

57. Center: $(2, -1)$; vertices: $(2, 4)$ and $(2, -6)$; transverse axis: $x = 2$; asymptotes: $y + 1 = \pm 3x - 2$

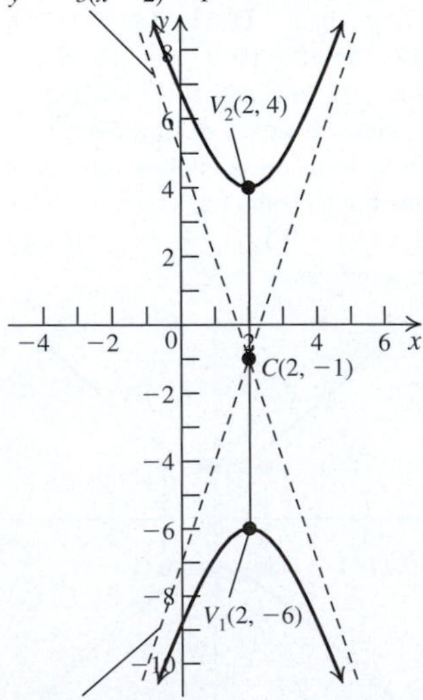

59. Center: $(-3, 0)$; vertices: $(-3 + 3\sqrt{5}, 0)$ and $(-3 - 3\sqrt{5}, 0)$; transverse axis: x-axis; asymptotes: $y = \pm(x + 3)$

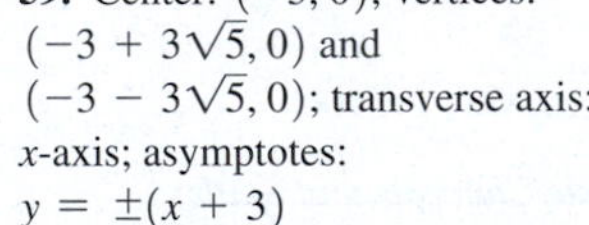

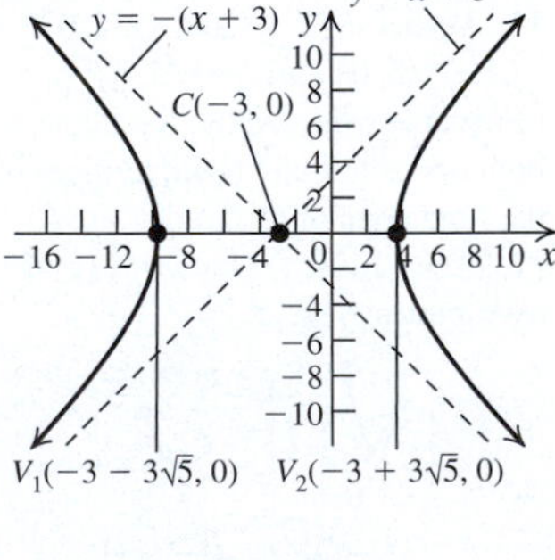

61. Center: $(2, 0)$; vertices: $(4, 0)$ and $(0, 0)$; transverse axis: x-axis; asymptotes: $y = \pm\frac{1}{2}(x - 2)$

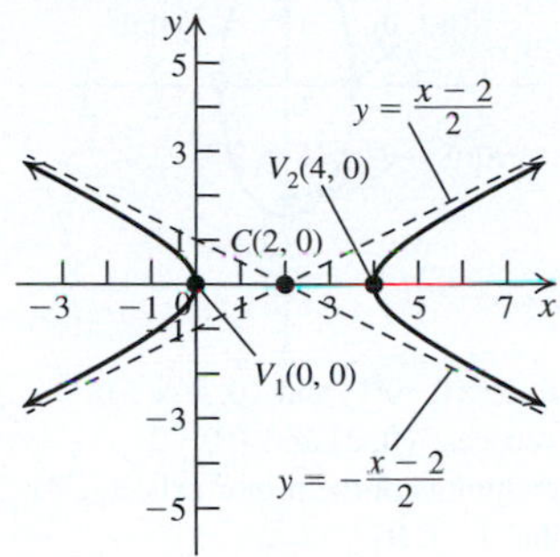

63. Center: $(-3, -4)$; vertices: $(-3, -3)$ and $(-3, -5)$; transverse axis: $x = -3$; asymptotes: $y + 4 = \pm\sqrt{2}(x + 3)$

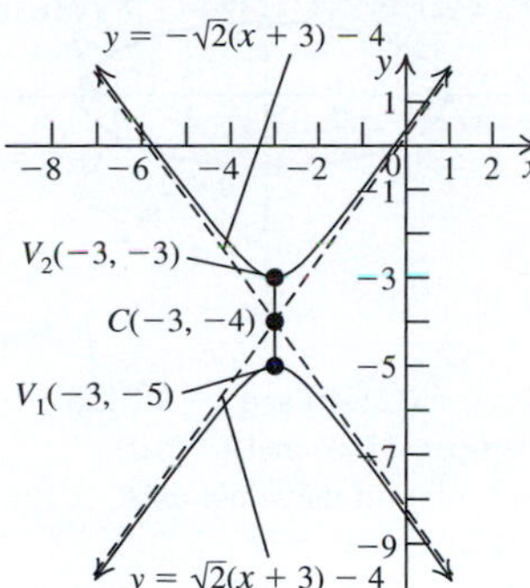

65. Center: $(3, -2)$; vertices: $(3 + \sqrt{6}, -2)$ and $(3 - \sqrt{6}, -2)$; transverse axis: $y = -2$; asymptotes:

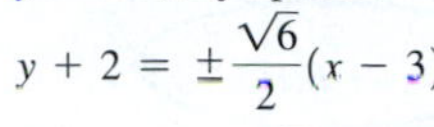

$y + 2 = \pm\frac{\sqrt{6}}{2}(x - 3)$

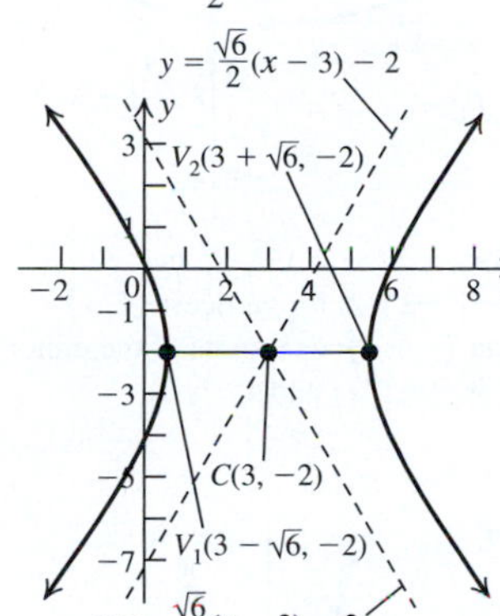

67. Center: $(\sqrt{2}, 0)$; vertices: $(\sqrt{2} + \sqrt{2\sqrt{2} + 1}, 0)$ and $(\sqrt{2} - \sqrt{2\sqrt{2} + 1}, 0)$; transverse axis: x-axis; asymptotes: $y = \pm(x - \sqrt{2})$

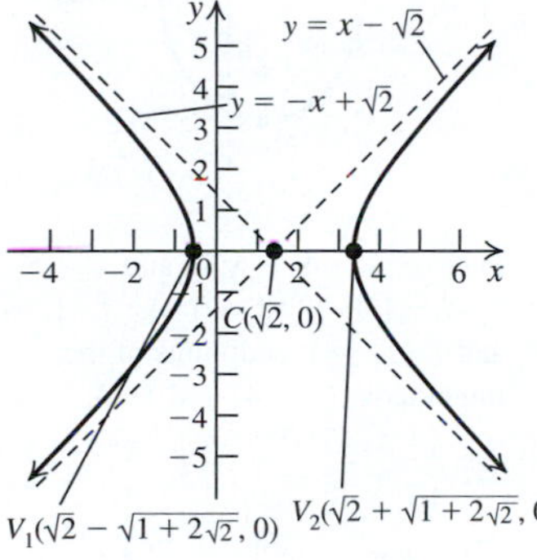

69. Parabola

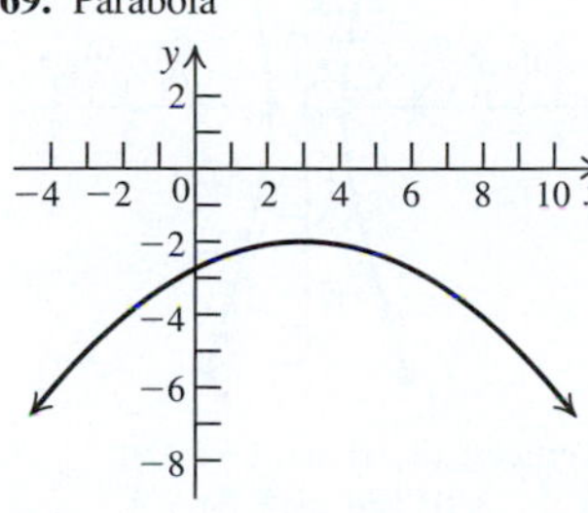

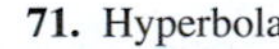

71. Hyperbola

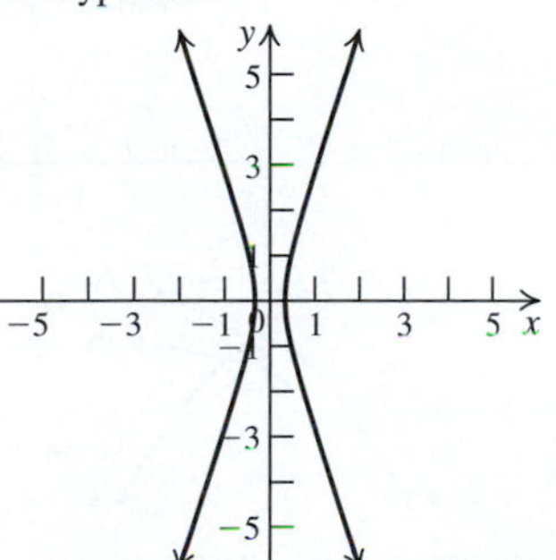

73. Circle

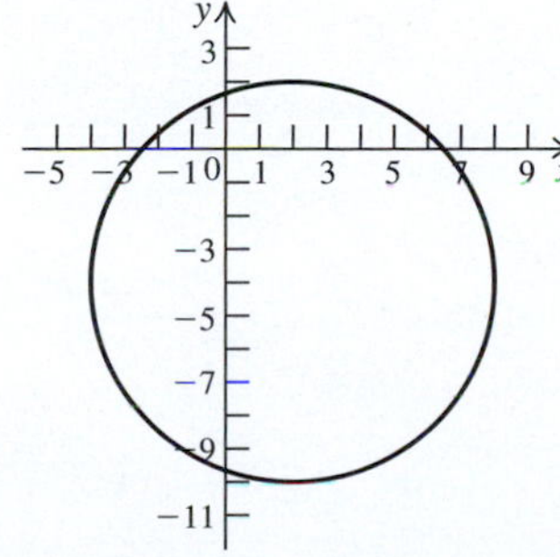

75. Parabola

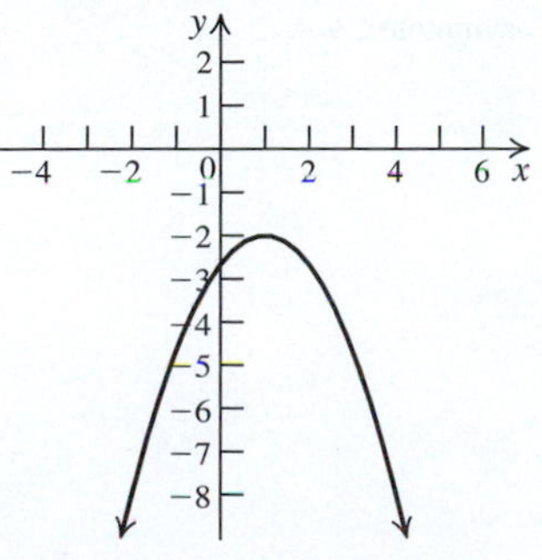

77. Ellipse

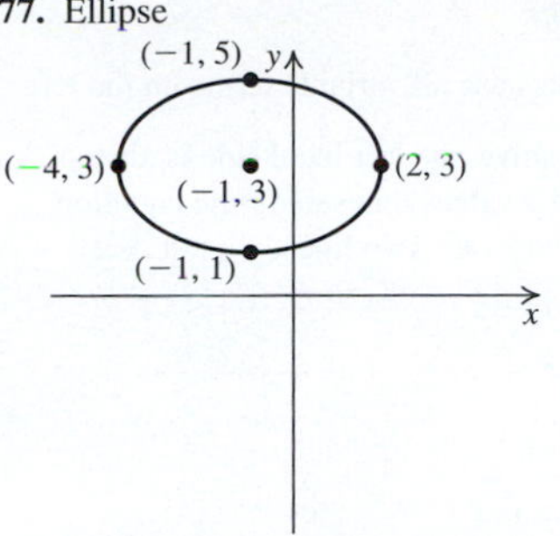

Applying the Concepts:

79. The difference in the distances of the location of the explosion from A and B is given 600 m. Therefore, the location of the explosion is restricted to points on a hyperbola, the length of whose transverse axis is 600 m. Equation: $\frac{x^2}{90{,}000} - \frac{y^2}{160{,}000} = 1$

81. Let the coordinates of Nicole be $(-4000, 0)$ and those of Juan be $(4000, 0)$. The lightening is 3300 ft closer to Nicole than Juan. The location of the lightening is on a hyperbola with equation $\frac{x^2}{2{,}722{,}500} - \frac{y^2}{13{,}277{,}500} = 1$.

83. Let the coordinates of A be $(-150, 0)$ and the coordinates of B be $(150, 0)$. Equation: $\frac{x^2}{5625} - \frac{y^2}{16{,}875} = 1$

85. $\frac{x^2}{193{,}600} - \frac{y^2}{446{,}400} = 1$ **87.** $(-68.5748, 44.2719)$

Beyond the Basics:

89. a. $x^2 - \frac{y^2}{24} = 1$ **b.** $\frac{x^2}{4} - \frac{y^2}{21} = 1$ **c.** $\frac{x^2}{16} - \frac{y^2}{9} = 1$

91. a. $y^2 - \frac{x^2}{35} = 1$ **b.** $y^2 - \frac{x^2}{15} = 1$ **c.** $y^2 - \frac{x^2}{8} = 1$

97. When $e = 1$, the graph is the union of two rays. **99.** $e = \frac{\sqrt{61}}{5}$; length of the latus rectum: $\frac{72}{5}$ **101.** $e = 3$; length of the latus rectum: 8

107. Points of intersection: $\left(-\frac{91}{20}, \frac{109}{10}\right)$

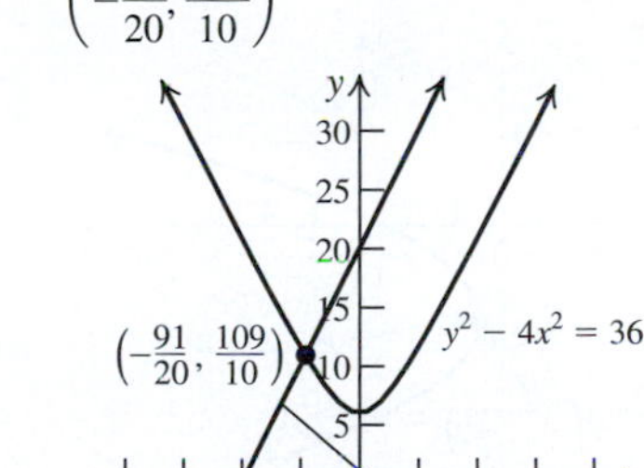

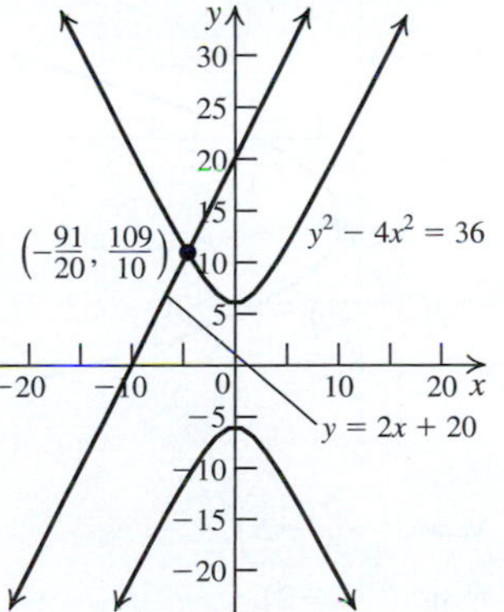

109. Points of intersection: $(-2\sqrt{2}, -\sqrt{7})$, $(-2\sqrt{2}, \sqrt{7})$, $(2\sqrt{2}, -\sqrt{7})$, $(2\sqrt{2}, \sqrt{7})$

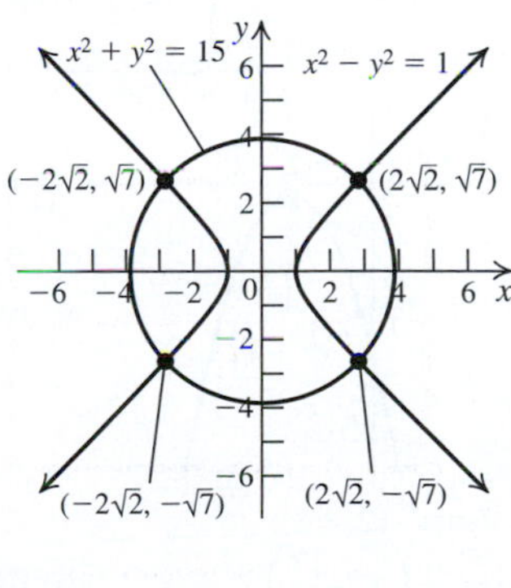

111. Points of intersection: $(-2, -3)$, $(-2, 3)$, $(2, -3)$, $(2, 3)$

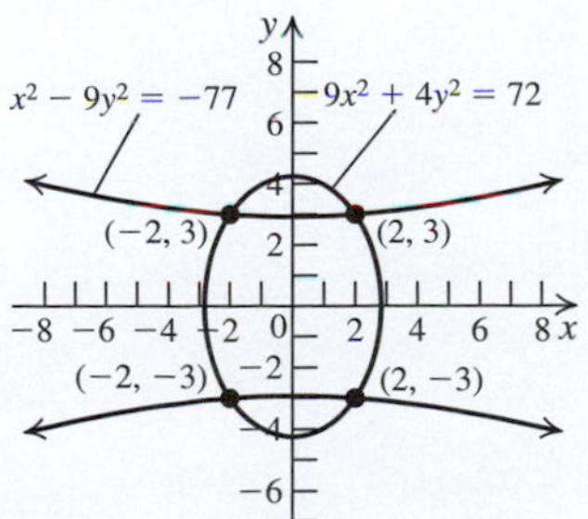

113. $y = \frac{6}{7}x - \frac{5}{7}$

Critical Thinking/Discussion/Writing:

115. a. Two lines, $y = \pm\frac{2}{3}x$ **b.** Because all variable terms on the left-hand side are nonnegative and 6 is positive, the left-hand side is always positive. Therefore, there are no x and y values that satisfy the equation. **c.** A point, $(0, 0)$ **d.** A point, $(2, -4)$ **e.** Two lines, $y = \pm\sqrt{2}x$

117.

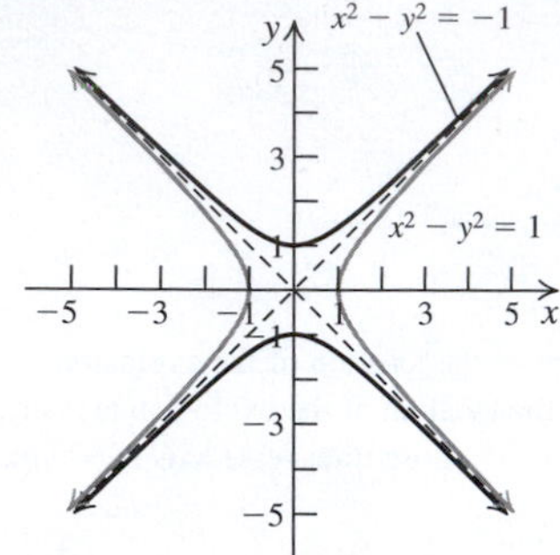

Maintaining Skills:

119. 21 **121.** 2 **123.** $\frac{1}{9}$ **125.** −16 **127.** 48 **129.** true

Review Exercises

1.

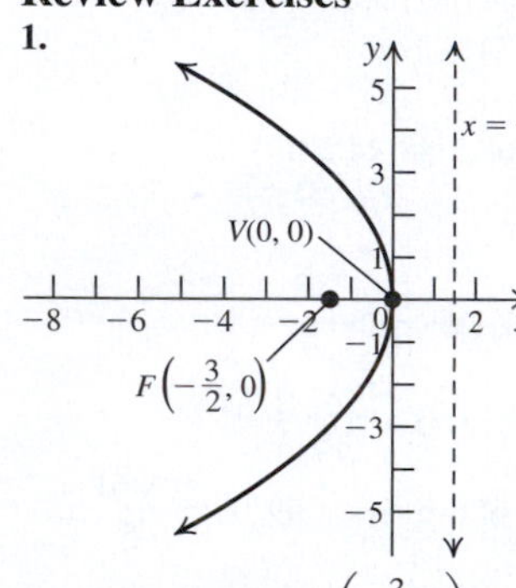

Vertex: $(0, 0)$; focus: $\left(-\frac{3}{2}, 0\right)$; axis: x-axis; directrix: $x = \frac{3}{2}$

3.

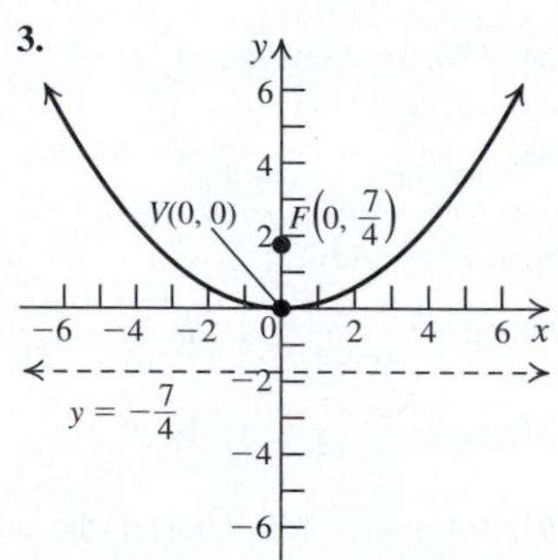

Vertex: $(0, 0)$; focus: $\left(0, \frac{7}{4}\right)$; axis: y-axis; directrix: $y = -\frac{7}{4}$

5.

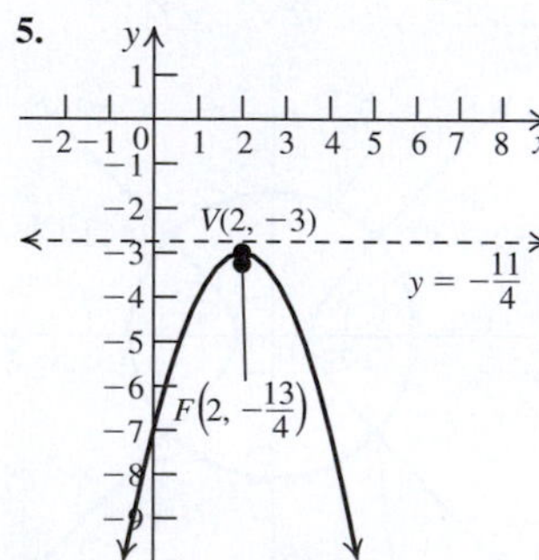
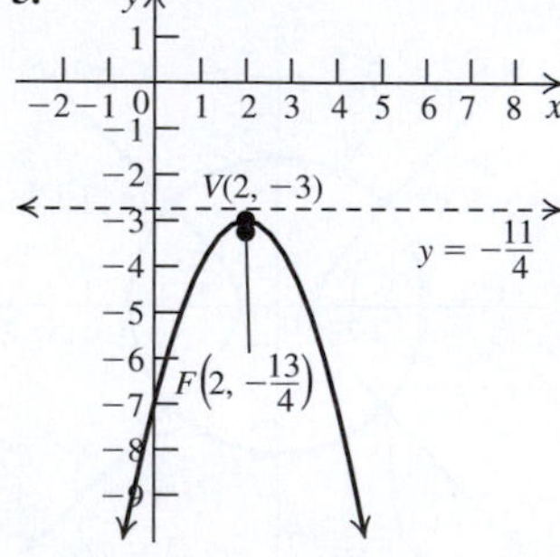

Vertex: $(2, -3)$; focus: $\left(2, -\frac{13}{4}\right)$; axis: $x = 2$; directrix: $y = -\frac{11}{4}$

7.

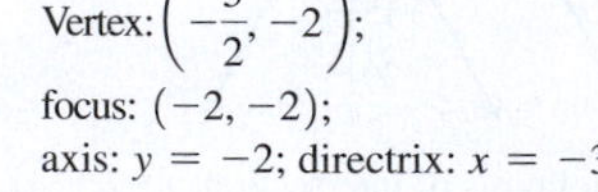

Vertex: $\left(-\frac{5}{2}, -2\right)$; focus: $(-2, -2)$; axis: $y = -2$; directrix: $x = -3$

9. $y^2 = -12x$ **11.** $x^2 = 16y$

13.

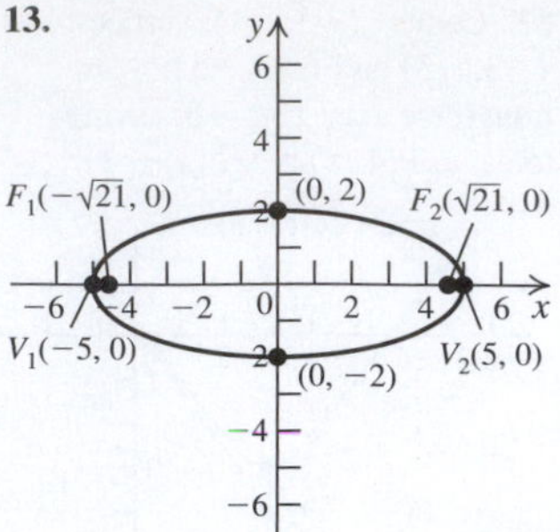

Foci: $(\sqrt{21}, 0)$ and $(-\sqrt{21}, 0)$; vertices: $(5, 0)$ and $(-5, 0)$; endpoints of the minor axis: $(0, 2)$ and $(0, -2)$

15.

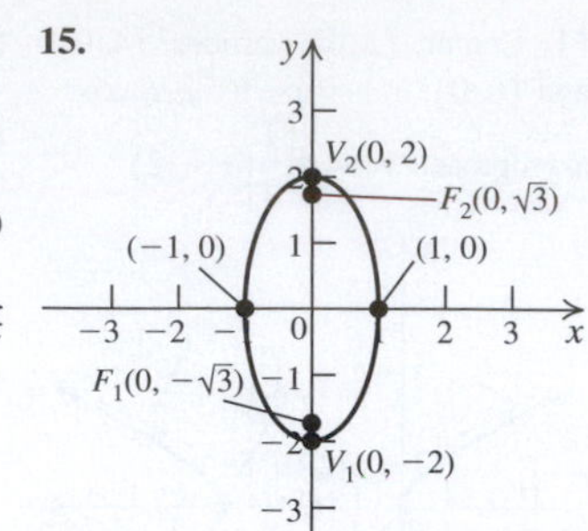

Foci: $(0, \sqrt{3})$ and $(0, -\sqrt{3})$; vertices: $(0, 2)$ and $(0, -2)$; endpoints of the minor axis: $(1, 0)$ and $(-1, 0)$

17.

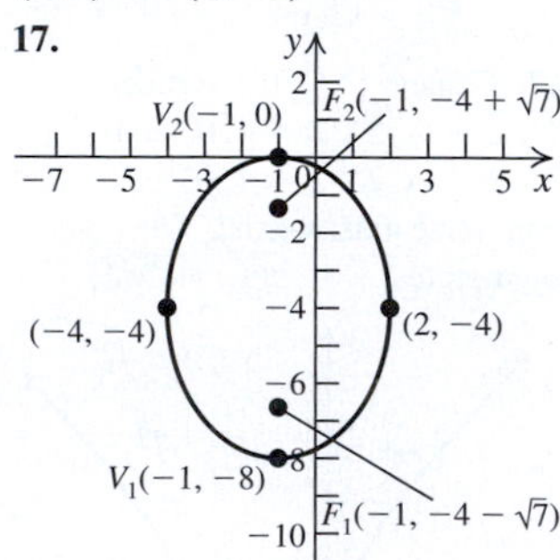

Foci: $(-1, -4 + \sqrt{7})$ and $(-1, -4 - \sqrt{7})$; vertices: $(-1, 0)$ and $(-1, -8)$; endpoints of the minor axis: $(2, -4)$ and $(-4, -4)$

19.

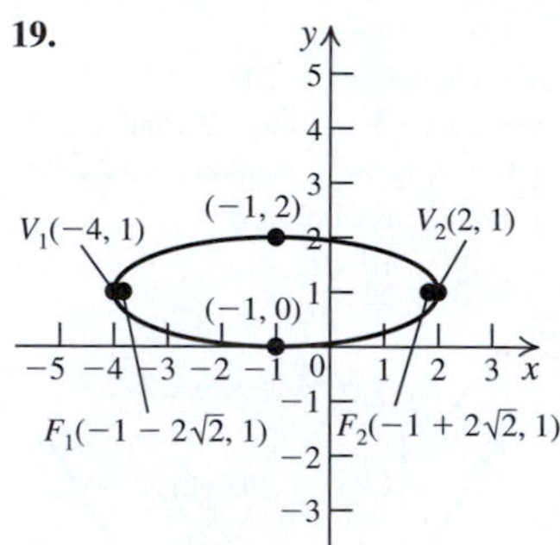

Foci: $(-1 + 2\sqrt{2}, 1)$ and $(-1 - 2\sqrt{2}, 1)$; vertices: $(2, 1)$ and $(-4, 1)$; endpoints of the minor axis: $(-1, 2)$ and $(-1, 0)$

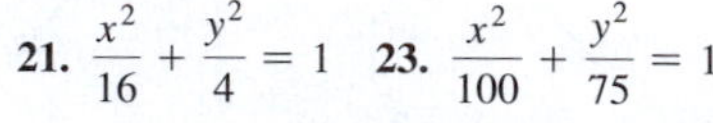

21. $\frac{x^2}{16} + \frac{y^2}{4} = 1$ **23.** $\frac{x^2}{100} + \frac{y^2}{75} = 1$

25.

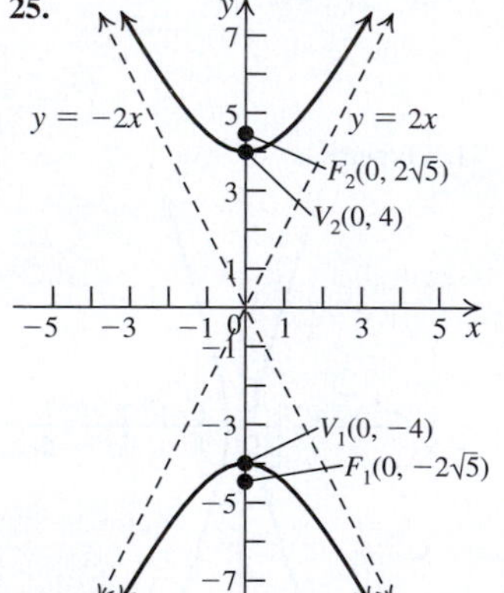

Vertices: $(0, 4)$ and $(0, -4)$; foci: $(0, 2\sqrt{5})$ and $(0, -2\sqrt{5})$; asymptotes: $y = \pm 2x$

27.

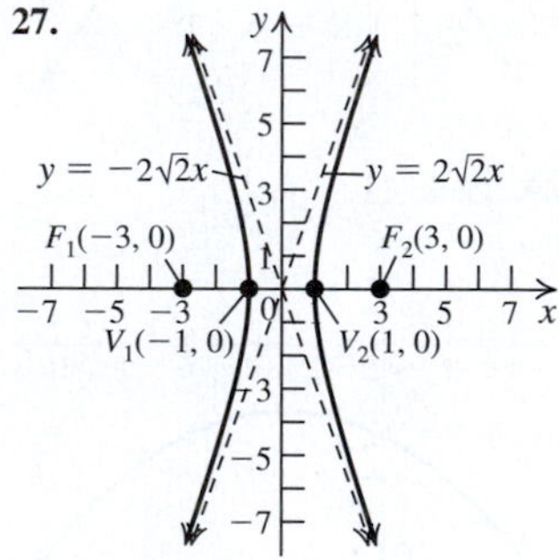

Vertices: $(1, 0)$ and $(-1, 0)$; foci: $(3, 0)$ and $(-3, 0)$ asymptotes: $y = \pm 2\sqrt{2}x$

29.

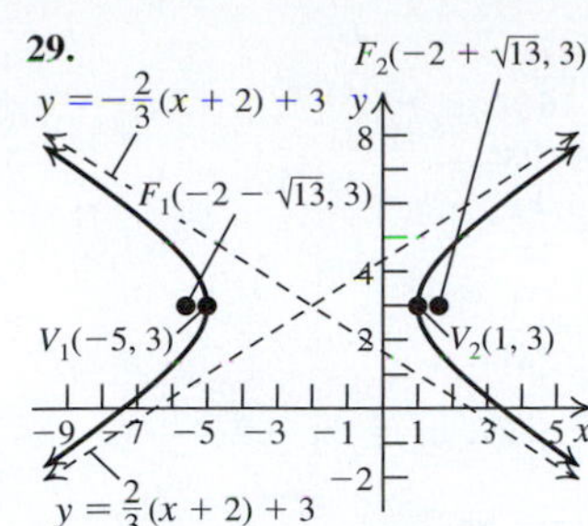

Vertices: $(1, 3)$ and $(-5, 3)$; foci: $(-2 + \sqrt{13}, 3)$ and $(-2 - \sqrt{13}, 3)$; asymptotes: $y - 3 = \pm\frac{2}{3}(x + 2)$

31.

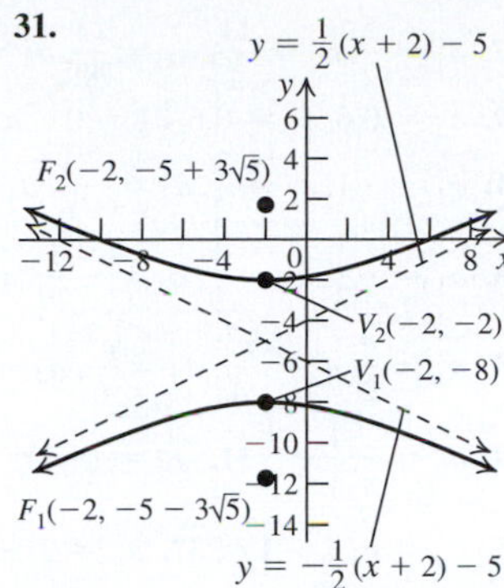

Vertices: $(-2, -2)$ and $(-2, -8)$; foci: $(-2, -5 + 3\sqrt{5})$ and $(-2, -5 - 3\sqrt{5})$; asymptotes: $y + 5 = \pm\frac{1}{2}(x + 2)$

33. $x^2 - \frac{y^2}{3} = 1$ **35.** $\frac{x^2}{4} - \frac{y^2}{36} = 1$

37. Hyperbola

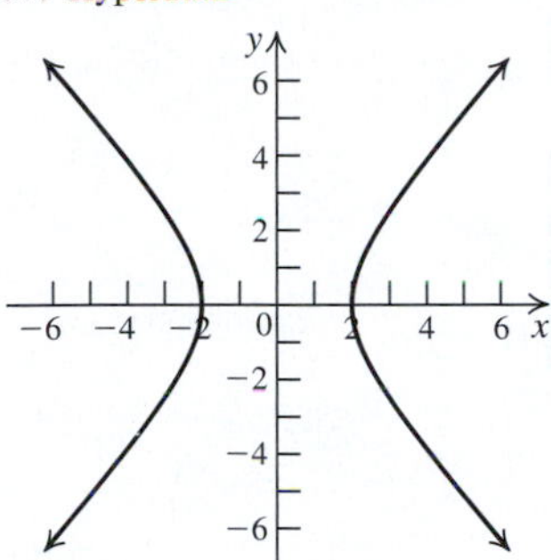

39. Ellipse

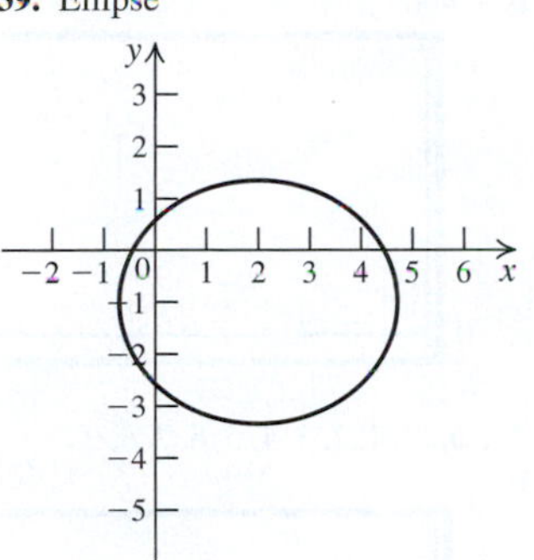

41. Circle

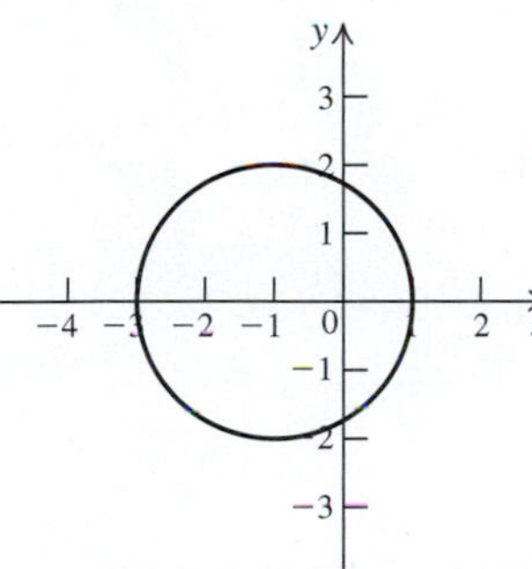

43. Hyperbola

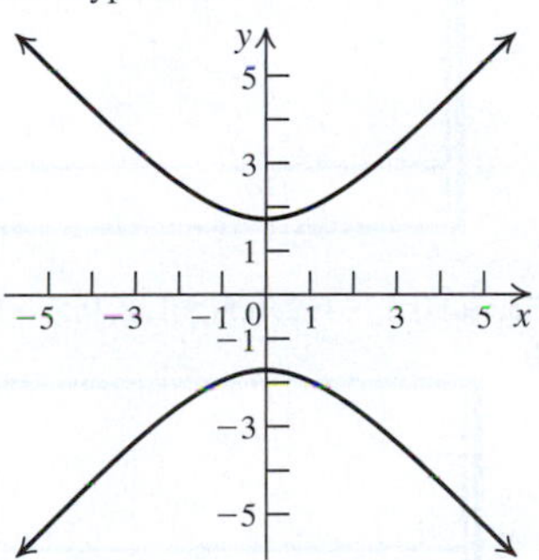

45. Circle

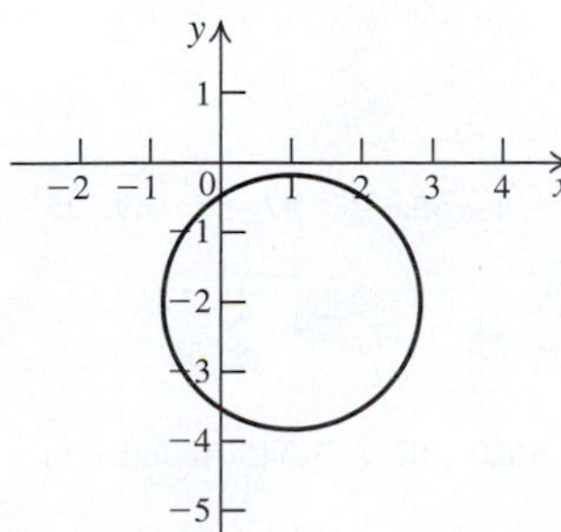

47. Ellipse

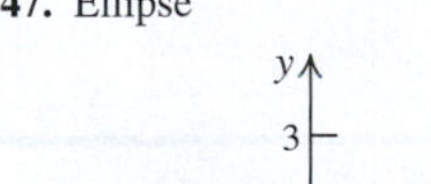

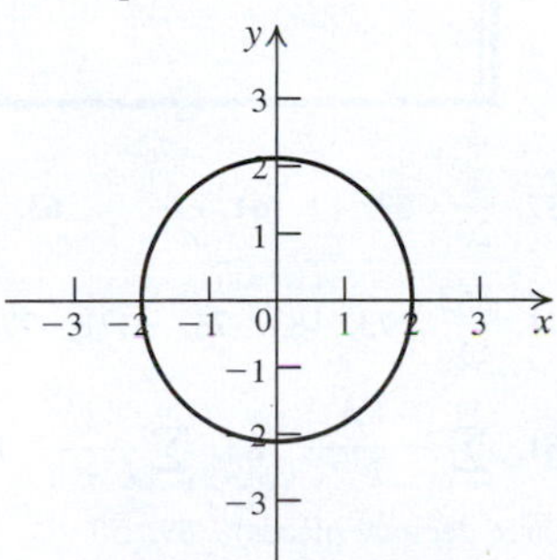

49. $\frac{x^2}{5} - \frac{y^2}{4} = 1$

51. Point of intersection: $\left(\frac{109}{10}, -\frac{91}{20}\right)$

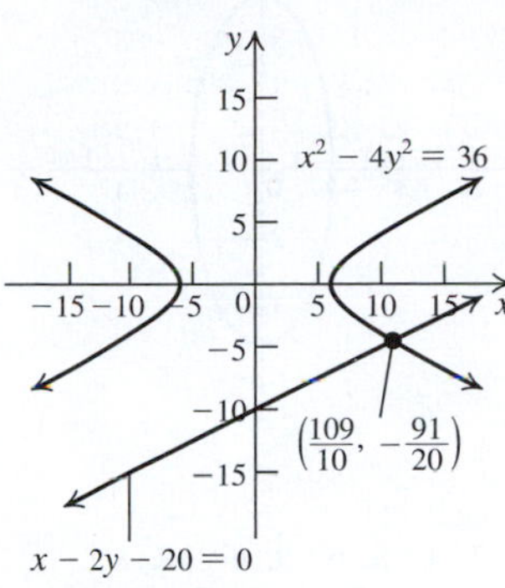

53. Points of intersection: $(2, 1), (2, -1), (-2, 1), (-2, -1)$

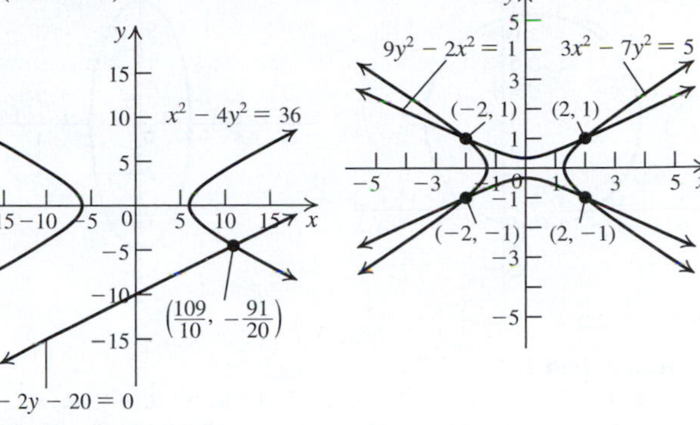

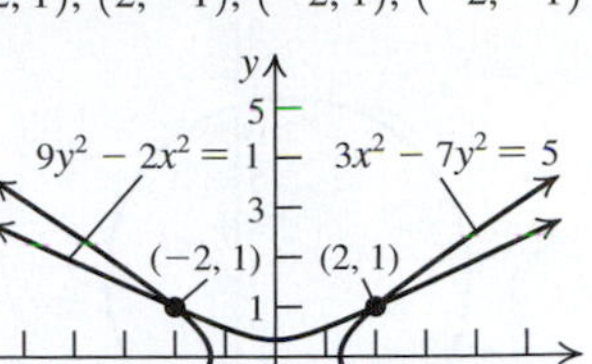

55. 14.6 ft **57.** 16.4 in.

Practice Test A

1. $(y - 12)^2 = 24(x + 6)$ **2.** $(y - 1)^2 = -8(x + 3)$ **3.** Focus: $\left(0, \frac{9}{4}\right)$; directrix: $x = \frac{9}{4}$

4.

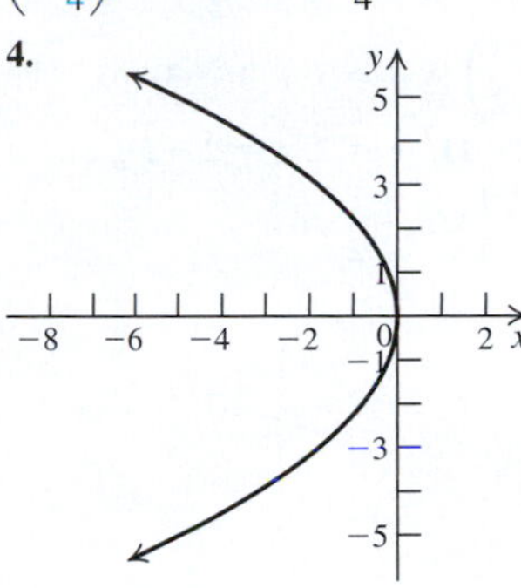

5. Vertex: $(-2, 1)$; focus: $(-2, -1)$; directrix: $y = 3$
6. 2/9 ft
7. $(y + 1)^2 = 24(x - 3)$
8. $\frac{x^2}{12} + \frac{y^2}{16} = 1$
9. $\frac{x^2}{29} + \frac{y^2}{25} = 1$
10. $\frac{x^2}{81} + \frac{y^2}{4} = 1$

11.

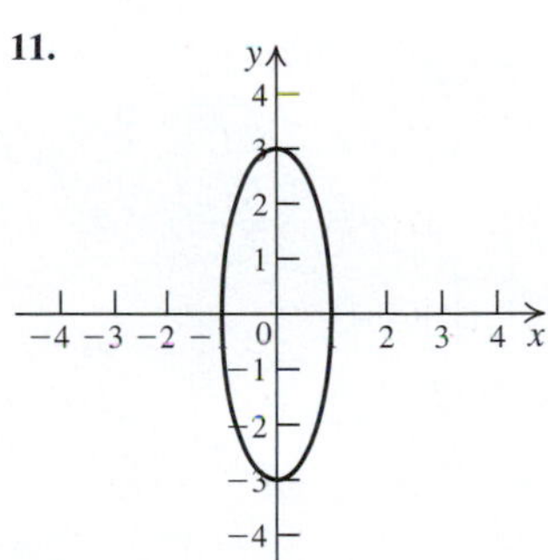

12. Vertices: $(0, 1)$ and $(0, -1)$, foci: $(0, 5\sqrt{2})$ and $(0, -5\sqrt{2})$

13.

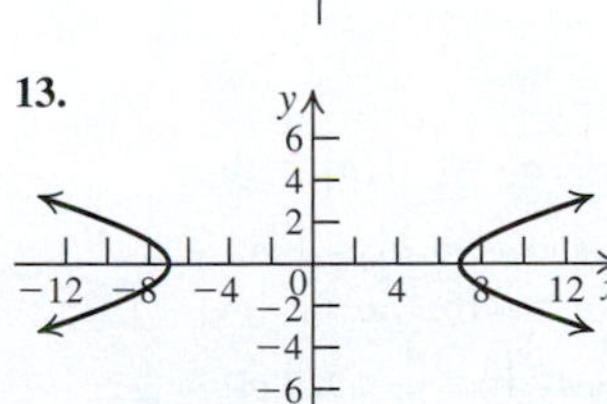

14. $\frac{y^2}{36} - \frac{x^2}{9} = 1$
15. $y^2 - (x - 1)^2 = 1$
16. $\frac{(x - 1)^2}{4} - (y + 2)^2 = 1$

17. Hyperbola

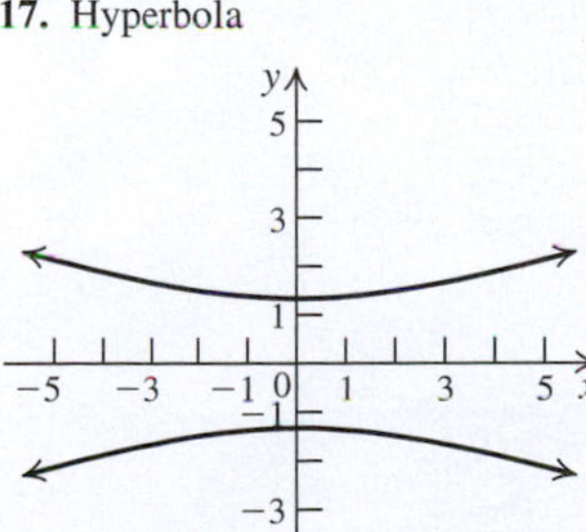

18. Parabola

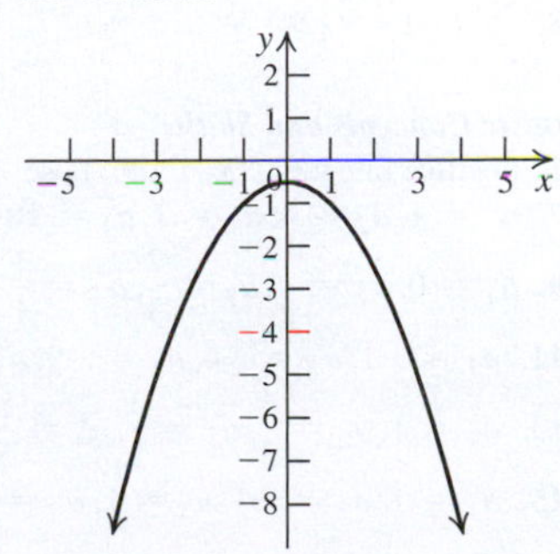

19. Circle

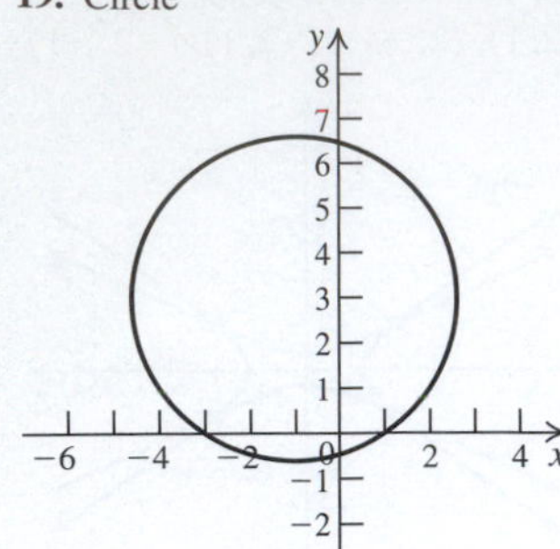

20. Ellipse

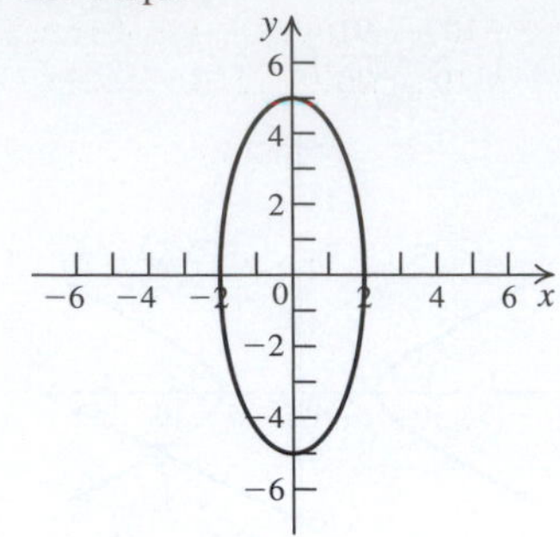

Practice Test B

1. b **2.** b **3.** a **4.** a **5.** a **6.** d **7.** a **8.** b **9.** d **10.** a **11.** a **12.** c **13.** b **14.** b **15.** c **16.** c **17.** c **18.** a **19.** d **20.** d

Cumulative Review Exercises (Chapters P–10)

1. $\dfrac{f(x+h)-f(x)}{h} = 2x + h - 3$ **3.** $f^{-1}(x) = \dfrac{1}{2}x + \dfrac{3}{2}$

$f(f^{-1}(x)) = f\left(\dfrac{1}{2}x + \dfrac{3}{2}\right) = 2\left(\dfrac{1}{2}x + \dfrac{3}{2}\right) - 3 = x + 3 - 3 = x$

5. $x = 4$ **7.** $(2, \infty)$ **9.** $x = 2, y = 3$ **11.** $x = 2, y = 2 - \log 2$

13. 0 **15.** $A^{-1} = \begin{bmatrix} 2 & 1 \\ \frac{5}{2} & \frac{3}{2} \end{bmatrix}$ **17.** $1, -1, \dfrac{1}{2}\sqrt{6}, -\dfrac{1}{2}\sqrt{6}$

19. Circle

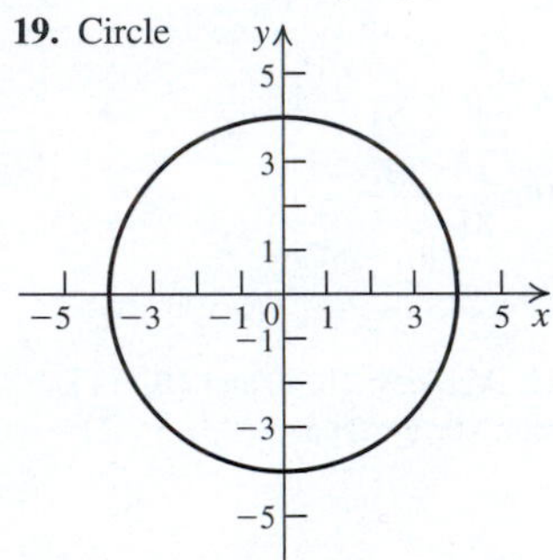

CHAPTER 11

Section 11.1

Answers to Practice Problems:

1. $a_1 = -2, a_2 = -4, a_3 = -8, a_4 = -16$

2. $a_n = (-1)^{n+1}\left(1 - \dfrac{1}{n}\right)$ **3.** $a_1 = -3, a_2 = -1, a_3 = 3, a_4 = 11, a_5 = 27$ **4.** $a_1 = 2, a_2 = 3$, and $a_k = a_{k-2} + a_{k-1}$, for all $k \ge 3$ **5. a.** 13 **b.** $n(n-1)(n-2)$

6. $a_1 = -2, a_2 = 2, a_3 = -\dfrac{4}{3}, a_4 = \dfrac{2}{3}, a_5 = -\dfrac{4}{15}$ **7.** −4

8. $\displaystyle\sum_{k=1}^{7} (-1)^{k+1}(2k)$

Basic Concepts and Skills:

1. positive integers **3.** 1 **5.** false

7. $a_1 = 1, a_2 = 4, a_3 = 7, a_4 = 10$

9. $a_1 = 0, a_2 = \dfrac{1}{2}, a_3 = \dfrac{2}{3}, a_4 = \dfrac{3}{4}$

11. $a_1 = -1, a_2 = -4, a_3 = -9, a_4 = -16$

13. $a_1 = 1, a_2 = \dfrac{4}{3}, a_3 = \dfrac{3}{2}, a_4 = \dfrac{8}{5}$

15. $a_1 = 1, a_2 = -1, a_3 = 1, a_4 = -1$

17. $a_1 = \dfrac{5}{2}, a_2 = \dfrac{11}{4}, a_3 = \dfrac{23}{8}, a_4 = \dfrac{47}{16}$

19. $a_1 = 0.6, a_2 = 0.6, a_3 = 0.6, a_4 = 0.6$

21. $a_1 = -1, a_2 = \dfrac{1}{2}, a_3 = -\dfrac{1}{6}, a_4 = \dfrac{1}{24}$

23. $a_1 = -\dfrac{1}{3}, a_2 = \dfrac{1}{9}, a_3 = -\dfrac{1}{27}, a_4 = \dfrac{1}{81}$

25. $a_1 = \dfrac{e}{2}, a_2 = \dfrac{e^2}{4}, a_3 = \dfrac{e^3}{6}, a_4 = \dfrac{e^4}{8}$ **27.** $a_n = 3n - 2$

29. $a_n = \dfrac{1}{n+1}$ **31.** $a_n = (-1)^{n-1}(2)$ **33.** $a_n = \dfrac{3^{n-1}}{2^n}$

35. $a_n = n(n+1)$ **37.** $a_n = 2 - \dfrac{(-1)^n}{n+1}$ **39.** $a_n = \dfrac{3^{n+1}}{n+1}$

41. $a_1 = 2, a_2 = 5, a_3 = 8, a_4 = 11, a_5 = 14$

43. $a_1 = 3, a_2 = 6, a_3 = 12, a_4 = 24, a_5 = 48$

45. $a_1 = 7, a_2 = -11, a_3 = 25, a_4 = -47, a_5 = 97$

47. $a_1 = 2, a_2 = \dfrac{1}{2}, a_3 = 2, a_4 = \dfrac{1}{2}, a_5 = 2$

49. $a_1 = 25, a_2 = -\dfrac{1}{125}, a_3 = -25, a_4 = \dfrac{1}{125}, a_5 = 25$

51. a. 2, 11, 26, 47, 107, 146, 191, 242, 299

b. 400

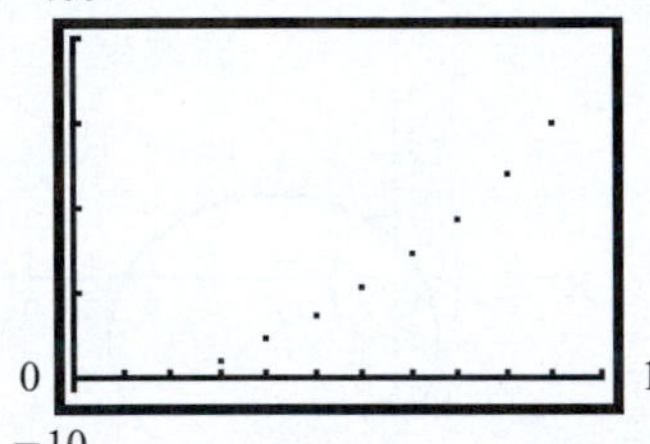

−10

53. a. 0, 1, 2, 3, 4, 5, 6, 7, 8, 9

b. 12

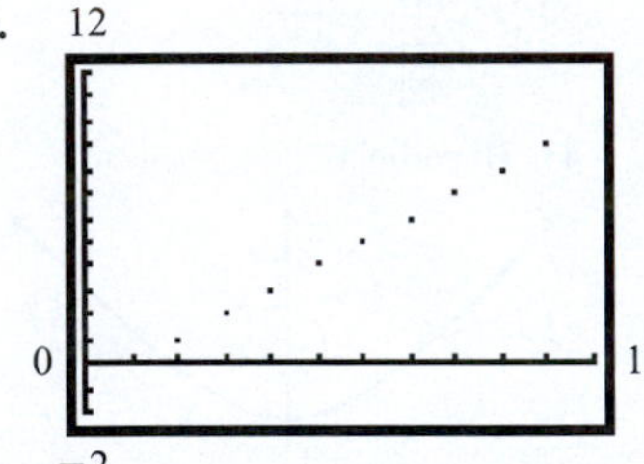

−2

55. a. 0.5, −1, 2, 0.5, −1, 2, 0.5, −1, 2, 0.5

b. 5

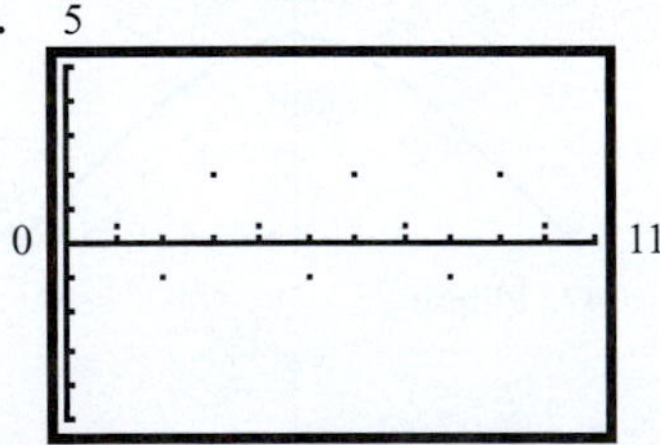

−5

57. $\dfrac{1}{20}$ **59.** 12 **61.** $\dfrac{1}{n+1}$ **63.** $2n + 1$ **65.** 35 **67.** 55 **69.** 25

71. $\dfrac{853}{140}$ **73.** 183 **75.** −118 **77.** $\displaystyle\sum_{k=1}^{51} (2k - 1)$ **79.** $\displaystyle\sum_{k=1}^{11} \frac{1}{5k}$

81. $\displaystyle\sum_{k=1}^{50} \frac{(-1)^{k+1}}{k}$ **83.** $\displaystyle\sum_{k=1}^{10} \frac{k}{k+1}$ **85.** 4620 **87.** −1.345 (rounded to three decimal places) **89.** 50

Applying the Concepts:

91. a. 208 ft **b.** $(16 + (n-1)32)$ ft **93.** \$250 **95.** 19,200 min **97. a.** $A_1 = 10{,}300, A_2 = 10{,}609, A_3 = 10{,}927.27, A_4 = 11{,}255.09, A_5 = 11{,}592.74, A_6 = 11{,}940.52$ **b.** \$16,047.10 **99.** 1st year: \$105,000; 2nd year: \$110,250; 3rd year: \$115,762.50; 4th year: \$121,550.63; 5th year: \$127,628.16; 6th year: \$134,009.56; 7th year: \$140,710.04; formula: $a_n = (100{,}000)(1.05^n)$

Beyond the Basics:

101. $a_n = 2^{1-(1/2^n)}$ **103. a.** $a_1 = 1, a_2 = \frac{1}{2}, a_3 = \frac{1}{4}, a_4 = \frac{1}{8},$ $a_5 = \frac{1}{16}$ **b.** $a_n = \frac{1}{2^{n-1}}$ **105.** $a_1 = 1, a_2 = 1, a_3 = 2, a_4 = 2,$ $a_5 = 3, a_6 = 4, a_7 = 4, a_8 = 4, a_9 = 5, a_{10} = 6$
107. $a_m = (m+1)^2$ **109.** $p = 2, q = 12$ **113.** $x = 49$

Critical Thinking/Discussion/Writing:
115. $k = 5$

Maintaining Skills:
119. 4, 4, 4, 4 **121.** −5, −5, −5, −5
123. $a_2 = 7, a_3 = 11, a_4 = 15, a_5 = 19$ **125.** $a_2 = 4, a_3 = 1,$ $a_4 = -2, a_5 = -5$ **127.** 3, 3, 3, 3 **129.** −2, −2, −2, −2
131. $a_{n+1} = 2n - 3, a_{n-1} = 2n - 7$

Section 11.2

Answers to Practice Problems:

1. −5 **2.** $a_n = 4n - 7$ **3.** $d = -3, a_n = 53 - 3n$ **4.** $\frac{85}{6}$
5. 2000 ft

Basic Concepts and Skills:
1. 5 **3.** −3 **5.** false **7.** Arithmetic; $a_1 = 1, d = 1$
9. Arithmetic; $a_1 = 2, d = 3$ **11.** Not arithmetic **13.** Not arithmetic
15. Arithmetic: $a_1 = 0.6, d = -0.4$ **17.** Arithmetic; $a_1 = 8, d = 2$
19. Not arithmetic **21.** $a_n = 3n + 2$ **23.** $a_n = 16 - 5n$
25. $a_n = \frac{3-n}{4}$ **27.** $a_n = -\frac{2n+1}{5}$ **29.** $a_n = 3n + e - 3$
31. $d = \frac{13}{2}, a_n = \frac{13}{2}n - 5$ **33.** $d = -2, a_n = 22 - 2n$
35. $d = \frac{1}{2}, a_n = \frac{n+11}{2}$ **37.** 1275 **39.** 2500 **41.** 15,150
43. −208 **45.** $\frac{121}{3}$ **47.** 6225 **49.** 460 **51.** 1340 **53.** $n = 38$
55. $n = 25$ **57.** $n = 25$

Applying the Concepts:
59. 1170 **61.** $12,375 **63.** $16.75 **65.** 1100 **67.** 408 **69.** 8

Beyond the Basics:
71. $d = \log\left(\frac{a}{b}\right)$ **73.** $d = 6$ **75.** 1368 **77.** Yes, it is harmonic because the reciprocals form an arithmetic sequence with $d = -\frac{1}{3}$.

Critical Thinking/Discussion/Writing:
81. $a_n = \frac{21-n}{2}$ **83.** 79

Maintaining Skills:
85. 2, 2, 2, 2 **87.** −3, −3, −3, −3
89. $a_2 = 6, a_3 = 18, a_4 = 54, a_5 = 162$
91. $a_2 = -2, a_3 = 4, a_4 = -8, a_5 = 16$ **93.** 2, 2, 2, 2
95. −3, −3, −3,−3 **97.** $a_{n+1} = -3 \cdot 2^{n+1}, a_{n-1} = -3 \cdot 2^{n-1}$

Section 11.3

Answers to Practice Problems

1. 3 **2.** It is geometric; $a_1 = \frac{3}{2}, r = \frac{3}{2}$ **3. a.** $a_1 = 2$ **b.** $r = \frac{3}{5}$
c. $a_n = 2\left(\frac{3}{5}\right)^{n-1}$ **4.** $7(1.5)^{17} \approx 6896.8288$ **5.** 8 **6.** ≈ 2
7. $91,510.60 **8.** 9 **9.** $\sum_{n=0}^{\infty}(10{,}000{,}000)(0.85)^n \approx \$66{,}666{,}666.67$

Basic Concepts and Skills:
1. 5 **3.** 3 **5.** true **7.** Geometric; $a_1 = 3, r = 2$
9. Not geometric **11.** Geometric; $a_1 = 1, r = -3$ **13.** Geometric; $a_1 = 7, r = -1$ **15.** Geometric; $a_1 = 9, r = \frac{1}{3}$ **17.** Geometric; $a_1 = -\frac{1}{2}, r = -\frac{1}{2}$ **19.** Geometric; $a_1 = 1, r = 2$ **21.** Not geometric
23. Geometric; $a_1 = \frac{1}{3}, r = \frac{1}{3}$ **25.** Geometric; $a_1 = \sqrt{5}, r = \sqrt{5}$
27. $a_1 = 2, r = 5, a_n = 2(5)^{n-1}$ **29.** $a_1 = 5, r = \frac{2}{3}, a_n = 5\left(\frac{2}{3}\right)^{n-1}$
31. $a_1 = 0.2, r = -3, a_n = 0.2(-3)^{n-1}$
33. $a_1 = \pi^4, r = \pi^2, a_n = \pi^{2n+2}$ **35.** $a_7 = 320$ **37.** $a_{10} = -1536$
39. $a_6 = \frac{243}{16}$ **41.** $a_9 = -\frac{390{,}625}{256}$ **43.** $a_{20} = -\frac{125}{131{,}072}$ **45.** 11
47. 10 **49.** 9 **51.** $S_{10} = \frac{1{,}220{,}703}{5}$ **53.** $S_{12} = -\frac{40{,}690{,}104}{25}$
55. $S_8 = \frac{109{,}225}{16{,}384}$ **57.** $\frac{31}{16}$ **59.** $\frac{6305}{729}$ **61.** 255 **63.** $\frac{3(5^{20} - 2^{20})}{35(2)^{19}}$
65. 0.397 **67.** 9841.333 **69.** 2.250 **71.** $\frac{1}{2}$ **73.** $-\frac{1}{3}$ **75.** $\frac{32}{5}$
77. $\frac{15}{2}$ **79.** $\frac{4}{5}$

Applying the Concepts:
81. 23,185 **83.** $3933.61 **85.** 1024 **87.** $1,190,374.35
89. 189.09 cm **91.** 20 m

Beyond the Basics:
93. $\frac{2}{3}$ **97.** common ratio $= \frac{1}{r}$

Critical Thinking/Discussion/Writing:
103. b_{1001} is larger **105.** $n = 15, k = 16$ **107.** 2, 5, and 8; or 26, 5, and −16

Maintaining Skills:
109. P_3: $3(3+1)$ is even; *true* P_4: $4(4+1)$ is even; *true*
111. P_3: $1 + 2 + 3 = \frac{3(3+1)}{2}$; *true*
P_4: $1 + 2 + 3 + 4 = \frac{4(4+1)}{2}$; *true*
113. P_{n+1}: $(n+1)^2 + (n+1)$ is divisible by 2.
115. P_{n+1}: $3^{n+1} > 5(n+1)$

Section 11.4

Answer to Practice Problem:
1. $P_{k+1}: (k+4)^2 > (k+1)^2 + 9$

Basic Concepts and Skills:
1. natural numbers **3.** P_{k+1} is true **5.** false
7. $P_{k+1}: (k+2)^2 - 2(k+1)$ **9.** $P_{k+1}: 2^{k+1} > 5(k+1)$

Applying the Concepts:
37. a. The number of sides of the nth figure is $3(4^{n-1})$. **b.** The perimeter of the nth figure is $3\left(\frac{4}{3}\right)^{n-1}$. **39.** The smallest number of moves to accomplish the transfer is $2^n - 1$.

Maintaining Skills:
59. n

Section 11.5

Answers to Practice Problems:
1. $(3y - x)^6 = 729y^6 - 1458y^5x + 1215y^4x^2 - 540y^3x^3 + 135y^2x^4 - 18yx^5 + x^6$ **2. a.** $\binom{6}{2} = 15$ **b.** $\binom{12}{9} = 220$

3. $(3x - y)^4 = 81x^4 - 108x^3y + 54x^2y^2 - 12xy^3 + y^4$
4. 220 **5.** $1{,}863{,}680x^3a^{12}$

Basic Concepts and Skills:
1. $n + 1$ **3.** alternating signs **5.** false **7.** 120 **9.** 12 **11.** 15
13. 1 **15.** 7 **17.** $(x + 2)^4 = x^4 + 8x^3 + 24x^2 + 32x + 16$
19. $(x - 2)^5 = x^5 - 10x^4 + 40x^3 - 80x^2 + 80x - 32$
21. $(2 - 3x)^3 = -27x^3 + 54x^2 - 36x + 8$
23. $(2x + 3y)^4 = 16x^4 + 96x^3y + 216x^2y^2 + 216xy^3 + 81y^4$
25. $(x + 1)^4 = x^4 + 4x^3 + 6x^2 + 4x + 1$
27. $(x - 1)^5 = x^5 - 5x^4 + 10x^3 - 10x^2 + 5x - 1$
29. $(y - 3)^3 = y^3 - 9y^2 + 27y - 27$
31. $(x + y)^6 = x^6 + 6x^5y + 15x^4y^2 + 20x^3y^3 + 15x^2y^4 + 6xy^5 + y^6$
33. $(1 + 3y)^5 = 243y^5 + 405y^4 + 270y^3 + 90y^2 + 15y + 1$
35. $(2x + 1)^4 = 16x^4 + 32x^3 + 24x^2 + 8x + 1$
37. $(x - 2y)^3 = x^3 - 6x^2y + 12xy^2 - 8y^3$
39. $(2x + y)^4 = 16x^4 + 32x^3y + 24x^2y^2 + 8xy^3 + y^4$
41. $\left(\frac{x}{2} + 2\right)^7 = \frac{1}{128}x^7 + \frac{7}{32}x^6 + \frac{21}{8}x^5 + \frac{35}{2}x^4 + 70x^3 + 168x^2 + 224x + 128$
43. $\left(a^2 - \frac{1}{3}\right)^4 = a^8 - \frac{4}{3}a^6 + \frac{2}{3}a^4 - \frac{4}{27}a^2 + \frac{1}{81}$
45. $\left(\frac{1}{x} + y\right)^3 = y^3 + 3\frac{y^2}{x} + 3\frac{y}{x^2} + \frac{1}{x^3}$ **47.** $120x^7y^3$
49. $-112.640x^3$ **51.** $16{,}128x^6y^2$ **53.** $-704{,}000x^2y^9$ **55.** 2.48832

Beyond the Basics:
57. $-\frac{8064\sqrt{x}}{x^8}$ **59.** $924\frac{x^{12}}{y^{18}}$ **65.** 16 **67.** x^5 **69.** $k = \pm 2$

Critical Thinking/Discussion/Writing:
77. $\frac{8!}{4!}$ **79.** $2^6 \cdot 6!$ **81.** 132 **83.** 56

Section 11.6

Answers to Practice Problems:
1.

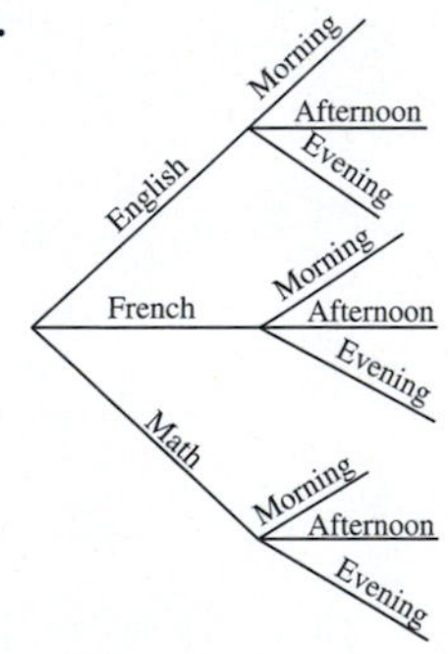

There are $(3)(3) = 9$ ways to choose a course. **2.** 35
3. 1,000,000 **4.** 5040 **5.** 3024 **6. a.** 72 **b.** 1
7. {bears, bulls, lions}, {bears, bulls, tigers}, {bears, lions, tigers}, {bulls, lions, tigers}; $C(4, 3) = 4$
8. a. 36 **b.** 1 **c.** 1 **9.** 220 **10.** 90

Basic Concepts and Skills:
1. permutation **3.** combination **5.** true **7.** 6 **9.** 56
11. 362,880 **13.** 1 **15.** 56 **17.** 126 **19.** 1 **21.** 1 **23. a.** 676
b. 650 **25.** 2450 **27.** 24 **29.** 60 **31.** 10 **33.** 55 **35.** 330
37. 20,160 **39.** 120 **41. a.** 160 **b.** 800 **43. a.** 12,144 **b.** 13,824
45. 256 **47.** Yes **49. a.** 5040 **b.** 5040 **51. a.** 10 **b.** 10 **c.** 5
53. 105 **55.** 3,628,800 **57.** The maximum number of times each child will go to the circus is 20. The father will go 35 times. **59.** 111,007,923,832,370,565 **61.** 80 **63.** 15 **65.** 360 **67.** 420

Beyond the Basics:
69. 6 **71.** 435 **73.** $m = 3$ **75.** $n = 6$ **77.** $n = 6$ **79.** $m = 9$

Critical Thinking/Discussion/Writing:
81. 25

Maintaining Skills:
83. 3 **85.** 1 **87.** 4, 5 **89.** $\{3, 5\}, 2$ **91.** 3, 4 **93.** $\varnothing, 0$

Section 11.7

Answers to Practice Problems:
1. a. $E_1 = \{1, 3, 5\}, P(E_1) = \frac{1}{2}$ **b.** $E_2 = \{5, 6\}, P(E_2) = \frac{1}{3}$
2. $\frac{1}{2}$ **3.** $\frac{1}{6}$ **4.** $\frac{1}{15{,}890{,}700}$ **5.** $\frac{2}{13}$ **6.** $\frac{3}{10}$ **7.** 0.69

Basic Concepts and Skills:
1. equally likely **3.** mutually exclusive
5. true (if "between 0 and 1" includes 0 and 1)
7. $S =$ {(Wendy's, McDonald's, Burger King), (Wendy's, Burger King, McDonald's), (McDonald's, Wendy's, Burger King), (McDonald's, Burger King, Wendy's), (Burger King, Wendy's, McDonald's), (Burger King, McDonald's, Wendy's,)}
9. $S =$ {{*Thriller, The Wall*}, {*Thriller, Eagles: Their Greatest Hits*}, {*Thriller, Led Zappelin IV*}, {*The Wall, Eagles: Their Greatest Hits*}, {*The Wall, Led Zappelin IV*}, {*Eagles: Their Greatest Hits, Led Zappelin IV*}}
11. $S =$ {(white, male), (white, female), (African American, male), (African American, female), (Native American, male), (Native American, female), (Asian, male), (Asian, female), (other, male), (other, female), (multiracial, male), (multiracial, female)}
13. 0 **15.** 1 **17.** experimental **19.** theoretical **21.** experimental
23. theoretical **25.** experimental
27. An event that is very likely to happen has probability 0.999.
An event that will surely happen has probability 1.
An event that is a rare event has probability 0.001.
An event that may or may not happen has probability 0.5.
An event that will never happen has probability 0.
29. $\frac{1}{3}$ **31.** $\frac{1}{3}$ **33.** $\frac{1}{2}$ **35.** $\frac{1}{13}$ **37.** $\frac{1}{4}$ **39.** $\frac{3}{13}$ **41.** $\frac{1}{10}$
43. $\frac{1}{10}$ **45.** $\frac{1}{2}$

Applying the Concepts:
47. 0.7 **49.** $\frac{1}{5}$ **51.** $\frac{47}{50}$ **53.** The probability of winning one prize is $\frac{2}{25}$. The probability of winning both is $\frac{9}{1550}$. **55. a.** $\frac{1}{4}$ **b.** $\frac{1}{4}$ **c.** $\frac{1}{2}$
57. a. $\frac{79}{148}$ **b.** $\frac{33}{40}$ **c.** 0 **d.** 1 **e.** $\frac{119}{120}$ **59.** The probability that a non-Hispanic white American has no lactose intolerance is $\frac{4}{5}$. The probability for the second group is $\frac{1}{4}$. **61. a.** $\frac{1}{10}$ **b.** $\frac{1}{10}$
63. 90; 135; 10; 765 **65.** Three of one gender and one of the other is more likely. **67. a.** $\frac{2}{5}$ **b.** $\frac{89}{180}$

Beyond the Basics:
69. $\frac{468{,}559}{1{,}000{,}000}$ **71.** $\frac{1}{12}$

Critical Thinking/Discussion/Writing:
73. $\frac{1}{3}$

Review Exercises

1. $a_1 = -1, a_2 = 1, a_3 = 3, a_4 = 5, a_5 = 7$

3. $a_1 = \frac{1}{3}, a_2 = \frac{2}{5}, a_3 = \frac{3}{7}, a_4 = \frac{4}{9}, a_5 = \frac{5}{11}$ **5.** $a_n = 32 - 2n$

7. 9 **9.** $n + 1$ **11.** 100 **13.** $\frac{1343}{140}$ **15.** $\sum_{k=1}^{50} \frac{1}{k}$

17. Arithmetic; $a_1 = 11, d = -5$ **19.** Not arithmetic **21.** $a_n = 3n$ **23.** $a_n = x + n - 1$ **25.** $d = 2, a_n = 2n + 1$ **27.** 352 **29.** 4020 **31.** Geometric: $a_1 = 4, r = -2$ **33.** Not geometric

35. $a_1 = 16, r = -\frac{1}{4}, a_n = \frac{(-1)^{n-1}}{4^{n-3}}$ **37.** $a_{10} = 39{,}366$

39. $S_{12} = \frac{819}{2}$ **41.** $\frac{3}{4}$ **43.** $\frac{3}{2}$ **47.** 792

49. $(x - 3)^4 = x^4 - 12x^3 + 54x^2 - 108x + 81$ **51.** $101{,}376x^5$ **53.** 24 **55.** 5040 **57.** 120 **59.** 120 **61.** 3696 **63.** 15,120

65. $\frac{1}{2}$ **67.** $\frac{2}{5}$ **69.** $\frac{15}{28}$ **71.** $\frac{1}{17}$ **73. a.** $\frac{1}{9}$ **b.** $\frac{4}{9}$ **c.** $\frac{5}{9}$

Practice Test A

1. $a_1 = 3, a_2 = -9, a_3 = -21, a_4 = -33, a_5 = -45$; arithmetic sequence **2.** $a_1 = -6, a_2 = -12, a_3 = -24, a_4 = -48, a_5 = -96$; geometric sequence **3.** $a_1 = -2, a_2 = -1, a_3 = 2, a_4 = 11, a_5 = 38$

4. $\frac{1}{n}$ **5.** $a_7 = 21$ **6.** $a_8 = -\frac{13}{128}$ **7.** 590 **8.** $\frac{93}{128}$ **9.** $\frac{2}{11}$ **10.** 1

11. $16x^4 - 32x^3 + 24x^2 - 8x + 1$ **12.** $8x$ **13.** 1024 **14.** 72

15. 21 **16.** 60 **17.** 6,760,000 **18.** $\frac{5}{12}$ **19.** $\frac{5}{6}$ **20.** $\frac{7}{51}$

Practice Test B

1. c **2.** d **3.** a **4.** c **5.** a **6.** b **7.** d **8.** a **9.** b **10.** b **11.** a **12.** d **13.** c **14.** a **15.** d **16.** b **17.** d **18.** c **19.** a **20.** d

Cumulative Review Exercises (Chapters P–11)

1. $\{2, 4\}$ **3.** $x = 2$ **5.** $x = 4$ **7.** $x = 4$

9. $(-\infty, -3 - 3\sqrt{3}] \cup [-3 + 3\sqrt{3}, \infty)$ **11.** $x = 10, y = 7, z = \frac{1}{2}$

13. $\frac{-\sqrt{6} - \sqrt{2}}{4}$

15. Shift the graph of $g(x) = \sqrt{x}$ three units to the right.

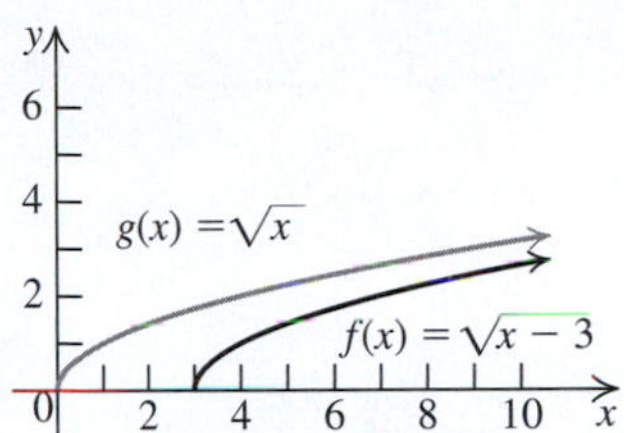

17. $\ln\left(\frac{x^7}{x - 5}\right)$ **19.** 12

CREDITS

Text Credits

Page 16 Table, Problem No: 154, Standard Car Features. Yahoo! Inc. **16** Problem No: 156, Population Projections. U.S. Census Bureau. **157** Table, Prevalence of Smoking Among Adults 18 Years and Older in United States. Centers for Disease Control and Prevention, National Health Interview Survey. **163** Problem No.47, Population Table. U.S. Census Bureau. **163** Problem No.49–52, Population Statistics Table. U.S. Census Bureau. **163** Problem No.53, Murders in United States. Disaster Center. **163** Problem No.54, Spending on Prescription Drugs. Centers for Disease Control and Prevention. **163** Problem No.55, Defense Spending. Usgovernmentspending.com **176** Problem No.90, Spending on Prescription Drugs. U.S. Census Bureau. **176** Problem No.92, Female Students in Colleges. United States Census Bureau. **192** Problem No.103, Health Insurance Coverage. Michigan Department of Community Health. **192** Problem No.104, Oil Consumption. Central Intelligence Agency. **194** The Ups and Downs of Drug Levels. Adapted from The Ups and Downs of Drug Levels, by Bob Munk. **237** Table, Problem No. 50, Federal Income Tax. Internal Revenue Service. **287** Table, Problem No. 105, Waste Disposal. Sioux Falls Health Department. **321** Problem No. 76. Source:www.aa.com. **321** Table, Problem No. 79, Deaths from FDA Approved Drugs. fda.com/drugs/GuidancecomplianceRegulatoryInformation/Surveillance/AdverseDrugEffects/ucm070461.html. **331** Example 08, Crude Oil and Petroleum Consumption. Source: www.eia.gov. **333** Problem No. 55, Total Energy Consumption. Source:www.eia.gov. **333** Problem No. 56, Lottery Sales. Source: www.floridafiscalportal.state.fl.us. **333** Problem No. 57, Marine Corps. Source: www.census.gov. **333** Problem No. 58, Unemployment. www.multpl.com. **441** Table, Example 05, Solving a Population Growth Problem. Central Intelligence Agency. **449** Problem No: 98, Facebook Users. BenPoster. **453** Example 03, Intensity of an Earthquake. U.S. Geological Survey. **454** Table, Major Earthquakes (M >= 7.5) of the 20th & 21st Century. National Earthquake Information Center, U.S. Geological Survey. **498** Problem No: 82, Height of a Building. Wikipedia. **534** Table, Problem No: 73, Average Temperatures. Weatherbase. **534** Table, Problem No: 74, Average Temperatures. ShapeFit.COM. **738** Table, Problem No:93, McDonald's Breakfast. ShapeFit.COM. **740** CAT Scans. Reprinted with permission from Mathematics Teacher, May 1996. © 1996 by the National Council of Teachers of Mathematics. **747** Table, CAT Scanner Ranges. Reprinted wiith permission from Mathematics Teacher, May 1996. © 1996 by the National Council of Teachers of Mathematics. **747** Figure, Three Grid Cells, Three X-rays. Reprinted wiith permission from Mathematics Teacher, May 1996. © 1996 by the National Council of Teachers of Mathematics. **750** Figure (a), Problem No:74, Four Grid Cells and Four X-rays. Reprinted wiith permission from Mathematics Teacher, May 1996. © 1996 by the National Council of Teachers of Mathematics. **750** Figure (b), Problem No:74, Four Grid Cells and Six X-rays. Reprinted with permission from Mathematics Teacher, May 1996. © 1996 by the National Council of Teachers of Mathematics. **920** Problem No. 105. http://el.media.mit.edu/logo-foundation/pubs/papers_as_11223.html. **970** Table, Example 07, Using Data to Estimate Probabilities. www.areaconnect.com. **972** Problem No. 49, Gender and the Airforce. www.af.mil/news. **972** Problem No. 51, Gender and the Marines. DMDC-June 2005, TFDW-June 2005.

Photo Credits

Chapter P: p. 1, Mykhaylo Palinchak/Shutterstock **Chapter 1: p. 76,** Caroline Letrange/Fotolia **p. 88,** Topdeq/Fotolia **p. 88,** Xalanx/Fotolia **p. 91,** Michele Burgess/Corbis **p. 91,** Corbis/SuperStock **p. 91,** Portrait of Lisa Gherardini (The Mona Lisa; La Gioconda) (ca. 1506), Leonardo da Vinci. Oil on poplar, 77 cm × 53 cm (30 in × 21 in). Acquired by François I, 1518 [Inv. 779]. Musée du Louvre/Andrew Howe/Getty Images **p. 101,** Jonelle Weaver/Photodisc/Getty Images **p. 104,** Library of Congress prints and Photographs Division [LC-USZC4-4940] **p. 115,** Library of Congress prints and Photographs Division/Doris Ulmann [LC-USZC4-4940] **p. 129,** Enno Kleinert/Alamy **p. 140,** Fuse/Thinkstock **p. 141,** JupiterImages/Thinkstock **p. 142,** Margo Harrison/Fotolia **p. 149,** Lynn Clayton Photography/Getty Images **p. 152,** Rolf Bruderer/Blend Images/Getty Images **Chapter 2: p. 155,** Bloomberg/Getty Images **p. 156,** Beth Anderson/Pearson Education, Inc. **p. 157,** Mark Yuill/Fotolia **p. 165** Ruszkiewicz/Fotolia **p. 178,** Newscom **p. 194,** Steve Cukrov/Shutterstock **p. 205,** George Doyle/Stockbyte/Getty Images **p. 207,** Moviestore Collection, Ltd./Alamy **p. 214,** Selecstock/Fotolia **p. 226,** Mj007/Shutterstock **p. 226,** David P. Stephens/Shutterstock **p. 239,** Piotr Marcinski/Fotolia **p. 257,** US Coast Guard/AP Images **p. 270,** Chris Ridley/Alamy **Chapter 3: p. 291,** Image Source/Getty Images **p. 292,** Rudi1976/Fotolia **p. 304,** Robert Glusic/Getty Images **p. 306,** Nickolae/Fotolia **p. 323,** David Carillet/Shutterstock **p. 331,** Bomboman/Fotolia **p. 335,** Alamy **p. 347,** Mary Evans Picture Library/Alamy **p. 355,** Eldon Doty/Getty Images **p. 374,** J. L. Charmet/Science Source/Photo Researchers, Inc. **Chapter 4: p. 391,** Buckwheat/Fotolia **p. 392,** Dmitrijs Gerciks/Fotolia **p. 399,** Keith Brofsky/Thinkstock **p. 402,** Georgios Kollidas/Fotolia **p. 405,** Inna Felker/Fotolia **p. 410,** Volodymyr Goinyk/Shutterstock **p. 417,** Pearson Education, Inc. **p. 420,** Gsmith/Fotolia **p. 425,** Andrew Holt/Alamy **p. 433,** Jules le Baron/Courtesy AIP Emilio Segre Visual Archives **p. 438,** Fotolia **p. 445,** Pearson Education, Inc. **p. 450,** Brian Lasenby/Shutterstock **p. 451,** SPL/Science Source/Photo Researchers, Inc. **p. 452,** AP Images **p. 456,** Drx/Fotolia **p. 459,** NASA **Chapter 5: p. 470,** Klaus Lang/All Canada Photos/Corbis **p. 471,** Pearson Education, Inc. **p. 485,** Robert Hardholt/Shutterstock **p. 500,** Paul Rapson/Alamy **p. 512,** Ekaterina V. Borisova/Shutterstock **p. 514,** HP Photo/Fotolia **p. 536,** Everett Collection **p. 547,** Anton Gvozdikov/Shutterstock **Chapter 6: p. 567,** Kerenby/Fotolia **p. 568,** North Wind Picture Archives **p. 582,** Corbis **p. 595,** David De Lossy/Photodisc/Getty Images **p. 614,** Lucidio Studio, Inc./Flickr/Getty Images **p. 618,** Nmarques74/Fotolia **p. 624,** Oriontrail/Shutterstock **p. 632,** Christian Wheatley/Shutterstock **Chapter 7: p. 639,** NASA **p. 640,** Jagdish Agarwal/Dinodia/Corbis **p. 652,** FloridaStock/Shutterstock **p. 654,** MaxFX/Shutterstock **p. 672,** Kirk Geisler/Shutterstock **p. 685,** Design Pics/Newscom **p. 696,** Oliver Sved/Shutterstock **p. 711,** John Foxx/Stockbyte/Getty Images **Chapter 8: p. 727,** Starush/Fotolia **p. 728,** Juulijs/Fotolia **p. 740,** Stockbyte/Getty Images **p. 752,** Mary Evans Picture Library/Alamy **p. 763,** xPACIFICA/Alamy **p. 782,** Pearson Education, Inc. **p. 786,** 20th Century Fox Film Corp/Everett Collection **Chapter 9: p. 797,** Yaacov Dagan/Alamy **p. 798,** AP Images **p. 814,** 20th Century Fox/Everett Collection **p. 815,** AKG Images/Newscom **p. 827,** Nomad Soul/Shutterstock **p. 828,** Science Source/Photo Researchers, Inc. **p. 842,** Pearson Education, Inc. **p. 847,** Science Source/Photo Researchers, Inc. **Chapter 10: p. 860,** Peter Barritt/Robert Harding World Imagery/Alamy **p. 861,** Xtuv Photography/Shutterstock **p. 863,** NASA **p. 863,** AP Images **p. 872,** Jess Alford/Getty Images **p. 875,** Kevin Beebe Custom Medical Stock Photo/Newscom **p. 882,** Interfoto/Alamy **p. 884,** NASA **p. 884,** Photo by Jill Britton/Table by Dan Bergerud/Camosun College, Victoria

BC/Courtesy of author **p. 887,** Pearson Education, Inc. **p. 905,** Photolink/Getty Images **Chapter 11: p. 909,** Andrew Zarivny/Shutterstock **p. 910,** Georgette Douwma/Getty Images **p. 912,** Pearson Education, Inc. **p. 921,** Chris Butler/Science Source/Photo Researchers, Inc. **p. 924,** Pearson Education, Inc. **p. 929,** ZUMA Press, Inc./Alamy **p. 941,** Stockbyte/Getty Images **p. 942,** Pearson Education, Inc. **p. 945,** Beth Anderson/Pearson Education, Inc. **p. 946,** Georgios Kollidas/Shutterstock **p. 946,** Dave King/Dorling Kindersley, Ltd. **p. 953,** John Paul Endress/Pearson Education, Inc. **p. 964,** David Gould/Getty Images **p. 968,** Beth Anderson/Pearson Education, Inc. **Front matter: p. xxi,** Sturti/E+/Getty Images. **FEP: p. 2,** Andrew Rich/E+/Getty Images **p. 3,** Shutterstock **p. 4,** Mike Flippo/Shutterstock.

INDEX

D

M

N

T

U

V

W

X

Y

Z

Algebra—Formulas and Definitions

Real-Number Properties

$a + b = b + a$ $\quad$ $ab = ba$

$(a + b) + c = a + (b + c)$ $\quad$ $(ab)c = a(bc)$

$a(b + c) = ab + ac$ $\quad$ $(a + b)c = ac + bc$

$a + 0 = 0 + a = a$ $\quad$ $a \cdot 1 = 1 \cdot a = a$

$a + (-a) = 0$ $\quad$ $a\left(\frac{1}{a}\right) = 1, a \neq 0$

$a \cdot 0 = 0 \cdot a = 0$

If $a \cdot b = 0$, then $a = 0$ or $b = 0$.

Absolute Value

$|a| = a$ if $a \geq 0$ and $|a| = -a$ if $a < 0$.

Assume that $a > 0$ and u is an algebraic expression:

$|u| = a$ is equivalent to $u = a$ or $u = -a$.

$|u| < a$ is equivalent to $-a < u < a$.

$|u| > a$ is equivalent to $u < -a$ or $u > a$.

Exponents

$a^n = \underbrace{a \cdot a \cdots a}_{n \text{ factors}}$ $\quad$ $a^0 = 1, a \neq 0$

$a^{-n} = \frac{1}{a^n}, a \neq 0$

$a^m a^n = a^{m+n}$ $\quad$ $\frac{a^m}{a^n} = a^{m-n}$

$(a^m)^n = a^{mn}$ $\quad$ $(a \cdot b)^n = a^n \cdot b^n$

$\left(\frac{a}{b}\right)^n = \frac{a^n}{b^n}$ $\quad$ $\left(\frac{a}{b}\right)^{-n} = \left(\frac{b}{a}\right)^n = \frac{b^n}{a^n}$

Logarithms

$y = \log_a x$ means $a^y = x$

$\log_a a^x = x$ $\quad$ $a^{\log_a x} = x$

$\log_a 1 = 0$ $\quad$ $\log_a a = 1$

$\log x = \log_{10} x$ $\quad$ $\ln x = \log_e x$

$\log_a MN = \log_a M + \log_a N$

$\log_a \frac{M}{N} = \log_a M - \log_a N$

$\log_a M^r = r \log_a M$

$\log_b x = \frac{\log_a x}{\log_a b} = \frac{\log x}{\log b} = \frac{\ln x}{\ln b}$ (change-of-base formula)

Radicals

$\sqrt{a^2} = |a|$

$\sqrt[n]{a^n} = |a|$ if n is even

$\sqrt[n]{a^n} = a$ if n is odd

$a^{\frac{1}{n}} = \sqrt[n]{a}$ $\quad$ $a^{\frac{m}{n}} = (\sqrt[n]{a})^m = \sqrt[n]{a^m}$

$\sqrt[n]{ab} = \sqrt[n]{a}\sqrt[n]{b}$ $\quad$ $\sqrt[n]{\frac{a}{b}} = \frac{\sqrt[n]{a}}{\sqrt[n]{b}}$

Rational Expressions

(Assume all denominators not to be zero.)

$\frac{A}{B} \cdot \frac{C}{D} = \frac{A \cdot C}{B \cdot D}$

$\frac{\frac{A}{B}}{\frac{C}{D}} = \frac{A}{B} \div \frac{C}{D} = \frac{A}{B} \cdot \frac{D}{C} = \frac{A \cdot D}{B \cdot C}$

$\frac{A}{C} \pm \frac{B}{C} = \frac{A \pm B}{C}$ $\quad$ $\frac{A}{B} \pm \frac{C}{D} = \frac{A \cdot D \pm B \cdot C}{B \cdot D}$

Common Products and Factors

$A^2 - B^2 = (A + B)(A - B)$

$A^2 + 2AB + B^2 = (A + B)^2$

$A^2 - 2AB + B^2 = (A - B)^2$

$A^3 - B^3 = (A - B)(A^2 + AB + B^2)$

$A^3 + B^3 = (A + B)(A^2 - AB + B^2)$

$A^3 + 3A^2B + 3AB^2 + B^3 = (A + B)^3$

$A^3 - 3A^2B + 3AB^2 - B^3 = (A - B)^3$

Distance Formula

The distance between $P(x_1, y_1)$ and $Q(x_2, y_2)$ is

$$d(P, Q) = \sqrt{(x_2 - x_1)^2 + (y_2 - y_1)^2}.$$

Midpoint Formula

The midpoint $M(x, y)$ of the line segment with endpoints $P(x_1, y_1)$ and $Q(x_2, y_2)$ is

$$M(x, y) = \left(\frac{x_1 + x_2}{2}, \frac{y_1 + y_2}{2}\right).$$

Lines

Slope	$m = \frac{\text{rise}}{\text{run}} = \frac{y_2 - y_1}{x_2 - x_1}$
Point-slope form	$y - y_1 = m(x - x_1)$
Slope-intercept form	$y = mx + b$
Horizontal line	$y = k$
Vertical line	$x = k$
General form	$ax + by + c = 0$

Quadratic Formula

The solutions to the equation $ax^2 + bx + c = 0$, $a \neq 0$, are

$$x = \frac{-b \pm \sqrt{b^2 - 4ac}}{2a}.$$

Standard Form of a Circle

$(x - h)^2 + (y - k)^2 = r^2$ $\quad$ center: (h, k)

radius: r

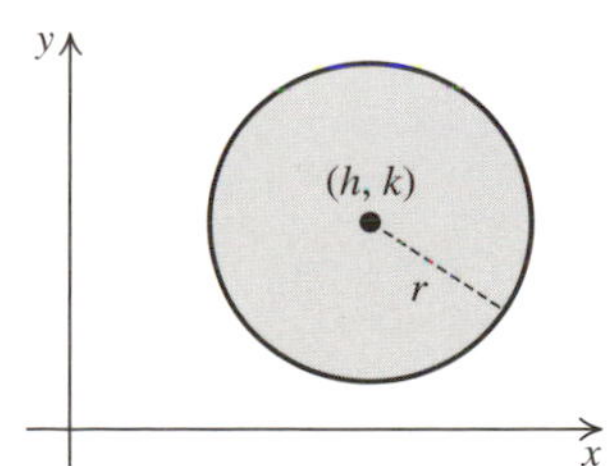

Trigonometry

Trigonometric Functions

Acute Angles

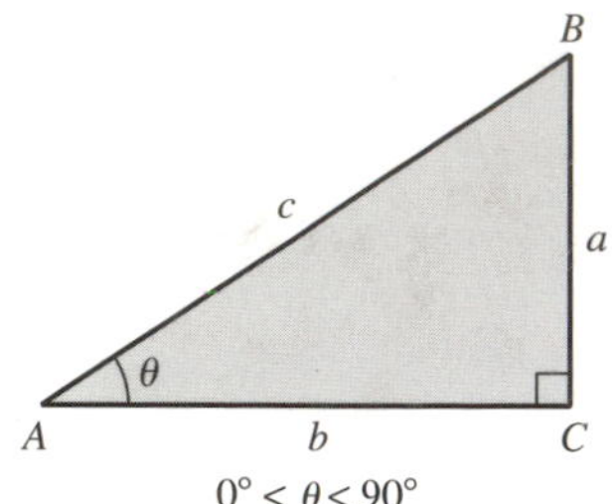

$0° < \theta < 90°$

a = length of the side opposite θ
b = length of the side adjacent to θ
c = length of the hypotenuse

$$\sin\theta = \frac{\text{opposite}}{\text{hypotenuse}} = \frac{a}{c} \qquad \csc\theta = \frac{\text{hypotenuse}}{\text{opposite}} = \frac{c}{a}$$

$$\cos\theta = \frac{\text{adjacent}}{\text{hypotenuse}} = \frac{b}{c} \qquad \sec\theta = \frac{\text{hypotenuse}}{\text{adjacent}} = \frac{c}{b}$$

$$\tan\theta = \frac{\text{opposite}}{\text{adjacent}} = \frac{a}{b} \qquad \cot\theta = \frac{\text{adjacent}}{\text{opposite}} = \frac{b}{a}$$

Any Angle

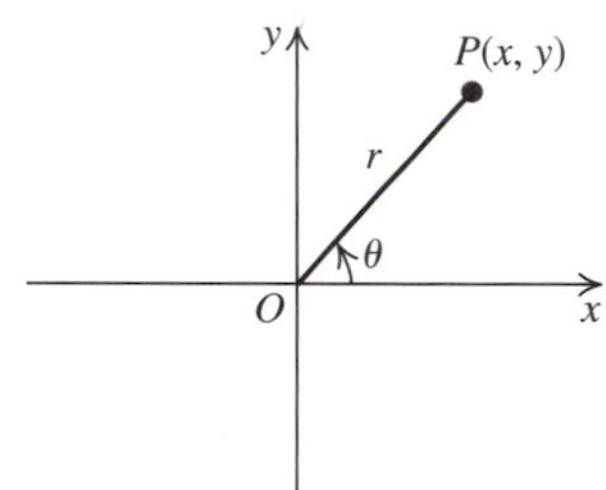

$$\sin\theta = \frac{y}{r}, \quad \csc\theta = \frac{r}{y},$$

$$\cos\theta = \frac{x}{r}, \quad \sec\theta = \frac{r}{x},$$

$$\tan\theta = \frac{y}{x}, \quad \cot\theta = \frac{x}{y}$$

Real Numbers

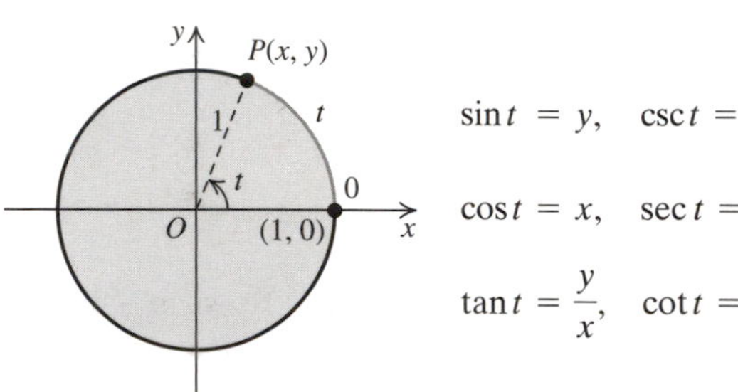

$$\sin t = y, \quad \csc t = \frac{1}{y},$$

$$\cos t = x, \quad \sec t = \frac{1}{x},$$

$$\tan t = \frac{y}{x}, \quad \cot t = \frac{x}{y},$$

Trigonometric Identities

Fundamental Identities

$$\tan\theta = \frac{\sin\theta}{\cos\theta} \qquad \cot\theta = \frac{\cos\theta}{\sin\theta}$$

$$\csc\theta = \frac{1}{\sin\theta} \qquad \sec\theta = \frac{1}{\cos\theta} \qquad \cot\theta = \frac{1}{\tan\theta}$$

$$\sin^2\theta + \cos^2\theta = 1$$

$$\tan^2\theta + 1 = \sec^2\theta$$

$$\cot^2\theta + 1 = \csc^2\theta$$

Even-Odd Identities

$$\sin(-\theta) = -\sin\theta \qquad \csc(-\theta) = -\csc\theta$$

$$\cos(-\theta) = \cos\theta \qquad \sec(-\theta) = \sec\theta$$

$$\tan(-\theta) = -\tan\theta \qquad \cot(-\theta) = -\cot\theta$$

Sum and Difference Formulas

$$\sin(u \pm v) = \sin u\cos v \pm \cos u\sin v$$

$$\cos(u \pm v) = \cos u\cos v \mp \sin u\sin v$$

$$\tan(u \pm v) = \frac{\tan u \pm \tan v}{1 - \tan u\tan v}$$

Cofunction Identities

$$\sin\theta = \cos(90° - \theta) \qquad \cos\theta = \sin(90° - \theta)$$

$$\tan\theta = \cot(90° - \theta) \qquad \cot\theta = \tan(90° - \theta)$$

$$\sec\theta = \csc(90° - \theta) \qquad \csc\theta = \sec(90° - \theta)$$

If θ is measured in radians, replace 90° with $\frac{\pi}{2}$.

Trigonometric Function Graphs

Sine and cosine function graphs and properties

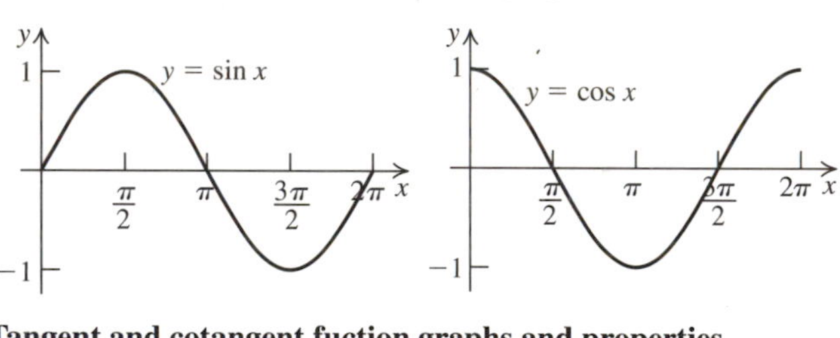

Tangent and cotangent fuction graphs and properties

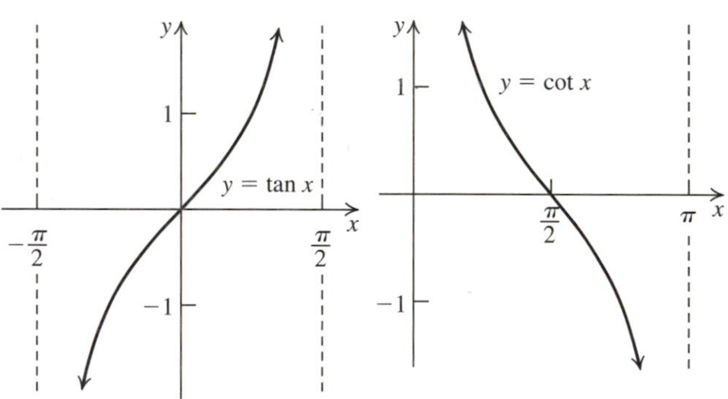

Secant and cosecant fuction graphs and properties

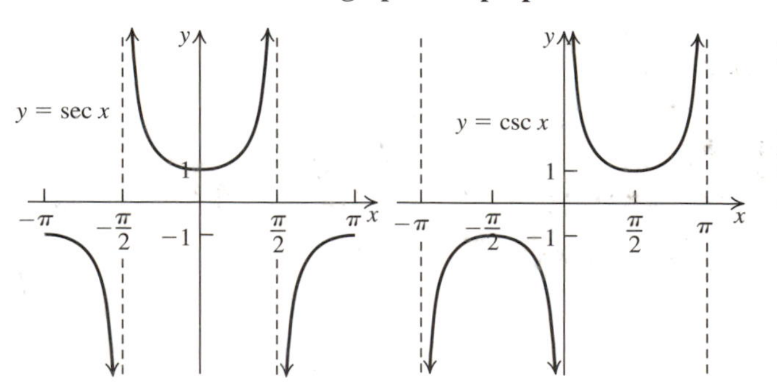

Solving Triangles

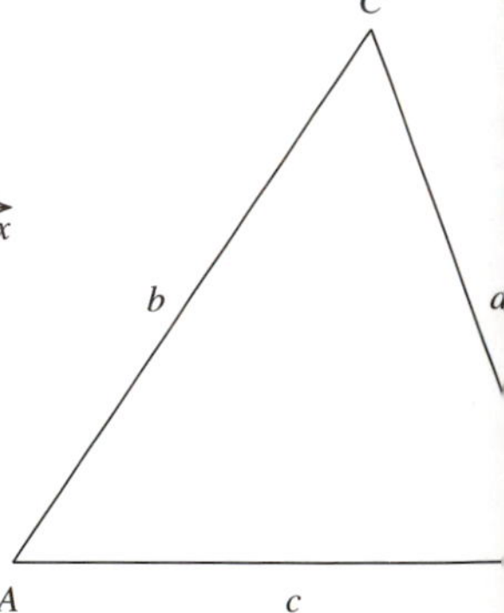

Law of Sines

$$\frac{\sin A}{a} = \frac{\sin B}{b} = \frac{\sin C}{c}$$

Law of Cosines

$$a^2 = b^2 + c^2 - 2bc\cos$$

$$b^2 = a^2 + c^2 - 2ac\cos$$

$$c^2 = a^2 + b^2 - 2ab\cos$$

Double-Angle Formula

$$\sin 2x = 2\sin x\cos x$$

$$\cos 2x = \cos^2 x - \sin^2$$

$$\cos 2x = 1 - 2\sin^2 x$$

$$\cos 2x = 2\cos^2 x - 1$$

$$\tan 2x = \frac{2\tan x}{1 - \tan^2 x}$$